METAMORPHISM AND CRUSTAL EVOLUTION OF THE WESTERN UNITED STATES

W. W. Rubey (1898–1974)

W.G. ERNST

Editor

METAMORPHISM AND CRUSTAL EVOLUTION OF THE WESTERN UNITED STATES

Rubey Volume VII

Prentice Hall, Englewood Cliffs, New Jersey 07632

Library of Congress Cataloging-in-Publication Data

Metamorphism and crustal evolution of the western
    United States.

    (Rubey volume ; 7)
    Papers presented at the Rubey Colloquium, held at
the University of California, Los Angeles, Jan. 1986.
    Includes bibliographies and index.
    1. Metamorphism (Geology)—West (U.S.)—Congresses.
2. Earth—Crust—Congresses. I. Ernst, W. G. (Wallace
Gary) (date). II. Rubey Colloquium (1986 :
University of California, Los Angeles) III. Series.
QE475.A2M47468   1988   551.8'0978   87-32896
ISBN 0-13-577719-4

Editorial/production supervision
    and interior design: *Denise Gannon*
Manufacturing buyer: *Lorraine Fumoso*

© 1988 by Prentice-Hall, Inc.
A Division of Simon & Schuster
Englewood Cliffs, New Jersey 07632

Printed in the United States of America

10 9 8 7 6 5 4 3 2 1

ISBN 0-13-577719-4

Prentice-Hall International (UK) Limited, *London*
Prentice-Hall of Australia Pty. Limited, *Sydney*
Prentice-Hall Canada Inc., *Toronto*
Prentice-Hall Hispanoamericana, S.A., *Mexico*
Prentice-Hall of India Private Limited, *New Delhi*
Prentice-Hall of Japan, Inc., *Tokyo*
Simon & Schuster Asia Pte. Ltd., *Singapore*
Editora Prentice-Hall do Brasil, Ltda., *Rio de Janeiro*

CONTENTS

PREFACE W. G. Ernst *xi*

1 TECTONIC SETTING AND VARIATIONS WITH DEPTH OF SOME
   CRETACEOUS AND CENOZOIC STRUCTURAL AND MAGMATIC
   SYSTEMS OF THE WESTERN UNITED STATES *1*
      Warren Hamilton

2 TECTONIC SIGNIFICANCE OF HIGH-PRESSURE PLUTONIC
   ROCKS IN THE WESTERN CORDILLERA OF NORTH
   AMERICA *41*
      E-an Zen

3 Pb ISOTOPIC EVIDENCE FOR THE FORMATION OF
PROTEROZOIC CRUST IN THE SOUTHWESTERN UNITED
STATES *69*
    J. L. Wooden, J. S. Stacey, K. A. Howard, B. R. Doe, and
    D. M. Miller

4 ISOTOPE GEOCHEMISTRY OF MESOZOIC AND TERTIARY
IGNEOUS ROCKS IN THE WESTERN UNITED STATES AND
IMPLICATIONS FOR THE STRUCTURE AND COMPOSITION
OF THE DEEP CONTINENTAL LITHOSPHERE *87*
    G. Lang Farmer

5 MESOZOIC CONTACT METAMORPHISM IN THE WESTERN
UNITED STATES *110*
    Mark D. Barton, Denise A. Battles, Gray E. Bebout, Rosemary
    C. Capo, John N. Christensen, Samuel R. Davis, R. Brooks
    Hanson, Carl J. Michelsen, and Heather E. Trim

6 TECTONIC AND METAMORPHIC EVOLUTION OF THE
NORTH CASCADES: AN OVERVIEW *179*
    Peter Misch

7 METAMORPHIC AND STRUCTURAL HISTORY OF THE
NORTHWEST CASCADES, WASHINGTON AND BRITISH
COLUMBIA *196*
    E. H. Brown

8 EVOLUTION OF THE CRYSTALLINE CORE OF THE NORTH
CASCADES RANGE *214*
    R. S. Babcock and Peter Misch

9 METAMORPHISM, STRUCTURAL PETROLOGY, AND
REGIONAL EVOLUTION OF THE OKANOGAN COMPLEX,
NORTHEASTERN WASHINGTON *233*
    Vicki L. Hansen and John W. Goodge

10 REGIONAL METAMORPHISM, STRUCTURE, AND TECTONICS
OF NORTHEASTERN WASHINGTON AND NORTHERN
IDAHO *271*
    Brady P. Rhodes and D. W. Hyndman

11 METAMORPHISM, STRUCTURAL DEVELOPMENT, AND AGE
OF THE CONTINENT: ISLAND ARC JUNCTURE IN WEST-
CENTRAL IDAHO *296*
    Karen Lund and L. W. Snee

12 POST-ARCHEAN METAMORPHIC AND TECTONIC EVOLUTION
OF WESTERN MONTANA AND NORTHERN IDAHO *332*
D. W. Hyndman, D. Alt, and J. W. Sears

13 METAMORPHIC PETROLOGY OF THE NORTHERN ARCHEAN
WYOMING PROVINCE, SOUTHWESTERN MONTANA:
EVIDENCE FOR ARCHEAN COLLISIONAL TECTONICS *362*
D. W. Mogk and Darrell J. Henry

14 A REVIEW OF THE GEOCHEMISTRY AND GEOCHRONOLOGY
OF THE ARCHEAN ROCKS OF THE NORTHERN PART OF
THE WYOMING PROVINCE *383*
J. L. Wooden, P. A. Mueller, and D. W. Mogk

15 THE ACCRETION OF PROTEROZOIC CRUST IN COLORADO:
IGNEOUS, SEDIMENTARY, DEFORMATIONAL, AND
METAMORPHIC HISTORY *411*
M. E. Bickford

16 EVOLUTION AND EARLY PROTEROZOIC HISTORY OF THE
MARGIN OF THE ARCHEAN CONTINENT IN UTAH *431*
Bruce Bryant

17 A SUMMARY OF PROTEROZOIC METAMORPHISM IN
NORTHERN AND CENTRAL NEW MEXICO: THE REGIONAL
DEVELOPMENT OF 520°C, 4-KB ROCKS *446*
Jeffrey A. Grambling

18 GEOLOGIC SETTING OF MESOZOIC AND CENOZOIC
METAMORPHISM IN ARIZONA *466*
Stephen J. Reynolds, Stephen M. Richard, Gordon B. Haxel,
Richard M. Tosdal, and Stephen E. Laubach

19 CORE COMPLEXES OF THE MOJAVE-SONORAN DESERT:
CONDITIONS OF PLUTONISM, MYLONITIZATION, AND
DECOMPRESSION *502*
J. Lawford Anderson

20 PROTEROZOIC HIGH-GRADE METAMORPHISM IN THE
COLORADO RIVER REGION, NEVADA, ARIZONA, AND
CALIFORNIA *526*
Warren M. Thomas, H. Steve Clark, Edward D. Young, Suzanne
E. Orrell, and J. Lawford Anderson

21 LATE CRETACEOUS REGIONAL METAMORPHISM IN SOUTHEASTERN CALIFORNIA *538*
Thomas D. Hoisch, Calvin F. Miller, M. T. Heizler, T. M. Harrison, and Edward F. Stoddard

22 PHANEROZOIC TECTONIC EVOLUTION OF THE GREAT BASIN *572*
R. Speed, M. W. Elison, and F. R. Heck

23 METAMORPHIC AND TECTONIC HISTORY OF THE NORTHEASTERN GREAT BASIN *606*
Arthur W. Snoke and David M. Miller

24 METAMORPHIC HISTORY OF THE EAST-CENTRAL BASIN AND RANGE PROVINCE: TECTONIC SETTING AND RELATIONSHIP TO MAGMATISM *649*
Elizabeth L. Miller, Phillip B. Gans, James E. Wright, and John F. Sutter

25 TECTONICS OF THE WALKER LANE BELT, WESTERN GREAT BASIN: MESOZOIC AND CENOZOIC DEFORMATION IN A ZONE OF SHEAR *683*
John H. Stewart

26 METAMORPHISM AND TECTONICS OF THE DEATH VALLEY REGION, CALIFORNIA *714*
Theodore C. Labotka and Arden L. Albee

27 METAMORPHISM AND TECTONICS OF THE NORTHERN SIERRA NEVADA *737*
Howard W. Day, Peter Schiffman, and E. M. Moores

28 TECTONISM AND METAMORPHISM IN THE NORTHERN SIERRA TERRANE, NORTHERN CALIFORNIA *764*
David S. Harwood

29 DEFORMATIONAL AND METAMORPHIC HISTORY OF PALEOZOIC AND MESOZOIC BASEMENT TERRANES IN THE WESTERN SIERRA NEVADA METAMORPHIC BELT *789*
Richard A. Schweickert, Charles Merguerian, and Nicholas L. Bogen

30 PRE-CRETACEOUS CRUSTAL EVOLUTION IN THE SIERRA NEVADA REGION, CALIFORNIA *823*
Warren D. Sharp

31 GEOLOGY AND PETROTECTONIC SIGNIFICANCE OF
CRYSTALLINE ROCKS OF THE SOUTHERNMOST SIERRA
NEVADA, CALIFORNIA *865*
David B. Sams and Jason B. Saleeby

32 METAMORPHIC AND TECTONIC EVOLUTION OF THE
NORTHERN PENINSULAR RANGES BATHOLITH, SOUTHERN
CALIFORNIA *894*
V. R. Todd, B. G. Erskine, and D. M. Morton

33 METAMORPHIC HISTORY OF THE SALINIAN BLOCK: AN
ISOTOPIC RECONNAISSANCE *938*
Eric W. James and James M. Mattinson

34 INVERTED METAMORPHIC GRADIENTS IN THE
WESTERNMOST CORDILLERA *953*
Simon M. Peacock

35 STRUCTURE, METAMORPHISM, AND TECTONIC
SIGNIFICANCE OF THE PELONA, OROCOPIA, AND RAND
SCHISTS, SOUTHERN CALIFORNIA *976*
Carl E. Jacobson, M. Robert Dawson, and Clay E. Postlethwaite

36 TECTONOMETAMORPHIC SIGNIFICANCE OF THE BASEMENT
ROCKS OF THE LOS ANGELES BASIN AND THE INNER
CALIFORNIA CONTINENTAL BORDERLAND *998*
Sorena Sorensen

37 CONSTRAINTS ON THE TIMING OF FRANCISCAN
METAMORPHISM: GEOCHRONOLOGICAL APPROACHES
AND THEIR LIMITATIONS *1023*
James M. Mattinson

38 METAMORPHIC AND TECTONIC EVOLUTION OF THE
FRANCISCAN COMPLEX, NORTHERN CALIFORNIA *1035*
M. C. Blake, Jr., A. S. Jayko, R. J. McLaughlin, and M. B.
Underwood

39 TECTONIC AND REGIONAL METAMORPHIC FRAMEWORK
OF THE KLAMATH MOUNTAINS AND ADJACENT COAST
RANGES, CALIFORNIA AND OREGON *1061*
R. G. Coleman, C. E. Manning, N. Mortimer, M. M. Donato,
and L. B. Hill

40 METAMORPHISM OF TRIASSIC-PALEOZOIC BELT ROCKS: A GUIDE TO FIELD AND PETROLOGIC RELATIONS IN THE OCEANIC MELANGE, KLAMATH AND BLUE MOUNTAINS, CALIFORNIA AND OREGON *1098*
  M. A. Kays, M. L. Ferns, and H. C. Brooks

INDEX *1143*

CONTENTS

PREFACE

The Department of Earth and Space Sciences of the University of California, Los Angeles conducts an annual lecture series entitled "The Rubey Colloquium." Specialists from research institutions around the country are invited to present lectures on a central theme of broad interest to earth and/or space scientists. As appropriate, written and elaborated summaries of these presentations subsequently have been gathered together as a book; collectively, this series is known as the Rubey Volumes. Both colloquia and book series are named in honor of the late W. W. Rubey (1898–1974), career geologist with the U.S. Geological Survey and Professor of Geology and Geophysics at UCLA. A brief sketch of Rubey's geologic accomplishments is contained in the Preface to Rubey Volume I, *The Geotectonic Development of California*, W. G. Ernst (ed.), 1981.

Rubey Volume No. VII, *Metamorphism and Crustal Evolution of the Western United States*, presents the results of a three-day colloquium that took place at UCLA in January 1986, followed by field trips to three contrasting metamorphic terranes in southern

California. The colloquium involved the largest number of speakers in the history of the series — as is reflected in the fact that Rubey Volume No. VII consists of 40 individual chapters.

The petrologic and structural evolution of the continental crust of the western conterminous United States is exceedingly complex, with constructional and thermal events ranging in age from Early Archean to the Cenozoic. Because of the magnitude of the subject, this collection of summary articles is necessarily incomplete. In aggregate, it focuses on the metamorphic recrystallization accompanying the various accretionary events responsible for continental growth, but the specific thermal and structural regimes cannot be divorced from either regional plate tectonic processes or local igneous histories. Furthermore, rifting, drifting, and later-stage petrotectonic events have obliterated, obscured, and overprinted most earlier lithic assemblages; accordingly, we see backwards through time progressively less distinctly. As Rubey Volume No. VII abundantly documents, many gaps and imperfections exist in our understanding of the metamorphism and crustal evolution of the western United States. If the reader is stimulated to clarify some of these unknowns, the collection of review papers will have served a useful purpose.

The volume is organized into five introductory overview chapters dealing with specific aspects of the isotopic and petrotectonic evolution of the American Cordillera, followed by 35 more detailed, regionally specific contributions, starting in the Pacific Northwest and working more or less clockwise around the western conterminous United States.

I thank UCLA for sponsoring and supporting both colloquium and volume. To all the authors, I extend my appreciation for their contributions, and trust that they, as well as the readers, will regard Rubey Volume VII as even better than the sum of its parts. My hope is that it will set the stage for future work.

Los Angeles   W. G. ERNST

THE RUBEY COLLOQUIUM SERIES

W. G. ERNST, ED.   *The Geotectonic Development of California*

W. G. ERNST and J. G. MORIN, EDS.   *The Environment of the Deep Sea*

R. L. PERRINE and W. G. ERNST, EDS.   *Energy: For Ourselves and Our Posterity*

MARGARET G. KIVELSON, ED.   *The Solar System: Observations and Interpretations*

DONALD CARLISLE ET AL., EDS.   *Mineral Exploration: Biological Systems and Organic Matter*

RAYMOND V. INGERSOLL and W. G. ERNST, EDS.   *Cenozoic Basin Development of Coastal California*

W. G. ERNST, ED.   *Metamorphism and Crustal Evolution of the Western United States*

Note added in proof: On July 23, 1987, Peter Misch, contributor to Chapters 6 and 8 of this volume, died at his home in Seattle after a brief illness. We deeply regret that Peter, who spent a lifetime studying the intricate interplay of metamorphic petrology and regional tectonics in mountain belts did not live to see Rubey Volume No. VII in print, for the book in aggregate epitomizes his career interests. In appreciation, we dedicate this volume to his memory.

Warren Hamilton
Branch of Geophysics
U. S. Geological Survey
Denver, Colorado 80225

1

TECTONIC SETTING AND VARIATIONS WITH DEPTH OF SOME CRETACEOUS AND CENOZOIC STRUCTURAL AND MAGMATIC SYSTEMS OF THE WESTERN UNITED STATES

ABSTRACT

Convergence of North American and eastern Pacific plates was slow during most of Cretaceous time; the subducting slab sank steeply below the advancing continent, and the magmatic arc was near the coast. Convergence was rapid during latest Cretaceous and early Paleogene time; the continent overrode the slab and was eroded against it, and the magmatic arc migrated far inland. Convergence slowed in late Paleogene time, and the slab again sank steeply. A strike-slip continental margin has evolved since middle Tertiary time.

Where rising arc magmas reacted with old continental crust, this vertical crustal column is seen to be typical: about 30 km depth, migmatites and granitic rocks of low-pressure-granulite facies; about 25–15 km, upper-amphibolite-facies migmatites and granitic rocks, many of them peraluminous, crystallized from melts cooler and wetter than those either deeper or shallower; about 15–4 km, crosscutting plutons of hornblende-biotite granitic rocks; about 4–0 km, calc-alkalic volcanic rocks, mostly as caldera complexes and ignimbrites, and as voluminous ash dispersed to mudstones and shales.

Accreted materials were metamorphosed beneath the leading part of the overriding continental plate when convergence was slow, and high-pressure rocks were cycled up into the gap above the sinking oceanic slab. During subsequent rapid convergence, accretionary-wedge complexes, metamorphosed at greenschist to lower amphibolite facies and locally blueschist facies, were underplated against tectonically eroded continental crust, beneath which they are now exposed.

Much of the foreland thrusting of preexisting stratal wedges was synchronous with steady-state arc magmatism. Gravitational spreading from the zone of great crustal thickening by magmatism may explain the thrusting. Syntectonic metamorphism deep in the thrust wedge records in part ambient geothermal gradients and in part magmatically perturbed gradients.

The western part of the continent was slowed by differential drag on the subducted slab during rapid convergence, and rotation of the Colorado Plateau relatively clockwise toward the continental interior produced Laramide crustal shortening. The great Laramide basement thrusts are seen on reflection profiles to flatten downward into anastomosing zones of ductile shear that define mid-crust lenses. Analogous mid-crust shear zones of the same age outline lenses exposed by south-central California.

Severe back-arc extension, orthogonal to the continental margin, affected magmatically heated Basin and Range crust during middle Tertiary slow convergence. Late Cenozoic oblique extension developed as the strike-slip continental margin evolved. Both orthogonal and oblique extensional regimes produced similar deformation. The middle crust was extended as lenses slid apart along ductile shear zones. The aggregate top — "detachment faults" — of these lenses increased in area as deep lenses emerged from beneath shallow ones. The brittle upper crust responded by block rotations. Range-sized fault blocks are allochthonous above the detachments, against which steeply rotated basin fills are truncated and new fill is deposited directly. Active range-front and strike-slip faults end downward against detachments. Most detachments originate at mid-crustal depths, although some surface as range-front faults. Some detachments remain active even after tectonic denudation has brought them to the surface, but others are themselves broken by new steep faults related to deeper detachments. The lower crust appears on reflection profiles to be extended by more pervasive ductile flow. The Basin and Range Mohorovicic discontinuity has been magmatically remade.

Similar extension affected Idaho and northeast Washington during Eocene time. Neogene uplift of the western United States largely records mantle heating, inferred to be due in part to convection consequent on extension and in part to conductive heating of subducted lithosphere.

The plate-tectonic settings and the geometric and temporal relationships of many Cretaceous and Cenozoic structural and petrologic assemblages in the western United States can be inferred with considerable confidence. Geologic and geophysical studies can be integrated to determine the vertical variations in these assemblages and thus to permit definition of the variations with depth of tectonic, magmatic, and metamorphic processes. This essay deduces such relationships in some of these systems.

Plate Motions

The plate regime affecting most of the western United States during Cretaceous and Paleogene time was one of subduction of eastern Pacific oceanic lithosphere of the Farallon (or the Kula) plate beneath North America. Convergence was slow during most of Cretaceous time, subduction was steep, and the continental magmatic arc — represented by the Peninsular, Sierra Nevada, Idaho, and other batholiths — lay within 300 km or so of the plate boundary. Convergence accelerated late in Cretaceous time, and slowed again in the Eocene (Engebretson *et al.*, 1985; Jurdy, 1984; Rea and Duncan, 1986). During the period of rapid convergence, from about 75 to 40 m.y., the subducting slab could not sink out of the way of the advancing continent, and continental tectonic and magmatic styles were very different from those of earlier Cretaceous time. The belt of arc magmatism migrated rapidly far to the east during latest Cretaceous time (Coney and Reynolds, 1977; Snyder *et al.*, 1976). The base of the continental lithosphere was eroded tectonically against the subducting plate (Cross and Pilger, 1978); metamorphosed oceanic sedimentary and crustal rocks are exposed in windows through the continental crust in southern California. At the same time, the western interior of the United States was deformed compressively, and the Laramide Rocky Mountains were formed. When convergence slowed in middle Paleogene time, the subducting slab again sank steeply, trench rollback was more rapid than the advance of the continent, and severe extension affected much of the interior West (Molnar and Atwater, 1978); voluminous accompanying magmatism was dominantly silicic, intermediate petrologically between arc and extensional types. Shortly thereafter, the [western] Pacific plate came in direct contact with North America, and a complex boundary dominated by strike-slip motion and oblique extension lengthened progressively along the continental margin as the subduction boundary shortened (Atwater, 1970; Dickinson, 1979; Dickinson and Snyder, 1979). Oblique late Neogene Basin and Range extension has been a response to the boundary change and has been accompanied by basaltic or bimodal basaltic-and-rhyolitic volcanism (Snyder *et al.*, 1976).

Mechanism of Subduction

An understanding of the mechanism of subduction is critical to tectonic analysis of western North America, as of any region of convergent-plate tectonics. Much published paleotectonic analysis incorporates the misconception that plate shortening necessarily accompanies plate convergence. Subduction too often is visualized as the sliding of oceanic lithosphere down a slot fixed in the mantle beneath an overriding plate — and this indeed would require shortening in the advancing upper plate if it occurred. The common case, however, is that the hinge rolls back, away from the overriding plate, with a horizontal velocity equal to or greater than the velocity of advance of the overriding plate. Among those who have documented evidence of many types for this are Carlson and Melia (1984), C. G. Chase (1978), Dewey (1980), Garfunkel *et al.* (1986), Hamilton (1979), Malinverno and Ryan (1986), Molnar and Atwater (1978), and Uyeda and Kanamori

(1979). Subduction commonly occurs at an angle steeper than the inclination of the Benioff seismic zone, which marks a position, not a trajectory, of the subducting plate.

A consequence of such subduction is that the dominant regime in an overriding plate above a sinking slab is one of extension or neutrality, not of shortening. I know of no active magmatic arcs, oceanic or continental, involving subduction of oceanic lithosphere without associated collision of light crustal masses, across which crustal shortening can be demonstrated. The Oligocene ashflows that form the great plateau of the Sierra Madre Occidental of Mexico must overlie a batholith dimensionally like the Cretaceous Sierra Nevada batholith of California, yet are subhorizontal. The smaller Oligocene batholith of the San Juan Mountains of Colorado is similarly capped by subhorizontal volcanic rocks. The late Oligocene and early Miocene arc magmatism of the southwestern United States was synchronous with severe extensional faulting. The migration of oceanic island arcs and the opening of back-arc basins behind them are manifestations of the generally extensional regime in overriding plates (Hamilton, in press *e*).

Nonshortening explanations should, in my view, generally be sought for deformation in a plate above a steeply subducting slab. This includes the formation of the Sevier belt, the foreland thrust belt of Cretaceous age in the western interior States, but not the Laramide belt of true crustal shortening within what is now the Rocky Mountain region but had previously been part of the craton.

Accretionary Wedge and Fore-Arc Basin

Extreme shear in a compressional regime is recorded by the Cretaceous and Paleogene Franciscan Complex of accretionary-wedge melange in coastal California. Such an accretionary wedge, however, represents weak surficial debris pushed in front of an advancing plate (Fig. 1-1), and does not require shortening of the lithosphere plate itself. The leading edges of overriding continental plates typically bear fore-arc basins of little-deformed strata, the presence of which precludes shortening of those thin, tapering parts of the overriding plates (Fig. 1-1; Hamilton, 1978, 1979). Such undeformed basins are present atop the thin leading edge even of the Pacific margin of South America (Coulbourn and Moberly, 1977), which most proponents of crustal shortening of overriding plates mistakenly assume to be the site of severe crumpling. The analogous Cretaceous and Paleogene fore-arc basin of California, which underlies the Great Valley and crops out in the eastern and medial Coast Ranges and in the western Transverse Ranges and regions to the south, similarly was little deformed during the long period of deposition of its fill. I have argued elsewhere (Hamilton, 1978, in press *e*) that such a fore-arc basin forms, after arc reversal, atop oceanic lithosphere which was earlier generated behind a migrating island-arc complex that collided with a continent — in the case of California, in Late Jurassic time. Such a collision is followed by inauguration of new subduction landward beneath the seaward side of the aggregate, leaving a narrow strip of back-arc basin crust attached to the continental plate as its leading edge. The front of this oceanic strip is raised as accretionary-wedge melange is stuffed beneath it, forming the basin, within which depocenters of successively younger deposits are displaced successively landward.

PRODUCTS OF CONTINENTAL ARC MAGMATISM

The long Cretaceous episode of steady-state subduction of oceanic lithosphere beneath North America was a period also of steady-state arc magmatism, the products of which we see now as a belt of great batholiths exposed at upper- and mid-crustal levels. Younger arc-magmatic assemblages are exposed at various crustal levels farther east. Variations

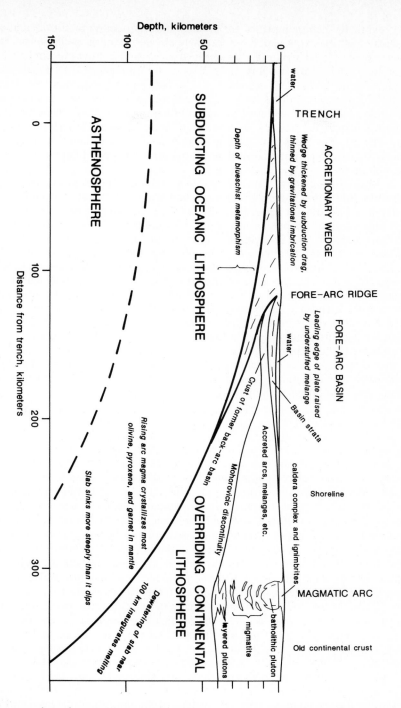

FIG. 1-1 Cross section of a continental-margin subduction system. The dimensions of the model are derived from modern Sumatra (Hamilton, 1979) but match those of the middle Cretaceous components of California.

5

with depth in these complexes are broadly similar to those in continental arc-magmatic terrains of all ages around the world. There are global consistencies in the dominant rock assemblages of different levels in continental crust (*cf*. Fountain and Salisbury, 1981; Hamilton, 1981a; Kay and Kay, 1981; Tarney and Windley, 1977; B. L. Weaver and Tarney, 1981; P. R. A. Wells, 1980, Windley, 1981). No two crustal sections are identical, but the similarities are more impressive than the differences because there is a mainline of crustal evolution. Analyses of western United States batholiths in these terms have been made by Hamilton (1981a, in press *d*), Hamilton and Myers (1967, 1974), Hyndman (1981, 1983), Ross (1985), and Sams and Saleeby (Chapter 31, this volume). Understanding of systematic variations permits integration into tectonic analysis of depth of formation of rock assemblages.

Magmatic arcs now form, along either oceanic island arcs or active continental margins, primarily above those parts of subducted slabs of oceanic lithosphere whose tops are at depths of about 100 km (Fig. 1-1). The initial melting at depth presumably is caused by water expelled by breakdown of hydrous metamorphic minerals in crust, partly serpentinized mantle, and sedimentary rocks of subducted oceanic lithosphere; the primary magmas must be equilibrated with overriding lithospheric mantle and hence be very mafic basalts. The volcanic rocks erupted at the surface, however, vary systematically in average composition with the character of the crust and mantle through which the magmas have risen. Magmas reaching the surface in a young oceanic island arc are dominantly basaltic; in a mature island arc, andesitic; and in a continental arc, rhyodacitic or quartz latitic. Such variations occur in arcs of the contrasted types, and they occur also as a progression wherever a continuous magmatic arc changes along trend from oceanic to continental (Hamilton, 1979, in press *e*). Lavas erupted even in the most primitive oceanic arcs must be products of series of partial or complete equilibrations by fractionation, zone refining, assimilation, and secondary melting in the mantle and crust. Fractionates must be complementary to voluminous more mafic materials.

Facies and Equilibria

Facies fields (Fig. 1-2) apply equally to metamorphic and igneous rocks. Igneous rocks do not have the compositions of melts, for fusibles are lost during crystallization, and solids equilibrated elsewhere commonly are contained in melts. Much of igneous petrology represents an attempt to demonstrate equilibria between solids and liquids. Confirmation is too often interpreted as indicating that liquid plus solid equal a source rock whose partial melting produced the liquid, or equal a magma whose fractionation produced the solid. Such simplistic concepts scarcely begin to account for the observed variations of crystalline rocks with depth of exposure. O'Hara and Mathews (1981) demonstrated the futility of most conventional geochemical modeling even for the simplest systems in which no wallrock reactions are involved and little variation occurs with depth. Walker (1983) reviewed complexities of magmatic evolution that are receiving increasing attention.

Deep continental crust consists largely of rocks in various granulite facies. There is no standard nomenclature for such rocks, which have received far less study than have the rocks of the middle crust; my own suggestions are summarized in Fig. 1-2. Reactions from amphibolite to granulite facies are dominated by dehydration, whereas reactions between the several granulite facies represent primarily decreasing stability of plagioclase (successively with olivine, orthopyroxene, and clinopyroxene), and increasing density, in the direction of increasing ratios of pressure to temperature. A garnet-amphibolite facies, reflecting intermediate activity of water, often intervenes between amphibolite and middle-pressure granulite; the *P-T* field of garnet amphibolite is poorly constrained, and it is omitted from the figure.

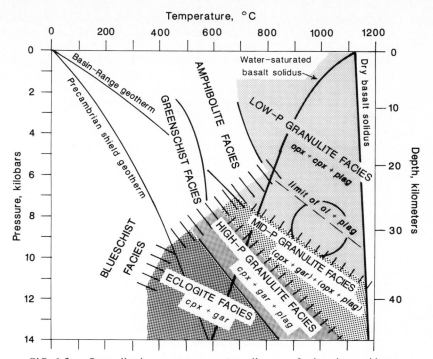

FIG. 1-2. Generalized pressure-temperature diagram of mineral assemblages relevant to the lower continental crust. Boundaries approximate those for mafic and intermediate rocks but vary with bulk composition; coexisting minerals vary in composition across each facies. The boundary between amphibolite and granulite facies at pressures greater than 5 or 6 kb shifts greatly with activity of H_2O and CO_2, and a garnet-amphibolite facies (not shown) often intervenes. Abbreviations: cpx, clinopyroxene; gar, garnet; ol, olivine; opx, orthopyroxene; plag, plagioclase. Rocks of similar bulk compositions become progressively more dense going from low-pressure granulite to eclogite as plagioclase reacts with ferromagnesian minerals to produce progressively denser phases; plagioclase reacts out successively with olivine, orthopyroxene, and clinopyroxene (although albite is stable in the higher-T/P part of the blueschist facies, and sanidine is stable in high-T eclogite). Adapted from many published papers, including Hansen (1981), Johnson et al. (1983), Newton and Perkins (1982), and papers referred to by each. Geotherms from Sclater et al. (1980) and Lachenbruch and Sass (1978).

Volcanism

Active magmatic arcs developed on mature continental lithosphere above subducting oceanic lithosphere, as in Sumatra and the Andes, are characterized by fields of silicic volcanic rocks, dominantly ashflow sheets erupted from large calderas atop granitic plutons, and by stratovolcanoes of rocks of intermediate composition. Little-eroded fossil systems include the vast belt of middle Tertiary ignimbrite fields of the San Juan Mountains of Colorado, Mogollon Plateau and other tracts in New Mexico and Arizona, and the Sierra Madre Occidental in Mexico. In the San Juan Mountains, calderas — which have diameters to 40 km, and from each of which great ashflow sheets were erupted — are contiguous or overlapping and, with geophysical data, define an underlying shallow, composite batholith (Lipman, 1984). Where erosion has penetrated such volcanic superstructures, the ashflows frequently are seen to cap plutons from which they were erupted. An example is the Late Cretaceous Boulder batholith of Montana (Hamilton and Myers, 1967, 1974). Most large "epizonal" and "mesozonal" plutons probably breached to erupt voluminous volcanic products; many plutons must form capped largely by their own volcanic ejecta, and be in effect gigantic mantled lava flows. The largest pluton yet mapped in the Cretaceous Sierra Nevada batholith, the Mount Whitney pluton, is about the same size, 25 × 80 km, as the largest known young caldera in an active magmatic arc, the late Pleistocene Lake Toba caldera of Sumatra.

Silicic magma chambers are zoned thermally and compositionally (Hildreth, 1981). Eruptions are generally from the volatile-rich tops of chambers and drain further volatiles from deeper in the chambers. The most highly fractionated magma is expelled first and hence is deposited stratigraphically lowest within ignimbrite sheets, which are produced primarily by the radial outflows of series of collapsing magma fountains. Volcanic rocks typically are more silicic and alkalic than are the plutons left beneath. Relatively high-pressure minerals (e.g., muscovite) in uncommon ignimbrites require that some chambers vent abruptly from depths as great as 18 km, but most ignimbrites have shallow sources (*cf*. Fig. 1-3).

Volcanic Ash, Hydrocarbons, and Uranium

Many accumulations of hydrocarbons and of uranium may be indirect by-products of arc magmatism. Volcanigenic shales appear to be major hydrocarbon source rocks, as C. E. Weaver (1960) early argued. Clay altered from silicic volcanic ash is dominated by smectite (interlayered illite and montmorillonite-beidellite). Such clays absorb organic molecules, hence tend to be rich in organic material and thus to be potential source rocks. Dehydration of the water-rich montmorillonitic layers to more layers of illite during burial metamorphism occurs in steps, which in part are at conditions appropriate for the maturation of hydrocarbons, and this probably is a major factor both in geopressuring shales and in facilitating migration of hydrocarbons once formed (Bethke, 1986; Bruce, 1984; Burst, 1969; C. E. Weaver, 1960).

An enormous volume of ash is dispersed into the atmosphere during major ignimbritic eruptions. The volume of airfall ash on the Indian Ocean floor from the late Pleistocene Toba caldera eruption of Sumatra is about 1000 km^3, equal to the volume of the comagmatic on-land ignimbrites (Ninkovich *et al.*, 1978). The shales and mudstones of the western interior of the United States from the Upper Triassic section upward are dominantly volcanigenic, and represent the air- and water-transported equivalents of the batholiths and ignimbrite fields farther west. Thus, altered volcanic ash comprises about 75 percent, by volume, of the extensive Upper Cretaceous shales of the Rocky Mountains and Great Plains (Schultz, 1978). The ash has been recycled repeatedly, and the shales and muds of the Gulf Coast and Gulf of Mexico also are dominantly volcanigenic. By contrast, the shales of the Atlantic margin of North America are dominated by clays produced by weathering of nonvolcanic materials. The great contrast in hydrocarbon productivity of Gulf and Atlantic provinces may be due to the contrast in mineralogy, hence sources, of clays.

The uranium deposits in sedimentary rocks in the western States mostly are associated with nonmarine mudstones rich in silicic ash, and similarly may be largely by-products of magmatism far to the west. (Most authorities on the deposits have sought more local sources.) Uranium is readily leached from such ash, and transported through nearby sandstones, by oxidizing water; it is deposited where fluids encounter reducing conditions, as at concentrations of organic matter.

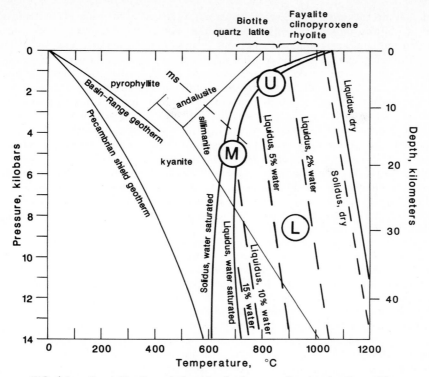

FIG. 1-3. Crystallization relationships for leucogranite, showing the melting interval for water-saturated granite and contours on the liquidus surface for undersaturated magma, after Huang and Wyllie (1981) and C. R. Stern and Wyllie (1981). Fields of the aluminum-silicate polymorphs are from Holdaway (1971); the andalusite-sillimanite boundary probably should incline more gently to the left. Primary muscovite can crystallize in granitic magma only deeper than the intersection of the stability limit of muscovite plus quartz (line ms; after Tracy, 1978; precise position varies with bulk composition) with the granite solidus. Typical fields are shown by circles for late-stage magmas of the upper (U), middle (M), and lower (L) crust; these could form granites bearing, respectively, biotite, muscovite, and hypersthene. Typical eruption temperatures of biotite quartz latite (commonly, a magmatic-arc magma) and of fayalite-clinopyroxene rhyolite (commonly, an extensional-setting magma) follow Hildreth (1981) and others. Geotherms from Sclater *et al*. (1980) and Lachenbruch and Sass (1978).

Upper and Middle Crust

Beneath the volcanic rocks of continental magmatic-arc complexes are upper-crustal plutons and composite batholiths of calc-alkalic granitic rocks. Among such complexes are those of the great magmatic arc, formed during a span that included most of Cretaceous time, that lies along the Cordillera of North America. The arc setting is shown by the relationship to the coeval accretionary wedge and fore-arc basin to the west; the geometry of wedge, basin, and magmatic arc is strikingly like that of modern Sumatra (Hamilton, 1979; Fig. 1-1). The Mesozoic magmatic complexes have been variably uplifted and eroded, and are exposed at various crustal levels that show systematic variations with depth.

Sierra Nevada batholith The Sierra Nevada batholith of California is a composite of hundreds of plutons, mostly of Cretaceous age, dominantly tonalite and granodiorite in the west and more felsic granodiorite and monzogranite in the east (Bateman, 1979; Bateman *et al.*, 1963; Noyes *et al.*, 1983). Plutons mostly are steep sided, show by their flow structures that they spread outward from rising, replenished medial zones, and have contact-metamorphic aureoles. The eastward compositional progression of the granitic rocks broadly parallels that of the petrologic maturity of pre-batholitic rocks, from the accreted island-arc and accretionary-wedge materials of the west to miogeoclinal assemblages (and, at depth, Precambrian basement?) of the east. The ratio of continental-crustal to mantle components in the granitic rocks increases correspondingly eastward (DePaolo and Farmer, 1984; Kistler and Peterman, 1978). The compositional changes also represent an age progression, for the Cretaceous plutons become in general younger eastward, from 125 m.y. in the west to 80–90 m.y. in the east (Chen and Moore, 1982; T. W. Stern *et al.*, 1981). The early assumption by some of us (e.g., Hamilton, 1969) that the eastward increase in such variables as the K_2O/SiO_2 ratio represented increasing height above the subducting slab of the time is not valid.

Depths of crystallization of central Sierran plutons can be inferred from broad petrologic features, although little petrobarometry has been attempted of either granitic or metamorphic rocks. Crystallization at depths of only 3-10 km is indicated for medial and eastern granites by the few detailed petrologic studies (e.g., Noyes *et al.*, 1983), by such regional petrological features as the absence of primary muscovite in granitic rocks and the prevalence of andalusite (often with muscovite but not biotite) in aluminous metasedimentary rocks (*cf.* Fig. 1-3), and by the local preservation of volcanic roof rocks. In general, western Sierran plutons crystallized deeper — no volcanic roof rocks remain, primary muscovite occurs in leucogranites, biotite is ubiquitous with sillimanite, although kyanite has not been reported — but still markedly shallower than the depth (13 km?; Fig. 1-3) of the aluminum-silicate triple point. Individual plutons are bounded sharply, and generally steeply, against each other and against the contact-metamorphosed rocks that separate many of them. Large plutons crystallized over spans of several million years each as new magma was introduced into their interiors, and typically are broadly zoned, the margins representing the hottest and driest magmas (Bateman and Chappell, 1979; Bateman and Nokleberg, 1978; Chen and Moore, 1982; Noyes *et al.*, 1983). Refractory plagioclase and hornblende were carried upward with the melts of many plutons of intermediate composition (Bateman and Nokleberg, 1978; Noyes *et al.*, 1983). whereas other plutons, mostly leucogranites, crystallized from nearly complete high-level melts (Bateman and Chappell, 1979). Migmatites are developed only as local contact phases except in the most deeply eroded tracts in the far west.

Pre-Cretaceous rocks form a broad, discontinuous belt along the east edge of the central part of the batholith. Cambrian to Permian sedimentary rocks are overlain by Upper Triassic, Jurassic, and lowermost Cretaceous volcanic rocks, which are dominantly the rhyodacitic ejecta of arc-magmatic plutons. Although much deformed also in the White and Inyo Mountains, just to the east of the Sierra Nevada, the section is there mostly upright and gently to moderately dipping; but the section rolls downward to the west to steep to vertical west dips in and adjacent to the Sierra Nevada, where it is contact metamorphosed and much more highly deformed. Large Late Cretaceous plutons of the main Sierra Nevada batholith (the Cathedral Peak pluton: Fig. 1-4; the similar Mono Recesses and Mount Whitney plutons; and the Mount Givens and other plutons) have complex contacts but in a general way lie with steep intrusive contacts against the west edge and stratigraphic top of this faulted pre-Cretaceous secion (*cf.* Bateman, 1965, Bateman and Moore, 1965, Bateman *et al.*, 1983, Fiske and Tobisch, 1978, Rinehart and Ross, 1964, and Tobisch *et al.*, 1986, none of whom made this point). I regard these contacts

FIG. 1-4. East contact of Cathedral Peak Granite, east-central Sierra Nevada. The subvertical contact is regarded as the depressed and outward-pushed *floor* of the pluton. Abbreviations: Kg, Cretaceous Cathedral Peak Granite, 88 m.y.; Jv, hornfelsed silicic and intermediate volcanic rocks of Jurassic age; Pz, hornfelsed sedimentary rocks, mostly clastic, of late Paleozoic age. Metasedimentary rocks dated by fossil collections, and granite by U-Pb determinations (Bateman *et al.*, 1983); metavolcanic rocks dated along strike to southeast by fossils and U-Pb determinations (Fiske and Tobisch, 1978; Bateman *et al.*, 1983, reported Late Triassic ages from whole-rock Rb-Sr determinations, but likely these are too old because of ^{87}Sr assimilated from crustal rocks). Contact between units Jv and Pz is marked for skyline ridge; contact lies at extreme right end of ridge just below skyline. Older Paleozoic metasedimentary rocks lie farther right, out of the picture. View north-northwestward over Steelhead Lake to Shepherd Crest.

as the depressed and outward-pushed *floors* of essentially extrusive, bathtub-shaped plutons, not as the "roof pendants" of popular assumption. Elongate semiconcordant Jurassic plutons within the metavolcanic section were rotated with the metavolcanic rocks in Late Cretaceous time. Some older and more deeply eroded large plutons farther west in the batholith probably also formed with extrusive relationships to flanking rocks.

Volcanic rocks, mostly ignimbrites, comagmatic with Cretaceous plutons are preserved primarily between plutons but cap them locally. A caldera complex atop the granitic pluton whence it came was documented by Fiske and Tobisch (1978).

The middle-crust substrate of the batholith is exposed at the south end of the Sierra Nevada batholith, which was tilted and eroded obliquely during latest Cretaceous or early Paleogene time (Elan, 1985; Ross, 1985, in press; Sams and Saleeby, Chapter 31, this volume; Sharry, 1981). Depth of formation of rocks now at the surface increases, over a distance of about 75 km, from less than 10 km to about 30 km. The eastern Sierran belt of felsic plutons trends southward into its substrate of mostly leucocratic granodiorite. Widespread primary muscovite in the granodiorite and the abundance of sillimanite and absence of andalusite in the wall rocks (although kyanite has not been reported) are indicative of middle-crust crystallizaiton (*cf.* Fig. 1-3). Wall rocks are migmatitic or gneissic: a broad terrain was heated to granite-solidus temperature at a pressure great enough to permit richly hydrous melts. The medial and western Sierran belts of mostly granodioritic and tonalitic plutons trend southward into a substrate of variably migmatitic hornblende tonalite and on into tonalitic and more mafic rocks containing brown hornblende, orthopyroxene, clinopyroxene, reddish biotite (which presumably has relatively high ratios of Ti and Mg to Fe), and, with hornblende or orthopyroxene (but not clinopyroxene), abundant pink or red garnet. Mineral assemblages are those of the garnet-amphibolite and low-pressure-granulite facies (Fig. 1-2). Metapelites in the transitional zone appropriately include the assemblage cordierite + sillimanite + almandine + K-feldspar (Elan, 1985), indicative of the middle-pressure part of the low-pressure-granulite facies, whereas cordierite was not found by Ross (in press) in the more deeply eroded southern rocks.

Peninsular batholith A southern sector of the Cretaceous composite arc-magmatic batholith is exposed in the Peninsular Ranges of southwestern California and northern Baja California (Silver *et al.* 1979; Todd and Shaw, 1979). Like the Sierra Nevada batholith, the southern batholith becomes both younger and more silicic, potassic, and radiogenic eastward, thus showing increasing incorporation of old crustal rocks, or of terrigenous strata proxying for them, in the magmas in that direction. Unlike the Sierra Nevada, much deeper rocks are exposed in the east than in the west. Coeval calc-alkalic volcanic rocks are widely preserved atop the shallow, crosscutting plutons of the far western part of the batholith. In the east, by contrast, mid-crustal migmatites and gneisses of uppermost amphibolite facies, and concordant granites including many crystallized from richly hydrous melts and containing primary muscovite, are exposed; dips of foliation and layering are mostly gentle to moderate.

Salinia A belt of batholithic rocks of middle and Late Cretaceous age is displaced 350 km relatively northwestward from the southern Sierra Nevada along the Neogene San Andreas fault. Upper-crust granites and mid-crust amphibolite-facies migmatites are both widely exposed, and deeper-seated rocks of garnet-amphibolite and low-pressure-granulite facies are exposed in the west-central part of the terrain (Compton, 1960; Ross, 1976, in press). The plutonic rocks are equivalent petrologically and chronologically to the medial parts of the Sierra Nevada and Peninsular batholiths (Kistler and Peterman, 1978; Mattinson, 1978; Ross, 1978). This "Salinian block" is now bounded on both sides by the coeval Franciscan accretionary-wedge complex, from which the plutonic rocks are separated by the San Andreas fault on the east and the Sur-Nacimiento megathrust on the west. Apparently, the batholith was sliced longitudinally by the San Andreas fault after its western part and the other belts that initially intervened between it and the Franciscan had been removed by subduction-related tectonic erosion, as discussed subsequently.

Idaho batholith The Idaho sector of the Cretaceous batholithic belt is exposed mostly at mid-crustal levels (R. B. Chase *et al.*, 1983; Hyndman, 1980, 1981, 1983). Shallow features of Sierra Nevada type — steep, sharply bounded plutons of massive rocks, septa of contact-metamorphic rocks, preserved coeval volcanic rocks — are largely lacking. The migmatitic floor of the batholith is exposed in the broad Salmon River arch, which trends westward across the center of the batholith. The part of the batholith, the Bitterroot lobe, north of this arch is rimmed by deep-seated gneisses and migmatites of upper amphibolite facies, the petrology of which (e.g., aluminum-silicate assemblages that include kyanite) indicates a general depth of formation of about 12–20 km, and is in broad aspect a synformal mass, the remnant of a great batholithic sheet (Hyndman, 1983). The base of the batholith is a zone of intercalated granitic, magmatitic, and gneissic rocks, in which dips are more commonly steep or moderate than gentle, and in which the proportion of granitic rocks decreases downward through a vertical range of 2 km as exposed in the great canyons of central Idaho (Cater *et al.*, 1973). Granitic rocks are characterized by primary muscovite as well as biotite, hence crystallized from magmas too hydrous to have risen to upper-crust levels (Hyndman, 1983; Fig. 1-3). Migmatitic rocks near the west margin of the batholith include such indicators of moderately high pressures as garnet amphibolite and, in granitic rocks, magmatic epidote (Zen, Chapter 2, this volume).

Also present in the medial and eastern parts of the Idaho batholith are Eocene igneous rocks: deep-seated plutons and gneisses on the one hand, and large epizonal plutons and ignimbrites and other volcanic rocks erupted from them on the other hand (Cater *et al.*, 1973; R. B. Chase *et al.*, 1983). Tectonic denudation accompanying Paleogene crustal extension may have produced the juxtapositions of deep and shallow rocks, as discussed in a subsequent section.

Arc rocks east of the Sierra Nevada In part of Jurassic time, and again during latest Cretaceous and Paleogene time, North American and eastern Pacific lithospheric plates converged more rapidly than they did during Early and early Late Cretaceous time, and the new magmatic arc formed to the east of the Idaho–Sierra Nevada–Peninsular batholitic belt, in the Basin and Range and other regions. Extremely variable uplift and erosion has affected these eastern magmatic complexes, which now can be seen at all levels of exposure from ashflows down through shallow batholiths and the hydrous-magma bases of batholiths atop migmatites, to middle-crest migmatites and concordant granitic sheets. Much of the erosion was tectonic, for middle-crust rocks are exposed primarily beneath extensional Tertiary detachment faults. Most upper-crust plutons within layered rocks appear to be sills inflated thickly to mushroom shapes. Thus, the small batholiths, mostly Jurassic, of the Inyo Mountains in eastern California in general have steep, outward-dipping contacts, which follow limited stratigraphic zones within the thick Paleozoic section for long distances (Ross, 1967). Best known of the batholiths that reached the surface over a broad area is the composite Late Cretaceous Boulder batholith of Montana, which is 100 km long and 50 wide and which spread laterally over a floor of all premagmatic rocks, Middle Proterozoic to Upper Cretaceous, pushing them down so that they steepen toward and dip beneath it; the batholith is overlain only by coeval volcanic rocks, and is in effect a gigantic extrusive mass (Hamilton and Myers, 1967, 1974). Some Sierra Nevada plutons are analogous, as noted previously.

The exposed middle-crust granites, including many as young as middle Tertiary, commonly contain primary muscovite and in many cases garnet, and have isotopic ratios of lead, strontium, and neodymium indicative of generally much more inclusion of crustal material in their melts than do the shallow granites (DePaolo and Farmer, 1984; Haxel *et al.*, 1984; Kistler *et al.*, 1981; C. F. Miller and Bradfish, 1980). Some of these plutons have contact-metamorphic aureoles but many others lie in rocks metamorphosed regionally at amphibolite facies. That these granites crystallized in the middle crust, and hence have depth significance, is indicated by such features as kyanite in wall rocks near some of them, and the richly hydrous character of the magmas recorded by the granites, migmatites, and voluminous pegmatites.

Lower Crust

Exposures of late Phanerozoic magmatic terrains in the western United States represent depths of formation no greater than the upper part of the lower crust, but lower continental crust is exposed in many other parts of the world beneath migmatite and low-pressure-granulite terrains such as those of the deeply eroded parts of the Cretaceous batholiths. Typical lower crust, which I assume to belong mostly to the arc-magmatic mainline, consists of distinctive igneous and metamorphic rocks equilibrated under low- or middle-pressure granulite conditions and typically retrograded, particularly in middle-pressure granulite facies but locally to high-pressure granulite and even eclogite facies (Hamilton, 1981a, in press *d*). Such retrogression can accompany isobaric cooling (Fig. 1-2), although many authors assume tectonic settings in which pressure increases with time. Exposed sections of deep crust, calibrated petrologically as representing depths of

30–45 km, generally display magmatic differentiates of granodioritic, tonalitic, and gabbroic bulk compositions, intruded from the mantle, and also the products of granitic magmas melted from wallrocks by the heat of those intrusions. The characteristic differentiates — gabbro, norite, anorthosite, and hypersthene granite (charnockite) — occur either as great layered complexes of denser rocks beneath lighter ones or as thinner intercalations in other rock types. Preexisting rocks often include metasedimentary rocks, such as quartzite, marble, and metapelite, which have lost most of their initial combined water. A major problem for interpretation is that exposed lower crust is dominantly of Precambrian age, and it is unclear to what extent younger deep crust is similar.

Structure

Contacts between and within granitic and metamorphic rocks typically are steep in shallow complexes and undulating with gentle dips in deep ones. The tendency toward gentle dips in the middle and deep crust is accentuated by ductile transposition accompanying retrograde metamorphism in many regions. Much such metamorphism and deformation probably records postmagmatic extension or shortening, but likely much also records synmagmatic gravitational flow of heated rocks that are displaced outward and downward by rising magmas and then flow subhorizontally beneath shallow spreading batholiths. Depression to deep crustal levels of supracrustal rocks may be due primarily to the repeated injection and eruption of magma above them rather than to the compressive or thrust tectonics commonly inferred.

Wallrock Deformation

Contact-metamorphic wallrocks of Sierra Nevada plutons commonly display superposed folds, more or less coaxial, which often are taken as evidence for successive compressive deformations of widely separated ages (e.g., Nokleberg and Kistler, 1980). Implicit in such analyses of polyphase folding, which mostly are unconstrained by data on changes in external shape of lithologic units, is the assumption that folding in metamorphic rocks records shortening in accordion fashion, and hence that refolding requires a new shortening. I see synmetamorphic folding, by contrast, as recording primarily a combination of pure and simple shear — of flattening plus discontinuous laminar flow — and any number of folds can be superposed within a continuum of deformation by such a mechanism. I described from southeastern California thin, distinctive units of metasedimentary rocks that are complexly deformed internally, with superimposed isoclinal and recumbent folds, yet that have plane-parallel external contacts which preclude accordion folding and require layer-parallel laminar flow with extreme flattening (Hamilton, 1982). Although multiple episodes of deformation must indeed often be recorded in metamorphic rocks, the geometric evidence commonly presented for them — the superposition of coaxial folds — is insufficient to demonstrate their reality. The dominant cause of deformation in contact-metamorphic wallrocks of upper-crust plutons is, I infer, the gravitational rise and spreading of the plutons and complementary downward flow of wallrocks, not regional shortening. Tobisch *et al.* (1986) argued on other grounds that the synmetamorphic deformation of eastern Sierran wallrocks, for which Nokleberg and Kistler (1980), among others, deduced sequential older deformations, is of Cretaceous age.

Heat and Water

The mainline of crustal evolution involves heating much of the deep and middle crust to the temperature of granite magma, which temperature is controlled by water content

(Fig. 1-3; Hamilton, 1981a). Mature magmatic-arc continental crust consists of largely magmatic rocks in upper sections, of generally more preexisting rock than new magmatic material in middle sections, and of dominantly magmatic rocks in deep sections. Most of such a crustal column was heated 300–500°C above steady-state geotherms. The enormous influx of mantle heat must have been carried upward in hot material which I deduce to be, in general, magma of arc origin.

Water in micas and other hydrous minerals of wallrocks increases upward in the crust, whereas the solubility of water in melts increases downward. Only in the middle crust is water generally abundantly available in wall rocks and is pressure adequate to maintain a high water content in magma, which hence typically is warm and wet, and does the setting permit accumulation of heat needed to raise large volumes of rock to solidus temperatures. There is little water in the deep crust, so magmas there are dry and hot.

Ordering of complexes by depth of formation permits general interpretation of the evolution of upper mantle and crust in terms of magmas that originate as very mafic basaltic liquids at depths near 100 km (Fig. 1-1). The evolution of crustal magma is profoundly influenced by the availability and behavior of water (Fig. 1-3). The rising magma evolves by zone refining, exothermically crystallizing refractories and endothermically assimilating fusibles. Little mantle material need be present in the final melts, but mantle heat carried by magma is the primary source of warming of the crust above ambient geotherms.

Olivine, pyroxenes, and garnet are mostly crystallized from the magmas within the subcontinental mantle: arc magmas rising to depths of 40 km in continental lithosphere mostly have evolved to compositions in the range from gabbro to granodiorite, and are relatively anhydrous. Deep-crustal crystallization, assimilation, and metamorphism produce norite-anorthosite-charnockite magmatic complexes and metamorphic rocks typically in middle-pressure-granulite facies. Magmas that pass through or rise from this zone produce relatively dry migmatites, in the deeper part of the middle crust, in low-pressure-granulite or garnet-amphibolite facies. Magmas rising higher commonly assimilate much water released by breakdown of heated wallrock micas and other hydrous minerals, or produce secondary melts by heating of hydrous rocks, so wetter migmatites, of upper amphibolite facies, are formed in the upper part of the middle crust. Magma once hydrated, and hence cooled, cannot rise appreciably above the level at which water pressure exceeds load pressure (Fig. 1-3), and it crystallizes in the middle crust. The upper part of the middle crust is characterized by migmatites that enclose large and small sheets of pegmatites and low-temperature granites, the latter typically peraluminous (having a molecular proportion of Al_2O_3 greater than that of $K_2O + Na_2O$; this is expressed mineralogically by the presence of muscovite, garnet, or other aluminum-rich minerals).

Enrichment in water and alumina is greater for magmas in metapelitic wallrocks than for those in old continental crust, and in one bit of popular terminology sediment-equilibrated granites are referred to as "S-type granites" (A. J. R. White *et al.*, 1986). Such analysis is often accompanied by the unwarranted assumption — as by Clemens (1984) — that equilibration requires crustal melting by heat mysteriously conducted into the crust and unrelated to mantle magmatism. As C. F. Miller (1986) emphasized, the "S-type" terminology appears to have outlasted its usefulness and to have degenerated into an exercise in picking nits.

The peraluminous granites of the western United States mostly were equilibrated with older plutonic rocks rather than metapelites. Many of these granites are temporally and spatially related to the exceptional subduction regime operating during latest Cretaceous and Paleogene time, as discussed in a subsequent section.

Magmas that cross the middle crust without equilibrating there can continue to rise because they remain relatively hot and dry. Voluminous magmas that reach the upper crust spread as shallow, steep-sided batholiths above the deeper migmatites and concordant granites, and erupt as capping ashflow sheets and as dispersed airborne ash when their own rise and crystallization produce water saturation. Most magmas of upper crustal batholiths probably contain less than 1.5% water and reach water saturation only at shallow depth after considerable crystallization, with resultant expulsion of a vapor-rich phase in pegmatites and volcanic eruptions (Maaloe and Wyllie, 1975). Granites in the middle crust, by contrast, have mostly crystallized from magmas with 3–5% water (Clemens, 1984; Green, 1976), hence reach saturation at much greater depth, expelling the voluminous pegmatites that characterize migmatite terrains; forced crystallization, and stopping of the rise of the magmas, necessarily result.

Such a scenario is compatible with experimental data (e.g., Wyllie, 1984) and also with the general requirement of trace-element studies that most granitic magmas have been equilibrated at depth with great volumes of rock rich in pyroxenes, hornblende, and calcic plagioclase (Tarney and Saunders, 1979). The required mafic and calcic differentiates or restites cannot generally lie in the middle crust, which is widely exposed, and presumably are in anorthositic and gabbroic complexes of the lower crust and in largely ultramafic differentiates of the upper mantle.

FORELAND THRUST BELT

Deformation and magmatism are related spatially and temporally, so structural and magmatic aspects of evolution of continental crust must be considered together. Until only 15 years or so ago, compression was widely viewed as a necessary precursor of most magmatism and metamorphism; and now it is widely believed that overriding plates are necessarily shortened. I disagree, and believe further that crustal shortening is not required even by the Sevier foreland thrust belt that formed to the east of the belt of Cretaceous batholiths.

Great shortening is recorded, of course, *within* the upper-crustal wedge represented by the broad fold-and-thrust belt. The belt was produced by the tectonic thickening, by imbrication relatively eastward toward the craton, of a preexisting, westward-thickening wedge of miogeoclinal and other strata. Relationships are best defined in southern Alberta, in the eastern 100 km of the belt, where there has been the least disruption by subsequent magmatism and extensional deformation and where geophysical and drilling data add critical constraints (Price, 1981).

The lithostatic head produced by tectonic thickening in the west caused the deforming wedge to flow gravitationally eastward, incorporating new foreland strata as it did so. Thrust sheets in the deep part of the wedge moved upward to the east because the top of the wedge sloped downward in that direction; uninvolved basement dips westward beneath the wedge. Rocks formed progressively deeper in the tectonic wedge are exposed progressively westward. Deformation styles of rocks exposed at the surface change correspondingly from imbricate thrusting plus flexural-slip folding in the east to pervasive synmetamorphic slip in the west. The latter deformation produced typically slate and phyllite of low greenschist facies, recording temperatures of 300–400°C and hence a likely depth in the wedge of the time of about 15 km.

The east front of the thrust belt trends irregularly southeastward and southward through Montana, Wyoming and eastern Idaho, and northern Utah, thence southwestward through southern Utah and southern Nevada into eastern California. The front was controlled throughout the east limit of thick, shearable strata, primarily miogeoclinal but including strata of the Cretaceous foreland basin and, in Montana, of the Proterozoic Belt basin. The thrust belt ends in California, where the stratal wedge trends obliquely into the batholithic belt.

The deep structure of the western part of the Sevier belt is exposed in the many "metamorphic core complexes" raised by extensional tectonic denudation during Eocene time in British Columbia and the northwestern States and during middle and late Tertiary time farther south. (These are discussed in a later section). Interpretations of these tracts are much debated, and they are complicated by deformation and magmatism of both pre- and post-thrusting ages; I see the thrust-belt component of the exposed rocks as representing synmetamorphic top-to-the-east slip, broadly involving basement rocks, at depths as great as 28 km, followed by severe extensional faulting.

I see no requirement that any part of the deformed wedge moved more than 100 km relatively eastward above the deep basement. By contrast, Price (1981) argued for more than 200 km of overthrusting on the basis of his quite different inferences regarding the amount and structural manifestations of Tertiary extension.

The thrust belt developed during the late Early Cretaceous (Aptian and Albian), Late Cretaceous, and early Paleogene (Heller *et al.*, 1986), and thus imbrication began soom after inauguration of major Cretaceous arc magmatism to the west, and continued 25 m.y. or so after magmatism had migrated relatively eastward away from the main Cretaceous belt in the southwestern United States. (It should be noted that dating of Cretaceous batholithic rocks is most complete in California, whereas dating of the thrust belt is most complete from northern Utah through southern Alberta.) At the time of completion of thrusting, the east edge of the thrust belt probably lay only some 150 km east of the composite Cretaceous batholiths, the distance having been much increased by subsequent Cenozoic extension (Hamilton, 1978), and compressive Cretaceous structures were continuous from batholiths to eastern thrust faults. Thrusting and batholiths were coupled mechanically: the buttress against which the thrust wedge was thickened lay in the region of the batholiths. Hamilton (1978) and Smith (1981) are among the few who have argued that arc-magmatic growth of the crust produced the topographic high that drove the thrusting. The dominant current view is probably that expressed by Price (1981): eastern cratonic basement was subducted relatively westward beneath the thrust belt and cycled at depth into the main east-dipping subduction system.

The ratio between extension across the arc and the growth of the crust by the mantle component of arc magmatism will determine the altitude of the top of the magmatic arc, and hence the gravitational potential for driving foreland thrusting. Foreland thrusting is now active along the east foot of that part of the Andes which has crestal altitudes generally above 5 km; perhaps thrusting in the Cretaceous foreland belt of western North America was a result of similar altitudes of the associated magmatic arc. No thrusting accompanied formation of the middle Tertiary arc of the Sierra Madre Occidental, the altitude of which (now mostly less than 3 km) may have been below the threshold needed for major gravitational spreading.

LATEST CRETACEOUS AND EARLY PALEOGENE TECTONICS

Many unusual tectonic and petrologic features of the western United States date from latest Cretaceous and earliest Paleogene time and appear to be related products of the rapid overriding of oceanic lithosphere. North American lithosphere dragged on subducting materials and was truncated by erosion against them, and this drag produced shortening of the cratonic continental interior.

Subcrustal Erosion

When North America rapidly overrode subducting Pacific lithosphere in latest Cretaceous and early Paleogene time, the continental plate was eroded tectonically from beneath and oceanic rocks were plated against its truncated base, which is exposed in uplifts, in interior southern California, that include the Tehachapi Mountains (the southernmost Sierra Nevada), the Rand Mountains in the northwest Mojave Desert, the central Transverse Ranges, and the Orocopia and Chocolate Mountains in southeastern California. The subjacent oceanic rocks — the Rand, Pelona, and Orocopia Schists — are mostly metamorphosed terrigenous quartzofeldspathic and pelitic strata but include much oceanic-crustal metabasalt and minor associated (pelagic?) metachert and ferromanganiferous metalimestone; serpentinite is present locally (Jacobson *et al.*, Chapter 35, this volume). The rocks are extremely transposed and have been metamorphosed at greenschist and lower amphibolite facies, and locally at transitional blueschist-greenschist facies (Ehlig, 1981; Haxel and Dillon, 1978; Jacobson *et al.*, Chapter 35, this volume; Postlethwaite and Jacobson, in press; Ross, in press; Sharry, 1981). Coherent high-amplitude reflections at a depth of 5-10 km beneath the thrust bounding the truncated continental crust in southeastern California may image the top of subducted oceanic lithosphere beneath the schist (Morris *et al.*, 1986).

The contact system least disturbed by subsequent deformation is probably that of the San Gabriel Mountains in the central Transverse Ranges (Ehlig, 1981; Jacobson *et al.*, Chapter 35, this volume). Mesozoic and Proterozoic middle- and lower-crust migmatites, granites, and granulites are truncated downward against Pelona Schist and are retrograded to mylonitic gneiss in a section hundreds of meters thick concordant to the fault. This metamorphism is isofacial with the metamorphism in the subjacent schist and records a temperature in excess of $500°C$, but temperature of metamorphism in the schist decreased downward (Jacobson *et al.*, Chapter 35, this volume). Integration of K-Ar and fission-track dating indicates that the lower part of the upper plate cooled from $>500°C$ to $<100°$ during the interval between about 70 and 57 m.y. (Mahaffie and Dokka, 1986): the truncation of the base of the continent occurred in latest Cretaceous or early Paleocene time. The mylonitic fabric records a relatively northwestward overriding of schist by gneiss (Carol Simpson, written communication, 1986); as the range has undergone net Neogene clockwise rotation (Terres and Luyendyk, 1985), the initial direction of relative overthrusting was westward.

As the surface of the upper-plate Cretaceous magmatic arc was 100 km or so above the slab being subducted at the time of its formation, the juxtaposition of rocks formed about 30 km deep in that arc against subducted oceanic rocks requires the tectonic erosion of the lower 70 km of continental lithosphere. The age of this erosion is bracketed between the early Late Cretaceous age of truncated granites and the Eocene age of marine strata that lie upon deeply eroded rocks. The upward increase in temperature of syntectonic metamorphism recorded by the subjacent schists (Jacobson *et al.*, Chapter 35, this volume) presumably represents heating of cool subducted rocks by heat retained in the rapidly truncated continental crust.

The thrust system both in the Rand Mountains and in southeastern California has been much disrupted by Tertiary extensional faulting. Mylonitic fabrics where studied in the latter region record overthrusting of gneiss and granulite relatively northeastward over Orocopia Schist (Haxel and Dillon, 1978). This has led to the inference that the sedimentary protoliths of the Orocopia Schist were deposited in a marginal basin and were overridden from the west by the plutonic crust of a minicontinent, which is bounded by a yet-unrecognized suture against autochthonous North American crust. Further study is essential to determine whether or not this east-directed fabric formed synchronously with the downward truncation of the overriding continental crust or represents instead later deformation, as by middle Tertiary midcrustal extensional deformation (see the discussion by Jacobson *et al.*, Chapter 35, this volume). As the southern Sierra Nevada is certainly part of both the continent and the overriding plate and as overriding by the San Gabriel Mountains (which share distinctive early Mesozoic granitic rock types with undoubted continental regions to the north) was relatively westward, I presently infer that the subduction recorded by the Rand, Pelonia, and Orocopia Schists was relatively eastward beneath the continent. (See note added in proof, p. 40.)

The truncated base of the continental plate may be exposed also in the southern Coast Ranges of California, where middle Cretaceous "Salinian" plutonic rocks lie structurally above coeval Franciscan accretionary-wedge melange on the Sur-Nacimiento megathrust. The plutonic rocks are equivalent to those of the medial part of the Sierra Nevada and Peninsular batholiths, as noted previously. The terrains equivalent to the western parts of those batholiths, to the pre-Cretaceous accreted assemblages next farther west, and to the fore-arc basin all have been removed tectonically, as has the lower 70 km of continental lithosphere that underlay the remaining plutonic rocks in middle Late Cretaceous time (*cf.* Fig. 1-1). This Coast Ranges complex lay west of the Rand-Pelona-Orocopia windows through the continental lithosphere before Neogene slip on the San Andreas fault. Page (1982) argued for subduction erosion of the missing Coast Ranges rocks, and I agree; he dated the event as probably Paleocene. Dickinson (1983) argued that, on the contrary, the missing terrains were moved by left-slip faulting.

Water Release

Many hydrous-magma muscovite and two-mica granites in the southwestern United States formed during latest Cretaceous and early Paleogene time, and hence during the general period in which oceanic lithosphere was being rapidly overridden by the continent, in the Basin and Range region. These were discussed earlier in terms of their middle-crust depths of origin. T. D. Hoisch[1] (Hoisch *et al.*, Chapter 21, this volume) has found several of these wet-granite terrains in southeastern California to be associated with latest Cretaceous or early Paleocene regional metamorphism characterized by the passage of enormous volumes of high-temperature water. This is indicated by the reaction of calcite plus quartz to wollastonite through voluminous terrains metamorphosed at middle-crust pressures. Hoisch suggested that much of the water recorded by both the granites and the metamorphism was liberated by dehydration metamorphism of sedimentary rocks subducted at shallow depth beneath the tectonically thinned continental lithosphere and that the water carried with it the heat needed for melts and metamorphism.

If voluminous sedimentary rocks were indeed subducted at shallow depth beneath the continent and there dehydrated, magma genesis might have been quite different from the usual arc-magmatic sort, in which it appears that deep-crustal magmas contain little water. Much of the eastward sweep of magmatism in late Late Cretaceous time, away from the earlier Cretaceous site of the magmatic arc in the Sierra Nevada and kindred batholiths, may not have been simply tracking the decreasing depth to the subducting slab with time, but may have recorded qualitatively different processes.

Peninsular Ranges Mylonites

The Cretaceous mid-crustal plutonic rocks of the east side of the Peninsular batholith of south-central California are cut by thick zones of greenschist- to amphibolite-facies

[1] T. D. Hoisch, written communication, 1986.

mylonite and mylonitic gneiss that dip gently to moderately eastward and outline large lenses of less-deformed rocks (Anderson, 1983; Sharp, 1979; C. Simpson, 1984, 1985). The deformed plutonic rocks are probably of middle Late Cretaceous age (*cf.* Silver *et al.*, 1979), and mylonitization was completed before 60 m.y. (references in C. Simpson, 1985). Sense of slip was of westward overthrusting (C. Simpson, 1984).

The deformation presumably belongs to the family of structures which record drag of the continental lithosphere on oceanic materials subjucted at a gentle angle beneath it. I infer the discontinuously ductile style of deformation — of lenses of little deformed rock separated by thick zones of ductile shear — to be common in the middle crust under both extensional and shear regimes, as emphasized subsequently.

Laramide Deformation

The Rocky Mountain region of the continental craton, from southern New Mexico to northern Montana, was distorted by shortening during the Laramide event of latest Cretaceous, Paleocene, and Eocene time, apparently as another manifestation of drag of the fast-moving continent on subducted materials beneath. The typical products of Laramide deformation are large, asymmetric anticlines of Precambrian rocks, bounded on one or both sides by reverse faults or steep monoclines against deep basins that subsided and received sediments concurrent with rise of uplift (Fig. 1-5). Major uplifts and basins have lengths to 300 km and structural reliefs exceeding 10 km. Although minor shortening affected the Colorado Plateau and Great Plains, the major structures lie in a belt trending northward through New Mexico to Colorado, where they splay out in successive arcs to the west, west-northwest, and northwest. Surface geology, industry drilling, and industry and COCORP reflection profiling all prove great basement shortening.

Deep reflection profiling (Sharry *et al.*, 1985; Smithson *et al.*, 1979) shows the basement thrusts to flatten downward and possibly to splay into anastomosing ductile shear zones that break the deeper part of the middle crust into lenses. Analogous and contemporaneous shear-bounded lenses exposed in southern California were discussed in the preceding section. Other analogs are exposed in the Proterozoic Grenville province in Ontario, where the crust has been thickened by northwest-verging imbrication of great lenses in deep-crustal rocks between thick mylonites of amphibolite and middle-pressure granulite facies (Davidson, 1984). Basement thrusts perhaps generally give way down-dip to subhorizontal complexes of lenses separated by thick ductile shear zones. The means by which such shortening is transmitted through still deeper crust and mantle is not yet clear, although pervasive deformation can be inferred on rock-mechanic grounds. Arc-magmatic rocks formed in parts of the Laramide province during the general period of shortening, so the lithosphere there presumably was 100 km or so thick at the time of shortening.

The Colorado Plateau and the continental interior each behaved as a relatively undeformed plate during this episode of crustal shortening; thus it must be possible to express the shortening distributed between them in terms of a rotation of the plateau relative to the interior. There is little deformation of any sort within the Laramide belt in central New Mexico; substantial right-lateral strike-slip plus minor shortening in northern New Mexico; and shortening across the belt that increases systematically northward and northwestward from there to a maximum in northeast Utah and Wyoming. I (Hamilton, in press *c*; *cf.* 1981b) deduced from this geometry of the compressive structures that the Laramide rotation amounted to about 4° clockwise relative to an Euler pole in or near central New Mexico (Fig. 1-6). This specific amount of rotation incorporates the assumption that the shortening is equal to half the width of exposed basement uplifts along small circles to the Euler pole, in reasonable accord with the best-constrained surface and sub-

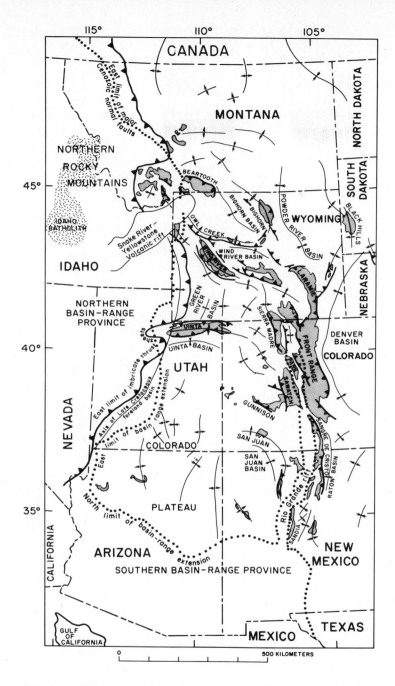

FIG. 1-5. Selected structural elements of the Rocky Mountain region. Most of the uplifts and basins indicated are of Laramide age, latest Cretaceous and early Paleogene. Outcrops of Precambrian rocks east of the foreland fold-and-thrust belt and north of the southern Basin and Range province are shaded.

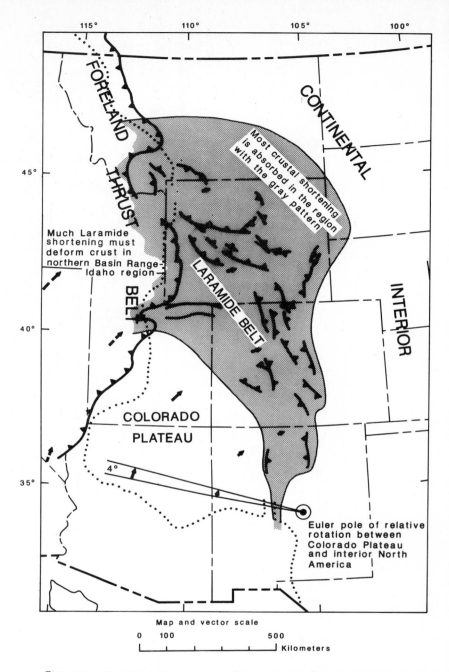

FIG. 1-6. Map illustrating rotation of the Colorado Plateau relative to interior North America, as the cause of the Laramide deformation of the Rocky Mountain region. The vectors depict, to scale, the relative rotation of the plateau by 4° as though about the inferred Euler pole in New Mexico. The motion was absorbed by crustal shortening in the Laramide belt. The southern boundary of the plateau plate is not defined here.

surface data. Bird (1984) showed that such rotation accords with sublithosphere drag patterns predicted from analysis of the geometry of the subducted plate.

A subsequent clockwise rotation of about 3°, relative to an Euler pole in central Colorado, is required by the middle Tertiary opening of the Rio Grande rift. The total of Laramide plus Rio Grande rotations is equivalent to a rotation of about 6° relative to an Euler pole in southern Colorado.

Steiner (1986) analyzed the available high-quality paleomagnetic data for correlative strata of the Colorado Plateau and the continental interior and concluded that the Late Cretaceous and Cenozoic relative rotation of the plateau had amounted to about 11°. Bryan and Gordon (1986) arued for only about 4° of relative rotation, but they incorporated statistically poor paleomagnetic data, poorly constrained ages of some units, and data sets from rocks of very different ages on and off the plateau.

CENOZOIC CRUSTAL EXTENSION

I infer both from analysis of the extensional structures themselves and from attempts to reconstruct palinspastically the Mesozoic components of the Cordillera (Hamilton, 1978) that Cenozoic extension has approximately doubled the width of the Basin and Range province and related terrains in the western United States. The structures on which this deformation was accomplished in the upper and middle crust are widely exposed.

Basin and Range Province

Subduction steepened after the westward advance of North America over Pacific lithosphere slowed in Eocene time. It is unclear whether the old, gently inclined slab sank, or was left at shallow depth and underrun by a new slab. From late Eocene through early Miocene time, a wave of intermediate and, mostly, silicic magmatism broadly affected, and was largely limited to, the region which was extended in middle and late Cenozoic time. The thermal softening of the crust that accompanied this magmatism presumably made possible the uncommon breadth, amount, and duration of accompanying and subsequent extension. The continental margin evolved during the period of extension from an early one of continuous subduction of Pacific lithosphere to one of strike-slip in the south and subduction along a progressively shortening sector in the north (Atwater, 1970; Dickinson, 1979). Early extension was approximately perpendicular to the continental margin and was synchronous with subduction along most of it, hence was back-arc spreading — severe extension of an overriding plate. Late extension has been mostly in a northwesterly direction relative to the continental interior, approximately parallel to the San Andreas plate-boundary direction but much complicated by adjustments of the shape of the continental plate to fit that boundary.

Middle Cenozoic spreading affected much of the Great Basin; the Basin and Range southern parts of California, Arizona, and New Mexico; the Rio Grande Rift system; and regions north of the eastern Snake River Plain in Idaho, western Montana, and southeasternmost British Columbia. (Eocene extension of south-central British Columbia, northeastern Washington, and Idaho is discussed subsequently.) Late Cenozoic extension of Basin and Range type has affected most strongly the Great Basin, though with much areal variation; the southern margin of the Colorado Plateau; the region around the eastern Snake River Plain; and eastern Oregon and adjacent areas. The late Cenozoic extension was in part superimposed on earlier extension but also affected new regions, and at any one time extension was rapid in some areas and slow or dormant elsewhere. Exposed structures of comparable crustal depths of formation are similar regardless of age and orientation. Regional altitude has increased during the long period of extension.

I have discussed these matters in other papers (Hamilton, in press *a, b*) from which this summary is adapted.

The structural style of deformation varies as functions of crustal depth and of amount of local extension. Upper-crust structures begin as rotating blocks and panels separated by brittle normal faults and, more locally, strike-slip faults. These shallow structures initially end downward at undulating zones of ductile shear. Slip on the undulating zones continues, but in increasingly brittle mode, as extension proceeds and tectonic denudation brings the evolving *detachment faults* progressively closer to the surface. The temporal and depth progression is shown by superimposed structures in zones of long-continued slip. Upon the products of early ductile slip — mylonite, formed at temperature >300°C, or even mylonitic gneiss, >500° — are superimposed progressively narrower zones of progressively more brittle fabrics, ending as low-temperature brittle gouge and breccia. Thermal-gradient considerations indicate that such ductile faults commonly began to form at depths greater than 10 km for mylonites and 15 km for gneisses, even in high-heat-flow regimes (Figs. 1-2, 1-3, and 1-7), and this is compatible with the general presence beneath detachment faults of crystalline rocks equilibrated at middle-crust conditions before such late deformation.

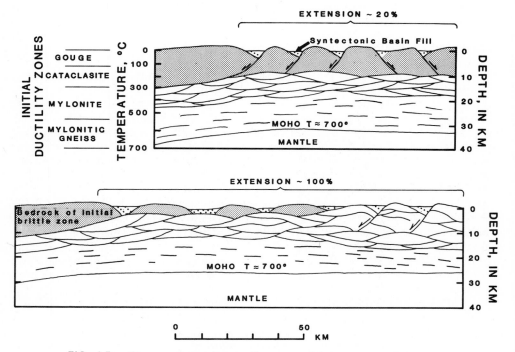

FIG. 1-7.   Cross sections of extending crust. Brittle upper-crust blocks rotate and separate. Middle-crust lenses slide apart along ductile shear zones; composite upper surface of lenses forms detachment faults that increase in total area with time. Lower crust flattens pervasively. Structural styles are superimposed as components rise toward surface with continuing attenuation. Crust is partly rebuilt by magmatism, so crust is thinned by a factor less than the extension ratio. Ductility zones after Sibson (1983). (See note added in proof, p. 40.)

TECTONIC SETTING AND VARIATIONS

Detachment faults that evolved with time from middle-crustal to near-surface depths are exposed in scores of ranges in the southern part of the Basin and Range province and in lesser numbers of them in the Great Basin. Upper plates are mostly of upper-crust rocks rotated to moderate to steep dips and truncated sharply at their bases, and abundantly include sedimentary and volcanic rocks deposited in basins synchronously with extension. Lower plates consist of crystalline rocks, of any age, that were formed in the middle crust. A crustal thickness of 10–15 km is missing across many detachment faults, and 20 km across a few of them. Detachment faults are connected to the surface by normal faults. Whether detachment faults at the edges of extended terrains end as shallow normal faults or emerge from beneath nonextended terrains is not yet established. The depth to active detachment faults varies with extensional maturity in the Great Basin, and ranges from 15 km in central Utah to 0–4 km in the Death Valley region.

Detachment faults crop out with undulating or domiform shapes. Slip on detachments commonly is in the same sense over several adjacent domes, up one side and down the other of each, although very late, near-surface slip is down-dip only. The common term "metamorphic core complexes" for subdetachment assemblages is often used with the incorrect connotation that detachment faults are related to local hot spots; any middle-crust rocks, including low-temperature slates, occur beneath the faults.

In the region of most severe Quaternary extension, that of Death Valley in eastern California, some gently dipping structures that originated in the middle crust have remained active after exposure at the surface and display even Holocene slip (Fig. 1-8), whereas others have been broken by steep faults, related presumably to new detachment faults at depth.

FIG. 1-8.   Truncation of normal fault against active detachment fault, Black Mountains, Death Valley, California. Interbedded Pliocene clastic strata, slide breccias, and volcanic rocks (lower left) dip obliquely back and right toward late Quaternary normal-fault scarp along high-standing plutonic rocks (left and center rear). Hanging wall, footwall, and fault are all truncated downward against the active Copper Canyon turtleback (detachment) fault, which forms the smooth face dipping obliquely left through the center of the view. Light-colored rocks on turtleback are a discontinuous carapace of mylonitized dolomite left by the Panamint Mountains block as it slid by early in the detachment history. An active range-front normal fault truncates both upper and lower plates at the top of the alluvial fan.

Half of the upper-plate rocks exposed against the late Cenozoic detachment faults in the ranges near Death Valley are synextensional sedimentary and volcanic strata, rotated down to marked truncations against the faults. Half of the upper-plate materials did not exist when extension began. The pre-Cenozoic rocks do not form mere rotated panels, one in contact with the next at the subsurface part of a range-front fault, but rather are completely allochthonous above the detachments. I regard the structural-geologic evidence as strong that central Death Valley and northern Panamint Valley record the Pliocene and Quaternary separations of the flanking bedrock ranges by the full widths of the basins, above detachment faults at two levels. The older and higher detachment faults in each case are exposed widely in the ranges flanking the valleys, whereas the younger, lower, and presently active major detachments are hidden and emerge in the subsurface from beneath the ranges on the east sides of each basin and floor the modern basin fills at depths of only a few kilometers.

De Voogd *et al.* (1986) interpreted COCORP seismic-reflection profiles to indicate that active imbricate normal faults dip moderately downward to a subhorizontal detachment fault deep within crystalline rocks 15 km beneath Death Valley. I have studied these profiles and also relevant shot-point gathers and the processing employed. Severe frequency-wave number (*f-k*) filtration applied before stacking produced in each of the COCORP profile displays a variable pervasive artifact fabric, inclined about 30° toward the vibrators along the one-sided geophone spread (regardless of the orientation of the line, and whether the spread was "pushed" or "pulled") and extending between about 2 and 5 sec on the displays; further, severe editing out of traces was done before stacking. The imbricate faults inferred by de Voogd *et al.* (1986) are, in my view, primarily artifacts introduced by this processing. The great inclined possible fault given major emphasis by de Voogd *et al.* (1986, Fig. 2, line 11, reflector C) represents this artifact in its shallow part, 2–4 sec. Although the part of this apparent reflector between 4–5 sec appears to be real on a shot-point gather, the lack of a pullup effect for an abrupt change of 1.1 sec of thickness of surficial basin sediments "above" it suggests to me that this part of the apparent reflector is a side echo, not a deep fault. Similarly, the lack of pullup variations for the "detachment fault" (including the "bright spot") near 6.5 sec on line 11 of de Voogd *et al.* (1986), and the lack of confirmation of the reflection on nearby and partly coincident line 9, permits the interpretation that this apparent structure also is a side echo.

The ductile shear zones that evolve into detachment faults from carapaces on crystalline rocks which otherwise largely retain their prefaulting fabric and mineralogy. Where carbonate rocks were present during the ductile stage in either upper or lower plates, they commonly were smeared out as lubricating veneers of foliated marble, 3–50 m thick, that now define the carapaces.

Crustal structure beneath detachment faults is characterized by ductile shear zones that outline large and small lenses that retain preextension fabrics. Anastomosing ductile shear zones have been recognized in the field, and can be inferred on many reflection profiles (e.g., Frost and Okaya, 1986; Hamilton, 1982), beneath detachment faults. Although the manner in which extension is transmitted through the middle crust, beneath the detachment faults that represent the base of early, brittle deformation, is much disputed, I infer that the middle crust is extended by the sliding apart of lenses bounded by gently dipping zones of ductile shear (Fig. 1-7). This can be likened to the sliding apart of a pile of wet fish — halibut, to make the analogy dimensionally apt. The brittle faulting of the upper crust is a response to the resulting increase in area of the composite top of the lenses separating at depth. I see the common domiform shapes of exposed detachment faults as due to the interaction of separating lenses; many investigators see them instead as products of postdetachment deformation.

The lower crust in the Basin and Range province is required by heat-flow data to be so hot that deep deformation must be accommodated by pervasive ductile flow (*cf.* Figs. 1-2 and 1-3). Extensional fault ramps cutting completely through the crust have been postulated by some investigators but could form only with extremely high strain rates; the subhorizontal, unbroken character of the Basin and Range Moho disproves the ramp concept and is discussed in a later section. Lower-crust extensional structures are not exposed within the region but can be inferred from the fabric displayed by reflection profiles to record pervasive ductile flattening. The style of deformation thus changes downward in the crust from brittle to discontinuously ductile to pervasively ductile.

The structures widely regarded as typical of the Basin and Range province — tilted panels of bedrock rotated against one another at range-front faults, the down-dropped part of each panel being covered by basin deposits — are present only where extension has been minor, hence commonly early in the extensional history of any one area. (Symmetrical horsts and grabens are uncommon at any stage.) Wherever detachment faults are exposed, it can be seen that basin fills have dragged directly against them, and that pre-extensional bedrock occurs as tilted blocks that are separated completely from one another atop the detachment-fault lenses. Extension has in many places corresponded to most of the distance between the bedrock of adjacent ranges. Strike-slip faults are limited to assemblages above detachment faults, and range-sized blocks formed by early extension perpendicular to the continental margin are now being further separated by oblique, northwestward slip closer to the San Andreas direction. (See note added in proof, p. 40.)

Paleomagnetic evidence for motion of the Sierra Nevada My tectonic analysis requires that Basin-Range extension have been accompanied by a counterclockwise rotation of the Sierra Nevada by 20 or 30° relative to the Colorado Plateau, and by little change in relative latitude, during middle and late Cenozoic time. Frei (1986) argued that, on the contrary, the paleomagnetism of Cretaceous granites of the central Sierra Nevada indicates that the Sierra has not rotated relative to the Colorado Plateau, although it has shifted northward about 600 km. Her data from the granites are suspect because neither the age of magnetization nor the lack of tilting can be proved. I earlier argued that the batholith has been eroded 5 km or so deeper in the west than in the east, and this, if applicable to the tracts studied by Frei, would calculate out to a postmagnetization counterclockwise rotation of the Sierra Nevada by about 5° from her data. Frei also overlooked the latest Cretaceous and Cenozoic rotation of the Colorado Plateau, discussed previously, by approximately 8° clockwise relative to the continental interior; this is equivalent to a counterclockwise rotation of the Sierra Nevada by 8° relative to the plateau. Further, the Cretaceous paleomagnetic reference pole Frei used for North America east of the Rocky Mountains is not based on reliable data. Her reference pole comes from three studies — mid Cretaceous diabase sheets of Ellef Ringnes Island, Arctic Canada; Upper Cretaceous strata of the Niobrara Formation in eastern Colorado and Kansas; and mid Cretaceous dikes in Arkansas. Map patterns, the attitudes of nearby rocks, and the magnetic data themselves all indicate to me that the Ellef Ringnes diabases sampled are in gently plunging open folds, although they were assumed to be horizontal (wall rocks and contacts of diabases are not exposed at the sample sites); moreover, Ellef Ringnes is separated from mainland Canada by the Northwest Passage plate boundary of poorly constrained character. The Niobrara and Arkansas studies yielded results so scattered as to be of minimal value and were included in the reference pole only because their mean paleomagnetic poles are close to that of the Ellef Ringnes study; the lack of a single reversely magnetized specimen among more than 300 Niobrara samples that spanned most of Late Cretaceous time makes the Niobrara study further suspect. Frei's reference pole requires a Cretaceous polar-wander curve for which there is no explanation in known plate motions.

Eocene Extension in the Northwest

Severe extension affected central and northern Idaho, northeastern Washington, and south-central British Columbia during Eocene time. Widely exposed detachment faults outline domiform masses of mid-crustal crystalline rocks that include the Okanogan, Kettle, and Priest River-Spokane domes in northeastern Washington and adjacent Idaho, and the many more of the Shuswap and allied families in British Columbia. The style of exposed structures is similar to that of the Basin and Range province, although deeper erosion has exposed subdetachment rocks and faults much more widely in the Eocene terrain. A subsurface relief of about 10 km on the composite detachment faults is indicated by gravity surveys (Cady and Fox, 1984). Rocks rotated down to truncations against domiform and undulating detachment faults include synextensional lower and middle Eocene sedimentary, volcanic, and granitic rocks, and sedimentary, metamorphic, and granitic rocks variously of Proterozoic, Paleozoic, and Mesozoic ages, whereas subdetachment rocks similarly range from Proterozoic to Eocene in age but are entirely of mid-crustal crystalline types (Brown and Read, 1983; Cheney, 1980; F. K. Miller, 1971; F. K. Miller and Engels, 1975; Rhodes, 1985; Rhodes and Cheney, 1981). Lower-plate rocks commonly have Eocene K-Ar cooling (uplift) ages regardless of their primary ages, whereas K-Ar ages of upper-plate pre-Tertiary rocks are erratically older (Mathews, 1981; F. K. Miller and Engels, 1975). The general direction of extension, as indicated by stretching lineations in mylonites and by axes of rotation of upper-plate Eocene strata, was westward to west-northwestward relative to the continental interior. The last slip on the detachment faults was brittle, but commonly lower plates carry carapaces recording shear that progressed with time from ductile to brittle, from early mylonitic gneiss to late chloritic microbreccia, and early Paleogene granites are among the mylonitized rocks (Carr, 1985; Lane, 1984; F. K. Miller, 1971).

Subdetachment ductile shear zones, with or without superimposed brittle structures, outline many lenses of mid-crustal rocks beneath the capping detachment faults (e.g., Spokane Dome mylonitic zone of Rhodes and Hyndman, 1984; and most faults shown by Journeay and Brown, 1986). The deep-reflection profile presented by Cook (1986) carries from near-outcrop to depth an exposed detachment fault (Standfast Creek fault) and an exposed fault (Columbia River fault) that bounds the next-lower structural lens and that flattens downward. Beneath these faults, as I read Cook's profile, is a stack, as thick as 15 km, of large lenses outlined by rocks transposed in ductile shear zones; and beneath those, the basal continental crust displays subhorizontal reflections that I infer to reflect rocks more pervasively transposed by ductile flattening (*cf.* Fig. 1-7). Similarly, I see lenses in the domiform and gently dipping reflectors of the COCORP profiles across the extended terrain of northern Washington (Potter *et al.*, in press; they assumed all west-dipping reflectors to be Mesozoic thrust faults, and outcropping west-dipping detachment faults to be either thrusts or minor normal faults).

Published interpretations of these complexes vary widely. Many geologists (Brown and Read, 1983; Cook, 1986; Journeay and Brown, 1986; Monger *et al.*, 1985) assign the ductile carapaces and deeper ductile faults largely to pre-Eocene overthrusting and infer relatively minor Eocene extension. The inference by Price (1981) of more than 200 km of foreland-thrustbelt shortening, referred to earlier, incorporates such an interpretation. The collective features of the Eocene terrain, as mapped and described by many geologists, so strikingly resemble those of the Basin and Range region, however, that I interpret them in similar terms of rise of mid-crust lenses to the surface as the result of great tectonic denudation accompanying extension, and of slip between the rising lenses that evolves from ductile at depth to brittle near the surface.

Other Eocene detachment faults atop mylonitic carapaces on mid-crustal rocks are exposed in the Idaho batholith region. Best known is that of the east flank, exposed along the east side of the Bitterroot Mountains, of a domiform carapace of mylonitic gneisses that affects middle-crust granites and migmatites of Eocene and Cretaceous ages (R. B. Chase *et al.*, 1983; Hyndman, 1980). I have seen a spectacularly exposed, west-dipping detachment fault farther south, in the canyon of the Middle Fork of the Salmon River at Cradle Creek, where upper-plate gneiss dips steeply east to a truncation against a west-dipping detachment fault capping a mylonitic carapace. I expect that detailed study in central Idaho — which mostly is both inaccessible and poorly exposed — will show widespread detachment faulting. Undulating or domiform contacts of epizonal Eocene granites above deep-seated plutonic rocks should be accorded particular scrutiny (e.g., in the canyon of the Middle Fork in the Camas Creek area; see Cater *et al.*, 1973). Farther south in central Idaho is the domiform Pioneer detachment fault (Wust, 1986).

Rifting and Magmatism

Discriminating cause and effect in rifting and magmatism is difficult. As rifting commonly is much too rapid to permit continuous equilibration by thermal conduction, geothermal gradients are steepened; the lithosphere is thinned both by extension and by incipient melting, due to depressurizing, of its lower part; induced mantle convection (Steckler, 1985) likely moves additional mantle heat upward. The widespread synextensional magmatism and high heat flow of the province both appear to be products, not causes, of the extension (Lachenbruch and Sass, 1978). Nevertheless, the initial middle Tertiary spreading of the Basin and Range province was in a back-arc mode and was accompanied by voluminous magmatism intermediate in character between the types commonly regarded as representing extensional and arc-magmatic settings (*cf.* Best, 1986), and the inauguration of the broad region of spreading may have been due to arc-magmatic heating of the lithosphere.

The late Neogene magmatism of the Basin and Range province has by contrast been largely bimodal, although the ratio of basalt to rhyolite has varied widely both areally and temporally. Silicic magmatism dominates early stages of rapid rifting of the continental in nonarc settings, as, for active examples, the Yellowstone–Snake River province of southern Idaho and northwest Wyoming, and the Owens Valley province of eastern California. A general model (Christiansen and McKee, 1978) for such magmatism, integrating isotopic and other data from the rock assemblages, is that basalt melted in the mantle, because of depressurizing by rapid extension, rises into the lower crust. The heat introduced by the mafic magmas partially melts lower-crustal rocks, and the resulting silicic magmas rise toward the surface. "Hot spots" are invoked for some such provinces, but in my view the concept they express is generally invalid, the melting being a product, not a cause, of extension.

New Volcanic Crust

Much of the interior Northwest exposes only Cenozoic basaltic and mafic-intermediate volcanic rocks and associated sedimentary rocks: northeastern California, the western Snake River Plain, all of Oregon except for the Klamath and Blue Mountains, most of southern and coastal Washington. I have argued (Hamilton, 1969, 1978; Hamilton and Myers, 1966) that these volcanic terrains represent Cenozoic magmatic and tectonic additions to the continental crust, and that pre-Cenozoic continental crust is lacking beneath much of them and is severely attenuated beneath the rest. Palinspastic reconstructions of Cretaceous tectonic and magmatic provinces in my view require such additions of new crust. Cenozoic rotations of western terrains shown by paleomagnetic data (e.g., Beck, 1980; Magill *et al.*, 1982; R. E. Wells, 1985) broadly fit the predicted motions. The volcanic rocks are variously of magmatic-arc, rift, and mixed types, but are petrologically primitive and show little incorporation of continental crust or of sediments derived from it (McKee *et al.*, 1983; C. M. White and McBirney, 1978).

THE MANTLE AND THE MOHO

Mantle Rocks

Upper mantle material, mostly of Proterozoic age, sampled as inclusions in alkali basalts and kimberlites of the western United States (Wilshire *et al.*, in press) shows the general variety typical of such occurrences elsewhere. Peridotite is the dominant rock type and consists of olivine, subordinate orthopyroxene, and minor clinopyroxene. Spinel is important particularly in uppermost-mantle lherzolite (two-pyroxene peridotite), and garnet is important in rocks from deeper in the mantle; plagioclase occurs in lherzolite recording pressures appropriate for lower continental crust rather than upper mantle (*cf.* Fig. 1-2). Garnetiferous granulites may represent either lower continental crust or upper mantle. All of the relevant rock types are also known within lowest crustal and uppermost mantle rocks exposed *in situ* in some parts of the world, although not in the United States. Metamorphic fabrics typify the mantle samples, but the primary assemblages mostly formed in equilibrium with mafic melts or were traversed by such melts (Wilshire *et al.*, in press). Such mantle rocks must include residues after partial melting, precipitates of refractory phases from melts, and products of zone refining, in widely varied sequences, and are quite different from the severely depleted harzburgites and dunites of oceanic upper mantle as exposed in ophiolite sheets.

Mohorovicic Discontinuity

As arc magmas must originate as mafic basalts yet have evolved to mafic and intermediate compositions by the time they reach the lower continental crust, the bulk of the olivine and pyroxene components of the protomagmas must crystallize in the mantle. The continental Mohorovicic discontinuity represents the shallow limit of crystallization of voluminous rocks dominated by olivine, pyroxenes, and garnet as well as the common deep limit of crystallization of silicic and intermediate rocks. In a region of active magmatism or tectonism, the Moho is a dynamic, evolving boundary, not a passive, fossil contact.

Magmatic Basin and Range Moho Middle and late Cenozoic extension has approximately doubled the width of the Great Basin, yet surface altitudes are now higher than they were early in the extension period (e.g., Axelrod, 1985), and the Moho is subhorizontal at a depth that is semiconstant in reflection time, 9–11 sec, beneath all of the diverse tectonic and petrologic assemblages of the upper crust (Klemperer *et al.*, 1986). Presumably, this crust is being rebuilt by magma from mantle sources, representing partial melting in response to extension, and also is being smoothed by gravitational flattening of the deep crust. Spreading of basaltic sheets atop dense mantle is inferred. Temperature low in the crust of much of the Basin and Range province must be high enough to melt granite where a modest amount of water is available (Fig. 1-3).

The Mohorovicic discontinuity appears as a strong reflector on many COCORP profiles in the Cenozoic extensional terrains of the western United States but not generally elsewhere, so a distinctive type of lower crust built by mafic magmatism as a by-product of extension may characterize extended terrains.

Neogene root of the Sierra Nevada The crestal region of the great tilted block — 4 km high, 100 km wide, and 600 km long — of the Sierra Nevada of California may now stand high because the Moho is 50-60 km deep beneath it (Pakiser and Brune, 1980); but refraction and gravity data can alternatively be integrated to infer that the crust is of normal thickness, whereas the mantle is of low density. Most of the rise from low uplands of middle Tertiary time has occurred in the past 10 m.y. (Huber, 1981), so either the crust has been thickened by 15 km or so, or the density of the mantle has been decreased, within that period. The surface rocks of the Sierra crest region are primarily Cretaceous granites and older metamorphic rocks and there is little late Neogene magmatism displayed to which the growth of a crustal root might be attributed. Volcanism has, however, been a locally intense accompaniment of the rapid late Neogene extension in the Owens Valley, which now bounds the Sierra Nevada on the east, and perhaps asthenospheric diapirs consequent on extension decreased the density of the lithospheric mantle and caused the uplift (*cf.* Buck, 1986; Mavko and Thompson, 1983; Steckler, 1985). Heat flow in the Sierra Nevada remains low because slow conduction through the crust has not yet produced a heat-flow signal at the surface.

A contrary argument was made by C. G. Chase and Wallace (1986), who accepted a Cretaceous age for a thick crustal root and suggested that this root was compensated isostatically on a regional scale until the breaking of the crust by late Neogene rifting permitted the operation of isostasy on the scale of the mountain range. This concept is reasonable qualitatively but not, it appears to me, quantitatively. The concept requires an enormously strong prerift elastic lithosphere and contains the implicit prediction that the Sierra Nevada of the early Neogene would have had a subregional isostatic residual gravity anomaly of something like —200 mgal. Such anomalies probably do not exist (*cf.* R. W. Simpson *et al.*, 1986).

Uplift of the Western Interior

Colorado Plateau, Laramide Rocky Mountains, and Great Plains were all near sea level — marine deposits are widespread — very late in Cretaceous time. Local relief was much increased by Laramide shortening, but the rise of the entire region to its present general altitudes of 1-3 km above sea level occurred during Neogene time. Crustal thickness (Allenby and Schnetzler, 1983) is mostly near 40 km, a standard stable-continental value; it locally exceeds 50 km, presumably as a result of Laramide shortening, although at the present state of knowledge the overthick regions correlate only in part with those of shortening and of current altitude. Uppermost mantle velocites are near normal, 8.0-8.2 km/sec (Allenby and Schnetzler, 1983), so the regional uplift must be largely compensated deeper in the mantle. The broad correlation between the region of uplift and the region beneath which Pacific lithosphere was earlier subducted at shallow depth (Bird, 1984) suggests that the cause of the Neogene uplift lies in some delayed effect of that plate interaction, perhaps in slow heating of the subducted slab or in sinking of the lower part of the continental lithosphere that had been underplated by Laramide subduction (*cf.* Bird, 1979).

OVERVIEW

In this essay I have sought integrative explanations for some of the observed vertical, horizontal, and temporal relationships between structural and petrologic assemblages formed during the past 100 m.y. in the western United States. Whatever the merit of these specific interpretations, unifying explanations must be possible for processes operating simultaneously in adjacent or subjacent sites; and those explanations are not likely to be viable unless the predictions implicit in them are compatible with knowledge of analogs exposed at different crustal levels or in other parts of the world.

REFERENCES

Allenby, R. J., and Schnetzler, C. C., 1983, United States crustal thickness: *Tectonophysics*, v. 93, p. 13–31.

Anderson, J. R., 1983, Petrology of a portion of the eastern Peninsular Ranges mylonite zone, southern California: *Contrib. Mineralogy Petrology*, v. 84, p. 253–271.

Atwater, Tanya, 1970, Implications of plate tectonics for the Cenozoic tectonic evolution of western North America: *Geol. Soc. America Bull.*, v. 81, p. 3513–3536.

Axelrod, D. I., 1985, Miocene floras from the Middlegate basin, west-central Nevada: *Univ. Calif. Publ. Geol. Sci.*, v. 129, 279 p.

Bateman, P. C., 1965 Geologic map of the Blackcap Mountain quadrangle, Fresno County, California: *U.S. Geol. Survey Geol. Quad. Map GQ-428.*

——, 1979, Cross section of the Sierra Nevada from Madera to the White Mountains, central California: *Geol. Soc. America Map Chart Series MC-28E.*

——, and Chappell, B. W., 1979, Crystallization, fractionation and solidification of the Tuolumne intrusive series, Yosemite National Park, California: *Geol. Soc. America Bull.*, v. 90, pt. 1, p. 465–482.

——, and Moore, J. G., 1965, Geologic map of the Mount Goddard quadrangle, Fresno and Inyo Counties, California: *U.S. Geol. Survey Geol. Quad. Map GQ-429.*

——, and Nokleberg, W. J., 1978, Solidification of the Mount Givens Granodiorite, Sierra Nevada, California: *J. Geology*, v. 86, p. 563–579.

——, *et al.*, 1963, The Sierra Nevada batholith — A synthesis of recent work across the central part: *U.S. Geol. Survey Prof. Paper 414-D*, 46 p.

——, Kistler, R. W., Peck, D. L., and Busacca, Alan, 1983, Geologic map of the Tuolumne Meadows quadrangle, Yosemite National Park, California: *U.S. Geol. Survey Geol. Quad. Map GQ-1570.*

Beck, M. E., Jr., 1980, Paleomagnetic record of plate-margin tectonic processes along the western edge of North America: *J. Geophys. Res.*, v. 85, p. 7115–7131.

Best, M. G., 1986, Some observations on Late Oligocene–Early Miocene volcanism in Nevada and Utah: *Geol. Soc. America Abstr. with Programs*, v. 18, p. 86.

Bethke, C. M., 1986, Inverse hydrologic analysis of the distribution and origin of Gulf Coast-type geopressured zones: *J. Geophys. Res.*, v. 91, p. 6535–6545.

Bird, Peter, 1979, Continental delamination and the Colorado Plateau: *J. Geophys. Res.*, v. 84, p. 7561–7571.

——, 1984, Laramide crustal thickening event in the Rocky Mountain foreland and Great Plains: *Tectonics*, v. 3, p. 741–758.

Brown, R. L., and Read, P. B., 1983, Shuswap terrane of British Columbia — a Mesozoic "core complex": *Geology*, v. 11, p. 164–168.

Bruce, C. H., 1984, Smectite dehydration — Its relation to structural development and hydrocarbon accumulation in northern Gulf of Mexico Basin: *Amer. Assoc. Petrol. Geologists Bull.*, v. 68, p. 673–683.

Bryan, Phillip, and Gordon, R. G., 1986, Rotation of the Colorado Plateau — An analysis of paleomagnetic data: *Tectonics*, v. 5, p. 661–667.

Buck, W. R., 1986, Small-scale convection induced by passive rifting — The cause for up-lift of rift shoulders: *Earth Planet. Sci. Lett.*, v. 77, p. 362–372.

Burst, J. F., 1969, Diagenesis of Gulf Coast clayey sediments and its possible relation to petroleum migration: *Amer. Assoc. Petrol. Geologists Bull.*, v. 53, p. 73–93.

Cady, J. W., and Fox, K. F., Jr., 1984, Geophysical interpretation of the gneiss terrane of northern Washington and southern British Columbia, and its implications for uranium exploration: *U.S. Geol. Survey Prof. Paper 1260*, 29 p.

Carlson, R. L., and Melia, P. J., 1984, Subduction hinge migration: *Tectonophysics*, v. 102, p. 399–411.

Carr, S. D., 1985, Ductile shearing and brittle faulting in the Valhalla gneiss complex, southeastern British Columbia: *Geol. Survey Canada Paper 85-1A*, p. 89–96.

Cater, F. W., *et al.*, 1973, Mineral resources of the Idaho Primitive Area and vicinity, Idaho: *U.S. Geol. Survey Bull. 1304*, 431 p.

Chase, C. G., 1978, Extension behind island arcs and motions relative to hot spots: *J. Geophys. Res.*, v. 83, p. 5385–5387.

——— .and Wallace, T. C., 1986, Uplift of the Sierra Nevada of California: *Geology*, v. 14, p. 730–733.

Chase, R. B., Bickford, M. E., and Arruda, E. C., 1983, Tectonic implications of Tertiary intrusion and shearing within the Bitterroot dome, northeastern Idaho batholith: *J. Geology*, v. 91, p. 462–470.

Chen, J. H., and Moore, J. G., 1982, Uranium-lead isotopic ages from the Sierra Nevada batholith, California: *J. Geophys. Res.*, v. 87, p. 4761–4784.

Cheney, E. S., 1980, Kettle dome and related structures of northeastern Washington: *Geol. Soc. America Mem. 153*, p. 463–483.

Christiansen, R. L., and McKee, E. H., 1978, Late Cenozoic volcanic and tectonic evolution of the Great Basin and Columbia Intermontane regions: *Geol. Soc. America Mem. 152*, p. 283–311.

Clemens, J. D., 1984, Water contents of silicic to intermediate magmas: *Lithos*, v. 17, p. 273–287.

Compton, R. R., 1960, Charnockitic rocks of Santa Lucia Range, California: *Amer. J. Sci.*, v. 158. p. 609–636.

Coney, P. J., and Reynolds, S. J., 1977, Cordilleran Benioff zones: *Nature*, v. 270, p. 403–406.

Cook, F. A., 1986, Seismic reflection geometry of the Columbia River fault zone and east margin of the Shuswap metamorphic complex in the Canadian Cordillera: *Tectonics*, v. 5, p. 669–685.

Coulbourn, W. T., and Moberly, R., 1977, Structural evidence of the evolution of fore-arc basins of South America: *Can. J. Earth Sci.*, v. 14, p. 102–116.

Cross, T. A., and Pilger, R. H., Jr., 1978, Constraints on absolute motion and plate inter-action inferred from Cenozoic igneous activity in the western United States: *Amer. J. Sci.*, v. 278, p. 865–902.

Davidson, Anthony, 1984, Identification of ductile shear zones in the southwestern Grenville Province of the Canadian Shield, *in* Kroner, A., and Greiling, R., eds., *Precambrian Tectonics Illustrated*: Stuttgart, West Germany, Schweizerbart'sche Verlagsbuchhandlung, p. 263–279.

DePaolo, D. J., and Farmer, G. L., 1984, Isotopic data bearing on the origin of Mesozoic and Tertiary granitic rocks in the western United States: *Philos. Trans. Roy. Soc. London*, v. A310, p. 743–753.

de Voogd, Beatrice, Serpa, Laura, *et al.*, 1986, Death Valley bright spot — A midcrustal magma body in the southern Great Basin, California? *Geology*, v. 14, p. 64–67.

Dewey, J. F., 1980, Periodicity, sequence, and style at convergent plate boundaries: *Geol. Assoc. Canada Spec. Paper 20*, p. 553–573.

Dickinson, W. R., 1979, Geometry of triple junctions related to San Andreas transform: *J. Geophys. Res.*, v. 84, p. 561–572.

——, 1983, Cretaceous sinistral strike slip along Nacimiento fault in coastal California: *Amer. Assoc. Petrol. Geologists Bull.*, v. 67, p. 624–645.

——, and Snyder, W. S., 1979, Geometry of subducted slabs related to San Andreas transform: *J. Geology*, v. 87, p. 609–627.

Ehlig, P. L., 1981, Origin and tectonic history of the basement terrane of the San Gabriel Mountains, central Transverse Ranges, *in* Ernst, W. G., ed., *The Geotectonic Development of California* (Rubey Vol. I): Englewood Cliffs, N. J., Prentice-Hall, p. 252–283.

Elan, Ron, 1985, High grade contact metamorphism at the Lake Isabella north shore roof pendant, southern Sierra Nevada, California: M.Sc. thesis, Univ. Southern California, Los Angeles, Calif., 202 p.

Engebretson, D. C., Cox, Allan, and Gordon, R. G., 1985, Relative motions between oceanic and continental plates in the Pacific basin: *Geol. Soc. America Spec. Paper 206*, 59 p.

Fiske, R. S., and Tobisch, O. T., 1978, Paleogeographic significance of volcanic rocks of the Ritter Range pendant, central Sierra Nevada, California; *in* Howell, D. G., and McDougall, K. A., eds., *Mesozoic Paleogeography of the Western United States*: Pacific Section, Soc. Econ. Paleontologists Mineralogists, Pacific Coast Paleogeography Symp. 2, p. 209–221.

Fountain, D. M., and Salisbury, M. H., 1981, Exposed cross-sections through the continental crust — implications for crustal structure, petrology and evolution: *Earth Planet. Sci. Lett.*, v. 56, p. 263–277.

Frei, L. S., 1986, Additional paleomagnetic results from the Sierra Nevada — Further constraints on Basin and Range extension and northward displacement of the western United States: *Geol. Soc. America Bull.*, v. 97, p. 840–849.

Frost, E. G., and Okaya, D. A., 1986, Geologic and seismologic setting of the CALCRUST Whipple line — Geometries of detachment faulting in the field and from reprocessed seismic-reflection profiles donated by CGG: *Geol. Soc. America Abstr. with Programs*, v. 18, p. 107.

Garfunkel, Z., Anderson, C. A., and Schubert, G., 1986, Mantle circulation and the lateral migration of subducted slabs: *J. Geophys. Res.*, v. 91, p. 7205–7223.

Green, T. H., 1976, Experimental generation of cordierite- or garnet-bearing granitic liquids from a pelitic composition: *Geology*, v. 4, p. 85–88.

Hamilton, Warren, 1969, Mesozoic California and the underflow of Pacific mantle: *Geol. Soc. America Bull.*, v. 80, p. 2409–2430.

——, 1978, Mesozoic tectonics of the Western United States, *in* Howell, D. G., and McDougall, K. A., eds., *Mesozoic Paleogeography of the Western United States*: Pacific Section, Soc. Econ. Paleontologists Mineralogists, Pacific Coast Paleogeography Symp. 2, p. 33–70.

——, 1979, Tectonics of the Indonesian region: *U.S. Geol. Survey Prof. Paper 1078*, 345 p. (Reprinted with minor corrections, 1981 and 1985.)

——, 1981a, Crustal evolution by arc magmatism: *Philos. Trans. Roy. Soc. London*, v. A302, p. 279–291.

——, 1981b, Plate-tectonic mechanism of Laramide deformation: *Univ. Wyo. Contrib. Geology*, v. 19, p. 87–92.

——, 1982, Structural evolution of the Big Maria Mountains, northeastern Riverside County, southeastern California, *in* Frost, E. G., and Martin, D. L., eds., *Mesozoic-Cenozoic Evolution of the Colorado River Region, California, Arizona, and Nevada:* San Diego, Calif., Cordilleran Publishers, p. 1–127.

——, in press *a,* Crustal extension in the Basin and Range Province, western United States, *in* Coward, M. P., Dewey, J. F., and Hancock, P. L., eds., *Continental Extensional Tectonics:* Geol. Soc. London Spec. Publ. 28.

——, in press *b,* Detachment faulting in the Death Valley region, California and Nevada, *in* Carr, M. D., ed., *Structural Geology of the Nevada Test Site and Vicinity:* U.S. Geol. Survey Bull. 1790.

——, in press *c,* Laramide crustal shortening, *in* Perry, W. J., and Schmidt, C. J., eds., *Laramide Tectonics of the Rocky Mountain Foreland:* Geol. Soc. America Mem.

——, in press *d,* Crustal geology of the United States, *in* Pakiser, L. C., and Mooney, W. D., eds., *Geophysical Framework of the United States:* Geol. Soc. America Mem.

——, in press *e,* Convergent-plate tectonics viewed from the Indonesian region, *in* Sengor, A. M. C., ed., *Tectonic Evolution of the Tethyan Domain:* Dordrecht, The Netherlands, D. Reidel.

——, and Myers, W. B., 1966, Cenozoic tectonics of the western United States: *Rev. Geophysics,* v. 4, p. 509–549.

——, and Myers, W. B., 1967, The nature of batholiths: *U.S. Geol. Survey Prof. Paper 554-C,* 30 p.

——, and Myers, W. B., 1974, Nature of the Boulder batholith of Montana: *Geol. Soc. America Bull.,* v. 85, p. 365–378, 1958–1960.

Hansen, Birger, 1981, The transition from pyroxene granulite facies to garnet clinopyroxene granulite facies — Experiments in the system CaO-MgO-Al$_2$O$_3$-SiO$_2$: *Contrib. Mineralogy Petrology,* v. 76, p. 234–242.

Haxel, Gordon, and Dillon, John, 1978, The Pelona-Orocopia Schist and Vincent-Chocolate Mountains thrust system, southern California; *in* Howell, D. G., and McDougall, K. A., eds., *Mesozoic Paleogeography of the Western United States:* Pacific Section, Soc. Econ. Paleontologists Mineralogists, Pacific Coast Paleogeography Symp. 2, p. 453–469.

——, Tosdal, R. M., May, D. J., and Wright, J. E., 1984, Latest Cretaceous and Early Tertiary orogenesis in south-central Arizona — Thrust faulting, regional metamorphism, and granitic plutonism: *Geol. Soc. America Bull.,* v. 95, p. 631–653.

Heller, P. L., *et al.,* 1986, Time of initial thrusting in the Sevier orogenic belt, Idaho-Wyoming and Utah: *Geology,* v. 14, p. 388–91.

Hildreth, Wes, 1981, Gradients in silicic magma chambers — Implications for lithospheric magmatism: *J. Geophys. Res.* v. 86, p. 10153–10192.

Holdaway, M. J., 1971, Stability of andalusite and the aluminum silicate phase diagram: *Amer. J. Sci.,* v. 271, p. 97–131.

Huang, W. L., and Wyllie, P. J., 1981, Phase relationships of S-type granite with H$_2$O to 35 Kbar — Muscovite granite from Harney Peak, South Dakota: *J. Geophys. Res.,* v. 86, p. 10515–10529.

Huber, N. K., 1981, Amount and timing of late Cenozoic uplift and tilt of the central Sierra Nevada, California — Evidence from the upper San Joaquin River basin: *U.S. Geol. Survey Prof. Paper 1197,* 28 p.

Hyndman, D. W., 1980, Bitterroot dome-Sapphire tectonic block, an example of a plutonic-core gneiss-dome complex with its detached suprastructure: *Geol. Soc. America Mem. 153,* p. 427–443.

—— , 1981, Controls on source and depth of emplacement of granitic magma: *Geology*, v. 9, p. 244–248.

—— , 1983, The Idaho batholith and associated plutons, Idaho and western Montana: *Geol. Soc. America Mem. 159*, p. 213–240.

Johnson, C. A., Bohlen, S. R., and Essene, E. J., 1983, An evaluation of garnet-clino-pyroxene geothermometry in granulites: *Contrib. Mineralogy Petrology* v. 84, p. 191–198.

Journeay, Murray, and Brown, R. L., 1986, Major tectonic boundaries of the Omineca Belt in southern British Columbia — A progress report: *Geol. Survey Canada Paper 86-1A*, p. 81–88.

Jurdy, D. M., 1984, The subduction of the Farallon plate beneath North America as derived from relative plate motions: *Tectonics*, v. 3, p. 107–113.

Kay, R. W., and Kay, S. M., 1981, The nature of the lower continental crust — Inferences from geophysics, surface geology, and crustal xenoliths: *Rev. Geophysics Space Physics*, v. 19, p. 271–297.

Kistler, R. W., and Peterman, Z. E., 1978, Reconstruction of crustal blocks of California on the basis of initial strontium isotopic compositions of Mesozoic granitic rocks: *U.S. Geol. Survey Prof. Paper 1071*, 17 p.

Kistler, R. W., Ghent, E. D., and O'Neil, J. R., 1981, Petrogenesis of garnet two-mica granites in the Ruby Mountains, Nevada: *J. Geophys. Res.*, v. 86, p. 10591–10606.

Klemperer, S. L., *et al.*, 1986, The Moho in the northern Basin and Range Province, Nevada, along the COCORP 40°N seismic reflection transect: *Geol. Soc. America Bull.*, v. 97, p. 603–618.

Lachenbruch, A. H., and Sass, J. H., 1978, Models of an extending lithosphere and heat flow in the Basin and Range province: *Geol. Soc. America Mem. 152*, p. 209–250.

Lane, L. S., 1984, Brittle deformation in the Columbia River fault zone near Revelstoke, southeastern British Columbia: *Can. J. Earth Sci.*, v. 21, p. 584–598.

Lipman, P. W., 1984, The roots of ash flow calderas in western North America — Windows into the tops of granitic batholiths: *J. Geophys. Res.*, v. 89, p. 8801–8814.

Maaloe, Sven, and Wyllie, P. J., 1975, Water content of a granite magma deduced from the sequence of crystallization determined experimentally with water-undersaturated conditions: *Contrib. Mineralogy Petrology*, v. 52, p. 175–191.

Magill, J. R., Wells, R. E., Simpson, R. W., and Cox, A. V., 1982, Post 12-m.y. rotation of southwest Washington: *J. Geophys. Res.*, v. 87, p. 3761–3776.

Mahaffie, M. J., and Dokka, R. K., 1986, Thermochronologic evidence for the age and cooling history of the upper plate of the Vincent thrust, California: *Geol. Soc. America Abstr. with Programs*, v. 18, p. 153.

Malinverno, Alberto, and Ryan, W. B. F., 1986, Extension in the Tyrrhenian Sea and shortening of the Apennines as result of arc migration driven by sinking of the lithosphere: *Tectonics*, v. 5, p. 227–245.

Mathews, W. H., 1981, Early Cenozoic resetting of potassium-argon dates and geothermal history of north Okanagan area, British Columbia: *Can. J. Earth Sci.*, v. 18, p. 1310–1319.

Mattinson, J. M., 1978, Age, origin, and thermal histories of some plutonic rocks from the Salinian block of California: *Contrib. Mineralogy Petrology*, v. 67, p. 233–245.

Mavko, B. B., and Thompson, G. A., 1983, Crustal and upper mantle structure of the northern and central Sierra Nevada: *J. Geophys. Res.*, v. 88, p. 5874–5892.

McKee, E. H., Duffield, W. A., and Stern, R. L., 1983, Late Miocene and early Pliocene basaltic rocks and their implications for crustal structure, northeastern California and south-central Oregon: *Geol. Soc. America Bull.*, v. 94, p. 292–304.

Miller, C. F., 1986, Comment on "S-type granites and their probable absence in south-western North America": *Geology*, v. 14, p. 804–805.

——, and Bradfish, L. J., 1980, An inner Cordilleran muscovite-bearing plutonic belt: *Geology*, v. 8, p. 412–416.

Miller, F. K., 1971, The Newport fault and associated mylonites, northeastern Washington: *U.S. Geol. Survey Prof. Paper 750-D*, p. 77–79.

——, and Engels, J. C., 1975, Distribution and trends of discordant ages of the plutonic rocks of northeastern Washington and northern Idaho: *Geol. Soc. America Bull.*, v. 86, p. 517–528.

Molnar, Peter, and Atwater, Tanya, 1978, Interarc spreading and Cordilleran tectonics as alternates related to the age of subducted oceanic lithosphere: *Earth Planet. Sci. Lett.*, v. 41, p. 330–340.

Monger, J. W. H., *et al.*, 1985, Juan de Fuca plate to Alberta plains: *Geol. Soc. America Continent-Ocean Transect 7 (B-2)*.

Morris, R. S., Okaya, D. A., Frost, E. G., and Malin, P. E., 1986, Base of the Orocopia Schist as imaged on seismic reflection data in the Chocolate and Cargo Muchacho Mtns. region of SE Calif., and the Sierra Pelona region near Palmdale, Calif.: *Geol. Soc. America Abstr. with Programs*, v. 18, p. 160.

Newton, R. C., and Perkins, D., III, 1982, Thermodynamic calibration of geobarometers based on the assemblages garnet-plagioclase-orthopyroxene (clinopyroxene)-quartz: *Amer. Mineralogist*, v. 67, p. 203–222.

Ninkovich, D., Sparks, R. S. J., and Ledbetter, M. T., 1978, The exceptional magnitude and intensity of the Toba eruption, Sumatra — An example of the use of deep-sea tephra layers as a geological tool: *Bull. Volcanologique*, v. 41, p. 1–13.

Nokleberg, W. J., and Kistler, R. W., 1980, Paleozoic and Mesozoic deformations in the central Sierra Nevada, California: *U.S. Geol. Survey Prof. Paper 1145*, 24 p.

Noyes, H. J., Wones, D. R., and Frey, F. A., 1983, A tale of two plutons — Petrographic and mineral constraints on the petrogenesis of the Red Lake and Eagle Peak plutons, central Sierra Nevada, California: *J. Geology*, v. 91, p. 353–379.

O'Hara, M. J., and Mathews, R. E., 1981, Geochemical evolution in advancing, periodically replenished, continuously fractionated magma chamber: *J. Geol. Soc. London*, v. 138, p. 237–277.

Page, B. M., 1982, Migration of Salinian composite block, California, and disappearance of fragments: *Amer. J. Sci.*, v. 282, p. 1694–1734.

Pakiser, L. C., and Brune, J. N., 1980, Seismic models of the root of the Sierra Nevada: *Science*, v. 210, p. 1088–1094.

Postlethwaite, C. E., and Jacobson, C. E., in press, Early history and reactivation of the Rand thrust, southern California: *J. Struct. Geology*.

Potter, C. J., *et al.*, in press, COCORP deep seismic reflection traverse of the interior of the North American Cordillera, Washington and Idaho — Implications for orogenic evolution: *Tectonics*.

Price, R. A., 1981, The Cordilleran foreland thrust and fold belt in the southern Canadian Rocky Mountains: *Geol. Soc. London Spec. Publ. 9*, p. 427–448.

Rea, D. K., and Duncan, R. A., 1986, North America plate convergence — A quantitative record of the past 140 m.y.: *Geology*, v. 14, p. 373–376.

Rhodes, B. P., 1985, Metamorphism of the Spokane dome mylonitic zone, Priest River complex — Constraints on the tectonic evolution of northeastern Washington and northern Idaho: *J. Geology*, v. 94, p. 539–566.

——, and Cheney, E. S., 1981, Low-angle faulting and the origin of Kettle dome, a metamorphic core complex in northeastern Washington: *Geology*, v. 9, p. 366–369.

____, and Hyndman, D. W., 1984, Kinematics of mylonites in the Priest River "metamorphic core complex," northern Idaho and northeastern Washington: *Can. J. Earth Sci.*, v. 21, p. 1161–1170.

Rinehart, C. D., and Ross, D. C., 1964, Geology and mineral deposits of the Mount Morrison quadrangle, Sierra Nevada, California: *U.S. Geol. Survey Prof. Paper 385*, 106 p.

Ross, D. C., 1967, Generalized geologic map of the Inyo Mountains region, California: *U.S. Geol. Survey Misc. Geol. Inv. Map I-506.*

____, 1976, Maps showing distribution of metamorphic rocks and occurrences of garnet, coarse graphite, sillimanite, orthopyroxene, clinopyroxene, and plagioclase amphibolite, Santa Lucia Range, Salinian block, California: *U.S. Geol. Survey Misc. Field Studies Map MF-751.*

____, 1978, The Salinian block — A Mesozoic granitic orphan in the California Coast Ranges; *in* Howell, D. G., and McDougall, K. A., eds., *Mesozoic Paleogeography of the Western United States:* Pacific Section, Soc. Econ. Paleontologists Mineralogists, Pacific Coast Paleogeography Symp. 2, p. 509–522.

____, 1985, Mafic gneissic complex (batholithic root?) in the southernmost Sierra Nevada, California: *Geology*, v. 13, p. 288–291.

____, in press, Metamorphic and plutonic rocks of the southernmost Sierra Nevada, California, and their tectonic framework: *U.S. Geol. Survey Prof. Paper*, in press.

Schultz, L. G., 1978, Mixed-layer clay in the Pierre Shale and equivalent rocks, northern Great Plains region: *U.S. Geol. Survey Prof. Paper 1064-A*, 28 p.

Sclater, J. G., Jaupart, C., and Galson, D., 1980, The heat flow through oceanic and continental crust and the heat loss of the Earth: *Rev. Geophysics Space Physics*, v. 18, p. 269–311.

Sharp, R. V., 1979, Some characteristics of the eastern Peninsular Ranges mylonite zone: *U.S. Geol. Survey Open File Rpt. 79-1239*, p. 258–267.

Sharry, John, 1981, The geology of the western Tehachapi Mountains, California: Ph.D. thesis, Massachusetts Inst. Technology, Cambridge, Mass., 215 p.

____, *et al.*, 1985, Enhanced imaging of the COCORP seismic line, Wind River Mountains: *Amer. Geophys. Un. Geodynamics Ser.*, v. 13, p. 223–236.

Sibson, R. H., 1983, Continental fault structure and the shallow earthquake source: *J. Geol. Soc. London*, v. 140, p. 741–767.

Silver, L. T., Taylor, H. P., Jr., and Chappell, Bruce, 1979, Some petrological, geochemical and geochronological observations of the Peninsular Ranges batholith near the international border of the U.S.A. and Mexico, *in* Abbott, P.L., and Todd, V. R., eds., *Mesozoic Crystalline Rocks — Peninsular Ranges Batholith and Pegmatites, Point Sal Ophiolite:* Guidebook Geol. Soc. America Ann. Meet., San Diego State Univ., San Diego, Calif., p. 83–110.

Simpson, Carol, 1984, Borrego Springs-Santa Rosa mylonite zone, a Cretaceous west-directed thrust in southern California: *Geology*, v. 12, p. 8–11.

____, 1985, Deformation of granitic rocks across the brittle-ductile transition: *J. Struct. Geology*, v. 7, p. 503–511.

Simpson, R. W., Jachens, R. C., Blakely, R. J., and Saltus, R. W., 1986, A new isostatic residual gravity map of the conterminous United States with a discussion of the significance of isostatic residual anomalies: *J. Geophys. Res.*, v. 91, p. 8348–8372.

Smith, A. G., 1981, Subduction and coeval thrust belts, with particular reference to North America: *Geol. Soc. London Spec. Publ. 9*, p. 111–124.

Smithson, S. B., *et al.*, 1979, Structure of the Laramide Wind River uplift, Wyoming, from COCORP deep reflection data and from gravity data: *J. Geophys. Res.*, v. 84, p. 5955–5972.

Snyder, W. S., Dickinson, W. R., and Silberman, M. L., 1976, Tectonic implications of space-time patterns of Cenozoic magmatism in the western United States: *Earth Planet. Sci. Lett.*, v. 32, p. 91–106.

Steckler, M. S., 1985, Uplift and extension at the Gulf of Suez — Indications of induced mantle convection: *Nature*, v. 317, p. 135–139.

Steiner, M. B., 1986, Rotation of the Colorado Plateau: *Tectonics*, v. 5, p. 649–660.

Stern, C. R., and Wyllie, P. J., 1981, Phase relationships of I-type granite with H_2O to 35 kilobars — The Dinkey Lakes biotite-granite from the Sierra Nevada batholith: *J. Geophys. Res.*, v. 86, p. 10412–10422.

Stern, T. W., *et al.*, 1981, Isotopic U-Pb ages of zircon from the granitoids of the central Sierra Nevada, California: *U.S. Geol. Survey Prof. Paper 1185*, 17 p.

Tarney, John, and Saunders, A. D., 1979, Trace element constraints on the origin of Cordilleran batholiths, *in* Atherton, M. P., and Tarney, J., eds., *Origin of Granite Batholiths — Geochemical Evidence:* Nantwick, Cheshire, England, Shiva, p. 90–105.

——, and Windley, B. F., 1977, Chemistry, thermal gradients, and evolution of the lower continental crust: *J. Geol. Soc. London*, v. 134, p. 153–172.

Terres, R. R., and Luyendyk, B. P., 1985, Tectonic rotation of the San Gabriel region, California, suggested by paleomagnetic vectors: *J. Geophys. Res.*, v. 90, p. 12467–12484.

Tobisch, O. T., Saleeby, J. B., and Fiske, R. S., 1986, Structural history of continental volcanic arc rocks, eastern Sierra Nevada, California — A case for extensional tectonics: *Tectonics*, v. 5, p. 65–94.

Todd, V. R., and Shaw, S. E., 1979, Structural, metamorphic and intrusive framework of the Peninsular Ranges batholith in southern San Diego County, California, *in* Abbott, P. L., and Todd, V. R., eds., *Mesozoic Crystalline Rocks — Peninsular Ranges Batholith and Pegmatites, Point Sal Ophiolite:* Guidebook Geol. Soc. America Ann. Meet., San Diego State Univ., San Diego, Calif., p. 178–231.

Tracy, R. J. 1978, High grade metamorphic reactions and partial melting of pelitic schist, west-central Massachusetts: *Amer. J. Sci.*, v. 278, p. 150–178.

Uyeda, Seiya, and Kanamori, Hiroo, 1979, Back-arc opening and the mode of subduction: *J. Geophys. Res.*, v. 84, p. 1049–1061.

Walker, David, 1983, New developments in magmatic processes: *Rev. Geophysics Space Physics*, v. 21, p. 1372–1384.

Weaver, B. L., and Tarney, John, 1981, Andesitic magmatism and continental growth, *in* Thorpe, R. S., ed., *Andesites:* New York, Wiley, p. 639–661.

Weaver, C. E., 1960, Possible uses of clay minerals in search for oil: *Amer. Assoc. Petrol. Geologists Bull.*, v. 44, p. 1505–1518.

Wells, P. R. A., 1980, Thermal models for the magmatic accretion and subsequent metamorphism of continental crust: *Earth Planet. Sci. Lett.*, v. 46, p. 253–265.

Wells, R. E., 1985, Paleomagnetism and geology of Eocene volcanic rocks of southwest Washington, implications for mechanisms of tectonic rotation: *J. Geophys. Res.*, v. 90, p. 1925–1947.

White, A. J. R., *et al.*, 1986, S-type granites and their probable absence in southwestern North America: *Geology*, v. 14, p. 115–118, 805–806.

White, C. M., and McBirney, A. R., 1978, Some quantitative aspects of orogenic volcanism in the Oregon Cascades: *Geol. Soc. America Mem. 152*, p. 369–388.

Wilshire, H. G., *et al.*, in press, Mafic and ultramafic xenoliths from volcanic rocks of the western United States: *U.S. Geol. Survey Prof. Paper*, in press.

Windley, B. F., 1981, Phanerozoic granulites: *J. Geol. Soc. London*, v. 138, p. 745–751.

Wust, S. L., 1986, Tertiary extensional detachment system of westward vergence, exposing brittle-to-ductile transition, Pioneer structural complex, central Idaho: *Geol. Soc. America Abstr. with Programs*, v. 18, p. 201.

Wyllie, P. J., 1984, Constraints imposed by experimental petrology on possible and impossible magma sources and products: *Philos. Trans., Roy. Soc. London*, v. A-310, p. 439–456.

Notes added in proof.

Hypothetical Orocopia suture in southeastern California (p. 18–19): The distinctive Paleozoic and lower Mesozoic stratigraphic section diagnostic of the southwest corner of the North American craton is now known southwest of this postulated suture as well as northeast of it, so no suture is present. In the Chocolate Mountains, the tectonic contact between continental rocks above and oceanic rocks beneath is not a synmetamorphic megathrust but instead is a northeast-directed middle Tertiary detachment fault marked by thin mylonite and involving also Tertiary rocks. The northeastward vergence recognized by Dillon and Haxel (1978) in retrograded rocks near the detachment, which cut out the megathrust, records middle Tertiary slip. (Existence of the conjectural Jurassic "Mojave-Sonora megashear" of Anderson and Silver is also precluded by the distribution of the cratonic strata.)

Crustal extension in Basin and Range province (p. 24–27): I argued in this paper (and in Hamilton, in press *a*) that detachment faults are the tops of spread-apart crustal lenses. An early-1987 fieldtrip across the southern Great Basin with Brian Wernicke, and my own fieldwork in the Death Valley region, have since convinced me of the validity of his contrary inference (as, in Can. J. Earth Sci., v. 22, p. 108–125, 1985) that detachment faults originate as great gently-dipping normal faults that are raised, flattened, and domed as their lower plates are denuded tectonically; but I see much more deformation than does Wernicke beneath the ramp faults. I regard the reflection profiles available from the southern Basin and Range Province, and rock-mechanic and thermal considerations, as indicating that flattening by spreading of lenses does occur in the deeper middle crust, and pervasive ductile flattening in the lower crust, and that the Wernicke simple-shear model does not apply at those depths. My current views are expressed in the final version of Hamilton (in press, *b*).

E-an Zen
U. S. Geological Survey
Reston, Virginia 22092

2

TECTONIC SIGNIFICANCE OF HIGH-PRESSURE PLUTONIC ROCKS IN THE WESTERNS CORDILLERA OF NORTH AMERICA

ABSTRACT

The petrology and mineral chemistry of magmatic epidote-bearing Late Cretaceous plutons, between Idaho and southeastern Alaska in the western Cordillera, indicate their consolidation at depths of 20–30 km. Their sialic sources and environments of emplacement imply that at the time of intrusion, the crust was thick (45–60 km) and the geothermal gradients were on the order of $20°$-$30°$C/km. The crust was subsequently uplifted at rates of about 0.2–0.7 mm/yr averaged over a few tens of million years. The plutons are located within the Mesozoic accreted terranes and are postaccretion; these inferred crustal properties indicate thick-skinned accretion processes.

Active plate margins showing supracrustal thickening and low geothermal gradients are commonly where blueschist-facies metamorphism is found. Are the two petrologic associations compatible? Both associations occur east of the Straight Creek fault in north-central Washington; they could reflect different residence times and depths of burial of the crust in the bural-uplift cycle rather than be due to the juxtapostition of terranes, as indicated by modeling of burial-thermal systems that invokes the initial presence of a thick (30 km) pile of cool supracrustal thrust cover.

Most existing geologic means to estimate paleo-thicknesses of crust refer to the amount of crust *removed*, not the amount remaining. Improved understanding of the tectonic evolution of sialic crusts at plate margins requires that we gain better knowledge of the paleo-thicknesses of these sialic crusts, as well as the way these thicknesses changed during orogenic evolution and are reflected in the styles of tectonic deformation.

INTRODUCTION

Interest in the constitution and evolution of the sialic crust has led to considerable ongoing effort in determining the pressures of formation of crystalline rocks. This effort is aided by improved knowledge of phase equilibria among metamorphic and igneous mineral assemblages as functions of temperature, pressure, and other external variables. The effort also provides new insights on the nature of tectonic processes at plate margins, including accretion of microplates as well as of sedimentary prisms. This paper will attempt to show that some magmatic epidote-bearing plutons found within the Mesozoic active margin of the western Cordillera, between western Idaho and southeastern Alaska, imply high pressures of consolidation. The information is then used to infer that these regions experienced low to normal paleogeothermal gradients, thick sialic crust, and high rates of crustal uplift during early Tertiary time. It is hoped that these preliminary conclusions, exploring means to infer tectonic environments from plutonic rocks, will stimulate further studies.

Carmichael (1978) summarized the concept of using metamorphic isograds in pelitic rocks to define pressure ranges, or "bathozones." Recent reports on mineral assemblages in the Canadian part of the Coast Range plutonic complex (e.g. Crawford and Hollister, 1982; Hollister, 1982; Lappin and Hollister, 1980; Selverstone and Hollister, 1980; Woodsworth *et al.*, 1983a) provide examples of the innovative use of careful petrographic observations to reconstruct the *evolution* of metamorphic terranes, from their early record of "high-grade" events through cooling and decompression. These efforts to decipher the *P-T* paths of rocks are aided by the so-called Gibbs method (Spear and Selverstone, 1983; Spear *et al.*, 1982, 1984), which under optimal conditions allows detailed determination of pressure-temperature-time history of rocks.

Most determinations in the literature on the depth, or pressure, of plutonic emplacement depended either on the nature of the contact metamorphic assemblage, or on stratigraphic reconstruction, that is, by using extrinsic data. Textural features of the igneous rocks themselves provide useful, albeit qualitative information, e.g., miarolitic

TECTONIC SIGNIFICANCE OF HIGH-PRESSURE PLUTONIC ROCKS

cavities and graphic intergrowth of quartz and feldspar indicating shallow intrusions, and association with migmatite indicating deep instrusions. Magmatic muscovite has been used to indicate pressures of emplacement, but the nonstoichiometry of the muscovite causes problems (see reviews by Miller *et al.*, 1981; Speer, 1984; Zen, 1985a). Another indirect method is the use of fluid-inclusion compositions to deduce pressures, which has met some success (e.g., Selverstone and Hollister, 1980; Sisson, 1985).

Hammarstrom and Zen (1985, 1986) proposed a geobarometer that applies to tonalitic to granodioritic plutonic rocks having the mineral assemblage quartz, plagioclase, potassic feldspar, calcic amphibole, biotite, sphene, an oxide phase (commonly magnetite), ± apatite, and a former melt phase. This barometer measures the total aluminum content of the amphibole equilibrated with quartz, and can distinguish pressures to within ±2 kb. A second, "on-off," geobarometer, also applicable to tonalitic and granodioritic plutonic rocks, is the presence of magmatic epidote (Zen and Hammarstrom, 1984b, 1986) in addition to the same mineral assemblage as above, except that the amphibole may be either stable or relict. Certain textural features (e.g., partially resorbed, embayed calcic amphibole overgrown by epidote; euhedral epidote in contact with xenomorphic igneous biotite; wormy intergrowth of epidote with unaltered igneous plagioclase) are used to show that the epidote is magmatic, and by comparison with experimental phase-equilibrium data, the mineralogical relations provide information on the depths of consolidation.

Naney (1983) showed that at 8 kb pressure and f_{O_2} values between the nickel-bunsenite (NB) and hematite-magnetite (HM) buffers, epidote can be stable in a synthetic granodiorite melt at temperatures just above the solidus. The mineral chemistry of Naney's experimental products is very close to that of the natural material except that his epidote is slightly more pistacite rich (*cf.* Zen and Hammarstrom, 1984a). Moreover, as was also pointed out by Crawford and Hollister (1982), appropriate to conditions of the NB buffer (pistacite content ~25%, about equal to or slightly more pistacite-rich than natural magmatic epidote), the stability curve or epidote + quartz (Liou, 1973) intersects the solidus of a H_2O-saturated tonalite-to-granodiorite melt at a pressure of 5–6 kb (Wyllie, 1977). This is the minimum pressure for stable epidote + quartz in a magma of this composition range and f_{O_2} value because both less H_2O in the melt and presence of additional phases which could react with epidote + quartz increase the minimum pressure of this intersection (Fig. 2-1). For this study, therefore, I accept a value of 6 kb as the minimum pressure for magmatic epidote in a tonalite-to-granodiorite magma containing the mineral assemblage specified above.

According to Naney's experimental results, at least in the 6–8 kb pressure range, magmatic epidote should only crystallize out of magmas that are volumetrically heavily dominated by crystals. Such mushes cannot be expected to move far, especially to regions of lower temperature, without completely freezing; therefore, the presence of late magmatic epidote in a rock of appropriate bulk composition and mineral assemblage means that the pressure of *consolidation* was not less than 6 kb. For the purpose of this paper, magmatic epidote is taken to imply a pressure of 6–8 kb. This is in excellent agreement with the pressure independently deduced from the synplutonic metamorphic assemblages of the country rock in the area immediately surrounding the Ecstall pluton in British Columbia (Crawford *et al.*, 1987; Woodsworth *et al.*, 1983a), 8 ± 1 kb, the one place where such direct comparison is available.

Another geobarometer is the presence of high-pressure garnet in several plutons from southeast Alaska. These plutons carry partially resorbed garnet that is nearly a 1:1 solid solution of grossular and almandine (Zen and Hammarstrom, 1984b, and references therein). The garnets are entirely sheathed by magmatic plagioclase, and the "matrix" mineral assemblage includes both high-aluminum amphibole and magmatic epidote. The composition of the garnet indicates pressures sufficiently high that the anorthite-pyroxene

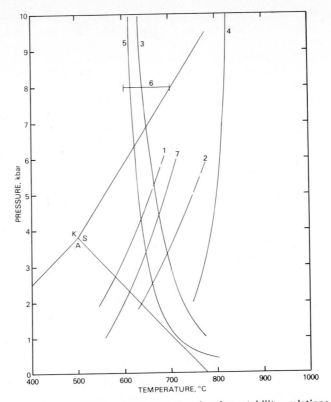

FIG. 2-1. Pressure-temperature diagram showing stability relations of epidote + quartz in various magmatic compositions. Curves 1 and 2, stability limit of epidote + quartz corresponding, respectively, to the nickel-bunsenite (NB) buffer and the magnetite-hematite (HM) buffer (pistacite content of epidote, respectively, 25% and 33%) (Liou 1973). Curve 3, solidus of H_2O-saturated tonalitic melt; curve 4, solidus of the same tonalite melt having just enough H_2O to ensure stability of hornblende and biotite; and curve 5, solidus of H_2O-saturated muscovite-granitic melt, all from Wyllie (1977). Bracket 6, temperature range in which epidote is stable in a H_2O-saturated synthetic granodioritic melt (Naney, 1983). Curve 7, stability limit of muscovite + quartz, (Chatterjee and Johannes, 1974). The stability boundaries of the aluminum-silicate polymorphs (K, kyanite; S, sillimanite; A, andalusite) are from Holdaway (1971).

assemblage is not stable. Presence of these garnets as sheathed relics therefore provides an independent argument for very high pressure ($\geq$13 kb) genesis of the magma in addition to high-pressure setting of its consolidation.

ENVIRONMENTS OF INTRUSION
AND INFERRED PALEOGEOGRAPHIC GRADIENTS

I will review the data (and reasons for surmises) on the local temperature and pressure conditions of intrusion of four groups of epidote-bearing plutons, previously described (Zen and Hammarstrom, 1984a); the reader should refer to that paper for data on mineral assemblages and chemistry. These plutons (Fig. 2-2) are in west-central Idaho, in north-central Washington, near Prince Rupert in British Columbia, and in southeastern Alaska. The estimated temperatures and pressures of intrusion lead to estimates of local paleogeothermal gradients at that time and to postintrusion uplift rates.

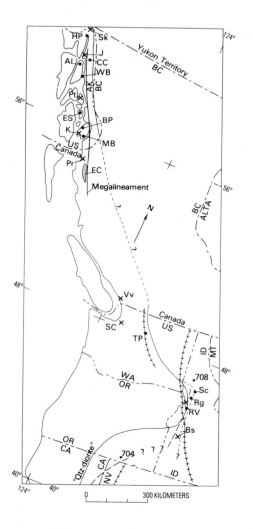

FIG. 2-2. Index map showing localities discussed in the text. Solid dots, geological features: RV, Round Valley, ID. SC, tonalite pluton at the head of Slate Creek, ID. TP, Tenpeak pluton, WA. EC, Ecstall pluton, BC. MB, Moth Bay pluton, AK. BP, Bushy Point pluton, AK. ES, Etolin Island, AK. CC, Carlson Creek pluton, AK. HP, Haines Point, AK. Crosses, communities: Bs, Boise, ID. Rg, Riggins, ID. Se, Seattle, WA. Vv, Vancouver, BC. Pr, Prince Rupert, BC. K, Ketchikan, AK. Pt, Petersburg, AK. J, Juneau, AK. Sk, Skagway, AK. Heavy solid line, the Coast Range megalineament (Brew and Ford, 1978), including its continuation in British Columbia as the Work Channel lineament. Light dashed and dotted lines, initial $^{87}Sr/^{86}Sr$ boundary lines for 0.704 and 0.708, respectively (Fleck and Criss, 1985; Kistler, 1974). Light solid line, the "quartz-diorite line" of Moore (1959) and of Moore *et al.* (1963). Line showing railroad-track pattern, boundaries of lead-isotope provinces (Zartman, 1974). Dash-dot lines, state or provincial boundaries; dash-triple dot lines, international boundary. Other features referred to in the text are shown on the local geologic maps, Figs. 2-3, 2-5 to 2-8.

Round Valley Pluton, Idaho

This medium-coarse-grained, flow-foliated pluton (Fig. 2-3) has a hornblende argon-spectrum age of about 110 m.y. (L. Snee, written communication, 1985). It is best exposed along U.S. highway 95 north of the Round Valley road junction. The pluton is cut by several generations of mildly peraluminous plagiogranite aplites and coarser-grained dikes. It also contains numerous disoriented xenoliths, a few centimeters to many meters in size, that are the local country rock of amphibolite and more felsic rocks, some showing "lit-par-lit injection" or partial fusion features (Fig. 2-4a). These xenoliths increase in

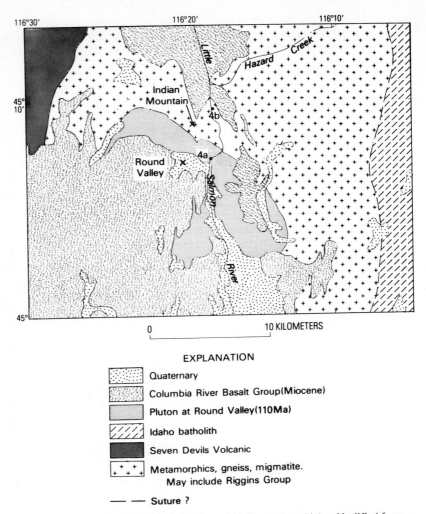

FIG. 2-3.   Geologic map of the Round Valley region, Idaho. Modified from Hamilton (1969). Localities of photographs shown in Fig. 2-4 are indicated.

abundance toward the contact with the country rocks, which are migmatitic gneisses having a fabric like some of the xenolith (Fig. 2-4b) and a mineral assemblage like that of the pluton, except for the local addition of garnet and/or muscovite. The country rock must have been thermally equilibrated with the pluton; these country rocks, mapped by Hamilton (1969) as trondhjemite gneiss immediately adjacent to the pluton and as amphibolite a few kilometers away, could be metamorphosed volcanic rocks of the Riggins Group, as he suggested (see also Kuntz, 1985).

The pluton probably originated by local partial melting. The temperature of a partial melt in coexistence with minerals is buffered by the low-variance heterogeneous equilibria (Presnall, 1969). Existing experimental data (Fig. 2-1) indicate that, for a H_2O-

FIG. 2-4.   (a) Outcrop of the tonalite of the Round Valley pluton showing details of migmatitic xenolith of gneissic meta-volcanic rock; U.S. highway 95, 250 m north of Round Valley road junction, SE1/4 Sec. 26, R.1E., T.21N., Indian Mountain quadrangle. (b) Country rock of Round Valley pluton, U.S. highway 95 roadcut 1.7 km south of Hazard Creek confluence, SW1/4 Sec. 11, R.1E., T.21N., Indian Mountain quadrangle. Shows banded but locally intrusive relation of gneiss (possibly metamorphosed Riggins Group) containing mineral assemblage similar to intrusive rock. Compare texture of rock with that of the xenolith in (a).

saturated tonalitic melt,[1] this temperature must have been between 650 and 700°C for pressures of 6–8 kb. However, for a tonalitic melt hydrous enough to form biotite and hornblende but not enough to generate a vapor phase, the temperature would be about 800°C (Wyllie, 1977; see also Naney, 1983). The paleogeothermal gradient during intrusion thus was probably no more than about 30 ± 5°C/km, which is a reasonable sialic crustal geothermal gradient (Lachenbruch and Saas, 1978).

The evidence for local melt formation suggests that barring major inversions of the crustal temperature profile, there is little likelihood of development of local anatectic magma bodies, as mafic as those now exposed, at higher tectonic levels.

The Tenpeak Area, Washington

The Tenpeak pluton (Fig. 2-5) has highly discordant K-Ar ages on hornblende (92.8 ± 3.1 m.y.) and biotite (77.3 ± 2.4 m.y.; Engels *et al.*, 1976. All ages corrected for new constants by R. W. Tabor, written communication, 1985). It is one of many synchronous plutons in the area; from the northwest to southeast they are the Chaval (Boak, 1977), the Sloan Creek, the Sulfur Mountain, the Tenpeak, and the large Mount Stuart batholith (Cater, 1982; Tabor *et al.*, in press). The Chaval and Tenpeak, as well as the apparently younger nearby Clark Mountain pluton (58.6- and 60.7-m.y. K-Ar ages on biotite and muscovite, respectively; Engels *et al.*, 1976), carry magmatic epidote.

Regional metamorphism of the country rocks surrounding the Cretaceous plutons, including the Jurassic(?) Chiwaukum Schist, has been studied by many people (Berti, 1983; Evans and Berti, 1986; Kaneda, 1980; Plummer, 1980; Tabor *et al.*, in press). The Chiwaukum Schist has undergone Barrovian metamorphism and contains kyanite-sillimanite-staurolite schists and amphibolite. Boak (1977) described a garnet schist from a roof pendant next to the contact with the Chaval pluton that includes kyanite in the assemblage; this rock was presumably equilibrated under intrusive conditions (Boak interpreted the andalusite and cordierite in the country rock near the Chaval pluton as the result of contact effect of the Tertiary Snowking pluton). The assemblage suggests temperatures of about 600°C and pressures of at least 4 kb (the pressure for the triple point of the aluminum-silicate polymorphs). On the basis of mineral assemblage and mineral chemistry relations of the Chiwaukum Schist, Evans and Berti (1986) suggest that the pressure in the area north of the Mount Stuart batholith, shortly after its intrusion and possibly synchronous with the intrusion of the Tenpeak pluton, was 6–7 kb. Detailed pressure-temperature-time paths in the area are being studied by Ralph A. Haugerud; for now I adopt 600°C and 6 kb for the environment of emplacement of the Tenpeak pluton. A maximum paleogeothermal gradient of 30°C/km is indicated, similar to that of the Round Valley, Idaho area.

About 15 km west of the Tenpeak pluton, the apparently local Evergreen fault separates the Chiwaukum Schist from the Tonga Schist of Yeats (1958) that has been metamorphosed in the blueschist facies. Both rock groups are east of the regionally important Straight Creek Fault. A possible explanation for the juxtaposition of the two types of metamorphic rocks will be discussed later in this paper.

---

[1] The magmatic epidote-bearing plutons are quartz diorite, tonalite, and granodiorite (Brew and Morrell, 1983; Hutchison, 1982; E. Zen and J. M. Hammarstrom, unpublished data). I will use the dominant tonalite composition in the discussion. During consolidation, the melt fraction of the magma could have been no more mafic than the bulk composition, so the temperature of the tonalite solidus is the upper bound to the host rock temperature during intrusion.

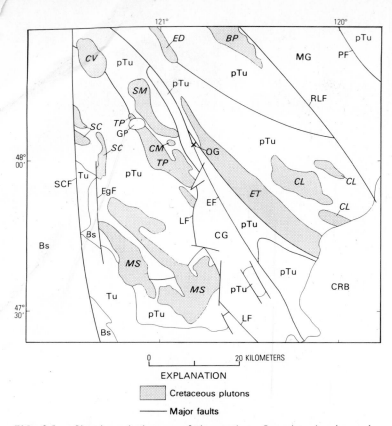

FIG. 2.5. Sketch geologic map of the northern Cascades, showing major faults and Cretaceous plutons, including the Tenpeak pluton. Geology from Tabor *et al.* (in press) and modified from Cater and Crowder (1967). Abbreviations: CG, Chiwaukum graben; MG, Methow graben; CRB, Columbia River Basalt Group; Tu, undifferentiated Tertiary rocks; Bs, undifferentiated rocks west of the Evergreen and the Straight Creek faults, including blueschist-facies metamorphic rocks (Tonga Schist of Yeats, 1958, and the Shuksan metamorphic suite of Misch, 1966; Brown *et al.*, 1982); pTu, undifferentiated pre-Tertiary metamorphic and igneous rocks, except where otherwise identified. Cretaceous and Paleocene intrusions: ED, Eldorado pluton; BP, Black Peak batholith; CL, Chelan batholith; ET, Entiat pluton; SM, Sulfur Mountain pluton; CV, Chaval pluton; SC, Sloan Creek plutons, CM, Clark Mountain plutons; TP, Tenpeak pluton; MS, Mount Stuart batholith. Selected high-angle faults: PF, Pasayten fault; RLF, Ross Lake fault; EF, Entiat fault; LF, Leavenworth fault; EgF, Evergreen fault; SCF, Straight Creek fault. OG, Old Gib volcanic neck; GP, Glacier Peak.

The Ecstall Pluton, British Columbia

The Ecstall pluton (Fig. 2-6) has a 98-m.y. zircon Pb/U age (Woodsworth *et al.*, 1983b). Crawford and Hollister (1982). Woodsworth *et al*, (1983a), and Crawford *et al.* (1987; see also Hutchison, 1970) described field evidence for the synchroneity of intrusion, thrusting, and regional metamorphism, and described the mineral assemblages of the country rocks from which they inferred peak pressure-temperature values of 8 ± 1 kb

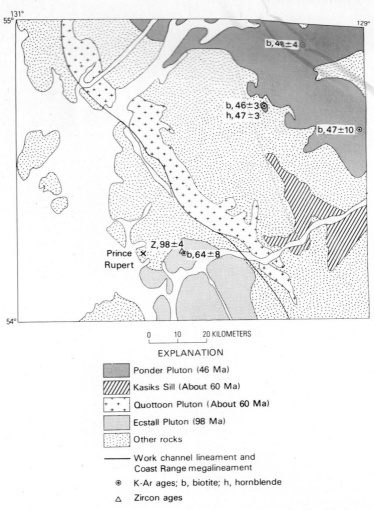

FIG. 2-6. Geologic map of Prince Rupert area, showing the Ecstall pluton, the Quottoon pluton, the Kasiks sill, and the Ponder pluton. The Work Channel lineament and the location of Khtada Lake are shown. The locations of samples for biotite (b), and hornblende (h) K-Ar ages are those reported by Hutchison (1982). The zircon sample location is from Woodsworth *et al.* (1983b). Geology from Hollister (1982).

(25 km) and 625 ± 25°C. This pressure is in excellent agreement with the stability of magmatic epidote. The paleogeothermal gradient is about 25°C/km. If, as suggested by Woodsworth *et al.* (1983a) and by Crawford *et al.* (1987), hotter rocks were emplaced over cooler during regional thrusting, then beneath the present level of exposure rocks metamorphosed at lower temperature might exist, so the average geothermal gradient might prove to be cooler yet.

TECTONIC SIGNIFICANCE OF HIGH-PRESSURE PLUTONIC ROCKS

Ketchikan-Petersburg Area, Southeast Alaska

Magmatic epidote-bearing plutons occur in an elongate belt extending from near Ketchikan north-northwestward, through the northern end and west half of Revillagigedo Island, into the eastern part of the Petersburg 1 X 2° quadrangle (Brew and Morrell, 1983; Burrell, 1984), and farther on north to the vicinity of Juneau. Comparable rocks apparently also occur at Haines Point near Skagway. I will, however, discuss only the 96–97-m.y. Moth Bay pluton (Sutter and Crawford, 1985) east of Ketchikan on the south shore of Revillagigedo Island, and certain *ca.* 90-m.y. foliated tonalite, granodiorite, and quartz diorite in the eastern part of the Petersburg map area to the north-northwest (Douglass *et al.*, 1985).

Rocks surrounding the Moth Bay pluton include metasedimentary rocks and Miocene plutonic rocks (Fig. 2-7). Some of the metasedimentary rocks are of low regional metamorphic grade; Berg *et al.* (1978) described these rocks (their unit MzPzs) near the Moth Bay pluton as "dark gray and silvery-gray phyllite, and fine-grained semischist and schist, and subordinate layers of green phyllite and schist [that contain] quartz, feldspar, biotite, garnet, muscovite, and pyrite, and minor graphite, hornblende or actinolite, and chlorite." The metamorphism is not younger than the Moth Bay pluton. Lack of signs of

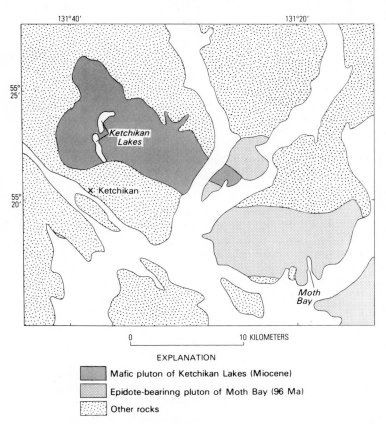

EXPLANATION

■ Mafic pluton of Ketchikan Lakes (Miocene)

▨ Epidote-bearinng pluton of Moth Bay (96 Ma)

▦ Other rocks

FIG. 2-7   Geologic map of the Moth Bay pluton area. From Koch and Elliott (1984) and Berg *et al.* (1978).

partial melting in the country rock, consisting of the plagioclase + quartz + muscovite assemblage, indicates a *maximum* temperature of about 650°C for an assumed pressure of 4 kb; this maximum temperature is lower for either higher or lower pressures (Wyllie, 1977; Zen, 1985a). Thus, I assign $T \leqslant 500°C$ and $P \geqslant 6$ kb for the environment of plutonism. The geothermal gradient is at most 25°C/km, but may be considerably less, possibly as low as 20°C/km.

A second area for which estimate is possible is the kyanite-staurolite-garnet-plagioclase-muscovite-biotite-quartz contact aureole assemblage from the east part of the Petersburg area (Douglass *et al.*, 1985). Once again, using the magmatic epidote to provide a pressure of at least 6 kb, the mineral assemblage indicates a temperature no more than 600°C in the aureole, and less away from the aureole. The maximum geothermal gradient is about 30°C/km.

CRUSTAL SIGNIFICANCE OF EPIDOTE-BEARING PLUTONS

Presence of epidote-bearing and related high-pressure plutons in an orogenic belt provides evidence for the postintrusion evolution of the sialic crust, some aspects of which were summarized previously (Zen, 1985b).

Crustal Thickness

Exposure of the high-pressure plutons on the earth's surface means that about 20–30 km of sialic crust has been removed, either by tectonic denudation or by erosion, since the plutons were emplaced. Yet the country rocks are invariably supracrustal — sedimentary and volcanic rocks metamorphosed from the greenschist facies, as in southeast Alaska, to the amphibolite facies, as in British Columbia, Washington, and Idaho. How much sialic crust might have underlain these rocks at the time of intrusion of the high-pressure plutons? The following thoughts may be helpful.

1. Regional temperatures of 500–600°C (southeast Alaska) correspond to mid-crustal levels. For the Alaska country rocks, the temperature difference between the country rock and the solidus of the tonalite, which probably had a crustal source, may be as much as 250°C. Extrapolating the regional geothermal gradient, one could infer a minimum of about 10 km of sialic crust between the exposed plutons of southeast Alaska and the level of magma generation; this level is actually exposed at Round Valley in Idaho.

2. The present thickness of the crust is around 38 km in the area of the Ecstall pluton (Roddick and Hutchison, 1974; Woodsworth *et al.*, 1983a) and in southeast Alaska (Barnes, 1977). In the Cascades of Washington, it is about 40 km (Rohay, 1982). Consideration of regional isostasy shows that barring major postintrusive regional crustal thickening or thinning, the modern thickness plus the amount denuded as indicated by the estimated pressure should approximate the thickness of the crust at the time of intrusion.

I conclude that areas where the high-pressure, magmatic epidote-bearing plutons occur had abnormally thick sialic crust during the plutonism. The thickness could have been as much as 50–60 km (the thickness of crust removed to expose the epidote-bearing plutons *plus* the thickness of crust presumed to underlie the plutons). A major source of uncertainty is possible subsequent crustal addition by underplating or other mechanisms. For example, Hollister (1982) and Crawford *et al.* (1987) estimated possible 8 km of added crust in the area east of the Work Channel lineament in British Columbia that resulted from Tertiary plutonic activities. This conclusion depends on assuming that these tabular bodies were originally subhorizontal and acquired the present dips later. However, even with a postintrusion addition of 10 km of crust, a crustal thickness of 40–50 km during intrusion of the epidote-bearing plutons seems ineluctable.

A crustal thickness of about 50 km is not unlike the thickness of the modern sialic crust in the central Peruvian Andes (James, 1971), the Himalaya region (see summary in Zeitler, 1985), or the modern central Alps (Müller, 1983).

Uplift Rates

Uplift rates are derived from pairs of depth and age relations for a given rock mass. Temperature of the rock mass enters into the relations only indirectly, for example if the depth must be estimated from the geothermal gradient, or if a time interval for cooling of the rock body to the blocking temperature of a particular isotopic clock must be added to the apparent radiometric age. Table 2-1 summarizes the calculated uplift rates based on the discussions below. More than one set of figures is given for each area, based on possible ranges of depths of burial of the epidote-bearing plutons, or on different bracketing data, as discussed below.

Round Valley area, Idaho Initial conditions for the uplift of this pluton is taken to be a depth of 25 km at 110 m.y.b.p. The younger event used to compute the uplift rate is the 17 m.y.-old Imnaha Basalt (Hooper, 1982; Hooper *et al.*, 1984; McKee *et al.*, 1981) of the Columbia River Basalt Group, which directly overlies the Round Valley pluton. The average uplift rate is therefore $(25-0)\text{km}/(110-17)$ m.y. $= 0.27$ mm/yr. Differences in the elevation of the land surface at the two bracketing ages are ignored.

TABLE 2-1. Minimum Uplift Rates Determined from Epidote-Bearing Plutons and Younger Igneous Events[a]

Area		Pre-uplift Depth (km)	Age (m.y.)	Post-uplift Depth (km)	Age (m.y.)	Average Rate of Uplift (mm/yr)
Round	A.	~30	~110	0	17	0.32
Valley	B.	~25	~110	0	17	0.27
Idaho	C.	~20	~110	0	17	0.22
Tenpeak	A.	~30	93	~1	30	0.46
Washington	B.	~25	93	~1	30	0.38
	C.	~20	93	~1	30	0.30
Ecstall	A.	~30	98	10	72	0.77[b]
British	B.	~25	98	10	72	0.58[b]
Columbia	C.	~20	98	10	10	0.38[b]
Moth Bay	A.	~30	96	~5	24	0.35[c]
Alaska	B.	~25	96	~5	24	0.28[c]
	C.	~20	96	~5	24	0.21[c]
Etolin	A.	~30	~90	~2	22	0.41
Island,	B.	~25	~90	~2	22	0.34
Alaska	C.	~20	~90	~2	22	0.26

[a]Rates assume original depth of A, 30 km; B, 25 km; C, 20 km. Entries in *italics* correspond to those given in the text.
[b]Based on cooling age of biotite, assuming cooling was through a regional geothermal gradient.
[c]Based on dated mafic intrusion of Ketchikan Lakes.

Sutter *et al.* (1984) proposed that the argon spectrum for hornblende from a magmatic epidote-bearing pluton at Slate Creek, about 30 km north-northeast of Riggins (Fig. 2-3; Lund, 1984), indicates cooling from ~525 to ~125°C between 84 and 79 m.y. "during a period of uplift along high-angle faults." If a geothermal gradient of 30°C/km is assumed, an *initial* uplift rate of 2.7 mm/yr would result. This is an order of magnitude greater than the Imnaha Basalt-bracketed rate, but could be real. The rate, however, does depend on interpretation of events that occurred very close to the time of intrusion; the cooling might not have been entirely the result of regional uplift but could be partly due to static cooling of a pluton.

Tenpeak area, Washington The initial conditions for uplift in the Tenpeal pluton area are at 20 km depth and 93 m.y.b.p. A younger bracket is provied by the Old Gib volcanic neck (Fig. 2-5; Cater and Crowder, 1967), located about 20 km to the east. The Old Gib has a published biotite K-Ar age of 45 m.y. (corrected for new constants from Engels *et al.*, 1976), but a better hornblende K-Ar age of ~30 m.y. (R. W. Tabor, written communication, 1985). Calculations are made using the hornblende age (Table 2-1). The Old Gib must have solidified virtually at the land surface, and I assigned to it a depth of 1 km. The uplift rate is $(20 - 1)/(93 - 30) = 0.30$ mm/yr. The existence of faults in the region counsels caution in regional application of the uplift rate; however, at an age of 30 m.y. ago, the areas occupied by the Old Gib and by the Tenpeak were presumably welded together.

Ketchikan-Petersburg area, southeast Alaska The Moth Bay pluton is 96 m.y. old (Sutter and Crawford, 1985) and was emplaced at a depth estimated at 20 km. It, or a satellite, is intruded by the mafic pluton of Ketchikan Lakes (Fig. 2-8; Koch and Elliott, 1984) that has nearly concordant K-Ar ages of 24.9 ± 0.8 m.y. on hornblende and 23.2 ± 0.7 m.y. on biotite. Assigning an arbitrary but seemingly generous value of 5 km to the depth of the mafic intrusion, we get for the uplift rate $(20 - 5)/(96 - 24) = 0.21$ mm/yr. Even a change of this depth to 8 km changes the uplift rate only by 20 percent. A tighter bracket based on other rocks west of the Coast Range megalineament (Fig. 2-2; Brew and Ford, 1978, 1981), however, is not currently available.

A second area providing data on uplift rate is on Etolin Island (Fig. 2-8). Hunt (1984) described a composite shallow intrusion near Burnett Inlet that contains abundant miarolitic cavities as well as extensive graphic intergrowth of quartz and feldspar. This composite pluton has a K-Ar age of 20–22 m.y. (nature of dated sample unspecified; Brew *et al.*, 1984). I arbitrarily but probably generously assigned a depth of 2 km to the body. The Burnett Inlet body cuts magmatic garnet-epidote bearing plutons, both with and without hornblende, assigned to the "90 m.y. group" by Brew and Morrell (1983) and Burrell (1984). These relations indicate an uplift rate of approximately $(20 - 2)/(90 - 22) = 0.27$ mm/yr, close to the estimates for the Moth Bay pluton.

Ecstall area, British Columbia Discussion of uplift rates of the Ecstall pluton is left to the last because the regional metamorphism, depth of intrusion of other nearby plutons, and the tectonic context of these events have been described. Following Woodworth *et al.* (1983a), I use a value of 25 km for the initial depth of the pluton now exposed at the surface. The Ecstall pluton has a K-Ar biotite age of 72 m.y. and a K-Ar hornblende age of 76 m.y. (Harrison *et al.*, 1979; all K-Ar ages corrected for revised decay constants). If the cooling resulted from regional uplift, if argon blocking temperature in hornblende is 525°C and in biotite is 250°C, and if the regional paleogeothermal gradient during intrusion, 25°C/km, is used, then the pluton was at a depth of 21 km at 76 m.y. and at a depth

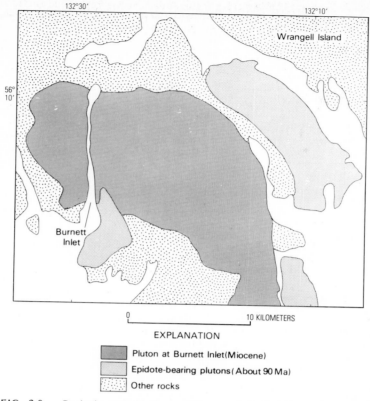

0   10 KILOMETERS

EXPLANATION

▨ Pluton at Burnett Inlet(Miocene)

▨ Epidote-bearing plutons (About 90 Ma)

▨ Other rocks

FIG. 2-8. Geologic map of the central part of Etolin Island, from Hunt (1984).

of 10 km at 72 m.y.b.p. The three uplift rates that can be calculated from pairs of these figures are shown in Tables 2-1 and 2-2; the overall rate between 98 and 72 m.y. is $(25 - 10)/(98 - 72) = 0.58$ mm/year. The large time lapse between intrusion and the biotite K-Ar age suggests that cooling might have been caused by regional uplift rather than by static cooling of a pluton. The possiblity of an inverted thermal profile, discussed earlier, suggests possibly a lower regional thermal gradient, therefore a higher uplift rate, if a larger packet of the crust is considered.

It is interesting to compare these rates with uplift rates based on the ages and depths of emplacement of the Quottoon (and the Kasiks) and the Ponder plutons in the area east of the Work Channel lineament (Hutchinson, 1982). Zircon from the Quottoon gave an age of 60 m.y. (Armstrong and Runkle, 1979) based on the Pb-U method. Kenah (1978) and Hollister (1982) estimated the depth of intrusion of this pluton at ~17 km, based on *P-T* data of Selverstone and Hollister (1980). This value agrees with the depth of intrusion for the Carlson Creek pluton east of Juneau, Alaska, that, like the Quottoon, is a unit in the tonalite sill complex (Grissom *et al.*, 1985). Hollister (1982) inferred that the Ponder pluton is about 49 m.y. old on the basis of nearly concordant K-Ar ages on hornblende and on biotite from correlative plutons just to the north in Alaska (Smith and Diggles,

TABLE 2-2. Incremental Uplift Rates for the Area near the Ecstall Pluton:[a]

| | Ecstall Pluton | | | | | |
	Total	98-76 m.y.	76-72 m.y.	72-40 m.y.	40-10 m.y.	10-0 m.y.
ΔZ (km)	25	$25-21=4$	$21-10=11$	$10-5=5$	$5-2=3$	2
Δt (m.y.)	98	22	4	32	30	10
$\Delta Z/\Delta t$	0.26	0.18	2.8	0.16	0.10	0.20

| | Quottoon/Kasiks Plutons | | | | |
	Total	60-47 m.y.	47-40 m.y.	40-10 m.y.	10-0 m.y.
ΔZ (km)	17	$17-6=9$	$6-5=1$	$5-3=2$	3
Δt (m.y.)	60	13	7	30	10
$\Delta Z/\Delta t$	0.28	0.69	0.14	0.07	0.30

| | Ponder Pluton | | | |
	Total	47-40 m.y.	40-10 m.y.	10-0 m.y.
ΔZ (km)	6	$6-5=1$	$5-2=3$	2
Δt (m.y.)	47	7	30	10
$\Delta Z/\Delta t$	0.13	0.14	0.10	0.20

[a] ΔZ, change in depth of burial resulting from the uplift during the time increment; Δt, time increment, in m.y. The two columns headed "40-10 m.y." and "10-0 m.y." are interpolated from the fission-track data of Parrish (1983). See the text for details.

1981). For the Ponder pluton itself, K-Ar ages on hornblende, 47 ± 3 m.y., and on biotite, 46 ± 3 m.y. (both ages corrected for revised decay constants), were reported from the same sample; other biotite K-Ar ages from the Ponder gave 48 ± 4 and 47 ± 10 m.y. (Fig. 2-6; Hutchison, 1982). So I use 47 m.y. for the Ponder. The depth of intrusion of the Ponder pluton was estimated at 6 km (Grissom et al., 1985; Sisson, 1985).

These two plutons yield an incremental uplift rate of 0.69 mm/yr (Table 2-2), which is close to the figure for the pre-72-m.y. uplift rate of the Ecstall area. Table 2-2 also includes the uplift calculations of Parrish (1983) based on fission track ages of minerals in the Central Gneiss Complex. Parrish's approach does not assume values of the paleogeothermal gradient. Instead, it assumes that isothermal surfaces are fixed relative to sea level and are parallel to one another, so that the quenched-in fission-track ages at different elevations for a given mineral and area directly measure regional uplift rates. Different minerals, having different fission-track quenching temperatures, allow estimating these rates over increments of time. Even though the Prince Rupert–Skeena River area is near the northern limit of Parrish's data, his results can be compared with mine. The data of Table 2-2 show a peak uplift rate of about 0.6 mm/yr in the Late Cretaceous and early Tertiary (Paleocene) on the two sides of the Work Channel lineament. Uplift slowed down during the late Eocene and Oligocene, but, based on Parrish's data (1983), resurged in late Miocene to Recent times.

The uplift rates presented in this section are within the range of Cenozoic and modern rates in orogenic systems, summarized in Table 2-3. The similarities are striking when the uncertainties and assumptions such as geothermal gradients, argon-closing temperatures for specific mineral samples, and even possible effects of glacial rebound in some

TABLE 2-3. Uplift Rates of Modern Orogens Estimated by Various Methods (Contemporary Except where Specified)

Area	Rate (mm/yr)	Methods	Reference
Central Alps	0.4–1	Denudation rate; based on radiometric age and heat flow	Clark and Jaeger, 1969
	0.2–1.3 (since Neogene)	Radiometric cooling ages	Wagner *et al.*, 1977
	0.8–2.6	Precise leveling	Gubler *et al.*, 1981
Himalayas	0.8 (Tertiary)	Comparison of various radiometric methods	Sharma *et al.*, 1980
	5 (Quaternary)	Dated stratigraphic and erosion events	Gansser, 1983
	0.5 (late Tertiary)	Fission-track ages	Zeitler, 1985
Taiwan	5.5	Denudation rate and terrace uplift rate	Li, 1976
Meridian Andes, Venezuela	5 0.8 (both Neogene)	Terrace elevation Apatite fission-track age	Giegengack *et al.*, 1984
Central Andes	3–5 (Neogene)	Various morphologic analyses, including unroofing of dated batholith	Gansser, 1983
Southern Alps, New Zealand	1.5–22	Sedimentation rate, sediment load, geomorphic analysis	Adams, 1980
	5–9 2–4 (before 140,000 b.p.)	Dated marine terraces	Bull and Bull, 1984
Japan	0.6–2.2 (average 1.3)	Denudation rate and precise leveling data	Yoshikawa, 1974
Coast Range, British Columbia	0.4 (last 10 m.y.)	Fission-track dating	Parrish, 1983

areas are taken into account. Post-tectonic regional uplift rates between 0.1 and 1mm/yr averaged over a few tens of million years may be typical of orogens that have undergone crustal thickening, plutonism, and metamorphism.

CAN BLUESCHIST AND EPIDOTE-BEARING PLUTONS BE PRODUCED IN A SINGLE TERRANE AND DURING THE SAME THERMAL EVENT?

R. G. Coleman once asked how it is possible to form "high pressure-high temperature epidote-bearing magmas where blueschists also sometimes form! The problem here is how do you refrigerate accretional terranes to produce blueschists and in the same setting produce magmatic rocks that require temperatures of 800–600°C? After formation, to preserve and expose these rocks (blueschists or epidote tonalites) we need exceptionally high erosion rates" (see comments in Zen, 1985b). Areas of supracrustal thickening and low geothermal gradients are commonly associated with development of high-pressure, low-temperature metamorphic rocks, including blueschists. Thus, even though the epidote-bearing pluton and blueschist association is uncommon, Coleman's concern requires serious consideration. In the Tenpeak area in Washington, in fact, these two rock types do come to within about 15 km across the apparently local Evergreen fault (Fig. 2-5).

I propose that the two "end-member" geological environments envisioned by Coleman are not necessarily incompatible. The starting point of my argument is a modeling of pressure-temperature-time paths for the burial and uplift of large orogenic masses. Even though the models are necessarily idealized and the initial and boundary conditions may not correspond to those envisioned for the rocks discussed in this paper, still, the model results counsel caution in assigning apparently disparate geologic terranes to different tectonic origins. [The thermal structure and subsequent evolution during uplift and erosion of such overthrust plates is also treated by Peacock (Chapter 34, this volume).]

The model in Fig 2-9, based on a computer program of Haugerud (1986), is chosen to illustrate the point. Unlike the initial conditions postulated by the model of England and Thompson (1984) that assumes an instantaneous duplication of the entire 35-km-thick crust by thrusting, our model assumes the instantaneous thrusting of 6-km-thick sheets of sedimentary material at the land surface. Five such sheets were emplaced at 0.5-m.y. intervals, during which heating and cooling occurred. A 5-m.y. incubation period followed the period of the thrusting, after which erosion and uplift at a rate of 0.4 km/m.y. proceeded for 75 m.y. to remove the thrust pile. Other modeling parameters are given in the figure caption.

The thermal history of four representative burial levels are shown in Fig. 2-9. Given the constraints of initial and boundary conditions, the rock masses buried to different depths evolved in sharply contrasting ways. Rock masses buried to 25 and 30 km (within the lowest thrust sheet and at the original land surface, respectively) would spend 20–30 m.y. in the blueschist-glaucophane schist facies, but upon continued uplift would heat to a maximum temperature of 450–500°C, and so might develop a greenschist facies overprint before the mineral assemblage is quenched-in near the surface. These rock masses would not, or would barely, pass through the argon blocking temperature for hornblende; hornblende formed during metamorphism presumably would largely record the time of beginning of its crystallization, at about 20–30 m.y. after uplift began. However, any biotite would record much younger cooling ages.

In contrast, rock masses initially buried to 40–45 km depth at the same moment, incubated for the same length of time, and uplifted at the same rate, would be heated to 600–650°C at a depth of about 30 km. Even though during the early stages, the rock masses resided in the higher part of the glaucophane schist-eclogite region, during uplift they would pass through the Barrovian metamorphic grades corresponding to upper amphibolite to pyroxene-hornfels facies. For rocks of appropriate composition such as andesite and related volcanogenic sediments, as well as pelitic sediments, partial melting would occur, producing magmas of metaluminous granodioritic/tonalitic compositions and of peraluminous granitic to granodioritic compositions, respectively. The less silica-rich portions of the melts should consolidate at depths of about 25 km, precisely the range of the epidote-bearing magmas. The more silica-rich melts, having lower solidus temperatures, would consolidate later and at shallower depths. It is probable that any preexisting eclogitic metamorphic assemblages and textures would be destroyed during this anatectic and high-grade metamorphic episode. The system would cool through the argon blocking temperature for hornblende about 55–65 m.y. after the beginning of uplift.

The two pairs of rock masses, therefore, would present an age discrepancy of several tens of millions of years, in addition to sharp contrasts in the metamorphic style (blue-

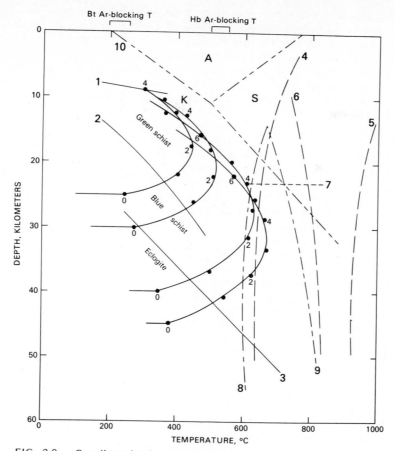

FIG. 2-9. One-dimensional calculation of pressure-temperature-time paths for burial-uplift history. Instantaneous thrusting of five sedimentary sheets each 6 km thick at 0.5-m.y. intervals, followed by a 5-m.y. incubation period. Uplift and erosion then proceeds at 0.4 mm/yr for 75 m.y. The heavy lines are trajectories of the *P-T* paths of rocks buried initially at 25-, 30-, 40-, and 45-km depths. Dots on these lines indicate position of the samples at 20-m.y. intervals since beginning of uplift (right end of the horizontal line segments). Labeled lines: 1, laumontite = lawsonite + quartz + H_2O (Nitsch, 1968); 2, Calcite = aragonite (Johannes and Puhan, 1971); 3, Albite = jadeite + quartz (Boettcher and Wyllie, 1969); 4, 5, and 6, tonalite solidi, respectively, for a H_2O-saturated melt, for a dry melt, and for a melt sufficiently hydrous to allow hornblende and biotite to crystallize (Wyllie, 1977); 7, 8-kb melting interval for H_2O-saturated synthetic granodiorite (Naney, 1983) from which epidote crystallized at $T < 700°C$; 8 and 9, wet and dry solidi for pelite melting (England and Thompson, 1984); 10, the stability fields of the polymorphs of aluminum silicate minerals (Holdaway, 1971). The argon-blocking temperature ranges for hornblende and for biotite are shown along the top margin of the diagram. Modeling parameters: Thermal conductivity, 2.25 W/(M-K); density, 2780 kg/m³; heat capacity, 900 J/(K-kg); heat generation, 2 μW/m³; restricted to upper 15 km of initial crust and overlying thrust sheets; zero elsewhere; surface temperature, 0°C; basal flux, 30 mW/m² at a depth of 150 km. Calculated with time steps of 0.25 m.y. for first 5 m.y. and of 1 m.y. thereafter, and space steps of 1 km. The locations of metamorphic facies are indicated.

schist with greenschist overprint versus rocks in amphibolite to pyroxene-hornfels facies, including kyanite-staurolite-sillimanite schist and anatectic plutons). Yet both rock assemblages would be developed within the same terrane.

It might be noted that the *average* model geothermal gradient at maximum temperatures of burial is considerably greater than the *equilibrium* gradients. For example, a rock mass initially buried to 45 km has an *average gradient from the surface* at the moment of maximum T (depth of 32 km) of 22°C/km. Because cooling lags behind uplift, the average gradient actually increases to 25°C/km at 25 km, and to nearly 28°C/km at 16 km (these values may be compared with the values derived by petrologic relations for the areas discussed, for instance, at the Ecstall pluton). By contrast, the equilibrium average gradient to 10 km is 22°C/km, and to 30 km it is only 17°C/km.

One possible actual example of this combination is in the area of the Tenpeak pluton of Washington, where the two rock types occur across the apparently local Evergreen fault (Fig. 2-5). Just east of the fault, the Tonga Schist (Yeats, 1958), which is a blueschist with a greenschist overprint, has been correlated with the blueschist of the Shuksan metamorphic suite of Misch (1966) that occurs farther west across the Straight Creek fault and that has yielded K-Ar and Rb-Sr isochron ages around 130 m.y. (Brown *et al.*, 1982). The 40-m.y. age difference between the blueschist and the epidote-bearing plutons is consistent with the age differences one might expect from the model described above. The early andalusite-cordierite contact assemblage around the Mount Stuart batholith, reported by Evans and Berti (1986), might pose a problem. However, geological and age data from the area do not yet adequately constrain the thermal history; the thermal model can be modified in various ways to fit the data. The Mount Stuart is apparently a few million years older than the Tenpeak, as indicated by available age data (Engels *et al.*, 1976). The post-Mount Stuart regional pressure increase reported by Evans and Berti and implied by the high-pressure plutons could be achieved by the emplacement of a thick (~15 km) layer of magma derived from a deeper crustal level; the Tenpeak pluton could be a part of that slightly younger magmatic history. Such magmatic events would not be recorded laterally in the cooler, more shallowly buried parts of the system, which include blueschist.

DISCUSSIONS AND SUMMARY

Many late Mesozoic plutons of the western Cordillera, in a corridor between western Idaho and southeastern Alaska, consolidated at depths between 20 and 30 km. Yet these plutons intruded supracrustal country rocks, some of which recorded low metamorphic grades. Thus at these places the paleogeothermal gradients were not abnormally high, and the sialic crust was thick (45–60 km). By dating the plutons and younger geologic events that took place at known depths, we can calculate the rates of regional uplift averaged over millions to tens of millions of years since intrusion. These rates are on the order of 0.2–0.6 mm/yr, comparable with or less than the reported uplift rates for several modern mountain systems.

The tectonic style in the areas that contain high-pressure plutons includes extensive plutonism and regional metamorphism, in addition to thickening of the sialic crust. This style contrasts with that of other areas of reported terrane accretion, such as in south-central Nevada, where early Mesozoic accretion of sedimentary sequences was not known to be accompanied by plutonic rocks, metamorphism, or crustal thickening (see, e.g., Speed, 1978, 1982). The difference between the "thick-skinned" and "thin-skinned" accretion may reflect the relative importance, respectively, of subduction-related versus transform-related mechanisms of terrane transport prior to terrane welding.

High rate of uplift implies high rate of erosion and/or tectonic denunation; the denuded material may be recorded elsewhere either in the arrival of thrust sheets, or in the nature of concurrent sedimentation. Additional tectonic constraints may be obtained by matching of sedimentary rocks of the same age that we know must have been derived from areas of rapid erosion, with presumed source areas. Such matching might reveal concurrent or later tectonic transport of the sedimentary deposits away from source areas (see Hollister, 1982), or provide the basis to refine the history of uplift by the provenance and the rate of supply of the sediments.

Are the phenomena described in this paper peculiar to the western Cordillera? Perhaps not. High-pressure plutons may be more widespread than is generally realized. The Andes have the excess sialic crust, but the system may be too young for exposure of the associated deeper plutonic features. Müller (1983) recently summarized the contemporary dynamics of the Swiss Alps. He showed that areas of highest uplift rates (1–2 mm/yr) coincide with areas where the sialic crust is more than 50 km thick. These are also areas of relatively low geothermal gradient and high-pressure Alpine metamorphism (Werner, 1980; Frey *et al.*, 1980). Associated synmetamorphic plutons are sparse, but it will be interesting to see if any of these carries evidence of a high-pressure environment of consolidation.

Sutter and Hatch (1985) calculated a 28-m.y. average uplift rate of 0.12 mm/yr for the Acadian orogeny in west-central Massachusetts, using a geothermal gradient of 30°C/km. This rate is similar to those given here. The Acadian crust in southern New England was locally thick; pressures as much as 7–8 kb have been determined based on metamorphic mineral assemblages east of the area studied by Sutter and Hatch (Tracy and Robinson, 1980) and supported by the composition of the hornblende in the Hardwick pluton (Hammarstrom and Zen, 1985, 1986; Shearer, 1983). The present crustal thickness of the same tectonic zone in southern New Hampshire is interpreted to be about 30 km (Ando *et al.*, 1983). This area is a good candidate for epidote-bearing high-pressure plutons.

As we try to define past tectonic processes at plate margins, we need reliable information on the local paleo-thicknesses of sialic crusts. Unfortunately, available data are scanty and are often not much better than guesses. Knowledge of depth of plutonism and metamorphism tells us how much crust had been *removed* since some reference time, rather than the amount of crust remaining. Vibroseismic data at subduction zones, combined with information on subduction rates, might yield data on crustal thickness in the recent past by backward projection of the subduction process. Knowledge of regional relief, reconstructed from the sedimentary record, might provide insight on crustal thickness (see Hsü, 1983); similarly, knowledge of the volume of sediments produced during increments of time could provide minimal estimates of erosion and uplift rates, and indirectly of crustal thickness. Knowledge of the vertical regional geothermal gradient could lead to estimates of minimum depth to some particular isotherm that might be associated with the Moho. Successful determination of a sialic source terrane for some plutonic body, coupled with knowledge of the temperature of the intruded country rock, could lead to minimum estimates of the thickness of the crust remaining. Other strategies may be discovered. However, few if any of these methods are free of *ad hoc* assumptions or of circular reasoning. To devise good means to estimate paleo-thicknesses of crusts at any stage of orogenic processes is a challenge to geologists and geophysicists alike that may gain importance as we study the dynamics and evolution of continental crusts.

ACKNOWLEDGMENTS

This study has benefited from help from many sources. I thank Joe Arth, Fred Barker, Hank Berg, Dave Brew, Bob Coleman, Bob Criss, Weecha Crawford, Sue Douglass, Bob Fleck, Warren Hamilton, Ralph Haugerud, Mary Lou Hill, Linc Hollister, Art Ford, Mel Kuntz, Peter Lipman, Karen Lund, Randy Parrish, George Plafker, Jinny Sisson, Larry Snee, Harold Stowell, John Sutter, Rowland Tabor, Bill Taubeneck, Glenn Woodsworth, and Bob Zartman, for sharing their samples, knowledge, and enthusiasm. Brew, Ford, Crawford, Hollister, Snee, Tabor, and Taubeneck have been especially generous with their data and ideas. Ralph Haugerud allowed me to use his computer program to model uplift histories, instructed me on its use, and shared with me his tectonic ideas. Collaboration with Jane Hammarstrom has been both fruitful and rewarding. This paper was reviewed by Jim Smith, Weecha Crawford, Ray Elliott, Ralph Haugerud, Warren Hamilton, Dave Brew, Gary Ernst, Harold Stowell, and Rowland Tabor; they have greatly improved my thinking but are in no sense responsible for my follies.

REFERENCES

Adams, J., 1980, Contemporary uplift and erosion of the southern Alps, New Zealand: *Geol. Soc. America Bull.*, v. 90, pt. II, p. 1–114.

Ando, C. J., Cook, F. A., Oliver, J. E., Brown, L. D., and Kaufman, Sidney, 1983, Crustal geometry of the Appalachian orogen from seismic reflection studies; *in* Hatcher, R. D., Jr., Williams, Harold, and Zietz, Isidore, eds., *Contributions to the Tectonics and Geophysics of Mountain Chains:* Geol. Soc. America Mem. 158, p. 83–101.

Armstrong, R. L., and Runkle, Dita, 1979, Rb-Sr geochronometry of the Ecstall, Kitkiata, and Quottoon plutons and their country rocks, Prince Rupert region, Coast Plutonic Complex, British Columbia: *Can. J. Earth Sci.*, v. 16, p. 387–399.

Barnes, D. F., 1977, Bouguer gravity map of Alaska: *U.S. Geol. Survey Geophys. Inv. Map GP-913*, scale 1:2,500,000.

Berg, H. C., Elliott, R. L., Smith, J. G., and Koch, R. D., 1978, Geologic map of the Ketchikan and Prince Rupert quadrangles, Alaska: *U.S. Geol. Survey Open File Rpt. 78-73A*, 1 sheet, scale 1:250,000.

Berti, J. W., 1983, Barrovian-type metamorphism superimposed on low-pressure metamorphism in the Chiwaukum Schist, Stevens Pass, Washington: *Geol. Soc. America Abstr. with Programs*, v. 15, p. 437.

Boak, J. L., 1977, Geology and petrology of the Mount Chaval area, north Cascades, Washington: M.S. thesis, Univ. Washington, Seattle, Wash., 87 p.

Boettcher, A. L., and Wyllie, P. J., 1969, Phase relationships in the system $NaAlSiO_4$-SiO_2-H_2O to 35 kilobars pressure: *Amer. J. Sci.*, v. 267, p. 875–909.

Brew, D. A., and Ford, A. B., 1978, Megalineament of southeastern Alaska marks southwest edge of Coast Range batholithic complex: *Can. J. Earth Sci.*, v. 15, p. 1763–1772.

____, and Ford, A. B., 1981, The coast plutonic complex sill, southeastern Alaska, *in* Albert, N. R. D., and Hudson, T., eds., *The United States Geological Survey in Alaska — Accomplishments during 1979:* U.S. Geol. Survey Circ. 823-B, p. B96–99.

____, and Morrell, R. P., 1983, Intrusive rocks and plutonic belts of southeastern Alaska, U.S.A., *in* Roddick, J. A., ed., *Circum-Pacific Plutonic Terranes:* Geol. Soc. America Mem. 159, p. 171–193.

——, Ovenshine, A. T., Karl, S. M., and Hunt, S. J., 1984, Preliminary reconnaissance geologic map of the Petersburg and parts of the Port Alexander and Sumdum 1:250,000 quadrangles, southeastern Alaska: *U.S. Geol. Survey Open File Rpt. 84-405*, 43 p., 2 sheets, scale 1:250,000.

Brown, E. H., Wilson, D. L., Armstrong, R. L., and Harakal, J. E., 1982, Petrologic, structural, and age relations of serpentinite, amphibolite, and blueschist in the Shuksan suite of the Iron Mountain–Gee Point area, north Cascades, Washington: *Geol. Soc. America Bull.*, v. 93, p. 1087–1098.

Bull, W. B. and Bull, R. P., 1984, Tectonic significance of rounded quartz pebbles on the southern Alps of New Zealand: *Geol. Soc. America Abstr. with Programs*, v. 16, p. 458.

Burrell, P. D., 1984, Map and table describing the Admiralty-Revillagigedo intrusive belt plutons in the Petersburg 1:250,000 quadrangle, southeastern Alaska: *U.S. Geol. Survey Open File Rpt. 84-171*, 6 p., scale 1:250,000.

Carmichael, D. M., 1978, Metamorphic bathozones and bathograds – A measure of the depth of post-metamorphic uplift and erosion on the regional scale: *Amer. J. Sci.*, v. 278, p. 769–797.

Cater, F. W., 1982, Intrusive rocks of the Holden and Lucerne quadrangles, Washington – The relation of depth zones, composition, texture, and emplacement of plutons: *U.S. Geol. Survey Prof. Paper 1220*, 108 p.

——, and Crowder, D. F., 1967, Geologic map of the Holden quadrangle: *U.S. Geol. Survey Geol. Quad. Map GQ-646.*

Chatterjee, N. D., and Johannes, W., 1974, Thermal stability and standard thermodynamic properties of synthetic $2M_1$-muscovite $KAl_2[AlSi_3O_{10}(OH)_2]$: *Contrib. Mineralogy Petrology*, v. 48, p. 89–114.

Clark, S. P., Jr., and Jaeger, E., 1969, Denudation rate in the Alps from geochronologic and heat flow data: *Amer. J. Sci.*, v. 267, p. 1143–1160.

Crawford, M. L., and Hollister, L. S., 1982, Contrasts of metamorphic and structural histories across the Work Channel lineament, coast plutonic complex, British Columbia: *J. Geophys. Res.*, v. 87, p. 3849–3860.

——, Hollister, L. S., and Woodsworth, G. L., 1987, Crustal deformation and regional metamorphism across a terrane boundary – Coast Plutonic Complex, British Columbia: *Tectonics*, v. 6, p. 343–361.

Douglass, S. L., Brew, D. A., and Lanphere, M. A., 1985, Polymetamorphism in the eastern part of the Petersburg map area, southeastern Alaska: *Geol. Soc. America Abstr. with Programs*, v. 17, p. 352.

Engels, J. C., Tabor, R. W., Miller, F. K., and Obradovich, J. D., 1976, Summary of K-Ar, U-Pb, Pb-alpha, and fission track ages from Washington State prior to 1975 (exclusive of Columbia Plateau basalts): *U.S. Geol. Survey Misc. Field Studies Series MF-710*, 2 sheets.

England, P. C., and Thompson, A. B., 1984, Pressure-temperature-time paths of regional metamorphism – I. Heat transfer during the evolution of regions of thickened continental crust: *J. Petrology*, v. 25, p. 894–928.

Evans, B. W., and Berti, J. W., 1986, A revised metamorphic history for the Chiwaukum Schist, North Cascades, Washington: *Geology*, v. 14, p. 695–698.

Fleck, R. J., and Criss, R. E., 1985, Strontium and oxygen isotopic variations in Mesozoic and Tertiary plutons of central Idaho: *Contrib. Mineralogy Petrology*, v. 90, p. 291–308.

Frey, Martin, Bucher, Kurt, Frank, Erik, and Mullis, Josef, 1980, Alpine metamorphism along the geotraverse Basel-Chiasso – A review: *Eclogae Geol. Helvetiae*, v. 73, p. 527–546.

Gansser, A., 1983, The morphogenic phase of mountain building, *in* Hsü, K, J., ed., Mountain Building Processes: *New York; Academic*, p. 221–228.

Giegengack, Robert, Kuhn, B. P., and Shagam, Reginald, 1984, Late Cenozoic rapid uplift of the Merida Andes, Venezuela: *Geol. Soc. America Abstr. with Programs*, v. 16, p. 518.

Grissom, G. C., Peters, E. K., Sisson, V. B., Stowell, H. H., and Hollister, L. S., 1985, Pressure of crystallization of plutons in the Coast Mountains, B.C. and Alaska, based on aluminum content of hornblende: *Geol. Soc. America Abstr. with Programs*, v. 17, p. 358.

Gubler, E., Kahle, H. G., Klingele, E., Müller, Stephan, and Oliver, R., 1981, Recent crustal movements in Switzerland and their geophysical interpretation: *Tectonophysics*, v. 71, p. 125–152.

Hamilton, Warren, 1969, Reconnaissance geologic map of the Riggins quadrangle, west-central Idaho: *U.S. Geol. Survey Misc. Geol. Inv. Map I-579*, scale 1:125,000.

Hammarstrom, J. M., and Zen, E-an, 1985, An empirical equation for igneous calcic amphibole geobarometry: *Geol. Soc. America Abstr. with Programs*, v. 17, p. 602.

——, and Zen, E-an, 1986, Aluminum in hornblende — An empirical igneous geobarometer: *Amer. Mineralogist*, 71, p. 1297–1313.

Harrison, T. M., Armstrong, R. L., Naeser, C. W., and Harakal, J. E., 1979, Geochronology and thermal history of the Coast Plutonic Complex, near Prince Rupert, British Columbia: *Can. J. Earth Sci.*, v. 16, p. 400–410.

Haugerud, R. A., 1986, 1DT — An interactive, screen-oriented microcomputer program for simulation of 1-dimensional geothermal histories: *U.S. Geol. Survey Open File Rpt 86-511*, 18 p. & diskette.

Holdaway, M. J., 1971, Stability of andalusite and the aluminum silicate phase diagram: *Amer. J. Sci.*, v. 271, p. 97–131.

Hollister, L. S., 1982, Metamorphic evidence for rapid (2 mm/yr) uplift of a portion of the central gneiss complex, Coast Mountains, B.C.: *Can. Mineralogist*, v. 20, p. 319–332.

Hooper, P. R., 1982, The Columbia River Basalt: *Science*, v. 215, no. 4539, p. 1463–1468.

——, Kleck, W. D., Knowles, C. R., Reidel, S. P., and Thiessen, R. L., 1984, Imnaha Basalt, Columbia River Basalt Group: *J. Petrology*, v. 25, p. 473–500.

Hsü, K. J., 1983, Geosynclines in plate-tectonic settings — Sediments in mountains, *in* Hsü, K. J., ed., *Mountain Building Processes:* New York; Academic, p. 3–12.

Hunt, S. J., 1984, Preliminary study of a zoned leucocratic-granite body on central Etolin Island, southeastern Alaska, *in* Coonrad, W. L., and Elliott, R. L., eds., *The United States Geological Survey in Alaska — Accomplishments during 1981:* U.S. Geol. Survey Circ. 868, p. 128–130.

Hutchinson, W. W., 1970, Metamorphic framework and plutonic styles in the Prince Rupert region of the Central Coast Mountains, British Columbia: *Can. J. Earth Sci.*, v. 7, p. 376–405.

——, 1982, Geology of the Prince Rupert–Skeena map area, British Columbia: *Geol. Survey Canada Mem. 394*, 116 p.

James, D. E., 1971, Plate tectonic model for the evolution of the central Andes: *Geol. Soc. America Bull.*, v. 82, p. 3325–3346.

Johannes, W., and Puhan, D., 1971, The calcite-aragonite transition, reinvestigated: *Contrib. Mineralogy Petrology*, v. 31, p. 28–38.

Kaneda, B. K., 1980, Contact metamorphism of the Chiwaukum Schist near Lake Edna, Chiwaukum Mountains, Washington: M.S. thesis, Univ. Washington, Seattle, Wash., 207 p.

Kenah, Christopher, 1978, Generation of the Quottoon pluton (British Columbia) by crustal anatexis: *Geol. Soc. America Abstr. with Programs*, v. 10, p. 433.

Kistler, R. W., 1974, Phanerozoic batholiths in western North America — A summary of some recent work on variations in time, space, chemistry, and isotopic compositions: *Ann. Rev. Earth Planet. Sci.*, v. 2, p. 403–418.

Koch, R. D., and Elliott, R. L., 1984, Late Oligocene gabbro near Ketchikan, southeastern Alaska, *in* Coonrad, W. L., and Elliott, R. L., eds., *The United States Geological Survey in Alaska — Accomplishments during 1981:* U.S. Geol. Survey Circ. 868, p. 126–128.

Kuntz, M. A., 1985, Rocks and structures at the western margin of the Idaho batholith near McCall, Idaho — A review: *Geol. Soc. America Abstr. with Programs*, v. 17, p. 250.

Lachenbruch, A. H., and Saas, J. H., 1978, Models of an extending lithosphere and heat flow in the Basin and Range province: *Geol. Soc. America Mem. 152*, p. 209–250.

Lappin, A. R., and Hollister, L. S., 1980, Partial melting in the central gneiss complex near Prince Rupert, British Columbia: *Amer. J. Sci.*, v. 280, p. 518–545.

Li, Y. H., 1976, Denudation of Taiwan island since the Pliocene Epoch: *Geology*, v. 4, p. 105–107.

Liou, J. G., 1973, Synthesis and stability relations of epidote, $Ca_2Al_2FeSi_3O_{12}(OH)$: *J. Petrology*, v. 14, p. 381–413.

Lund, Karen, 1984, Tectonic history of a continent-island arc boundary — West-central Idaho: Ph.D. thesis, Pennsylvania State Univ., University Park, Pa., 207 p.

McKee, E. H., Hooper, P. R., and Kleck, W. D., 1981, Age of Imnaha basalt — Oldest basalt flows of Columbia River Basalt Group, northwest United States: *Isachron/West*, v. 31, p. 31–33.

Miller, C. F., Stoddard, E. F., Bradfish, L. J., and Dollase, W. A., 1981, Composition of plutonic muscovite — Genetic implications: *Can. Mineralogist*, v. 19, p. 25–34.

Misch, Peter, 1966, Tectonic evolution of the northern Cascades, *in Symposium on the Tectonic History and Mineral Deposits of the Western Cordillera in British Columbia and Neighbouring Parts of the United States:* Montreal, Can. Inst. Mining Metallurgy Spec. Vol. 8, p. 101–148.

Moore, J. G., 1959, The quartz diorite boundary line in the western United States: *J. Geology*, v. 67, p. 198–210.

——, Grantz, Arthur, and Blake, M. C., Jr., 1963, The quartz diorite line in northwestern North America — Geological Survey Research 1962, Short papers in geology, hydrology, and topography: *U.S. Geol. Survey Prof. Paper 450-E*, p. E89–E93.

Müller, Stephan, 1983, Deep structure and recent dynamics in the Alps, *in* Hsü, K. J., ed., *Mountain Building Processes:* New York; Academic, p. 181–199.

Naney, M. T., 1983, Phase equilibria of rock-forming ferromagnesian silicates in granitic systems: *Amer. J. Sci.*, v. 283, p. 993–1033.

Nitsch, K.-H., 1968, Die Stabilität von Lawsonit: *Naturwissenschaften*, v. 55, p. 388.

Parrish, R. R., 1983, Cenozoic thermal evolution and tectonics of the Coast Mountains of British Columbia — 1, Fission track dating, apparent uplift rates, and patterns of uplift: *Tectonics*, v. 2, p. 601–631.

Plummer, C. C., 1980, Dynamothermal contact metamorphism superposed on regional metamorphism in the pelitic rocks of the Chiwaukum Mountains area, Washington Cascades: *Geol. Soc. America Bull.*, v. 91, pt. II, p. 1627–1668.

Presnall, D. C., 1969, The geometrical analysis of partial fusion: *Amer. J. Sci.*, v. 267, p. 1178–1194.

Roddick, J. A., and Hutchinson, W. W., 1974, Setting of the Coast Plutonic Complex, British Columbia: *Pacific Geology*, v. 8, p. 91–108.

Rohay, A. C., 1982, Crust and mantle structure of the north Cascades Range, Washington: Ph.D. thesis, Univ. Washington, Seattle, Wash., 174 p. [See *Dissertation Abstr. Internat.*, Sect. B, p. 661–B.]

Selverstone, J., and Hollister, L. S., 1980, Cordierite-bearing granulites from the Coast Ranges, British Columbia – *P-T* conditions of metamorphism: *Can Mineralogist*, v. 18, p. 119–129.

Sharma, K. K., Bal, K. D., Parshad, Rajinder, Lal, Nand, and Nagpaul, K. K., 1980, Paleo-uplift and cooling rates from various orogenic belts of India, as revealed by radiometric ages: *Tectonophysics*, v. 70, p. 135–158.

Shearer, C. K., 1983, Petrography, mineral chemistry, and geochemistry of the Hardwick Tonalite and associated igneous rocks: Ph.D. thesis, Univ. Massachusetts, Amherst, Mass., 265 p.

Sisson, V. B., 1985, Contact metamorphism and fluid evolution associated with the intrusion of the Ponder pluton, coast plutonic complex, British Columbia, Canada: Ph.D. thesis, Princeton Univ., Princeton, N.J., 345 p.

Smith, J. G., and Diggles, M. F., 1981, Potassium-argon determinations in the Ketchikan and Prince Rupert quadrangles, southeastern Alaska: *U.S. Geol. Survey Open File Rpt 78–73N*, 16 p., 1 sheet.

Spear, F. S., and Selverstone, J., 1983, Quantitative *P-T* paths from zoned minerals – Theory and tectonic applications: *Contrib. Mineralogy Petrology*, v. 83, p. 348–357.

――, Ferry, J. M., and Rumble, Douglas, III, 1982, Analytical formulation of phase equilibria – The Gibbs' method; *in* Ferry, J. M., ed., *Characterization of Metamorphism through Mineral Equilibria*: Mineral. Soc. America Rev. Mineralogy, v. 10, p. 105–152.

――, Selverstone, J., Hickmott, D., Crowley, P., and Hodges, K. V., 1984, *P-T* paths from garnet zoning – A new technique for deciphering tectonic processes in crystalline terranes: *Geology*, v. 12, p. 87–90.

Speed, R. C., 1978, Paleogeographic and plate tectonic evolution of the early Mesozoic marine province of the western Great Basin, *in* Howell, D. G., and McDougall, K. A., eds., *Mesozoic Paleogeography of the Western United States:* Pacific Section, Soc. Econ. Paleontologists Mineralogists, Pacific Coast Paleogeography Symp. 2 p. 253–270.

――, 1982, Evolution of the sialic margin of the central western United States, *in* Watkins, J. S., and Drake, C. L., eds., *Studies in Continental Margin Geology:* Amer. Assoc. Petrol. Geologists Mem. 34, p. 457–470.

Speer, J. A., 1984, Micas in igneous rocks, *in* Bailey, S. W., ed., *Micas:* Mineral. Soc. America Rev. Mineralogy, v. 13, p. 299–356.

Sutter, J. F., and Crawford, M. L., 1985, Timing of metamorphism and uplift in the vicinity of Prince Rupert, British Columbia and Ketchikan, Alaska: *Geol. Soc. America Abstr. with Programs*, v. 17, p. 411.

――, and Hatch, N. L., Jr., 1985, Timing of metamorphism in the Rowe-Hawley Zone, western Massachusetts: *Geol. Soc. America Abstr. with Programs*, v. 17, p. 65.

——, Snee, L. W., and Lund, Karen, 1984, Metamorphic, plutonic, and uplift history of a continent-island arc suture zone, west-central Idaho: *Geol. Soc. America Abstr. with Programs*, v. 16, p. 670–671.

Tabor, R. W., Zartman, R. E., and Frizzell, V. A., Jr., in press, Possible tectonostratigraphic terranes in the North Cascades crystalline core, Washington, *in* Schuster, E., ed., *Geology of Washington* (Symposium volume): Wash. State Div. Mines Earth Resources.

Tracy, R. J., and Robinson, Peter, 1980, Evolution of metamorphic belts: Information from detailed petrologic studies, *in* Wones, D. R., ed., *The Caledonides in the U.S.A. – IGCP Proc., Project 27:* Dept Geol. Sci. Va. Polytechnic Inst. Mem. 2, p. 189–195.

Wagner, G. A., Reimer, G. M., and Jaeger, E., 1977, Cooling ages derived by apatite fission-track, mica Rb-Sr and K-Ar dating — the uplift and cooling history of the central Alps: *Memorie Inst. Geologia Mineralogia Univ. Padova*, v. 30, 27 p.

Werner, Dietrich, 1980, Probleme der Geothermik im Bereich der Schweizer Zentralalpen: *Eclogae Geol. Helvetiae*, v. 73, p. 513–525.

Woodsworth, G. L., Crawford, M. L., and Hollister, L. S., 1983a, Metamorphism and structure of the coast plutonic complex and adjacent belts, Prince Rupert and Terrace areas, British Columbia: Geol. Assoc. Canada/Mineral. Assoc. Canada/Can. Geophys. Un. Annual Meet., Victoria, B.C., *Guidebook for Field Trip 14*, 62 p.

——, Loveridge, W. D., Parrish, R. R., and Sullivan, R. W., 1983b, Uranium-lead dates from the central gneiss complex and Ecstall pluton, Prince Rupert map area, British Columbia: *Can. J. Earth Sci.*, v. 20, p. 1475–1483.

Wyllie, P. J., 1977, Crustal anatexis — An experimental review: *Tectonophysics*, v. 43, p. 41–71.

Yeats, R. S.,1958, Geology of the Skykomish area in the Cascade Mountains of Washington: Ph.D. thesis, Univ. Washington, Seattle, Wash., 249 p. [See also *Dissertation Abstr.*, v. 19, no. 4, p. 775.]

Yoshikawa, Torao, 1974, Denudation and tectonic movement in contemporary Japan: *Univ. Tokyo Dept Geography Bull. 6*, p. 1–14.

Zartman, R. E., 1974, Lead isotopic provinces in the cordillera of the western United States and their geologic significance: *Econ. Geology*, v. 69, p. 792–803.

Zeitler, P. K., 1985, Cooling history of the NW Himalaya, Pakistan: *Tectonics*, v. 4, p. 127–151.

Zen, E-an, 1985a, *Muscovite* — McGraw-Hill Yearbook of Science and Technology: New York, McGraw-Hill, p. 283–287.

——, 1985b, Implications of magmatic epidote-bearing plutons on crustal evolution in the accreted terranes of northwestern North America: *Geology*, v. 13, p. 266–269.

——, and Hammarstrom, J. M., 1984a, Magmatic epidote and its petrologic significance: *Geology*, v. 12, p. 515–518.

——, and Hammarstrom, J. M., 1984b, Mineralogy and a petrogenetic model for the tonalite pluton at Bushy Point, Revillagigedo Island, Ketchikan 1° × 2° quadrangle, southeastern Alaska, *in* Reed, K. M., and Bartsch-Winkler, Susan, eds., *The United States Geological Survey in Alaska — Accomplishments during 1982:* U.S. Geol. Survey Circ. 939, p. 118–123.

——, and Hammarstrom, J. M., 1986, Reply to discussions by Tulloch and Moench: *Geology*, v. 14, p. 188–189.

J. L. Wooden, J. S. Stacey, and K. A. Howard
U.S. Geological Survey
Menlo Park, California 94025

B. R. Doe and D. M. Miller
U.S. Geological Survey
Reston, Virginia 22092

3

Pb ISOTOPIC EVIDENCE FOR THE FORMATION OF PROTEROZIC CRUST IN THE SOUTHWESTERN UNITED STATES

ABSTRACT

The common Pb isotopic system is a powerful tool for understanding crustal history. Pb isotopic studies in the western United States during the 1960s and 1970s determined the extent of Precambrian crust in the region, separated the Archean crust from the Proterozoic, and distinguished the difference between the Precambrian and Phanerozoic terranes. Recent Pb isotopic studies in southeastern California, Arizona, and New Mexico indicate (1) that the Pb in the Mesozoic and Cenozoic ore deposits and silicic intrusives of this area is largely derived from the Proterozoic basement, and (2) that systematic differences exist in the isotopic signature of the Proterozoic rocks within this area. The Proterozoic crust in SE California is characterized by $^{207}Pb/^{204}Pb$ ratios elevated above those associated with average crust and is similiar in this respect to the crust in parts of Utah and Nevada. The elevated $^{207}Pb/^{204}Pb$ ratios require that Pb from an Archean reservoir be involved in the genesis of this crust during the Proterozoic. In contrast, the Proterozoic crust in most of Arizona, New Mexico, and southern Colorado has a Pb isotopic signature compatible with derivation from a MORB/oceanic-like mantle reservoir. In addition, the $^{206}Pb/^{204}Pb$ ratios of rocks of all ages in SE California have a lower average value than those in Arizona. This feature is related to the higher metamorphic grade of the oldest (1.7-2.0 b.y. old) Proterozoic rocks in southeastern California which apparently lost U during a low granulite-grade event. Thus Pb isotopic data establish that two crustal terranes are juxtaposed in southeastern California and western Arizona. The boundary between these two crustal terranes runs roughly parallel to and east of the Colorado River in western Arizona. The nature of this boundary is unknown and is complicated by Phanerozoic tectonics. Chronologic data imply that the two terranes were joined by 1.7 b.y. ago. Nd isotopic data for the southwestern United States indicate approximately the same crustal terranes and crustal evolution model as the Pb isotopic data. Both isotopic models are supported by petrologic and geochemical data for the 1.4-b.y.-old Proterozoic intrusives that occur in both terranes.

INTRODUCTION

An extensive amount of crust was added to the North American craton during the Proterozoic. For the southwestern Unites States most of this crust was produced between 1.8 and 1.6 b.y. ago (Nelson and DePaolo, 1985; Silver, 1968; Silver *et al.*, 1977; Van Schmus and Bickford, 1981). Geochronologic data indicate that the Proterozoic rocks in the southwestern United States (Fig. 3-1) can be divided into two age provinces, 1.78–1.69 b.y. in the north (Colorado, southern Utah and Nevada, northwestern Arizona, southeastern California) and 1.68–1.61 b.y. in the south (southern Arizona and New Mexico). The tectonic mechanisms for this crustal addition are poorly understood, but an environment involving magmatic arcs that developed marginal to, or on, an Archean cratonic edge is often suggested (Condie, 1982, 1986). In much of the southwestern United States, this Proterozoic terrane has been involved in Phanerozoic orogenic events that have obscured the original relationships among the present exposed fragments of Proterozoic rocks (Burchfiel and Davis, 1972, 1975). Fortunately, many of the Phanerozoic silicic igneous rocks in this region were produced either by melting of the Proterozoic basement or had their isotopic characteristics strongly influenced by the Proterozoic crust through which their parental magmas ascended. This feature has allowed Sr, Nd, and Pb isotopic tracer studies to determine the presence or absence of Proterozoic crust in an area. When Proterozoic basement rocks are present, their isotopic characteristics can be determined, even when the Proterozoic rocks are not exposed at the surface.

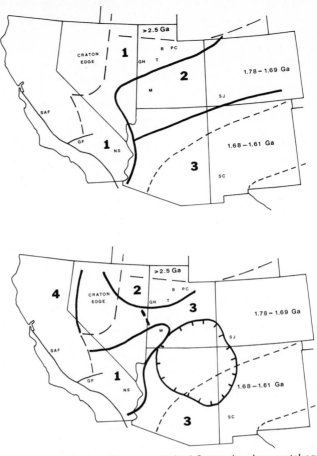

FIG. 3-1. Maps of the southwestern United States showing crustal age and isotopic provinces. Elements of the map were modified from Farmer and DePaolo (1984). Symbols refer to locations of Pb isotopic studies used in text and in Fig. 3-2 and are: PC, Park City; B, Bingham; T, Tintic; GH, Gold Hill; M, Milford; SJ, San Juan; SC, Silver City, New Mexico; NS, Needles Sheet; SAF, San Andreas fault, GF, Garlock fault. Crustal provinces based on Nd isotopic data (see Bennett and DePaolo, in press, for a complete discussion) are superposed on the upper map. The crustal provinces based on the Pb isotopic data discussed in this paper are shown on the lower map. The area enclosed by a hachured line in the lower map is a schematic representation of the extent of the Colorado Plateau for which there is essentially no isotopic or geochronologic data available because Proterozoic rocks are covered. The Pb isotopic provinces are: 1, the Mojave terrane; 2, an area with variable but very high contributions of Pb from an Archean age source; 3, a terrane with a MORB-like mantle Pb isotopic signature; a dotted line along the Utah and Nevada border separates cratonized from uncratonized crust; 4, the group III terrane of Zartman (1974).

Pb isotopic studies in the 1960s and 1970s recognized differences in the basement rocks of the western United States, the extent of the Proterozoic and Archean basement, and the isotopic contrast between crust formed in the Phanerozoic versus that formed in the Precambrian (Doe, 1967; Doe and Delevaux, 1973; Doe and Zartman, 1979;

1985; Doe *et al.*, 1979; Stacey and Zartman, 1978; Stacey *et al.*, 1968 Zartman, 1974; Zartman and Stacey, 1971). Sr isotopic studies in the 1970s confirmed and refined these concepts for the crust of the western United States (Kistler and Peterman, 1973, 1978). Some of the Pb isotopic data available for this area were recently summarized (Stacey and Hedlund, 1983) and differences in the Pb isotopic signature of the Proterozoic basement in Utah versus that in southern Colorado and New Mexico were described. Stacey and Hedlund (1983) discussed the implications of these data for early Proterozoic crustal evolution in the western United States. Some of their important points were that the early Proterozoic crust accreted around an Archean craton (the Wyoming Province) and that the crust that formed adjacent to the Archean craton incorporated a high percentage of Pb from this older crust. Crust that formed progressively farther from the Archean craton contained less of this radiogenic Pb until the crust in Arizona, New Mexico, and southern Colorado formed with a Pb isotopic signature approximately that of a MORB-like mantle (Fig. 3-2). Recent Nd isotopic studies of the western United States have confirmed the general nature of this model, refined the position of the western edge of the Proterozoic craton, and established that the older (2.0–1.4 b.y.) Proterozoic rocks are divisible into three major provinces (Fig. 3-1; Bennett and DePaolo, in press; Bennett *et al.*, 1984; DePaolo, 1981a, b; Farmer and DePaolo, 1983, 1984; Nelson and DePaolo, 1984, 1985).

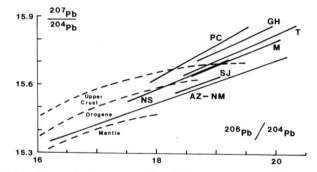

FIG. 3-2.   Summary diagram of $^{206}Pb/^{204}Pb$ versus $^{207}Pb/^{204}Pb$ for several areas within the southwestern United States. Locations are in Fig. 3-1. Dashed lines are average growth curves from the model of Zartman and Doe (1981). The orogene curve is approximately the same as the Stacey and Kramers (1975) uranogenic Pb model curve.

Pb and Nd isotopic data are equally good for identifying crustal provinces, although the two isotopic systems may not necessarily define the same provinces. This seeming paradox is a result of the very different geochemical properties of Pb and Nd and their parent elements U, Th, and Sm. The petrogenetic behavior of these elements results in a strong contrast between mantle and crustal concentrations of these elements, with Pb being enriched in the crust by a factor of about 100 while Nd is enriched by a factor of only about 20–30. A consequence of this concentration contrast is that it is difficult for a mantle-derived melt to retain its primary Pb isotopic composition after any contact with the continental crust. In addition, Sm/Nd ratios in the great majority of crustal rocks show little variation, whereas U/Pb ratios in crustal rocks can vary by more than an order of magnitude.

The Pb isotopic system is very sensitive to the crustal history of an area. This sensitivity is a result of the differing half-lives of ^{235}U, ^{238}U, and ^{232}Th and the fact that different regions of the crust and different types of rocks have strong variations in their U/Pb, Th/Pb, and Th/U ratios. The short half-life of ^{235}U means that the major production of radiogenic ^{207}Pb occurred during the first half of earth history. Any major differences that existed in U/Pb ratios during this period would have generated large differences in ^{207}Pb/^{204}Pb ratios. In contrast, during the second half of earth history, reservoirs with different U/Pb ratios would have developed large changes only of radiogenic ^{206}Pb which results from the decay of ^{238}U. (The present-day ratio of ^{238}U/^{235}U is 137.88.) The half-life of ^{232}Th is long relative to either of the U isotopes and changes in ^{208}Pb/^{204}Pb ratios are relatively small compared to those in ^{207}Pb/^{204}Pb and ^{206}Pb/^{204}Pb ratios. However, major differences in Th/U ratios do develop in rocks (the crustal average is about 4), and ^{208}Pb/^{204}Pb versus ^{206}Pb/^{204}Pb variations will indicate differences in Th/U ratios when those differences are sufficiently long-lived.

High-grade metamorphism can produce significant changes in the Th/U ratio (see Rudnick *et al.*, 1985 and references therein). During high-grade metamorphism, U, especially when in the +6 ionic state, may be lost from a rock as the result of loss of fluids or melts. Th only exists in the +4 ionic state and is not as mobile as U. Th may be bound or trapped in minerals with high Th/U ratios that are more stable under these conditions than those minerals with high U/Th ratios; consequently, the Th/U ratio of high-grade metamorphic rocks may be increased from that of their protoliths. Th, however, may also be lost at the very highest metamorphic conditions. Because crust below 20 km resides at temperatures and pressures associated with high-grade metamorphism, it is generally assumed to have low U/Pb ratios and high Th/U ratios. The timing of changes in U/Pb, Th/Pb, and Th/U ratios is also critical. If the crust experiences a general high-grade metamorphic event shortly after it is formed, it will evolve along a low U/Pb path below that of the Stacey and Kramers (1975) curve on a ^{207}Pb/^{204}Pb versus ^{206}Pb/^{204}Pb diagram. If Th is not lost at this time, the evolution path will tend to lie above that of the Stacey and Kramers (1975) curve on a ^{208}Pb/^{204}Pb versus ^{206}Pb/^{204}Pb diagram. These types of evolutionary paths will be most obvious in crust of Precambrian age since recent changes in U or Th content will not be reflected in the Pb isotopic ratios because of the lack of time for differences in these ratios to develop.

A word of caution is required about the use of average model curves for the U-Th-Pb system in the crust (Doe and Zartman, 1979; Stacey and Kramers, 1975; Zartman and Doe, 1981). Although those curves provide a very useful reference for Pb isotopic evolution, they can accurately predict Pb isotopic compositions only for those crustal terranes that are produced by mixing Pb from all the major mantle and crustal reservoirs and then evolve with U/Pb, Th/Pb, and Th/U ratios approximately equal to those of the average crust. However, because each crustal terrane actually has a fairly unique geologic history, it can follow Pb evolution paths widely divergent from the model curves. The evolution paths may be composed of several discrete stages that mark changes in U, Th, and Pb concentrations or mixing between markedly different Pb isotopic reservoirs. Pb isotopic compositions that appear similar today may be produced by very different evolution paths. An understanding of what a particular isotopic composition means will be gained only when all the geologic components of a particular crustal terrane are understood. The crust of the southwestern United States is an excellent example of this problem, and the initial attempts to solve this problem are presented below.

Pb ISOTOPIC DATA FOR SOUTHEASTERN CALIFORNIA

Recent results from Pb isotopic studies of Mesozoic intrusive rocks in the Needles 2° sheet (Fig. 3-1) and adjoining areas (Wooden *et al.*, 1986a) and of Early and Middle Proterozoic rocks in southeastern California (J. L. Wooden, D. M. Miller, and J. L. Anderson, unpublished results) and western Arizona (J. L. Wooden, B. Bryant, K. C. Condie, and J. L. Anderson, unpublished results) provide new constraints on the nature and history of the crust in this area. Pb isotopic data for the Mesozoic intrusive rocks of the Needles sheet have an unusually wide range of $^{206}Pb/^{204}Pb$ values (Fig. 3-3). In addition, these data lie along a linear trend on a plot of $^{207}Pb/^{204}Pb$ versus $^{206}Pb/^{204}Pb$ that has a slope equal to an age of about 1.6 b.y. A plot of $^{208}Pb/^{204}Pb$ versus $^{206}Pb/^{204}Pb$ (Fig. 3-3)

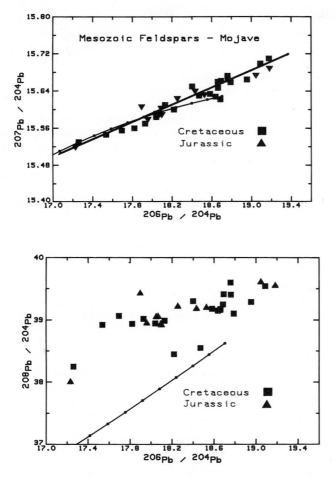

FIG. 3-3.   Plots of $^{206}Pb/^{204}Pb$ versus $^{207}Pb/^{204}Pb$ and $^{206}Pb/^{204}Pb$ versus $^{208}Pb/^{204}Pb$ for feldspars from Jurassic (triangles) and Cretaceous (squares) intrusive rocks in the Needles sheet, SE California. Straight line represents a 1600-m.y.-old reference isochron. Reference curves in this and succeeding diagrams are from Stacey and Kramers (1975) and represent average crust.

shows that most of these samples have $^{208}Pb/^{204}Pb$ ratios equal to 39.2 ± 0.2 over almost the entire range of $^{206}Pb/^{204}Pb$ ratios. On the plot of $^{208}Pb/^{204}Pb$ versus $^{206}Pb/^{204}Pb$ these data lie consistently above the Stacey and Kramers (1975) curve which represents average crust. Such a wide and continuous range of $^{206}Pb/^{204}Pb$ values is generally not found in suites of rocks produced by subduction at plate margins (e.g., see data in Church, 1976; Kay *et al.*, 1978) or by magmatism within oceanic basins, but are similar to those found in suites of rocks formed by reworking Precambrian, continental crust (Doe and Zartman, 1979; Moorbath and Taylor, 1981).

A great variety of compositional types are represented among the rocks analyzed for Pb isotopic composition (gabbros, monzodiorites to monzogranites, tonalites, granodiorites, and granites, including corundum normative ones), and there is no correlation between any compositional feature and the $^{206}Pb/^{204}Pb$ ratios. All these factors indicate a source region very heterogenous in bulk composition and Pb isotopic composition but with an average age of about 1.6. b.y. The wide range in $^{206}Pb/^{204}Pb$ ratios indicates strong variations in the U/Pb ratios of these rocks. The relatively constant $^{208}Pb/^{204}Pb$ ratio in most of these rocks suggests that their source rocks have about the same Th/Pb ratio. About one-half of the $^{206}Pb/^{204}Pb$ ratios are less than 18.4–18.6, which are the model values from the Stacey and Kramers (1975) curve for Jurassic to Cretaceous age rocks. Therefore, the sources for some of the Mesozoic intrusive rocks had U/Pb ratios distinctly less than the crustal average of about 9. Low U/Pb ratios are often attributed to U loss during a high-grade metamorphic event. This is not, however, the only way for rocks to acquire a low U/Pb ratio because many unfractionated basalts and some unmetamorphosed granitic rocks also have this characteristic[1] (Doe, 1967). If any of the source rocks did acquire their low U/Pb ratios during metamorphism, that metamorphic event must have occurred close to the time of formation of those rocks or otherwise there would have been insufficient time for major differences in $^{206}Pb/^{204}Pb$ ratios to develop. An additional feature of the Mesozoic Pb isotopic data is that they lie slightly above the Stacey and Kramers (1975) uranogenic Pb model curve (Fig. 3-3). These data therefore indicate a source that has a higher $^{207}Pb/^{204}Pb$ ratio than average crust. Reservoirs enriched in ^{207}Pb are most likely to have been created during the Archean. Therefore, rocks with $^{207}Pb/^{204}Pb$ ratios higher than the average crustal growth curve usually have had Archean-age material involved in their evolution.

The Pb isotopic and bulk compositional characteristics of the Mesozoic intrusives in the Needles sheet clearly indicate that they were derived from an Early to Middle Proterozoic basement terrane of variable composition and metamorphic grade. The exposed Precambrian basement of southeastern California into which the Mesozoic intrusives were emplaced has these general characteristics. U-Pb zircon geochronologic and common Pb isotopic studies of some of these basement rocks (Wooden et al., 1986b) provide evidence that they have exactly the right features to be the sources for the Mesozoic intrusives. Recent work has shown that some of the Proterozoic basement in the Kingman 2° sheet has undergone a granulite-grade metamorphic event (Elliott et al., 1986; Thomas et al., 1986; Young et al., 1986). U-Pb zircon dating in the New York and Providence Mountains (Wooden et al., 1986b) and Halloran Hills (DeWitt et al., 1984) provide strong evidence that this metamorphism was a regional event that occurred 1710–1700 m.y. ago. This high-grade metamorphic complex was intruded by granodioritic to granitic plutons between 1700 and 1650 m.y. ago (Wooden et al., 1986b) and later by the widespread, approximately 1400-m.y.-old anorogenic granitic suite (Anderson, 1983, 1984, 1986). Thus these rocks provide a compositionally heterogeneous source with an average age of about 1.6 b.y., and part of it has experienced a high-grade metamorphic event that could cause a loss of U. The common Pb isotopic characteristics of these basement rocks are shown in Fig. 3-4. The whole-rock data cover about the same range of $^{206}Pb/^{204}Pb$ ratios as seen in the Mesozoic intrusives. In addition, on the plot of $^{207}Pb/^{204}Pb$ versus $^{206}Pb/^{204}Pb$ (Fig. 3-4) the two data sets can be seen to define the same linear arrays. Figure 3-4 shows that the $^{208}Pb/^{204}Pb$ versus $^{206}Pb/^{204}Pb$ systematics of the Proterozoic rocks vary widely,

---

[1] Also J. L. Wooden and P. A. Mueller, unpublished data.

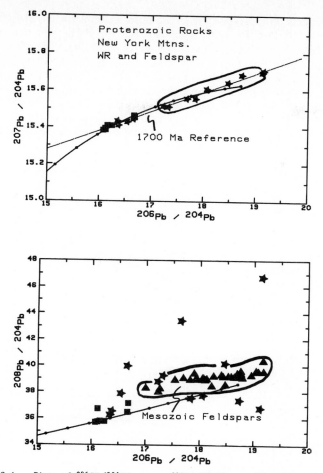

FIG. 3-4. Plots of $^{206}Pb/^{204}Pb$ versus $^{207}Pb/^{204}Pb$ and $^{206}Pb/^{204}Pb$ versus $^{208}Pb/^{204}Pb$ for feldspar (squares) and whole-rock (stars) samples of Proterozoic rocks from the New York Mountains, southeastern California. Outlined areas on both plots represents the range of the Mesozoic feldspar data from the Needles sheet (see Fig. 3-3). Actual Mesozoic data are shown as triangles on the lower plot. Straight line in upper plot represents a 1700-m.y.-old reference isochron.

but that the average Pb isotopic compositions of these Proterozoic rocks can provide the $^{208}Pb/^{204}Pb$ features of the Mesozoic rocks.

Pb ISOTOPIC DATA FOR ARIZONA

Published common Pb isotopic data for Arizona are still relatively scarce. However, we have analyzed, in cooperation with a number of USGS geologists and mine geologists, a number of the young ore deposits in southern Arizona. In additon, common Pb isotopic data are available for the rocks associated with the ore deposits at Silver Bell (Sawyer and Zartman, 1985) and Ajo (J.S. Stacey, J. L. Wooden, and Cox, unpublished data). These data are distinctly different from the common Pb isotopic data of the Mojave Desert region in two respects. For a given $^{206}Pb/^{204}Pb$ ratio, a sample from Arizona will generally have both a lower $^{207}Pb^{204}Pb$ ratio and a lower$^{208}Pb/^{204}Pb$ ratio than that from the Mojave Desert (Fig. 3-5). The Arizona data are like those from the Mojave in that they scatter along a reference line with a slope equal to an age of about 1.6 b.y. The range of $^{206}Pb/^{204}Pb$ ratios is about the same as that of the Mojave region data which, along with the 1.6-b.y. age, indicates that the Proterozoic basement is again the probable source of the Pb. However, the Arizona basement rocks must have had an initial $^{207}Pb/^{204}Pb$ ratio lower than that of the Mojave basement, and the Th/U ratios of the Arizona basement rocks must be about 4 (approximately the crustal average), although these rocks have variable U/Pb and Th/Pb ratios.

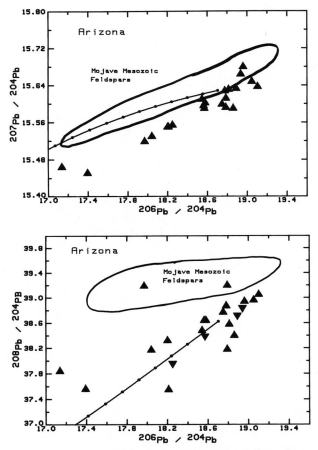

FIG. 3-5. Plots of $^{206}Pb/^{204}Pb$ versus $^{207}Pb/^{204}Pb$ and $^{206}Pb/^{204}Pb$ versus $^{208}Pb/^{204}Pb$ for sulfides and feldspars from Larimide and younger ores and rocks of southern Arizona. Outlined area represents the range of the Mesozoic feldspars from the Needles sheet, southeastern California.

Systematic studies of the common Pb isotopic characteristics of the basement rocks of Arizona are just starting. Pb isotopic analyses of galenas from the Early Proterozoic ore deposits at Jerome and Bagdad, Arizona, are available from earlier USGS work (Stacey *et al.*, 1976). Galenas from major ore deposits that formed early in the crustal history of an area provide a reasonable approximation of the initial Pb isotopic ratios of rocks because they have very low U/Pb ratios and represent Pb extracted from a large volume of rock. The Bagdad and Jerome data have $^{207}Pb/^{204}Pb$ ratios appropriate to lie on the extension of the southern Arizona sulfide trend, which has lower $^{207}Pb/^{204}Pb$ ratios than those associated with the Mojave region rocks.

Pb isotopic data for Early Proterozoic metavolcanic rocks (basalt to rhyolite; Vance and Condie, 1986) from the Jerome, Arizona, area are shown in Fig. 3-6. The $^{207}Pb/^{204}Pb$

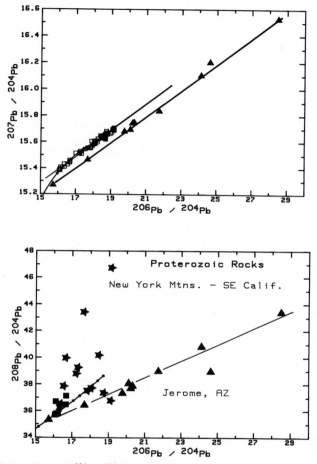

FIG. 3-6.   Plots of $^{206}Pb/^{204}Pb$ versus $^{207}Pb/^{204}Pb$ and $^{206}Pb/^{204}Pb$ versus $^{208}Pb/^{204}Pb$ for metavolcanic rocks from Jerome, Arizona (triangles). Data from SE California representing both Mesozoic and Proterozoic rocks are denoted by squares and/or stars. Straight lines in the upper plot are reference isochrons with ages of 1600 m.y. (Arizona) and 1700 m.y. (California). Straight line through Arizona data on the lower plot represents a Th/U ratio of 2.

versus $^{206}Pb/^{204}Pb$ data for these rocks define a slope equal to an age of about 1.6 b.y., and the trend of these data can be extended through the Jerome galena data. In addition, the $^{208}Pb/^{204}Pb$ versus $^{206}Pb/^{204}Pb$ data for these volcanics indicate an average Th/U ratio of about 2 (Fig. 3-6). This is an unusually low Th/U ratio, but it is in the range of that observed in modern-day MORB basalts and not too different from that observed for the San Juan Mountains of southern Colorado (Lipman *et al.*, 1978). Metavolcanic rocks similar to those at Jerome are a common component in the Early Proterozoic sequences of Arizona. If they all have similiarly low Th/U ratios, this could help explain why magmas formed by melting this basement, or ore fluids that leached the basement (both processes would average fairly large volumes of rock), also have low $^{208}Pb/^{204}Pb$ ratios.

IMPLICATIONS OF THE Pb AND Nd DATA

The new Pb isotopic data for Proterozoic and Mesozoic rocks in southeastern California indicate a strong similarity between the uranogenic Pb isotopic signature of this area and that of southern Nevada and Utah (Fig. 3-2). The Pb isotopic characteristics of both of these areas are distinct from those of the New Mexico–Arizona–southern Colorado region (Fig. 3-2). The geographic provinces indicated by the Pb isotopic data are similar to those defined by the Nd isotopic data (Fig. 3-1). Unfortunately, neither data set is complete enough or cross-correlated well enough to locate province boundaries precisely. However, there are enough data to show that neither of the isotopic provinces is correlated with any geochronologic province that is presently recognized within the Proterozoic terrane (Fig. 3-1).

The genetic implications of the Pb and Nd data are similar. The Arizona–New Mexico–southern Colorado region has an isotopic signature that requires a source area with long-term low U/Pb and Nd/Sm ratios (Nelson and DePaolo, 1984; Stacey and Hedlund, 1983). These characteristics are compatible with those that a modern MORB-like source would have had in the Proterozoic. The Pb isotopic data lie close to, but slightly above, the version II model mantle growth curve of Zartman and Doe (1981; Fig 3-2). This particular Pb isotopic signature is present in other Proterozoic rocks in North America (Ashwal *et al.*, 1986; Tilton, 1983; Zartman and Stacey, 1971) and for some of these rocks, Nd isotopic data have already indicated a MORB-like source (Ashwal and Wooden, 1983; Frost and O'Nions, 1984). The lack of any unquestioned Pb isotopic data for Proterozoic rocks with $^{207}Pb/^{204}Pb$ lower than that typical of the New Mexico-Arizona area ($^{206}Pb/^{204}Pb = 15.80$, $^{207}Pb/^{204}Pb = 15.30$) argues that this value be tentatively accepted as representing the MORB reservoir at approximately 1.8–1.4 b.y. ago. The environment in which these rocks formed must have been isolated from the contamination effects of the Archean cration. Environments such as an oceanic island arc, an arc well off the continental edge, or one screened from the craton by another arc are possible explanations.

It is not clear if the Pb and Nd data are in as good agreement concerning the geographic boundaries of the crustal provinces located to the west and northwest of the New Mexico–Arizona region (Fig. 3-1). The Pb isotopic data place southeastern California and southern Nevada and Utah in the same province, but the Nd data (Bennett and DePaolo, 1984, in press) indicate a distinction between these two areas. Qualitatively, however, both data sets require a significantly older crustal component to be mixed with the MORB-like reservoir. The Archean crust of the Wyoming Province is the obvious candidate for the high $^{207}Pb/^{204}Pb$ and Nd/Sm material needed. The rapid increase in $^{207}Pb/^{204}Pb$ for the Proterozoic basement (Stacey and Hedlund, 1983) as the Archean-Proterozoic craton boundary in northern Utah is approached (Figs. 3-1 and 3-2) is a clear indication that there was major interaction between the Archean craton and the new crust formed during the Proterozoic. The northern Utah picture is complicated, however, by the presence of Proterozoic rocks 2.0 b.y. old and older whose extent in the area is unknown (Hedge *et al.*, 1983; Stacey and Zartman, 1978; Stacey *et al.*, 1968).

An additional difference between crustal boundaries as indicated by Pb and Nd data is particularly apparent along the western edge of the Precambrian craton in Utah and Nevada (Fig. 3-1). The Phanerozoic plutons intruding the craton in Utah are characterized by low $^{206}Pb/^{204}Pb$ ratios which probably reflect a source region within the lower crust. Lead isotopic data for galenas from individual mining districts in Utah exhibit linear arrays with large ranges in $^{206}Pb/^{204}Pb$ values (Fig. 3-2) that are interpreted to result from mixing Pb from low U/Pb lower crust with Pb from high U/Pb upper crust (Stacey and Zartman, 1978; Stacey et al., 1968). In contrast, Pb isotopic data from mining districts in parts of Nevada exhibit comparatively small ranges of, and high average values for, $^{206}Pb/^{204}Pb$. This is consistent with the idea that these regions are underlain by sequences of uncratonized (low metamorphic grade) sediments and not by Precambrian crystalline basement (Zartman, 1974). The ability to differentiate clearly between both lower and upper crustal components is unique to the Pb system. In Nevada, Nd isotopic data indicate that the material was derived from Proterozoic and older crust, but cannot detect whether that crust was cratonized. Therefore, the terrane boundary defined by Nd isotopic data in Nevada, which lies to the west of the boundary defined by Pb isotopic data, may indicate only the extent of older sediments, not necessarily the edge of the craton itself.

The nature of the interaction between the Archean and Proterozoic Pb reservoirs is uncertain; however, Pb isotopic systematics can place some restrictions on possible processes. There are chronologic data that establish the existence of 2.0–2.3-b.y.-old rocks in Utah (see above and Girty et al., 1985) and southeastern California (Elliott et al., 1986; Wooden et al., 1986b). The more radiogenic nature of the Pb isotopic compositions in these areas cannot be the result of forming crust at about 2.0 b.y. ago from a MORB-like mantle and letting it evolve until the present because the $^{207}Pb/^{204}Pb$ versus $^{206}Pb/^{204}Pb$ trend of this crust now would lie below, not above, that observed for Arizona–New Mexico (Fig. 3-7). For crust formation ages older than 2.0 b.y., crust evolving from the model mantle curve would lie even farther below that of Arizona–New Mexico data array. These are simple two-stage models where stage 1 is the mantle and stage 2 is the crust. Three-stage models involving a period of crustal reworking (to reset mineral and rock ages and homogenize isotopic compositions) offer the possibility of producing crust with higher $^{207}Pb/^{204}Pb$ ratios (Fig. 3-8). The length of stage 2, however, is controlled by the average U/Pb ratio of the crustal material involved. The shorter the time span used for stage 2, the higher the average U/Pb ratio must be. An example is illustrated in Fig. 3-8. For crust produced from a MORB-like mantle at 2.0 b.y. ago, an average $^{238}U/^{204}Pb$ ratio (u) of 23 is required to generate the $^{206}Pb/^{204}Pb$ ratios found in the feldspars of the 1.7-b.y.-old Mojave crust. This u value is high, about 2.6 times higher than the crustal average, and in any case, could not produce a $^{207}Pb/^{204}Pb$ ratio as high as that observed in the Mojave Proterozoic feldspars (Fig. 3-8). Alternatively, if the second stage starts earlier at 2.8 b.y. ago, the average u value of the crust can be 12 and produce the 1.7-b.y.-old feldspar compositions. Note that this model does not allow the addition of younger crustal material derived from a MORB-like reservoir before reworking because that would lower $^{207}Pb/^{204}Pb$ ratios. Although late Archean crust with a reasonable average u value could have provided acceptable Pb isotopic compositions for the Mojave crust, it is difficult to believe that all the zircons incorporated in this crust could be reset totally to the younger age (starting time of the third stage). The problem presented by the absence of zircons older than 2.0 b.y. in Mojave crust also places constraints on models involving the physical mixing of sediments from an Archean craton into crust being formed 2.0–1.7 b.y. ago. An acceptable Pb isotopic mix could be produced this way, but it would require that the Archean sediments involved in the mixing have few if any zircons or that the Archean zircons be totally reset during reworking of the mixture.

The region covered by Proterozoic rocks with the Mojave, or more radiogenic, Pb

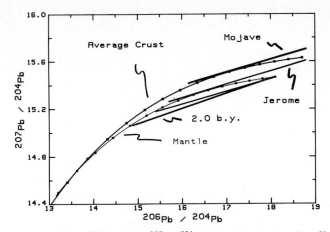

FIG. 3-7. A ^{206}Pb/^{204}Pb versus ^{207}Pb/^{204}Pb diagram showing the effect of deriving crust from a model MORB-like mantle reservoir at 1.7 and 2.0 b.y. ago and letting this crust evolve to the present. Two-stage models of this type cannot produce the Pb isotopic signature seen in the Mojave area.

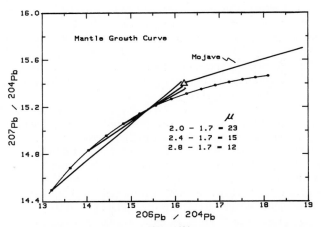

FIG. 3-8. A ^{206}Pb/^{204}Pb versus ^{207}Pb/^{204}Pb diagram showing three-stage Pb evolution models. Stage 1 in all cases is a MORB-like mantle reservoir. The start of the second stage varies in age from 2.8 to 2.0 b.y. and represents crust formed from this mantle reservoir. The third stage starts at 1.7 b.y. ago and represents a homogenization of the Pb isotopic composition of crust produced in the second stage. The average μ (^{238}U/^{204}Pb) values required to reach the initial Pb isotopic composition of the Mojave area at 1.7 b.y. are shown on the figure.

isotopic signature is large (10^5 km^2), and a major geologic process is required to produce such a large volume of crust with a seemingly homogeneous isotopic character. There are major problems with all the models previously discussed. An alternative, and preferred model is essentially the same process that is proposed to produce the isotopic features of many modern arc series rocks (Barreiro, 1983; Kay *et al.*, 1978; Woodhead and Fraser, 1985). This model involves subduction of continental detritus from an Archean craton

Pb ISOTOPIC EVIDENCE FOR THE FORMATION OF PROTEROZOIC CRUST

during the early Proterozoic, dewatering of the subducted material, and movement of fluids carrying Pb, Sr, and Nd into the overlying mantle. Because the mantle has very low concentrations of Pb, its Pb isotopic composition is easily changed by mixing with a small amount of crustal Pb (Fig. 3-9). This model also eliminates the problem of relict zircons that all reworking models face.

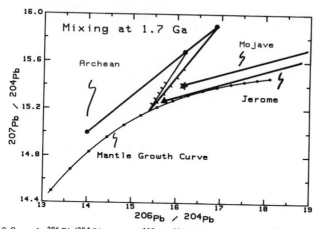

FIG. 3-9. A $^{206}Pb/^{204}Pb$ versus $^{207}Pb/^{204}Pb$ diagram illustrating mixing of Pb isotopic compositions between a 2.8-b.y.-old Archean craton and a MORB-like mantle reservoir at 1.7 b.y. ago. The Archean line shows the array of data as it would have looked for evolution between 2.8 and 1.7 b.y. and is marked by dots at its upper end for μ values of 9 and 12. A mix by weight of 70% mantle and 30% Archean Pb could produce the isotopic composition of the Mojave area.

Major partial melting of the mantle may not have occurred at the time of contamination, which could have been any time between approximately 2.0 and 1.7 b.y. ago. It is probable, however, that some fusion would have occurred during subduction and in the presence of these fluids. When this "crustally contaminated" mantle experienced significant partial melting, as it must have in a major crust forming event, it produced basaltic melts with relatively homogeneous isotopic compositions that were distinctly different from a MORB-like mantle. This mafic material must then have remelted to form the dominantly felsic rocks that are seen in the exposed Proterozoic of the southwestern United States (Condie, 1982; Van Schmus and Bickford, 1981). In fact, the remelting or reworking of the primary arc material would have helped to homogenize the isotopic signature of the region, and could have occurred a number of times after the arc was initially formed. Any local isotopic heterogeneities would have been averaged for the volume of crust that was involved in the partial melting. It is possible that some of the Proterozoic mafic arc rocks are still present in the lower crust, and that they served as a source for some of the intrusives that were produced when the Proterozoic craton was reacted during the Mesozoic. This situation would have resulted in the production of low- to moderate-K, calcalkaline series rocks which carry contaminated or evolved isotopic signatures (C. F. Miller et al., 1984). This process is suggested as an alternative to the mixing within the crust of mantle melts and crustal components during the Mesozoic (Farmer and DePaolo, 1983, 1984) to produce rocks with intermediate isotopic characteristics.

CONCLUSIONS

Pb isotopic data for Mesozoic rocks and ore deposits in southeastern California and Arizona strongly indicate that the Pb in these rocks is derived from the Proterozoic basement rocks of these regions. These data, together with Pb isotopic data for the Proterozoic basement rocks themselves, establish that the basement rocks in southeastern California have higher $^{207}Pb/^{204}Pb$ ratios relative to $^{206}Pb/^{204}Pb$ ratios than do those in central and southern Arizona. The boundary between these two areas runs approximately north-south through western Arizona. The nature of this boundary is uncertain, but there is a strong possibility that the two areas were joined at about 1.7 b.y. because U-Pb geochronology indicates a strong similarity in the timing of events in each area at and after this time. Nd isotopic data (Bennett and DePaolo, in press) for these areas indicate approximately the same crustal provinces. A model consistent with the combined isotopic signatures of these provinces involves multiple magmatic arcs. At least one of these arcs formed close enough to the Archean craton edge that its source region became uniformly contaminated with crustal material. An additional arc or arcs formed in an environment where no, or very little, Archean crustal material could enter the source regions, and therefore these rocks carry the isotopic signature of the Proterozoic mantle. These arcs collided with each other and the Archean craton, so that three or more terranes of different isotopic character were joined together.

The Pb isotopic data also indicate several general characteristics of the crust in these areas. The crust in both terranes has had very heterogeneous U/Pb ratios since the Middle Proterozoic. Time-integrated model values range from $^{238}U/^{204}Pb = 2$ to 30 for the Proterozoic rocks. The Early and Middle Proterozoic crust in the Mojave has the great majority of the low U/Pb ratios, and few ratios above the crustal average of 9. This is consistent with the higher metamorphic grade of most of the Early Proterozoic rocks in the Mojave (Anderson, 1986), and the information from Nd isotopic data (Bennett and DePaolo, in press) that the 1.4-b.y. granitoids were produced from the Early Proterozoic rocks. Th/U ratios in most of the Arizona crustal rocks stay close to or below the crustal average of 4 over the entire range of U/Pb ratios, whereas those in the Mojave are generally above the crustal average. These features are in part related to the "oceanic" character of the Arizona terrane, the loss of U from the Mojave terrane, and an absolute enrichment of Th in the Mojave terrane.

ACKNOWLEDGMENTS

Vicky Bennett generously shared her unpublished Nd data, and she and Lawford Anderson have shared their thoughts on the crustal provinces and Proterozoic geology of the southwestern United States. The reviews of John Aleinikoff and Dick Tosdal of early versions of this paper and of Gary Ernst and Clark Johnson of the final version of the paper are appreciated. Dennis Sorg, Cindy Benson, and Geoff Elliott prepared mineral separates and whole-rock powders. R. K. Vance and K. C. Condie provided the whole-rock powders for the Jerome, Arizona, metavolcanics, and Lawford Anderson provided samples from the 1.4-b.y.-old granitoids. JLW's research in southeastern California and Arizona is part of the USGS's Pacific to Arizona Crustal Experiment (PACE).

Anderson, J. L., 1983, Proterozoic anorogenic granite plutonism of North America, *in* Medaris, L. G., Mickelson, D. M., Byers, C. W., and Shanks, W. C., eds., *Proterozoic Geology:* Geol. Soc. America Mem. 161, p. 133–154.

——, 1984, Evolution and crystallization of Proterozoic anorogenic plutons of the southwestern U.S.: *Geol. Soc. America Abstr. with Programs*, v. 16, p. 429.

——, 1986, Proterozoic anorogenic granites of the southwestern U.S.: *Arizona Geol. Soc.*, in press.

Ashwal, L. D., and Wooden, J. L., 1983, Isotopic evidence from the eastern Canadian shield for geochemical discontinuity in the Proterozoic mantle: *Nature*, v. 306, p. 679–680.

——, Wooden, J. L., and Emslie, R. F., 1986, Sr, Nd, and Pb isotopes in Proterozoic intrusives astride the Grenville Front in Labrador — Implications for crustal contamination and basement mapping: *Geochim. Cosmochim. Acta*, v. 50, p. 2571–2586.

Barreiro, B., 1983, Lead isotopic compositions of South Sandwich Island volcanic rocks and their bearing on magmagenesis in intra-oceanic island arcs: *Geochim. Cosmochim. Acta*, v. 47, p. 817–822.

Bennett, V. C., and DePaolo, D. J., 1984, The definition of crustal provinces in the southern Rocky Mountain region using Sm-Nd isotopic characterizatics: *Geol. Soc. America Abstr. with Programs*, v. 16, p. 214.

——, and DePaolo, D. J., in press, Proterozoic crustal history of the western United States as determined by neodymium isotopic mapping: *Geol. Soc. America Bull.*, v. 99, in press.

Burchfiel, B. C., and Davis, G. A., 1972, Structural framework and evolution of the southern part of the Cordilleran orogen, western United States: *Amer. J. Sci.*, v. 272, 97–118.

——, and Davis, G. A., 1975, Nature and controls of the Cordilleran orogenies, western United States — Extensions of an earlier synthesis: *Amer. J. Sci.*, v. 275-A, p. 363–396.

Church, S. E., 1976, The Cascade Mountains revisited — A re-evaluation in light of new lead isotopic data: *Earth Planet. Sci. Lett.*, v. 29, p. 175–188.

Condie, K. C., 1982, Plate-tectonic model for Proterozoic continental accretion in the southwestern United States: *Geology*, v. 10, p. 37–42.

——, 1986, Geochemical constraints on Early Proterozoic tectonic setting in the Southwest: *Geol. Soc. America Abstr. with Programs*, v. 18, p. 347.

DePaolo, D. J., 1981a, A neodymium and strontium isotopic study of Mesozoic calc-alkaline granitic batholiths of the Sierra Nevada and Peninsular Ranges, California: *J. Geophys. Res.*, v. 86, p. 10470–10488.

——, 1981b, Neodymium isotopes in the Colorado Front Range and crust-mantle evolution in the Proterozoic: *Nature*, v. 291, p. 193–196.

DeWitt, E., Armstrong, E. L., Sutter, J. F., and Zartman, R. E., 1984, U-Th-Pb, Rb- Sr, and Ar-Ar mineral and whole-rock isotope systematics in a metamorphosed granitic terrane, southeastern California: *Geol. Soc. America Bull.*, v. 95, p. 723–739.

Doe, B. R., 1967, The bearing of lead isotopes on the source of granitic magma: *J. Petrology*, v. 8, p. 51–83.

____, and Delevaux, M. H., 1973, Variations in lead-isotopic compositions in Mesozoic granitic rocks of California — A preliminary investigation: *Geol. Soc. America Bull.*, v. 84, p. 3513--3526.

____, and Zartman, R. E., 1979, Plumbotectonics — I. The Phanerozoic, *in* Barnes, H., ed., *Geochemistry of Hydrothermal Ore Deposits*, 2nd Ed: New York, Wiley-Interscience, p. 22–70.

____, Steven, T. A., Delevaux, M. H., Stacey, J. S., Lipman, P. W., and Fisher, F. S., 1979, Genesis of ore deposits in the San Juan volcanic field, southwestern Colorado — Lead evidence: *Econ. Geology*, v. 74, p. 1–26.

Elliott, G., Wooden, J., Miller, D., and Miller, R., 1986, Lower-granulite grade Early Proterozoic supracrustal rocks, New York Mountains, SE California: *Geol. Soc. America Abstr. with Programs*, v. 18, p. 352.

Farmer, G. L., and DePaolo, D. J., 1983, Origin of Mesozoic and Tertiary granite in the western United States and implications for pre-Mesozoic crustal structure — I. Nd and Sr isotopic studies in the geocline of the northern Great Basin: *J. Geophys. Res.*, v. 88, p. 3379–3401.

____, and DePaolo, D. J., 1984, Origin of Mesozoic and Tertiary granite in the Western United States and implications for Pre-Mesozoic structure — 2. Nd and Sr isotopic studies of unmineralized and Cu- and Mo- mineralized granite in the Precambrian craton: *J. Geophys. Res.*, v. 89, p. 10141–10160.

Frost, C. D., and O'Nions, R. K., 1984, Nd evidence for Proterozoic crustal development in the Belt-Purcell Supergroup: *Nature*, v. 312, p. 53–56.

Girty, G. H., Reiland, D. N., and Wardlaw, M. S., 1985, Provenance of the Silurian Elder sandstone, north-central Nevada: *Geol. Soc. America Bull.*, v. 96, p. 925–930.

Hedge, C. E., Stacey, J. S., and Bryant, B., 1983, Geochronology of the Farmington Canyon Complex: *Geol. Soc. America Mem. 157*, p. 37–44.

Kay, R. W., Sun, S. S., and Lee-Hu, C. N., 1978, Pb and Sr isotopes in volcanic rocks from the Aleutian Islands and Pribilof Islands, Alaska: *Geochim. Cosmochim. Acta*, v. 42, p. 263–273.

Kistler, R. W., and Peterman, Z. E., 1973, Variations in Sr, Rb, K, Na, and initial $^{87}Sr/^{86}Sr$ in Mesozoic granitic rocks and intruded wall rocks in central California: *Geol. Soc. America Bull.*, v. 84, p. 3489–3518.

____, and Peterman, Z. E., 1978, Reconstruction of crustal blocks of California on the basis of initial strontium isotopic compositions of Mesozoic granitic rocks: *U.S. Geol. Survey Prof. Paper 1071*, 17 p.

Lipman, P. W., Doe, B. R., Hedge, C. E., and Steven, T. A., 1978, Petrologic evolution of the San Juan volcanic field, southwestern Colorado — Pb and Sr isotope evidence: *Geol. Soc. America Bull.*, v. 89, p. 59–82.

Miller, C. F., Bennett, V., Wooden, J. L., Solomon, G. C., Wright, J. E., and Hurst, R. W., 1984, Origin of the composite metaluminous/peraluminous Old Woman-Piute batholith, SE Calif.: *Geol. Soc. America Abstr. with Programs*, v. 16, p. 596.

Moorbath, S., and Taylor, P. N., 1981, Isotopic evidence for continental growth in the Precambrian, *in* Kroner, A., ed.: *Developments in Precambrian Geology, Vol. 4 — Precambrian Plate Tectonics;* New York, Elsevier, p. 491–525.

Nelson, B. K., and DePaolo, D. J., 1984, 1700-Myr greenstone volcanic successions in southwestern North America and isotopic evolution of Proterozoic mantle: *Nature*, v. 312, p. 143–146.

———, and DePaolo, D. J., 1985, Rapid production of continental crust 1.7 to 1.9 b.y. ago — Nd isotopic evidence from the basement of the North American mid-continent: *Geol. Soc. America Bull.*, v. 96, p. 746–754.

Rudnick, R. L., McLennan, S. M., and Taylor, S. R., 1985, Large ion lithophile elements in rocks from high-pressure granulite facies terrains: *Geochim. Cosmochim. Acta*, v. 49, p. 1645–1656.

Sawyer, D. A., and Zartman, R. E., 1985, Lead isotopes in continental arc magmas and origin of porphyry Cu deposits in Arizona: *Geol. Soc. America Abstr. with Programs*, v. 17, p. 708.

Silver, L. T., 1968, Precambrian batholiths of Arizona: *Geol. Soc. America Spec. Paper 121*, p. 558–559.

———, Anderson, C. A., Crittenden, M., and Robertson, J. M., 1977, Chronostratigraphic elements of the Proterozoic rocks of the southwestern and far western United States: *Geol. Soc. America Abstr. with Programs*, v. 9, p. 1176.

Stacey, J. S., and Hedlund, D. C., 1983, Lead-isotopic compositions of diverse igneous rocks and ore deposits from southwestern New Mexico and their implications for early Proterozoic crustal evolution in the western United States: *Geol. Soc. America Bull.*, v. 94, p. 43–57.

———, and Kramers, J. D., 1975, Approximation of terrestrial lead isotope evolution by a two-stage model: *Earth Planet. Sci. Lett.*, v. 26, p. 207–221.

———, and Zartman, R. E., 1978, A lead and strontium study of igneous rocks and ores from the Gold Hill mining district of Utah: *Utah Geology*, v. 5, p. 1–15.

———, Zartman, R. E., and Nkomo, I. T., 1968, A lead isotope study of galenas and selected feldspars from mining districts in Utah: *Econ. Geology*, v. 63, p. 796–814.

———, Doe, B. R., Silver, L. T., and Zartman, R. E., 1976, Plumbotectonics — IIA. Precambrian massive sulfide deposits: *U.S. Geol. Survey Open File Rpt.* 76–476.

Thomas, W. M., Clarke, H. S., Young, E. D., Orrell, S. E., and Anderson, J. L., 1986, High-grade Proterozoic metamorphism in the Colorado River region, Nevada, Arizona, and California: *Geol. Soc. America Abstr. with Programs*, v. 18, p. 418.

Tilton, G. R., 1983, Archean crust-mantle geochemical differentiation; *in* Ashwal, L. D., and Card, K. D., eds., *Workshop on a Cross Section of Archean Crust*, LPI tech. Rpt. 83-03: Lunar Planet. Inst., Houston, Tex., p. 92–94.

Vance, R. K., and Condie, K. C., 1986, Geochemistry and tectonic setting of the Early Proterozoic Ash Creek Group, Jerome, Arizona: *Geol. Soc. America Abst. with Programs*, v. 18, p. 419.

Van Schmus, W. R., and Bickford, M. E., 1981, Proterozoic chronology and evolution of the midcontinent region, North America, *in Developments in Precambrian Geology, Vol. 4 — Precambrian Plate Tectonics*, Kroner, A., ed.: New York, Elsevier, p. 261–296.

Wooden, J. L., Stacey, J., Howard, K., and Miller, D., 1986a, Crustal evolution in south-eastern California — Constraints from Pb isotopic studies: *Geol. Soc. America Abstr. with Programs*, v. 18, p. 200.

———, Miller, D., and Elliott, G., 1986b, Early Proterozoic geology of the northern New York Mountains, SE California: *Geol. Soc. America Abstr. with Programs*, v. 18, p 424.

Woodhead, J. D., and Fraser, D. G., 1985, Pb, Sr, and [10]Be isotopic studies of volcanic rocks from the northern Mariana Islands — Implications for magma genesis and crustal recycling in the western Pacific: *Geochim. Cosmochim. Acta*, v. 49, p. 1925–1930.

Young, E. D., Clarke, H. S., and Anderson, J. L., 1986, Proterozoic low-pressure horn-blende-granulite facies migmatites, McCullough Range, southern Nevada: *Geol. Soc. America Abstr. with Programs*, v. 18, p. 202.

Zartman, R. E., 1974, Lead isotope provinces in the Cordilleran of the western United States and their geologic significance: *Econ. Geology*, v. 69, p. 792–805.

Zartman, R. E., and Doe, B. R., 1981, Plumbotectonic — The model: *Tectonophysics*, v. 75, p. 135–162.

Zartman, R. E., and Stacey, J. S., 1971, Lead isotopes and mineralization ages in Belt Supergroup rocks, northwestern Montana and northern Idaho: *Econ. Geology*, v. 66, p. 849–860.

G. Lang Farmer
Department of Geological Sciences
University of Colorado
Boulder, Colorado 80309

4

ISOTOPE GEOCHEMISTRY OF MESOZOIC AND TERTIARY IGNEOUS ROCKS IN THE WESTERN UNITED STATES AND IMPLICATIONS FOR THE STRUCTURE AND COMPOSITION OF THE DEEP CONTINENTAL LITHOSPHERE

ABSTRACT

Combined Nd and Sr isotopic data from Mesozoic and Tertiary igneous rocks in the continental interior of the western United States can be used to infer aspects of the age, composition, and structure of deep, and otherwise inaccessible portions of the Precambrian continental lithosphere. The Nd isotopic compositions of Tertiary basalts in the western United States vary as a function of geographic position, with the lowermost ϵ_{Nd} values for pristine (noncrustally contaminated) basalts varying from -6 in southern Idaho, to -10 in southern Nevada, to $+2$ in northern New Mexico. The basalt ϵ_{Nd} values are lower than values for most oceanic basalts and require mantle sources that have experienced ancient light rare-earth element (LREE) enrichments. The basalt isotopic provinces also correspond spatially to discrete Nd isotopic provinces in the Precambrian continental crust, a correspondence that suggests that the basalts were derived from LREE-enriched "continental mantle" that has been coupled to the continental crust since the time each crustal segment formed. Basalts with the highest initial ϵ_{Nd} ($+8$) and lowest $^{87}Sr/^{86}Sr$ ratios (0.7030; $\epsilon_{Sr} = -21$) occur in regions of crustal extension, such as the central Great Basin and southern Rio Grande Rift, and are considered to have been derived from isotopically homogeneous, Rb- and LREE-depleted "oceanic" asthenosphere upwelling through the extending continental lithosphere. Late Cretaceous to early Tertiary peraluminous granites also have ϵ_{Nd} values that vary as a function of geographic position, but the ϵ_{Nd} values correspond exactly to the Nd isotopic compositions measured for felsic Precambrian crust. Therefore, the peraluminous granites must have been derived primarily from anatexis of old felsic crust. As a result, the distribution of peraluminous granites can be used to establish the subsurface distribution of Precambrian basement in the Great Basin and southern Arizona, regions where there are no basement exposures. The Nd isotopic compositions of Mesozoic and Tertiary metaluminous granites in the northern Great Basin suggest that these granites were also derived primarily from preexisting crust. The abrupt west to east decrease in initial $^{87}Sr/^{86}Sr$ displayed by these rocks in western Utah, from values >0.709 ($\epsilon_{Sr} = +60$) to values <0.709, marks a major discontinuity in the Sr isotopic characteristics of the lower continental crust in this region, possibly related to tectonic thinning of the continental craton during Late Precambrian continental rifting. Overall, the igneous rock isotopic data reveal an isotopically heterogenous continental crust, a variably LREE-enriched continental mantle, and an underlying asthenosphere with the isotopic characteristics of oceanic mantle. The origin of the Precambrian continental lithosphere is not well constrained, but the isotopic data are consistent with a model where the generation of the continental crust and the LREE enrichments in the upper mantle are related to fluid and/or magma production in ancient subduction zones.

INTRODUCTION

One key to deciphering the Mesozoic and Tertiary thermal and tectonic evolution of the upper continental crust in the western United States is an understanding of the deep structure and composition of the continental lithosphere, since varying thermal or tectonic conditions in the upper crust generally result from processes taking place deeper in the lithosphere. However, little direct information exists regarding the age, structural, and chemical characteristics of the lower continental crust or the underlying upper mantle. One indirect method of studying the deeper portions of continental lithosphere is through the Nd, Sr, and Pb isotopic compositions of Mesozoic and Tertiary igneous rocks. Since these igneous rocks have source regions within, or below, the continental lithosphere, they preserve a record of the isotopic compositions of those regions involved in the generation of their parental magmas. Any variations in the isotopic compositions of the magma source(s) inferred in this fashion can then be used to deduce aspects of the lithosphere, or sublithosphere, age, composition, and structure.

The western United States is an ideal region to apply such isotopic techniques of probing the continental lithosphere, first, because there was abundant Mesozoic and Tertiary magmatism and, second, because this magmatism was not confined to a narrow band parallel to the continental margin but instead occurred up to 1500 km inland (Coney and Reynolds, 1977; Cross and Pilger, 1978). One advantage of studying continental interior magmatism (as opposed to continental margin magmatism, such as in the Idaho, Sierra Nevada, and Peninsular Ranges batholiths) is that continental-interior magmatism in the western United States was much less voluminous than that at the continental margin, minimizing the possibility that the magmatism itself erased the isotopic record of the preexisting lithosphere (*cf.* Farmer and DePaolo, 1983). A more important advantage, however, is the fact that the continental-interior magmatism occurred in regions underlain by Precambrian continental crust and, as a result, may have tapped portions of the Precambrian lithosphere. The age, composition, and structure of the Precambrian lithosphere are particularly amenable to study using the radiogenic isotopic compositions of younger igneous rocks since the range of isotopic compositions in the continental lithosphere depends not only on the range of lithosphere parent/daughter ratios (i.e., Sm/Nd, Rb/Sr, U/Pb), but also on the amount of time that has passed since fractionations in the parent/daughter ratios occurred. Domains with contrasting Nd, Sr, or Pb isotopic compositions are therefore more likely to exist within old lithosphere (recording ancient parent/daughter fractionations related, for example, to crust formation), even though both old and newly formed lithosphere may show similar internal variations in Sm/Nd, Rb/Sr, or U/Pb.

In the following discussion, the available Nd and Sr isotopic data from Mesozoic Tertiary igneous rocks in the western United States will be reviewed in terms of the implications these data have regarding the composition of the deep Precambrian continental lithosphere. In this context, both basaltic extrusive rocks, and more silicic intrusive rocks will be considered, since the basaltic rocks provide information regarding the subcrustal mantle while the silicic rocks can be used as probes of the lower continental crust.

ISOTOPE SYSTEMATICS

Systematic variations in the Nd and Sr isotopic compositions exist within the earth due to the radioactive decay of ^{147}Sm to ^{143}Nd ($t_{1/2} = 106$ b.y.) and ^{87}Rb to ^{87}Sr ($t_{1/2} = 49$ b.y.) combined with fractionation of Sm from Nd, and Rb from Sr, during planetary differentiation processes such as the generation of continental crust from partial melting of the upper mantle. The basic systematics of combined Nd and Sr isotope geochemistry have been reviewed by DePaolo (1981a,b, 1983). Since Sm and Nd are both light rare-earth elements (LREE), their geochemical behavior is very similar (Henderson, 1984), resulting in a relatively narrow range of ^{147}Sm/^{144}Nd in most rock types (0.1–0.3; Hawkesworth and van Calsteren, 1984). The restricted range of ^{147}Sm/^{144}Nd, combined with the long half-life of ^{147}Sm, have allowed only small ^{143}Nd/^{144}Nd variations to evolve in present-day terrestrial rocks ($\sim$1%). In contrast, Rb and Sr belong to different chemical families (the alkali and alkali-earth elements, respectively), so that the Rb/Sr ratios, and ^{87}Sr/^{86}Sr, vary in the earth over a much wider range (Faure and Powell, 1972).

A generalized diagram showing the expected ranges of Nd and Sr isotopic compositions for oceanic and continental lithosphere is given in Fig. 4-1. On this diagram the Nd and Sr isotopic compositions are given in terms of ϵ values, which are defined as

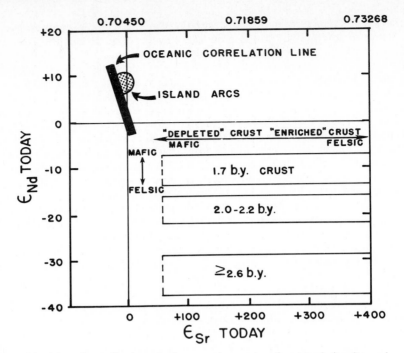

FIG. 4-1. Generalized range of ϵ_{Nd} and ϵ_{Sr} values for oceanic basalts, and for Precambrian continental crust in the western United States. Data sources given in text.

$$\epsilon_{Nd}(T) = \left[\frac{(^{143}\mathrm{Nd}/^{144}\mathrm{Nd})_{rock}(T)}{(^{143}\mathrm{Nd}/^{144}\mathrm{Nd})_{CHUR}(T)} - 1 \right] \times 10^4$$

and

$$\epsilon_{Sr}(T) = \left[\frac{(^{87}\mathrm{Sr}/^{86}\mathrm{Sr})_{rock}(T)}{(^{87}\mathrm{Sr}/^{86}\mathrm{Sr})_{UR}(T)} - 1 \right] \times 10^4$$

The $^{143}\mathrm{Nd}/^{144}\mathrm{Nd}_{CHUR}(T)$ refers to the $^{143}\mathrm{Nd}/^{144}\mathrm{Nd}$ ratio in a model chondritic reservoir (CHUR), at time T, and is defined as

$$^{143}\mathrm{Nd}/^{144}\mathrm{Nd}_{CHUR}(T) = {}^{143}\mathrm{Nd}/^{144}\mathrm{Nd}_{CHUR}(0) - 0.1967 \left[\exp(\lambda_{Sm}T) - 1 \right]$$

Unfortunately, Nd isotopic compositions reported in the literature are corrected for mass fractionation (during mass spectrometric analyses) relative to several different normalization schemes. The two most common normalization schemes assume either (1) $^{146}\mathrm{Nd}/^{142}\mathrm{Nd} = 0.63615$ (DePaolo and Wasserburg, 1976, 1977) or (2) $^{146}\mathrm{Nd}/^{144}\mathrm{Nd} = 0.7219$ (cf. Menzies et al., 1983), with two schemes yielding somewhat different absolute $^{143}\mathrm{Nd}/^{144}\mathrm{Nd}$ values for a given sample (by about 1 per mil). However, data acquired using either method can be directly compared, to a good approximation, using ϵ_{Nd} values [expressed relative to either $^{143}\mathrm{Nd}/^{144}\mathrm{Nd}_{(CHUR)}(0) = 0.511836$ or 0.51264, for normalization schemes 1 and 2, respectively].

The $^{87}Sr/^{86}Sr_{UR}(T)$ refers to the $^{87}Sr/^{86}Sr$ ratio in a model reservoir (UR; DePaolo and Wasserburg, 1976), where

$$^{87}Sr/^{86}Sr(T) = 0.7045 - 0.0827 \, [\exp(\lambda_{Rb}T) - 1]$$

In the following discussion, ϵ_{Nd} and ϵ_{Sr} will always refer to initial isotopic compositions (time $= T$).

A compilation of Nd and Sr isotopic data from oceanic basalts available up to 1984 is given by White (1985). In general, oceanic basalts plot in a narrow range of isotopic compositions (oceanic correlation line; Fig. 4-1), with the highest ϵ_{Nd} ($\approx +12$) and lowest ϵ_{Sr} (-25) corresponding to the most LREE- and Rb-"depleted" mantle reservoir(s). The low Rb/Sr, high Sm/Nd of the depleted mantle reservoirs are considered to be the result of continuous or semicontinuous extraction over time of magma enriched in LREE and Rb. In contrast, continental crust has lower ϵ_{Nd}, and generally higher ϵ_{Sr}, reflecting the higher Rb/Sr and lower Sm/Nd of the continental crust relative to the average upper mantle. Data summarized by Farmer and DePaolo (1984) suggest that the ϵ_{Nd} of Precambrian crust in the western United States depends on the crustal age and bulk composition, with mafic crustal rocks generally having higher ϵ_{Nd} and higher Sm/Nd ratios than felsic crust (see also Nelson and DePaolo, 1984). Studies by DePaolo (1981b), Farmer and DePaolo (1983, 1984), Nelson and DePaolo (1985), Leeman *et al.* (1985), and Bennett and DePaolo (in press) have also demonstrated that the Precambrian crust in the western United States can be divided into provinces with discretely different Nd model ages (T_{DM} ages), with the model ages in any given region (Fig. 4-2) being independent of

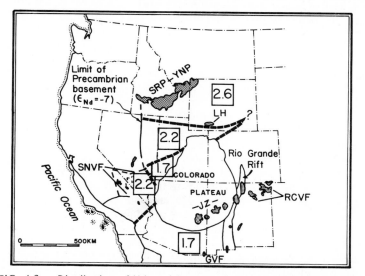

FIG. 4-2. Distribution of Nd model age provinces in the Precambrian continental crust of the western United States relative to selected regions of late Cenozoic basaltic and ultrapotassic volcanism. Dashed lines mark approximate boundaries between crustal provinces (references given in text) and blocked numbers are the average crustal Nd model ages (in b.y.). Note that Nd model age provinces 2 and 3 of Bennett and DePaolo (in press) have been combined into the $T_{DM} = 1.7$ b.y. old province. Western limit of Precambrian basement in the Great Basin from Farmer and DePaolo (1983). Abbreviations: RCVF, Raton-Clayton volcanic field; GVF, Geronimo volcanic field; SNVF, Southern Nevada volcanic field; JZ, Jemez volcanic zone; SRP-YNP, Snake River Plain-Yellowstone National Park volcanic fields; LH, Leucite Hills; Garlock and San Andreas faults shown in central and southern California.

FARMER

crustal bulk composition. The crustal ϵ_{Sr} values depend on the crustal age and bulk composition, and metamorphic grade, with rocks of similar bulk composition showing generally lower Rb/Sr and lower ϵ_{Sr} with increasing metamorphism, although the latter observations are based on a small data set from crustal xenoliths (*cf.* James *et al.*, 1980; see also Reid *et al.*, 1985).

The crustal isotopic compositions described above provide a framework in which the Nd and Sr isotopic compositions of Mesozoic and Tertiary igneous rocks in the continental interior of the western United States can be interpreted. The igneous rock isotopic data can be used to identify the magmatic source regions and to better characterize the isotopic compositions of the source regions. It should be emphasized, however, that the igneous rocks retain the isotopic compositions, but not necessarily the model ages of their sources. Fractionation in the parent/daughter ratio (Sm/Nd or Rb/Sr) can take place during partial melting or during the subsequent magma differentiation, resulting in deviations in the magma model age from that of its source (see the discussion of T_{DM} models ages in Bennett and DePaolo, in press). Therefore, in the following discussion the igneous rock initial isotopic compositions, but not their Nd (or Sr) model ages, will be used to infer the isotopic characteristics of their source regions.

BASALTIC MAGMATISM

Lead isotopic studies of Cenozoic basalts in the interior regions of the western United States have revealed that the Pb isotopic compositions of these basalts define secondary isochrons, the ages of which correspond closely to the age of the Precambrian basement through which the basalts erupted (Lipman, 1980). The primitive bulk chemical compositions of many of these basalts, and their regular Pb isotopic compositions, have been interpreted as evidence that the basalt Pb isotopic characteristics were not inherited from the continental crust, for example, through crustal assimilation by the basaltic parental magmas, but instead directly reflect the isotopic characteristics of the upper mantle directly underlying the Precambrian continental crust (Menzies *et al.*, 1983). As a cautionary note, it should be pointed out that the low Pb concentrations (<3 ppm) of primitive basaltic magmas make the Pb isotopic compositions of such magmas particularly susceptible to alteration by interaction with small amounts of felsic continental crust (with >20 ppm of Pb; Zartman and Doe, 1981). Therefore, the secondary Pb isochrons observed in the basalts could reflect crustal isotopic compositions even in undifferentiated rocks if the basalt magma interacted with a crustal source with comparatively high Pb concentrations. In any case, a model for the continental lithosphere of the western United States, based in part on the Pb isotopic data, has been proposed (*cf.* Lipman, 1980) that suggests that each segment of Precambrian continental crust is underlain by a "keel" of upper mantle which was isolated from the upper mantle as a whole at the time each segment of continental crust was formed. The continental lithosphere (crust and mantle) may extend to at least the depth of the continental Moho (~100 km), with asthenospheric mantle, similar isotopically to oceanic mantle, underlying the continental lithosphere as a whole.

Nd isotopic data from Cenozoic basalts and ultrapotassic rocks in the western United States can be used to test further the lithosphere model given above. Data are available from three regions of Cenozoic basaltic volcanism: the Snake River Plain and Yellowstone National Park, southwestern Nevada and eastern California, and the Rio Grande rift and vicinity (Fig. 4-2). Basalts in these three regions intruded through T_{DM} = 2.6-, 2.2-, and 1.7-b.y.-old crust, respectively, and so provide an opportunity to study the Nd and Sr isotopic characteristics of the subcrustal mantle beneath each of the main Nd crustal provinces.

T_{DM} = 1.7-b.y. Crust: Rio Grande Rift

The Rio Grande rift (RGR; Fig. 4-2) is a complex set of asymmetrical grabens located from central Colorado southward to northern Chihuahua (Baldridge et al., 1984). The largest amount of extension took place in the southern portions of the rift, beginning about 32 m.y. ago, while rifting in the northernmost portion of the rift began about 5 m.y. later (Chapin, 1979). Basaltic volcanism associated with rifting is primarily late Pliocene to Quaternary in age (<5 m.y.), with the southern rift (south of Socorro) dominated by alkali olivine basalts, the northern rift by aluminous olivine tholeiites (the Taos Plateau volcanic field; Dungan et al., 1984; Lipman and Mehnert, 1979), and the central rift by both tholeiitic and alkalic composition basalts (Baldridge, 1979).

Published Nd and Sr isotopic data from basalts both within and adjacent to the RGR are summarized in Fig. 4-3. Within the RGR, basalts show a north to south increase

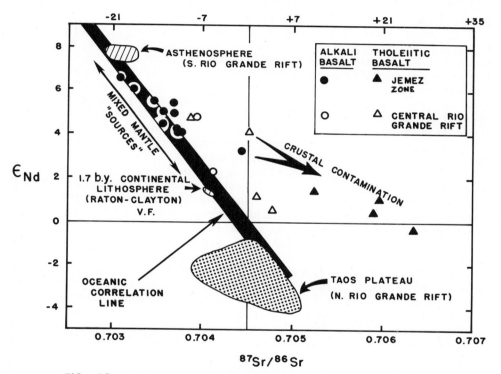

FIG. 4-3. ϵ_{Nd} versus ϵ_{Sr} plot for alkali and tholeiitic basalts in the Rio Grande rift and vicinity (T_{DM} = 1.7 b.y. crust). Data compiled from Perry et al. (1985), Williams and Murthy (1979), Menzies et al. (1983), Crowley et al. (1985), Phelps et al. (1983), and F. Perry (personal communication, 1985). Asthenosphere values inferred from Nd and Sr data from basalts in S. Rio Grande Rift and adjacent Basin and Range.

in ϵ_{Nd}, from values as low as −4 for the Taos Plateau tholeiitic basalts (Williams and Murthy, 1979), to values between 0 and +5 in the central RGR (for both tholeiitic and alkalic basalts; Perry et al., 1985), to values of +7 for alkali basalts in the southern rift (Crowley et al., 1985). The ϵ_{Sr} values for the RGR alkali basalts decrease regularly from north to south within the rift, from values of +10 to −20, with the combined Nd and Sr isotopic compositions generally plotting on the oceanic correlation line (Fig. 4-3; Crowley *et al.*, 1985). Adjacent to the central RGR, in the Jemez volcanic zone (JZ; Fig. 4-2), alkali basalts show a similar, but wider range of ϵ_{Nd} and ϵ_{Sr} values (+7 to +3, and −8 to −20, respectively) to basalts in the central RGR, but JZ tholeiitic basalts have low ϵ_{Nd} (−0.3 to +1.4) and high ϵ_{Sr} values (+7 to +25) that plot well off the oceanic correlation line (Fig. 4-3).

Some of the isotopic variations observed for the RGR and JZ basalts can be accounted for by crustal contamination. For example, the relatively evolved chemical compositions of the JZ and Taos Plateau tholeiitic basalts suggest a crustal component in these rocks (Dungan *et al.*, 1984, 1986; Perry *et al.*, 1985), with their isotopic compositions reflecting interaction with either high Rb/Sr upper crust, for the JZ basalts, or low Rb/Sr lower crust, for the Taos Plateau basalts. Both the JZ and RGR alkali basalts, however, are considered to have retained the isotopic compositions of their mantle sources and so reveal a considerable heterogeneity in the isotopic characteristics of the subcrustal mantle in these regions.

Following an earlier model proposed by Lipman and Mehnert (1975), Perry *et al.*[1] (1985) interpret the range of alkali basalt Nd isotopic compositions as being the result of the involvement of two discrete mantle sources in the basalt generation: continental lithosphere and underlying asthenosphere. In the southern rift and adjacent Basin and Range (Geronimo volcanic field; Menzies *et al.*, 1983) basalts were derived exclusively from upwelling asthenospheric mantle, which according to available geophysical data (Baldridge *et al.*, 1984) has reached the base of the continental crust and totally displaced the preexisting mantle lithosphere. The high ϵ_{Nd} and low ϵ_{Sr} values of the southern rift basalts require a LREE- and Rb-"depleted" asthenospheric source similar isotopically to the mantle sources of ocean island and island-arc basalts (DePaolo and Johnson, 1979; White, 1985). The Pb isotopic compositions of these basalts also require a source with oceanic mantle affinities (Everson, 1979). In contrast, the degree of lithosphere extension is significantly less in the central RGR and adjacent JZ (Baldridge *et al.*, 1984), correlating with the decrease in basalt ϵ_{Nd} and increase in ϵ_{Sr}. Perry *et al.* (1985) interpret the change in basalt isotopic compositions as the result of an increased involvement of "enriched" subcontinental mantle in the basalt generation, the former having remained at least partially intact in this region due to the lesser amount of lithospheric extension. The range of ϵ_{Nd} and ϵ_{Sr} for the central RGR and JZ alkali basalts then corresponds to variations in the relative amounts of asthenosphere and lithosphere components in the parental basalt magmas (Fig. 4-3), although the exact mechanism whereby the two sources "mix" is not known. The isotopic composition of that portion of the continental mantle involved in the basalt generation can be estimated from the isotopic compositions of basanites and nephelinites in the Raton-Clayton volcanic field (RCVF; Fig. 4-2), volcanics that erupted in a stable portion of the 1.7-b.y.-old craton and are considered to have crystallized from primary mantle-derived magmas (Phelps *et al.*, 1983). These rocks have a restricted range of $\epsilon_{Nd} = +1.0$, and $\epsilon_{Sr} = -6.0$ to −8.0 (Fig. 4-3), values that correspond closely to the lowest ϵ_{Nd} and highest ϵ_{Sr} values observed for the alkali basalts in the central RGR.

The exact Sm/Nd ratio of the continental mantle, and the time interval over which the continental mantle has maintained regions with lower Sm/Nd than the oceanic mantle, cannot be determined solely from the basalt Nd isotopic data. However, alkali basalts from the Geronimo volcanic field contain peridotite xenoliths which may represent direct samples of the $T_{DM} = 1.7$-b.y.-old continental mantle (Menzies, 1984). Menzies has described diopsides (which represent the major reservoir for rare earth elements in

---

[1] Also F. V. Perry, W. S. Baldridge, and D. J. DePaolo, unpublished data.

such nodules) with measured $\epsilon_{Nd} = -1$ to $+8$. The type "lb" diopsides have the lowest ϵ_{Nd} and lowest $^{147}Sm/^{144}Nd$ ratios (with average values of about 0.11). Approximate Nd model ages calculated relative to the depleted mantle evolution curve of Nelson and DePaolo (1984) for the type "lb" diopsides are on the order of 1.5–1.8 b.y., similar to values measured for exposed continental crust in this region (Nelson and DePaolo, 1984). Therefore, mantle "enrichment" and continental crust formation may have been largely contemporaneous and, possibly, genetically related.

$T_{DM} = 2.6$-b.y. Crust: Snake River Plain and Vicinity

Pliocene olivine tholeiites from the Snake River Plain (SRP; Fig. 4-2) show secondary Pb isochrons of about 2.7 b.y. (Leeman, 1975), suggesting that these basalts were derived exclusively from continental mantle associated with the Archean crust. The ϵ_{Nd} and ϵ_{Sr} values for these basalts, and for tholeiitic basalts in the vicinity of the Yellowstone National Park (YNP), range from -3 to -6, and $+20$ to $+50$, respectively (Fig. 4-4;

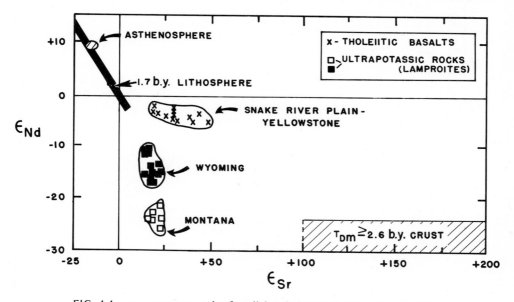

FIG. 4-4. ϵ_{Nd} versus ϵ_{Sr} plot for olivine tholeiitic basalts, and mafic ultra-potassic rocks in regions underlain by Archean ($T_{DM} > 2.6$ b.y.) crust. Data compiled from Fraser *et al.* (1985), Vollmer *et al.* (1984), and Menzies *et al.* (1984). Range of isotopic compositions for the Archean crust from Leeman *et al.* (1985).

Menzies *et al.*, 1983, 1984), which, presumably, reflects the isotopic compositions of portions of the continental mantle underlying the 2.6-b.y. continental crust. The lower ϵ_{Nd} and higher ϵ_{Sr} values for the SRP-YNP basalts, relative to the RGR basalts, are consistent with the derivation of the SRP-YNP basalts from a continental mantle reservoir with Sm/Nd and Rb/Sr ratios similar to, but with an age considerably older than, the mantle source of the RGR rift basalts. This relationship is illustrated for the Nd isotopic data in Fig. 4-5. In this diagram, the SRP-YNP and RGR basalt source regions are modeled as having formed via LREE enrichment of "depleting" mantle 2.6 and 1.7 b.y. ago, respectively. Within this framework, the present-day ϵ_{Nd} values for the basalt source

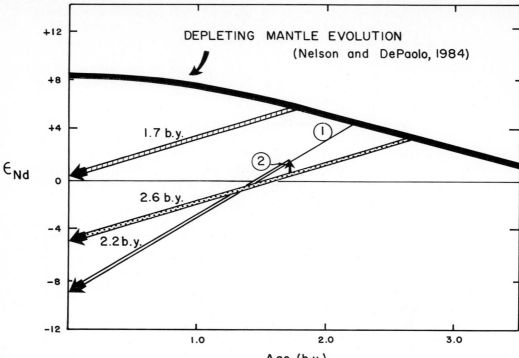

FIG. 4-5.   Idealized Nd isotopic evolution of basalt source regions within the T_{DM} = 1.7, 2.2, and 2.6 b.y. continental lithosphere. The 1.7- and 2.6-b.y. reservoirs can be modeled as having similar $^{147}Sm/^{144}Nd$ ratios (0.10 to 0.12), with the difference in present-day ϵ_{Nd} values arising from the difference in the times of segregation from the "depleting" mantle, the latter considered to be represented today by oceanic mantle. The 2.2-b.y. mantle source could represent a discrete lithosphere forming event 2.2 b.y. ago (1), but with a greater LREE enrichment (lower Sm/Nd) than the 1.7- or 2.6-b.y. sources, or, more likely, a segment of Archean lithosphere enriched in LREE via addition of "depleted" mantle derived magma or fluids during the formation of the 1.7-b.y. lithosphere (2).

regions in each mantle segment can be generated if both source regions had $^{147}Sm/^{144}Nd$ values of about 0.10–0.12. Therefore, basalt source regions within both the T_{DM} = 2.6-b.y. and 1.7-b.y. continental mantle can be considered to have undergone similar degrees of LREE enrichment at or near the time that each segment of continental lithosphere was formed.

Mafic, ultrapotassic volcanic rocks (lamproites) within the Archean craton have virtually identical ϵ_{Sr} values to the SRP-YNP basalts (Fig. 4-4) but much lower and more variable ϵ_{Nd} values, ranging from −10 to −17 in Wyoming (Leucite Hills; Vollmer *et al.*, 1984) and from −22 to −26 in Montana (Smoky Butte; Fraser *et al.*, 1985). The low ϵ_{Nd} values, which are also characteristic of Group II kimberlites in Australia and South Africa (McCulloch *et al.*, 1983; Smith, 1983), have been interpreted as reflecting an extreme LREE enrichment in the deepest levels of the Archean continental lithosphere (Fraser *et al.*, 1985). The relatively low and uniform ϵ_{Sr} values of the ultrapotassic rocks could be an indication that the LREE enrichment in the deep lithosphere was not accompanied by an increase in Rb/Sr, or that high Rb/Sr mineral phases were not stable in the region of ultrapotassic magma generation (Carlson *et al.*, 1985; Vollmer *et al.*, 1984).

Although the Nd isotopic data from both the basalts and ultrapotassic rocks imply that gross heterogeneities may exist in the amount of LREE enrichment in the Archean continental mantle lithosphere, few conclusions can be drawn from these data regarding the detailed characteristics of this enrichment. For example, if a greater depth of magma generation is required for the ultrapotassic rocks relative to the SRP-YNP basalts (*cf.* Ringwood, 1975), the lower ϵ_{Nd} values for the ultrapotassic rocks may imply that portions of the Archean mantle become increasingly LREE enriched with increasing depth. However, the Archean mantle could also consist throughout of relatively LREE-depleted (high-ϵ_{Nd}) periodotite, with regions or veins that are K-rich, hydrous, and LREE enriched (low-ϵ_{Nd}). In this case, the ultrapotassic rocks could represent small degrees of partial melting of the "veined" mantle which sampled only the most LREE-enriched and lowest-ϵ_{Nd} material. Greater degrees of partial melting of this same source would involve a larger proportion of the LREE-depleted peridotite and could yield basaltic magmas with higher ϵ_{Nd} values (F. V. Perry, W. S. Baldridge, and D. J. DePaolo, unpublished data).

The wide range of lamproite ϵ_{Nd} values may also imply that gross heterogeneities exist in the amount of LREE enrichment even in the deepest portions of the Archean lithosphere (Vollmer *et al.*, 1984). It is interesting to note, however, that secondary Pb isochrons for the YNP basalts in Wyoming are significantly younger than those for basalts associated with the Independence volcano in Montana (2.5–2.8 versus 3.8 b.y.; Carlson *et al.*, 1985; Doe *et al.*, 1982; Peterman *et al.*, 1970), suggesting that segments of the Archean lithosphere may have been isolated from the upper mantle at significantly different times. Therefore, the extremely low ϵ_{Nd} values for the Montana lamproites, relative to those in Wyoming, may reflect the older lithosphere age in this region, with the LREE enrichment in the lamproite source regions in Montana and Wyoming being related to crust forming events at 3.8 and 2.8 b.y. b.p., respectively. Such a correlation between lithosphere "age" and lamproite ϵ_{Nd} values, if fully confirmed by additional data, would be a strong argument for the derivation of these rocks from old, LREE-enriched continental lithosphere, rather than from enriched portions of the underlying asthenosphere (*cf.* D. L. Anderson, 1985).

T_{DM} = 2.2-b.y. Crust: SW Nevada

Late Miocene and younger transitional alkali basalts (hawaiites) in southern Nevada (Vaniman *et al.*, 1982) have long been recognized to have unusually radiogenic Sr (Fig. 4-6; Hedge and Noble, 1971; Leeman, 1970) and nonradiogenic Pb (Everson, 1979) relative to alkali basalts found elsewhere in the Basin and Range (Leeman, 1982). The latter basalts, such as those in the central Great Basin (Foland *et al.*, 1983), have ϵ_{Nd} values (+8), similar to the southern RGR basalts, suggesting a source in depleted oceanic asthenosphere upwelling in this region in conjunction with Basin and Range extension (Eaton *et al.*, 1978). However, within the southern Nevada isotopic province (the "Sierran" province of Leeman, 1982), hawaiite Nd isotopic compositions are much lower (ϵ_{Nd} = −6 to −9), correlating with the higher ϵ_{Sr} values (fig. 4-6). On the basis of bulk and trace element data, the hawaiite isotopic compositions are considered to be representative of the mantle sources of these basalts and not the result of crustal interaction (Hedge and Noble, 1971; Semken, 1984; Vaniman *et al.*, 1982). Therefore, the hawaiites seem to record evidence of an unusually LREE-enriched continental mantle beneath southern Nevada. In this context, it is significant that the geographic limits of the southern Nevada isotopic province correspond closely to the limits of T_{DM} = 2.2 b.y. Precambrian basement in this region (Fig. 4-2), as inferred from the Nd isotopic compositions of Tertiary

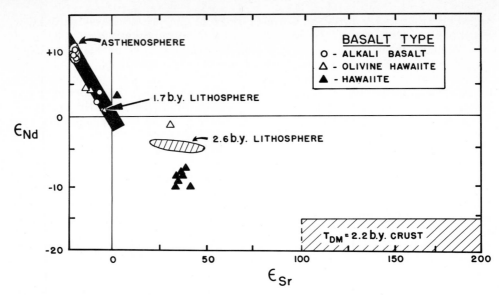

FIG. 4-6. ϵ_{Nd} versus ϵ_{Sr} plot for basalts in southern Nevada ($T_{DM} = 2.2$ b.y crust). Data from Semken (1984). Range of isotopic compositions for the $T_{DM} = 2.2$ b.y. crust estimated from the data of Farmer and DePaolo (1983).

peraluminous granite and exposed Precambrian basement rocks in the southern Great Basin and vicinity (Bennett and DePaolo, in press). Therefore, the hawaiite parental magmas could have been derived from the continental mantle associated with the $T_{DM} = 2.2$-b.y. continental lithosphere.

The origin of the $T_{DM} = 2.2$-b.y. lithosphere is not well constrained, but the available Nd isotopic data favor an origin via mixing of magma derived from "depleted" mantle (i.e., oceanic asthenosphere?) 1.7 b.y. ago with preexisting Archean continental lithosphere, rather than an origin as a discrete segment of continental lithosphere that formed 2.2 b.y. ago (Bennett and DePaolo, in press). In any model, however, the unusually low ϵ_{Nd} values for the southern Nevada hawaiites require that the source for these basalts have a lower Sm/Nd ratio than the sources of either the RGR or SRP-YNP basalts. One possible explanation for the extreme LREE enrichment of the continental mantle in southern Nevada is that this segment of mantle represents LREE enriched Archean mantle that was LREE enriched again during the 1.7-b.y. crust-forming event (Fig. 4-5).

SILICIC INTRUSIVE ROCKS

Combined Nd and Sr isotopic data from Mesozoic and Tertiary granite in the western United States (Farmer and DePaolo, 1983, 1984) have revealed that continental interior granites, herein defined as those intrusive bodies located inland of the Idaho, Sierra Nevada, and Peninsular Ranges batholiths, were derived primarily from preexisting middle to lower continental crust. As a result, the granite isotopic compositions can be used to investigate the chemical and structural characteristics of deep portions of the continental crust. The following discussion will consider separately peraluminous and metaluminous granites (molecular $Al_2O_3/(CaO + K_2O + Na_2O) > 1$, and < 1, respectively) because the isotopic data suggest that distinctly different source regions were involved in the generation of the two different granite types.

Peraluminous Granite

Late Cretaceous to early Tertiary peraluminous granite in the western United States (Lee et al., 1981; C. F. Miller and Bradfish, 1980) have ϵ_{Nd} values that vary regularly from north to south across the western United States (Farmer and DePaolo, 1984). As illustrated in Fig. 4-7, the peraluminous granite initial ϵ_{Nd} values are identical to the ϵ_{Nd} values

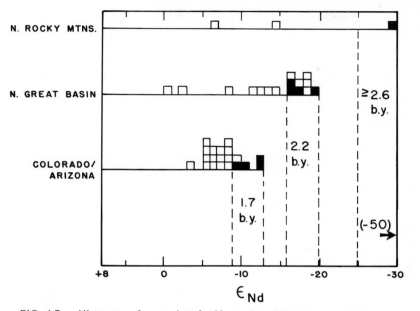

FIG. 4-7. Histogram of ϵ_{Nd} values for Mesozoic and Tertiary metaluminous (open squares) and peraluminous (closed squares) granite in the western United States relative to estimated range of ϵ_{Nd} values for felsic Precambrian basement (measured at the time of granite formation). Data from Farmer and DePaolo (1984).

for exposed, felsic composition, Precambrian basement in each region (as measured at the time of peraluminous granite formation). These data suggest that throughout the western United States, the majority of peraluminous granites were derived exclusively from partial or complete anatexis of felsic basement. In contrast, ϵ_{Sr} values for the peraluminous granites are considerably lower than values measured for exposed, felsic, Precambrian basement rocks, suggesting a lower Rb/Sr ratio for the crustal source of the peraluminous granites relative to the exposed felsic crust (Fig. 4-8). This observation implies either that the peraluminous granites were derived from deep-seated felsic crust which had undergone an ancient Rb-depletion event, possibly associated with devolatilization of the deep crust during regional metamorphism near the time of crust formation, or that variations in the bulk composition of the crust exist at depth which are reflected in changes in Rb/Sr but not in Sm/Nd.

Given that the peraluminous granites were derived exclusively from preexisting continental crust, the spatial distribution of these granites can then be used to delineate the boundaries of Precambrian basement terranes in regions where basement is not exposed. As demonstrated by Farmer and DePaolo (1984), the presence of semiautochthonous peraluminous granite with ϵ_{Nd} values of −9.9 to −12 in south-central Arizona (Fig. 4-9)

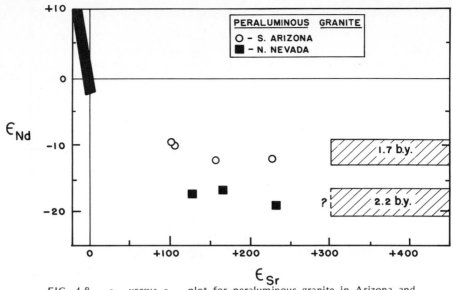

FIG. 4-8. ϵ_{Sr} versus ϵ_{Nd} plot for peraluminous granite in Arizona and northern Nevada relative to estimated isotopic compositions for felsic Precambrian basement in these two regions.

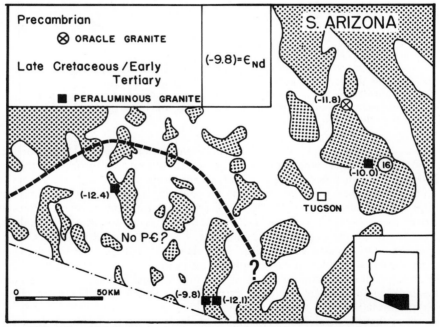

FIG. 4-9. Distribution and ϵ_{Nd} values of Late Cretaceous and early Tertiary peraluminous granite in southern Arizona. Autochthonous Precambrian basement rocks are not exposed south of the dashed line (Haxel *et al.*, 1984; personal communication, 1984). Modified from Farmer and DePaolo (1984, Fig. 3).

requires that this region be underlain by $T_{DM} = 1.7$-b.y. crust, despite the lack of exposed Precambrian basement rocks (Haxel *et al.*, 1984). In northeastern Nevada, the spatial distribution and isotopic compositions of peraluminous granites ($\epsilon_{Nd} = -16$ to -19) can be used to delineate the western limit of Precambrian basement concealed beneath the miogeoclinal sedimentary rocks of the Great Basin (Fig. 4-10), and to demonstrate that $T_{DM} = 2.2$-b.y. basement underlies much of the northern Great Basin (Farmer and DePaolo, 1983). Note from Fig. 4-10 that the $^{87}Sr/^{86}Sr = 0.706$ line, which has been taken to represent the edge of the Precambrian basement (Kistler and Peterman, 1973), lies some 50–100 km to the west of the edge of the Precambrian basement as defined by the distribution of the peraluminous granites.

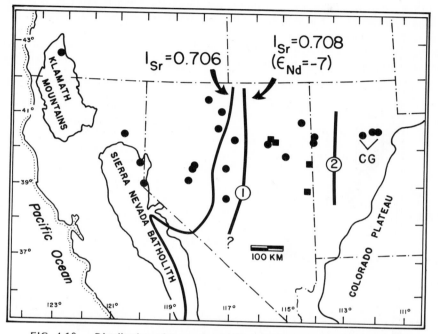

FIG. 4-10. Distribution of lower crustal Sr and Nd isotopic provinces in the Great Basin as defined by Mesozoic and Tertiary granites. Filled squares and filled circles are locations for peraluminous and metaluminous granites, respectively, studied by Farmer and DePaolo (1983). The $I_{Sr} = 0.706$ contour is from Farmer and DePaolo (1983) and Kistler and Peterman (1973). Granites west of this line have initial $^{87}Sr/^{86}Sr$ ratios less than 0.706. Line 1, where $I_{Sr} = 0.708$, is considered by Farmer and DePaolo (1983) to represent the western edge of Precambrian basement underlying the miogeoclinal sedimentary rocks in the Great Basin. Line 2 is the discontinuity in granite Sr isotopic compositions in western Utah (see the text). Abbreviation: CG, "cratonal" metaluminous granites. Modified from Farmer and DePaolo (1983, Fig. 9).

Metaluminous Granite

Unlike the peraluminous granites, Mesozoic and Tertiary metaluminous granite from the continental interior show a wide range of initial Nd isotopic compositions (Fig. 4-7), ranging from $\epsilon_{Nd} = -3.4$ to -8.6 in southern Arizona and Colorado ($T_{DM} = 1.7$-b.y.-old crust), and from $+0.7$ to -18.7 in the eastern Great Basin ($T_{DM} = 2.2$-b.y.-old crust). In general, these ϵ_{Nd} are greater than the ϵ_{Nd} values for peraluminous granite in each region. The trend toward higher ϵ_{Nd} values and lower ϵ_{Sr} values could be due to (1) anatexis of a mafic composition continental crust, with higher Sm/Nd and lower Rb/Sr than the crustal source of the peraluminous granites, or (2) the presence of a mantle-derived ($\epsilon_{Nd} \geqslant 0$) component in the metaluminous granites, which implies that these granites evolved from mantle-derived, mafic composition magmas that interacted with lower continental crust. In southern Arizona, the Nd isotopic compositions for metaluminous granite correspond closely to the range of Nd isotopic compositions of lower crustal amphibolites xenoliths found in potassic andesite lavas in central Arizona (-2 to -9; Esperanca *et al.*, 1985), suggesting that amphibolite lower crust could have been the dominant source for the Arizona metaluminous granites. In the eastern Great Basin, on the other hand, the range of Nd isotopic compositions is much larger than in southern Arizona. The range of Nd and Sr isotopic compositions can be interpreted as a mixing curve between high ϵ_{Nd} ($\geqslant 1$), low ϵ_{Sr} ($\leqslant 0$) mantle-derived magmas with a low ϵ_{Nd}, high ϵ_{Sr} lower crustal source, similar isotopically to the source region from which the peraluminous granite in this region were derived (Fig. 4-11; Farmer and DePaolo, 1983).

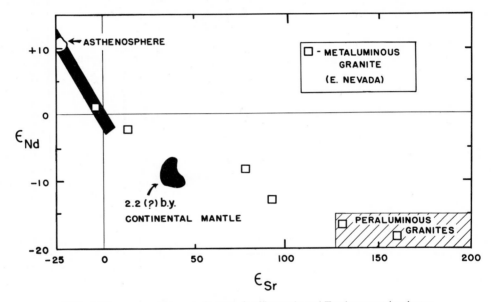

FIG. 4-11. ϵ_{Nd} versus ϵ_{Sr} diagram for Mesozoic and Tertiary metaluminous granite in eastern Nevada.

The parental magmas appear to have originated in "depleted" asthenosphere underlying the continental lithosphere, since the metaluminous granites trend toward higher ϵ_{Nd} and lower ϵ_{Sr} than expected for mafic magmas derived from the $T_{DM} = 2.2$-b.y. continental mantle (Fig. 4-11). Mixing between sublithospheric mantle and continental mantle could account for the range of granite ϵ_{Nd} values between 1 and -8, but those granites with $\epsilon_{Nd} < -8$ (and $\epsilon_{Sr} > 50$) clearly have interacted with felsic continental crust.

Metaluminous granites in the eastern Great Basin can also be divided into two geographic groups on the basis of their ϵ_{Sr} values; granites in the miogeocline of eastern Nevada and western Utah, with $\epsilon_{Sr} > +60$, and granites in, or near, the cratonal regions of central Utah, with $\epsilon_{Sr} < +60$ (labeled "CG" on Fig. 4-10). The lowermost ϵ_{Nd} values for metaluminous granite in both regions are similar (-19), suggesting that both granite sets were derived primarily from deep-seated, felsic, Precambrian basement with $T_{DM} = $

2.2 b.y. Therefore, the spatial variations in the granite ϵ_{Sr} values (for granites with $\epsilon_{Nd} = -19$) are considered to represent spatial variations in the Sr isotopic composition of the lowermost portions of the Precambrian basement (Farmer and DePaolo, 1983). The Sr isotopic variations could have originated through the tectonic removal of Rb-depleted (or mafic?), low-ϵ_{Sr} lower crust from the basement during the continental rifting event that formed the western margin of the North American craton 800 m.y. ago (Fig. 4-12). Alternatively, "basement" underlying the miogeoclinal sedimentary rocks could represent an ancient example of metasedimentary crust formed at the margin of rifted continental crust, in the transition zone between continental and newly formed oceanic crust (Nicolas, 1985). Such crust may be primarily composed of metamorphosed detritus derived from high-Rb/Sr, high-ϵ_{Sr} upper continental crust, accounting for the elevated ϵ_{Sr} inferred for the lower crust in eastern Nevada.

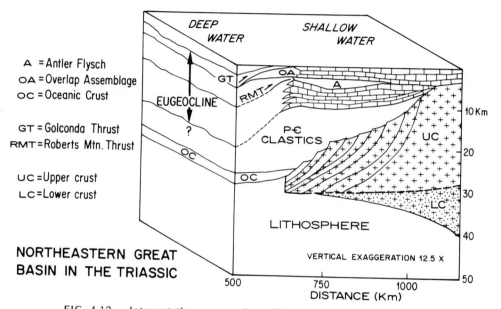

A = Antler Flysch
OA = Overlap Assemblage
OC = Oceanic Crust

GT = Golconda Thrust
RMT = Roberts Mtn. Thrust

UC = Upper crust
LC = Lower crust

NORTHEASTERN GREAT BASIN IN THE TRIASSIC

FIG. 4-12.   Interpretative cross section of the northwestern Great Basin in Triassic time. As depicted, tectonically thinned Precambrian basement (T_{DM} = 2.2 b.y.) is shown underlying the miogeocline, with the crust having non uniformly stretched during the rifting event at 800 m.y. The lower portion of the crust (LC) extended in a ductile fashion and was greatly attenuated near the continental margin. The upper crust (UC) behaved in a brittle fashion and extended by a smaller factor than the LC. Reprinted from Farmer and DePaolo (1983) by permission from the American Geophysical Union.

CONCLUSIONS

In summary, the Nd and Sr isotopic data from young igneous rocks in the western United States suggest that each segment of Precambrian continental crust is coupled to a discrete segment of continental mantle, with those segments of the continental mantle involved in basalt or ultrapotassic magma generation being LREE enriched relative to oceanic mantle. The regional variations in mafic magma isotopic compositions roughly correlate with the crustal Nd isotopic provinces, suggesting that the LREE enrichments in the deep continental lithosphere are long-lived, possibly dating back to the time of lithosphere formation. The contrast between basalt and ultrapotassic rock Nd isotopic compositions in the Archean lithosphere further suggests that the continental mantle may be vertically zoned in LREE. Asthenosphere underlying the continental lithosphere, at least in the Basin and Range, has the LREE- and Rb-depleted characteristics of oceanic mantle. Finally, the granite isotopic data identify regional isotopic variations in the lowermost portions of the Precambrian continental crust, which may relate to regional variations in the bulk composition and/or structural characteristics of the lower crust.

A speculative model for the formation of the ancient continental lithosphere which accommodates many of the observations above follows a uniformitarian model in which the continental crust forms via partial melting of the oceanic mantle wedge above a segment of subducting oceanic lithosphere (Fig. 4-13). The enriched zones within the con-

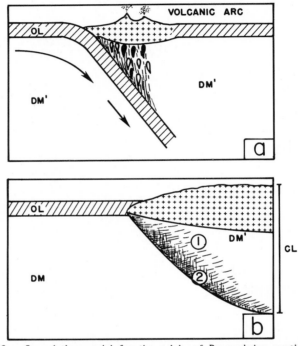

FIG. 4-13. Speculative model for the origin of Precambrian continental lithosphere in the western United States. In (a), proto-continental lithosphere (crust and mantle) are formed above a segment of subducting oceanic lithosphere (OL) in an island-arc environment. Shaded areas represent regions of mantle metasomatism and LREE enrichment, by fluids derived from dehydration of the subducting clab. Silicate magmas produced within the mantle wedge either rise and differentiate into proto-continental crust (open blebs), or solidify as discrete magma bodies or interstitial material (filled blebs), producing additional regions of LREE enrichment within the mantle wedge. DM' corresponds to the "depleting" mantle (Nelson and DePaolo, 1984) at the time of active subduction. DM' constitutes both the mantle wedge, and the ultimate source of both the oceanic and continental lithospheres. When subduction ceases (b), some portion of the old mantle wedge, consisting of DM', ①, and enriched DM', ②, becomes isolated from the depleting mantle (DM) by the insulating effect of the overlying continental crust. The continental mantle may preserve vertical zonations in the degree of LREE enrichment (shaded regions) inherited from the period of active subduction. CL = continental lithosphere (crust and mantle), which may extend to a depth of 150–200 km in cratonal regions (cf. Boyd et al., 1985).

tinental lithosphere originated in zones in the mantle wedge intensely metasomatized by LREE-enriched fluids, or magma, derived from the subducting slab (Hawkesworth et al., 1984; Nicholls and Ringwood, 1973; Sekine and Wyllie, 1982). The intensity of the metasomatism (and LREE enrichment) decreased upward and laterally away from the subducting slab. With the cessation of subduction, the enriched segments of the mantle wedge, as well as unmetasomatized segments of the oceanic mantle, became partially preserved in the mantle isolated beneath the continental crust due to thermal insulation by the crust itself (Oxburgh and Parmentier, 1978). Each age segment of continental crust thereby acquires a unique, chemically zoned, continental mantle, which was isolated from the "depleting" oceanic mantle (cf. Nelson and DePaolo, 1984) at the time of crust formation. This simple model would also allow for ancient periods of subduction beneath preexisting continental lithosphere, producing additional LREE enrichment in the lithosphere and continental crust with "mixed" isotopic characteristics, similar to the $T_{DM} = 2.2$-b.y. lithosphere in southern Nevada.

This model clearly oversimplifies the mechanisms whereby the cratonal segments of the western United States were formed and modified by subsequent tectonic and magmatic processes. For example, there is considerable debate regarding whether modern plate tectonics can be applied to models of the formation of the Precambrian continental crust (Anhaeusser, 1975; Baer, 1981). In any case, the igneous rock isotopic data provide information regarding the age, composition, and structure of the continental lithosphere in the western United States that cannot be obtained from other techniques. This information is not only critical to the understanding of the metamorphic and tectonic evolution of the upper continental crust in this region, but also provides important contraints regarding the origin and evolution of the continental lithosphere as a whole.

ACKNOWLEDGMENTS

Thanks go to F. Perry for the use of his unpublished data and for comments and discussions of earlier versions of this manuscript. Comments on the manuscript by B. Crowe, J. Wooden, V. Bennett, K. Tegtmeyer, R. Ratliff, and T. Scambos were also appreciated.

REFERENCES

Anderson, D. L., 1985, Hotspot magmas can form by fractionation and contamination of mid-ocean ridge basalts: *Nature*, v. 318, p. 145–149.

Anhaeusser, C. R., 1975, Precambrian tectonic environments: *Ann Rev. Earth Planet. Sci.*, v. 3, p. 31–54.

Baer, A. J., 1981, Geotherms, evolution of the lithosphere and plate tectonics: *Tectonophysics*, v. 72, p. 203–228.

Baldridge, W. S., 1979, Petrology and petrogenesis of Plio-Pleistocene basaltic rocks from the central Rio Grande rift, New Mexico, and their relation to rift structure, *in Rio Grande Rift — Tectonics and Magmatism:* Washington, D.C., Amer. Geophys. Un., p. 323–353.

——, Olsen, K. H., and Callender, J. F., 1984, *in Rio Grande Rift — Problems and Perspectives:* N. Mex. Geol. Soc. Guidebook 35, p. 1–12.

Boyd, F. R., Gurney, J. J., and Richardson, S. M., 1985, Evidence for a 150-200 km thick Archean lithosphere from diamond inclusion thermobarometry: *Nature*, v. 315, p. 387–389.

Carlson, R. W., Dudas, F. O., Meen, J. K., and Eggler, D. H., 1985, Formation and evolution of the Archean sub-continental mantle beneath the northwestern U.S.: *Trans. Amer. Geophys. Un.*, v. 66, p. 1109.

Chapin, C. E., 1979, Evolution of the Rio Grande rift — A summary, *in Rio Grande Rift — Tectonics and Magmatism:* Washington, D.C., Amer. Geophys. Un., p. 1-5.

Coney, P. J., and S. J. Reynolds, 1977, Cordilleran Benioff Zones: *Nature*, v. 270, p. 403–405.

Cross, T. A., and Pilger, R. H., Jr., 1978, Constraints on absolute plate motion and plate interaction inferred from Cenozoic igneous activity in the western United States: *Amer. J. Sc.*, v. 278, p. 865–902.

Crowley, J. C., Niemeyer, S., and Shaw, H. F., 1985, Nd isotopic evidence for multiple components in the generation of Rio Grande rift basalts: *Trans. Amer. Geophys. Un.*, v. 66, p. 1138.

DePaolo, D. J., 1981a, Nd isotopic studies — Some new perspectives on earth structure and evolution: (EOS) *Trans. Amer. Geophys. Un.*, v. 62, p. 137–140.

____, 1981b, Nd in the Colorado Front Range and implications for crust formation and mantle evolution in the Proterozoic: *Nature*, v. 291, p. 193–196.

____, 1983, Geochemical evolution of the crust and mantle: *Rev. Geophysics Space Physics*, v. 21, p. 1347–1358.

____, and Johnson, R. W., 1979, Magma genesis in the New Britain island-arc — Contributions from Nd and Sr isotopes and trace-element patterns: *Contrib. Mineralogy Petrology*, v. 70, p. 367–379.

____, and Wasserburg, G. J., 1976, Nd isotopic variations and petrogenetic models: *Geophys. Res. Lett.*, v. 3, p. 249–252.

____, and Wasserburg, G. J., 1977, The sources of island arcs as indicated by Nd and Sr isotopic studies: *Geophys. Res. Lett.*, v. 4, p. 465–368.

Doe, B. R., Leeman, W. P., Christiansen, R. L., and Hedge, C. E., 1982, Lead and strontium isotopes and related trace elements as genetic tracers in the upper Cenozoic rhyolite-basalt association of the Yellowstone Plateau volcanic field: *J. Geophys. Res.*, v. 87, p. 4785–4806.

Dugan, M. A., Muehlberger, W. R., Leininger, L., Peterson, C., McMillan, N. J., Gunn, G., Lindstrom, M., and Haskin, L., 1984, Volcanic and sedimentary stratigraphy of the Rio Grande gorge and the late Cenozoic geologic evolution of the Southern San Luis Valley, *in Rio Grande Rift — Problems and Perspectives:* N. Mex. Geol. Soc. Guidebook 35, p. 157–170.

____, Lindstrom, M. M., McMillan, N. J., Moorbath, S., Hoefs, J., and Haskin, L. A., 1986, Open system magmatic evolution of the Taos Plateau volcanic field, northern New Mexico — 1. The petrology and geochemistry of the Servilleta Basalt: *J. Geophys. Res.*, v. 91, p. 5999–6028.

Eaton, G. P., Wahl, R. R., Prostka, H. J., Mabey, Dr. R., and Kleinkopf, M. D., 1978, Regional gravity and tectonic patterns — Their relation to late Cenozoic epeirogeny and lateral spreading in the western Cordillera, *in* Smith, R. B., and Eaton, G. P., eds., *Cenozoic Tectonics and Regional Geophysics of the Western Cordillera:* Geol. Soc. America Mem. 152, p. 51–92.

Esperanca, S., Carlson, R. W., and Shirey, S. B., 1985, Isotopic characteristics of the lower crust under central Arizona — Evidence from xenoliths in the Camp Creek high-K latites: *Trans. Amer. Geophys. Un.*, v. 66, p. 1110.

Everson, J. E., 1979, Regional variations in the lead isotopic characteristics of late Cenozoic basalts from the southwestern U.S.: Ph.D. thesis, Calif. Inst. Technology, Pasadena, Calif., 454 p.

Farmer, G. L., and DePaolo, D. J., 1983, Origin of Mesozoic and Tertiary granite in the western United States and implications for pre-Mesozoic crustal structure — 1. Nd and Sr isotopic studies in the geocline of the northern Great Basin. *J. Geophys. Res.*, v. 88, p. 3379–3401.

——, and DePaolo, D. J., 1984, Origin of Mesozoic and Tertiary granite in the western United States and implications for pre-Mesozoic crustal structure — 2. Nd and Sr isotopic studies of unmineralized and Cu- and Mo- mineralized granite in the Precambrian craton: *J. Geophys. Res.*, v. 89, p. 10141–10160.

Faure, G., and Powell, J. L., 1972, *Strontium Isotope Geology*: Berlin, Springer-Verlag, 188 p.

Foland, K. A., Bergman, S. C., Hofmann, A. W., and Raczek, I., 1983, Nd and Sr isotopic variations in alkali basalts and megacrysts from the Lunar Crater volcanic field: *Trans. Amer. Geophys. Un.*, v. 64, p. 338.

Fraser, K. J., Hawkesworth, C. J., Erlank, A. J., Mitchell, R. H., and Scott-Smith, B. H., 1985/86, Sr, Nd and Pb isotope and minor element geochemistry of lamproites and kimberlites: *Earth Planet. Sci. Lett.*, v. 76, p. 57–70.

Haxel, G., Tosdal, R. M., May D. J., and Wright, J. E., 1984, Late Cretaceous and early Tertiary orogenesis in south-central Arizona — Thrust faulting, regional metamorphism, and granite plutonism: *Geol. Soc. America Bull.*, v. 95, p. 631–653.

Hawkesworth, C. J., and van Calsteren, P. W. C., 1984, Radiogenic isotopes — Some geological applications, *in* Henderson, P., ed., *Rare Earth Element Geochemistry*: Amsterdam, Elsevier, p. 375–421.

——, Rogers, N. W., van Calsteren, P. W. C., and Menzies, M. A., 1984, Mantle enrichment processes: *Nature*, v. 311, p. 311–335.

Hedge, C. E., and Noble, D. C., 1971, Upper Cenozoic basalts with high $^{87}Sr/^{86}Sr$ and Sr/Rb ratios, southern Great Basin, western United States: *Geol. Soc. Amer. Bull.*, v. 82, p. 3503–3510.

Henderson, P., 1984, General geochemical properties and abundances of the rare earth elements, *in* Henderson, P., ed., *Rare Earth Element Geochemistry*: Amsterdam, Elsevier, p. 1–32.

James, D. E., Padovani, E. R., and Hart, S. R., 1980, Preliminary results on the oxygen isotopic composition of the lower crust, Kilbourne Hole Maar, New Mexico: *Geophys. Res. Lett.*, v. 7, p. 321–324.

Kistler, R. W., and Peterman, Z. E., 1973, Variations in Sr, Rb, K, Na, and initial $^{87}Sr/^{86}Sr$ in Mesozoic granitic rocks and intruded wall rocks in central California: *Geol. Soc. America Bull.*, v. 84, p. 3489–3512.

Lee, D. E., Kistler, R. W., Friedman, I., and von Loenen, R. E., 1981, Two-mica granites of northeastern Nevada: *J. Geophys. Res.*, v. 86, p. 10607–10616.

Leeman, W. P., 1970, The isotopic composition of strontium in late-Cenozoic basalts from the Basin-Range province, western United States: *Geochim. Cosmochim. Acta*, v. 34, p. 857–872.

——, 1975, Radiogenic tracers applied to basalt genesis in the Snake River Plain-Yellowstone National Park region — Evidence for a 2.7 b.y.-old upper-mantle keel: *Geol. Soc. Amer. Abstr. with Programs*, v. 7, p. 1165.

——, 1982, Tectonic and magmatic significance of strontium isotopic variations in Cenozoic volcanic rocks from the western United States: *Geol. Soc. Amer. Bull.*, v. 93, p. 478–503.

——, Menzies, M. A., Matty, D. J., and Embree, G. F., 1985, Strontium, neodymium and lead isotopic compositions of deep crustal xenoliths from the Snake River Plain — evidence for Archean basement: *Earth Planet. Sci. Lett.*, v. 75, p. 354–368.

Lipman, P. W., 1980, Cenozoic volcanism in the western United States — Implications for continental tectonics, *in Continental Tectonics*: Washington, D.C., National Academy of Science, p. 161–174.

——, and Mehnert, H. H. 1975, Late Cenozoic basaltic volcanism and development of the Rio Grande depression in the southern Rocky Mountains: *Geol. Soc. America Mem.* 144, p. 119–154.

_____, and Mehnert, H. H., 1979, The Taos Plateau volcanic field, northern Rio Grande rift, New Mexico, *in Rio Grande Rift — Tectonics and Magmatism:* Washington, D.C., Amer. Geophys. Un.

McCulloch, M. T., Jaques, A. L., Nelson, D. R., and Lewis, J. D., 1983, Nd and Sr isotopes in kimberlites and lamproites from Western Australia — enriched mantle origin: *Nature,* v. 302, p. 400–403.

Menzies, M. A., 1984, Mantle ultramafic xenoliths in alkaline magmas — Evidence for mantle heterogeneity modified by magmatic activity *in* Hawkesworth, C. J., and Norry, M. J., eds., *Continental Basalts and Mantle Xenoliths:* Shiva Geology Series: Shiva, Nantwich, Cheshire, England, p. 111–138.

_____, Leeman, W. P., and Hawkesworth, C. J., 1983, Isotope geochemistry of Cenozoic volcanic rocks reveals mantle heterogeneity below western USA: *Nature,* v. 303, p. 205–209.

_____, Leeman, W. P., and Hawkesworth, C. J., 1984, Geochemical and isotopic evidence for the origin of continental flood basalts with particular reference to the Snake River Plain Idaho, U.S.A.: *Philos, Trans. Roy. Soc. London,* v. A310, p. 643–660.

Miller, C. F., and Bradfish, L. J., 1980, An inner Cordilleran belt of muscovite-bearing plutons: *Geology,* v. 8, p. 412–416.

Nelson, B. K., and DePaolo, D. J., 1984, 1,700-Myr greenstone volcanic successions in southwestern North America and isotopic evolution of Proterozoic mantle: *Nature,* v. 311, p. 143–146.

_____, and DePaolo, D. J., 1985, Rapid production of continental crust 1.7 to 1.9 b.y. ago — Nd isotopic evidence from the basement of the North American mid-continent, *Geol. Soc. Amer. Bull.,* v. 96, p. 746–754.

Nicholas, A., 1985, Novel type of crust produced during continental rifting: *Nature,* v. 315, p. 112–115.

Nicholls, I. A., and Ringwood, A. E., 1973, Effect of water on olivine stability in tholeiites and production of silica-saturated magmas in the island arc environment, *J. Geology,* v. 81, p. 285–300.

Oxburgh, E. R., and Parmentier, E. M., 1978, Thermal processes in the formation of continental lithosphere: *Philos. Trans. Roy. Soc.* London A, v. 288, p. 415–429.

Perry, F. V., Baldridge, W. S., and DePaolo, D. J., 1985, Asthenospheric upwelling beneath the Rio Grande rift region — Isotopic variations of basalts in central New Mexico: *Trans. Amer. Geophys. Union,* v. 66, p. 1109.

Peterman, Z. E., Doe, B. R., and Prostka, H. J., 1970, Lead and strontium isotopes in rocks of the Absaroka volcanic field, Wyoming, Contrib. Mineralogy Petrology, v. 27, p. 121–130.

Phelps, D. W., Gust, D. A., and Wooden, J. L., 1983, Petrogenesis of the mafic feldspathoidal lavas of the Raton-Clayton volcanic field, New Mexico: Contrib. Mineralogy Petrology, v. 84, p. 182–190.

Reid, M., Hart, S. R., and Padovani, E., 1985, Importance of sedimentary protoliths to the lower crust exemplified by the Kilbourne Hole paragneisses — Sr, Nd, Pb isotope geochemistry: *Trans. Amer. Geophys. Un.,* v. 66, p. 1110.

Ringwood, A. E., 1975, *Composition and Petrology of the Earths' Mantle:* New York, McGraw-Hill, 618 p.

Sekine, T., and Wyllie, P. J., 1982, Phase relationships in the system $KAlSiO_4$-Mg_2SiO_4-SiO_2-H_2O as a model for hybridization between hydrous siliceous melts and peridotite: *Contrib. Mineralogy Petrology,* v. 79, p. 368–374.

Semken, S. C., 1984, A neodymium and strontium isotopic study of late Cenozoic basaltic volcanism in the southwestern Basin and Range province: M.S. thesis, Univ. California, Los Angeles, Calif., 68 p.

Smith, C. B., 1983, Pb, Sr and Nd isotopic evidence for sources of southern African Cretaceous kimberlites: *Nature*, v. 304, p. 51–54.

Vaniman, D. T., Crowe, B. M., and Gladney, E. S., 1982, Petrology and geochemistry of hawaiite lavas from Crater Flat, Nevada: *Contrib. Mineralogy Petrology*, v. 80, p. 341–357.

Vollmer, R., Ogden, P., Schilling, J.-G., Kingsley, R. H., and Waggoner, D. G., 1984, Nd and Sr isotopes in ultrapotassic volcanic rocks from the Leucite Hills, Wyoming: *Contrib. Mineralogy Petrology*, v. 87, p. 359–368.

White, W. M., 1985, Sources of oceanic basalts — Radiogenic isotopic evidence: *Geology*, v. 13, p. 115–118.

Williams, S., and Murthy, V. R., 1979, Sources and genetic relationships of volcanic rocks from the northern Rio Grande rift — Rb-Sr and Sm-Nd evidence: *Trans. Amer. Geophys. Un.*, v. 60, p. 407.

Zartman, R. E., and Doe, B. R., 1981, Plumbotectonics — The model: *Tectonophysics*, v. 75, p. 135–162.

Mark D. Barton, Denise A. Battles, Gray E. Debout, Rosemary C. Capo,
John N. Christensen, Samuel R. Davis, R. Brooks Hanson, Carl J. Michelsen,
and Heather E. Trim

Department of Earth and Space Sciences
University of California, Los Angeles
Los Angeles, California 90024-1567

5

MESOZOIC CONTACT METAMORPHISM IN THE WESTERN UNITED STATES

Mesozoic magmatism caused widespread, varied types of metamorphism in the western United States. Metamorphism ranges from well-defined aureoles associated with isolated plutons to regionally extensive zones associated with batholiths. This paper reviews the major types of Mesozoic igneous-related metamorphism, and their temporal and spatial distributions; it concludes with discussion of the relationship between magmatism and regional metamorphism.

Contact metamorphism occurs abundantly with each of the three major pulses of Mesozoic granitic magmatism (240–200, 180–145, and 120–70 m.y.). Regional low-pressure facies series metamorphism is closely associated with areas containing >50% intrusions exposed at ≤10–14 km depth. Compiled dates are consistent with the contemporaneity of magmatism and metamorphism. Medium-pressure facies series terranes occur in some deeper batholithic belts, notably in the northwest, but generally flank the magmatic maxima on the east and west. Depth of exposure is the dominant control on the character of contact metamorphism. Pluton geometry and extent, and host and igneous compositions, are also important. The overall distribution of Mesozoic contact metamorphism can be systematized in terms of these parameters.

Shallow systems (<4–8 km) exhibit the most varied metamorphic effects. Aureoles range from huge to nonexistent; there is little correlation with intrusion exposure area. Shallow metamorphism is dominated by relatively static thermal and hydrothermal metamorphism, generally in discrete aureoles, but is regionally extensive some places where plutons are abundant. High levels have been preserved in Arizona, New Mexico, and Montana, and locally elsewhere, such as in parts of the Great Basin. Extensive Mesozoic hydrothermal metamorphism is prominent in the Arizona Laramide, and in Triassic and Jurassic rocks from the Mojave through the Great Basin to eastern Oregon. Correlative with this, magmatic-hydrothermal mineral deposits are strongly concentrated at shallow levels. Major copper deposits occur in the shallowest preserved levels (e.g., Arizona), whereas tungsten skarns occur where exposures are somewhat deeper (e.g., central California and western Nevada).

At moderate and deeper levels of emplacement (>4–8 km), aureoles are more nearly proportional to pluton size, and the plutons tend to be larger than shallow intrusions. Metasomatic aureoles become less abundant and decrease in size with depth; ore deposits are correspondingly sparse. Described thermal aureoles reach their maximum development at intermediate levels (6–12 km) in areas with moderate or low pluton densities (e.g., in the Great Basin). At greater depths or in areas with abundant intrusions, reported thermal aureoles narrow, commonly passing out into medium- to high-grade metamorphic rocks (e.g., along the batholiths from southern California to Washington). At deeper levels, penetrative deformation commonly accompanied pluton emplacement.

The characteristics of host and intrusive rocks play roles second only to depth in governing the types and distribution of contact metamorphism. Limited contact effects in Colorado, Arizona, and elsewhere reflect the local abundance of plutonic and high-grade metamorphic host rocks, which are little susceptible to obvious contact metamorphism. High-temperature rocks, however, commonly record later thermal events through cryptic changes in isotopic (Ar, H, etc.) compositions. Extensive Cretaceous resetting of K-Ar dates south of Idaho along the magmatic maximum demonstrates a regional greenschist-temperature event, which is only partly reflected by macroscopic metamorphism. Extensive hydrothermal alteration is most prevalent in volcanic rocks, reflecting their reactivity as well as their depth. Similarly, the distribution of skarns largely reflects the abundance of carbonate rocks along the miogeocline. Skarns are less common in the western eugeoclinal terranes and in the craton. Other things being equal, thermal aureoles decrease and metasomatic aureoles increase in size, and synintrusive deformation becomes more prominent with increasing silica content in the causative plutons. The distribution of types of mineralization parallels regional and temporal changes in pluton compositions.

Mesozoic contact metamorphism is clearly gradational with regional metamorphism in many cases. The close correspondence of low-pressure facies series regional metamorphism with areas containing abundant plutons demonstrates a connection. Simple thermal modeling coupled with kinetic considerations indicates that regionally extensive metamorphism can be produced by overlap of aureoles, regardless of intrusion rate. Alternative mechanisms fail to satisfy geologic constraints in the western United States. Effects characteristic of contact metamorphism, however, will be prominent only at shallow levels; at deeper levels igneous-related metamorphism can be quite similar to metamorphism produced by other processes. The model implies that the "high-temperature" metamorphic belt of the western United States reflects not a uniformly high geothermal gradient, but the time integration of a moderate regional gradient punctuated by local highs associated with intrusive centers. The "true" thermal maximum may have been east of the batholithic belt, consistent with independent data and models.

INTRODUCTION

Syntheses of Cordilleran Mesozoic history have emphasized tectonics, stratigraphy, and igneous activity (e.g., Burchfiel and Davis, 1972, 1975; Coney, 1978; Dickinson, 1981; Hamilton, 1978; Kistler, 1974) with relatively little attention paid to the metamorphism. The few general papers on metamorphism focus on particular aspects, such as skarn deposits (Einaudi, 1982a, b; Meinert *et al.*, 1981), the general distribution of regional metamorphism (Allmendinger and Jordan, 1981; Armstrong and Hansen, 1966; Hamilton, 1978), and metamorphic core complexes (e.g., Armstrong, 1982; Crittenden *et al.*, 1980). Reviews of contact metamorphism have concentrated on particular topics, such as skarn deposits (Einaudi *et al.*, 1981) or dominantly thermal aspects (Turner, 1981), not on regional relationships.

This paper examines the diversity of contact metamorphism in the western United States, reviews the distribution and timing of Mesozoic plutonism and metamorphism, and concludes with a discussion of possible relationships between igneous activity and the various types of metamorphism. The results and interpretations are based on an extensive literature review, including the development of CONTACT, a computer data base,[1] as well as some simple thermal modeling. Rather than presenting a comprehensive review of the literature, we adopt a process-oriented approach, using published data to examine the diversity of contact metamorphism and its relationship to broader petrotectonic processes. It is beyond the scope of this paper to provide a critical review of the bulk of the evidence for timing and conditions of regional metamorphism.

Regional tectonics have a profound control on the overall distribution of magmatism and host rocks, as well as subsequent histories (preservation, redistribution after formation). Tectonic factors do not directly affect contact metamorphism because the scale of contact metamorphism is generally quite small, about the scale of plutons. Therefore, large-scale tectonic issues are avoided here insofar as possible.

We define contact metamorphism to be a significant change in rock characteristics due to nearby emplacement or passage of magma. Changes can be mineralogical, textural,

[1]CONTACT contains information on Mesozoic intrusions and related metamorphism in the western United States. Data from over 3000 sources are incorporated; however, only key references are cited here. Figure 5-2 shows many of the 1600 locations included in the data base. The development of CONTACT was a primary goal of this continuing, collaborative project; many errors and inconsistencies remain in the data base at the time of writing.

MESOZOIC CONTACT METAMORPHISM IN THE WESTERN UNITED STATES

bulk compositional, or isotopic. The nature of the changes varies tremendously — from near-surface hydrothermal alteration caused by intrusion-driven groundwater circulation to deep, dynamothermal metamorphism associated with the emplacement of batholiths. Defined in this way, contact metamorphism will not always be clearly associated with an intrusion. Indeed, in many cases it is difficult to identify unambiguously the cause of metamorphism. In some places, contact and regional metamorphism are clearly inte-gradational. Identification of the underlying causes, however, is fundamental to understanding the nature of any metamorphic event; thus a genetic definition is preferred even though it leads to uncertainties in classification where insufficient data exist to identify the cause. An alternative term might be igneous metamorphism or igneous-related metamorphism. The definition includes regionally extensive metamorphism where magmatic activity is the cause. Although this relationship is implied by many writers, it is rarely stated explicitly. We emphasize, however, that in many areas, metamorphism is clearly polyphase, with igneous-related metamorphism superimposed on, or overprinted by, other metamorphic events. The point is that in *some* settings regional metamorphism can be directly attributed to magmatism. In contrast, Miyashiro (1961, p. 281) states that "metamorphism is to be regarded as regional providing the thermal structure of the resultant metamorphic terrain is independent of individual plutonic masses." This imputes a genetic meaning to "regional" that arbitrarily excludes some regionally extensive metamorphism.

Others commonly restrict contact metamorphism to features clearly correlated with intrusive contacts (e.g., Miyashiro, 1973; Turner, 1981). Turner specifically separates contact from regional metamorphism, a distinction that we think can be arbitrary. Although this observational approach may be a superior way of describing the metamorphism, it can lead to different classifications of nearly identical rocks formed by similar processes. For example, many pendants in the Sierra Nevada batholith comprise largely greenschist grade metamorphic rocks, yet described "contact metamorphism" is commonly restricted to narrow amphibolite-grade hornfelses or skarns occurring along intrusive contacts (Kerrick, 1970; Nokleberg and Kistler, 1980). In contrast, virtually identical greenschist-grade rocks, clearly in aureoles, occur adjacent to similar plutons just to the east of the Sierra Nevada in the White Mountains (Gastil *et al.*, 1967; Hanson, 1986; D. B. Nash, 1962). Although there is evidence for multiple metamorphic events in the Sierran pendants (Nokleberg and Kistler, 1980), most events can readily be interpreted to have occurred during several plutonic episodes over a period of 130 m.y. (Tobisch *et al.*, 1986). Thus, metamorphism which is clearly contact related in one area, and therfore spatially limited, is regionally extensive in another area. In other cases, regionally extensive metamorphism records apparent temperature gradients much higher than could be developed without advection of heat, even where an intrusive source is not obvious. For example, the Schell Creek Range, Nevada, lacks exposed plutons, but has a laterally extensive elevated Jurassic metamorphic gradient (E. L. Miller *et al.*, Chapter 24, this volume). However, in the Snake Range immediately to the east similar metamorphism clearly occurs around Jurassic plutons (see "Discussion").

THE DIVERSITY OF CONTACT METAMORPHISM

Before examining metamorphism in specific areas, it is useful to review briefly the diversity of Mesozoic contact metamorphism. Many transformations caused by pluton emplacement are not included in standard treatments of contact metamorphism. Metamorphism of any rock includes one or more changes in mineralogy, texture, and bulk composition. These changes may be either obvious (generally, macroscopic) or cryptic (requiring instrumental analysis). Such scale-independent changes, however, comprise only part of the story and do not necessarily connote igneous-related metamorphism. The elementary changes combine in many ways, with diverse distributions in space and time; they range, in space, from recrystallization about a single dike to polyphase metamorphism related to a composite batholith or, in time, from a simple cooling event associated with a small body to the history of a large region over tens of millions of years.

Several processes can lead to scale-independent changes: thermal perturbations, metasomatism, and deformation. Mineralogical changes result from thermal perturbations or metasomatism; deformation rarely causes mineralogical changes. Heat transfer to drive reactions takes place by conduction or by advection of fluids (Norton and Knight, 1977; Parmentier and Schedl, 1981). Metasomatism takes place by two analogous mechanisms: diffusion and infiltration (Korzhinskii, 1970). Thermal and metasomatic events do not always cause mineralogical changes; sometimes changes are cryptic (e.g., changes in structural state or partitioning of elements). Textural changes result from deformation, mineralogical changes, or simple recrystallization. With deformation, textures are generally anisotropic, whereas simple thermal recrystallization and mineralogical reconstitution lead to isotropic (granoblastic, hornfelsic) textures in low-strain environments. Bulk compositional changes, such as modification of elemental or isotopic compositions, most commonly result from metasomatism, but subtle changes, including the loss of volatile components (e.g., H_2O, CO_2, Ar), can occur by heating alone.

Minerals typical of prograde contact metamorphism (e.g., andalusite, cordierite; see Turner, 1981) belong to low-pressure facies series (Miyashiro, 1973); however, many other minerals also occur in the contact metamorphic environment, including some, such as kyanite, not normally considered contact related. The mineral assemblages themselves, of course, are not characteristic of the process by which they formed, only of the environment. Mineral assemblages produced during prograde thermal metamorphism typically have low variance, indicating that few components were mobile. Petrological and isotopic studies support this interpretation in several areas. Two notable Mesozoic examples include the Marysville, Montana, and Santa Rosa, Nevada, aureoles. At Marysville, progressive metamorphism of siliceous dolomite buffered H_2O-CO_2 fluid compositions (J. M. Rice, 1977a) and C and O isotopic variations fit a Rayleigh distillation model as expected for simple decarbonation (Lattanzi et al., 1980). In the Santa Rosa Range, successive mineral assemblages, bulk rock compositions and stable isotope data are compatible with simple loss of volatiles during progressive thermal metamorphism of pelites (Compton, 1960; Shieh and Taylor, 1969). Cryptic prograde variations include changes in the structural states of various minerals (e.g., K-feldspar in the aureole of the Eldora stock, Hart, 1964) and variations in the thermal maturity of various types of organic material, including condonts (e.g., in the central and eastern Great Basin, A. G. Harris, et al., 1980; Poole et al., 1983).

Well-described Mesozoic thermal "aureoles" range from distinct zones around individual plutons (e.g., the Santa Rosa Range, Compton, 1960; Notch Peak, Hover-Granath et al., 1983; Nabalek et al., 1984; Sierran pendants, Kerrick, 1970; the Boulder batholith and satellites, J. M. Rice, 1977a, b; Bowman and Essene, 1982, 1984), to a regionally extensive metamorphism (e.g., around the northern end of the Bitterroot lobe of the Idaho batholith, Hietanen, 1984; Hyndman, 1983; Lang and Rice, 1985; within pendants in the Sierra Nevada and western Nevada). In contrast to sedimentary rocks, intrusive rocks and greenschist and higher-grade metamorphic rocks show limited or absent prograde mineral reactions. For example, many plutons in the Sierra Nevada intrude granitic rocks with little or no mineralogical effect. On the other hand, igneous rocks (especially volcanics) with an earlier retrograde metamorphic history are susceptible to prograde thermal metamorphism. The best examples are the metamafic rocks (e.g., superposition of aureoles on high-pressure metamorphism in the Klamath Mountains, California; Coleman et al., Chapter 39, this volume).

Retrograde reactions are most obvious where the entire host had an earlier high-temperature history, such as fresh igneous rocks or previously metamorphosed rocks. They are also significant where prograde metamorphism took place as part of the same thermal event. Common examples of the latter include the replacement of prograde cordierite and andalusite by white mica- or chlorite-bearing assemblages (Turner, 1981) and the replacement of anhydrous skarn by hydrous skarn (Einaudi et al., 1981). Retrograde contact metamorphism is best developed at shallow to moderate depths where it is commonly associated with extensive fluid circulation. Classic examples include the widespread fluid-induced (hydrothermal) alteration of igneous assemblages in volcanic rocks to zeolite and greenschist grade assemblages (e.g., Jurassic Pony Trail volcanics, Muffler, 1964). This type of alteration may or may not be associated with significant changes in bulk composition. Such metamorphism is also common in previously metamorphosed rocks. For example, many shallow Laramide plutons in Arizona have extensive propylitic alteration aureoles where they intrude Precambrian metamorphic and igneous rocks (Lowell and Guilbert, 1970).

Like thermal metamorphism, metasomatism can be either prograde or retrograde depending on the nature of the fluids and the timing relative to other events. Metasomatic assemblages tend to be higher variance than thermal assemblages. Metasomatic changes in bulk composition ranges from small-scale diffusion-controlled exchange between unlike lithologies (e.g., reaction skarns formed between carbonates and igneous or clastic rocks) to extensive mass transfer due to infiltration or expulsion of fluids (Einaudi et al., 1981). These processes need not lead to mineralogical changes (e.g., isotope and alkali exchange in feldspars), although most do. The intensity of infiltration metasomatism reflects both the amount of material which passes through a rock and the degree to which that material departs from equilibrium with that rock. Accordingly, carbonate rocks generally contain the most conspicuous metasomatic effects. These include skarns and various replacement zones of other minerals, including quartz (jasperiods, etc.), sulfides (carbonate replacement mineral deposits), and other carbonates (e.g., hydrothermal dolomitization) (see Einaudi et al., 1981, for a review of the diversity and genesis of skarns and related ore deposits). For example, alkali-metasomatism in volcanic rocks is prominent in many places (e.g., albitization in the White Mountains, G. H. Anderson, 1937; Hanson, 1986). In aluminosilicate rocks (igneous, clastic, metamorphosed equivalents) metasomatism is best developed at shallow levels where pervasive alteration is generally related to addition of volatiles and exchange or removal of alkalies and alkaline earths with or without changes in silica content. Such alteration is widespread, although not necessarily pervasive, in Mesozoic volcanic rocks in the western United States. At greater depths (>4-6 km), pervasive metasomatism in aluminosilicate rocks is apparently less abundant; veins appear to dominate [an exception occurs along the northwest side of the Bitterroot lobe of the Idaho batholith, where Hietanen (1962) relates pervasive kilometer-scale Fe, Mg, Ca, Al metasomatism to the plutons]. Local segregation to form gneisses and migmatites is observed in the vicinity of many deeper and generally larger intrusions (e.g., around the Idaho batholith, Hyndman, 1983; Meyers, 1982).

Among the most widespread changes related to igneous activity are cryptic changes in isotopic compositions which are commonly accompanied by complementary mineralogical and textural transformations. Extensive exchange of O ahd H isotopes with circulating fluids at shallow crustal levels far beyond the limits of individual plutons has been throughly ocumented by H. P. Taylor and co-workers (e.g., see H. P. Taylor, 1974, 1977). Although most studies in the western United States have dealt with Tertiary events, good evidence exists that similar exchanges took place in the Mesozoic, commonly accompanied by hydrothermal metamorphism (e.g., low ^{18}O zones in the Mojave associated with shallow Jurassic plutons; Solomon, 1986). Recent evidence also suggests that extensive isotopic exchange may take place at greater depths associated with the emplacement of batholiths (Bitterroot lobe of the Idaho batholith; Fleck and Criss, 1985).

Another important cryptic change is the resetting of radiometric ages (K-Ar, Rb-Sr, U-Pb) by reequilibration or loss of isotopic components. The conditions of resetting depend on the isotopic system, and mineral(s) and times involved (Faure, 1986). Resetting occurs in many aureoles, and has been well documented near the Eldora stock, Colorado (Hart, 1964). The data in CONTACT demonstrate that regionally extensive Cretaceous resetting is important for Mesozoic K-Ar systematics, particularly in Nevada, California, and Arizona. Younger events, predominantly Eocene, reset K-Ar ages in Idaho and Washington (e.g., see Fox *et al.*, 1977; F. K. Miller and Engels, 1975), similar to the variations in the Canadian Cordillera (Armstrong, 1987).

Textural changes commonly give the clearest small-scale indication of contact metamorphism. Hornfelses are characteristic of the rapid heating rates (and thus reaction rates) in many contact systems. Although not definitive, granoblastic and other nonfoliated textures are suggestive of, and most common with, contact metamorphism. Many plutons, especially deeper ones, have foliated rather than hornfelsic aureoles. Intensity of foliation increases approaching some plutons but decreases near others. In some cases this is clearly a penetrative flattening parallel to the contact (e.g., the Papoose Flat aureole, Sylvester *et al.*, 1978). Tobisch *et al.* (1986) attribute regional flattening in the Ritter Range pendant to the effect of the emplacement of nearby Cretaceous plutons. Reactions almost always lead to textural change, but several other mechanisms can be significant as well. Annealing and/or recrystallization occurs in many rocks; formation of marbles is an important example. Metasomatically produced textures are commonly distinct from thermally produced textures. The former tend to be more coarse, fracture-controlled, or pseudomorphic than the latter.

The changes above are scale independent ("intensive") in the sense that they can be characterized without knowing anything about their spatial distribution; each can develop independently of magmatism. Equally diverse, but diagnostic of contact metamorphism, are the spatial and temporal distributions of metamorphic features relative to intrusions. Many of these scale-dependent ("extensive") aspects of igneous-related metamorphism can be related to the depth of exposure of the plutonic system. Buddington (1959) reviewed the distribution of igneous rocks on this basis and touched on a few related features in the host rocks. Some general observations are made here about the relationships between depths, extent, and type of metamorphism; more detailed correlations are discussed following the region-by-region description of the metamorphism.

Shallow intrusions (0–6 km) tend to have extensive hydrothermal aureoles. In volcanic, or other permeable, reactive rocks, hydrothermal systems are made obvious by widespread low-temperature alteration. The extent of this low-temperature metamorphism [zeolite to lower greenschist assemblages (= propylitic alteration) in mafic to intermediate aluminosilicate rocks] can be much greater than that of the causative intrusions at the level of exposure (e.g., Pony Trail volcanics, Muffler, 1964; Hidalgo volcanics, Lasky, 1947). These systems commonly grade laterally into unmetamorphosed equivalents. Little is published about the relative amount of hydrothermally altered material in volcanic complexes, but the extent and magnitude of isotopic anomalies suggest that a large fraction of older and proximal volcanic rocks are typically hydrothermally altered (e.g., Larson and Taylor, 1986; H. P. Taylor, 1977). Nonvolcanic rocks in the same areas commonly show less extensive obvious ingenous-related metamorphism; thermal aureoles are quite narrow (e.g., *cf.* the Pony Trail volcanics and the Cortez district; Gilluly and Mazursky, 1965; Muffler, 1964). Where plutonic or metamorphic rocks form the host, macroscopic changes are more restricted; they are rarely pervasive over large distances. When examined using appropriate methods (e.g., K-Ar, stable isotopes, detailed petrographic), however, even these rocks reveal extensive thermal or geochemical effects. A well-documented example is widespread Eocene O and H isotope exchange, K-Ar resetting, and alteration of some of the primary igneous minerals in the Atlanta lobe of the Idaho batholith (Criss and Taylor, 1983; H. P. Taylor and Magaritz, 1978). Moderate-grade metamorphic rocks commonly show hydrothermal overprints at shallow levels (Precambrian around stocks in Arizona, Colorado). Shallow intrusions are mostly discordant; where concordant (e.g., sills), the host rocks rarely show related penetrative deformation (Buddington, 1959). Mineralization is dominated by porphyry-type deposits, including associated skarns (generally, base-metal rich), plus epithermal (vein) and replacement deposits in aluminosilicate and carbonate rocks, respectively.

At moderate depths (4–14 km), many intrusions have well-developed thermal aureoles and may have extensive metasomatic aureoles, although they are generally not as extensive as in the epizonal intrusions. These rocks tend to be more equigranular, with fewer, but still abundant dikes. The extent of thermal metamorphism is generally quite large in low-grade hosts. Where host rocks had an earlier high-temperature history, thermal overprints are limited to narrow hornblende hornfels or higher-grade aureoles, although cryptic metamorphic events can be broadly distributed (*cf.* the described aureoles and the K-Ar resetting in western Nevada). Deformation and style of emplacement vary from concordant with strongly penetrative deformation to discordant with little deformation. Hornfels is well developed in the less deformed hosts. Overprinting of earlier metamorphic events is commonly reported, although in many cases it is not clear that the earlier events are unrelated to the pluton. In areas with isolated plutons, aureoles are quite distinct and reasonably extensive (e.g., central Great Basin). With increasing pluton density, described aureoles tend to be much more limited in their extent and generally of higher grade (e.g., the Sierran pendants). Alteration and mineralization in the intrusions is much less abundant, although ore deposits are fairly common outside the plutons. Skarns are widespread, but veins are somewhat less abundant; where present they tend to be limited in both number and absolute extent. Mineralization tends to Pb-Zn or W in skarns, with lesser amounts of lithophile metals and widespread, but relatively minor Ag and/or Au.

Deep plutonic suites (>12 km) rarely have distinct aureoles associated with single plutons. When present these aureoles are only of hornblende or pyroxene hornfels facies. Border zones are commonly migmatized or intensively intruded. Intrusive style tends to be concordant on both large and small scales, commonly with synintrusive penetrative deformation (Bateman, 1984; Buddington, 1959). Although aureoles about individual plutons are relatively rare, regional metamorphism usually reflects the shape of the larger composite bodies. For example, a prominent regional metamorphic gradient exists over 10–40 km around the Bitterroot lobe of the Idaho batholith (Hietanen, 1962, 1984). The relationship of this metamorphism to igneous activity is uncertain — it may be a consequence of the activity, or it may reflect shared underlying thermal processes (see, e.g., Turner, 1981, p. 460–468). Metasomatism, as recognized at shallower levels, is rare. Extensive zones of migmatization may record transfer of materials as might less obvious changes in bulk or isotopic composition (e.g., Bitterroot lobe, Idaho batholith, Fleck and Criss, 1985; Hietanen, 1962). Mineralization is generally absent or minor.

DISTRIBUTION AND TIMING OF MESOZOIC METAMORPHISM

Mesozoic contact metamorphism in the western United States is largely associated with the emplacement of the quartz dioritic to granitic batholiths and their satellitic plutons. Metamorphism also occurs with mafic intrusions that are less clearly associated with subduction and, in some cases, clearly associated with extension. This discussion is restricted to metamorphism that took place in a convergent margin environment, even though our definition includes spreading-center hydrothermal metamorphism as a type of contact metamorphism. Although important, such metamorphism is limited in variation and has been little described in the western United States.

Synopsis of Magmatism and Metamorphism

Figure 5-1 shows the apparent Mesozoic magmatic flux based on the data in CONTACT. Only those intrusions with ages known to better than ± 10 m.y. have been included.[2] These are apparent fluxes because of incomplete exposures, the vagaries of preservation (which favor younger rocks), and resetting of older dates by younger thermal events. Peaks in magmatic activity during the Cretaceous, Jurassic, and Triassic have long been recognized (Armstrong and Suppe, 1973; Armstrong *et al.*, 1977; Lanphere and Reed, 1973; F. K. Miller and Engels, 1975) and correspond closely with timing of Canadian magmatism (Armstrong, 1987). Major changes in magmatic flux appear to be correlated with significant variations in plate motions (Fig. 5-1; Engebretson *et al.*, 1985).

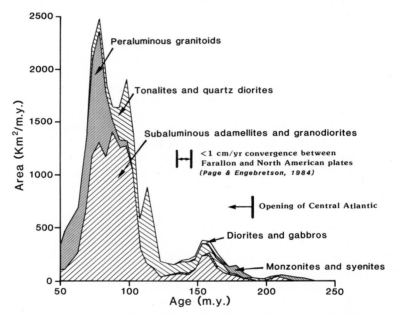

FIG. 5-1.   Apparent Mesozoic magmatic flux showing broad compositional trends for plutons with ≤±10 m.y. age uncertainty. The total apparent flux is about 20% higher.

In addition to broad cyclic variations in flux, distinct secular variations in magmatic compositions occur within each pulse (Fig. 5-1). These changes are best seen in the Cretaceous, but appear within each period of activity. Early plutons tend to be alkalic and/or less siliceous (low potash and silica to the west; alkalic, but distinctly low silica in the east). Subsequent plutonism is dominated by metaluminous granodiorites and granites.

[2]The dates used in this and subsequent figures are "best estimates," commonly obtained from comparison of multiple, different dates. Inconsistencies remain in the way age data are treated in the data base at the time of writing. Many plutons with uncertain ages (to our knowledge) have been included on Fig. 5-2. The 1983 DNAG geologic time scale is used throughout.

Later still, peraluminous magmatism becomes prominent in the Cretaceous; a hint of similar changes occurs in the Jurassic. To a degree, this distribution is biased because the major batholiths give the greatest control. For example, much of the Late Cretaceous peraluminous peak is due to the large exposures of the dominantly two-mica Idaho batholith. In detail, however, the pattern is observed elsewhere, in the southern California batholith (Taylor and Silver, 1978) and in the central Great Basin. These compositional changes are compatible with an increasing crustal component in the magmas over time in each pulse. Other, geographic factors, such as the distribution of old continental crust, are also clearly important (e.g., Farmer and DePaolo, 1983; Kistler, 1974; Taylor and Silver, 1978).

Variable levels of exposure might lead to similar effects if the mean depths of exposure vary systematically with age of intrusion and if compositions vary systematically with depth. For example, if peraluminous granitoids represent deeper levels of exposure than metaluminous granitoids (Hamilton, 1983; Hamilton and Myers, 1967), areas with Late Cretaceous plutonism must be more deeply eroded than those with older plutonism, a plausible, but somewhat surprising result. Alternatively, prolonged heating of the crust during each episode could lead to the observed systematic variation by promoting greater incorporation of crustal materials with time. The latter seems more likely and is consistent with the constraints on the timing of metamorphism (see below); however, there are also systematic variations with depth (see "Correlations" below).

Permo-Triassic magmatism occurs along the length of the western United States in a surprisingly continuous band (Fig. 5-2B) considering the diverse origins suggested for the host terranes (Rogers et al., 1974). These rocks form the first pulse of Mesozoic magmatism, but because of incomplete dating and serious problems with preservation, only the limits of distribution are shown on Fig. 5-2B. Only in California is there incontrovertible evidence for tying these to North America (Evernden and Kistler, 1970; E. L. Miller and Sutter, 1982). Triassic intrusions have a wide range of compositions, but are dominantly felsic granitoids.

Jurassic magmatism is widespread with an indistinct maximum toward the west (Fig. 5-2C). Earlier compilations (e.g., Kistler, 1974) have shown the Jurassic magmatic axis to lie distinctly to the west of the Cretaceous axis from central California northward. This may be reasonable for latest Jurassic and earliest Cretaceous time, but our weighted compilation (Fig. 5-2C) indicates that the total Jurassic arc was much wider. Dating in the last 15 years has demonstrated that Jurassic plutons are common as far east as western Utah. Again, problems with preservation, geochronology, and post-Jurassic structural events preclude a detailed analysis of the distribution of the magmatic flux. Part of the diffuseness of the Jurassic magmatism may result from distension by Cretaceous plutons. Evidence for distension is particularly good in central California, where abundant Jurassic plutons on the east and west sides of the Sierra Nevada are separated by a zone of >75% Cretaceous intrusions (cf. Fig. 5-2C and D). In contrast to the Triassic and Cretaceous, the proportion of quartz diorites and more mafic rocks is relatively high in the Jurassic. A distinct lull in the Early Cretaceous (Fig. 5-1) corresponds to a period of minimum convergence between North America and the Farallon plate (Page and Engebretson, 1984).

Cretaceous magmatism shows a rather simple, broad distribution (Fig. 5-2D). Embedded in the overall pattern are several systematic relationships. First, as has long been known in the Sierra Nevada, the main axis of pluton emplacement moved to the east with time (Evernden and Kistler, 1970), concurrent with increasing alkali and silica contents (e.g., Bateman and Dodge, 1970). The eastern limit of magmatism, however, remained in approximately the same place until the beginning of the Laramide orogeny at 70–80 m.y. Cretaceous magmatism to the east of the main arc in Nevada shows compositional variations sympathetic to those in the batholith. Early Cretaceous plutons in the

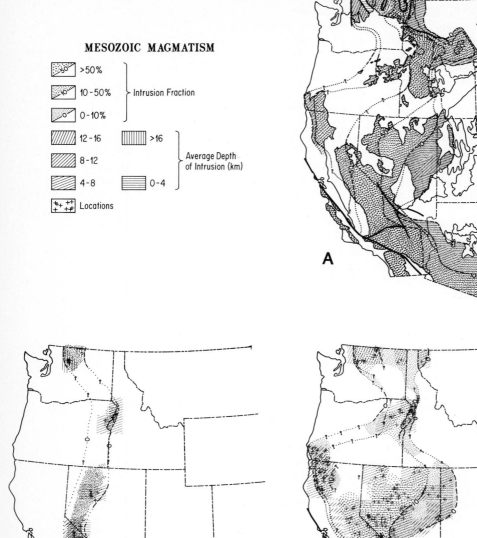

MESOZOIC MAGMATISM

Intrusion Fraction
- >50%
- 10-50%
- 0-10%

Average Depth of Intrusion (km)
- 12-16
- >16
- 8-12
- 4-8
- 0-4

Locations

A

B

C

FIG. 5-2

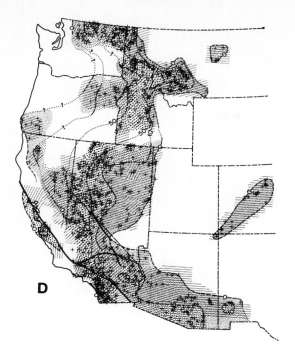

D

FIG. 5-2. Simplified modern distribution of Mesozoic magmatism showing intrusion fraction, average depth of intrusion, and locations. (A) Mesozoic intrusion fraction and average depth of intrusion. Cenozoic cover is shown. (B) Triassic intrusion fraction, average depth of intrusion, and locations. (C) Jurassic intrusion fraction, average depth of intrusion and locations. (D) Cretaceous intrusion fraction, average depth of intrusion, and locations. The intrusion fraction represent averages of proportion of pluton exposure after normalization by subtracting out the younger rocks (mostly Cenozoic, but also late Mesozoic when determining the Jurassic and Triassic). The data here and in Fig. 5-3 were originally averaged on a 1° grid, but have been modified to more closely reflect actual boundaries. Many details have been smoothed over. These figures show the modern distributions of like groups of rocks; they do not necessarily imply that those rocks formed contemporaneously or adjacent to one another. Palinspastic reconstructions are necessary for realistic views of Mesozoic distributions.

FIG. 5-2 Continued

Klamath Mountains and eastern Oregon considerably widen the Cretaceous arc; however, they are more appropriately included in the Jurassic event.

Figure 5-3 shows a much simplified and somewhat speculative compilation of Meszoic regional metamorphism. Rather than show specific types of metamorphism (e.g., as Ernst, 1974, did for California), which would be very complex, as well as spotty due to coverage, we have inferred the metamorphic facies series to which various areas belong. Metamorphic facies series (Miyashiro, 1961, 1973) provide a convenient way to indicate broad differences in mean metamorphic gradients (see Fig. 5-3A and Discussion), an advantage over simple facies which mostly reflect temperature (blueschist and eclogite excepted). Also shown in Fig. 5-3B-D is a speculative synthesis of the regional metamorphism by geological period along with the density of plutons of the same periods (from Fig. 5-2). Areas lacking Mesozoic granitoids (e.g., the Franciscan) are not discussed in the next section, however, their metamorphic history (grossly simplified) is included in Fig. 5-3.

Petrological studies away from the high-pressure belts are sparse; therefore, most constraints are derived from some combination of reported mineral associations, generalized rock descriptions, and various types of depth estimates. Medium-pressure facies series conditions are inferred in some areas lacking well-defined regional metamorphism, where it is clear that low-grade rocks had been at substantial depths. In the eastern and central Great Basin, for example, Cambrian and Upper Proterozoic argillaceous rocks are commonly chlorite grade or less (there are exceptions, see below). From stratigraphic control, these rocks must have been at 6–12-km depth (Armstrong, 1968; Gans and E. L. Miller,

Low Pressure	Metamorphic Facies Series (heavy lines indicate well documented metamorphism of appropriate age)
Medium Pressure	
High Pressure	

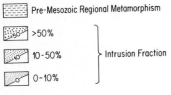

H — Areas of Documented Extensive Hydrothermal Metamorphism

Pre-Mesozoic Regional Metamorphism

>50%

10-50% } Intrusion Fraction

0-10%

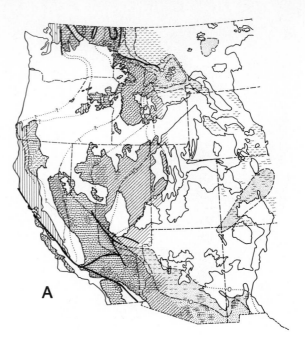

A

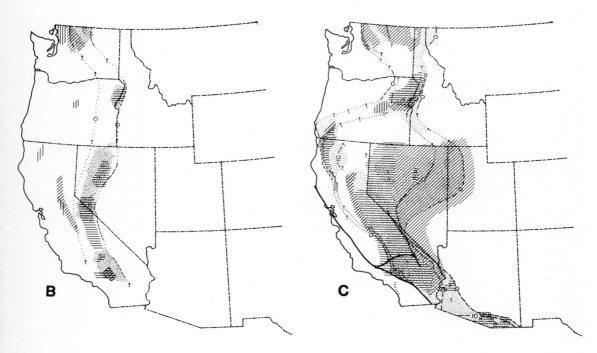

B

C

FIG. 5-3

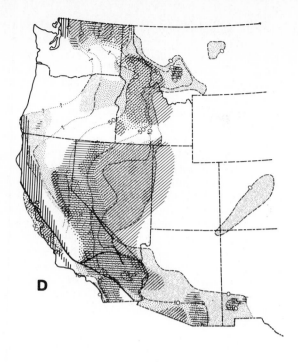

FIG. 5-3. Simplified modern distribution of Mesozoic regional metamorphic conditions (not necessarily metamorphism by metamorphic facies series with Mesozoic intrusion fractions (see the comments for Fig. 5-2). Heavier patterns in (B)–(D) indicate well-documented metamorphism of the appropriate age. (A) Mesozoic, with distribution of Cenozoic cover. (B) Triassic. (C) Jurassic. (D) Cretaceous.

D

FIG. 5-3   Continued

1983; Stewart, 1980). Therefore, thermal gradients likely never exceeded 25–30°C/km, easily within the medium-pressure range (*cf.* Fig. 5-7), and consistent with the local occurrence of kyanite in this area.

The timing of most metamorphism, although more poorly constrained than the timing of the igneous activity, is generally consistent with the periods of the igneous activity: Permo-Triassic, mid-Jurassic to earliest Cretaceous, and Late Cretaceous (extending into the early Tertiary in several regions). The relative geographic distributions of metamorphic facies series within each of the age groups (Fig. 5-3) appear broadly similar when one accounts for the poorer exposures of the older rocks. At epizonal and upper mesozonal levels, broad low-pressure facies series zones occur along the magmatic highs. Medium-pressure regimes occur on the flanks of the batholithic belts and in deeper exposures of the batholithic complexes (in Idaho and Washington). High-pressure metamorphism occurs discontinuously, but almost always to the west of plutonic rocks of similar age. We see no compelling evidence for a separate, eastern metamorphic culmination forming the "eastern Cordilleran metamorphic belt" (*cf.* Armstrong, 1982; C. F. Miller, 1978; C. F. Miller and Bradfish, 1980). The great width and diverse regimes seen in Nevada and eastern California undoubtedly reflect the large amount of Cenozoic extension in this region. Many areas, such as the Klamath Mountains, have complexly superposed magmatic and metamorphic events.

Recognized Permo-Triassic metamorphism (Fig. 5-3B) consists of scattered low- and high-pressure events. Moderate-pressure metamorphism in northwestern Washington

(Cater, 1982; Misch, 1966) and the Klamath Mountains (M. M. Donato, personal communication, 1986) may be partly Triassic, but dating is uncertain. Low-pressure rocks occur in eastern Oregon, western Nevada, and eastern California, where they are associated with Triassic plutons. These events include extensive hydrothermal metamorphism in arc-type volcanic rocks (e.g., Koipato Group), and poorly dated greenschist to lower amphibolite grade metamorphism of miogeoclinal materials in central and southern California. Exposures of Permo-Triassic blueschist facies metamorphism are scattered from northern California to northern Washington, always west of similar-aged intrusions.

Low-pressure metamorphism of definite or probable Jurassic age is widespread along the magmatic arc from eastern Oregon south to Mexico (Fig. 5-3C). Many Jurassic plutons in this region are relatively shallow. They commonly have extensive hydrothermal or thermal/metasomatic aureoles, sometimes in coeval volcanics. A variant of this type is regional hydrothermal metamorphism associated with emplacement of mafic intrusive complexes, which appears to be prominent only in the Jurassic. In some places these mafic intrusions are within the arc (e.g., the Humboldt Lopolith; Speed, 1976); others are in oceanic, marginal-basin, or back-arc settings (e.g., the Coast Range ophiolite, Hopson *et al.*, 1981). Jurassic medium-pressure metamorphism is best developed to the east of the Jurassic magmatic maximum south of Idaho. In Washington, poorly dated plutons and medium-pressure metamorphism are probably at least in part Jurassic. Jurassic moderate-pressure metamorphism also occurred along the west side of the several magmatic zones in California. Small terranes of dated Jurassic high-pressure rocks extend discontinuously from central California to central Oregon (Fig. 5-3C).

In California, Nevada, and parts of Oregon and Idaho, Cretaceous plutons appear to be exposed somewhat more deeply than Jurassic plutons (*cf.* Fig. 5-2C and D). In these areas Cretaceous volcanic rocks are rarer than Jurassic volcanic rocks, as is recognized regionally hydrothermal alteration. On the other hand, shallow Cretaceous igneous rocks are widely preserved in Arizona, New Mexico, and Montana. Igneous-related mineralization reflects this distribution; Nevada and California have significant amounts of deeper skarn and vein-type mineralization, in contrast to the shallower porphyry Cu type abundant in Arizona. Mineral assemblages, stratigraphy, and styles of emplacement all indicate low-pressure regional metamorphism (Fig. 5-3D) along the axis of the Cretaceous batholiths. Cretaceous cryptic metamorphism is further evidenced by extensive isotopic resetting (especially, K-Ar). In Idaho and Washington, Cretaceous resetting is less evident because of extensive Eocene plutonism (*cf.* F. K. Miller and Engels, 1975) and major Tertiary uplift. A few terranes in Washington, however, preserve evidence of a Cretaceous low-pressure event associated with plutonism.

Medium-pressure facies series conditions are indicated in the Cretaceous for much of the Great Basin east of the magmatic axis. Earlier metamorphism in most of Arizona and New Mexico apparently precluded extensive recognizable deep Cretaceous metamorphism there. Several areas in California plus much of Idaho and Washington underwent medium-pressure metamorphism during the Cretaceous within deep portions of the magmatic arc or on its eastern margin. Abundant Cretaceous high-pressure rocks extend from southern California to the Canadian border. These occur mostly to the west of similar-aged intrusions, although there are a few exceptions, such as the Rand-Orocopia Schist (Jacobson *et al.*, Chapter 35, this volume).

Specific Regions

This section summarizes the distribution and variation of magmatism, contact metamorphism, and regionally extensive metamorphism by area.

Colorado Although most Phanerozoic igneous activity in Colorado is Tertiary, magmatism began in the latest Cretaceous with emplacement of alkalic and calc-alkalic felsic intrusions (Armstrong, 1969; Obradovich *et al.*, 1969; Simmons and Hedge, 1978; Tweto, 1975). These bodies trend NE-SW along the Colorado mineral belt (Fig. 5-2D). Most plutons intruded Precambrian metamorphic and igneous rocks; Paleozoic and Mesozoic sediments in this region are thin to absent. Shallow emplacement of the intrusions is suggested by abundant dikes, porphyritic and chilled textures, petrologic data, proximity to the Precambrian-Phanerozoic unconformity, and the presence of coeval volcanic rocks (Tweto, 1975). No evidence exists for Mesozoic (or Tertiary) regional metamorphism. Metamorphism is restricted to the vicinity of intrusions, where it is commonly vein-related (e.g., the Central City District, C. M. Rice *et al.*, 1985) or inconspicuous (e.g., resetting of K-Ar ages and disordering of K-feldspar around the Eldora stock, Hart, 1964; Hart *et al.*, 1968). Where Phanerozoic rocks are cut, skarns, hornfelses, and marbles and locally developed, commonly with associated base-metal mineralization (e.g., Knees stock, Ekren and Houser, 1965; Rico stock, McKnight, 1974).

Arizona and New Mexico Southern and western Arizona and southwestern New Mexico underwent two periods of Mesozoic plutonic and volcanic activity. Jurassic magmatism in Arizona was restricted to the southern part of the state (Fig. 5-2C; also see T. H. Anderson and Silver, 1977). Late Cretaceous to early Tertiary (Laramide) intrusions occur throughout the region south of the Colorado Plateau (Fig. 5-2D). Magmatism migrated eastward in the Late Cretaceous to early Tertiary (Coney and Reynolds, 1977). Precambrian metavolcanic, metasedimentary, and plutonic rocks underlie most of the region. Paleozoic and Mesozoic sedimentary cover thins from greater than 3 km in southeastern New Mexico to less than 0.5 km(?) in northwestern Arizona. Incompletely preserved Mesozoic to early Tertiary volcanic and interbedded sedimentary rocks top the section (Titley, 1982a). Southwesternmost Arizona contains more complicated relations; it is discussed below with the Mojave.

Jurassic magmatism is represented in part by moderately large granitoid plutons (e.g., see Haxel *et al.*, 1984; Simons, 1973). Thick sequences of Jurassic volcanic rocks are locally preserved (Lipman and Sawyer, 1985), suggesting relatively shallow emplacement of nearby intrusions. Contact metamorphism associated with these intrusions has not been described in detail. A few smaller plutons, such as the Sacramento stock at Bisbee, have extensive hydrothermal aureoles in the Phanerozoic section with associated Cu mineralization (Bryant and Metz, 1966).

Laramide granitoids are abundant throughout the region (Fig. 5-2D). These rocks range from dikes to large, composite stocks (Titley, 1982a). Depths of emplacement average only a few kilometers (Fig. 5-2D) as indicated by associated caldera complexes (Lipman and Sawyer, 1985) and intrusions into the thin Phanerozoic sedimentary/volcanic sequence. The few plutons that are apparently deeper are incompletely described. Abundant hydrothermal alteration occurs within and around most of these intrusions (see, e.g., the articles in Titley and Hicks, 1966; Titley, 1982b). Extensive skarns and thermal aureoles formed in the carbonate-dominated Paleozoic and Mesozoic sedimentary rocks. Similarly, extensive hydrothermal aureoles occur in the Precambrian rocks (*cf.* Lowell and Guilbert, 1970). Mesozoic mineralization in Arizona is nearly all of porphyry Cu affinity. Most deposits are hosted by the igneous rocks, but alteration and some mineralization almost always extend into the country rocks (*cf.* Fig. 5-5). The economically important Cu skarn deposits of Arizona belong to this class (Einaudi, 1982a, b).

Away from the shallow intrusions, lack of reported Cretaceous metamorphism indicates that regional gradients did not exceed those achieved during the Precambrian metamorphism. Some of these areas, however, clearly underwent early Tertiary heating. Lastest Cretaceous to early Tertiary regional metamorphism of Mesozoic and older rocks is well developed in some metamorphic core complexes of southern Arizona (Haxel *et al.*, 1984; Keith *et al.*, 1980). In contrast to the restricted Mesozoic metamorphism in the older rocks, hydrothermal metamorphism extends through large portions of the volcanic rocks (e.g., the Hidalgo volcanics, Lasky, 1947; the Demetrie volcanics, Cooper, 1971; *cf.* Fig. 5-3D).

Mojave and southwestern Arizona Thin Phanerozoic sedimentary rocks overlie varied Proterozoic metamorphic rocks throughout this area, grading into thicker miogeoclinal rocks to the northwest (E. L. Miller and Cameron, 1981; Stone *et al.*, 1983). These rocks host several periods of magmatic activity, beginning in the Triassic or Late Permian (Burchfiel and Davis, 1981; Cox and Morton, 1984; E. L. Miller and Cameron, 1982; E. L. Miller and Sutter, 1981; J. D. Walker, 1984) and continuing to the Late Cretaceous in California (Burchfiel and Davis, 1981), or Eocene in Arizona (Haxel *et al.*, 1984; Fig. 5-2).

Sparse Early to Middle Triassic granitoids with primarily alkalic compositions occur in the western Mojave, where they intrude Precambrian to Permian rocks at variable depths in the Mojave and in the Transverse Ranges (C. F. Miller, 1978; E. L. Miller and Sutter, 1981; Silver, 1971). Permo-Triassic metamorphism and deformation probably accompanied emplacement of Early Triassic plutons (E. L. Miller and Cameron, 1982; J. D. Walker, 1985). The eastern Mojave was apparently unaffected by this Triassic metamorphic event (J. D. Walker, 1985).

Abundant Jurassic intrusive activity occurs across the central Mojave into southern Arizona and Sonora (C. M. Allen *et al.*, 1983; T. H. Anderson and Silver, 1978; Burchfiel and Davis, 1981). Associated volcanic rocks, extensive areas of hydrothermal alteration, discordant contacts, and low-pressure metamorphism suggest fairly shallow levels of emplacement. Contact metamorphism associated with many of these plutons is locally extensive, characterized by large metasomatic and thermal aureoles, especially in carbonate rocks (e.g., the Eagle Mountain Fe skarn, Dubois and Brummett, 1968). To the east, where Precambrian gneissic rocks host shallow Jurassic plutons, distinct aureoles are less obvious, but isotopic studies indicate resetting of Rb/Sr and K/Ar systems to Jurassic ages over large areas (Dewitt *et al.*, 1984; Lanphere, 1964). Oxygen isotope data indicate extensive overlapping hydrothermal aureoles around some of the shallow Jurassic plutons (Solomon, 1986). Mineralization associated with Jurassic intrusions is commonly Fe-rich.

Abundant Cretaceous batholiths lie out to the northeast, where deep Late Cretaceous (in California, Howard and Shaw, 1982) and pre-Late Cretaceous to early Tertiary (in Arizona, Haxel *et al.*, 1984; Reynolds *et al.*, Chapter 18, this volume) metaluminous to peraluminous plutons are closely associated with upper greenschist to amphibolite-grade regional metamorphism, thrusting, and nappe formation (elg., Old Woman Mountains, C. F. Miller *et al.*, 1982; Maria fold and thrust belt, Reynolds *et al.*, this volume, Chapter 18). Petrological data, the lack of associated volcanic rocks, and the generally concordant style of emplacement of most of the Cretaceous/early Tertiary intrusions are compatible with deeper levels of emplacement than in the Jurassic (C. M. Allen *et al.*, 1983; Howard and Shaw, 1982). Recognizable contact metamorphism is spatially restricted with mineral assemblages generally of the hornblende hornfels facies. Extensive cryptic igneous-related metamorphism in the central Mojave is expressed by Late Cretaceous regional resetting of K-Ar ages (*cf.* Fig. 5-3D). Hoisch *et al.* (Chapter 21, this volume) suggest that high fluxes of metamorphic waters caused regional metamorphism in the southeastern Mojave. Cretaceous tungsten mineralization occurs with a number of the plutons both in skarns (e.g., Old Woman Mountains) and in veins (e.g., New York Mountains, H. J. Brown, 1986).

Peninsular Range and Salinian block, California Relatively deep-seated (8–>20 km) Cretaceous plutons comprise the batholiths of the Peninsular Range and the Salinian block (James and Mattinson, Chapter 33, this volume; Silver and Taylor, 1978). Older intrusions are apparently absent. Host rocks include Proterozoic metamorphic rocks, probable miogeoclinal rocks, and Triassic to Cretaceous(?) arc-related sedimentary and volcanic rocks (D. C. Ross, 1983; Todd and Shaw, 1979).

Deep levels of exposure are indicated in the Salinian block by medium-pressure facies series (sillimanite is the only aluminum silicate), amphibolite-to granulite-grade metamorphism, and mesozonal to catazonal characteristics of the plutons (E. W. James and Mattinson, Chapter 33, this volume; D. C. Ross, 1983). Distinct-contact aureoles are difficult to recognize around individual plutons; where present, they are high grade and typically migmatitic. However, batholith-sized masses such as in the Santa Lucia Range contains zones of high-grade regional metamorphism (Compton 1966; D. C. Ross, 1983). Essentially no mineralization is associated with the intrusions (D. C. Ross, 1983).

Predominantly mesozonal levels are indicated for the Cretaceous Peninsular Ranges batholith, which grades from predominantly gabbros and tonalites on the west to predominantly tonalites, granodiorites, and two-mica granites on the east (Baird *et al.*, 1979; Larsen, 1948; Silver and Taylor, 1978; Todd and Shaw, 1979). Screens and pendants of primarily Mesozoic metasedimentary and metavolcanic andalusite-bearing rocks in the west typically have narrow hornblende-hornfels grade aureoles adjacent to plutonic contacts. Metamorphic grade and depth of exposure apparently increase to the east, where there are regions of sillimanite-only upper amphibolite-grade gneiss and prealuminous granitic rocks (Todd and Shaw, 1979). Metasomatism is locally prominent adjacent to the more felsic intrusions, particularly in carbonate rocks. A notable example is the boron-rich skarn at Crestmore, which formed adjacent to a pegmatite-bearing quartz monzonite (Burnham, 1954). Igneous-related mineralization is sparse; it is generally restricted to skarns or pegmatites within or adjacent to felsic plutons.

Sierra Nevada The composite Sierra Nevada batholith and satellitic plutons intrude diverse lithologies and exhibit multiple styles of contact metamorphism. Compositional and age variations across the batholith are well known — younger, more felsic, and more potassic to the east (Bateman, 1981; Bateman *et al.*, 1967; J. G. Moore, 1959). To the north and west, host rocks are a complex of previously metamorphosed Paleozoic and Mesozoic metavolcanic and metasedimentary rocks comprising several terranes of controversial affinity (Day *et al.*, 1984, Chapter 27, this volume; Schweikert, 1981; Schweikert *et al.*, Chapter 29, this volume). To the south and east, the batholith intrudes miogeoclinal rocks and overlying Triassic to Cretaceous volcanic and sedimentary rocks (J. N. Moore and Foster, 1980; Saleeby, 1978, 1981). Depth of exposure increases from north to south, going from epizonal in the northern and central portions to mesozonal and catazonal in the southern parts (Sharp, Chapter 30, this volume). Metamorphic grade increases with increasing pluton density (*cf.* Ernst, 1974). High- and medium-pressure facies series metamorphic rocks occur in areas with low pluton densities in the northern and western Sierra Nevada (Day *et al.*, Chapter 27, this volume).

Abundant, but typically separated, generally gabbroic to tonalitic granitoids of Jurassic and Cretaceous age dominate the west side of the range. Hornblende-hornfels thermal aureoles typically extend 1 km or more from plutonic contacts (e.g., Hietanen, 1981). Metasomatism is rare when compared with the more felsic, carbonate-hosted plutons in the central Sierra Nevada. Quartz dioritic to tonalitic plutons in the north-eastern Sierra are exposed at somewhat shallower depths. They have more extensive aureoles and associated copper mineralization (G. H. Anderson, 1937; A. R. Smith, 1970). Pre-Middle Jurassic medium- to locally high-pressure facies series regional metamorphism in the western Sierra preceded most of the magmatism (Saleeby, 1981; Saleeby and Sharp, 1980; Schweikert, 1981; Schweikert et al., 1984). Most rocks have greenschist-grade assemblages away from plutons. Gold mineralization in the Mother Lode (140–110 m.y.) postdates regional metamorphism but overlaps with the Cretaceous magmatism, although the relationship between the two, if any, is unclear (Bohlke and Kistler, 1986). Several intrusions in the northern and western Sierra Nevada have strongly foliated aureoles suggesting forceful emplacement (e.g., Bidwell Bar, Compton, 1955).

Cretaceous intrusions comprise most of the central and southern Sierra Nevada (Fig. 5-3D). Jurassic intrusions are abundant on both the west and the east, but absent in the southern Sierra. Known Triassic plutons (Chen and Moore, 1981; Stern et al., 1981) cluster in the east-central area where they are surrounded by younger plutons. Metamorphism in the pendants of the central and southern Sierra Nevada is of multiple types and multiple ages. In the central Sierra, associated volcanic rocks (Kistler and Swanson, 1981; Tobisch et al., 1986), and petrology (Bateman, 1981; Kerrick, 1970, 1977; D. C. Ross, 1985); abundant andalusite indicate epizonal to upper mesozonal levels of exposure. Cretaceous volcanic and sedimentary rocks are intruded by only slightly younger plutons. Thin, high-grade (hbl and rare px hornfels) aureoles, including abundant discontinuous skarns in carbonate rocks, are present along most contacts (e.g., see Kerrick, 1977; Krauskopf, 1953). Pendants are metamorphosed to upper greenschist to amphibolite grade, perhaps during multiple events (Bateman, 1983; Nokleberg and Kistler, 1980), which may have been contemporaneous with batholith emplacement (Saleeby, 1981; Tobisch et al., 1986). For example, plutons were emplaced adjacent to the Pine Creek pendant over a 130-m.y. span (Chen and Moore, 1981; Stern et al., 1981), each of which produced associated metamorphism (Bateman, 1965). Metamorphism within pendants usually cannot be attributed to any individual pluton, but probably represents the sum-mation of many thermal effects. Evidence for the timing of metamorphism is controversial: some workers suggest that one or more episodes of older (Devonian-Jurassic) regional metamorphism on which contact metamorphism is superimposed (Girty et al., 1984; Nokleberg and Kistler, 1980), in part based on a pronounced unconformity between upper Paleozoic miogeoclinal rocks and Mesozoic volcanic and sedimentary rocks. Others (e.g., Tobisch et al., 1986) argue that metamorphism and most deformation may be roughly synchronous with plutonism. Volcanic rocks typically show evidence of extensive hydrothermal metamorphism as in the Mineral King (Busby-Spera, 1984) and Ritter Range (Fiske and Tobisch, 1978) pendants. Although the plutons are rarely strongly altered or mineralized, tungsten skarns are abundant (Albers, 1981; Krauskopf, 1957; Newberry, 1982) K-Ar in particular, and Rb-Sr ages were widely reset in the Cretaceous within a broad band along the magmatic axis (Evernden and Kistler, 1980).

Wallrock fabrics are usually isotropic immediately adjacent to intrusive contacts; however, most pendants and wallrocks are foliated (e.g. Tobisch et al., 1986). Several plutons, particularly on the west side of the batholith, show clear evidence of forceful emplacement, locally, including deformation of adjacent plutons (Bateman et al., 1983). Modes of emplacement of Cretaceous plutons to the east are controversial (Nokleberg and Kistler, 1980; Schweikert, 1981; Tobisch et al., 1986); many intrusions and adjacent pendants are foliated and elongate in a northwest-southeast direction.

In the southern Sierra Nevada, deeper exposures are indicated by petrology and the catazonal nature of the intrusions (Sharry, 1981). Aureoles of any type are indistinct; where present, they tend to be discontinuous, high-grade, foliated, thermal aureoles adjacent to the more mafic intrusives (Elan, 1985; D. C. Ross, 1983). Mineral deposits are scarce in the southernmost part of the batholith north of the Garlock fault. Tungsten has been mined from skarns in Kernville Series rocks slightly north of the Garlock fault (Miller, 1940; Troxel and Morton, 1962). Metamorphism is amphibolite to granulite grade (medium-pressure facies series) with sillimanite as the sole aluminum silicate (D. C. Ross, 1985).

Western Great Basin The western Great Basin resembles the Sierra Nevada in having high pluton densities in a heterogeneous pile of sedimentary (Paleozoic to Jurassic) and volcanic (Permian to Cretaceous) rocks (Stewart, 1980). Cenozoic extension has exposed multiple structural levels as, for example, in the Yerington, Nevada, area (Proffett and Dilles, 1984). Emplacement depths average somewhat less than in the central Sierra Nevada (Figs. 5-3 and 5-9). Permo-Triassic magmatism is documented although sparse (Fig. 5-3B; Johnson, 1977). Triassic leucogranites are shallow, as evidenced by emplacement in roughly coeval volcanics (e.g., Koipato Group, M. G. Johnson, 1977). Jurassic and Cretaceous magmatism is abundant and well represented by felsic plutons and volcanic rocks (J. G. Smith et al., 1972; Fig. 5-3C and D). Shallow Jurassic mafic plutonism is widespread and locally voluminous; it is best represented by the Humboldt lopolith (Speed, 1976).

Similar to the Sierra Nevada, most described contact metamorphism is adjacent to plutons, but other, apparently igneous-related metamorphism is widespread. Extensive, low-temperature hydrothermal metamorphism is reported in volcanics from many areas. The Lower Triassic Koipato Group and related leucogranites are pervasively altered, generally by alkali (±hydrogen) exchange (Tatlock, 1961); alteration probably occurred during both the Triassic and the Late Cretaceous (Vikre, 1981). Areas with similar alteration include the White Mountains (in Jurassic metavolcanics, but perhaps caused by Cretaceous granites; G. H. Anderson, 1937; Crowder and Ross, 1973; Hanson, 1986) and the Yerington district (in the Jurassic Artesia Lake andesite, the volcanic cover of the Yerington batholith,[3] Proffett and Dilles, 1984). Regionally extensive greenschist-grade Mesozoic metavolcanic and metasedimentary rocks in western Nevada (e.g., Bingler, 1978; Garside, 1982) may owe part of their metamorphism to hydrothermal activity, but the geochemical data to prove this are lacking. A different, prominent type of hydrothermal metamorphism is extensive, shallow vein-related Fe- and Na-rich alteration associated with gabbroic to quartz dioritic intrusions such as the Humboldt lopolith and coeval volcanics (Reeves and Kral, 1955, Reeves et al., 1958; Speed, 1962).

Well-defined aureoles occur in a variety of protoliths at somewhat deeper levels. To the west, described contact metamorphism generally consists of narrow zones of hornblende-hornfels facies (e.g., in Washoe Co., Nevada, Bonham, 1969). To the east, much more extensive aureoles occur both in clastic rocks, where they are primarily hornfelses and veins, and in carbonates where marble, skarn, and replacement bodies are common. Excellent examples abound, including the Osgood Mountains (Hotz and Willden, 1964; B. E. Taylor and O'Neil, 1977) and the Santa Rosa Range (Compton, 1960). In the Yerington district, abundant hornfels, skarnoid, skarn, and associated Cu mineralization formed in Triassic-Lower Jurassic metavolcanic and metasedimentary rocks adjacent to the Yerington batholith (N. B. Harris and Einaudi, 1982). Albers and Stewart (1972) describe the large stocks in Esmeralda County, Nevada, as having well-defined aureoles up to 3 km wide, grading out into greenschist or subgreenschist Paleozoic metasedimentary rocks.

---

[3] Also M. T. Einaudi, personal communication, 1986.

Nearby in the White Mountains, a similar, extensive aureole occurs around the Jurassic Beer Creek pluton (Gastil *et al.*, 1967; D. B. Nash, 1962).

Mineralization of many types is prominent. Porphyry Cu is closely associated with Jurassic plutons (e.g., four deposits alone in the Yerington district; Proffett and Dilles, 1984); however, some is apparently also present in the Cretaceous (e.g., the Fish Creek prospect, B. W. Miller and Silberman, 1977). Copper skarns are common, but seemingly limited in extent. In the Yerington district, N. B. Harris and Einaudi suggest that this is due to the distance from the mineralizing intrusions to the nearest carbonate. Carbonate-hosted base-metal replacement deposits occur in the outer protions of some complex aureoles (e.g., Darwin, California; Hall and MacKevett, 1962; Rye *et al.*, 1974). Numerous tungsten skarns occur near granitic to granodioritic intrusions (Kerr, 1946). Tungsten-rich polymetallic skarns are also common; associated intrusions usually contain porphyry-type Mo ± Cu mineralization (e.g., New Boston–Blue Ribbon, K. S. Frost, 1984). The tungsten and polymetallic skarns are most often Late Cretaceous, as is rare epithermal vein mineralization associated with extensive hydrothermal metamorphism (e.g., the Rochester district, Vikre, 1981). Widespread Fe-rich veins and replacements occur with Jurassic mafic intrusives and cogenetic volcanic rocks (Reeves and Kral, 1955, 1956).

Deformation textures associated with the plutons are variable. In many areas, the aureoles have predominantly granoblastic textures, emplacement appears to have been passive. In other areas, however, the plutons and aureoles have weakly to prominently developed contact-parallel fabrics that die out going away from the plutons (Albers and Stewart, 1972; Bonham, 1969; Sylvester *et al.*, 1978; Willden, 1964). On a broad scale many plutons (particularly of Cretaceous age) are concordant with large-scale structures (e.g., see Albers and Stewart, 1972).

Regionally extensive greenschist to amphibolite-grade metamorphism is widespread in the western Great Basin, although many of the rocks are at the lower end of the grade spectrum (Albers and Stewart, 1972; Bingler, 1978; Bonham, 1969; Labotka and Albee, Chapter 26, this volume; Willden and Speed, 1974). Grades increase toward plutons; mean regional grades generally increase toward the west along with pluton density. In the Panamint Range, for example, a strong regional metamorphic gradient decreases away from a contemporaneous Middle Jurassic batholith (Labotka, 1985; Labotka and Albee, Chapter 26, this volume). To the east, as intrusion density decreases, mean regional grades are subgreenschist to greenschist. The Santa Rosa aureoles grade out into chlorite zone slates formed at about 7 km depth (Compton, 1960; Turner, 1981) in an area of relatively low pluton density. Abundant andalusite, the lack of reported kyanite, and stratigraphic control where available, along with the average greenschist-grade metamorphism, indicate low-pressure facies series conditions. Garside (1982) and Bingler (1978) describe regionally metamorphised greenschist-grade Mesozoic volcanic and sedimentary rocks which retain many primary structures and apparently lack penetrative fabrics (some have hornfelsic textures). In central Nevada, conodonts (A. G. Harris *et al.*, 1980) and organic maturation (Poole *et al.*, 1983) indicate greenschist-grade conditions in marine sediments as young as Triassic. Kyanite-bearing metasedimentary rocks in the intrusion-free Funeral Mountains (Labotka, 1980) are more characteristic of the medium-pressure metamorphic conditions of the central and eastern Great Basin.

Sparse radiometric dates indicate that some of the metamorphism is Jurassic; Jurassic low-pressure regional metamorphism is well documented in the Panamints (Labotka, 1985). Other dates on metamorphic rocks suggest Late Cretaceous metamorphism (e.g., Erickson *et al.*, 1978), as does extensive resetting of igneous biotite (and some hornblende) K-Ar dates to Late Cretaceous ages. Cretaceous ages (Speed and Kistler, morphism. Labotka also describes Late Cretaceous retrograde metamorphism contemporaneous with the intrusion of two-mica granites.

Central and Eastern Great Basin Carbonate and clastic rocks of the Cordilleran miogeocline host the sparse Jurassic (160 ± 15 m.y.) and Cretaceous (~120–70 m.y.) intrusions of the eastern and central Great Basin (Fig. 5-3; Stewart, 1980). The Upper Jurassic Pony Trail volcanics are the only Mesozoic volcanic rocks in this area and are intruded by contemporaneous shallow plutons (Muffler, 1964, J. F. Smith, 1972). The Cretaceous secular variation in pluton compositions resembles that for the whole western United States (Fig. 5-1).

Well-defined aureoles are recognized with most plutons. In most cases, the country rocks are low grade, and thermal aureoles can be distinguished for a considerable distance. The most extensive metasomatism and the widest relative thermal aureoles are associated with the tops of plutons (e.g., McCullough Butte, Barton, 1982; Ely, Westra, 1982, Monte Cristo, Sonnevil, 1979); whereas relatively narrower aureoles and considerably less extensive metasomatism are found adjacent to larger exposures such as the Tungstonia granite (Best *et al.*, 1974; E. L. Miller *et al.*, Chapter 24, this volume). Most of the obvious metasomatism occurs as skarns, a consequence of the carbonate-dominated section. Extensive isotopic aureoles arc have been documented at Notch Peak (Nabelek *et al.*, 1984) and McCullough Butte (Barton *et al.*, 1982). At McCullough Butte an extensive (>6 km) zone of meteric-signature veins surrounds an inner zone of magmatic-signature skarn and marble; similar relationships can be inferred around other intrusions. Regionally extensive hydrothermal metamorphism is prominent only in the Jurassic Pony Trail volcanics of central Nevada (Muffler, 1964). In the Upper Proterozoic/lower Cambrian clastic part of the section, aureoles are generally less pronounced with high-grade zones of limited relative extent passing into lower grade (generally greenschist) metamorphism of "regional" extent (*cf.* E. L. Miller *et al.* Chapter 24, this volume, Snoke and D. M. Miller, Chapter 23, this volume). K-Ar ages apparently approach emplacement ages except near younger (Cretaceous or Tertiary) plutons or detachment faults (D. E. Lee *et al.*, 1970), and in the deeper exposures (e.g., the Ruby Mountains).

For the most part, intrusions are discordant and the aureoles show little or no penetrative foliation. Concordant intrusions do occur however, and have strongly deformed margins and country rocks. Most of the Mesozoic intrusions in the Ruby Mountains and Snake Range have this character (Howard *et al.*, 1979; E. L. Miller *et al.*, Chapter 24, this volume; Snoke and D. M. Miller, Chapter 23, this volume). Another example is the Tungstonia two-mica granite in the Kern Mountains (Best *et al.*, 1974; E. L. Miller *et al.*, Chapter 24, this volume), which has a protoclastic border and highly flattened, mylonitized carbonate host rocks.

Most of the plutons have some associated mineralization, although it is commonly of minor economic and volumetric importance. Shallow (those intruding upper Paleozoic sediments) quartz diorite to quartz monzonite plutons contain copper mineralization and have nearby skarns (e.g., Ely, James, 1976; Westra, 1982; Dolly Vardon, Atkinson *et al.*, 1982). Similar plutons, but apparently deeper, have extensive nearby sulfide replacement deposits, generally within an extended thermal aureole, sometimes with closely associated skarns (e.g., Eureka, Nolan, 1962; Cortez, Gilluly and Mazursky, 1965). Deeper, granitic stocks contain Mo-Cu-type porphyry alteration and polymetallic W-rich skarns (e.g., Tempiute, Tschanz and Pampeyan, 1970; Monte Cristo, Sonnevil, 1979). Distinctive F- and Al-rich skarns are associated with deep (6–10 km) Late Cretaceous two-mica granites (Barton, 1987).

Regional metamorphism in the eastern and central Great Basin is variably developed

(Armstrong and Hansen, 1966). Greenschist- to amphibolite-grade rocks occur in many places (Snake, Schell Creek and Deep Creek Ranges, E. L. Miller *et al.*, Chapter 24, this volume; Ruby and East Humboldt Mountains, Snoke and Howard, 1984; Pilot Mountains, D. M. Miller, 1984; Toiyabe Range, Means, 1962; Grant Range, Cebull, 1970; Albion Range; Hodges and McKenna, 1986). Each of these areas exposes deep portions of the miogeocline and most contain a higher proportion of Mesozoic plutons than surrounding areas. Radiometric dating and the spatial distribution of metamorphism indicates that the higher grades are correlated with plutonic events in both the Jurassic and Cretaceous (Allmendinger *et al.*, 1984; D. E. Lee and Fisher, 1985; D. M. Miller, 1984; E. L. Miller *et al.*, Chapter 24, this volume; Snoke and Lush, 1984). Scattered kyanite occurrences indicate lower-temperature gradients than those recorded in rocks to the west. The plutons appear to be either part of this kyanite-producing metamorphic event (e.g., the Snake Range, E. L. Miller *et al.*, Chapter 24, this volume), or they appear to produce an andalusite-bearing overprint on an earlier event (e.g., the Ruby Range D. L. Johnson, 1981; Olson and Hinrichs, 1960). Moreover, areas in eastern Nevada lacking plutons have at most chlorite grade assemblages in the oldest pelitic rocks exposed (e.g., the Cambrian Pioche shale). As discussed above, this precludes regional thermal gradients in the low-pressure facies series regime. Although local highs exist, conodont color indices and organic maturation also indicate lower integrated thermal histories in eastern Nevada and western Utah than in central Nevada (A. G. Harris *et al.*, 1980; Poole *et al.*, 1983).

Klamath Mountains The Klamath Mountains province consists of a complex of metamorphosed Phanerozoic oceanic and island arc terranes that were juxtaposed in approximately their present relative configuration during the Jurassic (Irwin, 1981). Rare Paleozoic plutons are present in the eastern Klamaths (P. Irwin, 1985). Belts of Jurassic to Early Cretaceous quartz dioritic to tonalitic plutons (Fig. 5-3) pre- and postdate major thrust faults (Harper and Wright, 1984; Irwin, 1985). Magmatism may have migrated first west and then east across the Klamaths from the Middle Jurassic to Early Cretaceous (Harper and Wright, 1984). Alternatively, the various plutonic belts may represent juxtaposition of several arcs (Irwin, 1985). Within some belts, pluton densities exceed 50%, but the overall proportion of plutons is about 15%. The level of exposure deepens toward the Condrey Mountain window in the central Klamaths, as indicated by structural, petrologic, and paleomagnetic evidence (Barnes, 1984; Barnes and Rice, 1983; Coleman and Helper, 1983; Schultz and Levi, 1983).

Restricted contact aureoles are described around many of the plutons (*cf.* Bero, 1980; G. A. Davis, 1963; Godchaux, 1969). The type and extent of the contact metamorphism depends strongly on the host rock. For example, the Gibson Peak Pluton in the central Klamaths has a much wider aureole in ultramafics than it does in the amphibolite-grade pelites and calcareous rocks of the Grouse Ridge Formation (Lipman, 1963). At the other extreme, the Slinkard pluton lacks an obvious thermal aureole because it intruded the amphibolite-grade country rocks near their thermal maximum (Hill, 1985). C. M. Allen (1981), however, states that contact-metamorphosed ultramafic rocks around the Slinkard indicate postpeak intrusion.

Most aureoles lack mineralization or other evidence of significant hydrothermal activity. Oxygen isotopic data from metamafic rocks of the central portion of the western Tr-Pz terrane (Ernst, 1987), however, suggests large fluid flows, conceivably associated with plutonism. Ubiquitous quartz veins, many with precious and base metal mineralization, are at best tenuously related to the plutons (Irwin, 1960). A few tungsten skarns occur where carbonate is present (e.g., the Ashland pluton and nearby smaller bodies; Wolfe and White, 1951).

The Klamath region underwent multiple metamorphic and deformational events (see, e.g., Coleman *et al.*, Chapter 39, this volume; Hill, 1985; Mortimer, 1985). The igneous and metamorphic events in the eastern Klamaths and the central metamorphic belt are predominantly Paleozoic in age (Ernst, 1983). Lanphere *et al.*, (1968) document a mid-Jurassic or older greenschist overprint, roughly synchronous with plutonism, on Triassic high-pressure metamorphic assemblages. Thrusting and folding along with regional greenschist to amphibolite facies metamorphism has been dated at 165–170 m.y., overlapping in time with intermediate-composition volcanism (K/Ar hornblende 168–177) and plutonism (U/Pb zircon 163–174; Fahan and Wright, 1983). The greenschist overprint of blueschist in the Fort Jones terrane in the south is attributed to Jurrasic igneous activity by Jayko and Blake (1984). In the northeastern corner of the Klamaths, a region of low- to medium-pressure metamorphism is called regional contact metamorphism by Kays (1970). There, epidote-actinolite to sillimanite-grade metamorphism increase in grade toward earliest Cretaceous granitic plutons.

Despite the many events and terranes, it is apparent that plutonism, volcanism, and subduction were roughly contemporaneous in the Klamaths (Fahan and Wright, 1983; Harper and Wright, 1984), particularly in the western Hayfork terrane. Low-pressure facies series metamorphism is associated in a general way with the plutonic activity, whereas high-pressure metamorphism represented by the Condrey Mountain Schist is free of Jurassic plutons. The complex tectonometamorphic history obscures the precise relationship between the plutonism and metamorphism, which was clearly continued over a lengthy time. Interpretations vary; Coleman *et al.* (Chapter 39, this volume) interpret the low-pressure (Siskiyou) metamorphism as a preplutonic event perhaps related to rifting. The general overlap in timing of the volcanism, plutonism, and metamorphism and the apparently static nature of the low-pressure metamorphism, however, seem equally compatible with igneous-related metamorphism within an evolving arc as suggested by Fahan (1982) and Hill (1985).

Eastern Oregon and western Idaho Parts of one or more late-Paleozoic to mid-Mesozoic oceanic arcs occur in the east-northeast-trending Seven Devils and Huntington terranes of eastern Oregon and western Idaho. These arcs are separated by a melange terrane containing Devonian to Lower Triassic materials, and an overlying(?) Jurassic flysch terrane. These elements were assembled in the Late Triassic to Jurrasic (Brooks and Vallier, 1978; Mullen and Sarewitz, 1983; Oldow *et al.*, 1984). Plutonic activity is conveniently divided into four age groups: Permo-Triassic, Late Triassic to Early Jurassic, Middle Jurassic to Early Cretaceous, and Late Cretaceous. Late Cretaceous magmatism and metamorphism were apparently restricted to the Idaho batholith and are discussed in the next section.

Permian and Triassic gabbros, diorites, and tonalites (e.g., the Canyon Mountain and Sparta complexes) in the melange terrane in eastern Oregon are interpreted as the basement of one or two oceanic island arcs (Avé Lallèmant *et al.*, 1980; N. W. Walker, 1981; N. W. Walker and Mattinson, 1980). Arc volcanic rocks of similar age in the Seven Devils-Wallowa terrane may be genetically related to the intrusions (Gerlach *et al.*, 1981; Phelps and Avé Lallèmant 1980). Perhaps continuous with this magmatism are Late Triassic to Early Jurassic (220–200 m.y.) gabbro, diorite, and minor granodiorite plutons, intruding Permo-Triassic volcanic and volcanoclastic rocks and limestones in both the Seven Devils and Huntington terranes (Henricksen, 1974; N. W. Walker, 1981). Middle Jurassic to Early Cretaceous tonalitic (quartz diorite to granodiorite) plutons (e.g., the Wallowa and Bald Mountain batholiths) intrude metavolcanic and metasedimentary rocks and earlier plutons of the amalgamated terranes (Armstrong *et al.*, 1977; Henricksen, 1974; White, 1968). Late intrusions and stocks of the Wallowa batholith and Bald Mountain batholith are Early Cretaceous and include in the Wallowa Mountains a suite of cordierite trondhjemites (Taubeneck, 1964). Based on the absence of clearly coeval volcanics and local andalusite in contact aureoles, the middle to late Mesozoic plutons were probably emplaced at upper mesozonal to lower epizonal levels (Henricksen, 1974; White, 1973). Large, Early Cretaceous gneissic tonalitic plutons that intrude the eastern edge of the accreted terranes are discussed with the Idaho batholith.

Contact metamorphism has not been described with the Permo-Triassic intrusions; however, Vallier and Batiza (1978) document regional greenschist to amphibolite-grade hydrothermal metamorphism in the Permo-Triassic volcanic arc assemblages. Late Triassic/ Early Jurassic plutons in western Idaho typically have narrow hornblende-hornfels aureoles, commonly with andalusite in the pelitic rocks (e.g., at Cuddy and Iron Mountains). Metasomatism in these areas is expressed as skarns which in some places contain Fe, Cu, and Zn mineralization (Bruce, 1970; Henricksen, 1974). The more felsic intrusive phases contain porphyry Cu-Mo mineralization, a feature evidently absent in similar intrusions north of Cuddy and Iron Mountains in the Cuprum quadrangle (White, 1968).

Middle to Late Jurassic plutons exposed in western Idaho commonly have sharp contacts and have little evidence for contact metamorphism overprinting earlier(?) low-grade greenschist regional metamorphism of the Seven Devils volcanics (Hamilton, 1963; White, 1968; Wracher, 1969). In contrast, metamorphism around the Late Jurassic Bald Mountain and Wallowa batholiths of eastern Oregon is well defined, with extensive thermal aureoles containing local skarns. The batholiths are broadly concordant with generally increasing wall-rock foliation toward the contacts, but they also disrupt regional structural trends and are locally disordant (Krauskopf, 1943; Taubeneck, 1958). Extensive hornfelsic aureoles around scattered Early Cretaceous granodiorite and quartz diorite plutons (Deep Creek, Mann Creek, and Red Ledge stocks) clearly overprint both earlier intrusions and regional low-grade greenschist metavolcanics (Hamilton, 1963). Cu, W, and Mo skarn mineralization occurs locally in Triassic limestone (e.g., Deep Creek stocks, Hamilton, 1963; White, 1968).

Regional metamorphism events occurred throughout the Mesozoic. Dated Permo-Triassic high-pressure metamorphism is evidenced by scattered blueschist occurrences in the melange terrane (Hotz *et al.*, 1970; Mullen, 1980). Pre-Late Triassic greenschist/ epidote-amphibolite facies metamorphism occurred in arc-related rocks of the melange terrane (Mullen, 1980). Regionally extensive hydrothermal greenschists and amphibolites also formed in the Triassic (Vallier and Batiza, 1978). Regional low-pressure, lower greenschist-grade metamorphism in the accreted terranes is confined to the Middle to Late Jurassic and Cretaceous (Bruce, 1970; Brooks and Vallier, 1978; Hamilton, 1963; Henricksen, 1974; White, 1968). Widespread greenschist-grade metamorphism clearly affects Late Triassic/Early Jurassic plutons and their narrow aureoles (e.g., in the Cuddy Mountains; Henricksen, 1974). Middle to Late Jurassic plutons in the Cuprum quadrangle appear to be slightly younger or synchronous with regional metamorphism (White, 1968). The lack of thick cover and the overall grades suggests low-pressure facies series conditions. Metamorphism associated with the Late Jurassic Wallowa and Bald Mountain batholiths clearly overprints the regional greenschist metamorphism, limiting the regional greenschist event to Middle or Late Jurassic age. The presence of andalusite in argillites indicates a low-pressure origin for the metamorphic rocks spatially associated with the Late Jurassic Bald Mountain batholith of eastern Oregon (Prostka, 1962; Taubeneck, 1957, 1958). In western Idaho, the thermal effects of Early Cretaceous plutons overprint the Jurassic greenschist event. Cretaceous (~118 m.y.) greenschist to amphibolite facies metamorphism is confined to eastern exposures adjacent to the Idaho batholith. This metamorphism apparently took place at greater depths than the Jurassic metamorphism to the west (Hamilton, 1963; Lund and Snee, Chapter 11, this volume).

Idaho batholith Although traditionally separated into two lobes (the northern Bitterroot lobe and the southern Atlanta lobe; Armstrong, 1975a), the Late Cretaceous Idaho Batholith can be considered as one complex intruded over about 50 m.y. into variable lithologies (Hyndman, 1983). To the north and east, the batholith intrudes Proterozoic Belt greenschist- to amphibolite-grade metasedimentary rocks (and stratigraphic equivalents in central Idaho) and their underlying gneissic basement. Rocks of the Paleozoic miogeocline sediments abut the batholith on the southeast and in scattered inliers in the north-central portion of the batholith. The western margin of the batholith largely intrudes accreted Permian to Jurassic metasedimentary and metavolcanic rocks (Hyndman, 1983).

The west side of the batholith contains an extensive suite of foliated, largely concordant, tonalite/quartz diorite plutons (105-79 m.y., Armstrong, 1975b; Snee *et al.,* 1985; Toth and Stacey, 1985). A similar, but younger (73 ± 6 m.y.) and less extensive suite of tonalitic plutons comprise the northeastern portion of the batholith (Chase *et al.,* 1983). The cores of both the Atlanta and Bitterroot lobes consist of granodiorite, granite, and two-mica granite dated at 66 ± 10 m.y. for the Bitterroot lobe (Chase *et al.,* 1978) and 75–70 m.y. for the Atlanta lobe (Hyndman, 1983). Granodiorites and quartz diorites (e.g., Thompson Creek stock) dated at between 82 and 95 m.y. on the southeast flank of the batholith are roughly synchronous with the older tonalite suite exposed to the west (Bennett, 1984). Discordant isotopic dates within the batholith largely reflect post-intrusion uplift (especially in the Eocene Bitterroot core complex) or the effects of regional heating associated with the emplacement of epizonal Eocene granite and quartz syenite plutons (Criss and Fleck, 1985; Criss and Taylor, 1983). The western and northern margins of the batholith are generally concordant, gradational with common migmatitic zones, and heavily intruded. In contrast, on the east side, contacts are typically sharp and individual plutons clearly cross-cut regional structural trends.

On a regional scale, shallower levels of exposure occur to the east (Hyndman, 1983). In the Bitterroot lobe, coeval metamorphism is extensive and kyanite-bearing in the west, whereas on the eastern border low-pressure assemblages (andalusite and cordierite are found within narrow contact aureoles; Hyndman *et al.,* Chapter 12, this volume). Similarly, on the eastern flank of the Atlanta lobe metamorphism is clearly associated with individual plutons. There, skarn/hornfels aureoles occur around relatively high level (~5 km) satellitic granodioritic plutons that intrude the miogeoclinal carbonates and shales (White Cloud stock, Thompson Creek, Juliette Stock; Cavanaugh, 1979; C. P. Ross, 1937). Also on the east, deeper exposures in the Pioneer Mountains core complex ~10 km based on geobarometry; Silverberg, 1983) contain andalusite- and sillimanite-bearing schists that increase in grade toward a large, irregular gneissic granodioritic intrusion (Dover, 1983; Silverberg, 1986).

Little mineralization is associated with the Bitterroot lobe. In contrast, the Atlanta lobe contains a greater abundance of concomittant Au (associated with two-mica granites in its core (Lund *et al.,* 1986). Porphyry Mo, vein W, and skarn-related W, Pb, and Zn mineralization occurs mostly with satellitic plutons to the east. Furthermore, regionally the Atlanta lobe appears to be exposed at shallower levels than the Bitterroot lobe, because it intrudes presumably higher level miogeoclinal sediments and lacks associated kyanite. The greater abundance of concomitant mineralization may simply be due to the higher-level exposures (Hyndman, 1983).

Although the Bitterroot lobe contains little mineralization, it significantly affected its wallrocks. Roughly contact-parallel metamorphic zones, decreasing from sillimanite-orthoclase to greenschist assemblages, occur in the Belt Group rocks. Metamorphism extends up to 40 km away from the batholith (Hietanen, 1984; Hyndman, 1983; Lang and Rice, 1985). On the north west side of the batholith, a relatively static, deep-seated (based on stratigraphic and geobarometric estimates), high-temperature, medium-pressure event overprints an earlier (Precambrian?) synkinematic regional metamorphism (Hietanen, 1984; Hyndman, 1983; Lang and Rice, 1985). Dates of 64 m.y. on the metamorphic rocks suggest that metamorphism was associated with batholith emplacement (Hofmann, 1972). Mineralogical and isotopic evidence further suggests that extensive metasomatism up to 50 km away from the batholith was associated with postkinematic prograde metamorphism (Fleck and Criss, 1985; Hietanen, 1962). On the northeast side of the batholith a similar metamorphism has been dated as Late Cretaceous (Chase et al., 1978). There, two medium-pressure (sillimanite-kyanite) dynamothermal events occurred prior to intrusion of the earliest quartz diorite plutons (~73 m.y.), followed by one deformational event contemporaneous with lower-pressure (cordierite-producing) metamorphism and intrusion of the early quartz diorite plutons (Bickford et al., 1981; Chase, 1973; Chase et al., 1983). Thus, regional metamorphism is closely associated with the emplacement of the Bitterroot lobe, but a direct causal relationship is unclear. Several possibilities exist: (1) the batholith was a heat source leading to metamorphism on a regional scale; (2) the batholith may represent a metamorphic culmination, symptomatic of the metamorphic heat source but not the driving agent; (3) the emplacement of the batholith could have dragged the isograds upward; or (4) the two events are unrelated.

Cretaceous metamorphism around the Atlanta lobe resembles that around the Bitterroot lobe, with the added complication of metamorphism during accretion of terranes to the west. In the northern part of the lobe (e.g., in the Elk City and Buffalo Hump areas), sillimanite grade was locally attained in Proterozoic gneissic metasedimentary rocks prior to emplacement of the batholith (Lund et al., 1986; Reid et al., 1973). On the western margin, in the Harpster and Riggins areas, metamorphic grade increases from lower greenschist to sillimanite zone over 10-20 km approaching the batholith (Hamilton, 1963; Myers, 1982). This regional metamorphism may be due to the earliest phases of the batholith or juxtaposition of accreted terranes in the Late Cretaceous. Muscovite-biotite granite plutons of the core (75-70 m.y.) clearly postdate uplift and cooling of the earlier plutonic and metamorphic host rocks (Lund and Snee, Chapter 11; this volume). On the east side of the Atlanta lobe, metamorphism in the miogeoclinal sedimentary rocks is largely confined to aureoles about individual plutons at 5-10 km depth compatible with medium-pressure series background conditions.

Montana Mesozoic intrusions, almost entirely of Late Cretaceous age, occur in much of western Montana. In addition to rocks of the Bitterroot lobe of the Idaho batholith discussed above, intrusions are concentrated in southwestern Montana, west of the Montana disturbed belt (Smedes, 1968). These plutons include the composite Boulder batholith and many smaller granodioritic to granitic bodies (Hyndman, 1983; Tilling, 1973). Peraluminous granites occur in the Flint Creek Range and the Tobacco Root Mountains. The plutons intrude Belt sedimentary rocks, older Precambrian gneisses, and 3-5 km of Paleozoic and Mesozoic sedimentary rocks. Stratigraphic and petrologic evidence indicates that these rocks were intruded at relatively shallow levels (3-5 km; Hamilton and Myers, 1974; Hyndman, 1983; Klepper et al., 1971). Following minor mafic plutonism, the bulk of the Boulder batholith was emplaced into its own volcanic pile, the Elkhorn Mountain volcanics, at 78-68 m.y. (Hamilton and Myers, 1976; Robinson et al., 1968; Tilling et al., 1968). In northwestern Montana, several small, early Late Cretaceous plutons (Twenty-Odd, Ambassador, and Dry Creek stocks, Beaver Creek syenite), ranging in composition from quartz monzonite to syenite, intrude lower Belt series quartzite, argillite, and shale (Crowley, 1963; McDowell, 1971). Farther north, the Jurassic (184 m.y.; Fenton and Faure, 1969) Rainy Creek alkaline-ultramafic complex (Boettcher,

1967) is Montana's oldest reported Mesozoic intrusion. Alkalic laccoliths, sills, and plugs of latest Cretaceous and early Tertiary age (e.g., Judith and Moccasin Mountains, Little Rocky Mountains) extend into central Montana, where they were emplaced at very shallow levels (<1–4 km) into Paleozoic, Cretaceous, and lower Tertiary clastic sedimentary rocks (Lindsey, 1982; Marvin et al., 1980; R. N. Miller, 1959).

Extensive (0.5–2 km wide) hornfelsic aureoles in sedimentary and volcanic rocks surround many of the calc-alkaline intrusions from the Pioneer-Mount Torrey batholiths in the south (Zen et al., 1975) to the Boulder and Phillipsburg batholiths to the northwest (Emmons and Calkins, 1913; Holser, 1950; Hyndman, 1984). Detailed petrologic studies around the northern margin of the Boulder batholith (J. M. Rice, 1977b, 1979), the Black Butte stock (Bowman and Essene, 1982, 1984), and the Marysville stock (Knopf, 1950; Lattanzi et al., 1980; J. M. Rice, 1977a) indicate that calc-silicate and pelitic hornfelses formed by without appreciable infiltration by aqueous fluids. In sedimentary rocks, metasomatism is locally prominent, but tends to be much more restricted in space than the thermal effects (e.g., skarns around the Royal and Black Butte stocks extend less than 10 percent of the width of the thermal aureole; Allen, 1966; Mutch and McGill, 1962). The shallow alkalic plutons lack obvious aureoles; metamorphic effects generally are limited to fracture-controlled hydrothermal alteration (Lindsey, 1982).

Hydrothermal alteration is extensive in the Elkhorn Mountain volcanics (locally > 3 km thick) and associated hypabyssal intrusions; hornfels is restricted to the margin of the Boulder batholith (Smedes, 1966). The metamorphic rocks locally contain cordierite, dumortierite, and zeolites. Cross-cutting veinlets indicate at least four alteration events affected parts of the volcanic pile. The Kokoruda Ranch K-rich mafic rocks, the oldest part of the Boulder batholith, were converted to pyroxene hornfels by younger intrusive phases (Smedes, 1966). Alteration associated with late alaskite dikes includes sericitization, kaolinitization, and pyritization of the host quartz monzonite and of the dikes themselves, followed by late metalliferous quartz veins (Smedes and Thomas, 1965). Porphyry, vein, and skarn-type base-metal, precious-metal mineralization occurs within and adjacent to the calc-alkaline rocks of the Boulder batholith and related plutons (McClernan, 1983). The Butte district (Meyer et al., 1968) is the classic example, but significant production has occurred from skarns and veins, some related to porphyry-type mineralization, in many other areas, including the Phillipsburg region (Emmons and Calkins, 1913; Holser, 1950; Hyndman et al., 1982) and the Elkhorn Mountains (Knopf, 1913; Steefel and Atkinson, 1984). Minor F and Au vein mineralization occurs with the latest Cretaceous–early Tertiary alkalic complexes (Gardner, 1959; Lindsey, 1982; Marvin et al., 1980).

Regional metamorphism prior to Belt supergroup deposition resulted in gneisses and schists (Gilletti, 1966). Probable Precambrian age metamorphism to lower amphibolite facies affects Belt Prichard group rocks (Hyndman, 1983; Norwick, 1972, 1977). Regionally extensive Mesozoic metamorphism is restricted to areas around the larger intrusive bodies, including widespread hydrothermal metamorphism in the Elkhorn Mountain volcanics. Rocks marginal to the Bitterroot lobe record Jurassic or Cretaceous sillimanite-grade metamorphism and deformation (Hyndman, 1983).

Northern Idaho and Washington Northern Washington and northern Idaho have complex pre-Cenozoic geology essentially continuous with Canada (Coney et al., 1980; G. A. Davis et al., 1978; Okulitch, 1984). The region can be divided into five distinct metamorphic/plutonic provinces (from west to east): Northwest Cascades System (NWCS), Skagit Crystalline Core (SCC), Methow Graben Terrane (MGT), Okanogan Range (OR), and Northeastern Washington/Northern Idaho (NEWNI).

The NWCS is composed of Paleozoic and Mesozoic oceanic rocks that were subjected to regional high-P metamorphism in the Permo-Triassic (Vedder complex; Armstrong *et al.*, 1983) and Late Jurassic–Early Cretaceous (Shuksan suite; Brown, Chapter 7, this volume; Brown *et al.*, 1982). The Skagit Crystalline Core (SCC) consists largely of medium-P series amphibolite-facies, migmatitic Paleozoic, and Precambrian materials metamorphosed in the Cretaceous and perhaps also in the Jurassic and Precambrian (Babcock, and Misch, Chapter 8, this volume; Cater, 1982; Mattinson, 1972). Middle- and upper Mesozoic sedimentary and volcanic rocks comprise the fault-bounded MGT (Barksdale, 1973; G. A. Davis *et al.*, 1978). The OR is composed of weakly to moderately metamorphosed eugeoclinal rocks of Permian, Triassic, and Jurassic age and high-grade polymetamorphic rocks, at least some of which are considered to be more highly metamorphosed equivalents of the eugeoclinal rocks (Fox *et al.*, 1977; deformation and metamorphism of the Okanogan Gneiss Dome discussed by Hansen and Goodge, Chapter 9, this volume). Paleozoic miogeoclinal sedimentary rocks and Belt supergroup metasedimentary rocks comprise the NEWNI (F. K. Miller *et al.*, 1975). Fault-bounded metamorphic complexes (Okanogan Gneiss Dome and the Kettle/Lincoln and Priest River Complexes) in the OR and NEWNI are intepreted to represent exhumed roots of the Mesozoic magmatic arc (Okulitch, 1984). These complexes often have peripheral mylonitized zones and are considred to be the result of Late Cretaceous/early Teritary extension (for a discussion, see Okulitch, 1984; Rhodes and Hyndman, Chapter 10, this volume).

With the exception of the intrusion-free NWCS, all terranes contain abundant Mesozoic felsic intrusions and all but the MGT, where only Late Cretaceous–early Tertiary plutons invade Mesozoic sedimentary and volcanic rocks (Barksdale, 1973), demonstrate a history of magmatism that spanned the Mesozoic and continued into the early Tertiary. Systematic U-Pb dating has been reported only for rocks within the SCC (Mattinson, 1972) and there is therefore a strong bias toward younger ages imposed by regional resetting to Cretaceous or Eocene K-Ar and Rb-Sr ages (e.g., for the NEWNI, F. K. Miller and Engels, 1975; *cf*. Armstrong, 1986, for Canada). Igneous compositions became more felsic with time. In the SCC and OR, Triassic and Jurassic intrusions are quartz diorites to granodiorites; granodiorites and granites were emplaced in the Cretaceous and early Tertiary (Cater, 1982; Fox *et al.*, 1977). In the NEWNI, the sparse older intrusions are Jurassic granodiorites, whereas abundant Cretaceous and early Tertiary plutons are granodiorites and granites; in some cases the latter contain two igneous micas, indicating peraluminous compositions (F. K. Miller and Engels, 1975). The Paleogene(?) Colville batholith varies from early grandiorite to late granite; some of the latter are garnet-bearing and thus possibly prealuminous (Carlson, 1986; Colville Federated Tribes, 1984).

Triassic plutons occur in the SCC (Mattinson, 1972) and probably in the OR (Menzer, 1983). These tonalitic to granodioritic intrusions are relatively concordant with structural trends in surrounding amphibolite-grade metamorphic rocks. Examples include the Dumbell Pluton (Cater, 1982) and orthogneisses of the Marblemount Belt (Misch, 1966) of the SCC, and probably the Reed Creek, Leader Mountain, and Windy Hill Gneisses of the OR, which have been inferred on textural and petrologic grounds to be Late Triassic or Early Jurassic (Menser, 1983). The more abundant Jurassic plutons commonly show similar catazonal features (Cater, 1982; Menzer, 1983) or, more rarely, contact metamorphic assemblages diagnostic of relatively higher pressure (e.g., kyanite-bearing assemblages adjacent to the Flowery Trail Granodiorite of the NEWNI; F. K. Miller *et al.*, 1975). The relative timing of magmatism and tectonism is uncertain; some plutons appear to be pretectonic, others syntectonic (e.g., for the OR, Menzer, 1983).

Cretaceous to early Tertiary granitic intrusions are abundant in the SCC, OR, and NEWNI (Cater, 1982; Fox *et al.*, 1977; F. K.Miller and Engels, 1975; F. K. Miller *et al.*, 1975). Except where adjacent to gneiss domes (Cheney, 1980; Fox *et al.*, 1977; Okulitch, 1984; Rhodes and Hyndman, Chapter 10, this volume) or Tertiary plutons (F. K. Miller and Engels, 1975; F. K. Miller *et al.*, 1975), these rocks show concordant K-Ar behavior. They are inferred to have been emplaced at shallower levels (<10 km) than the earlier plutons, based largely on contact metamorphic assemblages. Emplacement into a realtively static environment is inferred from the lack of deformation textures and hornfelsic textures in their aureoles. The deepest exposures (>12 km) of plutonic/metamorphic rock are highly migmatized portions of the SCC, OR, and NEWNI. The shallowest exposures (<8 km) are inferred for plutons of the MGT, the Cretaceous plutons north of the Newport Fault in the NEWNI (which produced andalusite-bearing aureoles and porphyry Mo-W mineralization; F. K. Miller and Theodore, 1982), and the Cretaceous plutons of the OR (see Menzer, 1983). In each area, the Cretaceous intrusions are typically less concordant and foliated than earlier plutons (e.g., in the OR; Menzer, 1983; Rinehart and Fox, 1972).

Well-developed contact metamorphic aureoles occur around plutons exposed at shallow levels; these are typically Cretaceous in age or younger. Examples include the Blickensderfer, Galena Point, and Hall Mountain Plutons of the NEWNI (F. K. Miller, 1974; F. K. Miller and Theodore, 1982; F. K. Miller *et al.*, 1975); and the Aeneas Creek Pluton of the OR (Rinehart and Fox, 1972). These plutons also produced more prominent metasomatic alteration (e.g., Bayview pluton, Gillson, 1927; Kun, 1974). High-level plutons, especially those in unmetamorphosed or very slightly metamorphosed hosts (as in the MTG and parts of the NEWNI), produced spotted hornfelses containing andalusite and/or cordierite. Associated mineral deposits are typically Cu, Mo, W-rich and found both in hosts and plutonic rocks such as the porphyry-Mo-W deposits of the Hall Mountain intrusion (F. K. Miller and Theodore, 1982). Zn-Pb deposits in the Metaline District (Mills, 1977) and Late Cretaceous U deposits in Spokane County (J. T. Nash *et al.*, 1981) have been considered either Mesozoic hydrothermal deposits or earlier deposits reworked by the Mesozoic magmatic-hydrothermal processes.

Contact metamorphism related to the earlier, more mafic, foliated plutons differs from that around the later massive, felsic plutons. Thermal metamorphism is not apparent where plutons are exposed at deep levels in medium-*P* faces series terranes (e.g., the Triassic and Jurassic intrusives of the SCC, Cater, 1982). Exposures of medium-*P* facies series metamorphic rocks contain relatively minor quartz vein-hosted Cu, Ag, Au, Pb, and Zn mineralization, regardless of the temporal relationship between the metamorphism and the intrusions (e.g., adjacent to the Conconully Pluton of the OR, Moen, 1973).

The timing of magmatism relative to regional-scale metamorphism and the role of magmas in the thermal history of the region are often difficult to ascertain. For example, Babcock and Misch (Chapter 8, this volume) report a late Mesozoic to earliest Tertiary regional metamorphism superimposed on thermal metamorphism associated with well-dated 85–95-m.y. calc-alkaline tonalitic to granodioritic syntectonic intrusives in the SCC. In the Priest River complex, contact effects are probably superimposed on an earlier (Precambrian?), lower greenschist-grade regional metamorphism (Bennett *et al.*, 1975; J. E. Harrison and Jobin, 1965; Hobbs *et al.*, 1965). Areas of low-*P* facies series regional metamorphism that may have formed during emplacement of large volumes of magma include the Mount Stuart area of the SCC (Evans and Berti, 1986; Plummer, 1969), the upper plate of the Newport Fault Zone of the NEWNI (F. K. Miller, 1974; F. K. Miller and Theodore, 1982), the area around the Colville batholith (Colville Federated Tribes, 1984), and perhaps parts of the OR (for which Menzer, 1983, infers metamorphic gradients up to ~45°C/km). In the Mount Stuart area, Plummer (1969) interprets the well-foliated andalusite-cordierite Chiwaukum schist near the northern margin of the Mount Stuart batholith to have resulted from a Cretaceous contact-dynamothermal event associated with emplacement of batholith. This dynamic event was superimposed on an earlier "Barrovian" or medium-*P* facies series metamorphism of pre-Cretaceous, post-Devonian age that increases in grade to the north (Misch, 1966). For the same rocks, Evans and Berti (1986) concluded that a dynamic low-*P* facies series "contact metamorphism" was followed closely by a medium-*P* facies series "regional metamorphism." At the southern end of the Mount Stuart batholith, quartz diorite intrusions statically metamorphosed the ultramafic Ingalls complex to produce a 2-km-wide thermal/metasomatic aureole (R. B. Frost, 1975). Similarly, in the OR, the latest stage was seemingly static; Menzer (1983) and Hibbard (1971) indicate that the Salmon Schist records overprinting of an early dynamothermal event by a static thermal event; they suggest that either the two are distinct metamorphic events (with the latter related to magma emplacement), or that the thermal pulse of the metamorphic event outlasted the deformation event(s).

Correlations

From the preceding review the variety of metamorphism is apparent, but several themes stand out. The most important are correlations of distribution, process, and extent with depth, intrusion composition, and host rock composition. The extent, abundance, and distribution of metasomatic aureoles, thermal aureoles, and mineralization in the western United States can readily be interpreted in terms of these factors. These relationships are described further here; they are rationalized in the next section by a model for contact metamorphism about single plutons.

Depth Some general correlations with depth were discussed under "Diversity." Other key features vary with depth, including the average compositions and abundances of igneous rocks, the extent and character of thermal metamorphism, the extent and character of metasomatic metamorphism, the abundance and distribution of mineralization, and the extent and character of deformation.

Although not an explicit aspect of contact metamorphism, intrusion compositions correlate with many aspects of contact metamorphism; thus a few comments on their depth distribution are justified. Summing the areas of the broad classes of Mesozoic igneous rocks versus depth (rather than versus time, *cf.* Fig. 5-1) shows that mafic and alkalic rocks tend to be shallowest, ranging from about 0.5 to 8 km (median 4 km). Silica-rich metaluminous granitoids are slightly deeper on the average ranging from about 1.5 to 9 km (median 5.5 km). The deepest intrusions are the peraluminous granitoids, which are generally greater than 5 km, averaging around 9 km. This change is petrologically well grounded, with the deeper plutons forming from cooler, wetter melts (Wyllie, 1977). The relative abundance of peraluminous graitoids in Idaho and Washington is thus consistent with the generally deeper levels of exposure there (Fig. 5-2), although, as discussed earlier, many factors may be involved. Alternatively, igneous compositions in Idaho may have become peraluminous (crust dominated) in the latest Cretaceous/earliest Tertiary due to continued heating in the same area, whereas crustal heating ceased to the south when the locus of Laramide magmatism moved east.

The extents and distribution of thermal aureoles (Fig. 5-4A) vary in a complex way with depth. At shallow levels, thermal aureoles tend to be fairly narrow (<1 km, commonly much less), except where intrusions are quite large (e.g., the Butte quartz monzonite of the Boulder batholith) or hydrothermal fluid flow has been extensive (*cf.* Fig. 5-4A and C). Greenschist or lower-grade thermal aureoles are only reported at fairly shallow levels (<6-8 km), where pluton densities are low and the host rocks have not been previously metamorphosed. At greater depths, aureole extents vary considerably. Most are in the range 100 m-1 km; broader aureoles tend to occur about large, composite intrusions up to batholith size (e.g., the Idaho batholith). Deep plutons that are closely spaced or

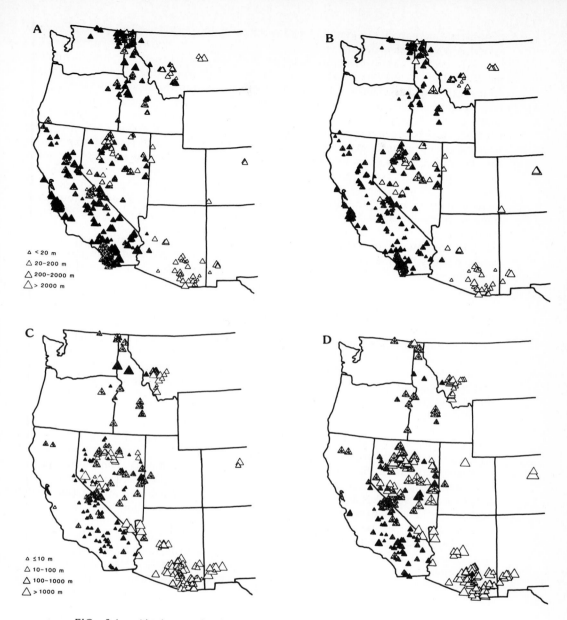

FIG. 5-4. Absolute and relative extents of thermal and metasomatic aureoles. These are only the locations where such metamorphism has been described in sufficient detail to quantify. Many other locations have similar relationships, but are not included. Open symbols indicate depth of emplacement <4 km; starred symbols indicate 4–8 km; solid symbols indicate >8 km. (A) Absolute extent of thermal aureoles. (B) Relative extent of thermal aureoles. The symbol size is proportional to log (metamorphic extent/igneous area). The range spans seven orders of magnitude. (C) Absolute extent of metasomatic aureoles. (D) Relative extent of metasomatic aureoles. The symbol size is proportional to log (metamorphic extent/igneous area). The range spans eight orders of magnitude.

that intrude amphibolite or granulite-grade rocks generally lack described aureoles or only have thin granulite-grade aureoles.

The relative extent of thermal aureoles (the width of the metamorphism divided by the half width of the intrusion) tends to decrease with depth (Fig. 5-4B), although the values scatter widely ($>10^2$–$<10^{-3}$). Less than one-third of the areas with reported contact metamorphism have extent data. About 70% of high-temperature (amphibolite or higher grade) aureoles for which at least semiquantitative extent data are available (>200 in CONTACT) have relative extents between 3 and 30%. The largest relative extents occur where little intrusion is exposed, either because of cover or high levels of exposure in the overall system (e.g., McCullough Butte, Barton, 1982). There are exceptions: many small, shallow intrusions (particularly dikes) have thin to absent macroscopic aureoles; some large, deeply exposed plutons have broad associated zones of regionally extensive metamorphism. Overall, the distribution of thermal aureoles reflects the distribution of plutons, although relatively few are described with catazonal plutons (e.g., in Idaho and Washington).

In contrast to the extent of thermal aureoles, the extent of metasomatic aureoles varies strongly with depth of formation. Where described, metasomatism tends to be much more extensive shallower than 8 km, particularly less than 4 km (Fig. 5-4C). Furthermore, metasomatism is more commonly reported at shallow than deep levels, another difference with thermal aureoles (cf. Fig. 5-4A and B with Fig. 5-4C and D). Metasomatism, however, is either unreported or absent from many areas, including many shallow plutons. The distribution of large metasomatic aureoles reflects average depths of emplacement (cf. Fig. 5-2 and Fig. 5-4C and D). Thus the scarcity of metasomatic aureoles in Idaho and Washington is consistent with the deeper exposures there. Another possibility is that large-scale metasomatism, as occurred around the northern end of the Idaho batholith (Fleck and Criss, 1985; Hietanen, 1962), has rarely been recognized because the changes are gradual or cryptic.

The relative extents of metasomatic aureoles vary from $>10^2$ to 10^{-5}. Also, the relative extent of metasomatism is generally much larger at shallow than deep levels (Fig. 5-4D). Where metasomatism is prominent, extents vary inversely with igneous exposure area, consistent with the idea that metasomatizing fluids concentrate near the tops of plutons regardless of fluid origin (magmatic, meteoric, connate). The most extensive shallow metasomatism (e.g., hydrothermal alteration in volcanic rocks), where commonly only isotopes, volatiles, and alkalies are transferred, is dominated by fracture-related exchange. These rocks need not have experienced high fluid flow even though large volumes may be pervasively altered (cf. H. P Taylor, 1977). Although relative extents of skarns vary widely, about two-thirds are in the range $10^{-0.5}$–$10^{-2.5}$. The largest skarns usually are associated with fairly small exposures of igneous rocks and commonly comprise major mineral deposits (e.g., Arizona Cu skarns, Einaudi, 1982). They almost certainly formed near the top of the magmatic systems. This is nicely illustrated by F-rich and related skarns associated with Late Cretaceous two-mica granites in the Great Basin (Barton, 1987).

Mineralization is a surprisingly common, important variety of metasomatism, but one which is commonly neglected in treatments of contact metamorphism. Mineralization is reported with more than half of the 1600+ locations in CONTACT. Of these, about 400 are significant deposits. Figure 5-5 shows the distribution of Mesozoic igneous-affiliated Cu and W deposits, including porphyry Cu deposits, Cu skarns, W skarns, and some other types. Excluded are lithophile-element skarns and greisens, and intrusion-hosted Mo-Cu deposits, even though they can contain significant Cu and W. Two features are of particular interest: First, mineralization is a strong function of emplacement level. Second, the overall distribution of mineralization reflects the average depths of exposure.

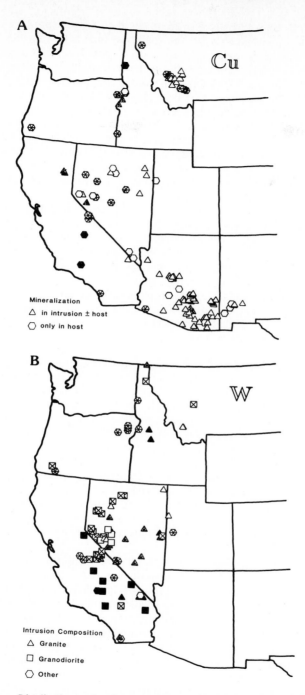

FIG. 5-5.   Distribution and estimated depths of Mesozoic Cu and W mineralization. Open symbols indicate depth of emplacement <4 km; starred symbols indicate 4–8 km; solid symbols indicate >8 km. Compare Fig. 5-2. (A) Cu mineralization. (B) W mineralization.

Virtually all copper deposits formed at less than 8 km depth (Fig. 5-A). Moreover, most economically important deposits are shallower than 4 km, including nearly all the intrusion-hosted mineralization (i.e., porphyry Cu) and related skarns. Between 4 and 8 km, lesser mineralization occurs, mostly in the form of skarns and replacement bodies. General depth relationships of this type have been suggested for some time (see Titley and Beane, 1981), but they have not been demonstrated on a regional scale. Instead, much has been made of the special geochemical or petrographic characteristics of Cu-rich provinces such as Arizona-New Mexico (cf. Titley and Beane, 1981). Close correspondence between Cu deposits and average emplacement depth is obvious from comparison of Fig. 5-5A and Fig. 5-2. Thus it seems clear that the modern distribution of Mesozoic Cu deposits is not so much a geological oddity as it is a consequence of unexceptional differences in geological history and the preservation potential of the deposits (cf. Mitchell and Garson, 1981; Sawkins, 1984).

In contrast to Cu mineralization, W mineralization in the western United States occurs predominantly in skarns. Although forming over a wider depth range, these skarns are best developed between 4 and 8 km depth. This is compatible with the suggestion of Newberry and Einaudi (1981) that some W skarns represent the deeper parts of magmatic systems that might have produced Cu mineralization. Many W skarns, however, are clearly related to the tops of deeper(?), granitic plutons commonly associated with porhyry Mo-Cu mineralization (E. Seedorff, personal communication, 1985). The regional distribution of W, like that of Cu, clearly reflects the average level of exposure. Deep exposures in Idaho and Washington and shallow ones in Arizona (where deep carbonates also happen to be absent) lack W deposits.

Intrusion composition The character of contact metamorphism varies in subtle ways with intrusion composition. Changes in metasomatic effects are most prominent, but systematic changes in the character of thermal effects and deformation can also be inferred. Metasomatism is better developed with felsic plutons. The most intense local hydrothermal alteration occurs with intermediate to felsic intrusions that produced reactive chlorine- or sulfur-rich fluids (e.g., porphyry Cu-type systems, Lowell and Guilbert, 1979). In shallow systems, however, where most widespread alteration results from circulation of nonmagmatic waters, metasomatic effects occur with all compositions (e.g., Fe, Na metasomatism with the Jurassic mafic intrusions in the Great Basin). At greater depths, intense, widespread metasomatism is restricted to the vicinity of felsic plutons. Deep (>4-6 km) gabbros, diorites, and most tonalites lack metasomatic aureoles. This is consistent with greater abundance of metasomatic rocks in the inner, more felsic portion of the arc (Fig. 5-4), although other factors, such as depth (above) and host composition (below), are also important.

Correlations among alteration types, metals, and compositions of associated intrusions have been documented by many investigators (e.g., see Keith, 1983; Mitchell and Garson, 1981). Worldwide, mafic rocks contain siderophile element enrichments of many types, but mineralization is rare with Mesozoic western U.S. mafic intrusions, with the exception of Fe skarns and veins. Similarly, mineralization associated with Mesozoic peraluminous granitoids appears limited to lithophile element and minor precious metal occurrences. The principal Mesozoic mineral deposits occur with metaluminous granitoids. They include porphyry Cu deposits and related oxidized skarns, Pb-Zn replacement deposits sometimes associated with porphyry-type alteration, shallow precious-metal base-metal veins in zones of extensive hydrothermal alteration, W skarn deposits associated with relatively unaltered intrusions, and polymetallic W skarns, commonly occurring with Mo-Cu porphyry mineralization (see Einaudi et al., 1981; Meinert et al., 1980).

Many variations occur within these categories. Tungsten skarns, for example, almost always occur with granodiorites and granites, at least where those rock types are abundant (Fig. 5-5B). Only on the west side of the Mesozoic magmatic arc are many W deposits associated with lower-silica or tonalitic intrusions and those deposits tend to be small. Even more restricted in space, greisen-type lithophile-element mineralization lies inboard of the batholiths along the locus of peraluminous granitoids (Barton, 1987; C. F. Miller and Bradfish, 1980).

Thermal aureoles appear to be more prominent adjacent to more mafic intrusions, except at shallow levels, where hydrothermal metamorphism dominates. Other factors being equal, granites, particularly forcefully emplaced two-mica granites, tend to have limited aureoles, whereas the intermediate and mafic plutons tend to have large aureoles. In addition, penetrative deformation is generally better developed with granitic plutons. For example, the two-mica granite of Papoose Flat has a narrow ($\leqslant$200 m) strongly foliated aureole of which only the inner part of hornblende hornfels facies (Sylvester et al., 1978). In contrast, the trondhjemitic Caribou Mountain and gabbroic Heather Lake plutons, which are of similar size and were emplaced at similar depths to Papoose Flat, have 450-m and 1-km-wide hornblende hornfels aureoles (Bero, 1980; G. A. Davis, 1963). Miller et al. (Chapter 24, this volume), describe another example of eastern Nevada, where Jurassic composite hornblende diorite to granite plutons have extensive dynamothermal aureoles with inner hornfelsic zones in miogeoclinal sedimentary rocks. Cretaceous two-mica granites intruded into the same stratigraphy at the same depth have narrow, lower-grade aureoles commonly with a contact parallel foliation.

Host composition The most obvious variations in contact metamorphism reflect the composition of the host rocks; most of these were discussed under "Diversity." The distribution of metasomatic and, to some degree, thermal aureoles in the western United States (Fig. 5-4C and D) can be rationalized by the distribution of appropriate host rocks. Concentrations of metasomatic aureoles in the southwest, Great Basin, and northwest reflect local abundance of reactive host rocks. These include the widely preserved Mesozoic volcanic rocks of the southwest, and the abundance of carbonate rocks of the miogeocline and the Belt basin in the other areas. Elsewhere, less reactive rock types comprise the majority of the hosts, restricting the widespread development of metasomatic aureoles. The scarcity of skarn deposits in eugeoclinal terranes reflects the paucity of carbonate, in addition to differences in igneous compositions and levels of exposure.

DISCUSSION

The systematic variations in contact metamorphism outlined so far can be qualitatively interpreted by a fairly simple model for contact metamorphism about individual plutons. From an analysis of the effects of metamorphism around a single pluton it is possible to evaluate the relationship between contact and regional metamorphism in terms of the effect of increasing pluton densities. In turn, this has implications for inferring the Mesozoic thermal structure of the western United States from the metamorphic record.

Contact Metamorphism around Single Plutons

Contact metamorphism results from the complete or partial dissipation of thermal, chemical, or mechanical disequilibria either caused or activated by magmatism. Contrasts in, and magnitudes of the physical and chemical properties of the intrusion and host govern the mechanisms by which this dissipation takes place. Geometric factors, including the size and shape of the pluton, and the distribution of rock types in the host, control the distribution and abundance of metamorphism. Straightforward physical and chemical reasoning in conjunction with geological information can be used to rationalize the type and extent of contact metamorphism in many areas. Because so many combinations of parameters are possible, and because these parameters are complexly coupled, any unified model would be exceedingly complicated. Fortunately, most of the effects can be separated for purposes of discussion, even though they intermingle in many aureoles.

Thermal contact metamorphism is influenced by many factors, including the thermal masses in a system, the initial and progressive temperature distributions, the mechanisms by which the heat is distributed, and the chemical and physical characteristics and distribution of the host rocks. Discussed here are qualitative aspects of these controls. For more thorough treatments of various theoretical aspects of thermal contact metamorphism, see the reviews by Turner (1981), Jaeger (1964, 1968), Lovering (1935, 1955), Parmentier and Schedl (1981), and Norton and Knight (1977). Their work forms the basis for the following interpretations.

Pure conduction is worth examining first, because in many ways it is the least complicated. In the simplest case, the extent of a conductive thermal aureole is linearly proportional to the size of the intrusion (Jaeger, 1968). Strictly, this extent is the distance at which the temperature (at a particular time, or, the maximum temperature) is a specified fraction of the difference between the initial temperatures of the intrusion and the host. The temperature profile is a more complicated, exponentially decreasing function of distance from the intrusion. Two immediate complications arise from the latent heat of fusion of the intrusion and convective movement within the intrusion. The higher liquidus temperatures and heats of fusion of mafic, when compared to felsic magmas (Carmichael *et al.*, 1974), are alone sufficient to cause observed differences in aureole extents. For example, a simplified calculation indicates that the 450°C isotherm would extend two-thirds times the intrusion width away from a gabbroic dike emplaced at 1200°C into rock originally at 200°C, whereas the distance would only be about one-sixth the width of a granitic dike intruded at 750°C. The principal effect of magmatic convection is to raise the contact temperature. Because convection is much more likely in intermediate to mafic magmas than it is in felsic magmas (Carmichael *et al.*, 1974), convection could increase the relative extent of mafic high-T thermal aureoles, but should not greatly change the overall width. The observed range of relative extents for most Mesozoic high-temperature thermal aureoles (3-30%) is consistent with conductive heat transport; however, advective terms (below) may also be important.

The geometry of the contact is important; where contacts dip shallowly the apparent thickness of an aureole is much greater than the true thickness. For many large plutons, contacts appear to be fairly steep (>45°), but this is clearly not the case near the tops (and bottoms?) of many intrusive systems, or possibly in areas with later tilting, where aureoles can have large relative extents. For example, many of the largest aureoles (especially in Arizona and Nevada) occur with dikes or small stocks; they almost certainly represent apical portions of larger masses. Hietanen (1984) attributes the unexpectedly large extent of metamorphism around the north end of the Idaho batholith to a shallowly dipping contact.

Three factors limit the size of observed aureoles: the intial temperature difference, the presence of background metamorphism, and the times available at elevated temperature and pressure for reaction and strain accumulation. Progressively deeper emplacement levels ultimately preclude low-T aureoles because of the increase in background temperatures with depth and the increasing probabilty of earlier metamorphism. This cannot be the only reason, however, because many shallow intrusions lack obvious low-T aureoles. Low background temperatures also limit aureole extent by reducing the part of the total temperature interval over which observable reactions take place. For example, consider two identical granitic plutons intruded into impure carbonate rocks, at 250°C for the first and 100°C for the other. Because decarbonation usually beings at 350–450°C (Kerrick, 1974), the extent of the aureole as measured by the outermost decarbonation reaction will be about four times greater for the pluton emplaced into 250°C rocks. Conversely, metamorphism will be more extensive the lower the threshold temperature for reaction. For example, consider interbedded shale and carbonate, where initial reactions plausibly take place (kinetics aside) at about 300°C and 400°C, respectively. For pure conduction, the 300°C maximum isotherm will be two to three times more extensive thatn the 400°C isotherm for a granite intruded into rock at 100°C. In very reactive rocks, such as volcanics with their highly metastable assemblages, reactions could take place at much lower temperatures, especially in the presence of water. Hydrothermal metamorphism takes place at temperatures well below 300°C in modern geothermal systems (Henley and Ellis, 1983). This kinetic effect helps account for the greater extent of metamorphism in volcanic rocks relative to many sedimentary and metamorphic rocks (e.g., in Arizona).

In contrast to the linearity of the distance scale for conduction, the time scale is inversely proportional to the square of the characteristic distance. That is, the larger the intrusion, the slower will be the heating and cooling at the same relative distance. This disproportionate narrowness of aureoles around many small intrusions probably is due to this effect (Turner, 1981, p. 21-22). Time scales are also extremely important in the generation of textures. With decreasing temperature, the exponential decline in growth, diffusion, and probably nucleation rates (see Lasaga, 1981; Walther and Wood, 1984) is not fully compensated by increasing time for reaction. In other words, below a certain temperature, rocks will heat and cool more rapidly than reactions can take place. This temperature limit is a function of many ill-known kinetic factors, but increasing the time of heating and cooling will lower the minimum temperature for reaction, thus increasing the relative width of the aureole. For short times, as around very shallow or small intrusions, reactions will not take place. In an intermediate time range, where heating and cooling are relatively rapid, granoblastic textures will develop due to high nucleation and growth rates relative to strain rates. In deep aureoles, especially those around large plutons, the time rate of change of temperature may be sufficiently slow that significant strain may be accumulated in the rock during metamorphism. In some cases, strain is induced by the intrusion itself during forceful emplacement. In other cases, because of the rapid heating and cooling rates near an intrusion, hornfelsic textures can develop in the inner aureole while anisotropic fabrics develop elsewhere.

Another reason for fairly sharp outer limits to recognizable contact metamorphism is the discontinuous initiation of many metamorphic reactions. If new reactions are spaced widely in temperature, the kinetic barrier will probably be overcome between reactions; consequently, only the next higher temperature reaction will take place. Continuous processes, such as simple coarsening or continuous reactions, would not show discontinuities. Rate factors, and the limited availability of some components such as volatiles, are of particular importance in inhibiting retrograde reaction.

Taken together, the kinetic and temperature effects predict that aureoles first would become larger with increasing depth and intrusion size, but then smaller. The increase in extent is seen at shallow and moderate depths in areas with fairly low pluton densities (e.g., the Great Basin). At greater depths (higher temperatures), the paucity of observed thermal aureoles results from small temperature gradients between magma and host, longer times and appropriate conditions for development of "regional" textures, and the increasing probability of earlier metamorphism.

These ideas are summarized in Fig. 5-6, which shows the results of a simple finite-difference thermal model for a sheetlike pluton intruding the crust with a uniform geothermal gradient of 30°C/km. For simplicity, the magma temperature was taken as 1000°C, which includes an approximation for the effect of latent heat (Jaeger, 1968).

The time required to dissipate 90% of the thermal anomaly is approximately 4 m.y. at 20 km depth but only about 2 m.y. near the top of the pluton. Note that the thermal anomaly remains extensive even at 20 km depth (a more realistic treatment of the heat of crystallization would lower the temperatures near the contact at depth, an effect opposed by convection within the magma or loss of magmatic fluids, but should not greatly decrease the width). The dotted area in Fig. 5-6 indicates a schematic limit to obvious contact metamorphism based on observed variations in extents and a rudimentary analysis of kinetic controls on texture development. The boundary between greenschist (low-T) and amphibolite grade (high-T) zones is near the 500°C isotherm (Turner 1981).

As can be seen, although intrusion-induced greenschist-grade metamorphism persists to considerable depths, low-T "aureoles" will be limited to shallower levels. Similarly, upgrading of high-T background metamorphic assemblages can take place beyond the observed aureole. At high levels, the cutoff will reflect lack of reaction. At deeper levels, metamorphism will take place to greater distances, but the combination of longer times and higher temperatures will facilitate the development of coarser and foliated textures (due to longer times at a constant strain rate), reducing the extent of the texturally distinct zone, and thus the apparent extent of the aureole.

The limiting effect of earlier metamorphism can also be qualitatively interpreted on Fig. 5-6. Consider an earlier metamorphic event that took place with a metamorphic gradient of about 40°C/km, rather than the 30°C/km value existing at the time of intrusion. The extent of heating above the old gradient would be restricted to a zone approximately limited by the intersection of the 500°C boundary with the 350°C base (dotted line) and the 650°C boundary with the 500°C base. Note that this is essentially the same as the schematic limits to "obvious" contact metamorphism in the "unmetamorphosed" case. Thus it may be difficult to distinguish the difference between contact metamorphism superimposed on earlier regional metamorphism and extensive metamorphic maxima due

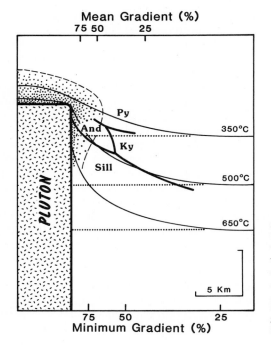

FIG. 5-6. Maximum temperatures achieved about a slab-shaped intrusion. The solid lines show the maximum temperatures obtained (which are not achieved simultaneously), the approximate limit of obvious contact metamorphism (see the text), the Al_2SiO_5 phase relations (And, andalusite; Ky, kyanite; Sill, sillimanite; Py, pyrophyllite), and positions for mean metamorphic gradients. The dotted lines indicate the background temperatures.

primarily to plutonism. In many cases, the principal evidence for earlier metamorphism is the development of hornfelsic textures on earlier fabrics. However, in a dynamic system, it seems plausible that hornfels might overprint earlier fabrics formed in the same overall event. If an intrusion is forcefully emplaced, early strain reflected by fabrics in lower-grade assemblages might be annealed or upgraded statically once the body is substantially in place.

Advection of aqueous fluids derived either from the intrusions or from the host can have a major effect on the development of thermal metamorphism (e.g., see Norton and Knight, 1977; Parmentier and Schedl, 1981). Loss of fluids from an intrusion could also raise the temperature near the contact at late stages (Hori, 1964), allowing an apparently static overprinting, but will not greatly change the overall dimensions of an aureole. Isotopic studies by Shieh and Taylor (1969a, b) suggest that such flow, if present, is not pervasive in some areas. Because most magmas saturate with fluid relatively late during crystallization, magmatic fluids will be most important in causing metasomatism and retrograde reactions. Thus the temperature near the contact will have passed through its maximum before exsolution of the bulk of the magmatic fluids. This role is compatible with that inferred for skarns (Einaudi *et al.*, 1981).

In contrast to movement of magmatic fluids, convection of connate or meteoric fluids can transfer heat on a large scale, as abundantly demonstrated by stable isotope studies. The principal effects of convection are to concentrate thermal anomalies near the top of intrusions and to extend the low-temperature metamorphic regimes. Fluid flow through intrusions enhances this upward redistribution of heat (Norton, 1982; Parmentier and Schedl, 1981). Permeabilities change with rock type, depth, and reaction; thus these factors markedly affect fluid and heat fluxes. Permeability generally decreases with depth due to a combination of higher-grade, less porous rocks, and confining pressure closing off connectivity. Fractures are easier to generate and maintain at shallow levels where rocks are brittle and tremendous energy is released by crystallization of hydrous magmas (Burnham and Ohmoto, 1980). With increasing depth, the change in volume during crystallization of a hydrous magma can be more readily accommodated by ductile or elastic rather than brittle behavior. These factors limit convective flow in high-grade or deep rocks, in comparison with shallow sedimentary and volcanic rocks. Primary porosity in many volcanic rocks is a major factor in localizing fluid flow for their metamorphism (Norton and Knapp, 1977). The extensive metamorphism in Mesozoic volcanics as compared to underlying rocks is probably a consequence both of their higher reactivity and higher permeability.

At depths greater than ~8 km, most reported aureoles occur in carbonate rocks. This can be attributed not only to the highly reactive nature of carbonates, but also to enhanced fluid flow by preferential development of secondary porosity during metamorphism (Rumble and Spear, 1983). For example, stable-isotope studies by Nabelek *et al* (1984) demonstrated that impure carbonate rocks around the Notch Peak pluton had much higher fluid to rock ratios than adjacent clean carbonate rocks which did not undergo decarbonation reactions. Although heat transport by fluids has been documented at moderate depths (Ferry, 1983), it is unclear how much this mechanism will change the overall thermal distribution. Fluid convection should strongly affect heat flow only in rather permeable (shallow) rocks; elsewhere the main effect will be metasomatic or catalytic.

Fluid infiltration is the dominant mechanism for metasomatism. Metasomatism also occurs by diffusion between contrasting lithologies, particularly carbonate rocks and intrusions or clastic rocks (forming bimetasomatic zones); however, the parabolic time dependence of diffusion distances precludes diffusional metasomatism over more than a few meters during contact metamorphic events. Many factors control the intensity and type of infiltration metasomatism. Chemical disequilibrium is a prerequisite for reaction, although the driving forces for infiltration are gradients in temperature and pressure. Both magmatic fluids and externally derived fluids cause contact metasomatism. At low pressure, saline magmatic fluids are effective carriers of metals, alkali, and alkaline earths. For several reasons, magmatic fluids become less saline with increasing pressure, but can contain larger quantities of other components such as silica. Systematic variations in mineralization with depth, in particular, and metasomatism, in general, reflect these changes. Mineralization is best developed at shallow levels where metals are most effectively transported. Similarly, the relatively intense metasomatism in this setting reflects the strong chemical and physical gradients between the magmatic fluids and the host. Not all magmas, however, produce such fluids; many appear to be dry (especially mafic magmas), whereas others either lacked the chlorine or evolved differently in their late stages. At deeper levels, metasomatism is less pronounced, in part reflecting smaller physical and chemical contrasts, and partly because of the reduced efficacy of transport of unusual components. A crude estimate of expected extents of skarns can be made from solubility data on silica in water, the water contents of magmas, and the assumption that most of the silica is removed from solution on encountering carbonate. Using 5 wt % water, 1 wt % silica in the water, the expected extent for skarn (containing 33% silica) equally distributed around the top half of a spherical pluton is 1% of the pluton radius. This is in approximate agreement with the range found in CONTACT (two-thirds between 0.3 and 30%) considering the vagaries of fluid flow and host distribution (the extent data in CONTACT are maximum values).

The concentration of metasomatic effects near the tops of intrusions results from the buoyancy of aqueous fluids relative to the surrounding magma or rock. Overpressures due to crystallization may drive magmatic fluids outward in many directions, not necessarily upward. In a convective system metasomatism is also concentrated on the cooling part of the fluid path because the highest integrated flux occurs there and the prograde solubility of many materials and decreasing pressure favor precipitation (J. W. Johnson and Norton, 1985). On the warming part of the fluid path, prograde solubility causes leaching over a large volume, with less conspicuous consequences. Exchange reactions, particularly isotopic exchanges, can occur throughout. By promoting fluid flow, higher permeabilities at shallow levels concentrate metasomatism there. In contrast, deep fluids, even though they can contain much dissolved material, tend to be neither physically concentrated nor pass through environments of rapidly changing physical conditions.

The variations in pluton emplacement style and the physical metamorphism caused thereby is a consequence of the thermal regime. At shallow levels where rocks are brittle, strain will be taken up within the intrusion and by brittle failure. Only where heating adjacent to the contact is significant prior to final emplacement will significant strain be accumulated in the host rocks. With increasing depth, however, the higher temperatures in the host rocks increases their ductility, thus reducing the ductility contrast. As it is primarily the ductility contrast that controls the relative accumulation of strain (Ramberg, 1970), the most deformed host rocks will be those that are most ductile (e.g., carbonates) adjacent to the most viscous plutons (e.g., felsic granitoids). This relationship is observed, for example, in the Great Basin. There two-mica granite plutons, which were the coolest and are among the most silica-rich, have highly deformed aureoles, especially in carbonate rocks. Other granitoids have less deformed or undeformed aureoles; more mafic intrustions rarely are described as having been forcefully emplaced.

In summary, metamorphism about a single pluton represents the time-integrated distribution of induced temperature and mass anomalies. Because physical and chemical contrasts between intrusions and host rocks decrease with depth, in the simplest case the extents of contact metamorphism are expected to decrease with depth. In reality, however, the time scale of the dissipative processes in an intrusive environment overlaps with time scales for magma-independant processes, thus the ultimate distribution is more complex. These interactions commonly obscure the effect of intrusion, resulting in readily distinguishable contact metamorphism being much more limited than the actual extent of igneous effects. Thermal metamorphism is generally more widespread than metasomatism, because intrusion introduces a large thermal pulse that must decay. In contrast, the transport constants for diffusive mass transfer preclude extensive exchange. It is only through coupling of mass flow driven by physical gradients with the chemical gradients resulting from the juxtaposition of materials of differing compositions that large-scale metasomatism can take place. Even with advection of fluids, the net effect is dominated by heat transfer, because fluids in the upper crust generally transfer heat more efficiently than solutes.

Relationship between Mesozoic Regional and Contact Metamorphism

Several lines of evidence indicate that much of the regionally extensive Mesozoic low-pressure facies series metamorphism and probably some of the medium-pressure facies metamorphism in the western United States formed as a direct consequence of magmatism. This metamorphism is a composite of multiple events; the present metamorphic-grade distribution may little reflect thermal surfaces extant at any given time. Constraints on the timing, although sparse, indicate that plutonism was often synchronous with metamorphism. Many geological and geochemical observations either point to, or are at least compatible with, igneous-related metamorphism. Furthermore, simple thermal calculations demonstrate that regionally extensive metamorphism is expected with the observed pluton densities, whereas alternative mechanisms appear ineffective for producing much of the observed metamorphism. Finally, heat flow measurements in modern arcs indicate that maximum geothermal gradients generally lie inboard of magmatic axes, not within them. Each of these lines of evidence is discussed below.

P-T-time relationships Consideration of the effect of superposing typical aureoles at observed pluton densities indicates that a regional metamorphic belt easily could be constructed by many "unrelated, minor" events, each associated with a separate, relatively short-lived intrusive center. A major consequence of this analysis is that an apparently simple thermal culmination indicated by metamorphism may be an artifact obtained by coalescence of many aureoles. In areas metamorphosed at less than 10–15 km depth, regional thermal surfaces may never have resembled the metamorphic surfaces observed today. The true thermal culmination could well have been significantly behind the major arc, compatible with heat flow in modern arcs (*cf.* Sugimura and Uyeda, 1973) and geophysical models of arc thermal structure (e.g., Andrews and Sleep, 1974). Previous interpretations of low-*P* metamorphic belts imply one or perhaps two relatively smooth thermal culminations (*cf.* Armstrong, 1982; Ernst, 1974; Oxburgh and Turcotte, 1971). Proposed here is something quite different for the magmatic arc — spotty thermal maxima associated with active magmatism. At any given time, a few areas undergoing active intrusion and metamorphism would have been separated by large areas near minimum (background) temperatures.

Figure 5-7 illustrates aspects of this idea. In Fig. 5-7A, emplacement of a pluton transiently increases the local thermal gradient, while permanently elevating the more regional metamorphic grade. Following decay of the first thermal anomaly, a second intrusion repeats the process (Fig. 5-7B). The result is a regionally elevated metamorphic field that does not reflect the temperature field at *any* time during the metamorphic

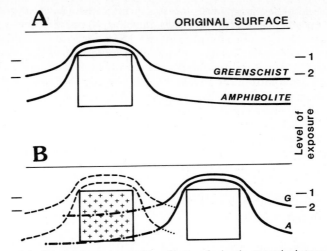

FIG. 5-7. Schematic metamorphic effects of closely spaced plutons. (A) Intrusion and associated metamorphism with first pluton. (B) Intrusion and metamorphism associated with a younger pluton. Heavy lines indicate approximate isothermal surfaces. Metamorphic upgrading takes place only where they are solid. Fine lines indicate metamorphic boundaries from (A). Note the development of "regional" greenschist grade metamorphism at level 2, whereas the aureoles are still discrete at level 1.

history. The metamorphic gradient is elevated above the background gradient, regardless of the magnitude of the background gradient; thus intrusions will *always* produce the highest-grade metamorphism in any thermal episode. Only earlier or later regional metamorphic events can lead to higher grades. Where plutons are widely spaced, thermal effects will not overlap. In such areas, the gradient imposed by other factors will be dominant and may well lead to independent regional metamorphism (e.g., parts of the eastern Cordilleran metamorphic belt, disussed below).

A necessary attribute of igneous-related metamorphism is contemporaneity of magmatism and metamorphism. This is difficult to establish unequivocally, but much circumstantial and a little direct evidence exists. Direct dating of low-pressure metamorphism is rare, but available dates in the western United States generally overlap (within 15 m.y.) with ages of magmatism (e.g., in the western Great Basin, Sierra Nevada). An additional demonstration of regional heating is the complete resetting of most biotite and partial resetting of many hornblende K-Ar dates in the Late Cretaceous along the Cretaceous magmatic axis south of Idaho (Evernden and Kistler, 1973; *cf.* Armstrong's 1986, compilation for Canada). Resetting of biotite requires minimum temperatures of ~280°C (T. M. Harrison and McDougall, 1980) near the lower boundary of the greenschist facies (Turner, 1981); hornblende resetting requires ~500°C. Another example is Cretaceous resetting of Rb-Sr whole-rock systematics in Triassic-Jurassic volcanic rocks of the Mineral King pendant contemporaneous with emplacement of the surrounding plutons (C. Busby-Spera, personal communication, 1985). This isotopic homogenization required extensive fluid circulation in addition to elevated temperatures. Regional metamorphism of Cretaceous volcanic and sedimentary rocks in western Nevada and the Sierra Nevada could plausibly have taken place only in the Late Cretaceous, concurrent with local putonism. Volcanic and volcanoclastic rocks in the western Hayfork terrane of the

Klamath Mountains, which were metamorphosed to grades ranging from prehnite-pumpellyite to amphibolite facies during the Siskiyou low-pressure event, are only a few million years older than the major plutons of the Klamaths (Coleman *et al.*, Chapter 39, this volume; Fahan, 1982). Although Coleman *et al.* interpret the Siskiyou event as pre-plutonic, perhaps related to an intra-arc rift, we see no compelling reason that metamorphism is not part of the same magmatic event as interpreted by Fahan (1982). Considerable heat must have been transferred during volcanism. Furthermore, the crust could not have completely cooled even in the maximum possible time between metamorphism and plutonism. The present plutons and their overprinting aureoles may simply record only the last stage of the development of a Middle Jurassic arc.

The regional association between high pluton densities and low-pressure facies series metamorphism (Fig. 5-3) strongly supports a cause-and-effect relationship. This is further supported by described "regional" (*not* "contact") metamorphis gradients toward large intrusive masses (e.g., northern Klamath Mountains, Kays, 1970; Panamint Range, Labotka and Albee, Chapter 26, this volume; Mount Stuart batholith, Evans and Berti, 1986; Plummer, 1969). Similar correlations with medium-pressure facies series metamorphism are commonly present, for example, as around the Idaho batholith (Hietanen, 1984). In other areas with medium-pressure facies series metamorphism, such as the eastern Great Basin, much of the lower-grade metamorphism appears unrelated to magmatic centers, although those centers are generally metamorphic maxima.

The paucity of penetrative deformation in some areas of greenschist and higher grade regional metamorphism is another argument for igneous-related metamorphism. Only advective processes for regional metamorphism are compatible with this observation. Moreover, the increase in penetrative deformation toward many plutons suggests a locally dominant role for plutonism in fabric development. Overprinting of hornfelsic textures on earlier anisotropic fabrics presents a serious problem of interpretation in some areas (e.g., Klamath Mountains, Sierra Nevada). For the other reasons outlined here, we think that peak regional grades were probably achieved during magmatic events, but we admit that in some cases the conventional interpretation of low-pressure metamorphism followed by plutonism associated with limited metamorphism is appealing. As discussed earlier, the secular evolution of textures might be one in which hornfelses develop relatively late during a prolonged magmatic/metamorphic history. E. L. Miller *et al.* (Chapter 24, this volume) describe late hornfelsic textures adjacent to plutons which have "localized deformation and upgraded the regional dynamothermal metamorphism to amphibolite facies." Another possibility is that static contact metamorphism could upgrade anisotropic fabrics formed during earlier, lower-grade events (G. Oertel, personal communication, 1986). In areas containing multiple plutons, a regional penetrative fabric might develop by continued strain during sustained heating by many plutons (Tobisch *et al.*, 1986). Consequently only the latest plutons, which will be the best defined and best preserved, will have hornfelsic aureoles overprinting earlier fabrics.

Widespread hydrothermal metamorphism provides another indication of regional igneous-related metamorphism. These changes are obvious in shallow Mesozoic volcanic sequences, where they are clearly related to magmatism. Volcanic rocks metamorphosed at greater depths are more ambiguous, although it seems only reasonable to attribute the addition of water, retrograde mineral reactions, and chemical and isotopic exchanges to regional contact metamorphism. The distribution and scale of metamorphism in these areas is consistent with overlapping hydrothermal aureoles (Norton and Knight, 1977; Parmentier and Schedl, 1981) given the intrusion abundances (see the discussion of thermal overlap below). Regional chemical and isotopic exchanges documented around the north end of the Bitterroot lobe (Fleck and Criss, 1985; Hietanen, 1962) are a further indication of a common process, although it is possible that they represent different aspects of a deeper process rather than the metasomatism resulting directly from the magmatism.

A model for regional igneous-related metamorphism The overlap idea is further developed in Figs. 5-6 and 5-8, where mean regional metamorphic gradients for various pluton densities are estimated. The mean metamorphic gradient is the gradient in degrees per kilometer which half the rocks in the region are above and half are below. This can be estimated in several ways. The method adopted here is to assume that each intrusion cools completely before the emplacement of the next, and that the plutons are evenly spaced in their final array. The mean gradients for planar plutons of half-width r and overall density $d(\%)$ is then the vertical component of the single pluton gradient at $r(100/d-1)$ from the edge of the intrusion (see Fig. 5-6). Model metamorphic gradients are plotted in Fig. 5-8 for various pluton densities with an assumed linear background geothermal gradient of 30°C/km. Pluton densities greater than about 25% can have a major effect on the regional gradient and can easily move "regional" metamorphic grades from medium-pressure to low-pressure facies series conditions (*cf.* the aluminum-silicate stability fields in Fig. 5-8 and Fig. 5-6).

This is generally a conservative model. It assumes a reasonable background gradient, single emplacement events for each pluton, and complete cooling between intrusive events. Higher background gradients could result from higher mantle heat flow or incomplete cooling between events, or both. Additionally, considerable heat can be transported through an intrusive system if volcanism takes place; this could considerably extend the thermal aureole. On the other hand, this model assumes heating to the base of the crust and neglects the effects of hydrothermal circulation. The means by which magmas are transported to their final resting places are poorly known (e.g., see Bateman, 1984; Marsh, 1982; Spera, 1980), but regardless of the specific mechanism, a significant amount of heat must be shared with deeper rocks. This probably requires multiple intrusions at least in the early stages of magma transport (*cf.* Marsh, 1982), although the amount of heat transferred is uncertain. As discussed above, hydrothermal advection of heat is qualitatively similar to thermal conduction, and thus unlikely to greatly reduce overall regional extents. Moreover, intrusion-driven fluid circulation is probably only important in the upper part of the crust. At greater depths, fluid advection caused by plutons is probably restricted to fluids either exsolved from magmas and expelled into the country rocks or

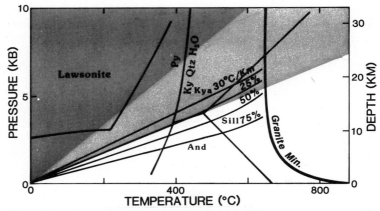

FIG. 5-8. Approximate low-, medium-, and high-pressure metamorphic facies series fields and mean elevated metamorphic gradients for the thermal model shown in Fig. 5-6. See the text for discussion.

produced from the host rocks themselves by devolatilization reactions. Movement of such fluids have a second-order effect on the overall heat flow because of their small relative volumes.

The spacing of plutons and their characteristic dimensions are also key factors. Regional heating is maximized when small plutons are evenly spaced. The thermal effects of clustered plutons are similar to larger single plutons up to a certain size if they are emplaced fairly close in time to one another. Pluton geometry is another important factor that has not been quantitatively evaluated. The general effect is that the larger the surface to volume ratio of the pluton, the smaller will be the increase in the mean metamorphic gradient. In the extreme case of spherical intrusions, this appears to be as much as a factor of nearly 2. Intrusion size becomes important when the width of the pluton approaches or exceeds the depth of emplacement. In this range, disproportionately large amounts of heat are lost through the roof; thus the flux out the sides is smaller than might otherwise be indicated (*cf.* Jaeger, 1968). Of course, in an area over a large body (or possibly beneath; *cf.* Hamilton and Myers, 1974), this would be reflected by high gradients across a much larger area than might be indicated by simple lateral heat flow from the observed intrusions (e.g., Hietanen's suggestion for Bitterroot lobe metamorphism). The rapid dropoff in grade on the western side of the Mesozoic batholiths reflects this effect — it is not possible to produce thermal aureoles much farther away than the batholith thickness. This is the problem with attributing metamorphism around Idaho batholith to heat flow from the batholith. In places like Idaho, and perhaps many metamorphic core complexes (Crittenden *et al.,* 1980), regional metamorphism probably reflects a deep source. However in many areas, such as the Snake, Schell, and Deep Creek Ranges, documented metamorphic highs are too localized and have gradients too steep to be associated with a distant (i.e., subcrustal) source.

Alternative mechanisms for producing regional low-pressure facies series metamorphism seem unsatisfactory for the Mesozoic western United States. These include homogeneous crustal thickening, inhomogenous crustal thickening, rifting, induced mantle flow, and aqueous fluid transport. Large structures (e.g., major nappes) with associated low-pressure regional metamorphism are absent, although they would be expected if regional metamorphism resulted from crustal thickening (*cf.* England and Thompson, 1984). Missing too are the tremendous volumes of sediments that would be expected to have been eroded off the top of a section of crust rising rapidly enough to move into the low-pressure field (England and Thompson, 1984). Structures roughly synchronous with metamorphism, such as the late Mesozoic thrust faults and nappes of the central and eastern Great Basin, are today exposed only at upper to mid-crustal levels. Compelling stratigraphic and structural arguments indicate that these rocks were never deeply ($>$20 km) buried by overriding structures (Armstrong, 1968; Gans and E. L. Miller, 1983) even though the whole crust in this area may have been quite thick during the late Mesozoic (Coney and Harms, 1984). Therefore, low-pressure regional metamorphism during thermal relaxation following crustal thickening seems implausible. On the west side of the batholiths and, to a certain extent, within them, crustal thickening by burial or structural thickening does appear to have been important for generation of prebatholithic moderate- to high-pressure facies series regional metamorphism.

Rifting can also produce high heat flow, possibly leading to low-pressure facies series metamorphism (Wickham and Oxburgh, 1985); however, rifting appears incompatible with geological observations for most of the Mesozoic. Rifting might explain some Late Jurassic metamorphism, particularly in some of the Mesozoic ophiolite complexes and related rocks; but even in these, the heat for metamorphism was introduced by magmas. Evidence for extensive rifting, such as voluminous mafic igneous rocks (excepting the Late Jurassic) and large sedimentary basins, is absent.

Subduction-induced asthenospheric upwelling appears to be an effective mechanism for producing higher heat flow, but calculations indicate that the region heated would be wide and behind the main magmatic arc (Andrews and Sleep, 1974). This is compatible with the observation of regional heat-flow highs behind, not within arcs (e.g., Japan, Sugimura and Uyeda, 1973). The induced flow, however, would have to be strongly concentrated at the corner of the mantle wedge in order to produce the inferred paleothermal gradients. Geophysical models suggest that this happens only for a few million years after inception of subduction (Andrews and Sleep, 1974). A good candidate for metamorphism caused by subduction-induced flow is the medium-pressure facies series metamorphism east of the magmatic axis (Fig. 5-3). Armstrong (1982) makes a persuasive argument for this; however, Farmer (Chapter 4, this volume) suggests that Sr and Nd isotope data from Tertiary basalts indicate that at least the eastern part of this area is underlain by "undisturbed" Precambrian mantle. Farmer's data present a problem for this otherwise attractive scenario because this old subcontinental mantle might be expected to have been dispersed by the induced flow.

Hoisch *et al.* (1986, Chapter 21, this volume) suggest that regional low-pressure metamorphism in parts of southeastern California may have formed by high volumes of fluid flow (similar ideas were suggested by Menzer, 1983, for the Okanogan Range, and by Oxburgh and Turcotte, 1971). Although this can certainly be the case on a small scale (= our hydrothermal metamorphism), passing large volumes of fluids through the crust encounters two problems. First, partial melting would take place (Wyllie, 1977), preventing fluid transit, unless the whole lithosphere is below 650–750°C, which seems unreasonably low in an area where large volumes of hotter granitic magma were being emplaced concurrently. Second, the volume of water needed to heat the entire region is enormous; seemingly much too large to be realistic. The petrologic evidence for high fluid/rock ratios presented by Hoisch *et al.* is compelling for some lithologies, but we suggest that the fluids were locally derived and channelized, compatible with a relatively low overall (regional) integrated fluid/rock radio.

Insofar as the observed pluton areas can be used to estimate magmatic fluxes, it is possible to estimate the overall elevation of the geothermal gradient in the arc due to advection of magmatic heat. Taking the Cretaceous arc as the best example, approximately a 1-km² area of pluton was emplaced per kilometer of arc per million years (*cf*. Fig. 5-1). If this represents a 25-km height of pluton plus heated crust and it is distributed over an original width of 100 km, the average increase in heat flow would have been about 0.5 heat-flow unit. This corresponds to an increase in the average geothermal gradient of only 5–10°C/km. The uncertainty in this calculation is likely to be no more than about a factor of 2. Thus, the average increase in the geothermal gradient is minor in comparison with the upgrading achieved in the metamorphic gradient by even moderate densities of plutons (Fig. 5-8). As pointed out by Oxburgh and Turcotte (1971), the magmatic flux needed to achieve increase geothermal gradients by 20–50°C/km over the entire region requires crustal dilation on the order of 5–10%/m.y.. This can be achieved only by back-arc spreading for which there is no evidence in the Cretaceous. Furthermore, the distribution of intrusive activity at any given time was quite localized (e.g., *cf*. the pluton distribution in the central Sierra Nevada, Bateman, 1983; and the spacing of volcanic centers in modern arcs, Gill, 1981). Thus local areas undoubtedly had high heat fluxes, whereas large intervening areas must have been elevated just slightly above the background gradient. With progressive emplacement of the batholiths, coalescence of the aureoles formed about local short-lived centers would give the appearance of uniformly high gradients. To the east of the batholiths, this effect would be exacerbated by the lower proportions of plutons.

Figure 5-9 summarized our view of the relationship between regional metamorphism

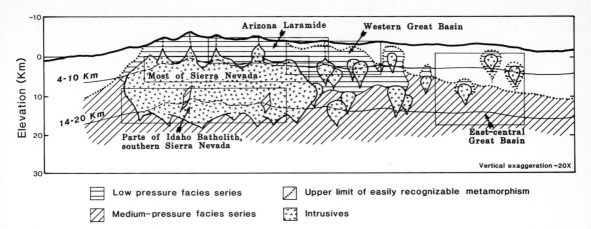

FIG. 5-9. Cross section showing the idealized distribution of metamorphic facies series in "unmetamorphosed" host rocks across the Mesozoic magmatic arc. This most closely resembles the Cretaceous; the older arcs clearly differ in some ways.

and magmatism with examples drawn from the western United States. This cartoon is most appropriate for the effects of a "single" arc like the Cretaceous western United States or the modern Andes. Low-pressure facies series metamorphism forms above, within, and marginal to the batholiths. Shallow metamorphism is dominated by relatively static thermal and hydrothermal metamorphism associated with the development of intrusive centers. Synintrusive structures develop within the batholith as it is emplaced; deeper and more deformed country rocks are preserved as inliers. Progressive emplacement of the batholith results in low-pressure facies series metamorphism over a broad band. On the west side, the relatively abrupt decrease in pluton density, correlative in a general way with the volcanic front, leads to an equally abrupt decrease in the metamorphic gradient. At deeper levels within the batholith, and possibly beneath it, increased total pressure puts all metamorphism in the medium-pressure facies series regime, even though the thermal drive is still magmatism. Deep regional metamorphism around the margins of batholiths (e.g., the Idaho batholith), however, could also result from deeper crustal or subcrustal heat sources and not the exposed batholith. With gradually decreasing pluton densities, low-pressure facies series metamorphism becomes intermixed with, and finally subordinate to, medium-pressure facies series metamorphism to the east. The overall trend is one of monotonically decreasing metamorphic gradient, with local maxima related to intrusive centers. In this area, the regional grades are almost certainly controlled by heat flow from depth. Thus, we interpret the eastern Cordilleran metamorphic belt as a composite of deep batholithic regimes (central Idaho and north) and deeply exposed areas east of the batholiths (e.g., central and eastern Great Basin), not as the result of a separate thermal culmination, as suggested by others.

This simple model ignores important factors such as earlier metamorphic history, vertical and horizontal tectonic movements, erosion of the superstructure, and variations in the timing and amounts of heat and material transfer from the mantle. Alone or combined, each of these factors was significant in the Mesozoic western United States. However, many of the complications of the overall distribution and timing of contact and regional metamorphism discussed in this paper can be rationalized by combined consideration of the present model, the principle of superposition of metamorphism, and the broader tectonic history of the western United States.

The ideas presented here contrast with the commonly presented notion (e.g., Miyashiro, 1973; Turner, 1981) that although regional metamorphism and magmatism may sometimes reflect the same underlying process, magmatism never causes regionally extensive metamorphism. Zwart (1967) pointed out that there at least two characteristic types of low-pressure metamorphic terranes: (1) broad zones without corresponding high-pressure belts ("Hercynotype"), perhaps the result of continental collison (England and Thompson, 1984); and (2) narrower zones belonging to paired metamorphic belts, related to subduction, and primarily preserved around the Pacific ocean (e.g., Ernst, 1974, 1984; Miyashiro, 1961). The phenomena observed and inferred in the Mesozoic low-pressure metamorphic belt in the western United States may be generally applicable to the second type.

Problems

Field observations Although there are many excellent topical studies related to contact metamorphism, few localities have been fully described. Complete characterizations are needed of all facets (mineralogical, compositional, and textural) of the metamorphic systems from intrusions to distant host. In particular, aspects that have not received much attention, such as the distribution and form of cryptic metamorphism or the emplacement and crystallization history of the intrusions (as related to metamorphism) need to be examined. Evaluation of fabric development in different areas may be a key to resolving timing relationships between different metamorphic assemblages.

Timing In spite of extensive geochronological studies, the ages of much Mesozoic plutonism and metamorphism are not well known. This is a particular problem in areas with complex histories, especially for the intepretation of older events. Detailed Ar-Ar and U-Pb studies (e.g., Labotka and Albee, Chapter 26, this volume; E. L. Miller *et al.*, Chapter 24, this volume) are required to establish the timing of magmatism, metamorphism, and deformation. Thermal histories need to be evaluated in well-defined systems by combined field observations, petrology, and suitable temperature-sensitive geochronometry (e.g., Ar-Ar, fission-track methods).

Petrology Although detailed metamorphic petrology has been done on many contact metamorphic systems, such studies rarely extend either outward into the regionally extensive materials or inward into the plutons. Furthermore, large areas of low-pressure metamorphism in the western United States lack any significant petrological study. Large- and small-scale petrological studies of regional metamorphism in conjunction with geochronology are needed to evaluate the relationship between magmatism and metamorphism. Determinations of regional *P-T*-time arrays are needed to test the magma-as-heat-source models.

Geochemistry Increasing evidence indicates that extensive fluid flow takes place during regional metamorphism. Examination of metavolcanic and metasedimentary sequences for cryptic hydrothermal metamorphism by geochemical techniques is needed to determine to what extent such metamorphism is related to magmatism. Metavolcanics, for exmaple, had to acquire water at some time. To what degree is hydration related to surface processes, shallow and contact metamorphism, or later, deeper inflow?

Theory Observations and interpretations need to be tied together through more realistic thermal modeling. Thermal modeling should not only treat heat flow realistically, but it should also involve consideration of chemical equilibration, texture development, and mass transfer. More studies like the pioneering study of Norton and Taylor (1979) are needed on the major types of contact metamorphism. In turn, theory should be used to guide further field studies.

ACKNOWLEDGMENTS

As in all reviews, the content of this paper is based on the work of many others. Because of the broad scope and limited space, we are unable to credit adequately the many geologists on whose work this is based. We hope to remedy this partially in the future with publication of the CONTACT data base.

A number of people have been of special help in one way or another. Richard Marvin generously provided much useful geochronological information from the Radiometric Age Data Bank (USGS). Steven Reynolds provided unpublished compilations for Arizona. Gail Barton entered most of the data. Sara Howe and Nantawan McLeod, UCLA Geology librarians, were tremendously helpful and patient in dealing with our many needs. Reviews by Gary Ernst and Elizabeth Miller improved the manuscript.

Areas within the Cordillera we handled as follows: Trim (Arizona and New Mexico), Michelsen (Idaho), Hanson (California), Davis (Utah), Christensen (Oregon and Klamath Mountains in California), Capo (Montana), Bebout (Washington), Battles (Colorado), and Barton (Nevada). The computing (data base development and thermal modeling) were done by Barton and Hanson. All shared in various tasks, but Barton did most of the organization, coordination, and writing. This work was partially supported by grants to MDB from the Committee on Research (Academic Senate, UCLA) and the NSF (EAR84-08388), and written while MDB was on appointment to the White Mountain Research Station (University of California).

REFERENCES

Albers, J. P., 1981, A lithologic-tectonic framework for the metallogenic provinces of California: *Econ. Geology*, v. 76, p. 765–790.

——, and Stewart, J. H., 1972, Geology and mineral deposits of Esmeralda County, Nevada: *Nev. Bur. Mines Geology Bull.*, v. 78, 80 p.

Allen, C. M., 1981, Intrusive relations of the Slinkard Pluton to the Western Paleozoic and Triassic belt, Klamath Mts. northern California [abstract]: *Geol. Soc. America Abstr. with Programs*, v. 13, no. 7. p. 395.

——, Miller, D. M., and Howard, K. A., 1983, Field petrologic, and chemical characteristics of Jurassic intrusive rocks, eastern Mojave Desert, Southeastern California: *Geol. Soc. America Abstr. with Programs*, v. 15, p. 410.

Allen, J. C., Jr., 1966, Structure and petrology of the Royal Creek stock, Flint Creek Range, central western Montana: *Geol. Soc. America Bull.*, v. 77, p. 291–302.

Allmendinger, R. W., and Jordan, T. E., 1981, Mesozoic evolution, hinterland of the Sevier orogenic belt: *Geology*, v. 9, p. 308–313.

——, Miller, D. M., and Jordan, T. E., 1984, Known and inferred Mesozoic deformation in the Hinterland of the Sevier belt, northwest Utah: *Utah Geol. Assoc. Publ.*, v. 13, p. 21–34.

Anderson, C. A., 1931, The geology of the Engels and Superior mines, Plumas County, California with a note on the ore deposits of the Superior Mine: *Univ. Calif. Geol. Publ.*, v. 20, p. 293–330.

Anderson, G. H., 1973, Granitization, albitization and related phenomena in the northern Inyo Mountains of California-Nevada: *Geol. Soc. America Bull.*, v. 48, p. 1–74.

Anderson, T. H., and Silver, L. T., 1977, U-Pb isotopic ages of granitic plutons near Cananea Sonora: *Econ. Geology*, v. 72, p. 827–836.

Andrews, D. J., and Sleep, N. H., 1974, Numerical modeling of tectonic flow behind island arcs: *Geophy. J. Roy. Astronom. Soc.*, v. 38, p. 237–251.

Armstrong, R. L., 1968, Sevier orogenic belt in Nevada and Utah: *Geol. Soc. America Bull.*, v. 79, p. 429–458.

———, 1969, K-Ar dating of laccolithic centers of the Colorado plateau and vicinity: *Geol. Soc. America Bull.*, v. 80, p. 2081–2086.

———, 1975a, Precambrian (1500 m.y. old) rocks of Central Idaho — The Salmon River Arch and its role in Cordilleran sedimentation and tectonics: *Amer. J. Sci.*, v. 275-A, p. 437–467.

———, 1975b, The geochronometry of Idaho: *Isochron/West*, v. 14, p. 1–50.

———, 1982, Cordilleran metamorphic core complexes — from Arizona to southern Canada: *Ann. Rev. Earth Planet. Sci.*, v. 10, p. 119–154.

1987, Mesozoic and Early Cenozoic magmatic evolution of the Canadian Cordillera, *in* Rodgers Symposium Volume, *Amer. J. Sci.* in press.

———, and Hansen, E., 1966, Cordilleran infrastructure of the Great Basin: *Amer. J. Sci.*, v. 264, p. 112–127.

———, and Suppe, J. L., 1973, Potassium-argon geochronology of Mesozoic igneous rocks in Nevada, Utah, and southern California: *Geol. Soc. America Bull.*, v. 84, p.. 1275–1392.

———, Taubeneck, W. H., and Hales, P. O., 1977, Rb-Sr and K-Ar geochronometry of Mesozoic granitic rocks and their Sr isotopic compositions, Oregon, Washington and Idaho: *Geol. Soc. America Bull.*, v. 86, p. 397–411.

———, Harakal, J. E., Brown, E. H., Bernardi, M. L., and Rady, P. M., 1983, Late Paleozoic high-pressure metamorphic rocks in northwestern Washington and southwestern British Columbia: *Geol. Soc. America Bull.*, v. 94, p. 451–458.

Atkinson, W. W., Jr., Kaczmarowski, J. H., and Erickson, A. J., Jr., 1982, Geology of a skarn-breccia orebody at the victoria Mine, Elko County, Nevada: *Econ. Geology*, v. 77, p. 899–918.

Avé Lallèmant, H. G., Phelps, D. W., and Sutter, J. F., 1980, ^{40}Ar-^{39}Ar ages of some pre-Tertiary plutonic and metamorphic rocks of eastern Oregon and their geologic relationships: *Geology*, v. 8, p. 371–374.

Baird, A. K., Baird, K. W., and Welday, E. E., 1979, Batholithic rocks of the northern Peninsular and Transverse Ranges, southern California, chemical composition and variation; *in* Abbott, P. L., and Todd, V. R., *Mesozoic Crystalline Rocks — Peninsular Ranges Batholith and Pegmatites, Point Sal Ophiolite:* Guidebook Geol. Soc. America Ann. Meet., San Diego State Univ., San Diego, Calif., p. 111–132.

Barksdale, J. D., 1975, Geology of the Methow Valley, Okanogan County, Washington: *Wash. Div. Geology Earth Resources Bull.*, v. 68, p. 72.

Barnes, C. G., and Allen, C. M., 1984, Open- and closed-system fractionation in the Wooley Creek batholith and Slinkard pluton magmatic systems, Klamath Mountains, northern California: *Geol. Soc. America Abstr. with Programs*, v. 16, p. 437.

———, and Rice, S. M., 1983, Tilted plutons in the Klamath Mountains: *Geol. Soc. America Abstr. with Programs*, v. 15, no. 5, p. 314.

Barrell, J., 1907, Geology of the Marysville Mining District, Montana: *U.S. Geol. Survey Prof. Paper 57*, 178 p.

Barton, M. D., 1982, Some aspects of the geology and mineralogy of the fluorine-rich skarn at McCullough Butte, Eureka County, Nevada: *Carnegie Inst. Wash. Yearbook 81*, p. 324–328.

———, 1987, Lithophile-element mineralization associated with Late Cretaceous two-mica granites in the Great Basin: *Geology*. v. 15, p. 337–340.

_____ , Ruiz, J., Ito, E., and Jones, L., 1982, Tracer studies of the fluorine-rich skarn at McCullough Butte, Eureka County, Nevada: *Carnegie Inst. Wash. Yearbook 81*, p. 328–331.

Bateman, P. C., 1965, Geology and tungsten mineralization of the Bishop district, California: *U.S. Geol. Survey Prof. Paper 470*, 208.

_____ , 1981, Geological and geophysical constraints on models for the origin of the Sierra Nevada batholith, California, *The Geotectonic Development of California:* Rubey Volume I (W. G. Ernst, ed.), Prentice Hall, Englewood Cliffs, N. J.: p. 71–86.

_____ , 1984, A summary of critical relations in the central part of the Sierra Nevada batholith, California, USA, *in* Roddick, J., ed., *Circum-Pacific Plutonic Terranes:* Geol. Soc. America Mem. 159, p. 241–254.

_____ , and Dodge, F. C. W., 1970, Variations of major chemical constituents across the central Sierra Nevada batholith: *Geol. Soc. America Bull.*, v. 81, p. 409–420.

_____ , Clark, L. C., Huber, N. K., Moore, J. G., and Rinehart, C. D., 1963, The Sierra Nevada batholith − A synthesis of recent work across the central part: *U.S. Geol. Survey Prof. Paper 414*, p. D1–D46.

_____ , Busacca, A. J., and Sawka, W. N., 1983, Cretaceous deformation in the western foothills of the Sierra Nevada, California: *Geol. Soc. America Bull.*, v. 94, p. 30–42.

Bennett, E. H., 1984, The general geology of the Idaho Batholith [abstract], *in* Beaver, P., ed., *Geology, Tectonics and Mineral Resources of Western and Southern Idaho:* Tobacco Root Geol. Soc., 9th Ann. Field Conf., p. 13.

Bennett, E. H., II, Kopp, R. S., and Galbraith, J. H., 1975, Reconnaissance geology and geochemistry of the Mt. Pend Oreille Quadrangle and surrounding areas: *Idaho Bur. Mines Geology Pamphlet 163*.

Bero, D. A., 1980, Petrology of the Heather Lake Pluton, Klamath Mountains California: M.S. thesis, California State Univ., Fresno., Calif.

Best, M. G., Armstrong, R. L., Graustein, W. C., Embrey, G. F., and Ahlborn, R. C., 1974, Mica granites of the Kern Mountains pluton, eastern White Pine County, Nevada − Remobilized basement of the Cordilleran miogeocline? *Geol. Soc. America Bull.*, v. 85, p. 1277–1286.

Bickford, M. E., Chase, R. B., Nelson, B. K., Shuster, R. D., and Arruda, E. C., 1981, U-Pb studies of zircon cores and overgrowths and monazite − implications for age and petrogenesis of the northeastern Idaho Batholith: *J. Geology*, v. 89, p. 433–457.

Bingler, E. C., 1978, Geologic map of the Schurz Quadrangle: *Nev. Bur. Mines Geol. Mapping*, v. 60.

Boettcher, A. L., 1967, The Rainy Creek alkaline-ultramafic igneous complex near Libby, Montana − 1, Ultramafic rock and finite: *J. Geology*, v. 75, p. 526–553.

Bohike, J. K., and Kistler, R. W., 1986, Rb-Sr, K-Ar, and stable isotope evidence for the ages and sources of fluid components in the northern Sierra Nevada foothills metamorphic belt, California: *Econ. Geol.*, v. 81, p. 296–322.

Bonham, H. F., 1969, Geology and mineral deposits of Wahoe and Storey Counties, Nevada − with a section on industrial rock and mineral deposits by Keith G. Papke: *Nev. Bur. Mines Bull.* 70, 140 p.

Bowman, J. R., and Essene, E. J., 1982, P-T-XCO_2 conditions of contact metamorphism in the Black Butte aureole, Elkhorn, Montana: *Amer. J. Sci.*, v. 282, p. 311–340.

_____ , and Essene, E. J., 1984, Contact skarn formation at Elkhorn, Montana. I − P-T component activity conditions of early skarn formation: *Amer. J. Sci.*, v. 284, p. 597–650.

Brooks, H. C., and Vallier, T. L., 1978, Mesozoic rocks and tectonic evolution of eastern Oregon and western Idaho, *in* Howell, D. G., and McDougall, K. A., eds., *Meso-*

zoic Paleogeography of the Western United States: Pacific Section, Soc. Econ. Paleontologists Mineralogists, Pacific Coast Paleogeography Symp. 2, p. 133–145.

Bruce, W. R., 1970, Geology, mineral deposits and alteration of parts of the Cuddy Mountain District, western Idaho: Ph.D. dissertation, Oregon State Univ. Corvallis, Oreg.

Bryant, D. G., and Metz, H. E., 1966, Geology and ore deposits of the Warren Mining District, *in* Titley, S. R., and Hicho, C. L., eds., *Geology of Porphyry Copper Deposits Southwestern North America:* Univ. Arizona Press, Tucson, Ariz., p. 189–204.

Buddington, A. F., 1959, Granite emplacement with special reference to North America: *Geol. Soc. America Bull.*, v. 70, p. 671–747.

Burchfiel, B. C., and Davis, G. A., 1972, Structural framework and evolution of the southern part of the Cordilleran orogen, western United States: *Amer. J. Sci.*, v. 272, p. 97–118.

―――, and Davis, G. A., 1975, Nature and controls of Cordilleran orogenesis, western United States – Extensions of an earlier synthesis: *Amer. J. Sci.*, v. 275A, p. 363–396.

―――, and Davis, G. A., 1981, Mojave Desert and environs, *in* Ernst, W. G., ed., *The Geotectonic Development of California* (Rubey Vol. I): Englewood Cliffs, N.J., Prentice-Hall, p. 217–252.

Burnham, C. W., 1954, Contact metamorphism at Crestmore, California, *in* Jahns, R. H., ed., *Geology of Southern California:* Calif. Div. Mines Bull. v. 270, p. 61–70.

―――, and Ohmoto, H., 1980, Late-state processes of felsic magmatism: *Mining Geology (Japan) Spec. Issue*, v. 8, p. 1–11.

Busby-Spera, C. J., 1984, The lower Mesozoic continental margin and marine intra-arc sedimentation at Mineral King, California, *in* Crouch, J. K., and Bachman, S., eds., *Tectonics and Sedimentation along the California Margin*: Pacific Section, Soc. Econ. Paleontologists Mineralogists, v. 38, p. 135–156.

Carlson, D. H., 1986, REE Chemistry and evolution of intrusive rocks associated with the Colville Batholith near Grand Coulee Dam, Washington: *Geol. Soc. America Abstr. with Programs*, v. 18, no. 2, p. 93.

Carmichael, I. S. E., Turner, F. J., and Verhoogen, J., 1974, *Igneous Petrology:* New York, McGraw-Hill, 739 p.

Cater, F. W., 1982, Intrusive rocks of the Holden and Luceren quadrangles, Washington; the relation of depth zones, composition, textures, and emplacement of plutons: *U.S. Geol. Survey Prof. Paper 1220*, 108 p.

Cavanaugh, P. C., 1979, The geology of the Little Boulder Creek molybdenum deposit, Custer County, Idaho: M.S. thesis, Univ. Montana, Missoula, Mont.

Cebull, S. E., 1970, Bedrock geology and organic succession in Southern Grant Range, Nye Co., Nevada: *Amer. Assoc. Petroleum Geol. Bull.*, 54, p. 1828–1842.

Chase, R. B., 1973, Petrology of the northeastern border zone of the Idaho batholith, Bitterroot Range, Montana: *Mont. Bur. Mines Geology, Mem. 43*, 28 p.

―――, Bickford, M. E., and Tripp, S. E., 1978, Rb-Sr and U-Pb isotopic studies of the northeastern Idaho batholith and border zone: *Geol. Soc. America Bull.*, v. 89, p. 1325–1334.

―――, Bickford, M. E., and Arruda, E. C., 1983, Tectonic implications of Tertiary intrusion and shearing within the Bitterroot Dome, northeastern Idaho Batholith: *J. Geology*, v. 91, p. 462–470.

Chen, J., and Moore, J., 1981, Uranium-lead isotopic ages from the Sierra Nevada batholith, California: *J. Geophy. Res.*, v. 87, p. 4761–4784.

Cheney, E. S., 1980, Kettle Dome and related structures of northeastern Washington, *in* Crittendon, M. D., Jr., Coney, P. J., and Davis, G. H., eds., *Cordilleran Metamorphic Core Complexes:* Geol. Soc. America Mem. 153. p. 463–484.

Coleman, R. G., and Helper, M. D., 1983, The significance of the Condrey Mountain Dome in the evolution of the Klamath Mountains, California and Oregon (abstract) in *Geol. Soc. America Abstr. with Prog.* 15, p. 294.

——, Helper, M. D., and Donato, M. M., 1983, Geologic Map, Condray Mtn. roadless area, California: *U. States Geol. Survey Misc. Field Studies Map MF-1540-A.*

Colville Confederated Tribes Geology Department, 1984, *Revised Geology and Mineral Potential of the Colville Indian Reservation, Washington* (Text and Map Folio): Nespelem, Wash., Colville Confederated Tribes Geol. Dept., Don Aubertin, Director.

Compton, R. R., 1955, Trondhjemite batholith near Bidwell Bar, California: *Geol. Soc. America Bull.*, v. 66, p. 9–44.

——, 1960, Contact metamorphism in the Santa Rosa Range, Nevada: *Geol. Soc. America Bull.*, v. 71, p. 1383–1416.

Compton, R. R., 1966, Granitic and metamorphic rocks of the Salinian block, California Coast Ranges: *Cal. Div. Mines and Geol. Bull.* 190, p. 277–287.

Coney, P. J., 1978, Mesozoic-Cenozoic Cordilleran plate tectonics *in* Smith, R. B., and Eaton, G. P., eds., *Cenozoic Tectonics and the Regional Geophysics of the Western Cordillera:* Geol. Soc. America Mem. 152, p. 33–50.

——, 1980, Cordilleran metamorphic core complexes – An overview, *in* Crittenden, M. D., Jr., Coney, P. J., and Davis, G. H., eds., *Cordilleran Metamorphic Core Complexes:* Geol. Soc. America Mem. 153, p. 4–34.

——, and Harms, T. A., 1984, Cordilleran metamorphic core complexes: Cenozoic relics of Mesozoic compression: *Geology*, v. 12, p. 550–554.

Coney, P. J., and Reynolds, S. J., 1977, Cordilleran benioff zones: *Nature* 270, p. 403–406.

Cooper, J. R., 1971, Mesozoic stratigraphy of the Sierrita Mountains, Pima Colorado, Arizona: *U.S. Geol. Survey*, v. 658–D, 42 p.

Cox, B. F., and Morton, J. L., 1980, Late Permian plutonism, El Paso Mountains, California: *Geol. Soc. America Abstr. with Programs*, v. 12, p. 103.

Criss, R. E., and Fleck, R. J., 1985, Isotopic characteristics of granitic rocks from the north half of the Idaho batholith: *Geol. Soc. America Abstr. with Programs*, v. 25, no. 6, p. 550.

——, and Taylor, H. P., Jr., 1983, Au 180/160 and D/H study of Tertiary hydrothermal systems in the southern half of the Idaho batholith: *Geol. Soc. America Bull.*, v. 94, p. 640–663.

Crittenden, M. D., Jr., Coney, P. J., and Davis, G. H., 1980, eds., Cordilleran metamorphic core complexes: *Geol. Soc. America Mem. 153*, 490 p.

Crowder, D. F., and Ross, D. C., 1973, Petrography of some granitic bodies in the northern White Mountains, California-Nevada: *U.S. Geol. Survey Prof. Paper 775.*

Crowley, F. A., 1963, Mines and mineral deposits, Sanders County, Montana: *Mont. Bur. Mining Geology Bull. 34*, 58 p.

Davis, G. A., 1963, Structure and mode of emplacement of Caribou Mt. pluton, Klamath Mts., California: *Geol. Soc. America Bull.*, v. 74, no. 3, p. 331–348.

——, Monger, J. W. H., and Burchfiel, B. C., 1978, Mesozoic construction of the Cordilleran "collage," central British Columbia to central California, *in* Howell, D. G., and Douglass, K. A., eds., *Mesozoic Paleogeography of the Western United States:* Pacific Section, Soc. Econ. Paleontologists Mineralogists, Pacific Coast Paleogeography Symp. 2, p. 1–31.

Day, H. W., Moore, E.M., and Tuminas, A. C., 1985, Structure and tectonics of the northern Sierra Nevada: *Geol. Soc. America Bull.*, v. 96, p. 436–450.

Dewitt, G., Armstrong, R. L., Sutter, J. F., and Zartman, R. E., 1984, U-Th-Pb, Rb-Sr and Ar-Ar mineral and whole rock isotopic systematics in a metamorphosed granitic terrane, southeastern, California: *Geol. Soc. America Bull.*, v. 95, p. 723–739.

Dickinson, W. R., 1981, Plate Tectonics and the Continental Margin of California, *in* Ernst, W. G., ed., *The Geotectonic Development of California* (Rubey Vol. I): Englewood Cliffs, N.J., Prentice-Hall, p. 1–28.

Dover, J. H., 1983, Geologic map and sections of the central Pioneer Mountains, Blaine and Custer Counties, central Idaho: *U.S. Geol. Survey Misc. Geol. Inv. Map I-1319.*

Dubois, R. L., and Brummett, 1968, Geology of the Eagle Mountain mine area. *in* Ridge, J. D., ed., *Ore Deposits of the United States, 1933-1967* (Graton-Sales vol.): New York, Amer. Inst. Min. Metal. Petrol. Engineers, p. 1592–1606.

Einaudi, M. T., 1982a, Description of skarns associated with porphyry copper plutons – Southwestern North America, *in* Titley, S. R., ed., *Advances in Geology of the Porphyry Copper Deposits:* Univ. Arizona Press, Tuscon, Ariz., p. 139–189.

_____ , 1982b, General features and origin of skarns associated with porphyry copper plutons, *in* Titley, S. R., ed., *Advances in Geology of the Porphyry Copper Deposits:* Univ. Arizona Press, Tuscon, Ariz., p. 185–209.

_____ , 1984, Yerington skarns, *in* Johnson, J. L., ed., *Exploration for Ore Deposits in the North American Cordillera:* Assoc. Explor. Geochemists Field Trip Guidebook, Field Trip 10; p. 31–39.

_____ , Meinert, L. D., and Newberry, R. J., 1981, Skarn deposits: *Econ. Geology*, 75th Ann. Vol., p. 327–391.

Ekren, E. B., and Houser, F. N., 1965, Geology and petrology of the Ute Mountains, Colorado: *U.S. Geol. Survey Prof. Paper 481*, 74 p.

Elan, R., 1985, High grade contact metamorphism at Lake Isabella North shore roof pendant, southern Sierra Nevada, California: M.S. thesis, Univ. Southern California, Los Angeles, Calif., 202 p.

Elder, J., 1981, *Geothermal Systems:* London, Academic, 508 p.

Emmons, W. H., and Calkings, F. C., 1913, *Geology and ore deposits of the Philipsburg Quadrangle, Montana:* U.S. Geol. Survey, 78 p.

Engebretson, D. C., Cox, A., and Gordon, R. G., 1985, Relative motions between oceanic and continental plates in the Pacific basin: *Geol. Soc. America Spec. Paper 206*, 59 p.

England, P. C., and Thompson, A. B., 1984, Pressure-temperature-time paths of regional metamorphism – I. Heat transfer during the evolution of regions of thickened continental crust: *J. Petrology*, v. 25, p. 894–928.

Erickson, R. L., Silberman, M. L., and Marsh, S. P., 1978, Age and composition of igneous rocks, Edna Mountain Quadrangle, Humboldt County, Nevada: *J. Res. U.S. Geol. Survey*, v. 6, p. 727–743.

Ernst, W. G., 1974, Metamorphism and ancient convergent continental margins, *in* Burk, C. A., and Drake, C. L., eds, *The Geology of Continental Margins:* New York, Springer, p. 907–919.

_____ , 1983, Phanerozoic continental accretion and the metamorphic evolution of northern and central California: *Tectonophysics*, v. 100, p. 287–320.

_____ , 1987, Mafic meta-igneous arc rocks of apparent komatiitic affinities, Sawyers Bar area, central Klamath Mountains, northern California *in* Mysen, B. O., ed., *Magmatic Process: Physicochemical Principles:* Geochem. Soc. Special Pub. No. 1, p. 191–208.

Evans, B. W., and Berti, J. W., 1986, Revised metamorphic history for the Chiwaukum Schist, North Cascades, Washington: *Geology*, v. 14, p. 695–698.

Evernden, J. F., and Kistler, R. W., 1970, Chronology of emplacement of Mesozoic batholith complexes in California and western Nevada: *U.S. Geol. Survey Prof. Paper 623*, 42 p.

Fahan, M. R., 1982, Geology and geochronology of a part of the Hayfork terrane, Klamath Mountains, California [M.S. thesis]: Santa Barbara, University of California.

Fahan, M. R., and Wright, J. E., 1983, Plutonism, volcanism, folding, regional metamorphism and thrust faulting — Contemporaneous aspects of a major middle Jurassic orogenic event within the Klamath Mountains, northern California: *Geol. Soc. America Abstr. with Programs*, v. 15, p. 272.

Farmer, G. L., and DePaolo, D. J., 1983, Origin of Mesozoic and Tertiary granite in the western United States and implications for pre-Mesozoic crustal structure — 1. Nd and Sr isotopic studies in the geocline of the Northern Great Basin: *J. Geophys. Res.*, v. 88 p. 3379–3401.

Faure, G., 1986, *Principles of Isotope Geology:* New York, Wiley, 589 p.

Fenton, M. D., and Faure, G., 1969, The age of the igneous rocks of the Stillwater Complex of Montana: *Geol. Soc. America Bull.* 80, p. 1599–1604.

Ferry, J. M., 1983, On the control of temperature, fluid composition, and reaction progress during metamorphism: *Amer. J. Sci.*, v. 283-A, p. 201–232.

Fiske, R. S., and Tobisch, O. T., 1978, Paleogeographic significance of volcanic rocks of the Ritter Range pendant, central Sierra Nevada, California, *in* Howell, D. G., and McDougall, K. A., eds., *Mesozoic Paleogeography of the Western United States:* Pacific Section, Soc. of Econ. Paleontologists Mineralogists, Pacific Coast Paleogeography Symp. p. 209–219.

Fleck, R. J., and Criss, R. E., 1985, Strontium and oxygen isotopic variations in Mesozoic and Tertiary plutons in central Idaho: *Contrib. Mineralogy Petrology*, v. 90, p. 291–308.

Fox, K. F., Rinehart, C. D., and Engels, J. C., 1977, Plutonism and orogeny in north-central Washington — Timing and regional context: *U.S. Geol. Survey Prof. Paper 989.* 27 p)

Frost, K. S., 1984, The New–Boston–Blue Ribbon polymetallic skarn, *in* Johnson, J. L., ed., *Exploration for Ore Deposits in the North American Cordillera:* Assoc. Explor. Geochemists Field Trip Guidebook, Field Trip 10, p. 25–29.

Frost, R. B., 1975, Contact metamorphism of serpentinite, chloritic blackwall and rodingite at Paddy-Go-Easy Pass, Central Cascades, Washington: *J. Petrology*, v. 16, p. 272–313.

Gans, P. B., and Miller, E. L., 1983, Style of mid-Tertiary extension in east-central Nevada, *in Geologic Excursions in the Overthrust Belt and Metamorphic Core Complexes of the Intermountain Region, Nevada:* Geol. Soc. America Field Trip Guidebook, Utah Geol. Min. Survey Spec. Studies 59, p. 107–160.

Gardner, L. S., 1959, Geology of the Lewiston area, Fergus County, Montana: *U.S. Geol. Sur. Map OM-199.*

Garside, L. J., 1982, Geologic map of the Moho Mountain Quadrangle, Nevada: *Nev. Bur. Mines Geology Map 74.*

Gastil, R. G., Delisle, M., and Morgan, J. R.: 1967, Some effects of progressive metamorphism on zircons: Geol. Soc. America Bull. v. 78, p. 879–905.

Gerlach, D. C., Avé Lallèmant, H. G., and Leeman, W. R., 1981, An island arc origin for the Canyon Mountain ophiolite complex, Eastern Oregon, U.S.A.: *Earth Planet. Sci. Lett.*, v. 53, p. 255–265.

Giletti, B. J., 1966, Isotopic ages from southwestern Montana: *J. Geophys. Res.*, v. 71, p. 4029-4036.

Gill, J. B., 1981, *Orogenic Andesites and Plate Tectonics:* Berlin, Springer-Verlag, 390 p.

Gillson, J. L., 1927, Granodiorites in the Pend Oreille district of northern Idaho: *J. Geology*, v. 35, p. 1-31.

Gilluly, J., and Masursky, H., 1965, Geology of the Cortez quadrangle, Nevada: *U.S. Geol. Survey Bull. 1175*, 117 p.

Girty, G. H., Wardlaw, M. S., Schweikhert, R. A., Hanson, R. E., and Bowring, S. A., 1984, Timing of pre-Antler deformation in the Shoo Fly Complex, Sierra Nevada, California: *Geology* 12, p. 673-676.

Godchaux, M. M., 1969, Petrology of the Greyback intrusive complex and contact aureole, southwestern Oregon: *Geol. Soc. America Abstr. 1969*, pt. 3, p. 18-19.

Hall, W. E., and MacKevett, E. M., 1962, Geology and ore deposits of the Darwin Quadrangle Inyo County, California: *U.S. Geol. Survey Prof. Paper 368*, 87 p.

Hamilton, W., 1963, Metamorphism in the Riggins region, western Idaho: *U.S. Geol. Survey Prof. Paper 436*, 95 p.

____, 1978, Mesozoic tectonics of the western United States, *in* Howell, D. G., and McDougall, K. A., eds., *Mesozoic Paleogeography of the Western United States:* Pacific Section, Soc. Econ. Paleontologists Mineralogists, Pacific Coast Paleogeography Symp. 2, p. 33-70.

____, 1983, Depth-related contrast between Idaho and Sierra Nevada batholiths: *Geol. Soc. America Abstr.* with Programs, v. 15, p. 334.

____, and Myers, W. B., 1967, The nature of batholiths: *U.S. Geol. Survey Prof. Paper 554-C*, 30p.

____, and Myers, W. B., 1974, Nature of the Boulder Batholith of Montana: *Geol. Soc. America Bull.*, v. 85, p. 365-378.

Hanson, R. B., 1986, Geology of Mesozoic metavolcanic and metasedimentary rocks, northern White Mountains, California: Ph.D. thesis, University of California, Los Angeles, 230 p.

Harper, G. D., 1983, A depositional contact between the Galice Formation and a Late Jurassic ophiolite in northwestern California and southwestern Oregon: *Oreg. Geology*, v. 45, p. 3-9.

____, and Wright, J. E., 1984, Middle to Late Jurassic tectonic evolution of the Klamath Mountains, California-Oregon: *Tectonics*, v. 3, no. 7, p. 759-772.

Harris, A. G., Warlan, B. R., Rust, C. C., and Merrill, G. K., 1980, Maps for assessing thermal maturity (conodont color alteration index maps) in Ordovician through Triassic rocks in Nevada and Utah and adjacent parts of Idaho and California: *U.S. Geol. Survey Misc. Geol. Inv. Map I-1249.*

Harris, N. B., and Einaudi, M. T., 1982, Skarn deposits in the Yerington District, Nevada — Metasomatic skarn evolution near Ludwig: *Econ. Geology*, v. 77, p. 877-898.

Harrison, J. E., and Jobin, D. A., 1965, Geologic map of the Packsaddle Mountain Quadrangle, Idaho: *U.S. Geol. Survey Geol. Quad. Map GQ-375.*

Harrison, T. M., and McDougall, I. M., 1980, Investigation of an intrusive contact, northwest Nelson, New Zealand — I. Thermal, chronological and isotopic constraints: *Geochim. Cosmochim. Acta*, v. 44, p. 1985-2003.

Hart, S. R., 1964, The petrology and isotopic-mineral age relations of a contact zone in the Front Range, Colorado: *J. Geology*, v. 72, p. 493-525.

____, Davis, G. L., Steiger, R. H., and London, G. R., 1968, A comparison of the isotopic mineral age variations and petrologic changes induced by contact metamorphism, *in* Hamilton, E. I., and Farquhar, R. M., eds., *Radiometric Dating for Geologists:* New York, Interscience, p. 73-110.

Hayes, P. T., and Drewes, H., 1978, Mesozoic depositional history of southeastern Arizona: *N.M. Geol. Soc. Guidebook*, 29th field conference, p. 201–207.

Hexel, G. B., Tosdal, R. M., May, D. J., and Wright, J. E., 1984, Latest Cretaceous and Early Tertiary orogenesis in south central Arizona — Thrust faulting, regional metamorphism and granitic plutonism: *Geol. Soc. America Bull.*, v. 95, p. 631–53.

Henley, R. W., and Ellis, A. J., 1983, Geothermal systems ancient and modern — A geochemical review: *Earth-Sci. Rev.*, v. 19, p. 1–50.

Henricksen, T. A., 1974, Geology and mineral deposits of the Mineral-Iron Mountain District, Washington County, Idaho, and of a metallized zone in western Idaho and eastern Oregon: Ph.D. dissertation, Oregon State Univ., Corvallis, Oreg., 260 p.

Hibbard, M. J., 1971, Evolution of a plutonic complex, Okanogan Range, Washington: *Geol. Soc. America Bull.*, v. 82, p. 3013–3046.

Hietanen, A., 1962, Metasomatic metamorphism in western Clearwater County, Idaho: *U.S. Geol. Survey Prof. Paper 344-A*, p. A1–A116.

——, 1984, Geology along the northwest border zone of the Idaho batholith, northern Idaho: *U.S. Geol. Survey Bull. 1608*, 17 p.

Hill, B. L., 1985, Metamorphic, deformational and temporal constraints on terrane assembly, northern Klamath Mountains, California, *in* Howell, D. G., ed., *Tectonostratigraphic Terranes of the Circum-Pacific Region*: Circum-Pacific Council Energy Min. Resources, Earth Sci. Series, no. 1, p. 173–186.

Hobbs, S. W., Griggs, A. B., Wallace, R. E., and Campbell, A. B., 1965, Geology of the Coeur d'Alene district, Shoshone County, Idaho: *U.S. Geol. Survey Prof. Paper 478*, 139 p.

Hodges, K. V., and McKenna, L., 1986, Structural and metamorphic characteristics of the Raft River—Quartzite assemblage juxtaposition, Albion Mountains, southern Idaho: *Geol. Soc. America Abstr. with Programs*, v. 18, p. 117.

Hofmann, A., 1972, Effect of regional metamorphism on the behavior of Rb and Sr in micas and whole-rock systems of the Belt Series, northern Idaho: *Carnegie Inst. Wash. Yearbook* 71, p. 559–563.

Holser, W. T., 1950, Metamorphism and associated mineralization in the Phillipsburg region, Montana: *Geol. Soc. America Bull.*, v. 61, p. 1053–1090.

Hopson, C. A., Mattinson, J. M., and Pessagmo, E. A., Jr., 1981, Coast Range ophiolith, western California, *in* Ernst, W. G., ed., *The Geotectonic Development of California:* Englewood Cliffs, N.J., Prentice Hall, p. 418–510.

Hori, F., 1964, On the role of water in heat transfer from a cooling magma: *Coll. Gen. Education, Univ. Tokyo Scient. Papers*, v. 14, p. 121–127.

Hotz, P. E., and Willden, R., 1964, Geology and mineral deposits of the Osgood Mountains quadrangle, Humboldt County Nevada: *U.S. Geol. Survey Prof. Paper 431.*, 128 p.

——, Lanphere, M. A., and Swanson, P., 1977, Triassic blueschist from northern California and north-central Oregon: *Geology*, v. 5, p. 659–663.

Hover-Granath, V. C., Papike, J. J., and Labotka, T. C., 1983, The Notch Peak contact metamorphic aureole, Utah — Petrology of the Big Horse Limestone Member of the Orr Formation: *Geol. Soc. of America Bull.*, v. 94, p. 889–906.

Howard, K. A., and Shaw, S. E., 1982, Mesozoic plutonism in the eastern Mojave desert, California: *Geol. Soc. America Abstr. with Programs*, v. 14, p. 174.

——, Kistler, K. W., Snoke, A. W., and Willden, R., 1979, Geological map of the Ruby Mountains, Nevada: *U.S. Geol. Survey Misc. Geol. Inv. Map I-1136*, scale 1:125,000.

Hyndman, D. W., 1983, The Idaho batholith and associated plutons, Idaho and western Montana, *in* Roddick, J. A., ed., *Circum-Pacific Plutonic Terranes: Geol. Soc. America Mem. 159*, p. 213–240.

_____ , 1984, A petrographic and chemical section through the Idaho batholith: *J. Geology*, v. 92, p. 83–102.

_____ , Silverman, A. J., Ehinger, R., Benoit, W. R., and Wold, R., 1982, The Phillipsburg batholith, western Montana, petrology, internal variation, and evolution: *Montana Bur. Mines and Geol.* Memoir 49, 37 p.

Irwin, W. P., 1960, Geologic reconnaissance of the northern Coast Ranges and Klamath Mountains, California, with a summary of the mineral resources: *Cal. Div. Mines Bull.*, 179, 80 p.

_____ , 1981, Tectonic accretion of the Klamath Mountains, *in* Ernst, W. G., ed., *The Geotectonic Development of California* (Rubey Vol. I): Englewood Cliffs, N. J., Prentice Hall, p. 29–49.

_____ , 1985, Age and tectonics of plutonic belts in accreted terranes of the Klamath Mountains, California and Oregon, *in* Howell, P. G., ed., *Tectonostratigraphic Terranes of the Circum-Pacific Region:* Circum-Pacific Council Energy Min. Resources, Earth Sci. Series, no. 1, p. 187–199.

James, L. P., 1976, Zoned alteration in limestone at porphyry copper deposits, Ely, Nevada: *Econ. Geology*, v. 71, p. 488–512.

Jaeger, J. C., 1964, Thermal effects of intrusions: *Rev. Geophysics*, v. 2, p. 443–466.

_____ , 1968, Colling and solidification of igneous rocks, *in* Hess, H. H., and Poldervaart, A., eds., *Basalts: the Poldervaart Treatise on Rocks of Basaltic Composition*, Vol. 2: New York, Interscience, p. 504–536.

Jayko, A. S., and Blake, M. C., Sr., 1984, Geologic Map, Orleans Mountain roadless area, Cul: *U.S. Geol. Survey Misc. Field Studies Map MF-1600-A.*

Johnson, D. L., 1981, Two-mica granites and metamorphic rocks of the east-central Ruby Mountains, Elko County, Nevada: M.S. thesis, Stanford Univ., Stamford, Calif., 145 p.

Johnson, J. W., and Norton, D., 1985, Theoretical prediction of hydrothermal conditions and chemical equilibria during skarn formation in prophyry copper systems: *Econ. Geology*, v. 80, p. 1797–1823.

Johnson, M. G., 1977, Geology and mineral deposits of Pershing County, Nevada: *Nevada Bureau of Mines and Geology Bulletin*, v. 89, 115 p.

Kays, M. A., 1970, Mesozoic metamorphism, May Creek schist belt, Klamath Mountains Oregon: *Geol. Soc. America Bull.*, v. 91, p. 2743–2758.

Keith, S. B. 1983, Distribution of fossil metallogenic systems and magma geochemical belts within the Great Basin and vicinity from 145 m.y. ago to present, *in* The role of heat in the development of energy and mineral resources in the northern Basin and Range province: *Geothermal Resources Council Special Report no. 13*, p. 285–286.

_____ , Reynolds, S. J., Damom, P. E., Snafiqullah, M., Livingston, D. E., and Pushkan, P. D., 1980, Evidence for multiple intrusion and deformation within the Santa Cataline-Rincon-Tortolita crystalline core complex, southeast Arizona: *Geol. Soc. America Mem. 153*, p. 217–267.

Kerr, P. F., 1946, Tungsten mineralization in the United States: *Geol. Soc. America Mem. 15*, 241 p.

Kerrick, D. M., 1970, Contact metamorphism in some areas of the Sierra Nevada, California: *Geol. Soc. America Bull.*, v. 81, p. 2913–2938.

_____ , 1974, Review of metamorphic mixed-volatile (H_2O-CO_2) equilibria: *American Mineral. 59*, p. 729–762.

_____ , 1977, The genesis of zoned skarns in the Sierra Nevada, California: *J. Petrology*, v. 18, p. 144–181.

Kistler, R. W., 1974, Phanerozoic batholiths in western North America − A summary of some recent work on variations in time, space, chemistry and isotopic compositions: *Ann. Rev. Earth Planet. Sci.*, v. 2, p. 403–418.

——, and Swanson, S. E., 1981, Petrology and geochronology of metamorphose volcanic rocks and a Middle Cretaceous volcanic neck in the east-central Sierrra Nevada, California: *J. Geophys. Res.*, v. 86, p. 10489–10501.

Klepper, M. R., Ruppel, G. T., Freeman, V. L., and Weeks, R. A., 1971, Geology and mineral deposits east flank of the Elkhorn Mountains, Broadwater County, Montana: *U.S. Geol. Survey Prof. Paper 665*, 66 p.

Knopf, A., 1913, Ore deposits of the Helena mining region, Montana: *U.S. Geol. Survey Bull. 527*, 127 p.

——, 1950, The Marysville granodiorite stock, Montana: *Amer. Mineralogist*, v. 35, p. 834–845.

Korzhinskii, D. S., 1970, *Theory of Metasomatic Zoning:* Oxford, Clarendon Press, 162 p.

Krauskopf, K. B., 1943, The Wallowa batholith: *Amer. J. Sci.*, v. 241, p. 607–628.

——, 1953, Tungsten deposits of Madera, Fresno and Tulare Counties, California: *Calif. Div. Mining Spec. Rpt.*, v. 35, 83 p.

——, 1968, A tale of ten plutons: *Geol. Soc. America Bull.*, v. 79, p. 1–18.

Kun, Peter, 1974, Geology and mineral resources of the Lakeview Mining District, Idaho: *Idaho Bur. Mines Geology Pamplet 156.*

Labotka, T. C., 1980, Petrology of a medium-pressure regional metamorphic terrane, Funeral Mountains, CA: *Amer. Mineral.*, v. 65, p. 670–689.

——, 1985, Polymetamorphism in the Panamint Mountains, California: an $^{39}AR-^{40}Ar$ study: *J. Geophys. Res.*, v. 90, p. 10359–10373.

Lang, H., and Rice, J. M., 1985, Metamorphism of pelitic rocks in the Snow Peak area, northern Idaho — Sequence of events and regional implications: *Geol. Soc. America Bull.* v. 96, no. 6, p. 731–736.

Lanphere, M. A., 1964, Geochronologic studies in the eastern Mojave Desert, California: *J. Geology*, v. 72, p. 381–399.

——, and Reed, B. L., 1973, Timing of Mesozoic and Cenozoic plutonic events in Circum-Pacific North America: *Geol. Soc. America Bull.*, v. 84, p. 3773–3782.

——, Irwin, W. P., and Hotz, P. E., 1968, Isotopic age of the Nevadan orogeny and older plutonic and metamorphic events in the Klamath Mountains, California: *Geol. Soc. America Bull.*, v. 79, p. 1027–1052.

Larsen, E. S., Jr., 1948, Batholith and associated rocks of Corona, Elsmore and San Luis Rey Quadrangles southern California: *Geol. Soc. America Mem. 29*, 182 p.

Larson, P. B., and Taylor, H. P., Jr., 1986, An oxygen-isotope study of water-rock interaction in the granite of Cataract Gulch, western San Juan Mountains, Colorado: *Geol. Soc. America Bull.*, v. 97, p. 505–515.

Lasaga, A. C., 1981, Rate laws of chemical reactions, *in* Lasaga, A. C., and Kirkpatrick, R. J., eds., *Kinetics of Geochemical Processes: Mineral. Soc. America Rev. Mineralogy*, v. 8, p. 1–68.

Lasky, S. G., 1947, Geology and ore deposits of the Little Hatchet Mountains, Hildapo and Grant Counties, New Mexico: *U.S. Geol. Survey Prof. Paper 208*, 101 p.

Lattanzi, P., Rye, D. M., and Rice, J. M., 1980, Behavior of ^{13}C and ^{18}O carbonates during contact metamorphism at Marysville Mountain—Implications for isotope systematics in impure dolomite limestones: *Amer. J. Sci.*, v. 280, p. 890–960.

Lee, D. E., and Fischer, L. B., 1985, Cretaceous metamorphism in the northern Snake Range, Nevada, a metamorphic core complex: *Isochron/West*, v. 42, p. 3–7.

_____ , Marvin, R. F., Stern, T. W., and Peterman, Z. E., 1970, Modification of potassium-argon ages by Tertiary thrusting in the Snake Range, White Pine County, Nevada: *U.S. Geol. Survey Prof. Paper 700-D*, p. 92–102.

Lindsey, D. A., 1982, Geologic map and discussion of selected mineral resources of the north and south Mocasin Mountains, Fergus County, Montana: *U.S. Geol. Survey Misc. Inv. Map I-1362.*

Lipman, P. W., 1963, Gibson Peak pluton—A discordant composite intrusion in the southeastern Trinity Alps, northern California: *Geol. Soc. America Bull.*, v. 74, no. 10, p. 1259–1280.

_____ , and Sawyer, D., 1985, Mesozoic ash-flow caldera fragments in southeast Arizona and their relation to porphyry copper deposits: *Geology*, v. 13, p. 652–656.

Lovering, T. S., 1935, Theory of heat conduction applied to geological problems: *Geol. Soc. America Bull.* v. 46, p. 69–94.

_____ , 1955, Temperatures in and near intrusions: *Econ. Geology*, v. 50, p. 249–281.

Lowell, J. D., and Guilbert, J. M., 1970, Lateral and vertical alteration-mineralization zoning in porphyry ore deposits: *Econ. Geology*, v. 65, p. 373–408.

Lund, K., Snee, L. W., and Evans, K. V., 1986, Age and genesis of precious-metals deposits, Buffalo Hump district, central Idaho: *Econ. Geology*, v. 91, p. 990–996.

Marsh, B. D., 1982, On the mechanics of igneous diapirism, stoping and zone melting: *Amer. J. Sci.*, v. 282, p. 808–855.

Marvin, R. F., Hearn, B. C., Jr., Mehnert, H. H., Naeser, C. W., Zartman, R. E., and Lindsey, D. A., 1980, Late Cretaceous–Paleocene–Eocene igneous activity in north-central Montana: *Isochron/West*, v. 29, p. 5–25.

Mattinson, J. M., 1972, Ages of zircons from the northern Cascade Mountains, Washington: *Geol. Soc. America Bull.*, v. 83, p. 3769–3783.

McClernan, H. G., 1983, Metallic mineral deposits of Lewis and Clark County: *Mon. Bur. Mines and Geology, Mem. 52*, 72 p.

McDowell, F. W., 1971, K-Ar ages of igneous rocks from the western United States: *Isochron/West*, v. 2, p. 1–16.

McKnight, E. T., 1974, Geology and ore deposits of the Rico District, Colorado: *U.S. Geol. Survey Prof. Paper 723*, 100 p.

Means, W. D., 1962, Structure and stratigraphy in the central Toiyabe Range, Nevada: *Cal. Univ. Publ., Geol Sci.* 42, p. 71–110.

Meinert, L. D., Newberry, R. and Einaudi, M. T., 1981, An overview of tungsten, copper and zinc-bearing skarns in western North America, *in* Silberman, M. L., ed., *Proceedings of the symposium on mineral deposits of the Pacific Northwest* U.S. Geol. Sur. Open File Report 81-0355, p. 303–327.

Menzer, F. J., 1983, Metamorphism and plutonism in the central part of the Okanogan Range, Washington: *Geol. Soc. America Bull.*, v. 94, p. 471–498.

Meyer C., Shea, F. P., and Goddard, C. C., Jr., 1968, Ore deposits at Butte, Montana, *in* Ridge, J. D., ed., *Ore Deposits of the United States, 1933–1967: Trans. AIME*, v. 2, p. 1373–1415.

Meyers, P. E., 1982, Geology of the Harpster area, Idaho County, Idaho: *Idaho Bur. Mines and Geology Bull.*, v. 25, 46 p.

Miller, B. W., and Silberman, M. L., 1977, Cretaceous K-Ar age of hydrothermal alteration at the North Fish Creek Porphyry Cu deposit, Fish Creek Mountains, Lander County, Nevada: *Isochron/West*, v. 18, p. 7.

Miller, C. F., 1978, Monzonitic plutons, California, and a model for generation of alkali-rich, near silica-saturated magmas: *Contrib. Mineralogy Petrology*, v. 67, p. 349–355.

_____, and Bradfish, L. J., 1980, An inner Cordilleran belt of muscovite-bearing plutons: _Geology_, v. 8, p. 412–416.

_____, Howard, K. A., and Hoisch, T. D., 1982, Mesozoic thrusting, metamorphism, and plutonism, Old Woman–Piute Range, southeastern California, _in_ Frost, E. G., and Martin, D. L., _Mesozoic-Cenozoic Tectonic Evolution of the Colorado River Region, California, Arizona, and Nevada:_ San Diego, Calif., Cordilleran Publishers, p. 561–581.

Miller, D. M., 1984, Sedimentary and igneous rocks of the Pilot Range and vicinity, Utah and Nevada: _Utah Geol. Assoc. Publ._, v. 13, p. 45–63.

Miller, E. L., and Cameron, C. S., 1982, Late Precambrian to Late Cretaceous evolution of the southwest Mojave Desert, California, _in_ Cooper, J. D., _Geology of Selected Areas in San Bernadino Mountains, Western Mojave Desert and Southern Great Basin, California, Field Guide:_ Geol. Soc. America Field Trip Guidebook, p. 21–34.

_____, and Sutter, J. F., 1981, ^{40}Ar/^{39}Ar age spectra for biotite and hornblende from plutonic rocks in the Victorville region, California: _Geol. Soc. America Bull._, v. 92, p. 164–169.

Miller, F. K., 1974, Preliminary geologic map of the Newport Number 1 quadrangle, Pend Oreille County, Washington, and Bonner County, Idaho: _Washington Div. Geology Earth Resources Geol. Map, GM-7._

_____, and Engels, J. C., 1975, Distribution and trends of discordant ages of the plutonic rocks of northeastern Washington and northern Idaho: _Geol. Soc. America Bull._, v. 86, p. 817–828.

_____, and Theodore, T. G., 1982, Molybdenum and tungsten mineralization associated with two stocks in the Harvey Creek area, northeastern Washington: _U.S. Geol. Survey Open-File Rpt., 82-295_, p. 31.

_____, Clark, L. D., and Engels, J. C., 1975, Geology of the Chewelah-Loon Lake area, Stevens and Spokane Counties, Washington: _U.S. Geol. Survey Prof. Paper 806_, 74 p.

Miller, R. N., 1959, Geology of the south Moccasin Mountains, Fergus Co., Montana: _Montana Bur. Mining Geologists Mem. 37_, 44 p.

Miller, W. J., and Webb, R. W., 1940, Descriptive geology of the Kernville Quadrangle, California: _Calif. J. Mines Geology_, v. 36, p. 343–378.

Mills, J. W., 1977, Zinc and lead ore deposits in carbonate rocks, Stevens County, Washington: _Wash. Div. Geology Earth Resources Bull._, v. 70, p. 171.

Misch, P., 1966, Tectonic evolution of the northern Cascades of Washington State, _in_ Gunning, H. C., ed., _Tectonic History of Mineral Deposits of the Western Cordillera:_ Can. Inst. Mining Metallurgy Spec. Vol. 8, p. 101–148.

_____, 1968, Plagioclase compositions and non-anatectic origin of migmatitic gneisses in northern Cascade Mountains of Washington State: _Contrib. Mineralogy Petrology_, v. 17, p. 1–70.

Mitchell, A. H. G., and Garson, M. S., 1981, _Mineral Deposits and Global Tectonic Settings:_ London, Academic 405 p.

Miyashiro, A., 1961, Evolution of metamorphic belts: _J. Petrology_, v. 2, p. 277–311.

_____, 1973, _Metamorphism and Metamorphic Belts:_ New York, Wiley, 492 p.

Moen, W. S., 1973, Conconully mining district of Okanogan County, Washington: _Wash. Div. Mines Geology Information Circ._, v. 49, p. 42.

Moore, J. G., 1959, The quartz diorite boundary line in the western United States: _J. Geology_, v. 67, p. 197–210.

Moore, J. N., and Foster, C. T., Jr., 1980, Lower Paleozoic metasedimentary rocks in the east-central Sierra Nevada, California—Correlation with Great Basin formations: _Geol. Soc. America Bull._, v. 91, p. 37–43.

Mortimer, N., 1985, Structural and metamorphic aspects of Middle Jurassic terrane juxtaposition, northeastern Klamath Mountains, California, *in* Howell, D. G., ed., *Tectonostratigraphic Terranes of the Circum-Pacific Region*: Circum-Pacific Council Energy Min. Resources, Earth Sci. Series, no. 1, p. 201–214.

Muffler, L. J. P., 1964, Geology of the Frenchie Creek Quadrangle north-central Nevada: *U.S. Geol. Survey Bull. 1179*, 99 p.

Mullen, E. D., 1980, Temperature-pressure progression in high pressure Permo-Triassic metamorphic rocks of northeastern Oregon: *Tran. Amer. Geophys. Un.*, v. 61, no. 16, p. 71.

_____, and Sanewitz, D., 1983, Paleozoic and Triassic terranes of the Blue Mountains, northeast Oregon; discussion and field trip guide—Pt. 1. A new consideration of old Problems: *Oreg. Geology*, v. 45, no. 6, p. 65-68.

Mutch, T. A., and McGill, G. E., 1962, Deformation in host rocks adjacent to an epizonal pluton (the Royal Stock, Montana): *Geol. Soc. America Bull.*, v. 73, p. 1541–1544.

Myers, P. E., 1982, Geology of the Harpster area, Idaho County, Idaho: *Idaho Bur. Mines Geology Bull. 25*, 46 p.

Nabalek, P. I., Labotka, T. C., O'Neil, J. R., and Papike, J. J., 1984, Contrasting fluid/rock interaction between the Notch Peak granitic intrusion and argillites and limestones in western Utah — Evidence from stable isotopes and phase assemblages: *Contrib. Mineralogy Petrology*, v. 86, p. 25-34.

Nash, D. B., 1962, Contact metamorphism at Birch Creek, Blanco Mountain quadrangle, Inyo County, California: M.S. thesis, Univ. California, Berkeley, Calif., 53 p.

Nash, J. T., Granger, H. C., and Adams, S. S., 1981, Geology and concepts of genesis of important types of uranium deposits: *Econ. Geology*, 75th Anniversary Vol., p. 63–116.

Newberry, R. J., 1979, Systematic variations in W-Mo-Cu skarn formation in the Sierra Nevada—An overview: Stanford Univ., *Dept. Geol., Geol. Soc. America, 92nd Ann. meet., SanDiego, Calif.* (November 5-8, 1979), *Geol. Soc. America Abstr. with Programs 11*, v. 7, 486 p.

_____, 1982, Tungsten-bearing skarns of the Sierra Nevada—I. The Pine Creek mine, California: *Econ. Geology*, v. 77, p. 823–844.

_____, and Einaudi, M. T., 1981, Tectonic and geochemical setting of tungsten skarn mineralization in the Cordillera, *in* Dickinson, W. R., and Payne, W. D., eds., *Relations of Tectonics to Ore Deposits in the Southern Cordillera*: Arizona Geol. Soc. Digest, v. 14, p. 99–112.

Nokleberg, W. J., and Kistler, R. W., 1980, Paleozoic and Mesozoic deformations in the central Sierra Nevada, California: *U.S. Geol. Survey Prof. Paper 1145*, 24 p.

Nolan, T. P., 1962, The Eureka mining district, Nevada: *U.S. Geol. Survey Prof. Paper 406*, 78 p.

Norton, D., 1982, Fluid and heat transport phenomena typical of copper-bearing pluton environments, southeastern Arizona, *in* Titley, S. R., ed., *Advances in Geology of Porphyry Copper Deposits: Southwestern North America*, Univ. of Ariz. Press, Tucson, p. 59–72.

_____, and Taylor, H. P., Jr., 1979, Quantitative simulation of the hydrothermal systems of crystallizing magmas on the basis of transport theory and oxygen isotope data; an analysis of the Skaergaard Intrusion: *J. Petrology 20*, p. 421–486.

_____, and Knapp, R., 1977, Transport phenomena in hydrothermal systems—The nature of porosity: *Amer. J. Sci.*, v. 277, p. 913–936.

_____, and Knight, J. 1977, Transport phenomena in hydrothermal systems—Cooling plutons: *Amer. J. Sci.*, v. 277, p. 937–981.

Norwick, S. A., 1972, The regional Precambrian metamorphic facies of the Prichard Formation of western Montana and Northern Idaho: Ph.D. thesis, Univ. Montana, Missoula, Mont., 129 p.

――――, 1977, Precambrian amphibolite facies metamorphism in the Belt rocks of northern Idaho: *Geol. Soc. America Abstr. with Programs*, v. 9, p. 753.

Obradovich, J. D., Mutschler, F. E., and Bryant, B., 1969, Potassium-Argon ages bearing on the igneous and tectonic history of the Elk Mountains and vicinity, Colorado – A preliminary report: *Geol. Soc. America Bull.*, v. 80, p. 1749-1756.

Okulitch, A. V., 1984, The role of the Shuswap Metamorphic Complex in Cordilleran tectonism—A review: *Can. J. Earth Sci.*, v. 21, p. 1171-1193.

Oldow, J. S., Avé Lallèmant, H. G., and Schmidt, W. J., 1984, Kinematics of plate convergence replaced from Mesozoic structures in the western Cordillera: *Tectonics*, v. 3, p. 201-227.

Olson, J. C., and Hinrichs, E. N., 1960, Beryl-bearing pegmatites in the Ruby Mountains and other areas in Nevada and northwestern Arizona: *U.S. Geol. Survey Bull. 1082-D*, 200 p.

Oxburgh, E. R., and Turcotte, D. L., 1971, Origin of paired metamorphic belts and crustal dilation in island arc regions: *J. Geophy. Res.*, v. 76, p. 1315-1327.

Page, B. M., and Engebretson, D. C., 1984, Correlation between the geologic record and computed plate motions in central California: *Tectonics*, v. 3, p. 133-135.

Parmentier, E. M., and Schedl, A., 1981, Thermal aureoles of igneous intrusions—Some possible indications of hydrothermal convective cooling: *J. Geology*, v. 89, p. 1-22.

Phelps, P., and Avé Lallèmant, H. G., 1980, The Sparta ophiolite complex, northeast Oregon—A plutonic equivalent to low K_2O island-arc volcanism: *Amer. J. Sci.*, v. 280A, p. 345-358.

Plummer, C., 1969, Geology of the crystalline rocks, Chiwaukum Mountains and vicinity, Washington Cascades: Unpublished Ph.D. thesis, Univ. Washington, Seattle, Wash., 160 p.

Poole, F. G., Claypool, G. E., and Fouch, T. D., 1983, Major episodes of petroleum generation in part of the northern Great Basin: *Geothermal Resources Council Spec. Rpt. 13*, p. 207-213.

Proffett, J. M., Jr., and Dilles, J. H., 1984, Geologic map of the Yerington District, Nevada: *Nev. Bur. Mines Geology Map 77*.

Prostka, H. S., 1962, Geology of the Sparta Quadrangle, Oregon: *Oregon Dept. Geology Min. Industries Map GMS1*.

Ramberg, H., 1970, Model studies in relation to intrusion of plutonic bodies, *in* Newall, G., and Rast, N., eds., *Mechanism of Igneous Intrusion*: p. 261-272.

Reeves, R. G., and Kral, V. E., 1955, Iron ore deposits of Nevada—part A. Geology and iron ore deposits of the Buena Vista Hills, Churchill and Pershing Counties, Nevada: *Nev. Bur. Mines Bull. 53*.

――――, Shawe, F. R., and Kral, V. E., 1958, Iron ore deposits of Nevada—Pt. B. Iron ore deposits of west-central Nevada: *Nev. Bur. Mines Bull. 53*.

Reid, R. R., Morrison, D. A., and Greenwood, W. R., 1973, The Clearwater orogenic zone—A relict of Proterozoic orogeny in central and northern Idaho: *Belt Symposium*, v. 2, Dept. Geology, Univ. Idaho, p. 10-56.

Rice, C. M., Lux, D. R., and Macintyre, R. M., 1982, Timing of mineralization and related intrusive activity near Central City, Colorado: *Econ. Geology*, v. 77, p. 1655-1666.

――――, Harmon, R. S., and Shepard, C. J., 1985, Central City, Colorado—The upper part of an alkaline prophyry molybdenum system: *Econ. Geology*, v. 80, p. 1769-1796.

Rice, J. M., 1977a, Progressive metamorphism of impure dolomitic limestone in the Marysville aureole, Montana: *Amer. J. Sci.*, v. 277, p. 1-24.

_____, 1977b, Contact metamorphism of impure dolomitic limestone in the Boulder aureole, Montana: *Contrib. Mineralogy Petrology*, v. 59, p. 237–260.

_____, 1979, Petrology of clintonite-bearing marbles in the Boulder aureole, Montana: *Amer. Mineralogist*, v. 64, p. 519–526.

Rinehart, C. D., and Fox, K. F., Jr., 1972, Geology of the Loomis quadrangle, Okanogan County, Washington: *Washington Div. Mines Geology Bull.*, v. 64, 124 p.

Robinson, G. D., Klepper, M. R., and Obradovich, J. D., 1968, Overlapping plutonism, volcanism and tectonism in the Boulder Batholith region, western Montana: *Geol. Soc. America Mem. 16*, p. 557–576.

Rogers, J. J. W., Burchfiel, B. C., Abbott, E. W., Anepohl, J. K., Ewing, A. H., Koehnken, P. J., Novitsky-Evans, J. M., and Talukdar, S. C., 1974, Paleozoic and Lower Mesozoic volcanism and continental growth in the western United States: *Geol. Idaho: U.S. Geol. Survey Bull. 877*, 161 p.

Ross, D. C., 1983, The Salinian block—A structurally displaced granitic block in the California Coast Ranges: *Geol. Soc. America Mem. 159*, 316 p.

Ross, D. C., 1983, The Salinian block — A structurally displace granitic block in the California Coast Ranges: *Geol. Soc. America Mem. 159*.

_____, 1985, Mafic gneissic complex in the southernmost Sierra Nevada: *Geology*, v. 13, p. 288–291.

Rumble, D., III, and Spear, F. S., 1983, Oxygen isotope equilibration and permeability enhancement during regional metamorphism: *J. Geol. Soc. London*, v. 140, p. 619–628.

Rye, R. O., Hall, W. E., and Ohnoto, H., 1974, Carbon, hydrogen, oxygen and sulfur isotope study of the Darwin lead-silver-zinc deposit, southern California: *Econ. Geology*, v. 69, p. 468–481.

Saleeby, J. B., 1981, Ocean floor accretion and volcanoplutonic arc evolution of the Mesozoic Sierra Nevada, *in* Ernst, W. G., ed., *The Geotectonic Development of California* (Rubey Vol. I): Englewood Cliffs, N.J., Prentice-Hall, p. 132–181.

_____, 1982, Time relations and structural-stratigraphic patterns in ophiolite accretion, west central Klamath Mountains, California: *Jour. Geophys. Res.* v. 87, p. 3831–3848.

_____, and Sharp, W., 1980, Chronology of the structural and petrologic development of the southwest Sierra Nevada foothills, California: *Geol. Soc. America Bull. 9*, v. 91, pt. I, summary, p. 327–320, pt. 2, p. 1416–1535.

_____, Goodin, S. E., Sharp, W. D., and Busby, C. J., 1978, Early Mesozoic paleotectonic-paleogeographic reconstruction of the southern Sierra Nevada region, *in* Howell, D. G., and McDougall, K. A., eds., *Mesozoic Paleogoegraphy of the Western United States:* Pacific Section, Soc. Econ. Paleontologists Mineralogists, Pacific Coast Paleogeography Symp. 2, p. 311–336.

Sawkins, F. J., 1984, *Metal Deposits in Relation to Plate Tectonics:* New York, Springer-Verlag, 325 p.

Schweikert, R. A., 1981, Tectonic evolution of the Sierra Nevada Range, *in* Ernst, W. G., ed., *The Geotectonic Development of California* (Ruby Vol. I): Englewood Cliffs, N.J., Prentice-Hall, p. 87–131.

_____, Bogen, N. L., Girty, G. H., Hanson, R. E., and Merguerian, C., 1984, Timing and expression of the Nevada orogeny, Sierra Nevada, California: *Geol. Soc. America Bull.*, v. 95, p. 967–979.

Sharry, J., 1981, The geology of the western Tehachapi Mountains, California: Ph.D. thesis, Massachusetts Inst. Technology, Cambridge, Mass., 215 p.

Shieh, Y. N., and Taylor, H. P., Jr., 1969, Oxygen and hydrogen isotope studies of contact metamorphism in the Santa Rosa Range, Nevada and other areas: *Contrib. Mineralogy Petrology*, v. 20, p. 306–356.

———, and Taylor, H. P., Jr., 1969a, Oxygen and carbon isotopic studies of contact metamorphism of carbonate rocks: *J. Petrology 10*, p. 307–331.

Schultz, K. L., and Levi, S., 1983, Paleomagnetism of Middle Jurassic plutons of the north-central Klamath Mountains: *Geol. Soc. America Abst.*, v. 15, p. 427.

Silver, L. T., 1971, Problems of crystalline rocks of the Transverse Ranges [abstract]: *Geol. Soc. America Abstr. with Programs*, v. 3, p. 193–194.

———, Taylor, H. P., Jr., and Chappell, B., 1979, Some petrologic geochemical and geochronologic observations of the Penninsular Range batholith near the international border of the United States of America and Mexico, *in* Abbott, P. L., and Todd, V. R., eds., *Mesozoic Crystalline Rocks — Peninsular Ranges Batholith and Pegmatites, Point Sal Ophiolite:* Guidebook Geol. Soc. America Ann. Meet., San Diego State Univ., San Diego, Calif., p. 83–116.

Silverberg, D. S., 1986, Metamorphic petrology and structure of the Hyndman and East Fork formations of the Pioneer Core Complex, Idaho: *Geol. Soc. America Abstr. with Programs*, v. 18, no. 2, p. 185.

Simmons, E. C., and Hedge, C. E., 1978, Minor-element and Sr-isotope geochemistry of Tertiary stocks, Colorado Mineral Belt: *Contrib. Mineralogy Petrology*, v. 67, p. 379–397.

Simons, F. S., 1973, Geologic map and sections of Nogales and Lochiel quads, Santa Cruz County, Arizona: *U.S. Geol. Survey Misc. Geol. Inv. Map I-762.*

Smedes, H. W., 1966, Geology and igneous petrology of the northern Elkhorn Mountains, Jefferson and Broadwater counties, Montana: *U.S. Geol. Survey Prof. Paper 510*, 116 p.

———, and Thomas, H. H., 1965, Reassignment of the Lowland Creek volcanics to Eocene age: *J. Geology*, v. 73, p. 508–510.

Smith, A. R., 1970, Trace elements in the Plumas copper districts, Plumas County, California: *Calif. Div. Mines Geology*, Spec. Rpt. 103, 26 p.

Smith, J. F., Jr., 1972, *Age of the Pony Trail Group in the Cortez Mountains, Eureka County, Nevada, in* Cohee, G. V., and Wright, W. B., eds., Changes in stratigraphic nomenclature: *U.S. Geol. Sur. Bull.*, 1934-A, p. A 83.

Smith, J. G., McKee, E. H., Tatlock, D. B., and Marvin, R. F., 1971, Mesozoic granitic rocks in northwestern Nevada — A link between the Sierra Nevada and Idaho batholiths: *Geol. Soc. America Bull.*, v. 82, no. 16, p. 2933–2944.

Snee, L. W., Lund, K., and Hoover, A. L., 1985, Discrimination of Idaho batholith plutons by structure and age: *Geol. Soc. America Abstr. with Programs*, v. 17, p. 409.

Snoke, A. W., and Howard, K. A., 1984, Geology of Ruby Mountains–East Humboldt Range, Nevada–A Cordilleran metamorphic core complex, *in* Lintz, J., Jr., ed., *Western Geological Excursions, Vol. 4:* Geol. Soc. America Ann. Meet., Mackay Sch. Mines, Reno, Nev., p. 260–303.

———, and Lush, A. P., 1984, Polyphase Mesozoic-Cenozoic deformational history of the northern Ruby Mountains-east Humboldt Range, Nevada, *in* Lintz, J., Jr., ed., *Western Geological Excursions, Vol. 4:* Geol. Soc. America Ann. Meet., Mackay School of Mines, Reno, Nev., p. 232–260.

Soloman, G. C., 1986, $^{18}O/^{16}O$ studies of Mesozoic-Cenozoic granites and their bearing on crustal evolution in southern California: *Geol. Soc. America Abstr. with Programs*, v. 18, p. 188.

Sonnevil, R. A., 1979, Evolution of skarn at Monte Cristo, Nevada: M.S. thesis, Stanford Univ., Stanford, Calif., 84 p.

Speed, R. C., 1962, Scapolitized gabbroic complex, west Humboldt Range, Nevada: Ph.D. thesis, Stanford Univ., Stanford, Calif., 255 p.

_____, 1976, Geologic map of the Humboldt Lopolith and surrounding terrain, Nevada: *Geol. Soc. America Map Chart Series MC-14.*

_____, and Jones, T. A., 1969, Synorogenic quartz sandstone in the Jurassic mobile belt of western Nevada—Boyer Ranch Formation: *Geol. Soc. America Bull.*, v. 80, p. 2551–2584.

_____, and Kistler, R. W., 1980, Cretaceous volcanism, Excelsior Mountains, Nevada: *Geol. Soc. America Bull.* 91, p. 392–398.

Steefel, C. I., and Atkinson, W. W., Jr., 1984, Hydrothermal andalusite and corundum in the Elkhorn district, Montana: *Econ. Geology*, v. 79, p. 573–579.

Spera, F. J., 1980, Aspects of magma transport, *in* Hargraves, R. B., ed., *Physics of Magmatic Processes*, Princeton University Press, Princeton, p. 265–324.

Stern, T. W., Bateman, P. C., Morgan, B. A., Newell, M. F., and Peck, D. L., 1981, Isotopic U-Pb ages of zircon from the granitoids of the central Sierra Nevada, California: *U.S. Geol. Survey Prof. Paper 1185*, 17 p.

Stewart, J. H., 1980, Geology of Nevada: *Nev. Bur. Mines Geology Spec. Pub.. 4*, 136 p.

Stone, P., Howard, K. A., and Hamilton, W. D., 1983, Correlation of Paleozoic strata of southeastern Mojave Desert region, California and Arizona: *Geol. Soc. America Bull.*, v. 94, p. 1135–1147.

Sugimura, A., and Uyeda, S., 1973, *Island Arcs: Japan and its Environs:* Amsterdam, Elsevier, 247 p.

Sylvester, A. G., Oertel, G., Nelson, C. A., and Christie, J. M., 1978, Papoose Flat pluton — A granitic blister in the Inyo Mountains, California: *Geol. Soc. America Bull.*, v. 89, 1205–1219.

Tatlock, D. B., 1961, Redistribution of K, Na, and Al in some felsic rocks in Nevada and Sweden: *Mining Engineering*, v. 13, no. 11, p. 1256.

Taubeneck, W. H., 1957, Geology of the Elkhorn Mountains, northeastern Oregon: *Geol. Soc. America Bull.*, v. 68, p. 181–238.

_____, 1958, Argillites in contact aureole of Bald Mountain batholith, Elkhorn Mountains, northeastern Oregon: *Geol. Soc. America Abstr.*, v. 69, no. 12, p. 1650.

Taubeneck, W. H., 1964, Cornucopia stock, Wallowa Mountains, northeastern Oregon-field relations: *Geol. Soc. America Bull.*, v. 75, p. 1093–1115.

Taylor, B. E., and O'Neil, J. R., 1977, Stable isotope studies of metasomatic Ca-Fe-Al-Si skarns and associated metamorphic and igneous rocks, Osgood Mountains, Nevada: *Contrib. Mineralogy Petrology*, v. 63, p. 1–49.

_____, 1974, Oxygen and hydrogen isotope evidence for large-scale circulation and interaction between ground waters and igneous intrusions, with particular reference to the San Juan volcanic field, Colorado, *in* Hofman, A. W., Giletti, B. J., Yoder, H. S., Jr., and Yund, R. A., eds., *Geochemical Transport and Kinetics:* Carnegie Inst. Wash. Publ. 634, p. 299–324.

_____, 1977, Water/rock interactions and the origin of H_2O in granitic batholiths: *J. Geol. Soc. London*, v. 133, p. 509–558.

_____, and Silver, L. T., 1978, Oxygen isotope relationships in plutonic rocks of the Peninsular Ranges Batholith, southern and Baja California, *in* Zartman, R. E., ed., *Short papers of the 4th international conference on geochronology, cosmochronology, and isotope geology: U.S. Geol. Sur.* Open-File Report #78-701, p. 423–426.

_____, 1978, Oxygen and hydrogen isotope studies of plutonic granitic rocks: *Earth Planet. Sci. Lett.*, v. 38, p. 177–210.

_____, and Magaritz, M., 1978, Oxygen and hydrogen isotope studies of the Cordilleran batholiths of western North America: *DSIR Bull. 210*, 151–173.

Tilling, R. I., 1973, Boulder batholith, Montana — A product of two contemporaneous but chemically distinct magma series: *Geol. Soc. America Bull.*, v. 84, p. 3879–3900.

_____, Klepper, M. R., and Obradovich, J. D., 1968, K-Ar ages and time span of emplacement of the Boulder Batholith, Montana: *Amer. J. Sci.*, v. 266, p. 671–689.

Titley, S. R., 1982a, *Advances in the Geology of Porphyry Copper Deposits, Southwestern North America:* Univ. Arizona Press., Tucson, Ariz.

_____, 1982b, Some features of tectonic history and ore genesis in the Pima Mining District, Pima County, Arizona, *in* Titley, S. R., ed., *Advances in Geology of the Porphyry Copper Deposits Southwestern North America:* Univ. Arizona Press, Tucson, Ariz., Chap. 19.

_____, and Beane, R. E., 1981, Porphyry copper deposits — Pt. 1. Geologic settings, petrology and tectogenesis: *Econ. Geology*, 75th Anniversary Vol., p. 214–234.

_____, and Hicks, 1966, *Geology of the Porphyry Copper Deposits, Southwest North America:* Univ. Arizona Press, Tucson, Ariz.

Tobisch, O. T., Saleeby, J. B., and Fiske, R. S., 1986, Structural history of continental volcanic arc rocks, eastern Sierra Nevada, California — A case for extensional tectonics: *Tectonics*, v. 5, p. 65–94.

Todd, V. R., and Shaw, S. E., 1979, Metamorphic, intrusive and structural framework of the Peninsular Ranges batholith in southern San Diego County California, *in* Abbott, P. L., and Todd, V. R., eds., *Mesozoic Crystalline Rocks — Peninsular Ranges Batholith and Pegmatites, Point Sal. Ophiolite:* Guidebook Geol. Soc. America Ann. Meet. San Diego State Univ., San Diego, Calif., p. 177–232.

Toth, M. I., and Stacey, J. S., 1985, Uranium-lead geochronology and lead isotope data for the Bitterroot lobe of the Idaho batholith: *Geol. Soc. America Abstr. with Programs*, v. 17, no. 4, p. 268.

Troxel, B. W., and Morton, P. K., 1962, Mines and mineral deposits of Kern County, California: *Calif. Div. Mining Geology County Rpt. 1*, 369 p.

Tschanz, C. M., and Pampeyan, E. H., 1970, Geology and mineral deposits of Lincoln County, Nevada: *Nev. Bur. Mines Bull.*, v. 73, 188 p.

Turner, F. J., 1981, *Metamorphic Petrology: Mineralogical, Field, and Tectonic Aspects:* 2nd Ed., New York, McGraw-Hill, 524 p.

Tweto, O., 1975, Laramide (late Cretaceous-early Tertiary) orogeny in the southern Rocky Mountains: *Geol. Soc. America Mem. 144*, p. 1–44.

Vallier, T. L., and Batiza, R., 1978, Petrogenesis of spilite of kerotophyre from a Permian and Triassic volcanic arc terrane, eastern Oregon and western Idaho, U.S.A.: *Canadian J. Earth Sci.*, v. 15, p. 1356–1369.

Vikre, P. G., 1981, Silver mineralization in the Rocherster Mining District, Pershing County, Nevada: *Econ. Geology*, v. 76 no. 3, p. 580–609.

Walawender, M. J., 1979, Basic plutons of the Peninsular Ranges batholith, southern California, *in* Abbott, P. L., and Todd, V. R., eds., *Mesozoic Crystalline Rocks — Peninsular Ranges Batholith and Pegmatites, Point Sal Ophiolite:* Guidebook Geol. Soc. America Ann. Meet., San Diego State Univ., San Diego, Calif., p. 151–162.

Walker, J. D., and Burchfiel, B. C., 1983, Permo-Triassic rocks in the northern Mojave Desert region and their tectonic significance, *in* Gurgel, K. D., ed., *Geologic excursions in stratigraphy and tectonics from southeastern Idaho to the southern Inyo Mountains, CA, via Canyonlands and Arches National Parks*, Utah: Guidebook, Part II, Special Studies, Utah Geological and Mineralogical Survey, v. 60, p. 48–50.

___, 1985, Permo-Triassic paleogeography and tectonics of the southwestern United States: Ph.D. thesis, Massachusetts Inst. Technology, Cambridge, Mass.

Walker, N. W., 1981, U-Pb geochronology of ophiolite and volcanic-plutonic arc terranes, northeastern Oregon and westernmost central Idaho [abstract]: (EOS) *Trans. Amer. Geophys. Un.*, v. 62, p. 1087.

___, and Mattinson, J. M., 1980, The Canyon Mountain complex, Oregon — U-Pb ages of zircons and possible tectonic correlations: *Geol. Soc. America Abstr. with Programs*, v. 12, p. 544.

Walther, J. V., and Wood, B. J., 1984, Rate and mechanism in prograde metamorphism: *Contrib. Mineralogy Petrology*, v. 88, p. 246–259.

Westra, G., 1982, Alteration and mineralization in the Ruth porphyry copper deposit near Ely, Nevada: *Econ. Geology*, v. 77, p. 950–970.

White, W. H., 1968, Plutonic rocks of the southern Seven Devils Mountains, Idaho: Ph.D. dissertation, Oregon State Univ., Corvallis, Oreg., 200 p.

___, 1973, Flow structure and form of the Deep Creek stock, southern Seven Devils Mountains, Idaho: *Geol. Soc. America Bull.*, v. 84, p. 199–210.

Wickham, S. M., and Oxburgh, E. R., 1985, Continental rifts as a setting for regional metamorphism: *Nature*, v. 318, p. 330–333.

Wilkinson, W. H., Jr., Vega, L. A., and Titley, S. R., 1982, Geology and ore deposits at Mineral Park Mojave County, Arizona, *in* Titley, S. R., ed., *Advances in Geology of the Porphyry Copper Deposits:* Univ. Arizona Press, Tucson, Ariz. Chap. 26.

Willden, R., 1964, Geology and mineral deposits of Humboldt County, Nevada: *Nev. Bur. Mines Bull.*, v. 59, 154 p.

___, and Speed, R., 1974, Geology and mineral deposits of Churchill County, Nevada: *Nev. Bur. Mines Geology Bull.*, v. 83, 95 p.

Wolfe, H. D., and White, D. S., 1951, Preliminary report on tungsten in Oregon: *Oreg. Dept. Geology Mining Ind. Short Paper 22*, 40 p.

Wracher, D. A., 1969, The geology and mineralization of the Peck Mountain area, Hornet Quadrangle, Idaho: MS. thesis, Oregon State Univ., Corvallis, Oreg.

Wylie, P. J., 1977, Crustal anatexis—An experimental review: *Tectonophysics*, v. 43, p. 41–71.

Zen, E-an, Marvin, R. F., and Mehnert, H. H., 1975, Preliminary petrographic, chemical and age data on some intrusive and associated contact metamorphic rocks, Pioneer Mountains, southwestern Montana: *Geol. Soc. America Bull.*, v. 86, p. 367–370.

Zwart, H. J., 1967, The duality of metamorphic belts: *Geologie Mijnbouw*, v. 46, p. 283–309.

Peter Misch*

Department of Geological Sciences
University of Washington
Seattle, Washington 98195

6

TECTONIC AND METAMORPHIC EVOLUTION OF THE NORTH CASCADES: AN OVERVIEW

*deceased

ABSTRACT

The Straight Creek fault on the west and the Ross Lake fault zone on the east divide the North Cascades into three tectonic units: the Northwest Cascades Thrust system; the Skagit Metamorphic Core; and the Methow-Pasayten Belt, flanked on the west by the Hozameen Belt.

Within the Skagit Core, the northwest-trending Barrovian Skagit Metamorphic Suite *s.str.* comprises the isochemical Cascade River Schist, the migmatitic Skagit Gneiss, and two major orthogneiss belts whose primary ages are 220 and 90 m.y. Main-stage Skagit metamorphism is dated at 90–60 m.y., but some late-metamorphically intruded orthogneisses are as young as 45 m.y. South of the Skagit Suite *s.str.*, the same Barrovian regional metamorphism is displayed by other northwest-trending belts of schists, migmatites, and orthogneisses (Skagit Metamorphic Suite *s.l.*). The southernmost belt consists of the Chiwaukum Schist, which is equivalent to the Settler Schist north of the Canadian Border and west of the Straight Creek fault.

The Northwest Cascades Thrust System displays mid-Cretaceous large-scale westward overthrusting and comprises, from above, the Skuksan thrust plate, the Church Mountain thrust plate and its associated Imbricate Zone, and the "Autochthon," the lowest unit exposed on the mainland. The highest plate consists of the blueschist-facies Shuksan Metamorphic Suite comprising the metabasaltic Shuksan Greenschist and the Darrington Phyllite. The Shuksan metamorphism dates back to 150 m.y.b.p., although much of it is reported to have occurred about 130 m.y. ago. The Shuksan metamorphism is unrelated to the mid-Cretaceous thrusting. The orogeny associated with this high-P/T metamorphism must have involved rapid loading by thrust masses no longer in evidence. Prior to mid-Cretaceous telescoping, the Shuksan belt was several 100 km inboard from the outer margin of the system. The Church Mountain thrust plate consists of upper Paleozoic and subordinate Mesozoic strata. These rocks, and those of the "Autochthon," display mid-Cretaceous synorogenic prehnite-pumpellyite-facies metamorphism. The Church Mountain plate includes the Imbricate Zone, which contains slices of various older crystallines. Most widespread is the continental-crustal Yellow Aster Complex, whose metaplutonic rocks record a middle Proterozoic granulite-facies(?) and a "Caledonian" amphibolite-facies cycle. Ubiquitous tectonically emplaced metaperidotites inhabit thrust contacts. The "Autochthon" mainly consists of Mesozoic strata.

The Upper Jurassic and Cretaceous clastics of the Methow-Pasayten Belt are strongly folded. Prehnite-pumpellyite-facies metamorphism is well displayed in the western part of the belt and westward increases to greenschist facies in the Jack Mountain Phyllite, which grades westward into the Buchan-type Elijah Ridge Schist adjacent to the Ross Lake fault. Within the Ross Lake fault zone, the belt of the Permian-Triassic Hozameen Group displays prehnite-pumpellyite-facies metamorphism and forms the Jack Mountain thrust prong at its southern end. Early movement along the Ross Lake fault predates intrusion at 90 m.y.b.p.

INTRODUCTION

In the course of the author's North Cascades Project initiated in 1948 and pursued since then by him and his students, it was found that, to quote from an earlier summary (Misch, 1971),[1] "North Cascades metamorphics greatly vary in grade, type of facies, and

---

[1] The metamorphic-facies map of the North Cascades of Washington and southernmost British Columbia which the author presented at the time was incorporated in the larger map published by Monger and Hutchison (1971).

age. Facies types generally are separated by major faults though superposition occurs also. Paleothermal regimes thus may be estimated for different belts." Closely tied to type of regional metamorphism, tectonic history is prolonged and highly complex, and it involves a variety of structural patterns characteristic of intensely deformed mobile belts.

The range is divided into three large tectonic units by two major early Tertiary fault systems, the Straight Creek fault (Misch, 1966; Vance, 1957) on the west, and the Ross Lake fault zone (Misch, 1966) on the east. Large-scale dextral slip is certain for the former and postulated for the latter (Misch, 1977a). Initiation of faulting prior to intrusion 90 m.y.b.p. is indicated for the Ross Lake fault. Major dextral slip along the Straight Creek fault largely predated local Eocene deposition. Later movements along both faults mainly involved dip-slip displacements. The Ross Lake fault ceased to be regionally active prior to intrusion 48 m.y. ago, whereas the Straight Creek fault was sealed by intrusion 35 m.y. ago.

The regional tectonic units bordered by the two fault systems are, from west to east (Fig. 6-1): The Northwest Cascades Thrust System; the Skagit Metamorphic Core; and the Methow-Pasayten Belt, whose northern part is flanked on the west by the Hozameen Belt located within the Ross Lake fault zone (bounded by the Ross Lake fault proper on the west and by the Jack Mountain-Hozameen fault on the east).

Within and near the Skagit Metamorphic Core, structural continuity is broken by Tertiary granitoid plutons. The largest of these bodies is the early to late Tertiary Chilliwack composite batholith (Daly, 1912; Misch, 1966; Richards and McTaggart, 1976). The most remarkable of the plutons is the Eocene Golden Horn batholith with its alkaline riebeckite granite (Misch, 1965; Stull, 1969). Late Cretaceous granitoid plutons in part likewise break structural continuity, but in part they participated in deformation and metamorphism of the Skagit Core. The postmetamorphic plutons are not discussed further in this paper.

SKAGIT METAMORPHIC CORE

The fault-bounded Skagit Core is of Barrovian type, ranging from chlorite to sillimanite zone in the type region, where the northwest-trending Skagit Metamorphic Suite *s.str.* (Misch, 1952, 1966, 1968, 1977b, 1979) comprises the isochemically metamorphosed Cascade River Schist on the southwest and the mostly migmatized rocks of the Skagit Gneiss on the northeast. The Cascade River Schist ranges from chlorite zone on the west to middle kyanite zone on the east and is derived from a variety of mostly immature clastics with subordinate basaltic and andesitic volcanics and minor limestones. The clastics include a locally prominent, thick graywacke member thought to be derived from adjacent 200-m.y.-old quartz diorite (see below). Very minor conglomerates include quartz-pebble conglomerate. Some of the phyllites and mica schists are highly quartzose. Locally, garbenschiefer is developed.

The Skagit Gneiss ranges from middle kyanite to sillimanite zone (Misch, 1971) and is predominantly derived from originally supracrustal rocks which include all of the lithologies seen in the Cascade River Schist, although there is a larger proportion of immature graywackes; at least partial stratigraphic equivalence is indicated. In the gneiss complex, most of these rocks are variously migmatized. Migmatization was attributed by Misch (1968) to combinations of metasomatism and metamorphic differentiation; a detailed geochemical study of the Skagit Gorge type section by Babcock (1970) led to a similar conclusion; a nonanatectic origin by metamorphic differentiation alone was proposed by Yardley (1978) for one common type of pegmatitic leucosome.

Both Skagit Gneiss and Cascade River Schist contain numerous small bodies of

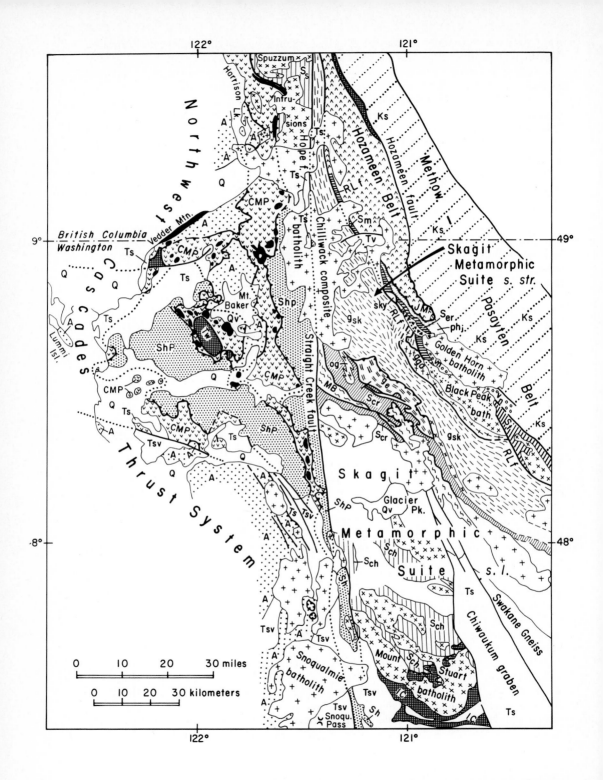

TECTONIC AND METAMORPHIC EVOLUTION OF THE NORTH CASCADES

LEGEND

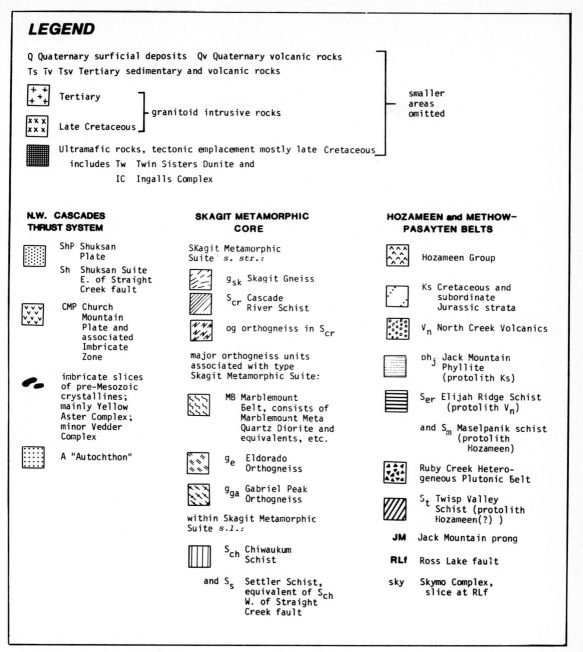

Q Quaternary surficial deposits   Qv Quaternary volcanic rocks
Ts Tv Tsv Tertiary sedimentary and volcanic rocks

⊞ Tertiary ⎤
          ⎬ granitoid intrusive rocks
⊠ Late Cretaceous ⎦

smaller
areas
omitted

▦ Ultramafic rocks, tectonic emplacement mostly late Cretaceous
     includes Tw Twin Sisters Dunite and
            IC Ingalls Complex

N.W. CASCADES THRUST SYSTEM

ShP Shuksan Plate

Sh Shuksan Suite E. of Straight Creek fault

CMP Church Mountain Plate and associated Imbricate Zone

imbricate slices of pre-Mesozoic crystallines; mainly Yellow Aster Complex; minor Vedder Complex

A "Autochthon"

SKAGIT METAMORPHIC CORE

SKagit Metamorphic Suite $s.$ $str.$:

g_{sk} Skagit Gneiss

S_{cr} Cascade River Schist

og orthogneiss in S_{cr}

major orthogneiss units associated with type Skagit Metamorphic Suite:

MB Marblemount Belt, consists of Marblemount Meta Quartz Diorite and equivalents, etc.

g_e Eldorado Orthogneiss

g_{ga} Gabriel Peak Orthogneiss

within Skagit Metamorphic Suite $s.l.$:

S_{ch} Chiwaukum Schist

and S_s Settler Schist, equivalent of S_{ch} W. of Straight Creek fault

HOZAMEEN and METHOW-PASAYTEN BELTS

Hozameen Group

Ks Cretaceous and subordinate Jurassic strata

V_n North Creek Volcanics

ph_j Jack Mountain Phyllite (protolith Ks)

S_{er} Elijah Ridge Schist (protolith V_n)

and S_m Maselpanik schist (protolith Hozameen)

Ruby Creek Heterogeneous Plutonic Belt

S_t Twisp Valley Schist (protolith Hozameen(?))

JM Jack Mountain prong

RLf Ross Lake fault

sky Skymo Complex, slice at RLf

FIG. 6-1.   Generalized map of the North Cascades, showing major structural and metamorphic units. Explanations in the text. Simplified from the following sources: Huntting *et al.* (1961), Misch (1966, 1977b), G. M. Miller (1979), Roddick *et al.* (1979), Tabor *et al.* (1980, 1982), Vance *et al.* (1980), Cater (1982), Haugerud (1985) (individual sources are not listed here if they have been fully incorporated in published compilations).

tectonically introduced metaperidotite, and of locally abundant metamorphosed intrusives ranging from stocks to thin dikes in size, from early to latest metamorphic in relative age, from fully to imperfectly recrystallized in texture, and from gabbroic to predominantly silicic in composition.

Zircon from a trondhjemitic leucosome of the main-stage migmatite in type Skagit Gneiss was dated by Mattinson (1972) as between 90 and 60 m.y. old (with a Precambrian "memory"). Ubiquitous, although volumetrically minor granitic orthogneiss dikes intruded main-stage Skagit Gneiss at the end of the metamorphic cycle (Misch 1966, 1968); they were dated at 45 m.y. by R. L. Armstrong (Babcock *et al.*, 1985). Hoppe (1985) obtained a date of 50 m.y. on a body of well-recrystallized orthogneiss, which the author had mapped within the southeastern portion of the Skagit Gneiss, not far west of unaltered dike swarms of almost 50-m.y.-old Golden Horn granite on the west side of the Ross Lake fault. The thermal history of Skagit metamorphism must have been prolonged, as well as variable in terms of cooling ages. Even so, these dates are not easily reconciled.

The Skagit Suite *s.str.* includes two large orthogneiss belts. The Marblemount Meta-Quartz Diorite (Misch, 1952, 1966, 1979), which predates Skagit metamorphism, is thought to represent basement to the Cascade River Schist within which it forms an anticlinal belt — the extensive Marblemount structural belt, which farther southeast also contains rocks other than the meta-quartz diorite (Grant, 1966; Tabor, 1961). The Marblemount has a primary age of 220 m.y., determined by Mattinson (1972) both in the type area and in the Holden area over 50 km to the southeast. The Marblemount equivalents near Holden were first reported by DuBois (1954; *cf.* Misch, 1966). Cater (1982) presented a different interpretation for these rocks.

The Eldorado Orthogneiss belt (Misch, 1966), likewise quartz-dioritic and recrystallized by Skagit metamorphism, is located between the Cascade River Schist and the Skagit Gneiss and has an intrusive age of 90 m.y. (Mattinson, 1972). This age is shared by some other quartz-dioritic orthogneiss units farther south, as well as by two batholiths, namely, the predominantly quartz-dioritic Black Peak batholith (Misch, 1952, 1966), which straddles the Ross Lake fault on the east side of the Skagit Gneiss, and the Mount Stuart batholith (Smith, 1904), which is at the southern end of the Skagit Core. Both Marblemount Meta-Quartz Diorite and Eldorado Orthogneiss possess faulted contacts and are strongly gneissose, and locally even schistose in marginal portions, whereas interior portions are only weakly gneissose, and locally directionless (especially, the Marblemount), with some igneous textures surviving despite usually complete mineralogical adjustment.

On the northeast side of the Skagit Gneiss and faulted against it, the Gabriel Peak Orthogneiss (Misch, 1966), dated at 65 m.y. by Hoppe (1985), appears to have formed from 90-m.y.-old Black Peak quartz diorite into which it grades to the northeast. Most of the Gabriel Peak is in albite-epidote amphibolite facies and blastomylonitic, whereas adjacent Skagit Gneiss is of higher grade. Local roof pendants of Skagit Gneiss were retrograded when involved in decollement thrusting during conversion of quartz diorite to orthogneiss. Subsequently, some of the diaphthoritic Skagit was downfolded into the orthogneiss. This complex tectonic history next to the Ross Lake fault postdates main-stage Skagit metamorphism.

The Skagit Suite *s.str.* is characterized by intense paracrystalline compressional deformation of considerable complexity in patterns and in history, which varies in details from place to place. Large structures in the type region include syn-metamorphic northeastward thrusts of the Eldorado and Marblemount orthogneiss belts (Eldorado and Magic Mountain thrusts, the latter emplaced during a retrogressive stage of Skagit metamorphism and subsequently downfolded into the structurally underlying Cascade River

Schist). At the southern end of the Skagit Core, R. B. Miller (1985) found metaperidotite correlated with the ophiolitic Ingalls Complex farther south to be thrust northward over Chiwaukum Schist (see below).

South of the Skagit Suite *s.str.*, the same Barrovian regional metamorphism is displayed by other northwest-trending belts of schists, migmatites, and orthogneisses (Skagit Metamorphic Suite *s.l.*), all truncated on the west by the Straight Creek fault zone. The southwesternmost belt consists of the pelitic Chiwaukum Schist (Page, 1939), restudied by several of the author's students, including Plummer (1969, 1980), and recently by Evans and Berti (1986).

Next to the Ross Lake fault, upper kyanite-grade Skagit Gneiss locally displays a Buchan-type overprint (andalusite and cordierite rims on Barrovian minerals); this overprint is mineralogically identical with assemblages characterizing the Elijah Ridge Schist across the fault (Misch, 1968, p. 31). Obviously, the Skagit Gneiss block here was uplifted along the fault and unloaded by erosion so as to enter the *P-T* field of the adjacent Buchan-type belt. Nowhere within the Skagit Suite *s.str.* was any record of pre-Barrovian metamorphism encountered. Armored relics of kyanite-zone minerals (kyanite and staurolite in almandine, pistacite in ferrohastingsite) have been found at a few places in sillimanite-zone Skagit Gneiss, attesting to the progressive nature of the Barrovian event. A record of progression from a lower grade is locally present in kyanite-grade Cascade River Schist also. Where Cascade River Schist and Marblemount Meta-Quartz Diorite are in greenschist facies, many samples indicate that the low-grade Skagit metamorphism was directly superposed on sedimentary and plutonic minerals and textures (Misch, 1966, p. 109).

In contrast, andalusite pseudomorphosed by Barrovian phases had been noted by some of the author's students at a few places in northern portions of the area underlain by Chiwaukum Schist (listed by Evans and Berti, 1986), and been attributed by Heath (1971) to local early contact metamorphism preceding the Barrovian regional Skagit metamorphism. Evans and Berti (1986) found this metamorphic history to be characteristic of the type Chiwaukum Schist, in contrast to the view (Plummer, 1969, 1980) that dynamic contact metamorphism by the 90-m.y.-old Mount Stuart batholith postdated the regional Barrovian event. As reported above, in the type area of the Skagit Metamorphic Suite, the Barrovian event involved the Eldorado Orthogneiss whose intrusive age is 90 m.y. (Mattinson, 1972).

Where the Straight Creek fault emerges from the Chilliwack composite batholith north of the Canadian border as the Hope fault, the Settler Schist (Lowes, 1972) duplicates the Chiwaukum Schist in lithology and structure, in metamorphic history as interpreted by Lowes (1972) and by Plummer (1969), and in associated plutonic history; however, the Settler is west of the Straight Creek-Hope fault. The Chiwaukum-Settler correlation was the basis for Misch's (1977a) proposal that dextral displacement on the fault was on the order of 120 miles (190 km). The reinterpretation of the metamorphic history of the Chiwaukum by Evans and Berti (1986) is matched by the reinterpreted history of the Settler given by Pigage (1976), Bartholomew (1979), and Gabites (1985). As pointed out by Evans and Berti (1986), this match further strengthens the Chiwaukum-Settler correlation. On the west, the Settler Schist is bordered by a tectonic zone which, in agreement with Lowes (1972), is here interpreted as the northern continuation of the root zone of the Shuksan thrust (see below), although a different rock unit is involved; unfortunately, the Chilliwack batholith has destroyed the evidence at this junction where the relationship between the Skagit Metamorphic Core and the Northwest Cascades Thrust System might have been established.

NORTHWEST CASCADES THRUST SYSTEM

The Northwest Cascades Thrust System (Misch, 1966), emplaced in mid-Cretaceous time, consists of large sheets overthrust to the southwest and west (to the northwest in a short salient straddling the Canadian border, the system being arcuate – see Misch, 1966, Fig. 7-17). This is documented by vergence indicators throughout the system; only non-ambiguous vergence indicators are used here – chiefly, overturned folds on all scales, formed during the mid-Cretaceous event. Oceanward overthrusting means continentward underthrusting. The thrust pile exposed on the mainland was well above the main sub-duction zone defined by the ocean plate underthrusting the mobile continental margin. The surface trace of this zone must have been west of the San Juan Islands.

In descending order, the system comprises the Shuksan thrust plate, the Church Mountain thrust plate and its associated Imbricate Zone, and the "Autochthon," which merely is the lowest unit exposed on the mainland; undoubtedly, lower thrusts exist in the subsurface, related to structural units exposed in the San Juan Islands (Brandon *et al.*, 1983). The highest units in the thrust pile are of the most easterly, farthest inboard derivation. Mid-Cretaceous telescoping was preceded by only indirectly indicated Late Jurassic (see below) and probably still earlier thrusting events. Prior to telescoping, the system was many times its present width. After emplacement, the Northwest Cascades Thrust System was subjected to several periods of folding, at least two being of Tertiary age. Thus, originally subhorizontal thrust planes at many places, but by no means every-where, now are broadly to tightly folded. No evidence was found that in mid-Cretaceous time overthrusting was combined with strike-slip faulting (except, of course, for local transverse tears due to differential overfolding or overthrusting). Superposed Tertiary, mainly high-angle faulting, however, is demonstrated at many places by involvement of Tertiary strata.

The Shuksan Thrust Plate

The Shuksan thrust plate consists of the Shuksan Metamorphic Suite, comprising the metabasaltic Shuksan Greenschist and the Darrington Phyllite (Misch, 1966). The syn-kinematic regional Shuksan metamorphism was assigned by Misch (1959) to a subfacies of broadly defined blueschist facies, transitional to greenschist facies. In this subfacies, tholeiitic metabasalts have produced blue amphiboles only in subordinate highly oxidized layers, whereas the less ferric bulk of the rocks is represented by actinolitic greenschists, Al-glaucophane being unstable (Misch, 1959; 1969, p. 45–47). The inter-bedding of greenschists and blueschists in the Shuksan was confirmed by Brown (1974; Brown *et al.*, 1982), and recently discussed by Dungan *et al.* (1983). Regionally, the Shuksan Greenschist is structurally above the Darrington Phyllite and conformable with it. As found by Jones (1959), a stratigraphic interval intervenes locally in which Shuksan and Darrington lithologies alternate (discounting repetition by tight folding, which is also common), and in which minor metamorphosed ironstones are present. Whether the order of superposition means that the Darrington is older than the metabasalt, or whether the sequence perhaps is regionally inverted, is debatable (*cf.* Dungan *et al.*, 1983).

The Darrington consists of pelitic and semipelitic, commonly graphitic phyllites associated with plagioclase-arkosic schistose metapsammites. In part of the region, the metapsammites contain lawsonite (Jones, 1959), as do some greenschists (Vance, 1957), whereas all blueschists carry epidote. The metapsammites indicate a sialic source rich in quartz-dioritic rocks. It needs to be emphasized that the Shuksan Suite as defined here is stratigraphically intact and structurally coherent throughout the large region in which it is preserved, in contrast to dismembered blueschist terranes and melanges, such as characterize the Franciscan Complex (see Blake *et al.*, Chapter 38, this volume).

Shuksan metamorphism dates back to 150 m.y. (Armstrong and Misch, in preparation), although, on the basis of a number of younger dates (including many measured on Darrington Phyllite), Armstrong (1980) and Brown *et al.* (1982) suggested that major metamorphism occurred about 130 m.y. ago. At any rate, the Shuksan metamorphism is unrelated to the mid-Cretaceous emplacement of the thrust plates which was accompanied by prehnite-pumpellyite-facies metamorphism. Intense paracrystalline compressional deformation displayed by the Shuksan Suite indicates that metamorphism was associated with major orogeny. This event must have involved rapid loading by thick thrust masses no longer in evidence, to account for P having reached at least 7 kb in the Shuksan rocks. The rapid burial to a depth in excess of 20 km is postulated to have been succeeded by rapid uplift and erosion to account for removal of the overlying thrust masses. All of this took place in a belt well inboard, prior to mid-Cretaceous telescoping not less than several 100 km from the outer margin of the thrust system defined by the surface trace of the main subduction zone (see above). Perhaps the tectonic setting of deposition of the Shuksan Suite was that of a back-arc basin or of a marginal basin. Whether the subsequent loading by Late Jurassic thrusting postulated here may have had the structural geometry of an internal subduction zone is a matter of speculation, and depends partly on the stratigraphic interpretation of the Shuksan Suite. During mid-Cretaceous thrust emplacement into their present position, the Darrington, and at thrust contacts also the Shuksan, were cataclastically redeformed and locally mylonitized.

The interesting, melangelike assemblage serpentinite-amphibolite-blueschist-minor eclogite discovered by Brown *et al.* (1982) at one locality, while structurally associated with normal Shuksan Suite, differs from it in major structural, stratigraphic, lithologic and facies aspects and perhaps represents a separate, though related terrane. Possibly, this unique occurrence represents a faulted fragment of a higher structural unit not preserved elsewhere, or of the Imbricate Zone below the sole of the Shuksan plate (see below); there is evidence for much Tertiary high-angle faulting near this locality (Jones, 1959).

The Church Mountain Thrust Plate

The Church Mountain thrust plate (Misch, 1966) consists of volcanic and sedimentary rocks of the Devonian to Permian Chilliwack Group (Daly, 1912; Danner, 1957, 1966; Misch, 1952, 1966; Monger, 1966), with lesser amounts of Mesozoic strata that recall units in the "Autochthon." In addition, this thrust sheet contains the complex Imbricate Zone, which tends to occupy the upper part of the sheet, but ranges from representing all of it to being nearly absent. Apart from tectonic slices of various previously metamorphosed units in the Imbricate Zone, both Church Mountain plate and "Autochthon" display a high-P variant of prehnite-pumpellyite facies (aragonite and rare lawsonite), reminiscent of what Vance (1968) found in the San Juan Islands. Neither Chilliwack Group nor Mesozoic strata have been affected by the Shuksan metamorphism, which obviously happened far away. Their prehnite-pumpellyite metamorphism accompanied the mid-Cretaceous thrusting event.

Among tectonic slices of pre-Mesozoic crystallines within the Imbricate Zone, the Yellow Aster Complex, which represents ancient continental crust (Misch, 1966) is the most prominent. Its gabbroic to leucotrondhjemitic metaplutonic rocks record an earlier granulite(?)-facies and a later amphibolite-facies cycle. A tentative Precambrian- + "Caledonian" age assignment was confirmed by Mattinson (1972), who dated the first cycle at 1452–2000 m.y. and the second at about 430 m.y. The "Caledonian" amphibolite-facies metamorphism occurs both as an overprint in Precambrian pyroxene gneisses, and in gneisses and amphibolites that lack any record of the first cycle. There also are "late Caledonian" intrusives that have not participated in the amphibolite-facies event but yield a similar isotopic age.

The "Caledonian" portion of the Yellow Aster may be correlated with the Turtle-back Complex of the San Juan Islands, first recognized as metaplutonic basement by J. A. Vance in 1958 (see Danner, 1966; Vance, 1968, 1975), and dated at about 430 m.y. by Mattinson (1972). The Turtleback is at a much lower structural level (Brandon et al., 1983) than the imbricate slices of Yellow Aster, which represent an originally more easterly belt. Within the "Autochthon" of the Northwest Cascades, a crystalline fault wedge on Vedder Mountain at the Canadian Border contains, in addition to late Paleozoic metamorphics (see below), some metaplutonic rocks that are doubtful Turtleback equivalents (McMillan, 1966; Misch, 1966). East of the Skagit Metamorphic Core, a slice of undated granulite-facies rocks at the Ross Lake fault (Skymo complex, Misch, 1966; Wallace, 1976) perhaps correlates with the Precambrian portion of the Yellow Aster.

Near the Canadian border, Yellow Aster slices have been squeezed up along the steep root of the Shuksan thrust and then transported westward below the sole of the Shuksan plate, becoming imbricated with strata of the Church Mountain plate. South of the Skagit River, the north-northwest-trending root zone, as interpreted by the author, is cut off by the north-south-trending Straight Creek fault, as shown by mapping by Vance (1957).

In the central Cascades, the Shuksan Suite is exposed east of the Straight Creek fault zone and was first reported by Smith (1903) as part of his Easton Schist. Ashleman (1979) recognized the continuation of the Shuksan thrust in this southern region. In the north, where the presumed Shuksan root zone emerges from the Chilliwack batholith as the western border of the Settler Schist (see above), the zone also contains metaplutonic slices that are considered probable Yellow Aster Complex (Lowes, 1972), an assignment confirmed by dating by Gabites (1985).

Other previously metamorphosed but post-Yellow Aster rocks occur in minor amounts in the Imbricate Zone. They comprise a few slices of albite-epidote amphibolite (Misch, 1977b), dated as Late Paleozoic by Armstrong et al. (1983), and small slices of Permian blueschists, including the first dated blueschist from the region (dated by R. L. Kulp, see Misch, 1966; confirmed by Armstrong et al., 1983). Some albite-epidote amphibolites display an incipient to advanced blueschist overprint, but some blueschists lack an albite-epidote-amphibolite antecedent. Locally, albite-epidote amphibolite facies occurs as an overprint on upper amphibolite-facies Yellow Aster rocks. The albite-epidote amphibolite slices in the Imbricate Zone are equivalent to similar amphibolite exposed in the crystalline fault wedge of Vedder Mountain (see above) within the "Autochthon" (Armstrong, et al., 1983; Bernardi, 1977; McMillan, 1966; Misch, 1977b), dated as Permian by McMillan and by Armstrong et al. Presumably, these rocks are equivalent to the Garrison Schist (Vance, 1975, 1977) of the San Juans.

Locally, the highest part of the Imbricate Zone contains slices of Darrington Phyllite detached from the base of the overriding Shuksan plate. Ubiquitous constituents of the Imbricate Zone and of the Shuksan root zone are slices of mostly tectonized and serpentinized alpine metaperidotites, usually inhabiting thrust contacts. Their mechanism of tectonic emplacement resembles that of the Yellow Aster Complex, but their source is thought to be in the upper mantle — suggesting that the root zone reached to great depths. The largest of these is the remarkably unaltered Twin Sisters Dunite (Christensen, 1971; Ragan, 1963; Thompson and Robinson, 1975). Northwest of the Twin Sisters, the Imbricate Zone contains slices of cataclastic clinopyroxenites grading into wehrlites. These rocks are possibly related to adjacent dated Yellow Aster slices.

The "Autochthon"

The "Autochthon," defined as the lowest structural unit exposed on the mainland, in the Mount Baker Window south of the Canadian border, comprises the at least partly submarine Middle Jurassic Wells Creek Volcanics and the Upper Jurassic and Lower Cretaceous marine clastics of the Nooksack Group (Misch, 1952, 1966), all in prehnite-pumpellyite facies (see above). The Wells Creek Volcanics are a typical arc deposit; they consist of flows, breccias, and tuffs of andesite-keratophyre and dacite-quartz keratophyre with some marine slate. The Nooksack consists of partly flyschlike graywacke and siltstone largely derived from predominantly andesitic arc volcanics, with local volcaniclastic andesite breccia and subordinate conglomerate, including a discontinuous, thin, basal conglomerate, and thick channel conglomerate. The latter is rich in volcanic and chert clasts and has a volcaniclastic matrix identical with ordinary Nooksack graywacke; some of the clasts, and local limestone boulders, beyond much doubt are derived from the upper Paleozoic Chilliwack Group, which now overrides the Nooksack along the Church Mountain thrust. Conglomerates contain rare, well-rounded quartz-dioritic pebbles. Locally, wildflyschlike conglomerates carry huge boulders both of Chilliwack Group greenstones and of Nooksack clastics. Perhaps some of the tectonic unrest suggested by this type of sediment was synchronous with the orogeny associated with the Shuksan metamorphism; nevertheless, in the Nooksack belt, marine deposition was continuous while that orogeny was taking place in the Shuksan belt.

Wells Creek and Nooksack rocks range from massive to slaty. The Nooksack tends to become phyllitic and complexly folded toward the top (Misch, 1966), near the overriding thrust pile, and toward the root zone. The Mount Baker Window represents a structural culmination controlled by Tertiary longitudinal and transverse folding.

Outside the Mount Baker Window, the "Autochthon" is represented in southernmost British Columbia by Daly's (1912) Upper Triassic Cultus Formation, now known to include both uppermost Triassic and Lower Jurassic strata (Monger, 1966). Its marine clastics are slates and graywackes of somewhat flyschlike aspect. North of the Fraser Valley, the "Autochthon" contains the Harrison Lake Volcanics, which correlate with the Wells Creek Volcanics, and clastic formations equivalent to the Nooksack Group, all described by Crickmay (1930a, b). He also recognized that on the east these rocks are in steep thrust contact with older strata; this fault is the northern continuation of the root zone of the Church Mountain thrust.

In front of the Northwest Cascades Thrust System of the mainland, west-verging strata assigned to the Nooksack Group were reported from Lummi Island in the easternmost San Juans by G. M. Miller and Misch (1963; see also Misch, 1966). Farther west, clastic strata correlative with the Nooksack contain intercalations of pillow basalt and chert (Brandon et al., 1983; Vance, 1975). The same holds about 80 km south of the type area of the Nooksack Group (Danner, 1957). Both occurrences appear to represent sites farther offshore than the type Nooksack. The arc-type depositional belt characterized by the Nooksack originally was located far outboard of the Shuksan belt, but it was still well inboard of the frontal foredeep, where the main subduction zone must have surfaced — a belt that is no longer in evidence.

Viewed as a whole, prior to mid-Cretaceous telescoping, the Northwest Cascades Thrust System must have contained a complex array of multiple basins, arcs, and continental-crustal fragments, differing in stratigraphic and tectonic-metamorphic histories, in age and chemistry of volcanism, in types of basement and of clastic source areas, and so on. Perhaps, in earlier Mesozoic times, the terranes making up this system resembled portions of the west Pacific rim of today. However, ancient continental crust — now thoroughly dismembered — was present at least in sizable parts of the system. Most of the multiple sequences present fail to meet the criteria for juvenile, as opposed to recycled, crust. Whether the Shuksan Suite is juvenile crust — as inferred by Dungan *et al.* (1983), who reported that the Shuksan tholeiites have MORB affinity — depends on the stratigraphic interpretation of this suite (see above). At any rate, this terrane once occupied the most internal belt within the Northwest Cascades System, and its clastics indicate a nearby sialic source. None of the sequences in the system represents an ocean-ridge-related abyssal environment far from continental crust.

METHOW-PASAYTEN AND HOZAMEEN BELTS

East of the Skagit Metamorphic Core, the Upper Jurassic and Cretaceous clastic strata of the Methow-Pasayten Belt (Barksdale, 1975; Daly, 1912) generally are strongly, although at most places not complexly, folded. In the western part of the belt (Misch, 1966), prehnite-pumpellyite-facies assemblages are well displayed by Lower Cretaceous graywackes resembling Nooksack lithologies. Associated conglomerates also are of Nooksack type and contain rare, well-rounded quartz-dioritic pebbles, some of which are gneiss. Toward the Ross Lake fault, metamorphism increases from prehnite-pumpellyite to greenschist facies in the Jack Mountain Phyllite, and fold patterns become highly complex. The phyllite grades westward into the higher-grade Elijah Ridge Schist, which is of Buchan type (Misch, 1966), carrying andalusite and the rare but critical assemblage almandine-cordierite-microcline-biotite-quartz. This metamorphically defined unit forms a narrow band along the Ross Lake fault. To the southeast, its protolith is seen to be represented by andesites and clastics of the pre-Upper Jurassic North Creek Volcanics (Misch, 1966). North of the Canadian Border, the protolith of structurally equivalent schist is Hozameen Group ("schists of Maaselpanik Creek" of Haugerud, 1985).

The volcanic and sedimentary rocks of the Hozameen Group (Daly, 1912; Haugerud, 1985; McTaggart and Thompson, 1967; Misch, 1966) are now known to include both Permian (Tennyson *et al.*, 1982) and Triassic (Haugerud, 1985) strata. They form the Hozameen Belt within the Ross Lake fault zone, between the Ross Lake fault proper on the west and the Hozameen fault on the east. There is a possibility that, south of the Canadian border, the stratigraphic sequence within the Hozameen Belt is inverted. At the southern end of the belt is the Jack Mountain thrust prong of Permian Hozameen greenstone overriding Cretaceous Jack Mountain Phyllite (Misch, 1952, 1966); the floating prong is folded into a north-plunging syncline. To the north, the Hozameen Belt becomes rooted. It seems that this peculiar structural geometry is best explained by a combination of local eastward thrusting and of regional dextral slip (Misch, 1977a), but problems remain. The Hozameen regionally displays prehnite-pumpellyite facies, but north of the Canadian border it rapidly grades into high-grade schists at its western margin (Haugerud, 1985; McTaggart and Thompson, 1967).

The Ross Lake fault zone had a long and apparently diverse history. The Ross Lake fault proper was established before the 90-m.y.-old Black Peak batholith that straddles the fault was emplaced. Any large-scale dextral displacement must have taken place mostly during this early stage. Evidence for local activity of the Ross Lake fault at the time of emplacement of the batholith, presumably in the form of compressional dip-slip, was presented by Adams (1964). Significant uplift of the Skagit Gneiss block, at the time when the eastern block was undergoing Buchan-type metamorphism, is locally recorded by metamorphic history (see above). Renewed activity in the fault zone is thought to have occurred when the Gabriel Peak Orthogneiss was formed 65 m.y. ago; at this time, a western splay of the fault zone seems to have played a major role, but the mechanics of this deformation are poorly understood. Final movement along the Ross Lake fault in this area involved reverse dip-slip; Gabriel Peak Orthogneiss and Elijah Ridge Schist were locally truncated at this stage. Fault activity had ceased in this area when the Golden Horn batholith was emplaced about 49 m.y. ago. From north of the Canadian border, Haugerud (1985) reported dextral strain along the Ross Lake fault to have affected rocks only 45 m.y. old.

One point needs to be reiterated (*cf.* Misch, 1977a). Whereas large-scale dextral slip is well documented for the younger Straight Creek fault, displacement of comparable magnitude merely is a matter of speculation in the case of the Ross Lake fault zone.

SUMMARY

Although the North Cascades represent a relatively small area by Cordilleran — but not by Alpine — standards, they display a remarkable array of highly varied tectonic and meta-morphic features: Paleozoic to Eocene metamorphism encompasses several facies types and virtually all grades; migmatization is superbly displayed in the Skagit Core; multiple belts record a wide range of depositional and orogenic histories; a variety of structural patterns, representing response to regional compressive stress, include spectacular examples of folded thrusts, of imbrication, of overfolding, of flow folding, and other forms of ductile deformation in metamorphic settings; large-scale strike-slip faulting is well documented for relatively late stages of the tectonic evolution of the region.

The common theme behind these features is a long history of high mobility and prevalence of major crustal shortening. In regard to plutonism, the overall predominance of quartz-dioritic rocks from Middle Proterozoic to late Cenozoic times is remarkable. Some features deserve special note: this region contains the only well-documented example of ancient continental crust in the westernmost Cordillera north of the San Gabriel Mountains; much of "the action" in terms of compressional tectonics was not in the main subduction zone (no longer preserved at the surface), but inboard from it, involving mobile crust which, prior to telescoping, was several 100 km wide; in the Northwest Cascades, blueschist metamorphism developed in an internal belt far inboard from the main subduction zone where the oceanic plate underthrust the mobile continental margin.

ACKNOWLEDGMENTS

The author thanks Darrel Cowan, Gary Ernst, Bernard Evans, and Joseph Vance for constructive critical reviews; Floyd Bardsley for drafting the map; and Sue Bolssen for typing the final manuscript.

REFERENCES

Adams, J. B., 1964, Origin of the Black Peak Quartz Diorite, Northern Cascades, Washington: *Amer. J. Sci.*, v. 262, p. 290–306.

Armstrong, R. L., 1980, Geochronology of the Shuksan Metamorphic Suite, North Cascades, Washington: *Geol. Soc. America Abstr. with Programs*, v. 12, p. 94.

_____, Harakal, J. E., Brown, E. H., Bernardi, M. L., and Rady, P. M., 1983, Late Paleozoic high-pressure metamorphic rocks in northwestern Washington and southwestern British Columbia — The Vedder Complex: *Geol. Soc. America Bull.*, v. 94, p. 451–458.

Ashleman, J. C., 1979, The geology of the western part of the Kachess Lake quadrangle: M. Sc. thesis; Univ. Washington, Seattle, Wash., 88 p.

Babcock, R. S., 1970, Geochemistry of the main-stage migmatitic gneisses in the Skagit gneiss complex: Ph.D. thesis; Univ. Washington, Seattle, Wash., 147 p.

____ , Armstrong, R. L., and Misch, P., 1985, Isotopic constraints on the age and origin of the Skagit Metamorphic Suite and related rocks: *Geol. Soc. America Abstr. with Programs*, v. 17, p. 339.

Barksdale, J. D., 1975, Geology of the Methow Valley, Washington: *Wash. Div. Geology Earth Resources Bull. 68*, 72 p.

Bartholomew, P. R., 1979, Geology and metamorphism of the Yale Creek Area, B. C.: M.Sc. thesis; Univ. British Columbia, Vancouver, B.C., 105 p.

Bernardi, M. L., 1977, Petrology of the crystalline rocks of Vedder Mountain, British Columbia: M.Sc. thesis; Western Washington Univ., Bellingham, Wash., 137 p.

Brandon, M. T., Cowan, D. S., Muller, J. E., and Vance, J. A., 1983, Pre-Tertiary geology of San Juan Islands, Washington, and southeast Vancouver Island, British Columbia: *Geol. Assoc. Canada Ann. Meet., Field Trip 5*, 65 p.

Brown, E. H., 1974, Comparison of the mineralogy and phase relations of bluschists from the North Cascades, Washington, and greenschists from Otago, New Zealand: *Geol. Soc. America Bull.*, v. 85, p. 333–344.

____ , Wilson, D. L., Armstrong, R. L., and Harakal, J. E., 1982, Petrologic, structural, and age relations of serpentinite, amphibolite, and blueschist in the Shuksan Suite of the Iron Mountain–Gee Point area, North Cascades, Washington: *Geol. Soc. America Bull.*, v. 93, p. 1087–1098.

Cater, F. W., 1982, Intrusive rocks of the Holden and Lucerne quadrangles, Washington — The relation of depth zones, composition, textures, and emplacement of plutons: *U.S. Geol. Survey Prof. Paper 1220*, 108 p.

Christensen, N. L., 1971, Fabric, seismic asymmetry, and tectonic history of the Twin Sisters Dunite, Washington: *Geol. Soc. America Bull.*, v. 82, p. 1681–1694.

Crickmay, C. H., 1930a, The structural connection between the Coast Range of British Columbia and the Cascade Range of Washington: *Geol. Mag.*, v. 67, p. 482–491.

____ , 1930b, Fossils from the Harrison Lake area, British Columbia: *Natl. Mus. Canada Geol. Ser. 51, Bull. 63*, p. 33–66.

Daly, R. A., 1912, Geology of the North American Cordillera at the Forty-Ninth Parrallel: *Geol. Survey Canada Mem. 38*, 857 p.

Danner, W. R., 1957, Stratigraphic reconnaissance in Northwestern Cascades and San Juan Islands of Washington: Ph.D. thesis; Univ. Washington, Seattle, Wash., 561 p.

____ , 1966, Limestone resources of western Washington: *Wash. Div. Mines Geology Bull. 52*, 474 p.

DuBois, R. L., 1954, Petrology and genesis of ores of Holden Mine area, Chelan County, Washington: Ph.D. thesis; Univ. Washington, Seattle, Wash., 222 p.

Dugan, M. A., Vance, J. A., and Blanchard, D. P., 1983, Geochemistry of the Shuksan greenschists and blueschists, North Cascades, Washington — Variably fractionated and altered metabasalts of oceanic affinity: *Contrib. Mineralogy Petrology*, v. 82, p. 131–146.

Evans, B. W., and Berti, J. W., 1986, A revised metamorphic history for the Chiwaukum Schist, North Cascades, Washington: *Geology*, v. 14, p. 695–698.

Gabites, J. E., 1985, Geology and geochronometry of the Cogburn Creek-Settler Creek area, northeast of Harrison Lake, B.C.: M.Sc. thesis; Univ. British Columbia, Vancouver, B.C., 153 p.

Grant, A. R., 1966, Geology and petrology of the Dome Peak area, Chelan, Skagit and Snohomish Counties, Northern Cascades, Washington: Ph.D. thesis; Univ. Washington, Seattle, Wash., 270 p.

Haugerud, R. A., 1985, Geology of the Hozameen Group and the Ross Lake shear zone, Maselpanik area, North Cascades, southwest British Columbia: Ph.D. thesis; Univ. Washington, Seattle, Wash., 263 p.

Heath, M. T., 1971, Bedrock geology of the Monte Cristo area, Northern Cascades, Washington: Ph.D. thesis; Univ. Washington, Seattle, Wash., 194 p.

Hoppe, W. J., 1985, Origin and age of the Gabriel Peak Orthogneiss, North Cascades, Washington: M.Sc. thesis); Univ. Kansas, Lawrence, Kans., 79 p.

Huntting, M. T., Bennett, W. A. G., Livingston, V. E., Jr., and Moen, W. S., compilers, 1961, *Geologic Map of Washington:* Wash. Div. Mines Geology.

Jones, R. W., 1959, Geology of the Finney Peak area, Northern Cascades, Washington: Ph.D. thesis; Univ. Washington, Seattle, Wash., 185 p.

Lowes, B. E., 1972, Metamorphic petrology and structural geology of the area east of Harrison Lake, British Columbia: Ph.D. thesis; Univ. Washington, Seattle, Wash., 207 p.

Mattinson, J. M., 1972, Ages of zircons from the Northern Cascade Mountains, Washington: *Geol. Soc. America Bull.,* v. 83, p. 3769–3784.

McMillan, W. J., 1966, Geology of Vedder Mountain, near Chilliwack, British Columbia: M.Sc. thesis; Univ. British Columbia, Vancouver, B.S., 59 p.

McTaggart, K. C., and Thompson, R. M., 1967, Geology of part of the Northern Cascades In southern British Columbia: *Can. J. Earth Sci.,* v. 4, p. 1199–1228.

Miller, G. M., 1979, Western extent of the Shuksan and Church Mountain Thrust Plates in Whatcom, Skagit, and Snohomish Counties, Washington: *Northwest Sci.,* v. 53, p. 229–241.

_____, and Misch, P., 1963, Early Eocene angular unconformity at western front of Northern Cascades, Whatcom County, Washington: *Amer. Assoc. Petrol. Geologists Bull.,* v. 47, p. 163–174.

Miller, R. B., 1985, The ophiolitic Ingalls Complex, north-central Cascade Mountains, Washington: *Geol. Soc. America Bull.,* v. 96, p. 27–42.

Misch, P., 1952, Geology of the Northern Cascades of Washington — Seattle: *Mountaineer,* v. 45, p. 3–22.

_____, 1959, Sodic amphiboles and metamorphic facies in Mount Shuksan belt, Northern Cascades, Washington [abstract] : *Geol. Soc. America Bull.,* v. 70, p. 1736–1737.

_____, 1965, Alkaline granite amidst the calc-alkaline intrusive suite of the Northern Cascades, Washington [abstract] : *Geol. Soc. America Spec. Paper 87,* p. 216–217.

_____, 1966, Tectonic evolution of the Northern Cascades of Washington State: A West-Cordilleran case history, in Gunning, H. C., ed., *A Symposium on the Tectonic History and Mineral Deposits of the Western Cordillera:* Can. Inst. Mining Metallurgy Spec. Vol. 8, p. 101–148.

_____, 1968, Plagioclase compositions and non-anatectic origin of migmatitic gneisses in Northern Cascade Mountains of Washington State: *Contrib. Mineralogy Petrology,* v. 17, p. 1–70.

_____, 1969, Paracrystalline microboudinage of zoned grains and other criteria for synkinematic growth of metamorphic minerals: *Amer. J. Sci.,* v. 267, p. 43–63.

_____, 1971, Metamorphic facies types in North Cascades: *Metamorphism in the Canadian Cordillera,* Cordilleran Section, Geol. Assoc. Canada, Symposium, Programs Abstr., p. 22–23.

_____, 1977a, Dextral displacements at some major strike faults in the North Cascades: *Geol. Assoc. Canada Programs with Abstr.*, v. 2, p. 37.

_____, 1977b, Bedrock geology of the North Cascades, *in* Brown, E. H., and Ellis, R. C., eds., *Geological Excursions in the Pacific Northwest:* Guidebook Geol. Soc. America *Ann. Meet., Western Washington Univ.*, Bellingham, Wash., p. 1–62.

_____, 1979, Geologic map of the Marblemount quadrangle: *Washington Div. Geology Earth Resources Geol. Map GM-23*.

Monger, J. W. H., 1966, The stratigraphy and structure of the type area of the Chilliwack Group: Ph.D. thesis; Univ. British Columbia, Vancouver, B.C., 173 p.

_____, and Hutchison, W. W., 1971, Metamorphic map of the Canadian Cordillera: *Geol. Survey Canada Paper 70-33*, 61 p.

Page, B. M., 1939, The geology of the southeast quarter of the Chiwaukum quadrangle: Ph.D. thesis; Stanford Univ., Stanford, Calif., 203 p.

Pigage, L. C., 1976, Metamorphism of the Settler Schist, southwest of Yale, British Columbia: *Can. J. Earth Sci.*, v. 13, p. 405–421.

Plummer, C. C., 1969, Geology of the crystalline rocks, Chiwaukum Mountains and vicinity, Washington Cascades: Ph.D. thesis; Univ. Washington, Seattle, Wash., 137 p.

_____, 1980, Dynamothermal contact metamorphism superposed on regional metamorphism in the pelitic rocks of the Chiwaukum Mountains area, Washington Cascades: *Geol. Soc. America Bull.*, v. 91, pt. II, p. 1627–1668.

Ragan, D. M., 1963, Emplacement of the Twin Sisters dunite, Washington: *Amer. J. Sci.*, v. 261, p. 549–565.

Richards, T. A., and McTaggart, K. C., 1976, Granitic rocks of the southern Coast Plutonic Complex and northern Cascades of British Columbia: *Geol. Soc. America Bull.*, v. 87, p. 935–953.

Roddick, J. E., Muller, J. A., and Okulitch, A. V., compilers, 1979, Fraser River Sheet, British Columbia-Washington: *Geol. Survey Canada Geol. Atlas Map 1386A* sheet 92, scale 1:1,000,000.

Smith, G. O., 1903, Contributions to the geology of Washington — The geology and physiography of Central Washington: *U.S. Geol. Survey Prof. Paper 19*, p. 9–39.

_____, 1904, Description of the Mount Stuart quadrangle: *U.S. Geol. Survey Atlas, Mount Stuart Folio 106*, 10 p.

Stull, R. J., 1969, The geochemistry of the southeastern portion of the Golden Horn Batholith, Northern Cascades, Washington: Ph.D. thesis; Univ. Washington, Seattle, Wash., 127 p.

Tabor, R. W., 1961, The crystalline geology of the area south of Cascade Pass, Northern Cascade Mountains, Washington: Ph.D. thesis; Univ. Washington, Seattle, Wash., 205 p.

_____, Frizzell, V. A., Jr., Whetten, J. T., Swanson, D. A., Byerly, G. R., Booth, D. B., Hetherington, M. J., and Waitt, R. B., Jr., 1980, Preliminary geologic map of the Chelan 1:100,000 quadrangle, Washington: *U.S. Geol. Survey Open File Map 80-841*, 43 p.

_____, Frizzell, V. A., Jr., Booth, D. B., Whetten, J. T., Waitt, R. B., Jr., and Zartman, R. E., 1982, Preliminary geologic map of the Skykomish River 1:100,000 quadrangle, Washington: *U.S. Geol. Survey Open File Map OF-82-747*, 25 p.

Tennyson, M. E., Jones, D. L., and Murchey, B., 1982, Age and nature of chert and mafic rocks of the Hozameen Group, North Cascade Range, Washington: *Geol. Soc. America Abstr. with Programs*, v. 14, p. 239–240.

Thompson, G. A., and Robinson, R., 1975, Gravity and magnetic investigation of the Twin Sisters Dunite, northern Washington: *Geol. Soc. America Bull.*, v. 86, p. 1413-1422.

Vance, J. A., 1957, Geology of the Sauk River area in North Cascades of Washington: Ph.D. thesis; Univ. Washington, Seattle, Wash., 312 p.

——, 1968, Metamorphic aragonite in the prehnite-pumpellyite facies, northwest Washington: *Amer. J. Sci.*, v. 266, p. 299–315.

——, 1975, Bedrock geology of San Juan County, *in* Russell, R. H., ed., *Geology and Water Resources of the San Juan Islands:* Wash. Dept. Ecology Water Supply Bull. 46, p. 3–19.

——, 1977, The stratigraphy and structure of Orcas Island, San Juan Islands, *in* Brown, E. H., and Ellis, R. C., ed., *Geological Excursions in the Pacific Northwest:* Guidebook Geol. Soc. Amer. Ann. Meet., Western Washington Univ., Bellingham, Wash., p. 170–203.

——, Dungan, M. A., Blanchard, D. P., and Rhodes, J. M., 1980, Tectonic setting and trace element geochemistry of Mesozoic ophiolitic rocks in western Washington: *Amer. J. Sci.*, v. 280A, p. 359–388.

Wallace, W. K., 1976, Bedrock geology of the Ross Lake fault zone in the Skymo Creek area, North Cascades National Park, Washington: M.Sc. thesis; Univ. Washington, Seattle, Wash., 111 p.

Yardley, B. W. D., 1978, Genesis of the Skagit Gneiss migmatites, Washington, and the distinction between possible mechanisms of migmatization: *Geol. Soc. America Bull.*, v. 89, p. 941–951.

E. H. Brown
Department of Geology
Western Washington University
Bellingham, Washington 98225

7

METAMORPHIC AND STRUCTURAL HISTORY OF THE NORTHWEST CASCADES, WASHINGTON AND BRITISH COLUMBIA

Pre-Tertiary metamorphic rocks of the northwestern Cascades, Washington and British Columbia, Misch's (1966) Northwest Cascades System (NWCS), are derived for the most part from Paleozoic and Mesozoic protolith materials of oceanic origin, apparently representing spreading ridge, arc/fan, and ocean-island settings. Metamorphism of these rocks is characterized by the high-P/T-index minerals lawsonite, Na-amphibole, and aragonite, but not jadeite + quartz. Two episodes of high-pressure metamorphism are represented: (1) Permian, in small, fault-bounded rock fragments of the Vedder Complex; and (2) Late Jurassic to Early Cretaceous, affecting most of the remaining rock in the region and best developed and most accurately dated in the Shuksan Suite.

A petrologic anomaly in the region is the Yellow Aster Complex, which occurs as widely dispersed, small fault-bounded fragments of lower Paleozoic gneiss and calc-alkaline intrusives of apparent continental derivation.

The earliest record of movement history is a set of northeast-southwest-oriented metamorphic stretching lineations in the Shuksan blueschist, which suggests Early Cretaceous convergence at a high angle to the continental margin. Mylonitic lineations in the Late Cretaceous fault zones along which the diverse units have been imbricated are NNW-SSE trending, and suggest movement parallel to the continental margin during uplift and emplacement. Wrangellia was apparently amalgamated with more easterly terranes by Late Cretaceous time and may have served as a "backstop" during northward translation and emplacement of the NWCS.

INTRODUCTION

The North Cascades of Washington and British Columbia are underlain by an extensive tract of pre-Tertiary metamorphic rock produced by processes of continental margin tectonism. Large topographic relief and good alpine exposures allow collection of structural and metamorphic data pertinent to solving problems of the mechanisms of this orogenesis. In this review paper, the focus is on rocks in a portion of the Northwest Cascades System (NWCS) of Misch (1966). These are high-P/T, thrust-imbricated metamorphic rocks on the western flank of the Cascades (Figs. 7-1 and 7-2). The NWCS is bordered to the west by Wrangellia and to the north by the Coast Plutonic Complex. High-grade crystalline rocks in the core of the Cascades, the Skagit Metamorphic Suite, lie across the Straight Creek fault to the east of NWCS and are the subject of a companion paper in this volume by Babcock and Misch (Chapter 8).

In outlining the geologic elements of this region, the metamorphic rocks are analyzed in terms of their preserved record of orogeny: that is, the age and tectonic setting of the protolith; the age and pressure-temperature conditions of metamorphism; the kinematics of metamorphic deformation and its relation to crystallization events; and the emplacement history of the rocks. Finally, these events are tentatively related to specific plate-tectonic interactions and the accretion of Wrangellia.

PETROLOGY

Yellow Aster Complex

The oldest rock of the region is the Yellow Aster Complex of Misch (1966), which consists of two components: (1) a quartzose, well-layered gneiss in which premetamorphic (detrital?) zircons are of Precambrian age (Mattinson, 1972); and intrusive into this rock,

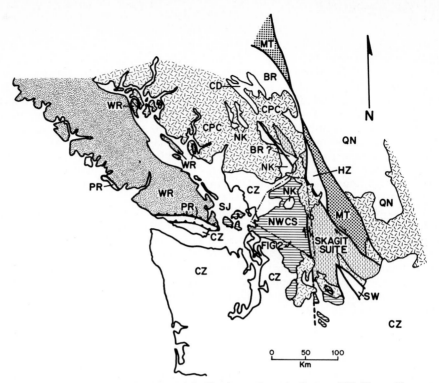

FIG. 7-1.   Regional setting of the Northwest Cascades System. WR, Wrangellia terrane; PR, Pacific Rim terrane; CPC, Coast Plutonic Complex; NK, Nooksack terrane; NWCS, Northwest Cascades System; CZ, Cenozoic; BR, Bridge River terrane; MT, Methow terrane; HZ, Hozameen terrane; QN, Quesnellia terrane; SJ, San Juan Islands; SW, Swakane Gneiss; CD, Cadwallader terrane. References: Roddick *et al*. (1979), Monger and Berg (1984), Silberling *et al*. (1984), Monger, (1985), Yorath *et al*. (1985).

(2) a suite of calc-alkaline dikes ranging from gabbro to trondhjemite and dated (Mattinson, 1972) as Ordovician. The Yellow Aster Complex occurs as small tectonic slices (<2 km wide) widely scattered in fault zones across the area. The felsic nature of the rock and its complex history suggest a continental origin (*cf*. Misch, 1966).

A Silurian amphibolite facies metamorphism is recorded in the quartzose gneiss (U/Pb sphene; Mattinson, 1972) by development of the assemblage quartz + plagioclase + clinopyroxene + epidote + garnet. The intrusive rocks are also in part gneissic, and thus may be pre- or synkinematic with respect to the amphibolite facies metamorphism. The bulk of the intrusive rocks, however, do not bear a tectonite fabric. All Yellow Aster rocks are overprinted by a greenschist facies assemblage: albite + epidote + chlorite + actinolite, which exhibits directionless textures. Later retrograde minerals include prehnite and pumpellyite in veins, and mats of lawsonite in plagioclase. The age of the greenschist facies and later minerals is not known; however Mattinson (1972) interpreted discordant zircon data to indicate a 90-m.y. retrograde event.

Chilliwack Group and Cultus Formation

The Chilliwack Group, and paraconformably overlying Cultus Formation, comprise an assemblage of predominantly andesitic pyroclastic deposits and volcanilithic sedimentary

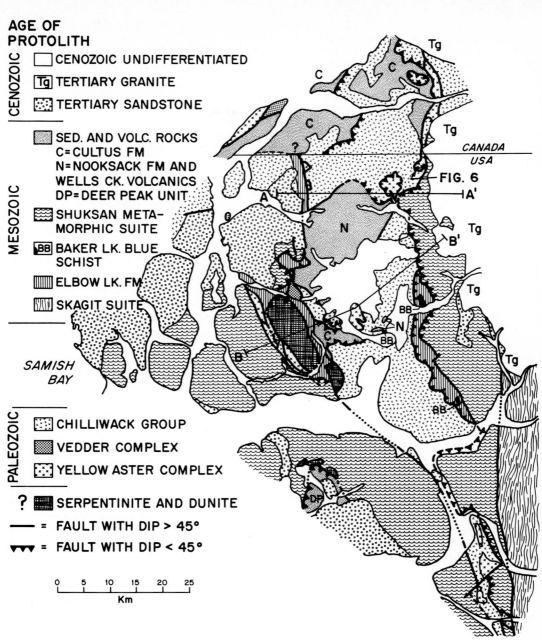

AGE OF PROTOLITH

CENOZOIC

☐ CENOZOIC UNDIFFERENTIATED

[Tg] TERTIARY GRANITE

▦ TERTIARY SANDSTONE

MESOZOIC

▨ SED. AND VOLC. ROCKS
C=CULTUS FM
N=NOOKSACK FM AND
WELLS CK. VOLCANICS
DP=DEER PEAK UNIT

▨ SHUKSAN META-MORPHIC SUITE

[BB] BAKER LK. BLUE SCHIST

▥ ELBOW LK. FM

▩ SKAGIT SUITE

PALEOZOIC

▦ CHILLIWACK GROUP

▦ VEDDER COMPLEX

▦ YELLOW ASTER COMPLEX

? ▦ SERPENTINITE AND DUNITE

—— = FAULT WITH DIP > 45°

▼▼▼ = FAULT WITH DIP < 45°

0 5 10 15 20 25
Km

FIG. 7-2. Geologic map of the northwest Cascades, Washington and British Columbia. Modified from Brown (1986). Cross sections A-A′ and B-B′ given in Figure 7-7. See Figure 7-1 for regional setting.

rocks, with lesser amounts of limestone, chert, and conglomerate. Coherent stratigraphy and mappable formations are recognized in the Chilliwack Valley, British Columbia (Monger, 1966, 1970) but generally not in the United States, where fault disruption is more intense. Fossils in the Chilliwack Group range in age from Devonian to Permian and in the Cultus from Triassic to Jurassic (Danner, 1966; Monger, 1966). Origin of the succession in an island-arc setting is suggested by the combination of lithologies and chemical composition of igneous rocks (Blackwell, 1983).

Metamorphism has only partially overprinted the igneous and sedimentary prololiths. Foliation is well developed in some areas, where metagreywackes are of textural grade II (following the scheme of Blake et al., 1967). Other rocks show no metamorphic fabric. Diagnostic mineral assemblages include epidote, pumpellyite, lawsonite, and aragonite. Amphiboles are exceedingly rare, although both actinolite and crossite/riebeckite have been reported (Sevigny, 1983; Smith, 1985). The timing of metamorphism is constrained by the youngest rocks affected, Late Jurassic age Cultus, and the Late Cretaceous post-metamorphic faults (see below) along which the units have been juxtaposed. The deformational fabric, which is best developed in proximity to faults and which is in part manifested by broken metamorphic minerals, appears to postdate the high-pressure metamorphism and to be related to the faulting event (Silverberg, 1985; Smith, 1985).

Vedder Complex

The Vedder Complex of Armstrong et al. (1983) consists of greenschists, blueschists, amphibolites, and quartzose mica schists of Permian metamorphic age. Chemical analyses of the amphibolites indicate a tholeiitic basalt protolith (Rady, 1980), which together with the associated quartzose, pelitic metasediment suggests an oceanic depositional site. The rock occurs as widely distributed small tectonic fragments. It is similar in age, metamorphism and protolith to the Garrison Schist, which occurs in fault zones in the San Juan Islands (Vance, 1975).

Metamorphic assemblages in the basic rocks contain quartz + albite + epidote + chlorite + muscovite + amphibole ± garnet. The amphiboles range from crossite to actinolite to barroisite to hornblende. Garnet occurs only in hornblende-bearing rocks. Metamorphic grade is variable from blueschist to high-pressure amphibolite facies, but zonal mapping is not practical due to the fault-disrupted nature of the rock.

Elbow Lake Formation

The Elbow Lake Formation (Elbow Lake unit of Blackwell, 1983) is dominantly meta-ribbon chert, but includes associated metapillow basalt and metavolcanilithic arenite. Poorly preserved radiolaria have been dated as Pennsylvania (D. Jones and W. R. Danner, personal communication, 1984) and Permian to Jurassic (R. Tabor, personal communication, 1985). High Ti-content of the basaltic rocks suggests an alkaline protolith and this evidence together with a presumed oceanic origin of the radiolarian cherts may indicate an ocean floor/seamount environment as the protolith site of deposition.

Metamorphism of the Elbow Lake Formation is characterized by epidote, pumpellyite, and actinolite in basic rocks and epidote, pumpellyite, and lawsonite in metasedimentary rocks. Aragonite and crossite occur also, but are uncommon (Blackwell, 1983; Ziegler, 1985). Metamorphic foliation is well developed in the metasedimentary rocks and virtually absent in the igneous rocks. The timing of metamorphism is not well constrained due to uncertainty about the age of the protolith, but is probably Late Jurassic–Early Cretaceous.

Baker Lake Blueschist

The Baker Lake blueschist occurs as tectonic slices up to 1 km wide within the Elbow Lake chert unit in the vicinity of Baker Lake. The protolith is basalt with minor but notable calcareous veins and lenses, locally intercalated marble, and rare meta-ribbon chert. An oceanic origin is suggested.

Fine-grained, well-foliated interlayered greenschist and blueschist are the dominant rock types of the unit. The abundance of lawsonite and aegirine in the basic schist distinguishes these rocks from other high-P schists of the region. Typical assemblages (including quartz, albite, and sphene) are: lawsonite + pumpellyite + glaucophane + aegirine, and lawsonite + epidote + actinolite + chlorite. The age of the protolith and that of metamorphism are unknown at this time.

Wells Creek Volcanics and Nooksack Group

The Jurassic Wells Creek Volcanics and overlying Upper Jurassic to Lower Cretaceous Nooksack Group are a calc-alkaline volcanic suite and related volcanilithic clastic sequence, respectively. Marine fossils abound in the Nooksack Formation. The protolith setting is interpreted to be a volcanic island arc and related fan (e.g., Franklin, 1985; Sondergaard, 1979).

Metamorphism is weak in these rocks. A penetrative foliation is generally absent, although a fracture cleavage inclined to bedding is common. Observed metamorphic assemblages (including quartz, albite, muscovite, epidote, sphene) are pumpellyite + epidote + calcite, and pumpellyite + prehnite (Sondergaard, 1979). Lawsonite, seen rarely, occurs as fine-grained brown mats replacing plagioclase crystals (Sevigny, 1983).

Deer Peak Unit

The Deer Peak unit, occurring in the southwest part of the area of Fig. 7-1, consists predominantly of andesitic to rhyolitic tuffaceous rock, with minor chert and argillite (Reller, in press). Metamorphic fabric is variably developed, being virtually absent in some areas, and represented by well-formed schistosity and stretching lineations elsewhere. Mineral assemblages include pumpellyite + actinolite + epidote in more mafic rocks and lawsonite in felsic rocks. Aragonite is widely distributed. The protolith and metamorphic ages are not yet known. This unit has been previously correlated with the Chilliwack Group (Misch, 1966, 1977), but the metamorphic facies is of distinctly higher grade. If radiometric dating of the protolith (in progress) proves this rock to be the same age as the Chilliwack, then the correlation is strengthened and a metamorphic facies change within the Chilliwack would be suggested.

Shuksan Metamorphic Suite

The Shuksan Metamorphic Suite of Misch (1966) consists of a varied assemblage of protolith materials which have mostly been thoroughly reconstituted to phyllites, phyllonites, and schists. Mappable units include the Shuksan Greenschist and the Darrington Phyllite. Chemical analyses of the Shuksan Greenschist indicate a tholeiitic basalt protolith with MORB affinity (Dungan et al., 1983; Street-Martin, 1981). Relict pillows, strongly deformed, are present in many places. The Darrington Phyllite is apparently derived from a quartzose, carbonaceous mud. It has rare interbeds of sedimentary ultramafic material. Fe-Mn-rich sediments of presumed oceanic hydrothermal origin (Street-Martin, 1981) occur commonly as relict beds between the greenschist and phyllite units, suggesting a preserved ocean-floor stratigraphy with the metabasalt on the bottom (Haugerud et al., 1981).

West of the Twin Sisters, Shuksan rocks are dominantly gray to black quartzose phyllite. Interbeds of phyllonite derived from volcanilithic sandstone are common. Other, relatively rare rocks are interbeds of green phyllite (metatuff) and lenses of metadiorite, gabbro, pillow basalt, and serpentinite which were emplaced (faulted?) into the Shuksan protolith prior to regional metamorphism.

The protolith age of the Shuksan suite was inferred to be Jurassic by Armstrong (1980) on the basis of the Early Cretaceous metamorphic age combined with low initial Sr isotope ratios. More recently, a Middle Jurassic age (163 m.y.) of igneous zircon in diorite involved in Shuksan metamorphism has been determined by Nick Walker (personal communication, 1984).

A number of metamorphic events are recorded in the Shuksan Suite. An early episode, yielding K/Ar ages from 144 to 160 m.y., produced a suite of coarse-grained barroisitic amphibolites, mica schists, and rare eclogite in an apparent contact aureole around periodotite, and is interpreted to represent initiation of subduction (Brown *et al.*, 1982). Lower-grade minerals of the regional blueschist metamorphism overprint these relatively high grade rocks. K/Ar and Rb/Sr ages of the regional metamorphism are mostly in the range 120–130 m.y. Mineral assemblages in the regionally metamorphosed Shuksan are the following (all including quartz + albite + chlorite + muscovite + epidote + sphene ± calcite/aragonite): actinolite + pumpellyite in greenschists, crossite + magnetite and/or hematite in blueschists, and lawsonite + paragonite (rare) in pelitic schists (Brown, 1986). Synmetamorphic deformation produced pervasive foliation, an associated northeast-trending stretching lineation defined by elongate relict clasts and phenocrysts, and northwest-trending late metamorphic flexural folds which deform the foliation surface and lineations. Axes of late folds are roughly normal to the stretching lineations. The late folds and earlier lineations are interpreted to have formed perpendicular to and parallel to, respectively, a more-or-less uniform metamorphic transport direction (Brown, in press).

Summary of Ages

Figure 7-3 summarizes the known protolith and metamorphic ages discussed above. Three metamorphic events have been delineated: (1) Silurian amphibolite facies of the Yellow Aster Complex, (2) Permian blueschist to high-pressure amphibolite facies of the Vedder Complex, and (3) Late Jurassic to Early Cretaceous blueschist and lower-grade facies in the Shuksan, Chilliwack, Cultus, Elbow Lake, and Wells Creek-Nooksack units. (See Table 7-1 for an explanation of symbols). An additional high-pressure metamorphic event, of Late Cretaceous age, has been recognized in the San Juan Islands (Brandon, 1982; Brandon *et al.*, 1983). Other, as yet undated metamorphic assemblages may represent additional

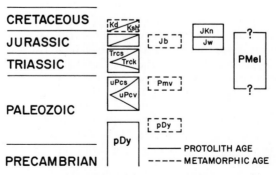

FIG. 7-3. Age and correlation of NWCS units. See Table 7-1 for explanation of symbols.

TABLE 7-1. Symbols Used in Figs. 7-3 and 7-7

Qs	Quaternary sedimentary deposits
Qv	Quaternary volcanic rocks
Tg	Tertiary granite, Chilliwack batholith
Ts	Tertiary sandstone, Chuckanut Formation
dpmv	Deer Peak metavolcanics
Kd	Darrington phyllite ⎫ Shuksan
Ksh	Shuksan Greenschist ⎬ Metamorphic
Jb	Barroisite schist ⎭ Suite
JKn	Nooksack Group
Jw	Wells Creek volcanics
Kbb	Baker Lake blueschist
Trcs	Sedimentary rocks ⎫ Cultus Formation
Trck	Keratophyre ⎭
PMel	Elbow Lake Formation
Pmv	Vedder Complex
uPcs	Sedimentary rocks ⎫ Chilliwack Group
uPcv	Volcanic rocks ⎭
pDy	Yellow Aster Complex
um	Ultramafic rocks

events, e.g., the overprint of greenschist and prehnite-pumpellyite minerals in the Yellow Aster Complex, and the Baker Lake blueschist. Some notable overlaps of protolith and metamorphic ages are (1) the sediments of the Chilliwack Group and contemporaneous amphibolite metamorphism of the Vedder Complex, and (2) the Nooksack sediments and Shuksan blueschists. Notable also is the fact that the Yellow Aster Complex does not bear a blueschist overprint which could be related to Vedder or Shuksan metamorphism, and the Vedder does not contain a Shuksan overprint. These relationships led Brown *et al.* (1981) to speculate that the various units of the region, all of which are mutually fault bounded, have developed in separate geotectonic settings, and were later tectonically juxtaposed.

Summary of Phase Relations

Phase assemblages of the various units are summarized and compared on Fig. 7-4. The triangular phase diagram represents a projection from a constant subassemblage of quartz + albite + chlorite + H_2O + CO_2 (*cf.* Brown, 1977). Pressure-temperature estimates are approximate and are based on the experimental equilibria shown on the diagram as well as oxygen-isotope temperatures of 330–400°C for the Shuksan Suite determined for quartz-magnetite pairs (Brown and O'Neil, 1982).

Inferred *P-T* conditions correspond to a low geothermal gradient (10°C/km). Phase assemblages are in part comparable to those in the South Fork Mountain Schist of California (*cf.* Brown and Ghent, 1983) and in the lower-grade parts of the Sanbagawa belt, Japan (Banno, 1964). They represent lower pressures, however, than the jadeite + quartz-bearing assemblages of the Franciscan Complex.

Several reaction relations among metamorphic index minerals are suggested by comparison of the phase assemblages. In a general form, and with reference to the diagrams of Fig. 7-4, these are:

$$\begin{array}{ccc} \textit{Chilliwack} & & \textit{Shuksan} \\ CaCO_3 + \text{Fe oxide} + & = & \text{epidote or pumpellyite} + \\ \text{chlorite} + \text{quartz} & & \text{amphibole} + H_2O + CO_2 \end{array}$$

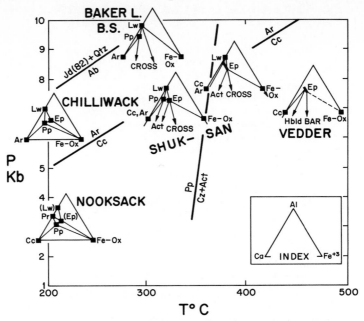

FIG. 7-4. Graphical display of mineral assemblages, modified from Brown *et al.*, 1981. Phases projected from quartz, albite, chlorite, H_2O, CO_2. Experimentally determined reaction lines from $Jd(82) = Qtz + Ab$ (Newton and Smith, 1967); $Ar = Cc$ (Crawford and Hoersch 1972; Johannes and Puhan, 1971); $Pp = Cz + Act$ (Nitsch, 1971).

| *Lower-Grade Shuksan* | | *Higher-Grade Shuksan* |
| pumpellyite + chlorite + quartz | = | epidote + actinolite + H_2O |

| *Baker Lake Blueschist* | | *Shuksan* |
| lawsonite + Na-amphibole | = | epidote + quartz + albite + chlorite + H_2O |

Details of these equilibria are discussed by Brown (1977), Brown *et al.* (1981), and Brown and Ghent (1983).

FAULT EMPLACEMENT STRUCTURES

Map units are bounded by a complex array of faults, the geometry and kinematics of which are presented in papers by Brown *et al.* (in press) and Brown (in press). All faults are postmetamorphic in the sense that they cross-cut fabrics and disrupt minerals related to the main episode of metamorphism in each unit. Minerals recrystallized in the fault zones include chlorite, calcite, stilnomelane, and muscovite, but not the blueschist index minerals, in contrast to the San Juan Islands where lawsonite and aragonite overprint fault structures (Brandon, 1982; Brandon *et al.*, 1983).

Deformation style in fault zones varies from brittle to ductile, and probably represents movement of different ages at different levels. The ductile zones, manifested by mylonites, are considered here to represent the deepest and oldest faults and those of most importance in reconstructing the postmetamorphic uplift and emplacement history.

The age of the ductile faulting is younger than Early Cretaceous Rb/Sr and K/Ar metamorphic ages of the Shuksan Suite (120–130 m.y.) and coeval fossil ages of the Nooksack (Valanginian). The faulting predates the 55–60-m.y. (Johnson, 1984) base of the Chuckanut Formation, and may predate the 75–110-m.y.-old Spuzzum batholith (Gabites, 1983; Richards and McTaggart, 1976) in British Columbia, although tracing of the faults that far north is problematic. Faulting in the San Juan Islands, which is probably related to that in the Cascades, involves Cenomanian (95 m.y.) strata and at least in part predates Santonian beds (86 m.y.), which contain metamorphic detritus related to faulting (Brandon, 1982).

The attitudes of the ductile faults vary from horizontal to north-northwest-striking and vertical. Misch (1966) suggested that high-angle faults along the eastern margin of the area (Fig. 7-1) constitute a root zone for a thrust plate of Shuksan Suite that moved out to the west, and that some more westerly high-angle faults represent folded thrusts. High-angle faults bounding the Twin Sisters dunite were interpreted by Ragan (1963) to be post-Chuckanut (post-Eocene) dip-slip faults. Recent mapping by E. H. Brown (unpublished) and co-workers have confirmed the importance of both low- and high-angle structures in the area, and have shown from analysis of mylonite fabrics that the movement direction along all faults is NNW-SEE (Fig. 7-5). Thus, the high-angle faults are of strike-slip nature. From cross-cutting relations, the high- and low-angle faults appear to have moved approximately contemporaneously: high-angle faults cut low-angle faults,

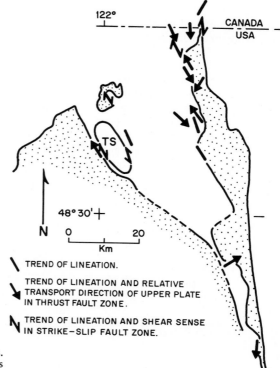

FIG. 7-5. Lineations in mylonite zones. TS, Twin Sisters. Stipled pattern represents the Shuksan Suite.

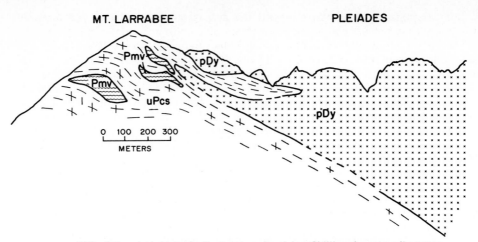

MT. LARRABEE   PLEIADES

FIG. 7-6.   Imbricate fault structure involving Chilliwack metasedimentary
rocks (uPcs), Vedder Complex amphibolites (Pmv) and Yellow Aster Complex
gneiss (pDy). View north. Fault movement sense is upper block toward viewer.
Location given on Fig. 7-2. See also Misch (1966, Plate 7-III) for a photograph
of geology on this mountainside.

and vice versa, and high-angle faults bend abruptly into low-angle faults. Evaluation of
the effects of postfault folding is difficult.

Development of tectonite fabric in the Chilliwack Group appears to be related to
this faulting; the degree of strain increases with proximity to the imbricate structures,
and foliation and stretching lineations are parallel to these same fabric elements in
mylonite zones (Smith, 1985).

Shear sense, as deduced from asymmetric augen and s-c intersections, is not uniform. In mylonites, 8 of 10 indicators in low-angle faults show movement of the upper
plate south; in the Chilliwack tectonites, 10 of 14 indicators show that the upper plate
moved north (Smith, 1985). From this disparity, one must conclude that the pattern of
regional shear is inhomogeneous. High-angle faults record a more uniform movement
sense: shear sense indicators in four of five faults are right-lateral.

Profound imbrication has occurred along this fault system, particularly in a zone
some 1-2 km thick adjacent to the Shuksan Suite (cf. Misch, 1966; Ziegler, 1985). Here
exotic blocks (i.e., of no regional extent) of dunite (including the Twin Sisters body),
Yellow Aster Complex, Vedder Complex, and Baker Lake blueschist are immersed in,
and imbricated with, Elbow Lake chert and Chilliwack Group (Figs. 7-6 and 7-7), form-

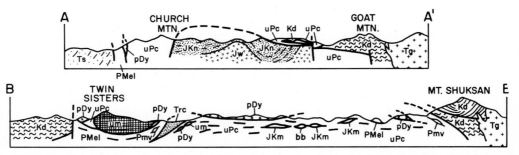

FIG. 7-7.   Cross sections through the northwest Cascades. See Table 7-1 for explanation of symbols and
Figure 7-2 for section lines. From Brown (in press).

ing a tectonic melange. Large displacements are suggested by the diversity of the fragments and their depths of origin.

TECTONIC HISTORY

With the exception of the Yellow Aster Complex, all pre-Tertiary rock units in the northwestern Cascades are ocean-derived and were subjected to high P/T metamorphism. These rocks are the products of sedimentation and volcanism in arc, trench, and ocean basin settings. One can imagine them to have been rafted to the continental margin, subducted to varying depths, and uplifted. The origin of the Yellow Aster Complex is problematic. Its long history of formation and lack of an oceanic signature suggest a continental origin, and its metamorphism is not indicative of the subduction environment. There is no evidence that the Yellow Aster Complex represents subjacent craton, particularly as the Paleozoic cratonal margin is recognized to lie well to the east[1] (e.g., Armstrong, et al., 1977; Monger et al., 1982).

Lineations in the Shuksan Suite suggest an ENE-WSW direction of transport during the high-pressure metamorphism (120–130 m.y.), whereas lineations related to emplacement of the units (80–100 m.y.?) trend NNW-SSE, at a high angle to the metamorphic lineations. Thus, a major change in transport direction appears to have taken place from metamorphism to emplacement. The uplift of the high-P metamorphic rocks from a subduction environment was possibly caused by this change in motion.

The possibility of correlating transport directions inferred from lineations with plate motions (cf. Shackelton and Ries, 1984) deduced from oceanic magnetic anomaly patterns (cf. Coney, 1978; Engebretson et al., 1985) is intriguing. However, a major problem in this exercise is the finding of post-Cretaceous rotations in the region of interest by paleomagnetic research: examples include 55–75° clockwise rotation of Cretaceous batholiths in southwestern British Columbia (Irving et al., 1985); and a 42° clockwise rotation of the Mount Stuart batholith in the central Washington Cascades (Beck et al., 1981, 1982). The magmatic and structural belts and internal fabric elements (fold axes, lineations) established during late Mesozoic orogeny define a structural grain parallel to the continental margin running the length of the Cordillera which does not appear to have suffered rotation. Much of the rotation could be restricted to the batholiths themselves and not be regional according to M. E. Beck, Jr., (personal communication, 1985). This question needs further attention.

Discounting possible rotations and other conceivable complexities (cf. Avé Lallemant, 1985), the transport directions indicated by the lineations correlate rather well with inferred plate motions (Fig. 7-8). High-angle convergence of the Farallon and North American plates during Early Cretaceous (Engebretson et al., 1985) corresponds to the timing and direction of metamorphic stretching lineations in the Shuksan Suite. Late Cretaceous fault translation in a direction roughly parallel to the continental margin agrees with an inferred change of plate interaction at this time to oblique convergence or transform motion.

REGIONAL CONSIDERATIONS

The petrotectonic framework in which the Northwest Cascades System (NWCS) is situated is as follows (Fig. 7-1): To the west is the high-P/T metamorphic thrust plate assemblage of the San Juan Islands. Many aspects of the age, lithology, and metamor-

---

[1] Also from R. L. Armstrong, unpublished data.

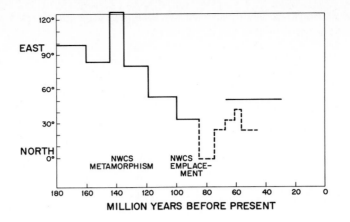

FIG. 7-8. Correlation of the ages of metamorphism and emplacement of the NWCS with direction of convergence of oceanic plates relative to North America (at Cape Mendocino). Solid line, Farallon; dashed line, Kula. Figure constructed by D. Engebretson, based on data in Engebretson *et al.* (1985).

phism of these rocks (*cf.* Brandon *et al.*, 1983; Vance, 1975) are similar to that in the NWCS and they are regarded here as a continuation of this system. To the west of the San Juan Islands is Vancouver Island, which has a fundamentally different geologic record; it is part of the broken-up exotic terrane Wrangellia (Jones *et al.*, 1977). Wrangellia gives way eastward to several small terranes lying between it and Quesnellia: Nooksack, Cadwallader, Bridge River–Hozameen, and Methow (Monger, 1985; Monger and Berg, 1984). The characterization and boundaries of these terranes are subjects of current research, complicated by the fact that they are intruded by batholiths of the Coast Plutonic Complex. East of the NWCS is the Skagit Suite (see Babcock and Misch, Chapter 8, this volume), and east of that a southerly continuation of the Methow terrane. No pre-Tertiary rocks are exposed south of the NWCS.

A significant question is how the metamorphism and deformation of the NWCS relate to accretion of the neighboring terranes, and in particular Wrangellia. The time at which Wrangellia was amalgamated to more easterly terranes is debatable. Armstrong (1985) interprets the distribution of plutons across the terranes to indicate closure of all sutures inboard of Wrangellia by 130 m.y.b.p. Woodsworth (personal communication, 1985) and Hill (1985) find overlap of sedimentary facies in the northern Coast Plutonic Complex similarly to indicate a pre-Early Cretaceous accretion of Wrangellia. However, closer to the NWCS, sedimentary facies distribution and provenance of sedimentary rocks in the Methow basin indicate that open ocean lay to the west of that area until the end of the Early Cretaceous, at which time a westerly sediment source became available (Tennyson and Cole, 1978). On this evidence, J. W. H. Monger (personal communication, 1985) regards the accretion of Wrangellia to have been completed at the end of the Early Cretaceous (~100 m.y.), but not before.

Imbricate faulting of the NWCS occurred sometime in the interval 60–120 m.y., and most likely at the same time as similar faults in the San Juan Islands, 85–95 m.y.b.p. This deformation therefore apparently postdates the accretion of Wrangellia.

Blueschist metamorphism, ranging in age from 160 to 120 m.y. in the Shuksan Suite, is probably contemporaneous with closure between Wrangellia and North America and may have formed in this setting. However, the timing and motion of the postmetamorphic imbricate faulting suggests an alternative interpretation for the metamorphic setting. The faults have undergone NNW-SSE motion, parallel to the continental margin. Large movement is suggested by the diversity of unrelated rock types involved: exotic gneisses of continental derivation, dunite mantle fragments, and scraps of Permian blueschist tectonic blocks, among others. The high-angle faults underwent right-lateral strike-slip, a sense of movement indicating translation of the NWCS northward with respect to inboard materials. These relations suggest formation of the NWCS by tectonic mixing of rock units during large northward translation along the continental margin. The end of this process came with impingement of the NWCS on the southern end of Wrangellia and inboard terranes, at which point the northward-moving materials were thrust under and over the already amalgamated terranes. The reentrant or "corner" in the continental margin created by the position of Wrangellia in effect trapped the NWCS. This corner was further filled in by a similar process during the Tertiary, when rocks of the Olympic Peninsula moved northward against, and were thrust under, Wrangellia (Beck and Engebretson, 1982; Yorath et al., 1985). If this hypothesis is correct, the blueschist metamorphism occurred well south of its present setting with respect to Wrangellia and therefore is not related to accretion of that terrane.

The Nooksack Group may not have fully participated in the deformational events of the other parts of the NWCS. Deformation and metamorphism are less intense in this unit. Misch (1966) regarded the Nooksack as being autochthonous. The Nooksack is probably correlative with rocks west of Harrison Lake (cf. Misch, 1966), intruded by the Coast Plutonic Complex, and may also be related to the Gambier Group (Woodsworth, personal communication, 1985), which extends north into the Coast Mountains of British Columbia. The Nooksack, Harrison Lake, and Gambier rocks are grouped together as the Nooksack terrane by Monger and Berg (1984; Fig. 7-2). If this relation is valid, the Nooksack Group was structurally incorporated into the NWCS during final emplacement, after major previous imbrication and transport of other parts of the complex.

Combining elements of the interpretations above, the NWCS (excluding the Nooksack?) is inferred to have been metamorphosed in the Early Cretaceous during high-angle Farallon–North America convergence at some more southerly locale. Shift of plate motion to a more oblique angle in Late Cretaceous exhumed the subducted rocks and moved them northward along the continental margin, imbricating them along the way with other, older metamorphic crustal and mantle rocks, and finally to thrust emplacement against the southern end of Wrangellia.

ACKNOWLEDGMENTS

This paper is in part a report of work in progress supported by a grant from the National Science Foundation (EAR85-08000). Discussions with J. W. H. Monger and R. S. Babcock concerning regional geology and their reviews of the manuscript, as well as that of R. A. Haugerud, are gratefully acknowledged.

REFERENCES

Armstrong, R. L., 1980, Geochronology of the Shuksan Metamorphic Suite, North Cascades, Washington: Geol. Soc. America Abstr. with Programs, v. 12, no. 3, p. 94.

——, 1985, Mesozoic-Early Cenozoic plutonism in the Canadian Cordillera — Distribution in time and space: Geol. Soc. America Abstr. with Programs, v. 17, p. 338.

_____, Taubeneck, W. H., and Hales, P. O., 1977, Rb-Sr and K-Ar geochronology of Mesozoic granitic rocks and their Sr isotopic composition, Oregon, Washington, and Idaho: *Geol. Soc. America Bull.*, v. 88, p. 387–411.

_____, Harakal, J. E., Brown, E. H., Bernardi, M. L., and Rady, P. M., 1983, Late Paleozoic high-pressure metamorphic rocks in northwestern Washington and southwestern British Columbia — The Vedder Complex: *Geol. Soc. America Bull.*, v. 94, p. 451–458.

Avé Lallèmant, H. G., 1985, Major Late Triassic strike-slip displacement in the Seven Devils terrane, Oregon and Idaho — A result of left-oblique plate convergence? *Tectonophysics*, v. 119, p. 299–328.

Banno, S., 1964, Petrologic studies on Sanbagawa crystalline schists in the Bessi-Ino district, central Sikoku, Japan: *J. Faculty Sci. Univ. Tokyo*, Sec. II, v. 15, p. 203–319.

Beck, M. E., Jr., and Engebretson, D. C., 1982, Paleomagnetism of small basalt exposures in the west Puget Sound area, Washington, and speculations on the accretionary origin of the Olympic Mountains: *J. Geophys. Res.*, v. 87, p. 3755–3760.

_____, Burmester, R. F., and Schoonover, R., 1981, Paleomagnetism and tectonics of Cretaceous Mt. Stuart batholith of Washington—Translation or tilt *Earth Planet. Sci. Lett.*, v. 56, p. 336 – 342.

Blackwell, D. L., 1983, Geology of the Park Butte–Loomis Mountain area, Washington: M.S. thesis, Western Washington Univ., Bellingham, Wash., 253 p.

Blake, M. C., Jr., Irwin, W. P., and Coleman, R. G., 1967, Upside-down metamorphic zonation, blueschist facies, along a regional thrust in California and Oregon: *U.S. Geol. Survey Prof. Paper 575-C*, p. C1–C9.

Brandon, M. T., 1982, Structural geology of middle Cretaceous thrust faulting on southern San Juan Island, Washington: M.S. thesis, Univ. Washington, Seattle, Wash., 130 p.

_____, Cowan, D. S., Muller, J. E., and Vance, J. A., 1983, Pre-Tertiary geology of San Juan Islands, Washington and southwest Vancouver Island, British Columbia: Field Trip Guide, Trip 5, Geol. Assoc. Canada, Mineral. Assoc. Canada, Can. Geophys. Un., 65 p.

Brown, E. H., 1977, Phase equilibria among pumpellyite, lawsonite, epidote, and associated minerals in low grade metamorphic rocks: *Contrib. Mineralogy Petrology*, v. 64, p. 123–136.

_____, 1986, Geology of the Shuksan Suite, North Cascades, Washington, USA; in Evans, B. W., and Brown, E. H., eds., *Blueschists and Eclogites*, Geol. Soc. America Mem. No. 164, p. 143–154.

_____, in press, Structural geology and accretionary history of the northwest Cascades, Washington and British Columbia: *Geol. Soc. America Bull.*

_____, and Ghent, E. D., 1983, Mineralogy and phase relations in the blueschist facies of the Black Butte and Ball Rock areas, northern California Coast Ranges: *Amer. Mineralogist*, v. 68, p. 365–372.

_____, and O'Neil, J. R., 1982, Oxygen isotope geothermometry and stability of lawsonite and pumpellyite in the Shuksan Suite, North Cascades, Washington: *Contrib. Mineralogy Petrology*, v. 80, p. 240–244.

_____, Bernardi, M. L., Christenson, M. L., Cruver, J. R., Haugerud, R. A., Rady, P. M., and Sondergaard, J. N., 1981, Metamorphic facies and tectonics in part of the Cascade Range and Puget Lowland of northwestern Washington: *Geol. Soc. America Bull.*, v. 92, p. 170–178.

_____, Wilson, D. L., Armstrong, R. L., and Harakal, J. E., 1982, Petrologic, structural, and age relations of serpentinite, amphibolite, and blueschist in the Shuksan Suite of the Iron Mountain-Gee Point area, North Cascades: *Geol. Soc. America Bull.*, v. 93, p. 1087–1098.

____, Blackwell, D. L., Christenson, B. W., Frasse, F. I., Haugerud, R. A., Jones, J. T., Leggi, P. A., Reller, G. J., Rady, P. M., Sevigny, J. H., Silverberg, D. S., Smith, M., Sondergaard, J. N., and Ziegler, C. B., in press, Geologic map of the northwestern Cascades, Washington: *Geol. Soc. America Map Chart Series.*

Coney, P. J., 1978, Mesozoic-Cenozoic Cordilleran plate tectonics; *in* Smith, R. B., and Eaton, G. P., eds., Cenozoic Tectonics and Regional Geophysics of the Western Cordillera: *Geol. Soc. America Mem. 152*, p. 33–50.

Crawford, W. A., and Hoersch, A. L., 1972, Calcite-aragonite equilibrium from 50°C to 150°C: *Amer. Mineralogist*, v. 57, p. 995–998.

Danner, W. R., 1966, Limestone resources of western Washington: *Wash. Div. Mines Geology Bull.* 52, 474 p.

Dungan, M. A., Vance, J. A., and Blanchard, D. P., 1983, Geochemistry of the Shuksan greenschists and blueschists, North Cascades, Washington — Variably fractionated and altered meta-basalts of oceanic affinity: *Contrib. Mineralogy Petrology*, v. 82, p. 131–146.

Engebretson, D. C., Gordon, R. G., and Cox, A., 1985, Relative motions between oceanic and continental plates in the Pacific Basin: *Geol. Soc. America Spec. Paper 206*, 59 p.

Franklin, R. J., 1985, Geology and mineralization of the Great Excelsior Mine, Whatcom County, Washington: M.S. thesis, Western Washington Univ., Bellingham, Wash., 119 p.

Gabites, J. E., 1983, Geology and geochronology east of Harrison Lake, British Columbia: *Geol. Assoc. Canada Abstr. with Programs*, v. 8, p. A25.

Haugerud, R. A., Morrison, M. L., and Brown, E. H., 1981, Structural and metamorphic history of the Shuksan Metamorphic Suite in the Mount Watson and Gee Point areas, North Cascades, Washington: *Geol. Soc. America Bull.*, v. 92, p. 374–383.

Hill, M. L., 1985, The Coast Plutonic Complex near Terrace, B.C. — A metamorphosed western extension of Stikinia: *Geol. Soc. America Abstr. with Programs*, v. 17, p. 362.

Irving, E. E., Woodsworth, G. J., Wynne, P. J., and Morrison, A., 1985, Paleomagnetic evidence for displacement from the south of the Coast Plutonic Complex, British Columbia: *Can. J. Earth Sci.*, v. 22, p. 584–598.

Johannes, W., and Puhan, D., 1971, The calcite-aragonite transition reinvestigated: *Contrib. Mineralogy Petrology*, v. 31, p. 28–38.

Johnson, S. Y., 1984, Stratigraphy, age and paleogeography of the Eocene Chuckanut Formation, northwest Washington: *Can. J. Earth Sci.*, v. 21, no. 1, p. 92–106.

Jones, D. L., Silberling, N. J., and HIllhouse, J., 1977, Wrangellia — A displaced terrane in northwestern North America: *Can. J. Earth Sci.*, v. 14, p. 2565–2577.

Mattinson, J. M., 1972, Ages of zircons from the northern Cascade Mountains, Washington: *Geol. Soc. America Bull.*, v. 83, p. 3769–3784.

Misch, P., 1966, Tectonic evolution of the North Cascades of Washington State — A west Cordilleran case history: *Can. Inst. Mining Metallurgy Spec. Vol. 8*, p. 101–148.

Misch, P., 1977, Bedrock geology of the North Cascades *in* Brown, E. H., and Ellis, R. C., eds., *Geological Excursions in the Pacific Northwest:* Guidebook Geol. Soc. America Ann. Meet., Western Washington Univ., Bellingham, Wash., p. 1–62.

Monger, J. W. H., 1966, Stratigraphy and structure of the type area of the Chilliwack Group, southwestern British Columbia: PhD. dissertation, Univ. British Columbia, Vancouver, B.C., 158 p.

____, 1970, Hope map-area, west half (92H W1/2), British Columbia: *Geol. Survey Canada Paper 69-47*, 75 p.

_____, 1985, Terranes in the southeastern Coast Plutonic Complex and Cascade fold belt: *Geol. Soc. America Abstr. with Programs*, v. 17, p. 371.

_____, and Berg, H. C., 1984, Lithotectonic terrane map of western Canada and south-eastern Alaska: *U. S. Geol. Survey Open File Rt.*

_____, Price, R. A., and Tempelman-Kluit, D. J., 1982, Tectonic accretion and the origin of the two major metamorphic and plutonic welts in the Canadian Cordillera: *Geology*, v. 10, p. 70–75.

Newton, R. C., and Smith, J. V., 1967, Investigations concerning the breakdown of albite at depth in the earth: *J. Geology*, v. 75, p. 268–286.

Nitsch, K.-H., 1971, Stabilitatsbeziehungen von prehnite- und pumpellyite-haltigen para-genesen: *Contrib. Mineralogy Petrology*, v. 30, p. 240–260.

Rady, P. M., 1980, Structure and petrology of the Groat Mountain area, North Cascades, Washington: M.S. thesis, Western Washington Univ. Bellingham, Wash., 132 p.

Ragan, D. M., 1963, Emplacement of the Twin Sistern Dunite, Washington: *Amer. J. Sci.*, v. 261, p. 549–565.

Reller, G. J., in press, Structure and petrology of the Deer Peak area, North Cascades, Washington: M.S. thesis, Western Washington Univ. Bellingham, Wash.

Richards, T. H., and McTaggart, K. D., 1976, Granitic rocks of the southern Coast Plutonic Complex, and northern Cascades of British Columbia: *Geol. Soc. America Bull.*, v. 87, p. 935–953.

Roddick, J. A., Muller, J. E., and Okulitch, A. V., 1979, Fraser River: *Geol. Survey Canada Map 1386A*, sheet 92, scale 1:1,000,000.

Sevigny, J. H., 1983, Structure and petrology of the Tomyhoi Peak area, North Cascade Range, Washington: M.S. thesis, Western Washington Univ. Bellingham, Wash., 203 p.

Shackleton, R. M., and Ries, A. C., 1984, The relation between regionally consistent stretching lineations and plate motions: *J. Struct. Geology*, v. 6, p. 111–117.

Silberling, N. J., Jones, D. L., Blake, M. C., and Howell, D. G., 1984, Lithotectonic terrane map of the western conterminous United States: *U.S. Geol. Survey Open File 84-523.*

Silverberg, D. L., 1985, Structure and petrology of the White Chuck Mountain–Mt. Pugh area, North Cascades, Washington: M.S. thesis, Western Washington Univ., Bellingham, Wash., 173 p.

Smith, M. T., 1985, Structure and petrology of the Grandy Ridge–Lake Shannon area, North Cascades, Washington: M.S. thesis, Western Washington Univ., Bellingham, Wash., 156 p.

Sondergaard, J. N., 1979, Stratigraphy and petrology of the Nooksack Group in the Glacier Creek-Skyline Divide area, North Cascades, Washington: M.S. thesis, Western Washington Univ., Bellingham, Wash., 103 p.

Street-Martin, L. V., 1981, The chemical composition of the Shuksan Metamorphic Suite in the Gee Point-Finney Creek area, North Cascades, Washington: M.S. thesis, Western Washington Univ., Bellingham, Wash., 91 p.

Tennyson, M. E., and Cole, M. R., 1978, Tectonic significance of upper Mesozoic Methow-Pasayten sequence, northeastern Cascade Range, Washington and British Columbia; *in* Howell, D. G., and McDougall, K. A., eds., *Mesozoic Paleogeography of the Western United States:* Pacific Section, Soc. Econ. Paleontologists Mineralogists, Pacific Coast Paleogeography Symp. 2, p. 499–508.

Vance, J. A., 1975, Bedrock geology of the southeastern San Juan Islands: *Washington Dept. Ecology Water Supply Bull. 46*, p. 3-19.

Yorath, C. J., Green, A. G., Clowes, R. M., Sutherland Brown, A., Brandon, M. T., Kanase-
wich, E. R., Hyndman, R. D., and Spencer, C., 1985, Lithoprobe, southern Vancouver Island — Seismic reflection sees through Wrangellia to the Juan de Fuca
plate: *Geology*, v. 13, p. 759–762.

Ziegler, C. B., 1985, Structure and petrology of the Swift Creek area, North Cascades,
Washington: M.S. thesis, Western Washington Univ. Bellingham, Wash., 191 p.

R. S. Babcock
Department of Geology
Western Washington University
Bellingham, Washington 98225

Peter Misch*
Department of Geological Sciences
University of Washington
Seattle, Washington 98195

8

EVOLUTION OF THE CRYSTALLINE CORE OF THE NORTH CASCADES RANGE

*deceased

ABSTRACT

The Skagit Crystalline Core (SCC) of the North Cascades Range is a fault-bounded wedge of mostly pre-Tertiary rocks that show evidence of Late Cretaceous to early Tertiary regional metamorphism and deformation (Skagit metamorphism). The SCC can be sub-divided into four major rock groups (1) metasediments and metavolcanics, (2) migmatites, (3) orthogneisses of Triassic intrusive age, and (4) partly syntectonic intrusives of Late Cretaceous to Eocene age. These occur mainly as northwest-trending belts bounded by faults or intrusive bodies elongated parallel to the structural trend.

The supracrustal rocks include a broad range of lithologies, with pelitic schists and metagreywackes most common. Metavolcanic rocks range from rhyolitic to basaltic with mafic types dominant. Protolith ages may be as old as Precambrian, but the main supra-crustal sequence, represented by the Chiwaukum, Settler, and Cascade River Schists, has so far yielded inferred depositional ages of about 210 m.y. This is close to the 220 m.y. intrusive ages obtained for pretectonic quartz-dioritic orthogneisses which constitute the Marblemount belt. In contrast, syntectonic granitic intrusives and migmatites, associated with Skagit metamorphism, yield ages that center around 90 m.y.

Geochemical, petrologic, and paleomagnetic data are consistent with the interpreta-tion that the SCC was deformed, intruded, and metamorphosed during a tectonic event related to the arrival of Wrangellia in the vicinity of the northern Baja California, most probably between 110 and 90 m.y.b.p. Subsequent northward movement resulted in dextral shear represented by movements on the Straight Creek–Fraser River and Ross Lake fault systems. The arrival time of the SCC at its present geographic position is uncer-tain, but it appears to have been in place prior to intrusion of the Chilliwack batholith in the middle to late Eocene.

INTRODUCTION

The Skagit Crystalline Core (SCC) is a term originally used by Misch (1966) to describe the greenschist to upper amphibolite facies rocks of the North Cascades Range which lie between the Straight Creek fault and the Ross Lake fault zone (Fig. 8-1).

The SCC is composed of numerous metaigneous and metasedimentary rock units that are grouped by Tabor *et al.* (1980, in press) and Siberling *et al.* (1984) into several different lithotectonic terranes (Fig. 8-2). However, Misch (1966, 1977, chapter 6, this volume) emphasizes the broadly correlative nature of rocks which occur in different "terranes" and points out that the common feature of the SCC as a whole is an episode of deformation, metamorphism, and syntectonic intrusion, mainly occurring during the middle Late Cretaceous. This orogenic event is termed Skagit metamorphism by Misch (1966).

Maps of the northern part of the SCC are found in reports by Misch (1952, 1966, 1977a, 1979). Also, see Misch (Chapter 6, this volume) for a comprehensive map of the SCC as a whole. Students of Misch have provided numerous thesis maps, which are listed in Misch (1966, 1977a, Chapter 6, this volume). U.S. Geological Survey maps within the SCC are as follows: Glacier Peak 15 min quadrangle (Crowder and Wright, 1966), Holden 15 min quadrangle (Cater and Crowder, 1967), Lucerne 15 min quadrangle (Cater and Wright, 1967), Skykomish 1:100,000 quadrangle (Tabor *et al.*, 1982), Eagle Rocks and Glacier Peak Roadless areas (Tabor *et al.*, 1982b), Wenatchee 1:100,000 quadrangle (Tabor *et al.*, 1982c), the Snoqualmie Pass 1:100,000 quadrangle (Frizzell *et al.*, 1984), various portions of the southern SCC (Tabor *et al.*, 1986), and a small part of the North Cascades (Staatz *et al.*, 1972).

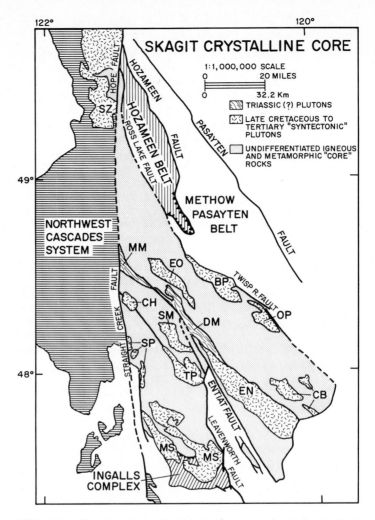

FIG. 8-1. Map showing major lithologies and structures in the Skagit Crystalline Core. Compiled mainly from Misch (1966, 1977a, b) and Tabor *et al.* (in press), plus U.S. Geological Survey maps and theses listed in the text. SZ, Spuzzum pluton; MM, Marblemount belt; EO, Eldorado Orthogneiss; BP, Black Peak batholith; CH, Cheval pluton; SM, Sulphur Mountain pluton; DM, Dumbell Mountain plutons; OP, Oval Peak pluton; SP, Sloan Peak plutons; TP, Ten Peak pluton; EN, Entiat pluton; CB, Chelan batholith; MS, Mount Stuart batholith. *Note:* Post-tectonic Tertiary plutons have been omitted for clarity.

The petrology and phase relationships of the Skagit Suite are described by Misch (1968), Misch and Rice (1975), Misch and Onyeagocha (1976), and Yardley (1978). The petrology and geochemistry of rocks in the Holden and Lucerne quadrangles are discussed by Crowder (1959) and Cater (1982). Erickson (1977a, b) provides data on the petrology and geochemistry of the Mount Stuart batholith, whereas the phase relationships and structural evolution of metamorphic rocks in the Mount Stuart area are interpreted by Van Diver (1967), Frost (1975, 1976), Miller and Frost (1977), Plummer

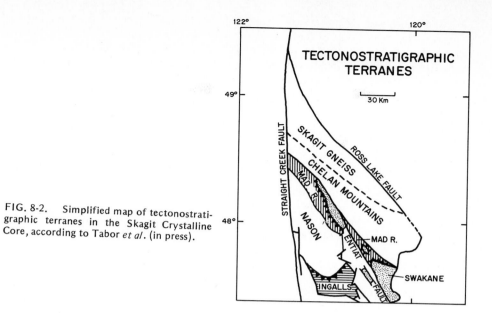

FIG. 8-2. Simplified map of tectonostratigraphic terranes in the Skagit Crystalline Core, according to Tabor *et al.* (in press).

(1980), Magloughlin (1986) and Evans and Berti (1986). Geochronologic data pertinent to the SCC are provided by Mattinson (1972), Engels *et al.* (1976), and Tabor *et al.* (1980, 1982a, in press). A very limited isotopic tracer study of SCC rocks is found in G. L. Davis *et al.* (1966). Regional tectonic reconstructions involving the SCC are discussed by Fox *et al.* (1977). G. A. Davis (1977), G. A. Davis *et al.* (1978), Monger *et al.* (1982), Tabor *et al.*, in press). Coney *et al.* (1980) consider the SCC to be part of a "suspect terrane" that they term the "North Cascades (composite)."

EVOLUTION OF THE SKAGIT CRYSTALLINE CORE

In this paper we subdivide the SCC into four major rock groups defined on the basis of age, lithology, and origin. These are (1) supracrustal metasedimentary and metavolcanic rocks, (2) heterogeneous migmatitic rocks, (3) orthogneisses of Triassic intrusive age (the Marblemount belt), and (4) partly syntectonic intrusives of Late Cretaceous to Eocene age.

Supracrustal Rocks

The oldest supracrustal unit is the Swakane Biotite Gneiss, which corresponds to the Swakane Terrane defined by Tabor *et al.* (in press). This is a remarkably uniform granofelsic gneiss that includes a Precambrian zircon component (Mattinson, 1972) and yields a tentative Rb/Sr protolith age of about 690 m.y. (R. J. Fleck and A. B. Ford, cited in Tabor *et al.*, in press). The Swakane Gneiss is in fault contact with rocks of the Napeequa River schist belt and correlatives, which comprise the Mad River Terrane of Tabor *et al.* (in press). In contrast to the Swakane, the Napeequa belt is dominated along its entire 80-km length by an "oceanic assemblage" of amphibolite and micaceous quartzite with numerous ultramafic pods interspersed (R. W. Tabor, personal communication). The age of the protolith is uncertain, but is interpreted as Late Paleozoic by Tabor *et al.* (in press). The Napeequa schist may correlate with rocks mapped as Cascade River Schist which crop out southwest of the Marblemount Belt (R. W. Tabor, personal communication). However, the main body of Cascade River Schist is quite different in lithology and has been included with the Chelan Mountains Terrane by Tabor *et al.* (in press). The Cascade River Schist and its probable equivalent, the Martin River Schist of Dubois (1954) is dominated by metagreywackes and quartzofeldspathic schists with minor amphibolites. This lithology is similar to the lower section of the Chiwaukum Schist of Page (1939). However, the Chiwaukum Schist and equivalent units also include an upper section that is dominated by pelitic schists (*cf.* Getsinger, 1978; Heath, 1971; Plummer, 1980). Whole-rock Rb-Sr analyses of the Chiwaukum (Magloughlin, 1986) and the Cascade River Schist (Babcock *et al.*, 1985) suggest a protolith age of 210–215 m.y. A similar Rb/Sr age is given for the Settler Schist (Gabites, 1985), which lies on the opposite (west) side of the Straight Creek (Hope) fault in SE British Columbia. This age correspondence is significant because lithologic and structural similarities of the Chiwaukum and Settler Schist (Lowes, 1972) led Misch (1977b) to propose that the units were originally contiguous but were offset by about 190 km of dextral slip on the Straight Creek Fault during post-Late Cretaceous time, following Skagit metamorphism.

Migmatitic Rocks

U/Pb zircon dating by Mattinson (1972) and geochronology reported by Tabor *et al.* (in press) suggest that the development of heterogeneous migmatitic gneisses in the SCC was related to Skagit Metamorphism in the Late Cretaceous. The Skagit Gneiss (Misch, 1966) and the Chelan Complex of Hopson and Mattinson (1971) show an apparent gradation from migmatitic gneisses into tracts of essentially isochemical supracrustal rock, which suggests a derivative relationship. The Spire Point Migmatitic Complex of Grant (1966) and the Totem Pass Unit of Ford (1959) are analogous and perhaps broadly correlative with the Skagit migmatites. However, neither of these units is coextensive with the Skagit-Chelan belt and both are dominated by orthogneisses, with only subordinate to minor amounts of supracrustal material.

Orthogneiss of Triassic Intrusive Age

The Marblemount belt of Misch (1966) consists predominantly of a series of elongate intrusives that yield nearly concordant U/Pb zircon ages of 220 m.y. (interpreted as the primary crystallization age by Mattinson, 1972). The northwest end of the belt consists of the Marblemount Meta-Quartz Diorite, which can be traced southeast into the Le Conte–Magic Mountain Gneiss of Tabor (1961) and is almost certainly correlative with the Bonanza Gneiss of Dubois (1954) and Dumbell Mountain plutons of Cater and Crowder (1967). In the Skagit region Misch (1952, 1966) interprets the Marblemount to be the basement on which the Cascade River Schist was deposited. During a late stage of Skagit metamorphism, the Marblemount was thrust over the supracrustal sequence and the two were metamorphosed together under greenschist facies conditions. In the Magic Mountain area this metamorphism was retrogressive. Conversely, in the Holden region, Cater (1982) argues that the Dumbell Mountain plutons are intrusive into, rather than thrust over their wallrocks, and that their internal structure is protoclastic rather than metamorphic. If this is true, then the wallrocks (the "Younger Gneisses of the Holden Quadrangle") must be older than the Cascade River Schist or the basement interpretation for the Marblemount is incorrect. However, data given by Dubois (1954) and Tabor *et al* (in press) make a protoclastic origin for the fabric in the Dumbell Mountain plutons unlikely.

"Syntectonic" Intrusives

The final assemblage in the SCC consists of numerous granitic intrusives of Late Cretaceous to Eocene age that show evidence of deformation during or just after intrusion (Fig. 8-1). U/Pb crystallization ages are mainly concordant and center around 90 m.y. for several of the intrusives. K/Ar biotite, muscovite, and hornblende ages tend to be concordant in the range 80–90 m.y.; but discordant younger ages are also common, reflecting a complex uplift and cooling history (*cf.* Tabor *et al.*, in press). The latest documented syntectonic plutonic activity in the SCC is represented by lineated dikes of the Skagit region (Misch, 1968). Estimated depths of emplacement are highly variable, ranging from about 10 km for the Mount Stuart batholith (Kaneda, 1980) to greater than 25 km for the Ten Peak and Sulphur Mountain plutons (Zen and Hammarstrom, 1984). Tabor *et al.* (in press) suggest that the migmatitic aspect of plutons in the Chelan Mountains may indicate an even deeper level of intrusion. In the Mount Stuart area, low-P/T metamorphism, probably related to intrusion, occurred before higher-pressure regional metamorphism (Evans and Berti, 1986; Heath, 1971; Oles, 1956). A similar relationship has been demonstrated for the supracrustal wallrocks of the Spuzzum pluton in southeastern British Columbia (Bartholomew, 1979; Gabites, 1985; Pigage, 1976). However, in the type section of the Skagit Gneiss, low-pressure metamorphism overprints the intermediate-pressure regional event (Misch, 1968). On a regional scale, it is clear that intrusion occurred before, during, and just after the culmination of Skagit metamorphism.

PETROCHEMISTRY OF THE TYPE AREA

The type area of the SCC is in the Skagit River region and is dominated by the Skagit Metamorphic Suite (SMS) as defined by Misch (1966). In the area shown by Fig. 8-3, the SMS can be subdivided into five lithologic units: the Marblemount Meta-Quartz Diorite (MMQD), the Cascade River Schist (CRS), the Skagit Gneiss, the Eldorado Orthogneiss, and the Gabriel Peak Orthogneiss (GPO). Details on the field, petrographic, and structural relationships of these rocks and their relationship to other units in the SCC can be found in Misch (1966, 1968, 1977a, b, chapter 6, this volume) and in Tabor *et al.* (in press).

Marblemount Meta-Quartz Diorite

The segment of the Marblemount belt in the vicinity of the Cascade River Valley comprises the Marblemount Meta-Quartz Diorite (Misch, 1952, 1966), which consists predominantly of plutonic rocks that have been recrystallized to a greenschist facies mineral assemblage of quartz-albite-epidote-chlorite-sericite. The metamorphic grade progressively increases toward the southeast, where amphibolite facies rocks are found in the Magic Mountain–Le Conte Gneiss (Bryant, 1955; Misch, 1966). Metamorphic deformation in the MMQD varies from blastomylonitic along the margins of the belt to faintly gneissose in the interior, with some sections showing well-preserved igneous textures despite development of a metamorphic mineralogy.

In the type area the MMQD consists predominantly of quartz diorite and tonalite, but minor diorite and rare gabbro also occur (Fig. 8-4). The metaluminous to slightly peraluminous intrusives of the main stage are cut by foliated muscovite-bearing dikes, as well as an unnamed metatrondhjemite on the northernmost end of the exposed MMQD (*cf.* Misch, 1979). Because these intrusives do not appear to cut across the contact of the

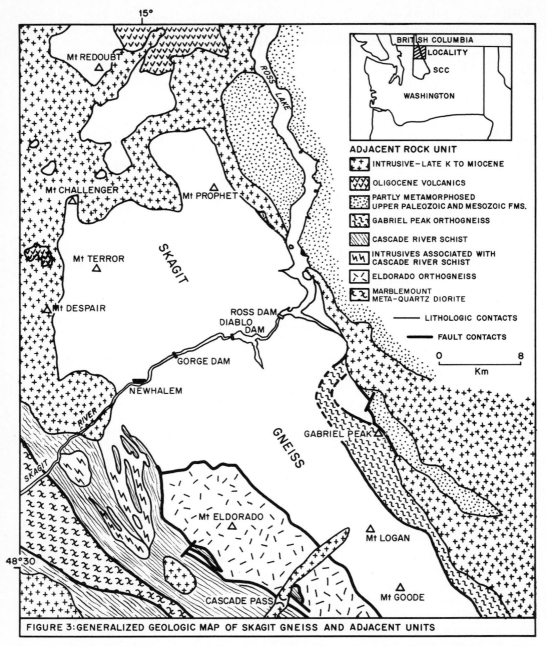

FIGURE 3: GENERALIZED GEOLOGIC MAP OF SKAGIT GNEISS AND ADJACENT UNITS

FIG. 8-3. Generalized geologic map of the type area of the Skagit Crystalline Core, modified from Misch (1966, 1977a, b).

adjacent Cascade River Schist, it can be inferred that a structural or erosional discontinuity (or both) developed between the two units prior to deposition of the CRS protolith. This evidence, along with conglomerates containing clasts of MMQD lithology in the

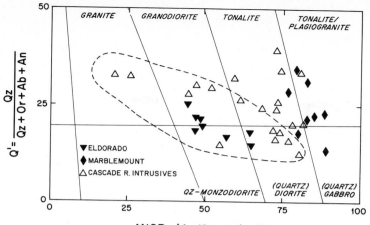

FIG. 8-4. Q′versus ANOR diagram after Streckeisen and Le Maitre (1979),
Dashed line encloses field of Sierra Nevada (I-type) granitic rocks (Bowden
et al., 1984).

lower part of the CRS, supports the conclusion of Misch (1966) that the MMQD represents the basement upon which the CRS was deposited.

Eldorado Orthogneiss

Like the MMQD, the Eldorado Orthogneiss of Misch (1966) is a homogeneous metaplutonic rock that is exposed in a fault-bounded belt running northwest-southeast, bounded by the Skagit Gneiss on the northeast and the CRS on the southwest. It also shows a foliation that varies from faint in the interior to intense along the margins. The Eldorado, however, is much younger, giving a concordant U/Pb zircon age of 92 m.y. (Mattinson, 1972). It is also more granodioritic in composition (Fig. 8-4) and mafic layers, lenses, and xenoliths are common.

The typical metamorphic assemblage is quartz-oligoclase-hornblende-epidote biotite. The same middle-amphibolite grade of metamorphism also appears in the adjacent Skagit Gneiss and Cascade River Schist. Tectonic emplacement of the Eldorado massif must have occurred prior to the later stages of Skagit metamorphism, because the rocks are fully recrystallized and an injection complex of foliated trondhjemite and pegmatite intrudes the fault boundary, where it overlies the Skagit Complex.

Cascade River Schist

As described by Misch (1952, 1966, 1977a, 1979), the Cascade River Schist is a northwest-southeast-trending belt of sedimentary and minor volcanic rocks that have been recrystallized under greenschist to middle amphibolite facies conditions. Among the metasediments, mica schists and quartzofeldspathic schists predominate; but also present are marbles, calc-silicates, para-amphibolites, quartzitic schists, local garbenshiefer, and metaconglomerates. Metavolcanic rocks include amphibolites (metabasalts and meta-andesites) as well as more siliceous metatuffs. In the type area, the grade of metamorphism increases from chlorite zone on the southwest margin to kyanite zone on the northeast.

Assuming metamorphism of the CRS to be essentially isochemical (Misch, 1966, 1968), the geochemistry of the metasediments suggests an immature provenance, probably

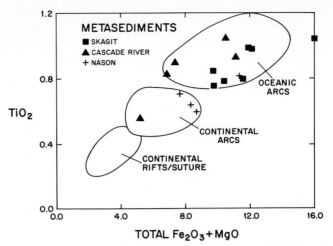

FIG. 8-5. Geochemistry of metagreywackes in the Skagit Crystalline Core. Fields of petrotectonic types from Condie and DeMalas (1985). Nason data from Ort and Tabor (1985).

a magmatic arc (Fig. 8-5). Amphibolite geochemistry (Fig. 8-6) shows an arc to oceanic affinity.

Orthogneisses constitute a minor, but petrogenetically important lithology in the CRS. The timing of these intrusives ranges from premetamorphic to late syntectonic (*cf.* Misch, 1966). On a normative Q' versus ANOR diagram (Streckeisen and Le Maitre, 1979), the orthogneisses range from granite to quartz diorite in composition (Fig. 8-7); however, the modal lithologies actually vary from leucotonalite to leucotrondhjemite.

Geochronological studies of the CRS have so far been inconclusive. Babcock *et al.* (1985) report that Rb-Sr isotopic analyses of mica schists and amphibolites fall into three groups. The first forms a scattered array with an early Mesozoic slope and an initial-

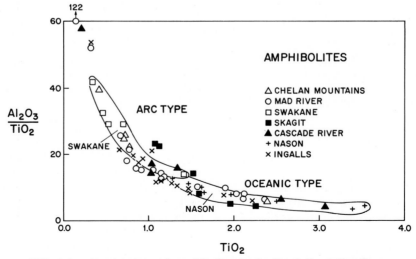

FIG. 8-6. Geochemistry of amphibolites in the Skagit Crystalline Core on an Al_2O_3/TiO_2 versus TiO_2 diagram adapted from Sun (1979). All data except Skagit and Cascade River from Ort and Tabor (1985).

Sr ratio of about 0.7038. The second group of samples is more radiogenic and either represents a hybrid suite containing a Precambrian component or interaction with Late Cretaceous to Tertiary granitic rocks. The latter hypothesis is preferred. Finally, there are samples with very high and uncorrelated Rb/Sr ratios that probably represent secondary isotopic reequilibration.

Skagit Gneiss

The Skagit Gneiss of Misch (1966) is a complex mixture of supracrustal metasediments and subordinate metavolcanics interspersed with migmatites and tracts of more homogeneous orthogneiss. The supracrustal rocks consist chiefly of biotite schist, quartzofeldspathic schist, and amphibolite that occur as migmatite "paleosomes" as well as larger layers and lenses intercalated with the predominant gneisses. The range of minor lithologies among the supracrustal rocks is the same as in the CRS, including metamorphosed ultramafic rocks. This coincidence of lithology, combined with the results of detailed petrographic work (Misch, 1968) and a study of field relationships along the boundary of the CRS with the Skagit Gneiss, led Misch (1966, 1968) to the conclusion that the Skagit migmatites were largely derived from rocks equivalent to the CRS.

The metamorphic grade of the Skagit Gneiss increases from middle-amphibolite facies at the southwest margin to upper-amphibolite facies in the interior of the belt (Misch, 1968, 1977a). Migmatites are prevalent even in the epidote-amphibolite facies and there is no significant increase in the degree of migmatization with metamorphic grade, so partial melting in place is unlikely to be a possible mechanism for migmatite formation. Misch (1968) presents petrographic arguments for metasomatic replacement coupled with metamorphic segregation as the mechanism of migmatization. However, isotopic relationships are complicated and the exact mechanism is still under investigation.

Petrologic and geochemical studies of the Skagit complex show that the orthogneisses are metaluminous quartz diorites and tonalites (Figs. 8-7 and 8-8). Migmatite leucosomes are mainly peraluminous and trondhjemitic and there is no significant difference between those associated with biotite schist versus amphibolitic melanosomes. Foliated pegmatite dikes which cut the main stage migmatites are also trondhjemitic.

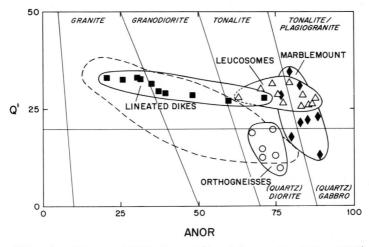

FIG. 8-7.   Q' versus ANOR diagram (Streckeisen and Le Maitre, 1979) for rocks of the Skagit Gneiss Complex. Fields of the MMQD and Sierra Nevada granites shown for comparison.

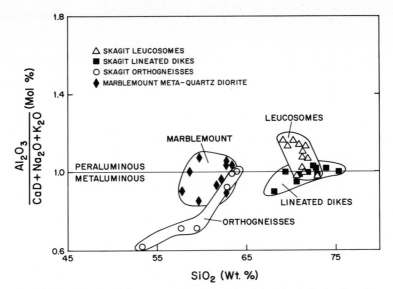

FIG. 8-8. Alumina index versus SiO$_2$ for rocks of the Skagit Gneiss Complex.

The late dikes and sills which share the lineation of their wallrocks are leucotonalite to true granite in composition. These intrusives have been interpreted by Misch (1968) as the products of partial melting at depths below presently exposed levels.

Geochronology of the Skagit Gneiss is indeterminate except for the late lineated dikes, which provide a fairly well defined whole-rick Rb/Sr isochron age of 45 m.y. with a surprisingly low initial-Sr ratio of 0.7040 (Babcock and Misch, 1984; Babcock *et al.*, 1985). Within the limits of error, this is the same as reset K/Ar ages on biotite schist and paragneiss (Engels *et al.*, 1976), indicating a widespread thermal event. It also overlaps K/Ar ages on diorites mapped as part of the Chilliwack composite batholith. Highly discordant U/Pb data from the Skagit Gneiss are interpreted by Mattinson (1972) to indicate a Precambrian age for detrital zircons, with a 60–90-m.y. age of regional metamorphism and migmatization. However, this conclusion assumes a correlation between the Swakane Gneiss and the Skagit Gneiss that is not supported by their different lithologies. A U/Pb age of 48 m.y. on an orthogneiss in the southeastern part of the Skagit belt is reported by Miller *et al.* (1985). Rb-Sr isotopic data for migmatites and orthogneisses of the Skagit complex are highly scattered and apparently uncorrelated, except that migmatite leucosomes are generally more radiogenic than melanosomes (Babcock *et al.*, 1985).

Gabriel Peak Orthogneiss

The Gabriel Peak Orthogneiss of Misch (1966) is a narrow belt of quartz-dioritic blasto-mylonite (*cf.* Misch, Chapter 6, this volume) that has recrystallized mainly in albite-epidote-amphibolite facies, in contrast to the upper middle and upper amphibolite facies conditions in the adjacent Skagit Gneiss (Misch, 1968). Miller *et al.* (1985) observe that a subhorizontal mineral lineation in the GPO is suggestive of strike-slip faulting and infer that the GPO has formed from a part of the Black Peak batholith affected by renewed movement along the Ross Lake fault zone (*cf.* Misch, 1966). Miller *et al.* (1985) also report U/Pb and Rb/Sr ages of about 65 m.y. for the GPO.

Paleomagnetic data from the SCC are difficult to interpret because they have been obtained almost entirely from granitic rocks that tend to be magnetically unstable and lack definitive geochronology and paleohorizontal control. Furthermore, the Late Cretaceous to Tertiary thermal events which have reset K/Ar ages are likely to have caused magnetic overprinting as well. Despite these shortcomings, the evidence discussed below suggests that Late Cretaceous to earliest Tertiary plutons in the SCC have been tectonically rotated and transported poleward since their intrusion, whereas younger plutonic rocks have been emplaced essentially *in situ*.

The pioneering work of Beck and Noson (1972), followed by Beck *et al.* (1981a), show that the Mount Stuart batholith has a well-defined paleopole that is rotated clockwise about 40° and has an inclination 28° shallower than that expected for the Late Cretaceous age of acquisition. Although this anomaly could have several causes, Beck *et al.* (1981b) prefer translation from a more southerly latitude. Nevertheless, as observed by McDougall (1980), beds of Eocene sandstone near Mount Stuart have moderate dips to the southwest which would partly compensate for the inclination anomaly. Also, the reported magmatic epidote in the Ten Peak pluton (Zen and Hammarstrom, 1984) suggests, but does not prove, an erosion level as much as 15 km deeper than the Mount Stuart batholith which crops out only 15-20 km to the southwest. Thus it appears that regional and local tilting could account for at least some of the discordance of the Mount Stuart paleopole.

Other discordant Late Mesozoic to Early Tertiary paleopoles are reported by Stauss (1982), Strickler (1982), and Harrison (1984), but the pattern of rotation and translation does not match that of the Mount Stuart block. Harrison (1984) defines a panel of rock which seems to have consistently discordant poles of Eocene age (45-50 m.y.) This panel is bounded on the east by the Ross Lake fault and on the west by the Entiat fault, which suggests an allochthonous relationship. However, the data presently available are too sketchy to establish this area as a separate tectonic domain. Elsewhere, Irving *et al.* (1985) provide evidence for northward displacement and rotation of the Late Cretaceous Spuzzum pluton, which apparently intrudes the north end of the SCC. Both Beck *et al.* (1981b) and Irving *et al.* (1985) emphasize the fact that Cretaceous batholiths throughout the North American Cordillera consistently show anomalously low paleolatitudes, suggesting a general pattern of poleward tectonic transport. Irving *et al.* (1985) also point out that the Mount Stuart paleopole is divergent from the main grouping of Cretaceous batholiths, unless a tilt correction of about 17° is applied. A similar correction is required to bring the Mount Stuart results into accord with an inferred paleopole from the Midnight Peak Formation of the Methow terrane (Granirer, 1985).

The possibility of a systematic relationship between oceanic plate movement and Cordilleran terrane history is discussed by Debiche *et al.* (1987). Taking the best estimates of the motion of the Kula and Farallon plates with respect to North America (Engebretson *et al.*, 1986), a terrane trajectory for Wrangellia can be calculated. This trajectory (Fig. 8-9) indicates a collision with the North American craton in the vicinity of northern Baja California at about 90-110 m.y.b.p. Subsequent northward movement, tangential to the continental margin, could have been driven by the Farallon plate (90-85 m.y.) and the Kula plate (85-55 m.y.).

In contrast to the allochthonous nature of the Mesozoic plutons, paleopoles of Tertiary granites measured by Beck *et al.* (1982) show no evidence of significant rotation or translation. Two of these, the Hope plutonic complex and the Chilliwack batholith, lie within the SCC and their K/Ar ages of 30-41-m.y. place a lower limit on the timing

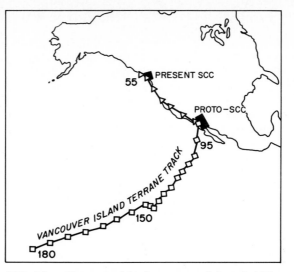

FIG. 8-9. Vancouver Island terrane track from Debiche *et al.*, 1987, and location of the SCC at about 90 m.y. based on paleomagnetic data for the Mount Stuart batholith and Spuzzum intrusives. A tilt correction of 17° has been applied to the Mount Stuart paleopole to accommodate regional dips shown by Eocene sediments (see the text).

of any possible terrane transport. However, for the North Cascades as a whole, northward motion must have ceased before about 43 m.y. (Beck *et al.*, 1982).

If the paleomagnetic data for northward displacement of the SCC (Beck *et al.*, 1981b; Irving *et al.*, 1985) are accepted, it is reasonable to assume that the proto-SCC at 90 m.y. existed along the margin of North America in the vicinity of northern Baja California. Then, if the terrane track for Wrangellia proposed by Debiche *et al.* (1987) is correct, the SCC could have evolved by collisional telescoping related to the arrival of Wrangellia (*cf.* G. A. Davis *et al.*, 1978). Such a collisional event might have resulted in syntectonic intrusion, metamorphism, and deformation, as envisioned by Monger *et al.* (1982) for the origin of the Coast Plutonic Complex and the Omineca Crystalline Belt in British Columbia. According to Beck *et al.* (1981b), northward movement could have been accomplished by imbricate faulting related to oblique subduction (Sunda type) or transport as a lithospheric fragment attached to the Kula and/or Farallon plates (California type).

SUMMARY AND CONCLUSIONS

The tectonic evolution of the SCC is difficult to interpret with the data presently available. It is clear that all of the rocks comprising the SCC participated in Skagit metamorphism during the Late Cretaceous and that intrusion and deformation continued until at least 45 m.y. ago. Orthogneisses of the Marblemount belt also have a well-established Triassic intrusive age and it appears that these rocks formed the basement upon which at least part of the supracrustal sequence was deposited. Rb/Sr scatterchrons suggest a Triassic depositional age for the Chiwaukum, Cascade River, and Settler Schists, but these may represent resetting during later metamorphism. Similarly, the Precambrian depositional age of the Swakane Gneiss is based on a single unpublished Rb/Sr date, plus a Precambrian U/Pb age on zircons that could well be detrital.

The tectonic environment in which the supracrustal sequence formed is uncertain. Based on geochemical and isotopic evidence, one of the authors (RSB) prefers a complex arc system similar to the present-day Phlipine Archipelago (*cf.* Hawkins *et al.*, 1985). The other author (PM) emphasizes the primary cohesion of the depositional system, however complex it may have been in detail (*cf.* Misch, Chapter 6, this volume).

The paleomagnetic data also are somewhat inconsistent but, taken as a whole, strongly suggest that between 90 and 105 m.y. ago the SCC existed at a latitude about equivalent to what is now northern Baja California. This position coincides with the calculated terrane track of Wrangellia, so it can be inferred that Late Cretaceous deformation, intrusion, and regional metamorphism are related to a collision between Wrangellia and the margin of North America (*cf.* G. A. Davis *et al.*, 1978; Silver and Smith, 1983).

After the inferred collision, the SCC must have been translated northward, roughly parallel to the continental margin, following the calculated trajectory of the Kula plate. Some of this movement may be represented by right-lateral shear distributed on the Straight Creek–Fraser River and Ross Lake fault systems. However, the total offset on these faults seems to be only on the order of a few hundreds of kilometers. Poleward translation of as much as 1400 km suggested by the paleomagnetic data would require a master fault system, somewhere to the east, which has not yet been recognized. The arrival time of the SCC at its present geographic position is constrained only by post-tectonic plutons of the Chilliwack batholith which show paleomagnetic poles concordant with the North American craton. The oldest of these has a K/Ar cooling age of about 43 m.y., but Beck (1987) argues for cessation of relative northward motion by 50 m.y. ago.

Note that the actual displacement is quite sensitive to interpretation of the paleomagnetic data. According to Beck (1987), "paleomagnetic Euler poles," first derived by Gordon *et al.* (1984) and modified by May and Butler (1986), suggest that most displaced terranes moved *southward* relative to North America prior to about 125 m.y. and rapidly returned northward during the late Cretaceous and early Tertiary. As an extreme case, the data available allow the SCC to have formed only a few hundred kilometers south of its present latitude (or less, within $a95$ error) and to have remained essentially in situ while most of the north-south absolute motion occurred in the North American plate. Another complicating factor is the possibility of as much as $14°$ of "true" polar wander during the Cretaceous reported by Gordon and Livermore (1987). However, this should not affect estimates of displacement based on a comparison with the apparent polar wander path for the North American craton, which would include any component of true polar wander.

REFERENCES

Babcock, R. S., and Misch, P., 1984, Geochemistry of the Crystalline Core of the North Cascades Range: (EOS) *Trans. Am. Geophys. Un.*, v. 65, p. 328–329.

—— , Armstrong, R. L., and Misch, P. 1985, Isotopic constraints on the age and origin of the Skagit Metamorphic Suite and related rocks: *Geol. Soc. America Abstr. with Programs*, v. 17, p. 339.

Bartholomew, P. R., 1979, Geology and metamorphism of the Yale Creek Area, B.C.: M.S. thesis, Univ. British Columbia, Vancouver, B.C.

Beck, M. E., Jr., and Noson, L., 1972, Anomalous paleolatitudes in Cretaceous granitic rocks: *Nature*, v. 235, p. 11–13.

_____, Burmester, R. F., and Schoonover, R., 1981a, Paleomagnetism and tectonics of Cretaceous Mt. Stuart batholith of Washington — Translation or tilt? *Earth Planet. Sci. Lett.*, v. 56, p. 336–342.

_____, Burmester, R. F., Engebretson, D. C., and Schoonover, R., 1981b, Northward translation of Mesozoic batholiths of western North America — Paleomagnetic evidence and tectonic significance: *Geofisica Internac.*, v. 20, p. 144–162.

_____, Burmester, R. F., and Schoonover, R., 1982, Tertiary paleomagnetism of the North Cascade Range, Washington: *Geophys. Res. Lett.*, v. 9, p. 515–518.

_____, 1987, Paleomagnetism of Continental North America: Implications for displacement of crustal blocks with the western Cordillera, Baja California to British Columbia. *in* Pakiser, L. C. and Mooney, W. D., editors: Geophysical Framework of the Continental United States: *Geol. Soc. America*, in press.

Bowden, P., Batchelor, R. A., Chappell, B. W., Didier, J., and Lameyre, J., 1984, Petrological, geochemical and source criteria for the classification of granitic rocks — A discussion: *Physics Earth Planet. Interiors*, v. 35, p. 1–11.

Cater, F. W., Jr., 1982, The intrusive rocks of the Holden and Lucerne quadrangles, Washington — The relation of depth zones, composition, textures, and emplacement of plutons: *U.S. Geol. Survey Prof. Paper 1220*, 108 p.

Cater, F. W., and Crowder, D. F., 1967, Geologic map of the Holden quadrangle, Snohomish and Chelan Counties, Washington: *U.S. Geol. Survey Geol. Quad. Map GQ-646.*

_____, and Wright, T. L., 1967, Geologic map of the Lucerne quadrangle, Chelan County, Washington: *U.S. Geol. Survey Geol. Quad. Map GQ-647.*

Condie, K. C., and DeMalas, J. P., 1985, The Pinal Schist — An early Proterozoic quartz wacke association in southeastern Arizona: *Precambrian Res.*, v. 27, p. 337–356.

Coney, P. J., Jones, D. L., and Monger, J. W. H., 1980, Cordilleran suspect terranes: *Nature*, v. 288, p. 329–333.

Crowder, D.F., 1959, Granitization, migmatization and fusion in the northern Entiat Mountains, Washington: *Geol. Soc. America. Bull.*, v. 70, p. 827–877.

_____, Tabor, R. W., and Ford, A. B., 1966, Geologic map of the Glacier Peak quadrangle, Snohomish and Chelan Counties, Washington: *U.S. Geol. Survey Geol. Quad. Map GQ-473.*

Davis, G. A., 1977, Tectonic evolution of the Pacific Northwest — Precambrian to present: Washington Public Power Supply System, Nuclear Project no. 1, subapp. 2RC, PSAR, amendment 23, p. i2R-C-46.

_____, Monger, J. W. H., and Burchfiel, B. C., 1978, Mesozoic reconstruction of the Cordilleran "collage," central British Columbia to central California; *in* Howell, D. G., and McDougall, K. A., eds., *Mesozoic Paleogeography of the Western United States:* Pacific Section, Soc. Econ. Paleontologists Mineralogists, Pacific Coast Paleogeography Symp. 2, p. 33–70.

Davis, G. L., Tilton, G. R., Aldrich, L. T., Hart, S. R., and Steiger, R. H., 1966, Isotopic composition of lead and strontium in crystalline rocks from the Northern Cascade Range, United States: *Carnegie Inst. Wash. Yearbook 64*, 1964–1965, p. 171–177.

Debiche, M. G., Cox, A., and Engebretson, D. C., 1987, The motion of allochthonous terranes across the Pacific Basin: *Geol. Soc. America spec. Paper 207*, 50 p.

Dubois, R. L., 1954, Petrology and ore deposits of the Holden mine area, Chelan County, Washington: Ph.D. thesis, Univ. Washington, Seattle, Wash.

Engebretson, D. C., Cox, A., and Gordon, R. G., 1986, Relative motions between oceanic and continental plates in the Pacific Basin: *Geol. Soc. America Spec. Paper 206*, 59 p.

Engels, J. C., Tabor, R. W., Miller, F. K., and Obradovich, J. D., 1976, Summary of K/Ar, Rb-Sr, U-Pb, and fission-track ages of rocks from Washington State prior to 1975 (exclusive of Columbia Plateau basalts): *U.S. Geol. Survey Misc. Field Studies Map MF-710.*

Erickson, E. H., Jr., 1977a, Petrology and petrogenesis of the Mount Stuart batholith — Plutonic equivalent of the high-alumina basalt association? *Contrib. Mineralogy Petrology*, v. 60, p. 183–207.

———, 1977b, Structure, stratigraphy, plutonism and volcanism of the central Cascades, Washington — Pt. II. General geologic setting of the Stevens Pass–Leavenworth–Swauk Pass area; *in* Brown, E. H., and Ellis, R. C., eds., *Geological Excursions in the Pacific Northwest:* Guidebook Geol. Soc. America Ann. Meet., Western Washington Univ., Bellingham, Wash., p. 276–282.

Evans, B. W., and Berti, J. W., 1986, A revised metamorphic history for the Chiwaukum Schist, Washington Cascades: *Geology*, v. 14, p. 695–698.

Ford, A. B., 1959, Geology and petrology of the Glacier Peak quadrangle, northern Cascades, Washington: Ph.D. thesis, Univ. Washington, Seattle, Wash.

———, Nelson, W. H., Sonnevil, R. A., Loney, R. A., Huie, C., Haugerud, R. A., and Garwin, S. L., 1986, Geologic map of the Glacier Peak Wilderness and vicinity, Chelan, Skagit, and Snohomish Counties, Washington: *U.S. Geol. Survey Misc. Field Inv. Map MP-1652-B*, in press.

Fox, K. F., Jr., Rinehart, C. D., and Engels, J. C., 1977, Plutonism and orogeny in north-central Washington — Timing and regional context: *U.S. Geol. Survey Prof. Paper 989*, 27 p.

Frizzell, V. A., Jr., Tabor, R. W., Booth, D. B., Ort, K., and Waitt, R. B., Jr., 1984, Preliminary geologic map of the Snoqualmie Pass 1:100,000 quadrangle, Washington: *U.S. Geol. Survey Open File Rt 84-693.*

Frost, B. R., 1975, Contact metamorphism of serpentinite, chlorite blackwall and rodingite at Paddy-Go-Easy Pass, Central Cascades, Washington: *J. Petrology* v. 16, p. 272–313.

———, 1976, Limits to the assemblage forsterite-anorthite as inferred from peridotite hornfels, Icicle Creek, Washington: *Amer. Mineralogist*, v. 61, p. 732–750.

Gabites, J., 1985, Geology and geochronometry of the Cogburn Creek-Settler Creek area, northeast of Harrison Lake, B.C.: M.Sc. thesis, Univ. British Columbia, Vancouver, B.C.

Getsinger, J. S., 1978, A structural and petrologic study of the Chiwaukum Schist on Nason Ridge, northeast of Stevens Pass, North Cascades, Washington: M.S. thesis, Univ. Washington, Seattle, Wash.

Gordon, R. G., and Livermore, R. A., 1987, Apparent polar wander of the mean-lithosphere reference frame: *Geo. J. Roy. Astronomical Soc.*, in press.

Granirer, J. L., 1985, Paleomagnetic evidence for northward transport of the Methow-Pasayten belt, North Central Washington: M.S. Thesis, Western Washington Univ., Bellingham, Wash.

Grant, A. R., 1966, Bedrock geology and petrology of the Dome Peak area, Chelan, Skagit and Snohomish counties, northern Cascades, Washington: Ph.D. thesis, Univ. Washington, Seattle, Wash.

Harrison, W., 1984, Paleomagnetism of the Hidden Lake, Sulphur Mountain, Ten Peak and Oval plutons: M.S. thesis, Western Washington Univ. Bellingham, Wash.

Hawkins, J. W., Moore, G. F., Villamor, R., Evans, C., Wright, E., 1985, Geology of the composite terranes of East and Central Mindanao; *in* Howell, D. G., ed., *Tectono-stratigraphic Terranes of the Circum-Pacific Region:* Circum-Pacific Council Energy Min. Resources, Earth Sci. Series, p. 437–463.

Heath, M. T., 1971, Bedrock geology of the Monte Cristo area, North Cascades, Washington: Ph.D. thesis, Univ. Washington, Seattle, Wash.

Hopson, C. A., and Mattison, J. M., 1971, Metamorphism and plutonism, Lake Chelan Region, Northern Cascades, Washington, *in Metamorphism in the Canadian Cordillera:* Cordilleran Section, Geol. Assoc. Canada Program Abstr., p. 13.

Irving, E., Woodsworth, G. J., Wynne, P. J., and Morrison, A., 1985, Paleomagnetic evidence for displacement from the south of the Coast Plutonic Complex, British Columbia: *Can. J. Earth Sci.*, v. 22, p. 584–598.

Kaneda, K., 1980, Contact metamorphism of the Chiwaukum Schist near Lake Edna, Chiwaukum Mountains, Washington: M.S. thesis, Univ. Washington, Seattle, Wash.

Lowes, B. E., 1972, Metamorphic petrology and structural geology of the area east of Harrison Lake, British Columbia, Canada: Ph.D. thesis, Univ. Washington, Seattle, Wash.

Magloughlin, J., 1986, Metamorphic petrology, structural history, geochronology, tectonics, and geothermometry/geobarometry in the Wenatchee Ridge area, North Cascades, Washington: M.S. thesis, Univ. Washington, Seattle, Wash.

Mattinson, J. M., 1972, Ages of zircons from the northern Cascade Mountains, Washington: *Geol. Soc. America Bull.*, v. 83, p. 3769–3784.

May, S. R., and Butler, R. F., 1986, North American Jurassic apparent polar wander: Implications for plate motion, paleogeography, and Cordilleran tectonics: *J. Geophys. Research*, v. 91, p. 11519–11544.

McDougall, J. W., 1980, Geology and structural evolution of the Foss River-Deception Creek area, Cascade Mountains, Washington: M.S. thesis, Oregon State Univ., Corvallis, Oreg.

McTaggart, K. C., 1970, Tectonic history of the northern Cascade Mountains, *in* Wheeler, J. O., ed., *Structure of the southern Canadian Cordillera:* Geol. Assoc. Canada Spec. Paper 6, p. 137–148.

——, and Thompson, R. M., 1967, Geology of part of the northern Cascades in southeast British Columbia: *Can. J. Earth Sci.*, v. 4, p. 1199–1228.

Miller, R. B., 1985, The ophiolitic Ingalls Complex, north-central Cascade Mountains, Washington: *Geol. Soc. America Bull.*, v. 96, p. 27–42.

——, and Frost, B. R., 1977, Structure, stratigraphy, plutonism, and volcanism of the central Cascades, Washington — Pt. III. Geology of the Ingalls Complex and related pre-Late Cretaceous rocks of the Mount Stuart uplift, Central Cascades, Washington, *in* Brown, E. H., and Ellis, R. C., eds., *Geological Excursions in the Pacific Northwest:* Guidebook Geol. Soc. America Ann. Meet. Western Washington Univ., Bellingham, Wash., p. 283–291.

——, Misch, P., and Hoppe, W. J., 1985, New observations on the central and southern segments of the Ross Lake Fault Zone, N. Cascades, Washington: *Geol. Soc. America Abstr. with Program*, v. 17, p. 370.

Misch, P., 1952, Geology of the northern Cascades of Washington: *Mountaineer*, v. 45, no. 12, p. 4–22.

——, 1966, Tectonic evolution of the northern Cascades of Washington State, *in* Gunning, H. C., ed., *A symposium on the Tectonic History and Mineral Deposits of the Western Cordillera:* Can. Inst. Mining Metallurgy Spec. Vol. 8, p. 101–148.

———, 1968, Plagioclase compositions and non-anatectic origin of migmatitic gneisses in Northern Cascade Mountains of Washington State: *Contri. Mineralogy Petrology*, v. 17, p. 1–70.

———, 1977a, Bedrock geology of the North Cascades, *in* Brown, E. H., and Ellis, R. C., eds., *Geological Excursions in the Pacific Northwest:* Guidebook Geol. Soc. America Ann. Meet. Western Washington Univ., Bellingham, Wash. p. 1–62.

———, 1977b, Dextral displacements at some major strike-slip faults in the North Cascades: *Geol. Assoc. Canada Programs with Abstr.*, p. 37.

———, 1979, Geologic map of the Marblemount Quadrangle, Washington: Washington Dept. Natl. Resources, *Div. Geology Earth Resources Geol Map GM-23.*

———, and Onyeagocha, A. C., 1976, Symplectite breakdown of Ca-rich almandines in upper amphibolite-facies Skagit Gneiss, North Cascades, Washington: *Contrib. Mineralogy Petrology*, v. 54, p. 189–224.

———, and Rice, J. M., 1975, Miscibility of tremolite and hornblende in progressive Skagit Metamorphic Suite, North Cascades, Washington: *J. Petrology*, v. 26, p. 1–21.

Monger, J. W. H., Price, R. A., and Templeman-Kluit, D. J., 1982, Tectonic accretion and the origin of the two major metamorphic and plutonic welts in the Canadian Cordillera: *Geology*, v. 10, p. 70–75.

Oles, K. F., 1956, The geology and petrology of the crystalline rocks of the Beckler River-Nason Ridge area, Washington: Ph.D. thesis, Univ. Washington, Seattle, Wash.

Ort, M. H., and Tabor, R. W., 1985, Major and trace-element composition of greenstones, greenschists, amphibolites and gneisses from the North Cascades, Washington: U.S. Geol. Survey, Open-file Rept. 85-434.

Page, B. M., 1939, Geology of part of the Chiwaukum quadrangle, Washington: Ph.D. thesis, Stanford Univ. Stanford, Calif.

Pigage, L. C., 1976, Metamorphism of the Settler Schist, southwest of Yale, British Columbia: *Can. J. Earth Sci.*, v. 13, p. 405–421.

Plummer, C. C., 1980, Dynamothermal contact metamorphism superposed on regional metamorphism in the pelitic rocks of the Chiwaukum Mountains area, Washington Cascades — Summary: *Geol. Soc. America Bull.*, v. 91, pt. 1, p. 386–388, pt. 2, 1627–1668.

Siberling, N. J., Jones, D. L., Blake, M. C., Jr., and Howell, D. G., 1984. Lithotectonic terrane map of the western conterminous United States — Pt. C, *in* Siberling, N.J., and Jones, D. L., eds., *Lithotectonic Terrane Maps of the North American Cordillera:* U.S. Geol. Survey Open File Rt. 84-523, p. C1–C43.

Silver, E. A., and Smith, R. B., 1983, Comparison of terrane accretion in modern southeast Asia and the Mesozoic North American Cordillera: *Geology*, v. 11, p. 198–202.

Staatz, M. H., Tabor, R. W., Weis, P. L., Robertson, J. F., Van Noy, R. M., and Patee, E. C., 1972, Geology and mineral resources of the northern part of North Cascades National Park, Washington: *U.S. Geol. Survey Bull. 1359*, 132 p.

Stauss, L. D., 1982, Anomalous paleomagnetic direction in Eocene dikes of Central Washington: M.S. thesis, Western Washington Univ. Bellingham, Wash.

Streckeisen, A., and Le Maitre, R. W., 1979, A chemical approximation to the modal QAPF classification of the igneous rocks: *Neues Jahrb. Abh.*, v. 136, p. 169–206.

Strickler, D. L., 1982, Paleomagnetism of three Late Cretaceous granitic plutons, North Cascades, Washington: M.S. thesis, Western Washington Univ., Bellingham, Wash.

Sun, S.-S., and Nesbitt, R. W., 1978, Geochemical regularities and genetic significance of ophiolitic basalts: *Geology*, v. 6, p. 689–693.

Tabor, R. W., 1961, The crystalline geology of the area south of Cascade Pass, northern Cascade mountains, Washington: Ph.D. thesis, Univ. Washington, Seattle, Wash.

——, Frizzel, V. A., Jr., Whetan, J. T., Swanson, D. A., Byerly, G. R., Booth, D. B., Hetherington, M. J., and Waitt, R. B., Jr., 1980, Preliminary geologic map of the Chelan 1:100,000 quadrangle, Washington: *U.S. Geol. Survey Open File Map 80–841*, 43 p.

——, Frizzell, V. A., Jr., Booth, D. A., Whetten, J. T., Waitt, R. B., Jr., and Zartman, R. E., 1982a, Preliminary geologic map of the Skykomish 1:100,000 quadrangle, Washington: *U.S. Geol. Survey Open File Map OF-82-747*, 31 p.

——, Frizzell, V. A., Jr., Yeats, R. S., and Whetten, J. T., 1982b, Geologic map of the Eagle Rock-Glacier Peak Roadless Areas: *U.S. Geol. Survey Misc. Field Studies Map MF-1380-A*.

——, Waitt, R. B., Jr., Frizzell, V. A., Jr., Swanson, D. A. and Byerly, G. R., 1982c, Geologic Map of the Wenatchee 1:100,000 quadrangle, Washington: *U.S. Geol. Survey Misc. Geol. Inv. Map I-1311*.

——, Zartman, R. E., Frizzell, V. A., Jr., in press, Possible tectonostratigraphic terranes in the North Cascades Crystalline Core, Washington, *in* Schuster, E., ed., *Geology of Washington Symposium Volume:* Wash. State Dept. Nat. Resources, in press.

Van Diver, B. B., 1967, Contemporaneous faulting-metamorphism in Wenatchee Ridge area, northern Cascades, Washington: *Amer. J. Sci.*, v. 265, p. 132–150.

Yardley, B. W. D., 1978, Genesis of the Skagit migmatites, Washington, and the distinction between possible mechanisms of migmatization: *Geol. Soc. America Bull.*, v. 89, p. 941–951.

Zen, E-an and Hammarstrom, J. M., 1984, Magmatic epidote and its petrologic significance: *Geology*, v. 12, p. 515–518.

Vicki L. Hansen and John W. Goodge
Department of Earth and Space Sciences
University of California
Los Angeles, California 90024

9

METAMORPHISM, STRUCTURAL PETROLOGY, AND REGIONAL EVOLUTION OF THE OKANOGAN COMPLEX, NORTHEASTERN WASHINGTON

ABSTRACT

We discuss models of Okanogan complex evolution in light of constraints imposed by combined study of metamorphic mineral assemblages and structural fabrics within rocks of the complex. Mineral assemblages formed during three mutually distinct metamorphic episodes are correlative with three contrasting styles of deformation: (1) mineral assemblages indicative of upper-amphibolite-facies metamorphism occur in association with regionally distributed metamorphic foliation and similar-style folds; (2) middle-greenschist-facies assemblages occur in mylonitic shear zone-type fabrics; and (3) zeolite to lowest-greenschist-facies assemblages are associated with cataclastic breccias along the complex margin.

By themselves, the metamorphic mineral assemblages define a simple loop in *P-T* space and therefore support neither a continuous nor a discontinuous model of Okanogan complex evolution. However, the structural styles and, more important, structural kinematics of the first two metamorphic/deformation episodes are quite dissimilar; vergence directions of folds associated with high-grade metamorphism are nearly opposite to the tectonic transport direction indicated in younger mylonites, thereby demonstrating tectonic and chronologic dissociation. Kinematics of structures formed by mylonitization and brecciation appear to be similar, and therefore the transition from middle greenschist to zeolite facies metamorphism is most easily interpreted as resulting from a continuous evolving process of decreasing metamorphic gradient.

Based on the structural petrology of upper- and lower-plate rocks, we favor a discontinuous multistep model for the evolution of the Okanogan complex. Foliated and folded high-grade metasedimentary rocks of both the upper and lower plates reflect an early period of crustal shortening at mid-crustal depths of approximately 15-25 km. Concurrent deformation and metamorphism are of probable Jurassic to Cretaceous age if correlation with isotopically dated rocks of similar lithology, structure, and metamorphic assemblages in southern British Columbia is valid; their effects may record accretion of the Quesnel terrane during the Jura-Cretaceous Columbian orogeny. During Late Cretaceous to Tertiary time, the regional stress regime changed from that of bulk crustal compression to crustal tension. Extension during this last period is recorded by widespread emplacement of granitic plutons, formation of a continuous west-dipping normal shear zone for over 250 km from Omak Lake in Washington to Vernon, British Columbia, marking the western margin of the Okanogan complex, and westward movement along imbricate, brittle, normal faults in the upper plate as far east as the western margin of the complex.

INTRODUCTION

Study of Mesozoic crustal evolution in the North American Cordillera began a shift in emphasis from foreland deformation in the fold-and-thrust belt to that in the so-called hinterland with the pioneering work of Misch (1960), Reesor (1965), and Armstrong and Hansen (1966). More recently, rocks within the hinterland region have gained widespread attention with the introduction of the concept of metamorphic core complexes, reviewed in Crittenden *et al.* (1980), and by Armstrong (1982). These isolated metamorphic terrains extend the length of the Cordillera from southwestern British Columbia to northwestern Mexico (Fig. 9-1). Metamorphic core complexes share a number of geologic features which suggest a common origin, although numerous attempts to present a unifying model have met with only partial success, owing to local and regional variation in structural, petrologic, and geochronologic relationships.

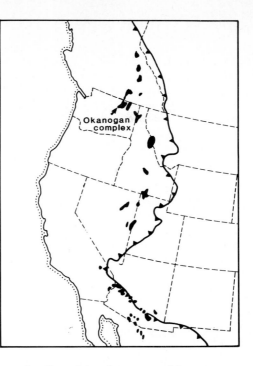

FIG. 9-1. Distribution of Cordilleran metamorphic core complexes (in black), modified after Coney (1980) and Armstrong (1982). Barbed line shows eastern limit of Cordilleran miogeocline, coincident in most areas with major thrusts of the fold-and-thrust belt.

Among the most critical problems concerning the origin of metamorphic core complexes are the timing of different styles of deformation, the conditions of deformation, and the vergence of structures ranging from folds to mylonite shear zones to brittle fault zones. In many complexes, mylonitization preceded brittle deformation in a relative sense, but only quite recently have efforts been made which firmly establish the absolute ages of deformation in a small number of these terrains (Bickford *et al.*, 1985; Davis *et al.*, 1982; E. L. Miller *et al.*, 1983; Parrish, 1983, 1985). Furthermore, the kinematic significance of mylonite shear fabrics common to the complexes remains controversial, as does an understanding of the strain state during ductile shear. Problems such as these leave unsettled how the deformation in these terrains relates to mechanisms of hinterland crustal deformation, whether by overall extension, shortening, or some combination of both. Despite an abundance of metamorphic rocks in these complexes, only minor study has been devoted to the details of synkinematic metamorphism and what it reveals of the orogenic history.

Studies of core complexes need to address tectonic evolution in terms of continuous and discontinuous processes, whether in regard to timing, metamorphic conditions, or mode of deformation. We here discuss how metamorphic assemblage relationships, combined with structural analysis, provide constraints for tectonic models of a metamorphic core complex in northeastern Washington. Metamorphic relations in the Okanogan complex, as well as other similar terrains, can contribute directly to our understanding of core complex evolution and constrain models invoking deep-level overthrusting, plutono-metamorphic diapirism, or bulk crustal extension. The Okanogan complex is suited for such study because critical areas are well mapped, rock compositions are diverse, metamorphic parageneses are well preserved, and characteristic structural elements of metamorphic core complexes are superposed.

A number of interpretations have been proposed to explain geologic relationships within the Okanogan complex. Based on geologic observation and few age constraints, Waters and Krauskopf (1941) postulated an origin of the Okanogan complex involving a sequence of tectonically linked events related to the rise, emplacement, and marginal deformation of a composite batholith. With a host of new radiometric age data, Fox *et al.* (1976, 1977) elucidated those interpretations in a similar model of somewhat larger scope; they reinforced earlier ideas that the Okanogan complex evolved through penecontemporaneous plutonometamorphic mechanisms resulting in a structural culmination related to diapiric emplacement of high-grade into lower-grade metamorphic rocks.

Snook (1965), operating without the benefit of radiometric age data, arrived at the contrary conclusion that the Okanogan complex evolved through a series of discrete steps, summarized as follows: (1) high-grade regional metamorphism and deformation, (2) penetrative mylonitization along a subhorizontal shear zone, (3) regional-scale warping of the mylonitic fabric, and (4) high-angle block faulting along the southern and western complex margins, elevating the crystalline core relative to rocks to the south and west. Cheney (1980) proposed a fourth hypothesis similar to that of Snook; he contended that Cenozoic thrusting separated metamorphic and plutonic basement rocks from upper-plate Precambrian and Tertiary rocks, followed by regional doming and post-Eocene folding.

The model presented in this paper combines aspects of each of the models outlined above. Relationships between metamorphic assemblages and structures presented herein are best explained by accretion of the Quesnel terrane and formation of the Quesnel suture in mid-Jurassic time, followed first by (and perhaps the cause for) an episode of high-grade regional metamorphism and associated deformation, and second by early Tertiary crustal extension. The Tertiary-age extension, marked principally by penetrative mylonitization and brecciation along an intracontinental shear zone, cross-cuts earlier-formed metamorphic and deformation patterns to place high-grade metamorphic rocks in both the upper and lower plates. These extensional structures and associated silicic magmas are the principal features that characterize the Okanogan complex as a metamorphic core complex.

REGIONAL TECTONIC SETTING

The Okanogan complex is one of several semicontinuous structural and metamorphic culminations in British Columbia, Washington, and Idaho referred to as "metamorphic core complexes" or "gneiss domes." Fox and Rinehart (1971) introduced the term "Okanogan gneiss dome" to describe crystalline rocks of the Okanogan highlands, and use of this term was maintained later by Fox *et al.* (1976, 1977). To avoid confusion with the term "mantled gneiss dome," which carries strong genetic connotations, and in keeping with the use of "metamorphic core complex" as a purely descriptive nomenclature, we refer to this terrain simply as the Okanogan complex.

Metamorphic core complexes in the northwestern United States and southwestern Canada are associated with the Shuswap Metamorphic Complex in the Omineca Crystalline Belt (Fox *et al.*, 1977; Monger *et al.*, 1972; Okulitch, 1984; Wheeler, 1965), of which the Okanogan complex is a part (Fig. 9-2). Okanogan complex rocks in Washington are continuous with rocks to the north in British Columbia referred to as the "Okanagan Plutonic and Metamorphic Complex" by Okulitch (1984) ("Okanogan" is spelled with an "o" south of the 49th parallel and with three "a"'s north of it). They are likely correlative to crystalline rocks within the Kettle complex to the east, being separated from them by en echelon fault grabens (Cheney, 1980). The Okanogan complex, although comprised of high-grade crystalline rocks, is flanked to the north and west by mostly unmetamorphosed pre-Upper Triassic sedimentary rocks of uncertain regional context. However, as suggested by Okulitch (1984), they may be the southernmost extension of the Quesnel

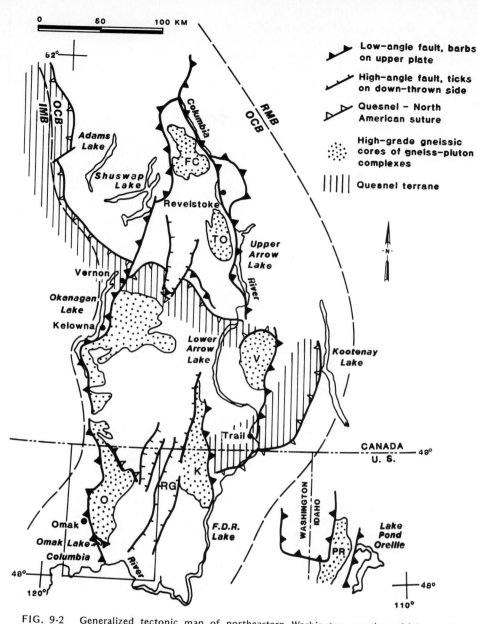

FIG. 9-2 Generalized tectonic map of northeastern Washington, northern Idaho, and southeastern British Columbia, showing distribution of metamorphic core complexes. Only the high-grade gneissic core areas of these complexes are shown (shaded), although they also include sedimentary, plutonic, and volcanic rocks. Low-angle faults are mostly of normal displacement and generally concordant with lower-plate mylonite zones (not shown). Quesnel-North America suture zone is approximately located from Okulitch (1984). Tectonic belts are from Monger *et al.* (1972): IMB, Intermontane Belt; OCB, Omineca Crystalline Belt; RMB, Rocky Mountain Belt, Core complexes are as follows: FC, Frenchman Cap; K, Kettle; O, Okanogan; PR, Priest River; TO, Thor-Odin, V, Valhalla, RG, Republic Graben. Data from Tipper *et al* (1981), Read and Brown (1981), Atwater and Rinehart (1984), Okulitch (1984), and Rhodes and Hyndman (1984). Area of Fig. 9-3 is outlined by box.

HANSEN, GOODGE

terrane, an allochthonous assemblage of marine sediments believed to overlie North American cratonal rocks in a thrust relationship (Fig. 9-2). Much of northeastern Washington, as well as adjacent parts of Idaho and British Columbia, consists of large masses of late Mesozoic and early Tertiary plutonic rock. The youngest granites in this general group are a series of syn- to post-tectonic plutons genetically related to coeval silicic volcanic rocks occupying north-northwest-trending grabens (see Atwater and Rinehart, 1984).

The oldest rocks of the region are compositionally varied gneisses which form the central and structurally lowest portions of the metamorphic core complexes, as well as the areally larger Shuswap Metamorphic Complex. Sparse evidence in British Columbia indicates rocks of the Shuswap Complex were involved in middle and late Paleozoic orogenesis. Devonian to Mississippian formation of east-vergent nappes associated with regional metamorphism and plutonism is of uncertain extent (Brown and Tippett, 1978; Fyson, 1970; Okulitch et al., 1975; Read, 1975; Read and Brown, 1981). Evidence for late Paleozic to earliest Mesozoic deformation and metamorphism is seen in eugeosynclinal sedimentary sequences in several areas adjacent to high-grade rocks of the Shuswap terrain (Fox et al., 1977; Okulitch, 1973; Read and Okulitch, 1977), but it is likely that much of the record for this event in the core complexes themselves is overprinted by later Mesozoic events.

Jurassic to Early Cretaceous deformation of the Shuswap Metamorphic Complex is well documented in Washington and British Columbia; this deformation and its associated high-grade metamorphism are attributed by many to the onset of eastward obduction of the Quesnel terrane (Brown and Read, 1983; Monger et al., 1982; Okulitch, 1984), termed the Columbian orogeny. Many workers recognize synmetamorphic deformation in core complexes of the Shuswap Complex, as well as deformation recorded in Triassic cover rocks (Fyles, 1970; Hyndman, 1968; Read and Brown, 1981; Read and Klepacki, 1981; Reesor, 1970; Rinehart and Fox, 1972, 1976; Snook, 1965; Wheeler, 1965). Characteristic aspects of this mid-Jurassic deformation are strongly appressed folds, often isoclinal and recumbent, which indicate a regime of crustal shortening.

Mesozoic orogeny in the entire region continued through Cretaceous time with a change in style during initial movement on east-directed thrust sheets in the foreland belt (Price, 1981; Price and Mountjoy, 1970), and shear along broad lower crustal décollement zones which brought high-grade "basement" rocks to higher structural levels outboard, or west, of the foreland (Brown and Read, 1983; Mattauer et al., 1983; Okulitch, 1984). Deformation of this age is believed to represent the latest stages in Quesnel–North American convergence. As with earlier Mesozoic structural episodes, the period is marked by severe crustal shortening. Mylonite zones exposed on the flanks of several core complexes are interpreted to represent a deep-seated basement décollement (Journeay and Brown, 1985; Okulitch, 1984; Read and Brown, 1981; Rhodes and Hyndman, 1984), although younger movement along certain of these zones may have extended into Tertiary time.

The familiar pattern of Mesozoic convergence and shortening changed dramatically in the early Tertiary, with onset of extension expressed by tectonic denudation and uplift of high-grade rocks in the core complexes, extensive normal faulting, and prolific plutonism and volcanism. Eocene extension was accommodated not only along normal faults, but at deeper levels in mylonite zones (Harms and Price, 1983; Rhodes and Hyndman, 1984; Tempelman-Kluit and Parkinson, 1986). Crustal extension has been related by Price (1979) to Eocene and younger translational motion on major strike-slip faults. Crustal extension of great magnitude slowed by Miocene time, as shown, for example, by flow of Columbia Plateau basalts across major normal faults along the southwest margin of the Okanogan complex.

Although the Okanogan complex crosses the international boundary, present studies do not. In this paper we review geologic relations only within the U.S. portion of the Okanogan complex and outline the constraints these relations place on evolution of the complex. In addition, we examine these constraints in light of recent work in southern British Columbia. Most of the detailed structural and metamorphic constraints presented herein are extracted from geologic relations within the southwest portion of the Okanogan complex.

The north-south trend of the elongate Okanogan complex is produced by bounding faults of the Republic Graben to the east and along the Okanogan River valley to the west (Fig. 9-2). The southern border becomes less well defined to the southeast, where it parallels Omak Lake. To the north, the Monashee décollement and the Quesnel suture separate the Okanagan Plutonic and Metamorphic Complex from the structurally lower Monashee complex (Okulitch, 1984). Rocks within the Okanogan complex everywhere consist of high-grade, layered paragneiss, orthogneiss, and Tertiary plutonic rocks.

The Okanogan complex is structurally asymmetric, and therefore, each of its margins is characterized by different geologic relations (Fig. 9-3). The upper plate to the west is composed of low-grade pre-Upper Triassic metasedimentary strata of the Anarchist Group, which are intruded by Jurassic and Cretaceous plutonic rocks (Rinehart and Fox, 1972, 1976). Low-grade eugeosynclinal strata of the Anarchist Group are considered to be a part of the allochthonous Quesnel terrane (Monger et al., 1982; Tipper, 1981). Farther south, high-grade metamorphic and associated plutonic rocks form the upper plate (Atwater and Rinehart, 1984; Goodge and Hansen, 1983; Menzer, 1983). A 1 to 2 km-thick zone of west-dipping mylonite gneiss, overprinted by a zone of chloritic breccia, separates the upper and lower plates along the western and southwestern margins of the complex. Along the northeast margin, a 1 km-wide gently dipping zone of scaly serpentinite marks the boundary of the complex between mylonitic gneiss of the lower plate and nonmylonitic, low-grade, Anarchist rocks of the upper plate. In contrast to the western margin of the complex, neither a metamorphic discontinuity nor a breccia zone disrupt the northeastern Anarchist-Okanogan complex contact, although it is truncated by Eocene high-angle faults locally (Orr, 1985). The eastern and southeastern margins of the Okanogan complex are marked by high-angle faults of the Republic Graben, which juxtapose upper-plate Anarchist Group and Eocene volcanic rocks against lower-plate, nonmylonitic Eocene granitic rocks (Atwater and Rinehart, 1984). Rocks within both upper and lower plates lack mylonitic fabric along the eastern margin.

STRUCTURES

Rocks of the Okanogan complex contain structures reflecting several periods of deformation and metamorphism, which appear correlative to similar events interpreted in rocks of the Shuswap Metamorphic Complex (see Brown and Read, 1983; Brown and Tippett, 1978; Fyles, 1970; Hyndman, 1968; McMillen, 1973; Read, 1980; Read and Brown, 1981; Read and Klepacki, 1981; Reesor, 1965; Reesor and Moore, 1971; Wheeler, 1965). Structures recognized in the Okanogan complex include polyphase folds, penetrative mylonitic fabrics, breccia zones, joints, and regional-scale tectonic warps (J. W. Goodge and V. L. Hansen, unpublished data). The relative timing and chronologic overlap of the tectonic events which formed these structures may be unraveled through study of the variously deformed rocks, their structural styles, orientation, and kinematics. High-grade

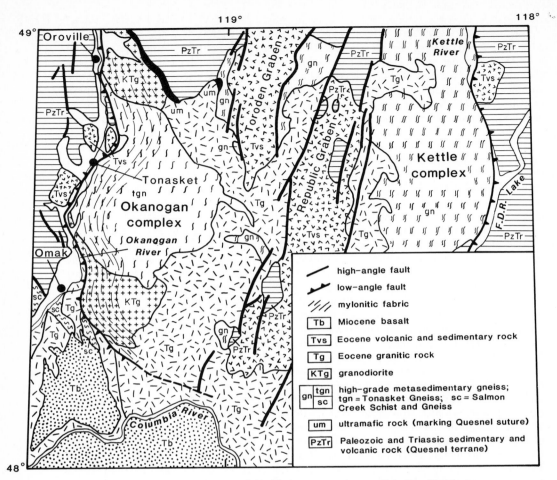

FIG. 9-3. Simplified geologic map of the Okanogan complex, northeastern Washington.

paragneiss records the entire structural evolution of the Okanogan complex from ductile-style folding to brecciation and warping. Granodiorite, which intrudes the paragneiss, displays evidence of penetrative mylonitization and brecciation, and lacks earlier-formed folds. Tertiary granitic plutons emplaced at shallow levels contain incipient to well-developed mylonitic shear zones and commonly display a primary (i.e., magmatic) mineral lineation parallel to the mylonitic lineation (Atwater, 1985). Each of the rock units above is also involved in brecciation at the complex margin locally (Atwater, 1985).

The Tonasket Gneiss (Fox *et al.*, 1976) hosts the oldest structures recognized within the U.S. portion of the Okanogan complex (Goodge, 1983a; Snook, 1965). Poly-phase folds, associated with high-grade regional metamorphism, are overprinted by a well-developed mylonitic fabric. These structures are interpreted to represent two distinct deformation events, an earlier deformation accompanied by amphibolite-grade metamorphism formed east-trending folds, and later subhorizontal ductile shearing deformed the previously formed folds (Goodge, 1983a, b; Snook, 1965).

A well-developed foliation and mineral lineation define the deformation fabric within these rocks related to high-grade regional metamorphism. Foliation consists of compositional layering and schistosity. Recrystallized biotite, hornblende, and where present, sillimanite form a variably developed mineral lineation, distinguished from a younger mylonitic lineation by its even distribution on schistosity surfaces, the undeformed character of the elongate biotite and hornblende, and its lack of stretched quartz rods or plagioclase augen. Blastic metamorphic textures indicate that high-grade metamorphic recrystallization occurred at a rate at least as rapid as deformation.

The Tonasket Gneiss metamorphic fabric is folded throughout the Okanogan complex (Atwater and Rinehart, 1984; Goodge, 1983b; Ross, 1981; Ross and Christie, 1979; Snook, 1965); textural evidence suggests that this deformation occurred synchronously with regional metamorphism. Axial planes of folds within the Tonasket trend easterly and dip steeply north and south (Fig. 9-4a; Goodge, 1983b; Ross, 1981; Ross and Christie, 1979). The age of deformation and associated high-grade metamorphism, interpreted to be Late Triassic to Early Jurassic (Snook, 1965), has not been substantiated by isotopic data. Regardless, there is strong evidence that deformation and associated metamorphism substantially predated mylonitization, as is discussed below.

The Salmon Creek Schist and Gneiss of Menzer (1983) southwest of the Okanogan complex margin displays structures similar to those preserved in the Tonasket Gneiss (Menzer, 1983). Although map-scale folds are difficult to recognize due to few and discontinuous marker units, tight similar-style folds deform an earlier metamorphic foliation. These small-scale, locally recumbent folds trend northwest with steeply northeast- and southwest-dipping axial planes. Relict mylonitic textures are present locally, although they are largely annealed (Menzer, 1983). Late syn- to postfolding granodiorite plutons intrude Salmon Creek rocks parallel to the preexisting metamorphic foliation (Menzer, 1983); however, these rocks contain no penetrative mylonitic fabrics, in contrast to similar lithologies in the Okanogan complex.

Within the Okanogan complex, granodioritic to dioritic plutons concordantly intrude the Tonasket Gneiss (Fig. 9-3). Intrusion of these bodies postdates folding, as shown by the absence of folds, as well as planar aplite and pegmatite dikes; however, intrusion predates the onset of mylonitization. Mineralogically and texturally similar granodiorite plutons which intrude the Salmon Creek rocks contain no mylonitic fabrics; whole-rock major element analyses of granodiorite units within Tonasket and Salmon Creek gneisses reveal no significant geochemical differences (J. W. Goodge and V. L. Hansen, unpublished data). Similarly, other mylonitic rocks within the complex closely resemble nonmylonitized rocks southwest of the complex in lithology and outcrop characteristics.

A zone of penetrative ductile mylonite, as defined by Bell and Etheridge (1973), dominates the form and structural fabric of the Okanogan complex. It is not present in rocks southwest or west of the complex. This broad 1 to 1.5 km-thick zone of mylonitic gneiss, concordant with the complex perimeter, forms "flatiron"-like aprons slanting up eastward from the Okanogan Valley. Intensity of mylonitic fabric increases structurally upward (westward) toward the border of the complex and decreases toward the complex interior. We believe the mylonite zone flattens to the west below upper-plate rocks exposed west of the Okanogan Valley, and flattens eastward if projected up structural dip.

The mylonitic fabric is best defined in granodiorite near the western border of the complex. The mylonitic fabric consists of a moderately to well-developed foliation defined by planar compositional segregations of quartz, feldspar, and biotite, and a well-developed, penetrative elongation lineation, L_e, marked by quartz rods, streaks of biotite, and feldspar augen. In more detail, the mylonitic foliation is a composite fabric comprised of S and C planes (Berthé et al., 1979a, b; Lister and Snoke, 1984; Simpson and Schmid, 1983). The geometric relationship between the concurrently formed S and C planes is used to interpret movement directions during mylonitic deformation (Berthé

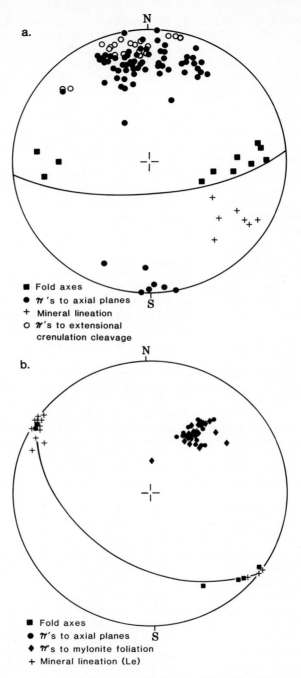

FIG. 9-4. (a) Structural features of Tonasket Gneiss below the mylonitic front. (b) Structural features of Tonasket Gneiss within the mylonite zone. Data were collected (Goodge, 1983b) in area northeast of Omak (Figure 9-3).

et al., 1979a, b; Lister and Snoke, 1984; Simpson and Schmid, 1983); regional tectonic transport directions are believed to parallel stretching lineations formed during ductile shear (Mattauer *et al.*, 1983).

L_e consistently trends approximately N60°W throughout the Okanogan complex from Omak Lake northward to Kelowna and Vernon in British Columbia; similarly, kinematic interpretation of S and C planes consistently indicates top-to-the-west, or west-side-down, movement within the plane of mylonitic foliation parallel to L_e, along the entire western border of the Okanogan complex (Fig. 9-3; Bardoux, 1985a, b; Hansen, 1983a, b; Parkinson. 1985). The amount of tectonic displacement within the mylonite zone is unknown.

Mylonitic deformation formed a penetrative zone along the entire western margin of the Okanogan complex, affecting both orthogneiss and paragneiss lithologies. The mylonitic fabric within the Tonasket Gneiss is similar to that in granodiorite, although previously formed folds are preserved within the paragneiss. Folds within mylonitized Tonasket Gneiss are similar in form to those observed in nonmylonitic paragneiss below the mylonitic front, although within the zone of mylonitic deformation, paragneiss displays a tightening of fold form, and an axial plane schistosity (mylonitic foliation) not present in nonmylonitized paragneiss (Goodge, 1983a, b). Axial planes of folds with such a schistosity are parallel to mylonitic foliation within mylonitized paragneiss; similarly, fold axes parallel L_e (Fig. 9-4b). Goodge (1983a, b) interprets the parallelism of these planar and linear structures as evidence of progressive reorientation of preexisting folds during later mylonitization. Axial planes that trend easterly and dip steeply south below the mylonitic front are progressively rotated into parallelism with northwest-trending, southwest-dipping mylonitic foliation. The sense of fold axial plane rotation is consistent with top-to-the-west shear, parallel to L_e, within the mylonite zone (Fig. 9-5a). Fold axes similarly are reoriented from steeply plunging east-west trends below the mylonitic front, to close parallelism with L_e within the shear plane, indicating bulk flattening in the mylonite zone associated with sinistral shear as viewed to the north. Although Goodge (1983a) originally interpreted fold reorientation trends as an indication of top-to-the-east shear in the mylonite zone because of observed initial clockwise fold axis rotation, progressive reorientation of the preexisting fold axes and axial planes is more simply consistent with a top-to-the-west sense of shear (Fig. 9-5b).

Progressive reorientation during mylonitization of previously formed folds, as well as a lack of folds in mylonitized granodiorite, provides strong evidence for a distinct hiatus in the structural evolution of the Okanogan complex between formation of folds and mylonitic deformation. Kinematic study of fold and mylonite structures also indicates a distinct change in the structural framework between episodes of folding and mylonitization. Fold trends indicate dominantly north-south vergence and shortening, whereas mylonitic fabrics record top-to-the-west displacement.

The breccia zone which defines the western and southern limits of the complex unquestionably disrupts the mylonitic fabric. This zone is well preserved in the Omak Lake valley as well as in several locations along the Okanogan River. Within the southwest-dipping 200 to 1000 m-thick zone of brecciation in the Omak Lake valley, the degree of cataclastic reduction varies; blocks of massive, undeformed rock as large as 1 km long and 0.25 km wide crop out locally within a thoroughly brecciated matrix of indistinguishable parent lithology. Slickensides preserved on randomly oriented planar dislocation surfaces within the northern Omak Lake valley are similarly random in their orientation and indicate no preferred movement direction within the breccia zone.

The zone of brecciation and low-grade metamorphism is poorly preserved in the axis of the broad synform which forms the drainage of Omak Creek (Snook, 1965). Very little of the breccia zone is preserved along the western boundary of the complex in this

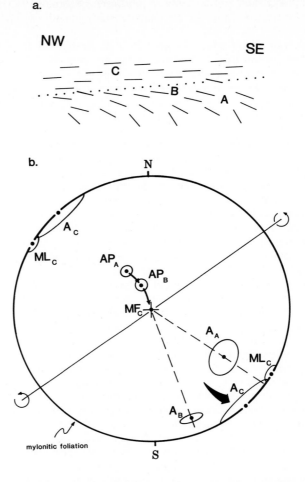

a.

NW

SE

C

B

A

b.

N

A_C

ML_C

AP_A

AP_B

MF_C

A_A

ML_C

A_C

A_B

mylonitic foliation

S

FIG. 9-5. Top-to-the-west reorientation of axial planes within mylonite zone. (a) Schematic cross section is drawn parallel to ~N60°W trend of mylonitic lineation. Short lines represent axial plane traces; dotted line is lowest structural limit of observed mylonitic fabrics. (b) Stereonet construction of fold reorientation in the mylonite zone (lower-hemisphere projection). MF, poles to mylonitic foliation; ML, mylonitic lineation; AP, poles to fold axial planes; A, fold axes. Subscripts refer to structural domains in the folded Tonasket gneiss; domain C is located within the mylonite zone, and domains B and A are located structurally deeper below the mylonite zone, as illustrated in (a). Ellipses show projected shape of circles of 95% confidence about calculated mean orientations for each structural element. NE-SW line shows general orientation of axis of rotation associated with reorientation of preexisting structures in the shear zone. Data are rotated such that mylonitic foliation lies in a horizontal reference plane.

area; however, small remaining exposures suggest that the breccia zone dips at a moderate to gentle angle westward below the Okanogan Valley. Within the synform, rocks grade structurally upward from coherent mylonitic orthogneiss to microfractured, brecciated, and chloritized rock of identical parent lithology. The zone of brecciation parallels mylonitic foliation and dips gently westward. Slickenside lineations preserved on extremely well polished surfaces of microbreccia within the axis of the synform trend N60°W, parallel to L_e, indicating latest movement in the microbrecca in a general northwest direction.

The Okanogan complex looses structural definition eastward along the southern boundary; the intensity of mylonitic fabrics and the occurrence of brecciated fault zones gradually decrease to the east, and lithologies which to the west are truncated by fault zones merge as coherent units across a projection of the Okanogan complex structural boundary. The mass of Tertiary granitic plutons which comprise the southern core of the complex intrude mylonitized granodiorite units. At the extreme southwestern margin of the complex, these felsic granitic rocks are penetratively deformed; however, within the complex they typically display thin nonpenetrative shear zones with a lineation parallel to L_e (Atwater and Rinehart, 1984). Parallelism of the lineation within Tertiary plutons to L_e indicates that the plutons may have been emplaced under the same stress regime which led to the formation of the mylonitic lineation. The nonpenetrative nature of the lineation within the Tertiary plutons may be either a result of crystallization below the zone of penetrative ductile shear, or emplacement during the wane of top-to-the-west tectonic denudation. It is possible that this entire suite of granitic rocks was emplaced during brittle deformation in the complex, and responded in a ductile manner due to their higher temperature and increased plasticity relative to neighboring rocks.

Brecciation clearly postdates mylonitization throughout the complex (Atwater, 1985; Bardoux, 1985a, b; Goodge and Hansen, 1983; Orr, 1985; Snook, 1965); however, the time interval separating ductile and brittle deformation may not be similar everywhere. For example, in the northeast portion of the complex, Orr (1985) describes high-angle normal faults which cut gently dipping mylonitic foliation and juxtapose mylonitic rocks against Eocene volcanic rocks. In contrast, Bardoux (1985a, b) describes an apparent gradation in structural features, from dominantly ductile mylonites, to extensional shear bands, to brittle normal faults, in which each records similar kinematic displacement during deformation. This gradation from ductile to brittle structures provides evidence of an evolving ductile to brittle shear zone along portions of the Okanogan complex. Atwater (1985) proposes that intrusion of a 48–50 m.y.-old pluton occurred before final cessation of ductile deformation and brecciation. Goodge and Hansen (1983) describe texturally isotropic Tertiary microdiorite dikes which postdate penetrative mylonitization of granodiorite units they intrude in the south-central portion of the complex, yet Parkinson (1985) documents a U/Pb zircon age of 52 m.y. from a ductilely deformed rhyolite porphyry dike.

Unraveling details of the timing of ductile versus brittle deformation promises to be a difficult task. One must consider not only the spatial relation of mylonite to breccia, but also the rheology of each unit at the assumed or determined time of deformation. Temperature, strain rate, composition, and volatile content each play a role as to whether a rock responds brittlely or plastically to tectonic stresses. As an example of the uncertainties, it is not clear whether mylonitic shear zones within granitic plutons formed at the same time as the penetrative ductile fabrics in the granodiorite units, or whether the nonpenetrative shear zones formed, perhaps, in a warmer, more volatile-rich pluton during concurrent brecciation of colder and drier adjacent rocks. It may be difficult to sort out this problem of timing as existing data[1] (Bardoux, 1985a, b) indicate that ductile mylonitic and brittle fault deformation occurred, at least locally, with the same kinematic sense. Although workers agree that brecciation postdates mylonitization (in the strict sense), it is a more difficult task to place precise ages on the termination of mylonitization and the onset of brecciation.

In addition to possibly large displacements within Okanogan mylonites and breccia zones, other features indicate perhaps modest extension. Within the complex, steeply dipping joints oriented perpendicular to L_e in the penetrative mylonitic fabrics increase in abundance and decrease in spacing toward the margin of the complex; these joints probably represent the latest stages of deformation associated with extension parallel to L_e, and did not form significantly during the peak of mylonitization because the nearly vertical joints, where present in the mylonite zone, are bounded by unsheared walls. Many north- to northwest-trending normal faults and zones of brecciation, commonly lacking clear large-scale displacements, dissect the upper plate west of the complex (Atwater and Rinehart, 1984; Menzer, 1983; Rinehart and Fox, 1976).

In summary, structural evolution of the Okanogan complex as constrained by geologic relations along the western and southwestern margins of the complex appears to consist of a series of distinct structural events. Early north-vergent folds formed during high-grade regional metamorphism predate the intrusion of granodiorite plutons. Mylonitic fabrics formed during top-to-the-west movement within a probably subhorizontal zone of ductile shear postdate the intrusion of granodiorite, to form orthogneiss, and appear to be gradationally overprinted by similarly top-to-the-west breccia zones formed along low-angle normal faults active parallel to the zone of mylonitization. Earlier folding probably reflects crustal shortening, whereas later mylonitization and brecciation record a period of crustal extension.

Although this sequence of structural events is documented along the western margin of the Okanogan complex from Omak Lake north to Vernon, British Columbia, Orr (1985) interprets a quite different structural history for the northeastern margin of the Okanogan complex in the Tenas Mary Creek area. According to Orr (1985), the northeastern margin of the complex contains a 1 km-wide zone of scaly serpentinite which marks the boundary of the complex between high-grade mylonitic paragneiss of the lower-plate and upper-plate Anarchist rocks. Zones of mylonitic foliation in the complex which are parallel to metamorphic foliation are also parallel to the boundary of the complex and increase in frequency of occurrence and intensity structurally upward toward the serpentinite. Parker and Caulkins (1964) identified rare intrafolial folds which they interpreted to have formed during Jurassic amphibolite facies metamorphism. In addition, Orr (1985) suggests that this mid-Mesozoic metamorphism was also accompanied by mylonitization of rocks structurally beneath the serpentinite.

No metamorphic discontinuity is present at the Anarchist-Okanogan complex contact; rather, there is a gradual decrease in metamorphic grade upward from lower-plate amphibolites to upper-plate Anarchist Group. High-angle faults, marked by breccia and low-grade alteration, locally disrupt the margin of the complex and justapose lower-plate Okanogan complex rocks with Eocene volcanic rocks. These faults truncate mylonitic foliation at a high angle, showing the faults and mylonitic fabrics to be structurally unrelated features (Orr, 1985). Based on these structural and metamorphic features, Orr interpreted that the internal fabric of the Okanogan complex formed as a result of Mesozoic top-to-the-east accretion of the allochthonous Anarchist Group, part of the Quesnel terrane. Kinematic determinations in the Tenas Mary Creek area have not been undertaken, but Ross (1981)

---

[1] Also V. L. Hansen and J. W. Goodge, unpublished data.

also interpreted top-to-the-east displacement within synmetamorphic mylonitic paragneiss, interpreted to have formed during Early Jurassic time. Although the apparent tectonic history of the northeastern margin of the Okanogan complex differs markedly from the tectonic history interpreted from study of the western margin, inferred transitions in general strain patterns in the crust are strikingly similar; along both margins, a period of regional shortening (Jura-Cretaceous age?) preceded by an undetermined time interval a period of deformation in an extensional regime (Eocene age). A tectonic model of evolution encompassing general features of each margin of the complex is discussed in the synthesis and conclusion section below.

METAMORPHISM

Study of metamorphism in the Okanogan complex is necessary in order to evaluate the evolution of the complex for the variety of reasons outlined at the beginning of this paper. Not only are the events which formed the high-grade crystalline rocks of the interior associated with metamorphic processes, but so too are those which brought about rise of this crystalline mass to shallow crustal levels.

Gneissic rocks of the Okanogan complex are varied in their whole-rock composition and include mafic to intermediate gneiss, pelitic schist, and calc-silicate gneiss. Not included in this suite are granodioritic orthogneisses which were emplaced as plutons after the main phase of metamorphism and folding (Goodge and Hansen, 1983). These metamorphic rocks are the principal constituents of both the Tonasket Gneiss, which occurs as a large mass mostly within the western portions of the Okanogan complex, and the Salmon Creek Schist and Gneiss, which lies within the upper plate in the vicinity of the Okanogan River (Fig. 9-3).

Characteristic mineral parageneses, which reflect conditions during specific periods in the evolution of the complex, are defined by their association with structural fabrics formed during different deformation episodes on the following basis: (1) high-grade regional metamorphism resulted in the formation of medium- to coarse-grained foliated, compositionally layered gneisses and schists, showing well-recrystallized hypidiomorphic textures and a mineral lineation; (2) mylonitization is recorded by equally well-foliated and layered rocks, commonly of fine- to medium-grain size and showing a strong unidirectional lineation, which display effects of elevated strain rate in the form of strained or broken porphyroclasts, and recrystallized subgrains of various minerals; and (3) brecciation and alteration, which occurred by cataclastic brittle failure, is marked by the formation of mineral assemblages in fractures and rock fragment interstices. Only on such textural grounds may mineral parageneses be distinguished.

All rocks in the Tonasket and Salmon Creek assemblages are affected by high-grade metamorphism, and the compositional diversity of these rocks provides an excellent means of gauging the peak conditions of metamorphism. Mylonitization is not expressed in Salmon Creek rocks, an important point in itself, but is evident in Tonasket intermediate to mafic gneisses and calc-silicates. Tonasket pelites, where mapped, occur structurally below the Okanogan mylonite zone, and have not been identified within the structurally higher zone of mylonite. For the same reason, only Tonasket mafic to intermediate gneisses contain evidence of brecciation; however, Salmon Creek rocks locally contain brecciation textures and alteration mineral assemblages similar to the Tonasket. In the remainder of the discussion on metamorphism, we describe mineral assemblages where the data permit, discuss probable metamorphic conditions, and infer time-dependent trends in the conditions of metamorphism.

Assemblages

Mineral assemblage and composition diagrams are shown for Tonasket and Salmon Creek metamorphic rocks in Figs. 9-6 to 9-9. Data for these diagrams are compiled from Snook (1965), Goodge and Hansen (1983, and unpublished data), and Menzer (1983).

Regional metamorphism of Tonasket and Salmon Creek rocks is regarded as the oldest tectonic event in the Okanogan area and was accompanied by strong folding (Goodge, 1983a; Goodge and Hansen, 1983; Menzer, 1983; Snook, 1965). Mineral assemblages formed during this event are shown in Figs. 9-6 and 9-7. Mafic metamorphic rocks, most commonly amphibolites, contain a characteristic mineral assemblage of plagioclase, hornblende, and diopside, although locally in Tonasket rocks they contain, in addition, garnet and biotite. Plagioclase is typically of intermediate andesine to labradorite composition, and calcic amphibole is pargasitic (Snook, 1965), both indicating relatively high temperature metamorphism. Quartz is commonly absent and sphene is the most common accessory mineral. Snook (1965) observed olivine, spinel, and corundum in some of the mafic gneisses, and we infer these to be possible relict igneous minerals of refractory composition which resisted complete recrystallization. Snook's interpretation, to the contrary, is that these phases are relict from an early episode of granulite facies metamorphism. It seems unlikely to us that granulite assemblages were later thoroughly recrystallized as lower-grade hydrous amphibolite assemblages.

An intermediate composition gneiss containing quartz, plagioclase, hornblende, and biotite, with minor sphene, is also present as a major lithology in the Tonasket; it is interpreted to be metaigneous in origin largely because of a silica-saturated and widely homogeneous composition. Minor amounts of pelitic schist and impure quartzite occur in both Tonasket and Salmon Creek assemblages, and these rocks characteristically contain quartz, plagioclase, orthoclase, muscovite, biotite, garnet, and sillimanite. Micas and sillimanite show strong mineral-preferred orientations; garnet contains minor inclusions but shows little evidence of porphyroblast rotation. Muscovite mostly is present as a relict phase with irregular outlines and is commonly overgrown by well-formed sillimanite; sillimanite also coexists with orthoclase. These petrographic relations are contrary to those described earlier by Goodge and Hansen (1983), in which sillimanite and orthoclase were believed not to coexist because of compositional layering. Fox *et al.* (1977, p. 15) comment that aluminosilicate phases are rare and that toward the interior of the complex, presumably pelitic rocks contain cordierite in metamorphic assemblages. They do not state, however, whether cordierite is present only in the vicinity of numerous plutons or is regionally extensive, and we do not include this phase in the regional metamorphic assemblages.

As with the pelitic rocks, calc-silicate gneiss occurs in thin bodies interlayered with mafic gneiss. The calc-silicates contain quite varied mineral assemblages, reflecting compositional inhomogeneity, leading to subdivision by Snook (1965) on this basis. In general, calc-silicate gneiss contains a metamorphic assemblage of calcite, quartz, plagioclase, and diopside; Tonasket calc-silicates also contain scapolite and idocrase, whereas Salmon Creek rocks may contain wollastonite, grossular-rich garnet, and/or idocrase (Menzer, 1983).

Finally, Salmon Creek rocks include small lenticular bodies of dunite containing olivine and minor orthopyroxene which is partly replaced by thin veinlets of serpentine (antigorite?) and clinochlore. For reasons outlined by Goodge and Hansen (1983), these dunite bodies probably are slivers of serpentinite emplaced tectonically prior to high-grade metamorphism; thus, olivine and orthopyroxene are metamorphic and not igenous phases.

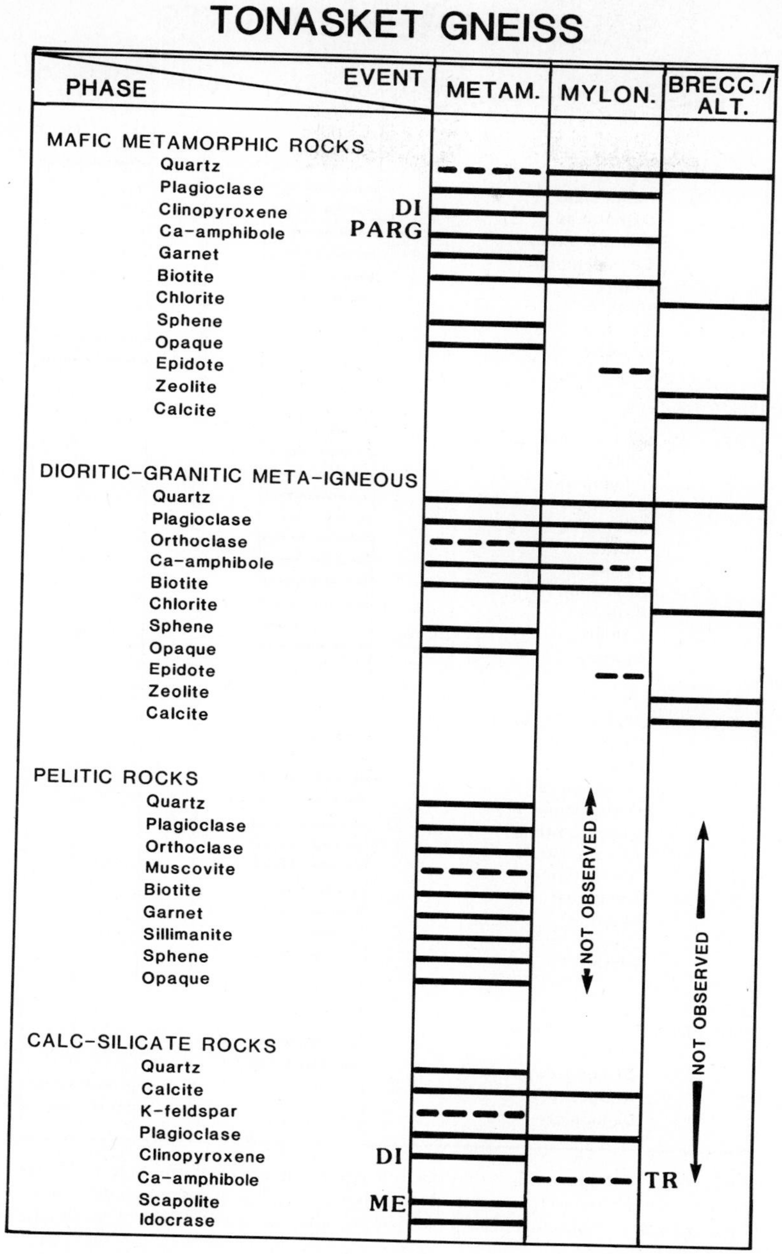

TONASKET GNEISS

FIG. 9-6. Metamorphic assemblage diagram for rocks of the Tonasket Gneiss (data in part from Snook, 1965). Parageneses shown are inferred equilibrium assemblages for recognized tectonic events as recorded by diagnostic structures; volumetrically minor pelitic and calc-silicate rocks do not occur in mylonite and breccia zones. Solid lines for phases always present; dashed lines for phases present locally due to compositional inhomogeneity.

SALMON CREEK SCHISTS

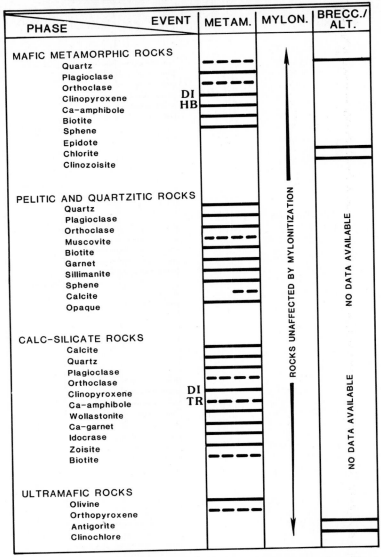

FIG. 9-7. Metamorphic assemblage diagram for rocks of the Salmon Creek Schist and Gneiss (data in part from Menzer, 1983). Parageneses shown are for high-grade metamorphism in all rocks, and for brecciation in mafic and ultramafic gneisses only. Salmon Creek rocks do not contain mylonitic fabrics. Solid lines for phases always present; dashed lines for phases present locally due to compositional inhomogeneity.

The following criteria indicate Tonasket and Salmon Creek rocks were metamorphosed in the sillimanite-orthoclase zone of the amphibolite facies: a plagioclase-hornblende-dominated assemblage in amphibolite, intermediate plagioclase compositions in all rocks, pargasitic hornblende compositions, the breakdown of muscovite to form

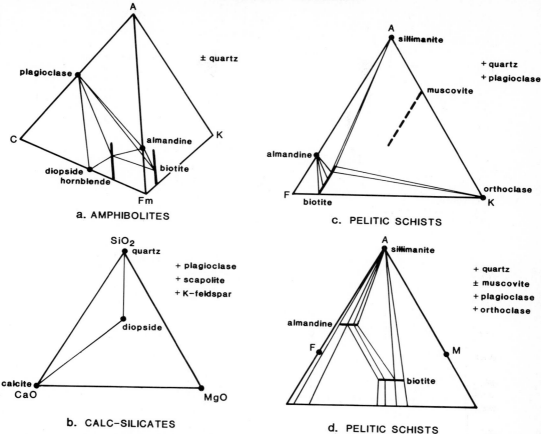

FIG. 9-8. Composition diagrams for amphibolite facies rocks of the Tonasket Gneiss. Compositions of phases showing solid solution are not determined. AFK diagram for pelitic rocks in (c) shows muscovite as a relict phase locally present following near-complete breakdown to form sillimanite + orthoclase.

sillimanite and orthoclase in pelites, and the formation of diopside, scapolite, and wollastonite in calc-silicate rocks.

Mylonitization is reflected by diagnostic fabrics only in lower-plate Tonasket rocks of the Okanogan complex, and mineral assemblages formed during this episode are shown in Fig. 9-6. Mafic to intermediate composition gneisses and some calc-silicates are the only paragneisses affected by mylonitization. The stable mineral assemblage formed in mafic to intermediate rocks at this time was quartz, plagioclase, biotite, and hornblende, with or without epidote, all of which show evidence of dynamic recrystallization during mylonitization. In rocks of compositions trending toward granodiorite, K-feldspar is also part of the mylonite assemblage. Calc-silicate rocks show evident syntectonic recrystallization of calcite and plagioclase, as well as the formation of tremolite after diopside. Scapolite formed during earlier metamorphism is described by Snook (1965, p. 764) as "rounded clear porphyroclasts," indicating that it was not a stable phase during mylonitization and was undergoing grain-size reduction. Snook (1965) suggests that Okanogan mylonites were formed under lower-grade conditions than the same rocks during earlier

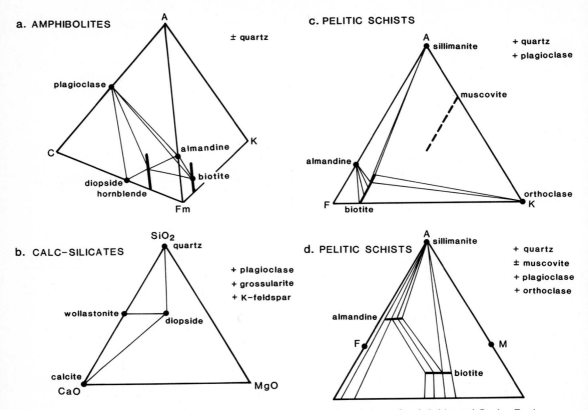

FIG. 9-9.   Composition diagrams for amphibolite facies rocks of the Salmon Creek Schist and Gneiss. Explanation same as for Fig. 9-8.

metamorphism. Based on the presence of biotite as the principal ferromagnesian phase and epidote in mafic rocks, and the formation of tremolite in calc-silicates, we interpret conditions of mylonitization to be in the general range of greenschist facies metamorphism. The presence of quartz in mylonitic rocks and the probably low X_{CO_2} in the fluid phase during mylonitization would not permit extension of epidote and tremolite stability to high temperatures; thus, the mylonites probably formed under conditions below that of the uppermost greenschist facies.

Brecciation textures and associated alteration mineral assemblages occur in the vicinity of fault zones which juxtapose upper- and lower-plate rocks of the Okanogan complex. In the case of Tonasket and Salmon Creek rocks, only mafic gneiss is in close enough proximity to these zones to have formed mineral assemblages associated with conditions of brecciation. The characteristic assemblage found within brecciated mafic gneiss is quartz, chlorite, calcite, and a zeolite mineral of unknown composition (Figs. 9-6 and 9-7). These minerals occur together and as individual phases in breccia fractures and interstices of rock fragments altered during this cataclastic episode. It is possible that zeolite precipitation also occurred during hydrothermal alteration following the cataclastic deformation. As noted above, dunite in the Salmon Creek assemblage contains antigorite(?) and clinochlore replacing metamorphic olivine and orthopyroxene; it is possible that these alteration phases formed during brecciation with accompanying hydration, or during emplacement of nearby plutons or hypabyssal dikes. The deposition of phases including quartz, chlorite, calcite, and zeolite indicates brecciation took place under conditions in the zeolite to possibly lowest greenschist facies of metamorphism.

Metamorphic mineral assemblages are plotted on a set of composition diagrams for common Tonasket and Salmon Creek rocks involved in high-grade metamorphism (Figs. 9-8 and 9-9). In amphibolites, the typical stable assemblage is plagioclase + hornblende + diopside (Figs. 9-8a and 9-9a), although bulk compositional variation provides for stabilization of plagioclase + hornblende + garnet + biotite. Calc-silicate assemblages (Figs. 9-8b and 9-9b) vary considerably from nearly pure marbles to silicate-rich rocks, and include diopside + scapolite + plagioclase + K-feldspar + quartz + calcite, and diopside + wollastonite + plagioclase + K-feldspar + quartz + calcite ± grossular. The appearance of wollastonite in Salmon Creek rocks may reflect higher temperatures of metamorphism than Tonasket rocks, as is discussed below. Metamorphic assemblages are essentially identical in Tonasket and Salmon Creek pelitic schists (Figs. 9-8c and d, and 9-9c and d). The common pelitic assemblage is quartz + plagioclase + orthoclase + biotite + garnet + sillimanite, where muscovite remains as a relict phase from incomplete breakdown in a reaction to form sillimanite and orthoclase. Menzer (1983) notes the presence in quartzitic rocks of muscovite, mostly as sericite, which has formed at the expense of biotite, sillimanite, and alkali feldspar, and attributes this second generation of white mica to retrograde metamorphism resulting from nearby granitic intrusion.

Although high-grade metamorphic assemblages may be observed in all rock compositions in the Okanogan gneisses, only mafic gneisses and calc-silicates show progressive phase changes during succeeding mylonitization, as well as brecciation in the case of mafic gneiss. Figures 9-10 and 9-11 illustrate these phase transitions on appropriate composition diagrams. Mylonitized amphibolites contain a stable assemblage of plagioclase + hornblende + biotite + quartz ± epidote (Fig. 9-10b), an assemblage we interpret to have involved no metasomatic or other changes of components following the previous metamorphic episode. Alteration assemblages in these mafic gneisses related to brecciation contain quartz + chlorite + calcite + zeolite (shown as laumontite in Fig. 9-10c). No potassic phases (such as white mica) are observed as part of this assemblage, indicating either a possible loss of K_2O during movement of fluids related to alteration, or simply the introduction of components from a fluid-rich source outside that of the host rock. Thus, metamorphism and mylonitization may have been relatively closed-system processes, whereas brecciation may have involved addition of material as part of open system influx. Snook (1965) postulated that sodium and potassium metasomatism of "migmatitic feldspathic gneisses" accompanied regional metamorphism; however, it is likely that granodiorite mylonitic gneisses rich in these components are metaplutonic rocks, thus discounting the necessity of alkali influx to the Tonasket rocks during metamorphism or later events.

Calc-silicate rocks show only minor phase changes between metamorphism and mylonitization, in which diopside-bearing metamorphic assemblages react to form a mylonitic assemblage containing tremolite + plagioclase + calcite (Fig. 9-11b and c). Snook (1965, p. 764) states that tremolite is "retrogressive after diopside," where deformation is the strongest, and generally absent in only weakly deformed rocks. Thus, calc-silicate rocks locally retrogress from a diopside-bearing metamorphic assemblage, to a similar mylonitic assemblage where the principal ferromagnesian mineral is tremolite.

TONASKET MAFIC GNEISS

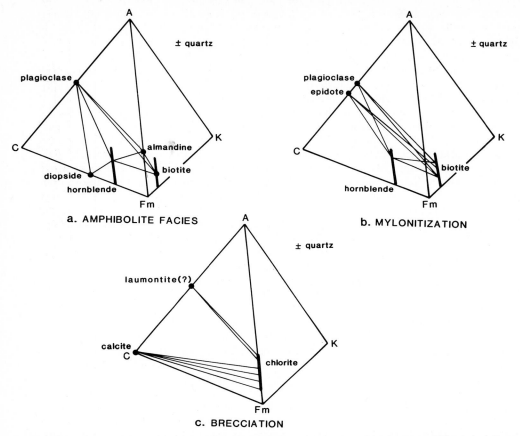

FIG. 9-10. ACFmK diagrams for Tonasket Gneiss mafic gneisses, showing phase assemblages in (a) amphibolite facies metamorphic rocks, (b) mylonite zone, and (c) breccia zone. Compositions of phases showing solid solution are not determined.

Conditions

Metamorphic assemblages such as those outlined above are useful in estimating *P-T* conditions prevailing during distinguishable tectonic events. In Fig. 9-12, relevant reaction boundaries are plotted and regions in *P-T* space of likely assemblage stability are shown for Okanogan amphibolites, calc-silicates, and pelites. (The figure captions provide complete reaction equations.)

P-T fields for amphibolite facies metamorphism are shown in Fig. 9-12a. A metamorphic assemblage of plagioclase + hornblende + diopside ± biotite ± garnet in mafic gneisses indicates temperatures surpassed those governing epidote stability, but were insufficiently high to form orthopyroxene in addition to diopside. Metamorphic pressures are unconstrained by the reaction boundaries shown for amphibolites, and the assemblages in these rocks could have formed at pressures greater than about 5 kb.

TONASKET CALC-SILICATE GNEISS

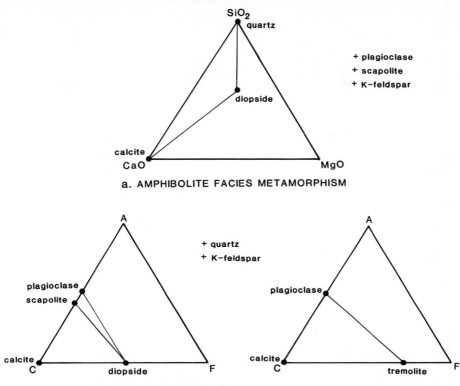

a. AMPHIBOLITE FACIES METAMORPHISM

b. AMPHIBOLITE FACIES METAMORPHISM   C. MYLONITIZATION

FIG. 9-11. SiO_2-CaO-MgO composition and ACF diagrams for Tonasket gneiss calc-silicates, showing phase assemblages for (a) and (b) high-grade metamorphism, and (c) mylonitization.

Metamorphic conditions for the formation of calc-silicate assemblages are limited by the diopside and wollastonite "in" boundaries, although as mentioned earlier, wollastonite is present only in Salmon Creek rocks, whereas diopside is stable in both. An upper temperature limit for wollastonite-bearing calc-silicates is unconstrained. Tonasket calc-silicates do contain scapolite, however, which indicates quite high metamorphic temperatures. Pure meionite (Ca end-member scapolite) is stable only above 875°C (Goldsmith and Newton, 1977), and coexisting calcic scapolite and plagioclase are also stable at high temperatures. Increased Na content, however, stabilizes scapolite at lower temperatures depending on X_{CO_2}, X_{H_2O}, and X_{NaCl}; Orville (1975) showed that Na-rich scapolite formed at temperatures as low as 750°C at 4 kb pressure. Scapolite in Tonasket rocks contains a high meionite component (Snook, 1965), suggesting that metamorphic temperatures on the order of 800°C are possible.

The clearest indicator of metamorphic conditions in Okanogan pelites is shown by the breakdown reaction of muscovite to form orthoclase and sillimanite; minor, possibly remnant, amounts of muscovite indicate incomplete reaction and metamorphic conditions near the muscovite "out" boundary. The absence of staurolite in the garnet-bearing

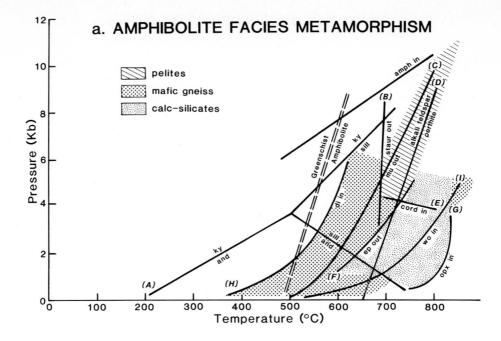

a. AMPHIBOLITE FACIES METAMORPHISM

FIG. 9-12. *P-T* diagrams showing regions of inferred mineral assemblage stability for rocks of the Tonasket Gneiss; lithologies include pelitic schist, mafic gneiss, and calc-silicate gneiss. (a) Amphibolite facies metamorphism. Reaction boundaries include: (*A*) alumino-silicate stability (Holdaway, 1971); (*B*) staur + qtz → alm + sill + H_2O (Ganguly, 1972; Richardson, 1968); (*C*) musc + qtz → sill + or + H_2O (Chatterjee and Johannes, 1974; Kerrick, 1972); (*D*) alkali feldspar → perthite/antiperthite (Bowen and Tuttle, 1950; Luth, 1976); (*E*) Fe-bi + sill + qtz → Fe-cord + or + H_2O (Holdaway and Lee, 1977); (*F*) ep + qtz → gross + mt + H_2O (Liou, 1973); (*G*) amph + pl + cpx + ilm → amph + pl + cpx + opx + ilm (Spear, 1981); (*H*) dol + qtz → di + CO_2 (Turner, 1981); (*I*) cc + qtz → wo + CO_2 (Greenwood, 1967; Harker and Tuttle, 1955); (*J*) ab + di → omph + qtz (Boettcher and Wyllie, 1968; Liou, 1971). Approximate location of greenschist-amphibolite facies boundary from Hyndman (1985). (b) Mylonitization. Reaction boundaries include: (*K*) Fe-chlor + qtz + mt → alm + H_2O (Hsu, 1968); (*L*) stilp + pheng → bi + chlor + qtz + H_2O (probably does not strictly apply because Tonasket amphibolites are not in every case silica saturated) (Haas and Holdaway, 1973); (*M*) trem + cc + qtz → di + CO_2 + H_2O (Turner, 1981). Box outlines temperature conditions estimated by Hansen (1983b) for Okanogan granodiorite mylonites based on feldspar geothermometer of Stormer (1975), calibrated at $P = 5$ kb. (c) Brecciation associated with brittle faulting. Reaction boundaries include: (*N*) heul → laum + qtz + H_2O (Coombs *et al.*, 1959); (*O*) laum → an + qtz + H_2O (Bird and Helgeson, 1981); (*P*) laum → lw + qtz + H_2O (Liou, 1971). Approximate location of zeolite-greenschist-prehnite-pumpellyite facies boundaries from Hyndman (1985).

assemblages indicates temperatures above the staurolite "out" boundary, and absence of cordierite indicates metamorphic pressures higher than that of the cordierite "in" reaction. Snook (1965) reports antiperthite exsolution textures in K-felspar-rich Tonasket "migmatitic" rocks, indicating temperatures above the feldspar solvus, although these textures may be related to crystallographic transitions which occurred during cooling of

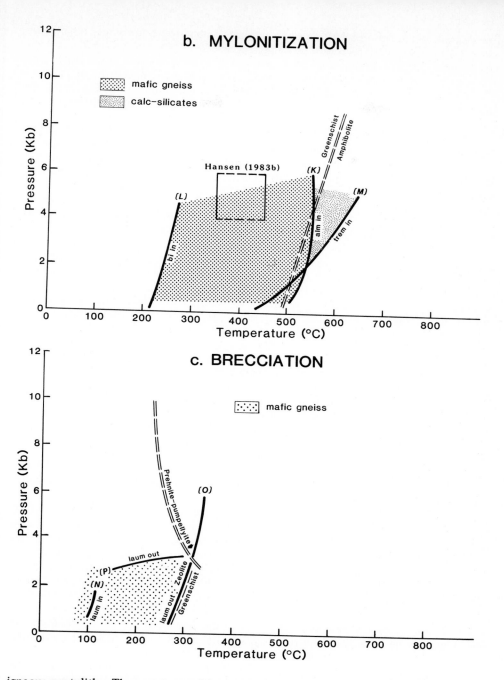

b. MYLONITIZATION

mafic gneiss

calc-silicates

Hansen (1983b)

(L) (K) (M)

Greenschist Amphibolite

bi in alm in trem in

Pressure (Kb)

Temperature (°C)

c. BRECCIATION

mafic gneiss

(O)

Prehnite-pumpellyite

laum out

(P)

(N) Zeolite

laum in laum out Greenschist

Pressure (Kb)

Temperature (°C)

igneous protoliths. Thus, metamorphic temperatures in Okanogan gneisses appear to have reached a minimum of ~750°C in order to satisfy phase assemblage constraints outlined above. Temperatures may not have exceeded ~825°C, based on an absence of ortho-pyroxene in amphibolites. Further, a lack of unambiguous evidence for regional anatectic phenomena suggests that conditions were overall rather dry, perhaps resulting from earlier abundant devolatilization reactions. Metamorphic pressures, nonetheless, are poorly constrained; pressures probably exceeded ~4 kb, but very little evidence exists which indicates an upper limit. An absence of kyanite in pelitic schist and jadeitic pyroxene in mafic rocks suggests pressures less than 10 kb at temperatures in the range cited above. Thus, we suggest peak metamorphic conditions for rocks within, and in the upper plate overlying, the Okanogan complex probably in the range 700-800°C and 4-8 kb.

P-T fields for mylonitization of Tonasket Gneiss mafic gneiss and calc-silicate are shown in Fig. 9-12b. Mylonitic amphibolites typically contain the assemblage plagioclase + biotite + hornblende + epidote, indicating temperatures greater than that required to form biotite, and below the upper stability limit of epidote as given by a reaction to form garnet. It should be noted here that the biotite "in" reaction, labeled "L" in Fig. 9-12b, assumes the assemblage contains quartz, which is not always the case for Tonasket mylonitic mafic gneisses. Thus, this lower temperature limit is not strictly applicable. Tremolite forming from diopside in calc-silicates indicates metamorphic temperatures below that of the tremolite "in" boundary. The breakdown of tremolite to form dolomite and quartz at lower temperatures requires a CO_2-rich volatile phase, which is not likely to be present after devolatilization during pro-grade metamorphism; thus, this reaction boundary is not shown. Hansen (1983b) analyzed coexisting plagioclase and orthoclase subgrains in well-recrystallized mylonitic orthogneisses of the Okanogan complex and applied their compositions to the two-feldspar geothermometer of Stormer (1975). Composition data were plotted on Stormer's temperature curves calibrated at 5 kb; this pressure was chosen based on Sibson's (1977) classification of fault rocks and mechanisms. Coexisting feldspar pairs plot near Stormer's 400°C temperature curve at 5 kb. Although feldspar geothermometers commonly give low temperatures compared to other methods, a temperature on this order is consistent with data from Kerrich *et al.* (1980) on mylonitized feldspars. In addition, temperatures obtained by this method reflect conditions during the latest stages of mylonitic recrystallization, not peak temperatures. As in the case of metamorphism, pressure estimates taken directly from Okanogan rocks are unconstrained, but we believe pressures in the range 4-6 kb to be not unreasonable. Thus, we estimate *P-T* conditions during Okanogan mylonitization to lie approximately in a range 350-525°C and 4-6 kb. Hansen's (1983b) data for Okanogan mylonites indicate that temperatures may actually lie in the more restricted range of 350-450°C during the latest stages of mylonitization.

A *P-T* field estimated for alteration assemblages in brecciated rocks is shown in Fig. 9-12c. Many of the phases present in the brecciated rocks have quite wide stability ranges (i.e., quartz, chlorite, calcite), and as such provide little information on metamorphic conditions during brecciation. Although zeolite compositions are not known, their presence places constraints on metamorphic temperatures and pressures. Laumontite is stable above a minimum temperature of approximately 100°C and may persist in an assemblage to temperatures as high as about 300°C (Fig. 9-12c). Stability below its dehydration reaction to form lawsonite and quartz indicates pressures less than about 3 kb. Therefore, we estimate conditions during the formation of brecciation/alteration assemblages to lie broadly in the range 100-300°C and 1-3 kb, conditions quite compatible with brittle deformation in the crust.

In summary, analysis of metamorphism in the Okanogan complex points to relationships which to this time have not been considered in models of evolution of the complex. First, not only are lithologic associations in Tonasket and Salmon Creek rocks strikingly similar, but so are the metamorphic assemblages which formed in these rocks during amphibolite facies metamorphism. High-grade metamorphic rocks occur in both the upper and lower plates, separated from one another by the Okanogan mylonite and brittle shear zones. Without much question, these units were once part of the same metamorphic sequence which has been dismembered by later deformation. Such relationships indicate that high-grade metamorphism of this type is not restricted to the lower plate of the Okanogan complex, and that metamorphism is therefore not directly related to origin of the complex itself through a process of crustal mobilization and upwelling. Similar conclusions were voiced by workers in the Kettle complex to the east (Rhodes and Cheney, 1981).

Second, the conditions which prevailed during periods of high-grade metamorphism and mylonitization are distinctly different. Although it could be argued that conditions of metamorphism and mylonitization lie within the same *P-T* loop, on structural grounds these events are quite dissimilar; vergence of folds associated with regional metamorphism is virtually opposed to the tectonic transport direction in the mylonitic gneisses, and mylonitic fabrics overprint those formed during metamorphism. Metamorphic conditions of mylonitization and brecciation are gradational with one another, as are the structures and kinematics interpreted for these two events. The gradation in metamorphic conditions, as well as the similar kinematic directions, argues in favor of truly gradational structural evolution from mylonitization to brecciation.

RADIOISOTOPIC AGE DATA

P-T Evolution and Deformation through Time

Combining the constraints on metamorphic conditions as outlined above with available geochronologic data, we may describe quite generally the evolution of tectonic processes through time. Figure 9-13 is a simplified *P-T-t* diagram showing what we believe to be the best available data concerning both conditions during which different processes operated as well as the age of the events which produced them. *P-T* conditions are taken from the previous section, and age data have been extracted from a variety of sources cited in the figure.

Most reliable constraints on the age of regional metamorphism in the region, including the amphibolite facies episode in the Okanogan complex, come from analysis of rocks in the British Columbia portion of the Shuswap Metamorphic Complex. Ages of regional metamorphism and deformation are indicated by the emplacement of postmetamorphic Middle Jurassic and older batholiths (Armstrong, 1982; Pigage, 1977), which show Rb-Sr ages as old as 150 m.y. Syn-tectonic plutons, believed to have been emplaced during peak metamorphism and deformation, have ages which fall in the range 170–165 m.y. (Archibald *et al.*, 1983); zircon data from the syn-kinematic Kuskanax batholith indicate a deformation age of 173 m.y. (Parrish and Wheeler, 1983). In contrast to our indirect correlations, Fox *et al.* (1977) rely on a number of K-Ar age determinations of Tonasket rocks in the Okanogan complex to conclude that high-grade metamorphism of these rocks occurred in early Tertiary time, as late as about 50 m.y. ago. However, this interpretation does not appear valid in light of documentation of a Tertiary-age thermal and/or uplift event reflected by widespread resetting of K-Ar ages (Mathews, 1981; F. K. Miller and Engels, 1975). It is indeed likely that this thermal event is coeval with Eocene extension responsible for the final uplift of the Okanogan complex.

As discussed above, reported ages of mylonitization vary widely, although much of this disparity may result from a lack of careful discrimination between mylonite zones of various types. It may be that at least two episodes of mylonitization of different ages, and different kinematics, are preserved in the Okanogan complex, as has been shown to be true in the Priest River complex to the east in Idaho and Washington (Rhodes and

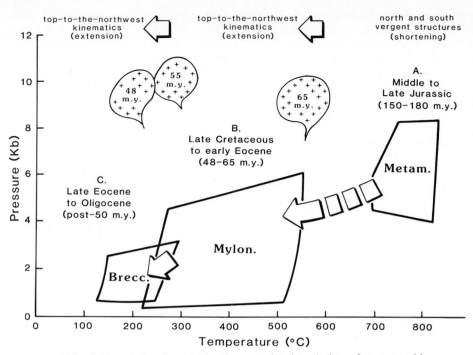

FIG. 9-13. *P-T-t* diagram showing postulated evolution of metamorphic conditions as Okanogan rocks became sequentially involved in high-grade metamorphism, mylonitization, and brecciation and alteration. Conditions taken from Fig. 9-12. Age data from (A) Pigage (1977), Read and Brown (1981), and Armstrong (1982); (B) Fox *et al.* (1977), Bickford *et al.* (1985), Parkinson (1985), and Okulitich (1984); (C) Fox *et al.* (1976, 1977), and Rinehart and Atwater (1984). Box arrows show direction of chronologic younging of Okanogan complex crystalline rocks as constrained by metamorphic conditions and structural relations (see the text for a complete discussion). Structural elements and cross-cutting intrusive relations for each of the three principal tectonic regimes are indicated at the top of the diagram; pluton locations imply nothing regarding *P-T* conditions of emplacement.

Hyndman, 1984). Mylonites related to top-to-the-east crustal shortening are of mid-Jurassic ages, such as the Monashee décollement, which ended movement by 157 m.y. (Brown and Read, 1983), and younger Cretaceous to Paleocene ages (Bickford *et al.*, 1985; Okulitich, 1984; Parrish, 1985); mylonite zones related to crustal extension reveal younger Eocene ages (Parrish, 1985; Tempelman-Kluit and Parkinson, 1986).[2]

Mylonites contemporaneous with high-grade metamorphism described by Orr (1985) along the northeastern portion of the complex are probably of Jura-Cretaceous age and may have formed during, or shortly after, the accretion of the Quesnel terrane. We agree with Orr that the 1 km-thick zone of sheared serpentinite along this portion of the Okanogan complex margin probably represents the Quesnel–North American suture and is of probable Early Jurassic age (Okulitich, 1984). Kinematic studies of the mylonites along the northeastern complex margin are in progress and will help to distinguish timing of mylonitization.

Top-to-the-west mylonites which formed after high-grade metamorphism along the

[2] Also from D. Parkinson, oral description, 1985.

western margin of the Okanogan complex are Eocene in age, as indicated by a 52-m.y. porphyry dike which intrudes mylonitic host rock, yet is itself mylonitized (D. Parkinson, oral presentation, 1985). Similarly, Mission Creek and French Valley granodiorite orthogneisses, which are penetratively mylonitized in the southwest portion of the complex, yield preliminary U/Pb crystallization ages of about 65 m.y. (D. Parkinson, personal communication, 1986). If these dates are correct, mylonitic deformation of the western margin of the complex is post-65 m.y. and may have continued as late as 48 m.y., as shown by incipient mylonitic shear zones within the Coyote Creek pluton (Atwater, 1985). Taken together, the available isotopic data, scanty though they are, indicate a broad time span between high-grade metamorphism and mylonitization (Fig. 9-13).

SYNTHESIS AND CONCLUSIONS

Two dominant models of core complex evolution have emerged in recent years — those describing continuous tectonic evolution versus those favoring a discontinuous origin. Models advocating a continuous evolution involve reconstruction of a protracted tectonic cycle from high-grade regional metamorphism and associated ductile deformation to successively younger mylonitization and brecciation as a largely crystalline crustal mass rises through progressively shallower crustal levels. Such a cycle is believed by most to occur during bulk crustal stretching (see Fox *et al.*, 1977; E. L. Miller *et al.*, 1983; Reynolds and Rehrig, 1980; Wernicke, 1981). The second type of model derives the same structural juxtaposition of contrasting geologic terrains as the first but evokes a polyphase, stepwise model of evolution. In these models, geologic terrains which are deformed and regionally metamorphosed at medium to high grades are later, in an apparently unrelated event, sheared and mylonitized during crustal shortening. During still younger crustal extension, the terrain is locally mylonitized and brecciated, arched and tectonically denuded to juxtapose high-grade core rocks with lower-grade upper-plate rocks (see Brown and Read, 1983; Read and Brown, 1981; Rhodes and Hyndman, 1984).

Rocks of the Okanogan complex record a tectonic history punctuated by regional metamorphism and associated folding, followed by intrusion, mylonitization, normal faulting, brecciation, and doming. Each sector of the margin of the complex displays distinctly different structural, metamorphic, and timing relationships, and therefore each margin outwardly appears to suggest an equally distinct model of Okanogan complex evolution. For example, geologic relations along the western margin of the complex seem to be most easily explained by an early period of crustal compression and resultant synchronous deformation and high-grade metamorphism in the form of east-directed crustal shortening and thickening, followed by much younger (100 m.y. or more) extension, which exposed the high-grade core of the complex. Tectonic denudation during this phase began with the formation of a low-angle, crustal-scale normal shear zone accommodating top-to-the-west movement which was later overprinted by a down-to-the-west complex-bounding breccia zone (see evolving shear zone model of Wernicke, 1985). In this model, extensional tectonics are held responsible for the majority of observed geologic relations. Older deformed and metamorphosed core rocks merely reflect the character of the crust prior to regional-scale extension and do not greatly affect the extensional history. Nonetheless, although this general model of early crustal shortening followed by later crustal extension appears to fit structural relations along the western margin of the complex from Omak Lake in Washington to Vernon, British Columbia, it does not adequately explain geologic relationships documented along the northeastern margin of the complex. Here Orr (1985) calls upon a shortening-dominated evolutionary model to explain the major relationships observed along this margin, with later minor extension merely acting to dissect the previously formed complex.

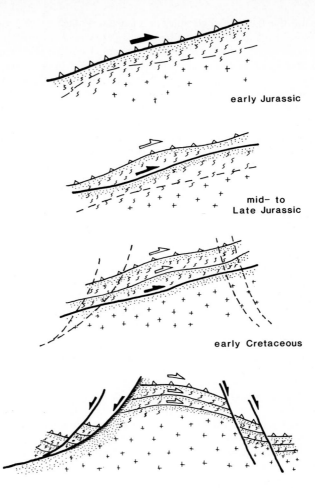

early Jurassic

mid- to
Late Jurassic

early Cretaceous

early Tertiary

FIG. 9-14.   Sequential tectonic evolution of the Okanogan
complex, from Early Jurassic to middle Tertiary time. See the
text for explanation. Light dashed lines indicate faults which
will become active in the following frame; heavy lines indicate
active faults in specific time periods. Wavy lines, high-grade
metamorphic rocks; crosses, plutonic rocks; stippling, mylonite;
hatchures, breccia. Model modified from Okulitch (1984).

    Rather than apply only one of these models to the entire Okanogan complex, we
may view each model as a tenable explanation of the relations observed along a particular
(and only that) portion of the proverbial elephant. With this approach in mind, Okulitch
(1984) outlined a tectonic model of evolution for the entire Shuswap Metamorphic Complex. Although Okulitch's model covers a much larger region, it adheres to the major
constraints required by each segment of the Okanogan complex margin. We have taken
Okulitch's model and modified it slightly to conform more closely to specific geologic
patterns of the Okanogan complex. Our model, illustrated in a sequence of cartoons
(Figs. 9-14 and 9-15), is discussed briefly below.
    The tectonic model begins with Early Jurassic accretion of the Quesnel terrane

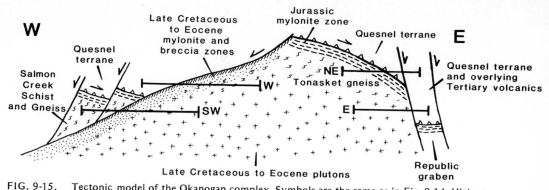

FIG. 9-15. Tectonic model of the Okanogan complex. Symbols are the same as in Fig. 9-14. Highest structural levels are exposed to the northeast, structurally lowest levels to the southwest. Observed geologic relations along each of the complex margins are shown by the horizontal bars; other relations are inferred in the model. Quesnel-North America suture shown by open-barbed line. Jurassic mylonites formed during eastward displacement of Quesnel terrane; Late Cretaceous to Eocene mylonites and breccias formed during westward tectonic denudation related to crustal extension.

against the continental margin and formation of the Quesnel suture during initial phases of the Columbian orogeny. Deformation and high-grade regional metamorphism closely postdate formation of the suture. The northeastern Okanogan complex margin preserves a portion of this early suture and post-suture history (Orr, 1985). The lower plate, consisting of transitional and North American continental crust, responded by synchronous regional deformation and high-grade metamorphism. By mid-Jurassic time, lower-plate rocks began to deform by a mode of ductile shear and became imbricated along the Monashee décollement, presently marked by synmetamorphic mylonites. Locally inverted Jurassic-age metamorphic gradients are preserved along the Monashee décollement (Journeay, 1985), reflecting significant amounts of tectonic throw along these east-directed shear zones. Orr's (1985) high-grade synmetamorphic mylonites along the northeastern margin of the Okanogan complex may be associated with this event. Similarly, mid-Jurassic metamorphism and deformation of the Tonasket Gneiss, prior to later penetrative mylonitization, within the complex and the Salmon Creek Schist and Gneiss southwest of the complex may be related to Quesnel-type tectonism.

With widening of the Atlantic rift, North America continued its westward drift and therefore maintained a convergent mode along its western margin relative to the Pacific plate system through Cretaceous and Paleocene time (Engebretson et al., 1984; Monger, 1977; Monger and Irving, 1979). The Monashee décollement and related crustal shear zones became inactive as easterly directed movement was accommodated along structurally lower Cretaceous- and Paleocene-age shear zones (Brown and Journeay, 1985; see also Bickford et al., 1985; Parrish, 1985; Rhodes and Hyndman, 1984). To date, there are no documented Cretaceous or Paleocene ages of mylonitization or metamorphism within the Okanogan complex. It is possible that events of such age may be identified with future work, but it is also possible that structures and metamorphic assemblages associated with this event are not present in the Okanogan complex because of insufficiently deep exposed structural levels, or because such zones originally formed farther east toward the continental margin.

Finally, in Late Cretaceous or early Tertiary time, the crustal stress regime changed from one of compression to tension. This change in crustal stress regime may have resulted from an overthickened crust, or as a result of slower or more oblique plate convergence, or both. In either case, the crust responded to the change by thinning. Within the Okanogan complex region, the crust was extended by at least three major mechanisms:

(1) a continuum of deformation ranging from deep-seated ductile flow within a low-angle shear zone involving normal-type down-to-the-west movement along the present western margin, to brittle cataclastic faulting in zones subparallel to the shear planes within the mylonite; (2) high-angle normal faulting associated with opening of the Republic and other structural grabens along the eastern margin; and (3) widespread intrusion of Late Cretaceous to Eocene granitic plutons and eruption of silicic volcanics dominantly in the eastern portions of the complex. Extension within the complex in early Tertiary time was attended by extension of upper-plate rocks along numerous high-angle normal faults which presumably sole into the principal ductile-brittle structural discontinuity which marks the margin of the complex.

Within this model, then, the Okanogan complex records tectonic evolution of the crust from Early Jurassic to Eocene time, an evolution which is preserved at various structural levels throughout the complex (Fig. 9-15). The structurally deepest level is preserved along the southwest margin of the complex where high-grade core rocks are juxtaposed against similarly high-grade rocks of the upper plate in an evolved extensional shear zone marked by penetratively mylonitized lower-plate rocks, and brecciated lower- and upper-plate rocks. Northward along the western margin, higher structural levels are observed as reflected in low-grade, allochthonous, upper-plate rocks which lie structurally upon lower-plate crystalline rocks. To the northeast, the structurally highest levels of the complex are preserved where serpentinite marks the Quesnel suture along the boundary of the complex. Core rocks here include high-grade synmetamorphic mylonite of probable mid-Jurassic age and upper-plate Anarchist allochthonous rocks which display a normal metamorphic gradient upward from the serpentinite zone. The upper and lower plates along the northeast margin were originally juxtaposed in Jura-Cretaceous time and were together dissected at a high angle by steep, Tertiary-age normal faults associated with opening of the Republic Graben. Along the easternmost boundary, nonmylonitic, dominantly Tertiary-age core granites are also juxtaposed against upper-plate Anarchist rocks and partly coeval volcanic rocks along high-angle normal faults of the Republic Graben (Atwater and Rinehart, 1984).

Therefore, although each margin of the Okanogan complex expresses seemingly very different structural, metamorphic, and timing constraints, the constraints imposed by geologic relations along each margin are reconciled in a coherent model of Okanogan complex evolution in which geologic features of the deepest structural levels are preserved in the southwest, and features reflecting the same tectonic events at shallower structural levels are preserved to the northeast.

ACKNOWLEDGMENTS

Conversations with B. F. Atwater, R. L. Brown, K. F. Fox, D. W. Hyndman, J. M. Journeay, A. V. Okulitch, B. P. Rhodes, C. D. Rinehart, and D. J. Tempelman-Kluit have greatly aided our understanding of metamorphic core complex relations in Washington, Idaho, and British Columbia, although they may well not endorse our interpretations. We also thank W. G. Ernst, D. W. Hyndman, and B. P. Rhodes for helpful reviews of an earlier draft. Research grants received from the Geological Society of America and Sigma Xi provided support to the authors for their field study in the Okanogan complex. Field work upon which portions of this paper is based was conducted while the authors were employed by the U.S. Geological Survey. We gratefully thank B. F. Atwater for providing us with additional samples of Tonasket Gneiss.

Archibald, D. A., Glover, J. K., Price, R. A., Farrar, E., and Carmichael, D. M., 1983, Geochronology and tectonic implications of magmatism and metamorphism, southern Kootenay Arc and neighbouring regions, southeastern British Columbia — Pt. I: Jurassic to mid-Cretaceous: *Can. J. Earth Sci.*, v. 20, p. 1891–1913.

Armstrong, R. L., 1982, Cordilleran metamorphic core complexes — From Arizona to southern Canada: *Ann. Rev. Earth Planet. Sci.*, v. 10, p. 129–154.

_____, and Hansen, E., 1966, Cordilleran infrastructure in the eastern Great Basin: *Amer. J. Sci.*, v. 264, p. 112–127.

Atwater, B. F., 1985, Contemporaneity of the Republic Graben and Okanogan gneiss dome — Evidence from the Coyote Creek pluton, southern Okanogan County, Washington: *Geol. Soc. America Abstr. with Programs*, v. 17, no. 6, p. 338.

_____, and Rinehart, C. D., 1984, Preliminary geologic map of the Colville Indian Reservation, Ferry and Okanogan Counties, Washington: *U.S. Geol. Survey Open File Rpt. 84-389*, 20 p.

Bardoux, M., 1985a, Tertiary tectonic denudation in the hinterland of the Canadian Cordillera — Initial results from Kelowna, B. C.: *Geol. Soc. America Abstr. with Programs*, v. 17, no. 6, p. 339.

_____, 1985b, The Kelowna detachment zone, Okanagan Valley, south-central British Columbia: *Geol. Survey Canada Current Res. 85-1A*, p. 333–348.

Bell, T. H., and Etheridge, M. A., 1973, Microstructure of mylonites and their descriptive terminology: *Lithos*, v. 6, p. 337–348.

Berthé, D., Choukroune, P., and Gapais, D., 1979a, Orientations préférentielles du quartz et orthogneissification progressive en régime cisaillant — l'example du cisaillement sud-armoricain: *Bull. Mineralogy*, v. 102, p. 265–272.

_____, Choukroune, P., and Jegouzo, P., 1979b, Orthogneiss, mylonite and noncoaxial deformation of granite — The example of the South Armorican shear zone: *J. Struct. Geology*, v. 1, p. 31–42.

Bickford, M. E., Rhodes, B. P., and Hyndman, D. W., 1985, Age of mylonitization in the southern Priest River complex, northern Idaho and northeastern Washington: *Geol. Soc. America Abstr. with Programs*, v. 17, p. 341.

Bird, D. K., and Helgeson, H. C., 1981, Chemical interpretation of aqueous solutions with epidote-feldspar mineral assemblages in geologic systems — II. Equilibrium constraints in metamorphic/geothermal processes: *Amer. J. Sci.*, v. 281, p. 576–614.

Boettcher, A. L., and Wyllie, P. J., 1968, Jadeite stability measured in the presence of silicate liquids in the system $NaAlSiO_4$-SiO_2-H_2O: *Geochim. Cosmochim. Acta*, v. 32, p. 999–1012.

Bowen, N. L., and Tuttle, O. F., 1950, The system $NaAlSi_3O_8$-$KAlSi_3O_8$-H_2O: *J. Geology*, v. 58, p. 498–511.

Brown, R. L., and Journeay, J. M., 1985, A duplex model of crustal evolution, Omineca Belt, southern British Columbia: *Geol. Soc. America Abstr. with Programs*, v. 17, no. 6, p. 344.

_____, and Read, P. B., 1983, Shuswap terrane of British Columbia — A Mesozoic "core complex:" *Geology*, v. 11, p. 164–168.

_____, and Tippett, C. R., 1978, The Selkirk fan structure of the southeastern Canadian Cordillera: *Geol. Soc. America Bull.* v. 89, p. 548–558.

Chatterjee, N. D., and Johannes, W., 1974, Thermal stability and standard thermodynamic properties of 2M1-muscovite, $KAl_2[AlSi_5O_{10}(OH)_2]$: *Contrib. Mineralogy Petrology*, v. 48, p. 89–114.

Cheney, E. S., 1980, Kettle dome and related structures of northeastern Washington, *in* Crittenden, M. D., Jr., Coney, P. J., and Davis, G. H., eds., *Cordilleran Metamorphic Core Complexes:* Geol. Soc. America Mem. 153, p. 463–484.

Coney, P. J., 1980, Cordilleran metamorphic core complexes — An overview, *in* Crittenden, M. D., Jr., Coney, P. J., and Davis, G. H., eds., *Cordilleran Metamorphic Core Complexes:* Geol. Soc. America Mem. 153, p. 7–31.

Coombs, D. S., Ellis, A. J., Fyfe, W. S., and Taylor, A. H., 1959, The zeolite facies with comments on the interpretation of hydrosynthesis: *Geochim. Cosmochim. Acta*, v. 17, p. 53–107.

Crittenden, M. D., Jr., Coney, P. J., and Davis, G. H., eds., 1980, Cordilleran Metamorphic Core Complexes. *Geol. Soc. America Mem. 153*, 490 p.

Davis, G. A., Anderson, J. L., Martin, D. L., Krummenacher, D., Frost, E. G., and Armstrong, R. L., 1982, Geologic and geochronologic relations in the lower plate of the Whipple detachment fault, Whipple Mountains, southeastern California: A progress report, *in* Frost, E. G., and Martin, D. L., eds., *Mesozoic-Cenozoic Tectonic Evolution of the Colorado River Region, California, Arizona and Nevada:* San Diego, Calif., Cordilleran Publishers, p. 408–432.

Engebretson, D. C., Cox, A., and Thompson, G. A., 1984, Correlation of plate motions with continental tectonics — Laramide to Basin-Range: *Tectonics*, v. 3, p. 115–119.

Fox, K. F., Jr., and Rinehart, C. D., 1971, Okanogan gneiss dome [abstract], *in Metamorphism in the Canadian Cordillera:* Cordilleran Section, Can. Geol. Assoc. Meet., Vancouver, B.C., Programs and Abstr., p. 10.

——, Rinehart, C. D., Engels, J. C., and Stern, T. W., 1976, Age of emplacement of the Okanogan gneiss dome, north-central Washington: *Geol. Soc. America Bull.*, v. 87, p. 1217–1224.

——, Rinehart, C. D., and Engels, J. C., 1977, Plutonism and orogeny in north-central Washington — Timing and regional context: *U.S. Geol. Survey Prof. Paper 989*, 27 p.

Fyles, J. T., 1970, Structure of the Shuswap metamorphic complex in the Jordan River area, northwest of Revelstoke, British Columbia, *in* Wheeler, J. O., ed., *Structure of the Southern Canadian Cordillera:* Geol. Assoc. Canada Spec. Paper 6, p. 87–98.

Fyson, W. K., 1970, Structural relations in metamorphic rocks, Shuswap Lake area, British Columbia, *in* Wheeler, J. O., ed., *Structure of the Southern Canadian Cordillera:* Geol. Assoc. Canada Spec. Paper 6, p. 107–122.

Ganguly, J., 1972, Staurolite stability and related parageneses: theory, experiments, and applications: *J. Petrology*, v. 13, p. 335–365.

Goldsmith, J. R., and Newton, R. C., 1977, Scapolite-plagioclase stability relations at high pressures and temperatures in the system $NaAlSi_3O_8$-$CaAl_2Si_2O_8$-$CaCO_3$-$CaSO_4$: *Amer. Mineralogist*, v. 62, p. 1063–1081.

Goodge, J. W., 1983a, Reorientation of folds by progressive mylonitization, Okanogan dome, north-central Washington: *Geol. Soc. America Abstr. with Programs*, v. 15, no. 5, p. 323.

——, 1983b, Fold reorientation and quartz microfabric in the Okanogan dome mylonite zone, Washington — Kinematic and tectonic implications: M.Sc. thesis, Univ. Montana, Missoula, Mont., 65 p.

——, and Hansen, V. L., 1983, Petrology and structure of rocks in the southwest portion of Okanogan dome, north-central Washington: *Northwest Geology*, v. 12, p. 13–24.

Greenwood, H. J., 1967, Wollastonite – Stability in H_2O-CO_2 mixtures and occurrence in a contact metamorphic aureole near Salmo, British Columbia, Canada: *Amer. Mineralogist*, v. 52, p. 1669–1680.

Haas, H., and Holdaway, M. J., 1973, Equilibria in the system Al_2O_3-SiO_2-H_2O involving the stability limits of pyrophyllite: *Amer. J. Sci.*, v. 273, p. 449–464.

Hansen, V. L., 1983a, Kinematic interpretation of mylonitic rocks in Okanogan dome, north-central Washington: *Geol. Soc. America Abstr. with Programs*, v. 15, p. 323.

——, 1983b, Kinematic interpretation of mylonitic rocks in Okanogan dome, north-central Washington, and implications for dome evolution: M.Sc. thesis, Univ. Montana, Missoula, Mont., 47 p.

Harker, R. I., and Tuttle, O. F., 1955, Studies in the system CaO-MgO-CO_2 – Pt. I: *Amer. J. Sci.*, v. 253, p. 209–224.

Harms, T. A., and Price, R. A., 1983, The Newport fault – Eocene crustal stretching, necking and listric normal faulting in NE Washington and NW Idaho: *Geol. Soc. America Abstr. with Programs*, v. 15, p. 309.

Holdaway, M. J., 1971, Stability of andalusite and the aluminum silicate phase diagram: *Amer. J. Sci.*, v. 271, p. 97–131.

——, and Lee, S. M., 1977, Fe-Mg cordierite stability in high-grade pelitic rocks based on experimental, theoretical and natural observations: *Contrib. Mineralogy Petrology*, v. 63, p. 175–198.

Hsu, L. C., 1968, Selected phase relations in the system Al-Mn-Fe-Si-O-H – A model for garnet equilibria: *J. Petrology*, v. 9, p. 40–83.

Hyndman, D. W., 1968, Mid-Mesozoic multiphase folding along the border of the Shuswap metamorphic complex: *Geol. Soc. America Bull.*, v. 79, p. 575–588.

——, 1985, *Petrology of Igneous and Metamorphic Rocks*, 2nd Ed.: New York, McGraw-Hill, 786 p.

Journeay, J. M., 1985, Thermo-tectonic evolution of the Monashee complex; a basement duplex, Omineca belt, B.C.: *Geol. Soc. America Abstr. with Programs*, v. 17, p. 364.

——, and Brown, R. L., 1985, Internal strain and kinematic evolution of the Monashee Complex; a basement duplex, B.C.: *Geol. Soc. America Abstr. with Programs*, v. 17, p. 364.

Kerrich, R., Allison, I., Barnett, R. L., Moss, S., and Starkey, J., 1980, Microstructural and chemical transformations accompanying deformation of granite in a shear zone at Miéville, Switzerland; with implications for stress corrosion cracking and superplastic flow: *Contrib. Mineralogy Petrology*, v. 73, p. 221–242.

Kerrick, D. M., 1972, Experimental determination of muscovite + quartz stability with $P_{H_2O} < P_{total}$: *Amer. J. Sci.*, v. 272, p. 946–958.

Liou, J. G., 1971, Synthesis and stability relations of prehnite, $Ca_2Al_2Si_3O_{10}(OH)_2$: *Amer. Mineralogist*, v. 56, p. 507–531.

——, 1973, Synthesis and stability relations of epidote, $Ca_2Al_2FeSi_3O_{12}(OH)$: *J. Petrology*, v. 14, p. 381–413.

Lister, G. S., and Snoke, A. W., 1984, S-C mylonites: *J. Struct. Geology*, v. 6, p. 617–638.

Luth, W. C., 1976, Granitic rocks, *in* Bailey, D. K., and Macdonald, R., eds., *The Evolution of the Crystalline Rocks:* New York, Academic, p. 335–417.

Mathews, W. M., 1981, Early Cenozoic resetting of potassium-argon dates and geothermal history of north Okanagan area, British Columbia: *Can. J. Earth Sci.*, v. 18, p. 1310–1319.

Mattauer, M., Collot, B., and Van den Driessche, J., 1983, Alpine model for the internal metamorphic zones of the North American Cordillera: *Geology*, v. 11, p. 11–15.

McMillen, W. J., 1973, Petrology and structure of the west flank, Frenchman Cap dome, near Revelstoke, British Columbia: *Geol. Survey Canada Paper 71-29*, 88 p.

Menzer, F. J., Jr., 1983, Metamorphism and plutonism in the central part of the Okanogan Range, Washington: *Geol. Soc. America Bull.*, v. 94, p. 471–498.

Miller, E. L., Gans, P. B., and Garing, J., 1983, The Snake Range decollement — An exhumed mid-Tertiary ductile-brittle transition: *Tectonics*, v. 2, p. 239–263.

Miller, F. K., and Engels, J. C., 1975, Distribution and trends of discordant ages of the plutonic rocks of northeastern Washington and northern Idaho: *Geol. Soc. America Bull.*, v. 86, p. 517–528.

Misch, P., 1960, Regional structural reconnaissance in central-northeast Nevada and some adjacent areas — Observations and interpretations: *Intermountain Assoc. Petrol. Geologists 11th Ann. Field Conf. Guidebook*, p. 17–42.

Monger, J. W. H., 1977, Upper Paleozoic rocks of the western Canadian Cordillera and their bearing on Cordilleran evolution: *Can. J. Earth Sci.*, v. 14, p. 1831–1859.

——, and Irving, E., 1979, The Canadian Cordilleran collage: *Geol. Soc. America Abstr. with Programs*, v. 11, p. 482.

——, Souther, J. G., and Gabrielse, H., 1972, Evolution of the Canadian Cordillera — A plate tectonic model: *Amer. J. Sci.*, v. 272, p. 577–602.

——, Price, R. A., and Tempelman-Kluit, D. J., 1982, Tectonic accretion and the origin of the two major metamorphic and plutonic welts in the Canadian Cordillera: *Geology*, v. 10, p. 70–75.

Okulitch, A. V., 1973, Age and correlation of the Kobau Group, Mount Kobau, British Columbia: *Can. J. Earth Sci.*, v. 10, p. 1508–1518.

——, 1984, The role of the Shuswap Metamorphic Complex in Cordilleran tectonism — A review: *Can. J. Earth Sci.*, v. 21, p. 1171–1193.

——, Wanless, R. K., and Loveridge, W. D., 1975, Devonian plutonism in south-central British Columbia: *Can. J. Earth Sci.*, v. 12, p. 1760–1769.

Orr, K. E., 1985, Structural features along the margin of Okanogan dome, Tenas Mary Creek area, NE Washington: *Geol. Soc. America Abstr. with Programs*, v. 17, no. 6, p. 398.

Orville, P. M., 1975, Stability of scapolite in the system Ab-An-NaCl-CaCO$_3$ at 4 kb and 750°C: *Geochim. Cosmochim. Acta*, v. 39, p. 1091–1105.

Parker, R. L., and Caulkins, J. A., 1964, Geology of the Curlew quadrangle, Ferry County, Washington: *U.S. Geol. Survey Bull. 1169*, 95 p.

Parkinson, D., 1985, Geochronology of the western side of the Okanogan metamorphic core complex, southern B.C.: *Geol. Soc. America Abstr. with Programs*, v. 17, no. 6, p. 399.

Parrish, R. R., 1983, Pb-U zircon dates reflecting Late Cretaceous-Early Tertiary plutonism, deformation and istopic resetting, Valhalla Complex, southeast British Columbia: *Geol. Assoc. Canada Program with Abstr.*, v. 8, p. A-53.

——, 1985, Metamorphic core complexes of southern B.C. — Distinctions between extensional or compressional origins: *Geol. Soc. America Abstr. with Programs*, v. 17, no. 6, p. 399.

——, and Wheeler, J. O., 1983, A U-Pb zircon age from the Kuskanax batholith, southeastern British Columbia: *Can. J. Earth Sci.*, v. 20, p. 1751–1756.

Pigage, L. C., 1977, Rb-Sr dates for granodiorite intrusions on the northeast margin of the Shuswap Metamorphic Complex: *Can. J. Earth Sci.*, v. 14, p. 1690–1695.

Price, R. A., 1979, Intracontinental ductile crustal spreading linking the Fraser River and northern Rocky Mountain Trench transform fault zones, south-central British

Columbia and northeast Washington: *Geol. Soc. America Abstr. with Programs*, v. 11, p. 499.

—, 1981, The Cordilleran foreland thrust and fold belt in the southern Canadian Rocky Mountains, *in* McClay, K. R., and Price, N. J., eds., *Thrust and Nappe Tectonics:* Geol. Soc. London Spec. Publ. 9, p. 427–488.

—, and Mountjoy, E. W., 1970, Geologic structure of the Canadian Rocky Mountains between Bow and Athabaska Rivers — A progress report, *in* Wheeler, J. O., ed., *Structure of the Southern Canadian Cordillera:* Geol. Assoc. Canada Spec. Paper 6, p. 7–26.

Read, P. B., 1975, Lardeau Group, Lardeau map-area, west half (82 K west half), British Columbia, *in Report of Activities* — Pt. A: Geol. Survey Canada Paper 75-1A, p. 29–30.

—, 1980, Stratigraphy and structure — Thor-Odin to Frenchman Cap "domes," Vernon east-half map area, southern British Columbia, *in Current Research* — Pt. A: Geol. Survey Canada Paper 80-1A, p. 19–25.

—, and Brown, R. L., 1981, Columbia River fault zone — Southeastern margin of the Shuswap and Monashee Complexes, southern British Columbia: *Can. J. Earth Sci.*, v. 18, p. 1127–1145.

—, and Klepacki, D. W., 1981, Stratigraphy and structure — Northern half of Thor-Odin nappe, Vernon east-half map area, southern British Columbia, *in Current Research:* Geol. Survey Canada Paper 81-1A, p. 169–173.

—, and Okulitch, A. V., 1977, The Triassic unconformity of south-central British Columbia: *Can. J. Earth Sci.*, v. 14, p. 606–638.

Reesor, J. E., 1965, Structural evolution and plutonism in Valhalla gneiss complex, British Columbia: *Geol. Survey Canada Bull. 129*, 128 p.

—, 1970, Some aspects of structural evolution and regional setting in part of the Shuswap Metamorphic Complex, *in* Wheeler, J. O., ed., *Structure of the Southern Canadian Cordillera:* Geol. Assoc. Canada Spec. Paper 6, p. 73–86.

—, and Moore, J. M., 1971, Petrology and structure of Thor-Odin gneiss dome, Shuswap metamorphic complex, British Columbia: *Geol. Survey Canada Bull. 195*, 149 p.

Reynolds, S. J., and Rehrig, W. A., 1980, Mid-Tertiary plutonism and mylonitization, South Mountains, central Arizona, *in* Crittenden, M. D., Jr., Coney, P. J., and Davis, G. H., eds., *Cordilleran Metamorphic Core Complexes:* Geol. Soc. America Mem. 153, p. 159–176.

Rhodes, B. P., and Cheney, E. S., 1981, Low-angle faulting and the origin of Kettle dome, a metamorphic core complex in northeastern Washington: *Geology*, v. 9, p. 366–369.

—, and Hyndman, D. W., 1984, Kinematics of mylonites in the Priest River "metamorphic core complex," northern Idaho and northeastern Washington: *Can. J. Earth Sci.*, v. 21, p. 1161–1170.

Richardson, S. W., 1968, The stability of Fe-staurolite + quartz: *Ann. Rpt. Geophys. Laboratory, 1966–67*, p. 398–402.

Rinehart, C. D., and Fox, K. F., Jr., 1972, Geology of the Loomis quadrangle, Okanogan County, Washington: *Wash. Div. Mines Geology Bull. 64*, 124 p.

—, and Fox, K. F., Jr., 1976, Bedrock geology of the Conconully quadrangle, Okanogan County, Washington: *U.S. Geol. Survey Bull. 1402*, 58 p.

Ross, J. V., 1981, A geodynamic model for some structures within and adjacent to the Okanagan Valley, southern British Columbia: *Can. J. Earth Sci.*, v. 18, p. 1581–1598.

_____ , and Christie, J. S., 1979, Early recumbent folding in some westernmost exposures of the Shuswap complex, southern Okanagan, British Columbia: *Can. J. Earth Sci.*, v. 16, p. 877–894.

Sibson, R., 1977, Fault rocks and fault mechanisms: *J. Geol. Soc. London*, v. 133, p. 191–213.

Simpson, C., and Schmid, S. M., 1983, An evaluation of the sense of movement in sheared rocks: *Geol. Soc. America Bull.*, v. 94, p. 1281–1288.

Snook, J. R., 1965, Metamorphic and structural history of "Colville batholith" gneisses, north-central Washington: *Geol. Soc. America Bull.*, v. 76, p. 759–776.

Spear, F. S., 1981, An experimental study of hornblende stability and compositional variability in amphibolite: *Amer. J. Sci.*, v. 281, p. 697–734.

Stormer, J. C., Jr., 1975, A practical two-feldspar geothermometer: *Amer. Mineralogist*, v. 60, p. 667–674.

Tempelman-Kluit, D. J., and Parkinson, D., 1986, Extension across the Eocene Okanagan crustal shear in southern British Columbia: *Geology*, v. 14, p. 318–321.

Tipper, H. W., 1981, Offset of an upper Pliensbachian geographic zonation in the North American Cordillera by transcurrent movement: *Can. J. Earth Sci.*, v. 18, p. 1788–1792.

_____ , Woodsworth, G. J., and Gabrielse, H., 1981, Tectonic assemblage map of the Canadian Cordillera and adjacent parts of the United States of America: *Geol. Survey Canada Map 1505A*, scale 1:2,000,000.

Turner, F. J., 1981, *Metamorphic Petrology – Mineralogical, Field, and Tectonic Aspects*, 2nd Ed.: New York, McGraw-Hill, 524 p.

Waters, A. C., and Krauskopf, K., 1941, Protoclastic border of the Colville batholith: *Geol. Soc. America Bull.*, v. 52, p. 1355–1418.

Wernicke, B., 1981, Low-angle normal faults in the Basin and Range province – Nappe tectonics in an extending orogen: *Nature*, v. 291, p. 645–648.

_____ , 1985, Uniform-sense normal simple shear of the continental lithosphere: *Can. J. Earth Sci.*, v. 22, p. 108–125.

Wheeler, J. O., 1965, Big-Bend map area, British Columbia (82M east half): *Geol. Survey Canada Paper 64-32*, 37 p.

Brady P. Rhodes
Department of Geological Sciences
California State University, Fullerton
Fullerton, California 92634

D. W. Hyndman
Department of Geology
University of Montana
Missoula, Montana 59812

10

REGIONAL METAMORPHISM, STRUCTURE, AND TECTONICS OF NORTHEASTERN WASHINGTON AND NORTHERN IDAHO

ABSTRACT

The northwestern U.S. Cordillera consists of three major tectonic provinces. The easternmost of these, the northern Rocky Mountain fold-and-thrust belt, consists of the strongly compressed Proterozoic and Paleozoic North American miogeocline and is unaffected by Phanerozoic, regional, dynamothermal metamorphism. To the west across the Purcell Trench is the 125-km-wide Kootenay Arc, which represents the westward continuation of the shortened miogeocline where it is pervaded by a dynamotheral metamorphism ranging in grade from the lower greenschist to upper amphibolite facies. Farther west across a major tectonic contact lies the accreted terrane of Quesnellia, a mosaic of eugeoclinal, metasedimentary, metavolcanic, and minor mafic metaplutonic rocks. Although it surrounds high-grade metamorphic complexes, Quesnellia itself is dominated by greenschist-facies metamorphism. Both the Kootenay Arc and Quesnellia are intruded by voluminous Jurassic-to-Tertiary batholitic granitic rocks.

Amphibolite facies metamorphic rocks occur within the Kettle, Lincoln, and Priest River complexes. The boundaries of these complexes are generally either low-angle to high-angle normal faults or zones of gradually increasing metamorphic grade. The Priest River complex is surrounded on three sides by low-grade rocks of the Belt Supergroup, and on a fourth side by the Purcell Trench. Its interior consists of metamorphosed Prichard Formation and undated, perhaps pre-Belt, sillimanite-grade para- and orthogneiss. Mylonitic rocks occur in the southern part of the complex within a 4-km-thick, domal, eastward-verging zone that cuts an older, synmetamorphic, north-northeastward-striking crystalloblastic foliation. The age of metamorphism may be Late Jurassic or Cretaceous, and the age of mylonitization is likely latest Cretaceous to Paleocene.

The Kettle complex is virtually surrounded by eugeoclinal rocks but consists predominantly of sillimanite-zone marble, quartzite, and pelitic gneiss, that may represent the metamorphosed miogeocline and/or cratonic basement of North America that was overriden by Quesnellia. On the east flank of the complex, these rocks are mylonitic within an upward-intensifying, gentle eastward dipping zone that is truncated by the Eocene Kettle River detachment fault. The west side of the complex is bounded by high-angle faults of the Republic and Keller grabens. The metamorphism in the Kettle complex may be Jurassic, the age of mylonitization is unknown. The Lincoln complex consists of a region of mylonitic granitic rocks south of the Kettle complex and the two likely join beneath intervening low-grade rocks.

The areal distribution and relative ages of metamorphic, plutonic, and deformational events in the northwestern United States, combined with sparse and sometimes equivocal radiometric data are consistent with the following tectonic sequence. Mid-Jurassic, west-dipping(?) subduction resulted in the collision of North America and Quesnellia, which may contain an associated, pre-Middle Jurassic, oceanic volcanic arc. During collision, Quesnellia overrode the North American craton, but slid beneath the deformed and thickened miogeocline of the Kootenay Arc. Collision caused subduction to jump to the west of Quesnellia and switch dip to the east. A magmatic arc, now represented by about 150–100-m.y.-old granitic plutons, intruded the Kootenay Arc and adjacent Quesnellia. Thickening of the crust during collision along with heat generated during the subsequent arc magmatism caused metamorphism of the miogeocline, the underlying cratonic basement, and perhaps the eastern part of Quesnellia. During the Late Cretaceous to Paleocene, the Rocky Mountain fold-and-thrust belt was active and perhaps resulted from eastward gravitational spreading of the previously thickened crust. By the middle Eocene (50 m.y.), thrusting had stopped and widespread extension caused the uplift and cooling of the metamorphic complexes, low-angle and high-angle normal faulting, and widespread intermediate volcanism.

Northeastern Washington and northern Idaho contain metamorphic rocks of many grades that, until recently, were largely unstudied. This area forms the southern extension of the relatively well studied Omineca Crystalline Belt and Kootenay Arc of British Columbia. Recent work includes mapping of the Colville Indian Reservation and adjacent areas of northeastern Washington (Atwater *et al.*, 1984; Mills, 1985), structural studies in the Kootenay Arc (Ellis, 1984), mapping, and petrologic structural studies in the Priest River complex (Miller, 1982a, b, 1983; Rhodes, 1986; Rhodes and Hyndman, 1984). In this review we summarize this recent work and outline the distribution, thermal history, structure, and tectonic setting of the metamorphic rocks that lie between the Republic graben of north-central Washington and the Rocky Mountain fold-and-thrust belt (Fig. 10-1).

Geologic maps now exist for much of the area of this review, but data are very scant regarding the metamorphism. Much of the data that we present are from U.S. Geological Survey reports published over the last 25 years. Our interpretation of the distribution and grade of the metamorphic rocks in this area draws heavily from such reports, which were not focused primarily on the metamorphism. In many cases, we made educated guesses of metamorphic grade based on brief published petrologic descriptions of our own field reconnaissance. One of our goals is to attempt to fit the metamorphic and structural history into an appropriate tectonic picture; it is our belief that in so doing we might better educate those unfamiliar with the area, and better understand where the greatest problems exist and where future research should focus.

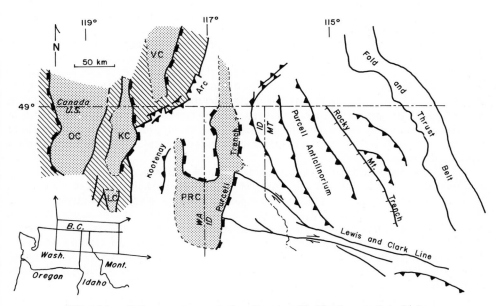

FIG. 10-1.   Index map to a part of northeastern Washington, northern Idaho, northwestern Montana, and southern British Columbia. Shaded regions are the high-grade metamorphic complexes: PRC, Priest River complex; KC, Kettle complex; OC, Okanogan complex; LC, Lincoln complex; VC, Valhalla complex. Diagonal lines represent the region underlain by Quesnellia.

Northeastern Washington, northern Idaho, and northwestern Montana contain four distinct tectonic elements. Three of these are, from east to west, the northern Rocky Mountain fold-and-thrust belt, the Kootenay Arc, and the composite accreted terrane of Quesnellia (Fig. 10-1). A fourth element is the high-grade metamorphic/plutonic Priest River, Kettle, and Lincoln complexes. These so-called "metamorphic core complexes" are superimposed on both the Kootenay Arc and Quesnellia.

The Rocky Mountain fold-and-thrust belt extends from the eastern Rocky Mountain front in northwestern Montana to the eastern edge of the Priest River complex in northern Idaho. Rocks in this zone include Precambrian crystalline basement of the craton overlain by the thick section (up to 20 km) of the fine-grained clastic rocks of the Proterozoic Belt basin (Harrison, 1972). Overlying the Belt rocks are sedimentary rocks of the eastward-tapering upper Propterozoic (<850 m.y.) and lower Paleozoic miogeoclinal wedge that accumulated on the trailing edge of North America (Stewart, 1972, 1976).

The Kootenay Arc consists of a roughly north-south-trending belt of predominantly low-grade, regionally metamorphosed rocks that extend from the Purcell Trench in northern Idaho to the eastern edge of Quesnellia. Most of the rocks within the Kootenay Arc are part of the western Belt basin. Locally particularly to the west, small enclaves of younger, upper Precambrian and lower Paleozoic miogeoclinal rocks are preserved within the arc. The area west of the Purcell Trench is also marked by voluminous granitic rocks, including the Kaniksu batholith of northern Idaho, and the Loon Lake batholith of northeastern Washington (Fig. 10-2). Locally, Eocene andesitic to rhyolitic volcanic and sedimentary rocks are preserved within structural lows.

The 118°W meridian in Washington marks the approximate location of an abrupt boundary between rocks formed as a part of the miogeocline of North America to the east and rocks of eugeoclinal character to the west. During the last 10 years, work to the north in central British Columbia shows that the rocks west of this boundary are part of a tectonic collage whose constituents formed outboard from and at a later time accreted to the North American craton (Coney et al., 1980; G. A. Davis et al., 1978). These allochthonous rocks have not been studied in detail in the northwestern United States east of the Cascade Mountains, but they probably belong to Quesnellia (Monger et al., 1982; Price et al., 1981). In northwestern Washington the rocks of Quesnellia range in age from Ordovician(?) to Jurassic and include eugeoclinal metagreywacke, argillite, and metavolcaniclastic rocks, plus minor marble, greenstone, and locally, small bodies of serpentinite. Locally, postaccretionary, unmetamorphosed Eocene volcanic and sedimentary rocks overlie Quesnellia.

High-grade metamorphic rocks in northeastern Washington and northern Idaho are confined to the Kettle, Lincoln, and Priest River complexes. These complexes form a part of a long chain of "metamorphic core complexes" (Coney, 1980) that extend from the Yukon Territory to Mexico. They contain a variety of metamorphic rocks, including orthogneiss, paragneiss, schist, amphibolite, calc-silicate gneiss, quartzite, and marble. Metamorphic grade is consistently within the sillimanite stability field in the interior of the complexes, although grades are locally lower near their boundaries. Domal mylonitic zones are present within all the complexes and low-angle normal faults commonly juxtapose high-grade against low-grade rocks at the border of the complexes.

Figures 10-2 and 10-3 summarize the geology from the Republic graben in northeastern Washington, almost to the Rocky Mountain Trench in northwestern Montana. Figure 10-4 outlines the extent and grade of regionally metamorphosed rocks over the same area as Fig. 10-2. To the east of the Priest River complex, regional metamorphism is confined to the pre-Belt crystalline basement, with the exception of regional load metamorphism

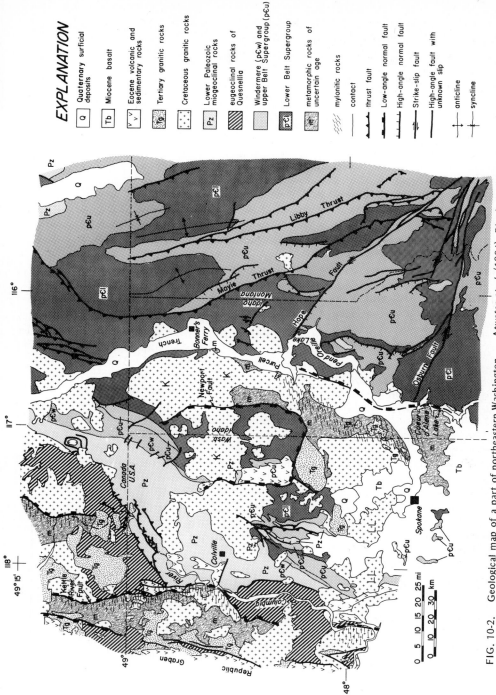

FIG. 10-2. Geological map of a part of northeastern Washington, northern Idaho, northwestern Montana, and southern British Columbia. Abbreviations: K, Kaniksu batholith; LL, Loon Lake batholith. Note that on this line of section, the Purcell Trench probably does not contain a major normal fault, although elsewhere it may. Compiled from Aadland and Bennett (1979), Griggs, (1973), Atwater et al. (1984), Rhodes and Hyndman (1984), Harrison et al. (1981), Cheney et al. (1982), Harms (1982), Cambell and Raup (1964), Cowan et al. (1986), Little (1957, 1960, 1982), Huntting et al. (1961), Becraft and Weis (1963), Rice (1941), Preto (1970), Leech (1960), and Parrish et al. (1985).

275

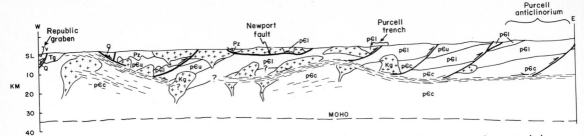

FIG. 10-3.   East-west cross section along latitude 48° 30' of Fig. 10-2. Q, Quesnellian rocks; p€c, cratonic base-ment; p€l, lower Belt Supergroup; p€u, upper Belt Supergroup and Winderemere Group; Pz, Paleozoic mio-geoclinal sedimentary rocks; Kg, Cretaceous granitic rocks; Tg, Tertiary granitic rocks; Tv, Eocene volcanic rocks. No vertical exaggeration.

of the Belt Supergroup, which is omitted from Fig. 10-4. West of the Purcell Trench, nearly all of the rocks show a regional, dynamothermal metamorphism ranging from lower greenschist to upper amphibolite grade.

STRUCTURE AND METAMORPHISM

The Rocky Mountain Fold-and-Thrust Belt

In northwestern Montana, east-verging thrust faults are concentrated in a band less than 50 km wide at the east margin of the Cordillera (Fig. 10-1). Behind this zone of thrust faults is a plate consisting of a section of up to 20 km of the Proterozoic Belt Supergroup that is cut by only a few, widely spaced thrust faults (Fig. 10-1). This plate is broadly warped into the Purcell anticlinorium (Fig. 10-2), with the lower Belt Prichard Formation exposed in the core. Harrison *et al.* (1980) mapped additional thrusts in northwestern Montana and northern Idaho; thus the thrust belt extends at least as far west as the Purcell Trench. East of the Rocky Mountain Trench, shortening appears to involve only the sedimentary cover overlying the gently westward-dipping craton. West of the Rocky Mountain Trench the available geophysical data are equivocal. Harrison *et al.* (1980) suggest that thrust faults rooted in the basement within and to the west of the Purcell anticlinorium explain the observed positive gravity anomalies. However, Price (1981) suggests that the continuity of magnetic anomalies across the Purcell Trench indicates that the basement is intact perhaps as far west as the Purcell Trench.

The age of thrusting is loosely bracketed by the Upper Jurassic, western-derived clastic rocks in the foreland, and lower Eocene sedimentary rocks that overlap compressional structures (Price, 1981). However, the Jurassic clastic rocks may document a thickening of the crust associated with subduction or accretion to the west and not necessarily the onset of thrusting. Direct dating of individual thrust faults implies that most of the shortening occurred from the mid-Cretaceous to early Eocene (Price, 1981).

The other major structure in the Rocky Mountain province is the Lewis and Clark line (Fig. 10-1 and 10-2), a zone of west-northwest-trending strike-slip faults. This zone experienced a long and complicated history. The latest movement was right lateral (Harrison *et al.*, 1974); however, the apparent offset of the continental cratonic margin, as defined by a zone of granitic batholiths or the $^{87}Sr/^{86}Sr = 0.706$ line, and left-lateral drag of Late Cretaceous fold axes (Hyndman *et al.*, Chapter 12, this volume) points to an earlier, large-scale left-lateral movement. The latest right-lateral movement likely occurred during Eocene extension (Sheriff *et al.*, 1984).

REGIONAL METAMORPHISM, STRUCTURE, AND TECTONICS

FIG. 10-4. Metamorphic map of the same region as Fig. 10-2. Thin solid lines are geologic contacts from Fig. 10-2. Thin dashed lines are approximate metamorphic facies boundaries, and dotted lines are inferred boundaries. Blank areas are unmetamorphosed but include rocks that have undergone burial metamorphism to the east of the Purcell Trench. See Fig. 10-2 for geologic and geographic names. Compiled from the same sources as Fig. 10-2 and Miller (1982a, b), Monger and Hutchison (1971), Archibald *et al.* (1983), Harms (1982), Clark (1973), and unpublished reconnaissance by the authors.

Legend:
- upper amphibolite facies = sillimanite zone
- amphibolite facies undifferentiated
- lower amphibolite facies = staurolite & kyanite zones
- upper greenschist facies = biotite & garnet zones
- greenschist facies undifferentiated
- lower greenschist facies = chlorite zone
- unmetamorphosed

Metamorphism in the Rocky Mountains is restricted to static load metamorphism which affected all units of the Belt Supergroup (Norwich, 1972). This metamorphism increases progressively with depth; the lower Belt Prichard Formation is everywhere in the upper greenschist facies, and the upper Belt Missoula Group is very low grade. Textures of the Prichard Formation are characterized by unoriented biotite porphyroblasts evenly scattered in sedimentary layers of uniform composition. The matrix is commonly very fine-grained white mica, metamorphic oligoclase, and quartz with only a faint preferred orientation, probably mimetic from the compaction of clay-rich sediments. Metamorphic grade appears to be essentially unchanged near Proterozoic diabase sills; thus the sills may be a regional but insignificant local source of heat.

The Priest River Complex

The Priest River complex is defined by a region of amphibolite-grade metamorphic rocks and associated granitic plutons that extend from the Purcell Trench westward into eastern Washington and from southern British Columbia southward into the southern panhandle of Idaho (Fig. 10-2). Metamorphic rocks within the complex include pelitic and semipelitic schist and gneiss, granodioritic to granitic orthogneiss, augen gneiss, and locally, amphibolite and/or thin layers of quartzite. The complex is abundantly intruded by pre-, syn-, and postkinematic granitic plutons. Miller and Engels (1975) divided the granitic rocks into a hornblende-biotite suite, and a two-mica suite. Within the high-grade rocks, most of the syn- and prekinematic granites are of the two-mica suite. The southern part of the complex contains a 4-km-thick domal zone of mylonitic rocks.

The Purcell Trench forms the eastern boundary of the Priest River complex over a distance of >200 km, from the south end of Coeur d'Alene Lake, where it appears to fade farther to the south, northward to southeastern British Columbia (Fig. 10-2). The Purcell Trench is a topographic low and, except locally, is filled by surficial deposits. Much of the trench contains an inferred fault that juxtaposes the sillimanite-grade rocks of the Priest River complex against Beltian sedimentary rocks that lack a regional dynamothermal metamorphism. This fault may be a low-angle, eastward-dipping detachment fault, but nowhere is it actually exposed. However, in the vicinity of Sandpoint, at the north end of Pend Oreille Lake, Harrison *et al.* (1972) mapped migmatites on the east side of the Purcell Trench. Our own reconnaissance indicates that the metamorphism in this area fades gradually to the east within the Prichard Formation, the oldest unit within the Belt Supergroup. Apparently, the Purcell Trench in this region does not contain a fault of significant offset (Fig. 10-3). Along the northern part of the Purcell Trench, the boundary is once again apparently faulted, but as the Canadian border is approached, this fault appears to fade again into a zone of gradational metamorphism (Archibald *et al.*, 1983).

Much of the north-central part of the Priest River complex is overlain by the synformal Newport fault. The fault itself is a zone of generally brittle deformation characterized by cataclastite and breccia; however, locally, a thin (<200 m) zone of greenschist facies mylonites is preserved beneath the breccias (Harms, 1982; Miller, 1971). The Newport fault separates tilted blocks of listrically(?) faulted lower Belt rocks intruded by granite, against high-grade rocks of the Priest River complex. On its northeastern limb, displacement appears to fade across the fault, and the metamorphic grade across it decreases, until it is lost within low-grade rocks of the Prichard Formation just south of the Canadian border (Miller, 1982b). The northwest limb likewise appears to fade within granitic rocks. The Newport fault cuts the 48-m.y. Silver Point pluton and thus is Eocene or younger. Microstructures from mylonitic rocks within the Newport fault zone indicate a normal sense on both limbs of the fault, although the data from the western limb are sparse (Harms, 1982).

The western boundary of the Priest River complex is difficult to locate because of the abundance of undeformed granitic rocks. Approximately 25 km northeast of Colville (Fig. 10-4), metamorphism within the Prichard Formation increases to the west from middle (?) greenschist facies to upper amphibolite facies (sillimanite zone). This region may mark the gradational boundary of the Priest River complex. Similarly, about 40 km southeast of Colville, a nearly complete section of Belt, Windermere, and lower Paleozoic rocks dips to the west and increases in metamorphic grade to the east and at stratigraphically lower levels. Coincident with this increase in grade is an increase in the intensity of mineral alignment, from well-developed fracture cleavage to the west to coarse schistosity to the east. This increase in grade cannot be due solely to burial metamorphism, but more likely represents the southward continuation of the gradational western boundary of the Priest River complex.

Much of the southern boundary of the Priest River complex is buried beneath the Miocene Columbia River Basalt. Along the southeast extension of the complex, just west of Coeur d'Alene Lake, lower Belt(?) greenschist-facies phyllite crops out 10 km south of sillimanite-grade paragneiss that protrudes through the Columbia River Basalt. Either the fault within the southern Purcell Trench curves to the west and juxtaposes the high- and low-grade rocks, or displacement on this fault diminishes to the south into a gradational boundary. South of Spokane, scattered outcrops of both upper and lower Belt rocks project through the basalt cover. Although exposures are poor, the metamorphic grade within these rocks is variable and seems to increase to the west. The entire southern boundary of the Priest River complex is probably predominantly gradational.

The earliest fabric within the Priest River complex is a crystalloblastic foliation within the metasedimentary rocks. Although the data are meager, this foliation in general dips at a high angle with an overall north-northeast trend throughout the complex[1] (Miller, 1974d, 1982a, b, 1983; Weis, 1968). This foliation developed during the peak of metamorphism in the upper amphibolite facies (Rhodes, 1986).

Mylonitic rocks occur in the 4-km-thick Spokane Dome mylonitic zone in the southern part of the Priest River complex (Fig. 10-2). The gently domal mylonitic foliation cuts and transposes the older crystalloblastic foliation at the edges of the zone. The eastern limb of the dome is cut by normal faults within the Purcell Trench. A nearly unidirectional mylonitic lineation trending N70°E defines the slip line within the zone. Microscopic and mesoscopic fabrics within the mylonitic rock, including S and C surfaces, and asymmetric porphyroclasts, indicate a top-to-the-east sense of shear on both limbs of the dome (Rhodes and Hyndman, 1984). Preliminary U-Pb data from the Mount Spokane Granite are consistent with a intrusive age of about 70 m.y.; thus much of the mylonitization may be younger. Dikes of the late-synkinematic Mount Rathdrum granite give complex, discordant U-Pb data which suggest that the zircons contain an approximately 1800-m.y. xenocrystic component, and that the intrusive age is no younger than approximately 55 m.y. Thus, mylonitization apparently occurred in the latest Cretaceous, Paleocene, or earliest Eocene[2] (Bickford *et al.*, 1985).

The fact that mylonitic rocks occur both within the Priest River complex and along the Newport fault has caused some confusion. The Newport fault is unique from other detachment faults in the Northwest, in that it is not superimposed on an older, thick

---

[1] Also B. P. Rhodes, unpublished data.

[2] M. E. Bickford personal communications, 1985.

amphibolite facies mylonitic zone at the border of the complex; thus the high-grade mylonites and low-grade chloritic breccias and mylonites are not necessarily related. The kinematics and temporal relationship between these two shearing events can be distinguished here more easily. The Spokane dome mylonitic zone projects under the Newport fault with an intervening region of nonmylonitic rocks (Fig. 10-3). The mylonitic zone is not overlain by a regional zone of semiconcordant cataclasis and breccia. Furthermore, the Spokane dome mylonitic zone is cut by widely spaced, discrete, very thin (1–5 cm) mylonitic zones that clearly cut the main mylonitic foliation, dip steeply to the west, and show a normal sense of displacement. These discrete zones probably represent younger deformation related to movement on the Newport fault.

The Newport fault is clearly extensional, but its tectonic mechanism, kinematics, and crustal geometry are still not clear. Harms and Price (1983) proposed that the Newport fault is representative of crustal megaboudinage, a concept developed by G. H. Davis and Coney (1979) for the metamorphic core complexes of southern Arizona. More recently, Wernicke (1981, 1985) suggested that detachment faults in the Basin and Range might root into the middle crust as mylonitic zones. The Newport fault could fit the geometry of his model if its kinematics were uniformly top-to-the-east, and if it was continous across an antiform with a similar fault buried in the Purcell Trench. However, Harms (1982) presents data for top-to-the-west movement on the east limb of the Newport fault.

Within the interior of the Priest River complex, the peak regional metamorphic grade is comparatively uniform (Fig. 10-4). Much of the south-central part of the complex, northeast of Spokane and south of the Newport fault, is underlain by the semipelitic Hauser Lake Gneiss (Weis, 1968; Weissenborn and Weis, 1976). The predominant mineral assemblage within the Hauser Lake Gneiss is quartz + plagioclase (An_{25-32}) + K-feldspar + biotite + sillimanite. The abundance of K-feldspar and the absence of muscovite suggests that the metamorphism reached the sillimanite-orthoclase zone of the amphibolite facies. In this same area, geobarometry indicates that pressure reached 6–8 kb during the peak of regional metamorphism (Rhodes, 1986).

Locally within the Hauser Lake Gneiss, kyanite occurs in possible equilibrium with sillimanite. Kyanite also occurs in similar pelitic rocks to the north along the Pend Oreille River (Clark, 1973). The occurrences of kyanite lie along the axis of a structural dome (see below) and thus may represent slightly deeper levels of exposure. Southeast of Spokane and south of the Spokane River Valley (Fig. 10-2), the Hauser Lake gneiss is in apparent fault contact with unnamed sillimanite- and muscovite-bearing schist and gneiss[3] (Weis, 1968); this region may represent a down-dropped block of a slightly higher level of the complex.

Northwest of Bonner's Ferry, a narrow finger of the Priest River complex is sandwiched between the northeastern limb on the Newport fault and the Purcell Trench (Fig. 10-4). The metamorphism south of the Canadian border is largely unstudied, but Miller (1983) reports variable metamorphism within the Prichard Formation with sillimanite-bearing schist and garnet-bearing amphibolite locally developed. In southern British Columbia, this finger continues to the north for more than 100 km (Archibald et al., 1983; Hoy, 1977), with the highest grade being in the sillimanite-orthoclase zone. North of the border, movement along the Purcell Trench and Newport fault is negligible and the complex has gradational boundaries. This finger of sillimanite-grade rocks appear to be largely confined to the topographically low valley surrounding Kootenay Lake. Perhaps this

---

[3] Also B. P. Rhodes, unpublished data.

REGIONAL METAMORPHISM, STRUCTURE, AND TECTONICS

narrow finger represents the combined effects of topography and a gentle, crustal scale antiform where erosion has exposed a narrow window into higher-grade rocks.

One of the major unsolved problems in the Priest River complex is the age of the protoliths of the high-grade metasedimentary rocks. Many workers (e.g., Archibald *et al.*, 1983; Griggs, 1973; Miller, 1947a, b, c, d) suggest that at least some of these high-grade rocks represent the metamorphosed equivalent of the Prichard Formation. This is certainly true along the western margin of the complex, where low-grade Prichard rocks grade into sillimanite-bearing schist. On the other hand, Weissenborn and Weis (1976) and Weis (1968) mapped the Hauser Lake Gneiss in the southern part of the complex as pre-Belt basement. Likewise, Harrison *et al.*, (1972) mapped similar high-grade rocks at the north end of Pend Oreille Lake as pre-Belt basement. Rb-Sr data from the Hauser Lake Gneiss is compatible with a pre-Belt origin (R. L. Armstrong, personal communication, 1985). Preliminary U-Pb data (Bickford *et al.*, 1985) from zircons in the granitic rocks that cut the Hauser Lake Gneiss suggest that these zircons contain a xenocrystic component that is approximately 1800 m.y. old; thus, Precambrian zircons, older than the Prichard, must occur nearby. It is possible that the as-yet-undiscovered unconformity between the Prichard Formation and its basement lies somewhere within the Priest River complex.

The age of many of the plutonic rocks within the Priest River complex is also largely unknown. The problem is compounded by the fact that nearly all the K-Ar data from plutonic rocks within the complex give ages that cluster around 50 m.y. and reflect the age of cooling of the rocks as they were uplifted and unroofed rather than the age of emplacement and crystallization. On the west side of the complex, granitic rocks of the Loon Lake batholith crop out continuously across the projected boundary of the Priest River complex. Within the complex, the K-Ar ages of biotite and hornblende are reset to approximately 50 m.y. Just to the west of the complex, K-Ar ages cluster around 170 m.y. and around 93–101 m.y. (Miller and Clark, 1976). The 50-m.y. ages presumably represent the time of cooling of deeper portions of the batholith as it rose.

The age of regional metamorphism within the Priest River complex is poorly constrained. Within the U.S. portion, it is bracketed as being younger than the Belt Supergroup and older than the postmetamorphic 50-m.y.-old Silver Point pluton (Miller and Engels, 1975). The best data come from the northeastern extension of the complex into southeastern British Columbia. Hoy (1977) describes metamorphosed rocks of the lower Paleozoic Lardeau and Proterozoic Windermere Groups in that area. Near the international border, Arichbald *et al.* (1983), describe plutons that are ringed by concentric, regional metamorphic zones and are thus presumed to be synmetamorphic. These plutons yield nearly concordant K-Ar, Ar-Ar, and U-Pb data that suggest an emplacement age in the range 156–171 m.y. (mid-Jurassic). This is supported by a similar post-Early Jurassic age of metamorphism in the Schuswap terrane of south-central British Columbia (Hyndman, 1968).

A second, less widespread metamorphism is preserved in, and occurred during, deformation within the mylonitic zone of the southern Priest River complex. This second metamorphism included the reactions of kyanite and sillimanite to form andalusite, orthoclase + sillimanite to form muscovite, and a retrograde garnet-biotite Mg-Fe exchange. Mineral assemblages and geothermobarometry suggest that this second metamorphism occurred at lower pressures (<3.5 kb) and lower temperatures (<600°C) than the peak metamorphism (Rhodes, 1986). The age of this second metamorphism and of mylonitization is probably Late Cretaceous as discussed below.

Figure 10-5 summarizes the metamorphic history of the southern part of the Priest River complex. The peak of metamorphism in the upper amphibolite facies is most likely

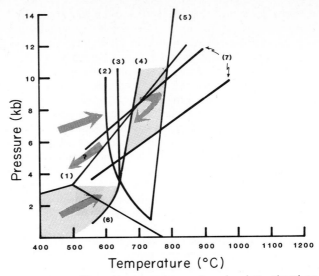

FIG. 10-5. Plot of pressure-temperature brackets placed on the high-grade and retrograde metamorphism in the Priest River complex by the mineral assemblages. See the text for a discussion. (1) Aluminosilicate triple point from Holdaway (1971); (2) quartz + Na-plagioclase + K-feldspar + water → liquid; (3) muscovite + quartz + Na-plagioclase + water → liquid + aluminosilicate; (4) muscovite + Na-plagioclase + quartz → K-feldspar + sillimanite + liquid; (5) biotite + quartz + Na-plagioclase → orthopyroxene + K-feldspar + liquid; (6) quartz + muscovite → aluminosilicate + K-feldspar + water. (2)–(6) from Clemens and Wall (1981). Curves (1)–(5) drawn for Fe/(Fe + Mg) for biotite = 0.4 and $P_{water} = P_{total}$. (7) Two curves that bound the cluster of curves derived from the equilibrium anorthite = grossular + sillimanite + quartz. Shaded areas define the conditions of the two metamorphisms discussed in the text. Thick arrows represent one possible P-T path.

Jurassic or Early Cretaceous. A Late Cretaceous (?) middle amphibolite facies metamorphism is localized within the mylonitic rocks. This event probably affected a much greater area, but the lower-grade assemblages grew only where intense shearing facilitated reaction. The exact path between the two metamorphic events is now known, but a large degree of uplift and denudation is required. Metamorphism ended by 50 m.y.b.p., the age of the final cooling of the Priest River complex.

The Kootenay Arc

The. U.S. part of the Kootenay Arc contains rocks of the Belt Supergroup, overlain by miogeoclinal rocks of late Proterozoic (Windermere) and early Paleozoic age, which are intruded by the 100-m.y.-old Loon Lake batholith. On a regional scale, the Kootenay Arc is a homocline that dips gently to the west, with Paleozoic rocks exposed mostly in the western part of the arc (Fig. 10-2). At a smaller scale, the internal structure of the arc is characterized by folds and thrust faults. The largest and best known of the thrusts is the west-dipping, eastward-verging Jump Off Joe thrust (Fig. 10-2). On the hanging wall of the thrust, the Belt-correlative Deer Trail Group is thrust to the east over lower Paleozoic miogeoclinal rocks (Miller and Engels, 1975). The age of this thrust is not known but it appears to cut granitic rocks of the approximately 100-m.y.-old Loon Lake batholith. Folds in the surrounding region verge both to the east and to the west, a characteristic common to much of the U.S. section of the arc (Ellis, 1984). Farther north, in the Canadian portion of the arc, the Selkirk fan structure contains westward-verging Jurassic structures overprinted by younger, steeply eastward-verging folds and thrust faults (Brown, 1978; Brown and Tippett, 1978). In the southern portion of the arc, the sequence of events, by analogy, may be similar.

Although the Kootenay Arc contains many of the same rocks as the Rocky Mountain fold-and-thrust belt to the east, it differs from that province in that nearly all of the rocks show the effects of regional, dynamothermal metamorphism. If the high-grade rocks of the Priest River complex are excluded, the entire Kootenay Arc lies within the greenschist facies (Fig. 10-4). Except very locally, the original sedimentary character of the rocks is preserved. Ellis (1984) describes biotite retrograded to chlorite in Precambrian and Cambrian slates in northeastern Washington. Lithologic descriptions from other reports (e.g., Miller and Clark, 1976) are consistent with a peak grade in the biotite zone.

The age of metamorphism is only very broadly bracketed within the Kootenay Arc. Metamorphism probably predates intrusion of the approximately 100-m.y. Loon Lake batholith, and no metamorphosed rocks younger than Ordovician exist within the arc. The fact that in several places this low-grade metamorphism can be traced upgrade into the Priest River complex suggests that this lower-grade event is the same Jurassic or Late Cretaceous metamorphism.

Quesnellia

The suture between Quesnellia and the Kootenay Arc can be traced as a tectonic contact from northeastern Washington into southern British Columbia, where it swings to the east in a large reentrant. In the southern portion of the arc, the suture appears as a linear, probably high-angle fault of uncertain displacement (Campbell and Raup, 1964). To the north, near Colville, the suture has not been well mapped because phyllites of the Quesnellian(?) Covada Group are juxtaposed against similar phyllitic rocks of the lower Paleozoic miogeocline (Mills, 1985). However, the suture is apparently cut by the Kettle River fault on the eastern flank of the Kettle complex. Northwest of Colville, the suture reemerges as a northeast-trending complex of northwestward-verging thrust faults that carry plates of lower Paleozoic miogeoclinal rocks over plates containing eugeoclinal rocks (Yates, 1971); this style and trend continue into southern British Columbia (Little, 1982).

The age of suturing between Quesnellia and North America is not well dated either stratigraphically or radiometrically. Monger *et al.* (1982) summarize evidence that the suture is mid-Jurassic. In the U.S. portion of Quesnellia, the youngest, clearly eugeoclinal rocks are the greenstones, metagreywackes, and mafic meta-intrusions of the Early to Middle Jurassic Rossland Group. The petrology and geochemistry of the Rossland Group suggest that it formed in or near an ocean volcanic arc, perhaps in a back-arc environment (Beddoe-Stephens and Lambert, 1981); thus the suture must be younger. Along the southern portion of the suture, west of Chewelah, Washington (Fig. 10-2), granitic plutons from the western part of the Loon Lake batholith sharply cut the high-angle suture. The granites here are undated but are similar to 100-m.y.-old granites farther east; thus suturing was probably complete by at least the mid-Cretaceous.

The interal structure of the southern part of Quesnellia is poorly exposed. Faults bound many of the stratigraphic units. As discussed below, the continental character of the protoliths within the Kettle complex suggests that Quesnellia overrode the Precambrian cratonic basement but slid under the miogeoclinal cover during suturing, producing west-directed thrusts within the Kootenay Arc, but perhaps only locally deforming the eastern edge of the accreting block. Many of the compressional structures within Quesnellia, including eastward-verging thrust faults (Cheney *et al.*, 1982; McMillen, 1979), might be related to postaccretionary, eastward-verging shortening in the Kootenay Arc. The latest structures involve north-northeast-trending normal faults related to Eocene extension and volcanism throughout the region.

In the United States Quesnellia consists of a variety of rock types as old as Ordovician. How many distinct tectonostratigraphic terranes actually make up Quesnellia is unknown. Virtually all the rocks of Quesnellia are metamorphosed within the greenschist facies and apparently no major metamorphic discontinuities exist (excluding the Kettle and Lincoln complexes). Furthermore, no distinct change in metamorphic grade occurs across Quesnellia's suture with the Kootenay Arc. If all of Quesnellia was metamorphosed together, the metamorphism can be no older than Jurassic. However, many elements of Quesnellia may have inherited a preaccretionary metamorphism.

The Kettle Complex

The Kettle complex is a north-south-trending 120-km by 30-km belt of high-grade metamorphic rocks nearly surrounded by Quesnellia (Fig. 10-1 and 10-2). The >100-km-long Kettle River detachment fault bounds the Kettle complex on its eastern flank. This fault separates the high-grade Kettle complex from greenschist-grade rocks of Quesnellia and unmetamorphosed Eocene volcanic and sedimentary rocks. A gently eastward-dipping, sillimanite-grade, upward-intensifying, 2–3-km-thick mylonitic zone underlies and is cut at a low angle by the Kettle River fault. This zone extends along the entire eastern flank of the complex. At its southern end, the Kettle River fault disappears within the composite Daisy Hill granitic suite (Atwater *et al.*, 1984). The Daisy Hill granites are gradational with high-grade paragneiss to the north and cut greenschist facies rock of the Quesnellian Covada Group to the south. Apparently, the Kettle complex plunges to the south beneath an essentially intrusive contact with Quesnellia. The Kettle River fault is either cut by the Daisy Hill granites, or displacement along it fades to the south.

The west side of the Kettle complex is truncated by high-angle normal faults of the Republic graben. The graben contains voluminous Eocene volcanic and sedimentary rocks and is floored by eugeoclinal rocks of Quesnellia. The northern Kettle complex extends approximately 40 km into southern British Columbia, where it narrows and is eventually lost within Tertiary(?) granitic rocks.

The internal, pre-mylonite structure of the Kettle complex is mostly unknown. The west side of the complex is characterized by crystalloblastic fioliations and weak or absent lineations in the metasedimentary rocks, and very weak foliations and/or lineations in the orthogneisses. Toward the east, the younger mylonitic fabric gradually intensifies. The mylonitic foliation dips gently to the east, contains a strong east-west mylonitic lineation, and contains S and C fabrics that indicate a top-to-the-east or normal sense of shear (A. J. Watkinson, personal communication, 1986; B. P. Rhodes, unpublished data). Rocks on the west side of the complex are not mylonitic. The age of this mylonitization is not known but it must predate the Kettle River fault, which cuts Eocene volcanic rocks.

Hurich *et al.* (1985) ran a short seismic reflection profile over the Kettle River mylonitic zone. Their profile shows several gently eastward-dipping shallow reflections with a set of deeper low-angle reflections, that extend to a depth of approximately 20 km. The uppermost reflections correlate with the Kettle River fault and/or the upper part of the mylonitic zone and can be traced on their short profile to a depth of approximately 3 km. They suggest that the deeper reflections could represent layered rocks within large nappes at mid-crustal depths. An alternative explanation is that they represent deeper, subparallel mylonitic zones. The Kettle complex was also traversed as a part of a COCORP transect across northeastern Washington and northern Idaho and the existence of several deep reflections on their records implies that the crust in this area is subhorizontally layered, possibly by numerous mylonitic zones (Chris Potter, personal communication, 1985).

The age of mylonitization is not known, nor is it known whether the mylonites developed in a pre-Tertiary compressional regime (with later doming), or in the same Eocene extensional regime that formed the Kettle River fault. The eastward vergence of the mylonites can be explained by compression or extension; only a detailed knowledge of the age of mylonitization will provide a conclusive answer. The most likely answer may be that the Kettle mylonitic zone experienced two episodes of movement, the first along a thick, high-grade zone in the middle crust that resulted from Jurassic to early Tertiary compression, and a second episode during Eocene extension, with the same eastward vergence, that partially reactivated the upper portions of the zone and culminated in brecciation and movement along the Kettle River detachment fault.

The Kettle complex contains a diverse collection of metamorphic rocks, including, in approximate order of decreasing abundance, granitic orthogneiss, quartzite, pelitic paragneiss, and marble with minor amounts of calc-silicate gneiss and amphibolite. The most common assemblage in the pelitic rocks of the northeastern section of the complex, south of the Canadian border, is sillimanite + K-feldspar + biotite + plagioclase + quartz, indicating that metamorphism reached the upper amphibolite facies (Rhodes and Cheney, 1981). Much of the rest of the complex contains rocks of similar grade (Cheney, 1980; Pearson, 1977; Preto, 1970; Wilson, 1981). One exception is the western portion of the complex just south of the international boundary. Here, high-grade rocks are overlain by a concordant sheetlike body of biotite-garnet schist, which is in turn structurally overlain by biotite-zone phyllite (Pearson, 1977) which may be a part of Quesnellia (Orr, 1985). This arrangement is suggestive of a metamorphic gradation, at least locally, between the Kettle complex and Quesnellia. Slightly lower-grade rocks may also exist on the eastern flank of the complex near the confluence of the Columbia and Kettle rivers. Here, muscovitic quartzite and greenish amphibolite may represent a small area of lower-amphibolite facies metamorphism.

Retrograde metamorphism is confined to a narrow belt adjacent to and just below the Kettle River fault. Within this zone, strong chloritization affected the granitic rocks and accompanied brecciation and cataclasis. However, locally, sillimanite-bearing mylonitic quartzite immediately adjacent to the fault is unaffected by the retrograde metamorphism.

The protoliths of the metasedimentary rocks of the Kettle complex were a sequence of interbedded shales, limestone, and clean, slightly feldspathic quartzose sandstones. No similar rocks occur within Quesnellia. These rocks may represent part of the metamorphosed Paleozoic miogeoclinal section, including the basal Cambrian Gypsy Quartzite and overlying pelitic rocks (Cheney, 1980). Another possiblity is that the Kettle complex contains remobilized Precambrian crystalline rocks of the craton. Precambrian ages have been obtained from the Shuswap complex to the north, but this hypothesis has not been tested in the Kettle complex. A third possibility, also untested, is that the Kettle complex is the crystalline basement of Quesnellia and not a part of North America. However, it seems most likely to us that the Kettle complex is a part of North America and Quesnellia overrode the craton. This is supported by structural relationships between the Okanogan complex and Quesnellia farther to the west in north-central Washington (Goodge and Hansen, 1983; Hansen and Goodge, Chapter 9, this volume; Orr, 1985).

The granitic gneisses in the Kettle complex are clearly metaplutonic, but their age is not known. Away from the mylontic zone on the east side of the complex, these gneisses are only very weakly foliated or massive; they appear to be unaffected by the Jurassic(?) sillimanite-grade dynamothermal event that affected the nearby metasedimentary rocks, and thus may be Cretaceous.

The age of metamorphism within the Kettle complex is not known. Rhodes and Cheney (1981) outline the circumstantial evidence for a Cretaceous or older age. In the Valhalla complex of southern British Columbia, Hyndman (1968) recognized a Jurassic metamorphism, and by extrapolation the metamorphism in the Kettle complex may also be Jurassic. Metamorphism ended by the Eocene, except for local greenschist-facies retrograde alteration adjacent to the Kettle River fault.

The Lincoln complex Approximately 30 km south of the Kettle complex, lineated orthogneiss occurs within granitic rocks in the Lincoln complex (Atwater *et al.*, 1984). The east side of the Lincoln complex is bounded by mylonitic rocks underlying a low-angle fault, and the west side is truncated by a high-angle normal fault of the Keller graben. This Lincoln complex probably represents the southern extension of the Kettle complex, and the low-angle fault on its east side may be the southern extension of the Kettle River fault.

STRUCTURAL, METAMORPHIC, AND PLUTONIC EVOLUTION

Table 10-1 summarizes the major events that occurred within Quesnellia, the Kootenay Arc, and the Rocky Mountains between the Middle Jurassic and Eocene. The age constraints on many of these events are circumstantial at best and as additional geochronology becomes available, this table will evolve. Before these events can be put into a tectonic framework, several important questions must be addressed: (1) What events led to the present surface distribution of high-grade metamorphic rocks? (2) Did the accretion of Quesnellia cause regional high-grade metamorphism? (3) Did accretion cause eastward thrust-faulting and/or mid-crustal ductile shearing in the Kootenay Arc and Rocky Mountains? and (4) Is there any relationship between the history of convergence rates of the North American plate with the Farallon plate and the geologic history of the northwestern Cordillera?

The present distribution of high-grade metamorphic rocks in the Kettle, Lincoln, and Priest River complexes is due largely to Tertiary extension. Wherever a clear relationship is present, the structures on the borders of the complexes are Eocene or younger. Furthermore, K-Ar ages of biotite within the complexes commonly yield Eocene cooling ages (Archibald *et al.*, 1984; Miller and Engels, 1975); thus, the complexes were uplifted and exhumed at that time. High-grade rocks probably continue beneath the structurally low Kootenay Arc and Republic graben, and are exposed only in the structural highs represented by the metamorphic complexes. This idea is supported by regional gravity modeling, which indicates the presence of high-density rocks under the Kootenay Arc (Cady and Fox, 1984).

The apparent Jurassic age of both the accretion of Quesnellia and the high-grade regional metamorphism in the Priest River and Kettle/Lincoln complexes suggests a correlation between the two events. Monger *et al.* (1982) propose such a correlation in the Shuswap complex of south-central British Columbia. They argue that the relatively narrow axis of the Schuswap metamorphic belt nearly coincides with the suture between North America and a large composite terrane that included Quesnellia, and that both accretion and metamorphism are mid-Jurassic. In the United States the suture lies within low-grade rocks of the Kootenay Arc; metamorphic rocks in the Priest River complex occur 50 km east of the suture (Fig. 10-2); in the Okanogan complex (Fig. 10-1) high-grade metamorphic rocks of possibly the same metamorphic age occur far to the west of the suture. If accretion caused metamorphism in the United States, it did so over a broad area and would seem to require the obduction of Quesnellia onto the broad western edge

TABLE 10-1. Jurassic-Eocene Structural, Metamorphic, and Plutonic Evolution of Northeastern Washington, Northern Idaho, and Northwestern Montana

Age	Structural Events	Metamorphic/Plutonic Events
Middle(?) Jurassic	Accretion and obduction(?) of Quesnellia onto the North American craton Crustal thickening Westward verging folds and thrusts in the Kootenay Arc Formation of crystalloblastic foliations in middle crust Possible mylonitization	High-grade to low-grade regional metamorphism Possible local partial melting and intrusion of early granitic plutons
Middle Jurassic to Late Cretaceous	Possible isostatic uplift of thickened crust Beginning of eastward thrusting in the Kootenay Arc(?) and Rocky Mountains with mylonitic roots in middle crust	Metamorphism continues beyond peak Abundant granitic intrusion in Quesnellia and Kootenay Arc (Loon Lake/Kaniksu batholiths), possibly in an Andean-type arc
Late Cretaceous to earliest Eocene	Eastward thrusting and thick, high-grade mylonitization peaks	In Priest River complex, metamorphism in middle amphibolite grade accompanies mylonitization In Kettle complex regional metamorphism continues in sillimanite zone Continued synkinematic granitic intrusions
Middle Eocene	Thrusting ends, extension throughout region High-angle normal faulting and down-dropping of the Republic and Keller Grabens Low-angle normal detachment faulting involving widespread cataclasis and possibly rooting as mylonitic zones in the middle crust Uplift and erosion of metamorphic complexes	Cooling of the metamorphic complexes and final setting of K-Ar ages Widespread volcanism and related granitic plutonism in wide magmatic arc Retrograde metamorphism along detachment faults

of North America. Such a low-angle accretion is supported by the existence of high-grade quartzites and marbles of a continental character as far west as the northeastern margin of the Okanogan complex (Orr, 1985), within the Tenas Marys Creek sequence of Parker and Calkins (1964). If this sequence is truly part of North America, then Quesnellia moved a minimum of 75 km over the margin of North America. As Quesnellia obducted, perhaps it slid beneath or bulldozed against the miogeoclinal sediments that were draped over the margin as suggested by Archibald *et al.* (1983), thus explaining the local development of west-virging folds and thrust faults along the suture. The deformation and thickening of the Kootenay Arc in front of the suture may have in part caused metamorphism in the Priest River complex, well to the east of the edge of Quesnellia.

It is kinematically consistent to suggest that the accretion of Quesnellia caused eastward thrust faulting in the Rocky Mountains. However, the available data suggest that most of the movement in the Rocky Mountains occurred well after the suturing of

Quesnellia and North America. Clearly, if Quesnellia did obduct onto North America, it must have done so along one or more large, probably mylonitic, shears. One possible candidate for such a shear is the mylonitic Columbia River fault zone of south-central British Columbia, which is intruded by a 157-m.y. (Late Jurassic) pluton (Brown and Murphy, 1982; Read and Brown, 1981). Another possibility is the mylonitic zone in the Kettle complex; the age of this zone is not known, but it is pre-Eocene and perhaps a part of its history is related to Jurassic accretion. The abundance of low-angle seismic reflectors beneath the Kootenay Arc and Priest River complex perhaps represent additional shear zones, some of which may be related to accretion.

Recent attention has focused on the correlation between plate motions and continental tectonics (e.g., Beck, 1984). The Jurassic to Eocene history of most of the United States corresponds to a time of convergence between the Farallon and the North American plates. Changes in the convergence rate, age of subducted lithosphere, and angle of subduction are used elsewhere in the Cordillera to explain such phenomena as Laramide thick-skinned thrust faulting in the foreland and the distribution and chemistry of magmatic arcs throughout the Cordillera.

How does the convergence history match up with the history of the northwestern U.S. Cordillera outlined in Table 10-1? Data from seafloor magnetic anamolies for the convergence of the Farallon plate with North America go back to 180 m.y. But for the northwestern United States, a major problem arises because of the uncertain location of a ridge separating the Kula and Farallon plates (Page and Engebretson, 1984). It is likely that for much of the earlier history of the northwest, this part of North America was interacting with the Kula plate and data for the Kula plate are relatively scarce.

Figure 10-6 is adapted from Page and Engebretson (1984) and shows the normal component of convergence between North America and both the Kula and Farallon plates. Superimposed are the approximate age ranges of major events of the northwestern Cordillera. The most striking feature of the convergence profile is the large peak in convergence, rate of the Farallon plate and to a lesser degree the Kula plate between 75 and 40 m.y. During this interval of increased convergence, at approximately 50 m.y., the deformational regime in the northwest apparently changed from compression to extension. This transition coincides with an abrupt increase in convergence rate with the Kula plate but does not correlate with any distinct changes in convergence with the Farallon

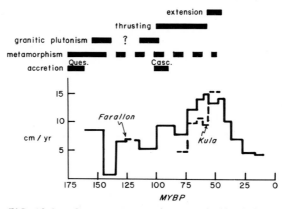

FIG. 10-6.   Convergence rates between the North American and Kula and Farallon plates and major tectonic events of the northwestern United States, after Page and Engebretson (1984).

plate. Thrust faulting appears to occur during a period of increasing convergence, although the age of the onset of thrusting is poorly constrained. The accretion of Cascadia had little apparent affect on convergence, although it could be suggested that the drop in convergence rate between 145 and 135 m.y. was caused by reorganization following the accretion of Quesnellia. Any other correlations are obscure. Apparently, accretionary events played a more significant role than convergence rate in controlling tectonic events in the northwestern United States.

TECTONIC HISTORY

The following plate tectonic history is an attempt to unify the known metamorphic, structural, and magmatic events into an internally consistent sequence. Figure 10-7 depicts this history diagrammatically. These diagrams represent an approximate section line located along a latitude of 48°, 30'N. An important factor not considered in this history is any strain and/or displacement parallel to the continental margin. Paleomagnetic evidence from the western portion of Quesnellia in south-central British Columbia indicates approximately 2400 km of northerly transport between the mid-Cretaceous (90 m.y.) and mid-Tertiary (Irving et al., 1985). However, we know of no evidence for

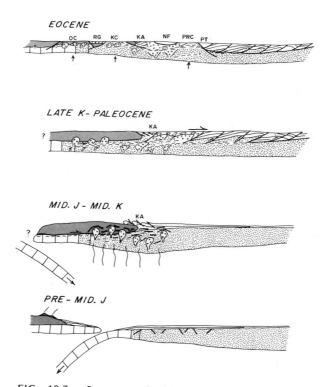

FIG. 10-7. Summary of plate tectonic events for the northwestern United States from mid-Jurassic to Eocene. Shaded area, Quesnellia; random dashing, cratonic basement; plus signs, granitic rocks; v's, Eocene volcanic rocks; vertical lines, oceanic crust; wavey lines, regional metamorphism. See the text for a discussion.

large-scale dextral shear along the suture between Quesnellia and the Kootenay Arc, and this displacement may have occurred along faults within western Quesnellia.

Pre-Middle Jurassic

In the northwestern United States, prior to the Middle Jurassic, cratonic North America was essentially a passive margin. An eastward-tapering wedge of miogeoclinal sediments was draped over the rifted and thinned western edge of the craton. The transition from old craton to oceanic crust was probably gradational and located west of the present-day Republic graben. No evidence exists for magmatic activity within cratonic North America prior to the Middle Jurassic, thus convergence with the Farallon (or Kula) plate was probably accommodated by west-dipping subduction, with old ocean crust subducting beneath Quesnellia. Such a geometry is supported by the existence of Early and Middle Jurassic, arc-related basaltic and basaltic-andesitic volcanic rocks within Quesnellia (e.g., Rossland Group and lower Nicola Group), although these could be related to eastward-dipping subduction farther to the west. The lack of andesites and more felsic rocks and the eugeoclinal nature of the sedimentary rocks within Quesnellia suggests that Quesnellia was part of an oceanic plate.

Middle Jurassic to Mid-Cretaceous

In southern British Columbia and the northwestern United States, Middle Jurassic plutons suture Quesnellia to North America by approximately 160 m.y.b.p. The existence of cratonic rocks in the Kettle and northeastern Okanogan complexes is evidence that Quesnellia obducted onto the thinned cratonic margin, overriding the transition between oceanic and continental crust. The existence of westward-verging structures in the Kootenay Arc suggests that the front of Quesnellia wedged between the craton and overlying sediments. The obduction of Quesnellia was probably accompanied by ductile shearing in the underlying crust.

Following accretion, the polarity of subduction switched to an eastward dip and moved to the western margin of Quesnellia. Plutonism during this interval apparently occurred during two distinct intervals (Armstrong, 1985). The first, >155 m.y., is much less widespread in the northwestern United States and was perhaps caused in part by thickening and heating related to obduction of Quesnellia, and in part by the development of a new magmatic arc. An inexplicable lull in magmatic activity occurred between 125 and 135 m.y. (Armstrong, 1985). However, by 100 m.y., a voluminous magmatic arc was established in a broad region across both the Kootenay Arc and Quesnellia.

Metamorphism of rocks at depth, beginning in the Middle Jurassic, was probably continuous from the eastern Kootenay Arc (i.e., the Priest River complex) well into Quesnellia (i.e., the Okanogan complex). This metamorphism required deep burial and thickening of the continental crust, perhaps caused by the obduction of Quesnellia and the thickening of a bulldozed wedge immediately to the east. Water released by subducted sediments may have induced the partial melting of the overlying mantle, forming mafic and/or intermediate magma, which carried heat upward into the continental crust, causing the *in situ* formation of granitic magma and simultaneous regional metamorphism. Metamorphism probably continued throughout this interval, with a peak during the Middle Jurassic to Early Cretaceous. Following the peak of metamorphism, uplift began with slow cooling.

REGIONAL METAMORPHISM, STRUCTURE, AND TECTONICS

Late Cretaceous to Paleocene

By the Late Cretaceous, magmatism ceased across the entire region, and shortening of the Kootenay Arc and Rocky Mountains resulted in eastward thrust faulting, apparently rooted in the middle crust as ductile shears. Lower-pressure conditions, in the middle amphibolite facies, existed during mylonitization in the Priest River complex. Thus, shortening in the Rocky Mountains occurred at the same time as uplift and eastward shearing in the Priest River complex. Perhaps isostatic uplift in the Priest River complex was caused by unloading of this thickened part of the crust as the overlying upper crust gravitationally spread toward the east (Price, 1973).

Eocene

By the late Paleocene or early Eocene, shortening gave way to extension, resulting in thinning of the crust. Extension was accommodated by low-angle detachment faulting, high-angle faulting, and local granitic plutonism and andesitic volcanism, perhaps caused by pressure-release melting. The end result was widespread tectonic and erosional denudation and exposure of midcrustal rocks in the Priest River, Kettle, and Lincoln complexes. The role of extensional, midcrustal ductile shears is uncertain. Extensional shearing may have locally reactivated earlier compressional mylonitic zones (e.g., the Kettle complex), but in the Priest River complex, the two events appear to be distinct. By 50 m.y.b.p., metamorphism had ceased and the rocks of the metamorphic complexes had cooled and set the K-Ar ages of biotite.

ACKNOWLEDGMENTS

Our research was partially supported by National Science Foundation grant EAR83-07690 to D. W. Hyndman. Acknowledgment is made to the Donors of The Petroleum Research Fund, administered by the American Chemical Society, for partial support of this research to B. P. Rhodes. We thank Gary Ernst and John Goodge for their thorough reviews of our manuscript.

REFERENCES

Aadland, R. K., and Bennett, E. H., 1979, *Geologic map of the Sandpoint quadrangle, Idaho and Washington*, scale 1:250,000: Idaho Bur. Mines Geology.

Archibald, D. A., Glover, J. K., Price, R. A., Farrar, E., and Carmichael, D. M., 1983, Geochronology and tectonic implications of magmatism and metamorphism, southern Kootenay Arc and neighboring regions, southeastern British Columbia — Pt. I. Jurassic to mid-Cretaceous: *Can. J. Earth Sci.*, v. 21, p. 567–583.

——, Krogh, T. E., Armstrong, R. L., and Farrar, E., 1984, Geochronology and tectonic implications of magmatism and metamorphism, southern Kootenay Arc and neighboring regions, southeastern British Columbia — Pt. II. Mid-Cretaceous to Eocene: *Can. J. Earth Sci.*, v. 21, p. 567–583.

Armstrong, R. L., 1985, Mesozoic — Early Cenezoic plutonism in the Canadian cordillera — Distribution in time and space: *Geol. Soc. America Abstr. with Programs*, v. 17, no. 6, p. 338.

Atwater, B. F., Rinehart, C. D., and Fleck, R. J., 1984, Preliminary geologic map of the Colville Indian Reservation, Ferry and Okanogan Counties, Washington: *U.S. Geol. Survey Open File Rpt. 84-389.*

Beck, M. E., 1984, Introduction to the special issue on correlations between plate motions and Cordilleran tectonics: *Tectonics*, v. 3, p. 103–106.

Becraft, G. E., and Weis, P. L., 1963, Geology and mineral deposits of the Turtle Lake quadrangle, Washington: *U.S. Geol. Survey Bull. 1131*, 73 p.

Beddoe-Stevens, B., and Lambert, R. St. J., 1981, Geochemical, mineralogical and isotopic data relating to the origin and tectonic setting of the Rossland volcanic rocks, southern British Columbia: *Can. J. Earth Sci.*, v. 18, p. 858–868.

Bickford, M. E., Rhodes, B. P., and Hyndman, D. W., 1985, Age of mylonitization in the southern Priest River Complex, northern Idaho and northeastern Washington: *Geol. Soc. America Abstr. with Programs*, v. 17, no. 6, p. 341.

Brown, R. L., 1978, Structural evolution of the southeast Canadian Cordillera — A new hypothesis: *Tectonophysics*, v. 48, p. 133–151.

——, and Murphy, D. C., 1982, Kinematic interpretation of mylonitic rocks in part of the Columbia River fault zone, Shuswap terrane, British Columbia: *Can. J. Earth Sci.*, v. 19, p. 456–466.

——, and Tippett, C. R., 1978, The Selkirk fan structure of the southeast Canadian Cordillera: *Geol. Soc. America Bull.*, v. 89, p. 548–558.

Cady, J. W., and Fox, K. F., 1984, Geophysical interpretation of the gneiss terrane of northern Washington and Southern British Columbia, and its implications for uranium exploration: *U.S. Geol. Survey Prof. Paper 1260*, 29 p.

Campbell, A. B., and Raup, O. B., 1964, Preliminary geologic map of the Hunters quadrangle, Stevens and Ferry Counties, Washington: *U.S. Geol. Survey Misc. Field Studies Map MF-276.*

Cheney, E. S., 1980, The Kettle dome and related structures in northeastern Washington, *in* Crittenden, M. D., Jr., Coney, P. J., Davis, G. H., eds., *Cordilleran Metamorphic Core Complexes:* Geol. Soc. America Mem. 153, p. 463–484.

——, Rhodes, B. P., Wilson, J. R., and Mcmillen, D. D., 1982, Metamorphic core complexes and low-angle faults of northeastern Washington, *in* Roberts, S., and Fountain, D., eds., *Tobacco Root Geol. Soc. 1980 Field Conf. Guidebook:* Univ. Montana, Missoula, Mont., p. 45–54.

Clark, S. H. B., 1973, Interpretation of a high-grade Precambrian terrane in northern Idaho: *Geol. Soc. America Bull.*, v. 84, p. 1999–2003.

Clemens, J. D., and Wall, V. J., 1981, Origin and crystallization of some peraluminous (S-type) granitic magmas: *Can. Mineralogist*, v. 19, p. 111–131.

Coney, P. J., 1980, Cordilleran metamorphic core complexes — An overview, *in* Crittenden, M. D., Jr., Coney, P. J., and Davis, G. H., eds., *Cordilleran Metamorphic Core Complexes:* Geol. Soc. America Mem. 153, p.

——, Jones, D. L., and Monger, J. W. H., 1980. Cordilleran suspect terranes: *Nature*, v. 288, p. 329–333.

Cowan, D. S., Potter, C. J., Brandon, M. T., Fountain, D. M., Hyndman, D. W., Johnson, S. Y., Lewis, B. T. R., McClain, K. J., and Swanson, D. A., 1986, Continent-Ocean Transe B3 — Transect #9/Be, Juan de Fuca spreading ridge to Montana thrust belt: *Geol. Soc. America Map.*

Davis, G. A., Monger, J. W. H., and Burchfiel, B. C., 1978, Mesozoic construction of the Cordilleran "collage," central British Columbia to central California, *in* Howell, D. G., and McDougall, K. A., eds., *Mesozoic Paleogeography of the Western United States:* Pacific Section, Soc. Econ. Paleontologists Mineralogists, Pacific Coast Paleogeography Symp. 2, p. 1–32.

Davis, G. H., and Coney, P. J., 1979, Geologic development of the Cordilleran metamorphic core complexes: *Geology*, v. 7, p. 120–124.

Ellis, M. A., 1984, Structural morphology and associated strain within parts of the U.S. section of the Kootenay arc, N.E. Washington: Ph.D. dissertation, Washington State Univ., Pullman, Wash., 198 p.

Goodge, J. W., and Hansen, V. L., 1983, Petrology and structure of rocks in the southwest portion of Okanogan dome, north-central Washington: *Northwest Geology*, v. 12, p. 13–24.

Griggs, A. B., 1973, Geological map of the Spokane quadrangle, Washington, Idaho, and Montana: *U.S. Geol. Survey Misc. Geol. Inv. Map I-768*, scale 1:250,000.

Harms, T. A., 1982, The Newport fault — Low-angle normal faulting and Eocene extension, northeast Washington and northwest Idaho: M.S. thesis, Queens Univ., Kingston, Ont., 157 p.

——, and Price, R. A., 1983, The Newport fault, Eocene crustal stretching, necking and listric normal faulting in NE Washington and NW Idaho: *Geol. Soc. America Abstr. with Programs*, v. 15, p. 309.

Harrison, J. E., 1972, Precambrian Belt basin of the northwestern United States — Its geometry, sedimentation, and copper occurrences: *Geol. Soc. America Bull.*, v. 83, p. 1215–1240.

——, Kleinkopf, M. D., and Obradovich, J. D., 1972, Tectonic events at the intersection of the Hope fault and Purcell Trench, northern Idaho: *U.S. Geol. Survey Prof. Paper 719*, 24 p.

——, Griggs, A. B., and Wells, J. D., 1974, Tectonic features of the Precambrian Belt basin and their influence on post-Belt structures: *U.S. Geol. Survey Prof. Paper 866*, 15 p.

——, Kleinkopf, M. D., and Wells, J. D., 1980, Phanerozoic thrusting in Proterozoic belt rocks, northwestern United States: *Geology*, v. 8, p. 407–411.

——, Griggs, A. B., and Wells, J. D., 1981, Generalized geologic map of the Wallace 1° × 2° quadrangle, Montana and Idaho: *U.S. Geol. Survey Misc. Field Studies Map MF-1354-A*, scale 1:250,000.

Holdaway, M. J., 1971, Stability of andalusite and the aluminum silicate phase diagram: *Amer. J. Sci.*, v. 271, p. 97–131.

Hoy, T., 1977, Stratigraphy and structure of the Kootenay arc in the Riondel area, southeastern British Columbia: *Can. J. Earth Sci.*, v. 14, p. 2301–2315.

Huntting, M. T., Bennett, W. A. G., Livingstone, V. E., and Moen, W. S., 1961, *Geologic map of Washington*, scale 1:500,000: Wash. Div. Mines Geology.

Hurich, C. A., Smithson, S. B., Fountain, D. M., and Humphreys, M. C., 1985, Seismic evidence of mylonite reflectivity and deep structure in the Kettle dome metamorphic core complex, Washington: *Geology*, v. 13, p. 577–580.

Hyndman, D. W., 1968, Mid-Mesozoic multiphase folding along the border of the Shuswap Metamorphic Complex: *Geol. Soc. America Bull.*, v. 79, p. 575–588.

Irving, E., Woodsworth, G. J., Wynne, P. J., and Morrison, A., 1985, Paleomagnetic evidence for displacement from the south of the Coast Plutonic Complex, British Columbia: *Can. J. Earth Sci.*, v. 22, p. 584–598.

Leech, G. B., 1960, Fernie west half, British Columbia: *Geol. Survey Canada Map 11-1960*.

Little, H. W., 1957, Kettle River (east half), British Columbia: *Geol. Survey Canada Map 6-1957*, scale 1:253,440.

——, 1960, Nelson map-area, west half, British Columbia: *Can. Geol. Survey Mem. 308*, 205 p.

_____,1982, Geology of the Rossland-Trail map-area, British Columbia: *Geol. Survey Canada Paper 79-26*, 38 p.

McMillen, D. D., 1979, The structure and economic geology of Buckhorn Mountain, Okanogan County, Washington: M.S. thesis, Univ. Washington, Seattle, Wash., 68 p.

Miller, F. K., 1971, The Newport fault and associated mylonites, northeastern Washington: *U.S. Geol. Survey Prof. Paper 750-D*, p. D77–D79.

_____,1974a, Preliminary geological map of the Newport number 1 quadrangle: *Wash. Div. Geology Earth Resources Map GM-10*, scale 1:62,500.

_____,1974b, Preliminary geologic map of the Newport number 2 quadrangle: *Wash. Div. Geology Earth Resources Map GM-8*, scale 1:62,500.

_____,1974c, Preliminary geological map of the Newport number 3 quadrangle: *Wash. Div. Geology Earth Resources Map GM-10*, scale 1:62,500.

_____,1974d, Preliminary geological map of the Newport number 4 quadrangle: *Wash. Div. Geology Earth Resources Map GM-10*, scale 1:62,500.

_____,1982a, Geologic map of the Salmo-Priest wilderness study area (RARE E6-981 Al-981), Pend Oreille County, Washington, and Boundary County, Idaho: *U.S. Geol. Survey Misc. Field Studies Map MF-1192-A*, scale 1:48,000.

_____,1982b, Preliminary geologic map of the Continental Mountain area, Idaho: *U.S. Geol. Survey Open File Rpt. 82-1062.*

_____,1983, Geologic map of the Selkirk Roadless area, Boundary County, Idaho: *U.S. Geol. Survey Misc. Field Studies Map MF-1447-A*, scale 1:48,000.

_____,and Clark, L. D., 1976, Geology of the Chewelah-Loon Lake area, Stevens and Spokane Counties, Washington: *U.S. Geol. Survey Prof. Paper 806*, 76 p.

_____,and Engels, J. C., 1975, Distribution and trends of discordant ages of the plutonic rocks of northeastern Washington and northern Idaho: *Geol. Soc. America Bull.*, v. 86, p. 517–528.

Mills, J. W., 1985, Geologic maps of the Marcus and Kettle Falls quadrangles, Stevens and Ferry Counties, Washington: *Wash. Dept. Nat. Resources Map GM-32*, scale 1:24,000.

Monger, J. W. H., and Hutchinson, W. W., 1971, Metamorphic map of the Canadian Cordillera: *Geol. Survey Canada Map 1322A*, scale 1:5,000,000.

_____,Price, R. A., and Tempelman-Kluit, D. J., 1982, Tectonic accretion and the origin of the two major metamorphic and plutonic welts in the Canadian Cordillera: *Geology*, v. 10, p. 70–75.

Norwick, S. A., 1972, Burial metamorphism of the Proterozoic Prichard Formation in northern Idaho and northwestern Montana: Ph.D. dissertation, Univ. Montana, Missoula, Mont.

Orr, K. E., 1985, Structural features along the margin of Okanogan dome, Tenas Mary Creek area, northeastern NE Washington: *Geol. Soc. America Abstr. with Programs*, v. 17, no. 6, p. 398.

Page, B. M., and Engebretson, D. C., 1984, Correlation between the geologic record and computed plate motions for central California: *Tectonics*, v. 3, p. 133–156.

Parrish, R., Carr, S. D., and Parkinson, D., 1985, Metamorphic complexes and extentional tectonics, southern Shuswap complex, southeastern British Columbia: *Cordilleran Section, Geol. Soc. America, Field Trip Guidebook, Trip 12*, 15 p.

Parker, R. L., and Calkins, J. A., 1964, Geology of the Curlew quadrangle, Ferry County, Washington: *U.S. Geol. Survey Bull. 1169*, 95 p.

Pearson, R. C., 1977, Preliminary geologic map of the Togo Mountain quadrangle, Ferry County, Washington: *U.S. Geol. Survey Open File Rpt. 77-371*, scale 1:62,500.

Preto, V. A., 1970, Structure and petrology of the Grand Forks Group, British Columbia: *Geol. Survey Canada Paper 69-22*, 80 p.

Price, R. A., 1973, Large scale gravitational flow of supracrustal rocks, southern Canadian Rockies, *in* De Jong, K. A., and Scholten, R., eds., *Gravity and Tectonics:* New York, Wiley, p. 491–502.

——, 1981, The Cordilleran foreland thrust and fold belt in the southern Canadian Rocky Mountains, *in* Price, N. J., and MacClay, K., eds., *Thrust and Nappe Tectonics:* Geol. Soc. London Spec. Publ. 9, p. 427–448.

——, Monger, J. W. H., and Muller, J. E., 1981, Cordilleran cross-section, Calgary to Victoria, *in* Thompson, R. I., and Cook, D. G., eds., *Field Guides to the Geology and Mineral Deposits:* Geol. Assoc. Canada Guidebook, 1981 Meet., Galgary, p. 261–334.

Read, P. B., and Brown, R. L., 1981, Columbia River fault zone — Southeastern margin of the Shuswap and Monashee complexes, southeastern British Columbia: *Can J. Earth Sci.*, v. 18, p. 1127–1145.

Rhodes, B. P., 1986, Metamorphic history of the Spokane dome mylonitic zone, Priest River Complex, northeastern Washington and northern Idaho and constraints on the crustal history: *J. Geology,* in press.

——, and Cheney, E. S., 1981, Low-angle faulting and the origin of Kettle dome, a core complex in northeastern Washington: *Geology*, v. 9, p. 366–369.

——, and Hyndman, D. H., 1984, Kinematics of mylonites in the Priest River "metamorphic core complex," northern Idaho and northeastern Washington: *Can. J. Earth Sci.*, v. 21, p. 1161–1170.

Rice, H. M. A., 1941, Nelson map-area, east half, British Columbia: *Geol. Survey Canada Mem. 228.*

Sheriff, S. D., Sears, J. W., and Moore, J. W., 1984, Montana's Lewis and Clark fault zone, an intracratonic transform fault system: *Geol. Soc. America Abstr. with Programs*, v. 16, no. 6, p. 653.

Stewart, J. H., 1972, Initial deposits in the Cordilleran geosyncline, evidence of a late (<850 m.y.) Precambrian continental separation: *Geol. Soc. American Bull.*, v. 83, p. 1345–1360.

——, 1976, Late Precambrian evolution of North America — Plate tectonic implications: *Geology*, v. 4, p. 11–15.

Weis, P. L., 1968, Geologic map of the Greenacres quadrangle, Washington and Idaho: *U.S. Geol. Survey Geol. Quad. Map GQ-734*, scale 1:62,500.

Weissenborn, A. E., and Weis, P. L., 1976, Geologic map of the Mount Spokane quadrangle, Spokane County, Washington and Kootenai and Bonner Counties, Idaho: *U.S. Geol. Survey Geol. Quad. Map GQ-1336*, scale 1:62,600.

Wilson, J. R., 1981, Structural development of the Kettle gneiss dome in the Boyds and Bangs Mountain quadrangles, northeastern Washington: M.S. thesis, Washington State Univ., Pullman, Wash., 156 p.

Wernicke, B., 1981, Low-angle normal faults in the Basin and Range Province — Nappe tectonics in an extending orogen: *Nature*, v. 291, p. 645–648.

——, 1985, Uniform-sense normal simple shear of the continental lithosphere: *Can. J. Earth Sci.*, v. 22, p. 108–125.

Yates, R. G., 1971, Geologic map of the Northport quadrangle, Washington: *U.S. Geol. Survey Misc. Geol. Inv. Map I-603.*

Karen Lund and L. W. Snee
Central Mineral Resources
U.S. Geological Survey
Denver, Colorado 80225

11

METAMORPHISM, STRUCTURAL DEVELOPMENT, AND AGE OF THE CONTINENT— ISLAND ARC JUNCTURE IN WEST-CENTRAL IDAHO

Deep-seated metamorphism and deformation, which developed along a convergent strike-slip orogenic belt, can be observed along the Salmon River suture, an accretionary plate boundary in west-central Idaho. Results of stratigraphic, metamorphic, and structural analysis, combined with $^{40}Ar/^{39}Ar$ age-spectrum dating of rocks from adjacent metamorphic terranes, elucidate timing and dynamothermal effects of juxtaposition of allochthonous island arc rocks of the western terrane against North American continental rocks of the eastern terrane.

Beginning at about 118 m.y.b.p., consanguineous regional metamorphism and deformation produced greenschist to upper amphibolite facies rocks with bedding-parallel fabric. From 118 to about 93 m.y., multiple growths of metamorphic minerals and fabrics were associated with pulses of megascopic deformation in both terranes. Structures show diverging lateral transport outward from the juncture and vertical transport near the juncture. The deformation resulted in inverted metamorphic sequences; in both terranes, metamorphic grade decreases structurally downward and spatially away from the suture zone. Deeper crustal levels are exposed toward the suture.

A final dynamothermal event, which was accompanied by emplacement of tonalitic plutons of the Idaho batholith, occurred from 93 to about 88 m.y.b.p. A secondary schistosity, which parallels the juncture and truncates previous structural trends, is superimposed both on the plutons and on roof pendants within them. A few kilometers from the suture zone, unfoliated tonalites cross-cut metamorphic and structural fabrics. Rapid uplift, cooling and erosion of plutons and host rocks from about 88 m.y.b.p. preceded emplacement of Idaho batholith muscovite-biotite granite plutons from 75 to 70 m.y.b.p.

The suture zone is an abrupt, nearly vertical juncture between basement blocks, as suggested by geologic features and corroborated by an abrupt change in geochemical provinces. The suture was formed by a convergent transcurrent fault, which had a probable right-lateral sense, along which the edge of the continent may have been removed before slices of exotic oceanic terrane were emplaced. Metamorphism and deformation of both terranes near their juncture are the result of movement along the Salmon River suture during the late Early to Late Cretaceous.

Earlier metamorphism, deformation, and plutonism occurred in parts of both terranes away from the Salmon River suture zone. In the continental terrane, several previous periods of deformation and plutonism are indicated but not well dated. In the western terrane outboard of the suture zone, both Late Triassic and Late Jurassic orogenic events are preserved. Although tectonic processes responsible for the early events have not been identified, it is now apparent that pre-118-m.y.-old features in the interiors of the plates do not reflect collision of exotic terranes with North America. Outboard of the suture, Cretaceous conglomeratic rocks are the only known deposits associated with events along the Salmon River suture.

INTRODUCTION

The present western extent of sialic crust in the northwestern United States extends northward through Nevada (Kistler and Peterman, 1978) to west-central Idaho and from there steps westward across Washington (Armstrong et al., 1977). This boundary has been considered to be the pre-Cretaceous Pacific margin of the North American plate (Coney et al., 1980; Davis et al., 1978; Hamilton, 1969). Features of the boundary are best exposed in west-central Idaho, where oceanic rocks of island arc origin are juxtaposed directly alongside older continental rocks. Stratigraphic trends of the Middle Proterozoic

LUND, SNEE

Belt basin and the Phanerozoic miogeocline in this region are thought to terminate as a result of having been cut out by rifting (Burchfiel and Davis, 1975; Hamilton, 1976) or as a result of never having been deposited (Armstrong, 1975; Dickinson, 1981; J. E. Harrison *et al.*, 1974). Initial strontium isotope ratios, used to separate continental and oceanic geochemical provinces, indicate a sharp break (less than 5 km wide) between continental values ($^{87}Sr/^{86}Sr$ = 0.706 or greater) and oceanic values (0.704 or less; Armstrong *et al.*, 1977; Fleck and Criss, 1985).

Previous conclusions about the geometry and geology of the juncture in west-central Idaho have been inferred from incomplete geologic understanding of the area. Few studies have been made of either side of the ocean-continent boundary in the region; no studies, which compared terranes, have been undertaken since this major tectonic feature was identified (Armstrong *et al.*, 1977; Hamilton, 1963a, b) or since complex models were advanced that consider the Cordilleran to be a "collage" of terranes accreted to the continent and rearranged by tectonic processes (Davis *et al.*, 1978; Hamilton, 1978; Helwig, 1974).

This study presents results of an investigation of stratigraphy, metamorphism, structure, and plutonism of both sides of the continent-ocean boundary, herein named the Salmon River suture zone, in part of west-central Idaho (Fig. 11-1). Rocks of the continent are referred to herein as belonging to the continental or eastern metamorphic terrane, whereas oceanic rocks are referred to as belonging to the oceanic, island arc, or western metamorphic terrane. Both of these terranes are really superterranes in the sense of Monger *et al.* (1982).

The area chosen for the field study includes the Gospel-Hump Wilderness Area and the drainage area of Slate Creek, which together comprise the best exposed cross section of the Salmon River suture zone and of rocks on either side. Combined with the field study are $^{40}Ar/^{39}Ar$ age-spectrum data of a more regional nature. The purpose of the work is (1) to synthesize and contrast the geologic history of both sides of the suture zone, (2) to produce an understanding of timing of regional tectonism, and (3) to formulate models for orogenic processes.

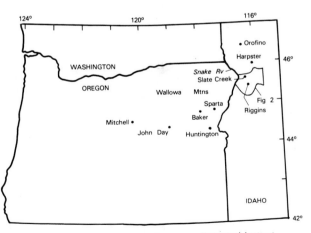

FIG. 11-1.   Location map for areas discussed in text.

Eastern Terrane

In the study area, rocks east of the Salmon River suture are metasedimentary rocks that represent protoliths derived from terrigenous sources. These have been divided into five tectonostratigraphic packages (Fig. 11-2) that have been subdivided into 14 map units (Lund, 1984). The metamorphic rocks are preserved as roof pendants and xenoliths in the Idaho batholith.

The structurally lowest and probably oldest tectonostratigraphic metamorphic unit [the Middle Proterozoic(?) "basal metasedimentary unit"] is composed of mica schist and feldspathic gneiss with interleaved calc-silicate and quartzitic gneiss. No consistent stratigraphic horizons have been identified in these rocks, possibly because of multiple phases of metamorphism and deformation (Reid, 1959) and because of incomplete regional mapping. Other than the assurance that these rocks produce continental strontium signatures (Fleck and Criss, 1985), their origin is problematic. These rocks are part of the terrain called the Salmon River Arch by Armstrong (1975) — an idea disputed by K. V. Evans (1981) and K. V. Evans and Lund (1981). The "basal metasedimentary unit" was intruded by igneous rocks that are now amphibolite and augen gneiss. The U/Pb zircon age of these igneous rocks, although complicated by inheritance, is probably 1370 m.y. (K. V. Evans and Fischer, 1986). Similar intrusive rocks 80 km east, near Shoup, Idaho, which have been dated more conclusively by U/Pb zircon methods at 1370 ± 10 m.y. (K. V. Evans, 1981; K. V. Evans and Fischer, 1986), intruded low-grade, deformed Belt-equivalent metasedimentary rocks (A. L. Anderson, 1956; K. V. Evans, 1981). Thus although not well constrained, the "basal metasedimentary unit" is possibly equivalent to part of the Belt Supergroup.

Two informally named Middle Proterozoic (?) tectonostratigraphic units with consistent internal stratigraphy were mapped in rocks exposed at intermediate structural levels and are presumed to be younger than the "basal metasedimentary unit" (Lund, 1984). The Concord (lower) unit is composed of three map units of calc-silicate gneiss, whereas the Quartzite Butte (upper) unit is composed of two quartzitic map units (Fig. 11-2). The Concord and Quartzite Butte units have a combined exposed thickness of about 3000 m. Despite upper greenschist to amphibolite facies metamorphism, the rocks retain a remarkable array of primary sedimentary structures. These include cross-bedding and oscillation ripple marks in quartzite and ball and pillow structures, climbing ripples, soft sediment folding, flame structures, and laminar bedding in calc-silicate gneisses. Premetamorphic amphibolite dikes and sills of unknown age intruded the units. The stratigraphy reflects slow, cyclic deposition of continent-derived sediments.

The informally named Paleozoic (?) Moores and Umbrella Butte tectonostratigraphic units (Fig. 11-2) were mapped in rocks exposed at highest structural levels overlying and probably younger than the other three tectonostratigraphic units (Lund, 1984). The Moores and Umbrella Butte units total about 6500 m thick and are composed of a distinctive sequence of quartzite, quartzite-pebble quartzose conglomerate, calc-silicate gneiss, mica schist, and marble lithologies. The Moores unit was subdivided into eight map units based on continuous stratigraphic packages, whereas a single map unit comprises the Unbrella Butte unit. The combination of amphibolite facies metamorphism and strong cleavage development has destroyed most primary sedimentary structures except for cross-bedding in quartzite-bearing units. Compositional variety and relatively abrupt facies

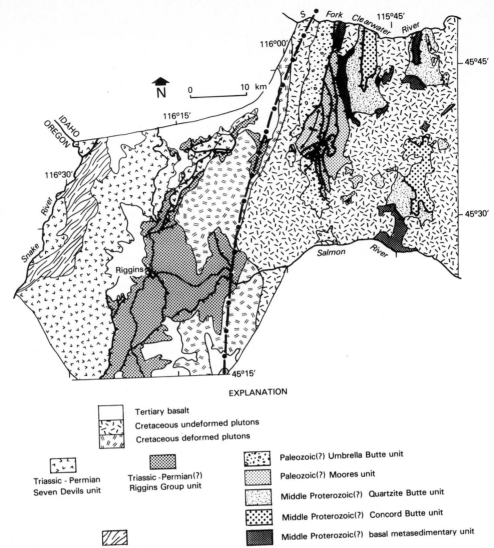

EXPLANATION

Tertiary basalt

Cretaceous undeformed plutons

Cretaceous deformed plutons

Triassic - Permian Seven Devils unit

Triassic - Permian(?) Riggins Group unit

Paleozoic(?) Umbrella Butte unit

Paleozoic(?) Moores unit

Middle Proterozoic(?) Quartzite Butte unit

Middle Proterozoic(?) Concord Butte unit

Middle Proterozoic(?) basal metasedimentary unit

Paleozoic basement (?)

FIG. 11-2. Geologic map of the study area along the island arc–continent suture zone in west-central Idaho. Includes the Gospel-Hump Wilderness and Slate Creek areas from this report and mapping after Hamilton (1963b), Onasch (1977), Vallier (1977), Lund (1984), and K. Lund and K. V. Evans (U.S. Geol. Survey, unpublished mapping). Heavy dash-dot line is initial strontium ratio isopleth; < 0.704 to west and > 0.706 to east, from Fleck and Criss (1985).

changes in the Moores and Umbrella Butte tectonostratigraphic units represent a variety of continental sources and rapidly changing depositional environments.

The four upper tectonostratigraphic units (Concord, Quartzite Butte, Moores, and Umbrella Butte) of the eastern terrane contain enough primary sedimentary features to determine that the rocks are generally right-side-up (Lund, 1984). However, metamorphism and deformation have destroyed evidence for sedimentary contact relationships, so the original stratigraphic superposition of tectonostratigraphic units has not been determined. Despite a lack of protolith age data, generalized correlations for these tectonostratigraphic units can be made on the basis of stratigraphy, thickness, primary sedimentary features, and regional continuity. Rocks of the Concord and Quartzite Butte units possibly represent Middle Proterozoic sediments equivalent to some part of the Belt Supergroup. Rocks of the Moores and Umbrella Butte units seem to correlate best with some of the Upper Proterozoic or Paleozoic sedimentary rocks of the Cordilleran miogeocline. Middle Proterozoic to Paleozoic sedimentary deposits were previously assumed to be missing in this area (Armstrong, 1975; Burchfiel and Davis, 1975; Dickinson, 1981; Hamilton, 1976; J. E. Harrison et al., 1974).

Western Terrane

West of the suture zone, metamorphic country rocks comprise felsic to mafic volcanic, volcaniclastic, intrusive, volcanogenic sedimentary, and calcareous rocks (Fig. 11-2; Hamilton, 1963b; Hietanen, 1962; Lund, 1984; Myers, 1982; Onasch, 1977; Vallier, 1977). Because unequivocal sedimentary younging indicators have not been found near the suture zone, stratigraphic superposition and thickness are even more difficult to determine than in the eastern terrane. This difficulty presents a problem with mapping and regional correlation. Based on differences in metamorphic grade, on possible differences in structural history, and on the position of serpentinite bodies, Hamilton (1963b) separated the oceanic terrane of west-central Idaho into two island arc sequences (Fig. 11-2). Rocks of the Riggins Group (named by Hamilton, 1963b) of unknown pre-Tertiary age are of low, medium, and high metamorphic grades according to Hamilton (1963b). The Riggins Group structurally overlies more regionally exposed Permian to Triassic rocks of the Seven Devils island arc sequence (Dickinson, 1979; also known as the Wallowa terrane of Silberling and Jones, 1984) which "are of low metamorphic grade throughout" according to Hamilton (1963b). For the purposes of this paper and consistency with the eastern terrane nomenclature, these two sequences are treated as tectonostratigraphic units.

Rocks of the Permian to Triassic Seven Devils island arc are exposed in northeastern Oregon and west-central Idaho (Brooks and Vallier, 1978; Hamilton, 1963b; Vallier et al., 1977). Vallier (1977) subdivided the primarily volcanic Seven Devils Group into four rock-stratigraphic units. The lower two formations of Early Permian age are separated from the upper two formations of Middle to Late Triassic age by a major unconformity. Chemistry of the volcanic rocks is described by Hamilton (1963b) and Vallier and Batiza (1978).

In this study area, rocks of the Seven Devils Group are greenschist to lower amphibolite facies metamorphosed volcanic rock with minor volcaniclastic conglomerate and sedimentary layers. Tonalite plutons intruded rocks of the Seven Devils Group prior to regional metamorphism. In the absence of age information, low-grade volcanic rocks in west-central Idaho have generally been correlated with the Triassic formations in the upper Seven Devils Group (Myers, 1982) because of association of volcanic rocks with sedimentary rocks that have been correlated with Triassic rocks of Oregon (Hamilton, 1963b; Vallier, 1977). The upper lithologies of the Seven Devils tectonostratigraphic unit (rocks overlying the Seven Devils Group) appear to be a stratigraphically interfingered sequence of volcanic-derived sedimentary, calcareous, and minor volcanic layers (Lund, 1984; Lund et al., 1983a; McCollough, 1984) as has also been found in northeastern Oregon (Follo and Siever, 1984). This suggests a gradual change upward in the section from protoliths with volcanic predominance (the Seven Devils Group) to carbonaceous-calcareous shale and carbonate lithologies (and in eastern Oregon, younger noncalcareous shale and conglomerate lithologies). Therefore, stratigraphic section reflects a complex history of volcanic island arc formation through late-stage quiescence (Follo and Siever, 1984; Lund et al., 1983a, b). In western Idaho, previous descriptions of these upper units seem to have been too simplified, thereby evoking a complex tectonic intercalation of separate lithologies in order to explain map patterns. A thorough study and possible revision of regional stratigraphy is warranted.

Because the Riggins Group is only described from structurally and metamorphically complex exposures in west-central Idaho, a complete stratigraphy is unknown. Map units are defined on the basis of gross compositional changes (Hamilton, 1963b) and minor conglomerate layers (Lund, 1984; Onasch, 1977), which are present as enveloping surfaces, are parallel or subparallel to the planar metamorphic fabric, and are the only characteristics that can be used to indicate a relationship between metamorphic layering and bedding. The preserved stratigraphy is composed of two lower volcanic units and a structurally upper carbonaceous-calcareous phyllite to schist and minor carbonate unit. Sheared serpentinite bodies are located along faults (Hamilton, 1963b; Onasch, 1977). Chemistry of the volcanic rocks of the Riggins Group was described by Hamilton (1963b) to be "similar in many respects to the Seven Devils volcanics" and, in addition, he noted that "the distribution of minor elements (in the Seven Devils Group) is . . . identical with that in the meta-volcanic rocks of the Riggins Group." Based on more detailed mapping (Fig. 11-2), Onasch (1977), Lund (1984), and McCollough (1984) have traced units of the Riggins Group northward along strike and across metamorphic isograds into areas where the Riggins Group had been mapped as a (higher grade) part of the Seven Devils island arc in earlier reconnaissance mapping (Hamilton, 1963b). In Slate Creek, as elsewhere in the region, the Riggins Group crops out closer to the eastern margin of the oceanic plate than does the Seven Devils arc.

The origin of the Riggins Group rocks has been problematic because the age is unknown. Early descriptive study of the group resulted in separating these rocks from other island arc sequences in the region (Hamilton, 1963b). With the advent of terrane analysis, separation of these rocks from those of the Seven Devils island arc has been even more codified (Brooks and Vallier, 1978; Coney et al., 1980; Silberling and Jones, 1984), whereas similarities have been largely ignored. Based on stratigraphy, petrography, and preliminary geochemistry, there is good reason to suggest that the Riggins Group is a higher-grade metamorphic equivalent of rocks of the Seven Devils island arc and more detailed study should be conducted to test the validity of this hypothesis.

Stratigraphic Differences across the Suture

The age and lithologic differences between the eastern and western sequences indicate that these rocks are unrelated. Based on contrasting lithologies, neither sequence could have formed from the other. There are no interfingered or transitional relationships between the sequences. Indeed, the eastern rocks are probably considerably older than the western. No pre-Tertiary strata overlap the boundary between the terranes (Fig. 11-2).

Stratigraphic information on an abrupt change from terriginous to oceanic compositions is corroborated by geochemical information. Initial strontium isotope ratios from preaccretionary and postaccretionary plutons of both terranes (Armstrong et al., 1977; Fleck and Criss, 1985) also demonstrate a sharp break (less than 5 km of intermediate or mixed values) between terranes. Unlike the conclusions of large-scale crustal contamination of magmas presented by those authors, we think this indicates that the change in rock types seen at the surface corresponds to a geochemical change at depth, probably a change from oceanic to continental basement types.

Previous Work

Previous workers were in agreement that metamorphism in the vicinity of the Salmon River suture was caused by magmatic events, but there were differences of opinion with regard to both mode and timing of metamorphism. In the western terrane, metamorphism was described as "contact metamorphism on a regional scale" caused by the Idaho batholith (Hamilton, 1963a). Metamorphism in the eastern terrane was described as (1) metasomatic metamorphism that was genetically related to Cretaceous igneous activity (Hietanen, 1962) or (2) a complex sequence combining a Late Proterozoic metamorphism, Triassic metamorphism, and processes of granitization to form the Idaho batholith (Reid, 1959). These workers concluded that formation of the metamorphic fabric and most of the structure was caused by, but preceded final emplacement of, the Idaho batholith. Because of the spatial relationship between large plutonic intrusions and regional metamorphic rocks, alternative explanations for metamorphism, which were not related to regional contact metamorphism, were not sought. Such views of passive metamorphism and dynamic batholith emplacement have dominated work in the region since then (Armstrong *et al.*, 1977; D. W. Hyndman, 1979; D. W. Hyndman and Talbot, 1976; Myers, 1982; Onasch, 1977).

Little recent work has been published on the stratigraphy or metamorphism of roof pendants in the central and western Idaho batholith. Early studies of the eastern terrane described the stratigraphy and regional metamorphic nature of some of the roof pendant rocks (Beckwith, 1928; Lindgren, 1904; Shenon and Reed, 1934) but did not adequately portray regional metamorphic isograds. The only comprehensive study of metamorphism in metasedimentary roof pendants of the eastern terrane near the Salmon River suture is in an area where the tectonic setting is still poorly understood (Hietanen, 1962, 1963a, b, c, 1968).

Metamorphism in rocks of the western terrane near the suture zone has been the subject of several specific studies. These studies documented that metamorphic isograds generally trend north-northeast and metamorphic grade increases toward the east (Hamilton, 1963a; Myers, 1982; Onasch, 1977). The patterns of the isograds are disrupted by faulting to the extent that telescoping of metamorphic grade and termination of isograds can be seen macroscopically. However, because structures, rock packages, and plutonic events were commonly correlated according to associated metamorphic grade, there is confusion about timing of events and relationship of metamorphism to tectonic events.

Eastern Terrane

The grade of eastern terrane metasedimentary rocks ranges from lower to upper amphibolite facies. The highest-grade rocks are sillimanite-bearing metapsammitic rocks, garnet- and sillimanite-bearing metapelitic rocks, and diopside-bearing calcsilicate gneiss. These upper amphibolite facies rocks occur in a band about 6 km wide adjacent to, and east of, the suture zone. Metamorphic grade decreases to the east away from the suture zone and was disrupted by structural activity. Because of stacking by thrust faulting, metamorphic facies are in an inverted sequence, with higher grade rocks structurally overlying lower grade rocks (Fig. 11-3).

The "basal metasedimentary unit" does not follow patterns seen in other tectonostratigraphic units; rocks of this unit maintain a homogeneous medium grade of metamorphism across 30 km of continuous outcrop north of the area of Fig. 11-2. Both from

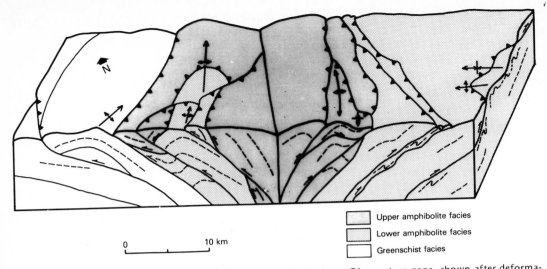

Upper amphibolite facies

Lower amphibolite facies

Greenschist facies

0   10 km

FIG. 11-3.   Reconstruction of metamorphic grades near the Salmon River suture zone, shown after deformation and prior to intrusion.

previous work (Reid, 1959) and from this study[1] (Lund, 1984), it is apparent that these rocks have undergone more intense metamorphism and deformation than the rest of the terrigenous rocks in the study area. The possibility of successive metamorphic events separated by long periods of geologic time as suggested by Reid (1959) cannot be evaluated without detailed work. However, Middle Proterozoic intrusive rocks, which cross-cut metamorphosed and deformed Belt-equivalent rocks in the region, give evidence of a regional Middle Proterozoic compressional event (Evans, 1981). At present, there is not enough information to delimit metamorphic events or regional changes in metamorphic characteristics of this complex "basal metasedimentary unit."

The mineralogy of gneisses of the Concord tectonostratigraphic unit is heterogeneous (Lund, 1984). Two main mineral assemblages recur as layers. Muscovite-biotite quartzofeldspathic gneiss is interlayered with calcsilicate-bearing layers. Calcsilicate layers contain plagioclase (An^{10}), epidote, actinolite, biotite, quartz, and microcline. Minor pods and stringers of dolomitic marble occur with calcsilicate gneiss in some places. Garnet-epidote-actinolite skarn found along late-stage quartz veins and fault zones is the only metamorphic assemblage not related to regional metamorphism.

Metapsammitic rocks of the Quartzite Butte tectonostratigraphic unit, about 28 km east of the Salmon River suture, contain about 90% quartz with a few percent tourmaline and muscovite; biotite, apatite, plagioclase, orthoclase, epidote, pyrite, and graphite are present in some layers. Higher grade equivalents of the Quartzite Butte unit that crop out 5 km east of the suture zone contain sillimanite that replaces muscovite and forms fibrolite in quartz grains.

Rocks of the Moores tectonostratigraphic unit crop out in a wide area on the east side of the suture zone. The metamorphic grade of calcsilicate gneiss in this unit is higher than calcsilicate lithologies in units farther east (Concord and Quartzite Butte units). Two common diagnostic associations for layers in the westernmost calcsilicate gneiss rocks of the Moores tectonostratigraphic unit are: (1) diopside, orthoclase, tremolite/actinolite, quartz; and (2) diopside, quartz, tremolite/actinolite, plagioclase (An^{60}),

[1] Also K. Lund, unpublished mapping.

biotite. Quartz-pebble muscovite-biotite quartzitic conglomerate in the upper (western) part of the Moores unit contains sillimanite needles in muscovite and quartz. Because the diverse lithologies of the Moores tectonostratigraphic unit are repeated by thrust faulting, metamorphic mineral assemblages can be compared between the same units in different structural plates (Fig. 11-2). In the eastern, structurally lower of these plates, the lowest map unit is composed of muscovite-biotite schist; on the western side of the structurally higher plate, this same unit is a sillimanite-muscovite-biotite schist. In contrast, the mineral assemblages of muscovite-bearing quartzite and calcsilicate units, which are repeated in both plates, do not appear to change with structural or geographic position.

The quartzitic, quartzite-pebble conglomeratic, and schistose rocks of the Umbrella Butte tectonostratigraphic unit (the structurally highest and nearest to the suture zone of the terrigenous units) contain ubiquitous sillimanite. Sillimanite occurs as fibrolite in quartz and as needles with muscovite in quartzitic rocks. Schist layers are composed of tourmaline, garnet, sillimanite, muscovite, biotite, plagioclase, and quartz. The mineral assemblage of calcsilicate and marble layers is similar to rocks of the same composition in the Moores unit.

Except for the "basal metasedimentary unit," rocks of the eastern terrane generally record a simple metamorphic history. Metamorphic minerals are formed in the plane of schistosity. Overprinting of metamorphism is seen only in circumstances of minor thermal effects related to magmatic activity and as formation of secondary fabrics or retrogradation near fault zones.

Western Terrane

Along the Snake River (Vallier, 1977) and near Riggins (Hamilton, 1963b), the Seven Devils Group consists of volcanic and volcaniclastic greenstones. Metamorphic grade in these rocks increases without any known faulting to the northeast from the Riggins area. In the Slate Creek area, the Seven Devils Group is greenschist facies rocks composed of metavolcanic clasts in a metatuffaceous matrix (Lund, 1984). Recrystallized quartz, untwinned plagioclase, chlorite, tremolite, epidote, biotite, and muscovite comprise the mineralogically variable fine-grained groundmass. Plagioclase, garnet, and chlorite porphyroblasts are common. Early plutonic rocks that cut the Seven Devils Group have been metamorphosed to the same degree as the volcanic rocks.

The mineralogy of the upper sedimentary part of the Seven Devils island arc also changes northeast of Riggins. At lowest grade, these rocks are fine-grained calcareous and dolomitic marbles and orange-weathering, dark gray, carbonaceous-calcareous phyllite with siderite porphyroblasts (Hamilton, 1963b). Northeastward, the rocks change to blue-gray pyrite-phlogopite carbonaceous marble and dark gray, graphitic calcareous schist (Lund, 1984). The mineral assemblage of the schists is calcite, plagioclase, biotite, fine rounded quartz, epidote, and siderite. In addition, abundant sphene, garnet, and amphibole are present in highest grade, schistose rocks nearer to the suture.

The three formations of the Riggins Group are at lowest metamorphic grade near Riggins and increase in grade toward the northeast, east, and south (Hamilton, 1963b; Onasch, 1977). At lowest grades, the lower, volcanic and volcaniclastic formations of the Riggins Group are mineralogically similar to rocks of the Seven Devils Group at equivalent metamorphic grades (Lund, 1984). Greenschist facies rocks are muscovite and/or chlorite schist, with plagioclase, quartz, and epidote in the groundmass and, at slightly higher grade, with porphyroblasts of biotite flakes, unoriented hornblende needles, and carbonate rhombs (Hamilton, 1963b). Toward the suture zone, these rocks become garnet-muscovite schist, biotite schist, garnet-hornblende-biotite schist, and garnet-amphibolite gneiss, depending on the original composition. Metavolcaniclastic conglomerate layers are present in volcanic rocks of all metamorphic grades. Leucotonalite intruded into the Riggins Group in several places and was metamorphosed and deformed together with Riggins Group country rocks (Hamilton, 1963b; Lund, 1984; Onasch, 1977).

Metamorphosed volcaniclastic sedimentary rocks of the upper Riggins Group have not been conclusively located outside of the type section (west-central part of Fig. 11-2), where a range in metamorphic grade was described by Hamilton (1963b). Low-grade rocks are gray carbonaceous-calcareous phyllite and minor dark-colored marble. At higher metamorphic grades, the rocks are garnet-biotite-calcsilicate schist, garnet-biotite-amphibole schist, and pyrite-phlogopite carbonaceous marble[2] (Hamilton, 1963b).

Rocks of the western terrane in west-central Idaho, which lie at least 10 km west of the suture zone, record a fairly simple metamorphic history. Only one foliation is seen in these rocks. At low metamorphic grade, porphyroblasts (especially hornblende) are randomly oriented in the plane of foliation; at higher metamorphic grades, porphyroblasts are well aligned. These single-fabric rocks underwent a single metamorphic event.

Along a narrow belt parallel to the suture zone and spatially related to suture zone plutons (see below), oceanic rocks contain clear evidence of more than one dynamothermal event. This high-grade overprinted zone is about 10 km wide in the northern part of Fig. 11-2; the zone widens slightly to the south. In these rocks, garnet, hornblende, and other porphyroblasts were rolled during formation of a secondary schistosity that cross-cuts the original, bedding-parallel schistosity. Growth of new minerals is indicated in many places by the presence of two phases of the same mineral that have different orientations and/or characteristics (e.g., rolled hornblende porphyroblasts surrounded by smaller, oriented hornblende laths). More than two episodes of metamorphism and associated deformation are recorded in some places. Rocks south of Slate Creek, which contain rotated garnet and deformed biotite porphyroblasts wrapped by a deformed muscovite fabric, give evidence of two high-grade syndeformational events (wherein porphyroblasts were formed and then rolled) and one lower grade metamorphic overprint (wherein biotite was chloritized and euhedral garnet rims were formed around syntectonic garnet cores; McCollough, 1984).[3] In addition, evidence of low-grade metamorphic overprinting of higher grade events is found outside this belt. East of Riggins, oriented chlorite flakes grew in the axial planar cleavage associated with macroscopic folds that formed later than amphibolite facies metamorphism (Onasch, 1977).

A range from greenschist to upper amphibolite-facies metamorphic rocks is found in both of these western terrane tectonostratigraphic units. The metamorphic grades of the Seven Devils island arc rocks observed in Slate Creek (Lund, 1984) are higher than those observed in the Riggins area (Hamilton, 1963b) or in the South Fork of the Clearwater River canyon[4] (Myers, 1982). The amount of regional change of metamorphic grade within these units was not fully appreciated by previous workers because of gaps in mapping. Comparison of lithologically similar rocks of the Seven Devils and Riggins Group tectonostratigraphic units at similar metamorphic grades (both high and very low) suggests that there is little difference between the mineralogy on a regional basis. As much internal lithologic variation appears to exist within the two tectonostratigraphic units as between them.

As described by all recent workers (Hamilton, 1963b; Lund, 1984; Myers, 1982; Onasch, 1977; Wagner, 1945), metamorphic grade in the western terrane increases eastward and is disrupted by faults. A change in metamorphic grade is observed across postmetamorphic thrust faults and specifically across the thrust fault (Rapid River thrust)

[2]Ibid.

[3]Ibid.

[4]Ibid.

that separates the Seven Devils tectonostratigraphic unit from the overlying Riggins Group tectonostratigraphic unit in Slate Creek (Lund, 1984) as in the Riggins area (Hamilton, 1963b; Onasch, 1977). Because of nearly equivalent high metamorphic grades of the rocks, the change in metamorphic grade across the thrust fault in Slate Creek is not as dramatic as to the south. As in the eastern terrane, because of structural stacking and subsequent erosion as well as metamorphic processes, the metamorphic facies appear to be in an inverted sequence (Fig. 11-3).

Comparison of Metamorphism across Suture

Metamorphic grade decreases symmetrically away from the Salmon River suture in both terranes (Fig. 11-3). Although peak metamorphic grade is approximately equal in both terranes, the metamorphic belt is wider on the continental side. Low-grade metamorphic rocks are found within 15 km of the suture zone on the oceanic side of the juncture, whereas the westernmost Proterozoic to Paleozoic low-grade rocks in the continent near this latitude are about 65 km east of the suture (B. F. Leonard, unpublished mapping).

Although isotopic dating is in progress on metamorphic rocks of the eastern terrane in an attempt to understand their complex thermal history, many factors inhibit the development of an hypothesis about the regional geologic history of the continental rocks through time. Evidence of Middle Proterozoic metamorphism in some of the continental rocks of central Idaho (Evans, 1981) generally supports the idea that rocks of the continent may have been regionally overprinted by multiple metamorphic events of different ages. The scarcity of detailed regional mapping and description of metasedimentary rocks of central Idaho makes it difficult to distinguish among multiple events of similar grade and extent. In addition, original rock compositions and resulting metamorphic mineral assemblages are commonly neither diagnostic of metamorphic conditions nor appropriate for dating metamorphic events. The effects of voluminous plutonic activity in the eastern terrane complicate the thermal history of dynamothermal metamorphic events. Because of telescoping of metamorphic grade and relative ease of correlating metamorphic and tectonic events in the western terrane, the relative and absolute timing of metamorphic history in the oceanic rocks is easier to evaluate on a regional basis.

AGE OF METAMORPHISM

$^{40}Ar/^{39}Ar$ age-spectrum dating of hornblende, biotite, muscovite, and microcline, which formed during metamorphism, is currently the best method for determining thermal history and age of complex metamorphic terranes. Ideally, a mineral date determined by this method marks the time when that mineral became closed to diffusion of argon. Closure of a particular mineral to diffusion is controlled chiefly by temperature, to a smaller extent by cooling rate (Dodson, 1973), and possibly by chemical composition or strain. Each mineral, which is appropriate for argon dating, has a characteristic closure temperature that is known with a precision of about $\pm20°C$. The closure temperature is higher for minerals that cooled rapidly, and conversely. Commonly accepted closure-temperature ranges that span from rapid cooling (1000°C/m.y.) to slow cooling (5°C/m.y.) are 580–480°C for hornblende (T. M. Harrison, 1981), 325–270°C for muscovite (Snee et al., in press), 300–260°C for biotite (T. M. Harrison and McDougall, 1980; Snee, 1982), and 160–100°C for microcline (T. M. Harrison and McDougall, 1982). For simplicity, intermediate closure temperatures (i.e., for intermediate cooling rates of 100–500°C/m.y.) are assumed in this study (530, 300, 280, and 130°C for hornblende, muscovite, biotite, and microcline, respectively).

Because argon-closure temperatures are faily well known, $^{40}Ar/^{39}Ar$ age-spectrum dating has valuable applications not only for determination of the age of metamorphism but also for evaluation of the postmetamorphic thermal history of the region. In addition, an age spectrum of a mineral is a record of distribution of argon within the mineral because argon is released from the mineral in progressively increasing temperature steps and an age is calculated for each increment. If the mineral has a relatively simple thermal history, argon is evenly distributed, the age of the majority of released argon is the same within error, and a "plateau" is defined. Alternatively, if the mineral has had a complex thermal history, a disturbed spectrum is commonly exhibited. Disturbed spectra typically result because radiogenic argon has been lost or gained with respect to the amount that corresponds to the real age. Samples that have gained radiogenic argon produce "excess-argon" spectra characterized by anomalously old first, and in some cases last, steps; samples that have lost radiogenic argon produce "argon-loss" spectra characterized by an increase in the age of increasing temperature steps. Argon spectra of samples in this study are both simple and complex and many exhibit argon loss. A brief summary, including character of the age spectra, is presented in Table 11-1. Sample locations are shown on Fig. 11-4. Raw data are available in Snee *et al.* (1986).

Even though $^{40}Ar/^{39}Ar$ age spectra of metamorphic minerals from the western terrane are complex, some general observations are evident. Age spectra of the majority of 14 dated hornblendes exhibit some argon loss. Plateaus are produced by only two of the samples; one of these displays minor argon loss. Hornblende dates form at least three statistically distinct groups at about 118, 109, and 101 m.y. These age groups show a clear geographic relationship; the 118-m.y. group is located south and west of Riggins, the 109-m.y. group forms an apparent belt to the northeast, east, and south of the 118-m.y. group, and the 101-m.y. group is located farther east along the west side of the suture zone. This geographic distribution is shown schematically in Fig. 11-4. The age groups also exhibit a relationship, albeit less clearly defined, to metamorphic grade described by Hamilton (1963b), Onasch (1977), and Lund (1984). The 118-m.y. group occurs in greenschist facies, garnet-zone rocks but also in upper amphibolite facies rocks; the 109-m.y. group occurs in amphibolite facies, oligoclase-, and andesine-zone rocks; and the 101-m.y. group occurs in upper amphibolite facies, sillimanite-zone rocks. Although argon loss from hornblende is present in all age groups, it is found only in rocks of

TABLE 11-1. Summary of $^{40}Ar/^{39}Ar$ Age-Spectrum Data[a]

Sample	Mineral	Zone	Unit	Age and Characteristics of $^{40}Ar/^{39}Ar$ Spectrum[b, c]
			Metamorphic Rocks	
R7	Hb	Gt	Riggins Gp	Tmp = 117.0 ± 0.5 m.y. minor excess argon
R30	Hb	(Hi)	Riggins Gp?	Tmp = 118.0 ± 0.6 m.y. Ar loss to 114 m.y.
R3	Mu	Ch	Seven Devils island arc	Ar loss: 104–100 m.y.
R28	Hb	(Hi)	Riggins Gp?	Ar loss: 108–100 m.y.
R29	Hb	(Hi)	Riggins Gp?	Tp = 108.1 ± 0.5 m.y. Ar loss to 107 m.y.
R18	Hb	O1	Riggins Gp	Tmp = 106.5 ± 1.4 m.y. disturbed
	Bi		Riggins Gp	Excess argon
R17	Hb	O1	Riggins Gp	Ar loss: 107–105 m.y.
R34	Hb	Gt	Riggins Gp	Tp = 109.1 ± 0.6 m.y.

TABLE 11-1. Summary of ^{40}Ar/^{39}Ar Age-Spectrum Data (continued)

Sample	Mineral	Zone	Unit	Age and Characteristics of ^{40}Ar/^{39}Ar Spectrum[b,c]
R27	Hb	O1	Riggins Gp	Tmp = 108.9 ± 0.6 m.y. minor excess Ar
R26	Hb	An	Riggins Gp	Ar loss: 109–94 m.y.
	Bi		Riggins Gp	Tp = 89.9 ± 0.5 m.y.
R35	Mu	Bi	Riggins Gp	Ar loss: 101–97 m.y.
R16	Hb	An	Riggins Gp	Ar loss: 100.5–93 m.y.
	Bi		Riggins Gp	Tp = 88.2 ± 0.5 m.y.
R14	Hb	Si	Riggins Gp	Ar loss: 101–89 m.y.
R25	Bi	An	Seven Devils island arc	Tp = 84.1 ± 0.4 m.y.
Suture-Zone Rocks				
R24	Hb	(Hi)	Seven Devils island arc	Ar loss: 92.5–88 m.y.
	Mu	–	Deformed pegmatite in R24 gneiss	Tp = 83.4 ± 0.4 m.y.
	Mi	–	Deformed pegmatite in R24 gneiss	Ar loss: 79–69.5 m.y.
R23	Hb	(Hi)	Riggins Gp inclusion	Ar loss: 99.5–93 m.y.
R22	Hb	(Hi)	Riggins Gp inclusion	Ar loss: 90–88 m.y.
R12	Hb	–	Riggins Gp	Ar loss: 93.5–88 m.y.
	Bi	–	Riggins Gp	Tp = 82.5 ± 0.4 m.y.
R11	Hb	–	Deformed tonalite	Tp = 85.1 ± 0.4 m.y.
Undeformed Plutons				
R21	Hb	–	Tonalite	Tp = 83.9 ± 0.4 m.y.
	Bi	–	Tonalite	Tp = 81.1 ± 0.4 m.y.
	Mi	–	Tonalite	Ar loss: 78–75 m.y.
R10	Mu	–	Pegmatite in tonalite	Tp = 76.7 ± 0.4 m.y.
R8	Mu	–	Muscovite-biotite granite	Tp = 75.3 ± 0.4 m.y.
	Bi	–	Muscovite-biotite granite	Tp = 74.7 ± 0.4 m.y.
	Mi	–	Muscovite-biotite granite	Ar loss: 69–47 m.y.

[a]Hb, hornblende; Mu, muscovite; Bi, biotite; Mi, microcline; Gt, garnet; Ol, oligoclase; An, andesine; Ch, chlorite; Si, sillimanite; (Hi), upper amphibolite facies or higher; Ar, argon; Tp, plateau age; Tmp, preferred age of sample.

[b]All errors are 1 standard deviation.

[c]Decay constants and isotopic abundances used in this study are those recommended by Steiger and Jäger (1977).

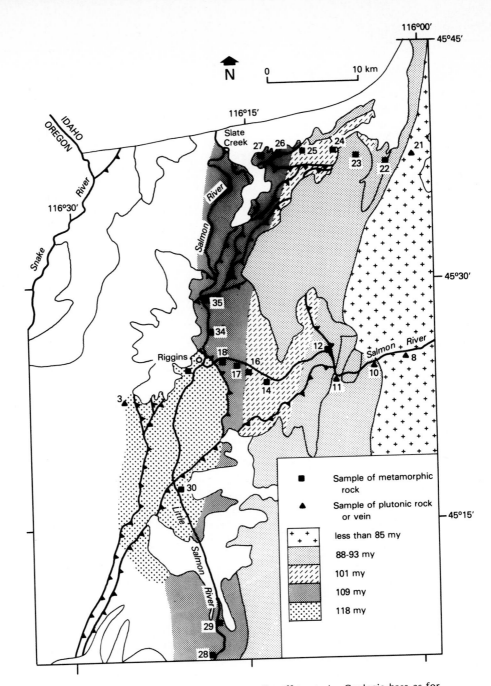

FIG. 11-4.　Sample locality map for $^{40}Ar/^{39}Ar$ study. Geologic base as for Fig. 11-2. Age distribution pattern is based on hornblende ages and is grossly generalized. Sample 23 is from an inclusion of metamorphic rock within foliated tonalite.

oligoclase-zone or higher; argon loss is most severe in areas where the rock contains two hornblende growths or multiple foliations. Two dated muscovites from greenschist facies, biotite-zone rocks exhibit argon loss. Three biotites from andesine-zone rocks display distinctly younger ages than coexisting hornblendes.

An important conclusion from these data is that the oldest metamorphism associated with the Salmon River suture occurred at 118 m.y. One sample (R7), which records this age, is a garnet-zone rock with a single foliation defined by nonlineated hornblende. This hornblende probably formed within the lower limits of the stability field of hornblende based on the mineral assemblage and probably at a temperature near 500°C or lower (uppermost greenschist facies; Winkler, 1979). Because the argon-closure temperature of hornblende is about 500°C even at slow cooling rates, this hornblende probably closed to diffusion of argon upon formation (i.e., the age does not reflect any significant period of cooling). Hornblende (R30) from upper amphibolite-facies rock also has a single hornblende-defined foliation; this sample, which formed above 500°C (well above the hornblende argon-closure temperature), also gives an age of 118 m.y. Therefore, it is unlikely that an extended period of cooling could have occurred after metamorphism in either upper amphibolite-facies or greenschist-facies rocks of this age.

The ^{40}Ar/^{39}Ar age-spectrum data show that a second period of hornblende growth and argon closure occurred at about 109 m.y. Samples that best show this are two post-kinematic hornblendes (R27 and R34); these hornblendes formed after deformation that produced strong foliation in garnet-zone rocks. No argon loss is exhibited by either spectrum, and sample R34 has a well-defined plateau. Like the 118-m.y. garnet-zone sample, these younger garnet-zone hornblendes probably formed at a temperature below that for hornblende argon closure. Thus, the dates closely approximate the age of a period of static hornblende formation. That this thermal activity was accompanied by deformation is supported by five other synkinetmatic hornblendes from middle to upper amphibolite facies rocks (R17, 18, 26, 28, and 29), which show variable amounts of argon loss superposed on a maximum age of about 109 m.y. Argon loss recorded by these five hornblende samples probably resulted from one or more overprinting event(s).

Sillimanite-zone samples (R23, 16, and 14), which clearly show either two foliations, a lineation-dominated fabric, or two different hornblende growth patterns, have age spectra with minor argon loss from high-temperature ages between 99.5 and 101 m.y. to low-temperature ages between 89 and 93 m.y. Although it is difficult to interpret confidently the high-temperature ages of these three samples, these data indicate another pulse of thermal activity at about 101 m.y.b.p. Argon loss exhibited by the low-temperature ages of these three samples probably occurred during emplacement and deformation of suture zone plutons (discussed below).

STRUCTURE OF METAMORPHIC ROCKS

Eastern Terrane

In part because earlier workers in the continental terrane were not able to identify a traceable stratigraphy, macroscopic folds were not described previously. However, high-angle normal faults and mesoscopic folds have been described by most previous workers (Greenwood and Morrison, 1973; Kopp, 1973; Myers, 1982; Reid, 1959) without interpretation of kinematic history or correlation with tectonic events. Farther north, Hietanen (1962) described a sequence of folds and faults in metasedimentary rocks and correlated formation of structures to tectonism during the Nevadan orogeny.

In the eastern terrane, the bedding-parallel metamorphic fabric of terrigenous metasedimentary rocks is multiply deformed. Folding events related to the fabric-forming metamorphic event (intrafolial folds with axial planar schistosity) are found in the structurally (and stratigraphically?) lowest-layered rocks ("basal metasedimentary unit") in the northeast corner of Fig. 11-2 (Reid, 1959). At present, this unit has been insufficiently studied for an interpretation to be made on vergence or chronology of structural elements.

Macroscopic folds and thrust stacking of different structural domains and metamorphic regimes are common in the four upper tectonostratigraphic units (Concord, Quartzite Butte, Moores, and Umbrella Butte units) of the eastern terrane. North-plunging, east-vergent mesoscopic folds and mineral lineations in the Concord unit formed during east-directed thrust faulting of the Quartzite Butte tectonostratigraphic unit over the Concord tectonostratigraphic unit. After assembly of the plates, these units were folded together into northwest-plunging folds paralleled by another set of mineral lineations.

Farther west, rocks of the Moores and Umbrella Butte units were deformed into generally southwest-plunging macroscopic folds that are associated with mesoscopic folds and mineral lineations. Concomitant thrust faults formed internally in large-scale folds. The faults progressively cut up-section to the east, thereby shortening the radius of folds. Segments of these folds were rotated out of parallelism during stacking by thrust faults (Fig. 11-2). This process resulted in juxtaposition of rocks of higher metamorphic grades above rocks of lower grades (Fig. 11-3) and formation of east-vergent, tight to isoclinal mesoscopic folds near the bases of overriding plates. These folds and thrust faults were refolded by a generation of gently west-plunging, upright macroscopic cross-folds.

The structural vergence of these phases of deformation is toward the east (Fig. 11-5). Folds on the eastern side of the study area are strongly asymmetric and indicate eastward transport. Folds nearest to the terrane boundary are more upright and exhibit only minor eastward asymmetry. Thrust faults are steep near the suture zone and more shallow to the east. Cross-folds superimposed on early asymmetric folds and thrust faults indicate continued compression but perhaps with a change in stress orientation.

Western Terrane

Mesoscopic and macroscopic folds in the island arc terrane of western Idaho were mentioned by Hamilton (1963b) and by Myers (1982). Onasch (1977, 1978) described multi-

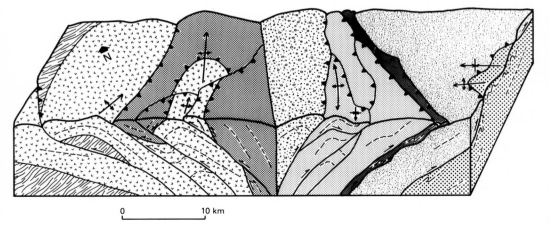

0   10 km

FIG. 11-5. Reconstruction of the macroscopic structures in the layered metamorphic rocks near the Salmon River suture. Patterning for metamorphic units same as Fig. 11-2.

ple phases of deformation and contrasted kinematic information from rocks of both the Riggins Group and Seven Devils island arc.

Early structural events (including intrafolial folds) described from lower greenschist facies rocks (Onasch, 1977, 1978) have orientations and vergence that form inconsistent patterns and cannot be related to formation of later structures. In higher-grade rocks, none of the fold events can be related confidently to formation of metamorphic fabric. Early structures are sparse in Slate Creek; no consistent chronology or orientation was determined (Lund, 1984). Rotation by later events, the probable presence of unrecognized macroscopic structures, and overprinting of metamorphic and deformational events make it difficult to identify and interpret domains of the early structural events. Because of this, early structures in the Slate Creek area could not be correlated confidently with those of Onasch.

The first structure, which can be correlated regionally, is a west-directed thrust fault that juxtaposes rocks of different metamorphic grades. Near Riggins, this fault, called the Rapid River thrust (Hamilton, 1963b), was described as separating greenschist facies metamorphic rocks of the Seven Devils island arc from overlying greenschist to amphibolite facies rocks of the Riggins Group. A major postmetamorphic thrust fault in the Slate Creek area is the northern extension of the Rapid River thrust fault.

In the Slate Creek area, the Rapid River thrust fault juxtaposes upper amphibolite facies metamorphic rocks correlated with the Riggins Group above lower amphibolite facies rocks correlated with the Seven Devils island arc. The Rapid River thrust was subsequently deformed into a macroscopic, upright, open antiform of both the Seven Devils and Riggins Group rocks (Fig. 11-5). This fold, the Slate Creek antiform, and associated minor folds have steep east-dipping axial planes, shallow northeast plunges, and west-directed asymmetry. An imbricate system of later west-moving reverse faults further disrupted metamorphic patterns, disrupted the Slate Creek antiform, over-steepened parts of the eastern limb, and offset segments of the antiformal axis (Lund, 1984). Minor folds associated with late reverse faults trend north-northeast and have west vergence.

Other macroscopic folds described near Riggins expose only Riggins Group rocks (Hamilton, 1963b; Onasch, 1977) and do not demonstrate a relationship between folding and juxtaposition of metamorphic grade along thrust faults such as the Rapid River thrust. Because of the depth of exposure in the northern fold, the Slate Creek antiform, it is possible to subdivide macroscopic deformational events into three nearly coaxial stages of faulting, folding, and later faulting.

Syn- to postmetamorphic deformation in greenschist and amphibolite facies rocks near the suture zone is west-directed (Fig. 11-5) as indicated by east-over-west vergence of both minor folds related to faulting and major folds. The nearly coaxial trends of these folds suggest prolonged deformation and consistent orientation of the regional stress fields. Folds near the suture are upright; those farther west exhibit west-vergent asymmetry. Near the juncture, steep reverse faults are common; shallower thrust faults predominate to the west.

Age of Deformation

Argon age-spectrum data constrain the timing of structural activity. Primary foliation, which in places partially transposes bedding, is defined (in the appropriate lithologies) by 118-m.y. hornblende that is associated with the onset of metamorphism (Table 11-1). Thus, structures that deform metamorphic fabric in these rocks are all 118 m.y. old or younger. Although we feel that the Rapid River thrust is only one of several important postmetamorphic fault systems within the western terrane, it is a well-known feature for which an age can be derived from the argon data. Because this fault system cuts 118- and

109-m.y. metamorphic rocks that have short cooling histories, the thrust formed after 109 m.y.b.p.

SUTURE ZONE PLUTONS
AND SUPERPOSED DEFORMATION

A series of hornblende-biotite tonalite plutons, some of which contain magmatic epidote, intruded along the suture zone after the metamorphic and structural activity described above (Fig. 11-6). The tectonic significance of such epidote-bearing granitoids is described by Zen (Chapter 2, this volume). These plutons cut rocks of both terranes and truncate metamorphic and structural trends in country rocks of both terranes; emplacement of these first-phase Idaho batholith plutons marks the time when the eastern and western terranes were contiguous in the near-present-day geometry. Tonalitic plutons and roof pendants and xenoliths within them were deformed together by a postemplacement event(s) that was confined to a zone about 5 km wide along the suture. The fabrics, which formed during this deformation in plutons and roof pendants that occupy the suture zone, are manifested by steep schistosity that parallels the suture zone and especially by steep southeast-plunging mineral lineations in plutonic rocks on the west side of the juncture (Lund, 1984); fabric in some of these rocks is lineation-dominated. The metamorphic event that accompanied deformation in xenolith and roof pendant rocks is manifested by retrogression of minerals or formation of second-generation minerals. Thus, during emplacement and deformation of the suture zone tonalitic plutons, amphibolite-facies country rocks locally underwent an additional metamorphic event of approximately equal magnitude to the earlier event(s) and were deformed into a new fabric within and parallel to the suture zone. The age of these plutons (see below) provides a minimum age for macroscopic deformation (described above).

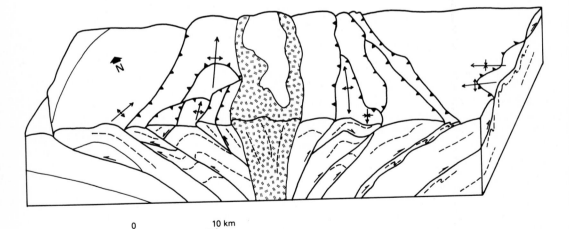

0   10 km

FIG. 11-6. Reconstruction of the intrusion and deformation of foliated tonalite plutons along the suture zone (accompanied by the latest metamorphic event and a cross-cutting fabric in the roof rocks in the suture zone area). Patterning for deformed plutonic units of the Idaho batholith same as Fig. 11-2.

Age of Suture Zone Plutons and Superposed Deformation

Determination of the age of tonalitic plutons within the suture zone by the $^{40}Ar/^{39}Ar$ age-spectrum technique is difficult because minerals may remain open to argon diffusion during postcrystallization cooling and deformation. Because the emplacement temperature of tonalite is about 850°C (Naney, 1983) and because the argon-closure temperature for hornblende is near 550°C, hornblende ages should provide a minimum estimate that, unless the plutons had an extended early cooling history, should be close to the age of crystallization. The $^{40}Ar/^{39}Ar$ age spectrum for a sample of hornblende taken from a metasedimentary xenolith (R23) in a suture zone pluton (Table 11-1; Fig. 11-4) provides important constraints on the emplacement history of the enclosing pluton. The argon-loss pattern preserved in this age spectrum is strong evidence that the metamorphic rock from which the hornblende was taken had cooled below the hornblende argon-closure temperature at, or before, 99.5 m.y.b.p.; this hornblende formed during an earlier metamorphic event. Argon loss preserved in the spectrum probably occurred during emplacement of the enveloping tonalite. Importantly, because this hornblende was not completely reset during pluton emplacement, it seems likely that the heat of the pluton was lost quickly. This suggests a short cooling history and strongly implies a high crustal emplacement level for the tonalite; therefore, the 93.5-m.y. date, an estimate of the time of the argon loss, may be close to the age of emplacement.

The age spectrum of hornblende sample R12 from the suture zone tonalite supports a 93.5-m.y. age of emplacement. Hornblende in this sample, which contains no epidote, has been deformed by a strong postemplacement event that produced a steep southeast-dipping lineation defined by deformed hornblendes. Argon loss exhibited by the age spectrum apparently occurred at 88 m.y., probably during deformation; the date of 93.5 m.y. for the highest-temperature extraction step is a minimum estimate of the age of emplacement of the tonalite.

The virtually identical data of hornblende samples R24 and R22, which were collected from the suture zone over 25 km north of R12 (Fig. 11-4), lends additional support to the interpretation that at least some tonalites within the suture zone were emplaced at a relatively shallow level at about 93 m.y. and that a major episode of deformation caused argon loss at about 88 m.y.b.p. Therefore, in the Slate Creek area, where the Rapid River thrust is folded and cut by later thrust faults, the original Rapid River thrusting event plus the later folding and rethrusting all occurred between 109 and 93 m.y. ago. Tonalite was emplaced at about 93 m.y.b.p. and before the 88-m.y. event formed a new fabric. This younger fabric cut off the 118- and 109-m.y.-old metamorphic fabrics near the suture zone.

The new fabric, whether planar or lineation-dominated, probably formed by means of significant vertical movement during the 88-m.y. event. The plateau age of 85.1 ± 0.4 m.y. of deformed hornblende from sample R11 is evidence that deformation in this area continued after 88 m.y. Plateau ages ranging from 78.8 ± 0.4 to 80.8 ± 0.5 m.y. on samples of deformed hornblende from tonalitic plutons about 50 km south of R11 (L. W. Snee, unpublished data) show that deformation, which was probably produced by vertical movement, occurred along some parts of the suture zone until about 79 m.y.b.p.

The emplacement, uplift, and cooling history of deformed plutons in the Riggins area is complex; however, the argon data apparently support the contention that epidote-bearing plutons crystallized at a significant depth. Within the suture zone itself where both epidote-bearing and epidote-free plutons occur side by side, the cooling data are apparently contradictory. Considering the complex, long-lived, vertical and horizontal movement on this zone, it would not be surprising for deep-seated rocks to be juxtaposed against shallow rocks.

UNDEFORMED PLUTONS

Away from the suture zone, hornblende-biotite tonalite and biotite granodiorite plutons intruded both terranes (Fig. 11-7). Epidote is present as a primary magmatic mineral in some of the undeformed tonalitic rocks. The plutons are mostly homogeneous in texture, although faint, inconsistently oriented foliation is present in places. The magmas intruded by stoping their roof rocks. These may be equivalent to, or comagmatic with, tonalitic rocks that intruded into, and were deformed along, the suture zone.

In the eastern terrane, the tonalite, granodiorite, and metamorphic country rocks were intruded by muscovite-biotite granite plutons. Muscovite is primary and these rocks have no directional fabrics. The granite, like the tonalite, intruded passively by stoping the roof. Country rock xenoliths, which are prevalent in the upper parts of the plutons, were not rotated during stoping and much of the mapping of structure and stratigraphy of the wall rocks was facilitated by continuity of stratigraphy and structure from one xenolith block to the next.

Both groups of undeformed plutons, as well as deformed suture zone tonalitic rocks, comprise the central Idaho batholith. Emplacement of these plutons documents a time when the terranes were juxtaposed in the approximate present-day configuration and most of the activity along the suture zone between terranes had ceased.

Age of Emplacement and Cooling of Undeformed Plutons

Table 11-1 summarizes ^{40}Ar/^{39}Ar age-spectrum data for undeformed plutons east of the suture zone. Sample R21, tonalite that contains magmatic epidote, is from the oldest undeformed phase of the Idaho batholith. The ^{40}Ar/^{39}Ar age-spectrum dates for hornblende, biotite, and microcline define the cooling history of this sample from about 530°C at 84 m.y., to 280°C at 81 m.y., to 130°C at 78 m.y. (see also Fig. 11-9). The hornblende plateau-date of 83.9 ± 0.4 m.y. is a minimum estimate for the age of emplace-

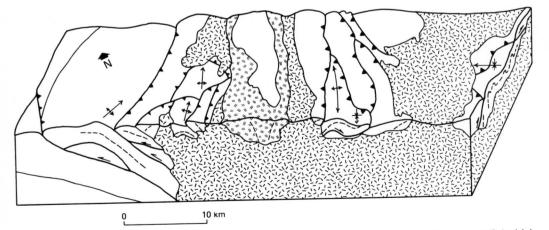

0   10 km

FIG. 11-7.   Reconstruction of the intrusion of unfoliated tonalite-granodiorite and granite plutons of the Idaho batholith into previously metamorphosed and deformed rocks of both terranes. Patterning for Idaho batholith plutonic units same as Fig. 11-2.

ment. Because an unpublished zircon U-Pb date (K. V. Evans, USGS, personal communication, 1984) of 89 ± 5 m.y. is similar to the hornblende ^{40}Ar/^{39}Ar date and because the presence of primary epidote suggests deep emplacement (>25 km; Zen, 1985), age of emplacement of this pluton is greater than 84 m.y.; possibly, these unfoliated tonalites were emplaced at the same time as chemically similar 93-m.y.-old suture-zone tonalites.

Sample R8 is a representative sample of the voluminous Idaho batholith muscovite-biotite granites. Muscovite and biotite plateau dates of 75.3 ± 0.4 and 74.7 ± 0.4 m.y., respectively, are statistically indistinguishable and are good approximations for age of crystallization, providing there was no extended cooling that affected the pluton. The suggestion that this is an emplacement age is supported by the 76.7 ± 0.4 m.y. date on a sample of pegmatite dike (R10) that intruded undeformed tonalite and that may be part of the same muscovite-biotite granite intrusive episode. However, the strongest evidence against prolonged cooling is that muscovite-biotite granite magmas were emplaced into country rock that includes the above-discussed tonalite, which had cooled below 130°C by 78 m.y. Assuming a minimum geothermal gradient of 20°C/km (see Zen, 1985 for arguments about geothermal gradient) at the time of emplacement, the maximum depth of emplacement of muscovite-biotite granite was 6.5 km (i.e., 130°C, microcline closure temperature, divided by 20°C/km). This depth is consistent with the conclusion of Lund *et al.* (1986) that 74-m.y.-old muscovite-biotite granite located 40 km to the east was emplaced at a depth of less than 9 km, and the conclusion of Toth (1985) that other muscovite-biotite granites of the Idaho batholith may have been emplaced as shallowly as 5 km. In general, the argon data support contentions of C. F. Miller *et al.* (1981) and of J. L. Anderson and Rowley (1981) that primary muscovite of impure composition can form at relatively low pressure. Additional ^{40}Ar/^{39}Ar muscovite dates from muscovite-biotite granite throughout the central and southern parts of the Idaho batholith (Snee, unpublished data) range from 70.1 ± 0.3 to 76.7 ± 0.4 m.y. Muscovite dates become younger toward the center of the large area of muscovite-biotite granite. This trend may be a result of multiple intrusive pulses, differential cooling, or exposure of deeper levels of emplacement.

Uplift History of the Eastern Terrane

The last pre-Miocene structural event in the region is manifested by sets of north-northeast trending, high-angle normal faults that are parallel to the Salmon River suture and along which uplift occurred (Fig. 11-8). The uplift event began during emplacement of unfoliated tonalitic plutons of the Cretaceous Idaho batholith (Lund *et al.*, 1986) and continued at least until after emplacement of Eocene epizonal granite magmas in the same region (Lund *et al.*, 1983b).

The rate of this uplift can be estimated from cooling rates. The cooling rate for the undeformed tonalite from 84 to 81 m.y. is about 95°C/m.y.; from 81 to 78 m.y., the rate is about 45°C/m.y. This rapid cooling rate is consistent with either high-level emplacement or rapid uplift following an unknown period of slow cooling after deep emplacement. The presence of magmatic epidote suggests emplacement at a depth of greater than 25 km (Zen and Hammarstrom, 1984) if the epidote crystallized at the level of pluton emplacement (B. W. Evans and Vance, 1985). Assuming that the tonalite was emplaced at a depth of 25 km or deeper and assuming a minimum geothermal gradient of 20°C/km based on a synthesis of studies by Carmichael (1978), R. D. Hyndman *et al.* (1979), Hollister (1982), and Zen (1985), the uplift rate from 84 to 81 m.y. was about 4 mm/yr; from 81 to 78 m.y., the rate was about 2 mm/yr. If a higher geothermal gradient were assumed, uplift would have been faster. These rates are only slightly less

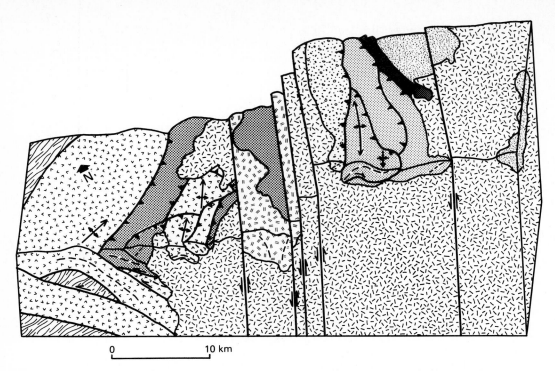

FIG. 11-8. Reconstruction of uplift that occurred along and parallel to the suture zone after emplacement of the Idaho batholith plutons (see Figs. 11-6 and 11-7). Patterning for units same as Fig. 11-2.

than the rapid, recent uplift of 5 mm/yr of the Nanga Pargat massif in the Himalaya (Zeitler, 1985) and 2.5–14 mm/yr along the Alpine fault in New Zealand (Adams, 1981).

After most of the uplift had occurred, basalt of the Clearwater Embayment of the Columbia River Plateau (Bond, 1963) covered pre-Tertiary rocks of the western terrane but did not overlap onto the eastern terrane. The basalt flows may have been moated against scarps along steep uplift-fault zones and then subsequently cut and displaced by post-Miocene faults that are parallel to earlier steep faults (Lund, 1984).

TECTONISM OUTBOARD
OF THE SALMON RIVER SUTURE

Previous models for both timing and style of accretion of exotic rocks to North America have been based on evidence for Late Triassic and Late Jurassic metamorphic, structural, and plutonic events outboard of the Salmon River suture. Because the Salmon River suture trends north and the effects of suturing also appear to trend north in belts parallel to the boundary, it is most effective for the purposes of this study to summarize the history of the outboard rocks with respect to time and geographic location (Fig. 11-9) and without regard to west-trending, internal terrane boundaries as defined by Brooks (1979), Dickinson (1979), Silberling and Jones (1984), and Vallier *et al.* (1977).

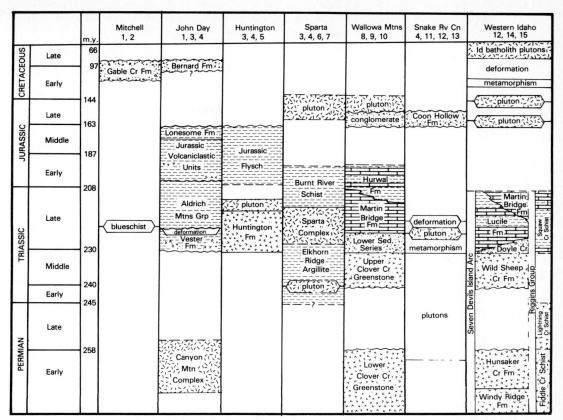

FIG. 11-9. Generalized correlation diagram of oceanic rocks from west-central Idaho to central Oregon after (1) Dickinson (1979); (2) Hotz *et al.* (1977); (3) Mullen and Sarewitz (1983); (4) Avé Lallemant *et al.* (1980); (5) Brooks (1979); (6) Walker (1981); (7) Walker (1983); (8) Ross (1938); (9) Smith and Allen (1941); (10) Follo and Siever (1984); (11) Balcer (1980); (12) Vallier (1977); (13) Avé Lallemant *et al.* (1985); (14) this study; and (15) Hamilton (1963b). See Fig. 11-1 for locations.

Late Triassic Tectonism

During the Late Triassic, metamorphism (at Mitchell and the Snake River Canyon), deformation (near John Day and in the Snake River Canyon), and plutonism (at Sparta and the Snake River Canyon) were contemporaneous with the volcanism associated with the Huntington island arc (near Huntington, also called the Olds Ferry terrane by Silberling and Jones, 1984) and with deposition of conglomeratic rocks in the upper part of the Seven Devils island arc (near John Day, in the Wallowa Mountains, and in western Idaho). As suggested by Brooks and Vallier (1978) and Dickinson and Thayer (1978), these events may be indicative of a Late Triassic tectonic event of major regional consequence. However, the geologic setting of this tectonism is not yet understood. Earliest workers suggested that these events reflected accretion of the exotic rocks to North America (Brooks and Vallier, 1978; Dickinson, 1979). Recent workers have suggested that these events are related to preaccretionary subduction (Avé Lallèmant *et al.*, 1980;

Hillhouse *et al.*, 1982; Oldow *et al.*, 1984). More specifically, Avé Lallèmant *et al.* (1985) interpreted a Late Triassic shear zone in the Snake River basement rocks to be an intra-arc left-lateral fault related to Late Triassic left-lateral plate convergence. Whatever the origin of these Late Triassic features, our study shows that these events are only mani-fested in the exotic rocks and are not a result of suturing these rocks to North America in their present position.

Late Jurassic Tectonism

Tonalitic plutons with poorly constrained Rb/Sr ages, which average about 150 m.y. old, intruded all central Oregon terranes (Armstrong *et al.*, 1977). Because the plutons cut terranes that have no other known connection, the emplacement of these plutons has been taken as the minimum time of amalgamation of the central Oregon terranes and has been interpreted as the time of accretion of the combined terranes to the North American continent (Avé Lallèmant *et al.*, 1980; Brooks and Vallier, 1978; Dickinson and Thayer, 1978; Hillhouse *et al.*, 1982). Paleomagnetic studies using samples from these presumed postaccretionary plutons have resulted in models for postcollisional tectonic rotation of blocks in the northwestern United States (Hillhouse *et al.*, 1982; Wilson and Cox, 1980). Because our study shows that accretion occurred between 118 and 93 m.y.b.p., these plutons do not represent the time of final accretion of the superterrane to North America. Rotation exhibited by these approximately 150-m.y.-old plutons may be pre-, syn-, or postaccretionary.

In the Wallowa Mountains, Late Jurassic plutons cut overturned folds in Lower Jurassic to Upper Triassic rocks of the Seven Devils island arc[5] (Prostka, 1962). The plutons do not represent the time of final accretion of the superterrane to North America. Mountains (Follo and Siever, 1984) and in the Snake River Canyon area (Morrison, 1964; Vallier, 1977). Thus, during the Late Jurassic, there was tectonism in the oceanic terrane that was not related to accretion of these rocks to the continent along the Salmon River suture.

Upper Cretaceous Deposits

In central Oregon, conglomeratic rocks of the Albian to Cenomanian Gable Creek Forma-tion crop out in the Mitchell inlier (Wilkinson and Oles, 1968). Partly correlative conglom-eratic rocks of the Cenomanian Bernard Formation crop out in the John Day inlier (Dickinson *et al.*, 1979). Detrital grains in sandstone layers include quartz, mica, and potassium feldspar. Conglomeratic layers include granitic and metaquartzite clasts as well as more common greenstone, volcanic, and phyllitic clasts. These sediments reflect a mixed provenance of metamorphic, plutonic, and volcanic rocks that was thought to be "a mature and dissected arc terrane" (Dickinson *et al.*, 1979).

Based on our study, it seems likely that the provenance contained both oceanic and continental sources. The origin of the Albian to Cenomanian conglomeratic rocks of the Gable Creek and Bernard Formations can be directly linked in time to metamorphic, deformational, plutonic, and uplift events along the Salmon River Suture in west-central Idaho. The Cretaceous orogenic belt along the Salmon River suture provides a source for lithic clasts and detrital grains that were not entirely derived from the oceanic rocks upon which they were deposited. At present, the only suspected outboard effect of accretion of the exotic terrane to North America is formation of these Cretaceous conglomeratic deposits.

[5] Also M. F. Follo, personal communication, 1985.

Geometry of Suture Zone

Stratigraphic information (Lund, 1984) and geochemical data (Armstrong *et al.*, 1977; Criss and Fleck, 1985) from the suture zone area indicate that the two juxtaposed terranes formed independently of each other. Depositional ages and compositions of layered rocks in the terranes preclude that either sequence formed by erosion of the other. Because the change in rock types at the surface corresponds to a change in basement type at depth, the Salmon River suture appears to be a steep crustal boundary (Lund, 1984). Such a near-vertical suture and lack of transitional rocks is dissimilar to what would be expected for subduction-produced boundaries.

The Salmon River suture zone lacks most features that are commonly associated with subduction zones (Ernst, 1984). Along the suture, there are no melange zones that consist of either forearc sedimentary deposits or forearc structures. Blueschist facies metamorphic assemblages are absent. Transitional deposits, which might indicate an ancient back-arc setting or a preaccretionary (pre-Cretaceous) connection of any type between the arc terrane and the North American continent, have not been demonstrated. There are also no ophiolitic rocks at the juncture to indicate the remnants of oceanic and transitional crust that must have once separated the terranes. The absence of these features along the terrane boundary indicates that subduction was not the major mechanism that formed the juncture.

Thermal History

The directional metamorphic fabric in both terranes was developed and multiply deformed between 118 and 93 m.y.b.p. prior to the intrusion of the main, undeformed plutons of the Idaho batholith at about 75 m.y.b.p.; the Idaho batholith was not the cause of regional metamorphism. Metamorphic isograds extend along the suture zone and metamorphic grade decreases away from the zone in both directions. This bilateral symmetry of metamorphic facies across the juncture is due to a temporal overlap of metamorphism and deformation. Deeper, more highly metamorphosed rocks were juxtaposed against lower grade rocks along thrust faults that had vergence outward from the suture zone. The resulting metamorphic and structural geometries are used to elucidate movement between the terranes.

Up to four episodes of metamorphism and/or deformation, two periods of plutonism, and a time of rapid uplift in the suture zone area are revealed by the combination of field and argon data. A summary of the thermal history is displayed on a time-temperature diagram (Fig. 11-10). The first event was clearly dynamothermal and was marked by a rapid rise to maximum metamorphic conditions at 118 m.y. The next two events also may have been dynamothermal, or the ages may reflect nonthermal structural activity that exposed two deeper, hotter crustal levels to cooling at 109 and 101 m.y.b.p., respectively. The maximum temperature during any of these three events has not yet been defined. The last stage of metamorphism and deformation at 93 m.y. included emplacement of tonalitic plutons. Subsequent cooling after each event is only constrained by a few samples and is approximated on Fig. 11-10. Nonetheless, the metamorphic rocks affected by the first three events had cooled below about 280°C (biotite argon-closure temperature) before the 93-m.y.-old event; heat associated with this latest event (an emplacement temperature of 850°C is estimated from Naney, 1983), although restricted to the immediate area of the suture zone, was great enough to disturb hornblende age spectra of minerals in country rocks that had formed during earlier events.

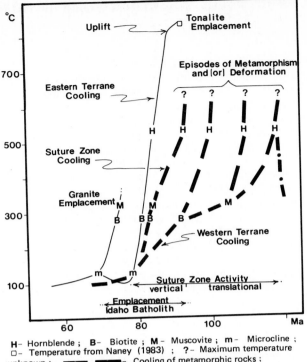

FIG. 11-10. Thermal history of the Salmon River suture zone, Idaho, based on time and temperature relationships derived from $^{40}Ar/^{39}Ar$ age-spectrum data.

Emplacement of tonalitic plutons stitched the terranes together and the western terrane became part of the stable North American continent. Deformation that was accompanied by emplacement of foliated tonalites may have been caused either by plate interaction or by the beginning stages of uplift. The eastern terrane began to be uplifted no later than 88 m.y.b.p. Late-stage, steep brittle faults formed at high crustal levels during uplift. Emplacement of muscovite-biotite granites took place at shallower crustal levels after much of the uplift was complete.

Sense of Movement on Suture Zone

Fold vergence, sense of displacement on faults, and attitudes of structural fabrics described in this study area are like those formed along large-scale strike-slip faults at other active plate margins (Badham, 1982; Harland, 1971; Lowell, 1972; Page et al., 1979; Sylvester and Smith, 1976). The structures are also similar to those produced in experimental models for strike-slip faults (Lowell, 1972; Odonne and Vialon, 1983; Wilcox et al., 1973). Presently active strike-slip zones such as the San Andreas and Alpine faults appear to be mostly one-sided. Along this cross section of the Salmon River suture zone, there is also some asymmetry. The metamorphic belt is wider on the east side, there has been larger scale thrust faulting eastward (well into western Montana; Snee, 1982), and we have documented much more uplift on the continental side. However, possibly because of the depth of exposure of the Salmon River suture, the 45-km-wide cross sections shown on Figs. 11-2, 11-3, and 11-5 depict a gross symmetry much like what is expected from experimental work.

Based on the foregoing geologic constraints, we conclude that the Salmon River suture formed as the result of oblique strike-slip accretion of the island arc superterrane to the North American continent. This convergent strike-slip faulting (or transpression; Harland, 1971) probably involved a complex series of events that is responsible for many of the enigmatic features of this suture. This mechanism is the simplest explanation for the direct juxtaposition of island arc rocks against continental rocks along a vertical boundary and for the concomitant absence of ocean basin (transitional) rocks. Although other processes may have been active as well, the Paleozoic shelf and slope deposits, which are largely missing in central Idaho, were probably cut out by transcurrent faulting in the same manner as other transitional rocks.

Because of subsequent plutonism and late-stage uplift processes, no shear indicators have been found along the suture zone that could directly give the relative movement between terranes. From regional information on Proterozoic and Paleozoic continental shelf sedimentary rocks in northern Washington and southern British Columbia (Price et al., 1985) and on the sense of displacement on other major Late Cretaceous transcurrent faults in the region (Gabrielse, 1985; Monger, 1984), it is most plausible to suggest that movement along the Salmon River suture was right-lateral. In addition, paleomagnetic data on 150-m.y.-old plutons that intruded several of the terranes outboard (in central Oregon) indicate post-150-m.y. and pre-Eocene clockwise rotation (see below; Hillhouse et al., 1982; Wilson and Cox, 1980); if the rotation is found to be the same age as orogenic events in the western terrane, this sense of rotation would be consistent with right-lateral faulting in much the same way as postulated for more recent events by Beck (1976).

SUMMARY

Exotic rocks of eastern Oregon and west-central Idaho were sutured to the North American continent during the interval 118 to 93 m.y.b.p. by transpressional movement. During this 25-m.y. interval, compression and crustal thickening resulted in metamorphic pulses. At about 93 m.y.b.p., metamorphic processes gave way to melting and the first plutonic phases were emplaced during the last stage of metamorphism and ductile deformation. After about 90 m.y.b.p., compression related to suturing relaxed, and consanguineous uplift, melting, and emplacement of voluminous magmas concluded the suturing process.

DISCUSSION

A transpressive model for the tectonic setting of the Salmon River suture zone leads to a better understanding of the processes of metamorphism, deformation, and plutonism that were active in the region. Deeply eroded cores of compressive strike-slip fault zones are not commonly recognized. Most examples of regional continental strike-slip faulting cited in the literature are active systems in which young or surficial rocks are disrupted. Minor crystalline components have been described along only a few major strike-slip faults (Howell, 1954; Jennings and Troxel, 1954; Lowell, 1972); even fewer are thought to be associated with an accompanying metamorphic event (Badham, 1982), and none have been directly related to plutonic activity. Therefore, several features of the geology of the Salmon River suture zone are enigmatic or controversial in the context of the model for transpressive fault zones. Because metamorphism, deformation, and plutonism can be related in time and space to transpressive motion along the suture zone in this case, the Salmon River suture may be one model for explaining features of exposed orogens and for understanding processes occurring at depth within compressional strike-slip faults.

The metamorphic and structural model for the Salmon River suture has ramifications for the origin of the enigmatic rocks of the Riggins Group. Detailed work on rocks of both the Seven Devils island arc and the Riggins Group has demonstrated similarities in lithology, chemistry, and protolith of these rocks. When observed at like metamorphic grades (both high and low grades), rocks of similar lithology contain the same mineralogy. Combined with this, regional metamorphic and structural geometries indicate that the Riggins Group may be a sliver of high-grade rocks that has been thrust over lower grade, equivalent rocks of the Seven Devils island arc.

Another problematic feature of the Salmon River suture zone is the string of ultramafic slivers that are recognized in the western terrane near the future zone (A. L. Anderson, 1930, 1931; Bonnichsen, 1985; Hamilton, 1963b; Onasch, 1977; Tullis, 1944). Sheared ultramafic bodies were emplaced along faults (Hamilton, 1963b; Onasch, 1977) that are west-vergent reverse faults parallel to the Salmon River suture. Although the west-central Idaho ultramafic slivers have been ascribed to subduction-related melange zones (Bonnichsen, 1985; Hamilton, 1963b), rocks that fit the accepted description of melange (Silver and Beutner, 1980) are not found (Onasch, 1977). In northern Sumatra, Page et al. (1979) described deformed ultramafic slices along faults parallel to an active, convergent strike-slip plate boundary and concluded that basement slivers had been brought up from depth by fault movement; ultramafic rocks in west-central Idaho may be analogous. Sheared serpentinite bodies in west-central Idaho may have been mobilized from oceanic basement deep within the orogenic belt by means of transpressive faulting. These ultramafic slivers may have been emplaced into the west side of the suture zone area by the same upwelling structural style that brought high-grade metamorphic rocks (the Riggins Group) up and over rocks from lower temperature-pressure regimes, and that brought deep-seated plutons up from great crustal depths.

Metamorphic patterns observed in the west-central Idaho suture zone are similar to those along the Alpine fault in New Zealand (Sibson et al., 1981) and the Work Channel lineament in western British Columbia (Crawford and Hollister, 1982). However, in both regions, the asymmetric geometry of metamorphic isograds is ascribed to differential uplift across the fault.

In the case of the Alpine fault, metamorphic grade increases toward the fault; metamorphic rocks form a band on the east side of the fault. Metamorphic isograds are parallel to the fault. The movement history along the Alpine fault before 20 m.y.b.p. is the subject of debate, but Sibson et al. (1981) and Adams (1981) have documented that 10 km of uplift and at least 120 km of transcurrent movement have occurred during the last 25 m.y. Thermal activity resulting from fault movement caused (1) formation of pseudotachylyte, (2) crystallization of micas in blastomylonite and adjacent schists (i.e., greenschist metamorphism), and (3) argon loss from micas in schists (temperatures of at least 300°C amounting to greenschist facies overprint are necessary; Snee et al., in press) up to 15 km away from the fault but at less than 10 km depth (Adams, 1981; Sibson et al., 1981).

This thermal information on a geometrically better constrained fault is significant in that greenschist-facies metamorphism occurred at shallow depths over a wide area. More heat would be expected at greater crustal depths. This strongly suggests that higher grade metamorphic rocks would occur below present exposure levels along the Alpine fault just as is seen in the more deeply eroded Salmon River suture.

In the case of the Work Channel Lineament, a suture zone between the Alexander and Stikine terranes in the Coast Ranges plutonic complex of British Columbia is characterized by a geometry and sequence of events that are similar to those in Idaho, except that plutonic activity was involved at an earlier stage in the British Columbian example. Inverted metamorphic sequences occur where high-grade metamorphic rocks were moved by thrust faults outward from the center of a metamorphic belt over lower-grade rocks (Crawford and Hollister, 1982; Hollister and Crawford, 1986). In addition, the highly deformed, high-grade metamorphic core was intruded at a later stage by elongated tonalite plutons that were ductilely deformed together with the host rocks during uplift of the central part of the orogenic belt. The origin of the Work Channel lineament is not explained, but metamorphism, deformation, and plutonism are thought to be concurrent processes that occurred in "surges." The direct cause of formation of plutons is said to be thickening of the crust during compression and underthrusting related to subduction (Hollister and Crawford, 1986).

The fold-thrust orogenic belt recognized in west-central Idaho indicates large-scale compression and crustal thickening that were caused by transpression rather than by subduction. Although it is possible that some as-yet-undiscovered subduction zone (coupled with the transpressive zone) did exist outboard of the study area or that at about 100 m.y.b.p. transpressive motion switched to an outboard subduction zone, we think it is likely that the processes of compression and crustal thickening may have been sufficient to have caused metamorphism and, ultimately, melting.

CONCLUSIONS

Metamorphic and structural fabrics in rocks of both oceanic and continental terranes in west-central Idaho along the deeply eroded Salmon River suture formed as the direct result of oblique transcurrent movement between terranes. Accretion of the exotic rocks to the North American continent took place during a 25-m.y. interval in the Cretaceous from 118 to 93 m.y.b.p. Although direct observations of the sense of movement along this transpressive fault are obscured by plutonic rocks and no offset of units can be documented at present, the movement was probably right-lateral. The first tonalite plutons of the Idaho batholith were emplaced at depths of about 25 km during the last stages of transcurrent movement at about 93 m.y. Emplacement of these plutons tacked the two terranes together by intruding both terranes. The last stage of metamorphism and ductile deformation occurred at about 88 m.y.b.p. in suture zone tonalites and their roof rocks. Large-scale uplift and erosion of the eastern terrane began at, or before, 88 m.y.b.p.; emplacement of muscovite-biotite granites took place around 75 m.y. at shallow crustal levels of about 6.5 km.

Outboard of the Salmon River suture, the Cretaceous collisional effects can be separated from older fabrics that formed during earlier events. Those structural and metamorphic features that are related to earlier events must be identified and investigated separately. This study shows that any future tectonic models to explain the history of interior plate processes in the region must take into account information on tectonic interactions at the plate margins.

Observations presented in this study demonstrate that major metamorphic and fabric elements near the juncture between these two terranes are related to events that brought the terranes together. Compression and crustal thickening caused by transpressive accretion first resulted in high-grade metamorphism and ultimately in formation of voluminous magmas.

LUND, SNEE

ACKNOWLEDGMENTS

This paper was made possible through the guidance and inspiration of Robert Scholten, E-an Zen, and Anna Hietanen. Careful, critical reviews by T. L. Vallier, J. A. D'Allura, W. W. G. Ernst, and W. B. Hamilton greatly improved the manuscript. The ^{40}Ar/^{39}Ar analyses were performed in J. F. Sutter's laboratory at the U.S. Geological Survey. We thank Sutter for the use of the facilities and for his continued involvement in this project. However, the authors alone are responsible for interpretations presented in this paper and for any errors in ideas or presentation.

REFERENCES

Adams, D. J., 1981, Uplift rates and thermal structure in the Alpine Fault Zone and Alpine schists, southern Alps, New Zealand, *in* McClay, K. R., and Price, N. J., eds., *Thrust and Nappe Tectonics:* Boston, Blackwell Scientific Publications, p. 211–222.

Anderson, A. L., 1930, The geology and mineral resources of the region about Orofino, Idaho: *Idaho Bur. Mines Geology Pamphlet 34*, 63 p.

____, 1931, Genesis of the anthophyllite deposits near Kamiah, Idaho: *J. Geology*, v. 39, p. 68–81.

____, 1956, Geology and mineral resources of the Salmon Quadrangle, Lemhi County, Idaho: *Idaho Bur. Mines Geology Pamphlet 106*, 102 p.

Anderson, J. L., and Rowley, M. C., 1981, Synkinematic intrusion of peraluminous and associated metaluminous granitic magmas, Whipple Mountains, California: *Can. Mineralogist*, v. 19, p. 83–101.

Armstrong, R. L., 1975, Precambrian (1500 m.y. old) rocks of central Idaho — The Salmon River Arch and its role in Cordilleran sedimentation and tectonics: *Amer. J. Sci.*, v. 275-A, p. 437–467.

____, Taubeneck, W. H., and Hales, P. O., 1977, Rb-Sr and K-Ar geochronometry of Mesozoic granitic rocks and their Sr isotopic composition, Oregon, Washington, and Idaho: *Geol. Soc. America Bull.*, v. 88, p. 397–411.

Avé Lallèmant, H. G., Phelps, D. W., and Sutter, J. F., 1980, ^{40}Ar/^{39}Ar ages of some pre-Tertiary plutonic and metamorphic rocks of eastern Oregon and their geologic relationships: *Geology*, v. 8, p. 371–374.

____, Schmidt, W. J., and Kraft, J. L., 1985, Major Late Triassic strike-slip displacement in the Seven Devils terrane, Oregon and Idaho — A result of left-oblique plate convergence? *Tectonophysics*, v. 119, p. 299–328.

Badham, J. P. N., 1982, Strike-slip orogens — An explanation for the Hercynides: *J. Geol. Soc. London*, v. 139, p. 493–504.

Balcer, D. E., 1980, ^{40}Ar/^{39}Ar ages and REE geochemistry of basement terranes in the Snake River Canyon, northeastern Oregon — western Idaho: M.S. thesis. Ohio State Univ. Columbus, Ohio, 111 p.

Beck, M. E., Jr., 1976, Discordant paleomagnetic pole positions as evidence of regional shear in the western Cordillera of North America: *Amer. J. Sci.*, v. 276, p. 694–712.

Beckwith, R. H., 1928, The geology and ore deposits of the Buffalo Hump district: *N.Y. Acad. Sci. Annals*, v. 30, p. 263–296.

Bond, J. G., 1963, Geology of the Clearwater embayment: *Idaho Bur Mines Geology Pamphlet 128*, 83 p.

Bonnichsen, B., 1985, Dunite at New Meadows, Idaho — An accreted fragment of oceanic crust: *Geol. Soc. America Abstr. with Programs*, v. 17, p. 209.

Brooks, H. C., 1979, Plate tectonics and the geologic history of the Blue Mountains: *Oreg. Geology*, v. 41, p. 71–80.

——, and Vallier, T. L., 1978, Mesozoic rocks and tectonic evolution of eastern Oregon and western Idaho, *in* Howell, D. G., and McDougall, K. A., eds., *Mesozoic Paleogeography of the Western United States:* Pacific Section, Soc. Econ. Paleontologists Mineralogists, Pacific Coast Paleogeography Symp. 2, p. 133–146.

Burchfiel, B. C., and Davis, G. A., 1975, Nature and controls of Cordilleran orogenesis, western United States — Extensions of an earlier synthesis: *Amer. J. Sci.*, v. 275-A, p. 363–396.

Carmichael, D. M., 1978, Metamorphic bathozones and bathograds — A measure of the depth of post-metamorphic uplift and erosion on the regional scale: *Amer. J. Sci.*, v. 278, p. 769–797.

Coney, P. J., Jones, D. L., and Monger, J. W. H., 1980, Cordilleran suspect terranes: *Nature*, v. 288, p. 329–333.

Crawford, M. L., and Hollister, L. S., 1982, Contrast of metamorphic and structural histories across the Work Channel lineament, Coast Plutonic Complex, British Columbia: *J. Geophys. Res.*, v. 87, p. 3849–3860.

Davis, G. A., Monger, J. W. H., and Burchfiel, B. C., 1978, Mesozoic construction of the Cordilleran "collage," central British Columbia to central California, *in* Howell, D. E., and McDougall, K. A., eds., *Mesozoic Paleogeography of the Western United States:* Pacific Section, Soc. Econ. Paleontologists Mineralogists, Pacific Coast Paleogeography Symp. 2, p. 1–32.

Dickinson, W. R., 1979, Mesozoic forearc basin in central Oregon: *Geology*, v. 7, p. 166–170.

——, 1981, Plate tectonics and the continental margin of California, *in* Ernst, W. G., ed., *The Geotectonic Development of California* (Rubey Vol. I): Englewood Cliffs, N.J., Prentice-Hall, p. 1–28.

——, and Thayer, T. P., 1978, Paleogeographic and paleotectonic implications of Mesozoic stratigraphy and structure in the John Day inlier of central Oregon, *in* Howell, D. G., and McDougall, K. A., eds., *Mesozoic Paleogeography of the Western United States:* Pacific Section, Soc. Econ. Paleontologists Mineralogists, Pacific Coast Paleogeography Symp. 2, p. 147–161.

——, Helmold, K. P., and Stein, J. A., 1979, Mesozoic lithic sandstones in central Oregon: *J. Sediment. Petrology*, v. 49, p. 501–516.

Dodson, M. H., 1973, Closure temperature in cooling geochronological and petrological systems: *Contrib. Mineralogy Petrology*, v. 40, p. 259–274.

Ernst, W. G., 1984, California blueschists, subduction, and the significance of tectono-stratigraphic terranes: *Geology*, v. 12, p. 436–440.

Evans, B. W., and Vance, J. A., 1985, Properties of truly magmatic epidote: *Geol. Soc. America Abstr. with Programs*, v. 17, p. 576.

Evans, K. V., 1981, Geology and geochronology of the eastern Salmon River Mountains, Idaho, and implications for regional Precambrian tectonics: Ph.D. thesis, Pennsylvania State Univ., University Park, Pa., 222 p.

——, and Fischer, L. B., 1986, U-Pb geochronology of two augen gneiss terranes, Idaho — New data and tectonic implications: *Can. J. Earth Sci.*, v. 23, p. 1919–1927.

——, and Lund, Karen, 1981, The Salmon River "Arch"? *Geol. Soc. America Abstr. with Programs*, v. 13, p. 448.

Fleck, R. J., and Criss, R. E., 1985, Strontium and oxygen isotopic variations in Mesozoic and Tertiary plutons of central Idaho: *Contrib. Mineralogy Petrology*, v. 90, p. 291–308.

Follo, M. F., and Siever, Raymond, 1984, Conglomerates as clues to the evolution of a suspect terrane: Wallowa Mountains, Oregon: *Geol. Soc. America Abstr. with Programs*, v. 16, p. 510.

Gabrielse, H., 1985, Major dextral transcurrent displacements along the Northern Rocky Mountain Trench and related lineaments in north-central British Columbia: *Geol. Soc. America Bull.*, v. 96, p. 1–14.

Greenwood, W. R., and Morrison, D. A., 1973, Reconnaissance geology of the Selway — Bitterroot Wilderness Area: *Idaho Bur. Mines Geology Pamphlet 154*, 30 p.

Hamilton, W. R., 1963a, Overlapping of late Mesozoic orogens in western Idaho: *Geol. Soc. America Bull.*, v. 74, p. 779–788.

____, 1963b, Metamorphism in the Riggins region, western Idaho: *U.S. Geol. Survey Prof. Paper 436*, 95 p.

____, 1969, Mesozoic California and the underflow of Pacific mantle: *Geol. Soc. America Bull.*, v. 80, p. 2409–2430.

____, 1976, Tectonic history of west-central Idaho: *Geol. Soc. America Abstr. with Programs*, v. 8, p. 378.

____, 1978, Mesozoic tectonics of the western United States, *in* Howell, D. G., and McDougall, K. A., eds., *Mesozoic Paleogeography of the Western United States:* Pacific Section, Soc. Econ. Paleontologists Mineralogists, Pacific Coast Paleogeography Symp. 2, p. 33–70.

Harland, W. B., 1971, Tectonic transpression in Caledonian Spitsbergen: *Geol. Mag.*, v. 108, p. 27–42.

Harrison, J. E., Griggs, A, B., and Wells, J. D., 1974, Tectonic features of the Precambrian Belt basin and their influence on post-Belt structures: *U.S. Geol. Survey Prof. Paper 866*, 15 p.

Harrison, T. M., 1981, Diffusion of ^{40}Ar in hornblende: *Contrib. Mineralogy Petrology*, v. 78, p. 324–331.

____, and McDougall, I., 1980, Investigations of an intrusive contact, northwest Nelson, New Zealand — I. Thermal, chronological, and isotopic constraints: *Geochim. Cosmochim. Acta*, v. 44, p. 1985–2003.

____, and McDougall, I., 1982, The thermal significance of potassium feldspar K-Ar ages inferred from ^{40}Ar/^{39}Ar age spectrum results: *Geochim. Cosmochim. Acta*, v. 46, p. 1811–1820.

Helwig, James, 1974, Eugeosynclinal basement and a collage concept of orogenic belts, *in* Dott, R. H., Jr., and Shaver, R. H., eds., *Modern and Ancient Geosynclincal Sedimentation:* Soc. Econ. Paleontologists Mineralogists Spec. Publ. 19, p. 359–376.

Hietanen, A., 1962, Metasomatic metamorphism in western Clearwater County Idaho: *U.S. Geol. Survey Prof. Paper 344-A*, 116 p.

____, 1963a, Anorthosite and associated rocks in the Boehls Butte quadrangle and vicinity, Idaho: *U.S. Geol. Survey Prof. Paper 344-B*, 78 p.

____, 1963b, Metamorphism of the Belt Series in the Elk River-Clarkia area, Idaho: *U.S. Geol. Survey Prof. Paper 344-C*, 49 p.

____, 1963c, Idaho batholith near Pierce and Bungalow, Clearwater County, Idaho: *U.S. Geol. Survey Prof. Paper 344-D*, 42 p.

____, 1968, Belt series in the region around Snow Peak and Mallard Peak, Idaho: *U.S. Geol. Survey Prof. Paper 344-E*, 34 p.

Hillhouse, J. W., Gromme, C. S., and Vallier, T. L., 1982, Paleomagnetism and Mesozoic tectonics of the Seven Devils volcanic arc in northeastern Oregon: *J. Geophys. Res.*, v. 87, no. 85, p. 3777–3794.

Hollister, L. S., 1982, Metamorphic evidence for rapid (2 mm/yr) uplift of a portion of the Central Gneiss Complex, Coast Mountains, B.C.: *Can. Mineralogists*, v. 20, p. 319–332.

——, and Crawford, M. L., 1986, Melt-enhanced deformation — A major tectonic process: *Geology*, v. 14, p. 558–561.

Hotz, P. E., Lanphere, M. A., and Swanson, D. A., 1977, Triassic blueschist from northern California and north-central Oregon: *Geology*, v. 5, p. 659–663.

Howell, B. F., Jr., 1954, Geology in the Little Tujunga area, Los Angeles County, *in* Jahns, R. H., ed., *Geology of Southern California:* Calif. Div. Mines Bull. 170, map sheet 10.

Hyndman, D. W., 1979, Major tectonic elements and tectonic problems along the line of section from northeastern Oregon to west-central Montana: *Geol. Soc. America Map Chart Series MC-28C*.

——, and Talbot, J. L., 1976, The Idaho batholith and related subduction complex: *Field Guide 4, Cordilleran Section, Geol. Soc. America*, 15 p.

Hyndman, R. D., Jessop, A. M., Judge, A. S., and Rankin, D. S., 1979, Heat flow in the Maritime Provinces of Canada: *Can. J. Earth Sci.*, v. 16, p. 1154–1165.

Jennings, C. W., and Troxel, B. W., 1954, Geologic guide through the Ventura basin and adjacent areas, southern California, *in* Jahns, R. H., ed., *Geology of Southern California:* Calif. Div. Mines Bull. 1970, Geol. Guide 2, p. 15–19.

Kistler, R. W., and Peterman, Z. E., 1978, Reconstruction of crustal blocks of California on the basis of initial strontium isotopic compositions of Mesozoic granitic rocks: *U.S. Geol. Survey Prof. Paper 1071*, 17 p.

Kopp, R. S., 1973, Stratigraphy, primary sedimentary structures, and depositional environment of metasedimentary rocks of the Gospel Peak area, Idaho County, Idaho, *in Belt Symposium 1973:* Idaho Bur. Mines Geology, v. 1, p. 206–207.

Lindgren, W., 1904, A geological reconnaissance across the Bitterroot Range and Clearwater Mountains in Montana and Idaho: *U.S. Geol. Survey Prof. Paper 27*, 123 p.

Lowell, J. D., 1972, Spitsbergen Tertiary orogenic belt and the Spitsbergen fracture zone: *Geol. Soc. America Bull.*, v. 83, p. 3091–3102.

Lund, K., 1984, Tectonic history of a continent-island arc boundary — West-central Idaho: Ph.D. thesis, Pennsylvania State Univ., University Park, Pa., 207 p.

——, Scholten, R., and McCollough, F. M., 1983a, Consequences of interfingered lithologies in the Seven Devils island arc: *Geol. Soc. America Abstr. with Programs*, v. 15, p. 284.

——, Rehn, W. R., and Holloway, C. D., 1983b, Geologic map of the Blue Joint Wilderness study area, Ravalli County, Montana, and the Blue Joint roadless area, Lemhi County, Idaho: *U.S. Geol. Survey Misc. Field Studies Map MF-1557-B*.

——, Snee, L. W., and Evans, K. V., 1986, Age and genesis of precious-metals deposits, Buffalo Hump district, central Idaho: Implications for depth of emplacement of quartz veins: *Econ. Geology*, v. 81, p. 990–996.

McCollough, W. F., 1984, Stratigraphy, structure, and metamorphism of Permo-Triassic rocks along the western margin of the Idaho batholith, John Day Creek, Idaho: M.S. thesis, Pennsylvania State Univ., University Park, Pa., 141 p.

Miller, C. F., Stoddard, E. F., Bradfish, L. J., and Dollase, W. A., 1981, Composition of plutonic muscovite: genetic implications: *Can. Mineralogist*, v. 19, p. 25–34.

Monger, J. W. H., 1984, Cordilleran tectonics: A Canadian perspective: *Bull. Soc. Geol. France*, v. 27, no. 2, p. 255–278.

——, Price, R. A., and Tempalman-Kluit, D. J., 1982, Tectonic accretion and the origin of the two major metamorphic and plutonic welts in the Canadian Cordillera: *Geology*, v. 10, p. 70–75.

Morrison, R. F., 1964, Upper Jurassic mudstone unit named in Snake River Canyon, Oregon-Idaho boundary: *Northwest Sci.*, v. 38, p. 83–87.

Mullen, E. D., and Sarewitz, D., 1983, Paleozoic and Triassic terranes of the Blue Mountains, northeast Oregon — Discussion and field trip guide, Pt. 1. A new consideration of old problems: *Oreg. Geology*, v. 45, p. 65–68.

Myers, P. E., 1982, Geology of the Harpster area, Idaho County, Idaho: *Idaho Bur. Mines Geology Bull. 25*, 46 p.

Naney, M. T., 1983, Phase equilibria of rock-forming ferromagnesian silicates in granitic systems: *Amer. J. Sci.*, v. 283, p. 993–1033.

Odonne, F., and Vialon, P., 1983, Analogue models of folds above a wrench fault: *Tectonophysics*, v. 99, p. 31–46.

Oldow, J. S., Avé Lallèmant, H. G., and Schmidt, W. J., 1984, Kinematics of plate convergence deduced from Mesozoic structures in the western Cordillera: *Tectonics*, v. 3, p. 201–227.

Onasch, C. M., 1977, Structural evolution of the western margin of the Idaho batholith in the Riggins, Idaho area: Ph.D. thesis, Pennsylvania State Univ., University Park, Pa., 196 p.

——, 1978, Multiple folding along the western margin of the Idaho batholith in the Riggins, Idaho area: *Northwest Geology*, v. 7, p. 34–38.

Page, B. G. N., Bennett, J. D., Cameron, N. R., Bridge, D. McC., Jeffery, D. H., Keats, W., and Thaib, J., 1979, A review of the main structural and magmatic features of northern Sumatra: *J. Geol. Soc. London*, v. 136, p. 569–579.

Price, R. A., Monger, J. W. H., and Roddick, J. A., 1985, Cordilleran cross-section; Calgary to Vancouver: *Geol. Soc. America Fieldtrip Guide*, 85 p.

Prostka, H. J., 1962, Geology of the Sparta quadrangle, Oregon: *Oreg. Dept. Geology Miner. Industries Geol. Map Series GMS-1*.

Reid, R. R., 1959, Reconnaissance geology of the Elk City region, Idaho: *Idaho Bur. Mines Geology Pamphlet 121*, 74 p.

Ross, C. P., 1938, The geology of part of the Wallowa Mountains, Oregon: *Oreg. Dept. Geology Miner. Industries Bull., 3*, 74 p.

Shenon, P. J., and Reed, J. C., 1934, Geology and ore deposits of the Elk City, Orogrande, Buffalo Hump, and Tenmile districts, Idaho County, Idaho: *U.S. Geol. Survey Circ. 9*, 89 p.

Sibson, R. H., White, F. H., and Atkinson, B. K., 1981, Structure and distribution of fault rocks in the Alpine Fault Zone, New Zealand, *in* McClay, K. R., and Price, N.J, eds., *Thrust and Nappe Tectonics:* Boston, Blackwell Scientific, p. 197–210.

Silberling, N. J., and Jones, D. L., 1984, Lithotectonic terrane maps of the North American Cordillera: *U.S. Geol. Survey Open File Rpt. 84-523*, pt. C, 43 p.

Silver, E. A., and Beutner, E. C., 1980, Melanges: *Geology*, v. 8, p. 32–34.

Smith, W. D., and Allen, J. D., 1941, Geology and physiography of the northern Wallowa Mountains, Oregon: *Oreg. Dept. Geology Mining Industries Bull. 12*, 65 p.

Snee, L. W., 1982, Emplacement and cooling of the Pioneer batholith, southwestern Montana: Ph.D. thesis, Ohio State Univ. Columbus, Ohio, 320 p.

——, Sutter, J. F., Lund, K., Balcer, D. E., and Evans, K. V., 1986, ^{40}Ar/^{39}Ar age-spectrum data for metamorphic and plutonic rocks from west-central Idaho: *U.S. Geol. Survey Open File OF 87-0052*, 20 p.

——, Sutter, J. F., and Kelly, W. C., in press, Thermochronology of economic mineral deposits — Dating the stages of mineralization at Panasqueira, Portugal by high-precision ^{40}Ar/^{39}Ar age-spectrum techniques on muscovite: *Econ. Geology*.

Steiger, R. H., and Jäger, E., 1977, Subcommission on geochronology — Convention on the use of decay constants in geo- and cosmo-chronology: *Earth Planet. Sci. Lett.*, v. 36, p. 359–362.

Sylvester, A. G., and Smith, R. R., 1976, Tectonic transpression and basement-controlled deformation in San Andreas fault zone, Salton trough, California: *Amer. Assoc. Petrol. Geologists Bull.*, v. 60, p. 2081–2102.

Toth, M. I., 1985, Geology of the Idaho batholith: *Geol. Soc. America Abstr. with Programs*, v. 17, p. 269.

Tullis, E., 1944, Contributions to the geology of Latah County, Idaho: *Geol. Soc. America Bull.*, v. 55, p. 131–164.

Vallier, T. L., 1977, The Permian and Triassic Seven Devils Group, western Idaho and northeastern Oregon: *U.S. Geol. Survey Bull. 1437*, 58 p.

——, and Batiza, Rodey, 1978, Petrogenesis of spilite and keratophyre from a Permian and Triassic volcanic arc terrane, eastern Oregon and western Idaho, U.S.A.: *Can. J. Earth Sci.*, v. 15, p. 1356–1369.

——, Brooks, H. C., and Thayer, T. P., 1977, Paleozoic rocks of eastern Oregon and western Idaho, *in* Stewart, J. H., Stevens, C. H., and Fritsche, A. E., eds., *Paleozoic Paleogeography of the Western United States:* Pacific Section, Soc. Econ. Paleontologists Mineralogists, Pacific Coast Paleogeography Symp. 1, p. 455–466.

Wagner, W. R., 1945, A geological reconnaissance between the Snake and Salmon Rivers north of Riggins, Idaho: *Idaho Bur. Mines Geology Pamphlet 74*, 16 p.

Walker, N. W., 1981, U-Pb geochronology of ophiolitic and volcanic-plutonic arc terranes, northeastern Oregon and westernmost-central Idaho [abstract]: (EOS) *Trans. Amer. Geophys. Un.*, v. 62, p. 1087.

——, 1983, Pre-Tertiary evolution of northeastern Oregon and west-central Idaho — Constraints based on U/Pb ages of zircons: *Geol. Soc. America Abstr. with Programs*, v. 15, p. 371.

Wilcox, R. E., Harding, T. P., and Seely, D. R., 1973, Basic wrench tectonics: *Amer. Assoc. Petrol. Geologists Bull.*, v. 57, p. 74–96.

Wilkinson, W. D., and Oles, K. F., 1968, Stratigraphy and paleoenvironments of Cretaceous rocks, Mitchell quadrangle, Oregon: *Amer. Assoc. Petrol. Geologists Bull.*, v. 52, p. 129–161.

Wilson, D., and Cox, A., 1980, Paleomagnetic evidence for tectonic rotation of Jurassic plutons in Blue Mountains, eastern Oregon: *J. Geophys. Res.*, v. 85, no. 87, p. 3681–3689.

Winkler, H. G. F., 1979, *Petrogenesis of Metamorphic Rocks:* New York, Springer-Verlag, 348 p.

Zeitler, P. K., 1985, Cooling history of the NW Himalaya, Pakistan: *Tectonics*, v. 4, p. 127–151.

Zen, E., 1985, Implications of magmatic epidote-bearing plutons on crustal evolution in the accreted terranes of northwestern North America: *Geology*, v. 13, p. 266–269.

——, and Hammarstrom, J. M., 1984, Magmatic epidote and its petrologic significance: *Geology*, v. 12, p. 515–518.

D. W. Hyndman, D. Alt, and J. W. Sears
Department of Geology
University of Montana
Missoula, Montana 59812

12

POST-ARCHEAN METAMORPHIC AND TECTONIC EVOLUTION OF WESTERN MONTANA AND NORTHERN IDAHO

ABSTRACT

Post-Archean evolution of the northern Rocky Mountains began during Proterozoic time with deposition of enormously thick sedimentary sequences and the establishment, through rifting, of a new continental margin. When that margin became a collision boundary during Late Jurassic and Cretaceous time, the former continental shelf was crushed against the edge of the continent, along with formerly offshore island arcs and their associated sediments. Simultaneous strong heating of the western part of the continent above the sinking oceanic slab regionally metamorphosed the Proterozoic sedimentary pile and melted part of it, along with its basement, to form the Bitterroot and Atlanta batholiths, the two major parts of the Idaho batholith complex.

These batholiths lie within a broad band of regionally metamorphosed rocks that range in grade from sillimanite-orthoclase zone assemblages near the contact to greenschist facies rocks 50 km or more away. Regionally metamorphosed rocks near the deeper, western part of the Bitterroot batholith probably crystallized at a depth of at least 25 km. They locally contain kyanite with superimposed andalusite and cordierite. Andalusite and cordierite appear in the mineral assemblages farther southeast as the zone of regionally metamorphosed rocks narrows and finally disappears in favor of relatively minor contact haloes around shallow plutons. Three main stages of deformation are associated with the primary high-grade metamorphism, the rise of granitic magmas, and their emplacement.

Regional metamorphism preceded large-scale thrusting and folding, which became progressively younger eastward. Granitic magmas rose between 5 and 10 km into the metamorphosed Proterozoic section, then crystallized at a depth of 10–15 km to form the main part of the Bitterroot batholith. Associated plutons east of the main complex were emplaced at much shallower crustal levels, presumably because the magmas were drier there. Emplacement of the batholith complex continued from about 90 to 70 m.y. ago, with the eastern parts being younger. Large parts of the complex were tectonically unroofed, probably during the latest stages of emplacement.

A second episode of igneous and tectonic activity developed about 50 m.y. ago. Numerous, widely scattered granitic intrusions were emplaced at extremely shallow depth. Many erupted. Simultaneous tectonic activity involved emplacement of a large dike swarm on a north-northeast trend and strike-slip faulting along the nearly perpendicular Lewis and Clark fault zone, and several parallel trends.

INTRODUCTION

Daniel and Berg (1981) compiled radiometric age dates on Montana rocks, and plotted a histogram of frequency versus age. That graph shows distinct peaks at 2700, 1600, approximately 1100, 70, and 50 m.y. ago. In this review we very briefly consider Precambrian events, then dwell at length on the tectonic and metamorphic development of the northern Rocky Mountains during late Mesozoic time. We conclude with a brief review of Eocene tectonism and igneous activity.

THE ARCHEAN BASEMENT

High-grade metamorphic rocks of the continental basement are widely exposed in southwestern Montana, as well as in parts of central and northern Idaho. Basement rock in most areas is dominantly of middle to high amphibolite grade, assemblages in the sillimanite-orthoclase zone being most widespread. Low granulite assemblages occur locally in rocks of mafic or ultramafic composition that probably descended from essentially anhydrous predecessors. Many basement rocks of southwestern Montana contain retrograde greenschist mineral assemblages, most notably in the large talc deposits of southwestern Montana, which apparently formed through alteration of high-grade dolomitic marbles.

The Archean basement of southwestern Montana contains numerous large diabase dikes. Wooden *et al.* (1978) provided a good general description of those in the southern Tobacco Root Range, which are probably typical. The relatively few dates that Daniel and Berg (1981) cite for those dikes cluster in the ranges from 1000 to 1130 and from 1430 to 1455 m.y. years, so they may record at least two tectonic events.

PROTEROZOIC SEDIMENTATION, DIABASE INTRUSION, AND THE CONTINENTAL RIFT

Daniel and Berg (1981) cite a large group of dates on Proterozoic sedimentary rocks that spreads from 800 to 1400 m.y. Those dates refer to deposition and burial metamorphism of the sedimentary section and the emplacement into it, and into the basement, of swarms of diabase sills and dikes.

Proterozoic Deposition

The Middle Proterozoic Belt Supergroup of western Montana, northern Idaho, and easternmost Washington is the most voluminous and widely distributed sedimentary section in the northern Rocky Mountains. Other Lower or Middle Proterozoic sedimentary sections include the Lemhi Group and Yellowjacket Formation of central Idaho and nearby southwestern Montana. The Upper Proterozoic Windermere Group and its equivalents extend discontinuously from British Columbia south at least to Death Valley, with a noteable gap between eastern Washington and central Idaho.

Total thickness of the Proterozoic section is unknown and doubtless varies from one area to another, but is certainly very great. Harrison (1972), and C. Wallace (personal communication, 1979) for example, used stratigraphic reasoning to estimate the total thickness of the Belt Supergroup at about 20 or 21 km; D. Winston (personal communication, 1979) based his estimate of about 15 km on similar reasoning.

Diabase Sills and Dikes

Throughout western Montana and northern Idaho, the Proterozoic formations contain numerous thick sills and less numerous dikes of diabase. Several authors, including Sears and Price (1978), interpreted the diabase as evidence of continental rifting. The abrupt disappearance of Proterozoic formations in westernmost Idaho and easternmost Washington also suggests rifting, as does the absence of those formations in the region farther west. Figure 12-1 shows the geographic relationships. Daniel and Berg (1981) cite relatively few radiometric ages of diabase intrusions in the Proterozoic sections of western Montana and northern Idaho. Those range from 750 to 1430 m.y. but tend to cluster around 1400, 1100, and 800 m.y.

The Windermere Group and its equivalents include a variety of rock types, such as pillow basalts and graywacke, that can plausibly be associated with a newly established continental margin. The linear distribution of the Windermere formations near the continental margin established in Proterozoic time also suggests an association between

Windermere deposition and rifting. Radiometric ages on diabase sills in the Windermere section cluster near 800 m.y., so it seems reasonable to accept the inference of Crittenden *et al.* (1972) and Stewart (1972) of a Late Proterozoic date for establishment of a continental margin that was to last until late Mesozoic time. Figure 12-1 shows the approximate present course of that margin.

Precambrian Metamorphism of the Proterozoic Section

Many of the Proterozoic formations, especially those low in the section, contain greenschist facies mineral assemblages, apparently formed through burial metamorphism. Maxwell and Hower (1967) showed that the grade of metamorphism decreases with increasing height in the Belt section. Norwick (1972) found that upper greenschist assemblages are nearly ubiquitous in the enormously thick Prichard Formation of northwestern Montana and northern Idaho.

Potassium-argon ages on greenschist mineral assemblages in Belt rocks range from 900 to 1400 m.y., leaving little doubt that burial metamorphism accompanied sedimentation. The rocks typically lack foliation and in large areas remain only slightly deformed. Indeed, many appear quite unmetamorphosed unless examined in thin section. More locally, the secondary minerals form conspicuous biotite clots that resemble those in a contact hornfels. One of the chief effects of burial metamorphism was to enhance bedding-plane anisotropy, which may well have influenced Mesozoic deformation.

Heat from the numerous diabase intrusions in the Belt section certainly contributed to burial metamorphism. However, except for very narrow contact zones, there is no generally clear association between growth of secondary minerals and proximity to diabase intrusions. Norwick's (1972) suggestion that the geothermally driven heat flux was the major agent of metamorphism seems reasonable.

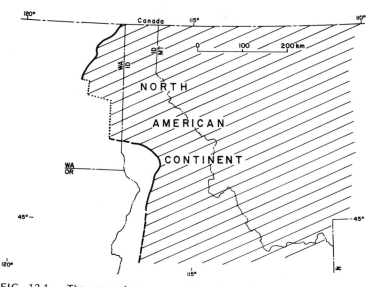

FIG. 12-1. The approximate present location of the continental margin established during the Late Proterozoic rifting event and modified by later events.

PALEOZOIC DEPOSITION

Considerable sections of marine Paleozoic formations survive east and south of the Idaho batholith and in northeastern Washington. If the Antler orogeny affected the part of the northern Rocky Mountains under discussion here, the evidence of its occurrence remains unrecognized.

THE EARLY MESOZOIC COLLISION

Coney (1972) related opening of the Atlantic Ocean during Jurassic and Early Cretaceous time to collision between the Pacific Ocean floor and the continental margin. That encounter telescoped the continental shelf that had accumulated since Late Proterozoic time against the continental margin, metamorphosed and partially melted the rocks, and added exotic oceanic terranes. Remnants of that tectonic accumulation exist in discontinuous exposure in northeastern Washington, western Idaho, and northeastern Oregon, as shown in Fig. 12-2. Subduction associated with that collision created a broad zone of metamorphic and igneous activity, the Idaho batholith complex, near the old western margin of the continent. For discussion of portions of these terranes in eastern Washington and west-central Idaho, see Hansen and Goodge, Rhodes and Hyndman, and Lund and Snee, Chapters 9, 10 and 11 respectively, this volume.

The Fold Belt of Northeastern Washington

A strong fold belt continuous with the Kootenay arc of British Columbia trends generally south through northeastern Washington, then disappears beneath Miocene basalts of the Columbia Plateau. It consists of tightly folded and metamorphosed Proterozoic, Paleozoic, and lower Mesozoic sedimentary formations extensively invaded by granitic intrusions. This fold belt strongly resembles the northern Sierra Nevada, which may indeed be its southern continuation.

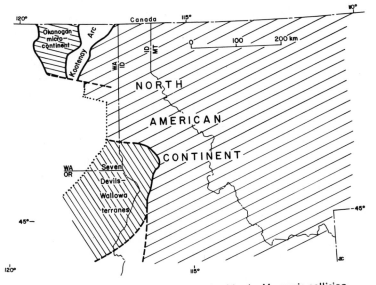

FIG. 12-2.   Distribution of rocks involved in the Mesozoic collision.

POST-ARCHEAN METAMORPHIC AND TECTONIC EVOLUTION

Many of the formations in the northeastern Washington fold belt have the general aspect of sediments deposited in relatively shallow water. The rocks strongly suggest a tightly deformed continental shelf and coastal plain crushed against the continental margin on which it formed. The accumulation also includes lesser volumes of apparent abyssal sediments such as ribbon cherts and turbidites. The youngest deformed sedimentary units contain marine fossils of Jurassic age. Miller and Engels (1975) reported radiometric dates on granitic intrusions that range from about 50 to 100 m.y.

THE WESTERN BORDER DEFORMED BELT

Rocks and structures in the western part of the Idaho batholith complex, illustrated in Fig. 12-3, record subduction and accretion west of the old continental margin as well as metamorphism and emplacement of the batholith complex in the old continent. To a large and still incompletely known extent, the structures formed in those two environments overprint each other.

The Former Western Border of the Continent

Hamilton (1963, 1969) showed that the Riggins Group of westernmost Idaho consists largely of strongly deformed volcanic rocks and oceanic sediments, locally metamorphosed to the amphibolite facies and penetrated by granitic plutons. The Rapid River thrust fault separates the Riggins Group from the less metamorphosed Seven Devils Complex that underlies it on the west. The Seven Devils Complex of the Hell's Canyon area contains large volumes of severely deformed Permian and Triassic volcanic rocks of generally intermediate composition, along with Triassic and Lower Jurassic sedimentary rocks. At least some of the sedimentary rocks appear to be reef facies with faunas exotic to North America. Lund and Snee (1985) interpreted the Seven Devils Complex and Riggins Group as two parts of a single accreted terrane.

Armstrong et al. (1977) and Fleck and Criss (1985) showed that the boundary between initial strontium isotope ratios typical of continental (>0.704) and oceanic (<0.704) regions coincides with a westward transition from granitic to dioritic rocks in the western part of the Idaho batholith. P. E. Meyers (personal communication, 1985) mapped a largely mylonite zone*, the Western Border mylonite of Fig. 12-3, that closely approximates the strontium isotope boundary. Lund and Snee (1985) and Lund et al. (1985) reported dates between 85 and 95 m.y. for dioritic rocks in the western part of the Idaho batholith. Toth and Stacey (1985) found ages between 80 and 105 m.y. for similar rocks, and Fleck and Criss (1985) reported ages between 80 and 95 m.y. These dates are distinctly older than those typical of the more granitic parts of the Bitterroot batholith farther east.

We suggest that the southern extension of the Kootenay arc in northeastern Washington and the more mafic igneous rocks of the western part of the Idaho batholith are parts of the same fold belt. The former western edge of North America probably coincides with the transition from granitic to dioritic rocks in the western part of the Idaho batholithic complex, and lies close to the Western Border mylonite. This segment of the margin may be a transform fault that truncated the old continental margin before the Seven Devils terrane and Riggins Group docked.

---

*Strayer et al. (1987) demonstrate that collision to the Seven Devils — Wallowa terrane formed a major mylonite zone with movement plunging about 50° NE.

The Salmon River Arch

Armstrong (1975) reported radiometric dates on rocks from the Salmon River arch in a range around 1500 m.y. If Mesozoic granite formerly extended across the Salmon River arch (Hyndman, 1983), it is possible that the rocks there were reheated, and their ages partially reset. If valid, the 1500-m.y. dates are barely old enough to permit interpretation of rocks in the Salmon River arch as basement for the Proterozoic Belt sedimentary section.

THE CENTRAL PLUTONIC AND METAMORPHIC ZONE

The Idaho batholith (Figs. 12-3 and 12-4) includes two major bodies of late Mesozoic granitic rocks north and south of the Salmon River arch. Armstrong (1975) called them the Bitterroot and Atlanta lobes of the Idaho batholith, our Bitterroot and Atlanta batholiths. This review focuses on the Bitterroot batholith and its environs.

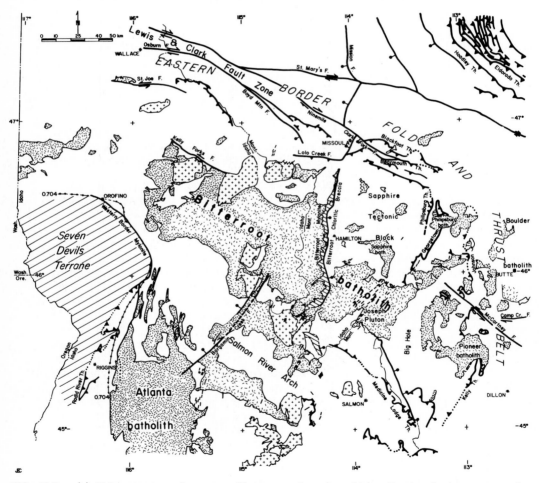

FIG. 12-3. (a) Major structures in western Montana and northern Idaho. Random hachure pattern, Late Cretaceous plutons; plus pattern, Early Tertiary plutons. (b) Main regional subdivisions.

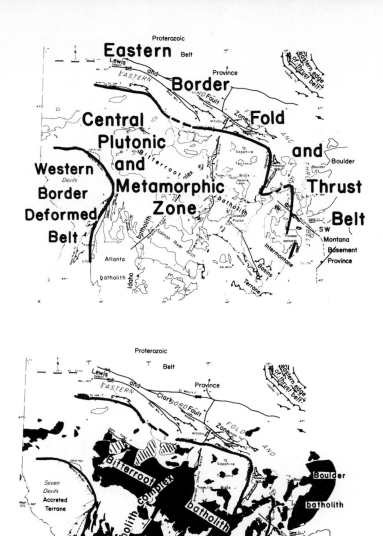

The Batholith Complex
and Its Associated Metamorphic Rocks

The geologic map of Idaho (Bond, 1978) and Fig. 12-4 show the country rocks surrounding the Bitterroot batholith becoming generally older with increasing distance from the granite. The granite therefore lies in the trough of a regional syncline, which, like the batholith, trends and plunges southeast.

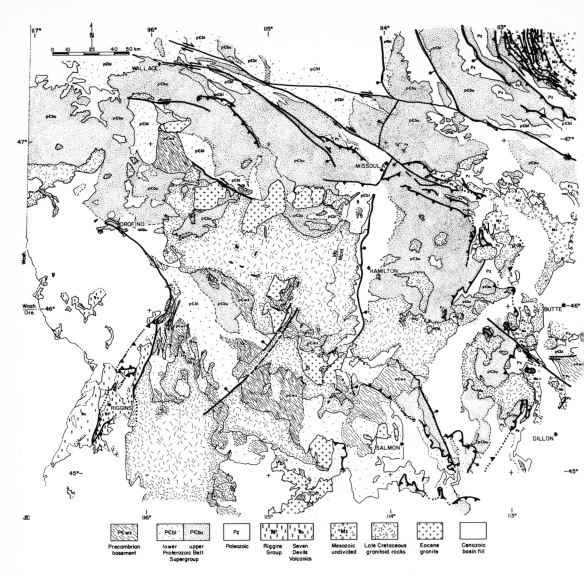

FIG. 12-4. Major geological subdivisions of the Idaho batholith region.

Figure 12-5 shows metamorphic grades reaching the sillimanite zone adjacent to the granite contact around most of the northwestern two-thirds of the Bitterroot batholith. A broad zone of regionally metamorphosed rocks with strongly developed foliation extends outward from the contact for variable distances, locally as great as about 70 km. Their foliation reliably distinguishes Proterozoic rocks regionally metamorphosed during Mesozoic time from those subjected only to burial metamorphism during Proterozoic time. Metamorphism around the southeastern end of the batholith and its satellite plutons is limited to contact aureoles generally less than 1 km wide. Metamorphic mineral assemblages also indicate that the depth of granite emplacement shallows southeastward.

In general, the metamorphic zones form a pattern crudely concentric to the contacts

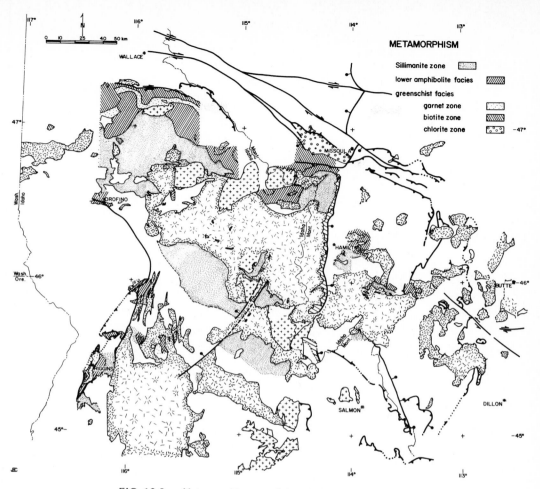

FIG. 12-5. Metamorphic map of the Idaho batholith region.

of the northwestern two-thirds of the Bitterroot batholith. They decline in grade with increasing distance from the granite. Many metamorphic zones in the range from high amphibolite to greenschist facies rocks are missing from the map of Fig. 12-5 because the original Proterozoic sedimentary section contains relatively few rocks with aluminous compositions that might recrystallize into diagnostic mineral assemblages. Chase and Johnson (1975) pointed out that the granite locally cuts at a low angle across the trend of foliation and the isograds of its metamorphic envelope.

Hietanen (1956, 1961) found a Barrovian sequence of mineral assemblages northwest of the Bitterroot batholith that requires reaction of staurolite and quartz to produce kyanite and garnet before the appearance of sillimanite — see field 1 of Fig. 12-6. In nearby areas, Childs (1982) reported assemblages that include quartz, muscovite, biotite, staurolite, kyanite, and sillimanite, which correspond to field 2 of Fig. 12-6. Both assemblages suggest that the rocks recrystallized at a depth greater than 25 km. On the northeast side of the Bitterroot batholith, in the northern Bitterroot Range, Nold (1968, 1974), Chase (1973), Wehrenberg (1972), and Cheney (1975) described migmatites with mineral assemblages that include quartz, muscovite, biotite, sillimanite, and orthoclase. The

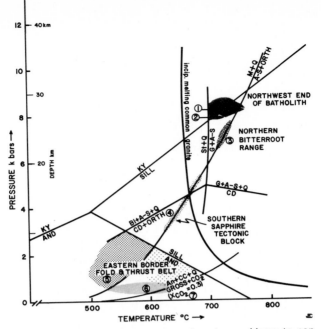

FIG. 12-6.   *P-T* diagram relating mineral assemblages to conditions of metamorphism around the Bitterroot batholith.

position of those assemblages in Fig. 12-6 implies crystallization at depths between 20 and 25 km.

Farther east, in the southern part of the Sapphire tectonic block, Presley (1970) and LaTour (1974) found assemblages of sillimanite, muscovite, and orthoclase in rocks that do not contain migmatites; that situation corresponds to field 4 of Fig. 12-6, and indicates a depth of less than 15 to 18 km. Still farther east, in the eastern fold and thrust belt, Stuart (1966) reported a foliated assemblage of quartz, muscovite, biotite, and cordierite that corresponds to field 5 of Fig. 12-6 in contact aureoles around the Royal stock in the Flint Creek Range. In a nearby area, Hyndman *et al.* (1982) reported calc-silicate assemblages that correspond to field 6 of Fig. 12-6. Alonso and Friberg (1985) reported an assemblage that includes andalusite and microcline in contact rocks at McCartney Mountain in southwestern Montana. Grossularite is abundant nearby. That corresponds to field 7 of Fig. 12-6. Those associations, typical of those around the eastern fringes of the Idaho batholith complex, suggest crystallization at shallow depth. The presence of miarolitic cavities in the granite and its emplacement in Paleozoic sedimentary formations confirm that inference.

Generation of Granite Magma

Hietanen (1956, 1961, 1963, 1984) found that the metamorphic terrane northwest of the Bitterroot batholith includes rocks that contain quartz, muscovite, kyanite, and sillimanite. Hyndman (1981) showed that an assemblage that contains muscovite, quartz, and oligoclase melts in the pressure and temperature conditions that define field 1 in Fig. 12-7, and that the melt would contain enough water to cause it to crystallize at constant temperature

POST-ARCHEAN METAMORPHIC AND TECTONIC EVOLUTION

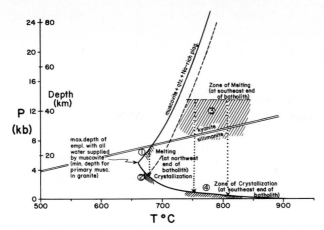

FIG. 12-7.   Melting relations near the northwestern end of the
Bitterroot batholith.

western Bitterroot batholith melted at a depth of between 20 and 25 km. Then it rose to a depth of approximately 15 km, where the northwestern part crystallized.

Granite rose to higher levels that correspond to field 4 of Fig. 12-7 near the southeastern end of the Bitterroot batholith, presumably because those magmas melted at deeper levels. The, source, perhaps in the lower crust, must have been initially undersaturated in water to permit crystallization at the shallow depths inferred from field 7 of Fig. 12-6. The large volume and felsic nature of the granite indicate that it was generated within the continental crust. Assuming a present crustal thickness of about 40 or 45 km and a shallow level of emplacement and thus minimal erosion, the granite magma would have been generated at depths less than about 45 km.

Compositionally plausible source rocks for the granite magma include the basement and some lower parts of the Proterozoic section, most notably the immensely thick Prichard Formation of the Belt Supergroup. Dexter *et al.* (1979) and Schuster and Bickford (1985) found that zircons from the Bitterroot batholith have uranium-lead upper intercept ages of about 1700 m.y. Their results strongly suggest that at least some of the magma formed through melting of basement rocks.

Deformation Associated with Regional Metamorphism

Table 12-1 contains our synthesis of the literature on the higher-grade metamorphic rocks around the Bitterroot batholith. Three generally recognizable phases of deformation overlap in time throughout the strongly metamorphosed rocks around the northwestern two-thirds of the batholith. The phases of deformation apparently accompanied the initial high-grade metamorphism that accompanies anatexis and emplacement of the Bitterroot batholith. The association between deformation and intrusion, shown in the sequence of cross sections of Fig. 12-8, is generally clear near the batholith, much less so near the western border deformed belt.

Lund and Snee (1985) contended that transform faulting ceased along the former western border of the continent about 105 m.y. ago. Folding and thrust faulting began then, and continued until about 80 m.y. ago, when the suture between the old continental margin and the rocks to the west was closed and stitched by emplacement of early plutons of the batholith complex (Fleck and Criss, 1985). The metamorphism that created the batholith was contemporaneous with the early, more mafic igneous rocks associated with its western border, and probably began about 100 m.y. ago.

First Stage of Deformation

The first phase of deformation, D1, probably began about 105 m.y. ago during emplacement of early tonalitic plutons in the western part of the Idaho batholith complex. Those rocks are foliated, and their contacts tend to conform with the foliation of their country rocks. We attribute that foliation to the D1 stage of deformation.

The D1 deformation produced tight isoclinal folds with very strong axial plane schistosity that generally parallels compositional layering. Although D1 folds are typically relatively small, large structures have been reported in several areas. Flood (1974), for example, found large D1 folds in the Anaconda Range near the eastern end of the Bitterroot batholith. In a nearby area, Heise (1983) found foliation of the D1 deformation lying parallel to the original bedding in Paleozoic deformations. Childs (1982) described other large folds along the northwestern margin of the Bitterroot batholith, as did Reid *et al.* (1973). In the northern Bitterroot Range, Hyndman and Chase (1979) showed a large overturned fold in the Proterozoic section, probably with a core of basement rock, and R. E. Kell (unpublished data) found a large refolded nappe.

Metamorphism associated with D1 folding is generally in the middle to upper part of the amphibolite facies. Childs (1982) and Hietanen (1956, 1961, 1963) reported that the sequence of isograds northwest of the Bitterroot batholith goes to the sillimanite-orthoclase zone, which corresponds to the onset of melting in Figs. 12-6, and 12-7. Chase (1973), Nold (1974) and Cheney (1975) all reported sillimanite-muscovite-orthoclase assemblages in the northeast border zone of the batholith. Inasmuch as D1 is the only phase of metamorphism that generally produced mineral assemblages and anatectic migmatites corresponding to the onset of melting, it seems reasonable to associate it with generation of the magma that became the Bitterroot batholith.

Second Stage of Deformation

The second, D2, phase of deformation produced open to nearly isoclinal folds that deform the axial plane schistosity developed during the D1 stage. Near the batholith, D2 folds typically plunge nearly vertically, and show close association with thrust faulting. This stage of deformation probably accompanied rise of the magma and final emplacement of the Bitterroot batholith, the second diagram of Fig. 12-8. A new axial plane schistosity, locally strong but generally weak, developed along with the D2 folding. It typically involved growth of muscovite and biotite in the new schistosity. Childs (1982) described development of sillimanite during D2 folding near the northwestern end of the Bitterroot batholith.

Third Stage of Deformation

The third, D3, stage of deformation involved concentric and flexural slip folding of the schistosities formed during the D1 and D2 stages of deformation. The D3 folds are typically upright and relatively open, generally near the margins of the batholith. Their formation appears to correspond to the final stages of forceful emplacement of the Bitterroot batholith. No new axial plane schistosity developed in them.

Some authors describe what may be another stage of deformation still too poorly defined to warrant formal designation. These last folds are very broad warps typically

TABLE 12-1. The Sequence and Times of Major Events in the Development of the Bitterroot Batholith and Associated Structures

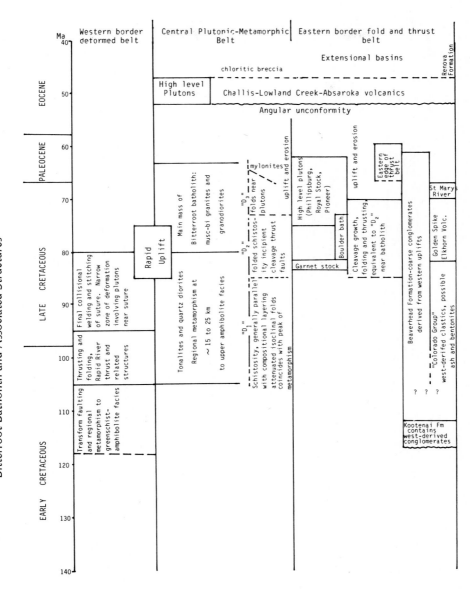

345

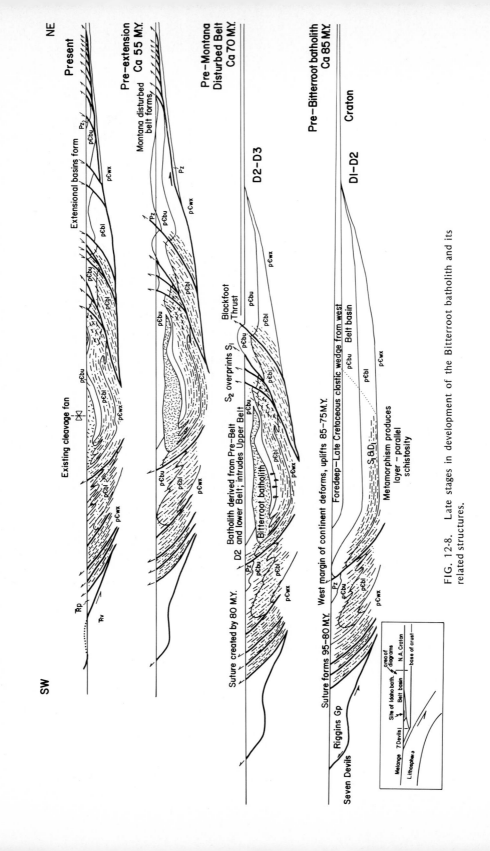

FIG. 12-8. Late stages in development of the Bitterroot batholith and its related structures.

several kilometers across and generally parallel to the margins of the Bitterroot batholith. They probably formed during final uplift of the batholith complex.

Eastern Border Fold-and-Thrust Belt

Greenschist facies rocks farther from the batholith typically show well-developed slaty cleavage that appears to merge into the D2 schistosity closer to the batholith. The slaty cleavage parallels the axial planes of folds, and involves sedimentary formations deposited as recently as mid-Cretaceous time. Folds are associated with east- and northeast-directed thrust faults that cut latest Cretaceous rocks along the margins of the Sapphire and Pioneer detachment blocks in western Montana. Eocene volcanic rocks overlap the faults.

EMPLACEMENT OF THE BITTERROOT BATHOLITH

Figure 12-9 shows schematically the tendency of the Bitterroot batholith to become shallower southeastward. Granite magma that melted within the zone of most intense regional metamorphism rose as the surrounding metamorphosed country rocks subsided to form the regional syncline. Rising magma dragged older country rock upward near the contacts of the batholith. In most areas those older rocks were the metamorphosed Lower Proterozoic section, but locally they may include Archean basement.

The highly calcareous Wallace Formation, now metamorphosed to calc-sillicate gneisses, and highly quartzose formations of the Ravalli Group surround most of the

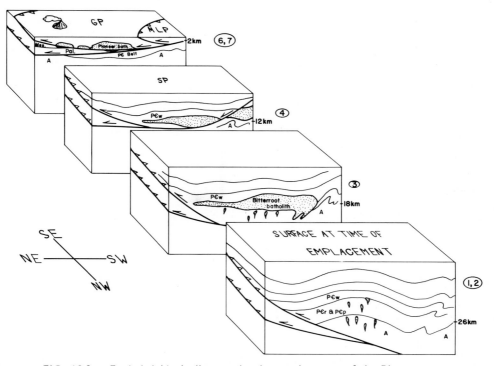

FIG. 12-9.   Exploded block diagram showing emplacement of the Bitterroot batholith, looking south. The numbers correspond to those of the numbered fields of Fig. 12-6.

northwestern part of the Bitterroot batholith. Neither includes compositions that might have melted to form granitic magma. The Wallace Formation and Ravalli Group lie several kilometers above the Prichard Formation, which is dominantly semipelitic or quartzofeldspathic, so the magma rose at least that far before it crystallized.

In a study of migmatites associated with the Bitterroot batholith, Johnson (1975) showed that the overwhelming majority of the granitic dikes and sheets were emplaced by injection. Very few at the present level of exposure formed through partial melting. We infer that the injection migmatites formed as the magma rose to the level where both are now exposed. Migmatites formed through partial melting exist only in gneisses apparently derived from the Prichard Formation. Rising magma probably dragged those rocks up from the zone of melting.

Berg (1968) and Cheney (1975) described anorthosites in the northeastern border zone of the Bitterroot batholith, as did Hietanen (1956) in the northwestern border zone. All appear to be basement rock. Jens (1972) described strongly metamorphosed mafic rocks in the northeastern border zone, which could be part of a layered mafic intrusion dragged up from the basement, or a large diabase sill injected during Proterozoic time.

Wiswall (1979) showed that part of one pluton in the western Bitterroot batholith is a thin sheet of granite with its basal intrusive contact locally exposed. Evidently, the rising magma spread laterally after it rose above its zone of melting. The substratum of that part of the batholith is a little-investigated complex of high-grade metamorphic rocks, basement or metamorphosed Proterozoic formations.

The southeastern end of the Bitterroot batholith and its satellite plutons locally intrude Paleozoic and Mesozoic sedimentary formations. Associated contact hornfels and skarns rarely extend more than 1 km from the igneous contact, and exhibit little foliation. Mineral assemblages locally include andalusite and cordierite, and reach the lower amphibolite facies. The general absence of a strongly developed metamorphic environment suggests that the magma of the southeastern end of the Bitterroot batholith rose well above its zone of melting. The close association in that region between granite and thrust faults suggests that at least some magma may have moved some distance east of its zone of melting.

The tendency for the depth of granite emplacement to become shallower eastward indicates that the water content of the magma in the eastern part of the region was lower than in that farther west. Either the proportion of drier basement rock that melted in the metamorphic environment was higher in the eastern part of the region than farther west, or less steam was rising from the deeper parts of the sinking oceanic slab.

Mafic Dikes

Hyndman (1985) found large numbers of mafic dikes throughout the Bitterroot batholith. Foster (1986) measured in detail a 10-km-long traverse across part of the northern Bitterroot batholith, in which mafic dikes account for 20% of the rock. Less systematic observation reveals that similar dikes abound generally in the Bitterroot and Atlanta batholiths.

Field observation and compositional data suggest that the mafic dikes were emplaced while the granite was at least partly molten. Dikes and granite typically display mutually intrusive contact relations. Virtually all the dikes are metamorphosed, with the foliation generally oriented obliquely to the contact. Some dikes break up into pillow structures within the granite. The most mafic dikes are calc-alkaline basaltic andesites. Compositions of others fall on a linear trend toward that of the granite, which evidently reflects combination of primary granite and calc-alkaline basaltic andesite. Some dikes contain portions of intermediate composition, which locally include discrete areas of mafic and granitic composition that are mechanically mixed, but not chemically combined.

Hyndman (1985) argued that the simultaneous existence of large volumes of granitic and basaltic magma suggests that heat transfered through rising basaltic magma was a major contributor to regional metamorphism and anatexis. Intrusion of swarms of dikes into the granitic magma certainly means that extension accompanied emplacement of the batholith. It is not clear whether the entire crust was spreading, or merely the rising batholith. Resolution of that question depends in large part on knowing whether the dike swarm crosses the Salmon River arch.

THE LEWIS AND CLARK FAULT ZONE

Figure 12-3 shows an enigmatic but obviously major structure, the Lewis and Clark fault zone, which passes north of the Bitterroot batholith. It is a swarm of faults that clearly extends at least from the vicinity of Spokane southeast to the northern margin of the Boulder batholith. Structures somehow associated with the Lewis and Clark fault zone may continue southeast in the parallel trends of the Lake Basin, Nye Bowler, Cat Creek, and similar fault zones of central Montana. It may also extend northwest along the boundary between the Columbia Plateau and the Okanogan microcontinent to the Okanogan Valley of central Washington.

Weidman (1965) provided general descriptions of the zone, and Reynolds and Kleinkopf (1977) interpreted it as a continental transform structure. Sheriff *et al.* (1984) related movement on the zone to Tertiary crustal extension. Many authors have interpreted the Lewis and Clark fault zone as a swarm of large right-lateral transcurrent faults. Nevertheless, several lines of evidence converge to persuade us that major displacement was left lateral, and happened during Late Cretaceous time. Some of the faults apparently moved again in a right-lateral direction during Tertiary time.

The Geologic Map of Idaho (Bond, 1978) shows that the northerly trend of batholiths is offset approximately 90 km to the west, from the Bitterroot batholith south of the Lewis and Clark fault zone to the Kaniksu batholith farther north. Meyers (1985) mapped the Western Border mylonite zone curving sharply west as it approaches the Lewis and Clark fault zone from the south. We interpret both the offset of the batholith trend and curvature of the mylonite zone as evidence of left-lateral displacement on the Lewis and Clark fault zone. R. E. Kell (unpublished data) found that the Kelly Fork and Osburn faults, elements of the Lewis and Clark fault zone, offset thrust faults north of the Bitterroot batholith in a pattern that suggests left-lateral displacement.

It is extremely difficult to assign an age to movement along faults of the Lewis and Clark zone because few of the rocks involved in its western reach are younger than Proterozoic. The offset thrust faults north of the Bitterroot batholith are associated with emplacement of the granite, so they are probably between 70 and 80 m.y. old. Large east-trending shears in the Boulder batholith appear to have formed soon after the granite crystallized, about 70 m.y. ago. We suggest that Late Cretaceous movement continued until at least this time.

DETACHMENT BLOCKS AND THRUST FAULTING

Two large blocks of the upper crust, the Sapphire and Pioneer tectonic blocks, Fig. 12-10, apparently detached from the Bitterroot batholith. Large parts of the Bitterroot and Atlanta batholiths well beyond the former positions of those two detached blocks were certainly exposed before mid-Eocene time, when volcanic rocks blanketed an erosion surface developed on them. It seems likely that those areas were also tectonically exposed, and that other large detachments remain to be recognized.

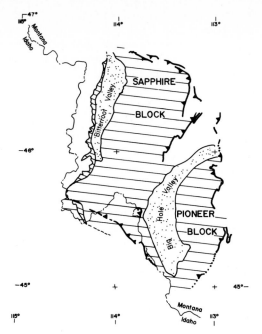

FIG. 12-10. The Sapphire and Pioneer detachment blocks. Both apparently moved off the Idaho batholith complex, leaving the Bitterroot Valley and the Big Hole behind their trailing edges.

Exposures of the Bitterroot batholith consistently reveal numerous, nearly horizontal slickensides and veins of pegmatite. Examination of the granite in thin section invariably reveals strained and broken mineral grains. Nearby Tertiary granites contain neither horizontal pegmatites nor evidence of internal strain. We infer that the pervasive internal deformation of the Bitterroot batholith occurred during Cretaceous time, during unloading of the Sapphire and Pioneer detachment blocks.

Hietanen (1956) reported metamorphic assemblages in the area northwest of the Bitterroot batholith that contain kyanite, sillimanite, and andalusite. Hyundman and Alt (1972) showed that the andalusite is clearly later than the rest of the assemblage. Cheney (1975) found similarly superimposed andalusite in rocks near the northeastern border of the Bitterroot batholith. We believe the andalusite probably formed in response to the drop in pressure that accompanied unloading of the detachment blocks.

One major effect of the detachment of two large blocks of the upper crust, and other large-scale thrust faulting east of the Bitterroot batholith, was to distribute granite over a large area east of the main mass of the Bitterroot batholith. Contact metamorphism, locally on a large scale, is associated with those intrusions.

The Sapphire Tectonic Block

Hyndman *et al.* (1975) and Hyndman (1980, 1983) argued that the Sapphire tectonic block detached from the Bitterroot batholith and moved east gravitationally, some 70 km into western Montana, while large volumes of granite magma reduced the shear strength of the crust. The broad arc open to the west of the Garnet, Flint Creek,

POST-ARCHEAN METAMORPHIC AND TECTONIC EVOLUTION

and Anaconda-Pintlar ranges defines the northern, eastern, and southern borders of the Sapphire block. The Bitterroot Valley appears to be a gap, now deeply filled with sediment, that opened behind its trailing margin.

Hyndman (1980, 1983) described the Bitterroot mylonite zone that truncates the eastern margin of the Bitterroot batholith and its associated regionally metamorphosed rocks for a distance of at least 80 km. The mylonite contains strongly sheared mineral assemblages in the sillimanite zone, a grade consistent with crystallization at temperatures approaching the magmatic range. LaTour and Barnett (1987) and Kerrich and Hyndman (1986) reported detailed mineral equilibria and oxygen isotope studies that suggest the same conclusion. An intense mylonitic lineation is oriented slightly south of east throughout the entire length of the zone. Mylonite fabrics such as S and C surfaces and mica "fish" consistently indicate a top-to-the-east sense of movement.

Childs (1982) found a similar mylonite zone along the northern edge of the Bitterroot batholith, with lineations that point in the same direction as those in the Bitterroot batholith. That directional similarity suggests that the two mylonites may be remnants of a zone that formerly extended completely across the Bitterroot batholith. Perhaps the entire roof of the batholith may have sheared off along a continuous mylonite zone.

The east flank of the Bitterroot mylonite grades abruptly upward into a chloritic breccia nearly lacking preferred orientation of mineral grains. The low-greenschist facies mineral assemblage in the chloritic breccia, and oxygen isotope studies by Kerrich and Hyndman (1986) both indicate that it developed at relatively low temperature and in a shallow crustal environment that involved circulation of meteoric water. The chloritic breccia may have formed during Eocene crustal extension, after displacement of the Sapphire tectonic block.

Bounding ranges around the Sapphire block consist essentially of very tightly folded and intricately thrust sedimentary formations. The deformation apparently occurred in the leading edges of the detachment block, and in the sedimentary section bulldozed ahead of it. Large masses of granite intrude those ranges.

Hyndman et al. (1975) and Hyndman (1980) suggested that the detachment block sheared through the partially molten top of the batholith, and moved on a sole of granite magma, now partially exposed in the large and incompletely mapped Joseph pluton (Desmarais, 1983). If so, the Late Cretaceous granite in the bounding ranges may well be displaced from the Bitterroot batholith.

Sears (1985) found that the contact aureole of the Garnet stock, a granite pluton in the northern margin of the Sapphire block, cuts foliation in its country rock. Ruppel et al. (1981) reported dates of 79 and 81 m.y. for the intrusion. Evidently, the Sapphire block was moving when the granite was emplaced. Hyndman et al. (1982) found that the older part of the Philipsburg batholith in the Flint Creek Range is slightly deformed, presumably because the detachment block was moving when it was emplaced. The younger, eastern part of the batholith is undeformed, and was emplaced along one of the thrust faults at the leading edge of the block. Radiometric ages on the granite suggest that the detachment block was still moving about 75 m.y. ago.

The Pioneer Tectonic Block

Alt and Hyndman (1977) suggested that the Pioneer Range of southwestern Montana appears to be another detachment block that moved off the Idaho batholith while the granite was still partly molten. Like the bounding ranges around the Sapphire block, the Pioneer Range consists of very tightly folded and thrust-faulted sedimentary formations heavily engorged with Late Cretaceous granite — the Pioneer batholith and its satellites. The Big Hole, a large basin west of the Pioneer Range, apparently opened behind the detachment block as it moved east. Figure 12-3 shows no mylonite bounding the Big Hole on the west. It remains to be seen whether further mapping in that region will reveal a continuous mylonite zone comparable to the Bitterroot mylonite.

Zen *et al.* (1975, 1980) reported ages of 70 and 71 m.y. for gneiss recrystallized from the Cambrian Silver Hill and the Cretaceous Kootenai formations in the contact zone of the Pioneer batholith. An assemblage of andalusite, cordierite, and biotite in the Kootenai Formation indicates emplacement of the granite at shallow depth, as does the country rock. Snee and Sutter (1979) reported potassium-argon radiometric dates of 67 m.y. for andesites from southwestern Montana. That age permits an inference that they erupted from the Pioneer batholith.

An Alternative Interpretation

Bickford *et al.* (1981), and Chase *et al.* (1978, 1983) concluded on the basis of lead-uranium zircon dates that the Bitterroot mylonite zone involves intrusive rocks that range in age from 48 to 60 m.y. They further concluded that the major shearing occurred since about 50 m.y. ago, and is therefore related to early Tertiary igneous activity. Argon 40/39 dates led Garmezy and Sutter (1983) to conclude that mylonitization began 45.5 m.y. ago, and continued for 2 million years. If, as Hyndman *et al.* (1975) suggested, chloritic brecciation followed mylonitization as a continuum, and if the argon dates were not reset during hydrothermal alteration around the shallow plutons, or by the intense greenschist facies alteration that marks the chloritic breccias, the Bitterroot mylonite might conceivably have formed during Eocene extension.

If mylonite formation did indeed accompany early Tertiary volcanism, an alternative tectonic model becomes necessary. In this model, Late Cretaceous thrusts that emerge along the edges of the Sapphire block would root into deeper levels of the crust, and pass beneath the Bitterroot batholith. Continued movement on the thrust faults after it crystallized would make the batholith at least somewhat allochthonous. That interpretation presents several problems.

If mylonite development is to be interpreted as an Eocene event, then more than 10 km of rock must have disappeared along the abrupt contact at the top of the mylonite during progressive formation of the ductile mylonite and overlying brittle breccia during Eocene time. And if Eocene dates in the deep-seated mylonites do in fact indicate the time of mylonitization, rather than resetting during formation of chloritic breccias, they are inconsistent with the environment and time of emplacement of Eocene plutonic and volcanic rocks. The miarolitic cavities in the undeformed Eocene granites, and the eruption of many of them, show that they were emplaced at extremely shallow depth. Holloway (1980) found dikes associated with the Eocene volcanic rocks cutting the Bitterroot mylonite.

Brecciated Eocene volcanic rocks rest on mylonitized granite of the Bitterroot batholith in the southern part of the Bitterroot Valley. The simplest interpretation, which we favor, would have the volcanic rocks deposited on a much older Bitterroot mylonite already exposed during Eocene time. Those large volumes of Eocene volcanic rocks lie on an erosion surface of considerable local relief developed on the Bitterroot and Atlanta batholiths. If the batholith was exposed about 50 m.y. ago through Eocene extension, insufficient time was available for erosion to carve a rugged landscape on the granite before eruption of the volcanic rocks, also about 50 m.y. ago. Furthermore, the 45- m.y. age Garmezy and Sutter (1983) reported for the Bitterroot mylonite is distinctly younger than the 50-m.y. age of the Eocene volcanic rocks that unconformably overlie it.

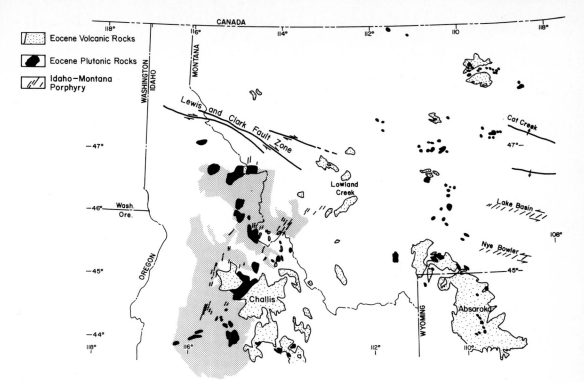

FIG. 12-11. Distribution of Eocene igneous rocks and clearly associated tectonism in the Northern Rocky Mountains.

central Montana. Each displaces the Paleocene Fort Union Formation. Each also has a component of vertical offset, with the block south of the fault zone uplifted. The high Beartooth Plateau, for example, rose south of the Nye-Bowler fault zone.

Early Tertiary crustal spreading may also have reversed the direction of movement on some of the faults in the western part of the Lewis and Clark zone. That could explain Childs' (1982) discovery of apparent right-lateral offset along the Kelly Forks fault of a dike swarm associated with the Tertiary Bungalow pluton of central Idaho. Many other authors, notably Harrison *et al.* (1981), have inferred right-lateral offset along faults in the western part of the Lewis and Clark system.

The Absaroka Volcanic Center

Also about 50 m.y. ago, the Absaroka volcanic center erupted enormous volumes of andesite in an area centered in the eastern part of Yellowstone Park. Like the Cretaceous Elkhorn Mountains volcanic pile erupted from the Boulder batholith, the Absaroka volcanos produced large volumes of andesite in an area remote from an active trench. We interpret the Absaroka volcanic center as an oversized resurgent caldera.

Eocene Granite and Volcanic Rocks

Figure 12-11 shows that Eocene granitic rocks abound along two broad zones that trend generally north-south. One passes through the Idaho batholith complex, the other through northwestern Wyoming and west-central Montana. Plutons range in size from small stocks to batholiths. Tertiary granites are typically massive, and commonly contain abundant gas cavities. Many contain small amounts of fluorite, topaz, and smokey quartz. High-level emplacement of the early Tertiary granites indicates that the water content of the magmas was extremely low, presumably because the magmas melted deep in the crust.

Most of the Tertiary granites erupted, many of them voluminously. They are the source of the large Lowland Creek and Challis volcanic fields, as well as of a number of smaller volcanic piles. In general, the erupted material is dominantly rhyolite. Both the volume of volcanic rocks and the proportion of andesite increase to the southwest along the trend from the Lowland Creek to the Challis volcanic piles. Basalt is rare.

Except in central Montana north of the projected trend of the Lewis and Clark fault zone, early Tertiary igneous rocks of mafic composition are extremely rare. Most are shonkinite, a rock composed of essential potassium feldspar, augite, and some nepheline, that appeared in four large centers. Other igneous centers widely scattered through central Montana produced intrusions of syenite, monzonite, and granite, many greatly enriched in sodium or potassium.

The regional distribution of early Tertiary granite, alkalic intrusive rocks, and volcanic rocks in the northern Rocky Mountains shown in Fig. 12-11 reveals little apparent internal pattern and no clear relationship to tectonic activity. The only clear compositional trend is the tendency for alkalic magmas to appear in central Montana north of the Lewis and Clark fault zone.

POST-EOCENE TECTONISM

All but the youngest intermontane basins of western Montana contain deep deposits of valley fill sediment, the Renova Formation, which accumulated between latest Eocene and mid-Miocene time. Although the Renova Formation is nearly everywhere deformed, available information does not support any generalization about the pattern of its deformation. It is clear only that tectonic activity continued into mid-Tertiary time.

From latest Miocene through the end of Pliocene time, a second generation of basin fill sediment, the Six Mile Creek Formation, accumulated in the intermontane basins of the northern Rocky Mountains. It is equivalent in time and origin to the Flaxville and Oglalla gravels of the High Plains.

Deformation of the Six Mile Creek Formation is apparent within 250 km of the Yellowstone region and within the Intermountain seismic zone that extends north from the Yellowstone resurgent caldera, then branches northwest and northeast near Helena, Montana. Elsewhere in the northern Rocky Mountains, upper Tertiary gravels remain essentially undisturbed. Faults that offset the Six Mile Creek Formation typically trend northwest, less commonly northeast. All are normal faults, and most are marginal to new ranges and intermontane basins. Strong topographic scarps mark those faults, and most are seismically active. There is little associated volcanism.

Attempts to propose a cause-and-effect relationship between post-Cretaceous tectonism and igneous activity in Idaho and Montana and the subduction then current west of the Okanogan microcontinent inflict gross violence on the standard plate tectonics models. New insights are needed.

SUMMARY

Late Jurassic to Cretaceous accretion of offshore island arcs in the western border zone of the Idaho batholith complex was accompanied and followed by Cretaceous regional metamorphism in the greenschist and amphibolite facies.

Alt, D., and Hyndman, D. W., 1977, Tectonic problems in the northern Rockies: *Geol. Soc. America Abstr. with Programs*, v. 9, p. 706.

Alonso, M., and Friberg, L. M., 1985, Petrology of the contact metamorphic aureole around McCartney Mountain, southwestern Montana: *Geol. Soc. America Abstr. with Programs*, v. 17, p. 206.

Armstrong, R. L., 1975, Precambrian (1500 m.y. old) rocks of central Idaho — The Salmon River arch and its role in Cordilleran sedimentation and tectonics: *Amer. J. Sci.*, v. 275-A, p. 437–467.

——, Taubeneck, W. H., and Hales, P. O., 1977, Rb-Sr and K-Ar geochronometry of Mesozoic granitic rocks and their Sr isotopic composition, Oregon, Washington, and Idaho: *Geol. Soc. America Bull.*, v. 88, p. 397–441.

Badley, R. H., 1978, Petrography and chemistry of the East Fork dike swarm, Ravalli County, Montana: M.S. thesis, Univ. Montana, Missoula, Mont., 54 p.

Berg, R. B., 1968, Petrology of anorthosites of the Bitterroot Range, Montana, *in* Isachsen, Y. W., ed., *Origin of Anorthosite and Related Rocks:* N.Y. State Mus. Sci. Series, Mem. 18, p. 387–398.

Bickford, M. E., Chase, R. B., Nelson, B. K., Schuster, R. D., and Arruda, E. C., 1981, U-Pb studies of zircon cores and overgrowths, and monazite — Implications for age and petrogenesis of the northeastern Idaho batholith: *J. Geology*, v. 89, p. 433–547.

Bond, J. G., 1978, *Geologic Map of Idaho:* Moscow, Idaho, Idaho Dept. Lands, Bur. Mines Geology, scale 1:500,000.

Brandon, W. R., 1984, An origin for the McCartney's Mountain salient of the southwestern Montana fold and thrust belt: M.S. thesis, Univ. Montana, Missoula, Mont., 128 p.

Chase, R. L., 1973, Petrology of the northeast border zone of the Idaho batholith, Bitterroot Range, Montana: *Mont. Bur. Mines Geology Mem. 43*, 28 p.

——, and Johnson, B., 1975, Border zone relationships of the northern Idaho batholith: *Northwest Geology*, v. 4, p. 38–50.

——, Bickford, M. E., and Tripp, S. E., 1978, Rb-Sr and U-Pb isotopic studies of the northeastern Idaho batholith and border zone: *Geol. Soc. America Bull.*, v. 89, p. 1235–1325.

——, Bickford, M. E., and Arruda, E. C., 1983, Tectonic implications of Tertiary intrusion and shearing within the Bitterroot dome, northeastern Idaho batholith: *J. Geology*, v. 91, p. 462–470.

Cheney, J. T., 1975, Kyanite, sillimanite, phlogopite, cordierite layers in the Bass Creek anorthosites, Bitterroot Range, Montana: *Northwest Geology*, v. 4, p. 77–82.

Childs, J. F., 1982, Geology of the Precambrian Belt Supergroup and the northern margin of the Idaho batholith, Clearwater County, Idaho: Ph.D. dissertation, Univ. California, Santa Cruz, Calif. p. 136.

Coney, P. J., 1972, Cordilleran tectonics and North America plate motion: *Amer. J. Sci.*, v. 272, p. 603–628.

Crittenden, M. D., Jr., Stewart, J. H., and Wallace, C. A., 1972, Regional correlation of upper Precambrian strata in western North America: *Proc. 24th Internat. Geol. Cong., Montreal, Sec. 1*, p. 334–341.

Csejtey, B., 1962, Geology of the southeast flank of the Flint Creek Range, western Montana: Ph.D. dissertation, Princeton Univ., Princeton, N.J., 175 p.

Daniel, F. and Berg, R. B., 1981, Radiometric dates of rocks in Montana: *Mont. Bur. Mines Geology Bull. 114*, 136 p.

Desmarais, N. R., 1983, Geology and geochronology of the Chief Joseph plutonic-metamorphic complex, Idaho: Ph.D. dissertation, Univ. Washington, Seattle, Wash., 150 p.

Dexter, J. J., Chase, R. B., and Bickford, M. E., 1979, U/Pb zircon ages and crustal contamination of the northeastern Idaho batholith: *Geol. Soc. America Abstr. with Programs*, v. 11, p. 270.

Doe, B. R., Tilling, R. I., Hedge, C. E., and Klepper, M. R., 1968, Lead and strontium isotope studies of the Boulder batholith, southwestern Montana: *Econ. Geology*, v. 63, p. 884–906.

Emmons, W. H., and Calkins, F. C., 1913, Geology and ore deposits of the Philipsburg Quadrangle, Montana: *U.S. Geol. Survey Prof. Paper 78*, 271 p.

Fleck, R. J., and Criss, R. E., 1985, Strontium and oxygen isotope variations in Mesozoic and Tertiary plutons of central Idaho: *Contrib. Mineralogy Petrology*, v. 90, p. 291–308.

Flood, R. E., Jr., 1974, Structural geology of the upper Fishtrap Creek area, central Anaconda Range, Deer Lodge County, Montana: M.S. thesis, Univ. Montana, Missoula, Mont., 71 p.

Foster, D., 1986, Synplutonic dikes of the Idaho batholith, Idaho and western Montana and their relationship to the generation of the batholith; M.S. thesis, Univ. Montana, Missoula, Mont., 79 p.

Fraser, G. D., and Waldrop, W. A., 1972, Geologic map of the Wise River quadrangle, Silver Bow and Beaverhead Counties, Montana: *U.S. Geol. Survey Geol. Quad. Map GQ-988*.

Garmezy, L. and Sutter, J. F. 1983, Mylonitization coincident with uplift in an extensional setting, Bitterroot Range, Montana-Idaho: *Geol. Soc. America Abstr. with Programs*, v. 15, p. 578.

Hamilton, W., 1963, Metamorphism in the Riggins region, western Idaho: *U.S. Geol. Survey Prof. Paper 436*, 95 p.

———, 1969, Reconnaissance geologic map of the Riggins Quadrangle, west-central Idaho: *U.S. Geol. Survey Misc. Geol. Inv. Map I-579*.

———, and Myers, W. B., 1967, The nature of batholiths: *U.S. Geol. Survey Prof. Paper 554-C*, p. C1–C30.

Harrison, J. E., 1972, Precambrian Belt basin of northwestern United States — Its geometry, sedimentation, and copper occurrences: *Geol. Soc. America Bull.*, v. 83, p. 1215–1240.

———, Griggs, A. B., and Wells, J. D., 1981, Generalized geologic map of the Wallace 1 X 2 quadrangle, Montana and Idaho: *U.S. Geol. Survey Misc. Field Studies Map MF-1354-A*.

Heise, B. A., 1983, Structural geology of the Mt. Haggin area, Deer Lodge County, Montana: M.S. thesis, Univ. Montana, Missoula, Mont., 77 p.

Hietanen, A., 1956, Kyanite, andalusite, and sillimanite in the schist in Boehls Butte Quadrangle, Idaho: *Amer. Mineralogist*, v. 41, p. 1–27.

———, 1961, Relationship between deformation, metamorphism, metasomatism, and intrusion along the northwest border zone of the Idaho batholith, Idaho: *U.S. Geol. Survey Prof. Paper 424-D*, p. 161–164.

———, 1963, Metamorphism of the Belt Series in the Elk River–Clarkia area, Idaho: *U.S. Geol. Survey Prof. Paper 344-C*, 49 p.

———, 1984, Geology along the northwest border zone of the Idaho batholith, northern Idaho: *U.S. Geol. Survey Bull. 1608*, 17 p.

Hyndman, D. W., 1980, Bitterroot dome-Sapphire tectonic block, an example of a plutonic-core gneiss-dome complex with its detached superstructure, *in* Crittenden, M. D., Jr., Cooney, P. J., and Davis, G. H., eds., *Cordilleran Metamorphic Core Complexes:* Geol. Soc. America Mem. 153, p. 427–443.

——, 1981, Controls on source and depth of emplacement of granitic magma: *Geology*, v. 9, p. 244–249.

——, 1983, The Idaho batholith and associated plutons, Idaho and western Montana; *in* Roddick, J. A., ed., *Circum-Pacific Batholiths:* Geol. Soc. America Mem. 159, p. 213–240.

——, 1985, Source and formation of the Idaho batholith, *in Symposium on the Geology of the Idaho Batholith:* Geol. Soc. America Abstr. with Programs, v. 17, p. 226.

——, and Alt, D., 1972, The kyanite-andalusite-sillimanite problem on Goat Mountain, Idaho: *Northwest Geology*, v. 1, p. 42–46.

——, Talbot, J. L., and Chase, R. B., 1975, Boulder batholith, a result of emplacement of a block detached from the Idaho batholith infrastructure? *Geology*, v. 3, p. 401–404.

——, and Chase, R. B., 1979, Major tectonic elements and tectonic problems along the line of section from northeastern Oregon to west-central Montana: *Geol. Soc. America Map Chart Series MC-28c*, 11p + 1:250,000 strip map and cross-section.

——, Badley, R., and Rebal, D., 1977, Northwest trending dike swarm in central Idaho and western Montana: *Geol. Soc. America Abstr, with Programs*, v. 9, p. 734–735.

——, Silverman, A. J., Ehinger, R., Benoit, R., and Wold, R., 1982, The Philipsburg batholith, western Montana — Field, petrographic, chemical relationships and evolution: *Mont. Bur. Mines Geology Mem. 49*, 37 p.

Johnson, B. R., 1975, Migmatites along the northern border of the Idaho batholith: Ph.D. dissertation, Univ. Montana, Missoula, Mont., 120 p.

Jens, J. C., 1972, Petrology of a mafic layered intrusion near Lolo Pass, Idaho: M.S. thesis, Univ. Montana, Missoula, Mont., 85 p.

Kerrich, R., and Hyndman, D. W., 1986, Thermal and fluid regimes in the Bitterroot lobe-Sapphire Block: *Geol. Soc. America Bull..*, v. 97, p. 147–155.

Klepper, M. R., Robinson, G. D., and Smedes, H., 1974, On the nature of the Boulder batholith of Montana: *Geol. Soc. America Bull.*, v. 82, p. 1563–1580.

Knopf, A., 1957, The Boulder batholith of Montana: *Amer. J. Sci.,* v. 255, p. 81–103.

LaTour, T., 1974, An examination of metamorphism and scapolite in the Skalkaho region, southern Sapphire Range, Montana: M.S. thesis, Univ. Montana, Missoula, Mont., 95 p.

——, and Barnett, R. L., 1987, Mineralogical changes accompanying mylonitization in the Bitterroot lobe of the Idaho batholith — Implication for timing of deformation: *Geol. Soc. America Bull*, v. 98, p. 356–363.

Lund, K., and Snee, L. W., 1985, Structural and metamorphic setting of the central Idaho batholith: *Geol. Soc. America Abstr. with Programs*, v. 17, p. 253.

——, Snee, L. W., and Sutter, J. F., 1985, Style and timing of suture related deformation in island arc rocks of western Idaho: *Geol. Soc. America Abstr. with Programs*, v. 17 p. 367.

Luthy, S., 1981, Petrology of Cretaceous and Tertiary intrusive rocks, Red Mountain–Bull Trout Point area, Boise, Valley, and Custer Counties, Idaho: M.S. thesis, Univ. Montana, Missoula, Mont., 109 p.

Maxwell, D., and Hower, J., 1967, High grade diagenesis and low grade metamorphism of illite in the Precambrian Belt Series: *Amer. Mineralogist*, v. 52, p. 843–857.

Miller, F. K., and Engels, J. C., 1975, Distribution and trends of discordant ages of the plutonic rocks of northeastern Washington and northern Idaho: *Geol. Soc. America Bull.*, v. 86, p. 517–528.

Nold, J. L., 1968, Geology of the northeastern border zone of the Idaho batholith, Montana and Idaho: Ph.D. dissertation, Univ. Montana, Missoula, Mont., 159 p.

_____ , 1974, Geology of the northeastern border zone of the Idaho batholith: *Northwest Geology*, v. 3, p. 47–52.

Norwick, S., 1972, The regional Precambrian metamorphic facies of the Prichard formation of western Montana and northern Idaho: Ph.D. dissertation, Univ. Montana, Missoula, Mont., 129 p.

Olson, H. J., 1968, The geology and tectonics of the Idaho porphyry belt from the Boise basin to the Casto Quadrangle: Ph.D. dissertation, Univ. Arizona, Tucson, Ariz., 154 p.

Pearson, R. C., and Zen, E-An, 1985, Geologic map of the eastern Pioneer Mountains, Beaverhead County, Montana: *U.S. Geol. Survey Misc. Field Studies Map MF-1806-A.*

Presley, M. W., 1970, Igneous and metamorphic geology of the Willow Creek drainage basin, southern Sapphire Mountains, Montana: M.S. thesis, Univ. Montana, Missoula, Mont., 64 p.

Reid, R. R., Morrison, D. A., and Greenwood, W. R., 1973, The Clearwater orogenic zone — A relic of Proterozoic orogeny in central and northern Idaho: *Belt Symposium*, Univ. Idaho, Moscow, Idaho, p. 10–56.

Reynolds, M. W., and Kleinkopf, M. D., 1977, The Lewis and Clark line, Montana-Idaho — A major intraplate tectonic boundary: *Geol. Soc. America Abstr. with Programs*, v. 9, p. 1140–1141.

Ruppel, E. T., and Lopez, D. A., 1984, The thrust belt in southwest Montana and east-central Idaho: *U.S. Geol. Survey Prof. Paper 1278*, 41 p.

_____ , Wallace, C. A., Schmidt, R. G., and Lopez, D. A., 1981, Preliminary interpretation of the thrust belt in southwest and west-central Montana and east-central Idaho: *Mont. Geol. Soc. Field Conf. Guidebook*, p. 139–159.

Schmidt, C. J., and Garihan, J. M., 1983, Laramide tectonic development of the Rocky Mountain foreland in southwestern Montana, *in* Cowell, J. D., ed., *Rocky Mountain Foreland Basins and Uplifts:* Rocky Mountain Assoc. Geologists Symp., p. 271–294.

Schuster, R. D., and Bickford, M. E., 1985, chemical and isotopic evidence for the petrogenesis of the northeastern Idaho batholith: *J. Geology*, v. 93, p. 727–742.

Sears, J. W., 1985, Implications of small-scale structures under the Sapphire plate, west-central Montana: *Geol. Soc. America Abstr. with Programs*, v. 17, p. 264.

_____ , and Price, R. A., 1978, The Siberian connection — A case for Precambrian separation of the North American and Siberian cratons: *Geology*, v. 6, p. 267–270.

Sheriff, S. D., Sears, J. W., and Moore, J. N., 1984, Montana's Lewis and Clark fault zone — A relic of Proterozoic orogeny in central and northern Idaho: *Belt Symposium*, Univ. Idaho, Moscow, Idaho, p. 10–56.

Smedes, H. W., Klepper, M. R., and Tilley, R. I., 1968, Boulder batholith — A description of geology and road log: *Rocky Mountain Section, Geol. Soc. America, Field Trip 3, Bozeman, Montana*, p. 21.

Snee, L. W. and Sutter, J. F., 1979, K-Ar geochronology and major element geochemistry of plutonic and associated volcanic rocks from the southeastern Pioneer Mountains, Montana: *Geol. Soc. America Abstr. with Programs*, v. 11, p. 302.

Stewart, J. H., 1972, Initial deposits in the Cordilleran geosyncline — Evidence of a late

Precambrian (850 m.y.) continental separation: *Geol. Soc. America Bull.*, v. 83, p. 1345–1360.

Strayer, L. M., Hyndman, D. W., and Sears, J. W., 1987, Movement direction and displacement estimate in the Western Idaho Suture Zone mylonite: Dworshak Dam/Orofino area, west-central Idaho: *Geol. Soc. America Abstr. with Programs*, v. 19, p. 857.

Stuart, C. J., 1966, Metamorphism in the central Flint Creek Range, Montana: M.S. thesis, Univ. Montana, Missoula, Mont. 103 p.

Toth, M., and Stacey, J. S., 1985, Uranium-lead geochronology and lead isotope data for the Bitterroot lobe of the Idaho batholith: *Geol. Soc. America Abstr. with Programs*, v. 17, p. 268.

Wehrenberg, J. P., 1972, Geology of the Lolo Peak area, northern Bitterroot Range, Montana: *Northwest Geology*, v. 1, p. 25–32.

Weidman, R. W., 1965, The Montana lineament: *Billings Geol. Soc. 16th Ann. Field Conf. Guidebook*, p. 137–143.

Wiswall, G., 1979, Field and structural relationships below the Bitterroot lobe of the Idaho batholith: *Northwest Geology*, v. 8, p. 18–28.

Wooden, J. L., Vitaliano, C. J., Koehler, S. W., and Ragland, P. C., 1978, The late Precambrian mafic dikes of the southern Tobacco Root Mountains, Montana — Geochemistry, Rb-Sr geochronology, and relationship to Belt tectonics: *Can. J. Earth Sci.*, v. 15, p. 467–479.

Zen, E-An, Marvin, R. F., and Mehnert, H. H., 1975. Preliminary petrographic, chemical, and age data on some intrusive and associated contact metamorphic rocks, Pioneer Mountains, southwestern Montana: *Geol. Soc. America Bull.*, v. 86, p. 367–370.

——— , Arth, J. G., and Marvin, R. F., 1980, Petrology, age, and some isotope geochemistry of the Cretaceous and Paleocene intrusive rocks, Pioneer batholith, southwestern Montana: *Geol. Soc. America Abstr. with Programs*, v. 12, p. 309.

D. M. Mogk
Department of Earth Sciences
Montana State University
Bozeman, Montana 59717

Darrell J. Henry
Department of Geology and Geophysics
Louisiana State University
Baton Rouge, Louisiana 70803

13

METAMORPHIC PETROLOGY OF THE NORTHERN ARCHEAN WYOMING PROVINCE, SOUTHWESTERN MONTANA: EVIDENCE FOR ARCHEAN COLLISIONAL TECTONICS

The Archean basement of southwestern Montana exhibits a variety of metamorphic and structural styles that developed over the first billion years of continental growth in North America. In the western portion of the Beartooth Mountains, a mobile belt marks a fundamental discontinuity in the nature of the Archean crust: Archean terrains to the east are dominated by Late Archean plutons with some older supracrustal xenoliths, whereas terrains to the west are dominated by high-grade metasedimentary sequences.

In the eastern Beartooth Mountains, large xenoliths of platform-type supracrustal rocks from the Quad Creek and Hellroaring Plateau areas were metamorphosed to granulite grade at about 3.4–3.2 b.y. ago. This sequence must have been originally derived from an older sialic crust, and was deposited during a period of relative quiescence. Subsequent burial to at least 20 km, presumably by means of tectonic thickening ended this orogenic cycle. At ~2.8 b.y.b.p., a second orogenic cycle began with the generation of large volumes of andesitic rocks. Soom after emplacement, these andesitic rocks were affected by a pervasive amphibolite grade metamorphism. Finally, all of the older rocks were intruded by the voluminous calc-alkaline granitoids that dominate the central and eastern Beartooth Mountains. This granitoid emplacement at ~2.75 b.y. ago is the culmination of the second orogenic cycle and represents the final crustal stabilization of this area.

In contrast, the North Snowy Block of the western Beartooth Mountains contains abundant platformal sedimentary rocks with a paucity of Late Archean granitoids. The area is characterized by tectonic juxtaposition of lithologic sequences with different metamorphic grades, structural styles, and isotopic ages. Emplacement of these units was along transcurrent and thrust faults and represents tectonic thickening of the crust penecontemporaneous with the magmatic growth observed to the east.

The Late Archean basement to the west of the Beartooth Mountains, including most of the rocks in the Gallatin, Madison, Tobacco Root, Ruby and Blacktail Ranges, is characteristically of supracrustal origin. A working model calls for deposition of coarse clastics and platform-type sediments in an ensialic or Tethyan-type basin, followed by collapse of the basin, isoclinal folding and nappe emplacement, and generation of granulite-migmatite associations at about 2.75–2.70 b.y.b.p.

INTRODUCTION

The Archean basement of southwestern Montana comprises the northernmost part of the Wyoming Province as defined by Condie (1976). Archean rocks are currently exposed in a series of foreland block uplifts across this terrain (e.g., Foose et al., 1961). A major discontinuity in the nature of this Archean continental crust is exhibited in a mobile belt in the North Snowy Block, western Beartooth Mountains (Reid et al., 1975; D. W. Mogk, P. A. Mueller, and J. L. Wooden, unpublished data; Fig. 13-1). To the east of the North Snowy Block mobile belt, exposures in the central and eastern Beartooth Mountains, Bighorn Mountains, and samples from deep-drill cores in eastern Wyoming and Montana of Archean rocks are comprised dominantly of Late Archean granitoids with inclusions of older supracrustal rocks (Barker et al., 1979, Henry et al., 1982; Mueller et al., 1985; Peterman, 1981; Timm, 1982). To the west of the North Snowy Block, the Archean terrain consists dominantly of ~2.75-b.y.-old (James and Hedge, 1980) high-grade metasedimentary rocks as exposed in the Gallatin (Spencer and Kozak, 1975), Madison (Erslev, 1983), Ruby (Garihan, 1979), and Tobacco Root Mountains (Vitaliano et al., 1979) (Fig. 13-1). These rocks characteristically have attained upper amphibolite to

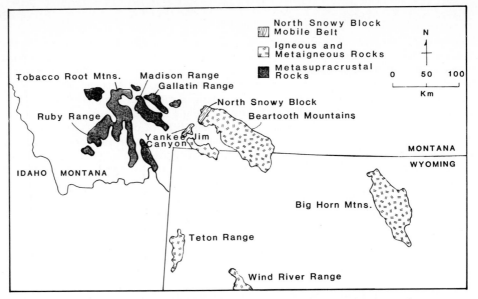

FIG. 13-1. Sketch map of the northern Wyoming Province showing major exposures of Archean rocks. The North Snowy Block mobile belt delineates a major discontinuity in the nature of the continental crust in this area, with dominantly rocks of igneous origin to the east and rocks of supracrustal origin to the west.

granulite grades of metamorphism, have been isoclinally folded, and have been locally emplaced as nappes.

The metamorphic history of each of these areas provides an important framework for understanding the tectonic evolution of this Archean continental crust. The determination of the physical conditions of metamorphism, using heterogeneous phase equilibria and mineralogical geothermobarometers, provides the basis for construction of *P-T* trajectories, allows the interpretation of tectonic and geochemical processes, and aids in the recognition of allochthonous units. Specific areas in the Beartooth Mountains and adjacent ranges will be described in detail to characterize their lithologic associations, metamorphic grade, and structural style. This information will be combined with geochemical and geochronological information (Wooden *et al.*, Chapter 14, this volume) to present a working model for the evolution of this continental crust.

BEARTOOTH MOUNTAINS

The Beartooth Mountains have been divided into four distinct domains: the eastern Beartooth Block, the Stillwater Block, the North Snowy Block, and the South Snowy Block (Wilson, 1936; Fig. 13-2). A major project conducted by Arie Poldervaart and his students in the eastern and central Beartooth Mountains provided the first field and petrologic studies (Bently, 1967; Butler, 1966, 1969; Casella, 1964, 1969; Eckelmann and Poldervaart, 1957; Harris, 1959; Larsen *et al.*, 1966; Prinz, 1964; Skinner *et al.*, 1969; Spencer, 1959). Their initial interpretation of this area called for granitization of a sequence of openly folded supracrustal rocks. Recent petrologic, geochemical, and geochronologic studies (Mueller *et al.*, 1985, and references) have demonstrated that the Beartooth Mountains consist predominantly of voluminous Late Archean granitoids with inclusions

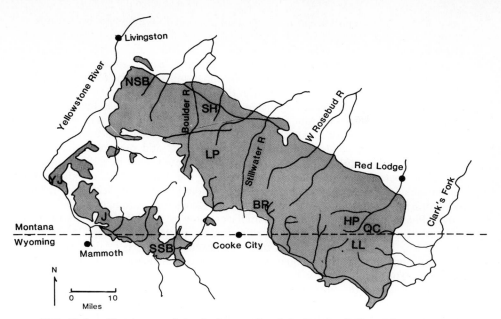

FIG. 13-2. Sketch map of the Archean rocks of the Beartooth Mountains, showing the location of the areas discussed in the text: QC, Quad Creek; HP, Hellroaring Plateau; LL, Long Lake; BR, Broadwater River; LP, Lake Plateau; SH, Stillwater hornfels aureole; NSB, North Snowy Block; YJ, Yankee Jim Canyon; J, Jardine; SSB, South Snowy Block.

of supracrustal rocks which exhibit wide ranges in composition, metamorphic grade, and isotopic age. The following sections describe those areas in the Beartooth Mountains where detailed petrologic studies have been done.

Quad Creek and Hellroaring Plateau, Eastern Beartooth Block

The easternmost portion of the Beartooth Mountains is composed predominantly of granitic to tonalitic granitoids, gneisses, and migmatites whose Rb-Sr isotopic systematics indicate an age of formation of 2.8 b.y. (Wooden *et al.*, Chapter 14, this volume). Within these felsic rocks, inclusions of various supracrustal lithologies range in size from a few centimeters to several kilometers. These oldest lithologies include quartzite, felsic gneiss, pyribolite, amphibolite, iron formation, pelitic schist, and ultramafite. Similar to many high-grade terrains in the Archean (Windley, 1977), this assemblage of lithologies is more reminiscent of a shelf environment than a eugeoclinal or typical greenstone belt setting. This is supported by bulk compositions and REE abundances of the lithologic units (Mueller *et al.*, 1982, 1985).

The structural features of the rocks of the eastern Beartooth Mountains are dominated by two phases of deformation. The first major phase of deformation produced a series of south-southwest-plunging isoclinal and intrafolial folds that formed during passive flow folding associated with a high-grade metamorphic event (Rowan, 1969). The second major deformation produced upright, typically nonisoclinal folds which formed about south-southwest- to south-plunging axes and in which flexure is the dominant fold style. The common large antiformal and synformal structures with wavelengths of several kilometers are a manifestation of the second deformation (Casella, 1969).

Minor subsequent deformations produced broad, symmetric, open folds and relatively minor development of mineral lineations parallel to the axial planes of the folds (Rowan and Mueller, 1971; Skinner *et al.*, 1969). The supracrustal lithologic units are commonly boudin-shaped, are separated by mylonitic zones, and probably underwent a significant amount of tectonic thinning. Isoclinal and intrafolial folding are common in these units indicating their involvement in the first deformation.

Most of these earlier lithologies display mineral assemblages that are indicative of granulite-grade metamorphism [M1] (Table 13-1). Application of a series of geothermometers and geobarometers (Ellis and Green, 1979; Ferry and Spear, 1978; Newton and Perkins, 1982) yield temperatures of 750–800°C and pressures of 5–6 kb (Henry *et al.*, 1982). Some lithologies are partially to completely reset by a subsequent amphibolite grade metamorphism [M2] (Table 13-1). In the pelitic schists coexisting sillimanite and muscovite are locally stable, and place the stability field for M2 below the muscovite breakdown curve (Fig. 13-3). Application of geothermobarometers (Ferry and Spear, 1978; Plyusnina, 1982) to those assemblages showing amphibolite grade reequilibration indicate that this metamorphism developed at temperatures of 575–625°C and pressures of 3–5 kb. The degree of retrogression of a given lithology is related to the size of the xenolith and rock type (the quartzites and ironstones being the least subject to retrogression). The Rb-Sr isotopic systematics of the earlier supracrustal rocks are very different from those of the felsic intrusive rocks involved in the 2.8-b.y. event. Most samples define a 3.4-b.y. isochron that has been interpreted as the approximate time of granulite-grade metamorphism, with a few samples partially reset toward a 2.8-b.y. isochron (Henry *et al.*, 1982).

We believe that the granulite-grade supracrustal lithologies are restricted to the eastern portion of the central Beartooth Block. Consequently, this must have been a site of deposition of a series of platform-type supracrustal rocks that suffered large-scale horizontal shortening and deep burial to about 20 km probably about 3.4 b.y. ago. A continent-continent collisional mechanism can most easily explain such deep burial of

TABLE 13-1. Typical Granulite and Amphibolite (Retrogressed) Mineral Assemblages of the Supracrustal Lithologies[a]

Lithology	Granulite Assemblage	Amphibolite Assemblage
Quartzite	Qz + biot + opx + plag (1) Qz + biot + cord + sill	Qz + biot + hb + plag
Pelitic schist	Qz + biot + plag + ksp + sill Qz + biot + gar + cord + sill + plag Qz + biot + plag + sill + cord	Qz + biot + plag + mu + sill
Pyribolite	Qz + biot + plag + hb + gar + opx + cpx Qz + biot + plag + hb + opx Qz + biot + plag + cord + opx	
Amphibolite		Qz + biot + hb + plag Hb + plag Qz + biot + plag + cord + anth
Iron formation	Qz + mt + gar + opx + cpx + hb Qz + mt + gar + opx	Qz + mt + grun
Ultramafite	Hb + opx + gr sp ol + opx + hb + gr sp	Anth + biot + hb

[a]anth, anthophyllite; biot, biotite; cord, cordierite; cpx, clinopyroxene; gar, garnet; gr sp, green spinel; grun, grunerite; hb, hornblende; ksp, K feldspar; mt, magnetite; mu, muscovite; ol, olivine; opx, orthopyroxene; plag, plagioclase; Qz, quartz; sill, sillimanite.

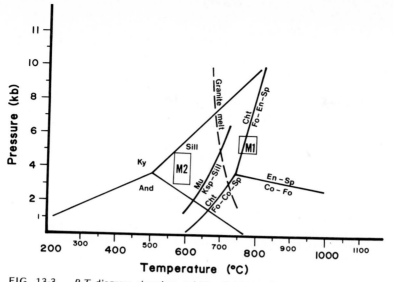

FIG. 13-3.   *P-T* diagram showing stability fields for [M1] and [M2] metamorphism based on several geothermobarometers. Reference reaction curves
are shown for aluminosilicates (Holdaway, 1971), muscovite breakdown
(Chatterjee and Johannes, 1975), granite melting (Thompson and Algor, 1971),
Mg-chlorite and Mg-cordierite breakdown (Evans, 1977).

supracrustal rocks and implies that this Early Archean crust locally doubled to at least
40 km thickness (Mueller *et al.*, 1985; Newton, 1983).

Long Lake and Broadwater River Areas, Eastern Beartooth Block

The Long Lake portion of the central Beartooth Block is also dominated by granitoids,
gneisses, and migmatites. The most voluminous rock type is a tonalitic-to-granitic pluton
[the Long Lake granite (LLG)] that is weakly foliated, retains its igneous texture, and
surrounds the rest of the major rock units (Warner *et al.*, 1982; Wooden *et al.*, Chapter
14, this volume). The LLG intrudes a ubiquitous, strongly lineated gneiss of dominantly
granodioritic composition (LLGd). Both of these units include blocks (1 m to 1 km) of
a strongly lineated and foliated amphibolite of basaltic andesite to andesite (AA). This
amphibolite shows no evidence of having previously reached granulite-grade metamorphism. There are also minor amounts of quartz-microcline pegmatitic bodies that cut all
of these units. The structural features that affect this area are similar to those of the
eastern Beartooth Mountains (Casella, 1969; Harris, 1959; Khoury and Ghaly, 1973).
　　The typical mineral assemblages of these lithologies are given in Table 13-2. Although no well-calibrated geothermobarometers for these mineral assemblages exist,
some indication of the *P-T* conditions can be obtained from the hornblende-plagioclase
relations (Nabelek and Lindsley, 1985; Plyusnina, 1982; Spear, 1980). In the amphibolites, temperatures of 550-600°C are obtained. The geobarometric calibration of
Plyusnina (1982) yields pressures of 4.5-6 kb. However, the high tetrahedral Al in the
hornblende may indicate metamorphism of even higher pressures (Spear, 1981). The
minor amount of pegmatitic dikes suggest a relatively low P_{H_2O}.
　　Rb-Sr and Sm-Nd isotopic systematics of these units suggest that the AA and

TABLE 13-2. Mineral Assemblages of the Lithologic Units of the Long Lake Area

Unit	Mineral Assemblage[a]
Long Lake granite	Qz + biot + plag + mt + ksp
Long Lake granodiorite	Qz + biot + plag + mt ± ksp
Andesitic amphibolite	Hb + plag + Qz + biot + mt + epid + sph ± ksp
Pegmatite	Qz + ksp ± plag ± mu

[a]biot, biotite; epid, epidote; hb, hornblende; ksp, K feldspar; mt, magnetite; mu, muscovite; plag, plagioclase; Qz, quartz; sph, sphene.

possibly LLGd were formed and underwent metamorphism at 2.8 b.y.b.p. (Wooden *et al.*, Chapter 14, this volume).

In addition to these units, minor amounts of supracrustal xenoliths (including pelitic schist, quartzite, and iron formation) are found to the west of Long Lake in the Lonesome Mountain and Broadwater River areas (Larsen *et al.*, 1966; Timm, 1982). Unlike the Quad Creek area, these supracrustal lithologies contain metamorphic assemblages of medium-to-upper amphibolite grade, with some indications of an earlier lower *P-T* metamorphism (Timm, 1982). Preliminary U-Pb analyses of zircons from some of the metasedimentary xenoliths suggest that the source region for the sediment was about 3.1 b.y. old (Montgomery *et al.*, 1984).

These data are interpreted as signaling a major crust-forming event beginning at 2.8 b.y.b.p. with the generation of large amounts of andesitic magma as extrusive or shallow intrusive rocks in a convergent plate margin (Mueller *et al.*, 1985). A compressional regime may have continued until about 2.75 b.y. ago, resulting in amphibolite-grade metamorphism and culminating in generation of voluminous granitoid magmas that also entrapped older metasedimentary units.

Lake Plateau Area

The Lake Plateau area is similar in many respects to the eastern Beartooth Mountains; however, it does appear to have formed at much deeper crustal levels. Regional mapping was originally done by Page *et al.* (1973a, b) and Butler (1966) and detailed petrologic studies of this area have been reported by Richmond and Mogk (1985). Large volumes of pegmatite-rich granitic to granodioritic magmas have been intruded as sheetlike bodies into a series of metasupracrustal rocks (Wooden *et al.*, Chapter 14, this volume). To the east, the inclusions are dominantly andesitic amphibolites, similar to those reported by Butler (1966, 1969) in the Cathedral Peak and Long Lake areas (as described above). The west side of the area has meter- to kilometer-sized inclusions of pelitic and psammitic schists. Both types of inclusions have a strong crystallization schistosity developed parallel to the axial surfaces of isoclinal folds. Injection of the granites appear to be in lit-par-lit fashion into this earlier schistosity.

Amphibolite facies metamorphism of the supracrustal inclusions predates intrusion of the granites. The dominant assemblage in the pelitic schist is garnet-biotite-plagioclase-quartz ± cordierite ± sillimanite; staurolite is present in the Boulder River area. Application of the garnet-biotite geothermometer (Ferry and Spear, 1978) yields temperatures in the range 570-620°C. The garnet-cordierite-sillimanite-quartz barometer (Lonker, 1981; Newton, 1983; Thompson, 1976) yields pressures of 7-8 kb (assuming that $P_{H_2O} = P_{total}$) and the garnet-plagioclase-sillimanite-quartz barometer (Ghent, 1976;

Newton, 1983) yields similar pressure estimates. The andesitic amphibolites contain the assemblage hornblende-plagioclase-biotite-quartz ± epidote.

The granites of the Lake Plateau are similar in many respects to Caledonian-type granites as described by Pitcher (1982). They have a restricted range of compositions, from granite to granodiorite (tonalite is conspicuously absent). Rb-Sr whole-rock systematics indicate an age of 2698 ± 86 m.y. and a U-Pb zircon age of 2748 ± 20 m.y. has been determined (LaFrenz *et al.*, in press). These granites are pegmatite-rich and contain magmatic muscovite (wt % TiO2 > 0.7) (Speer, 1984) suggesting P_{H_2O} in excess of 4 kb. The occurrence of magmatic epidote (euhedral grains against hornblende and biotite) indicates that P_{H_2O} may have been as high as 8 kb (Zen and Hammarstrom, 1984), which is consistent with the pressure estimates based on the pelitic assemblages. The granites were emplaced into amphibolite-grade supracrustal rocks at the culmination of this Late Archean orogenic cycle. Generation of the Lake Plateau granites is interpreted as the result of postcollisional, adiabatic melting (Richmond and Mogk, 1985; Wooden *et al.*, Chapter 14, this volume).

North Snowy Block

The North Snowy Block (NSB) is an Archean mobile belt characterized by tectonic juxtaposition of both metaigneous and metasedimentary rocks. The NSB consists of four lithologically and metamorphically distinct linear belts, separated by transcurrent faults and overlain by two east-verging thrust sheets (Mogk, 1984; Fig. 13-4). These six units are distinguished by abrupt discontinuities in metamorphic grade, structural style, and isotopic age. The evolution of this continental crust is dominated by tectonic thickening as opposed to the magmatic growth that occurred in the eastern and central Beartooth areas. The following summarizes the characteristics of these units as they occur in an east-west cross section (Table 13-3).

1. The Paragneiss Unit consists of a wide variety of quartzofeldspathic gneisses, pelitic schists, and amphibolites. Anastomosing shear zones are responsible for abrupt discontinuities of lithology on a meter to tens-of meters scale; a strong subhorizontal lineation is defined by mineral streak lineations. Metamorphism is generally upper amphibolite facies with garnet-biotite pairs yielding temperatures of 700°C; tectonic mixing of the individual units occurred under epidote amphibolite or greenschist facies conditions.

2. The Mount Cowen Augen Gneiss is a late- to postkinematic, sill-like granitic body which has been emplaced along a postulated fault between the Paragneiss Unit and the Davis Creek Schist. The augen texture is defined by an anastomosing foliation of ragged chlorite and muscovite, which wraps around microcline porphyroclasts. The presence of epidote and chlorite suggest recrystallization in the greenschist facies. The Mount Cowen Augen Gneiss is the youngest major rock unit in the NSB (2737 ± 52 m.y.) (D. W. Mogk, P. A. Mueller, and J. L. Wooden, unpublished data).

3. The Davis Creek Schist is a phyllitic metapelite with subordinate quartzite layers. The dominant metamorphic assemblage is chlorite-muscovite-albite-quartz. Rare intrafolial isoclinal folds are overprinted by asymmetric kink folds. Contacts with the overlying Trondhjemitic Gneiss-Amphibolite Complex are strongly mylonitic and occur as wispy intercalations.

4. The Trondhjemitic Gneiss-Amphibolite Complex is a ductile shear zone characterized by blastomylonitic texture, passive flow folding, and subhorizontal mineral streak lineation in the trondhjemitic gneiss. The amphibolites occur as rigid bodies that have been rotated into conformity with the ductile shear foliation in the

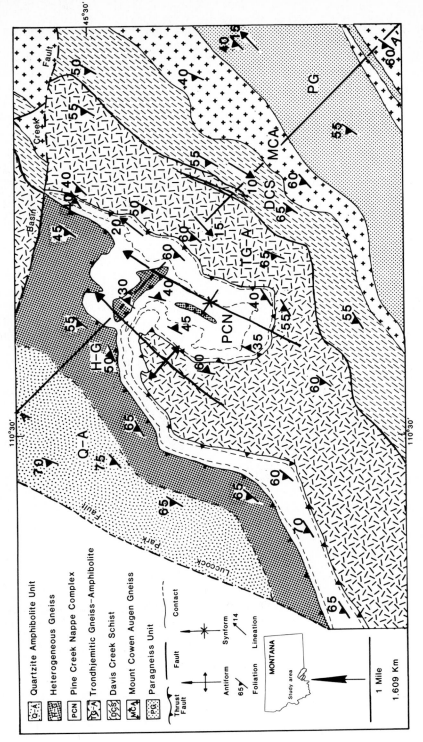

FIG. 13-4. Sketch map of part of the North Snowy Block, showing distribution of major units: (1) heterogeneous gneiss; (2) Pine Creek nappe complex; (3) trondhjemitic gneiss-amphibolite complex; (4) Davis Creek Schist; (5) Mount Cowen Augen Gneiss; (6) paragneiss.

370

TABLE 13-3. Characteristics of the Lithologic Units of the North Snowy Block

Unit	Metamorphic Grade	Structural Style	Isotopic Ages
Heterogeneous Gneiss	Upper amphibolite 650–700°C (gar-bio)	Transposition foliation, intrafolial isoclinal folds	3.4 b.y. Rb-Sr whole rock on injected migmatites
Pine Creek Nappe Complex	Mid-Upper-Amphibolite 600–650°C (gar-bio)	Isoclinal folding on all scales	3.2 b.y. Sm-Nd chondritic model age on amphibolite
Trondhjemite-Amphibolite Complex	Epidote-oligoclase zone 500°C, coexisting albite-oligoclase	Ductile shear zone, blastomylonitic, passive flow folds	3.55 and 3.26 b.y. Sm-Nd chondritic model age ~3.4 b.y. Rb-Sr whole rock
Davis Creek Schist	Greenschist Facies chlor-musc-albite-qtz	Phyllitic, local isoclinal fold, late-stage kinks	
Mount Cowen Augen Gneiss	Greenschist Facies	Granitic augen gneiss	2.74 b.y. Rb-Sr whole-rock isochron
Paragneiss	Upper amphibolite 700°C (gar-bio)	Anastomosing shear zones, high degree of internal tectonic mixing	~2.8 b.y. on quartzofeldspathic gneiss in Yankee Jim Canyon

trondhjemitic gneiss. They include fine-grained, well-lineated varieties as well as coarse-grained metagabbros and anorthositic gabbros, metamorphosed in the epidote-bearing and epidote-free amphibolite facies. Hornblende-plagioclase relations (Spear, 1980) are consistent with recrystallization in the lower to middle amphibolite facies. Individual layers yield a variety of *P-T* conditions (e.g., 520–540°C, 6-7 kb; 560–580°C, 4-6 kb; 620–650°C, 4-5 kb; using the phase relations of Plyusnina, 1982), suggesting tectonic mixing of different levels of mafic crust during ductile shearing. Dynamic recrystallization of both albite and oligoclase neoblasts around oligoclase porphyroclasts in the trondhjemitic gneiss indicates ductile shearing occurred at less than 500°C (below the crest of the peristerite solvus), and postdates the peak amphibolite metamorphism recorded in the amphibolites.

5. The Pine Creek Nappe Complex is an isoclinally folded thrust nappe consisting of amphibolite (core), with symmetrically disposed marble and quartzite on the upper and lower limbs. The lower limb is strongly attenuated and mylonites are well developed in the quarzite at the lower contact. Isoclinal folding occurs on all scales and is contemporaneous with a crystallization schistosity which is parallel to the axial surfaces. Metamorphism is in the middle to upper amphibolite facies, as indicated by hornblende-plagioclase relations in the amphibolites (Spear, 1980) with average temperatures of 580–620°C and pressures of 4-6 kb (Plyusnina, 1982), garnet-biotite temperatures of 600-650°C in rare pelitic layers in the quartzite, and calcite-dolomite-diopside-phlogopite-retrograde tremolite assemblages in the marble.

6. The Heterogeneous Gneiss unit consists of a migmatitic complex associated with a high-grade metasupracrustal sequence which has been thrust over the Pine Creek Nappe Complex. In the lower part of the thrust sheet, lit-par-lit injections of granite to tonalite invade the country rock. Locally, foliation is truncated, and partial assimilation of the supracrustal rocks is common. Higher in the thrust sheet, the igneous rocks are absent and the supracrustal assemblage includes orthoquartzite, feldspathic quartzite, amphibolite, and minor pelite. Metamorphism at the base of the thrust sheet is in the upper amphibolite facies, as demonstrated by the occur-

rence of sillimanite in pelitic rocks; hornblende-plagioclase relations and garnet-biotite temperatures are in the range 650–700°C.

Within the North Snowy Block, the wide variety of rock types and the abrupt discontinuities in metamorphic grade, structural style and isotopic ages suggest significant tectonic displacements. The Paragneiss Unit, Davis Creek Schist, and Trondhjemitic Gneiss-Amphibolite Complex probably were emplaced along transcurrent faults, as evidenced by their subhorizontal lineations. These units cannot be readily restored to a prefaulting stratigraphy, as would be expected if they currently represent a series of stacked thrust faults. In addition, the Paragneiss Unit is a chaotic mixture of tectonically juxtaposed, diverse rock types; the style of deformation of this unit is analogous to the tectonic mixing associated with wrench faults rather than thrust faults. The minimum age for the transcurrent faulting is 2.75 b.y. based on the age of the protolith of the Mount Cowen Augen Gneiss. The emplacement of the two thrust sheets marks a significant change in the tectonic style of the NSB and appears to postdate the transcurrent faulting. The ductile shearing observed in the Trondhjemitic Gneiss is absent in the overlying thrust sheets. It is also significant that the metamorphic grade increases discontinuously up-section from the Trondhjemitic Gneiss through the overlying thrust sheets.

Yankee Jim Canyon

The Yankee Jim Canyon area is a ductile shear zone (Burnham, 1982) which is similar in all respects to the Paragneiss Unit in the North Snowy Block. A wide variety of lithologies, including quartzofeldspathic gneisses, quartzite, pelites, amphibolites, and ultramafites, are tectonically mixed on a meter to tens-of-meters scale. Common assemblages include:

1. K-spar-plag-qtz-bio ± musc (quartzofeldspathic gneisses)
2. Bio-gar-sill-plag-qtz (pelitic rocks)
3. Hornblende-plagioclase-quartz ± biotite (metabasites)
4. Opx-hornblende (ultramafites)

Although diagnostic index minerals are rare, these assemblages indicate that peak metamorphism is in the upper amphibolite facies. Garnet-biotite temperatures are in the range 680–740°C. Retrograde minerals occur in anastomosing shear zones in this area and include chlorite, epidote, actinolite, and sericite. Calculated rim temperatures on garnet-biotite pairs of 480–520°C reflect retrograde metamorphic temperatures. The age of metamorphism is about 2.8 b.y. (Rb-Sr whole-rock age by Paul Mueller, unpublished data) and there is some evidence of zircons as old as 3.6 b.y. in some of the units (Guy and Sinha, 1985). The mixture of lithologies, metamorphic grade, and structural style of the Yankee Jim Canyon area are similar to those observed in the Paragneiss Unit of the North Snowy Block.

Jardine-South Snowy Block

The Jardine area is characterized by a sequence of fine-grained detrital sediments and associated iron formation. Included in this sequence are quartz-biotite schists, chlorite-muscovite schists, and grunerite-bearing ironstones (Hallager, 1984). This sequence is interpreted as the distal fan facies associated with a rifted continental margin (Thurston, 1986). Sedimentary structures such as graded bedding and cut-and-fill channel structures are locally preserved; penetrative deformation is conspicuously absent. Toward the margins of this sequence, the grade apparently increases, as evidenced by the occurrence of andalusite, garnet, and staurolite in the pelitic units. These low-grade metasedimentary rocks are rare in the Archean basement of southwestern Montana. The Jardine sequence has been tectonically juxtaposed along ductile shear zones in the Yankee Jim Canyon area to the west. In the Broadwater River area to the east, 2.6-b.y.-old quartz monzonite stocks have been emplaced in this sequence.

The bulk of the South Snowy Block consists of an extensive metasedimentary sequence which was intruded by Late Archean granites (Casella *et al.*, 1982). The dominant rock type is biotite schist with varying modal abundances of quartz, plagioclase, muscovite, garnet, staurolite, and andalusite. Grunerite-bearing ironstones are also present in this area. On the eastern margin, the schists are upgraded to upper amphibolite facies as indicated by the presence of sillimanite and injection-type migmatites. Isoclinal folding accompanies amphibolite-grade metamorphism and a second open folding event is also recorded.

Stillwater Complex Hornfels Aureole

The Stillwater Complex aureole has been described by Page (1977). Vaniman *et al.* (1980), Page and Zientek (1985) and Labotka (1985). The aureole consists of layered, fine-grained clastic sediments, iron formation, blue quartzite, and diamictite. Sedimentary structures such as cross bedding, graded bedding, and cut-and-fill structures are locally preserved. The diagnostic assemblages in pelitic rocks are orthopyroxene-cordierite and anthophyllite-cordierite with associated quartz, plagioclase, spinel, biotite, ilmenite, and sulfide minerals. The iron formation contains the assemblage: quartz-magnetite-orthopyroxene-grunerite; clinopyroxene, olivine, and chlorite are locally present. The estimated conditions of metamorphism are $P = 2$-3 kb based on the composition of pigeonite in the iron formations (Vaniman *et al.*, 1980) and $T = 710°C$ (garnet-biotite) to 825°C (minimum temperature of pigeonite stability at 3 kb) (Labotka, 1985). The hornfels aureole is truncated by a splay of the Mill Creek–Stillwater Fault Zone (Geissman and Mogk, in press). Metasediments in the Boulder River area (Weeks, 1981) on the south side of this fault have mineral assemblages, penetrative deformation, calculated pressures and temperatures, and whole-rock chemistries more characteristic of the regionally metamorphosed schists described in the Lake Plateau area (Geissman and Mogk, in press). Labotka (1985) has recently reported fine-grained garnet, staurolite, and chlorite in this area. The Stillwater Complex and its associated aureole have been tectonically emplaced against the Beartooth massif shortly after crystallization of the complex (2700–2720 m.y.b.p.) and prior to the emplacement of the 2700-m.y.-old Mouat Quartz Monzonite (Geissman and Mogk, in press; Mogk and Geissman, 1984).

Summary of the Archean Geology
of the Beartooth Mountains

The Beartooth Mountains have evolved through at least two orogenic cycles separated by a long period of tectonic quiescence during the Mid to Late Archean (Mogk *et al.*, 1984; Mueller *et al.*, 1985). Platform-type sediments in the Quad Creek and Hellroaring Plateau areas have been metamorphosed to granulite grade and are the only remnants of the first cycle, which occurred ~3.4 b.y. ago. Deep burial of these sediments is interpreted as the result of continental collision. A second orogenic cycle involved both magmatic and tectonic growth of the crust. Generation of andesitic rocks occurred at about 2.8 b.y. and voluminous granitoids were emplaced at 2.75 b.y. in the eastern and central Beartooth Mountains. These rocks are interpreted to be subduction related (Mueller *et al.*, 1985). Tectonic thickening in the North Snowy Block is roughly contemporaneous with the magmatic activity of the central Beartooth Mountains. The metasupracrustal rocks of the Beartooth Mountains exhibit a broad range of metamorphic grade, structural style, and whole-rock chemistry, suggesting that they were derived from a variety of source areas and experienced independent geologic histories[1] (Mueller *et al.*, 1984).

OVERVIEW OF THE WESTERN WYOMING PROVINCE

The mobile belt in the North Snowy Block marks a major discontinuity in the nature of the Archean continental crust of the northern Wyoming Province (Fig. 13-1). Rocks to the west of this mobile belt include vast expanses of quartzofeldspathic gneiss with inter-layered mafic granulite and extensive platform-type metasedimentary rocks. Metamorphism is dominantly in the upper amphibolite to granulite facies. Isoclinal folding on all scales and nappe emplacement characterize the regional structural style. Recognizable plutonic rocks are restricted to one narrow belt in the Spanish Peaks area of the northern Madison Range. Rocks with oceanic affinity are also absent in the western Wyoming Province; there are no greenstone belt associations similar to those observed in the southern Wyoming Province and the sedimentary rocks are distinctly of platform type as opposed to a eugeoclinal assemblage. A model is proposed that calls for early development of a rift-bounded ensialic, or Tethyan-type basin with only limited generation of oceanic crust, followed by deep burial of sediments by means of tectonic thickening (A-type subduction). The following summarizes the major features of the exposures of Archean basement in an east-west progression.

Northern Gallatin Range

The northern Gallatin Range consists dominantly of a variety of quartzofeldspathic gneisses with subordinate layers of orthoquartzite, pelitic schists, and meter-scale boudins of mafic granulite. Compositional layering on a centimeter scale is defined by wide variations of modal abundances of microcline and plagioclase and by color index. This fine-scale layering and the association of quartzites with pelites strongly suggests that these rocks are derived from supracrustal assemblages. Metamorphism is in the upper amphibolite to hornblende granulite facies. The dominant assemblage in mafic layers is clinopyroxene-garnet-hornblende-plagioclase-quartz ± scapolite. Garnet-clinopyroxene temperatures in mafic granulites (Dahl 1980; Ellis and Green, 1979) yield temperatures of 700–750°C. Orthopyroxene is also recognized in centimeter-scale mafic layers in the quartzofeldspathic gneisses. Garnet-biotite pairs in the gneissic and schistose layers also yield temperatures of 700–750°C. Numerous layers of both concordant and discordant remobilized granitic leucosome occur within the quartzofeldspathic gneisses. These are interpreted as anatectic melts, produced in response to vapor-absent melting during granulite facies metamorphism. The granulite facies assemblages are overprinted by amphibolite facies assemblages; no new structural elements are recognized in this area. The amphibolite event is therefore interpreted as part of the retrograde path of this metamorphic cycle.

Spanish Peaks Area, Northern Madison Range

The Spanish Peaks area has many of the same components recognized in the Gallatin Range. Migmatitic gneiss-granulite associations are present in both the Gallatin and Madison River Canyons. The lithologic sequences, metamorphic grades, and structural styles

[1] Also D. W. Mogk, P. A. Mueller, and J. L. Wooden, unpublished data.

in these areas are identical to those reported above. However, in addition, a mesozonal batholithic complex is present in the central part of the range (Salt and Mogk, 1985). The batholithic terrain is separated from the high-grade metasupracrustal terrains by northeast-trending ductile shear zones. The intrusive phases follow a differentiation trend from gabbro-quartz diorite-monzodiorite-granodiorite-granite; only the late-stage granites are unfoliated. Magmatic epidote has been observed and suggests intrusion of these plutonic rocks under 8-kb P_{H_2O}. The host rock is dominantly tonalitic gneiss with subordinate kyanite-bearing pelitic schists and mafic granulites. Garnet-biotite and clinopyroxene-garnet geothermometry yields temperatures of 650–700°C; garnet-plagioclase-quartz-kyanite geobarometry indicates pressures in the range of 7–8 kb.

West of the batholithic terrain, the paragneisses are characteristically K-rich. In the pelitic rocks, sillimanite replaces kyanite as the stable aluminosilicate polymorph, and amphibolite assemblages replace the mafic granulite. Calculated temperatures are somewhat higher in this terrane, in the range 680–750°C and pressures are lower, in the range 6–7 kb. There is a conspicuous absence of plutonic rocks in this terrane, as well. The apparent break in metamorphic grade and lithologic sequences suggests large-scale displacements along the ductile shear zones; dominant down-dip lineations suggest that juxtaposition occurred along thrust faults.

Tobacco Root and Ruby Ranges

Extensive platform-type metasedimentary rocks that include marble, pelitic schists, banded iron formation, and orthoquartzites are exposed in these ranges. Quartzofeldspathic gneisses, both K-spar- and plagioclase-rich, are the most abundant rocks; mafic granulites and metaultramafites are also present (e.g., Cordua, 1976; Desmarais, 1980; Garihan, 1979; Vitaliano *et al.*, 1979). Isoclinal folding is present on all scales, and detachment along attenuated limbs accompanied nappe emplacement. This sequence has also been metamorphosed in the upper amphibolite to granulite facies. Diagnostic metamorphic assemblages include:

1. Olivine-diopside-scapolite-phlogopite-calcite-dolomite (marble)
2. Garnet-biotite-kyanite-sillimanite-plagioclase (pelitic schist)
3. Garnet-orthopyroxene-quartz-magnetite-retrograde grunerite (BIF)
4. Olivine-orthopyroxene-spinel (ultramafites)
5. Garnet-clinopyroxene-plagioclase-hornblende (metabasites)

Garnet-biotite temperatures are in the range 700–800°C (D. W. Mogk, unpublished data) as are garnet-clinopyroxene temperatures (Dahl, 1979, 1980). Pressure estimates based on the garnet-aluminosilicate-quartz-plagioclase barometer are on the order of 7 kb.

Blacktail Range

West of the platform-type assemblages exposed in the Tobacco Root and Ruby Ranges, the dominant rock type is again quartzofeldspathic gneiss with a wide range of plagioclase/K-spar ratios and color index. These gneisses characteristically have attained a granoblastic texture and the following mineral assemblages have been recognized in discrete centimeter-scale layers:

1. Plag-K-spar-qtz
2. Biotite-garnet-K-spar-qtz
3. Biotite-cordierite-sillimanite-plagioclase-qtz

These gneisses are interpreted as a supracrustal sequence originally comprised of arkoses, graywackes, mafic sills or flows, and possibly felsic volcaniclastic sediments (Clark and Mogk, 1985, 1986).

EVOLUTION OF THE WESTERN WYOMING PROVINCE

To the west of the Beartooth Mountains, during the interval 2.75–2.70 b.y., crustal growth was dominantly through tectonic thickening. A working model for the evolution of this terrain includes early formation of a rift-bounded ensialic basin, with possible formation of a small amount of oceanic crust, followed by collapse of this basin and ultimate deep burial through stacking of nappes. The quartzofeldspathic gneisses have been interpreted as paragneisses based on fine-scale compositional layering and the intimate interlayering of pelitic and quartzitic units. It is significant to note that these are the oldest K-rich gneisses (sediments) recognized in this area. Presumably, they were once arkoses derived from an orogenic terrain similar to those found in the Beartooth Mountains, which were rich in Late Archean silicic volcanic and plutonic rocks. The mafic rocks have been interpreted as continental quartz tholeiites (e.g., Hanley, 1976) and probably occurred originally as flows or sills in the supracrustal pile. The platform sediments must have been deposited in the basin during a long period of quiescence.

The structural style and high metamorphic grade now observed in these rocks requires large-scale horizontal shortening and deep burial of these supracrustal rocks. Continental collision is the most reasonable process, as suggested by Newton and Perkins (1982). No direct evidence of ancient oceanic crust is preserved in the northern Wyoming Province. There is a distinct lack of large volumes of mafic rocks or eugeoclinal sediments preserved in this area (in contrast to characteristic greenstone belt assemblages). However, the batholithic rocks in the Spanish Peaks area may be a remnant of a magmatic arc generated in response to consumption of a small oceanic basin. The absence of platform sediments in the ranges in the westernmost part of the Wyoming Province (and dominance of quartzofeldspathic gneisses) suggests that there was another sialic source of clastics to the west. This phantom sialic source area could simply represent the western margin of the basin, another colliding continental mass (in a Tethyan-type setting), or a colliding island arc (e.g., Cheyenne Belt, Wyoming; Karlstrom and Houston, 1984). Figure 13-5 presents an interpretive east-west cross section of the northern Wyoming Province, showing a model for the proposed basin prior to the Late Archean orogenic cycle. Figure 13-6 is a schematic cross section of this terrain highlighting the salient properties of each of the major ranges in their present geologic setting.

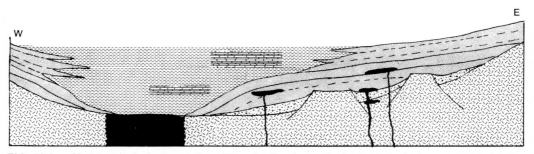

FIG. 13-5.   Model of basin in the northern Wyoming Province prior to the Late Archean orogeny. A rift-bounded basin formed adjacent to the ancient crust formed in the Beartooth Mountains. Attenuation of the crust may have led to generation of a small oceanic basin. Thick clastic sequences are deposited on the east and west sides of the basin and stable platform sediments are deposited in the interior. Basaltic material is emplaced as either sills or flows into this sequence. This sequence of rocks best represents the protoliths for the high-grade rocks observed in the northern Wyoming Province.

Within the Archean basement of southwestern Montana, crustal evolution has occurred by means of both tectonic and magmatic processes. Discrete metamorphic/deformational cycles have been followed by long periods of quiescence and deposition of platform-type sediments. The best documented orogenic events recorded in the Archean basement of southwestern Montana include:

1. Granulite-grade metamorphism of platform-type sediments at a time of 3.4 b.y. in the Quad Creek and Hellroaring Plateau areas of the Beartooth Mountains. Deep burial of these sediments is interpreted as the result of tectonic thickening.
2. Generation of 2.8-b.y.-old andesites and voluminous 2.75-b.y.-old granitoids in the eastern and central Beartooth Mountains, and contemporaneous tectonic thickening by means of thrust and transcurrent faulting in the North Snowy Block.
3. Deposition of thick supracrustal sequences in the northwestern Wyoming Province, followed by large-scale crustal shortening and granulite-grade metamorphism of platform-type sediments in the latest Archean (*ca.* 2.70-2.75 b.y.). This style of orogeny is interpreted to be the result of continental collision.

Based on these observations, we propose that continental growth in the latest Archean occurred through processes analogous with those of contemporary-style plate tectonics. The most ancient rocks in the Beartooth Mountains, including the 3.6-b.y.-old trondhjemitic gneiss in the North Snowy Block and the 3.4-b.y.-old metasupracrustal sequences in the Quad Creek and Hellroaring Plateau areas, are remnants of the oldest continental material recognized in the Wyoming Province. The orogeny that occurred between 2.8 and 2.75 b.y. in the Beartooth Mountains produced large volumes of igneous

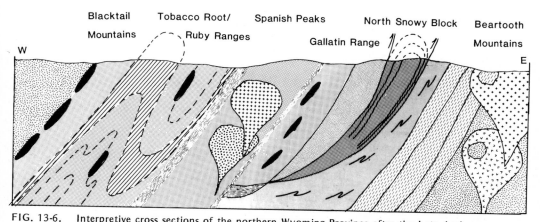

FIG. 13-6. Interpretive cross sections of the northern Wyoming Province after the Late Archean orogeny (2.8–2.7 b.y.b.p.). The prominant features of this terrain are featured from east to west: Beartooth Mountains: voluminous 2.75-b.y.-old calc-alkaline rocks with inclusions of older, high-grade metasupracrustal rocks; North Snowy Block: linear belts of tectonically juxtaposed units, ductile shearing of trondhjemite-amphibolite complex, emplacement of thrust sheets; Gallatin Range: ductile shear zones and mafic granulites in quartzofeldspathic gneiss matrix; Spanish Peaks: batholithic complex emplaced into granulite-grade meta-supracrustal rocks, separated from other units by ductile shear zones; Tobacco Root/Ruby Ranges: large scale isoclinal folds and granulite-grade metamorphism of stable platform-type sedimens; Blacktail Range; granulite grade metamorphism of paragneisses. The large-scale crustal shortening of this terrain and the associated high-grade metamorphism of supracrustal rocks is interpreted as the result of a Late Archean continental collision.

rocks through subduction-related processes (Mueller *et al.*, 1985), and the North Snowy Block has many aspects of a modern Cordilleran-type continental margin (D. W. Mogk, P. A. Mueller, and J. L. Wooden, unpublished data). Orogenesis continued to the west of the Beartooth Mountains in the latest Archean, and continental collision is believed to be the culminating event in this cycle. The evolution of the continental crust of the northern Wyoming Province occurred through Late Archean magmatic and tectonic accretion to an ancient sialic continental nucleus.

The metamorphic petrology and structural relations observed in the northern Wyoming Province have implications for the nature of the Archean continental crust in general. The pressures and temperatures calculated from the granulite-grade mineral assemblages in supracrustal rocks requires deep burial of these rocks, on the order of 20-25 km as far back as 3.4 b.y. ago. This requires that thick continental crust must have been developed, at least locally, early in the geologic record. The widespread occurrence of platform-type sediments in these high-grade terrains requires long periods of tectonic quiescence, calling into question the concept of the Archean "permobile" tectonic regime. Finally, the temperatures and pressures of metamorphism recorded in these Archean rocks are similar to those observed in modern orogens, suggesting that the mechanism for dissipation of the earth's internal heat has been fundamentally the same throughout the history of the earth.

ACKNOWLEDGMENTS

This research has been supported by the NSF-EPSCOR program and the NASA Early Crustal Genesis project (DWM). DJH would like to acknowledge support by the National Research Council for the early stages of this study. Paul A. Mueller and Joseph L. Wooden have contributed substantially to the ideas presented in this paper through numerous years of collaboration. Critical reviews by Gary Ernst and Paul Mueller greatly improved the quality of this manuscript.

REFERENCES

Barker, F., Arth, J. G., and Millard, H. T., Jr., 1979, Archean trondhjemites of the southwestern Bighorn Mountains, Wyoming — A preliminary report, *in* Barker, F., ed., *Trondhjemites, Dacites and Related Rocks:* Amsterdam, Elsevier, p. 401–414.

Bentley, R. D., 1967, Geologic evolution of the Beartooth Mountains, Montana and Wyoming — Pt. 9. The Cloverleaf Lakes area: Ph.D. dissertation, Columbia Univ., New York, N.Y., 153 p.

Burnham, R. L., 1982, Mylonitic basement rocks in the Yankee Jim Canyon and Sixmile Creek areas, Park County, Montana: *Rocky Mountain Section, Geol. Soc. America Abstr. with Programs*, p. 305.

Butler, J. R., 1966, Geologic evolution of the Beartooth Mountains, Montana and Wyoming — Pt. 6. The Cathedral Peak area, Montana: *Geol. Soc. America Bull.*, v. 77, p. 45–64.

——, 1969, Origin of Precambrian granite gneisses in the Beartooth Mountains, Montana and Wyoming, *in* Larsen, L. H., ed., *Igneous and Metamorphic Geology:* Geol. Soc. America Mem. 115, p. 73–101.

Casella, C. J., 1964, Geologic evolution of the Beartooth Mountains, Montana and Wyoming — Pt. 4. Relationship between Precambrian and Laramide structures in the Line Creek area: *Geol. Soc. America Bull.*, v. 75, p. 969–984.

_____, 1969, A review of the Precambrian geology in the eastern Beartooth Mountains, Montana and Wyoming, *in* Larsen, L. H., ed., *Igneous and Metamorphic Geology:* Geol. Soc. America Mem. 115, p. 53–71.

_____, Levay, E. E., Hirst, B., Huffman, K., Lahti, V., and Metzger, R., 1982, Precambrian geology of the southwestern Beartooth Mountains, Yellowstone National Park, Montana and Wyoming, *in* Mueller, P. A., and Wooden, J. L., eds., *Precambrian Geology of the Beartooth Mountains, Montana and Wyoming:* Mont. Bur. Mines Geology Spec. Publ. 84, p. 1–24.

Chatterjee, N. D., and Froese, E., 1975, A thermodynamic study of the pseudobinary join muscovite-paragonite in the system KA1Si308-NaA1Si308-A1203-Si02-H2O: *Amer. Mineralogist*, v. 60, p. 985–993.

Clark, M. L., and Mogk, D. W., 1985, Development and significance of the Blacktail Mountains Archean metamorphic complex, Beaverhead County, Montana: *Rocky Mountain Section, Geol. Soc. America Meet., Boise, Idaho, Abstr. with Programs*, p. 212.

_____, and Mogk, D. W., 1986, Evolution of the Archean Metasupracrustal Sequence, Blacktail Mountains, Montana: *Rocky Mountain Section, Geol. Soc. America Meet., Flagstaff, Ariz., Abstr. with Programs,* p. 346.

Condie, K. C., 1976, The Wyoming Archean Province in the western United States, *in* Windley, B. F., ed., *The Early History of the Earth:* London, Wiley, p. 499–511.

Cordua, W. S., 1976, Precambrian geology of the southern Tobacco Root Mountains, Madison County, Montana: Ph.D. dissertation, Indiana Univ., Bloomington, Ind., 247 p.

Dahl, P. S., 1979, Comparative geothermometry based on major element and oxygen isotope distributions in Precambrian metamorphic rocks from southwestern Montana: *Amer. Mineralogist*, v. 64, p. 1280–1293.

_____, 1980, The thermal-compositional dependence of Fe^2-Mg distributions between co-existing garnet and pyroxene — Applications to geothermometry: *Amer. Mineralogist*, v. 65, p. 852–866.

Desmarais, N. R., 1980, Metamorphosed Precambrian ultramafic rocks in the Ruby Range, Montana: *Precambrian Res.*, v. 16, p. 67–101.

Eckelmann, F. D., and Poldervaart, A., 1957, Geologic evolution of the Beartooth Mountains, Montana and Wyoming — Pt. 1. Archean history of the Quad Creek area: *Geol. Soc. America Bull.*, v. 68, p. 1225–1262.

Ellis, D. J., and Green, D. H., 1979, An experimental study of the effect of Ca upon garnet-clinopyroxene exchange equilibria: *Contrib. Mineralogy Petrology*, v. 71, p. 13–22.

Erslev, E. A., 1983, Pre-Beltian geology of the southern Madison Range, southwestern Montana: *Mont. Bur. Mines Geology Mem. 55*, p. 1–26.

Evans, B. W., 1977, Metamorphism of alpine periodotite and serpentinite: *Ann. Rev. Earth Planet. Sci.*, v. 5, p. 397–447.

Ferry, J. M., and Spear, F. S., 1978, Experimental calibration of the partitioning of Fe and Mg between biotite and garnet: *Contrib. Mineralogy Petrology*, v. 66, p. 113–117.

Foose, R. M., Wise, D. U., and Garbarini, G. S., 1961, Structural geology of the Beartooth Mountains, Montana and Wyoming: *Geol. Soc. America Bull.*, v. 72, p. 1143–1172.

Garihan, J. M., 1979, Geology and structure of the central Ruby Range, Madison County, Montana: *Geol. Soc. America Bull.*, Pt. II, v. 90, p. 695–788.

Geissman, J. W., and Mogk, D. W., in press, Late Archean tectonic emplacement of the Stillwater Complex along reactivated basement structures, northern Beartooth Mountains, southern Montana, USA: *Proceedings VI International Conference on Basement Tectonics.*

Ghent, E. D., 1976, Plagioclase-garnet-Al$_2$SiO$_5$-quartz — A potential geobarometer-geothermometer: *Amer. Mineralogist*, v. 61, p. 309–340.

Guy, R. E., and Sinha, A. K. 1985, Petrology and isotopic geochemistry of the Archean basement lithologies near Gardiner, Montana: *Geol. Soc. America Abstr. with Programs*, v. 17, p. 601.

Hallager, W. S., 1984, Geology of gold-bearing metasediments near Jardine, Montana, *in* Foster, R. P., ed., *Gold 82 — The Geology, Geochemistry and Genesis of Gold Deposits:* Rotterdam, Balkema, p. 191–218.

Hanley, T. B., 1976, Stratigraphy and structure of the central fault block, northwestern Tobacco Root Mountains, Madison County, Montana: *Tobacco Root Geol. Soc. 1976 Field Conf. Guidebook*, Mont. Bur. Mines Geology Spec. Publ. 73, p. 7–14.

Harris, R. L., Jr., 1959, Geologic evolution of the Beartooth Mountains, Montana and Wyoming — Pt. 3. Gardner Lake Area, Wyoming: *Geol. Soc. America Bull.*, v. 70, p. 1185–1216.

Henry, D. J., Mueller, P. A., Wooden, J. L., Warner, J. L., and Lee-Berman, R., 1982. Granulite grade supracrustal assemblages of the Quad Creek area, eastern Beartooth Mountains, Montana, *in* Mueller, P. A., and Wooden, J. L., eds., *Precambrian Geology of the Beartooth Mountains, Montana and Wyoming:* Mont. Bur. Mines Geology Spec. Publ. 84, p. 147–159.

Holdaway, M. J., 1971, Stability of andalusite and the aluminum silicate phase diagram: *Amer. J. Sci.*, v. 271, p. 97–131.

James, H. L., and Hedge, C. E., 1980, Age of the basement rocks of southwest Montana: *Geol. Soc. America Bull.*, v. 91, p. 11–15.

Karlstrom, K. E., and Houston, R. S., 1984, The Cheyenne Belt — Analysis of a Proterozoic suture in southern Wyoming: *Precambrian Res.*, v. 25, p. 415–446.

Khoury, S. G., and Ghaly, T. S., 1973, Geological evolution of the Archean basement in the Long Lake area, Beartooth Mountains: *Earth Sci.*, v. 1, p. 1–15.

Labotka, T. C., 1985, Petrogenesis of the metamorphic rocks beneath the Stillwater Complex — Assemblages and conditions of metamorphism, *in* Czamanske, G. K., and Zientek, M. L., eds., *The Stillwater Complex, Montana — Geology and Guide:* Mont. Bur. Mines Geology Spec. Publ. 92, p. 70–76.

Larsen, L. H., Poldervaart, A., and Kirchmeyer, M., 1966, Geologic evolution of the Beartooth Mountains, Montana and Wyoming — Pt. 7. Structural homogeneity of gneisses in the Lonesome Mountain area: *Geol. Soc. America Bull.*, v. 77, p. 1277–1292.

Lonker, S. W., 1981, The P-T-X relations of the cordierite-garnet-sillimanite-quartz equilibrium: *Amer. J. Sci.*, v. 281, p. 1056–1090.

Mogk, D. W., 1984, Petrology, geochemistry and structure of an Archean terrane in the North Snowy Block, Beartooth Mountains, Montana: Ph.D. dissertation, Univ. Washington, Seattle, Wash., 440 p.

——, and Geissman, J. W., 1984, The Stillwater Complex is allochthonous: *Geol. Soc. America Ann. Meet. Reno, Nev., Abstr. with Programs*, p. 598.

——, Mueller, P. A., and Wooden, J. L., 1984, Secular variation in Archean tectonic style, Beartooth Mountains, Montana: (EOS) *Trans. Amer. Geophys. Un.*, v. 65, p. 230.

Montgomery, C. W., Kirsling, T. J., and Gray, B. A., 1984, Ages and Sr and O isotope systematics of Archean granitic gneisses of the south-central Beartooth Mountains: *Geol. Soc. America Abstr. with Programs*, v. 16, p. 599.

Mueller, P. A., Wooden, J. L., Odom, A. L., and Bowes, D. R., 1982, Geochemistry of the Archean rocks of the Quad Creek and Hellroaring Plateau areas of the eastern Beartooth Mountains, *in* Mueller, P. A., and Wooden, J. L., eds., *Precambrian Geology of the Beartooth Mountains, Montana and Wyoming:* Mont. Bur. Mines Geology Spec. Publ. 84, p. 69–82.

____, Mogk, D. W., Wooden, J. L., Henry, D. J., and Bowes, D. R., 1984, Archean metasedimentary rocks from the Beartooth Mountains — Evidence for accreted terrane? *Geol. Soc. America Abstr. with Programs*, v. 16, p. 602.

____, Wooden, J. L., Henry, D. J., and Bowes, D. R., 1985, Archean crustal evolution of the eastern Beartooth Mountains, Montana and Wyoming, *in* Czamanske, G. K., and Zientek, M. L., eds., *The Stillwater Complex, Montana — Geology and Guide:* Mont. Bur. Mines Geology Spec. Publ. 92, p. 9–20.

Nabelek, C. R., and Lindsley, D. R., 1985, Tetrahedral A1 in amphibole — A potential thermometer for some mafic rocks: *Geol. Soc. America Abstr. with Programs*, v. 17, p. 673.

Newton, R. C., 1983, Geobarometry of high-grade metamorphic rocks: *Amer. J. Sci.*, v. 283-A, p. 1–28.

____, and Perkins, D., 1982, Thermodynamic calibration of geobarometers based on the assemblages garnet-plagioclase-orthopyroxene (clinopyroxene)-quartz: *Amer. Mineralogist*, v. 67, p. 203–222.

Page, N. J., 1977, Stillwater Complex, Montana — Rock succession, metamorphism and structure of the complex and adjacent rocks: *U.S. Geol. Survey Prof. Paper 999*, 79 p.

____, and Zientek, M. L., 1985, Petrogenesis of the metamorphic rocks beneath the Stillwater Complex — Lithologies and structures *in* Czamanske, G. K., and Zientek, M. L., eds., *The Stillwater Complex, Montana — Geology and Guide:* Mont. Bur. Mines Geology Spec. Publ. 92, p. 55–69.

____, Simons, F. S., and Dohrenwand, J. C., 1973a, Reconnaissance geologic map of the Mount Douglas quadrangle, Montana: *U.S. Geol. Survey Misc. Field Studies Map MF-488*, scale 1:62,500.

____, Simons, F. S., and Dohrenwand, J. C., 1973b, Reconnaissance geologic map of the Mount Wood Quadrangle, Montana: *U.S. Geol. Survey Misc. Field Studies Map MF-491*, scale 1:62,500.

Peterman, Z. E., 1981, Dating of Archean basement in northeastern Wyoming and southern Montana: *Geol. Soc. America Bull.*, Pt. I, v. 92, p. 139–146.

Pitcher, W. S., 1982, Granite type and tectonic environment, *in* Hsü, K. J., ed., *Mountain Building Processes*, London, Academic, p. 19–40.

Plyusnina, L. P., 1982, Geothermometry and geobarometry of plagioclase-hornblende bearing assemblages: *Contrib. Mineralogy Petrology*, v. 80, p. 140–146.

Prinz, M. J., 1964, Geologic evolution of the Beartooth Mountains, Montana and Wyoming — Pt. 5. Mafic dike swarms of the southern Beartooth Mountains: *Geol. Soc. America Bull.*, v. 75, p. 1217–1248.

Reid, R. R., McMannis, W. J., and Palmquist, J. C., 1975, Precambrian geology of the North Snowy Block, Beartooth Mountains: *Mont. Geol. Soc. America Spec. Paper 157*, p. 1–135.

Richmond, D. P., and Mogk, D. W., 1985, Archean Geology of the Lake Plateau Area, Beartooth Mountains, Montana: *Rocky Mountain Section, Geol. Soc. America Meet., Boise, Idaho, Abstr. with Programs*, p. 262.

Rowan, L. C., 1969, Structural geology of the Quad–Wyoming–Line Creeks area, Beartooth Mountains, Montana: *Geol. Soc. America Mem. 115*, p. 1–18.

____, and Mueller, P. A., 1971, Relations of folded dikes and Precambrian polyphase deformation, Gardner Lake area, Beartooth Mountains, Wyoming: *Geol. Soc. America Bull.*, v. 82, p. 2177–2186.

Salt, K. J., and Mogk, D. W., 1985, Archean geology of the Spanish Peaks area, southwestern Montana: *Rocky Mountain Section, Geol. Soc. America Meet., Boise, Idaho, Abstr. with Programs*, p. 263.

Skinner, W. R., Bowes, D. R., and Khoury, S. G., 1969, Polyphase deformation in the Archean basement complex, Beartooth Mountains, Montana and Wyoming: *Geol. Soc. America Bull.*, v. 80, p. 1053–1060.

Spear, F. S., 1980, NaSi=CaAl exchange equilibrium between plagioclase and amphibole: *Contrib. Mineralogy Petrology*, v. 72, p. 33–41.

———, 1981, An experimental study of hornblende stability and compositional variability in amphibole: *Amer. J. Sci.*, v. 281, p. 697–734.

Speer, J. A., 1984, Micas in igneous rocks, *in* Bailey, S. W., ed., *Micas:* Mineral. Soc. America Rev. Mineralogy, v. 13, p. 299–356.

Spencer, E. W., 1959, Geologic evolution of the Beartooth Mountains, Montana and Wyoming — Pt. 2. Fracture patterns: *Geol. Soc. America Bull.*, v. 7o, p. 467–508.

———, and Kozak, S. V., 1975, Precambrian evolution of the Spanish Peaks, Montana: *Geol. Soc. America Bull.*, v. 86, p. 785–792.

Thompson, A. B., 1976, Mineral reactions in pelitic rocks — II. Calculation of some P-T-X (Fe-Mg) phase relations: *Amer. J. Sci.*, v. 276, p. 425–454.

———, and Algor, J. B., 1977, Model systems for anatexis of pelitic rocks — I. Theory of melting reactions in the system $KAlO_2$-$NaAlO_2$-Al_2O_3-SiO_2-H_2O: *Contrib. Mineralogy Petrology*, v. 63, p. 247–269.

Thurston, P. B., 1986, Geochemistry and provenance of Archean metasedimentary rocks in the southwestern Beartooth Mountains, Montana: M.S. thesis, Montana State University, Bozeman, Montana, 74 p.

Timm, R. W., 1982, Mineralogy and petrology of some metasedimentary xenoliths in granitic gneisses of the Broadwater River area, Beartooth Mountains, Montana, *in* Mueller, P. A., and Wooden, J. L., eds., *Precambrian Geology of the Beartooth Mountains, Montana and Wyoming:* Mont. Bur. Mines Geology Spec. Publ. 84, p. 25–40.

Vaniman, D. T., Papike, J. J., and Labotka, T. C., 1980, Contact metamorphic effects of the Stillwater Complex, Montana — The concordant iron formation: *Amer. Mineralogist*, v. 65, p. 1087–1102.

Vitaliano, C. J., Cordua, W. S., Hess, D. F., Burger, H. R., Hanley, T. B., and Root, F. K., 1979, Explanatory text to accompany geologic map of southern Tobacco Root Mountains, Madison County, Montana: *Geol. Soc. America Map Chart Series MC-31.*

Warner, J. L., Lee-Berman, R., and Simonds, C. H., 1982, Field and petrologic relations of some Archean rocks near Long Lake, eastern Beartooth Mountains, Montana and Wyoming, *in* Mueller, P. A., and Wooden, J. L., eds., *Precambrian Geology of the Beartooth Mountains, Montana and Wyoming*: Mont. Bur. Mines Geology Spec. Publ. 84, p. 57–68.

Weeks, G., 1981, Precambrian geology of the Boulder River area, Beartooth Mountains, Montana: M.S. thesis, Univ. Montana, Missoula, Mont., 58 p.

Wilson, J. T., 1936, Geology of the Mill Creek-Stillwater area, Montana: Ph.D. dissertation, Princeton Univ., Princeton, N.J., 202 p.

Windley, B. F., 1977, *The Evolving Continents*: London, Wiley, 385 p.

Zen, E-An, and Hammarstrom, J. M., 1984, Magmatic epidote and its petrologic significance: *Geology*, v. 12, p. 515–518.

J. L. Wooden
U.S. Geological Survey
Menlo Park, California 94025

P. A. Mueller
Department of Geology
University of Florida
Gainesville, Florida 32611

D. W. Mogk
Department of Earth Science
Montana State University
Bozeman, Montana 59717

14

A REVIEW OF THE GEOCHEMISTRY AND GEOCHRONONOLOGY OF THE ARCHEAN ROCKS OF THE NORTHERN PART OF THE WYOMING PROVINCE

ABSTRACT

The major exposures of Archean rocks in the western United States are found in the Wyoming Province, which encompasses Wyoming, southwestern Montana, and parts of Idaho, Utah, and South Dakota. Archean exposures are confined to crustal blocks uplifted during Laramide time and which, therefore, represent only a small fraction of the surface area of this region. This lack of continuous exposure prevents widespread correlations and means that the exposed rocks are not necessarily representative of the province as a whole. The major exposures in the northern part of the Wyoming Province occur in the Beartooth, Bighorn, Owl Creek, and Tobacco Root Mountains, and the Gallatin, Madison, and Ruby Ranges.

The Beartooth Mountains have been more extensively studied than the other areas and can be divided into several distinct terranes that have general counterparts in other ranges. The eastern and central Beartooth Mountains are dominated by a Late Archean suite of andesitic amphibolites and granitoids that have ages in the range 2.79–2.74 b.y. A period of deformation and middle amphibolite-grade metamorphism occurred in the earlier part of this time span. Many of the foliated granitoids in the Bighorn Mountains are of a similar age and were synkinematically emplaced. Thus the eastern Beartooth and Bighorn Mountains may be part of a major, Late Archean magmatic province. In both areas Late Archean plutonic suites intruded older rocks. In the Bighorn Mountains these rocks form a trondhjemitic, tonalitic, and amphibolitic gneiss suite that is 2.9–3.1 b.y. old. A suite of mafic amphibolites and dacitic gneisses similar in age and composition to the older Bighorn Mountains suite is present in the Owl Creek Mountains. The Owl Creek suite is the closest approximation to a greenstone belt type of assemblage in this part of the Wyoming Province. In the eastern Beartooth Mountains, the older rocks appear to be a varied supracrustal suite that is 3.2–3.4 b.y. old. Supracrustal rocks 2.9–3.1 b.y. old are present in the southwestern Beartooth Mountains and immediately south of the 2.70-b.y.-old Stillwater Complex in the northcentral Beartooth Mountains.

Rocks of the northwestern Beartooth Mountains represent a different type of terrane. It is proposed that this area was assembled from as many as six different crustal segments by tectonic processes like those that operated in the Cordilleran orogeny. Some of the crustal segments in this terrane may be as old as 3.4 b.y. (a trondhjemitic and amphibolitic gneiss complex), whereas others are Late Archean in age. The Late Archean event that formed this terrane may be the same one that was responsible for generation of the 2.8-b.y.-old magmatic province to the east. The northern part of the southern Madison Range also is cored by a 3.4-b.y.-old tonalitic gneiss complex that is intruded by Late Archean gneissic granitoids. Thus the oldest rocks presently recognized in the northern Wyoming Province are found in a limited area encompassing the Beartooth Mountains and the southern Madison Range.

The northern Madison and Gallatin Ranges are a transitional area composed dominantly of strongly deformed, high-grade quartzofeldspathic gneisses that were intruded by a Late Archean granitoid suite. This area lies between terranes to the east that contain Early Archean rocks and/or are dominated by Late Archean granitoids, and terranes to the west that contain rocks of a definite continental shelf affinity and seem to be only Late Archean in age. Thus the region from the Ruby Range and Tobacco Root Mountains on the west to the northwestern Beartooth Mountains on the east may represent a Late Archean continental margin that was shaped by tectonic processes like those operating in the Phanerozoic that produce areas of strong deformation and high metamorphic grade, juxtaposition of terranes, and magmatic arcs. The number of discrete events that affected this margin is not resolved at this time.

INTRODUCTION

There are two major exposures of Archean rocks in the United States. These are (1) the southern extension of the Superior Province into Minnesota, northern Michigan, and Wisconsin; and (2) the Wyoming Province of Wyoming, southwestern Montana, and minor parts of Idaho, Utah, and western South Dakota. This paper will concentrate on the northern part of the Wyoming Province. The reader is referred to the papers of Peterman (1979) and Condie (1976a) for more comprehensive reviews of the geology and geochronology of the Wyoming Province.

Exposures of Archean rocks in the Wyoming Province (Fig. 14-1) differ from those of most other shield areas in that they consist of discrete blocks uplifted during the Laramide orogeny. Unlike the Superior Province, where there is little relief and exposures are almost continuous, the Archean rocks of the Wyoming Province are characterized by more than 1 km of topographic relief and as much as 10 km of structural relief. In the Wyoming Province, less than one-third of the Precambrian rock is exposed. The wide separation of exposures (Fig. 14-1) poses a major problem for interpretation of Archean history. It prevents widespread correlation of structural or lithologic trends and may make it impossible to know if the area is divisible into subprovinces such as those proposed for the Superior Province. The detailed lithologic character of the province also is impossible to know. As presently exposed, the province would seem to be dominated by gneissic and granitoid rocks and to contain only minor exposures of greenstone belt-type rocks. However, Peterman (1979) has pointed out, by superposing a map of the Wyoming Province exposure pattern on the Superior Province, that the distribution of lithologic types in the Wyoming Province could easily be derived from an area with alternating greenstone-granite and granite-gneiss terranes such as characterize the Superior Province.

The Archean rocks of the Wyoming Province have not been studied in detail. Other Archean shields have received more extensive study for two major reasons. The first of these reflects interest in economic deposits which, in Archean rocks, are concentrated in greenstone-volcanic belts. The previously mentioned paucity of exposed greenstone belt-like rocks in the Wyoming Province can be directly related to the minor occurrence of major economic deposits in the area. The second reason for major interest in a particular Archean area is the presence of the earliest Archean rocks (3.8 b.y. old), which provide information about the earliest geologic history of the earth. Although there is growing evidence that rocks 3.4 b.y. or older must exist in the Wyoming Province, the oldest unambiguously dated rocks in the Wyoming Province are 3.2 b.y. old.

The northern part of the Wyoming Province, as discussed in detail in this paper, consists of the Beartooth, Bighorn, Owl Creek, Madison, Gallatin, Ruby, and Tobacco Root Mountains. The geochemistry and geochronology of each area is considered separately. A special emphasis on the rocks of the Beartooth Mountains results from the authors' research experience there, the large number of studies by previous investigators, and the availability of Archean rocks with a greater range of ages and lithologies than is recognized in other exposures of Archean rocks in this province. Many of the problems of geologic interpretation that are detailed for the rocks of the Beartooth Mountains are valid for the other areas discussed and for Archean terranes in general. The reader is referred to Mogk and Henry (Chapter 13, this volume) for a complementary study of the

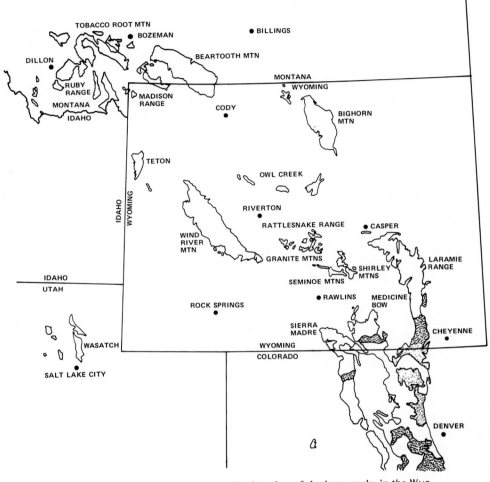

FIG. 14-1.   Sketch map showing the location of Archean rocks in the Wyoming Province.

metamorphic petrology and tectonic evolution of the Archean rocks of southwestern Montana.

BEARTOOTH MOUNTAINS

The Beartooth Mountains of Montana and Wyoming (Fig. 14-2) lie immediately north and northeast of Yellowstone National Park as a roughly west-northwest-trending block

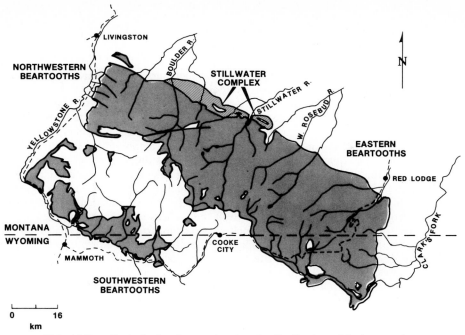

FIG. 14-2.   Geologic sketch map showing the distribution of Archean rocks in the Beartooth Mountains of Montana and Wyoming.

that contains over 5000 km² of Archean rocks. The Beartooth block can be divided into four areas: (1) the late Archean granitoid rocks of the eastern area that contain kilometer-size inclusions of older supracrustal rocks (Mueller and Wooden, 1982; Mueller *et al.*, 1982a, b, 1985a, b;Wooden *et al.*, 1982a, b); (2) the metasedimentary rocks of the north-central area into which the Stillwater Igneous Complex and associated granitoid rocks were intruded (Page, 1977); (3) the metasedimentary and minor granitoid rocks of the southwestern area (Casella *et al.*, 1982); and (4) the metamorphosed supracrustal and igneous rocks of the northwestern area that have been tectonically interleaved by thrusting (Mogk, 1982, 1984; Mogk *et al.*, 1986; Reid *et al.*, 1975). Because the geology of each area is significantly different, individual areas will be discussed separately. It is not clear what the relationships of individual areas are to each other: it is possible that fundamentally different areas were tectonically juxtaposed in the Late Archean.

Eastern Beartooth Mountains

The eastern Beartooth Mountains were studied in detail in the late 1950s and early 1960s by Arie Poldervaart and his students (Casella, 1969; Eckelmann and Poldervaart, 1957; Harris, 1959; Larsen *et al.*, 1966). The dominant granitic rocks of this area, as well as the minor associated metasedimentary rocks, were explained originally as the result of the granitization of a sequence of folded sedimentary rocks. Later syntheses from this group, however, recognized the probable roles of igneous processes and multiple deformation and metamorphism. Work during the last 10 years by Mueller, Wooden, Bowes, and co-workers (see Mueller *et al.*, 1985b for a summary) has led to a model involving the intrusion of a suite of Late Archean rocks into a Middle to Early Archean supracrustal sequence, parts of which had experienced a granulite facies metamorphic event (Henry *et al.*, 1982).

Geochemistry of the Late Archean Rocks

There are at least three distinctive compositional members of the Late Archean suite (Fig. 14-3a). These members have been given the following informal names: andesitic amphibolite, Long Lake granodiorite, and Long Lake granite. Field relationships indicate that the andesitic amphibolite is the oldest member of this group and that these rocks experienced an amphibolite-grade metamorphism before the other members of the suite were emplaced. The present mineral assemblage of these amphibolites is biotite, horn-blende, plagioclase, and quartz. The andesitic amphibolite is found as meter- to kilometer-size inclusions in the younger granitoid rocks. The Long Lake granodiorite is intermediate in age, being found in some places as weakly foliated inclusions in the Long Lake granite. The Long Lake granodiorite is compositionally distinctive but very difficult to distinguish in the field from the Long Lake granite. Both these rocks are leucocratic, medium- to coarse-grained, two-feldspar and quartz rocks having biotite as the only mafic mineral (Warner *et al.*, 1982; Wooden *et al.*, 1982a, b). It is difficult, therefore, to estimate the relative abundance of the granodiorite versus the granite in the field. Geochemical sampling indicates, however, that the granite is much more abundant.

The major element composition of the andesitic amphibolites is restricted to the andesitic or dioritic field (Fig. 14-3a) and is not unusual with respect to modern or Archean andesites (Condie, 1976a, b, 1982; Gill, 1981). There is, however, a good deal of variety in major element abundances in this group, and several subgroups can be distinguished within which there are regular elemental changes that may be related to fractionation processes. According to the classification developed by Gill (1981) for modern orogenic andesites, the andesitic amphibolites have major element compositions that fall into both the calc-alkaline and tholeiitic fields. The calc-alkaline types dominate and are found over a wide geographic area in the eastern and central Beartooth Mountains. The tholeiitic types seem to be restricted to the eastern Beartooth Mountains. Because these rocks have been recrystallized in the middle amphibolite facies, it is not possible to reliably classify them according to their alkali contents.

The trace element concentrations of the andesitic amphibolites deserve special mention. Whereas the ranges and abundances of the compatible trace elements in these rocks are within the normal ranges found in andesitic rocks, those of the incompatible trace elements are not. Sr concentrations range from 200 to 1000 ppm, Zr from 40 to 300 ppm, Ce from 20 to 300 ppm, and Ba 200 to 1500 ppm. In addition, the abundances of these incompatible elements are well correlated with each other (Fig. 14-3b). This is particularly unusual for Sr, which typically shows little variation within an andesitic suite even when fractionation has produced obvious variations in other elements. The best explanation for these chemical characteristics is that an incompatible element-rich fluid was involved with partial melting of the mantle (Mueller *et al.*, 1983). Variable degrees of partial melting can explain the range of major element compositions and some of the trace-element variation. A fluid rich in incompatible elements is needed to produce the unusually high concentrations of these elements in many of the rocks. Because Sr acts as an incompatible element, these processes must be occurring in a plagioclase-free environment, probably the mantle.

The Long Lake granodiorite has the major element composition (Fig. 14-3a and c) of a typical calc-alkaline granodiorite (63–70% SiO_2). Although it overlaps the Long Lake granite in SiO_2 content, it maintains a lower Na_2O concentration. In keeping with its lower average SiO_2 content, the granodiorite is generally more mafic than the granite,

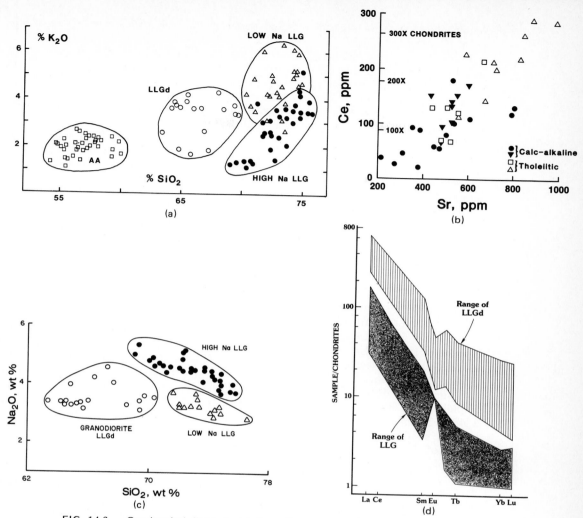

FIG. 14-3. Geochemical diagrams showing salient compositional features of the late Archean suite of the eastern Beartooth Mountains. Figure a shows K_2O versus SiO_2 relationships for the andesitic amphibolites (AA), the Long Lake granodiorite (LLGd), and the two groups of the Long Lake granite (high Na and low Na LLG). Figure b shows ppm Ce versus ppm Sr for the andesitic amphibolites. Figure c illustrates Na_2O versus SiO_2 for the Long Lake diorite and granite. This diagram shows the fields for the high- and low-Na groups of the granite series. Figure d is a composite REE diagram for the Long Lake granodiorite and granite. The higher REE contents of the LLGd distinguish it from the LLG.

having higher FeO, MgO, and CaO concentrations. However, the trace-element contents of the granodiorite, especially its higher Sr, Ba, and REE concentrations, distinguish it from the granite. The REE pattern of the granodiorite is particularly distinctive, being higher in both LREE and HREE and having a noticeable negative Eu anomaly (Fig. 14-3d). The high concentrations of incompatible trace elements in the granodiorite suggest that it also could have had a trace element-rich fluid involved in its genesis. The negative Eu anomaly, however, indicates that plagioclase was important either in the source or as a fractionating mineral and therefore that this rock's genesis was accomplished at crustal P and T.

The Long Lake granite is the volumetrically dominant rock type in the eastern Beartooth Mountains. It is divisible into at least two subgroups on the basis of Na_2O versus SiO_2 relationships (Fig. 14-3c). The high-Na group is volumetrically more important and compositionally more coherent than the low-Na group. Although the silica content of the granite suite is restricted, the variation in Na_2O and K_2O concentrations (and modal plagioclase and K feldspar) mean that the rock types vary from high-silica tonalite to typical granite. The negative correlation between Na and Si in both subgroups is unusual in modern calc-alkaline rocks but is typical of Archean tonalite and trondhjemite suites. The genesis of tonalite and trondhjemite suites is best explained as the result of partial melting and/or fractionation of a basaltic parent. The lack of sodic suite rocks of intermediate composition strongly favors an origin by partial melting of a mafic source (basaltic and/or mafic andesite) that contained residual garnet, amphibole, or clinopyroxene to hold the HREE concentrations at 4X chondrites or lower. The strong LREE versus HREE fractionation seen in these rocks, along with the relatively low Sr and Rb contents, the low Rb/Sr ratios, and the high average K/Rb ratio of 350 with respect to other granitic rocks, are compatible with derivation from a mafic source.

Geochronology and Isotopic Systematics of the Late Archean Rocks

A range of isotopic data is now available for the Late Archean rocks of the eastern Beartooth Mountains. New U-Pb zircon data (P. A. Mueller, R. D. Shuster, M. A. Graves) provide the best chronologic information for this group. These data are in agreement with the previously discussed field relationships and give the following ages: andesitic amphibolite, 2789 ± 5 m.y.; Long Lake granodiorite, 2782 ± 3 m.y.; Long Lake granite, 2748 ± 25 m.y. The imprecise age of the Long Lake granite results from these zircons being at least 60% discordant. Within confidence limits this suite of rocks covers a time period of 15 to 65 m.y. Previously available Rb-Sr whole-rock data (Wooden *et al.*, 1982a, b) produced a composite isochron for all three major groups with an age of 2790 ± 45 m.y. (Fig. 14-4a) and an initial Sr ratio of 0.7022 ± 0.0002. Common Pb isotopic data (Fig. 14-5) for whole rocks and feldspars from the same three groups give an age of 2770 ± 60 m.y. (Wooden and Mueller, in press). The feldspar separates provide a good estimate of the initial Pb isotopic ratio in these rocks at the time of their formation. These ratios are 13.86 for $^{206}Pb/^{204}Pb$, 14.97 for $^{207}Pb/^{204}Pb$, and 34.06 for $^{208}Pb/^{204}Pb$. Sm-Nd chondritic model ages (Fig. 14-4c) vary from 2.88 to 3.02 b.y., and initial ϵ_{Nd} values calculated for an age of 2.78 b.y. range from −1.5 to −3.1 for five samples.

The chronologic and isotopic data for the Late Archean suite clearly show that they are restricted to a small time interval and are remarkably homogeneous in their intial Sr, Nd, and Pb isotopic ratios. A time interval of 10–50 m.y. for andesitic volcanism, deformation, and metamorphism, and additional major plutonism is not remarkable for the Archean or the Phanerozoic. It is unusual that such a compositionally diverse suite of rocks would have the same initial isotopic ratios. These ratios are not what would be expected for new crust that was forming from primitive or depleted mantle. Initial Sr, Nd, and Pb ratios in this case would be approximately 0.7010, $\epsilon_{Nd} = +2$, and $^{206}Pb/^{204}Pb = 13.40$, $^{207}Pb/^{204}Pb = 14.58$, $^{208}Pb/^{204}Pb = 33.16$, respectively. The difference between these values and those observed in the Late Archean suite suggest either an enriched mantle source or contamination of the suite by older Archean crust. The very high Pb ratios strongly favor the involvement of older Archean crust in whatever process was

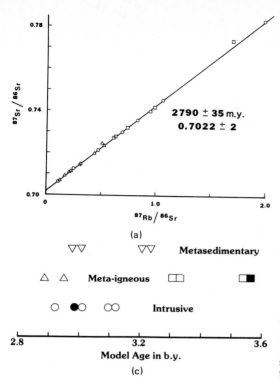

(a)

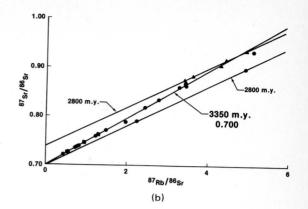

(b)

FIG. 14-4. Composite Rb-Sr isochron diagrams for the (a) Late and (b) Early Archean rocks of the eastern Beartooth Mountains. Symbols in (a) are circles for the Long Lake granite, squares for the Long Lake granodiorite, and triangles for the andesitic amphibolites. In (b) reference isochrons are given for 3350 m.y. and 2800 m.y. The 2800-m.y. lines are interpreted as a time of resetting of the older rocks either by rehomogenization or loss of radiogenic Sr. A summary of Sm-Nd chondritic model ages for the same rocks is given in (c).

responsible, because it is only in crustal environments that the necessary high $^{238}U/^{204}Pb$ ratios could be produced to allow the $^{206}Pb/^{204}Pb$ and $^{207}Pb/^{204}Pb$ ratios to grow as high as needed. If crustal contamination occurred during emplacement of the Late Archean suite into an older crust, it is difficult to understand how the necessary homogeneity was achieved. It is suggested that a portion of the mantle itself was contaminated, possibly by introduction of crustal material through subduction, dewatering of the slab, and penetration of the overlying mantle wedge by fluids carrying Pb, Sr, and Nd with a crustal signature (Wooden and Mueller. in press). Crust made from this contaminated (but relatively homogeneous) mantle would have the necessary isotopic values and be isotopically homogeneous across a whole spectrum of compositional types, especially if the crust-forming cycle was confined to a short time period that limited further radiogenic growth.

Geochemistry of the Older Archean Rocks

Enclosed in the Late Archean rocks of the eastern Beartooth Mountains are meter- to several-kilometer-size inclusions of metamorphosed supracrustal rocks. At least some of these rocks have experienced both a granulite-grade metamorphic event (Henry *et al.*, 1982) and the Late Archean amphibolite-grade event discussed above. This granulite event was characterized by temperatures of about 800°C and pressures of about 6 kb, which suggest a geothermal gradient similar to those of modern, high heat-flow orogenic zones (about 40–45°C/km). These rocks are strongly deformed and isoclinally folded, and many lithologic layers exist only as boudins that are a few to tens of meters in

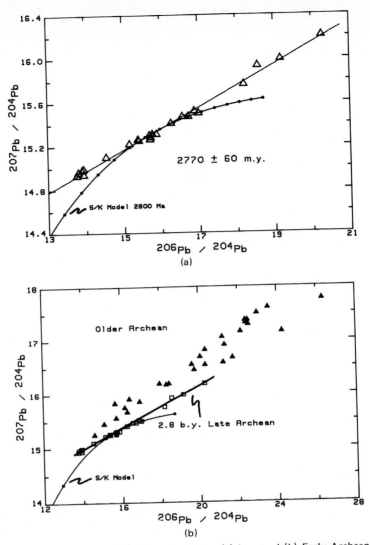

FIG. 14-5. Pb-Pb isotopic data for the (a) Late and (b) Early Archean rocks of the eastern Beartooth Mountains. S/K model stands for the Stacey and Kramers (1975) model for Pb isotopic growth in average crust. The scatter in the older Archean data (b) is in part related to the multistage evolution of this Pb isotopic system. The very radiogenic nature of the older Archean data requires at least Middle Archeran ages and high U/Pb ratios.

length. The metamorphic equivalents of ironstones, basalts and ultramafic rocks, pelites, wackes, quartzites, and felsic volcanics can all be found in the supracrustal assemblages. At present no older plutonic rocks have been unequivocally identified.

In spite of the deformation and metamorphism that has affected these rocks, they appear to have largely retained their original bulk chemistries. The variation diagrams shown in Fig. 14-6a have no unusual features. If the obvious samples with strongly fractionated sedimentary compositions are not considered (high-Si quartzites, low-Na

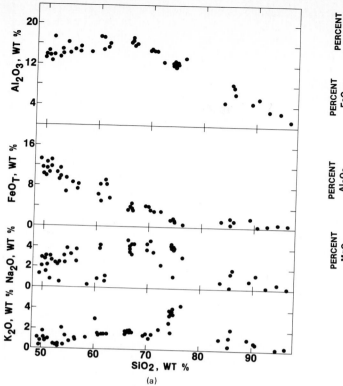

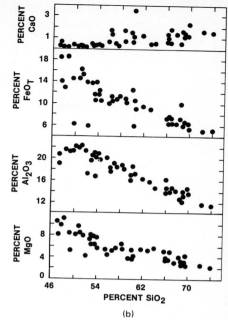

(b)

(a)

FIG. 14-6.   Compositional diagrams for the older Archean, supracrustal rocks of the (a) eastern Beartooth Mountains and (b) Stillwater Complex area.

pelites, high-Fe ironstones), the remaining samples show the expected compositional variations of an igneous calc-alkaline sequence. Most of this sequence has moderate K_2O contents (2 wt % or less), and K_2O/Na_2O less than 1 (weight ratios). Only the samples with wt % SiO_2 in the mid-70s have K_2O/Na_2O of about 1.

Specific compositional features of some of the rock types in the older Archean sequence are worth emphasizing. The basaltic amphibolites are close to average basalt in composition and show little to only moderate Fe enrichment. Rare earth patterns are generally unfractionated and less than 20X chondrites, Sr contents are about 100 ppm, and Cr and Ni contents are about 250 and 100 ppm, respectively. The samples with SiO_2 contents between 65 and 73 wt % are comparable to medium-K dacites. These samples have strongly fractionated REE patterns and no Eu anomalies. A group of rhyolitic composition rocks with $K_2O/Na_2O = 1$ have fractionated REE patterns and strong negative Eu anomalies. The quartzites range in SiO_2 from 80 to 97 wt %, are locally fuchsitic, and contain as much as 100 ppm Cr and obvious detrital zircons. Iron-rich rocks vary in SiO_2 from 47 to 61 wt % and in total Fe as FeO from 30 to 40 wt %. Their compositional characteristics favor a continental shelf rather than a volcanic depositional environment. Pelitic rocks in the section have wt % SiO_2 contents in the low 60s, and very low CaO and NaO_2 contents. The combination of quartzites, iron-rich rocks, and pelitic rocks strongly indicate that at least part of the history of this sequence involved deposition on a mature continental shelf.

Geochronology of the Older Archean Rocks

The age of the older Archean sequence is problematical. Zircon U-Pb data (Mueller *et al.*, 1982a, b) indicate a minimum age ($^{207}Pb/^{206}Pb$ age) of 3295 m.y. for a quartzite and 3220 m.y. for a granitic migmatite from Hellroaring Plateau. These data were produced by acid leaching of zircons that fall on a discordia line that intersects the concordia curve at about 3100 m.y. The 3100-m.y. age may have no geological meaning because it may be a point along an episodic cord between the true age of the zircons and the time of new zircon growth and/or Pb loss. The lower end of this episodic cord is represented by clear, acicular zircons giving an age of about 2800 m.y. The 2800-m.y. age is consistent with the amphibolite facies metamorphic event that is recorded by the andesitic amphibolites, although it could also represent an event that preceded the crystallization of the igneous precursors of the andesitic amphibolites at 2790 m.y. ago.

Rb-Sr, Sm-Nd, and common Pb data also provide information about the age of the older Archean rocks. A Rb-Sr isochron for the granitic migmatite discussed above gives an age of 2830 ± 130 m.y. and an initial ratio of 0.738 ± 0.007. The high initial ratio of the isochron clearly shows that it represents the time at which a much older rock had its Rb-Sr system reset. A model age based on the average Rb and Sr contents and Sr isotopic values of this rock is approximately 3500 m.y. if an initial Sr ratio of 0.700 is assumed. Rb-Sr data for other rocks in the older complex also indicate an early Archean age. Henry *et al.* (1982) reported data for many of the lithologic types discussed above that fall along a 3350-m.y. reference isochron with an initial ratio of 0.700 (Fig. 14-4b). Additional work (P. A. Mueller and J. L. Wooden, unpublished data) continues to confirm this trend. These data do not define an isochron because samples scatter both above and below the reference line. This scatter may be caused by many factors related to the complicated geologic history of these rocks. Possibilities include Rb or ^{87}Sr loss during high-grade metamorphism, Rb addition during metamorphism or later plutonism, mixing of different age rocks during plutonism or deformation, or improper identification of younger rocks included in the older complex. Although the lower Rb/Sr ratio samples could represent mixing between Late and Early Archean materials, the samples with high

Rb/Sr ratios on the reference line provide clear evidence that some Early Archean material must be present.

The implication for ages of 3.3 b.y. from the Rb-Sr system is supported by Sm-Nd and Pb-Pb data. Four samples have Sm-Nd chondritic model ages between 3.3 and 3.5 b.y. (Fig. 14-4c). Three other samples have model ages between 3.1 and 3.2 b.y. Common Pb studies show that the same samples that have high Rb-Sr and Sm-Nd model ages have very radiogenic Pb compositions that are consistent with a minimum age of 3.3 b.y. and growth in a high U/Pb environment (Fig. 14-5b). The high initial Pb ratios of the Late Archean suite (see above) also require the presence of older Archean material. Therefore, there is both strong inferential and reasonable direct evidence that rocks at least 3.3 b.y. old exist in the eastern Beartooth Mountains.

Northcentral Beartooth Mountains

The area surrounding the Stillwater Complex has been of special interest to geologists studying the origin of this famous layered mafic igneous complex. Unfortunately, field relationships in this area are complicated by major faults that separate the Stillwater Complex and its metasedimentary wallrocks from the main exposures of Archean rocks in the rest of the Beartooth Mountains. Butler (1966) published one of the first studies to consider the transition from the Stillwater Complex and its contact metamorphosed border rocks into the dominantly crystalline Archean rocks to the south. Reports by Page (1977), Wooden et al. (1982a), and Czemanske and Zientek (1985) contain recent information on the area. The following discussion also utilizes unpublished data from current studies by Wooden, Mueller, and Mogk and students.

The Stillwater Complex intruded a sequence of metasedimentary rocks (Page, 1977; Page and Zientek, 1985) 2700 m.y. ago (DePaolo and Wasserburg, 1979; Mueller and Wooden, 1976). These metasedimentary rocks are variable in composition having wt % SiO_2 ranging between 45 and 80% (Fig. 14-6b). Low Na and Ca contents throughout this range indicate that all of these rocks went through a strong weathering cycle. Ubiquitous, high Fe, Mg, Cr, and Ni contents indicate that these rocks formed by mixing between quartz-rich and high-Mg sources. Nunes and Tilton (1971) reported U-Pb zircon ages for these rocks of 3060 and 3090 m.y. (Fig. 14-7a and b). DePaolo and Wasserburg (1979) reported a single chondritic model age of 3130 m.y. for a hornfels sample. These rocks, or their source terrane, must have a minimum age of about 3100 m.y. At present there is no compositional and geochronological equivalent to these rocks known elsewhere in the Beartooth Mountains.

These metasedimentary rocks are separated from the crystalline rocks of the Beartooth Mountains by faults. This crystalline complex is similar in many ways to the Late Archean complex of the eastern Beartooth Mountains. The granitoid rocks are dominated by high-SiO_2 members that have variable K_2O/Na_2O ratios (Fig. 14-8). These rocks contain numerous inclusions of amphibolitic and schistose rocks that have dioritic/andesitic bulk compositions. The style of intrusion for the granitic rocks is one of numerous thin sheets that produce a lit-par-lit appearance. Pegmatite and aplite veins are the latest intrusions and account for about 20% of the volume. A composite Rb-Sr isochron for the granitoid rocks gives an age of 2700 ± 100 m.y. with an initial Sr ratio of 0.7023 (Fig. 14-8). A U-Pb zircon age for one of the granitoids is 2752 ± 14 m.y. Thus, these rocks are roughly equivalent to those of the eastern Beartooth Mountains in both age and composition.

A small body of coarse-grained granite occupies the boundary between the Stillwater Complex and the metasedimentary rocks. This granite may be a member of a major suite of medium-grained granitoids (Page and Nokleberg, 1972). Nunes and Tilton (1971)

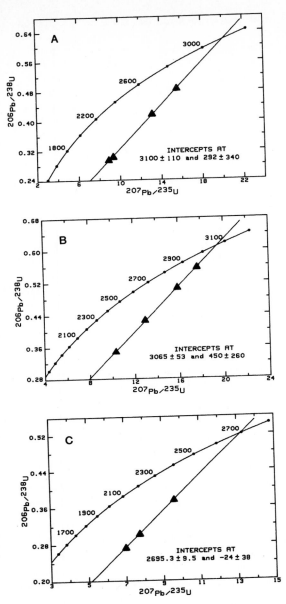

FIG. 14-7. U-Pb concordia diagrams for (a) schists south of the Stillwater complex, (b) hornfels produced by the complex, and (c) granitoids at the base of the Stillwater Complex. All ages in m.y. Data from Nunes and Tilton (1971).

obtained a zircon age of 2700 m.y. on the coarse-grained granite (Fig. 14-7c), the youngest reliable age for granitoids in the Beartooth Mountains.

Southwestern Beartooth Mountains

The Archean geology of the southwestern Beartooth Mountains is known from the work of Casella *et al.* (1982). Metasedimentary rocks dominate this section. These units are intruded by a variety of synkinematic granitoids producing migmatitic zones. The metasedimentary rocks occur as thinly bedded layers of schist, quartzite, metaconglomerate,

GEOCHEMISTRY AND GEOCHRONOLOGY OF ARCHEAN ROCKS

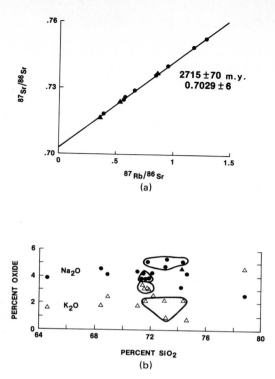

FIG. 14-8. Rb-Sr isochron and compositional diagrams for granitoids of the central
Beartooth Mountains. Different symbols on
the isochron diagram represent different
localities. Enclosed data on compositional
diagrams represent multiple samples from a
single sampling site.

and rare iron formation. Sedimentary structures including graded bedding, cross-bedding,
and channel cut-and-fill have been preserved in spite of multiple periods of deformation
and metamorphism. The major deformation produced isoclinal folds and was accompanied
by amphibolite-grade metamorphism. A second period of amphibolite-grade metamorphism is associated with only minor deformation and the intrusion of two-mica granites.
The latter have a minimum Rb-Sr age of 2740 m.y. and a high initial Sr ratio, indicating
that they are partial melts of older crustal rocks (Wooden, 1979). Bulk compositional
data for the metasedimentary rocks suggest their protoliths were graywackes with a minor
shale component. The age of the metasedimentary sequence is uncertain but seems to be
in the range of 2.9–3.1 b.y. based on model Rb-Sr data and a single model Sm-Nd age[1]
(Montgomery and Lytwyn, 1984).

The metasedimentary sequence is intruded along the eastern edge of its exposure by
granitoids. The earliest of these granitoids is a quartz-hornblende diorite that may be the
equivalent of the andesitic amphibolites of the eastern Beartooth Mountains. Intrusive
into these rocks is a composite batholith with phases varying from tonalite to granite,
but with granodiorite dominating. The bulk compositions of some of these rocks are
similar to the granitoids of the eastern Beartooth Mountains. However, the majority of
these rocks have higher K_2O/Na_2O ratios, and a greater number of the samples have wt %
SiO_2 contents in the 60s in keeping with their granodioritic modal classification. Rb-Sr
and U-Pb zircon data (Montgomery, 1982; Montgomery and Lytwyn, 1984; Wooden,
1979) are consistent with these rocks being the same age as the granitoids of the eastern
Beartooth Mountains, but a significant reheating of this area in the Proterozoic has complicated the Rb-Sr systematics of these rocks.

[1] Also J. L. Wooden, unpublished data.

WOODEN, MUELLER, MOGK

397

Northwestern Beartooth Mountains

The northwestern part of the Beartooth Mountains is commonly referred to as the North Snowy block (Fig. 14-9a). This area was first described in detail by Reid *et al.* (1975). Subsequent work by Mogk (1982, 1984) has added to the structural and geochemical knowledge of the area. Mogk *et al* (in press) describe this area as a series of lithologically

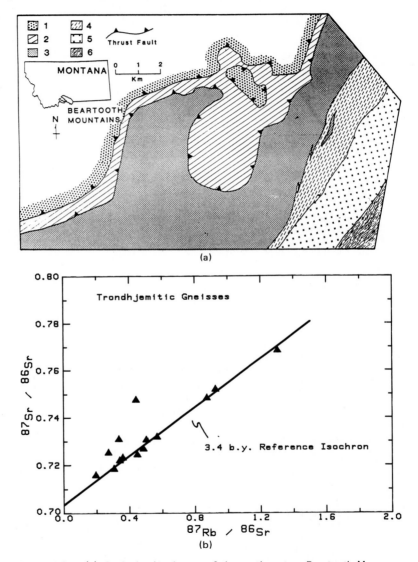

FIG. 14-9. (a) Geologic sketch map of the northwestern Beartooth Mountains (see Mogk and Henry, this vol. Chapter 13, for more information) and (b) a Rb-Sr isochron diagram for the trondhjemitic gneiss of the trondhjemitic gneiss-amphibolite complex.

and metamorphically distinct packages that were juxtaposed by tectonic processes. Six major units can be defined from east to west:

1. A paragneiss unit consisting of a heterogeneous assemblage of supracrustal rocks, including quartzofeldspathic gneisses, pelitic schists, amphibolites, and banded iron formation.

2. The Mount Cowen augen gneiss (Fig. 14-10) is a granitic, sill-like body. Unlike the granitic rocks of the east and central Beartooth Mountains, this unit has K_2O/Na_2O ratios greater than 1 and higher Rb/Sr ratios. A Rb-Sr isochron for this rock gives an age of 2740 ± 50 m.y. and an initial ratio of 0.7023, both of which are within error of the data for the eastern and central granitoids.

3. The Davis Creek schist is a phyllitic metapelite containing minor layers of quartzite.

4. A trondhjemite-amphibolite complex consists of trondhjemitic gneisses interlayered with a variety of amphibolites that range in composition from basaltic to anorthositic gabbroic. The trondhjemitic gneisses are like those found in Archean terranes elsewhere. They are high in wt % SiO_2 (68-76) and Na_2O (8-4), low in wt % K_2O (<2.5), total Fe (<2), and MgO (<1). Unfortunately, these rocks are in a ductile shear zone metamorphosed in the epidote-oligoclase facies. This may explain why the Rb-Sr whole-rock data for these rocks are scattered (Fig. 14-9b). The majority

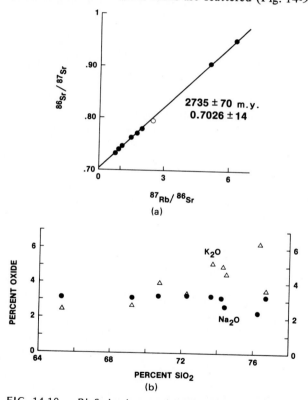

FIG. 14-10. Rb-Sr isochron and compositional diagrams for the Mt. Cowen granitic gneiss of the northwestern Beartooth Mountains.

WOODEN, MUELLER, MOGK

of the data lie close to a 3.4-b.y. reference line, but a significant number of samples lie to the left of this line, indicating impossibly old ages. An Early to Mid-Archean age for this unit is supported by two Sm-Nd chondritic model ages of 3.26 and 3.59 b.y.

5. The Pine Creek nappe complex is cored by amphibolite that is partly of andesitic composition and has outer, symmetrically disposed quartzite and marble. A single chondritic Sm-Nd model age of 3.2 b.y. is the only chronologic information available for this unit.

6. The heterogeneous gneiss consists of a supracrustal package of quartzites, amphibolites, and minor schists bearing injected(?) gneisses of granitic to tonalitic composition. Rb-Sr whole-rock data for gneissic samples indicate that these rocks are approximately 3.4 b.y. old. The data lie along the reference isochron for the trondhjemitic gneiss discussed above, and there are compositional similarities between some members of the heterogeneous gneiss and the trondhjemitic gneiss.

SOUTHERN MADISON RANGE

The southern Madison Range lies southwest of the Beartooth Mountains and has been studied in detail by Erslev (1981, 1982, 1983). The field-oriented studies of Erslev are now being extended by chronologic studies involving Erslev, P. Mueller, and J. Sutter. The ages used in this section should be considered preliminary until these studies are complete and formally published. The major features of this range are a northern gneissic and migmatitic terrane. This terrane is separated from a southern sequence of metapelitic to psammitic schists and marbles called the Cherry Creek metamorphic suite (Erslev, 1983) by a thick sequence of mylonites called the Madison mylonite zone (Erslev, 1982).

Two units in the northern terrane have been analyzed by the Rb-Sr whole-rock technique. A tonalitic gneiss has a limited spread of Rb/Sr ratios and somewhat disturbed systematics, but the data for this unit clearly fall along a 3400-m.y. reference isochron Fig. 14-11). This age is similar to that of the trondhjemitic gneiss of the northwestern Beartooth Mountains and the supracrustal sequence of the eastern Beartooth Mountains. The composition of the tonalitic gneiss is distinct from any member of these other units inasmuch as it is a typical calc-alkaline tonalite with wt % SiO_2 in the low to middle 60s. The other unit examined is a granitic augen gneiss that is the main phase of a gneiss dome that penetrated the tonalitic gneiss. Limited Rb-Sr data for this unit indicate that it is approximately 2700 m.y. old and support the observations of Erslev (1983) that the tonalitic gneiss is older than the granitic gneiss. The granitic gneiss is richer in K and Rb and has higher Rb/Sr ratios than the typical granitoid of the eastern and central Beartooth Mountains. The apparent similarity in age between this part of the southern Madison Range and the Beartooth Mountains is additional evidence for an Early Archean craton being involved in a Late Archean orogenic event.

The only chronologic data available for the southern terrane consists of limited Rb-Sr and U-Pb zircon data that indicate an age of about 2500 m.y. for a granodioritic augen gneiss. This gneiss was syntectonically intruded into the metasedimentary sequence and provides a minimum age for the time of the last deformation and for the depositional age of the sediments in this area. This age is distinctly younger than those presently known from other parts of the northern Wyoming Province. The younger age, the lithologic distinction of this supracrustal sequence from those in the Beartooth Mountains, and the supracrustal versus orthogneiss contrast between northern and southern terranes suggest that the Madison mylonite zone may represent a significant crustal discontinuity.

GEOCHEMISTRY AND GEOCHRONOLOGY OF ARCHEAN ROCKS

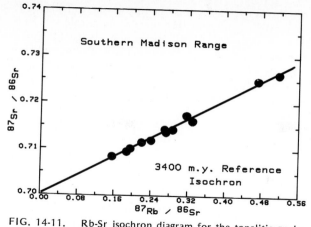

FIG. 14-11. Rb-Sr isochron diagram for the tonalitic gneiss of the southern Madison Range.

NORTHERN MADISON AND GALLATIN RANGES

Earlier work in the northern Madison Range by Spencer and Kozak (1975) described an extensive terrane of quartzofeldspathic gneiss that had been multiply deformed and metamorphosed in the amphibolite facies. This work is currently being extended by D. Migk and students. Salt and Mogk (1985, and unpublished data) have recognized three distinctive terranes. The first is a granulite and migmatite association in the Gallatin and Madison River valleys. A previously unrecognized batholitic complex that contains quartz-diorite, monzodiorite, granodiorite, and granite comprises the second terrane. The third terrane consists of metasupracrustal rocks in the upper amphibolite to granulite facies. The northern Gallatin Range appears to be an extension of the granulite and migmatite terrane mentioned above (May and D. W. Mogk, unpublished data). James and Hedge (1980) included three samples from this area in their regional Rb-Sr study that produced a composite isochron age of about 2750 m.y. for the Ruby, Tobacco Root, and northern Gallatin ranges. This is an important area because it represents a transition between the metamorphosed shelf sequences to the west and the granitoid and older supracrustal sequences to the east.

TOBACCO ROOT AND RUBY RANGES

Vitaliano *et al.* (1979) and Garihan (1979) have described the Archean geology of the Tobacco Root and Ruby Ranges, respectively. These two areas are similar and consist of a heterogeneous assemblage of quartzofeldspathic and mafic gneiss, para- and ortho-amphibolite, metamorphosed ultramafic rock, marble, quartzite, pelitic schist, and iron formation. These areas have experienced at least one period of deformation that resulted in isoclinal folds. Two periods of metamorphism are possible, with an amphibolite facies event overprinting a granulite facies event. Both areas contain a major and distinctive metasedimentary component indicative of a continental shelf environment — marble, quartzite, schist, and iron formation.

Two Rb-Sr whole-rock studies supply the only geochronologic information for these areas. Mueller and Cordua (1976) obtained an age of 2670 m.y. for quartzofeldspathic gneisses of the Horse Creek area in the southern Tobacco Root Mountains. An initial ratio of about 0.704 indicates that these metamorphic rocks could not have had a long crustal history before this time. James and Hedge (1980) analyzed a suite of quartzofeldspathic gneisses from the Tobacco Root, Ruby, and Gallatin Ranges. These samples gave an age of 2760 ± 115 m.y. by themselves or an age of 2730 ± 85 m.y. when combined with the data of Mueller and Cordua (1976). No comprehensive geochemical data are available for these samples, but their Rb and Sr contents and Rb/Sr ratios are consistent with evolved granitic rocks or arkosic sediments. There is no evidence from either of these studies that these units existed before approximately 2800 m.y. ago; however, the data base is small and only quartzofeldspathic samples have been examined.

BIGHORN MOUNTAINS

The Bighorn Mountains are located in north-central Wyoming (Fig. 14-1), east-southeast of the Beartooth Mountains, and contain major exposures of Archean rocks (about 2800 km^2). The Bighorn Mountains can be divided into a northern terrane of granitoids ranging in composition from tonalite to granite and a southern terrane consisting of orthogneisses and foliated granitoids (Barker et al., 1979; Heimlich et al., 1972). K-Ar dating throughout the range (Condie and Heimlich, 1969; Heimlich and Armstrong, 1972; Heimlich and Banks, 1968) showed that all the country rocks were Archean, and a careful K-Ar biotite age study (Heimlich and Armstrong, 1972) showed that ages in the northern terrane averaged 2.73 b.y., whereas those in the southern terrane averaged 2.51 b.y. The reasons for the difference in age are unclear, but it shows that the two areas have more than lithologic differences. U-Pb studies in the northern terrane gave ages between 2840 and 2865 m.y., while those in the southern terrane gave ages in the range 2890–2905 m.y. with one younger age of 2710 m.y. (Banks and Heimlich, 1976; Heimlich and Banks, 1968). A Rb-Sr whole-rock study of granitoids and gneisses from both terranes gave an age of 2805 ± 60 m.y. (Steuber and Heimlich, 1977).

The only integrated field, geochemical, and geochronologic study in the Bighorn Mountains is that of Barker et al. (1979) and Arth et al. (1980) in the Lake Helen area of the southern terrane. These studies established that the rocks were produced in two events — an older E-1, and a younger E-2. The older E-1 assemblage consists of trondhjemitic and tonalitic gneiss, basaltic amphibolite, and hornblende-biotite gneiss. The sequence of events was intrusion of trondhjemitic magmas, deformation and metamorphism, synkinematic intrusions of tonalitic magmas, and very late synkinematic intrusion of andesitic (hornblende-biotite gneiss) magmas. The basaltic amphibolites are associated with the younger trondhjemitic rocks and may represent inclusions or later dikes of mafic compositions. These trondhjemitic and tonalitic rocks (Fig. 14-12) have similar major element contents with SiO_2 = 69–72 wt %, Al_2O_3 = 15–16 wt %, and NaO_2/K_2O ratios between 3 and 4. REE patterns are strongly fractionated, with minimal Eu anomalies and moderate LREE abundances (30–80X chrondrites) and low HREE abundances (1–4X). The amphibolites show a range of basaltic compositions for both major and trace elements. A composite Rb-Sr whole-rock isochron (Fig. 14-13) for E-1 trondhjemitic, tonalitic, and amphibolitic samples gave an age of 3007 ± 68 m.y. (I_{SR} = 0.7001 ± 1), and a U-Pb zircon age for the trondhjemitic-tonalitic gneisses was 2947 ± 100 m.y. The compositions of these rocks are similar to those of the trondhjemitic-amphibolitic complex of the western Beartooth Mountains, but apparently they are distinct in age. The andesitic gneisses of the Bighorn Mountains are so similar in major and trace element contents to those of the andesitic amphibolites of the eastern and central Beartooth Mountains that some genetic relationship must exist (Mueller et al., 1983). The Bighorn

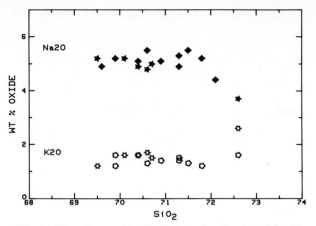

FIG. 14-12. Compositional diagram for the trondhjemitic and tonalitic rocks of the Lake Helen area (Barker *et al.*, 1979). E-1 rocks represented by diamonds and pentagons and E-2 rocks by stars.

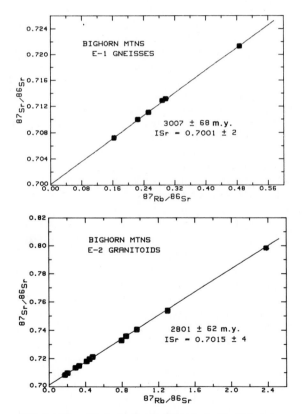

FIG. 14-13. Rb-Sr isochron diagrams for the E-1 gneisses and E-2 granitoids of the Lake Helen area, southern Bighorn Mountains (Arth *et al.*, 1980).

andesitic rocks have not been directly dated, but field relations place them between the 3.0- and 2.8-b.y.-old events.

The younger E-2 event in the Lake Helen area starts with synkinematic intrusion of a trondhjemitic to leucogranodioritic pluton that sharply cuts structures of the E-1 rocks. This pluton was followed by synkinematic intrusion of hornblende-biotite tonalite, biotite tonalite, biotite granodiorite, and biotite granite. The trondhjemitic rocks are compositionally similar to the older E-1 trondhjemitic-tonalitic rocks (Fig. 14-12) in all respects except for slightly higher LREE contents. Compositional data for the later intrusions is not published except for information that these rocks range in SiO_2 from 57 to more than 75 wt % and are typically calc-alkaline. A Rb-Sr whole-rock isochron (Fig. 14-13) that includes samples from all the E-2 rock types gave an age of 2801 ± 62 m.y. with an initial ratio of 0.7015 ± 0.0002.

OWL CREEK MOUNTAINS

The Owl Creek Mountains are located in central Wyoming and represent several small areas of exposure. The work that has been done there is concentrated in Wind River Canyon which transects one of the areas of exposure. This area has been described by Condie (1967), Granath (1975), Mueller *et al.* (1985a), and Stuckless *et al.*, (1986). The rocks exposed in the canyon represent a multiply folded, amphibolite grade, supracrustal sequence that consists of interlayered gneiss, amphibolite, and minor schist. The supracrustal sequence is cut by a potassic granite of Late Archean age.

Mueller *et al.* (1985a) have shown that the supracrustal sequence is largely of igneous origin. As presently exposed, the sequence is composed mostly of two types of amphibolite — one of tholeiitic basaltic composition with flat REE patterns, and another of basaltic-andesitic composition with LREE-enriched REE patterns. A gneiss of dacitic composition is found interlayered with the tholeiitic amphibolites. These dacitic rocks have SiO_2 between 66 and 72 wt %, average $K_2O/Na_2O = 0.65$, and strongly fractionated REE patterns (La = 100X and HREE = 3X). A U-Pb zircon age (Fig. 14-14) for a dacitic sample is 2906 ± 10 m.y. and is interpreted as the time of igneous crystallization. A Rb-Sr isochron (Fig. 14-14) for dacitic samples gives an age of 2755 ± 96 m.y. and is interpreted as the time of metamorphism. Samples of basaltic andesite fall on this isochron but tholeiitic samples fall slightly below, suggesting a genetic link between the dacitic and basaltic andesitic samples. Model initial Sr ratios in the range 0.702–0.704 for the dacitic-basaltic andesitic rocks between 2.75 and 2.90 b.y. ago indicate that the genesis of these rocks involved the crust in some way.

SYNTHESIS

The amount of information available about the geochemistry and geochronology of the northern part of the Wyoming Province allow only first-order hypotheses about the origin of this area to be made. The authors wish to make it clear that the data available for this province remain fragmentary, even in the better studied parts such as the Beartooth and Bighorn Mountains. Therefore, the ideas presented here may change a great deal as more information becomes available. This review should convince readers that the Archean terranes of the western United States contain much information of value about early crustal evolution and are worthy of additional study.

There is a growing body of data that strongly suggest the existence of rocks 3.3 b.y. old and older in the northern Wyoming Province. Rb-Sr data from the eastern and north-western Beartooth Mountains and the southern Madison Range indicate ages of 3.3–3.4

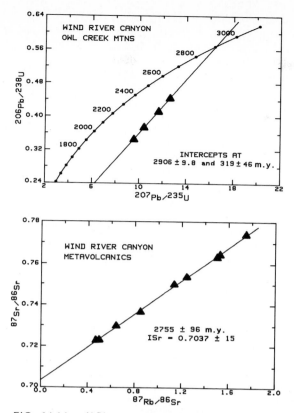

FIG. 14-14. U-Pb concordia and Rb-Sr isochron diagrams for the dacitic gneisses of the Wind River Canyon, Owl Creek Mountains (Mueller *et al.*, 1985a).

b.y. Sm-Nd and Pb-Pb data support these ages in the Beartooth Mountains. Unfortunately, the meager zircon data available for these rocks tend to provide only minimum ages of about 3.1 b.y.; however, the data contain obvious indications of complex systematics, and modern analytical techniques should provide more information for these systems. These older rocks are trondhjemites and tonalites in the southern Madison Range and northwestern Beartooth Mountains. This type of rock is known to be the most common component of Early Archean terranes (Barker and Peterman, 1974; Wooden *et al.*, 1980). However, the older rocks of the eastern Beartooth Mountains are dominantly potassic gneisses and a lithologically varied supracrustal sequence. Trace-element and common Pb isotopic data support the evolved nature of these rocks. The Early Archean rocks of the eastern Beartooth Mountains require that crustal processing of the more "primitive," low-K rocks suites more common in the Early Archean took place. The nature and timing of the earliest metamorphic events that affected these rocks is uncertain, largely because they were strongly overprinted by Late Archean events. A period of granulite-grade metamorphism in the eastern Beartooth Mountains may date from this period (Henry *et al.*, 1982).

Middle Archean rocks (2.9–3.1 b.y. old) are common in the northern Wyoming Province. Metamorphosed supracrustal rocks of a dominantly sedimentary origin in the northcentral and southwestern Beartooth Mountains and trondhjemitic and dacitic gneisses and basaltic amphibolites in the Bighorn and Owl Creek Mountains are all in this age range. It is perhaps significant that the low-K suites of this age are common in the areas where there is no evidence of older Archean crust. It seems likely that these Middle Archean rocks experienced at least an amphibolite grade metamorphic event before they became involved in major Late Archean events. Both the eastern and central Beartooth Mountains and the Bighorn Mountains contain major suites of magmatic rocks that are 2.75–2.80 b.y. old. In the Beartooth Mountains, the earliest members of this suite are andesitic amphibolites that record a middle amphibolite grade metamorphic event. This period of metamorphism preceded the intrusion of a dominantly high-silica and sodic granitoid suite. Activity in the Bighorn Mountains started with trondhjemitic rocks that are intruded by a synkinematic calc-alkaline suite. The geology of both these areas is consistent with the development of a major magmatic arc similar to those associated with modern-day convergent plate tectonics.

The northwestern Beartooth Mountains and Archean terranes to the west represent a very different late Archean environment (Mogk and Henry, Chapter 13, this volume). There is evidence for plutonic activity only in the northern and southern Madison Ranges and the northwestern Beartooth Mountains. The plutonic rocks are subordinate to sequences of quartzofeldspathic gneisses and/or supracrustal rocks containing quartzites and carbonates. The rock associations in the Tobacco Root Mountains and the Ruby Range are indicative of a continental shelf environment. Many of the rocks in this western area have high amphibolite- or granulite-grade metamorphic assemblages that could be produced in crustal doubling events of the Himalayan-Tibetan type (Newton and Perkins, 1982). A good case can be made that the northwestern Beartooth Mountains are an assemblage of at least six terranes of variable metamorphic grade that were tectonically juxtaposed during the Late Archean. Thus this area may have experienced major tectonic activity akin to that seen along modern continental margins undergoing convergent plate tectonics, whereas the area to the east was the site of a magmatic arc. The major activity in the magmatic arc may be limited to a time span of about 50 m.y., but chronologic constraints are not available for the western area. Its history may involve several distinct events over a 200-m.y. time span.

ACKNOWLEDGMENTS

We wish to thank the many scientists who share our interest in the Archean rocks of the Wyoming Province. We hope that we have presented their work fairly in this review and apologize to those whose work was not referenced because of lack of space or our ignorance of its existence or importance. A special thanks goes to F. Barker and W. G. Ernst for critically reviewing and editing the manuscript. The authors' work in this area has been supported by grants from the National Science Foundation, the U.S. Geological Survey, Research Corporation, NASA's Crustal Genesis Program, and NASA RTOPs to L. N. Nyquist, which supported JLW between 1976 and 1983.

REFERENCES

Arth, J. G., Barker, F., and Stern, T. W., 1980, Geochronology of Archean gneisses in the Lake Helen area, southwestern Bighorn Mountains, Wyoming: *Precambrian Geology*, v. 11, p. 11–22.

Barker, F., 1982, Geologic map of the Lake Helen quadrangle, Big Horn and Johnson Counties, Wyoming: *U.S. Geol. Survey Geol. Quad. Map GQ-1563*.

——, and Peterman, Z. E., 1974, Bimodal tholeiitic-dacitic magmatism and the Early Precambrian crust: *Precambrian Res.*, v. 1, p. 1-12.

——, Arth, J. G., and Millard, H. T., 1979, Archean trondhjemites of the southwestern Bighorn Mountains, Wyoming — A preliminary report, *in* Barker, F., ed., *Trondhjemites, Dacites, and Related Rocks:* Amsterdam, Elsevier, p. 401-414.

Butler, J. R., 1966, Geologic evolution of the Beartooth Mountains, Montana and Wyoming — Pt. 6. Cathedral Peak area, Montana: *Geol. Soc. America Bull.*, v. 77, p. 45-64.

——, 1969, Origin of Precambrian granitic gneiss in the Beartooth Mountains, Montana and Wyoming: *Geol. Soc. America Mem. 115*, p. 73-101.

Casella, C. J., 1964, Geologic evolution of the Beartooth Mountains, Montana and Wyoming — Pt. 4. Relationship between Precambrian and Laramide structures in the Line Creek area: *Geol. Soc. America Bull.*, v. 75, p. 969-986.

——, 1969, A review of the Precambrian geology of the eastern Beartooth Mountains, Montana and Wyoming: *Geol. Soc. America Mem. 115*, p. 53-71.

——, Levay, J., Eble, E., Hirst, B., Huffman, K., Lahti, V., and Metzer, R., 1982, Precambrian geology of the southwestern Beartooth Mountains, Montana and Wyoming, *in* Mueller, P. A., and Wooden, J. L., eds., *Precambrian Geology of the Beartooth Mountains, Montana and Wyoming:* Mont. Bur. Mines Geology Spec. Publ. 84, p. 1-24.

Clark, M. L., and Mogk, D. W., 1985, Development and significance of the Blacktail Mountains Archean metamorphic complex, Beaverhead County, Montana: *Geol. Soc. America Abstr. with Programs*, v. 17, p. 212.

Condie, K. C. 1967, Petrologic reconnaissance of the Precambrian rocks in the Wind River Canyon, Central Owl Creek Mountains, Wyoming: *Univ. Wyo., Contrib. Geology*, v. 6, p. 123-129.

——, 1976a, The Wyoming Archean province in the western United States, *in* Windley, B. F., ed., *The Early History of the Earth:* New York, Wiley, p. 499-510.

——, 1976b, Trace element geochemistry of Archean greenstone belts: *Earth Sci. Rev.*, v. 112, p. 393-417.

——, 1982, Plate Tectonics and Crustal Evolution: Amsertdam, Elsevier, 434 p.

——, and Heimlich, R. A., 1969, Interpretation of Precambrian K-Ar biotite dates in the Bighorn Mountains, Wyoming: *Earth Planet. Sci. Lett.*, v. 6, p. 209-212.

Czemanske, G. K., and Zientek, M. L., 1985, eds., *The Stillwater Complex, Montana: Geology and Guide:* Mont. Bur. Mines Geology Spec. Publ. 92.

DePaolo, D. J., and Wasserburg, G. J., 1979, Sm-Nd age of the Stillwater Complex and the mantle evolution curve for neodynium: *Geochim. Cosmochim. Acta*, v. 43, p. 999-1008.

Eckelmann, F. D., and Poldervaart, A., 1957, Geologic evolution of the Beartooth Mountains, Montana and Wyoming — Pt. 1. Archean history of the Quad Creek area: *Geol. Soc. America Bull.*, v. 68, p. 1225-1262.

Erslev, E. A., 1981, Petrology and structure of the Precambrian metamorphic rocks of the southern Madison Range, southwestern Montana: Ph.D. thesis, Harvard Univ., Cambridge, Mass., 133 p.

——, 1982, The Madison mylonite zone — A major shear zone in the Archean basement of southwestern Montana: *33rd Ann. Wyo. Geol. Assoc. Guidebook*, p. 213-221.

——, 1983, Pre-Beltian geology of the southern Madison Range, southwestern Montana: *Mont. Bur. Mines Geology Mem. 55*, 26 p.

Garihan, J. M., 1979, Geology and structure of the central Ruby Range, Madison County, Montana: *Geol. Soc. America Bull.*, v. 90, Pt. II, p. 695-788.

Gill, J. B., 1981, *Orogenic Andesites and Plate Tectonics:* New York, Springer-Verlag, 390 p.

Granath, J. W., 1975, Wind River Canyon — An example of a greenstone belt in the Archean of Wyoming, U.S.A.: *Precambrian Res.*, v. 2, p. 71–91.

Harris, R. L., 1959, Geologic evolution of the Beartooth Mountains, Montana and Wyoming — Pt. 3. Gardner Lake area, Wyoming: *Geol. Soc. America Bull.*, v. 70, p. 1185–1216.

Heimlich, R. A., and Armstrong, R. L., 1972, Variance of Precambrian K-Ar biotite dates, Bighorn Mountains, Wyoming: *Earth Planet. Sci. Lett.*, v. 14, p. 75–78.

——, and Banks, P. O., 1968, Radiometric age determinations, Bighorn Mountains, Wyoming: *Amer. J. Sci.*, v. 266, p. 180–192.

——, Nelson, G. C., and Malcuit, R. J., 1972, Mineralogy of Precambrian gneiss from the Bighorn Mountains, Wyoming: *Geol. Mag.*, v. 109, p. 215–230.

Henry, D. J., Mueller, P. A., Wooden, J. L., Warner, J. L., and Lee-Berman, R., 1982, Granulite grade supracrustal assemblages of the Quad Creek area, eastern Beartooth Mountains, Montana, *in* Mueller, P. A., and Wooden, J. L., eds., *Precambrian Geology of the Beartooth Mountains, Montana and Wyoming:* Mont. Bur. Mines Geology Spec. Publ. 84, p. 147–156.

James, H. L., and Hedge, C. E., 1980, Age of the basement rocks of southwest Montana: *Geol. Soc. America Bull.*, v. 91, p. 11–15.

Larsen, L. H., Poldervaart, A., and Kirchmayer, M., 1966, Geologic evolution of the Beartooth Mountains, Montana and Wyoming — Pt. 7. Structural homogeneity of gneisses in the Lonesome Mountain area: *Geol. Soc. America Bull.*, v. 77, p. 1277–1292.

Mogk, D. W., 1982, The nature of the trondhjemitic gneiss-amphibolite basement complex in the North Snowy Block, Beartooth Mountains, Montana, *in* Mueller, P. A., and Wooden, J. L., eds., *Precambrian Geology of the Beartooth Mountains, Montana and Wyoming:* Mont. Bur. Mines Geology Spec. Publ. 84, p. 83–90.

——, 1984, Petrology, geochemistry, and structure of an Archean terrane in the North Snowy block, Beartooth Mountains, Montana: Ph.D. thesis, Univ. Washington, Seattle, Wash.

——, Muller, P. A., and Wooden, J. L., in press, Archean tectonics of the North Snowy Block, Beartooth Mountains, Montana: *J. Geology*.

Montgomery, C. W., 1982, Preliminary U-Pb dating of biotite granodiorite from the South Snowy Block, Beartooth Mountains, *in* Mueller, P. A., and Wooden, J. L., eds., *Precambrian Geology of the Beartooth Mountains, Montana and Wyoming;* Mont. Bur. Mines Geology Spec. Publ. 84, p. 41–44.

——, and Lytwyn, J. N., 1984, Rb-Sr systematics and ages of principal Precambrian lithologies in the South Snowy Block, Beartooth Mountains: *J. Geology*, v. 92, p. 103–112.

Mueller, P. A., and Cordua, W. S., 1976, Rb-Sr whole rock age of gneisses from the Horse Creek area, Tobacco Root Mountains, Montana: *Isochron/West*, v. 16, p. 33–36.

——, and Wooden, J. L., 1976, Rb-Sr whole rock age of the contact aureole of the stillwater igneous complex, Montana: Earth Planet. Sci. Lett., v. 29, p. 384-388.

——, Wooden, J. L., and Bowes, D. R., 1982a, Precambrian evolution of the Beartooth Mountains, Montana and Wyoming, U.S.A.: *Rev. Bras. Geociencias*, v. 12, p. 215–222.

——, Wooden, J. L., Odom, A. L., and Bowes, D. R., 1982b, Geochemistry of the Archean rocks of the Quad Creek and Hellroaring Plateau areas of the eastern Beartooth Mountains, *in* Mueller, P. A., and Wooden, J. L., eds., *Precambrian Geology of the*

Beartooth Mountains, Montana and Wyoming: Mont. Bur. Mines Geology Spec. Publ. 84, p. 69–82.

———, Wooden, J. L., Schulz, K., and Bowes, D. R., 1983, Incompatible element rich andesitic amphibolites from the Archean of Montana and Wyoming: Evidence for mantle metasomatism: *Geology*, v. 11, p. 203–206.

———, Peterman, Z. E., and Granath, J. W., 1985a, A bimodal Archean volcanic series, Owl Creek Mountains, Wyoming: *J. Geology*, v. 93, p. 701–712.

———, Wooden, J. L., Henry, D. J., and Bowes, D. R., 1985b, Archean crustal evolution of the eastern Beartooth Mountains, Montana and Wyoming; *in* Czamanske, G. K., and Zientek, M. L., eds., *The Stillwater Complex: Montana – Geology and Guide:* Mont. Bur. Mines and Geology Spec. Publ. 92, p. 9–20.

Newton, R. C., and Perkins, D., 1982, Thermodynamic calibration of geobarometers based on the assemblages garnet-plagioclase-orthopyroxene (clinopyroxene)-quartz: *Amer. Mineralogist*, v. 67, p. 203–222.

Nunes, P. D., and Tilton, G. R., 1971, Uranium-lead ages of minerals from the Stillwater igneous complex and associated rocks, Montana: *Geol. Soc. America Bull.*, v. 82, p. 2231–2250.

Page, N. J., 1977, Stillwater complex, Montana – Rock succession, metamorphism and structure of the complex and adjacent rocks: *U.S. Geol. Survey Prof. Paper 999.*

———, and Nokleberg, W. J., 1972, Genesis of mesozonal granitic rocks below the base of the Stillwater Complex, in the Beartooth Mountains, Montana: *U.S. Geol. Survey Prof. Paper 800D*, p. D127–D141.

———, and Zientek, M. L., 1985, Petrogenesis of the metamorphic rocks beneath the Stillwater Complex – Lithologies and structures, *in* Czemanske, G. K., and Zientek, M. L., eds., *The Stillwater Complex, Montana – Geology and Guide:* Mont. Bur. Mines Geology Spec. Publ. 92, p. 55–69.

Peterman, Z. E., 1979, Geochronology and the Archean of the United States: *Econ. Geology*, v. 74, p. 1544–1562.

———, 1981, Dating of Archean basement in northeastern Wyoming and southern Montana: *Geol. Soc. America Bull.*, v. 92, Pt. 1, p. 139–146.

Reed, J. C., and Zartman, R. E., 1973, Geochronology of the Precambrian rocks of the Teton Range, Wyoming: *Geol. Soc. America Bull.*, v. 84, p. 561–582.

Reid, R. R., McMannis, W. J., and Palmquist, J. C., 1975, Precambrian geology of the North Snowy block, Beartooth Mountains, Montana: *Geol. Soc. America Spec. Paper 157*, 135 p.

Richmond, D. P., and Mogk, D. W., 1985, Archean geology of the Lake Plateau area, Beartooth Mountains, Montana: *Geol. Soc. America Abstr. with Programs*, v. 17, p. 262.

Salt, K. J., and Mogk, D. W., 1985, Archean geology of the Spanish Peaks area, southwestern Montana: *Geol. Soc. America Abstr. with Programs*, v. 17, p. 263.

Spencer, E. W., and Kozak, S. V., 1975, Precambrian evolution of the Spanish Peaks, Montana: *Geol. Soc. America Bull.*, v. 86, p. 785–792.

Stuckless, J. S., Meisch, A. T., and Wenner, D. B., 1986, Chemical and isotopic studies of granitic rocks, Owl Creek Mountains, Wyoming: *U.S. Geol. Survey Prof. Paper*, in press.

Stueber A. M., and Heimlich, R. A., 1977, Rb-Sr isochron age of the Precambrian basement complex, Bighorn Mountains, Wyoming: *Geol. Soc. America Bull.*, v. 88, p. 441–444.

Vitaliano, C. J., Cordua, W. S., Hess, D. F., Burger, H. R., Hanley, T. B., and Root, F. K., 1979, Explanatory text to accompany geologic map of southern Tobacco Root

Mountains, Madison County, Montana: *Geol. Soc. America Map Chart Series MC-31.*

Warner, J. L., Lee-Berman, R., and Simonds, C. H., 1982, Field and petrologic relations of some Archean rocks near Long Lake, eastern Beartooth Mountains, Montana and Wyoming, *in* Mueller, P. A., and Wooden, J. L., eds., *Precambrian Geology of the Beartooth Mountains, Montana and Wyoming:* Mont. Bur. Mines Geology Spec. Publ. 84, p. 57–68.

Wooden, J. L., 1979, Geochemistry and geochronology of Archean gneissic rocks of the southwestern Beartooth Mountains — A preliminary study, *in* Mueller, P. A., and Wooden, J. L., eds., *Guidebook. U.S. Natl. Comm. Archean Geochemistry 1979 Field Conf.,* p. 10–17.

_____ , Goldich, S. S., and Suhr, H. H., 1980, Origin of the Morton gneiss, southwestern Minnesota — Pt. II. Geochemistry: *Geol. Soc. America Spec. Paper 182,* p. 57–76.

_____ , Mueller, P. A., and Bowes, D. R., 1982a, An informal guidebook for the Precambrian rocks of the Beartooth Mountains, Montana-Wyoming, *in* Magmatic processes of early planetary crusts — Magma oceans and stratiform layered intrusions: *Lunar Planet. Inst. Tech. Rt. No. 82–01,* p. 195–234.

_____ , Mueller, P. A., Hunt, D. K., and Bowes, D. R., 1982b, Geochemistry and Rb-Sr geochronology of Archean rocks from the interior of the southeastern Beartooth Mountains, Montana and Wyoming, *in* Mueller, P. A., and Wooden, J. L., eds., *Precambrian Geology of the Beartooth Mountains,* Mont. Bur. Mines Geology Spec. Publ. 84, p. 45–56.

_____ , and Mueller, P.A., in press, Pb, Sr, and Nd isotopic compositions of a suite of Late Archean igneous rocks, eastern Beartooth Mountains: Implications for crust-mantle evolution: *Earth Planet. Sci. Lett.*

M. E. Bickford
Department of Geology
University of Kansas
Lawrence, Kansas 66045

15

THE ACCRETION OF PROTEROZOIC CRUST IN COLORADO: IGNEOUS, SEDIMENTARY, DEFORMATIONAL, AND METAMORPHIC HISTORY

ABSTRACT

The crystalline crust of Colorado consists mostly of Lower Proterozoic rocks that are separated from the Archean Wyoming Craton by the Cheyenne Belt, a major northeast-trending shear zone. The Lower Proterozoic crust is composed of bimodal metavolcanic assemblages, thick accumulations of associated metagreywacke, and metapelite-to-quartzite that represent sedimentary deposits of nonvolcanic origin. All have been intruded by a variety of granitic plutons. Volcanic rocks and related metasediments mostly occur in the central and southwestern region, whereas metapelites and related metasediments occur in the Front Range and the Wet Mountains. Thick argillite and quartzite of the middle Proterozoic Uinta Mountains Group occurs in the northwestern part of the state. The buried basement of eastern Colorado is evidently underlain by Lower Proterozoic rocks similar to those in the exposed areas except for the Las Animas Formation, an accumulation of Upper Proterozoic metasedimentary and volcanic rocks known in southeastern Colorado from drill hole data.

All of the Proterozoic rocks in Colorado except the Uinta Mountains Group and the Las Animas Formation are strongly deformed and metamorphosed. Metamorphic grade ranges from upper greenschist to granulite facies, but most rocks are in middle to upper amphibolite facies; migmatite is common in metasedimentary rocks of the Wet Mountains. Variations in grade seem to be related to the level of crust exposed through differential uplift and erosion of crustal blocks during Ancestral Rockies and Laramide tectonism.

The oldest rocks known in Colorado are metavolcanics of the Irving Formation of the Needle Mountains and granite from the Powderhorn area west of Gunnison; these rocks were formed 1780 m.y. ago. A thick sequence of basalt, rhyolite, and associated mafic sedimentary rocks was formed in the Powderhorn area 1770–1760 m.y. ago; these rocks were intruded by granitic plutons about 1755 m.y. ago. A second period of bimodal volcanism and accumulation of graywacke turbidite occurred 1740–1730 m.y. ago in the area south of Gunnison. Following at least two episodes of intense deformation, including tight isoclinal folding and metamorphism in upper greenschist to lower amphibolite facies, these rocks were intruded by granitic plutons 1725–1700 m.y. ago. To the east, in the Front Range and Wet Mountains, the accumulation, and probably initial deformation and metamorphism, of pelitic rocks and graywacke occurred before the intrusion of granitic plutons about 1700 m.y. ago. Late granitic plutons whose ages are about 1670 m.y. were emplaced throughout Colorado and at least as far east as western Kansas.

Mostly nonfoliated anorogenic granitic plutons were emplaced throughout Colorado 1420–1470 m.y. ago. The amount of deformation accompanying this igneous event is not well understood, but the widespread occurrence of pegmatite swarms with similar ages suggests that preexisting crust reached the melting point locally and extensive metamorphism may have occurred at this time. A second period of anorogenic intrusion, occurring about 1360 m.y. ago, is represented by the San Isabel batholith of the Wet Mountains. By analogy with the extensive 1450- and 1370-m.y.-old granite-rhyolite terranes of the midcontinent region, intrusives of both episodes may have been accompanied by surface eruptives that were removed by erosion following uplift. Each successive period of igneous activity seems to have caused resetting of the Rb-Sr system in the older rocks, perhaps by driving hydrothermal systems that removed radiogenic Sr.

The geologic record described is one of successive accretion of essentially juvenile, volcanogenic terranes on the southern margin of the Wyoming craton. The juvenile nature of the volcanogenic rocks is supported by their primitive initial Sr and Pb isotopic compositions and the fact that their Sm-Nd mantle separation ages are about 1800 m.y., essentially the same as their U-Pb crystallization ages. Granitic rocks and plutons of the 1450- and 1370-m.y.-old anorogenic suites, however, also have mantle separation ages of about 1800 m.y., indicating that they were derived through remelting of preexisting crust. This probably occurred during widespread rifting following accretion and stabilization of the earlier volcanogenic terranes.

INTRODUCTION

It is difficult to discuss evolution of the crust in Colorado except in the context of crustal evolution in all of the southwestern United States and at least across parts of the midcontinent area. However, in this paper I will concentrate on the geochronology and petrogenesis of Proterozoic crust in Colorado while keeping the events discussed in the broader context.

PROTEROZOIC CRUSTAL ROCKS IN COLORADO

The Cheyenne Belt and Its Associated Rocks

The crustal province of which Colorado is part is fundamentally bounded to the north by the Cheyenne Belt, a major northeast-trending shear zone which separates gneissic Archean and miogeoclinal lower Proterozoic rocks to the north in Wyoming from mostly volcanogenic, eugeoclinal lower Proterozoic rocks in Colorado (Divis, 1976; Hills and Houston, 1979; Houston et al., 1968, 1979; Fig. 15-1). The Cheyenne Belt is commonly interpreted as a suture between Archean rocks of the stable Wyoming Craton and its lower Proterozoic passive margin cover rocks against a complex of somewhat younger oceanic arc complexes that were accreted against a southwest-dipping subduction zone (Hills and Houston, 1979; Karlstrom et al., 1983). The mylonite zone of the Cheyenne Belt has apparently experienced lateral movement, perhaps totaling 7 km of left-lateral offset (Houston and Parker, 1963; Houston et al., 1975). Major movement on the Cheyenne Belt is constrained between 1780 and 1750 m.y. ago by the U-Pb ages of sheared and unsheared rocks, respectively (Premo, 1984). Whether the volcanogenic rocks to the south of the Cheyenne Belt are outboard arcs formed during subduction or are exotic oceanic arcs accreted laterally is still debated. However, it now seems clear that there are no Archean rocks to the south of the Cheyenne Belt and that the lower Proterozoic rocks of the southwest represent accretion of mostly juvenile material to the continent.

Lower Proterozoic Assemblages South of the Cheyenne Belt

The lower Proterozoic crust south of the Cheyenne Belt is a complex of volcanogenic and nonvolcanogenic metamorphic rocks, most at high grades, that have been intruded by a variety of granitic plutons. The volcanogenic rocks occur chiefly in the central and southern parts of Colorado, although they also occur just south of the Cheyenne Belt in north-central Colorado and adjacent south-central Wyoming (Fig. 15-1; Premo, 1984). These rocks are typified by the exposed sequences in the Gunnison area (Fig. 15-2; Afifi, 1981; Bickford and Boardman, 1984; Condie and Nuter, 1981; Hedlund and Olson, 1981).

Gunnison region The Gunnison region is underlain by sequences of bimodal volanic rocks intercalated with thick turbiditic metasedimentary rocks. Metamorphic grade is upper greenschist to lower amphibolite facies, and although the rocks have been isoclinally folded, preservation of primary features is good to excellent. Most of the basaltic rocks were evidently flows, but breccias and debris-flow deposits also occur. Many of the basalt

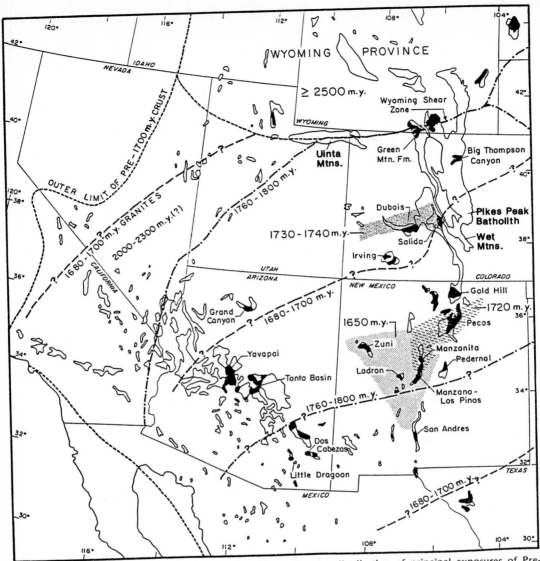

FIG. 15-1. Sketch map of southwestern United States showing distribution of principal exposures of Precambrian rocks. Major exposures of supracrustal rocks are shown in black and most areas discussed in the text are named. Modified after Condie (1986).

flows have spectacularly preserved pillows and pillow-breccias (Fig. 15-3), indicating submarine eruption. The felsic volcanic rocks include both ash-flow tuff and apparent debris-flow deposits. In the former, eutaxitic texture is preserved despite deformation and metamorphism, although phenocrysts are commonly polycrystalline and metamorphic micas may have developed in old compaction foliation planes. The metasedimentary rocks are typically quartz-feldspar wackes in which the matrix is largely chlorite and white micas.

These rocks initially may have contained felsic volcanic rock fragments, but they

414   **ACCRETION OF PROTEROZOIC CRUST IN COLORADO**

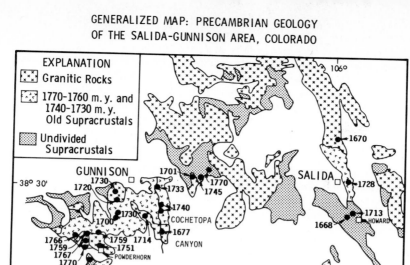

EXPLANATION
Granitic Rocks
1770-1760 m. y. and
1740-1730 m. y.
Old Supracrustals
Undivided
Supracrustals

FIG. 15-2. Sketch geologic map of the Gunnison-Salida area of Colorado.
Numbers indicate the locations of dated samples and their ages. Modified from
Bickford and Boardman (1984).

have not been preserved. Graded bedding is common, as are such features as flame structures, basal scours, and convolute bedding; these features suggest deposition by subaqueous gravity flows, perhaps in submarine fans.

Prior to deformation, these rocks were intruded by both granitic plutons and gabbroic sills. Like the basalts, the gabbroic sills are now amphibolites, but primary ophitic texture is locally preserved despite complete recrystallization. Deformation was extreme, and the entire section was folded at least twice (Cummings *et al.*, 1984) and metamorphosed. Following metamorphism and deformation, granitic plutons were again intruded. The chronology of these events has been determined by U-Pb zircon analysis and will be discussed below.

Chemical studies (Bennett, 1984; Bickford and Boardman, 1984; Condie and Nuter, 1981; Tobin, 1982) indicate that the basalts are rather typical ocean-floor tholeiites, whereas the felsic volcanic rocks range from high-silica rhyolite to dacite.

FIG. 15-3. Basaltic pillow-breccia, Salida area. From Bickford and Boardman, 1984.

Bennett (1984) reported a few analyses in the basaltic andesite range, and Afifi (1981) observed an andesitic unit, but it is clear that andesite is rare and the volcanic assemblage is bimodal (Fig. 15-4). Bennett (1984) analyzed numerous samples of metasedimentary rocks; many of these are of essentially andesitic composition, but it is not clear whether they were originally andesitic debris, perhaps from andesitic stratovolcanos that were rapidly eroded, or simply mixtures of debris from the end members of the bimodal volcanic rocks. The only plutons studied extensively in this area are those associated with an annular complex about 5 km south of Gunnison (Fig. 15-2). These were studied by Vance (1984) and Tobin (1982). Tobin showed that the rocks range from quartz diorite through tonalite to granite, and that the suite has a calc-alkaline chemical trend.

Salida area Proterozoic rocks in the Salida area have been studied by Boardman (1976, 1986), Bickford and Boardman (1984), and by Boardman and Condie (1986). These rocks are exposed in a large, gently tilted block north of Salida and, although metamorphosed in upper amphibolite facies, are mostly not penetratively deformed. Mafic volcanogenic rocks are largely volcaniclastic and include: thick breccia or conglomeratic units; coarsely fragmental units with crude bedding or cross-bedding; and laminated or cross-laminated, fine-grained units. Pillowed basaltic flows and pillow breccias are less common. These rocks suggest accumulation in shallow water near volcanic sources. Felsic volcanogenic rocks include submarine debris-flow deposits that grade upward into tuffaceous sedimentary units. Thick tholeiitic sills have intruded these sequences. It is clear that the Salida Proterozoic section is quite similar to that exposed in the Gunnison area, although metamorphic grade is higher. In a later section it will be shown that these rocks are coeval with the younger volcanic-plutonic sequence of the Gunnison area.

Needle Mountains A major volcanogenic succession in the Needle Mountains of south-western Colorado (Fig. 15-1) is the Irving Formation (Cross *et al.*, 1905). These high-grade felsic gneisses and amphibolites have been interpreted as a sequence of basalt and felsic volcanic rocks (Barker, 1969). They were intruded by granitic plutons, eroded, and then covered by a sequence of sandstone, shale, and graywacke (the Uncompahgre Series;

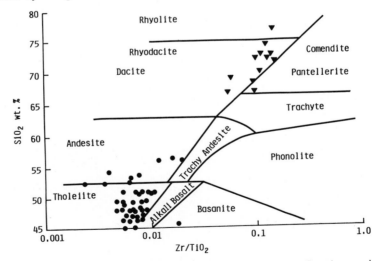

FIG. 15-4. Data for basaltic and gabbroic rocks from Gunnison and Salida areas and felsic rocks from Gunnison area only, plotted on SiO_2 versus Zr/TiO_2 diagram of Winchester and Floyd (1977). After Bickford and Boardman (1984).

ACCRETION OF PROTEROZOIC CRUST IN COLORADO

now metamorphosed to slate and quartzite). Subsequently, a younger series of plutons was intruded. The volcanogenic sequence of the Irving Formation is clearly similar to the rocks exposed in the Gunnison and Salida areas and, as will be shown later, is coeval with the older part of these volcanic successions.

Sierra Madre Range, southeastern Wyoming The Sierra Madre Range is the northern extension of the Park Range of Colorado. Because rocks there have been studied and because this region lies south of the Cheyenne Belt, it is included in this discussion. The major volcanogenic unit is the Green Mountain Formation (Condie and Shadel, 1984; Divis, 1976); as in the Gunnison area, metamorphism is in greenschist to amphibolite facies, and primary structures and textures are commonly well preserved. The exposed volcanogenic rocks occur in deformed pendants in granitic plutons and comprise a bimodal volcanic series with compositions clustering in the tholeiite and dacite to rhyolite fields in Zr/TiO_2 versus SiO_2 diagrams (Fig. 15-5). According to Condie and Shadel (1984), mafic volcanic rocks include tuff breccias and agglomerates, probably accumulated as mudflows and associated mafic sediments, as well as pillowed basaltic lava flows. Felsic volcanic rocks are entirely fragmental and include tuff breccias, agglomerates, and tuffaceous sedimentary rocks as well as ashflow tuff with well-preserved eutaxitic textures. The Green Mountain Formation is intruded by the Encampment River granodiorite and by the Sierra Madre granite.

Front Range and Wet Mountains The oldest Proterozoic rocks in these eastern ranges of Colorado are those of the Idaho Springs formation, mostly biotite quartzofeldspathic gneiss. Minor amphibolite, evidently derived from volcanic rocks, occurs also, as do local occurrences of quartzite and muscovite-quartz rocks interbedded with muscovitic schist. However, the dominant lithology suggests that this region was the site of sedimentary basins in which thick sequences of shale and graywacke were deposited, in contrast to the dominantly volcanogenic terranes to the west (Gunnison area; Needle Mountains) and northwest (Sierra Madre Range). Metamorphism of these rocks was in upper

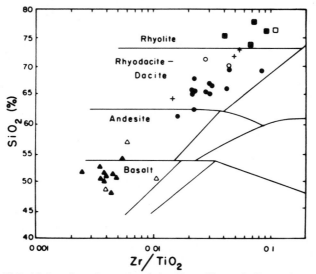

FIG. 15-5.   Data for rocks of the Green Mountain Formation, Sierra Madre Range, Wyoming, plotted on SiO_2 versus Zr/TiO_2 diagram of Winchester and Floyd (1977). After Condie and Shadel (1984).

amphibolite facies throughout much of the Front Range, and migmatite, indicating metamorphic conditions in sillimanite-orthoclase facies at the granite solidus, is common in much of the Wet Mountains. Charnockite and charnockitic gneiss, indicating metamorphism in granulite facies, also occur in the Mount Tyndall area of the western Wet Mountains (Brock and Singewald, 1968).

Major granitic plutons were emplaced into the Idaho Springs Formation over a significant part of the Proterozoic. Swarms of pegmatite, such as the Eight Mile Pegmatite of the northern Wet Mountains, are also characteristic of this terrane. The chronology of these events is presented in a later section of this paper.

Uinta Mountains, northwest Colorado A thick sequence of middle to upper Proterozoic sedimentary rocks occurs in the Uinta Mountains of northwestern Colorado. These rocks, the Uinta Mountain Group, include quartzite, arkose, and shale and are virtually unmetamorphosed. Because the present exposures in the Uinta Mountains are bounded by faults, and by major basins on both north (Green River Basin) and south (Uinta Basin), the limits of the formation and its relation with other Proterozoic rocks are poorly known. The age of the Uinta Mountain Group is provisionally put at 950 m.y. near the top and, indirectly, at 1400 m.y. at the base (Crittenden and Peterman, 1975); it is thus broadly coeval with the Belt Supergroup of Montana and Idaho.

Buried basement of eastern Colorado Little is known about the buried basement of eastern Colorado because the number of basement penetrations by drilling there is limited. Most of what is known has been summarized by Tweto (1980). The basement appears to consist of rocks mostly like those exposed in the Front Range and the Wet Mountains, that is, metasedimentary gneisses and granitic plutons with minor amounts of metavolcanic rocks. The buried basement of Nebraska and northern Kansas and Missouri includes similar rocks (Bickford *et al.*, 1981, 1986; Lidiak, 1972; Van Schmus *et al.*, 1986), and it appears that this terrane extends at least as far east as northern Missouri.

Tweto (1983) has described a sequence of only slightly metamorphosed slate, phyllite, graywacke, and chert with carbonate rocks and volcanics in its upper part from the subsurface of southeastern Colorado. According to Tweto (1983), who called these rocks the Las Animas Formation, they are probably rift sediments and are similar to the Tillman Metasedimentary Group (Ham *et al.*, 1964) of the Wichita Mountains region of southern Oklahoma. Although the age of these rocks has not been determined by radiometric methods, Tweto has assigned them to the late Precambrian.

DEFORMATION AND METAMORPHISM

The deformational and metamorphic history of most of the Proterozoic rocks in Colorado is not well known. Taylor (1975) and Braddock (1970) recognized three folding periods in Proterozoic rocks of the Front Range and it is clear from the mapping of Hedlund (1974), Olson (1974, 1976), and Hedlund and Olson (1974, 1975) that rocks in the Gunnison area experienced at least two periods of folding. Recent structural studies by Afifi (1981) and Cummings *et al.* (1984) in the northern part of the Iris Quadrangle southeast of Gunnison (Fig. 15-2) indicate three deformational periods. The first two resulted in folding and the third produced a pervasive cleavage. Metamorphism in upper greenschist facies accompanied each of these deformational phases.

To my knowledge there has not been a serious regional study or metamorphism in Colorado except perhaps that of Boos and Boos (1957). The metamorphic grade of Proterozoic rocks in Colorado ranges from greenschist to granulite facies, but in most parts of the state, rocks are in amphibolite facies. As the geologic and tectonic relationships in the Proterozoic unfold, significant patterns in the deformation and metamorphism of the rocks may emerge; in fact, studies in the Gunnison area, where both geochronological and geochemical studies have recently been done, will probably remain at a standstill until a major structural study is done.

As a first approximation, the distribution of metamorphic grades is determined by the crustal level exposed, which in turn is related to the amount of vertical uplift of crustal blocks during late Paleozoic (Ancestral Rockies) and Cenozoic (Laramide) tectonism. Thus, rocks in the Front Range and the Park Range are mostly in middle to upper amphibolite facies. In the Wet Mountains migmatite is common and rocks are mostly in upper amphibolite facies; as noted above, however, a region of charnockite and charnockitic gneiss occurs in the Mount Tyndall area of the western Wet Mountains. Rocks in the northern Sawatch Range are in middle to upper amphibolite facies, but much of the region south of Gunnison is in upper greenschist or lower amphibolite facies.

GEOCHRONOLOGY

Orogenic Assemblages

Geochronological study in Colorado began almost as soon as the application of radiometric techniques became widespread. Most of the early measurements, which were done in the late 1950s and early 1960s, employed the Rb-Sr method, commonly on separated minerals such as muscovite and biotite. Application of the Rb-Sr whole-rock method began with the work of Wetherill and Bickford (1965) in the Gunnison area, of Hedge (1967, 1970), Hedge *et al.* (1967), and Peterman *et al.* (1968) in the Front Range, and Bickford *et al.* (1969) in the Needle Mountains. As discussed in a later section, Rb-Sr ages are commonly reset by later thermal events, and most geochronologists now consider U-Pb analysis of zircons a more reliable method for obtaining primary crystallization ages. Published zircon data are sparse, although Silver and Barker (1968) reported zircon ages from the Needle Mountains in an abstract. What follows is based largely on new zircon ages (Bickford *et al.*, 1984; Bickford and Boardman, 1984; Premo, 1984) and some of the reliable Rb-Sr whole-rock age results.

The earliest known igneous activity in Colorado was the formation of the Irving Formation in the Needle Mountains about 1780 m.y. ago (Barker *et al.*, 1969; Silver and Barker, 1968). Premo (1984) has reported an age of 1790 m.y. for a volcanic unit in the Green Mountain Formation of the Sierra Madre Range, southern Wyoming; this unit is intruded by the 1780-m.y.-old Encampment granodiorite. Bickford and Boardman (1984) reported an age of 1780 m.y. for a fine-grained granite south of Powderhorn in the southwestern part of the Gunnison area (Fig. 15-2).

The thick sequence of basalt, rhyolite-to-dacite, and related mafic sedimentary rocks in the Powderhorn area (Fig. 15-2) were formed between 1770 and 1760 m.y. ago (Bickford and Boardman, 1984). These rocks were subsequently intruded by granitic plutons about 1755 m.y. ago. Somewhat to the east, in the area south of Gunnison and east to Cochetopa Canyon (Fig. 15-2), a quite similar sequence of bimodal volcanic rocks with thick interbedded turbidite was formed between 1740 and 1730 m.y. ago. These rocks were intruded by syntectonic granitic plutons 1730 to 1725 m.y. ago and by posttectonic granite about 1715 m.y. ago. The extreme deformation and metamorphism of the volcanogenic sequence is thus bracketed between 1730 and 1715 m.y. ago. The history of deformation and metamorphism of the older volcanogenic sequence to the west is not as well constrained; these rocks were likely deformed and metamorphosed during the 1730–1715-m.y.-old event, but because of their greater age they may have been deformed and metamorphosed earlier. Careful structural studies, alluded to above, will be required to sort out this history.

The volcanic assemblage in the Salida region (Fig. 15-2) accumulated at least in part during the 1740–1730-m.y. period, for that is the zircon age of a rhyolitic ashflow or debris-flow unit there (Bickford and Boardman, 1984). Only further age determinations will reveal whether rocks formed during the earlier (1770–1760 m.y.) event are present there. To the southeast, near Howard, Colorado, metarhyolite has yielded ages of 1713 and 1668 m.y. These rocks were recrystallized to upper amphibolite facies mineral assemblages and it is unclear whether the zircon data represent crystallization ages or metamorphic ages.

Farther east, in the Front Range and the Wet Mountains, the accumulation, and probably the initial deformation and metamorphism, of the thick sequences of pelite and graywacke that formed the protolith of the Idaho Springs Formation occurred before 1700 m.y. ago, for that is the age of numerous granitic plutons that intrude the earlier rocks. These include the Boulder Creek batholith (Stern *et al.*, 1971) of the northern Front Range and plutons exposed in the Arkansas River Canyon at the northern end of the West Mountains (Bickford *et al.*, 1984). Plutons of this age occur in the Needle Mountains (Bickford *et al.*, 1969; Silver and Barker, 1968), in the Gunnison area (Bickford and Boardman, 1984), and a granitic pluton of this age (1720 m.y.) is known from the basement of southwestern Kansas (Van Schmus *et al.*, 1987). Late granitic plutons whose ages are 1660–1680 m.y. were also emplaced over a broad area, occurring in the Wet Mountains, the Salida area, the Gunnison area, and across the midcontinent as far east as central Missouri (Bickford *et al.*, 1981, 1986; Van Schmus *et al.*, 1987). It seems likely that these late granites were formed inboard of a major orogenesis that occurred to the south in central and southern New Mexico and Arizona and in the subsurface across the southern midcontinent (Bickford *et al.*, 1986; Condie and Bowring, 1984; Silver, 1968; Van Schmus *et al.*, 1987). In this area, volcanism occurred between about 1680 and 1700 m.y. ago and these rocks were intruded by mostly granitic plutons between 1630 and 1650 m.y. ago. Alternatively, these late plutons may have formed during terminal collision of a collage of volcanic arc and back-arc terranes with the Wyoming and Superior cratons.

Anorogenic Assemblages

A series of mostly nonfoliated and therefore "anorogenic" plutons was intruded into preexisting crustal rocks in Colorado during the period 1450–1420 m.y. ago and about 1360 m.y. ago. These rocks are almost exclusively granitic in composition, although one small 1460-m.y.-old gabbro body, the Gabbro of Electra Lake, occurs in the Needle Mountains (Bickford *et al.*, 1969; Ghuma, 1971). This suite of rocks has been extensively discussed in the literature (e.g., Anderson, 1983; Bickford *et al.*, 1986; Emslie, 1978; Silver *et al.*, 1977; Van Schmus and Bickford, 1981; Van Schmus *et al.*, 1987), for rocks of this age and type are known as intrusives into older Proterozoic crust from Labrador, across the midcontinent region (Fig. 15-6), and into Colorado, New Mexico, Arizona, and California. Indeed, Silver *et al.* (1977) called this event the "Proterozoic Anorogenic Perforation of North America." In the eastern midcontinent region (Ohio, Indiana, Illinois, and Missouri; Fig. 15-6) plutonic rocks about 1480 m.y. old are accompanied by extensive coeval rhyolite. Originally, it was believed that this volcanic and intrusive event occurred exclusively between about 1480 and 1420 m.y. ago, but recent studies of exposed and basement rocks in Missouri, southern Kansas, Oklahoma, and the Texas Panhandle have shown that a second granite-rhyolite terrane was formed between 1340

ACCRETION OF PROTEROZOIC CRUST IN COLORADO

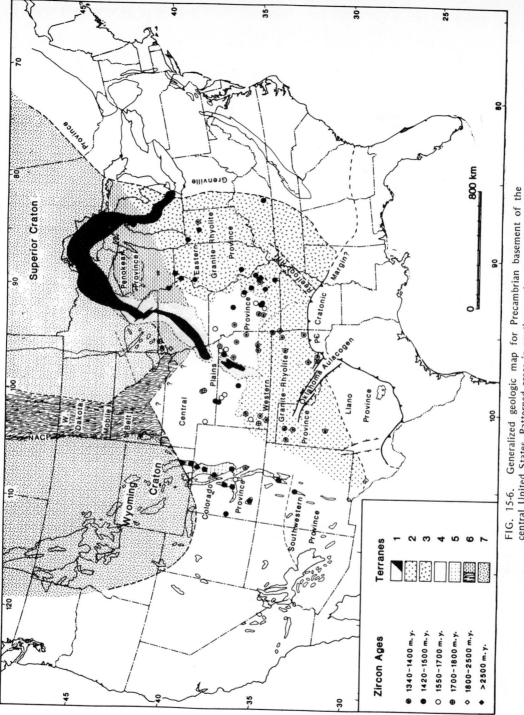

FIG. 15-6. Generalized geologic map for Precambrian basement of the central United States. Patterned areas in southern and eastern midcontinent are granite-rhyolite terranes. After Bickford *et al.* (1986).

421

and 1400 m.y. ago in these areas (Bickford and Van Schmus, 1985b; Bickford *et al.*, 1981, 1986; Thomas *et al.*, 1984; Fig. 15-6).

In Colorado, anorogenic plutons of the 1480- to 1420-m.y.-old group are best known, and include the Silver Plume, Sherman, and Log Cabin granites of the northern Front Range and southern Wyoming (Peterman and Hedge, 1968); several unnamed plutons in the southern Front Range and Wet Mountains (Bickford *et al.*, 1984); the St. Kevin granite of the northern Sawatch Range (Doe and Pearson, 1969); the Curecanti pluton of the Black Canyon of the Gunnison River (Hansen and Peterman, 1968); and plutons in the Unaweap Canyon of extreme western Colorado (Bickford and Cudzilo, 1975; Mose and Bickford, 1969). In the Needle Mountains, the Eolus granite, the Granite of Florida River, and the Gabbro of Electra Lake are all about 1460 m.y. old (Bickford *et al.*, 1969) and clearly intrude the Irving Formation, older granite plutons (about 1720 m.y. old), and the metasedimentary Uncompahgre Series. No volcanic rocks of this age are known in Colorado.

The principal representative of the 1340- to 1400-m.y.-old anorogenic rocks is the San Isabel batholith of the southern Wet Mountains, although at least one other small pluton of this age is also known from the Wet Mountains (Bickford *et al.*, 1984; Thomas *et al.*, 1984). Bickford *et al.* (1969) reported a Rb-Sr age of 1350 m.y. for the Trimble granite of the Needle Mountains. As with the older anorogenic plutons, no volcanic rocks with ages in the range 1340 to 1400 m.y. are known in Colorado, but because extensive regions of the buried basement to the east are underlain by volcanics of this age, it seems likely that volcanism probably accompanied this event in Colorado as well; presumably, the volcanic rocks were removed by erosion during late Paleozoic and Cenozoic uplifts.

It is interesting that the major pegmatite swarms in Colorado have ages of about 1450 m.y. (e.g., Quartz Creek Pegmatite District, Sawatch Range; Wetherill and Bickford, 1965; Eight Mile Pegmatite, northern Wet Mountains; M. E. Bickford and E. C. Simmons, unpublished data) but occur within country rocks whose ages are greater than 1700 m.y. It seems likely that the formation of these pegmatites was related to heating of the crust attending emplacement of the anorogenic plutons. This event probably also produced metamorphism and deformation.

The latest occurrence of an anorogenic pluton in Colorado is the 1060-m.y.-old Pikes Peak batholith (Barker *et al.*, 1975; Hedge, 1976; Hutchinson, 1960). This major intrusive occurs in the southern Front Range (Fig. 15-1), where it cross-cuts all previously formed rocks and structures. The Pikes Peak batholith is coeval with rocks of the Llano Province of central Texas (Zartman, 1964, 1965), with the formation of the midcontinent rift system (Van Schmus and Hinze, 1985), and with much of the metamorphism and igneous activity of the Grenville Province of eastern North America; however, the relationship of these widespread magmatic and tectonic events is not understood.

RESETTING OF Rb-Sr AGES

It has been known for a number of years that Rb-Sr ages, both those determined from separated minerals and those determined from whole-rock samples, commonly yield ages that are less than U-Pb ages determined from zircons from the same rocks. Bickford and Van Schmus (1985a) have recently pointed out that there is an apparent systematic relationship between Rb-Sr ages and the age of subsequent igneous activity. In numerous cases it can be shown that the Rb-Sr system is evidently reset by the next period of igneous activity, perhaps because the emplacement of new magma drives hydrothermal systems that pervasively alter preexisting crustal rocks.

It seems likely that this phenomenon has occurred in Colorado and adjacent New Mexico. A suite of rocks from the Gunnison area studied by Wetherill and Bickford

ACCRETION OF PROTEROZOIC CRUST IN COLORADO

(1965) yielded a Rb-Sr whole-rock isochron age of 1650 m.y. Although this suite included rocks from several plutons with zircon ages ranging from 1725 to 1700 m.y., four of the samples are from the Granite of Wood Gulch (Olson, 1976), which has a zircon age of 1701 ± 10 m.y. (M. E. Bickford, unpublished data); the Rb-Sr isochron age of these four samples is 1580 m.y. Thus, the Rb-Sr system in these rocks was evidently affected by events occurring between 1650 and 1580 m.y. ago. It should be recalled that orogenesis and the emplacement of plutons occurred to the south in New Mexico and Arizona between about 1630 and 1650 m.y. ago, and that plutons with zircon ages as young as 1670 m.y. occur in both the Gunnison and Salida areas. In the Grand Canyon of Arizona, the Phantom pluton yielded a Rb-Sr isochron age of 1635 m.y. (Brown *et al.*, 1979), whereas the zircon age is about 1700 m.y. (Pasteels and Silver, 1965).

Wetherill and Bickford (1965) also reported that minerals from rocks of the Gunnison area, including muscovite, biotite, K-feldspar, plagioclase, and apatite, define isochrons which include the whole-rock analysis but indicate ages ranging from 1380 to 1450 m.y. These results, and the presence of the 1450-m.y.-old pegmatites of the Quartz Creek District (Staatz and Trites, 1955), strongly suggest that this region was affected by the emplacement of at least the 1450-m.y.-old anorogenic plutons and perhaps the 1370-m.y.-old plutons as well. Similar results can be shown from New Mexico, where the Magdalena pluton yields a zircon age of 1650 m.y. (Bowring *et al.*, 1983) but has Rb-Sr whole-rock isochron ages of 1326 ± 138 and 1247 ± 63 m.y. (White, 1977), and where metarhyolite from the Zuni Mountains yielded a Rb-Sr whole-rock age of 1385 ± 40 m.y. (Brookins *et al.*, 1978), whereas its zircon age is 1650 m.y. (Bowring and Condie, 1982).

These results suggest that Rb-Sr ages most commonly reflect the most recent thermal event capable of altering the Rb-Sr system. As noted above, this is most likely magmatism that drives hydrothermal activity, removing radiogenic Sr from minerals and probably whole-rock systems as well. Thus, Rb-Sr age data most likely give information about the timing of subsequent igneous phenomena and their related tectonism rather than primary crystallization history.

CRUSTAL ACCRETION DURING THE PROTEROZOIC

Early Proterozoic: Petrologic and Geochemical Data

The geologic record described above for the early Proterozoic is clearly one of successive accretion of essentially juvenile, volcanogenic terranes on the southern margin of the Archean craton. Rocks occurring south of the Cheyenne Belt are the kind of volcanic and immature sedimentary rocks to be expected in an island arc, fore-arc basin, or back-arc basin environment, and strongly suggest that the mechanism was one of tectonic addition of these terranes to the edge of the craton. The lower Proterozoic sedimentary rocks found north of the Cheyenne Belt (Libby Creek Group; Karlstrom *et al.*, 1983) have been interpreted as a passive margin sequence, indicating that the craton edge was rifted; age constraints (Karlstrom *et al.*, 1983; Premo, 1984) indicate that rifting occurred prior to 2100 m.y. ago and that deposition of the Libby Creek Group was essentially complete by 2000 m.y. ago.

The record of accretion of volcanogenic rocks evidently spans the period between about 1790 m.y. ago (metafelsite in the Green Mountain Formation) and about 1630 m.y. ago (youngest plutons in the volcanogenic portion of the Proterozoic of New Mexico and Arizona). It seems clear that these events occurred more-or-less episodically, beginning with accretion of volcanic terranes in the Sierra Madre Mountains, the Needle Mountains, the western parts of the Gunnison area, and parts of northern Arizona between 1790 and 1760 m.y. ago. Similar rocks were formed in the eastern part of the Gunnison area and in the Salida area between 1740 and 1730 m.y. ago, and rocks of this age may occur in northern New Mexico. There is a record of volcanism about 1710 m.y. ago (and possibly as recently as 1670 m.y. ago) in the northern Sangre de Cristo Range near Howard, Colorado. Volcanism occurred broadly across southern New Mexico and southern Arizona between about 1700 and 1680 m.y. ago. Each of these volcanic terranes was intruded by granitic plutons about 10 to 20 m.y. after volcanism, but true subvolcanic (and thus coeval) plutons are rarely recognized.

Geochemically, the outstanding characteristic of these volcanic rocks is that the suite is bimodal. Basaltic compositions are common and their major and trace element chemistry suggests that they are similar to modern ocean floor or continental rift basalts (Bennett, 1984; Bickford and Boardman, 1984; Boardman and Condie, 1986; Condie and Nuter, 1981). Felsic volcanic rocks are dacitic to rhyolitic in composition; andesite, the characteristic volcanic rock of most modern volcanic arcs, is rare to absent. Both Condie and Shadel (1984) and Bickford and Boardman (1984) have addressed this problem, pointing out that bimodal volcanism does occur in some primitive modern arcs. Condie and Shadel also suggest that steeper thermal gradients in the early Proterozoic may have produced a higher proportion of basalt. A more difficult problem is the origin of the felsic volcanic rocks; Yoder (1973) suggested that fractional melting of mantle rocks could initially yield rhyolitic composition, but most workers believe that rhyolites represent melting of felsic to intermediate lower crust. As will be shown below, however, isotopic data suggest that both basaltic and felsic volcanic rocks were derived from mantle sources at essentially the time of crystallization of the magmas.

Bennett (1984) and Bickford and Boardman (1984) have pointed out that the great abundance of fragmental material, in both volcanic and sedimentary rocks, is suggestive of island arc environments (Garcia, 1978). Condie and Nuter (1981) have suggested that at least the younger volcanic sequence of the Gunnison area was formed in a back-arc environment upon somewhat older continental crust. The ancient environments in which these rocks accumulated is, therefore, still obscure, but all workers seem agreed that the observed assemblages record accretion of essentially juvenile, mantle-derived rocks to the southern margin of the Wyoming craton. Much more detailed geochemical, structural, and sedimentological studies must be done before these questions can be answered to complete satisfaction.

Early Proterozoic: Isotopic data Besides the obvious contribution of geochronological data to understanding of the petrogenesis of the early Proterozoic volcanic terranes of Colorado and the southwest, the application of isotope systematics to petrogenetic problems has produced significant constraints on the origin of these rocks; further study has even greater potential. DePaolo (1981a) and Nelson and DePaolo (1985) have obtained initial $^{87}Sr/^{86}Sr$ ratios for some rocks from the Gunnison area and have calculated Sm-Nd mantle-separation ages for numerous samples from Colorado and across the midcontinent.

The initial $^{87}Sr/^{86}Sr$ ratios were all measured on samples of basalt; mostly they are low, in the range 0.702–0.704 and indicate derivation from mantle sources as expected. Unfortunately, Nelson and DePaolo did not determine this isotopic parameter for suites of felsic volcanic rocks or for any of the granitic plutons, and it is these felsic rocks whose source is so much in question. This work is currently under way in our laboratory at the University of Kansas.

The Sm-Nd mantle separation ages determined by DePaolo (1981a) and Nelson and DePaolo (1983), however, offer an important constraint upon interpretation of the felsic volcanogenic rocks. The determined mantle separation ages are about 1800 ± 100 m.y. and, within analytical uncertainty, are the same as the crystallization ages of the rocks themselves. These data preclude derivation of these rocks from continental crust that is much older than 1800 m.y. and greatly strengthen the conclusion that these volcanic terranes are juvenile, essentially oceanic, additions to the continental crust. Stacey and Hedlund (1983) have studied Pb isotopic compositions in igneous rocks and ore minerals in southwestern New Mexico and similarly concluded that the data do not allow derivation from Archean crust.

Middle Proterozoic: The Anorogenic Plutons

As noted above, the suite of 1420–1480- and 1340–1400-m.y.-old anorogenic plutons and, in the eastern and southwestern midcontinent, the coeval granite-rhyolite terranes (Fig. 15-6), have been the subject of a fairly extensive literature. These characteristically granitic plutons seem always to be emplaced within older Proterozoic crust; nowhere are they known as intrusives into Archean rocks. Anderson (1983) has attempted to categorize the chemical characteristics of this suite, but aside from the fact that they are overwhelmingly granitic and commonly show rapakivi texture, the rocks are rather variable. Emslie (1978) noted that the anorogenic suite of granitic rocks is associated with anorthosite in eastern North America; this association also occurs in the Wolf River batholith of central Wisconsin (Anderson *et al.*, 1980; Van Schmus *et al.*, 1975) and in the Laramie Anorthosite-Sherman Granite association of south-central Wyoming and northern Colorado (Peterman and Hedge, 1968; Subbarayudu *et al.*, 1975).

Nelson and DePaolo (1985) have studied the Sm-Nd systematics of a number of granitic rocks of the anorogenic suite and find that regardless of their crystallization ages, they have mantle separation ages (i.e., the time when their $^{143}Nd/^{144}Nd$ ratios were those of the CHUR or depleted mantle curve; DePaolo, 1981b) of about 1800 ± 100 m.y. Thus, the anorogenic plutons and their volcanic equivalents in the basement of the midcontinent appear to have been derived from remelting of Proterozoic crust, presumably the volcanogenic crust accreted to the continent between 1630 and 1790 m.y. ago.

The anorogenic plutonic suite and its associated volcanic rocks are overwhelmingly granitic in composition. Mafic rocks of this age are rare, although diabase dikes intrude the 1480-m.y.-old granite and rhyolite of the Saint Francois Mountains of southeastern Missouri (Bickford and Mose, 1975) and the Gabbro of Electra Lake intrudes older Proterozoic rocks in the Needle Mountains. Rocks of intermediate composition, specifically andesite, are notably absent except for a minor occurrence in the Saint Francois Mountains and in the subsurface of southwestern Missouri and adjacent parts of Kansas and Oklahoma (Denison, 1966). Moreover, sedimentary or metamorphic rocks of this age are rare and the entire suite is characteristically undeformed; the plutons are typically nonfoliated.

These characteristics seem to rule out formation of these rocks in a continental arc, although this was suggested by Nelson and DePaolo (1985). Van Schmus and Bickford (1981) and Bickford *et al.* (1986) have suggested that these rocks are the response of crust accreted during the orogenic episodes 1630–1790 m.y. ago to at least two periods of crustal extension and consequent melting. The rarity of mafic rocks and the broad occurrence of the granitic or rhyolitic phases suggests that extension was widely distributed and probably relatively slow, so that the mafic magmas which presumably carried heat from the mantle into the newly formed volcanogenic crust reached the surface or near-surface only rarely.

ACKNOWLEDGMENTS

Many people have helped shape the ideas in this paper. I would like to mention especially K. C. Condie, S. J. Boardman, S. A. Bowring, G. H. Girty, L. T. Silver, and my close colleague W. R. Van Schmus. R. D. Shuster and W. R. Premo did many of the zircon age determinations in the Gunnison, Salida, and Wet Mountains areas. Chemical data for Gunnison area rocks were largely obtained by my students R. J. Tobin, G. R. Bennett, and Amare Astatke. Financial support was provided by NSF grants EAR80-25257 and EAR82-18463.

REFERENCES

Afifi, A. M., 1981, Precambrian geology of the Iris area, Gunnison and Saguache Counties, Colorado: Unpublished M.S. thesis, Colorado Sch. Mines, Golden, Colo., 197 p.

Anderson, J. L., 1983, Proterozoic anorogenic granite plutonism of North America: *Geol. Soc. America Mem. 161*, p. 133–154.

____, Cullers, R. L., and Van Schmus, W. R., 1980, Anorogenic metaluminous and peraluminous granite plutonism in the mid-Proterozoic of Wisconsin, USA: *Contrib. Mineralogy Petrology*, v. 74, p. 311–328.

Barker, F., 1969, Precambrian geology of the Needle Mountains, southwestern Colorado: *U.S. Geol. Survey Prof. Paper 644–A*, 35 p.

____, Peterman, Z. E., and Hildreth, R. A., 1969, A rubidium-strontium study of the Twilight Gneiss, West Needle Mountains, Colorado: *Contrib. Mineralogy Petrology*, v. 23, p. 271–283.

____, Wones, D. R., Sharp, W. N., and Desborough, G. A., 1975, The Pikes Peak batholith, Colorado Front Range, and a model for the origin of the gabbro-anorthosite-syenite-potassic granite suite: *Precambrian Res.*, v. 2, p. 97–160.

Bennett, G. S., 1984, Geochemistry of bimodal volcanic and volcaniclastic metasedimentary rocks, Cochetopa Canyon area, central Colorado: Unpublished M.S. thesis, Univ. Kansas, Lawrence, Kans., 82 p.

Bickford, M. E., and Boardman, S. J., 1984, A Proterozoic volcano-plutonic terrane, Gunnison and Salida areas, Colorado: *J. Geology*, v. 92, p. 657–666.

____, and Cudzilo, T. F., 1975, U-Pb ages of zircons from Vernal Mesa-type quartz monzonite, west-central Colorado: *Geol. Soc. America Bull.*, v. 86, p. 1432–1434.

____, and Mose, D. G., 1975, Geochronology of Precambrian rocks in the St. Francois Mountains, southeastern Missouri: *Geol. Soc. America Spec. Paper 165*, 48 p.

____, and Van Schmus, W. R., 1985a, Resetting of whole-rock and mineral Rb-Sr ages by subsequent Proterozoic orogenies: *Geol. Soc. America Abstr. with Programs*, v. 17, p. 523.

____, and Van Schmus, W. R., 1985b, Discovery of two Proterozoic granite-rhyolite terranes in the buried midcontinent basement — The case for shallow drill holes: *Proceedings of the First International Symposium on Continental Drilling*, New York, Springer-Verlag, p. 355–364.

____, Wetherill, G. W., Barker, F., and Lee-Hu, C.-N., 1969, Precambrian Rb-Sr geochronology in the Needle Mountains, southwestern Colorado: *J. Geophys. Res.*, v. 74, p. 1660–1676.

____, Harrower, K. L., Hoppe, W. J., Nelson, B. K., Nusbaum, R. L., and Thomas, J. J., 1981, Rb-Sr and U-Pb geochronology and distribution of rock types in the Precambrian basement of Missouri and Kansas: *Geol. Soc. America Bull.*, v. 92, Pt. 1, p. 323–341.

———, Cullers, R. L., and Van Schmus, W. R., 1984, U-Pb zircon chronology of early and middle Proterozoic igneous events in the Gunnison, Salida, and Wet Mountains areas, Colorado [abstract]: *Geol. Soc. America, Abstr. with Programs*, v. 16, p. 215.

———, Van Schmus, W. R., and Zietz, I., 1986, Proterozoic history of the midcontinent region of North America: *Geology*, v. 14, p. 492–496.

Boardman, S. J., 1976, Geology of the Precambrian metamorphic rocks of the Salida area, Chaffee County, Colorado: *Mountain Geologist*, v. 13, p. 89–100.

———, 1986, Early Proterozoic bimodal volcanic rocks in central Colorado, U.S.A. Pt. 1: Petrography, stratigraphy, and depositional history: *Precambrina Res.*, v. 34, p. 1–36.

———, and Condie, K. C., 1986, Early Proterozoic subaqueous bimodal volcanic rocks in central Colorado, U.S.A. Pt. II: Geochemistry, petrogenesis, and tectonic setting: *Pre-cambrian Res.*, v. 34, p. 37–68.

Boos, C. M., and Boos, M. F., 1957, Tectonics of eastern flank and foothills of Front Range, Colorado: *Amer. Assoc. Petrol. Geologists Bull.*, v. 41, p. 2603–2676.

Bowring, S. A., and Condie, K. C., 1982, U-Pb zircon ages from northern and central New Mexico: *Geol. Soc. America Abstr. with Programs*, v. 8, p. 304.

———, Kent, S. C., and Sumner, W., 1983, Geology and U-Pb geochronology of Proterozoic rocks in the vicinity of Socorro, New Mexico, *in* Chapin, C. E. and Callender, J. F., eds.: *Socorro Region II, N. Mex. Geol. Soc. Guidebook 34*, p. 137–142.

Braddock, W. A., 1970, Origin of slaty cleavage — Evidence from Precambrian rocks in Colorado: *Geol. Soc. America Bull.*, v. 81, p. 589–600.

Brock, M. R., and Singewald, Q. D., 1968, Geologic map of the Mount Tyndall quadrangle, Custer County, Colorado: *U.S. Geol. Survey Geol. Quad. Map GQ-596*.

Brookins, D. G., Della Valle, R. S., and Lee, M. J., 1978, Rb-Sr geochronologic investigation of Precambrian silicic rocks from the Zuni Mountains, New Mexico: *Mountain Geologist*, v. 15, p. 67–71.

Brown, E. H., Babcock, R. S., Clark, M. D., and Livingston, D. E., 1979, Geology of the older Precambrian rocks of the Grand Canyon — Pt. I. Petrology and structure of the Vishnu Complex: *Precambrian Res.*, v. 8, p. 219–241.

Condie, K. C., 1986, Geochemistry and tectonic setting of early Proterozoic supracrustal rocks in the southwestern United States: *J. Geology*, v. 94, p. 845–864.

———, and Bowring, S. A., 1984, Early Proterozoic supracrustal associations in the southwest — An update [abstract]: *Geol. Soc. America Abstr. with Programs*, v. 16, p. 218.

———, and Nuter, J. A., 1981, Geochemistry of the Dubois greenstone succession — An early Proterozoic bimodal volcanic association in west-central Colorado: *Precambrian Res.*, v. 15, p. 131–155.

———, and Shadel, C. A., 1984, An early Proterozoic volcanic arc succession in southeast Wyoming: *Can. J. Earth Sci.*, v. 21, p. 415–427.

Crittenden, M. D., Jr., and Peterman, Z. E., 1975, Provisional Rb/Sr age of the Precambrian Uinta Mountain Group, northeastern Utah: *Utah Geology*, v. 2, p. 75–77.

Cross, W., Howe, E., Irving, J. D., and Emmons, W. H., 1905, Needle Mountain folio: *U.S. Geol. Survey Folio 131*.

Cummings, D. O., Coffman, L. E., and Girty, G. H., 1984, A synthesis of deformational events in south-central Colorado: *Geol. Soc. America Abstr. with Programs*, v. 16, p. 219.

Denison, R. E., 1966, Basement rocks in adjoining parts of Oklahoma, Kansas, Missouri, and Arkansas: unpublished Ph.D. dissertation, Univ. Texas, Austin, Tex., 328 p.

DePaolo, D. J., 1981a, Neodymium isotopes in the Colorado Front Range and crust-mantle evolution in the Proterozic: *Nature*, v. 291, p. 193–196.

_____, 1981b, Nd isotopic studies – Some new perspectives on earth structure and evolution: (EOS) *Trans. Amer. Geophys. Un.*, v. 62, p. 137–140.

Divis, A. F., 1976, Geology and geochemistry of the Sierra Madre Range, Wyoming: *Colo. Sch. Mines Q.*, v. 71, 127 p.

Doe, B. R., and Pearson, R. C., 1969, U-Th-Pb geochronology of zircons from the St. Kevin granite northern Sawatch Range, Colorado: *Geol. Soc. America Bull.*, v. 80, p. 2495–2502.

Emslie, R. F., 1978, Anorthosite massifs, rapakivi granites, and late Proterozoic rifting of North America: *Precambrian Res.*, v. 7, p. 61–98.

Garcia, M. O., 1978, Criteria for the identification of ancient volcanic arcs: *Earth Sci. Rev.*, v. 14, p. 147–165.

Ghuma, M. A., 1971, Petrology of the gabbro of Electra Lake, Needle Mountains, southwestern Colorado: Unpublished M.S. thesis, Univ. Kansas, Lawrence, Kans., 48 p.

Ham, W. E., Denison, R. W., and Merritt, C. A., 1964, Basement rocks and structural evolution of southern Oklahoma: *Okla. Geol. Survey Bull 95*, 302 p.

Hansen, W. R., and Peterman, Z. E., 1968, Basement-rock geochronology of the Black Canyon of the Gunnison, Colorado, *in Geological Survey Research 1968:* U.S. Geol. Survey Prof. Paper 660-C, p. C80–C90.

Hedge, C. E., 1967, The age of the gneiss at the bottom of the Rocky Mountain Arsenal well: *Mountain Geologist*, v. 4, p. 115–117.

_____, 1970, Whole-rock Rb-Sr age of the Pikes Peak batholith, Colorado, *in Geological Survey Research, 1970:* U.S. Geol. Survey Prof. Paper 700-B, p. B86–B89.

_____, Peterman, Z. E., and Braddock, W. A., 1967, Age of the major Precambrian regional metamorphism in the northern Front Range, Colorado: *Geol. Soc. America Bull.*, v. 78, p. 551–558.

Hedlund, D. C., 1974, Geologic map of the Big Mesa Quadrangle, Gunnison County, Colorado: *U.S. Geol. Survey Geol. Quad. Map GQ-1153.*

_____, and Olson, J. C., 1974, Geologic map of the Iris NW Quadrangle, Gunnison and Saguache Counties, Colorado: *U.S. Geol. Survey Geol. Quad. Map GQ-1134.*

_____, and Olson, J. C., 1975, Geologic map of the Powderhorn Quadrangle, Gunnison and Saguache Counties, Colorado: *U.S. Geol. Survey Geol. Quad. Map GQ-1178.*

_____, and Olson, J. C., 1981, Precambrian geology along part of the Gunnison uplift of southwestern Colorado, *in* Epis, R. C., and Callender, J. F., eds., *Western Slope Colorado; Western Colorado and Eastern Utah:* N. Mex. Geol. Soc. Guidebook 32, p. 267–272.

Hills, F. A., and Houston, R. S., 1979, Early Proterozoic tectonics of the central Rocky Mountains, North America: *Univ. Wyo. Contrib. Geology*, v. 17, p. 89–109.

Houston, R. S., and Parker, R. B., 1963, Structural analysis of a folded quartzite, Medicine Bow Mountains, Wyoming: *Geol. Soc. America Bull.*, v. 74, p. 197–202.

_____, et al., 1968, A regional study of rocks of Precambrian age in that part of the Medicine Bow Mountains lying in southeastern Wyoming, *with a chapter on* The relationship between Precambrian and Laramide structure: *Wyo. Geo. Survey Mem. 1*, 167 p.

_____, Schuster, J. E., and Ebbett, B. E., 1975, Preliminary report on the distributions of copper and platinum group metals in mafic igneous rocks of the Sierra Madre, Wyoming: *U.S. Geol. Survey Open File Rt. 75-85*, 129 p.

_____, Karlstrom, K. E., Hills, F. A., and Smithson, S. B., 1979, The Cheyenne Belt – A major Precambrian crustal boundary in the western United States [abstract]: *Geol. Soc. America Abstr. with Programs*, v. 11, p. 446.

Hutchinson, R. M., 1960, Structure and petrology of the north end of the Pikes Peak batholith, Colorado, *in* Weimer, R. J., and Haun, J. D., eds., *Guide to the Geology of Colorado:* Rocky Mountain Assoc. Geologists, p. 170–180.

Karlstrom, K. E., Flukey, A. J., and Houston, R. S., 1983, Stratigraphy and depositional setting of the Proterozoic Snowy Pass Supergroup, southeastern Wyoming — Record of an Early Proterozoic Atlantic-type cratonic margin: *Geol. Soc. America Bull.*, v. 94, p. 1257–1274.

Lidiak, E. G., 1972, Precambrian rocks in the subsurface of Nebraska: *Nebr. Geol. Survey Bull. 26,* 41 p.

Mose, D. G., and Bickford, M. E., 1969, Precambrian geochronology in the Unaweep Canyon, west-central Colorado: *J. Geophys. Res.*, v. 74, p. 1677–1687.

Nelson, B. K., and DePaolo, D. J., 1984, 1700-Myr greenstone volcanic successions in southwestern North America and isotopic evolution of the Proterozoic mantle: *Nature*, v. 312, p. 143–146.

_____ , and DePaolo, D. J., 1985, Rapid production of continental crust 1.7-1.9 b.y. ago — Nd isotopic evidence from the basement of the North American continent: *Geol. Soc. America Bull.*, v. 96, p. 746–754.

Olson, J. C., 1974, Geologic map of the Rudolph Hill Quadrangle, Gunnison, Hinsdale, and Saguache Counties, Colorado: *U.S. Geol. Survey Geol. Quad. Map GQ-1177.*

_____ , 1976, Geologic map of the Iris Quadrangle, Gunnison and Saguache Counties, Colorado: *U.S. Geol. Survey Geol. Quad. Map GQ-1286.*

Pasteels, P., and Silver, L. T., 1965, Geochronological investigations in the crystalline rocks of the Grand Canyon, Arizona [abstract]: *Geol. Soc. America Ann. Meet. Program*, p. 122.

Peterman, Z. E., and Hedge, C. E., 1968, Chronology of Precambrian events in the Front Range, Colorado: *Can. J. Earth Sci.*, v. 5, p. 749–756.

_____ , Hedge, C. E., and Braddock, W. A., 1968, Age of Precambrian events in the northeastern Front Range, Colorado: *J. Geophys. Res.*, v. 73, p. 2277–2296.

Premo, W. R., 1984, U-Pb zircon geochronology of Early Proterozoic plutonism in N. Colorado and SE. Wyoming: *Geol. Soc. America Abstr. with Programs*, v. 16, p. 251.

Silver, L. T., 1968, Precambrian batholiths of Arizona [abstract]: *Geol. Soc. America Abstr. for 1968, Spec. Paper 121*, p. 558–559.

_____ , and Barker, F., 1968, Geochronology of Precambrian rocks of the Needle Mountains, southwestern Colorado — Pt. I. U-Pb results [abstract]: *Geol. Soc. America Abstr. for 1967, Spec. Paper 115*, p. 204–205.

_____ , Bickford, M. E., Van Schmus, W. R., Anderson, J. L., Anderson, T. H., and Medaris, L. G., 1977, The 1.4-1.5 b.y. transcontinental anorogenic perforation of North America: *Geol. Soc. America Abstr. with Programs*, v. 9, p. 1176–1177.

Staatz, M. M., and Trites, A. F., 1955, Geology of the Quartz Creek pegmatite district, Gunnison County, Colorado: *U.S. Geol. Survey Prof. Paper 265.*

Stacey, J. S., and Hedlund, D. C., 1983, Lead-isotopic compositions of diverse igneous rocks and ore deposits from southwestern New Mexico and their implications for early Proterozoic crustal evolution in the western United States: *Geol. Soc. America Bull.*, v. 94, p. 43–57.

Stern, T. W., Phair, G., and Newell, M. F., 1971, Boulder Creek batholith, Colorado — Pt. 2. Isotopic age of emplacement and morphology of zircon: *Geol. Soc. America Bull.*, v. 82, p. 1615–1634.

Subbarayudu, G. V., Hills, A. F., and Zartman, R. E., 1975, Age and Sr isotopic evidence for the origin of the Laramie anorthosite and syenite complex, Laramie Range, Wyoming: *Geol. Soc. America Abstr. with Programs*, v. 7, p. 1287.

Taylor, R. B., 1975, Geologic map of the Blackhawk Quadrangle, Colorado: *U.S. Geol. Survey Geol. Quad. Map GQ-1248.*

Thomas, J. J., Shuster, R. D., and Bickford, M. E., 1984, A terrane of 1350-1400 m.y. old silicic volcanic and plutonic rocks in the buried Proterozoic of the midcontinent and in the Wet Mountains, Colorado: *Geol. Soc. America Bull.*, v. 95, p. 1150–1157.

Tobin, R. J., 1982, Petrology and geochronology of a Proterozoic annular complex near Gunnison, Colorado: Unpublished M.S. thesis, Univ. Kansas, Lawrence, Kans., 62 p.

Tweto, Ogden, 1980, Precambrian geology of Colorado, *in Colorado Geology:* Rocky Mountain Assoc. Geologists, p. 37–46.

_____, 1983, Las Animas Formation (Upper Precambrian) in the subsurface of southeastern Colorado — U.S. Geological Survey contributions to stratigraphy, *Geol. Survey Bull. 1529-G*, 14 p.

Vance, R. K., 1984, Geology and geochemistry of the Gunnison intrusive complex, Gunnison County, Colorado: Unpublished M.S. thesis, Univ. Kentucky, Lexington, Ky., 261 p.

Van Schmus, W. R., and Bickford, M. E., 1981, Proterozoic chronology and evolution of the midcontinent region, North America, *in* Kroner, A., ed., *Precambrian Plate Tectonics:* Amsterdam, Elsevier, p. 261–296.

_____, and Hinze, W. J., 1985, The Midcontinent Rift System: *Ann. Rev. Earth Planet. Sci.*, v. 13, p. 345–383.

_____, Medaris, L. G., and Banks, P. O., 1975, Geology and age of the Wolf River batholith, Wisconsin: *Geol. Soc. America Bull.*, v. 86, p. 907–914.

_____, Bickford, M. E., and Zietz, I., 1987, Early and middle Proterozoic provinces in the central United States: *Amer. Geophys. Un. Geodynamics Series*, v. 17, p. 43–68.

Wetherill, G. W., and Bickford, M. E., 1965, Primary and metamorphic Rb-Sr chronology in central Colorado: *J. Geophys. Res.*, v. 71, p. 4669–4689.

White, D. L., 1977, A Rb-Sr isotopic study of the Precambrian intrusives of south-central New Mexico: Unpublished Ph.D. dissertation, Miami Univ., Oxford, Ohio, 88 p.

Winchester, J. A., and Floyd, P. A., 1977, Geochemical discrimination of different magma series and their differentiation products using immobile elements: *Chem. Geology*, v. 20, p. 325–343.

Yoder, H. S., Jr., 1973, Contemporaneous basaltic and rhyolitic magmas: *Amer. Mineralogist*, v. 58, p. 153–171.

Zartman, R. E., 1964, A geochronologic study of the Lone Grove pluton from the Llano Uplift, Texas: *J. Petrology*, v. 5, p. 359–408.

_____, 1965, Rubidium-strontium age of some metamorphic rocks from the Llano Uplift, Texas: *J. Petrology*, v. 6, p. 28–36.

Bruce Bryant
U.S. Geological Survey
Denver Federal Center
Denver, Colorado 80225

16

EVOLUTION AND EARLY PROTEROZIC HISTORY OF THE MARGIN OF THE ARCHEAN CONTINENT IN UTAH

ABSTRACT

The southern margin of the Archean continent trends east-west across northern Utah. Only a few areas of Precambrian crystalline rocks are available for study in the region, so data to constrain interpretations are sparse. In the Wasatch Mountains, the Farmington Canyon Complex represents crystalline rocks at the margin of the Archean continent; mapping, petrographic, chemical, and isotopic data from that unit suggest the following history. Detritus containing materials as old as 3600 m.y. was shed from the Archean craton and deposited on oceanic crust, and later, intruded by tholeiitic dikes and sills. About 2600 m.y.b.p. the sediments and intrusives were metamorphosed, perhaps to granulite facies; Sr isotopes were only partly reequilibrated. Mafic intrusives possibly were emplaced after this metamorphic event. To the west of the Wasatch Mountains, evidence suggests that new crust may have formed subsequently along the margin of the craton at 2000–2300 m.y.b.p. About 1800 m.y. ago, the Farmington Canyon Complex underwent migmatization and low-temperature melting, and was metamorphosed to amphibolite facies during a metamorphic event that was widespread along the southern margin of the craton. New crust was formed over a large area to the south during the Early Proterozoic.

INTRODUCTION

Metamorphic and plutonic basement rocks crop out in only a few areas in Utah. The largest lies in the northwestern part of the state and extends into Idaho. It includes outcrops in the Raft River Mountains, Grouse Creek Mountains, and Albion Range. The second largest area is in the Wasatch Mountains north of Bountiful, and the third is on Antelope Island in Great Salt Lake. Crystalline rocks are exposed in small areas in the Wasatch Mountains southeast of Salt Lake City, in the southern Wasatch Mountains east of Santaquin, in Cottonwood Canyon just east of the Wasatch Mountains, in the Mineral Mountains in the southwest part of the state, and on the north flank of the Uinta Mountains in the northeast corner (Fig. 16-1). These limited exposures preclude definitive knowledge of the origin and history of the crust of this region. However, a synthesis of recent work suggests some hypotheses concerning the evolution of the crust in the region during Precambrian time. (For syntheses of Precambrian petrotectonics in northern New

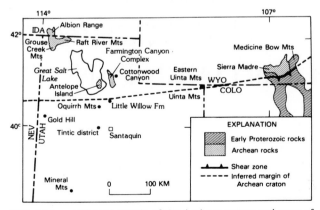

FIG. 16-1. Inferred margin of the Archean craton and areas of exposed Precambrian crystalline rock in Utah and adjacent states. Margin of craton in Sierra Madre and Medicine Bow Mountains from Hills and Houston (1979) and in Utah from Bryant (1985).

Mexico and the northern Wyoming province, see the contributions to this volume by Grambling, Chapter 17, Mogk and Henry, Chapter 13, and Wooden *et al.*, Chapter 14.)

THE MARGIN OF THE ARCHEAN CONTINENT

Isotopic studies in recent decades show that the margin of the Archean craton trends westerly across northern Utah (Fig. 16-1). Rocks of Archean age have been documented in the Raft River and Grouse Creek Mountains (Compton *et al.*, 1977), the Albion Range (Armstrong, 1976; Small and Slauson, 1968), the northern Wasatch Mountains (Hedge *et al.*, 1983), and the eastern Uinta Mountains (Sears *et al.*, 1982). The principal evidence for Proterozoic crust just south of these exposures of Archean rock is indirect. Studies of lead isotopes in galena and feldspar show that Proterozoic crust underlies Gold Hill, the Oquirrh Mountains, and areas to the south (Stacey and Zartman, 1978; Stacey *et al.*, 1968); and studies of neodynium isotopes in Mesozoic and Tertiary granite show that Proterozoic crust extends into eastern Nevada (Farmer and DePaolo, 1983, 1984). Published isotopic studies of areas of exposed Proterozoic crust in Utah are lacking except in the Mineral Mountains well to the south of the Archean cratonic margin (Aleinikoff *et al.*, 1986).

THE FARMINGTON CANYON COMPLEX

Crystalline rocks exposed in the northern Wasatch Mountains were named the Farmington Canyon Complex by Eardley and Hatch (1940) on the basis of reconnaissance studies. Detailed study of the area between Farmington and Weber Canyons (Bell, 1952) and reconnaissance study of the entire Farmington Canyon Complex south of the Ogden River have provided additional data on the unit[1] (Bryant, 1984).

Here I separate a unit of metamorphosed plutonic rock of Early Proterozoic age from the main body of layered metamorphic rocks of Archean age, which I call the Farmington Canyon Complex (Fig. 16-2). The younger plutonic rock occurs north of Weber Canyon and consists of fine- to medium-grained, biotite-hornblende quartz monzonite gneiss containing lenticular inclusions of amphibolite and a few inclusions of quartzite. It has generally sharp contacts with country rock on the south, but two smaller outlying bodies of granitic gneiss (Bryant, 1984) have partly gradational contacts and contain some layered and partly migmatitic gneiss. South of the pluton between Weber and Farmington Canyons, migmatitic rocks are dominant in the Farmington Canyon Complex. They consist of interlayered and intergradational biotite-hornblende quartz monzonite gneiss, biotite-garnet-quartz-feldspar gneiss and schist, garnet-quartz-feldspar gneiss, garnet-biotite schist, and sillimanite-garnet-biotite schist. Southward, migmatite is progressively less abundant, and the dominant rocks are biotite-feldspar-quartz gneiss, garnet-biotite-quartz-feldspar gneiss, sillimanite-biotite schist, and quartzite. In one area east of Bountiful, the schist and gneiss contain numerous thick layers of quartzite. Pegmatite is abundant throughout, and where concordant with foliation, commonly has indistinct contacts, but sharp contacts where discordant. The discordant pegmatite bodies are as much as several meters thick and 100 long, and are more abundant south of the migmatite zone. Internal stratigraphy and structure of the layered metamorphic rocks are poorly known because no facing criteria or marker horizons have been identified. Amphibolite, metadiabase, and metagabbro typically form lenses 1–3 m thick and a few meters

[1] Also B. Bryant, unpublished data.

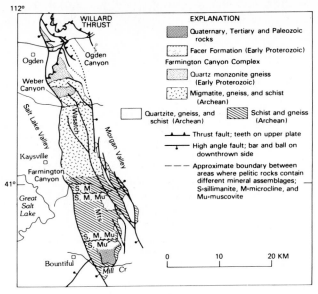

FIG. 16-2. Generalized geologic map of the Farmington Canyon Complex showing major rock units and areas of metamorphic mineral assemblages in pelitic rocks.

long throughout the complex, although a few bodies of amphibolite are as much as 10 m thick and 500 m long.

PROTOLITHS OF THE FARMINGTON CANYON COMPLEX

Layered Gneiss

Modal and chemical compositions of the layered metamorphic rocks suggest that they were derived predominantly from a metasedimentary sequence, as recognized by Eardley and Hatch (1940). Layers of sillimanite schist and mica schist have been derived from aluminous shales, quartzite from quartz sandstone, and feldspar-quartz gneiss from feldspathic sandstone. Comparison of modes of the feldspar-quartz gneiss from the schist and gneiss unit (Fig. 16-3) with mean detrital modes of modern sea-floor sands from different tectonic settings (Dickinson and Valloni, 1980, Table 2, left triangle of Fig. 3) suggests that these gneisses could have been derived from sandstone from an ancient craton and orogenic belt, and deposited along a rifted continental margin. The most important conclusion is that sedimentary protoliths of the Farmington Canyon Complex were derived from preexisting continental crust.

Many layers in the migmatite unit resemble those in the schist and gneiss unit, but others are more feldspathic and richer in microcline (Fig. 16-4). Some of the layers of quartz-feldspar gneiss in the migmatite unit resemble the rock in the pluton of Proterozoic age to the north, but samples of gneiss analyzed from the migmatite unit have a lower alkali content than samples from that pluton (Fig. 16-5). The quartz-feldspar gneiss layers probably originally contained more detrital feldspar than most of the gneiss layers in the nonmigmatitic unit to the south. Concentrations of mafic minerals that could be interpreted as restite occur only as thin partings, lenses, and clots. Textural variations and indistinct contacts may be due to local metamorphic differentiation during metamorphic

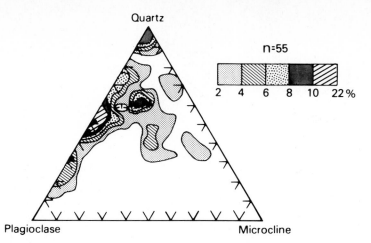

FIG. 16-3. Proportions of quartz, plagioclase, and microcline in samples of quartz feldspar gneiss from the schist and gneiss unit that do not show extensive effects of retrogressive metamorphism. Dots represent analyzed samples shown on Fig. 16-5.

recrystallization, but the feldspathic gneiss layers are probably not the result of large-scale metamorphic differentiation. Those layers could represent immature sediments derived from a craton, or could have a component of felsic volcanic material in their protolith, or both.

Amphibolites

Chemical analyses of amphibolites indicate that they were derived from rocks of basaltic composition. SiO_2 ranges from 47 to 52%. Alkali content is variable, and inclusions in the quartz monzonite gneiss and from migmatitic Farmington Canyon Complex have higher alkali contents than samples from nonmigmatitic Farmington Canyon Complex. Samples

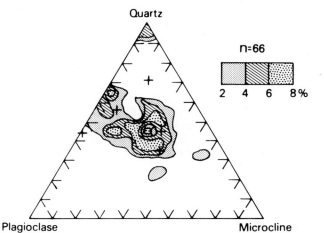

FIG. 16-4. Proportions of quartz, plagioclase, and microcline in samples of quartz feldspar gneiss from the migmatite unit that do not show extensive effects of retrogressive metamorphism. Crosses represent analyzed samples shown in Fig. 16-5.

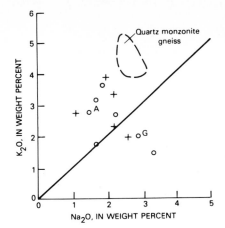

FIG. 16-5. K₂O/Na₂O diagram for samples of quartz feldspar gneiss from schist and gneiss unit (circles) and from migmatite unit (crosses). Line represents a ratio of 1.0. A, Mean composition of arkose; G, mean composition of graywacke (Pettijohn, 1963).

from the migmatitic Farmington Canyon Complex and from inclusions in the Proterozoic quartz monzonite gneiss have olivine in their norms, whereas samples from the nonmigmatitic Farmington Canyon Complex have normative quartz. The TiO_2 content of the amphibolites is less than that of most basalts, but is within the range given for tholeiites (Winchester and Floyd, 1977) and close to the average of Nockolds' (1954) "central basalt." Zirconium and cerium contents are low, but in the range of values obtained from tholeiites (Winchester and Floyd, 1977).

On a TiO_2-K_2O-P_2O_5 diagram (Fig. 16-6A), amphibolites from the Farmington Canyon Complex fall into two groups. Analyses from the migmatitic Farmington Canyon Complex and inclusions in the quartz monzonite gneiss fall well within the field designated as continental basalt, whereas those from nonmigmatitic Farmington Canyon Complex lie in the field of oceanic basalt. T. H. Pearce *et al.* (1975) suggest that metamorphism will tend to enrich basalts in K_2O, and thus tend to move them toward the field of continental basalt. In the Farmington Canyon Complex, that is indeed what may have happened, for samples from the migmatitic Farmington Canyon Complex and inclusions in the quartz monzonite contain more K_2O than samples from nonmigmatitic Farmington Canyon Complex. In addition to being from areas of migmatization and melting, the rocks containing more K_2O have more secondary actinolite and chlorite, due to later retrogressive alteration, than samples from the south. This alteration is associated with later Proterozoic or Phanerozoic events not discussed here. The hypothesis that the samples of amphibolite from the quartz monzonite gneiss and migmatitic Farmington Canyon Complex differ from those in nonmigmatitic Farmington Canyon Complex due to metamorphic effects, is supported by their tight grouping on a FeO^*-MgO-Al_2O_3 diagram (Fig. 16-6B; T. H. Pearce *et al.*, 1977), even though some of the analyses do not pass the criteria used in that study (51-56% SiO_2 content and a tholeiitic designation based on the classification of Macdonald and Katsura, 1964). FeO, MgO, and Al_2O_3 are immobile relative to K_2O during metamorphism in the amphibolite facies, although one or more of these components could be mobile in greenschist-facies metamorphism during which numerous epidote segregations may form in mafic igneous rocks. The analyses plot in the field of oceanic basalt, but whether the detailed division of tectonic environments given, based on Cenozoic igneous rocks, should be applied to Archean rocks is uncertain (T. H. Pearce *et al.*, 1977).

The analyses of the amphibolites from the nonmigmatitic Farmington Canyon Complex fall in the ocean-floor basalt field in the Ti-Zr diagram (Fig. 16-6C). That field is not particularly diagnostic, inasmuch as it includes all three possible rock categories. However, three of the four samples that fall in the field of calc-alkali basalts are from the

EVOLUTION AND EARLY PROTEROZOIC HISTORY

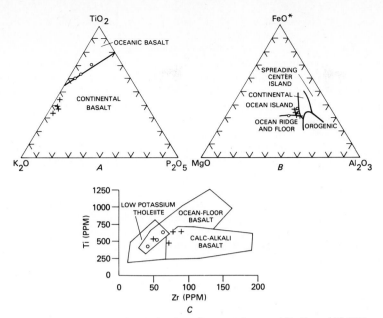

FIG. 16-6. Chemical discrimination diagrams for amphibolites. (A) TiO₂-
K₂O-P₂O₅ diagram (after T. H. Pearce et al., 1975). (B) FeO*-MgO-Al₂O₃ diagram (after T. H. Pearce et al., 1977; FeO* = total iron oxide calculated as
FeO. (C) Ti-Zr diagram (after J. A. Pearce and Cann, 1973). Circles, samples
from schist and gneiss unit; crosses, samples from migmatite unit and inclusions
in the quartz monzonite gneiss.

quartz monzonite gneiss and migmatite, so that the difference between the amphibolite
from regions of migmatization and melting, and from those from areas lacking evidence
for action of these processes, is apparent also in that diagram.

The compositions of the amphibolites indicate that they were derived from igneous
rocks of basaltic composition. Relict textures in a few bodies show that they were originally gabbro or diabase. The lenticular shape and small size of many of the amphibolites
suggest that they may be tectonically disrupted sills or dikes emplaced before deformation
and metamorphism. Amphibolites in the quartz monzonite gneiss are interpreted as
undigested relics from country rock that was melted to form the quartz monzonite. As in
many parts of the Archean crust, the mafic intrusives may be of several ages, but cross-
cutting relations between amphibolites were not found. Based on minor element contents,
basaltic protoliths of the samples analyzed may have erupted through oceanic crust
underlying the sedimentary sequence forming the protoliths of the layered gneisses.

AGE OF THE ROCKS AND METAMORPHIC EVENTS

Rb-Sr and Sm-Nd studies of rocks of the Farmington Canyon Complex indicate that the
layered metamorphic rocks are Archean (Hedge et al., 1983). Whole-rock Rb-Sr data, on a
total of 11 samples from migmatite in Weber Canyon and from gneiss and amphibolite in
the gneiss and schist unit south of Bountiful Peak, do not define a simple line on an
isochron diagram. Instead, they define a triangular area with the lower boundary much
sharper than the upper boundary (Hedge et al., 1983, Fig. 3). The lower boundary
corresponds to an age of 2600 m.y. and the upper boundary to an age of 3600 m.y. The

favored explanation for this (Hedge *et al.*, 1983) is that regional metamorphism equilibrated the Rb-Sr ratios in most, but not all, of the rocks 2600 m.y. ago, and that the older apparent ages are inherited from the premetamorphic rocks.

Sm-Nd data from migmatites in Weber Canyon give model ages ranging from 2740 m.y. for amphibolite to 3430 m.y. for a gneiss layer. The age of the amphibolite may be close to its true age because the amphibolite was originally intruded into a pile of stratified rocks as a basalt dike derived from the mantle. The older Sm-Nd age could be inherited from detritus derived from older crust and deposited as sediments.

The metamorphic event producing the main mineral assemblage presently in the rocks occurred during or after the emplacement of the quartz monzonite gneiss in the Early Proterozoic, because foliation and lineation trends in the main body of the Farmington Canyon Complex and the quartz monzonite gneiss of the complex are similar. U-Pb data from zircons from the Farmington Canyon Complex yield Pb^{207}/Pb^{206} ages ranging from 1770 to 2271 m.y., indicating that they were severely reset during the Early Proterozoic. The fact that the Sr. isotopic ratios in the Farmington Canyon Complex were not severely disturbed by the event supports the conclusion that a high-temperature mineral assemblage relatively resistant to later resetting was created during a Late Archean metamorphic event (Hedge *et al.*, 1983).

METAMORPHIC HISTORY

The main metamorphic assemblages of the rocks of the Farmington Canyon Complex in the Wasatch Mountains are of amphibolite grade. Layers of sillimanite-garnet-biotite schist are found throughout the major rock units. Some of the schist layers contain microcline, but they generally lack muscovite in the northern part of the area (Fig. 16-2). In the central part of the area, rocks containing sillimanite and microcline have small quantities of muscovite that appears to have been in equilibrium with the sillimanite and microcline. To the south, contemporaneous muscovite and sillimanite occur together, and microcline is absent in the sillimanite-bearing rocks (Fig. 16-2). A similar zone, in which pelitic rocks contain muscovite, sillimanite, and microcline in equilibrium, has been described by Evans and Guidotti (1966) and Lundgren (1966). Evans and Guidotti point out that this ideally univariant assemblage is actually divariant and can be explained by assuming that during dehydration, P_{H_2O} slowly increased from initial values less than lithostatic, under conditions of low permeability, and was buffered by the assemblage muscovite + sillimanite + potassic-feldspar + plagioclase + quartz, controlled by local values of P and T. Subsequent mineralogical studies by Cheney and Guidotti (1979) support this conclusion. Application of this explanation to the wide zone of muscovite + sillimanite + potassic-feldspar facies rocks in the Farmington Canyon Complex is uncertain because of probable major metamorphism of these rocks during the Archean. Consequently, the Early Proterozoic metamorphism probably did not involve dehydration and could even have involved hydration. The isograd of Evans and Guidotti (1966) would correspond to the boundary between rocks to the south containing sillimanite and muscovite but lacking microcline, and those to the north containing all three minerals, whereas the isograd of Lundgren (1966) would correspond to the boundary at Farmington Canyon between rocks to the south containing sillimanite, microcline, and muscovite in apparent equilibrium and those to the north in which muscovite is lacking in the main assemblage.

The rocks north of Farmington Canyon are migmatitic like those in the sillimanite-microcline zone of Connecticut (Lundgren, 1966). South of Farmington Canyon, the rocks just barely reached the temperature of about 650°C necessary for the beginning of anatexis at pressure of 3 kb or more, assuming that $P_{H_2O} \approx P$ total (Winkler, 1979, p. 247), producing liquids rich in feldspar and quartz components; the latter segregated, or were injected into, planes of weakness in the schist and gneiss where they crystallized to form pegmatites. To the north, a somewhat larger proportion of the rock was melted to produce numerous pegmatites in the migmatitic gneiss and schist. Experiments show that only a very small increase in temperature can lead to a great increase in the proportion of rock melted (Winkler, 1979, p. 310). Also, diffusion rates increase exponentially with temperature, so that diffusion may be an important mechanism to form segregation pegmatite at high temperature (Lindh and Wahlgren, 1985). The formation of pegmatite evidently took place over a considerable interval of time during the main metamorphism, for concordant pegmatites are locally folded and cut by discordant ones.

Bell (1952) stated that the highest-grade rocks in the Farmington Canyon Complex contain hypersthene, and thus are in the granulite facies, but he gave no details on the occurrence. I found little definitive evidence for granulite-facies rocks in the Farmington Canyon Complex. A few rocks do contain orthorhombic pyroxene as relict grains, not in equilibrium with the main amphibolite facies mineral assemblage. Although the complex contains layers of rocks rich in garnet and lacking in biotite, biotite is very widespread. Pyroxene, even monoclinic pyroxene, is rare in the rocks of the Farmington Canyon Complex. Relict hypersthene was found in one sample of migmatitic gneiss adjacent to the Early Proterozoic quartz monzonite gneiss, in an inclusion of garnet-plagioclase-quartz rock, and in an inclusion of metamorphosed ultramafic rock in the quartz monzonite gneiss. These few relict grains are the only mineralogical evidence for an earlier metamorphic event, which isotopic studies suggest occurred about 2600 m.y. b.p. That metamorphic event may have taken place under granulite-facies conditions.

AGE AND ORIGIN OF THE QUARTZ MONZONITE GNEISS

Normative compositions of analyzed samples of quartz monzonite gneiss, plotted on a projection of isotherms on cotectic surfaces in the Qz-Ab-Or-An tetrahedron (Fig. 16-7), show that the quartz monzonite gneiss plots near the cotectic lines in both diagrams and thus near the intersection of the cotectic surfaces, indicating that the rocks were derived from a relatively low-temperature melt.

Whole-rock Rb-Sr determinations on the quartz monzonite gneiss suggest an age of 1808 ± 34 m.y. and U-Pb determinations on zircon indicate an age of 1780 ± 20 m.y.; the ages were combined to assign an age of 1790 ± 20 m.y. to the unit (Hedge *et al.*, 1983). Well-developed metamorphic foliation and texture, and generally concordant and locally gradational contacts indicate that the quartz monzonite was emplaced during deformation that produced the amphibolite facies mineral assemblage in the Farmington Canyon Complex. The initial Sr^{87}/Sr^{86} ratio of 0.769 for the quartz monzonite gneiss is very high, indicating that it probably formed by partial melting of crustal rocks similar to the surrounding migmatite, gneiss, and schist. Feldspars from the gneiss have lead isotopic values indicative of derivation from Archean crustal material (J. S. Stacey, oral communication, 1985).

Figure 16-8 is a strontium-evolution diagram showing how the high initial ratio of the quartz monzonite gneiss might have originated in Early Proterozoic time from Archean crustal material similar to that now exposed. The $^{87}Sr/^{86}Sr$ ratios of layered metamorphic rock were calculated for 1800 and 2600 m.y. based on the measured Rb/Sr ratios. The samples from the quartz monzonite gneiss all had ratios 1800 m.y. ago close to the calculated 0.769 initial ratio of the rock. That initial ratio falls in the upper range of the values at 1800 m.y. calculated for samples of the country rock. Those values range from

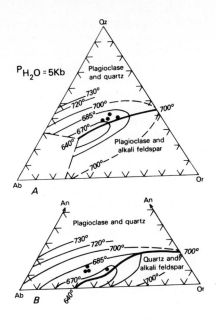

FIG. 16-7. Normative composition of quartz monzonite gneiss (dots) compared to experimental data for the system quartz (Qz)–albite (Ab)–orthoclase (Or)–anorthite (An) at 5 kb ($P_{H_2O} \approx P_{total}$; Winkler, 1979). Isotherms show configuration of liquidus surfaces; solid line shows cotectic composition between 700 and 640°C. (A) Data projected on Qz-Ab-Or; (B) data projected on Ab-Or-An.

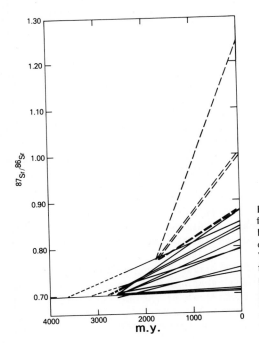

FIG. 16-8   Strontium-evolution diagram for samples from the main body of the Farmington Canyon Complex and from the quartz monzonite gneiss of the complex. Values calculated for 1800 and 2600 m.y.b.p. for main body (solid lines) and for quartz monzonite gneiss (dashed lines). Dotted lines extend Farmington Canyon Complex lines to mantle values. Heavy line shows composition of average chondritic mantle through time (Peterman, 1979).

0.704 to 0.776. Recalculation of the ratios in the country rock samples at 2600 m.y. still shows scatter, reflecting the fact that those samples do not form an isochron and that crustal materials older than 2600 m.y. were incorporated in the metamorphic rocks.

Ages of minerals from the Farmington Canyon Complex reflect cooling after the metamorphism and emplacement of the quartz monzonite gneiss 1790 m.y. ago and resetting to varying degrees by subsequent events. K-Ar ages of hornblende range from 1364 to 1700 m.y. b.p. (Hashad, 1964). A Rb-Sr age of 1580 m.y. was obtained on muscovite from a pegmatite in rock above the Ogden thrust (Giletti and Gast, 1961). K-Ar ages of biotite are younger, due to various degrees of resetting during Phanerozoic thermal events.

SUMMARY OF CRUSTAL HISTORY BASED ON STUDY OF THE FARMINGTON CANYON COMPLEX

Before 2600 m.y.b.p., continental crustal rocks possibly as old as 3600 m.y. and probably with a complex history, existed in the region and shed detritus onto adjacent oceanic crust, as inferred by the minor element ratios in the amphibolites. The trend and location of the margin of this older continent in relation to the Farmington Canyon Complex and to the margin of the Archean continent which was established at least 1800 m.y. ago is unknown. Possibly volcanos near the margin of the continent erupted and added a component of tuff to the sediments to produce the high feldspar content of some layers. The sedimentary prism was intruded by numerous small dikes and sills of diabase and gabbro. About 2600 m.y.b.p., the sediments were metamorphosed, and the mafic intrusives disrupted. Metamorphism only partly equilibrated the Sr-isotopic composition of sediments and intrusives. Granulite-facies mineral assemblages may have been produced during the metamorphism. Between 2600 and 1800 m.y., we lack records of events. Possibly more mafic intrusives were emplaced, and by 1800 m.y. the present east-west trend of the Archean craton was established, perhaps by rifting of the Archean continent at some time during that long interval between 2600 and 1800 m.y.b.p. An amphibolite-facies metamorphism led to production of anatectic magma from relatively dehydrated crustal material that already had a high-temperature mineral assemblage, and perhaps to hydration of much of the country rock at present levels of exposure. Inasmuch as no isotopic evidence of Archean continental crust to the south has been found (Farmer and DePaolo, 1984; Stacey *et al.*, 1968), the Early Proterozoic continental crust therefore must have been produced by processes acting on oceanic crust. These processes will probably have to be inferred from studies of Early Proterozoic crust where it is more extensively exposed in Colorado.

OTHER AREAS

Other evidence for Early Proterozoic metamorphism in the region is found in the Facer Formation, a sequence of quartzite, muscovite, and chlorite schist, thin sills of amphibolite, and a few beds of limestone from just above the Willard thrust, north of Ogden Canyon (Fig. 16-1) (Sorensen and Crittenden, 1976, Fig. 1). Muscovite from the schist has an Rb-Sr age of 1660 ± 50 m.y. (Crittenden and Sorensen, 1980). Hornblende from a diorite cutting the Facer Formation has a K-Ar age of 1681 ± 12 m.y. (Crittenden and Sorensen, 1980). These rocks of lower metamorphic grade have been thrust tens of kilometers to the east onto the Farmington Canyon Complex.

On Antelope Island, about 25 km west of the Wasatch Mountains, exposed basement rocks may have been transported eastward in relation to those in the Wasatch Mountains along part of the thrust system of the Sevier orogenic belt. Structural trends are northward on Antelope Island (Larsen 1957), compared to west or northwest trends in the Farmington Canyon Complex (Bryant, 1984).

West of the Wasatch Mountains, fewer details are available concerning the history of the crystalline basement rocks. On Antelope Island, migmatitic schist, gneiss, and granite gneiss contain concordant lenses and discordant dikes of amphibolite (Bryant and Graff, 1980; Larsen, 1957). A pluton of well-foliated granite, containing monoclinic pyroxene as the principal mafic mineral includes zircon with a U-Pb age of about 2020 m.y. (Hedge *et al.*, 1983). No other isotopic data are available for comparison with that in the Wasatch Mountains. However, a 2000-m.y.-old metamorphic event has been postulated to explain Sr isotopic data in the Albion Range (Armstrong, 1976); moreover, Nd isotopic data suggest that some crust may have formed along the edge of the Archean craton west of the Wasatch Mountains 2100 m.y.b.p. (Farmer and DePaolo, 1983). These ages may represent an extension of a 2000–2300-m.y.-old crustal province recognized in southern California and Nevada by Nd isotopic studies (Farmer and DePaolo, 1984). Pb isotopic data from Gold Hill, Tintic, and the Oquirrh Mountains were interpreted as indicating mixing of Archean and 1650-m.y.-old crustal material (Stacey and Zartman, 1978; Stacey *et al.*, 1968), but the data also may reflect an earlier 2000–2300-m.y. crust-forming event. Such an event might have affected the apparent age of the zircons in the Farmington Canyon Complex on which the oldest Pb^{207}/Pb^{206} age is 2271 m.y.

Galena from the Park City area of Utah about 20 km southeast of the Farmington Canyon Complex has lead-isotope ratios that indicate a basement age of 2345 ± 50 m.y. (Stacey *et al.*, 1968). Perhaps this indicates another metamorphic event younger than the one postulated at 2600 m.y. along the margin of the Archean craton, or it represents a mixture of the 2600- and 1800-m.y.-old crustal components.

The best exposures of the margin of the Archean craton are in the Medicine Bow Mountains and Sierra Madre in south-central Wyoming, where the boundary is along a shear zone. The following data are summarized from Hills and Houston (1979) and Karlstrom and Houston (1984). North of the boundary in Wyoming, an ensialic sedimentary and volcanic sequence 10 km thick of Early Proterozoic age overlies Archean granitic rock, migmatite, and gneiss, which have Rb-Sr ages as great as 2960 m.y. This sequence was deposited during and after rifting of the Archean continent about 2300 m.y. b.p. South of the boundary, an ensimatic sequence of Early Proterozoic age (see Fig. 16-1) of unknown thickness occurs. The northern sequence is metamorphosed to upper greenschist facies assemblages and the southern to upper amphibolite facies assemblages. No isotopic evidence of Archean crust is known south of the shear zone (DePaolo, 1981). Major displacement on the shear zone occurred between 1730 and 1645 m.y.b.p., and some displacement may have occurred earlier. Hills and Houston (1970) and Karlstrom and Houston (1984) interpret these relations as the result of an Atlantic-type continental margin, developed during the Early Proterozoic, being overridden by a volcanic arc from the southeast along a shear zone that developed from a south-dipping subduction zone, forming a suture between Archean and Proterozoic crust.

In the eastern Uinta Mountains in northeastern Utah, the Red Creek Quartzite resembles part of the ensialic sequence of the Medicine Bow Mountains. It was thrust north over the Owiyakuts Complex of Archean age along a ductile deformation zone during an Early Proterozoic amphibolite facies metamorphic event (Sears *et al.*, 1982). The thrust is not entirely analogous to the major shear zone separating Archean from

Proterozoic rocks in the Medicine Bow Mountains, but it may be a thrust subsidiary to, and north of, a westward extension of that shear zone according to Sears *et al.* (1982). Work in progress suggests that rocks called Owiyakuts complex are part of the Red Creek quartzite (Swayze and Holden, 1986).

Farther west in Utah, the boundary between Proterozoic and Archean crust is not exposed, nor are Lower Proterozoic stratified sequences known to unconformably overlie Archean rocks. Gneiss and amphibolite of the Little Willow Formation, exposed in a small area in the Big Cottonwood uplift southeast of Salt Lake City, have not been dated but may be part of an Early Proterozoic ensialic sequence. The Facer Formation in the Willard thrust sheet north of the Farmington Canyon Complex originally may have been part of an Early Proterozoic ensialic sequence deposited on the Archean craton, but it since has been detached and moved by thrusting associated with the Sevier orogeny.

Rocks along the margin of the Archean craton comprising the main body of the Farmington Canyon Complex in central Utah were metamorphosed in the Early Proterozoic at temperatures sufficient to partially melt the crust. Either a deeper level in the crust is exposed in central Utah than in the Medicine Bow Mountains and Sierra Madre, or the Early Proterozoic metamorphic event was much more intense in Utah. Farther west, in the Raft River Mountains, Albion Range, and Grouse Creek Mountains, evidence for a Proterozoic metamorphic overprint is somewhat obscured by a Tertiary metamorphic overprint. However, one pluton has a Rb-Sr whole-rock isochron of 2134 ± 190 m.y. similar, within the limits of error, to the U-Pb age of zircon from granite gneiss on Antelope Island, and it has an initial $^{87}Sr/^{86}Sr$ ratio of 0.764 (Compton *et al.*, 1977), very similar to that obtained from the quartz monzonite gneiss in the Wasatch Mountains; this suggests a similar origin by melting of upper crust Archean rocks.

CONCLUSIONS

Evidence from Sr and Nd isotopes suggests that the Archean craton had a long and complex history before a Late Archean metamorphic event about 2600 m.y. The trend of the craton during Archean time is unknown; it may have been different from the east-west trend established by the Early Proterozoic.

A metamorphic and plutonic event about 1800 m.y.b.p. almost destroyed evidence of the earlier history along the edge of the craton in the Wasatch Mountains. It was the initial stage of a process that led to the formation of new crust over a wide area to the south of the Archean craton during the Early Proterozoic 1800–1650 m.y. ago. Some evidence from isotopic and geochronologic studies suggests a somewhat older Early Proterozoic crust-forming event along the margin of the Archean craton west of the Wasatch Mountains.

Because of the limited outcrops of Precambrian crystalline rock in Utah, the history and processes of crust formation in the region are highly speculative. Isotopic tracers such as Sr, Nd, and Pb, and studies of the exposed crystalline rock allow only general speculation on the history of the concealed basement.

ACKNOWLEDGMENTS

I thank Zell Peterman for help in calculating the $^{87}Sr/^{86}Sr$ ratios for times in the past from data measured in the present, and Jim Cole and Jack Reed for thorough and helpful reviews of this manuscript.

REFERENCES

Aleinikoff, J. N., Nielson, D. L., Hedge, C. E., and Evans, S. H., Jr., 1986, Geochronology of Precambrian and Tertiary rocks in the Mineral Mountains, south-central Utah, *in:* Peterman, Z. E., and Schuabel, D. C., eds., *Shorter Contributions to Isotope Research:* U.S. Geol. Survey Bull. 1622, p. 1–12.

Armstrong, R. L., 1976, The geochronometry of Idaho: *Isochron/West*, v. 15, pt. 2, p. 1–33.

Bell, G. L., 1952, Geology of the northern Farmington Mountains, *in* Marsell, R. E., ed., *Geology of the Central Wasatch Mountains — Guidebook to the Geology of Utah no. 8:* Utah Geol. Soc., p. 38–51.

Bryant, Bruce, 1984, Reconnaissance geologic map of the Precambrian Farmington Canyon Complex and surrounding rocks in the Wasatch Mountains between Ogden and Bountiful, Utah: *U.S. Geol. Survey Misc. Geol. Inv. Map I-1447*, scale 1:50,000.

____ , 1985, Structural ancestry of the Uinta Mountains, *in* Picard, M. D., ed., Geology and energy resources, Uinta Basin, Utah: *Utah Geol. Assoc. Publ. 12*, p. 115–120.

____ , and Graff, Paul, 1980, Metaigneous rocks on Antelope Island *Geol. Soc. America Abstr. with Programs*, v. 12, no. 6, p. 269.

Cheney, J. T., and Guidotti, C. V., 1979, Muscovite-plagioclase equilibria in sillimanite + quartz-bearing metapelites, Puzzle Mountain area, northwest Maine: *Amer. J. Sci.*, v. 279, p. 411–434.

Compton, R. R., Todd, V. R., Zartman, R. E., and Naeser, C. W., 1977, Oligocene and Miocene metamorphism, folding, and low-angle faulting in northwestern Utah: *Geol. Soc. America Bull.*, v. 88, p. 1237–1250.

Crittenden, M. D., Jr., and Sorensen, M. L., 1980, The Facer Formation, a new early Proterozoic unit in northern Utah: *U.S. Geol. Survey Bull. 1482-F*, p. F1–F28.

DePaolo, D. J., 1981, Neodynium isotopes in the Colorado Front Range and crust-mantle evolution in the Proterozoic: *Nature*, v. 291, p. 193–196.

Dickinson, W. R., and Valloni, Renzo, 1980, Plate settings and provenance of sands in modern ocean basins: *Geology*, v. 8, p. 82–86.

Eardley, A. J., and Hatch, R. A., 1940, Pre-cambrian crystalline rocks of north-central Utah: *J. Geology*, v. 48, p. 58–72.

Evans, B. W., and Guidotti, C. V., 1966, The sillimanite-potash feldspar isograd in western Maine, USA: *Contrib. Mineralogy Petrology*, v. 12, p. 25–62.

Farmer, G. L., and DePaolo, D. J., 1983, Origin of Mesozoic and Tertiary granite in the western United States and implications for pre-Mesozoic crustal structure — 1. Nd and Sr isotopic studies in the geocline of the northern Great Basin: *J. Geophys. Res.*, v. 88, no. B4, p. 3379–3401.

____ , and DePaolo, D. J., 1984, Origin of Mesozoic and Tertiary granite in the western United States and implications for pre-Mesozoic crustal structure — 2. Nd and Sr isotopic studies of unmineralized and Cu-and Mo-mineralized granite in the Precambrian craton: *J. Geophys. Res.*, v. 29, no. B12, p. 10141–10160.

Giletti, B. J., and Gast, P. W., 1961, Absolute age of Precambrian rocks in Wyoming and Montana: *N.Y. Acad. Sci. Annals*, v. 91, p. 454–458.

Hashad, A. H., 1964, Geochronologic studies in the central Wasatch Mountains, Utah: Unpublished Ph.D. thesis, Univ. Utah, Salt Lake City, Utah, 98 p.

Hedge, C. E., Stacey, J. S., and Bryant, Bruce, 1983, Geochronology of the Farmington Canyon Complex, *in* Miller, D. M., Todd, V. R., and Howard, K. A., eds., *Tectonic and Stratigraphic Studies in the Eastern Great Basin:* Geol. Soc. America Mem. 157, p. 33–44.

Hills, F. A., and Houston, R. S., 1979, Early Proterozoic tectonics of the central Rocky Mountains, North America: *Univ. Wyo. Contrib. Geology*, v. 17, no. 2, p. 89–110.

Karlstrom, K. E., and Houston, R. S., 1984, The Cheyenne belt — Analysis of a Proterozoic suture in southern Wyoming: *Precambrian Res.*, v. 25, p. 415–446.

Larsen, W. N., 1957, Petrology and structure of Antelope Island, Davis County, Utah: Ph.D. thesis, Univ. Utah, Salt Lake City, Utah, 142 p.

Lindh, Anders, and Wahlgren, Carl-Herric, 1985, Migmatite formation at subsolidus conditions — An alternative to anatexis: *J. Metamorphic Petrology*, v. 3, p. 1–12.

Lundgren, L. W., Jr., 1966, Muscovite and partial melting in southeastern Connecticut: *J. Petrology*, v. 7, p. 421–453.

Macdonald, G. A., and Katsura, T., 1964, Chemical composition of Hawaiian lavas: *J. Petrology*, v. 5, p. 82–133.

Nockolds, S. R., 1954, Average chemical composition of some igneous rocks: *Geol. Soc. America Bull.*, v. 65, no. 10, p. 1007–1032.

Pearce, J. A., and Cann, J. R., 1973, Tectonic setting of basic igneous rocks determined using trace element analyses: *Earth Planet. Sci. Lett.*, v. 19, p. 290–300.

Pearce, T. H., Gorman, B. E., and Birkett, T. C., 1975, The TiO_2-K_2O-P_2O_5 diagram — A method of discrimination between oceanic and non-oceanic basalts: *Earth Planet. Sci. Lett.*, v. 24, p. 419–426.

——, Gorman, B. E., and Birkett, T. C., 1977, The relationship between major element chemistry and tectonic environment of basic and intermediate volcanic rocks: *Earth Planet. Sci. Lett.*, v. 36, p. 121–132.

Peterman, Z. E., 1979, Strontium isotope geochemistry of Late Archean to Late Cretaceous tonalites and trondhjemites, *in* Barker, Fred, ed., *Trondhjemites Dacites, and Related Rocks:* Amsterdam, Elsevier, p. 133–147.

Pettijohn, F. J., 1963, Chemical composition of sandstones — excluding carbonate and volcanic sands, *in* Data of Geochemistry 6th Ed.: U.S. Geol. Survey Prof. Paper 440-S, 21 p.

Sears, J. W., Graff, P. J., and Holden, G. S., 1982, Tectonic evolution of Proterozoic rocks, Uinta Mountains, Utah and Colorado: *Geol. Soc. America Bull.*, v. 93, no. 10, p. 990–997.

Small, W. D., and Slauson, W. F., 1968, Lead isotopic abundances in selected regions of Montana, Wyoming, and Idaho [abstract]: *Geol. Soc. America Spec. Paper 121*, p. 638.

Sorensen, M. L., and Crittenden, M. D., Jr., 1976, Preliminary geologic map of the Mantua quadrangle and part of the Willard quadrangle, Boxelder, Weber, and Cache Counties, Utah: *U.S. Geol. Survey Misc. Field Studies Map MF-720*, scale 1:24,000.

Stacey, J. S., and Zartman, R. E., 1978, A lead and strontium isotopic study of igneous rocks and ores from the Gold Hill mining district, Utah: *Utah Geology*, v. 5, p. 1–15.

——, Zartman, R. E., and Nkomo, I. T., 1968, A lead isotope study of galenas and selected feldspars from mining districts in Utah: *Econ. Geology*, v. 63, p. 796–814.

Swayze, G. A., and Holden, G. S., 1986, Metamorphic conditions, Beaver Creek block, Precambrian Red Creek terrain, N.E. Utah: *Geol. Soc. America Abstr. with Programs*, v. 18, no. 5, p. 417.

Winchester, J. A., and Floyd, P. A., 1977, Geochemical discrimination of different magma series and their differentiation products using immobile elements: *Chem. Geology*, v. 20, p. 325–343.

Winkler, H. G. F., 1979, *Petrogenesis of Metamorphic Rocks*, 5th ed.: Heidelberg, Springer-Verlag, 348 p.

Jeffrey A. Grambling
Department of Geology
University of New Mexico
Albuquerque, New Mexico 87131

17

A SUMMARY OF PROTEROZOIC METAMORPHISM IN NORTHERN AND CENTRAL NEW MEXICO: THE REGIONAL DEVELOPMENT OF 520°C, 4-KB ROCKS

Regionally metamorphosed Proterozoic rocks crystallized near 520°C, 4 kb, across a large area in central and northern New Mexico. Relict metamorphic isograds and isobars are subhorizontal, but rock fabrics and mineral zoning document that deformation accompanied metamorphism. The thermal peak came during and after a young deformational event (D_3). During this event, the growth of zoned porphyroblasts records P-T conditions increasing from 450°C, 3.4 kb to 520°C, 4.1 kb, documenting that synmetamorphic deformation thickened the local, upper continental crust by at least 20%. Deformation of the exposed units took place by ductile horizontal shortening and vertical thickening. The magnitude of this ductile thickening is sufficient to account for the 20% pressure increase recorded by the zoned minerals.

The rocks cooled isobarically at a depth near 14 km. Cooling rates estimated from 15–90-μm-thick retrograde rims on garnet were less than 1°C/m.y. for the first 25°C of cooling. Isobaric cooling may have been a major factor in preserving the remarkable degree of metamorphic equilibrium shown by these rocks. The cooling path may also indicate that the premetamorphic crust was fairly thin, perhaps 25–30 km.

Postmetamorphic erosion has removed 11–15 km of overburden from the metamorphic terrane, which is now exposed discontinuously across an area of 75,000 km². The unconformity between Paleozoic (generally Mississippian) and Proterozoic rocks is unfolded, and the 4-kb rocks lie immediately below this unconformity. The deep regional erosion must have occurred prior to 350 m.y.b.p.

INTRODUCTION

Isolated mountain ranges in northern and central New Mexico contain abundant exposures of Proterozoic metamorphic rocks. These rocks record a highly unusual feature: across an area of 250 km by 300 km, metamorphic mineral assemblages formed in a narrow range of P-T conditions. Most rocks preserve temperatures near 520°C and pressures near 4 kb. Mapped isograds are subhorizontal, giving the impression that a widespread regional metamorphism occurred in the interior of a tectonically stable craton.

This paper documents the P-T conditions of metamorphism across northern New Mexico and interprets the causes of the widespread, uniform conditions. Results are based on detailed geologic and petrologic studies in the Truchas and Rio Mora Ranges (Fig. 17-1), extensive sampling in the Picuris and Tusas Mountains, and reconnaissance mapping, combined with locally detailed work, in 11 other mountain ranges. The study area encompasses most Precambrian exposures in the northern half of the state. In most cases, the metamorphic data come from isolated, fault-bounded uplifts flanking the Rio Grande rift.

This paper is an abbreviated version of a manuscript to be published elsewhere. Readers interested in the full-length paper are referred to J. A. Grambling (unpublished data).

REGIONAL GEOLOGY

Proterozoic supracrustal rocks are best studied in the northern half of the area shown in Fig. 17-1. There, the oldest rocks consist of metavolcanic units, both mafic and felsic, with minor amounts of pelitic schist, banded iron formation, and meta-arkose (Reed, 1984; Robertson and Moench, 1979; Wobus, 1985). Massive quartzite and pelitic schist of the Ortega Group form a thick sequence overlying the metavolcanics. Farther to the

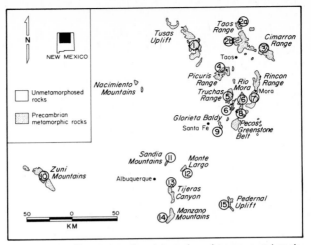

FIG. 17-1.　Exposures of metamorphosed supracrustal rocks in northern and central New Mexico. Inset shows general location.

south the stratigraphic relations are less well understood, but supracrustal lithologies consist of pelitic, quartzose, and arkosic metasediments interlayered with felsic and rare mafic metavolcanics. U-Pb zircon ages of supracrustal rocks across the study area range from 1750 to 1650 m.y. (Bowring and Condie, 1982; Bowring et al., 1983, 1984). In general, U-Pb zircon ages become younger toward the south, but there are a number of local exceptions to this trend.

The supracrustal rocks are intruded by plutons ranging from mafic to felsic in composition. Most intrusive rocks are deformed, metamorphosed, and define the same age range as the 1750–1650-m.y. supracrustal rocks (Bowring and Condie, 1982; Bowring et al., 1983, 1984), with the only exceptions being the 1400–1500-m.y.-old anorogenic granites (Brookins, 1982; Silver et al., 1977) exposed in and south of the Picuris Range (Fig. 17-1).

Precambrian metamorphic rocks appear to share a common structural history across the entire 75,000-km^2 study area. The earliest deformation produced a bedding-parallel foliation in most rocks. Locally, this S_1 foliation is mylonitic. Asymmetric recumbent folds with amplitudes up to 3 km have been mapped in the Rio Mora and Taos uplifts and appear to be related to this early deformation. Quartzites in areas as widely separated as the Pedernal, Rio Mora, and Tusas uplifts have bladed kyanite crystals aligned parallel to the north-south D_1 extension direction. This indicates that early deformation occurred at temperatures above 400°C and pressures above 2.5 kb, where kyanite is stable with quartz (Haas and Holdaway, 1973; Holdaway, 1971).

The D_1 event was overprinted by several younger deformations which also seem to have been widespread. D_2 structures are tight, upright to overturned, northeast-to-northwest trending. The third generation of folds trends easterly. These upright structures formed during the thermal peak, as indicated by porphyroblast-matrix fabric relationships. Some peak metamorphic porphyroblasts show "millipede" inclusion geometry (Bell and Rubenach, 1983), others show the youngest fabric deflected around them, and still others have overgrown east-west crenulations.

Peak metamorphic conditions were determined from samples collected in 15 separate mountain ranges. *P-T* conditions were estimated from the following calibrations: the Al_2SiO_5 minerals (Grambling and Williams, 1985; Holdaway, 1971) the equilibrium chloritoid + Al_2SiO_5 = staurolite + quartz + H_2O (Ganguly, 1972; Richardson, 1968); the transition from greenschist to epidote-amphibolite facies in metabasalt (Apted and Liou, 1983; Moody *et al.*, 1983); the equilibrium serpentine = forsterite + talc + H_2O (Bowen and Tuttle, 1949; Yoder, 1966; it should be noted that these experiments were not reversed); and garnet-biotite geothermometry (Ferry and Spear, 1978). The precision of the *P-T* estimates range from ±15°C, ±0.3 kb, to ±50°C, ±1-2 kb.

The *P-T* estimates cited below are meaningful only if the metamorphic minerals crystallized in chemical equilibrium. This is a critical assumption: many of the *P-T* estimates rely heavily on occurrences of two or three aluminum silicate minerals, yet these minerals are notorious for their metastable behavior. There are several reasons to suspect that the assemblages do preserve an arrested state of chemical equilibrium. These reasons, documented in Grambling (1981) and Grambling and Williams (1985), are listed below.

1. The spatial distribution of Al_2SiO_5 minerals is consistent with the attainment of equilibrium. Actual triple-point occurrences occupy relatively small areas surrounded by zones with one or two polymorphs.
2. Al_2SiO_5 isograds, where carefully mapped, are sharp. Transitions generally occur over distances less than 100 m, implying little metastable coexistence.
3. The aluminum silicate minerals typically contain minor amounts of Fe_2O_3 and Mn_2O_3, which are distributed systematically among coexisting polymorphs.
4. Al-silicate thermobarometry gives results compatible with other mineral thermometers and barometers.

The following discussion considers mineral assemblages and *P-T* conditions within the individual uplifts, presented in geographic order from north to south. Results are summarized in Table 17-1. The discussion concerns only rocks south of latitude 36°40'N, the latitude of Questa, New Mexico.

Tusas Mountains

All three polymorph of Al_2SiO_5 occur in the Tusas Range (Grambling and Williams, 1985, Fig. 5). Kyanite coexists with andalusite in the northwest. Near Kiowa Mountain in the central Tusas Range, M. L. Williams (personal communication, 1985) has recognized kyanite, andalusite, and sillimanite within a 1-km² area. Kyanite coexists with sillimanite farther south, with no andalusite present.

The aluminum silicates are most abundant in the massive Ortega Quartzite, the uppermost Proterozoic unit in the Tusas stratigraphy. Pelitic schists in the underlying Vadito Group locally have garnet coexisting with biotite. M. L. Williams has shown that garnet-biotite pairs from rocks 4 km north of Kiowa Mountain give temperatures near 480°C. Geothermometry at Kiowa Mountain yields a *T* of 500°C, near the area with all three Al-silicate minerals, implying pressures near 3.8 kb. Garnet-biotite temperatures rise to 525°C in rocks 8 km south of Kiowa Mountain.

TABLE 17-1. Locations, Mineral Assemblages, *P-T* Conditions, and Abbreviations Used in Text

Location	Mineral Assemblages[a]	P (kb)	T (°C)	Distance below Phanerozoic-Precambrian Unconformity (m)	References
Tusas Mountains	KAS; KS-gar-bi	4.0	515	100	Grambling and Williams, 1985
Taos Range	KAS; KS-gar-bi	3.8	520	120	This study; J. Reed, Jr., personal communication, 1984
Cimarron Range	S-ctd	3.9	520	150	Leyenberger, 1983
Picuris Range	KAS-ctd-staur; gar-bi	3.8	505	?	Holdaway, 1978; Grambling and Williams, 1985
Truchas Range	KAS-ctd-staur; gar-bi; K-cord-bi-chl	4.2	530	~400	Grambling, 1981; Grambling and Williams, 1985
Rio Mora	KS-Mn A; gar-bi	4.6	540	190	Grambling and Williams, 1985
Rincon Range	S-gar-bi; A[b]	3.8	535	100	This study
Pecos Greenstone Belt	staur-gar-bi	?	515	130	This study (no pressure estimate)
Glorieta Baldy	S-gar-bi	3.8	500	?	Renshaw, 1984
Zuni Uplift	hbl-epid-plag-chl; serp-talc	?	525	50	Lambert, 1983 (no pressure estimate)
Monte Largo	AS-ctd; pseudomorphs of S after K	3.7	520	30	This study
Tijeras Canyon	AS; A-staur; KA[c]	3.8	500?	125	J. Connolly, personal communication, 1983; W. Cavin, personal communication, 1984
Manzano Mountains	KAS-ctd-staur; S-staur-gar-bi	3.8	500	?	This study
Pedernal Hills	K-staur; greenschist metabasalts	?	500	50	This study (no pressure estimate)

[a] A, andalusite; bi, biotite; cord, cordierite; ctd, chloritoid; epid, epidote; gar, garnet; hbl, hornblende; K, kyanite; plag, plagioclase; S, sillimanite; serp, serpentine; staur, staurolite. All samples except those in locality 10 contain both quartz and muscovite.

[b] Andalusite occurs 5 km east of S-gar-bi.

[c] AS, KA occur within a 5 × 5 km area on opposite sides of a small(?) fault.

Taos Range

The metamorphic data from the southern Taos Range, south of Questa, come from approximately 100 specimens collected by the author and by J. C. Reed, Jr. Mineral assemblages are summarized in Fig. 17-2. The southern Taos Range contains one mapped occurrence of coexisting kyanite, andalusite, and sillimanite. Surrounding rocks have various combinations of one or two Al-silicate minerals. Chloritoid and/or staurolite coexist with kyanite and/or andalusite in a number of outcrops.

The aluminous assemblages occur in massive quartzite, probably equivalent to the Ortego Quartzite, and in underlying pelitic schists. They indicate metamorphic conditions near 500°C, 3.8 kb across a large part of the southwestern Taos Range. *P-T* conditions in the southeastern Taos Range are poorly constrained, but the local occurrence of rocks transitional between upper greenschist and lower amphibolite facies, together with a single occurrence of sillimanite, suggests a similar *P* and *T*.

Cimarron Mountains

The Cimarron Mountains lie directly east of the southern Taos Range and have been studied by Leyenberger (1983) and Wobus (1984). Metamorphic interpretations are based largely on a single specimen, hence must be considered preliminary. The specimen is a quartzite with chloritoid, fibrolite, and coarse sillimanite, together with minor muscovite and FeTi oxides. The specimen, collected by Leyenberger from outcrops south of U.S. Highway 64 west of the Fowler Pass fault, appears to record textural equilibrium. It indicates *P-T* conditions of 3.2–4.7 kb, 500–550°C. Such conditions are consistent with the presence of blue-green hornblende and calcic plagioclase in nearby metabasalts (Leyenberger, 1983).

Picuris Range

Holdaway (1978) reported the occurrence of kyanite-andalusite-sillimanite and sillimanite-chloritoid in the Picuris Range. Grambling and Williams (1985) presented addi-

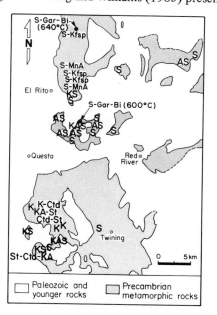

FIG. 17-2. Metamorphic assemblages in the Taos Range. Data from this study and J. Reed, Jr. (personal communication, 1984). St = staurolite, Kfsp = K feldspar, other symbols in Table 17-1.

tional data pertinent to metamorphism in that area. Their Fig. 4 is used as a basis for the following discussion.

The three aluminum silicates coexist in a number of localities along Hondo Canyon in the northern Picuris Range. South of Hondo Canyon, the topography rises gradually toward Picuris Peak. Sillimanite disappears 300 m above Hondo Canyon, where kyanite and andalusite can both be found. Andalusite disappears at 2900 m elevation. Between 2900 m and the 3300-m summit of Picuris Peak, only kyanite has been found.

Seven kilometers southwest of Hondo Canyon, the three Al-silicate minerals occur singly in samples collected in a 1-km^2 area. Four kilometers farther south, near Copper Hill, rocks within a 2-km^2 area have kyanite-andalusite and kyanite-sillimanite.

The distribution of Al-silicates indicates peak P-T conditions of 500–530°C, 3.6–4 kb across the Picuris Range. Garnet-biotite temperatures range from 500 to 525°C.

Truchas Range

The Truchas Range also contains all three polymorphs of Al_2SiO_5. In the extreme northern Truchas Range, sillimanite is the only Al_2SiO_5 polymorph. Kyanite appears first near the summit of 3993-m-high Truchas Peak, where kyanite-bearing rocks overlie the topographically lower sillimanite-bearing rocks. The kyanite-sillimanite isograd forms a planar surface which dips gently to the south (Grambling, 1981).

In the southernmost Truchas Range, kyanite occurs on mountain summits and andalusite on hillslopes. A subhorizontal kyanite-andalusite isograd follows topographic contours 200 m below the rugged summits of Pecos and East Pecos Baldy mountains. The kyanite-andalusite and kyanite-sillimanite isograds intersect in the central Truchas Range, where aluminum silicate triple-point assemblages are common.

Mineral assemblages and compositions indicate that metamorphism peaked at 4.2 kb, 530°C, consistent with the 500–530°C garnet-biotite temperatures in the area.

Rio Mora Uplift

Metamorphic relations are best exposed along the escarpment east of Rio Mora, where elevations fall 850–1000 m in slightly over 2 km. In this area, kyanite occurs along the crest of the escarpment. The first sillimanite appears at slightly lower elevations, where kyanite and sillimanite coexist. The two polymorphs occur together over a vertical distance of less than 180 m. This narrow zone of coexistence can be followed along the entire 15-km outcrop trace of the isograd. The planar isograd dips west; sillimanite is the only Al_2SiO_5 mineral at lower elevations.

Andalusite also occurs in this region, but only along a single stratigraphic layer which lies on the contact between Ortega Quartzite and underlying metavolcanic rocks. The crystals are all enriched in Mn_2O_3 (0.3–12 wt %), which stabilized the andalusite to unusually high P and T.

The rocks are highly folded. Wherever the Mn^{3+}-rich layer crosses the kyanite-sillimanite isograd, rocks with kyanite-sillimanite-Mn andalusite have been found. Compositions of the triple-point minerals may be used to calculate the P and T of metamorphism. These calculations (Grambling and Williams, 1985) yield conditions near 540°C, 4.6 kb. Garnet-biotite geothermometry gives comparable temperatures in rocks collected near the kyanite-sillimanite isograd.

Rincon Range

Proterozoic rocks have been mapped in only two small portions of this uplift, at its southern end south of the town of Mora and at its northern end near Guadalupita (Fig.

SUMMARY OF PROTEROZOIC METAMORPHISM IN NEW MEXICO

17-1). Baltz and O'Neill (1980a, b) reported sillimanite near Mora and andalusite about 10 km south of Mora. A sample of garnet-biotite-sillimanite schist was analyzed for this study, collected from rocks 15 km southwest of Mora. It yielded a temperature of 535°C. The presence of sillimanite implies a pressure of 3.3–4.4 kb. Because Baltz and O'Neill (1980a, b) identified andalusite but no kyanite in nearby rocks, pressures may have been in the lower half of the reported range.

Preliminary mapping in the northern Rincon Range, 35 km north of Mora, has documented chlorite-epidote-hornblende-plagioclase assemblages in amphibolites. Inter-layered schists have staurolite-biotite-kyanite-garnet and kyanite-staurolite-chloritoid, both with muscovite and quartz. Small andalusite crystals have been identified in two hand samples, one with kyanite. These preliminary data suggest that P-T conditions peaked near 3.5–3.8 kb, 500°C.

Pecos Greenstone Belt

Epidote-amphibolite and upper greenschist facies metabasalts characterize the Pecos greenstone belt (Klich, 1983). Assemblages include hornblende-calcic plagioclase-epidote-biotite-chlorite-quartz and actinolite-plagioclase-chlorite-epidote, with the two assem-blages commonly occurring side by side. Pelitic rocks are rare. One analyzed sample with staurolite, muscovite, and quartz gave a garnet-biotite temperature of 520°C. The assem-blages do not constrain peak metamorphic pressure.

Glorieta Baldy

A large fault-bounded slice of rocks near Glorieta Baldy consists of metasedimentary and metavolcanic material (Renshaw, 1984). One sample of garnet-biotite-staurolite schist indicated a temperature of 500°C. The occurrence of sillimanite in several rocks, com-bined with the garnet-biotite temperatures, implies pressure near 3.8 kb.

Zuni Uplift

Lambert (1983) described mafic and ultramafic rocks in a part of the Zuni Mountains. She described epidote-amphibolite assemblages in metabasalts and serpentine-talc mineralogy in metamorphosed ultramafic rocks. The assemblages suggest that peak meta-morphic temperatures were in the range 480–550°C, but the data do not constrain pressures.

Sandia Mountains

Metamorphosed supracrustal rocks are exposed near Placitas, Tijeras Canyon, and the Monte Largo Hills (Fig. 17-3). The outcrops partially surround exposures of the 1445 ± 40-m.y.-old Sandia Granite (Brookins, 1982) and are the only outcrops in the study area that show evidence of contact metamorphism.

Berkley and Callender (1979) mapped zones of chlorite phyllite, andalusite schist, and sillimanite schist surrounding the Sandia pluton near Placitas. The 500-m-wide sillimanite zone has abundant relict andalusite except within 50 m of the intrusion, where sillimanite coexists with potassium feldspar. The map pattern is complicated by a large number of faults. Preliminary garnet-biotite geothermometry indicates that tem-peratures increased toward the granite, peaking at 650–700°C in the sillimanite-K feldspar zone. Further work has focused on rocks near the first appearance of sillimanite. Here, embayed relict kyanite has been found in two samples. Garnet-biotite pairs have been sampled, but their significance is not obvious because of the extreme zoning shown by garnet. The highly zoned garnets have rims up to 100 μm thick which are compositionally distinct from the nearly unzoned garnet cores. If the rim zoning formed during cooling, a reasonable temperature may be obtained by comparing matrix biotite to garnet cores or to garnet just inward of the rim. Such an approach gives 640°C. Alternatively, the rim composition could indicate regional metamorphism superimposed on the contact event. Garnet rim-biotite geothermometry gives a temperature of 525°C. The two approaches, when coupled with associated andalusite-sillimanite occurrences, indicate pressure of 1.8–3.4 kb. Relict kyanite suggests the higher pressure may be more reasonable. However, because of the large uncertainty in metamorphic P and T, and the contact-metamorphic nature of the exposures, this area has been omitted from subsequent discussions.

Along Tijeras Canyon (Fig. 17-3), the southeastern edge of the Sandia Granite is in tectonic contact with andalusite-sillimanite rocks (R. H. Vernon, personal communication, 1985). The Tijeras fault truncates the andalusite-sillimanite zone. Immediately south of the fault, rocks contain andalusite and staurolite. A second fault lies southeast of the Tijeras fault, separating the andalusite rocks from kyanite-bearing units farther south. Interpretation of P-T conditions depends critically on the nature of the two faults. If fault offsets are small, the rocks record near-triple-point pressures, in a contact-metamorphic environment. Conversely, some authors have suggested substantial offsets on these faults (Lisenbee *et al.*, 1979), which would remove constraints on the metamorphic conditions.

Metamorphic rocks near Monte Largo (Fig. 17-3) are dominated by a massive, hematitic quartzite with abundant aluminous interlayers (the Ortega Quartzite?). The rocks have sillimanite with rare andalusite. Andalusite crystals are embayed and appear to be relict. Sillimanite commonly occurs in pseudomorphs after kyanite, but no kyanite

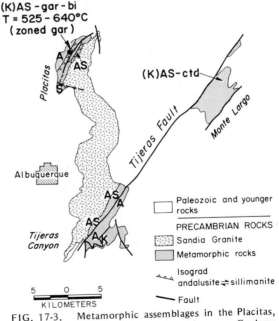

FIG. 17-3. Metamorphic assemblages in the Placitas, Tijeras Canyon, and Monte Largo study areas. Geology from Cavin *et al.* (1982).

SUMMARY OF PROTEROZOIC METAMORPHISM IN NEW MEXICO

remains. Chloritoid has been identified in one sample of sillimanitic quartzite. The mineralogy suggests peak *P-T* conditions of 3.2-3.8 kb, 500-540°C, and further implies a paragenetic sequence of kyanite(?) → andalusite → sillimanite.

Manzano Mountains

Bauer (1983) collected several garnet-biotite rocks from the southern Manzano Mountains. These were analyzed in the present study and gave peak metamorphic temperatures clustering at 500°C (Fig. 17-4). One of the specimens has sillimanite and two have staurolite. A fifth sample was collected from aluminous quartzite in the same area; it contains kyanite-andalusite-sillimanite-chloritoid-staurolite in apparent textural equilibrium. Metamorphism in the southern Manzano Mountains apparently peaked near 3.8 kb, 500°C.

Pedernal Hills

The northern Pedernal Hills consist largely of massive hematitic quartzite, with kyanite concentrated in some layers. Gonzales (1968) recognized staurolite in one sample of quartzite. Immediately south of the quartzite, mafic metavolcanic rocks have greenschist-facies assemblages with actinolite-chlorite-albite-biotite-clinozoisite and calc-silicate rocks have epidote, diopside, and garnet (Armstrong and Holcombe, 1982). The mineral assemblages broadly constrain metamorphic conditions: pressures must have exceeded 3.4 kb, at temperatures of 475-525°C.

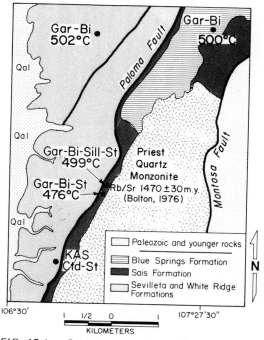

FIG. 17-4.   Geology and metamorphic assemblages in the southern Manzano Mountains. Geology from Bauer (1983).

Summary of Peak P-T Conditions

Nearly all metamorphic rocks across the area of Fig. 17-1 recrystallized between 3.2 and 4.6 kb, 475–540°C (Table 17-1). Exceptions occur only in the contact aureole of the Sandia Granite. The depth of postmetamorphic erosion varies only slightly, from 11 to 15 km, across the 75,000-km² study area. It is this limited erosional depth, centered around the depth of the kyanite-andalusite-sillimanite triple point, that accounts for the common occurrence of the Al-silicate triple-point assemblages.

METAMORPHIC ISOTHERMAL-ISOBARIC SURFACES

In several ranges, vertical relief is sufficiently great that Proterozoic peak metamorphic isobaric and isothermal surfaces can be mapped in three dimensions. This can be done with the most confidence in the Rio Mora, Truchas, and Picuris Ranges (Fig. 17-1), where the density of metamorphic data is greatest.

At Rio Mora, the kyanite-sillimanite isograd forms a planar surface that dips west at 7°. Compositions of kyanite, sillimanite, and Mn-andalusite at several places along the isograd indicate that the isograd is both isobaric and isothermal (Grambling and Williams, 1985).

Aluminum silicate compositions and isograds in rocks from the Truchas Range define a 4.2-kb isobaric surface dipping 10° to the west-northwest. Al-silicate mineralogy and garnet-biotite temperatures show that relict isotherms dip gently southwest.

The 3.8-kb and 500°C isothermal-isobaric surfaces are coincident in the Picuris Range. They dip approximately 3° to the west.

Each of these regions, then, contains planar isobaric surfaces that dip gently to the west. This gentle dip is remarkably similar to the 8–15° westerly dip on the Precambrian-Paleozoic unconformity, where it is exposed along Rio Mora. The metamorphic and geologic data imply that little tilting occurred between metamorphism and 350 m.y. ago, but a gently westward tilting occurred, at least east of the Rio Grande rift, after the Paleozoic.

Specimens cited in Table 17-1 were generally collected from outcrops within 500 m of the overlying Paleozoic-Precambrian unconformity. The Proterozoic isobars trend roughly parallel to this contact. This implies that the 11–15 km of erosion occurred prior to Mississippian time, the age of the basal Paleozoic sediments (Baltz and O'Neill, 1980 a, b). Because the metamorphic isobars remained nearly horizontal during the post-metamorphic uplift and erosion, it seems likely that metamorphism accompanied the final stabilization of this part of the southwestern North American craton, and was followed by extremely slow nonorogenic rise toward the surface.

The extreme topographic relief of the southern Rocky Mountains results mainly from Cenozoic tectonism related to opening of the Rio Grande rift (Manley, 1984). Cenozoic structures have disrupted the Proterozoic isobaric surfaces, so that 4-kb rocks now occur, in various uplifts, at elevations varying from 1800 to 3650 m above sea level. Because of the complexity of these young structures (Baldridge et al., 1984), it is difficult to project the geology downward from present exposures, and there is no significance to estimates of synmetamorphic crustal thickness based on combining present-day crustal thickness with the amount of material removed since the thermal peak.

METAMORPHIC P-T PATH

The assemblage garnet-biotite-plagioclase-muscovite occurs in a number of samples from the study area. Ghent and Stout (1981) proposed two empirical geobarometers for this assemblage. Their equation lb (the Mg reaction) has been used to calculate pressures: lines of constant K_{1b} are nearly independent of temperature, so small errors in estimated temperatures have virtually no effect on calculated pressures.

Nine specimens from the Truchas and Rio Mora uplifts have been analyzed for the compositions of the coexisting minerals. Because biotite and plagioclase are included in garnet, and garnet and plagioclase crystals are chemically zoned, the rocks preserve a chemical record of changes in P-T conditions during mineral growth.

Garnet Zoning

Typically, X_{Mn} decreases from the centers to the edges of garnet porphyroblasts. Only one sample (81-234S) shows different Mn zoning; it contains 3-5-cm crystals in contrast to the 2-10-mm crystals found in other rocks. Several garnets also show thin outer rims, with X_{Mn} increasing abruptly in the outermost 15-50 μm.

Garnets are also zoned in calcium. Most have X_{Ca} falling slightly from cores to near the edges of the crystals. Outer rims show reversals in this pattern, with X_{Ca} increasing in the outermost 15-50 μm.

All samples except 81-234S are zoned in Fe/(Fe + Mg). This ratio gradually decreases from garnet cores to zones 15-90 μm from garnet rims. The outer rims show abrupt zoning reversals, with large increases in Fe/(Fe + Mg).

The zoning is interpreted to be a prograde feature, preserving changes in garnet composition with progressive garnet growth, except in the narrow outer rims. The narrow rims are interpreted as retrograde features, formed by reaction among garnet, biotite, and plagioclase after the thermal peak. Several observations support these interpretations of garnet zoning.

1. Biotite grains are included in garnet cores in four samples (Table 17-2). These included biotite-garnet pairs yield temperatures 20-80°C lower than those calculated from peak mineral compositions, and are interpreted as prepeak assemblages.
2. The highest temperatures are obtained by comparing matrix biotite to garnet just inside the narrow, apparently retrograde rims. Such "peak" temperatures are compatible with other mineral geothermometers.
3. Temperatures calculated from garnet retrograde rims and matrix biotite are 15-60°C lower than peak temperatures. Biotite is modally abundant in nearly all samples, the garnet retrograde rims are narrow, and volumetric ratios of biotite to garnet retrograde rims exceed 100. Therefore, it seems probable that biotite compositions changed negligibly during the retrograde event.
4. Garnets are included inside staurolite in several samples. These garnets show no retrograde rims. Retrograde rims in isolated garnets are poorly developed on garnet-quartz contacts and best developed on garnet-biotite contacts.

Plagioclase Zoning

Groundmass and porphyroblastic plagioclase crystals are zoned in nearly all specimens (Table 17-2). The zoning is erratic in several samples but it is quite regular in the rest: X_{Ca} decreases from plagioclase cores to zones 7-40 μm from their rims. In the thin outer rims, mineral compositions change dramatically. Some samples show abrupt increases in X_{Ca}, others abrupt decreases, and in one specimen (81-71b), plagioclase shows no edge zoning.

Crystals of plagioclase are included in garnet in several samples. The cores of these inclusions are more sodic than matrix plagioclase grains in the same specimen. However,

TABLE 17-2. Mineral Compositions Used in Calculating P-T Paths, and Pressures Calculated from Equation 1b of Ghent and Stout (1981)

Sample and Location	Garnet[a] X_{Fe}	X_{Mg}	X_{Mn}	X_{Ca}	t^{b} (μm)	Plagioclase[a] X_{Ca}	t^{b} (μm)	Biotite[c,d] Fe	Mg	Muscovite[d] X_K	X^{VI}_{Al}	ln K, from Ghent and Stout (1981)[a]	P (bars)[e]
77-39 near Pecos Baldy, Truchas Range	0.7847	0.0532	0.0799	0.0822	15	0.34	11	0.4080	0.3190	0.81	0.98	9.884	3881
	0.8564	0.0759	0.0233	0.0434		0.20		0.4710	0.3120	0.76	0.96	9.180	4417
	0.8565	0.0696	0.0226	0.0513		0.29						10.053	3693
77-41 near Pecos Baldy, Truchas Range	0.7863	0.0528	0.0753	0.0856	26	0.37	30	–	–	–	–	10.184	3633
	0.8464	0.0798	0.0198	0.0541		0.31		0.4498	0.3362	0.82	0.98	9.791	3878
	0.8541	0.0727	0.0191	0.0540		0.30						9.977	3758
77-42 near Pecos Baldy, Truchas Range	0.8166	0.0676	0.0785	0.0373	51	0.22	7	0.4867	0.3033	–	–	10.116	3689
	0.8359	0.0735	0.0411	0.0494		0.20		0.4825	0.2988	0.78	0.98	8.691	4850
	0.8486	0.0676	0.0379	0.0459		0.12						7.630	5765
82TRU-2 near Truchas Peak, Truchas Range	0.4540	0.1099	0.3763	0.0599	54	0.34	5	0.2917	0.5154	–	–	10.301	3536
	0.4908	0.1527	0.3133	0.0432		0.30		0.2848	0.5304	0.80	0.89	10.060	3642
	0.4976	0.1468	0.3003	0.0553		0.27						9.067	4537
81-71a near Gascon, east of Rio Mora	0.5298	0.0535	0.2396	0.1771	63	0.94	14	–	–	0.98	0.84	11.961	2164
	0.5412	0.1019	0.1834	0.1735		0.93		0.3188	0.5215	0.98	0.95	10.058	3547
	0.5541	0.1041	0.1986	0.1431		0.90						10.473	3334
81-71b near Gascon, east of Rio Mora	0.4815	0.0621	0.2883	0.1681	33	0.93		–	–	–	–	11.920	2198
	0.5236	0.0964	0.1705	0.2094		0.93		0.3185	0.5276	0.98	0.84	9.941	3657
	0.5247	0.1003	0.1943	0.1806		0.93						10.266	3512
81-234S near Pacheco Lake, east of Rio Mora	0.7155	0.2116	0.0359	0.0698	50	0.79	7	0.2450	0.6394	–	–	10.804	3121
	0.6894	0.2173	0.0296	0.0636		0.51		0.3156	0.5065	0.94	0.93	8.991	4550
	0.6751	0.1740	0.0870	0.0640		0.35						8.510	5104
82-12 near Pyramid Peak, north of Rio Mora	0.3705	0.1758	0.4129	0.0408	35	0.23	40	–	–	–	–	9.497	4201
	0.3644	0.1841	0.4106	0.0409		0.23		0.1837	0.6370	0.79	0.90	9.351	4212
	0.3498	0.1373	0.4666	0.0463		0.16						8.770	4791
82-62 near Pyramid Peak, north of Rio Mora	0.3760	0.1317	0.4026	0.0897	90	0.38	38	–	–	–	–	9.391	4288
	0.3783	0.1637	0.3845	0.0735		0.35		0.2175	0.6263	0.86	0.89	9.090	4458
	0.3824	0.1655	0.3859	0.0662		0.38						9.618	4066

[a] Data presented in order: mineral core, "peak," retrograde rim.

[b] Thickness of retrograde rim.

[c] Fe, Mg as cations per 11/3 oxygens.

[d] Data presented in order: inclusion in staurolite or garnet; matrix.

[e] Pressures calculated from Ghent and Stout (1981), eq. 1b, in order: core (at $T = 450°C$), "peak" (at $T = 500–550°C$; see the text), and retrograde rim (at $T = 475°C$). Pressures are not sensitive to small changes in T.

most inclusions are concentrically zoned. X_{Ca} in plagioclase increases toward the surrounding garnet in such a way that X_{Ca} at the inclusion-garnet contact is the same as X_{Ca} in the cores of matrix plagioclase grains.

The sodic cores of included plagioclase are interpreted as relics formed before the nucleation of garnet. The calcic rims on plagioclase inclusions, and the regular decrease in X_{Ca} across matrix plagioclase crystals, are interpreted to preserve a record of changes in plagioclase composition during porphyroblast growth. The thin outer rims (7–40 μm) on matrix grains appear to be retrograde features. These may have developed by reactions among plagioclase, garnet, and micas. The latter interpretation is supported by an additional observation: plagioclase crystals included within staurolite lack the thin retrograde rims.

Changes in *P-T* During Porphyroblast Growth

Temperatures derived from inclusions of biotite within garnet average 450°C. This is assumed to be the temperature at which garnet nucleated in all nine specimens. Peak metamorphic temperatures varied slightly across the area from which the samples were collected. They were approximately 500°C in the southern Truchas Range, 530°C in the northern Truchas Range, and 540°C near Rio Mora (Grambling, 1981; Grambling and Williams, 1985).

Temperatures derived from retrograde garnet rims fall between 470 and 530°C. For the purposes of calculating metamorphic *P-T* paths, it was assumed that all retrograde rims formed as the rocks cooled to 475°C. The exact temperature used in this calculation is not critical because metamorphic pressures, calculated from isopleths of $\ln K_1$ (Ghent and Stout, 1981), are insensitive to temperature.

Metamorphic pressures calculated from mineral core, "peak," and retrograde rim compositions are given in Table 17-2. Each sample records an increase in pressure during progressive metamorphism, followed by nearly isobaric cooling (Fig. 17-5). The nine samples yield an average "core" pressure of 3.4 kb, peak pressure of 4.1 kb, and retrograde pressure of 4.3 kb. The data indicate that the rocks followed a metamorphic path that plots as a counterclockwise loop on a *P-T* diagram.

Rate of Retrograde Cooling

The 15–90-μm thickness of retrograde rims on garnets can be used to estimate postmetamorphic cooling rates. Cygan and Lasaga (1985) presented experimental data for Mg self-diffusion in garnet, $E_A = 239 \pm 16$ kJ/mol and log $D_0 = -8.01 \pm 0.75$ m^2/sec. They further suggested, from studies of Lasaga *et al.* (1977), that these diffusion parameters may apply to Fe-Mg interdiffusion at temperatures as low as 500°C. Their experimental results are used in the following calculation.

Crank (1975) and Cygan and Lasaga (1985) presented the equation

$$x^2 = 4Dt$$

to relate diffusion rate, D, to the approximate penetration depth, x, of a diffusing species. In this relation t represents the duration of diffusion. An implicit assumption necessary to apply experimental Mg self-diffusion to Fe-Mg interdiffusion is that Mg diffusion is the rate-limiting step in the coupled diffusion.

Garnet Fe-Mg interdiffusion rates estimated from Cygan and Lasaga (1985) are 7×10^{-21} cm^2/sec at 500°C and 2×10^{-21} cm^2/sec at 475°C. The former is the maximum temperature at which the retrograde diffusion occurred, because 500°C is the maximum

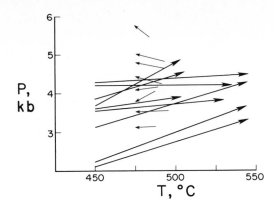

FIG. 17-5. Metamorphic *P-T* paths determined for nine samples from Truchas and Rio Mora areas, based on interpretations of zoning and inclusion relationships in garnet, plagioclase, biotite, and muscovite. Constructed from data in Table 17-2. Large, heavy arrows show the evolution of *P-T* conditions during garnet growth; small, light arrows indicate *P-T* changes during retrograde alteration of garnet rims.

temperature attained by rocks in the southern Truchas Range, and garnets from those rocks preserve retrograde-zoned rims averaging 30 μm in thickness. The duration of retrograde Fe-Mg interdiffusion can be estimated using two extreme, limiting approaches.

The shortest time required to develop the 30-μm retrograde rims, assuming that the experimental data are relevant, is 10 m.y. This assumes that all retrograde diffusion occurred at 500°C, the peak metamorphic temperature. The longest time considered reasonable is 99 m.y., calculated assuming that all retrograde diffusion occurred at 475°C, the approximate final temperature of garnet-biotite equilibration. Because temperature undoubtedly fell gradually during retrograde Fe-Mg interdiffusion in garnet, an intermediate cooling time of 40–50 m.y. seems reasonable. If the above interpretations of mineral zoning are correct, postmetamorphic cooling rates were slightly less than 1°C/m.y. for the first 25°C of cooling.

These inferred slow cooling rates are consistent with the *P-T* path shown by the rocks. One would expect slow cooling in rocks buried by 11–15 km of overburden. The slow cooling rates for the New Mexico samples contrast sharply with the 13°C/m.y. cooling rates determined from a rapidly uplifted metamorphic sequence in British Columbia (Harrison *et al.*, 1979).

Isobaric cooling may be a major reason why the New Mexico rocks preserve such a remarkable degree of metamorphic equilibration. During isobaric cooling, rocks attain their maximum entropy state at the highest *P* and *T* they experience. In contrast, during isothermal decompression (a "clockwise" *P-T* path), maximum entropy is achieved after maximum pressures and at or even after maximum temperatures. A clockwise *P-T* path may enhance the development of disequilibrium metamorphic assemblages.

THE RELATIONS BETWEEN METAMORPHISM AND DEFORMATION

The regional uniformity of metamorphic conditions argues strongly that plutons contributed minimally to the metamorphic heat budget, except in local areas such as near the Sandia Granite. The regional similarity in structural fabrics can be interpreted to mean that the entire map area underwent the same deformational history. The regionally consistent peak metamorphic conditions indicate that all rocks may have experienced similar thermal histories as well. This interpretation is supported by the observation that garnets in the southern Manzano Mountains, near the southern edge of the study area, preserve 50–80-μm-thick retrograde rims very similar to those in the Truchas-Rio Mora samples. It is thus inferred that the *P-T* path calculated above, based upon mineral chemistry in the Truchas and Rio Mora uplifts, may be applicable to the entire study area. The horizontal nature of metamorphic isograds, with isothermal surfaces cutting across even the youngest folds (Grambling and Williams, 1985, Fig. 3), requires that heat from the thermal event outlasted the deformation.

Deformation and the *P-T* Path

Garnet and plagioclase grew during crenulation of the early schistosity. The 0.7-kb (21%) pressure increase recorded by mineral zoning thus records a pressure change during the young deformation. Minor folds associated with the crenulation event record minimum shortening strains of 20–50%. The magnitude of this shortening strain is sufficient to account for the entire recorded increase in metamorphic pressure.

A plausible mechanism for metamorphism and deformation is that the rocks were shortened horizontally during the young deformation. In response, the rocks thickened, and at least part of the thickening occurred by ductile processes. This thickening resulted in the *P-T* path recorded by the zoned minerals. The regional uniformity in metamorphic conditions and deformational history supports the interpretation that such a process was distributed across a tremendous area.

Three different "types" of regional metamorphic *P-T* paths have been described in the literature. The first is the "clockwise" *P-T* path proposed by England and Richardson (1977), England and Thompson (1984), Thompson and England (1984), and Day (1985). Clockwise *P-T* paths, inferred to result from extreme continental thickening followed by rapid uplift and erosion, have been documented by Hollister *et al.* (1979), Spear and Selverstone (1983), and several other workers. The second type of *P-T* path is isobaric cooling after (presumed) isobaric heating. This *P-T* path has been atrributed to cooling following magmatic heating, as described by Vernon (1982), Bohlen *et al.* (1983), and several additional studies. A third type of *P-T* path is described in this report: a counterclockwise *P-T* path with initial increases in both *P* and *T*, followed by nearly isobaric cooling.

This third type of *P-T* path may be limited to situations where only minor amounts of synmetamorphic crustal thickening took place. The *P-T* path may occur only where deformation occurred by ductile mechanisms, or where overthrust sheets were extremely thin. Thick overthrust sheets presumably result in a thick, isostatically elevated crust which enhances rapid erosion, thus creating a clockwise *P-T* path. A minor amount of crustal thickening may not trigger rapid erosion.

An interesting speculation concerns the thickness of the premetamorphic crust in New Mexico. The *P-T* data presented in this study record 2–3 km of crustal thickening during the D_3 event. D_2 may have caused additional thickening, but at present the D_2 deformation is poorly understood. Isoclinal recumbent folds and (at least local) nappes associated with D_1 record a significant amount of crustal thickening. If one accepts that isobaric cooling can occur only with a crustal thickness of 40 km or less, and subtracts a reasonable amount of D_1 and D_2 thickening, it is conceivable that the premetamorphic crust in this area was as thin as 25–30 km.

CONCLUSIONS

Proterozic regional metamorphism occurred during deformation in northern and central New Mexico. Rock mineralogy records a metamorphic *P-T* path in which pressures and temperatures both increased, followed by very slow cooling at depth. Metamorphic peak temperatures were probably attained during the final stages of deformation, but isograds overprint mapped folds, presumably because deformational stresses dissipated faster than metamorphic heat.

The deformation and metamorphism resulted in an unusual distribution of metamorphic *P-T* conditions: across an area of approximately 75,000 km², metamorphic regional peak conditions cluster near 520°C and 4 kb. The metamorphic *P-T* conditions are reflected in the widespread development of rocks with coexisting kyanite, andalusite, sillimanite, and related minerals that preserve textural, spatial, and chemical evidence consistent with the attainment of chemical equilibrium.

ACKNOWLEDGMENTS

Research was supported by NSF grants EAR83-09503, EAR84-06572, and EAR82-01211, which partially supported purchase of the electron microprobe. Additional support came from the New Mexico Bureau of Mines and Mineral Resources and Anaconda Minerals, Inc. In addition to these organizations, I thank Tim Bell, Bill Cavin, Kent Condie, Jim Connolly, Rod Holcombe, Steve Phipps, Jack Reed, Jr., Jamie Robertson, Mike Williams, and Lee Woodward for contributions of samples and discussions.

REFERENCES

Anderson, J. L., 1983, Proterozoic anorogenic granite plutonism of North America, *in* Medaris, L. G., Jr., *et al.*, eds., *Proterozoic Geology:* Geol. Soc. America Mem. 161, p. 133–154.

Apted, M. J., and Liou, J. G., 1983, Phase relations among greenschist, epidote amphibolite, and amphibolite in a basaltic system: *Amer. J. Sci.*, v. 283–A, p. 328–354.

Armstrong, D. G., and Holcombe, R. J., 1982, Precambrian rocks of a portion of the Pedernal Highlands, Torrance County, New Mexico; in Grambling, J. A., and Wells, S. G., eds.: *Albuquerque County II:* N.M. Geol. Soc. Guidebook 33, p. 203–210.

Baldridge, W. S., Olsen, K. H., and Callender, J. F., 1984, Rio Grande rift — Problems and perspectives; *in* Baldridge, W. S., Dickerson, P. W., Riecker, R. E., and Zidek, J., eds., *Rio Grande Rift — Northern New Mexico:* N.M. Geol. Soc. Guidebook 35.

Baltz, E. H., and O'Neill, J. M., 1980a, Preliminary geologic map of the Mora River area, Sangre de Cristo Mountains, Mora County, New Mexico: *U.S. Geol. Survey Open File Rpt. 80-374.*

——, and O'Neill, J. M., 1980b, Preliminary geologic map of the Sapello River area, Sangre de Cristo Mountains, Mora and San Miguel Counties, New Mexico: *U.S. Geol. Survey Open File Rt. 80-398.*

Bauer, P., 1983, Geology of Precambrian rocks of the southern Manzano Mountains, New Mexico: M.S. thesis; Univ. New Mexico, Albuquerque, N. Mex., 133 p.

Bell, T. H., and Rubenach, M. J., 1983, Sequential porphyroblast growth and crenulation cleavage development during progressive deformation: *Tectonophysics*, v. 92, p. 171–194.

Berkley, J. L., and Callender, J. F., 1979, Precambrian metamorphism in the Placitas-Juan Tabo area, northwestern Sandia Mountains, New Mexico; *in* Ingersoll, R. V., and Woodward, L. A., eds., *Guidebook of Santa Fe Country:* N.M. Geol. Soc. Guidebook 30, p. 181–188.

Bohlen, S. R., Wall, V. J., and Boettcher, A. L., 1983, Experimental investigations and geological applications of equilibria in the system $FeO-TiO_2-Al_2O_3-SiO_2-H_2O$: *Amer. Mineralogist*, v. 68, p. 1049–1058.

Bolton, W. R., 1976, Precambrian geochronology of the Sevilleta metarhyolite and the Los Pinos, Sepultura, and Priest plutons of the southern Manzano uplift, New Mexico: M.S. thesis; New Mexico Inst. Mining Technol., Socorro, N. Mex., 45 p.

Bowen, N. L., and Tuttle, O. F., 1949, The system MgO-SiO_2-H_2O: Geol. Soc. America Bull., v. 60, p. 439–460.

Bowring, S. A., and Condie, K. C., 1982, U-Pb zircon ages from northern and central New Mexico: Geol. Soc. America Abstr. with Programs, v. 14, p. 304.

——, Kent, S. C., and Sumner, W., 1983, Geology and U-Pb geochronology of Proterozoic rocks in the vicinity of Socorro, New Mexico; in Chapin, C. E., and Callender, J. F., eds., Socorro Region II: N.M. Geol. Soc. Guidebook 34, p. 137–142.

——, Reed, J. C., Jr., and Condie, K. C., 1984, U-Pb geochronology of Proterozoic volcanic and plutonic rocks, Sangre de Cristo Mountains, New Mexico: Geol. Soc. America Abstr. with Programs, v. 16, p. 216.

Brookins, D. G., 1982, Radiometric age of Precambrian rocks from central New Mexico; in Grambling, J. A., and Wells, S. G., eds., Albuquerque Country II: N.M. Geol. Soc. Guidebook, v. 33, p. 187–190.

Calvin, W. J., Connolly, J. R., Woodward, L. A., Edwards, D. L., and Parchman, M., 1982, Precambrian stratigraphy of Manzanita and North Manzano Mountains, New Mexico; in Grambling, J. A., and Wells, S. G., eds., Albuquerque Country II: N.M. Geol. Soc. Guidebook 33, p. 191–196.

Crank, J., 1975, The Mathematics of Diffusion: Oxford University Press, Oxford, 414 p.

Cygan, R. T., and Lasaga, A. C., 1985, Self-diffusion of magnesium in garnet at 750° to 900°C: Amer. J. Sci., v. 285, p. 328–350.

Day, H. W., 1985, Distinguishing among major controls on the apparent thermal and barometric structure of metamorphic belts: Geol. Soc. America Abstr. with Programs, v. 17, p. 559.

England, P. C., and Richardson, S. W., 1977, The influence of erosion upon mineral facies of rocks from different metamorphic environments: Geol. Soc. London J, v. 134, p. 201–213.

——, and Thompson, A. B., 1984, Pressure-temperature-time paths of regional metamorphism — I. Heat transfer during the evolution of regions of thickened continental crust: J. Petrology, v. 25, p. 894–928.

Ferry, J. M., and Spear, F. S., 1978, Experimental calibration of the partitioning of Fe and Mg between garnet and biotite: Contrib. Mineralogy Petrology, v. 66, p. 113–117.

Ganguly, J., 1972, Staurolite stability and related parageneses: J. Petrology, v. 13, p. 335–365.

Ghent, E. D., and Stout, M. Z., 1981, Geobarometry and geothermometry of plagioclase-biotite-garnet-muscovite assemblages: Contrib. Mineralogy Petrology, v. 76, p. 92–97.

Gonzales, R. A., 1968, Petrography and structure of the Pedernal Hills, Torrance County, New Mexico: M.S. thesis; Univ. New Mexico, Albuquerque, N. Mex., 78 p.

Grambling, J. A., 1981, Kyanite, andalusite, sillimanite and related mineral assemblages in the Truchas Peaks region, New Mexico: Amer. Mineralogist, v. 66, p. 702–722.

——, and Williams, M. L., 1985, The effects of Fe^{3+} and Mn^{3+} on aluminum silicate phase relations in north-central New Mexico, U.S.A.: J. Petrology, v. 26, p. 324–354.

Haas, H., and Holdaway, M. J., 1973, Equilibria in the system Al_2O_3-SiO_2-H_2O involving the stability limits of pyrophyllite, and thermodynamic data of pyrophyllite: Amer. J. Sci., v. 273, p. 449–464.

GRAMBLING

Harrison, T. M., Armstrong, R. L., Naeser, C. W., and Harakal, J. E., 1979, Geochronology and thermal history of the Coast Plutonic Complex, near Prince Rupert, British Columbia: *Can. J. Earth Sci.*, v. 16, p. 400–410.

Holdaway, M. J., 1971, Stability of andalusite and the aluminum silicate phase diagram: *Amer. J. Sci.*, v. 271, p. 97–131.

____, 1978, Significance of chloritoid-bearing and staurolite-bearing rocks in the Picuris Range, New Mexico: *Geol. Soc. America Bull.*, v. 89, p. 1404–1414.

Hollister, L. S., Burruss, R. C., Henry, D. L., and Hendel, E., 1979, Physical conditions during uplift of metamorphic terranes, as recorded by fluid inclusions: *Bull. Mineralogie*, v. 102, p. 555–561.

Klich, I., 1983, Precambrian geology of the Elk Mountain–Spring Mountain area, San Miguel County, New Mexico: M.S. thesis; New Mexico Inst. Mining and Technology, Socorro, N. Mex., 147 p.

Lambert, E. E., 1983, Geology and petrochemistry of ultramafic and orbicular rocks, Zuni Mountains, Cibola County, New Mexico: M.S. thesis; Univ. New Mexico, Albuquerque, N. Mex., 166 p.

Lasaga, A. C., Richardson, S. M., and Holland, H. D., 1977, The mathematics of cation diffusion and exchange between silicate minerals during retrograde metamorphism; *in* Saxena, S. K., and Bhattacharji, S., eds., *Energetics of Geological Processes:* New York, Springer-Verlag, p. 353–388.

Leyenberger, T. L., 1983, Precambrian geology of Cimarron Canyon, Colfax County, New Mexico: M.S. thesis; Univ. New Mexico, Albuquerque, N. Mex., 93 p.

Lisenbee, A. L., Woodward, L. A., and Connolly, J. R., 1979, Tijeras-Cañoncito fault system — A major zone of recurrent movement in north-central New Mexico; *in* Ingersoll, R. V., and Woodward, L. A., eds., *Guidebook of Santa Fe Country:* N.Mex. Geol. Soc. Guidebook 30, p. 89–99.

Manley, K., 1984, Brief summary of the Tertiary geologic history of the Rio Grande rift in northern New Mexico; *in* Baldridge, W. S., Dickerson, P. W., Riecker, R. E., and Zidek, J., eds., *Rio Grande Rift — Northern New Mexico:* N.Mex. Geol. Soc. Guidebook 35, p. 63–66.

Moody, J. B., Meyer, D., and Jenkins, J. E., 1983, Experimental characterization of the greenschist/amphibolite boundary in mafic systems: *Amer. J. Sci.*, v. 283, p. 48–96.

Reed, J. C., Jr., 1984, Proterozoic rocks of the Taos Range, Sangre de Cristo Mountains, New Mexico; *in* Baldridge, W. S., Dickerson, P. W., Riecker, R. E., and Zidek, J., eds., *Rio Grande Rift — Northern New Mexico:* N.Mex. Geol. Soc. Guidebook, 35, p. 179–185.

Renshaw, J. L., 1984, Precambrian geology of the Thompson Peak area, Santa Fe County, New Mexico: M.S. thesis; Univ. New Mexico, Albuquerque, N. Mex., 197 p.

Richardson, S. W., 1968, Staurolite stability in a part of the system Fe-Al-Si-O-H: *J. Petrology*, v. 9, p. 467–488.

Robertson, J. M., and Moench, R. H., 1979, The Pecos greenstone belt: Proterozoic volcano-sedimentary sequence in the southern Sangre de Cristo Mountains, New Mexico; *in* Ingersoll, R. V., and Woodward, L. A., eds.: *Guidebook of Santa Fe Country:* N.Mex. Geol. Soc. Guidebook, 30, p. 165–174.

Silver, L. T., Bickford, M. E., Van Schmus, W. R., Anderson, J. L., and Medaris, L. G., 1977, The 1.4–1.5 b.y. transcontinental anorogenic plutonic perforation of North America: *Geol. Soc. America Abstr. with Programs*, v. 9, p. 1176–1177.

Spear, F. S., and Selverstone, J., 1983, Quantitative P-T paths from zoned minerals: theory and tectonic applications: *Contrib. Mineralogy Petrology*, v. 83, p. 348–357.

Thompson, A. B., and England, P. C., 1984, Pressure-temperature-time paths of regional metamorphism — II. Their inference and interpretation using mineral assemblages in metamorphic rocks: *J. Petrology*, v. 25, p. 929–955.

Vernon, R. H., 1982, Isobaric cooling of two regional metamorphic complexes related to igneous intrusions in southeastern Australia: *Geology*, v. 10, p. 76–81.

Wobus, R. A., 1984, Proterozoic supracrustal rocks and plutons of the Cimarron Canyon area, north-central New Mexico: *Geol. Soc. America Abstr. with Programs*, v. 16, p. 260.

———, 1985, Changes in the nomenclature and stratigraphy of Proterozoic metamorphic rocks, Tusas Mountains, north-central New Mexico: *U.S. Geol. Survey Bull. 1571*, 19 p.

Yoder, H. S., 1966, Spilites and serpentinites: *Carnegie Inst. Wash. Yearbook 65*, p. 269–283.

Stephen J. Reynolds
Arizona Geological Survey
Tucson, Arizona 85719

Stephen M. Richard
Department of Geological Sciences
University of California
Santa Barbara, California 93106

Gordon B. Haxel
U.S. Geological Survey
Flagstaff, Arizona 86001

Richard M. Tosdal
U.S. Geological Survey
Menlo Park, California 94025

Stephen E. Laubach
Texas Bureau of Economic Geology
Austin, Texas 78713

18

GEOLOGIC SETTING OF MESOZOIC AND CENOZOIC METAMORPHISM IN ARIZONA

ABSTRACT

Rocks affected by Mesozoic and Cenozoic regional metamorphism in Arizona are restricted to the Basin and Range Province, which has undergone a complex history of post-Paleozoic tectonism. Evidence of such metamorphism is lacking, at exposed crustal levels, in the other two physiographic-geologic provinces, the Colorado Plateau and Transition Zone. The contrasting thermal history of the Basin and Range Province is reflected by higher conodont color-alteration indices of Paleozoic rocks and by widespread Late Cretaceous to middle Tertiary K-Ar and fission-track cooling ages of Proterozoic and younger rocks.

Most prograde metamorphism occurred in Mesozoic to early Cenozoic time and was associated with regional deformation, plutonism, or both. Metamorphism associated with regional deformation but not plutonism occurs along major Cretaceous thrust faults in west-central and southwestern Arizona. This metamorphism produced mylonitic fabrics in preexisting crystalline rocks within the thrust zones or along subsidiary ductile shear zones. Paleozoic and Mesozoic supracrustal rocks beneath the thrust sheets have been variably converted into crystalloblastic schists or metamorphic tectonites that locally increase in textural and metamorphic grade upward toward major thrusts.

Metamorphism associated with plutonism but not thrusting occurred in Triassic and Late Jurassic time in southwestern Arizona and southeastern California and in Late Cretaceous to early Tertiary time in west-central and southeastern Arizona. This type of metamorphism is characterized by structurally downward increases in textural and metamorphic grade and, at relatively deep levels, by associated synkinematic orthogneiss and pegmatite. Metamorphism associated with both thrusting and plutonism is best represented in south-central Arizona, where Late Cretaceous to early Tertiary metamorphism occurred concurrently with emplacement of peraluminous granites and movement along the Baboquivari thrust and related faults.

A distinct episode of dominantly retrograde metamorphism occurred in middle Tertiary time and was associated with mylonitization and brittle deformation along regional, normal-displacement shear zones, referred to as metamorphic core complexes and detachment faults. Mylonitization along deeper segments of the shear zones was accompanied by incomplete retrogression of plutonic and preexisting metamorphic protoliths. Detachment faults and associated breccias that formed along shallower segments of the shear zones are associated with widespread chlorite-epidote alteration. Normal displacement on the shear zones resulted in progressive unroofing of lower-plate rocks with fabrics formed both by Tertiary mylonitization along deeper levels of the shear zone and by earlier, unrelated episodes of prograde metamorphism. Tertiary mylonitization and detachment faulting occurred within a regime of continuously decreasing temperature and pressure, and therefore produced fabrics that are distinct from those formed during earlier prograde metamorphism.

INTRODUCTION

Arizona includes three physiographic-geologic provinces: the Colorado Plateau, the Transition Zone, and the Basin and Range Province (Fig. 18-1). The Colorado Plateau of northeastern Arizona is characterized by gently dipping Paleozoic and Mesozoic sedimentary rocks deposited on a basement of Proterozoic metamorphic and granitic rocks. In the adjacent Transition Zone, late Mesozoic to early Cenozoic erosion has largely removed the Paleozoic and Mesozoic cover, resulting in widespread exposure of the underlying Proterozoic rocks (Peirce, 1986). The Basin and Range Province of southern and western Arizona has experienced a more complicated tectonic history, culminating in middle to late Tertiary extensional faulting that outlined the present-day mountain blocks and basins.

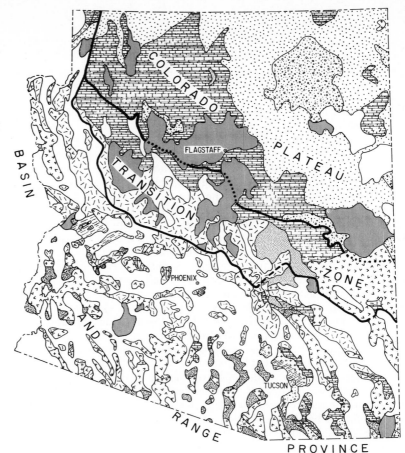

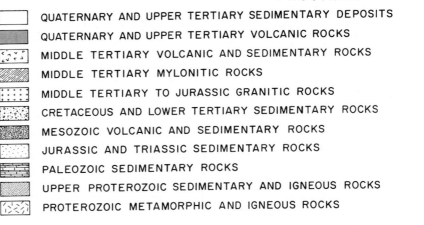

QUATERNARY AND UPPER TERTIARY SEDIMENTARY DEPOSITS

QUATERNARY AND UPPER TERTIARY VOLCANIC ROCKS

MIDDLE TERTIARY VOLCANIC AND SEDIMENTARY ROCKS

MIDDLE TERTIARY MYLONITIC ROCKS

MIDDLE TERTIARY TO JURASSIC GRANITIC ROCKS

CRETACEOUS AND LOWER TERTIARY SEDIMENTARY ROCKS

MESOZOIC VOLCANIC AND SEDIMENTARY ROCKS

JURASSIC AND TRIASSIC SEDIMENTARY ROCKS

PALEOZOIC SEDIMENTARY ROCKS

UPPER PROTEROZOIC SEDIMENTARY AND IGNEOUS ROCKS

PROTEROZOIC METAMORPHIC AND IGNEOUS ROCKS

FIG. 18-1. Generalized geologic map of Arizona showing boundaries of physiographic-geologic provinces. This map, in conjunction with Fig. 18-2 and Table 18-1, illustrates rock units affected by Mesozoic and Cenozoic metamorphism. All faults have been omitted.

Mesozoic and Cenozoic regional metamorphism has affected much of the Basin and Range Province (Fig. 18-2), but is not manifest at the surface in the other two provinces. Prograde metamorphism in the Basin and Range Province was episodic and primarily occurred during Mesozoic and early Tertiary plutonism and crustal shortening. A distinct episode of retrograde metamorphism and alteration occurred during middle Tertiary time and was associated with normal displacement on regional, low-angle shear zones, com-

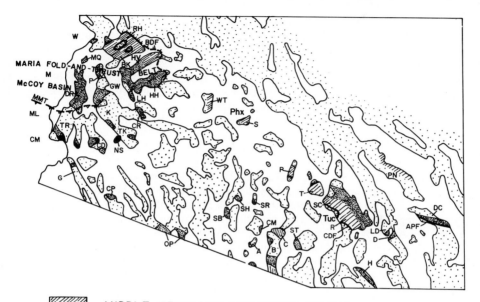

◫ MIDDLE TERTIARY MYLONITIC FABRIC

■ OROCOPIA SCHIST

▨ MESOZOIC – EARLY CENOZOIC METAMORPHIC FABRIC

▦ ALL OTHER BEDROCK

FIG. 18-2. Metamorphic map of southern Arizona. Hatchured lines are middle Tertiary detachment faults and barbed lines are Mesozoic to early Tertiary thrust faults. Labeled features are as follows (by region): (1) southeastern Arizona: APF, Apache Pass fault zone; CDF, Catalina detachment fault; D, Dragoon Mtns.; DC, Dos Cabezas Mtns.; H, Huachuca Mtns.; LD, Little Dragoon Mtns.; P, Picacho Mtns.; PN, Pinaleno Mtns.; R, Rincon Mtns.; SC, Santa Catalina Mtns.; T, Tortolita Mtns.; Tuc, Tucson; (2) south-central Arizona: A, Artesa Mtns.; B, Baboquivari Mtns.; C, Coyote Mtns.; CM, Comobabi Mtns.; OP, Organ Pipe National Monument and Quitobaquito Hills; SB, Sierra Blanca; SH, Sheridan Mtns.; SR, Santa Rosa Mtns.; ST, Sierrita Mtns.; and (3) western and central Arizona: BDF, Buckskin detachment fault; BK, Buckskin Mtns.; CD, Castle Dome Mtns.; CM, Chocolate Mtns.; CP, Copper Mtns. and Baker Peaks; CR, Cemetary Ridge; DR, Dome Rocks Mtns.; G, Gila Mtns.; GW, Granite Wash Mtns.; HH, Harquahala Mtns.; HV, Harcuvar Mtns.; K, Kofa Mtns.; LH, Little Harquahala Mtns.; M, McCoy Mtns.; ML, Mule Mtns.; MMT, Mule Mountains thrust; MO, Mesquite Mtns.; NS, Neversweat Ridge; P, Plomosa Mtns.; Phx, Phoenix; RH, Rawhide Mtns.; S, South Mtns.; TK, Tank Mtns.; TR, Trigo Mtns.; W, Whipple Mtns.; WT, White Tank Mtns.

monly referred to as metamorphic core complexes or detachment faults. Metamorphic and structural fabrics produced during this middle Tertiary extension are distinct from those formed during the previous metamorphic episodes and will be discussed separately.

THERMAL HISTORY OF ARIZONA

The three provinces of Arizona have experienced different Phanerozoic thermal histories. The continental crust that underlies all three provinces was formed between 1.8 and 1.4 b.y. ago. This Proterozoic basement was subsequently covered by a relatively uniform thickness of less than 1.5 km of Paleozoic cratonic strata and a more variable thickness of Mesozoic and Cenozoic rocks. Some variations in thermal history might be expected, therefore, because of uneven Mesozoic and Cenozoic burial, but more extreme lateral and vertical variations in temperature would have resulted from Mesozoic and Cenozoic plutonism and tectonism.

The cumulative thermal history of any area is reflected in K-Ar and fission-track cooling ages of the oldest rocks. Proterozoic rocks in the Colorado Plateau and the Transition Zone yield K-Ar cooling ages in excess of 1000–1500 m.y. (Fig. 18-3), which demonstrates that these rocks were shielded from significant Phanerozoic heating and were never buried deeply enough to fully reset the ages. In contrast, Proterozoic rocks within the Basin and Range Province yield K-Ar ages that range from greater than 1500 m.y. to less than 25 m.y. Areas of the Basin and Range Province that have retained Proterozoic cooling ages typically lack large Mesozoic and Cenozoic plutons and structures. For example, the presence of Proterozoic cooling ages in the Hualapai and Cerbat Mountains in the Basin and Range Province of northwestern Arizona indicates that these rocks were not subjected to major Phanerozoic thermal events, and therefore have a history akin to that of the adjacent Transition Zone (Fig. 18-3). These mountain ranges appear to represent an area where Neogene block faulting has cut across the northwestward continuation of a preexisting, late Mesozoic to early Tertiary uplift, termed the Mogollon Highlands (Peirce, 1986). Other areas of the Basin and Range Province with preserved Proterozoic or partially reset Paleozoic cooling ages include (1) areas immediately north and south of the Santa Catalina-Rincon metamorphic core complex, (2) extensive exposures of relatively undeformed Proterozoic granite south and southwest of Phoenix, and (3) the southern Pinaleno Mountains and western Dos Cabezas Mountains of southeastern Arizona. These areas cannot have been significantly affected by Mesozoic and Cenozoic plutonism, regional metamorphism, or deep burial beneath thrust sheets.

Rapid lateral variations in cooling ages within a small area largely reflect the localized thermal effects of Mesozoic and Cenozoic plutons. An excellent example of this occurs in east-central Arizona (near closely spaced data points on Fig. 18-3), where emplacement of a large Late Cretaceous to early Tertiary granitic pluton has variably reset K-Ar ages in the Proterozoic host rocks (Creasey, 1980). Local variations in cooling ages within other areas of Arizona are due to juxtaposition of different crustal levels across faults with significant vertical separation. Notable examples of this include (1) the Catalina detachment fault in the Santa Catalina and Rincon Mountains, where middle Tertiary normal displacement has juxtaposed upper-plate rocks bearing Proterozoic K-Ar ages against mylonitic lower-plate rocks that yield late Oligocene and early Miocene cooling ages; and (2) the Apache Pass fault zone in the Dos Cabezas Mountains of southeastern Arizona, which separates Proterozoic rocks with early Tertiary cooling ages to the northeast from those with Proterozoic cooling ages to the southwest (Fig. 18-3; see also Fig. 18-6).

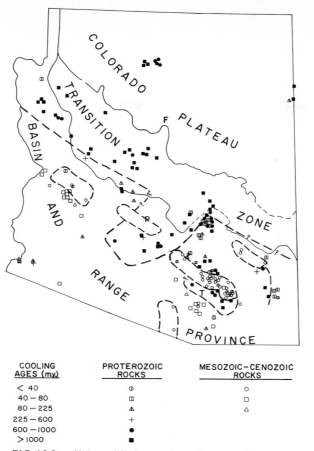

COOLING AGES (m.y.)	PROTEROZOIC ROCKS	MESOZOIC-CENOZOIC ROCKS
< 40	⊕	○
40 – 80	⊞	□
80 – 225	▲	△
225 – 600	+	
600 – 1000	●	
> 1000	■	

FIG. 18-3.   K-Ar and fission-track cooling ages of Proterozoic, Mesozoic, and Cenozoic crystalline rocks in Arizona (Reynolds *et al.*, 1986c). Dashed lines outline areas with similar cooling ages or metamorphic histories.

Younger rock units, such as Mesozoic and Cenozoic plutonic and metamorphic rocks, also locally yield cooling ages that are significantly younger than the plutonic or metamorphic event that formed the dated mineral (Fig. 18-3). This is especially true in southwestern and west-central Arizona, where rock units of various ages yield Late Cretaceous to early Tertiary cooling ages. These ages reflect regional uplift that accompanied or followed Mesozoic to early Tertiary plutonism, metamorphism, and crustal shortening. Middle Tertiary cooling ages occur within the northwest-trending zone of metamorphic core complexes and locally in south-central Arizona, where detachment faulting has tectonically exhumed lower-plate rocks from moderate depths.

In addition to K-Ar dates, regional and local thermal histories can be partially discerned by using the color alteration index (CAI) of conodonts, apatitic microfossils that systematically change color when heated (Epstein *et al.*, 1977). CAI values, which are a function of temperature and the time period over which that temperature persisted, have been determined for Paleozoic localities throughout much of Arizona (Fig. 18-4; Wardlaw and Harris, 1984). Conodonts from rocks in the Colorado Plateau generally have CAI values of 1–1.5, corresponding to maximum paleotemperatures of approximately

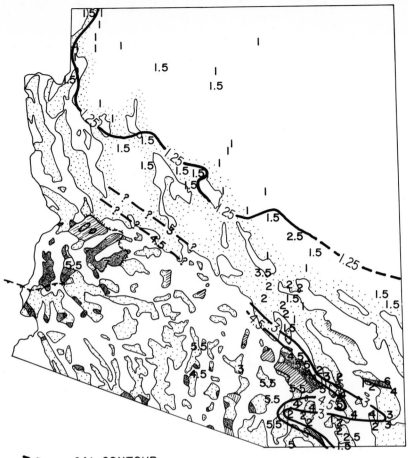

FIG. 18-4. Map showing the lowest conodont-color-alteration index (CAI) within subregions of Arizona. CAI data are from Wardlaw *et al.* (1983). The lowest CAI value within any area is most representative of the regional thermal history, whereas higher CAI values within a region of predominantly low CAI values probably reflect localized thermal effects of plutons (Wardlaw and Harris, 1984).

50°C, assuming a 100–200 m.y. duration of burial beneath the thin Paleozoic and Mesozoic cover. Maximum paleotemperatures as high as 75°C are indicated for parts of the Transition Zone by CAI values of 1.5–2. These slightly higher plaeotemperatures were probably due, in part, to an increased thermal input during Mesozoic and Cenozoic magmatism, since the total Paleozoic and Mesozoic cover was probably less on the Transition Zone than on the Colorado Plateau. Nevertheless, the consistently low CAI values indicate that both provinces escaped significant Mesozoic and Cenozoic metamorphism, at least at the present level of exposure. Evidence exists, however, that rocks deep beneath the southwestern edge of the Transition Zone were affected by metamorphism and ductile shearing in both Mesozoic and Cenozoic time (Reynolds and Spencer, 1985).

CAI values in the Basin and Range Province are much more variable and generally much higher, with values less than 2 being restricted to a few scattered locations (Wardlaw and Harris, 1984). These higher values probably reflect: (1) contact metamorphism and hydrothermal circulation adjacent to Mesozoic and Cenozoic plutons; (2) regional metamorphism, in part due to burial beneath thrust sheets; and (3) increased stratigraphic burial in west-central and southern Arizona, where the cumulative Paleozoic and Mesozoic section is thickest. CAI values that vary widely within a small area indicate either pluton-related thermal effects or the presence of major faults, such as the Apache Pass fault zone in the Dos Cabezas Mountains, that have juxtaposed Paleozoic rocks with contrasting thermal histories (Fig. 18-4; see also Fig. 18-6). In contrast, CAI values that are consistently high over larger areas reflect regional metamorphism. Areas of Arizona with regionally high CAI values of 5 or greater, such as in south-central and west-central Arizona, are those areas affected by Mesozoic to early Cenozoic prograde metamorphism. CAI values of 5 suggest maximum paleotemperatures of approximately 325°C for a heating event of 10 m.y. duration (Epstein *et al.*, 1977).

The conodont CAI values, in conjunction with the cooling ages discussed earlier, can be used to define regions with similar *cumulative* thermal histories (Fig. 18-5). The region with the lowest paleotemperatures includes all of the Colorado Plateau and parts of the Transition Zone. This zone is defined by CAI values of 1.5 or less and by Proterozoic cooling ages of Proterozoic rocks. The next paleothermal zone to the southwest comprises the remainder of the Transition Zone and parts of the Basin and Range Province, including the Hualapai and Cerbat Mountains in northwestern Arizona. This zone is characterized by CAI values of 1.5–2 and by relatively unperturbed cooling ages on Proterozoic rocks, which together suggest maximum paleotemperatures of approximately 75°C. The Basin and Range Province contains the following additional paleothermal zones: (1) a zone in southeastern Arizona with interspersed high and low CAI values and variably reset to unperturbed cooling ages; (2) a broad zone in west-central, southwestern, and south-central Arizona with uniformly high CAI values and strongly perturbed, but pre-middle Tertiary cooling ages; and (3) a northwest-trending, discontinuous belt with middle Tertiary cooling ages, corresponding to the zone of metamorphic core complexes.

MESOZOIC AND EARLY CENOZOIC METAMORPHISM

Rocks metamorphosed in Mesozoic to early Cenozoic time are exposed within a broad, northwest-trending belt across southern Arizona (Fig. 18-2; Table 18-1). This region was part of the relatively stable North American craton during Paleozoic and earliest Mesozoic time, but began to evolve into a complex orogen in Late Triassic or Early Jurassic time (Coney, 1978; Haxel *et al.*, 1980; Hayes and Drewes, 1978; Reynolds *et al.*, 1986b; Tosdal *et al.*, 1987). Silicic and intermediate volcanism and plutonism were widespread during most of Jurassic time and were accompanied by regional metamorphism at deeper crustal levels in at least some areas of southern Arizona (Tosdal, 1986a; Tosdal *et al.*, 1987). Postmagmatic faulting formed basins in which thick sequences of uppermost Jurassic and Cretaceous nonmarine clastic rocks were deposited (Bilodeau, 1978; Harding and Coney, 1985; Tosdal *et al.*, 1987).

Major crustal shortening, expressed by large-scale folding, thrust and high-angle reverse faulting, and cleavage formation, occurred during Cretaceous and early Tertiary

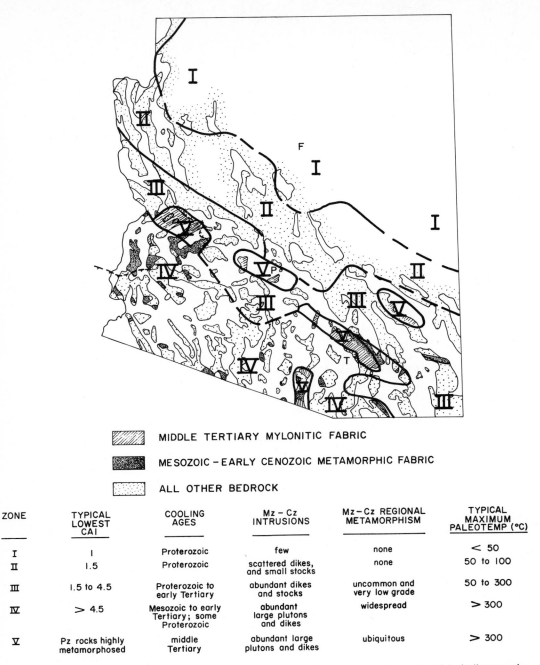

ZONE	TYPICAL LOWEST CAI	COOLING AGES	Mz – Cz INTRUSIONS	Mz–Cz REGIONAL METAMORPHISM	TYPICAL MAXIMUM PALEOTEMP (°C)
I	1	Proterozoic	few	none	< 50
II	1.5	Proterozoic	scattered dikes, and small stocks	none	50 to 100
III	1.5 to 4.5	Proterozoic to early Tertiary	abundant dikes and stocks	uncommon and very low grade	50 to 300
IV	> 4.5	Mesozoic to early Tertiary; some Proterozoic	abundant large plutons and dikes	widespread	> 300
V	Pz rocks highly metamorphosed	middle Tertiary	abundant large plutons and dikes	ubiquitous	> 300

FIG. 18-5.   Map of interpreted paleothermal zones of Arizona. Each zone contains rocks with similar cumulative Phanerozoic thermal histories or with similar Mesozoic and Cenozoic metamorphic histories. Zones I and II have largely escaped Phanerozoic thermal events areas, whereas zones III and IV contain areas affected by Cretaceous and Tertiary thermal disturbances. Rocks in Zone V, the zone of metamorphic core complexes, have been overprinted by middle Tertiary heating and uplift-related cooling so that little geochronologic record of previous thermal events is preserved.

time (G. H. Davis, 1979; Drewes, 1981; Haxel *et al.*, 1984; Reynolds *et al.*, 1986b). This deformation was accompanied by metamorphism of sub-greenschist to lower-amphibolite facies. Significant variations in metamorphic grade take place over distances of 1-10 km and reflect differences in exposed crustal levels, metamorphic gradients beneath thrust faults, and proximity to syn- or late-metamorphic granitic plutons.

In addition to the regional metamorphism described in this paper, contact metamorphism is common around many Jurassic, Cretaceous, and Tertiary plutons. Contact metamorphism is most obvious where such plutons have intruded relatively unmetamorphosed Paleozoic and Mesozoic supracrustal rocks, such as in the Sierrita, Santa Rita, Empire, and Dragoon Mountains of southeastern Arizona (Fig. 18-6). The thermal effects of pluton emplacement and attendant hydrothermal circulation are reflected by rapid lateral variations in the CAI values of the host rocks. Further discussion of contact metamorphism is contained in Barton *et al.* (Chapter 5, this volume).

Southeastern Arizona

Southeastern Arizona, the region east of the Santa Cruz River (Fig. 18-6), contains extensive exposures of Paleozoic and Mesozoic supracrustal rocks that appear in the field to be unmetamorphosed or weakly metamorphosed. These rocks are locally folded and cut by Late Cretaceous and early Tertiary (Laramide) thrust and reverse faults (G. H. Davis, 1979; Drewes, 1981), but generally lack mesoscopic metamorphic minerals and penetrative cleavage. Regionally metamorphosed Paleozoic and Mesozoic rocks are most widely exposed in a northwest-trending belt through the Santa Catalina-Rincon-Tortolita metamorphic core complex. This belt of metamorphic rocks is coincident with an area of consistently high CAI values that continues southeastward into the adjacent Little Dragoon and Dragoon Mountains. The belt of metamorphic rocks contains large Late Cretaceous, early Tertiary, and middle Tertiary plutons and major, basement-involved Laramide thrusts, reverse faults, and folds.

In the Santa Catalina, Rincon, and Tortolita Mountains, strongly tectonized and metamorphosed Paleozoic and Mesozoic rocks are exposed in the lower plate of the middle Tertiary Catalina detachment fault. Normal slip along this regionally southwest-dipping shear zone has resulted in complete tectonic and erosional denudation of tectonized metasedimentary and metaplutonic rocks. These tectonized rocks contain metamorphic fabrics formed during Late Cretaceous to early Tertiary deformation (Keith *et al.*, 1980; Thorman and Drewes, 1981), in addition to the widespread mylonitic fabric formed during middle Tertiary normal displacement on the shear zone (G. H. Davis, 1983). A major episode of metamorphism in the eastern Santa Catalina and Rincon Mountains occurred between emplacement of Late Cretaceous granodioritic to quartz dioritic plutons and Eocene garnet-two-mica granites (Keith *et al.*, 1980). Metamorphism and deformation affected sedimentary rocks as young as the Lower Cretaceous Bisbee Group and was associated, in large part, with the emplacement of east-vergent thrust sheets of Proterozoic rocks over Cretaceous and Paleozoic metasedimentary rocks (Bykerk-Kaufmann, 1986; Drewes, 1981; Thorman and Drewes, 1981). The thrust zones represent a moderately deep structural level because they are marked by variably developed mylonitic fabrics in the plutonic rocks and by metamorphic tectonite fabrics in the metasedimentary rocks. Dramatic changes in the degree of synkinematic metamorphism between adjacent thrust faults (e.g., the eastern Rincon Mountains versus the Johnny Lyon Hills) demonstrate either rapid lateral variations in thermal regime during thrusting, probably due to plutonism, or that the thrusts formed at different crustal levels. We suggest that the most highly metamorphosed Paleozoic and Mesozoic rocks, such as those in the eastern Rincon Mountains, were the most deeply buried beneath thrust sheets of

TABLE 18-1. Summary of Mesozoic–Early Cenozoic Metamorphic Mineral Assemblages[a]

Mountain Range	Affected Rocks	Mineral Assemblage[b,c]	Timing of Metamorphism	References[d]
Southeast Arizona				
Dos Cabezas Mountains Apache Pass fault zone	Paleozoic sedimentary rocks, Bisbee group	Pelite: and (after ky)-biot (after stau)-chld-tour-graphite	Early Tertiary K-Ar cooling ages	Drewes, 1984
Rincon Mountains below Catalina fault	Upper Proterozoic, Paleozoic, and Mesozoic sedimentary rocks	Pelite: q-cord-biot-andesine-cc to q-plag-ksp-musc-biot-cc; rare stau and sill. Calc-silicate: cc-dolo-talc-trem to act-epi-diop-ido-gar to epi-amph-gar	Post-Leatherwood quartz diorite (74 m.y.) pre-peraluminous granitoid (50–44 m.y.)	Keith et al., 1980; Bykerk-Kaufman, 1986; Lindgrey, 1983
Santa Catalina Mountains below Catalina Fault	Paleozoic sedimentary rocks, Proterozoic granitoids	Calc-silicates: act-epi-diop-ido-gar. Psammite: q-fs-epi-amph-gar-hem-mag	Post-Leatherwood quartz diorite (74 m.y.) pre-peraluminous granitoid (50–44 m.y.)	Keith et al., 1980; Kykerk-Kaufman, 1986
Tortolita Mountains	Proterozoic and lower Paleozoic sediments; Proterozoic granitoids	Pelite: rare sillimanite	Post-Late Cretaceous Chirreon Wash granodiorite; Middle Tertiary K-Ar cooling ages	Keith et al., 1980
Western Arizona				
Buckskin Mountains upper plate of detachment fault	Mesozoic clastic rocks	Sub-GS to lower GS	Cretaceous(?)	S. J. Reynolds and J. E. Spencer, unpublished data
Castle Dome Mountains	Rocks of Slumgullion Orocopia Schist	q-fs-biot-hbld-w. mica aA-lA	Jurassic(?) Late Cretaceous	Haxel et al., 1985; M. Grubensky, written communication, 1986
Dome Rock Mountains Southern	Jurassic silicic to intermediate volcanic rocks, McCoy Mountains Formation	Semi-pelites: w. mica-chl-q-brn carb to w. mica-biot-q-chl-epi. Calc-silicates: brn carb-q-epi-w. mica-chl to q-act-epi-biot	Post-McCoy Mountains Formation, pre-Late Cretaceous (Mule Mountains thrust)	Tosdal, 1986a
	Mule Mountains thrust zone	Pelites: biot-q-w. mica-chl. Meta-diorite: cc-q-chl	Late Cretaceous musc and biot K-Ar cooling ages from thrust zone	Tosdal, 1986a

Location	Rock units	Assemblage	Age/Description	References
Northern	Paleozoic and lower Mesozoic sedimentary rocks; Tung Hill shear zone; Meta-diorite in Proterozoic or Mesozoic schist	Pelites: q-biot-musc-chl; Semi-pelites: q-fs-w. mica-chl-epi; Green amph, peimontite; Blue-green amph (to act)-plag	Jurassic(?); intruded by plutonic suite resembling Trigo Peaks superunit; Unknown	Yeats, 1985; Yeats, 1985
Gila Mountains	Paleozoic sedimentary rocks, Proterozoic gneisses, Mesozoic plutonic rocks	GS to 1A	Jurassic or Cretaceous intruded by Gunnery Range batholith or granite of Sierra Prieta	S. J. Reynolds, unpublished data
Harcuvar Mountains	Proterozoic gneisses, Mesozoic plutons	q-fs-biot-hbld	K-Ar, synkinematic hbld, 70 m.y.; biot infoliated and unfoliated granitoid 44–51 m.y.	Shafiqullah et al., 1980; Reynolds, 1982
Harquahala Mountains	Paleozoic and Mesozoic sedimentary rocks; Proterozoic igneous and metamorphic rocks	Pelites: q-fs-w. mica-chl to q-fs-biot-gar-sta-(ky); Calc-silicates: dolo-q to dolo-talc-trem	K-Ar, musc, 62 m.y., biot 56 m.y.; from Proterozoic schist; K-Ar biot 22 m.y., hbld 28 m.y., post metamorphic dikes	Hardy, 1984; S. M. Richard, unpublished data; Shafiqullah et al., 1980
Kofa Mountains Northwestern	McCoy Mountains Formation	GS	Jurassic or Cretaceous	
Southwestern	Mesozoic clastic rocks	Unmeta. to A	Late Jurassic	Tosdal et al., 1987
Little Harquahala Mountains	Paleozoic and Mesozoic supracrustal rocks	Semi-pelites: q-alb-w. mica-chl-cc-(biot); Meta-andesite: q-alb-chl-epi-w. mica	Pre-Granite Wash granodiorite (biot K-Ar 66 and 71 m.y.)	Reynolds et al., 1986a, b, c; Shafiqullah et al., 1980
Middle Mountains	Orocopia Schist	aA-1A	Late Cretaceous	Haxel and Dillon, 1978
Muggins Mountains	Proterozoic granitoid, Paleozoic and Mesozoic sedimentary rocks	q-chl-w. mica	Jurassic(?)	R. M. Tosdal, published data
Neversweat Ridge	Orocopia Schist	uGS-1A		
Plomosa Mountains Southern	Paleozoic and Mesozoic sedimentary rocks	Semi-pelites: q-w. mica-chl (CAI7)	Deformation and very low grade metamorphism correlated with that in Little Harquahala Mountains	Reynolds et al., 1986a, b, c; Wardlaw and Harris, 1984
Northern	Paleozoic and Mesozoic sedimentary rocks; Proterozoic(?) igneous and metamorphic rocks	GS(?)	Intruded by Late Cretaceous Muddersback granite	Stoneman, 1985; Keith and Wilt, 1986

TABLE 18-1. Summary of Mesozoic–Early Cenozoic Metamorphic Mineral Assemblages[a] (continued)

Mountain Range	Affected Rocks	Mineral Assemblage[b,c]	Timing of Metamorphism	References[d]
Trigo Mountains	Winterhaven Formation (volcanic member)	q-chl-epi-biot-w. mica-cc (act)	Late Jurassic, latest Cretaceous	Haxel et al., 1985
	Orocopia Schist	Semi-pelite: alb-q-ksp-musc-biot-clz to q-musc-biot-gar-olig; Basalt: hbld-alb-epi-(gar)	Late Cretaceous	Haxel and Dillon, 1978; Haxel and Tosdal, 1986
	Orocopia schist and Winterhaven Formation near granite of Marcus Wash	Intermediate volcanics: Semi-pelite:musc-biot-epi-chl-cc	Late Cretaceous–early Tertiary	Haxel et al., 1985; Frost and Martin, 1983
	Triassic plutons	Metagabbro: biot-green hbld-andesine	Triassic: affects Lowe granodiorite, intruded by Trigo Peaks super-unit	Tosdal, 1986a
	Jurassic plutons (Trigo Peaks superunit)	Metadiorite: biot-q-alb-epi	Jurassic: older phases of suite metamorphosed; younger phases not affected	Tosdal, 1986a
South-Central Arizona				
Artesa	Jurassic supracrustal rocks	Unmetamorphosed to 1A?	Late Jurassic	Haxel et al., 1984; Tosdal et al., 1987
Baboquivari Mountains	Jurassic granitoids and supracrustal rocks	Psammite: q-fs-epi; q-fs-biot-musc; Meta-volcanics: q-fs-biot-epi-sph-act-tour	Late Cretaceous–early Tertiary	Haxel et al., 1984
Bates Well Hills	Paleozoic sediments	Calc-silicates: cc-diop-gar-clz-wol		M. Grubensky, written communication, 1986
North Comobabi Mountains	Jurassic granitoids and supracrustal rocks	Meta-volcanics: q-w. mica-epi-ksp-plag-chl-carb	Late Cretaceous–early Tertiary	Haxel et al., 1984
South Comobabi Mountains	Jurassic and Cretaceous supracrustal rocks	Meta-andesite: q-w. mica-sph-epi-carb	Late Cretaceous–early Tertiary	Haxel et al., 1984
Coyote Mountains	Paleozoic sediments	Calc-silicates: woll-diop-gar-plag; q-trem; cc-diop-clh or phlogopite	Late Cretaceous–early Tertiary	Wright and Haxel, 1982
Gunsight Hills	Late Cretaceous granitoids in shear zone	Granodiorite: biot-musc-q-alb to biot-musc-q-andesine-hbld	K-Ar synkinematic hbld 58 m.y.; biot 32 m.y.	Haxel et al., 1984; Tosdal et al., 1986

Puerta Blanco Mountains	Rocks of La Abra (Jurassic supracrustal and granitic rocks)	Meta-rhyolite: olig-ksp-q-musc-epi-(hbld) Psammite: olig-biot-epi-q-chl	Late Cretaceous–early Tertiary or Jurassic(?)	Haxel et al., 1984
Quijotoa Mountains	Mesozoic supracrustal rocks	biot-musc-q-fs; q-epi-mica		Rytuba et al., 1978
Quitoboquito Hills	Rocks of La Abra (Jurassic supracrustal and granitic rocks)	q-plag-biot-w. mica-epi-chl	Late Cretaceous–early Tertiary or Jurassic(?)	Haxel et al., 1984
Sierrita Mountains	Mesozoic supracrustal rocks	Meta-rhyolite: q-fs-chl-epi-cc-w. mica Meta-andesite: act-q-alb-w. mica-chl-biot-cc Semi-pelite: chl-biot-w. mica-q-fs		Cooper, 1971
Sheridan Mountains	Rocks of Gu Achi (Mesozoic clastic rocks)	GS-1A(?)	Late Cretaceous–early Tertiary	Haxel et al., 1984
Sierra Blanca	Jurassic sedimentary rocks	GS, A	Late Cretaceous–early Tertiary	Haxel et al., 1984

[a] Geographic division between southeast, south-central, and western Arizona follows usage in the text.

[b] Metamorphic grade (used where detailed mineral assemblage information unavailable): sub-GS, subgreenschist; GS, greenschist; A, amphibolite; aA, albite amphibolite; l, lower; u, upper.

[c] Minerals: act, actinolite; alb, albite; amph, amphibole; and, andalusite; biot, biotite; carb, carbonate mineral; cc, calcite; chl, chlorite; chld, chloritoid; clh, clinohumite; clz, clinozoisite; cord, cordierite; diop, diopside; dolo, dolomite; epi, epidote; fs, feldspar; gar, garnet; hem, hematite; hbld, hornblende; ido, idocrase; ksp, potassium feldspar; ky, kyanite; mag, magnetite; musc, muscovite; olig, oligoclase; plag, plagioclase; pyx, pyroxene; q, Quartz; sill, sillimanite; sph, sphene; stau, slaurolite; tour, tourmaline; trem, tremolite; w. mica, white mica; wol., wollastonite.

[d] References are made to review papers, including more complete reference lists, when possible.

479

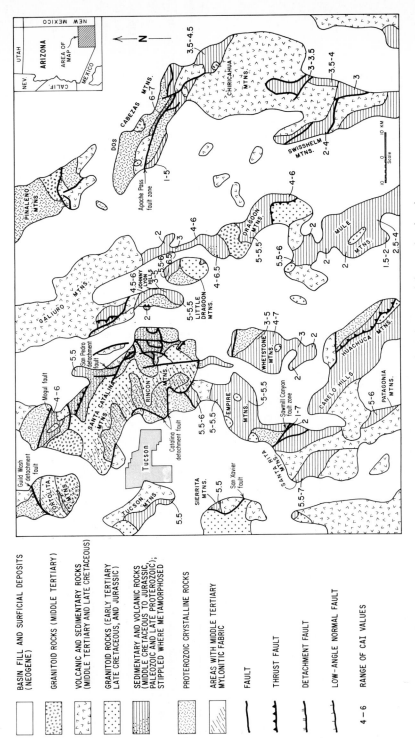

FIG. 18-6. Generalized geologic map of southeastern Arizona, showing the distribution of Mesozoic and Cenozoic metamorphic fabrics and CAI values on Paleozoic rocks. CAI values are from Wardlaw et al. (1983).

BASIN FILL AND SURFICIAL DEPOSITS (NEOGENE)

GRANITOID ROCKS (MIDDLE TERTIARY)

VOLCANIC AND SEDIMENTARY ROCKS (MIDDLE TERTIARY AND LATE CRETACEOUS)

GRANITOID ROCKS (EARLY TERTIARY LATE CRETACEOUS, AND JURASSIC)

SEDIMENTARY AND VOLCANIC ROCKS (MIDDLE CRETACEOUS TO JURASSIC, PALEOZOIC AND LATE PROTEROZOIC); STIPPLED WHERE METAMORPHOSED

PROTEROZOIC CRYSTALLINE ROCKS

AREAS WITH MIDDLE TERTIARY MYLONITIC FABRIC

FAULT

THRUST FAULT

DETACHMENT FAULT

LOW-ANGLE NORMAL FAULT

4 – 6   RANGE OF CAI VALUES

Proterozoic crystalline rocks. These relatively deep-level rocks were brought to high crustal levels by tectonic denudation during middle Tertiary detachment faulting.

In addition to the thrust-related metamorphism described above, evidence exists in the forerange of the Santa Catalina Mountains for another metamorphic event that appears to be related to plutonism rather than thrusting. This previously unrecognized metamorphic event is represented by northeast- to east-striking foliation and gneissic banding that is concordant with, and partly defined by, near-vertical pegmatite dikes related to an Eocene peraluminous granite. This steep fabric is not mylonitic in character and is clearly crosscut by middle Tertiary mylonitic fabric. A similar, well-developed, steep, nonmylonitic foliation and gneissic banding occurs in the peraluminous granite and adjacent metasedimentary rocks in the western Santa Catalina Mountains. This non-mylonitic fabric is discordantly crosscut by apophyses of middle Tertiary granite (Keith *et al.*, 1980, p. 251) and is interpreted here as developing synchronously with emplacement of the Eocene peraluminous granite. Correlative metamorphic fabrics may be present in the Tortolita Mountains to the west, where steep, northeast-trending foliation in a Late Cretaceous granodiorite is cut by subhorizontal mylonitic fabrics of middle Tertiary age.

The presence of metamorphosed and ductilely deformed Paleozoic and Mesozoic rocks in the Santa Catalina, Rincon, and Tortolita Mountains within a larger terrane of mostly unmetamorphosed rocks poses some interesting questions. Was ductile deformation restricted to this area because of the large, associated plutons, or are such tectonized rocks present at depth below much of the region and only exposed where middle Tertiary denudation has brought midcrustal rocks to the surface? The presently available data suggest that ductilely deformed rocks are indeed more widespread at depth than at the surface, but that such rocks are localized along major, basement-involved thrusts or near large, synkinematic plutons.

South-Central Arizona

In contrast to southeastern Arizona, much of south-central Arizona was affected by Late Cretaceous and early Tertiary regional metamorphism (Fig. 18-7; Haxel *et al.*, 1980, 1984). This metamorphic episode was associated with thrust faulting and with intrusion of syntectonic peraluminous plutons. The thrust faults, typically marked by mylonitic zones several or many meters thick, place Proterozoic and Mesozoic crystalline rocks over Mesozoic supracrustal rocks of lower metamorphic grade. There are two principal thrust fault systems: the Baboquivari thrust system, including the Window Mountain Well thrust, and the Quitobaquito thrust and associated subsidiary ductile shear zones (Fig. 18-7).

The lower plate of the Baboquivari thrust consists largely of variably metamorphosed Jurassic supracrustal rocks intruded by locally metamorphosed Jurassic, Late Cretaceous, and early Tertiary granitoids. Textural and mineralogical metamorphic grades in lower-plate rocks generally increase structurally upward toward exposed or inferred thrust faults. Within these metamorphic gradients, unmetamorphosed sandstones and siltstones are progressively converted to phyllite and schist ranging in grade from greenschist to lower-amphibolite facies. Metamorphism is also manifest by progressive recrystallization, development of foliation and lineation, obscuring of bedding and other primary structures or textures, and local isoclinal folding and transposition of bedding. In most mountain ranges, foliation and lineation are folded by mesoscopic kink or crenulation folds.

Lower-plate supracrustal and plutonic rocks below the Quitobaquito thrust, and rocks along thrust faults in the western Sierrita Mountains (Drewes and Cooper, 1973), are cut by imbricate blastomylonitic shear zones marked by intensification of the

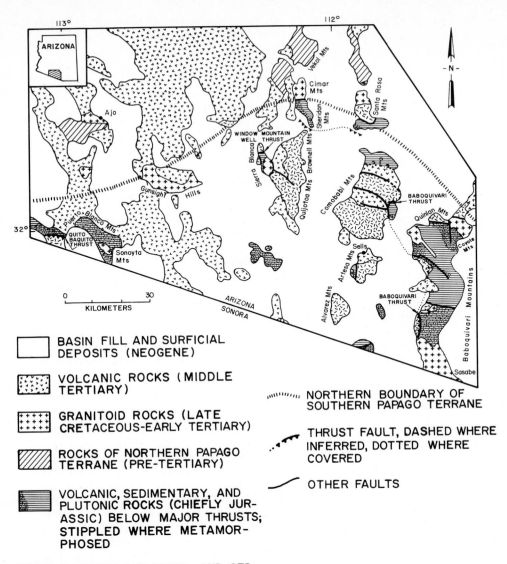

FIG. 18-7.   Generalized geologic map of south-central Arizona (from Haxel *et al.*, 1984). As discussed by Haxel *et al.* (1984), the Northern Papago terrane contains Paleozoic and Proterozoic rocks typical of southeastern and central Arizona, whereas the Southern Papago terrane lacks these rocks, but contains Mesozoic supracrustal rocks and Mesozoic and early Tertiary plutonic rocks.

Map legend:

- BASIN FILL AND SURFICIAL DEPOSITS (NEOGENE)
- VOLCANIC ROCKS (MIDDLE TERTIARY)
- GRANITOID ROCKS (LATE CRETACEOUS-EARLY TERTIARY)
- ROCKS OF NORTHERN PAPAGO TERRANE (PRE-TERTIARY)
- VOLCANIC, SEDIMENTARY, AND PLUTONIC ROCKS (CHIEFLY JURASSIC) BELOW MAJOR THRUSTS; STIPPLED WHERE METAMORPHOSED
- PLUTONIC, VOLCANIC, AND SEDIMENTARY ROCKS (CHIEFLY JURASSIC) ABOVE MAJOR THRUSTS

- NORTHERN BOUNDARY OF SOUTHERN PAPAGO TERRANE
- THRUST FAULT, DASHED WHERE INFERRED, DOTTED WHERE COVERED
- OTHER FAULTS

regional metamorphic fabric (Haxel *et al.*, 1984). A several-kilometer-thick, earliest Tertiary shear zone in the Gunsight Hills consists of amphibolite-facies orthogneiss and orthoschist that are in part blastomylonitic (Tosdal *et al.*, 1986).

Plutonic and supracrustal rocks in the upper plate of the Baboquivari thrust system show considerably less evidence of Late Cretaceous and early Tertiary regional metamorphism than do those in the lower plate. Upper-plate Mesozoic supracrustal rocks are unmetamorphosed except in a few ranges where they have been affected by Jurassic regional metamorphism (Tosdal *et al.*, 1987). Upper-plate Paleozoic strata generally lack penetrative metamorphic fabrics, but locally reached lower-greenschist temperatures, as indicated by CAI values of 4.5–5.5 (Wardlaw and Harris, 1984). In most areas of south-central Arizona, such temperatures are probably due to Jurassic and/or Late Cretaceous plutonism. Low-grade metamorphism of the Paleozoic strata may also locally reflect tectonic burial beneath thrust sheets structurally higher than the Baboquivari thrust system; a possible example is the Copperosity thrust in the Vekol Mountains (Dockter and Keith, 1978).

Regional metamorphism in south-central Arizona was accompanied by intrusion of leucocratic, high-silica granites containing various combinations of the accessory minerals biotite, primary muscovite, and garnet. These peraluminous granites, based on their instrusive and fabric relations with metamorphic host rocks, are syn- or late-kinematic to postkinematic. Such variations in the apparent relations between metamorphism and plutonism largely reflect differences in crustal level, rather than significant differences in the relative ages of metamorphism and plutonism — the synkinematic granites represent somewhat deeper crustal levels than the postkinematic granites.

U-Pb zircon ages indicate that peraluminous granites in the Coyote Mountains (Wright and Haxel, 1982) and similar granites in Sonora (T. H. Anderson *et al.*, 1980) are about 58 m.y. old. K-Ar ages of metamorphic rocks and field relations of the granites indicate that the later stages, and probably the culmination, of regional metamorphism occurred about 58–60 m.y. ago (Haxel *et al.*, 1984).

Field relations and geochronologic data indicate that thrust faulting, regional metamorphism, and granitic plutonism were closely related. Late Cretaceous and early Tertiary orogenesis in south-central Arizona evidently was due to the conjunction of (1) crustal compression and thickening, and (2) anomalous heat flux into the crust from the underlying mantle (Haxel *et al.*, 1984). The high heat flux was probably largely due to mantle-derived metaluminous magmas ascending off the flattening subduction zone. Thrusting within this regime of high heat flux and corresponding steep geothermal gradients resulted in regional metamorphism of supracrustal rocks that were tectonically buried beneath the thrust sheets. Local partial melting of the middle crust produced silicic granitic magmas (Farmer and DePaolo, 1984), which rose to intrude the upper crust.

Western Arizona

Western Arizona and parts of adjacent southeastern California have a more complicated history of Mesozoic and early Tertiary metamorphism and deformation than the rest of Arizona. In addition to middle Tertiary mylonitization and detachment faulting, at least five major metamorphic episodes, ranging in age from Triassic to early Tertiary, have affected the region. Most metamorphic episodes were associated with regional deformation, including large-scale thrusting, and some were accompanied by synkinematic plutonism. In several ranges, large-scale transport on Mesozoic thrust faults and middle to late Tertiary low-angle normal faults resulted in juxtaposition of rocks with different metamorphic histories.

Two major regional thrust faults divide the western Arizona region into three tectonic domains (Fig. 18-8). The sotherly and structurally lowest domain is the late Mesozoic Orocopia Schist, exposed in windows through the Chocolate Mountains thrust (Haxel and Dillon, 1978). The second metamorphic domain comprises Mesozoic metamorphic rocks above the Chocolate Mountains thrust and south of the south-dipping

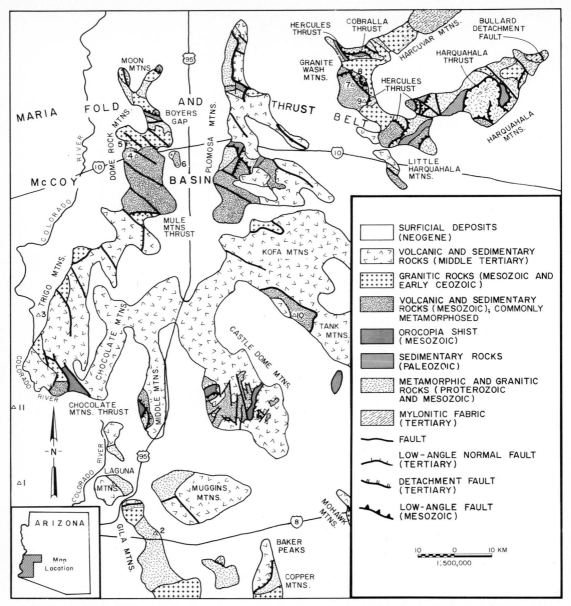

FIG. 18-8. Generalized geologic map of west-central and southwestern Arizona. Occurrences of aluminous metasomatic rocks are numbered in accordance with Table 18-2. Occurrences in southeastern California include Cargo Muchacho Mountains (no. 1) and Indian Pass (west of map area). Precise location of occurrence in Gila Mountains is unknown.

Mule Mountains thrust. The third domain includes regionally metamorphosed Mesozoic, Paleozoic, and Proterozoic rocks that lie north of the Mule Mountains thrust, within the McCoy Basin (Harding and Coney, 1985) and along the Maria fold-and-thrust belt (Reynolds *et al.*, 1986b). Some metamorphic rocks in this northern domain have a metamorphic history similar to those in the second domain.

Orocopia Schist and Chocolate Mountains thrust The pre-Tertiary structural architecture of southwesternmost Arizona and adjacent southeastern California is dominated by the Late Cretaceous Chocolate Mountains thrust, which places Proterozoic and Mesozoic crystalline rocks over the late Mesozoic Orocopia Schist (Haxel and Dillon, 1978; Jacobson *et al.*, Chapter 35, this volume). Latest Cretaceous and middle Tertiary low-angle normal faulting and middle to late Tertiary folding have resulted in exposure of the Orocopia Schist in a belt of east-trending antiformal culminations (Fig. 18-8; Haxel *et al.*, 1985). The Orocopia Schist and Chocolate Mountains thrust have been strongly disrupted and altered by middle to late Tertiary faulting and magmatism in southwestern Arizona, but are better preserved in southeastern California (Dillon, 1976; Haxel, 1977).

The protolith of schist was Jurassic and Cretaceous(?) sedimentary and volcanic rocks, largely graywacke with subordinate to rare semipelite, basalt, ferromanganiferous chert, and siliceous marble; rare ultramafic rocks are also present (Haxel and Tosdal, 1986; Mukasa *et al.*, 1984).

Metamorphic assemblages in the Orocopia Schist are those of the upper-greenschist facies, albite-epidote amphibolite facies, and lower-amphibolite facies. The related Pelona Schist of the central Transverse Ranges of California (Ehlig, 1981) evidently was metamorphosed at depths of 20–30 km (Graham and Powell, 1984); several mineralogic features qualitatively suggest relatively high-pressure metamorphism of the Orocopia Schist as well (Haxel and Dillon, 1978). Orocopia metagraywacke and semipelite are two-mica, quartzofeldspathic schists. Gray to black porphyroblasts of graphitic albite are characteristic and fuchsite (Cr-bearing mica) is sporatic but widespread. Orocopia metabasalts typically are fine-grained, epidote-bearing albite amphibolites, some of which also contain garnet. In a few areas, metabasalts are greenschist rather than amphibolite. Metacherts contain spessartine, magnetite, and a variety of accessory minerals, locally including riebeckite. Rare bodies of talc-actinolite rocks or antigorite serpentinite occur in the schist.

Metamorphism of the Orocopia Schist occurred concurrently with Late Cretaceous movement on the overlying Chocolate Mountains thrust. Metamorphic grade in the schist increases upward toward the thrust zone, where metagraywackes commonly contain garnet, and metabasalts are oligoclase or andesine amphibolites.

The tectonic regime in which the Orocopia and Pelona Schists were metamorphosed, however, is unresolved (Burchfiel and Davis, 1981; Crowell, 1981). The tectonic setting and relatively high-pressure metamorphism suggest that the schist was metamorphosed as it was subducted easterly or northeasterly beneath southwestern North America. However, structural evidence from the Chocolate Mountains thrust zone indicates the opposite sense of motion — northeastward or northward overthrusting (J. T. Dillon, G. B. Haxel, and R. M. Tosdal, unpublished data). Such structural data are consistent with a scenario in which the protolith of the schist was trapped in a convergence zone between the North American mainland and a smaller crustal block represented by crystalline rocks of the central Transverse Ranges (Crowell, 1981; Haxel and Tosdal, 1986). An alternative interpretation is that structures indicating continentward overthrusting along the Chocolate Mountains and related thrusts reflect uplift following subduction (Jacobson *et al.*, Chapter 35, this volume).

Upper plates of the Chocolate Mountains and Mule Mountains thrusts At least two episodes of Mesozoic regional metamorphism and deformation affected rocks in the upper plates of the Chocolate Mountains and Mule Mountains thrusts (Tosdal, 1986a). Although the Chocolate Mountains thrust dips northward, toward the subsurface projection of the south-dipping Mule Mountains thrust, the two thrusts are probably not the same fault.

Late Triassic and Triassic(?) plutonic complexes in several ranges on either side of the Colorado River north of Yuma were deformed and metamorphosed at amphibolite-facies conditions during their emplacement (Tosdal, 1986a). Parts of these intrusive and metamorphic complexes can be correlated with the Late Triassic Mount Lowe plutonic suite (Ehlig, 1981), which was intruded synkinematically at depths greater than 26 km[1] (Barth, 1982). These metamorphic and deep-level plutonic rocks may have been uplifted in Late Triassic or Early Jurassic time to account for their proximity to lower-grade Jurassic supracrustal rocks. Such an uplift could represent the source of clasts of Proterozoic and Triassic igneous rocks (Peirce, 1986) that were shed onto the Colorado Plateau in Late Triassic time.

In a number of ranges in southwestern Arizona and adjacent California, greenschist- to amphibolite-facies regional metamorphism was coeval with intrusion of Late [and Middle(?)] Jurassic leucocratic granite and, in some areas, pegmatite. This granite is the youngest unit of a widespread, compositionally expanded suite of Middle and Late Jurassic granitoids (Tosdal, 1986a; Tosdal *et al.*, 1987). This Late Jurassic metamorphic episode affected Proterozoic crystalline rocks, Paleozoic and lower Mesozoic cratonic strata, and Jurassic supracrustal and plutonic rocks.

Contrasts in metamorphic grade and intrusive style indicate that the Late Jurassic metamorphic terranes represent a considerable range of crustal levels. Particularly instructive progressions in crustal level are exposed in the southwestern Kofa Mountains and in the Cargo Muchacho Mountains of southeastern California. At the highest exposed structural level, metamorphosed supracrustal rocks are argillitic and phyllitic or, near contacts with granite, hornfelsic. The granites are unfoliated, cross-cutting, and lack associated pegmatites. At the deepest exposed Late Jurassic crustal level, regional metamorphic rocks are amphibolite-facies schists, and the granites are orthogneisses that are concordant with and share the fabric of their metamorphic country rocks. Associated with the granites are concordant to discordant sheets and pods of intrusive pegmatite. Even at the deepest exposed crustal levels, the rocks are not migmatitic.

Late Jurassic regional metamorphism was accompanied by movement along ductile shear zones, some of which cut the leucocratic orthogneisses and are in turn cut by synkinematic pegmatites (Dillon, 1976; Tosdal, 1986b). Fluid flow and metasomatism were important aspects of this metamorphism and deformation, as evidenced by the presence of highly aluminous metasomatic rocks (described below) and by the conversion of biotite-poor, hornblende-bearing rocks into biotite schist.

In Late Cretaceous time, the Triassic and Jurassic metamorphic and plutonic rocks were locally mylonitized and retrograded along ductile shear zones related to the Chocolate Mountains thrust and Mule Mountains thrust (Haxel and Tosdal, 1986; Tosdal 1986a). Uplift associated with or following thrusting resulted in widespread latest Cretaceous and early Tertiary K-Ar cooling ages obtained on a variety of metamorphic and plutonic rocks (Fig. 18-3).

McCoy basin and Maria fold-and-thrust belt North of, and structurally below, the northeast-vergent Mule Mountains thrust lies the McCoy Basin, an east-west-trending belt composed of thick sequences of Jurassic or Cretaceous McCoy Mountains Formation and its substrate of Jurassic volcanic and hypabyssal rocks (Harding and Coney, 1985). The basin is flanked and partially overlapped to the north by the Maria fold-and-thrust belt, an east-west-trending zone of major overturned folds and thrust sheets with generally southwest, south, and southeast transport (Reynolds *et al.*, 1986b). Rocks within the McCoy Basin and Maria thrust belt have been subjected to multiple episodes of deformation and associated metamorphism. In the Dome Rock Mountains (Fig. 18-8), the McCoy Mountains Formation was affected by a complex sequence of deformations accompanied

---

[1] Also A. P. Barth, written communication, 1986.

by regional metamorphism of lower- to middle-greenschist facies at higher structural levels and upper-greenschist facies at depth (Tosdal, 1986a). These fabrics are Cretaceous in age and are overprinted by thin, imbricate ductile shear zones and more localized greenschist-facies metamorphism associated with the latest Cretaceous Mule Mountains thrust (Tosdal, 1986a).

Metamorphism and associated deformation also occur along the northern margin of the McCoy Basin, within the Maria fold-and-thrust belt (Reynolds et al., 1986b). In the northern Dome Rock Mountains, greenschist-facies Paleozoic metasediments have been interfolded with their original crystalline basement, cut by northeast- and southwest-vergent ductile shear zones, and intruded by variably deformed Mesozoic granites (Yeats, 1985). Structurally higher Mesozoic metasedimentary rocks locally contain green amphibole and piemontite beneath a thrust sheet of mylonitic granite. Similar thrust faults and folded Paleozoic and Lower Mesozoic metasedimentary rocks are present in the Plomosa Mountains, the next range to the east (F. K. Miller, 1970; Scarborough and Meader, 1983; Stoneman, 1985).

Detailed structural mapping and metamorphic petrology in the Granite Wash Mountains to the east (Fig. 18-8) have demonstrated a close relationship between metamorphism and deformation (Laubach, 1986; Laubach et al., 1986; Reynolds et al., 1986b). In the central part of the range, the southwest-vergent Hercules thrust has emplaced Proterozoic and Jurassic crystalline rocks over a south- to southeast-facing sequence of cratonic Paleozoic units, Jurassic volcanics, and Jurassic or Cretaceous sedimentary rocks of the McCoy Mountains Formation. Rocks below the thrust contain structures produced by three phases of folding and cleavage formation, designated D1, D2, and D3. The first deformation, D1, produced a texturally uniform, fine-grained slaty cleavage that formed synchronously with a large south- to southeast-vergent fold that upturned the Paleozoic-Mesozoic section. This deformation was accompanied by chlorite-grade, greenschist-facies metamorphism at temperatures of less than 450°C and pressures of less than 4 kb.

A second, unrelated episode of metamorphism and cleavage formation (D2) occurred as Proterozoic and Jurassic crystalline rocks above the Hercules thrust were emplaced southwestward across the upturned section. The effects of D2 deformation and metamorphism vary markedly with increasing distance below the thrust. At deepest structural levels, Mesozoic sedimentary rocks largely retain their primary sedimentary structures and are cut by spaced, disjunctive cleavage or crenulation, reflecting pressure-solution effects. At higher levels, the rocks are phyllitic with greenschist-facies mineral assemblages. Directly beneath the thrust, the Mesozoic rocks have been converted into schists with assemblages transitional from upper greenschist to lowest amphibolite facies. There is, therefore, an apparent inverted metamorphic gradient that is directly related to emplacement of crystalline rocks above the Hercules thrust. In addition, the thrust zone contains a thrust-bounded sliver of lithologically distinct Mesozoic metasedimentary rocks that contain abundant postkinematic andalusite, which is absent from rocks below the thrust zone. Elsewhere, the thrust zone contains kyanite-quartz rocks associated with altered metavolcanic units. Subsequent to thrusting and D2 deformation, the rocks were cut by a disjunctive, pressure-solution-type cleavage (D3) and by two postkinematic Late Cretaceous plutons.

Two distinct episodes of deformation and metamorphism (D1 and D2) also affected rocks in the northern Granite Wash Mountains (Cunningham, 1986; Reynolds et al., 1986b). The first episode, correlated with D1 in the central part of the range, formed a south-southeast-vergent, ductile thrust that emplaced Proterozoic crystalline rocks over an overturned, south-southeast-facing section of tremolite-bearing Paleozoic metasedimentary rocks that have been attenuated to less than 5% of their original thickness. South-southeast-vergent D1 structures are truncated along strike by D2 structures associated with the southwest-vergent Hercules thrust, which places Proterozoic gneiss over metamorphosed and attenuated Paleozoic and Mesozoic rocks.

A similar sequence of structures and metamorphic events is present in the Little Harquahala and Harquahala Mountains to the south and east (Reynolds *et al.*, 1986b). Both ranges contain regional thrust faults and ductile shear zones that juxtapose Proterozoic and Jurassic crystalline rocks with southeast-facing, commonly overturned Paleozoic and Mesozoic sections. The structurally lowest rocks in both ranges are Jurassic volcanic rocks and sedimentary and volcanic rocks assigned to the Jurassic or Cretaceous McCoy Mountains Formation (Spencer *et al.*, 1985). These rocks are at subgreenschist to lower-greenschist facies and mostly retain their sedimentary features. They have been over-ridden by Proterozoic and Jurassic crystalline rocks along the Hercules thrust. The crystalline rocks are in turn overlain along the structurally higher Centennial thrust by a northeast-striking, southeast-facing section of Proterozoic granite, Paleozoic cratonic strata, Jurassic volcanic rocks, and McCoy Mountains Formation. These rocks appear relatively unmetamorphosed, but nevertheless contain conodonts with a CAI value of 7 (Wardlaw and Harris, 1984).

The Paleozoic and Mesozoic sequence becomes increasingly metamorphosed and deformed along strike to the northeast into the central Harquahala Mountains[2] (Reynolds *et al.*, 1986b). At the northeast end of the belt, the stratigraphic section is inverted (Hardy, 1984) and metamorphosed to lower-amphibolite facies, with rare kyanite and staurolite in Mesozoic metasedimentary rocks (B. Bryant, written communication, 1984). These rocks are overlain by the south-vergent Harquahala thrust, which places Proterozoic metamorphic and granitic rocks discordantly across the inverted Phanerozoic section. Randomly oriented crystals of tremolite and other metamorphic minerals indicate that peak metamorphic conditions persisted after movement on the Harquahala thrust. The northeastward increase in grade and the amount of recrystallization from the Little Harquahala Mountains to the Harquahala Mountains may reflect (1) an increased burial beneath the Harquahala thrust; (2) a deeper level of exposure due to tectonic denudation on the middle Tertiary Bullard detachment fault; or (3) proximity to a post-thrusting, garnet-two-mica granite on the northeastern end of the range.

Another regionally important metamorphic-deformational event is represented in the northeastern Harquahala Mountains and Harcuvar Mountains to the north. This event occurred under amphibolite-facies conditions and produced a generally subhorizontal fabric in migmatized, remobilized Proterozoic gneiss and in Late Cretaceous granites (Rehrig and Reynolds, 1980; Reynolds, 1982). Peak metamorphism occurred in Late Cretaceous time, prior to intrusion of crosscutting, muscovite-bearing early Tertiary(?) pegmatites. This Late Cretaceous metamorphic event is probably also present in the Buckskin and Whipple Mountains to the north, and may correlate with high-grade metamorphism of Paleozoic and Mesozoic rocks in the Big Maria and Old Woman Mountains (Hoisch *et al.*, Chapter 21, this volume).

Aluminous metasomatic rocks Highly aluminous metasomatic rocks occur in at least 11 localities in western Arizona and adjacent southeastern California (Table 18-2). The distribution and geologic setting of these unusual rocks is not widely appreciated, chiefly because half of the known occurrences have been discovered within the last 5 years.

The most striking of the aluminous metasomatic rocks are those composed almost entirely of quartz and kyanite; other common combinations are quartz-muscovite and quartz-pyrophyllite. Locally, these grade into quartz-rich rocks formed by pre- or syn-metamorphic metasomatism or hydrothermal activity. Accessory minerals or uncommon essential minerals are andalusite, sillimanite, tourmaline, rutile, ilmenite, biotite, lazulite

---

[2]Also S. M. Richard, unpublished data.

TABLE 18-2. Aluminous Metasomatic Rocks in Western Arizona and Southeastern California (Fig. 18-8)

Locality	Mineralogy (+qtz)[a]	Geologic Setting and References[b]
1. American Girl Canyon, Cargo Muchacho Mountains	ky, ru, mu, tour, laz, py, py op, dum, mag	Pods and irregular replacements in metamorphic rocks adjacent to Jurassic granite (Henshaw, 1942; Dillon, 1976)
2. Gila Mountains[c]	ky	In metamorphic (?) rocks adjacent to Jurassic granite (Anthony et al., 1977)
3. Clip Wash, Trigo Mountains	ky, dum, mu	Boulders of quartzose schist in terrace gravels along Colorado River (Wilson, 1933)
4. Dome Rock Mountains	ky, ru, mu	Meter-thick lens enclosed within foliated, porphyritic Jurassic granite
5. Dome Rock Mountains	ky, hyp, mu, tour	Near contact of metamorphosed Triassic (?) clastic rocks, Jurassic quartz-porphyry volcanics, and alaskitic Jurassic granite (Crowl, 1979)
6. Dome Rock Mountains	ky, and, sill, mu, dum, pyrop, hem, mag, ru, py	Schistose zone in Jurassic metavolcanic rocks near a Jurassic granite (Wilson, 1929)
7. Granite Wash Mountains	ky, mu, py, tour	Foliated and granofelsic masses within a schistose quartzite near the stratigraphic top of altered Jurassic metavolcanics in lower plate of Hercules thrust
8. Granite Wash Mountains	ky, mu, hem, cov	Irregular pods in ferruginous quartzite layer within altered Jurassic metavolcanic rocks below Hercules thrust
9. Granite Wash Mountains	and, mu, py, ky	Andalusite-muscovite-pyrite schist with rare kyanite within Jurassic metavolcanic rocks; associated with ferruginous quartzite
10 Kofa Mountains	pyrop, mu, tour, ky	Within Jurassic metasedimentary rocks near a Jurassic granite
11. Indian Pass, Chocolate Mountains	ky, mu, dum, tour	Boulders in Neogene fanglomerate (Haxel, 1977)

[a] and, andalusite; cov, covelite; dum, dumortierite; hem, hematite; hyp, hypersthene; ky, kyanite; laz, lazulite; mag, magnetite; mu, muscovite; py, pyrite; pyrop, pyrophyllite; ru, rutile; sill, sillimanite; tour, tourmaline.
[b] Additional data from unpublished studies of authors in all areas except numbers 2 and 3.
[c] Precise location in Gila Mountains unknown.

(hydrous Mg-Fe aluminophosphate), apatite, dumortierite, staurolite, K-feldspar, magnetite, and pyrite[3] (Dillon, 1976; Henshaw, 1942; Wilson, 1929). The most common assemblage comprises quartz, an aluminosilicate, a Ti-bearing mineral, and a P-bearing mineral. Pyrophyllite is typically retrograde after higher-grade aluminosilicate minerals, generally kyanite.

The aluminous metasomatic rocks commonly occur as gneiss or schist in regional metamorphic terranes and as granofels in undeformed plutonic rocks. Granofelsic and gneissic or schistose aluminous rocks also locally occur together.

[3] Also S. J. Reynolds, G. B. Haxel, S. M. Richard, R. M. Tosdal, and S. E. Laubach, unpublished data.

The aluminous metasomatic rocks occur in at least three different structural settings (Table 18-2): (1) along intrusive contacts between Late Jurassic leucocratic granite or pegmatite and quartzofeldspathic metavolcanic or metaplutonic country rocks; (2) along low-angle synplutonic structural contacts that place Late Jurassic leucocratic granite over quartzofeldspathic lower-plate rocks; and (3) entirely within metamorphosed and altered Triassic(?) clastic strata and Jurassic silicic volcanic rocks.

The quartz-kyanite and other aluminous rocks of the Cargo Muchacho Mountains in southeastern California (Henshaw, 1942) are good examples of types 1 and 2. The aluminous rocks were derived both from metamorphosed volcanic and volcaniclastic rocks and from granitic orthogneiss, and range in texture from well-foliated gneiss and schist to granofels. They are intruded by synkinematic pegmatite and were produced by metasomatic alteration coeval with plutonism, deformation, and amphibolite-facies metamorphism (Dillon, 1976; Tosdal, 1986b). From field evidence and thermodynamic reasoning, it is inferred that the aluminous rocks were produced by hydrogen-ion metasomatism of quartzofeldspathic rocks during circulation of Cl-rich, acidic fluids at high water-to-rock ratios (Dillon, 1976; Wise, 1975). During metasomatism, the rocks were leached of relatively mobile elements and left enriched in the relatively immobile elements Al, Ti, and P.

The type 3 aluminous rocks are apparently not associated with granitic rocks, but were still probably formed by hydrogen-ion metasomatism in a hydrothermal environment. Aluminous rocks of this type occur in the Granite Wash Mountains near the top of a sequence of Jurassic silicic volcanic and volcaniclastic rocks. These quartz-kyanite rocks are interpreted as the metamorphosed products of a shallow-level hydrothermal system; synvolcanic near-surface argillic alteration produced clay-rich rocks and hydrothermal sinter that were subsequently converted into quartz-kyanite rocks during Cretaceous thrust-related metamorphism and tectonic burial.

Controls of Mesozoic and Early Cenozoic Regional Metamorphism

Despite the spatial and temporal variability of Mesozoic to early Cenozoic regional metamorphism in southern and western Arizona, certain broad patterns are apparent. In particular, most metamorphism is associated with thrust faulting, granitic plutonism, or both.

Regional metamorphism associated with plutonism but not thrust faulting is typified by Late Jurassic metamorphism in the Cargo Muchacho and Kofa Mountains, and by Late Cretaceous to early Tertiary metamorphism in the Harcuvar and Harquahala Mountains, and Santa Catalina Mountains. Characteristics of this type of metamorphism are structurally downward increases of mineralogic and textural metamorphic grade and, at relatively deep structural levels, association with synkinematic orthogneiss and with sheets of late-kinematic to postkinematic pegmatite and granite intruded parallel to foliation.

Metamorphic rocks associated with thrust faulting are of two different types. The first type includes mylonitic rocks localized along the thrusts themselves or within ductile shear zones related to the thrusts. These mylonitic rocks are fault rocks (Sibson, 1977), rather than regional metamorphic rocks, and are most common where thrust-related fabrics are superimposed on earlier plutonic or high-grade metamorphic rocks, such as in the upper plates of the Hercules and Chocolate Mountains thrusts.

The second type of thrust-related metamorphic rocks are crystalloblastic schists in which mineralogic and textural metamorphic grades increase structurally upward, through distances on the order of hundreds of meters to several kilometers, toward the overlying thrust fault. Notable examples are schists below the Baboquivari thrust, the Orocopia Schist underneath the Chocolate Mountains thrust, and metamorphic rocks underlying the Hercules and related thrusts. Metamorphism of these rocks was apparently in part consequent to tectonic burial beneath a crystalline thrust sheet (England and Thompson, 1984, Haxel et al., 1984; Reynolds et al., 1986b). The effects of metamorphism were most pronounced near the thrusts, where metamorphic reactions were aided by dynamically induced recrystallization and by the availability of fluid derived from prograde, dehydration reactions. The inverted metamorphic gradients may reflect heat introduced by fluids ascending the permeable thrust zones or by the emplacement of hot upper-plate rocks over cooler lower-plate rocks.

The distinction between mylonitic fault rocks and regional metamorphic rocks, however, is not always clear, especially where (1) thrust-zone mylonitic fabrics grade downward into crystalloblastic fabrics in lower-plate schists or gneisses, or (2) a mylonitic thrust zone has been overprinted by static recrystallization, as in the Harquahala Mountains.

The metamorphic terranes beneath thrust faults in southwestern and west-central Arizona evidently are not associated with granitic plutonism, at least at exposed crustal levels. The clearest example is the Orocopia Schist, which is intruded by only a single, small epizonal granitic pluton that is distinctly younger than and unrelated to the metamorphism of the schist (Haxel et al., 1985). In contrast, Late Cretaceous and early Tertiary metamorphic rocks in south-central Arizona are associated with both synmetamorphic thrust faults and syn- to postmetamorphic granites; thrusting, regional metamorphism, and granitic plutonism were closely related to one another.

Regional metamorphic rocks associated with granitic plutons in southwestern Arizona and those in south-central Arizona have several interesting contrasts. Metamorphic grade at exposed crustal levels increases structurally downward in southwestern Arizona, but increases upward toward thrust faults in south-central Arizona. Synmetamorphic plutons in southwestern Arizona are the granitic end member of a compositionally diverse, calc-alkaline suite (Tosdal et al., 1987), whereas those in south-central Arizona are compositionally restricted, high-silica peraluminous granites. In southwestern Arizona, synmetamorphic granites are typically orthogneisses, whereas those in south-central Arizona are widely foliated but only locally metamorphosed.

Some of the contrasts in metamorphic style between pluton-related and thrust-related metamorphism reflect expected differences between pressure-temperature paths of the rocks in the two settings. Metamorphism associated with deep-level plutonism may be characterized by an approximately isobaric rise in temperature, whereas thrusting may produce a more complex pressure-temperature history that includes (1) initial increase in pressure and, to a lesser degree, temperature during thrust loading; (2) further increases in temperature as isotherms become reestablished after thrusting; and (3) a final period of decompression and cooling due to erosional or tectonic denudation of the over-thickened crustal welt (England and Thompson, 1984). This more complicated history may help explain why metamorphic fabrics formed during thrusting locally have an over-print of static recrystallization, such as in the Granite Wash and Harquahala Mountains of western Arizona.

The highest-grade Mesozoic to early Cenozoic metamorphic rocks in Arizona are generally situated in the lower plate of middle Tertiary detachment faults. For example, the highest-grade Mesozoic metamorphic rocks in west-central Arizona are in the Harquahala and Harcuvar Mountains, beneath the Bullard detachment fault. Likewise, the highest-grade metamorphic rocks in southeastern Arizona are below the Catalina detachment fault. These deep-seated rocks (J. L. Anderson, Chapter 19, this volume) were exhumed by Tertiary detachment faulting and provide a sample of what the middle crust, or at least the lower levels of the upper crust, contained at the end of Mesozoic and early Cenozoic tectonism. Judging from these exposures, the middle crust has a pronounced low-dipping or subhorizontal layering defined by polyphase metamorphic fabrics, thrust faults and ductile shear zones, thrust-bounded slivers of rock, and sheet-like intrusions of granite and pegmatite. Low-dipping reflectors observed in many seismic reflection profiles (Frost and Okaya, 1986) probably image both this pre-middle Tertiary structural and plutonic layering and middle Tertiary detachment faults and mylonitic fabrics.

MIDDLE TERTIARY MYLONITIZATION AND DETACHMENT FAULTING

Middle Tertiary crustal extension in the southwestern United States was accommodated by the formation of regional, gently dipping normal faults (Fig. 18-9), referred to as detachment faults (G. A. Davis *et al.*, 1980, 1986; Reynolds and Spencer, 1985; Spencer and Reynolds, 1987). Rocks in the upper plate of detachment faults have been extensively faulted, brittlely distended, and rotated within tilted fault blocks that commonly include steeply dipping middle Tertiary volcanic and sedimentary rocks. In contrast, rocks below detachment faults consist of plutonic and metamorphic rocks that are generally unfaulted, except within a zone of chloritic breccia and microbreccia at high structural levels directly beneath the detachment fault.

Beneath some detachment faults, lower-plate crystalline rocks are overprinted by middle Tertiary mylonitic fabrics; such areas are referred to as metamorphic core complexes (Crittenden *et al.*, 1980). These mylonitic fabrics were largely formed by ductile, laminar flow within shear zones that represent the deeper-level, down-dip continuations of the detachment faults (G. A. Davis *et al.*, 1986; G. H. Davis, 1983; Reynolds, 1985; Wernicke, 1985). During detachment faulting, lower-plate rocks became less deeply buried as they were drawn out from beneath the normal-faulted upper plate (Fig. 18-9). As lower-plate rocks along the shear zone were progressively unroofed and transported upward through the brittle-ductile transition, they ceased ductile flow and were overprinted by progressively more brittle deformation. Given enough normal displacement on a detachment fault, lower-plate rocks that originated at mid-crustal depths below the brittle-ductile transition can be juxtaposed directly against shallow-level, upper-plate rocks, such as syntectonic volcanic and clastic rocks. Detachment faulting therefore presents an opportunity to compare styles of deformation at different crustal levels and provides a sample, represented by the lower plate, of the rock types and structures of the middle crust or lower parts of the upper crust (Anderson, Chapter 19, this volume; G. A. Davis *et al.*, 1986).

There have been few studies of metamorphism associated with metamorphic core complexes and detachment faults in Arizona, largely because most workers have focused instead on geochronologic or structural aspects (Armstrong, 1982; Crittenden *et al.*, 1980). Middle Tertiary mylonitic fabrics within the complexes have been imprinted on a wide variety of protoliths, including Proterozoic crystalline rocks, Paleozoic and Mesozoic metasedimentary rocks, and Jurassic to middle Tertiary plutons. The protoliths of mylonitic rocks are generally still discernible, because many minerals, such as plagioclase and amphibole, are relics that survived mylonitization with little or no recrystallization.

Minerals that formed during mylonitization, such as biotite in pressure shadows, indicate that most mylonitization occurred under middle-greenschist to lower-amphibolite-facies conditions. Temperatures during mylonitization increased with structural depth and ranged from as high as 550°C at deeper structural levels (J. L. Anderson, 1981,

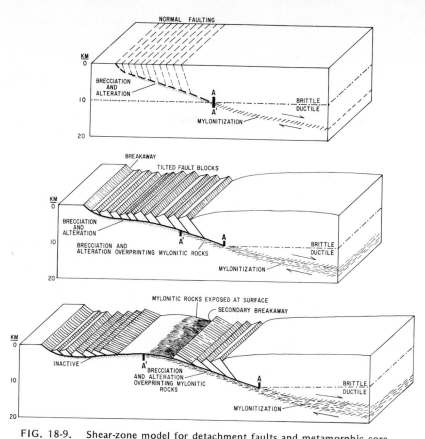

FIG. 18-9.   Shear-zone model for detachment faults and metamorphic core complexes. Normal displacement along the low-angle shear zone produces mylonitic fabrics at depth and more brittle structures along shallower segments of the fault. As the lower plate becomes less deeply buried with time due to tectonic denudation of the upper plate, mylonitic rocks originally below the brittle-ductile transition are overprinted by successively more brittle structures. Complete tectonic denudation, in conjunction with isostatic uplift, brings the brittlely overprinted mylonitic rocks to the surface (see G. A. Davis *et al.*, 1986; Spencer and Reynolds, 1986, 1987; Wernicke, 1985).

Chapter 19, this volume) to probably less than 400°C. Such temperatures are consistent with available geochronologic data and with the common, but by no means universal, association of mylonitization with synkinematic middle Tertiary plutons (Reynolds *et al.*, 1986a). In most complexes, there is a clear pattern of decreasing metamorphic grade with time, as indicated by thin, late-stage mylonitic zones that cut and retrogress thicker zones of more penetrative mylonitic fabric (Reynolds and Lister, 1987). The latest developed mylonitic fabrics occur as thin selvages along cross-cutting quartz veins and fractures. The increased localization and decrease in metamorphic grade of mylonitization with time reflects the ascent of the rocks toward shallower levels due to tectonic denudation and isostatic uplift during detachment faulting.

Pressure during mylonitization varied among adjacent complexes and during the evolution of a single complex. Pressures during mylonitization have been estimated at 2-5 kb from mineralogic geobarometers (J. L. Anderson, Chapter 19, this volume), but were probably less in complexes such as the South Mountains, where mylonitization was associated with hypabyssal plutonism (Reynolds, 1985). In addition, a progressive drop in pressures during mylonitization must have occurred within each complex as the lower-plate rocks were unroofed.

Variations in pressure between different complexes can be viewed as a spectrum between "shallow" and "deep" end members. In the shallow type of core complex, typified by the South Mountains, mylonitization was restricted to a relatively thin zone less than several hundred meters thick within or near large, synkinematic Tertiary plutons; mylonitization occurred at relatively shallow crustal levels because of heat conveyed into the upper crust by the plutons. Metamorphic complexes of this shallow type include, in addition to the South Mountains, the Tortolita, and Picacho Mountains.

In contrast, the relatively deep type of core complex, exemplified by the Santa Catalina, Rincon, and Buckskin Mountains, is characterized by a relatively thick mylonitic zone more than 1 km across. Mylonitization in this type of complex affected a variety of older rocks, commonly many kilometers from Tertiary plutons. Rocks in these complexes were hot enough to deform by mylonitization because they were at substantial depths, perhaps 10 km or more. Other metamorphic complexes, such as the Coyote Mountains, are intermediate in character between the shallow and deep types.

As the mylonitic and nonmylonitic rocks of the complexes were uplifted to shallower levels, they were extensively overprinted by brecciation, fracturing, and chlorite-epidote alteration in the footwall of the overlying detachment fault. This zone of brecciation and alteration, referred to as the chloritic breccia zone, ranges in thickness from 0 to 300 m and commonly contains four rock types. The most common rock type is chloritic breccia, a highly fractured and brecciated rock that is strongly affected by fracture-controlled and void-filling chlorite-epidote-hematite alteration (G. A. Davis et al., 1980; Rehrig and Reynolds, 1980; Reynolds, 1985). At highest structural levels near the detachment fault, the chloritic breccia is capped by or locally intermixed with zones less than 1 m thick of microbreccia or cataclasite, a very fine-grained, flinty rock composed of microscopic angular fragments. The chloritic breccia zone and underlying, less fractured rocks also locally contain thin selvages of psuedotachylite along small fault surfaces and in wedge-shaped injection veins adjacent to the faults. The fourth rock type within the chloritic breccia zone includes fine-grained foliated and lineated rocks that are intermediate in structural character between mylonite and layered cataclasite, and that appear to represent the transition from ductile to brittle deformation (Reynolds, 1985).

The pronounced change in both mesoscopic and microscopic appearance due to the successive development of mylonite, chloritic breccia, and microbreccia is not necessarily accompanied by major chemical changes. In the South Mountains, a complete gradation from middle Tertiary granodiorite to chloritic breccia and microbreccia derived from the granodiorite occurs within any large changes in major-element geochemistry or in oxygen and hydrogen isotopic composition[4] (Smith and Reynolds, 1985).

It is important to emphasize that almost all mineralogic changes that occurred during mylonitization and subsequent formation of chloritic breccia were retrograde in character. Mylonitic fabrics almost invariably overprint either plutonic rocks or metamorphic rocks that have an older, higher-grade metamorphic assemblage. The only places where mylonitization was a prograde event are near large Tertiary plutons, such as in pendants within synkinematic plutons in the Tortolita Mountains. The dominance of retrograde metamorphic reactions is consistent with progressive transport of the rocks to shallower crustal levels during detachment faulting. The widespread preservation of pre-existing higher-grade assemblages probably reflects a limited supply of aqueous fluids,

_____
[4] Also B. M. Smith and S. J. Reynolds, unpublished data.

since retrograde metamorphic reactions commonly involve hydration and require sufficient fluid (Fyfe *et al.*, 1978). Retrogression during mylonitization was most pronounced within zones of high strain and along thin mylonite zones that formed during the latest stages of mylonitization (Reynolds and Lister, 1987). There is also a close spatial and temporal correspondence between widespread retrogression and pervasive fracturing within the chloritic breccia zone. Unfractured lower-plate rocks below the chloritic breccia zone retain their pre-mylonitic plutonic or high-grade metamorphic mineral assemblages and are not retrogressed, yet must have undergone the same *P-T* evolution as the strongly retrogressed rocks within the breccia zone. These relationships emphasize the importance of fluid pathways and availability of fluid in determining where retrogression occurred.

COMPARISON OF MESOZOIC TO EARLY TERTIARY METAMORPHISM AND MIDDLE TERTIARY MYLONITIZATION

In the late 1970s, there was confusion and controversy about the distinctions between middle Tertiary core-complex mylonitization and earlier Mesozoic and early Tertiary metamorphism (Crittenden *et al.*, 1980). In retrospect, this confusion arose from inadequate geochronology and from the widespread imposition of middle Tertiary mylonitic fabrics on rocks with prior metamorphic histories. Recent geologic mapping and geochronology have clarified these issues in most core complexes and it is now possible to characterize the differences between middle Tertiary and older fabrics.

Fabrics produced during middle Tertiary mylonitization and detachment faulting differ in several respects from those produced during the Mesozoic and early Tertiary metamorphic episodes. Mylonitization and detachment faulting were accompanied by retrograde metamorphism. In contrast, Mesozoic and early Tertiary metamorphic episodes were commonly prograde, because they affected previously unmetamorphosed Paleozoic and Mesozoic supracrustal rocks or synmetamorphic plutonic rocks. Extensive Mesozoic to early Tertiary retrogression is confined to those areas where metamorphism affected older higher-grade metamorphic rocks, such as Proterozoic gneiss, or premetamorphic plutonic rocks; these relationships are especially clear along crystalline overthrusts, where lower-plate supracrustal rocks become increasingly metamorphosed upward toward the thrust, whereas upper-plate crystalline rocks are retrogressed in the thrust zone.

These differences in metamorphic style are reflected by contrasts in the character of the associated fabrics. Fabrics produced during Mesozoic and early Tertiary prograde metamorphism are generally schistose rather than mylonitic, probably because prograde metamorphism is dominated by dehydration reactions that provide abundant fluids to promote diffusional mass transfer, recrystallization, and neomineralization. In contrast, middle Tertiary mylonitic fabrics are typically more strongly lineated, less schistose, and characterized by evidence for crystal-plastic deformation of quartz (Lister and Snoke, 1984). These fabrics and the limited development of retrograde (rehydration) reactions indicate that mylonitization occurred under relatively dry conditions, as is suggested by the restriction of retrogression to thin shear zones that contain vein quartz or other indications of high levels of fluid activity. This low level of fluid activity does not necessarily imply that fluid pressures were low, since the presence of dilatant fractures indicates that fluid pressures occasionally exceeded the least principle compressive stress (Reynolds and Lister, 1987).

The better development of foliation than lineation in most Mesozoic and early Tertiary metamorphic terranes may also indicate that strain away from major thrust zones or adjacent to intrusions was dominated by flattening rather than rotational shear. In contrast, lineation is better developed than foliation in many middle Tertiary mylonites, which reflects the formation of these rocks in zones of large shear strain.

Another distinction between the two types of metamorphic fabrics is that Mesozoic to early Tertiary metamorphic fabrics are locally overprinted by a static recrystallization, due either to thrust loading or to posttectonic plutonism. In contrast, middle Tertiary mylonitic fabrics are overprinted only by successively more brittle fabrics, in large part because deformation persisted past typical metamorphic temperatures and because no subsequent episode of plutonism and metamorphism has affected the rocks.

The differences in style of metamorphism between the two ages of fabrics are due, in part, to differences in tectonic setting. Mesozoic to early Cenozoic metamorphism generally occurred within a context of crustal shortening, commonly accompanied by magmatism. Triassic and Late Jurassic metamorphism occurred within or near a continental-margin arc that trended northwest across southwestern Arizona and adjacent areas. The main pulse of thrusting and metamorphism in west-central Arizona, however, was not accompanied by significant magmatism, probably because deformation occurred within a back-arc setting, behind the Late Jurassic to Cretaceous coastal batholiths. Major magmatism, namely emplacement of garnet-two-mica granites, did accompany Late Cretaceous to early Tertiary deformation and metamorphism in south-central Arizona, the Harcuvar Mountains of west-central Arizona, and the Santa Catalina and Rincon Mountains in southeastern Arizona. This metamorphic episode occurred while magmatism migrated eastward across Arizona, presumably in response to a decreasing eastward dip of the subducted slab beneath the region (Coney and Reynolds, 1977). This decrease in dip may also have caused much of the crustal shortening during this time period (Dickinson and Snyder, 1978).

In contrast, Tertiary mylonitization and detachment faulting occurred during a westward, or return, sweep of magmatism across Arizona between 35 and 15 m.y. ago. This magmatic episode, which included eruption of voluminous silicic ash and lava flows, is interpreted to reflect a steepening and collapse of the east-dipping subduction zone beneath the region (Coney and Reynolds, 1977). Crustal extension, represented by the mylonitic zones and detachment faults, locally both preceded and outlasted magmatism, but occurred during the same overall time period (Spencer and Reynolds, 1987). Mylonitization may have been facilitated by the large thermal input of middle Tertiary magmas, but also occurred in areas away from large plutons.

Finally, there has been much controversy about the possible relationship between mylonitization and earlier episodes of metamorphism (Crittenden *et al.*, 1980). The presence of Tertiary mylonitic fabrics in complexes that lack any evidence of pre-existing Mesozoic to early Cenozoic metamorphism, such as in the South Mountains, indicates that older metamorphism was not a prerequisite for subsequent mylonitization. Instead, Tertiary mylonitization is commonly overprinted on earlier metamorphic rocks because these are the types of rocks present in the crustal levels where mylonites form. Such pre-mylonite metamorphic fabrics range in age from early Tertiary to Proterozoic and have no direct genetic relationship to subsequent mylonitization.

ACKNOWLEDGMENTS

Some of the concepts and data presented in this paper were developed through cooperative research with G. Lister, E. DeWitt, J. Dillon, D. May, J. Spencer, and J. Wright. We also appreciate past cooperation and discussions with T. Anderson, P. Coney, G. A. Davis, G. H. Davis, W. Dickinson, C. Jacobson, S. Keith, M. Roddy, L. Silver, B. Smith, and

J. Welty. M. Grubensky provided unpublished petrographic data for south-central and southwestern Arizona. The manuscript was improved by reviews of J. L. Anderson, M. Donato, W. G. Ernst, H. W. Peirce, and J. Spencer. The figures were drafted by J. LaVoie and P. Corrao.

REFERENCES

Anderson, J. L., 1981, Conditions of mylonitization in a Cordilleran metamorphic core complex, Whipple Mountains, California, *in* Howard, K. A. *et al.*, eds., *Tectonic Framework of the Mohave and Sonoran Deserts, California and Arizona:* U.S. Geol. Survey Open File Rpt. 81-503, p. 73-75.

Anderson, T. H., Silver, L. T., and Salas, G. A., 1980, Distribution and U-Pb isotope ages of some lineated plutons, northwestern Mexico, *in* Crittenden, M. D., Jr., *et al.*, eds., *Cordilleran Metamorphic Core Complexes:* Geol. Soc. America Mem. 153, p. 269-283.

Anthony, J. W., Williams, S. A., and Bideaux, R. A., 1977, *Mineralogy of Arizona:* University of Arizona Press, Tucson, Ariz., 254 p.

Armstrong, R. L., 1982, Cordilleran metamorphic core complexes — From Arizona to southern Canada: *Ann. Rev. Earth Planet. Sci.*, v. 10, p. 129-154.

Barth, A. P., 1982, Petrology of the Mount Lowe intrusion, San Gabriel Mountains, California: M.S. thesis, California State Univ. Los Angeles, Calif., 82 p.

Bilodeau, W. L., 1978, The Glance Conglomerate, a Lower Cretaceous syntectonic deposit in southeastern Arizona, *in* Callender, J. F., *et al.*, eds., *Land of Cochise — Southeastern Arizona:* N.Mex. Geol. Soc. Guidebook 29, p. 209-214.

Burchfiel, B. C., and Davis, G. A., 1981, Mojave desert and environs, *in* Ernst, W. G., ed., *The Geotectonic Development of California* (Rubey Vol. I): Englewood Cliffs, N.J., Prentice-Hall, p. 217-252.

Bykerk-Kaufmann, Ann, 1986, Multiple episodes of ductile deformation within the lower plate of the Santa Catalina metamorphic core complex, southeastern Arizona, *in* Beatty, Barbara, and Wilkinson, P. A. K., eds., *Frontiers in Geology and Ore Deposits of Arizona and the Southwest:* Ariz. Geol. Soc. Digest, v. 16, p. 460-463.

Coney, P. J., 1978, Plate tectonic setting of southeastern Arizona, *in* Callender, J. F., *et al.*, eds., *Land of Cochise — Southeastern Arizona:* N.Mex. Geol. Soc. Guidebook 29, p. 285-290.

——, and Reynolds, S. J., 1977, Cordilleran Benioff zones: *Nature*, v. 270, p. 403-406.

Cooper, J. R., 1971, Mesozoic stratigraphy of the Sierrita Mountains, Pima County, Arizona: *U.S. Geol. Survey Prof. Paper 658-D*, 42 p.

Creasey, S. C., 1980, Chronology of intrusion and deposition of porphyry copper ores, Globe-Miami district, Arizona: *Econ. Geology*, v. 75, p. 830-844.

Crittenden, M. D., Jr., Coney, P. J., and Davis, G. H., eds., 1980, Cordilleran metamorphic core complexes: *Geol. Soc. America Mem. 153*, 490 p.

Crowl, W. J., 1979, Geology of the central Dome Rock Mountains, Yuma County, Arizona: M.S. thesis, Univ. Arizona, Tucson, Ariz., 76 p.

Crowell, J. C., 1981, An outline of the tectonic history of southeastern California, *in* Ernst, W. G., ed., *The Geotectonic Development of California* (Rubey Vol. I): Englewood Cliffs, N.J., Prentice-Hall, p. 583-600.

Cunningham, W. D., 1986, Superimposed thrusting in the northern Granite Wash Mountains, La Paz County, Arizona: M.S. thesis, Univ. Arizona, Tucson, Ariz., 112 p.

Davis, G. A., Anderson, J. L., Frost, E. G., and Shackelford, T. J., 1980, Mylonitization and detachment faulting in the Whipple–Buckskin–Rawhide Mountains terrane, southeastern California and western Arizona, *in* Crittenden, M. D., Jr., *et al.*, eds., *Cordilleran Metamorphic Core Complexes:* Geol. Soc. America Mem. 153, p. 79–129.

____, Lister, G. S., and Reynolds, S. J., 1986, Structural evolution of the Whipple and South Mountains shear zones, southwestern United States: *Geology*, v. 14, p. 7–10.

Davis, G. H., 1979, Laramide folding and faulting in southeastern Arizona: *Amer. J. Sci.*, v. 279, p. 543–569.

____, 1983, Shear-zone model for the origin of metamorphic core complexes: *Geology*, v. 11, p. 342–347.

Dickinson, W. R., and Snyder, W. S., 1978, Plate tectonics of the Laramide orogeny: *Geol. Soc. America Mem. 151*, p. 355–366.

Dillon, J. T., 1976, Geology of the Chocolate and Cargo Muchacho Mountains, southeasternmost California: Ph.D. dissertation, Univ. California, Santa Barbara, Calif., 405 p.

Dockter, R. D., and Keith, W. J., 1978, Reconnaissance geologic map of Vekol Mountains quadrangle, Arizona: *U.S. Geol. Survey Misc. Field Studies Map MF-931.*

Drewes, H. D., 1981, Tectonics of southeastern Arizona: *U.S. Geol. Survey Prof. Paper 1144*, 96 p.

____, 1984, Geologic map and section of the Bowie Mountain North quadrangle, Cochise County, Arizona: *U.S. Geol. Survey Misc. Geol. Inv. Map I-1492*, scale 1:24,000.

____, and Cooper, J. R., 1973, Reconnaissance geologic map of the west side of the Sierrita Mountains, Palo Alto Ranch quadrangle, Pima County, Arizona: *U.S. Geol. Survey Misc. Field Studies Map MF-538*, scale 1:24,000.

Ehlig, P. L., 1981, Origin and tectonic history of the basement terrane of the San Gabriel Mountains, central Transverse Ranges, *in* Ernst, W. G., ed., *The Geotectonic Development of California* (Rubey Vol I): Englewood Cliffs, N.J., Prentice-Hall, p. 253–283.

England, P. C., and Thompson, A. B., 1984, *P-T-t* paths of regional metamorphism — I. Heat transfer during the evolution of regions of thickened crust: *J. Petrology*, v. 25, p. 894–928.

Epstein, A. G., Epstein, J. B., and Harris, L. D., 1977, Conodont color alteration — An index to organic metamorphism: *U.S. Geol. Survey Prof. Paper 995*, 27 p.

Farmer, G. L., and Depaolo, D. J., 1984, Origin of Mesozoic and Tertiary granite in the western United States and implications for pre-Mesozoic crustal structure — 2. Nd and Sr studies of unmineralized and Cu- and Mo-mineralized granite in the Precambrian craton: *J. Geophys. Res.*, v. 89, p. 10141–10160.

Frost, E. G., and Martin, D. L., 1983, Overprint of Tertiary detachment deformation on the Mesozoic Orocopia Schist and Chocolate Mountains Thrust: *Geol. Soc. America Abstr. with Programs*, v. 15, p. 577.

Frost, E. G., and Okaya, D. A., 1986, Application of seismic reflection profiles to tectonic analysis in mineral exploration, *in* Beatty, Barbara, and Wilkinson, P.A.K., eds., Frontiers in geology and ore deposits of Arizona and the Southwest: Tucson, Arizona Geological Society Digest, v. 16, p. 137–151.

Fyfe, W. S., Price, N. J., and Thompson, A. B., 1978, *Fluids in the Earth's Crust:* New York, Elsevier, 383 p.

Graham, C. M., and Powell, Roger, 1984, A garnet-hornblende geothermometer — Calibration, testing, and application to the Pelona Schist, southern California: *J. Metamorphic Geology*, v. 2, p. 12–31.

Harding, L. E., and Coney, P. J., 1985, The geology of the McCoy Mountains Formation, southeastern California and southwestern Arizona: *Geol. Soc. America Bull.*, v. 96, p. 755–769.

Hardy, J. J., Jr., 1984, The structural geology, tectonics and metamorphic geology of the Arrastre Gulch window, south-central Harquahala Mountains, Maricopa County, Arizona: M.S. thesis, Northern Arizona Univ., Flagstaff, Ariz., 99 p.

Haxel, G. B., 1977, The Orocopa Schist, and the Chocolate Mountains thrust, Picacho-Peter Kane Mountain area, southeasternmost California: Ph.D. dissertation, Univ. California, Santa Barbara, Calif., 277 p.

——, and Dillon, John, 1978, The Pelona-Orocopia schist and Vincent–Chocolate Mountain thrust system, southern California, *in* Howell, D. G., and McDougall, K. A., eds., *Mesozoic Paleogeography of the Western United States:* Pacific Section, Soc. Econ. Paleontologists Mineralogists, Pacific Coast Paleogeography Symp. 2, p. 453–469.

——, and Tosdal, R. M., 1986, Significance of the Orocopia Schist and Chocolate Mountains thrust in the late Mesozoic tectonic evolution of the southeastern California–southwestern Arizona region, *in* Beatty, Barbara, and Wilkinson, P. A. K., eds., *Frontiers in Geology and Ore Deposits of Arizona and the Southwest:* Ariz. Geol. Soc. Digest, v. 16, p. 52–61.

——, Wright, J. E., May, D. J., and Tosdal, R. M., 1980, Reconnaissance geology of the Mesozoic and Cenozoic rocks of the southern Papago Indian Reservation, Arizona — A preliminary report, *in* Jenney, J. P., and Stone, Claudia, eds., *Studies in Western Arizona:* Ariz. Geol. Soc. Digest, v. 12, p. 17–29.

——, Tosdal, R. M., May, D. J., and Wright, J. E., 1984, Latest Cretaceous and early Tertiary orogenesis in south-central Arizona — Thrust faulting, regional metamorphism, and granitic plutonism: *Geol. Soc. America Bull.*, v. 95, p. 631–653.

——, Tosdal, R. M., and Dillon, J. T., 1985, Tectonic setting and lithology of the Winterhaven Formation, a new Mezozoic stratigraphic unit in southeasternmost California and southwestern Arizona: *U.S. Geol. Survey Bull. 1599*, 19 p.

Hayes, P. T., and Drewes, Harald, 1978, Mesozoic depositional history of southeastern Arizona, *in* Callender, J. F., *et al.*, eds., *Land of Cochise — Southeastern Arizona:* N.Mex. Geol. Soc. Guidebook 29, p. 201–207.

Henshaw, P. C., 1942, Geology and mineral resources of the Cargo Muchacho Mountains, Imperial County, California: *Calif. Div. Mines Rt. 38*, p. 147–196.

Keith, S. B., and Wilt, J. C., 1986, Laramide Orogeny in Arizona and adjacent regions — A strato-tectonic synthesis, *in* Beatty, Barbara, and Wilkinson, P. A. K., eds., Frontiers in geology and Ore Deposits of Arizona and the Southwest: Ariz. Geol. Soc. Digest, v. 16, p. 502–554.

——, Reynolds, S. J., Damon, P. E., Shafiqullah, M., Livingston, D. E., and Pushkar, P. D., 1980, Evidence for multiple intrusion and deformation within the Santa Catalina–Rincon–Tortolita crystalline complex, southeastern Arizona, *in* Crittenden, M. D., Jr., *et al.*, eds., *Cordilleran Metamorphic Core Complexes:* Geol. Soc. America Mem. 153, p. 217–267.

Laubach, S. E., 1986, Polyphase deformation, thrust-induced strain and metamorphism, and Mesozoic stratigraphy of the Granite Wash Mountains, west-central Arizona: Ph.D. dissertation, Univ. Illinois, Urbana-Champaign, Ill., 180 p.

——, Reynolds, S. J., Spencer, J. E., and Richard, S. M., 1986, Thrust-related metamorphism, Granite Wash Mountains, west-central Arizona: *Geol. Soc. America Abstr. with Programs*, v. 18, p. 126.

Lindgrey, Steven, 1983, Kinematic analysis of a mid-Tertiary shear zone, eastern Rincon Mountains, southern Arizona: *Geol. Soc. America Abstr. with Programs*, v. 15, p. 314.

Lister, G. S., and Snoke, A. W., 1984, S-C mylonites: *J. Struct. Geology*, v. 6, p. 617–638.

Miller, F. K., 1970, Geologic Map of the Quartzsite quadrangle, Yuma County, Arizona: *U.S. Geol. Survey Geol. Quad. Map GQ-841*, scale 1:62,500.

Mukasa, S. B., Dillon, J. T., and Tosdal, R. M., 1984, A late Jurassic minimum age for the Peolona-Orocopia Schist Protolith, southern California: *Geol. Soc. America Abstr. with Programs*, v. 16, p. 323.

Peirce, H. W., 1986, Paleogeographic linkage between Arizona's physiogeographic provinces, *in* Beatty, Barbara, and Wilkinson, P. A. K., eds., *Frontiers in Geology and Ore Deposits of Arizona and the Southwest:* Ariz. Geol. Soc. Digest, v. 16, p. 74–82.

Rehrig, W. A., and Reynolds, S. J., 1980, Geologic and geochronologic reconnaissance of a northwest-trending zone of metamorphic core complexes in southern and western Arizona, *in* Crittenden, M. D., Jr., *et al.*, eds., *Cordilleran Metamorphic Core Complexes:* Geol. Soc. America Mem. 153, p. 131–157.

Reynolds, S. J., 1982, Multiple deformation in the Harcuvar and Harquahala Mountains, west-central Arizona, *in* Frost, E. G., and Martin, D. L., eds., *Mesozoic-Cenozoic Tectonic Evolution of the Colorado River Region, California, Arizona, and Nevada:* San Diego, Calif., Cordilleran Publishers, p. 137–142.

——, 1985, Geology of the South Mountains, central Arizona: *Ariz. Bur. Geology Mineral Technology Bull. 195,* 61 p.

——, and Lister, G. S., 1987, Structural aspects of fluid-rock interactions within detachment zones: *Geology*, v. 15, p. 362–366.

——, and Spencer, J. E., 1985, Evidence for large-scale transport on the Bullard detachment fault, west-central Arizona: *Geology*, v. 13, p. 353–356.

——, Shafiqullah, M., Damon, P. E., and DeWitt, Ed, 1986a, Early Miocene mylonitization and detachment faulting, South Mountains, central Arizona: *Geology*, v. 14, p. 283–286.

——, Spencer, J. E., Richard, S. M., and Laubach, S. E., 1986b, Mesozoic structures in west-central Arizona, *in* Beatty, Barbara, and Wilkinson, P. A. K., eds., *Frontiers in Geology and Ore Deposits of Arizona and the Southwest:* Ariz. Geol. Soc. Digest, v. 16, p. 35–51.

——, *et al.*, 1986c, Compilation of radiometric age determinations in Arizona: *Ariz. Bur. Geology Miner. Technology Bull. 197.*

Rytuba, J. J., Till, A. B., Blair, Will, and Haxel, Gordon, 1978, Reconnaissance geologic map of the Quijotoa Mountains quadrangle, Pima County, Arizona: *U.S. Geol. SurveyMisc. Field Studies Map MF-937*, scale 1:62,500.

Scarborough, R., and Meader, N., 1983, Reconnaissance geology of the northern Plomosa Mountains, La Paz County, Arizona: *Ariz. Bur. Geology Mineral Technology Open File Rpt. 83–24*, 35 p.

Shafiqullah, M., Damon, P. E., Lynch, D. J., Reynolds, S. J., Rehrig, W. A., and Raymond, R. H., 1980, K-Ar geochronology and geologic history of southwestern Arizona and adjacent areas, *in* Jenney, J. P., and Stone, Claudia, eds., *Studies in Western Arizona:* Ariz. Geol. Soc. Digest, v. 12, p. 201–260.

Sibson, R. H., 1977, Fault rocks and fault mechanisms: *J. Geol. Soc. London*, v. 133, p. 191–213.

Smith, B. M., and Reynolds, S. J., 1985, Oxygen and hydrogen isotope study of mylonitization and detachment faulting, South Mountains, central Arizona: (EOS) *Trans. Amer. Geophys. Un.*, v. 66, p. 1138.

Spencer, J. E., and Reynolds, S. J., 1986, Some aspects of the middle Tertiary tectonics of Arizona and southeastern California, *in* Beatty, Barbara, and Wilkinson, P. A. K., eds., *Frontiers in Geology and Ore Deposits of Arizona and the Southwest:* Ariz. Geol. Soc. Digest, v. 16, p. 102–107.

————, and Reynolds, S. J., 1987, Middle Tertiary tectonics of Arizona and the southwest, *in* Jenney, J. P., and Reynolds, S. J., eds., *Geologic Evolution of Arizona:* Ariz. Geol. Soc. Digest, v. 17, in press.

————, Richard, S. M., and Reynolds, S. J., 1985, Geologic map of the Little Harquahala Mountains: *Ariz. Bur. Geology Miner. Technology Open File Rpt. 85-9*, scale 1:24,000.

Stoneman, D. A., 1985, Structural geology of the Plomosa Pass area, northern Plomosa Mountains, La Paz County, Arizona: M.S. thesis, Univ. Arizona, Tucson, Ariz., 99 p.

Thorman, C. H., and Drewes, Harald, 1981, Geology of the Rincon wilderness study area, Pima County, Arizona, *in Mineral Resources of the Rincon Wilderness Study Area, Pima County, Arizona:* U.S. Geol. Survey Bull. 1500, p. 5–37.

Tosdal, R. M., 1986a, Mesozoic ductile deformations in the southern Dome Rock Mountains, northern Trigo Mountains, Trigo Peak and Livingston Hills, southwestern Arizona, and Mule Mountains, southeastern California, *in* Beatty, Barbara, and Wilkinson, P. A. K., eds., *Frontiers in Geology and Ore Deposits of Arizona and the Southwest:* Ariz. Geol. Soc. Digest, v. 16, p. 62–71.

————, 1986b, Gneissic host rocks to gold mineralization in the Cargo Muchacho Mountains, southeastern California, *in Geological Setting of Gold and Silver Mineralization in Southeastern California and Southwestern Arizona:* Cordilleran Section, Geol. Soc. America Field Guide 1986 Meet., v. 5, 6, p. 139–142.

————, Peterson, D. W., May, D. J., LeVeque, R. A., and Miller, R. J., 1986, Reconnaissance geologic map of the Mount Ajo and part of the Pisinimo quadrangles, Pima County, Arizona: *U.S. Geol. Survey Misc. Field Studies Map MF-1820.*

————, Haxel, G. B., and Wright, J. E., 1987, Jurassic geology of the Sonoran Desert region, southern Arizona, southeast California, and northernmost Sonora — Construction of a continental-margin magmatic arc, *in* Jenney, J. P., and Reynolds, S. J., eds., *Geologic Evolution of Arizona:* Ariz. Geol. Soc. Digest, v. 17, in press.

Wardlaw, B. R., and Harris, A. G., 1984, Conodont-based thermal maturation of Paleozoic rocks in Arizona: *Amer. Assoc. Petrol. Geologists Bull.*, v. 68, p. 1101–1106.

————, Harris, A. G., and Schindler, K. S., 1983, Thermal maturation values (conodont color alteration indices) for Paleozoic rocks in Arizona: *U.S. Geol. Survey Open File Rpt. 83-819*, 13 p.

Wernicke, Brian, 1985, Uniform-sense normal simple shear of the continental lithosphere: *Can. J. Earth Sci.*, v. 22, p. 108–125.

Wilson, E. D., 1929, An occurrence of dumortierite near Quartzsite, Arizona: *Amer. Mineralogist*, v. 14, no. 10, p. 373–381.

————, 1933, Geology and mineral deposits of southern Yuma County, Arizona: *Ariz. Bur. Mines Bull. 134*, Geol. Ser. 7, 236 p.

Wise, W. S., 1975, The origin of the assemblage: quartz+Al-silicate+rutile+Al-phosphate: *Fortschr. Mineralogie*, v. 52, p. 151–159.

Wright, J. E., and Haxel, G. B., 1982, A garnet-two-mica granite, Coyote Mountains, southern Arizona: geologic setting, uranium-lead isotopic systematics of zircon, and nature of the granite source region: *Geol. Soc. America Bull.*, v. 93, p. 1176–1188.

Yeats, K. J., 1985, Geology and structure of the northern Dome Rock Mountains, La Paz County, Arizona: M.S. thesis, Univ. Arizona, Tucson, Ariz., 123 p.

J. Lawford Anderson
Department of Geological Sciences
University of Southern California
Los Angeles, California 90089-0741

19

CORE COMPLEXES OF THE MOJAVE-SONORAN DESERT: CONDITIONS OF PLUTONISM, MYLONITIZATION, AND DECOMPRESSION

The Late Cretaceous to Tertiary core complexes of the Mojave-Sonoran Desert contain a strikingly common history of petrotectonic development, including (1) inflation by multiple intrusion of two-mica and other granitic magmas spanning tens of m.y., (2) flattening and shear by protracted mylonitization, and (3) a final period of detachment faulting and related extension. Recent studies (Anderson, 1985) have shown that decompression is another important attribute, evidenced by the deep-seated emplacement of the early intrusions, apparent shallower levels of mylonitization, and the near-surface nature of detachment faulting. Thermobarometric estimation of the intrusive and mylonitization events are herein presented utilizing existing calibrations for GAR-PL-BIO-MU, MU-BIO-KSP-QZ, and KSP-PL equilibria and the *P-T* limits of alumina solubility in hornblende.

In the Whipple Mountains of southeastern California, plutonism involved more than 14 intrusions separated into four episodes: 89 ± 3 m.y., 73 ± 3 m.y., 26 ± 5 m.y., and 19 m.y. (Wright *et al.*, 1986). Mylonitization was a protracted event, being synkinematic with intrusions of 26 m.y. age and ending prior to 19 m.y. The two earliest intrusive episodes involved emplacement of two mica ± garnet granodiorites and biotite-hornblende quartz diorite, respectively, all at midcrustal depths. Crystallization conditions for these two Late Cretaceous intrusive episodes are estimated at 746 ± 55°C at 9.2 ± 0.9 kb, and 785 ± 7°C at 7.5 ± 0.4 kb, respectively. The differing pressures are more reflective of calibration uncertainty than absolute variations. The major mylonitization occurred from 458 ± 35 to 535 ± 44°C, increasing with depth, at an average estimated pressure of 4.4 ± 1.1 kb.

The ages of plutonism and mylonitization in the nearby Sacramento Mountains of California is presently unknown, but both pluton emplacement and subsequent deformation appear to have taken place at shallower levels relative to those of the Whipple Mountains. Estimated crystallization conditions are 783 ± 6°C at 4.2 ± 0.7 kb for a biotite-hornblende granodiorite and 736 ± 38°C at 2.9 ± 0.5 kb for a younger garnet, two-mica granite. Mylonitization is estimated to have occurred at 484 ± 26°C and less than 2.0 kb.

The Santa Catalina Mountains of southern Arizona contain at least 12 plutons separated into three intrusive epochs (60–75 m.y., 44–50 m.y., and 25–29 m.y.; see Keith *et al.*, 1980), each followed by or coeval with mylonitization. Work presented here pertains only to the 44–50-m.y.-old, two-mica, garnet-bearing Wilderness granite. Emplacement and deformation are estimated to have occurred at 676 ± 42°C and 494 ± 10°C, respectively, both at about the same depth, which is loosely constrained between 3.8 ± 0.9 and 5.2 ± 1.0 kb. The lesser figure seems more realistic, based on stratigraphic constraints. A much deeper emplacement (>6 kb) for the 60–75-m.y.-old Leatherwood quartz diorite is indicated by the probable occurrence of magmatic epidote in this intrusive.

Compared to these other complexes, the Iron Mountains of southeastern California have a much different character. All plutonism and mylonitization is Late Cretaceous in age[1] (>67 m.y.; D. M. Miller and Howard, 1985) and exposures lack any record of detachment faulting or related extension. Moreover, magma intrusion and mylonitization (*ca.* 440 ± 20°C) appear to have occurred at the uniformly shallow level of 1.7 ± 0.2 kb.

INTRODUCTION

One of the several geologic features common to nearly every "core complex" mylonitic/detachment terrane of the Cordillera is the spatial association with late Mesozoic to

---

[1] Also J. Wright, unpublished data.

Tertiary granitic magma emplacement. Regions or exposed crustal levels lacking such plutonism usually also lack any development of these terranes, now regarded as a significant structural element of Cordilleran tectonics. As further testimony to the amount of magmatism involved, it is common for 30% or more of the mylonitic gneiss protolith to be the crystallized products of this phase of pre-core-complex history. Moreover, granitic magmas continued to pervade the level of crust affected by the mylonitization throughout the deformation phase, which also was apparently a protracted episode. Most typically, the granites are peraluminous, bearing two micas ± garnet. Lesser amounts of calc-alkaline, metaluminous (biotite-sphene±hornblende) granites were also involved.

Recent work on estimating the conditions of emplacement and subsequent mylonitization (Anderson, 1985) within a few complexes of the southwest United States has indicated a moderately deep crustal origin for the initial stages of core-complex development. Moreover, significant decompression (lowering of confining pressure) with time is implied, with deep-seated early plutonism giving way to less deep levels of mylonitization in transition to the shallow crustal levels that characterize these terranes during their final evolutionary phases of detachment faulting and brittle extension.

This paper reports results for four mylonitic terranes of the Mojave-Sonoran Desert region, specifically, the Whipple, Sacramento, and Iron Mountains of southeastern California and the Santa Catalina Mountains of southern Arizona (Fig. 19-1). The principal effort has been in the Whipple terrane; work in the others is preliminary and part of an ongoing regional study.

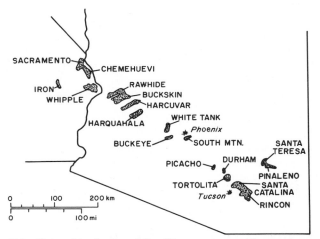

FIG. 19-1. Distribution of Cordilleran metamorphic complexes and associated terranes of southeastern California and southern Arizona.

GEOLOGIC SETTING

As noted by Dewitt (1980), variations in timing, structural style, and overall complexity characterize core complexes of the Cordillera. The common features are well known (Coney, 1980) and include (1) a low-angle detachment or décollement fault of extensional origin, often with subsidiary splays, separating two plates of fundamentally con-

CORE COMPLEXES OF THE MOJAVE-SONORAN DESERT

trasting structural style; (2) a lower plate or "core" of metamorphosed crystalline rocks, bearing a shallow-dipping mylonitic foliation and uniform trending (usually northeast-southwest) lineation truncated by the overlying detachment fault and associated chloritic breccias; and (3) an upper plate of crystalline and supracrustal rocks which lack the mylonitic deformation and metamorphism of the lower plate, and which occur extended, imbricated, and uniformally rotated by an assemblage of normal faults that have a sense of motion consistent with that of the detachment fault. Exclusive of the Iron Mountains, which lack a detachment fault, the statement above characterizes the Whipple, Sacramento, and Santa Catalina complexes and most others of the Mojave-Sonoran Desert, including that of the Chemehuevi (John, 1982), Rawhide (Shackelford, 1980), and Harcuvar and Harquahala Mountains (Reynolds and Rehrig, 1980).

In the context of Cordilleran metamorphism for this region, it appears that the evolution of the core complexes was the latest stage or culmination of thermal and deformational events that originated in the Mesozoic. On a regional scale, the thermal evolution of this area of the Southwest was heterogeneous. Some ranges lack any evidence of Mesozoic or younger metamorphism; others contain only elements of a Mesozoic metamorphism and show none of the distinctive features of core-complex deformation. Examples of the latter case include the Jurassic (~140-170 m.y.) metamorphism of the Halloran Hills and the Death Valley region (DeWitt et al., 1984; Labotka and Albee, Chapter 26, this volume), and the Cretaceous metamorphism (Hoisch et al., Chapter 21, this volume) that occurs in the Big Maria and Old Woman Mountains. This older metamorphism is most readily recognized where the Paleozoic cratonic section has been preserved. Ductile deformation, recumbent nappe formation, and thrust faulting are common and clearly indicative of a compressional setting. In the Whipple, Sacramento, and other complexes, this supracrustal section, although once regionally existent, is absent below the upper plate Tertiary nonconformity.

Recognition of pre-mylonitization Mesozoic metamorphism in the crystalline rocks of core complexes is made difficult by the abundance of younger plutonism and the obliteration of older fabric by the mylonitization. However, an older (pre-70 m.y.) metamorphism has been recognized in the Harcuvar Mountains (Rehrig and Reynolds, 1980), and the Whipple Mountains contain at least two premylonization events, one between 73(?) and 89 m.y. and another in excess of 130 m.y. (G. A. Davis et al., 1982). Both geologic relations and isotopic dating show that core-complex mylonitization is younger than any of these variably aged Mesozoic metamorphisms. For most complexes, the age of mylonitization is Tertiary; exceptions include that of the Chemehuevi (John, 1982) and Iron (D. M. Miller and Howard, 1985) Mountains which reportedly contain Late Cretaceous mylonitic fabrics.

The deformation history of core complexes is distinctive, with the semibrittle/semiductile character of mylonitization giving way in time to brittle structures associated with detachment faulting; both stages of deformation, however, appear to exhibit the same kinematic history as evidenced by the similar orientation of planar and linear structures and the same sense of shear. The relationship between mylonitization and detachment faulting has been controversial. Some workers (G. A. Davis et al., 1980; Frost, 1981; Howard, 1980) have suggested that the two phenomena are unrelated. However, an emerging consensus from a spectrum of studies (G. A. Davis et al., 1986; G. H. Davis, 1980, 1983; E. L. Miller et al., 1983; Rehrig and Reynolds, 1980; Reynolds, 1985; Wernicke, 1981) is that mylonitization and detachment faulting are cogenetic and, in part, coeval, representing ductile and brittle manifestations, respectively, of crustal extension.

Whipple Mountains

The geology of the Whipple Mountains has been described previously by Anderson *et al.* (1979), G. A. Davis *et al.* (1980, 1982, 1986), and Anderson and Rowley (1981). The lower plate contains in excess of 3.9 km of mylonitic gneisses, the bottom of which is not exposed. Commonly, these rocks are upwardly truncated at the Whipple detachment fault except in the southwest portion of the range, where they pass into nonmylonitic rocks through a 3- to 30-m-thick zone of structural transition, or "mylonitic front." The style of mylonitization gradually changes with depth, being more cataclastic at upper levels and involving more recrystallization at lower levels. The lineation uniformly trends N45° ± 10°E or S45° ± 10°W with a shallow plunge.

More than 14 separate prekinematic, synkinematic, and postkinematic plutons occur below the mylonitic front. Recent U-Pb (zircon) dating by Wright *et al.* (1986) has constrained the ages of most, if not all, to four intrusive epochs. Having an age of 89 ± 3 m.y. are two generations of peraluminous sheetlike intrusions that include biotite granodiorite, two-mica monzogranite, garnet-two-mica granodiorite, and two-mica tonalite. A 73 ± 3-m.y. hornblende-biotite quartz diorite is the structurally deepest, occurring as four sheetlike intrusions exposed about the central high peaks of the range. All of these rocks are intensively mylonitized and lineated, although the early generation of the peraluminous suite also contains an older, steeper fabric. Anderson and Rowley (1981) presented field evidence indicating that some of these intrusions are synkinematic to the mylonitization, but new isotopic and field data now suggest a prekinematic origin. Numerous, thin biotite tonalite and trondhjemitic aplite sheets or sills synkinematically pervade the complex at high structural levels. These intrusions locally cross-cut the mylonitic fabric of the older granitic rocks, but are mylonitized in their own plane of intrusion and have the same lineation. Wright *et al.* (1986) have dated the tonalites at 26 ± 5 m.y. Recent mapping by G. A. Davis (personal communication, 1984) in the northwest portions of the range has revealed two postkinematic plutons, the older consisting of a hornblende-biotite-clinopyroxene diorite and the younger, a biotite-hornblende granodiorite. Wright *et al.* (1986) report an age of 19 m.y. for the diorite.

The age of the mylonitization in the Whipples has remained uncertain due largely to the disturbed nature of the isotopic systems (K-Ar, Rb-Sr) used in earlier dating (G. A. Davis *et al.*, 1980, 1982; Martin *et al.*, 1980). Previous suggestions of a Late Cretaceous to early Tertiary age were based on a variety of older dates, including (1) 83 ± 8 m.y. (fission track, sphene) and 79 ± 6 m.y. (K-Ar, biotite) from mylonitic clasts in upper plate sedimentary units of late Oligocene or Miocene age, (2) Rb-Sr ages with wide uncertainties, (3) K-Ar (hornblende) ages of 42 ± 1, 50 ± 5, and 51 ± 4 m.y. from deep structural levels, and (4) semiconcordant K-Ar hornblende and biotite ages of 57 and 52 m.y., respectively, reported by Shackelford (1980) for the nearby Rawhide Mountains. Most K-Ar ages, however, are Tertiary and young structurally upward toward the detachment fault, the last movement of which is bracketed between 16 to 19 m.y. From the work of Wright *et al.* (1986), it is now certain that the mylonitization is principally, if not exclusively, Tertiary in age. As indicated above, the mylonitization was coeval with intrusions dated at 26 m.y. and was postdated by intrusions having a 19-m.y. age. Although Cretaceous mylonites are apparently exposed in nearby ranges (Howard *et al.*, 1982; John, 1982; their existence in the crystalline complex of the Whipple Mountains has not been unequivocally demonstrated.

Sacramento Mountains

The geology of the Sacramento Mountains has most recently been studied by McClelland (1982) and Spencer and Turner (1982). The dominant structure of the range is the Sacramento Mountains detachment fault which separates a crystalline lower plate from three allochthonous upper plates, each bounded by low-angle detachment faults. Crystalline units throughout most of the lower plate contain a mylonitic foliation and southwest-plunging lineation below a southwest-dipping mylonite front that McClelland mapped in the western part of the range. Three prekinematic(?) to synkinematic(?) plutons intrude strongly mylonitized gneisses of presumed Proterozoic age. The oldest pluton is a multiply intruded biotite-hornblende granodiorite that possesses a weak to moderately developed mylonitic fabric. Garnet, two-mica monzogranite intrudes the granodiorite and older gneiss as large sheets (to 50 m thick) concordant with the mylonitic foliation. The mylonitic foliation in the monzogranite is variably developed, ranging from weak to strong. Intrusive into the granodiorite and monzogranite are thin sills of biotite tonalite and aplite, both strongly foliated and lineated. None of these plutons has been dated, but the intrusive style and composition of the monzogranite and tonalite are similar to the 89- and 26-m.y.-old intrusive units, respectively, of the Whipple Mountains.

Iron Mountains

The mylonitic gneisses of the Iron Mountains (Fig. 19-1) represent the uppermost portion and the oldest plutonic units of the Cretaceous Cadiz Valley batholith (Howard *et al.*, 1982; D. M. Miller and Howard, 1985; D. M. Miller *et al.*, 1981). Three sills, with inter-layered Proterozoic(?) gneisses, make up the 1.2-km-thick mylonitic section. Mylonitiza-tion, most intense at the top, decreases downsection into the batholith proper which is largely undeformed except for localized zones of shear. The low-angle fabric consistently contains an E-W-trending lineation. The principal rock type is two-mica monzogranite to granodiorite with minor amounts of garnet. D. M. Miller *et al.* (1981) report a number of K-Ar (mica) dates, all between 62 and 67 m.y., on both mylonitic and postkinematic phases. J. E. Wright (in Howard *et al.*, 1982) has reported a Late Cretaceous U-Pb (zircon) age from elsewhere in the batholith.

Santa Catalina Mountains

The complex exposed in the Santa Catalina Mountains (Fig. 19-1) covers an extensive area that also includes the Rincon Mountains to the southeast and the Tortolita Moun-tains to the northwest. Over the past three decades, the range has received a considerable amount of study, some of which (G. H. Davis, 1973, 1975, 1977; G. H. Davis and Coney, 1979) laid the foundation for our basic understanding of the general features of core complexes. Papers by G. H. Davis (1980), Banks (1980), and Keith *et al.* (1980) give geologic descriptions of the complex. Most pertinent to this study is the work of Keith *et al.* (1980), who have presented a case for multiple intrusion and mylonitization begin-ning in the Late Cretaceous or early Tertiary and ending in the early Miocene, based on a variety of isotopic systems, including K-Ar, Rb-Sr, and U-Pb. Three intrusive epochs involving 12 plutons are each interpreted to be coeval with or followed by mylonitiza-tion leading to the final crystalline construction of this core complex. The 60- to 75-m.y.-old Leatherwood suite includes three plutons (Chirreon Wash, Leatherwood, Rice Peak) of biotite-hornblende granodiorite to quartz diorite composition. Five plutons (Derrio Canyon, Wilderness, Youtcy, Espirito Canyon, Wrong Mountain) of the Wilderness suite all have an age of about 44–50 m.y. and are composed of garnet, two-mica monzogranite. Forming the youngest intrusions of the complex is the Catalina suite, four plutons of biotite-sphene ± hornblende granite (monzogranite to syenogranite) having an age be-tween 25 and 29 m.y. Bykerk-Kaufmann and Janecke (1986) have shown that the post-Leatherwood, pre-Wilderness mylonitic fabric is characterized by an east-west lineation

ANDERSON

507

and is compressional in origin, while the mid-Tertiary, post-Wilderness mylonitic fabric, characterized by a northeast-southwest lineation, formed in an extensional stress regime.

DETAILED MINERALOGY

The deformation is termed mylonitic, owing to the brittle behavior exhibited by several minerals, including plagioclase, garnet, and amphibole, which occur both as broken porphyroclasts and finely comminuted material in a variably recrystallized matrix. Other minerals exhibit crystal-plastic behavior, including quartz, alkali feldspar, and the micas. Due to ductility contrast during deformation, all of the minerals responded differently and, as a consequence, occur segregated in submillimeter-wide monomineralic bands defining the mylonitic foliation (Fig. 19-2). Mineral reactions were sparse and largely confined to the fine-grained mylonitic matrix. Many of the minerals have retained original compositions and typically have a bimodal spread of composition following the textural division of porphyroclast and matrix. Garnet appears to have been the most resistive to compositional change; although in part replaced by biotite, detailed profiling has revealed no secondary garnet composition. In contrast, plagioclase, biotite, muscovite, and hornblende all exhibit partial reequilibration to new compositions. Alkali feldspar and magnetite have thoroughly reequilibrated to near end-member compositions.

The sections below describe the compositional nature of primary (igneous) and secondary (from mylonitization) mineral phases in the peraluminous and metaluminous granitic intrusives of the ranges studied. The gneissic host units of some of these granites were also investigated but were found not to be amenable to systematic petrologic study even though some of the bulk compositions involved are theoretically more useful. For example, kyanite and sillimanite-bearing pelitic gneisses of presumed Proterozoic age occur below the mylonite front of the Whipple Mountains. Their phase relations, however, were found to be complex, apparently due to a more complicated thermal history

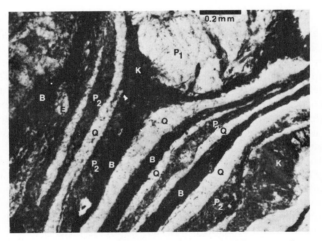

FIG. 19-2. Photomicrograph of typical mylonitic texture from the Whipple Mountains. Note monomineralogic nature of laminar banding and disruption of banding about porphyroclasts. Mineral symbols: P_1, porphyroclast plagioclase; P_2, matrix plagioclase; K, alkali feldspar; B, biotite; M, muscovite; Q, quartz; and E, epidote.

CORE COMPLEXES OF THE MOJAVE-SONORAN DESERT

involving one Proterozoic high-grade metamorphism and one or two Mesozoic metamorphisms prior to the onset of core-complex mylonitization.

Feldspars

Due to the inherent sluggishness of the CaAl = NaSi exchange, plagioclase reequilibration to the metamorphic conditions of mylonitization was limited. Porphyroclasts commonly preserve delicate oscillatory and normal zoning patterns of the igneous parentage, despite the penetrative fabric development that has affected the rocks. Relict compositions also exist in the matrix where they are mechanically mixed with reequilibrated compositions. The by-product of the retrogression is the formation of epidote, which, as the only porphyroblast phase, occurs as large crystals deflecting and in part rolled in the laminar structure of the matrix foliation. The compositional gap between relict and more albitic reequilibrated plagioclase varies with bulk composition and metamorphic grade accompanying the deformation. In general, the reequilibrated compositions range from An_{03-19} at higher structural levels to An_{21-27} at deeper levels.

Alkali feldspar responded much differently to the deformation, being more prone to recrystallization and local displacement by pressure solution. Commonly, new, coarse, homogeneous crystals occur in strain shadows at the ends of porphyroclasts and in porphyroclast pullaparts, as well as discontinuous laminae extending from alkali feldspar porphyroclasts. All occurrences have about the same composition, ranging from Or_{88-92} at shallow levels to Or_{82-89} at deeper levels, with a concomitant change in structual state from a relatively disordered structural state approaching orthoclase at greater depths to maximum or intermediate microcline at shallow levels.

Muscovite

The distinctive chemical feature of plutonic white mica is the substitution of Fe^{2+}, Fe^{3+}, Mg, and Ti, making its composition much different from end-member muscovite (Anderson and Rowley, 1981; C. F. Miller et al., 1981). Terms like phengitic or celadonitic are not appropriate as several components are involved. Compositions observed in this study can be described by six components (W. Thomas and Anderson, unpublished data), muscovite (mu; KAl_2Si_3Al), paragonite (pg; $NaAl_2Si_3Al$), Fe celadonite (fce; $KFe^{2+}AlSi_4$), Mg celadonite (mce; $KMgAlSi_4$), ferrimuscovite (fe; $KFe_2^{3+}Si_3Al$), and a titanium end member (Ti; $KTiAlSi_2Al_2$), plus Fl and Cl substitution in the hydroxyl site. Assuming no octahedral or interlayer vacancy, charge balance constraints allow estimation of minimum Fe^{3+} and maximum Fe^{2+}. Uniformly, most of the iron is ferric, represented here as the end member ferrimuscovite. In order of abundance, the most important components in primary muscovite are Mg celadonite, ferrimuscovite, and "Ti," their sum being the highest in the Whipple muscovites whose average total X_{mu} is only 0.58. Secondary muscovite, largely restricted in occurrence to the mylonitic matrix, differs in composition by significant loss of Ti (Fig. 19-3) and ferrimuscovite components with retrogression. Textural features indicate the by-product to be biotite.

Biotite

In all rocks studied, biotite coexisting with muscovite was found to have higher Al^{IV}, Al^{VI}, and Fe/Fe + Mg relative to that coexisting with sphene ± hornblende. Despite the intensity of mylonitization, much the biotite retained its original igneous composition, particularly that occurring as porphyroclasts and some in the matrix. The major change in composition of retrogressed matrix biotite is a depletion in Ti (Fig. 19-4) balanced by

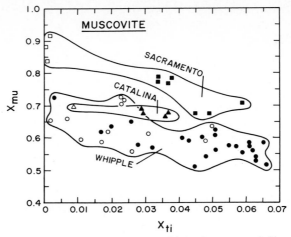

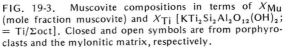

FIG. 19-3. Muscovite compositions in terms of X_{Mu} (mole fraction muscovite) and X_{Ti} [$KTi_2Si_2Al_2O_{12}(OH)_2$; $= Ti/\Sigma oct$]. Closed and open symbols are from porphyroclasts and the mylonitic matrix, respectively.

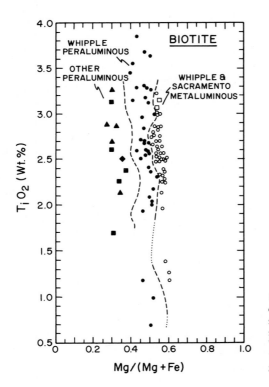

FIG. 19-4. Biotite compositions in terms of wt % TiO_2 and Mg/(Mg + Fe). Closed symbols from metaluminous granites; open symbols from peraluminous granites; circles, Whipple plutons; squares, Sacramento plutons; triangles, Santa Catalina Wilderness granite; diamonds, Iron Mountains.

an increase in Mg by the exchange mechanism $Ti + []^{VI} = 2Mg$. Similar depletions in Ti for sheared biotite have been reported by Kerrich *et al.* (1981) and it is well known that the solubility of Ti in biotite decreases with lower temperature (Guidotti *et al.*, 1976; Robert, 1976). Texturally, the low-Ti biotite appears to have formed from magmatic biotite (with magnetite, rutile, and alkali feldspar as by-products), from high-Ti muscovite in the two-mica granites, and from hornblende in the biotite hornblende granodiorites and quartz diorites. Chlorite did not form during mylonitization, although this phase does occur replacing biotite in rocks (chloritic breccias) affected by alteration associated with detachment faulting.

Amphibole

Hornblende occurs as a primary phase in the structurally deep 73-m.y.-old quartz diorite of the Whipple Mountains and in the undated granodiorite of the Sacramento Mountains. The hornblendes are pargasitic, but the Whipple hornblende is markedly more aluminous (12.2–13.1 verus 8.4–9.9 wt % Al_2O_3), a consideration that is important in the barometry calculations.

Mylonitization modified the rim compositions of deformed, primary grains and produced new amphiboles of differing compositions in the recrystallized and mylonitic matrix and in late-forming tension gashes. Tension gashes are widespread throughout the 73-m.y.-old Whipple Mountains quartz diorite pluton. Oriented perpendicular to the mylonitic lineation, the gashes are up to 0.5 cm wide and filled with crack-seal assemblages of untwinned orthoclase, oligoclase, and prismatic crystals of actinolitic hornblende aligned parallel to the lineation. The variable textures of deformed, recrystallized, and new amphibole exhibit a range of composition involving a nonlinear loss of Al^{IV}, Al^{VI}, Ti, Fe, Na, K, and enrichment in Si, Ca, and Mg (Fig. 19-5). The end result is the actinolitic hornblende of the tension gashes with 6.0–8.1 wt % Al_2O_3. The principal solution mechanisms involved are (1) $Ti + 2Al^{IV} = Mg + 2Si$, (2) $Al^{VI} + Al^{IV} = Mg + Si$, (3) $(Na + K)^A + Al^{IV} = []^A + Si$, and (4) $Fe^{2+} = Mg$.

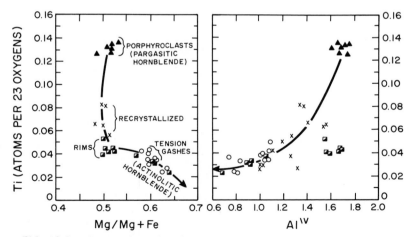

FIG. 19-5.   Range of hornblende chemistry from the mylonitized 73-m.y.-old hornblende-biotite quartz diorite, Whipple Mountains. Porphyroclasts are considered primary, all other textural occurrences interpreted to result from reequilibration during deformation.

Garnet

Porphyroclasts of garnet are common in the peraluminous granitoids of all four ranges studied. Those of the Sacramentos, Santa Catalinas, and Irons are like most plutonic garnets, being principally an almandine-spessartine solution (Fig. 19-6). Garnets from the Whipple plutons are unusual due to a high grossular component, mole fractions of which range up to 0.24.

Texturally, the garnet in all rocks studied appear to be prekinematic, being variably broken and extended by the mylonitic deformation. Retrogressed garnets are expected to have Mn-enriched rims (Hodges and Royden, 1984). Those of the Sacramento Mountains have minor Mn enrichment from core to rim (change in X_{sp} less than 0.04). The Santa Catalina (Wilderness granite), and the Whipple garnets have Mn-depleted rims, a feature more commonly associated with igneous crystallization (Anderson and Rowley, 1981). Hence, the interpretation presented here is that garnet did not retrogress during mylonitization due to the inhibited nature of reaction kinetics common to almost all assemblages in these rocks.

GEOTHERMOMETRY AND GEOBAROMETRY

Mylonites are petrologically complicated inasmuch as their mineral assemblages do not represent a complete reequilibration to the metamorphic conditions of the deformation event. Broad ranges of mineral chemistry exist due to a mixture of reequilibrated, partially(?) reequilibrated, and protolith mineral compositions. There is much to be gained from a study of such variable compositions, for they are indicative of the multiple history experienced by these rocks. The premise taken here is that the protolith mineral chemis-

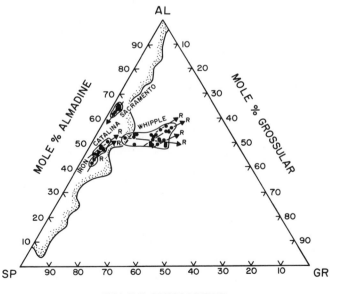

FIG. 19-6. Garnets from various two-mica granites in terms of almandine, spessartine, and grossular mole proportions. R, direction of zoning toward garnet rims. The stippled field represents compositions of most plutonic garnets.

try (that of the porphyroclasts) can be utilized to retrieve igneous conditions of these rocks, and the range of markedly different compositions exhibited by phases within the mylonitic matrix represent an approach to equilibration during mylonitization. In application, this requires every mineral phase in each sample to be thoroughly characterized in all of its textural occurrences. Only then can appropriate phase compositions of probable equilibrium assemblages be identified. This is the tack taken in this study and several thermobarometric calibrations have been found to be applicable, including garnet-biotite, garnet-muscovite-plagioclase-biotite, muscovite-biotite-alkali feldspar-quartz, and two feldspars.

Garnet-Muscovite-Plagioclase-Biotite Equilibria

Thermobarometry for most granitic rocks is limited, due to their high-variance mineral assemblages. Garnet, two-mica granites are an exception because the assemblage garnet-muscovite-plagioclase-biotite allows the application of three thermometers and barometers:

$$\underset{\text{almandine}}{Fe_3Al_2Si_3O_{12}} + \underset{\text{phlogopite}}{KMg_3AlSi_3O_{10}(OH)_2} = \underset{\text{pyrope}}{Mg_3Al_2Si_3O_{12}} + \underset{\text{annite}}{KFe_3AlSi_3O_{10}(OH)_2} \quad (1)$$

$$\underset{\text{anorthite}}{3CaAl_2Si_2O_8} + \underset{\text{annite}}{KFe_3AlSi_3O_{10}(OH)_2} = \underset{\text{almandine}}{Fe_3Al_2Si_3O_{12}} + \underset{\text{grossular}}{Ca_3Al_2Si_3O_{12}}$$
$$+ \underset{\text{muscovite}}{KAl_2Si_3AlO_{10}(OH)_2} \quad (2)$$

$$\underset{\text{anorthite}}{3CaAl_2Si_2O_8} + \underset{\text{phlogopite}}{KMg_3AlSi_3O_{10}(OH)_2} = \underset{\text{pyrope}}{Mg_3Al_2Si_3O_{12}} + \underset{\text{grossular}}{Ca_3Al_2Si_3O_{12}}$$
$$+ \underset{\text{muscovite}}{KAl_2Si_3AlO_{10}(OH)_2} \quad (3)$$

At face value, simultaneous solution of any two of the three can theoretically give a unique fluid-independent determination of P and T.

The temperature-sensitive exchange of Mg and Fe between garnet and biotite has had a history of study, including efforts by Thompson (1976), Goldman and Albee (1977), Ferry and Spear (1978), Hodges and Royden (1984), Ganguly and Saxena (1984), and Indares and Martignole (1985). The latter three present modifications of the earlier calibrations through use of more appropriate activity models for garnet and biotite with the aim of correcting for the influence of additional components, an important consideration for their application to granitic rocks.

The results of five calibrations for a representative sample from each complex are depicted in Fig. 19-7. Inasmuch as the garnets appear not to have reequilibrated under the conditions of mylonitization, the aim here is to determine crystallization conditions for the pre-mylonitization intrusions by careful choice of plagioclase, muscovite, and biotite compositions, which, as described earlier, usually are those from porphyroclasts. Convincingly, all but one of the Whipple and Sacramento samples yield magmatic temperatures, although there is a spread, with the Ferry and Spear (FS) calibration giving a lower result and Hodges and Royden (HR), Ganguly and Saxena (GSX), and Indares and Martignole (IM) calibrations yielding higher values. The results of the Goldman and Albee (GA) calibration are less uniform but also appear to run low except for the Whipple garnets (Table 19-1). For the Whipple plutons, which ubiquitously have high grossular

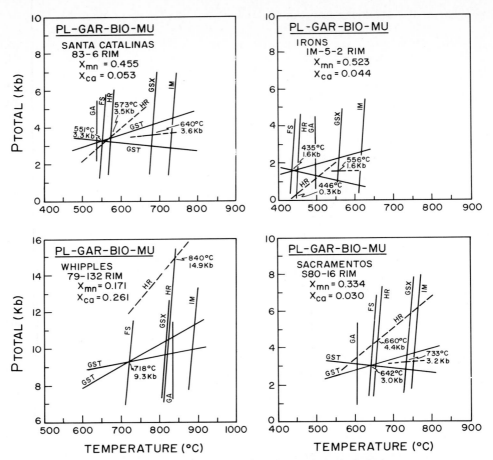

FIG. 19-7. Results of garnet-biotite thermometry and plagioclase-garnet-biotite-muscovite barometry for representative samples of the Whipple, Sacramento, Santa Catalina, and Iron Mountains. The various calibration labels are: FS, Ferry and Spear (1978); GA, Goldman and Albee (1977); HR, Hodges and Royden (1984); GSX, Ganguly and Saxena (1984); IM, Indares and Martignole (1985); GST, Ghent and Stout (1981).

garnet, all of the calibrations except FS give results seemingly too high in temperature for granodiorites. The other suites have garnets low in Ca, but markedly higher Mn. The Wilderness granite (Santa Catalina) and that of the Iron Mountains have very high Mn, and there appears to be a general trend toward lower calculated temperatures with decreasing Ca and increasing Mn in the garnet. No one calibration seems suitably adapted to the wide variations in garnet compositions found in this study. The FS calibration seems best for the high-Ca, moderate-Mn garnets (Whipples) and the GSX calibration yields acceptable results for garnets with X_{Mn} up to 0.48. Only samples from the Iron Mountains give submagmatic temperatures by all methods of garnet-biotite thermometry; presumably, this could be due to the high Mn of garnets in those plutons or to resetting during mylonitization.

More fundamental to this study are estimations of pressure or depth. Equilibria 2 and 3 are pressure sensitive, basically involving the transfer of Ca to the grossular com-

TABLE 19-1. Input Variables and Results for PLAG-GAR-BIO-MU Thermobarometry

	Whipple Mountains							Sacramento Mountains				Iron Mountains		Santa Catalina Mountains			
Sample:	WW6-C	183-R	183-C	241-R	241-C	132-R	132-C	15-R	15-C	16-R	16-C	5-2-R	5-2-C	836-R	836-C	837-R	837-C
Plagioclase no.	(09-1)	(I2)	(A1)	(A1)	(A1)	(A1)	(B1)	(G3)	(G1)	(L2)	(L3)	(A-1)	(A-1)	(N-1)	(N-3)	(I-3)	(I-4)
Xan	.219	.195	.237	.239	.241	.229	.257	.169	.187	.204	.221	.211	.211	.190	.205	.149	.154
Biotite no.	(07)	(E1)	(F1)	(A3)	(A-1)	(I2)	(I1)	(H3)	(H2)	(P1)	(P1)	(B1)	(B1)	(O-1)	(0-2)	(E-1)	(D-1)
XMg	.440	.425	.418	.384	.378	.350	.347	.257	.246	.269	.301	.300	.300	.221	.232	.257	.231
XFe	.423	.430	.423	.431	.438	.459	.498	.580	.563	.504	.504	.537	.537	.572	.535	.534	.528
XTi	.073	.047	.051	.058	.061	.053	.064	.052	.066	.045	.045	.053	.053	.056	.053	.060	.067
XA1^{vi}	.055	.086	.096	.107	.101	.123	.077	.098	.111	.159	.159	.157	.157	.123	.153	.127	.151
Muscovite no.	(20)	(M1)	(M2)	(A1)	(A2)	(C3)	(C2)	(H1)	(F2)	(P1)	(O1)	(a)	(a)	(O1)	(O2)	(C1)	(C1)
XK	.974	.993	.954	.977	.977	.991	.990	.986	.977	.966	.968	.98	.98	.980	.978	.974	.974
XA1^{vi}	.731	.750	.757	.746	.721	.675	.690	.839	.876	.823	.826	.75	.75	.776	.772	.768	.768
Garnet no.	(07)	(F1)	(F2)	(H1)	(H4)	(I-1)	(G1)	(A4)	(A2)	(G2)	(G3)	(E1)	(C1)	(11)	(14)	(12)	(14)
XCa	.279	.234	.246	.246	.167	.261	.239	.019	.025	.030	.031	.044	.048	.053	.044	.052	.051
XMg	.064	.128	.120	.087	.099	.089	.089	.056	.064	.066	.068	.026	.024	.030	.029	.035	.039
XFe^{2+}	.438	.470	.458	.455	.458	.479	.424	.593	.596	.570	.573	.408	.420	.463	.437	.485	.460
XMn	.218	.168	.176	.212	.277	.171	.249	.332	.315	.334	.328	.523	.508	.455	.490	.428	.451
XA1^{vi}	.955	.955	.958	.975	.953	.954	.938	1.000	.933	.988	.987	1.000	1.000	1.000	1.000	1.000	1.000

Pressure-Temperature Solutions[b]

	WW6-C	183-R	183-C	241-R	241-C	132-R	132-C	15-R	15-C	16-R	16-C	5-2-R	5-2-C	836-R	836-C	837-R	837-C
1																	
P(kb)	7.4	10.6	9.8	8.7	7.9	9.3	9.0	2.0	2.9	3.0	3.0	1.6	1.7	3.3	2.4	3.9	4.3
T(°C)	515	785	762	665	725	718	822	632	698	642	651	435	411	551	524	518	602
2																	
P(kb)	8.2	11.3	10.5	9.5	8.5	10.1	9.9	2.1	3.1	3.2	3.1	1.6	1.8	3.6	2.5	4.3	4.7
T(°C)	618	857	841	768	820	817	948	730	792	733	741	556	526	680	653	636	733
3																	
P(kb)	9.8	17.2	15.9	12.9	11.9	14.9	15.1	3.0	4.9	4.4	4.4	0.3	0.2	3.5	2.2	3.8	5.2
T(°C)	624	902	882	775	808	840	941	645	717	660	670	446	424	573	541	538	626

Comparative Thermometry (°C)[c]

	WW6-C	183-R	183-C	241-R	241-C	132-R	132-C	15-R	15-C	16-R	16-C	5-2-R	5-2-C	836-R	836-C	837-R	837-C
FS	517	772	754	662	725	713	817	641	703	646	655	443	418	554	530	518	601
HR	618	863	849	756	791	813	910	649	713	658	668	458	435	575	547	539	622
GSX	618	842	830	761	818	808	939	738	797	736	745	563	533	681	658	635	731
IM	674	947	928	829	891	875	1036	750	784	756	763	621	585	731	712	666	759
GA	608	913	916	784	803	833	949	545	588	606	613	498	470	537	555	510	599

[a] Assumed value.

[b] Simultaneous solution of (1) Ghent and Stout (1981) and Ferry and Spear (1978), (2) Ghent and Stout (1981) and Ganguly and Saxena (1984), (3) Hodges and Royden (1984).

[c] FS, Ferry and Spear (1978); HR, Hodges and Royden (1984); GSX, Ganguly and Saxena (1984); IM, Indares and Martignole (1985); GA, Goldman and Albee (1977); results are at $P = 8$ kb for Whipple Mountains, 4 kb for others.

ponent in garnet at the expense of anorthite component in plagioclase with increasing pressure. Ghent and Stout (1981) presented an empirical calibration for both equilibria using natural samples from which P and T had been independently determined from garnet-biotite and garnet-plagioclase-aluminosilicate-quartz equilibria. Hodges and Spear (1982) found that the Ghent and Stout calibrations yield consistent results with other thermobarometers, but that equilibria 2 was more sensitive to analytical uncertainty. Garnets in that study had X_{Mn} up to 0.26 and X_{Ca} up to 0.16. Garnet activity models, modified from that of Newton and Haselton (1981), encompass terms to account for the effect of Mn and Ca on the solution.

A comparison of the Hodges and Royden (HR) and Ghent and Stout (GST) calibrations are given in Fig. 19-7. By either method, the high-Ca garnets of the Whipples point to moderate crustal pressures, yet it seems certain that the HR calibration yields pressures much too high. Figure 19-8 gives a compilation for all four complexes using simultaneous solution of GST-FS for the Whipple Mountains and GST-GSX for the Sacramento, Santa Catalina, and Iron Mountains. The average results for the two-mica granites are 9.2 ± 0.9 kb (at 746 ± 55°C) for the Whipple Mountains, 2.9 ± 0.5 kb (at 736 ± 38°C) for the Sacramento Mountains, 3.8 ± 0.9 kb (at 676 ± 42°C) for the Santa Catalina Mountains, and 1.7 ± 0.2 kb (at 541 ± 15°C) for the Iron Mountains. Results using rim or near-rim (R) and core (C) compositions are also designated in Fig. 19-8 as most plagioclase and garnet porphyroclasts exhibit some zonation. It is emphasized that these absolute values of P and T must be considered preliminary, due to the fact that both the garnet and the muscovite in these granites are unlike the compositions for which these calibrations were originally intended. Admittedly, the 9.2-kb estimate for the Whipple Mountains seems oddly high and the low pressure calculated for the Iron Mountains (1.7 kb) may be in part a function of the low calculated temperature. On a relative basis, however, the calculations indicate deep levels of emplacement of the Whipple two-mica plutons and shallower levels for the Santa Catalina, Sacramento, and Iron Mountain plutons.

Alumina Solubility in Hornblende

Hammerstrom and Zen (1985) have recently developed a geobarometer for calcalkaline granites based on total Al in hornblende. Largely empirical, the barometer is based on

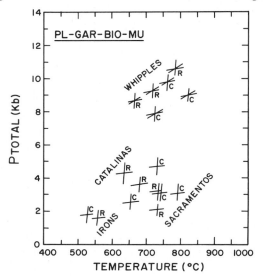

FIG. 19-8. Compilation of plagioclase-garnet-biotite-muscovite thermobarometry for all complexes. C, core; R, rim compositions.

data for natural hornblendes in granitic rocks for which P and T had been independently estimated, plus a limited number of experimental determinations. There is much to indicate that Al solubility in hornblende does increase with pressure, but the absence of a temperature term in the calibration (Fig. 19-9) points to the preliminary nature of the barometer. Moreover, reanalysis of the same data, enlarged to include other analyses from the Mojave-Sonoran region (Anderson, unpublished data) show that Al hornblende solubility is also affected by the $Fe/(Fe + Mg)$ ratio. There is a good positive correlation of $Fe/(Fe + Mg)$ and Al_{total} (E. Young, unpublished data); thus, iron rich hornblendes will yield erroneously high apparent pressures. Most of the data of Hammerstrom and Zen have $Fe/(Fe + Mg)$ ratios from 0.30 to 0.70, and until a proper correction can be made, it seems best to restrict its application. Primary hornblendes from the Whipple and Sacramento Mountains have similar $Fe/(Fe + Mg)$ ratios, ranging from 0.45 to 0.51.

Hammerstrom and Zen (1985) recognized a temperature dependence of Al^{IV} in hornblende and Nabelek and Lindsley (1985) have developed an approximate empirical calibration for cases where the coexisting plagioclase is An_{40} or greater. The plagioclase porphyroclasts in the 73-m.y.-old hornblende-bearing quartz diorite pluton of the Whipples and the hornblende-bearing granodiorite of the Sacramentos have An contents up to An_{42} and An_{46}, respectively.

Although both calibrations are as yet untested, simultaneous solution of each can yield a unique $P-T$ determination. Again, the objective is to determine the emplacement level of the plutons. The result is shown in Fig. 19-9. The primary Whipple hornblendes are far more aluminous than any other that this writer has observed in the Mojave-Sonoran region. Thus, it is not surprising that the estimated conditions, 7.5 ± 0.4 kb (at $785 \pm 7°C$), are at a moderate crustal level. Much different results occur for the biotite-hornblende granodiorite of the Sacramento Mountains. Although the estimated temperature ($783 \pm 6°C$) is similar to that of the Whipple Mountains, the pressure, at 4.2 ± 0.7 kb, is much lower.

An independent check is a comparison with emplacement pressures for the associated two-mica granites given above. Given the uncertainties of both calibrations, the results (7.5 versus 9.2 kb for the Whipple Mountains and 4.2 versus 2.9 kb for the Sacramento Mountains) are broadly similar. For each complex, the higher estimate is derived from the older plutonic suite; however, the differences are insufficiently distinct to be conclusive. What is apparent is that the highly calcic nature of the Whipple garnets and the aluminous nature of the hornblendes, although in different magma systems, are

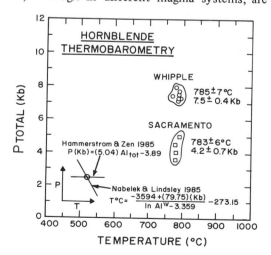

FIG. 19-9.   Results of hornblende thermobarometry for quartz diorite and granodiorite of the Whipple and Sacramento Mountains. From simultaneous solution of Hammerstrom and Zen (1985) and Nabelek and Lindsley (1985) calibrations.

indicative of a common theme, one of deep pluton emplacement relative to the other complexes of this study.

Mylonitization Feldspar Thermometry

Due to the fact that none of the alkali feldspar in these mylonitized granitic intrusives has retained an igneous composition, the application of two-feldspar thermometry can only be used for the metamorphic conditions of mylonitization. Nearly every sample contains a range, commonly bimodal, of plagioclase compositions with the less calcic members restricted to the mylonitic matrix. Coupling the more sodic plagioclase with the rather uniform composition of the alkali feldspar forms a basis for thermometry of the fabric-forming mylonitization event.

Several calibrations exist (Hazelton *et al.*, 1982; Price, 1985; Stormer, 1975; Whitney and Stormer, 1977); yet only that of Whitney and Stormer is appropriate for low-temperature feldspar pairs. All others are developed for high structural state feldspars (sanidine-high albite). Some of the alkali feldspar in these mylonitic rocks is orthoclase (only in the deep structural levels of the Whipples) but most is an ordered feldspar, closer to intermediate or maximum microcline. After employing all calibrations (the sanidine models indicate temperatures lower by as much as 80°C), the choice made here is to use that of Whitney and Stormer for all of the data so that the use of relative thermometry is maximized.

The study of the Whipple complex has been far more extensive (85 reequilibrated pairs) and the results for that range are given in Fig. 19-10. Although there is a spread in the data, it is certain that the grade of metamorphism attending mylonitization increases with depth. This is consistent with the mylonitization involving greater amounts of recrystallization (less cataclasis) at deeper levels and the depth-dependent grade change as deduced from mineral assemblages, that being upper greenschist (biotite + albite + Or_{91} alkali feldspar) at shallow levels to lower amphibolite (biotite + actinolitic hornblende + oligoclase + Or_{86} alkali feldspar) in the structurally deeper portions of the range.

Muscovite-Biotite-Alkali Feldspar-Quartz Equilibria

Powell and Evans (1983) formulated a geobarometer for the equilibrium

$$3KMgAlSi_4O_{10}(OH)_2 = KMg_3AlSi_3O_{10}(OH)_2 + 2KAlSi_3O_8 + 3SiO_2 + 2H_2O$$
Mg celadonite   phlogopite   orthoclase   quartz   fluid

based on the study of Velde (1965). Velde reversed this reaction at 375–390°C and 2.0 kb and at 440–480°C and 4.5 kb for muscovite containing 20–30 mol % celadonite. Because the alkali feldspar in the mylonitic rocks of this study have been thoroughly reequilibrated, this barometer can only be applied to mylonitization conditions. Appropriately, the temperature of mylonitization (above section) and the amount of Mg celadonite in the secondary muscovites is reasonably close to that of the experimental conditions. However, one notable concern, as pointed out by T. Hoisch (personal communication, 1986) is that the Powell and Evans formulation uses thermodynamic data for sanidine, not orthoclase or microcline.

In natural rocks, this equilibrium takes the form of a continuous reaction where muscovite becomes increasingly celadonitic with increasing pressure or decreasing temperature at the expense of components of the other phases (biotite, feldspar, quartz). Using the results of feldspar thermometry to give temperature brackets for each sample, the

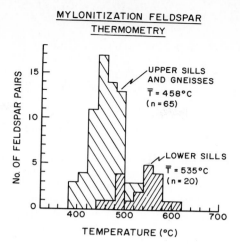

FIG. 19-10. Summary of feldspar thermometry at 4 kb for the Whipple Mountains mylonitization. Calibration is that of Whitney and Stormer (1977).

results are depicted in Fig. 19-11. The Whipple and Santa Catalina assemblages yield similar pressures of 4.5 ± 1.1 and 5.2 ± 1.0 kb. The Sacramentos, which have muscovites with much less Mg celadonite, yield a much lower pressure of 1.4 ± 0.4 kb.

Because this barometer is relatively new, absolute results of these estimates are viewed with reservation. The 5.2-kb estimate for the post-Wilderness mylonitization of the Santa Catalina Mountains seems high given the stratigraphic levels of supracrustal rocks intruded by the Wilderness granite (Keith *et al.*, 1980). Relative barometry is more informative. The two earlier described barometers indicated a major contrast in crustal depth for the Whipple and Sacramento terranes at the time of pluton emplacement. It appears consistent that a similar difference characterizes the structural level of mylonitization.

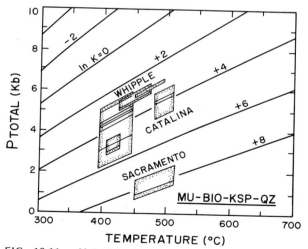

FIG. 19-11. Mylonitization barometry for the assemblage muscovite-biotite-alkali feldspar-quartz. In $K = (a_{phl})(a_{or}^2)/(a_{cel})^3$ of Powell and Evans (1983).

CONCLUSION

A systematic petrologic study of core complex evolution is a challenging exercise for several reasons: (1) the rocks are cataclastically disturbed, constituting a mixture of protolith and reequilibrated mineral phases; (2) the major minerals in the various granitic rocks have compositions that are difficult to describe rigorously with current activity models; and (3) most of the applicable thermometers and barometers were developed for other bulk compositions.

Yet, despite these reservations, the thermobarometric results show notable internal agreement. Mylonitization thermometry is consistent with what can be deduced from metamorphic grade of the mineral assemblages and the style of cataclastic strain. The constraints placed on crystallization pressure have been evaluated by two independent approaches. At face value, the overall scheme of the thermobarometry is provocative (Fig. 19-12). The Iron Mountains, already recognized to have an evolution different from most core complexes (e.g., no detachment history), appear to have formed at quite shallow levels. The Whipple and Sacramento Mountains complexes exhibit similar histories of decompression originating from middle crustal levels. Both terranes, from stratigraphic reasoning, were at shallow levels during detachment faulting (<5 km) and, in fact, part of the Whipple lower plate had been unroofed during the middle Miocene (G. A. Davis *et al.*, 1980). The tack then becomes one of determining the crustal levels during mylonitization and premylonitization plutonism.

At less than an estimated 8 km, the Sacramento mylonitization may not have been deep, but then the plutonic emplacement (~11 to 16 km) was also comparatively shallow relative to that of the Whipples. The Whipple terrane, with its more calcic garnets and aluminous hornblendes, must have originated at moderate crustal levels (~27 km) during emplacement of its 89- and 73-m.y.-old plutons. Subsequent mylonitization was at a shallower level (~16 km), although it was deeper than mylonitization in the Sacramento

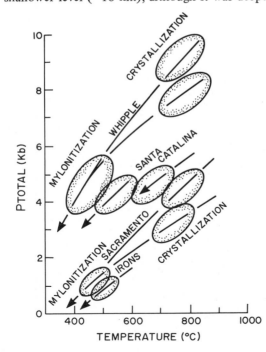

FIG. 19-12. Representation of estimated conditions for pluton emplacement and subsequent mylonitization for the Whipple, Sacramento, Santa Catalina, and Iron Mountains complexes. Results for the Santa Catalinas include only that for the Wilderness granite.

CORE COMPLEXES OF THE MOJAVE-SONORAN DESERT

Mountains as indicated by the more celadonitic nature of the retrogressed muscovite in the Whipple mylonites. The fact that younger plutons in each of the two complexes appear to have been emplaced at slightly shallower levels than the older plutons is interesting but not well constrained.

The Santa Catalina Mountains terrane deserves more work. Only one pluton was studied in some detail. Emplacement and deformation of the Eocene Wilderness granite appears, with limited data, to have occurred at about the same level, broadly constrained between 9 and 17 km. The older Leatherwood and younger Catalina suites are obvious targets for more study. Both contain hornblende. Preliminary study shows that the Laramide-aged Leatherwood also contains conspicuous amounts of epidote of at least two generations. Some of the epidote is secondary, yet an earlier generation includes large euhedra enclosed, or partly enclosed, in biotite. If this epidote is magmatic, as its texture implies, it is certain that the Leatherwood originated at even deeper levels (>6 kb or >22 km) based on the magmatic stability of this phase (Naney, 1983; Zen and Hammerstrom, 1984). Thus, the Santa Catalina terrane may have experienced a decompression rivaling that of the Whipple Mountains.

Core-complex decompression has not yet been widely recognized in other complexes of the Cordillera. Recently, however, Rhodes (1986) has described shallowing of metamorphic conditions in the Priest River complex of the Spokane dome, where Cretaceous, midcrustal mylonitization (>13 km) was superimposed on Jurassic metamorphic rocks that originated from depths in excess of 20 km. Exclusive of the core complexes, the various crystalline terranes of the Mojave-Sonoran Desert do not represent deep crustal levels. Mesozoic plutons in some mountain ranges intrude high into the Paleozoic section and limited barometric evidence for other granitic suites (Brand and Anderson, 1982, Beckerman et al, 1982) indicate upper crustal crystallization ranging from 7-11 km.

It is evident from this study that the premylonitization Mesozoic history of some core complexes originated at much deeper levels, specifically that of the middle crust. Moreover, they owe their present surface exposure to profound, although localized, uplift and/or doming, which in turn removed large sections of crust by the middle to late Tertiary. This accounts for the missing Paleozoic supracrustal section below the Tertiary nonconformities of several core complexes of this region, including that of the Whipple and Sacramento Mountains. Relative to Mesozoic metamorphic terranes, such as those found in the thrust-faulted Big Maria and Old Woman Mountains (Hoisch et al., Chapter 21, this volume), the deformation history of core complexes was younger, with pre-core-complex Mesozoic roots that were deeper. The kinematic nature of the decompression is at present only speculative. Certainly, erosion as a consequence of tectonic uplift is important. Crustal thinning due to flattening and shear during protracted mylonitization is another possible mechanism. Recently, G. A. Davis (1986) and G. A. Davis et al. (1986) have described the mechanics of rapid upward transport during a coupled mylonitization-detachment process.

ACKNOWLEDGMENTS

This research, initiated in 1977 in the Whipple Mountains, could not have been accomplished without the very beneficial and close collaboration with Gregory A. Davis. Several of our past students have aided by their own fundamental research, including Valerie Krass, Mark Rowley, Eric Frost, Linda Thurn, and Tom Hoisch. In fact, Tom was the first to warn me that the Whipple "story" may have originated at deeper crustal levels than I originally envisioned. Keith Howard and the U.S. Geological Survey provided us with logistic support at times during this study. Keith and Dave Miller personally escorted me through their work in the Iron Mountains. Likewise, Steve Reynolds and Stan Keith led me to choice outcrops of the Wilderness granite in the Santa Catalina Mountains. Finally, I am indebted to the Earth Sciences Division of the National Science Foundation, which has supported this work through research grants EAR77-09695, EAR84-17017 (both with G. A. Davis), and EAR81-20880. Additional support has been given by the Foss Foundation for Mineralogic Research (USC). I thank Andrew Barth, Greg Davis, Bill Dickinson, Gary Ernst, Tom Hoisch, Sue Orrell, Warren Thomas, and Ed Young for thoughtful comments and/or reviews of this manuscript.

REFERENCES

Anderson, J. L., 1985, Contrasting depths of "core complex" mylonitization — Barometric evidence: *Geol. Soc. America Abstr. with Programs*, v. 17, p. 337.

——, and Rowley, M. C., 1981, Synkinematic intrusion of two-mica and associated metaluminous granitoids, Whipple Mountains, California: *Can. Mineralogist*, v. 19, p. 83–101.

——, Davis, G. A., and Frost, E. G., 1979, Field guide to regional Miocene detachment faulting and early Tertiary(?) mylonitization, Whipple-Buckskin-Rawhide Mountains, southeastern California and western Arizona, *in* Abbott, P. L., ed., *Geologic Excursions in the Southern California Area:* Geol. Soc. America Guidebook, p. 109–133.

Banks, N. G., 1980, Geology of a zone of metamorphic core complexes in southeastern Arizona, *in* Crittenden, M. D., Coney, P. J., and Davis, G. H., eds., *Cordilleran Metamorphic Core Complexes:* Geol. Soc. America Mem. 153, p. 177–215.

Beckerman, G. M., Robinson, J. P., and Anderson, J. L., 1982, The Teutonia batholith — A large intrusive complex of Jurassic and Cretaceous age in the eastern Mojave Desert, California, *in* Frost, E. G., and Martin, D. L., eds., *Mesozoic-Cenozoic Tectonic Evolution of the Colorado River Region, California, Arizona, and Nevada:* San Diego, Calif., Cordilleran Publishers, p. 205–221.

Brand, H., and Anderson, J. L., 1982, Mesozoic alkalic monzonites and peraluminous adamellites of the Joshua Tree National Monument, southern California: *Geol. Soc. America Abstr. with Programs*, v. 14, p. 151–152.

Coney, P. J., 1980, Cordilleran metamorphic core complexes — An overview, *in* Crittenden, M. D., Jr., Coney, P. J., and Davis, G. H., eds., *Cordilleran Metamorphic Core Complexes:* Geol. Soc. America Mem. 153, p. 7–34.

Davis, G. A., 1986, Upward transport of mid-crustal mylonitic gneisses in the footwall of a Miocene detachment fault, Whipple Mountains, southern California: *Geol. Soc. America Abstr. with Programs*, v. 18, p. 98–99.

Anderson, J. L., Frost, E. G., and Shackelford, T. J., 1980, Mylonitization and detachment faulting of the Whipple-Buckskin-Rawhide Mountain terrane, California-Arizona, *in* Crittenden, M. D., Jr., Coney, P. J., and Davis, G. H., eds., *Cordilleran Metamorphic Core Complexes:* Geol. Soc. America Mem. 153, p. 79–130.

——, Anderson, J. L., Martin, D. L., Krummenacher, Daniel, Frost, E. G., and Armstrong, R. L., 1982, Geologic and geochronologic relations in the lower plate of the Whipple detachment fault, Whipple Mountains, southeastern California — A progress report; *in* Frost, E. G., and Martin, D. L., eds., *Mesozoic-Cenozoic Tectonic Evolution of the Colorado River Region, California, Arizona, and Nevada:* San Diego, Calif., Cordilleran Publishers, p. 408–432.

CORE COMPLEXES OF THE MOJAVE-SONORAN DESERT

_____, Lister, G. S., and Reynolds, S. J., 1983, Interpretation of Cordilleran core complexes as evolving crustal shear zones in an extending orogen: *Geol. Soc. America Abstr. with Programs*, v. 15, p. 311.

_____, Lister, G. S., and Reynolds, S. J., 1986, Structural evolution of the Whipple and South Mountains shear zones, southwestern United States: *Geology*, v. 14, p. 7–10.

Davis, G. H., 1973, Mid-Tertiary gravity-glide folding near Tucson, Arizona: *Geol. Soc. America Abstr. with Programs*, v. 5, p. 592.

_____, 1975, Gravity-induced folding of a gneiss dome complex, Rincon Mountains, Arizona: *Geol. Soc. America Bull.*, v. 86, p. 979–990.

_____, 1977, Characteristics of metamorphic core complexes, southern Arizona: *Geol. Soc. America Abstr. with Programs*, v. 9, p. 944.

_____, 1979, Laramide folding and faulting in southeastern Arizona: *Amer. J. Sci.*, v. 279, p. 543–569.

_____, 1980, Structural characteristics of metamorphic core complexes, southern Arizona, *in* Crittenden, M. D., Jr., Coney, P. J., and Davis, G. H., eds., *Cordilleran Metamorphic Core Complexes:* Geol. Soc. America Mem. 153, p. 35–77.

_____, 1983, Shear-zone model for the origin of metamorphic core complexes: *Geology*, v. 11, p. 342–347.

_____, and Coney, P. J., 1979, Geologic development of the Cordilleran metamorphic core complexes: *Geology*, v. 7, p. 120–124.

DeWitt, E., 1980, Comment on "Geologic development of the Cordilleran metamorphic core complex": *Geology*, v. 8, p. 6–9.

_____, Armstrong, R. L. Sutter, J. F., and Zartman, R. E., 1984, U-Th-Pb, Rb-Sr, and Ar-Ar mineral and whole rock isotopic systematics in a metamorphosed granitic terrane, southeastern California: *Geol. Soc. America Bull.*, v. 95, p. 723–739.

Ferry, J. M., and Spear, F. S., 1978, Experimental calibration of the partitioning of Fe and Mg between biotite and garnet: *Contrib. Mineralogy Petrology*, v. 66, p. 113–117.

Frost, E. G., 1981, Structural style of detachment faulting in the Whipple Mountains, California, and Buckskin Mountains, Arizona: *Ariz. Geol. Soc. Digest*, v. 13, p. 24–29.

Ganguly, J., and Saxena, S., 1984, Mixing properties of aluminosilicate garnets — Constraints from natural and experimental data, and applications to geothermobarometry: *Amer. Mineralogist*, v. 69, p. 88–97.

Ghent, E. D., and M. Z. Stout, 1981, Geobarometry and geothermometry of plagioclase-biotite-garnet-muscovite assemblages: *Contrib. Mineralogy Petrology*, v. 76, p. 92–97.

Goldman, D. S., and Albee, A. L., 1977, Correlation of Mg/Fe partitioning between garnet and biotite with $^{18}O/^{16}O$ partitioning between quartz and magnetite: *Amer. J. Sci.*, v. 277, p. 750–767.

Guidotti, C. V., Cheney, J. T., and Guggenheim, S., 1976, Distribution of titanium between coexisting muscovite and biotite in pelitic schists from northwestern Maine: *Amer. Mineralogist*, v. 62, p. 438–448.

Hammerstrom, J. M., and Zen, E-an, 1985, An empirical equation for igneous calcic amphibole geobarometry: *Geol. Soc. America Abstr. with Programs*, v. 17, p. 602.

Hazelton, H. T., Hovis, G. L., Hemingway, B. S., and Robie, R. A., 1982, Calorimetric investigation of the excess entropy of mixing in analbite-sanidine solid solutions — Lack of evidence for Na, K short range order and implications for two-feldspar thermometry: *Amer. Mineralogist*, v. 68, **p.** 398–413.

Hodges, K. V., and Royden, L., 1984, Geologic thermobarometry of retrograded metamorphic rocks — An indication of uplift trajectory of a portion of the northern Scandinavian Caledonides: *J. Geophys. Res.* v. 89, p. 7077–7090.

——, and F. S. Spear, 1982, Geothermometry, geobarometry and the Al_2SiO_5 triple point at Mt. Moosilauke, New Hampshire: *Amer. Mineralogist*, v. 67, p. 1118–1134.

Howard, K. A., 1980, Metamorphic infrastructure in the northern Ruby Mountains, Nevada, *in* Crittenden, M. D., Jr., Coney, P. J., and Davis, G. H., eds., *Cordilleran Metamorphic Core Complexes:* Geol. Soc. America Mem. 153, p. 335–346.

——, Miller, D. M., and John, B. E., 1982, Regional character of mylonitic gneiss in the Cadiz Valley area, southeastern California; *in* Frost, E. G., and Martin, D. L., eds., *Mesozoic-Cenozoic Tectonic Evolution of the Colorado River region, California, Arizona, and Nevada:* San Diego, Calif., Cordilleran Publishers, p. 441–447.

Indares, A., and Martignole, 1985, Biotite-garnet geothermometry in the granulite facies — The influence of Ti and Al in biotite: *Amer. Mineralogist*, v. 70, p. 272–278.

John, B. E., 1982, Geologic framework of the Chemehuevi Mountains, southeastern California; *in* Frost, E. G., and Martin, D. L., eds., *Mesozoic-Cenozoic Tectonic Evolution of the Colorado River Region, California, Arizona, and Nevada:* San Diego Calif., Cordilleran Publishers, p. 317–325.

Keith, S. B., Reynolds, S. J., Damon, P. E., Shafiqullah, M., Livingston, D. E., and Pushkar, P. D., 1980, Evidence for multiple intrusion and deformation within the Santa Catalina–Rincon–Tortolita metamorphic core complex, *in* Crittenden, M. D., Jr., Coney, P. J., and Davis, G. H., eds., *Cordilleran Metamorphic Core Complexes:* Geol. Soc. America Mem. 153, p. 217–267.

Kerrich, R., LaTour, T. E., and Barnett, R. L., 1981, Mineral reactions participating in intragranular fracture propagation — Implications for stress corrosion cracking: *J. Struct. Geology*, v. 3, p. 77–87.

Martin, D. L., Barry, W. K., Krummenacher, D., and Frost, E. G., 1980, K-Ar dating of mylonitization and detachment faulting in the Whipple Mountains, San Bernardino County, California and the Buckskin Mountains, Yuma County, Arizona: *Geol. Soc. America Abstr. with Programs*, v. 12, p. 118.

McClelland, W. C., 1982, Structural geology of the central Sacramento Mountains, San Bernardino County, California; *in* Frost, E. G., and Martin, D. L., eds., *Mesozoic-Cenozoic Tectonic Evolution of the Colorado River Region, California, Arizona, and Nevada:* San Diego, Calif., Cordilleran Publishers, p. 401–407.

Miller, C. F., Stoddard, E. F., Bradfish, L. J., and Dollase, W. A., 1981, Composition of plutonic muscovite — Genetic implications: *Can. Mineralogist*, v. 19, p. 25–34.

Miller, D. M., and Howard, K. A., 1985, Bedrock geologic map of the Iron Mountains Quadrangle, San Bernardino and Riverside Counties, California: *U.S. Geol. Survey Misc. Field Studies Map MF-1736*, 1:62,500.

——, Howard, K. A., and Anderson, J. L., 1981, Mylonitic gneiss related to emplacement of a Cretaceous batholith, Iron Mountains, southern California, *in* Howard, K. A., Carr, M. D., Miller, D. M., eds., *Tectonic Framework of the Mojave and Sonoran Deserts, California and Arizona:* U.S. Geol. Survey Open File Rpt. 81–503, p. 73–75.

Miller, E. L., Gans, P. B., and Garing, J., 1983, The Snake Range decollement — An exhumed mid-Tertiary brittle-ductile transition: *Tectonics*, v. 2, p. 239–263.

Nabelek, C. R., and Lindsley, D. H., 1985, Tetrahedral Al in amphibole — A potential thermometer for some mafic rocks: *Geol. Soc. America Abstr. with Programs*, v. 17, p. 673.

Naney, M. T., 1983, Phase equilibria of rock forming ferromagnesian silicates in granitic systems: *Amer. J. Sci.*, v. 283, p. 993–1033.

Newton, R. C., and Haselton, H. T., 1981, Thermodynamics of the garnet-plagioclase-

Al$_2$SiO$_5$-quartz geobarometer; *in* Newton, R. C., *et al.*, eds., *Thermodynamics of Minerals and Melts:* Springer-Verlag, New York, p. 131–147.

Powell, R., and Evans, J. A., 1983, A new geobarometer for the assemblage biotite-muscovite-chorite-quartz: *J. Metamorphic Geology*, v. 1, p. 331–336.

Price, J. G., 1985, Ideal site mixing in solid solutions, with an application to two-felspar geothermometry: *Amer. Mineralogist*, v. 70, p. 696–701.

Rehrig, W. A., and Reynolds, S. J., 1980, Geologic and geochronologic reconnaissance of a northwest-trending zone of metamorphic core complexes in southern and western Arizona, *in* Crittenden, M. D., Jr., Coney, P. J., and Davis, G. H., eds., *Cordilleran Metamorphic Core Complexes:* Geol. Soc. America Mem. 153, p. 131–157.

Reynolds, S. J., 1985, Geology of the South Mountains, central Arizona: *Ariz. Bur. Geology Mineral Technology Bull. 195*, 61 p.

——, and Rehrig, W. A., 1980, Mid-Tertiary plutonism and mylonitization, South Mountains, central Arizona, *in* Crittenden, M. D., Jr., Coney, P. J., and Davis, G. H., eds., *Cordilleran Metamorphic Core Complexes:* Geol. Soc. America Mem. 153, p. 159–175.

Rhodes, B. P., 1986, Metamorphism of the Spokane Dome mylonitic zone, Priest River complex — constraints on the tectonic evolution of northeastern Washington and northern Idaho: *J. Geology*, v. 94, p. 539–556.

Robert, J. L., 1976, Titanium solubility in synthetic phlogopite solid solutions: *Chem. Geology*, v. 17, p. 213–227.

Shackelford, T. J., 1980, Tertiary tectonic denudation of a Mesozoic-Early Tertiary(?) gneiss complex, Rawhide Mountains, western Arizona: *Geology*, v. 8, p. 190–194.

Spencer, J. E., and Turner, R. D., 1982, Dike swarms and low-angle faults, Homer Mountain and north-western Sacramento Mountains, southeastern California; *in* Frost, E. G., and Martin, D. L., eds., *Mesozoic-Cenozoic Tectonic Evolution of the Colorado River Region, California, Arizona, and Nevada:* San Diego, Calif., Cordilleran Publishers, p. 97–107.

Stormer, J. C., Jr., 1975, A practical two feldspar geothermometer: *Amer. Mineralogist*, v. 60, p. 667–674.

Thompson, A. B., 1976, Mineral reactions in pelitic rocks: Prediction of P-T-X (Fe-Mg) phase relations: *Amer. J. Sci.*, v. 276, p. 401–424.

Velde, B., 1965, Phengitic micas: synthesis, stability, and natural occurrence: *Amer. J. Sci.*, v. 263, p. 886–913.

Wernicke, B., 1981, Low-angle faults in the Basin and Range Province — Nappe tectonics in an extending orogen: *Nature*, v. 291, p. 645–648.

Whitney, J. A., and Stormer, J. C., Jr., 1977, The distribution of NaAlSi$_3$O$_8$ between coexisting microcline and plagioclase and its effect on geothermometric calculations: *Amer. Mineralogist*, v. 62, p. 687–691.

Wright, J. E., Anderson, J. L., and Davis, G. A., 1986, Timing of plutonism, mylonitization, and decompression in a metamorphic core complex, Whipple Mountains, CA: *Geol. Soc. America Abstr. with Programs*, v. 18 p. 201.

Zen, E-an, and Hammarstrom, J. M., 1984, Magmatic epidote and its petrologic significance: *Geology*, v. 12, p. 515–518.

Warren M. Thomas, H. Steve Clark, Edward D. Young,
Suzanne E. Orrell, and J. Lawford Anderson
Department of Geological Sciences
University of Southern California
Los Angeles, California 90089-0741

20

PROTEROZOIC HIGH-GRADE METAMORPHISM IN THE COLORADO RIVER REGION, NEVADA, ARIZONA, AND CALIFORNIA

ABSTRACT

The distribution of grade in Proterozoic metamorphic rocks in the Colorado River region of western Arizona and adjacent California and Nevada exhibits a northward increase from high amphibolite to granulite facies, all significantly higher grade than that of metamorphic terranes to the east in central and southeastern Arizona.

The granulite-grade rocks of the Gold Butte area and McCullough Range of southern Nevada include muscovite-absent pelitic gneisses containing garnet + cordierite + biotite + sillimanite + plagioclase + K-feldspar + quartz + ilmenite + hercynite and mafic rocks of hornblende + biotite + clinopyroxene + orthopyroxene + plagioclase ± cummingtonite. Thermobarometric calculations for the metapelites indicate minimum temperatures of 650–725°C at 2–4 kb. Higher peak temperatures in excess of 750°C are suggested by two-pyroxene and hornblende-plagioclase thermometry for interbedded mafic rocks. The occurrence of cummingtonite and lack of garnet in the mafic rocks is consistent with metamorphism at low pressure. Relatively flat compositional profiles in garnets from Gold Butte suggest that these rocks cooled slowly enough for diffusion to modify any original growth profiles. Retrograde effects are minor and local, although two-feldspar temperatures have been reset to the range 450–550°C.

Representative of Proterozoic metamorphic terranes farther south along the Colorado River, high amphibolite-grade (K-feldspar-sillimanite zone) rocks of the Whipple Mountains, California, contain muscovite, in part replaced by fibrolitic sillimanite, and lack cordierite, hercynite, and prismatic sillimanite. In contrast to the granulite-grade rocks to the north, thermobarometric calculations yield lower temperatures of 620–680°C at the same range of low pressure.

The entire high-grade terrane represents a low-pressure facies series, the high grade being the result of increased temperature, not greater depth, relative to lower-grade rocks to the east. Calculated average metamorphic gradients are greater than 50°C/km.

INTRODUCTION

Proterozoic metamorphic rocks are exposed throughout the southern Basin and Range in southern Nevada, eastern California, and western to southeastern Arizona. The age of these rocks varies from 1.78 to 1.69 b.y. in the northwest to 1.68–1.61 b.y. in the southeast (Silver, 1968). The grade of metamorphism ranges from greenschist in south and southeastern Arizona to low-granulite facies in southern Nevada, with a conspicuous area of high-grade rocks located in the Colorado River region. Figure 20-1 shows a preliminary compilation of the grade of Proterozoic metamorphic rocks in this region. It must be remembered that the boundaries between areas of similar grade are not isograds inasmuch as the metamorphism of these rocks is by no means coeval. In a very general way, the lower-grade rocks of southeastern Arizona correspond to the younger ages of metamorphism.

The present study focuses on three exposures of Proterozoic metamorphic rocks along the thermal high in the Colorado River region. The most northern is that of the Gold Butte region of southeastern Nevada, in which Early Proterozoic metamorphic rocks have been intruded by the anorogenic, Middle Proterozoic Gold Butte granite. Farther south in the McCullough Range, Proterozoic rocks are similar to those of Gold Butte, but include many Proterozoic foliated granites and lack any representatives of the anorogenic granitic episode. Both the Gold Butte and McCullough Range rocks are metamorphosed to low-pressure low granulite facies. High amphibolite-grade assemblages occur to the south along the Colorado River, including exposures in the Whipple Mountains, the third area of study. Rock types include Proterozoic semipelites, amphibolites, and meta-

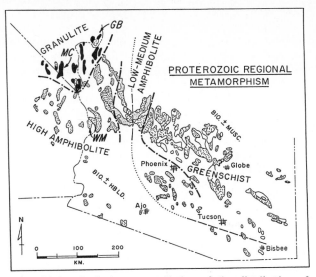

FIG. 20-1. Preliminary compilation of the distribution of Proterozoic metamorphic grade in the southwestern United States, after Anderson (1986). The heavy dot-dash lines indicate approximate changes in metamorphic grade but are not isograds inasmuch as the age of metamorphism ranges from ~1.75 b.y. in the west to 1.65 b.y. in the east (Silver, 1968). The symbols are: solid dark, granulite facies; heavy stipple, high amphibolite facies; light stipple, low to medium amphibolite facies; blank, greenschist facies; short line patterns, Proterozoic anorogenic granites. The light dotted and light dashed line separates anorogenic granites with predominately biotite ± hornblende from those with biotite ± muscovite. Location of the present study areas are given by GB, Gold Butte; MC, McCullough Range; WM, Whipple Mountains.

plutonics now located in the upper plate of the Tertiary Whipple detachment fault. All three localities will be examined in terms of their conditions of metamorphism in order to characterize the Proterozoic thermal high in the Colorado River region.

GOLD BUTTE

The Gold Butte area comprises the central and southern portion of the South Virgin Mountains of southern Nevada and has been the subject of a number of previous studies (Dexter *et al.*, 1983; Volborth, 1962) that have dealt with the metamorphic rocks in a reconnaissance fashion. The Gold Butte region consists of a Proterozoic metamorphic complex of schists, gneisses, gneissic granite, charnockite, and metaultramafics intruded by the 1.45 b.y.-old anorogenic Gold Butte granite (Kwok, 1983; Volborth, 1962). Migmatite, anorthosite, aplite, and pegmatite of Proterozoic age are also present. These metamorphic and plutonic rocks form the basement on which Paleozoic platform sediments, exposed to the north and east, were deposited. Stewart (1980) believed that the sedimentary and volcanic precursors were deposited from 1.77 to 1.73 b.y. ago and were intruded by plutons and metamorphosed from 1.74 to 1.70 b.y. ago. Gneissic granite lenses within the metamorphic complex have yielded a U-Pb zircon age of 1.74 b.y. (Silver, 1973, oral communication, quoted in Stewart, 1980). The metaultramafic rocks occur in small (less than 0.4 km) but abundant discordant and concordant pods and lenses and appear to be younger than the metamorphic complex, but older than the Gold Butte granite (Dexter *et al.*, 1983). The Gold Butte granite has been dated by U-Pb at 1.45 ± 0.025 b.y. (Silver, 1973, oral communication, quoted in Stewart, 1980) and its emplacement is part of a 1.4–1.5-b.y. trans-continental anorogenic magmatic event extending from the southwestern United States to Labrador (Anderson, 1983, 1986).

Gneisses in the Gold Butte region contain the distinctive assemblage garnet + cordierite + biotite + sillimanite + plagioclase + K-feldspar + quartz + ilmenite + hercynite. In hand sample, they are commonly banded with prominent garnet porphyroblasts up to 5 mm in diameter. In thin section, textures typical of high-grade metamorphic rocks are often conspicuous: coarse perthites and antiperthites, polygonal grain boundaries between like minerals, and rounded, meniscus-like boundaries between quartz and enclosing garnet poikiloblasts. Low-temperature retrograde effects are minor, but higher-temperature reaction textures, both prograde and retrograde, are common.

Interbedded mafic rocks have the assemblage hornblende + biotite + clinopyroxene + Fe-Ti oxides + plagioclase + orthopyroxene + quartz. The mineralogy of the ultramafic pods is variable, but can be characterized by Mg-rich hornblende + biotite or phlogopite + orthopyroxene + olivine. The metaultramafics are locally altered, with vermiculite occurring in some. Charnockites have the classic mineralogy quartz + K-feldspar + antiperthitic plagioclase + orthopyroxene + garnet. The mineral assemblages in the Gold Butte area are clearly indicative of a low-pressure granulite-facies metamorphism.

The effect of the intrusion of the Gold Butte granite on the metamorphic complex is unclear. Although well-developed aureoles are uncommon around epizonal anorogenic granitic plutons (Anderson, 1983), the presence of orthopyroxene and several distinctive reaction textures in the metapelites within 0.3 km of the pluton's contact suggests that the granite may have had some effect on the previously metamorphosed rocks. These textures preserve presumed *prograde* reactions, which may be inconsistent with the temperatures that granite magmas, even hot and dry ones, can impose on a contact aureole.

The phase assemblage of the pelitic gneisses in the Gold Butte area permits the application of a number of geothermometers. The garnets in the pelitic gneisses are almandine-rich with relatively flat compositional profiles and significant zoning only at the rim. A typical core composition is $Alm_{75}Pyr_{20}Sp_{02}Gr_{03}$; the corresponding rim is $Alm_{76}Pyr_{18}Sp_{03}Gr_{03}$. Garnet rims are slightly more extreme in composition when they are in contact with biotite, suggesting even lower-temperature reequilibration over short distances. A typical composition of a garnet rim in contact with biotite is $Alm_{81}Pyr_{13}Sp_{04}Gr_{02}$. Biotite compositions are relatively uniform from sample to sample and are not significantly zoned except close to a contact with garnet, again suggesting low-temperature reequilibration. Plagioclase compositions are typified by $An_{25}Ab_{72}Or_{03}$, but in one sample, irregular zoning ranges from $An_{21}Ab_{77}Or_{01}$ to $An_{35}Ab_{64}Or_{01}$. In another sample, a coarse antiperthite exhibits a host of $An_{23}Ab_{76}Or_{01}$ exsolving lamellae of $An_{01}Ab_{13}Or_{86}$. Cordierite is not zoned.

Garnet-biotite temperatures were calculated using the calibrations of Ferry and Spear (1978) and Indares and Martignole (1985). Results of the former are higher temperatures and are preferred as being consistent with the mineralogy of the rocks. Calculatins were made using garnet core and matrix biotite compositions, as has been justified by Tracy *et al.* (1976) and Chamberlain and Lyons (1983) for cases in which biotite is much more abundant modally than garnet, or equivalently, in which retrograde zoning in garnet is limited to a narrow rim. The relatively flat compositional profiles of the garnets suggests that any previously existing growth zoning was obliterated by diffusion above about 650°C (Tracy, 1982, and references therein). The two garnet-biotite Fe-Mg exchange calibrations were solved simultaneously with the Ghent *et al.* (1979) and Newton and Haselton (1981) calibrations of the garnet-plagioclase-sillimanite-quartz equilibrium, yielding four solutions ranging from 610°C and 1.9 kb to 690°C and 3.2 kb. The Indares and Martignole (1985) calibration yielded temperatures lower by 90°C and Newton and Haselton (1981) lower pressures by about 0.2 kb. Inasmuch as the Indares and Martignole (1985) calibration appears to give results too low to be compatible with phase equilibria, the Ferry and Spear (1978) results are preferred and a pressure of 3.2 kb is assumed for further thermometric calibrations. Temperatures for all samples range at this pressure from 690 to 770°C. Garnet-cordierite Fe-Mg exchange thermometry was also employed using the calibration of Holdaway and Lee (1977). Results from this pair range from 675 to 745°C at 3.2 kb. Finally, the pressure-independent calibration of Fe-Mg exchange between cordierite and hercynite (Vielzeuf, 1983) yields a temperature of 675°C.

Most metamorphic cordierite contains a small amount of H_2O in the channel site which is thought to be a function of the activity of H_2O during metamorphism. From microprobe data, water must be estimated by taking the difference of the total from 100%, which results in values with large uncertainties. Using this approach, cordierite from Gold Butte has approximately 0.5 mol of H_2O per 18-oxygen formula unit. With this value and the compositions of coexisting garnet and cordierite, the graphical calibration of Martignole and Sisi (1981) indicates parameters of metamorphism of 740-760°C, 5.2-5.6 kb and $f_{H_2O} = 1.3$-1.45 kb. This corresponds to an ideal activity of H_2O of about 0.25. The pressure, however, is markedly higher than that calculated from other thermobarometers, a result consistent with the experience of other workers (among them St-Onge, 1984; and Elan, 1985). Nonetheless, a qualitative indication of low a_{H_2O} is obtained, similar to conditions proposed for other granulite facies terranes in India (Janardhan *et al.*, 1982), Greenland (Wells, 1980), the Australia (Philips, 1980).

In contrast to the pelites, the interbedded mafic rocks record higher temperatures. Tie lines between Na(M4) in amphibole and mole percent An in plagioclase are parallel with those from other granulite terranes (Engel and Engel, 1962; Spear, 1980). Using Spear's (1980) thermometer from NaSi = CaAl exchange between amphibole and plagioclase, a wide range of temperatures is permitted, from 510 to greater than 725°C, depending on how much Fe^{3+} is estimated to be in the amphibole. Assumption of the minimum ferric content yields the lower temperature; assumption of the maximum Fe^{3+} gives the higher temperature. Inasmuch as the composition of the plagioclase falls within a miscibility gap for plagioclase compositions which probably closes at higher temperatures, the upper end of the range (>725°C) is preferred.

Orthopyroxene and clinopyroxene in the mafic rocks are unzoned and coexist in apparent textural equilibrium. Application of Lindsley's (1983) graphical two-pyroxene thermometer indicates that the pairs equilibrated at 825 ± 50°C. Other two-pyroxene thermometers yield unreasonably high temperatures of 909°C (Wood and Banno, 1973) and 957°C (Wells, 1977), similar to the results found for hornblende granulites from Quairanding, Australia (Davidson, 1968; Lindsley, 1983).

The interpretation of the higher temperatures for the mafic hornblende granulite is unclear, inasmuch as the sample analyzed occurs within 0.3 km of the contact with the Gold Butte granite. It is not clear whether these higher temperatures reflect more nearly peak conditions of a regional metamorphism than do the pelites or whether they represent a contact metamorphic overprint.

Figure 20-2 shows the results of the thermobarometric calculations plotted on a petrogenetic grid. The calculated conditions fall between Kerrick's (1972) curves for the dehydration of muscovite and minimum melting, both at a $X_{H_2O}^{fl} = 0.5$, consistent with inferred low a_{H_2O}, and the absence of muscovite. A number of samples contain coarse matrix hercynite, formerly in contact with quartz, now separated by a narrow selvage of

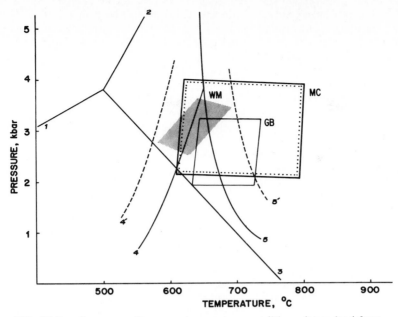

FIG. 20-2. Summary of pressure-temperature conditions determined from thermobarometric calculations for Gold Butte (GB), the McCullough Range (MC) and the Whipple Mountains (WM). Curves are: (1), (2), and (3) andalusite = kyanite, kyanite = sillimanite, and andalusite = sillimanite (Holdaway, 1971); (4) and (4') mus + qtz = als + ksp + vap at $X_{H_2O} = 1.0$ and 0.5 respectively (Kerrick, 1972); and (5) and (5) granite melting at $X_{H_2O} = 1.0$ and 0.5, respectively (Kerrick, 1972).

cordierite. The minimum temperature for the coexistence of hercynite + quartz is 775°C at 2.9 kb (Bohlen and Dollase, 1983). Inasmuch as the hercynite contains about 5.5 wt % ZnO, this assemblage may have been stable at slightly lower temperatures.

MCCULLOUGH RANGE

The McCullough Range is a 56 km-long, north-trending range in southern Nevada forming the northern extension of the New York Mountains of California. The geology of the McCullough Range has been described on a reconnaissance basis by Hewitt (1956), Longwell *et al.* (1965), and Bingler and Bonham (1972). The northern half of the range consists of volcanic and volcanoclastic rocks of Tertiary age, whereas the southern portion is a Precambrian gneissic and plutonic complex.

The southern portion of the McCullough Range has been studied recently by Anderson *et al.* (1985) and by Clarke (1985). The oldest rocks are metasedimentary garnet-cordierite-sillimanite gneisses and a variety of younger foliated granitic plutons. Leucogranitic amphibolitic and ultramafic layers occur interbedded with the gneisses. The crystalline rocks of the McCullough Range are considered to be Proterozoic based on U/Pb zircon dates reported from similar and probably correlative rocks from the immediately adjacent New York Mountains (Elliott *et al.*, 1986; Wooden *et al.*, 1986). The oldest rock recognized by these authors is a garnet-biotite-plagioclase-quartz gneiss with an age of 1975 m.y.; granitic orthogneisses from the same complex have minimum zircon ages of 1700-1710 m.y. Together with granitic gneisses from the Providence Mountains farther south, these orthogneisses form a composite discordia line with an age of 1710 ± 10 m.y., which is interpreted as the age of metamorphism and deformation of the complex (Elliott *et al.*, 1986; Wooden *et al.*, 1986). That the McCullough Range rocks have been thermally undisturbed since the Proterozoic is suggested by the uniform antiquity of K-Ar biotite and hornblende cooling ages, all greater than 708 m.y. (J. L. Anderson, unpublished data). In contrast to the Gold Butte region, 1.4-b.y.-aged anorogenic granites are absent in the McCullough Range. Mylonitic shear zones crosscut the older foliation and are indicative of postmetamorphic cataclasis. Together these rocks represent the oldest units in the area and are similar to other high-grade gneisses in the region. The gneisses are treated in the remainder of this study.

The most abundant preplutonic gneisses range from migmatitic to nonmigmatitic and contain biotite ± garnet ± sillimanite ± cordierite + plagioclase + K-feldspar + quartz + Fe-Ti oxides ± hercynite. Gray alkali feldspar poikiloblasts are common and readily distinguished in hand samples. Parallel alignment of biotite, and sillimanite and cordierite where present, defines the predominant foliation. Other minerals include quartz, plagioclase (An_{23-31}), hercynite, and Fe-Ti oxides. Migmatitic leucosomes generally range from a few to 40 mm thick and coalesce to form mappable units in some parts of the range. Non-migmatitic gneisses have the same biotite + garnet + cordierite mineralogy as the migmatitic varieties. Small-scale isoclinal and multiple folding is common.

The amphibolite and metaultramafic rocks also contain granulitic assemblages. Major minerals in the amphibolite include hornblende + clinopyroxene + plagioclase ± quartz ± biotite ± orthopyroxene ± Fe-Ti oxides ± minor cummingtonite. Texturally, the rocks range from granoblastic to having a foliation defined by the parallel alignment of hornblende. The metaultramafic rocks are of an unusual composition and consist of actinolitic hornblende (6–8 wt % Al_2O_3), orthopyroxene, and biotite in apparent textural equilibrium. They occur as layers up to 3 m wide associated with amphibolite. It is unclear whether they originated from metamorphic segregation from the amphibolite or are premetamorphic intrusives.

Thermobarometry of the garnet-biotite-plagioclase-sillimanite-quartz-cordierite-hercynite assemblage yields high temperature but variable results. Simultaneous solutions of garnet-biotite (Ferry and Spear, 1978) and garnet-plagioclase-sillimanite-quartz (Ghent *et al.*, 1979; Newton and Haselton, 1981) equilibria give results ranging from 600°C and 2.2 kb to 690°C and 3.6 kb. Calculations using garnet rim compositions may be up to 100°C lower. A few garnet-biotite temperatures range up to 775°C at 2 kb. In contrast, garnet-cordierite Mg-Fe exchange (Holdaway and Lee, 1977; Martignole and Sisi, 1981) consistently yields higher temperatures and pressures, ranging from 607°C and 2.0 kb to 730°C and 3.8 kb. Utilization of Vielzeuf's (1983) calibration of Fe-Mg exchange between hercynite and cordierite gives pressure-independent temperatures of 670–690°C.

As in the Gold Butte region, the mafic rocks record systematically higher temperatures than the pelites. Plagioclase-amphibole thermometry (Spear, 1980) suggests temperatures in excess of 725°C, which is higher than the calibrated range. Nabelek and Lindsley's (1985) empirical hornblende thermometer yields 855°C at an assumed pressure of 2.5 kb. Although both orthopyroxene and clinopyroxene are present in some of the mafic rocks, their compositions do not seem to represent equilibrium pairs, with the orthopyroxene having reequilibrated to lower-temperature, more iron-rich compositions and exhibiting thick exsolution lamellae. In contrast, clinopyroxene compositions seem to be similar to those from other granulite facies rocks (Lindsley, 1983). Temperatures calculated from these apparently disequilibrium pairs are unreasonably high, ranging from 850 to 1010°C using the calibrations of Wood and Banno (1973) and Wells (1977).

Cordierite from the McCullough gneisses exhibits less than 0.17 mol of H_2O per formula unit, suggesting a low a_{H_2O} during equilibration. Furthermore, the equilibrium

PROTEROZOIC HIGH-GRADE METAMORPHISM

biotite + sillimanite + quartz = garnet + K-feldspar + $_{H_2O}$ at temperatures between 600 and 700°C and pressure from 2 to 4 kb indicates that a_{H_2O} was between about 0.03 and 0.26 (Phillips, 1980). This is in contrast to the qualititative indications of higher P_{H_2O} given by the nearly ubiquitous *in situ* migmatites and moderate X_{Fe} in cordierite. It is possible that the activity of H_2O may have been high during an early stage of metamorphism but was reduced by incorporation of water into the migmatitic melt. The cordierite thus represents equilibration under the later, more dehydrated conditions.

In summary, the gneisses the McCullough Range record a granulite-facies metamorphic event with minimum temperatures ranging from 600 to 700°C and minimum pressures from 2.5 to 3 kb. Peak conditions may have been as high as 750°C and 3-4 kb. Figure 20-20-2 shows a summary of the thermobarometric results plotted on a petrogenetic grid.

WHIPPLE MOUNTAINS

The Whipple Mountains of California lie approximately 150 km south of the McCullough Range and expose Proterozoic metamorphic rocks both in the upper and lower plates of the Tertiary Whipple detachment fault. The Proterozoic metamorphic rocks of both plates consist of quartzofeldspathic metasedimentary gneisses and amphibolite, intruded by several foliated to nonfoliated granitic plutons (Davis *et al.*, 1980). Representative of Proterozoic metamorphic terranes in this portion of the Colorado River, these rocks are of high amphibolite grade, as demonstrated by the diagnostic mineral assemblage quartz + K-feldspar + plagioclase (An_{20-34}) + biotite + garnet + muscovite + fibrolitic sillimanite + Fe-Ti oxides. These metasedimentary rocks are medium- to fine-grained, granoblastic to strongly foliated, with sedimentary structures such as graded bedding, cross-bedding, and channels locally preserved in metaquartz arenites. A near-vertical tectonic foliation parallels the bedding, and small-scale isoclinal folds with fold axes parallel to foliation are common (Davis, *et al.*, 1980).

In this study, samples were examined from the Proterozoic upper plate metasedimentary gneisses of the eastern Whipple Mountains. These rocks have been intruded by 1.74-b.y. foliated granites and two large plutons having a U-Pb zircon age of 1.41 b.y. (J. L. Anderson and J. E. Wright, unpublished data). Some retrogression may have been caused by the later intrusions or by a Jurassic(?) thermal event that has reset K-Ar ages in Precambrian units (Davis *et al.*, 1980). The gneisses differ from those in the Gold Butte area and the McCullough Range in that coarse muscovite is present, and cordierite and prismatic sillimanite are absent. All sillimanite occurs as fibrolitic clots within coarse muscovite. Furthermore, these rocks lack the exsolution and myrmekitic textures typical of the granulite-grade rocks to the north.

Garnets are zoned with Mn increasing the Mg decreasing toward the rim; Fe shows a less consistent decrease toward the rim and Ca shows no systematic radial change. Rims on all garnets are nearly the same composition, whereas cores are more variable. Typical core compositions range from $Alm_{74}Pyr_{08}Gr_{02}Sp_{16}$ to $Alm_{69}Pyr_{06}Gr_{03}Sp_{22}$; rims are represented by $Alm_{68}Pyr_{08}Gr_{03}Sp_{24}$. Biotite is unzoned.

Garnet-biotite thermometry is somewhat complicated, owing to the high Mn content of the garnets, which may render the Ferry and Spear (1978) calibration unsuitable. Among the available calibrations, Ganguly and Saxena (1984) allow for the effect of Mn on Mg-Fe mixing in garnet, but apparent temperatures for these rocks calculated with the latter calibration are too high to be compatible with the observed phase assemblage. Temperatures and pressures calculated by simultaneous solution of the Ferry and Spear (1978) garnet-biotite and the Ghent *et al.* (1979) garnet-plagioclase-sillimanite-quartz equations using garnet core compositions yields 685°C and 3.4 kb to 618°C and 3.1 kb. Utilization

THOMAS ET AL.   533 of the Indares and Martignole (1985) calibration yields temperatures about 20°C lower for a given pressure; Newton and Haselton's (1981) garnet-plagioclase-sillimanite-quartz calibration yields pressures about 0.3 kb lower for a given temperature. Calculations using garnet rims return temperatures as much as 120°C lower and are considered to represent retrograde reequilibration. Temperatures were also calculated using the muscovite-garnet-biotite-plagioclase calibrations of Ghent and Stout (1981) and Hodges and Royden (1984). These results may be less reliable inasmuch as the muscovite is probably metastable. Results employing the Ghent and Stout (1981) calibration range from 625°C and 2.4 kb to 675°C and 2.5 kb.

Figure 20-2 summarizes the results of thermobarometric calculations on a petrogenetic grid. The presence of fibrolite is consistent with an overstepping of the muscovite dehydration curve by less than 50°C (Kerrick and Heninger, 1984), whereas all calculated pressures seem to be too low to be consistent with the stable coexistence of garnet + biotite + sillimanite. Nevertheless, equivalent thermobarometers yield lower temperatures for the Whipple Mountains rocks than for the Gold Butte and McCullough Range samples, consistent with the differences in the mineral assemblages.

CONCLUSIONS

Comparison of thermobarometric results from the three areas studied suggests that temperatures and pressures calculated for the McCullough Range and Gold Butte may be minima. Although the two northern areas have definitively higher temperature assemblages than the Whipple Mountain rocks, there is a marked overlap between the calculated conditions for all three areas. This could result from the temperatures in fact being similar, but with a_{H_2O} being lower in the granulite terranes. Alternatively, the temperatures recorded could represent values reset during slow cooling with diffusion becoming ineffective below approximately 650°C.

Simple calculations of linear metamorphic gradients are provocative. Values for Gold Butte, assuming a pressure of 3.2 kb, gives gradients from 62 to 67°C/km for determinations from the pelitic rocks. McCullough Range and Whipple Mountain pelites yield similar values. These gradients are higher than usually reported for regionally metamorphosed terranes (Miyashiro, 1973). Syntectonic to late- or post-tectonic plutons occur in all three ranges but distinct contact aureoles are not definable. Solving the thermometric equations for a pressure of 4 kb results in temperatures slightly higher, and lower metamorphic gradients of around 50°C/km. Assumption of such higher pressures is consistent only with the results of cordierite-garnet Fe^{2+}-Mg-H_2O thermobarometry, but as noted above, this thermobarometer has systematically yielded higher pressures in other studies (Elan, 1985; St-Onge, 1984).

Data on pressure-temperature paths is limited, but calculations on zoned garnet, zoned plagioclase, and unzoned biotite from a sample from Gold Butte suggest a path from 638°C and 2.6 kb to 593°C and 4.3 kb. Such a trajectory is consistent with these rocks being part of a hot lower nappe or plate over which colder rocks were emplaced (Spear *et al.*, 1984), an interpretation that is consistent with the apparent orogenic nature of the approximately 1.7-b.y.-aged metamorphism.

Despite the difficulties in constraining the pressures of metamorphism, it is apparent from both thermobarometric calculations and mineral assemblages that the low granulite to high amphibolite facies rocks of the Colorado River region represent hot, relatively shallow crust. Like some other granulite terranes they are not samples of the deep crust.

ACKNOWLEDGMENTS

The authors gratefully acknowledge support from NSF grants EAR81-20880 (JLA) and EAR84-17017 (JLA and G. A. Davis), U.S. Geological Survey grant 052790-85 (WMT), and the Foss Foundation for Mineralogical Research at the University of Southern California.

REFERENCES

Anderson, J. L., 1983, Proterozoic anorogenic granite plutonism of North America, *in* L. G. Medaris, *et al.*, eds., *Proterozoic Geology:* Geol. Soc. America Mem. 161, p. 133–154.

——, 1986, Proterozoic anorogenic granites of the southwestern U.S.: *Ariz. Geol. Soc. Digest*, in press.

——, Young, E. D., Clarke, H. S., Orrell, S. E., Winn, M., Schmidt, C. S., Weber, M. E., and Smith, E. I., 1985, The Geology of the McCullough Range Wilderness Area, Clark County, Nevada: *Final Techn. Rpt. U.S. Geol. Survey*.

Bingler, E. C., and Bonham, H. F., 1972, Reconnaissance geologic map of the McCullough Range and adjacent areas, Clark County, Nevada: *Nev. Bur. Mines Map 45*, scale 1:125,000.

Bohlen, S. R., and Dollase, W. A., 1983, Calibration of hercynite-quartz stability: *Geol. Soc. America Abstr. with Programs*, v. 15, p. 529.

Chamberlain, C. P., and Lyons, J. B., 1983, Pressure, temperature and metamorphic zonation studies of pelitic schists in the Merrimack Synclinorium, south-central New Hampshire: *Amer. Mineralogist*, v. 68, p. 530–540.

Clarke, H. S., 1985, Proterozoic granulite terrain, southern Nevada: *Geol. Soc. America Abstr. with Programs*, v. 17, p. 348.

Davidson, L. R., 1968, Variations in ferrous-iron magnesium distribution coefficients of metamorphic pyroxenes from Quairading, Western Australia: *Contrib. Mineralogy Petrology*, v. 80, p. 239–259.

Davis, G. A., Anderson, J. L., Frost, E. G., and Shackelford, T. J., 1980, Mylonitization and detachment faulting in the Whipple-Buckskin-Rawhide Mountains terrane, southeastern California and western Arizona; *in* Crittenden, M. D., Jr., Coney, P. J., and Davis, G. H., eds., *Cordilleran Metamorphic Core Complexes:* Geol. Soc. America Mem. 153, p. 79–129.

——, Anderson, J. L., Martin, D. L., Krummenacher, D., Frost, E. G., and Armstrong, R. L., 1982, Geologic and geochronologic relations in the lower plate of the Whipple detachment fault, Whipple Mountains, southeastern California — A progress report; *in* Frost, E. G., Martin, D. L., and Cameron, T., eds., *Mesozoic-Cenozoic Tectonic Evolution of the Colorado River Region, California, Arizona, and Nevada:* San Diego, Calif., Cordilleran Publishers, p. 409–432.

Dexter, J. J., Goodknight, C. S., Dayvault, R. D., and Dickson, R. E., 1983, Mineral evaluation of part of the Gold Butte district, Clark County, Nevada: *Dept. Energy Open File Rpt. GJBX-18*, 31 p.

Elan, R., 1985, High grade contact metamorphism at the Lake Isabella north shore roof pendant, southern Sierra Nevada, California: M.S. thesis, Univ. Southern California, Los Angeles, Calif., 202 p.

Elliott, G. S., Wooden, J. L., and Miller, C. M., 1986, Lower-granulite grade early Proterozoic supracrustal rocks, New York Mountains, SE California: *Geol. Soc. America Abstr. with Programs*, v. 18, p. 353.

Engel, A. E. J., and Engel, C. G., 1962, Hornblendes formed during progressive metamorphism of amphibolites, northwest Adirondack Mountains, New York: *Geol. Soc. America Bull.*, v. 73, p. 1499–1514.

Ferry, J. M., and Spear, F. S., 1978, Experimental calibration of the partitioning of Fe and Mg between biotite and garnet: *Contrib. Mineralogy Petrology*, v. 66, p. 113–117.

Ganguly, J., and Saxena, S., 1984, Mixing properties of aluminosilicate garnets — Constraints from natural and experimental data, and applications to geothermobarometry: *Amer. Mineralogist*, v. 69, p. 88–97.

Ghent, E. D., and Stout, M. Z., 1981, Geobarometry and geothermometry of plagioclase-biotite-garnet-muscovite assemblages: *Contrib. Mineralogy Petrology*, v. 76, p. 92–97.

——, Robbins, D. B., and Stout, M. T., 1979, Geothermometry, geobarometry and fluid compositions of metamorphised calc-silicates and pelites, Mica Creek, British Columbia: *Amer. Mineralogist*, v. 64, p. 874–885.

Hewitt, D. F., 1956, Geology and mineral resources of the Ivanpah Quadrangle, California and Nevada: *U.S. Geol. Survey Prof. Paper 275*, 172 p.

Hodges, K. V., and Royden, L., 1984, Geologic thermobarometry of retrograded rocks — An indication of the uplift trajectory of a portion of the northern Caledonides: *J. Geophys. Res.*, v. B89, 7077–7090.

Holdaway, M. J., 1971, Stability of andalusite and the aluminum silicate phase diagram: *Amer. J. Sci.*, v. 271, p. 97–131.

——, and Lee, S. M., 1977, Fe-Mg cordierite stability in high-grade pelitic rocks based on experimental, theoretical and natural observations: *Contrib. Mineralogy Petrology*, v. 63, p. 175–198.

Indares, A., and Martignole, J., 1985, Biotite-garnet geothermometry in the granulite facies — The influence of Ti and Al in biotite: *Amer. Mineralogist*, v. 70, p. 272–278.

Janardhan, A. S., Newton, R. C., and Hansen, E. C., 1982, The transformation of amphibolite facies gneiss to charnockite in southern Karnataka and northern Tamil Nadu, India: *Contrib. Mineralogy Petrology*, v. 79, p. 130–149.

Kerrick, D. M., 1972, Experimental determination of muscovite + quartz stability with $PH_{20} < P_{total}$: *Amer. J. Sci.*, v. 272, p. 946–958.

——, and Heninger, S. G., 1984, The sillimanite-andalusite equilibrium revisited: *Geol. Soc. America Abstr. with Programs*, v. 16, p. 558.

Kwok, K., 1983, Petrochemistry and mineralogy of anorogenic granites of the southwestern U.S.A.: M.S. thesis, Univ. Southern California, Los Angeles, Calif., 149 p.

Lindsley, D. H., 1983, Pyroxene thermometry: *Amer. Mineralogist*, v. 68, p. 477–494.

Longwell, C. R., Pampeyan, E. H., Bowyer, B., and Roberts, R. J., 1965, Geology and mineral deposits of Clark County, Nevada: *Nev. Bur. Mines Bull. 62*, 218 p.

Lonker, S. W., 1981, The P-T-X relations of the cordierite garnet-sillimanite-quartz equilibrium: *Amer. J. Sci.*, v. 281, p. 38–46.

Martignole, J., and Sisi, J. C., 1981, Cordierite-garnet-H_2O equilibrium — A geological thermometer, barometer and water fugacity indicator: *Contrib. Mineralogy Petrology*, v. 77, p. 38–46.

Miyashiro, A., 1973, *Metamorphism and Metamorphic Belts:* George Allen & Unwin, Ltd., London, 492 p.

Nabelek, C. R., and Lindsley, D. H., 1985, Tetrahedral Al in amphibole — A potential thermometer for some mafic rocks: *Geol. Soc. America Abstr. with Programs*, v. 17, p. 673.

Newton, R. C., and Haselton, H. T., 1981, Thermodynamics of the garnet-plagioclase-Al2SiO5-quartz geobarometer; *in* Newton, R. C., Nevrotsky, A., and Wood, B. J., eds., *Thermodynamics of Minerals and Melts:* Springer-Verlag, New York, p. 129–145.

Phillips, G. N., 1980, Water activity changes across an amphibolite-granulite facies transition, Broken Hill, Australia: *Contrib. Mineralogy Petrology*, v. 75, p. 377–386.

Silver, L. T., 1968, Precambrian batholiths of Arizona: *Geol. Soc. America Spec. Paper 121*, p. 558–559.

———, Anderson, C. A., Crittenden, M., and Robertson, J. M., 1977, Chronostratigraphic elements of the Precambrian rocks of the southwestern and far western United States: *Geol. Soc. America Abstr. with Programs*, v. 9, p. 1176.

Spear, F. S., 1980, NaSi=CaAl exchange equilibrium between plagioclase and amphibole — An empirical model: *Contrib. Mineralogy Petrology*, v. 73, p. 22–41.

———, Selverstone, J., Hickmott, D., Crowley, P., and Hodges, K. V., 1984, P-T paths from garnet zoning — A new technique for deciphering tectonic processes in crystalline terranes: *Geology*, v. 12, p. 87–90.

Stewart, J. H., 1980, Geology of Nevada: *Nev. Bur. Mines Geology Spec. Publ. 4*, 136 p.

St.-Onge, M. R., 1984, Geothermometry and geobarometry in pelitic rocks of north-central Wopmay Orogen (Early Proterozoic), Northwest Territories, Canada: *Geol. Soc. America Bull.*, v. 95, p. 196–208.

Tracy, R. J., 1982, Compositional zoning and inclusions in metamorphic minerals: *Rev. Mineralogy*, v. 10, p. 255–397.

———, Robinson, P., and Thompson, A. B., 1976, Garnet composition and zoning in the determination of temperature and pressure of metamorphism, central Massachusetts: *Amer. Mineralogist*, v. 61, p. 762–775.

Vielzeuf, D., 1983, The spinel and quartz associations in high grade xenoliths from Tallante (S. E. Spain) and their potential use in geothermometry and barometry: *Contrib. Mineralogy Petrology*, v. 82, 301–311.

Volborth, A., 1962, Rapakivi-type granites in the Precambrian complex of Gold Butte, Clark County, Nevada: *Geol. Soc. America Bull.*, v. 73, 813–832.

———, 1973, Geology of the granite complex of the Eldorado, Newberry, and northern Dead Mountains, Clark County, Nevada: *Nev. Bur. Mines Geology Bull. 80*, 40 p.

Wells, P. R. A., 1977, Pyroxene thermometry in simple and complex systems: *Contrib. Mineralogy Petrology*, v. 62, p. 129–139.

———, 1980, Thermal models for the magmatic accretion and subsequent metamorphism of continental crust: *Earth Planet. Sci. Lett.*, v. 46, p. 253–265.

Wood, B. J., and Banno, S., 1973, Garnet-orthopyroxene and orthopyroxene-clinopyroxene relationships in simple and complex systems: *Contrib. Mineralogy Petrology*, v. 42, p. 109–124.

Wooden, J., Miller, D., and Elliott, G., 1986, Early Proterozoic geology of the northern New York Mountains, SE California: *Geol. Soc. America Abstr. with Programs*, v. 18, p. 424.

Thomas D. Hoisch
Department of Geological Sciences
University of Southern California
Los Angeles, California 90089-0741

M. T. Heizler and T. M. Harrison
Department of Geological Sciences
State University of New York
Albany, New York 12222

Calvin F. Miller
Department of Geology
Vanderbilt University
Nashville, Tennessee 37235

Edward F. Stoddard
Department of Geosciences
North Carolina State University
Raleigh, North Carolina 27650

21

LATE CRETACEOUS REGIONAL METAMORPHISM IN SOUTHEASTERN CALIFORNIA

In southeastern California, ductile nappes formed within the ancient craton are exposed in several mountain ranges. Paleozoic cratonal sediments involved in the structures were metamorphosed syntectonically at middle greenschist to upper amphibolite facies. Metamorphic mineral assemblages in the Piute Mountains, Big Maria Mountains, and Old Woman Mountains indicate pressures corresponding to depths of about 9 km, 9-12 km, and >12 km, respectively. $^{40}Ar/^{39}Ar$ spectra of a variety of minerals from these ranges indicate post-80-m.y. cooling from peak metamorphic temperatures, suggesting that peak metamorphism and synmetamorphic nappe formation were Late Cretaceous events. In nearby uplifted areas which lack Late Cretaceous nappes or thrusts, such as the Turtle Mountains, removal of the cratonal sedimentary sequence by erosion explains its absence. Post-Late Cretaceous uplift of areas within southeastern California is an important component of the tectonic history. $^{40}Ar/^{39}Ar$ data indicate rapid cooling at the end of the Cretaceous in the Old Woman–Piute Range, probably associated with rapid uplift.

Distributions of Late Cretaceous regional metamorphism and deformation in southeastern California imply a link between metamorphism and nappe development. Granitic plutons of similar age, however, intrude not only some regionally metamorphosed areas but also some unmetamorphosed areas, making a direct link between plutonism and metamorphism difficult to establish. In the Big Maria Mountains, no Late Cretaceous plutons are exposed; metamorphism by hot infiltrating fluids is indicated by huge volumes of vesuvianite-wollastonite rocks for which fluid/rock ratios >17:1 are estimated. Warping of the ancient craton which led to nappe development may have been localized where hot infiltrating fluids softened the crust.

INTRODUCTION

Southeastern California records intense Late Cretaceous tectonism involving large-scale ductile nappes, regional metamorphism, and the anatectic derivation of large volumes of granitic magmas. Nappes involved both the Paleozoic to lower Mesozoic cratonal sedimentary assemblage and its underlying Precambrian crystalline basement. Within nappes, the cratonal assemblage was carried to mid-crustal depths and metamorphosed at lower greenschist to upper amphibolite facies. Analysis of metamorphic fabrics, mineral assemblages, and mineral chemistries from the exposed rocks permit characterization of metamorphic pressures and temperatures, assessment of P-T paths, and determination of fluid/rock ratios. These data, combined with field, geochronologic, and other data, constrain possible heat sources and possible causes of crustal anatexis.

The present study stems mostly from field, petrologic, and geochronologic work in two large mountain ranges in southeastern California, the Old Woman-Piute Range (which includes the Old Woman Mountains and Piute Mountains) and the Big Maria Mountains (Fig. 21-1).

Tectonic Setting of Southeastern California

The crust of southeastern California has experienced multiple episodes of tectonism of diverse styles since its initial formation in Early Proterozoic time. After a major orogenic episode about 1.7 b.y. ago, a period of general stability lasted through early Mesozoic time. During this time, platform facies sediments bearing a close resemblance to the Grand Canyon sequence were deposited (e.g., Stone *et al.*, 1983). A transition from platform-facies sediments to shelf-facies sediments occurs along a roughly northeast-trending line

FIG. 21-1. Index map of mountain ranges in southeastern California.

within southeastern California, crossing just west of the Old Woman Mountains (e.g., Burchfiel and Davis, 1972, 1981).

Granitic plutonism affected the area of southeastern California shown in Fig. 21-1 in Jurassic and Cretaceous time. This was accompanied in Late Cretaceous time by widespread deformation and metamorphism, expressed as ductile thrusts and nappes that developed under greenschist to amphibolite facies conditions. Geochronology from this study establishes that tectonism within southeastern California is similar in timing to Sevier thrusting in southern Nevada. It is, however, much different in character. In southern Nevada, Sevier thrusting was initiated along the Paleozoic miogeoclinal hinge and did not involve crystalline basement, whereas in southeastern California structures were developed within the ancient craton. Northwest of the area shown in Fig. 21-1, metamorphism and deformation of Jurassic age has been documented (Dewitt *et al.*, 1984; Labotka *et al.*, 1985b).

Intense extensional deformation affected much of the southwestern United States, including southeastern California, during Cenozoic time. Basin and Range normal faults and low-angle detachment faults formed in response to extension. In southeastern

California, mountain ranges flanked by detachment faults commonly expose rocks that were deeply buried in Late Cretaceous time, as documented by exposed metamorphic mineral assemblages. Major vertical crustal movements therefore were likely associated with this extension.

Previous Work

Previous field, petrologic, geochronologic, and structural work in the Old Woman–Piute Range was summarized in C. F. Miller *et al.* (1982). A synthesis of field data and of structural evolution in the Big Maria Mountains was presented by Hamilton (1982, 1984).

Few studies of metamorphic petrology in southeastern California have been conducted. Investigations include those of Hoisch (1982, 1985) in the Old Woman–Piute Range and Big Maria Mountains, and Shklanka (1963) in the Little Maria Mountains. Studies of regional metamorphism have been conducted in the Funeral and Panamint Mountains (Labotka, 1980, 1981; Labotka and Albee, Chapter 26, this volume; Labotka *et al.*, 1985b) about 240 km north-northwest of the Old Woman Mountains.

OLD WOMAN–PIUTE RANGE

General Geology

The Old Woman–Piute Range comprises mainly an 800-km^2 pre-Tertiary complex. This complex is subequally divided between Late Cretaceous batholithic rocks and Proterozoic through early Mesozoic metamorphic rocks. Intensely, ductilely deformed and metamorphosed Cambrian through lower Mesozoic cratonal sedimentary rocks are the best structural markers in the area (Stone *et al.*, 1983). They reveal a great northeast-trending overturned anticline or nappe cored by crystalline basement (Howard *et al.*, 1980). Conformable overturned Cambrian strata mark the lower limb, which is thrust-faulted with a southeast vergence over a generally upright middle Paleozoic through lower Mesozoic section. This section in turn overlies a thrust fault which separates it from a package of Precambrian and Mesozoic(?) metamorphic rocks, dominantly orthogneisses (Howard *et al.*, 1980; C. F. Miller *et al.*, 1982). The Paleozoic metasedimentary rocks which define the nappe and thrust complex extend for nearly 50 km from the Little Piute Mountains across the Old Woman Mountains to the Kilbeck Hills, but constitute only a few percent of the exposed area of prebatholithic metamorphic rocks. Ductilely deformed Paleozoic metasedimentary rocks are also exposed in the Piute Mountains, but their relation to the main nappe remains problematical.

All pre-Cretaceous lithologies in the Old Woman Mountains were injected by Late Cretaceous granitoid dikes and sills. In the lower-grade Piute and Little Piute Mountains, the Late Cretaceous granitoids are confined to discrete plutons and associated dikes. Late Cretaceous intrusive rocks are generally undeformed and unmetamorphosed, but in a few areas record mild to intense ductile shearing.

The most clearly preserved metamorphism and deformation in most of the range is Late Cretaceous in age and is recorded in the Cambrian through lower Mesozoic section. Precambrian rocks share fabric features with and commonly have mineral assemblages consistent with the grade of the Cambrian through lower Mesozoic section (Hoisch, 1982; C. F. Miller *et al.*, 1982). Early Proterozoic tectonism can also be recognized in the area, as indicated by almost undeformed Proterozoic diabase dikes which cut foliated 1.7-b.y.-old orthogneiss in the northern Piute Mountains (K. A. Howard, personal communication, 1985), by local preservation of relic assemblages in Precambrian rocks that are higher in grade than those in Paleozoic rocks, and by lead isotope ratios in feldspars of Cretaceous granitoids which indicate very high-grade Precambrian metamorphism of underlying crustal source materials (J. Wooden, unpublished data). The area has also been overprinted, although generally mildly, by brittle features associated with Miocene extensional tectonism.

Metamorphism

In the Old Woman–Piute Range, a series of loosely constrained isograds representing discontinuous AFM or AKNa reactions can be drawn from mineral assemblages observed in the Cambrian Bright Angel Shale and metapelites of the Precambrian basement (Fig 21-2). The recent discovery of kyanite in the northern Old Woman Mountains suggests that the initial interpretations of Hoisch (1982) and C. F. Miller *et al.* (1982) should be modified to distinguish a different "metamorphic field gradient" (usage of Spear *et al.*, 1984) in the Piute Mountains than in the Old Woman Mountains. In the Piute Mountains, the metamorphic field gradient traverses the andalusite-sillimanite boundary, whereas in the Old Woman Mountains, it traverses the kyanite-sillimanite boundary (Fig. 21-3). AFM and AKNa projections showing observed mineral assemblages for the two areas are given in Fig. 21-4. Reactions which delineate isograds are shown below (after A. B. Thompson, 1976a):

Piute Mountains

$$\text{chlorite} + \text{staurolite} + \text{muscovite} = \text{biotite} + \text{andalusite} + \text{quartz} + H_2O \qquad (1)$$
$$\text{ecstaurolite} + \text{muscovite} + \text{quartz} = \text{andalusite} + \text{biotite} + \text{garnet} + H_2O \qquad (2)$$
$$\text{andalusite} = \text{sillimanite} \qquad (3)$$

Old Woman Mountains

$$\text{kyanite} = \text{sillimanite} \qquad (4)$$
$$\text{staurolite} + \text{muscovite} + \text{quartz} = \text{sillimanite} + \text{biotite} + \text{garnet} + H_2O \qquad (5)$$
$$\text{muscovite} + \text{albite} = \text{K-feldspar} + \text{sillimanite} + H_2O \qquad (6)$$

The isograds distinguish four metamorphic zones in the Piute Mountains increasing in grade toward the southwest: The Staurolite Zone (STZ), Andalusite Zone I (ANZ I), Andalusite Zone II (ANZ II), and Upper Sillimanite Zone (USZ). The lowest-grade area is located in the northern portion of the western Piute Mountains, where chlorite + muscovite + biotite + quartz assemblages are present, but neither staurolite nor andalusite is found in metamorphosed Bright Angel Shale. The stable AFM assemblage here may be staurolite + chlorite + biotite as indicated by metastable inclusions in andalusite porphyroblasts in the immediately adjacent Andalusite Zone I. The appearance of andalusite marks the transition to the Andalusite Zone I, and the occurrence of reaction 1. In the eastern Piute Mountains, the occurrence of the AFM univariant assemblage staurolite + muscovite + garnet + quartz + andalusite marks the transition to Andalusite Zone II and the attainment of conditions corresponding to reaction 2. The appearance of fibrolitic sillimanite coincides with the disappearance of staurolite and garnet and marks the transition to the Upper Sillimanite Zone by the crossing of reaction 3.

Isograds in the Old Woman Mountains also distinguish four metamorphic zones: Kyanite Zone (KZ), Lower Sillimanite Zone (LSZ), Upper Sillimanite Zone (USZ), and Sillimanite-K feldspar Zone (SKZ), increasing in grade southward (Fig. 21-2b), Kyanite occurs as inclusions in cordierite together with phlogopite and quartz (no muscovite or K-feldspar) in the northern Old Woman Mountains, an area where abundant staurolite and garnet-bearing AFM assemblages are found in Precambrian metapelites; this defines

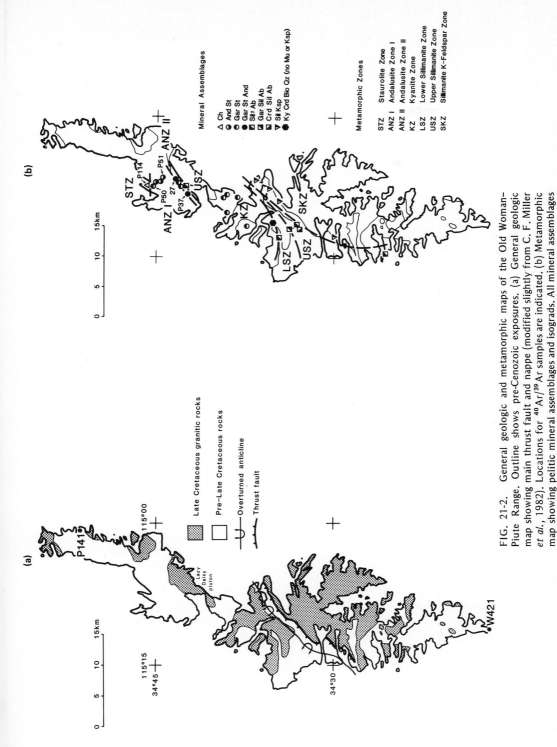

FIG. 21-2. General geologic and metamorphic maps of the Old Woman–Piute Range. Outline shows pre-Cenozoic exposures. (a) General geologic map showing main thrust fault and nappe (modified slightly from C. F. Miller *et al.*, 1982). Locations for $^{40}Ar/^{39}Ar$ samples are indicated. (b) Metamorphic map showing pelitic mineral assemblages and isograds. All mineral assemblages include muscovite, biotite and quartz, except where indicated. See Fig. 21-3 for mineral abbreviations. Localities for samples mentioned in the text and tables are indicated.

Mineral Assemblages

△ Ch
◔ And St
◑ Gar St
● Gar St And
◩ Sil± Ab
◪ Gar Sil Ab
◹ Crd Sil Ab
▽ Sil Ksp
● Ky Crd Bio Qz (no Mu or Ksp)

Metamorphic Zones

STZ   Staurolite Zone
ANZ I Andalusite Zone I
ANZ II Andalusite Zone II
KZ   Kyanite Zone
LSZ   Lower Sillimanite Zone
USZ   Upper Sillimanite Zone
SKZ   Sillimanite K-Feldspar Zone

Late Cretaceous granitic rocks

Pre-Late Cretaceous rocks

Overturned anticline

Thrust fault

543

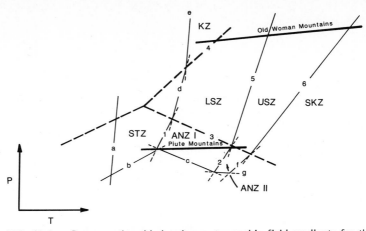

FIG. 21-3. Petrogenetic grid showing metamorphic field gradients for the Old Woman Mountains and Piute Mountains, and reactions delineating metamorphic zones. Reactions are based on Hess (1973) and A. B. Thompson (1976b). Reactions 1–6 are given in the text. Reactions a, b, d, e, and f are written with the high-temperature assemblage on the right, and reactions c and g are written with the high-pressure assemblage on the right (abbreviations: Ab, albite; And, andalusite; Bio, biotite; Ch, chlorite; Crd, cordierite; Gar, garnet; Ksp, K-feldspar; Ky, kyanite; Mu, muscovite; Qz, quartz; Sil, sillimanite; St, staurolite): a: $Gar + Ch + Mu = St + Bio + Qz + H_2O$; b: $Ch + And + Qz = Crd + St + H_2O$; c: $St + Crd + Mu = And + Bio + Qz + H_2O$; d: $Mu + St + Ch + Qz = Bio + Sil + H_2O$; e: $Mu + St + Ch + Qz = Bio + Ky + H_2O$; f: $Mu + Ab + Qz = And + Ksp + H_2O$; g: $Mu + Crd + Gar = And + Bio + Qz + H_2O$.

the Kyanite Zone. Progressing southward, the loss of staurolite and garnet corresponds roughly to the appearance of sillimanite; this denotes the crossing of reaction 4 and entry into the Lower Sillimanite Zone. The loss of staurolite and garnet may be explained as resulting from AFM continuous reactions (e.g., A. B. Thompson, 1976a). The transition from the Lower Sillimanite Zone to the Upper Sillimanite Zone is recognized by the terminal stability of staurolite in AFM assemblages (reaction 5), which cannot be delineated precisely from the present set of data, owing to its earlier disappearance by continuous reactions. Somewhat Fe-richer bulk compositions are needed to preserve staurolite within the Lower Sillimanite Zone and to delineate precisely its terminal stability. Entry into the Sillimanite-K Feldspar Zone is indicated by the appearance of K-feldspar in pelitic schist, reflecting the crossing of AKNa discontinuous reaction 6.

Piute Mountains: Detailed Study

The four metamorphic zones delineated in the central Piute Mountains are based in large part on mineral assemblages observed in the Bright Angel Shale contained within a 5-km northeast-southwest-trending band of Paleozoic metasedimentary rocks. The temperature of metamorphism increases about 150-200°C in the southwestward direction, resulting in mineralogical, textural, and chemical changes within the metamorphosed Bright Angel Shale. Primary sedimentary stratification (S_0) can be recognized along the entire band as thin beds (1-10 cm) of variable color and composition. At the lowest-grade (northernmost) point, the Bright Angel Shale is phyllitic, with phyllosilicates (major muscovite and biotite, minor chlorite) crystallizing parallel to S_0, forming an incipient schistosity (S_1).

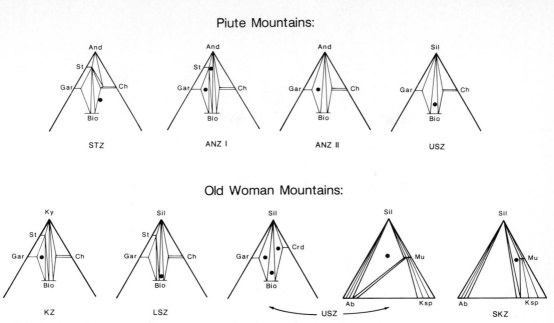

Piute Mountains:

STZ   ANZ I   ANZ II   USZ

Old Woman Mountains:

KZ   LSZ   USZ   SKZ

FIG. 21-4   AFM and AKNa representations for the Piute Mountains and Old Woman Mountains, with observed mineral assemblages indicated by dots. AFM projections (open at the bottom) represent the system $K_2O-FeO-MgO-Al_2O_3-SiO_2-H_2O$ projected through muscovite, quartz and H_2O (J. B. Thompson, 1957), and AKNa projections (closed triangles) represent the system $K_2O-Na_2O-Al_2O_3-SiO_2-H_2O$ projected through quartz and H_2O (J. B. Thompson, 1961).

Progressing southwestward to the central part of the band, the rocks coarsen to schist, and a second schistosity (S_2) is developed along which slip cleavages are locally formed (Fig. 21-5). Staurolite, garnet, and andalusite appear, but never all three together. Staurolite and second-generation biotite parallel S_2 (Fig. 21-6).

At the extreme southeastern part of the band, the Paleozoic sequence is in contact with the Lazy Daisy pluton, an unmetamorphosed Late Cretaceous two-mica granite. The Bright Angel Shale at this locality is locally hornfelsic and contains large poikiloblastic amoeboid andalusite with inclusions of early-formed biotite (Fig. 21-7). All of S_0, S_1, and S_2 are easily recognizable in some samples. Earlier-formed garnet and staurolite are also present, the garnet occurring both as inclusions in andalusite porphyroblasts and within the matrix. Staurolite is ragged and contained within mica pods (Fig. 21-8), suggesting a decomposition reaction. Similar staurolite textures have been described in metapelites by Foster (1983) and Guidotti (1968), who suggest that they result from prograde reactions.

Garnet-biotite geothermometry for samples from the Piute Mountains is shown in Table 21-1. Locations for these samples are shown in Fig. 21-2b. In applying garnet-biotite thermometry to zoned garnets, a single composition for biotite was used. The modal ratio of biotite to garnet within the samples analyzed is generally greater than 3:1, requiring that any change in the distribution of Fe and Mg between garnet and biotite be reflected primarily in changes of garnet composition.

FIG. 21-5. Photomicrograph of sample P114 (Piute Mountains). Long edge equals approximately 2 mm. A schistosity (S_1) which developed parallel to original sedimentary stratification (S_0) is deformed into microfolds. Second-generation biotite parallels axial surfaces of microfolds (S_2).

P-T paths experienced by the andalusite-bearing samples may be determined using the "Gibbs method" as described by Spear *et al.* (1982) and Spear and Selverstone (1983). The matrix of equations which describe the relationship between mineral compositional variations and variations in pressure and temperature for the assemblage quartz + muscovite + biotite + andalusite + garnet is shown in Table 21-2. This system of equations treats quaternary garnets, in contrast to Table 2 of Spear and Selverstone (1983), in which binary garnets (almandine-pyrope) are treated. In applying the Gibbs method, it is necessary to know how both biotite and garnet changed composition during crystal growth or metamorphic reactions. Garnet preserves its compositional variation through zoning, but the biotite was found to be largely homogeneous within a given sample. Variations in biotite composition may have been lost through homogenization. For the system of interest as represented in Table 21-2, a change of $+0.01X_{ph1}$ (mole fraction phlogopite

FIG. 21-6. Photomicrograph of sample P50, Piute Mountains. Long edge equals approximately 2 mm. A foliation which developed parallel to original sedimentary stratification is deformed into slip cleavages, with second-generation biotite and staurolite paralleling the axial surfaces.

FIG. 21-7. Photomicrograph of sample 27, Piute Mountains. Long edge equals approximately 2 mm. All but the lower right portion of the photo is a poikiloblastic andalusite porphyroblast with inclusions of earlier-formed biotite. Garnet (high relief, round) and staurolite (high relief, ragged) are also visible.

in biotite) starting with biotite from sample 27 at constant garnet composition (average of rim and core of garnet 27-1, Table 21-4) would denote a temperature drop of 13°C and a pressure increase of 1.5 bar. For the purposes of applying the Gibbs method, biotite compositions were assumed to have remained constant during metamorphism. These calculations indicate, however, that variations in biotite composition are important.

Calculated *P-T* paths for andalusite-bearing samples are shown in Table 21-3. Rims of two garnets from sample 27 record substantially higher temperatures than cores, with significant increases in pressure (Fig. 21-7). This is interpreted as resulting from contact heating by the Lazy Daisy pluton at a deeper crustal level than the early stage of metamorphism. This sequence of events is also suggested by textures in the Bright Angel Shale at the contact, where andalusite porphyroblasts which resulted from a prograde, staurolite-consuming reaction clearly postdate the development of both S1 and S2. The very large values of ΔP (Table 21-3) may, however, be artifacts of the improperly applied assumption of constant biotite composition.

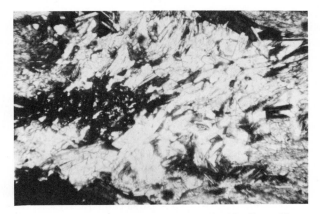

FIG. 21-8. Photomicrograph of sample 27, Piute Mountains. Long edge equals approximately 2 mm. Photo shows ragged staurolite within pod of micas.

TABLE 21-1. Garnet-Biotite Thermometry (°C)

Garnet	Ganguly and Saxena (1984)[a]	Ferry and Spear (1978)[a]	Holdaway and Lee (1977)[a]	Goldman and Albee (1977)[b]	Perchuck (1977)[b]	A. B. Thompson (1976b)[a]
P37E2 core	519	518	535	498	529	538
P37E2 rim	501	495	519	480	515	519
P37D core	516	531	544	500	542	548
P37D rim	499	491	516	495	515	517
P51-1 core	512	470	501	502	481	500
P51-1 rim	564	540	551	539	540	556
P51-2 core	569	546	555	554	539	560
P51-2 rim	579	570	571	558	560	579
27-1 core	508	460	493	499	475	491
27-1 rim	608	600	591	583	575	602
27-2 core	568	538	549	542	528	554
27-2 rim	596	574	574	566	557	582

[a]A pressure of 2.5 kb is assumed in the calculation.
[b]The effect of pressure is disregarded in the formulation of the geothermometer.

In the Piute Mountains, the presence of andalusite, biotite, and staurolite, and the absence of cordierite in AFM assemblages constrains pressures to be within specific limiting reactions on an AFM petrogenetic grid corresponding to the ANZ I field on Fig. 21-3. Pressures of metamorphism throughout the Piute Mountains must therefore have been toward the high-pressure part of the andalusite stability field. An estimate of pressure corresponding to the pressure of intrusion of the Lazy Daisy pluton may be obtained by consideration of its contact effect on adjacent pelitic schist. Intrusion of the Lazy Daisy pluton caused the temperature at the contact to increase until the conditions of AFM discontinuous reaction 2 were reached. The pressure at the contact of the pluton at the time of intrusion may therefore be constrained from the location of reaction 2 as depicted on Fig. 21-3. It lies at pressures somewhat below the Al_2SiO_5 triple point, but does not extend far in the low-pressure direction. Considering the Al_2SiO_5 stabilities to be correctly determined in the study of Holdaway (1971), a pressure of 2.5 ± 0.5 kb is reasonable, although this is not a deterministic estimate.

Confirmation of reaction 2 may be obtained by modeling it from mineral compositions in sample 27. Using the garnet rim composition of garnet 27-1, muscovite, biotite, and staurolite (Table 21-4), and assuming stoichiometric quartz, ilmenite, and rutile (all present), a system of eight equations in eight unknowns was formulated. Because Ca occurs in significant abundance only within garnet and Zn only within staurolite, Ca and Zn were omitted in the calculations. It is likely that these elements either concentrate in their respective hosts with increasing temperature via continuous net transfer reactions or are evolved into the fluid phase. For garnet, the remaining eightfold-coordinated cations (Mn, Fe, Mg) were renormalized to three, and for staurolite, the remaining fourfold coordinated R^{2+} cations (Fe, Mg) were renormalized to the sum of Zn + Fe + Mg. Reaction 2 is then calculated to be

(2a)   1.46quartz + 0.66muscovite + 1.00staurolite + 2.27rutile
= 0.67biotite + 0.02garnet + 9.40andalusite + 2.31ilmenite + H_2O (2a)

This reaction is very similar to reaction 6 of Pigage (1982) who derived it through a linear regression of mineral compositions, but differs in that sillimanite was the stable Al_2SiO_5

TABLE 21-2. System of Equations Describing the Relationship among Variations in Pressure (P), Temperature (T), and Composition (X) for the Assemblage Andalusite + Quartz + Muscovite + Biotite + Garnet.[a,b]

The system has the form $\mathbf{M}\cdot\mathbf{v} = \mathbf{0}$, where $\mathbf{v} = [\,dT,\ dP,\ d\mu_{qz},\ d\mu_{mus},\ d\mu_{and},\ d\mu_{py}^{gar},\ d\mu_{al}^{gar},\ d\mu_{gr}^{gar},\ d\mu_{sp}^{gar},\ d\mu_{phl}^{bio},\ d\mu_{ann}^{bio},\ dX_{sp}^{gar},\ dX_{py}^{gar},\ dX_{gr}^{gar},\ dX_{phl}^{bio}\,]^{T}$

dT	dP	$d\mu_{qz}$	$d\mu_{mus}$	$d\mu_{and}$	$d\mu_{py}^{gar}$	$d\mu_{al}^{gar}$	$d\mu_{gr}^{gar}$	$d\mu_{sp}^{gar}$	$d\mu_{phl}^{bio}$	$d\mu_{ann}^{bio}$	dX_{sp}^{gar}	dX_{py}^{gar}	dX_{gr}^{gar}	dX_{phl}^{bio}		
$\bar{S}_{qz}$	$-\bar{V}_{qz}$	1	0	0	0	0	0	0	0	0	0	0	0	0	=	0
$\bar{S}_{mus}$	$-\bar{V}_{mus}$	0	1	0	0	0	0	0	0	0	0	0	0	0	=	0
$\bar{S}_{and}$	$-\bar{V}_{and}$	0	0	1	0	0	0	0	0	0	0	0	0	0	=	0
$\bar{S}_{gar}$	$-\bar{V}_{gar}$	0	0	0	X_{py}^{gar}	X_{al}^{gar}	X_{gr}^{gar}	X_{sp}^{gar}	0	0	0	0	0	0	=	0
$\bar{S}_{bio}$	$-\bar{V}_{bio}$	0	0	0	0	0	0	0	X_{phl}^{bio}	X_{ann}^{bio}	0	0	0	0	=	0
0	0	2	-2	4	0	-2	0	0	0	2	0	0	0	0	=	0
0	0	2	-2	4	-2	0	0	0	2	0	0	0	0	0	=	0
$-(\bar{S}_{py}^{gar}-\bar{S}_{al}^{gar})$	$(\bar{V}_{py}^{gar}-\bar{V}_{al}^{gar})$	0	0	0	-1	1	0	0	0	0	$\dfrac{\delta^2\bar{G}^{gar}}{\delta X_{sp}\,\delta X_{py}}$	$\dfrac{\delta^2\bar{G}^{gar}}{\delta X_{py}^2}$	$\dfrac{\delta^2\bar{G}^{gar}}{\delta X_{gr}\,\delta X_{py}}$	0	=	0
$-(\bar{S}_{sp}^{gar}-\bar{S}_{al}^{gar})$	$(\bar{V}_{sp}^{gar}-\bar{V}_{al}^{gar})$	0	0	0	0	1	0	-1	0	0	$\dfrac{\delta^2\bar{G}^{gar}}{\delta X_{sp}^2}$	$\dfrac{\delta^2\bar{G}^{gar}}{\delta X_{sp}\,\delta X_{py}}$	$\dfrac{\delta^2\bar{G}^{gar}}{\delta X_{sp}\,\delta X_{gr}}$	0	=	0
$-(\bar{S}_{gr}^{gar}-\bar{S}_{al}^{gar})$	$(\bar{V}_{gr}^{gar}-\bar{V}_{al}^{gar})$	0	0	0	0	1	-1	0	0	0	$\dfrac{\delta^2\bar{G}^{gar}}{\delta X_{sp}\,\delta X_{gr}}$	$\dfrac{\delta^2\bar{G}^{gar}}{\delta X_{py}\,\delta X_{gr}}$	$\dfrac{\delta^2\bar{G}^{gar}}{\delta X_{gr}^2}$	0	=	0
$-(\bar{S}_{phl}^{bio}-\bar{S}_{ann}^{bio})$	$(\bar{V}_{phl}^{bio}-\bar{V}_{ann}^{bio})$	0	0	0	0	0	0	0	-1	1	0	0	0	$\dfrac{\delta^2\bar{G}^{bio}}{\delta X_{phl}^2}$	=	0

[a] al, almandine; and, andalusite; ann, annite; bio, biotite; gar, garnet; gr, grossular; mus, muscovite; phl, phlogopite; py, pyrope; qz, quartz; sp, spessartine.

[b] $\bar{S}$, molar entropy; $\bar{V}$, molar volume; $\bar{G}$, molar Gibbs free energy; μ, chemical potential.

TABLE 21-3. Pressure-Temperature Paths from Garnet Zoning in Andalusite-Bearing Pelitic Schist from the Piute Mountains

Garnet Core-to-Rim	ΔT (°C)	ΔP (Bar)
27-1	148	2219
27-2	42	605
P37D	−40	−888

TABLE 21-4. Microprobe Analyses of Minerals from Sample 27, Piute Mountains

Muscovite		Biotite		Staurolite		Garnet 27-1		
						3 core	2 rim	— average
No. pts:	7		4		4			
SiO_2	45.80	SiO_2	35.36	SiO_2	27.61	SiO_2	37.91	38.42	38.17
Al_2O_3	35.33	Al_2O_3	19.49	Al_2O_3	54.88	Al_2O_3	21.29	21.41	21.35
FeO	1.11	FeO	21.97	FeO	13.54	FeO	31.17	34.03	32.60
MgO	0.51	MgO	7.73	MgO	1.31	MgO	1.33	2.34	1.84
CaO	0.00	CaO	0.00	CaO	0.00	CaO	4.53	3.01	3.77
Na_2O	0.51	Na_2O	0.26	TiO_2	0.47	TiO_2	0.08	0.04	0.06
K_2O	10.55	K_2O	9.28	MnO	0.24	MnO	5.41	3.09	4.25
TiO_2	0.55	TiO_2	2.34	ZnO	0.71				
MnO	0.08	MnO	0.10	F	0.01				
F	0.22								
Cl	0.01								
Total	94.58	Total	96.53	Total	98.77	Total	101.72	102.34	102.03
Sum O_x	11.00	Sum O_x	11.00	Sum O_x	47.00	Sum O_x	12.00	12.00	12.00
Si	3.07	Si	2.69	Si	7.77	Si	3.01	3.02	3.02
Al	2.80	Al	1.74	Al	18.21	Al	1.99	1.99	1.99
Fe^{2+}	0.06	Fe^{2+}	1.40	Fe^{2+}	3.19	Fe^{2+}	2.07	2.24	2.16
Mg	0.05	Mg	0.88	Mg	0.55	Mg	0.16	0.27	0.22
Ca	0.00	Ca	0.00	Ca	0.00	Ca	0.39	0.25	0.32
Na	0.07	Na	0.04	Ti	0.10	Ti	0.00	0.00	0.00
K	0.90	K	0.90	Mn	0.06	Mn	0.36	0.21	0.28
Ti	0.03	Ti	0.13	Zn	0.15				
Mn	0.00	Mn	0.01	F	0.01				
F	0.05								
Cl	0.00								

polymorph. Thus, the determined reaction stoichiometry is consistent with textural and temporal interpretations.

General Geology

The Big Maria Mountains are similar in many respects to the Old Woman Mountains, comprising a middle greenschist to middle amphibolite facies, Late Cretaceous, regional metamorphic terrane approximately 200 km^2 in area which lies structurally beneath a Tertiary detachment fault (Hamilton, 1982, 1984). The Paleozoic to lower Mesozoic crational sedimentary assemblage is folded and attenuated within a large basement-involved northwest-trending nappe (Hamilton, 1982, 1984). Equilibrium metamorphic textures in lineated rocks in which mineral lineations parallel fold axes indicate that metamorphism and deformation were coeval.

The area was intruded in Jurassic time by granitic plutons ranging in composition from gabbro to monzogranite and mostly emplaced within the Precambrian complex. Two U-Pb dates of about 160 m.y. confirm the age of the plutonism (dates by L. T. Silver, cited in Hamilton, 1982, 1984). Evidence of contact metamorphism associated with this event is not preserved through the strong Late Cretaceous metamorphism and deformation.

The Tertiary history of the area is characterized by faulting related to extension and to right-lateral shear. Low-angle detachment faults of extensional origin are present in the northern part of the range (Hamilton, 1982, 1984) and are inferred to extend beneath alluvium to separate the Big Maria Mountains terrane from the upper plate of the Riverside Mountains to the north. Pervasive cataclasis affects the high structural levels of the range near the detachment fault. Several major normal and right-lateral faults cut and therefore postdate the detachment faults (Hamilton, 1982, 1984).

An important difference between the geology of the Big Maria Mountains and the Old Woman Mountains is that no Late Cretaceous plutons are exposed in the Big Maria Mountains. In the northwestern Big Maria Mountains, abundant late-metamorphic, muscovite-bearing pegmatite dikes of Late Cretaceous(?) age may be indicative of an underlying pluton.

Regional Metamorphism

Metamorphism in the Big Maria Mountains was studied by examining a number of different rock types in order to determine the physical conditions and to define isograds. The metamorphosed siliceous dolomites in the area were particularly useful for these purposes. The Cambrian Muav, Mississippian Redwall, and Permian Kaibab Formations and an unnamed middle Paleozoic dolomite marble are each in whole or in part siliceous dolomites, and together comprise about a third of the Paleozoic section. Mineral assemblages developed within these units are sensitive to P-T-X conditions, and were interpreted together with T-X topologies to draw isograds. Equilibrium assemblages of calcite and dolomite were utilized for calcite-dolomite thermometry. Pressure for the northwestern part of the area was estimated from the location in P-T-X space of the observed isobaric invariant assemblage forsterite + diopside + tremolite + calcite + dolomite.

The metamorphosed Supai Formation is a siliceous limestone which developed mineral assemblages consisting of about 90% wollastonite throughout most of the range. These rocks occur in great volume and record extremely high fluid/rock ratios (discussed below). A sharp isograd was mapped within this unit based on the presence or absence of wollastonite.

Jurassic metagranites from the southern part of the range contain two plagioclases in equilibrium about the peristerite solvus. In these, the composition of the albite was used as an indicator of pressure, following Maruyama *et al.* (1982). Temperatures were obtained from Jurassic metagranites in the southern and central part of the range using two-feldspar thermometry.

Isograds from Siliceous Dolomites

Approaches to mapping isograds in siliceous dolomites have been discussed by Skippen (1974), Greenwood (1975), and Rice (1977). Isograds mapped by the first occurrence of minerals in siliceous dolomites are inadequate because such isograds assume external control of the fluid phase composition (J. B. Thompson, 1957). When the fluid-phase composition is internally controlled, as is often thought to be the case with siliceous dolomites, the critical assemblage will occur over a range of temperatures at constant pressure (Guidotti, 1974). In this case, isograds corresponding to the traces of isobaric univariant thermal maxima ("singular points" of Rice, 1977) should appear as sharp lines in the field, even when total pressure varies widely. Figure 21-9 shows the general isobaric T-X_{CO_2} topology for the siliceous dolomite system. In the study area, isograds corresponding to the trace of invariant points I and II and the thermal maximum of the reaction labeled "6" in Fig. 21-9 are mappable from the available mineral-assemblage data.

A general northwestward increase in metamorphic grade is indicated by two-feldspar and calcite-dolomite thermometry (discussed below). A traverse northwestward across the area should therefore yield changes in mineral associations that follow an upward traverse across Fig. 21-9. Starting at the southeastern edge of the range and moving to the northwest, the last occurrences of the stable associations talc + quartz + calcite and talc + quartz + calcite + dolomite mark the crossing of invariant point I. Continuing,

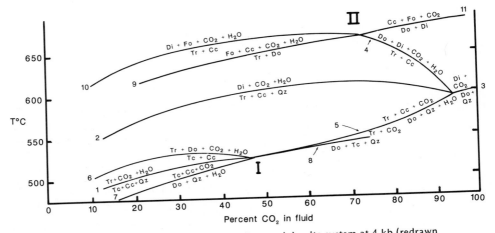

FIG. 21-9. T-X topology for the siliceous dolomite system at 4 kb (redrawn from Eggert and Kerrick, 1981, with slight modification to reaction 6). Numbers for reactions correspond to those of Eggert and Kerrick (1981). Cc, calcite; Dol, dolomite; Di, diopside; Fo, forsterite; Qz, quartz; Tc, talc; Tr, tremolite.

the last occurrence of stable talc in the associations tremolite + talc + calcite, talc + calcite + dolomite, and talc + quartz + dolomite + calcite mark the crossing of the thermal maximum of reaction "6" (Fig. 21-9). Invariant point II is crossed at the occurrence of the isobaric invariant assemblage forsterite + tremolite + diopside + calcite + dolomite. Two localities actually contain the isobaric invariant assemblage, but textures suggest disequilibrium; it is likely that the forsterite-forming reactions were overstepped. Thus, ths isograd representing the trace of invariant point II is drawn at slightly lower temperatures than prevailed at these localities. Figure 21-10 shows mineral assemblages and the resultant isograd map. The three isograds delineate four metamorphic zones which are designated Lower Talc Zone (LTZ), Upper Talc Zone (UTZ), Diopside-Tremolite Zone (DTZ), and Forsterite Zone (FZ), in prograde order. The isograd which separates the Lower Talc Zone from the Upper Talc Zone is particularly well constrained, whereas the others are only approximately located. The assemblage talc + tremolite + calcite occurs at two localities at grades higher than the Upper Talc Zone and is interpreted to be the product of retrograde reequilibration.

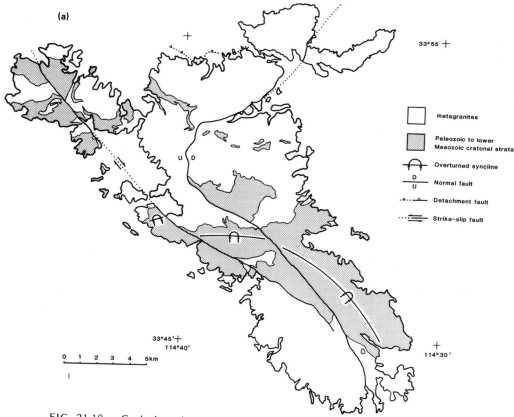

(a)

33°55

metagranites

Paleozoic to lower
Mesozoic cratonal strata

Overturned syncline

Normal fault

Detachment fault

Strike–slip fault

33°45'
114°40'

0 1 2 3 4 5km

33°30'

FIG. 21-10.   Geologic and metamorphic maps for the Big Maria Mountains (a) General geologic map (after Hamilton, 1984). (b) Metamorphic map showing isograds and mineral assemblages from siliceous dolomites. Numbers indicate temperatures (°C) derived from calcite-dolomite and two-feldspar thermometry. See Fig. 21-9 for mineral abbreviations. Localities for $^{40}Ar/^{39}Ar$ samples are indicated (SW2-2-1 and SW2-2-2).

(b)

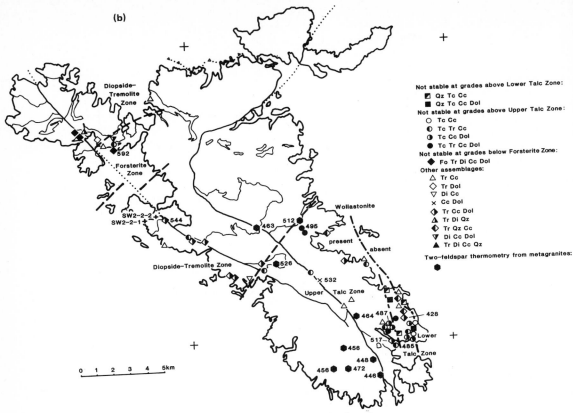

Not stable at grades above Lower Talc Zone:
- ◩ Qz Tc Cc
- ◼ Qz Tc Cc Dol

Not stable at grades above Upper Talc Zone:
- ○ Tc Cc
- ◔ Tc Tr Cc
- ◑ Tc Cc Dol
- ● Tc Tr Cc Dol

Not stable at grades below Forsterite Zone:
- ◆ Fo Tr Di Cc Dol

Other assemblages:
- △ Tr Cc
- ◇ Tr Dol
- ▽ Di Cc
- × Cc Dol
- ◈ Tr Cc Dol
- ◮ Tr Di Qz
- ◆ Tr Qz Cc
- ▼ Di Cc Dol
- ▲ Tr Di Cc Qz

Two-feldspar thermometry from metagranites:
- ⬣

FIG. 21-10.   Continued

Geothermometry

Temperatures of metamorphism in the Big Maria Mountains were estimated from calcite-dolomite thermometry in siliceous dolomites and two-feldspar geothermometry in meta-morphosed Jurassic metagranites.

Calcite-dolomite thermometry is based on the calcite-dolomite solvus. Only the calcite composition varies sensitively with temperature, hence most formulations of the calcite-dolomite geothermometer consider only the calcite and not the dolomite composition. Variation with pressure is slight (Goldsmith and Newton, 1969). In the present study, the formulation of Rice (1977) is used.

Common complications in the determination of the calcite composition at the peak of metamorphism are the exsolution of dolomite, and concentric zoning within calcite grains (Bowman and Essene, 1982; Hutcheon and Moore, 1973; Nesbitt and Essene, 1982; Perkins *et al.*, 1982; Puhan, 1976; Wada and Suzuki, 1983). Neither was observed in any of the samples analyzed in the present study. Moderate inhomogeneity within calcite grains was generally observed; different compositions were distributed erratically except that somewhat more magnesian compositions occurred commonly in the cores of large grains. Erratic compositional variation within calcite grains coexisting with dolomite was also noted by Essene (1983), but how it should be treated with regard to the application of calcite-dolomite geothermometry is unclear. In the present study, as many as 30 microprobe points per sample were taken from the cores of large calcite grains and averaged to obtain the calcite composition used in the thermometry calculation (Hoisch, 1985). A standard deviation of about 35°C is derived from uncertainty in the composition.

To apply two-feldspar thermometry to the Jurassic metagranites, the Whitney and Stormer (1977) formulation of the two-feldspar thermometer was used. This formulation assumes microcline as the stable structural state of K-feldspar, in contrast to other formulations which assume a high-structural state (e.g., Ghiorso, 1984; Haselton et al., 1983; Price, 1985; Stormer, 1975). The Whitney and Stormer (1977) formulation inherits considerable uncertainty from the high-temperature solvus calibration of Bachinski and Muller (1971) on which it is based. Nesbitt and Essene (1982) found that it gives temperatures as much as 80°C lower than calcite-dolomite thermometry. In the present study, 50°C have been added to the results as this appears to yield temperatures which conform to trends seen in calcite-dolomite temperatures. A pressure of 4 kb was assumed in the calculations.

The microcline-albite solvus shows strong variation in microline composition with temperature, and only slight variation in albite composition. To ascertain the K-feldspar composition to be used in the thermometry calculations, as many as 27 microprobe points were taken per sample from several grains. Because K-feldspar grains comprise heavily exsolved, metamorphosed igneous phenocrysts, determination of the K-feldspar composition by integration would likely have yielded strongly albitic igneous compositions. In this study, exsolved K-feldspar phenocrysts are considered to represent closed systems of coexisting K-feldspar and albite. The most albitic K-feldspar composition found from the analyses within a sample was taken to reflect the composition at the peak of metamorphism.

Evidence for chemical and textural equilibration of the Jurassic metagranites during metamorphism is considerable. The biotite changed color from its initial igneous color of reddish-brown to an even greenish brown while also changing composition from titaniferous to nontitaniferous. Titanium from the biotite formed intricate grids of rutile inclusions. Epidote and sphene also formed, and plagioclase underwent exsolution about the peristerite solvus (discussed below). Felsic minerals interstitial to porphyroblasts (formerly igneous phenocrysts) formed granoblastic-polygonal textures (Fig. 21-11). The results of two-feldspar and calcite-dolomite thermometry are depicted on Fig. 21-10b. Both indicate a general trend of temperature increase from the southeast to the northeast.

FIG. 21-11. Photomicrograph of a Jurassic metagranite from the southern Big Maria Mountains showing well-developed granoblastic-polygonal texture. Long edge equal approximately 1 mm.

Geobarometry: Siliceous Dolomites

Many investigators have discussed the effect of total pressure on the siliceous dolomite system (e.g., Eggert and Kerrick, 1981; Metz 1977, 1983; Skippen, 1974; Slaughter et al., 1975; Trommsdorf, 1972). Metz (1977, 1983) suggested the use of the isobaric invariant assemblage calcite + dolomite + tremolite + forsterite + diopside as a pressure indicator (point II on Fig. 21-9). In P-T-X_{CO_2} coordinates, its stability is represented as a line, which has been calibrated experimentally by Metz (1983). Knowing any one variable — pressure, temperature, or X_{CO_2} — yields unique values for the other two. Pressure and X_{CO_2} thus may be determined by using calcite-dolomite thermometry to ascertain the temperature.

The invariant assemblage has been found in two areas in the northwestern Big Maria Mountains. The samples containing this assemblage have complex textures that are difficult to interpret, but suggest disequilibrium. All silicates have amoeboid shapes and are extremely poikiloblastic. The samples are composed of about 70% silicates by volume and all dolomite occurs as large porphyroblasts 4–5 mm across. Samples from these areas also contain clinohumite, phlogopite, and clinochlore. Rather than being the result of equilibration at the precise conditions of the invariant point, it seems more likely that the assemblage results from reactions in which forsterite was produced by the overstepping of forsterite producing reactions in either the high-temperature or high-water-content direction.

Calcite-dolomite thermometry has been applied to one sample. The determined temperature of 592 ± 35°C indicates a pressure of 1.5–3.0 kb according to Metz (1983, Fig. 5). Pressure in this area is further constrained by the occurrence of muscovite in late-metamorphic pegmatite dikes 2 km east of this locality. Calculations by A. B. Thompson (1982) for the system K_2O-FeO-Na_2O-Al_2O_3-SiO_2-H_2O place the minimum pressure at about 3.5 kb. Uncertainties and additional components could lower this pressure (Anderson and Rowley, 1981; C. F. Miller et al., 1981). The Lazy Daisy pluton (Piute Mountains) is a two-mica granite which is interpreted to have intruded at a pressure of 2.5 ± 0.5 kb in the present study, as judged from its contact effect on pelitic schist. This is likely to be close to the minimum pressure for muscovite stability in granites. Pressure in the northwestern Big Maria Mountains is therefore approximately 2.5–3.0 kb.

Extra components in the system, particularly fluorine substitution into tremolite, will significantly affect the P-T-X_{CO_2} location of the isobaric invariant point. At constant total pressure, the invariant point rises in temperature with increasing fluorine in tremolite; the tremolite was not analyzed, so this possible effect cannot be evaluated. Other possible problems are uncertainties in the experimental calibrations, and too low a temperature caused by retrograde reequilibration of the calcite composition. Errors in the temperature estimate would result in errors in the pressure estimate, with higher temperatures corresponding to higher pressures.

Geobarometry: Peristerite Gap

Four approximate calibrations of the peristerite solvus from field studies have recently been compiled by Maruyama et al. (1982, Fig. 7). These data show marked pressure sensitivity and temperature insensitivity of the composition of the albite limb. An equation expressing albite composition versus pressure may be regressed from the four field calibrations. Using the albite compositions at 400°C and the middle of the estimated pressure range for each solvus calibration, a least-squares fit to an equation in the form $P = m \ln X_{An} + b$ was carried out:

$$P(\text{kb}) = -2.17 \ln X_{An} - 3.60 \tag{7}$$

This equation reproduces the average pressures within 1 kb for each of the four field calibrations.

The fit equation used is strictly empirical, having no particular theoretical basis. Choosing an appropriate theoretical model for the peristerite gap is difficult, as indicated by the disparity of views held on the subject. Some have interpreted it as a simple solvus (Crawford, 1966; Laves, 1954; Nord et al., 1978; Ribbe, 1962), some as a binary loop involving a first-order Al/Si order/disorder transformation in albite (Orville, 1974; Smith, 1972, 1983), and Carpenter (1981) has interpreted it as a conditional solvus involving a conditional spinodal.

A compilation of plagioclase analyses taken from five samples of metaluminous Jurassic metagranites in the southern part of the range is shown in Fig. 21-12. This shows two distinct peaks at An_{04} and An_{18}, with analyses from each of the five samples contributing to both peaks. A composition of An_{04} yields a pressure of 3.4 kb according to equation (7).

Compositions which scatter away from the peaks may be explained as the combination of several effects. Because grain boundaries separating exsolved albite from oligoclase could not be observed in transmitted light, some of the analyzed points probably overlapped grain boundaries. Some of the scatter may also be due to partial equilibration during cooling or incomplete equilibration at the peak of metamorphism.

According to Maruyama et al. (1982), determination of the equilibrium compositions in a sample must be carried out by looking at the compositions of adjacent albite and oligiclase within 5 μm of a mutual contact. They base their conclusion on observations made from rocks within two contact metamorphic aureoles. This method was not applied in the other studies used to compile their Fig. 21-7, which are from higher-pressure regionally metamorphosed areas. In rocks from a contact aureole, it might be expected that equilibrium would be achieved only at the immediate contacts because of

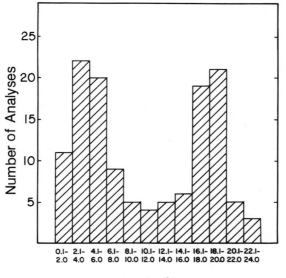

FIG. 21-12. Compilation of plagioclase compositions from five samples of Jurassic metagranites from the southern Big Maria Mountains. All five samples have compositions which contribute to both peaks.

problems with ion diffusion in plagioclase, which is made exceedingly difficult by the coupled substitution. The protracted thermal histories of regionally metamorphosed areas, however, might tend to cause a more complete equilibration at the peak of metamorphism.

Wollastonite Isograd in the Supai Formation, and Evidence for Large-Scale Fluid Flux

The Supai Formation comprises about 30% of the volume of Paleozoic metasedimentary rocks preserved in the Big Maria Mountains. The metamorphosed Supai Formation contains the assemblage wollastonite + grossular + diopside + vesuvianite + sphene ± calcite ± quartz ± K-feldspar throughout the range except at the extreme southeastern edge, where assemblages lack wollastonite. The isograd that separates wollastonite-bearing from wollastonite-absent regions (Fig. 21-10b) may be interpreted as representing a temperature slightly lower than that of the isograd separating the Lower Talc Zone from the Upper Talc Zone.

The volume of fluid relative to the volume of rock necessary to explain the petrogenesis of wollastonite in the metamorphosed Supai Formation can be determined using the method of Rice and Ferry (1982, p. 291). Mineral reactions which buffer the composition of a H_2O-CO_2 fluid will progress to an extent which is dependent on the composition and quantity of infiltrating fluids. At a specific pressure and temperature, a unique fluid composition is in equilibrium with both reactants and products. When the composition of the fluid introduced differs from the equilibrium composition, the reaction will proceed if volatiles liberated by it modify the fluid composition toward the equilibrium composition. If infiltrating fluids continuously impose disequilibrium fluid compositions, reactions will progress until either infiltration stops or at least one of the reactants is spent. The method of Rice and Ferry (1982) permits calculation of the amount of fluid needed to explain the observed reaction progress as judged by the abundances of product phases. It assumes that during infiltration, the fluid composition was spontaneously buffered to that of the equilibrium composition, and that all fluid which flowed through the rock interacted chemically. Errors in these assumptions will cause the fluid/rock ratio to be underestimated.

Needed for the calculation are the equilibrium fluid composition and its molar volume, the composition and molar volume of the fluid introduced, and modal abundances of the final rock. The P-T-X relations of the wollastonite-forming reaction (quartz + calcite = wollastonite + CO_2) at conditions of 500°C at 3 kb indicate a fluid composition of about $X_{H_2O} = 0.97$. The presence of vesuvianite confirms the very water-rich fluid composition (Valley et al., 1985). Assuming these P-T-X conditions for the reaction, a composition of pure water for the infiltrating fluid, and a final rock consisting of 90% wollastonite, a volume of fluid 17 times the volume of the final rock is required (data of Burnham et al., 1969, were used for fluid molar volumes; the molar volumes of fluid with composition $X_{H_2O} = 0.97$ is assumed to be that of pure water). If the fluid introduced contained even a slight CO_2 component, much higher fluid/rock ratios would be calculated. Thus, the ratio of 17:1 is a minimum. The calculation is rather insensitive to uncertainties in estimated temperature.

The fluid compositions indicated by mineral assemblages in siliceous dolomites within the Paleozoic section are far less water-rich than recorded in the Supai Formation, and fluid/rock ratios required to explain their petrogenesis are more than an order of magnitude less. A similar disparity of fluid evolution between these two different rock types has also been observed in the Notch Peak contact metamorphic aureole, Utah (Hover-Granath et al., 1983; Labotka et al., 1985A; Nabelek et al., 1984). There, studies have suggested that fluid flow was channelized within the wollastonite-rich unit. In the

Big Maria Mountains, channelization may have occurred locally, but discontinuities of units and complex structures make it improbable that fluids could have reached the Supai Formation without first passing through siliceous dolomites. Locally developed quartz veins within the siliceous dolomites may represent channels through which water passed without interacting chemically. Walther and Orville (1982) have also interpreted quartz veins as channels which accommodate the escape of volatiles during metamorphism. No such veins are observed in wollastonite-bearing Supai Formation; the high degree of volume reduction (about 33%) during the wollastonite-forming reaction may have enhanced permeability to the extent that fluid flux could be accommodated without fracturing.

AGE OF MESOZOIC METAMORPHISM

Old Woman–Piute Range

In the Old Woman–Piute Range, the presence of widespread metamorphosed Cambrian through lower Mesozoic strata intruded by unmetamorphosed Late Cretaceous plutons indicates a Mesozoic timing for the metamorphism. C. F. Miller *et al.* (1982) suggested that this metamorphism took place primarily during the Late Cretaceous, with peak temperatures occurring for the most part before batholith emplacement. This interpretation was based on the following observations:

1. Late Cretaceous plutonic rocks are predominantly unmetamorphosed and cut foliation and major structures, indicating that, in general, they postdate fabric development.
2. Plutons lack chilled margins and generally impose no distinct metamorphic aureole on their surroundings, suggesting that country rock was hot at the time of emplacement.
3. Pressure at the time of granite emplacement, estimated by aplite compositions in the system Qz-Ab-Or-An (Hay, 1981) was roughly 4 kb, about the same as that estimated for peak metamorphic conditions (Hoisch, 1982; C. F. Miller *et al.*, 1982). Note, however, that interpretations of metamorphic pressures in the present study differ from the earlier interpretations.
4. Plutons appear to intrude isograds. Only in a few areas was granite emplacement contemporaneous with or earlier than the peak of metamorphism and deformation.

Big Maria Mountains

Field evidence and previous isotopic dating constrain the age of metamorphism in the Big Maria Mountains to the Middle Jurassic to Late Cretaceous interval. The metamorphism fully affected plutonic rocks dated by U-Pb methods at about 160 m.y. (dates by L. T. Silver, cited in Hamilton, 1982, 1984) and also affected Lower Jurassic sedimentary rocks of the cratonal assemblage. The upper age constraint is provided by late-metamorphic pegmatite dikes, loosely dated by K-Ar methods as Late Cretaceous in age (Martin *et al.*, 1982).

^{40}Ar/^{39}Ar Thermochronology

The lines of evidence above do not prove conclusively a Late Cretaceous timing for metamorphism in the Old Woman–Piute Range or Big Maria Mountains. Regional metamorphism of Jurassic age has been documented in areas about 240 km northwest of the Old Woman Mountains (Dewitt *et al.*, 1984; Labotka *et al.*, 1985b). In nearby western Arizona, based a Late Cretaceous age for regional metamorphism has been suggested on K-Ar dating (Rehrig and Reynolds, 1980).

As part of a study aimed at elucidating the thermal history of parts of southeastern California, we have initiated a program of $^{40}Ar/^{39}Ar$ dating of samples from the Big Maria Mountains and Old Woman–Piute Range (Fig. 21-13). The preliminary data strongly suggest high temperatures and subsequent rapid cooling near the end of the Cretaceous. In both ranges, minerals analyzed (hornblende, micas, K-feldspar), regardless of blocking temperature, record cooling through argon-retention intervals in latest Cretaceous or earliest Tertiary time. None clearly records earlier events, although a few spectra (M. T. Heizler and T. M. Harrison, unpublished data) suggest complex thermal histories. Rapid cooling in the Old Woman–Piute area was probably related to very rapid uplift and unroofing (*cf.* Knoll *et al.*, 1985).

Our current interpretation of data acquired to date is that a major thermal episode in these areas occurred during Late Cretaceous time. We also infer, based on both fabric development and the high degree of ductility, that deformation was synmetamorphic, probably occurring at least in part near the metamorphic peak. Possible earlier Mesozoic metamorphic and deformational events may not be discernible through the strong Late Cretaceous effects.

CRUSTAL ANATEXIS: A LINK TO METAMORPHISM?

We inferred earlier that the Late Cretaceous plutons intruded the Old Woman–Piute Range shortly after the thermal peak of metamorphism. Plutons were emplaced throughout the range, but are most abundant in the higher-grade parts. Two distinct magma series were involved, one strongly peraluminous and granitic, the other primarily metaluminous and granodioritic. Their isotopic compositions (Sr, Nd, Pb, O) indicate that both were largely or entirely produced by crustal anatexis (C. F. Miller *et al.*, 1984). The initial $^{87}Sr/^{86}Sr$ ratio and ϵ_{Nd} of peraluminous granites were about 0.718 and −16, respectively, and of metaluminous granodiorites about 0.710 and −12. We interpret the granitic magmas to have been derived from intermediate to felsic crust, and the granodioritic magmas from more mafic crustal material. Source regions of both were variably affected by pre-1.0-b.y.-old high-grade metamorphism, as indicated by variable but typically low feldspar-lead isotope ratios ($^{206}Pb/^{204}Pb$ as low as 17.2).

An anatectic event of this magnitude must be a result of a major change in crustal conditions. Tectonic loading, implied by the development of nappes, is a possible mechanism for inducing crustal anatexis in a water-rich environment, but the chronology of events in the Old Woman–Piute Range makes this unworkable. According to our present interpretation, only a short period of time separates nappe development from granite intrusion — much less than needed to induce anatexis from tectonic loading. Alternatively, rapid unloading may induce generation of water-undersaturated melts. Rapid cooling of the Old Woman–Piute Range at the end of the Cretaceous implies rapid unloading, but magmas in the Old Woman–Piute Range intruded at about the time cooling began. Magmas generated by uplift at an earlier stage, however, are not precluded. Underplating of the crust by mafic magmas (e.g., Haxel *et al.*, 1984; C. F. Miller, 1982) or erosion of the lithosphere by hot rising asthenosphere (Farmer and De Paolo, 1983) may act to enhance the supply of heat to the middle crust, promoting anatexis and metamorphism. These may have played roles in metamorphism and plutonism in southeastern California.

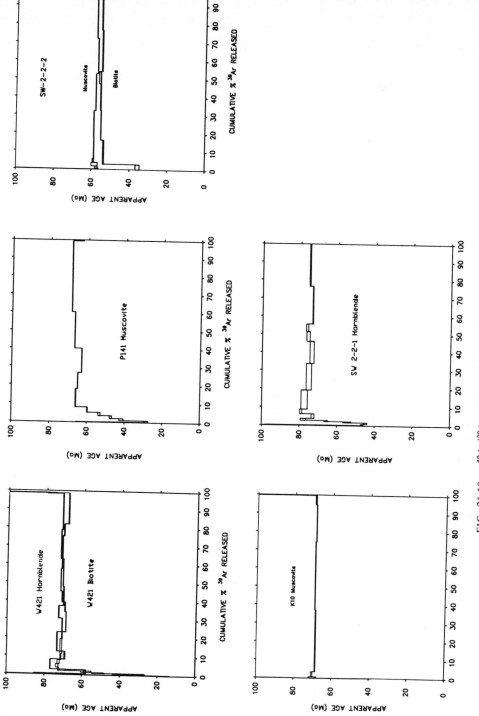

FIG. 21-13. $^{40}Ar/^{39}Ar$ spectra. Samples W421 (Proterozoic? gneissic tona-lite) and P141 (Proterozoic gneissic granite) are from the Old Woman–Piute Range (Fig. 21-2); K10 (metamorphosed Bright Angel Shale) is from the Kilbeck Hills (Fig. 21-1); SW2-2-1 (Jurassic metadiorite) and SW2-2-2 (lower Mesozoic metapelite) are from the Big Maria Mountains (Fig. 21-10b).

Neither heat conducted from plutons nor temperatures associated with steady-state geotherms are adequate to explain the observed metamorphic grades in the Big Maria Mountains and Old Woman–Piute Range. In the Big Maria Mountains, no Late Cretaceous plutons are exposed; pegmatite dikes in the area clearly postdate the peak of metamorphism. Using the stable craton geotherm of Lachenbruch and Sass (1977), a pressure of 3.0 kb corresponds to a temperature of 200°C, which is some 230°C lower than the lowest determined temperature for metamorphism in the Big Maria Mountains. Thermal models which consider the effects of crustal thickening followed by uplift (England and Thompson, 1984) produce maximum temperatures higher than the prethickening steady-state geotherm, but still inadequate to explain these conditions. These models are also inadequate in that maximum temperatures are achieved after crustal thickening — in both the Big Maria Mountains and the Old Woman–Piute Range, maximum temperatures were probably achieved during crustal thickening.

In the Piute Mountains, metamorphism is both shallower and cooler than in the Old Woman Mountains. In the western Piute Mountains, the Lazy Daisy pluton was emplaced adjacent to rocks undergoing regional metamorphism at ANZ I grade. Heat from the pluton caused a temperature increase of about 140°C at the contact, as judged from garnet zoning (garnet 27-1, Table 21-1). Textures in metamorphosed Bright Angel Shale show that the contact effect is superimposed upon regional fabrics formed earlier. At depths corresponding to a pressure of 2.5 kb, the steady-state geotherm for the stable craton indicates a temperature of 180°C, which is some 260°C too low to explain the preintrusion temperature.

Hoisch (1982) and C. F. Miller *et al.* (1982) recognized an association between the distribution of plutons and metamorphic grade in the Old Woman–Piute Range. This suggests a causal relationship between plutons and metamorphism, but geochronology and field relations suggest that the plutons were not responsible for the observed pattern of temperature distribution. They intruded shortly after the peak of metamorphism (based on geochronology), they appear to have intruded isograds, and they imposed little or no contact aureoles on their surroundings. The association between metamorphic grade and the distribution of plutons may result from the vertical distribution of plutons, in that pluton emplacement concentrated at the level of the core of the Old Woman Mountains and both metamorphic grade and granite volume generally diminish upward. It is possible that heat derived from ascending magmas which were emplaced above the current level of exposure contributed to the overall pattern of temperature distribution, but there is no specific evidence for this.

The introduction of heat by hot infiltrating fluids most easily explains the observed metamorphic grades. Huge volumes of wollastonite-vesuvianite rocks in the Big Maria Mountains document a major event of fluid infiltration during metamorphism. Large-scale fluid infiltration during regional metamorphism is also indicated in the Old Woman Mountains by the local development of massive wollastonite within the Supai Formation.

If hot infiltrating fluids were important in controlling the thermal structure of the area, then channelization should have resulted in lateral thermal gradients. A strong lateral thermal gradient is observed in the central Old Woman Mountains, where metamorphic grade drops northwestward from the Sillimanite-K feldspar Zone to the Kyanite Zone over a distance of about 3 km. The isograd defining the Sillimanite-K feldspar Zone closely parallels the main thrust fault in the region. Fluids channelized along the thrust fault could possibly explain why the grade of metamorphism was highest there, as well as explaining the high lateral thermal gradient away from it. This cannot be explained by heat derived from granitic magmas (see the discussion above).

Generalized maps of southeastern California showing the distributions of Paleozoic cratonal sedimentary rocks and Late Cretaceous plutons, nappes, and metamorphic grade are shown in Fig. 21-14. Mineral assemblages which developed during regional metamorphism in the Piute Mountains, Big Maria Mountains, and Old Woman Mountains require post-Late Cretaceous uplift of 9 km, 9–12 km, and >12 km, respectively. This is probably typical of other similar nearby areas such as the Little Maria and Palen Mountains. Within these areas, the cratonal strata were carried to mid-crustal depths by the development of thrusts and nappes. In uplifted areas where such structures are lacking, the cratonal strata were removed by erosion. This explains their absence in the Turtle, Sacramento, Chemehuevi, and Whipple Mountains. Post-Late Cretaceous uplift in most of these areas is indicated by exposed mid-crustal Late Cretaceous plutons emplaced within Precambrian basement.

In the Turtle Mountains, K-Ar dates from Precambrian basement rocks (Howard *et al.*, 1982) were only partially reset by heat from Late Cretaceous plutons or other sources. This is also true of the upper-plate Precambrian crystalline rocks of the Whipple Mountains (Anderson and Frost, 1981). In these areas, metamorphism associated with the resetting, as indicated by textural or mineralogic changes, is not observed. Evidently, the large-scale fluid flux associated with metamorphism in the Big Maria Mountains and Old Woman–Piute Range did not affect these areas at the depths presently exposed.

The distribution of features shown in Fig. 21-14 and the findings of the present study can be explained by a simple scenario which begins with the localized upward migration of hot infiltrating fluids. Such fluids, rising rapidly through the crust, should cause melting near the source, where conditions are hottest. Above the zone of melting, the enhanced supplies of heat and volatiles facilitated metamorphic reactions and textural equilibration. The hydrous melts which formed, segregated and then intruded the overlying area while still hot.

This proposed sequence of events is strongly dependent on the rates of magma generation, segregation, and ascent, and the rate at which hot infiltrating fluids ascended and caused metamorphism. It seems reasonable to expect that of these processes, fluid ascent and associated metamorphism would be fastest. Anatexis should have been coincident in timing with metamorphism, and segregation and ascent of magmas should have logically followed. In southeastern California, warping of the ancient craton leading to nappe development may have been localized along zones weakened by infiltrating fluids, ultimately subjecting the cratonal assemblage to mid-crustal conditions.

The origin of a large-scale volatile flux and possible reasons for localization can only be speculated upon at present. Large-scale volatile production may originate from a dehydrating subducting slab (Fyfe and Kerrick, 1985), from a reduction in porosity of tectonically buried rocks (Fyfe and Kerrich, 1985), or from the crystallization of underplating hydrous magmas. Localization could be attributed to a source which is local in nature, or to channeling by deep penetrative structures or anisotropies within the crust.

ACKNOWLEDGMENTS

We would like to express our appreciation to Warren Hamilton and Keith Howard for insightful discussions on southeastern California geology. Thanks are owed to them and to W. G. Ernst for helpful reviews of this paper. Support for this project came from the National Science Foundation (grants EAR78-23694 to Calvin F. Miller and EAR81-

HOISCH ET AL.

120880 to J. Lawford Anderson), the U.S. Geological Survey (contract 137877-82, and other support), the Geological Society of America, Sigma Xi, the Vanderbilt University Research Council, and the U.S.C. Geological Sciences Graduate Student Research Fund. This project was completed while TDH held a National Research Council — USGS Research Associateship.

(a)

FIG. 21-14. Southeastern California maps of Late Cretaceous regional metamorphism, unmetamorphosed Late Cretaceous plutons, nappes, and Paleozoic cratonal strata. Sources include Bishop (1963), Stone *et al.* (1983), Stone and Howard (1979), Pelka (1973), Shklanka (1963), Hoisch (1982, 1985, unpublished data), C. F. Miller *et al.* (1982), G. A. Davis *et al.* (1980), Howard *et al.* (1982), John (1981, 1982), and McClelland (1982), W. Hamilton (personal communication, 1985), K. A. Howard (personal communication, 1986), and J. L. Anderson (personal communication, 1985). (a) Map showing distribution of major overturned folds (nappes) and Paleozoic cratonal strata (black). (b) Map showing approximate Late Cretaceous metamorphic grade.

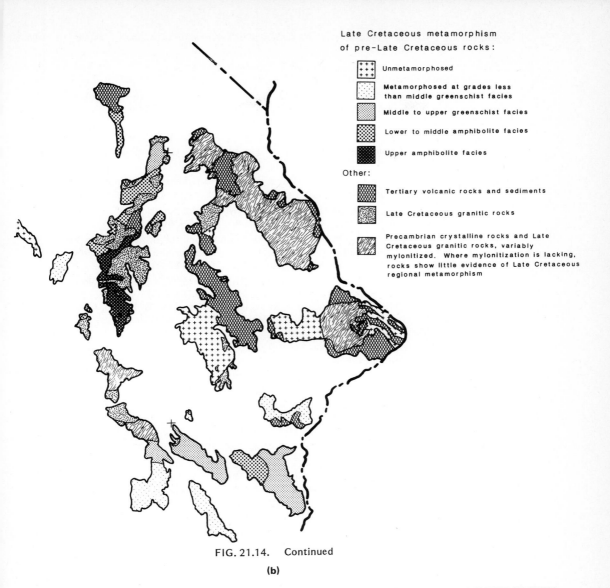

Late Cretaceous metamorphism
of pre-Late Cretaceous rocks:

- Unmetamorphosed
- Metamorphosed at grades less than middle greenschist facies
- Middle to upper greenschist facies
- Lower to middle amphibolite facies
- Upper amphibolite facies

Other:

- Tertiary volcanic rocks and sediments
- Late Cretaceous granitic rocks
- Precambrian crystalline rocks and Late Cretaceous granitic rocks, variably mylonitized. Where mylonitization is lacking, rocks show little evidence of Late Cretaceous regional metamorphism

FIG. 21.14.   Continued

(b)

REFERENCES

Anderson, J. L., and Frost, E. J., 1981, Petrologic, geochronologic, and structural evaluation of the allochthonous crystalline terrane in the Copper Basin area, eastern Whipple Mountains, California: *Univ. Southern California Final Tech. Rpt. U.S. Geol. Survey Contract* (unpublished), 63 p.

____ , and Rowley, M. C., 1981, Synkinematic intrusion of peraluminous and associated metaluminous granitic magmas, Whipple Mountains, California: *Can. Mineralogist*, v. 19, p. 83–101.

Bachinski, S. W., and Muller, G., 1971, Experimental determinations of the microline-low albite solvus: *J. Petrology*, v. 12, p. 329–256.

Bishop, C. C., 1963, *Geologic map of California, Needles sheet*; scale 1:250,000: Calif. Div. Mines Geology.

Bowman, J. R., and Essene, E. J., 1982, P-T-X(CO_2) conditions of contact metamorphism in the Black Butte aureole, Elkhorn, Montana: *Amer. J. Sci.*, v. 282, p. 311–340.

Burchfiel, B. C., and Davis, G. A., 1972, Structural framework and evolution of the southern part of the Cordilleran orogen, western United States: *Amer. J. Sci.*, v. 272, p. 97–118.

_____ , and Davis, G. A., 1981, Mojave Desert and environs, *in* Ernst, W. G., ed., *The Geotectonic Development of California*, (Rubey Vol. I): Englewood Cliffs, N.J. Prentice-Hall, p. 217–252.

Burnham, C. W., Holloway, J. R., and Davis, N. F., 1969, Thermodynamic properties of water to 1000°C and 10,000 bars: *Geol. Soc. America Spec. Paper 132*, 96 p.

Carpenter, M. A., 1981, A "conditional spinodal" within the peristerite miscibility gap of plagioclase feldspars: *Amer. Mineralogist*, v. 66, p. 553–560.

Crawford, M. L., 1966, Composition of plagioclase and associated minerals from Vermont, U.S.A., and south Westland, New Zealand, with inferences about the peristerite solvus: *Contrib. Mineralogy Petrology*, v. 13, p. 269–294.

Davis, G. A., Anderson, J. L., Frost, E. G., and Shackelford, T. J., 1980, Mylonitization and detachment faulting in the Whipple-Buckskin-Rawhide Mountains terrane, southeastern California and western Arizona, *in* Crittenden, M. D., Jr., Coney, P. J., and Davis, G. H., eds., *Cordilleran Metamorphic Core Complexes:* Geol. Soc. America Mem. 153, p. 79–129.

Dewitt, E., Armstrong, R. L., Sutter, J. F., and Zartman, R. E., 1984, U-Th-Pb, Rb-Sr, and Ar-Ar mineral and whole-rock isotopic systematics in a metamorphosed granitic terrane, southeastern California: *Geol. Soc. America Bull.*, v. 95, p. 723–739.

Eggert, R. G., and Kerrick, D. M., 1981, Metamorphic equilibria in the siliceous dolomite system — 6 kbar experimental data and geologic implications: *Geochim. Cosmochim. Acta*, v. 45, p. 1039–1049.

England, P. C., and Thompson, A. B., 1984, Pressure-temperature-time paths of regional metamorphism — I. Heat transfer during the evolution of thickened continental crust: *J. Petrology*, v. 25, p. 894–928.

Essene, E. J., 1983, Solid solutions and solvi among metamorphic carbonates with applications to geologic thermobarometry, *in* Reeder, R. J., *Carbonates — Mineralogy and Chemistry:* Min. Soc. America Rev. Mineralogy, v. 11, p. 77–96.

Farmer, G. L., and De Paolo, D.J., 1983, Origin of Mesozoic and Tertiary granite in the western United States and implications for pre-Mesozoic crustal structure — I. Nd and Sr. isotopic studies in the geocline of the northern Great Basin: *J. Geophys. Res.*, v. 88, p. 3379--3401.

Ferry, J. M., and Spear, F. S., 1978, Experimental calibration of the partitioning of Fe and Mg between biotite and garnet: *Contrib. Mineralogy Petrology*, v. 66, p. 113–117.

Foster, C. T., Jr., 1983, Thermodynamic models of biotite pseudomorphs after staurolite: *Amer. Mineralogist*, v. 68, p. 389–397.

Fyfe, W. S., and Kerrich, R., 1985, Fluids and thrusting: *Chem. Geology*, v. 49, p. 353–362.

Ganguly, J., and Saxena, S. K., 1984, Mixing properties of aluminosilicate garnets: constraints from natural and experimental data, and applications to geothermobarometry: *Amer. Mineralogist*, v. 69, p. 88–97.

Ghiorso, M. S., 1984, Activity/composition relations in the ternary feldspars: *Contrib. Mineralogy Petrology*, v. 87, p. 282–296.

Goldman, D. J., and Albee, A. L., 1977, Correlation of Mg-Fe partitioning between garnet and biotite with $^{18}O/^{16}O$ partitioning between quartz and magnetite: *Amer. J. Sci.*, v. 277, p. 750–761.

Goldsmith, J. R., and Newton, R. C., 1969, *P-T-X* relations in the system $CaCO_3$-$MgCO_3$ at high temperatures and pressures: *Amer. J. Sci.*, v. 267-A, p. 160–190.

Greenwood, H. J., 1975, Buffering of pore fluids by metamorphic reactions: *Amer. J. Sci.*, v. 275, p. 573–594.

Guidotti, C. V., 1968, Prograde muscovite pseudomorphs after staurolite in the Rangely-Oquossoc area, Maine: *Amer. Mineralogist*, v. 53, p. 1368–1376.

———, Transition from the staurolite to sillimanite zone, Rangely quadrangle, Maine: *Geol. Soc. America Bull.*, v. 85, p. 475–490.

Hamilton, W., 1982, Structural evolution of the Big Maria Mountains, northwestern Riverside County, southeastern California, *in* Frost, E. G., and Martin, D. L., eds., *Mesozoic-Cenozoic Tectonic Evolution of the Colorado River Region, California, Arizona, and Nevada* (Anderson-Hamilton volume): San Diego, Calif., Cordilleran Publishers, p. 1–28.

———, 1984, Generalized geologic map of the Big Maria Mountains region, northeastern Riverside County, southeastern California: *U.S. Geol. Survey Open File Rpt. 84-407*.

Haselton, H. T., Jr., Hovis, G. L., Hemingway, B. S., and Robie, R. A., 1983, Calorimetric investigation of the excess entropy of mixing in analbite-sanidine solid solutions — Lack of evidence for K-Na short-range order and implications for two-feldspar thermometry: *Amer. Mineralogist*, v. 68, p. 398–413.

Haxel, G. B., Tosdal, R. M., May, D. J., and Wright, J. E., 1984, Latest Cretaceous and early Tertiary orogenesis in south-central Arizona — Thrust faulting, regional metamorphism, and granitic plutonism: *Geol. Soc. America Bull.*, v. 95, p. 631–653.

Hay, D. E., 1981, Structural, plutonic, and metamorphic history of Paramount Wash., eastern Mojave Desert: Unpublished M.S. thesis, Vanderbilt Univ., Nashville, Tenn., 194 p.

Hess, P. C., 1973, Figure 3-9, *in* Miyashiro, A., *Metamorphism and Metamorphic Belts:* London, Allen & Unwin, p. 81.

Hoisch, T. D., 1982, The metamorphism of Bright Angel Shale, Old Woman-Piute Range, Southeastern California: Unpublished M.S. thesis, Vanderbilt Univ., Nashville, Tenn., 109 p.

———, 1985, Metamorphism in the Big Maria Mountains, southeastern California: Unpublished Ph.D. thesis, Univ. Southern California, Los Angeles, Calif., 264 p.

Holdaway, M. J., 1971, Stability of andalusite and the aluminum silicate phase diagram: *Amer. J. Sci.*, v. 271, p. 97–131.

———, and Lee, S. M., 1977, Fe-Mg cordierite stability in high grade pelitic rocks based on experimental, theoretical, and natural observations: *Contrib. Mineralogy Petrology*, v. 63, p. 175–198.

Hover-Granath, V. C., Papike, J. J., and Labotka, T. C., 1983, The Notch Peak contact metamorphic aureole, Utah — Petrology of the Big Horse Limestone member of the Orr Formation: *Geol. Soc. America Bull.*, v. 94, p. 889–906.

Howard, K. A., Miller, C. F., and Stone, P., 1980, Mesozoic thrusting in the eastern Mojave Desert, California [abstract]: *Geol. Soc. America Abstr. with Programs*, v. 12, p. 112.

———, Stone, P., Pernokas, M. A., and Marvin, R. F., 1982, Geologic and geochronologic reconnaissance of the Turtle Mountains area, California — West border of the Whipple Mountains detachment terrane, *in* Frost, E. G., and Martin, D. L., eds.,

Mesozoic-Cenozoic Tectonic Evolution of the Colorado River Region, California, Arizona, and Nevada (Anderson-Hamilton volume): San Diego, Calif., Cordilleran Publishers, p. 341–355.

Hutcheon, I., and Moore, J. M., 1973, The tremolite isograd near Marble Lake, Ontario: *Can. J. Earth Sci.*, v. 10., 936–947.

John, B. E., 1981, Reconnaissance study of Mesozoic plutonic rocks in the Mojave Desert Region, *in* Howard, K. A., Carr, M. D., and Miller, D. M., eds., *Tectonic Framework of the Mojave and Sonoran Deserts, California and Arizona:* U.S. Geol. Survey Open File Rpt. 81-503, p. 48–50.

——, 1982, Geologic framework of the Chemehuevi Mountains, southeastern California, *in* Frost, E. G., and Martin, D. L., eds., *Mesozoic-Cenozoic Tectonic Evolution of the Colorado River Region, California, Arizona, and Nevada* (Anderson-Hamilton volume): San Diego, Calif., Cordilleran Publishers, p. 317–325.

Knoll, M. A., Harrison, T. M., Miller, C. F., Howard, K. A., Duddy, I. R., and Miller, D. S., 1985, Pre-Peach Springs tuff (18 m.y.) unroofing of the Old Woman Mountains, crystalline complex, southeastern California – Implications for Tertiary extensional tectonics: *Geol. Soc. America Abstr. with Programs*, v. 17, p. 365.

Labotka, T. C., 1980, Petrology of a medium pressure metamorphic terrane, Funeral Mountains, California: *Amer. Mineralogist*, v. 65, p. 670–689.

——, 1981, Petrology of an andalusite-type regional metamorphic terrane, Panamint Mountains, California: *J. Petrology*, v. 22, p. 261–296.

——, Papike, J. J., and Nabelek, P. I., 1985a, Fluid evolution in the Notch Peak aureole [abstract]: *Trans. Am. Geophys. Un.*, v. 66, p. 389.

——, Warasila, R. L., and Spangler, R. R., 1985b, Polymetamorphism in the Panamint Mountains, California – ^{39}Ar-^{40}Ar study: *J. Geophys. Res.*, v. 90, p. 10359–10372.

Lachenbruch, A. H., and Sass, J. H., 1977, Heat flow in the United States and the thermal regime of the crust, *in* Heacock, J. G., ed., *The Nature and Physical Properties of the Earth's Crust:* Geophys. Monogr. Ser., v. 20, Washington, D.C., Amer. Geophy. Un., p. 626–675.

Laves, F., 1954, The coexistance of two plagioclases in the oligiclase composition range: *J. Geology*, v. 62, p. 409–411.

Martin, D. L., Krummenacher, D., and Frost, E. G., 1982, K-Ar geochronologic record of Mesozoic and Tertiary tectonics of the Big Maria-Little Maria-Riversides Mountains terrane, *in* Frost, E. G., and Martin, D. L., eds., *Mesozoic-Cenozoic Tectonic Evolution of the Colorado River Region, California, Arizona, and Nevada* (Anderson-Hamilton volume): San Diego, Calif., Cordilleran Publishers, p. 518–550.

Maruyama, S., Liou, J. G., and Susuki, K., 1982, The peristerite gap in low-grade metamorphic rocks: *Contrib. Mineralogy Petrology*, v. 81, p. 268–276.

McLelland, W. C., 1982, Structural geology of the central Sacramento Mountains, San Bernardino County, California, *in* Frost, E. G., and Martin, D. L., eds., *Mesozoic-Cenozoic Tectonic Evolution of the Colorado River Region, California, Arizona, and Nevada* (Anderson-Hamilton volume): San Diego, Calif., Cordilleran Publishers, p. 401–406.

Metz, P., 1977, Temperature-pressure, and H_2O-CO_2 gas composition during metamorphism of siliceous dolomitic limestones deduced from experimentally determined equilibria of forsterite-forming reactions: *Tectonophysics*, v. 43, p. 163–167.

——, 1983, Experimental investigation of the stability conditions of petrologically significant calc-silicate assemblages observed in the Damara Orogen, *in* Martin H., and Eder F. W., eds., *Intracontinental Fold Belts:* Berlin, Springer-Verlag, p. 785–793.

Miller, C. F., 1982, Cordilleran tectonics and crustal anatexis [abstract]: *Geol. Soc. America Abstr. with Programs*, v. 14, p. 216.

——, Stoddard, E. F., Bradfish, L. J., and Dollase, W. A., 1981, Composition of plutonic muscovite: genetic implications: *Can. Mineralogist*, v. 19, p. 25–34.

——, Howard, K. A., and Hoisch, T. D., 1982, Mesozoic thrusting, metamorphism, and plutonism, Old Woman–Piute Range, southeastern California, *in* Frost, E. G., and Martin, D. L., eds., *Mesozoic-Cenozoic Tectonic Evolution of the Colorado River Region, California, Arizona, and Nevada* (Anderson-Hamilton volume): San Diego, Calif., Cordilleran Publishers, p. 561–581.

——, Bennett, V., Wooden, J. L., Solomon, G. C., Wright, J. E., and Hurst, R. W., 1984, Origin of the composite metaluminous/peraluminous Old Woman-Piute batholith, S.E. California — Isotopic constraints [abstract]: *Geol. Soc. America Abstr. with Programs*, v. 16, p. 596.

Nabelek, P. I., Labotka, T. C., O'Neil, J. R., and Papike, J. J., 1984, Contrasting fluid/rock interaction between the Notch Peak granitic intrusion and argillites and limestones in western Utah — Evidence from stable isotopes and phase assemblages: *Contrib. Mineralogy Petrology*, v. 86, p. 25–34.

Nesbitt, B. E., and Essene, E. J., 1982, Metamorphic thermometry and barometry of a portion of the southern Blue Ridge province: *Amer. J. Sci.*, v. 282, p. 701–729.

Nord, G. L., Jr., Hammarstrom, J. L., and Zen, E-An, 1978, Zoned plagioclase and peristerite formation in phyllites from southwestern Massachusetts: *Amer. Mineralogist*, v. 63, p. 947–955.

Orville, P. M., 1974, The peristerite gap as an equilibrium between ordered albite and disordered plagioclase solid solution: *Bull. Soc. Fr. Mineralogie Cristallographie*, v. 97, p. 386–392.

Pelka, G. J., 1973, Geology of the McCoy and Palen Mountains, southeastern California: Unpublished Ph.D. thesis, Univ. California, Santa Barbara, Calif., 162 p.

Perchuk, L. L., 1977, Thermodynamic control of geological processes, *in* Saxena, S. K., and Batacharji, eds., *Energetics of Geologic Processes:* New York, Springer-Verlag, p. 285–352.

Perkins, D., III, Essene, E. J., and Marcotty, L. A., 1982, Thermometry and barometry of some amphibolite-granulite facies rocks from the Otter Lake area, southern Quebec: *Can. J. Earth Sci.*, v. 19, p. 1759–1774.

Pigage, L. C., 1982, Linear regression analysis of sillimanite-forming reactions at Azure Lake, British Columbia: *Can. Mineralogist*, v. 20, p. 349–378.

Price, J. G., 1985, Ideal site mixing in solid solutions, with an application to two-feldspar geothermometry: *Amer. Mineralogist*, v. 70, p. 696–701.

Puhan, D. 1976, Metamorphic temperature determined by means of the dolomite-calcite solvus geothermometer — Examples from the central Damara orogen (southwest Africa): *Contrib. Mineralogy Petrology*, v. 58, p. 23–28.

Rehrig, W. A., and Reynolds, S. J., 1980, Geologic and geochronologic reconnaissance of a northwest-trending zone of metamorphic core complexes in southern and western Arizona, *in* Crittenden, M. D., Jr., Coney, P. J., and Davis, G. H., eds., *Cordilleran Metamorphic Core Complexes:* Geol. Soc. America Mem. 153, p. 131–151.

Ribbe, P. H., 1962, Observations on the nature of unmixing in peristerite plagioclases: *Norsk Geol. Tidsskr.*, v. 42, p. 138–151.

Rice, J. M., 1977, Progressive metamorphism of impure dolomitic limestone in the Marysville aureole, Montana: *Amer. J. Sci.*, v. 277, p. 1–24.

——, and Ferry, J. M., 1982, Buffering, infiltration, and the control of intensive variables during metamorphism, *in* Ferry, J. M., ed., *Characterization of Metamorphism through Mineral Equilibria:* Min. Soc. America, Rev. Mineralogy, v. 10, p. 263–326.

Shklanka, R., 1963, Repeated metamorphism and deformation of evaporite-bearing sediments, Little Maria Mountains, California: Unpublished Ph.D. thesis, Stanford Univ. Stanford, Calif., 156 p.

Skippen, G., 1974, An experimental model for low pressure metamorphism of siliceous dolomitic marble: *Amer. J. Sci.*, v. 274, p. 487–509.

Slaughter, J., Kerrick, D. M., and Wall, V. L., 1975, Experimental and thermodynamic study of equilibria in the system $CaO-MgO-SiO_2-H_2O-CO_2$: *Amer. J. Sci.*, v. 275, p. 143–162.

Smith, J. V., 1972, Critical review of synthesis and occurrence of plagioclase feldspars and a possible phase diagram: *J. Geology*, v. 80, p. 505–525.

——, 1983, Phase equilibria of plagioclase, *in* Ribbe, P.H., ed., *Feldspar Mineralogy*, 2nd Ed.: Min. Soc. America, Rev. Mineralogy v. 2, p. 223–240.

Spear, F. S., and Selverstone, J., 1983, Quantitative P-T paths from zoned minerals — Theory and tectonic applications: *Contrib. Mineralogy Petrology*, v. 83, p. 348–357.

——, Ferry, J. M., and Rumble, D., III, 1982, Analytical formulation of phase equilibria — the Gibbs' method, *in*, Ferry, J. M., ed., *Characterization of Metamorphism through Mineral Equilibria:* Min. Soc. America Rev. Mineralogy, v. 10, p. 105–152.

——, Selverstone, J., Hickmott, D., Crowley, P., and Hodges, K. V., 1984, P-T paths from garnet zoning — A new technique for deciphering tectonic processes in crystalline terranes: *Geology*, v. 12, p. 87–90.

Stone, P., and Howard, K. A., 1979, Compilation of geologic mapping in the Needles 1° by 2° sheet, California and Arizona: *U.S. Geol. Survey Open File Rpt. 79-388.*

——, Howard, K. A., and Hamilton, W., 1983, Correlation of metamorphosed Paleozoic strata of the southwestern Mojave Desert region, California and Arizona: *Geol. Soc. America Bull.*, v. 94, p. 1135–1147.

Stromer, J. C., Jr., 1975, A practical two-feldspar geothermometer: *Amer. Mineralogist*, v. 60, p. 667–674.

Thompson, A. B., 1976a, Mineral reactions in pelitic rocks — I. Prediction of P-T-X (Fe-Mg) phase relations: *Amer. J. Sci.*, v. 276, p. 401–424.

——, 1976b, Mineral reactions in pelitic rocks — II. Calculation of some P-T-X (Fe-Mg) phase relations: *Amer. J. Sci.*, v. 276, p. 401–424.

——, 1982, Dehydration partial melting of pelitic rocks and the generation of H20-undersaturated granitic liquids: *Amer. J. Sci.*, v. 282, p. 1567–1595.

Thompson, J. B., Jr., 1957, The graphical analysis of mineral assemblages in pelitic schists: *Amer. Mineralogist*, v. 42, p. 842–858.

——, 1961, Mineral facies in pelitic schists (in Russian with English summary), *in* Sokolov, G. A., ed., *Physico-chemical Problems of the Formation of Rocks and Ores:* Moscow, Akad. Nauk USSR, p. 313–325.

Trommsdorf, V., 1972, Change in T-X during metamorphism of siliceous dolomitic rocks of the central Alps: *Schweiz. Mineral. Petrog. Mitt.*, v. 52, p. 567–571.

Valley, J. W., Bowman, J. R., Peacor, D. R., Essene, E. J., and Allard, M. J., 1985, Crystal chemistry of a Mg-vesuvianite and implications of phase equilibria in the system $CaO-MgO-Ai_2O_3-SiO_2-H_2O-CO_2$: *J. Metamorphic Geology*, v. 3, p. 137–153.

Wada, H., and Suzuki, K., 1983, Carbon isotope thermometry calibrated by dolomite-calcite solvus temperatures: *Geochim. Cosmochim. Acta*, v. 47, p. 697–706.

Walther, J. V., and Orville, P. M., 1982, Volatile production and transport in regional metamorphism: *Contrib. Mineralogy Petrology*, v. 79, p. 252–257.

Whitney, J. A., and Stormer, J. R., 1977, The distribution of NaAlSi308 between coexisting microcline and plagioclase and its effect on geothermometric calculations: *Amer. Mineralogist*, v. 62, p. 687–691.

R. Speed, M. W. Elison, and F. R. Heck
Department of Geological Sciences
Northwestern University
Evanston, Illinois 60201

22

PHANEROZOIC TECTONIC EVOLUTION OF THE GREAT BASIN

ABSTRACT

The Phanerozoic tectonic evolution of western North America is more completely recorded in the Great Basin than in other reaches of the continental margin because the Great Basin has apparently not undergone substantial tectonic attrition in which tracts of sialic continent plus earlier accreted terranes have been removed. Between the Sierra Nevada and Utah in the Great Basin, rocks at the surface can be assigned to seven major tectono-stratigraphic units. The development of these units in their current form began with rifting and drifting during the formation of a passive margin in late Precambrian time. The Precambrian edge of the sialic continent so created probably persists to the present. A thick sedimentary cover was laid down through Paleozoic and Triassic times on the wide, subsiding shelf of the passive margin, and allochthons of oceanic strata overrode the outer shelf in Mississippian and Early Triassic times. Deep shelf cover was probably subjected to greenschist facies metamorphism, and below the allochthons, possibly blueschist facies during thrusting.

An active margin to North America began in the Great Basin in Middle or Late Triassic time with the development of a continental magmatic arc. The persistence of the arc in position and time suggests a constant configuration of downgoing slab and continent for about 150 m.y. The region between the Sierran arc and central Utah was the foreland of North America that underwent contraction in Jurassic and Cretaceous times. Plutonism in the foreland, however, was probably related to foreland deformation and not directly to a downgoing slab. Contraction was taken up in three parallel-surface thin-skinned thrust belts (Winnemucca, Eureka, Sevier), each governed in position and vergence by inherited crustal declivities. Between the eastern two belts, there was a large enclave capped by little-deformed cover strata that we interpret as a huge thrust sheet. Displacements, mainly horizontal shear, were transmitted in lower strata of the enclave between adjacent surface thrust belts. The only region of apparently substantial unroofing and exposure of deep-seated metamorphic rocks in Mesozoic time was the Toiyabe uplift zone of the central Great Basin. This zone may be a chain of pluton-cored domes of Late Jurassic and probably later uplift. Spreading off the western flank of the Edna-Osgood dome led to the extensive, west-verging Willow Creek allochthon. A model of thermal diapirism as a contractile mechanism is presented.

The Mesozoic Great Basin is distinctive among contracting forelands of the world in its greater width, lack of extensive unroofing, and absence of overriding on the ocean-ward side by basement sheets or hot nappes of deep-seated rocks. The differences may indicate that unlike other regions, Mesozoic western North America in or west of the Great Basin was never subject to collision or overriding by bouyant basement terranes of microcontinental or greater size.

INTRODUCTION

We address the sequence and nature of major tectonic events that caused deformation and metamorphism in the outer reaches of western North America in what is now the Great Basin. The duration considered is the Phanerozoic, which extends from late Precambrian, approximately 700 m.y.b.p., to the present. Our topical emphasis is Jurassic and Cretaceous deformation in the Great Basin, when it was a contracting foreland between a continental arc on the west and the craton in central Utah on the east.

The Great Basin province (Fig. 22-1) is our geographic focus because it may be the only region of western North America that preserves in place a full record of pre-Cenozoic continental margin tectonics. The Great Basin evidently escaped strong tectonic attrition

SPEED, ELISON, HECK

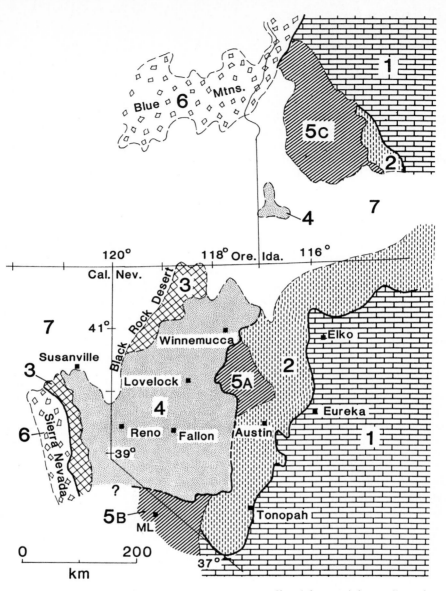

FIG. 22-1. Regional tectonostratigraphic units at the surface in the western Great Basin and surrounding regions. Heavy lines are tectonic boundaries, light lines, depositional. Figure 22-2 shows a stacking diagram and some unit subdivisions.

1) North America: Precambrian crust plus parautochthonous cover

2) displaced Paleozoic oceanic terranes overlying N. America; Early Triassic and older attachment

3) displaced Paleozoic terranes; probable Early Triassic attachment to North America

4) parautochthonous Mesozoic cover to and intrusions in terrane 3

5) mainly autochthonous Mesozoic cover to and intrusions in terranes 1 and 2

6) displaced terranes; Late Triassic or younger attachment

7) Quaternary to Upper Cretaceous cover to all other units

that elsewhere removed sizable tracts of sialic continent and early accreted terranes from the continental margin. The Great Basin, therefore, provides an almost unique glimpse into the past of marginal western North America.

PHANEROZOIC TECTONOSTRATIGRAPHY

Figures 22-1 and 22-2 identify generalized affiliations of outcrops in the western Great Basin and vicinity on the basis of allochthoneity and timing of emplacement. Such units are employed later in a model of the Phanerozoic tectonic evolution of the region. Different tectonostratigraphic organizations of this region are given by Silberling *et al.* (1984) and Oldow (1984a).

Unit 1 (Fig. 22-1) is Precambrian sialic North America and its sedimentary cover of upper Precambrian and higher strata that are as young as Triassic in the eastern Great Basin and Devonian in the western. Unit 1 has probably maintained near-coherency in Phanerozoic time, although it was deformed by probable major extension in late Precambrian, contraction in Mesozoic, and extension in Cenozoic time. The western 150 km or so of Precambrian sialic North America are overlain by displaced Paleozoic oceanic sedimentary and basaltic rocks of unit 2 (Fig. 22-2). Unit 2 includes the Roberts Mountains allochthon that was emplaced early in Mississippian time (Roberts *et al.*, 1958; J. F. Smith and Ketner, 1968) and the Golconda allochthon that was emplaced in Early Triassic time (Silberling, 1973; Speed, 1977). The existence of unit 1 below unit 2 is indicated by erosional windows through unit 2 that expose Paleozoic carbonates that almost certainly belong to unit 1.

Deformed Paleozoic rocks that were probably of arc and oceanic origin exist in the Black Rock Desert region and the northern Sierra Nevada (unit 3, Fig. 22-1). At both sites, they are overlain and intruded by Mesozoic rocks of unit 4. The affiliation of Paleozoic rocks at the two exposure areas of unit 3 and the subsurface extent of unit 3 are unknown. In one model (Speed, 1977, 1979), rocks of unit 3 constitute a microplate or group of microplates called Sonomia that is basement to, and approximately coextensive with, the Mesozoic cover in unit 4. Thus, unit 3 is probably greatly displaced from its sites of accumulation. It arrived in Early Triassic time together with the structurally discrete Golconda allochthon. Other models are given by Harwood (1983) and E. L. Miller *et al.* (1984).

Units 4 and 5 (Figs. 22-1 and 22-2) contain Mesozoic sedimentary and igneous

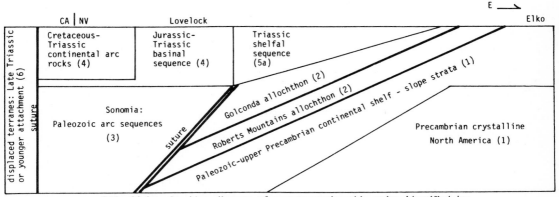

FIG. 22-2.   Stacking diagram of tectonostratigraphic units, identified by number as in Fig. 22-1, in a section approximately along 40°N between the Sierra Nevada and Elko, Nevada.

SPEED, ELISON, HECK

rocks that were deposited or intruded near their present sites. Unit 4 includes Triassic and Jurassic strata of mainly basinal but locally shelfal affiliation (Oldow, 1981; Speed, 1978b) and magmatic rocks, including the Sierran batholith, of probable arc affinity (Russell, 1984) and latest Triassic through Late Cretaceous ages. Scattered meager exposures of its base imply that unit 4 formed as cover to unit 3 after the attachment of unit 3 to North America. Unit 4, however, is now everywhere allochthonous with respect to North America on the Fencemaker and Luning thrusts, due to Mesozoic contractional tectonics at the edge of the North American foreland (Oldow, 1981, 1984a; Speed, 1978a). Displacements of unit 4 are probably small (<300 km) because some distinctive strata of unit 4 can be tied to continental sediment delivery systems preserved in autochthonous unit 5 that lies above North America (unit 1) (Lupe and Silberling, 1985; Silberling and Wallace, 1969; Speed, 1978a).

Unit 5 is composed of mainly autochthonous Mesozoic rocks that probably or certainly formed above or in North America (unit 1). Unit 5a contains chiefly shelfal strata (Nichols and Silberling, 1977) and rhyolite of Triassic ages that were deposited on unit 2 (Golconda allochthon), above subjacent but unexposed North America. Unit 5b consists of mainly, volcanogenic and intrusive Mesozoic rocks that formed on or in unit 2 and farther south, directly on or in North America. In Idaho, unit 5c consists mainly of the Late Cretaceous Idaho batholith. Although the Cretaceous phase of the Sierran and Idaho batholiths are probably continuous below Cenozoic cover, the two exposure regions are included in different tectonostratigraphic units (4 and 5c, respectively). This is because the wallrocks and strontium isotopic composition (Armstrong *et al.*, 1977; Kistler and Peterman, 1973, 1978) imply the Idaho batholith was generated from and emplaced in Precambrian sialic North America, whereas the Cretaceous northern Sierran batholith had an ensimatic source and was emplaced in Sonomia (unit 3) after Sonomia was attached to North America.

Unit 6 consists of displaced and suture-related terranes of Mesozoic melange or of arc affiliation that attached at their current position between Late Triassic and Late Cretaceous times (H. C. Brook and Vallier, 1977; Dickinson and Thayer, 1978; Saleeby, 1983; Saleeby and Sharp, 1980; Schweickert and Cowan, 1975; Schweickert *et al.*, 1984).

Unit 7 covers all the preceding units. It consists of sedimentary and volcanic rocks that are as old as Late Cretaceous in California and Oregon and Tertiary farther east.

Sections that follow give interpretations of the major events from older to younger in the Phanerozoic evolution of the Great Basin that created the tectonostratigraphy and structures that exist today (Figs. 22-1 and 22-2). Figure 22-3 outlines a timetable of events. In general, the Phanerozoic was governed first by passive, then active margin regimes. The switch occurred in Middle or Late Triassic time.

PASSIVE MARGIN FORMATION

The first event, beginning shortly before Cambrian time, created a passive margin within sialic North America (Burchfiel and Davis, 1972; Stewart, 1972, 1976). Such tectonics presumably caused the rifting and stretching of sialic crust west of a hinge with the cratonal platform in central Utah. This was followed by the drifting away of a part of the continent of unknown size (Fig. 22-3) and the growth of oceanic lithosphere progressively away from the new sialic edge. The rifted and stretched crust is nowhere exposed. It presumably includes a brittlely extended upper zone and ductilely stretched lower zone.

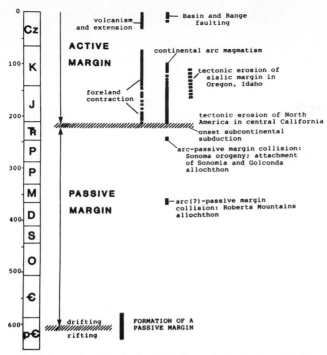

FIG. 22-3. Phanerozoic timetable for major tectonic events in the Great Basin and nearby regions; full lines indicate probable continuous activity; short dashes indicate possible activity or uncertain range of activity.

Passive Margin Subsidence and Sedimentation

Following late Precambrian rifting, a passive continental margin existed in the Great Basin until Middle or Late Triassic time. The thinned and probably diked sialic crust west of the cratonal hinge in central Utah became a subsiding shelf on which a great thickness of sediments accumulated as the cover within unit 1 (Figs. 22-1 and 22-2). In the eastern Great Basin, sedimentation was virtually continuous from late Precambrian to Early Triassic times, yielding a 10–12-km stratal thickness (Stewart and Suczek, 1977). Upper Precambrian to Middle Cambrian terrigenous clastic strata form the lower half of the section and shoalwater carbonate the upper half.

In the central Great Basin, cover thickness in unit 1 is less certain because of less Cenozoic structural relief and because shelf and slope sedimentation was interrupted by the emplacement onto the outer continental shelf of the oceanic allochthons of unit 2. The thickness of the upper Precambrian-Middle Cambrian clastic wedge, however, probably increased west to a maximum at the toe of slope and exceeded the 5 km of the eastern Great Basin. The succeeding lower Paleozoic carbonate strata in the central Great Basin include subtidal outershelf facies, and at the westernmost exposures (Fig. 22-4), local shoalwater, outer ridge, and continental slope facies (Kay and Crawford, 1964; Matti and McKee, 1977; Rowell *et al.*, 1979; Stewart and Poole, 1974).

Given at least a 12-km thickness of cover in unit 1 across the Great Basin and a normal continental temperature gradient of 30°C/km, the lower sixth of the cover sediments would have been in greenschist metamorphic facies ($\geqslant$250°C, $\geqslant$3 kb) by the end of the passive margin phase (Fig. 22-3).

Edge of Precambrian North America

Precambrian sialic crust, exposed at places in the eastern Great Basin (Stewart, 1980), presumably extends west in the subsurface below autochthonous cover (unit 1) and overriding allochthons (unit 2) (Fig. 22-2). The present locus of the buried edge of sialic crust can be approximated by the 0.706 contour of initial strontium isotopic ratios (Fig. 22-4) in plutons that are samples of the lithosphere in which they are emplaced, according to data and arguments of Kistler and Peterman (1973, 1978), and Armstrong *et al.* (1977). Location of sialic edge by strontium is generally corroborated by spatial transitions in neodymium isotope ratios and pluton mineralogy (Farmer and DePaolo, 1983; C. F. Miller and Bradfish, 1980). West of the 0.706 contour in the Great Basin, the lithosphere is more ensimatic and is thought to belong to a collided microplate (Sonomia, unit 3, Fig. 22-2).

The present sialic edge is approximately coincident with the westernmost limit of exposures of autochthonous or parautochthonous Paleozoic cover to the sialic crust (Fig. 22-4). Because such strata are interpreted as outer shelf, outer ridge, and slope facies, they probably represent deposits at or near the boundary between continental and oceanic crust at the late Precambrian passive margin. Two important points follow this relation. First, the age of the edge of sialic crust in the Great Basin may be late Precambrian and the edge may be that created during late Precambrian drifting. Later discussion will indi-

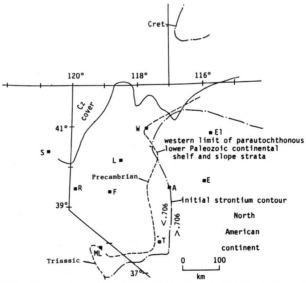

FIG. 22-4. Approximate position of present buried edge of Precambrian sialic continental North America defined by (1) initial strontium isotopic ratio of 0.706 in Mesozoic plutons from Kistler and Peterman (1978), and (2) western limit of outcrops of parautochthonous lower Paleozoic continental shelf and slope strata. Age symbols give probable dates of formation of present edge of sialic continent after Speed (1983).

cate that this possibility is unlikely elsewhere in western North America. Second, there has not been significant detachment and horizontal displacement of cover relative to basement at the western edge of Precambrian North America (unit 1). The second relation forms a constraint in reconstruction of thin-skinned foreland displacements during Mesozoic contraction.

Passive Margin Collisions

During the period in which North America had a passive margin in the Great Basin, there were two collisions of outboard terranes with the sialic edge. These resulted in the transport of allochthons of oceanic sediments and lava (unit 2, Fig. 22-2) over 100 km continentward of the sialic edge and the attachment of an ensimatic lithosphere (unit 3, Fig. 22-2) to the sialic continent at or near its Precambrian edge.

Discussion below and Fig. 22-5 illustrate a conceptual process of how passive margin collisions might have worked, assuming the colliding terranes are relatively far displaced arc systems (Brueckner and Snyder, 1985; Speed, 1977, 1979, 1983). Recent alternative models are given by Harwood (1983) and E. L. Miller et al. (1984). A basic difference in models is whether the strata in the allochthons were accreted into a forearc

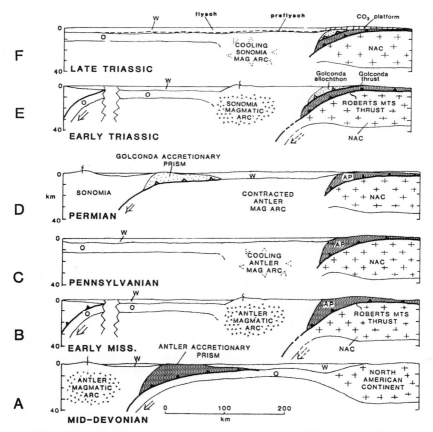

FIG. 22-5. Model of sequential arc-passive margin collisions in the Great Basin, emplacement of Roberts Mountains and Golconda allochthons as accretionary prisms; modified after Speed (1979) and Speed and Sleep (1982).

over a long reach of oceanic lithosphere fronting western North America before collision, or whether the allochthons (at least the Golconda, Fig. 22-2) are stratigraphic successions that accumulated at the toe of the continent in the Great Basin before disruption and tectonic transport. In Fig. 22-5D, Sonomia is shown as a Permian magmatic arc system migrating with closing and lateral components along coastal North America. The arc surmounted a subduction zone in which the downgoing slab was oceanic lithosphere attached to the passive margin of North America. Closure ceased when continental lithosphere started down below Sonomia, by which time the forearc, the Golconda allochthon, was almost fully emplaced on the outer continental shelf (Fig. 22-5E). If the oceanic slab attached to Paleozoic North America broke off and sank upon cessation of closure, a finite but probably minor width of sialic continental edge may have been taken into the mantle with it. Upon welding to North America, a new ocean-facing convergent zone probably developed somewhere west of Sonomia, and the Sonomian magmatic arc underwent thermal contraction and subsidence due to loss of subduction-related heating. The origin of the deep-water Triassic successor basin that is approximately coextensive with Sonomia is thus explained (Fig. 22-5F).

Because of the tectonic similarity of the emplacements of the Roberts Mountains and Golconda allochthons, a similar collisional process is envisioned for both allochthons. The Roberts Mountains allochthon is also regarded as an accretionary prism. (Dickinson, 1977; Speed, 1977; Speed and Sleep, 1982) that was underrun by the continental shelf before final collision between the continent and a migrating island-arc system (Figs. 22-A-C). The Antler magmatic arc, however, is a postulate because it is nowhere exposed. The Antler arc is assumed to have thermally contracted, like Sonomia, because upon collision, the subduction zone jumped west and the arc lost its heat source (Speed and Sleep, 1982). The vertically contracted Antler magmatic arc may have been completely subducted below Sonomia (Fig. 22-5D).

A major effect of the Mississippian collision was the generation of an asymmetric foreland basin with amplitude $\leqslant 3.5$ km that rimmed the continentward edge of the Roberts Mountains allochthon (Poole, 1974). The basin probably reflects elastic or elasticoviscous downflexing of strong continental lithosphere during vertical loading by the allochthon and broadening during progressive sedimentation which widened the region of loading (Speed and Sleep, 1982).

The principal effects of the accretion of Sonomia to the passive sialic margin were the addition of large girth to the morphologic continent and the generation of a deep-water basin west of the Sonomia suture and the edge of sialic North America. The Triassic basin was successor to and approximately coextensive with Sonomia and accumulated early pelagic and hemipelagic sediments followed by great thicknesses of Upper Triassic terrigenous flysch (Speed, 1978b). The flysch forms the major component of the Jurassic Fencemaker allochthon.

Metamorphism during Passive Margin Collisions

What was the metamorphic effect of emplacement of allochthons of unit 2 upon the Precambrian-Paleozoic shelf cover of North American continental crust? By analogy with modern accretionary prisms, each allochthon may have been initially very thick, perhaps as great as 20 km, and cool, with temperature gradients as low as $10°C/km$. This implies that metamorphism of the overrun strata may have been high P/T, at least initially, because heating of the autochthon by the allochthon would have been minor. In fact, the existence of brittle deformation and insignificant metamorphism in shelf cover just below and within each allochthon testify that the allochthons were cool throughout.

Observations are less clear on metamorphism due to passive margin collisions at deeper levels of the 12-km-thick shelf section, however, because any such metamorphism has not been studied closely nor discriminated from that due to later active margin thermotectonics. To investigate further possible effects of metamorphism by passive margin collisions, we present a couple of models in Fig. 22-6. The main variables include (1) the initial thickness and rate of erosion of the allochthons, (2) the thickness and initial temperature gradient of the autochthonous cover, and (3) the emplacement rate of the allochthon. The first variable is the major uncertainty, whereas values of 12 km and 30°C/km are reasonable for the second. Regarding the third variable, the autochthon can be assumed to have been buried almost isothermally if the allochthon closed at rates ≥0.5 cm/yr (Brewer, 1981).

In Fig. 22-6, curve 1 indicates a steady-state geotherm of 30°C/km in the Precambrian-Paleozoic shelf section before a passive margin collision. Curve 2 shows that rapid emplacement of a 20-km-thick allochthon on the shelf cover causes most of the cover to be initially in the blueschist facies. In the ensuing 50 m.y. or so, the buried shelf cover is heated, mainly from below, to restablish a steady-state geotherm of 30°C/km. The heating path of points on curve 2 in temperature-depth space (Fig. 22-6) are controlled by the rate and depth of erosion of the allochthon. Vectors attached to the base of the cover (curve 2) illustrate the possible paths: (a) for no erosion, (c) for instantaneous complete erosion, and (b) an intermediate rate of unroofing. Paths near 2a imply transformation of the initial blueschist in most of the cover to amphibolite facies and probable melting of wet siliciclastic rocks. Paths near 2c, in contrast, would follow nearly isothermal trajectories into lower-pressure facies. Such retrogradation might permit relics of the earlier blueschist metamorphism to be preserved, particularly in the middle third of the cover.

Under the same conditions, a 9-km-thick allochthon would cause none of the shelf cover to reach blueschist grade (curve 3, Fig. 22-6). Upon subsequent reequilibration to a 30°C/km geotherm, only the base of the cover could possibly melt, and progradational metamorphism to amphibolite facies could occur only in the lower half of the cover.

The greatest stratigraphic unroofing in the region of passive margin collision of which we are aware is in the Osgood Mountains (see Fig. 22-8; Hotz and Willden, 1964). There, Lower Cambrian and/or upper Precambrian siliciclastic strata are the lowest exposed beds of unit 1 (Fig. 22-2) and may be about 6 km below the top of the shelf cover. Such rocks are at greenschist grade. Assuming the greenschist is of prograde origin

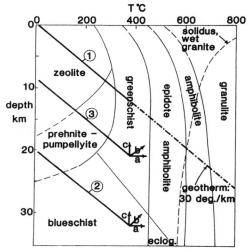

FIG. 22-6. Possible temperature-depth relationships in late Precambrian-Paleozoic shelf cover during passive margin phase; metamorphic facies boundaries from Ernst (1976) converted to depth using an average density of 2.6 g/cm³. Line 1 for 12-km-thick cover before allochthon emplacement. Lines 2 and 3 for same cover after very rapid emplacement of 20-km- and 9-km-thick allochthons, respectively. Heating paths a, b, and c explained in the text.

and not superposed on blueschist, 9 km appears to have been a maximum thickness of the allochthon to keep the upper 6 km of cover in the greenschist field if the early erosion rate was small (path 3a). If the initial erosion rate were very high (path 3c), however, the allochthon could have been as thick as 12 km and only progradational greenschist metamorphism would have occurred in the shelf cover.

The models of Fig. 22-6 assume only one allochthon above the shelf cover, whereas there were two sequential allochthons separated by about 100 m.y. To satisfy the condition that the greenschist grade was never surpassed midway down in the cover, the aggregate thickness of the two allochthons did not exceed 9 km if erosion were nil. In contrast, each allochthon could initially have been much thicker (12 km) if early erosion rates of each were extremely rapid upon emplacement.

MESOZOIC ACTIVE MARGIN TECTONICS

An active margin developed on western North America (Burchfiel and Davis, 1972; Hamilton, 1969) in Middle or Late Triassic time, as indicated by the oldest dates of silicic volcanic and plutonic rocks of a continental magmatic arc (C. A. Brook, 1977; Dilles *et al.*, 1983; Evernden and Kistler, 1970; Kistler, 1974). The subduction trace between an east-dipping oceanic slab and the morphologic continent of North America then existed at a suture in the Sierra Nevada foothills at the eastern side of unit 6 (Fig. 22-1; Sharp, chapter 30, this volume). At least part of Sonomia and perhaps other previously accreted terranes thus were incorporated into Triassic North America.

During Jurassic and Cretaceous time in the Great Basin, three principal phenomena were associated with active margin tectonism: magmatism and deformation in a continental arc, tectonic attrition, and deformation of the continental foreland which lay between the arc and the earlier craton-shelf hinge (Fig. 22-7). Each phenomenon is discussed below.

Mesozoic Continental Arc

The position of a continental magmatic arc in the westernmost Great Basin is defined by the existence of volcanogenic rocks and by a high proportion ($\sim$0.5) of plutonic rocks among exposed Mesozoic rocks (Fig. 22-7). Dating indicates that magmatism may have occurred more or less continuously in westernmost Nevada from Late Triassic to Late Cretaceous times, except for a possible lull in the Early Cretaceous (Chen and Moore, 1982; Dilles *et al.*, 1983; Evernden and Kistler, 1970; Oldow, 1981, 1984a; Russell, 1984; Speed, 1978a, 1983; Speed and Kistler, 1980; Stern *et al.*, 1981). The arc crosses the suture between Precambrian North America and Sonomia at about 38°N (Fig. 22-7). The Cretaceous phase of the arc was probably continuous from Nevada to Idaho, where it rejoins Precambrian North America. The older part of the arc, however, is apparently absent from Idaho. The main locus of Mesozoic continental arc magmatism in western Nevada north of about 37°N seems not to have varied with time. The apparent constancy of continental arc magmatism in space and time in westernmost Nevada and eastern California implies that the active margin of North America and underriding slabs maintained a relatively steady configuration over as much as 150 m.y. in the Mesozoic.

The main effects of arc development on the continent were crustal thickening and intra-arc deformation. Where continental arc rocks invade the sialic continent (area of unit 5B and south, Fig. 22-1), crustal thickness is 5-15 km greater than that of the normal craton (Eaton, 1963; Prohdehl, 1979). Moreover, the crust below Sonomia thickens substantially below the continental arc (Eaton, 1963; Prohdehl, 1979; Stauber, 1980).

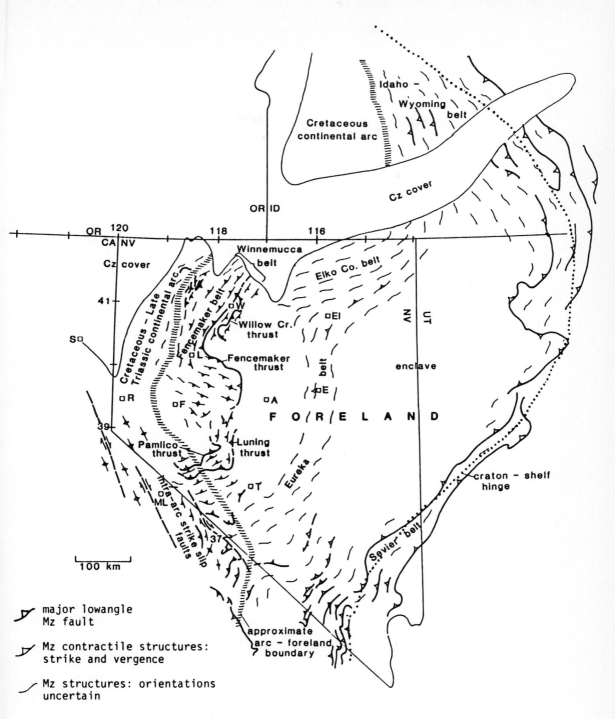

FIG. 22-7. Active margin features in the Mesozoic Great Basin and nearby regions. Belts are thin-skinned fold and thrust belts. Mz, Mesozoic.

Chemical and isotopic studies imply that major arc magma additions to the crust were mantle-derived but reflective of the different lithospheres in which they were generated or contaminated (Farmer and DePaolo, 1983; Kistler and Peterman, 1973, 1978).

Mesozoic Magmatism and Metamorphism in the Foreland

Magmatism at deep crustal levels also occurred during Mesozoic time within the continental foreland (Fig. 22-7) from the continental arc nearly as far east as the Paleozoic craton-shelf hingeline. Peaks of plutonism occurred at 155–165 and 70–80 m.y.b.p. (Allmendinger and Jordan, 1981; Armstrong and Suppe, 1973; Lee *et al.*, 1981). Thermal phenomena of the foreland differ from those of the contemporaneous continental arc by (1) the restriction to relatively deep levels and absence of evident volcanism (except for one site, the Upper Jurassic Pony Trail Group of Muffler (1964); (2) the smallness and discreteness of individual plutons; and (3) the compositions of foreland plutons indicating crustal or sedimentary precursors (mainly S type versus I type of arc). Mesozoic metamorphism and deformation of Precambrian and lower Paleozoic strata occurred deep within the continental shelf succession, but such effects are apparently absent in the upper few kilometers of Paleozoic strata (Armstrong, 1972; Howard, 1971, 1980; Snoke, 1980). Mesozoic magma generation and metamorphism in the foreland are likely to have been related processes. Metamorphic minerals indicate depths deeper than the undisturbed, expected stratigraphic depth of the metamorphosed strata. Thus, either metamorphism occurred in strata of anomalous depositional thickness or where strata were thickened tectonically. In the latter case, a small amount of thickening may reflect early contraction at deep levels in the foreland during the Mesozoic (Armstrong, 1972).

It is a fundamental question whether the elevation of the geotherms at depth in the Mesozoic foreland was directly related to subduction or whether it was a product of motions within the continental lithosphere and/or subjacent asthenosphere. An appeal to foreland heating by the same slab that caused magmatism in the continental arc would require the unlikely condition that melting occurred across a zone about 600 km wide in two pulses of short period in Late Jurassic and Late Cretaceous times. Foreland plutons are not contemporaneous with the possible Early Cretaceous lull in continental arc magmatism, a time when a shallowly dipping slab could have existed. Therefore, we argue that foreland plutonism was not due to melting above a downgoing slab and was not related to the long-lived continental arc of western Nevada and California. Armstrong's (1972) hypothesis of anatexis due to crustal thickening seems to account best for the broad reach of foreland thermal phenomena and their restriction to deeper crustal levels in the Mesozoic.

Tectonic Attrition

Figure 22-1 indicates a marked difference in the array of tectonostratigraphic units of the Great Basin in Nevada and those of Idaho, notably the absence or paucity of units 2, 3, 4, and 5A in Idaho. This implies either that accretion and continental margin sedimentation of Paleozoic and Triassic age were vastly less rich in Idaho or that the products of such processes in Idaho have been removed. Tectonic attrition of the sialic edge and earlier accretants in Idaho in Jurassic or Cretaceous time is the most plausible explanation for this difference (Hamilton, 1976; Speed, 1983). First, some 150 m.y. of progressive tectonostratigraphic development in Nevada is almost completely missing on the edge of sialic North America in Idaho only 300–400 km away; although tectonics certainly can vary rapidly along margins, it is unlikely that there would not be some correspondence in record over such a long time in such a short distance. Second, the sialic edge in Idaho intersects the Paleozoic craton-shelf hinge (Fig. 22-4), an unlikely primary arrangement. Third, displaced terranes of the Blue Mountains (Fig. 22-1) with large late Mesozoic rotations (Hillhouse *et al.*, 1982), which are absent from rocks in Nevada, were sutured during the Cretaceous against the sialic edge in Idaho (Lund and Snee, Chapter 11, this volume; Speed, 1983).

We suggest that a large raft that included a chunk of sialic North America and units 2–5A (Fig. 22-1) was extracted from the region of eastern Oregon and western Idaho in Jurassic or Cretaceous time. The withdrawal of the raft may have resulted from the propagation of a rift-transform system inboard of the sialic margin during highly oblique convergence in mid-Mesozoic time. The formation of the Gulf of California and northwestward rafting of Baja California may be a modern analog. The fragment removed from Oregon and Idaho was presumably replaced by oceanic lithosphere that has since been consumed below the incoming terranes of the Blue Mountains (Fig. 22-1) and covered by later deposits. The fragment or fragments of the raft may now be lodged in southern British Columbia and perhaps other places to the north of Idaho (Speed, 1983). Therefore, the age of the edge of Precambrian North America in Idaho is Mesozoic, unlike its greatly older age in Nevada (Fig. 22-4).

Foreland Deformation

The region east of the continental arc to and including the Sevier thrust belt in Utah and southern Nevada (Fig. 22-7) was the foreland of western North America that underwent mainly contractile deformation from Jurassic to Paleocene times. Within the Great Basin, foreland deformation was heterogeneous in surface distribution and depth, apparently unlike that in its northerly prolongation, the Idaho-Wyoming belt (Fig. 22-7). The Paleozoic-Triassic strata that were laid down before the onset of active margin tectonism record thin-skinned tectonics in a braided pattern (Speed, 1983). The three main thrust belts in which horizontal displacements ramped to the surface are the Sevier belt (Allmendinger and Jordan, 1981, 1984; Armstrong, 1968, 1972; Royce *et al.*, 1975), the Eureka belt (Speed, 1983), and the Winnemucca belt[1] (Oldow, 1981, 1984a; Speed, 1978a, 1983). Between the Sevier and Eureka belts, there is a tectonic enclave of little shallow deformation, and between the Eureka and Winnemucca belts (Fig. 22-7) occurs a possible belt of moderate unroofing and exposure of formerly deep-seated, metamorphosed rocks, the Toiyabe uplift zone. Sections that follow consider each of these features and conclude in a new kinematic model of Mesozoic foreland contraction across the Great Basin.

SEVIER AND EUREKA BELTS AND TECTONIC ENCLAVE

Both the Sevier and Eureka thrust belts (Fig. 22-7) are zones of displacement climb to the free surface with maximum recognized throw ≤7 km, syntectonic sedimentation, eastward vergence, and little evident involvement of crystalline basement or metamorphism, reflecting emplacement of nappes from deep sources. The Sevier belt lies above the Paleozoic craton-shelf hinge and a rapidly eastward thinning cover section. A crystalline slice within the Sevier belt probably is due to intersection of the frontal footwall ramp with the top of the west-facing late Precambrian hinge in the sialic crust. The Sevier belt and regions immediately west of it record cratonward piggyback propagation of thrust sheets from Late Jurassic through Paleocene times with a total displacement of about

[1] Also F. Heck, M. Elison, and R. Speed, unpublished data.

80 km (Allmendinger and Jordan, 1981; Armstrong, 1968; Heller *et al.*, 1986; Royce *et al.*, 1975). The Eureka belt is less well known as to timing, propagation, and total displacement but includes Lower Cretaceous(?) sediments (Newark Canyon Group) that are probably syntectonic.

The tectonic enclave (Fig. 22-7) is defined by an upper interval of cover strata of variable thickness from the Triassic down to Mississippian or deeper that underwent little or no deformation or stripping in Mesozoic time (Armstrong, 1972; Gans and E. L. Miller, 1983). Below this upper interval, however, Mesozoic deformation and metamorphism are known in deeper cover strata of unit 1 (Fig. 22-1) (mainly upper Precambrian and Cambrian) that are now exposed in areas unroofed in Cenozoic core complexes (Allmendinger and Jordan, 1984; Armstrong, 1972; Howard, 1971, 1980; D. M. Miller, 1982; E. L. Miller *et al.*, Chapter 24, this volume Snoke, 1980; Snoke and D. M. Miller, Chapter 23, this volume). The Mesozoic structures in the lower stratal interval include recumbent fold nappes, thrusts, and low-angle normal faults. Such relations indicate that Mesozoic displacements were probably continuous between the Sevier and Eureka belts on an extensive detachment through the lower interval of the tectonic enclave (see Fig. 22-15B; Speed, 1983). The detachment was a flat thrust, whereas the two thrust belts were ramp zones. The upper interval in the enclave was thus a coherent thrust sheet of unusual dimensions. It was either decoupled from the more deformable lower interval that underwent horizontal contraction, or more likely, was the top layer in a system of horizontal heterogeneous simple shear. The occurrence of recumbent Mesozoic fold nappes in exhumed Precambrian strata in the Ruby Mountains (Snoke, 1980) indicates that horizontal shear was at least part of the deep-seated motion. Moreover, the preservation of the Paleozoic-Triassic sedimentary cover across the enclave into mid-Cenozoic time (Armstrong, 1972) argues that large uplift and tectonic or erosional stripping at the surface did not occur in the Mesozoic and that horizontal simple shear was more significant than contraction in transmitting displacements through the lower interval. We employ this deduction in a model, given later, of foreland motions in the Great Basin.

The positioning of theSevier and Eureka thrust belts relative to the huge rigid cap of the tectonic enclave is easily explained. The belts both lie above major declivities in crustal layering. The Sevier lies above the west-facing craton-shelf hinge which has a 10-km amplitude on the top of the crystalline basement. The Eureka lies at the toe of the Roberts Mountains allochthon (Fig. 22-2), which caused a substantial west-facing declivity in the Mississippian, as recorded by foreland basin strata (Poole, 1974). Theory predicts that the maximum slope of the deflection of crustal layers should lie below the toe of the allochthon (Speed and Sleep, 1982). By contrast, layering in the enclave may have been more plane parallel, thus without cause for displacement climb.

WINNEMUCCA DEFORMATION BELT

The Winnemucca deformation belt (Fig. 22-7) is flanked on the west by the Mesozoic continental arc and on the east with poorly defined boundary by a possible zone of Mesozoic thermal doming and unroofing that we describe later. The Winnemucca belt probably lies astride the edge of Proterozoic sialic North America.

The northern half of the Winnemucca belt contains four regional structures (Figs. 22-8, 22-9, and 22-10), from the west to east: Fencemaker allochthons A and B, autochthon, and the Willow Creek allochthon. The Fencemaker allochthons are composed of mainly basinal lower Mesozoic sediments that were shortened and thrust shelfward on the Fencemaker thrusts that approximately follow the basin to shelf transition (unit 4-5 break, Fig. 22-2) at the edge of Triassic North America (Speed, 1978b). The Fence-

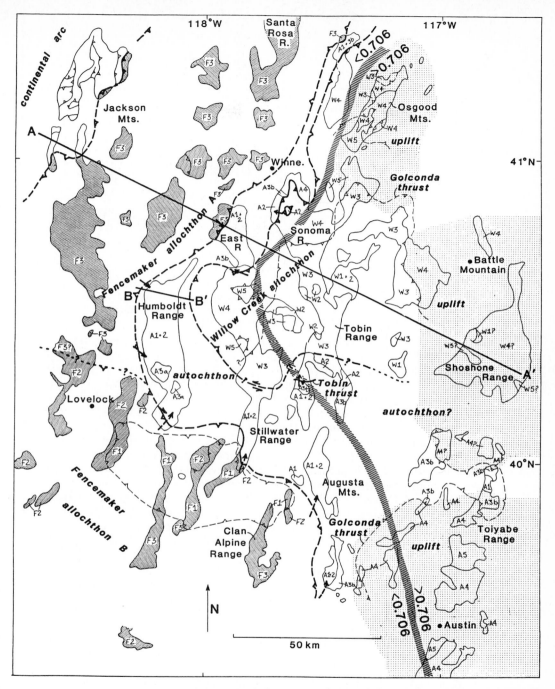

FIG. 22-8. Northern Winnemucca deformation belt showing five major structural features, Fencemaker allochthons A and B, autochthon, Willow Creek allochthon, and pluton-cored domes; letter-number symbols within features refer to tectonostratigraphic units explained in Fig. 22-9; arrows indicate allochthon transport trend; cross section AA' in Fig. 22-10.

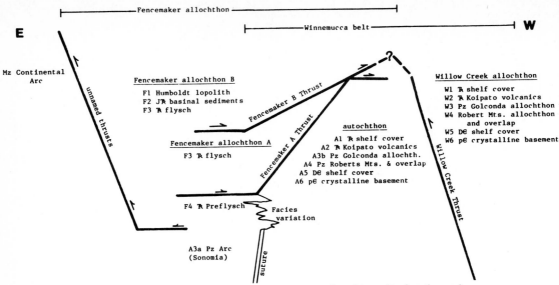

FIG. 22-9 Stacking diagram of tectonostratigraphic units for the region shown in Fig. 22-8, showing the three major thrusts of thw Winnemucca belt and the bivergent nature of the Fencemaker allochthon.

maker was originally thought to be a single allochthon and to be continuous around a salient in the Triassic shelf margin near Lovelock (Figs. 22-7 and 22-8). We now believe that two allochthons, A and B (Fig. 22-8), exist. They lie discretely on the northwest- and southwest-facing flanks of the salient, contain partly different stratigraphic successions, and probably moved in different directions at different times (F. Heck, M. Elison, and R. Speed, unpublished data). Fencemaker allochthon A moved east-southeast relative to the shelf in today's coordinates. Transport occurred after about 215 m.y.b.p., the age of youngest strata in the allochthon, and before the undated final emplacement of the Willow Creek allochthon (Fig. 22-8), whose movements imposed structures that overprint east-verging structures of Fencemaker A. Structures within the eastern part of Fencemaker A are uniformly east-vergent and suggest minimum horizontal contraction of 70%. The western flank of Fencemaker allochthon A is at the eastern flank of the Mesozoic continental arc (Russell, 1984). There, Fencemaker structures verge west, and some at least may have formed in Early Cretaceous and perhaps later times. Thus, Fencemaker allochthon A is bivergent (Fig. 22-10) and may have undergone east to west propagation of deformation. The eastern two-thirds of the allochthon contain only Upper Triassic flysch. Such strata probably detached from older basinal beds (preflysch, Figs. 22-9 and 22-10) and were imbricated against the apparently less deformable shelf and arc margins as those margins closed toward one another. We postulate that the sediments and Sonomia basement below the flysch detachment underrode the continental arc to the west in the Jurassic and/or Cretaceous (Fig. 22-10). The choice of west versus east underriding by the Sonomia basement stems from the constraint that the continental crust and its Phanerozoic tectonostratigraphic cover appear tied together in the Winnemucca autochthon (Figs. 22-9 and 22-10).

Fencemaker allochthon B (Fig. 22-8) was transported north-northeast relative to the shelf edge. Structures generated within the autochthon by Fencemaker B overprint those caused by Fencemaker allochthon A. The age of final emplacement of Fencemaker allochthon B is uncertain, although contraction in Fencemaker B was ongoing between

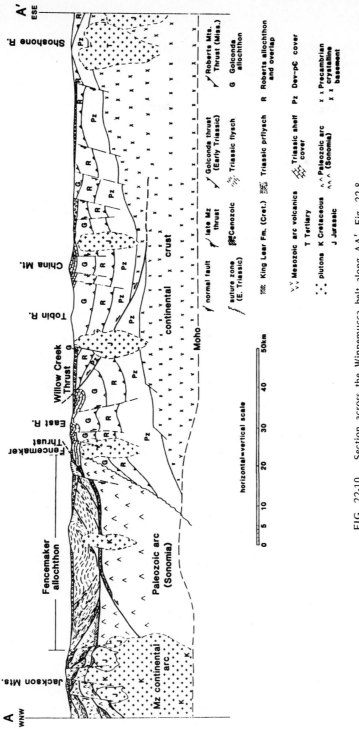

FIG. 22-10. Section across the Winnemucca belt along AA', Fig. 22-8.

589

150 and 165 m.y., as indicated by syntectonic sedimentation and magmatism (Speed and Jones, 1969).

The Winnemucca autochthon to the Fencemaker and Willow Creek thrusts (Figs. 22-8, 22-9, and 22-10) is capped almost completely by Triassic shelf and shelf edge strata (Nichols and Silberling, 1977). The tectonostratigraphic succession of the autochthon is only slightly disrupted by Mesozoic low-angle faults, both repetitional and omissional, that appear to have displacements ≤2 km. The Tobin thrust (Fig. 22-8) is an example of an intra-autochthon structure. Deformation in Triassic rocks of the autochthon, estimated by cleavage spacing and orientation with respect to bedding, fold tightness, and width of and distance between shear zones is proportional to proximity to the major Mesozoic thrusts (Fig. 22-8).

The Willow Creek allochthon (Figs. 22-8, 22-9, and 22-10) contains a moderately disrupted tectonostratigraphic succession of rocks that lay above North America before thrusting. The lowest exposed unit that is clearly within the allochthon is Ordovician carbonate of unit 1 in the northern East Range (M. Elison, unpublished data; conodont data courtesy of D. Whitebread). The allochthon's initial tectonostratigraphic thickness may have been 10 km or much more; the uncertainty depends on the initial structural thicknesses of the Roberts Mountains and Golconda allochthons (Fig. 22-2) which are known only to have exceeded the 2–3 km present exposed thicknesses. The Willow Creek allochthon was emplaced above the Triassic shelf cover and locally, lower rocks of the autochthon. Transport in the East and Sonoma Ranges was west-northwest, and the minimum displacement is certainly 10 km and more likely, 30 km. Folds in the East Range (Fig. 22-8) related to motions of the Willow Creek allochthon are cut by an undeformed pluton (Yellowstone Canyon) that is dated at 151 m.y. (K-Ar, hornblende with slightly discordant biotite). Thus, we assume that at least some transport of the Willow Creek allochthon was before 151 m.y.b.p., but the age of final emplacement is unknown. We suspect that Fencemaker B as well as A preceded Willow Creek emplacement because open folds with northerly axial traces are widespread in the autochthon and deform the Fencemaker B thrust. Such folds are appropriately oriented to have been generated by motions related to Willow Creek transport.

As indicated in Fig. 22-8, we are as yet uncertain as to the southern terminus, if any, of the Willow Creek allochthon and the discrimination of it from autochthon in the Toiyabe and Shoshone Ranges. In Fig. 22-10 we portray the sole of the Willow Creek allochthon as a listric thrust that flattens east into sialic crust. This interpretation is based on the idea that the Willow Creek allochthon was caused by spreading away from a chain of thermal domes that lay immediately east, to be discussed presently.

At the southern end of the Winnemucca belt (Fig. 22-7), the Luning and Pamlico allochthons were driven southeast with respect to the sialic shelf edge (Oldow, 1981, 1984a). Early motions in this area were probably Jurassic but thrusting occurred as late as Late Cretaceous, between 70 and 100 m.y. (Speed and Kistler, 1980). Controversy exists whether the deflection of the Luning thrust is due to strike slip faulting (Albers, 1967; Speed, 1978a) or whether the deflection is due to an inherited bend in the Precambrian edge of North America (Kistler, 1978; Oldow, 1984b).

To conclude, the Mesozoic Winnemucca deformation belt in the western Great Basin may have begun motions in Early or Middle Jurassic time. However, current information indicates that thereafter, movement directions and times were heterogeneous and sporadic, perhaps occurring at one place or another throughout the Jurassic and Cretaceous. The cause of the heterogeneity is probably twofold: (1) buttresses and declivities in the tectonostratigraphic succession with varied initial orientations that caused varied directions of flattening in rocks and displacement climb in thrust sheets (Speed, 1983);

and (2) the occurrence of major wrench faulting (Oldow, 1984a) in zones parallel to strike-slip faults of the continental arc (Fig. 22-7).

TOIYABE UPLIFT ZONE

The region of the western Great Basin between the Winnemucca and Eureka thrust belts (Fig. 22-7) exposes moderately deep-seated tectonostratigraphic units and moderate metamorphic grades with unroofing ages that are at least partly, if not entirely, Mesozoic. We call this region the Toiyabe uplift zone with distribution shown in Fig. 22-11[2] (based on Albers and Stewart, 1972; Means, 1962; D. L. Smith, 1984). The zone may be a continuous belt of moderate unroofing, or more likely, it is a chain of domes, at least some of which are thermal and diapiric, as discussed below. The Toiyabe uplift zone lies above Proterozoic North America (unit 1, Fig. 22-1) with evident geographic proximity to the sialic edge (0.706 line, Fig. 22-11) and to the region of North America overrun by oceanic allochthons (unit 2, Figs. 22-1 and 22-2).

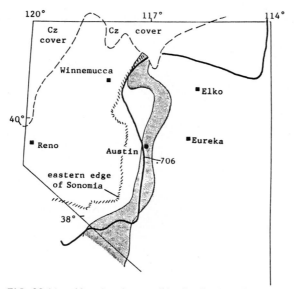

FIG. 22-11. Map showing possible distribution of Toiyabe uplift zone (stipple).

Within the northern Toiyabe uplift zone (Fig. 22-8), we resolve three unroofed areas, from north to south: Edna-Osgood Mountains, Shoshone Range, and Toiyabe Range centered near Austin. Demarcation of the unroofed areas is based on exposure of Paleozoic shelf cover (unit 1, Fig. 22-2; Figs. 22-8, 22-9, and 22-10) and on conodont alteration indices (CAI) of high value (≥ 4.5, $\simeq 350°C$) in all Jurassic units at a given site (Fig. 22-12). In the Edna-Osgood Mountain area, quartzite of Lower Cambrian or older age of the sialic cover (unit 1, Fig. 22-2) is exposed. This implies at least 13 km of unroofing, given 6 km for the supraquartzite Paleozoic shelf section, 3 km each for the Roberts Mountains and Golconda allochthons, and 1 km of Triassic shelf strata; all these

[2]R. Speed, unpublished data from Toquima and Toiyabe Ranges, Osgood and Edna Mountains.

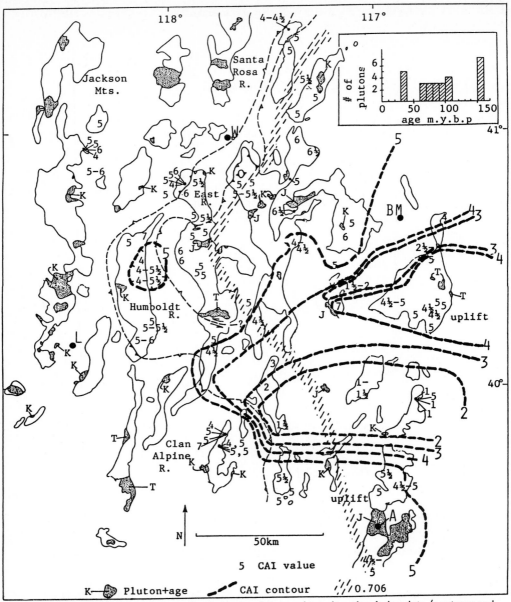

FIG. 22-12. Winnemucca deformation belt, showing conodont alteration index data (contours and site values) in rocks from Ordovician to Triassic age; data from Harris *et al.* (1980) and F. Heck, M. Elison, and R. Speed (unpublished data); stippled areas are plutons: J, Jurassic; K, Cretaceous, T, Tertiray.

figures for thickness are minima. The Shoshone Mountains expose several windows through the Roberts Mountains allochthon (Fig. 22-8), and the Toiyabe Range south of Austin is probably unroofed as deeply as the Edna-Osgood area.

In two of the three deeply unroofed areas, Jurassic plutons intrude the lowest tectonostratigraphic units. The plutons yield K-Ar dates on hornblende between 150 and

160 m.y. (Fig. 22-12). In the third area, the Shoshone Mountains, a large Jurassic pluton is exposed just east of Fig. 22-8, and we presume it underlies the Shoshone Mountains as well (Fig. 22-10). The more deeply seated rocks of the Edna-Osgood and Toiyabe Range unroofed areas are at chlorite or biotite grade, even away from exposed plutons.

The areas of unroofing and high conodont alteration index (CAI) values are separated by narrow zones in which shallow pre-Jurassic units are preserved, and in which CAIs are low, ≤3 and as low as 1 (Fig. 22-12). Such geometry suggests that the unroofed areas are discrete, at least at shallow levels, and are individually domical rather than part of a two-dimensional belt.

A major point is that east of the Winnemucca deformation belt (east of the Willow Creek and Fencemaker thrust) in the Toiyabe uplift zone, rocks from Triassic to as old as Ordovician have approximately the same CAI, either high or low, at a given position. This implies that metamorphism did not occur before the tectonostratigraphic succession was completely built up in Late Triassic time, not before about 215 m.y.b.p. Moreover, the localized heating throughout the succession suggests (1) advective heat sources, such as plutons; (2) strong vertical attenuation of the tectonostratigraphic succession, greatly increased temperature gradient, and a surficial thermal blanket; or (3) both (1) and (2). The occurrence of plutons in or near the cores of each of the three domes certainly supports the first idea. Moreover, the occurrence implies that uplift and unroofing were contemporaneous with magma rise, thus older than or at the time of argon locking recorded by the Late Jurassic hornblende ages of the plutons.

Metamorphism through tectonic attenuation was probably less effective than magma advection because no thermal blanket above the Triassic is recognized and because a steep metamorphic gradient with depth in a tectonostratigraphic succession should probably be more apparent than it is.

There is, however, evidence for tectonostratigraphic attenuation of Mesozoic age in the northernmost dome of the Toiyabe uplift belt (Osgood-Edna Mountains, Fig. 22-8). There, subhorizontal brittle fault slices of sedimentary rocks of units 1 and 2 (Fig. 22-2) are interleaved, mainly with omission but also with repetition[3] (Erickson and Marsh, 1974; Hotz and Willden, 1964). Some of these faults are cut by plutons or their contact metamorphic zones with Late Cretaceous K-Ar mineral ages. In proposing a Mesozoic age for the mainly omissional faulting in the Edna-Osgood dome, we assume but have not yet proven that the base of the massive Permian Edna Mountain Formation above the Iron Point thrust in the Edna Mountains is a fault, not depositional as inferred by Erickson and Marsh (1974). D. L. Smith (1984) found pre-Oligocene low-angle normal faulting near Austin in what we believe to be the southernmost of the three domes in the area of Fig. 22-8. Smith's pre-Oligocene faults may have formed during Mesozoic unroofing of that dome.

MODEL OF THERMAL DOMING

There are evident relationships between the Edna-Osgood dome and the Willow Creek allochthon. The two features appear to be geographically continuous (Figs. 22-8, 22-10, and 22-12) and may have undergone movements at the same times, in the Jurassic and probably, into the Cretaceous. The Willow Creek allochthon seems to maintain the tectonostratigraphic succession but with contractile structures, whereas the Edna-Osgood dome is more disrupted, apparently chiefly by brittle extensional faults. We therefore suggest that the Willow Creek allochthon is a product of spreading west from the Edna-Osgood dome.

[3] Also R. Speed and M. Elison, unpublished work.

Figure 22-13 presents a model that may depict the development of the Edna-Osgood dome and possibly other domal unroofed areas of the Toiyabe uplift zone (Fig. 22-11). The model assumes that the Eureka thrust belt (Fig. 22-7) represents spreading from the east side of the Toiyabe uplift zone. The model further assumes that the Toiyabe uplift zone is a response to regional contraction because Mesozoic movements in the thrust belts on either side of the uplift zone indicate compression across the foreland in the Great Basin in that era.

The model (Fig. 22-13) postulates that crustal contraction takes place in a laterally narrow zone (±200 km wide) by volume transfer and doming of the free surface. The dome (Fig. 22-13, 2) spreads laterally on surface thrusts, and extension occurs in the central region due to lateral flow and stretching by progressive uparching. Magmas generated at the base of the sialic crust rise diapirically through the contracting part of the dome, increasing its ductility and the rate of local horizontal shortening. The plutons also provide a mechanism of mass transfer to higher levels. Because the unroofing is greater on the west side of the dome, we assume that uplift and diapirism went on for a longer time there than on the east side (Fig. 22-13, 3).

Given an average of 10 km unroofing across the model dome and assuming no depression of the sialic crust below the dome for compensation of mountains, the mechanism shown in Fig. 22-13 could accommodate 33% horizontal shortening relative to an initial 30-km-thick crust. For a 200-km initial width of the domed region, about 66 km of contraction could have been taken up.

Why did thermal domes develop in what seems to be a belt just inboard of the sialic edge and not evidently in other regions of the Mesozoic foreland of the Great Basin? We suppose that the pre-Jurassic subsidence of the region of the Toiyabe uplift zone by the Roberts Mountains and Golconda allochthons (unit 2, Figs. 22-1 and 22-2) caused two geographically local responses: first, a declivity that caused displacement climb at the Eureka belt, and second, deeper and warmer sialic crust that was more prepared to respond ductilely and/or melt during initial homogeneous foreland contraction in the Jurassic than crust east of unit 2.

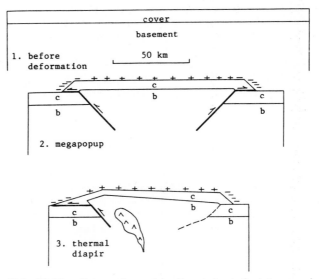

FIG. 22-13. Schematic model of evolution of pluton-cored dome during regional crustal contraction: c, cover; b, basement; −, horizontal contractile strain; +, horizontal extensional strain.

PHANEROZOIC TECTONIC EVOLUTION OF THE GREAT BASIN

The model of Fig. 22-13 suggests that if in fact Mesozoic thermal domes extend in a belt south of the Edna-Osgood dome, west-moving allochthons like the Willow Creek may exist at other places to the south (Fig. 22-11). We are unaware of evidence to support or deny such a prediction.

CENOZOIC EXTENSION

The onset of extension so evident from modern structures and seismicity in the Great Basin began apparently late in Oligocene time. This phase of active margin tectonism has been addressed extensively in the recent literature (Allmendinger *et al.*, 1983; Bartley and Wernicke, 1984; Gans and E. L. Miller, 1983; McDonald, 1976; E. L. Miller *et al.*, 1983; R. B. Smith, 1978; Snoke, 1980; Stewart, 1978; Wernicke *et al.*, 1982; Zoback *et al.*, 1981). Our discussion below focuses only on amounts of Cenozoic extension that need be considered for Mesozoic reconstructions, with the conclusion that it is large in the eastern half of the Great Basin but probably small in the western.

Widespread large Neogene extension in the eastern half of the Great Basin is indicated by surface and seismic reflection structures: extensive low-angle detachment faults with probable large displacement (Allmendinger, *et al.*, 1983; McDonald, 1976; Wernicke *et al.*, 1982), sequential sets of omissional faults with brittle rotation of strata and early faults (Gans and Miller, 1983), and exhumed Neogene mylonite zones (core complexes) (E. L. Miller *et al.*, 1983; Snoke, 1980) East-west Neogene elongations as great as 200% occur in some areas, but the average elongation across the eastern Great Basin is probably 80-100%[4] (Wernicke, 1981), indicating major heterogeneity.

In contrast, evidence for large Neogene extension in the western half of the Great Basin is sparse except at Yerington[5] (Proffett, 1977) and possibly, in the central Toiyabe Range (Smith, 1984). Shallow detachment faults of large strike and dip length have not been resolved in seismic records. Omissional faults with strongly rotated hanging walls are rare in the western Great Basin, and where they occur, many are confined to detached Tertiary sections. Such faults commonly do not penetrate below the Tertiary. Normal fault sets that are currently active appear planar to depths of 12-15 km (Vetter and Ryall, 1983). Thus, the western Great Basin seems to differ in Neogene structure and kinematics from the eastern.

We estimate the average EW elongation of the western half of the Great Basin to be between 0 and 20% with the following rationale. Bogen and Schweickert (1985) and Frei *et al.* (1984) found no significant paleomagnetic declination anomaly of the late Mesozoic Sierra Nevada relative to the Colorado Plateau. This means that extension of the Great Basin included little or no rotation, that is, little or no north-south gradient in total east-west displacement. Using the structurally derived estimates for total extension in the southern Great Basin of Wernicke *et al.* (1982) and a range of counterclockwise rotation of the Sierra Nevada of 0 to 6° permitted by the paleomagnetic data, Bogen and Schweickert (1985) figured a total EW extension of 178 ± 33 km across the northern Great Basin at 40°N. Such displacement is evidently more concentrated in the eastern than the western Great Basin. Figure 22-14 shows how average strains in the two halves of the Great Basin may be related; assuming that 80% is a good average for the eastern half, strains in the western half are 0-20%. It is noteworthy that a value of 15-20% fits well with extension due solely to the horst and graben structure with steeply dipping normal faults that characterizes the province today. The 100% extension at Yerington found by Profett (1977) in the western Great Basin testifies to at least local heterogeneity of Neogene strain

[4] Also P. B.Gans, personal communication, 1986.
[5] Also J. M. Proffett, oral communication, 1984.

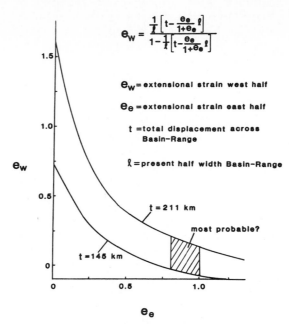

$$e_w = \frac{\frac{1}{l}\left[t - \frac{e_e}{1+e_e}\ell\right]}{1 - \frac{1}{l}\left[t - \frac{e_e}{1+e_e}\ell\right]}$$

e_w = extensional strain west half

e_e = extensional strain east half

t = total displacement across Basin-Range

ℓ = present half width Basin-Range

t = 211 km

most probable?

t = 145 km

FIG. 22-14. Proportionation of horizontal extensional strain in present eastern and western halves of Great Basin; parameter is total Cenozoic extensional displacement across Great Basin; see the text.

magnitude at constant orientation of principal components. This implies that at other local areas, the extension is less than the average and conceivably, is negative.

The Great Basin has probably been a province of very high heat flow throughout Neogene time. The average modern heat flow in the western region is probably 1.5–1.7 times that of the same region in Paleogene time (Lachenbruch and Sass, 1978). This means that the top of greenschist grade (250°C) is now about 5 km deep and moved up a couple of kilometers in early Neogene time. Regional greenschist facies rocks are not widely exposed in the western Great Basin, however, indicating that Cenozoic erosional or tectonic stripping has not been generally effective in that region.

KINEMATIC MODEL OF FORELAND CONTRACTION ACROSS THE GREAT BASIN

We propose a general model of sites of major displacement in the heterogeneously contracting foreland of the Mesozoic Great Basin in Fig. 22-15. Figure 22-15A illustrates a possible pre-Jurassic configuration with restoration for Cenozoic horizontal extensional strain of nil in the western half and 100% in the eastern. These are close to figures arrived at in the previous section and are satisfactory for present purposes. The titled fault blocks in the continental crust represent effects of Precambrian passive margin development.

The structure of the foreland after Mesozoic contraction (Fig. 22-15B) is drawn with the craton-shelf hinge fixed in position and with the following conditions derived from previous discussion: (1) the Sevier thrust belt is a zone of displacement climb above rigid basement at the craton-shelf hinge with about 80 km of cratonward displacement within the imbricate stack; (2) the sole fault of the Sevier thrust belt extends west below the

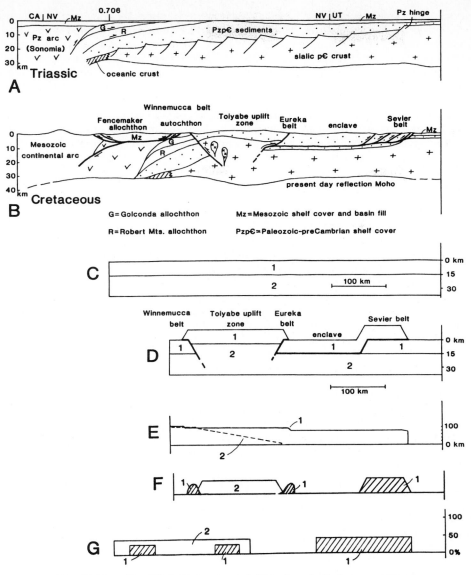

FIG. 22-15. Simplified kinematic model for Jurassic-Cretaceous deformation in the Great Basin at 40–41°N. (A) Generalized section across the Great Basin in the Late Triassic prior to the onset of deformation. (B) Generalized section across the Great Basin after Jurassic-Cretaceous deformation. (C) Two-layer model of undeformed Great Basin; layers 1 and 2 are arbitrary crustal divisions. (D) Two-layer geometry following Jurassic-Cretaceous deformation. (E) Easterly horizontal displacement of layer 1 (solid line) and layer 2 (dashed line) relative to the craton (right margin). (F) Contractile strain in layers 1 and 2 relative to undeformed length shown qualitatively as a function of position in deformed foreland. (G) Average contractile strain vs. original lengths for layers 1 and 2; areas of the curves for the two layers are equal assuming they stay attached at the western end (Winnemucca autochthon) and eastern end (craton) of the section.

tectonic enclave as a subhorizontal detachment within cover strata; some of the eastward displacement in the Sevier thrust belt may have been taken up in horizontal simple shear in lower layers of the enclave, but we assume the enclave was translated as a rigid sheet; (3) the western end of the detachment below the tectonic enclave is an east-verging thrust that branches with faults of the Eureka thrust belt and roots below the Toiyabe uplift zone; (4) major contraction occurs across the Toiyabe uplift zone by conjugate rooted thrusts (the Willow Creek thrust and Eureka thrust belt) and advection of mass to positions of higher elevation; the cover is assumed to remain attached to sialic crust in the uplift zone; the problem of crust-mantle interaction is discussed below; (5) the autochthon of the Winnemucca belt migrates toward the craton and maintains its pre-Jurassic tecto-nostratigraphy without serious detachment (and thus contains a pinline); (6) rocks below the western two-thirds of the Fencemaker allochthon underride the Mesozoic continental arc; and (7) movements on the master faults of the system (Fig. 22-15B) (sub-Fencemaker fault, conjugate rooted thrusts, subenclave detachment) occurred episodically or con-tinously in Jurassic and Cretaceous times.

Figures 22-15C and D show a balanced two-layer kinematic model of movement zones from the Winnemucca autochthon to the Seiver thrust belts (Fig. 22-15B). Transla-tion of a detached upper layer (layer 1) in the eastern half of the foreland is compensated by differential displacements of attached upper and lower layers in the western half (Fig. 22-15E). This leads to heterogeneous strain as a function of layer and position (Fig. 22-15F). If we are correct that pre-Mesozoic vertical sequences stay attached during foreland con-traction in the craton east of the Sevier belt and in the Winnemucca autochthon (Fig. 22-15B), strains integrated across the foreland in each layer should be similar, assuming constant volume. This is depicted in Fig. 22-15G for our two-layer model, where areas under the curves for layers 1 and 2 are equal. The total horizontal contraction across the region considered (autochthon of Winnemucca belt to eastern toe of Sevier belt) is 18% (Figs. 22-15C and D).

Because the contraction model (Fig. 22-15D) includes vertical as well as horizontal displacement components, gravity may affect layer boundary configurations and hence, the amount of horizontal contraction. We assumed in Fig. 22-15 that the base of layer 2 is a detachment and that compensation is either taken up later or is attained below layer 2. An alternative choice is that the lower boundary of layer 2 deflects below uplifts so as to maintain mass balance (e.g., Airy isostatic model). The alternative would cause far more shortening than that in the model we depict. In fact, there appears to be a maximum in depth to the seismic reflection Moho below the Toiyabe uplift zone (Fig. 22-15B; Klemperer et al., 1986). This may reflect Mesozoic contraction or, alternatively, may be due to Neogene processes (Klemperer et al., 1986) that have obscured the older crustal structures at depth.

FORELAND COMPARISONS

Mesozoic foreland tectonism of western North America in the Great Basin differed sub-stantially from that of other contracting forelands of the world. The foreland in the Great Basin was unusually broad and widely retained its pretectonic cover except above the relatively narrow Toiyabe uplift zone (Fig. 22-11). Parautochthonous or exotic nappes of strongly metamorphosed or crystalline basement rocks were not driven to shallow levels anywhere in the Great Basin. Thus, foreland tectonism did not create large regional uplift and unroofing. There was no systematic cratonward progagation of contraction except across the inherited passive margin hinge where the Sevier thrust belt exists. There was no general oceanward increase in depth of unroofing and grade of metamorphism as exists in many foreland contractile zones such as the southern Appalachians, Alps, and Canadian Rocky Mountains. The differences may indicate that Mesozoic North America in the Great Basin experienced long-term active margin tectonics without the collision against — or without being overridden by — large buoyant terranes of the type that affected other foreland belts.

CONCLUSIONS

1. Thick sedimentary and tectonic successions emplaced on the subsiding shelf of sialic North America in the Great Basin may have caused low-grade metamorphism at places in the upper Precambrian to Triassic passive margin cover. At normal continental temperature gradients (30°C/km), the lower sixth of the cover succession may have been in greenschist facies at the end of Paleozoic time by progressive sedimentation. Below oceanic allochthons (Roberts Mountains and Golconda), intervals of shelf cover may have been initially in blueschist facies if the accumulated thickness of the allochthons exceeded 12 km. Relics of such blueschist metamorphism may exist in deeper levels of the shelf cover if large uplift and erosion of the allochthons rapidly followed their emplacements.

2. The buried edge of sialic North America in the Great Basin may be the preserved late Precambrian boundary formed during the drift phase of passive margin formation. The cover to the sialic crust at this edge is apparently autochthonous.

3. The continental magmatic arc existed during active margin tectonism of western North America approximately in place in the western Great Basin from Late Triassic to Late Cretaceous times except for a possible lull in the Early Cretaceous. This implies that the downgoing slab maintained a more or less constant configuration with respect to the continent for 150 m.y. Plutonic pulses within the Great Basin as far east as western Utah in Late Jurassic and Late Cretaceous times were not directly related to continental arc magmatism.

4. Jurassic and Cretaceous foreland contraction in the Great Basin was heterogeneously distributed at the surface and with depth. Three thrust belts form braided strands of displacement climb to the surface. Between the eastern two belts, a large tectonic enclave probably connected displacements between the two belts at depth below a rigid sheet of unusually great size. Each of the thrust belts is thin-skinned.

5. The Toiyabe uplift zone in central Nevada may be a chain of pluton-cored Mesozoic domes which record the only significant uproofing and exposure of deep-seated metamorphic rocks during Mesozoic time in the Great Basin. The position of this zone may be controlled by crustal declivities and greater depth of continental crust caused by the earlier overriding of allochthons of oceanic sediments. Spreading off the western flank of the Edna-Osgood dome in this uplift zone may have caused the extensive, west-verging Willow Creek allochthon.

6. Mesozoic contraction across the Great Basin can be modeled between pinlines in the craton on the east and the Winnemucca autochthon on the west as taken up on rooted conjugate thrusts in the western region and on a thin-skinned, east-verging detachment in the eastern, with an average horizontal shortening of 18%.

7. Mesozoic foreland tectonism in the Great Basin differed substantially from that of other contracting forelands of the world by the lack of extensive regional uplift and unroofing, and lack of overriding by hot basement nappes. This may indicate that the edge of Mesozoic North America bordering the Great Basin did not experience collision with large buoyant terranes.

8. Regional Cenozoic extension in the western half of the Great Basin may be substantially less than that of the eastern half.

ACKNOWLEDGMENTS

This work was supported by National Science Foundation grants EAR83-06242 and EAR85-19966 and National Aeronautics and Space Administration grant NAS5-27238A004.

REFERENCES

Albers, J. P., 1967, Belt of sigmoidal bending and right-lateral faulting in the western Great Basin: *Geol. Soc. America Bull.*, v. 78, p. 143–156.

———, and Stewart, J. H., 1972, Geology and mineral deposits of Esmeralda County, Nevada: *Nev. Bur. Mines Geology Bull.*, v. 78, 80 p.

Allmendinger, R. W., and Jordan, T. E., 1981, Mesozoic evolution, hinterland of the Sevier orogenic belt: *Geology*, v. 9, p. 308–314.

———, and Jordan, T. E., 1984, Mesozoic structure of the Newfoundland Mountains, Utah — Horizontal shortening and subsequent extension of the hinterland of the Sevier Belt: *Geol. Soc. America Bull.*, v. 95, p. 1280–1292.

———, Sharp, J. W., Von Tish, D., Serpa, L., Brown, L., Kaufman, S., Oliver, J., and Smith, R. B., 1983, Cenozoic and Mesozoic structure of the eastern Basin and Range province, Utah: *Geology*, v. 11, p. 532–536.

———, Miller, D. M., and Jordan, T. E., 1985, Known and inferred Mesozoic deformation in the hinterland of the Sevier belt, northwest Utah, *in* Kerns, G. J., ed., *Geology of Northwest Utah, Southern Idaho and Northeast Nevada*: Utah Geol. Assoc. Publ. 13, p. 21–34.

———, Sharp, J. W., Von Tish, D., Serpa, L., Brown, L., Kaufman, S. Oliver, J., and Smith, R. B., 1983, Cenozoic and Mesozoic structure of the eastern Basin and Range province, Utah: *Geology*, v. 11, p. 532–536.

Armstrong, R. L., 1968, Sevier orogenic belt in Nevada and Utah: *Geol. Soc. America Bull.*, v. 79 p. 439–458.

———, 1972, Low-angle (denudation) faults, hinterland of the Sevier orogenic belt, eastern Nevada and western Utah: *Geol. Soc. America Bull.*, v. 83, p. 1729–1754.

———, and Suppe, J., 1973, Potassium-argon geochronometry of Mesozoic igneous rocks in Nevada, Utah, and southern California: *Geol. Soc. America Bull.*, v. 84, p. 1375–1392.

———, Taubeneck, W. H., and Hales, P. O., 1977, Rb-Sr and K-Ar geochronometry of Mesozoic granitic rocks and their Sr isotopic composition, Oregon, Washington, and Idaho: *Geol. Soc. America Bull.*, v. 88, p. 321–331.

Bartley, J. M. and Wernicke, B. P., 1984, The Snake Range decollement interpreted as a major extensional shear zone: *Tectonics*, v. 3, p. 647–657.

Bogen, N. L. and Schweickert, R. A., 1985, Magnitude of crustal extension across the northern Basin and Range province: constraints from paleomagnetism: *Earth Planet. Sci. Lett.*, v. 75, p. 93–100.

Brewer, J., 1981, Thermal effects of thrust faulting: *Earth Planet. Sci. Lett.*, v. 56, p. 233–244.

Brook, C. A., 1977, Stratigraphy and structure of the Saddlebag Lake roof pendant, Sierra Nevada: *Geol. Soc. America Bull.*, v. 88, p. 321–331.

Brook, H. C., and Vallier, T. L., 1977, Mesozoic rocks and tectonic evolution of eastern Oregon and western Idaho, *in* Howell, D. G., and McDougall, K. A., eds., *Mesozoic Paleogeography of the Western United States:* Pacific Section, Soc. Econ. Paleontologists Mineralogists, Pacific Coast Paleogeography Symp. 2, p. 133–146.

Brueckner, H. K., and Snyder, W. S., 1985, Structure of the Havallah sequence, Golconda allochthon, Nevada — Evidence for prolonged evolution in an accretionary prism: *Geol. Soc. America Bull.*, v. 96, p. 1113–1130.

Burchfiel, B. C., and Davis, G. A., 1972, Structural framework and evolution of the southern part of the Cordilleran orogen, western United States: *Am. J. Sci.*, v. 272, p. 97–118.

Chen, J. H., and Moore, J. G., 1982, U-Pb isotopic ages from the Sierra Nevada batholith, California: *J. Geophys. Res.*, v. 87, p. 4761–4785.

Cogbill, A. H., 1979, Relationships of crustal structure and seismicity, western Great Basin: Ph.D. thesis, Northwestern Univ., Evanston, Ill., 289 p.

Dickinson, W. R., 1977, Paleozoic plate tectonics and the evolution of the Cordilleran continental margin, *in* Stewart, J. H., Stevens, C. H., and Fritsche, A. E., eds., *Paleozoic Paleogeography of the Western United States:* Pacific Section, Soc. Econ. Paleontologists Mineralogists, Pacific Coast Paleogeography Symp. 1, p. 137–155.

—— , and Thayer, T. P., 1978, Paleogeographic and paleotectonic implications of Mesozoic stratigraphy and structure in the John Day inlier of central Oregon, *in* Howell, D. G., and McDougall, K. A., eds., *Mesozoic Paleogeography of the Western United States:* Pacific Section, Soc. Econ. Paleontologists Mineralogists, Pacific Coast Paleogeography Symp. 2, p. 395–408.

Dilles, J. H., Wright, J. E., and Proffett, J. M., 1983, Chronology of early Mesozoic plutonism and volcanism in the Yerington District, NV: *Geol. Soc. America Abstr. with Programs*, v. 15, p. 383.

Eaton, J. P., 1963, Crustal structure from San Francisco, California, to Eureka, Nevada from seismic refraction measurements: *J. Geophys. Res.*, v. 68, p. 5789–5806.

Erickson, R. L., and Marsh, S. P., 1974, Paleozoic tectonics in the Edna Mountain quadrangle, Nevada: *U.S. Geol. Survey J. Res.*, v. 2, p. 331–337.

Ernst, W. G., 1976, *Petrologic Phase Equilibria:* San Francisco, W. H. Freeman Co., 333 p.

Evernden, J. F., and Kistler, R. W., 1970, Chronology of emplacement of Mesozoic batholithic complexes in California and western Nevada: *U.S. Geol. Survey Prof. Paper 623*, 67 p.

Farmer, G. L., and DePaolo, D. J., 1983, Origin of Mesozoic and Tertiary granite in the western United States and implications for pre-Mesozoic crustal structure — I. Nd and Sr isotopic studies in the geosyncline of the northern Great Basin: *J. Geophys. Res.*, v. 88, p. 3379–3401.

Frei, L. S., Magill, J. R., and Cox, A., 1984, Paleomagnetism results from the central Sierra Nevada: constraints on reconstruction of the western United States: *Tectonics*, v. 3, p. 157–177.

Gans, P. B., and Miller, E. L., 1983, Style of mid-Tertiary extension in east-central Nevada: *Utah Geol. Min. Survey Spec. Studies 59*, p. 107–139.

Hamilton, W., 1969, Mesozoic California and the underflow of the Pacific mantle: *Geol. Soc. America Bull.*, v. 80, p. 2409–2430.

—— , 1976, Tectonic history in the Riggins region, western Idaho: *Geol. Soc. America Abstr. with Programs*, v. 8, p. 378.

Harris, A. G., Wardlaw, B. R., Rust, C. C., and Merrill, G. K., 1980, Maps for assessing thermal maturity (conodont color alteration index maps) in Ordovician through Triassic rocks in Nevada and Utah and adjacent parts of Idaho and California: *U.S. Geol. Survey Misc. Inv. Series Map I-1249*.

Harwood, D. S., 1983, Stratigraphy of upper Paleozoic volcanic rocks and regional unconformities in part of the northern Sierra Nevada terrane, California: *Geol. Soc. America Bull.*, v. 94, p. 413–422.

Heller, P. L., Bowdler, S. S., Chambers, H. P., Coogan, J. C., Hagen, E. S., Shuster, M. W., Winslow, N. S., and Lawton, T. F., 1986, Time of initial thrusting in Sevier orogenic belt, Idaho-Wyoming and Utah: *Geology*, v. 14, p. 388–391.

Hillhouse, J. W., Gromme, C. S., and Vallier, T. L., 1982, Paleomagnetism and Mesozoic tectonics of the Seven Devils volcanic arc, northeastern Oregon: *J. Geophys. Res.*, v. 87, p. 3777–3794.

Hotz, P. E., and Willden, R., 1964, Geology and mineral deposits of the Osgood Mountains quadrangle, Humboldt County, Nevada: *U.S. Geol. Survey Prof. Paper 431*, 128 p.

Howard, K. A., 1971, Paleozoic metasediments in the northern Ruby Mountains, Nevada: *Geol. Soc. America Bull.*, v. 82, p. 259–264.

——, 1980, Metamorphic infrastructure in the northern Ruby Mountains, Nevada; *in* Crittenden, M. D., Jr., Coney, P. J. and Davis, G. H., eds., *Cordilleran Metamorphic Core Complexes:* Geol. Soc. America Mem. 153, p. 335–347.

Kay, M., and Crawford, J. P., 1964, Paleozoic facies from the miogeosynclinal to the eugeosynclinal belt in the thrust slices, central Nevada: *Geol. Soc. America Bull.*, v. 75, p. 425–454.

Kistler, R. W., 1974, Phanerozoic batholiths in western North America: a summary of some recent work on variations in time, space, chemistry, and isotopic composition: *Ann. Rev. Earth Planet. Sci.*, v. 2, p. 403–418.

——, 1978, Mesozoic paleogeography of California, *in* Howell, D. G., and McDougall, K. A., eds., *Mesozoic Paleogeography of the Western United States:* Pacific Section, Soc. Econ. Paleontologists Mineralogists, Pacific Coast Paleogeography Symp. 2, p. 75–85.

——, and Peterman, Z. E., 1973, Variations in Sr, Rb, K, Na, and initial Sr87/Sr86 in Mesozoic granitic rocks and intruded wall rocks in central California: *Geol. Soc. America Bull.*, v. 84, p. 3489–3512.

——, and Peterman, Z. E., 1978, Reconstruction of crustal blocks of California on the basis of initial strontium isotopic composition of Mesozoic granitic rocks: *U.S. Geol. Survey Prof. Paper 1071*, 17 p.

Klemperer, S. L., Hauge, T. A., Hauser, E. C., Oliver, J. E., and Potter, C. J., 1986, The Moho in the northern Basin and Range province, Nevada, along the COCORP 40°N seismic reflection transect: *Geol. Soc. America Bull.*, v. 97, p. 603–618.

Lachenbruch, A. H., and Sass, J. H., 1978, Models of an extending lithosphere and heat flow in the Basin and Range province: *Geol. Soc. America Mem. 152*, p. 209–251.

Lee, D. E., 1984, Analytical data for a suite of granitoid rocks from the Basin and Range province: *U.S. Geol. Survey Bull.*, v. 1602, 54 p.

——, Kistler, R. W., Friedman, I., and Van Loenen, R. E., 1981, Two mica granites of northeastern Nevada: *J. Geophys. Res.*, pt. B, v. 86, p. 10607–10616.

Lupe, R., and Silberling, N. J., 1985, Genetic relationship between lower Mesozoic continental strata of the Colorado Plateau and marine strata of the western Great Basin: significance for accretionary history of Cordilleran lithotectonic terrane, *in* Howell, D. G. ed., *Tectonostratigraphic Terranes of the Circum-Pacific Region:* Circum-Pacific Council Energy Min. Resources, Earth Sci. Series, no. 1, pp. 263–271.

Matti, J. C., and McKee, E. H., 1977, Silurian and Lower Devonian paleogeography of the outer continental shelf of the Cordilleran miogeocline, central Nevada, *in* Stewart, J. H., Stevens, C. H., and Fritsche, A. E., eds., *Paleozoic Paleogeography of the*

Western United States: Pacific Section, Soc. Econ. Paleontologists Mineralogists, Pacific Coast Paleogeography Symp. 1, p. 181–217.

McDonald, R. E., 1976, Tertiary tectonics and sedimentary rocks along the transition, Basin and Range province to plateau and thrust belt province, Utah: *Proc. Rocky Mountain Assoc. Geologists Symp.*, p. 281–317.

Means, W. D., 1962, *Structure and Stratigraphy in the Central Toiyabe Range, Nevada:* Berkeley, Calif., University of California Press.

Miller, C. F., and Bradfish, L. J., 1980, An inner Cordilleran belt of muscovite-bearing plutons: *Geology,* v. 8, p. 412–416.

Miller, D. M., 1982, Relations between younger-on-older and older-on-younger low-angle faults, Pilot Range, Nevada and Utah: *Geol. Soc. America Abstr. with Programs,* v. 14, p. 216.

Miller, E. L., Gans, P. B., and Garing, J., 1983, The Snake Range décollement — An exhumed mid-Tertiary ductile-brittle transition: *Tectonics,* v. 2, p. 239–263.

——, Holdsworth, B. K., Whiteford, W. B., and Rodgers, D., 1984, Stratigraphy and structure of the Schoonover complex, northeastern Nevada — Implications for Paleozoic plate-margin tectonics: *Geol. Soc. America Bull.,* v. 95, p. 1063–1076.

Muffler, L. J. P., 1964, Geology of the Frenchie Creek quadrangle, north-central Nevada: *Geol. Soc. America Bull.,* v. 47, p. 241–251.

Nichols, K. M., and Silberling, N. J., 1977, Stratigraphy and depositional history of the Star Peak Group (Triassic), northwestern Nevada: *Geol. Soc. America Spec. Paper 178,* 73 p.

Oldow, J. S., 1981, Kinematics of late Mesozoic thrusting, Pilot Mountains, Nevada: *J. Struct. Geology,* v. 3, p. 39–51.

——, 1984a, Evolution of a late Mesozoic back-arc fold and thrust belt, northwestern Great Basin, USA: *Tectonophysics,* v. 102, p. 245–274.

——, 1984b, Spatial variability in the structure of the Roberts Mountains allochthon, western Nevada: *Geol. Soc. America Bull.,* v. 95, p. 174–185.

Poole, F. G., 1974, Flysch deposits of the Antler foreland basin: *Soc. Econ. Paleontologists Mineralogists Spec. Publ. 22,* p. 58–82.

Proffett, J. M., 1977, Cenozoic geology of the Yerington district, Nevada, and implications for origin of basin-range faulting: *Geol. Soc. America Bull.,* v. 88, p. 247–266.

Prohdehl, C., 1979, Crustal structure of the western United States: *U.S. Geol. Survey Prof. Paper 1034,* 74 p.

Roberts, R. J., Hotz, P. E., Gilluly, J., and Purgusan, H. G., 1958, Paleozoic rocks of northcentral Nevada: *Amer. Assoc. Petrol. Geologists Bull.,* v. 42, p. 2813–2857.

Rowell, A. J., Rees, M. N., and Suczek, C. A., 1979, Margin of the North American continent in Nevada during Late Cambrian time: *Am. J. Sci.,* v. 279, p. 1–18.

Royce, F., Warner, M. A., and Resse, D. L., 1975, Thrust belt geometry and related stratigraphic problems Wy-Ida-Ut: *Proc. Rocky Mountain Assoc. Geologists Symp.,* p. 41–54.

Russell, B. J., 1984, Mesozoic geology of the Jackson Mountains, northwestern Nevada: *Geol. Soc. America Bull.,* v. 95, p. 313–323.

Saleeby, J. B., 1983, Accretionary tectonics of the North American Cordillera: *Ann. Rev. Earth Planet. Sci.,* v. 11, 45–73.

——, and Sharp, W., 1980, Chronology of structural and petrologic development of the SW Sierra Nevada foothills, California — Pt. I.: *Geol. Soc. America Bull.,* v. 91, p. 317–320.

Schweickert, R. A., and Cowan, D. S., 1975, Early Mesozoic evolution of the western Sierra Nevada, California: *Geol. Soc. America Bull.*, v. 86, p. 1329–1336.

——, Bogen, N. L., Girty, G. H., Hanson, R. E., and Merguerian, C., 1984, Timing and structural expression of the Nevadan orogeny, Sierra Nevada, California: *Geol. Soc. America Bull.*, v. 95, p. 967–979.

Sharp, W. D., 1988 Pre-Cretaceous, crustal evolution in the Sierra Nevada region, California: this volume, Chapt. 30.

Silberling, N. J., 1973, Geologic events during Permian-Triassic time along the Pacific margin of the U.S.: *Alberta Soc. Petrol. Geologists Mem. 2*, p. 345–362.

——, and Wallace, R. E., 1969, Stratigraphy of the Star Peak Group (Triassic) and overlying lower Mesozoic rocks, Humboldt Range, Nevada: *U.S. Geol. Survey Prof. Paper 592*, 50 p.

——, Jones, D. L., Blake, M. C., Jr., and Howell, D. G., 1984, Lithotectonic terrane map of the North American Cordillera: *U.S. Geol. Survey Misc. Field Studies Map C*.

Smith, D. L., 1984, Effects of unrecognized Oligocene extension in central Nevada on the interpretation of older structures: *Geol. Soc. America Abstr. with Programs*, v. 16, p. 660.

Smith, J. F., and Ketner, K. B., 1968, Devonian and Mississippian rocks and the date of the Roberts Mountains thrust in the Carlin–Pinon Range area, Nevada: *U.S. Geol. Survey Bull. 1251-I*, p. 11–118.

Smith, R. B., 1978, Seismicity, crustal structure and intraplate tectonics of the interior of the western Cordillera; *in* Smith, R. B., and Eaton, G. P., eds., *Cenozoic Tectonics and Regional Geophysics of the Western Cordillera:* Geol. Soc. America Mem. 152, p. 111–144.

Snoke, A. W., 1980, Transition from infrastructure to suprastructure in the northern Ruby Mountains, Nevada, *in* Crittenden, M. D., Jr., Coney, P. J., and Davis, G. H., eds., *Cordilleran Metamorphic Core Complexes:* Geol. Soc. America Mem. 153, p. 287–333.

Speed, R. C., 1977, Island arc and other paleogeographic terranes of late Paleozoic age in the western Great Basin, *in* Stewart, J. H., Stevens, C. H., and Fritsche, A. E., eds., *Paleozoic Paleogeography of the Western United States:* Pacific Section, Soc. Econ. Paleontologists Mineralogists, Pacific Coast Paleogeography Symp. 1, p. 349–362.

——, 1978a, Paleogeographic and plate tectonic evolution of the early Mesozoic marine province of the western Great Basin, *in* Howell, D. G., and McDougall, K. A., eds., *Mesozoic Paleogeography of the Western United States:* Pacific Section, Soc. Econ. Paleontologists Mineralogists, Pacific Coast Paleogeography Symp. 2, p. 253–270.

——, 1978b, Basinal terrane of the early Mesozoic marine province of the western Great Basin, *in* Howell, D. G., and McDougall, K. A., eds., *Mesozoic Paleogeography of the Western United States:* Pacific Section, Soc. Econ. Paleontologists Mineralogists, Pacific Coast Paleogeography Symp. 2, p. 237–252.

——, 1979, Collided Paleozoic microplate in the western United States: *J. Geology*, v. 87, p. 279–292.

——, 1983, Evolution of the sialic margin in the central-western United States, *in* Watkins, J. S., and Drake, C. L., eds., *Studies in Continental Margin Geology:* Amer. Assoc. Petrol. Geologists Mem. 34 (Hedberg Volume), p. 457–468.

——, and Jones, T. A., 1969, Synorogenic quartz sandstone in the Jurassic mobile belt of western Nevada — Boyer Ranch Formation: *Geol. Soc. America Bull.*, v. 80, p. 2551–2584.

——, and Kistler, K. W., 1980, Cretaceous volcanism, Excelsior Mountains, Nevada: *Geol. Soc. America Bull.*, v. 91, p. 392–398.

____ , and Sleep, N. H., 1982, Antler orogeny and foreland basin — A model: *Geol. Soc. America Bull.*, v. 93, p. 815–828.

Stauber, D. A., 1980, Crustal structure in the Battle Mountain heat flow high in northern Nevada from seismic refraction profiles and Rayleigh wave phase: Ph.D. thesis, Stanford Univ., Stanford, Calif., 316 p.

Stern, T. W., Bateman, P. C., Morgan, B. A., Newell, M. F., and Peck, D. L., 1981, Isotopic U-Pb ages of zircon from the granitoids of the central Sierra Nevada, California: *U.S. Geol. Survey Prof. Paper 1185*, 17 p.

Stewart, J. H., 1972, Initial deposits of the Cordilleran geosyncline — Evidence of a late Precambrian (850 my) continental separation: *Geol. Soc. America Bull.*, v. 83, p. 1345–1360.

____ , 1976, Late Precambrian evolution of North America — Plate tectonics implication: *Geology*, v. 4, p. 11–15.

____ , 1978, Basin and Range structure in western North America — A review, *in* Smith, R. B., and Eaton, G. P., eds., *Cenozoic Tectonics and Regional Geophysics of the Western Cordillera:* Geol. Soc. America Mem. 152, p. 1–32.

____ , 1980, Geology of Nevada, a discussion to accompany the geologic map of Nevada: *Nev. Bur. Mines Geology Spec. Publ. 4*, 136 p.

____ , and Poole, F. G., 1974, Lower Paleozoic and uppermost Precambrian Cordilleran miogeocline, Great Basin: *Soc. Econ. Paleontologists Mineralogists Spec. Publ. 22*, p. 27–57.

____ , and Suczek, C. A., 1977, Cambrian and latest Precambrian paleogeography and tectonics in the western United States, *in* Stewart, J. H., Stevens, C. H., and Fritsche, A. E., eds., *Paleozoic Paleogeography of the Western United States:* Pacific Section, Soc. Econ. Paleontologists Mineralogists, Pacific Coast Paleogeography Symp. 1, p. 1–17.

Vetter, U. R., and Ryall, A. S., 1983, Systematic change of focal mechanism with depth in the western Great Basin: *J. Geophys. Res.*, v. 88, p. 8237–8250.

Wernicke, B., 1981, Low angle normal faults in the Basin and Range province — Nappe tectonics in an extending orogen: *Nature*, v. 291, p. 645–648.

____ , Spencer, J. E., Burchfiel, B. C., and Guth, P. L., 1982, Magnitude of crustal extension in the southern Great Basin: *Geology*, v. 10, p. 499–502.

Zoback, M. L., Anderson, R. E., and Thompson, G. A., 1981, Cenozoic evolution of state of stress and style of tectonism of the Basin and Range province of western U.S.: *Philos. Trans. Roy. Soc. London*, v. A300, p. 407–434.

Arthur W. Snoke
Department of Geology and Geophysics
University of Wyoming
Laramie, Wyoming 82071

David M. Miller
U. S. Geological Survey
National Center
Reston, Virginia 22092

23

METAMORPHIC
AND TECTONIC
HISTORY OF
THE NORTHEASTERN
GREAT BASIN

Metamorphic rocks in terranes ranging from lower greenschist facies to migmatitic amphibolite facies underlie parts of many ranges within the northeastern Great Basin in Nevada, Utah, and Idaho. Although these metamorphic terranes are discontinuously exposed, we infer that they represent parts of a once-continuous metamorphosed region that was exhumed to different crustal levels by Cenozoic crustal extension. The high-grade terranes are tectonic windows of the Cenozoic and Mesozoic Cordilleran infrastructure, whereas low-grade terranes provide important structural and stratigraphic links to unmetamorphosed cover rocks, also extensively exposed in the northeastern Great Basin.

The metamorphic terranes of the northeastern Great Basin occur along the margin of the Precambrian continent, where it was thinned by Late Proterozoic rifting; they lie east of the Paleozoic Antler orogenic belt but mainly west of the locus of late Mesozoic and early Cenozoic overthrusting. During middle and late Mesozoic time, this part of the hinterland of the Sevier belt underwent deformation, plutonism, and regional metamorphism associated with lateral shortening and crustal thickening. Structural elements associated with Mesozoic metamorphism were pre- and synmetamorphic thrust faults, large-scale recumbent folds displaying variable vergence, and polyphase metamorphic fabrics. During middle Tertiary time, the Mesozoic hinterland metamorphic terranes were tectonically denuded of their upper-crustal cover rocks by crustal extension along low-angle faults. This exhumation of the high-grade terranes was relatively rapid, with mid-crustal mylonitic shear zones and upper-crustal low-angle normal faults as the key structural manifestations.

The protoliths of the hinterland metamorphic rocks include Precambrian crystalline basement and supracrustal rocks, and sedimentary rocks of the Cordilleran miogeocline. In many of the metamorphic terranes, these rocks were intruded by Late Jurassic and Tertiary (Eocene and Oligocene) granitoid rocks, including peraluminous granitoids that are especially characteristic of the higher-grade terranes. These peraluminous granitoids commonly yield initial $^{87}Sr/^{86}Sr$ ratios greater than 0.710, indicating the prominent role of continental crust in the melted source terrane. Because these granitoids were principally derived by the anatexis of crustal rocks, their intrusive ages must closely approximate a time of deep crustal metamorphism in the region.

The polymetamorphic terranes in the northeastern Great Basin display varying degrees of Mesozoic and Cenozoic imprints. Low-grade metamorphic terranes in the Black Pine Mountains and the Pilot Range area chiefly display a Mesozoic metamorphic history, whereas the high-grade terrane in the Ruby Mountains and East Humboldt Range only locally preserves Mesozoic structural features, which were strongly overprinted by Tertiary ductile fabrics and pervasively intruded by Tertiary granitoid magmas. The difficulty with distinguishing timing of metamorphism and accompanying tectonic styles has contributed much uncertainty to tectonic models for the area. On the basis of our review of metamorphism and tectonics of the region, characteristic tectonic styles emerge for the two periods of metamorphism. We advocate tectonic wedging/overthrusting and crustal-scale detachment to account for Mesozoic and Cenozoic metamorphism, respectively.

INTRODUCTION

In northeast Nevada, northwest Utah, and southeast Idaho, metamorphic rocks ranging from low-grade, greenschist facies terranes to high-grade, amphibolite facies migmatitic terranes underlie parts of many mountain ranges within the late Cenozoic Basin and Range province. Most of the metamorphic rocks are Upper Proterozoic and Paleozoic miogeoclinal strata that were metamorphosed during Mesozoic and Cenozoic time, but in a few ranges Archean and Lower Proterozoic rocks that were metamorphosed during Precambrian time occur. Although the Mesozoic metamorphic terranes are discontinuously exposed, we infer that they represent parts of a once-continuous metamorphosed region that was segmented and exhumed to different crustal levels in Cenozoic time. This hypothesis is supported by data from drill holes through unmetamorphosed near-surface rocks, located between the metamorphic terranes, that intersected metamorphosed strata at moderate depth. The high-grade terranes in most cases provide windows into the Mesozoic and Cenozoic Cordilleran infrastructure, whereas the low-grade terranes provide important structural and stratigraphic links to the unmetamorphosed cover rocks, also extensively exposed throughout the northeastern Great Basin (Fig. 23-1).

   The metamorphic terranes of the northeastern Great Basin occur in a rifted zone of the Precambrian continent; they lie east of the middle Paleozoic Antler orogenic belt and west of the late Mesozoic and early Cenozoic Sevier fold-and-thrust belt (Fig. 23-1). The terranes are conveniently referred to as "hinterland metamorphic complexes" with regard to their spatial relation to the Sevier fold-and-thrust belt, although much of the hinterland metamorphism apparently predated major movement on thrusts in the fold-and-thrust belt.

   This synthesis of metamorphism and tectonism encompasses a region greater than 100,000 km² (Fig. 23-1) in the northern part of the hinterland of the Sevier orogenic

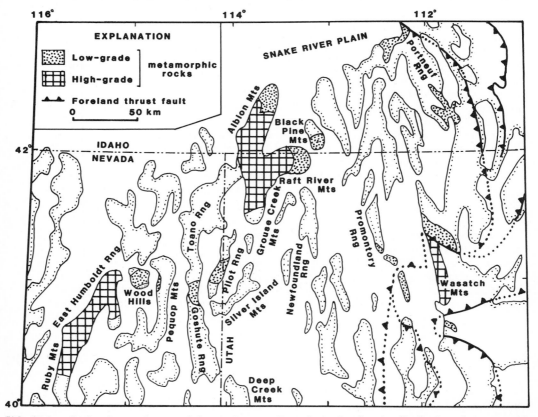

FIG. 23-1.   Regional tectonic map of the northeastern Great Basin showing the distribution of metamorphic terranes. Just west of the area shown are the Sr and Nd ratio lines referred to in the text.

belt, as defined by Armstrong (1968a). We present case studies for five subareas within this region to document the character of metamorphism and associated tectonic events at various crustal levels: (1) Black Pine Mountains; (2) Pilot Range; (3) Wood Hills and northern Pequop Mountains; (4) Albion, Raft River, and Grouse Creek Mountains; and (5) Ruby Mountains and East Humboldt Range (Fig. 23-1). With this data base along with ancillary data from other metamorphosed terranes in the region, we address several tectonic questions critical in our regional synthesis: (1) timing of Mesozoic metamorphism and deformation in the hinterland and its relation to shortening in the Sevier fold-and-thrust belt, (2) possible causes of hinterland metamorphism and P-T constraints, (3) role of crystalline basement in the tectonic evolution of the metamorphic terranes, and (4) significance of mylonitic rocks in the metamorphic terranes. Finally, we synthesize the main events of the region, describe tectonic models for Mesozoic shortening and Cenozoic extension, and enumerate several problems requiring further study.

Our studies form much of the basis for this paper, but the ideas presented herein represent an evolution from earlier concepts, particularly those from influential syntheses of Mesozoic and Cenozoic tectonics by Misch (1960), Misch and Hazzard (1962), Armstrong and Hansen (1966), and Armstrong (1968a, 1972, 1982).

GEOLOGIC FRAMEWORK

The area of our synthesis lies east of the $I_{Sr} = 0.706$ line considered by many investigators to mark the rifted west edge of Precambrian crust (Armstrong et al., 1977; Kistler and Peterman, 1978; Speed, 1982). Farmer and DePaolo (1983) placed the basement edge in northeastern Nevada 100 km farther east, as marked by the $I_{Sr} = 0.708$ and $\epsilon_{Nd} = -7$ contour; this boundary is also west of the region under consideration.

A complex suture between Archean rocks of the Wyoming Province and Proterozoic rocks of the Colorado Front Ranges, exposed in the Medicine Bow Mountains of southeastern Wyoming (Duebendorfer and Houston, 1986; Karlstrom and Houston, 1984), extends west to eastern Utah (Sears et al., 1982) and perhaps northeastern Nevada (Condie, 1982). Nelson and DePaolo (1985) identified two distinct Precambrian crustal provinces based on Nd model ages: northernmost Utah, characterized by Nd model ages ⩾2.5 b.y.; and parts of eastern Nevada and western Utah, characterized by model ages of 2.0-2.3 b.y. (Nelson et al., 1983).

Precambrian basement rocks are sparsely exposed west of the Colorado, Utah, and Wyoming Rocky Mountains; three exposures are noteworthy: (1) Archean and Proterozoic Farmington Canyon Complex, Wasatch Mountains (Bryant, 1979, Chapter 16, this volume); (2) Archean rocks of the Albion, Raft River, and Grouse Creek Mountains (Armstrong, 1976; Armstrong and Hills, 1967; Compton et al., 1977); and (3) Archean(?) rocks in a recumbent fold in the northern East Humboldt Range (Snoke and Lush, 1984).

During the early Middle Proterozoic and again during the Late Proterozoic, the Precambrian craton of western North America was rifted (Gabrielse, 1972; Sears and Price, 1978; Stewart, 1972, 1976). The Middle Proterozoic Belt and Purcell Supergroups and equivalent strata are the supracrustal rocks associated with the early rifting event, whereas the Middle(?) and Upper Proterozoic Windermere Group and equivalent strata are the sedimentary rocks deposited during the later rifting. Possible Belt equivalent strata are sparse in the area of our synthesis but probable Windermere equivalents, such as the Upper Proterozoic Huntsville sequence of Crittenden et al. (1971) and McCoy Creek Group of Misch and Hazzard (1962), are widespread and represent the earliest sediments associated with the establishment of the late Precambrian and early Paleozoic miogeocline of western North America (Stewart, 1972).

Marine and nonmarine Upper Proterozoic and Lower Cambrian quartzite, siltstone, shale of the McCoy Creek Group and Huntsville sequence, and overlying quartzite account for at least half the thickness of the late Precambrian and lower Paleozoic Cordilleran miogeocline. These clastic rocks thicken irregularly westward from the Wasatch Mountains to over 6000 m in eastern Nevada (Misch and Hazzard, 1962; Stewart and Poole, 1974). During Early Cambrian to Late Devonian time, these clastic rocks were overlain by a heterogeneous sequence of shallow-marine intertidal and supratidal carbonate and detrital strata. This shallow-marine deposition was terminated in eastern Nevada with the establishment of a Mississippian foreland basin east of orogenic highlands associated with the emplacement of the Roberts Mountain allochthon, the principal event of the Late Devonian and Early Mississippian Antler orogeny (Dickinson *et al.*, 1983; Poole, 1974; Roberts *et al.*, 1958). Depositional patterns in western Utah were less affected by the Antler orogeny, but a thick shale wedge formed in the middle and Late Mississippian.

During the Pennsylvanian and Permian, shallow carbonate depositional environments returned to the Antler foreland region, but much of the area experienced a deformational event manifested by a Middle Pennsylvanian regional unconformity (Marcantel, 1975; Steele, 1960) and an increase in coarse detritus in sedimentary basins east of the Antler belt (Ketner, 1977). Approximately coeval with this deformational event in northeastern Nevada was the development of the Oquirrh basin of northern Utah and southeastern Idaho (Jordan and Douglass, 1980; Roberts *et al.*, 1965). These tectonostratigraphic events in the miogeocline coincide with the development of the Ancestral Rockies within the craton, which may represent intracratonic response to the continental collision manifested in the Ouachita orogenic belt (Kluth and Coney, 1981).

The Early Triassic Sonoma orogeny ended Paleozoic depositional patterns (Silberling and Roberts, 1962) by eastward thrusting of Pennsylvanian and Permian oceanic rocks (the Havallah sequence), as the Golconda allochthon, over rocks of the Antler belt. Remnants of the Golconda allochthon lie west of our synthesis area, but a widespread disconformity between mid-Permian and Lower Triassic rocks in eastern Nevada and western Utah may be a manifestation of this orogenic event (Collinson *et al.*, 1976).

Despite the episodic tectonism from the middle Paleozoic to the Triassic, marine sedimentation essentially was continuous from Late Proterozoic or Early Cambrian to Early Triassic time in the area of the hinterland under discussion (Armstrong, 1968a). However, in the Late Triassic, the eastern Great Basin experienced a major change from marine to nonmarine sedimentation. The stratigraphic record following marine withdrawal is very sparse in the eastern Great Basin, but isolated exposures of Upper Triassic and Lower Jurassic continental sedimentary rocks near Currie, Nevada, suggest that similar rocks may have covered the region and perhaps were coextensive with the Upper Triassic and Lower Jurassic continental sequence of the Colorado Plateau (Stewart, 1980). The Cretaceous Newark Canyon Formation (conglomerate, sandstone, siltstone, mudstone, shale, and limestone) is exposed at scattered localities in east-central Nevada (Fouch, 1979), and depositional facies indicate the development of isolated basins, possibly of extensional origin, in at least part of the hinterland of the Sevier orogenic belt.

A broad belt of eastward-directed thin-skinned thrust faults developed near the miogeocline-craton hinge during Cretaceous to early Eocene time (Armstrong and Oriel, 1965). Deeper, westward extensions of the thrusts, in the hinterland of the thrust belt (Armstrong, 1968a) are poorly understood, but middle to late Mesozoic thrusts, folds, and metamorphism point to regional deformation in the hinterland (Allmendinger and Jordan, 1981; Allmendinger *et al.*, 1984; Armstrong and Hansen, 1966; Coats and Riva, 1983; Kistler *et al.*, 1981; Misch, 1960; Snoke and Lush, 1984). The thrust faults and metamorphism in the hinterland appear to be older in most cases than the thin-skinned structures in the fold-and-thrust belt (Allmendinger *et al.*, 1984).

Both metaluminous and peraluminous Mesozoic plutons are widely scattered in the eastern Great Basin. In the region of Fig. 23-1, the Mesozoic plutons are largely, if not wholly, of Late Jurassic age (Moore and McKee, 1983) but Late Cretaceous plutons are widespread to the south and west. The Late Jurassic plutons in many places cut, or are associated with, faults and folds (Allmendinger *et al.*, 1984).

Paleocene strata are not documented within or near our study area, although the Cretaceous Newark Canyon Formation could include Paleocene rocks according to Fouch (1979), and Eocene alluvial fan-lacustrine sequences may also include older Tertiary strata (J. F. Smith and Ketner, 1976; Fouch, 1979). However, by the middle Eocene, a series of broad, shallow nonmarine basins, possibly caused by upper crustal extension accommodated by detachment faults,[1] were established in the northeastern Great Basin (Fouch, 1979; Solomon *et al.*, 1979). Beginning perhaps as early as 43 m.y., calc-alkaline magmatism began with widespread silicic ash-flow tuff inundating the landscape, burying tilted and folded Eocene rocks (J. F. Smith and Ketner, 1976). Deformation that affected these Eocene rocks must have immediately preceded the massive ash-flow tuff eruptions; it perhaps was caused by large-scale, low-angle normal faulting associated with the initiation of the middle Tertiary magmatism. Plutons of late Eocene and early Oligocene age that represent the magma chambers that fed the tuff eruptions are now widely exposed in the northeastern Great Basin. During late Oligocene and Miocene time, extensive landscape disruption established numerous smaller basins that served as depocenters for the widespread Humboldt and Salt Lake Formations (Van Houten, 1956). After a brief middle Miocene magmatic lull (McKee *et al.*, 1970), basaltic dikes and plugs, and rhyolite flows and shallow intrusive masses, were widely emplaced. In several areas, low-angle normal faults (detachments) cut these middle Miocene igneous rocks, indicating low-angle faulting at least until 8 m.y. ago. Faulting that outlined the present ranges is poorly dated, but probably began after 10 m.y. and was perhaps much later (Zoback *et al.*, 1981); many of these range-bounding faults show geomorphic evidence for Quaternary movement.

TECTONIC STYLE
OF UNMETAMORPHOSED ROCKS

Essentially unmetamorphosed rocks crop out in all of the mountain ranges of the eastern Great Basin, and until a few decades ago were regarded as virtually the sole constituent of these ranges. Misch (1960) and Armstrong and Hansen (1966) first highlighted the metamorphic rocks and granitoids scattered throughout the northeastern Great Basin, and subsequent work has shown additional tracts of metamorphic rocks. However, most of the pre-Cenozoic strata west of the foreland fold-and-thrust belt and east of approximately the Nevada-Utah border are unmetamorphosed, and greater than 50% of the strata cropping out in ranges from that state border west to the Roberts Mountain allochthon are unmetamorphosed.

We regard rocks with no obvious textural or mineralogic criteria for at least greenschist facies metamorphism as being "essentially unmetamorphosed." Included are widespread tracts of subgreenschist facies, slightly metamorphosed rocks outlined by Christensen (1975), and carbonate rocks that locally achieved fairly high temperatures based on conodont color alteration index but that show no dynamothermal fabrics or

---

[1]*Detachments* or *detachment faults* we define as low-angle normal faults that contain upper crustal rocks, commonly in tilted fault blocks, in the hanging wall, and rocks derived from greater depth in the footwall. The footwall rocks commonly were metamorphosed and ductilely deformed. Other authors have used "detachment" in a different context (e.g., Dahlstrom, 1970; Pierce, 1963).

crystallization to new mineral assemblages. Upper Proterozoic rocks, once deeply buried, are commonly of lower greenschist facies; in those ranges where they grade up stratigraphic section to unmetamorphosed rocks they are included as "essentially unmetamorphosed." However, these rocks show modest dynamothermal effects associated with metamorphism and in some cases are strongly folded and cleaved (Link *et al.*, 1985). Contact metamorphism is not considered.

Direct information on the temperature and pressure conditions experienced by the essentially unmetamorphosed rocks are sparse. Conodont alteration indexes in these ranges are generally low (Harris *et al.*, 1980), corroborating textural and compositional data. Higher alteration indexes in places indicate that rocks locally reached higher temperatures than suggested by textures, perhaps because of hydrothermal alteration or contact metamorphism. Regionwide, Upper Proterozoic to Triassic strata are some 10 km thick, and thus it is not surprising that the oldest strata exhibit the slight metamorphism expected from burial within crust with typical geothermal gradients. Following early Mesozoic time, varying amounts of upper Paleozoic and lower Mesozoic strata were eroded (Armstrong, 1968a) as indicated by middle and upper Cenozoic deposits unconformably overlying Permian and Carboniferous strata. Thus, "burial" metamorphism peaked during the early or middle Mesozoic, before substantial erosion.

The complex pattern of faulting of unmetamorphosed rocks in many ranges eluded synthesis until detailed studies and new techniques were employed. In all cases, observed or inferred late Cenozoic faults border the mountain ranges; they are generally north-striking, moderately dipping normal faults of the Basin and Range event. Related high-angle normal faults occur at places within the ranges, but in general, older faults predominate. Older faults are at high and low angles to horizontal and to bedding. In some cases bedding and faults have been rotated during faulting (E. L. Miller *et al.*, 1983), and thus both geographic and bedding-plane reference frames must be used. Examples of thrust faults, normal faults at low angles to bedding, low-angle (nearly horizontal) normal faults, and high-angle normal and reverse faults are reported in numerous ranges (e.g., Allmendinger and Jordan, 1985; Coats and Riva, 1983; Link, *et al.*, 1985; Schneyer, 1984). All of these fault styles have been documented as Mesozoic in places (Allmendinger *et al.*, 1984), and all styles except thrust faults have been documented in other places as Cenozoic (E. L. Miller *et al.*, 1983; Schneyer, 1984). Mesozoic and Cenozoic faults are commonly difficult to distinguish, but in some cases granitoids and volcanic rocks provide time constraints.

DETAILED CASE STUDIES OF AREAS OF CONTRASTING METAMORPHIC GRADE

Throughout the northeastern Great Basin, mountain ranges contain rocks exhumed from contrasting crustal levels. Many of the ranges contain upper-crustal, fossiliferous rocks that are essentially unmetamorphosed, whereas others contain large tracts of the same strata that are regionally metamorphosed. Low-grade metamorphic terranes chiefly display a Mesozoic metamorphic history with only a weak Cenozoic imprint, whereas high-grade terranes display locally preserved Mesozoic structural features that were overprinted strongly by Cenozoic ductile fabrics and intruded pervasively by Cenozoic granitoid magmas of crustal derivation.

As used in the ensuing five case studies, "low-grade" metamorphism refers to predominantly greenschist facies, and "high-grade" metamorphism to predominantly amphibolite facies. The areas are described in order of increasing metamorphic grade.

Black Pine Mountains, Idaho

The Black Pine Mountains lie 10 km northeast of the high-grade rocks of the Raft River Mountains (Fig. 23-1), and are bordered on the north and east by ranges in which upper Paleozoic unmetamorphosed strata crop out. J. F. Smith (1982, 1983) mapped the Black Pine Mountains, and Allmendinger *et al.* (1984) outlined further detailed studies. As described by Smith, the range consists of a nested sequence of low-angle fault plates of low-grade, middle and upper Paleozoic strata.

The Black Pine Mountains are divided by a northeast-striking high-angle(?) reverse fault into north and south blocks (Fig. 23-2) containing rocks of similar age but dissimilar stratigraphy. The north block consists of Upper Pennsylvanian and Lower Permian strata in an upright position. In this block the younger, structurally higher strata were inferred by J. F. Smith (1982) to rest in low-angle fault contact on older rocks because bedding in underlying rocks is truncated. Rocks in the north block attained low to moderate temperatures, as indicated by conodont color alteration indices (CAI) of 3 and 4, roughly corresponding to 100 to 300°C (J. F. Smith, 1983). Neither ductile deformation textures nor minor structures are reported for rocks in this block.

In contrast, rocks in the south block are cut by numerous low-angle faults, are folded and cleaved, and widely attained CAI values of 5 and higher (greater than 300°C). The lower plate (Fig. 23-2) consists of faulted Devonian, Mississippian, and Lower Pennsylvanian rocks; silty rocks are cleaved and metamorphosed to muscovite-chlorite and chloritoid-andalusite assemblages (Christensen, 1975). Mica beards and quartz pressure shadows on quartz grains indicate synkinematic mineral growth. Structurally overlying the lower plate rocks are Pennsylvanian limestone and sandstone of the Oquirrh Formation, of which at least 2400 m are overturned. The axis of the large recumbent fold associated with the overturned strata trends approximately north; smaller chevron folds also occur in parts of the plate. Parts of the middle plate are internally imbricated by older-over-

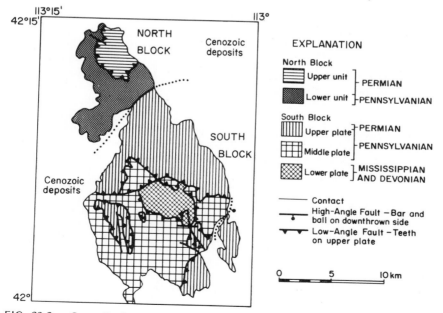

FIG. 23-2. Generalized tectonic map of the Black Pine Mountains, Idaho, after J. F. Smith (1982, 1983).

younger (thrust) faults (Allmendinger *et al.*, 1984). The upper plate consists of Upper Pennsylvanian and Lower Permian sandstone and siltstone that are unlike age-equivalent strata in the north block. The upright strata in this plate rest in structural discordance on overturned strata of the middle plate and cleaved rocks of the lower plate.

Abundant minor structures in rocks of the south block include minor folds, cleavage in silty rocks, and closely spaced calcite veins (J. F. Smith, 1982). Folds are more common near some low-angle faults, but are not confined to faulted zones; they range from open to isoclinal and apparently are not systematically oriented. Veins near low-angle faults have been elongated locally by as much as 100% parallel to the faults (Allmendinger *et al.*, 1984).

Metamorphism of lower-plate rocks in the south block occurred after or during the low-angle faulting because metamorphic grade is similar in all structural plates, but cleavage composed of recrystallized micas indicates strain in the rocks during metamorphism (Allmendinger *et al.*, 1984). Metamorphism is partly dated by nine whole-rock K-Ar ages ranging from 76.6 to 111 m.y. on slates of the lower plate (J. F. Smith, 1982). The high CAI values for rocks in several plates support the interpretation that the K-Ar ages record Cretaceous cooling following low-grade metamorphism, probably during middle and late Mesozoic time. The only firm upper bound on the timing of metamorphism is provided by unexposed range-bounding normal faults that cut upper Miocene strata (Covington, 1983).

The simplest interpretation of the nested sequence of low-angle faults that attenuate section is that they are extensional. However, the overturned strata in the middle plate most plausibly represent a dismembered thick overturned limb of an east-vergent(?) recumbent fold; this structure, along with the associated imbricate thrust faults, is more consonant with the kinematics of shortening (thrusting).

Pilot Range Area, Nevada and Utah

The Pilot Range and adjacent Silver Island Mountains and northern Toano Range contain a rock record of Mesozoic metamorphism, ductile deformation, faulting, and plutonism; and Cenozoic faulting and igneous activity. A gently dipping detachment fault — the Pilot Peak décollement — occurs in all three ranges. Strata in the Pilot Range area belong to three tectonostratigraphic groups distinguished by age and depositional facies (Fig. 23-3). In the footwall of the décollement are Upper Proterozoic, Cambrian, and Ordovician metamorphosed clastic and carbonate rocks. The hanging-wall rocks are comprised of Upper Cambrian to Permian strata that are predominantly unmetamorphosed carbonate rocks. Cambrian strata in the footwall were deposited on a continental shelf margin, whereas age-equivalent strata in the hanging wall were deposited on a carbonate platform shelf (McCollum and McCollum, 1984). Structurally and depositionally above the Paleozoic strata in the hanging wall are Oligocene and Miocene volcanic rocks and terrigeneous sedimentary rocks.

Greenschist and amphibolite facies metamorphism affected rocks in the footwall of the décollement. Rocks containing retrogressed staurolite and probable garnet occur east of Pilot Peak and locally in the Toano Range (Fig. 23-3), and contrast with the remainder of footwall rocks, which are typically metamorphosed to the biotite or chlorite zone of greenschist facies. The higher-grade rocks are crudely spatially associated with small, foliated Jurassic two-mica granite bodies in the Pilot Range (D. M. Miller, *et al.*, in press) and Mesozoic two-mica granite in the Toano Range (L. L. Glick, unpublished mapping, 1985). Although the hanging-wall rocks generally show no signs of regional

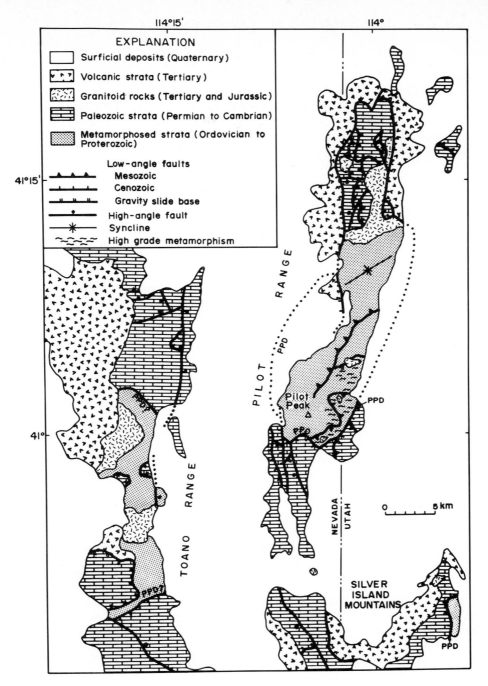

FIG. 23-3. Generalized geologic map of the Pilot Range area, Utah and Nevada. Pilot Peak décollement is labeled "PPD."

metamorphism, color alteration values for conodonts in these rocks indicate that they are heated to temperatures ranging from 170°C to greater than 400°C. These rocks show wide contact-metamorphic aureoles near Mesozoic and Cenozoic granitoid plutons.

Relics of metamorphic minerals such as wollastonite, tremolite, actinolite, garnet, staurolite, cordierite, and hornblende occur in the high-grade rocks. Pseudomorphs of garnet, staurolite, and cordierite contain retrograde assemblages such as epidote, hornblende, white mica, chlorite, and biotite. Pseudomorph outlines in Pilot Range rocks have strong preferred orientations, but retrogressive minerals do not, indicating that amphibolite facies metamorphism accompanied deformation, but retrogression was most likely a thermal event accompanied by little strain. Although amphibolite facies minerals and synkinematic quartz-feldspar veins in the Pilot Range are best developed in the area of exposed Jurassic two-mica granite, they are not completely restricted to exposures associated with the pluton. Thus, the metamorphism may not have been caused by pluton emplacement. Hornblende in the Pilot Range is a product of retrogression adjacent to Tertiary granodiorite dikes and small plutons, suggesting that Tertiary plutonism caused the retrogression.

The sequence of structural development within each tectonostratigraphic group is well established, but major structures bounding the groups make correlation of structures among the groups more difficult. The Pilot Peak décollement separates footwall metamorphic rocks and hanging-wall sedimentary rocks. The décollement in the northern Pilot Range apparently was intruded by an early Oligocene pluton (Fig. 23-3). Tertiary strata in the Pilot Range are underlain by one or more low-angle faults that cut and have caused the rotation of strata as young as 8 m.y. (D. M. Miller, 1985); in adjacent ranges the Tertiary strata commonly rest unconformably on older strata.

The styles of structures and amounts of strain affecting rocks in the footwall of the Pilot Peak décollement are strongly dependent on metamorphic grade and lithology. Upper Proterozoic quartzite and siltstone strata that are metamorphosed to chlorite zone are little strained but form large southeast-overturned folds. Proterozoic rocks that contain biotite and, locally, garnet are more tightly folded, penetratively cleaved, and quartzite shows moderate strain. Here, mylonite occurs in quartzite deformed along bedding-plane fault zones. Textures in the metamorphic rocks indicate that strain accompanied recrystallization. East of Pilot Peak (Fig. 23-3), where Upper Proterozoic and Cambrian clastic and carbonate strata of amphibolite facies are complexly folded and foliated, three sets of folds formed following low-angle faulting within the metamorphosed rocks. The foliation and metamorphic fabrics formed after intrusion of a pluton dated at 155 to 165 m.y. (D. M. Miller, 1984); metamorphic hornblende yields a $^{40}Ar/^{39}Ar$ plateau age of 150 m.y., which indicates that peak metamorphic temperatures existed for a short time in the Late Jurassic. Temperatures at peak metamorphism were 520 to 580°C, and pressure less than 6 kb (D. M. Miller and Lush, 1981).

Bedding-plane faults cutting footwall strata formed prior to or during synmetamorphic folding are generally younger-over-older; that is, they attenuate stratigraphic section. However, a fault east of Pilot Peak duplicates part of the Upper Proterozoic and Cambrian section, and thus apparently represents shortening. The youngest low-angle normal faults beneath the Pilot Peak décollement cut foliation and bedding at moderate angles but are cut by the décollement.

Structures in rocks above the décollement generally show brittle styles. Bedding-plane faults that generally attenuate section are the earliest structures, and are cut and rotated by high-angle faults. Generally, east-striking faults functioned as tear faults for the bedding-plane faults and also were reactivated as later north-striking high-angle faults formed. The north-striking faults are normal, down to the west, and rotate bedding and the bedding-plane faults. All of these faults in the hanging wall are cut by the underlying décollement.

The Pilot Peak décollement, as exposed south of Pilot Peak, lies generally parallel to bedding in Cambrian schist, marble, and phyllite of the footwall, and discordantly beneath moderately dipping Cambrian and Ordovician limestone. In the Silver Island Mountains, it occupies a similar structural position, but in the Toano Range it appears to lie on low-grade Middle Ordovician strata, with Devonian rocks apparently comprising the hanging wall. Although the décollement generally displays ductile structures in the foot-wall, the last movement in the Pilot Range brittlely affected hanging-wall and footwall rocks in a zone less than 1 m wide. Bedding and cleavage in rocks of the footwall are folded into parallelism with the fault where it cuts across strata, and drag folds indicate an east-ward component of movement of the décollement after metamorphism. In contrast, tilted fault blocks of hanging-wall rocks and offset depositional facies suggest westward translation of hanging-wall rocks. The décollement cuts virtually all faults and folds in hanging-wall and footwall strata but is intruded by granodiorite sills about 40 m.y. old that show only minor local brecciation.

The structural development of the Pilot Range area encompasses Mesozoic and Cenozoic faulting and folding. Southeast-facing folds indicate southeast-directed motion on related bedding-plane faults that attenuated the metamorphosed rocks (D. M. Miller and Lush, 1981) during the Late Jurassic and Early Cretaceous interval of metamorphism. Although related minor folds deformed granite dikes that were intruded 155–165 m.y. ago, the rocks had cooled to Ar blocking temperatures for hornblende by 150 m.y. and for muscovite and biotite by 83–64 m.y.[2] (D. M. Miller, 1984), constraining metamorphism to the Late Jurassic and Late Cretaceous. Low-angle, bedding-plane faults in unmetamor-phosed strata were kinematically similar to those in metamorphosed strata and, therefore, also may have been Mesozoic structures. All metamorphic rocks were relatively cool by 40–36 m.y., when plutons and dikes with chilled margins were intruded widely in the Pilot Range. The plutons cut the Pilot Peak décollement, indicating that the detachment fault and associated hanging-wall normal faults had emplaced unmetamorphosed on metamorphosed strata by latest Eocene time and possibly during the Late Cretaceous cooling recorded by K-Ar mica ages. Subsequent low- and high-angle faulting, also in an extensional upper-crustal setting, rotated blocks with strata as young as 8 m.y. above de-tachment faults at a structural level much higher than the Pilot Peak décollement (Fig. 23-4). These younger detachment faults may have been reactivated by faults blocking out the present ranges, which cut deposits as young as Pleistocene.

Wood Hills and Northern Pequop Mountains, Nevada

The Wood Hills and northern Pequop Mountains lie east of the high-grade migmatitic metamorphic terrane of the northern East Humboldt Range, which includes metamor-phosed Paleozoic strata of the Cordilleran miogeocline, and west of the low-grade terrane cropping out in the Pilot and Toano Ranges (Fig. 23-1). Metamorphosed Paleozoic miogeoclinal carbonate, siliciclastic, and pelitic rocks in the Wood Hills and northern Pequop Mountains, described by Thorman (1970), lie structurally below unmetamorphosed carbonate miogeoclinal strata in the hanging wall of a regionally extensive low-angle fault (Fig. 23-5). The protoliths of the footwall metasedimentary rocks range from Late Cambrian through Late Devonian in age, whereas ages of unmetamorphosed hanging-wall

[2] Also D. M. Miller, *et al.*, in press.

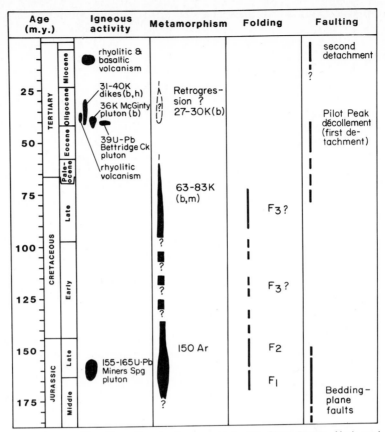

FIG. 23-4. Time-line diagram for events in the Pilot Range, Utah and Nevada. Events (in m.y.) are dated by the following techniques: Ar, $^{40}Ar/^{39}Ar$; K, K-Ar; U-Pb, U-Pb on zircon. Materials dated are: b, biotite; m, muscovite; h, hornblende. Time scale used is that of Palmer (1983).

rocks range from Ordovician through Permian, including much of the same stratigraphic succession.

Although Thorman (1970) referred to the low-angle fault that separates metamorphosed and unmetamorphosed Paleozoic rocks in the Pequop Mountains as the "Wood Hills thrust," it is a detachment fault according to our usage. In the northern Wood Hills, bedding in the hanging wall dips chiefly eastward into the detachment, consistent with a low-angle normal fault interpretation, but in the southern part of the Wood Hills and the northern Pequop Mountains, bedding in the hanging wall is crudely parallel to the low-angle fault. Thorman (1970) suggested that these relations indicate regional thrusting of décollement type; however, the emplacement of unmetamorphosed rocks on metamorphosed rocks is more characteristic of crustal thinning and extensional tectonics as developed in subsequent models.

Structurally above the Wood Hills "thrust" and cutting unmetamorphosed strata is the Valley View thrust (Fig. 23-5), an older-on-younger low-angle fault (Thorman, 1970). This fault is truncated by the Wood Hills "thrust," in accordance with a model that the Valley View fault is a Mesozoic thrust truncated by a Cenozoic low-angle normal fault.

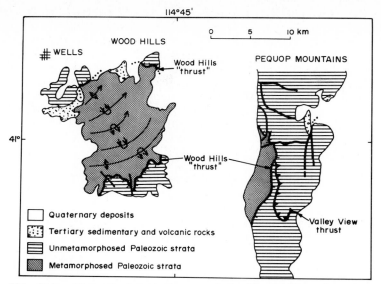

FIG. 23-5. Generalized geologic map of the Wood Hills and northern Pequop Mountains, Nevada, after Thorman (1970). Inferred Mesozoic low-angle faults shown by barbs; inferred Cenozoic low-angle faults shown by hachures.

Large-scale northwest-vergent folds (Fig. 23-5) deform the metamorphic rocks below, and are cut by, the Wood Hills "thrust." According to Thorman (1970), these folds are postmetamorphic; the orientations of foliation and bedding shown on his map suggest that the main phase foliation is deformed by the large-scale folds. However, boudinage and attenuation is evident near fold hinges and in the limbs of folds, indicating ductile flowage during fold development.

Maximum metamorphic conditions in the Wood Hills and northern Pequop Mountains may be estimated because Thorman (1970) reported synkinematic staurolite and local kyanite, as well as postkinematic andalusite in schist, and diopside and tremolite in siliceous metacarbonate rocks. These minerals indicate lower amphibolite facies conditions; the local coexistence of staurolite and kyanite indicates that the schist crystallized at greater than $T = 550°C$ and $P = 4.5$ kb (Greenwood, 1976; Wernicke, 1982). However, most of the metamorphic rocks are greenschist facies.

The age of metamorphism in the Wood Hills is partly constrained by K-Ar biotite dates of 42.4 m.y. on the metamorphosed Cambrian Dunderberg Shale, and 29.8 m.y. and 41.5 m.y. on pegmatites (Thorman, 1965). Thorman interpreted these as either cooling ages or ages reset by Tertiary magmatism, and suggested that the age of metamorphism was middle Mesozoic by comparison with other metamorphic terranes in the region.

Large-scale overturned folds in the metamorphosed terrane and an older-on-younger low-angle fault in the unmetamorphosed terrane are suggestive of tectonic shortening. In contrast, the low-angle detachment fault that emplaced unmetamorphosed on metamorphosed rocks indicates upper crustal extension. Because the detachment fault (Wood Hills "thrust") separates correlative middle Paleozoic metamorphosed and unmetamorphosed rocks, the fault must have lateral displacement greater than the width of exposed metamorphosed rocks — about 10 km. Detailed depositional facies and structural studies are required to determine the directions of movement during overthrusting, and during subsequent juxtaposition of metamorphosed and unmetamorphosed sequences by detachment faulting.

In comparison with the Pilot-Toano terrane to the east and the Ruby-East Humboldt terrane to the west, the Wood Hills–Pequop Mountains terrane may represent an intermediate Mesozoic crustal position. Rocks are consistently higher grade and show more ductile strain than those in the Pilot-Toano terrane, but are not migmatitic like those in the East Humboldt Range. It is notable that the Wood Hills and Pequop Mountains lack granitoids other than rare pegmatite dikes, in sharp contrast with most metamorphic terranes of the northeastern Great Basin. Timing constraints for metamorphic and deformational episodes are too meager in the Wood Hills to permit rigorous comparison with other terranes, but the data are compatible with timing of events in the Pilot Range: middle Mesozoic metamorphism, thrusting, and folding; and early Tertiary uplift and cooling by detachment faulting.

Albion, Raft River, and Grouse Creek Mountains, Idaho and Utah

A sprawling metamorphic terrane is widely exposed in the Raft River and Grouse Creek Mountains of Utah, and the Albion Mountains and Middle Mountain of Idaho. Medium- to high-grade rocks of the terrane occur as close as 10 km southwest of low-grade rocks in the Black Pine Mountains and 45 km northeast of low-grade rocks in the Pilot Range (Fig. 23-1). Extensive studies of the Albion–Raft River–Grouse Creek Mountains metamorphic terrane during the last two decades have revealed many details of its metamorphic and tectonic development; chief among these studies are those by Armstrong (1968b, 1976, 1982), Armstrong and Hills (1967), Compton (1972, 1975, 1983), Compton *et al.* (1977), D. M. Miller (1980, 1983a), Todd (1980, 1983), and D. M. Miller *et al.* (1983). The data from these studies are employed in the following synthesis without citation in many cases.

The Albion–Raft River–Grouse Creek metamorphic terrane consists of upper greenschist to upper amphibolite facies metamorphic rocks ranging in age from Archean to Mississippian that are cut by pre- to early-metamorphic low-angle faults attenuating section. Lower-grade upper Paleozoic rocks rest on high-grade rocks along a middle Cenozoic detachment fault, and upper Miocene rocks rest on a variety of underlying rocks along a late Cenozoic detachment (Fig. 23-6). Gently dipping foliation, prominent lineation, broad domes, and detachment faults mark this terrane as a "metamorphic core complex" according to the usage outlined by Coney (1979).

Archean basement rocks widely exposed in the metamorphic terrane and termed the Green Creek Complex by Armstrong (1968b), and the "older schist" unit and "adamellite" unit by Compton *et al.* (1977), consist of granodiorite to monzogranite plutons that are in most places strongly foliated, and biotite schist, amphibolite, and trondhjemite. Unconformably overlying the Green Creek Complex are alternating quartzite and schist strata of uncertain age; ages of Early Proterozoic, Late Proterozoic, and early Paleozoic have been suggested. These metamorphosed rocks are conformable in some places, but are in fault contact in many others, with overlying marble representing metamorphosed Ordovician miogeoclinal strata and succeeding black phyllite, quartzite, and schist representing metamorphosed Mississippian strata. Structurally overlying these metamorphic rocks are slightly metamorphosed Pennsylvanian and Permian rocks similar to those in the southern block of the Black Pine Mountains. In the western Grouse Creek and northernmost Albion Mountains, Triassic strata also occur. Remarkably, this same stratigraphic sequence, with some facies and structural variations, occurs throughout the entire terrane. A structurally higher and lithologically distinct sequence of strata crops out along the west side of the Albion Mountains and Middle Mountain. This thick sequence of metamorphosed quartzite and schist, the quartzite assemblage of D. M. Miller (1983a), may consist of Upper Proterozoic strata. Part of the sequence is overturned; throughout the

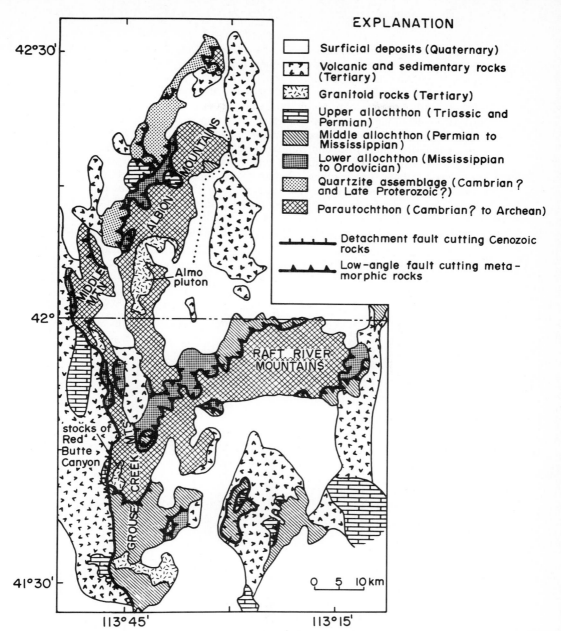

EXPLANATION

☐ Surficial deposits (Quaternary)

Volcanic and sedimentary rocks (Tertiary)

Granitoid rocks (Tertiary)

Upper allochthon (Triassic and Permian)

Middle allochthon (Permian to Mississippian)

Lower allochthon (Mississippian to Ordovician)

Quartzite assemblage (Cambrian? and Late Proterozoic?)

Parautochthon (Cambrian? to Archean)

Detachment fault cutting Cenozoic rocks

Low-angle fault cutting metamorphic rocks

FIG. 23-6.   Generalized geologic map of the Albion, Raft River, and Grouse Creek Mountains area, Idaho and Utah, after Compton *et al.* (1977) and D. M. Miller (1983a, b).

mountains it rests on metamorphosed Mississippian and older strata. Lake beds, volcaniclastic rocks, tuff, and rhyolite flows of late Miocene age rest unconformably upon and are faulted against the older rocks.

Middle Tertiary plutonic rocks occur in part of the terrane, and other plutons possibly were intruded and deformed during the Mesozoic. Three Tertiary plutons are well dated by the whole-rock Rb-Sr technique (Fig. 23-7): Immigrant Pass pluton (38 m.y.), stocks of Red Butte Canyon (25 m.y.), and Almo pluton (28 m.y.). The possible Mesozoic plutonic rocks apparently assimilated much sedimentary rock, rendering the isotopic systems exceedingly complex (gneiss of Camel Rock, Armstrong, 1976; Vipont Mountain intrusion, Compton *et al.*, 1977).

Patterns of metamorphic assemblages reveal generally lower-grade rocks in the northern and eastern Albion Mountains, southern Grouse Creek Mountains, and eastern Raft River Mountains, and the highest-grade rocks near the intersection of the three ranges. Highest-grade assemblages are in the sillimanite zone; pelitic rocks contain synkinematic cordierite, andalusite, and sillimanite in roof pendants of the Almo pluton in the southwestern Albion Mountains. Other high-grade mineral assemblages in pelitic rocks in the central Albion and northern Grouse Creek Mountains include kyanite-staurolite-garnet, and carbonate rocks include talc or tremolite in the presence of quartz and calcite. Greenschist facies rocks contain albite-epidote-chlorite in pelitic rocks and dolomite-muscovite-quartz in carbonate rocks. Armstrong (1968b; updated by Hodges and McKenna, 1986) pointed out that widespead high-grade assemblages in the west-

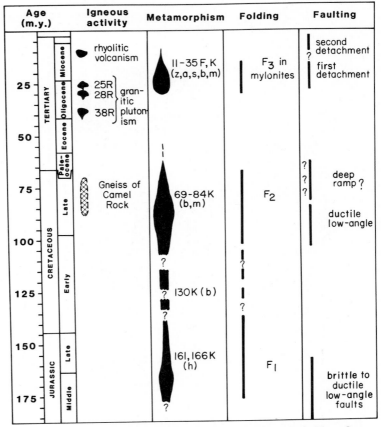

FIG. 23-7.   Time-line diagram for events in the Albion–Raft River–Grouse Creek Mountains metamorphic terrane, Idaho and Utah. Dating techniques are: F, fission track; R, Rb-Sr; K, K-Ar. Materials dated are: b, biotite; m, muscovite; h, hornblende; z, zircon; s, sphene; a, apatite. Time scale used is that of Palmer (1983).

ern Albion Mountains indicate higher pressure and moderate temperature conditions (P = 4.8–5.4 kb; T between 480 and 580°C), compared to the assemblages near the Almo pluton, which indicate lower pressure (less than 4.2 kb) and higher temperature. Armstrong suggested that these metamorphic assemblages developed at different times.

Major low-angle faults cut strata of the terrane at low angles to bedding, in most places attenuating strata and disrupting metamorphic isograds. Compton *et al.* (1977) divided the faulted rocks in the Utah portion of the terrane into three major fault sheets and a parautochthon, each of which is internally cut by faults with lesser separations. A similar subdivision is possible in the Albion Mountains, but the additional fault-bounded sheet of Upper Proterozoic(?) quartzite assemblage adds uncertainty concerning correspondence of the faulted sheets. In general, the parautochthon and lower allochthonous sheet are most highly metamorphosed, with metamorphic contrasts only rarely occurring across the intervening fault. Rocks in these two tectonic packages bear similar sets of minor structures and the faults are metamorphosed, indicating that faulting of the lower sheet predated or accompanied metamorphism and folding. The Upper Proterozoic(?) quartzite assemblage in the western Albion Mountains is faulted over lower allochthon rocks of lower metamorphic grade (Hodges and McKenna, 1986), requiring that the fault separating those tectonic packages was late- to postmetamorphic or intervened between two metamorphic events. The middle allochthonous sheet is generally low grade to unmetamorphosed, and was brittlely emplaced over highly metamorphosed rocks of the lower sheet and the parautochthon. Last movement on this fault is indicated by deformation in the border of a 25-m.y.-old stock emplaced near the fault. The upper allochthonous sheet in the Grouse Creek Mountains consists of unmetamorphosed Permian and Tertiary strata, locally faulted onto Miocene sedimentary and volcanic rocks (Compton, 1983). Faulted Miocene strata east of the Albion Mountains perhaps should be included in this sheet.

The low-angle faults in the metamorphic terrane generally attenuate section and place less metamorphosed rocks on more metamorphosed rocks, resulting in a remarkable thinning of strata and truncation of temperature profiles. However, important exceptions exist: higher-grade parts of the middle sheet locally rest on low-grade parautochtonous rocks in the Raft River Mountains, allochthonous quartzite assemblage strata are more highly metamorphosed than underlying lower sheet rocks in the Albion Mountains, and local imbricate thrusts in the western Raft River Mountains repeat strata.

Widespread ductile minor structures in the parautochthon and lower allochthon define two regionally developed sets of folds, lineations, and foliations. The foliations are subhorizontal, and lineations and parallel fold axes lie within the foliation planes and are locally consistently oriented, but regionally describe arcs. The first lineation trends in the northeast quadrant, and the second lineation in the northwest quadrant. Shear indicators associated with the second set indicate top-to-the-east shear zone deformation for a broad zone of rocks above the Green Creek Complex (Sabisky, 1985). Where minerals that grew during the first and second deformations are dated by K-Ar in low-to medium-grade parts of the Albion Mountains, they give Cretaceous ages for micas (69–84 m.y.) and Jurassic ages for hornblende[3] (161, 166 m.y.; Armstrong, 1976), indicating Cretaceous cooling of one or more middle Mesozoic metamorphic events. However, low-grade rocks in the eastern Raft River Mountains yielded early Tertiary K-Ar ages (Armstrong and Hansen, 1966) and even younger fission track ages for sphene (Compton *et al.*, 1977).

A third set of ductile minor structures is locally well developed in the region of low-P, high-T metamorphism near the Almo pluton and elsewhere near the intersection of the three main ranges, where it is manifested by west-northwest-trending lineation and

[3] Also R. L. Armstrong, written communication, 1977.

minor folds, and subhorizontal foliation. Textures accompanying the third-phase structures are mylonitic in places such as at the border of the Red Butte Canyon stocks, parts of the Almo pluton, and at Middle Mountain. In these areas, K-Ar and fission-track ages of metamorphic minerals are Oligocene and Miocene, and the structures are superposed on Oligocene granitic rocks. Kinematic indicators in mylonitic rocks of Middle Mountain indicate top-to-the-west shear for this Tertiary deformation (Saltzer and Hodges, 1986).

In summary, early low-angle faulting predated and (or) accompanied metamorphism during eastward shearing of strata above the Archean basement. This metamorphism and its associated first and second phases of folding occurred during the middle Mesozoic and ended by mid-Cretaceous time (Fig. 23-7). During Oligocene time, several plutons were emplaced, and migmatites, mylonitic rocks, and metamorphic assemblages in the region of the plutons indicate westward shearing under lower pressures and locally higher temperatures than for the Mesozoic event. A detachment fault shear zone coincided with the Oligocene event (Todd, 1980), locally deforming plutons, and emplacing little-metamorphosed rocks on high-grade rocks. A second detachment fault during the late Miocene (or Pliocene?) time emplaced strata as young as 8 m.y. onto metamorphic rocks (Compton, 1983; Covington, 1983; D. M. Miller et al., 1983; Todd, 1983).

Although the Mesozoic deformation accomplished a remarkable attenuation of stratigraphic section by low-angle faulting and plastic thinning, major overturned sections of strata, juxtaposition of higher-pressure facies over lower-pressure facies, and a few large recumbent folds indicate that much regional strain was shortening. This strain was concentrated in strata above the Archean basement rocks in a thick top-to-the-east shear zone.

Detachment faults apparently resulted from two Tertiary periods of upper crustal extension. An Oligocene deep to intermediate-level fault is manifested by ductile deformation accompanied by metamorphism near Oligocene plutons and brittle normal faulting throughout much of the terrane, for which westward sense of shear contrasts sharply with that for earlier ductile deformation. Younger detachment faults cut and rotate upper Miocene rocks and assorted Paleozoic and Mesozoic strata; the younger detachment plates are inferred from internal structures in the southern Grouse Creek Mountains and east of the Albion Mountains to have moved east or northeast after 10–8 m.y. (Compton, 1983; Covington, 1983; Todd, 1983).

Ruby Mountains and East Humboldt Range, Nevada

The Ruby Mountains and East Humboldt Range are a late Cenozoic system of west-tilted horsts over 120 km long in northeast Nevada (Fig. 23-1). Widely exposed metamorphic rocks exhibiting gently dipping foliation and persistent lineation form a terrane within the horst system that is commonly cited as an example of a Cordilleran metamorphic core complex (Crittenden et al., 1980).

The bedrock within these ranges can be divided into three broad categories: (1) igneous and metamorphic complex, (2) low-grade to unmetamorphosed miogeoclinal rocks (Cambrian to Triassic), and (3) Tertiary sedimentary, volcanic, and plutonic rocks (Fig. 23-8). They also are broadly divisible into three structural terranes, from deep to shallow: migmatitic core, mylonitic zone, and cover. These three structural terranes developed during polyphase Mesozoic and Cenozoic deformation (Snoke and Lush, 1984).

The age and stratigraphic affiliation of the rocks in the igneous and metamorphic complex are fairly well established (Howard, 1971), although not accepted by some workers. A low- to medium-grade, locally fossiliferous, metasedimentary sequence in the southeast East Humboldt Range that represents metamorphosed Middle Cambrian to

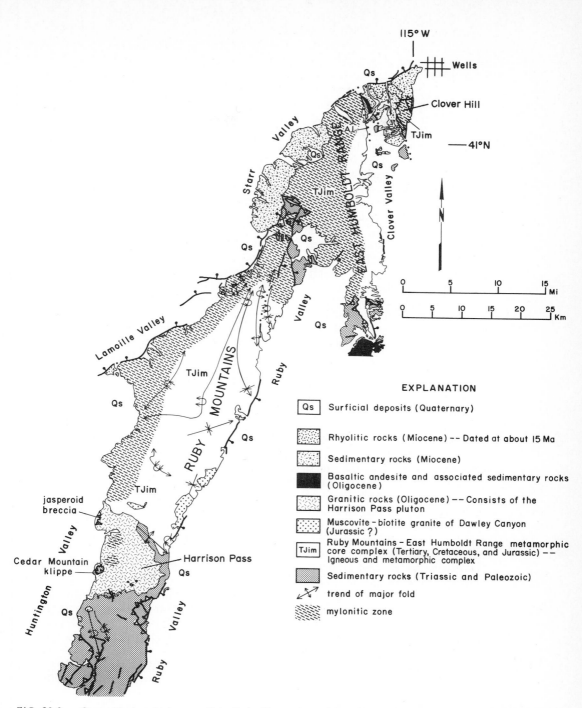

FIG. 23-8. Generalized geologic map of the Ruby Mountains and East Humboldt Range, Nevada, after Howard *et al.* (1979) and Snoke and Lush (1984). A. L. = Angel Lake.

Upper Devonian miogeoclinal rocks (Hope and Coats, 1976; Taylor, 1984) is lithologically similar to metamorphosed miogeoclinal rocks mapped by Thorman (1970) in the Wood Hills. A sequence of metasedimentary metaquartzite, pelitic schist, gneiss, and marble that includes the same metasedimentary stratigraphic units as well as Upper Proterozoic and Lower Cambrian miogeoclinal strata, but is extensively intruded by granitic rocks, forms the migmatitic core of the northern Ruby Mountains (Howard, 1966, 1971, 1980; Snoke, 1980). The same metasedimentary sequence continues into the high country of the East Humboldt Range, but there is dominated by impure quartzite and schist of the Upper Proterozoic and Lower Cambrian metaclastic sequence. However, Kistler et al. (1981) argued that the sequence is chiefly Precambrian (>1450 m.y. old) and was deformed during Precambrian time, but also was intensely metamorphosed at about 550 m.y.

In the northern East Humboldt Range, a large-scale recumbent fold encloses in its core a distinctive rock suite (metamorphic suite of Angel Lake) that is characterized by an association of orthogneiss, paragneiss, and amphibolite. Snoke and Lush (1984) interpreted the suite of Angel Lake as Precambrian crystalline basement, partly based on similarity with the Archean Green Creek Complex of Armstrong (1968b) in the Albion Mountains. Recent U-Pb zircon data on orthogneiss of the suite of Angel Lake indicate an upper concordia intercept greater than 2.5 b.y. (J. E. Wright, personal communication, 1985), supporting the correlation.

Structurally above the igneous and metamorphic complex is a cover terrane that includes low-grade to unmetamorphosed miogeoclinal rocks and thick sections of Oligocene and Miocene sedimentary and volcanic rocks. The cover terrane consists chiefly of a fault-bounded mosaic of rock units; faults dip at low to high angles and are chiefly of Tertiary age. Pre-Tertiary cover rocks commonly display a more complex structural history than the Tertiary strata, presumably reflecting both Mesozoic and Cenozoic deformations (Snoke and Lush, 1984; Taylor, 1984). Some of the lithostratigraphic units of the cover terrane also occur in the mylonitic zone and migmatitic core, where they are intensely metamorphosed and polydeformed (Snoke and Lush, 1984; Taylor, 1984).

The metasedimentary rocks of the northern Ruby Mountains–East Humboldt Range are ubiquitously invaded by dikes and complex concordant sheets ranging from quartz gabbro to granite. The most conspicuous intrusive rock is leucocratic, commonly coarse-grained to pegmatitic, and spans the granite to trondhjemite modal fields. Muscovite and biotite are common, and garnet, sillimanite, zircon and monazite are accessory. These peraluminous granitoids commonly yield initial $^{87}Sr/^{86}Sr$ ratios greater than 0.710, indicating the prominent role of continental crust in their melted source rocks[4] (Kistler et al., 1981). Inasmuch as these granitic rocks were principally derived by the anatexis of crustal rocks, their age of igneous emplacement closely approximates a time of deep crustal metamorphism in the region.

Kistler et al. (1981) dated high initial $^{87}Sr/^{86}Sr$, peraluminous granitoids in the southern Ruby Mountains that are intimately interlayered with adjacent amphibolite facies metamorphic rocks as 160 ± 3 m.y. and 83 ± 1.3 m.y. old. The 160-m.y. age is derived from a suite of whole-rock Rb-Sr samples; the 83-m.y. age is principally constrained by Rb-Sr muscovite-whole-rock pairs (Kistler et al., 1981, their Fig. 9 and Table 9). The 160-m.y. age may be close to the age of initial crystallization, and a Late Jurassic age is also indicated for a muscovite granite porphyry dike intruded into the cover sequence near Secret Valley, southwestern East Humboldt Range[5] (Dallmeyer et al., 1986). The 83-m.y. age may be a cooling age, reflecting when the ^{87}Sr closure temperature for mus-

[4] Also A. W. Snoke and P. D. Fullagar, unpublished data, 1986.

[5] Ibid.

covite was achieved during Late Cretaceous cooling. In support of this interpretation, similar Late Cretaceous U-Pb monazite dates have been obtained from pelitic schist from the northern Ruby Mountains (Snoke *et al.*, 1979).

Peraluminous granitoids were intruded during the Tertiary near Harrison Pass in the southern Ruby Mountains (Snoke and Howard, 1984; Snoke and Lush, 1984; Willden and Kistler, 1969), including two rock units that have been dated at about 36 m.y. (U-Pb zircon, Wright, and Snoke, 1986): porphyritic biotite ± hornblende granodiorite to monzogranite and biotite ± muscovite granite.

Widespread sillimanite in metapelitic rocks and diopside in impure metacarbonate rocks indicate that the migmatitic core reached upper amphibolite facies conditions (Howard, 1966; Snelson, 1957). The distribution of aluminosilicate minerals suggests that rocks of the southern Ruby Mountains were metamorphosed at lower pressure than those of the northern East Humboldt Range 80 km away. In the southern Ruby Mountains, sillimanite ± staurolite has been widely reported (Johnson, 1981; Olson and Hinrichs, 1960; Willden and Kistler, 1969), and petrographic reconaissance of the area indicates that andalusite is also common at some localities[6] (Johnson, 1981; Olson and Hinrichs, 1960).

At Clover Hill, northeastern East Humboldt Range, the assemblage kyanite + garnet + biotite + sillimanite + muscovite + quartz + plagioclase + ilmenite + rutile is common in metapelitic rocks. In the nearby migmatitic core of the northern East Humboldt Range, Lush (1982) found sillimanite-rich pelitic schist with sparse kyanite relics. In neither of these localities is staurolite present, and consequently, the sillimanite paragenesis probably reflects the kyanite = sillimanite inversion, the breakdown of muscovite in the presence of quartz, or perhaps the simultaneous role of both reactions. Regardless of the exact petrogenesis, the widespread occurrence of sillimanite ± kyanite indicates *minimum* temperatures of 550–600°C and pressures greater than 4.5 kb were achieved during the peak of regional metamorphism (Greenwood, 1976) in the northern East Humboldt Range. The terrane is extensively migmatitic, suggesting that higher temperatures (*ca.* 700°C) are probably a more realistic estimate of the thermal regime during metamorphism. Also, epidote is a common phase in hornblende-biotite quartz dioritic orthogneisses of the East Humboldt Range, and possibly is of magmatic origin indicative of moderately high pressures (Zen, Chapter 2, this volume; Zen and Hammarstrom, 1984). Although more textural and mineralogical work is required to test the magmatic petrogenesis of the epidote, results are in harmony with mineral assemblages indicating higher-pressure metamorphism at Clover Hill than in the southern Ruby Mountains.

Rocks of the northern Ruby Mountains–East Humboldt Range record a complex polyphase Mesozoic and Cenozoic deformational history, summarized in Fig. 23-9. Premetamorphic thrust faulting is required by duplicated upper Precambrian to Ordovician miogeoclinal strata in the Ruby Mountains (Howard, 1966). Similarly, in the northern East Humboldt Range, Precambrian crystalline basement rocks were faulted into the miogeoclinal supracrustal sequence prior to metamorphism and associated polyphase ductile deformation (Lush, 1982; Snoke and Lush, 1984). These premetamorphic thrust faults possibly were early manifestations of crustal shortening.

First-phase pervasive foliation and associated compositional layering were folded by second-phase folds ranging from huge nappes to microfolds. Various late to postmetamorphic folds locally deformed these earlier structures.

Superimposed on the Mesozoic (?) polyphase structures is a locally pervasive mylonitic shear zone deformation (Fig. 23-8) that is interpreted as extensional, analogous in part to the rooted, extensional shear zones hypothesized by Wernicke (1981). Physical conditions of the mylonitization are poorly constrained but must have been upper greenschist facies

---

[6] Also A. W. Snoke, unpublished data.

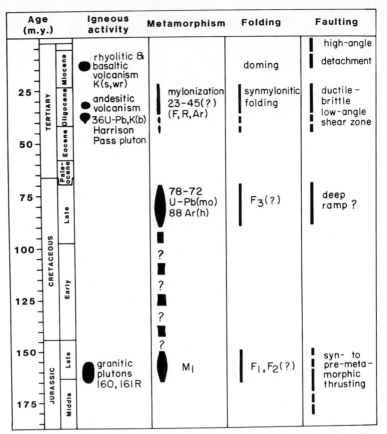

FIG. 23-9. Time-line diagram for events in the Ruby Mountains–East Humboldt Range metamorphic terrane. Dating techniques are: F, fission track; K, K-Ar; R, Rb-Sr; Ar, $^{40}Ar/^{39}Ar$. Materials dated are: b, biotite; m, muscovite; z, zircon; sp, sphene; ap, apatite; wr, whole-rock. Time scale used is that of Palmer (1983).

or above, for biotite was stable. The mylonitic deformation stretched and rotated the first-phase foliation, and second-phase folds are locally preserved as highly appressed isoclines whose fold hinges have been stretched parallel to the mylonitic elongation lineation. Other folds in the shear zone deform the mylonitic foliation and apparently represent manifestations of progressive simple shear. A WNW-ESE lineation and well-developed foliation are the most characteristic features of the mylonite zone, and abundant kinematic indicators as described by Simpson and Schmid (1983) and by Lister and Snoke (1984) indicate that the sense of shear was top to the west-northwest throughout most of the zone. However, S-C relations in thin mylonitic ductile shear zones in the Harrison Pass pluton indicate top to the east-southeast. These shear zones may be conjugate to the main zone, resulting from unequal strain partitioning. Alternatively, the small-scale shear zones in the pluton may be late zones antithetic to the main shear zone, such as those described in Arizona core complexes (Reynolds *et al.*, 1986).

Studies of the mylonite zone in the northern Ruby Mountains and East Humboldt Range (Dallmeyer *et al.*, 1986; Dokka *et al.*, 1986; Howard, 1966, 1980; Lister and Snoke, 1984; Snelson, 1957; Snoke, 1980; Snoke and Lush, 1984; Snoke *et al.*, 1984;

Wright and Snoke, 1986) have demonstrated the following: (1) the mylonitic fabrics were superposed on rocks as young as Oligocene granitoids; (2) a west-northwest mylonitic elongation lineation is regionally extensive; (3) sense of shear criteria in the mylonitic rocks chiefly indicate top to the west-northwest; (4) the physical conditions of mylonitization were apparently variable, but dynamically recrystallized biotite suggests at least upper greenschist facies conditions; (5) most mylonitization postdates a 36-m.y.-old pluton and predates 23 m.y., based on an Rb-Sr (wr-muscovite) cooling age for mylonitic two-mica granite gneiss; (6) superposed on the mylonitic shear zone are numerous brittle deformational features that suggest progressive cooling through unroofing of the shear zone during displacement along a large-scale, low-angle normal fault; and (7) the mylonitic shear zone and associated brittle detachment fault system is truncated by high-angle, Miocene to Quaternary Basin and Range normal faults.

Apparently, as the shear zone evolved and ambient physical conditions changed in response to uplift and tectonic denudation of cover rocks, the extensional deformation changed from ductile to brittle, and low-angle normal faults were localized in the mylonitic shear zone. These low-angle faults further reworked and dismembered early macroscopic folds related to the Mesozoic compressional regime, and juxtaposed unmetamorphosed to low-grade metamorphic rocks with high-grade metamorphic rocks.

Mylonitic rocks describe a domical foliation surface in portions of the Ruby Mountains–East Humboldt Range, especially along the western flanks of the range system. The domical shape may partly result from an initial convex-upward mylonitic shear zone geometry (possibly the top of a mid-crustal lens — Hamilton, 1982). Late uplift, probably related to the regional tectonic denudation of upper crustal rocks that formerly overlay the crystalline complex, deformed the shear zone and complex. This late uplift in part may have been an isostatic response as suggested by Rehrig and Reynolds (1980) and Spencer (1984). The uplift may have coincided with low- and high-angle faulting that cut volcanic rocks as young as 15 m.y. old (Fig. 23-9). Basin and Range normal faults cutting the foliation domes at the margin of the ranges are as young as Pleistocene; they represent the final manifestation of a complex history of Cenozoic crustal extension.

REGIONAL TECTONIC PROBLEMS

Timing of Metamorphism and Deformation

Constraints that can be placed on the timing of events described in the preceding case studies are summarized in Fig. 23-10, in which foreland fold-and-thrust belt events from southwestern Wyoming (Armstrong and Oriel, 1965) are included for comparison. Although timing constraints vary considerably, in many places a middle Mesozoic metamorphic event during Middle and Late Jurassic time was followed by a widespread cooling recorded by Late Cretaceous closure of various isotopic systems. Whether this represents one protracted regional metamorphic event or two or more pulses in the Jurassic and Cretaceous is uncertain, although the lack of documented metamorphic overprints and the essentially continuous deformational fabrics point to the former hypothesis. Low-angle faults and/or ductile shear zones preceded or accompanied metamorphism in Middle Jurassic time. The Jurassic metamorphism is constrained mainly by relations with dated Jurassic plutons of 155–165 m.y. age, leading to uncertainties for the time of beginning and ending of metamorphism shown in Fig. 23-10. The metamorphism at exposed crustal levels may be related to regional heating associated with a pulse of plutons derived from deeper areas undergoing metamorphism.

It is still uncertain whether deformation and metamorphism in the hinterland accom-

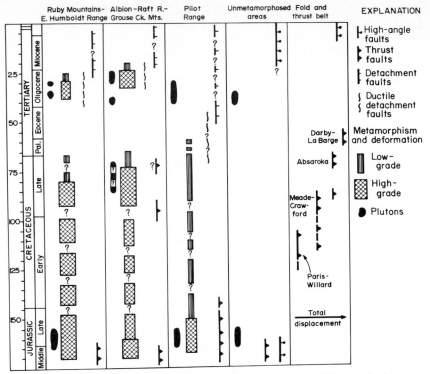

FIG. 23-10.   Time-line summary diagram for well-constrained events in the northeastern Great Basin. Time scale used is that of Palmer (1983).

panied the first well-documented thrusting in the foreland (the Paris-Willard system) in Early Cretaceous time. However, region-wide waning of metamorphic temperatures coincided with major thrusting on the Absaroka thrust system (Fig. 23-10). Later thrusts are difficult to correlate with structures in the hinterland. Fragmented thrust systems west of the well-established Sevier thrusts are probably older than the Paris-Willard system (Allmendinger *et al.*, 1984), and may represent a thrust system temporally and spatially intermediate between Sevier thin-skinned thrusts and hinterland metamorphism, folding, and thrusting. The middle Mesozoic hinterland metamorphism is thus mostly or entirely older than Sevier belt thrusting and contemporaneous with or younger than Mesozoic backarc thrusting and metamorphism in western Nevada and eastern California (e.g., Oldow, 1984; Schweickert *et al.*, 1984).

Ductile to brittle extensional deformation within shear zones and detachment faults are documented for Late Cretaceous to Oligocene time in several areas, and yet later brittle detachments and high-angle normal faults are documented in middle Miocene to Quaternary time. Brittle detachment faulting is chiefly dated by relations with Tertiary volcanic and plutonic rocks, which are of early Oligocene and middle and late Miocene age. It is, therefore, uncertain whether extensional structures developed more or less continuously, or episodically. Evidence such as 17-m.y.-old basalt dikes with chilled margins that cut foliated Archean to Oligocene rocks in the Ruby Mountains and East Humboldt Range suggest that the majority of unroofing, and extension, occurred during the first half of the period of Tertiary extension. Well-documented Cenozoic metamorphism associated with ductile shear zones in the Middle Mountain and Grouse Creek Mountains area shows a direct association with Oligocene plutons, suggesting that plutons emplaced near the shear zone created thermal nodes in the middle crust.

P-T Constraints for Hinterland Metamorphism

No detailed quantitative studies on the physical conditions of hinterland metamorphism in the northeastern Great Basin are presently available. Other than the preliminary data recently reported by Hodges and McKenna (1986) from the Albion Mountains, the constraints on metamorphic conditions have been determined by a comparison of petrographic data with experimentally determined P-T stability curves for various mineral assemblages.

A summary of the physical conditions of peak metamorphism (Table 23-1) indicates estimated pressures were 4.5 kb or higher for parts of many metamorphic terranes. The maximum inferred temperatures of peak metamorphism vary greatly, but the migmatitic parts of the Albion, Grouse Creek, and Ruby Mountains and the East Humboldt Range probably exceeded 700°C, judging from the locally pervasive nature of granitic leucosome within these areas. Other areas show 450-500°C maximum temperatures. These geothermobaric constraints suggest that rocks from 15 (to 20?) km deep in the crust have been exhumed. These data, coupled with present crustal thicknesses of ~20 to 30 km (R. B. Smith, 1978) suggest that the Mesozoic crust was once 40-50 km thick, supporting the concept of an overthickened crustal welt (Coney and Harms, 1984) in the hinterland during the Mesozoic prior to the extensional regime of the middle Tertiary. The present crustal thickness probably partly results from addition of material from the mantle to the lower crust during Cenozoic extension to account for heat flow and voluminous magma production. However, this process we consider unlikely to fully account for the present crustal thickness following major extension.

The mechanisms for developing overthickened crust in the Mesozoic hinterland are obscure, because only relatively small duplications of strata are represented by documented Mesozoic thrust faults, and ductile strain apparently largely occurred in broad shear zones parallel to stratigraphic units. A possible reason for not recognizing the required major duplications of strata is that many faults were reactivated during Cenozoic extension, obscuring or eliminating Mesozoic overlaps. Tectonic thickening of at least 10 km is required by metamorphic assemblages in Upper Devonian to Upper Mississippian strata in the Albion and Ruby Mountains and Wood Hills, since these strata were buried at least 15 km and yet the miogeoclinal strata deposited on them were at most 5-6 km thick. This tectonic thickening requires emplacing roughly the entire thickness of the miogeoclinal section on Mississippian shales, a degree of thickening too extreme to be accounted for by local structure. Furthermore, major thrust faults did not break the surface in the eastern hinterland, because Tertiary rocks uniformly rest on upper Paleozoic strata (Armstrong, 1968a). This relation requires that the uppermost crust of the hinterland was not significantly faulted or folded on a regional scale prior to Cenozoic time.

One model that is compatible with the crustal thickening we infer from metamorphic assemblages and the lack of large upper-crustal structures is that of a gigantic crustal duplex. Upper crustal rocks in the roof of the duplex were not cut by large structures, rocks at lower levels underwent large amounts of translation in both layer-parallel shortening and even local extensional settings, and deep-seated rocks served as the parautochthon. However, major thrust faults cutting to the surface are not recognized in the eastern hinterland, and overthrust belt structures are too young. The Samaria Mountain thrust, located just west of the overthrust belt, is probably middle and (or) late Mesozoic in age, but it moved only about 15 km at its front (Allmendinger et al., 1984). The possibility that major Mesozoic thrusts surfaced west of the hinterland remains.

TABLE 23-1. Summary of Maximum Temperature, Maximum Pressure, and Sense of Shear Data for Metamorphic Terranes in the Northeastern Great Basin

	Ruby–East Humboldt[1]	Albion–Raft River–Grouse Creek[2]	Wood Hills[3]	Pilot–Toano[4]	Black Pine[5]	Base of Willard Thrust Sheet[6]
Mesozoic						
P(kb)	>4.5	>4.5	>4.5	<6	?	~4
T(°C)	>700° in migmatitic core	550–580°	>550°	Locally 520–580°	>300°	300–350°
Shear sense	Variable	Variably east	Northwest(?)	East and southeast	East	East
Cenozoic						
Eocene and Oligocene						
P(kb)	<4.5	<4.2				
T(°C)	>350°, locally >600°C	Locally >600°	Greenschist facies(?)	None(?)	None	None
Shear sense	West-northwest	West-northwest	Southeast(?)	East and west(?)	East(?)	—
Miocene						
Shear sense	West	East	?	West	Normal faults	Normal faults

Sources: (1) Howard (1966, 1980), Kistler *et al.* (1981), Snoke and Lush (1984); (2) Armstrong (1986b), Compton *et al.* (1977), D. M. Miller *et al.* (1983), Saltzer and Hodges (1986); (3) Sabisky (1985); (4) Thorman (1970); (4) D. M. Miller (1984), D. M. Miller *et al.* (in press); (5) J. F. Smith (1983); (6) Beck (1982), Naeser *et al.* (1983).

Another model is that of a tectonic wedge emplacing a thick allochthon on Mississippian shales, but causing little regional-scale structure in the hanging wall of the wedge. Supporting such a model is the widespread faulting in Mississippian shales, and the fact that high-grade metamorphism is always restricted to Mississippian and older rocks, whereas younger rocks are rarely metamorphosed to even low grades. The Samaria Mountain thrust could represent a minor frontal breakthrough. Similar models have been invoked for parts of the Canadian Cordillera (Brown *et al.*, 1986; Price, 1986). Passive uplift of an areally extensive hanging-wall section, forming a high, arid plateau, could explain the erosion of only 1–2 km of the youngest Paleozoic and Mesozoic strata.

Geologic processes other than tectonic burial may have contributed to the metamorphism of supracrustal rocks. The spatial association of high-grade metamorphic rocks with Mesozoic granitoids in the southern Ruby Mountains, Pilot Range, and Toano Range suggest that emplacement of magmas may have elevated the regional geothermal gradients.

Role of Crystalline Basement

Precambrian basement rocks played key roles in both the geochemical and structural evolution of hinterland metamorphic complexes. The geochemical role is manifested by Mesozoic and Cenozoic magma genesis; the structural role is manifested by varying mechanical behavior, including ductile shortening and relatively stiff, unyielding basement during Mesozoic compressional tectonics.

Initial $^{87}Sr/^{86}Sr$ ratios greater than 0.710 have been reported from many granitoids in hinterland metamorphic complexes of the northeastern Great Basin[7] (Armstrong, 1976; Best *et al.*, 1974; Compton *et al.*, 1977; Kistler *et al.*, 1981), indicating anatexis of crustal rocks (probably Precambrian crystalline basement ± metasedimentary rocks) during melting. Neodymium isotopic analyses on some of these granitoids substantiate this role for Precambrian crustal materials during magma genesis (Farmer and DePaolo, 1983). Geochronological data indicate that the processes critical in the evolution of high initial $^{87}Sr/^{86}Sr$ granitoids occurred during Jurassic, Cretaceous(?), and Oligocene time. Consequently, the geochemical recycling of Precambrian crustal materials was a process that occurred during both Mesozoic compressional and Cenozoic extensional tectonic regimes.

The mechanical role of crystalline basement shows variations in the hinterland, suggesting some intriguing regional models for crustal strain. However, knowledge is currently fragmentary because (1) basement crops out in only a few ranges, hence many critical structural relations are in the subsurface; (2) few detailed structural studies of the basement rocks exist; and (3) the structural histories of basement rocks are especially complex, having had components of deformation during three or more periods.

Studies by Royse *et al.* (1975), Bruhn and Beck (1981), and Beck and Bruhn (1983) of the Archean and Early Proterozoic Farmington Canyon Complex in the Wasatch Mountains provided insights into basement behavior during crustal shortening (Bryant, Chapter 16, this volume). The Farmington Canyon Complex occurs as a basement culmination above a major thrust belt ramp system (Bruhn *et al.*, 1983; Royse *et al.*, 1975). Migmatitic rocks and widespread distribution of sillimanite indicate Proterozoic amphibolite facies metamorphic conditions. The complex is laced with retrograde shear zones that Bell (1952) interpreted as an anastomosing set of Cretaceous thrust faults. These zones developed under greenschist facies conditions and involved hydrothermal reactions (Bell, 1952; Hollet *et al.*, 1978). The basement culmination in the Wasatch Mountains was shown by Schirmer (1985) to be a result of the stacking of several basement-cored thrust sheets that ramp eastward and connect with regional thrusts in the overlying sedimentary

---

[7] Also A. W. Snoke and P. D. Fullagar, unpublished data, 1986.

rock sequence. Schirmer suggested that imbrication of basement in the Wasatch Mountains is related to the location of the Precambrian rifted margin in the footwall of the regional décollement of the Sevier orogenic belt. It is also possible that the faulted(?) margin of the Pennsylvanian and Permian Oquirrh basin localized basement-involved thrusts. The critical factors that apparently controlled the heterogeneous Mesozoic deformation of the Farmington Canyon Complex are structural anisotropies inherited from Proterozoic or late Paleozoic rifting of the basement and the migration and localization of fluids in the basement terrane during Mesozoic shortening.

The style of deformation of the crystalline basement of the Farmington Canyon Complex contrasts markedly with the penetrative deformation of Lower and Upper Proterozoic strata at the base of the structurally overlying Willard thrust sheet (Beck and Bruhn, 1983; Crittenden, 1972). The pre- and early-miogeoclinal rocks that form the lower 3 km of the Willard sheet experienced lower greenschist-facies metamorphism during the early movement history of the allochthon (Beck, 1982), and the Lower Proterozoic strata previously underwent Proterozoic upper greenschist-facies metamorphism (Crittenden and Sorensen, 1980). The rocks are foliated and lineated and contain tight to isoclinal folds (Beck and Bruhn, 1983), some of which apparently were rotated subparallel to the common east-northeast elongation lineation. The estimated physical conditions during the Mesozoic greenschist-facies metamorphism were $T = 300–350°C$ and $P = 4$ kb (Beck, 1982).

Palinspastic restoration of the Willard thrust sheet suggests an initial location near the Raft River Mountains metamorphic terrane (Beck and Bruhn, 1983). Crittenden (1979) suggested that the units at the base of the Willard thrust sheet, such as the Proterozoic Facer Formation (Crittenden and Sorensen, 1980), may correlate with metasedimentary and metavolcanic rocks unconformably overlying Archean basement in the Raft River–Albion area. Furthermore, Crittenden (1979) argued that a major tectonic break in the Raft River Mountains exists below the metamorphosed Ordovician Pogonip Group but above the units he correlated with the Proterozoic Facer Formation. This hypothesis permits deriving the Willard thrust sheet from the Raft River–Albion area, but remains unproven.

The Raft River, Grouse Creek, and Albion Mountains contain the most extensive tract of crystalline basement rocks in the hinterland of the Sevier orogenic belt. These Archean rocks are described as an "autochthon" (Compton et al., 1977) because they are less deformed than overlying rocks. D. M. Miller (1980) noted that based on textural and strain data, the maximum flow of ductile rocks was concentrated within carbonate and pelitic units above the Archean parautochthon, and that Archean basement shows downward decreasing strain gradients, suggesting that strain occurred in thick shear zones controlled by lithology-caused mechanical anistropy. Local remobilization of the Precambrian basement rocks is indicated by rhemorphic offshoots of the older granitoid into overlying younger metasedimentary rocks (Compton et al., 1977; Todd, 1980).

The allochthonous nature of the Farmington Canyon Complex in the Wasatch Mountains (Bruhn et al., 1983; Royse et al., 1975; Schirmer, 1985) suggests that the Archean rocks in the Raft River, Grouse Creek, and Albion Mountains are also in the hanging wall of a large Mesozoic thrust complex. Surface geologic data does not allow this hypothesis to be tested; however, late Mesozoic K-Ar dates in the Albion Mountains (Armstrong, 1976) indicate cooling possibly caused by uplift roughly coincident with the maximum movement along the Absaroka thrust, the regional thrust fault considered to have caused the imbrication of the Farmington Canyon Complex (D. M. Miller, 1983b).

Further to the west, remobilized basement rocks recently identified in the northern East Humboldt Range are probably Archean in age. An Archean(?) granitoid intruded a heterogeneous paragneiss wallrock suite, and amphibolitic rocks (metamorphosed mafic bodies) occur within both paragneiss and granitoid. These basement rocks occur in a basement-cored nappe in the hanging wall of a low-angle fault that emplaced the nappe onto an upright sequence of metamorphosed lower Paleozoic rocks, and Precambrian basement rocks have not been recognized below the fault.

If the three exposures of Archean basement — from thrust belt to the heart of the hinterland — are near their respective Mesozoic positions, they represent fragments of an east-west traverse across the Mesozoic orogen. The basement rocks display progressively greater ductile strain westward into the hinterland, from localized, fluid-controlled retrograde shear zones cutting rocks otherwise only slightly deformed during the Mesozoic, to moderately developed penetrative foliation, and to recumbently folded migmatitic gneiss. Mesozoic thrusts apparently existed as narrow fault zones in the east, were broader simple shear zones predominantly parallel to mechanical anisotropies induced by basement and cover lithologies farther west, and perhaps rooted into mushrooming highly ductile middle crust even farther west. These zones of different ductile behavior may have changed position with time, as suggested by Burchfiel and Davis (1975), although their movement was not necessarily progressively eastward. Setting of fission track ages in basement of the western thrust belt (Naeser *et al.*, 1983), and K-Ar mica (D. M. Miller, 1983b) and U-Pb monazite (Snoke *et al.*, 1979) ages in metamorphosed cover rocks in much of the hinterland during middle Late Cretaceous time suggests that cooling was caused by a widespread tectonic event such as the ramping of basement rocks inferred in the Wasatch Mountains. Such ramping apparently cooled eastern basement rocks and hinterland cover rocks sufficiently to cause brittle conditions, whereas many basement rocks in the hinterland remained in the ductile realm, probably at upper greenschist to lower amphibolite facies conditions, until being exhumed under ductile to brittle conditions during middle Cenozoic time.

Significance of Mylonitic Rocks in the Metamorphic Terranes

Extensive tracts of mylonitic rocks were recognized by Misch (1960) in the northeastern Great Basin; he interpreted them as dynamically deformed by Mesozoic décollement faulting. Armstrong and Hansen (1966) developed a stockwork tectonic model for the metamorphic terranes of the northeastern Great Basin, and hypothesized that the mylonitic rocks developed in an *Abscherungszone* between mobile upwelling infrastructure and more rigid, overlying suprastructure. Later studies (Howard, 1980; Snoke, 1980) in the northern Ruby Mountains, where mylonitic rocks are extensively exposed, adopted the stockwork model, relating the mylonitic deformation to Mesozoic tectonics. Reappraisal of Mesozoic timing for mylonitic fabric development in the northeastern Great Basin was caused by discovery of mylonitic textures in the Oligocene stocks of Red Butte Canyon in the central Grouse Creek Mountains (Compton *et al.*, 1977). This report of Tertiary mylonitic fabrics was supported by reports of extensive Tertiary mylonitic deformation in the metamorphic core complexes of southern Arizona (Coney, 1979; Reynolds and Rehrig, 1980).

These early reports of Tertiary mylonitization led to several tectonic models (G. H. Davis and Coney, 1979; Hamilton, 1982; E. L. Miller *et al.*, 1983; Rehrig and Reynolds, 1980; Wernicke, 1981) for Cenozoic ductile-brittle extension in Cordilleran metamorphic core complexes. These models form three overlapping groups: (1) crustal-scale lenses and boudins (G. H. Davis and Coney, 1979; Hamilton, 1982); (2) rooted, low-angle, normal-sense ductile-brittle shear zone (Wernicke, 1981); and (3) *in situ* pure shear deformation associated with batholith emplacement (E. L. Miller *et al.*, 1983). A recent modification to the models resulted from recognizing ductile shear zones in the deeper parts of detachment fault systems (G. H. Davis, 1983; G. A. Davis et al., 1986; Snoke and Lush, 1984). These ductile shear zones show a consistent sense of shear within individual complexes and indicate the regional slip line during Tertiary crustal extension (Wust, 1986).

In the northeastern Great Basin, the most extensive studies on the mylonitic rocks of the metamorphic terranes have been undertaken in the Ruby Mountains and East Humboldt Range, where Oligocene mylonites show a regional west-northwest sense of shear. Superposition of the mylonitic deformation on previously deformed metamorphic tectonites can be seen at a variety of scales. The base of the kilometer-scale mylonitic zone is a transition from massive migmatite to flaggy, well-foliated mylonitic rocks. This lower boundary cuts across large-scale, nappelike folds that can be traced from the migmatitic core into the mylonitic zone (Howard, 1968, 1980), where they were stretched and flattened within the mylonitic zone and in some cases the hinge lines were rotated crudely parallel to the mylonitic elongation lineation. These folds with curvilinear hinge lines are gigantic sheath folds similar to those described by Henderson (1981). Centimeter-scale ductile shear zones, common in the migmatitic core terrane near the transition into the mylonitic zone, are superposed on amphibolite facies metamorphic rocks, and have structural fabrics that are discordant to structures in the core terrane. Streaky mineral lineations in the zones are essentially coincident with the elongation lineation pervasively developed in the main mylonitic shear zone. The small-scale shear zones in the migmatitic core terrane are interpreted as a manifestation of progressive, inhomogeneous simple shear.

In the Middle Mountain area, southern Idaho, mylonitization is superposed on previously deformed metamorphic tectonites. Structural studies from this mylonitic zone indicate a prominent N60–70°W lineation, and sense of shear criteria that indicate top to the west-northwest. Detailed geochronological data do not tightly constrain the age of mylonitization, although Armstrong (1976) reported numerous K-Ar (mica) ages ranging from 21 to 16 m.y. Combined with three whole-rock-muscovite Rb-Sr dates of 23 ± 2, 26 ± 2, 34 ± 2 m.y. (Armstrong, 1976), these ages probably represent cooling following mylonitic deformation. Similar fabrics are locally developed within the 25-m.y.-old stocks of Red Butte Canyon and the 28-m.y.-old Almo pluton in nearby parts of the metamorphic terrane.

The west-northwest shear sense in, and middle Tertiary age for, mylonitic rocks from the northern Ruby Mountains–East Humboldt Range and the Middle Mountain area strongly argue against tectonic models that relate the mylonitic deformation with the east-directed shortening of the Mesozoic Sevier fold-and-thrust belt. The consistency for the direction of the middle Tertiary shear sense in each area and the progressive transition from ductile to brittle deformation is compatible with a model that employs a rooted, low-angle, normal-sense ductile-brittle shear zone (G. A. Davis et al., 1986; Wernicke, 1981, 1985). The mylonitic rocks represent ductile deformation that accommodated the development of a middle crustal, normal-sense shear zone that at a higher level presumably connected with a brittle detachment fault system. Displacement along ductile-brittle shear zones was responsible for the middle Tertiary exhumation of the Ruby Mountains–East Humboldt and Albion–Raft River–Grouse Creek Mountains metamorphic terranes and setting of Oligocene and Miocene K-Ar and fission track cooling ages in the terranes.

TECTONIC MODELS

The previous discussions of major tectonic problems in the northeastern Great Basin have synthesized much of the available data for this complex region of overlapping Mesozoic and Cenozoic deformation and metamorphism. In this final section we present a two-part cartooned developmental model that appears to be consonant with most of the stratigraphic, structural, magmatic, and metamorphic data for the region. This developmental model (Fig. 23-11) incorporates the following aspects of the history of the region:

1. Precambrian rifting that thinned the basement by normal faulting.

2. Deposition of a thick (>10 km) miogeoclinal sedimentary succession.

3. Widespread Late Jurassic metamorphism in the hinterland of the Sevier orogenic belt and the development of Cordilleran infrastructure, coincident with tectonic crustal thickening

4. Shear zones between supracrustal rocks and basement developed in the Mesozoic hinterland.

5. Cretaceous folding and thrusting of miogeoclinal sedimentary rocks in the Sevier orogenic belt, involving Precambrian basement rocks in the hanging wall of some thrust faults along the western margin of the fold-and-thrust belt.

6. Upramping of portions of the hinterland and Sevier belt during the late Late Cretaceous Absaroka thrusting.

7. Widespread middle Tertiary extension involving the development of west- and east-dipping, ductile-brittle shear zones that exhumed middle-crustal rocks during the Oligocene.

8. Continued displacement along brittle shear zones principally manifested by upper-level detachment fault systems that displaced Miocene rocks.

9. Superposition of Basin and Range listric and planar normal faults on the middle Tertiary extensional structures.

Complexly overlapping Mesozoic and Cenozoic metamorphic assemblages, structures, and plutons are now demonstrated in several ranges in the northeastern Great Basin, but within these and numerous other ranges, many structures and metamorphic fabrics still cannot be assigned unambiguously to Mesozoic or Cenozoic deformational regimes. The difficulty of assigning ages and kinematic interpretations for structures has created overlapping views of the regional metamorphic and tectonic development of the northeastern Great Basin. These include assigning the metamorphism entirely to the Mesozoic (Armstrong and Hansen, 1966; Misch, 1960; Thorman, 1970) or emphasizing the Cenozoic

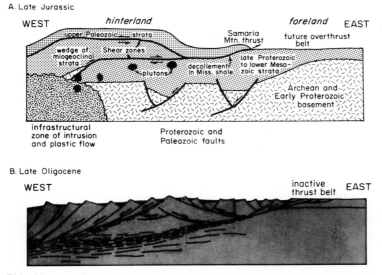

FIG. 23-11. Mesozoic and Cenozoic tectonic models for the northeastern Great Basin and adjacent overthrust belt area.

metamorphism (Compton *et al.*, 1977; E. L. Miller *et al.*, 1983), and attributing the attendant deformation to either shortening or extension modes. The divergent data and interpretations seem to require multiphase models (Armstrong, 1972, 1982; Snoke and Lush, 1984), but geologic relations demonstrating superposed Mesozoic and Cenozoic deformation are few (Allmendinger *et al.*, 1984).

Factors contributing to the difficulty of sorting out kinematic and thermal regimes include (1) difficulty in placing precise age constraints on events, (2) vague models that cannot be rigorously applied as inductive tests, (3) structures and metamorphism that are different in age having in some cases remarkably similar attributes, and (4) reactivated structures and prolonged or overprinting metamorphic events that are difficult to identify.

Superposition of Mesozoic and Cenozoic metamorphism and ductile deformation in metamorphic core complexes is not surprising, and in fact is predicted by models of crustal extension invoking shear zones at several depths in the crust. In the regions that have undergone the greatest extension, extensional processes are predicted to have exhumed the deepest rocks, those showing Cenozoic metamorphism prior to and during early extension. These same deep-seated rocks were likely even deeper during middle and late Mesozoic time before the effects of erosion and tectonic denudation late in the history of the Cordilleran orogen. Thus, superposition of Cenozoic metamorphism on Mesozoic metamorphic rocks is expected, and crustal extension models lead to the conclusion that it would be unexpected to find Cenozoic metamorphism that was *not* superposed on earlier metamorphism.

Mesozoic Model

As has been documented in preceding case study descriptions, a wide variety of structures have been attributed to Mesozoic deformation. Unmetamorphosed terranes exhibit Jurassic thrust faults, normal faults at low angles to bedding, flat normal faults, and high-angle normal and reverse faults (Allmendinger and Jordan, 1985). Ductile deformation is primarily manifested as attenuation of strata by plastic flow accompanied by ductile or brittle low-angle normal faults. Major overturned folds and fault-bounded overturned sequences indicate shortening parallel to strata in many places. Kinematic indicators give locally varying sense of shear, but generally with an eastward component. Maximum pressures given by metamorphic assemblages are greater than 4.5 kb, and maximum temperatures are about 550–600°C except in migmatitic terranes where temperatures greater than 700°C were likely. The great pressures require several kilometers of tectonic thickening.

Inferences drawn from the above synthesis yield an apparent contradiction: widespread overturned sequences, large-scale overturned folds, scattered thrust faults, and crustal thickening require overall shortening in the region during Mesozoic metamorphism, but widespread attenuation of strata by plastic and brittle processes implies regional extensional processes. It is possible that the two kinematic regimes alternated in time or space, but models relying on this scenario are difficult to apply on a regional basis. For example, gravity-driven extension in upper parts of a thickened crust could contrast with deeper plastic shortening, but shortening and extensional structures should not be consistently superposed, and a progression from shortening to extension with time should be evident. In contrast, documented Mesozoic structures indicate that attenuation of strata and shortening each occurred throughout the duration of ductile deformation, and shortening and attenuation are represented in rocks ranging from unmetamorphosed to highly metamorphosed.

We advocate that attenuation was a feature of regional shortening. One example is progressive simple shear flow with a component of extension parallel to the flow direction — the extending flow of Nye (1952) and Price (1972) and extending simple shear seen in parts of glaciers (Hudleston, 1983). Sabisky (1985) hypothesized flow of this sort of metamorphic rocks of the Raft River Mountains. In a second case, some attenuation may represent thinning in shear zones that formed at low angles to bedding. A third scenario that permits attenuation in an overall shortening regime is that of a tectonic wedge, in which widespread omission of strata might occur in the hanging-wall rocks and within the wedge if uplift and crustal thickening are significant. Large-scale omission of strata may occur in the roof thrust zone if it cuts up-section in the direction of transport.

A model of middle Mesozoic tectonic wedging best explains the available data (Fig. 23-11a) that require tectonic thickening without a surface breakout and widespread emplacement of younger rocks on older, but the model requires further testing.

Cenozoic Model

Cenozoic metamorphic rocks in the northeastern Great Basin are poorly constrained with respect to pressure, temperature, and age. They appear to have formed under lower pressures than highest-grade Mesozoic metamorphic rocks. Senses of Cenozoic shear given by various kinematic indicators are both eastward and westward (Table 23-1), perhaps for extension at different times. Extreme attenuation during Cenozoic time led to extensive mylonite development, best exemplified in the Ruby-East Humboldt terrane. In the Albion and Grouse Creek Mountains, metamorphism and deformation may have been spatially associated with Oligocene granitoids. At higher crustal levels, early Cenozoic detachment faults were ductile to brittle structures such as in the Pilot Range.

A general model of detachment tectonics incorporating aspects of the Wernicke (1981) and Hamilton (1982) models accounts for the Cenozoic extensional tectonic development of the northeastern Great Basin. At shallow crustal levels, faulted blocks were tilted and shuffled above a brittle detachment fault. At deeper levels, the fault zone was a broader normal-sense ductile shear zone. Metamorphic rocks were transported to shallower levels beneath unmetamorphosed rocks along the shear and fault system. At yet deeper levels, it may have formed anastomosing shear zones or connected to even deeper levels where large-scale lower crustal flow occurred. Allochthonous rocks moved both east and west with respect to rocks beneath detachments in the various ranges in the eastern Great Basin, suggesting that detachments dipped both east and west, which requires either that they merged in a broad plastic zone or that they cut one another. The several levels of detachment faults/shear zones are exposed widely.

Questions and New Directions

Uncertainties abound in these tectonic models, the most important of which include:

1. What is the age and the number of Mesozoic metamorphic events experienced by the hinterland? Was the 160–150 m.y.-old metamorphic event overprinted by Cretaceous metamorphism, or were the Jurassic metamorphic rocks maintained at high temperatures until Late Cretaceous cooling of the hinterland?

2. Does the Precambrian crystalline basement exposed in the Albion–Raft River–Grouse Creek terrane represent the hanging wall of a west-dipping Mesozoic ductile thrust zone but the footwall of west- and east-dipping Cenozoic low-angle normal faults?

3. What are the precise *P-T* constraints for the Mesozoic and Cenozoic metamorphic events? Are moderately high pressures (5–6 kb) required by the Mesozoic metamorphic assemblages?

4. If tectonic thickening of the hinterland was an important mechanism to facilitate burial of the miogeoclinal rocks, how was it accomplished? What role did Mesozoic thrust

faulting play, and how can these faults be recognized given the widespread Cenozoic low-angle fault overprint? Was Mesozoic extension widespread in addition to shortening?

5. Was Cenozoic extension episodic or continuous? Do active range-bounding faults represent surface expressions of the extensional fault systems operating throughout the Cenozoic, or has there been an evolution of extensional fault geometries?

6. What transfer mechanisms accommodated different amounts of extension in different areas? Were the highly extended metamorphic complexes connected by lateral ramp systems?

The answers to these questions may be achieved as more sophisticated quantitative tools are applied to the rocks and structures of the region. Several techniques could be especially important to solving these problems:

1. Geochronology will continue to be needed to constrain the ages of plutonic, deformational, and metamorphic events as well as for detailing the cooling histories of igneous and metamorphic terranes.

2. Stratigraphic studies, especially facies analyses, will be critical for identifying separations on low-angle faults that juxtapose dissimilar sedimentary packages.

3. Quantitative metamorphic petrology is needed to establish P-T-time relations in the metamorphic terranes.

4. Detailed examinations of Tertiary strata are needed to determine basin developmental histories, including magma source regions for volcanic rocks.

5. New methods for studying P-T-time relations in fault materials could complement detailed kinematic analyses of faults.

6. Deep crustal seismic reflection studies could help for understanding the crustal fabric of the hinterland and perhaps for discriminating deep-seated Mesozoic compressional structures from Cenozoic extensional shear zones. If seismic reflection studies reveal drill targets, deep boreholes could provide clues to the evolution of the region.

ACKNOWLEDGMENTS

Numerous geologists contributed to the data base from which this paper was derived; the reference list acknowledges these contributions. However, we both owe special gratitude to the late Max D. Crittenden, Jr., who introduced us to much of the geology discussed in this regional summary. The geologic studies of Snoke have been supported by National Science Foundation research grants EAR79-04204, EAR82-14236, and EAR84-18390. We thank L. L. Glick, K. A. Howard, T. E. Jordan, S. J. Reynolds, and J. E. Spencer for helpful comments that improved earlier versions of this paper.

REFERENCES

Allmendinger, R. W., and Jordan, T. E., 1981, Mesozoic evolution, hinterland of the Sevier orogenic belt: *Geology*, v. 9, p. 308–313.

――, and Jordan, T. E., 1985, Mesozoic structure of the Newfoundland Mountains, Utah–Horizontal shortening and subsequent extension in the hinterland of the Sevier belt: *Geol. Soc. America Bull.*, v. 95, p. 1280–1292.

――, Miller, D. M., and Jordan, T. E., 1984, Known and inferred Mesozoic deformation in the hinterland of the Sevier belt, northwest Utah, *in* Kerns, G. J., and Kerns, R. L., eds., *Geology of Northwest Utah, Southern Idaho and Northeast Nevada:* Utah Geol. Assoc. Publ. 13, p. 21–34.

Armstrong, F. C., and Oriel, S. S., 1965, Tectonic development of Idaho-Wyoming thrust belt: *Amer. Assoc. Petrol. Geologists Bull.*, v. 49, p. 1847–1866.

Armstrong, R. L., 1968a, Sevier orogenic belt in Nevada and Utah: *Geol. Soc. America Bull.*, v. 79, p. 429–458.

_____, 1968b, Mantled gneiss domes in the Albion Range, southern Idaho: *Geol. Soc. America Bull.*, v. 79, p. 1295–1314.

_____, 1972, Low-angle (denudation) faults, hinterland of the Sevier orogenic belt, eastern Nevada and western Utah: *Geol. Soc. America Bull.*, v. 83, p. 1729–1754.

_____, 1976, The geochronometry of Idaho — Pt. II: *Isochron/West*, v. 15, p. 1–22.

_____, 1982, Cordilleran metamorphic core complexes — From Arizona to southern Canada: *Ann. Rev. Earth Planet. Sci.*, v. 10, p. 129–154.

_____, and Hansen, E., 1966, Cordilleran infrastructure in the eastern Great Basin: *Amer. J. Sci.*, v. 264, p. 112–127.

_____, and Hills, F. A., 1967, Rb-Sr and K-Ar geochronologic studies of mantled gneiss domes, Albion Range, southern Idaho, U.S.A.: *Earth Planet. Sci. Lett.*, v. 3, p. 114–124.

_____, Taubeneck, W. H., and Hales, P. O., 1977, Rb-Sr and K-Ar geochronometry of Mesozoic granitic rocks and their Sr isotopic composition, Oregon, Washington, and Idaho: *Geol. Soc. America Bull.*, v. 88, p. 397–411.

Beck, S. L., 1982, Deformation in the Willard thrust plate in northern Utah and its regional implications: M.S. thesis; Univ. Utah, Salt Lake City, Utah, 79 p.

_____, and Bruhn, R. L., 1983, Ductile deformation and basement involvement contemporaneous with a Mesozoic thrust fault in the Sevier orogenic belt, northern Utah: *Geol. Soc. America Abstr. with Programs*, v. 15, p. 377.

Bell, G. L., 1952, Geology of the northern Farmington Mountains, *in* Marsell, R. E., ed., *Guidebook to the Geology of Utah, no. 8, Geology of the Central Wasatch Mountains, Utah:* Utah Geol. Min. Survey, p. 38–51.

Best, M. G., Armstrong, R. L., Graustein, W. C., Embree, G. F., and Ahlborn, R. C., 1974, Mica granites of the Kern Mountains pluton, eastern White Pine County, Nevada — Remobilized basement of the Cordilleran miogeosyncline? *Geol. Soc. America Bull.*, v. 85, p. 1277–1286.

Brown, R. L., Journeay, J. M., Murphy, D. C., and Rees, C. J., 1986, Obduction, backfolding and piggyback thrusting in the metamorphic hinterland of the southeast Canadian Cordillera: *J. Struct. Geology*, v. 8, p. 255–268.

Bruhn, R. L., and Beck, S. L., 1981, Mechanics of thrust faulting in crystalline basement, Sevier orogenic belt, Utah: *Geology*, v. 9, p. 200–204.

_____, Picard, M. D., and Beck, S. L., 1983, Mesozoic and early Tertiary structure and sedimentology of the central Wasatch Mountains, Uinta Mountains and Uinta Basin, *in* Gurgel, K. D., ed., *Geologic Excursions in the Overthrust Belt and Metamorphic Core Complexes of the Intermountain Region, Nevada:* Geol. Soc. America Field Trip Guidebook, Utah Geol. Min. Survey Spec. Studies 59, p. 63–105.

Bryant, B., 1979, Reconnaissance geologic map of the Precambrian Farmington Canyon Complex and surrounding rocks in the Wasatch Mountains between Ogden and Bountiful, Utah: *U.S. Geol. Survey Open File Rpt. 79-709.*

Burchfiel, B. C., and Davis, G. A., 1975, Nature and controls of Cordilleran orogenesis, western United States — Extensions of an earlier synthesis: *Amer. J. Sci.*, v. 275-A, p. 363–396.

Christensen, O. D., 1975, Metamorphism of the Manning Canyon and Chainman Formations: Ph.D. dissertation; Stanford Univ., Stanford, Calif., 194 p.

Coats, R. R., and Riva, J. F., 1983, Overlapping overthrust belts of Late Paleozoic and Mesozoic ages, northern Elko County, Nevada, *in* Miller, D. M., Todd, V. R., and

Howard, K. A., eds., *Tectonic and Stratigraphic Studies in the Eastern Great Basin:* Geol. Soc. America Mem. 157, p. 305–327.

Collinson, J. W., Kendall, C. G. St. C., and Marcantel, J. B., 1976, Permian-Triassic boundary in eastern Nevada and west-central Utah: *Geol. Soc. America Bull.*, v. 87, p. 821–824.

Compton, R. R., 1972, Geologic map of the Yost quadrangle, Box Elder County, Utah, and Cassia County, Idaho: *U.S. Geol. Survey Misc. Geol. Inv. Map I-672.*

——, 1975, Geologic map of the Park Valley quadrangle, Box Elder County, Utah, and Cassia County, Idaho: *U.S. Geol. Survey Misc. Geol. Inv. Map I-873.*

——, 1983, Displaced Miocene rocks on the west flank of the Raft River–Grouse Creek core complex, Utah, *in* Miller, D. M., Todd, V.R., and Howard, K. A., eds., *Tectonic and Stratigraphic Studies in the Eastern Great Basin:* Geol. Soc. America Mem. 157, p. 271–280.

——, Todd, V. R., Zartman, R. E., and Naeser, C. W., 1977, Oligocene and Miocene metamorphism, folding, and low-angle faulting in northwestern Utah: *Geol. Soc. America Bull.*, v. 88, p. 1237–1250.

Condie, K. C., 1982, Plate-tectonics model for Proterozoic continental accretion in the southwestern United States: *Geology*, v. 10, p. 37–42.

Coney, P. J., 1979, Tertiary evolution of Cordilleran metamorphic core complexes, *in* Armentrout, J. M., Cole, M. R., and TerBest, H., Jr., eds., *Cenozoic Paleogeography of the Western United States:* Pacific Section, Soc. Econ. Paleontologists Mineralogists, Pacific Coast Paleogeography Symp. 3, p. 15–28.

——, and Harms, T. A., 1984, Cordilleran metamorphic core complexes — Cenozoic extensional relics of Mesozoic compression: *Geology*, v. 12, p. 550–554.

Covington, H. R., 1983, Structural evolution of the Raft River Basin, Idaho, *in* Miller, D. M., Todd, V. R., and Howard, K. A., eds., *Tectonic and Stratigraphic Studies in the Eastern Great Basin:* Geol. Soc. America Mem. 157, p. 229–237.

Crittenden, M. D., Jr., 1972, Willard thrust and the Cache allochthon, Utah: *Geol. Soc. America Bull.*, v. 83, p. 2871–2880.

——, 1979, Oligocene and Miocene metamorphism, folding, and low-angle faulting in northwestern Utah — Discussion: *Geol. Soc. America Bull.*, v. 90, p. 305–306.

——, and Sorensen, M. L., 1980, The Facer Formation, a new Early Proterozoic unit in northern Utah: *U.S. Geol. Survey Bull. 1482-F*, p. F1–F28.

——, Schaeffer, F. E., Trimble, D. E., and Woodward, L. A., 1971, Nomenclature and correlation of some upper Precambrian and basal Cambrian sequences in western Utah and southern Idaho: *Geol. Soc. America Bull.*, v. 82, p. 581–602.

——, Coney, P. J., and Davis, G. H., eds., 1980, Cordilleran metamorphic core complexes: *Geol. Soc. America Mem. 153*, 490 p.

Dahlstrom, C. D. A., 1970, Structural geology in the eastern margin of the Canadian Rocky Mountains: *Bull. Can. Petrol. Geology*, v. 18, p. 332–406.

Dallmeyer, R. D., Snoke, A. W., and McKee, E. H., 1986, The Mesozoic-Cenozoic tectonothermal evolution of the Ruby Mountains–East Humboldt Range, Nevada — A Cordilleran metamorphic core complex: *Tectonics*, v. 5, p. 931–954.

Davis, G. A., Lister, G. S., and Reynolds, S. J., 1986, Structural evolution of the Whipple and South Mountains shear zones, southwestern United States: *Geology*, v. 14, p. 7–10.

Davis, G. H., 1983, Shear-zone model for the origin of metamorphic core complexes: *Geology*, v. 11, p. 342–347.

——, and Coney, P. J., 1979, Geologic development of the Cordilleran metamorphic core complexes: *Geology*, v. 7, p. 120–124.

Dickinson, W. R., Harbaugh, D. W., Saller, A. H., Heller, P. L., and Snyder, W. S., 1983, Detrital modes of Upper Paleozoic sandstones derived from Antler orogen in Nevada — Implications for nature of Antler orogeny: *Amer. J. Sci.*, v. 283, p. 481–509.

Dokka, R. K., Mahaffie, M. J., and Snoke, A. W., 1986, Thermochronologic evidence of major tectonic denudation associated with detachment faulting, northern Ruby Mountains-East Humboldt Range, Nevada: *Tectonics*, v. 5, p. 995–1006.

Duebendorfer, E. M., and Houston, R. S., 1986, Kinematic history of the Cheyenne belt, Medicine Bow Mountains, southeastern Wyoming: *Geology*, v. 14, p. 171–174.

Farmer, G. L., and DePaolo, D. J., 1983, Origin of Mesozoic and Tertiary granite in the western United States and implications for pre-Mesozoic crustal structure — 1. Nd and Sr isotopic studies in the geocline of the northern Great Basin: *J. Geophys. Res.*, v. 88, p. 3379–3401.

Fouch, T. D., 1979, Character and paleogeographic distribution of Upper Cretaceous(?) and Paleogene nonmarine sedimentary rocks in east-central Nevada, *in* Armentrout, J. M., Cole, M. R., and TerBest, H., Jr., eds., *Cenozoic Paleogeography of the Western United States:* Pacific Section, Soc. Econ. Paleontologists Mineralogists, Pacific Coast Paleogeography Symp. 3, p. 97–111.

Gabrielse, H., 1972, Younger Precambrian of the Canadian Cordillera: *Amer. J. Sci.*, v. 272, p. 521–536.

Greenwood, H. J., 1976, Metamorphism at moderate temperatures and pressures, *in* Bailey, D. K., and MacDonald, R., eds., *The Evolution of the Crystalline Rocks:* London, Academic, p. 187–259.

Hamilton, W. B., 1982, Structural evolution of the Big Maria Mountains, northeastern Riverside County, southeastern California, *in* Frost, E. G., and Martin, D. L., eds., *Mesozoic-Cenozoic Tectonic Evolution of the Colorado River Region, California, Arizona, and Nevada* (Anderson-Hamilton volume): San Diego, Calif. Cordilleran Publishers, p. 1–27.

Harris, A. G., Wardlaw, B. R., Rust, C. C., and Merrill, G. K., 1980, Maps for assessing thermal maturity (conodont color alteration index maps) in Ordovician through Triassic rocks in Nevada and Utah and adjacent parts of Idaho and California: *U.S. Geol. Survey Misc. Geol. Inv. Map I-1249*, scale 1:2,500,000.

Henderson, J. R., 1981, Structural analysis of sheath folds with horizontal X-axes, northeast Canada: *J. Struct. Geology*, v. 3, p. 203–210.

Hodges, K. V., and McKenna, L., 1986, Structural and metamorphic characteristics of the Raft River-Quartzite Assemblage juxtaposition, Albion Mountains, southern Idaho: *Geol. Soc. America Abstr. with Programs*, v. 18, p. 117.

Hollet, D. W., Bruhn, R. L., and Parry, W. T., 1978, Physiochemical aspects of thrust faulting in Precambrian gneiss, Sevier orogenic belt, Utah: *Geol. Soc. America Abstr. with Programs*, v. 10, p. 424.

Hope, R. A., and Coats, R. R., 1976, Preliminary geologic map of Elko County, Nevada: *U.S. Geol. Survey Open File Rpt. 76-779*, scale 1:100,000.

Howard, K. A., 1966, Structure of the metamorphic rocks of the northern Ruby Mountains, Nevada: Ph.D. thesis, Yale Univ., New Haven, Conn., 170 p.

——, 1968, Flow direction in triclinic folded rocks: *Amer. J. Sci.*, v. 266, p. 758–765.

——, 1971, Paleozoic metasediments in the northern Ruby Mountains, Nevada: *Geol. Soc. America Bull.*, v. 82, p. 259–264.

——, 1980, Metamorphic infrastructure in the northern Ruby Mountains, Nevada *in* Crittenden, M. D., Jr., Coney, P. J., and Davis, G. A., eds., *Cordilleran Metamorphic Core Complexes:* Geol. Soc. America Mem. 153, p. 335–347.

_____, Kistler, R. W., Snoke, A. W., and Willden, R., 1979, Geologic map of the Ruby Mountains, Nevada: *U.S. Geol. Survey Misc. Geol. Inv. Map I-1136*, scale 1:125,000.

Hudleston, P. J., 1983, Strain patterns in an ice cap and implications for strain variations in shear zones: *J. Struct. Geology*, v. 5, p. 455–463.

Johnson, D. L., 1981, Two-mica granites and metamorphic rocks of the east-central Ruby Mountains, Elko County, Nevada: M.S. Thesis; Stanford Univ., Stanford, Calif., 145 p.

Jordan, T. E., and Douglass, R. C., 1980, Paleogeography and structural development of the Late Pennsylvanian to Early Permian Oquirrh Basin, northwestern Utah, *in* Fouch, T. D., and Magathan, E. R., eds., *Paleozoic Paleogeography of West-Central United States:* West-Central United States Rocky Mountain Section, Soc. Econ. Paleontologists Mineralogists, Paleogeography Symp. 1, p. 217–238.

Karlstrom, K. E., and Houston, R. S., 1984, The Cheyenne belt — Analysis of a Proterozoic suture in southern Wyoming: *Precambrian Res.*, v. 25, p. 415–446.

Ketner, K. B., 1977, Late Paleozoic orogeny and sedimentation, southern California, Nevada, Idaho, and Montana, *in* Stewart, J. H., Stevens, C. H., and Fritsche, A. E., eds., *Paleozoic Paleogeography of the Western United States:* Pacific Section, Soc. Econ. Paleontologists Mineralogists, Pacific Coast Paleogeography Symp. 1, p. 363–369.

Kistler, R. W., and Peterman, Z. E., 1978, Reconstruction of crustal blocks of California on the basis of initial strontium isotopic compositions of Mesozoic granitic rocks: *U.S. Geol. Survey Prof. Paper 1071*, 17 p.

_____, Ghent, E. D., and O'Neil, J. R., 1981, Petrogenesis of garnet two-mica granites in the Ruby Mountains, Nevada: *J. Geophys. Res.*, v. 86, p. 10591–10606.

Kluth, C. F., and Coney, P. J., 1981, Plate tectonics of the ancestral Rocky Mountains: *Geology*, v. 9, p. 10–15.

Link, P. K., LeFebre, G. B., Pogue, K. R., and Burgel, W. D., 1985, Structural geology between the Putnam thrust and the Snake River Plain, southeastern Idaho, *in* Kerns, G. J., and Kerns, R. L., eds., *Orogenic Patterns and Stratigraphy of North-Central Utah and Southeastern Idaho:* Utah Geol. Assoc. Publ. 14, p. 97–117.

Lister, G. S., and Snoke, A. W., 1984, S-C mylonites: *J. Struct. Geology*, v. 6, p. 617–638.

Lush, A. P., 1982, Geology of part of the northern East Humboldt Range, Elko County, Nevada: M.S. thesis, Univ. South Carolina, Columbia, S.C., 138 p.

Marcantel, J., 1975, Late Pennsylvanian and Early Permian sedimentation in northeast Nevada: *Amer. Assoc. Petrol. Geologists Bull.*, v. 59, p. 2079–2098.

McCollum, L. B., and McCollum, M. B., 1984, Comparison of a Cambrian medial shelf margin sequence with an outer shelf margin sequence, northern Great Basin, *in* Kerns, G. J., and Kerns, R. L., eds., *Geology of Northwest Utah, Southern Idaho and Northeast Nevada:* Utah Geol. Assoc. Publ. 13, p. 35–44.

McKee, E. H., Noble, D. C., and Silberman, M. L., 1970, Middle Miocene hiatus in volcanic activity in the Great Basin area of the western United States: *Earth Planet. Sci. Lett.*, v. 8, p. 93–96.

Miller, D. M., 1980, Structural geology of the northern Albion Mountains, south-central Idaho, *in* Crittenden, M. D., Jr., Coney, P. J., and Davis, G. H., eds., *Cordilleran Metamorphic Core Complexes:* Geol. Soc. America Mem. 153, p. 399–423.

_____, 1983a, Allochthonous quartzite sequence in the Albion Mountains, Idaho, and proposed Proterozoic Z and Lower Cambrian correlatives in the Pilot Range, Utah and Nevada, *in* Miller, D. M., Todd, V. R., and Howard, K. A., eds., *Tectonic and Stratigraphic Studies in the Eastern Great Basin:* Geol. Soc. America Mem. 157, p. 191–213.

———, 1983b, Mesozoic metamorphism and low-angle faults in the hinterland of Nevada linked to Sevier-belt thrusts: *Geol. Soc. America Abstr. with Programs*, v. 15, p. 644.

———, 1984, Sedimentary and igneous rocks of the Pilot Range and vicinity, Utah and Nevada, *in* Kerns, G. J., and Kerns, R. L., eds., *Geology of Northwest Utah, Southern Idaho and Northeast Nevada:* Utah Geol. Assoc. Publ. 13, p. 45–63.

———, 1985, Geologic map of the Lucin quadrangle, Box Elder County, Utah: *Utah Geol. Min. Survey Map 78*, 10 p., scale 1:24,000.

———, and Lush, A. P., 1981, Preliminary geologic map of the Pilot Peak and adjacent quadrangles, Elko County, Nevada, and Box Elder County, Utah: *U.S. Geol. Survey Open File Rpt. 81-658*, 18 p., scale 1:24,000.

———, Armstrong, R. L., Compton, R. R., and Todd, V. R., 1983, Geology of the Albion-Raft River–Grouse Creek Mountains area, northwestern Utah and southern Idaho: *Utah Geol. and Min. Survey Spec. Studies 59*, p. 1–62.

———, Hillhouse, W. C., Zartman, R. E., and Lanphere, M. A., in press, Geochronology of intrusive and metamorphic rocks the Pilot Range, Utah and Nevada, and comparison with regional patterns. Geol. Soc. America Bull., v. 100.

Miller, E. L., Gans, P. B., and Garing, J., 1983, The Snake Range décollement — An exhumed mid-Tertiary ductile-brittle transition: *Tectonics*, v. 2, p. 239–263.

Misch, P., 1960, Regional structural reconnaissance in central northeast Nevada and some adjacent areas — Observations and interpretations, *in Geology of East-Central Nevada:* Intermountain Assoc. Petrol. Geologists 11th Ann. Field Conf. Guidebook, p. 17–42.

———, and Hazzard, J. C., 1962, Stratigraphy and metamorphism of late Precambrian rocks in central northeastern Nevada and adjacent Utah: *Amer. Assoc. Petrol. Geologists Bull.*, v. 46, p. 289–343.

Moore, W. J., and McKee, E. H., 1983, Phanerozoic magmatism and mineralization in the Tooele 1° X 2° quadrangle, Utah, *in* Miller, D. M., Todd, V. R., and Howard, K. A., eds., *Tectonic and Stratigraphic Studies in the Eastern Great Basin:* Geol. Soc. America Mem. 157, p. 183–190.

Naeser, C. W., Bryant, B., Crittenden, M. D., Jr., and Sorensen, M. L., 1983, Fission-track ages of apatite in the Wasatch Mountains, Utah — An uplift study, *in* Miller, D. M., Todd, V. R., and Howard, K. A., eds., *Tectonic and Stratigraphic Studies in the Eastern Great Basin:* Geol. Soc. America Mem. 157, p. 29–36.

Nelson, B. K., and DePaolo, D. J., 1985, Rapid production of continental crust 1.7 to 1.9 b.y. ago — Nd isotopic evidence from the basement of the North American mid-continent: *Geol. Soc. America Bull.*, v. 96, p. 746–754.

———, Farmer, G. L., Bennett, V. C., and DePaolo, D. J., 1983, Sm-Nd evidence for a possible Penokean-correlative basement terrane in the eastern Great Basin: (EOS) *Trans. Amer. Geophys. Un.*, v. 64, p. 331.

Nye, J. F., 1952, The mechanics of glacier flow: *J. Glaciology*, v. 2, p. 82–93.

Oldow, J. S., 1984, Evolution of a Late Mesozoic back-arc fold and thrust belt, northwestern Great Basin, U.S.A.: *Tectonophysics*, v. 102, p. 245–274.

Olson, J. C., and Hinrichs, E. N., 1960, Beryl-bearing pegmatites in the Ruby Mountains and other areas in Nevada and northwestern Arizona: *U.S. Geol. Survey Bull. 1082-D*, p. 135–200.

Palmer, A. R., compiler, 1983, The Decade of North American Geology 1983 geologic time scale: *Geology*, v. 11, p. 503–504.

Pierce, W. G., 1963, Reef Creek detachment fault, northwestern Wyoming: *Geol. Soc. America Bull.*, v. 74, p. 1225–1236.

Poole, F. G., 1974, Flysch deposits of Antler foreland basin, western United States, *in*

Dickinson, W. R., ed., *Tectonics and Sedimentation:* Soc. Econ. Paleontologists Mineralogists Spec. Publ. 22, p. 58–82.

Price, R. A., 1972, The distinction between displacement and distortion in flow, and the origin of diachronism in tectonic overprinting in orogenic belts: *Proc. 24th Internat. Geol. Cong., Sect. 3*, p. 545–551.

——, 1986, The southeastern Canadian Cordillera — Thrust faulting, tectonic wedging, and delamination of the lithosphere: *J. Struct. Geology*, v. 8, p. 239–254.

Rehrig, W. A., and Reynolds, S. J., 1980, Geologic and geochronologic reconnaissance of a northwest-trending zone of metamorphic core complexes in southern and western Arizona, *in* Crittenden, M. D., Jr., Coney, P. J., and Davis, G. H., eds., *Cordilleran Metamorphic Core Complexes:* Geol. Soc. America Mem. 153, p. 131–157.

Reynolds, S. J., and Rehrig, W. A., 1980, Mid-Tertiary plutonism and mylonitization, South Mountains, central Arizona, *in* Crittenden, M. D., Jr., Coney, P. J., and Davis, G. H., eds., *Cordilleran Metamorphic Core Complexes:* Geol. Soc. America Mem. 153, p. 159–175.

——, Shafiqullah, M., Damon, P. E., and DeWitt, E., 1986, Early Miocene mylonitization and detachment faulting, South Mountains, central Arizona: *Geology*, v. 14, p. 283–286.

Roberts, R. J., Hotz, P. E., Gilluly, J., and Ferguson, H. G., 1958, Paleozoic rocks of north-central Nevada: *Amer. Assoc. Petrol. Geologists Bull.*, v. 42, p. 2813–2857.

——, Crittenden, M. D., Jr., Tooker, E. W., Morris, H. T., Hose, R. K., and Cheney, T. M., 1965, Pennsylvanian and Permian basins in northwestern Utah, northeastern Nevada, and south-central Idaho: *Amer. Assoc. Petrol. Geologists Bull.*, v. 49, p. 1926–1956.

Royse, F., Jr., Warner, M. A., and Reese, D. L., 1975, Thrust belt structural geometry and related stratigraphic problems, Wyoming-Idaho-northern Utah, *in* Bolyard, D. W., ed., *Deep Drilling Frontiers of the Central Rocky Mountains:* Rocky Mountain Assoc. Geologists Symp., p. 41–54.

Sabisky, M., 1985, Ductile flow and folding in the Raft River Mountains, northwestern Utah: (EOS) *Trans. Amer. Geophys. Un.*, v. 66, p. 1089.

Saltzer, S. D., and Hodges, K. V., 1986 Mylonitic fabric analysis at Middle Mountain, southern Idaho: *Geol. Soc. America Abstr. with Programs*, v. 18, p. 180.

Schirmer, T. W., 1985, Basement thrusting in north-central Utah — A model for the development of the northern Utah highland, *in* Kerns, G. J., and Kerns, R. L., eds., *Orogenic Patterns and Stratigraphy of North-Central Utah and Southeastern Idaho:* Utah Geol. Assoc. Publ. 14, p. 129–143.

Schneyer, J. D., 1984, Structural and stratigraphic complexities within an extensional terrain: Examples from the Leppy Hills area, southern Silver Island Mountains, near Wendover, Utah, *in* Kerns, G. L., and Kerns, R. L., Jr., eds., Geology of Northwest Utah, Southern Idaho and Northeast Nevada: Utah Geological Association Publication 13, p. 93–116.

Schweickert, R. A., Bogen, N. L., Girty, G. H., Hanson, R. E., and Merguerian, C., 1984, Timing and structural expression of the Nevadan orogeny, Sierra Nevada, California: *Geol. Soc. America Bull.*, v. 95, p. 967–979.

Sears, J. W., and Price, R. A., 1978, The Siberian Connection — A case for Precambrian separation of the North American and Siberian cratons: *Geology*, v. 6, p. 267–270.

——, Graff, P. J., and Holden, G. S., 1982, Tectonic evolution of lower Proterozoic rocks, Uinta Mountains, Utah and Colorado: *Geol. Soc. America Bull.*, v. 93, p. 990–997.

Silberling, N. J., and Roberts, R. J., 1962, Pre-Tertiary stratigraphy and structure of northwestern Nevada: *Geol. Soc. America Spec. Paper 72*, 58 p.

Simpson, C., and Schmid, S. M., 1983, An evaluation of criteria to deduce the sense of movement in sheared rocks: *Geol. Soc. America Bull.*, v. 94, p. 1281–1288.

Smith, J. F., Jr., 1982, Geologic map of the Strevell 15 minute quadrangle, Cassia County, Idaho: *U.S. Geol. Survey Misc. Geol. Inv. Map I-1403*, scale 1:62,500.

———, 1983, Paleozoic rocks in the Black Pine Mountains, Cassia County, Idaho: *U.S. Geol. Survey Bull. 1536*, 36 p.

———, and Ketner, K. B., 1976, Stratigraphy of post-Paleozoic rocks and summary of resources in the Carlin-Pinon Range area, Nevada *with a section on* Aeromagnetic survey by D. R. Mabey: *U.S. Geol. Survey Prof. Paper 867-B*, 48 p.

Smith, R. B., 1978, Seismicity, crustal structure, and intraplate tectonics of the interior of the western Cordillera, *in* Smith, R. B., and Eaton, G. P., eds., *Cenozoic Tectonics and Regional Geophysics of the Western Cordillera*, Geol. Soc. America Mem. 152, p. 111–144.

Snelson, S., 1957, The geology of the northern Ruby Mountains and the East Humboldt Range, Elko County, Nevada: Ph.D. thesis, Univ. Washington, Seattle, Wash., 268 p.

Snoke, A. W., 1980, Transition from infrastructure to suprastructure in the northern Ruby Mountains, Nevada, *in* Crittenden, M. D., Jr., Coney, P. J., and Davis, G. H., eds., *Cordilleran Metamorphic Core Complexes*, Geol. Soc. America Mem. 153, p. 287–333.

———, and Howard, K. A., 1984, Geology of the Ruby Mountains-East Humboldt Range, Nevada — A Cordilleran metamorphic core complex, *in* Lintz, Joseph, Jr., ed., *Western Geological Excursions, Vol. 4*; Geol. Soc. America Ann. Meet., MacKay Sch. Mines, Reno, Nev., p. 260–303.

———, and Lush, A. P., 1984, Polyphase Mesozoic-Cenozoic deformational history of the northern Ruby Mountains-East Humboldt Range, Nevada, *in* Lintz, Joseph, Jr., ed., *Western Geological Excursions, Vol. 4*; Geol. Soc. America Ann. Meet., Mackay Sch. Mines, Reno, Nev., p. 232–260.

———, McKee, E. H., and Stern, T. W., 1979, Plutonic, metamorphic, and structural chronology in the northern Ruby Mountains, Nevada — A preliminary report: *Geol. Soc. America Abstr. with Programs*, v. 11, p. 520–521.

———, Dallmeyer, R. D., and Fullagar, P. D., 1984, Superimposed Tertiary mylonitization on a Mesozoic metamorphic terrane, Ruby Mountains-East Humboldt Range, Nevada: *Geol. Soc. America Abstr. with Programs*, v. 16, p. 662.

Solomon, B. J., McKee, E. H., and Anderson, D. W., 1979, Paleogene rocks near Elko, Nevada, *in* Armentrout, J. M., Cole, M. R., and TerBest, H., Jr., eds., *Cenozoic Paleogeography of the Western United States:* Pacific Section, Soc. Econ. Paleontologists Mineralogists, Pacific Coast Paleogeography Symp. 3, p. 75–89.

Speed, R. C., 1982, Evolution of the sialic margin in central western United States, *in* Watkins, J. S., and Drake, C. L., eds., *Studies in Continental Margin Geology:* Amer. Assoc. Petrol. Geologists Mem. 34, p. 457–468.

Spencer, J. E., 1984, Role of tectonic denudation in warping and uplift of low-angle normal faults: *Geology*, v. 12, p. 95–98.

Steele, G., 1960, Pennsylvanian-Permian stratigraphy of east-central Nevada and adjacent Utah, *in Geology of East-Central Nevada:* Intermountain Assoc. Petrol. Geologists 11th Ann. Field Conf. Guidebook, p. 91–113.

Stewart, J. H., 1972, Initial deposits in the Cordilleran geosyncline — Evidence of a late Precambrian (<850 m.y.) continental separation: *Geol. Soc. America Bull.*, v. 83, p. 1345–1360.

———, 1976, Late Precambrian evolution of North America — Plate tectonics implication: *Geology*, v. 4, p. 11–15.

———, 1980, Geology of Nevada — A discussion to accompany the Geologic Map of Nevada: *Nev. Bur. Mines Geology Spec. Publ. 4*, 136 p.

———, and Poole, F. G., 1974, Lower Paleozoic and uppermost Precambrian Cordilleran miogeocline, Great Basin, western United States, *in* Dickinson, W. R., ed., *Tectonics and Sedimentation:* Soc. Econ. Paleontologists Mineralogists Spec. Publ. 22, p. 28–57.

Taylor, G. K., 1984, Stratigraphy, metamorphism, and structure of the southeastern East Humboldt Range, Elko County, Nevada: M.S. thesis, Univ. South Carolina, Columbia, S.C., 148 p.

Thorman, C. H., 1965, Mid-Tertiary K-Ar dates from late Mesozoic metamorphosed rocks, Wood Hills and Ruby–East Humboldt Range, Elko County, Nevada: *Geol. Soc. America Spec. Paper 87*, p. 234–235.

———, 1970, Metamorphosed and nonmetamorphosed Paleozoic rocks in the Wood Hills and Pequop Mountains, northeast Nevada: *Geol. Soc. America Bull.*, v. 81, p. 2417–2448.

Todd, V. R., 1980, Structure and petrology of a Tertiary gneiss complex in northwestern Utah, *in* Crittenden, M. D., Jr., Coney, P. J., and Davis, G. H., eds., *Cordilleran Metamorphic Core Complexes:* Geol. Soc. America Mem. 153, p. 349–383.

———, 1983, Late Miocene displacement of pre-Tertiary and Tertiary rocks in the Matlin Mountains, northwestern Utah, *in* Miller, D. M., Todd, V. R., and Howard, K. A., eds., *Tectonic and Stratigraphic Studies in the Eastern Great Basin:* Geol. Soc. America Mem. 157, p. 239–270.

Van Houten, F. B., 1956, Reconnaissance of Cenozoic sedimentary rocks of Nevada: *Amer. Assoc. Petrol. Geologists Bull.*, v. 40, p. 2801–2825.

Wernicke, B., 1981, Low-angle normal faults in the Basin and Range province — Nappe tectonics in an extending orogen: *Nature*, v. 291, p. 645–648.

———, 1982, Mesozoic evolution, hinterland of the Sevier orogenic belt — Comment: *Geology*, v. 10, p. 3–5.

———, 1985, Uniform-sense normal simple shear of the continental lithosphere: *Can. J. Earth Sci.*, v. 22, p. 108–125.

Willden, R., and Kistler, R. W., 1979, Precambrian and Paleozoic stratigraphy in central Ruby Mountains, Elko County, Nevada, *in* Newman, G. W., and Goode, H. D., eds., *Basin and Range Symposium and Great Basin Field Conference:* Rocky Mountain Assoc. Geologists Utah Geol. Assoc., p. 221–243.

Wright, J. E., and Snoke, A. W., 1986, Mid-Tertiary mylonitization in the Ruby Mountains–East Humboldt Range metamorphic core complex, Nevada: *Geol. Soc. America Abstr. with Programs*, v. 18, p. 795.

Wust, S. L., 1986, Regional correlation of extension directions in Cordilleran metamorphic core complexes: *Geology*, v. 14, p. 828–830.

Zen, E-an, and Hammarstrom, J. M., 1984, Magmatic epidote and its petrologic significance: *Geology*, v. 12, p. 515–518.

Zoback, M. L., Anderson, R. E., and Thompson, G. A., 1981, Cainozoic evolution of the state of stress and style of tectonism of the Basin and Range province of the western United States: *Philos. Trans. Roy. Soc. London*, v. A300, p. 407–434.

Elizabeth L. Miller, Phillip B. Gans, and James E. Wright
Department of Geology
Stanford University
Stanford, California 94305

John F. Sutter
U. S. Geological Survey
Reston, Virginia 22092

24

METAMORPHIC HISTORY OF THE EAST-CENTRAL BASIN AND RANGE PROVINCE: TECTONIC SETTING AND RELATIONSHIP TO MAGMATISM

ABSTRACT

Studies in the northern and southern Snake Range, Schell Creek Range, Kern Mountains, and Deep Creek Range coupled with published and new U-Pb and $^{40}Ar/^{39}Ar$ dates on igneous and metamorphic rocks permit some broad generalizations about the metamorphic history of the miogeoclinal succession of this region. Three distinct metamorphic events, of Jurassic, Cretaceous, and Tertiary age, all increase in grade with depth, were accompanied by deformation, and temporally overlapped with the intrusion of granitic plutons. Synkinematic aureoles of plutons all contain approximately the same assemblages: and + mu + bi ± stt ± sil ± cord. These three events affect about the same crustal levels (7–15 km), creating polyphase fabrics in many areas. Lateral and vertical temperature gradients were high, locally greater than 100°C/km, suggesting that the emplacement of plutons was probably the primary cause of heating at these structural levels.

Jurassic metamorphism was accompanied by the formation of an east-dipping, regionally developed cleavage (but no folding) which begins at about 7–8 km depth and increases to amphibolite facies at depth and toward plutons dated in the southern Snake Range as 160 m.y. old. Bedding-cleavage angles and their variation with depth suggest strain is a result of *top-to-the-west* bedding-parallel shear; ductile strain at depth either dies upward into folds or may reflect west-vergent thrusting.

Cretaceous metamorphism is associated with a well-developed cleavage that cuts the older Jurassic fabric and is west-dipping with respect to bedding. Decreasing bedding-cleavage angles and increasing strain with depth indicate *top-to-the-east* bedding-parallel shear, and suggest that exposures of deep-seated rocks in east-central Nevada represent the upper part of the (ductile) root zone for faults of the Cretaceous Sevier thrust belt to the east and/or lie in the upper plate of their basal thrust. Dating of fabrics and plutons indicate that Cretaceous magmatism and deformation may have been protracted, and its timing is compatible with the history of thrust faulting to the east. Tonalitic to granitic plutons as old as 110 m.y. predate the intrusion of Late Cretaceous muscovite-bearing granites. The latter have been interpreted as the products of crustal anatexis related to basement shortening beneath the region. The intrusion of the older plutons may have elevated isotherms and aided this process.

Tertiary strain was many orders of magnitude greater than that which accompanied metamorphism in the Mesozoic. High-angle normal faults cut upper crustal rocks to depths that varied from 7 to 15 km. Below, ductile fabrics formed in response to vertical shortening and WNW-ESE extension. Zones of ductile simple shear and brittle high-angle faults overprint and uplift older deep-seated extensional fabrics. This event began with the intrusion of plutons in the Oligocene about 36 m.y. ago; seismic data across the region suggest that the process of extensional ductile flow at depth may continue to the present in concert with Basin and Range faulting at the surface.

Magmatism, metamorphism, and deformation appear to be intimately related in the upper 15 km of crust across east-central Nevada. At the levels exposed today, metamorphic and deformational events are distinct, and rocks were able to cool between these events; at deeper levels we would expect these distinctions to be less clear and to find evidence for partial to wholesale remobilization of the crust and more complete overprinting of older events by younger events.

INTRODUCTION

The "hinterland" of the Mesozoic Sevier fold-and-thrust belt in the northern Basin and Range province as defined by Armstrong (1968, 1972) exposes deep seated rocks that were deformed, metamorphosed, and intruded in the Mesozoic. Most workers believe that this deformation and metamorphism is geometrically and genetically linked to supracrustal thrust faulting to the east, but isolated exposure due to basin and range faulting, complicated isotopic systems, and locally intense Tertiary deformational and metamorphic overprint have hindered our understanding of the relationship of these events to foreland deformation.

This paper describes the deformational, metamorphic, and intrusive history of deep-seated rocks exposed in the Schell Creek Range, Snake Range, and Deep Creek Range portion of the hinterland (Figs. 24-1 and 24-2), addresses the question of the timing of these events with respect to the development of the thrust belt, and discusses the nature of superimposed Tertiary fabrics and metamorphism. Our study represents the first comprehensive overview of metamorphism in this region since the pioneering work and papers of Misch (1960) and his students, Misch and Hazzard (1962), and the later compilation of Hose and Blake (1976).

Although the details of the style as well as the exact timing of events might be ex-

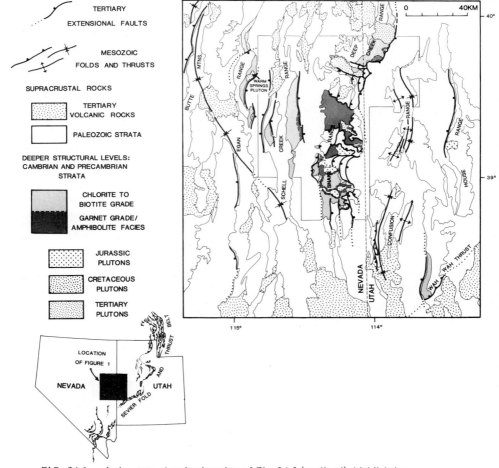

FIG. 24-1.   Index map showing location of Fig. 24-2 (outlined), highlighting the exposures and approximate metamorphic grade of deep-seated rocks uplifted by Tertiary extensional faulting in east-central Nevada and adjacent Utah. Modified after Stewart and Carlsen (1978) and Hintze (1980).

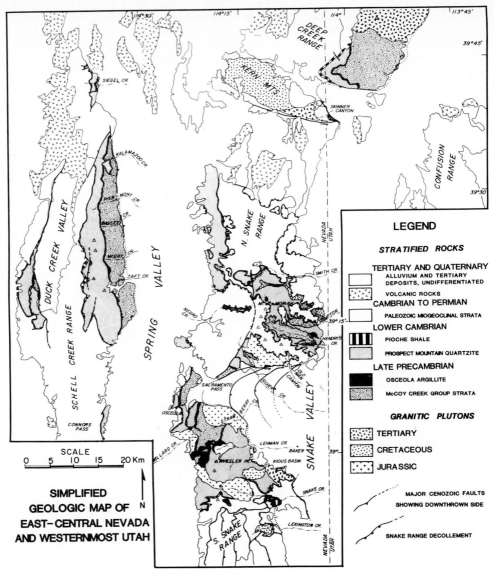

FIG. 24-2. Simplified geologic map of east-central Nevada and part of westernmost Utah with emphasis on the distribution of upper Precambrian-Cambrian miogeoclinal strata and the plutons that intrude these rocks. The Cambrian Pioche Shale and the upper Precambrian Osceola Argillite (Misch and Hazzard, 1962) are highlighted inasmuch as these pelitic units are particularly useful for comparing the metamorphic and deformational history of the miogeoclinal succession both within mountain ranges and between mountain ranges. Geologic base map modified from Hose and Blake (1976) and Stokes (1963).

pected to vary somewhat up and down the hinterland of the thrust belt, our general observations regarding the interrelationship of metamorphism, deformation, and magmatism should be representative of this belt in general. As such, we hope that this study will provide a good case history with which to compare evolving stories from similar regions along strike in the Cordillera such as the Albion-Raft River Range region and the Ruby Mountains in the northeastern Basin and Range (Snoke and D. M. Miller, Chapter 23, this volume) and the Death Valley region in the southern part of the Basin and Range province (Labotka and Albee, Chapter 26, this volume).

GEOLOGIC SETTING

East-central Nevada was part of a broad, subsiding continental shelf from the latest Precambrian through at least the Triassic (Stewart, 1980; Stewart and Poole, 1974). The shelf or miogeoclinal section in this region is greater than 15 km thick and consists of regionally persistent carbonate, quartzite, and shale units (Fig. 24-3).

Both Mesozoic and Cenozoic deformational and metamorphic events have affected the eastern Great Basin, creating polyphase fabrics in many areas and leading to much confusion as to the relative importance of "Mesozoic compressional" versus "Cenozoic extensional" structures. This problem has been particularly acute in east-central Nevada, where some early workers (e.g., Drewes, 1967; Fritz, 1968; Misch, 1960; Nelson, 1966) identified numerous Mesozoic "thrust" faults, whereas workers in adjacent areas (e.g., Kellogg, 1964; Moores *et al.*, 1968; Young, 1960) found little evidence for significant pre-Tertiary faulting. Armstrong (1972) summarized this problem and argued that many of the low-angle faults previously mapped as Mesozoic "thrusts" were actually Tertiary normal faults. He emphasized that early Oligocene rocks were generally deposited unconformably on upper Paleozoic rocks, suggesting that no major thrust faults breached the surface in this area. The recent compilation and palinspastic reconstruction of the Tertiary unconformity by Gans and E. L. Miller (1983) support Armstrong's observations and indicate that Mesozoic supracrustal deformation across this region was limited to gentle folds and small-displacement thrust faults. Conformable stratigraphic sections that span the entire late Precambrian to late Paleozoic can be pieced together beneath this early Tertiary unconformity, thereby precluding any regionally extensive Mesozoic low-angle faults at any of the exposed structural levels. The simple picture that emerges is that prior to Tertiary extension, the miogeoclinal succession in east-central Nevada was approximately flat-lying and lay to the west of, or in the upper plate of, the highest thrust fault of the Sevier thrust belt, the Wah Wah–Canyon Range thrust (Fig. 24-1).

Beginning in the early Oligocene, east-central Nevada was the site of voluminous

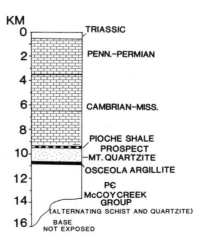

FIG. 24-3. Simplified stratigraphic column of the miogeoclinal succession in east-central Nevada showing the approximate structural depths during Mesozoic time represented by the Lower Cambrian and upper Precambrian pelitic and quartzite units. Stratigraphic thicknesses from Hose and Blake (1976) and Armstrong (1963).

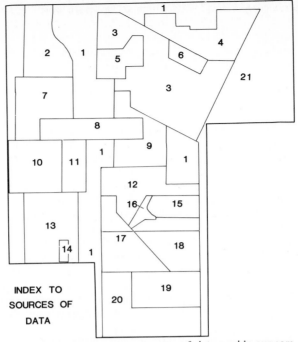

FIG. 24-4. Index map to sources of data used in our compilation: (1) Hose and Blake (1976); (2) Dechert (1967); (3) P. B. Gans and E. L. Miller (unpublished data); (4) D. W. Rodgers (1987); (5) Clark (1985); (6) E. L. Miller, P. B. Gans and Stanford Geol. Survey (SGS) (unpublished data); (7) Young (1960), P. B. Gans (1987); (8) Gans *et al*. (1985); (9) J. Lee (unpublished data); (10) P. B. Gans and SGS (unpublished data); (11) P. B. Gans and E. L. Miller (unpublished data); (12) E. L. Miller *et al*. (1983, 1987), Rowles (1982), J. Lee *et al*. (1987), R. Geving (1987); (13) Drewes (1967); (14) P. B. Gans and E. L. Miller (unpublished data); (15) E. L. Miller (unpublished data); (16) P. B. Gans (unpublished data); (17) E. L. Miller (unpublished data); (18) Grier (1984); (19) A. McGrew (1987); (20) Whitebread (1969); (21) Stokes (1963).

magmatism and large magnitude extension. Gans and E. L. Miller (1983) have delineated a northeast-trending belt of about 250% WNW-ESE-directed extension that includes the northern Egan, Schell Creek, and Snake Ranges. It is the large differential uplift and block rotations across this belt that allow us to see sufficiently deep structural levels to study both Mesozoic and Tertiary ductile fabrics and metamorphism (Fig. 24-1).

Three distinct metamorphic events, Jurassic, Cretaceous, and Tertiary, affected the thick miogeoclinal sedimentary prism across this region. All three of these metamorphic events increased in grade with depth, were accompanied by deformation and, at least in part, temporally overlapped with intrusions of granitic plutons. These three events affected about the same crustal levels (7–15 km), creating polyphase fabrics in many areas. Lateral and vertical temperature gradients were high, locally greater than 100°C/km, suggesting that the emplacement of plutons was probably the primary cause of heating at these relatively high levels of the crust.

We have shown all pertinent structural, metamorphic, and geochronologic data on maps that represent three different time spans, Jurassic, Cretaceous, and Tertiary, and referenced our sources of data in Fig. 24-4. On our simplified geologic map (Fig. 24-2), which is the base for our metamorphic maps, we have highlighted the lowest exposed units of the miogeoclinal succession: the Lower Cambrian Pioche Shale, the Prospect Mountain Quartzite, and the upper Precambrian McCoy Creek Group, of which the Osceola Argillite is the youngest unit (Fig. 24-3). Our knowledge of the metamorphic grade of these rocks is based on mineral assemblages present in the pelitic units; the interbedded quartzites and the overlying Paleozoic carbonate succession in general do not contain diagnostic mineral assemblages or well-developed metamorphic fabrics.

Tertiary faulting has disrupted the map pattern of metamorphic terranes that were once more continuous, such that relationship between events depicted in individual ranges is not always obvious. Field and petrographic data on the orientation, nature, and metamorphic grade of deformational fabrics have been extensively utilized in the correlation of deformational events within and across mountain ranges. Although this approach is generally valid, the reader should be aware that fabrics of similar geometry and orientation may well have developed diachronously across the study area.

Our current knowledge of the timing of events in east-central Nevada relies on both previously published and our new ^{40}Ar/^{39}Ar and U-Pb geochronologic studies. Many standard K-Ar ages on a wide variety of rock types exist for this region, thanks to the work of D. E. Lee *et al.* (1970, 1980), but show a bewildering range in ages from Precambrian to Miocene. As discussed below, this range probably reflects the composite effect of at least two and sometimes three superimposed metamorphic events and/or slow cooling after these events. Thus, except under unusual circumstances, K-Ar ages do not provide tight constraints on the age of metamorphism and deformation. In some places, extensive new work has still not resolved the exact timing of metamorphism. We have been most successful in dating regional events by mapping these fabrics into the contact aureoles of plutons and carefully documenting the relationship of deformation to metamorphism as fabrics are upgraded adjacent to the plutons. Multiple fractions of zircon from most of the plutons in the study area have now been dated by the U-Pb method.

Based on our Tertiary reconstructions, we have assumed that stratigraphic depths within the miogeoclinal succession are approximately equivalent to structural depths except where noted otherwise. An uncertain amount of erosion occurred between Triassic time and the development of the latest Eocene–early Oligocene unconformity; thus, when discussing Jurassic and Cretaceous metamorphism, we have incorporated the thickest reported sections of Permian and Triassic rocks in our depth estimates (Fig. 24-3). There is apparently no indication that thick successions of post-Triassic and pre-Tertiary rocks ever covered this region, inasmuch as these lithologies are not present either as preserved remnants in late Mesozoic and early Tertiary grabens, or as clasts beneath or within the base of the Tertiary section.

JURASSIC

Metamorphic fabrics and plutons of known or inferred Jurassic age are well exposed in the Schell Creek Range and southern Snake Range (Fig. 24-5). Pelitic units in the northern Snake Range were also locally metamorphosed and deformed in the Jurassic but were so strongly overprinted by both Cretaceous and Tertiary metamorphism and deformation that study of older fabrics is impractical. Jurassic-age fabrics and plutons have not been

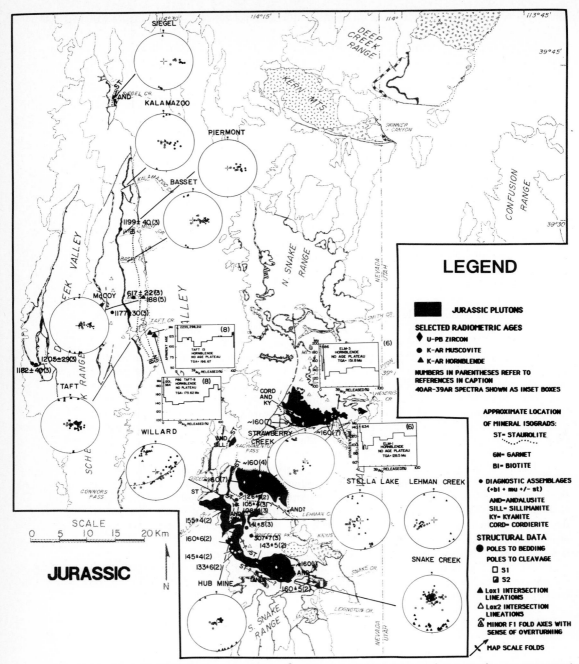

FIG. 24-5. Map showing distribution of Jurassic plutons, metamorphism, and deformation in east-central Nevada and part of westernmost Utah with selected published and new radiometric age data. Numbers in parentheses by radiometric dates refer to the following references: (1) D. E. Lee *et al.* (1968); (2) D. E. Lee *et al.* (1970); (3) D. E. Lee *et al.* (1980); (4) D. E. Lee *et al.* (in press *b*); (5) Hose and Blake (1976); (6) J. Lee *et al.* (in press); (7) J. E. Wright and E. L. Miller (unpublished data); (8) this paper. K-Ar dates have been recalculated using the new constants (Dalrymple, 1979) where necessary.

described from either the Deep Creek Range or from the Kern Mountains, where Cretaceous age deformation and metamorphism is important (Gans *et al.*, 1986; Rodgers, 1987).

Schell Creek Range

The Schell Creek Range is a west-tilted structural block bound by the east-dipping Schell Creek fault in Spring Valley (Gans *et al.*, 1985). Within the range, bedding dips consistently 30–50° westward. The Pioche Shale at the crest of the range appears to be largely unmetamorphosed, whereas the next lower pelitic unit, the Osceola Argillite, is chlorite grade and possesses two weakly developed cleavages. The earlier cleavage (S_1) is a penetrative slaty cleavage that is shallowly dipping to subhorizontal. It is crenulated by a younger spaced cleavage (S_2) that presently dips steeply to moderately westward. For reasons discussed below, we believe the older cleavage was developed during the Jurassic, whereas the younger cleavage is Cretaceous. Metamorphism associated with both cleavages increases in grade down-section, an observation first made by Misch and Hazzard (1962). Metamorphism synchronous with S_1 increases downward to biotite grade and, finally, garnet grade near the base of the section in the vicinity of Taft Creek (Fig. 24-5). In the area just south of Siegel Creek (Fig. 24-5), the Osceola Argillite contains retrograded staurolite(?) porphyroblasts that grew synchronously with S_1. Here the McCoy Creek Group is intruded by an undated (Jurassic?) granite that contains andalusite in its immediate aureole and which may have provided the heat necessary to locally upgrade this regional cleavage.

Strain related to cleavage formation in the Schell Creek Range was largely accommodated in pelitic horizons. Interbedded quartzites show little or no evidence for penetrative deformation in outcrop or in thin section. Mesoscopic folds are extremely rare but where present, the ones associated with the first cleavage are overturned toward the west and have axes that are subparallel to north-south bedding-cleavage intersection lineations (Fig. 24-5). Cross-bedded quartzites throughout the section consistently indicate that bedding is upright.

The angle between bedding and cleavage for both cleavages in the Schell Creek Range varies from approximately 45° to 0° and appears to be largely a function of the total strain in the rock. In general, this angle decreases down-section and with increasing metamorphic grade (Fig. 24-6). Thus, the Osceolla Argillite with its incipient cleavage has the highest bedding-to-cleavage angles, whereas bedding and cleavage are nearly parallel in strongly deformed marbles and schists near the base of the McCoy Creek Group. On a local scale, bedding-to-cleavage angles vary dramatically depending on lithology, with the smallest angles in the more highly strained pelitic and carbonate horizons, and progressively larger angles (or nonexistent) in the less deformed, more quartzose intervals (Fig. 24-7A).

The relationships described above between cleavage, bedding, finite strain, lithology, and stratigraphic depth lend themselves to a straightforward kinematic interpretation. As cleavage represents the XY or flattening plane of the finite strain ellipsoid (see the discussion in Ramsay and Huber, 1983, p. 181–184), the progressive increase of strain and the decreasing angles between bedding and cleavage with structural depth in the Schell Creek Range represent a pattern similar to that developed within the upper part of a ductile shear zone as modeled by Ramsay and Huber (p. 35) and shown schematically in Fig. 24-6. The relationships above suggest that the first cleavage in the Schell Creek Range is related to top-to-the-west layer-parallel ductile shear, rather than to layer-parallel shortening. The direction of movement during deformation is approximately east-west or perpendicular to rare west-vergent folds and the bedding-cleavage intersection lineation. This interpretation explains the increasing strain and the decreasing bedding-to-cleavage angles with structural depth even though bedding maintains approximately the same attitude. The lack of strain in interbedded quartzites is compatible with layer-parallel shear but would pose serious strain incompatibility problems in a layer-parallel shortening model. Finally, this interpretation suggests that west-vergent, bedding-parallel synmetamorphic faults may be present at depth beneath the Schell Creek Range. Well-documented examples of similar variation of strain with depth and toward ductile thrust faults exist in the Helvetic nappes in the western Alps as described by Ramsay (1981) and in other mountain ranges as summarized by Sanderson (1982).

Throughout most of the Schell Creek Range, peak metamorphism was synchronous with the older cleavage. We have attempted to date this metamorphism directly using the ^{40}Ar/^{39}Ar technique[1] but have not been successful. Hornblende was dated from an amphibolite sill and from a hornblende-bearing quartzite near the base of the exposed section of the McCoy Creek Group just south of Taft Creek (Fig. 24-5). Neither sample yielded a plateau age; instead, both contain large amounts of excess argon in the low-temperature steps and exhibit complex, somewhat saddle-shaped spectra which are characteristic of minerals that contain excess argon. The complexities of these spectra indicate that the total gas ages of these hornblendes (175 and 196 m.y.), which are analogous to standard K-Ar ages, are not geologically significant and cannot be interpreted as minimum ages for the time of metamorphism. Our total gas ages are similar to the conventional K-Ar ages of 188 and 232 m.y. obtained from the same amphibolite sill farther north in McCoy Creek by Hose and Blake (1976) and by Armstrong *et al.* (1976), who also suggested that the anomalously old age of 232 m.y. might indicate the presence of excess argon. The highly disturbed spectra obtained from these hornblendes may be the result of diffusional loss accompanied by excess argon gain during subsequent metamorphism in the Cretaceous (see below), or may be due to the fact that the rocks, transitional between upper greenschist and amphibolite facies, may have crystallized at temperatures below the hornblende blocking temperature.

D. E. Lee *et al.* (1980) have dated muscovite from Precambrian pelitic rocks and from Cambrian quartzites in the Schell Creek Range (Fig. 24-5). To the west of the Schell Creek Range, in the Duck Creek Range, where the metamorphic grade of Precambrian and Cambrian rocks is perhaps lowest in the entire study area, detrital muscovite yields ages of 1182 and 1205 m.y. The Cambrian Prospect Mountain Quartzite near the crest of the Schell Creek Range yields ages of 1177 and 1199 m.y. Down-section, rocks that possess two cleavages yielded an age of 617 m.y. Although fine-grained white mica and chlorite were growing in these rocks during the time of formation of the first cleavage, larger, detrital muscovite grains apparently still retained much of their Precambrian argon. Further down-section, both cleavages were accompanied by new growth of muscovite and biotite such that the latest Cretaceous conventional K/Ar and ^{40}Ar/^{39}Ar ages obtained in this area (see below) reflect complete degassing (followed by slow cooling?) of both the

[1] A description of the analytical techniques and the isotopic analyses for samples that are discussed in this paper but have not been published elsewhere are available by request from the authors because they are too lengthy to be included here. The use of the ^{40}Ar/^{39}Ar geochronology technique on hornblende, biotite, and muscovite has been well documented (see, e.g., Dallmeyer and Sutter, 1976; Lanphere and Dalrymple, 1971) and is a technique well suited to deciphering the thermal history of a region that has suffered multiple heating events (see Sutter *et al.*, 1985). Our use of the term "plateau age" follows that of Fleck *et al.* (1977). If no age plateau exists, the age spectrum is said to be disturbed or discordant and may or may not be interpretable depending on its complexity. Our interpretation of complex spectra indicating loss and/or excess argon follows that of Harrison and McDougall (1980b) and Miller and Sutter (1982). The argon closure temperatures for hornblende, muscovite, and biotite are considered to be about 530 ± 40°C, 320 ± 40°C, and 280 ± 40°C, respectively (Harrison and McDougall, 1980a).

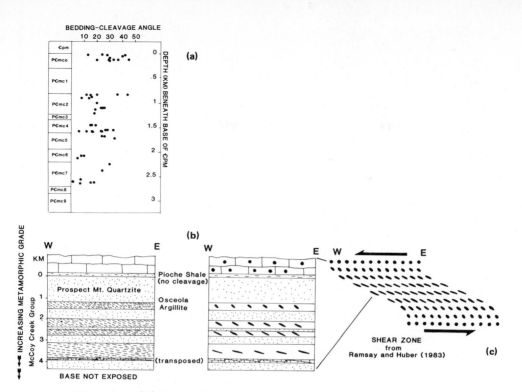

FIG. 24-6.   (A) Relationship of bedding to cleavage angles with structural depth in the Schell Creek Range. (B) Attitude of the long axis (X) and short axis (Z) of strain ellipsoids as inferred from the attitude of cleavage (the XY or flattening plane) in the Schell Creek Range. View is perpendicular to bedding-cleavage intersection lineations (Y). The relative lengths of strain ellipsoid axes shown is schematic. (C) Comparison of (B) to the idealized representation of the strain pattern developed in rocks across a zone of ductile simple shear from Ramsay and Huber (1983).

Precambrian detrital muscovite and the inferred Jurassic muscovite during metamorphism associated with the second cleavage. Thus, within the Schell Creek Range, the older east-dipping cleavage can presently only be constrained as post-latest Precambrian and pre-latest Cretaceous. However, we infer a Jurassic age for it based on constraints from the southern Snake Range described below.

Southern Snake Range

Precambrian and Cambrian rocks in the southern Snake Range are disposed in a broad, doubly plunging antiform whose apex lies close to Wheeler Peak. Everywhere these rocks possess a cleavage which dips variably, depending on the attitude of bedding. If bedding is rotated to horizontal, this cleavage is approximately north-south striking and gently to moderately east-dipping, like the early (S_1) cleavage in the Schell Creek Range. Locally, it is crenulated by a younger, gently to moderately west-dipping cleavage similar to S_2 in the Schell Creek Range, further strengthening the correlation of these cleavages between the two ranges. S_1 in the southern Snake Range occurs in rocks as high as the Middle Cambrian carbonates, but is developed primarily in Lower Cambrian and older pelitic

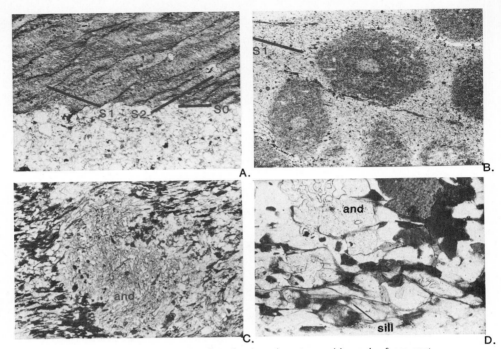

FIG. 24-7. Photomicrographs of selected metamorphic rocks from east-central Nevada. Field of view is 2.5 mm across. (A) Oppositely dipping S_1 and S_2 cleavages in interbedded phyllite (strained) and quartzite (unstrained) of the McCoy Creek Group, Schell Creek Range. (B) Chlorite-white mica pseudo-morphs of idioblastic staurolite in aureole of Jurassic Willard Creek pluton, southern Snake Range. (C) Andalusite indicating syngrowth rotation in biotite-muscovite schist in the immediate contact aureole of the Jurassic Willard Creek pluton, southern Snake Range. (D) Biotite-muscovite-andalusite-sillimanite schist in the immediate contact aureole of the Jurassic Willard Creek Pluton, southern Snake Range, Nevada.

units. As in the Schell Creek Range, both the strain and metamorphic grade associated with this cleavage increase down-section. Throughout the southern Snake Range, peak metamorphism was synchronous with development of S_1. In addition to the gradual downward increase in metamorphic grade, dramatic lateral metamorphic gradients exist as well. These gradients are defined by the successive appearance of first chlorite, then biotite porphyroblasts, then staurolite ($\pm$garnet), and finally andalusite$\pm$sillimanite (Figs. 24-5 and 24-7B–D). Because biotite is almost everywhere retrograded to chlorite, we have not been able to map a biotite-in isograd. Nearly everywhere, staurolite porphyroblasts have been retrograded to chlorite, white mica and opaque oxides, but their identity is clear from pseudomorphs of idioblastic sections (Fig. 24-7). The staurolite-in isograds clearly transect sedimentary units and are roughly concentric about the three Jurassic plutons in the range (Fig. 24-5).

Of critical importance is the relationship of the "contact" metamorphic porphyro-blasts in the southern Snake Range to the more "regional" S_1 cleavage. In some areas, porphyroblasts of staurolite and andalusite grew either synchronously with, or postdated, S_1 cleavage in that they include the cleavage and occasionally indicate syngrowth rotation into the plane of cleavage (Fig. 24-7C). Close to the intrusive contacts of the plutons, polyphase fabrics and coaxial refolding of the first cleavage are common and postdated growth of porphyroblasts, indicating that locally, deformation clearly outlasted the thermal peak of metamorphism (Fig. 24-5). These relationships suggest that (1) the S_1 cleavage began to develop prior to the emplacement of the Jurassic plutons to their present crustal levels, (2) heating of the country rocks during pluton emplacement localized deformation and upgraded the regional dynamothermal metamorphism to amphibolite facies, and (3) deformation continued during their cooling. There is no evidence for forceful emplacement of these plutons inasmuch as rocks immediately adjacent to their contacts generally exhibit a hornfels texture and are cut by randomly oriented aplite and pegmatite dikes. In two localities (Sacramento Pass and south of Willard Creek), units are seen to "roll over" from subhorizontal to steeply dipping inward, parallel to the margin of the plutons, as if country rocks had flowed downward, parallel to the steeply inclined intrusive contacts instead of being bowed upward.

All three Jurassic plutons in the southern Snake Range are well dated, and the Snake Creek pluton has been studied in detail (D. E. Lee and Christiansen, 1983). U-Pb ages for zircon of approximately 160 m.y. have been obtained by D. E. Lee et al. (1968, in press b) from the Snake Creek and Strawberry Creek–Osceola plutons. Additional U-Pb dating of multiple fractions of zircon from these same plutons by J. E. Wright and E. L. Miller (unpublished data) further confirm a 160-m.y. age. In addition, the Snake Creek pluton has yielded a hornblende K-Ar age of 160 ± 5 m.y. (D. E. Lee et al., 1970). Late-stage east to northeast-striking quartz-muscovite veins in the Snake Creek and Willard Creek plutons have yielded K-Ar muscovite ages ranging from 133 to 160 m.y. (Fig. 24-5; D. E. Lee et al., 1970; Whitebread, 1969), suggesting that the Jurassic plutons cooled to below the argon retention temperature for muscovite (320 ± 40°C) soon after their emplacement. Argon loss from Precambrian detrital micas during the Jurassic is evident as one moves down-section near Wheeler Peak (Fig. 24-5). The upper part of the Prospect Mountain Quartzite underlying Wheeler peak yielded a K-Ar age of 307 m.y., whereas metamorphic muscovite nearly 1000 m deeper in the underlying Osceola Argillite at Stella Lake yields an age of 141 m.y. (D. E. Lee et al., 1980; Fig. 24-5). North of Wheeler Peak, K-Ar ages from the Strawberry Creek–Osceola pluton and from metamorphic country rocks are variably younger (Fig. 24-5; D. E. Lee et al., 1970, 1980), perhaps due to argon loss during reheating in either the Cretaceous or Tertiary.

Northern Snake Range

The southern flank of the northern Snake Range is underlain by gneissic plutonic rocks that have yielded only Tertiary K-Ar mica ages (Fig. 24-5). Our unpublished mapping of this area indicates that two distinct intrusions are present. The older and easternmost of the two, informally called the Old Man pluton, grades from a hornblende diorite in its easternmost exposures to a biotite tonalite and granite on the west. The Old Man pluton is intruded by the younger Silver Creek granite, a two-mica granite that underlies the Silver Creek region. New U-Pb ages of several zircon fractions from both the Old Man and Silver Creek plutons suggest they are both about 160 m.y. old, the same age as the three plutons in the southern Snake Range (J. E. Wright and E. L. Miller, unpublished data). Hornblende from the Old Man pluton has yielded disturbed ^{40}Ar-^{39}Ar spectra that are compatible with a Jurassic crystallization age but indicate severe thermal disturbance during younger Cretaceous and/or Tertiary metamorphic events (Fig. 24-5; J. Lee et al., 1987).

Textures and mineral assemblages related to Jurassic metamorphism in the northern Snake Range have been largely obliterated or retrograded by younger events. The record of Jurassic metamorphism is preserved only within a narrow septum of metamorphosed McCoy Creek Group and Prospect Mountain quartzite, sandwiched between the Old Man and Silver Creek plutons (Fig. 24-5). These rocks record two superimposed metamorphic events that are spatially related to the two Jurassic plutons. Muscovite, biotite, staurolite, garnet, and andalusite grew during the first metamorphism (M_1) associated with the intrusion of the Old Man pluton. In most places these minerals were severely retrograded during the second metamorphism (M_2), which was synchronous with the development of a much stronger foliation. Muscovite, biotite, staurolite, andalusite, and cordierite grew during this second metamorphism. Andalusite and cordierite are developed in close proximity to the Silver Creek pluton contact. Rare kyanite is present in the more distal aureole of the Silver Creek Pluton as small, euhedral crystals growing in retrograded M_1 andalusite and in M_2 cordierite. The presence of andalusite and cordierite in the contact aureole of this composite Jurassic pluton suggests that the depth of intrusion of these plutons is similar to that in the southern Snake Range and is compatible with the stratigraphic levels intruded. Neoblastic kyanite suggests cooling at constant pressures at or below the aluminum silicate triple point. Jurassic fabrics in the Silver Creek region are cut by a younger subhorizontal foliation of interred Tertiary age.

CRETACEOUS

Cretaceous age metamorphic and deformational event(s) affect most of the ranges in the map area. The pervasive S_2 crenulation cleavage in the Schell Creek Range was probably produced during Late Cretaceous deformation. Cretaceous regional metamorphism increases from greenschist to amphibolite facies northward in the northern Snake Range and affects deep-seated strata adjacent to the Late Cretaceous Tungstonia Granite in the Kern Mountains and Deep Creek Range (Fig. 24-8). Although Cretaceous plutons are present in the southern Snake Range, their contact aureoles are narrow and Cretaceous metamorphism and deformation in this area is quite localized at the structural levels presently exposed. In contrast to the Jurassic plutons, which are all the same age, Cretaceous plutons range in age from about 110 to 75 m.y. Existing and new geochronologic data suggest that Cretaceous metamorphic and deformational fabrics may also vary in age in different parts of the map area (Fig. 24-8). In fact, we suspect that east-central Nevada had a protracted history of deep-seated metamorphism, deformation, and plutonism throughout much of Cretaceous time, but more data are needed to substantiate this.

Schell Creek Range

A strong second cleavage (S_2) crenulates the older Jurassic(?) penetrative cleavage in the Schell Creek Range. This second cleavage dips west with respect to bedding, and bedding cleavage intersection lineations trend north-south (Fig. 24-8). Rare, small-scale, asymmetric folds parallel the bedding-cleavage intersection lineations and are overturned toward the east. Mineral elongation or stretching lineations within the S_2 cleavage plane are approximately perpendicular to fold axes and trend east-west. These define the long axis of the strain ellipsoid, X, for this deformational event and are inferred to be parallel to the flow direction or tectonic transport direction during deformation (Fig. 24-8).

The second cleavage in the Schell Creek Range decreases in intensity westward, and like the older Jurassic cleavage, dies out up-section. In the Duck Creek Range, to the west of the Schell Creek Range, it is developed locally within the Osceola Argillite as a single, weak fissility. Farther east, near the crest of the Schell Creek Range, this cleavage becomes a spaced crenulation cleavage that cuts the older Jurassic(?) cleavage. No new mineral growth is associated with the second cleavage at these higher stratigraphic levels. At the deepest structural levels exposed, syntectonic muscovite, biotite, and garnet are

FIG. 24-8. Map showing distribution of Cretaceous plutons, metamorphism, and deformation in east-central Nevada and part of westernmost Utah together with selected published and new radiometric age data. Numbers in parentheses by radiometric dates refer to the following references: (1) D. E. Lee *et al.* (1980); (2) D. E. Lee *et al.* (1981); (3) D. E. Lee *et al.* (in press a); (4) J. Lee *et al.* (1987); (5) Best *et al.* (1974); (6) D. E. Lee *et al.* (in press b); (7) D. E. Lee *et al.* (1970); (8) U-Pb zircon data (this paper); (9) ^{40}Ar/^{39}Ar data (this paper). K-Ar dates have been recalculated using the new constants where necessary.

MILLER, GANS, WRIGHT, SUTTER

663

present. The metamorphism associated with the second cleavage first retrogrades and then overprints the older cleavage.

The pattern of decreasing bedding-to-cleavage angles, increasing strain, and increasing metamorphic grade with depth for S_2 in the Schell Creek Range is similar to that associated with the older Jurassic cleavage and can be interpreted in a similar fashion. The younger cleavage, however, dips *west* with respect to bedding and is associated with rare east-vergent folds, suggesting that it formed in response to *top-to-the-east* layer-parallel shear. The strains associated with the second cleavage were also lithologically controlled, and near the base of the section, strain was sufficiently high to transpose bedding in marble and schist units. Hose and Blake (1976) mapped a thrust fault near the marbles at the base of the McCoy Creek Group section. Our mapping shows that the marbles are in stratigraphic contact with both underlying and overlying units such that duplication of the section is precluded; however, we would certainly agree that this could be a zone of large-magnitude, bedding-parallel, top-to-the-east shear.

As with the inferred Jurassic metamorphism, it has been difficult to date this younger metamorphism directly. D. E. Lee *et al.* (1980) report K-Ar muscovite ages from the base of the McCoy Creek Group section that range from 61 to 63.6 m.y. near McCoy Creek (Fig. 24-8). A $^{40}Ar/^{39}Ar$ date on muscovite from nearby Taft Creek yields a similar total gas age of 61.8 m.y. but does not exhibit a plateau. The general increase of ages with higher-temperature steps in this muscovite suggest that it may represent an argon-loss pattern. As all of these ages were obtained from rocks that had new growth of coarse-grained muscovite and biotite in the second cleavage, we believe that they represent valid minimum ages for the second metamorphism. We may have obtained a better approximation of the age of this metamorphism by dating phlogopite in marbles of the Middle Cambrian Lincoln Peak Formation at Connor's Pass (Fig. 24-8). The metamorphism at Connor's Pass affects unusually high stratigraphic levels and dies out very quickly both up-section and laterally, so it is likely that these rocks cooled quickly after the growth of metamorphic minerals. Phlogopite sample MP-5 does not exhibit a plateau but shows a pattern that may indicate cooling through its argon retention temperature at some time prior to about 82 m.y. ago, the age of the highest-temperature step, as well as diffusive loss of argon at a time younger than 80 m.y. D. E. Lee *et al.* (1980) report a K-Ar age of 66.9 m.y. from the underlying Pioche Shale near this locality (Fig. 24-8), which supports the interpretation of younger heating and argon loss from rocks initially metamorphosed in the Cretaceous. Harrison *et al.* (1985) have shown that ^{40}Ar diffusivity in biotites is strongly dependent on the Fe/Mg ratios, and that phlogopite has a closure temperature to argon diffusion that is closer to that of muscovite than biotite (320°C versus 280°C). Thus, the 82-m.y. age of MP-5 is probably a good minimum age for this metamorphism. Unfortunately, the relationship between the synmetamorphic fabrics at Connor's Pass and west-dipping S_2 elsewhere in the Schell Creek Range is not entirely clear. The phlogopite we have dated grew synchronously with the development of a weak foliation and a N50°–60°W lineation that is subparallel to bedding in these rocks (Fig. 24-8).

K-Ar muscovite ages of 90.2 and 87.4 m.y. from farther north along the eastern flank of the Schell Creek Range and from the western flank of the Snake Range were obtained from the Osceola Argillite and the Prospect Mountain Quartzite, respectively (D. E. Lee *et al.*, 1980; Fig. 24-8). In both of these localities, peak metamorphic conditions occurred during the formation of the first cleavage, and no new mineral growth was associated with the second, spaced cleavage. Thus it is unclear if these ages represent Cretaceous cooling and/or metamorphic ages or if they represent younger, Cretaceous and/or Tertiary, argon loss from older (Jurassic) micas.

Southern Snake Range

The southern Snake Range appears to be the least affected by Cretaceous regional metamorphism. At about the latitude of the town of Osceola (Fig. 24-8), however, a west-dipping cleavage is evident in slates and phyllites of the McCoy Creek Group and Pioche Shale. Near the town of Osceola is the only cleavage present in the rocks. This west-dipping cleavage forms approximately north-south bedding-cleavage intersection lineations and dies out quickly up-section (Fig. 24-8). In the northernmost part of the southern Snake Range, and on its northeastern flank, this cleavage strongly crenulates the older, Jurassic age cleavage. The younging of mica ages toward the northernmost part of the southern Snake Range may reflect superimposed Cretaceous on Jurassic metamorphism (Fig. 24-8).

Two Cretaceous plutons in the southern Snake Range have been dated by D. E. Lee et al. (1970, 1980, in press b). These are unusual, muscovite phenocryst-bearing granites. Muscovite K-Ar ages from these plutons indicate a minimum age of 79.7 m.y. for the Pole Canyon pluton and 86.3 m.y. for the Lexington Creek pluton (Fig. 24-8). These two plutons have very narrow contact metamorphic aureoles and commonly a foliation is developed subparallel to their contacts. No andalusite has been found in their contact aureoles.

Northern Snake Range

Precambrian and Cambrian quartzite and schist are exposed beneath the northern Snake Range décollement in the deeply incised canyon walls along the southeast side of the range (Fig. 24-8). These units have been metamorphosed to upper greenschist and amphibolite facies and geochronologic data discussed below suggest that throughout most of the range, peak metamorphism occurred in the Cretaceous. Although the minerals related to Cretaceous metamorphism are largely well preserved, the superimposed Tertiary strain has unfortunately transposed and/or obliterated any fabrics related to the older metamorphic event (Rowles, 1982).

Metamorphic grade appears to vary dramatically in the northern Snake Range, both vertically and laterally. In general, metamorphic grade increases with depth and to the north. On the east side of the range, the highest pelitic unit, the Pioche Shale, contains biotite and muscovite and locally garnet at its base. The next lower unit, the Osceola Argillite, contains biotite, muscovite, and locally kyanite (Hampton Creek) or staurolite + garnet (north of Hampton Creek), and lower units in the McCoy Creek Group all contain coarse biotite, muscovite, garnet, and staurolite. Within the pelitic units of the McCoy Creek Group on the east side of the range, we have mapped a series of "mineral-in" isograds that trend approximately east-west, transect layering, and indicate increasing metamorphic grade to the north (Fig. 24-8). The Pioche Shale in the southernmost part of the range apparently lacks garnet, whereas farther north the Pioche Shale and Prospect Mountain Quartzite are everywhere garnet-muscovite-biotite bearing (Fig. 24-8). Northward, kyanite first appears in the Osceola Argillite between Hendry's and Hampton Creek, and then disappears further north where garnet and staurolite are stable (Fig. 24-8). Garnet, then staurolite, also make their appearance northward within individual pelitic units lower in the McCoy Creek Group (Geving, 1987).

D. E. Lee and Fischer (1985) have used the U-Pb method to date five size fractions of detrital zircon from Hampton Creek, which together with a U-Pb date on metamorphic monazite (D. E. Lee et al., 1981), define a chord whose lower intercept with concordia is 78 ± 9 m.y. (the inferred age of metamorphism) and upper intercept is 1726 ± 26 m.y. (the inferred age of the detrital zircons). J. Lee *et al.* (1987) have dated hornblende from a metadiorite dike in Hampton Creek with the $^{40}Ar/^{39}Ar$ technique (Fig. 24-8). The age spectrum from sample JL-219 shows the incorporation of small amounts of excess argon and loss of argon at some time younger than 54 m.y. and then climbs and levels out at about 75 m.y., although it does not strictly plateau. The flat part of the spectrum at about 75 m.y. may be reasonably close to the age of cooling of the hornblende and is thus in good agreement with D. E. Lee and Fischer's (1985) data, indicating a Late Cretaceous age of metamorphism.

A very large tonalitic to granitic plutonic complex underlies the central part of the northern Snake Range and is exposed continuously along the bottoms of Smith Creek, Deadman Creek, and Horse Canyon (Fig. 24-8). Abundant coarse-grained pegmatite dikes are also present and intrude the tonalite and overlying rocks. The pluton and the pegmatite dikes were involved in Tertiary age deformation and to date have yielded only Tertiary K-Ar ages. Two separate localities (Fig. 24-8) were sampled for U-Pb dating. Although reconnaissance mapping suggests that these two sample localities may be part of a single pluton, the zircon data discussed below suggest that more detailed studies are needed to confirm the preliminary field relations. The two zircon fractions analyzed from each sample locality are strongly discordant (Table 24-1) and on the concordia diagram (Fig. 24-9) display isotopic systematics indicative of mixing between magmatic and older Precambrian inherited or xenocrystic zircon components (i.e., Wright and Haxel, 1982 and references therein). The two zircon size fractions analyzed from each of the sample localities define essentially parallel chords with lower concordia intercepts (magmatic ages) of 108 ± 2 (sample 20-210-2 from the east side of the range; Table 24-1, Fig. 24-9). A combined regression analysis of all four zircon analyses yields an age of 100 ± 8 m.y. Without more detailed mapping or analyzing more zircon fractions from the two sites, we cannot unambiguously conclude whether more than one pluton is present. Although the zircon data suggest that two discrete plutons may be present, it is possible that mixing of a magmatic zircon

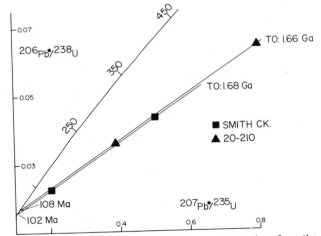

FIG. 24-9. Concordia diagram of zircon size fractions from the Cretaceous tonalitic-to-granitic complex that underlies the central part of the northern Snake Range. Smith Creek is from the headwaters of Smith Creek near the crest of the range, 20–210 is from the east flank of the range, north of Hampton Creek (Fig. 24-8).

TABLE 24-1. Uranium-Lead Isotopic Data[a]

Sample	Wt (mg)	U (ppm)	206Pb[b] (ppm)	Measured Ratios			Atomic Ratios			Apparent Ages (Ma)		
				$\frac{206Pb}{204Pb}$	$\frac{207Pb}{206Pb}$	$\frac{208Pb}{206Pb}$	$\frac{206Pb[b]}{238U}$	$\frac{207Pb[b]}{235U}$	$\frac{207Pb[b]}{206Pb[b]}$	$\frac{206Pb[b]}{238U}$	$\frac{207Pb[b]}{235U}$	$\frac{207Pb[b]}{206Pb[b]}$
Cretaceous												
Smith Creek C	14.8	661.2	24.54	611	0.10774	0.15020	0.04321	0.50526	0.08481	272.7	415.2	1311
Smith Creek VF	17.6	570.6	11.11	6920	0.06740	0.07387	0.02266	0.20408	0.06532	144.4	188.6	785
20-210C	16.2	409.5	22.65	12,315	0.09141	0.11644	0.06439	0.80149	0.09028	402.2	597.7	1432
20-210VF	7.8	438.2	13.49	9259	0.07990	0.10442	0.03584	0.38722	0.07837	226.9	332.3	1156
Tertiary												
Ibapah VC	16.7	756.6	5.92	1156	0.07760	0.15812	0.00911	0.08183	0.06516	58.4	79.9	780
Ibapah F	16.3	893.9	4.97	757	0.06922	0.23937	0.00647	0.04449	0.04984	41.6	44.2	187
Skinner Can. C	17.1	1868	8.78	2174	0.05358	1.24180	0.00547	0.03531	0.04680	35.2	35.2	39
Skinner Can. VF	17.8	1399	6.51	1623	0.05595	0.28712	0.00541	0.03499	0.04687	34.8	34.9	43
Young Can. C	18.4	2418	12.82	1572	0.06145	0.09528	0.00617	0.04439	0.05214	39.7	44.1	292
Young Can. VF	15.0	3130	15.22	3658	0.05148	0.09095	0.00566	0.03704	0.04746	36.4	36.9	72
Egan Range Total	27.1	2383	15.41	3254	0.06485	0.11170	0.00753	0.06271	0.06040	48.4	61.8	618

[a]Sample dissolution and ion exchange chemistry were modified from Krogh (1973). Mass spectrometry was carried out on a Finnagan MAT 261 multicollector mass spectrometer. All measured isotopic ratios were corrected for a mass fractionation of 0.125% per mass unit. Uranium and lead concentrations were determined on solution aliquots from each sample. Measured $^{208}Pb/^{206}Pb$ ratios were precise to at least 0.1%, and $^{206}Pb/^{204}Pb$ ratios were precise to 1–7%. Precisions (2-sigma) of $^{206}Pb[b]/^{238}U$ ratios are about 0.2% based upon replicate analyses of two "standard" zircon fractions. Total lead blanks were in the 0.1 to 0.5 ng range. Concordia intercept errors were calculated according to the method of Ludwig (1984). Common Pb correction: $^{206}Pb/^{204}Pb = 18.6$; $^{207}Pb/^{204}Pb = 15.6$; decay constants: $^{235}U = 0.98485 \times 10\text{-}9\text{yr-}1$; $^{238}U = 0.155125 \times 10\text{-}9\text{yr-}1$; VC, C, F, VF refers to size fractions >100, >200, <200, and <325 mesh, respectively.

[b]Radiogenic Pb.

667

population with an isotopically heterogeneous older included zircon population could produce essentially parallel chords from the two separate sample localities, with slightly different apparent magmatic ages. Nonetheless, the isotopic data indicate a Cretaceous emplacement age, and we conservatively choose the 100 ± 8 m.y. age calculated from all the isotopic data as approximating the actual age of the intrusive rocks.

Although it appears as if the mineral isograds mapped in the northern Snake Range may be spatially related to this pluton (Fig. 24-8), the 25-m.y. difference in the time between the crystallization age of the plutonic rocks and the time of cooling to below 540°C (closure T of hornblende) as dated in the Hampton Creek area presents a problem. More work is needed to determine if the pegmatites are younger and if the pluton itself was involved in a younger Late Cretaceous metamorphism.

Kern Mountains–Deep Creek Range

Cretaceous metamorphism and deformation have strongly affected the Kern Mountains–Deep Creek Range region. The Deep Creek Range is a west-tilted fault block that exposes an almost continuous 13-km-thick crustal section of miogeoclinal strata. The structurally deepest exposures occur on the east side of the range, where they have been studied in detailed by Nelson (1966, 1969) and more recently by Rodgers (1987).

Deep structural levels in the Deep Creek Range exhibit two well-developed cleavages. The earlier one is a penetrative, west-dipping slaty to schistose cleavage that cuts bedding at low angles. This cleavage increases in metamorphic grade both down-section and southward toward the Tungstonia Granite in the adjacent Kern Mountains (Fig. 24-8; Nelson, 1969; Rodgers, 1987). Metamorphic amphibole from a mafic sill within the region of amphibolite facies metamorphism on the south flank of the range has yielded a $^{40}Ar/^{39}Ar$ plateau age of 73 m.y., indistinguishable from the inferred age of the Tungstonia Granite in the adjacent Kern Mountains (Rodgers, 1987).

Only minor offset faults separate the Deep Creek Range structural block from Paleozoic metamorphic rocks in the Kern Mountains that are intruded by the Tungstonia and Skinner Canyon Granite (Fig. 24-8). The Tungstonia Granite is a large muscovite-phenocryst-bearing granite that has been mapped and described by Best *et al.* (1974) and more recently remapped by Gans *et al* (1986). As described by Best *et al.* (1974), Clark (1985) and Gans *et al.*, (1986), the Tungstonia Granite has a strongly developed protoclastic border and narrow, strongly foliated contact aureole, where it intrudes higher Paleozoic units on the west side of the Kern Mountains. Folding, possibly related in part to the forceful emplacement of the pluton to shallow levels of the crust (4 km or less), occurs on the southwest side of the intrusion (Fig. 24-8; Clark, 1985). Published K-Ar ages obtained from the Tungstonia vary from 69 to 47 m.y., and young eastward toward the Tertiary Skinner Canyon Granite (Fig. 24-8), a trend that is documented in greater detail by P. B. Gans, and J. F. Sutter (unpublished data). Three different size fractions of zircons from the Tungstonia granite yield U-Pb ages that lie on a chord whose lower intercept is 75 ± 9 m.y. (D. E. Lee *et al.*, in press *b*).

TERTIARY

Tertiary structures have shaped the present basins and ranges in east-central Nevada, tilting, uplifting, and exposing the deep-seated Mesozoic metamorphic rocks described above. In addition, several of these ranges expose rocks that have been intruded and regionally metamorphosed during the Tertiary (Fig. 24-10). The structural levels metamorphosed in the Tertiary are similar to those affected by Mesozoic metamorphism, and lateral temperature gradients were similar to, and locally greater than, Mesozoic gradients. For example, the Schell Creek Range exposes paleodepths of up to 10 km, yet even the deepest structural levels do not appear to have been significantly metamorphosed or ductilely deformed during the Tertiary, whereas equivalent stratigraphic levels in the adjacent Snake Range were metamorphosed to greenschist facies and mylonitized during the Tertiary. On a regional scale, Tertiary deformation or strain was probably an order of magnitude greater than that which accompanied metamorphism during the Mesozoic. Ductile fabrics that formed at deeper structural levels during the Tertiary record this strain and show evidence for progressive ductile to brittle deformation reflecting uplift and cooling. Unlike their Mesozoic counterparts, plutons intruded during the early stages of Tertiary extension vented thick sequences of volcanic rocks to the surface, allowing us to compare the timing of extensional deformation affecting both shallow and deep-seated levels of the crust.

Southern Snake Range

The southern Snake Range décollement mapped by Whitebread (1969) is a Tertiary fault that separates an upper plate of faulted Paleozoic rocks from a lower plate of Cambrian and upper Precambrian quartzite and schist intruded by plutonic rocks (Fig. 24-10). It presently dips about 15° to the east, has normal-sense offset, cuts 36-m.y.-old plutonic rocks, and contains moderately west-dipping (30–40°) Tertiary conglomerate in its hanging wall (McGrew, 1986). Tertiary metamorphism and deformation affects only lower plate rocks along the easternmost flank of the range (Fig. 24-10).

Along Snake Creek, the Prospect Mountain Quartzite has been ductilely attenuated and possesses a well-developed, gently east-dipping foliation and N75°W mineral elongation lineation. These fabrics are parallel to, and are presumably related to the development of the southern Snake Range décollement (A. McGrew, unpublished data). The Young Canyon pluton intrudes these rocks north of Snake Creek (Fig. 24-10), and older Jurassic-age metamorphic minerals are retrograded in the vicinity of the pluton. Andalusite + biotite ± garnet are locally present in its contact aureole (McGrew, 1986). Although the Young Canyon pluton is not itself penetratively deformed, its margins and aplite dikes are crushed and brecciated and appear to be deformed together with the country rocks along the easternmost flank of the range.

Two size fractions of zircon separated from the Young Canyon pluton are slightly discordant (Table 24-1) and on the concordia diagram (Fig. 24-11) display isotopic systematics indicative of mixing between magmatic and older Precambrian inherited or xenocrystic zircon components. The Young Canyon data define a chord with a lower concordia intercept (emplacement age) of 36 ± 1 m.y. We have plotted the Young Canyon zircon data together with data from a single zircon fraction collected from a similar pluton, the Warm Springs pluton in the northern Egan Range described by Gans (1982), whose locality is shown on Fig. 24-1. The Egan Range zircon data fall approximately on the Young Canyon chord, and we also interpret its emplacement age to be about 36 m.y., coeval with dikes emplaced at higher structural levels and volcanic rocks extruded at the surface (P.B. Gans, G. A. Mahood, and E. Schermer, unpublished data).

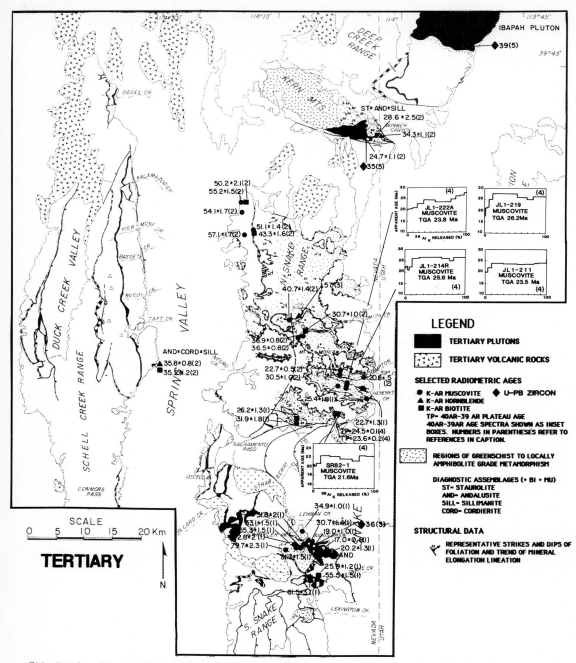

FIG. 24-10.   Map showing distribution of Tertiary plutons, metamorphism, and deformation in east-central Nevada and part of westernmost Utah, together with selected published radiometric age dates. Numbers in parentheses by radiometric dates refer to the following references: (1) D. E. Lee *et al.* (1970); (2) D. E. Lee *et al.* (1980); (3) Hose and Blake (1976); (4) J. Lee *et al.* (1987); (5) U-Pb zircon data (this paper). K-Ar dates have been recalculated using the new constants where necessary.

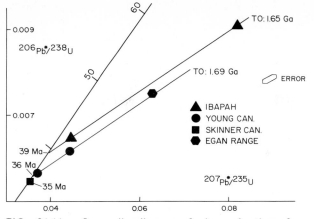

FIG. 24-11. Concordia diagram of zircon fractions from Tertiary plutons in east-central Nevada and westernmost Utah.

The K-Ar ages of 61.5, 55.5, and 25.9 m.y. obtained from deformed Cambrian schist and quartzite units exposed along Snake Creek in the southern Snake Range (D. E. Lee *et al.*, 1980) probably reflect either resetting of, or argon loss from, Jurassic or Cretaceous micas during Tertiary metamorphism and intrusion. A very systematic eastward younging of K-Ar ages from 79.7 m.y. on the west to 31.8 m.y. on the east has been documented by D. E. Lee *et al.* (1980) in the Cretaceous Pole Canyon pluton, which the Young Canyon pluton intrudes on the east (Fig. 24-10). D. E. Lee *et al.* (1970) have argued that the younging of ages in the southern Snake Range reflects progressive resetting in a vertical direction due to thermal (frictional) heating by movement along the Snake Range décollement, which they refer to as a Tertiary "thrust" fault. We suggest, instead, that the younging of ages along the east flank of the range reflects argon loss and resetting in an *eastward*, not upward direction and is due in part to the intrusion of the Young Canyon pluton during the Oligocene. This argument does not, however, explain the even younger progression of mica ages from about 31 to 17 m.y. observed within the Young Canyon pluton itself (Fig. 24-10; D. E. Lee *et al.*, 1980). Although these ages could represent differential cooling of the Young Canyon pluton due to relative uplift of the footwall of the southern Snake Range décollement, the extreme variation of ages over very short distances and the fact that the youngest ages from this area were apparently obtained from cataclastic rocks that have anomalously low $\delta^{18}O$ and δD values (D. E. Lee *et al.*, 1984) suggest that either hydrothermal white mica was dated and/or argon loss occurred in Oligocene micas due to this younger hydrothermal alteration.

Northern Snake Range

Tertiary metamorphism and ductile deformation affect a broad portion of the northern Snake Range and have been described in detail by J. Lee *et al* (1987). Ductile strain produced a penetrative subhorizontal foliation characterized by a WNW-ESE-trending mineral elongation lineation and resulted in remarkable thinning of Cambrian and upper Precambrian units beneath the northern Snake Range décollement. The progression of deformational fabrics in lower-plate rocks has led J. Lee *et al.* (1987) to conclude that an earlier history of coaxial stretching and thinning was followed by a component of noncoaxial, down-to-the-east ductile simple shear along the east flank of the range. Fabrics and structures developed in the lower-plate rocks record the transition from ductile to brittle through time.

Metamorphism accompanying this deformation was generally low to middle green-schist in grade and increases with structural depth. The deepest exposures of lower-plate rocks contain syntectonic biotite and muscovite, whereas chlorite was stable during deformation at higher structural levels (J. Lee et al., 1987). Temperatures during deformation for at least the eastern half of the range can be inferred to have been higher than 320°C, the argon blocking temperature of muscovite, but below 530°C, the blocking temperature of hornblende (J. Lee et al., 1987). Palinspastic reconstruction of the well-known stratigraphy in the upper plate of the Snake Range décollement indicates that lower-plate rocks were at depths ranging from 7–10 km prior to Tertiary faulting (Gans and E. L. Miller, 1983), suggesting an average gradient in the northern Snake Range at the beginning of extension of about 40°C/km. Stratigraphic units at comparable or even greater structural depths in the southern Snake Range and Schell Creek Range were not significantly metamorphosed during the Tertiary, suggesting that the average Tertiary thermal gradient in these places may have been somewhat lower than that in the northern Snake Range.

Micas from lower plate rocks in the northern Snake Range mostly yield Tertiary K-Ar ages. A pronounced eastward younging of K-Ar ages is apparent in the northern Snake Range, but the distance over which this younging occurs is much greater than that in the southern Snake Range (Fig. 24-10). The reason for this eastward progression of ages is not well understood, and at least two explanations are possible, given the present data base: (1) the pattern of ages could indicate differential uplift of deeper structural levels on the east side of the range, provided that isotherms were horizontal; and (2) the pattern of ages may represent progressive argon loss due to lateral thermal gradients related to one or more heating events in the Tertiary. In addition, the youngest dates on the easternmost flank of the range may be in part due to hydrothermal activity as argued by D. E. Lee et al. (1984) for the southern Snake Range. As an example of this problem, the Eocene mica ages in the northwest corner of the range may represent either Cretaceous metamorphic micas that were partially degassed during mid-Tertiary metamorphism, or micas that did not cool through their blocking temperatures until the early Tertiary. Additional unpublished ^{40}Ar/^{39}Ar data from this same region unfortunately do not resolve this ambiguity.

Preliminary ^{40}Ar/^{39}Ar data on lower plate muscovites by J. Lee et al. (in press; Fig. 24-10) indicate that ages not only decrease eastward but may decrease with structural depth as well. Total gas ages of muscovite samples from a vertical transect on the east flank of the range vary from 26 to 23 m.y. from shallow to deeper levels, with the exception of the highest sample, which yields a younger 23.8-m.y. total gas age (Fig. 24-10). These preliminary data are compatible with our observations of increasing metamorphic grade with depth. The data are in part incompatible with the interpretation of D. E. Lee et al. (1970), who invoked progressive resetting of mica ages upward toward the Snake Range décollement, but the age of the highest sample may in fact reflect loss of argon due to hydrothermal activity in proximity to the décollement.

The observation of increasing metamorphic grade with depth, when coupled with regional geological considerations, suggest that rocks in the northern Snake Range underwent extensive ductile flow because of localized heat and water input during the Tertiary. Textures in lower-plate rocks indicate that ductile deformation of quartz outlasted metamorphism. Based on the flow laws of "wet" quartzite, ductile deformation of quartz is possible at the argon closure T values of micas (about 300°C; Koch, 1983), suggesting that lower-plate rocks were probably still deforming in earliest Miocene times (J. Lee et al., 1987).

In concert with lower-plate deformation and metamorphism, extensional deformation of the upper plate of the northern Snake Range occurred by brittle faulting. At least two sets of down-to-the-southeast normal faults cut the Paleozoic stratified sequence above the northern Snake Range décollement in response to WNW-ESE-directed extension. Geometric auguments necessitate that these faults soled or bottomed into the décollement at depth (Gans and E. L. Miller, 1983). The age of extensional deformation at supracrustal levels is best constrained by the syntectonic relationship between faulting and volcanism. Normal faults at the northern end of the Snake Range cut volcanic rocks as old as 35 m.y. old, and a 24-m.y.-old vitric tuff is moderately tilted by the youngest set of normal faults (Gans, 1987), indicating that upper-plate deformation continued into the Miocene. As in the southern Snake Range, no stratigraphic relations have been found that place a younger age limit on brittle faulting in the uppermost levels of the crust.

Kern Mountains and Southern Deep Creek Range

Tertiary deformation and metamorphism has affected the entire eastern Kern Mountains and has been described by Gans et al. (1986). Paleozoic rocks are polydeformed and metamorphic grade increases to amphibolite facies toward the Skinner Canyon Granite, where andalusite and sillimanite are developed in its immediate contact aureole. Deformation associated with peak metamorphism was synchronous with the intrusion of the Skinner Canyon Granite and greatly thinned Paleozoic units, transposing them into parallelism with a steep northwest-southeast-trending foliation (Fig. 24-10). During later stages of deformation, this steeply dipping foliation was recumbently folded about subhorizontal axial planes. Both deformations involve offshoots and aplite dikes of the Skinner Canyon Granite, but this later deformation postdated the peak of metamorphism. The main body of the Skinner Canyon Granite is undeformed.

K-Ar ages of the Cretaceous Tungstonia Granite become progressively younger toward the Skinner Canyon Granite (Fig. 24-10). The Skinner Canyon Granite, suspected to be Tertiary in age by Best et al. (1974), has been dated using the U-Pb method on zircon. The isotopic analysis of two zircon fractions separated from the Skinner Canyon Granite indicate an emplacement of 35 ± 1 m.y., with no evidence of an older inherited or xenocrystic zircon component (Table 24-1, Fig. 24-11). The much younger (24.7-28 m.y.) K-Ar ages reported from the Skinner Canyon Granite and its country rocks in the easternmost part of the Kern Mountains (Fig. 24-10) may be related to hydrothermal activity, as suggested by D. E. Lee et al. (1984).

In contrast to the eastern Kern Mountains, the adjacent Deep Creek Range and the western part of the Kern Mountains are little affected by Cenozoic metamorphism as described by Rodgers (1987). A narrow zone of contact metamorphism is developed along the border of the Ibapah pluton, which has been dated using the U-Pb method on zircon. The isotopic systematics of two zircon fractions separated from the Ibapah pluton are readily interpreted as representing a mixing line between an older Precambrian inherited or xenocrystic zircon population and a 39 ± 1 m.y. magmatic population (Table 24-1, Fig. 24-11).

DISCUSSION AND SUMMARY

Utilizing the data presented in this paper, we have constructed a series of simplified cross sections that schematically show the evolution of the upper part of the crust across easternmost Nevada from Mesozoic to present times (Fig. 24-12).

Three district metamorphic events of Jurassic, Cretaceous, and Tertiary age affected the upper 13–15 km of crust exposed in east-central Nevada. All three metamorphic events increase in grade downward and were accompanied by penetrative deformation. In all cases, deformation and metamorphism can be shown to have overlapped at least partially in time with the intrusion of granitic plutons. About the same levels of the crust were metamorphosed during each event and, based on the mapping of "mineral-in" isograds, it appears that both lateral and vertical temperature gradients were significant and locally greater than 100°C/km. This suggests that the emplacement of plutons was probably the primary cause of heating at these relatively high levels of the crust. As expected, local temperature gradients related to the intrusion of plutons become less pronounced with structural depth (see the discussion in Barton *et al.*, Chapter 5, this volume), where deformational and metamorphic fabrics are more "regional" in nature and appearance.

New ^{40}Ar/^{39}Ar geochronology together with published K-Ar ages by D. E. Lee *et al.* (1970, 1980) across the study area shows that the variation in ages is a function of, and is compatible with, the observed increase in metamorphic grade with structural depth as well as with the observed sequence of overprinting of metamorphic events. In addition, these data suggest that peak metamorphic conditions within the upper crustal column across most of east-central Nevada were attained during the Mesozoic, prior to Tertiary normal faulting. These relationships support the views held by Misch (1960) and Misch and Hazzard (1962) on the nature of Mesozoic metamorphism in east-central Nevada but are not compatible with the interpretation of D. E. Lee *et al.* (1970, 1980), who proposed that argon loss from dated minerals is related to frictional heating due to movement on nearby Tertiary "thrust" faults.

During the Jurassic, metamorphic and deformational fabrics developed at depth across east-central Nevada have characteristics that are both "regional" and "contact" in nature. Fabrics typical of "regional" metamorphism are best developed in the Schell Creek Range, where chlorite, biotite, and garnet are stable with progressively increasing depths. As detailed above, strain associated with Jurassic regional metamorphism is best explained by top-to-the-west bedding-parallel simple shear (Fig. 24-12A). At the crustal levels presently exposed, the strain represented by this fabric is not great, but nevertheless may indicate the presence of west-vergent, synmetamorphic thrust faults at depth. There are not sufficient data to say whether mid-Jurassic west-vergent structures are common elsewhere in the eastern Great Basin or whether this represents a localized phenomenon. For example, a detailed study further to the northeast in the Newfoundland Mountains, Utah, has outlined evidence for minor *east-directed* thrusting prior to intrusion of a 150-m.y.-old (minimum age) pluton (Allmendinger and Jordan, 1984).

The precise age of the earliest metamorphism and deformation is best dated in the southern Snake Range, where "regional" fabrics and metamorphism are upgraded in the contact aureoles of plutons that are approximately 160 m.y. old (Fig. 24-12A). The regional cleavage affects the highest stratigraphic levels adjacent to the plutons, and contact metamorphic porphyroblasts can be shown to be syn- to posttectonic with cleavage formation. Immediately adjacent to the plutons, deformation locally outlasted metamorphism. These relations suggest that the earliest metamorphism and deformation occurred

JURASSIC

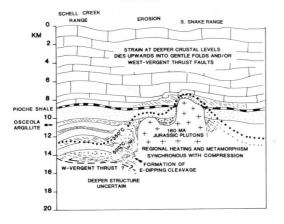

SCHELL CREEK RANGE   EROSION   S. SNAKE RANGE

KM

STRAIN AT DEEPER CRUSTAL LEVELS
DIES UPWARDS INTO GENTLE FOLDS AND/OR
WEST-VERGENT THRUST FAULTS

PIOCHE SHALE

OSCEOLA ARGILLITE →

350°C

600°C

160 MA
JURASSIC PLUTONS

REGIONAL HEATING AND METAMORPHISM
SYNCHRONOUS WITH COMPRESSION

FORMATION OF
E-DIPPING CLEAVAGE

W-VERGENT THRUST ?

DEEPER STRUCTURE
UNCERTAIN

CRETACEOUS

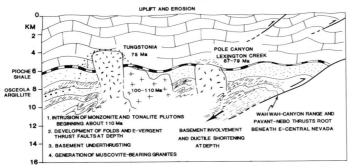

UPLIFT AND EROSION

KM

TUNGSTONIA
75 Ma

POLE CANYON
LEXINGTON CREEK
87-79 Ma

PIOCHE SHALE

OSCEOLA ARGILLITE

100-110 Ma

1. INTRUSION OF MONZONITE AND TONALITE PLUTONS
BEGINNING ABOUT 110 Ma

2. DEVELOPMENT OF FOLDS AND E-VERGENT
THRUST FAULTS AT DEPTH

3. BASEMENT UNDERTHRUSTING

4. GENERATION OF MUSCOVITE-BEARING GRANITES

BASEMENT INVOLVEMENT
AND DUCTILE SHORTENING
AT DEPTH

WAH WAH-CANYON RANGE AND
PAVANT-NEBO THRUSTS ROOT
BENEATH E-CENTRAL NEVADA

OLIGOCENE-MIOCENE

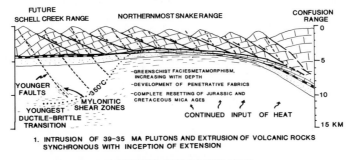

FUTURE
SCHELL CREEK RANGE   NORTHERNMOST SNAKE RANGE   CONFUSION RANGE

-GREENSCHIST FACIES METAMORPHISM,
INCREASING WITH DEPTH

-DEVELOPMENT OF PENETRATIVE FABRICS

-COMPLETE RESETTING OF JURASSIC AND
CRETACEOUS MICA AGES

YOUNGER
FAULTS

350°C

MYLONITIC
SHEAR ZONES

YOUNGEST
DUCTILE-BRITTLE
TRANSITION

CONTINUED INPUT OF HEAT

15 KM

1. INTRUSION OF 39-35 MA PLUTONS AND EXTRUSION OF VOLCANIC ROCKS
SYNCHRONOUS WITH INCEPTION OF EXTENSION

2. HEATING AND DUCTILE YIELDING OF DEEP-SEATED ROCKS,
FAILURE OF UPPER CRUST ALONG NORMAL FAULTS

-WHOLESALE REMOBILIZATION OF MIDDLE AND LOWER CRUST-

FIG. 24-12.   Highly simplified and schematic cross sections of east-central
Nevada, illustrating the evolution of the upper part of the crust from Mesozoic
to present times utilizing the data presented in this paper. Scales are only
approximate, and for clarity, older features have been omitted in successively
younger cross sections.

in the mid-Jurassic, about 160 m.y. ago, and was accompanied by, and related to, the emplacement of synkinematic plutons. It is important to point out that not only is the mid-Jurassic deformational event in east-central Nevada west-vergent, but also substantially predates the onset of thrusting directly in the Sevier Belt at this latitude (see discussion below), and thus cannot be genetically related to it.

A significant hiatus in magmatism of about 40 m.y. appears to have occurred both here and elsewhere in Nevada between the time of intrusion of Middle to Late Jurassic plutons and the oldest Cretaceous plutons. A 100 ± 8-m.y.-old tonalite to granite intrusive complex in the northern Snake Range is the oldest Cretaceous pluton in the study area (Fig. 24-12B) and is about the same age as a series of hornblende-biotite-bearing monzonite porphyry stocks near Ely, Nevada, dated by the K-Ar method as 110 m.y. (McDowell and Kulp, 1967). It is unclear whether any of these late Early Cretaceous plutons were accompanied by deformation and regional metamorphism at depth, because the Ely stocks are exposed at too high a level to tell, and most pre-Tertiary fabrics in the northern Snake Range have been obliterated by superimposed Tertiary strain.

Several compositionally distinct muscovite-bearing plutons were emplaced in the Late Cretaceous, including the Lexington Creek pluton (87 m.y., minimum age), the Pole Canyon pluton (about 79 m.y.), and the Tungstonia Granite (75 ± 9 m.y.). The dating of these intrusions is not very precise; they could be coeval or could span more than a 10-m.y. interval. East-vergent fabrics that increase in intensity with depth were developed at this time, but without further work it is unclear if these developed synchronously or diachronously across the area. The regionally developed, west-dipping cleavage that flattens downward with increasing metamorphic grade and strain is compatible with top-to-the-east bedding-parallel ductile shear (Fig. 24-12B). At higher structural levels, deformation was probably limited to open folding and small-displacement thrust faults. Structures that are likely but not demonstrated to be Cretaceous are the Butte and Confusion synclinoria of Hose and Blake (1976) and associated structures, including several minor displacement thrust faults in the southern Confusion Range mapped by Hintze et al. (unpublished mapping, Brigham Young University field studies) (Figs. 24-1 and 24-12B).

Deep-seated Late Cretaceous deformation in the study area spans a time interval similar to that of known movement on thrust faults directly east of this area in the foreland belt. Synthrusting deposits of the Indianola Group, deposited in the foredeep of the Sevier thrust belt, indicate that thrusting probably began in the Albian and occurred episodically until the late Campanian or early Maestrichtian (a time span from about 100 to 75 m.y.) (Heller et al., 1986; Lawton, 1985). The highest major thrust fault in the Sevier belt, the Canyon Range/Wah Wah thrust (Fig. 24-1; Armstrong, 1963; Misch, 1960) cuts down-section to the west and must underlie or root beneath the study area. We suggest that the Late Cretaceous east-vergent fabrics in east-central Nevada are related to movement on the underlying, deep-seated part of the Sevier thrust system (Fig. 24-12B), although a more precise chronology is needed on both the thrust faults and the fabrics developed in deep-seated rocks before a more accurate comparison can be made. Our reconstructed metamorphic gradients suggest that the entire hinterland region must have undergone Cretaceous amphibolite facies metamorphism at depths greater than about 15 km, thus basement shortening at depth, required to accommodate the brittle displacement on thrusts to the east, must have occurred by ductile yielding of both sedimentary and underlying crystalline basement rocks.

Available field, chemical, and isotopic data support the present view that muscovite-bearing peraluminous granitic rocks which occur within the hinterland of the Sevier thrust belt were generated largely, if not entirely, by crustal melting (Best *et al.*, 1974; C. F. Miller and Bradfish, 1980; Wright and Haxel, 1982; Farmer and DePaolo, 1983; 1984; Haxel *et al.*, 1984). Haxel *et al.* have suggested that the heat necessary for Late Cretaceous–early Tertiary crustal anatexis in south-central Arizona was provided principally by a combination of crustal thickening during thrust faulting and the intrusion of mantle-derived magmas into the lower crust, as indicated by the presence there of Late Cretaceous and earliest Tertiary metaluminous granitic plutons with a mantle component. We suggest that a similar set of circumstances may account for the origin of Late Cretaceous muscovite-bearing peraluminous plutons within east-central Nevada. Specifically, we suggest that crustal anatexis was driven by heat due to the emplacement of mantle-derived magmas into the lower crust and by crustal thickening during basement shortening that accompanied movement on the Sevier thrust belt. Available chemical and isotopic data suggest that the approximately 90–110-m.y.-old metaluminous Cretaceous plutons in the Great Basin, which predated the emplacement of Late Cretaceous peraluminous plutons, were derived in large part from a mantle source[7] (see the discussion in Barton, *et al.*, Chapter 5, this volume, and Barton, 1987), and likely aided heating of the lower to midcrust.

At the inception of Tertiary faulting, the 15-km-thick miogeoclincal stratigraphy was still relatively intact across east-central Nevada (Gans and E. L. Miller, 1983). Figure 24-12C, modified from Gans *et al.* (1985), shows in a schematic fashion how the crust in east-central Nevada was modified by Tertiary extensional deformation. Tertiary plutons that are exposed to view today were emplaced early in the history of extension and, based on new U-Pb isotopic data, are as old as 39 m.y. but are mostly 35–36 m.y. old. The extrusion of large volumes of coeval intermediate to silicic volcanic rocks at the surface indicates that even larger batholiths must be present at depth (Gans, 1987).

Approximately 250% extension has occurred across east-central Nevada and began synchronously with volcanism. Even at the scale of our study area, extensional strain at supracrustal levels is remarkably heterogeneous (Gans and E. L. Miller, 1983). This extension was accommodated in the brittle part of the crust by successive generations of imbricate, high-angle normal faults. These faults were approximately planar in the brittle part of the crust, but must have bottomed or curved into subhorizontal surfaces at depth because both faults and strata rotated and tilted as faults moved (Gans and E. L. Miller, 1983). We infer that all normal faults interacted at depth with a complex zone of brittle-to-ductile transition, beneath which rocks extended in a plastic fashion. Temperatures necessary for this transition depend largely on rock composition and strain rates, but are probably on the order of 350°C for quartzose rocks (Sibson, 1977). Based on the relationships of Tertiary metamorphism to deformation described in this paper, the transition downward to ductile behavior occurred at fairly shallow depths of about 7 km in the northern Snake Range and along the easternmost flank of the southern Snake Range and Kern Mountains, but lower heat flow and a deeper transition zone characterized the Schell Creek Range and the Deep Creek Range regions. We argue that plastic flow at depth in the Tertiary was localized and driven by the input of heat and water into the crust, and that as faulting continued and the cover thinned, additional heat must have been added to the upper crust in order to keep temperatures locally elevated until the early Miocene (Fig. 24-12C).

Younger faults cut to deeper structural levels in the cooler parts of the crust, becoming ductile at depths ranging from 10 to 15 km (Fig. 24-12C). Differential uplift due to bending, offset, and rotation along these younger faults ultimately uplifted and exposed the rocks we see in the present-day ranges of east-central Nevada. In view of the style of progressive extensional deformation depicted in Fig. 24-12C, all variations in the ductile strain history of rocks, from brittle offset faults to ductile shear zones and to subhorizontal pure shear stretching fabrics, are to be expected, depending on temperatures, strain rate, percent strain, and the local boundary conditions for ductile flow. Pure shear stretching at depth would be expected in an extensional tectonic setting where the least principal stress was horizontal and heat flow high. Pure shear stretching does not present a strain incompatibility problem, provided that strain is continuous throughout the crustal column or is ultimately compensated by the intrusion of magmas or by the lateral and vertical flow of material at depth. Seismic reflection data from east-central Nevada, reviewed and discussed by McCarthy (1986), support wholesale deformation and remobilization of the middle and lower crust during Tertiary regional extension.

Much more work is needed to address the specific details of the interplay between magmatism, metamorphism, and deformation in the Great Basin. Based on the presently available data from east-central Nevada and westernmost Utah, however, it appears as if these three processes are intimately associated, particularly in the upper 15 km of the crust. There is good evidence that increased heat flow due to the intrusion of plutons during the Jurassic, Cretaceous, and Tertiary apparently triggered ductile flow of surrounding rocks because they were under a state of deviatoric stress. Fabrics formed during these events indicate horizontal shortening during the Jurassic and Cretaceous and vertical shortening and horizontal extension in the Tertiary. We would imagine that the documented pattern, progression, and interrelationship of plutonism, metamorphism, and deformation is very much a function of the crustal levels exposed. At the particular levels studied, we can document three distinct metamorphic and deformational events, and data suggest that, to a large extent, rocks were able to cool between these events. If somewhat deeper levels of the crust were exposed, these distinctions would be less clear and the times necessary for cooling between events longer. Deeper still, we would expect evidence for partial to wholesale reconstitution and remobilization of the crust during these three events, and a more complete overprinting of older events by younger ones. Certainly, if one went any deeper in the crust today, one would find negligible evidence for older Mesozoic events and instead find a regional metamorphic, migmatitic, and igneous terrane of largely Tertiary age.

ACKNOWLEDGMENTS

This work has been supported by NSF grant EAR84-18678 awarded to E. L. Miller, NSF grant EAR83-13733 awarded to J. E. Wright, and by Sohio Petroleum Company, which provided support for Stanford Geological Survey mapping projects. Most of the previously unpublished structural and metamorphic data discussed in this paper represent the work of P. B. Gans and E. L. Miller, but we would like to acknowledge the important contributions and comments by E. L. Miller's graduate students D. Clark, J. Lee, A. McGrew, D. Rodgers, and L. Rowles.

REFERENCES

Allmendinger, R. W., and Jordan, T. E., 1984, Mesozoic structure of the Newfoundland Mountains, Utah — Horizontal shortening and subsequent extension in the hinterland of the Sevier Belt: *Geol. Soc. America Bull.*, v. 95, p. 1280–1292.

Armstrong, R. L., 1963, Geochronology and geology of the eastern Great Basin in Nevada and Utah: Ph.D. thesis, Yale Univ., New Haven, Conn.

—— , 1968, Sevier orogenic belt in Nevada and Utah: *Geol. Soc. America Bull.*, v. 79, p. 429–458.

—— , 1972, Low angle (denudation) faults, hinterland of the Sevier orogenic belt, eastern Nevada and western Utah: *Geol. Soc. America Bull.*, v. 83, p. 1729–1754.

—— , Speed, R. C., Graustein, W. C., and Young, A. Y., 1976, K-Ar dates from Arizona, Montana, Nevada, Utah, and Wyoming: *Isochron/West* v. 16, p. 1–6.

Barton, M. D., 1987, Cretaceous magmatism, mineralization and metamorphism in the east-central Great Basin: *Geol. Soc. America Abstr. with Programs*, v. 19, p. 357.

Best, M. G., Armstrong R. L., Graustein, W. C., Embre, G. F., and Ahlborn, R. C., 1974, Two mica granites of the Kern Mountains pluton, eastern White Pine County, Nevada — Remobilized basement of the Cordilleran miogeosyncline? *Geol. Soc. America Bull.*, v. 85, p. 1277–1286.

Clark, D. H., 1985, Tectonic evolution of the Red Hills–southwestern Kern Mountains area, east-central Nevada: M.S. thesis, Stanford Univ., Stanford, Calif., 51 p.

Dallmeyer, R. D., and Sutter, J. F., 1976, ^{40}AR/^{39}Ar incremental release ages of biotite and hornblende from variably retrograded basement gneisses of the northeastern-most Reading Prong, New York — Their bearing on Early Paleozoic metamorphic history: *Amer. J. Sci.*, v. 276, p. 731–747.

Dalrymple, G. B., 1979, Critical tables for conversion of K-Ar ages from old to new constants: *Geology*, v. 7, p. 558–560.

—— , and Lanphere, M. A., 1971, ^{40}Ar/^{39}Ar technique of K-Ar dating — A comparison with the conventional technique: *Earth Planet. Sci. Lett.*, v. 12, p. 300–308.

Dechert, C. P., 1967, Bedrock geology of the northern Schell Creek Range, White Pine County, Nevada: Ph.D. thesis, Univ. Washington, Seattle, Wash.

Drewes, H., 1967, Geology of the Connor's Pass quadrangle, Schell Creek Range, east-central Nevada: *U.S. Geol. Survey Prof. Paper 557.*

Farmer, G. L., and DePaolo, D. J., 1983, Origin of Mesozoic and Tertiary granite in the western United States and implications for pre-Mesozoic crustal structure — 1. Nd and Sr isotopic studies in the geocline of the northern Great Basin: *J. Geophys. Res.*, v. 88, p. 3379–3401.

—— , and DePaolo, D. J., 1984, Origin of Mesozoic and Tertiary granite in the western United States and implications for pre-Mesozoic crustal structure — 2. Nd and Sr studies of unmineralized and Cu- and Mo- mineralized granite in the Precambrian craton: *J. Geophys. Res.*, v. 89, p. 10141–10160.

Fleck, R. J., Sutter, J. F., and Elliot, D. H., 1977, Interpretation of discordant ^{40}Ar/^{39}Ar age spectra of Mesozoic tholeiites from Antarctica: *Geochim. Cosmochim. Acta*, v. 41, p. 15–32.

Fritz, W. H., 1968, Geologic map and sections of the southern Cherry Creek and northern Egan Ranges, White Pine County, Nevada: *Nev. Bur. Mines Map 35.*

Gans, P. B., 1987, Cenozoic extension and magmatism in the eastern Great Basin: Ph.D. thesis, Stanford Univ., Stanford, CA., 174 p.

—— , Clark, D. H., Miller, E. L., Wright, J. E., and Sutter, J. F., 1986, Structural development of the Kern Mtns. and N. Snake Range: *Geol. Soc. America Abstr. with Programs*, v. 18, p. 108.

—— , and Miller, E. L., 1983, Style of mid-Tertiary extension in east-central Nevada, *in* Gurgel, K. D., ed., *Geologic Excursions in the Overthrust Belt and Metamorphic Core Complexes of the Intermountain Region, Nevada:* Geol. Soc. America Field Trip Guidebook, Utah Geol. Min. Survey Spec. Studies 59, p. 107–160.

_____, and Miller, E. L., 1985, Comment on "The Snake Range decollement interpreted as a major extensional shear zone" by J. M. Bartley and B. P. Wernicke: *Tectonics*, v. 4, p. 411–415.

_____, Miller, E. L., McCarthy, J., and Ouldcott, M. L., 1985, Tertiary extensional faulting and evolving ductile-brittle transition zones in the northern Snake Range and vicinity – New insights from seismic data: *Geology*, v. 13, p. 189–193.

Gering, R. L., 1987, A study of the metamorphic petrology of the northern Snake Range, east-central Nevada; M.S. thesis, Southern Methodist Univ., Dallas, TX., 75 p.

Grier, S. P., 1983, Tertiary stratigraphy and geologic history of the Sacramento Pass area *in* Gugel, K. D., ed., *Geologic Excursions in the Overthrust Belt and Metamorphic Core Complexes of the Intermountain Region, Nevada:* Geol. Soc. America Field Trip Guidebook, Utah Geol. Mineral Survey Spec. Studies 59, p. 139–144.

_____, 1984, Alluvial fan and lacustrine carbonate deposits in the Snake Range – A study of Tertiary sedimentation and associated tectonism: M.S. thesis, Stanford Univ., Stanford, Calif., 61 p.

Harrison, T. M., and McDougall, I. M. 1980a, Investigations of an intrusive contact, northwest Nelson, New Zealand – I. Thermal, chronological and isotopic constraints: *Geochim. Cosmochim. Acta*, v. 44, p. 1985–2003.

_____, and McDougall, I. M., 1980b, Investigations of an intrusive contact, northwest Nelson, New Zealand – II. Diffusion of radiogenic and excess ^{40}Ar in hornblende revealed by ^{40}Ar/^{39}Ar age spectrum analysis: *Geochim. Cosmochim. Acta*, v. 44, p. 2005–2020.

_____, Duncan, I., and McDougall, I., 1985, Diffusion of ^{40}Ar in biotite – Temperature, pressure and compositional effects: *Geochim. Cosmochim. Acta*, v. 49, p. 2461–2468.

Haxel, G. B., Tosdal, R. M., May, D. J., and Wright, J. E., 1984, Latest Cretaceous and early Tertiary orogenesis in south-central Arizona – Thrust faulting, regional metamorphism, and granitic plutonism: *Geol. Soc. America Bull.*, v. 95, p. 631–653.

Heller, P. L., Bowdler, S. S., Chambers, H. P., Coogan, J. C., Hagen, E. S., Shuster, M. W., Winslow, N. S., and Lawton, T. F., 1986, Time of initial thrusting in the Sevier Orogenic belt, Idaho-Wyoming and Utah: *Geology*, v. 14, p. 388–391.

Hintze, L. F., 1980, *Geologic Map of Utah*; scale 1:500,000: Utah Geol. Min. Survey.

Hose, R. K., and Blake, M. C., Jr., 1976, Geology and mineral resources of White Pine County, Nevada – Pt. I. Geology: *Nev. Bur. Mines and Geology Bull. 85*, p. 1–35.

Kellogg, H. E., 1964, Cenozoic stratigraphy and structure of the southern Egan Range, Nevada: *Geol. Soc. America Bull.*, v. 75, p. 949–968.

Koch, P. S., 1983, Rheology and microstructures of experimentally deformed quartz aggregates: Ph.D. thesis, Univ. California, Los Angeles, Calif., 464 p.

Krough, T. E., 1973, A low contamination method for hydrothermal decomposition of zircon and extraction of U and Pb for isotopic age determinations: *Geochim. Cosmochim. Acta*, v. 37, p. 485–494.

Lanphere, M. A., and Dalrymple, G. B., 1971, a test of the ^{40}Ar/^{39}Ar age spectrum technique on some terrestrial materials: *Earth Planet. Sci. Lett.*, v. 12, p. 359–372.

Lawton, T. F., 1985, Style and timing of frontal structures, thrust belt, central Utah: *Amer. Assoc. Petrol. Geologists Bull.*, v. 69, p. 1145–1159.

Lee, D. E., and Christiansen, E. H., 1983, The granite problem as exposed in the southern Snake Range, Nevada: *Contrib. Mineralogy Petrology*, v. 83, p. 99–116.

_____, and Fischer, L. B., 1985, Cretaceous metamorphism in the northern Snake Range, Nevada, a metamorphic core complex: *Isochron/West*, v. 42, p. 3–7.

____, Marvin, R. F., Stern, T. W., Mays, R. E., and Van Loenen, R. E., 1968, Accessory zircon from granitoid rocks of the Mount Wheeler mine area, Nevada, *in Geological Survey Research:* U.S. Geol. Survey Prof. Paper 600D, p. D197–D203.

____, Marvin, R. F., Stern, T. W., and Peterman, Z. E., 1970, Modification of K-Ar ages by Tertiary thrusting in the Snake Range, White Pine County, Nevada, *in Geological Survey Research:* U.S. Geol. Survey Prof. Paper 700-D, p. 93–102.

____, Marvin, R. F., and Mehnert, H. H., 1980, A radiometric study of Mesozoic-Cenozoic metamorphism in eastern White Pine County, Nevada and adjacent Utah: *U.S. Geol. Survey Prof. Paper 1158C*, p. C17–C28.

____, Stern, T. W., and Marvin, R. F., 1981, Uranium-thorium lead isotopic ages of metamorphic monazite from the northern Snake Range, Nevada: *Isochron/West*, v. p. 23.

____, Friedman, I., and Gleason, J. D., 1984, Modification of delta D values in eastern Nevada granitoid rocks spatially related to thrust faults: *Contrib. Mineralogy Petrology*, v. 88, p. 288–298.

____, Kistler, R. W., and Robinson, A. C., in press *a*, The strontium isotope composition of granitoid rocks from the southern Snake Range, Nevada, *in Shorter Contributions to Isotope Research*, U.S. Geol. Survey Prof. Paper Chap. P, p.

____, Stacey, J. S. D., Fischer, L., in press *b*, Muscovite phenocrystic two-mica granites of northeastern Nevada are Late Cretaceous in age, *in Shorter Contributions to Isotope Geology:* U.S. Geol. Survey Prof. Paper, Chap. D, p.

Lee, J., Miller, E. L., and Sutter, J. F., 1987, Ductile strain and metamorphism in an extensional tectonic setting – A case study from the northern Snake Range, Nevada, U.S.A.: *J. Geol. Soc. London, Spec. Publication no. 28: Continental Extensional Tectonics*, p. 267–298.

Ludwig, K. R., 1984, Plotting and regression programs for isotope geochemists, for use with HP-86/87 microcomputers: *U.S. Geol. Open File Rpt. 83-849*.

Marvin, R. F., 1968, Transcontinental Geophysical Survey (35°–39°N), Radiometric age determinations of rocks: *U.S. Geol. Survey Misc. Geol. Inv. Map I-537*.

McCarthy, J., 1986, Reflection profiles from the Snake Range metamorphic core complex – A window into the mid-crust, *in Geodynamics Series vol. 14, Reflection Seismology – The Continental Crust*, Washington, D.C., Amer. Geophys. Un., p. 281–292.

McDowell, F. W., and Kulp, J. L., 1967, Age of intrusion and ore deposition in the Robinson mining district of Nevada: *Econ. Geology*, v. 62, p. 905–909.

McGrew, A. J., 1986, Deformation history of the southern Snake Range, Nevada, and the origin of the southern Snake Range décollement: M.S. thesis, Stanford Univ., Stanford, CA., 46 p.

Miller, E. L., and Sutter, J. F., 1982, Structural geology and $^{40}Ar/^{39}Ar$ geochronology of the Goldstone-Lane Mountain area, Mojave Desert, California: *Geol. Soc. America Bull.*, v. 93, p. 1191–1207.

____, Gans, P. B., and Garing, J., 1983, The Snake Range decollement – An exhumed mid-Tertiary ductile-brittle transition: *Tectonics*, v. 2, p. 239–263.

____, Gans, P. B., and Lee, J., 1987, The Snake Range décollement, eastern Nevada. *Geol. Soc. America Centennial Field Guide* Cordilleran Section, p. 77-82.

Misch, P., 1960, Regional structural reconnaissance in central-northeast Nevada and some adjacent areas – Observations and interpretations: *Intermountain Assoc. Petrol. Geologists 11th Ann. Field Conf. Guidebook*, p. 17–42.

____, and Hazzard, J. C., 1962, Stratigraphy and metamorphism of the Late Precambrian rocks in central northeastern Nevada and adjacent Utah: *Amer. Assoc. Petrol. Geologists Bull.*, v. 46, p. 289–343.

Moores, E. M., Scott, R. B., and Lumbsden, W. W., 1968, Tertiary tectonics of the White Pine-Grant Range regions, east-central Nevada, and some regional implications: *Geol. Soc. America Bull.*, v. 79, p. 1703–1726.

Nelson, R. B., 1966, Structural development of the northernmost Snake Range, Kern Mountains and Deep Creek Range, Nevada and Utah: *Amer. Assoc. Petrol. Geologists Bull.*, v. 50, p. 921–951.

_____ , 1969, Relation and history of structures in a sedimentary succession with deeper metamorphic structures, eastern Great Basin: *Amer. Assoc. Petrol. Geologists Bull.*, v. 53, p. 307–339.

Ramsay, J. G., 1981, Tectonics of the Helvetic Nappes, *in* McCoy, K. R., and Price, N. J., eds., *Thrust and Nappe Tectonics:* Geol. Soc. London Spec. Publ., p. 293–309.

_____ , and Huber, M. I., 1983, *The Tecniques of Modern Structural Geology, Vol. 1 – Strain Analysis:* New York, Academic, 307 p.

Rodgers, D. W., 1987, Thermal and structural evolution of the southern Deep Creek Range, west-central Utah and east-central Nevada: Ph.D. thesis, Stanford Univ., Stanford, CA., 149 p.

Rowles, L. D., 1982, Deformational history of the Hampton Creek area, northern Snake Range, Nevada: M.S. thesis, Stanford Univ., Stanford, Calif., 80 p.

Sanderson, D. J., 1982, Models of strain variation in nappes and thrust sheets – A review: *Tectonophysics*, v. 88, p. 201–233.

Sibson, R. H., 1977, Fault rocks and fault mechanisms: *J. Geol. Soc. London*, v. 133, p. 191–213.

Stewart, J. H., 1980, Geology of Nevada: *Nev. Bur. Mines and Geology Spec. Publ. 4*, 136 p.

_____ , and Carlson, J. E., 1978, *Geologic Map of Nevada*; scale 1:500,000: Geol. Survey.

_____ , and Poole, F. G., 1974, Lower Paleozoic and uppermost Precambrian Cordilleran miogeocline, Great Basin, Western United States, *in* Dickinson, W. R., ed., *Tectonics and Sedimentation:* Soc. Econ. Paleontologists Mineralogists Spec. Publ. 22, p. 28–57.

Stokes, W. L., 1963, *Geologic Map of Northwestern Utah*, scale 1:250,000 Coll. Mines Min. Industries, Univ. Utah, Salt Lake City, Utah.

Sutter, J. F., Ratcliffe, N. M., and Mukasa, S. B., 1985, ^{40}Ar/^{39}Ar and K-Ar date bearing on the metamorphic and tectonic history of western New England: *Geol. Soc. America Bull.*, v. 96, p. 123–136.

Whitebread, D. H., 1969, Geologic map of the Wheeler Peak and Garrison quadrangles, White Pine County, Nevada, and Milford County, Utah: *U.S. Geol. Survey Misc. Geol. Inv. Map I-578.*

Wright, J. E., and Haxel, G., 1982, A garnet-two mica granite, Coyote Mountains, southern Arizona – Geologic setting, uranium-lead isotopic systematics of zircon, and nature of the granite source region: *Geol. Soc. America Bull.*, v. 93, p. 1176–1188.

Young, J. C., 1960, Structure and stratigraphy in the north central Schell Creek Range, eastern Nevada: Ph.D. thesis, Princeton Univ., Princeton, N.J.

John H. Stewart
U.S. Geological Survey
Menlo Park, California 94025

25

TECTONICS OF THE WALKER LANE BELT, WESTERN GREAT BASIN: MESOZOIC AND CENOZOIC DEFORMATION IN A ZONE OF SHEAR

ABSTRACT

The Walker Lane was originally defined as a narrow northwest-trending zone of right-lateral faulting in the western Great Basin, separating areas of north-northeast-trending basin-range topography to the east from areas of diverse topography to the west. More recently, the term "Walker Lane belt" has been used to describe a broad northwest-trending zone of diverse topography and strike-slip faulting east of the Sierra Nevada. Defined in this way, the belt is about 700 km long and 100–300 km wide.

Rather than consisting of a single throughgoing right-lateral fault or a system of right-lateral faults as indicated by many workers, the Walker Lane belt is a complex zone which can be subdivided into nine regional blocks of diverse character. Three of these blocks contain major northwest-trending right-lateral faults, three contain major northeast-trending left-lateral faults, two are locally bounded by strike-slip faults but contain no major internal strike-slip faults, and one (Excelsior-Coaldale block) contains major east-west faults with sizable Mesozoic right-lateral offset and minor left-lateral Cenozoic offset. Strike-slip faults in any one structural block rarely, if ever, extend into an adjacent block. Lateral offset on individual major faults ranges from a few to perhaps 100 km.

The Walker Lane belt may have been initiated in Late Triassic or Jurassic time by oblique subduction that produced right-lateral faulting along and in back of a magmatic arc. North-south compression, perhaps related to oblique subduction, produced a "kink" in this system and resulted in the east-west cross-faults of the Excelsior-Coaldale block. These east-west faults as well as proposed strike-slip faults in adjacent parts of the Sierra Nevada are considered to be Mesozoic in age because of cross-cutting relations of Mesozoic intrusive rocks.

Although Mesozoic initiation of the Walker Lane belt seems likely, latest movement on most of the faults in the belt is Cenozoic in age. Some Cenozoic faults may be reactivated Mesozoic structures, but others were apparently initiated by Cenozoic deformation that includes major right-lateral and left-lateral faults, detachment faults and core complexes produced in highly extended areas, and basin-range blocks that locally follow the trends of older northwest-trending strike-slip faults. Diverse Cenozoic structures are related to repeated changes in the stress field that produced alternating extensional and strike-slip deformation.

INTRODUCTION

The Walker Lane belt is an important northwest-trending zone about 700 km long and 100–300 km wide in the western Great Basin. It is characterized by mountain ranges with unusual shapes or trends lying between the Sierra Nevada on the west and areas of long north-northwest-trending mountains and valleys typical of basin-range topography on the east. This article is a summary of the Walker Lane belt in which characteristic structures of the belt are described and analyzed. The article emphasizes a possible Mesozoic initiation of the zone and a reactivation of Mesozoic structures, as well as the development of new structures in the Cenozoic. The geology of the Walker Lane belt is important in understanding the Mesozoic and Cenozoic history of the western United States as well as in revealing the style of Cenozoic strike-slip deformation in a region well inland from the present transform plate boundary.

DEFINITION

The topographic contrast (Fig. 25-1) between areas of diverse trends of mountain ranges in the western Great Basin and areas of typical north-northwest-trending basin-range

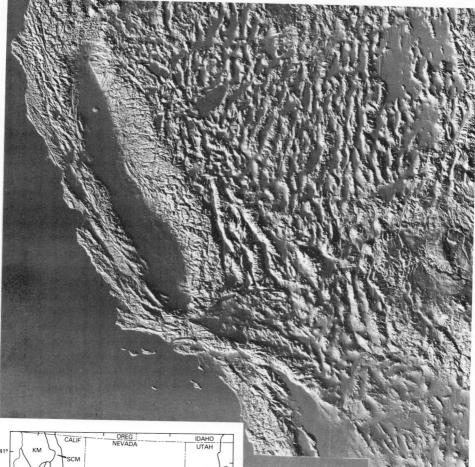

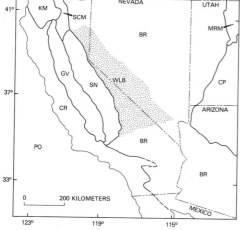

FIG. 25-1. Computer-generated shadded-relief mosaic of the western United States produced by methods described by Batson *et al.* (1975). Courtesy of Gerald Schaber, U.S. Geological Survey, Flagstaff, Arizona. Note Walker Lane belt of diverse topography between Sierra Nevada block on west and areas of typical basin-range topography on east. BR, Basin and Range Province; CP Colorado Plateau; CR, Coast Ranges; GV, Great Valley of California; KM, Klamath Mountains; MRM, Middle Rock Mountains; PO, Pacific Ocean; SCM, Southern Cascade Mountains; SN, Sierra Nevada; WLB, Walker Lane Belt.

STEWART

685

topography to the east was noted by Giannella and Callaghan (1934) in a paper describing surface rupture during the Cedar Mountain earthquake in 1932. They noted that the change occurred along a fairly narrow zone that they suggested might be the physiographic expression of a right-lateral fault comparable to the San Andreas fault. Billingsley and Locke (1939, 1941) referred to this zone as the "Walker line" and later Locke *et al.* (1940) changed the name to the "Walker Lane." Chiefly, they described the Walker Lane as a line separating regions of contrasting topographic style.

The term "Walker Lane" has also been used to describe a broad zone of diverse topography of which the Walker Lane or line (as defined by Billingsley and Locke, 1939, 1941; Locke *et al.*, 1940) is the eastern boundary. This dual usage of the name is confusing and led me to suggest (Stewart, 1980) that the name "Walker belt" be used to describe the narrow to broad zone of diverse topography west of the Walker Lane or line. Other geologists (e.g., Carr, 1984a, b) have used the term "Walker Lane belt" for this zone. I will use the name Walker Lane belt to describe all of the region lying between the Sierra Nevada on the west and the region of typical basin and range topography on the east, although some other geologists (e.g., Hardyman, 1984) have restricted the Walker Lane belt to a narrower zone.

STRUCTURAL BLOCKS, FAULT PATTERNS, AND DISPLACEMENTS

The Walker Lane belt is subdivided into nine structural blocks (Figs. 25-2 and 25-3), each of which, for the most part, acted independently of adjacent blocks. Three of these blocks contain major northwest-trending right-lateral faults, three contain major northeast-trending left-lateral faults, two contain no major internal strike-slip faults, although they are bounded locally by strike-slip faults, and one (Excelsior-Coaldale block) contains east-west faults that had sizable Mesozoic right-lateral movement and minor left-lateral Cenozoic movement. Strike-slip faults in any one block rarely extend into an adjacent block.

Pyramid Lake Section

The Pyramid Lake section (E. J. Bell, 1984), the northernmost segment of the Walker Lane belt, extends from the Honey Lake area of California to the Reno area, Nevada. This section may continue into northwest California, where numerous faults, some on line with faults of the Pyramid Lake section, trend northwest.

The Pyramid Lake section is marked by five major northwest-trending faults (Grizzly Valley, Last Chance, Honey Lake, Warm Springs Valley, and Pyramid Lake) and many minor faults, primarily with northwest and north trends. Segments of several of these faults were active in the late Cenozoic, and some show Holocene offset (E. J. Bell, 1984; E. J. Bell and Slemmons, 1979; J. W. Bell, 1984a, b; Bonham, 1969). The Pyramid Lake fault is associated with numerous en-echelon northwest- and northeast-striking faults interpreted as Riedel and conjugate Riedel shears developed during right-lateral movement (E. J. Bell and Slemmons, 1979). The fault zone is also characterized by sag ponds, elongate depressions and troughs, offset stream channels, transcurrent buckles, rhombohedral and wedge-shaped enclosed depressions, and recent scarps, all indicative of Holocene strike-slip (L. W. Anderson and Hawkins, 1984; E. J. Bell, 1984; E. J. Bell and Slemmons, 1979). Bonham (1969) suggested a cumulative right-lateral offset of 32 km, based on the apparent offset of the northern outcrop limit of a group of Cenozoic tuffs (Fig. 25-4), although he indicates that this estimate is highly conjectural. Offset on the major northwest-trending faults in the Pyramid Lake section, however, not only

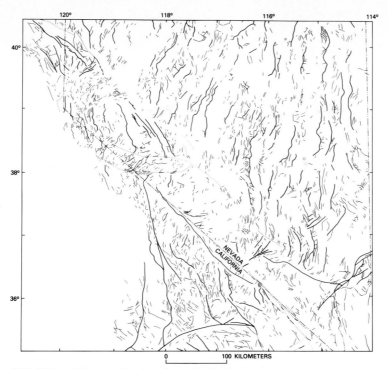

FIG. 25-2. Map showing late Cenozoic faults in western Great Basin. Compare with Figs. 25-1 and 25-3 for location of Walker Lane belt. Heavy lines are major faults, dotted where inferred; fine lines are minor faults. Based on many sources, including Jennings (1975), Stewart and Carlson (1978), Stewart *et al.* (1982), J. W. Bell (1984a), Carr (1984b), and Stewart (1985).

includes segments that are strike-slip and characterized by little topographic relief across the fault zone, but other segments interpreted to be oblique-slip or dip-slip and characterized by faults bounding high, uplifted mountain ranges (E. J. Bell, 1984; Bonham, 1969).

Surface faulting may have occurred along the Pyramid Lake fault during an earthquake in 1852 (estimated Richter magnitude of 7; E. J. Bell, 1984; Slemmons *et al.*, 1965), although the location of the epicenter for this earthquake is not precise and only tentatively related to the Pyramid Lake fault. In 1950, a small amount of surface rupture (Giannella, 1957) occurred along a north-trending normal fault during a 5.6 magnitude earthquake (Slemmons, 1967) centered a short distance north of the Honey Lake fault near the California-Nevada state line.

Carson Section

The Carson section is characterized by zones of northeast-trending faults of known or presumed left-lateral displacement that interrupt the dominant northwest grain of the Walker Lane belt. Three such fault zones from north to south are recognized, the Olinghouse fault zone on the north, the Carson Lineament in the middle, and the Wabuska Lineament on the south.

The Olinghouse system includes a group of northeast-trending faults that extend from about 10 km east of Reno to the Pyramid fault zone (Fig. 25-3). Surface rupture

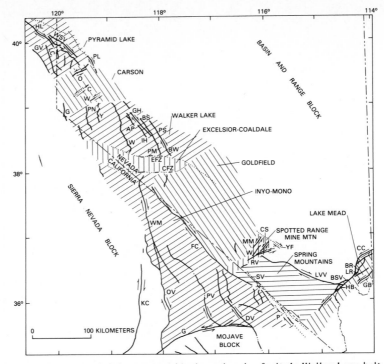

FIG. 25-3. Regional structural blocks and major faults in Walker Lane belt.
Arrows indicate relative movement on strike-slip faults. Major faults or fault
zone listed by structural blocks: PYRAMID LAKE BLOCK: HL, Honey Lake;
GV, Grizzly Valley; LC, Last Chance; WSV, Warm Springs Valley; PL, Pyra-
mid Lake. CARSON BLOCK: O, Olinghouse; C, Carson; W. Wabuska. WALKER
LAKE BLOCK: G, Genoa; PN, Pine Nut; Y, Yerington; W, Wassuk; AP, Agai
Pah Hills; GH, Gumdrop Hills; IH, Indian Head; BS, Benton Spring; PS, Petri-
field Springs; PM, Pilot Mountains; BW, Bettles Well. EXCELSIOR-COALDALE
BLOCK: EFZ, Excelsior; CFZ, Coaldale. INYO-MONO BLOCK: KC, Kern
Canyon; I, Independence; WM, White Mountain; OV, Owens Valley; FC,
Furnace Creek; PV, Panamint Valley; DV, Death Valley; G, Garlock; SV,
Stewart Valley; P, Pahrump. SPOTTED RANGE–MINE MOUNTAIN BLOCK:
MM, Mine Mountain; W, Wahmonie; RV, Rock Valley; CS, Cane Spring; YF,
Yucca-Frenchman. SPRING MOUNTAINS BLOCK: LVV, Las Vegas Valley.
LAKE MEAD BLOCK: BSV, Bitter Spring Valley; HB, Hamblin Bay; CC,
Cabin Canyon; BR, Bitter Ridge; LR, Lime Ridge; GB, Gold Butte.

(Sanders and Slemmons, 1979) is inferred to have occurred along the Olinghouse fault,
the most important fault in the Olinghouse system, during an earthquake of about
magnitude 7 in 1869. Maximum vertical displacement was 3 m and maximum left-slip
displacement was 3.65 m (Sanders and Slemmons, 1979), suggesting overall left-
oblique slip.

The Carson Lineament (Fig. 25-3) was defined by D. K. Rogers (1975) as a structural zone expressed by near-alignment of valleys and mountain ranges, and numerous
northeast faults. It has been considered to be a zone of left slip (J. W. Bell, 1981,
1984a, b; D. K. Rogers, 1975; Slemmons *et al.*, 1979) based on its northeast trend
parallel to the left- or left-oblique-slip Olinghouse zone. D. K. Rogers (1975) suggested
that initial movement on faults of the Carson Lineament occurred sometime prior to the

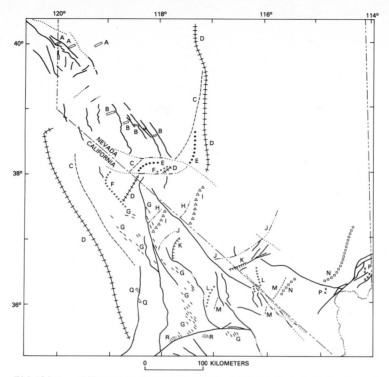

FIG. 25-4.   Offset geologic trends on faults in Walker Lane belt. Faults same as on Fig. 25-3. A, Northern outcrop limit of a group of Cenozoic tuffs (Bonham, 1969); B, southern limit of a group of Cenozoic tuffs (Ekren and Byers, 1984; Ekren *et al.*, 1980; Hardyman, 1984; Hardyman *et al.*, 1975); C, eastern limit of Upper Triassic and Lower Jurassic rocks, from Fig. 25-5; D, $^{87}Sr/^{86}Sr = 0.706$, after R. W. Kistler, written communication, 1986; E, eastern limit of Permian Mina Formation (Stewart, 1985); F, eastern limit of a Devonian sandy limestone (Stewart, 1985); G, Jurassic Independence dike swarm (Chen and Moore, 1979); J. G. Moore and Hopson, 1961); H, facies boundary in Lower Cambrian Harkless Formation (Stewart, 1970); I, generalized facies trends in Lower Cambrian Poleta Formation (J. N. Moore, 1976a, b); J, western limit of Lower Mississippian limestone unit (F. G. Poole *in* Stewart *et al.*, 1968); K, 100-m isopach for Upper Ordovician (Caradocian and Ashgillian) rocks (Miller and Walch, 1977, Fig. 14); L, 150-m isopach of Lower Cambrian Zabriskie Quartzite (slightly modified from Stewart, 1970, Fig. 25); M, 600-m isopach of Late Proterozoic Stirling Quartzite (Stewart, 1970); N, southwest limit of Upper Mississippian (late Chesterian) shale unit (modified from F. G. Poole *in* Stewart *et al.*, 1968); P, Mesozoic and Tertiary strata (Bohannon, 1979, 1984); Q, boundary of Mesozoic pluton (J. G. Moore and du Bray, 1978); R. Paleozoic metasedimentary rocks (G. I. Smith and Ketner, 1970).

mid-Cretaceous because the lineament follows the pre-mid-Cretaceous fabric of metamorphic rocks, although displacements have continued until the Holocene.

The name Wabuska Lineament (Fig. 25-3) is proposed here for a zone of northeast-trending faults and mountain ranges south of the Carson Lineament. The zone interrupts the conspicuous northwest grain of faults that approach the zone from the south.

Walker Lake Section

The Walker Lake section contains conspicuous subparallel north-northwest-trending faults, some of which are considered to be right-lateral faults and others basin-range normal faults. The trends of these faults are at an angle of about 40° to the typical north-northeast trend of basin-range faults to the east of the Walker Lane belt (Fig. 25-2).

Basin-range normal faults include the Genoa, Pine Nut, Yerington, and Wassuk faults (Fig. 25-3), which have irregular map traces and bound major, west-tilted mountain blocks. Striae on the Wassuk fault have rakes of about 60° and plunge to the south, where the fault strike is about N20°W and the dip is 45° east (R. C. Bucknam *in* Zoback, 1986). These basin-range faults are subparallel to right-slip faults to the east (Ekren *et al.*, 1980; Ferguson and Muller, 1949; Nielsen, 1965) that are characterized by generally straight map traces, little topographic relief across the fault, steep fault planes, and low-angle to horizontal striae on fault surfaces (Ekren and Byers, 1984; Hardyman, 1978, 1984). These strike-slip faults include the Agai Pah Hills, Indian Pass, Gumdrop Hills–Pilot Mountains, and Petrified Spring–Bettles Well faults. Displacements on these faults have been estimated from the apparent offset of the outcrop limits of Oligocene tuffs that suggest from a few kilometers to perhaps as much as 18 km of offset on individual faults, and 48–60 km of cumulative offset across the entire fault zone (Ekren and Byers, 1984; Ekren *et al.*, 1980; Hardyman, 1984; Hardyman *et al.*, 1975). Offset of Mesozoic granitic rocks (Hardyman, 1978, 1984) and of Triassic sedimentary rocks (Ekren and Byers, 1984) on the Gumdrop Hills fault is comparable in magnitude to the estimates of offset on that fault based on the southern outcrop limit of the Oligocene tuffs.

Overlap of landslide deposits, considered to be related to strike-slip movement, by 24-m.y.-old tuffs and deposition of a 22–24-m.y.-old tuff against a preexisting fault surface, has been cited as evidence (Ekren and Byers, 1984) that movement on strike-slip faults in the Walker Lake section commenced about 24 m.y. ago. Offset of younger Tertiary rocks and of Quaternary alluvial deposits indicates that normal to strike-slip displacement has continued into the Holocene. Surface rupture occurred in the southern part of the Walker Lake section, a short distance east of the southern part of the Petrified Spring–Bettles Well fault, during the 7.2–7.3-magnitude Cedar Mountain earthquake of 1932 (Giannella and Callaghan, 1934; Molinari, 1984). Surface faulting was right-normal oblique slip and occurred along a system of generally north-northwest-trending faults.

Excelsior-Coaldale Section

The Excelsior-Coaldale section is a structurally important zone of east-west-trending faults cutting across the dominant north-northwest grain of the Walker Lane belt. The faults appear to have had major right slip in the Mesozoic and the same faults to have had local left slip in the Cenozoic. Understanding this section involves a larger problem of a change, or a disruption, in the trend of Precambrian, Paleozoic, and Mesozoic thicknesses, facies, and structural features. These trends (e.g., C in Fig. 25-4) are generally south-southwest in central Nevada, but in the Excelsior-Coaldale section are either offset right-laterally (Fig. 25-4) or, as interpreted by other geologists (Albers, 1967; Oldow, 1984), curve sharply westward. Farther west in eastern California, the trends of these features are generally south-southeast. This Z-shaped pattern has been interpreted in different ways. Several geologists (Ferguson and Muller, 1949; Geissman *et al.*, 1984; Oldow, 1984) consider the trends to indicate original depositional patterns rather than tectonically disrupted patterns. They have suggested that the pattern reflects the original shape of the Precambrian to Mesozoic continental margin. Kistler and Peterman (1978), in a unique idea, suggested that the pattern reflects the position of the continental margin but related it to two rifting events, one in the Precambrian and another from about 600 to 350 m.y. ago when a piece of continental crust was removed from the North American continent forming an indent (the Z-shaped pattern) in the continental margin. Other geologists have ascribed the pattern to major tectonic distortion that has disrupted originally more linear trends. Albers (1967) proposed that the distortion was produced by large-scale bending or drag (oroflexural folding) in an area of right-lateral shear. Wetterauer (1977) related the distortion (which he referred to as the "Mina deflection") to a crustal-scale fold produced by northeast-southwest compression. In an earlier article, I proposed (Stewart, 1985) that the distortion is related to right-lateral movement on two major east-trending fault zones, the Excelsior and Coaldale fault zones (Fig. 25-4). Right-lateral offset of 45–55 km on the Excelsior fault zone and 60–80 km on the Coaldale fault zone is indicated by apparent disruption (Stewart, 1985) of the trends of Precambrian through Mesozoic rocks and of the line of initial $^{87}Sr/^{86}Sr = 0.706$ (R. W. Kistler, written communication, 1986).

The right-lateral disruption of facies trends may extend into the Sierra Nevada. Here, Upper Triassic and Lower Jurassic rocks of the King sequence (Saleeby *et al.*, 1978) may be displaced from lithologically similar and generally age-equivalent sequences in western Nevada (Fig. 25-5). The apparent disruption of these rocks can be accounted for by right-slip on the Excelsior and Coaldale fault zones and on a system of northwest-trending faults in the eastern Sierra Nevada proposed by Kistler *et al.* (1980) to account for apparent disruption in the line of iniital $^{87}Sr/^{86}Sr = 0.706$ (Fig. 25-5).

Offset on the Excelsior and Coaldale fault zones appears to be late Mesozoic in age, because the Lower Jurassic to Cretaceous(?) Dunlap Formation appears to be offset and mid-Cretaceous and younger plutons appear to intrude the faults (Stewart, 1985). The faults locally cut Cenozoic rocks, but lateral displacements, if any, are small compared to Mesozoic displacements. Oligocene and Miocene tuff units may have been displaced 20–25 km left-laterally (Stewart, 1985, Fig. 3) along the Coaldale fault zones, indicating a reversal in the sense of movement from Mesozoic to Cenozoic. Speed and Cogbill (1979) also indicate a component of left-lateral offset on the Cenozoic Candelaria fault zone that lies between and subparallel to the Excelsior and Coaldale fault zones, and a component of left-lateral movement was also reported on surface rupture near the Excelsior fault zone during the 6.3-magnitude Excelsior Mountains earthquake in 1934 (Callaghan and Giannella, 1935).

The trend of the Excelsior and Coaldale fault zones is unique for a system of its size in the Walker Lane belt, and the fault zones truncate other faults in the Walker Lane belt. A major system of normal and right-lateral northwest-trending faults in the Walker Lake section terminates southward at the Excelsior fault zone, and the northwest-trending Owens Valley–White Mountain fault system in eastern California, and the Furnace Creek fault zone in eastern California and western Nevada, terminates northward at the Coaldale fault zone. This relation is difficult to relate to the proposed Mesozoic age of the Excelsior and Coaldale faults, and the late Cenozoic age of the northwest-trending faults that they terminate. Conceivably, the northwest-trending faults had a long history and were initiated in the Mesozoic prior to the Excelsior and Coaldale faults and were reactivated as major faults in the late Cenozoic. If the northwest-trending faults did initiate prior to the Excelsior and Coaldale faults, the curve of some of the northwest-trending faults as they approach the Excelsior and Coaldale zones can be explained as right-lateral drag on the Excelsior and Coaldale faults. In addition, this concept can explain the northeast trends of faults between the originally northwest-trending faults in a zone of right-lateral shear between the Excelsior and Coaldale fault zones. Alternatively, if displacement on the northwest-trending faults is entirely younger than the Excelsior and Coaldale fault zones,

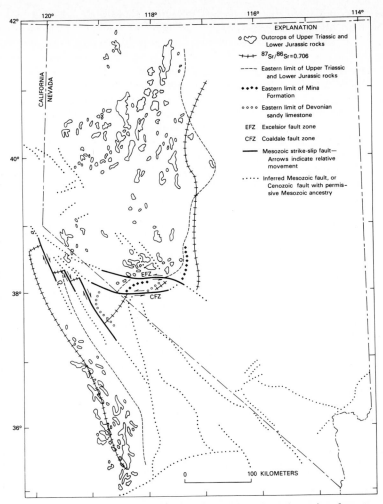

EXPLANATION

○ ◠ Outcrops of Upper Triassic and
  Lower Jurassic rocks

+++ $^{87}Sr/^{86}Sr = 0.706$

--- Eastern limit of Upper Triassic
  and Lower Jurassic rocks

•••• Eastern limit of Mina
  Formation

∘∘∘∘ Eastern limit of Devonian
  sandy limestone

EFZ Excelsior fault zone

CFZ Coaldale fault zone

— Mesozoic strike-slip fault—
  Arrows indicate relative
  movement

··· Inferred Mesozoic fault, or
  Cenozoic fault with permis-
  sive Mesozoic ancestry

FIG. 25-5.   Offset of Upper Triassic and Lower Jurassic rocks of western
Great Basin and the King sequence of the Sierra Nevada. Distribution of
Upper Triassic and Lower Jurassic rocks based mainly on Stewart (1980) and
Nokleberg (1983).

the northwest-trending faults may have abutted against and terminated at the cross structures of the Excelsior and Coaldale block.

Inyo-Mono Section

A large triangular section of the Walker Lane belt lying south of the Excelsior-Coaldale
section, east of the Sierra Nevada region, and west of the Furnace Creek fault zone is
referred to here as the Inyo-Mono section, following the usage of Carr (1984b). This
section, not commonly included in the Walker Lane belt, is included here because of the
presence of major north-northwest-trending faults similar to those in the Pyramid Lake
and Walker Lake sections of the belt. The Inyo-Mono section is characterized by high
relief, abrupt linear mountain fronts, and major northwest-trending right-lateral faults.

The Owens Valley–White Mountains fault zone lies along the western side of Inyo-Mono section. It extends from the Coaldale fault on the north to the Garlock fault on the south, a total distance of 300 km. It is a complex system of aligned or subparallel dip-slip and strike-slip faults that in the southern segment includes major normal faults along the east side of the Sierra Nevada, in the middle segment includes high-angle right-lateral faults within Owens Valley, and in the northern segment includes high-angle normal faults on the west side of the White Mountains. In the middle segment, right-lateral faults in Owens Valley have undergone recurrent Holocene offset with a ratio of strike slip to dip slip averaging 6 to 1 (Sarah Beanland, oral communication, 1986). Major surface rupture occurred along the Owens Valley fault system in the 1872 earthquake that has an estimated magnitude of about 8 and right-lateral displacements averaging 6 m (Sarah Beanland, oral communication, 1986; D. B. Slemmons, *quoted in* Hill, 1972; Oakeshott *et al.*, 1972). In contrast to the right-lateral Owens Valley fault, the close-by and subparallel Independence fault that bounds the east side of the Sierra Nevada has offset Pleistocene glacial deposits 30 m vertically, yet has a negligible component of stike slip (M. M. Clark, oral communication, 1986). Normal movement on the Independence fault and parts of the Owens Valley-White Mountains fault system produced the deep Owens Valley rift graben in the late Cenozoic (Sheridan, 1978) and the high eastern escarpment of the Sierra Nevada.

Investigations and speculations on the importance of the Owens Valley–White Mountains fault zone prior to the late Cenozoic have led to a variety of interpretations. J. G. Moore and Hopson (1961) noted that a major system of dikes (the Independence dike swarm) of Late Jurassic age (148 m.y., Chen and Moore, 1979) crosses Owens Valley and the Owens Valley fault zone without obvious offset, limiting strike-slip movement to a few kilometers at most. Schweickert (1981) indicates, however, that about 19 km of left-lateral offset of the dike swarm is possible across Owens Valley. The conclusion that little lateral offset has occurred is supported by recognition of a 170-m.y.-old (R. W. Kistler, oral communication, 1985) pluton that is also apparently not offset across Owens Valley (Ross, 1962). Pre-Late Jurassic offset is possible and has been postulated by Nokleberg (1983) on the basis of different facies of Paleozoic rocks on either side of the valley, although J. N. Moore and Foster (1980) using some of the same data on the Paleozoic rocks indicated that no major strike-slip offset is possible across Owens Valley.

The north-trending Kern Canyon fault which is subparallel to the Owens Valley–White Mountains fault zone and lies within the Sierra Nevada west of the Walker Lane belt is a right-lateral fault with as much as 13 km of offset (J. G. Moore and du Bray, 1978). This fault cuts plutonic rocks as young as 80 m.y. old (J. G. Moore and du Bray, 1978), although Schweickert (1981) believes that most movement occurred prior to 80 m.y. ago.

The north- to northwest-trending Panamint Valley fault in the central part of the Owens Valley–Death Valley section of the Walker Lane belt probably includes major segments of normal, as well as segments of right-lateral, offset (Jennings, 1975). Holocene right-slip of 20 m and somewhat less dip-slip have been reported (R. S. U. Smith, 1979).

The Death Valley fault is a right-lateral fault with less than 8 km of offset, based on Precambrian stratigraphic trends across southern Death Valley (Davis, 1977; Wright and Troxel, 1967). However, Brady (1984) estimates possibly 20 km of right-lateral offset since late Tertiary time, based on provenance studies of gravels and conglomerates.

The Furnace Creek fault zone extends for about 250 km with a fairly straight north-northwest trend. Its southern two-thirds lies within a valley separating high mountain ranges. North of this segment it cuts through bedrock for a short distance, and its northern third bounds the east side of the high and rugged White Mountains. The fault cuts Quaternary alluvium along much of its length, and offset of drainages and the presence of left-stepping faults demonstrate Quaternary right-lateral displacement in some segments, although dip-slip components occur in many segments and constitute the main form of Quaternary displacement in some areas (Brogan, 1979). Pre-Quaternary right-lateral offset has been estimated at 40-100 km on the basis of apparent offsets of stratigraphic trends of upper Precambrian and Paleozoic rocks (Fig. 25-4) and the distribution of granitoids (Cooper, *et al.*, 1982; McKee, 1968; Miller and Walch, 1977; J. N. Moore, 1976a, b; Pelton, 1966; Poole and Sandberg, 1977; Poole *et al.*, 1967; 1977; Stewart, 1967; Stewart *et al.*, 1968). No consensus has developed on the structural history of the Death Valley region and the Furance Creek fault zone (see the discussions by Hamilton, 1986; Stewart *et al.*, 1968; 1970; Wernicke, 1981; Wright and Troxel, 1967; 1970), although most workers agree that the area has undergone considerable extension during the late Cenozoic. The southern part of the Furnace Creek fault may bound a region of greater extension on the southwest relative to the northeast (Stewart, 1983b; Wright and Troxel, 1970).

The Pahrump fault zone (Liggett and Childs, 1973), which corresponds in the northern part to the Stewart Valley fault zone (Burchfiel *et al.*, 1983; Stewart *et al.*, 1968), extends for about 130 km near the California-Nevada state line along the southeastern side of the Inyo-Mono block. The zone cuts bedrock and locally displaces Tertiary and Quaternary basin sediments. Right-lateral displacement of 16-19 km is indicated by offset of Precambrian and Paleozoic rocks and by consistent indication of drag on either side of the mapped fault zone (R. L. Christiansen, *in* Stewart *et al.*, 1968).

The left-lateral Garlock fault is the southern boundary of the Owens Valley–Death Valley section of the Walker Lane belt. Left-lateral offset of between 48 and 68 km has been estimated on the basis of the apparent offset of the Independence dike swarm (G. I. Smith, 1962), of Paleozoic rocks (G. I. Smith and Ketner, 1970), and of a Mesozoic thrust fault (Davis and Burchfiel, 1973). The Garlock fault has been interpreted as an intracontinental transform structure that accommodates greater Cenozoic extension to the north than to the south (Davis and Burchfiel, 1973).

Goldfield Section

The Goldfield section is unusual because of the lack of major throughgoing northwest-trending strike-slip faults and the scarcity of major basin-range faults (Figs. 25-2 and 25-3). It is characterized by mountain ranges that have irregular, in places arcuate, shapes and that have been uplifted in places by folding rather than faulting (Albers and Stewart, 1972, p. 46). Albers (1967) proposed that the arcuate shape of some of the mountain ranges reflect trends of tilted beds, fold axes, and, in places, elongate plutons. He applied the name "oroflexes" to such mountain ranges and related the arcuate structural trends to tectonic bending of the crust. The evidence for and against such structural bending is discussed under the section on characteristic structures of the Walker Lane belt.

Spotted Range–Mine Mountain Section

The Spotted Range–Mine Mountain section (Carr, 1984b) consists of a zone about 30-60 km wide containing northeast-trending left-lateral faults. The faults cut bedrock in places but characteristically are inferred to lie beneath alluvial cover in major northeast-trending valleys. The faults locally cut Quaternary alluvial deposits. Displacements are 1-2 km or less on individual faults (Carr, 1984b).

Spring Mountains Section

The Spring Mountains section consists of a relatively intact block composed of the large, regionally anomalous, mountain mass of the Spring Mountains. It is bounded on the southwest by the Pahrump fault zone (described under the Mono-Inyo section), on the northwest by the Spotted Range–Mine Mountain block (described above), on the northeast by the Las Vegas Valley shear zone (described here), and on the southeast by the Lake Mead fault zone (described below).

The Las Vegas Valley shear zone is a complex structure with 40–67 km of right-lateral offset based on trends of Paleozoic strata and Mesozoic structures (Longwell, 1960, 1974; Stewart et al., 1968; Wernicke et al., 1982, 1984). The faults are buried by alluvium in Las Vegas Valley, but drag is evident from the curving of structural features as they approach the fault zone (Longwell, 1960). Fleck (1970) and R. E. Anderson et al. (1972) indicate that much of the displacement may have occurred between 15 and 11 m.y. ago, because 15-m.y.-old strata adjacent to the Las Vegas Valley shear zone are structurally disrupted in the same style as Paleozoic strata, whereas 11-m.y.-old volcanic rocks are relatively undeformed. Large-scale extension, as much as 100%, has occurred during the Cenozoic on the north side of the fault zone, and the offset on the shear zone has been related to the relatively large extension north of the shear zone relative to that to the south (Guth, 1981; Wernicke et al., 1982, 1983, 1984). On the other hand, Royse (1983) has argued that the shear zone was developed as a lateral ramp during Cretaceous thrusting.

Lake Mead Section

The Lake Mead section is a northeast-trending zone as much as 30 km wide containing the left-lateral Bitter Spring Valley, Hamblin Bay, Cabin Canyon, Bitter Ridge, Lime Ridge, and Gold Butte faults (R. E. Anderson, 1973; Bohannon, 1979, 1984). Offset across the entire zone is estimated to be 65 km, based on the apparent displacement of a distinctive facies of lower and middle Miocene rocks, the apparent offset of the line defined by the Tertiary unconformable truncation of Mesozoic formations, and the apparent offset of a source terrane for distinctive rapakivi granite clasts in Tertiary rocks (R. E. Anderson, 1973; Bohannon, 1979, 1984). On individual faults, R. E. Anderson (1973) has described clear evidence of 20 km of offset on the Hambin Bay fault based on offset of the 11–13-m.y.-old Hamblin-Cleopatra volcano and its radial dike swarms.

Major offset on the Lake Mead fault zone is clearly late Tertiary in age based on the offset of the late Tertiary Hamblin-Cleopatra volcano and offset of distinctive facies of lower and middle Miocene rocks. Bohannon (1984) has outlined a history of strike-slip movement on the Lake Mead fault zone occurring between 17 and perhaps 10 m.y. ago that is reflected in the facies patterns of Miocene sedimentary rocks.

CHARACTERISTIC STRUCTURES OF THE WALKER LANE BELT

The Walker Lane belt is characterized by a variety of structural features, some of which appear to distinguish the belt from adjacent regions and other of which are less distinctive. The characteristic features discussed below are (1) strike-slip faults, (2) regional structural blocks, (3) basin-range blocks, (4) oroflexural folds, and (5) areas of large-scale extension, detachment faults, and metamorphic core complexes.

1. Strike-slip faults The most distinctive structures of the Walker Lane belt are strike-slip faults that include both northwest-trending right-lateral systems and northeast-trending left-lateral systems. These faults are characterized by straight map traces, vertical or near-vertical attitude of the fault planes, low-angle or horizontal-slip striae, and little, if any, topographic relief across the fault zones. The strike-slip faults do not extend unbroken along the entire length of the belt and thus are not analogous, as once proposed, to the major throughgoing right-lateral faults of the San Andreas system of California. The strike-slip faults of the Walker Lane belt are restricted to individual structural blocks (the Pyramid Lake, Carson, and other such blocks described above) or form the boundaries between such blocks, but they rarely, if ever, extend from one block to another.

2. Regional structural blocks The Walker Lane belt is characterized by distinct regions of contrasting structural and faulting style (Fig. 25-3). Such regional and subregional blocks apparently characterize much of the western United States inland of the San Andreas transform system (Fig. 25-6). This blocklike character is in part responsible for the diverse physiography that originally led to the recognition of the Walker Lane belt as distinctively different from adjacent regions.

3. Basin-range faults Parts of the Walker Lane belt contain mountain blocks bounded on one or both sides by normal faults similar to basin-range faults found elsewhere in the Basin and Range Province. These faults are characterized by irregular or wavy map traces, moderate to steep dip of the fault planes, striae with high rake angles, and positions bounding major mountain ranges. Basin-range faults are particularly well defined in

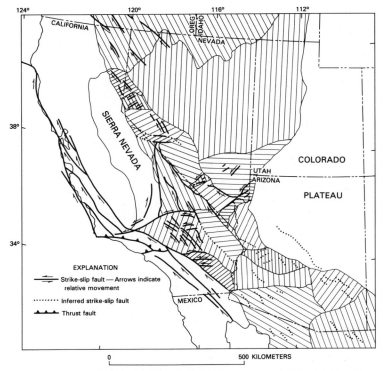

FIG. 25-6. Distribution of major Cenozoic faults in western United States and regional structural blocks (patterned) in the Basin and Range province.

the Walker Lane section, where the Pine Nut, Yerington, and Wassuk faults bound major blocklike ranges (Dohrenwend, 1982).

4. Oroflexural folds Albers (1967) suggested that the Walker Lane belt is characterized by gigantic dextral drag in a right-lateral shear system. The dextral drag, in his interpretation, is defined by a sigmoidal pattern of facies boundaries and structural trend lines, and by arcuate mountain ranges that result from tectonic bending of the crust. He proposed the term "oroflex" to describe such a tectonically bent mountain range.

The best-defined drag features (oroflexural folds) occur along the Las Vegas Valley shear zone, where structures hook westward on the north side and eastward on the south side of this right-lateral fault system (Albers, 1967; Longwell, 1960). Curving structural trends near the north end of the Las Vegas shear zone have also been related to oroflexural folding (Albers, 1967; Carr, 1984b).

Albers (1967) also proposed major oroflexural folds in what are here called the Walker Lake, Excelsior-Coaldale, and Goldfield sections of the Walker Lane belt. These folds were based on apparent curving patterns of facies boundaries and structural-trend lines, as well as arcuate mountain ranges. Other interpretations relate the arcuate pattern to original depositional trends along an irregular continental margin originally formed by Precambrian rifting (Oldow, 1984), to patterns formed in two stages of rifting (Kistler and Peterman, 1978), to crustal-scale folding produced by northeast-southwest compression (Wetterauer, 1977), or to displacements on the Excelsior and Coaldale fault zones (Stewart, 1985, and this report).

Paleomagnetic studies in the north-central part of the Walker Lake section (Geissman *et al.*, 1982, 1984) indicate no structural rotation of Middle Jurassic, and younger plutonic rocks and Oligocene and Pliocene volcanic rocks. Similar studies in the southeastern part of the section indicate no structural rotation of mid-Cretaceous plutonic rocks. Thus, either no oroflexural folds are present in the Walker Lake section or such folds are pre-Middle Jurassic in age. In addition, the generally uniform northwest trend of Jurassic and Cretaceous plutons in the Sierra Nevada and western Great Basin (Fig. 25-7) indicates an apparent lack of oroflexural folding. Any oroflexural folding apparently is older than these plutonic rocks.

Albers (1967) based his interpretation of oroflexural folds in Esmeralda County partly on the changing trends of structures in pre-Tertiary rocks from northeast-trending in the eastern part of the Goldfield section to north-northwest-trending in the western part. Some of the mountain ranges in the Goldfield section are arcuate, further supporting Albers' (1967) interpretation. Alternatively, the northeast trends in the eastern part of the Goldfield section may reflect Paleozoic structural belts, which are generally northeast-trending in Nevada (Stewart, 1980), whereas the north-northwest trends may reflect superimposed Mesozoic structural trends which are characteristically north-northwest in the granitic terrane and wallrocks of the Sierra Nevada located directly west of the Walker Lane belt. In this view, the curving mountains are the result of intersecting structural trends rather than oroflexural folding.

A further complication in determining the validity of the oroflexes in the Goldfield section is the presence of major late Cenozoic extensional structures. Some anticlinal features, such as Mineral Ridge, that were considered by Albers (1967) to be part of the generally curving pattern of structural features in the Goldfield section are clearly related to late Cenozoic structural uplift and detachment faulting (Kirsch, 1968, 1971) rather than to Mesozoic deformation that is supposed to define the oroflexural folds in the Goldfield section.

5. Large-scale extension, detachment faults, and metamorphic core complexes Local and regional areas (Fig. 25-8) of large-scale extension occur within the Walker Lane belt, as

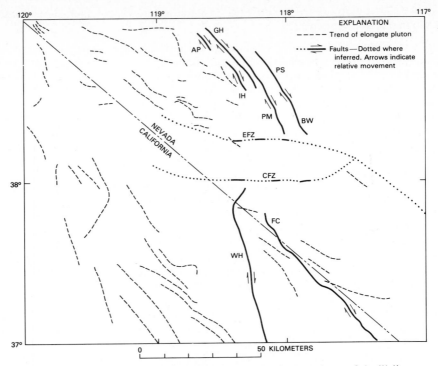

FIG. 25-7.   Trends of elongate Mesozoic plutons in central part of the Walker Lane belt and adjacent Sierra Nevada. Based mainly on John (1983), Stern *et al*. (1981), and Albers and Stewart (1972). Symbols for faults same as Fig. 25-3.

well as in many other parts of the Basin and Range Province. Extension may be greater in the Walker Lane belt, however, than in the Basin and Range Province as a whole, judging from the concentration of mapped low-angle extensional faults and the greater-than-normal dips of Tertiary rocks (Fig. 25-9). Such relatively high dips of Tertiary rocks have been interpreted to develop by rotation of blocks above detachment surfaces in areas of large-scale extension (Morton and Black, 1975; Stewart, 1979).

Large-scale extension in the Walker Lane belt has been recognized in the Yerington area (Proffett, 1977), Paradise Range-Royston Hills area (D. A. John and D. H. White-bread, oral communication, 1985), the Mineral Ridge-Weepah region (J. H. Stewart, geologic mapping, 1983–1985), the Bullfrog Hills area (M. D. Carr, written communication, 1985), the Death Valley area (Hamilton, 1986; Stewart, 1983b; Wright and Troxel, 1973); the Sheep and Desert Ranges (Guth, 1981), and the El Dorado Mountains (R. E. Anderson, 1971). Regional detachment faults are recognized or proposed in the Mineral Ridge-Weepah area (J. H. Stewart, geologic mapping, 1983–1985), the Bullfrog Hills-Funeral Mountains area (M. D. Carr, written communication, 1985; Labotka, 1980), and in the Death Valley area (M. D. Carr, written communication, 1985; Hamilton, 1986; Stewart, 1983b).

Extension, tectonic denudation, and uplift has exposed midcrust amphibolite-facies metamorphic rocks (Hamilton, 1986; Kirsch, 1968, 1971; Labotka, 1980) in the Mineral Ridge-Weepah area, Trappman Hills, Bullfrog Hills, Funeral Mountains, and Panamint Mountains (Fig. 25-8). In places, these metamorphic rocks occur below a

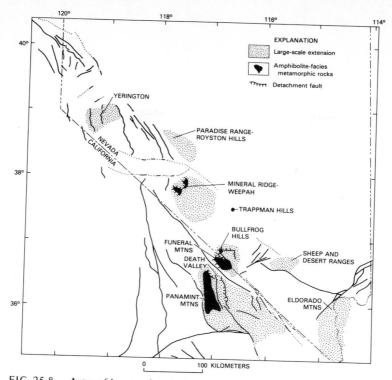

FIG. 25-8. Areas of large-scale extension, detachment faults, and amphibolite-facies metamorphic rocks in Walker Lane belt. High-angle Cenozoic faults same as on Fig. 25-3. Based on many sources, including Proffett (1977), Kirsch (1968; 1971), Labotka (1980), Wright and Troxel (1973), Stewart (1983a, b), Hamilton (1986), Guth (1981), R. E. Anderson (1971), McKee (1983), Armstrong (1982), and unpublished information of M. D. Carr, D. A. John, D. H. Whitebread, and J. H. Stewart.

carapace of relatively unmetamorphosed rocks in domal uplifts referred to as meta-morphic core complexes (Coney, 1980). The metamorphic rocks reach the garnet and, rarely, the sillimanite grade (Kirsch, 1968, 1971; Labotka, 1980). Three of the areas of metamorphic rocks yield late Miocene, reset K-Ar ages (McKee, 1983).

DEVELOPMENT OF MESOZOIC STRUCTURES

The time of initial development of the Walker Lane belt is a matter of conjecture. A Cenozoic origin is commonly indicated, as Giannella and Callaghan (1934) stated when they first recognized the Walker Lane and suggested that it was a structure similar to the late Cenozoic San Andreas fault. Subsequent workers have elaborated on this concept, emphasizing the similarity in the trends, directions, and timing of offset of the two systems. This concept was stressed by Atwater (1970), who indicated that late Cenozoic faults in the western United States that parallel the San Andreas system may be part of a wide, soft boundary between two rigid moving crustal plates.

Other geologists have indicated a Mesozoic initiation for the Walker Lane belt. Locke *et al.*, (1940), for example, in their definition of the Walker Lane, suggested a

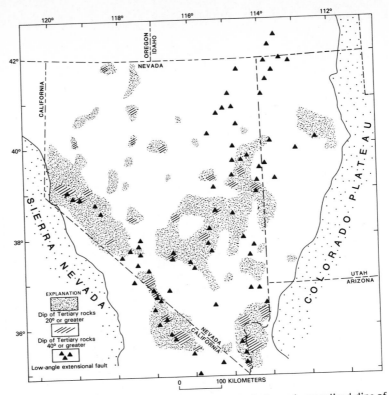

FIG. 25-9. Distribution of Cenozoic low-angle faults and generalized dips of Tertiary rocks in Great Basin. After Stewart (1983a).

close relationship between Mesozoic granitic structures and the Walker Lane. Albers (1967) also indicated a Mesozoic initiation of strike-slip faulting and oreflexural folding in the belt, perhaps in the Jurassic, with subsequent faulting lasting until the late Cenozoic. Royse (1983) has suggested that the Las Vegas Valley shear zone is a ramp structure developed during Cretaceous thrusting. Speed (1978) proposed that the Walker Lane belt is, at least in part, coextensive with an inferred belt of Mesozoic right-lateral distortion and has suggested that the Cenozoic deformation in the Walker Lane belt utilized preexisting faults and zones of weakness in this older belt of deformation (R. C. Speed, oral communication, 1974). In an earlier article (Stewart, 1985), I indicated that the Excelsior and Coaldale fault zones are Mesozoic in age based on intrusive relations of mid-Cretaceous granitic rocks and the lack of major lateral offset of Tertiary rocks. In addition, Kistler et al. (1980) indicated that the apparent right-lateral offset of the $^{87}Sr/^{86}Sr$ 0.706 line in the eastern Sierra Nevada is Jurassic in age. The evidence from these two areas (the Excelsior and Coaldale fault zones, and the eastern Sierra Nevada) indicate a Mesozoic initiation of strike-slip deformation in parts of the Walker Lane belt, although the extent of this Mesozoic deformation, or even whether it generally follows the trend of the Cenozoic Walker Lane belt, is not clear.

The Walker Lane belt may have originated as a system of northwest-trending strike-slip faults within or inboard of a Late Triassic and Jurassic arc system (Fig. 25-10) in a tectonic setting, similar to that in Sumatra, where strike-slip faults are intra-arc and interpreted to be related to oblique subduction (Beck, 1983; Fitch, 1972). Such a possibility

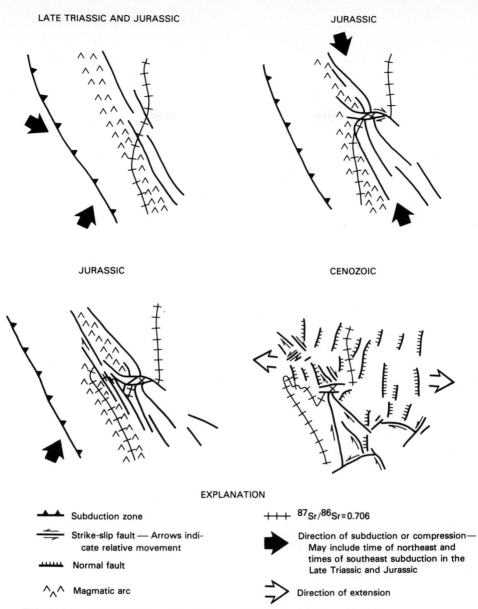

LATE TRIASSIC AND JURASSIC

JURASSIC

JURASSIC

CENOZOIC

EXPLANATION

▲▲ Subduction zone	+++ $^{87}Sr/^{86}Sr = 0.706$
⇌ Strike-slip fault — Arrows indicate relative movement	▶ Direction of subduction or compression— May include time of northeast and times of southeast subduction in the Late Triassic and Jurassic
⊥⊥⊥⊥ Normal fault	
∧∧∧ Magmatic arc	⇒ Direction of extension

FIG. 25-10.   Model of Mesozoic and Cenozoic tectonic development of the Walker Lane belt.

is suggested by the parallelism of the Walker Lane belt and outcrops of Late Triassic and Jurassic granitoid and volcanic rocks (Fig. 25-11) that are interpreted to be part of a magmatic arc.

The continuation of the magmatic arc into Arizona (Fig. 25-11) may indicate that the strike-slip belt also extended into Arizona. Kluth (1983), for example, has indicated a relationship between Mesozoic volcanism and strike-slip faulting in southern Arizona

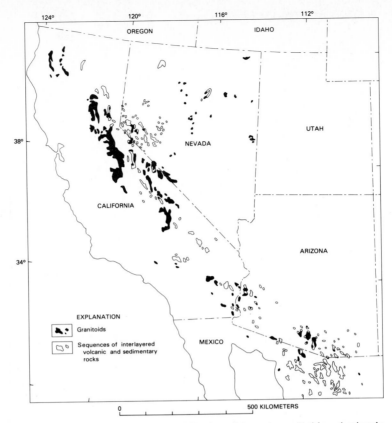

FIG. 25-11. Distribution of Late Triassic and Jurassic granitoids and volcanic rocks, western United States and northern Mexico. Based on Evernden and Kistler (1970), Jennings (1977), Stewart (1980), and Stewart *et al.* (1986).

that is analogous, in part, to the concept presented here for the Walker Lane belt. Continuation of the Walker Lane belt of northwest-trending strike-slip faults into Arizona also has been suggested by Warner (1978, Fig. 5). Finally, Mesozoic strike-slip faults in southern Arizona have been described by Drewes (1981), as well as by Titley (1976), who suggests a pattern of northwest-trending Mesozoic linear tectonic features related to trancurrent faulting.

The Mesozoic history of the Walker Lane belt undoubtedly was complex. Some authors have stressed right-lateral Mesozoic displacements on northwest-trending structures in the Walker Lane belt (Albers, 1967; Speed, 1978), whereas other workers (T. H. Anderson and Schmidt, 1983; T. H. Anderson and Silver, 1979; Oldow, 1983; Saleeby, 1981; Silver and Anderson, 1974) have proposed left-lateral Mesozoic displacements either within the Walker Lane belt or along the western Mesozoic North America continental margin which they interpreted as lying a short distance to the west. Following the initial strike-slip movement, whether right- or left-lateral, north-south compression appears to have been necessary to produce the east-west faults of the Excelsior-Coaldale fault zones. These fault zones may represent a "kink" in the shear system (Fig. 25-10).

DEVELOPMENT OF CENOZOIC STRUCTURES

Although a Mesozoic initiation of the Walker Lane belt seems likely, most, if not all, of the Walker Lane faults have been active in the Cenozoic. A Cenozoic age is clear because major faults in the Walker Lane cut Cenozoic rocks and locally, as in the Walker Lake and Lake Mead sections, displacements are measured on apparent offset of Cenozoic rocks. Some Cenozoic faults may be reactivated Mesozoic structures, but others apparently were initiated by Cenozoic movements.

Cenozoic structural movements in the Walker Lane belt are diverse and involve lateral, normal, or oblique slip on high-angle faults as well as local areas of large-scale extension on low-angle normal faults. Strike-slip faults, characterized mainly by their straight map traces and by offset geologic features, are commonly parallel or subparallel to dip-slip faults, characterized by their irregular or wavy map traces and by bounding major mountain blocks. The side-by-side occurrence of these two styles of faults is particularly evident in the Walker Lake section, where the oblique-normal-slip Wassuk fault is subparallel to strike-slip faults to the east (see the description under Walker Lake section), as well as in the Owens Valley area, where Quaternary displacement on the Owens Valley fault is largely strike-slip and Quaternary displacement on the close-by and subparallel Independence fault is largely normal dip-slip (see the description under the Inyo-Mono section). Displacement on individual faults also appears to vary significantly along the trend of the fault. Movement on the Benton Spring–Pilot Mountain fault in the Walker Lake section and the Owens Valley–White Mountains and Furnace Creek faults in the Inyo-Mono section, for example, appears in some segments to be mainly strike-slip, whereas in other segments of the same fault it appears to be mainly dip- or oblique-slip. Finally, the segmentation of the Walker Lane belt into distinct regional blocks each of which have undergone deformation independent of adjacent blocks indicates that any shear that affects the entire Walker Lane belt is accommodated by internal deformation of individual blocks or by displacement between blocks, but not by throughgoing strike-slip faults extending from one block into the next.

A model (Fig. 25-12) is proposed here to explain these diverse movements in the Walker Lane belt based on concepts proposed by Donath (1962), Wright (1976), Angelier et al. (1985), R. E. Anderson and Bernhard (1984), and R. E. Anderson (1984). The model relates deformation to repeated changes in the stress field from an extension mode to a strike-slip mode. In detail, the deformation is related to a stress field in which the extension direction (least principal stress, S_3) is west-northwest in a direction similar to the presumed S_3 direction for north-northeast-trending basin-range faults in areas of typical basin-range faulting east of the Walker Lane belt. This direction of S_3 is the same as the present-day S_3 in the Walker Lane belt based on focal mechanisms of earthquakes (A. M. Rogers et al., 1983; R. B. Smith and Lindh, 1978; Vetter and Ryall, 1983; Zoback and Zoback, 1980). Within a stress field with S_3 oriented west-northwest, changes from the extensional mode to the strike-slip mode have been explained in two different ways. In the first explanation (Angelier et al., 1985; Wright, 1976), the maximum principal stress (S_1) and intermediate principal stress (S_2) are close in magnitude relative to S_3. Relatively minor changes in the stress field will cause permutations in S_1/S_2 and result in alternating and interfering patterns of strike slip, when S_1 is horizontal, to dip slip, when S_1 is vertical. In the second explanation (Zoback and Beanland, 1986), changes from extensional mode to strike-slip mode are caused by large temporal fluctuations in the relative magnitude of the maximum horizontal stress. Normal dip slip occurs when the maximum horizontal stress is close in magnitude to the minimum horizontal stress, and

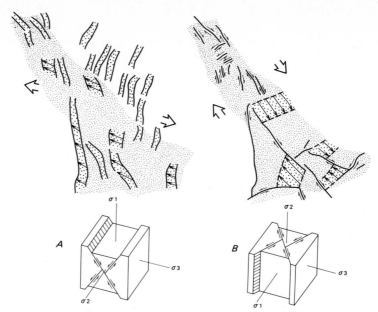

FIG. 25-12. Model of late Cenozoic tectonics of the Walker Lane belt.
(A) extensional mode. (B) Strike-slip mode. Fine stipple is Walker Lane belt.
Coarse stipple indicates area of major extension. Arrows in coarsely stippled
areas indicate direction of extension. Single-headed arrows indicate relative
displacement on strike-slip faults. Large arrows in (A) indicate direction of
extension. Large arrows in (B) indicate direction of shear.

strike slip when the maximum horizontal stress and vertical stress are approximately equal.

The model is based in part on studies of paleostress indicators in the Lake Mead area and elsewhere in or near the Great Basin (R. E. Anderson, 1984; R. E. Anderson and Bernhard, 1984; Angelier *et al.*, 1985) where both strike-slip and dip-slip modes of deformation are apparently related to the same tectonic regime and to permutations in S_1/S_2. The model (Fig. 25-12) does not consider changes in the direction of S_3 with time, although a clockwise rotation of the stress axis from southwest to west northwest has been proposed (R. E. Anderson and Ekren, 1977; Angelier *et al.*, 1985; Zoback and Thompson, 1978; Zoback *et al.*, 1981). Such a change in the stress field seems likely and would undoubtedly create even more complex structures than those accounted for by the relatively simple model presented here.

A model of repeated changes from an extension mode to a strike-slip mode accounts for the presence in the same region of structures related to extension, such as basin-range blocks and major grabens of Owens and Death Valley, and of structures related to shear, such as major strike-slip faults. In this scheme, one set of structures may have been active during the extension mode and another set during the strike-slip mode. In addition, some faults, or parts of some faults, that originated as either dip-slip or strike-slip may have been reactivated as strike-slip or dip-slip, respectively, after a permutation in S_1/S_2. Such a change in faulting style may account for the lateral variability along individual faults from segments of dip-slip, formed during a time of extension, to segments of strike-slip, formed during a time of shear. Furthermore, the Pine Nut, Yerington, and Wassuk faults that have trends similar to closely right-lateral faults and are predominantly dip- or oblique-slip may have originated as strike-slip faults and changed to dip- or oblique-slip normal faults during extension.

In the model presented here, some areas of large-scale extension (Fig. 25-8) formed as a consequence of extensional stress (Fig. 25-12A), and other areas formed as a consequence of shear stress (Fig. 25-12B). In the latter interpretation, shear stress produced the 40–100 km of right-lateral displacement on the northwest-trending Furnace Creek fault zone, as well as the associated large-scale northwest extension on detachment faults (Amargosa and related systems) in the Death Valley area and the corresponding southwest extension of a similar magnitude on detachment faults (Mineral Ridge–Weepah and related systems) in the Goldfield section (Fig. 25-12B). Relating extension to shear differs from other models in which the strike-slip faults are viewed as transform faults that accommodate large extension on one side of a fault relative to the other side (Davis and Burchfiel, 1973; Guth, 1981; Liggett and Childs, 1977; Wernicke *et al.*, 1984). On transform faults, by analogy to oceanic systems, the axis of least principal stress (S_3) is parallel to the extension direction and thus parallel to strike-slip faults. Such a transform fault model is difficult to relate to the Walker Lane belt, where northwest-trending right-lateral faults and northeast-trending left-lateral faults require, in the transformal fault model, that S_3 be parallel at times to the northwest right-lateral faults and at other times parallel to the northeast left-lateral faults. Such a radical shift in the direction of S_3 seems unlikely. A model relating extension to shear is supported by the presence of large-scale drag features on the Las Vegas Valley shear zone that indicates a somewhat compressional boundary along the fault, rather than a passive boundary as might be expected if the fault surface is parallel to S_3. Ron, *et al.* (1986), furthermore, have proposed on the basis of paleomagnetic studies that rocks adjacent to the Hamblin Bay fault on the Lake Mead fault system have been rotated about a vertical axis, compatible with deformation in a strike-slip system, rather than about a horizontal axis as would be expected by listric rotation of extensional blocks. The cause of large-scale extension related to shear is somewhat of a mystery. One explanation is that the gridlock blocks of the Walker Lane belt have caused shear strain to be absorbed both by strike-slip faulting and by large-scale extension.

ACKNOWLEDGMENTS

I wish to thank Sarah Beanland, R. G. Bohannon, and M. L. Zoback for their thoughtful reviews that greatly improved the quality of the manuscript.

REFERENCES

Albers, J. P., 1967, Belt of sigmoidal bending and right-lateral faulting in the western Great Basin: *Geol. Soc. America Bull.*, v. 78, p. 143–156.

——, and Stewart, J. H., 1972, Geology and mineral deposits of Esmeralda County, Nevada: *Nev. Bur. Mines Geology Bull., v. 78*, 80 p.

Anderson, L. W., and Hawkins, F. F., 1984, Recurrent Holocene strike-slip faulting, Pyramid Lake fault zone, western Nevada: *Geology*, v. 12, p. 681–684.

Anderson, R. E., 1971, Thin skin distension in Tertiary rocks of southeastern Nevada: *Geol. Soc. America Bull.*, v. 82, no. 1, p. 43–58.

——, 1973, Large-magnitude Late Tertiary strike-slip faulting north of Lake Mead, Nevada: *U.S. Geol. Survey Prof. Paper 794*, 18 p.

——, 1984, Strike-slip faults associated with extension in and adjacent to the Great Basin: *Geol. Soc. America Abstr. with Programs*, v. 16, no. 6, p. 429.

——, and Barnhard, T. P., 1984, Extensional and compressional paleostresses and their relationship to paleoseismiscity and seismicity, central Sevier Valley, Utah: *U.S. Geol. Survey Open File Rpt. 84-763*, p. 515–546.

____, and Ekren, E. B., 1977, Comment on "Late Cenozoic fault patterns and stress fields in the Great Basin and westward displacement of the Sierra Nevada block:" *Geology*, v. 5, no. 7, p. 388–389.

____, Longwell, C. R., Armstrong, R. L., and Marvin, R. F., 1972, Significance of K-Ar ages of Tertiary rocks from the Lake Mead region, Nevada-Arizona: *Geol. Soc. America Bull.*, v. 83, p. 273–287.

Anderson, T. H., and Schmidt, V. A., 1983, The evolution of Middle America and the Gulf of Mexico — Caribbean Sea region during Mesozoic time: *Geol. Soc. America Bull.*, v. 94, p. 941–966.

____, and Silver, L. T., 1979, The role of the Mojave-Sonora megashear in the tectonic evolution of northern Sonora, *in* Anderson, T. H., and Roldan-Quintana, Jaime, eds., *Geology of Northern Sonora:* Geol. Soc. America Ann. Meet., Guidebook, p. 59–68.

Angelier, Jacques, Colleta, Bernard, and Anderson, R. E., 1985, Neogene paleostress changes in the Basin and Range — A case study at Hoover Dam, Nevada-Arizona: *Geol. Soc. America Bull.*, v. 96, p. 347–361.

Armstrong, R. L., 1982, Cordilleran metamorphic core complexes: *Ann. Rev. Earth Planet. Sci.*, v. 10, p. 129–154.

Atwater, Tanya, 1970, Implications of plate tectonics for the Cenozoic evolution of North America: *Geol. Soc. America Bull.*, v. 81, p. 3513–3536.

Batson, R. M., Edwards, Kathleen, Eliason, E. M., 1975, Computer-generated shaded-relief images: *U.S. Geol. Survey J. Res.*, v. 3, no. 4, p. 401–408.

Beck, M. L., 1983, On the mechanism of tectonic transport in zones of oblique subduction: *Tectonophysics*, v. 93, p. 1–11.

Bell, E. J., 1984, Overview of Late Cenozoic tectonics of the Walker Lane, *in* Lintz, Joseph, Jr., *Western Geological Excursions, Vol. 4:* Geol. Soc. America Ann. Meet., Mackay Sch. Mines, Reno, Nev., p. 407–413.

____, and Slemmons, D. B., 1979, Recent crustal movement in the central Sierra Nevada-Walker Lane region of California-Nevada — Pt II. The Pyramid Lake right-slip fault zone segment of the Walker Lane: *Tectonophysics*, v. 52, no. 1-4, p. 571–583.

Bell, J. W., 1981, Quaternary fault map of the Reno 1 by 2 degree quadrangle: *U.S. Geol. Survey Open File Rpt. 81-982*, scale, 1:250,000.

____, 1984a, Quaternary fault map of Nevada, Reno sheet: *Nev. Bur. Mines Geology Map 79*, scale, 1:250,000.

____, 1984b, Holocene faulting in western Nevada and recurrence of large-magnitude earthquakes, *in* Lintz, Joseph, Jr., *Western Geological Excursions, Vol. 4:* Geol. Soc. America Ann. Meet., Mackay Sch. Mines, Reno, Nev., p. 388–402.

Billingsley, Paul, and Locke, Augustus, 1939, *Structure of ore districts in the continental framework* [pamphlet]: New York. Amer. Inst. Mining Metall. Engineers. 51 p.

____, and Locke, Augustus, 1941, Structure of ore districts in the continental framework: *Amer. Inst. Mining Metall. Engineers Trans.*, v. 144, p. 9–59.

Bohannon, R. G., 1979, Strike-slip faults of the Lake Mead region of southern Nevada, *in* Armentrout, J. M., Cole, M. R., and TerBest, Harry, Jr., eds., *Cenozoic Paleogeography of the Western United States:* Pacific Section, Soc. Econ. Paleontologists Mineralogists, Pacific Coast Paleogeography Symp. 3, p. 129–139.

____, 1984, Nonmarine sedimentary rocks of Tertiary age in the Lake Mead region, southeastern Nevada and northwestern Arizona: *U.S. Geol. Survey Prof. Paper 1259*, 72 p.

Bonham, H. F., 1969, Geology and mineral deposits of Washoe and Storey Counties, Nevada: *Nev. Bur. Mines Bull. 70*, 140 p.

Brady, R. H., III, 1984, Neogene stratigraphy of the Avawaltz Mountains between the Garlock and Death Valley fault zones, southern Death Valley, California — Implications as to late Cenozoic tectonism: *Sediment. Geology*, v. 38, p. 127–157.

Brogan, G. E., 1979, Late Quaternary faulting along the Death Valley-Furnace Creek fault system, California and Nevada: *U.S. Geol. Survey Contract Rpt.*, Contract No. 14-08-0001-17801, scale 1:24,000.

Burchfiel, B. C., Hamill, G. C., and Wilhelms, D. E., 1983, Structural geology of the Montgomery Mountains and the northern half of the Nopah and Resting Spring Ranges, Nevada and California: *Geol. Soc. America Bull.*, v. 94, no. 11, p. 1359–1376.

Callaghan, E., and Gianella, V. P., 1935, The earthquake of January 30, 1934, at Excelsior Mountains, Nevada: *Seismol. Soc. America Bull.*, v. 25, p. 161–168.

Carr, W. J., 1948a, Timing and style of tectonism and localization of volcanism in the Walker Lane belt of southwestern Nevada: *Geol. Soc. America Abstr. with Programs*, v. 16, no. 6, p. 464.

—— , 1984b, Regional structural setting of Yucca Mountain, southwestern Nevada, and Late Cenozoic rates of tectonic activity in part of the southwestern Great Basin, Nevada and California: *U.S. Geol. Survey Open File Rpt. 84-854*, 109 p.

Chen, J. H., and Moore, J. G., 1979, Late Jurassic Independence dike swarm in eastern California: *Geology*, v. 7, p. 129–133.

Coney, P. J., 1980, Cordilleran metamorphic core complexes — An overview *in* Crittenden, M. D., Jr., Coney, P. J., and Davis, G. H., eds., *Cordilleran Metamorphic Core Complexes:* Geol. Soc. America Mem. 153, p. 7–31.

Cooper, J. D., Miller, R. H., and Sundberg, F. A., 1982, Environmental stratigraphy of the lower part of the Nopah Formation (Upper Cambrian), southwestern Great Basin, *in* Cooper, J. D., Troxel, B. W., and Wright, L. A., eds., *Geology of selected areas in the San Bernardino Mountains, western Mojave Desert, and southern Great Basin, California:* Cordilleran Section, Geol. Soc. America, Guidebook, Field trip 9, p. 97–114.

Davis, G. A., 1977, Limitations on displacement and southeastward extent of the Death Valley fault zone, California, *in Short Contributions to California Geology:* Calif. Div. Mines and Geology Spec. Rpt. 129, p. 27–33.

—— , and Burchfiel, B. C., 1973, Garlock fault: — An intracontinental transform structure, southern California: *Geol. Soc. America Bull.*, v. 84, p. 1407–1422.

Dohrenwend, J. C., 1982, Map showing Late Cenozoic faults in the Walker Lake 1° by 2° quadrangle, Nevada-California: *U.S. Geol. Survey Misc. Field Studies Map MF-1382-D.*

Donath, F. A., 1962, Analyses of basin-range structure, south-central Oregon: *Geol. Soc. America Bull.*, v. 73, p. 1–16.

Drewes, H., 1981, Tectonics of southern Arizona: *U.S. Geol. Survey Prof. Paper 1144*, 96 p.

Ekren, E. B., and Byers, F. M., Jr., 1984, The Gabbs Valley Range — A well-exposed segment of the Walker Lane in west-central Nevada, *in* Lintz, Joseph, Jr., *Western Geological Excursions, Vol. 4:* Geol. Soc. America Ann. Meet., Mackay Sch. Mines, Reno, Nev., p. 203–215.

—— , Byers, F. M., Jr., Hardyman, R. F., Marvin, R. F., and Silberman, M. L., 1980, Stratigraphy, preliminary petrology, and some structural features of Tertiary volcanic rocks in the Gabbs Valley and Gillis Ranges, Mineral County, Nevada: *U. S. Geol. Survey Bull. 1464*, 54 p.

Evernden, J. F., and Kistler, R. W., 1970, Chronology of emplacement of Mesozoic batholitic complexes in California and western Nevada: *U.S. Geol. Survey Prof. Paper 623*, 42 p.

Ferguson, H. G., and Muller, S. W., 1949, Structural geology of the Hawthorne and Tonopah quadrangles, Nevada: *U.S. Geol. Survey Prof. Paper 216*, 55 p.

Fitch, T. J., 1972, Plate convergence, transcurrent faults, and internal deformation adjacent to Southeast Asia and the Western Pacific: *J. Geophys. Res.*, v. 77, p. 4432–4461.

Fleck, R. J., 1970, Age and possible origin of the Las Vegas Valley shear zone, Clark and Nye Counties, Nevada: *Geol. Soc. America Abstr. with Programs*, v. 2, p. 333.

Geissman, J. W., Van Der Voo, Rob, and Howard, K. L., Jr., 1982, A paleomagnetic study of the structural deformation in the Yerington District, Nevada: *Amer. J. Sci.*, v. 282, p. 1042–1109.

——, Callian, J. T., and Oldow, J. S., 1984, Paleomagnetic assessment of oroflexural deformation in west-central Nevada and significance for emplacement of allochthonous assemblages: *Tectonics*, v. 3, p. 179–200.

Gianella, V. P., 1957, Earthquake and faulting, Fort Sage Mountains, California, December, 1950: *Seismol. Soc. America Bull.*, v. 47, p. 173–177.

——, and Callaghan, Eugene, 1934, The earthquake of December 20, 1932, at Cedar Mountain, Nevada, and its bearing on the genesis of Basin Range structure: *J. Geology*, v. 42, p. 1–22.

Guth, P. L., 1981, Tertiary extension north of the Las Vegas Valley shear zone, Sheep and Desert Ranges, Clark County, Nevada: *Geol. Soc. America Bull.*, v. 92, Pt. I, p. 763–771.

Hamilton, Warren, 1986, Crustal extension in the Basin and Range province, southwestern United States: *Geol. Soc. London, Spec. Paper*, in press.

Hardyman, R. F., 1978, Volcanic stratigraphy and structural geology of Gillis Canyon quadrangle, northern Gillis Range, Mineral County: Ph.D. thesis, Univ. Nev., Reno, Nev., 248 p.

——, 1984, Strike-slip, normal, and detachment faults in the northern Gillis Range, Walker Lane of west-central Nevada; *in* Lintz, Joseph, Jr., *Western Geological Excursions, Vol. 4:* Geol. Soc. America Ann. Meet., Mackay Sch. Mines, Reno, Nev., p. 184–199.

——, Ekren, E. B., and Byers, F. M., Jr., 1975, Cenozoic strike-slip, normal and detachment faults in northern part of the Walker Lane, west central Nevada [abstract]: *Geol. Soc. America Abstr. with Programs*, v. 7, no. 7, p. 1100.

Hill, M. R., 1972, Owens Valley earthquake of 1872: *Calif. Geology*, v. 25, no. 3, p. 51–54.

Jennings, C. W., 1975, Fault map of California: *Calif. Div. Mines Geology, Calif. Geol. Data Map Series, Map 1, Faults, volcanoes, thermal springs, and wells*, scale 1:750,000.

——, 1977, Geologic map of California: Calif. Div. Mines and Geology, Calif. Geol. Data Map Series, scale 1:750,000.

John, D. A., 1983, Map showing distribution, ages, and petrographic character of Mesozoic plutonic rocks in the Walker Lake 1° by 2° quadrangle, California and Nevada: *U.S. Geol. Survey Misc. Field Studies Map MF-1382-B*, scale 1:250,000.

Kirsch, S. A., 1968, Structure of the metamorphic and sedimentary rocks of Mineral Ridge, Esmeralda County, Nevada: Ph.D. thesis; Univ. Calif., Berkeley, Calif., 79 p.

——, 1971, Chaos structure and turtleback dome, Mineral Ridge, Esmeralda County, Nevada: *Geol. Soc. America Bull.*, v. 82, p. 3169–3176.

Kistler, R. W., and Peterman, Z. E., 1978, Reconstruction of crustal blocks of California on the basis of initial strontium isotopic compositions of Mesozoic granitic rocks: *U.S. Geol. Survey Prof. Paper 1071*, 17 p.

——, Robinson, A. C., and Fleck, R. J., 1980, Mesozoic right-lateral fault in eastern California: *Geol. Soc. America Abstr. with Programs*, v. 12, p. 115.

Kluth, C. F., 1983, Geology of the northern Canelo Hills and implications for the Mesozoic tectonics of southeastern Arizona, *in* Reynolds, M. W., and Dolly, E. D., eds., *Mesozoic Paleogeography of West-Central United States:* Rocky Mountain Section, Soc. Econ. Paleontologists Mineralogists, Rocky Mountain Paleogeography Symp. 2, p. 159–171.

Labotka, T. C., 1980, Petrology of a medium-pressure regional metamorphic terrane, Funeral Mountains, California: *Amer. Mineralogist*, v. 65, nos. 7, 8, p. 670–689.

Liggett, M. A., and Childs, J. F., 1973, Evidence of a major fault zone along the California-Nevada state line, 35°30' to 36°30' N. latitude: Argus Exploration Company, *NASA Rpt. Inv. CR-133140, E73-10773*, 13 p.

——, and Childs, J. F., 1977, An application of satellite imagery to mineral exploration *in* Woll, P. W., and Fischer, W. A., Proceedings of the first annual William T. Pecora Memorial Symposium, October 1975, Sioux Falls, S.D.: U.S. Geol. Survey Prof. Paper 1015, p. 253–270.

Locke, A., Billingsley, P. R., and Mayo, E. B., 1940, Sierra Nevada tectonic patterns: *Geol. Soc. America Bull.*, v. 51, p. 513–540.

Longwell, C. R., 1960, Possible explanation of diverse structural patterns in southern Nevada: *Amer. J. Sci.*, v. 258-A, p. 192–203.

——, 1974, Measure and date of movement on Las Vegas Valley shear zone, Clark County, Nevada: *Geol. Soc. America Bull.*, v. 85, p. 985–990.

McKee, E. H., 1968, Age and rate of movement of the northern part of the Death Valley-Furnace Creek fault zone, California: *Geol. Soc. America Bull.*, v. 29, p. 509–512.

——, 1983, Reset K-Ar ages — Evidence for three metamorphic core complexes, western Nevada: *Isochron/West*, v. 38, p. 17–20.

Miller, R. H., and Walch, C. A., 1977, Depositional environments of Upper Ordovician through Lower Devonian rocks in the southern Great Basin, *in* Stewart, J. H., Stevens, C. H., and Fristche, A. E., eds., *Paleozoic Paleogeography of the Western United States:* Pacific Section, Soc. Econ. Paleontologists Mineralogists, Pacific Coast Paleogeography Symp. 1, p. 165–180.

Molinari, M. P., 1984, Late Cenozoic structural geology of Stewart and Monte Cristo Valleys, Walker Lane of west central Nevada, *in* Lintz, Joseph, Jr., *Western Geological Excursions, Vol. 4:* Geol. Soc. America Ann. Meet., Mackay Sch. Mines, Reno, Nev., p. 219–231.

Moore, J. G., and du Bray, E., 1978, Mapped offset on the right-lateral Kern Canyon fault, southern Sierra Nevada, California: *Geology*, v. 6, p. 205–208.

——, and Hopson, C. A., 1961, The Independence dike swarm in eastern California: *Amer. J. Sci.*, v. 259, p. 241–259.

Moore, J. N., 1976a, The Lower Cambrian Poleta Formation — A tidally dominated, open coastal and carbonate bank depositional complex, western Great Basin: Ph.D. thesis, Univ. California, Los Angeles, Calif. 284 p.

——, 1976b, Depositional environments of Lower Cambrian Poleta Formation and its stratigraphic equivalents, California and Nevada: *Brigham Young Univ. Geol. Stud.*, v. 23, pt. 2, p. 23–38.

——, and Foster, C. T., Jr., 1980, Lower Paleozoic metasedimentary rocks in the east-central Sierra Nevada, California — Correlation with Great Basin formations: *Geol. Soc. America Bull.*, v. 91, pt. I, p. 37–43.

Morton, W. H., and Black R., 1975, Crustal alteration in Afar, *in* Pilger, A., and Rosler, A., eds., *Afar Depression of Ethiopia, Inter-Union Commission on Geodynamics — International Symposium on the Afar Region and Related Rift Problems*, Proceedings, Scientific Report No. 14: Stuttgart, West Germany, Schweizerbartische Verlagsbuchhandlung, p. 55–56.

Nielsen, R. L., 1965, Right-lateral strike-slip faulting in the Walker Lane, west-central Nevada: *Geol. Soc. America Bull.*, v. 76, p. 1301–1308.

Nokleberg, W. J., 1983, Wallrocks of the central Sierra Nevada batholith, California — A collage of accreted tectono-stratigraphic terranes: *U.S. Geol. Survey Prof. Paper 1255*, 28 p.

Oakeshott, G. B., Greensfelder, R. W., and Kahle, J. E., 1972, Owens Valley Earthquake of 1872 — One hundred years later: *Calif. Geology*, v. 25, no. 3, p. 55–61.

Oldow, J. S., 1983, Tectonic implications of a Late Mesozoic fold and thrust belt in northwestern Nevada: *Geology*, v. 11, p. 542–546.

———, 1984, Spatial variability in the structure of the Roberts Mountains allochthon, western Nevada: *Geol. Soc. America Bull.*, v. 92, p. 174–185.

Pelton, P. J., 1966, Mississippian rocks of the southwestern Great Basin, Nevada and California: Ph.D. thesis, Rice Univ., Houston, Tex., 99 p.

Poole, F. G., and Sandberg, C. A., 1977, Mississippian paleogeography and tectonics of the western United States, *in* Stewart, J. H., Stevens, C. H., and Fritsche, A. E., eds., *Paleozoic Paleogeography of the Western United States:* Pacific Section, Soc. Econ. Paleontologists Mineralogists, Pacific Coast Paleogeography Symp. 1, p. 67–85.

———, Baars, D. L., Drewes, Harald, Hayes, P. T., Ketner, K. B., McKee, E. D., Teichert, Curt, and Williams, J. S., 1967, Devonian of the southwestern United States, *in* Oswald, D. H., ed., *International Symposium on the Devonian System, Vol. 1:* Calgary, Alta., Alberta Soc. Petrol. Geologists, p. 879–912.

———, Sandberg, C. A., and Boucot, A. J., 1977, Silurian and Devonian paleogeography of the western United States, *in* Stewart, J. H., Stevens, C. H., and Fritsche, A. E., eds., *Paleozoic Paleogeography of the Western United States:* Pacific Section, Soc. Econ. Paleontologists Mineralogists, Pacific Coast Paleogeography Symp. 1, p. 39–65.

Proffett, J. M., Jr., 1977, Cenozoic geology of the Yerington district, Nevada, and implications for the nature and origin of Basin and Range faulting: *Geol. Soc. America Bull.*, v. 88, no. 2, p. 247–266.

Rogers, A. M., Harmsen, S. C., Carr, W. J., and Spence, W. J., 1983, Southern Great Basin seismological data report for 1981, and preliminary data analysis: *U.S. Geol. Survey Open File Rpt. 83-669*, 240 p.

Rogers, D. K., 1975, The Carson Lineament — Its influence on recent left-lateral faulting near Carson City, Nevada: *Geol. Soc. America Abstr. with Programs*, v. 7, no. 7, p. 1250.

Ron, H., Aydin, A., and Nur, A., 1986, Strike-slip faulting and block rotation in the Lake Mead system: *Geology*, v. 14, p. 1020–1023.

Ross, D. C., 1962, Correlation of granitic plutons across faulted Owens Valley, California *in Geological Survey Research 1962:* U.S. Geol. Survey Prof. Paper 450-D, p. D86–D88.

Royse, Frank, 1983, Comment on "Magnitude of crustal extension in the southern Great Basin:" *Geology*, v. 11, p. 495–496.

Saleeby, J. B., 1981, Ocean floor accretion and volcanoplutonic arc evolution of the Mesozoic Sierra Nevada *in* Ernst, W. G., ed., *The Geotectonic Development of California* (Rubey Vol. I): Prentice-Hall, Inc., Englewood Cliffs, New Jersey, p. 132–181.

_____ , Gooding, S. E., Sharp, W. D., and Busby, C. J., 1978, Early Mesozoic paleotectonic-paleogeographic reconstruction of the southern Sierra Nevada *in* Howell, D. G., and McDougall, K. A., eds., *Mesozoic Paleogeography of the Western United States:* Pacific Section, Soc. Econ. Paleontologists Mineralogists, Pacific Coast Paleogeography Symp. 2, p. 311–336.

Sanders, C. O., and Slemmons, D. B., 1979, Recent crustal movements in the central Sierra Nevada-Walker Lane region of California-Nevada — Pt. III, The Olinghouse fault zone: *Tectonophysics*, v. 52, p. 585–597.

Schweickert, R. A., 1981, Tectonic evolution of the Sierra Nevada range, *in* Ernst, W. G., ed., *The Geotectonic Development of California* (Rubey Vol. I): Englewood Cliffs, N.J., Prentice-Hall, p. 87–131.

Sheridan, M. F., 1978, Owens Valley — A major rift between the Sierra Nevada batholith and basin and range province, USA, *in* Ramberg, I. B., and Neuman, E.-R., eds., *Tectonics and Geophysics of Continental Rifts, Vol. 2:* Dordrecht, The Netherlands, D. Reidel, p. 81–88.

Silver, L. T., and Anderson, T. H., 1974, Possible left-lateral Early to Middle Mesozoic disruption of the southwestern North America cratonic margin: *Geol. Soc. America Abstr. with Programs*, v. 6, no. 7, p. 955–956.

Slemmons, D. B., 1967, Pliocene and Quaternary movements of the Basin-and-Range province USA: *J. Geosci., Osaka City Univ.*, v. 10, p. 91–103.

_____ , Jones, A. E., and Gimlett, J. I., 1965, Catalog of Nevada earthquakes, 1852–1960: *Seismol. Soc. America Bull.*, v. 55, no. 2, p. 519–566.

_____ , Van Wormer, Douglas, Bell, E. J., and Silberman, M. L., 1979, Recent crustal movements in the Sierra Nevada-Walker Lane region of California-Nevada — Pt. I. Rate and style of deformation: *Tectonophysics*, v. 52, p. 561–570.

Smith, G. I., 1962, Large lateral displacement on Garlock fault, California, as measured from offset dike swarms: *Amer. Assoc. Petrol. Geologists Bull.*, v. 46, p. 85–104.

_____ , and Ketner, K. B., 1970, Lateral displacement on the Garlock fault, southeastern California, suggested by offset sections of similar metasedimentary rocks: *U.S. Geol. Survey Prof. Paper 700-D*, p. D1–D9.

Smith, R. B., and Lindh, A. G., 1978, Fault-plane solutions of the western United States — A compilation; *in* Smith, R. B., and Eaton, G. P., eds., *Cenozoic Tectonics and Regional Geophysics of the Western Cordillera:* Geol. Soc. America Mem. 152, p. 107–109.

Smith, R. S. U., 1979, Holocene offset and seismicity along the Panamint Valley fault zone, western Basin-and-Range Province, California: *Tectonophysics*, v. 52, p. 411–415.

Speed, R. C., 1978, Paleogeographic and plate tectonic evolution of the Early Mesozoic marine province of the western Great Basin, *in* Howell, D. G., and MacDougall, K. A., eds., *Mesozoic Paleogeography of the Western United States:* Pacific Section, Soc. Econ. Paleontologists Mineralogists, Pacific Coast Paleogeography Symp. 2, p. 253–270.

_____ , and Cogbill, A. H., 1979, Candelaria and other left-oblique slip faults of the Candelaria region, Nevada: *Geol. Soc. America Bull.*, v. 90, pt. 1, p. 149–163.

Stern, T. W., Bateman, P. C., Morgan, B. A., Newell, M. F., and Peck, D. L., 1981, Isotopic U-Pb ages of zircon from the granitoids of the central Sierra Nevada, California: *U.S. Geol. Survey Prof. Paper 1185*, 17 p.

Stewart, J. H., 1967, Possible large right-lateral displacement along fault and shear zones in Death Valley–Las Vegas area, California and Nevada: *Geol. Soc. America Bull.*, v. 78, p. 131–142.

____, 1970, Upper Precambrian and Lower Cambrian strata of the southern Great Basin, California and Nevada: *U.S. Geol. Survey Prof. Paper 620*, 206 p.

____, 1979, Regional tilt patterns of Late Cenozoic basin-range fault blocks, western United States: Geol. Soc. America Bull., v. 91, pt. 1, p. 460–464.

____, 1980, Geology of Nevada: *Nev. Bur. Mines Geology Spec. Publ. 4*, 136 p.

____, 1983a, Cenozoic structure and tectonics of the northern Basin and Range province, California, Nevada, and Utah *in The Role of Heat in the Development of Energy and Mineral Resources in the Northern Basin and Range Province:* Geothermal Resources Council, Spec. Rpt. 13, p. 25–39.

____, 1983b, Extensional tectonics in the Death Valley area, California — Transport of the Panamint Range structural block 80 km northwestward: *Geology*, v. 11, p. 153–157.

____, 1985, East-trending dextral faults in the western Great Basin — An explanation for anomalous trends of pre-Cenozoic strata and Cenozoic faults: *Tectonics*, v. 4, no. 6, p. 547–564.

____, and Carlson, J. E., 1978, *Geologic Map of Nevada*; scale 1:500,000: *U.S. Geol. Survey.*

____, Albers, J. P., and Poole, F. G., 1968, Summary of regional evidence for right-lateral displacement in the western Great Basin: *Geol. Soc. America Bull.*, v. 79, p. 1407–1413.

____, Albers, J. P., and Poole, F. G., 1970, Reply to "Summary of regional evidence for right-lateral displacement in the western Great Basin:" *Geol. Soc. America Bull.*, v. 81, p. 2175–2180.

____, Carlson, J. E., and Johannesen, Dann C., 1982, Geologic map of the Walker Lake 1° by 2° quadrangle, California and Nevada: *U.S. Geol. Survey Misc. Field Studies Map MF-1382-A*, scale 1:250,000.

____, Anderson, T. H., Haxel, G. B., Silver, L. T., and Wright, J. E., 1986, Late Triassic paleogeography of the southern Cordillera — The problem of a source for voluminous volcanic detritus in the Chinle Formation of the Colorado Plateau region: *Geology*, v. 14, p. 567–570.

Titley, S. R., 1976, Evidence for a Mesozoic linear tectonic pattern in southeastern Arizona: *Ariz. Geol. Soc. Digest*, v. 10, p. 71–101.

Vetter, U. R., and Ryall, A. S., 1983, Systematic change of focal mechanisms with depth in western Great Basin: *J. Geophys. Res.*, v. 88, no. B10, p. 8237–8250.

Warner, L. A., 1978, The Colorado Lineament — A Middle Proterozoic wrench fault system: *Geol. Soc. America Bull.*, v. 89, p. 161–171.

Wernicke, Brian, 1981, Low-angle normal faults in the Basin and Range Province — Nappe tectonics in an extending orogen: *Nature*, v. 291, p. 645–648.

____, Spencer, J. E., Burchfiel, B. C., and Guth, P. L., 1982, Magnitude of crustal extension in the southern Great Basin: *Geology*, v. 10, p. 499–502.

____, Spencer, Jon, and Guth, P. L., 1983, Reply on "Magnitude of crustal extension in the southern Great Basin": *Geology*, v. 11, p. 496–497.

____, Guth, P. L., and Axen, G. J., 1984, Tertiary extensional tectonics in the Sevier thrust belt of southern Nevada, *in* Lintz, Joseph, Jr., *Western Geological Excursions, Vol. 4:* Geol. Soc. America Ann. Meet., Mackay Sch. Mines, Reno, Nev., p. 473–495.

Wetterauer, R. H., 1977, The Mina deflection — A new interpretation based on the history of the Lower Jurassic Dunlap Formation, western Nevada: Ph.D. thesis; Northwestern Univ., Evanston, Ill., 155 p.

Wright, L. A., 1976, Late Cenozoic fault patterns and stress fields in the Great Basin and western displacement of the Sierra Nevada block: *Geology*, v. 4, p. 489–494.

——, and Troxel, B. W., 1967, Limitations on right-lateral, strike-slip displacement, Death Valley and Furnace Creek fault zones, California: *Geol. Soc. America Bull.*, v. 78, p. 933–949.

——, and Troxel, B. W., 1970, Discussion on "Summary of regional evidence for right-lateral displacement in the western Great Basin:" *Geol. Soc. America Bull.*, v. 81, p. 2167–2174.

——, and Troxel, B. W., 1973, Shallow-fault interpretation of Basin and Range structure, southwestern Great Basin, *in* De Jong, K. A., and Scholten, Robert, eds., *Gravity and Tectonics:* New York, Wiley, p. 397–407.

Zoback, M. L., and Beanland, Sarah, 1986, Temporal variations in stress magnitude and style of faulting along the Sierran frontal fault system: *Geol. Soc. America Abstr. with Programs*, v. 18, p. 801.

——, and Thompson, G. A., 1978, Basin and Range rifting in northern Nevada — Clues from a mid-Miocene rift and its subsequent offset: *Geology*, v. 6, no. 2, p. 111–116.

——, and Zoback, Mark, 1980, State of stress in the conterminous United States: *J. Geophys. Res.*, v. 85, no. B11, p. 6113–6156.

——, Anderson, R. E., and Thompson, G. A., 1981, Cenozoic evolution of the state of stress and style of tectonism of the Basin and Range province of the western United States, *in* Vine, F. J., and Smith, A. D., eds., *Extensional Tectonics Associated with Convergent Plate Boundaries:* London, Royal Society, p. 189–216.

Theodore C. Labotka
Department of Geological Sciences
University of Tennessee
Knoxville, Tennessee 37996

Arden L. Albee
Division of Geological and Planetary Sciences
California Institute of Technology
Pasadena, California 91125

26

METAMORPHISM AND TECTONICS OF THE DEATH VALLEY REGION, CALIFORNIA AND NEVADA

The Panamint and Funeral Mountains, in the Death Valley area, California, comprise some of the largest exposures of the regional metamorphic terrain marginal to the Sierra Nevada batholith. The metamorphic rocks are exposed in northwest-trending anticlines and have protoliths of Middle Proterozoic gneissic basement, Upper Proterozoic Pahrump Group, and Upper Proterozoic Noonday Dolomite, Johnnie Formation, and Stirling Quartzite. Textures indicate that prograde metamorphism occurred prior to folding and that folding was accompanied by retrograde metamorphism and/or secondary foliation development. Mineral assemblages indicate that metamorphic conditions in the Panamint Mountains were vastly different from those in the Funeral Mountains. Pelitic schists in the Panamint Mountains are characterized by andalusite and cordierite assemblages, by the local occurrence of chloritoid + biotite, and by the persistence of staurolite in the sillimanite zone. In the Funeral Mountains, kyanite was the stable aluminosilicate and cordierite is absent, chloritoid broke down to garnet + staurolite + chlorite, and staurolite broke down to kyanite + garnet + biotite. Quantitative estimates of physical conditions give $P = 3 \pm 1$ kb and $400 < T \leqslant 650°C$ in the Panamint Mountains and $P = 6.5 \pm 1$ kb and $450 < T \leqslant 700°C$ in the Funeral Mountains. The pressure attained in the Panamint Mountains can be ascribed to stratigraphic overburden, but the overburden in the Funeral Mountains must have been twice the normal stratigraphic section. This extra overburden was probably derived from the emplacement of a nappe complex rooted in the Inyo Range. Despite the difference in conditions between the Panamint and Funeral Mountains, metamorphic grade in both ranges increased westward. The age of prograde metamorphism in the Panamint Mountains is 170–150 m.y.b.p. and in the Funeral Mountains $\geqslant 135$ m.y.b.p. Retrograde metamorphism occurred 80–70 m.y. ago in the Panamint Mountains and extensive resetting of K-Ar ages in the Funeral Mountains occurred at this time. The ages of metamorphism postdate periods of thrusting in the Inyo and Argus ranges to the west and in the foreland thrust belt to the east. Analysis of heat conduction indicates that metamorphism could not have resulted simply from the emplacement of the nappes, in the manner suggested for continental collision zones. Rather, the source of heat for metamorphism must have been related to the Sierran magmatic arc. A mantle heat flux that decreased with distance from the arc accounts for the essentially isobaric nature of metamorphism in both the Panamint and Funeral Mountains. Although the role that the development of the regional metamorphic terrain plays in the formation of the foreland thrust belt is obscure, the metamorphic belt is clearly a natural consequence of the emplacement of the Sierran batholith.

INTRODUCTION

The region in the southwestern Great Basin around Death Valley, California and Nevada, comprises one of the largest exposures of regionally metamorphosed rocks in the Cordillera of the western United States. The central Death Valley region is the culmination of an anticlinorium in which Middle Proterozoic gneisses and schists and Upper Proterozoic and lower Phanerozoic metasedimentary rocks crop out in the rugged mountain ranges surrounding Death Valley. East of this region, lower to middle Paleozoic rocks were thrust over middle Paleozoic to lower Mesozoic rocks along faults associated with the Sevier Orogenic Belt. West of Death Valley lie the White and Inyo Ranges and the Sierra Nevada, which consist of Mesozoic granitic rocks of the Sierra Nevada batholith. The metasedimentary rocks in the Death Valley region underwent at least two periods of metamorphism during Mesozoic time and are considered to be part of a belt of metamorphic rocks marginal to the batholith and interior to the foreland fold-and-thrust belt. This batholithic marginal belt includes metamorphic rocks in the Silver Peak area, the Snake and Schell Creek Ranges, Ruby Range, Grouse Creek and Albion Ranges, the Belt Supergroup rocks on the northern margin of the Idaho batholith, and the Shuswap metamorphic complex. All these areas have undergone Mesozoic regional metamorphism, which in many areas has been overprinted by Tertiary thermal events associated with uplift and extension in the Great Basin.

A generalized geologic map of the Death Valley region is shown in Fig. 26-1. The greatest areal distribution of regionally metamorphic rocks occurs in the Panamint and Funeral Mountains, although small exposures occur in the Black Mountains, Bare Mountain and Bullfrog Hills, and roof pendants occur in the Inyo Range and White Mountains. The regionally metamorphosed rocks occur in northwest-trending anticlines and domes and have protoliths of Upper Precambrian rocks of the Pahrump Group, Noonday Dolomite, Johnnie Formation, Stirling Quartzite, and Lower Cambrian Wood Canyon Formation. The metamorphic terrains contain minor amounts of granitic rocks in comparison with the ranges west of Panamint Valley; the Panamint Valley Fault Zone appears to separate the batholithic terrain from the regional metamorphic complex.

The petrology of the metamorphic rocks places strong constraints on the tectonic development of the batholithic marginal terrain. Pressure and temperature during metamorphism, calculated from phase equilibria or determined from the sequence of prograde reactions, are a measure of the heat flux and thermal gradient. The pressure also locates the terrain within the crust during the peak temperature of metamorphism. These data, combined with isotopic evidence for the age of metamorphism, provide a basis for the thermal history of the terrain and for the relations between metamorphism and the emplacement of the Sierran batholith and the development of the foreland thrust belt.

The primary focus of this discussion is the petrology of the metamorphic rocks in the Panamint and Funeral Mountains. These areas make up the majority of the exposed metamorphic terrain and have been most thoroughly studied. The geology, petrology, and age of metamorphism of each area are described to determine the thermal history and the tectonic environment of metamorphism during Mesozoic time.

METAMORPHISM IN THE DEATH VALLEY AREA

Panamint Mountains

The Panamint Mountains bound the western side of central Death Valley (Fig. 26-1), forming an imposing range with a relief of almost 3500 m. The large amount of relief provides a deep section into the upper part of the crust in which the Mesozoic metamorphic terrain is well exposed. Figure 26-2 shows a generalized geologic map of the Telescope Peak quadrangle, which is located in the central part of the range. The geology is taken from Albee *et al.* (1981) and is described in greater detail by Labotka *et al.* (1980). The structure of the Panamint Mountains is dominated by a north-northwest-trending anticline, which contains Middle Proterozoic granite, gneiss, and schist in the core, and Upper Proterozoic metasedimentary rocks on the flanks. The Proterozoic rocks include augen gneiss, mica schist, and amphibolite; the age of the augen gneiss has been dated by Lanphere *et al.* (1964) as ~1750 m.y. In the southern part of the quadrangle, the augen gneiss was intruded by a porphyritic quartz monzonite ~1400 m.y. ago; the augen gneiss and quartz monzonite make up the World Beater Dome. The Upper Proterozoic rocks lie nonconformably upon the older gneissic complex and comprise the Pahrump Group, Noonday Dolomite, Johnnie Formation, and Stirling Quartzite. The Pahrump Group

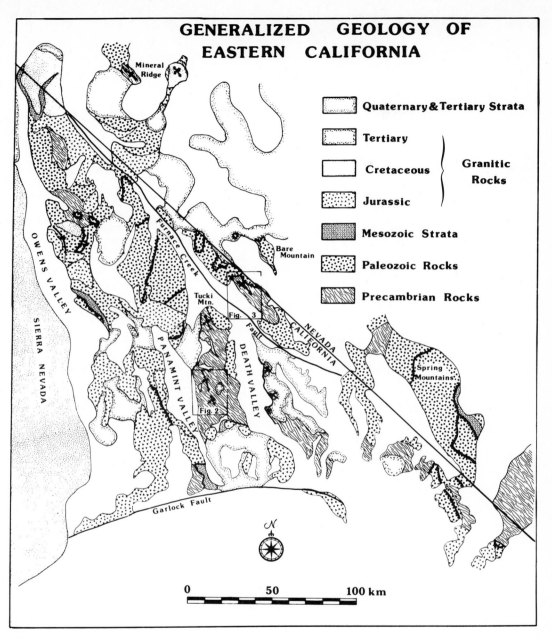

GENERALIZED GEOLOGY OF EASTERN CALIFORNIA

Mineral Ridge

Bare Mountain

Tucki Mtn.

Fig. 3

Fig. 2

OWENS VALLEY

SIERRA NEVADA

PANAMINT VALLEY

DEATH VALLEY

Furnace Creek Fault

NEVADA
CALIFORNIA

Spring Mountains

Garlock Fault

	Quaternary & Tertiary Strata
	Tertiary
	Cretaceous
	Jurassic
	Mesozoic Strata
	Paleozoic Rocks
	Precambrian Rocks

Granitic Rocks

N

0   50   100 km

FIG. 26-1. Generalized geology of eastern California showing the distribution of batholithic rocks, Proterozoic and younger sedimentary rocks, and major faults. Locations of the Telescope Peak quadrangle (Fig. 26-2) and the Chloride Cliff quadrangle (Fig. 26-3) are indicated.

consists of three formations: the Crystal Spring Formation, Beck Spring Dolomite, and Kingston Peak Formation. The Crystal Spring Formation contains dolomitic marble, muscovite-biotite schist, quartzite, and metadiabase sills and dikes. The Beck Spring Dolomite consists of massive dolomitic marble and minor interbedded argillite near the

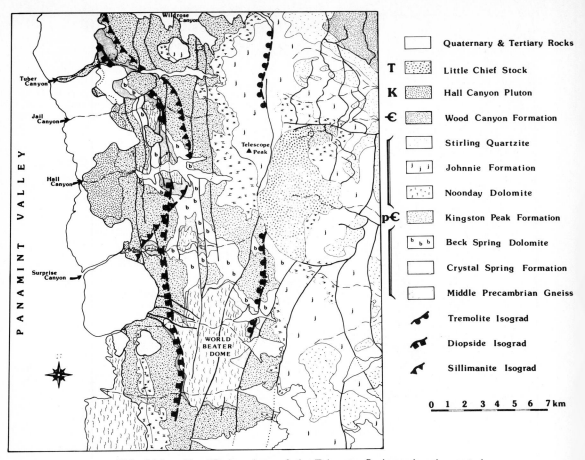

FIG. 26-2. Simplified geology of the Telescope Peak quadrangle, central Panamint Mountains. The geology is modified from Albee *et al.* (1981), and the isograds are taken from Labotka (1981). The tremolite isograd represents the first occurrence of tremolite + calcite assemblages in dolomitic rocks. The diopside isograd represents the first appearance of diopside in calcic schists and carbonate rocks. The sillimanite isograd represents the transition from andalusite to sillimanite.

top. The Kingston Peak Formation consists mostly of calcic argillites interbedded with conglomerates, massive conglomerate and diamictite, and minor amounts of marble. The Pahrump Group is overlain by the Noonday Dolomite, which consists mostly of thin-bedded to massive dolomitic marble with minor argillite, the Johnnie Formation, consisting of argillite, pelitic schist, and minor marble, and the Stirling Quartzite, which contains quartzite, feldspathic quartzite, and minor pelitic schist. The Upper Proterozoic rocks are overlain by Lower Cambrian Wood Canyon Formation, which is essentially unmetamorphosed in the Panamint Mountains.

Deformation of the rocks in the central Panamint Mountains occurred primarily during Mesozoic and Cenozoic time, although there is evidence for Precambrian tectonic events in the development of the augen gneiss in World Beater dome and in the stratigraphy of the Pahrump Group (Labotka and Albee, 1977; J. M. Miller, 1983). Folding along north-northwest-trending axes resulted in the anticline, although much of the strain during the folding event was taken up by displacement across preexisting north-trending, high-angle faults. In the Pleasant Canyon area, the Precambrian rocks were domed, producing the World Beater Dome with a core of 1400-m.y.-old quartz monzonite. At this time, the muscovite-bearing granodiorite (Hall Canyon pluton) was emplaced along the western margin of the Panamint Mountains. The age of folding appears to be Late Cretaceous, as discussed below, but Wright *et al.* (1981) suggested that the anticlinorium that culminates in the Panamint and Black mountains formed prior to the emplacement in the Middle Jurassic part of the Sierran batholith. J. M. Miller (1983) also described evidence from the southern part of the Panamint Mountains, where the development of the major folds occurred before the emplacement of the Middle or Late Jurassic Manly Peak pluton. There may indeed have been several periods of Mesozoic deformation with the last occurring during Late Cretaceous time.

Extension began during middle to late Tertiary time, and perhaps continues today. The extensional structures in the central Panamint Mountains are shallow-dipping normal faults, which have westward displacement. These faults are clearly younger than the folds; in many places, hanging-wall rocks cut across the core of the anticline and the high-angle faults. Several low-angle faults cut the rocks on the east side of the Panamint Mountains (Hunt and Mabey, 1966), which, in turn, are intruded by dikes and sills and are cut by structures associated with the emplacement of the Little Chief granitic stock (McDowell, 1974) 12 m.y. ago (Stern *et al.*, 1966). The stock has a very narrow contact metamorphic aureole and appears to have been emplaced very near the surface on the Earth. Its emplacement appears to postdate all rocks, with the possible exception of the monolithic breccias that form the western front of the Panamint Mountains.

The most impressive structures related to the uplift of the Panamint Mountains in the Telescope Peak quadrangle are the massive mono-lithologic breccias along the western margin of the range. The breccias have easily recognized parent lithologies, but the lithologies are extremely fractured and brecciated and, in some places, finely ground like fault gouge. The displacement of the breccia across the west-dipping fault surface is approximately 2–3 km. The breccias probably formed in the same state of stress as the other low-angle faults in the Panamint Mountains, but under much less cover and appear to be gravity slides developed during uplift of the range. The age of the slide mass breccias is Pliocene or Quaternary. North of Wildrose Canyon, Pliocene fanglomerates, which contain interbedded basalt flows, are in fault contact with the slide mass. The slide breccias locally contain clasts of the basalt, indicating that at least some of the slide mass development occurred after deposition of the Pliocene fanglomerate.

The metamorphic rocks in the Panamint Mountains show evidence for at least two periods of post-Precambrian regional metamorphism. The earlier period of prograde metamorphism produced rocks with a weakly developed foliation. Pelitic schists contain porphyroblastic andalusite and cordierite in a fine-grained, granoblastic matrix. The second period of metamorphism was generally retrograde, produced rocks with well-developed foliation and lineation, and recrystallized the prograde assemblages most thoroughly in the vicinity of the axis of the anticline. Prograde porphyroblasts in pelitic schist commonly show incipient to complete replacement by white mica or chlorite. In the core of the anticline, sillimanite-grade schists were nearly completely recrystallized to garnet-grade assemblages. The isotopic systematics of the Precambrian rocks in the World Beater Dome were extensively reset during doming. Field relations place few constraints on the ages of metamorphism, except that retrograde metamorphism appears to have occurred during the latest folding event. Isotopic data regarding the ages of metamorphism are discussed below.

The petrology of the Panamint metamorphic complex is described in detail by

Labotka (1981).[1] Grade during prograde metamorphism increases westward and is indicated by three isograds on Fig. 26-2. The lowest grade of metamorphism is marked by the tremolite isograd, which represents the first coexistence of calcite and tremolite in dolomitic marbles. East of this isograd, carbonate rocks can contain dolomite and quartz. Farther west, diopside became stable in calc-silicate rocks; its eastern stability limit is indicated by the diopside isograd. The sillimanite isograd separates low-grade, andalusite-bearing pelitic rocks from high- grade, sillimanite-bearing rocks to the west. Other pelitic isograds could not be mapped because of the paucity of pelitic schists in the higher-grade rocks. The diopside and sillimanite isograds intersect near Surprise Canyon. This relation resulted from a stratigraphic variation in CO_2/H_2O ratios of the fluid phase in the rock types that are distributed over the Telescope Peak quadrangle. North of the intersection, diopside first appears in dolomitic marbles in equilibrium with CO_2-rich fluid. South of the intersection, dolomitic marbles are scarce, and diopside formed in dolomite-clast conglomerates in equilibrium with H_2O-rich fluid. Thus, the diopside isograd is polythermal.

Although not very abundant, the pelitic schists are extremely useful for characterizing the metamorphic facies series because of the numerous possible reactions. Pelitic rocks from the west side of the range crest contain a variety of assemblages because of a range in Fe/(Fe + Mg); these include garnet + staurolite + chloritoid + biotite, andalusite + staurolite + biotite, andalusite + biotite + chlorite, and biotite + chlorite + cordierite. These assemblages appear to have been formed from the nearly complete breakdown of chlorite over a narrow temperature range. In the higher grade portions of the andalusite zone, chlorite is absent, and the assemblage andalusite + biotite + cordierite occurs. In the sillimanite zone, the only pelitic assemblage found is sillimanite + staurolite + biotite; retrograde metamorphism in the Tuber Canyon area greatly affected the sillimanite-zone rocks and replaced the assemblage with garnet + chlorite + biotite or garnet + chlorite + chloritoid.

Temperatures and pressure during the peak of metamorphism were determined by reconstructing the overlying stratigraphic column and by calculations based on phase equilibria (Labotka, 1981). Barometric calculations have been unsuccessful because appropriate mineral assemblages are generally lacking or equilibrium constants are poorly known. For example, calculations based on an empirical relation for a reaction among garnet, biotite, muscovite, plagioclase, and quartz (Ghent and Stout, 1981) give a wide range in pressure from 2.4 to 6.0 kb. The only reliable pressure estimates were based on the reconstructed stratigraphy. The estimates range from 2.3 kb at the base of the Stirling Quartzite to 3.0 kb at the base of the Pahrump Group. These are probably minimum values because additional overburden from thrust sheets or from volcanic rocks could have been present during metamorphism, but later removed by erosion. Nevertheless, pressures in the range 2.3–3.0 kb are consistent with the observed mineral assemblages. Temperatures were calculated from several phase equilibria, including the partitioning of Fe and Mg between garnet and biotite (Goldman and Albee, 1977), and are consistent with the westward increase in grade and with the position of the sillimanite isograd at 2.75 kb. The temperature ranged from <415° near the tremolite isograd to ~650°C in the sillimanite zone. At 2.75 kb, the sillimanite isograd represents an isotherm of 575°C. Temperature during retrograde metamorphism was 450 ± 50°C. Equilibrium fluid compositions depended on the rock type. Fluids in equilibrium with dolomitic marbles were CO_2-rich, those in pelitic schists had $X_{CO_2} = 0.2$, and fluids in calcic argillites were nearly pure H_2O.

The age of metamorphism was first recognized to be middle to late Mesozoic by

---

[1] Also T. C. Labotka, unpublished data.

METAMORPHISM AND TECTONICS OF THE DEATH VALLEY REGION

Lanphere (1964) and Lanphere *et al.* (1964). Labotka *et al.* (1985) have investigated the effects of polymetamorphism on ^{39}Ar-^{40}Ar systematics of metamorphic and granitic rocks from the Panamint Mountains. Their results clearly show the two metamorphic episodes, with the earlier, prograde metamorphism occurring 170–150 m.y. ago, and the retrograde metamorphism occurring 80–70 m.y. ago. These results imply that the last compressional deformation in the Panamint Mountains occurred during Late Cretaceous time.

Funeral Mountains

The Funeral Mountains lie on the east side of Death Valley, separated from the Panamint Mountains by the Furnace Creek–Death Valley fault zone and by 5–50 km of right-lateral displacement. The geology of the Funeral Mountains is similar to that of the Panamint Mountains, but the nature of the metamorphism is quite different.

The geology of the Funeral Mountains has been mapped by Wright and Troxel (personal communication), and a simplified version of their map is shown in Fig. 26-3. The oldest rocks in the Funeral Mountains are migmatitic gneisses that occur in the floor of Monarch Canyon. The gneissic rocks are overlain by pelitic schists, amphibolites, marbles, and metaconglomerates that appear to be equivalent to the Pahrump Group. These metamorphic rocks are, in turn, overlain by pelitic schists and quartzites of the Johnnie Formation and by quartzite and minor pelitic schist of the Stirling Quartzite. The metamorphic rocks form a northwest-trending, doubly plunging anticline with a culmination near Chloride Cliff. The metamorphic rocks in the core of the range are separated from weakly metamorphosed to unmetamorphosed, lower Paleozoic rocks by the gently undulating, north-dipping Boundary Canyon fault. Similar north-dipping faults also displace the metamorphic rocks. The low-angle faults displace Oligocene rocks, indicating that faults of this type, like those in the Panamint Mountains, are middle to late Tertiary in age.

The textures of the metamorphic rocks are much different from those of rocks in the Panamint Mountains. The rocks are schistose and generally have two foliation surfaces. The first foliation is defined by the micas and is generally parallel to compositional layering. The second is a slip cleavage oriented at a high angle to compositional layering. Most porphyroblasts grew prior to the development of the slip cleavage.

Metamorphic grade increases westward or northwestward, ranging from biotite grade in the southeast to sillimanite grade near Monarch Canyon. There appears to have been only one regionally extensive metamorphic event in the Funeral Mountains, although retrograde metamorphic effects are locally significant. Rocks near the Keane Wonder mine and along the sole of the Boundary Canyon fault are thoroughly altered to chlorite schist. Unlike the Panamint metamorphic complex, folding of the Funeral metamorphic terrain appears to have been accompanied only by the development of the slip cleavage and by little retrograde metamorphism.

The petrology of the metamorphic rocks has been studied by Labotka (1980). Unlike the low-pressure Panamint metamorphic complex, the metamorphic rocks in the Funeral Mountains formed under Barrovian facies series conditions. Kyanite, rather than andalusite, is the aluminosilicate mineral stable over nearly the entire range in grade. Pelitic schists are more abundant in the Funeral Mountains than in the Panamint Mountains because of the wider distribution of metamorphosed Johnnie Formation and the greater abundance of pelitic schists in the Pahrump Group in the Funeral Mountains. This wide distribution permits the mapping of the familiar Barrovian isograds, and three are indicated on Fig. 26-3. Near Indian Pass, in the southeastern part of the range, the garnet isograd represents the first appearance of garnet and the garnet-grade assemblage

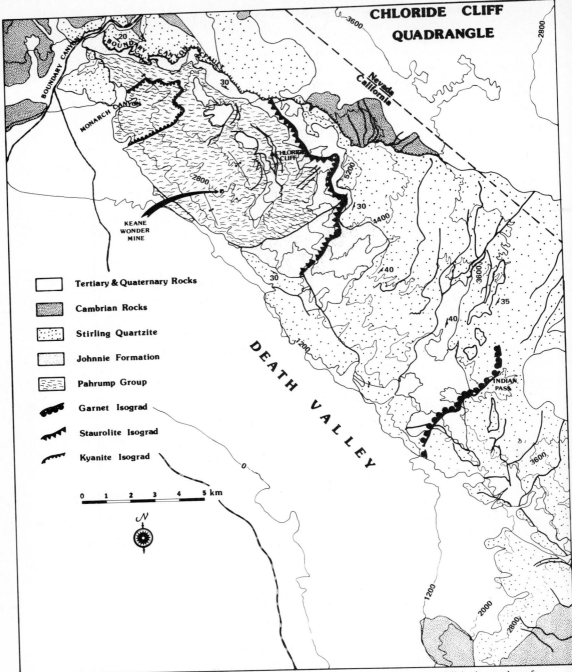

FIG. 26-3.   Geologic map of the Chloride Cliff quadrangle, simplified from unpublished mapping of
L. A. Wright and B. W. Troxel. Isograds are taken from Labotka (1980). The garnet zone is characterized by the coexistence of garnet and chlorite. In the staurolite zone, staurolite + biotite
assemblages occur. The kyanite zone contains rocks with the assemblage kyanite + garnet + biotite;
staurolite has broken down.

722   METAMORPHISM AND TECTONICS OF THE DEATH VALLEY REGION

garnet + biotite + chlorite. Other assemblages common at Indian Pass are garnet + chloritoid + chlorite and kyanite + chloritoid + chlorite. Kyanite has made its first appearance in the low-grade rocks near the garnet isograd. Northwest of Indian Pass, chloritoid was unstable and broke down to the assemblage garnet + staurolite + chlorite. The staurolite isograd was defined as the first occurrence of staurolite + biotite assemblages, including garnet + staurolite + biotite, staurolite + biotite + chlorite, and kyanite + staurolite + biotite. Near Monarch Canyon, the kyanite isograd marks the first occurrence of the assemblage kyanite + garnet + biotite and the last occurrence of stable staurolite. Sillimanite replaces kyanite in the highest-grade rocks in the floor of Monarch Canyon; a sillimanite isograd is not mapped because of the limited distribution of sillimanite-bearing assemblages.

Temperature during the peak of metamorphism, calculated from the partitioning of Fe and Mg between garnet and biotite, was in the range 450-500°C in the garnet zone, 500-550°C in the staurolite zone, and greater than 550°C in the kyanite zone. Considerable uncertainty is associated with these values, particularly the kyanite zone values, because of the wide range in compositions of garnet and biotite in some samples. Kyanite-zone samples give temperatures ranging from 500 to 750°C, with most samples giving values of 600 ± 30°C. Pressure during metamorphism was recalculated from the data of Labotka (1980) using recent experimental work of Bohlen et al. (1983) and Newton and Haselton (1981) and using the empirical relation of Ghent and Stout (1981). The results are reasonably consistent among the various methods, although imprecise. Values are 6.0 ± 0.5 kb for garnet- and staurolite-zone samples, and 6.6 ± 0.8 kb for kyanite-zone samples. The kyanite = sillimanite transition at 600 ± 30°C occurs at 5.7 ± 0.6 kb (Holdaway, 1971), which is consistent with the calculated values. The pressure estimate is about one and one-half times as great as that based on the amount of stratigraphic overburden. The high pressure must have resulted from duplication of the stratigraphic section by thrust faults.

The age of metamorphism is not precisely known. There are few geological constraints on the age. Correlation with the metamorphism in the Panamint Mountains suggests a Middle Jurassic age, but the nature of metamorphism in the Funeral Mountains is significantly different from that in the Panamint Mountains. Perhaps the two metamorphic terrains are not closely related. Recent ^{39}Ar-^{40}Ar data of E. DeWitt (personal communication, 1985) show very complicated systematics and give evidence for a protracted Mesozoic thermal history. Data from hornblende and some mica separates indicate that prograde metamorphism occurred $\geq$ 130 m.y. ago. Most samples from the high-grade portion of the terrain near Monarch Canyon, where numerous muscovite-bearing granites have intruded, give latest Cretaceous (65-85 m.y.b.p.) ages. Pegmatites that cut all rocks are mid-Tertiary (20-30 m.y.) in age[2] (Wasserburg et al., 1959). The age of prograde metamorphism is 20-40 m.y. younger in the Funeral Mountains than in the Panamint Mountains; this difference could have resulted from the greater depth of metamorphism and, therefore, a slower cooling rate in the Funeral Mountains than in the Panamint Mountains. The Late Cretaceous ages recorded in the Funeral Mountains are evident in samples from the core of the fold, indicating that the age of folding may also be Late Cretaceous. This is the same relation observed in the Panamint Mountains. Thus, both the Panamint and the Funeral metamorphic complexes are believed to have formed in the same tectonic environment, despite the Death Valley-Furnace Creek fault zone that separates them.

---

[2] Also E. DeWitt, personal communication, 1985.

Other Localities in the Vicinity of Death Valley

Tucki Mountain, at the north end of the Panamint Mountains, is constructed of highly deformed rocks of the Pahrump Group, which occur in the core of a dome. The dome is flanked by Upper Proterozoic and Paleozoic rocks; the contacts between the rocks in the core of the dome and the superjacent rocks are gently dipping fault surfaces with hanging walls displaced relatively westward. Proterozoic rocks in the core of the dome are weakly metamorphosed; pelitic rocks contain chloritoid, and some carbonate rocks have talc in the assemblage. Muscovite-bearing granite, the Skiddo pluton, intruded the Proterozoic rocks on the south side of Tucki Mountain and is similar in appearance to the Hall Canyon pluton. The geologic history of Tucki Mountain is like that of the Telescope Peak area, and the age of folding is probably Late Cretaceous, while the low-angle faults are probably middle to late Tertiary in age. Hodges *et al.* (1984), however, interpret the latest folding event to coincide with the development of the Tertiary extensional faults.

Bare Mountain and Bullfrog Hills are located northeast of the Funeral Mountains, near Beatty, Nevada. The pre-Tertiary rocks are predominantly early Paleozoic in age and generally dip homoclinally north or east (Cornwall and Kleinhampl, 1961). At the northwest end of the range, the rocks are folded into east- or southeast-plunging anticlines, and pelitic rocks from the Wood Canyon Formation contain the garnet-grade assemblage garnet + biotite + chlorite. These folded rocks are extensively cut by low-angle faults that appear to have normal displacement across them. Although the history of metamorphism and deformation is not well determined, the style of metamorphism is similar to that of the Funeral Mountains, and the metamorphic rocks are believed to be part of the Funeral metamorphic complex.

North of Death Valley, near Tonopah, Nevada, upper Precambrian metasedimentary rocks crop out in the core of a southeast-plunging anticline at Mineral Ridge (Albers and Stewart, 1972). The rocks were metamorphosed under lower amphibolite-facies conditions and intruded by numerous pegmatites. Assemblages reported by Bailly (1951) include garnet + biotite in schists, epidote + phlogopite in marbles, and garnet + hornblende + epidote in calcic schists. Textures indicate that metamorphic mineral growth was followed by the development of crenulation cleavage. The folding that produced the anticline appears to have occurred after metamorphism.

TECTONICS OF METAMORPHISM

The Sierran Batholith

Petrologic and isotopic studies of the metamorphic rocks in the Panamint Mountains indicate two periods of metamorphism during Mesozoic time (Labotka *et al.*, 1985). The evidence from the Funeral Mountains does not as clearly support two separate thermal events, but is consistent with a protracted thermal history contemporaneous with metamorphism in the Panamint Mountains. The ages of metamorphism closely correspond with the ages of emplacement of granitic rocks in the Argus, Inyo, and White ranges and with the ages of emplacement of batholithic rocks in the southeastern Sierra Nevada and the Mojave Desert. This correspondence is described in detail by Labotka *et al.* (1985), who show that the granitic rocks at Hunter Mountain and in the Argus Range were emplaced during the Middle to Late Jurassic period of metamorphism and that the Late Cretaceous period of metamorphism occurred during the emplacement of the numerous granitic rocks in the Argus Range and White Mountains and in the eastern Sierra Nevada.

The distribution of the granitic rocks in eastern California and the disposition of the metamorphic isograds in the Panamint and Funeral Mountains are shown in Fig. 26-4, which is a reconstruction of middle Mesozoic eastern California. Metamorphic grade in both ranges increases westward, toward the batholith. Because the ages of emplacement and metamorphism are the same and because metamorphic grade increases toward the granitic rocks, the cause of metamorphism could be ascribed to contact metamorphism around the batholith. However, the regional scale of metamorphism cannot be produced by simple conduction of heat from the batholith, even at a depth of 10 km where the ambient temperature is ~250°C. Liberal estimates of the thermal properties of the granitic and metamorphic rocks give calculated widths of the metamorphic terrain of less than 5 km, at which distance the temperature never rises above 450°C. Metamorphism, although coinciding with the emplacement of granitic magma, must have resulted from a high heat flux from the mantle or burial by thrust sheets at great depths.

Nappes

Metamorphism is thought to be intimately related to thrust faulting. The location of the Cordilleran regional metamorphic terrain and associated décollement structures in the hinterland of the foreland fold and thrust belt has led many workers to relate the development of the foreland belt to the uplift of the metamorphic infrastructure and gravity sliding of the suprastructure (e.g., Roberts and Crittenden, 1973; Scholten, 1973). Price (1973) suggested a variation of this idea in which the rising metamorphic terrain spread laterally, producing the foreland belt.

The Death Valley area lies close to the foreland thrust belt. The ages of thrust faults in this portion of the foreland belt have been well determined by Burchfiel and his co-workers (Burchfiel and Davis, 1977; Burchfiel et al., 1974) and can be related to the ages of metamorphism. Burchfiel and Davis (1977) describe four periods of thrusting and high-angle faulting prior to ~200 m.y. ago and major thrusting and high-angle faulting between 135 and 95 m.y. ago in the New York Mountains. Metamorphism in the Death Valley area appears to have postdated periods of thrusting in the foreland by 20-50 m.y.

West and north of Death Valley, numerous thrust faults crop out in the Argus, Inyo, Cottonwood, and Last Chance ranges. The geometric relations among the faults and associated folds are well summarized by Dunne et al. (1978), based on the geologic mapping of Hall, Moore, Nelson, and Ross. The faults generally place upper Precambrian and lower Paleozoic sedimentary rocks over middle to upper Paleozoic and lower Mesozoic rocks. Two major systems of faults occur in these ranges. The older system includes the Last Chance thrust of Stewart et al. (1966), the Race Track thrust of McAllister (1956), and the Lemoigne thrust of Hall (1971). The development of these faults appears to have coincided with the formation of folds in the northern Inyo and southern White Mountains, including a large-amplitude recumbent anticline. The distribution of rocks that lie on the upper plate of this thrust system, which we call the Inyo nappe, is sketched on Fig. 26-4. The Last Chance and related faults are intruded by granitic rocks of the Hunter Mountain batholith, which has an age of 185-170 m.y.p.b. (Chen and Moore, 1982). The younger system includes the Coso and Argus-Sterling thrusts (Moore, 1974) in the southern Inyo and Argus ranges. These faults place Paleozoic sedimentary rocks and Jurassic granitic rocks over upper Paleozoic to lower Mesozoic sedimentary and volcanic rocks. The youngest granitic rocks cut by this system are about 150 m.y. old; the upper age limit on this thrust is not constrained. West and north of the Panamint Mountains, the development of nappes appears to predate periods of metamorphism, as it does south and east of Death Valley.

The Panamint Mountain region, including the Funeral and Black Mountains, may lie on the upper plate of a thrust fault, but if so, the fault must be one of the foreland faults.

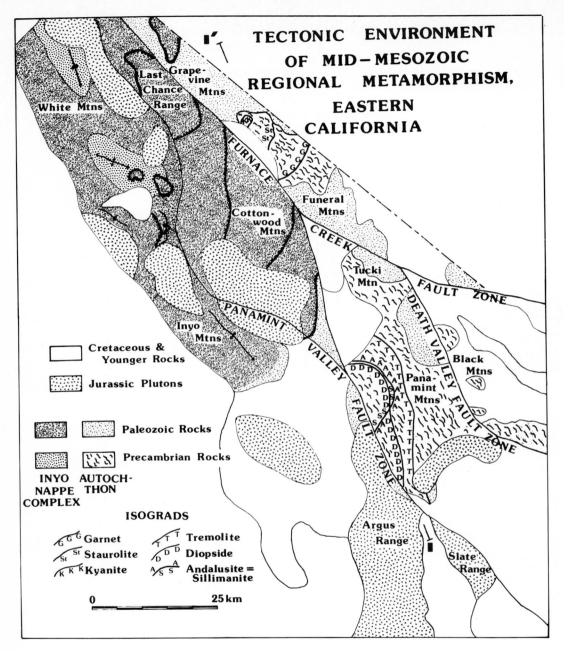

FIG. 26-4.   Tectonic map of eastern California showing the relative positions of the Panamint and Funeral Mountains after restoring an aggregate of 50 km of right-lateral displacement across the Death Valley–Furnace Creek fault zone and 20 km of displacement across the Panamint Valley fault zone. The generalized map shows the distributioons of Jurassic granitic rocks, regional metamorphic terrains and metamorphic isograds, and allochthonous rocks of the Inyo nappe complex. The indicated section line 1–1' is the section schematically shown in Fig. 26-7. In Fig. 26-7, the section, which is oblique to the trend of the batholith, is projected onto a northeast-trending line, approximately normal to the batholith.

The available geologic evidence is equivocal; there are no exposures of thrust faults within the Panamint block, nor any clear evidence for basement involvement in thrusting north of the Clark Mountains, California (Burchfiel and Davis, 1971). Thus, the Panamint block is an inlier either of autochthonous crust or of the structurally lowest nappe in the orogenic belt. Metamorphism in this block appears to have postdated periods of nappe emplacement.

Palinspastic Reconstruction

Middle to late Tertiary extension and late Tertiary to Recent right-lateral strike-slip faults have greatly modified the Mesozoic structures. Offset across the Furance Creek–Death Valley fault zone has been estimated to be as little as 8 km (Wright and Troxel, 1967) or as much as 90 km (Stewart, 1967). Stewart (1983) has estimated a northwestward displacement of the Panamint Mountains of 80 km along the Furnace Creek fault in a model that accounts for extension in central Death Valley across the Amargosa fault. The basis for the 80-km displacement is the apparent offset of isopachous lines across the Furnace Creek fault zone; upper Precambrian and lower Paleozoic strata appear to be thinner in the Panamint Mountains than expected from regional trends. The Panamint Mountains may have had positive relief during much of geologic time (Labotka and Albee, 1977), and the bending of the isopachous lines in the region of the Panamint Mountains may have partly resulted from the paleotopography. The amount of offset across the Furnace Creek fault zone is not well determined; 80 km is taken to be the upper limit. The effect of restoring the displacement across the fault zone is to line up the Funeral Mountains with the Panamint and Tucki Mountains into a single, north-trending range.

The great amount of extension in the Death Valley area resulted in the development of west-dipping listric normal faults and variably dipping low-angle detachment faults. Included among these are the system of faults that produced the Amargosa Chaos of Noble (1941) and the Amargosa thrust of Hunt and Mabey (1966), the detachment faults on the Turtlebacks in the Black Mountains, the Boundary Canyon fault in the Funeral Mountains, and the numerous low-angle faults around Tucki Mountain, and the east slope of the Panamint Mountains. The effect of restoring the displacement along these faults is to reduce the width of the valleys and perhaps overlap some mountain blocks. For example, the eastern part of the Panamint Mountains and the western part of the Black Mountains may have been closely nestled prior to the opening of central Death Valley when upper Precambrian sedimentary rocks in the Panamint Mountains were displaced westward along the Amargosa fault. The preextension geology appears to have consisted of a more compact, essentially contiguous metamorphic terrain in which temperature increased westward and the amount of overburden increased northward.

Figure 26-4 was constructed by removing the right-lateral displacement along the Panamint Valley, Death Valley, and Furnace Creek fault zones (labeled on Fig. 26-4). The amount of displacement removed was that necessary to close Panamint and Death Valleys and consists of 20 km along the Panamint Valley fault zone, 15 km along the Death Valley fault zone, and 35 km along the Furnace Creek fault zone. The aggregate of 50 km of right-lateral displacement across the Death Valley–Furnace Creek fault zone is less than that proposed by Stewart (1983), but is consistent with geologic evidence. The resulting configuration aligns the Paleozoic rocks at the southern end of the Funeral Mountains with those at Tucki Mountain, brings the Black Mountains along the Panamint Mountains, and creates a regular outline for the Hunter Mountain batholith.

The distribution of rocks belonging to the Inyo nappe complex, as shown on Fig. 26-4, is generalized and may include rocks from more than one thrust slice. The nappe appears to root in the Inyo Mountains where the structure culminates near Waucoba

Mountain. The greatest areal extent of the nappe complex is west of the reconstructed position of the Funeral Mountains. The folded rocks in the Inyo Mountains plunge south-ward, and the scattered occurrences of Paleozoic rocks in the Argus Range are shown as autochthonous rocks. This distribution is conjectural, but the great thickness of allo-chthonous rocks appears to be north of the Panamint Mountains and west of the Funeral Mountains. Thus, the greatest thickness of the nappe appears to have overridden the Funeral Mountains, whereas the Panamint Mountains were little affected by the thrust faults.

Thermal Environment during Metamorphism

The relative time relation of metamorphism and granitic batholith emplacement following thrust faulting indicates either that the formation of nappe structures induced meta-morphism and magma production or that there is no relation between thrusting and metamorphism. The thermal environment during metamorphism is reconstructed from the mineral assemblages to determine whether the emplacement of nappes could have caused metamorphism, aided the development of the metamorphic terrains, or simply provided a differential amount of overburden in the Death Valley area.

The mineral assemblages in pelitic schists are sensitive indicators of thermal gradients during metamorphism because of the numerous discontinuous reactions that can occur among the minerals in this rock type. This is well illustrated by the contrast in the mineral assemblages from the Panamint and Funeral Mountains. The sequences of mineral assem-blages and reactions for the two metamorphic terrains are illustrated in Fig. 26-5. The facies series for the two areas contrast in several ways. The most obvious is the occurrence of kyanite in the Funeral Mountains and andalusite in the Panamint Mountains. In the Funeral Mountains, chloritoid broke down prior to the instability of the garnet + chlorite tie line, whereas chloritoid persisted in the Panamint Mountains where the assemblage chloritoid + biotite occurs locally (Albee, 1965). Cordierite-bearing assemblages are common in the Panamint Mountains, but have not been found in the Funeral Mountains. And staurolite broke down in the Funeral Mountains while kyanite was still stable, but staurolite remained within the sillimanite zone in the Panamint Mountains.

These several differences can be related by the petrogenetic grid for pelitic schists, a portion of which is shown in Fig. 26-6. This particular version of the grid contains the experimentally determined P-T locations of a few pertinent reactions, including the stability of chlorite from Bird and Fawcett (1973), the aluminosilicate transformations from Holdaway (1971), and the stability of pyrophyllite from Kerrick (1968). The loca-tions of a few discontinuous reactions affecting pelitic schists, simplified after Albee (1965) and Labotka (1981), are shown as dashed lines and are placed to be consistent with natural assemblages. The ranges in estimated P and T during metamorphism in the Panamint and Funeral Mountains are also placed on Fig. 26-6. These ranges define the metamorphic field gradient; that is, the distribution of P and T exposed at the surface. These values do not necessarily correspond to a true geothermal gradient, as discussed by Richardson, England, Spear, and many other workers.

The field gradients show increases in temperature with only small, if any, changes in pressure, although there is a great difference in the recorded pressures between the Pana-mint and Funeral Mountains. A steady-state geotherm, labeled ∇T, is also shown on Fig. 26-6 to determine whether metamorphism could have occurred in a uniform gradient. This geotherm was calculated by assuming that radioactive heat production occurs in the 12 km of mostly carbonate rocks at a rate of 0.4 $\mu W/m^3$ and in the upper 15 km of the granitic portion of the crust at a rate of 2 $\mu W/m^3$, that the conductivity of the crust is 2.25 W/m/K, and that the mantle heat flux is 30 mW/m². These values are based on

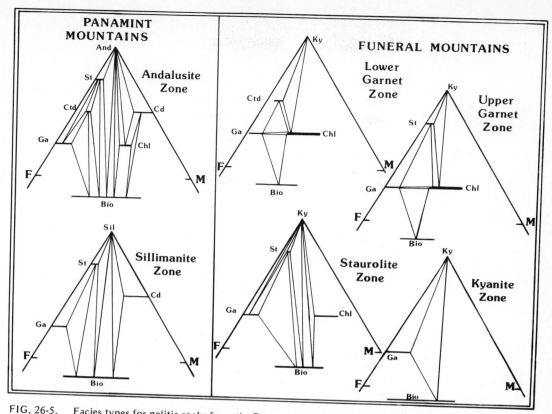

FIG. 26-5. Facies types for pelitic rocks from the Panamint Mountains (left) and the Funeral Mountains (right) illustrating the difference in the sequence of mineral assemblages resulting from the difference in pressure between the two ranges. All assemblages indicated by the three-phase triangles coexist with muscovite and quartz. Each facies type indicates all possible assemblages, insofar as can be determined from the actual assemblages, for a given temperature, pressure, and partial pressure of H_2O. The facies types shown do not constitute the complete set of types representing the progressive metamorphism of pelitic schists under the conditions of the Panamint and Funeral Mountains, but rather illustrate the characteristic or the most common assemblages in each zone. Mineral abbreviations: And, andalusite; Bio, biotite; Cd, cordierite; Chl, chlorite; Ctd, chloritoid; Gar, garnet; Ky, kyanite; Sil, sillimanite; St, staurolite.

intermediate values for the ranges given by England and Thompson (1984) and on the abudance of radioactive elements in carbonate rocks compiled by Clark (1965). The gradient (dP/dT) is steeper than the locus of $P\text{-}T$ values for either the Panamint or Funeral Mountains and cannot possibly explain the metamorphic conditions in both the Panamint and Funeral Mountains.

The concept of emplacement of hot thrust sheets has been successful in explaining metamorphism in the Alps (England and Richardson, 1976; Oxburgh and Turcotte, 1974). Thrust sheets were emplaced in the Death Valley region prior to metamorphism, although the sheets themselves were not metamorphosed. Even though the nappes were not hot, the thickening of the crust would have caused the crust to heat because the nappes provide an insulating cover and emplace additional radioactive material. A curve describing the $P\text{-}T\text{-}t$ path of rock at the basement-Pahrump contact is shown on Fig. 26-6. The path is calculated assuming the nappes doubled the thickness of the sedimentary cover and that the extra thickness was eroded at a constant rate of 1.5×10^{-4} m/yr. This rate is sufficient to expose the basement 160 m.y. after thrusting. The maximum tempera-

LABOTKA, ALBEE

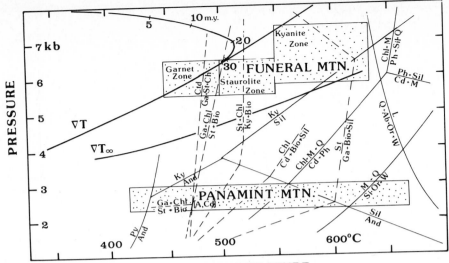

FIG. 26-6. Petrogenetic grid for the metamorphism of pelitic schists, showing some experimentally determined locations of reactions (solid lines) and the approximate locations of discontinuous reactions affecting the facies types of Fig. 26-5 (dashed lines). Mineral abbreviations as in Fig. 26-5 plus: Ab, albite; M, muscovite; Or, orthoclase; Ph, phlogopite; Q, quartz; Py, pyrophyllite; W, H_2O. The ranges in pressure and temperature calculated from mineral assemblages in the Panamint and Funeral Mountains are shown in the stippled boxes. The curve labeled ∇T represents the distribution of temperatures in a steady-state gradient for heat sources and flux appropriate for orogenic terrains. The curve labeled ∇T_∞ represents the steady-state distribution of temperatures in a terrain that has a doubled sedimentary section, a result of thrust faulting. The curved line with tick marks in millions of years represents the temperature-pressure-time path of a rock at the contact between basement and sedimentary cover after thrusting 12 km of additional section. The thickened section is eroded at a constant rate 1.5×10^{-4} m/yr.

ture of this rock would have been attained between 20 and 30 m.y. after thrusting and would never have exceeded the temperature of the steady-state geotherm. If erosion did not occur until after a new steady state was attained, the temperature could have attained values given by the curve labeled ∇T_∞.

In either case, a single thermal gradient could not have produced both the Panamint and Funeral metamorphic terrains. The effect of nappe emplacement appears to be primarily the addition of overburden; the thermal effect of emplacement was minor.

The relatively high temperatures at low pressure in the Panamint Mountains could not have resulted from convection of hot fluids from depth because the equilibrium fluid compositions determined by Labotka (1981) depend on the rock type. The fluids in marbles were CO_2-rich; those in pelitic schists were H_2O-rich; and those in calcic schists were nearly pure H_2O. Thus, the fluids appear to have been strata bound; there is no evidence for convection of fluids across lithologic boundaries. The difference in P/T between the Panamint and Funeral Mountains must have resulted from a regional variation in the mantle heat flux.

A schematic cross section that adequately describes the middle Mesozoic metamorphic event is shown in Fig. 26-7. This section extends from the Panamint Mountains,

METAMORPHISM AND TECTONICS OF THE DEATH VALLEY REGION

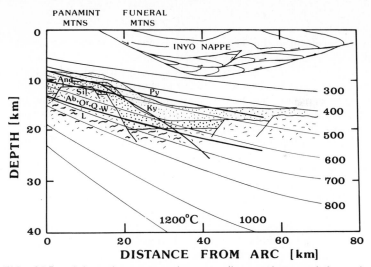

FIG. 26-7. Schematic cross section extending northeastward from the Jurassic arc. The line of section is shown on Fig. 26-4 and is projected onto the northeast trend. This projection allows the thermal environment to be treated as a two-dimensional problem. The effect of the projection makes the Inyo nappe appear to ride over the Panamints and load only the Funerals. This merely results from the northwest-southeast condensation of the section. See Fig. 26-4 for the true relation between the Inyo nappe complex and the regional metamorphic terrains. The hachured area represents Middle Proterozoic basement with a high in the Panamint Mountains and a basin in the Funeral Mountains, remnants of the later Proterozoic rift basin. The heavily stippled layer represents the Pahrump Group, and the lighter layer represents the uppermost Precambrian sedimentary section. Temperature contours were calculated for the steady-state condition with an exponentially decreasing mantle heat flux with distance from the arc. Heavy lines represent the aluminosilicate reactions with: Py, pyrophyllite; And, andalusite; Ky, kyanite; Sil, sillimanite. The granite minimum melting reaction albite + orthoclase + quartz + H_2O = liquid is also shown. This type of thermal environment is characterized by horizontal thermal gradients and gently east-dipping isograds; both features occur in the Panamint and Funeral Mountains.

adjacent to the Jurassic arc, to the Funeral Mountains, projected onto a line normal to the Jurassic arc (see Fig. 26-4). The Panamint Mountains are shown as a basement high, and the Funeral Mountains are buried by the Inyo nappe complex. Superimposed on the cross section are isotherms that were calculated for a steady state with an exponentially decreasing mantle heat flux with distance from the arc. There is no physical basis for this type, or any other type, of boundary condition in the vicinity of a magmatic arc. The time-varying distribution of heat sources is almost certainly more complex than the static distribution used to calculate the isotherms in Fig. 26-7. The chosen exponential decrease simply illustrates the type of thermal environment next to an arc. The aluminosilicate reactions, pyrophyllite stability, and the granite minimum-melting curve from Fig. 26-6 are also placed on the cross section. The isograds are gently dipping, with the andalusite-sillimanite transition occurring in the Panamint Mountains and the kyanite-sillimanite transition in the Funeral Mountains. This arrangement of isotherms is consistent with the essentially isobaric nature of metamorphism in both the Panamint and Funeral Mountinas.

SUMMARY, CONCLUSIONS, AND PROBLEMS

Regional metamorphism in the Sierran batholith marginal belt in the Death Valley area is characterized by a large range in thermal gradients from >55°C/km in the Panamint Mountains to <30°C/km in the Funeral Mountains. Within each block, temperature increased westward with little or no increase in pressure, other than that resulting from differences in stratigraphic depth. However, the difference in pressure of ~3–4 kb between the Panamint and Funeral Mountains indicates that the overburden in the Funeral Mountains was about twice that in the Panamint Mountains.

Mineral assemblages in pelitic schists are sensitive to the great difference in pressure. In the Panamint Mountains, andalusite was the stable aluminosilicate at low grades, cordierite-bearing assemblages are common, chlorite broke down to biotite + an aluminous mineral over nearly the entire range in $Fe/(Fe + Mg)$ over a narrow range in temperature, chloritoid + biotite assemblages occur, and staurolite remained stable in the sillimanite zone. In the Funeral Mountains, kyanite was the stable aluminosilicate, cordierite-bearing assemblages are absent, chlorite-bearing assemblages were stable over a wide range in grade, chloritoid broke down to garnet + staurolite + chlorite, and staurolite broke down to kyanite + garnet + biotite.

Prograde metamorphism occurred during the emplacement of the Middle Jurassic batholith, although the Funeral Mountain terrain did not cool until Late Jurassic or Early Cretaceous time. During Late Cretaceous time, the metamorphic complexes were folded along northwest-trending axes. At this time, retrograde metamorphism and intrusion of leucocratic granite occurred in the Panamint Mountains, and K-Ar ages were extensively reset near Monarch Canyon in the Funeral Mountains. Thermal effects of Tertiary extension appear to be limited to the intrusion of the high-level Little Chief stock and associated dikes in the Panamint Mountains. In the Funeral Mountains, numerous pegmatite dikes were intruded in the Monarch Canyon area during middle Tertiary time, and retrograde metamorphism occurred along the sole of the Boundary Canyon fault and near the Keane Wonder Mine at Chloride Cliff.

Thrust faulting northwest of Death Valley and in the foreland east and south of Death Valley appears to have preceded metamorphism and intrusion of granitic rocks. Emplacement of the pre-Middle Jurassic Inyo nappe accounts for the great amount of overburden indicated by metamorphism in the Funeral Mountains. This nappe plunges to the south and apparently did not cover the Panamint Mountains, probably because the Panamint Mountains, constructed of a large block of Middle Proterozoic basement and its cover, remained a topographic high throughout much of geologic time. Unlike continental collision zones, the emplacement of the nappe cannot be solely responsible for metamorphism in the Death Valley area; a regional variation in the mantle heat flux associated with generation of the batholith must have existed. This variation in flux implies that horizontal thermal gradients existed, which accounts for the essentially isobaric range in temperatures in both the Panamint and Funeral Mountains.

The thermal history of the Death Valley region during the Late Cretaceous folding event and associated retrograde metamorphism is not well known. Labotka (1981) argued that retrograde metamorphism occurred as a result of an influx of H_2O during folding of the slowly cooling prograde terrain. The isotopic results of Labotka *et al.* (1985) indicate that if retrograde metamorphism occurred in this manner, extremely slow cooling rates, ~1°C/m.y., and implausibly large horizontal thermal gradients, >50°C/km, persisting for ~80 m.y., are required. Rather, the Late Cretaceous metamorphism probably represents a separate thermal event associated with the emplacement of the eastern Sierra batholith because retrograde metamorphism occurred during intrusion of the leucocratic granite and because this second metamorphism was locally higher grade than the first

(Labotka, 1981). The Funeral Mountains did not undergo extensive retrogradation at this time, possibly because the Funeral Mountains are farther from the Sierran batholith than the Panamint Mountains.

The relation between the metamorphic complex and the thrust belts, particularly the foreland thrust belt, is yet not clear. There is no evidence that uplift of the metamorphic terrain is in some way the cause of the foreland thrust belt. Indeed, the root of the east-directed Inyo nappe complex lies west of the metamorphic terrain. The thrust faults must owe their development to convergence and uplift of the batholithic complex itself.

The metamorphic complex in the Death Valley area is probably part of a once-continuous metamorphic belt marginal to the batholith; the metamorphic belt formed as a result of the thermal environment adjacent to the arc. The metamorphic belt is now exposed in discontinuous structural highs that formed during Tertiary uplift and extension. Except where the crust was thickened by pre-batholithic thrusting, the high temperatures of metamorphism were confined to the Middle Proterozoic basement. Mineralogic evidence of Mesozoic metamorphism in the basement gneisses is obscure because the mineral assemblages in granitic gneisses are stable over a wide range in grade. Only in places like Death Valley, where Late Proterozoic basins contain thick accumulations of sedimentary rocks, are the extensive effects of Mesozoic metamorphism evident. Although the perplexing problem relating the metamorphic terrain to the foreland thrust belt remains, the development of the regional metamorphic complex is demonstrably and logically related to the emplacement of the batholith.

ACKNOWLEDGMENTS

Research in the Panamint and Funeral metamorphic complexes has been supported over the past 25 years by grants from the National Science Foundation, the Geological Society of America, the National Aeronautics and Space Administration, and the Discretionary Fund of the Department of Geological Sciences, the University of Tennessee. E. DeWitt, B. Troxel, and L. Wright kindly provided us with unpublished data and many hours of lively discussion about Death Valley geology. The comments of W. G. Ernst and the participants in Rubey Colloquium VII are gratefully appreciated and indicate that much has yet to be understood about the thermal environment of continental margin arcs.

REFERENCES

Albee, A. L., 1965, A petrogenetic grid for the Fe-Mg silicates of pelitic schists: *Amer. J. Sci.*, v. 263, p. 512–536.

——, Labotka, T. C., Lanphere, M. A., and McDowell, S. D., 1981, Geologic map of the Telescope Peak quadrangle, California: *U.S. Geol. Survey Map Geol. Quad. GQ 1532.*

Albers, J. P., and Stewart, J. H., 1972, Geology and mineral deposits of Esmeralda County, Nevada: *Nev. Bur. Mines Geology Bull.*, v. 78, 80 p.

Bailly, P. A., 1951, Geology of the southeastern part of Mineral Ridge, Esmeralda County, Nevada: Ph.D. dissertation, Stanford Univ., Stanford, Calif.

Bird, G. W., and Fawcett, J. J., 1973, Stability relations of Mg-chlorite, muscovite, and quartz between 5 and 10 kb water pressure: *J. Petrology*, v. 14, p. 415–428.

Bohlen, S. R., Wall, V. J., and Boettcher, A. L., 1983, Experimental investigations and geological applications of equilibria in the system $FeO-TiO_2-Al_2O_3-SiO_2-H_2O$: *Amer. Mineralogist*, v. 68, p. 1049–1058.

Burchfiel, B. C., and Davis, G. A., 1971, Clark Mountain thrust complex in the Cordillera of southeastern California – Geologic summary and field guide: *Campus Mus. Contrib. 1*, Univ. California, Riverside, Calif., p. 1–28.

____, and Davis, G. A., 1977, Geology of the Sagamore Canyon-Slaughterhouse Spring area, New York Mountains, California: *Geol. Soc. America Bull.*, v. 88, p. 1623–1640.

____, Fleck, R. J., Secor, D. T., Vincelette, R. R., and Davis, G. A., 1974, Geology of the Spring Mountains, Nevada: *Geol. Soc. America Bull.*, v. 85, p. 1013–1022.

Chen, J. H., and Moore, J. G., 1982, Uranium-lead isotopic ages from the Sierra Nevada batholith, California: *J. Geophys. Res.*, v. 87, p. 4761–4784.

Clark, S. P., Jr., ed. 1965, Handbook of physical constants: *Geol. Soc. America Mem. 97*.

Cornwall, H. R., and Kleinhampl, F. J., 1961, Geology of the Bare Mountain quadrangle, Nevada: *U.S. Geol. Survey Map Geol. Quad. GQ-157*.

Dunne, G. C., Gulliver, R. M., and Sylvester, A. G., 1978, Mesozoic evolution of rocks of the White, Inyo, Argus, and Slate Ranges, eastern California, in Howell, D. G., and McDougall, K., eds., *Mesozoic Paleogeography of the Western United States:* Pacific Section, Soc. Econ. Paleontologists Mineralogists, Pacific Coast Paleogeography Symp. 2, p. 189–207.

England, P. C., and Richardson, S. W., 1977, The influence of erosion upon the mineral facies of rocks from different metamorphic environments: *J. Geol. Soc. London*, v. 134, p. 201–213.

____, and Thompson, A. B., 1984, Pressure-temperature-time paths of regional metamorphism – I. Heat transfer during evolution of regions of thickened continental crust: *J. Petrology*, v. 25, p. 894–928.

Ghent, E. D., and Stout, M. A., 1981, Geobarometry and geothermometry of plagioclase-biotite-garnet-muscovite assemblages: *Contrib. Mineralogy Petrology*, v. 76, p. 92–97.

Goldman, D. S., and Albee, A. L., 1977, Correlation of Mg/Fe partitioning between garnet and biotite with $^{18}O/^{16}O$ partitioning between quartz and magnetite: *Amer. J. Sci.*, v. 277, p. 750–767.

Hall, W. E., 1971, Geology of the Panamint Butte quadrangle, Inyo County, California: *U.S. Geol. Survey Bull. 1299*.

Hodges, K. V., Walker, J. D., and Wernicke, B. P., 1984, Tertiary folding and extension, Tucki Mountain area, Death Valley region, CA [abstract]: *Geol. Soc. America Abstr. with Programs*, v. 16, p. 540.

Holdaway, M. J., 1971, Stability of andalusite and the aluminum silicate phase diagram: *Amer. J. Sci.*, v. 271, p. 97–131.

Hunt, C. B., and Mabey, D. R., 1966, Stratigraphy and structure, Death Valley, California: *U.S. Geol. Survey Prof. Paper 494-A*.

Kerrick, D. M., 1968, Experiments on the upper stability limit of pyrophyllite at 1.8 kilobars and 3.9 kilobars water pressure: *Amer. J. Sci.*, v. 266, p. 204–214.

Labotka, T. C., 1980, Petrology of a medium-pressure regional metamorphic terrain, Funeral Mountains, California: *Amer. Mineralogist*, v. 65, p. 670–689.

____, 1981, Petrology of an andalusite-type regional metamorphic terrain, Panamint Mountains, California: *J. Petrology*, v. 22, p. 261–296.

____, and Albee, A. L., 1977, Late Precambrian depositional environment of the Pahrump Group, Panamint Mountains, California: *Calif. Div. Mines and Geology Spec. Rpt.*, v. 129, p. 93–100.

————, Albee, A. L., Lanphere, M. A., and McDowell, S. D., 1980, Stratigraphy, structure, and metamorphism in the central Panamint Mountains (Telescope Peak quadrangle), Death Valley area, California: *Geol. Soc. Amer. Bull.*, v. 91, pt. I, p. 125–129, pt. II, p. 843–933.

————, Warasila, R. L., and Spangler, R. R., 1985, Polymetamorphism in the Panamint Mountains, California — A ^{39}Ar-^{40}Ar study: *J. Geophys. Res.*, v. 90, p. 10359–10371.

Lanphere, M. A., 1964, Geochronologic studies in the eastern Mojave Desert, California: *J. Geology*, v. 72, p. 381–399.

————, Wasserburg, G. J., Albee, A. L., and Tilton, G. R., 1964, Redistribution of strontium and rubidium isotopes during metamorphism, World Beater Complex, Panamint Range, California, *in* Craig, H., Miller, S. L., and Wasserburg, G. J., eds., *Isotopic and Cosmic Chemistry:* Amsterdam, North-Holland, p. 269–320.

McAllister, J. F., 1956, Geology of the Ubehebe Peak quadrangle, California: *U.S. Geol. Survey Geol. Quad. Map GQ 95.*

McDowell, S. D., 1974, Emplacement of the Little Chief stock, Panamint Range, California: *Geol. Soc. America Bull.*, v. 85, p. 1535–1546.

Miller, J. M., 1983, Stratigraphy and sedimentology of the Upper Proterozoic Kingston Peak Formation, Panamint Range, eastern California: Ph.D. dissertation, Univ. California, Santa Barbara, Calif.

Moore, S. C., 1974, Syn-batholithic thrusting of Jurassic (?) age in the Argus Range, Inyo County, California [abstract]: *Geol. Soc. America Abstr. with Programs*, v. 6, p. 223.

Newton, R. C., and Haselton, H. T., 1981, Thermodynamics of the garnet-plagioclase-Al$_2$SiO$_5$-quartz geobarometer, *in* Newton, R. C., Navrotsky, A., and Wood, B. J., eds., *Thermodynamics of Minerals and Melts:* New York, Springer, p. 129–145.

Noble, L. F., 1941, Structural features of the Virgin Spring area, Death Valley, California: *Geol. Soc. America Bull.*, v. 52, p. 941–1000.

Oxburgh, E. R., and Turcotte, D. L., 1974, Thermal gradients and regional metamorphism in overthrust terrains with special reference to the eastern Alps: *Schweiz. Mineral. Petrogr. Mitt.*, v. 54, p. 641–662.

Price, R. A., 1973, Large-scale gravitational flow of supracrustal rocks, southern Canadian Rockies, *in* De Jong, K. A., and Scholten, R., eds., *Gravity and Tectonics:* New York, Wiley-Interscience, p. 491–502.

Roberts, R. J., and Crittenden, M. D., Jr., 1973, Orogenic mechanisms, Sevier orogenic belt, Nevada and Utah, *in* De Jong, K. A., and Scholten, R., eds., *Gravity and Tectonics:* New York, Wiley-Interscience, p. 409–427.

Scholten, R., 1973, Gravitational mechanisms in the northern Rocky Mountains of the United States, *in* De Jong, K. A., and Scholten, R., eds., *Gravity and Tectonics:* New York, Wiley-Interscience, p. 473–489.

Stern, T. W., Newell, M. F., and Hunt, C. B., 1966, Uranium-lead and potassium argon ages of parts of the Amargosa thrust complex, Death Valley, California: *U.S. Geol. Survey Prof. Paper 550-B*, p. 142–147.

Stewart, J. H., 1967, Possible large right-lateral displacement along fault and shear zones in the Death Valley-Las Vegas area, California and Nevada: *Geol. Soc. America Bull.*, v. 78, p. 131–142.

————, 1983, Extensional tectonics in the Death Valley area, California — Transport of the Panamint Range structural block 80 kilometers northwestward: *Geology*, v. 11, p. 153–157.

————, Ross, D. C., Nelson, C. A., and Burchfiel, B. C., 1966, Last Chance thrust — A major fault in the eastern part of Inyo County, California: *U.S. Geol. Survey Prof. Paper 550-D*, p. 23–24.

Wasserburg, G. J., Wetherill, G. W., and Wright, L. A., 1959, Ages in the Precambrian tertain of Death Valley, California: *J. Geology,* v. 67, p. 702–708.

Wright, L. A., and Troxel, B. W., 1967, Limitations on strike-slip displacement along the Death Valley and Furnace Creek fault zones, California: *Geol. Soc. America Bull.,* v. 78, p. 933–950.

―――― , Troxel, B. W., Burchfiel, B. C., Chapman, R. H., and Labotka, T. C., 1981, Geologic cross section from the Sierra Nevada to the Las Vegas Valley, eastern California to southern Nevada: *Geol. Soc. America Map Chart Series MC-28M.*

Howard W. Day, Peter Schiffman, and E. M. Moores
Department of Geology
University of California
Davis, California 95616

27

METAMORPHISM
AND TECTONICS
OF THE NORTHERN
SIERRA NEVADA

ABSTRACT

The northern Sierra Nevada can be divided into four major lithotectonic belts: (1) the Eastern Belt of Ordovician(?) through Jurassic age, (2) the Feather River Peridotite Belt of Devonian(?) to Jurassic age, (3) the Central Belt of Permian to Jurassic age, and (4) the Western Belt of Jurassic age. All of these rocks were deformed during the Late Jurassic Nevadan "orogeny." Evidence for pre-Nevadan metamorphism is preserved in each of the four lithotectonic belts, but Nevadan and post-Nevadan metamorphism are difficult to identify unambiguously.

Five types of regional metamorphism are recognized: (1) Permian or older amphibolite facies metamorphism of mafic and ultramafic rocks in the Feather River Belt, (2) pre-Late Triassic low-pressure metamorphism of older ophiolitic rocks in the Central Belt, (3) Middle Jurassic or older blueschist facies metamorphism of volcanic and sedimentary rocks in the Feather River Peridotite Belt, (4) prehnite-pumpellyite facies burial metamorphism of Paleozoic and Mesozoic volcanic arcs, and (5) Late Jurassic, Nevadan or younger, greenschist facies metamorphism.

Low-grade metamorphism in the Smartville Complex and other arc complexes of the northern Sierra represents a common style of metamorphism. Their regional prehnite-pumpellyite facies assemblages probably were caused by deep burial during construction of the volcanic edifice. There is no evidence for unusual thermal gradients during the arc metamorphism. The alteration of the associated dikes and other hypabyssal and plutonic rocks is interpreted as autometamorphism during the cooling of the intrusives.

Nevadan metamorphism is expressed as a weak, greenschist facies overprint of pre-Nevadan assemblages in the Western and Eastern belts. It is best developed in penetratively deformed argillaceous rocks in the Feather River Belt and in the Central Belt, where Nevadan metamorphic grade may have been highest. There is no evidence for unusually high- or unusually low-pressure metamorphism during Nevadan deformation.

INTRODUCTION

The rocks of the northern Sierra Nevada record a geologic history extending back to the early Phanerozoic. In general, that record contains evidence for both oceanic and continental margin tectonics and contains stratigraphic evidence of abundant volcanism in oceanic, near-continental, and continental environments. Four main lithotectonic belts are separated by prominent faults (Fig. 27-1; after H. W. Day *et al.*, 1985): the Eastern, Feather River, Central, and Western Belts. The oldest known rocks within each belt are younger in each successive belt to the west, reflecting substantially different geologic histories.

The metamorphism of the northern Sierra has long been considered to consist primarily of a low-grade recrystallization associated with the Late Jurassic Nevadan "orogeny" followed by contact metamorphism during the intrusion of shallow, post-tectonic plutons and the Sierra Nevada batholith. Nevadan assemblages are best developed and best recognized where unambiguous Nevadan deformation has produced a penetrative fabric. Although such metamorphism is widespread and regionally important, large tracts of rocks contain pre-Nevadan prehnite-pumpellyite, lawsonite-blueschist, epidote-amphibolite, and amphibolite facies assemblages. Low greenschist facies Nevadan assemblages overprint such assemblages in most areas. Pre-Nevadan metamorphism occurred prior to Late Jurassic time and, therefore, is not related directly to either the Nevadan orogeny or to the emplacement of the Cretaceous Sierra Nevada batholith. Four episodes of pre-Nevadan metamorphism affected different terranes. In addition, a

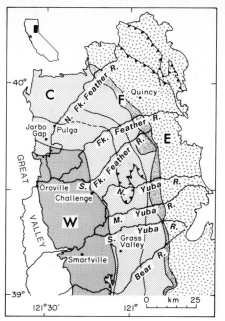

FIG. 27-1. Lithotectonic belts of the northern Sierra Nevada, California. After Day *et al.*, 1985. The Western, Central, Feather River Peridotite, and Eastern Belts are indicated by W, C, F, and E, respectively.

characteristic style of low-grade metamorphism associated with the construction of volcanic edifices may have occurred at different times in four volcanic terranes. At present, it is difficult to distinguish reliably among Nevadan and pre-Nevadan effects.

The purpose of this paper is to review the metamorphism of rocks in the northern Sierra and to evaluate the relationship of the metamorphism to the tectonic history of the region. An overview of the geology and structure of the four principal lithotectonic belts is followed by a summary of the metamorphism of the principal structural and lithologic units in the northern Sierra. An improved understanding of the metamorphic history will ultimately help constrain tectonic models for the evolution of the Sierra. This general region and that farther south are also treated by Harwood and by Schweickert *et al.* (Chapters 28 and 29, this volume).

The Western Belt

The Western Belt (Fig. 27-1) in the northern Sierra Nevada is occupied by the Smartville Complex (Beard, 1985; Cady, 1975; H. W. Day *et al.*, 1985; Moores, 1975; Schweickert and Cowan, 1975; Xenophontos, 1984), a Late Jurassic, rifted volcanic arc complex (Beard and Day, 1986). It is bounded on the north and east by the Big Bend–Wolf Creek fault zone (H. W. Day *et al.*, 1985; Hietanen, 1981; Tuminas, 1983) and is unconformably overlain on the west by Cretaceous and younger sedimentary rocks of the Great Valley sequence. The volcanic rocks (SV, Fig. 27-2) are basaltic to intermediate in composition and include a lower volcanic unit that is characterized by dominant pillowed flows, and an upper volcanic unit containing abundant pyroclastic volcanic rocks and volcanogenic, epiclastic, and volcanoclastic rocks (Xenophontos, 1984). In the vicinity of Lake Oroville, the upper volcanic unit corresponds in large part to the Bloomer Hill Formation of Hietanen (1977) and the Oregon City Formation mapped by Creely (1965). Included within the volcanic unit in Fig. 27-2 are the Monte De Oro Formation and related Jurassic sedimentary rocks that overlie the upper volcanic unit north of Oroville and may be the

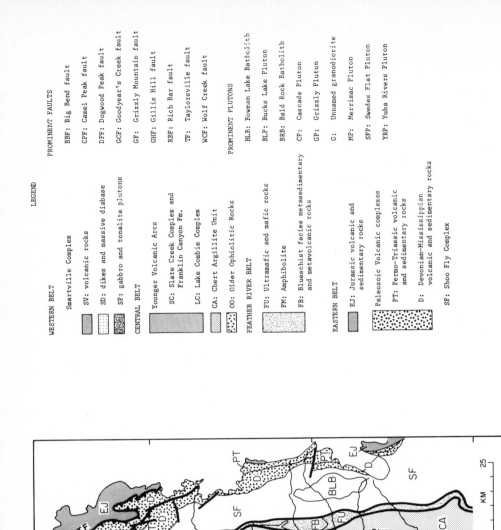

FIG. 27-2.　Simplified geologic map of the northern Sierra Nevada, California. Dots indicate the geographic localities shown in Fig. 27-1.

youngest Jurassic rocks in the Smartville Complex (Cole and McJunkin, 1979; Creely, 1965; Vaitl, 1980).

The volcanic rocks were intruded by an extensive suite of hypabyssal (SD) and plutonic rocks (SP) (Beard, 1985; Beard and Day, 1986). The oldest intrusive rocks are hypabyssal, massive diabase, and metagabbro. Those in turn were intruded by zoned, gabbro-diorite plutons, elongate bodies of tonalite, including "plagiogranite," and a unit of hypabyssal dikes that are mutually intrusive with the tonalite. Locally, the dike unit consists of 100% sheeted dikes and appears to record an episode of extension late in the history of the arc (Beard, 1985). Clasts of plutonic rocks occur in volcanic and epiclastic rocks of the uppermost(?) volcanic unit, suggesting that volcanism and plutonism are broadly coeval. The few fossils and isotopic measurements presently known suggest that the Smartville Complex is Oxfordian-Kimmeridgian or about 160 m.y. old (Creely, 1965; Marlette et al., 1979; McJunkin et al., 1979; Saleeby, 1981; Saleeby and Moores, 1979).

The basement on which the Smartville arc was deposited probably included already deformed ultramafic rocks. Serpentinized ultramafic rocks exposed in the Higgins Corner window (southeastern corner of the Smartville Complex, Fig. 27-2) were intruded by plutonic and hyabyssal Smartville rocks and are in fault contact with the chert-argillite unit of the Central Belt. North of the Complex, near Jarbo Gap, a dike that intrudes older ophiolitic rocks is lithologically similar to Smartville dikes and the same age[1] (Dilek, 1985). Plutonic or hypabyssal rocks of the Smartville Complex do not intrude the nearby chert-argillite unit in the Central Belt. Ricci (1983) specifically noted that dioritic dikes in the Bloomer Hill Formation, at the northernmost border of the Complex, are not found in the adjacent chert-argillite rocks of the Central Belt. South-southeast of the Complex, however, the Pine Hill gabbro intrudes and metamorphoses rocks that are probably equivalent to the chert-argillite unit of the Central Belt (Springer, 1980). The Pine Hill intrusion is lithologically similar to the gabbros in the Smartville Complex and the same age (Saleeby et al., 1981), but its relationship to the Smartville Complex is unknown.

The Central Belt

The Central Belt (Fig. 27-1) includes all those rocks west of the Feather River Belt and east of the Smartville Complex. It is separated from the Smartville Complex by the Big Bend–Wolf Creek fault zone (H. W. Day et al., 1985; Hietanen, 1977; BBF, WCF, Fig. 27-2) and from the Feather River Belt by the Rich Bar and Goodyears Creek faults (RBF, GCF, Fig. 27-2) and their poorly known extensions south of the North Yuba River (Hietanen, 1981). The Central Belt is stratigraphically and structurally complex. It is useful to distinguish three major units (H. W. Day et al., 1985): (1) an older unit of chert, argillite and undifferentiated metasedimentary rocks (CA, Fig. 27-2); (2) older ophiolitic rocks and undifferentiated mafic and ultramafic rocks that are tectonically interleaved with the chert argillite unit (00, Fig. 27-2); and (3) younger volcanic arc-ophiolite complexes that tectonically overlie the chert argillite unit (SC, LC, Fig. 27-2).

Chert-argillite unit The chert-argillite unit includes rocks mapped by Hietanen (1981) as Calaveras Formation, Horseshoe Bend Formation metasedimentary rocks, and Triassic metasedimentary rocks. South of the North Yuba River, these units were mapped by Chandra (1961) as Calaveras Group and by Tuminas (1983) as Clipper Gap Formation. The unit contains abundant chert and its metamorphic equivalents as well as diverse fine-grained metasedimentary rocks summarized as "argillite." The latter include siliceous argillite closely associated with chert, volcanogenic argillite containing relict volcanic

[1] Also J. B. Saleeby, personal communication, 1986.

phenocrysts, and pelitic argillites and phyllites containing abundant white mica. The three types of "argillite" are commonly difficult to distinguish in the field unless metamorphic minerals (e.g., actinolite) are present that reveal the bulk chemical composition.

Older ophiolitic rocks Underlying the younger volcanic arc complexes and tectonically interleaved with the chert-argillite unit are a few fault-bounded blocks of older ophiolitic rocks as well as numerous bodies of mafic and ultramafic rocks (OO, Fig. 27-2). Hietanen (*cf.* 1981 and earlier work) included these older ophiolitic rocks, some fragments of the younger volcanic arc, and parts of the chert-argillite unit within the Horseshoe Bend Formation. We interpret ultramafic and mafic rocks near Pulga, Jarbo Gap, and Challenge (Figs. 27-1 and 27-2) as fragments of older ophiolitic complexes (Dilek, 1985; Dilek and Moores, 1986; Mazaheri, 1982; Murphy, 1986; Murphy and Moores, 1985).

The complex south of Pulga (Hietanen, 1973; Mazaheri, 1982) consists of meta-basalts and a large serpentinized peridotite mass that was intruded by a distinctive, deformed clinopyroxene + hornblende gabbro. J. B. Saleeby (personal communication, 1983) determined an age of 202 m.y. (U-Pb zircon) for a small mass of silicic rock associated with the deformed gabbro, confirming that the complex is older than the Smartville Complex and related rocks. Dilek (1985) mapped ultramafic and mafic rocks south of Jarbo Gap that were considered by Hietanen (1977) to be Jurassic serpentinite and Permian(?) Horseshoe Bend Formation. He showed that serpentinite and mafic breccias were intruded by mafic dikes that are, in places, highly deformed. These rocks in turn were intruded by a younger, subhorizontal, undeformed plagiogranite dike. J. B. Saleeby (personal communication, 1983) determined a U-Pb age of 159 m.y. on zircon from the plagiogranite dike, suggesting that the deformed serpentinite and mafic dikes are older than the Smartville Complex. Murphy (1986) mapped a serpentinite matrix melange, informally named the Indian Creek melange near Challenge. The melange contains blocks of metalliferous chert, metavolcanic and hypabyssal rocks, metagabbro, pyroxenite, and serpentinized ultramafic rocks. Murphy (1986) argued that the melange formed prior to the Early Jurassic because (1) the melange is inferred to lie structurally and statigraphically beneath less disrupted, Upper Triassic–Lower Jurassic (?) meta-sedimentary rocks; (2) argillaceous sedimentary debris is completely absent in the ophiolitic melange, suggesting that the melange formed at some distance from the depo-centers of the adjacent rocks; and (3) the Indian Creek rocks are similar to known older ophiolitic rocks elsewhere in the Sierra.

Younger volcanic arc complexes Tectonically overlying the chert-argillite unit and older ophiolitic rocks are Jurassic ultramafic, plutonic, and volcanic rocks in the Slate Creek Complex and the Lake Combie Complex that are interpreted by H. W. Day *et al.* (1985) as fragments of volcanic arc-ophiolite. The Slate Creek Complex (SC) is bounded on the west by the Camel Peak fault zone (CPF) and on the east, in part, by the Dogwood Peak fault (DPF). In the region south of the South Fork of the Feather River, the volcanic rocks in the Slate Creek Complex correspond approximately to the Franklin Canyon Formation (Hietanen, 1981) and it seems reasonable to speculate that a large part of the Franklin Canyon Formation, as well as associated metagabbros and ultramafic rocks to the north, are directly related to the Slate Creek Complex. A similar sequence of ultra-mafic, plutonic, and volcanic rocks, called the Lost Creek block, lies just west of the Cascade pluton and is also considered part of the Slate Creek Complex (H. W. Day *et al.*, 1985; Murphy, 1986).

The Jurassic Lake Combie Complex (LC, Fig. 27-2; Tuminas, 1983) is bounded on the west by the Smartville Complex along the Wolf Creek fault zone (WCF) and on the east by chert argillite and undifferentiated metasedimentary rocks along the Gillis Hill faults (GHF). The lowermost rocks of the Lake Combie Complex are ultramafic tectonites that were intruded and overlain by a mafic plutonic unit. Hypabyssal dikes are common and locally form a unit of chilled but unsheeted dikes and massive diabase near the top of the plutonic unit. A thick sequence (>5 km) of mafic volcanic and epiclastic rocks overlies the plutonic and hypabyssal units. Also included in the Lake Combie Complex in Fig. 27-2 are Upper Jurassic metasedimentary rocks of the Colfax sequence (H. W. Day et al., 1985) that are probably equivalent to the Mariposa Formation farther south in the Sierra.

The oldest known rocks in the Central belt are Lower Permian. Standlee and Nestell (1985) reported well-preserved Early Permian fusulinids from marble in the chert-argillite unit between the Middle Fork and the South Fork of the Feather River (Hietanen, 1981). Watkins et al. (1986) also reported Early Permian fusulinids from the chert-argillite unit near Lake Oroville. The chert-argillite unit in which the fossiliferous limestone blocks and coherent beds is found, however, has so far produced only Mesozoic fauna and the marbles may be olistostromal blocks. Irwin et al. (1978) identified Late Triassic or Jurassic radiolaria from cherts along the North Fork of the Feather River. C. Blake (personal communication, 1986) has found late Paleozoic radiolaria in nearby rocks, but the extent of these rocks is unknown. Hietanen (1981) reported two localities near the North Yuba River in which Middle to Late Triassic and Late Triassic to Early Jurassic fauna were identified in chert. Near Challenge, Murphy (1986) found poorly preserved Mesozoic radiolaria in rocks assigned to the chert-argillite unit south of the South Fork of the Feather River, west of the Cascade Pluton.

The Feather River Belt

The Feather River Belt (Fig. 27-1) is a fault-bounded terrane that separates the relatively coherent sequence of Paleozoic and Mesozoic rocks in the Eastern Belt from highly deformed, primarily Mesozoic rocks in the Central Belt to the west for approximately 150 km along strike in the northern Sierra Nevada (H. W. Day et al., 1985). It clearly represents a major tectonic boundary.

The belt is dominated by metaperidotite, metagabbro, amphibolite, and serpentinite (FU, Fig. 27-2) in the northern part. Near the North Yuba River, it also contains multiply deformed blueschist, fine-grained quartzite, and fissile slate that is tectonically imbricated with serpentinite and associated mafic rocks (FB, Fig. 27-2). Near the intersection of the Rich Bar Fault (RBF) and the Goodyears Creek fault (GCF), on Slate Creek, Hietanen (1981) mapped a large body of amphibolite and associated metavolcanic rocks (FM, Fig. 27-2) that is in contact with serpentinite, with metasedimentary rocks assigned to the Shoo Fly Complex of the Eastern Belt and with metasedimentary rocks of the Central Belt. S. H. Edelman (unpublished data) found sheeted dikes in the amphibolite and interpreted it as an ophiolite fragment. Although its relationship to the Feather River Periodotite is not well established, for the purposes of this discussion, we have included the amphibolite in the Feather River Belt. Standlee (1978) reported an $^{40}Ar/^{39}Ar$ age of 278 m.y. (revised using new decay constants, L. A. Standlee, personal communication, 1986) on hornblende from this unit, confirming a Permian or older age.

The ages of the most abundant rocks in this belt are not well known. Standlee[2] (1978) reported an $^{40}Ar/^{39}Ar$ age of 395 m.y. on hornblende from a dike in the peridotite, suggesting that the peridotite is Devonian or older. Hornblende from a foliated mafic inclusion in the peridotite gave an $^{40}Ar/^{39}Ar$ cooling age of 236 m.y. (Weisenberg, 1979). Schweickert et al. (1980) reported K-Ar ages as young as 174 m.y. from white mica in the schists near the North Yuba River that must represent a minimum age for the blueschist metamorphism.

[2] Also L. A. Standlee, personal communication, 1986.

The Eastern Belt

The Eastern Belt (Fig. 27-1) is bounded on the west by a prominent fault zone that separates it from the Feather River Belt, and on the east by Tertiary volcanic rocks and Mesozoic granitic rocks of the Sierra Nevada batholith. The oldest known rocks in the Eastern belt are Ordovician(?) to Devonian(?) siliciclastic metasedimentary rocks of the Shoo Fly Complex (SF, Fig. 27-2). The Complex consists of quartz-rich metasedimentary rocks and shale matrix melange containing serpentinite, limestone, tuffaceous phyllite, and chert. Although variably recrystallized and deformed, these rocks still retain many of their primary textures, and the provenance and tectonic setting of the unit have been extensively debated (e.g., Bond and Devay, 1980; Girty and Wardlaw, 1984, 1985; Schweickert and Snyder, 1981).

Unconformably overlying the Shoo Fly Complex are volcanic rocks and associated intrusives that were formed during three major episodes of magmatic activity. The earliest volcanism is recorded in the Late Devonian to Early Mississippian extrusive and intrusive rocks in the Sierra Buttes, Taylor, and Peale Formations (D, Fig. 27-2); Later episodes are found in the Permian and Triassic volcanic rocks in the Arlington, Goodhue, and Reeve Formations (PT, Fig. 27-2) and in the Jurassic volcanic rocks and volcaniclastic sedimentary rocks in the Mount Jura area and Sailor Canyon Formation (EJ, Fig. 27-2). Overviews of the stratigraphy, structure, and tectonic significance of these arc complexes have been presented by McMath (1966), D'Allura *et al.* (1977), Schweickert and Snyder (1981), Varga and Moores (1981), and Harwood (1983). They probably represent volcanic arc rocks formed marginal to North America. All rocks were deformed in early Late Jurassic time by east-vergent folds and thrust faults.

STRUCTURE

Most of the Late Jurassic rocks in the Sierra Nevada were deformed during the "Nevadan orogeny" (Clark, 1960; Taliaferro, 1942), the essential expression of the Jurassic amalgamation of oceanic terranes and the Mesozoic continental margin (Moores, 1970, 1972; Saleeby, 1981; Schweikert and Cowen, 1975; Schweickert, 1981). Much of the deformation occurred in the interval 155 ± 3 m.y. (Schweickert *et al.*, 1984). All the major lithotectonic belts of the northern Sierra contain rocks that have also suffered significant pre-Nevadan deformation and, presumably, metamorphism.

Pre-Nevadan Deformation

All four of the major lithotectonic belts of the northern Sierra Nevada were probably deformed prior to the Nevadan orogeny. Commonly, however, the principal expression of the earlier deformations is a cleavage or metamorphic foliation that dips steeply and strikes NNW. In the absence of independent evidence, however, such foliations most often cannot be assigned unambiguously to Nevadan or pre-Nevadan episodes of deformation.

Beard (1985) argued that volcanic rocks in the Smartville Complex were folded prior to the intrusion of the sheeted dike complex. Neither that folding nor Nevadan deformation produced penetrative fabrics in the Smartville rocks, except in widely separated, through-going shear zones. No pre-Nevadan deformation has yet been documented in the equivalent younger volcanic arc complexes of the Central Belt, although the ultramafic basement of these complexes contains tectonites that are older than the igneous activity[3] (Murphy, 1986; Tuminas, 1983). Older ophiolitic rocks in the Central

---

[3] Also S. H. Edelman, unpublished data.

Belt, however, contain clear evidence for pre-Nevadan deformation at Pulga (Mazaheri, 1982), Jarbo Gap (Dilek, 1985; Dilek and Moores, 1986), and near Challenge (Murphy, 1986; Murphy and Moores, 1985). The Mesozoic chert-argillite unit in the Central Belt commonly contains a pronounced foliation of indeterminate age. Weisenberg and Ave Lallèment (1977) argued for a pre-Nevadan mantle origin for tectonite fabrics in the Feather River peridotite. Important pre-Nevadan deformation is implied by the 175-m.y. minimum ages of foliated blueschists (Schweickert et al., 1980) and the 278-m.y. age of foliated amphibolite (Standlee, 1978) in this belt. Regional reconstructions suggest that steep faults separating the Eastern, Feather River, and Central Belts, as well as subunits within the chert-argillite unit, must have accommodated profound movements of pre-Nevadan age (S. H. Edelman, unpublished data). Pre-Nevadan deformation in the Eastern Belt is evident in the record of Paleozoic unconformities and polyphase deformation (Harwood, 1983; Varga, 1985; Varga and Moores, 1981). Although the nature and extent of pre-Nevadan deformation is poorly understood, the possibility of multiple episodes of metamorphism accompanying these deformations must be addressed.

Nevadan Deformation

The most obvious Nevadan structures in the northern Sierra Nevada are the steep faults of the Foothills fault system (Clark, 1960) that bound the lithotectonic belts and control the major outcrop patterns. They deform Upper Jurassic rocks and are truncated by latest Jurassic and earliest Cretaceous plutons. The faults dip steeply east and record reverse motion in their latest movements. Mineral lineations in these fault zones plunge steeply in the down-dip direction of the associated foliations. Although the total displacement on these faults is unknown, the late Nevadan displacement was probably not profound because several Late Jurassic plutons and Jurassic arc-ophiolites in the Central and Western belts are deformed but not significantly offset by these steep faults. Some of these faults had a significant pre-Nevadan history but they deform early Nevadan, generally east-directed, overthrusts. In our view, the presently observed orientation and sense of movement on the steep faults are late Nevadan features.

Thrust faults are more common in the northern Sierra than previously recognized (H. W. Day et al. 1985; Moores and Day, 1984). In the Eastern belt, the Paleozoic section is repeated across the Grizzly fault (GF, Fig. 27-2) and lies in thrust contact over Jurassic rocks on the Taylorsville fault (TF) (Diller, 1892). Steep, late Nevadan faults deform earlier thrust faults in the Western and Central Belts (H. W. Day et al., 1985; Moores and Day, 1984). The principal evidence for those thrust faults is the presence of arc ophiolite fragments in the Central Belt (Slate Creek Complex, Lake Combie Complex, and Lost Creek block) that may correlate with the Smartville Complex to the west. These correlated arc-ophiolite suites then represent a marker unit that helps unravel the structure in a region that is strongly deformed by high-angle faults.

The evidence for thrust faulting includes several low- to moderate-angle faults that placed various units of the Lake Combie Complex over chert-argillite (H. W. Day et al., 1985; Tuminas, 1983), a window through the southeastern Smartville Complex exposing chert-argillite near Higgins Corner (Beard, 1985; H. W. Day et al., 1985; Tuminas, 1983), and a detailed gravity study of the northern contact (Ricci et al., 1985), which shows that the bottom of the Smartville dips at low angles to the west. Finally, the basal ultramafic units of the Slate Creek Complex and the Lost Creek block overlie the chert argillite unit and there is no evidence that magmas directly related to the remainder of the overlying arc-ophiolite units have intruded the metasedimentary rocks. This relationship is best explained as a thrust fault.

If the Smartville, Lake Combie, and Slate Creek Complexes are remnants of the same volcanic arc, at least several tens of kilometers of across-strike shortening is implied. How much more shortening might have been accommodated on these faults is presently indeterminate. However, regional geophysical studies suggest that the lower contact of the Western belt rocks can be traced for large distances to the south and west of the Smartville Complex. The gravity and magnetic data of Suppe (1979) and Griscom (personal communication, 1983) suggest that there is a dense, highly magnetic slab in the Foothills dipping west beneath the Great Valley, perhaps corresponding to the Smartville Complex and related rocks. Farther south, in the central Sierra, Zoback *et al.* (1985) have found prominent west-dipping reflectors beneath the Great Valley and the Central Sierra Foothills that may correspond to the top and bottom, respectively, of the volcanic and related plutonic rocks in the Western Belt.

METAMORPHISM

Systematic studies that specifically address the metamorphic petrology of rocks in the northern Sierra are rare. However, brief descriptions of metamorphic features are present in many regional studies focusing on structural, statigraphic, and sedimentologic problems. Ernst (1983) previously summarized metamorphism in the Sierra Nevada as part of a larger compilation of metamorphism in California. However, this report represents the first attempt at a metamorphic synthesis for the northern Sierra.

Western Belt: The Smartville Complex

Volcanic rocks The upper and lower volcanic units (SV, Fig. 27-2) contain massive flows, pillowed flows, and abundant volcaniclastic and epiclastic deposits of basaltic and intermediate composition. They are only weakly deformed in most places and original volcanic structures and textures are well preserved. Even the most delicate quench textures are commonly preserved in the glassy margins of pillows.

The rocks are also only very weakly recrystallized but contain widespread prehnite and pumpellyite[4] (Beiersdorfer, 1982; Xenophontos, 1984). Epidote, albite (An_{01}-An_{07}), and chlorite are virtually ubiquitous and actinolite is rare. Sphene, magnetite, pyrite, carbonate, and potassium feldspar are common accessory minerals. Actinolite is common, however, where the volcanic rocks are deformed in late Nevadan shear zones, for example, in the western Smartville Complex and the eastern margin of the Complex, adjacent to the Yuba Rivers pluton. Hietanen (1977) and Ricci (1983) report that actinolite is common in the Bloomer Hill Formation west of the Bald Rock batholith; this may reflect contact effects of the batholith. Beiersdorfer (1982) and Xenophontos (1984) also report minor actinolite in some samples of the upper volcanic unit. Zeolites are conspicuously absent in the Smartville volcanic rocks.

Both prehnite and pumpellyite are widespread in the volcanic rocks from Oroville to the Bear River. Both occur in veins and vesicles as well as in albitized plagioclase. Pumpellyite commonly occurs as acicular grains in radial aggregates and, rarely, as substantial prisms. It is typically a deep pleochroic green and may have been mistakenly identified in some studies as green clinozoisite or epidote. Some palagonite in formerly glassy rocks is also a similar deep green color and might be cryptocrystalline pumpellyite.

Figure 27-3 shows data on the compositions of pumpellyite + chlorite, prehnite + chlorite and actinolite + chlorite that coexist with albite, epidote, and quartz, projected from albite, epidote, quartz, and H_2O to the plane $(Al_2O_3 + Fe_2O_3)-(FeO + MnO)-MgO$ (Beiersdorfer, 1982; Beiersdorfer and Day, 1983). In iron-rich compositions, the associa-

---

[4]Also H. W. Day and P. Schiffman, unpublished data.

tion chlorite + pumpellyite is common, whereas rarer magnesium-rich rocks contain chlorite + amphibole. The prehnite + chlorite pair does not appear to be compatible with pumpellyite + chlorite in this projection, and may not represent the same equilibrium conditions. Alternatively, the projection used here or the recalculations of microprobe analyses may not adequately represent the role of ferric iron in the mineral assemblages.

An important point to be made using this diagram is that not all occurrences of albite + epidote + chlorite + actinolite necessarily correspond to the greenschist facies. Dashed lines connect the compositions of stilpnomelane, chlorite, and actinolite from the lowest-grade greenschists in Vermont analysed by Laird (1980). The chlorite + actinolite pair from the Smartville volcanics is equally compatible with the lower greenschist pairs from Vermont or the pumpellyite + chlorite tielines. The pumpellyite + chlorite pairs are clearly not compatible with the Vermont greenschists on this projection. It may not be possible to distinguish lower greenschist facies assemblages and the pumpellyite-bearing assemblages indicative of lower grades, in magnesium-rich compositions. Insufficient data are available to judge whether or not iron-rich compositions are generally more sensitive indicators of low-grade metamorphism in mafic rocks.

Dike unit The Smartville dike unit (SD, Fig. 27-2) contains fine-grained massive and diabasic rocks as well as abundant cross-cutting and sheeted dikes of basaltic and intermediate compositions (Beard, 1985; Xenophontos, 1984). Silicic dikes are much less common (S. D. Day, 1977). Scattered dikes intrude both volcanic units and rocks of both volcanic units are included as screens within the dike complex.

Dikes commonly contain the alteration assemblage albite + epidote + actinolite + chlorite ± quartz, sphene, magnetite, and carbonate. The alteration is pervasive in some dikes but is not restricted to the dike unit proper. Scattered dikes that intruded the upper and lower volcanic units also exhibit the alteration assemblage. The intrusion of scattered dikes creates barely discernible contact effects and does not appear to affect the regionally developed prehnite + pumpellyite alteration in the volcanic host rocks. The compositions of coexisting chlorite and amphibole in these dikes are compared to chlorite and pumpellyite in the volcanic rocks in Fig. 27-3. The crossing tielines confirm our suggestion that the alteration of the dikes occurred during cooling at higher temperatures than the pumpellyite assemblages. Late alteration in some plutons may also have formed in this way.

We interpret the alteration of the dikes as an autometamorphic phenomenon in which the dikes served as their own source of heat, and interaction with surrounding fluids permitted the recrystallization (Beiersdorfer and Day, 1983). No similar recrystallization of the host rocks has been found, presumably because they were not hot enough. The simplest interpretation of the available data is that the dikes and their associated alteration apparently postdate the regional prehnite + pumpellyite alteration.

Plutonic rocks Pre-Nevadan gabbro and tonalite plutons (SP, Fig. 27-2) created contact aureoles usually only a few hundred meters wide. They contain two-pyroxene and hornblende hornfelses that overprint the regionally developed prehnite + pumpellite facies metamorphism (Beard, 1985). Secondary amphiboles in the plutons are apparently more aluminous than those found in the dikes, consistent with higher temperatures of autometamorphism or contact effects of later plutons.

Central Belt

Younger volcanic arc complexes The Lake Combie Complex (LC, Fig. 27-2), the Slate Creek Complex, and the Lost Creek block, as well as the Franklin Canyon Formation

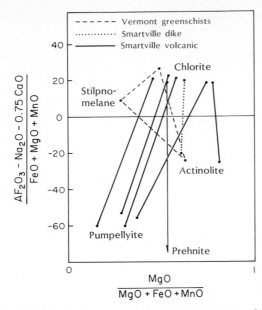

FIG. 27-3. Compositions of coexisting minerals from the Smartville Complex. Projection from albite, epidote, quartz, and H_2O to the plane $Al_2O_3 + Fe_2O_3$, FeO + MnO, and MgO. Smartville data from Beiersdorfer, 1982; Vermont greenschists from Laird, 1980.

(SC, Fig. 27-2), appear to represent fragments of a volcanic arc that are equivalent to the Smartville Complex. Actinolite and biotite are more widely reported in the younger volcanic arc complexes of the Central Belt than in the Smartville Complex and pumpellyite has not yet been confirmed (Hietanen, 1973, 1974, 1976, 1981). Hietanen (1973, 1981) reported the widespread occurrence of albite + epidote + actinolite ("rarely hornblende") + chlorite + quartz + magnetite in metatuffs and the groundmass of meta-andesites in the Franklin Canyon Formation. Metabasalts and meta-andesites commonly contain more than one epidote group mineral and more than one amphibole, even outside contact aureoles (Hietanen, 1974, 1976). The volcanic units of the Slate Creek Complex (S. H. Edelman, unpublished data) and the Lost Creek block (Murphy, 1986) are for the most part weakly recrystallized and mildly deformed. They contain widespread albite + epidote + actinolite ± chlorite ± calcite ± biotite. Delicate skeletal quench textures are preserved by plagioclase phenocrysts in pillow lavas. Tuminas (1983) reported the assemblage albite + epidote + actinolite + chlorite in metabasites from the Lake Combie Complex, and noted veins of calcite + quartz, epidote + quartz, and prehnite.

Chert-argillite unit The chert-argillite unit of the Central Belt, like the overlying arc complexes is also slightly higher in grade than the Smartville Complex. Pelitic rocks commonly contain quartz + albite + white mica + chlorite + biotite and rare garnet (Hietanen, 1973, 1981). Volcanogenic argillite contains unrecrystallized volcanic clasts and prehnite is an uncommon but widespread metamorphic mineral in the Yuba Rivers area (S. H. Edelman, unpublished). Hacker (1984) reported sporadic biotite in the argillites between the North and South Yuba Rivers. Siliceous argillite and chert contain primarily quartz, relics of micro-fossils that are nearly totally recrystallized, minor white mica, and opaque minerals.

Older ophiolitic rocks The older ophiolitic rocks (OO, Fig. 27-2) described in this paper were included by Hietanen (1973, 1977, 1981) within the Horseshoe Bend Formation and generally contain assemblages of the greenschist and epidote amphibolite facies. Mazaheri (1982) reported that massive serpentinites in the Pulga area contain antigorite + brucite + magnetite, but where the serpentinites are deformed in Nevadan fault zones, talc + tremolite + carbonate ± chlorite is the typical assemblage. Metagabbro contains hornblende + andesine (An_{33-47}) + clinozoisite ± epidote ± quartz in the core of the pluton and, in the foliated margins, hornblende + actinolite + oligoclase (An_{12-20}) + clinozoisite + chlorite ± quartz. These assemblages record deformation during late- and postmagmatic cooling of the pluton. Foliated metavolcanic rocks near Pulga contain hornblende + albite-oligoclase ($An_{06} - An_{15}$) + epidote ± quartz ± chlorite. In later foliations, near Nevadan faults, the assemblage is actinolite + albite (An_{04-10}) + epidote + chlorite + quartz. The earlier assemblage is similar to epidote-amphibolite facies assemblages where the later assemblage is a typical greenschist facies assemblage. Many of the rocks contain two plagioclases or optically determined plagioclase compositions that lie within the range of peristerite pairs. The extent to which these compositions represent equilibrium pairs or unrecognized peristerite intergrowths is not known.

Farther south in the Central Belt, near Jarbo Gap, dikes intruding mafic breccias and serpentinite contain similar greenschist facies assemblages (Dilek, 1985). Murphy (1986) found uncommon, isoclinally folded metavolcanic blocks in serpentinite matrix melange that contain quartz + "sodic plagioclase" + green hornblende or actinolite + chlorite. Feathery overgrowths of actinolite on the green hornblende are common. Other undifferentiated metavolcanic rocks interleaved with the chert-argillite formations contain the same assemblage in addition to epidote/clinozoisite ± biotite.

Feather River Belt

Ultramafic rocks Metaperidotites near North Fork of the Feather River (FU, Fig. 27-2) contain serpentine, talc, chlorite, tremolite, and anthophyllite that define a metamorphic foliation and replace primary phases (Ehrenberg, 1975; Weisenberg, 1979). Relict olivine + tremolite ± enstatite are preserved in peridotites, and are commonly replaced by serpentine and talc, suggesting that amphibolite facies metamorphism preceded the more ubiquitous greenschist facies recrystallization (Ehrenberg, 1975). Widespread occurrences of tremolite + olivine + chlorite schists within the peridotite predate the most recent serpentinization (Weisenberg, 1979).

Mafic rocks Foliated mafic inclusions within the peridotite body contain evidence for more than one stage of metamorphism. Hornblende + plagioclase (An_{15-36}) + diopside and quartz + hornblende + garnet + diopside are early amphibolite facies assemblages. Actinolite (uralite) + albite (An_{01-04}) + epidote is a later greenschist facies assemblage (Ehrenberg, 1975). Mafic and calcareous schists, which are faulted against the peridotite body along its western margin, have a similar recrystallization history. Most of those rocks contain green hornblende + albite ($Ab_{0.5-11}$) + epidote + chlorite. Garnet occurs together with the hornblende in several samples but diopside was found in only one sample (Ehrenberg, 1975). Calcareous metasediments contain quartz + oligoclase (An_{17-22}) + garnet (containing up to 10 wt % CaO) ± chlorite ± muscovite ± hornblende. Albite is common in samples containing garnet and hornblende. Hornblende is altered to actinolite ± chlorite; plagioclase is sericitized, and garnet is replaced by chlorite + stilpnomelane.

Hietanen (1981) described a large "intrusive body" of Permian amphibolite on the west margin of the Feather River Belt (FM, Fig. 27-2). S. H. Edelman (unpublished data) found sheeted dikes in exposures of this amphibolite and interpreted it as an ophiolite fragment. The amphibolite mass is strongly foliated and contains hornblende (ferroan pargasite) + plagioclase (An_{30}) + quartz + epidote + sphene. It is partially bounded by mafic volcanic and volcanogenic sedimentary rocks that contain epidote + hornblende + plagioclase (An_{28}) + chlorite + sphene + magnetite + quartz. These assemblages suggest lower amphibolite facies recrystallization.

Blueschists South of the amphibolite body, Hietanen (1981) described a 13-km-long zone of metasedimentary and metavolcanic rocks within the Feather River Belt, near the North Yuba River (FB, Fig. 27-2). The metasedimentary rocks include cherts and micaceous quartzites that contain phengite + chlorite ± stilpnomelane. Schweickert *et al.* (1980) report quartz + muscovite ± chlorite, stilpnomelane, albite, epidote, sphene, lawsonite, and blue amphibole in similar rocks.

Metavolcanic rocks exposed near the North Yuba River at Goodyears Bar, Rosassco Ravine, and Downieville have relict igneous textures but also contain metamorphic assemblages that appear to reflect different *P-T* conditions. Hietanen (1981) reported crossite + actinolitic hornblende + chlorite + epidote + albite; chlorite + actinolite + albite + quartz; albite + epidote + calcite; actinolite + chlorite + epidote + albite; lawsonite + chlorite + phengite + quartz + stilpnomelane; pumpellyite + crossite + chlorite + sphene. Schweickert *et al.* (1980) report (1) albite + epidote + chlorite + quartz + stilpnomelane in massive basaltic lava with fracture fillings of crossite + stilpnomelane + chlorite, (2) metabasaltic breccias with unfoliated clasts consisting of albite + epidote + chlorite + stilpnomelane + prehnite and foliated matrices also containing glaucophane-crossite + lawsonite + muscovite + quartz, and (3) strongly foliated metatuffs containing albite + epidote + chlorite + muscovite + quartz + glaucophane.

Sierran blueschist facies rocks were first recognized approximately 10 km south of the North Yuba River by Ferguson and Gannett (1932), who described "glaucophane schists" as part of the wallrock assemblage in Alleghany district gold deposits. The blue amphibole occurs as rims on hornblende or actinolite within foliated amphibolites (Coveney, 1981). These rocks record a complicated series of recrystallization events, including contact metamorphism by pre-Nevadan granitic plutons that are foliated and contain stilpnomelane (Coveney, 1981). Mica schists in the Alleghany district contain the assemblage white mica + chlorite + quartz + sphene + albite + lawsonite (S. E. Rosenbaum, unpublished data). Rosenbaum also found metadiabase/metagabbro containing pumpellyite + albite + chlorite + quartz + sphene, pumpellyite veins are cross-cut by Cretaceous quartz-sulfide-gold(?) veins. The relationship of blueschist facies rocks south of the North Yuba River to those farther north is unknown.

Eastern Belt

Shoo Fly Complex The Shoo Fly Complex (SF, Fig. 27-2) contains greenschist and sub-greenschist facies assemblages; most studies have concentrated on its sedimentology. The most common rocks of the Complex are deformed quartz-rich metasediments. Phyllites northwest of Quincy contain quartz + white mica ± chlorite, biotite, epidote, albite, and sphene (Hietanen, 1973). Chlorite and white mica are parallel to the dominant foliation; biotite transects and is apparently younger than this foliation. Reddish-brown stilpnomelane occurs locally instead of biotite. Adjacent to the Feather River Belt, some metasediments contain coarse-grained muscovite and chlorite lying in the dominant north-northwest-striking foliation and metamorphic biotite that lies in a north-northeast-striking, later crenulation cleavage (D. S. Harwood, personal communication, 1986). In the Quincy district, two foliations occur locally, each containing aligned muscovite and stilpnomelane (D'Allura, 1977). Detrital biotite is replaced by chlorite and sphene. Other metamorphic minerals include chlorite, calcite, epidote, actinolite, magnetite, and sphene. Devay (1981) described the petrology of quartzose metasandstones between the North Yuba River and the East Branch of the Feather River. Matrix (metamorphic?) phyllosilicates in these rocks include muscovite and/or phengite, chlorite, and stilpnomelane. Between the Middle and South Yuba Rivers, Shoo Fly metasediments locally record three phases of

Mesozoic deformation and are regionally metamorphosed to "chlorite-grade" greenschist assemblages (Girty and Schweickert, 1984).

Most workers propose that the Shoo Fly Complex was metamorphosed at greenschist facies conditions. We feel that the available data do not necessarily warrant this conclusion. The presence of biotite or stilpnomelane, if unambiguously authigenic, indicates greenschist facies metamorphism. However, the commonly reported assemblage, quartz + white mica + chlorite, is equally compatible with prehnite-pumpellyite facies metamorphism. In appropriate bulk compositions, actinolite also may be present in the transition between the prehnite-pumpellyite and greenschist facies. This transition is generally characterized by illite crystallinity and vitrinite reflectance techniques in argillaceous sediments (e.g., Kisch, 1983). More detailed metamorphic study of the Shoo Fly Complex, incorporating these techniques, is clearly warranted.

Paleozic-Mesozoic volcanic arc complexes The Shoo Fly Complex is overlain by three volcanic-arc complexes of Devonian-Mississipian (D, Fig. 27-2), Permo-Triassic (PT), and Jurassic (EJ) age. East of Quincy, the Devonian-Mississipian and Permo-Triassic arc sequences are repeated by thrusting on the Grizzly Mountain Fault (D'Allura, 1977; Hannah, 1980). On both sides of the fault, mafic to silicic volcanic rocks contain prehnite-pumpellyite or greenschist facies mineral assemblages. Much of the primary igneous fabric is preserved in these rocks although some deformational fabrics are locally superimposed.

Hannah (1980) mapped the Devonian-Mississipian and Permo-Triassic volcanic arc sequences between the Grizzly Mountain and Taylorsville faults along Keddie Ridge. In this region, the volcanic rocks are much less intensely deformed than correlative rocks west of the Grizzly Mountain fault. The former are moderately folded and most unfoliated, whereas the latter are strongly foliated, flattened, and isoclinally folded. On Keddie Ridge, volcanic rocks of the two arc complexes contain prehnite-pumpellyite or low-grade greenschist assemblages. Silicic rocks contain quartz + albite + sericite ± calcite and chlorite; mafic rocks contain quartz + albite + epidote + chlorite ± calcite. Prehnite and more rarely pumpellyite occur in the groundmass and within veins and amygdules. Zeolites are absent; rare yellow-brown palagonite is present with pumpellyite in some glassy volcanic rocks. Volcanic rocks in the Goodhue Formation (Permo-Triassic arc) lack pumpellyite and contain uralitized clinopyroxene and rare biotite and, locally, contain abundant actinolite. Thus, they may be slightly higher grade than similar lithologies within the Taylor Formation (Devonian-Mississipian arc) (D'Allura, 1977). However, volcanic rocks in the Goodhue Formation are much more magnesian than in the Taylor Formation (D'Allura, 1977) and the differences in mineralogy probably reflect the differences in chemistry (*cf.* Fig. 27-3).

In addition to the observed differences in deformation style and intensity, Hannah (1980) argued for a significant discontinuity in metamorphic grade across the Grizzly Mountain fault. Actinolite + ferrostilpnomelane are present on the west side of the fault and prehnite-pumpellyite on the east. The reason for these differences is not clear, however, and actinolite is known to be present locally on the east side of the fault. Such variations may be attributable to differences in bulk chemical composition, to nearby buried, hypabyssal plutons or to factors such as metastable equilibria or variations in fluid pressure, chemical activity, and oxygen fugacity resulting from permeability-controlled gradients, and are a not uncommon feature of burial metamorphism in volcanic arcs (Levi *et al.*, 1982). The presence of major differences in deformation style as well as the mineralogical differences suggests that small differences in grade may exist across the fault.

The Jurassic volcanic arc complex east of the Taylorsville Fault (EJ, Fig. 27-2) is composed of a 4–5-km-thick section of intermediate to silicic volcanogenic sediments (McMath, 1966) that is intruded to the east by the Sierra Nevada batholith. McMath (1958) reported that plagioclase phenocrysts from dacitic tuff breccias of the Jurassic lower Kettle Formation were replaced by authigenic albite and alkali feldspar; epidote was noted as being conspicuously absent in these rocks. Plagioclase phenocrysts in lithologically similar rocks from the upper part of the Kettle Formation are less pervasively albitized. These observations suggest that the Jurassic volcanic arc complex east of the Taylorsville Fault may have been metamorphosed at slightly lower grade than the Devonian-Mississippian and Permo-Triassic arcs exposed to the west.

The metamorphic grade in the Eastern Belt apparently does not increase appreciably toward the Sierra Nevada batholith across the Eastern Belt. In fact, the available data suggest that the metamorphic grade decreases from perhaps as high as biotite grade in the greenschist facies near the Feather River Belt, to prehnite-pumpellyite facies or lower east of the Taylorsville Fault. Reported variations in metamorphic grade within the Devonian-Mississippian and Permo-Triassic arc sequences may be controlled by differences in bulk chemical composition of the volcanic rocks or by local nonequilibrium effects consistent with a burial metamorphic origin. We see no evidence for any regional metamorphism related to batholithic emplacement. Rather, the observed patterns of metamorphism in this region are consistent with a model of pre-Nevadan static burial metamorphism with locally superimposed Nevadan dynamic metamorphism of similar grade.

Contact Metamorphism

Regionally metamorphosed Paleozoic through Jurassic rocks of the northern Sierra Nevada have locally undergone contact metamorphism adjacent to Jurassic and Cretaceous plutons. We have attempted to obtain geobarometric information from available mineral assemblages reported for northern Sierra contact metamorphic aureoles in order to constrain maximum burial depths of country rocks.

The metavolcanic rocks of the Smartville Complex were intruded by both pre-Nevadan and post-Nevadan plutons. The former are zoned gabbro-diorite and tonalite plutons that, as discussed above, produce contact aureoles less than a few hundred meters wide containing two-pyroxene and hornblende hornfels (Beard, 1985). Metabasaltic rocks adjacent to the post-Nevadan Bald Rock batholith (BRB, Fig. 27-2) contain amphibolite facies assemblages (hornblende + An_{40}) (Compton, 1955; Hietanen; 1973). Swanson (1970) reported the assemblage andalusite + biotite + muscovite + plagioclase + quartz + graphite in black slates intercalated with Jurassic volcanic rocks adjacent to the Early Cretaceous Rocklin pluton (south of the Bear River in the Western Belt). Mafic volcanic rocks in the aureole of the Rocklin pluton contain the amphibolite facies assemblage hornblende + plagioclase + diopside.

In the Central Belt, pelitic rocks in the contact aureoles of a number of Late Jurassic or Early Cretaceous plutons contain andalusite + staurolite (in the aureole of the Merrimac pluton, MP, Fig. 27-2), andalusite + corderite, and andalusite + anthophyllite, in addition to muscovite + biotite + quartz (Hietanen, 1973, 1976). Sillimanite, with andalusite + corderite, is reported only from the southern margin of the Bucks Lake pluton (BLP). Bobbitt (1982) reported staurolite + andalusite + garnet + biotite + muscovite + quartz from the contact aureole of the Yuba Rivers pluton (YRP, Fig. 27-2). The occurrence of staurolite + biotite + andalusite in these aureoles suggests pressures less than the aluminum silicate triple point but above the lower pressure stability of staurolite + quartz, perhaps about 3 kb.

The Eastern Belt contains a number of Paleozoic as well as Mesozoic plutons (Hannah, 1980; McMath, 1958). The contact aureole in quartz-rich metasediments of the Shoo Fly Complex, adjacent to the Devonian Bowman Lake batholith (BLB, Fig. 27-2), contains mineral assemblages indicative of the "albite-epidote hornfels" facies (Girty and Schweickert, 1984). Granitic plutons associated with the Cretaceous Sierra Nevada batholith also have andalusite-bearing contact aureoles. For example, black slates of the Jurassic Trail Formation, east of the Taylorsville fault, contain andalusite + chloritoid in contact aureoles (McMath, 1958) and farther south, in the Walker Mine district, similar Jurassic rocks contain contact metamorphic andalusite + garnet + muscovite + biotite + quartz (Don Pietz, personal communication, 1986).

The Northern Sierran contact metamorphic aureoles described above uniformily reflect high crustal levels for pluton emplacement as indicated by pelitic assemblages containing andalusite ± staurolite. The presence of andalusite in metapelitic assemblages constrains maximum depth of burial no more than about 14 km at the time of intrusion (Holdaway, 1971).

DISCUSSION

We have interpreted many of the mafic rocks of the Western and Central Belts of the northern Sierra Nevada as remnants of volcanic arc-ophiolites and older ultramafic-plutonic complexes formed in oceanic settings. It is natural to ask how much of the metamorphic mineralogy and fabric of these rocks was acquired in the environment where they formed, prior to their final amalgamation during the Late Jurassic Nevadan orogeny. We believe that rocks in the northern Sierra have a significant history of pre-Nevadan metamorphism and deformation. Our efforts to summarize the episodes and physical conditions of metamorphism are illustrated in Figs. 27-4 and 27-5, respectively.

Time of Metamorphism

Evidence for pre-Nevadan metamorphism is clear in the rocks of the Smartville Complex. The volcanic rocks contain prehnite and pumpellyite, are not penetratively deformed in most places, and were intruded by dikes and plutons during pre-Nevadan extension (Beard and Day, 1986). Although actinolite is not rare in the complex, it is most abundant in dikes and plutonic or hypabyssal diabase where it developed by autometamorphic alteration during cooling of the intrusions. Actinolite in the volcanic rocks apparently is limited to the contact aureoles of plutons (Bobbitt, 1982; Compton, 1955) or to shear zones interpreted to be Nevadan or younger. The contact aureoles and the alteration of the intrusive rocks are younger than the regional metamorphism of the volcanic rocks. J. B. Saleeby (personal communication, 1986) has determined U-Pb zircon ages of about 160 m.y. on both the volcanic rocks and the dikes. The prehnite and pumpellyite, therefore, appear to reflect deep burial metamorphism during the pre-Nevadan construction of a thick volcanic pile.

The metamorphism of the Smartville units taken together constitutes a distinctive style of metamorphism that might be characteristic of volcanic arc terranes. It is especially notable that there is no evidence for a steep metamorphic gradient through the volcanic section of the sort that is common in hydrothermally altered oceanic crust. The Smartville volcanic units may be as much as 6.5 km thick (Xenophontos, 1984) and there is little evidence for systematic mineral zonation with increasing depth. On the other hand, in remnants of the California Coast Range ophiolite, the metamorphic gradient, and probably also the geothermal gradients, attending submarine hydrothermal metamorphism exceeded 100°C/km. Zeolites are common at the top of the stratigraphic section and the index minerals pumpellyite, epidote, and actinolite appear successively through a stratigraphic interval of about 2 km (Schiffman et al., 1984, 1986). The Smartville

Age	WESTERN BELT	CENTRAL BELT	FEATHER RIVER BELT	EASTERN BELT
	Contact Aureoles	Contact Aureoles		Contact Aureoles
K — 143				
— 152		NEVADAN DEFORMATION AND METAMORPHISM		
	Greenschist Facies Shear Zones	Widespread Middle to Upper Greenschist Overprint	Lower Greenschist	Lower Greenschist Facies Overprint
J — 158				
	Burial Metamorphism of Smartville (?) Younger Arc-Ophiolites Prehnite-Pumpellyite Facies			Very low grade Burial Metamorphism of Jurassic Volcanic Rocks
— 160				
— 175			Blueschist Facies Met & Def (Pre-175 MA)	
— 212		Amphibolite Facies Metamorphism of Late Tr. (?) Older Ophiolite		Prehnite-Pumpellyite Facies Burial Metamorphism of Permo-Triassic Volcanic Rocks
Tr	Amphibolite Facies Metamorphism of Late Tr. (?) Older Ophiolite			
— 247			Permian or Older Amphibolite Facies Metamorphism of Mafic/Ultramafic Rocks	
P				
C				
D				Prehnite-Pumpellyite Facies Burial Metamorphism of Dev.-Miss. Volcanic Rocks

FIG. 27-4. Episodes of metamorphism in the northern Sierra Nevada. As outlined in the text, the Central Belt may also contain pre-Nevadan greenschist facies metamorphism. In the Eastern Belt, multiple episodes of burial metamorphism have been assigned to arc complexes primarily by analogy with the Smartville complex.

754

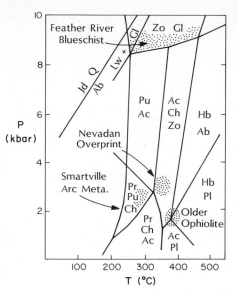

FIG. 27-5. Estimated stability fields of mineral assemblages in a model, iron-free metabasalt, after Liou et al., 1985. All assemblages coexist with quartz + albite + H_2O. Ab, albite; Ac, actinolite; Ch, chlorite; Gl, glaucophane; Hb, hornblende; Lw, lawsonite; Pl, oligoclase; Pr, prehnite; Pu, pumpellyite; Q, quartz; Zo, zoisite or epidote. Addition of iron to this model system would affect the location of the boundaries illustrated, but probably not their topology. In particular, the fields in which glaucophane appears would probably shift toward lower pressures.

volcanic rocks were metamorphosed under conditions more consistent with burial metamorphism. Zeolite facies metamorphism may have occurred in the stratigraphically highest levels of the Smartville volcanics, rocks now eroded or covered by Cretaceous sedimentary rocks in the Great Valley.

Nevadan metamorphism was probably important in the Central Belt, but there is also clear evidence for pre-Nevadan deformation and recrystallization. The older ophiolitic rocks near Pulga and Challenge (Mazaheri, 1982; Murphy, 1986) contain assemblages suggesting that the pre-Nevadan metamorphism led to assemblages transitional between epidote-amphibolite and amphibolite facies. The presence of actinolite + oligoclase in the deformed margin of a gabbro pluton suggests relatively low-pressure conditions (Fig. 27-5). Actinolite and albite occur in later foliations and as overgrowths on earlier minerals. This, together with the presence of biotite in undifferentiated mafic rocks, apparently represents greenschist facies metamorphism of Nevadan age.

The chert-argillite unit with which the older ophiolitic rocks are interleaved also contains actinolite and biotite in foliations that are interpreted as Nevadan(?). This suggests that greenschist facies metamorphism in the Central Belt was widespread and a consequence of the Nevadan deformation. Biotite is reported sporadically in the region between the South and North Yuba Rivers (Hacker, 1984); the reported occurrence of prehnite and relict volcanic features in these rocks may indicate a substantially lower grade of metamorphism than that farther north (Edelman, 1986).

Younger volcanic arc complexes in the Central Belt that are correlated with the Smartville Complex may also have suffered a similar burial metamorphism. Unambiguous burial metamorphic assemblages have not yet been recognized, although the presence of prehnite suggests such conditions. Actinolite and biotite are much more common in these rocks than in the Smartville Complex to the west, and they have sufferred a greenschist facies overprint of Nevadan or possibly post-Nevadan age. This conclusion is consistent with the deep burial (12–14 km) of the western part of the Central Belt during Nevadan deformation inferred from the presence of staurolite + andalusite in the aureoles of the late- or post-Nevadan Yuba Rivers and Merrimac plutons.

Aside from the paucity of detailed studies, several other factors must temper our view of this Nevadan(?) greenschist facies metamorphism. First, it is clear from our examination of the Smartville Complex that greenschist facies assemblages in fine-grained hypabyssal mafic rocks and associated plutons can be autometamorphic features acquired during the cooling of the igneous rocks. If a subsequent metamorphism is low grade, it may be very difficult to distinguish the regional metamorphic effects from earlier alteration. The most sensitive indicators of subsequent metamorphism are the volcanic rocks, which, like the Smartville volcanic units, may have suffered previously only very low-grade burial metamorphism. In many parts of the northern Sierra, however, it is difficult to distinguish whether the protoliths of fine-grained mafic rocks were volcanic or hypabyssal. Second, the chert-argillite unit is older than the Smartville rocks and equivalent younger volcanic arc complexes in the Central Belt, and probably suffered the older deformations that are recorded in the pre-Nevadan melanges (Murphy, 1986) and in the blueschists of the adjacent Feather River Belt (Schweickert *et al.*, 1980). The fabrics of these two deformations are nearly parallel and, for the most part, are difficult to distinguish. In much of the Central Belt, therefore, our present understanding does not permit a reliable distinction between pre-Nevadan and Nevadan metamorphism of the chert-argillite unit. Finally, our present understanding of the Central Belt does not permit us to ascertain the degree to which metamorphism is correlated with the abundance of syn- and post-tectonic plutons.

The Feather River Peridotite and associated rocks are the highest-grade regionally metamorphosed rocks in the northern Sierra, and clearly represent an exotic metamorphic belt as well as marking a fundamental tectonic break in the region. The argon "ages" on amphiboles (Hietanen, 1981; Standlee, 1978; Weisenberg and Avé Lallèment, 1977) suggest that the amphibolite or epidote amphibolite facies metamorphism was Permian or older (Fig. 27-4), but the data permit considerable latitude in interpretation. The relationship of the blueschists to the ultramafic and mafic rocks is clearly tectonic. Deformation and recrystallization are pre-Nevadan and no younger than 175 m.y. (Schweickert *et al.*, 1980). Nevadan greenschist facies metamorphism followed. Finally, serpentinization of peridotites occurred in response to large influxes of CO_2-rich fluids that were also responsible for quartz-gold veins (S. E. Rosenbaum, unpublished data) during the Early Cretaceous (e.g., K-Ar and Rb-Sr dates between 140 and 110 m.y. reported for hydrothermal minerals by Evans and Bowen (1977) and Bohlke *et al.* (1984)).

Greenschist facies assemblages in the peridotite and associated mafic rocks and in the quartz-mica schists may be related to Nevadan deformation, but the observation of blue amphibole rims on actinolite in the Alleghany district suggests that greenschist metamorphism also took place prior to 175 m.y. ago. The 175-m.y. argon ages on white mica from rocks of the Feather River Belt (Schweickert *et al.*, 1980) suggest that they last cooled below 350°C at about that time. Alternatively, subsequent burial and heating was neither sufficiently long nor sufficiently hot to reset older ages completely.

The Eastern Belt was regionally metamorphosed at low grades, but the relative contributions of Nevadan and Pre-Nevadan metamorphism cannot yet be distinguished. Several lines of evidence suggest that pre-Nevadan metamorphism may have been important. Varga and Moores (1981) found foliated clasts of Shoo Fly meta-argillite in the Devonian Sierra Buttes Formation suggesting pre-Devonian deformation at unknown metamorphic conditions. Such clasts typically contain the same mineral assemblage as the host Shoo Fly (D. S. Harwood, personal communication, 1986). The sporadic occurrence of biotite, especially in the western part of the Shoo Fly, chlorite after biotite, and stilpnomelane suggest that most of the Shoo Fly contains lower greenschist facies assemblages. The Devonian-Mississippian, Permian-Triassic, and Jurassic volcanic rocks all contain very low-grade prehnite-pumpellyite assemblages with sporadic actinolite and biotite. These observations are consistent with multiple episodes of burial metamorphism during formation of the volcanic arcs followed by Nevadan(?) overprinting of lower greenschist facies metamorphism on both the volcanic rocks and the Shoo Fly Complex. Nevadan thrust faults separate rocks that are more penetratively deformed and possibly slightly higher grade to the west, consistent with pre-Nevadan metamorphism and deformation.

Conditions of Metamorphism

If the Nevadan orogeny were directly related to a subduction event, we might expect to find a record of unusual metamorphic conditions during Nevadan metamorphism. A remarkable aspect of northern Sierran metamorphic rocks is the lack of evidence for unusual geothermal gradients (Fig. 27-5). Although contact aureoles reflect unusually high temperatures at the crustal level of intrusion, the regionally developed assemblages accompanying deformation during the Nevadan orogeny show no evidence for the unusually low temperatures and high pressures that are characteristic of subduction or the unusually high temperatures and low pressures that reflect regions of high heat flow in a paired metamorphic belt.

The coexisting prehnite and pumpellyite in the Smartville Complex suggests temperatures of about 275°C and pressures about 2.5 kb (Liou *et al.*, 1985), corresponding to about 9.5 km depth and an apparent thermal gradient of about 29°C/km. The pre-Nevadan metamorphism of the Jurassic arcs is therefore consistent with normal burial metamorphic gradients and there is no evidence for gradients approaching 100°C/km typical of hydrothermal alteration in the sea floor environment. The presence of actinolite + albite in shear zones through the Smartville suggests that temperatures may have been somewhat higher during Nevadan deformation, but recrystallization was incomplete because the rocks are for the most part not penetratively deformed.

The Central Belt appears to have been more deeply buried and hotter during Nevadan deformation than the other belts. In the rocks tectonically underlying the Jurassic arcs, the abundance of actinolite suggests that temperatures were about 375°C (Liou *et al.*, 1985); contact aureoles formed at about 12 km depth. Such conditions also represent gradients typical of normal burial. In the Nevadan shear zones, actinolite everywhere coexists with albite, nowhere with oligoclase, arguing against unusually low pressures of metamorphism.

The presence of anthophyllite in the Feather River metaperidotite indicates pre-Nevadan temperatures on the order of 650°C and pressures less than about 10 kb (H. W. Day *et al.*, 1985), consistent with amphibolite facies assemblages in the coexisting mafic rocks (Fig. 27-5). Hornblende coexists with albite in some mafic rocks suggesting epidote-amphibolite facies assemblages (Fig. 27-5), and a later greenschist facies overprint is ubiquitous. Currently available studies do not permit us to test whether or not the metamorphism of this major ultramafic body is similar to that expected during the emplacement of ophiolitic slabs in oceanic or near continental environments.

Conditions of metamorphism in the Eastern Belt are poorly constrained and no information suggests unusually shallow or unusually deep burial for these rocks at any time during their history. The presence of the pair prehnite + pumpellyite in the volcanic arc complexes suggests that conditions were similar to those in the Smartville, but in the Eastern Belt, assemblages more commonly contain actinolite or stilpnomelane and are transitional to the greenschist facies.

Evidence for unusual conditions of metamorphism comes primarily from pre-Nevadan metamorphism of rocks in the Central and Feather River Belts. The occurrence of actinolite + oligoclase in the deformed margins of a 200-m.y.-old gabbro near Pulga (Mazaheri, 1982) suggests that the older ophiolitic rocks in the Central Belt may have formed at very low pressures (Fig. 27-5).

The blueschists in the Feather River Belt contain the most important evidence of unusual conditions of metamorphism in the northern Sierra. They contain glaucophane + lawsonite, but the absence of jadeite suggests that the rocks did not reach pressures greater than about 10 kb (Fig. 27-5). The common occurrence of crossite or glaucophane with clinozoisite or epidote suggests, however, that pressures were at least 6-8 kb (Brown, 1977; Liou *et al.* 1985). The assemblages reported, and discussed previously, probably were not formed at the same physical conditions. The presence of lawsonite + glaucophane suggests that temperature may have been less than 250°C, based on the iron-free model phase diagram (Fig. 27-5). However, the presence of quartz, albite, and chlorite together with epidote + crossite, epidote + actinolite, or actinolite + crossite, as well as the occurrence of stilpnomelane, suggests metamorphic temperatures on the order of 350°C. The fact that white mica in these rocks preserves argon ages of about 175 m.y. suggests that temperatures have not substantially exceeded 350°C since that time (Schweickert *et al.*, 1980).

Tectonic Implications

Pre-Nevadan metamorphism is well preserved in each of the four lithotectonic belts of the northern Sierra Nevada. The highest-grade rocks in the northern Sierra are the ultramafic and mafic rocks in the Feather River Belt. Metamorphic grade in the Eastern as well as the Western and Central Belts increases toward the Feather River Belt. Clearly, the latter marks an important tectonic and metamorphic discontinuity.

The presence of blueschists strongly suggests that the Feather River Belt includes the remnants of a subduction complex. The subduction event was pre-Nevadan, however, and no evidence now exists for high-pressure metamorphism of the rocks on either side of the Feather River Belt. These observations are consistent with the proposal by S. H. Edelman (unpublished data) that the steeply east-dipping faults bounding the Feather River Belt and in the Central Belt represent cryptic pre-Nevadan faults that accommodated only relatively minor Nevadan movements. Those faults juxtapose units that were deposited far from one another and must have accommodated a large but unknown amount of shortening (subduction?) prior to Nevadan thrusting.

The occurrence of actinolite + oligoclase in the margins of a pluton at Pulga, in the Central Belt, suggests that the associated pre-Nevadan metamorphic rocks may have been metamorphosed at very low pressures. The metamorphic evidence is therefore consistent with the other geologic evidence, suggesting that some of the older ophiolitic rocks were formed in a fracture zone setting (Dilek and Moores, 1986).

Nevadan assemblages are very difficult to distinguish reliably from pre-Nevadan metamorphism in most belts. A weak greenschist facies overprint that is Nevadan or younger is probably present in the Western Belt, and Jurassic rocks in the Eastern Belt contain greenschist assemblages similar to those in pre-Jurassic rocks. Where rocks are more penetratively deformed in the Feather River and Central Belts, Nevadan greenschist facies assemblages are better developed and possibly higher grade, but still very difficult to distinguish from pre-Nevadan metamorphic assemblages. Aside from contact aureoles of plutons on the east side of the Eastern Belt, alteration assemblages associated with gold-bearing(?) veins in the Allegheny district of the Feather River Belt are the only known Cretaceous metamorphic effects that might be associated with the intrusion of the Sierra Nevada batholith.

We have proposed that the Nevadan deformation in the northern Sierra included early Nevadan, east-directed overthrusting in which the Smartville Complex was the highest known thrust nappe and that the thrust fault complex was deformed subsequently by steep east-dipping faults (H. W. Day *et al.*, 1985; Moores and Day, 1984). The inference that the Central Belt was more deeply buried and hotter than the Smartville Complex during the Nevadan deformation is consistent with its proposed burial by thrust nappes of Smartville and related rocks. There is no evidence that the Feather River Peridotite and Eastern Belts were buried more deeply than the Central Belt during or after the Nevadan orogeny. The Eastern Belt contains metamorphic assemblages similar to those of the Smartville Complex. It seems unlikely, therefore, that the Eastern Belt was ever buried significantly deeper than the Smartville Complex.

Pre-Nevadan metamorphism is a significant element in the history of the amalgamated northern Sierra Nevada. Burial metamorphism of the Smartville Complex, and possibly other arc complexes of the northern Sierra, is genetically related to the thermal and hydrothermal regimes that existed during the construction of the volcanic edifice. No evidence exists for unusual geothermal gradients during Nevadan deformation. Consequently, neither a subduction environment nor a high-heat-flow regime seems to have been present during the Nevadan orogeny. Because the data on which our suggestions are based are so sparse, however, our proposals are clearly a call for further work.

ACKNOWLEDGMENTS

We are grateful to W. G. Ernst for the opportunity to construct this review. It would not have been possible without the support and hard work of our colleagues at U.C. Davis, especially J. S. Beard, S. H. Edelman, J. Hannah, A. C. Tuminas, R. Varga, and C. Xenophontos. C. Blake, W. G. Ernst, D. S. Harwood, V. Hoeck, and S. Sorenson kindly provided useful and constructive reviews. Most of our work was carried out with the financial support of NSF grants EAR78-03640 and EAR80-19697 (HWD and EMM) and EAR85-16262 (HWD).

REFERENCES

Beard, J. S., 1985, The geology and petrology of the Smartville intrusive complex, northern Sierra Nevada foothills, California: Ph.D. dissertation; Univ. California, Davis, Calif., 356 p.

____ , and Day, H. W., 1986, Petrology and origin of gabbro pegmatite in the Smartville instrusive complex, northern Sierra Nevada: *Amer. Mineralogist*, v. 71, p. 1085–1099.

Beiersdorfer, R. E., 1982, Metamorphic petrology of the Smartville Complex, northern Sierra Nevada foothills, California: M.Sc. thesis; Univ. California, Davis, Calif., 152 p.

____ , and Day, H. W., 1983, Pumpellyite-actinolite facies metamorphism in the Smartville Complex, northern Sierra Nevada: *Geol. Soc. America Abstr. with Programs*, v. 15, p. 436.

Bobbitt, J. B., 1982, Petrology, structure and contact relations of part of the Yuba Rivers Pluton, northwestern Sierra Nevada foothills, California: M.Sc. thesis; Univ. California, Davis, Calif., 160 p.

Bohlke, J. K., Kistler, R. W., and McKee, E. H., 1984, K-Ar, Rb-Sr, and stable isotope data on the ages and fluid sources of gold-quartz veins in the Sierra Nevada foothills metamorphic belt, California: *Geol. Soc. America Abstr. with Programs*, v. 16, p. 448–489.

Bond, G. C., and Devay, J. C., 1980, Pre-upper Devonian quartzose sandstones in the Shoo Fly Formation northern California — Petrology, provenance, and implications for regional tectonics: *J. Geology*, v. 88, p. 285–308.

Brown, E. H., 1977, Phase equilibria among pumpellyite, lawsonite, epidote, and associated minerals in low grade metamorphic rocks: *Contrib. Mineralogy Petrology*, v. 64, p. 123–136.

Cady, J. W., 1975, Magnetic and gravity anomalies in the Great Valley and western Sierra Nevada metamorphic belt, California: *Geol. Soc. America Spec. Paper 168*, 56 p.

Chandra, D. K., 1961, Geology and mineral deposits of the Colfax and Foresthill quadrangles: *Calif. Div. Mines Geology Spec. Paper 67*, p. 50.

Clark, L. D., 1960, Foothills fault system, western Sierra Nevada, California: *Geol. Soc. America Bull.*, v. 71, p. 483–496.

Cole, K. A., and McJunkin, R. D., 1979, Geology of the Lake Oroville area, Butte County, California (Map): *Calif. Div. Mines Geology Bull. 203-78*, pl. 1.

Compton, R., 1955, Trondhjemite batholith near Bidwell Bar, California: *Geol. Soc. America Bull.*, v. 66, p. 9–44.

Coveney, R. M., 1981, Gold quartz veins and auriferous granite at the Oriental mines, Alleghany district, California: *Econ. Geology*, v. 76, p. 2176–2199.

Creely, R. S., 1965, Geology of the Oroville quadrangle, California: *Calif. Div. Mines Geology Bull. 184*, 86 p.

D'Allura, J. A., 1977, Stratigraphy, structure, petrology and regional correlations of metamorphosed Upper Paleozoic volcanic rocks in portions of Plumas, Sierra, and Nevada Counties, California: Ph.D. dissertation; Univ. California, Davis, Calif., 338 p.

―――, Moores, E. M., and Robinson, L. S., 1977, Paleozoic Rocks of the northern Sierra Nevada, their structural and paleogeographic implications, *in* Stewart, J. H., Stevens, C. H., and Fritsche, A. E., eds., *Paleozoic Paleogeography of the Western United States:* Pacific Section, Soc. Econ. Paleontologists Mineralogists, Pacific Coast Paleogeography Symp. 1; p. 396–408.

Day, H. W., Moores, E. M., and Tuminas, A. C., 1985, Structure and tectonics of the northern Sierra Nevada: *Geol. Soc. America Bull.*, v. 96, p. 436–450.

Day, S. D., 1977, Petrology and intrusive complexities of sheeted dikes in the Smartville ophiolite, northwestern Sierra foothills, California: M.Sc. thesis; Univ. California, Davis, Calif., 113 p.

Devay, J. C., 1981, The petrology and provenance of the Shoo Fly Formation quartzose sandstones, northern Sierra Nevada, California: M.Sc. thesis; Univ. California, Davis, Calif., 110 p.

Dilek, Y., 1985, Structure and petrology of the Big Bend fault and associated mafic dike complex, northern Sierra Nevada foothills, California: M.Sc. thesis; Univ. California, Davis, Calif., 276 p.

―――, and Moores, E. M., 1986, A possible allochthonous oceanic transform fault zone in the northwestern Sierra Nevada, California: *Geol. Soc. America Abstr. with Programs*, v. 18, p. 101.

Diller, J. S., 1892, Geology of the Taylorville region of California: *Geol. Soc. America Bull.*, v. 3, 369–394.

Ehrenberg, S. M., 1975, Feather River ultramafic body, northern Sierra Nevada, California: *Geol. Soc. America Bull.*, v. 86, p. 1235–1243.

Ernst, W. G., 1983, Phanerozoic continental accretion and metamorphic evolution of northern and central California: *Tectonophysics*, v. 100, p. 287–320.

Evans, J. R., and Bowen, O. E., 1977, Geology of the southern Mother Lode, Tuolomne, and Mariposa Counties, California (Map): *Calif. Div. Mines Geology Map Sheet 36*.

Ferguson, H. G., and Gannett, R. W., 1932, Gold quartz veins of the Alleghany district, California: *U.S. Geol. Survey Prof. Paper 172*, 139 p.

Girty, G. H., and Schweickert, R. A., 1984, The Culbertson Lake allochthon, a newly identified structure within the Shoo Fly Complex, California — Evidence for four phases of deformation and extension of the Antler orogeny to the northern Sierra Nevada: *Modern Geology*, v. 8, p. 181–198.

____, and Wardlaw, M. S., 1984, Was the Alexander terrane a source of feldspathic sandstones in the Shoo Fly Complex, Sierra Nevada, California? *Geology*, v. 12, p. 339–342.

____, and Wardlaw, M. S., 1985, Petrology and provenance of pre-Late Devonian sandstones, Shoo Fly Complex, northern Sierra Nevada, California: *Geol. Soc. America Bull.*, v. 96, p. 516–521.

Hacker, B. R., 1984, Stratigraphy and structure of the Yuba Rivers area, Central Belt, northern Sierra Nevada, California: M.Sc. thesis; Univ. California, Davis, Calif., 125 p.

Hannah, J. L., 1980, Stratigraphy, petrology, paleomagnetism, and tectonics of Paleozoic arc complexes, Northern Sierra Nevada, California: Ph.D. dissertation; Univ. California, Davis, Calif., 323 p.

Harwood, D. S., 1983, Stratigraphy of Upper Paleozoic rocks and regional unconformities in part of the northern Sierra terrane, California: *Geol. Soc. America Bull.*, v. 94, p. 413–422.

Hietanen, A., 1951, Metamorphic and igneous rocks of the Merrimac area. Plumas National Forest, California: *Geol. Soc. America Bull.*, v. 67, p. 565–607.

____, 1973, Geology of the Pulga and Bucks Lake quadrangles, Butte and Plumas counties, California: *U.S. Geol. Survey Prof. Paper 731*, 66 p.

____, 1974, Amphibole pairs, epidote minerals, chlorite and plagioclase in metamorphic rocks, northern Sierra Nevada, California: *Amer. Mineralogist*, v. 59, p. 22–40.

____, 1976, Metamorphism and plutonism around the middle and south forks of the Feather River, California: *U.S. Geol. Survey Prof. Paper 920*, 30 p.

____, 1977, Paleozoic-Mesozoic boundary in the Berry Creek quadrangle, northwestern Sierra Nevada, California: *U.S. Geol. Survey Prof. Paper 1027*, 22 p.

____, 1981, Petrologic and structural studies in the northwestern Sierra Nevada, California: *U.S. Geol. Survey Prof. Paper 1226*, 59 p.

Holdaway, M. J., 1971, Stability of andalusite and the aluminum silicate phase diagram: *Amer. J. Sci.*, v. 271, p. 97–131.

Irwin, W. P., Jones, D. L., and Kaplan, T. A., 1978, Radiolarians from pre-Nevadan rocks of the Klamath Mountains, California and Oregon, *in* Howell, D. G., and McDougall, K. A., eds., *Mesozoic Paleogeography of the Western United States:* Pacific Section, Soc. Econ. Paleontologists Mineralogists, Pacific Coast Paleogeography Symp. 2; p. 303–310.

Kisch, H., 1983, Mineralogy and petrology of burial diagenesis and incipient metamorphism in clastic rocks, *in* Larsen, G., and Chilinger, G. V., eds., Diagenesis in Sediment, and Sedimentary Rocks, 2. *Developments in Sedimentology:* Elsevier, Amsterdam, p. 289–493.

Laird, J., 1980, Phase equilibria in mafic schist from Vermont: *J. Petrology*, v. 21, p. 1–37.

Levi, B., Aguirre, L., and Nystrom, J. O., 1982, Metamorphic gradients in burial metamorphosed vesicular lavas — Comparison of basalt and spilite in Cretaceous basic flows from central Chile: *Contrib. Mineralogy Petrology*, v. 80, p. 49–59.

Liou, J. G., Maruyama, S., and Cho, M., 1985, Phase equilibria and mineral parageneses of metabasites in low grade metamorphism: *Mineral. Mag.*, v. 49, p. 321–334.

Marlette, J. W., Akers, R. J., Cole, K. A., and McJunkin, R. D., 1979, Geologic Investigations, *in The August 1, 1975 Oroville Earthquake Investigations: Calif. Div. Mines Geology Bull.* 203–78, p. 15–121.

Mazaheri, S. A., 1982, The petrology and metamorphism of ultramafic and mafic rocks near Pulga, Butte County, California: M.Sc. thesis; Univ. California, Davis, Calif., 139 p.

McJunkin, R. D., Davis, T. E., and Criscione, J. J., 1979, An isotopic age for Smartville ophiolite and the obduction of metavolcanic rocks in the northwestern Sierran foothills, California: *Geol. Soc. America Abstr. with Programs*, v. 11, p. 91.

McMath, V. E., 1958, The geology of the Taylorsville area, Plumas County, California: Ph.D dissertation; Univ. California, Los Angeles, Calif., 199 p.

———, 1966, Geology of the Taylorsville area, northern Sierra Nevada, California: *Calif. Div. Mines and Geology Bull.* 190, p. 173–183.

Moores, E. M., 1970, Ultramafics and orogeny, with models for the U.S. Cordillera and the Tethys: *Nature*, v. 228, p. 837–842.

———, 1972, Model for Jurassic island arc, continental margin collision in California: *Geol. Soc. America Abstr. with Programs*, v. 4, p. 202.

———, 1975, The Smartville terrain, northwestern Sierra Nevada, a major pre-Late Jurassic ophiolite complex: *Geol. Soc. America Abstr. with Programs*, v. 7, p. 352.

———, and Day, H. W., 1984, An overthrust model for the Sierra Nevada: *Geology*, v. 12, p. 416–419.

Murphy, T. P., 1986, Stratigraphy and structure of the Clipper Mills area, northern Sierra Nevada, California: M.Sc. thesis; Univ. California, Davis, Calif., 224 p.

———, and Moores, E. M., 1985, Two ophiolitic, tectonostratigraphic terranes in the western Central Belt, northern Sierra Nevada, California: *Geol. Soc. America Abstr. with Programs*, v. 17, p. 372.

Ricci, M. P., 1983, Geology, structure, and gravity of the northern margin of the Smartville Complex, northern Sierra Nevada foothills, California: M.Sc. thesis; Univ. California, Davis, Calif., 130 p.

———, Moores, E. M., Verosub, K. L., and McClain, J. S., 1985, Geologic and gravity evidence for thrust emplacement of the Smartville ophiolite: *Tectonics*, v. 4, p. 539–546.

Saleeby, J. B., 1981, Ocean floor accretion and volcano plutonic arc evolution in the Mesozoic Sierra Nevada, *in* Ernst, W. G., ed., *The Geotectonic Development of California* (Rubey Vol. I): Prentice-Hall, Englewood Cliffs, N.J., p. 132–181.

———, and Moores, E. M., 1979, Zircon ages on northern Sierra Nevadan ophiolite remnants and some possible regional correlations: *Geol. Soc. America Abstr. with Programs*, v. 11, p. 125.

Saleeby, J. B., Wright, W. H., III, Behrman, P. G., and Springer, R. K., 1981, Polygenetic ophiolitic assemblages of the lower American and Consumnes Rivers, western Sierra Nevada, California: *Geol. Soc. America Abstr. with Programs*, v. 13, p. 104.

Schiffman, P., Williams, A. E., and Evarts, R. C., 1984, Oxygen isotope evidence for submarine hydrothermal alteration of the Del Puerto ophiolite, California: *Earth Planet. Sci. Lett.*, v. 70, p. 207–220.

———, Bettison, L., and Williams, A. E., 1986, Hydrothermal metamorphism of the Point Sal Remnant, California Coast Ranges Ophiolite: *Proceedings of the Fifth International Symposium on Water Rock Interaction*, p. 489–492.

Schweickert, R. A., 1981, Tectonic Evolution of the Sierra Nevada Range, *in* Ernst, W. G., ed., *The Geotectonic Development of California* (Rubey Vol. I): Englewood Cliffs, N.J., Prentice-Hall, p. 87–131.

———, and Cowan, D. S., 1975, Early Mesozoic tectonic evolution of the western Sierra Nevada, California: *Geol. Soc. America Bull.*, v. 86, p. 1329–1336.

———, and Snyder, W. S., 1981, Paleozoic plate tectonics of the Sierra Nevada and adjacent regions, *in* Ernst, W. G., ed., *The Geotectonic Development of California* (Rubey Vol. I): Englewood Cliffs, N.J., Prentice-Hall, p. 182–201.

———, Armstrong, R. L., and Harakal, J. E., 1980, Lawsonite blueschist in the northern Sierra Nevada, California: *Geology*, v. 8, p. 27-31.

———, Bogen, N. I., Girty, G. H., Hanson, R. E., and Merguerian, C., 1984, Timing and structural expression of the Nevadan orogeny, Sierra Nevada, California: *Geol. Soc. America Bull.*, v. 95, p. 967–979.

Springer, R. K., 1980, Geology of the Pine Hill intrusive complex, a layered gabbroic body in the western Sierra Nevada foothills, California — Summary: *Geol. Soc. America Bull.*, v. 91, p. 381–385.

Standlee, L. A., 1978, Middle Paleozoic ophiolite in the Melones fault zone, northern Sierra Nevada, California: *Geol. Soc. America Abstr. with Programs*, v. 10, p. 148.

Standlee, L. A., and Nestell, M. K., 1985, Age and tectonic significance of terranes adjacent to the Melones fault zone, N. Sierra Nevada, California: *Geol. Soc. America Abstr. with Programs*, v. 17, p. 410.

Suppe, J., 1979, Structural interpretation of the southern part of the northern Coast Ranges and Sacramento Valley, California — Summary: *Geol. Soc. America Bull.*, v. 90, p. 327–330.

Swanson, S. E., 1970, Mineralogy and petrology of the Rocklin pluton, Placer and Sacramento Counties, California: M.Sc. thesis; Univ. California, Davis, Calif., 85 p.

Taliaferro, N. L., 1942, Geologic history and correlation of the Jurassic of southwestern Oregon and California: *Geol. Soc. America Bull.*, v. 53, p. 71-112.

Tuminas, A. C., 1983, Structural and stratigraphic relations in the Grass Valley-Colfax area of the northern Sierra Nevada foothills, California: Ph.D. dissertation; Univ. California, Davis, Calif., 415 p.

Vaitl, J., 1980, Geology of the Cherokee area, northern Sierra Nevada, California: M.Sc. thesis; Univ. California, Davis, Calif., 93 p.

Varga, R. J., 1985, Mesoscopic structural fabric of Early Paleozoic rocks in the northern Sierra Nevada, California and implications for Late Jurassic plate kinematics: *J. Struct. Geology*, v. 7, p. 667–682.

———, and Moores, E. M., 1981, Age, origin and significance of an unconformity that predates island-arc volcanism in the northern Sierra Nevada, California: *Geology*, v. 9, p. 512-518.

Watkins, R., Wallance, J. W., Reinheimer, C. E., and Nestell, M. K., 1986, Permian sedimentary megamictite at Oroville Lake, northern Sierra Nevada, California: *Geol. Soc. America Abstr. with Programs*, v. 18, p. 195.

Weisenberg, C. W., 1979, Structural development of the Red Hill portion of the Feather River ultramafic complex, Plumas County, California: Ph.D. dissertation; Rice Univ. Houston, Tex., 166 p.

———, and Avé Lallèmant, H., 1977, Permo-triassic emplacement of the Feather River ultramafic body, northern Sierra Nevada mountains, California: *Geol. Soc. America Abstr. with Programs*, v. 9, p. 525.

Xenophontos, C., 1984, Geology, petrology, and geochemistry of part of the Smartville Complex, northern Sierra Nevada, California: Ph.D. dissertation; Univ. California, Davis, Calif., 446 p.

Zoback, M. D., Wentworth, C. M., and Moore, J. G., 1985, Central California seismic reflection transect — II. The eastern Great Valley and western Sierra Nevada: (EOS), *Trans. Amer. Geophys. Un.*, v. 66, p. 1073.

David S. Harwood
U.S. Geological Survey
Menlo Park, California 94025

28

TECTONISM AND METAMORPHISM IN THE NORTHERN SIERRA TERRANE, NORTHERN CALIFORNIA

Regional unconformities separate rocks of the Northern Sierra terrane of northeastern Sierra Nevada into four lithotectonic sequences and record episodes of deformation in pre-Late Devonian, Late Pennsylvanian and Early Permian, Late Permian to Late Triassic, and Late Jurassic time. Greenschist-facies regional metamorphism accompanied the Late Jurassic deformation and masks equivalent or lower-grade metamorphism that may have occurred during earlier periods of deformation.

At some unknown time in the pre-Late Devonian, quartzose flysch, chert, submarine volcanic rocks, and serpentinite-bearing melange of the lower Paleozoic Shoo Fly Complex were internally folded and telescoped together in a series of imbricate thrust slices. The thrust slices were beveled by erosion and unconformably overlapped by Upper Devonian epiclastic deposits and submarine volcanic rocks of the Taylorsville sequence.

The Taylorsville sequence consists of two volcanic assemblages separated by chert and an erosional unconformity. The lower volcanic assemblage records a period of island-arc volcanism that lasted from the Late Devonian to the mid Mississippian. By late mid Mississippian time, that volcanic arc had subsided and the region received a widespread blanket of pelagic sediment now marked by the chert member of the Peale Formation. Pelagic sedimentation continued without dilution of other clastic debris at least to the Middle Pennsylvanian. Between the Middle Pennsylvanian and the late Early Permian, the terrane was uplifted, possibly along normal faults, differentially eroded, and unconformably overlapped by upper Lower Permian epiclastic deposits, pillow basalt, and andesite of the upper volcanic assemblage of the Taylorsville sequence. This period of arc volcanism extended into the early Late Permian, but its exact duration is unknown because the Permian volcanic rocks are unconformably overlain by Upper Triassic limestone as well as Lower and Middle Jurassic volcanic rocks that represent a third period of island-arc volcanism in the northern Sierra Nevada.

In the southeastern part of the terrane, Upper Devonian volcanic rocks of the Taylorsville sequence interfinger with pelite, quartzite, and conglomerate derived from an eastern source. These conglomeratic rocks, together with limestone and orthoquartzite that gradationally overlie them, make up the Tahoe sequence. The Tahoe sequence may have been derived from the Antler orogenic highland.

The dominant northwest-trending structural grain of the terrane was imposed during the Late Jurassic Nevadan orogeny. This period of compressive deformation produced east-directed thrust faults, major northwest-trending folds, a penetrative fabric in the pelitic rocks, and regional greenschist-facies metamorphism. Arenaceous and pelitic rocks throughout the terrane contain quartz, muscovite, chlorite, albite, tourmaline, and zircon. Biotite occurs locally in the Shoo Fly Complex east of the Melones fault zone. Volcanic rocks commonly contain quartz, plagioclase, actinolite, chlorite, epidote, calcite, and sericite. Prehnite and pumpellyite occur in volcanic rocks in the northeastern part of the terrane.

INTRODUCTION

The Northern Sierra terrane encompasses the rocks of the Sierra Nevada that lie generally north of the 39th parallel and east of the Melones fault zone (Fig. 28-1). It is composed of four belts of metamorphic rocks that include, from west to east, (1) the lower Paleozoic Shoo Fly Complex; (2) middle and upper Paleozoic volcanic and volcaniclastic rock of the Taylorsville sequence; (3) the newly recognized Tahoe sequence, which is composed of middle Paleozoic siliciclastic and carbonate rocks originally included with the Mesozoic

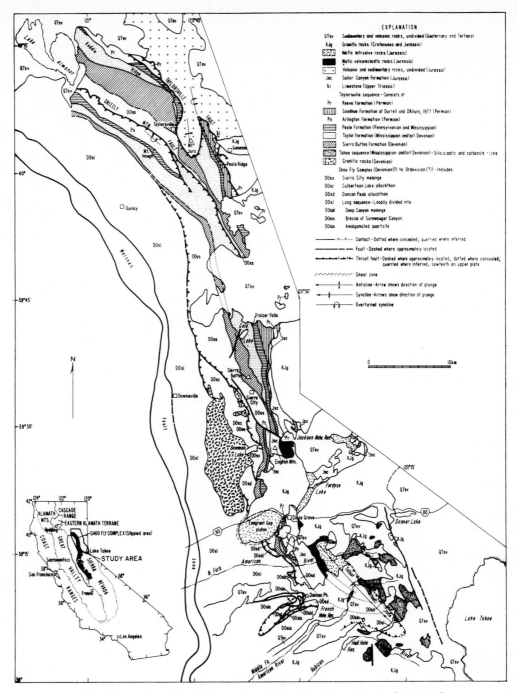

FIG. 28-1. Generalized geologic map of Northern Sierra terrane. Sources of data: D'Allura *et al.* (1977), Durrell and D'Allura (1977), Hannah (1980), D. S. Harwood (1983, unpublished data); Schweickert *et al.* (1984a, b).

TECTONISM AND METAMORPHISM IN THE NORTHERN SIERRA TERRANE

rocks by Lindgren (1897); and (4) Lower and Middle Jurassic volcanic and volcaniclastic rocks.

Over the past two decades, remarkable progress has been made in deciphering the stratigraphy and structure of the Northern Sierra terrane, primarily because fossils are relatively abundant in the Taylorsville sequence and the Mesozoic rocks and because low-greenschist-facies regional metamorphism has not obliterated primary volcanic and sedimentary textures. Regional metamorphism accompanied Late Jurassic compressive tectonism, which imparted the dominant northwest-trending structural grain to the northern Sierra Nevada. In addition, Paleozoic rocks in the Northern Sierra terrane contain a variety of folds and faults that attest to repeated episodes of deformation between the pre-Late Devonian and the Late Jurassic. Regional unconformities divide the rocks of the Northern Sierra terrane into tectonostratigraphic sequences and help date the episodes of deformation.

The Northern Sierra terrane, except for the Shoo Fly Complex, is truncated near the 39th parallel by Cretaceous plutons of the Sierra Nevada batholith, which separate the terrane from screens of metamorphic rocks in the central part of the range. For discussion of the southern part of this terrane, see Schweickert *et al.*, Chapter 29, this volume. To the north and east, the metamorphic rocks are buried by thick deposits of Tertiary and Quaternary volcanic rocks and alluvium, which obscure direct stratigraphic and structural ties between the Northern Sierra terrane and partly coeval rocks in the eastern Klamath Mountains and north-central Nevada. The western boundary of the terrane is the Melones fault zone and the steeply dipping Feather River peridotite belt that is marked along most of its 150-km length by serpentinized ultramafic rocks, meta-gabbro, amphibolite, schist, and phyllite. The kinematic history of the Feather River peridotite belt and its bounding faults is not well understood and is complicated by the presence of glaucophane-lawsonite-bearing blueschist rocks that occur locally (Schweickert *et al.*, 1980). However, it is generally agreed that this belt represents a major tectonic boundary related to early Mesozoic accretion along the western continental margin (Day *et al.*, 1985; Moores, 1972; Saleeby, 1981, 1982; Schweickert, 1978, 1981; Schweickert and Cowan, 1975). For discussion of Feather River peridotite belt and rocks west of it, see Day *et al.*, Chapter 27, this volume.

SHOO FLY COMPLEX

The Shoo Fly Complex is the oldest and most extensive unit in the Northern Sierra terrane. It has been traced for nearly 350 km along the western flank of the Sierra Nevada (Merguerian, 1985; Schweickert, 1981), but most of the detailed work has been done north of the 39th parallel, where the Shoo Fly is unconformably overlain by Upper Devonian rocks of the Taylorsville sequence. Mapping by D'Allura (1977), D'Allura *et al.* (1977), Standlee (1978), Varga (1980), Girty (1983), Hanson (1983), Harwood (1983), and Girty and Schweickert (1984) has defined four major lithotectonic units in the northern part of the Shoo Fly Complex. They are, from structurally lowest to highest, the Lang sequence, the Duncan Peak allochthon, the Culbertson Lake allochthon, and the Sierra City melange (Fig. 28-1). These four regionally extensive thrust-bounded units were internally folded and thrust together before the Late Devonian (Frasnian).

Lang Sequence

The informally named Lang sequence is composed predominantly of quartzite and pelite turbidites and lesser amounts of quartz-microcline grit, chert, and limestone. The turbidite beds are of variable thickness, are commonly graded, and locally contain flame structures and slump features (Schweickert *et al.*, 1984b). Bond and DeVay (1980) concluded from sedimentologic evidence that these rocks are deep-water submarine fan deposits that formed off the margin of a continental terrane composed of plutonic and metamorphic rocks.

In the vicinity of the North Fork of the American River (Fig. 28-1), two olisto-stromes lie disconformably on quartzite-pelite turbidites of the Lang sequence. The olistostromes are composed of large blocks of chert, feldspathic grit, and massive ortho-quartzite in a broken but unsheared matrix of pelite and quartzite. The tectonic significance of the olistostromes is uncertain and probably will remain so until the ages of the various lithotectonic units in the Shoo Fly are known. The olistostromes are large slump deposits that may be essentially coeval with the Lang sequence, or they may have originated during the period of pre-Late Devonian thrusting that amalgamated the Shoo Fly Complex. For reasons discussed in the section on pre-Late Devonian deformation, the second hypo-thesis is favored.

Duncan Peak Allochthon

Between the North Fork of the American River and Bowman Lake (Fig. 28-1), the Lang sequence is structurally overlain by the Duncan Peak allochthon, which is composed of radiolarian chert interbedded with variable amounts of black siliceous argillite and sparse quartzite. The contact between the Lang sequence and the Duncan Peak allochthon is a thrust fault locally marked by thick zones of chert breccia west of Duncan Peak and by interlayered slivers of chert, pelite, and quartzite north of the American River. In the vicinity of Duncan Peak, the allochthon consists of several overlapping thrust slices of chert locally separated by chert breccia, chaotic mixtures of chert and black pelite, and slivers of the Lang sequence. The lower thrust slices of chert were tightly folded either before or, more likely, during emplacement of the higher thrust slices in the allochthon.

Culbertson Lake Allochthon

East of Bowman Lake (Fig. 28-1), the northern part of the Duncan Peak allochthon is structurally overlain by rocks of the Culbertson Lake allochthon (Girty, 1983; Girty and Schweickert 1984), but to the north the Culbertson Lake allochthon completely overlaps the Duncan Peak allochthon and structurally overlies the Lang sequence (Schweickert *et al.*, 1984b).

The Culbertson Lake allochthon contains a lower lithologic assemblage composed of pillowed and brecciated basalt interlayered with chert and overlain by limestone, chert, and argillite, which Girty (1983) interpreted to be part of an intraplate seamount. The upper lithologic assemblage, which depositionally overlies the lower assemblage, consists primarily of subarkose and quartz arenite that contains a suite of 2.09-b.y.-old detrital zircons. Girty and Wardlaw (1984, 1985) interpreted the upper assemblage as a submarine fan complex deposited adjacent to a continental land mass composed of Precambrian plutonic, metamorphic, and probably supracrustal rocks.

Sierra City Melange

The structurally highest lithotectonic unit in the Shoo Fly Complex is the Sierra City melange (Schweickert *et al.*, 1984b). The melange is thrust over the Culbertson Lake allochthon east of Bowman Lake, but farther north it rests structurally above rocks that probably correlate with the Lang sequence (D'Allura *et al.*, 1977; Schweickert *et al.*, 1984b). The melange is composed of blocks of serpentinite, gabbro, pillowed and massive basalt, chert, sandstone, limestone, and dolostone in a sheared matrix of slate, sandstone, and chert. Blocks of arkosic sandstone in the melange contain volcanic lithic fragments, plutonic detritus, and a suite of detrital zircons dated at 506 m.y. (Girty and Wardlaw, 1984). Girty and Wardlaw (1985) also report a preliminary and unspecified Precambrian age for detrital zircons obtained from quartz arenite blocks in the melange. They infer a volcanic source terrane for the arkosic-lithic sandstone blocks with the lower Paleozoic detrital zircon suite and, following the original proposal by Jones *et al.* (1972), suggest that the Alexander terrane, presently lodged in southeastern Alaska, is a possible source (see Gehrels and Saleeby, 1984, for an alternative interpretation). Girty and Wardlaw (1985) suggested that the quartz arenite blocks with the Precambrian detrital zircon suite may have been derived from the underlying Culbertson Lake allochthon during thrusting.

PRE-LATE DEVONIAN DEFORMATION

Available data provide some constraints on the age of the Shoo Fly Complex, as a whole, and set a minimum age for its tectonic amalgamation. However, until the stratigraphy in each thrust slice is dated, the sequence and time of thrusting and, ultimately, the tectonic origin of the Shoo Fly will be matters of speculation.

At present, the most reliable minimum age for the Shoo Fly Complex and the thrusting is provided by Late Devonian (Frasnian) fossils in epiclastic rocks at Sierra Buttes (Hanson, 1983), which are correlative with the Grizzly Formation (Diller, 1908). Hanson *et al.* (in press), however, report a discordant Pb/U model age of 364–385 m.y. for zircon from trondhjemitic, granodiorite, and granitic phases of the Bowman Lake batholith, which intrudes the Shoo Fly Complex (Fig. 28-1). If the Bowman Lake batholith is as old as 385 m.y., the minimum age of thrusting could be Middle Devonian (Eifelian).

The maximum age of the Shoo Fly Complex is unknown. Varga and Moores (1981) reported Ordovician or Silurian radiolarians from a chert block in the Sierra City melange and D'Allura *et al.* (1977) reported Late Ordovician megafossils from limestone blocks in the melange south of Taylorsville (Fig. 28-1) (the Montgomery Limestone of Diller, 1908). Conodonts from the limestone block at Montgomery Creek provide an unequivocal Late Ordovician (middle Maysvillian through Gamachian) age and "the species association is indicative of North American Midcontinent Faunas 12 and 13 (relatively warm water biofacies) and deposition in generally warm shelfal or platformal settings" (A. Harris, written communication, 1985). A clast of coral-bearing limestone in Upper Devonian epiclastic rocks at Sierra Buttes, presumably derived from the underlying Sierra City melange, also yielded Late Ordovician conodonts (A. Harris, written communication, 1985). The Late Ordovician age of the limestone blocks does not provide a maximum age for the Sierra City melange, unfortunately, because the onset of melange formation may have preceded incorporation of the limestone blocks.

The thrust slices, which appear to stack up to the north, contain macroscopic and mesoscopic folds that formed during the thrusting. The form and orientation of these early folds differ significantly throughout the Shoo Fly Complex, but all have axial surfaces truncated by the unconformity at the base of the Upper Devonian rocks.

Near Bowman Lake, Girty (1983) mapped a set of early isoclinal folds in the Culbertson Lake allochthon that have northwest-trending, near-vertical axial surfaces that strike into and are truncated by the Upper Devonian rocks. Axial surfaces of these early folds are deformed about a pervasive N30°W, vertical cleavage that Girty (1983) related to the Late Jurassic Nevadan deformation. Early fold axes plunge northwest and southeast nearly in the plane of the Late Jurassic cleavage.

Early mesoscopic isoclinal folds that were not present in the Upper Devonian rocks were found in the Sierra City melange about 30 km north of Bowman Lake (Varga, 1980; Varga and Moores, 1982). The early folds in the Sierra City melange are geometrically similar to those in the Culbertson Lake allochthon, and they have also been rotated into the plane of the dominant northwest-trending Jurassic cleavage (Varga and Moores, 1982). No metamorphic fabric that predated or formed contemporaneously with the early period of folding was reported by Varga and Moores in the Sierra City melange. Conceivably, some of the macroscopic folds could be related to slumping in that chaotic unit. However, the widespread distribution of early folds in the coherent stratigraphic succession of the Culbertson Lake allochthon, the geometric similarity between early folds in the Sierra City melange and the Culbertson Lake allochthon, and the subparallel orientation of early-fold axial surfaces and the thrust fault between these lithotectonic units argues against folding by slumping and indicates an episode of crustal shortening at shallow crustal levels (Girty and Schweickert, 1984).

South of Interstate Highway 80 (I80), structural trends in the Shoo Fly Complex (Fig. 28-2) swing out of the northwest and strike eastward nearly perpendicular to the trend of the upper Paleozoic rocks and the Jurassic Sailor Canyon Formation. In this area, the Shoo Fly Complex is composed of the Lang sequence and the overlying Duncan Peak allochthon. Slices of these lithotectonic units are tectonically interlayered and locally both units are unconformably overlain by the Upper Devonian Sierra Buttes Formation north of the North Fork of the American River (Fig. 28-1).

The Duncan Peak allochthon is composed of numerous thin thrust slices of gray chert that are locally separated by lenses of chert-argillite breccia or slivers of well-bedded quartzite and pelite of the Lang sequence. In general, the thrust slices in the Duncan Peak allochthon trend northwest and dip northeast, but many local variations in strike and dip exist where slices of chert were folded independently of adjacent slices during the thrusting. The chert apparently was internally folded and interleaved with slices of the Lang sequence during emplacement of the Duncan Peak allochthon prior to deposition of the Upper Devonian rocks.

Near the North Fork of the American River, two mappable olistostromal units help define the internal structure in the Lang sequence. The lower olistostrome, referred to informally as the breccia of Screwauger Canyon (Fig. 28-2), contains blocks of chert, quartz-granule conglomerate, and massive, fine-grained gray quartzite incorporated in a matrix of black pelite and broken quartz-rich sandstone and siltstone. The upper olistostrome, shown on Fig. 28-2 as the Deep Canyon melange, contains the same blocks as the Screwauger Canyon breccia in a matrix of purple, green, and gray pelite that contains relatively continuous and disrupted beds of parallel- and cross-laminated, very fine-grained quartzose sandstone.

West of Duncan Peak (Fig. 28-2), the Screwauger Canyon breccia outlines the nose of an east-northeast-trending, steeply northeast-plunging anticline cored by thick-bedded amalgamated quartzite that contains thin zones of intrafolial slump folds. Sedimentary structures in the amalgamated quartzite define the trace of the axial surface of the anticline, but no pervasive foliation was found parallel to the axial surface in either the amalgamated quartzite or the Screwauger Canyon breccia. A second band of Screwauger Canyon breccia extends southwest from the base of Little Bald Mountain (Fig. 28-2) and, with the amalgamated quartzite to the south, defines the southeast limb of a syncline cored in part by the Deep Canyon melange. Sedimentary structures that indicate top-facing directions are not consistent with through-going axial traces of these folds, and it appears that the shared limb of the syncline and anticline is faulted out to the southwest by the thrust fault at the top of the Screwauger Canyon breccia.

Disrupted bedding in the Screwauger Canyon breccia defines a foliation that conforms to the trends of bedding in the core of the anticline. At the nose of the anticline, the foliation is crudely developed, and small bedding fragments tend to be equant, randomly oriented, and dispersed in the pelitic matrix. On the limbs of the anticline, where

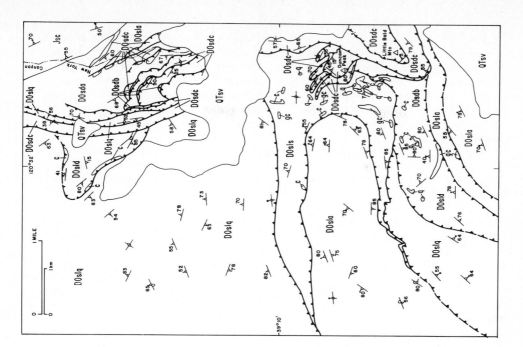

EXPLANATION

QTsv – Sedimentary and volcanic rocks, undivided (Quaternary and Tertiary)

Jsc – Sailor Canyon Formation (Jurassic)

Shoo Fly Complex (Devonian? to Ordovician?) – Includes:

    Duncan Peak allochthon – Includes:

DOsdc – Chert

DOsdb – Chert breccia

DOsda – Argillite, chert, and minor quartzite

    Long sequence – Includes:

DOsld – Deep Canyon mélange – Includes quartzite (q), chert (c), and granule conglomerate (gc)

DOsls – Breccia of Screwauger Canyon – Includes quartzite (q), chert (c), and granule conglomerate (gc)

DOslq – Quartzite: pelite turbidites, grit, and minor chert and limestone

DOsla – Amalgamated quartzite

———— CONTACT

⊥⊥⊥⊥ THRUST FAULT – Dashed where approximately located; queried where inferred; sawteeth on upper plate

STRIKE AND DIP OF BEDDING AND FOLIATION

⟍⁵⁸ Inclined   ⟍⁵⁸ Inclined, top known

⟂ Vertical   ⟂ Vertical, top known

⟍²³ Overturned

FIG. 28-2. Geologic map of part of Duncan Peak quadrangle, showing distribution of lithotectonic units in Shoo Fly Complex.

771

the breccia thins rapidly and eventually pinches out, the foliation is well developed and bedding fragments are elongate, flattened, and aligned. The changes in internal character of the breccia suggest that the anticline formed during or soon after the breccia was deposited when the breccia and surrounding rocks were semilithified. Where the Screwauger Canyon breccia is absent, the bounding faults cannot be identified, and their position is inferred from major lithologic changes and zones of soft-sediment deformation. It appears that folding was concentrated at the nose of the anticline, but to the southwest, strain release was distributed throughout a broad expanse of rocks by intrafolial soft-sediment slump and shearing.

To the northeast, the Screwauger Canyon breccia and Deep Canyon melange are truncated by folded thrust slices of the Duncan Peak allochthon. Early northeast-trending overturned folds in the lower chert slices at Duncan Peak are so dissimilar in style and orientation from the anticline outlined by the Screwauger Canyon breccia, it is concluded that they formed independently of the anticline in the underlying rocks, probably when the chert slices were thrust faulted. This conclusion is supported by the fact that a southeast-trending syncline, cored by the Deep Canyon melange, is truncated by northwest-trending slices of chert west of New York Canyon (Fig. 28-2).

The field relations indicate that the pre-Late Devonian deformation in these lithotectonic units of the Shoo Fly was diachronous and dissimilar. None of the blocks in the Screwauger Canyon breccia or the Deep Canyon melange are exotic, and all could have been derived from the Lang sequence and the Duncan Peak allochthon. The balance of evidence suggests that the Lang and Duncan Peak lithotectonic units were in reasonable proximity when the olistostromes were deposited and that each unit contributed to the chaotic mixtures. During or soon after deposition, the olistostromes and the Lang sequence were deformed and, subsequently, overridden by imbricate thrust slices of the Duncan Peak allochthon. This scenario is consistent with previous suggestions that the Shoo Fly is an accretionary complex formed at shallow depths by compressional or transpressional tectonism along a collapsing plate margin (Schweickert and Snyder, 1981; Schweickert et al., 1984b).

Mineral assemblages in the Shoo Fly Complex north of the 39th parallel suggest shallow-level tectonic amalgamation. Quartz, muscovite, chlorite, albite, tourmaline, and zircon make up the arenaceous and pelite rocks. Biotite occurs with this assemblage in rocks just east of the Melones fault zone (Standlee, 1978). Plagioclase, clinozoisite, calcite, chlorite, white mica, and actinolite are present in the sparse volcanic rocks of the Shoo Fly Complex and scattered lenses of limestone are composed primarily of calcite, quartz, and tremolite; dolomite occurs locally. Jurassic rocks in the terrane contain the same mineral assemblages, and available data are insufficient to differentiate Jurassic greenschist metamorphism from possible pre-Late Devonian greenschist metamorphism.

TAYLORSVILLE SEQUENCE

The Taylorsville sequence consists of submarine volcanic and volcaniclastic rocks, radiolarian chert, and locally prominent epiclastic deposits that range in age from Late Devonian (Frasnian) to the late Early Permian (Wordian). Many of the stratigraphic and structural problems that thwarted early regional syntheses (Diller, 1908; Turner, 1897) were successfully resolved by McMath (1958, 1966), and over the past 25 years the stratigraphy McMath determined in the Taylorsville area has been extended from Lake Almanor to the North Fork of the American River (D'Allura et al., 1977; Durrell and D'Allura, 1977; Hannah, 1980; Hannah and Moores, 1986; Hanson, 1983; Harwood, 1983). Unlike the Shoo Fly

Complex, the Taylorsville sequence is relatively fossiliferous; therefore, its depositional history and internal tectonic evolution are fairly well known.

D'Allura *et al.* (1977) first pointed out that the Taylorsville sequence is composed of two distinct volcanic assemblages separated by a widespread unit of chert that marked a prolonged period of volcanic quiescence. Schweickert (1981) and Harwood (1983) recognized an erosional unconformity separating the two volcanic assemblages in the central and southern parts of the sequence, respectively, and suggested that the hiatus spanned much of the Pennsylvanian and the Early Permian (Harwood, 1983). New stratigraphic data (Fig. 28-3) indicate that (1) the hiatus probably spanned, at most, the Late Pennsylvanian and the Early Permian; (2) erosion extended to significantly different levels along the length of the Taylorsville sequence; and (3) that basaltic volcanism of the Goodhue Formation (Durrell and D'Allura, 1977), which marks the onset of the younger volcanic phase in the Taylorsville sequence, is spatially, and probably tectonically, related to the areas of maximum erosion and inferred uplift and extension of the older volcanic assemblage.

Lower Volcanic Assemblage

The lower volcanic assemblage of the Taylorsville sequence consists of the Grizzly, Sierra Buttes, and Taylor Formations, and the lower member of the Peale Formation. The Grizzly Formation (Diller, 1908), which is present only locally along the profound

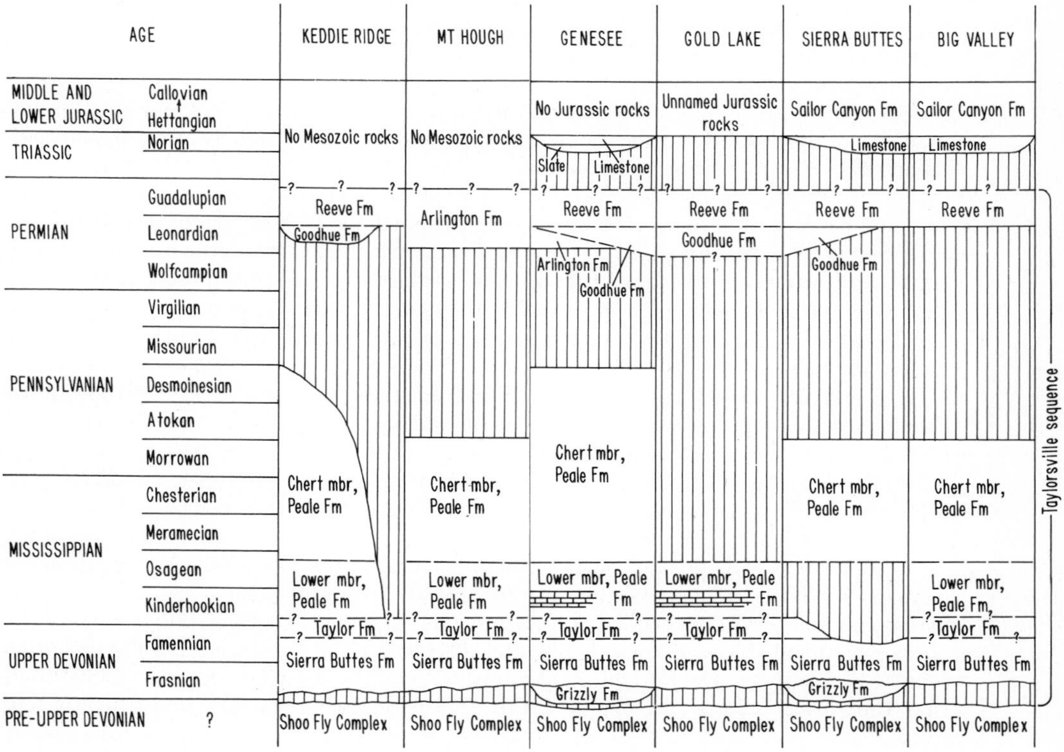

AGE		KEDDIE RIDGE	MT HOUGH	GENESEE	GOLD LAKE	SIERRA BUTTES	BIG VALLEY
MIDDLE AND LOWER JURASSIC	Callovian / Hettangian	No Mesozoic rocks	No Mesozoic rocks	No Jurassic rocks	Unnamed Jurassic rocks	Sailor Canyon Fm	Sailor Canyon Fm
TRIASSIC	Norian			Slate / Limestone		Limestone	Limestone
PERMIAN	Guadalupian	Reeve Fm	Arlington Fm	Reeve Fm	Reeve Fm	Reeve Fm	Reeve Fm
	Leonardian	Goodhue Fm			Goodhue Fm		
	Wolfcampian			Arlington Fm / Goodhue Fm	?	Goodhue Fm	
PENNSYLVANIAN	Virgilian						
	Missourian						
	Desmoinesian						
	Atokan						
	Morrowan	Chert mbr, Peale Fm	Chert mbr, Peale Fm	Chert mbr, Peale Fm		Chert mbr, Peale Fm	Chert mbr, Peale Fm
	Chesterian						
MISSISSIPPIAN	Meramecian						
	Osagean	Lower mbr, Peale Fm	Lower mbr, Peale Fm	Lower mbr, Peale Fm	Lower mbr, Peale Fm		Lower mbr, Peale Fm,
	Kinderhookian						
UPPER DEVONIAN	Famennian	Taylor Fm	Taylor Fm	Taylor Fm	Taylor Fm		Taylor Fm
	Frasnian	Sierra Buttes Fm	Sierra Buttes Fm	Sierra Buttes Fm / Grizzly Fm	Sierra Buttes Fm	Sierra Buttes Fm / Grizzly Fm	Sierra Buttes Fm
PRE-UPPER DEVONIAN	?	Shoo Fly Complex	Shoo Fly Complex	Shoo Fly Complex	Shoo Fly Complex	Shoo Fly Complex	Shoo Fly Complex

FIG. 28-3. Stratigraphic columns from the Northern Sierra terrane. Note erosional unconformity separating late Early Permian volcanic rocks from older volcanic assemblage of Taylorsville sequence.

unconformity between the Shoo Fly Complex and the Taylorsville sequence, is composed of conglomerate, breccia, quartz-rich and quartzofeldspathic sandstone, mudstone, argillite, and phosphatic-streaked black chert. Hanson (1983) noted that the conglomerate and breccia, which contain clasts derived from the Shoo Fly Complex, occur as lenticular channel deposits interbedded with quartzofeldspathic turbidites that locally form upward thinning and fining sequences. He concluded that the Grizzly Formation in the vicinity of Sierra Buttes was deposited as a channelized submarine fan complex in moderately deep water. Although McMath (1958) found beds of orthoquartzite associated with quartzo-feldspathic turbidites, pebble conglomerate lenses, and mudstone in the Grizzly Formation near Taylorsville, he also concluded that the Grizzly was not a shallow-water onlap deposit, but rather that it formed in a submarine slope-rise environment. Hanson (1983) found Late Devonian (Frasnian-Famennian) conodonts in chert interbedded in the Grizzly, and Frasnian corals and gastropods in thin-bedded sandstone and limestone that he inter-preted to be a large slump block in the submarine fan deposits. No current-direction data are available from the Grizzly, so the orientation of the paleoslope is unknown.

Epiclastic deposits of the Grizzly Formation grade abruptly into felsic volcaniclastic rocks of the Sierra Buttes Formation. Along its 120-km strike length, the Sierra Buttes Formation shows major variations in thickness and composition that are due primarily to the proximity of a given exposure to centers of submarine volcanism. From Hanson's (1983) detailed work, it seems certain that one source area for the formation was located in the vicinity of Sierra Buttes (see also D'Allura, 1977). The abundance of felsic and intermediate dikes in the Sierra Buttes Formation near Taylorsville (McMath, 1958) and on Keddie Ridge (Hannah, 1980) suggests that a second source area may have been located near the northern end of the belt.

Hanson (1983) recognized four major lithologic units in the near-source deposits of the Sierra Buttes Formation. The lower three units contain some andesitic detritus but are predominantly rhyolitic to dacitic in composition. The upper unit is primarily andesitic. Regardless of composition, all four units consist of thick, discontinuous beds of lapilli tuff, tuff breccia, and tuffaceous turbidites with carbonaceous, locally phosphatic chert and laminated tuff interlayered in variable proportions. Coarse-grained lapilli tuff and tuff-breccia beds are composed of felsic and andesitic lithic clasts, felsic long-tube pumice lapilli, and intraclasts of black chert and tuff. Hanson (1983) interpreted these rocks as submarine mass flow deposits produced directly by eruptions or by slumping of near-vent accumulations. Chert and fine-grained tuff accumulated during intervals of relative volcanic quiessence.

Coeval hypabyssal rocks, ranging from basaltic andesite to rhyolite, intrude all four lithologic units near Sierra Buttes. The intrusive rocks show a variety of quench textures and structures, described by Hanson (1983) and Brooks *et al.* (1982), that indicate they were intruded into wet volcaniclastic sediment. Hanson (1983) and Hanson *et al.* (in press) interpret the Bowman Lake batholith to be an epizonal pluton coeval with the Sierra Buttes Formation.

McMath (1958) recognized two lithologic units in the Sierra Buttes Formation near Taylorsville. Both units are composed primarily of felsic tuff breccia and lapilli tuff with variable amounts of interbedded tuffaceous slate, argillite, and black phosphate-streaked chert. Abundant felsic and mafic dikes and sills intrude the lower unit and the underlying Shoo Fly Complex. On Keddie Ridge northwest of Taylorsville, Hannah (1980) found that the lower part of the Sierra Buttes Formation consists predominantly of felsic hypabyssal rocks, including leuocratic biotite tonalite, whereas the upper part contains felsic tuff, tuff-breccia, quartz-bearing volcaniclastic sandstone, siltstone, and chert. Exact correlation of the lithologic units near Taylorsville with those mapped by Hanson (1983) near Sierra Buttes is impossible, but general equivalence of the formation is assured by the Late Devonian radiolarian fauna in chert from the lower lithologic unit on Keddie Ridge (Hannah, 1980).

South of I80, the Sierra Buttes Formation consists of a lower unit of felsic tuff-breccia with abundant clasts of black chert, and an upper unit of black tuffaceous slate that contains scattered thin lenses of chert-granule conglomerate (Harwood, 1983). The lower unit is probably equivalent to unit C of Hanson (1983), and the upper unit is a relatively fine-grained facies not present to the north. The formation south of the highway is about one-third as thick as it is to the north and it lacks hypabyssal intrusives. These observations suggest that a source area for the formation lay near Sierra Buttes and that the southern part of the formation represents a more distal apron of the Late Devonian submarine volcanic complex.

The Sierra Buttes Formation is conformably and abruptly overlain by andesitic submarine volcanic and volcaniclastic rocks of the Taylor Formation. The Taylor consists primarily of augite-bearing andesite breccia, massive and pillowed andesite and basaltic andesite flows, and andesitic volcaniclastic turbidites interbedded in variable proportions. South of I80, andesitic crystal-lithic tuff turbidites make up the entire formation.

The thickness of the Taylor Formation varies abruptly along its strike length. On Keddie Ridge, the formation thickens northward from 0 to 3 km over a distance of 20 km (Hannah, 1980). From Taylorsville south to Gold Lake, its thickness is fairly constant and ranges from 2.8 to 3 km (Durrell and D'Allura, 1977; McMath, 1958). However, between Gold Lake and Sierra City (Fig. 28-1), the thickness of the formation varies between 2.7 and 0.3 km, and the formation pinches out completely southeast of Sierra City (Schweickert et al., 1984b). South of I80, andesitic volcaniclastic rocks of the Taylor are 0.3 km thick. The variations in thickness probably reflect a combination of factors that include proximity to source area, differential compaction, syndepositional faulting, and postdepositional erosion. No source area has been identified for the formation, but the greater thickness and abundance of flows and coarse breccia at and north of Gold Lake suggest that the source or sources for the Taylor lay in the northern part of the terrane.

The Taylor Formation has not been dated directly. However, it can be no older than Late Devonian (Famennian), the age of the youngest fossils in the underlying Sierra Buttes Formation (Anderson et al., 1974; Hanson, 1983), and no younger than Early Mississippian (late Kinderhookian), the age of the oldest fossils in the overlying lower member of the Peale Formation.

The Taylor is overlain by the Peale Formation, which consists of a lower volcanic-volcaniclastic member and an upper chert member. The lower member contains massive trachytic to quartz latitic flows, tuff breccia, tuff, tuffaceous siltstone and sandstone, scattered lenses of chert-granule conglomerate, and minor limestone. The flows and clasts in the tuff breccia are characterized by pink alkali feldspar phenocrysts, granophyric intergrowths of alkali feldspar and quartz, and locally quartz phenocrysts. Flows and tuff breccia compose most of the lower member in the northern part of the region and extend as far south as the vicinity of Gold Lake (D'Allura et al., 1977; Durrell and D'Allura, 1977). In the area east and southeast of Sierra Buttes, however, the lower member is absent (Schweickert et al., 1984b) and the chert member rests directly on either the Taylor or Sierra Buttes Formations (Fig. 28-3, Sierra Buttes). South of I80, the lower member is present above andesitic tuff turbidites of the Taylor Formation and contains only minor amounts of alkali feldspar-phyric tuff and tuff breccia associated with chert-pebble conglomerate in its lower part. The tuff and tuff breccia grade upward into tuffaceous siltstone and sandstone that contain chert-granule conglomerate lenses. The upper part of the member is composed of thick grain-flow volcaniclastic units that are locally conglomeratic (Harwood, 1983). Therefore, like the underlying Taylor and Sierra Buttes Formations, the lower member of the Peale Formation shows a southward decrease in thickness and amount of volcanic flows and breccia and an increase in relatively fine-grained volcanic sedimentary deposits that indicate source areas in the northern part of the region.

A lens of crinoidal limestone located about 80 m above the base of the lower member of the Peale at Frazier Falls (Fig. 28-3, Gold Lake) yielded Early Mississippian (late Kinderhookian) conodonts (A. Harris, written communication, 1984). Another lens of limestone located about 50 m below the top of the lower member on Peale Ridge (Fig. 28-3, Genesee) produced the same late Kinderhookian conodont fauna (A. Harris, written communication, 1985). The lens of limestone on Peale Ridge is believed to be the one originally found by Diller (1908) from which McMath (1966) reported Early Mississippian brachiopods and trilobites.

Chert Member of the Peale Formation

The lower member of the Peale Formation, the youngest volcanic unit in the lower assemblage of the Taylorsville sequence, grades abruptly upward into the chert member of the Peale, which serves as a widespread marker unit occurring throughout the region except near Gold Lake (Fig. 28-3). The chert member of the Peale is composed of thin-bedded black, red, green, and gray radiolarian chert that locally contains minor amounts of purple and black argillite. D'Allura *et al.* (1977) report a megascopic gradation between the lower member and the chert member of the Peale. Chert from beds low in the member commonly contain volcanic detritus and bipyramidal quartz (B. Murchey, oral communication, 1985), which also suggests that the chert member lies gradationally above the lower member.

Radiolarians have been extracted from various stratigraphic levels of the chert member along its strike. The radiolarians, identified by B. Murchey and D. L. Jones (written communication, 1983–1985), indicate that the chert member ranges in age from late middle Mississippian (Osagean) to Middle Pennsylvanian (Desmoinesian) and that much of the region contains only the lower part of the chert section. The upper age of the chert member, however, may not be the minimum age of volcanic quiescence, subsidence, and silicious pelagic deposition because the chert member is separated from the overlying upper Lower Permian volcanic rocks of the upper assemblage of the Taylorsville sequence by an erosional unconformity.

In addition to the pronounced break in ages, the erosional unconformity is marked by locally thick accumulations of epiclastic rocks in the lower part of the Permian Arlington Formation and by thinner but widespread epiclastic deposits in the lower part of the Permian Reeve Formation. In spite of this pronounced erosional break, no folds or faults that can be related to a period of compressional tectonism have been recognized in the lower part of the Taylorsville sequence or in the Shoo Fly Complex. Instead, it appears that the Late Pennsylvanian and Early Permian unconformity represents a period of uplift probably associated with crustal heating and extension prior to late Early Permian volcanism. The pattern of differential erosion in the lower part of the Taylorsville sequence, shown in Fig. 28-3, and the fact that the epiclastic deposits of the Arlington Formation contain sparse quartzite clasts, apparently derived from the Shoo Fly Complex, suggest that the uplift was associated with block faulting, but normal faults have not been recognized, thus far, in the lower part of the section.

Upper Volcanic Assemblage

The upper volcanic assemblage of the Taylorsville sequence consists of the Arlington, Goodhue, and Reeve Formations. The Arlington is composed of epiclastic conglomerate and volcaniclastic sedimentary rocks that contain late Early Permain (Leonardian) mega-fossils near the base of the formation near Mount Hough (T. Dutro, written communication, 1986). The Goodhue is primarily pillow basalt, and the Reeve contains local polymict conglomerate, volcanic sandstone, and chert near its base that grade upward into andesitic breccia and flows locally intruded by hypabyssal plagioclase porphyry. The lower part of all three formations is late Early Permian in age, and each formation rests unconformably on the older volcanic assemblage at some point in the Northern Sierra terrane (Fig. 28-3). The Reeve is unconformably overlain by Upper Triassic (Norian) limestone and Jurassic volcaniclastic rocks in the southern part of the terrane and by Norian limestone and slate in the northern part of the area.

This new stratigraphic information suggests that some rocks traditionally included in the younger volcanic assemblage need reevaluation. The Robinson Formation of Diller (1908) is considered here to be a facies of the Arlington, as first suggested by McMath (1958). Rocks located southwest of Lake Almanor that were correlated with the Arlington by D'Allura *et al.* (1977) are excluded here from the Arlington, because they reportedly form a continuous section between Early Permian (Wolfcampian) and Late Triassic (Norian) limestones. If such is the case, the rocks southwest of Lake Almanor may have greater stratigraphic similarity to upper Paleozoic and lower Mesozoic rocks in the Eastern Klamath terrane (Albers and Robertson, 1961; Coney *et al.*, 1980) than to rocks in the Northern Sierra terrane with which they may be tectonically juxtaposed. Until more stratigraphic and structural data are available, it seems best to apply the name Arlington Formation only to those rocks southeast of Lake Almanor that are continuous with the type Arlington.

On Peale Ridge (Fig. 28-3, Genesee), the Arlington, Goodhue, and Reeve Formations occur in stratigraphic sequence and have gradational boundaries. Elsewhere, however, only one or two of these formations are present. Coarse basal conglomerate and finer grained volcaniclastic rocks of the Arlington Formation are the dominant Permian rocks in the upper Taylorsville thrust block (Fig. 28-3, Mt. Hough), where they are about 2 km thick. Small isolated patches of the Goodhue Formation occur disconformably above the chert member of the Peale on Keddie Ridge (Hannah, 1980), but the major mass of the Goodhue occurs in the central part of the terrane. The Goodhue thickens southeastward from the Genesee area (Fig. 28-1) and reaches a maximum thickness of about 2.3 km at Frazier Falls (Fig. 28-3, Gold Lake), where it rests unconformably on the lower member of the Peale Formation. The Goodhue pinches out southeast of Sierra Buttes, and in the southern part of the terrane (Fig. 28-3, Big Valley) only the Reeve Formation is present and composed of basal chert-pebble conglomerate and marble that grades up into fine-grained tuff and tuffaceous slate (Harwood, 1983). Except for the small patches of Goodhue mentioned above, the Reeve forms the bulk of the Permian section at the north end of the terrane (Fig. 28-3, Keddie Ridge), where it is composed of chert, volcaniclastic rocks, and chert-pebble conglomerate in the lower part and andesitic breccia, flows, and plagioclase porphyry in the upper part. On Keddie Ridge, the Reeve rests unconformably on either the chert member or the lower member of the Peale and locally on the underlying Taylor Formation.

The field relations and new radiolarian data from the chert member of the Peale Formation (B. Murchey, written communication, 1985) indicate that major differential erosion of the lower volcanic assemblage occurred prior to deposition of the upper Lower Permian rocks of the upper volcanic assemblage. Furthermore, the distribution of the Goodhue Formation appears to correlate with areas of deeper erosion in the subjacent rocks, and the Arlington Formation overlies some of the least eroded areas of the lower volcanic assemblage. These relations suggest that differential uplift and erosion of the lower volcanic assemblage and chert member of the Peale Formation may have been related to local crustal heating and warping by magma that eventually was extruded onto the sea floor and accumulated in the areas of maximum erosion as pillow basalt of the Goodhue Formation. Chert and volcanic rocks of the Peale Formation, which were eroded from areas of greatest thermal(?) uplift, accumulated as massive slump deposits of Arlington Formation and possibly as the epiclastic deposits of the Reeve in subsea areas that were uplifted to a lesser degree. Quartzite clasts in the Arlington, which probably were derived from the Shoo Fly Complex, indicate that the Shoo Fly was exposed to erosion, possibly by block faulting, somewhere outside the map area.

TAHOE SEQUENCE

Lindgren (1897) correlated metamorphic rocks exposed near the Sierra crest in the Donner Lake area with the Jurassic Sailor Canyon Formation, which he had mapped farther to the west. However, new stratigraphic, structural, and paleontological data indicate that the rocks near the Sierra crest are Paleozoic, no older than Late Devonian, and that they form a coherent stratigraphic assemblage of siliciclastic and carbonate rocks that is tectonically juxtaposed against the Mesozoic rocks to the west. This belt of rocks is referred to informally as the Tahoe sequence. Several of the stratigraphic units in the Tahoe sequence are discontinuous, and the stratigraphy has been pieced together with particular emphasis on exposures in the North Fork of the American River drainage and the Blackwood Creek basin where the fossil localities occur (Fig. 28-4).

The oldest unit in the Tahoe sequence is composed of polydeformed, interbedded quartzite and black pelite that contains lenses of chert-pebble and cobble conglomerate throughout the area, and mappable masses of felsic to mafic volcanic rocks in its lower, westernmost exposures. Black and white chert clasts form the bulk of the conglomerate lenses, but black pelite and white orthoquartzite clasts are also present. Conglomerate lenses are thicker (up to 2 m) and more abundant, and the clasts are larger in the eastern exposures, generally indicating an eastern source for the coarse detritus. Volcanic rocks, which are locally abundant along Five Lakes Creek and in the Middle Fork of the American River (Fig. 28-4), are predominantly fine-grained quartz-bearing volcaniclastic grain-flow deposits with scattered phosphatic-streaked black chert clasts. The thick-bedded grain-flow deposits are interspersed with thin beds of very fine-grained felsic tuff. A few thick beds of mafic tuff breccia occur in the felsic volcaniclastic rocks. The volcaniclastic rocks, which are lithologically identical to rocks in the Sierra Buttes Formation, are interbedded locally with quartzite, pelite, and conglomerate. The interbedded volcaniclastic and siliciclastic rocks provide an important stratigraphic link between the early deposits of the Taylorsville and Tahoe sequences.

Along the Rubicon River (Fig. 28-4), the quartzite-pelite-conglomerate unit is overlain gradationally by thin-bedded black pelite that contains interbeds of gray sandy limestone turbidites. Calcareous, feldspathic grain-flow units, interbedded with the sandy limestone turbidites along Blackwood Creek (Fig. 28-4), contain poorly preserved molds of small ammonites. The ammonites were studied by N. J. Silberling (written communication, 1984), who found one specimen having a "simple 8-lobed suture that appears to be goniatitic," indicating that the rock can be no older than latest Devonian and no younger than Jurassic. His "best guess" on the age of the material was latest Devonian or Mississippian, but because of the poor state of preservation he could not unequivocally rule out a Mesozoic age. Considered independently, therefore, the ammonites do little to date the Tahoe sequence except to restrict the maximum age of the enclosing rocks to the Late Devonian; but in their stratigraphic context and in association with the other fossil data, discussed below, that meager information is extremely valuable.

At Blackwood Creek, the ammonite-bearing unit is gradationally overlain by well-bedded white orthoquartzite that serves as a distinctive marker unit throughout the region. At Serena Creek (Fig. 28-4), the pelite and sandy limestone turbidite unit is missing, and the white orthoquartzite unit gradationally overlies the pelite-quartzite-conglomerate unit. The orthoquartzite, in turn, grades upward into massive gray limestone. Quartzite beds near the base of the unit contain scattered, poorly preserved bivalve fragments, which T. Dutro and J. Pojeta (written communication, 1984) concluded were Paleozoic (post-Early Ordovician) in age. Because the Paleozoic bivalve fragments occur in the orthoquartzite unit, which overlies the ammonite-bearing rocks to the south, they rule out a Jurassic age for the ammonites and strengthen the conclusion that the Tahoe sequence is middle Paleozoic.

In the higher and eastern parts of the Tahoe sequence, the orthoquartzite unit locally contains conglomerate beds composed of chaotic mixtures of quartzite and carbonate clasts in a quartzite matrix. The conglomerate beds probably represent slump deposits derived from a shallow-water shelf-type environment. Orthoquartzite clasts occur sparsely in the chert-rich conglomerate lenses in the pelite-quartzite-conglomerate unit, which also contains thick, discontinuous beds (or blocks?) of white orthoquartzite that lie generally above the volcaniclastic deposits.

Contrary to the earlier interpretation of Harwood and Fisher (1984), it is now clear that the basin where the Tahoe sequence was deposited received submarine volcaniclastic material from the older volcanic assemblage of the Taylorsville sequence and chert-quartz-rich detritus from a totally different, apparently eastern source during its early history. Sedimentation from these different source areas continued until the ammonite-bearing pelite and sandy limestone turbidite unit was deposited, probably in Late Devonian and/or Early Mississippian time. No volcaniclastic rocks have been recognized in the orthoquartzite unit or in the overlying limestone unit, suggesting that arc volcanism had ceased or that the locus of volcanism had shifted by the time those units were deposited.

The Tahoe sequence contains folds related to two separate periods of deformation. The early folds are isoclinal and the dip of their axial surfaces varies from near vertical to horizontal in different parts of the region, but their axial surfaces strike east or nearly so. The early folds are refolded by north-northwest-trending, upright, open folds that have a penetrative cleavage parallel to their axial surfaces. The late folds and the penetrative fabric predate the Cretaceous granitoid rocks of the Sierra Nevada batholith and probably are related to Late Jurassic deformation. The age of the early period of folding is unknown, but it must be middle Paleozoic or younger.

PERMIAN AND TRIASSIC DEFORMATION

A profound angular unconformity at the base of the Mesozoic metamorphic rocks indicates that tectonism occurred in the Northern Sierra terrane between the Late Permian and the Late Triassic, but little direct evidence is available on the nature of that deformation. South of I80, rocks of the Taylorsville sequence contain scattered minor folds that are tightly appressed and oriented roughly parallel to bedding. A recrystallization fabric also lies parallel to bedding, but it is impossible to prove that either the folds or the fabric are related to pre-Late Triassic rather than to Late Jurassic deformation throughout most of the area. At the North Fork of the American River, however, the Taylorsville sequence is folded into a broad northeast-trending fold and its axial trace is truncated by Upper Triassic limestone. At I80, a second macroscopic fold with a northeast-trending axial surface occurs in the Taylorsville sequence and is associated with a northeast-trending fault that does not offset the base of the Jurassic Sailor Canyon Formation (Harwood, 1983). No penetrative fabric is associated with these structures.

Nokelberg and Kistler (1980) reviewed the evidence for Paleozoic and Mesozoic

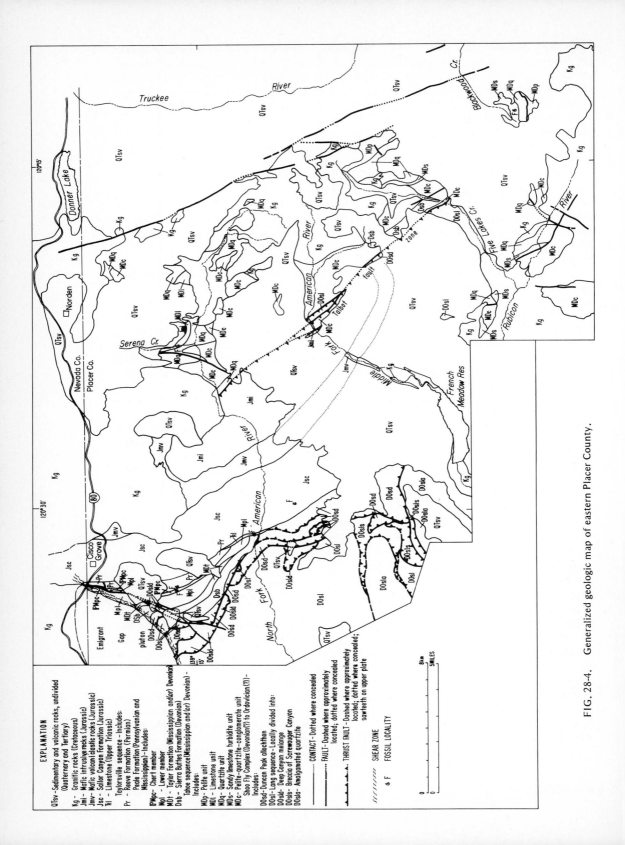

FIG. 28-4. Generalized geologic map of eastern Placer County.

deformations in metamorphic rocks of the central Sierra Nevada and concluded that a penetrative fabric formed in the Permian and/or Early Triassic. It is possible that the folding and normal faulting at the southern end of the Northern Sierra terrane represents the northern extent of that deformation. Certainly, the profound unconformity below the Upper Triassic and Jurassic rocks at the North Fork of the American River is exceptional, and, farther north, Mesozoic rocks lie disconformably above the upper Lower Permian rocks.

The Taylorsville sequence contains mineral assemblages typical of the greenschist facies. Felsic and intermediate volcanic rocks and volcaniclastic sandstones contain quartz, albite, muscovite, chlorite, actinolite, epidote and calcite. Quartz is present but rare in basaltic rocks of the Goodhue Formation, which is composed of chlorite, actinolite, epidote, calcite, talc or a serpentine-group mineral, and sphene(?) in addition to primary augite and plagioclase. On Keddie Ridge, Hannah (1980) found prehnite and pumpellyite in the groundmass and in veins cutting andesitic volcanic rocks. Actinolite is sparsely developed in the rocks on Keddie Ridge, suggesting somewhat lower conditions of metamorphism there than elsewhere in the terrane. Traditionally, metamorphism of the Taylorsville sequence has been related to the Jurassic Nevadan orogeny and no new evidence contradicts this conclusion.

TRIASSIC AND JURASSIC ROCKS

Triassic and Jurassic rocks occur above and below the Taylorsville fault (Fig. 28-1), and their lithologic composition and stratigraphy differ considerably depending on their structural position. From the vicinity of Genesee (Fig. 28-1) southeast to the Middle Fork of the American River, Mesozoic rocks occur above the Taylorsville fault. These rocks commonly rest either disconformably or unconformably on upper Lower Permian rocks of the Taylorsville sequence except southeast of Bowman Lake, where Schweickert *et al.* (1984b) report that Jurassic rocks rest unconformably on the chert member of the Peale Formation, and at the North Fork of the American River, where Upper Triassic and Jurassic rocks truncate all of the Taylorsville sequence and rest unconformably on the Shoo Fly Complex (Fig. 28-1; Harwood, 1983).

At the North Fork of the American River (Fig. 28-4), the Upper Triassic section contains a basal chert-pebble conglomerate that grades upward into thin-bedded limestone and calcareous chert-rich sandstone that is overlain by bioclastic limestone and gray to black massive limestone. This section represent a shallow-water onlap sequence. Andesitic lapilli tuff sharply overlies the limestone and it is overlain by pelite and feldspathic sandstone turbidites of the Sailor Canyon Formation, which contains Early and Middle Jurassic (Sinemurian to Bajocian) ammonites (Imlay, 1968). Coarse andesitic tuff breccia overlies the Sailor Canyon Formation and is intruded on the east by metamorphosed pyroxene diorite.

North of I80, conglomerate has not been reported at the base of the Mesozoic section that rests disconformably on the Taylorsville sequence. Discontinuous patches of limestone, probably of Late Triassic age, occur at the base of the section and are overlain by pelite and tuffaceous sandstone of the Sailor Canyon Formation. Andesitic tuff breccia occurs above the Sailor Canyon Formation as far north as English Mountain (Fig. 28-1; Schweickert *et al.*, 1984b; Stuart-Alexander, 1966).

Near Genesee, the Upper Triassic Hosselkus Limestone (Diller, 1908) appears to rest unconformably on upper Lower Permian rocks, but the area is structurally complex and McMath (1958) could not exclude the possibility of faulting along the unconformity. The Hosselkus Limestone grades upward into the Swearinger Slate (Diller, 1908), which contains argillaceous limestone, quartzose sandstone, and conglomerate that are also Late

Triassic (Norian) in age (McMath, 1958). No Jurassic rocks occur above the Taylorsville fault in the northern part of the terrane, where the Mesozoic section is only about 300 m thick. Mesozoic rocks gradually thicken toward the south, however, and are about 3 km thick at the North Fork of the American River where Lower and Middle Jurassic rocks make up all but the basal 100 m of the section.

Mesozoic rocks below the Taylorsville fault in the northeastern part of the terrane contrast sharply with the relatively thin section of Upper Traissic rocks above the fault. Below the fault, the Mesozoic rocks are at least 4 km thick and consist of Lower and Middle Jurassic (Imlay, 1961) volcanic and volcaniclastic rocks of dacitic and andesitic composition. Although the section consists predominantly of flows, tuff breccia, and tuff, it contains significant amounts of interbedded volcaniclastic conglomerate, sandstone, and mudstone. The upper part of the section contains polymict conglomerate, volcaniclastic- and quartz-rich graywacke, and locally plant-bearing slate and mudstone. No Triassic rocks have been identified below the fault and the basement of the thick Jurassic volcanic sequence is unknown, although it is generally assumed to be the Taylorsville sequence (McMath, 1958).

JURASSIC DEFORMATION AND METAMORPHISM

The dominant northwest-trending structural grain of the Northern Sierra terrane was imposed during the Late Jurassic Nevadan orogeny (Day et al., 1985; Schweickert, 1981; Schweickert et al., 1984a; Varga and Moores, 1981). This period of compressive deformation produced east-directed thrust faults, major northwest-trending folds, a penetrative fabric in the peltic rocks, and regional greenschist-facies metamorphism. Superposed structures indicate protracted and complex deformation possibly dating back to 179 m.y.

In the southern part of the terrane, a northeast-trending zone of ductile shearing that partly surrounds the Emigrant Gap pluton (Fig. 28-1) provides evidence for some of the earliest Jurassic deformation. Along the northwest margin of the pluton, where the shear zone is developed in the Shoo Fly Complex, mesoscopic isoclinal folds have axial surfaces that trend N25°E and dip steeply northwest. Fold axes plunge moderately to steeply northwest. Schweickert et al. (1984b) report that the shear zone deforms bedding and a poorly developed early fabric in a dextral sense. On the southeast side of the pluton, the Shoo Fly and overlying rocks of the Taylorsville sequence are deformed in a sinistral sense to produce macroscopic northeast-plunging folds against the margin of the pluton (Harwood, 1983). Intense ductile shearing dies out to the southwest, and at the southwest margin of the pluton, beds in the Shoo Fly strike northwest or northeast and are intensely crenulated by a northwest-trending, near-vertical slip cleavage. Biotite, muscovite, and chlorite are oriented in the slip cleavage within the contact aureole of the pluton. Feldspar porphyry dikes that contain mafic and ultramafic xenoliths are deformed in the shear zones along the northwest and southeast margins of the pluton. The dikes commonly strike northwest parallel to the slip cleavage near the southwest margin of the pluton. The spatial distribution of the deformation and its relation to dikes derived from the pluton suggest that the shearing was developed during forceful injection of the Emigrant Gap pluton (James, 1971). Ages on the pluton range from 179 to 152 m.y. (Snoke et al., 1982). Intrusion of the Emigrant Gap pluton apparently occurred during the waning stages of volcanism recorded by the upper part of the Jurassic volcanic rocks. The pluton may represent a deeper part of the Jurassic magmatic arc.

Intrusion of the Emigrant Gap pluton was followed by northeast-directed compressive tectonism that produced a variety of fold-thrust structures in the Paleozoic and Mesozoic rocks. The northern part of the terrane contains two southwest-dipping thrust faults shown on Fig. 28-1 as the Taylorsville and Grizzly Mountain faults. The Taylorsville fault (McMath, 1958) is a southwest-dipping thrust fault that places a west-dipping, east-facing inverted section of upper Paleozoic rocks over Jurassic rocks. The Jurassic rocks in the lower plate were folded into an east-verging syncline prior to or during the early phases of thrusting. Upper Paleozoic rocks above the Taylorsville fault overlie the overturned west limb and the axial trace of the syncline giving a minimum estimate of northeast displacement of about 8 km (McMath, 1958). The Grizzly Mountain fault, which only repeats the Paleozoic section and does not involve exposed Mesozoic rocks, appears to be a high-angle reverse fault or an ancillary thrust that splayed off the Taylorsville fault. Upper Paleozoic rocks southwest of the Grizzly Mountain fault contain a well-developed southwest-dipping shear fabric and actinolite-bearing mineral assemblages. To the northeast on Keddie Ridge, however, the upper Paleozoic rocks contain only a weakly developed spaced cleavage and prehnite-pumpellyite-bearing assemblages are more common than actinolite-bearing assemblages. Day *et al.* (1985) cite these structural and mineralogical differences as evidence for thrust movement on the Grizzly Mountain fault. It is possible that Cenozoic normal faulting on or near the trace of the Grizzly Mountain fault has obscured some of the original thrust movement.

A pronounced shear fabric is developed within a few tens of meters of the Taylorsville fault, and a similar fabric pervades all but the most massive volcanic rocks above the Grizzly Mountain fault. East-verging mesoscopic folds, southwest-plunging fault mullions, and rotated clasts in conglomerate beds indicate northeast transport on the Taylorsville and Grizzly Mountain faults. Actinolite and chlorite occur in the shear fabric of the Taylorsville fault and indicate that greenschist-facies metamorphism accompanied or closely followed the thrusting. The presence of prehnite and pumpellyite on Keddie Ridge suggests that the metamorphic grade may decrease slightly to the northeast, but the metamorphic gradient is not parallel to either the Grizzly Mountain or the Taylorsville faults.

The sinuous trace of the Taylorsville thrust around Mount Jura (Fig. 28-1) defines a broad, south-plunging anticline that formed after thrusting. This anticline is not reflected in the map pattern of the upper Paleozoic rocks above the thrust, nor is there a penetrative fabric in the Paleozoic rocks related to the post-thrust folding. Instead, the inverted sequence of upper Paleozoic rocks above the Taylorsville fault represents the overturned limb of a large east-verging recumbent anticline that formed early in the fold-thrust deformation. Structural data are still too sparse to indicate whether folding of the Taylorsville thrust is related to deformation that produced west-verging late folds in the northwest part of the terrane (Day *et al.*, 1985).

The pattern of Mesozoic deformation in the southern part of the terrane differs significantly from that to the north, and the paucity of critical outcrops in the intervening area makes tectonic synthesis speculative. At the North Fork of the American River, Mesozoic rocks rest unconformably on lower and upper Paleozoic rocks (Harwood, 1983), and they extend eastward to the Talbot fault zone (Fig. 28-4) without any significant tectonic break. The Talbot fault zone contains several vertical to steeply west-dipping faults that are marked by zones of flattening and ductile shearing a few tens of meters wide. In its northern exposures, the Talbot fault zone juxtaposes Jurassic metadiorite on the west against various rocks of the Tahoe sequence, but in its southern exposures at Five Lakes Creek (Fig. 28-4), it juxtaposes rocks of the Tahoe sequence against the Shoo Fly Complex to the west.

Mesozoic rocks west of the Talbot fault zone define a northwest-trending syncline that plunges moderately to the south in its northern exposures. The southern extent of the syncline is covered by Tertiary volcanic rocks, but it is assumed that the syncline is doubly plunging and that the Mesozoic unconformity intersects the Talbot fault zone north of the exposures of the Shoo Fly Complex at Five Lakes Creek. The Talbot fault zone apparently is also tightly folded in the south, because the Tahoe sequence in the Rubicon River canyon (Fig. 28-4) strikes west and dips north below the Shoo Fly Complex.

The Talbot fault zone cuts rocks at least as young as Middle Jurassic and is intruded by Cretaceous plutons about 100 m.y. old. The same time constraints limit movement on the Taylorsville fault, and it seems possible that the Talbot fault zone is either the southern extension of the Taylorsville fault or a fault related to the same period of deformation as the Taylorsville fault.

SUMMARY

Mesocopic structures and regional unconformities indicate four separate periods of deformation in the Northern Sierra terrane. The earliest and least understood deformation amalgamated oceanic rocks of the Shoo Fly Complex along regional thrust faults sometime prior to the Late Devonian. Serpentinite-bearing melange, major olistostromes, and isoclinal folds formed during the thrusting, but there is no indication that the Shoo Fly Complex was metamorphosed above low greenschist facies, if at all, during the pre-Late Devonian deformation.

From the Late Devonian to the Late Jurassic, three island-arc volcanic sequences were built unconformably on a deformed basement of Shoo Fly Complex. The periods of arc volcanism and the hiatuses that separate them are shown in Fig. 28-5. The cross-sectional areas of the volcanic units, taken from Fig. 28-1, are also shown on Fig. 28-5 in order to compare the relative abundance of the volcanic units. Ideally, the volumes of the volcanic units should be compared, but that is impossible in the Northern Sierra terrane, and I have assumed that the cross-sectional areas of the units are a reasonable approximation of their relative abundance. If such is the case, the Sierra Buttes and Taylor Formations contain about the same relative amounts of material and each was erupted over a period of about 10 m.y. The lower member of the Peale Formation, however, represents less than a quarter of the material in the Sierra Buttes or Taylor Formations, although it was erupted over a comparable length of time, about 10 m.y. If Fig. 28-1 is representative of the arc sequence as a whole, it appears that volcanism in the lower Taylorsville sequence waned dramatically in the Early Mississippian. By the late middle Mississippian, the arc had apparently cooled and subsided to form a basin that received siliceous pelagic sediment now represented by the chert member of the Peale. The period of early arc volcanism in the Northern Sierra terrane corresponds closely with the time of emplacement of the Roberts Mountain allochthon in northwestern Nevada (Johnson and Pendergast, 1981;

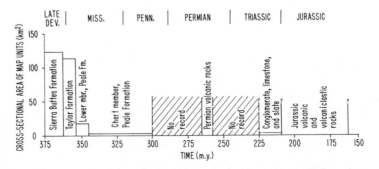

FIG. 28-5.   Graph of cross-sectional area of volcanic rocks versus time, showing relative abundance and times of volcanism in Northern Sierra terrane.

Roberts *et al.*, 1958; Speed and Sleep, 1982). Early deposits of the Tahoe sequence that are interbedded with volcanic rocks correlative with the Sierra Buttes Formation may represent material shed westward from the allochthons, and the younger quartzite and carbonate units of the Tahoe sequence may be deposits formed in relatively shallow water on the west slope of the Antler orogenic highland.

Pelagic sedimentation characterized the basin that formed over the submerged arc sequence for at least 40 m.y., from the late middle Mississippian to the Middle Pennsylvanian. Basin conditions may have lasted significantly longer, but the depositional record is missing between the Desmoinesian Stage of the Middle Pennsylvanian and the Leonardian Stage of the Lower Permian. During that time, the basin was uplifted and differentially eroded (Fig. 28-3). No compressional tectonic features have been found in the older arc sequence and basin uplift may have been related to deep crustal heating that gave way eventually to late Early Permian volcanism.

Permian volcanism in the Northern Sierra terrane produced about half the amount of material recorded in the Taylor or Sierra Buttes Formations. Volcanism lasted at least 12 m.y., but the upper age limit is not well constrained because of the profound unconformity between the upper Lower Permian volcanic rocks and overlying Upper Triassic rocks. During the Permian-Triassic hiatus, thrust sheets of the Golconda allochthon were emplaced in northwestern Nevada (Silberling and Roberts, 1962). Although the Upper Triassic rocks rest unconformably on all the rocks of the Taylorsville sequence and the Shoo Fly Complex, there is sparse evidence of compressional tectonism and no unequivocal evidence of metamorphism related to the Permian and Triassic deformation.

Upper Triassic conglomerate, limestone, and shale form an onlap sequence above the upper Lower Permian volcanic rocks. The Triassic rocks give way abruptly to volcanic and volcaniclastic rocks that range in age from Early Jurassic (Pliensbachian) to late Middle Jurassic (Callovian). The Jurassic volcanism clearly spanned a significantly longer period of geologic time than either of the earlier arc sequences, but the relative amount of material cannot be compared because of Cretaceous plutonism and cover by Tertiary and Quaternary rocks.

Late Jurassic compressive deformation produced northeast-verging folds above and below major thrust faults and regional, low-greenschist-facies metamorphism in the Jurassic rocks. If the older rocks in the Northern Sierra terrane were metamorphosed during the Paleozoic or early Mesozoic deformational events, that metamorphism was of low greenschist facies or less and is masked by the Late Jurassic metamorphic overprint.

ACKNOWLEDGMENTS

I am grateful to M. C. Blake, Jr., G. R. Fisher, G. H. Girty, A. Jayko, J. L. Hannah, R. Hanson, R. A. Schweickert, and B. J. Waugh for discussions that contributed to the ideas presented here. Thanks are extended to T. Dutro, M. Gordon, A. Harris, D. L. Jones, B. Murchey, J. Pojeta, and N. J. Silberling for identifying fossil collections from the Northern Sierra terrane. The manuscript benefitted from critical reviews by W. G. Ernst, A. Jayko, P. Stone, and H. Day.

REFERENCES

Albers, J. P., and Robertson, J. F., 1961, Geology and ore deposits of east Shasta copper-zinc district, Shasta County, California: *U.S. Geol. Survey Prof. Paper 338*, 107 p.

Anderson, T. B., Woodard, G. D., Strathouse, S. M., and Twichell, M. K., 1974, Geology of a Late Devonian fossil locality in the Sierra Buttes Formation, Dugan Pond, Sierra City quadrangle, California [abstract]: *Geol. Soc. America Abstr., with Programs*, v. 6, p. 139.

Bond, G. C., and DeVay, J. C., 1980, Pre-upper Devonian quartzose sandstones in the Shoo Fly Formation, northern California — Petrology, provenance, and implications for regional tectonics: *Geology*, v. 88, p. 285–308.

Brooks, E. R., Wood, M. M., and Garbutt, P. L., 1982, Origin and metamorphism of peperite and associated rocks in the Devonian Elwell Formation, northern Sierra Nevada, California: *Geol. Soc. America Bull.*, v. 93, p. 1208–1231.

Coney, P. J., Jones, D. L., and Monger, J. W. H., 1980, Cordilleran suspect terranes: *Nature*, v. 288, no. 5789, p. 329–333.

D'Allura, J. A., 1977, Stratigraphy, structure, petrology, and regional correlations of metamorphosed upper Paleozoic volcanic rocks in portions of Plumas, Sierra, and Nevada Counties, California: Ph.D. dissertation, Univ. California, Davis, Calif., 338 p.

____, Moores, E. M., and Robinson, L., 1977, Paleozoic rocks of the northern Sierra Nevada — Their structural and paleogeographic implications, *in* Stewart, J. H., et al., eds., *Paleozoic Paleogeography of the Western United States:* Pacific Section, Soc. Econ. Paleontologists Mineralogists, Pacific Coast Paleogeography Symp. 1, p. 395–408.

Day, H. W., Moores, E. M., and Tuminas, A. C., 1985, Structure and tectonics of the northern Sierra Nevada: *Geol. Soc. America Bull.*, v. 96, p. 436–450.

Diller, J. S., 1908, Geology of the Taylorsville region, California: *U.S. Geol. Survey Bull. 353*, 128 p.

Durrell, Cordell, and D'Allura, J. A., 1977, Upper Paleozoic section in eastern Plumas and Sierra Counties, Sierra Nevada, California: *Geol. Soc. America Bull.*, v. 88, p. 844–852.

Gehrels, G. W., and Saleeby, J. B., 1984, Paleozoic geologic history of the Alexander terrane in SE Alaska, and comparisons with other orogenic belts [abstract]: *Geol. Soc. America Abstr. with Programs*, v. 16, no. 6, p. 516.

Girty, G. H., 1983, The Culbertson Lake allochthon — A newly identified structural unit in the Shoo Fly Complex — Sedimentological, stratigraphic, and structural evidence for extension of the Antler orogenic belt to the northern Sierra Nevada, California: Ph.D. dissertation, Columbia Univ., New York, N.Y., 155 p.

____, and Schweickert, R. A., 1984, The Culbertson Lake allochthon, a newly identified structure within the Shoo Fly Complex, California — Evidence for four phases of deformation and extension of the Antler orogeny to the northern Sierra Nevada: *Modern Geology*, v. 8, p. 181–198.

____, and Wardlaw, M. S., 1984, Was the Alexander Terrane a source of feldspathic sandstones in the Shoo Fly Complex, Sierra Nevada, California?: *Geology*, v. 12, p. 339–342.

____, Wardlaw, M. S., 1985, Petrology and provenance of pre-Late Devonian sandstones, Shoo Fly Complex, northern Sierra Nevada, California: *Geol. Soc. America Bull.*, v. 96, p. 516–521.

____, Wardlaw, M. S., Schweickert, R. A., Hanson, R., and Bowring, S., 1984, Timing of pre-Antler deformation in the Shoo Fly Complex, Sierra Nevada, California: *Geology*, v. 12, p. 673–676.

Hannah, J. L., 1980, stratigraphy, petrology, paleomagnetism, and tectonics of Paleozoic arc complexes, northern Sierra Nevada, California: Ph.D. dissertation, Univ. California, Davis, Calif., 323 p.

____, and Moores, E. M., 1986, Age relations and depositional environments of Paleozoic strata, northern Sierra Nevada, California: *Geol. Soc. America Bull.*, v. 97, p. 787–797.

Hanson, R. E., 1983, Volcanism, plutonism and sedimentation in a Late Devonian submarine island-arc setting, northern Sierra Nevada, California: Ph.D. dissertation, Columbia Univ., New York, N.Y., 345 p.

——, Saleeby, J. B., Schweickert, R. A., in press, Composite Devonian island-arc batholith in the northern Sierra Nevada, California: *Geol. Soc. America Bull.*

Harwood, D. S., 1983, Stratigraphy of upper Paleozoic volcanic rocks and regional unconformities in part of the Northern Sierra terrane: *Geol. Soc. America Bull.*, v. 94, p. 413–422.

——, and Fisher, G. R., 1984, Paleozoic rocks along the Sierran crest west of Lake Tahoe, California [abstract]: *Geol. Soc. America Abstr. with Programs*, v. 16, no. 6, p. 531.

Imlay, R. W., 1961, Late Jurassic ammonites from the western Sierra Nevada, California: *U.S. Geol. Survey Prof. Paper 374-D*, 30 p.

——, 1968, Lower Jurassic (Pliensbachian and Toarcian) ammonites from eastern Oregon and California: *U.S. Geol. Survey Prof. Paper 593*, 51 p.

James, O. B., 1971, Origin and emplacement of the ultramafic rocks of the Emigrant Gap area, California: *J. Petrology*, v. 12, pt. 3, p. 532–560.

Johnson, J. G., and Pendergast, A., 1981, Timing and mode of emplacement of the Roberts Mountain allochthon, Antler orogeny: *Geol. Soc. America Bull.*, v. 92, pt. 1, p. 648–658.

Jones, D. L., Irwin, W. P., and Ovenshine, A. T., 1972, Southeast Alaska — A displaced continental fragment?: *U.S. Geol. Survey Prof. Paper 800-B*, p. B211–B217.

Lindgren, Waldamar, 1897, Description of the gold belt, description of the Truckee quadrangle, California: *U.S. Geol. Survey Geol. Atlas, Folio 39*, 8 p.

McMath, V. E., 1958, The geology of the Taylorsville area, Plumas County, California: Univ. California, Los Angeles, Calif., Ph.D. dissertation, 199 p.

——, 1966, Geology of the Taylorsville area, northern Sierra Nevada, *in* Bailey, E. H., ed., *Geology of Northern California:* Calif. Div. Mines and Geology Bull. 190, p. 173–183.

Merguerian, Charles, 1985, Stratigraphy, structural geology, and tectonic implications of the Shoo Fly Complex and the Calavaras–Shoo Fly thrust, central Sierra Nevada, California: Ph.D. dissertation, Columbia Univ., New York, N.Y., 255 p.

Moores, E. M., 1972, Model for Jurassic island arc, continental margin collision in California [abstract]: *Geol. Soc. America Abstr. with Programs*, v. 4, p. 202.

Nokleberg, W. J., and Kistler, R. W., 1980, Paleozoic and Mesozoic deformations in the central Sierra Nevada, California: *U.S. Geol. Survey Prof. Paper 1145*, 24 p.

Roberts, R. J., Hotz, P. E., Gilluly, J., and Ferguson, H. G., 1958, Paleozoic rocks of north-central Nevada: *Amer. Assoc. Petrol. Geologists Bull.*, v. 42, p. 2813–2857.

Saleeby, J. B., 1981, Ocean floor accretion and volcanoplutonic arc evolution in the Mesozoic Sierra Nevada, *in* Ernst, W. G., ed., *The Geotectonic Development of California*, (Rubey Vol. I): Englewood Cliffs, N.J., Prentice-Hall, p. 132–181.

——, 1982, Polygenetic ophiolite belt of the California Sierra Nevada, geochronological and tectonostratigraphic development: *J. Geophys. Res.*, v. 87, p. 1803–1824.

Schweickert, R. A., 1978, Triassic and Jurassic paleogeography of the Sierra Nevada and adjacent regions, California and western Nevada, *in* Howell, D. G., and McDougall, K. A., eds., *Mesozoic Paleogeography of the Western United States:* Pacific Section, Soc. Econ. Paleontologists Mineralogists, Pacific Coast Paleogeography Symp. 2, p. 361–384.

_____ , 1981, Tectonic evolution of the Sierra Nevada range, *in* Ernst, W. G., ed., *The Geotectonic Development of California*, (Rubey Vol. I): Englewood Cliffs, N.J., Prentice-Hall, p. 87–131.

_____ , and Cowan, D. S., 1975, Early Mesozoic tectonic evolution of the western Sierra Nevada, California: *Geol. Soc. America Bull.*, v. 86, p. 1329–1336.

_____ , and Snyder, W. S., 1981, Paleozoic plate tectonics of the Sierra Nevada and adjacent regions, *in* Ernst, W. G., ed., *The Geotectonic Development of California* (Rubey Vol. I): Englewood Cliffs, N.J., Prentice-Hall, p. 182–201.

_____ , Armstrong, R. L., and Harakal, J. E., 1980, Lawsonite blueschist in the northern Sierra Nevada, California: *Geology*, v. 8, p. 27–31.

_____ , Bogen, N. L., Girty, G. H., Hanson, R. E., and Merguerian, C., 1984a, Timing and structural expression of the Nevadan orogeny, Sierra Nevada, California: *Geol. Soc. America Bull.*, v. 95, p. 967–979.

_____ , Harwood, D. S., Girty, G. H., and Hanson, R. E., 1984b, Tectonic development of the Northern Sierra terrane — An accreted Late Paleozoic island arc and its basement, *in* Lintz, J., Jr., ed., *Western Geological Excursions; Vol. 4:* Geol. Soc. America Guidebook, p. 1–65.

Silberling, N. J., and Roberts, R. J., 1962, Pre-Tertiary stratigraphy and structure of northwestern Nevada: *Geol. Soc. America Spec. Paper 163*, 28 p.

Snoke, A. W., Sharp, W. D., Wright, J. E., and Saleeby, J. B., 1982, Significance of mid-Mesozoic peridotitic to dioritic intrusive complexes, Klamath Mountains-western Sierra Nevada, California: *Geology*, v. 10, p. 160–166.

Speed, R. C., and Sleep, N. H., 1982, Antler orogeny and foreland basin — A model: *Geol. Soc. America Bull.*, v. 93, p. 815–828.

Standlee, L. A., 1978, Geology of the northern Sierra Nevada basement rocks, Quincy-Downieville area, California: Ph.D. dissertation, Rice Univ., Houston, Tex., 176 p.

Stuart-Alexander, D. E., 1966, Contrasting deformation of Paleozoic and Mesozoic rocks near Sierra City, northern Sierra Nevada, California: Ph.D. dissertation, Stanford Univ., Stanford, Calif., 91 p.

Turner, H. W., 1897, Description of the Downieville quadrangle, California: *U.S. Geol. Survey Geol. Atlas, Folio 37*, 15 p.

Varga, R. J., 1980, Structural and tectonic evolution of the early Paleozoic Shoo Fly Formation, Sierra Nevada range, California: Ph.D. dissertation, Univ. California, Davis, Calif., 248 p.

_____ , and Moores, E. M., 1981, Age, origin, and significance of an unconformity that predates island-arc volcanism in the northern Sierra Nevada: *Geology*, v. 9, p. 512–518.

Richard A. Schweickert
Department of Geological Sciences
University of Nevada-Reno
Reno, Nevada 89557

Charles Merguerian
Geology Department
Hofstra University
Hempstead, New York 11550

Nicholas L. Bogen
Carlson and Sweatt-Monenco
275 7th Avenue
New York, New York 10001

29

DEFORMATIONAL AND METAMORPHIC HISTORY OF PALEOZOIC AND MESOZOIC BASEMENT TERRANES IN THE WESTERN SIERRA NEVADA METAMORPHIC BELT

ABSTRACT

Rocks of the western Sierra Nevada metamorphic belt contain evidence of several important deformational and metamorphic events that affected parts of the southwestern Cordillera in Paleozoic and Mesozoic time. An understanding of these events is important for relating the basement terranes of the Sierra Nevada with other terranes in the Cordillera. Detailed structural studies have been made of all of the major tectonostratigraphic units in the southern part of the belt, and a clear assessment can now be made of the relations between deformation, metamorphism, and faulting for all terranes, a situation that exists for few locations within the Cordillera. Although correlation of structural generations in polydeformed and polymetamorphic terranes is a difficult task, the use of style, overprinting, and orientation of structures, together with the use of synmetamorphic faults and several suites of igneous intrusions as structural markers, have enabled us to establish the following structural and metamorphic sequence. Age estimates are based on published isotopic and faunal ages.

Overall, there is an eastward increase in age, metamorphic grade, and structural complexity in the southern part of the western metamorphic belt. The lower Paleozoic Shoo Fly Complex (SFC) contains evidence of an early to mid-Paleozoic deformational/metamorphic event and a second, very intense, late Paleozoic(?) event. These two events are separated by the intrusion of a suite of gabbroic to granitic, and syenitic, intrusions. The third deformational/metamorphic event, shared by chaotic rocks of the Calaveras Complex (CC), in Early Triassic or later time, coincided with the juxtaposition of the two terranes along the Calaveras–Shoo Fly thrust (CSFT), and internal imbrication of the two terranes. During this event, both terranes attained amphibolite facies conditions. The SFC, CC, and CSFT were strongly folded by east-southeast-trending folds accompanied by metamorphic recrystallization prior to the Middle Jurassic. Post-tectonic, forcefully intruded plutons of Middle Jurassic age intruded SFC and CC, deforming the early fabrics. These intrusions were followed by east-west-trending mafic dikes of the Sonora dike swarm in the Late Jurassic. During the Late Jurassic Nevadan orogeny, the Don Pedro and Foothills terranes underthrust the CC along the Sonora fault. Nevadan deformation included development of crenulation cleavage in the SFC and CC, polyphase ductile deformation at upper greenschist facies in the Don Pedro terrane, thrusting along the Melones fault and development of large, upright folds with slaty cleavage at lower greenschist facies in the Foothills terrane.

We interpret the structural and metamorphic sequence in the southern part of the western metamorphic belt to reflect deep-seated pre-Carboniferous deformation and metamorphism of the SFC (during the Antler orogeny or earlier), juxtaposition of SFC and CC during the Triassic (Sonoma orogeny?), Early to Middle Jurassic deformation of both terranes, and juxtaposition of Foothills and Don Pedro terranes with CC and SFC during the Late Jurassic Nevadan orogeny.

INTRODUCTION

Rocks of the western Sierra Nevada metamorphic belt contain evidence of several important deformational and metamorphic events. The belt is significant because it provides an important comparison with Paleozoic and Mesozoic rocks of Nevada and also is an important link among coeval metamorphic terranes of the western Cordillera in British Columbia, Washington, Idaho, Oregon, the Klamath Mountains, and the Peninsular Ranges and Baja California. Furthermore, it provides an important glimpse of structure and metamorphism along the western perimeter of the Sierra Nevada batholith, and may be the key to deciphering the geologic history of terranes now obscured by the batholith (for a description of the northern part of this belt, see Harwood, Chapter 28, this volume).

Detailed structural studies in the southern part of the belt have encompassed all of the major lithotectonic units, and a clear assessment can be made of the relations between deformation, metamorphism, and faulting for the entire belt, a situation that exists for few locations within the southwestern Cordillera.

Some parts of the belt are polymetamorphic terranes characterized by a rather complex structural and metamorphic history. The aims of this paper are to discuss methods of structural analysis, to synthesize a large amount of structural data obtained from the various units over the past decade, to discuss the major structural events and their relations to metamorphism, and to summarize some interpretations of these relations that are consistent with available data.

Regional Setting

The Sierra Nevada batholith invaded the western Sierran metamorphic belt along its eastern and southeastern margins during Jurassic and Cretaceous time (Fig. 29-1). Cretaceous and Cenozoic deposits of the Great Valley overstep the metamorphic belt on its western edge.

The southern part of the metamorphic belt is subdivided into four main tectono-stratigraphic units that are separated by three ductile fault zones. The easternmost and structurally highest terrane is known as the Shoo Fly Complex. The Shoo Fly is an assemblage of highly deformed and metamorphosed sedimentary and igneous rocks, most of which are pre-Carboniferous in age. The western edge of the Shoo Fly is marked by the Calaveras–Shoo Fly thrust, which places the Shoo Fly structurally over the Calaveras Complex. The Calaveras is a highly deformed and moderately metamorphosed assemblage of chaotic chert and argillite, with lesser amounts of limestone, basalt, quartzose siltstone and sandstone, and is of late Paleozoic to early Mesozoic age. Both the Shoo Fly and Calaveras are intruded by a variety of felsic and mafic dikes and calcalkaline plutons that form important structural markers.

The Sonora fault is a ductile thrust zone that separates the Calaveras Complex from underlying Jurassic units to the west. Immediately west of the Sonora fault are strongly deformed, greenschist-grade volcanic and sedimentary rocks of the phyllite-greenschist belt (Don Pedro terrane of Blake *et al.*, 1982). These rocks in turn are in thrust contact along the more westerly Melones fault with less deformed and very weakly metamorphosed volcanic and sedimentary rocks of the slate belt (Foothills terrane of Blake *et al.*, 1982).

METHODS OF STRUCTURAL ANALYSIS

In the western metamorphic belt, where polyphase deformed and polymetamorphosed rocks are widespread, it is of considerable importance to attempt to establish a relative chronology of structural generations in each terrane and to relate these generations to episodes of metamorphic recrystallization. Ideally, with sufficient age constraints, the structural generations may then be used to correlate tectonic events from one terrane to another.

Problems of Structural Correlation

Structural correlation is not an easy task, and numerous authors have discussed problems of structural correlation within orogenic belts, from both conceptual and operational points of view. Roberts (1977), among others, discussed the fact that temperature and

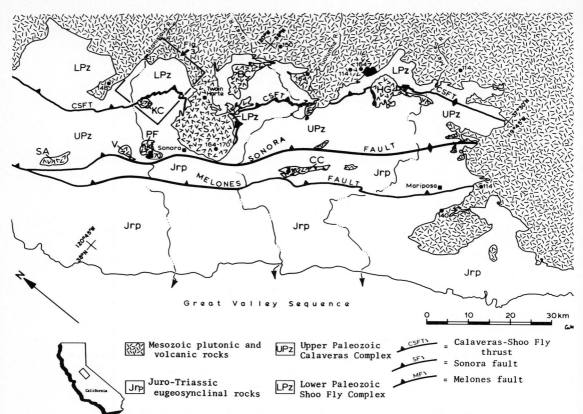

FIG. 29-1. Geologic map of the southern end of the western Sierra Nevada metamorphic belt in Calaveras, Tuolumne, and Mariposa Counties, California. Shoo Fly Complex (LPz), Calaveras Complex (UPz), Don Pedro and Foothills terranes (Jrp). Some map contacts from Bowen (1969), Merguerian (1981a, b), Morgan (1976), Rogers (1966), Schweickert and Bogen (1983), R. A. Schweickert (unpub. data), Strand (1967), Strand and Koenig (1965), Tobisch (1960), Turner and Ransome (1897, 1898), and Wagner et al. (1981). Middle Jurassic plutons indicated by V-pattern are SA, San Andreas; V, Vallecitos; PF, Parrotts Ferry; KC, Knight Creek; S, Standard; BC, Basin Creek; CC, Cobbs Creek; HG, Hazel Green. Mafic-ultramafic plutons and phases are solid black. Granitoids of the Sierra Nevada batholith are stippled. Isotopic dates from Sharp and Saleeby (1979) and Stern et al. (1981) are indicated by filled circles.

pressure will vary within a large volume of rock, and therefore, during a single deformational "event," different deformational mechanisms will be activated, resulting in different structural styles. In addition, structural generations are likely to develop diachronously over a large region and therefore should be expected to vary in age from place to place (Hobbs et al., 1976; Roberts, 1977; Williams, 1985). Finally, separate generations of structures are as likely to have formed during a continuum of progressive deformation as from discrete or separate events. The structures themselves cannot resolve this problem. Isotopic dating or other evidence is necessary to evaluate this question.

The establishment of a relative chronology of structural generations commonly is very difficult. Hobbs et al. (1976), Roberts (1977), Borradaile (1978), Williams (1985), and many others have discussed the use of such criteria as style, overprinting, and orientation for establishing structural generations, and have pointed out potential pitfalls.

Terranes in the Sierra Nevada metamorphic belt (and in a larger sense, tectonostratigraphic terranes in general) pose special problems for structural correlation because they have been tectonically juxtaposed during various intervals of their deformational histories and therefore probably were structurally independent of each other during parts of their history. Finally, terranes like the Calaveras and Shoo Fly Complexes lack information about sedimentary facing directions because of chaotic soft-sediment deformation in the Calaveras and tectonic transposition in the Shoo Fly.

Mindful of these difficulties, we have applied the following criteria as carefully and extensively as possible to arrive at a structural model for each tectonostratigraphic unit.

Conventional Methods

Exposure in the western metamorphic belt in general is poor. However, rivers and streams are deeply incised and provide excellent, semicontinuous exposures along their courses. Accordingly, most of our work has consisted of structural traverses along rivers and streams within the Mokelumne, Stanislaus, Tuolumne, and Merced River drainages (Fig. 29-1). These structural traverses have enabled us to correlate structural features over large areas, but the limited outcrop has severely limited our ability to map in detail large, map-scale structures.

Overprinting of foliations and folds by younger foliations and folds has been the principal criterion used in this study. Style and orientation of older and younger structures have then been used within the same outcrop to classify structures into generations. This procedure has been used in working from outcrop to outcrop along stream traverses to establish a working structural chronology. The structural chronology has thus been established from a continuous network of structural traverses.

Additional Methods

The Shoo Fly and Calaveras Complexes are intruded by numerous igneous intrusions, both irregular plutons and sheetlike dikes and sills. In areas where the structural chronology is relatively clear, intrusions of a given type and compositional range generally appear to have consistent structural relations. Based on this observation, we have assumed that the different igneous intrusions are the products of fairly discrete intrusive events. This has enabled us to use mafic dike swarms, felsic dikes, and certain calcalkaline plutons as structural markers, separating structures of different generations.

From our experience, certain generalizations can be made about the use of igneous intrusions as structural markers. Dike swarms, which probably reflect certain crustal stress conditions (Nakamura *et al.*, 1977; Suppe, 1984), are likely to give rather unambiguous results. For example, in the Shoo Fly and Calaveras, a clearcut distinction is possible between structures that predate and those that postdate (and therefore deform) the Sonora mafic dike swarm.

Plutons, in contrast, can give somewhat ambiguous results. For example, a concordant pluton, with foliation in its wallrocks paralleling its external contact, can represent at least three different situations. (1) The pluton may have caused the development of a strong foliation and associated folds in the wallrocks as a result of forceful intrusion. In this case, foliation dates from the time of intrusion. (2) The pluton may have been intruded after an episode of deformation of the wallrocks, and forced the preexisting foliation into parallellism with its walls. (3) Deformation may postdate the intrusion of the pluton, with greater strain taken up in the wallrocks, so that foliation is deflected around the margins of the pluton. It is important therefore to examine the interior parts of plutons to see whether deformational fabrics or shear zones occur that might indicate major postplutonic deformation. In addition, the relations between foliations near and far from the pluton must be studied to see whether evidence exists for deformation of a pre-existing wallrock fabric (case 2 above), or whether foliation in the wallrocks and in the pluton is the result of a later deformation (case 3 above). The presence of static contact metamorphic minerals overprinting a fabric in the contact aureole may be consistent with either case 1 or 2 above. A common development may be case 1 followed by case 3, making it very difficult to determine which is the correct situation.

The major faults that separate the terranes in the foothills provide a final technique for structural correlation: synmetamorphic fault zones can be used as structural markers. We have found it very useful first to examine and characterize fault or shear zone structures, and then to carry these structures outward into hanging-wall or footwall domains to observe the order of superimposition of fault-related structures with respect to those of the wallrocks. Using the Calaveras–Shoo Fly thrust as an example, it is apparent that the fault fabrics represent intense, mylonitic development of the earliest foliation observed in the Calaveras Complex in the footwall, while these same fault fabrics are superimposed on two earlier generations of structures in the Shoo Fly Complex in the hanging wall.

We emphasize that the results of our analysis should be viewed as structural models that are consistent with available outcrop data, style, overprinting, orientation, and relations to igneous intrusions and faults. During structural analysis, we have not assumed strict synchroneity of structural generations, nor have we assumed that different generations necessarily indicate discrete events. Some generations may be parts of sequences of progressive deformation, while others may be discrete. However, it is important to note that we have observed no evidence for repeated parallel deformational events wherein a single foliation or fold generation may represent multiple deformational events, as argued by Tobisch and Fiske (1982) for a locality in the eastern Sierra.

BASEMENT TERRANES OF THE WESTERN SIERRA NEVADA METAMORPHIC BELT

The Shoo Fly Complex

Lithology The Shoo Fly Complex in the southern part of the Sierra Nevada metamorphic belt is a lower amphibolite-grade sequence of polyphase-deformed quartzite, quartzofeldspathic gneiss, garnet schist, calc-silicate rock, and marble. The Shoo Fly was mapped in reconnaissance from the Stanislaus River southward beyond the Merced River (Fig. 29-2; Merguerian, 1981a), and was studied in detail in the Jupiter area of the Stanislaus drainage (Fig. 29-3; Merguerian, 1985a, b). Gneissic granitoids associated with the Shoo Fly indicate that the metasedimentary rocks were intruded by magmas ranging in composition from gabbro to granite, and syenite (Merguerian, 1985c; Merguerian and Schweickert, in press).

Original stratigraphic relations have been obscured by several episodes of folding and ductile shearing, but the overall lithologic assemblage suggests that the Shoo Fly represents a thick slope-rise sequence composed of continentally derived, mature clastic sediments.

Age Fossils are not preserved in the Shoo Fly, and age estimates are based on isotopic data from the orthogneisses and on regional comparisons of the lithologic assemblage.

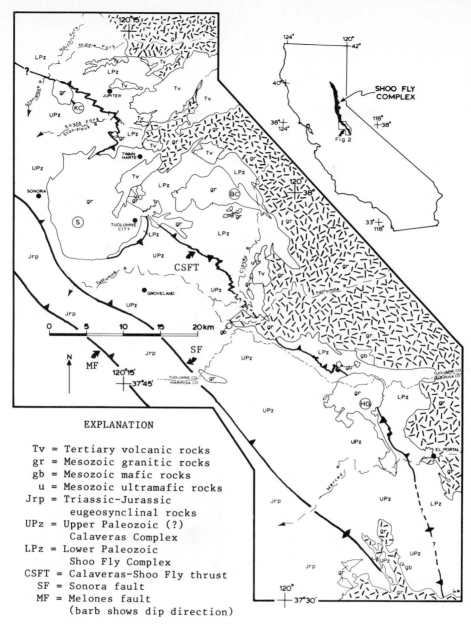

EXPLANATION

Tv = Tertiary volcanic rocks
gr = Mesozoic granitic rocks
gb = Mesozoic mafic rocks
 u = Mesozoic ultramafic rocks
Jrp = Triassic-Jurassic
        eugeosynclinal rocks
UPz = Upper Paleozoic (?)
        Calaveras Complex
LPz = Lower Paleozoic
        Shoo Fly Complex
CSFT = Calaveras-Shoo Fly thrust
 SF = Sonora fault
 MF = Melones fault
      (barb shows dip direction)

FIG. 29-2.   Geologic sketchmap of the Shoo Fly Complex and the Calaveras-
Shoo Fly thrust in Tuolumne and Mariposa Counties, California. The distribution of the Shoo Fly and geometry of the thrust zone are outlined from
detailed mapping at 1:24,000 scale (Merguerian, 1985a, b). Some map contacts are from Bowen (1969), Strand and Koenig (1965), Tobisch (1960),
R. A. Schweickert (unpublished data), Schweickert and Bogen (1983), Turner
and Ransome (1897, 1898, and Wagner *et al.* (1981). Labels on plutons are
the same as in Fig. 29-1.

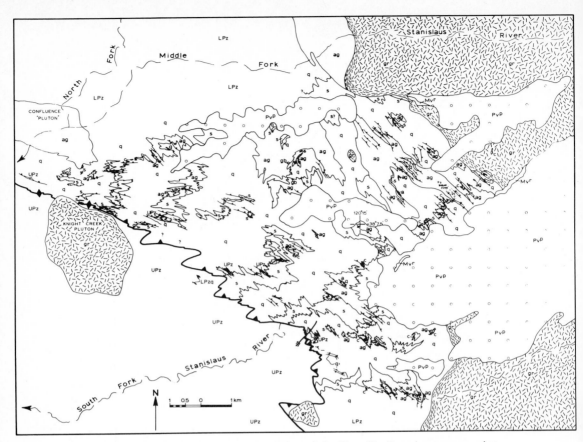

FIG. 29-3. Bedrock geologic map of the Shoo Fly Complex encompassing parts of the Stanislaus, Columbia SE, Crandall Peak, and Twain Harte 7.5-minute quadrangles. This map was compiled from 1:24,000 scale maps (Merguerian, 1985a, b). Lithologies in the Shoo Fly Complex (LPZ) are: q, quartzite and quartzofeldspathic gneiss; s, schist and phyllite; ag, gneissic granitooids; c, calc-silicate and marble; a, amphibolite; Pvp, Tertiary cover rocks; gr, Mesozoic granitic rocks; UPz, Upper Paleozoic(?) to lower Mesozoic Calaveras Complex. D3 ductile shear zones in the Shoo Fly appear as thin barbed lines. (b) Map of S3, S2 and S1 hinge surface traces in the Shoo Fly [to be used in conjunction with (a)]. The hinge surface traces are schematic in that they are projected across stream channels and are not regionally continuous as indicated. Note that S2 traces are truncated at a high angle by the Calaveras-Shoo Fly thrust.

Sharp *et al.* (1982, in press) obtained U-Pb data from zircons in the granitic orthogneisses that indicate Paleozoic igneous ages. The zircons show complex U-Pb behavior and evidence for a component of inherited Precambrian lead, so interpretations are somewhat tenuous, but model ages of 275 ± 10 m.y. and 370 m.y. have been obtained from different igneous bodies (Sharp *et al.*, 1982, in press). Rb/Sr data on one of the syenite bodies suggests a 363-m.y. igneous age (H. K. Brueckner, personal communication, 1983). These data strongly suggest that Shoo Fly protoliths are pre-Late Devonian. Based on the isotopic data and on regional correlations, Shoo Fly rocks are considered to be of early Paleozoic age.

DEFORMATIONAL AND METAMORPHIC HISTORY

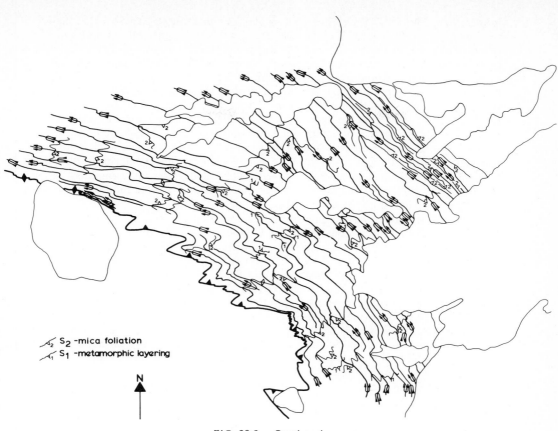

S_2 -mica foliation
S_1 -metamorphic layering

N

FIG. 29-3   Continued

Igneous intrusions The Shoo Fly Complex was intruded by Paleozoic granitoids mentioned above and by a number of Mesozoic sills, dikes, and plutons (Fig. 29-2). The oldest Mesozoic intrusions are late syntectonic foliated granitoid sills selectively intruded within the Calaveras–Shoo Fly thrust zone during its development. The thrust is also intruded by a Middle Jurassic calcalkaline pluton (the Standard pluton of Sharp and Saleeby, 1979; Schweickert, 1981; Sharp, 1984) which forms part of an extensive belt of similar plutons (the Jawbone intrusive suite of Stern *et al.*, 1981) developed across the trend of the metamorphic belt (Fig. 29-1).

The Jawbone plutons are spatially, temporally, and perhaps genetically associated with an extensive swarm of andesite, lamprophyre, and basalt dikes of the Sonora dike swarm (Merguerian, 1985a, 1986; see also Fig. 29-6). These dikes, which also intrude the Calaveras Complex and the Standard pluton, have yielded Middle Jurassic K-Ar hornblende ages (Sharp, 1980). They are important structural markers because they were intruded after an episode of regional folding of the Calaveras–Shoo Fly thrust, yet are deformed by northwest-trending Nevadan folds and cleavage.

Structural summary The Shoo Fly bears evidence for seven phases of superimposed deformation, including two phases of penetrative deformation that predate the formation of the Calaveras–Shoo Fly thrust. The thrust contact, as shown in Fig. 29-2, is a complicated zone of ductile deformation that truncates lithologic units and early deformational fabrics of the Shoo Fly. The earliest of these, D1, is represented by an S1 mica foliation in the hinge areas of F2 folds. The F1 folds are rare due to the penetrative nature of subsequent deformations, but the complicated map pattern in the Jupiter area (Fig. 29-3a) and contact relationships of the orthogneiss bodies (Fig. 29-4A) argue strongly for the existence of an intense D1 regional event. Subsequent to D1, the Shoo Fly was intruded by cal-calkaline to alkaline plutons that were protoliths of the gneissic granitoids.

D2 was the first regional event to have affected the protoliths of the gneissic granitoids (Fig. 29-4B). In the Shoo Fly, D2 is expressed as long-limbed isoclinal F2 folds (Fig. 29-5) locally with a mylonitic S2 fabric developed along their S2 hinge surfaces. Where observed, L2 stretching lineations parallel the F2 hingelines. S2, commonly a penetrative mica foliation, resulted from nearly complete recrystallization during D2. Internally, the orthogneisses were partly recrystallized and cut by an S2 amphibole + mica foliation (Fig. 29-4b). At the margins of the gneiss bodies, F2 folds and L2 lineations are remarkably well developed by comparison.

The D3 structures vary in intensity within the Shoo Fly, and are most intense in the Calaveras–Shoo Fly thrust zone. Near the Calaveras–Shoo Fly thrust, D3 transposition, mylonite, and coeval F3 folds with sheared out limbs and L3 stretching lineations that parallel F3 hingelines, resulted in obliteration of the oldest fabrics. The parallelism of S3 with the regional trace of the Calaveras–Shoo Fly thrust and the regional truncation of stratigraphic units and D1 + D2 structures (Fig. 29-3b) together indicate that the thrust and D3 regional deformation were coeval. From D3 onward, the Shoo Fly and Calaveras shared identical deformational histories.

The D4 event was responsible for regional folding of the Calaveras–Shoo Fly thrust and development of a spaced S4 mica schistosity parallel to the hinge surfaces of tight F4 folds. The F4 folds and related structures were later cut by the Sonora dike swarm and the Knight Creek pluton (Figs. 29-1 and 29-6).

D5 and D6 in the Shoo Fly are probably of Late Jurassic (Nevadan) age since they

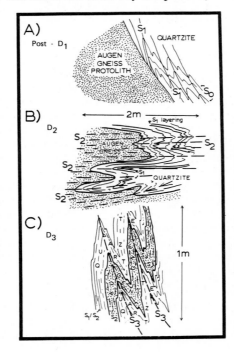

FIG. 29-4. Field drawings from various localities summarizing the relationship of the augen gneisses (stippled) to structural elements formed during D1, D2, and D3 in the Shoo Fly. (A) Contact relations of the augen gneiss protoliths indicate that they postdate an early D1 phase of folding and metamorphism in the Shoo Fly. (B) The Shoo Fly and the augen gneiss protoliths are deformed by F2 isoclinal folds and cut by a penetrative S2 foliation. (C) F3 folding and shearing along S3 hinge surfaces redeforms the foliated gneiss and Shoo Fly host rocks in response to the D3 regional deformation (and formation of the Calaveras–Shoo Fly thrust). Curvature of the S3 hinge surfaces is due to F4 and younger folding.

FIG. 29-5. F2 isoclinal folds of Shoo Fly laminated biotite quartzite deform a penetrative S1 mylonitic foliation. Note the sheath fold near the top of the photo. Millimeter-scale, spaced S3 ductile shears can be traced into the hinge surfaces of F3 folds nearby. Pencil for scale.

cross-cut all preexisting structures and the Middle Jurassic Standard pluton and the Sonora dikes (Figs. 29-1 and 29-6). S5 and S6 are northwest and northeast-trending crenulations and spaced cleavage that formed parallel to the hinge surfaces of sparse, asymmetric to open F5 and F6 folds.

Finally, sporadic west-northwest to east-west-trending spaced cleavage (S7) cuts the Nevadan structures (and all older structures) and is associated with open F7 folds of probably Cretaceous age.

Calaveras–Shoo Fly thrust The Calaveras–Shoo Fly thrust is a folded, east-dipping mylonitic shear zone that separates rocks of contrasting lithology, structure, and age. Numerous detailed traverses across the 1-2-km-wide thrust zone reveal mylonitic structures involving deformed and typically highly silicified Shoo Fly and Calaveras rocks intercalated on the scale of centimeters to tens of meters. The mapped thrust contact (Figs. 29-2 and 29-3) represents a form line that separates regions of >50% Shoo Fly from >50% Calaveras lithologies. During thrusting, the lower-plate Calaveras rocks acquired a flattening foliation (S1) related to long-limbed intrafolial folds, while the previously deformed Shoo Fly rocks developed a coeval S3 mylonitic foliation and discrete, anastomosing mylonitic shear zones. D3 in the Shoo Fly was contemporaneous with D1 in the Calaveras (Fig. 29-7).

D3 thrust zone structures in the Shoo Fly include rootless, isoclinal to tight F3 folds and S1 + S2 metamorphic layering transposed into parallelism with S3. The thrust zone is characterized by megascopic to microscopic, ellipsoidal slivers of foliated Shoo Fly rocks sheathed in S3 mylonite. The slivers, which attain long dimensions of 50 m, are flattened parallel to S3, show internal F3 folds of S1 and S2, and are elongate parallel to F3 hingelines and L3 stretching lineations that occur in their mylonitic envelopes. Except in these slivers, the intensity of ductile shearing in the thrust zone obliterated almost all trace of pre-D3 fabrics in the Shoo Fly, but away from the thrust zone, where the spacing between shear zones increases, vestiges of the pre-D3 folds and metamorphic fabrics are preserved. Detailed descriptions of various transects through the thrust zone are in Merguerian (1985a).

Field and petrographic evidence indicates that the Calaveras–Shoo Fly thrust experienced a protracted history of displacement. Late-D3 structures in the thrust zone include thin seams of cataclasite developed at acute angles to S3, branching veins of pseudotachy-

SCHWEICKERT, MERGUERIAN, BOGEN

799

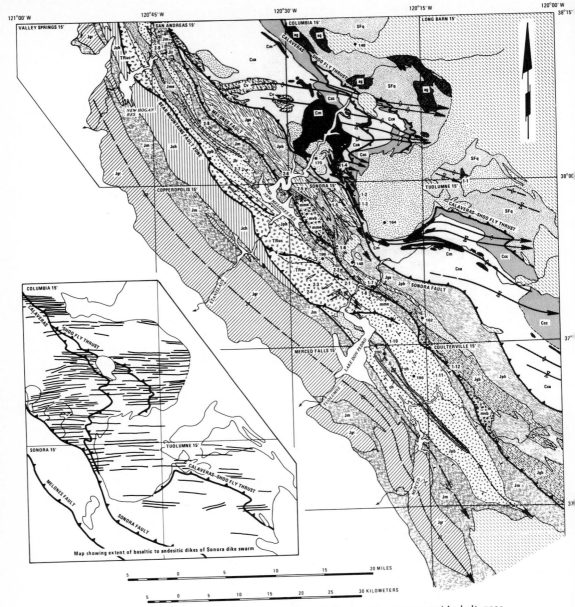

FIG. 29-6. Structural map of part of the western metamorphic belt near Sonora, California (Schweickert and Bogen, 1983), showing some of the principal structural features of the Calaveras, Don Pedro, and Foothills terranes. Thrust fault within Calaveras Complex north of Standard pluton is the American Camp thrust. In the center of the Sonora quadrangle, the Page Mountain pluton is incorrectly shown as post-Nevadan. Our field observations indicate it is a deformed, pre-Nevadan pluton, and this has been confirmed by new isotopic data of Sharp (1985). Inset map shows extent of the Sonora dike swarm. Note that these dikes are truncated by the Sonora fault.

DEFORMATIONAL AND METAMORPHIC HISTORY

EXPLANATION

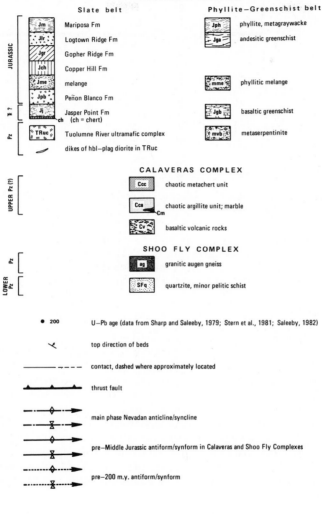

post—Nevadan granitoid

pre—Nevadan granitoid

FOOTHILLS TERRANE

Slate belt

Jm	Mariposa Fm
Jlr	Logtown Ridge Fm
Jgr	Gopher Ridge Fm
Jch	Copper Hill Fm
Jme	melange
Jpb	Peñon Blanco Fm
Ji	Jasper Point Fm (ch = chert)
TRuc	Tuolumne River ultramafic complex

dikes of hbl—plag diorite in TRuc

Phyllite—Greenschist belt

Jph	phyllite, metagraywacke
Jga	andesitic greenschist
mme	phyllitic melange
Jgb	basaltic greenschist
mvb	metaserpentinite

JURASSIC / ₮ ? / Pz

CALAVERAS COMPLEX

UPPER Pz (?)

Ccc	chaotic metachert unit
Cca / Cm	chaotic argillite unit; marble
Cv	basaltic volcanic rocks

SHOO FLY COMPLEX

g	granitic augen gneiss
SFq	quartzite, minor pelitic schist

LOWER Pz / Pz

● 200   U—Pb age (data from Sharp and Saleeby, 1979; Stern et al., 1981; Saleeby, 1982)

top direction of beds

——— — — — contact, dashed where approximately located

thrust fault

main phase Nevadan anticline/syncline

pre—Middle Jurassic antiform/synform in Calaveras and Shoo Fly Complexes

pre—200 m.y. antiform/synform

Sources of mapping:
1) Valley Springs 15′ quad: Wagner and others, 1981.
2) San Andreas 15′ quad: Clark and others, 1963; Schweickert, unpub.
3) Columbia 15′ quad: Baird, 1962; Schweickert and Merguerian, unpub.
4) Long Barn 15′ quad: Schweickert and Merguerian, unpub.
5) Copperopolis 15′ quad: Taliaferro and Solari, 1949.
6) Sonora 15′ quad: Eric and others, 1955; Morgan, 1976; Schweickert and Bogen, unpub.
7) Tuolumne 15′ quad: Schweickert, unpub.
8) Merced Falls 15′ quad: Bogen and Schweickert, unpub.
9) Coulterville 15′ quad: Bogen and Schweickert, unpub.; O.E. Bowen, unpub.

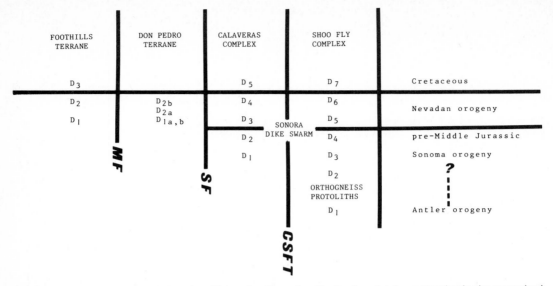

FOOTHILLS TERRANE	DON PEDRO TERRANE	CALAVERAS COMPLEX	SHOO FLY COMPLEX	
D3		D5	D7	Cretaceous
D2	D2b / D2a	D4	D6	Nevadan orogeny
D1	D1a,b	D3	D5	
		D2 SONORA DIKE SWARM	D4	pre-Middle Jurassic
		D1	D3	Sonoma orogeny
			D2 ORTHOGNEISS PROTOLITHS	?
			D1	Antler orogeny

MF SF CSFT

FIG. 29-7.   Structural correlation chart illustrating the regionally developed deformational episodes recognized in tectonostratigraphic units of the foothills metamorphic belt.

lyte and protocataclasite, minor crenulate post-F3 but pre-F4 asymmetric folds, and late-D3 foliated granitoid sills and bleblike injections. The transition from ductile to brittle fabrics during D3 is also indicated by minor retrograde recrystallization of the penetrative S3 mylonitic foliation and silicification of the Shoo Fly and Calaveras wallrocks.

Calaveras Complex

Lithology The Calaveras Complex contrasts markedly in lithology with the Shoo Fly Complex, and consists of two principal mappable units, both metamorphosed to upper greenschist to lower amphibolite facies (Fig. 29-6). Most extensive is a unit of chaotic argillite and siliceous argillite containing large olistoliths and tectonic slices of limestone and smaller blocks of chert, limestone, and basalt. Minor quartzose siltstone and sandstone occur locally. A more areally restricted unit consists principally of chert, varying from coherent to chaotic, with lesser chaotic argillite. Both units have previously been described in detail by Schweickert *et al.* (1977), who documented the chaotic nature of the Calaveras and discussed evidence that the chaotic rocks formed by soft-sediment deformation prior to lithification. A large, irregular mass of limestone near the Stanislaus River (Fig. 29-6) probably represents the sedimentary cover of a seamount that was tectonically accreted into the complex.

Age Horncorals of probable Permo-Carboniferous age were reported from a limestone olistolith southeast of the Standard pluton by Schweickert *et al.* (1977). Bateman *et al.* (1985) reported Lower Triassic conodonts from several limestone olistoliths along the Merced River. This evidence indicates that rocks of late Paleozoic to Early Triassic age occur in the Calaveras, and that deformation affecting the Complex must be Early Triassic or younger.

Igneous intrusions The Calaveras is intruded by several suites of igneous intrusions. Oldest is a suite of felsite dikes that predate the earliest deformational structures and postdate

DEFORMATIONAL AND METAMORPHIC HISTORY

the soft-sediment deformation. Next are concordant, forcefully intruded, calcalkaline plutons like the Standard, Parrotts Ferry, and Vallecitos plutons, with radiometric ages of about 165–170 m.y. (Sharp and Saleeby, 1979; Stern et al., 1981). These plutons postdate and deform second-generation structures in the Calaveras. Next youngest is a swarm of mafic dikes, part of the Sonora dike swarm (Fig. 29-6), that postdates the plutons and predates the youngest deformational features, the F_3 to F_5 folds and associated crenulation cleavages. A small, undated calcalkaline pluton near Rose Creek (the Knight Creek pluton, Figs. 29-1 and 29-2) intrudes the mafic dikes and also cuts both upper and lower plates of the Calaveras–Shoo Fly thrust.

Structural summary Earliest structures in the Calaveras Complex are those related to soft-sediment deformation (D_0). Most of the Calaveras is characterized by chaotic argillite consisting of lenses and fragments of chert, limestone, and rare basaltic rocks in a matrix of featureless argillite. In many areas, smears and streaks of siltstone are intimately mixed with the darker argillite matrix. Blocks of rhythmically bedded chert and of marble commonly make up large olistoliths in the argillite, and are about the only rocks that show relict bedding, except for extremely rare zones of stratified quartzose siltstone and sandstone. The large map unit of chert consists of very large masses of bedded chert that are closely packed in a matrix of chaotic argillite.

Superimposed upon the sedimentary chaos are D_1 structures, which include shape-fabric foliation and local lithological layering (S_1), extension lineation (L_1), and isoclinal folds (F_1). These structures are very strongly developed near thrust faults like the Calaveras-Shoo Fly thrust and the American Camp fault, but also occur throughout the Calaveras (Fig. 29-8a).

D_2 structures are very large-scale F_2 folds of the F_1 foliation, abundant F_2 minor folds, S_2 cleavage and schistosity, and L_2 extension lineation. The map pattern of lithologic units within the Calaveras to a large extent reflects the geometry of D_2 structures (Figs. 29-6 and 29-8). The hinge surfaces of major F_2 folds have been spectacularly deformed and reoriented by the forcefully intruded plutons (Fig. 29-8b). Following emplacement of the plutons, the entire complex was invaded by mafic dikes of the Sonora dike swarm, which extend from the Shoo Fly to the Sonora fault (Fig. 29-6).

Third- and fourth-generation structures cut all the preceding structures, plutons, and mafic dikes, and must have formed during the Late Jurassic Nevadan orogeny. These structures include northwest- and northeast-trending crenulations and spaced cleavages (S_3 and S_4), local asymmetrical folds (F_3 and F_4), and intersection lineations (L_3 and L_4). The youngest structures are sporadic east-west-trending crenulations and spaced cleavage (S_5) that may be of Cretaceous age.

The Sonora fault The Sonora fault separates the Calaveras Complex from underlying rocks of the Don Pedro terrane (Fig. 29-6). Greenschist-grade Upper Jurassic phyllite and greenschist, with perfect, continuous phyllitic cleavage, occur in the footwall. Calaveras metachert and meta-argillite with shape-fabric foliation and mica schistosity, of upper greenschist to amphibolite grade (with biotite) occur in the hangingwall. Significantly, the hanging wall structure and metamorphism predates the Sonora dike swarm, and F_2 folds in the Calaveras and the dikes themselves are highly deformed as they approach the fault. The dike swarm and F2 folds of the Calaveras are truncated at a moderate to large angle by the Sonora fault, as shown in Figs. 29-6 and 29-8b. Penetrative footwall structures postdate 160-m.y.-old plutons and Oxfordian-Kimmeridgian strata.

The fault is marked by a zone a few tens of meters wide of phyllonite and mylonite with rootless intrafolial folds and a down-dip stretching lineation. Adjacent to the fault, the older fabrics in the Calaveras, and Nevadan fabrics in the phyllitic rocks, parallel the

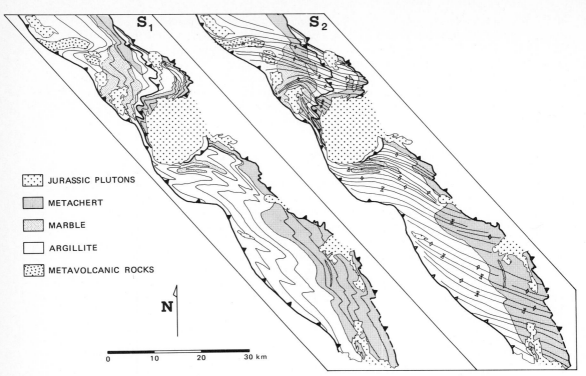

JURASSIC PLUTONS

METACHERT

MARBLE

ARGILLITE

METAVOLCANIC ROCKS

N

0   10   20   30 km

FIG. 29-8. (a) (left). Interpretive map of S1 form surfaces in the Calaveras Complex. The map was prepared using data on orientation of S1 fabrics in the Calaveras, lithologic contacts, and geometry of F2 folds. (b) (right). Interpretive map of S2 form surfaces in the Calaveras Complex. The map was prepared using data on geometry of F2 folds, lithologic contacts, and attitudes of S2 in the Calaveras Complex. Note discordance of S2 form lines to the Sonora fault, and deformation of S2 surfaces around the Standard and Parrots Ferry plutons.

fault trace. Earlier fabrics in hanging-wall Calaveras rocks are overprinted at a low angle by a new, penetrative schistosity related to the fault.

Don Pedro Terrane (Phyllite-Greenschist Belt)

Lithology Rocks of this terrane consists of five principal types (Fig. 29-6), listed below. (1) Phyllite, phyllitic sandstone, and conglomerate, the most widespread lithologies of the belt. Sandstone is volcaniclastic and chert-rich graywacke; conglomerate contains abundant chert and volcanic clasts. (2) Schistose augite- and plagioclase-bearing, andesitic tuff and tuff-breccia, and local rhyolitic to dacitic ash-flow tuff. These units are the predominant metavolcanic units north of latitude 37°45′. (3) Schistose basaltic pillow lava and pillow breccia, which form the main metavolcanic units south of latitude 37°45′. (4) Metamelange, commonly occurring in close proximity to the Melones fault, and containing large blocks of marble, chert, sandstone, and greenschist. (5) Metaserpentinite, varying from talc-antigorite schist to foliated metaperidotite, and metagabbro.

Age Units of the Don Pedro terrane appear to be correlative with those of the Foothills terrane. The phyllite unit is at least partly of Oxfordian-Kimmeridgian age, having yielded the ammonite *Perisphinctes* (Clark, 1964). The unit also bears a close lithologic resemblance to the Mariposa Formation of the Foothills terrane. The schistose augite- and plagioclase-bearing metavolcanic rocks probably are correlative with the Lower Jurassic Penon Blanco Formation and/or the Upper Jurassic Logtown Ridge and Gopher Ridge Formations. Bateman *et al.* (1985) reported a 187 ± 10 m.y. Rb-Sr isochron age from andesitic metavolcanic rocks on the Merced River, indicating these rocks are about the same as the Penon Blanco. The metamelange probably corresponds to a melange that lies stratigraphically between the Penon Blanco and Logtown Ridge Formations in the Foothills terrane. The pillow basalt and pillow breccia probably are correlative with the Jurassic-Triassic Jasper Point Formation to the west (Bogen, 1985; Schweickert and Bogen, 1983).

Igneous intrusions Rocks of the Don Pedro terrane are intruded by the Page Mountain and Cobbs Creek plutons (Fig. 29-6, central part of Sonora quad and southwest corner of Tuolumne quad, respectively). Our observations indicate that both plutons have been deformed together with their wallrocks, although Sharp (1985) stated that these plutons postdate the regional deformation of the wallrocks. Published U-Pb zircon ages on the Page Mountain pluton are 148 m.y. (Morgan, 1976; Stern *et al.*, 1981) and 166–170 m.y. (Sharp, 1985). Published ages on the Cobbs Creek pluton are 162 m.y. (U-Pb zircon; Morgan, 1975; Stern *et al.*, 1981) and 169 m.y. (40Ar/39Ar; Herzig *et al.*, 1985). Scattered northwest-trending andesitic dikes also occur within the belt.

Structural summary Rocks of the Don Pedro terrane contain evidence for two generations of penetrative, isoclinal folding, and in many places the dominant schistosity is a composite of S1a and S1b. Original stratigraphic relations have been obscured by this deformation and by numerous ductile shear zones that occur within the terrane. Recrystallization of these rocks has obliterated most primary sedimentary or igneous textures, except in uncommon areas where strains were smaller. The large mass of metavolcanic rocks in the north-central part of the Sonora quadrangle (Fig. 29-6) occupies the core of a large-scale F1a isoclinal fold that has been strongly refolded by F1b folds. Local northwest and northeast-trending crenulation cleavages deform the S1a X S1b schistosity.

   Inasmuch as all deformation in the terrane appears to postdate Oxfordian-Kimmeridgian rocks and Middle Jurassic plutons, it must postdate the penetrative deformation of the Calaveras Complex and is probably of Nevadan age (155 ± 3 m.y.) (Schweickert *et al.*, 1984a, 1985).

The Melones fault In the southern part of the western metamorphic belt, the Melones fault separates upper greenschist-grade phyllite and schist of the Don Pedro terrane from very low-grade greenschist rocks of the Foothills terrane (Fig. 29-6). Its geometry is that of a steeply east-dipping thrust fault and it was active during the main phase of the Nevadan orogeny. Unlike the Sonora fault, the Melones has suffered a complex history of reactivation, in both Cretaceous and Cenozoic time.

   We have made 16 detailed structural traverses across the Melones in an attempt to establish its structural geometry and displacement history (Schweickert *et al.*, unpublished data). Where the Late Jurassic Melones fault can be identified, it is marked by a narrow zone, commonly 5 m wide, of phyllonite showing complete obliteration of fabric elements in hanging-wall rocks. Strain-slip cleavage related to the fault has been observed up to about 20 m away from the fault in the hanging wall, where it deformed phyllite and greenschist of the Don Pedro terrane. Generally, in footwall rocks, intensity of slaty cleavage increases progressively over a few meters toward the fault, where it merges into the fault fabric. Locally, the relation of fault to footwall fabrics is more complex, and the fault appears to cut folds in the footwall at a low angle (Schweickert *et al.*, 1984b). In at least one locality, slaty cleavage is refolded and cut by a secondary spaced cleavage parallel to the fault fabric. We interpret these relationships to indicate that displacement on the fault both accompanied and outlasted cleavage development in footwall slates (Schweickert *et al.*, 1984a).

Lineations in the fault zone plunge down the dip of foliation, suggesting dip-slip displacement. Since more highly deformed and metamorphosed rocks lie in the hanging wall, we regard the present geometry as that of a high-angle reverse fault. Bogen *et al.* (1985) presented paleomagnetic data from the Sonora dike swarm that suggests 25–30° of rigid body rotation of the Calaveras about an axis trending about N60°W. If the Sonora and Melones faults rotated a like amount, original dips would have been about 50–60° northeast.

In several localities, the fault trace is folded into asymmetrical folds with north-northeast-trending vertical hinge surfaces which we interpret as late-phase Nevadan folds, indicating that the Melones fault was locked by the end of the Nevadan orogeny.

Early Cretaceous, brittle, high-angle reverse faults are localized near the Melones and root either along the Melones or in its hanging wall. These faults were brittle, seismic, and experienced small dip-slip displacement (Schweickert *et al.*, unpublished data). The Cretaceous faults are characterized by fault gouge, fault breccia, and quartz veins, and must represent a considerably shallower, and hence later, episode of faulting than the ductile Jurassic fault. Several isotopic ages have been reported from micas associated with the quartz veins and these all suggest Early Cretaceous faulting (Evans and Bowen, 1977; Kistler *et al.*, 1983).

Woodward-Clyde Consultants (1978) obtained evidence that Cenozoic normal displacements occurred on parts of the Melones fault zone. Based on their trench logs and our examination of the Melones fault, we believe the Cenozoic displacement was confined to segments of the fault that underwent Cretaceous reactivation.

Based on our structural study, we conclude that no evidence exists for significant strike-slip displacement on the Melones. Nor is there any evidence for eastward-vergent structures in the fault zone.

Foothills Terrane

Lithology The Foothills terrane consists predominantly of mafic to intermediate volcaniclastic rocks that commonly overlie tectonic melange and are overlain by flyschlike mudstone and sandstone (Fig. 29-6). Ultramafic rocks and mafic to intermediate plutons occur locally (Clark, 1964, 1976; Duffield and Sharp, 1975; Taliaferro and Solari, 1949). Rock types and ages are similar to those of the Don Pedro terrane, but primary features are well preserved and the stratigraphic succession is well known in the Foothills terrane.

An excellent reference section occurs in the southern part of the terrane, near the Merced River, where a main-phase Nevadan anticline exposes four formations with aggregate thickness in excess of 8 km (Fig. 29-6). At the base, the Jasper Point Formation (Bogen, 1985) includes massive pillowed and brecciated basalt more than 900 m thick overlain by varicolored, radiolarian chert as much as 100 m thick, with manganiferous and ferruginous lenses in the lower part. The unit resembles the upper layers of oceanic crust (Bogen, 1985; Schweickert, 1978, 1981).

The chert is intercalated with fine tuff of the Penon Blanco Formation, which consists of ankaramitic volcaniclastic and porphyritic rocks with a maximum preserved thickness of 5 km. Much of the unit appears to have been extruded from a single vent and may represent a submarine volcanic edifice developed upon oceanic crust (Bogen, 1985). Mafic, intermediate, and silicic volcaniclastic and porphyritic rocks of the Gopher Ridge Formation overlie the Penon Blanco locally. Flyschlike mudstone, sandstone, and volcaniclastic rocks of the Mariposa Formation concordantly overlie or interfinger with the volcaniclastic units within the terrane and are in excess of 2 km thick (Bogen, 1984). North of the Stanislaus River, the Mariposa also overlies and interfingers with up to 3 km of ankaramitic porphyry and volcaniclastic strata of the Logtown Ridge Formation (Fig. 29-6; Duffield and Sharp, 1975). A distinctive sedimentary melange bearing blocks of Triassic and Permian limestone with a Tethyan fauna occurs stratigraphically between the Penon Blanco and Logtown Ridge Formations north of the Stanislaus River (Schweickert and Bogen, 1983). This melange unit probably is correlative with the metamelange within the Don Pedro terrane. The volcanic units of the Foothills terrane (and the Don Pedro terrane) represent parts of a long-lived island-arc system (Schweickert and Cowan, 1975) that is lithologically similar to volcanic rocks of the New Hebrides and Mariana arcs.

Age The ages of most of the units within the Foothills terrane are known from fossils or radiometric dates. Best constrained is the youngest unit, the Mariposa Formation, which contains ammonities, ammonoids, and pelecypods diagnostic of late Oxfordian to early Kimmeridgian age (Clark, 1964; Imlay, 1961). The Logtown Ridge and Gopher Ridge Formations interfinger locally with the Mariposa and therefore are early Kimmeridgian or older. Most of the Logtown Ridge is known from ammonites to be Callovian, however (Clark, 1964; Duffield and Sharp, 1975; Imlay, 1961). The Penon Blanco Formation is unfossiliferous and the Jasper Point Formation has yielded no diagnostic fossils, but several dioritic igneous bodies that intruded one or both units have yielded Early Jurassic U-Pb ages (see below). The Penon Blanco Formation therefore is Early Jurassic or older and the Jasper Point Formation may be much older, especially the basalt unit that lies beneath thick chert deposits (Bogen, 1985).

Igneous intrusions Dioritic plutons, stocks, and dikes with two distinct groups of ages intrude the Foothills terrane. The diorite of Don Pedro and related stocks, and the diorite of Chinese Camp have yielded Early Jurassic U-Pb zircon ages ranging from about 180 to 200 m.y. (Morgan, 1976; Saleeby, 1982; Stern et al., 1981). These bodies intrude the Penon Blanco and Jasper Point Formations. Hypabyssal diorite that intrudes the Penon Blanco porphyries has a U-Pb zircon age of about 196 m.y. (Saleeby, 1982) and may be a feeder dike.

Younger, dioritic to granodioritic plutons intrude the Mariposa Formation and units throughout the Foothills terrane (Evernden and Kistler, 1970). K-Ar ages as old as 140–145 m.y. are known in the far northern foothills but the dated plutons cut only chaotic rocks. In the south, the Guadelupe pluton, which has a K-Ar biotite age of about 139 m.y. (Evernden and Kistler, 1970), intrudes the Mariposa Formation (Fig. 29-6).

Structural summary Outcrop patterns and steeply dipping strata in the Foothills terrane are the result of the single deformational phase that affected all the terranes of the western Sierra and is known as the Nevadan orogeny (Clark, 1964; Schweickert et al., 1984a; Taliaferro, 1942). Slaty cleavages are well developed only in the cores of major folds. Sandstone with spaced cleavage, examined in thin section, reveals evidence of pressure solution of quartz, predominantly, and crystallization of chlorite and white micas, locally; evidence of metamorphism is otherwise lacking. The presence of chlorite and albitized plagioclase in volcanic rocks may be a diagenetic feature. The bulk of the rock does not appear to have reached metamorphic equilibrium during deformation.

Both the Melones and the Bear Mountains faults were active as steep reverse or thrust faults during the Nevadan orogeny, and other faults developed near the cores of the tight Nevadan folds. The Melones, which has been studied in detail, was a ductile thrust zone (see above).

Younger structures reorient those of Nevadan age locally (Bogen, 1979). Vertical, kinklike asymmetric folds with northeast-striking hinge surfaces are the most widely noted (Schweickert *et al.*, 1984a).

Structures older than the Nevadan orogeny in the southern foothills are less well known. The thrust at the base of the Tuolumne River ultramafic complex (Fig. 29-6) is approximately 190–200 m.y. old, as indicated by K-Ar hornblende ages on amphibolite-grade fault blocks (Morgan, 1976). Some structures within the multiply deformed ultramafic body are probably of similar age but have not yet been studied in detail. Blocks of metamorphic rock with 190–200-m.y. ages described by Behrman (1978) also occur near the Cosumnes River to the north of Fig. 29-1, in part of the melange unit that separates the Penon Blanco and Logtown Ridge Formations.

METAMORPHISM AND DEFORMATION

In this section we describe the general metamorphic characteristics of the various terranes, together with available data on the relations between metamorphic recrystallization and the deformational events described above.

Paleozoic Metamorphism

Limited evidence exists for two Paleozoic amphibolite facies metamorphic events in the western Sierra Nevada metamorphic belt. As noted above, the Shoo Fly Complex is the only unit that underwent Paleozoic deformation and metamorphism. This deformation included the D1 and D2 events. The metamorphic grade during D1 is uncertain as the rocks were thoroughly recrystallized during D2 and D3, but relict mica, quartz, feldspar, and calc-silicate minerals parallel the S1 foliation, suggesting that lower-amphibolite-grade conditions existed.

Shearing and recrystallization of quartz, feldspar, and mica along S2 hinge surfaces are common in semipelitic units, and S2 diopside and biotite occur in calc-silicate rocks. In the gneissic granitoids, oriented hornblende and hastingsite, together with biotite, feldspar, and quartz, define the S2 foliation. These relations suggest amphibolite facies conditions also existed during D2 deformation in the Shoo Fly.

Mesozoic Metamorphism

Several Mesozoic metamorphic events affected the western metamorphic belt. Amphibolite facies metamorphism accompanied deformation in the Shoo Fly and Calaveras Complexes during the Triassic. Both terranes experienced a greenschist facies event prior to Middle Jurassic time. A number of Middle Jurassic plutons forcefully intruded the Calaveras and Shoo Fly Complexes, and developed narrow contact aureoles parallel to their margins. Late Jurassic Nevadan metamorphic events resulted in retrograde assemblages in the older terranes and greenschist facies assemblages in Jurassic rocks. Within the Foothills terrane, very localized evidence exists for Early Jurassic amphibolite facies metamorphism.

The oldest Mesozoic deformation was D3 and the Shoo Fly Complex and D1 in the Calaveras (hereafter D1/D3), which included the development of the Calaveras-Shoo Fly thrust and internal faults and shear zones within the Shoo Fly and Calaveras. D1/D3 clearly was post-earliest Triassic. Next was the D2/D4 event, affecting both the Calaveras and Shoo Fly, which was followed closely by numerous granitic intrusions in Middle Jurassic time. The age of the D2/D4 event is not tightly constrained, but may have been Early or Middle Jurassic. Significantly, no evidence of this pre-Nevadan event exists in the Don Pedro and Foothills terranes, although the Foothills terrane contains sparse evidence of a localized high-grade Lower Jurassic event that seems to have been unique to that terrane. During the Late Jurassic Nevadan orogeny, the Don Pedro and Foothills terranes were juxtaposed with the Shoo Fly and Calaveras, and all four terranes underwent deformation and metamorphism (D5, D6 in Shoo Fly; D3, D4 in Calaveras; D1a, D1b, D2a, D2b in Don Pedro terrane; and D1 and D2 in Foothills terrane).

D1/D3 metamorphic events In semipelitic and calc-silicate units of the Shoo Fly Complex, S3 is marked by recrystallization of mica, quartz, tremolite, calcite, and to a lesser extent feldspar. S3 is typically highly laminar and wraps around centimeter-to-meter scale augen of foliated quartzite and quartzofeldspathic gneiss as well as porphyroclasts of S2 biotite and amphibole. In the gneissic granitoids, recrystallized quartz, feldspar, mica, hastingsite, epidote, and retrograde stilpnomelane define the S3 foliation, which anastomoses about porphyroclasts of S2 feldspar and amphibole and domains of less-deformed granitoid rock. These minerals suggest that D3 occurred under epidote amphibolite facies conditions.

In the Calaveras, amphibolite-grade mineral assemblages are well developed in a variety of lithologies. S1 in meta-argillite is marked by planar preferred orientation of biotite, muscovite, and zones of opaque minerals. Typical assemblages are biotite-muscovite-quartz ± garnet. In metachert, S1 is commonly marked by stylolitic seams of opaque material; in some instances, these seams cut across randomly oriented biotite flakes, suggesting that some mineral growth predated development of S1. Mineral assemblages in metacherts are generally the same as those in meta-argillites. Marbles in the Calaveras commonly consist of calcite-quartz ± diopside ± tremolite. Metavolcanic rocks contain green hornblende-plagioclase-epidote ± biotite ± sphene ± clinozoisite. S1 foliation is defined by preferred orientation of hornblende needles and by hornblende-rich layers.

Regionally, metamorphic grade in the Calaveras appears to be somewhat higher in the vicinity of the Tuolumne River, where meta-argillites several kilometers from the margins of Middle Jurassic plutons locally contain biotite-muscovite-quartz ± sillimanite ± cordierite. Conversely, metamorphic grade seems to be somewhat lower in the vicinity of the Merced River, where it may approach epidote amphibolite facies (Bhattacharyya and Paterson, 1984; Schweickert *et al.*, 1985).

D2/D4 metamorphic events Typically, S4 in the Shoo Fly Complex is a spaced schistosity composed of aligned subidioblastic biotite and muscovite which can be traced into the hinge surfaces of F4 folds. In quartzitic rocks, S4 is a spaced cleavage. The dominantly retrograde metamorphic effects suggest that the Shoo Fly was elevated to shallower crustal levels after D3.

In the Calaveras, S2 schistosity, which parallels hinge surfaces of F2 folds, is widely developed, but commonly is much less conspicuous than S1 foliation. In thin section, S2 is defined by aligned biotite flakes in argillite and by dark, opaque-rich seams cutting S0/S1 layering in cherts. Locally, these seams trim the edges of biotite and garnet grains in the S1 foliation, indicating that S2 developed as a spaced, metamorphic pressure solution cleavage. Metamorphic grade was apparently greenschist or slightly higher.

Contact metamorphism related to Middle Jurassic forceful intrusions During the Middle Jurassic, and following the D2/D4 deformational event, the Standard, Parrotts Ferry, Vallecitos, and San Andreas plutons forcefully intruded the Calaveras Complex and parts of the Shoo Fly (Fig. 29-6). Figures 29-6 and 29-8 show the extent to which F2 folds and S2 hinge surfaces in the Calaveras were deformed and reoriented by the intrusion of these plutons. These relations are especially evident along the Stanislaus River north of the Parrotts Ferry pluton and along the Tuolumne River south of the Standard pluton, where west-northwest-trending F2 folds undergo up to 90° deflections adjacent to these plutons.

Locally, within the contact aureoles of the Standard and Vallecitos plutons, evidence of folds and foliations related to the intrusions has been found.

Baird (1962) indicated that minor contact metamorphic effects around the Vallecitos pluton involve the development of hornfels in a zone only about 8 m wide, and the coarsening of grain size in marble over a larger area. Similarly, our observations suggest that contact metamorphic effects along the margins of the Parrotts Ferry pluton are very minor. Baird (1962) interpreted the lack of pronounced contact metamorphic features to indicate that these plutons were syntectonically emplaced, although it is important to emphasize that field relations indicate that the plutons postdate the major D2/D4 deformation.

Contact metamorphism related to the Standard pluton is more pronounced. Kerrick (1970) described the contact aureole of the pluton on its southeastern margin along Turnback Creek, near the eastern edge of the Sonora 15-minute quadrangle (Fig. 29-6). In this area, wallrocks are parts of the Shoo Fly complex, the Calaveras–Shoo Fly thrust, and the Calaveras Complex. Kerrick (1970) recognized an inner zone near the contact, averaging about 300 m wide, characterized by the development of sillimanite, and containing the assemblage sillimanite-quartz-muscovite-plagioclase-K feldspar ± garnet. Beyond the sillimanite zone, schists contain quartz-biotite-muscovite ± plagioclase ± K feldspar. According to Kerrick (1970), the sillimanite zone was controlled by lithologic variations rather than by temperature, and this is consistent with the fact that the sillimanite zone is almost entirely confined to Shoo Fly rocks in the aureole, while the schists outside the sillimanite zone are part of the Calaveras Complex. We have observed the contact assemblage sillimanite-biotite-muscovite-quartz in Shoo Fly rocks near the northern and eastern margins of the Standard pluton. In addition, the assemblages muscovite-biotite-quartz ± andalusite ± tourmaline developed in Calaveras schists and calcite-diopside-tremolite-phlogopite-quartz developed in marbles near the western and southern margins of the pluton. These mineral assemblages indicate that a contact aureole of hornblende hornfels facies metamorphic rocks developed around the perimeter of the Standard pluton.

Nevadan metamorphic events Nevadan metamorphism resulted in a retrograde greenschist facies overprint on the Calaveras and Shoo Fly, and prograde greenschist facies minerals in the Don Pedro terrane. Metamorphic grade was very low greenschist facies in the Foothills terrane.

S5 and S6 in the Shoo Fly are defined by fine-grained chlorite, quartz, and opaque minerals. The D5 and D6 events are responsible for retrograde effects in most samples.

Main phase (D3) Nevadan folds and cleavages are widely but weakly developed in the Calaveras Complex. Crenulation cleavage (S3) is common in the more micaceous schists, while spaced cleavages (S3) occur infrequently in more siliceous rocks and in marble. The S3 cleavages parallel the hinge surfaces of minor asymmetric, kinklike F3 folds.

Late-phase, N20°E-trending, domainal spaced cleavage and crenulation cleavage (S4) occur even less commonly than the main-phase structures, but nevertheless are widely distributed. These parallel the hinge surfaces of open, asymmetric F4 folds in many areas. Minor growth of chlorite parallel to S3 and S4 crenulations is the only metamorphic effect noted.

In the Don Pedro terrane, greenschist minerals, including white mica, chlorite, stilpnomelane, and actinolite, typically define the S1a schistosity. Biotite and garnet occur locally in some areas. In the Foothills terrane, as noted by Schweickert et al. (1984a), cleavage in graywacke developed by flattening of incompetent volcanic rock fragments, dissolution of quartz, and crystallization of very fine-grained chlorite and white mica.

Metamorphic events unique to the Foothills terrane Evidence of deformation and amphibolite-grade metamorphism occurs in two areas in the Foothills terrane. There is no evidence that these events affected any of the other terranes. The more southerly occurrence includes blocks of garnet amphibolite and crossite-bearing chert within the fault zone at the base of the Tuolumne River ultramafic complex (Morgan, 1976; Fig. 29-6). Morgan (1976) obtained K-Ar hornblende ages between 190 and 200 m.y. from the amphibolite blocks. Displacement on the fault also appears to have occurred within this span as the fault cuts most, but not all, of the Penon Blanco Formation (Fig. 29-6). In addition, diorites of the Don Pedro and Chinese Camp plutons intruded the Penon Blanco at approximately the same time (Morgan, 1976). The Don Pedro intrusion appears to have risen to a level just beneath the fault, without piercing it, but the diorite of Chinese Camp is in a higher thrust sheet and must have pierced the fault locally (Fig. 29-6), probably before movement terminated. This fault is one of several thrusts imbricating the Penon Blanco and its ultramafic basement. Morgan (1978) documented the existence of highly magnesian olivine (greater than Fo_{97}) in the ultramafic complex. Laboratory studies indicate that the olivine, accompanied by diopside, probably formed during prograde metamorphism of previously serpentinized ultramafic rock (Morgan, 1978), but it is not known whether this prograde event accompanied the events discussed above. The amphibolite-grade metamorphism represented by the tectonic blocks in the fault zones may have resulted from the overthrusting of hot ultramafic rocks (Schweickert and Bogen, 1983).

A second area of (epidote) amphibolite grade metamorphism with ages of 190–200 m.y. occurs in the central part of the Foothills terrane, near the Cosumnes River. Behrman (1978) obtained K-Ar hornblende ages of 190 ± 2, 191 ± 2, and 179 ± 12 m.y. from isolated blocks of metavolcanic and metagabbroic rocks in melange. Their sources and sites of metamorphism are not known.

Post-Nevadan metamorphism In the Shoo Fly and Calaveras Complexes, sporadically developed spaced brittle crenulation cleavage (S5/S7), trending about N65°W to east-west is related to broad, open F5/F7 folds and warps. The S5/S7 fabric is weak and dominantly brittle in nature, with little recrystallization except for incipient chloritization of pre-existing metamorphic fabrics.

SUMMARY

Structural studies and bedrock mapping in the central Sierra Nevada foothills between the Great Valley and the Sierra Nevada batholith have delineated four ductile-fault-bounded tectonostratigraphic terranes with diachronous, but overlapping, structural and metamorphic histories. Immediately west of the batholith, the lower Paleozoic Shoo Fly Complex is a lower amphibolite grade sequence of quartzite, quartzofeldspathic gneiss, garnet schist, calc-silicate rocks, and marble intruded by Paleozoic granitoids (now orthogneisses). The Shoo Fly, which has experienced seven phases of superimposed deformation, is in mylonitic thrust contact with the upper Paleozoic to lower Mesozoic Calaveras Complex. Lithologic units and metamorphic fabrics of two early deformational events in the Shoo Fly are truncated and deformed along the pre-Middle Jurassic Calaveras–Shoo Fly thrust (Figs. 29-9, and 29-10).

The Calaveras Complex, a lower amphibolite-grade, chaotic assemblage of argillite, chert, marble, talc schist, metabasalt, and rare metasandstone and metasiltstone, experienced five deformational events. A penetrative, metamorphic flattening foliation formed during juxtaposition with the Shoo Fly, and subsequent east-west-trending, east-plunging folds deformed the thrust with the development of a schistosity parallel to hinge surfaces in both the Shoo Fly and the Calaveras. Subsequently, the Calaveras and Shoo Fly were forcefully intruded and contact metamorphosed by a suite of Middle Jurassic plutons. The two sets of early, pre-Middle Jurassic structures in the Calaveras, and a swarm of east-west mafic dikes, are truncated by the Sonora fault and are overprinted and retrograded by Nevadan structures (Figs. 29-9 and 29-10).

The Don Pedro terrane, west of the Sonora fault, consists of Jurassic phyllitic rocks and metavolcanic greenschists, and deformed plutons. During early phases of the Nevadan orogeny, these rocks were juxtaposed beneath the Calaveras along the Sonora fault, and were subjected to two intense phases of isoclinal folding in upper greenschist facies conditions. Later northwest and northeast-trending crenulations in the Don Pedro terrane probably correlate directly with main-phase and late-phase Nevadan structures in the Foothills terrane. The Melones fault truncates and overprints the penetrative structures of the Don Pedro terrane (Figs. 29-9 and 29-10).

The Foothills terrane, west of the Melones fault, consists of Upper Jurassic flysch and Lower to Upper Jurassic arc volcanic rocks resting on pre-Lower Jurassic oceanic crust, serpentinite, and melange. These rocks were tightly folded and metamorphosed to lowest greenschist facies during the Nevadan orogeny. Slaty cleavage developed widely in the fine-grained, pelitic rocks of the Foothills terrane, but cleavage development is very sparse in the coarse, massive volcaniclastic rocks. Northeast-trending, late Nevadan folds and crenulation cleavages locally deformed the folds and also warped the traces of the Sonora and Melones faults.

The southern part of the western metamorphic belt contains tectonostratigraphic

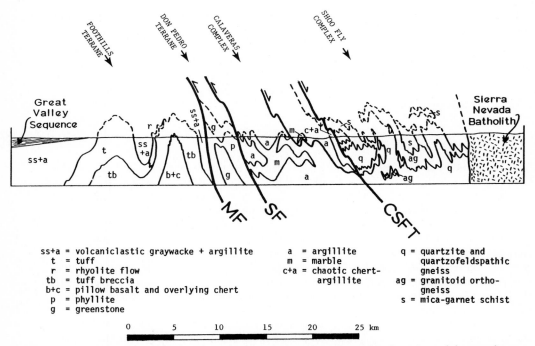

ss+a = volcaniclastic graywacke + argillite   a = argillite   q = quartzite and
t = tuff   m = marble   quartzofeldspathic
r = rhyolite flow   c+a = chaotic chert-   gneiss
tb = tuff breccia   argillite   ag = granitoid ortho-
b+c = pillow basalt and overlying chert   gneiss
p = phyllite   s = mica-garnet schist
g = greenstone

0   5   10   15   20   25 km

FIG. 29-9.   Structure section across the western metamorphic belt illustrating the westward decrease in age, structural complexity, and metamorphism of the tectonostratigraphic units. The section is composite; data from the Shoo Fly, Calaveras, and Don Pedro terranes come from the Stanislaus River but data from the Foothills terrane are from the Merced River. Mf, Melones fault; Sf, Sonora fault; CSFT, Calaveras–Shoo Fly thrust. Data from R. A. Schweickert (unpublished data), Bogen (1983), and Merguerian (1985a). No vertical exaggeration.

DEFORMATIONAL AND METAMORPHIC HISTORY

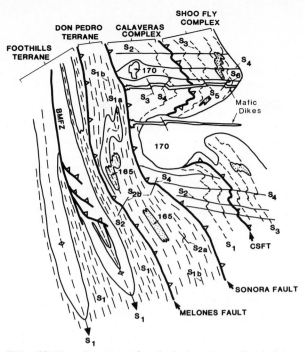

FIG. 29-10. Cartoon of principal structural relations among the terranes in the southern part of the western metamorphic belt. Consult Fig. 29-7 for correlation of structural symbols; labels on structural symbols correspond to their order of development in each terrane. For example, S2 in Calaveras = S4 in Shoo Fly. In the Don Pedro terrane, the Page Mountain and Cobbs Creek plutons are stippled. In the Calaveras Complex, the Parrotts Ferry and Standard plutons are shown with 170-m.y. ages. Paleozoic orthogneisses in the Shoo Fly are shown with dotted S2 fabrics predating S3 foliations. CSFT, Calaveras–Shoo Fly thrust.

units that reveal an eastward increase in age, metamorphic grade, and structural complexity, with abrupt transitions occurring at the Melones fault, Sonora fault, and Calaveras–Shoo Fly thrust (Figs. 29-9 and 29-10). Figure 29-10 is a cartoon showing schematically the principal structural relations across the metamorphic belt. It is noteworthy that in nearly all cases, deformation and metamorphic recrystallization occurred together. The Shoo Fly thus experienced three episodes of recrystallization at amphibolite facies, followed by several retrograde events. The Calaveras Complex experienced one episode of recrystallization at amphibolite facies, followed by several retrograde events. The Calaveras Complex experienced one episode of recrystallization at lower amphibolite facies, with as many as four minor retrograde events. The Don Pedro terrane underwent two prograde greenschist metamorphic events, and the Foothills terrane experienced one very weak, lower greenschist facies metamorphic event. Both the Don Pedro and Foothills terranes experienced minor retrograde metamorphism during weak post-Nevadan deformation. Definition of the structural and metamorphic sequence in the metamorphic belt and preliminary age data on cross-cutting igneous bodies should allow comparisons to be made with orogenic episodes documented elsewhere in the SW Cordillera.

DISCUSSION

In this section we comment on the significance of our structural data in regard to several controversial issues that have surfaced in the past few years. These issues include the following questions, discussed below: (1) Is there really a structural contrast between the Shoo Fly Complex and Calaveras Complex across the Calaveras–Shoo Fly thrust (Bhattacharyya and Paterson, 1985; Schweickert *et al.*, 1985)? (2) Is the Calaveras Complex a melange (Bateman *et al.*, 1985; Schweickert, 1981; Schweickert *et al.*, 1977)? (3) Do structural contrasts exist between the Calaveras Complex and Don Pedro terranes (Bogen and Schweickert, 1983; Schweickert *et al.*, 1984a; Sharp, 1985)? (4) Does the Sonora fault exist (Bhattacharyya and Paterson, 1985; Schweickert *et al.*, 1985; Sharp, 1985)? (5) Do the rocks of the western metamorphic belt face westward (Bateman *et al.*, 1985)? (6) What is the significance of conjugate crenulations in the belt (Bhattacharyya and Paterson, 1985; Schweickert *et al.*, 1984b, 1985; Tobisch and Fiske, 1976)? (7) Are the dominant Nevadan structures in the western metamorphic belt east-vergent folds and thrusts that were backfolded by west-vergent folds and faults (Day *et al.*, 1984; Saleeby *et al.*, 1986; Wentworth and Zoback, 1985)?

Item 1

Bhattacharyya and Paterson (1985) stated that along the Merced River, the Calaveras and Shoo Fly rocks both contain only one slaty cleavage. As described above and discussed by Schweickert and Bogen (1983), Merguerian (1985a), and Schweickert *et al.* (1985), structures related to the Calaveras–Shoo Fly thrust correlate directly with the earliest structural fabric (S1) in the Calaveras, but overprint two older, penetrative fabrics in the Shoo Fly, implying that the Shoo Fly contains two generations of structures that do not occur in the Calaveras. This fact is not readily evident in every outcrop of the Shoo Fly, but is best appreciated by carefully noting the style and structural overprinting in a number of contiguous outcrops, and by noting the relations of the structures to intrusions or to the Calaveras–Shoo Fly thrust. The Calaveras itself contains two generations of structures that predate mafic dikes which probably are related to the Sonora dike swarm (Schweickert *et al.*, 1985).

Item 2

Bateman *et al.* (1985) expressed doubt that the Calaveras is a melange, at least in the Merced River drainage, based on the lateral continuity of a large mass of limestone from which they obtained Lower Triassic fossils. Our structural observations along the North, Middle, and South Forks of the Merced River show conclusively that stratal continuity does not exist at outcrop scale. Most rocks are unbedded, chaotic argillite with olistoliths of chert and siltstone. Even chert-rich zones show bedding that is only traceable a few meters at best, and in numerous cases, lenses of bedded chert can be shown to be enclosed within chaotic argillite. The large limestone body described by Bateman *et al.* (1985) is remarkable, and we suggest that it is an unusually large olistolith.

Item 3

According to Sharp (1985), D2 structures in the Calaveras can be traced across the Sonora fault into the Don Pedro terrane where they predate 166–170-m.y. intrusions. Field evidence we have obtained does not support this interpretation. Figures 29-8a and b show our interpretation of F2 folds in the Calaveras that are discordantly truncated by the Sonora fault. Locally, discordance is nearly 90°. S2 in the Calaveras is oriented about

DEFORMATIONAL AND METAMORPHIC HISTORY

N70°W, even near the fault, but no such structures have turned up within the Don Pedro terrane. The trace of the Sonora fault shows no involvement in the D2 structures. Our observations in the Page Mountain and Cobbs Creek plutons demonstrate that the plutons are locally strongly foliated and lineated, and have experienced much or all of the deformation of their wallrocks, indicating that the deformation/metamorphism in the Don Pedro terrane postdates, rather than predates, the plutons. The penetrative D1 and D2 Calaveras structures predate 170-m.y. plutons, whereas those of the Don Pedro terrane postdate 165-m.y. plutons and strata containing Oxfordian-Kimmeridgian fossils. Hence the structural sequences in the two terranes are of distinctly different age, even though they are somewhat similar in style. As noted above, metamorphic grade associated with the two sets of structures differs also.

Item 4

Does the Sonora fault exist? Bhattacharyya and Paterson (1984) and Sharp (1985) have apparently had difficulty in locating this feature and have questioned its existence. This is an important question because the Sonora fault is the best candidate for a Nevadan suture in the southern part of the western metamorphic belt. We have mapped good exposures of the fault on Coyote Creek and in Sonora, and a fair exposure on the North Fork of the Merced River. Evidence for the Sonora fault includes the following: (1) As noted in item 3 above, structures on two sides of the fault are of dramatically different age. (2) A metamorphic contrast occurs across the fault. (3) A major lithologic contrast occurs across the fault. (4) F2 folds in the Calaveras are truncated by the fault, locally at angles approaching 90° (Fig. 29-8b). (5) The Sonora dike swarm in the Calaveras and Shoo Fly is deformed at the fault and does not extend into the Don Pedro terrane. (6) Folds and foliation of similar orientation to D2 Calaveras structures do not occur in the Don Pedro terrane.

Item 5

Bateman et al. (1985) inferred, on the basis of several radiometric and fossil ages, which young progressively westward, that the units in the southern part of the western metamorphic belt form a west-facing sequence. This seems highly unlikely for several reasons: (1) The Calaveras is a chaotic melange terrane, lacking stratigraphic order. (2) The Calaveras and adjacent terranes have been penetratively deformed and metamorphosed at least twice, and the Shoo Fly and Don Pedro terranes have undergone tectonic transposition. Even if they were coherent sequences, facing directions would be expected to be parallel to the layering, rather than perpendicular to it as Bateman et al. (1985) assumed. (3) Major faults separate the terranes. The stacking order of the terranes suggests that the eastern terranes now are structurally high relative to the western terranes.

Item 6

Tobisch and Fiske (1976) first described conjugate crenulations in rocks of the Shoo Fly and Don Pedro terranes, together with crenulations from some localities in the eastern Sierra Nevada. They inferred that the crenulations deform a Jurassic slaty cleavage in each terrane, and suggested that the crenulations formed as a result of "relaxation" and layer-parallel shortening at the end of the Nevadan orogeny. As discussed by Schweickert et al. (1984a, 1985), this interpretation is untenable for crenulations in the Shoo Fly and Calaveras Complexes, inasmuch as the penetrative fabrics in these complexes consist of several pre-Nevadan foliations, and because other pre-Nevadan fabrics commonly show up as crenulations. Schweickert *et al.* (1984a, 1985) suggested that crenulations of the type described by Tobisch and Fiske (1976) could have formed in the Calaveras and Shoo Fly either during the Nevadan orogeny (where they might represent main-phase and late-phase structures) or as a result of superimposition of any of the brittle fabrics noted in the Calaveras and Shoo Fly Complexes.

Item 7

Day *et al.* (1985) presented evidence from the northwestern Sierra Nevada that indicates the Nevadan orogeny was characterized by early, east-vergent thrusts and folds that were subsequently backfolded and deformed by west-vergent faults and folds. Saleeby *et al.* (1986) extended this interpretation to the southern part of the metamorphic belt, and inferred that major Nevadan structures beneath the Great Valley and the southern foothills were east-vergent. Significantly, Zoback *et al.* (1985) presented seismic data that shows a major west-dipping reflector beneath the Great Valley that projects toward the surface near the eastern edge of the western metamorphic belt.

During our structural examination of the Foothills and Don Pedro terranes we have found no evidence of the kind reported by Day *et al.* (1984) for east-vergent structures. Rather, all the major faults dip steeply east and appear to have reverse displacement, and the major F1 folds of the Foothills terrane appear to be upright to slightly west-vergent. We believe the surface structure requires the interpretation that the Foothills and Don Pedro terranes were thrust beneath the Calaveras and Shoo Fly Complexes in a west-vergent system during the Nevadan orogeny. Perhaps a transition occurs somewhere along the length of the belt between east-vergent structures and the west-vergent structures we have documented. As to the nature of the west-dipping reflector described by Zoback *et al.* (1985), we suspect this may represent structural grain in the basement of the island arc in the Foothills terrane, an interpretation similar to an option suggested by these authors.

TECTONIC INTERPRETATIONS

In this final section we present our views of the simplest tectonic interpretations of the deformational and metamorphic history that are consistent with the data discussed in this paper.

Paleozoic Deformation and Metamorphism of the Shoo Fly Complex

The Shoo Fly in the region described here apparently experienced a deformational and metamorphic history similar to that of the Shoo Fly in the northern Sierra, described by Schweickert *et al.* (1984b). In both regions, important deformation predated the emplacement of Devonian(?) granitoids, while the grade of metamorphism was higher in the southern part of the Shoo Fly. As suggested by many authors, this early deformation could have been a precursor to the Late Devonian–Early Mississippian Antler orogeny, within an island arc system that approached and then collided with the passive margin of western North America. The early D1 deformation recorded within the Shoo Fly would in such a scenario represent subduction and accretion of continentally derived sediments into the accretionary prism during approach of the arc system. The Devonian gneissic granitoids probably represent magmas derived by melting of this large accretionary prism just prior to the collision. The Antler orogeny *sensu stricto* in Nevada would represent the late stage, diachronous thrusting of the large accretionary prism onto the continental margin of North America. This scenario is consistent with the data we have presented, and although there is as yet no conclusive proof that the Shoo Fly and its overlying arc sequence in the northern Sierra were involved in the Antler orogeny, much circumstantial evidence places this terrane in a paleogeographic setting near the western margin of North America during early to mid-Paleozoic time (Coles and Snyder, 1985; Girty *et al.*, 1984; Hanson and Schweickert, 1986; Varga, 1982).

The age of the D2 event in the Shoo Fly is unknown, but it is not considered to have been progressive with D3 because of differences in style and the nearly orthogonal regional truncation of S2 hinge surface traces against the D3 Calaveras–Shoo Fly thrust (Fig. 29-3a and b). D2 may be of late Paleozoic age and, if so, may be due to late-stage Antler convergence.

Mesozoic Deformation and Metamorphism of the Calaveras and Shoo Fly Complexes

The age of the post-earliest Triassic thrusting of the Calaveras beneath the Shoo Fly and D1/D3 deformation and metamorphism of both complexes is not yet well constrained, but may possibly have coincided in time with the Sonoma orogeny in Nevada, which is bracketed as post-earliest Triassic and pre-latest Triassic (Schweickert and Lahren, 1984, 1987; Speed, 1984). The Shoo Fly is part of the basement of the arc massif known as Sonomia, which is inferred to have collided with the North American margin during the Sonoma orogeny (Burchfiel and Davis, 1972; Schweickert and Snyder, 1981; Silberling, 1973; Speed, 1979). The Calaveras Complex probably is a subduction complex formed by accretion and slumping of oceanic sediments as Sonomia approached the continental margin, and probably corresponds with the Golconda allochthon of Nevada, although it is more highly metamorphosed and probably represents a deeper level of the subduction complex. If deformation and metamorphism related to juxtaposition of the Calaveras and Shoo Fly were instead Early Jurassic events, accretion of the Calaveras must have been a separate, pre-Middle Jurassic event, an interpretation favored by Saleeby *et al.* (1986).

The major D2/D4 event which resulted in strong folding of both the Calaveras and Shoo Fly may have been a progressive event related to D1/D3. Alternatively, it may mark a distinctly younger Triassic or Jurassic event of cryptic tectonic significance. One possibility, suggested by Saleeby *et al.* (1986), is that these structures reflect a major dextral strike-slip component along the Calaveras–Shoo Fly thrust during or following thrust movements. However, we have observed no structures within the Calaveras–Shoo Fly thrust zone that are compatible with strike-slip displacements prior to or during the D2/D4 event. Another possibility is that these structures are related to oblique pre-Nevadan subduction along the approximate position of the Sonora fault. Structures related to this event have not been positively identified elsewhere in the Sierra Nevada region.

Nevadan Deformation and Metamorphism

Nevadan events, including D3/D5 and D4/D6 in the Calaveras and Shoo Fly, and all Late Jurassic deformations in the Don Pedro and Foothills terranes, can reasonably be interpreted as the result of the collision of an oceanic island arc or arcs in the Don Pedro and Foothills terranes with a convergent margin at the west edge of the North American continent, as discussed by Schweickert *et al.* (1984a) and many other authors, resulting in a west-vergent system of Nevadan folds and thrust faults. In this view, the Sonora fault represents the Nevadan suture, and the Melones fault is an important imbrication within the arc terrane. It is beyond the scope of this paper to discuss all the pros and cons of this interpretation of the Nevadan orogeny; our main point is that the structural and meta-morphic data we have presented are consistent with this hypothesis.

NOTE ADDED IN PROOF: Paterson et al., (1987) and Tobish et al., (1987), recently described evidence for post-Nevadan deformation in the Foothills terrane south of the Merced River. The extent and significance of this deformation remains to be determined.

ACKNOWLEDGMENTS

Our work in the Sierran Nevada foothills has been supported by the National Science Foundation (grants EAR76-10979 and EAR78-23567 to RAS), the U.S. Geological Survey (grants 14-08-0001-18376 and 14-08-0001-17724 to RAS), and California Division of Mines and Geology (field support to NLB in 1980, field support to CM in 1981). We are deeply grateful to each of these organizations for its generous support. M. Lahren kindly helped with the preparation of the manuscript.

REFERENCES

Baird, A. K., 1962, Superposed deformations in the central Sierra Nevada foothills east of the Mother Lode: *Univ. Calif., Publ. Geol. Sci. Bull.*, v. 42, p. 1–70.

Bateman, P. C., Harris, A. G., Kistler, R. W., and Krauskopf, K. B., 1985, Calaveras reversed — Westward facing is indicated: *Geology*, v. 13, p. 338–342.

Behrman, P. S., 1978, Pre-Calloviagn rocks, west of the Melones fault zone, central Sierra Nevada foothills, *in* Howell, D. G., and McDougall, K. A., eds., *Mesozoic Paleogeography of the Western United States:* Pacific Section, Soc. Econ. Paleontologists Mineralogists, Pacific Coast Paleogeography Symp. 2, p. 337–348.

Bhattacharyya, T., and Paterson, S. R., 1985, Timing and structural expression of the Nevadan orogeny, Sierra Nevada, California — Discussion: *Geol. Soc. America Bull.*, v. 96, p. 1346–1347.

Blake, M. C., Jones, D. L., and Howell, D. G., 1982, Preliminary tectonostratigraphic terrane map of California: *U.S. Geol. Survey Open File Rpt. 82-593.*

Bogen, N. L., 1979, Four sets of structures in the Upper Jurassic Mariposa Formation, Tuolumne County, California [abstract]: *Geol. Soc. America Abstr. with Programs*, v. 11, p. 70.

——, 1983, Studies of the Jurassic geology of the west-central Sierra Nevada of California: Ph.D. dissertation; Columbia Univ. New York, N.Y., 240 p.

——, 1984, Stratigraphy and sedimentary petrology of the Upper Jurassic Mariposa Formation, western Sierra Nevada, California, *in* Crouch, J. K., and Bachman, S. B., eds., *Tectonics and Sedimentation along the California Margin:* Pacific Section, Soc. Econ. Paleontologists Mineralogists, v. 38, p. 119–134.

——, 1985, Stratigraphic and sedimentologic evidence of a submarine island-arc volcano in the Lower Messozoic Penon Blanco and Jasper Point Formations, Mariposa County, California: *Geol. Soc. America Bull.*, v. 96, p. 1322–1331.

——, Kent, D. V., and Schweickert, R. A., 1985, Paleomagnetism of Jurassic rocks in the western Sierra Nevada metamorphic belt and its bearing on the structural evolution of the Sierra Nevada block: *J. Geophys. Res.*, v. 90, p. 4627–4638.

Borradaile, G. J., 1978, Transected folds — A study illustrated with examples from Canada and Scotland: *Geol. Soc. America Bull.*, v. 89, p. 481–493.

Bowen, O. E., 1969, Geology of the El Portal-Coulterville area: *Calif. Div. Mines and*

Geology open-file, Kinsley, Feliciana Mountain, El Portal, and Buckingham Mountain quadrangles, scale 1:24,000.

Burchfiel, B. C., and Davis, G. A., 1972, Structural framework and evolution of the southern part of the Cordilleran orogen, western United States: *Amer. J. Sci.*, v. 272, p. 97–118.

Clark, L. D., 1964, Stratigraphy and structure of part of the western Sierra Nevada metamorphic belt, California: *U.S. Geol. Survey Prof. Paper 410*, 70 p.

———, 1976, Stratigraphy of the north half of the western Sierra Nevada metamorphic belt, California: *U.S. Geol. Survey Prof. Paper 923*, 26 p.

———, Stromquist, A. A., and Tatlock, D. B., 1963, Geologic map of the San Andreas quadrangle, Calaveras County, California: *U.S. Geol. Survey Geol. Quad. Map GQ-222*, scale 1:62,500.

Coles, K. S., and Snyder, W. S., 1985, Significance of lower and middle Paleozoic phosphatic chert in the Toquima Range, central Nevada: *Geology*, v. 13, p. 573–577.

Day, H. W., Moores, E. M., and Tuminas, A. C., 1985, Structure and tectonics of the northern Sierra Nevada: *Geol. Soc. America Bull.*, v. 96, p. 436–450.

Duffield, W. A., and Sharp, R. V., 1975, Geology of the Sierra foothills melange and adjacent areas, Amador County, California: *U.S. Geol. Survey Prof. Paper 827*, 30 p.

Eric, J. H., Stromquist, A. A., and Swinney, C. M., 1955, Geology and mineral deposits of the Angels Camp and Sonora quadrangles, Calaveras and Tuolumne Counties, California: *Calif. Div. Mines Geology Spec. Rpt. 41*, 55 p.

Evans, J. R., and Bowen, O. E., 1977, Geology of the southern Mother Lode, Tuolumne and Mariposa Counties, California: *Calif. Div. Mines Geology Map Sheet 36*, scale 1:24,000.

Evernden, J. F., and Kistler, R. W., 1970, Chronology of emplacement of Mesozoic batholithic complexes in California and Nevada: *U.S. Geol. Survey Prof. Paper 623*, 42 p.

Girty, G. H., Wardlaw, M. S., Schweickert, R. A., Hanson, R. E., and Bowring, S. A., 1984, Timing of pre-Antler deformation in the Shoo Fly Complex, Sierra Nevada, California: *Geology*, v. 12, p. 673–676.

Hanson, R. E., and Schweickert, R. A., 1986, Stratigraphy of mid-Paleozoic island-arc rocks in part of the northern Sierra Nevada, Sierra and Nevada Counties, California: *Geol. Soc. America Bull.*, v. 97, p. 986–998.

Herzig, C. T., Sharp, W. D., and Spangler, R., 1985, New constraints on the nature and origin of the Melones fault zone: *Geol. Soc. America Abstr. with Programs*, v. 17, p. 361.

Hobbs, B. E., Means, W. D., and Williams, P. F., 1976, *An Outline of Structural Geology:* New York, 571 p.

Imlay, R. W., 1961, Late Jurassic ammonites from the western Sierra Nevada, California: *U.S. Geol. Survey Prof. Paper 374D*, p. D1–D30.

Kerrick, D. M., 1970, Contact metamorphism in some areas of the Sierra Nevada: *Geol. Soc. America Bull.*, v. 81, p. 2913–2938.

Kistler, R. W., Dodge, F. C. W., and Silberman, M. L., 1983, Isotopic studies of Mariposite-bearing rocks from the south-central Mother Lode, California: *Calif. Geology*, Sept. 1983, p. 201–203.

Merguerian, C., 1981a, *The Extension of the Calaveras-Shoo Fly Thrust (CSFT) to the Southern End of the Sierra Nevada Metamorphic Belt, California — Preliminary Geologic Maps of the Tuolumne, Duckwall Mountain, Groveland, Jawbone Ridge, Lake Eleanor SW, SE, and NE ¼, Buckhorn Peak, Kinsley, El Portal, and Buckingham Mountain 7.5' Quadrangles*, accompanied by 10-page report: San Francisco, Calif. Div. Mines and Geology.

_____, 1981b, Tectonic significance of the Calaveras-Shoo Fly thrust (CSFT), Tuolumne County, California: *Geol. Soc. America Abstr. with Programs*, v. 13, p. 96.

_____, 1985a, Stratigraphy, structural geology, and tectonic implications of the Shoo Fly Complex and the Calaveras–Shoo Fly thrust, central Sierra Nevada, California: Ph.D. dissertation, Columbia Univ., New York, N.Y., 255 p.

_____, 1985b, Bedrock geologic map of the Shoo Fly Complex in the Jupiter area, Stanislaus River drainage, Tuolumne County, California: *Calif. Div. Mines Geology Open-File Rpt. 85-11 SF*, scale 1:24,000.

_____, 1985c, Tectonic significance of mylonitic Paleozoic gneissic granitoids in the Shoo Fly Complex, Tuolumne County, California: *Geol. Soc. America Abstr. with Programs*, v. 17, p. 369.

_____, 1986, Geology of the Sonora dike swarm, Sierra Nevada, California: *Geol. Soc. America Abstr. with Programs*, v. 18, p. 157.

_____, and Schweickert, R. A., in press, Paleozoic gneissic granitoids in the Shoo Fly Complex, central Sierra Nevada, California: *Geol. Soc. America Bull.*

Morgan, B. A., 1976, Geology of Chinese Camp and Moccasin quadrangles, Tuolumne County, California: *U.S. Geol. Survey Misc. Field Studies Map MF-840*, scale 1:24,000.

_____, 1978, Metamorphic forsterite and diopside from the ultramafic complex at the Tuolumne River, California: *U.S. Geol. Survey J. Res.*, v. 6, p. 73–80.

Nakamura, K., Jacob, K. H., and Davies, J. N., 1977, Volcanoes as possible indicators of tectonic stress orientation — Aleutians and Alaska: *Pageoph.* v. 115, p. 87–112.

Paterson, S. R., Tobisch, O. T., and Radloff, J. K., 1987, Post-Nevadan deformation along the Bear Mountains fault zone. Implications for the Foothills terrane, central Sierra Nevada, California: *Geology*, v. 15, p. 513–516.

Roberts, J. L., 1977, The structural analysis of metamorphic rocks in orogenic belts, *in* Saxena, S. K., and Bhattacharji, S., eds., *Energetics in Geological Processes:* New York, Springer-Verlag, p. 151–168.

Rogers, T. H., 1966, San Jose sheet, *Geologic Map of California*; scale 1:250,000: Calif. Div. Mines Geology.

Saleeby, J. B., 1982, Polygenetic ophiolite belt of the California Sierra Nevada — Geochronological and tectonostratigraphic development: *J. Geophys. Res.*, v. 87, p. 1803–1824.

_____, *et al.*, 1986, Explanatory text for the Continent-ocean transect — Corridor C-2, central California offshore to Colorado Plateau: *Geol. Soc. America Centennial Continent-Ocean Transect 10.*

Schweickert, R. A., 1978, Triassic and Jurassic paleogeography of the Sierra Nevada and adjacent regions in California and Nevada, *in* Howell, D. G., and McDougall, K. A., eds., *Mesozoic Paleogeography of the Western United States:* Pacific Section, Soc. Econ. Paleontologists Mineralogists, Pacific Coast Paleogeography Symp. 2, p. 361–384.

_____, 1981, Tectonic evolution of the Sierra Nevada Range, *in* Ernst, W. G., ed., *The Geotectonic Development of California* (Rubey Vol. I): Englewood Cliffs, N.J. Prentice-Hall, p. 87–131.

_____, and Bogen, N. L., 1983, *Tectonic Transect of Sierran Paleozoic through Jurassic Accreted Belts:* Soc. Econ. Paleontologists Mineralogists, Pacific Section, 22 p.

_____, and Cowan, D. S., 1975, Early Mesozoic tectonic evolution of the Western Sierra Nevada, California: *Geol. Soc. America Bull.*, v. 86, p. 1329–1336.

_____, and Lahren, M. M., 1984, Extent of Antler and Sonoma belts, sutures, and transcurrent faults in eastern Sierra Nevada, California: *Geol. Soc. America Abstr. with Programs*, v. 16, p. 648.

Schweickert, R. A., and Lahren, M. M., 1987, Continuation of Antler and Sonoma orogenic belts to the eastern Sierra Nevada, California, and Late Triassic thrusting in a compressional arc: *Geology*, v. 15, p. 270–273.

——, and Snyder, W. S., 1981, Paleozoic plate tectonics of the Sierra Nevada and adjacent regions, *in* Ernst, W. G., ed., *The Geotectonic Development of California:* (Rubey Vol. I): Englewood Cliffs, N.J., Prentice-Hall, p. 182–201.

——, Saleeby, J. B., Tobisch, O. T., and Wright, W. H., 1977, Paleotectonic and paleogeographic significance of the Calaveras Complex, western Sierra Nevada, California *in* Stewart, J. H., Stevens, C. H., and Fritsche, A. E., eds., *Paleozoic Paleogeography of the Western United States:* Pacific Section, Soc. Econ. Paleontologists Mineralogists, Pacific Coast Paleogeography Symp. 1, p. 381–394.

——, Bogen, N. L., Girty, G. H., Hanson, R. E., and Merguerian, C., 1984a, Timing and structural expression of the Nevadan orogeny, Sierra Nevada, California: *Geol. Soc. America Bull.*, v. 95, p. 967–979.

——, Harwood, D. S., Girty, G. H., and Hanson, R. E., 1984b, Tectonic development of the northern Sierra terrane — An accreted late Paleozoic island arc and its basement, *in* Lintz, J., Jr., ed., *Western Geological Excursions, Vol. 4:* Soc. America Ann. Meet., Geol. Mackay Sch. Mines, Reno, Nev., p. 1–65.

——, Bogen, N. L., Girty, G. H., Hanson, R. E., and Merguerian, C., 1985, Timing and structural expression of the Nevadan orogeny, Sierra Nevada, California — Reply: *Geol. Soc. America Bull.*, v. 96, p. 1349–1352.

Sharp, W. D., 1980, Ophiolite accretion in the northern Sierra: *Trans. Amer. Geophys. Un.*, v. 61, p. 1122.

——, 1984, Structure, petrology, and geochronology of a part of the central Sierra Nevada foothills metamorphic belt: Ph.D. dissertation; Univ. California, Berkeley, Calif. 183 p.

——, 1985, The Nevadan orogeny of the Foothills metamorphic belt, California — A collision without a suture?: *Geol. Soc. America Abstr. with Programs*, v. 17, p. 407.

——, and Saleeby, J. B., 1979, The Calaveras Formation and syntectonic mid-Jurassic plutons between the Stanislaus and Tuolumne Rivers, California: *Geol. Soc. America Abstr. with Programs*, v. 11, p. 127.

——, Saleeby, J. B., Schweickert, R. A., Merguerian, C., Kistler, R. W., Tobisch, O. T., and Wright, W. H., 1982, Age and tectonic significance of Paleozoic orthogneisses of the Sierra Nevada foothills metamorphic belt, California: *Geol. Soc. America Abstr. with Programs*, v. 14, p. 233.

——, Saleeby, J. B., Schweickert, R. A., Merguerian, C., Kistler, R. W., Tobisch, O. T., and Wright, W. H., in press, Age and tectonic significance of Paleozoic orthogneisses of the Sierra Nevada foothills metamorphic belt, California: *Geol. Soc. America Bull.*

Silberling, N. J., 1973, Geologic events during Permian-Triassic time along the Pacific margin of the United States: *Alberta Soc. Petrol. Geologists Mem. 2*, p. 345–362.

Speed, R. C., 1979, Collided Paleozoic microplate in the western United States: *J. Geology*, v. 87, p. 279–292.

——, 1984, Paleozoic and Mesozoic continental margin collision zone features: Mina to Candelaria, NV, traverse, *in* Lintz, J., Jr., ed., *Western Geological Excursions, Vol. 4:* Geol. Soc. America Ann. Meet., Mackay Sch. Mines, Reno, Nev., p. 66–80.

Stern, T. W., Bateman, P. C., Morgan, B. A., Newell, M. F., and Peck, D. L., 1981, Isotopic

U-Pb ages of zircon from the granitoids of the central Sierra Nevada, California: *U.S. Geol. Survey Prof. Paper 1185*, 17 p.

Strand, R. G., 1967, *Geologic Map of California, Mariposa sheet*, scale 1:250,000: *Calif. Div. Mines Geology*.

——, and Koenig, J. B., 1965, *Geologic map of California; Sacramento sheet*, scale 1:250,000: *Div. Mines and Geology*., Calif.

Suppe, J., 1984, *Principles of Structural Geology:* Englewood Cliffs, N.J., Prentice-Hall, 537 p.

Taliaferro, N. L., 1942, Geologic history and correlation of the Jurassic of southwestern Oregon and California: *Geol. Soc. America Bull.*, v. 53, p. 71–112.

——, and Solari, A. J., 1949, Geology of the Copperopolis quadrangle, California: *Calif. Div. Mines Bull. 145*, plate 1.

Tobisch, O. T., 1960, Geology of the Crane Flat-Pilot Peak area, Yosemite district, California: M.A. thesis, Univ. California, Berkeley, Calif.

——, 1984, Development of foliation and fold interference patterns produced by sedimentary processes: *Geology*, v. 12, p. 441–444.

——, and Fiske, R. S., 1976, Significance of conjugate folds and crenulations in the central Sierra Nevada, California: *Geol. Soc. America Bull.*, v. 87, p. 1411–1420.

——, and Fiske, R. S., 1982, Repeated parallel deformation in part of the eastern Sierra Nevada, California, and its implications in the dating of structural events: *J. Struct. Geology*, v. 4, p. 177–195.

——, Paterson, S. R., Longiaru, S., and Bhattacharyya, T., 1987, Extent of the Nevadan orogeny, central Sierra Nevada, California: *Geology*, v. 15, p. 132–135.

Turner, H. W., and Ransome, F. L., 1897, Sonora, California: *U.S. Geol. Survey Geol. Atlas, Folio 41*, 7 p.

——, and Ransome, F. L., 1898, Big Trees, California: *U.S. Geol. Survey Geol. Atlas, Folio 51*, 8 p.

Varga, R., 1982, Implications of Paleozoic phosphorites in the northern Sierra Nevada range: *Nature*, v. 297, p. 217–220.

Wagner, D. L., Jennings, C. W., Bedrossian, T. L., and Bortugno, E. J., 1981, Geologic map of the Sacramento quadrangle, California: *Calif. Div. Mines Geology Map No. 1A*, scale 1:250,000.

Wentowrth, C., and Zoback, M. D., 1985, Central California seismic reflection transect – I. The eastern Coast Ranges and western Great Valley [abstract]: (EOS) *Trans. Amer. Geophys Un.*, v. 66, p. 1073.

Williams, P. F., 1985, Multiply deformed terrains – Problems of correlation: *J. Struct. Geology*, v. 7, p. 269–280.

Woodward-Clyde Consultants, 1978, SNPS Preliminary Safety Analysis Report: Unpublished report for Pacific Gas and Electric Company, San Francisco, Calif.

Zoback, M. D., Wentworth, C. M., and Moore, J. G., 1985, Central California seismic reflection transect – II. The eastern Great Valley and western Sierra Nevada [abstract]: (EOS) *Trans. Amer. Geophys. Un.*, v. 66, p. 1073.

Warren D. Sharp
Department of Earth and Space Sciences
State University of New York
Stony Brook, New York 11794

30

PRE-CRETACEOUS CRUSTAL EVOLUTION IN THE SIERRA NEVADA REGION, CALIFORNIA

ABSTRACT

The Sierra Nevada region spans the transition from North American Proterozoic sialic crust to Jurassic juvenile crust. Pre-Tertiary rocks of the region consist of six principal assemblages, including five basement terranes and widespread Middle to Late Jurassic arc and related ophiolitic rocks.

Parautochthonous North American sialic materials consisting of Proterozoic crystalline basement and derived clastic sedimentary rocks are inferred to underlie much of the Sierra Nevada batholith. These materials form the nucleus against which four additional basement terranes were accreted in Triassic to Early Jurassic time. From east to west these include (1) a Paleozoic continental borderland, (2) the Calaveras Complex, (3) the Sullivan Creek terrane, and (4) the Foothills ophiolite terrane.

The oldest rocks of the continental borderland are exposed in the northern Sierra. They consist of thrust slices of an early Paleozoic arc and the continentally derived fill of an ocean basin that have been telescoped by pre-Late Devonian, south-vergent thrusting. Unconformably overlying this composite basement is a well-preserved upper Paleozoic marine arc sequence. More deeply seated equivalents of these two assemblages are recognized in the central Sierra. The Paleozoic arcs represented by this terrane probably developed along the fringes of the North American margin, based on the presence of continental materials in its basement and consideration of faunal data. During the Early Triassic Sonoman orogeny, this terrane was accreted to western North America.

The Calaveras Complex consists of late Paleozoic to Early Triassic, uppermost oceanic crust that has been structurally thickened. This assemblage probably formed in a Middle to Late Triassic subduction complex located along the western margin of North America. A coeval marine arc in eastern California and Nevada may be the site of complementary magmatism, and its low-lying marine character is consistent with the paucity of terrigenous detritus within the Calaveras Complex.

The Sullivan Creek and Foothills ophiolite terranes include Lower to Middle Jurassic arc sequences that unconformably overlie disrupted polygenetic ophiolite. Recent age and isotopic data suggest correlation between the earliest elements of the Foothills ophiolite terrane and coeval ophiolite of the continental borderland of the northern Sierra and eastern Klamaths. This possibility, and the presence of continentally derived detritus in the Lower to Middle Jurassic arc strata, suggest earlier proximity of these terranes to the continental margin than previously appreciated. Early Jurassic, high-pressure tectonite melange blocks of the Foothills ophiolite are derived from supracrustal protoliths. These may reflect lithospheric thickening related to the early stages of accretion of these terranes.

Juxtaposition of the Calaveras Complex and Sullivan Creek terrane with the already accreted continental borderland occurred along west-vergent, pre-late Middle Jurassic thrusts. Shortly thereafter, a Middle to Late Jurassic arc was superposed across the accreted basement terranes and sialic North America.

This arc consists of coeval intrusions, supracrustal sequences, and related ophiolitic complexes. Voluminous, mantle-derived, parental magmas of the arc are a significant addition to the Sierran crust. Structural and stratigraphic evidence indicates that the arc was synorogenic throughout its history.

These relations indicate relatively early accretion of the Sierran terranes, and prolonged, postaccretionary orogenesis. This view of Sierran structural evolution differs from that of some earlier workers who relate Sierran orogenesis chiefly to a brief event termed the Nevadan orogeny, believed to be caused by a Late Jurassic arc-continent collision.

If the Sierran pattern may be generalized, continental crustal growth at Cordilleran-type margins may proceed by formation of intermediate-type crust (e.g., that of the modern Aleutians) in continent-fringing island arc systems, including accretionary wedges and arc-related ophiolitic complexes. Accretion of these assemblages at a long-lived, convergent continental margin may be followed by crustal thickening due to noncollisional shortening of the overriding plate and superposition of arc-type magmatism.

INTRODUCTION

Comparison of the present western margin of North America with its inferred early Paleozoic location reveals that there has been an extensive increase in the area of the continent (Fig. 30-1). The Sierra Nevada region includes the edge of Proterozoic North American sialic basement as well as rock assemblages that have been added to the continent in Mesozoic time. The region provides a well-studied example of Phanerozoic growth of continental crust.

To the southeast of the Sierra Nevada lies North American sialic crust, including Proterozoic crystalline basement and its sedimentary cover. This cover consists of terrigenous clastic rocks and carbonates, of late Precambrian to early Paleozoic age, formed on a passive continental margin that resulted from rifting at about 600 m.y.b.p. (Stewart, 1972; Stewart and Suczek, 1977). The shape, in map pattern, of the edge of North American sialic crust is largely inherited from this rifting event, although it has been modified by continentward thrusting and strike-slip faulting related to subsequent active continental margin tectonics. Palinspastic restorations of the rifted margin sequence show that it is characterized by great lateral continuity and gradual facies changes.

In contrast, north and west of the inferred edge of Proterozoic sial, the crust consists of a patchwork of fault-bounded fragments of diverse origin. Such fragments are termed tectonostratigraphic terranes. Simply defined, a terrane is a fault-bounded domain with a geological development that is distinct from that of its neighbors. Some terranes, or components within them, appear to have been displaced hundreds or thousands of kilometers (e.g., Coney et al., 1980; Jones et al., 1977). Primary relationships among terranes, and between a given terrane and cratonic North America, cannot be assumed nor established by direct continuity. Many terranes consist of fragments of oceanic lithosphere, primitive magmatic arcs, and accretionary wedges. In light of these relations, some major questions facing Cordilleran geologists are: In what tectonic environments were the accreted terranes formed? What are the sources for igneous and sedimentary rocks that comprise the accreted terranes? What is the role of collision of exotic terranes in producing orogenesis at a long-lived convergent margin? Answers to these questions will help us understand the role of Cordilleran-type plate margins in the spectrum of processes leading to the growth and development of continental crust.

A westward traverse across the central Sierra Nevada passes from parautochthonous, North American basement to predominantly ensimatic accreted terranes. This paper is structured about the premise that the central Sierra Nevada region (near latitude 38°N) can be viewed as consisting of six principal rock assemblages; five diverse basement terranes and widespread Middle to Late Jurassic arc rocks (Fig 30-2). The basement terranes include the faulted edge of the ancient North American continent (parautochthonous North American basement) and four accreted terranes. From east to west, in the informal nomenclature of this paper, these are the continental borderland, the Calaveras Complex, the Sullivan Creek terrane, and the Foothills ophiolite terrane (Fig. 30-2). These four terranes consist chiefly of Paleozoic to early Mesozoic rocks, many of which appear to have formed without continental influence. The general plate margin environments in which each terrane has evolved may be inferred from their lithologic assemblages, structures, and other features. However, comparison to modern plate margin assemblages

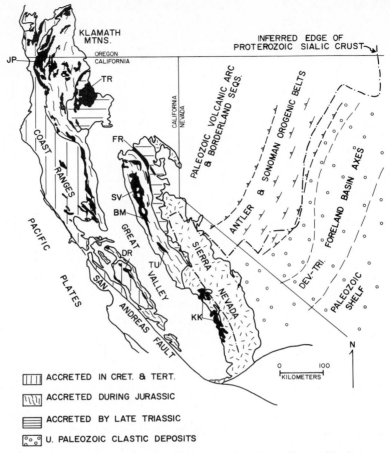

FIG. 30-1.   Regional map showing geological provinces discussed in the text. Ultramafic and related rocks are shown in black. Location of early Paleozoic continental margin is approximately given by the cryptic edge of Proterozoic sialic crust inferred from Sr. isotopic studies of Mesozoic granitoids (after Kistler, *in* Saleeby *et al.*, 1986). Ages of accretion given for the Sierra Nevada region, Klamath Mountains, and Coast Ranges refer to times when the respective belts may be directly linked to North America. Lettered localities are as follows: JP, Josephine peridotite; TR, Trinity ophiolite; FR, Feather River peridotite; SV, Smartville ophiolitic complex; BM, Bear Mountains area; TU, Tuolumne ultramafic complex; KK, Kings-Kaweah area; DR, Diablo Range occurrences of Coast Range ophiolite.

indicates that these terranes are only small fragments of the tectonic systems in which they formed (e.g., Burchfiel and Davis, 1981). Processes that produce such fragments are rifting and transform faulting (e.g., Saleeby, 1981, 1983). As a result, explicit paleogeographic reconstructions for these basement terranes are a matter of much inference.

Middle to Late Jurassic magmatic arc rocks occur in each basement terrane and are considered herein to have formed in a single arc. This arc was build across the five basement terranes that had previously been juxtaposed by faulting (Fig. 30-2b). Orogenesis, including faulting, folding, penetrative deformation, flysch-type sedimentation, and metamorphism, occurred throughout Middle to Late Jurassic time in the region of Fig.

30-2a and adjacent area. Orogenesis coincided spatially and temporally with arc magmatism.

Voluminous, mafic magmas played a key role in the Jurassic evolution of the Sierran crust. They constitute an important addition of mantle-derived material to the crust. Also, accompanying heat caused regional-scale metamorphism and may have caused thermal weakening of the subarc lithosphere. This effect may have reduced lithospheric strength to the point where yielding took place in response to regional stresses generated as a result of plate convergence. This explanation for Middle to Late Jurassic Sierran orogenesis is suggested as an alternative to collisional models (e.g., Ingersoll and Schweickert, 1986; Moores and Day, 1984; Schweickert and Cowan, 1975).

The geological development of each basement terrane will be summarized. Then stratigraphic and structural relations bearing on the initial juxtaposition and accretion of the basement terranes will be presented. Finally, the structural and petrologic development of the Middle to Late Jurassic arc-related rocks will be discussed.

PARAUTOCHTHONOUS NORTH AMERICAN BASEMENT

North American crystalline basement of Proterozoic (1.8 b.y.) age is widespread in the southwestern Cordillera (Barker *et al.*, 1969; Hanson and Peterman, 1968; Hedge *et al.*, 1967, 1968; Lanphere *et al.*, 1963; Peterman *et al.*, 1967; Silver, 1971; Silver *et al.*, 1977a, b). Such basement is not exposed in the Sierran region, but it is inferred to exist at depth beneath much of the batholith based on geochemical properties of the Mesozoic granitoids and xenoliths in upper Cenozoic volcanic rocks (Chen and Tilton, 1978; DePaolo, 1981; Domenick *et al.*, 1983; Kistler and Peterman, 1973, 1978; Saleeby *et al.*, 1986).

The $^{87}Sr/^{86}Sr = 0.706$ isopleth, shown in Figs. 30-1 and 30-2a, is inferred to approximately coincide with the northern and western limit of sialic crust, consisting of Proterozoic crystalline basement and its thick cover of Paleozoic terrigenous clastic sediment. This edge formed initially by late Precambrian rifting (Stewart, 1972; Stewart and Suczek, 1977). Within the Sierra Nevada, this edge is thought to have been modified by a family of prebatholithic, Jurassic or Cretaceous dextral-slip faults with an aggregate displacement of about 250 km (Kistler *et al.*, 1980; Saleeby *et al.*, 1986). Thus, North American sial in the region of the batholith is considered parautochthonous.

Crustal composition and structure within the batholith were profoundly affected by voluminous Cretaceous (mainly 100–80 m.y.) magmatism. In this event, discussed by Sams and Saleeby (Chapter 31, this volume), the entire crust appears to have been reworked by melting and high-temperature metamorphism. This event was driven by voluminous, mantle-derived, mafic magmas rising above a subduction zone whose surface trace lay to the west, and may be represented by early elements of the Franciscan Complex (Hamilton, 1969).

Pre-batholithic cover rocks have been largely eroded; however, local remnants are preserved as keel-like bodies between plutons (Tobisch *et al.*, 1985). Such rocks are generally steeply dipping, penetratively deformed, and metamorphosed to hornblende-hornfels facies. Three principal protolith assemblages are recognized within these metamorphic screens. The first consists of quartzose sandstone, pelite, marble, calc-silicate rock, and silicic volcanics of Mesozoic age (Bateman and Clark, 1974; Bateman and Moore, 1965; Christensen, 1963; Fiske and Tobisch, 1978; Jones and Moore, 1973; Kistler, 1966; Rinehart *et al.*, 1959; Tobisch *et al.*, 1985). These record sedimentation along the continental margin under shallow marine to subaerial conditions, in part within

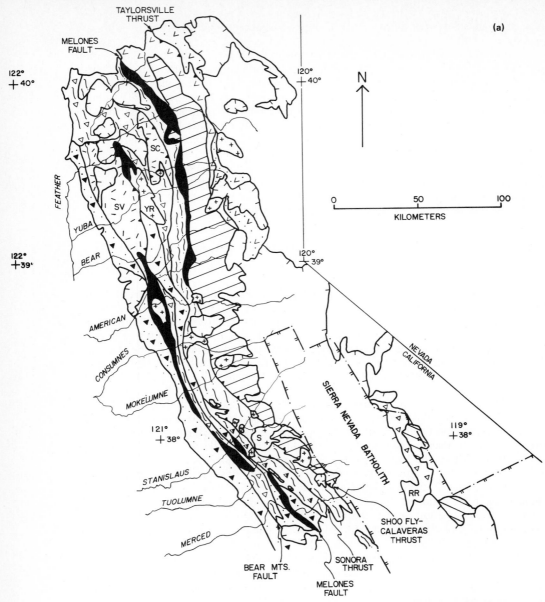

FIG. 30-2. (a) Map of the central and northern part of the Sierra Nevada region, showing the contiguous part of the foothills metamorphic belt. Symbols for units are given in (b). Units discussed in the text, major faults, and rivers are shown. Localities discussed in text are designated as follows: SC, Slate Creek ophiolitic complex; SV, Smartville ophiolitic complex; YR, Yuba River pluton; S, Standard pluton. Sources are discussed in the text. (b) Schematic diagram showing structural and stratigraphic relationships in the area of near latitude 38°N. This diagram serves as a key for (a). The diagram shows the age range of rocks in each of the six principal assemblages, including five basement terranes and the overlapping Middle to Late Jurassic arc. Large arrows indicate early links between basement terranes, based on provenance analysis and structural relations, as discussed in the text.

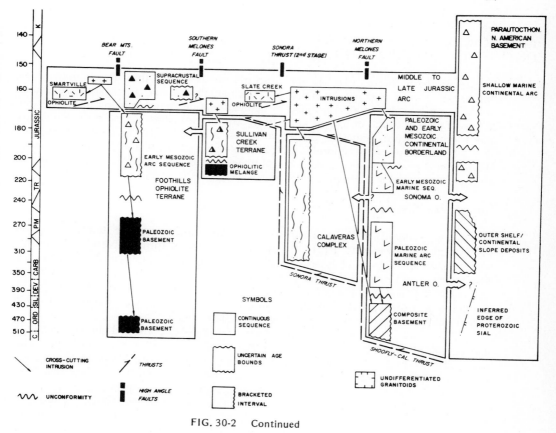

FIG. 30-2   Continued

an active ignimbrite field (Busby-Spera, 1984, 1985; Saleeby *et al.*, 1978). The second assemblage is comprised of outer shelf or continental slope deposits, in part of Mississippian to Permian age (Brook, 1977; Huber and Rinehart, 1965; Rinehart and Ross, 1964; Russell and Nokleberg, 1977). The third assemblage consists of lower Paleozoic quartzose sandstone and interbedded pelite, calcarenite, and thinly bedded limestone and pelite (Bateman and Moore, 1965; Kistler, 1966; Kistler and Bateman, 1966; Moore and Foster, 1980). These rocks yield Ordovician and Silurian fossils and represent chiefly shelf-facies rocks, probably correlatiive with those of southern Nevada (Moore and Foster, 1980).

An additional, local component of the batholithic screens are 100-m.y.-old, silicic ignimbrites of the Ritter Range, which along with coeval, shallow-level plutonic rocks, form a synbatholithic caldera collapse complex (Fiske and Tobisch, 1978; Fiske *et al.*, 1977; Peck, 1980).

PALEOZOIC TO EARLY MESOZOIC CONTINENTAL BORDERLAND

In the northern Sierra, two regional unconformities divide this terrane into three "packages" of rock (Fig. 30-2b). These are (1) a Late Proterozoic to early Paleozoic composite basement, (2) a Late Devonian to Permian marine arc assemblage, and (3) lower Mesozoic marine strata, in part of arc affinity. Southward along strike, in the Stanislaus and Tuolumne River drainages, this tripartite assemblage of little-metamorphosed rock gives way to polyphase metamorphic tectonites of upper greenschist to amphibolite grade. Mapping and geochronological study reveal these to be more deeply seated equivalents of some parts of "packages" 1 and 2 in the north.

Composite Basement

The oldest "package" in the northern Sierra is a structural composite, termed the Shoo Fly Complex (Schweickert, 1981). It consists of a stack of thrust slices of accretionary wedge materials, abyssal plain sediments, a seamount-subsea fan assemblage, and continental slope and rise deposits. Dated components range in age from Late Proterozoic to Silurian (Girty and Wardlaw, 1985; Saleeby *et al.*, 1986; Varga and Moores, 1981). Turbidite sandstones of the subsea fan are of continental block provenance and yield Early Proterozoic detrital zircons (Girty and Wardlaw, 1985).

The thrust slices represent part of an early Paleozoic arc and the continentally derived fill of an ocean basin that were telescoped by south-vergent thrusting. This occurred prior to Late Devonian time (Schweickert *et al.*, 1984), when the thrust stack was beveled and overlapped by marine arc strata and intruded by the Bowman Lake batholith dated at 370 ± 5 m.y. (Pb/U zircon age, Saleeby *et al.*, in press).

Plate configurations leading to these events are controversial. Briefly, the origins envisioned are as follows. A northwest-facing arc system, represented by rocks in the eastern Klamaths, separated from the North American continental shelf by a back-arc basin, collapsed by obduction of the basin onto the continental margin (Burchfiel and Davis, 1972, 1975).

Alternatively, a southeast-facing arc collided with the passive margin of North America due to consumption of intervening oceanic crust, ramping its accretionary wedge onto the continental shelf (Schweickert and Snyder, 1981).

In the plate models above, the pre-Late Devonian thrusting in the northern Sierra is correlated with the Late Devonian–Early Mississippian emplacement of ocean-basin fill onto the continental margin of central Nevada, termed the Antler Orogeny. This correlation is suggested by the similar vergence and approximately similar timing of thrusting in the northern Sierra and central Nevada. It is supported by lithologic similarity between elements of the Shoo Fly Complex and the Antler allochthon (Schweickert and Snyder, 1981). A third possibility, delineated by Speed (1979) and Speed and Sleep (1982), implies that pre-Late Devonian thrusting in the northern Sierra is unrelated to the Antler Orogeny. In this scenario, the Shoo Fly Complex is part of the basement of a Paleozoic microplate, termed Sonomia, that was not accreted to North America until Early Triassic time. All of these models seem feasible at present.

Late Devonian to Permian Marine Arc

The second "package" in the northern Sierra section has been described by McMath (1966), Harwood (1983), and Durell and D'Allura (1977). It begins with Upper Devonian, felsic, submarine pyroclastic debris flows of the Sierra Buttes Formation. These, along with tuffaceous slate, were deposited across the beveled thrust slices of the Shoo Fly Complex. Above these lie chert, followed by andesitic turbidites of the Taylor Formation. These are overlain by the Peale Formation, consisting of a lower tuff-siltstone member, and an upper chert member, and fine-grained tuff of the Lower Permian Reeve Formation.

These rocks represent the distal deposits of a submarine volcaniclastic apron formed around an island arc. The chert of the Peale Formation marks a period of regional volcanic quiescence and subsidence. Detailed correlation between this section and time equivalents in central Nevada (Murchey *et al.*, 1986) leave little doubt that these areas were part of the same plate in Early Permian time. Plate settings envisioned by various workers for the Paleozoic marine arc sequence are strongly dependent on that favored for the underlying composite basement. Such settings include the following. A northwest-facing island arc separated from North America by a back-arc basin is depicted by Burchfiel and Davis (1972, 1975), Poole (1974), Poole and Sandberg (1977), Silberling (1975), and Schweickert and Snyder (1981). Speed (1978), Schweickert and Snyder (1981). Speed (1978), 1979) envisions a southeast-facing arc, probably of "quasi-local" origin.

Unconformably overlying the Paleozoic marine arc sequence in the northern Sierra are strata as old as Late Triassic. Strata as young as Permian are preserved locally below the unconformity. Elsewhere, much section is missing and large angular discordance is observed (Harwood, 1983). This surface is generally regarded as a reflection of the Early Triassic Sonoman orogeny, first defined in central Nevada as thrusting of Paleozoic marine strata (the Golconda allochthon) southward onto the continental margin. This event mirrors the earlier Antler orogeny in terms of the vergence and rock types involved, but the associated foreland basin strata are not as voluminous.

A gamut of plate configurations similar to those for the Antler orogeny are proposed for the Sonoman event. These include closure of a back-arc basin (Burchfiel and Davis, 1972, 1975; E. L. Miller *et al.*, 1984; Schweickert and Snyder, 1981) and partial subduction of the continental margin beneath a southward-facing arc, causing obduction of that arc's accretionary wedge (Speed, 1978, 1979). Evidence for prolonged deformation in the Golconda allochthon, occurring in unlithified to lithified sediment, has been cited in favor of the obducted accretionary wedge model (Breuckner and Snyder, 1985). Apparent continuity of magmatism and deposition through the Permo-Triassic interval in the Klamath Mountains is consistent with a northwest-facing arc, implying the Sonoman event is a back-arc closure (Davis *et al.*, 1978). The facies relationships and provenance of rocks emplaced in the Sonoman orogeny also seem best reconciled with deposition in a back-arc basin (E. L. Miller *et al.*, 1984). Again, it is not possible to eliminate any of these alternatives at present. The significant point here with respect to Sierran evolution is that *all* models that seem broadly applicable agree that by Middle Triassic time, the Shoo Fly Complex and overlying Paleozoic strata were accreted to North America.

Lower Mesozoic Marine Strata

Unconformably overlying the Paleozoic marine arc assemblage are Upper Triassic conodont-bearing limestone and Lower to Middle Jurassic continental margin arc deposits. These consist of graywacke, andesitic to dacitic flows and volcaniclastic rock, and intrusions of the Sailor Canyon and Milton Formations and the Taylorsville sequence (Figs. 30-2; Clark, 1964; D'Allura *et al.*, 1977; Harwood, 1983; McMath, 1986).

Shoo Fly Complex of the Central Sierra

Paleozoic continental borderland rocks in the drainages of the Stanislaus and Tuolumne Rivers consist of schist and gneiss derived chiefly from quartzofeldspathic sandstone, pelite, and lesser limestone and chert, as well as orthogneiss derived from granite, lesser gabbro and syenite. This assemblage is included in the Shoo Fly Complex (Schweickert, 1981; Schweickert and Snyder, 1981), but is distinguished from the Shoo Fly Complex of the northern Sierra by its polyphase, penetrative deformation and upper greenschist to amphibolite-facies metamorphic grade.

Detailed mapping reveals that some orthogneiss is derived from intrusions that cross-cut lithologic layering and early isoclinal folds (Merguerian, 1985; Merguerian and Schweickert, 1987; Sharp, 1984). A granitic stock, the Confluence Pluton, exhibiting these contact relations has yeilded an emplacement age of 370 ± 5 m.y.[1] An earlier report (Sharp *et al.*, 1982) of an age of 440 m.y. for the Confluence was in error. Orthogneiss with an igneous crystallization age of 275 ± 5 m.y. and an inherited component of Early Proterozoic age has also been identified (Sharp *et al.*, 1982).

Folding and penetrative deformation have also affected this pluton (Merguerian, 1987). Metamorphic ages from several bodies of orthogneiss, including the two mentioned above, are in the range of 235 ± 15 m.y. based on Pb/U sphene and K/Ar hornblende analyses[2]. Conclusions based on these field and geochronological data are as follows. Sediment of largely terrigenous origin, perhaps deposited in a continental rise-slope environment, was folded prior to intrusion by Late Devonian granitoids. The section was again intruded by Permian granitoids and strongly deformed and metamorphosed in Triassic time.

Parallels with the northern Sierra are noteworthy. Late Devonian and Early Permian intrusions in the central Sierra are the same age as strong pulses of magmatism in the northern Sierra. Episodes of deformation in the central Sierra appear to correspond to the two major unconformities recognized in the northern Sierra. The pre-Late Devonian age of metasediments of the central Sierra is consistent with their correlation to quartzofeldspathic sandstones of the structurally lowest and southernmost plate in the Shoo Fly Complex of the northern Sierra (Schweickert, 1981; Sharp, 1984). These observations indicate the Shoo Fly of the central Sierra is a more deeply seated equivalent of part of the Complex (the Lang-Halford sequence of Schweickert *et al.*, 1984) and overlying Paleozoic marine arc rocks in the northern Sierra.

Western Boundary of Continental Borderland

The fault bounding the continental borderland terrane on its west side is designated the Calaveras–Shoo Fly thrust (Schweickert, 1981) and places oceanic-affinity rocks of the Calaveras Complex beneath the Shoo Fly Complex. In the northern Sierra, this fault zone broadens and contains peridotite, Paleozoic gabbro, and amphibolite comprising the

---

[1] Upper intercept age of abraded zircon fractions, W. D. Sharp, unpublished data.

[2] W. D. Sharp, unpublished data.

Feather River ophiolite (Ehrenberg, 1975) and related rocks, as well as Mesozoic blueschist. In the area between the Feather and Yuba Rivers the suture is up to 7 kms across and is underlain by dunite and harzburgite tectonite with high-temperature (mantle) flow fabrics. Associated mafic and cumulate ultramafic rocks are cut by a gabbro dike dated at 387 ± 7 m.y. (Ar/Ar plateau on hornblende, Standlee, 1978). The peridotite is bounded on either side by faults that are discordant to its mantle fabric, but concordant with secondary foliations in ultramafic and mafic schists formed by syntectonic metamorphism at amphibolite grade. Mafic schists along the Rick Bar fault on the west side of the peridotite are dated at 236 ± 4 m.y. (Ar/Ar plateau on hornblende) perhaps reflecting cooling related to emplacement of the peridotite (Weisenberg and Avé Lallèmant, 1977). Other amphibolites and metagabbro associated with the peridotite yield hornblende ages of 272 ± 6 m.y. (Ar/Ar plateau, Standlee, 1978) and 285 and 245 m.y. (K/Ar, Hietanen, 1981). The two younger ages come from a large fault-bounded slab of amphibolite and structurally concordant psammitic schist within and to the west of the fault. The schistose to gneissic amphibolite is derived in part from dikes and pillow lavas of high-Ti basalt of possible alkalic affinity, while the psammites are correlated with the Shoo Fly Complex (Fig. 30-2). The internal contact between amphibolite and psammitic schist apparently preserves an early contact between Shoo Fly Complex and Paleozoic metabasite, which has elsewhere been disrupted by later faulting.

Map-scale inclusions of amphibolite, schist, and cometamorphic quartzite within the fault zone in the Alleghany district near the Yuba River yield K/Ar hornblende ages as old as 345 ± 9 m.y. (Bohlke and McKee, 1984). These ages approach those of the Central Metamorphic Belt of the Klamaths (Lanphere et al., 1968). The mineralogy (brown hornblende + plagioclase ± clinopyroxene ± garnet) and protolith (high-Ti basalt with interleaved quartzite) of these metabasites are also similar to the Grouse Ridge Formation of the Central Metamorphic Belt, suggesting that the Alleghany slab is a Sierran analog of the CMB or its offset equivalent.

Also present within the fault zone in the Yuba Rivers area is an inclusion of metachert, mafic metavolcanic rock, and metagraywacke, bearing prograde lawsonite + glaucophane + albite + muscovite ± quartz ± epidote ± sphene assemblages indicative of blueschist facies metamorphism (Schweickert et al., 1980). Five K/Ar analyses of whole rocks and muscovite concentrates from the metacherts yield ages of 157–190 m.y. with three determinations clustering at 173–176 m.y. The 190 m.y. age is possible too old due to inherited argon, and so an age of 174 m.y. is taken as a *minimum* age of blueschist metamorphism. Both the younger age (157 m.y.) and the presence of two generations of structures which deform the prograde assemblages suggest that significant disturbance of the K/Ar system has occurred.

South of latitude 38°50', the zone narrows but is marked by lenses of serpentinite-matrix melange (Schweickert, 1977) and metabasite. Hornblende from gneissic metagabbro near latitude 38°15' N yields a minimum Ar/Ar age of 262 ± 5 m.y.[3] In the central Sierra, near latitutde 38°N, major displacement on the fault zone can be shown to be early Mesozoic (Middle Jurassic or older, but post-Permian), as discussed in a later section.

South of latitude 38°30', the boundary between the continental borderland and primitive oceanic rocks to the west cannot be directly traced due to overlap of younger strata and disruption by dominantly Cretaceous granitoids. However, it appears to continue along strike as a cryptic feature termed the Foothills suture (Saleeby, 1978, 1981). East of outcrops of Paleozoic ophiolitic melange in the Kings and Kaweah Rivers areas (Fig. 30-1), steep gradients in crustal thickness, heat flow, and the chemical and isotopic

[3] W. D. Sharp, unpublished data.

composition of the younger grantoids indicate a rapid eastward transition onto thicker, older, more evolved crust inferred to be the edge of sialic North America.

CALAVERAS COMPLEX

The Calaveral Complex (Schweickert *et al.*, 1977) lies west of the continental borderland and related rocks described above. Along its western boundary, the Calaveras is in contact with the Foothills ophiolite and Sullivan Creek terranes (Fig. 30-2).

Lithology

The Calaveras Complex consists dominantly of metamorphosed siliceous shale, siltstone, and chert that characteristically lack coherent bedding and form a matrix for inclusions of bedded chert, marble, ultramafic schist, greenstone, and rare sandstone. These inclusions consist of blocks of a single rock type ranging from millimeters to hundreds of meters on a side, and larger inclusions, termed slabs, with internal contacts between two or more component rock types. The largest slabs, consisting of marble, amphibolite, and chert, are several kilometers across and are continuous along strike for tens of kilometers. Large areas of the Calaveras Complex are melange formed by sedimentary mixing, tectonic mixing, or both.

For reasons discussed below, the Calaveras Complex is considered to be a remnant of an early Mesozoic accretionary complex. Subsequent to melange formation, the Calaveras was penetratively deformed and metamorphosed under primarily greenschist facies conditions during emplacement of a suite of Middle Jurassic peridotitic to dioritic plutons.

Marble-amphibolite-metachert slabs These slabs consist primarily of coarsely crystalline calcite marble. However, locally interleaved with the calcite marble are amphibolite, metachert, mica schist, and metamorphosed conglomerates. These conglomerates contain clasts of amphibolite, mica schist, limestone, and dolostone. Siliciclastic detritus coarser than silt is absent from the marble-amphibolite slabs, indicating isolation from terrigenous sources, as does the presence of pure, gray ribbon chert that lacks interbedded shale.

Amphibolite within the marble-amphibolite slabs is derived from gabbro and mafic volcanic rock. Relict primary features of the protoliths and their contact relations with the marble are locally preserved. Such features include (1) dikes of medium-grained gabbro and diorite that intrude marble, (2) basaltic pillow lava (and associated feeder dikes) overlain by marble, (3) sheets of fine-grained mafic tuff interbedded with marble, and (4) conglomerates with a tuffaceous matrix interbedded with marble.

The lack of intergradation of marble and quartzite-phyllite, combined with the absence of amphibolite outside the marble-amphibolite slabs, indicates that the slabs developed prior to inclusion in the metachert and phyllite matrix in which they are presently found. Thus, the marble-amphibolite slabs are considered exotic with respect to the matrix.

The major- and trace-element compositions of the amphibolite help to define the environment of formation of these slabs[4]. Ti, Zr, Y, and the light rare earths (LREEs) are relatively immobile during metamorphism, are not abundant in the enclosing carbonate, and may be used to determine the affinity of the amphibolite protoliths. The basaltic protoliths for the amphibolites are diverse, varying widely in their trace-element concentrations. However, the high TiO_2 and Zr contents of some are quite unlike those of magmas associated with Benioff zones and the enrichment of the LREEs distinguishes these samples from midocean ridge basalts. Compositionally similar alkali basalts are

[4]W. D. Sharp, unpublished data.

characteristic of oceanic islands (Carmichael *et al.*, 1974). Carbonate buildups consisting of reefs and related material form on seamounts derived from volcanic islands as the underlying lithosphere cools and sinks with age. Thus, considering the lithologic characteristics of the marble-amphibolite slabs, it is suggested that they originated as limestone and associated mafic igenous rocks capping seamounts.

Ultramafic schist Irregular bodies of metamorphosed serpentinite are structurally interleaved with quartzite and quartz mica schists of the Calaveras Complex. Individual ultamafic schist bodies mapped are up to 1 km long and hundreds of meters wide. Within these ultramafic schists, and in close association with them, occur blocks of metamorphosed pink, metalliferous chert, and amphibolite. These metacherts lack interbedded pelitic schists, hence are distinct from metacherts of the melange matrix. Amphibolite in these schists is derived from fine-grained mafic rock. The ultramafic rocks are talc-antigorite-tremolite schists derived from foliated and massive serpentinite and serpentinized harzburgite, and the blocks within them are rock types typical of ocean crust. Thus, the ultramafic schists and related rocks appear to be tectonic inclusions of disrupted oceanic crust and lithosphere.

Other inclusions Other rock types that occur as blocks in the Calaveras Complex include metamorphosed, interbedded chert and shale, marble, greenstones, and quartzofeldspathic gneisses. Blocks of interbedded chert and shale with well-preserved bedding bounded by chaotic mixtures of chert and shale are a very common feature. Such blocks range in size from part of a meter to the extent of the largest continuous outcrops, hundreds of meters. Isolated blocks of marble resembling that of the marble-amphibolite slabs are also common. Greenstones, usually small blocks derived from mafic volcanic rock (sometimes pillow lava), are a minor component, constituting 2 or 3% of the Calaveras Complex.

Matrix The matrix consists of mixtures of argillite or phyllite and quartzite. The relative portions vary from pure argillite to 80% quartzite and 20% argillite. Argillite is usually siliceous and often graphitic. Quartzite is derived either from siltstone or chert. Although both clastic textures (siltstone) and relict radiolaria (chert) may sometimes be recognized, recrystallization due to deformation and metamorphism frequently precludes distinction between the two in the field.

Coherent bedding is often lacking in the matrix of the Calaveras Complex. Rather, quarzite occurs as (1) flattened, discontinuous slabs, resembling relict bedding; (2) tabular and roughly equidimensional clasts; (3) rootless tight folds; and (4) irregularly bounded fragments of complex shape, all in a matrix of dark siliceous argillite or mica schist.

Age

Because the Calaveras Complex is a melange, it may be thought of as having two types of ages. The first type is the stratigraphic age of its constituents. Marble of the Calaveras Complex has yielded several poorly preserved individuals of *Caninia* sp., a Permo-Carboniferous horn coral, from a locality near the Tuolumne River (Schweickert *et al.*, 1977). The fossils were found in a single block of marble surrounded by metachert-phyllite matrix within a belt of similar blocks and marble-amphibolite slabs like those described above. Early Triassic conodonts have been recovered from small limestone blocks near the Merced River (Bateman *et al.*, 1985). Pennsylvanian or Permian conodonts and Early Permian fusulinids were reported from the Feather River area (Hietanen, 1981; Standlee and Nestell, 1985). Because the marble and marble-amphibolite are exotic blocks and slabs, their ages cannot be extended to other components of the complex.

A second sort of age for melanges is their time of assembly. In practice, formation of melange may be a protracted event, perhaps lasting tens of millions of years, and can only be approximately bracketed. An upper bound on the age of tectonic mixing of the Calaveras Complex is provided by the Old Gulch Pluton that intrudes the Complex, post-dating early structures, and has yielded a concordant U/Pb zircon age of 177 m.y.[5] Based on this relationship and the Early Triassic age of the youngest known component of the melange, assembly of the Calaveras Complex is bracketed as Middle Jurassic or older, and at least in part, post-Early Triassic. A schematic summary of these age relations is presented in Fig. 30-3.

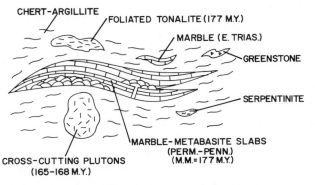

FIG. 30-3. Schematic summary of age relations among components of the Calaveras Complex. M. M. designates an Ar/Ar plateau age on hornblende from amphibolite from a marble-metabasite slab. Other ages given are concordant Pb/U zircon analyses (W. D. Sharp, 1985, unpublished data).

Origin

The Calaveras Complex is inferred to be a remnant of an early Mesozoic accretionary wedge. Observations leading to this conclusion include the following. No intact basement contacts for the sedimentary rocks have been observed, suggesting that the sedimentary cover has been detached from its substrate, although serpentinite and greenstone in the Calaveras Complex suggest that the basement for the sediments was oceanic. The steep dip of lithologic layering and broad outcrop indicate an apparent thickness of >10 km for the Calaveras Complex. Because pelagic and hemipelagic sediments are deposited very slowly, such a great thickness is unlikely to be a stratigraphic succession and probably results from structural repetition of a relatively thin stratigraphic section by folding and faulting. Intraformational conglomerate, breccia, and slump deposits are common, indicating that some deposition took place in a tectonically unstable environment. Marble-amphibolite was deposited in an area of active alkaline basaltic magmatism, perhaps on a seamount, prior to being encased in the chert-shale sequence. Its occurrence within metamorphosed chert and shale suggests that it is a fault-bounded slice within the Calaveras Complex. Thus the Calaveras Complex is a structural stack of seafloor sediments and seamount fragments that are detached from their basement. Such a deposit indicates shortening of a marine basin, most likely by consumption of oceanic crust and lithosphere in a subduction zone of Triassic to Middle Jurassic age.

[5] W. D. Sharp, unpublished data.

Correlation

The lithologic assemblage, along with its Early Triassic to Middle Jurassic age of mixing, forms the basis for correlation of the Calaveras Complex. Terranes consisting of this same distinctive lithologic assemblage and of similar age form a discontinuous belt extending from the Klamath Mountains to British Columbia (Davis *et al.*, 1978). These terranes include the Eastern Hayfork, North Fork, and Stuart Fork terranes of the Klamath Mountains (Ando *et al.*, 1983; Wright, 1982), the Canyon Mountain Complex of eastern Oregon, the Trafton Group of Washington (Danner, 1977) and the Cache Creek Group of British Columbia (Monger, 1977). Many of these other terranes lack the Middle Jurassic penetrative deformation and metamorphism superposed on the Calaveras Complex (see below) and preserve blueschist facies mineral assemblages dated as Late Triassic and Early Jurassic (Davis *et al.*, 1978).

The location of the subduction zone in which the Calaveras may have formed can be inferred only from indirect evidence. The presence of Upper Triassic to Lower Jurassic volcanic arc rocks in eastern California and Nevada (Chen and Tilton, 1978; Evernden and Kistler, 1970; Speed, 1978, 1979) indicates that a subduction zone existed along the western North American margin in about the appropriate time and place for the Calaveras Complex (Fig. 30-4). The paucity of sand-size or coarser terrigenous detritus in the Calaveras may be the result of the low-lying marine character of the continental margin at this time (e.g., western marine province of Speed, 1978) rather than indicating an intra-oceanic origin.

SULLIVAN CREEK TERRANE

Recently, it has been shown that an Early to Middle Jurassic arc sequence occurs *east* of the Melones Fault Zone. This sequence, which has been termed the Sullivan Creek terrane, (Herzig *et al.*, 1984; Sharp, 1983, 1984, 1985) and lies west of the Calaveras Complex, south of latitude 38°20' (Fig. 30-2a). These rocks have also been referred to as the greenschist-phyllite belt (Schweickert and Bogen, 1983; Schweickert *et al.*, 1984). The

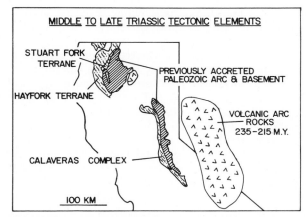

FIG. 30-4. Diagram of Middle to Late Triassic tectonic elements. Shows present outcrop of Calaveras Complex and Klamath Mountain equivalents, inferred to represent early Mesozoic subduction zones. Approximate extent of coeval arc magmatism in Nevada and eastern California are after Speed (1978) and Kistler (1978).

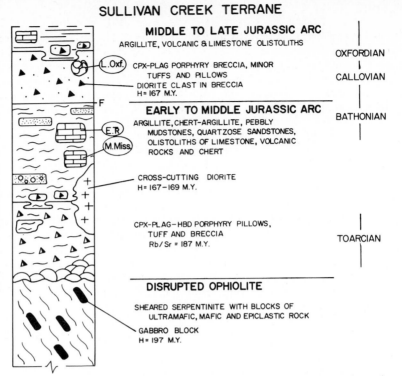

SULLIVAN CREEK TERRANE

MIDDLE TO LATE JURASSIC ARC

ARGILLITE, VOLCANIC & LIMESTONE OLISTOLITHS

OXFORDIAN

CPX-PLAG PORPHYRY BRECCIA, MINOR
TUFFS AND PILLOWS

CALLOVIAN

DIORITE CLAST IN BRECCIA
H = 167 M.Y.

EARLY TO MIDDLE JURASSIC ARC

BATHONIAN

ARGILLITE, CHERT-ARGILLITE, PEBBLY
MUDSTONES, QUARTZOSE SANDSTONES,
OLISTOLITHS OF LIMESTONE, VOLCANIC
ROCKS AND CHERT

CROSS-CUTTING DIORITE
H = 167-169 M.Y.

CPX-PLAG-HBD PORPHYRY PILLOWS,
TUFF AND BRECCIA
Rb/Sr = 187 M.Y.

TOARCIAN

DISRUPTED OPHIOLITE

SHEARED SERPENTINITE WITH BLOCKS OF
ULTRAMAFIC, MAFIC AND EPICLASTIC ROCK

GABBRO BLOCK
H = 197 M.Y.

FIG. 30-5. Schematic stratigraphic column of Sullivan Creek terrane and overlying Middle to Late Jurassic arc rocks. H designates Ar/Ar plateau ages on hornblendes (Herzig *et al.*, 1984), Rb/Sr designates whole rock isochron (Bateman *et al.*, 1985).

1.5–3-km-thick sequence (Fig. 30-5) consists of Early to Middle Jurassic greenstone derived from aphyric and plagioclase-phyric mafic pillow lavas and overlying plagioclase-augite ± hornblende-phyric mafic tuffs, and breccias. The greenstones may have been deposited on ophiolitic melange, although the original contact has been sheared. Depositionally above the greenstones is a heterogenous metasedimentary assemblage consisting of slate and phyllite derived from mudstone, pebbly mudstone and impure chert and containing clasts and disrupted beds of sandstone, Mississippian and Early Triassic limestone clasts, and chert. The greenstones are dated at 187 ± 10 m.y. (Rb/Sr isochron, Bateman *et al.*, 1985). Metavolcanic and metasedimentary rocks of the sequence are intruded and metamorphosed by gabbroic to quartz dioritic plutons that yield ages of 168–170 m.y. (Ar/Ar hornblende plateau ages, Herzig *et al.*, 1984. Modal analysis of disrupted beds and clasts of sandstone reveal a mixed provenance (Herzig, 1985). Some sandstones are rich in monocrystalline quartz grains (Q_{65-75}, F_{20-30}, L_{2-10}), indicating a continental source (Dickinson and Suczek, 1979), while others are rich in subangular to angular andesine and lithic fragments derived from local volcanic, metamorphic, and plutonic sources (Fig. 30-30-6).

Faunas from Mississippian limestones indicate warm, shallow-water depositional environments, but fusilinids are of cosmopolitan type, lacking provincial significance (M. Nestell, written communication, 1984). The Lower Triassic limestones are lithologically and faunally similar to a 3-km-long lens of limestone interbedded with chert found in the Calaveras Complex, and probably are olistoliths derived from the Calaveras (Bateman

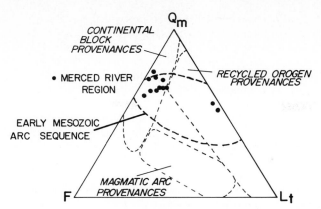

FIG. 30-6. Diagram showing framework clast composition of sandstones. Data from Merced River region of the Sullivan Creek terrane (Herzig, 1985) are compared to the field for sandstones of the early Mesozoic arc sequence from the Consumnes River area (Behrman, 1978) and general provenance types (italics) from Dickinson and Suczek (1979).

et al., 1985). If correct, this establishes a pre-170-m.y. depositional link between the Calaveras Complex and the Sullivan Creek terrane.

FOOTHILLS OPHIOLITE TERRANE

This terrane is the westernmost unit of the Foothills belt (Fig. 30-2). The oldest elements of this terrane consist of mafic and ultramafic rock with minor felsic differentiates, chert and limestone, comprising an ophiolitic assemblage. This assemblage is polygenetic and structurally complex. It yields a wide span of Paleozoic stratigraphic and metamorphic ages and is largely disrupted to form serpentinite-matrix melange. This Paleozoic basement is unconformably overlain by thick, submarine mafic volcanic piles of Triassic (?) to Early Jurassic age. Together with coeval mafic-ultramafic plutons and amphibolite, the volcanic rocks comprise a primitive, early Mesozoic arc assemblage.

Paleozoic Basement

Our knowledge of the Paleozoic basement is drawn primarily from studies of four localities. From south to north, these are the Kaweah River area (Saleeby, 1979; Saleeby and Sharp, 1980), the Kings River area (Saleeby, 1978; Saleeby and Sharp, 1980), the Tuolumne ultramafic complex (Morgan, 1976, 1978; Schweickert and Bogen, 1983), and the Bear Mountains area (Behrman, 1978; Duffield and Sharp, 1975). Saleeby (1982) summarizes Pb/U zircon data from all four areas and Shaw *et al.* (1984, 1987) present Sm/Nd isochron ages and Nd, Sr, and Pb isotopic data for the Kings River Area.

At the Kings River locality, the ophiolitic assemblage is a melange with slabs up to several kilometers in size separated by relatively narrow zones of schistose serpentinite. Two ensimatic magmatic systems are recognized within these slabs, which preserve primary igneous contacts in their interiors.

The older system is a cogenetic volcanoplutonic suite comprising an upward sequence of cumulate olivine- and clinopyroxene-rich peridotite, clinopyroxene gabbro, sheeted basaltic dikes, and pillowed, massive, or brecciated basaltic lava. Elements of this suite yield a Sm/Nd isochron age of 480 ± 25 m.y. Initial isotopic values of Sr, Pb, and

Nd reveal that the suite was derived from a source which was characterized by long-term depletion of Rb/Sr, U/Pb, and Nd/Sm relative to the bulk earth. Such values are characteristic of rocks formed by partial melting of the upper mantle. Dunite and harzburgite tectonite, inferred to be relict mantle depleted by partial melting, are spatially associated with the suite and presumably formed its basement. These features are consistent with formation of the suite at a spreading center such as those at modern midocean ridges or in arc-related extensional basins.

Within the 480-m.y. slabs at the Kings River occur cross-cutting mafic rocks which yield a Sm/Nd isochron age of 285 ± 45 m.y. Here again, Sr, Pb, and Nd isotopic values are consistent with upper mantle derivation. However, the source for the younger suite is distinct from that of the older one, and the isotopic values suggest arc affinity magmatism (Shaw *et al.*, 1984).

Other age data for the ophiolitic basement are provided by Pb/U analyses of zircon from metaplagiogranite. These felsic rocks occur as monolithologic melange blocks, as screens within slabs of mafic dike complex, and as concordant layers within mafic-ultamafic slabs. Plagiogranites from the Kaweah River and Bear Mountains areas yield inferred igneous ages of 270–310 m.y., apparently corresponding to the younger Kings River suite. Some samples from the Bear Mountains exhibit pronounced discordance and yield Pb/Pb ages of 350–370 m.y. These ages indicate that an inherited component of radiogenic Pb is present in the zircon, or that crystallization ages for these zircon exceed 350 m.y. The second interpretation is supported by an Ar/Ar plateau age of 368 ± 2 m.y. for hornblende.[6] obtained for a majic amphibolite melange block from the Bear Mountains (Fig. 30-2a).

In summary, two generations of Paleozoic magmatism are recognized in the ophiolitic assemblage on the basis of structural relations, isotopic compositions, and age data of the Foothills ophiolite terrane. The oldest of these is Ordovician oceanic lithosphere considered to have formed at a spreading center. This lithosphere subsequently was intruded by Permo-Pennsylvanian magmas of arc affinity. In addition, Late Devonian amphibolite and perhaps plagiogranite occurs within the ophiolitic melange, although their relation to other components of the melange is uncertain at present.

Origin

Previous concepts of the origin of Paleozoic basement of the Foothills ophiolite terrane were based on the inference that it represented far-traveled oceanic lithosphere. Such hypotheses include those of Schweickert and Cowan (1975), Davis *et al.* (1978), and Saleeby (1981 and earlier papers). In the Schweickert and Cowan hypothesis, the Foothills ophiolite terrane evolved separately from North America until its accretion in a Late Jurassic arc-continent collision. Davis *et al.* and Saleeby envision juxtaposition of far-traveled oceanic lithosphere, corresponding to the Foothills ophiolite terrane of this paper, with the tectonically truncated western margin of North America in Triassic time. In this view, initial emplacement of the ophiolite and the truncation of the westward projection of Paleozoic, northeast-southwest depositional and tectonic trends present in Nevada (Fig. 30-1) took place along a north-striking transform fault system lying within the current Sierra Nevada and Klamath Mountains. These hypotheses were developed prior to the application of the Sm/Nd and Ar/Ar techniques to dating of mafic igneous and metamorphic components of the ophiolitic basement, and must be reevaluated in light of new data which indicate previously unsuspected similarities between the Paleozoic basement of the Foothills ophiolite and rocks of the eastern Klamath province.

[6]C. Leighton and W. D. Sharp, unpublished data.

The eastern Klamath province consists of a Paleozoic ensimatic oceanic arc built on the Trinity ultramafic sheet (Burchfiel and Davis, 1981). This arc and related materials may also underlie the northern Sierra east of the Melones fault, as well as much of northern and western Nevada (Burchfiel and Davis, 1972; Davis et al., 1978; Speed, 1979). Ordovician, Devonian, and Permian faunal assemblages of the eastern Klamath Mountains have North American affinities (Boucot and Potter, 1977) and paleomagnetic data for Devonian rocks of the eastern Klamaths (Scott et al., 1985) yield North American expected directions, suggesting proximity of the eastern Klamath province to the continent during much of its development.

Petrogenesis of the Trinity ultramafic sheet has been dated as Ordovician based on a Sm/Nd isochron age of 472 ± 32 m.y. for plagioclase lherzolite (Jacobsen et al., 1984) and Pb/U zircon ages of 455–480 m.y. on related gabbro and diorite (Mattinson and Hopson, 1972). The initial Nd isotopic composition, $\epsilon_{Nd}(T)$, of the Trinity lherzolite is +10.4 ± 0.5. These values compare favorably with the Sm/Nd isochron age of 480 ± 25 m.y. and $\epsilon_{Nd}(T)$ of 10.7 ± 0.5 obtained for the Foothills ophiolitic basement at the Kings River locality (Shaw et al., 1984, 1987). Thus, Early Ordovician oceanic lithosphere of distinctive isotopic composition forms basement for both the eastern Klamath province and the Foothills ophiolite terrane, suggesting that the two may be correlative. The Feather River ophiolite, long thought to be correlative with the Trinity (Davis, 1969), is also shown as part of this basement terrane on Fig. 30-2a.

Since the eastern Klamath province lies *east* of the various loci proposed for the line of continental truncation and apparently developed along the North American margin, the correlation suggests a similar origin for the Foothills ophiolite terrane. The occurrence of correlative oceanic lithosphere east and west of the early Mesozoic accretionary complexes (Calaveras Complex and its Klamath Mountain equivalent, the blueschist-bearing Stuart Fork Formation) is anomalous with respect to existing concepts for the origin of the Foothills ophiolite. Assuming that the correlation is meaningful, two explanations seem possible. The occurrence of Paleozoic ophiolitic basement *west* of the Calaveras Complex may be the result of westward-directed thrusting. Such thrusting would have carried Ordovician oceanic lithosphere, over the early Mesozoic accretionary complex. Such thrusting is analogous to displacement on the Jurassic Siskyou thrust of the Klamath Mountains which places eastern Klamath basement and Central Metamorphic Belt rocks, now exposed as a klippe, over and west of the Stuart Fork Formation (Ando et al., 1983; Davis et al., 1978). In the Sierra, this thrust may have rooted beneath the west side of the Feather River ophiolite. In this view, COCORP data that show moderately eastward-dipping, midcrustal reflectors (Nelson et al., 1986) would be the down-dip continuation of the proposed thrust zone. Alternatively, the distribution of Ordovician ophiolite could reflect a paleogeographic setting in which an expanse of Ordovician oceanic lithosphere served as a common basement for one or more Mesozoic oceanic arc systems whose accreted remnants include the early Mesozoic arc sequences and the Calaveras Complex (J. B. Saleeby, personal communication, 1986).

Pre-Jurassic Tectonism

The Paleozoic basement was disrupted to form serpentinite-matrix melange, uplifted and deeply eroded prior to superposition of an early Mesozoic arc sequence described below. Such processes are evident at the Tuolumne ultramafic complex (Fig. 30-1), where in the hanging wall of a thrust fault, Paleozoic (Ordovician?) dunite and harzburgite tectonite are unconformably overlain by Lower Jurassic arc volcanic rocks (Morgan, 1976; Fig. 30-7). The tectonite probably represents oceanic mantle, implying that the entire expected thickness (6–10 km) of oceanic crust was removed prior to deposition of the Lower Jurassic

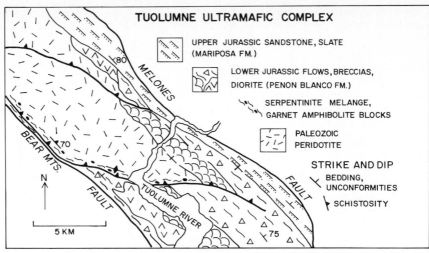

FIG. 30-7.   Map of Tuolumne ultramafic complex (labeled TU in Fig. 30-1) and unconformably overlying strata. Note serpentinite-matrix melange zone along the basal thrust of the peridotite, and steep, uniform dips in Jurassic strata. Modified after Morgan (1976), Schweickert and Bogen (1983), and W. D. Sharp (unpublished field mapping).

section. Garnet amphibolite, derived from chert and basalt, occurs as blocks within serpentinite-matrix melange along the thrust. Experimentally calibrated geobarometers (Bohlen and Liotta, 1986) applied to these rocks yield minimum pressures of 6 kb at inferred temperatures of $500°C$ (Sharp and Leighton, 1987). Thus, supracrustal rocks have been metamorphosed at ≥24 km and returned to the surface. Amphiboles from these blocks yield K/Ar and Ar/Ar ages of 200–178 m.y. (Morgan, 1976; Sharp and Leighton, 1987). Geologic relations and the oldest isotopic ages indicate that high-pressure metamorphism and uplift to shallow depths occurred prior to 200 m.y. Additional uplift took place in two stages, with the first at about 178 m.y. (Middle Jurassic). Final uplift and tilting occurred after deposition of the Upper Jurassic Mariposa Formation. The blocks may reflect thrusting of a thick (≥20 km) plate of oceanic lithosphere over supracrustal rocks during or prior to the Early Jurassic (Schweickert and Bogen, 1983).

In the Kaweah segment of the belt, lower Mesozoic pillow lavas were erupted across a substrate of serpentinite melange and ophiolitic debris (Saleeby, 1982). Overthrusting and internal disruption of the Foothills ophiolite at this time may have been related to its emplacement along the continental margin by transform faulting or, more likely, oblique convergence (Saleeby, 1981). This is consistent with the first appearance of continental detritus in strata unconformably overlying the disrupted ophiolitic basement, as discussed in the following section.

Early Mesozoic Arc Sequence

The composite Paleozoic lithosphere described above forms basement for thick mafic volcanic piles, related intrusions, and heterogeneous sediments constituting an Upper Triassic to lower Middle Jurassic arc sequence (Fig. 30-8). Stratigraphic relations within this sequence are complex and laterally variable, reflecting superposition of composite volcanic centers on a background of pelagic and epiclastic sedimentation. A well-preserved section through a submarine volcano, the Peñon Blanco Formation of Bogen (1985), occurs in

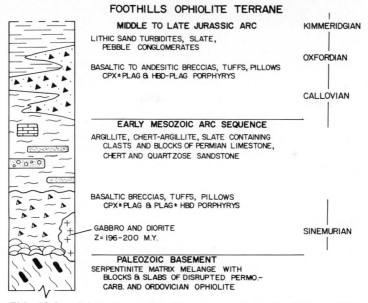

FOOTHILLS OPHIOLITE TERRANE

MIDDLE TO LATE JURASSIC ARC   KIMMERIDGIAN

LITHIC SAND TURBIDITES, SLATE,
   PEBBLE CONGLOMERATES

                                                          OXFORDIAN

BASALTIC TO ANDESITIC BRECCIAS, TUFFS, PILLOWS
   CPX±PLAG & HBD-PLAG PORPHYRYS

                                                          CALLOVIAN

EARLY MESOZOIC ARC SEQUENCE

ARGILLITE, CHERT-ARGILLITE, SLATE CONTAINING
   CLASTS AND BLOCKS OF PERMIAN LIMESTONE,
   CHERT AND QUARTZOSE SANDSTONE

BASALTIC BRECCIAS, TUFFS, PILLOWS
   CPX±PLAG & PLAG±HBD PORPHYRYS

GABBRO AND DIORITE   SINEMURIAN
   Z=196-200 M.Y.

PALEOZOIC BASEMENT
SERPENTINITE MATRIX MELANGE WITH
   BLOCKS & SLABS OF DISRUPTED PERMO.-
   CARB. AND ORDOVICIAN OPHIOLITE

FIG. 30-8.   Schematic stratigraphic column of the Foothills ophiolite terrane and overlying Middle to Late Jurassic arc rocks. Z designates concordant Pb/U zircon ages from Saleeby (1982). Relations shown are generalized from the segment of the terrane between the Stanislaus and American Rivers where the section is most complete and are after Duffield and Sharp (1975), Behrman (1978), and Schweickert and Bogen (1983).

the Tuolumne Rivers area. The Peñon Blanco Formation (PBF) is locally up to 5 km thick, consisting of a near-vent sequence of subaqueously deposited tuff, tuff-breccia, and massive to bedded porphyry. Rocks of the PBF are characteristically phenocryst-rich basalts and basaltic-andesites with plagioclase, augite, and hornblende occurring alone or in combination. Wehrlite, gabbro, and diorite stocks, sills, and dikes that intrude the PBF and its basement are believed to be its feeders, and have been dated at 196–200 m.y. (concordant Pb/U zircon ages, Saleeby, 1982), making the PBF Early Jurassic in age.

Mafic amphibolites, with the assemblage green hornblende-plagioclase-sphene ± rutile ± epidote ± garnet, occur as widely dispersed blocks within serpentinite-matrix melange of the Paleozoic basement. These amphibolites yield metamorphic ages of 195–212 m.y. (K/Ar hornblende ages, Behrman, 1978; Saleeby and Sharp, 1980[7]. The amphibolites are plateau ages, C. Leighton and W. D. Sharp, unpublished data). The amphibolites are inferred to be fragments of Paleozoic basement that have been metamorphosed by the heat from Early Jurassic magmatism. The wide dispersal of the blocks and their structurally and mineralogically discordant contacts with the melange matrix indicate that significant melange mixing postdates the Early Jurassic magmatic event.

In the area between the Merced and Tuolumne Rivers, the early Mesozoic arc sequence is truncated by an unconformity that places the Upper Jurassic Mariposa Formation directly on top of the Peñon Blanco Formation (Morgan, 1976). However, north along strike, in the vicinity of the Stanislaus River, additional section is present below the unconformity. There, the PBF is overlain by a chaotic assemblage of heterogeneous sediments including pebbly mudstone, conglomerate, slate and sandstone con-

[7]Ar/Ar hornblende plateau ages, C. Leighton and W. D. Sharp, unpublished data.

taining outcrop-scale blocks of Permian limestone, chert, metachert, and sandstone (Schweickert and Bogen, 1983). Fusilinacean faunas from the limestones are of Tethyan type (Douglas, 1967). See Ross and Ross (1983) for a discussion of faunal provinciality. This unit extends northward to the Consumnes River area and was mapped as melange by Duffield and Sharp (1975), Parkinson (1976), and Behrman (1978). Relations described by Schweickert and Bogen (1983) indicate that the melange was formed primarily by sedimentary rather than tectonic mixing since it is in stratigraphic contact with coherent units above and below. The depositional age of the melange in the Consumnes River area is Middle Jurassic or older (pre-Callovian), as associated strata of Callovian to Kimmer-idgian age are not mixed into the melange. Similarly, in the Stanislaus River area, melange lies depositionally below Callovian mafic volcaniclastic rocks of the Logtown Ridge Formation, although locally the melange occupies higher stratigraphic positions, inter-tonguing with the Oxfordian-Kimmeridgian Mariposa Formation (Schweickert and Bogen, 1983).

The melange contains poorly sorted, bedded sandstones and sandstone clasts in pebbly mudstone consisting of well-rounded to angular quartz grains, and subangular plagioclase and lithic clasts, including polycrystalline quartz, chert, and silicic volcanic rock. Modal analysis (Behrman, 1978) of 18 samples yielded a compositional range of Q_{71-31}, F_{18-2}, L_{8-64} (Fig. 30-6). The textural and modal diversity of the sandstones is inferred to reflect a mixed provenance including a continental source region contributing rounded, monocrystalline quartz grains, and a local source supplying subangular quartz, plagioclase, chert, and fragments of mafic volcanic rock.

Prior to deposition of the Early to Middle Jurassic melange, the Foothills ophiolite terrane was apparently ensimatic. Mafic and ultramafic igneous and metaigneous rocks of primitive isotopic composition were dominant and significant volumes of silicic rock were lacking. Pre-Jurassic sedimentary rocks, including chert and limestone, lack identifiable terrigenous detritus and reflect marine conditions free of continental influence. Thus, deposition of the melange, containing a component of well-rounded quartz sand and silicic volcanic debris, reflects continental influence in the Foothills ophiolite terrane for the first time.

ASSEMBLY OF THE FOOTHILLS TERRANES

Provenance and structural relations discussed below indicate that the four basement terranes of the central Sierra Foothills were mutually juxtaposed and accreted to North America by late Middle Jurassic time.

Epiclastic Rocks of the Foothills Ophiolite and Sullivan Creek Terranes

Lower to Middle Jurassic epiclastic rocks of the Foothills ophiolite terrane and the Sullivan Creek terrane share many distinctive features and are thought to be correlative (Figs. 30-5, and 30-8). Both assemblages are predominantly argillite and chert-argillite but contain olistoliths of Paleozoic limestone and sandstones of mixed provenance, including quartzose sandstones of continental affinity (Fig. 30-6). Both assemblages depositionally overlie thick piles of Lower Jurassic, mafic volcaniclastic rock, and form basement for upper Middle Jurassic (Callovian) magmatic rocks.

Likely sources for the externally derived detritus of the Foothills ophiolite and

Sullivan Creek Terranes are found in other Sierran-Klamath basement terranes and autochthonous North American strata. As mentioned earlier, Lower Triassic limestone, olistoliths in the Sullivan Creek terrane are probably derived from the Calaveras Complex. The Calaveras Complex also contains large masses of limestone which are in part Permo-Pennsylvanian and could be the source of Permo-Carboniferous limestone blocks. The full age range and faunal provinciality of Calaveras Complex carbonates is unknown because they were largely recrystallized during emplacement of late Middle Jurassic plutons. It should be noted that this recrystallization postdates deposition of the limestones in the Foothills ophiolite and Sullivan Creek terranes and so does not contradict the proposed relationship. Furthermore, the Hayfork and North Fork terranes of the Klamath Mountains, probably correlative with the Calaveras Complex, contain Permian, Tethyan limestones (Irwin et al., 1978; Ross and Ross, 1983, Fig. 30-4). As discussed below, Early to Middle Jurassic folding and thrust faulting affected these assemblages and it is inferred that this activity produced uplift and erosion at the surface, supplying detritus to the epiclastic rocks of the Foothills ophiolite and Sullivan Creek terranes.

Quartzose sandstone and silicic volcanic clasts in the Early to Middle Jurassic foothills terranes may have been derived from coeval autochthonous North American strata. Upper Triassic to Middle Jurassic sequences within the southern Sierra Nevada and Mojave regions record mixing of texturally mature quartzose sandstone, probably of the Dunlap-Aztec-Navajo lithosome, with abundant juvenile, intermediate to silicic volcanic debris in a shallow marine environment along the North American continental margin (Busby-Spera, 1984; Miller and Carr, 1978; Stanley et al., 1971). Thus, the continental margin was rich in the same materials that are prevalent in the foothills sequences, strongly suggesting that the two were part of a common depositional system. If this is correct, the early Mesozoic epiclastic strata indicate that the four basement terranes of the foothills metamorphic belt — the Foothills ophiolite, Sullivan Creek terrane, Calaveras Complex, and Shoo Fly Complex — were in mutual proximity and adjacent to North America by Early to Middle Jurassic time.

The plate configuration leading to the assembly of the foothills terranes is poorly constrained; however, the presence of high-pressure metamorphic blocks in fault zones that were initially shallowly dipping (e.g., garnet amphibolites of Fig. 30-7) indicates that a convergent component of motion was important. A second consideration is that the distribution of Early Jurassic magmatic arc rocks suggests that current map relations do not reflect the Early Jurassic paleogeography. Arc rocks of this age include those of the Foothills ophiolite and Sullivan Creek terranes as well as those of the continental borderland and parautochthonous North America (Fig. 30-2). Early Jurassic magmatic rocks are not known in the intervening Calaveras Complex, however, despite extensive field and geochronological study. Thus, Early to early Middle Jurassic arc assemblages occur both outboard and inboard of the Calaveras Complex that represents a synchronous or slightly older subduction zone. The absence of Early Jurassic magmatic rocks in the Calaveras indicates that magmatism did not migrate across the region of Fig. 30-2a at this time. Given the presence of continental-affinity detritus in the outboard Early to early Middle Jurassic arc sequences, as discussed above, it is likely that they formed in the same continental arc as the coeval inboard arc sequences, and owe their current position to faulting along the continental margin. If the distinctive Paleozoic basement of the outboard Early Jurassic arc sequence (e.g., Paleozoic basement of the Foothills ophiolite and Sullivan Creek terranes, Fig. 30-2) is indeed correlative to the Trinity ophiolite of the eastern Klamaths (see the discussion of origin of Foothills ophiolite above), then the Early Jurassic position of the Foothills ophiolite and Sullivan Creek terranes may have been inboard and north of their current positions.

Middle Jurassic Thrusting

The relationships just discussed strongly suggest that proximity of the foothills terranes to the North American margin extends at least as far back as Early Jurassic time. Direct contact between the Sullivan Creek terrane and the Calaveras Complex, and between the Calaveras Complex and the already accreted Shoo Fly Complex was established somewhat later. The Shoo Fly Complex, Calaveras Complex, and Sullivan Creek terrane are juxtaposed along westward-vergent thrust faults. Displacement on these thrusts is Middle Jurassic or older.

In the central Sierra, the Shoo Fly and Calaveras Complexes are juxtaposed along the Calaveras–Shoo Fly thrust (Schweickert, 1981; Fig. 30-2a). The thrust is intruded by the 166 ± 2-m.y.-old Standard pluton (concordant Pb/U zircon ages, Sharp, 1984; Fig. 30-2b). Pre-Standard, eastward-plunging, upright folds and associated schistosity and lineation deform and overprint the thrust. These structures are dated by amphibole from synkinematic albite-epidote amphibolite that yields an age of 176 ± 3 m.y.[8] Rocks as young as middle Early Triassic are present in the lower plate of the thrust (Bateman *et al.*, 1985) and metamorphism of the upper plate, dated at 235 m.y. (see above), predates thrusting. Thus, juxtaposition along the Shoo Fly–Calaveras thrust is bracketed as pre-176 m.y. and post-middle Early Triassic.

Along its western margins, between latitudes 37°30' and 38°15'N, the Calaveras Complex is in structural contact with the Sullivan Creek terrane (Schweickert *et al.*, 1984; Sharp, 1984, 1985, Fig. 30-2a). The intervening fault is termed the Sonora thrust. The age of displacement on the Sonora thrust is controversial. Schweickert *et al.* (1984) indicate that displacement is Late Jurassic. This conclusion is based on the observation that in the region west of the Standard pluton, a swarm of mafic dikes dated at 157–159 m.y. (Sharp, 1980) are locally truncated at the Calaveras–Sullivan Creek contact. In this area, the fault strikes northwest, paralleling Late Jurassic cleavage. However, to the south, in the vicinity of Big Oak Flat, the fault strikes nearly due east (see Fig. 30-9). In this area, Late Jurassic cleavage is poorly developed and crosses the fault at high angles. Another *earlier* schistosity may also be traced across the fault (Sharp, 1984, 1985). This earlier schistosity and a related lineation formed prior to emplacement of the Standard pluton as shown by overgrowth of these fabrics by garnet in the contact aureole of the pluton. These relations indicate that displacement along the eastward-striking segment of the Calaveras–Sullivan Creek contact *predates* emplacement of the Standard pluton and hence is >166 m.y. in age. The disparate structural relations and ages of displacement along the two segments of the fault imply a two-stage history for motion along the Calaveras–Sullivan Creek contact. Initial juxtaposition of the Calaveras Complex with the Sullivan Creek terrane took place in the Middle Jurassic, based on the relations described above and the presence of mafic volcanic rocks in the footwall dated at 187 ± 10 m.y. (Bateman *et al.*, 1985). Subsequently, this boundary was modified along much of its length by Late Jurassic faulting. This later stage corresponds to the Sonora thrust as envisioned by Schweickert *et al.* (1984).

In the northern Sierra, west-vergent thrusts juxtapose, in descending structural order, the Shoo Fly Complex, the Feather River ophiolite, the Calaveras Complex and related rocks and the Foothills ophiolite terrane[9] (Edelman, 1986). These west-vergent faults are truncated by the younger Canyon Creek thrust, which is in turn intruded by the 160 ± 5-m.y.-old Scales pluton (K/Ar hornblende age, Hietanen, 1981). Felsic metavolcanic rock of the Foothills ophiolite terrane in the footwall of one of the west-vergent

[8] Ar/Ar plateau on hornblende, W. D. Sharp, unpublished data.

[9] Also S. H. Edelman, written communication, 1985.

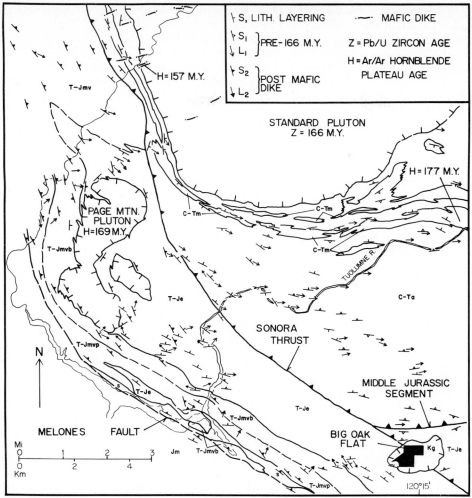

FIG. 30-9. Detailed map of the contact relations between the Calaveras Complex and Sullivan Creek terrane in the area of the Tuolumne River. Units are as follows: C-Ta, Calaveras Complex argillite; C-Tm, Calaveras Complex marble-metabasite slabs; T-Je, Sullivan Creek epiclastic rocks; T-Jmvb, Sullivan Creek metavolcanic breccias and tuffs; T-Jmvp, Sullivan Creek pillowed lava flows; Jm, Mariposa Formation. Hornblende age (H) near eastern edge of map is from amphibolite with synkinematic S1 fabric (age data from Sharp, 1984, 1985). Lithologic contacts after Eric *et al.* (1965), Morgan (1976), Hart (1959), and Sharp (1984). Structural age data from Sharp (1980, 1984, 1985). Compare with Fig. 30-2 of Schweickert and Bogen (1983).

thrusts has yielded an age of 203 m.y.[10], indicating that west-vergent thrusting here is early to Middle Jurassic.

The Foothills ophiolite has also been internally imbricated on several thrusts that repeat the basal contact of the Lower Jurassic Peñon Blanco Formation (Morgan, 1976; Schweickert and Bogen, 1983; e.g., Fig. 30-7). Faults of this group cut diorite dated at

[10] Pb/U zircon age, J. B. Saleeby, personal communication, 1986.

196 to 200 m.y. (concordant Pb/U zircon ages, Saleeby, 1982), but are overlapped by the Upper Jurassic Mariposa Formation, indicating Early to Middle Jurassic displacement. Garnet amphibolite blocks, brought up along the structurally lowest of these faults, yield ages of 178 ± 3 m.y. (Ar/Ar plateau ages on hornblende, Sharp and C. Leighton, 1986). These ages date cooling of the amphiboles due to upward thrasport along the thrust.

These relations, along with depositional links between the Foothills ophiolite and Sullivan Creek terranes described above, indicate that all pre-Jurassic basement terranes of the metamorphic belt were mutually juxtaposed and some were internally imbricated prior to or during the Middle Jurassic. Since the Shoo Fly Complex was already accreted at this time, the Foothills basement terranes are considered to be part of the North American plate as of late Middle Jurassic time (see Fig. 30-2b). This constraint on the timing of accretion is significant in two respects. First, accretion is older than previously believed. Pre-Late Jurassic assembly and internal deformation of the foothills basement terranes is inconsistent with models which relate Sierran accretion and deformation solely to Late Jurassic collision of an exotic terrane with the North American margin (e.g. Ingersoll and Schweickert, 1986; Moores and Day, 1984; Schweickert, 1981; Schweickert and Cowan, 1975; Schweickert et al., 1984). Second, along with Middle Jurassic thrusting delineated in the Klamath Mountains (Davis et al., 1978; Wright, 1982; Wright and Fahan, in press) and Nevada (E. L. Miller et al., 1984; Speed, 1978), the Sierran data reflect a major, pre-Late Jurassic crustal shortening event in the western Cordillera.

The amount of east-west shortening effected by west-vergent, Middle Jurassic thrusting in the Sierran-Klamath region is not fully known but is potentially quite large as shown by the following examples. Surface mapping of the Middle Jurassic Siskiyou thrust of the Klamaths reveals that the upper plate, consisting of the Trinity Ophiolite and amphibolite of the Central Metamorphic Belt, have been carried westward at least 40 km over the early Mesozoic accretionary complex (Stuart Fork Formation, Davis et al., 1965). Modelling of seismic refraction, gravity, and aeromagnetic data indicates that the Siskiyou thrust continues eastward at the base of an 8–9-km-thick plate of Trinity ophiolite for an additional 60 km (Zucca et al., 1986). Thus, a total of 100 km of westward displacement is indicated for the Siskiyou thrust. In the northern Sierra, I suggest that the Early to Middle Jurassic thrusts that juxtapose the Shoo Fly Complex, Feather River Ophiolite, and Calaveras Complex (Edelman, 1985) may be traced some 40 km eastward in the subsurface as moderately dipping (35°) seismic reflectors reported by Nelson et al. (1986). Note that this interpretation differs from that of Nelson et al., who considers the reflectors to be Late Jurassic (Nevadan) thrusts. The latter interpretation is difficult to reconcile with the steep dip of exposed Late Jurassic faults in the area (Day et al., 1985) and the demonstrably pre-Late Jurassic age of the Shoo Fly-Calaveras thrust in the central Sierra. I suggest that the Late Jurassic displacement at the projected surface trace of the reflector (e.g., the Melones fault zone) represents reactivation of earlier thrusts, as discussed by Day et al. (1985).

Significant westward transport of upper plate rocks of the Shoo Fly–Calaveras thrust of the central Sierra is also indicated. Deformed Paleozoic granitoids of the Shoo Fly Complex (e.g., the Confluence pluton) have major and trace element abundances, and Pb and Sr isotopic compositions indicative of an origin by melting of continental crustal materials[11] (Sharp, 1984), while adjacent Middle Jurassic and younger granitoids lack such signatures (Kistler and Peterman, 1973, 1978). The strong continental affinities of the Paleozoic granitoids indicate that they formed due to melting within continental crust of significant ($\geqslant 35$ km) thickness, and that therefore at the time of their petrogenesis the Shoo Fly Complex was underlain by such crust. I infer that west-directed

[11] Also W. D. Sharp, unpublished data.

thrusting detached the Shoo Fly Complex from its continental crustal roots and placed it in its current position over oceanic rocks of the Calaveras Complex. Subsequent to thrusting, Middle Jurassic and younger magmas emplaced into the Shoo Fly Complex ascended primarily through ensimatic rocks, below the thrust. Thus, the Standard pluton, which postdates thrusting, pierces the Shoo Fly Complex only at high levels, explaining its relatively primitive chemical and isotopic composition.

The relations discussed above indicate that the foothills metamorphic belt consists of a stack of rootless nappes assembled by west-directed Early to Middle Jurassic thrusting. Although these disparate basement terranes may have originated on different lithospheric plates, by Middle Jurassic time they were assembled into a single, composite plate which formed basement for a Middle to Late Jurassic magmatic arc.

Broadly synchronous folding and major thrust faulting occurred over large areas of the western Cordillera. The Klamath Mountains are a stacked sequence of thin, subhorizontal thrust plates separated by east-rooting faults (Davis, 1968; Irwin, 1972), several of which are Middle Jurassic (Davis *et al.*, 1978; Wright and Fahan, in press). In central and western Nevada, late Early and Middle Jurassic folding and thrusting caused destruction of a marine depositional province in which strata had accumulated since Early Triassic (Speed, 1978). Vergence there was generally to the east or south (Oldow, 1978, 1981; Speed, 1978). In the Canadian Cordillera, Early to Middle Jurassic thrusting and related crustal thickening are well known (Archibald *et al.*, 1983; Monger, 1977; Monger, *et al.*, 1982; R. A. Price, 1981; Read and Wheeler, 1976). Interestingly, vergence there is also mixed. Initial, east-directed thrusting was followed by formation of west-verging folds and thrusts in the western part of the belt. The transition between east- and west-vergent structures (e.g., Selkirk fan) occurs in the general vicinity of the suture zone linking allochthonous and suspect terranes to rocks that were deposited on or adjacent to North America Price, 1986). This region also corresponds to the locus of Middle Jurassic ductile deformation, high-pressure metamorphism, and crustal anatexis (Monger, 1982). Early to Middle Jurassic, west-vergent thrusting in the Sierra Nevada foothills and Klamath Mountains and the broadly coeval continentward thrusting in Nevada are also complementary aspects of a single, major crustal shortening event. This geometry, thrusts verging outward from the central locus of arc magmatism, is similar to the "two sided nature of the Cordilleran orogen" envisioned by Burchfiel and Davis (1972).

East-Directed Thrusting

Recently, attention has been drawn to the role of *east-vergent* thrust faults in Sierran structural evolution (Day *et al.*, 1985; Moores and Day, 1984). East-vergent thrusts include the Taylorsville thrust, which imbricates Paleozoic and Jurassic strata overlying the Shoo Fly Complex, and the thrusts beneath the Late Jurassic Smartville and Slate Creek ophiolitic complexes. In addition, deep seismic reflection data from a transect near the southern end of the metamorphic belt reveals well-defined, west-dipping reflectors in the middle to lower crust which may represent eastward-vergent faults (Zoback *et al.*, 1985). The basal thrust at Smartville affects rocks as young as 162 ± 2 m.y. (Smartville sheeted dikes) and is intruded by the 158 ± 1 m.y. Yuba Rivers pluton[12]. These dates closely bracket the age of thrusting at about 160 m.y. and limit the interval between genesis and emplacement of Smartville to a few millions of years, implying a quasi-local origin for it. Similar, though less precise, constraints apply to the Slate Creek ophiolitic plate. There, gabbro dated at 163 ± 8 m.y. and related ophiolitic rocks are emplaced on the east-directed Canyon Creek thrust (Edelman, 1985), which is intruded by the 160 ± 5-m.y.

[12] Pb/U zircon ages, J. B. Saleeby, unpublished data.

Scales pluton (K/Ar on hornblende, Hietanan, 1981). These data, and field relations described by Edelman (1985), lead to the conclusion that eastward thrusting of Smartville and Slate Creek occurred *after* juxtaposition of the foothills terranes along west-vergent thrusts.

MIDDLE TO LATE JURASSIC ARC

Middle to Late Jurassic intrusions and supracrustal sequences are voluminous components of the Sierra Nevada foothills belt (Fig. 30-2a). They have long been considered products of convergent margin magmatism (Bateman and Clark, 1974; Hamilton, 1969; Schweickert and Cowan, 1975). However, the plate configuration in which they formed is controversial. Previously, they have been regarded as (1) a single arc superposed across already accreted basement terranes (Davis *et al.*, 1978; Saleeby, 1981; Sharp, 1985), (2) a single arc repeated by Late Jurassic strike-slip faulting (Kistler, 1978), and (3) two oppositely facing arcs that collided in the Late Jurassic (Ingersoll and Schweickert, 1986; Schweickert, 1981; Schweickert and Cowan, 1975).

Based on relations discussed above, structural and depositional links predating the Middle to Late Jurassic arc rocks are recognized among the terranes that form their basement. Thus, these rocks are considered herein to have formed in a single arc superposed on basement terranes that had previously been accreted to the western margin of North America.

Intrusions

Intrusions, chiefly quartz diorite to granodiorite dated at 177–157 m.y., occur within each of the Sierran basement terranes (Fig. 30-2a, Evernden and Kistler, 1970; Herzig *et al.*, 1985; Saleeby, 1982; Sharp, 1984; Stern *et al.*, 1981; Wright and Sharp, 1982). These include, but are not restricted to, distinctive zoned plutons with spatially and temporarily related mafic dike swarms. Such plutons, for example, the Standard pluton, consist of (1) olivine- and clinopyroxene-rich ultramafic rocks with relict cumulate textures; (2) two-pyroxene bearing gabbro, diorite, and monzodiorite; and (3) hornblende ± pyroxene diorite and tonalite. This lithologic association is characteristic of Middle Jurassic intrusive complexes distributed for 400 km along strike throughout the Sierran-Klamath region (Snoke *et al.*, 1982; Wright and Fagin, in press; Wright and Sharp, 1982). These intrusions form the northern part of a belt of Middle Jurassic intrusions that extends southeastward well into the craton (Kistler, 1978).

In the western Klamaths, these complexes intrude thick, coeval, volcaniclastic sequences. Along with the linear, beltlike distribution of the complexes, this leads to the interpretation that they are the roots of a Middle to Late Jurassic magmatic arc (Harper and Wright, 1984; Sharp, 1984; Snoke *et al.*, 1982). Direct association between the plutons and volcaniclastics is not generally observed in the Sierra; however, many of the Sierran complexes are deeply eroded, as evidenced by development of high-temperature inner contact aureoles. For example, as the Standard pluton (Fig. 30-2a) is approached, regional fabrics in its wallrocks of Calaveras Complex chert-argillite are progressively rotated into parallelism with the plutonic contact and overprinted by high-temperature flow fabrics (Sharp, 1984). An inner aureole of pyroxene hornfels facies, 200–400 wide, characterized by assemblages bearing sillimanite and potassium feldspar is present along the southern and southwestern perimeter of the pluton. Quartz-mica schist-bearing Mn-rich garnet occurs up to 4 km from the pluton (Kerrick, 1970). These features suggest that the current surface lay at perhaps 10 km depth within the active arc, and suggest that direct extrusive equivalents of the Standard and other Sierran intrusions have been eroded away. Voluminous Middle to Late Jurassic volcaniclastic sequences are preserved in adjacent, less deeply eroded structural blocks, as discussed below.

Among the zoned complexes, intermediate rocks, typically with 57-60% SiO_2, predominate. These rocks have major and trace element compositions that correspond to high-potassium calc-alkaline andesite of Gill (1981)[13] The presence of large volumes of compositionally evolved rock in an arc built on a largely ensimatic basement is noteworthy. The repeated association of characteristic rock types in the Sierran-Klamath complexes implies a genetic relationship among the constituent rock types. Field and geochronological data are consistent with this premise, indicating close contemporaneity among the various rock types at each complex (Sharp, 1984; Wright and Sharp, 1982). Fo- and Ni-rich olivine and Cr-rich pyroxene in the ultramafic rocks suggest that the parental liquids for the plutons could be mantle derived[14] (Snoke et al., 1981). Coeval, mafic dikes around the Standard pluton consist, in part, of near-primitive, high-Mg basalt and andesite and may represent parental liquids for the pluton (Sharp and Stern, 1985). Variable Sr isotopic compositions of different rock types of the Standard pluton indicate that they were not derived by fractionation of melts from a single homogeneous source, and raise the possibility that modest assimilation of older crust has occurred. Nevertheless, based on the available data, it is likely that the complexes are largely mantle derived. This suggests that these complexes are examples of the generation of significant volumes of chemically evolved rock (e.g., high-K andesite) in an ensimatic crustal environment. As such, they may be analogous to some granitoids in Archean greenstone belts that lack isotopic evidence of involvement of older crust in their petrogenesis (e.g., Moorbath and Taylor, 1981; Shirey and Hanson, 1984).

Supracrustal Sequences

The supracrustal sequences contain Callovian and Oxfordian-Kimmeridgian fossils (Clark, 1964) indicating an age range of 169-152 m.y. according to the time scale of Palmer (1983). The sequences are 3-5 km thick and consist chiefly of complexly intertongued volcaniclastic and epiclastic sediments (Behrman and Parkinson, 1978; Schweickert and Bogen, 1983). Bedding features indicate deposition in submarine fan and slope environments. Pillow lavas, massive flows, and porphyritic dikes and sills occur locally. Volcanic and volcaniclastic rocks are primarily basaltic to andesitic and are characteristically rich in phenocrysts of clinopyroxene, hornblende, and plagioclase.

Detailed study of mafic volcanic and volcaniclastic rocks along the Consumnes River (Beard and Shervias, 1986) reveals that flows and breccia clasts include ankaramite, bearing olivine, Cr-diopside, and Cr-spinel phenocrysts. Phenocryst compositions suggest that the ankaramite may have been derived from the upper mantle with little or no modification, and show a depletion of Ti-group elements typical of arc basalts. More evolved magmas are also present at the Consumnes section, including augite basalt and hornblende andesite tuff-breccias and flows.

Epiclastic rocks consist of slate, siltstone, graywacke, and conglomerate. Their framework clast compositions have been studied near the Consumnes River by Behrman and Parkinson (1978), and by Bogen (1984) near the Merced River. Sandstones are characterized by a dominance of lithic clasts over quartz and feldspar. They range from those with lithic clast populations of chert and metamorphic rock $\gg$ argillite = volcanic rock $>$ limestone, to those with volcanic rock $>$ chert and metamorphics $>$ argillite =

---

[13] Also W. D. Sharp, unpublished data.

[14] Ibid.

limestone. This variability reflects a mixed provenance, with the Calaveras and Shoo Fly Complexes being likely sources for chert and metamorphic rock. Volcanic rock fragments were primarily derived from the synchronous arc, although contributions from the early Mesozoic arc sequence are likely but difficult to detect.

Ophiolites

Extensive Middle to Late Jurassic ophiolites are also considered part of this assemblage. Among these are the Middle to Late Jurassic Smartville, Lake Combie, and Slate Creek ophiolitic complexes (Day *et al.*, 1985). Their cover is arc-derived detritus, indicating that they formed within an active magmatic arc (Menzies *et al.*, 1980; Xenophotos and Bond, 1978). Although they are in west-dipping thrust contact with older Sierran basement (Edelman, 1985; Moores and Day, 1984; Fig. 30-2a) the interval between formation and thrust emplacement of these ophiolites is only a few m.y. (see the discussion above of east-directed thrusting). This indicates that they formed in relative proximity to the previously accreted Sierran basement terranes.

Middle to Late Jurassic ophiolite also occurs in the eastern Coast Ranges (Hopson *et al.*, 1981), where it tectonically overlies the Franciscan Complex. In the Diablo Range, Late Jurassic volcaniclastic turbidite overlies the ophiolite (Evarts, 1977; Sharp and Evarts, 1982). The abundance of intermediate composition rock within the ophiolite there (Evarts, 1977), and some of its geochemical properties (Shervais and Kimbrough, 1983), along with its arc-derived cover, lead to the conclusion that this segment of the Coast Range ophiolite formed within a rifting magmatic arc. Between the Coast Ranges and the foothills lies the Great Valley with its sedimentary fill of mudstone and turbidite sandstone of Late Jurassic to Cretaceous age, the Great Valley Sequence (GVS).

The GVS unconformably overlies the Coast Range ophiolite (CRO) and its volcaniclastic cover on the west side of the Valley, and folded Middle to Late Jurassic volcaniclastic and epiclastic strata on the east side. Thus, the extent of the CRO under the Great Valley Sequence and its contact relations with the foothills units cannot be directly observed. However, extensive geophysical data suggest that mafic crust underlies much of the Great Valley (Cady, 1975; Colburn and Moody, 1986; Suppe, 1979; Wentworth and Zoback, 1985; Wentworth *et al*, 1987; Zoback and Wentworth, 1985). The structure of this mafic crust is poorly known but is apparently complex based on the presence of velocity reversals and several sets of dipping reflectors (Wentworth and Zoback, 1985). Based on magnetic, drilling, and gravity data, Cady (1975) concluded that the Great Valley basement consisted of tectonically thickened ophiolitic crust. Suppe (1979) suggested that this ophiolite was equivalent to that of the Coast Ranges and was emplaced by east-directed thrusting. Previous workers have regarded the Smartville and Coast Range ophiolites, as well as inferred ophiolitic crust beneath the Great Valley, as crust formed in intra-arc rift zones within the Middle to Late Jurassic arc of the foothills (Harper *et al.*, 1985); Saleeby, 1981; Schweickert and Cowan, 1975; Sharp, 1984).

Alternative models in which the Coast Range and Smartville ophiolites are considered to have formed independently from the Sierran arc have been proposed by Hopson *et al.* (1981) and Moores and Day (1984), respectively.

Syn-Arc Orogenesis

Several lines of evidence suggest that the Middle to Late Jurassic arc is synorogenic, and that significant orogenic activity including uplift and related sedimentation, penetrative deformation and metamorphism, and faulting occurred throughout Middle to Late Jurassic time.

Sedimentation Nearly continuous synorogenic sedimentation in the Sierra Nevada and adjacent regions spans Middle to Late Jurassic time. Important characteristics of these sequences are: they contain detritus from uplifted underlying or adjacent basement terranes; and rapid sedimentation and/or tectonic instability in the depositional basin are sometimes indicated by slumping and soft-sediment deformation. Such units include (1) the epiclastic component of the early Mesozoic arc sequence of the Foothills ophiolite and Sullivan Creek terranes, which is pre-Callovian to Oxfordian-Kimmeridgian in age (Behrman, 1978; Duffield and Sharp, 1975; Schweickert and Bogen, 1983); (2) the Mariposa Formation, of Callovian to Kimmeridgian age (Behrman and Parkinson, 1978; Bogen, 1984); (3) the Kimmeridgian to Neocomian Stony Creek and Platina Formations of the Great Valley Group (Bertucci and Ingersoll, 1983; Ingersoll, 1983); and (4) the Kimmeridgian to Tithonian ophiolitic breccias of the basal Great Valley and Coast Range ophiolite (Bailey *et al.*, 1970; Hopson *et al.*, 1981; Phipps, 1984).

Penetrative deformation Middle to Late Jurassic plutons have internal fabrics varying from pristine igneous to those indicating much internal strain (Day *et al.*, 1985; Saleeby, 1978; Sharp, 1984, 1985). Contact relations with wallrock fabrics also indicate syn-, post-, and pretectonic emplacement (Baird, 1962; Ingersoll and Schweickert, 1986; Schweickert, 1981; Schweickert *et al.*, 1984; Sharp, 1985). Significantly, no simple correlation between age and relation to deformation is recognized on the scale of the Sierran region, although local chronologies can be established. This implies that deformation was diachronous on a regional scale, was broadly contemporaneous with intrusion, and hence spans the Middle to Late Jurassic life of the arc.

Late Jurassic deformation includes folding and formation of cleavage in rocks as young as Kimmeridgian (156–152 m.y., Palmer, 1983) and displacement of these strata on high-angle faults. Folds involving the Oxfordian-Kimmeridgian Mariposa Formation are of large amplitude with subhorizontal, north-trending axes, and steeply dipping axial planes. Post folding magnetization was locally acquired in the Late Jurassic (Bogen *et al.*, 1985) and cleavage in Late Jurassic rocks is cut by plutons as old as 154 ± 4 m.y. (Schweickert *et al.*, 1984), closely bracketing the age of folding. Associated metamorphism was prehnite-pumpellyite to lower greenschist facies. Biotite from widespread localities yield K/Ar ages of 146–154 m.y., dating cooling below $280°C$[15] (Behrman, 1978).

Faulting Pre-late Middle Jurassic (>166 m.y.), west-vergent thrusts, and early-Late Jurassic (about 160 m.y.) east-vergent thrusts have already been discussed. Younger faulting includes high-angle faults of the Foothills fault system (Clark, 1960), such as the Bear Mountains and Melones faults (Fig. 30-2). These faults truncate Kimmeridgian strata of the Mariposa Formation and related rocks and so are post-155 m.y. They strike north to northwest and dip steeply to the east, and are often marked by heightened fabric development, tectonic breccia, or schistose serpentinite in zones hundreds of meters or more wide.

Large displacements on western strands of the Foothills fault system predate intrusion of the 128-m.y.-old Rocklin pluton (Swanson, 1977). A significant component of dip slip with east side up is indicated for the Bear Mountains and Melones fault zones, where they lie within the Foothills ophiolite terrane, based on the presence of older, more deeply seated rocks on their east sides. Dip slip is also indicated by strong, down-dip lineations developed within the fault zones (Clark, 1960; Schweickert *et al.*, 1984).

Two aspects of this latest stage of deformation are emphasized here. First, though the Late Jurassic to early Cretaceous faults divide the Foothills metamorphic belt into the

---

[15] Also W. D. Sharp, unpublished data.

structural blocks most apparent today, these faults are superposed on the nappes assembled by earlier, chiefly Middle Jurassic thrusting.

Latest Jurassic faulting in the Sierra foothills produced differential uplift of the Middle to Late Jurassic arc and its basement. West of the Bear Mountains fault zone, only supracrustal levels of the arc are exposed. In the block between the Bear Mountains and Melones faults, sections through the arc to its basement are exposed. East of the Melones fault, the plutonic roots of the arc are exposed and only local patches of the supracrustal sequence remain.

A second important point is that the metamorphic belt acquired its steeply dipping structure late in its history. That tilting is a late-stage feature is most clearly shown by relations at the latitude of the Tuolumne River in the structural block between the Bear Mountains and Melones faults. There Paleozoic ultramafic tectonite is unconformably overlain by the Lower Jurassic Peñon Blanco Formation, which is in turn unconformably overlain by the Mariposa Formation (Morgan, 1976; Fig. 30-7). Both unconformities and bedding in the Mariposa are parallel and vertically dipping, indicating that the entire sequence was tilted as a unit in post-Mariposa time. This pattern characterizes the structural block lying between the Bear Mountains and Melones faults for at least 100 km along strike, based on detailed mapping (Behrman, 1978; Bogen, 1985; Duffield and Sharp, 1975). Recognition of this late-stage tilting indicates that pre-Late Jurassic melange fabrics in this segment of the Foothills ophiolite terrane (e.g., the zone bearing garnet amphibolite blocks in Fig. 30-7) were initially shallowly dipping. Locally, these melange fabrics can be shown to have formed at temperatures of 500°C, based on syn-kinematic growth of forsterite (Morgan, 1978). Such temperatures indicate that these fabrics formed at depths of 10 km or more. These observations are difficult to reconcile with concepts of Sierran structural evolution that emphasize strike-slip tectonics (e.g., the fracture zone and transpression models of Saleeby 1977, 1981), since deeply seated structures in strike-slip systems would be expected to have *steep* primary dips. Rather, early stages of motion along the thrust at Tuolumne provide evidence of Early Jurassic shortening.

In summary, the accreted terranes of the Sierran region appear to have been proximal to one another and the North American margin by Early Jurassic time and were in direct contact by the late Middle Jurassic. Shortening across the Sierran region, and related aspects of orogenesis, span much of Middle to Late Jurassic time and are associated in time and space with arc magmatism. These conclusions contrast with the widely held view that orogenesis in the Sierran region is largely due to a brief, Late Jurassic event termed the Nevadan orogeny (compare Bateman and Clark, 1974). Also, the relations described are not consistent with models of Sierran structural evolution which relate Nevadan deformation to a Late Jurassic collision between terranes of the foothills (compare Ingersoll and Schweickert, 1986; Moores and Day, 1984; Schweickert and Cowan, 1975). Several alternative causes of Jurassic Sierran orogenesis may be envisioned. It may have resulted from the prolonged effects of Early Jurassic collision between the Foothills ophiolite terrane and North America. Collision of an unknown entity in a more outboard position (e.g., west of the Coast Range thrust) also seems possible. Such a colliding entity is no longer present in the Sierran region, but then neither is the subduction complex complementary to the Middle to Late Jurassic Sierran arc. Yet another possibility is suggested by the Tertiary Andean margin of South America, where extensive shortening in the arc and back-arc regions is unrelated to collision of allochthonous terranes (e.g., Burchfiel, 1980). There, good coupling across the convergent plate margin is apparently favored by "absolute" movement of the South American plate into the subduction zone and perhaps by a young and therefore relatively buoyant down-going plate (Uyeda, 1982). These effects, plus thermal weakening of the subarc lithosphere by magmatic heating, may be sufficient to cause crustal shortening in the overriding plate without collision.

ACKNOWLEDGMENTS

Conversations and written communications with P. Bateman, S. Edelman, R. Drake, W. G. Ernst, G. Harper, C. Herzig, R. Kistler, C. Leighton, C. Merguerian, B. Morgan, S. Patterson, J. Saleeby, R. A. Schweickert, R. Spangler, J. E. Wright, and W. H. Wright III contributed significantly to the data and ideas presented in this paper. Some of the geochronological data were collected in the laboratories of G. Curtis (U. C., Berkeley), J. Saleeby (Cal. Tech.), and G. Tilton (U.S., Santa Barbara). I am grateful for financial support from the Geological Society of America, Penrose Fund, and the State University of New York, Stony Brook.

REFERENCES

Ando, C., Irwin, W. P., Jones, D. L., and Saleeby, J. B., 1983, The ophiolitic North Fork terrane in the Salmon River region, Central Klamath Mountains, California: *Geol. Soc. America Bull.*, v. 94, p. 236–252.

Archibald, D. A., Glover, J. K., Price, R. A., Fanar, E., and Carmichael, D. M., 1983, Geochronology and tectonic implications of magmatism and metamorphism, southern Kootenay Arc and neighboring regions, southeastern British Columbia — Pt. 1. Jurassic to mid-Cretaceous: *Can. J. Earth Sci.*, v. 20, p. 1891–1913.

Bailey, E. H., Blake, M. C., and Jones, D. L., 1970, Onland Mesozoic oceanic crust in California Coast Ranges: *U.S. Geol. Survey Prof. Paper 700-C*, p. C70–C81.

Baird, A. K., 1962, Superposed deformation in the central Sierra Nevada Foothills east of the Mother Lode: *Calif. Univ. Publ. Geol. Sci.*, v. 42, p. 1–70.

Barker, F., Peterman, Z. E., and Hildreth, R. A., 1969, A rubidium-strontium study of the Twilight Gneiss, West Needle Mountains, Colorado: *Contrib. Mineralogy Petrology*, v. 23, p. 271–282.

Bateman, P. C., and Clark, L. D., 1974, Stratigraphic and structural setting of the Sierra Nevada batholith, California: *Pacific Geology*, v. 8, p. 79–89.

——, and Moore, J. G., 1965, Geologic map of the Mount Goddard quadrangle, Fresno and Inyo Counties, California: *U.S. Geol. Survey Geol. Quad. Map GQ-429*, scale 1:62,500.

——, Harris, A. G., Kistler, R. W., and Krauskopf, K. B., 1985, Calaveras reversed — Westward younging is indicated: *Geology*, v. 13, p. 338–341.

Beard, J., and Shervais, J., 1986, Volcanic stratigraphy and petrology of a Jurassic island arc — The Logtown Ridge Formation, California: *Trans. Amer. Geophys. Un.*, v. 67, p. 402.

Behrman, P. G., 1978, Pre-Callovian rocks west of the Melones Fault Zone, central Sierra Nevada foothills; *in* Howell, D. G., and McDougall, K. A., eds., *Mesozoic Paleogeography of the Western United States:* Pacific Section, Soc. Econ. Paleontologists Mineralogists, Pacific Coast Paleogeography Symp. 2, p. 337–348.

——, and Parkison, G. A., 1978, Paleogeographic significance of the Callovian to Kimmeridgian strata, central Sierra Nevada foothills, California; *in* Howell, D. G., and McDougall, K. A., eds., *Mesozoic Paleogeography of the Western United States:* Pacific Section, Soc. Econ. Paleontologists Mineralogists, Pacific Coast Paleogeography Symp. 2, p. 349–360.

Bogen, N. L., 1984, Stratigraphy and sedimentary petrology of the Upper Jurassic Mariposa Formation, western Sierra Nevada, California, *in* Crouch, J. K., and Bachman, S. B., eds., *Tectonics and Sedimentation along the California Margin:* Pacific Section, Soc. Econ. Paleontologists Mineralogists, Los Angeles, v. 38, p. 119–134.

_____, 1985, Stratigraphic and sedimentologic evidence of a submarine island-arc volcano in the lower Mesozoic Penon Blanco and Jasper Point Formations, Mariposa Co., California: _Geol. Soc. America Bull._, v. 96, p. 1322–1331.

_____, Kent, D. V., and Schweickert, R. A., 1985, Paleomagnetic constraints on the structural development of the Melones and Sonora faults, central Sierran foothills, California: _J. Geophys. Res._, v. 90, p. 4627–4638.

Bohlen, S. R., and Liotta, J. J., 1986, A barometer for garnet amphibolites and garnet granulites: _J. Petrology_, v. 27, p. 1025–1034.

Bohlke, J. K., and McKee, E. H., 1984, K-Ar ages relating to metamorphism, plutonism, and gold-quartz vein mineralization near Allegheny, Sierra County, California: _Isochron/West_, no. 39, p. 3–7.

Boucot, A. J., and Potter, A. W., 1977, Middle Devonian orogeny and biogeographical relations in areas along the North American Pacific Rim: _Univ. Calif. Riverside Mus. Contrib. 4_, p. 210–219.

Breuckner, H. K., and Snyder, W. S., 1985, Structure of the Havallah sequence, Golconda allochton — Evidence for prolonged evolution of an accretionary prism: _Geol. Soc. America Bull._, v. 96, p. 1113–1130.

Brook, C. A., 1977, Stratigraphy and structure of the Saddlebag Lake roof pendant, Sierra Nevada, California: _Geol. Soc. America Bull._, v. 88, p. 321–331.

Burchfiel, B. C., 1980, Tectonics of Non-collisional regimes, _in_ Burchfiel, B. C., Oliver, J. E., and Silver, L. T., eds., _Continental Tectonics:_ Washington, D.C., _Nat. Acad. Sci._

_____, and Davis, G. A., 1972, Structural framework and evolution of the southern part of the Cordilleran orogen, Western United States: _Amer. J. Sci._, v. 272, p. 97–118.

_____, and Davis, G. A., 1975, Nature and controls of Cordilleran orogenesis, western United States — Extensions of an earlier synthesis: _Amer. J. Sci._, v. 275-A, p. 363–396.

_____, and Davis, G. A., 1981, Triassic and Jurassic tectonic evolution of the Klamath Mountains–Sierra Nevada geologic terrane; _in_ Ernst, W. G., ed., _The Geotectonic Development of California_ (Rubey Vol. I): Englewood Cliffs, N.J., Prentice-Hall, p. 50–70.

Busby-Spera, C. J., 1984, Large-volume ash flow eruptions and submarine caldera collapse in the Lower Mesozoic Sierra Nevada, California: _J. Geophys. Res._, v. 89, p. 8417–8427.

_____, 1985, A sand-rich submarine fan in the Lower Mesozoic Mineral King caldera complex, Sierra Nevada California: _J. Sediment. Petrology._ v. 55, p. 376–391.

Cady, J. W., 1975, Magnetic and gravity anomalies in the Great Valley and western Sierra Nevada metamorphic belt, California: _Geol. Soc. America Spec. Paper 168_, 56 p.

Carmichael, I. S. E., Turner, F. J., and Verhoogen, J., 1974, _Igneous Petrology:_ New York, McGraw-Hill, 739 p.

Chen, J. H., and Tilton, G. T., 1978, Lead and strontium isotopic studies of the southern Sierra Nevada batholith, California: _Geol. Soc. America Abstr. with Programs_, v. 10, p. 99–100.

Christensen, M. N., 1963, Structure of metamorphic rocks at Mineral King, California: _Univ. Calif. Publ. Geol. Sci._, v. 42, p. 159–198.

Clark, L. D., 1960, Foothills fault system, western Sierra Nevada, California: _Geol. Soc. America Bull._, v. 71, p. 483–496.

_____, 1964, Stratigraphy and structure of part of the western Sierra Nevada metamorphic belt, California: _U.S. Geol. Survey Prof. Paper 410_, 70 p.

Colburn, R. H., and Mooney, W. D., 1986, Two dimensional velocity structure along the synclinal axis of the Great Valley, California: *Seismol. Soc. America Bull.*, v. 76, p. 1305–1322.

Coney, P. J., Jones, D. L., and Monger, J. W. H., 1980, Cordilleran suspect terranes: *Nature*, v. 288, p. 329–333.

Danner, W. R., 1977, Paleozoic rocks of northwestern Washington and adjacent parts of British Columbia; *in* Stewart, J. H., *et al.*, eds., *Paleozoic Paleogeography of the Western United States:* Pacific Section, Soc. Econ. Paleontologists Mineralogists, Pacific Coast Paleogeography Symp. 1, p. 395–408.

Davis, G. A., 1968, Westward thrusting in the south-central Klamath Mountains, California: *Geol. Soc. America Bull.*, v. 79, p. 911–933.

——, 1969, Tectonic correlations, Klamath Mountains and western Sierra Nevada, California: *Geol. Soc. America Bull.*, v. 80, p. 1095–1108.

——, *et al.*, 1965, Structure, metamorphism, and plutonism in the south-central Klamath Mountains, California: *Geol. Soc. America Bull.*, v. 76, p. 933–966.

——, Monger, J. W. H., and Burchfiel, B. C., 1978, Mesozoic construction of the Cordilleran "collage," central British Columbia to central California; *in* Howell, D. G., and McDougall, K. A., eds., *Mesozoic Paleogeography of the Western United States:* Pacific Section, Soc. Econ. Paleontologists Mineralogists, Pacific Coast Paleogeography Symp. 2, p. 1–32.

Day, H. W., Moores, E. M., and Tuminas, A. C., 1985, Structure and tectonics of the northern Sierra Nevada: *Geol. Soc. America Bull.*, v. 96, p. 436–450.

DePaolo, D. J., 1981, A neodymium and strontium isotopic study of the Mesozoic calc-alklaline granitic batholiths of the Sierra Nevada and Peninsular ranges, California: *J. Geophys. Res.*, v. 86, p. 10470–10488.

Dickinson, W. R., and Suczek, C. A., 1979, Plate tectonics and sandstone compositions: *Amer. Assoc. Petrol. Geologists Bull.*, v. 63, p. 2164–2182.

Domenick, M. A., Kistler, R. W., Dodge, F. C. W., and Tatsumoto, M., 1983, Nd and Sr isotopic study of crustal and mantle inclusions from the Sierra Nevada xenoliths and implications for batholith petrogenesis: *Geol. Soc. America Bull.*, v. 94, p. 713–719.

Douglass, R. C., 1967, Permian Tethyan fusulinids from California: *U.S. Geol. Survey Prof. Paper 583-A*, p. 7–43.

Duffield, W. A., and Sharp. R. V., 1975, Geology of the Sierra foothills melange and adjacent areas, Amador County, California: *U.S. Geol. Survey Prof. Paper 827*, 30 p.

Durell, C., and D'Allura, J. A., 1977, Upper Paleozoic section in eastern Plumas and Sierra Counties, northern Sierra Nevada, California: *Geol. Soc. America Bull.*, v. 88, p. 844–852.

Edelman, S. H., 1985, Slate belt structures related to arc-continent collision and vergence reversal, northern Sierra Nevada, California: *Geol. Soc. America Abstr. with Programs*, v. 17, p. 353.

——, 1986, Structure across the northern Sierra Nevada Foothills suture zone, California: *Geol. Soc. America Abstr. with Programs*, v. 18, p. 103.

Ehrenberg, S. N., 1975, Feather River ultramafic body, northern Sierra Nevada, California: *Geol. Soc. America Bull.*, v. 86, p. 1235–1243.

Evarts, R. C., 1977, The geology and petrology of the Del Puerto ophiolite, Diablo Range, central California Coast Ranges; *in* Coleman, R. G., and Irwin, W. P., eds., *North American Ophiolites:* Oreg. Dept. Geology Miner. Industries Bull., v. 95, p. 121–139.

Evernden, J. F., and Kistler, R. W., 1970, Chronology of emplacement of Mesozoic batholithic complexes in California and western Nevada: *U.S. Geol. Survey Prof. Paper 623*, 67 p.

Fiske, R. S., and Tobisch, O. T., 1978, Paleogeographic significance of volcanic rocks of the Ritter Range pendant, central Sierra Nevada, California, *in* Howell, D. G., and McDougall, K. A., eds., *Mesozoic Paleogeography of the Western United States:* Pacific Section, Soc. Econ. Paleontologists Mineralogists, Pacific Coast Paleogeography Symp. 2, p. 209–222.

Fiske, R. S., Tobisch, O. T., Kistler, R. W., Stern, T. W., and Tatsuomoto, M., 1977, Minarets Caldera — A Cretaceous volcanic center in the Ritter Range pendant, central Sierra Nevada, California [abstract]: *Geol. Soc. America Abstr. with Programs,* v. 9, p. 975.

Gill, J. B., 1981, *Orogenic Andesites and Plate Tectonics:* New York, Springer-Verlag.

Girty, G. H., and Wardlaw, M. S., 1985, Petrology and provenance of pre-late Devonian sandstones, Shoo Fly Complex, northern Sierra Nevada, California: *Geol. Soc. America Bull.,* v. 96, p. 516–521.

Hamilton, W., 1969, Mesozoic California and the mantle: *Geol. Soc. America Bull.,* v. 80, p. 2409–2430.

Hanson, W. R., and Peterman, Z. E., 1968, Basement-rock geochronology of the Black Canyon of the Gunnison, Colorado, *in Geological Survey Research 1968:* U.S. Geol. Survey Prof. Paper 600-C, p. C80–C90.

Harper, G. D., and Wright, J. E., 1984, Middle to Late Jurassic tectonic evolution of the Klamath Mountains, California-Oregon: *Tectonics,* v. 3, p. 759–772.

——, Saleeby, J. B., and Norman, E., 1985, Geometry and tectonic setting of sea-floor spreading for the Josephine ophiolite, and implications for Jurassic accretionary events along the California margin; *in* Howell, D. G., ed., *Tectonostratigraphic Terranes of the Circum-Pacific Region:* Circum-Pacific Council Energy Min. Resources, Earth Sci. Series, no. 1, p. 239–257.

Harwood, D. S., 1983, Stratigraphy of Upper Paleozoic volcanic rocks and regional unconformities in part of the northern Sierra terrane, California: *Geol. Soc. America Bull.,* v. 94, p. 413–422.

Hedge, C. E., Peterman, Z. E., and Braddock, W. A., 1967, Age of the major Precambrian regional metamorphism in the northern Front Range, Colorado: *Geol. Soc. America Bull.,* v. 78, no. 4, p. 551–558.

——, Peterman, Z. E., Case, J. E., and Obradovich, J. D., 1968, Precambrian geochronology of the northwestern Unconpahgre Plateau, Utah and Colorado; *in Geological Survey Research 1968:* U.S. Geol. Survey Prof. Paper 600-C, p. C91–C96.

Herzig, C. T., 1985, Jurassic rocks east of the Melones Fault Zone, central Sierra Nevada foothills, California: Unpublished M.S. thesis, State Univ. New York, Stony Brook, N.Y., 88 p.

——, Sharp, W. D., and Spangler, R., 1985, New constraints on the origin and nature of the Melones Fault Zone: *Geol. Soc. America Abstr. with Programs,* v. 17, p. 361.

Hietanen, A., 1981, Geology west of the Melones Fault between the Feather and North Yuba Rivers, California: *U.S. Geol. Survey Prof. Paper 1226-A,* 35 p.

Hopson, C. A., Mattinson, J. M., and Pessagno, E. A., Jr., 1981, Coast Range ophiolite, western California, *in* Ernst, W. G., ed., *The Geotectonic Development of California* (Rubey Vol. I): Englewood Cliffs, N.J., Prentice-Hall, p. 418–510.

Huber, N. K., and Rinehart, C. D., 1965, Geologic map of the Shuteye Peak quadrangle, Sierra Nevada, California: *U.S. Geol. Survey Geol. Quad. Map GQ-728,* scale 1:62,500.

Ingersoll, R. V., 1983, Petrofacies and provenance of late Mesozoic forearc basin, northern and central California: *Amer. Assoc. Petroleum Geol. Bull.,* v. 67, p. 1125–1142.

Ingersoll, R. V., and Schweickert, R. A., 1986, A plate-tectonic model for Late Jurassic ophiolite genesis, Nevadan orogeny and forearc initiation, Northern California: *Tectonics,* v. 5, p. 901–912.

Irwin, W. P., Jones, D. L., and Kaplan, T. H., 1978, Radiolarians from pre-Nevadan rocks of the Klamath Mountains, California and Oregon, *in* Howell, D. G., and McDougall, K. A., eds., *Mesozoic Paleogeography of the Western United States:* Pacific Section, Soc. Econ. Paleontologists Mineralogists, Pacific Coast Paleogeography Symp. 2, p. 303–310.

Jacobsen, S. B., Quick, J. E., and Wasserburg, G. J., 1984, A Nd and Sr isotopic study of the Trinity peridotite — Implications for mantle evolution: *Earth Planet. Sci. Lett.*, v. 68, p. 361–378.

Jones, D. L., and Moore, J. G., 1973, Lower Jurassic ammonite from the south-central Sierra Nevada, California: *U.S. Geol. Survey J. Res.*, v. 1, p. 453–458.

——, *et al.*, 1977, Wrangellia — A displaced terrane in northwestern North America: *Can. J. Earth Sci.*, v. 14, p. 2565–2577.

Kerrick, D. M., 1970, Contact metamorphism in some areas of the Sierra Nevada, California: *Geol. Soc. America Bull.*, v. 81, p. 2913–2938.

Kistler, R. W., 1966, Structure and metamorphism in the Mono Craters quadrangle, Sierra Nevada, California: *U.S. Geol. Survey Bull., 1221-E*, p. E1–E53.

——, 1978, Mesozoic paleogeography of California — A viewpoint from isotope geology, *in* Howell, D. G., and McDougall, K. A., eds., *Mesozoic Paleogeography of the Western United States:* Pacific Section, Soc. Econ. Paleontologists Mineralogists, Pacific Coast Paleogeography Symp. 2, p. 75–84.

——, and Bateman, P. C., 1966, Stratigraphy and structure of the Dinkey Creek roof pendant in the central Sierra Nevada, California: *U.S. Geol. Survey Prof. Paper 524-G*, 14 p.

——, and Peterman, Z. E., 1973, Variations in Sr, Rb, K, Na, and initial $^{87}Sr/^{86}Sr$ in Mesozoic granitic rocks and intruded wall rocks in central California: *Geol. Soc. America Bull.*, v. 84, no. 11, p. 3489–3512.

——, and Peterman, Z. E., 1978, Reconstruction of crustal blocks of California on the basis of initial strontium isotopic compositions of mesozoic granitic rocks: *U.S. Geol. Survey Prof. Paper 1071*, 17 p.

——, Robinson, A. C., and Fleck, R. W., 1980, Mesozoic right-lateral fault in eastern California: *Geol. Soc. America with Programs*, v. 12, p. 115.

Lanphere, M. A., Wasserburg, G. J. F., Albe, A. L., and Tilton, G. R., 1963, Redistribution of strontium and rubidium isotopes during metamorphism, World Beater Complex, Panamint Range, California, *in* Craig, H., Miller, S. L., and Wasserburg, G. J., eds., *Isotopic and Cosmic Chemistry:* Amsterdam, North-Holland, p. 269–320.

——, *et al.*, 1968, Isotopic age of the Nevadan orogeny and older plutonic and metamorphic events in the Klamath Mountains, California: *Geol. Soc. America Bull.*, v. 79, p. 1027–1052.

Mattinson, J. M., and Hopson, C. A., 1972, Paleozoic ophiolite complexes in Washington and northern California: *Carnegie Inst. Wash. Yearbook 71*, p. 578–583.

McMath, V. E., 1966, Geology of the Taylorsville area, northern Sierra Nevada, California: *Calif. Div. Mines Geology Bull.*, v. 190, p. 173–183.

Menzies, M., Blanchard, D., and Xenophontos, C., 1980, Genesis of the Smartville arc-ophiolite, Sierra Nevada foothills, California: *Amer. J. Sci.*, v. 280-A, p. 329–344.

Merguerian, C., 1985, Stratigraphy, structural geology and tectonic implications of the Shoo Fly complex and the Calaveras-Shoo Fly thrust, central Sierra Nevada, California: Ph.D. thesis, Columbia Univ., New York, N.Y., 255 p.

Merguerian, C., and Schweickert, R. A., 1987, Paleozoic gneissic granitoids in the Shoo Fly Complex, central Sierra Nevada, California: *Geol. Soc. America Bull.*, v. 99, p. 699–717.

Miller, E. L., and Carr, M. D., 1978, Recognition of possible Aztec-equivalent sandstones and associated Mesozoic sedimentary deposits within the Mesozoic magmatic arc in the southwestern Mojave Desert, California, *in* Howell, D. G., and McDougall, K. A., eds., *Mesozoic Paleogeography of the Western United States:* Pacific Section, Soc. Econ. Paleontologists, Mineralogists, Pacific Coast Paleogeography Symp. 2, p. 283–290.

———, Holdsworth, B. K., Whiteford, W. B., and Rodgers, D., 1984, Straitgraphy and structure of the Schoonover complex, northeastern Nevada: *Geol. Soc. America Bull.*, v. 95, p. 1063–1076.

Monger, J. W. H., 1977, Upper Paleozoic rocks of the western Canadian Cordillera and their bearing on Cordilleran evolution: *Can. J. Earth Sci.*, v. 14, p. 1832–1859.

———, Price, R. A., and Tempelman-Kluit, D. J., 1982, Tectonic accretion and the origin of the two major metamorphic and plutonic welts in the Canadian Cordillera: *Geology*, v. 10, p. 70–75.

Moorbath, S., and Taylor, P. M., 1981, Isotopic evidence for continental growth in the Precambrian: p. 491–525 *in* Kröner, A., ed., *Precambrian Plate Tectonics*, Elsevier, Amsterdam, 781 p.

Moore, J. N., and Foster, C. J., 1980, Lower Paleozoic meta-sedimentary rocks in the east-central Sierra Nevada, California — Correlations with Great Basin formations: *Geol. Soc. America Bull.*, v. 91, p. 37–43.

Moores, E., and Day, H. W., 1984, Overthrust model for the Sierra Nevada: *Geology*, v. 12, p. 416–419.

Morgan, B. A., 1976, Geology of Chinese Camp and Moccasin quadrangles, Tuolumne County, California: *U.S. Geol. Survey Misc. Field Studies Map MF-840*, scale 1:24000.

———, 1978, Metamorphic forsterite and diopside from the ultramafic complex at the Tuolumne River, California: *U.S. Geol. Survey J. Res.*, v. 6, p. 73–80.

Murchey, B., Harwood, D. S., and Jones, D. L., 1986, Correlative chert sequences from the northern Sierra Nevada and Havallah sequence, Nevada: *Geol. Soc. America Abstr. with Programs*, v. 18, p. 162.

Nelson, K. D., Zhu, T. F., Gibbs, A., Harris, R., Oliver, J. E., Kaufmann, S., Brown, L., and Schweickert, R. A., 1986, COCORP deep deismic reflection profiling in the northern Sierra Nevada, California: *Tectonics*, v. 5, p. 321–333.

Oldow, J. S., 1978, Triassic Pamlico Formation — An allochtonous sequence of volcano-genic-carbonate rocks in west-central Nevada, *in* Howell, D. G., and McDougall, K. A., eds., *Mesozoic Paleogeography of the Western United States:* Pacific Section, Soc. Econ. Paleontologists Mineralogists, Pacific Coast Paleogeography Symp. 2, p. 223–235.

———, 1981, Structure and stratigraphy of the Luning allochthon and kinematics of allochton emplacement, Pilot Mountains, west-central Nevada: *Geol. Soc. America Bull.*, v. 92, Pt. I, p. 889–911, Pt. II, p. 1647–1669.

Palmer, A. R., 1983, The decade of North America geology, 1983 geologic time scale: *Geology*, v. 11, p. 503–504.

Parkison, G. A., 1976, Tectonics and sedimentation along a Late Jurassic (?) active continental margin, Western Sierra Nevada Foothills, California: M.S. thesis, Univ. California, Berkeley, Calif., 160 p.

Peck, D. L., 1980, Geologic map of the Merced Peak quadrangle, Yosemite National Park, California: *U.S. Geol. Survey Geol. Quad. Map GQ-1531*, scale 1:62, 500.

Peterman, Z. E., Hedge, C. E., Coleman, R. G., and Snarely, P.D., Jr., 1967, $^{87}Sr/^{86}Sr$ ratios in some eugeosynclinal sedimentary rocks and their bearing on the origin of granitic magma in orogenic belts: *Earth Planet. Sci. Lett.*, v. 2, p. 433–439.

Poole, F. G., 1974, Flysch deposits of the Antler United States, *in* Dickinson, W. R., ed.,

Tectonics and Sedimentation: Soc. Econ. Paleontologists Mineralogists Spec. Publ. 22, p. 58–82.

———, and Sandberg, C. A., 1977, Mississippian paleogeography and tectonics of the western United States, *in* Stewart, J. H., Stevens, C. H., and Fritsche, A. E., eds., *Paleozoic Paleogeography of the Western United States:* Pacific Section, Soc. Econ. Paleontologists Mineralogists, Pacific Coast Paleogeography Symp. 1, p. 67–85.

Price, J., 1986, The southeast Canadian Cordillera — Thrust faultering, tectonic wedging and delamination of lithosphere: *J. Struct. Geology*, v. 8, p. 239–254.

Price, R. A., 1981, The Cordilleran foreland thrust and fold belt in the southern Canadian Rocky Mountains, *in* Price N. J. and MacClay, K., eds., *Thrust and Nappe Tectonics:* Geol. Soc. London Spec. Publ. 9, p. 427–448.

Read, P. B., and Wheeler, J. O., 1976, Geology, Tardeau west-half, British Columbia: *Geol. Survey Can. Open File Map 432.*

Rinehart, C. D., and Ross, D. C., 1964, Geology and mineral deposits of the Mount Morrison quadrangle, Sierra Nevada, California, *with a section on* A gravity study of Long Valley, by L. C. Pakiser: *U.S. Geol. Survey Prof. Paper 385*, 106 p.

———, Ross, D. C., and Huber, N. K., 1959, Paleozoic and Mesozoic fossils in a thick stratigraphic section in the eastern Sierra Nevada, California: *Geol. Soc. America Bull.*, v. 70, p. 941–946.

Ross, C. A., and Ross, J. R. P., 1983, Late Paleozoic accreted terranes of western North America, *in* Stevens, C. H., ed., *Pre-Jurassic Rocks in Western North American Suspect Terranes:* Soc. Econ. Paleontologists Mineralogists, Pacific Section, p. 7–22.

Russell, S. J., and Nokleberg, W. J., 1977, Superimposition and timing of the deformations in the Mount Morrison roof pendant and in the central Sierra Nevada, California: *Geol. Soc. America Bull.*, v. 88, p. 335–345.

Saleeby, J. B., 1977, Fracture zone tectonics, continental margin fragmentation and emplacement of the Kings-Kaweah ophiolite belt, southwest Sierra Nevada, California, *in* Coleman, R. E., and Irwin, W. P., eds., *North American Ophiolite:* Oreg. Dept. Geology Miner. Industries Bull., v. 95, p. 141–160.

———, 1978, Kings River Ophiolite, southwest Sierra Nevada foothills, California: *Geol. Soc. America Bull.*, v. 89, p. 617–636.

———, 1979, Kaweah serpentinite melange, southwest Sierra Nevada, California: *Geol. Soc. America Bull.*, v. 90, p. 29–46.

———, 1981, Ocean floor accretion and volcano-plutonic arc evolution of the Mesozoic Sierra Nevada, California, *in* Ernst, W. G., *The Geotectonic Development of California* (Rubey Vol. I): Englewood Cliffs, N.J., Prentice-Hall, p. 132–181.

———, 1982, Polygenetic ophiolite belt of the California Sierra Nevada — Geochronological and tectonostratigraphic development: *J. Geophys. Res.*, v. 87, p. 1803–1824.

———, 1983, Accretionary tectonics of the North American Cordillera: *Ann. Rev. Earth Planet. Sci.*, v. 15, p. 45–73.

———, *et al.*, 1986, Corridor C-2, Central California offshore to Colorado Plateau: *Geol. Soc. America*, Centennial Continent-Ocean Transect #10.

———, and Sharp, W. D., 1980, Chronology of the structural and petrologic development of the southwest Sierra Nevada foothills, California — Pt. I: *Geol. Soc. America Bull.*, v. 91, p. 317–320.

———, and Sharp, W. D., 1980, Chronology of the structural and petrologic development of the southwestern Sierra Nevada foothills, California — Pt. II: *Geol. Soc. America Bull.*, v. 91, p. 1416–1535.

———, Goodin, S. E., Sharp, W. D., and Busby, C. J., 1978, Early Mesozoic paleotectonic-paleogeographic reconstruction of the southern Sierra Nevada region, *in* Howell, D. G., and McDougall, K. A., eds., *Mesozoic Paleogeography of the Western United*

States: Pacific Section, Soc. Econ. Paleontologists Mineralogists, Pacific Coast Paleogeography Symp. 2, p. 311–336.

Schweickert, R. A., 1977, Major pre-Jurassic thrust fault between the Shoo Fly and Calaveras complexes, Sierra Nevada, California: *Geol. Soc. America Abstr. with Programs*, v. 9, p. 497.

——, 1981, Tectonic evolution of the Sierra Nevada range, *in* Ernst, W. G., ed., *The Geotectonic Development of California* (Rubey Vol. I): Englewood Cliffs, N.J., Prentice-Hall, p. 87–131.

——, and Bogen, N. L., 1983, *Tectonic Transect of Sierran Paleozoic through Jurassic Accreted Belts:* Pacific Section, Soc. Econ. Paleontologists Mineralogists, 22 p.

——, and Cowan, D. S., 1975, Early Mesozoic tectonic evolution of the western Sierran Nevada, California: *Geol. Soc. America Bull.*, v. 86, p. 1329–1336.

——, and Snyder, W., 1981, Paleozoic plate tectonics of the Sierra Nevada and adjacent regions, *in* Ernst, W. G., ed., *The Geotectonic Development of California* (Rubey Vol. I): Englewood Cliffs, N.J., Prentice-Hall, p. 182–202.

——, Tobisch, O. T., Wright, W. T., III, and Saleeby, J. B., 1977, Paleotectonic and paleogeographic significance of the Calaveras Complex, western Sierra Nevada, California, *in* Stewart, J. H., et al., eds., *Paleozoic Paleogeography of the Western United States:* Pacific Section, Soc. Econ. Paleontologists Mineralogists, Pacific Coast Paleogeography Symp. 1, p. 381–394.

——, Armstrong, R. L., and Harakal, J. E., 1980, Lawsonite blueschist in the northern Sierra Nevada, California: *Geology*, v. 8, p. 27–31.

——, Bogen, N. L., Girty, G. H., Hanson, R. E., and Merguerian, C., 1984, Timing and structural expression of the Nevadan orogeny, Sierra Nevada, California: *Geol. Soc. America Bull.*, v. 95, p. 967–979.

Scott, G. R., Renne, P. R., Bazard, D. R., and Johnston, J. M., 1985, Paleomagnetism of a Permian Nosoni Fm. ignimbrite and implications to accretionary models of eastern Klamath Belt tectonics: *Geol. Soc. America Abstr. with Programs*, v. 17, p. 407.

Sharp, W. D., 1980, Ophiolite accretion in the northern Sierra: *Trans. Amer. Geophys. Un.*, v. 61, p. 1122.

——, 1983, Chronology of the structural development of the central Sierra Nevada Foothills, CA: *Geol. Soc. America Abstr. with Programs*, v. 15, p. 272.

——, 1984, Structure, petrology, and geochronology of a part of the central Sierra Nevada foothills metamorphic belt, California: Unpublished Ph.D. dissertation, Univ. California, Berkeley, Calif., 173 p.

——, 1985, The Nevadan Orogeny of the foothills metamorphic belt, California — A collision without a suture? *Geol. Soc. America Abstr. with Programs*, v. 17, p. 407.

——, and Evarts, R. C., 1982, New constraints on the environment of formation of the Coast Range ophiolite at Del Puerto Canyon, California: *Geol. Soc. America Abstr. with Programs*, v. 14, p. 233.

——, and Leighton, C. W., 1987, Accretion of the Foothills ophiolite, western Sierra Nevada Foothills, California: *Geol. Soc. America Abstr. with Programs*, v. 19, p. 450.

——, and Stern, R., 1985, The role of mantle-derived magmas in crustal growth in accreted terranes, California: *Trans. Amer. Geophys. Un.*, v. 66, p. 420.

——, *et al.*, 1982, Age and tectonic significance of Paleozoic orthogneisses of the Sierra Nevada Foothills Metamorphic Belt, California: *Geol. Soc. America Abstr. with Programs*, v. 14, p. 233.

Shaw, H. F., Chen, J. H., Wasserburg, J. G., and Saleeby, J. B., 1984, Nd-Sr-Pb systematics and age of the Kings-Kaweah ophiolite, California: *Trans. Amer. Geophys. Un.*, v. 65, p. 1147.

_____ ,Chen, J. H., Saleeby, J. B., and Wasserburg, J. G., 1987, Nd-Sr-Pb systematics and age of the Kings River ophiolite, California: implications for depleted mantle evolution: Contrib. *Mineralogy Petrology*, v. 96, p. 281–290.

Shervais, J. W., and Kimbrough, D. L., 1983, The Coast Range ophiolite of California — A composite island arc-oceanic crust terrane: *Geol. Soc. America Abstr. with Programs*, v. 15, p. 685.

Shirey, S. B., and Hanson, G. N., 1984, Mantle-derived Archaean monozodiorites and trachyandesites: *Nature*, v. 310, p. 222–224.

Silberling, N. J., 1975, Age relationships of the Golconda thrust fault, Sonoma Range, north-central Nevada: *Geol. Soc. America Spec. Paper 163*, 28 p.

Silver, L. T., 1971, Problems of crystalline rocks of the Transverse Ranges: Cordilleran Section, *Geol. Soc. America Abstr. with Programs*, v. 3, p. 193–194.

_____ , Anderson, C. A., Crittenden, M., and Robertson, J. M., 1977a, Chronostratigraphic elements of the Precambrian rocks of the southwestern and far western United States: *Geol. Soc. America Abstr. with Programs*, v. 9, p. 1176.

_____ , Bickford, M. E., Van Schmus, W. R., Anderson, J. L., Anderson, T. H., and Medaris, L. G., Jr., 1977b, The 1.4-1.5 b.y. transcontinental anorogenic plutonic performation of North America: Geol. Soc. America Ann. Meet., *Abstr. with Programs*, v. 9, p. 1176–1177.

Snoke, A. W., Quick, J. E., and Bowman, H. R., 1981, Bear Mountain igneous complex, Klamath Mountains, California — An ultrabasic to silicic calc-alklaline suite: *J. Petrology*, v. 22, p. 501–552.

_____ , Sharp, W. D., Wright, J. E., Saleeby, J. B., 1982, Significance of mid-Mesozoic peridotitic to dioritic intrusive complexes, Klamath Mountains-Western Sierra Nevada, California: *Geology*, v. 15, p. 160–166.

Speed, R. C., 1977, Island arc and other paleogeographic terranes of late Paleozoic age in the western Great Basin, *in* Stewart, J. H., *et al.*, eds., *Paleozoic Paleogeography of the Western United States:* Pacific Section, Soc. Econ. Paleontologists and Mineralogists, Pacific Coast Paleogeography Symp. 1, p. 349–362.

_____ , 1978, Paleogeographic and plate tectonic evolution of the early Mesozoic marine providence of the western Great Basin, *in* Howell, D. G., and McDougall, K. A., eds., *Mesozoic Paleogeography of the Western United States:* Pacific Section, Soc. Econ. Paleontologists Mineralogists Pacific Coast Paleogeography Symp. 2, p. 253–270.

_____ , 1979, Collided Paleozoic microplate in the western United States: *J. Geology*, v. 87, p. 279–292.

_____ , and Sleep, N. H., 1982, Antler orogeny, a model: *Geol. Soc. America Bull.*, v. 93, p. 815–828.

Standlee, L. A., 1978, Middle Paleozoic ophiolite in the Melones Fault Zone, northern Sierra Nevada, California: *Geol. Soc. America Abstr. with Programs*, v. 10, p. 148.

_____ , and Nestell, M. K., 1985, Age and tectonic significance of terranes adjacent to the Melones Fault Zone, N. Sierra Nevada, California. *Geol. Soc. America Abstr. with Programs*, v. 17, p. 410.

Stanley, K. O., Jordan, W. M., and Dott, R. H., Jr., 1971, New hypothesis of Early Jurassic paleogeography and sediment dispersal for western United States: *Amer. Assoc. Petroleum Geologists Bull.*, v. 55, p. 10-19.

Stern, T. W., Bateman, P. C., Morgan, B. A., Newell, M. F., and Peck, D. L., 1981, Isotopic U-Pb ages of zircon from granitoids of the central Sierra Nevada, California: *U.S. Geol. Survey Prof. Paper 1185*, 17 p.

Stewart, J. H., 1972, Initial deposits in the Cordilleran geosyncline — Evidence of a Late Precambrian (< 850 m.y.) continental separation: *Geol. Soc. America Bull.*, v. 83, p. 1345–1360.

volcanic arc rocks along part of the eastern Sierra Nevada, California — A case for extensional tectonics: *Tectonics*, v. 5, no. 1, p. 65–94.

Uyeda, S., 1982, Subduction zones, an introduction to comparative subductology: *Tectonophysics:* v. 81, p. 133–147.

Varga, R. J., and Moores, E. M., 1981, Age, origin and significance of an unconformity that predates island-arc volcanism in the northern Sierra Nevada: *Geology*, v. 9, p. 512–518.

Weisenberg, C. W., and Avé Lallèment, H., 1977, Permo-Triassic emplacement of the Feather River ultramafic body, northern Sierra Nevada Mountains, California: *Geol. Soc. America Abstr. with Programs*, v. 9, p. 525.

Wentworth, C. M., and Zoback, M. D., 1985, Central California seismic reflection transect — I. The eastern Coast Ranges and western Great Valley: *Trans. Amer. Geophys. Un.*, v. 66, p. 1073.

____ ,*et al.*, 1987, A transect across the Mesozoic accretionary margin of central California: Geophys. J. Roy. Astronom. Soc., v. 89, p. 105–110.

Wright, J. E., 1982, Permo-Triassic subduction complex, southern Klamath Mountains, northern California: *J. Geophys. Res.*, v. 87, p. 3805–3818.

____ , and Fahan, M. R., in press, Middle Jurassic (pre-Nevadan) orogenesis in the Klamath Mountains, northern California: Regional deformation/metamorphism and thrust faulting within an active arc environment: *Geol. Soc. America Bull.*

____ , and Miller, E. L., 1986, An expanded view of Jurassic orogenesis for the western U.S. Cordillera: *Geol. Soc. America Abstr. with Programs*, v. 18, p. 201.

____ , and Sharp, W. D., 1982, Mafic-ultramafic intrusive complexes of the Klamath-Sierra region, California — Remnants of a Middle Jurassic arc complex: *Geol. Soc. America Abstr. with Programs*, v. 14, p. 245–246.

Xenophotos, C., and Bond, G. C., 1978, Petrology, sedimentation, and paleogeography of the Smartville terrane (Jurassic) — bearing on the genesis of the Smartville ophiolite, *in* Howell, D. G., and McDougall, K. A., eds., *Mesozoic Paleogeography of the Western United States:* Pacific Section, Soc. Econ. Paleontologists Mineralogists, Pacific Coast Paleogeography Symp. 2, p. 291–302.

Zoback, M. D., Wentworth, C. M., and Moore, J. G., 1985, Central California seismic reflection transect — II. The eastern Great Valley and western Sierra Nevada: *Trans. Amer. Geophys. Un.*, v. 66, p. 1073.

Zucca, J. J., *et al.*, 1986, Crustal structure of northeastern California: *J. Geophys. Res.*, v. 91, p. 7359–7382.

David B. Sams and Jason B. Saleeby
Division of Geological and Planetary Sciences
California Institute of Technology
Pasadena, California 91125

31

GEOLOGY AND PETROTECTONIC SIGNIFICANCE OF CRYSTALLINE ROCKS OF THE SOUTHERNMOST SIERRA NEVADA, CALIFORNIA

ABSTRACT

Crystalline rocks of the southernmost Sierra Nevada north of the Garlock fault consist primarily of Early Cretaceous orthogneisses with subordinate paragneiss, and a mid-Cretaceous tonalite batholith complex with coeval gabbroic intrusives. Quartz-rich meta-sedimentary rocks and marble constitute the main framework into which the plutonic rocks were emplaced. The orthogneisses are predominantly tonalitic in composition with significant layers and domains of granodioritic to granitic, and dioritic to gabbroic gneiss. Field relations, isotopic Rb/Sr, O, and zircon U/Pb data demonstrate assimilation of meta-sedimentary material into the orthogneisses and magma mixing between mafic, tonalitic, and anatectic granitic magma derived from the metasedimentary material. The ortho-gneisses yield a family of internally and externally concordant zircon ages, indicating igneous crystallization between 110 and 120 m.y.b.p. A bulk-rock Rb/Sr isochron determined on the orthogneiss zircon sample suite yields a similar age of 117 ± 5 m.y. The tonalite batholith complex lies generally to the northeast of the gneiss complex with some phases intruding or grading into the gneisses. The batholith complex along with small (5-km-scale) cumulate gabbro bodies intruded into the gneisses yields numerous, mainly concordant, zircon ages of 100 ± 2 m.y. A series of late to postdeformational intrusives yield zircon ages of 90-95 m.y., thus putting a tight constraint on the cessation of ductile deformation under high-temperature conditions.

The orthogneisses exhibit a banding which developed under igneous to hot sub-solidus conditions by the transposition of magmatic segregations and/or igneous mineral assemblages. Blastomylonitic fabrics commonly overprint such banding and represent continued or superimposed deformation compared to tonalitic to mafic varieties, perhaps as the result of bulk material properties. Rocks of the batholith complex have a late-stage deformational fabric shown primarily in the tonalites as pervasive foliation and faint gneissic banding. Much of the tonalite batholith complex was deformed under igneous conditions, but late phases lack significant deformation features. Some gabbros and diorites show deep-level (6-8 kb) postcrystallization autometamorphic garnet growth. The major deformational fabrics exhibited in these suites of rocks may be related to intrusion of the tonalite batholith into the lower crust, and/or the result of intra-arc shearing that was preferentially concentrated in selected intrusive bodies. Alternatively, the deformational fabrics may represent mid- to lower-crustal flow and perhaps detachment related to the overall dynamics of batholithic magmatism.

Crystalline rocks of the southernmost Sierra Nevada represent the deepest exposed levels of the composite Sierra Nevada batholith. Several petrologic relations suggest a depth of formation corresponding to about 8 kb pressure. At such crustal levels, igneous and metamorphic processes and products cannot be clearly separated. The southwestern margin of the 100-m.y.-old tonalitic batholithic complex may represent the tilted floor zone of the batholith which originally spread laterally over its gneissic substrate. A number of high-grade metamorphic features produced in this environment appear to have formed on a retrogressing P/T trajectory upon descent from solidus conditions. Rapid uplift of the crystalline rocks occurred prior to late Eocene time, perhaps in response to major thrust tectonics documented throughout southern California. Crystalline rocks of the southernmost Sierra Nevada provide an excellent opportunity to study deep-level Phanerozoic batholithic processes, which in this case operated by the addition of mantle-derived magma that interacted with and reconstituted previously tectonically assembled crustal fragments into new sialic crust. The central Sierra Nevada exposes volcanic and epizonal levels of the mid-Cretaceous batholithic belt, and progressively deeper rocks are exposed southward. Thus, the southern half of the range offers a down-plunge view of the composite batholith, with the southernmost rocks sampling perhaps as deep as 20–25 km.

INTRODUCTION

The structure and composition of the lower levels of continental crust present some of the fundamental problems confronting modern geology. Most studies in rocks believed to have originated at deep continental levels have been conducted in Precambrian shields, with relatively few studies conducted in Phanerozoic mountainous terranes. This bias in sampling requires that we closely consider the question of long-term secular variations in deep crustal processes. Regardless of age, basic questions concern the interplay of igneous and metamorphic processes, and their interplay with the tectonic accretion of preexisting crustal fragments. The superpositioning of voluminous magmatic and related regional thermal metamorphic regimes over tectonically assembled crustal fragments is an expected and observed phenomenon in Phanerozoic and probably older active continental margin orogens (Hamilton, 1969; Saleeby, 1983). In most cases, such superpositioning probably results from the migration or initiation of an arc magmatic belt across an earlier subduction or collisional tectonic accretionary zone, and would conceivably lead to the production of annealed continental crustal columns consisting of both recycled crustal and juvenile mantle-derived materials.

Meaningful exploration of the questions above requires the identification of young (Phanerozoic), deep continental crustal assemblages, and the application of well-integrated geological, geochemical, and geophysical studies. The Sierra Nevada is known for its voluminous Cretaceous composite batholith of primarily mesozonal character (Bateman and Eaton, 1967; Evernden and Kistler, 1970; Presnall and Bateman, 1973). Metamorphic framework rocks of the batholith have recently received wide attention as representing an accumulation of tectonically accreted crustal and upper mantle fragments (Hamilton, 1969; Saleeby, 1981, 1983; Schweickert and Cowan, 1975). Most studies have considered batholithic processes from the perspective of numerous granitoid plutons being emplaced at high crustal levels with discrete contact aureoles, with regional metamorphism considered primarily in relation to prebatholithic tectonic accretion. This conceptual separation of magmatic and regional metamorphic processes arises from the general mesozonal character along >400 km of the composite batholith.

Recent studies conducted at the extreme southern end of the Sierra Nevada have revealed a considerably deeper level of exposure in the batholith and its metamorphic framework (Ross, 1985, 1986; Saleeby et al., 1986; Sams et al., 1983; Sharry, 1981). The transition from the mesozonal composite batholith to the deep-level batholith terrane occurs over a distance >100 km (Saleeby et al., 1986). Throughout the transition zone and within the deep-level terrane, batholithic and regional metamorphic processes cannot be separated. Furthermore, these processes are seen to transform tectonically accreted crustal fragments into a new sialic crustal column. In this report we provide a synthesis of field and petrologic data from this important deep-level terrane, and interpret the petrologic and structural observations in the context of regional magmatic and tectonic phenomena. Such interpretations lead to discussions of the structure, composition, and genesis of lower continental crust and to the tectonic processes that resulted in the uplift and exposure of this young, deep-level terrane. Indeed, a fundamental question concerning the exposure of any deep-level terrane is the uplift mechanism, and what constitutes the modern deep crustal underpinnings of the terrane.

Exposures of deep-level rocks occur primarily where the crystalline rocks of the Sierra Nevada diverge abruptly from a regional north-northwest to an east-northeast trend

(Fig. 31-1). This "tail," so termed by Ross (1980, 1985), constitutes the Tehachapi Mountains. The geomorphic anomaly in the exposure pattern of Sierran crystalline rocks may be closely related to the processes which resulted in the uplift of the deep-level terrane. Discussion begins with a synthesis of field and petrographic observations, and then proceeds into geochronology, petrogenesis, and tectonics. The observations are not intended to be

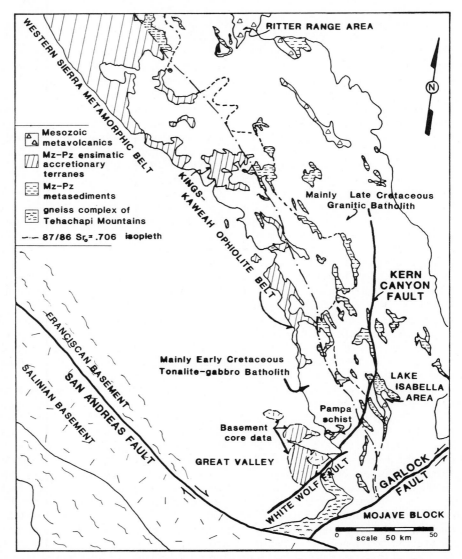

FIG. 31-1. Regional map showing location of study area and major tectonic and geographic features referred to in the text. Initial $^{87}Sr/^{86}Sr$ on Sierra Nevada batholithic rocks after Kistler and Peterman (1973, 1978). Generalized contacts of major lithologic assemblages after Saleeby (1981) and basement domains of the southeastern San Joaquin Valley based on petrographic data from basement cores (Ross, 1979) and comparisons with nearby exposures in the western Sierran foothills.

GEOLOGY AND PETROTECTONIC SIGNIFICANCE OF CRYSTALLINE ROCKS

systematic or complete, but rather to focus on attributes deemed most pertinent to the emphasis of this report. For a more systematic treatment, the reader is referred to Ross (1980, 1983b, 1986), Sharry (1981), Saleeby *et al.* (1987), and Sams (1986).

SYNTHESIS OF FIELD, RADIOMETRIC, AND PETROGRAPHIC DATA

Mountainous areas of the southernmost Sierra Nevada and Tehachapi Mountains (generalized geology shown in Fig. 31-2) are underlain almost entirely by crystalline rocks. Eocene to Holocene strata lap southeastward onto the crystalline rocks of the Sierran-Tehachapi block, and in general dip moderately northwest into the San Joaquin Valley.

Important faults of the study area include the White Wolf, Garlock, and eastern end of the Pleito thrust. The White Wolf fault is active, with the last major event of magnitude 7.7 in 1952 (Arvin-Tehachapi earthquake). Surface traces of the fault and the distribution of aftershock and microseismic foci suggest a southeast-dipping reverse fault (Ross, 1986), although a complex ground breakage pattern mapped by Buwalda and St. Amand (1955) following the 1952 event revealed significant components of strike as well as dip slip. The relationships between the White Wolf fault and the Breckenridge and Edison faults to the northeast are unknown. The Garlock fault is an east-northeast-trending set of mainly steep-dipping faults which in the study area encase an elongate block of allochthonous greenschist (Rand schist; see Jacobson *et al.*, Chapter 13, this volume) not found elsewhere in the Sierran-Tehachapi block. The Garlock fault is suggested to be an intracontinental transform structure separating the westward-extending Basin and Range–Sierra Nevada province from the more rigid Mojave block (Davis and Burchfiel, 1973). Such a kinematic pattern implies ~50 km of left slip within the study area, which is reported to be similar to apparent offsets within basement rocks of the region (Ross, 1986). Field observations by Sharry (1981) along the western termination of the north branch of the Garlock fault suggest a north-dipping low-angle thrust contact between the Rand schist and Tehachapi crystalline rocks. The Pleito thrust bounds the north edge of the western Tehachapi Mountains and the San Emigdio range to the west. It is an active south-dipping low-angle thrust that has not broken for >600 years, and which places Tehachapi and San Emigdio crystalline rocks northward over Cenozoic strata (Hall, 1984). Late Cenozoic movement on the White Wolf, Garlock, and Pleito faults has yielded the current relief and tectono-morphic grain in the study area. Early Cenozoic uplift of the Tehachapi crystalline rocks may be related to regional thrusting, represented in part perhaps by the north branch of the Garlock fault.

Batholithic rocks typical of the western and axial Sierra Nevada can be traced southward into the study area as far as Tejon Creek (Fig. 31-2). Within this drainage, and along the extreme southwest Sierran foothills, a gradation is observed between such batholithic rocks and heterogeneous gneissose rocks. The batholithic rocks are foliated for the most part, but not as intensely as the gneisses. Foliation surfaces parallel lithologic contacts. North of Cummings Mountain, northwest strikes and near-vertical dips prevail, although there is a large open Z-shaped bend in structural trends in the northeast part of the map area. South of Cummings Mountain, the outcrop patterns and some of the major contacts in the crystalline rocks change to east-northeast and foliation trends are more variable but still predominantly northwest.

The crystalline rocks are divided into map units on the basis of lithologic criteria. These map units can be grouped into three suites based on U/Pb geochronological data and gross structural relations. The gneiss complex of the Tehachapi Mountains is defined as heterogeneous mafic to felsic metaigneous rock with U/Pb zircon ages of range 120–110

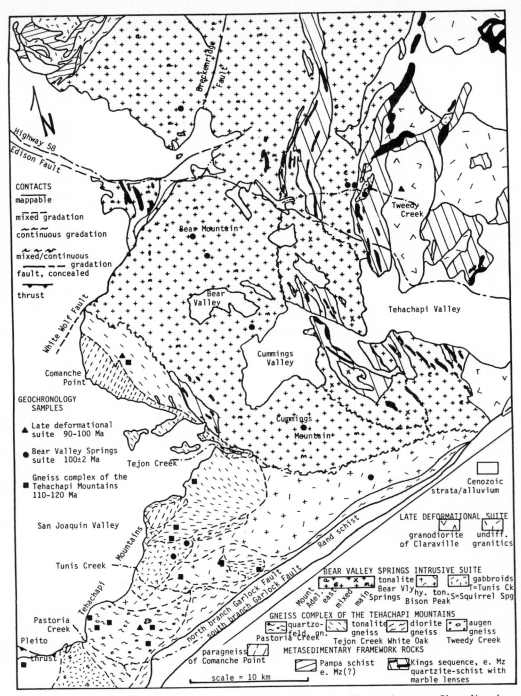

FIG. 31-2. Generalized geologic map showing crystalline rocks of the southernmost Sierra Nevada and U/PB zircon ages. Lithologic units taken primarily from Sams (1986) with modifications after Sharry (1981) and Ross (1980, 1986).

GEOLOGY AND PETROTECTONIC SIGNIFICANCE OF CRYSTALLINE ROCKS

m.y. The gneiss complex also includes paragneiss and rare marble and quartzite lenses. The Bear Valley Springs igneous suite consists of voluminous intrusive rocks with zircon ages clustered at 100 ± 2 m.y. This suite includes the tonalite of Bear Valley Springs, a batholith-scale pluton, and intrusive bodies, including the hypersthene tonalite of Bison Peak and the gabbroids of Tunis Creek that are intermingled with the gneiss complex. The third suite consits of non- to mildly deformed granitoid rocks, several of which yield zircon ages in the range 100–90 m.y. Such rocks are referred to as the late-deformational intrusive suite based on local structural relations.

Metasedimentary septa run discontinuously along the eastern border zone of the tonalite of Bear Valley Springs, and separate the tonalite in most places from the grano-diorite of Claraville. Contacts between batholithic rocks and such septa are typically sharp. The metasedimentary septa are distinguished from paragneiss enclaves of the gneiss complex by a lack or paucity of partial melt or extensive metamorphic segregation features. Margins of paragneiss enclaves are typically obscure, and many are gradational into metaigneous rock.

Contact relations between rocks of the Bear Valley Springs suite and those of the gneiss complex are commonly gradational. The southwestern margin of the tonalite of Bear Valley Springs is gradational into tonalite gneiss of the Tehachapi complex imme-diately south of the White Wolf fault (Fig. 31-2). A metasedimentary spectum lies along the gradation zone to the south. The Bear Valley Springs grades into the hypersthene tonalite of Bison Peak along its southern margin. The hypersthene tonalite and the gabbroids of Tunis Creek are intrusive into, and in places intergradational with, rocks of the gneiss complex. The various lithologic constituents of the gneiss complex are con-cordantly interlayered at centimeter-to-kilometer scales. Lithologic gradations are mixed and continuous. Most contacts appear to be transposed by extreme ductile flow. Most distinguishable intrusive contacts are imparted by members of the younger suites.

Metasedimentary Framework Rocks

These rocks constitute remnants of the structural-stratigraphic framework into which the igneous and metaigneous rocks of the study area were emplaced. Two distinct framework assemblages are present: the Kings sequence and Pampa schist. The Kings sequence forms a series of discontinuous septa from Tejon Creek to areas north of the study area. Similar rocks have been traced >200 km northward (Fig. 31-1), and constitute one of the main framework assemblages for the composite Sierra Nevada batholith (Saleeby *et al.*, 1978). The Pampa schist forms a cluster of small septa in the northwest corner of the study area, and is lithologically distinct from the Kings sequence.

Kings sequence septa of the study area are concordantly intruded by the grano-diorite of Claraville, an augen gneiss and small ovoid tonalitic stocks at Tweedy Creek, and the tonalite of Bear Valley Springs. The septa consist of lenses of impure quartzite, marble, and calc-silicate rock, encased within a quartz-mica schist matrix. Psammitic to quartzofeldspathic granofels and gneiss are locally present, especially in the vicinity of Cummings Valley. Quartzite lenses are typically massive to faintly banded, and grada-tional with schistose host rocks. Protolith compositions were orthoquartzite to subarkose with a possible subordinate tuffaceous component. The quartz-mica schist matrix was derived from impure quartz-rich sandstone with rare pelitic layers now containing sill-imanite.

Internal schistose fabrics and lithologic layering is concordant with the outcrop pattern of the septa and the major intrusive contacts. Tight to isoclinal folding is common in the marbles and calc-silicates, but rare in other rock types. Fold axes and stretching lineations plunge moderately to steeply southeast. The steep linear shape fabrics are commonly found without associated folds. Deformational fabrics within the septum at Tweedy Creek are both cut by and shared with an augen gneiss intrusive body. Here the lithologic layering of the metasedimentary rocks is cut obliquely by the intrusive contacts of the granodioritic orthogneiss, but the dominant shape fabrics of each are parallel. Textural expression of the fabric in the metasedimentary host rock is a coarse crystalloblastic schistosity and in the orthogneiss blastomylonitic banding. Structural relations at Tweedy Creek show that the framework rocks were deformed by folding and at least local disruption of bedding prior to intrusions of the orthogneiss protolith. Subsequently, the small intrusion and its metasedimentary host were strongly flattened and stretched along a moderate to steep southeast plunge. Folds within the metasedimentary rocks may have been tightened into an isoclinal geometry during the later phase of deformation. The augen gneiss of Tweedy Creek has yielded a zircon age of 114 ± 3 m.y., and thus the relations are important in the regional structural chronology.

The small septum along the southwest margin of the tonalite of Bear Valley Springs has the same lithologic and mineralogic assemblage as septa to the east, but possesses granofelsic to gneissic fabrics. This septum is distinguished from the nearby paragneiss of Comanche Point by the presence of marble, quartzite, and sillimanite-bearing pelite, and by fewer migmatitic structures. The septum also exhibits clear intruded contacts by adjacent plutonic rock, which is generally not the case with the paragneiss.

Stratigraphic relations and numerous protolith features are preserved in the framework septa of the Lake Isabella region to the north (Saleeby and Busby-Spera, 1986). Here the section is dominated by orthoquartzite, subarkosic quartzite and tuffaceous quartz sandstone, marble and calc-silicate rock, and psammitic to quartz-rich schist with several distinct intervals of pelite. Late Triassic to Early Jurassic bivalves have been recovered from several locations farther north in Kings sequence rocks (Saleeby *et al.*, 1978).

The Pampa schist consists of sillimanite and andalusite-bearing graphitic pelite interlayered with psammitic schist and local lenses of amphibolitic mafic to intermediate volcaniclastic rocks (Dibblee and Chesterman, 1953). This assemblage strongly resembles lower Mesozoic slaty rocks which lie depositionally above the Paleozoic Kings-Kaweah ophiolite belt ~50 km to the north (Saleeby, 1979). Blocks of quartzite and stratified sequences of quartz-rich turbidites are intermixed with the Kings-Kaweah slaty strata and are suggested to be a western facies of the Kings sequence (Saleeby *et al.*, 1978).

The paragneiss of Comanche Point is a homogeneous migmatitic unit of mainly psammitic composition with local calc-silicate rock. Granitic leucosomes pervade the paragneiss and thus protolith features and stratigraphic affinity are highly obscured. Rare lenses of marble and quartzite are scattered as lenses within Tehachapi orthogneisses as well as Comanche-type paragneiss lenses, suggesting that the Comanche protolith was at least in part coextensive with Kings sequence rocks prior to plutonism. Basement rocks for the Kings sequence are unknown, except for the apparent westward overlap relation with Paleozoic ophiolitic rocks. Thus, the only pre-batholithic framework rocks observed within the study area are of early Mesozoic age and consist of continent-derived clastic rocks with shallow-water carbonates, and ensimatic argillaceous and subordinate volcanic strata.

Bear Valley Springs Igneous Suite

This suite consists of the tonalite of Bear Valley Springs, the hypersthene tonalite of Bison Peak, and the gabbroids of Tunis Creek. The gabbroid of Squirrel Spring and smaller bodies scattered to the north are also considered members of this suite, although direct age data are lacking. Map relations which distinguish this suite from the gneiss complex include at least local intrusive relations with the complex, and a general lack of gneissic banding. However, rocks of this suite commonly possess a planar fabric representing ductile deformation under solidus to hot subsolidus conditions, or in many of the more mafic members, igneous flow and crystal accumulation processes.

Tonalite of Bear Valley Springs The tonalite of Bear Valley Springs is a composite batholith which extends at least 30 km north of the study area and may be divided into four phases. The main, mixed and eastern border phases are intergradational, whereas the Mount Adelaide phase intrudes the main and eastern border phases. Zircon ages on these four phases are all 100 ± 2 m.y. The intergradational phases constitute the main mass of the batholith in the study area. In general it is a coarse-grained, hypidiomorphic granular, biotite hornblende tonalite. It is pervasively foliated with steep fabrics of increasing intensity oriented subparallel to its margins. The foliation consists of aligned mafic minerals, and distorted mafic inclusions and schlieren. Hypersthene is a local, but important, early igneous phase and is most common in the southern part of the batholith where it is surrounded by hornblende ± biotite clusters. Quartz is ductilely deformed and partially recrystallized, particularly in the southern part of the batholith. Plagioclase shows weak zoning around Cummings Valley that becomes stronger to the north.

The tonalite retains an igneous texture throughout, although it is ductilely deformed, especially in the south. Mylonitic fabrics are locally present near metamorphic septa, particularly in the eastern border phase. The strong planar fabric developed most intensely in the southern part of the batholith, formed under mainly synplutonic conditions, as evidenced by the alignment of mafic tabular and platy igneous minerals, as well as mafic inclusions and schlieren, and the cross-cutting relations of the Mount Adelaide phase. The gradational contact of the main phase of the batholith with the tonalite gneiss of Tejon Creek and the parallel alignment of their respective deformation fabrics with high-grade metamorphic fabrics of the intervening metamorphic septum and nearby migmatitic paragneiss also indicate a synplutonic deformational regime. Retrograde granoblastic and schistose mineral domains and local cataclastic fabrics overprint the high-temperature fabrics.

The mixed phase of the batholith is distinguished as a generally north-trending domain containing high concentrations of mafic inclusions and schlieren, major zones of hypersthene-bearing tonalite, and lesser leucotonalite that locally grades into granodiorite. The mixed phase is localized along a highly dismembered framework septum. It grades into the eastern and main phases by a decrease in mafic inclusions, hypersthene-bearing rocks and leucotonalitic rocks. The zircon ages suggest that the eastern phase could be slightly younger (~1 m.y.?) than the main phase even though they are lithologically similar. The Mount Adelaide phase clearly intrudes the main phase north of the study area, whereas it appears to both cut and grade into the eastern phase. These relations are consistent with the zircon ages. The Mount Adelaide phase is distinguished by greater modal ratios of biotite to hornblende, large euhedral biotite books, fewer inclusions, and significantly less deformation. Age and structural relations suggest that the main and mixed phases were emplaced first, shortly followed by the eastern and then Mount Adelaide phases. Most of the deformation took place prior to Mount Adelaide emplacement. As discussed below, there may be significant structural relief between the northern and southern ends of the exposed batholith complex

The mafic inclusions and schlieren are considered significant in regard to the petrogenesis of the entire Bear Valley Springs igneous suite. Compositions range from mafic tonalite to gabbro, with diorite-quartz diorite predominating. Margins of the inclusions are both sharp and diffuse, with structural settings ranging from mafic grain clusters having interstices continuous with host tonalite, to well-defined elliptical bodies. Mineralogical equilibration with the tonalite host is pervasive. In the southern part of the batholith, some inclusions contain hypersthene in addition to hornblende and biotite; textural relations mimic those of the hypersthene-bearing host tonalite. Zircon U/Pb systematics of a representative inclusions from the main phase are nearly identical to more typical tonalite samples (Saleeby *et al.*, 1987). The inclusions clearly are not xenoliths derived from framework rocks. Bulk compositions are drastically different, and framework rock zircon populations are dominated by mid-Proterozoic grains.

The most reasonable interpretations for the mafic inclusions are the following: (1) they represent fragments of restite or residuum transported with the tonalite host from source regions; or (2) they are the products of magma mixing between tonalitic and mafic end members. The second interpretation is favored. Near and within the mixed phase of the batholith, main-phase tonalite undergoes mixed and continuous gradations into mafic tonalite and diorite which are similar in texture and mineralogy to the mafic inclusions dispersed throughout the batholith. Such rocks also exhibit compositions and textures that are similar to parts of the more mafic members of the Bear Valley Springs suite (Bison Peak and Tunis Creek units). The mafic inclusions also exhibit structural relations with tonalite host rocks suggestive of synplutonic conditions. These include disrupted dike-form and pillow-form structures that have been described in settings suggestive of magma mixing (Blake *et al.*, 1965; Marshall and Sparks, 1984). In general, such features are obscured by synplutonic flow deformation in the tonalite batholith.

Except for the framework septum west of Cummings Valley, the southwest margin of the tonalite batholith is gradational into adjacent map units (Fig. 31-1). The westward gradation into the tonalite gneiss of Tejon Creek is defined by an increase in K-feldspar, scattered appearance of garnet or muscovite, pronounced gneissosity and a reduction in average grain size. U/Pb zircon systematics are distinct between the tonalite batholith and the tonalite gneiss unit. The southward gradation of the tonalite batholith into the hypersthene tonalite of Bison Peak is more subtle. Modal biotite and quartz decrease and early igneous hypersthene is more common, but not pervasive. Hypersthene is also locally joined by clinopyroxene. The transition is also marked by a decrease in the intensity of planar deformation fabrics within the Bison Peak unit.

Hypersthene tonalite of Bison Peak The composition of this tonalite unit ranges from tonalite to quartz diorite. Inclusions, schlieren, and deformed dikes of dioritic to gabbroic rock of Tunis Creek unit affinity also occur within the Bison Peak unit. The main tonalite is a medium- to coarse-grained, hypidiomorphic granular to xenomorphic, weakly to moderately foliated rock. Gneissic banding is rare.

The main mineral assemblage is quartz, biotite, poikilitic hornblende, and weakly zoned plagioclase. Hypersthene is a common minor constituent. Garnet and clinopyroxene are less common accessory minerals, but are not found together. Hypersthene and clinopyroxene are locally found together and, where so, they are early igneous phases. Cummingtonite is also present as a minor constituent, mainly as a reaction zone between hypersthene and hornblende. The rock commonly shows faint ductile deformation fabrics with little annealing. Cataclastic textures are locally present near the Garlock fault and adajcent to the diorite gneiss of White Oak.

The Bison Peak unit appears to intrude both the tonalitic and dioritic gneisses of Tejon Creek and White Oak, although contact relations are poorly exposed, and superimposed tectonitic fabrics complicate relations with the White Oak unit. Several metasedimentary bodies (quartzite, marble, paragneiss, calc-silicate rock, and psammatic schist) are enclosed within the hypersthene tonalite. They represent large xenoliths that are compositionally distinct from mafic inclusions and schlieren. Compositions, textures, and structural settingss of the mafic inclusions grade into rocks which suggest synplutonic relations between the hypersthene tonalite and the gabbroidal rocks of Tunis Creek. These include continuous gradations between hornblende (± hypersthene) quartz diorite of Bison Peak with dioritic variants of the Tunis Creek unit near their mutual contact, dikes of Tunis Creek-type gabbro cutting the tonalite, and nearly meter-scale lenses of similar gabbro completely encased in tonalite. A 1–2-km thick selvage-like unit of Bison Peak-type tonalite borders the southern margin of the Tunis Creek unit (Fig. 31-2). The selvage intrudes the quartzofeldspathic gneiss of Pastoria Creek, but exhibits mixed and continuous gradations with Tunis Creek unit.

Gabbroids of Tunis Creek and Squirrel Spring The gabbroid of Tunis Creek is fine- to coarse-grained, hypidiomorphic to allotriomorphic granular, non- to weakly foliated rocks of mainly mafic but also ultramafic and dioritic composition. Most specimens consist of subequal amounts of hornblende and plagioclase, and lithologic compositions most commonly range from gabbro to diorite. Minor amounts of garnet, opaques, hypersthene and clinopyroxene are locally present. Hornblende locally contains cores of clinopyroxene and commonly poikilitically encloses plagioclase grains. Local olivine occurs as cores rimmed by orthopyroxene, while secondary cummingtonite locally occurs as a rim around orthopyroxene within hornblende oikocrysts. Calcic- to intermediate-plagioclase occasionally shows weak zoning. Almandine garnet porphyroblasts are euhedral to subhedral, and contain plagioclase inclusions. The garnet has grown granoblastically across earlier igneous textures and structures, and less commonly across ductile deformation fabrics in igneous mineral assemblages (Sams, 1986). In most instances, garnet has grown at the expense of hornblende. Garnet and two pyroxenes are not found together. The textural settings of the pyroxenes indicate an igneous origin as early liquidus phases or as olivine reaction products. Subordinate ultramafic rocks consist of centimeter-scale patches of hornblendite and meter-scale pods of cumulate wehrlite, hornblende-plagioclase clinopyroxenite, and hornblende orthopyroxenite.

Textures within the Tunis Creek unit are highly variable at both outcrop and map scale. Sharp contacts between different textural domains suggest numerous injections of mafic magma batches, each with a slightly different composition or crystallization history. Hornblende most clearly defines textural variation ranging from subhedral prisms to large centimeter-scale oikocrysts. Centimeter-to-meter-scale patches of hornblende pegmatite also occur locally. Occasional planar rhythmic variation in modal poikilitic hornblende suggests a heteradcumulate texture. Allotriomorphic textures with non- or weakly zoned plagioclase suggest adcumulate crystal growth. Crystal accumulation cannot be shown to have occurred by gravitational settling along a horizontal or subhorizontal magma chamber floor. Sedimentation features are rare and occur in pods of rock that appear to have been remobilized or engulfed by later magma batches. Sharp, commonly irregular contacts between texture domains suggest multiple injections with flow differentiation and/or filter pressing as an important crystal accumulation mechanism. Cumulate textures and structures may also have developed along static boundaries between largely crystalline and magma domains (McBirney and Noyes, 1979). However, field relations suggest a complex geometry for such domains with numerous disruptions by repeated magma injections.

Metamorphic textures indicating prograde mineral growth within the gabbroids are apparently absent. Ductile deformation fabrics are rare and most concentrated near marginal contacts or in enclaves or dikes within adjacent units. Such deformation textures show green hornblende clusters as tails on igneous hornblende porphyroclasts, and partially recrystallized tails on ductilely deformed plagioclase grains. Garnet is present locally as a disequilibrium phase, statically grown across deformational as well as igneous fabrics. Textures suggestive of complete metamorphic annealing are locally observed, but without garnet in an equilibrium texture. Such textures are likely to be adcumulate in origin since the phases are identical to those present in clear igneous textures.

The gabbroidal rocks of Squirrel Spring on the whole are less mafic than those of Tunis Creek with notable amounts of hornblende quartz diorite, a compositional variant only observed along boundaries of the Tunis Creek unit. Cummingtonite is locally observed in the Squirrel Spring unit, but neither pyroxenes nor olivine has been observed. Garnet is locally present, and at least locally occurs as a late static phase. The petrographic data base on this unit is not nearly as dense as that of the Tunis Creek unit. The inclusion of the Squirrel Spring unit into the Bear Valley Springs suite is based on its significantly less deformed state compared to the gneiss complex, at least local intrusive contacts into the complex, and petrographic similarities with parts of the Tunis Creek unit.

Contact relations between the Tunis Creek and Bison Peak units indicate synplutonic injection of tonalitic and mafic magma. Furthermore, the gradational relations between the Bison Peak unit and the tonalite of Bear Valley Springs indicate that Tunis Creek mafic magmatism accompanied the emplacement of the tonalite batholith. Such relations suggest that the myriad mafic inclusions within both the tonalite batholith and the tonalite of Bison Peak are the mineralogically equilibrated remnants of Tunis Creek affinity magma batches that mixed with the more voluminous tonalitic magmas.

Gneiss Complex of the Tehachapi Mountains

This heterogeneous suite of rocks consists of penetratively deformed gneissic rocks of mafic to felsic and psammitic to calc-silicate composition, with small ultramafic pods and local lenses of metaquartzite, highly recrystallized marble and granulite. Map units consist of the tonalite gneiss of Tejon Creek, quartzofeldspathic gneiss of Pastoria Creek, diorite gneiss of White Oak, and the paragneiss of Comanche Point.

The diorite gneiss of White Oak (after Sharry, 1981) forms a narrow strip that parallels the Garlock fault. It is pervasively sheared parallel to the fault, with brittle deformation and granulation increasing toward the fault. It is for the most part a biotite hornblende diorite with lesser amounts of quartz diorite. Pyroxenes and garnet occur rarely, but not together. It is distinguished from the hypersthene tonalite of Bison Peak and the quartzofeldspathic gneiss of Pastoria Creek by a lower quartz, biotite, and hypersthene content, and on its pervasive mylonitic to cataclastic foliation. Remnants of high-temperature ductile deformation fabrics occur locally, but in most places are obscured by the lower-temperature fabrics. The retrograde chlorite appears to postdate retrograde muscovite and greenish biotite ± amphibole that are also widespread, but not dominant, in much of the gneiss complex. White Oak unit contacts with other units are rare, and appear to be concordant with compositional and structural layering, although such structural relations may represent transposition by the late-stage deformation. Such late-stage deformation and retrograde metamorphism may be related to the initial uplift of the gneiss complex.

Gneiss units of Tejon Creek and Comanche Point These units are discussed together due to their intimate contact and map distribution relations. The Tejon Creek unit is for the most part a tonalitic orthogneiss, whereas the Comanche Point unit is psammitic to calc-silicate paragneiss. These units form a north-northwest-trending belt along the southwest margin of the tonalite of Bear Valley Springs, and turn abruptly to an east-west trend along the northernmost slopes of the Tehachapi Mountains. In the Tehachapi Mountains, these two units bear complex contact relations with the Bison Peak and Tunis Creek intrusive units, as well as quartzofeldspathic gneiss of Pastoria Creek affinity.

The paragneiss of Comanche Point consists mainly of quartzofeldspathic and biotite-rich layers which alternate at centimeter scales. Quartzofeldspathic layers consist mainly of quartz, K-feldspar, and plagioclase in approximate granite-minimum relative proportions, with subordinate biotite and local muscovite or garnet. In many areas, the banding is clearly the result of partial melting as shown by crosscutting vein and dike-type structures. However, in some locations the banding may mimic earlier (primary?) compositional layering; such layering may actually control leucosome development, as seen to the north in the Lake Isabella area (Saleeby and Busby-Spera, 1986). The overall composition of the paragneiss is psammitic. Calc-silicate rich layers are subordinate yet widely dispersed. These consist of the quartzofeldspathic assemblage but with notable appearance of green hornblende, diopside, and scapolite, and an increase in modal plagioclase and/or garnet. Some dark layers, with or without calc-silicate minerals, contain accessory graphite. Aluminosilicate minerals are apparently lacking as a result of composition. Textures within quartzofeldspathic and calc-silicate layers are granoblastic, with K-feldspar and occasionally garnet as distinct porphyroblasts. Biotite-rich layers are moderately to strongly schistose. Open to tight flexural flow folding of apparent random orientation accompanied the peak metamorphism. Unequivocal premigmatization tectonic structures have not been recognized in the paragneiss.

The tonalite gneiss of Tejon Creek is typically a medium-grained, allotriomorphic granular, well-foliated tonalite. The foliation is defined by feldspar and ductilely deformed quartz aggregates interleaved with aligned mafic grain aggregates. The gneiss resembles the tonalite of Bear Valley Springs, but with a stronger deformational fabric. U/Pb zircon ages, however, suggest a 10–20-m.y. age difference between the two tonalites. The Tejon Creek unit is interpreted to represent a distinctly earlier pulse of tonalitic magmatism than that of the Bear Valley Springs. Plagioclase is locally zoned, although most rims are recrystallized. K-feldspar, and clinopyroxene and orthopyroxene (mutually exclusive) that occur as cores in hornblende, opaques, and zircon are accessory minerals. The tonalite gneiss exhibits a relict igneous texture with a gneissic overprint. Mafic lenses and ovoid inclusions are common, but not with the density as that in the tonalite of Bear Valley Springs. The lenses and inclusions lie concordantly within the gneissic fabric of the host, and range in composition from mafic tonalite to diorite. Mineralogical equilibration with the host tonalite is widespread. The inclusions are clearly distinct from vestiges of the paragneiss enclaves, and are suspected to be of synplutonic origin.

The transition between the tonalite gneiss of Tejon Creek and the tonalite batholith of Bear Valley Springs was discussed above. The contact between the tonalite gneiss and the framework septum east of Cummings Valley is sharp yet concordant. The relative paucity of partial melt products in the septum versus the Comanche Point unit is apparently compositionally controlled with significantly more siliceous protoliths dominating in septum. In the vicinity of the migmatic paragneiss, the tonalitic gneiss grades into granodioritic (± garnet ± muscovite) gneiss by variations in K-feldspar/plagioclase modes. Contact relations between the tonalite gneiss of Tejon Creek and gabbroids of Tunis Creek are poorly exposed. However, within the contact zone, quartz diorite and mafic tonalite were observed, suggesting a gradational contact.

A number of interesting contact relations exist between the paragneiss of Comanche Point and the tonalite gneiss of Tejon Creek. Partially transposed apophyses of the tonalite are locally observed intruding the paragneiss, but in general the paragneiss lies concordantly against the tonalite gneiss. Concordant contacts are both sharp and diffuse, suggesting various degrees of assimilation of the paragneiss into the tonalitic magma. In several locations, leucosome veins and dikes appear to have been partly fused and transported out of the paragneiss into the tonalite. Such veinites are distinguished from the apophyses by scale (centimeters versus tens of meters), continuity with the paragneiss migmatitic veinwork, and a granitic (± garnet) versus tonalitic composition. An additional contact relation is exhibited within the tonalite gneiss where 10-meter-scale domains of more intensely banded gneiss appear to be nearly, yet incompletely digested vestiges of paragneiss. Analogous contact relations are exhibited between the paragneiss and gabbroic intrusions of the Tunis Creek unit along the northern edge of the Tehachapi Mountains. In one well-exposed traverse, hornblende gabbro passes rapidly into hornblende diorite, quartz diorite-mafic tonalite, inclusion-rich granodioritic gneiss, and then paragneiss over a distance of about 75 m. The granodioritic gneiss appears to represent nearly completely melted paragneiss, with the inclusions being vestiges of paragneiss. Numerous centimeter-scale granitic dikes occur as small swarms in the granodioritic-quartz diorite transition zone. These relations suggest that the gabbroic magma further facilitated melting of the paragneiss, and the resulting melt products contaminated the margin of the gabbroic intrusive. Isotopic data discussed below indicate widespread contamination of mafic and tonalitic magmas by metasedimentary melt products.

Quartzofeldspathic gneiss of Pastoria Creek Heterogeneous quartzofeldspathic gneisses occur as distinct 1–10-km bodies throughout the Tehachapi Mountains (Fig. 31-2). Such rocks are distinct from the gabbroidal, tonalitic, and paragneiss units discussed above in that they are more heterogeneous at map as well as mesoscales and possess a wider compositional range.

Main elements of the Pastoria Creek unit are granitic to granodioritic banded and augen gneiss, and tonalitic to dioritic banded gneiss. Much of the banding has resulted from the transposition of biotite ± hornblende grain aggregates into partially recrystallized domains commonly with strong linear shape fabrics. Granitic and tonalitic gneisses are present in subequal amounts. The granitic gneisses locally grade into banded gneiss and granofels of metasedimentary origin, some with granulite facies mineral assemblages. The subordinate granulites occur adjacent to the Tunis Creek and Bison Peak intrusives. Subordinate gabbroidal lenses and dikes, and small pods of ultramafic cumulates are also present within both granitic and tonalitic gneisses of the Pastoria Creek unit. Concordant to slightly discordant layers of granitic pegmatite (biotite + muscovite ± garnet) as well as rare lenses of metaquartzite and marble are scattered throughout the gneisses.

Granitic and tonalitic varieties of gneiss at least locally grade into one another by variation in K-feldspar/plagioclase modes. K-feldspar is commonly porphyroclastic, and plagioclase is typically weakly zoned, with stronger zonation occurring in more mafic varieties. Quartz is commonly ductilely deformed into undulose ribbons with local recrystallization, particularly in K-feldspar augen gneiss. Granofelsic textures with significant annealing are also common. The augen gneisses may represent domains with a more protracted deformation history. Granitic gneisses typically lack hornblende and commonly contain garnet. Tonalitic varieties contain subequal green hornblende and biotite. Biotite is deep red to brown pleochroic. Large (1–5 cm) red almandine garnets have locally grown statically across deformation fabrics in tonalitic to dioritic gneisses at the expense of hornblende as in the Tunis Creek unit. Finer garnets within granitic gneisses are more commonly set in apparent equilibrium textures.

Some tonalite gneisses contain euhedral to subhedral epidote as a replacement product of hornblende. Well-formed crystal faces on epidote grains from such samples are also observed embayed into biotite flakes, and vermicular contact zones between the epidote and plagioclase ± quartz are also present. Such textural features in tonalites are interpreted as indicators of magmatic crystallization of the epidote under high-pressure conditions (Zen, 1985; Zen and Hammarstrom, 1984). These authors also note the weak zonation of plagioclase in such deep-level tonalites as compared to shallower tonalites. It is possible that such "magmatic epidote" textures once pervaded the Pastoria Creek tonalites, but have been obscured by the ductile deformation fabrics.

High-grade metasedimentary enclaves within the Pastoria Creek unit include paragneiss similar to that of the Comanche Point unit, metaquartzite, marble, and rare quartzo-feldspathic granulite. The granulitic rocks occur as diffuse zones of massive to faintly gneissose rock within granitic gneiss. In addition to quartz, feldspars, red biotite, and hornblende, such granulites contain hypersthene + graphite or clinopyroxene + graphite ± garnet. A metasedimentary origin is based on rare intergradations with psammitic para-gneiss, the presence of graphite, and on heavy oxygen isotopic compositions (Ross, 1983c). The presence of these granulites raises the question of how much of the granitic gneiss could be ultrametamorphosed sedimentary material. As discussed below, isotopic data suggest that the typical granitic gneisses of the Pastoria Creek unit are contaminated with metasedimentary material, but that the constituent minerals formed under magmatic conditions.

In summary, the quartzofeldspathic gneiss of Pastoria Creek represents a mixed assemblage of magmatic and high-grade metasedimentary materials. Deep-level tonalites, diorites, and gabbros invaded a metasedimentary framework which underwent partial to complete melting and local granulite facies metamorphism. As more clearly observed with the Tejon Creek unit, most of the granitic gneiss is interpreted to be magmatic material contaminated by metasedimentary melt products. Continuous gradations between tonalitic and granitic compositions of Pastoria Creek also permit magmatic differentiation to have operated in addition to assimiliation. Penetrative, ductile flow extended through solidus and hot subsolidus conditions. Garnet grew as a static metamorphic phase in the hot subsolidus environment primarily within tonalitic and more mafic compositions and apparently as an igneous phase in granitic compositions. Scattered mylonitic fabrics indicate that domainal flow or shear persisted well into the subsolidus, or alternatively, an additional tectonic event was superimposed while the rocks were in medium-grade meta-morphic conditions.

Late- and Postdeformational Intrusives

In the discussions above, considerable attention is placed on the textural and structural expression of high-temperature deformation fabrics present in the gneiss complex of the Tehachapi Mountains and the Bear Valley Springs intrusive suite. It is thought that most of the high-temperature fabrics developed in conjunction with deep crustal magmatism which culminated at ~100 m.y.b.p. A number of intrusive bodies cross-cut these fabrics and are only moderately to weakly deformed. Major plutonic bodies exhibiting late-defor-mational structural relations are the tonalite of Mount Adelaide (latest phase of the Bear Valley Springs complex) and the granodiorite of Claraville, which cut the penetrative deformation fabric in the southern end of the tonalite batholith as well as the deformation fabric of the framework septum. Three small late-deformational intrusive bodies include small ovoid tonalitic stocks that cross-cut the main deformation fabrics in the augen gneiss of Tweedy Creek and its metasedimentary host, a swarm of non- to weakly deformed pegmatite dikes that cross-cut the main deformation fabrics in the tonalitic gneiss of Tejon Creek and the paragneiss of Comanche Point, and a foliated two-mica garnet granite dike that cuts the main deformation fabric of the quartzofeldspathic gneiss of Pastoria Creek.

Thus, the 90–100-m.y. U/Pb ages determined on selected cross-cutting intrusives provide important age constraints on the development of the high-grade deformation fabrics, provide a test for the synplutonic deformation interpretation, and are consistent with the field and petrographic observations that major ductile deformation of the region accompanies igneous and high-grade metamorphism recorded in the gneiss complex of the Tehachapi Mountains and in the Bear Valley Springs intrusive suite.

Igneous ages derived from U/Pb zircon data for the major crystalline rocks of the study area are shown on Fig. 31-2 (after Sams, 1986; Saleeby *et al.*, 1987). Such ages fall into a narrow range of 90–120 m.y., and consist of three main igneous suites. The first includes the gneiss complex of the Tehachapi Mountains and the augen gneiss of Tweedy Creek. Internally and externally concordant zircon ages of the orthogneisses indicate igneous crystallization between 110 and 120 m.y.b.p. A number of subtle discordances along with the physical properties of the gneiss zircon populations suggest minor inheritance of Mid- to Late Proterozoic zircon and limited open system behavior following the 120–110-m.y. igneous event in response to a major 100-m.y. plutonic event. The second age suite, the 100 ± 2-m.y.-old Bear Valley Springs intrusive suite (tonalites of Mount Adelaide, Bear Valley Springs, and Bison Peak; gabbroid of Tunis Creek) contains rocks which cross-cut the older, more deformed suite, but exhibit synplutonic deformation. The zircon systematics of this suite are for the most part internally and externally concordant, indicating igneous crystallization, with only local evidence of inheritance of Late Proterozoic zircon. The 90–94-m.y. suite of late to postdeformational intrusive rocks yields both concordant and discordant zircon ages. In general, strongly discordant samples are rare, and include the granodiorite of Claraville (concordia intercepts of 90/1900 m.y.), the paragneiss of Comanche Point (108/1450 m.y.), and a quartzite in the Kings sequence metasedimentary framework rocks (1700 m.y. upper intercept).

A summary concordia diagram for the U/Pb zircon dates is shown in Fig. 31-3. The limits of the inheritance trajectories are derived by projecting the upper intercepts from the granodiorite of Claraville, Kings sequence quartzite, paragneiss of Comanche Point, and metagabbro of Tunis Creek samples to concordia through the field of the gneiss complex data points. The upper intercepts (1900, 1700, 1450, and 1383 m.y., respectively) reflect the overall isotopic character of the contaminate zircon, but do not necessarily represent actual ages. The downward dispersion of the ages on the gneiss complex from about 115 ± 5 m.y. is probably due to lead loss during the extended mid- to lower-crustal residency of the complex. Thus, a lead evolution envelope is shown which is constrained by inheritance trajectories from Proterozoic to mid-Cretaceous ages, and superimposed open-system lead loss or uranium gain trajectories from the concordant gneiss complex data points toward the 100-m.y. Bear Valley thermal culmination.

It is suggested that the position of most or all the gneiss complex data points within the envelope reflects mainly igneous crystallization at 115 ± 5 m.y., with minor discordances related to both inheritance and lead loss in a multistage history. Magmas that produced the orthogneisses locally inherited minor amounts of Proterozoic zircon from partial to complete melts derived from the metasedimentary framework rocks. Some of the magmas have inherited small amounts of early Phanerozoic zircon from deeper levels, as suggested by some of the discordance patterns in the zircon systematics (Sams, 1986). The overall zircon population was isotopically dominated by the growth of new mid-Cretaceous igneous zircon. Such igneous zircon may have crystallized over an extended period of time, and/or acted as partially open systems in the high-temperature environment that lasted through the Bear Valley Springs intrusive event.

The field relations and zircon isotopic data presented above clearly demonstrate assimilation and inheritance of metasedimentary material into orogenic magmas. Further evidence for mixing can be seen in the strontium and oxygen isotopic data on the zircon sample suite (Saleeby *et al.*, 1987), which suggests simple two-component mixing of mantle-derived gabbroic to tonalitic magmas (modelled after the hypersthene tonalite of Bison Peak-metagabbro of Tunis Creek) with partial to complete melt products from metasedimentary framework rocks (Kings sequence quartzite-paragneiss of Comanche Point).

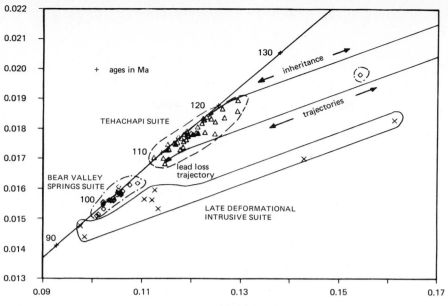

FIG. 31-3. Summary concordia diagram (after Wetherill, 1956) showing inheritance trajectories indicating Proterozoic contaminates in Cretaceous magmas, Cretaceous disturbance trajectories of Proterozoic detrital zircon populations, and Pb-evolution envelope for orthogneiss populations interpreted to have incorporated minor components of Proterozoic zircon during ~117-m.y.-old magmatism and then remained partially open until ~100 m.y.b.p.

Sedimentary admixtures for granitic rocks may be as high as 45% but no higher than about 15% for the tonalitic batholithic complex, with younger and more easterly samples of the Bear Valley Springs pluton or other rock units (granodiorite of Claraville) requiring a larger metasedimentary component. Modeled values of 10–20% are typical for orthogneisses.

A plot of $\delta^{18}O$ versus initial $^{87}Sr/^{86}Sr$ (denoted by Sr_0) for the sample suite is shown in Fig. 31-4. The quartzite and paragneiss of Comanche Point are shown with isotopic compositions at 100 to 108 m.y. Sr_0 correlates directly with $\delta^{18}O$, and generally correlates inversely with Sr content for most of the samples, supporting the mixing model postulated on the basis of the field and zircon studies. The data have been grouped into two sectors based on geographic distribution. The eastern sector samples are those that lie along or east of the line where $Sr_0 = 0.706$[1] (0.706 isopleth of Kistler and Peterman, 1973, 1978) (tonalite of Bear Valley Springs, granodiorite of Claraville, tonalite stock and augen gneiss of Tweedy Creek), and have quartzite as representative of their metamorphic framework. The western sector samples are those with $Sr_0 < 0.706$ (metagabbro of Tunis Creek, hypersthene tonalite of Bison Peak, gneiss complex of the Tehachapi Mountains), and have paragneiss as their representative metamorphic framework rock. The two sectors are distinctly separated on the plot with little overlap, and have trajectories that project toward their respective metasedimentary endmembers. This suggests that the rocks east and north of Tejon Creek were contaminated preferentially

[1] Also R.W. Kistler, unpublished data.

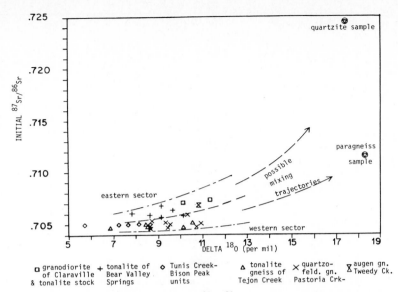

FIG. 31-4. Plot of $\delta^{18}O$ versus initial $^{87}Sr/^{86}Sr$ for all rocks of the southern-most Sierra Nevada, showing mixing trajectories from low $\delta^{18}O$-low initial $^{87}Sr/^{86}Sr$ toward sampled metasedimentary framework values.

by Kings sequence rocks, whereas those to the south and west interacted with rocks represented by the paragneiss.

Subtle complexities in the Sr isotopic data and zircon inheritance patterns (Sams, 1986; Saleeby *et al.*, 1987) suggest the involvement of a third, cryptic component. Such a component is thought to be early Phanerozoic ensimatic accretionary terranes which constitute the Sierra Nevada western metamorphic belt to the north. Such rocks are not observed in the study area, although regional geological relations (Saleeby, 1979) and basement core data in southernmost San Joaquin Valley (Ross, 1979), as shown on Fig. 31-1, suggest that such rocks may have extended into or beneath the observed metamorphic framework for the Cretaceous igneous complex. Thus, their potential contribution to the system should not be ignored.

A reconnaissance study of the Nd/Sm isotopic ratios of batholithic and metasedimentary rocks from the central Sierra Nevada and Peninsular Ranges batholiths east of the $Sr_0 = 0.706$ isopleth was performed by DePaolo (1981). Results showed an inverse correlation of ϵ_{Nd} to both Sr_0 and $\delta^{18}O$, with a model Nd/Sm age for the source of the metasedimentary rocks of 1.5-1.9 b.y. His model for the generation of the batholithic rocks involves a mixture of mantle-derived island-arc type magmas and craton-derived metasedimentary components. The approximate percentages are: modeled by Nd systematics, 70-80% upper mantle, 20-30% crustal; using $\delta^{18}O$, 65% mantle to 35% crustal. Chen and Tilton (1978) constructed Pb evolution diagrams from feldspar data on plutonic rocks of the south-central Sierra Nevada, and obtained a similar 1.8-b.y. model age for the source region for the batholithic magmas. They believed that the source region for the batholith was in part Proterozoic, sialic material. Kistler and Peterman (1978), based on Sr_0, constructed a limiting isochron of 1.7 b.y. ($Sr_0 = 0.7020$, slope equivalent to 1.6 b.y. at 100 m.y.) for intrusive rocks of the south-central Sierras. They believed that magmas formed by melting at depths that ranged from upper mantle (west of the $Sr_0 = 0.706$ isopleth) to the lower crust (east of the 0.706 isopleth), and that the source reservoirs formed at 1.7 b.y. Based on neodymium-samarium and rubidium-strontium studies of mid- to lower-crustal xenoliths in the central Sierra, Domenick *et al.* (1983) suggested that the covariation in Sr_0 with $^{87}Rb/^{86}Sr$ and $\delta^{18}O$ is not due to simple mixing of a two-end member system, and that the source region for Sierran magmas was heterogeneous. They concluded that the generation of magmas was a result of partial melting of an isotopically diverse source region in the lower crustal or upper mantle, with potential contamination and mixing in mid- to upper-crustal regions.

The data presented in this paper and in Saleeby *et al.* (1987) generally support the model for the source for the magmas that formed the southernmost Sierra Nevada batholith as requiring upper mantle derived magmas that incorporated varying amounts of cratonal derived metasedimentary material prior to crystallization. A third, cryptic early Phanerozoic ensimatic crustal component may be required to explain the subtle isotopic variations in a number of samples. Such a third component could be Paleozoic and early Mesozoic accretionary terranes exposed along the western Sierra metamorphic belt to the north.

INTERRELATIONS BETWEEN IGNEOUS AND METAMORPHIC PHENOMENA

The literature on the gneiss complex of the Tehachapi Mountains is confusing with respect to the separation of the products of igneous versus metamorphic phenomena. Sharry (1981, 1982) refers to the gneiss complex as a granulite terrane, and Ross (1985) discusses a number of equivocal criteria for distinguishing the complex as a granulite versus high-grade amphibolite metamorphic complex. The gabbroids of Tunis Creek and the tonalite of Bison Peak constitute the main granulite assemblage of Sharry (1981), and Ross (1985) refers to the "Tunis Creek hot spot" in terms of an area from which metamorphic grade appears to descend from granulite(?) facies. Both of these workers emphasized the local occurrence of hypersthene or garnet as criteria for designating the Tunis Creek and Bison Peak units as granulites or granulite-affinity rocks.

The Tunis Creek and Bison Peak units are considered here to be typical members of the western gabbroic to tonalitic phases of the Cretaceous Sierra Nevada batholith. The more mafic members of the western Sierran batholith consist of (1) hornblende-rich gabbros with various admixtures of olivine, olivine $\rightarrow$ hypersthene, and hypersthene $\pm$ cummingtonite $\pm$ clinopyroxene; (2) two-pyroxene gabbro-diorite with various admixtures of hornblende and reddish biotite; and (3) hypersthene-hornblende-biotite quartz diorite-tonalite with various admixtures of cummingtonite $\pm$ clinopyroxene (Best, 1963; Ehrreich, 1965; Mack *et al.*, 1979; Saleeby and Sharp, 1980; Saleeby and Williams, 1979). Subordinate pods of hornblende wehrlitic and orthopyroxenitic ultramafic rocks are scattered throughout this distinctive batholithic assemblage. In the more mafic gabbroic rocks, brown hornblende and calcic plagioclase dominate, and are commonly interlocked in an adcumulate texture which resembles an annealed metamorphic texture. An additional important textural relation is the poikilitic habit of hornblende enclosing numerous plagioclase grains. Olivine or hypersthene commonly occur with reaction rims (pyroxene or cummingtonite, respectively) within the hornblende oikocrysts. In the hypersthene or two-pyroxene quartz-bearing rocks, hypersthene $\pm$ clinopyroxene commonly occurs within green to brown hornblende oikocrysts. Based on regional field relations, petrography, and zircon ages, the Tunis Creek and Bison Peak units can be confidently designated gabbroic to tonalitic intrusives related to the western Sierra Nevada batholith.

An important difference between the gabbroidal phases of the western Sierra Nevada batholith north of the study area and the more mafic rocks of the study area is the local subsolidus growth of garnet in the latter. Reddish almandine-rich garnet porphyroblasts, some up to 5 cm in diameter, have grown at the expense of igneous hornblende. Vestiges of igneous plagioclase grains are commonly included within the garnet, and the localized hornblende depletion commonly leaves a "bleached" plagioclase-rich halo around garnet grains. Such garnet growth occurs most commonly in gabbroic to dioritic rocks, but is also observed in some of the more mafic tonalites. It is texturally distinct from garnet in the granitic rocks which is more typically set in an apparent equilibrium texture. The subsolidus growth of garnet in the gabbroids preferentially occurred near or along contacts with quartzofeldspathic assemblages. Chains of porphyroblasts are locally distributed along fractures and thin garnet-fill veins. Large and small garnet grains, "bleached halos," and fracture system clusters clearly cut and postdate igneous textures (Sams, 1986). Locally such features statically cut across gneissic banding that presumably formed by synplutonic or hot-subsolidus flow.

The local late-stage garnet growth in the more mafic rocks of the study area is interpreted to have resulted from a protracted hot-subsolidus residence time. Such extended hot-subsolidus conditions resulted from a deeper level of intrusion as compared to similar gabbroids to the north. The preferred localization of garnet growth along contacts with quartzofeldspathic assemblages, which show predominantly retrograde paragenetic relations (Sams, 1986), is perhaps related to the existence of pathways for the diffusion of water liberated by hornblende breakdown. Analogous hot-subsolidus garnet growth is reported from the border phases of the mafic Scourie dyke, Sutherland, which was emplaced into a high-grade metamorphic environment represented by the host Lewisian gneisses (O'Hara, 1961). The critical point concerning the metamorphic garnet growth in mafic rocks of the study area is that they represent a period of high-grade autometamorphism developed along a retrograde P-T trajectory which initiated with magmatic crystallization of wet magmas. The alternative explanation of the garnet representing prograde metamorphic mineral growth is unlikely inasmuch as the garnet occurs in the youngest intrusive phases of the complex that are in large part non- or weakly deformed, and by the fact that retrograde paragenetic sequences across igneous assemblages are the predominant pattern.

Mineral paragenetic sequences in most of the quartzofeldspathic gneisses are clearly retrograde as summarized above. The zircon ages along with the preservation of some intrusive contacts with igneous textures indicate that most of these gneisses were in an igneous state in the mid-Cretaceous. The clustering of K/Ar ages between 90–80 m.y. for rocks of the gneiss complex and the tonalite of Bear Valley Springs (Ross, 1983a) reflects cooling of the crystalline rocks through the biotite and hornblende argon-retention temperatures starting ~10 m.y. year following the Bear Valley Springs magmatic culmination. It is thus reasonable to suspect that most or all of the retrograde metamorphism followed a P-T trajectory that descended from igneous states. As discussed below, the retrogressing trajectory appears to have continued on through to a rapid tectonic uplift event in the early Tertiary.

Prograde paragenetic sequences appear to be restricted to metasedimentary framework remnants. Compositional members that would most clearly show granulite-facies mineral assemblages (pelitic, psammitic, quartzofeldspathic) were for the most part melted during the invasion of the much more voluminous magmatic bodies. However, as noted in the petrographic overview section, rare lenses of granulite facies paragneiss and granofels occur in the Pastoria Creek unit, in the vicinity of the Tunis Creek and Bison Peak intrusions. Much of the framework complex possesses bulk compositions that do not yield diagnostic metamorphic mineral assemblages (quartzite and marble). However, calc-silicate layers in the Comanche paragneiss unit contain scapolite, which is consistent with high-amphibolite to granulite facies metamorphism (Winkler, 1967). The subordinance of prograde paragenetic sequences is most clearly related to the fact that the bulk of the gneiss complex was in a magmatic state during the peak phases of high-grade metamorphism.

Paleodepth estimates for the genesis of the gneiss complex are important for both petrologic and tectonic considerations. Sharry (1981) performed thermobarometric calculations on which Ross (1985) based some of his major interpretations.The two-pyroxene geothermometer (Wells, 1980; Wood and Banno, 1973) on the Bison Peak unit yielded temperatures of 850–925°C, which were considered to be unrealistically high. But, as noted above, the pyroxenes are magmatic, not metamorphic, and the analyses reported in Sharry (1981) are within the range of two-pyroxene intrusive complexes from the western Sierra and Klamath Mountains[2] (Snoke et al., 1981). The two-pyroxene thermometer is thus interpreted here to represent the approximate temperature at the point of chemical isolation of the pyroxene grains during descent through the solidus. Thermobarometric calculations using garnet-biotite thermometry and garnet-biotite-plagioclase-muscovite barometry on the Pastoria Creek unit yielded considerable scatter about 850°C and a pressure of ∼8 kb. However, muscovite has not been observed in a metamorphic equilibrium texture. It occurs locally as an igneous phase in two-mica granitoids and more commonly as a retrograde phase in quartzofeldspathic gneisses. Thus, the validity of these calculations is uncertain.

Nevertheless, a deep-level genesis, perhaps ∼8 kb, is suggested for the gneiss complex on other petrologic grounds. Phase relations for wet basaltic to andesitic magmatic systems summarized by Green (1982) suggest that an isobaric cooling trajectory through the solidus at a pressure of 7–8 kb would encounter a garnet phase boundary. This is consistent with textural relations discussed above for the scattered garnet overprints in gabbroic to mafic tonalitic compositions. The highly scattered distribution of such garnet porphyroblasts and their settings in nonequilibrium textures is thought to reflect the descending temperature trajectory and resulting deceleration of reaction rates. An additional petrologic argument for a deep-level origin for the gneiss complex is the presence of magmatic epidote in tonalitic members of the Pastoria Creek unit. Zen and Hammarstrom (1984) present arguments for ⩾8-kb crystallization depths for such magmas, although substantial experimental work to confirm such a pressure has not been performed. Thus a number of independent lines of evidence suggest a pressure of formation for the gneiss complex of perhaps 8 kb. Accordingly, the depth of formation is tentatively taken as ∼25 km, considering a largely granitoid and silicic volcanic overburden in the mid-Cretaceous (Saleeby et al., 1986).

PETROTECTONIC SIGNIFICANCE OF THE GNEISS COMPLEX

The petrogenesis and tectonics of the gneiss complex are for the most part inseparable subjects. The tectonic position of the gneiss complex and the related batholith of Bear Valley Springs pose fundamental questions concerning the reconstitution of accretionary terranes into new sialic crust, and the mechanism(s) for the uplift and deep-level exposure of such relatively young crust. The isotopic and field data presented above show clearly that mantle-derived juvenile magmas mixed with melt products of craton-derived sediments to form hybrid magmatic rocks. The involvement of Phanerozoic ensialic accretionary terranes as an additional source component is hinted at by some of the isotopic patterns, and perhaps expected on regional tectonic grounds, but not directly supported by field observations.

---

[2] Also J. B. Saleeby, unpublished microprobe data.

Regional tectonic relations suggesting that such a third component should not be ignored are summarized in Fig. 31-1. Here, the southern end of the largely ensimatic western Sierra metamorphic belt is shown broken into smaller framework fragments to the south. The southern fragments constitute the Kings-Kaweah ophiolite belt. On this generalized map, the $Sr_0 = 0.706$ isopleth is shown generally separating the framework domain of Paleozoic ophiolite belt and its lower Mesozoic cover from the Kings sequence domain. This relation cannot be traced southward into the study area due to an apparent southern termination of the ophiolite belt. However, basement core data from the southern San Joaquin Valley (Fig. 31-1) suggest that ophiolitic rocks continue in the subsurface southward nearly to the White Wolf fault. Furthermore, the small septa of the Pampa schist (Dibblee and Chesterman, 1953) could be remnants of part of the lower Mesozoic section originally built on the ophiolite. These regional relations suggest that the boundary (suture) between the ophiolite belt and the Kings sequence once extended into the study area, but has been obliterated along with most of the framework rock. The prebatholith basement for the Kings sequence is unknown, but major alternatives include (1) Paleozoic shelfal strata with a possible attenuated Proterozoic crystalline substrate, (2) Paleozoic oceanic crust of possible Kings-Kaweah affinity, and (3) no basement, due to structural detachment during tectonic accretion during and shortly after deposition as a thick sedimentary prism. As noted above, an apparent western turbidite and slide block facies of the Kings sequence lies depositionally above the Kings-Kaweah ophiolite belt. The major point here is that there is reason to suspect that ophiolitic basement rocks and at least some of its cover rocks extended into the study area prior to batholith petrogenesis. Such basement rocks constitute one of the major components of the early Phanerozoic ensimatic accretionary complex of the western and northern Sierra. The interaction of mantle-derived magma, presumably generated in response to Franciscan subduction (Hamilton, 1969) with such ensimatic as well as Kings sequence framework rocks resulted in further growth and reconstitution of the tectonically assembled crust into new sialic crust.

Geobarometric estimates discussed in the preceding section along with the regional structure of the composite Sierra Nevada batholith suggest that progressively deeper levels of the batholith are exposed southward, and that the gneiss complex of the Tehachapi Mountains represents the deepest exposed levels (Saleeby et al., 1986). In this regard, the transition between the tonalite of Bear Valley Springs and the gneiss complex represents a key structural relation. Two alternative models are considered: (1) Uplift and rotation of the southernmost Sierra (discussed below) resulted in considerable tilting of the crystalline rocks such that the transition represents the floor of the tonalite batholith. In this model the increase in deformation toward the southwest margin of the batholith can be explained as a result of lateral spreading of the batholith over its gneissose substrate (similar to Hamilton and Meyers, 1967). The northern Mount Adelaide phase of the batholith represents a high-level magma batch that completely escaped the floor-level deformation. The Tunis Creek and Bison Peak units represent coeval subbatholithic magma batches that perhaps shared a common conduit system as the tonalitic batholith. (2) Regional patterns in Cretaceous batholithic activity of both the Sierra Nevada and Peninsular Ranges to the south suggest a reorganization in the locus of batholith emplacement at between 105 and 100 m.y.b.p. (Chen and Moore, 1982; Saleeby and Sharp, 1980; Silver et al., 1979; Stern et al, 1981). At this point in time, nearly the entire eastern edge of the Early Cretaceous magmatic belt was truncated by mid-Cretaceous plutons, and subsequent plutonism migrated continuously eastward. Such a regional change seems to require a tectonic event. The greater deformation observed along the southwestern margin of the tonalite batholith could represent a shear zone developed during the reorganization event which juxtaposed different crustal levels of adjacent magmatic age belts. The map distribution of the major units, particularly Tunis Creek and Bison Peak, and restoration models of Tertiary uplift, tilting, and rotation favor the geometry of the first model (Sams, 1986).

The structural continuity between the gneiss complex and more typical Sierran batholithic rocks to the north has important implications for the regional crustal structure and petrogenesis of the composite Sierran batholith. In addition to the paleodepth-age data discussed above for the southernmost Sierra Nevada, paleodepth-age relations of the Ritter Range and the Lake Isabella region (Fig. 31-1) clearly shows a southward increase in depth of exposure for crystalline rocks generated at ~100 m.y. In the Ritter Range area, a 100-m.y. resurgent caldera and subcaldera pluton are preserved, and record surface-level batholithic processes (Fiske and Tobisch, 1978; Tobisch et al., 1985). In the Lake Isabella area, metamorphic mineral assemblages equilibrated at ~3 kb pressure at ~100 m.y. (Elan, 1985; Saleeby and Busby-Spera, 1986). Thus, on a regional scale, the ~100-m.y. batholithic belt is exposed from silicic volcanic to sub- or deep batholithic levels between the Ritter Range area and the southernmost Sierra Nevada. Consideration of this regional relation offers a down-plunge view of the composite batholith which may be used to explore its gross crustal structure (Saleeby et al., 1986).

Consideration of the down-plunge view of the batholith along with structural relations discussed for the Ritter Range area (Tobisch et al., 1985), the Lake Isabella area (Saleeby and Busby-Spera, 1986), and those discussed above for the southernmost Sierra lead to some interesting implications. (1) Most of the crust consists of batholithic igneous material and metaigneous derivatives with only subordinate enclaves of preexisting framework rock. (2) Mixing of mafic magma (mantle-derived), tonalitic to quartz dioritic magma (derivation uncertain), and partial to complete melts from psammitic-pelitic framework rock is a major process at deep levels (~25 km). (3) Most metamorphic framework remnants at mesozonal to epizonal levels consist of steeply dipping, flattened and/or vertically stretched screens which typically separate different plutons. (4) Roof rocks consist primarily of caldera fill and outflow sequences which are nearly coeval with batholithic intrusives. (5) Fragments of such volcanic sequences may be penetratively deformed and transported downward shortly after eruption; such downflow may commence within a resurgent caldera cycle. A model for batholith kinematics based on these implications includes upward transport and high-level spreading of silicic magma derived from lower crustal mixing reservoirs represented in part by the gneiss complex of the Tehachapi Mountains. Upward magma transport may by dynamically linked to downward return flow of framework screens which occasionally trap synclinal or half-graben infolds of near-surface batholith-related volcanics (Tobisch et al., 1985). The paucity of prebatholithic framework rocks within roof domains suggests major extension during batholith emplacement and related volcanic venting (Tobisch et al., 1985). Such batholithic activity appears to have almost completely reconstituted the Sierran crustal column along the axis of the composite batholith. Critical questions are posed concerning the overall crustal extension resulting from such batholithic dynamics which ultimately must be in part constrained by mixing models, and a finer resolution of the contribution of mantle-derived materials to the reconstituted crustal column.

The rapid postbatholithic uplift of the southernmost Sierra Nevada is clearly documented by unconformable relations between the gneiss complex and Eocene marine and nonmarine strata in the western and southern Tehachapi Mountains, respectively (Nilsen and Clarke, 1975). Biotite and hornblende K/Ar ages on the crystalline rocks fall between 108 and 67 m.y. (Evernden and Kistler, 1970; Kistler and Peterman, 1978), and are interpreted to represent cooling from a 100-m.y. thermal peak, perhaps during the early stages of the uplift event. If the depth of the gneiss complex at 100 m.y.b.p. is taken as

~25 km, and the uplift was accomplished by 45 m.y. ago (mid-Eocene, Palmer, 1983), then an average uplift of ~0.5 mm/yr is implied. Such a rate is within the range measured for actively growing orogenic zones (Schumm, 1963).

The uplift event was accompanied or followed shortly by an apparent major rigid clockwise rotation event. Kanter and McWilliams (1982) report a clockwise rotation of 45 ± 15° with no resolvable latitudinal translation for the tonalite of Bear Valley Springs within the study area. They also report a reversed polarity for the batholith, which suggests rotation following the prolonged mid-Cretaceous normal polarity epoch (post- 84 m.y.; Harland *et al.*, 1982). These relations are consistent with the K/Ar dates and the cooling of the batholith through the Curie point. Kanter and McWilliams (1982) also report no resolvable rotation in a Miocene volcanic complex built on batholithic rocks northeast of the study area. McWilliams and Li (1983, 1985) and Plescia and Calderone (1986) reported that the tonalites of Bear Valley Springs and Mount Adelaide to the west of the White Wolf–Breckenridge–Kern Canyon fault system show no rotation or translation with respect to the craton since the Late Cretaceous. McWilliams and Li indicated that rocks from the gneiss complex near the mouth of Pastoria Creek had been rotated clockwise 59 ± 16°, whereas Plescia and Calderone working in the same area reported a clockwise rotation of 40 ± 10° since about 25 m.y. (K/Ar age on volcanic strata overlapping the crystalline basement). McWilliams and Li (1983) reported that the granodiorite of Lebec (south of the Garlock fault) has 90° + of clockwise rotation. These relations suggest that the southern end of the tonalite batholith along with the gneiss complex underwent a very sharp clockwise rotation between 80 and 20 m.y.b.p.

The northern branch of the Garlock fault (Fig. 31-2) has been suggested to be the modified remnants of a thrust fault between lower-plate greenschist facies Rand schist and the gneiss complex (Sharry, 1981). The Rand schist is a member of distinctive suite of lower-plate schists scattered throughout southern California, in all cases tectonically beneath both Mesozoic batholithic rocks and Proterozoic crystalline rocks (Ehlig, 1981; Haxel and Dillon, 1978). Zircon geochronological work reported in Saleeby *et al.* (1987) were initiated partly in search for Proterozoic rocks in the study area with these regional relations in mind. Regional age constraints on the thrusting of the southern California schist suite beneath the wide array of crystalline rocks converge on latest Cretaceous to early Tertiary time (Dillon, 1986; Ehlig, 1981; Haxel and Dillon, 1978; Silver and Nourse, 1986). Thus, it is not unreasonable to suspect that the rapid uplift of the southernmost Sierra and perhaps the clockwise rotation were related to the regional thrusting event. The structural consequence of coupled uplift, tilting, and rotation is the exposure of the sub-Bear Valley Springs batholith gneiss complex in the aberrant "tail" structure of the Tehachapi Mountains (Sams, 1986). Structural and textural features within the study area that are considered likely to have been related to this event include greenschist facies retrograde metamorphism developed primarily in the White Oak unit, cataclastic and mylonitic zones throughout this unit, and localized mylonite and cataclasite zones present in units north of the White Oak unit. Diffuse domains of incipient brittle granulation and undulose quartz fabrics present within much of the crystalline complex may also reflect this event or alternatively the modern tectonic regime of renewed uplift and White Wolf-Pleito-Garlock faulting.

Numerous workers have suggested that the allochthonous Salinia crystalline terrane of the California Coast Ranges (Fig. 31-1) was offset from the region adjoining the southern Sierra by the San Andreas fault (Page, 1981, 1982; Ross, 1984; Silver, 1984). Recent comparisons between the gneiss complex of the Tehachapi Mountains and the mid-Cretaceous charnockitic rocks of Salinia have been offered as possible substantiation of such offset (Ross, 1984, 1986). Derivation of Salinia from the region south of the study area is plausible, but features that uniquely tie the two crystalline complexes together are lacking. Mid-Cretaceous time saw peak conditions in batholithic magmatism for well over a 1000-km length of the Sierra Nevada-Peninsular Ranges belt (Chen and Moore, 1982; Evernden and Kistler, 1970; Saleeby and Sharp, 1980; Silver et al., 1979; Stern et al., 1981). Furthermore, and as a major conclusion of this work, crystalline rocks of the study area are perhaps typical of mid- to deep crustal levels in Phanerozoic batholithic belts developed across accretionary terranes, and thus tectonic regimes which disrupt and uplift such belts are likely to expose similar lithologic assemblages independent of detailed location along the belt.

ACKNOWLEDGMENTS

Field and zircon geochronological work were supported by NSF grants EAR80-18811, EAR82-18460, and EAR84-19731 awarded to Saleeby. Fieldwork by Sams was also supported by Geological Society of America Penrose grants. Special thanks to the Tejon Ranch Company for access to much of the study area. Discussions with D. C. Ross, R. W. Kistler, John Sharry, L. T. Silver, A. L. Albee, and H. P. Taylor, Jr. have been very helpful.

REFERENCES

Bateman, P. C., and Eaton, J. P., 1967, Sierra Nevada batholith: *Science*, v. 158, p. 1407–1417.

Best, M. G., 1963, Petrology and structural analysis of metamorphic rocks in the southwestern Sierra Nevada foothills, California: *Univ. Calif. Publ. Geol. Sci.*, v. 42, p. 111–159.

Blake, D. H., Elwell, R. W. D., Gibson, I. L., Skelhorn, R. R., and Walker, G. P. L., 1965, Some relationships resulting from the intimate association of acid and basic magmas: *J. Geol. Soc. London*, v. 121, p. 31–49.

Buwalda, J. P., and St. Amand, P., 1955, Geological effects of the Arvin-Tehachapi earthquake: *Calif. Div. Mines Geology Bull.*, v. 171, p. 41–56.

Chen, J. H., and Moore, J. G., 1982, Uranium-lead isotopic ages from the Sierra Nevada batholith: *J. Geophys. Res.*, v. 87, p. 4761–4784.

——, and Tilton, G. T., 1978, Lead and strontium isotopic studies of the southern Sierra Nevada batholith, California: *Geol. Soc. America Abstr. with Programs*, v. 10, p. 99–100.

Davis, G. A., and Burchfiel, B. C., 1973, Garlock fault — An intracontinental transform structure, southern California: *Geol. Soc. America Bull.*, v. 84, p. 1407–1422.

DePaolo, D. J., 1981, A neodymium and strontium isotopic study of the Mesozoic calc-alkaline granitic batholiths of the Sierra Nevada and Peninsular Ranges, California: *J. Geophys. Res.*, v. 86, p. 10470–10488.

Dibblee, T. W., Jr., and Chesterman, C. W., 1953, Geology of the Breckenridge Mountain quadrangle, California: *Calif. Div. Mines Geology Bull.*, v. 168, 56 p.

Dillon, J. T., 1986, Timing of thrusting and metamorphism along the Vincent-Chocolate Mountain thrust system, southern California: *Geol. Soc. America Abstr. with Programs*, v. 18, p. 101.

Domenick, M. A., Kistler, R. W., Dodge, F. C. W., and Tatsumoto, M., 1983, Nd and Sr isotopic study of crustal and mantle inclusions from the Sierra Nevada and implications for batholith petrogenesis: *Geol. Soc. America Bull.*, v. 94, p. 713–719.

Ehlig, P. L., 1981, Origin and tectonic history of the basement terrane of the San Gabriel Mountains, central Transverse Ranges, *in* Ernst, W. G., ed., *The Geotectonic*

Development of California (Rubey Vol. I): Englewood Cliffs, N.J., Prentice-Hall, p. 253–283.

Ehrreich, A. L., 1965, Metamorphism, migmatization, and intrusion in the foothills of the Sierra Nevada, Madera, Mariposa, and Merced counties: Ph.D. thesis, Univ. California, Los Angeles, Calif., 320 p.

Elan, R., 1985, High grade contact metamorphism at the Lake Isabella north shore roof pendant, southern Sierra Nevada, California: Ph.D. thesis, Univ. Southern California, Los Angeles, Calif. 202 p.

Evernden, J. F., and Kistler, R. W., 1970, Chronology of emplacement of Mesozoic batholithic complexes in California and western Nevada: *U.S. Geol. Survey Prof. Paper 623*, 42 p.

Fiske, R. S., and Tobisch, O. T., 1978, Paleogeographic significance of volcanic rocks of the Ritter Range pendant, central Sierra Nevada, California, *in* Howell, D. G., and McDougall, K. A., eds., *Mesozoic Paleogeography of the Western United States:* Pacific Section, Soc. Econ. Paleontologists Mineralogists, Pacific Coast Paleogeography Symp. 2, p. 209–222.

Green, T. H., 1982, Anatexis of mafic crust and high pressure crystallization of andesite, *in* Thorpe, R.S., ed., *Andesites:* New York, Wiley.

Hall, N. T., 1984, Recurrence interval and Late Quaternary history of the eastern Pleito thrust fault, northern Transverse Ranges, California: *Final Technical Report*, June 1984, U.S. Geol. Survey Earthquake Hazards Reduction Program, 89 p.

Hamilton, W., 1969, Mesozoic California and the underflow of the Pacific mantle: *Geol. Soc. America Bull.*, v. 80, p. 2409–2430.

——, and Myers, W. B., 1967, The nature of batholiths: *U.S. Geol. Survey Prof. Paper 554-C*, p. C1–C30.

Harland, W. B., Cox, A. V., Llewellyn, P. G., Pickton, C. A. G., Smith, A. G., and Walters, R., 1982, *A Geologic Time Scale:* Cambridge, Cambridge University Press.

Haxel, G. B., and Dillon, J. T., 1978, The Pelona-Orocopia schist and Vincent-Chocolate Mountain thrust system, southern California, *in* Howell, D. G., and McDougall, K. A., eds., *Mesozoic Paleogeography of the Western United States:* Pacific Section, Soc. Econ. Paleontologists Mineralogists, Pacific Coast Paleogeography Symp. 2, p. 453–469.

Kanter, L. R., and McWilliams, M. O., 1982, Rotation of the southernmost Sierra Nevada, California: *J. Geophys. Res.*, v. 87, p. 3819–3830.

Kistler, R. W., and Peterman, Z. E., 1973, Variations in Sr, Rb, K, Na and initial $^{87}RB/^{86}Sr$ in Mesozoic granitic rocks and intruded wall rocks in central California: *Geol. Soc. America Bull.*, v. 84, p. 3489–3512.

——, and Peterman, Z. E., 1978, Reconstruction of crustal blocks of California on the basis of initial strontium isotopic compositions of Mesozoic granitic rocks: *U.S. Geol. Survey Prof. Paper 1071*, 17 p.

Mack, S., Saleeby, J. B., and Ferrell, J. F., 1979, Origin and emplacement of the Academy pluton, Fresno County, California: *Geol. Soc. America Bull.*, v. 90, Pt. I, p. 321–323, Pt. II, p. 633–694.

Marshall, L. A., and Sparks, R. S. J., 1984, Origins of some mixed-magma and net-veined ring intrusions: *J. Geol. Soc. London*, v. 141, p. 171–182.

McBirney, A. R., and Noyes, R. M., 1979, Crystallization and layering of the Skaergaard intrusion: *J. Petrology*, v. 20, p. 487–554.

McWilliams, M., and Li, Y., 1983, A paleomagnetic test of the Sierran orocline hypothesis: (EOS) *Trans. Amer. Geophys. Un.*, v. 64, p. 686.

——, and Li, Y., 1985, Oroclinal bending of the southern Sierra Nevada batholith: *Science*, v. 230, p. 172–175.

Nilsen, T. H., and Clarke, S. H., Jr., 1975, Sedimentation and tectonics in the early Tertiary continental borderland of central California: *U.S. Geol. Survey Prof. Paper 925*, 64 p.

O'Hara, M. J., 1961, Petrology of the Scourie dyke, Sutherland: *Miner. Mag.*, v. 32, p. 848–865.

Page, B. M., 1981, The southern Coast Ranges, *in* Ernst, W. G., ed., *The Geotectonic Development of California* (Rubey Vol. I): Englewood Cliffs, N.J., Prentice-Hall, p. 329–417.

——, 1982, Migration of Salinian composite block, California, and disappearance of fragments: *Amer. J. Sci.*, v. 282, p. 1694–1734.

Palmer, A. R., 1983, The Decade of North America geologic time scale: *Geology*, v. 11, p. 503–504.

Plescia, J. B., and Calderone, G. J., 1986, Paleomagnetic constraints on the timing and extent of rotation of the Tehachapi Mountains, California: *Geol. Soc. America Abstr. with Programs*, v. 18, p. 171.

Presnall, D. C., and Bateman, P. C., 1973, Fusion relationships in the system $NaAlSi_3O_8$-$CaAl_2Si_2O_8$-$KAlSi_3O_8$-H_2O and generation of granitic magmas in the Sierra Nevada batholith: *Geol. Soc. America Bull.*, v. 84, no. 10, p. 3181–3202.

Ross, D. C., 1979, Summary of petrographic data of basement rock samples from oil wells in the southeast San Joaquin Valley, California: *U.S. Geol. Survey Open File Rpt. 79-400*, 11 p.

——, 1980, Reconnaissance geologic map of the southernmost Sierra Nevada (north to 35°30'N.): *U.S. Geol. Survey Open File Rpt. 80-307*, 22 p.

——, 1983a, Generalized geologic map of the southern Sierra Nevada, California, showing the location for which K-Ar radiometric age data and Rb/Sr data have been determined: *U.S. Geol. Survey Open File Rpt. 83-231*, scale 1:250,000.

——, 1983b, Petrographic (thin section) notes on selected samples from hornblende-rich metamorphic terranes in the southernmost Sierra Nevada, California: *U.S. Geol. Survey Open File Rpt. 83-587*, 36 p.

——, 1983c, Generalized geologic map of the southern Sierra Nevada, California, showing the location of basement samples for which whole rock [18]O has been determined: *U.S. Geol. Survey Open File Rpt. 83-904*, scale 1:250,000.

——, 1984, Possible correlations of basement rocks across the San Andreas, San Gregorio-Hosgri, and Rinconada-Reliz-King City faults, California: *U.S. Geol. Survey Prof. Paper 1317*, 37 p.

——, 1985, Mafic gneissic complex (batholithic root?) in the southernmost Sierra Nevada, California: *Geology*, v. 13, p. 288–291.

——, 1986, The metamorphic and plutonic rocks of the southernmost Sierra Nevada, California, and their tectonic framework: *U.S. Geol. Survey Prof. Paper*, in press.

Saleeby, J. B., 1979, Kaweah serpentine melange, southwest Sierra Nevada Foothills, California: *Geol. Soc. America Bull.*, v. 90, p. 29-46.

——, 1981, Ocean floor accretion and volcano-plutonic arc evolution of the Mesozoic Sierra Nevada, *in* Ernst, W. G., ed., *The Geotectonic Development of California* (Rubey Vol. I): Englewood Cliffs, N.J., Prentice-Hall, p. 132–181.

——, 1983, Accretionary tectonics of the North American Cordillera: *Ann. Rev. Earth Planet. Sci.*, v. 15, p. 45–73.

——, and Busby-Spera, C., 1986, Fieldtrip guide to the metamorphic framework rocks of the Lake Isabella area, southern Sierra Nevada, California, *in* Dunne, G. C., ed., *Mesozoic and Cenozoic Structural Evolution of Selected Areas, East-central California:* Cordilleran Section, Geol. Soc. America Guidebook, p. 81–94.

____, and Sharp, W. D., 1980, Chronology of the structural and petrologic development of the southwest Sierra Nevada Foothills, California: *Geol. Soc. America Bull.*, v. 91, pt. I, p. 317–320, pt. II, p. 1416–1535.

____, and Williams, H., 1979, Possible origin of California Great Valley gravity-magnetic anomalies: (EOS) *Trans. Amer. Geophys. Un.*, v. 59, p. 1189.

____, Goodin, S. E., Sharp, W. D., and Busby, C. J., 1978, Early Mesozoic paleotectonic-paleogeographic reconstruction of the southern Sierra Nevada region, *in* Howell, D. G., and McDougall, K. A., eds., *Mesozoic Paleogeography of the Western United States:* Pacific Section, Soc. Econ. Paleontologists Mineralogists, Pacific Coast Paleogeography Symp. 2, p. 311–336.

____, Sams, D. B. and Kistler, R. W., 1987, Geochronology and geochemistry of crystalline rocks of the southernmost Sierra Nevada, California: *J. Geophys. Res.*, v. 92, p. 10443–10466.

____, Sams, D. B., and Tobisch, O. T., 1986, A down-plunge view of the Sierra Nevada batholith — From resurgent caldera to ultrametamorphic levels: *Geol. Soc. America Abstr. with Programs*, v. 18, p. 179.

Sams, D. B., 1986, U/Pb zircon geochronology, petrology, and structural geology of the crystalline rocks of the southernmost Sierra Nevada and Tehachapi Mountains, Kern County, California: Ph.D. thesis, California Inst. Technology, Pasedena, Calif., 315 p.

Sams, D. B., Saleeby, J. B., Ross, D. C., and Kistler, R. W., 1983, Cretaceous igneous, metamorphic and deformational events of the southernmost Sierra Nevada, California: *Geol. Soc. America Abstr. with Programs*, v. 15, p. 294.

Schumm, S. A., 1963, The disparity between present rates of denudation and orogeny: *U.S. Geol. Survey Prof. Paper 454H*, p. H1–H13.

Schweickert, R. A., and Cowan, D. S., 1975, Early Mesozoic tectonic evolution of the western Sierra Nevada, California: *Geol. Soc. America Bull.*, v. 86, p. 1329–1336.

Sharry, J., 1981, The geology of the western Tehachapi Mountains, California: Ph.D. thesis, Massachusetts Inst. Technology, Cambridge, Mass., 215 p.

____, Minimum age and westward continuation of the Garlock fault zone, Tehachapi Mountains, California: *Geol. Soc. America Abstr. with Programs*, v. 14, p. 233.

Silver, L. T., 1983, Paleogene overthrusting in the tectonic evolution of the Transverse Ranges, Mojave and Salinian regions, California: *Geol. Soc. America Abstr. with Programs*, v. 15, p. 438.

____, and Nourse, J. A., 1986, The Rand "thrust" complex in comparison with the Vincent thrust-Pelona schist relationship, southern California: *Geol. Soc. America Abstr. with Programs*, v. 18, p. 185.

____, Taylor, H. P., Jr., and Chappell, B., 1979, Some petrological, geochemical, and geochronological observations of the Peninsular Ranges batholith near the international border of the U.S.A. and Mexico, *in* Abbott, P. L., and Todd, V. R., eds., *Mesozoic crystalline Rocks — Peninsular Ranges Batholith and Pegmatites, Point Sal Ophiolite*, Guidebook Geol. Soc. America Ann. Meet., San Diego Univ., San Diego, Calif., p. 83–110.

Snoke, A. W., Quick, J. E., and Bowman, H. R., 1981, Bear Mountain Igneous complex, Klamath Mountains, California — An ultrabasic to silicic calc-alkaline suite: *J. Petrology*, v. 22, p. 501–552.

Stern, T. W., Bateman, P. C., Morgan, B. A., Newell, M. F., and Peck, D. L., 1981, Isotopic U-Pb ages of zircon from granitoids of the central Sierra Nevada, California: *U.S. Geol. Survey Prof. Paper 1185*, 17 p.

Tobisch, O. T., Saleeby, J. B., and Fiske, R. S., 1985, Structural history of continental volcanic arc rocks, eastern Sierra Nevada, California — A case for extensional tectonics: *Tectonics*, v. 5, p. 65–94.

Wells, R. A., 1980, Thermal models for the magmatic accretions and subsequent metamorphism of continental crust: *Earth Planet. Sci. Lett.*, v. 46, p. 253–265.

Wetherill, G. W., 1956, Discordant uranium-lead ages — I: (EOS) *Trans. Amer. Geophys. Un.*, v. 37, p. 320–326.

Winkler, H. G. F., 1967, *Petrogenesis of Metamorphic Rocks:* New York, Springer-Verlag, 237 p.

Wood, B. J., and Banno, S., 1973, Garnet-orthopyroxene and orthopyroxene-clinopyroxene relationships in simple and complex systems: *Contrib. Mineralogy Petrology*, v. 42, p. 109--124.

Zen, E., 1985, Implications of magmatic epidote-bearing plutons on crustal evolution in the accreted terranes of northwestern North America: *Geology*, v. 13, p. 266–269.

____, and Hammarstrom, J. N., 1984, Magmatic epidote and its petrologic significance: *Geology*, v. 12, p. 515–518.

V. R. Todd and B. G. Erskine
U.S. Geological Survey
Menlo Park, California 94025

D. M. Morton
U.S. Geological Survey
Department of Earth Sciences
University of California, Riverside
Riverside, California 92521

32

METAMORPHIC AND TECTONIC EVOLUTION OF THE NORTHERN PENINSULAR RANGES BATHOLITH, SOUTHERN CALIFORNIA

Systematic lithologic, geochemical, geophysical, and structural variations across the Creta-ceous Peninsular Ranges batholith and its wallrocks indicate the existence of an east-dipping Early Cretaceous suture between a parautochthonous(?) Late Jurassic oceanic arc and a Mesozoic marginal flysch basin that lay adjacent to western North America. Pene-trative deformational fabrics in batholithic rocks indicate considerable east-northeast shortening. These fabrics and the spatial distribution of prebatholithic and plutonic rocks suggest that the island arc collided with, and was thrust beneath, the western edge of North America.

Synchronous with and following accretion of the island arc, the locus of magmatic activity moved slightly eastward, producing partial melts in arc-related rocks west of the suture, and in the Mesozoic sedimentary prism east of, and above it. The resulting plutons stitched across the suture, imparting a twofold nature to the new Cretaceous crust.

In the Late Cretaceous and early Tertiary, the magmatic arc migrated eastward into the eastern Peninsular Ranges, southern Arizona, and northwest Mexico, where it pro-duced widespread melting in a new, predominantly continental source region. As the Cretaceous arc moved eastward, the zone of most intense deformation also shifted to the east. Compressive, largely syn-intrusive deformation was concentrated in thermally weakened crust in the central part of the batholith in the Early Cretaceous and in the eastern part in the Late Cretaceous. In both areas, deep-seated batholithic rocks were thrust westward over shallower rocks.

In early and middle Tertiary time, as the western edge of North America underwent sweeping uplift and dilation, the compressive tectonic regime that had dominated the northern part of the batholith since at least the Early Cretaceous gave way to one of extensional tectonics marked by widespread east-directed low-angle faulting.

No evidence appears to exist for strike-slip tectonic transport of regional dimensions within the northern Peninsular Ranges batholith between Early Cretaceous and middle Tertiary time. The data do not rule out either large northward displacements of pre-batholithic rocks by transform faults, or the existence of a syn- or postbatholithic trans-form fault east of the presently exposed batholith.

INTRODUCTION

The Peninsular Ranges batholith of southern California, United States, and the Baja California peninsula, Mexico (Fig. 32-1) is one part of the chain of great Mesozoic Circum-Pacific batholiths. The development of plate tectonic theory over the past two decades has provided a framework in which the origin of the circum-Pacific batholiths and associ-ated supracrustal rocks can be understood. In particular, the formation of batholiths along continental margins by the subduction of oceanic lithosphere and arc magmatism is recognized as a major factor in the growth of continents (Hamilton, 1981). More recently, geologic, geophysical, and paleontologic studies indicate that some crustal blocks (tectonostratigraphic terranes) have been displaced from their sites of origin in the Pacific basin to collide with, and become accreted to, western North America (Coney et al., 1980; G. A. Davis et al., 1978; D. L. Jones et al., 1983). Paleomagnetic studies suggest that the northward translation and clockwise rotation of allochthonous terranes relative to the North American craton have been major elements in shaping the western Cordillera (Beck, 1980; Beck et al., 1981). In particular, paleomagnetic studies of rocks of the Peninsular Ranges of southern California and Baja California indicate that these rocks originated more than 1000 km south of their present location (Hagstrum et al., 1985, 1986).

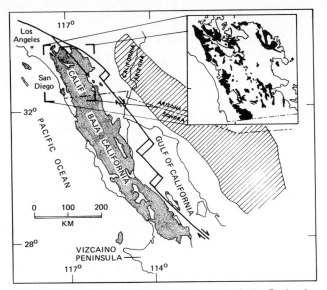

FIG. 32-1. Index map showing location of the Peninsular Ranges batholith and the study area. Stipple pattern indicates batholith and associated prebatholithic rocks (from Jahns, 1954). Line pattern indicates Jurassic-Cretaceous volcanic-plutonic ensialic arc; ensialic arc and Gulf of California–San Andreas rift-transform system from Hagstrum *et al.* (1985). Inset map shows study area, solid areas are prebatholithic screens, modified from Rogers (1965) and Strand (1962).

The application of plate tectonic and paleomagnetic data to hypotheses concerning the origin of the Peninsular Ranges batholith requires detailed knowledge of the geology of prebatholithic and batholithic rocks. Over the past two decades, many studies have contributed to our knowledge of the ages and tectonic settings of prebatholithic and plutonic rocks of the batholith. The purpose of this paper is to synthesize the results of recent geologic studies in the southern California segment of the batholith and to point out areas that need further study. Many problems remain to be solved, and our conclusions are working hypotheses in an evolving understanding of the origin and tectonic development of the batholith.

REGIONAL SETTING

The Peninsular Ranges batholith is well exposed from 34°N in southern California to 28°N in Baja California (Frizzell, 1984; Jahns, 1954; Fig. 32-1). Although batholithic rocks are largely covered by Cenozoic rocks south of the 28th parallel, the batholith is generally considered to continue to the tip of the peninsula. However, the petrologic affinities of Cretaceous granitoids in the southernmost part of the peninsula are uncertain (Frizzell, 1984).

Prebatholithic sedimentary and volcanic rocks in the Peninsular Ranges were subjected to synkinematic metamorphism and Cretaceous intrusion. These metamorphic rocks occur chiefly as screens and roof pendants partly or wholly surrounded by batholithic rocks (Fig. 32-1). Bodies of prebatholithic rocks range in size from pods a few meters long to screens that are as much as 3–5 km wide and 35–40 km long. In southern

PENINSULAR RANGES METAMORPHIC AND TECTONIC EVOLUTION

California, fossils have been found in prebatholithic rocks at several localities, but the correlation of rocks between screens is based chiefly on lithologic similarities. In the Baja California peninsula, detailed studies of prebatholithic rocks over the past decade have revealed a wide range of depositional environments, from miogeoclinal to eugeoclinal, and a variety of ages, from early Paleozoic to mid-Cretaceous (Gastil and Miller, 1984). Many screens in the batholith either have not been studied in detail or do not contain materials that are suitable for paleontologic or radiometric dating, and therefore their ages and relations to one another are poorly understood. Nevertheless, volcanic and epiclastic rocks of Mesozoic age generally occur on the wide side of the batholith, and epiclastic rocks of Paleozoic age occur on the east side (Gastil et al., 1975; Hill and Silver, 1983; Larsen, 1948).

The volcanic and volcaniclastic rocks on the west side of the batholith are deposits of an island arc that was active in latest Jurassic time and continued into the Early Cretaceous (Gastil et al., 1975; Silver et al., 1963, 1979). The island arc may have been paired with an eastern ensialic magmatic arc of similar age located in what is now southern Arizona and west-central Sonora (Fig. 32-1; Gastil et al., 1978, 1981). An alternative hypothesis suggests that the western magmatic arc was originally the southwestern part of a single Mesozoic arc built on continental crust in the north and oceanic crust in the south (Gastil et al., 1972; Gastil and Miller, 1983). In this case, the southern part of the arc was tectonically displaced to its present position by right-lateral transform faulting possibly associated with oblique subduction of the Farallon plate beneath western North America between the Late Cretaceous and late Miocene (Beck and Plumley, 1979; Erskine and Marshall, 1980; Hagstrum et al., 1985; Teissere and Beck, 1973). Whether or not the island arc was paired with an eastern ensialic arc, the exposure of blueschist-facies rocks throughout the continental borderland and at several onshore localities in the Peninsular Ranges is compelling evidence for the existence of a late Mesozoic subduction zone off the southwest Cordilleran margin (Ernst, 1984).

Prebatholithic rocks in southern California underwent regional dynamothermal metamorphism of low pressure–high temperature (Abukuma) type (Miyashiro, 1961) probably beginning in the Late Jurassic and continuing throughout the Early Cretaceous, synchronous with intrusion in the western part of the batholith (Gastil et al., 1975; Schwarcz, 1969; Silver et al., 1969; Todd and Shaw, 1979). Structures formed during metamorphism and intrusion include northwest-striking, east-dipping isoclinal folds and foliation, and mineral lineations that generally plunge down the dip of foliation. Metamorphic grade varies from greenschist facies in the western part of the batholith to upper amphibolite facies in the eastern part. The metamorphic gradient from greenschist to upper amphibolite facies is locally abrupt in the northern part of the batholith (Schwarcz, 1969), but it is more gradual in the batholith east of San Diego (Todd and Shaw, 1979). Contact metamorphism due to the emplacement of the batholith is generally absent, although its effects are recognized locally in areas of lowest-grade regional metamorphism (Gastil et al., 1975) and can be profound where batholithic rocks intruded carbonates under amphibolite facies conditions (Burnham, 1959). In general, the metamorphic grade increases to the east as shown by progressive development of partial melt leucosomes, metamorphic layering, transposition of bedding, and loss of recognizable primary structures.

The Peninsular Ranges batholith was emplaced between 140 and 80 m.y.b.p. Silver, quoted in Hill, 1984; Silver et al., 1979). In the area east of San Diego and in northern Baja California, U-Pb zircon ages of plutons on the west side of the batholith range from 120 to 105 m.y. with no systematic geographic distribution, while those of plutons on the east side range from 105 to 89 m.y. and decrease progressively eastward. This distribution of plutonic ages suggested to Silver et al. (1979) that a magmatic arc was fixed, or stationary, on the west side of the batholith for much of Early Cretaceous time and began to migrate eastward at about 105 m.y.b.p. However, this hypothesis may be too simplistic because preliminary data suggest the existence of plutons that are older than 105 m.y. in the eastern part of the batholith (R. G. Gastil, written communication, 1986).

The batholith exhibits systematic west-to-east variations in lithology, geochemistry, and structural character (Baird and Miesch, 1984; DePaolo, 1981; Gastil, 1975, 1983; Larsen, 1948; Silver *et al.*, 1979; Todd and Shaw, 1979). Plutonic compositions range broadly from gabbro and tonalite in the west, to granodiorite and granite with more alkaline compositions in the east; tonalite is probably the most abundant rock type in the batholith (Gastil, 1975, 1983). Both western and eastern intrusive rocks are calcic (Silver *et al.*, 1979; Todd and Shaw, 1986).

During the period when the greatest volume of plutonic rocks was emplaced (120–90 m.y.), the region was uplifted and unroofed, and large volumes of metamorphic and plutonic detritus were shed westward into marine basins along the continental margin (Gastil *et al.*, 1978; Kennedy and Peterson, 1975; Patterson, 1979; Schoellhamer *et al.*, 1981). By Eocene time, rivers carried gravels from the continental interior westward across the deeply eroded batholith to the continental shelf (Gastil *et al.*, 1981; Kennedy and Peterson, 1975; Minch, 1972). The present-day tectonic setting of peninsular California began to take shape about 37 m.y. ago in the late Eocene and early Oligocene (Berggren *et al.*, 1985), when the Pacific-Farallon ridge first encountered the North American trench and the San Andreas transform fault system was initiated (Engebretson *et al.*, 1985).

PREBATHOLITHIC ROCKS

Prebatholithic rocks in the northern part of the Peninsular Ranges batholith are divided into three zones in this report: volcanic and volcaniclastic rocks in the western part of the batholith (Zone I), submarine fan deposits in the central part (Zone II), and miogeoclinal rocks in the eastern part (Zone III) (Fig. 32-2). A group of rocks in the western part of Zone III may be transitional between Zones II and III.

Zone I: Volcanic and Volcaniclastic Rocks

Volcanic and volcaniclastic rocks assigned to the Santiago Peak Volcanics (Larsen, 1948) form a discontinuous coastal belt, typically less than 18 km wide, that was intruded by the Cretaceous batholith and is partly covered by Upper Cretaceous, Tertiary, and Quaternary sedimentary deposits (Fig. 32-2). The volcanic rocks are an alternating sequence of weakly metamorphosed andesitic to rhyolitic flows, tuff, agglomerate, and breccia with minor basalt (Adams, 1979; Buesch, 1984; Larsen, 1948). The intrusive contact between the batholith and the volcanic rocks is steep and irregular (Adams, 1979; Kennedy and Peterson, 1975). Small granodiorite and gabbro plutons, many with hypabyssal textures, intruded the volcanic rocks and are probably feeders (Kennedy and Peterson, 1975; Larsen, 1948). A sequence of poorly exposed interbedded volcaniclastic sandstone, siltstone, mudstone, and sedimentary breccia with sparse tuff layers is exposed about 5–10 km east of Del Mar (Fig. 32-2), west of the volcanic rocks Balch, 1981) Similar rocks are interbedded locally with the main volcanic sequence to the east and north (Buesch, 1984).

Andesitic and dacitic tuff and breccia about 11 km south of the international border are probably the extension of the Santiago Peak Volcanics (Hawkins, 1970). The Alisitos Formation (Allison, 1955; Santillan and Barrera, 1930), a sequence of metamorphosed andesitic volcanic and volcaniclastic rocks with minor interbedded limestone

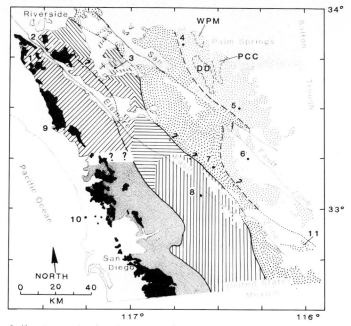

Sedimentary and volcanic deposits (Quaternary, Tertiary, and Late
   Cretaceous)

ZONE I Metamorphosed volcanic and volcaniclastic rocks

■ Santiago Peak Volcanics (latest Jurassic) and quartz latite
porphyry including Temescal Wash Quartz Latite Porphyry
(Larsen, 1948)

▦ Metavolcanic rocks of unknown age with scarce metasedimentary layers

ZONE II Metamorphosed submarine fan deposits

▨ Bedford Canyon Formation (Jurassic and Triassic) and rocks
generally regarded as equivalent; west of dashed line, includes significant metavolcanic component

▧ French Valley Formation (Triassic?)

▥ Julian Schist (Triassic?) and rocks generally regarded as
equivalent

▤ Undivided metasedimentary rocks

ZONE III Metamorphosed miogeoclinal and transitional(?) sedimentary rocks

▨ Metasedimentary rocks transitional between Zones II and
III (Mesozoic? and Paleozoic)

▨ Miogeoclinal metasedimentary rocks (Paleozoic? and late
Precambrian?); short dashed lines outline rock units cited
in text, DD=Desert Divide Group, WPM=Windy Point metamorphic rocks, PCC=Palm Canyon Complex

FIG. 32-2.   Prebatholithic zones in northern Peninsular Ranges batholith
generalized from distribution of rock types in screens; plutonic rocks not
shown. Solid lines are approximate boundaries between zones, long dashes
separate subzones that have different ages, lithologies, or depositional settings.
Data from Strand (1962), Rogers (1965), and sources cited in text, modified
by unpublished data of authors. Locations mentioned in text: (1) Santa Ana
Mountains, (2) Temescal Wash, (3) Winchester, (4) San Jacinto Mountains,
(5) Santa Rosa Mountains, (6) Borrego Valley, (7) Ranchita, (8) Julian,
(9) Camp Pendleton, (10) Del Mar, (11) Coyote Mountains.

in the coastal area of northern Baja California, is the probable southern continuation of the unit (Gastil *et al.*, 1981). Undated amphibolite-grade metavolcanic rocks occur in isolated screens in the eastern part of Zone I (Fig. 32-2, close stipple). These rocks generally have been assigned to the Santiago Peak Volcanics (Lillis *et al.*, 1979; Todd and Shaw, 1979; Walawender *et al.*, 1979), but they are lithologically dissimilar to that unit in its type area in the Santa Ana Mountains (D. M. Morton, unpublished mapping) and may therefore represent a new volcanic unit.

The age of a part of the Santiago Peak Volcanics is latest Jurassic (Tithonian) based on marine fossils in the volcaniclastic sequence east of Del Mar (Fig. 32-2; Balch *et al.*, 1984; Fife *et al.*, 1967; D. A. Jones and Miller, 1982; D. A. Jones, *et al.*, 1983). Adams (1979) described a volcanic subunit that unconformably overlies the Tithonian volcaniclastic rocks, suggesting that part of the formation in this area may be Early Cretaceous in age. Local gradations between the metavolcanic rocks in screens in the eastern part of Zone I (Fig. 32-2, close stipple) and adjacent Early Cretaceous plutons (Todd, 1983) suggest that these rocks may be Early Cretaceous in age. The age of the Santiago Peak Volcanics in the southern part of the study area is therefore tentatively considered to be latest Jurassic and Early Cretaceous(?). The Santiago Peak Volcanics have not been dated in the Santa Ana Mountains, but Schoellhamer *et al.* (1981) tentatively considered the formation to be Late Jurassic(?) and Early(?) Cretaceous on the basis of stratigraphic relations and a K-Ar radiometric determination on dacite tuff that gave an Early Cretaceous apparent age. The Alisitos Formation of northern Baja California contains biohermal limestone strata that yielded a mid-Cretaceous (Aptian-Albian) fauna (Silver *et al.*, 1963). A considerable part of the formation is unfossiliferous and may be of different age (Gastil *et al.*, 1975; Silver *et al.*, 1963).

The Santiago Peak Volcanics underwent greenschist facies metamorphism that varies from the lowest to slightly higher-grade parts of the chlorite zone (Adams, 1979; Hawkins, 1970; Jahns and Lance, 1950; Larsen, 1948; Turner, 1981; Fig. 32-3). The aggregate metamorphic mineral assemblage in volcanic rocks is chlorite-epidote-serpentine-calcite-albite-quartz-clinozoisite-stilpnomelane ± muscovite ± biotite ± actinolite (Table 32-1). The Santiago Peak Volcanics bear nonpenetrative micaceous cleavage and possible isoclinal folds (Adams, 1979). The metamorphic mineral assemblage chlorite-illite-sericite ± epidote ± calcite of the latest Jurassic volcaniclastic rocks east of Del Mar together with diagnenetic textures indicates a transition from highest-grade burial metamorphism to lowest-grade greenschist facies (Balch, 1981). These rocks were probably buried to depths of 3 km or more prior to uplift and local overturning (Balch, 1981).

The most likely provenance for the Santiago Peak Volcanics was an emerging andesitic volcanic arc located near an ocean basin (Adams, 1979; Balch, 1981; Buesch, 1984). Scarce occurrences of columnar jointing in flows and terrestrial plant fossils in volcanic breccia suggest that volcanic eruptions were in part subaerial and that the arc was located at, or near, the continental margin (Adams, 1979). The available geologic data do not constrain the direction of dip of the subduction zone that was associated with the volcanic arc. However, the scattered occurrences of blueschist facies and ophiolitic rocks in the continental Borderland suggest that the subducted slab dipped to the east. Rb/Sr isotopic data on metavolcanic rocks (Santiago Peak Volcanics?) in screens east of San Diego suggest that the volcanic arc was built on oceanic crust and/or sediments (V. R. Todd and S. E. Shaw, unpublished data).

The volcaniclastic rocks of the formation were derived from a nearby andesitic volcanic arc that was undergoing rapid erosion (Balch, 1981; Buesch, 1984). Volcanic clasts were deposited in deep ocean waters (trench or fore-arc basin?) by west- and south-southwest-directed turbidity currents and submarine pyroclastic flows. The volcanic

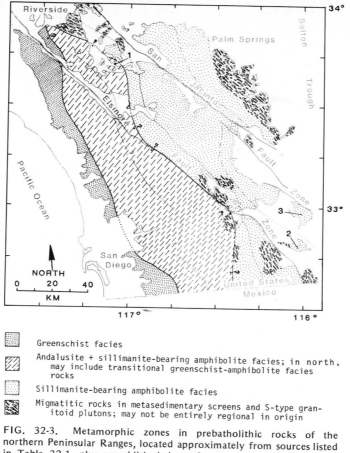

FIG. 32-3. Metamorphic zones in prebatholithic rocks of the northern Peninsular Ranges, located approximately from sources listed in Table 32-1, plus unpublished data of authors. Boundaries of prebatholithic zones (gray lines) and base from Fig. 32-2. Locations mentioned in text: (1) Winchester, (2) Coyote Mountains, (3) Fish Creek Mountains. Metamorphic rocks in parts of Coyote and Fish Creek Mountains may be of lower grade than rocks in northern part of zone of sillimanite-bearing amphibolite facies (Miller and Dockum, 1983; V. R. Todd, unpublished data).

Legend:
- Greenschist facies
- Andalusite + sillimanite-bearing amphibolite facies; in north, may include transitional greenschist-amphibolite facies rocks
- Sillimanite-bearing amphibolite facies
- Migmatitic rocks in metasedimentary screens and S-type granitoid plutons; may not be entirely regional in origin

terrane was intruded by volcanic plugs and hypabyssal plutons that may have contributed clasts to the volcaniclastic sediments as parts of the arc were unroofed.

In northern Baja California, the western prebatholithic volcanic and volcaniclastic belt is considered to be the high-level expression of the magmatism that produced the western plutons of the batholith (Gastil *et al.*, 1975; Hawkins, 1970; Silver *et al.*, 1963, 1969). East of San Diego, elliptical (in plan), steeply dipping, sheeted plutonic complexes in the western part of the batholith include screens of metamorphosed porphyritic and pyroclastic volcanic rocks that locally grade into plutonic rocks (Todd, 1983). These relations, and the westward decrease in the size and abundance of plutons and increase in the abundance of volcanic rocks (Silver *et al.*, 1979), suggest that the westernmost bedrock exposures were near the roof of the batholith. Rocks of the volcanic-plutonic arc prob-

TABLE 32-1. Characteristic Aggregate Metamorphic Mineral Assemblages in Prebatholithic Rocks

Protolith	Mineral Assemblage[a]	Metamorphic Facies	References
Volcanic rocks	Chlorite-albite-epidote/clinozoisite-serpentine-calcite-ankerite-stilpnomelane ± zeolites ± muscovite ± biotite ± actinolite ± pyrophyllite	Greenschist facies	Larsen, 1948; Jahns and Lance, 1950; Hawkins, 1970; Adams, 1979
Epiclastic and volcaniclastic rocks	Chlorite-calcite ± epidote ± albite ± sericite/muscovite ± biotite	Greenschist facies	Larsen, 1948; Balch, 1981
Semipelitic and pelitic rocks	Quartz-biotite-muscovite-andalusite-sillimanite-graphite-plagioclase-alkali feldspar ± cordierite ± garnet	Low-pressure (?) amphibolite facies	Schwarcz, 1969; Berggreen and Walawender, 1977; Todd and Shaw, 1979; Germinario, 1982
Semipelitic and pelitic rocks	Quartz-biotite-alkali feldspar-sillimanite ± plagioclase ± muscovite ± cordierite ± garnet ± sphene ± epidote	Intermediate-pressure (?) amphibolite facies	Theodore, 1970; Sydnor, 1975; Brown, 1981; Hill, 1984; Detterman, 1984; Erskine, 1986a; Grove, 1986
Calc-silicate rocks	Quartz-diopside-epidote/clinozoisite-plagioclase-tremolite/actinolite-calcite ± alkali feldspar ± grossularite ± wollastonite ± dolomite ± hornblende ± biotite ± phlogopite ± scapolite ± sphene ± muscovite ± graphite	Low- and intermediate-pressure amphibolite facies	Schwarcz, 1969; Theodore, 1970; Detterman, 1984; Erskine, 1986a

[a]Opaque oxide and sulfide minerals not listed.

ably continue for some distance to the west, where they are truncated by an offshore fault zone in the vicinity of the Coronado escarpment (Howell and Vedder, 1981).

Zone II: Submarine Fan Deposits

A north-northwest-trending zone that consists of screens of flysch-type metasedimentary rocks intruded and surrounded by plutonic rocks occurs in the central part of the batholith (Zone II, Fig. 32-2). The metasedimentary rocks include variably metamorphosed and deformed sandstone and shale that were deposited by turbidity currents on one or more west-facing Mesozoic submarine fan systems. Minor amounts of conglomerate, limestone, and volcanic rocks are interbedded with these rocks. While the tectonic setting of these submarine fan deposits, whether forearc, back-arc, or trench, is not known, they were deposited in a deep-water marine basin, or basins, marginal to North America. Three units are recognized in Zone II: the Bedford Canyon Formation and unnamed rocks generally correlated with it crop out in the northern part of the zone; the French Valley Formation crops out east of the Bedford Canyon Formation; and the Julian Schist occupies the southern part of the zone (Fig. 32-2). The rocks of Zone II are part of the shale-sandstone belt of Gastil *et al.* (1975), which extends southward into Baja California at least as far as the 28th parallel and includes a large number of prebatholithic rock sequences of varied lithology, age, and depositional setting (Gastil and Miller, 1984). The ages and correlations of the prebatholithic units are poorly known because detailed stratigraphic studies and reliable age data are lacking in many areas, and because metamorphism and intrusion have obscured contacts. Although the rocks of Zone II are generally considered to be Mesozoic in age based on rather scant paleontologic evidence, Paleozoic rocks may also be present (R. G. Gastil, written communication, 1986).

Bedford Canyon Formation The Bedford Canyon Formation was named by Larsen (1948) for exposures in the northern Santa Ana Mountains. As mapped by Larsen, the formation consists of weakly metamorphosed slate and argillite with minor quartzite, limestone, conglomerate, and scarce metaigneous layers that he correlated tentatively with the Santiago Peak Volcanics. Larsen and later workers included in the Bedford Canyon Formation more highly metamorphosed prebatholithic rocks that occur as screens in the batholith southeast and east of the northern Santa Ana Mountains. However, in many places, these rocks differ lithologically from the formation in the type area. Rocks mapped as Bedford Canyon Formation beyond the type area consist mainly of slate and phyllite east of the Elsinore fault zone (Fig. 32-2) and black siliceous quartzite and phyllite west of the fault zone (D. M. Morton, unpublished mapping). Rocks in the vicinity of Camp Pendleton (Fig. 32-2) that Larsen included in the Bedford Canyon Formation are considered by Buesch (1984) to be volcaniclastic subunits of the Santiago Peak Volcanics. In view of the lithologic variability of these prebatholithic rocks, the name Bedford Canyon Formation should be restricted to occurrences in the type area in the northern Santa Ana Mountains.

Moran (1976) described the Bedford Canyon Formation in the type area as an unmetamorphosed to weakly metamorphosed flysch-type sequence of sandstone and shale with minor tuffaceous strata, limestone, conglomerate, pebbly mudstone, and chert. He suggested that these sediments were deposited by turbidity currents in a bathyal marine basin into which scarce clastic carbonate rocks from an older, or penecontemporaneous, marine shelf were deposited periodically by submarine mass flows and local slumping (olistoliths). The most likely provenance of the sands and silts is terrestrial crystalline and (or) sedimentary rocks (Moran, 1976). This is supported by Precambrian lead-alpha ages of zircons from sandstone (Bushee *et al.*, 1963) and by Rb/Sr whole-rock isochron ages that indicate probable derivation from a Precambrian igneous and meta-morphic terrane (Criscione *et al.*, 1978).

The Bedford Canyon Formation is complexly folded and at least partly overturned (Criscione *et al.*, 1978; Moscoso, 1967; Schoellhamer *et al.*, 1981). Buckley *et al.* (1975) consider the formation to be the overturned limb of a nappe that was uplifted prior to deposition of the overlying Santiago Peak Volcanics. The plate tectonic setting of the depositional basin is not known. Gastil *et al.* (1978, 1981) consider the unit to be a clastic wedge derived from the North American craton in a fore-arc or back-arc setting, while the highly deformed nature of the formation and the occurrence in the northern Santa Ana Mountains of scattered serpentinite bodies led Criscione *et al.* (1978) to suggest that the formation was deposited in an oceanic trench and later accreted to the North American continent during subduction.

The Bedford Canyon Formation is unconformably overlain by the Santiago Peak Volcanics in the northern Santa Ana Mountains; its base is not exposed (Larsen, 1948; Schoellhamer *et al.*, 1981). Some workers consider that the two formations are entirely, or almost entirely, in fault contact (T. E. Davis, *quoted in* Buesch, 1984). The contact relations among the Bedford Canyon Formation and other rocks historically included with it, the French Valley Formation, and the Julian Schist are unknown. Schwarcz (1969) concluded that the French Valley Formation overlies and interfingers with the Bedford Canyon Formation in the area south of Winchester (Fig. 32-2). Here the rocks mapped as Bedford Canyon Formation are essentially all phyllite in contrast to the type Bedford Canyon Formation and thus may represent a different unit (D. M. Morton, unpublished mapping). Furthermore, the French Valley Formation is isoclinally folded and may not be upright. A large number of small screens occur between the type localities of the Bedford Canyon Formation and the Julian Schist. To our knowledge, these screens have not been studied in detail.

Some limestone bodies in the Bedford Canyon Formation yielded Bajocian and Callovian fossils, indicating a Middle Jurassic or younger age for part of the formation (Imlay, 1963, 1964; Silberling *et al.*, 1961) while others contained Paleozoic fossils (D. L. Jones, oral communication, 1986). Some chert bodies in the formation contained Triassic radiolarians (D. L. Jones, oral communication, 1986). The paleontologic data support whole-rock Rb/Sr isochron ages of fine-grained sediments in the formation that suggest two sedimentation and/or diagenetic events, one in Middle to Late Triassic time and a second in Middle Jurassic time (Criscione *et al.*, 1978). The istopic evidence for a Jurassic depositional event suggests that some limestone bodies were coeval with the flysch-type sediments.

Julian Schist and French Valley Formation Metasedimentary rocks in the vicinity of Julian, Calif. (Fig. 32-2) were named the Julian group by Merrill (1914) and later called the Julian Schist Series or Schist by Hudson (1922). The Julian Schist consists of pelitic and semi-pelitic schist and paragneiss, calc-silicate-bearing feldspathic quartzite that is interbedded with laminated, locally pebbly quartz-mica schist, and minor amphibolite, marble, metaconglomerate, and clean metaquartzite and quartzite-pebble conglomerate (Berggreen and Walawender, 1977; Detterman, 1984; Germinario, 1982, Grove, 1986; Todd and Shaw, 1979). The area shown as Julian Schist in Fig. 32-2 includes meta-sedimentary rocks that are lithologically similar to those in the vicinity of Julian, but some of these rocks may be either lateral variants of the formation or unrelated units of different age (R. G. Gastil, written communication, 1986). Most workers have considered amphibolite of the unit to be metamorphosed mafic flows, tuff, tuff-breccia, and/or sills (Detterman, 1984; Everhart, 1951; Grove, 1986; Hudson, 1922; Merriam, 1946; Schwarcz, 1969; Todd and Shaw, 1979). Todd and Shaw reported local occurrences of interlayered felsic and mafic metavolcanic rocks in the Julian Schist, and local gradation of amphibolite layers into pods of fine-grained gabbro.

The protolith and depositional setting of the Julian Schist were studied by Germinario (1982), Detterman (1984), and Grove (1986), who considered the original clastic rocks to be a predominantly fine-grained, thinly interbedded sequence of shale, argillaceous sandstone, calcareous arkosic sandstone, and quartz-rich graywacke. Limestone formed very scarce, thin, lenticular beds and large (up to 10 m or more) blocks of clean carbonate in this sequence. Sedimentary structures such as graded bedding, cross-bedding, and flame and rip-up structures preserved locally in the limbs of isoclinal folds suggest that these sediments were deposited by turbidity currents on a submarine fan, or a series of coalescing fans in a marginal basin (Detterman, 1984; Germinario, 1982). The rare occurrence of beds of limestone and clean quartzite may record infrequent incursions of shelf sediments, while the large limestone blocks suggest penecontemporaneous or later tectonic slumping and/or debris flows of shelf sediments. Intermittent volcanic activity to the west (?) apparently supplied small volumes of flows and pyroclastic rocks that were associated with shallow intrusions. Germinario (1982) and Detterman (1984) concluded that the most likely provenance of the Julian Schist was a cratonic plutonic and metamorphic complex that lay to the east of the marginal basin. This is supported by Precambrian Pb-alpha ages determined by Bushee *et al.* (1963) on detrital zircons from metaquartzite in the Julian Schist.

The age of the Julian Schist is not known. The only fossil found in the unit was an imprint on a piece of quartzite float of an ammonite identified by J. P. Smith as Triassic (Hudson, 1922). This fossil was later lost (Germinaro, 1982). Todd and Shaw (1979, 1985) suggested that the Julian Schist includes volcanic rocks derived from the Santiago Peak volcanic arc and that a part of the Julian Schist is therefore latest Jurassic in age. However, the volcanic rocks in the unit may have been derived from an older, unrelated arc or from an ancestral Santiago Peak arc. The nature of the contact between the Julian Schist and the Santiago Peak Volcanics is discussed in a later section.

The French Valley Formation described by Schwarcz (1969) from exposures near Winchester (Fig. 32-2) is correlated with the Julian Schist by most workers on the basis of lithologic similarities. Included in the French Valley Formation are minor occurrences of metaserpentinite, metamorphosed silica-carbonate rocks, and tremolite-anthophyllite schist[1] (Schwarcz, 1969). D. M. Morton (unpublished mapping) extended the unit to include exposures in the northernmost part of the batholith east of the Elsinore fault zone (Fig. 32-2). Schwarcz concluded that the original sediments were deposited rapidly by prograding deltaic and lagoonal systems into a shallow marginal basin. Some sedimentary structures suggest shallow turbidite deposition. M. A. Murphy (oral communication, 1966) collected poorly preserved mollusks from an outcrop of calc-silicate rocks within a sequence of phyllite, lithic graywacke, and interlayered quartzite-biotite schist, probably at least in part equivalent to the French Valley Formation[2] (Lamb, 1970). These mollusks are similar in appearance to Upper Triassic monotids (M. A. Murphy, oral communication, 1966; D. L. Jones, oral communication, 1976).

Conditions of metamorphism Metamorphic mineral assemblages and textures of the pre-batholithic rocks of Zone II indicate that the rocks were subjected to greenschist to amphibolite-grade metamorphism during regional deformation (Fig. 32-3, Table 32-1). The Bedford Canyon Formation in the Santa Ana Mountains was weakly metamorphosed to assemblages containing sericite, biotite, and chlorite (Larsen, 1948). Critical mineral

---

[1] Also D. M. Morton, unpublished mapping.

[2] Ibid.

assemblages in metapelitic rocks of the Julian Schist and its probable correlative, the French Valley Formation, include muscovite + biotite + sillimanite + alkali feldspar + quartz ± andalusite ± cordierite ± garnet (Table 32-1). Metamorphic reactions suggest peak pressure and temperature ranges of about 2–4 kb and 600–700°C, respectively. (Berggreen and Walawender, 1977; Detterman, 1984; Germinario, 1982; Schwarcz, 1969). Grove (1986) estimated peak metamorphic conditions of 4–5 kb and 650–700°C in the eastern part of Zone II.

In the northern part of the batholith, the low pressure–high temperature metamorphic gradient from greenschist to upper amphibolite facies is relatively steep (Fig. 32-3). A remarkably steep metamorphic gradient occurs in the Winchester area (Fig. 32-3), where the mineral assemblage in metapelitic rocks of the French Valley Formation varies eastward over a distance of 4 km from muscovite-biotite-quartz ± chlorite, to biotite-muscovite-andalusite-quartz, to sillimanite-biotite-K-feldspar-quartz-oligoclase ± cordierite (Schwarcz, 1969). Minor amounts of oligoclase and K-feldspar occur in the first two assemblages. A northwest orientation of metamorphic isograds in the Winchester area changes northward to a west-northwest orientation (D. M. Morton, unpublished mapping). The regional difference in the slope of the metamorphic gradient from steep in the northern part of the batholith to gentle in the batholith east of San Diego may reflect Late Cretaceous and early Tertiary eastward and northward regional tilting of the batholith and erosion to deeper crustal levels in the south. Local variations in the orientation of metamorphic isograds and in the erosional depth of plutons in the northernmost part of the batholith[3] (Morton, 1986) probably reflect this deformation.

In the batholith east of San Diego, recent studies of the Julian Schist by Detterman (1984), Grove (1986), and V. R. Todd (unpublished data) suggest that there may be a rather abrupt west-to-east transition from metapelitic assemblages that contain both andalusite and sillimanite to those with sillimanite only (Fig. 32-3). In some places, steep metamorphic gradients from andalusite- to sillimanite-bearing assemblages appear to be associated with the intrusion of high-temperature gabbroic magma (R. G. Gastil, written communication, 1986). However, the regional transition depicted in Fig. 32-3 suggests that amphibolite-grade metamorphism may have occurred at lower pressures in the western part of Zone II than in the eastern part (Zwart et al., 1967). The andalusite + sillimanite to sillimanite only transition may coincide approximately with west-to-east textural contrasts in the Julian Schist. Relict sedimentary structures are seen locally in the limbs of isoclinal folds in the largest screens throughout much of Zone II, but in a narrow eastern belt, bedding is rarely seen. There, the isoclinal folds and axial plane foliation of the western rocks have been refolded and anatectic veining is pervasive (Grove, 1986; Figs. 32-3 and 32-4).

Transitional subzone between Zones II and III Included in the western part of Zone III (Fig. 32-2) is a group of screens that contain undated high-grade metasedimentary rocks that appear to be transitional in lithology, and may be transitional in age and/or depositional environment, between the Mesozoic submarine fan deposits of Zone II and the largely Paleozoic miogeoclinal rocks of Zone III. It is also possible that these "transitional" rocks are unrelated to the rocks of Zones II and III. Although few of these screens have been mapped in detail, the transitional rocks probably include the typical feldspathic quartzite and semipelitic schist of the Julian Schist as well as significant amounts of orthoquartzite, bedded chert, marble, and amphibolite. Marble in the transitional subzone occurs in relatively thin, laterally continuous layers that were interbedded with the other lithologies prior to metamorphism and folding. Marble is notably more abundant in this subzone than it is in the Julian Schist, but is less abundant and thinner-

[3] Ibid.

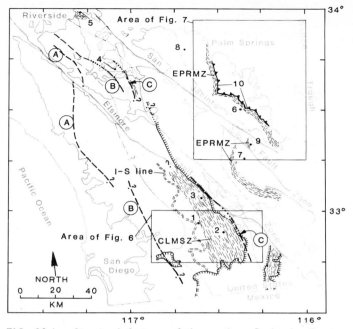

FIG. 32-4. Structural features of the northern Peninsular Ranges
batholith. Data from Sharp (1979), Todd and Shaw (1979, 1985),
Erskine (1986a), and unpublished data of authors. Boundaries of prebatholithic zones (gray lines) and base from Fig. 32-2. I-S line (Todd
and Shaw, 1985) shown by double dashes. Hachured lines indicate
western boundary of late- to post-tectonic plutons, tick marks point to
interiors of satellitic pluton and main intrusive complex. Dash pattern
indicates ductile shear zone: CLMSZ, Cuyamaca–Laguna Mountains
shear zone; EPRMZ, eastern Peninsular Ranges mylonite zone. Heavy
dash lines separate zones of different structural style: A, approximate
western limit of prebatholithic rocks with folded primary layering
(F_1, S_0); B, approximate western limit of prebatholithic rocks with
pervasive transposition of folds and primary layering (S_1); C, approximate western limit of prebatholithic rocks with pervasive transposition
of folded(?) transposed layering (S_2, F_2). Locations mentioned in the
text: (1) Cuyamaca Mountains, (2) Laguna Mountains, (3) Julian,
(4) Winchester, (5) Box Springs Mountains, (6) Santa Rosa Mountains,
(7) Borrego Springs, (8) San Jacinto Mountains, (9) Coyote Mountain,
(10) Palm Canyon fault, teeth on upper plate. Insets indicate areas of
Figs. 32-6 and 32-7.

bedded than marble in the miogeoclinal rocks to the east (D. M. Morton and V. R. Todd,
unpublished mapping). The volume of amphibolite varies throughout the transitional
subzone, but is probably greater in the southern part of the subzone than in the northern
part. Amphibolite commonly occurs in relatively thick layers that were originally interbedded with the sedimentary rocks and it grades laterally into hornblende schist (metatuff?) and pods of fine-grained gabbro (sills?).

Engel and Schultejann (1984) described mica-feldspar-quartz schists (argillaceous
sandstone) and lenses of marble in roof pendants and screens in the Ranchita area (Fig.
32-2) that may be equivalent to the rocks of the transitional subzone. According to these
authors, the metasedimentary rocks in the Ranchita area, which resemble the Julian
Schist, appear to grade eastward to a platform sequence of quartz-rich sandstone, shale,

and limestone. A number of prebatholithic sequences that contain turbidites, bedded chert, amphibolite, and minor carbonate occur in the shale-sandstone belt in the northern part of Baja California (Gastil and Miller, 1984). The ages and affinities of these sequences with one another, and with the transitional subzone in southern California, are not known at this time, but Gastil (1983) considered them to be possibly Carboniferous to Triassic in age.

Zone III: Miogeoclinal Rocks

The prebatholithic rocks of Zone III are quartz-rich, carbonate-bearing metasedimentary rocks that are probably deformed and metamorphosed remnants of a miogeoclinal prism shed westward from the North American craton. Exposures range in size from small isolated pods within plutons to relatively large screens that separate major intrusive units. Two groups of metasedimentary rocks are included within this zone: autochthonous rocks that occur west of the eastern Peninsular Ranges mylonite zone, and parautochthonous rocks that occur in the upper plate of the Palm Canyon fault east of the mylonite zone (Fig. 32-4). Except for a general lithologic similarity, evidence correlating the two groups is lacking and it is possible that they may differ in age and/or depositional setting.

Autochthonous metasedimentary rocks The autochthonous rocks of Zone III are well exposed in the San Jacinto and Santa Rosa Mountains, in the ranges west of Borrego Valley, and in the Coyote Mountains to the south (Fig. 32-2). The Coyote Mountains have been included within the transitional subzone in Fig. 32-2 because the range exposes both miogeoclinal and transitional (flysch, chert, carbonate rocks, and amphibolite) assemblages in unclear structural relations (R. G. Gastil, written communication, 1986). Among the most completely studied autochthonous rocks are metamorphic rocks located northwest of Palm Springs called the Windy Point metamorphic rocks by Sydnor (1975), and the southern part of the Desert Divide Group (Brown, 1981), which forms a large north-northwest-striking screen along the crest of the San Jacinto Mountains (Fig. 32-2). The Windy Point metamorphic rocks contain significant volumes of amphibolite and plagioclase-rich schist in addition to quartz- and carbonate-rich metasedimentary rocks, and may therefore be anomalous to Zone III (Hill, 1984). The Desert Divide Group is similar in lithology and metamorphic grade to most of the autochthonous prebatholithic rocks of the eastern part of the batholith, and the characteristics of this unit are therefore considered to be representative of Zone III.

The contact between the Desert Divide Group and adjacent plutonic rocks is intrusive except where it is faulted. The northernmost part of the screen exceeds 4.3 km in thickness, but parts may thin to 1.3 km. Brown (1981) divided the unit into two subunits: a large quartzite body on the east called the Ken Quartzite, and a diverse sequence of metacarbonate rocks, schist, and gneiss on the west called the Bull Canyon Formation. The Ken Quartzite consists of medium- to coarse-grained pure quartzite with minor biotite, plagioclase, alkali feldspar, muscovite, and ilmenite and minor lenses of marble, schist, and gneiss. The Bull Canyon Formation is composed of four lithologic units, or facies (Brown, 1981). The gneiss facies consists of moderately to well-foliated quartz-rich gneiss with minor or trace amounts of biotite, plagioclase, alkali feldspar, garnet, sillimanite, muscovite, calcite, ilmenite, and sphene. The quartz-rich gneiss grades across and along compositional banding into the sillimanite gneiss facies, which contains quartz, sillimanite, alkali feldspar, and biotite, with subsidiary plagioclase, muscovite, garnet, and ilmenite. The schist facies is composed of a quartz-rich biotite schist interlayered with quartzite. The marble facies commonly is composed of pure calcite; plagioclase, alkali feldspar, muscovite, diopside, tremolite, zoisite, scapolite, almandine, sphene, graphite, and ilmenite are common in calc-silicate interlayers.

Although no direct age determinations have been made on rocks of the Desert Divide Group or Windy Point metamorphic rocks, isotopic evidence from the San Jacinto Mountains and fossils from similar metasedimentary rocks in the Coyote Mountains to the south provide constraints on the age of sedimentation. Strontium isotopic analysis of samples from the Desert Divide Group are consistent with a late Precambrian age or a provenance from late Precambrian rocks (Hill, 1984). Carbonates from the Desert Divide Group have thus far yielded no datable fossil material (R. H. Miller and R. G. Gastil, oral communication, 1985), but metacarbonates from the Coyote Mountains yielded Ordovician conodonts demonstrating that at least some of the autochthonous prebatholithic rocks are Paleozoic in age (R. H. Miller and Dockum, 1983). Strontium isotopic analysis of samples from the Windy Point metamorphic rocks are consistent with a late Paleozoic or early Mesozoic depositional age for the unit (Hill, 1984).

The relatively thick quartzite sequence and interbedded quartz-rich sediments and carbonates of the Desert Divide Group are suggestive of a relatively mature clastic sequence derived from nearby continental rocks. Evidence discussed above suggests that Zone III rocks were derived largely from a Precambrian source terrane and deposited in a Paleozoic miogeoclinal clastic wedge along the cratonic margin. Since the Desert Divide Group was probably thinned during regional metamorphism and intrusion, and because mapping shows that lithologic sections are not repeated on a large scale (Erskine, 1986a), the section in the northern part represents a minimum thickness (4.3 km) for the miogeoclinal wedge.

Parautochthonous metasedimentary rocks Metasedimentary rocks that extend for 80 km from north to south in the Santa Rosa Mountains are the easternmost exposures of prebatholithic rocks in the northern Peninsular Ranges. Metasedimentary rocks in the Palm Springs area were called the Palm Canyon Complex by Miller (W. J. 1944) and this name was used by Erskine (1986a) for similar rocks in the central and southern Santa Rosa Mountains. The rocks occupy the upper plate of the Palm Canyon fault (Fig. 32-4) and are considered parautochthonous to the batholith.

The Palm Canyon Complex is a sequence of intensely deformed metasedimentary rocks, anatexites, and orthogneiss whose thickness ranges from 1.2 to 5.7 km. The structural and metamorphic complexity of the unit prohibit an accurate estimate of the thickness of the protolith. The age of the Palm Canyon Complex is unknown, and isotopic data are lacking. Only the northern and central parts of the complex have been mapped in detail (Erskine, 1986a; Matti *et al.*, 1983).

Except for the general absence of pure quartzites, the metasedimentary rocks of the Palm Canyon Complex are similar to those below the Palm Canyon fault. Marbles are common, and increase in relative abundance westward toward the trace of the Palm Canyon fault. Their compositions range from monomineralic marble to less pure carbonate rocks with diopside, epidote, garnet, ilmenite, muscovite, plagioclase, quartz, scapolite, sphene, and tremolite (Erskine, 1986a). Associated with these rocks are calc-silicate hornfels that consist of calcite, diopside, epidote, hornblende, plagioclase, quartz, sphene ± garnet ± hypersthene ± scapolite. Also common are a variety of pelitic schists and gneisses with alkali feldspar, biotite, cordierite, garnet, muscovite, quartz, and sillimanite. Amphibolite is present in relatively minor amounts. The Palm Canyon Complex appears to represent a quartz-rich clastic and carbonate-bearing miogeoclinal sequence deposited along a cratonic margin. The gross similarity in composition between this unit and the autochthonous metasedimentary rocks permits a tentative correlation between the two groups.

Conditions of metamorphism The general absence of andalusite in the metapelitic rocks of Zone III suggests that these rocks were metamorphosed under conditions of intermediate pressure amphibolite facies (Fig. 32-3, Table 32-1). On the basis of metamorphic mineral reactions and the compositions of coexisting biotite + cordierite, and biotite + garnet, Hill (1984) estimated pressures and temperatures in the Desert Divide Group at the peak of metamorphism to be in the range 3.2–4.2 kb and 620–800°C, respectively. Pressure and temperature estimates based on phase relations in the Palm Canyon Complex are probably at the upper ends of the ranges 2–4.5 kb and 650–800°C, respectively (Erskine, 1986a).

PLUTONIC ROCKS

Systematic variations in composition, size, and structural characteristics of plutons across the batholith have led workers to propose a number of classifications, all of which have common features. Larsen (1948) noted the geographic asymmetry in the distribution of plutonic rock types across the northern part of the batholith, pointing out that transitions between lithologic zones are sharp and trends are parallel to the long axis of the batholith. Gastil (1975, 1983) divided the plutonic rocks of the batholith into five compositional belts, four of which are present in southern California and the Baja California peninsula. These are, from west to east, (1) the Pacific Margin belt (western Transverse Ranges, Channel Islands of southern California; Isla Cedros and Vizcaino Peninsula, Baja California peninsula), which includes allochthonous oceanic plutonic rocks of Triassic, Jurassic, and Early Cretaceous age (Moore, 1984); (2) the gabbro belt (western part of the Peninsular Ranges), consisting chiefly of gabbro, tonalite, and granodiorite; (3) the leucotonalite belt (eastern Peninsular Ranges), which includes relatively large, homogeneous leucocratic plutons that vary from "potash feldspar-free tonalite to granodiorite"; and (4) the granodiorite-granite belt (easternmost part of Baja California peninsula and western Sonora, Mexico), which includes large adamellite plutons with well-formed, large, pink K-feldspar crystals.

On the west side of the batholith, Silver *et al.* (1979, p. 105) reported diverse lithologies that range from "gabbros and quartz gabbros to abundant tonalites (relatively mafic and inclusion-rich) with fairly abundant silica-rich leucogranodiorites and rare adamellites" and noted the prominence of hornblende throughout these rocks. On the east side, they found a relatively limited range of lithologies that are dominated by "sphene-hornblende-biotite-bearing tonalites and low K_2O granodiorites." Oxygen isotopic compositions of plutonic rocks reflect the west-to-east lithologic asymmetry noted above. Plutons on the west side of the batholith have "normal" igneous $\delta^{18}O$ values (+6 to +8.5), while those on the east side have values of +9 to +11 or higher, with a sharp discontinuity, or "step," between the two regions (Taylor and Silver, 1978; Fig. 32-5). The $\delta^{18}O$ step coincides approximately with the boundary between the older, western static arc and the younger, eastern migrating arc of Silver *et al.* (1979). Rare earth element (REE) abundances also vary systematically across the batholith (Gromet, 1979; Gromet and Silver, 1979). Slightly fractionated REE patterns, commonly with negative Eu anomalies in plutons in the western region of the batholith give way to middle and heavy REE fractionated and depleted patterns with subdued Eu anomalies, if any, in the central region of the batholith, and to strongly enriched patterns of light REE in eastern plutons.

Major elements of about 334 granitoid rocks of the northern Peninsular Ranges batholith were analyzed by statistical methods in order to characterize the source materials of the batholith (Baird and Miesch, 1984). The data indicate that the granitoid rocks formed from the mixing of basaltic and quartzofeldspathic end-member magmas whose

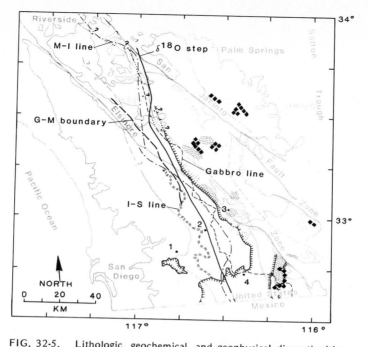

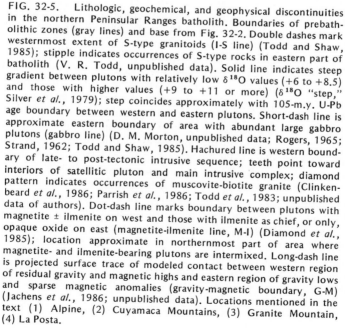

FIG. 32-5. Lithologic, geochemical, and geophysical discontinuities in the northern Peninsular Ranges batholith. Boundaries of prebatholithic zones (gray lines) and base from Fig. 32-2. Double dashes mark westernmost extent of S-type granitoids (I-S line) (Todd and Shaw, 1985); stipple indicates occurrences of S-type rocks in eastern part of batholith (V. R. Todd, unpublished data). Solid line indicates steep gradient between plutons with relatively low $\delta^{18}O$ values (+6 to +8.5) and those with higher values ($\delta^{18}O$ +9 to +11 or more) ($\delta^{18}O$ "step," Silver et al., 1979); step coincides approximately with 105-m.y. U-Pb age boundary between western and eastern plutons. Short-dash line is approximate eastern boundary of area with abundant large gabbro plutons (gabbro line) (D. M. Morton, unpublished data; Rogers, 1965; Strand, 1962; Todd and Shaw, 1985). Hachured line is western boundary of late- to post-tectonic intrusive sequence; teeth point toward interiors of satellitic pluton and main intrusive complex; diamond pattern indicates occurrences of muscovite-biotite granite (Clinkenbeard et al., 1986; Parrish et al., 1986; Todd et al., 1983; unpublished data of authors). Dot-dash line marks boundary between plutons with magnetite ± ilmenite on west and those with ilmenite as chief, or only, opaque oxide on east (magnetite-ilmenite line, M-I) (Diamond et al., 1985); location approximate in northernmost part of area where magnetite- and ilmenite-bearing plutons are intermixed. Long-dash line is projected surface trace of modeled contact between western region of residual gravity and magnetic highs and eastern region of gravity lows and sparse magnetic anomalies (gravity-magnetic boundary, G-M) (Jachens et al., 1986; unpublished data). Locations mentioned in the text (1) Alpine, (2) Cuyamaca Mountains, (3) Granite Mountain, (4) La Posta.

compositions reflected variations in their source materials. The boundary between the two different source regions is a north-northwest-striking discontinuity that lies near the center of the batholith and is interpreted as the western margin of major continental crustal contribution to the batholith. This discontinuity coincides approximately with the $\delta^{18}O$ and 105-m.y. U-Pb age steps discussed above.

Todd and Shaw (1979, 1985, 1986) divided intrusive rocks into a syntectonic sequence in the western part of the batholith and a late- to posttectonic sequence that occurs chiefly, but not exclusively, in the eastern part of the batholith. The syntectonic intrusive sequence includes both I-type granitoids, those with geochemical and isotopic characteristics primarily derived from partial melting of an igneous protolith, and S-type granitoids, those with characteristics primarily derived from partial melting of a crustal sedimentary, or metasedimentary, source (Chappell and White, 1974; A. J. R. White and Chappell, 1977). The western extent of S-type granitoids was considered to mark an I-S line in the batholith (Fig. 32-5).

Gabbroic plutons are confined largely to the western part of the batholith (Fig. 32-5), where they occur in coeval intrusive complexes with both I- and S-type granitoids. The syntectonic intrusive sequence corresponds approximately to the gabbro belt of Gastil (1975, 1983) and to the western region of Gromet (1979). The late- to post-tectonic intrusive sequence includes large, homogeneous, relatively leucocratic, strontium-rich plutons that vary from marginal hornblende-biotite-tonalite (tonalite of Granite Mountain; Todd, 1979), through biotite trondhjemite and granodiorite (La Posta Quartz Diorite; W. J. Miller, 1935), to garnetiferous muscovite-biotite granite. The late- to post-tectonic intrusive sequence is approximately equivalent to the leucotonalite belt of Gastil (1975, 1983) and to the central region of Gromet (1979). Although geochemical and isotopic characteristics generally indicate that these granitoids are I- type in character, their $^{87}Sr/^{86}Sr$ initial ratios and $\delta^{18}O$ values locally are somewhat higher than those of the western I-types. These data, and the occurrence of two-mica granites as part of the sequence, suggest some contribution of continental material in the source (Clinkenbeard et al., 1986; Gunn, 1986; Hill, 1984; Parrish et al., 1986; Shaw et al., 1986; Taylor and Silver, 1978).

Plutons of the syntectonic intrusive sequence are part of the western age group (120–105 m.y.) of Silver et al. (1979). The S-type plutons have not been dated by U-Pb zircon method, but are considered to be approximately coeval with the I-type plutons in the sequence on the basis of contacts and structural relations. The S-type plutons may be somewhat older than the associated I-types, as suggested by K-Ar replicate biotite ages of 115 and 122 m.y. for a sample from an S-type pluton in the Cuyamaca Mountains (W. C. Hillhouse, U.S. Geol. Survey). Biotite K-Ar ages in this region are as much as 20 m.y. younger than corresponding U-Pb zircon ages and probably represent the age of syn-kinematic recrystallization of plutons (Todd and Shaw, 1979). The 115- and 122-m.y. K-Ar ages are the oldest biotite ages obtained for any granitoid pluton in the Cuyamaca Mountains and thus suggest that the parent S-type pluton is older than 120 m.y.

Plutons of the late- to post-tectonic intrusive sequence are part of the eastern age group (105–89 m.y.) of Silver et al. (1979). A late-tectonic pluton at Granite Mountain (Fig. 32-5) has a U-Pb zircon age of 100 m.y., and the large post-tectonic La Posta pluton mentioned above has ages of 98 and 97 m.y. (L. T. Silver, oral communication, 1979). U-Pb zircon dating of the La Posta pluton by Clinkenbeard et al. (1986) gave consistent 95-m.y. ages. Late- to post-tectonic plutons that occur in the western part of San Diego County have not been dated by U-Pb zircon method, but they may be somewhat older than those in the eastern part. A late-tectonic pluton near Alpine yielded nearly concordant K-Ar hornblende and biotite ages of 103 and 101 m.y., respectively (W. C. Hillhouse, U.S. Geol. Survey), suggesting a slightly older age of emplacement than that of the pluton at Granite Mountain. We should note that the designation of the younger group of plutons as late- to post-tectonic may apply only to the southern part of the study area. Plutons in the northeast part of the batholith that are tentatively correlated with the late-to post-tectonic intrusive sequence on the basis of lithology and geochemistry locally are strongly deformed, as will be discussed below.

PREBATHOLITHIC SUTURE

Most workers consider the Santiago Peak Volcanics and the Alisitos Formation to represent the deposits of a fringing island arc that collided with, and was accreted to, western North America in late Mesozoic time (Gastil *et al.*, 1981). Todd and Shaw (1985) suggested that compressive structures in Early Cretaceous plutons and their metamorphosed wallrocks east of San Diego were caused by collision of the Santiago Peak island arc with the Triassic and Jurassic marginal basin. The exact location and nature of the contact between arc-related rocks and marginal-basin sediments in the northernmost part of the batholith are unclear, in large part because the distribution, ages, and affinities of metavolcanic rocks there are not fully known. In the batholith east of San Diego, the westernmost screens consist mainly of metavolcanic rocks (Fig. 32-2, close stipple). The southern part of the prebatholithic boundary between Zones I and II is considered to be the approximate midpoint of a 5- to 10-km-wide zone of interfingering metavolcanic and metasedimentary wallrocks (Todd and Shaw, 1985; Fig. 32-6). Metasedimentary rocks are rare in screens west of this interfingering zone, and, as mentioned above, minor orthoamphibolite layers are present in the Julian Schist well east of the zone. The location of the prebatholithic boundary between Zones I and II is uncertain in the area between the Santa Ana Mountains and 33°N (Fig. 32-2). The southwesternmost screens in this area contain metamorphosed felsic and mafic volcanic rocks (Todd, unpublished mapping), which suggests that the boundary strikes north-northwest, nearly parallel to the I-S line (Fig. 32-5). Further to the north, screens that lie west of the dashed line in Fig. 32-2 contain interstratified metasedimentary rocks ("Bedford Canyon Formation") and metavolcanic rocks[4] (R. Engel, 1959). This line, inferred to be the boundary between Zones I and II, marks the eastern limit of screens containing significant amounts of metavolcanic rocks. Minor amphibolite layers of possible volcanic origin occur east of this boundary in the French Valley Formation (Schwarcz, 1969). The relations above suggest that the Mesozoic marginal basin received detritus from both a western(?) volcanic arc and eastern cratonic highlands prior to collision.

Further evidence for the existence of a major prebatholithic tectonic contact, or suture, in the northern part of the batholith are the north-northwest-striking lithologic, geochemical, and geophysical discontinuities in plutons discussed in the preceding section (Fig. 32-5). Although their origins are not well understood, these discontinuities almost certainly reflect a fundamental west-to-east contrast in the composition of the prebatholithic crust that in part formed the source rocks for the plutons (Shaw *et al.*, 1986; Taylor and Silver, 1978; Todd and Shaw, 1985, 1986). The I-S line, gabbro line, and $\delta^{18}O$ step all reflect a change in source rock from primitive, oceanic crust on the west to continental crust and/or continentally derived sediments on the east. From the relative positions of the prebatholithic boundary between Zones I and II, the I-S line, and the gabbro line, Todd and Shaw (1985) suggested that the prebatholithic suture between the Santiago Peak volcanic arc and the Mesozoic marginal basin dipped to the east (Fig. 32-6). Thus, the suture was oriented subparallel to regional batholithic structures that imply northeast-southwest shortening (northwest-striking, northeast-dipping tight to isoclinal folds, axial plane foliation, and ductile shears). The eastward offset of the lithologic and geochemical discontinuities from this prebatholithic boundary (Fig. 32-6) is probably an artifact of the eastward dip of the suture. The geometry produced by the inferred east-

[4]Ibid.

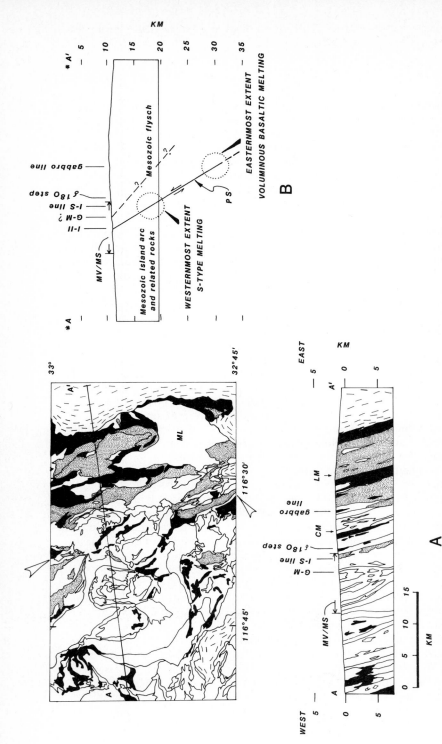

FIG. 32-6. Geologic map and cross sections of part of the Cuyamaca and Laguna Mountains. (A) Map shows contacts between Mesozoic wallrocks (black), Early Cretaceous I-type granitoid and gabbroic plutons (unpatterned), and Early Cretaceous(?) S-type granitoids (stipple); mid-Cretaceous, late- to post-tectonic plutons shown in dash pattern. Arrows indicate approximate western boundary of Cuyamaca-Laguna Mountains shear zone. Abbreviations: CM, Cuyamaca Mountains; LM, Laguna Mountains; ML, 105-m.y.-old tonalite pluton; MV/MS, zone of mixed metavolcanic

and metasedimentary rocks. (B) Schematic cross section *A-*A' showing location of prebatholithic suture (PS) during Early Cretaceous underthrusting of island arc and syntectonic intrusion and before substantial westward thrusting in Cuyamaca-Laguna Mountains shear zone; arrows indicate sense of shear. Location of suture inferred from relative positions of boundary between prebatholithic zones I and II (I-II), I-S line, and gabbro line; other plutonic and prebatholithic boundaries as in (A). Dashed line is inferred contact between western and eastern gravity-magnetic regions.

dipping suture and by regional structures that suggest northeast-southwest compression, and a pervasive, steep northeast-plunging lineation indicate that the Santiago Peak arc was thrust beneath the western edge of North America.

The gravity-magnetic boundary shown in Fig. 32-5 is the projected surface trace of the contact, inferred from geophysical modeling, between a western region of residual gravity and magnetic highs and an eastern region of gravity lows and sparse magnetic anomalies (Jachens *et al.*, 1986). This contact apparently marks the transition from oceanic crust intruded by relatively mafic, I-type plutons on the west to continental crust and sediments intruded by relatively leucocratic late- to post-tectonic plutons on the east. The boundary lies near, but is not coincident with, the magnetite-ilmenite line, which separates rocks containing magnetite ± ilmenite on the west from those with ilmenite as the chief opaque oxide on the east (Diamond *et al.*, 1985). The geophysical modeling implies the existence at depth of a remarkably regular contact that dips about 45° to the east and extends to a depth of 15 km or more. North of the segment of the gravity-magnetic boundary shown in Fig. 32-5, the gravity and magnetic fields are complex, and the contact, if present, cannot be identified (R. C. Jachens, oral communication, 1986). South of this segment, the contact is well defined, dips about 40° to the east, and extends to a minimum depth of 10 km (A. Griscom, oral communication, 1986). The relatively simple geometry of this contact suggested to Jachens *et al.* (1986) that it is tectonic rather than intrusive and is probably the subsurface location of the prebatholithic suture. Alternatively, the gravity-magnetic contact may be a syn- or postintrusive ductile shear zone that formed during an episode of Early cretaceous ductile deformation (discussed below).

MAJOR ZONES OF DUCTILE DEFORMATION IN THE BATHOLITH

Two major zones of ductile deformation have been recognized in the northern part of the batholith, the Early Cretaceous Cuyamaca–Laguna Mountains shear zone in the south-central part of the study area, and the Late Cretaceous eastern Peninsular Ranges mylonite zone in the eastern part (Fig. 32-4). The older zone strikes north-northwest and dips steeply to the east, while the younger zone strikes predominantly from north-northwest to northwest and dips moderately to the east. Intrusion probably preceded and accompanied deformation in both zones.

Cuyumaca–Laguna Mountains Shear Zone

Work in the southern part of the study area indicates that folding of the prebatholithic rocks began prior to and continued during intrusion of the batholith. During the Early Cretaceous, plutonic deformation was concentrated in a zone that extends north-northwest for at least 40 km from the southern end of the Laguna Mountains to the vicinity of Julian, California (Fig. 32-4). This zone is characterized by (1) strongly flattened and elongated, north-northwest-striking, east-dipping plutons; (2) strong foliation and local development of mylonite gneiss; (3) local hypabyssal textures in plutonic rock types that occur as hypidiomorphic- to allotriomorphic-granular plutons outside the shear zone; and (4) emplacement of intermediate and mafic dike swarms. The shear zone is intruded to the east and south by younger (100–95 m.y. old) plutons of the late- to post-tectonic intrusive sequence and therefore its original eastern and southern extents are not known. The zone probably extends for an unknown distance to the north, where it may be intruded by younger plutons and also partly covered by broad areas of

Quaternary alluvium. In the northern part of the study area, the steep metamorphic gradient in the Winchester area mentioned above and a northeast-dipping zone of tectonically thinned plutonic units on the southwest side of the Box Springs Mountains are on strike with the shear zone and may have formed by related(?) ductile shear (D. M. Morton, unpublished mapping). The western boundary of the shear zone is considered to be the transition from strongly deformed, north-northwest-trending plutons on the east to plutons having variable deformation fabrics and orientations on the west (Fig. 32-6).

The plutonic and metamorphic rocks of the Cuyamaca-Laguna Mountains shear zone are characterized by a steep (65-85°) east-dipping foliation and steeply plunging mineral lineation that lies within the plane of foliation. Isoclinal folds, ranging in amplitude from a few centimeters to several meters, are common in prebatholithic rocks, and folds with amplitudes up to several kilometers occur sporadically (Detterman, 1984; Schwarcz, 1969). Isoclinal folds are also seen locally in plutons, where dikes, inclusions, or schlieren serve as markers (Todd and Shaw, 1979). The axial surfaces of these folds are coplanar with batholithic foliation, and fold axes are generally steep and presumably collinear with mineral lineations in the prebatholithic and plutonic rocks[5] (Detterman, 1984; Germinario, 1982). Mafic inclusions in strongly deformed plutons are flattened triaxial ellipsoids whose greatest axes are oriented parallel to lineation. In the prebatholithic rocks, the fabric elements formed in a single prograde metamorphic event. The overall fabric, therefore, is considered to have formed during Early Cretaceous metamorphism and intrusion in a tectonic setting dominated by regional compression. The steep lineation and fold axes probably resulted from subvertical extension that accompanied regional compression as well as from the upward flow of highly crystallized, viscous magma. Plutons in the Cuyamaca–Laguna Mountains shear zone locally consist of mylonite gneiss containing thin isolated bands of mylonite that are oriented parallel to the north-northwest fabric of the zone. However, high-grade metamorphic rocks in the adjacent prebatholithic screens are not mylonitized, suggesting that deformation was concentrated in plutons that were not completely crystallized.

Todd and Shaw (1986) identified five granitoid suites, or magma types, each of which composes a large number of discrete plutons throughout the southern part of the study area. Within the Cuyamaca–Laguna Mountains shear zone, several of these suites are represented mainly by long, narrow, dikelike plutons with hypabyssal textures. Furthermore, a group of relatively high-alkali intermediate and mafic dikes were emplaced preferentially into the shear zone. The hypabyssal plutons and high-alkali dikes typically are strongly deformed. These relations suggest the occurrence in the shear zone of one or more periods of fracturing (relatively high strain rate?) at which times the zone was connected by fractures to the near-surface environment and magmas were cooled rapidly. These periods were closely followed by ductile deformation. If the shear zone represents the deeper parts of an Early Cretaceous fault zone, it may have undergone a gradual ductile-to-brittle transition due to regional uplift while intrusion was still taking place. Although the exact sequence of tectonic and intrusive events in the shear zone is not known, it seems clear that intrusion was contemporaneous with deformation.

Comparison of Figs. 32-4 and 32-5 shows that the Cuyamaca–Laguna Mountains shear zone lies just east of most of the lithologic, geochemical, and geophysical boundaries mentioned in the preceding section. In particular, the shear zone lies from about 6 to 20 km east of the prebatholithic boundary between Zones I and II. The shear zone approximately coincides with the zone of interfingering I- and S-type granitoid plutons that probably originated by partial melting of crustal rocks in the vicinity of the prebatholithic suture, where flysch lay structurally above rocks of the volcanic arc (Fig. 32-6).

---

[5] Also V. R. Todd, unpublished data.

The contrast in mechanical properties between I-type granitoid plutons that contain abundant refractory minerals such as hornblende, calcic plagioclase, and pyroxene, and S-type granitoids that contain abundant quartz, mica, and hydrous metasedimentary inclusions probably enhanced the zone of weakness created in the Early Cretaceous crust by the prebatholithic suture.

The timing of movements in the Cuyamaca–Laguna Mountains shear zone is broadly constrained by U-Pb zircon ages of granitoid plutons in the southern part of the study area, but constraints are only approximate, since relatively few plutons in and near the zone have been dated. Ages of the I-type granitoid plutons in and west of the shear zone range from 120 to 105 m.y. (Silver et al., 1979). Thus, deformation in the shear zone may have begun as early as 120 m.y. ago, or even earlier, if the S-type granitoid plutons are older than the I-types. The large I-type pluton in the southern part of the Laguna Mountains (labled ML in Fig. 32-6) yielded a U-Pb zircon date of 105 m.y. (L. T. Silver, oral communication, 1979). The margins of this pluton are deformed, as suggested in Fig. 32-6, by its "ears," but the core is relatively undeformed. These relations suggest that deformation in the shear zone was waning by 105 m.y.b.p. East of the shear zone, plutonic ages appear to decrease progressively eastward from 100 to 89 m.y. (Silver et al., 1979).

The younger (100–95 m.y.), late- to post-tectonic plutons intrude the older plutonic terrane along a remarkably linear north-northwest-trending contact located on the east side of the shear zone (Fig. 32-4). Enclaves of older plutonic rocks occur within these plutons east and south of the area of Fig. 32-6. To the south, the younger plutons cut across the shear zone, and a satellitic pluton of the younger intrusive sequence occurs within the older plutonic terrane. The parallelism of much of the western boundary of the younger plutonic terrane with the shear zone suggests that the oldest plutons of this group were emplaced during the last movements in the shear zone. Plutons that are correlated with the 100-m.y.-old tonalite of Granite Mountain (Todd, 1979) lie between the older and younger plutonic terranes along the length of their boundary (Fig. 32-4) and have lithologic and geochemical characteristics that are transitional between the two (Todd, 1980). Plutons of Granite Mountain type that lie on the east side of the shear zone locally underwent complex deformation, including the development of early mylonitic fabrics and high-grade metamorphic folds that probably were associated with movements in the shear zone (Grove, 1986). This deformation was followed by large-scale folding (amplitudes of several kilometers) of plutons and metasedimentary screens that apparently postdated movements in the shear zone (Grove, 1986). Granite Mountain-type plutons that occur west of the shear zone, and 98–95-m.y.-old La Posta-type plutons on both sides of the zone, are foliated parallel to their walls and have relatively massive cores. Thus, the latest movements in the Cuyamaca–Laguna Mountains shear zone probably took place at about 100 m.y., and movement had ceased by about 98 m.y.

The sense of shear in the Cuyamaca–Laguna Mountains shear zone has not been determined rigorously, but it can be inferred from the geometry of the zone and from regional geologic relations. The zone probably dips steeply to the east, as inferred from steep, east-dipping plutonic contacts and foliation (Fig. 32-6). Although the same prebatholithic and plutonic units are present on both sides of the shear zone, rocks on the east side appear to have formed at a deeper level of the batholith than those on the west side, implying that the eastern block was thrust over the western block. Grain size of mica in the Julian Schist and in S-type granitoids on the east side of the shear zone is notably coarser than it is in these units on the west side of the zone. As mentioned above, the Julian Schist on the east side of the shear zone shows a high degree of granitic veining (Fig. 32-3) that is probably the result of metamorphic segregation and incipient melting. Similarly, S-type rocks east of the shear zone are chiefly complex mixtures of partly melted schist inclusions, wispy S-type granitoid bodies, and rafts of refractory calc-silicate-bearing quartzite and amphibolite. This material is believed to have formed close to the zone of S-type melting. Similar anatectic migmatites probably occur below the discrete S-type plutons west of the shear zone (Todd and Shaw, 1985).

While injection migmatites are common in the batholith, the migmatitic Julian Schist and S-type granitoids in the eastern part of the shear zone appear to be regional in extent rather than related to specific intrusives, and are therefore more likely anatectic in origin. Approximately coincident with the anatectic migmatites east of the shear zone are an apparent increase in metamorphic pressures (Fig. 32-3), virtual obliteration of sedimentary structures, and development of polyphase ductile folds in the Julian Schist (Fig. 32-4). The apparent west-to-east increase in depth and the geometry of the shear zone suggest deep-seated east-over-west thrusting. The occurrence of the same rock types on both sides of the shear zone and the steep lineations in the rocks of the zone further suggest that movement was predominantly dip-slip. If strike-slip movements did occur, they were probably not of regional extent.

Eastern Peninsular Ranges Mylonite Zone

The eastern Peninsular Ranges mylonite zone (Sharp, 1979) is a 100-km-long belt of deformation located in the eastern part of the northern Peninsular Ranges batholith. The zone strikes southward from Palm Springs to the southern Santa Rosa Mountains, where it is offset to the west by two strands of the right-lateral San Jacinto fault (Sharp, 1967; Fig. 32-7). Southeast of Borrego Springs, the zone is buried beneath alluvium. The mylonite zone is probably part of a more extensive zone of deformation that lies beneath the Coachella and Imperial Valleys and the Gulf of California. Spatially associated with the mylonite zone are a series of east-dipping low-angle faults that strike subparallel to, but generally dip more gently than, the regional east-dipping mylonitic foliation. Although ductile mylonitic deformation and brittle low-angle faulting were probably parts of a single protracted episode of deformation, the chronologic and kinematic relations between the two are uncertain. The two types of deformation are discussed separately below.

Ductile deformation Rocks in, and associated with, the eastern Peninsular Ranges mylonite zone have been divided into three units, based on lithology, rock fabrics, and contact relations. From structurally lowest to highest these are (1) strongly deformed mylonitic rocks that define the mylonite zone itself, (2) metasedimentary tectonites and anatexites of the Palm Canyon Complex, and (3) "upper-plate" granitic rocks (Fig. 32-7). All of these rocks are linked structurally by the presence of a foliation and lineation that are uniform in attitude, and the three units probably represent different parts of a broad zone of deformation that have been juxtaposed by ductile shearing and later low-angle faulting. For this reason, the Palm Canyon Complex and "upper-plate" granitic rocks are considered parautochthonous rather than allochthonous. The critical features of each unit are summarized below.

The Mylonite Zone. The rocks of the mylonite zone are autochthonous to the batholith and have not been displaced by movements along low-angle faults (Fig. 32-7). As the mylonite zone is approached from the west, rocks of the northern Peninsular Ranges batholith grade rather abruptly into mylonite gneiss, mylonite, and ultramylonite characterized by the presence of an intense foliation and a well-developed stretching lineation. The foliation is regionally consistent, striking north-northwest and dipping 35–55° eastward. The attitude of the lineation is extremely uniform in a given region, but its trend changes systematically from eastward in the southern part of the mylonite zone to about N55°E in the San Jacinto and northern Santa Rosa Mountains. Classic shear indica-

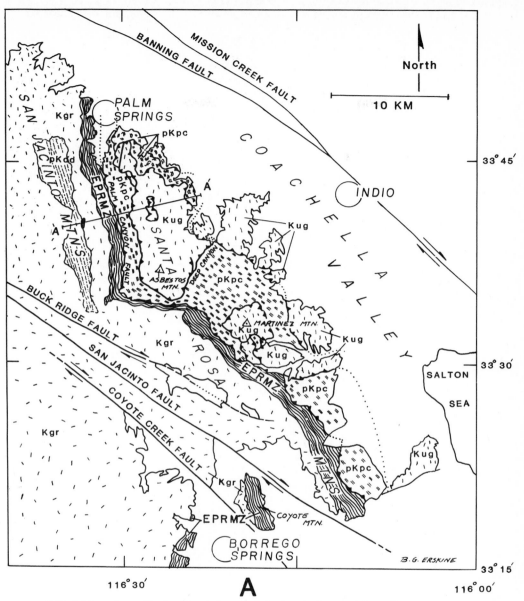

FIG. 32-7.   Generalized geology of part of the San Jacinto and Santa Rosa Mountains, modified from Erskine (1986a) and Sharp (1979). (A) Geologic map. (B) Enlarged cross section A-A', vertical and horizontal scales equal.

tors such as multiple foliations and asymmetric fabrics described by Platt and Vissers (1980), S. H. White *et al.* (1980), and Lister and Snoke (1984) indicate an east-over-west sense of shear (thrust) throughout the zone (Simpson, 1984). This sense of shear is confirmed by asymmetric quartz textures in mylonitic plutonic rocks and in quartz-rich leucocratic dikes that were transposed along the foliation (Erskine, 1986a). Sharp

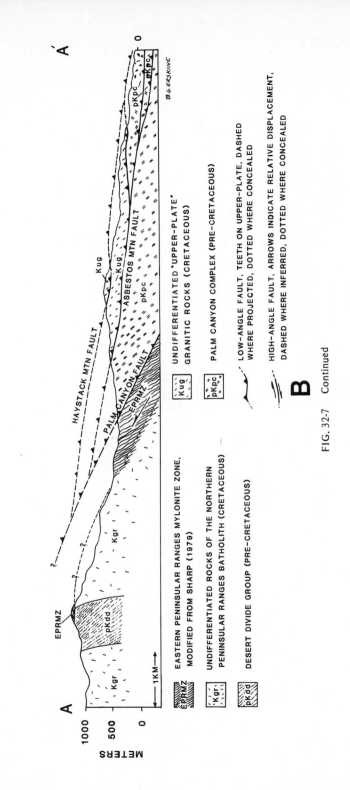

FIG. 32-7 Continued

(1979) proposed westward to southwestward thrusting in the mylonite zone based on regional geologic studies. Engel and Schultejann (1984) reached a similar conclusion based on studies of highly deformed metasedimentary rocks and gneiss in and adjacent to the eastern Peninsular Ranges mylonite zone. They described Late Cretaceous imbricate thrusts and nappes that first moved westward and later north-northwestward over the rising autochthonous core of the batholith.

The contact of the mylonite zone with underlying plutonic rocks dips less steeply (30–40°) than the mylonitic foliation, and the geometric relations between the two are consistent with the expected geometry of a postcrystalline shear zone (Ramsay, 1980). Based on examination of microstructures in the southern part of the eastern Peninsular Ranges mylonite zone, Simpson (1985) concluded that the mylonite zone represents a postintrusive shear zone. However, a number of structural relations between intrusive rocks both below and within the northern part of the mylonite zone suggest that deformation was, in part, contemporaneous with intrusion (Anderson, 1983; Erskine, 1986b). The concordance of foliation, lineation, and strained mafic inclusion fabrics in plutonic rocks below the mylonite zone with mylonitic foliation and lineation within the zone supports this view. In addition, at least one plutonic unit within the mylonite zone (leucogranite of Cactus Spring, Matti *et al.*, 1983) has the geometry and structure that would be expected of an intrusion that invaded an active deformation zone and crystallized before deformation ceased. It appears likely that rocks within the region were deformed prior to, during, and following intrusion of the eastern part of the batholith.

Although the age of mylonitization is not well constrained, it probably began at some time between 97 and 62 m.y.b.p. U-Pb ages of zircons from tonalite plutons in the San Jacinto Mountains indicate emplacement ages of 97 m.y. (Hill, 1981, 1984), providing a lower limit for the onset of deformation. The upper limit is constrained by concordant 62-m.y. apatite-sphene-zircon fission-track ages of mylonitic rocks in the northern part of the mylonite zone (Dokka, 1984). These ages suggest rapid uplift of the region at that time.

Palm Canyon Complex. The Palm Canyon Complex is the name given to a group of mylonitic metasedimentary rocks, anatexites, and orthogneiss located above the Palm Canyon fault (Fig. 32-7). The unit is well exposed from the mouth of Palm Canyon southward for at least 30 km to the area between Deep Canyon and Martinez Mountain, where about 5–6 km of section is exposed in a structural culmination. The structural top and bottom of the complex are truncated by low-angle faults. The most characteristic feature of the Palm Canyon Complex is the presence of substantial volumes of mylonitic, garnetiferous leucogranite anatexites that are associated lit-par-lit with the metasedimentary rocks and orthogneiss. The anatexites are strongly foliated and commonly carry a well-developed downdip lineation that has the same east-northeast plunge as lineation in the mylonitic rocks below the Palm Canyon fault. Metasedimentary rocks that lack anatexite textures are generally nonmylonitic, except for the calc-silicate and marble tectonites that are found throughout the Palm Canyon Complex. A variety of minor structures are present within the nonmylonitic metasedimentary rocks, but rarely do these rocks possess the well-developed northeast down-dip lineation characteristic of the anatexites and rocks of the mylonite zone.

The foliation and lineation in the Palm Canyon Complex appear to have formed during anatexis of the pelitic schists, and these structures therefore record deformation at high grades of metamorphism. The lack of significant anatexites in similar metasedimentary rocks in the Desert Divide Group to the west suggests that the Palm Canyon Complex represents the relatively deep part of a thick section of deformed rocks that was juxtaposed against shallower rocks by westward thrusting along the mylonite zone and later along the Palm Canyon fault.

"Upper-Plate" Granitic Rocks. The term "upper-plate" granitic rocks refers to all granitic rocks that occur as parautochthonous sheets and klippen above the Palm Canyon Complex (Fig. 32-7). The most completely studied unit of these granitic rocks is the granodiorite of Asbestos Mountain, which occupies a klippe above the Asbestos Mountain fault in the northern Santa Rosa Mountains. Foliation and lineation in the "upper-plate" granitic rocks are roughly parallel to the regional mylonitic structures, except where the parautochthonous sheets are folded. However, this foliation dips less steeply than that of rocks within the underlying mylonite zone and Palm Canyon Complex. The presence of east-dipping foliations and east- to northeast-trending lineations in the parautochthonous granitic rocks suggests that these rocks were emplaced during the episode of regional deformation that produced the mylonite zone.

The granodiorite of Asbestos Mountain is lithologically and geochemically similar to autochthonous plutonic rocks of the northern Peninsular Ranges batholith[6] (Anderson, 1983; Erskine, 1986a; Hill, 1984). These similarities suggest that the Asbestos Mountain unit and, by inference, the other "upper-plate" granitic rocks are not exotic to the batholith. However, the "upper-plate" rocks are anomalous in relation to the eastern part of the batholith in that they are magnetite-series granitoids, whereas the granitic rocks beneath the mylonite zone are ilmenite series. This suggests that either the source terrane of the upper-plate rocks lay some 60 km to the west, west of the magnetite-ilmenite line (Fig. 32-5), or as seems more likely, a belt of magnetite-series granitoids may once have occupied the eastern part of the batholith and been tectonically displaced.

Conditions of Metamorphism. The pressures and temperatures that existed during mylonitic deformation are not precisely known, but they can be inferred from the *P-T* conditions of metasedimentary rocks in and adjacent to the mylonite zone. As mentioned in the section on prebatholithic zones, the pressures and temperatures at the peak of metamorphism were probably similar for rocks below and above the mylonite zone: 3.2–4.5 kb and 650–800°C (Erskine, 1986a; Hill, 1984). Theodore (1970) reported a broader range of pressures (3.4–7.0 kb) and a relatively restricted range of temperatures (580–660°C) at the peak of metamorphism during mylonitization in rocks at Coyote Mountain. Following recalibration of the metamorphic reactions used by Theodore to estimate *P-T* conditions, these ranges were revised to 600–700°C and 3–5 kb (Hill, 1984).

Brittle deformation Spatially associated with the eastern Peninsular Ranges mylonite zone are imbricate low-angle faults that bound the packets of deformed rocks described in the sections above. It is unclear at present whether the low-angle faults are west-directed thrusts, east-directed normal faults, or a combination of the two. The low-angle faults have a complex movement history that probably involves several stages of reactivation. The termination of mylonitic deformation and the commencement of low-angle faulting may be related to the 62-m.y.b.p. uplift and cooling of the region documented in the northern Santa Rosa Mountains by fission-track dates (Dokka, 1984).

Two lines of evidence suggest that low-angle faulting is kinematically linked to ductile deformation. First, the fault system is spatially and geometrically related to the mylonite zone. Although the dips of the low-angle faults are generally shallower than that of the mylonitic foliation, the basal Palm Canyon fault is subparallel to mylonitic foliation and is parallel to the lower contact of mylonite zone (Fig. 32-7). Unlike low-angle faults that occur higher in the section, the Palm Canyon fault is neither truncated by, nor does it join with, any other low-angle fault. The Palm Canyon fault thus appears to be a basal décollement that is kinematically related to, but younger than, the mylonite

---

[6] Also S. E. Shaw, oral communication, 1983 and R. G. Gastil and M. J. Walawender, written communication, 1984.

zone. The second line of evidence relating the low-angle fault system and mylonitic deformation is that reasonable palinspastic reconstructions of the eastern Peninsular Ranges mylonite zone require west-directed slip (thrusting) along the low-angle faults. Westward thrusting explains the present structural position of the deep-seated anatexites in the Palm Canyon Complex, and the anomalous opaque mineralogy and implied eastern source region of the "upper-plate" granitic rocks (Baird and Miesch, 1984; Baird *et al.*, 1979). Furthermore, concordant fission-track ages above and below low-angle faults (Dokka, 1984) suggest that faulting occurred prior to 62 m.y.b.p., close to the time of mylonitic deformation.

In spite of this evidence linking the low-angle faults to mylonitic deformation and westward thrusting, kinematic indicators along faults generally suggest eastward displacement of the parautochthonous sheets (listric normal faulting). As the region was uplifted and unroofed in latest Cretaceous and early Tertiary time, east-directed normal faults probably formed by reactivation of the older thrust faults and propagated downward into the mylonite zone. At least one low-angle fault in the southern part of the eastern Peninsular Ranges mylonite zone placed Miocene(?) sedimentary rocks over mylonitic plutonic rocks (Schultejann, 1984), suggesting reactivation of the low-angle faults during mid-Cenozoic extension.

TECTONIC EVOLUTION OF THE NORTHERN PART OF THE PENINSULAR RANGES BATHOLITH

The earliest event, or events, that can be recognized in the tectonic history of the northern part of the Peninsular Ranges batholith was the deposition of one or more Paleozoic marine sedimentary prisms along the southern part of the Cordilleran continental margin (no. 1, Fig. 32-8). Evidence for the timing and plate tectonic setting of the disruption of the Paleozoic continental margin is scattered throughout peninsular California and the northwest part of mainland Mexico. Scattered occurrences of carbonate-rich Paleozoic rocks in the eastern part of the batholith were once thought to lie considerably west of the North American craton (Gastil *et al.*, 1978, 1981). Studies of newly recognized Paleozoic rock sequences in peninsular California and mainland Mexico (Gastil and Miller, 1984) suggest that the craton originally extended westward into what is now peninsular California, and that the Paleozoic miogeoclinal and deeper-water, or transitional, facies extended from southern Nevada through peninsular California to western Sonora (Dickinson, 1981; A. E. J. Engel and Schultejann, 1984; Gastil and Miller, 1984; Hill, 1984).

Although earlier tectonic events cannot be ruled out, the Paleozoic continental margin probably began to break up in late Paleozoic time. Upper Paleozoic arclike volcanic and plutonic rocks, and marginal basin or trench deposits at several localities in Baja California and mainland Mexico (no. 2, Fig. 32-8) suggest the existence of a late Paleozoic convergent margin (Gastil *et al.*, 1978, 1981). During Triassic and Jurassic time, a thick wedge of clastic rocks was shed westward from cratonic highlands into one or more deep marine basins marginal to the North American continent (nos. 3, 4, Fig. 32-8). The plate-tectonic setting of these basins, whether fore-arc, back-arc, or trench, is not known.

Studies of arc-related rocks in the California continental borderland indicate that a volcanic-plutonic island arc began to accumulate in the ocean basin off the southwest coast of North America in Late Triassic time, and extended along the length of the borderland by Middle Jurassic time (Gastil *et al.*, 1978; 1981; no. 5, Fig. 32-8). Ophiolitic rocks, spatially associated with this arc, were considered to be fragments of ancient oceanic crust formed during Triassic rifting (Gastil *et al.*, 1978, 1981; Rangin, 1978).

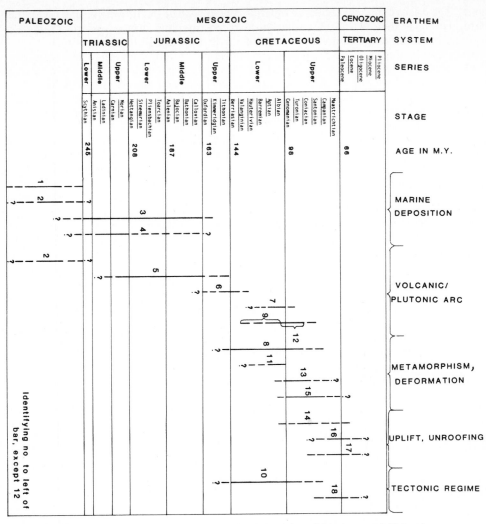

FIG. 32-8   Major events in the tectonic evolution of the northern Peninsular Ranges. Stages of Mesozoic systems and Jurassic stage boundaries from Palmer (1983).

1. Deposition of Paleozoic miogeoclinal and transitional rocks (Windy Point metamorphic rocks, Desert Divide Group, and Palm Canyon Complex).
2. Late Paleozoic arc magmatism and deposition in marginal basin or trench (Baja California and mainland Mexico).
3. Deposition of Bedford Canyon Formation in marginal basin.
4. Deposition of Julian Schist and French Valley Formation in marginal basin, or basins.
5. Construction of volcanic-plutonic island arc(s) (California Continental Borderland; includes Vizcaino terrane).
6. Deposition of Santiago peak Volcanics.
7. Deposition of Alisitos Formation.
8. Collision and underthrusting of Santiago Peak island arc.
9. Emplacement of western, syntectonic plutons (~120–105 m.y.).
10. Regional compression.
11. Cuyamaca–Laguna Mountains shear zone.
12. Emplacement of eastern, late- to post-tectonic plutons (~105–89 m.y.).
13. Northwest-southeast shortening (?).
14. Uplift of western part of batholith.
15. Eastern Peninsular Ranges mylonite zone.
16. Low-angle normal faulting.
17. Uplift of eastern part of batholith.
18. Regional extension.

However, recent studies by Moore (1983, 1984, 1986) indicate that ophiolitic rocks in the southern part of the continental borderland formed in more than one plate tectonic setting. There, the Vizcaino terrane is composed of three smaller paleo-oceanic terranes, two of which include ophiolitic rocks that formed in one or more Late Triassic marginal basin-island arc systems, and a third that includes pre-middle Mesozoic ophiolitic rocks that may have formed at a midocean ridge. Moore concluded that these terranes were probably not associated with a continental margin before the Late Jurassic. The Vizcaino terrane was probably accreted to the Baja California peninsula by Early Cretaceous time, and is separated from Cretaceous volcanic and plutonic rocks of the peninsula (including the Alisitos Formation) by a major structural discontinuity (Rangin, 1978).

By Late Jurassic time, a fringing volcanic island arc represented by the Santiago Peak Volcanics probably extended from the continental margin in what is now southern California southward to an oceanic island in northern Baja California (the Alisitos Formation) (Silver *et al.*, 1963, 1969; nos. 6, 7, Fig. 32-8). The Santiago Peak segment of the arc was probably emergent and located just inboard of a deep ocean basin (Adams, 1979; Balch, 1981; Buesch, 1984). Scarce metavolcanic layers in flysch-type metasedimentary rocks east of the Santiago Peak segment of the arc may represent incursions of arc detritus into the Late Jurassic marginal basin, or may include older arc-related rocks. If these volcanic layers are as old as Triassic or Early Jurassic, then a fringing arc was never far from the coast of North America.

Collision of the Santiago Peak–Alisitos arc with western North America in the Early Cretaceous resulted in folding of the continental-margin deposits and eventual underthrusting of the arc beneath the continental margin (no. 8, Fig. 32-8; Gastil *et al.*, 1978, 1981; Todd and Shaw, 1985). As indicated by the end of prebatholithic marine deposition and the onset of deformation, the arc apparently arrived first in the northern part of peninsular California. In the Santa Ana Mountains, the Triassic and Jurassic Bedford Canyon Formation was probably folded, overturned, uplifted, and eroded prior to depositional and/or tectonic emplacement of the Santiago Peak Volcanics. Even if the contact between the two units were entirely tectonic, the difference in intensity of deformation between them suggests that the Bedford Canyon Formation was deformed before it came in contact with the Santiago Peak Volcanics. In the batholith east of San Diego, the uppermost Jurassic and Lower Cretaceous(?) Santiago Peak Volcanics were folded and metamorphosed in Early Cretaceous time. In the eastern part of the study area, this deformation fabric may have been superposed upon a pre-Late Jurassic fabric. In Baja California, the Alisitos Formation of Aptian and Albian age had been deformed and uplifted by late Albian time (Gastil *et al.*, 1978, 1981). The staggered arrival of the island arc suggested to Gastil *et al.* (1981) that it was offset by a series of transform faults that connected paired western and eastern magmatic arcs. However, it is possible that the Santiago Peak and Alisitos arc segments represent fragments of an island arc that was dismembered by northward-directed transform faulting and that the older fragments traveled farther, and arrived earlier, than the younger ones. In this case the arc rocks of the continental borderland could have belonged to the same dismembered island arc.

By Early Cretaceous time, the locus of magmatic activity in the northern part of the batholith had moved eastward from the Santiago Peak volcanic arc to the vicinity of the Triassic and Jurassic marginal basin (no. 9, Fig. 32-8). The volcanic rocks that were superjacent to the Early Cretaceous plutons have been removed by erosion, but downfolded and foundered volcanic strata and/or subvolcanic pyroclastic rocks probably occur in plutonic ring complexes in the western part of the study area (Todd, 1983). Early Cretaceous plutons intruded the east-dipping suture between the accreted fringing island arc and the folded marginal basin. West-to-east variations in the compositions of these plutons reflect the composition of the prebatholithic crust above and below the suture:

partial melts of oceanic arc rocks in the structurally lower, western block produced I-type granitoid magmas, and melts of predominantly flysch-type sedimentary rocks in the structurally higher, eastern block produced predominantly S-type granitoid magmas. Melting in the mantle wedge above the subducted slab produced basaltic magmas that probably provided some of the heat for melting crustal materials and rose simultaneously with I- and S-type granitoid magmas into the overlying crust.

Regional compression continued during Early Cretaceous intrusion in the batholith east of San Diego (Todd and Shaw, 1979; no. 10, Fig. 32-8). Deformation was essentially penetrative in plutons and wallrocks in the western part of the area, and early fabrics were modified by diapiric intrusion. Deformation became concentrated in the east-dipping Cuyamaca–Laguna Mountains shear zone (no. 11, Fig. 32-8; Fig. 32-4), which is slightly east of and subparallel to the inferred position of the prebatholithic suture. The fabric in this part of the batholith indicates that the direction of maximum shortening was approximately east-northeast. If the predominant north-northwest orientation of the prebatholithic zones and the magmatic arc reflect the trend of the outboard trench and subduction zone, then convergence at this time was approximately normal to the continental margin. Steeply plunging lineations in both plutonic and metamorphic rocks of the shear zone indicate that movements were predominantly dip-slip. The apparent west-to-east depth contrast in S-type granitoids and metamorphosed flysch-type sedimentary wallrocks across the zone suggests that the east side was thrust over the west side (deep-seated thrust).

Approximately 100 m.y. ago, the locus of magmatism moved eastward into what had been the Paleozoic miogeocline, and a markedly different type of magma was produced (younger, late- to post-tectonic intrusive sequence) (Clinkenbeard et al., 1986; Hill, 1984; Silver et al., 1979; Todd and Shaw, 1986; no. 12, Fig. 32-8). Although plutons of the younger sequence are not penetratively deformed, strong fabrics did develop locally. In the Cuyamaca–Laguna Mountains shear zone, and in a 40-km-wide zone that lies east of, and parallel to the shear zone, the older plutons of this sequence (~100 m.y.) were locally intensely deformed. The younger plutons of the sequence (98–95 m.y.) are relatively massive or are weakly foliated parallel to their walls. Intrusion of these relatively undeformed plutons across the shear zone indicates that movements in it ended about 100–98 m.y. ago.

The eastward migration of magmatism may have coincided with a change from east-northeast shortening to northwest-southeast shortening in the western part of the batholith (no. 13, Fig. 32-8). Folds and faults that suggest northwest-southeast compression developed west of the Cuyamaca–Laguna Mountains shear zone during, and/or following, the last stages of westward thrusting in that zone (V. R. Todd, unpublished data). The youngest rocks that are involved in northwest-southeast shortening are plutons of Granite Mountain and La Posta type, one of which has nearly concordant hornblende and biotite K-Ar ages of 103 and 101 m.y., respectively. The relations above suggest that sometime between 102 and 98 m.y.p.b., shortening in this region changed from a direction approximately normal to the continental margin to one that was subparallel to it.

Uplift and erosion began in the western part of the batholith while the younger group of plutons was being emplaced (Gastil et al., 1981; no. 14, Fig. 32-8). In the northern Santa Ana Mountains, the oldest postbatholithic marine deposits are late Turonian (Schoellhamer et al., 1981), while to the south in San Diego County, postbatholithic continental deposition preceded Campanian marine deposition (Kennedy and Peterson, 1975). In northern Baja California, marine deposition was also taking place in Late Cretaceous time, just a few m.y. after intrusion of 90-m.y.-old plutons (Silver et al., 1979).

Sometime between 97 and 62 m.y., the eastern Peninsular Ranges mylonite zone developed in the eastern part of the batholith (no. 15, Fig. 32-8). The zone cuts mio-geoclinal wallrocks and plutonic rocks, the latter probably equivalent to the late-to-post-tectonic intrusive sequence of the southern part of the study area. In parts of the eastern Peninsular Ranges, deformation was contemporaneous with intrusion, high-grade meta-morphism, and anatexis. Although the structure of the mylonite zone is complex (Fig. 32-7), its overall geometry and fabrics are similar to those of the older Cuyamaca–Laguna Mountains shear zone with the exception that foliation in the younger shear zone dips less steeply than that in the older shear zone. Shear-sense indicators in the mylonite zone, the pervasive down-dip lineation, and the development of anatexites in the metasedi-mentary rocks that overlie the mylonite zone, suggest that the eastern side was thrust over the western side (A. E. J. Engel and Schultejann, 1984; Erskine, 1986a). The amount of shortening across the mylonite zone has been estimated to be on the order of 50–100 km (A. E. J. Engel and C. G. Engel, 1982; Silver, 1982, 1983). North-northwest-directed thrusting and tectonic transport are interpreted to have overlapped with, and followed, Late Cretaceous westward thrusting in parts of the eastern Peninsular Ranges mylonite zone (A. E. J. Engel and Schultejann, 1984). A problem with this interpretation is that the linear tectonic elements attributed by these authors to north-northwest tectonic transport are collinear with the east-northeast mylonitic linear fabric elements that are generally considered to have formed during westward tectonic transport (Erskine, 1986c).

In its early stages, the relatively deep-seated Peninsular Ranges mylonite zone prob-ably graded upward into a low-angle fault system. At some time prior to 62 m.y., the mylonite zone began to be displaced by low-angle faults that juxtaposed different struc-tural levels of the mylonitic terrane (no. 16, Fig. 32-8). This transition from ductile mylonitic deformation to brittle low-angle faulting probably accompanied uplift and unroofing of the eastern part of the batholith in the latest Cretaceous and early Tertiary (no. 17, Fig. 32-8). The oldest low-angle faults were probably west-directed thrusts (Erskine, 1986a, c). Uplift continued and by late Paleogene time the region was under-going extension (no. 18, Fig. 32-8). Low-angle normal faults displaced rocks eastward and structurally downward against high-grade metamorphic rocks of the mylonite zone (Dickinson, 1981; A. E. J. Engel and Schultejann, 1984). These Late Cretaceous and Paleogene structures were displaced and reactivated by the Neogene Gulf of California–San Andreas rift-transform system.

CONCLUSIONS AND SPECULATIONS

The fundamental tectonic feature in the northern part of the Peninsular Ranges batholith is a suture that formed when the Late Jurassic Santiago Peak island arc collided with the Mesozoic marginal basin in Early Cretaceous time. The existence of this suture is indi-cated by the change in prebatholithic lithology from volcanogenic rocks in the western part of the batholith to flysch-like sedimentary rocks in the central part. The probable occurrence of volcanic strata in these sedimentary rocks suggests that the island arc may never have been far from the North American continent.

The existence of the prebatholithic suture is further demonstrated by systematic west-to-east petrologic variations in the Cretaceous batholith. Plutonic zones in the batho-lith correspond closely with prebatholithic zones: Early Cretaceous I-type plutons in-truded metavolcanic rocks of Santiago Peak-type west of the suture, while apparently coeval S-type plutons were emplaced into flysch east of the suture. These plutons are interpreted as products of partial melting in the plutonic root of the island arc on the west side of the suture, and in the lower part of the Mesozoic sedimentary prism on the east side as the magmatic arc moved toward the North American continent in Early Cretaceous time. The distribution of the Early Cretaceous plutons relative to the continental margin and the inferred outboard trench suggests that the suture dipped to the east. This inferred dip, and the orientation of batholithic structures that indicate east-northeast shortening, further suggest that the island arc was thrust under the western edge of North America.

The eastward migration of the Cretaceous magmatic arc into southern Arizona and northwest Mexico in Late Cretaceous and early Tertiary time is generally attributed to a decrease in the dip of the subducted slab (Dickinson, 1981; Gastil *et al.*, 1981). Conditions favoring shallow subduction beneath western North America are thought to have prevailed during much of this time (Beck, 1986). In the northern part of the batholith, the flattening of the subducted slab coincided with widespread melting in a new source region as the magmatic arc began to penetrate eastward into continental crust. The compositions of the mid- and Late Cretaceous plutons appear to vary systematically with those of their wallrocks. Mid- and Late Cretaceous plutons that occur west of the prebatholithic suture have geochemical characteristics similar to those of the Early Cretaceous I-type plutons, whereas plutons that occur east of the suture have characteristics that suggest a significant contribution by continental material in the source region. The source of the mid- and Late Cretaceous magmas is not fully understood, but it seems likely that the magmas interacted with different kinds of crust on the east and west sides of the prebatholithic suture.

As the Cretaceous magmatic arc migrated eastward, the site of most intense deformation also shifted eastward. Synintrusive deformation was concentrated in the central part of the batholith (Cuyamaca–Laguna Mountains shear zone) in Early Cretaceous time and in the eastern part (eastern Peninsular Ranges mylonite zone) in Late Cretaceous time. Deep-seated batholithic rocks were thrust westward over shallower rocks in both shear zones. The decrease in dip of regional foliation from the older, western shear zone to the younger, eastern zone may reflect the flattening of the subduction plane in Late Cretaceous time. The locations of the two shear zones may also have been controlled by prebatholithic crustal structure (Fig. 32-4). The Cuyamaca–Laguna Mountains shear zone lies close to, and slightly east of, the prebatholithic suture and the I-S line, which together mark a major discontinuity in the Late Jurassic–Early Cretaceous crust. The eastern Peninsular Ranges mylonite zone lies just east of the inferred prebatholithic boundary between transitional deeper-water(?) sedimentary rocks and shallower shelf rocks. If this boundary approximates the position of the Paleozoic shelf-slope break, the mylonite zone may have formed in crust of anomalous thickness located between thin oceanic crust to the west and thicker continental crust to the east. The close temporal relation between intrusion and deformation suggests that deformation was enhanced in crust that was thermally weakened.

As the Pacific-Farallon spreading center approached the trench along the continental margin in early and middle Tertiary time, the western edge of North America underwent sweeping uplift and dilation as the result of subduction of progressively younger, more buoyant oceanic lithosphere and isostatic recovery following the emplacement of the great volume of Cretaceous plutonic rocks (Dickinson, 1981; Gastil *et al.*, 1981). In the northern part of the Peninsular Ranges batholith, a compressive tectonic regime that had begun by at least Early Cretaceous time with underthrusting of the Santiago Peak island arc gave way to an extensional regime dominated by east-directed low-angle normal faulting in early and middle Tertiary time.

Paleomagnetic studies indicate that the Peninsular Ranges batholith and its associated pre- and postbatholithic rocks have undergone northward displacement on the order of 1200 km between Late Cretaceous and late Miocene time (Erskine, 1982; Erskine and

Marshall, 1980; Hagstrum *et al.*, 1985, 1986; Patterson, 1984; Teissere and Beck, 1973). The paleomagnetic data challenge existing reconstructions of southern California paleogeography, most of which permit only relatively small displacements (~300 km) due to the northwestward movement of peninsular California along the San Andreas transform beginning in Neogene time (Gastil *et al.*, 1978, 1981; Rangin, 1978). Although the resolution of the problem of the total displacement of peninsular California is beyond the scope of this paper, the data of this study indicate that the inferred transform fault was not located within the part of the batholith that is presently exposed in the northern Peninsular Ranges. In this region, the dominant stress regime from Late Jurassic to Late Cretaceous time was compressive with a major component of east-northeast shortening. Thus, during this period, convergence was directed approximately perpendicular to the continental margin, a condition that would not favor the development of transform faults (Beck, 1986). Furthermore, the pervasive steeply to moderately plunging down-dip lineations in the major structural zones of the northern part of the batholith argue against significant strike-slip movements. Prebatholithic and plutonic rock units can be matched across the two batholithic shear zones, which makes large lateral displacements of batholithic rocks unlikely. Finally, as the batholith was uplifted and unroofed, the Cretaceous compressive tectonic episode was succeeded by regional extension that may have continued into Miocene time. The data of this study do not rule out large relative displacements within, or between, prebatholithic zones prior to Late Jurassic time or large displacements along the proto-San Andreas fault.

The transition in Late Cretaceous time from a relatively steep subduction angle to a shallow angle is considered to have favored north-oblique subduction and transform faulting (Beck, 1986). Limited evidence in the study area suggests a transition in the Late Cretaceous from convergence normal to the continental margin to tectonic transport parallel to the margin (oblique subduction?). Local occurrences of nonpenetrative folds and faults that suggest Late and/or post-Cretaceous northward tectonic transport may record responses in warm lithosphere to transform faulting in a zone located to the east of the batholith as presently exposed.

Several lines of evidence suggest that the southern part of the study area underwent greater uplift and erosion to deeper crustal levels than the northern part. Metamorphic gradients are more abrupt (Fig. 32-3), contact metamorphism is more widespread, and Early Cretaceous plutons may be less deformed in the northern part of the area. S-type granitoid plutons have not been recognized in the north, perhaps because they never rose into the upper crust. As mentioned earlier, the magnetic field is more complex in the northernmost part of the study area than in the southern part, probably due to the intermingling of magnetite- and ilmenite-bearing plutons in the north. The wider occurrence of magnetite-bearing plutons in the north may reflect a higher degree of oxidation in plutons emplaced at relatively higher levels in the crust. The cause of the apparent north-south depth gradient is unknown, but must predate development of the late Cretaceous and early Tertiary Peninsular Ranges erosion surface, or surfaces. Local west-northwest bending of batholithic structures (Fig. 32-6) and of prebatholithic and plutonic discontinuities (Fig. 32-5) suggests that the Cretaceous magmatic arc may have been offset left-laterally by widely spaced east-west-trending(?) ductile faults, as suggested by Gastil *et al.* (1981). Fault-bounded blocks having the dimensions of the study area may have been tilted during this deformation.

Although highly speculative, this scenario has features in common with a model proposed by May (1986) in which west-directed thrusts in the Transverse Ranges were offset from similar thrusts in the eastern Peninsular Ranges by an east-west-trending, left-lateral ductile fault system that was active during Late Cretaceous intrusion. Clearly, detailed structural studies of prebatholithic and plutonic rocks should be given the highest priority in ongoing research in the northern Peninsular Ranges batholith in order to evaluate the foregoing, and other, models.

ACKNOWLEDGMENTS

We are grateful for helpful reviews of the manuscript by D. S. Harwood, R. F. Yerkes, R. G. Gastil, and R. C. Jachens. We also thank A. K. Baird, B. F. Cox, R. G. Gastil, M. Grove, J. C. Matti, R. E. Powell, R. V. Sharp, and S. E. Shaw for numerous discussions and field trips that aided in the development of our ideas about the Peninsular Ranges batholith.

REFERENCES

Adams, M. A., 1979, Stratigraphy and petrography of the Santiago Peak Volcanics east of Rancho Santa Fe, California: unpublished M.S. thesis, San Diego State Univ., San Diego, Calif., 123 p.

Allison, E. C., 1955, Middle Cretaceous Gastropoda from Punta China, Baja California, Mexico: *J. Paleontology*, v. 29, p. 400–432.

Anderson, J. R., 1983, Petrology of a portion of the Eastern Peninsular Ranges mylonite zone, Southern California: *Contrib. Mineralogy Petrology*, v. 84, p. 253–271.

Baird, A. K., and Miesch, A. T., 1984, Batholithic rocks of southern California — A model for the petrochemical nature of their source materials: *U.S. Geol. Survey Prof. Paper 1284*, 42 p.

_____, Baird, K. W., and Welday, B. E., 1979, Batholithic rocks of the Northern Peninsular and Transverse Ranges, southern California, *in* Abbott, P. L., and Todd, V. R., eds., *Mesozoic Crystalline Rocks — Peninsular Ranges Batholith and Pegmatites, Point Sal Ophiolite:* Guidebook for Geol. Soc. America Ann. Meet., San Diego State Univ., San Diego, Calif. (Nov. 1979), p. 111-132.

Balch, D. C., 1981, Sedimentology of the Santiago Peak volcaniclastic rocks, San Diego County, California: unpublished M.S. thesis, San Diego State Univ., San Diego, Calif., 135 p.

_____, Bartling, S. H., and Abbott, P. L., 1984, Volcaniclastic strata of the Upper Jurassic Santiago Peak Volcanics, San Diego, California, *in* Crouch, J. K., and Bachman, S. B., eds., *Tectonics and Sedimentation along the California Margin:* Pacific Section, Soc. Econ. Paleontologists Mineralogists, v. 38, p. 157–170.

Beck, M. E., Jr., 1980, Paleomagnetic record of plate-margin processes along the western edge of North America: *J. Geophys. Res.*, v. 85, p. 7115–7131.

_____, 1986, Model for Late Mesozoic-Early Tertiary tectonics of coastal California and western Mexico and speculations on the origin of the San Andreas fault: *Tectonics*, v. 5, no. 1, p. 49–64.

_____, and Plumley, P. W., 1979, Late Cenozoic subduction and continental-margin truncation along the northern Middle America Trench — Discussion: *Geol. Soc. America Bull.*, v. 90, p. 792–794.

_____, Burmester, R. F., Engebretson, D. C., and Schoonover, R., 1981, Northward translation of Mesozoic batholiths, western North America — Paleomagnetic evidence and tectonic significance: *Geofisica Internac.*, v. 20-3, p. 143–162.

Berggreen, R. G., and Walawender, M. J., 1977, Petrography and metamorphism of the

Morena Reservoir roof pendant, southern California: *Calif. Div. Mines Geology Spec. Rpt. 129*, p. 61–65.

Berggren, W. A., Kent, D. V., Flynn, J. J., and Van Couvering, J. A., 1985, Cenozoic geochronology: *Geol. Soc. America Bull.*, v. 96, p. 1407–1418.

Brown, A. R., 1981, Structural history of the metamorphic, granitic and cataclastic rocks in southeastern San Jacinto Mountains, *in* Brown, A. R., and Ruff, R. W., eds., *Geology of the San Jacinto Mountains:* South Coast Geol. Soc., Santa Ana, Calif., Ann. Fieldtrip Guidebook, no. 9, p. 100–138.

Buckley, C. P., Condra, D. A., and Cooper, J. P., 1975, The Jurassic flysch of the Santa Ana Mountains – An example of obduction? *Geol. Soc. America Abstr. with Programs*, v. 7, p. 299–300.

Buesch, D. C., 1984, The depositional environment and subsequent metamorphism of the Santiago Peak Volcanic rocks, Camp Pendleton, California: Unpublished M.S. thesis, California State Univ., Los Angeles, Calif., 113 p.

Burnham, C. W., 1959, Contact metamorphism of magnesian limestones at Crestmore, California: *Geol. Soc. America Bull.*, v. 70, p. 879–920.

Bushee, J., Holden, J., Geyer, B., and Gastil, R. G., 1963, Lead-alpha dates for some basement rocks of southwestern California: *Geol. Soc. America Bull.*, v. 74, p. 803–806.

Chappell, B. W., and White, A. J. R., 1974, Two contrasting granite types: *Pacific Geology*, v. 8, p. 173–174.

Clinkenbeard, J. P., Walawender, M. J., Parrish, K. E., and Wardlaw, M. S., 1986, The geochemical and isotopic composition of the La Posta granodiorite, San Diego County, California. *Geol. Soc. America Abstr. with Programs*, v. 18, p. 95.

Coney, P. J., Jones, D. L., and Monger, J. W. H., 1980, Cordilleran suspect terranes: *Nature*, v. 288, p. 329–333.

Criscione, J. L., Davis, T. E., and Ehlig, P., 1978, The age of sedimentation/diagenesis for the Bedford Canyon Formation and the Santa Monica Formation in southern California – A Rb/Sr evaluation, *in* Howell, D. G., and McDougall, K. A., eds., *Mesozoic Paleogeography of the Western United States:* Pacific Section, Soc. Econ. Paleontologists Mineralogists, Pacific Coast Paleogeography Symp. 2, p. 385–396.

Davis, G. A., Monger, J. W. H., and Burchfiel, B. C., 1978, Mesozoic construction of the Cordilleran "collage," central British Columbia to central California, *in* Howell, D. G., and McDougall, K. A., eds., *Mesozoic Paleogeography of the Western United States:* Pacific Section, Soc. Econ. Paleontologists Mineralogists, Pacific Coast Paleogeography Symp. 2, p. 1–32.

DePaolo, D. J., 1981, A neodymium and strontium isotopic study of the Mesozoic calc-alkaline granitic batholiths of the Sierra Nevada and Peninsular Ranges, California: *J. Geophys. Res.*, v. 86, p. 10470–10488.

Detterman, M. E., 1984, Geology of the Metal Mountain district, In-ko-pah Mountains, San Diego County, California: unpublished M.S. thesis, San Diego State Univ., San Diego, Calif., 216 p.

Diamond, J. L., Knaack, C. M., Gastil, R. G., Erskine, B. G., Walawender, M. J., Marshall, M., and Cameron, G. J., 1985, The magnetite-ilmenite line in the Peninsular Ranges of southern and Baja California, U.S.A., and Mexico: *Geol. Soc. America Abstr. with Programs*, v. 17, p. 351.

Dickinson, W. R., 1981, Plate tectonic evolution of the southern Cordillera, *in* Dickinson, W. R., and Payne, W. D., eds., *Relations of Tectonics to Ore Deposits in the Southern Cordillera:* Ariz. Geol. Soc. Digest, v. 14, p. 113–135.

Dokka, R. K., 1984, Fission-track geochronologic evidence for Late Cretaceous mylonitization and early Paleocene uplift of the northeastern Peninsular Ranges, California: *Geophys. Res. Lett.*, v. 11, p. 46–49.

Engebretson, D. C., Cox, Allan, and Gordon, R. G., 1985, Relative motions between continental plates in the Pacific basin: *Geol. Soc. America Spec. Paper 206*, 59 p.

Engel, A. E. J., and Engel, C. G., 1982, Late Mesozoic and Early Cenozoic tectonics along the Salton Trough-Peninsular Range boundary *Geol. Soc. America Abstr. with Programs*, v. 14, p. 162.

——, and Schultejann, P. A., 1984, Late Mesozoic and Cenozoic tectonic history of south central California: *Tectonics*, v. 3, no. 6, p. 659–675.

Engel, R., 1959, Geology of the Lake Elsinore quadrangle, California: *Calif. Div. Mines Bull. 146*, p. 9–58.

Ernst, W. G., 1984, Californian blueschists, subduction, and the significance of tectono-stratigraphic terranes: *Geology*, v. 12, p. 436–440.

Erskine, B. G., 1982, A paleomagnetic and rock magnetic investigation of the northern Peninsular Ranges batholith, southern California: unpublished M.S. thesis, San Diego State Univ., San Diego, Calif., 194 p.

——, 1986a, Mylonitic deformation and associated low-angle faulting in the Santa Rosa mylonite zone, southern California: unpublished Ph.D. dissertation, Univ. California, Berkeley, Calif., 247 p.

——, 1986b, Syntectonic granitic intrusion and mylonitic deformation along the eastern margin of the northern Peninsular Ranges batholith, southern California *Geol. Soc. America Abstr. with Programs*, v. 18, p. 105.

——, 1986c, Metamorphic and deformation history of the eastern Peninsular Ranges mylonite zone — Implications on tectonic reconstructions of southern California *Geol. Soc. America Abstr. with Programs*, v. 18, no. 2, p. 105.

——, and Marshall, C. M., 1980, A paleomagnetic and rock magnetic investigation of the northern Peninsular Ranges batholith, southern California: *Trans. Am. Geophys. Union*, v. 61, p. 948.

Everhart, D. L., 1951, Geology of the Cuyamaca Peak quadrangle, San Diego County, California: *Calif. Div. Mines Bull. 159*, p. 51–115.

Fife, D. L., Minch, J. A., and Crampton, P. J., 1967, Late Jurassic age of the Santiago Peak Volcanics, California: *Geol. Soc. America Bull.*, v. 78, p. 299–304.

Frizzell, V. A., Jr., 1984, The geology of the Baja California Peninsula — An introduction, *in* Frizzell, V. A., Jr., ed., *Geology of the Baja California Peninsula:* Pacific Section, Soc. Econ. Paleontologists Mineralogists, v. 39, p. 1–7.

Gastil, R. G., 1975, Plutonic zones in the Peninsular Ranges of southern California and northern Baja California: *Geology*, v. 3, p. 361–363.

——, 1983, Mesozoic and Cenozoic granitic rocks of southern California and western Mexico: *Geol. Soc. America Mem. 159*, p. 265–275.

——, and Miller, R. H., 1983, Stratal-tectonic elements of the eastern Peninsular Ranges, Baja California, Mexico: *Geol. Soc. America Abstr. with Programs*, v. 15, p. 439.

——, and Miller, R. H., 1984, Prebatholithic paleogeography of peninsula California and adjacent Mexico, *in* Frizzell, V. A., Jr., ed., *Geology of the Baja California Peninsula:* Pacific Section, Soc. Econ. Paleontologists Mineralogists, v. 39, p. 9–16.

——, Phillips, R. P., and Rodriguez-Torres, R., 1972, The reconstruction of Mesozoic California: *Proc. 24th Internat. Geol. Congress, Section 3*, p. 217–229.

——, Phillips, R. P., and Allison, E. C., 1975, Reconnaissance geology of the state of Baja California: *Geol. Soc. America Mem. 140*, 170 p.

——, Morgan, G. J., and Krummenacher, D., 1978, Mesozoic history of peninsular

California and related areas east of the Gulf of California, *in* Howell, D. G., and McDougall, K. A., eds., *Mesozoic Paleogeography of the Western United States:* Pacific Section, Soc. Econ. Paleontologists Mineralogists, Pacific Coast Paleogeography Symp. 2, p. 107–116.

——, Morgan, G. J., and Krummenacher, D., 1981, The tectonic history of peninsular California and adjacent Mexico, *in* Ernst, W. G., ed., *The Geotectonic Development of California* (Rubey Vol. I): Englewood Cliffs, N.J., Prentice-Hall, p. 284–306.

Germinario, M. P., 1982, The depositional and tectonic environments of the Julian Schist, Julian, California: unpublished M.S. thesis, San Diego State Univ., San Diego, Calif., 95 p.

Gromet, L. P., 1979, Rare earths abundances and fractionations and their implications for batholithic petrogenesis in the Peninsular Ranges Batholith, California, U.S.A., and Baja California, Mexico: unpublished Ph.D. dissertation, California Inst. Technology, Pasadena, Calif., 337 p.

——, and Silver, L. T., 1979, Profile of rare earth element characteristics across the Peninsular Ranges batholith near the international border, *in* Abbott, P. L., and Todd, V. R., eds., *Mesozoic Crystalline Rocks – Peninsular Ranges Batholith and Pegmatites, Point Sal Ophiolite:* Guidebook Geol. Soc. America Ann. Meet. San Diego State Univ., San Diego, Calif. (Nov. 1979), p. 133–142.

Grove, M., 1986, Metamorphism and deformation in the Box Canyon area, eastern Peninsular Ranges, San Diego County, California: unpublished M.S. thesis, Univ. California, Los Angeles, Calif.

Gunn, S. H., 1986, Isotopic constraints on the origin of the Laguna Juarez pluton, Baja California, Mexico: *Geol. Soc. America Abstr. with Programs*, v. 18, p. 111–112.

Hagstrum, J. T., McWilliams, M., Howell, D. G., and Gromme, S., 1985, Mesozoic paleomagnetism and northward translation of the Baja California Peninsula: *Geol. Soc. America Bull.*, v. 96, p. 1077–1090.

——, McWilliams, M., Howell, D. G., and Gromme, S., 1986, Mesozoic paleomagnetism and northward translation of the Baja California peninsula: *Geol. Soc. America Abstr. with Programs*, v. 18, p. 112.

Hamilton, 1981, Crustal evolution by arc magmatism: *Philos. Trans. Roy. Soc. London,* v. A301, p. 279–291.

Hawkins, J. W., 1970, Metamorphosed Late Jurassic andesites and dacites of the Tijuana-Tecate area, California, *in Pacific slope geology of northern Baja California and adjacent Alta California:* Pacific Section, Amer. Assoc. Petrol. Geologists, Soc. Econ. Paleontologists Mineralogists, Soc. Explor. Geophysics Guidebook, p. 25–29.

Hill, R. I., 1981, Field, petrologic and isotopic studies of the intrusive complex of San Jacinto Mountains, *in* Brown, A. R., and Ruff, R. W., eds., *Geology of the San Jacinto Mountains:* South Coast Geol. Soc., Santa Ana, Calif., Ann. Field Trip Guidebook, no. 9, p. 76–89.

——, 1984, Petrology and petrogenesis of batholithic rocks, San Jacinto Mountains, southern California: unpublished Ph.D. dissertation, California Inst. Technology, Pasedena, Calif., 660 p.

——, and Silver, L. T., 1983, Contrasting west-east strontium isotope properties of wallrocks, northern Peninsular Ranges batholith, California: *Geol. Soc. America Abstr. with Programs*, v. 15, p. 439.

Howell, D. G., and Vedder, J. G., 1981, Structural implications of stratigraphic discontinuities across the southern California Borderland, *in* Ernst, W. G., ed., *The Geotectonic Development of California* (Rubey Vol. I): Englewood Cliffs, N.J., Prentice-Hall, p. 535–558.

Hudson, F. S., 1922, Geology of the Cuyamaca region of California, with special reference to the origin of nickeliferous pyrrhotite: *Univ. Calif. Dept. Geol. Sci. Bull.*, v. 13, p. 175–252.

Imlay, R. W., 1963, Jurassic fossils from southern California: *J. Paleontology*, v. 37, p. 97–107.

____, 1964, Middle and Upper Jurassic fossils from southern California: *J. Paleontology*, v. 38, p. 505–509.

Jachens, R. C., Simpson, R. W., Griscom, A., and Mariano, J., 1986, Plutonic belts in southern California defined by gravity and magnetic anomalies: *Geol. Soc. America Abstr. with Programs*, v. 18, p. 120.

Jahns, R. H., 1954, Geology of the Peninsular Range province, southern California and Baja California, *in Geology of Southern California:* Calif. Div. Mines Geology Bull. 170 (September 1954), p. 29–52.

____, and Lance, J. F., 1950, Geology of the San Dieguito pyrophyllite area, San Diego County, California: *Calif. Div. Mines Geology Spec. Rpt. 4*, 32 p.

Jones, D. A., and Miller, R. H., 1982, Jurassic fossils from the Santiago Peak Volcanics, San Diego County, California, *in* Abbott, P. L., ed., *Geologic Studies in San Diego:* San Diego Assoc. Geologists Publ., p. 93–103.

____, Miller, R. H., and Kling, S. A., 1983, Late Jurassic fossils from Santiago Peak Volcanics, San Diego County, California: *Geol. Soc. America Abstr. with Programs*, v. 15, p. 413.

Jones, D. L., Howell, D. G., Coney, P. J., and Monger, J. W. H., 1983, Recognition, character, and analysis of tectonostratigraphic terranes in western North America, *in* Hashimoto, M., and Uyeda, S., eds., *Accretion Tectonics in the Circum-Pacific Regions:* Tokyo, Terra Scientific Publishing Co., p. 21–35.

Kennedy, M. P., and Peterson, G. L., 1975, Geology of the eastern San Diego metropolitan area, California: *Calif. Div. Mines Geology Bull. 200*, p. 43–56.

Lamb, T. N., 1970, Fossiliferous Triassic (?) metasedimentary rocks near Sun City, Riverside County, California: *Geol. Soc. America Abstr. with Programs*, v. 2, p. 110–111.

Larsen, E. S., Jr., 1948, Batholith and associated rocks of Corona, Elsinore and San Luis Rey quadrangles, southern California: *Geol. Soc. America Mem. 29*, 182 p.

Lillis, P. G., Walawender, M. J., Smith, T. E., and Wilson, J., 1979, Petrology and emplacement of the Corte Madera gabbro pluton, southern California, *in* Abbott, P. L., and Todd, V. R., eds., *Mesozoic Crystalline Rocks — Peninsular Ranges Batholith and Pegmatites, Point Sal Ophiolite:* Guidebook Geol. Soc. America Ann. Meet., San Diego State Univ., San Diego, Calif. (Nov. 1979), p. 143–150.

Lister, G. S., and Snoke, A. W., 1984, S-C mylonites: *J. Struct. Geology*, v. 6, p. 617–638.

Matti, J. C., Cox, B. F., Powell, R. E., Oliver, H. W., and Kuizon, L., 1983, Mineral resource potential of the Cactus Spring Roadless Area, Riverside County, California: *U.S. Geol. Survey Misc. Field Studies Map MF-1650-A*, 10 p.

May, D. J., 1986, Amalgamation of metamorphic terranes in the southeastern San Gabriel Mountains, California: unpublished Ph.D. dissertation, Univ. California, Santa Barbara, Calif., 325 p.

Merriam, R. H., 1946, Igneous and metamorphic rocks of the southwestern part of the Ramona quadrangle, San Diego County, California: *Geol. Soc. America Bull.*, v. 57, p. 223–260.

Merrill, F. J. H., 1914, Geology and mineral resources of San Diego and Imperial Counties, California: *Calif. State Mining Bur. Rpt. 14*, p. 653–662.

Miller, R. H., and Dockum, M. S., 1983, Ordovician conodonts from metamorphosed carbonates of the Salton Trough, California: *Geology*, v. 11, p. 410–412.

Miller, W. J., 1935, A geologic section across the southern Peninsular Range of California: *Calif. J. Mines Geology*, v. 381, p. 115–142.

———, 1944, Geology of the Palm Spring-Blythe strip, Riverside County, California: *Calif. J. Mines and Geology*, v. 40, p. 11–72.

Minch, J. A., 1972, The Late Mesozoic-Early Tertiary framework of continental sedimentation, northern Peninsular Ranges, Baja California, Mexico: unpublished Ph.D. dissertation, Univ. California, Riverside, Calif., 192 p.

Miyashiro, A., 1961, Evolution of metamorphic belts: *J. Petrology*, v. 2, p. 277–311.

Moore, T. E., 1983, Geology, petrology, and tectonic significance of the Mesozoic paleo-oceanic terranes of the Vizcaino peninsula, Baja California Sur, Mexico: unpublished Ph.D. dissertation, Stanford Univ., Stanford, Calif., 376 p.

———, 1984, Tectonic significance of Mesozoic terranes of the Vizcaino peninsula, Baja California Sur, Mexico: *Pacific Section, Amer. Assoc. Petrol. Geologists, Soc. Econ. Paleontologists Mineralogists Ann. Meet., Program Abstr.*, p. 102.

———, 1986, Late Triassic ophiolites of the Vizcaino peninsula — remnants of an island arc/marginal basin system in Baja California? *Geol. Soc. America Abstr. with Programs*, v. 18, p. 159.

Moran, A. J., 1976, Allochthonous carbonate debris in Mesozoic flysch deposits in Santa Ana Mountains, California: *Amer. Assoc. Petrol. Geologists Bull.*, v. 60, no. 11, p. 2038–2043.

Morton, D. M., 1986, Contrasting plutons in the northern Peninsular Ranges batholith: *Geol. Soc. America Abstr. with Programs*, v. 18, p. 161.

Moscoso, B. A., 1967, A thick section of Jurassic "flysch" in the Santa Ana Mountains, southern California: unpublished M.S. thesis, San Diego State Univ., San Diego, Calif., 106 p.

Palmer, A. R., 1983, The Decade of North American Geology 1983 Geologic Time Scale: *Geology*, v. 11, p. 503–504.

Parrish, K. E., Walawender, M. J., Clinkenbeard, J. P., and Wardlaw, M. S., 1986, The Indian Hill granodiorite, a peraluminous, garnet-bearing granitoid, eastern Peninsular Ranges: *Geol. Soc. America Abstr. with Programs*, v. 18, p. 168.

Patterson, D. L., 1979, The Valle Formation — Physical stratigraphy and depositional model, southern Vizcaino peninsula, Baja California Sur, *in* Abbott, P. L., and Gastil, R. G., eds., *Baja California Geology, Field Guides and Papers:* San Diego State Univ., Dept. Geol. Sci., San Diego, Calif., p. 73–76.

———, 1984, Paleomagnetism of the Valle Formation and the Late Cretaceous paleogeography of the Vizcaino Basin, Baja California, Mexico, *in* Frizzell, V. A., Jr., ed., *Geology of the Baja California Peninsula:* Pacific Section, Soc. Econ. Paleontologists Mineralogists, v. 39, p. 173–182.

Platt, J. P., and Vissers, R. L. M., 1980, Extensional structures in anisotropic rocks: *J. Struct. Geology*, v. 2, p. 397–410.

Ramsay, J. G., 1980, Shear zone geometry — A review: *J. Struct. Geology*, v. 2, p. 83–99.

Rangin, C., 1978, Speculative model of Mesozoic geodynamics, central Baja California to northeastern Sonora, Mexico, *in* Howell, D. G., and McDougall, K. A., eds., *Mesozoic Paleogeography of the Western United States:* Pacific Section, Soc. Econ. Paleontologists Mineralogists, Pacific Coast Paleogeography Symp. 2, p. 85–106.

Rogers, T. H., 1965, *Geologic map of California, Santa Ana sheet*, scale 1:250,000: Calif. Div. Mines Geology.

Santillan, M., and Barrera, T., 1930, Los Posibilidades petroliferas en la costa occidental de la Baja California, entre los paralelos 30° y 32° de latitude norte: *Inst. Geol. Mexico Anales*, v. 5, p. 1–37.

Schoellhamer, J. E., Vedder, J. G., Yerkes, R. F., and Kinney, D. M., 1981, Geology of the northern Santa Ana Mountains: *U.S. Geol. Survey Prof. Paper 420-D*, 109 p., scale 1:24,000.

Schultejann, P. A., 1984, The Yaqui Ridge antiform and detachment fault — Mid-Cenozoic extensional terrane west of the San Andreas fault: *Tectonics*, v. 3, p. 677–691.

Schwarcz, H. P., 1969, Pre-Cretaceous sedimentation and metamorphism in the Winchester area, northern Peninsular Ranges, California: *Geol. Soc. America Spec. Paper 100*, 61 p.

Sharp, R. V., 1967, San Jacinto fault zone in the Peninsular Ranges of southern California: *Geol. Soc. America Bull.* v. 78, p. 705–730.

——, 1979, Some characteristics of the eastern Peninsular Ranges mylonite zone, *in Proceedings, Conference VIII, Analysis of Actual Fault Zones in Bedrock:* U.S. Geol. Survey Open File Rpt. 79-1239, p. 258–267.

Shaw, S. E., Cooper, J. A., O'Neil, J. R., Todd, V. R., and Wooden, J. L., 1986, Strontium, oxygen, and lead isotope variations across a segment of the Peninsular Ranges batholith, San Diego County, California: *Geol. Soc. America Abstr. with Programs*, v. 18, p. 183.

Silberling, N. J., Schoellhamer, J. E., Gray, C. H., Jr., and Imlay, R. W., 1961, Upper Jurassic fossils from Bedford Canyon Formation, southern California: *Amer. Assoc. Petrol. Geologists Bull.*, v. 45, no. 10, p. 1746–1765.

Silver, L. T., 1982, Evidence and a model for west-directed Early to mid-Cenozic basement overthrusting in southern California: *Geol. Soc. America Abstr. with Programs*, v. 14, p. 617.

——, 1983, Paleogene overthrusting in the tectonic evolution of the Transverse Ranges, Mojave and Salinian regions, California: *Geol. Soc. America Abstr. with Programs*, v. 15, p. 438.

——, Stehli, F. G., and Allen, C. R., 1963, Lower Cretaceous pre-batholithic rocks of northern Baja California, Mexico: *Amer. Assoc. Petrol. Geologists Bull.*, v. 47, p. 2054–2059.

——, Allen, C. R., and Stehli, F. G., 1969, Geological and geochronological observations in a portion of the Peninsular Ranges batholith of northwestern Baja California, Mexico: *Geol. Soc. America Spec. Paper 121*, p. 279–280.

——, Taylor, H. P., Jr., and Chappell, B., 1979, Some petrological, geochemical and geochronological observations of the Penincular Ranges batholith near the International Border of the U.S.A. and Mexico, *in* Abbott, P. L., and Todd, V. R., eds., *Mesozoic Crystalline Rocks — Peninsular Ranges Batholith and Pegmatites, Point Sal Orphiolite:* Guidebook Geol. Soc. America Ann. Meet., San Diego State Univ., San Diego, Calif. (Nov. 1979), p. 83–110.

Simpson, C., 1984, Borrego Springs-Santa Rosa mylonite zone — A Late Cretaceous west-directed thrust in southern California: *Geology*, v. 12, p. 8–11.

——, 1985, Deformation of granitic rocks across the ductile-brittle transition: *J. Struct. Geology*, v. 7, p. 503–511.

Strand, R. G., 1962, *Geologic map of California, San Diego-El Centro sheet*, scale 1:250,000: *Calif. Div. Mines Geology.*

Sydnor, R. H., 1975, Geology of the northeast border zone of the San Jacinto Pluton, Palm Springs, California: unpublished M.S. thesis, Univ. California, Riverside, Calif., 121 p.

Taylor, H. P., and Silver, L. T., 1978, Oxygen isotope relationships in plutonic igneous rocks of the Peninsular Ranges batholith, southern and Baja California, *in* Zartman, R. E., *Short Papers of the Fourth International Conference on Geochronology,*

Cosmochronology and Isotope Geology: U.S. Geol. Survey Open File Rpt. 78-701, p. 423-426.

Teissere, R. E., and Beck, M. E., Jr., 1973, Divergent Cretaceous paleomagnetic pole position for the southern California batholith, U.S.A.: *Earth Planet. Sci. Lett.*, v. 18, p. 296-300.

Theodore, T. G., 1970, Petrogenesis of mylonites of high metamorphic grade in the Peninsular Ranges of southern California: *Geol. Soc. America Bull.*, v. 81, p. 435-450.

Todd, V. R., 1979, Geologic map of the Mount Laguna quadrangle, San Diego County, California: *U.S. Geol. Survey Open File Rpt. 79-862*, 39 p., scale 1:24,000.

———, 1980, Geologic map of the Alpine quadrangle, San Diego County, California: *U.S. Geol. Survey Open File Rpt. 80-82*, 42 p., scale 1:24,000.

———, 1983, Geologic map of the El Cajon Mountain quadrangle, San Diego County, California: *U.S. Geol. Survey Open File Rpt. 83-781*, 20 p., scale 1:24,000.

———, and Shaw, S. E., 1979, Structural, metamorphic and intrusive framework of the Peninsular Ranges batholith in southern San Diego County, California, *in* Abbott, P. L., and Todd, V. R., eds., *Mesozoic Crystalline Rocks — Peninsular Ranges Batholith and Pegmatites, Point Sal Ophiolite:* Guidebook Geol. Soc. America Ann. Meet., San Diego State Univ., San Diego, Calif., (Nov. 1979), p. 177-231.

———, and Shaw, S. E., 1985, S-type granitoids and an I-S line in the Peninsular Ranges batholith, southern California: *Geology*, v. 13, p. 231-233.

———, and Shaw, S. E., 1986, Granitoid suites in the Peninsular Ranges batholith, San Diego County, California — Implications for a heterogeneous source region *Geol. Soc. America Abstr. with Programs*, v. 18, p. 193.

———, Learned, R. E., Peters, T. J., and Mayerle, R. T., 1983, Mineral resource potential of the Sill Hill, Hauser, and Caliente Roadless Areas, San Diego County, California: *U.S. Geol. Survey Misc. Field Studies Map MF-1547-A*, 12 p., scale 1:62,500.

Turner, F. J., 1981, *Metamorphic Petrology:* New York, McGraw-Hill, 524 p.

Walawender, M. J., Hoppler, H., Smith, T. E., and Riddle, C., 1979, Trace element evidence for contamination in a gabbronorite-quartz diorite sequence in the Peninsular Ranges batholith: *J. Geology*, v. 87, p. 87-97.

White, A. J. R., and Chappell, B. W., 1977, Ultrametamorphism and granitoid genesis: *Tectonophysics*, v. 43, p. 7-22.

White, S. H., Burrows, S. E., Carreras, J., Shaw, M. D., and Humphreys, E. J., 1980, On mylonites in ductile shear zones: *J. Struct. Geology*, v. 2, p. 175-187.

Zwart, H. J., Corvalan, J., James, H. L., Miyashiro, A., Saggerson, E. P., Sobolev, V. S., Subramaniam, A. P., and Vallance, T. G., 1967, A scheme of metamorphic facies for the cartographic representation of regional metamorphic belts: *International Union Geological Sciences Geol. Newsletter*, v. 1967, no. 2, p. 57-72.

Eric W. James and James M. Mattinson
Department of Geological Sciences
University of California, Santa Barbara
Santa Barbara, California 93106

33

METAMORPHIC HISTORY OF THE SALINIAN BLOCK: AN ISOTOPIC RECONNAISSANCE

ABSTRACT

Metamorphic rocks of the enigmatic Salinian block of western California fall into three suites: (1) banded and homogeneous gneiss in the southeastern part of the block at Barrett Ridge, Mount Abel, and Mount Pinos; (2) "Sur Series" amphibolite to lower granulite-grade pendants composed primarily of quartzofeldspathic schist and granofels, with subordinate interlayered pelitic, calcareous, and graphitic rocks; and (3) the schist of Sierra de Salinas, a homogeneous, amphibolite-grade quartzofeldspathic biotite schist, probably a metagraywacke. All three suites have been intruded by mid- to Late Cretaceous plutonic rocks.

Metamorphism and deformation of the "Sur Series" and schist of Sierra de Salinas were spatially and temporally associated with plutonism. Early synkinematic plutons with U/Pb ages between approximately 130 and 108 m.y. share east-northeast- to northeast-trending folds and lineations with their wall rocks. Refolding along north- to northwest-trending axes is associated with later intrusions of weakly deformed to undeformed plutons with U/Pb ages between 108 and 78 m.y.

Reconnaissance Rb/Sr isotopic analyses show the effects of Cretaceous metamorphism and plutonism on isotopic ratios controlled by provenance and diagenetic processes. Samples of the Sierra de Salinas schist have $^{87}Sr/^{86}Sr$ and $^{87}Rb/^{86}Sr$ ratios that are comparable to those from metamorphosed clastic rocks derived from Mesozoic protoliths. Rb and Sr isotopic ratios from a single "Sur Series" sample suggest a Paleozoic or Precambrian protolith. Published Rb/Sr isotopic work in the southeastern Salinian block supports a Precambrian age for gneisses near Mount Able and Mount Pinos.

Both the "Sur Series" and Sierra de Salinas schist contain metamorphosed Precambrian detrital zircons. These zircons fall on chords on concordia diagrams with roughly 100-m.y. lower intercepts and 1.7-b.y. upper intercepts. Samples of gneiss from Barrett Ridge have discordant U/Pb ages with an upper intercept of about 1.7 b.y. and a lower intercept of 78 m.y.

The metamorphic events documented in the Salinian block are not unique in the western Cordillera. Metamorphism identical in age, grade, and involvement of plutonic rocks is known from the southern Sierra Nevada, southeastern San Gabriel Mountains, and Peninsular Ranges of California.

INTRODUCTION

The high-grade metamorphic and plutonic rocks of the California Coast Ranges north of the Big Pine fault and between the San Andreas and Sur-Nacimiento faults comprise the crystalline basement of the Salinian block. These "continental" rocks are clearly out of place, juxtaposed against "oceanic" rocks of the Franciscan assemblage (Fig. 33-1). Although many correlations of Salinian basement rocks to provinces now on the other side of the San Andreas fault have been proposed (e.g., Hill and Dibblee, 1953; Ross, 1976b, 1985; Wiebe, 1970), no correlation has met with universal acceptance. In fact, recent paleomagnetic studies (Champion *et al.*, 1984; McWilliams and Howell, 1983; Page, 1982; Vedder *et al.*, 1982) suggest that it may be necessary to look much farther afield to find correlatives.

Metamorphic rocks contain the bulk of the geologic history of the Salinian block. Emplacement of plutonic rocks only spans the middle to Late Cretaceous (Mattinson and James, 1985), whereas parts of the Salinian basement have metamorphic histories that extend from the Precambrian (Kistler *et al.*, 1973; Mattinson, 1983) through Late Cretaceous.

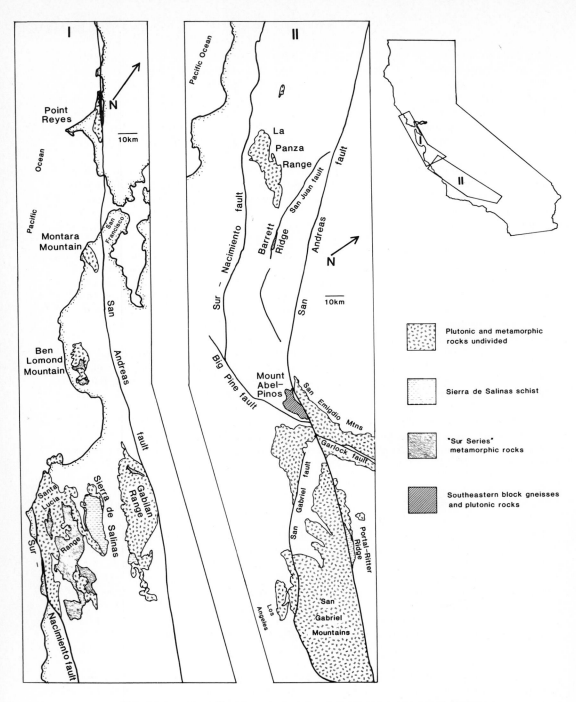

FIG. 33-1. Crystalline basement rocks of the Salinian block and adjacent areas of western California.

METAMORPHIC HISTORY OF THE SALINIAN BLOCK

Obviously, the geochronology and isotopic characteristics of the metamorphic rocks are important in any terrane analysis or attempt at correlation. However, an isotopic and geochronologic view of the metamorphic history of the Salinian block would be little more than descriptive without the petrologic and structural framework that comprises the conventional geologic history. Several comprehensive articles (Compton, 1966; Mattinson and James, 1985; Page, 1981, 1982; Ross, 1972, 1977, 1978, 1985) present the framework and highlight the problems of the region. These articles also present information about specific parts of the Salinian block, as do articles by Compton (1960), Leo (1961, 1967), Wiebe (1970a, b), Ross (1976 a, b), John (1981), and Frizzell and Powell (1982). The following description of the structure and petrology of the Salinian metamorphic rocks is based on these sources.

PETROLOGIC AND STRUCTURAL FRAMEWORK

Metamorphic rocks of the Salinian block can be grouped into three lithologically and geographically distinct suites (Ross, 1977) (Fig. 33-1). These are (1) high-grade gneisses of the southeastern sub-block of the Salinian block which crop out east of the Red Hills-San Juan–Chimeneas fault at Barrett Ridge and near Mount Abel and Mount Pinos; (2) "Sur Series" metamorphic rocks found throughout the block at Point Reyes, Ben Lomond, the Santa Lucia, Gabilan, and La Panza ranges, and possibly near Mount Able and Mount Pinos; and (3) the schist of Sierra de Salinas in the eastern Santa Lucia Range and the southern Gabilan Range. A brief description of each suite follows.

Rocks of the Southeastern Sub-block

Rocks north of the Big Pine fault in the Mount Pinos–Mount Able area and in a small area at Barrett Ridge are usually grouped with the Salinian block (Ross, 1977, 1978). Studies by Carman (1964), Lofgren (1967), Ross (1972), and Frizzell and Powell (1982) indicate that these rocks are high-grade migmatites and quartzofeldspathic gneisses. Local amphibolite, schist, marble, and quartzite may represent a separate terrane of "Sur Series" rocks (Frizzell and Powell, 1982). The metamorphic rocks are intruded on the south by gneissic granite and granodiorite and faulted against Pelona schist on the north. To the northwest, rocks at Barrett Ridge are thin-layered gneisses composed of plagioclase, quartz, K-feldspar, biotite, and hornblende. The gneiss includes minor amounts of quartzite and marble (Ross, 1972).

Although no specific lithologic correlations have been advanced, gneissic and migmatitic rocks in the southeastern subblock of the Salinian block may have general affinities with Precambrian rocks in the San Gabriel and San Bernardino Mountains (Ross, 1972, 1978).

"Sur Series" Rocks

The name "Sur Series" was first used by Trask (1926) "as a comprehensive name for the schist series of the Southern Coast Ranges." The name includes virtually all metamorphic pendants in the Salinian block and, as pointed out by Ross (1977), has no real stratigraphic meaning. The rocks are chiefly quartzofeldspathic schist, granofels, and gneiss. Common variants are aluminous, graphitic, or calcareous. Marble is conspicuous in the field but is volumetrically minor. Pure quartzites are thin-bedded and rare. There are no thick sequences of any one rock type. In general, the "Sur Series" protoliths are medium- to fine-grain clastic sedimentary rocks composed of feldspar and quartz mixed with varying proportions of argillaceous, organic, and limey material. There are no known meta-volcanic rocks — examples of amphibolite appear to have sedimentary protoliths. In aggregate, the section represents platform and shelf deposits.

Unfortunately, there are no well-preserved fossils known from the "Sur Series" or any other Salinian metamorphic rock. There are vague forms in limestones from the Gabilan Range suggestive of crinoid and coral fossils (Bowen and Grey, 1959), but the preservation is so poor and the status of the collections so confused (Page, 1981) that they are of no value for age determinations. Attempts to separate conodonts from the marbles have not been successful (E. W. James, unpublished data).

Metamorphic grade in "Sur-Series" rocks ranges from lower granulite to lower amphibolite. Metamorphism is associated with plutonism, and the highest metamorphic grade generally occurs near intrusions (Compton, 1966; DeCristofore, 1975, John, 1981; Leo, 1967; but also see DeCristoforo and Cameron, 1977). Dikeing and mixing of metamorphic and plutonic rocks is common, particularly in the Santa Lucia Range.

Comparison of studies by Compton (1960; 1966), Leo (1961, 1967), Wiebe (1970a, b), DeCristoforo (1975) DeCristoforo and Cameron (1977) and John (1981) shows that with the exception of the granulites in the Santa Lucia Range, conditions of metamorphism were similar across the Salinian block. Typical mineral assemblages in pelitic rocks are quartz + plagioclase + biotite + sillimanite ± K-feldspar ± garnet ± cordierite. The best estimate of peak metamorphic conditions (John, 1981) is $700 \pm 50°C$ and 4 ± 1 kb. This estimate probably applies to all areas except locally developed pockets of granulite-facies metamorphism. These higher-grade rocks are commonly related to loss of fluids along fracture systems (Compton, 1960, 1966).

The "Sur Series" metamorphic rocks have been folded along two axes. Folds of the first set are synmetamorphic, east- to northeast-trending, isoclinal, and plunge gently. The earliest plutonic rocks share this fabric. The first folds are refolded by north-south-to north-northwest-trending, open, steeply plunging folds. The refolding is most obvious near weakly deformed to undeformed younger plutons that make up the bulk of the Salinian basement (Compton, 1966). These relationships are important in determining the chronology of metamorphism and will be discussed again.

There are no metasediments outside of the Salinian block that are accepted correlatives of the "Sur Series." One possible exception (Powell, 1982) is the Placerita Formation. The Placerita Formation of the western and southern flank of the San Gabriel Mountains (Miller, 1934; Oakeshott, 1958) is composed of roof pendants of quartzite, marble, and pelitic, feldspathic, and graphitic schist. Identification of this terrane is important because it may tie together the Mount Able–Mount Pinos area, Transverse Ranges, Peninsular Ranges, and the rest of the Salinian block (Frizzell and Powell, 1982; Powell, 1982).

Schist of Sierra de Salinas

The schist of Sierra de Salinas (Ross, 1976a, b) makes up much of the eastern part of the Santa Lucia Range and extends into the southern Gabilan Range. The schist is encountered in the subsurface as far south as San Ardo and nearly as far east as the San Andreas fault. These rocks are homogeneous biotite quartzofeldspathic schists with very minor amounts of interbedded quartzite, marble, and amphibolite. On the basis of modal and chemical composition, Ross (1976b) suggests that the schist was derived from a thick, uniform pile of graywacke.

Metamorphic grade of the schist of Sierra de Salinas is similar to that of surrounding "Sur Series" rocks. Some samples contain garnet, sillimanite, and K-feldspar (Ross, 1976a, b). In contrast to the "Sur Series," the schist of Sierra de Salinas is *not* thoroughly intruded and intermixed with granitic rocks. There are intrusive contacts along the west side of the Sierra de Salinas and in the Gabilan Range, but in the main body of schist intrusions are limited to rare dikes and metamorphic pegmatites.

Structural data from the Sierra de Salinas schist are sparse. Reconnaissance mapping by Ross (1976a) illustrates the presence of northwest-trending foliation and folds similar to mapped trends in adjacent metamorphic and plutonic rocks.

Ross (1976b) correlates the Sierra de Salinas schist with schist at Portal and Ritter Ridge east of the San Andreas fault. His correlation rests on the modal and chemical similarity of the schists and corroborates established slip estimates of about 305 km for the San Andreas fault. Further evidence for correlation includes the similarity of pre-San Andreas structural elements in the Portal-Ritter Ridge schist (Evans, 1966) to structural elements in the Salinian block (Compton, 1966; John, 1981; Wiebe, 1970b). The correlation is currently the best link between Salinian basement and rocks east of the San Andreas fault.

The Sierra de Salinas–Portal–Ritter Ridge schist has been included with the Pelona-Orocopia-Rand schist by some authors (Haxel and Dillon, 1978 and references therein). However, evidence for this far-reaching correlation is contradictory. First, there are differences in lithology. Distinctive light and dark layering that characterizes some portions of the Pelona-Orocopia schist is not developed in the Sierra de Salinas. Fuchsite and piemontite, typical minor components of the Pelona-Orocopia and Rand schists, are absent from the Sierra de Salinas schist. The Sierra de Salinas and Portal–Ritter Ridge schists are metamorphosed to amphibolite grade, whereas lower amphibolite-grade rocks in the Pelona-Orocopia schist are found only near the overlying Vincent-Chocolate Mountain thrust. There are similarities and differences in the ages of intrusive materials. The Sierra de Salinas schist is intruded by 80–85-m.y.-old plutons (Mattinson, 1982; Mattinson and James, 1985). The Rand schist is intruded by a ~78-m.y. stock in the Rand Mountains (Silver et al., 1984) and a ~131-m.y.-old pluton in the San Emigdio Mountains (James, 1986). Intrusions into the Pelona-Orocopia schist are very rare; only one is known, a 163-m.y. metadiorite in the Chocolate Mountains (Mukasa et al., 1984). Overall, correlation of the Sierra de Salinas and Portal–Ritter Ridge schists appears compelling. Correlations of Sierra de Salinas and Rand schist are less convincing and correlation with the Pelona-Orocopia schist less likely.

GEOCHRONOLOGY

Age of the Southeastern Sub-block

Banded and migmatitic gneisses in the Mount Pinos–Mount Able area are Precambrian in age based on their proximity and lithologic similarities to dated Precambrian rocks of the San Gabriel Mountains (Silver, 1971) and on the basis of high $^{87}Sr/^{86}Sr$ isotopic ratios (Kistler et al., 1973).

Kistler et al. (1973) have also determined a two-point Rb/Sr isochron on granitic rocks intruding the Precambrian gneisses and propose an age of about 180 m.y. The strong possibility of differential contamination by Precambrian wall rocks makes this age assignment difficult to evaluate; a conservative approach suggests only a Mesozoic age.

Rocks at Barrett Ridge are separated from Mount Pinos and Mount Able and the rest of Salinia by major faults and Tertiary cover. However, they comprise both Precambrian and Mesozoic rocks similar to both the Mount Pinos–Mount Able gneisses and to intrusive rocks of the southern Salinian block. Barrett Ridge banded gneiss has discordant U/Pb zircon ages that yield an upper intercept age of about 1.7 b.y. and a lower intercept age of about 76 m.y. (Mattinson, 1983). The lower intercept age is the age of leucocratic granitic rock that intrudes the gneiss, which is also approximately the age of plutonic rocks in the nearby La Panza Range.

Present-day Sr isotopic ratios (measured on apatite, probably representing Cretaceous equilibration) of the Barrett Ridge gneiss are very high, ~0.76, in keeping with its Precambrian age (Mattinson, 1983).

Metamorphic Age of the "Sur Series"

Our understanding of the age of metamorphism of the schist of Sierra de Salinas and "Sur Series" rocks rests largely on the age of intrusive rocks and the relative ages of folds and lineations. In the Ben Lomond area and the Santa Lucia Mountains (Compton, 1966; DeCristoforo, 1975; Leo, 1961, 1967; Wiebe, 1970a, b) the plutonic rocks can be divided into two groups: (1) a deformed suite, and (2) a weakly deformed to undeformed suite.

The deformed plutons are synmetamorphic. They generally share the fabric with their wallrocks (northeast- and north-northwest-trending folds and lineations) but contacts are locally discordant — cross-cutting and crumpling foliation. The deformed plutons have U/Pb zircon ages of between about 108 and 130 m.y. (James, 1984; Mattinson, 1982). Unfortunately, zircon systematics are affected by inheritance and by lead loss, preventing an exact age determination. In some cases, the older plutons have a hornfelsic overprint imposed by later plutons.

The younger plutons are responsible for the refolding of earlier folds along steep north- to north-northwest-trending axes but have very weak fabrics, if any, themselves (Compton 1966; John, 1981; Wiebe, 1970a, b). They have concordant to near concordant U/Pb ages ranging from about 108 to 76 m.y. (James, 1984; Mattinson, 1978, 1982; Mattinson and James, 1985).

In the La Panza Range there is evidence that both periods of folding lasted until the end of Salinian plutonic activity. Compton (1966) reports the results of petrofabric studies on plutonic rocks that demonstrate the presence of both periods of folding. The weakly deformed sample is from one of the youngest plutonic units in the Salinian block (~78 m.y., J. M. Mattinson, unpublished data) yet it has both northeast- and north-west-trending lineations.

In the Santa Lucia range there is no evidence for appreciable time between development of northeast-trending folds and lineation in older plutons and their wall rocks, and the development of north-northwest-trending folds associated with the emplacement of weakly to undeformed deformed plutons (Compton, 1966; Wiebe, 1970a). Also the 100–105-m.y. age (Mattinson, 1978) of charnockitic tonalite in the Santa Lucia Range suggests that highest metamorphic conditions were reached in mid-Cretaceous time.

Provenance Age of the "Sur Series"

Ages of metamorphosed detrital zircon from schist and quartzite on Ben Lomond Mountain are consistent with a mid-Cretaceous metamorphism of sedimentary rocks containing Precambrian detritus (James, 1984). Interpretation is difficult because a simple concordia-intercept model is not strictly applicable. A detrital zircon sample could contain several zircon populations, each with its own predepositional history. At least one episode of metamorphism further complicates the interpretation in this case. However, it is surprising that the analyses show little scatter around a chord from 100 m.y. to 1.7 b.y on the concordia diagram (Fig. 33-2). It can be said with certainty that the minimum age for the oldest zircon population in the Santa Cruz schist is 1640 m.y., the $^{207}Pb*/^{206}Pb*$ (*denotes radiogenic Pb) age of the coarse zircon fraction. The Precambrian Pb/Pb date and upper intercept are consistent with a provenance age of at least 1.6 b.y., very probably about 1.7 m.y. This age of provenance requires deposition of the protolith after 1.6-1.7 b.y. and before mid-Cretaceous.

Metamorphic Age of Schist of Sierra de Salinas

The metamorphic age of the schist of Sierra de Salinas is constrained by the age of undeformed plutons that intrude it. In the Gabilan Range the schist is intruded by plutons with 80–85-m.y. U/Pb zircon ages (Mattinson, 1982). Deformed granitic rocks, probably slightly older than the Gabilan plutons, intrude the schist on the west side of the Sierra de Salinas. The schist appears to have the same fabric as these granitic rocks (Ross, 1976a). Rb/Sr data presented in the next section also bear on metamorphic age of the schist.

Provenance Age of the Schist of Sierra de Salinas

The age of metamorphosed detrital zircon from schist east of the Salinas Valley also shows the presence of a ~1.7-b.y.-old component that has been affected by lead loss at about 80 m.y. (Mattinson, 1982). The data are very similar (Fig. 33-2) to U/Pb data from metamorphosed detrital zircon separated from schist near Santa Cruz (James, 1984).

The 1.7-b.y. upper intercept age for detrital zircon in the Schist of Sierra de Salinas and "Sur Series" schist near Santa Cruz indicates that at least part of the sediment that comprises these rocks is Precambrian in age.

Reconnaissance work for a comprehensive Rb/Sr study of Salinian metamorphic rocks allows some general conclusions about their age. Isotopic data from whole-rock samples of the Sierra de Salinas schist from both sides of the Salinas Valley suggest, in a general way, an age of deposition. Samples have Rb/Sr ratios and Sr isotopic ratios that represent a composite of detrital components modified by seawater interaction, diagenesis, and metamorphism. These complicating factors plus the scatter and small spread of data on the isochron diagram (Figs. 33-3 and 33-4) precludes rigorous age interpretations, but the data still suggest some general age ranges.

The lower-most four points in Fig. 33-3 (three samples and one replicate analysis) are colinear and might be taken to suggest a 200-m.y. isochron. This is surely an over-optimistic interpretation of the data given the statistical uncertainties of the data and the complicated chemical history of the samples.

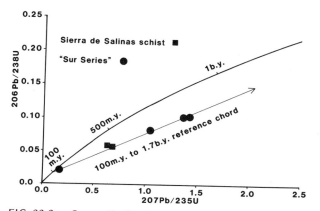

FIG. 33-2.   Concordia diagram showing ages of metamorphosed detrital zircons from Sierra de Salinas schist and "Sur Series" schist and quartzite.

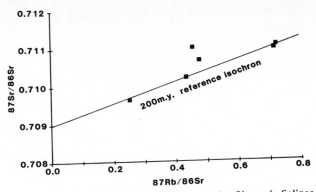

FIG. 33-3. Rb/Sr isochron diagram showing Sierra de Salinas schist whole-rock analyses.

Some simple model age calculations impose a realistic limit on the age of the detrital component and perhaps early metamorphism as well. Considering the zircon evidence for a Precambrian continental detrital component, it is probable that the sediments had a high initial $^{87}Sr/^{86}Sr$ ratio. Therefore, assuming a low initial ratio (0.704) should yield a maximum age of deposition or possibly the last metamorphism. Assuming an initial ratio of 0.704 yields maximum ages of between about 1550 and 680 m.y. These ages can only be considered crude maxima.

Perhaps a better approach is direct comparison of Rb/Sr isotopic ratios with other rocks of known depositional age and setting. Comparison with work by Hill (1984) and Hill and Silver (1983) in the northern Peninsular Ranges is particularly informative (Fig. 33-4). These rocks are similar to the schist of Sierra de Salinas in lithology, grade, and setting. Hill (1984) and Hill and Silver (1983) found that the metasediments in the western and central part of the Peninsular Ranges have $^{87}Sr/^{86}Sr$ ratios that range from 0.7040 to 0.710 at $^{87}Rb/^{86}Sr$ <0.1 to 0.750 at $^{87}Rb/^{86}Sr$ = 12 (Fig. 33-4). These metamorphic rocks contain Triassic and Jurassic fossils. The Sierra de Salinas Rb/Sr analyses fall within the same range of $^{87}Sr/^{86}Sr$ as the central suite of the Peninsular Ranges.

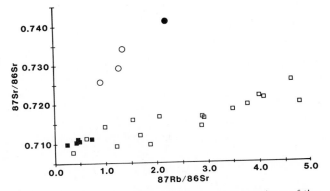

FIG. 33-4. Rb/Sr isochron diagram comparing analyses of the Sierra de Salinas schist (filled squares) and "Sur Series" schist (filled circle) with analyses of Mesozoic (?) metasediments (open squares) and Paleozoic or Precambrian (?) metasediments (open circles) of the Peninsular Ranges of southern California (Hill, 1984; Hill and Silver, 1983).

METAMORPHIC HISTORY OF THE SALINIAN BLOCK

Other Mesozoic "eugeoclinal" metasediments in the Peninsular Ranges (Criscione *et al.*, 1978), as well as Franciscan and Great Valley sequence rocks (Peterman *et al.*, 1967, 1981) have similar Sr isotopic characteristics, although the Sierra de Salinas samples have a smaller range of Rb/Sr.

Returning briefly to the "Sur Series" metasediments, a single determination of the $^{87}Sr/^{86}Sr$ ratio in schist from near Santa Cruz yields a ration of 0.74102 at a $^{87}Rb/^{86}Sr$ ratio of about 2.2. This $^{87}Sr/^{86}Sr$ ratio is higher than whole-rock ratios from the schist of Sierra de Salinas and may indicate a Paleozoic or Precambrian source or age (Fig. 33-4). The isotopic ratios are in the same range as samples from presumed Paleozoic rocks of the eastern Peninsular Ranges (Hill, 1984; Hill and Silver, 1983).

SUMMARY AND DISCUSSION

Similarities in Metamorphic Suites

Despite differences in lithology, age, and setting, the three suites of Salinian metamorphic rocks share some important general attributes. The three metamorphic suites all contain 1.7-b.y.-old zircons, either as detritus or as an original igneous component. They also have Sr isotopic ratios that are relatively radiogenic, indicating that they contain old "continental" materials.

The "Sur Series" and Sierra de Salinas rocks show evidence of a mid-Cretaceous high-grade metamorphism. Synmetamorphic deformation produced east- to northeast-trending folds followed by north- to northwest-trending folds. Metamorphism was associated with voluminous syn- to post-kinematic plutonism. The Barrett Ridge rocks show the effects of Cretaceous contact metamorphism (Mattinson, 1983), but have not developed an obvious new fabric.

All three suites are intruded by mid-Cretaceous plutons. Although the intruding plutons are not identical in age or lithology, they are linked by a northwest to southeast increase in radiogenic Sr (Kistler and Peterman, 1978; Kistler *et al.*, 1973; Mattinson and James, 1985), radiogenic Pb isotopes (Mattinson and James, 1985), and a decrease in age from mid- to Late Cretaceous (Mattinson and James, 1985; Silver, 1982, 1983). These zonations are typical of arc plutonic rocks developed on a wedge of continental crust.

Comparison with Other Terranes

The southernmost Sierra Nevada, the nearest crystalline terrane to the Salinian block in palinspastic reconstructions of the early Tertiary, has ages of metamorphism that match those of the Salinian block. In the Tehachapi Mountains, Sams *et al.* (1983) and Sams and Saleeby (1986) have found 110–120-m.y.-old augen gneiss intruded by less deformed 100-m.y.-old tonalite. Some of the U/Pb zircon analyses from the augen gneisses have discordance patterns identical to deformed plutons from the Salinian block. Cretaceous deformation in the Sierra Nevada is predominant in the south, apparently giving way to coaxial Jurassic fabrics to the north (Tobisch and Fiske, 1982; Tobisch and Saleeby, 1983).

May (1984) and Walker and May (1985) show that much of the deformation and granulite- to amphibolite-grade metamorphism in the southeastern San Gabriel Mountains is mid-Cretaceous. Cretaceous plutons in this area range from deformed to undeformed and contain ~1.7-b.y.-old inherited zircon, as in the Salinian block. The metasedimentary rocks intruded by the plutons are part of the Placerita terrane (Powell, 1982).

Farther south, in the Southern California batholith, work by Silver *et al.* (1979)

and Todd and Shaw (1979) documents a period of older, 120–105-m.y. plutonism, in part synmetamorphic. These plutons are followed by 105–90-m.y.-old undeformed intrusions. Similarity of the Peninsular Ranges with the Salinian block also includes Sr isotopic ratios of Mesozoic metamorphic rocks that parallel analyses of the Sierra de Salinas schist.

Although our Sr data for "Sur Series" rocks consist only of a single analysis, the similarity to the eastern Peninsular Ranges metamorphic rock data of Hill and Silver (1983) and Hill (1984) suggests that the "Sur Series" has a Paleozoic protolith. A comparison of rock types that comprise the "Sur Series" (Ross, 1977, Wiebe, 1970a) and metamorphic rock types in the eastern Peninsular Ranges (Todd and Shaw, 1979) shows considerable similarity. As mentioned earlier, Powell (1982) extends the Placerita Formation into the eastern Peninsular Ranges.

The 1.7-b.y. upper intercept of the Barrett Ridge Gneiss, 1.7-b.y.-old component in the "Sur Series" and Sierra de Salinas schist and the ubiquitous inherited zircon component in Salinian plutonic rocks provide a possible longstanding tie between these rocks. A comparison to Precambrian age provences in the Cordillera is also instructive. Basement ages of 1.7 b.y. are common east of the San Andreas fault and in the San Gabriel Mountains (Silver, 1971; Walker and May, 1986). A 1.7–1.8-b.y. basement age provence extends along the eastern side of the Gulf of California to north of the latitude of Hermosillo, Mexico (Anderson and Silver, 1981). The next outcrops of Precambrian rock are approximately 1500 km to the south. These are Grenville age (1.1 b.y.) granulite facies rocks of the Oaxaca Complex (Anderson and Silver, 1971, 1981). Interestingly, these rocks are in the area suggested as the source for the Salinian block by paleomagnetic studies (Champion *et al.*, 1984).

The age and isotopic characteristics of Salinian metamorphism show no striking differences from other high-temperature/moderate-pressure metamorphic terranes of California. Salinian metamorphic rocks show striking resemblances to metamorphic suites in the southern Sierra Nevada, southern and western San Gabriel Mountains, and central to eastern Peninsular Ranges. These similarities involve: age of preexisting basement; provenance age and type; metamorphic grade; the relationship of metamorphism to plutonism; and the timing of metamorphic, deformational, and intrusive events. All these similarities suggest that a careful reevaluation of the "orphan" status of the Salinian block is in order.

ACKNOWLEDGMENTS

We thank Dr. Donald Ross of the U.S. Geological Survey for providing samples of Sierra de Salinas schist. Research was supported by National Science Foundation grant EAR80-08215, a U.C. Santa Barbara faculty research grant, and Penrose grant 2848-81.

REFERENCES

Anderson, T. H., and Silver, L. T., 1981, An overview of Precambrian rocks in Sonora: *Univ. Nal. Auton. Mexico, Inst. Geologica Rev.*, v. 5, no. 2, p. 131–139.

____ , and Silver, L. T., 1971, Age of granulite metmorphism during the Oaxacan Orogeny, Mexico: *Geol. Soc. America Abstr. with Programs*, v. 4, p. 492.

Bowen, O. E., and Gray, C. H., 1959, Geologic and economic possibilities of the limestone and dolomite deposits of the northern Gabilan Range, California: *Calif. Div. Mines Geology Spec. Rpt. 466*, 40 p.

Carman, M. F., Jr., 1964, Geology of the Lockwood Valley area, Kern and Ventura counties, California: *Calif. Div. Mines Spec. Rpt. 81*, 62 p.

Champion, D. E., Howell, D. G., and Gromme, C. S., 1984, Paleomagnetic and geologic data indicating 2500 km of northward displacement for the Salinian and related Terranes, California: *J. Geophys. Res.*, v. 89, p. 7736–7752.

Compton, R. R., 1960, Charnockitic rocks of the Santa Lucia Range, California: *Amer. J. Sci.*, v. 258, p. 609–636.

——, 1966, Granitic and metamorphic rocks of the Salinian block, California Coast Ranges: *Calif. Div. Mines Geology Bull. 190*, p. 277–287.

Criscione, J. J., Joseph, J., Davis, T. E., and Ehlig, P., 1978, The age of sedimentation/diagenesis for the Bedford Canyon formation and the Santa Monica Formation in Southern California — A Rb/Sr evaluation, *in* Howell, D. G. and McDougall, K. A., eds., *Mesozoic Paleogeography of the Western United States:* Pacific Section, Soc. Econ. Paleontologists Mineralogists, Pacific Coast Paleogeography Symp. 2, p. 385–396.

DeCristoforo, D. T., 1975, Petrology of the U.C.S.C. campus pelitic schist: B.S. thesis, Univ. California Santa Cruz, Calif., 53 p.

——, and Cameron, K. L., 1977, Petrology of sillimanite-K-spar zone metapelites from Ben Lomond Mountain, central Coast Ranges, California: *Geol. Soc. America Abstr. with Programs*, v. 9, p. 411.

Evans, J. G., 1966, Structural analysis and movement of the San Andreas fault zone near Palmdale, southern California: Ph.D. thesis, Univ. California, Los Angeles, Calif., 186 p.

Frizzell, V. A., Jr., and Powell, R. E., 1982, Crystalline rocks near Frazier and Alamo Mountains, Western Transverse Ranges, California — A comparative study: *Geol. Soc. America Abstr. with Programs*, v. 14, p. 164.

Haxel, J., and Dillon, J. 1978, The Pelona-Orocopia Schist and the Vincent-Chocolate Mountain thrust system, southern California, *in* Howell, D. G. and McDougall, K. A. eds., *Mesozoic Paleogeography of the Western United States:* Pacific Section, Soc. Econ. Paleontologists Mineralogists, Pacific Coast Paleogeography Symp. 2, p. 453–469.

Hill, M. L., and Dibblee, T. W., Jr., 1953, San Andreas, Garlock and Big Pine faults, California — A study of the character, history and tectonic significance of their displacements: *Geol. Soc. America Bull.*, v. 64, p. 443–458.

Hill, R. I., 1984, Petrology and petrogenesis of batholithic rocks, Mt. San Jacinto, California: Ph.D. Thesis, California Inst. Technology, Los Angeles, Calif.

——, and Silver, L. T., 1983, Contrasting west-east strontium isotopic properties of wall rocks, no. Peninsular Ranges batholith, Calif.: *Geol. Soc. America Abstr. with Programs*, v. 15, p. 439.

James, E. W., 1984, U/Pb ages of plutonism and metamorphism in part of the Salinian block, Santa Cruz and San Mateo counties, CA: *Geol. Soc. America Abstr. with Programs*, v. 1, p. 291.

——, 1986, U/Pb age of the Antimony Peak tonalite and its relation to the Rand Schist in the San Emigdio Mountains, California: *Geol. Soc. America Abstr. with Programs*, v. 1, p. 291.

John, D. A., 1981, Structure and petrology of pelitic schist in the Fremont Peak pendant, northern Gabilan Range, California: *Geol. Soc. America Bull.*, v. 92, p. 237–246.

Kistler, R. W., and Peterman, Z. E., 1978, Reconstruction of crustal blocks of California on the basis of initial strontium isotopic compositions of Mesozoic granitic rocks: *U.S. Geol. Survey Prof. Paper 1071*, 17 p.

____ , Peterman, Z. E., Ross, D. C., and Gottfried, D., 1973, Strontium isotopes and San Andreas fault, *in* Kovach, R. L., and Nur, A., eds., *Proceedings of the Conference on Tectonic Problems of the San Andreas Fault System:* Stanford Univ. Publ. Geol. Sci., v. 13, p. 339–347.

Leo, G. W., 1961, The plutonic and metamorphic rocks of Ben Lomond Mountain, Santa Cruz County, California: Ph.D. dissertation, Stanford Univ., Stanford, Calif. 194 p.

____ , 1967, The plutonic and metamorphic rocks of the Ben Lomond Mountain area, Santa Cruz County, California: *Calif. Div. Mines Geology Spec. Rpt. 91*, p. 27–43.

Lofgren, G. E., 1967, Geology of the Mount Pinos basement complex [abstract] : *Geol. Soc. America Spec. Paper 115*, p. 337.

Mattinson, J. M. 1978, Age, origin and thermal histories of some plutonic rocks from the Salinian block of California: *Contrib. Mineralogy Petrology*, v. 67, p. 233–245.

____ , 1982, Granitic rocks of the Gabilan Range — U-Pb isotopic systematics and implications for age and origin: *Geol. Soc. America Abstr. with Programs*, v. 14, p. 184.

____ , 1983, Basement rocks of the southeastern Salinian Block, Califormia — U-Pb relationships: *Geol. Soc. America Abstr. with Programs*, v. 15, p. 414.

____ , and James, E. W., 1985, Salinian block U-Pb age and isotopic variations: implications for origin and emplacement of the Salinian terrane, *in* Howell, D. G., ed., *Tectonostratagraphic Terranes of the Circum-Pacific Region:* Circum-Pacific Council Energy Min. Resources, Earth Sci. Series, no. 1, p. 215–226.

May, D. J. 1985, Mylonite belts in the southeastern San Gabriel Mts., California — Remnants of a Late Cretaceous sinistral transcurrent shear zone: *Geol. Soc. America Abstr. with Programs*, v. 17, p. 368.

McWilliams, M. O., and Howell, D. G., 1982, Exotic terranes of western California: *Nature*, v. 297, p. 215–217.

Miller, W. J., 1934, Geology of the western San Gabriel Mountains of California: *Univ. Calif. Publ. Math. Phys. Sci.*, v. 1, p. 1–114.

Mukasa, S. B., Dillon, J. T., and Tosdal, R. M., 1984, A Late Jurassic minimum age for the Pelona-Orocopia schist protolith, southern California: *Geol. Soc. America Abstr. with Programs*, v. 16, p. 323.

Oakeshott, G. B., 1958, Geology and mineral deposits of San Fernando quadrangle, Los Angeles County, California: *Calif. Div. Mines. Bull. 172*, 147 p.

Page, B. M., 1981, The southern Coast Ranges, *in* Ernst, W. G., ed., *The Geotectonic Development of California* (Rubey Vol. I: Englewood Cliffs, N.J., Prentice-Hall, p. 329–417.

____ , 1982, Migration of Salinian composite block, California, and disappearance of fragments: *Amer. J. Sci.*, v. 282, p. 1694–1734.

Peterman, Z. E., Hedge, C. E., Coleman, R. G., and Snavely, P. D., Jr., 1967, $^{87}Sr/^{86}Sr$ ratios in some eugeosynclinal sedimentary rocks and their bearing on the origin of granitic magma in orogenic belts: *Earth Planet. Sci. Lett.*, v. 2, p. 433–439.

____ , Coleman, R. G., and Bunker, C. M., 1981, Provenance of Eocene graywackes of the Flournoy Formation near Agness, Oregon — A geochemical approach: *Geology*, v. 9, p. 81–86.

Powell, R. E., 1982, Prebatholithic terranes in the crystalline basement of the Transverse Ranges, southern California: *Geol. Soc. America Abstr. with Programs*, v. 14, p. 225.

Ross, D. C., 1972, Petrographic and chemical reconnaissance study of some granitic and gneissic rocks near the San Andreas fault from Bodega Head to Cajon Pass, California: *U.S. Geol. Survey Prof. Paper 698*, 92 p.

———, 1976a, Reconnaissance geologic map of the pre-Cenozoic basement rocks, northern Santa Lucia range, Monterey County, California: *U.S. Geol. Survey Misc. Field Inv. Map MF 750.*

———, 1976b, Metagraywacke in the Salinian block, central Coast Ranges, California- and a possible correlative across the San Andreas fault: *U.S. Geol. Survey J. Res.*, v. 4, p. 683–696.

———, 1977, Pre-intrusive metasedimentary rocks of the Salinian block, California — A paleotectonic dilemma, *in* Steward, J. H., Stevens, C. H., and Fritsche, A. E., eds., *Paleogeography of the Western United States:* Pacific Section, Soc. Econ. Paleontologists Mineralogists, Pacific Coast Paleogeography Symp. 1, p. 371–380.

———, 1978, The Salinian block — A Mesozoic granitic orphan in the California Coast Ranges, *in* Howell, D. G., and McDougall, K. A., eds., *Mesozoic Paleogeography of the Western United States:* Pacific Section, Soc. Econ. Paleontologists Mineralogists, Pacific Coast Paleogeography Symp. 2, p. 509–522.

———, 1985, Basement rocks of the Salinian block and southernmost Sierra Nevada and possible correlations across the San Andreas, San Gregorio-Hosgri, and Rinconada-Reliz-King City fault zones: *U.S. Geol. Survey Prof. Paper 1317.*

Sams, D. B., and Saleeby, J. B., 1986, U/Pb zircon ages, and implications on the structural and petrologic development of the southernmost Sierra Nevada: *Geol. Soc. America Abstr. with Programs*, v. 18, p. 180.

———, Saleeby, J. B., Kistler, R. W., and Ross, D. C., 1983, Cretaceous igneous, metamorphic and deformational events in the southern-most Sierra Nevada, California: *Geol. Soc. America Abstr. with Programs*, v. 15, p. 294–295.

Silver, L. T., 1971, Problems of crystalline rocks of the Transverse Ranges: *Geol. Soc. America Abstr. with Programs*, v. 3, p. 193–194.

———, 1982, Evidence and a model for west directed Early to Mid-Cenozoic basement overthrusting in southern California: *Geol. Soc. America Abstr. with Programs*, v. 14, p. 617.

———, 1983, Paleogene overthrusting in the tectonic evolution of the Transverse Ranges, Mojave and Salinian regions, California: *Geol. Soc. America Abstr. with Programs*, v. 15, p. 438.

———, Taylor, H. P., and Chappell, B., 1979, Some Petrological, geochemical and geochronological observations of the Peninsular Ranges batholith near the international border of the U.S.A. and Mexico, *in* Abbott, P. L., and Todd, V. R., eds., *Mesozoic Crystalline Rocks — Peninsular Ranges Batholith and Pegmatites, Point Sal Ophiolite:* Guidebook Geol. Soc. America Ann. Meet., San Diego State Univ., San Diego, Calif., p. 84–110.

———, Sams, D. B., Bursik, M. I., Graymer, R. W., Nourse, J. A., Richards, M. A., and Salyards, S. L., 1984, Some observations on the tectonic history of the Rand Mountains, Mojave Desert, California: *Geol Soc. America Abstr. with Programs*, v. 16, p. 333.

Tobisch, O. T., and Fiske, R. S., 1982, Repeated parallel deformation in part of the Sierra Nevada, California and its implications for dating structural events: *J. Struct. Geol.*, v. 2, p. 177–196.

———, and Saleeby, J. B., 1983, Nature and timing of late Mesozoic deformation in central and southern Sierra Nevada, California: *Geol. Soc. America Abstr. with Programs*, v. 15, p. 294.

Todd, V. R., and Shaw, S. E., 1979, Metamorphic, intrusive and structural framework of the Peninsular Ranges Batholith in southern San Diego County, California, *in* Abbott, P. L., and Todd, V. R., eds., *Mesozoic Crystalline Rocks — Peninsular Ranges Batholith and Pegmatites, Point Sal Ophiolite:* Guidebook Geol. Soc. America Ann. Meet., San Diego State Univ., San Diego, Calif., p. 177-231.

Trask, P. D., 1926, Geology of the Point Sur quadrangle, California: *Calif. Univ. Dept. Geol. Sci. Bull.*, v. 16, no. 6, p. 119-186.

Vedder, J. G., Howell, D. G., and McLean, H., 1983, Stratigraphy, sedimentation and tectonic accretion of exotic terranes, southern Coast Ranges, California: *Amer. Assoc. Petrol. Geologists Mem. 34*, p. 471-496.

Walker, N. W., and May D. J., 1986, U-Pb ages from the SE San Gabriel Mts., CA — Evidence for Cretaceous metamorphism, plutonism, and mylonitic deformation predating the Vincent thrust: *Geol. Soc. America Abstr. with Programs*, v. 18, p. 195.

Wiebe, R. A., 1970a, Relations of granitic and gabbroic rocks, northern Santa Lucia Range, California: *Geol. Soc. America Bull.*, v. 81, p. 105-116.

_____ , 1970b, Pre-Cenozoic tectonic history of the Salinian block, western California: *Geol. Soc. America Bull.*, v. 81, p. 1837-1842.

Simon M. Peacock
Department of Geology
Arizona State University
Tempe, Arizona 85287

34

INVERTED METAMORPHIC GRADIENTS IN THE WESTERMOST CORDILLERA

ABSTRACT

Metamorphic grade and intensity increase structurally *upward* in the lower plate of several thrust zones in the westernmost Cordillera. Such inverted metamorphic gradients (IMGs) include (1) the Late Cretaceous/early Tertiary Pelona-Orocopia Schist ($T = 400 \rightarrow 650°C$, $P = \sim9$ kb) beneath the Vincent thrust system; (2) the Cretaceous Catalina Schist ($350 \rightarrow 650°C$, ~9 kb) beneath serpentinite; (3) the Devonian Central Metamorphic Belt ($450 \rightarrow 650°C$, ~6 kb) beneath the Trinity thrust; and (4) the Early Cretaceous South Fork Mountain Schist ($<200 \rightarrow 300°C$, ~7 kb) beneath the northern Coast Range thrust. (2), (3), and (4) formed beneath ultramafic hanging walls, whereas (1) developed beneath "granitic" crystalline rocks. Structural thicknesses of the IMGs range from approximately 1 km (1,2,3) to 2-5 km (4), implying inverted metamorphic gradients on the order of $-100°C/km$. In Washington, the Jurassic/Cretaceous Shuksan Suite ($\sim450 \rightarrow 750°C$, ~8 kb) may represent a structurally disrupted IMG. The tectonic setting and high pressures of metamorphism suggest that the IMGs formed in paleosubduction zones.

Any realistic thermal model of IMGs must both *create* and *preserve* an IMG. Inverted thermal gradients induced by thrusting decay ~1 m.y. after movement ceases due to the rapid conduction of heat downward. Uplift/erosion rates would have to be implausibly high to quench such an IMG once thrusting ceased. Frictional heating in the thrust zone requires unreasonably high shear strengths (on the order of 1 kb) for rocks under metamorphic conditions.

In this paper, I present three one-dimensional models that approximate the subduction process. The results of these models show that an IMG can be created and preserved without requiring frictional heating or unreasonable erosion rates. (1) During the initiation of subduction, an IMG forms at the top of the subducting plate as it descends beneath a hot hanging wall. During descent, the lower plate is continuously heated from above, resulting in maximum temperatures in the IMG *equal* to hanging-wall temperatures, in contrast to instantaneous thrust emplacement models. The IMG then becomes welded to the hanging wall, perhaps because of the weakening effect of dehydration reactions occurring within the subducting slab. (2) Continued subduction for 5-20 m.y. removes most of the heat from the overlying mantle wedge, effectively refrigerating the upper plate and accreted IMG. (3) After subduction ceases, moderate uplift rates of less than 0.1 mm/yr will preserve the previously accreted, but now cooled, IMG.

The model predicts that the highest-grade rocks in the IMG form early in the subduction process and effectively date the initiation of subduction. Later, lower temperature, blueschist metamorphism should postdate high-grade metamorphism, as is observed in the Franciscan Complex (e.g., Coleman and Lanphere, 1971; McDowell, *et al.*, 1984) and in the Shuksan Suite (Brown *et al.*, 1982). Because millions of years of subduction is required to refrigerate the hanging wall of a subduction zone, IMGs mark sites of major plate consumption, where 1000 km or more of oceanic lithosphere has been subducted.

GENERAL OVERVIEW

In tectonically stable areas, one expects temperature to increase monotonically with depth in the earth. Prograde metamorphic rocks contain mineral assemblages that generally reflect the highest temperature the rock has experienced. Thus, in a metamorphic terrane, we might expect to observe rocks of increasing metamorphic grade with deeper exhumed structural level. In areas characterized by compressional tectonics, however, it is not uncommon to observe decreasing metamorphic grade with increasing depth — a phenomenon termed an "inverted metamorphic gradient" (IMG).

Inverted metamorphic gradients associated with major thrust faults occur in three different geologic settings: (1) metamorphic aureoles beneath ophiolites, (2) continent-continent collision zones, and (3) high-pressure terranes associated with subduction zones. Whereas the lithologies and conditions of metamorphism vary among the three different settings, all represent convergent plate boundaries and are characterized by decreasing metamorphic grade and deformation with increasing depth.

Metamorphic aureoles beneath ophiolites represent extreme examples of inverted metamorphic gradients. An aureole of high-grade metamorphic rocks, <300 m thick, is present at the base of the Bay of Islands and White Hills ophiolites in west Newfoundland (Williams and Smyth, 1973). Pyroxene-bearing amphibolites, in contact with peridotite, grade downward through lower amphibolite and greenschist facies metabasites to virtually unmetamorphosed volcanic and sedimentary rocks. Beneath the Samail ophiolite in Oman, upper amphibolite facies conditions next to the peridotite decrease to greenschist facies conditions 130 m below the peridotite/amphibolite contact (Ghent and Stout, 1981). Metamorphic aureoles are also observed beneath ophiolites in the "Tethyan" belt (e.g., Woodcock and Robertson, 1977), Quebec (Feininger, 1981), Cornwall (Green, 1964), and Scotland (Spray and Williams, 1980). The metamorphic aureoles are commonly considered to have a "dynamothermal" origin related to the obduction and early transport history of the ophiolite (e.g., Ghent and Stout, 1981; Williams and Smyth, 1973). Inverted metamorphic gradients, calculated from metamorphic temperatures based on mineral assemblages and phase equilibria, approach or exceed −1000°C/km!

Less steep, inverted metamorphic gradients are observed in regional metamorphic terranes characterized by nappe tectonics resulting from continent-continent collision. Peak metamorphic temperatures decrease ~27°C/km from the garnet zone to biotite zone beneath the Tay Nappe in the Scottish Dalradian (Watkins, 1985). In the Himalayas, an IMG from sillimanite zone grading down to chlorite/biotite zone occurs in the Main Central Thrust zone (Sinha-Roy, 1982). Other IMGs are observed in Norway (Andreasson and Lagerblad, 1980), Greenland (Talbot, 1979), and the Appalachians (Thompson *et al.*, 1968). In some of these areas, such as Sulitjelma, Norway (Mason, 1984) and parts of New England (Rosenfeld, 1968), preexisting metamorphic isograds were probably overturned during later nappe development.

A third type of IMG, associated with subduction zones, occurs in the westernmost U.S. Cordillera; the best documented example is developed in the Pelona Schist beneath the Vincent thrust system in southern California (e.g., Ehlig, 1968; Graham and England, 1976). The first part of this paper summarizes the evidence for inverted metamorphic gradients in five different areas of the westernmost Cordillera: the Pelona-Orocopia Schists, Catalina Schist, Central Metamorphic Belt, South Fork Mountain Schist, and the Shuksan Suite (Fig. 34-1). Each of these metamorphic terranes is interpreted to have formed in the lower plate of a paleosubduction zone. Second, I present an overview of previous thermal models of IMGs followed by a new model which illustrates how an IMG may be *created* and *preserved* without requiring unreasonable uplift rates or frictional heating. Finally, I discuss some of the implications of this model for Cordilleran tectonics.

DOCUMENTATION OF CORDILLERAN IMGs

Pelona-Orocopia Schists

The best documented IMG in the western Cordillera is preserved in the Pelona-type schists (including the Orocopia, Chocolate Mountain, and Rand schists) of southern California and southwestern Arizona (see also Jacobson *et al.*, Chapter 35, this volume). The Pelona-

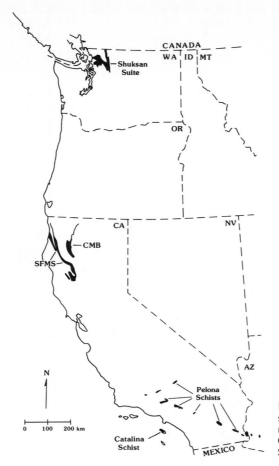

FIG. 34-1. Map of western United States showing the location of Cordilleran inverted metamorphic gradients. Abbreviations: CMB, Central Metamorphic Belt; SFMS, South Fork Mountain Schist.

Orocopia Schists are comprised predominantly of metagraywacke with lesser amounts of metabasite, Fe-Mn metachert, metaultramafic, marble and metapelite (Haxel and Dillion, 1978), suggesting a protolith of arc-derived sedimentary rocks and oceanic crust. Structurally, the Pelona-type schists lie in the lower plate of the Vincent thrust system (= Orocopia = Chocolate Mountain = Rand thrusts). Precambrian through Mesozoic gneisses and intrusive rocks comprise the upper plate of the thrust system. The Pelona Schists were metamorphosed during Paleocene and/or Late Cretaceous time (Ehlig, 1981); the age of the protoliths is unknown.

Metamorphic grade and grain size in the Pelona Schist increase structurally up toward the Vincent thrust in several well-studied areas, such as the Sierra Pelona (Ehlig, 1968). Cataclasis and retrograde metamorphism have affected rocks of the upper plate within a kilometer of the Vincent thrust (Haxel and Dillon, 1978). Peak metamorphic temperatures in the Pelona Schist are similar to the estimated temperatures of retrograde metamorphism (Graham and England, 1976), indicating that the thrust is *not* a metamorphic discontinuity. Metamorphic foliation and lineation are parallel between the schist and retrograde rocks. These observations strongly suggest that metamorphism of the Pelona Schist was genetically related to movement along the Vincent thrust system.

Whereas estimated temperatures and pressures of metamorphism of the Pelona Schist (discussed below) suggest formation in a subduction zone, there is a major controversy regarding the direction of underthrusting. Structural vergence data, including the famous overturned Narrows synform, suggest that the Pelona-type schists formed in a *west*-dipping subduction zone (model 1). The Pelona Schist protolith may have been deposited in a back-arc basin (1A) (Haxel and Dillon, 1978) or a marginal basin adjacent to North America (1B) (Ehlig, 1981). Closure of the back-arc basin (1A) or accretion of a relatively far-traveled microcontinent/arc (1B) resulted in westward subduction of the Pelona Schist protolith. On the other hand, metamorphism of the Pelona Schist occurred essentially contemporaneous with the Laramide orogeny, which is generally interpreted as a period of very shallow subduction. This has led several workers to propose that the Pelona Schist formed in an *east*-dipping subduction zone beneath continental crust (model 2) (e.g., Burchfiel and Davis, 1972). The lack of a well-defined suture zone marking the closure of the back-arc or marginal basin is cited as evidence in favor of model 2. The important point for this paper is that the Pelona Schist was metamorphosed in a subduction zone, regardless of the sense of vergence.

Ehlig (1968) first described an inverted metamorphic gradient in Pelona Schist from the Sierra Pelona area of the San Gabriel Mountains (Fig. 34-2). More detailed petrographic and microprobe studies by Graham and England (1976) and Graham and Powell (1984) have confirmed and quantified the IMG. The Pelona Schist is approximately 1000 m thick in the Sierra Pelona area (Ehlig, 1981). Structurally lower rocks were metamorphosed in the greenschist facies, as indicated by the mineral assemblage qtz + alb + epi + act + chl in mafic lithologies. The presence of barroisite instead of actinolite in more Fe-rich bulk compositions suggests high-pressure greenschist facies conditions. As one moves structurally up toward the Vincent thrust, metamorphic conditions increase from greenschist facies to albite-epidote amphibolite facies to lower amphibolite facies, separated by the garnet and oligoclase isograds, respectively. Schists in the amphibolite facies locally contain clinopyroxene, suggesting upper amphibolite facies conditions.

Graham and Powell (1984) estimated metamorphic pressures of 8–10 kb based on the phengite geobarometer (Powell and Evans, 1983) and the jadeite content of clinopyroxene. Peak metamorphic temperatures ranged from 480°C at the garnet isograd to 570°C at the oligoclase isograd to 620–650°C near the thrust, based on garnet-hornblende geothermometry (Graham and Powell, 1984) and observed mineral assemblages. The

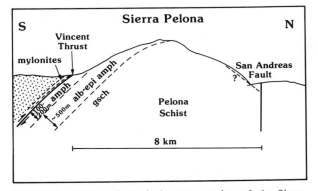

FIG. 34-2. Schematic geologic cross section of the Sierra Pelona area showing IMG preserved in the Pelona Schist after Graham and Powell (1984). Not to scale; maximum vertical relief is ~750 m.

garnet isograd lies 600–700 m below the Vincent thrust, which leads to an average inverted metamorphic gradient of −240°C/km close to the thrust.

Catalina Schist

The Catalina Schist exposed on Santa Catalina Island and the Palos Verdes Peninsula in the southern California Borderland consists of Mesozoic metamorphic rocks (see Sorensen, Chapter 36, this volume). On Santa Catalina Island, thrust sheets of blueschist, greenschist, and amphibolite lie structurally beneath serpentinite (Fig. 34-3) (Platt, 1975). High-grade tectonic blocks are present along postmetamorphic thrust faults that separate the different units. Metamorphism occurred in Cretaceous time based on K-Ar ages of 95–110 m.y. for the Catalina Amphibolite Unit reported by Suppe and Armstrong (1972). Platt (1975) considers the Catalina Schist to have formed beneath a hot hanging wall in a subduction zone.

The Catalina Schist represents an IMG ranging from amphibolite facies through high-pressure greenschist facies to blueschist facies. The Catalina Blueschist Unit consists of metagraywacke, metachert, mafic metavolcanic, and ultramafic rocks metamorphosed under blueschist facies conditions as indicated by the assemblage glauc + laws ± jad (Platt, 1975). Sedimentary textures, such as bedding planes and rare graded beds, are still preserved despite isoclinal folding and several subsequent deformation events (Platt, 1976).

The Catalina Greenschist Unit contains more mafic volcanics than the Blueschist Unit and metasedimentary rocks are dominantly pelitic (Platt, 1975). Metabasites contain assemblages characteristic of the epidote-amphibolite and greenschist facies (Sorensen, 1986); locally abundant glaucophane and crossite suggest metamorphism occurred in the high-pressure portion of these facies. Crossite is locally preserved in greenschist facies metasedimentary rocks. Intense deformation has largely destroyed original protolith textures. Lower greenschist facies retrograde metamorphism affected parts of the Greenschist Unit; locally pumpellyite is present (Platt, 1975).

The structurally highest Amphibolite Unit consists of green hornblende-zoisite schist, with minor brown hornblende-garnet schist, semipelitic schist, and garnet quartzite (Platt, 1975). All rock types are coarse-grained and thoroughly recrystallized; no original textures are preserved. Mineral assemblages suggest that amphibolite facies metamorphism was overprinted by a later retrograde metamorphism that produced sericite from plagioclase and chlorite from garnet and biotite.

A serpentinite unit consisting primarily of lizardite and chrysotile overlies the Amphibolite Unit. Platt (1975) considers the serpentinite to be part of the Amphibolite Unit because the contact between the two units is sharp, high-grade blocks are present in

Santa Catalina Island

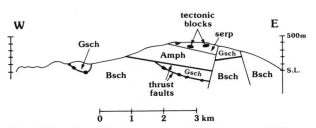

FIG. 34-3. Geologic cross section through Santa Catalina Island after Platt (1975). Serp, serpentinite; Amph, Amphibolite Unit; Gsch, Greenschist Unit; Bsch, Blueschist Unit.

the serpentinite, and the units share a common structural history. Serpentinization is presumed to have occurred later than amphibolite facies metamorphism because lizardite and chrysotile are unstable under such conditions. Locally, relatively high-grade assemblages containing anthophyllite are preserved.

Estimates of peak metamorphic conditions range from 300°C (Blueschist Unit) to 450°C (Greenschist Unit) to 600°C (Amphibolite Unit) under pressures of ~9 kb, based on observed mineral assemblages and garnet-clinopyroxene geothermometry (Platt, 1975). Recent work by Sorensen (1986) confirms Platt's (1975) general estimates of metamorphic conditions, with more recent phase equilibria studies suggesting peak temperatures ~50°C higher for each unit.

Metamorphic grade does not decrease systematically from amphibolite to blueschist facies, but rather the different units are separated by subhorizontal, postmetamorphic faults. It is possible that the Catalina Schist does not represent an IMG, but instead reflects a tectonic stacking of distinctly different metamorphic terranes formed far apart from one another. However, metamorphic grade does decrease downward *within* the Greenschist Unit (Sorensen, 1986) suggesting that displacements along the postmetamorphic faults may not be large. Estimating the thickness of the Catalina IMG is complicated by the thrust faults. Platt's (1975) cross section yields a thickness of 900 m from the serpentinite/amphibolite contact to the greenschist/blueschist contact suggesting a steep inverted metamorphic gradient of approximately −330°C/km. This estimate is considered a maximum because of the unknown thickness of rocks removed by postmetamorphic faulting.

Central Metamorphic Belt

The Central Metamorphic Belt (CMB) forms the lower plate of the Trinity (= Bully Choop) thrust system in the Klamath province, northern California (Davis *et al.*, 1965) (Fig. 34-4). Metabasalts and metasediments record greenschist/amphibolite facies metamorphism. Rb-Sr, K-Ar, and ^{40}Ar-^{39}Ar radiometric dates range from 370 to 399 m.y., suggesting Devonian metamorphism (Cashman, 1980; Hotz, 1977; Lanphere *et al.*, 1968). The Trinity peridotite and intruded gabbros overlie the CMB; both peridotite and intrusive gabbros yield Ordovician Rb-Sr and Sm-Nd crystallization ages (Jacobson *et al.*, 1984; Lanphere *et al.*, 1968; Mattinson and Hopson, 1972). The Trinity peridotite forms the structural base of the Eastern Klamath plate, an Ordovician through Jurassic section of dominantly volcaniclastic and volcanic rocks. Murray and Condie (1973) recognized four major pulses of volcanism in the Eastern Klamath plate, including a Late Silurian-Devonian pulse represented by the Copley Greenstone and Balaklala Rhyolite. The CMB has been interpreted by Burchfiel and Davis (1975), Potter *et al.* (1977), and Irwin (1977, 1981) to represent a crustal fragment accreted to the eastern Klamaths in an east-dipping Devonian subduction zone. Transitional blueschist/greenschist facies metamorphic rocks of the Permo-Triassic Stuart Fork Formation structurally underlie the CMB (Davis *et al.*, 1965; Lanphere *et al.*, 1968) and represent the adjacent, younger accretionary terrane.

The CMB consists of metabasalts (Salmon Hornblende Schist and Grouse Ridge Hornblende Gneiss) overlain by calc-schists/marbles, impure quartzites, and minor amphibolites (Grouse Ridge metasediments = Abrams Mica Schist) (Davis and Lipman, 1962; Irwin, 1960). Prior to metamorphism, the CMB may have been a fragment of oceanic crust composed of basalt overlain by carbonate and sandstone and/or chert. Metamorphic recrystallization completely destroyed all primary igneous and sedimentary textures in the CMB protoliths. Metamorphic foliation within the CMB and overlying Trinity peridotite generally parallels the Trinity thrust, where not affected by later north-south upright folding. Grain size within metabasalt increases toward the thrust. Extensive hydration of the Trinity peridotite along the Trinity thrust resulted from the infiltration of fluids

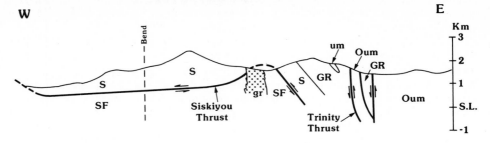

Oum - Trinity mafic/ultramafic complex (Ord.)   after Davis et al. (1979)

CMB { GR - Grouse Ridge Fm.
      S - Salmon Hornblende Schist } (Dev)

SF - Stuart Fork Fm. (Pm-T)

gr - granite

FIG. 34-4.   Geologic cross section through the eastern Klamath Mountains
after Davis *et al.* (1979).

derived from prograde metamorphic reactions occurring in the underlying CMB (Peacock, 1987). These observations strongly suggest that metamorphism of the CMB occurred contemporaneously with thrusting along the Trinity thrust.

Lipman (1964) and Davis *et al.* (1965) first recognized an inverted metamorphic gradient in the CMB, although at that time they included the lower grade and much younger Stuart Fork Formation in the CMB. Holdaway (1965) reported a poorly defined albite → oligoclase transition structurally upward from west to east. Metabasalt mineral assemblages reflect decreasing metamorphic grade from lower amphibolite facies at the Trinity thrust through albite-epidote amphibolite to upper greenschist facies at the base of the CMB. Peak metamorphic temperatures deduced from the observed mineral assemblages and very limited garnet-hornblende geothermometry decrease from $650 \pm 50°C$ at the Trinity thrust to $450 \pm 50°C$ at the base of the CMB (Peacock, 1985). Some of the greenschist facies metamorphism at the base of the CMB may be related to retrogression associated with the much younger underthrusting of the Stuart Fork Formation. Mineral assemblages and amphibole mineral chemistry constrain metamorphic pressures to have been 4–8 kb at 450–650°C (Peacock, 1985).

The original thickness of the CMB is difficult to determine because of later accretion events in the Klamath Mountains. Estimates of the present-day structural thickness range from 1 to 5 km (Davis *et al.*, 1965, 1979; Irwin, 1981). A semiquantitative estimate of the inverted metamorphic gradient is therefore on the order of $-100°C/km$.

South Fork Mountain Schist

The easternmost portion of the Franciscan Complex in northern California is termed the South Fork Mountain Schist (Fig. 34-5) (see also Blake *et al.*, Chapter 38, this volume. The Coast Range thrust separates the South Fork Mountain Schist from structurally higher Klamath Province and Great Valley sequence rocks to the east. K-Ar, $^{40}Ar/^{39}Ar$, and Rb-Sr dating summarized by Lanphere *et al.* (1978) suggests an Early Cretaceous (115–120 m.y.) metamorphism.

Blake *et al.* (1967) first defined the South Fork Mountain Schist on the basis of

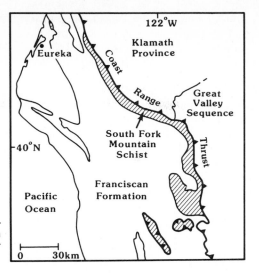

FIG. 34-5. Geologic map showing the distribution of the South Fork Mountain Schist in northern California modified after Bishop (1977).

greater textural development (schistosity) as compared to structurally lower Franciscan rocks. The contact between the South Fork Mountain Schist and "normal" Franciscan rocks is gradational in some places (Blake *et al.*, 1967; Wood, 1971) and a fault in others (Suppe, 1973). The South Fork Mountain Schist protolith contained more mudstone and mafic volcanics than the predominantly metagraywacke Franciscan Complex to the west (Bishop, 1977). The maximum thickness of the South Fork Mountain Schist is ~2–5 km in the North Yolla Bolly Mountain area (Blake *et al.*, 1967).

Blake *et al.* (1967) documented an upward increase in metamorphism within eastern Franciscan rocks based on mineralogical and textural changes. Incipiently metamorphosed graywacke, basalt, and chert of textural zone 1 becomes increasingly cataclasized and foliated toward the Coast Range thrust. Original sedimentary textures and pillow lavas are preserved in textural zone 1 rocks. Metagraywacke contains the assemblage qtz + alb + musc + chl + pump; metabasalt also contains pumpellyite. Calcite is common, although metamorphic aragonite is also observed.

The transition from textural zone 1 to textural zone 2 is marked by the appearance of platy cleavage caused by incipient cataclasis in metagraywacke. Metavolcanic rocks appear sheared in thin section, but massive in outcrop. Metagraywacke contains the diagnostic minerals lawsonite and aragonite, whereas only pumpellyite is found in metabasalt (Blake *et al.*, 1967).

Textural zone 3 rocks are distinguished by the complete destruction of original sedimentary textures and the development of metamorphic segregation banding (Blake *et al.*, 1967). Both metagraywacke and metabasalt appear schistose (to gneissose) in outcrop and thin section. Coarse-grained lawsonite and rare aragonite rimmed by calcite are observed in metagraywacke; garnet and epidote are observed in more mafic schists. Metabasalts contain alb + chl + act + epi with minor stilp, pump, musc, sph, and qtz. Local bluish layers contain crossite. Jadeitic pyroxene is observed in some samples (Brown and Ghent, 1983).

The IMG preserved in the South Fork Mountain Schist of northern California is quite distinct from the Cordilleran IMGs discussed above. Brown and Ghent (1983) estimate metamorphic conditions of 250–300°C at ~7 kb for blueschist facies rocks in the Black Butte and Ball Rock areas based on mineral equilibria. Textural zone 1 rocks may reflect lower temperatures, pressures, and/or differential stresses. In the previous IMGs, the metamorphic gradient developed primarily as a result of differences in peak metamorphic temperatures, whereas in the South Fork Mountain Schist, pressure gradients may have been important (e.g., Blake *et al.*, 1967; Ernst, 1971). Blake *et al.* (1967) consider the metamorphic zonations to result from fluid overpressures increasing toward the Coast Range Thrust. Such a model is considered unlikely by the author because cracks should form in a metamorphic rock, and fluids should drain, as fluid pressures cause the tensile strength of the rock to be exceeded. If Zone 1 rocks represent lower temperature conditions, the preserved inverted metamorphic gradient is on the order of $-30°C/km$.

The thermal model presented below does not attempt to describe the formation of the IMG preserved in the South Fork Mountain Schist. Its discussion here is included for the sake of completeness and because it is one of the earliest inverted metamorphic gradients described from the Cordillera.

Shuksan Suite

Brown *et al.* (1982) describe geologic relationships within the Shuksan Suite of the North Cascades that appear to represent a structurally disrupted IMG. The Iron Mountain–Gee Point area studied by Brown *et al.* (1982) is a part of a more than 150-km-long Mesozoic blueschist terrane exposed along the west flank of the North Cascades. Faults separate the Shuksan suite from older rocks to the west and east.

In the area studied by Brown *et al.* (1982), serpentinite, amphibolite, barroisite schist, rare eclogite, and "blackwall" metasomatic rock occur within regional blueschist facies rocks. Ultramafic lithologies occur as lenses up to hundreds of meters wide and 2 km long, consisting of antigorite with minor amounts of magnetite + brucite, or magnesite. Relict brown A1-chromite suggests an Alpine peridotite protolith.

The amphibolite unit is well foliated, generally gneissic, and consists of interlayered garnet amphibolite, garnet hornblende schist, coarse muscovite schist, and Fe-Mn quartzose schist (Brown *et al.*, 1982). Amphibolite facies mineral assemblages (hbl + gar + epi + musc + qtz + alb + rut ± sph) are commonly replaced by lower-grade blueschist facies minerals (cross, laws, pump, chl, alb, sph) associated with a secondary foliation. Amphibolite generally crops out next to serpentinite, with "blackwall" metasomatic rock formed between the two.

Barroisite schist consists of coarse Na-amphibole (glaucophane to crossite) + musc + epi with lesser amounts of qtz, alb, chl, rut, and/or sph (Brown *et al.*, 1982). Omphacite is common, but does not coexist with quartz. Retrogression resulted in coarse barroisite ⇝ cross or glauc and omphacite ⇝ pump and/or cross or glauc. Rare eclogitic rocks contain the assemblage omphacite + gar + Na-amph + alb + epi + musc.

Mineral assemblages and geothermometry (gar-cpx, gar-hbl) suggest peak metamorphic conditions ranging from 450°C (barroisite schists) to 700–800°C (amphibolites) at pressures of 7–14 kb (Brown *et al.*, 1982). K-Ar and Rb-Sr dating yields metamorphic ages of 148 ± 5 to 164 ± 6 m.y. for amphibolite and barroisite schist (Brown *et al.*, 1982). Oxygen isotope data and mineral compositions suggest peak metamorphic temperatures of 360–400°C at 7–8 kb for the regional blueschists (Brown *et al.*, 1982). Radiometric dating of blueschists yields metamorphic ages of 129 ± 5 m.y., substantially postdating the high-grade metamorphic event.

A complete section from serpentinite through amphibolite to blueschist is not preserved in the Shuksan Suite; therefore, no estimates of the thickness, and hence metamorphic gradient, can be made. The close proximity of the amphibolites to the serpentinite led Brown *et al.* (1972) to propose that the high-grade metamorphism occurred beneath a hot hanging wall in a subduction zone. Their model is supported by the younger radiometric ages for the regional blueschists as compared to the higher grade units. Such a model is very similar to that proposed by Platt (1975) for the formation of the Catalina Schist.

Summary of Cordilleran IMGs

Table 34-1 presents a summary of the metamorphic conditions preserved in each of the IMGs discussed above. With the exception of the South Fork Mountain Schist, conditions of metamorphism are quite similar among the IMGs. Each of the IMGs has been interpreted to represent the lower plate of a paleosubduction zone by previous workers. The high pressures of metamorphism (7–10 kb) for four out of five of the IMGs support this interpretation. The Catalina Schist, Central Metamorphic Belt, and South Fork Mountain Schist developed beneath ultramafic hanging walls and the highest-grade rocks in the Shuksan Suite occur adjacent to serpentinite. Only the Pelona-type schists developed beneath "granitic" crystalline rocks. Platt (1975) and Brown *et al.* (1982), among others, have presented models whereby an inverted metamorphic gradient forms early in the subduction process. My thermal model presented below supports such a scenario.

Thermal Modeling of IMGs

Before examining different thermal models for the formation of IMGs, let us discuss the relationship between a crustal geotherm and the observed metamorphic geotherm. In stable cratons, temperature increases monotonically with depth in the crust; the absolute temperatures are primarily a function of the thermal properties of the rock pile, the amount of radioactive heating, and the heat flux into the base of the crust from the mantle. In stable cratonic areas, temperature at a given depth is essentially constant over time; that is, there is a steady-state geotherm. The slope of the geotherm is called the geothermal gradient. In tectonically active areas, such as subduction zones, temperature varies over time and in general a steady-state geotherm is never achieved.

Consider the *P-T* path of a typical metamorphic rock shown in Fig. 34-6. At a given time and depth, the metamorphic rock is, by definition, at the temperture dictated by the geotherm *for that time.* The geotherm changes with time, however, reflecting different tectonic processes such as burial and erosion. Most metamorphic reactions are temperature sensitive; thus the mineral assemblage of a metamorphic rock generally reflects the highest temperature the rock has experienced. In general, the mineral assemblage will reflect the confining pressure at T_{max} (designated P_{Tmax}), which is substantially lower

TABLE 34-1. Estimated Metamorphic Conditions, Structural Thicknesses, and Calculated Metamorphic Gradients of Cordilleran Inverted Metamorphic Gradients[a]

Geologic Unit	T (°C)	P (kb)	Thickness	Metamorphic Gradient
Pelona Schist	400 → 650	8–10	1 km	−240°C/km
Catalina Schist	350 → 650	8–10	900 m	−330°C/km[b]
Central Metamorphic Belt	450 → 650	4–8	1–5 km	−100°C/km
South Fork Mountain Schist	200 → 300	~7	2–5 km	−30°C/km
Shuksan Suite	450 → 750	7–14	Disrupted	−

[a]See the text for references.
[b]Complicated by several thrust faults.

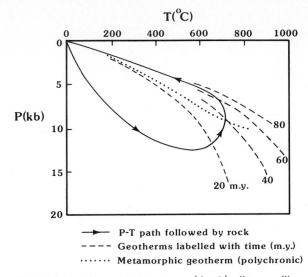

T(°C)

FIG. 34-6. Temperature-pressure (depth) diagram illustrating the relationship between a geotherm, a rock's *P-T* path, and the metamorphic geotherm (metamorphic field gradient), after England and Richardson (1977).

than the maximum pressure experienced by the rock. The observed metamorphic geotherm represents the locus of T_{max}-P_{Tmax} points for a suite of metamorphic rocks (Fig. 34-6) exposed at the earth's surface. In general, each rock reaches its T_{max} at a different time; thus a metamorphic geotherm is polychronic rather than representing a single geotherm (England and Richardson, 1977). The slope of a metamorphic geotherm is called the metamorphic gradient, or metamorphic field gradient, and can be calculated from the observed variation in *P-T* conditions within a metamorphic terrane (Spear *et al.*, 1984). In the discussion below, keep in mind that for an inverted thermal gradient to be *preserved* in the metamorphic record, temperatures in the rock column must not reach values greater than the original inverted thermal gradient, or else new mineral assemblages will form and the IMG will be modified or erased.

Thermal models of tectonic processes based on heat transfer theory place useful constraints on the formation of metamorphic terranes such as IMGs. While heat transfer theory has been known for many years, only with the development of mainframe computers has numerical modeling of complex geologic problems become possible. The heat transfer equation for a moving medium is

$$\frac{\partial T}{\partial t} = \kappa \, \nabla^2 T - \mathbf{v} \, \text{grad} \, T + \frac{A}{\rho C} \tag{1}$$

(Carslaw and Jaeger, 1959) which in one-dimension, with z = depth, reduces to

$$\frac{\partial T}{\partial t} = \kappa \frac{\partial^2 T}{\partial z^2} - v_z \frac{\partial T}{\partial z} + \frac{A}{\rho C} \tag{2}$$

$$\underset{\text{conduction}}{} \quad \underset{\text{advection}}{} \quad \underset{\substack{\text{heat} \\ \text{sources/} \\ \text{sinks}}}{}$$

where T = temperature (K)

κ = thermal diffusivity (m²/sec)

v_z = velocity of medium in the z-direction (m/sec)

A = volumetric heat production (W/m³)

ρ = density (kg/m³)

C = specific heat (J/kg-K)

The change in temperature with time depends on three different types of terms: conduction, advection, and heat sources/sinks. Examples of heat advection include erosion (denudation) and metamorphic fluid flow. Frictional heating, radioactive decay, and metamorphic reactions are examples of heat sources and sinks.

Equation (2) may be solved analytically for simple geometries. For more complex scenarios as in this paper, a numerical, finite-difference method is commonly employed. Either method requires that an initial condition and boundary conditions (B.C.s) be specified as shown in Fig. 34-7. In general, a constant-temperature (0°C) boundary condition is used at the surface and a constant heat flux across the basal boundary. [In the literature, heat flux (W/m²) is commonly expressed in heat flow units (H.F.U.), where 1 H.F.U. = 10^{-6} cal/cm²-sec = 0.042 W/m².] The shape of the initial geotherm [$T(z)$ at t_0] depends on the particular model under consideration. In this paper I am modeling a thrust fault 20 km below the surface, suitable for the pressure conditions of the Central Metamorphic Belt in the Klamath Mountains. I have used an array spacing (Δz) of 500 m, which requires time steps (Δt) of 2500 years to avoid numerical instabilities associated with the explicit finite-difference method.

Previous Work

Oxburgh and Turcotte (1974) used the Péclet number to demonstrate that thrusting causes a major perturbation of the geotherm. The Péclet number is a dimensionless ratio between the rate of heat emplacement (e.g., thrusting) and the rate of heat conduction, defined as

$$\text{Péclet number} = \frac{v \times l}{\kappa} \tag{3}$$

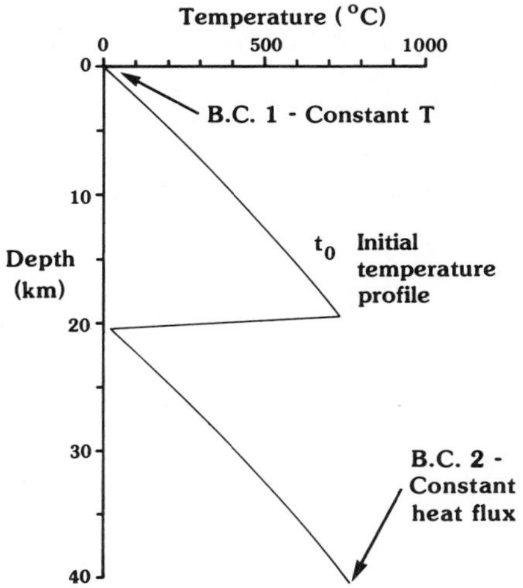

FIG. 34-7. Temperature-depth (T-z) diagram showing initial "sawtooth" geotherm for upper and lower plates separated by a thrust fault at 20 km depth. Initial geotherm: 10-m.y.-old oceanic geotherm of Turcotte and Schubert (1982, p. 164, Eq. 4-125) repeated above and below the thrust fault. Boundary conditions (B.C.): T(surface) = 0°C, constant heat flux (z = 40 km) = 0.089 W/m². Thermal diffusivity (κ) = 1 mm²/sec.

where v = velocity of thrusting (m/sec)

l = characteristic length scale (thickness of thrust sheet) (m)

κ = average thermal diffusivity (m^2/sec)

Péclet numbers much greater than 1 mean that heat is emplaced faster than it is conducted away. For plate tectonic rates of emplacement (3 cm/yr), a 20-km-thick thrust sheet, and $\kappa = 1$ mm^2/sec, the Péclet number is approximately 20. Thus, Oxburgh and Turcotte (1974) chose to model the thermal effects of thrust faulting by considering an instantaneously emplaced thrust sheet. The initial "sawtooth" geotherm shown in Fig. 34-7 is derived simply by stacking a 20-km thrust sheet, with its associated geotherm, on top of the original geotherm. The original geotherm shown is a 10-m.y. oceanic geotherm of Turcotte and Schubert (1982, p. 164, Eq. 4–125). The slight curvature of the geotherm results from ongoing cooling of the oceanic lithosphere rather than radioactive decay, which is assumed to be negligible.

Figure 34-8 illustrates the thermal relaxation of the "sawtooth" geotherm as a function of time. Initially, heat is conducted rapidly downward because of the steep inverted thermal gradients in the vicinity of the thrust. After a period of 1 m.y., there is no longer an inverted thermal gradient at the thrust; temperatures increase monotonically with depth. Over a period of tens of millions of years, the geotherm approaches the steady-state geotherm defined by the boundary conditions.

To preserve the inverted metamorphic gradient created in the lower plate within the first few hundred thousand years, geologically unreasonable uplift rates of greater than 10 mm/yr are required. So whereas the instantaneous thrusting model *creates* an IMG, rapid conduction of heat downward effectively *erases* the IMG within 1 m.y. In addition, note that the maximum temperature achieved in the lower plate is one-half the initial temperature of the base of the upper plate. Thus, unreasonably high upper-plate temperatures are required to produce the metamorphic temperatures >500°C recorded in many of the IMGs discussed above.

The inadequacy of the model above led Graham and England (1976) to propose that shear (frictional) heating along the Vincent thrust system was responsible for the formation of the Pelona Schist IMG. Their thermal model starts with a relatively cool "sawtooth" geotherm, as shown in Fig. 34-9. A heat source is then allowed to operate in the thrust zone for several million years. The three curves labeled t_f in Fig. 34-9 show the shape of the geotherm when the temperature at the thrust reached 600°C for different

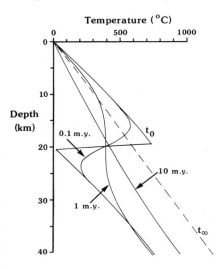

FIG. 34-8. Temperature-depth diagram for upper and lower plates illustrating the thermal relaxation of the "sawtooth" geotherm shown in Fig. 34-7. Individual curves are labeled with time elapsed after the end of thrusting. Also shown is the steady-state solution (t_∞) that is approached over tens of m.y.

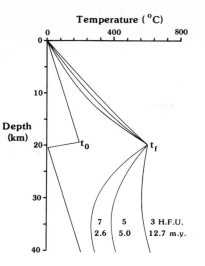

FIG. 34-9. Temperature-depth diagram for upper and lower plates showing effects of thrust heating on a cool initial "sawtooth" geotherm (t_0). Thrust heating was allowed to operate until the temperature at the thrust reached 600°C. Three different curves (labeled t_f) illustrate the shape of the geotherm resulting from 3, 5, and 7 H.F.U. of thrust heating. Each geotherm is labeled with the amount of thrust heating and the time required for the temperature of the thrust to reach 600°C. The thrust is at a depth of 20 km. Note that higher amounts of thrust heating result in more rapid heating and steeper inverted thermal gradients. Initial condition: "cool" sawtooth geotherm with $dT/dt = 10°C/km$. Boundary conditions (B.C.): $T(surface) = 0°C$, constant heat flux $(z = 40$ km$) = 0.030$ W/m². Thermal diffusivity $(\kappa) = 1$ mm²/sec.

magnitudes of thrust heating. Stronger heat sources result in more rapid heating and steeper inverted thermal gradients. After thrusting (and therefore thrust heating) ceases, temperatures relax toward the lower steady-state geotherm, thereby preserving the inverted metamorphic gradient created in the lower plate.

The major problem with this model is the nature of the heat source operating at the thrust. The most obvious heat source chosen by Graham and England (1976) and others is shear, or frictional, heating, which is defined as

$$q = u \times \tau \tag{4}$$

where $q =$ mean rate of heat production per unit area (W/m²)
$u =$ rate of movement along the surface (m/sec)
$\tau =$ shear stress across the surface (Pa $= 10^{-5}$ bar)
(Turcotte and Schubert, 1982, p. 189).

To generate a 5-H.F.U. heat source by frictional heating, a constant shear stress of 1.2 kb is required for a convergence rate of 5 cm/yr. Such a shear stress appears to be one to two orders of magnitude greater than estimates of rock strengths at elevated temperatures and pressures (e.g., Brune, 1970; Sleep, 1975), especially for rocks undergoing dehydration reactions (Heard and Rubey, 1966; Raleigh and Paterson, 1965).

Alternatively, fluids flowing up the thrust zone could provide the necessary heat source in the one-dimensional model. Thermal gradients parallel to the thrust are less than 10°C/km (Anderson *et al.*, 1978), much lower than those perpendicular to the thrust. Using 10°C/km, an up-dip Darcy velocity of 1.2×10^{-8} m³/m²-sec is required to produce a 5-H.F.U. heat source. This volumetric fluid flux per unit area was calculated for a 500-m-wide thrust zone (Δz in my model); narrower zones of fluid flow require correspondingly higher Darcy velocities. Integrated fluid/rock ratios of greater than 10^5 to 1 over a period of 1 m.y. are implied for rocks in the thrust zone, an incredibly large ratio even for a convecting system.

Thus, while thermal models invoking a heat source operating in the thrust zone can create an inverted metamorphic gradient, neither frictional heating nor upward fluid flow along the thrust is an adequate heat source based on reasonable estimates of the parameters involved.

Proposed Model

Ideally, a thermal model of inverted metamorphic gradients would be two-dimensional and have a grid size on the order of 100 m. Large amounts of computer time are required for such a model; timing studies are presently under way. Prior to the completion of a fully two-dimensional model, an understanding of the problem can be obtained by a set of one-dimensional models that approximate the two-dimensional process.

Figure 34-10 is a highly schematic cross section of a subduction zone just after the initiation of subduction. The geotherm on both sides of the initial subduction "break" is the same, a 10-m.y. oceanic geotherm (Oxburgh and Turcotte, 1974) in my modeling. At the onset of subduction, conduction of heat downward from the hanging wall will warm up the top portion of the descending slab. An inverted thermal gradient will be created in the lower plate that is a function of the angle of subduction, convergence rate, initial geotherm, and the thermal properties of the rocks. The IMG created in the top portion of the descending slab will be preserved *if* it is accreted to the upper plate. Continued subduction of lithosphere will remove heat from the upper plate, effectively refrigerating the hanging wall and any previously accreted IMG. After the cessation of subduction, temperatures in both plates will relax toward the steady-state geotherm. Below, I discuss three separate models which approximate (1) the warming of the lower plate by conduction of heat downward during the initiation of underflow, (2) the refrigeration of the upper plate by continued descent, and (3) the thermal relaxation toward the steady-state geotherm after subduction ceases.

Heating upon descent The heating of the lower plate during descent can be approximated by considering an initial oceanic geotherm with a changing upper boundary condition. Instead of a constant-temperature condition, the top of the lower plate will be in contact with rock of increasing higher temperature as it is subducted. During early stages of subduction, this temperature is simply the initial geotherm of the upper plate. With continued subduction upper-plate temperatures, at any given depth, will decrease because of the

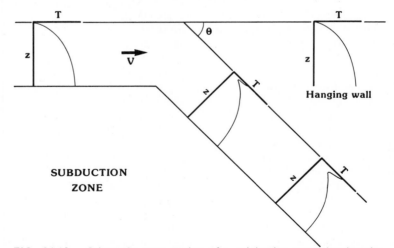

FIG. 34-10. Schematic cross section of a subduction zone showing the heating up of the subducting slab as it descends beneath a hot hanging wall during the initiation of subduction. The exact shape of the lower-plate geotherm will be a function of the initial upper- and lower-plate geotherms, the thermal properties of the rocks, the rate of plate convergence, and the angle of subduction (θ).

conduction of heat downward into previously subducted material. Thus, the schematic model shown in Fig. 34-10 is valid only during the initial stages of subduction, perhaps the first 1 m.y. or so.

Figure 34-11 illustrates the shape of the geotherm in the *lower* plate after 400,000 years of convergence at 10 cm/yr and a subduction angle of 30°. At this time 20 km of upper plate overlies the lower plate; that is, the thrust is now at a depth of 20 km. Only the top 10 km of the lower plate have been affected by heat conducted downward. Thermal gradients within a kilometer of the thrust are −160°C/km, the same order of magnitude as those observed in the Cordilleran IMGs discussed above. Slower convergence rates and shallower subduction angles result in less steep inverted thermal gradients. The temperature at the top of the lower plate is equal to the temperature at the base of the upper plate, in contrast to that predicted by the instantaneous ("sawtooth") model. In the proposed model, the lower plate is heated continuously as it subducts to a depth of 20 km, whereas in the "sawtooth" model heating commences after the lower plate is instantaneously emplaced at 20 km depth.

An inverted *metamorphic* gradient will be created in the lower plate upon its descent during the early stages of subduction. My model requires that this IMG is welded onto, or accreted, to the base of the upper plate as proposed by Brown *et al.* (1972) and Cloos (1985), among others. The accretion process is poorly understood. It must become easier to decouple the rock several kilometers into the lower plate rather than to continue movement along the preexisting thrust fault; hence, the thrust steps down with time. The lower temperature of 300–400°C preserved near the base of several IMGs suggests that dehydration reactions occurring at the onset of greenschist facies metamorphism may substantially decrease the strength of rocks *within* the lower plate.

Refrigeration Continued subduction removes heat from the upper plate because of the conduction of heat downward into the lower plate. A crude model of this process can be made by considering infinitely fast subduction. In this extreme limiting case, the upper plate will be always in contact with a 0°C lower plate, the temperature at the top of initial lower plate geotherm. This can be modeled by imposing a constant-temperature (0°C) basal boundary condition. Figure 34-12 illustrates the changing shape of the geotherm with time in a 20-km-thick upper plate. Most of the heat is removed from the upper plate in 1 m.y.; by 5 m.y. essentially all of the heat has been removed.

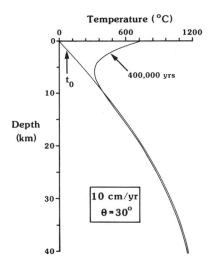

FIG. 34-11. Temperature-depth diagram for the *lower* plate showing the extent of heating during descent beneath a hot hanging wall. After 400,000 years the top of the lower plate is at a depth of 20 km, the thrust depth modeled in this paper. Thermal gradients within a kilometer of the thrust are −160°C/km. The two curves are not coincident below 10 km because of the ongoing cooling of the initial 10-m.y. oceanic geotherm. Initial condition: 10-m.y. oceanic geotherm of Turcotte and Schubert (1982). Boundary conditions: T(surface) changes along the 10-m.y. geotherm, constant heat flux ($z = 40$ km) = 0.034 W/m². Convergence rate = 10 cm/yr; angle of subduction (θ) = 30°; thermal diffusivity (κ) = 1 mm²/sec.

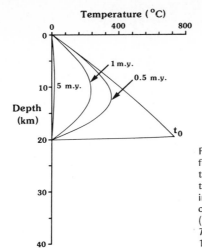

Temperature (°C)

FIG. 34-12. Temperature-depth diagram for the *upper* plate illustrating the refrigeration effect of continued underthrusting for the end-member case where subduction is infinitely fast. Initial condition: 10-m.y. oceanic geotherm of Turcotte and Schubert (1982). Boundary conditions: T(surface) = T(20 km) = 0°C. Thermal diffusivity (κ) = 1 mm²/sec.

A more accurate picture requires two-dimensional modeling, such as that reported by Dumitru and Cloos (1986). They found that the inboard region (15–35 km thick) of the overriding plate requires a few tens of m.y. to cool after the initiation of fast subduction. For a fast convergence rate of 10 cm/yr operating for 20 m.y., 2000 km of oceanic lithosphere will be consumed. More detailed two-dimensional thermal models of subduction are presently being developed. In the meantime, a reasonable estimate is that a minimum of ~1000 km of oceanic lithosphere needs to be subducted in order to refrigerate the overriding plate.

Postsubduction thermal relaxation After subduction ceases, the geotherm will resemble the initial geotherm illustrated in Fig. 34-13. A constant temperature of 0°C in the upper 20 km represents an upper plate that has had all of its heat removed by prior underthrusting. Temperatures in the lower plate are represented by the standard 10-m.y. oceanic geotherm. Temperatures in the model will steadily increase with time as the geotherm relaxes toward the steady-state geotherm defined by the boundary conditions. Inasmuch as we are modeling oceanic lithosphere, radioactive heat sources are not important in this model. Thus, all of the heat required to raise the geotherm flows into the base of the model from below.

The steady-state geotherm is approached over a period of tens of million years. It takes 20 m.y. for temperatures several kilometers beneath the thrust to reach 500°C. Only moderate uplift rates of less than 0.1 mm/yr are required to ensure that temperatures in the previously accreted IMG never reach 500°C. Such uplift rates are comparable to those required to preserve the high-pressure metamorphic signature (England and Richardson, 1977) recorded in many of the Cordilleran IMGs. The prior removal of heat from the upper plate effectively refrigerates the accreted IMG.

Summary and Implications

The model presented above *creates* and *preserves* an inverted metamorphic gradient without requiring unreasonable uplift rates or frictional heating. The IMG is formed during the initiation of subduction as the downgoing slab comes into contact with hot upper plate material. Through processes poorly understood at present, the IMG detaches from the downgoing slab and is welded (accreted) to the base of the upper plate. Continued

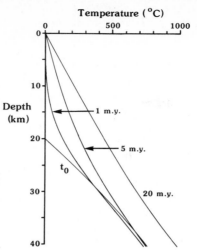

Temperature (°C)

FIG. 34-13. Temperature-depth diagram for both upper and lower plates illustrating postsubduction thermal relaxation. Initial condition: 0°C upper plate, 10-m.y. oceanic geotherm lower plate. Boundary conditions: T(surface) = 0°C, constant heat flux (z = 40 km) = 0.089 W/m². Thermal diffusivity (κ) = 1 mm²/sec.

subduction removes most of the heat from the upper plate and accreted IMG over several million years. After subduction ceases, only moderate uplift rates of less than 0.1 mm/year are required to ensure that the IMG is preserved.

The model presented above has two important consequences for Cordilleran IMGs. First, the highest-grade rocks of the IMG form during the earliest stages of subduction and therefore effectively date the initiation of subduction. Later blueschist metamorphism should postdate high-grade metamorphism by millions of years. In the Shuksan Suite, metamorphic ages of high-grade amphibolite and barroisite schist *predate* regional blueschist ages by 19–35 m.y. The same relationship appears to hold for the Franciscan Complex, where regional blueschist rocks yield younger metamorphic ages than "high-grade" blueschist and eclogite blocks (e.g., Coleman and Lanphere, 1971; McDowell *et al.*, 1984). Second, IMGs mark the sites of major lithospheric consumption where a minimum of ~1000 km of oceanic lithosphere must have been subducted in order to refrigerate the IMG and upper plate. Thus, if the Pelona-type schists formed in a back-arc basin, the arc was at least 1000 km offshore.

Future work will involve the development of a realistic two-dimensional model to better constrain the amount of subduction required to refrigerate the IMG. Research will also be directed toward a better understanding of the processes responsible for the detachment and welding of the IMG to the upper plate.

ACKNOWLEDGMENTS

The thermal modeling presented here is an outgrowth of stimulating formal and informal discussions with many people, including Mark Barton, Ed Bolton, Don DePaolo, Tom Drake, Gary Ernst, John Goodge, Vicki Hansen, Brooks Hanson, Bruce Nelson, John Rosenfeld, and Dave Williams. One day I hope to convince all of them that my model is both realistic and plausible. I thank Gary Ernst for a detailed review of the manuscript and Susan Selkirk for her help in drafting the figures. This research began as part of a Ph.D. dissertation at U.C.L.A. under the guidance of Gary Ernst (NSF grant EAR82-11625 to Ernst) and continued at Arizona State University with research "startup" funds.

REFERENCES

Anderson, R. N., DeLong, S., and Schwarz, W. M., 1978, Thermal model for subduction with dehydration in the down going slab: *J. Geology*, v. 86, p. 731–739.

Andreasson, P. G., and Lagerblad, B., 1980, Occurrence and significance of inverted metamorphic gradients in the western Scandinavian Caledonides: *J. Geol. Soc. London*, v. 137, p. 217–230.

Bishop, D. G., 1977, South Fork Mountain Schist at Black Butte and Cottonwood Creek, northern California: *Geology*, v. 5, p. 595–599.

Blake, M. C., Jr., Irwin, W. P., and Coleman, R. G., 1967, Upside-down metamorphic zonation, blueschist facies, along a regional thrust in California and Oregon: *U.S. Geol. Survey Prof. Paper 575-C*, p. C1–C9.

Brown, E. H., and Ghent, E. D., 1983, Mineralogy and phase relations in the blueschist facies of the Black Butte and Ball Rock areas, northern California Coast Ranges: *Amer. Mineralogist*, v. 68, p. 365–372.

—— , Wilson, D. L., Armstrong, R. L., and Harakal, J. E., 1982, Petrologic, structural, and age relations of serpentinite, amphibolite, and blueschist in the Shuksan Suite of the Iron Mountain — Gee Point area, North Cascades, Washington: *Geol. Soc. America Bull.*, v. 93, p. 1087–1098.

Brune, J. N., 1970, Tectonic stress and the spectra of seismic shear waves from earthquakes: *J. Geophys. Res.*, v. 75, p. 4997–5009.

Burchfiel, B. C., and Davis, G. A., 1972, Structural framework and evolution of the southern part of the Cordilleran orogen, western United States: *Amer. J. Sci.*, v. 272, p. 97–118.

—— , and Davis, G. A., 1975, Nature and controls of Cordilleran orogenesis, western United States — Extensions of an earlier synthesis: *Amer. J. Sci.*, v. 275–A, p. 363–396.

Carslaw, H. S., and Jaeger, J. C., 1959, *Conduction of Heat in Solids:* Oxford, Clarendon Press.

Cashman, S. M., 1980, Devonian metamorphic event in the northeastern Klamath Mountains, California: *Geol. Soc. America Bull.*, v. 91, p. 1453–1459.

Cloos, M., 1985, Thermal evolution of convergent plate margins — Thermal modeling and reevaluation of isotopic Ar-ages for blueschists in the Franciscan complex of California: *Tectonics*, v. 4, p. 421–433.

Coleman, R. G., and Lanphere, M. A., 1971, Distribution and age of high-grade blueschists, associated eclogites, and amphibolites from Oregon and California: *Geol. Soc. America Bull.*, v. 82, p. 2397–2412.

Davis, G. A., and Lipman, P. W., 1962, Revised structural sequence of pre-Cretaceous metamorphic rocks in the southern Klamath Mountains, California: *Geol. Soc. America Bull.*, v. 73, p. 1547–1552.

—— , Holdaway, M. J., Lipman, P. W., and Romey, W. D, 1965, Structure, metamorphism, and plutonism in the south-central Klamath Mountains, California: *Geol. Soc. America Bull.*, v. 76, p. 933–966.

—— , Ando, C. J., Cashman, P. H., and Goulland, L., 1979, Cross-section of the central Klamath Mountains, California: *Geol. Soc. America Map Chart Series MC-281*.

Dumitru, T., and Cloos, M., 1986, Subnormal subduction-period geothermal gradients in the Great Valley forearc basin, CA — Fission track analysis, computer modelling, and tectonic implications: *Geol. Soc. America Abstr. with Programs*, v. 18, p. 103.

Ehlig, P. L., 1968, Causes and distribution of Pelona, Rand, and Orocopia Schist along the

San Andreas and Garlock faults: *Stanford Univ. Publ. Geol. Sci.*, v. 11, p. 294–305.

——, 1981, Origin and tectonic history of the basement terrane of the San Gabriel Mountains, central Transverse Ranges, *in* Ernst, W. G., ed., *The Geotectonic Development of California* (Rubey Vol. I): Englewood Cliffs, N.J., Prentice-Hall, p. 253–283.

England, P. C., and Richardson, S. W., 1977, The influence of erosion upon the mineral facies of rocks from different metamorphic environments: *Geol. Soc. London J.*, v. 134, p. 201–213.

Ernst, W. G., 1971, Metamorphic zonations on the presumably subducted lithospheric plates from Japan, California, and the Alps: *Contrib. Mineralogy Petrology*, v. 34, p. 43–59.

Feininger, T., 1981, Amphibolite associated with the Thetford Mines ophiolite complex at Belmina Ridge, Quebec: *Can. J. Earth Sci.*, v. 18, p. 1878–1892.

Ghent, E. D., and Stout, M. Z., 1981, Metamorphism at the base of the Samail ophiolite, southeastern Oman Mountains: *J. Geophys. Res.*, v. 86, p. 2557–2571.

Graham, C. M., and England, P. C., 1976, Thermal regimes and regional metamorphism in the vicinity of overthrust faults — An example of shear heating and inverted metamorphic zonation from southern California: *Earth Planet. Sci. Lett.*, v. 31, p. 142–152.

——, and Powell, R., 1984, A garnet-hornblende geothermometer: calibration, testing, and application to the Pelona Schist, Southern California: *J. Metamorph. Geology*, v. 2, p. 13–31.

Green, D. H., 1964, The metamorphic aureole of the peridotite at the Lizard, Cornwall: *J. Geology*, v. 72, p. 543–563.

Haxel, G., and Dillon, J., 1978, The Pelona-Orocopia Schist and Vincent-Chocolate Mountain thrust system, southern California, *in* Howell, D. G., and McDougall, K. A. eds. *Mesozoic Paleogeography of the Western United States:* Pacific Section, Soc. Econ. Paleontologists Mineralogists, Pacific Coast Paleogeography Symp. 2, p. 453–469.

Heard, H. C., and Rubey, W. W., 1966, Tectonic implications of gypsum dehydration: *Geol. Soc. America Bull.*, v. 77, p. 741–760.

Holdaway, M. J., 1965, Basic regional metamorphic rocks in part of the Klamath Mountains, northern California: *Amer. Mineralogist*, v. 50, p. 953–977.

Hotz, P. E., 1977, Geology of the Yreka area, Siskiyou County, California: *U.S. Geol. Survey Bull. 1436*, 72 p.

Irwin, W. P., 1960, Relations between the Abrams Mica Schist and Salmon Hornblende Schist in Weaverville Quadrangle, California: *U.S. Geol. Survey Prof. Paper 400-B*, p. 315–316.

——, 1977, Review of Paleozoic rocks of the Klamath Mountains, *in* Stewart, J. H., *et al.*, eds., *Paleozoic Paleogeography of the Western United States:* Pacific Section, Soc. Econ. Paleontologists Mineralogists, Pacific Coast Paleogeography Symp. 1, p. 441–454.

——, 1981, Tectonic accretion of the Klamath Mountains, *in* Ernst, W. G., ed., *The Geotectonic Development of California* (Rubey Vol. I): Englewood Cliffs, N.J., Prentice-Hall, p. 29–49.

Jacobson, S. B., Quick, J. E., and Wasserburg, G. J., 1984, A Nd and Sr isotopic study of the Trinity peridotite — implications for mantle evolution: *Earth Planet. Sci. Lett.*, v. 68, p. 361–378.

Lanphere, M. A., Irwin, W. P., and Hotz, P. E., 1968, Isotopic age of the Nevadan orogeny and older plutonic and metamorphic events in the Klamath Mountains: *Geol. Soc. America Bull.*, v. 79, p. 1027–1052.

_____, Blake, M. C., and Irwin, W. P., 1978, Early Cretaceous metamorphic age of the South Fork Mountain Schist in the northern Coast Ranges of California: *Amer. J. Sci.*, v. 278, p. 798–815.

Lipman, P. W., 1964, Structure and origin of an ultramafic pluton in the Klamath Mountains, California: *Amer. J. Sci.*, v. 262, p. 199–222.

Mason, R., 1984, Inverted isograds at Sulitjelma, Norway — The result of shear zone deformation: *J. Metamoph. Geology*, v. 2, p. 77–82.

Mattinson, J. M., and Hopson, C. A., 1972, Paleozoic ophiolite complexes in Washington and northern California: *Carnegie Inst. Wash. Yearbook 71*, p. 578–583.

McDowell, F. W., Lehman, D. H., Gucwa, P. R., Fritz, D., and Maxwell, J. C., 1984, Glaucophane schists and ophiolites of the northern California Coast Ranges — Isotopic ages and their tectonic implications: *Geol. Soc. America Bull.*, v. 95, p. 1373–1382.

Murray, M., and Condie, K. C., 1973, Post Ordovician to Early Mesozoic history of the Eastern Klamath Subprovince, northern California: *J. Sediment. Petrology*, v. 43, p. 505–515.

Oxburgh, E. R., and Turcotte, D. L., 1974, Thermal gradients and regional metamorphism in overthrust terrains with special reference to the eastern Alps: *Schweiz. Mineral. Petrogr. Mitt.*, v. 54, p. 641–662.

Peacock, S. M., 1987, Serpentinization and infiltration metasomatism in the Trinity peridotite, Klamath Province, Northern California: Implications for subduction zones: *Contrib. Mineralogy Petrology*, v. 95, p. 55–70.

_____, 1985, Thermal and fluid evolution of the Trinity Thrust System, Klamath Province, Northern California — Implications for the effect of fluids in subduction zones: Ph.D. dissertation, Univ. California, Los Angeles, Calif., 328 p.

Platt, J. P., 1975, Metamorphic and deformational processes in the Franciscan Complex, California — Some insights from the Catalina Schist terrane: *Geol. Soc. America Bull.*, v. 86, p. 1337–1347.

_____, 1976, The petrology, structure, and geologic history of the Catalina Schist terrain, southern California: *Univ. Calif. Publ. Geol. Sci.*, v. 112, p. 1–111.

Potter, A. W., Hotz, P. E., and Rohr, D. M., 1977, Stratigraphy and inferred tectonic framework of Lower Paleozoic rocks in the eastern Klamath Mountains, northern California, *in* Stewart, J. H., *et al.*, eds., *Paleozoic Paleogeography of the Western United States:* Pacific Section, Soc. Econ. Paleontologists Mineralogists, Pacific Coast Paleogeography Symp. 1, p. 421–440.

Powell, R., and Evans, J., 1983, A new geothermometer for the assemblage biotite-muscovite-chlorite-quartz: *J. Metamorph. Geology*, v. 1, p. 331–336.

Raleigh, C. B., and Paterson, M. S., 1965, Experimental deformation of serpentinite and its tectonic implications: *J. Geophys. Res.*, v. 70, p. 3965–3985.

Rosenfeld, J. L., 1968, Garnet rotation due to the major Paleozoic deformations in southeast Vermont, *in* Zen, E., White, W. S., Hadley, J., and Thompson, J. B., eds., *Studies of Appalachian Geology:* New York, Wiley-Interscience, p. 185–202.

Sinha-Roy, S., 1982, Himalayan Main Central Thrust and its tectonic implications for Himalayan inverted metamorphism: *Tectonophysics*, v. 84, p. 197–224.

Sleep, N. H., 1975, Stress and flow beneath island arcs: *Geophys. J. Roy. Astronom. Soc.*, v. 42, p. 827–857.

Sorensen, S. S., 1986, Petrologic and geochemical comparison of the blueschist and greenschist units of the Catalina Schist terrane, southern California: *Geol. Soc. America Mem. 164*, p. 59–75.

Spear, F. S., Selvestone, J., Hickmott, D., Crowley, P., and Hodges, K. V., 1984, P-T paths from garnet zoning — A new technique for deciphering tectonic processes in crystalline terranes: *Geology*, v. 12, p. 87–90.

Spray, J. G., and Williams, G. D., 1980, The sub-ophiolite metamorphic rocks of the Ballantrae Igneous Complex, SW Scotland: *Geol. Soc. London*, v. 137, p. 359–368.

Suppe, J., 1973, Geology of the Leech Lake Mountain-Ball Mountain region, California: *Calif. Univ. Publ. Geol. Sci.*, v. 107, 82 p.

—— , and Armstrong, R. L., 1972, Potassium-argon dating of Franciscan metamorphic rocks: *Amer. J. Sci.*, v. 272, p. 217–233.

Talbot, C. J., 1979, Infrastructural migmatitic upwelling in East Greenland interpreted as thermal convective structures: *Precambrian Res.*, v. 8, p. 77–93.

Thompson, J. B., Robinson, P., Clifford, T. N., and Trask, N. J., 1968, Nappes and gneiss domes in west-central New England, *in* Zen, E., White, W. S., Hadley, J., and Thompson, J. B., eds., *Studies of Appalachian Geology:* New York, Wiley-Interscience, p. 203–218.

Turcotte, D. L., and Schubert, G., 1982, *Geodynamics: Application of Continuum Physics to Geological Problems:* New York, Wiley, 450 p.

Watkins, K. P., 1985, Geothermometry and geobarometry of inverted metamorphic zones in the west central Scottish Dalradian: *J. Geol. Soc. London*, v. 142, p. 157–165.

Williams, H., and Smyth, W. R., 1973, Metamorphic aureoles beneath ophiolite suites and alpine peridotites — Tectonic implications with west Newfoundland examples: *Amer. J. Sci.*, v. 273, p. 594–621.

Wood, B. L., 1971, Structure and relationships of Late Mesozoic schists of northwest California and southwest Oregon: *New Zealand J. Geology Geophysics*, v. 14, p. 219–239.

Woodcock, N. H., and Robertson, A. H. F., 1977, Origin of some ophiolite-related metamorphic rocks of the 'Tethyan' belt: *Geology*, v. 5, p. 373–376.

Carl E. Jacobson, M. Robert Dawson, and Clay E. Postlethwaite
Department of Earth Sciences
Iowa State University
Ames, Iowa 50011

35

STRUCTURE, METAMORPHISM, AND TECTONIC SIGNIFICANCE OF THE PELONA, OROCPIA, AND RAND SCHISTS, SOUTHERN CALIFORNIA

The Pelona, Orocopia, and Rand schists (POR schists) of southern California and south-western Arizona are assemblages of metagraywacke, mafic schist, and associated lithologies that were metamorphosed at relatively high P/T. The schists are overlain by Precambrian to Mesozoic igneous and metamorphic rocks of continental affinity along the Vincent, Chocolate Mountains, Orocopia, and Rand thrusts (VCM thrust), which are thought to be of Late Cretaceous age. A controversy exists as to whether metamorphism of the schists occurred in an eastward-dipping subduction zone at the western margin of North America, or in a westward-dipping subduction zone between North America and an outboard island arc or continental sliver. The former model is inconsistent with structural evidence that the latest phase of movement of the VCM thrust involved transport of the upper plate to the northeast. The latter model is inconsistent with the absence of a suture zone inboard of the outcrop belt of POR schists. The POR schists have undergone a complex structural history and exhibit at least two sets of isoclinal folds and two of open-to-tight folds. Metamorphism of the schists is predominantly of greenschist to oligoclase amphibolite facies, but local relics of crossite indicate an early metamorphism transitional to the blue-schist facies. Both early blueschists and later greenschists to oligoclase amphibolites apparently are related to subduction, in contrast to some other orogenic belts where greenschist to amphibolite overprints of blueschists are attributed to uplift. The preservation of high-P/T subduction assemblages, geochronologic evidence of a Late Cretaceous-early Tertiary thermal disturbance, and the recent discovery in southeasternmost California of a pre-60-m.y., low-angle normal fault that transects the VCM thrust are all evidence of major uplift of the POR schists during or shortly after subduction. This, in turn, may indicate that the northeast vergence presently observed for the VCM thrust reflects only the uplift of the POR schists rather than their initial subduction.

INTRODUCTION

One of the most elusive problems of southern California geology is that of determining the regional paleogeography at the time of formation of the Pelona, Orocopia, and Rand Schists (POR schists) (Fig. 35-1). These units were derived from sedimentary and volcanic protoliths deposited in an oceanic environment, exhibit metamorphism of glaucophanic-greenschist, greenschist, albite-epidote amphibolite, and oligoclase-amphibolite facies and are generally thought to have formed in a subduction zone (Burchfiel and Davis, 1981; Crowell, 1968, 1981; Dickinson, 1981; Ehlig, 1981; Graham and England, 1976; Graham and Powell, 1984; Haxel and Dillon, 1978; Jacobson, 1980, 1983a; Jacobson and Sorensen, 1986; Yeats, 1968). The schists are located interior to the continental margin and lie structurally beneath the Mesozoic magmatic arc of western North America along a regionally extensive system of low-angle faults known locally as the Vincent, Orocopia, Chocolate Mountains, and Rand thrusts (VCM thrust).

Two basic models have been developed to explain the intracontinental setting of the POR schists. (1) Yeats (1968), Crowell (1968, 1981), Burchfield and Davis (1981), and Dickinson (1981) concluded that the POR schists constitute portions of the Franciscan Complex that were underthrust beneath the western margin of North America, possibly at several different times, in a low-angle, east-directed subduction zone. (2) Haxel and Dillon (1978), however, noted that eastward subduction is inconsistent with structural evidence (Dillon, 1976; Ehlig 1958, 1981; Haxel, 1977) that the upper plate of the VCM thrust was transported northeastward. Instead, they proposed that the protoliths of the schist were deposited either in a marginal basin between North America and an

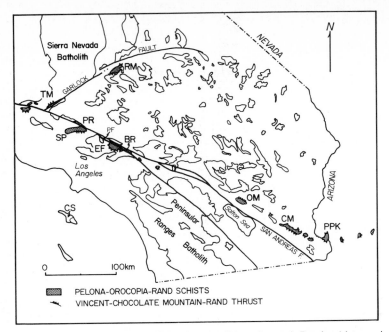

FIG. 35-1. Distribution of the Pelona, Orocopia, and Rand schists and Vincent, Chocolate Mountains, Orocopia, and Rand thrusts. Mesozoic magmatic rocks indicated by unpatterned outlined areas. BR, Blue Ridge area of San Gabriel Mountains; CM, Chocolate Mountains; CS, Catalina Schist; EF, East Fork area of San Gabriel Mountains; OM, Orocopia Mountains, PF, Punchbowl fault; PPK, Picacho–Peter Kane Mountain area; PR, Portal and Ritter Ridges; RM, Rand Mountains; SP, Sierra Pelona; TM, Tehachapi Mountains.

outboard island arc or in an ensimatic basin similar to the present Gulf of California and located at the southeast termination of the Late Cretaceous–early Tertiary proto-San Andreas fault. Deformation and metamorphism of the POR schists were attributed to closing of the basin along a southwest-dipping subduction zone.

Haxel and Dillon (1978) made no judgment as to whether their hypothesized island arc was formed by rifting of North America (Japanese-type margin) or whether it was exotic. Later, an exotic origin was proposed by Ehlig (1981), Vedder *et al.* (1983), and Harding and Coney (1985). Haxel, *et al.* (1985), however, concluded that certain units in the upper plate of the VCM thrust are correlative with rocks of cratonic North America, thus arguing against an exotic origin.

Recent work also indicates that at least part of the protolith of the POR schists is pre-Late Jurassic in age and is thus older than both the protolith of the Franciscan Complex and the inferred proto-San Andreas fault (Mukasa *et al.*, 1984). This has caused Tosdal *et al.* (1984) to replace the "proto-San Andreas" model by an analogous one in which the protoliths of the POR schists were deposited in a Gulf of California-like basin related to the Mojave-Sonora megashear (Silver and Anderson, 1974).

Although models involving SW-directed subduction are consistent with northeast transport of the upper plate, they require a suture zone inboard of the POR schist exposures. As emphasized by Burchfiel and Davis (1981), such a suture has not been observed. Furthermore, some workers have argued that the evidence of northeast thrusting is equivocal. This direction of movement was first proposed by Ehlig (1958, 1981) in a pioneering study of the Pelona Schist and Vincent thrust in the eastern San Gabriel Mountains. Ehlig noted that foliation, fold axes, and stretching lineations in the Pelona Schist were parallel to equivalent structures in mylonites at the base of the upper plate, that maximum temperature of metamorphism of the Pelona Schist increased toward the upper plate (inverted metamorphic zonation), and that mylonites in the upper plate were of the same metamorphic facies as directly subjacent Pelona Schist (metamorphic convergence of upper and lower plates). From these relations, Ehlig concluded that deformation and metamorphism of the Pelona Schist and mylonitization of the upper plate were due to thrusting. Northeastward transport was postulated from the sense of overturning of a northwest-southeast-trending macroscopic synform (Narrows synform) in the Pelona Schist directly beneath the upper plate. This fold is superposed upon isoclinal folds with axial-plane schistosity that also trend northwest-southeast. Both generations of folds were considered by Ehlig to have their axes perpendicular to movement direction.

Ehlig's (1958, 1981) analysis of thrusting direction has been criticized on several grounds. First, Burchfiel and Davis (1981) argued that the Narrows synform has folded the thrust-related schistosity and thus might postdate thrusting. Second, as summarized by Bryant and Reed (1969), Escher and Watterson (1974), and Bell and Hammond (1984), lineations and axes of tight folds in deep-seated thrust zones tend to be parallel rather than perpendicular to thrusting direction. Finally, the northwest-southeast trend of lineations and isoclinal folds in the San Gabriel Mountains is anomalous, because in most of the POR schists similar structures are oriented north-south to northeast-southwest (Fig. 35-2). It is not clear whether the divergent nature of structures in the San Gabriel Mountains is due to primary inhomogeneities among the POR schists or to rotations along Tertiary high-angle faults (Jacobson, 1983a).

Although the evidence of northeastward vergence in the San Gabriel Mountains may be problematic, the same movement direction has been inferred by Haxel and Dillon (1978) for the segment of the VCM thrust in the Chocolate Mountains and Picacho-Peter Kane Mountain areas (Fig. 35-1). Haxel and Dillon derived their thrusting direction by application of the separation-arc method of Hansen (1967) to postschistosity folds (style 2 folds) near the thrust contact in both the upper and lower plates (Fig. 35-3A). The separation-arc method does not depend on movement being perpendicular to fold axes and lineations. In fact, in the area analyzed by Haxel and Dillon (1978), lineations and axes of isoclinal folds (style 1 folds) define a broad maximum that is nearly parallel to inferred direction of transport (compare Figs. 35-2C and 35-3A). Style 2 folds in this region range from parallel to perpendicular to the presumed transport direction. Haxel and Dillon's (1978) analysis of movement direction of the Chocolate Mountains thrust using folds has lately been confirmed by microstructural studies (Simpson, 1986)[1] None theless, as discussed below, these movement indicators reflect only the latest history of the thrust and it is controversial as to whether or not they formed during the original subduction event (Burchfiel and Davis, 1981; Crowell, 1981; Jacobson, 1983c, 1984).

Despite considerable debate regarding the significance of style 2 folds, little attention has been paid to the origin of style 1 folds. Since the initial work of Ehlig (1958), a universal assumption has been that the latter folds and their associated schistosity and lineation were produced during *under*thrusting of the POR schists (equated to subduction with the advent of plate tectonics) (Harvill, 1969; Haxel and Dillon, 1978; Jacobson, 1980, 1983a; Raleigh, 1958; Sharry, 1981; Vargo, 1972). Recent interpretations of the metamorphic evolution of orogenic belts, however, indicate that this assumption needs further justification. The style 1 folds developed during metamorphism of greenschist to oligoclase-amphibolite facies. In several areas, relict crossite indicates that the greenschists

---

[1] Also J. T. Dillon, unpublished data.

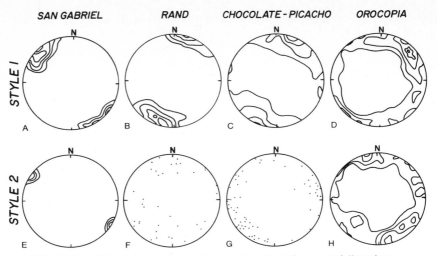

SAN GABRIEL   RAND   CHOCOLATE - PICACHO   OROCOPIA

STYLE 1

STYLE 2

FIG. 35-2.   Orientations of style 1 and style 2 fold axes and lineations.
Contour intervals: (A) 3, 6, 9, and 12% per 1% area; (B) 4 sigma (Kamb
method); (C) 2, 6, 10, and 14% per 1% area; (D) 2 sigma; (E) 15, 30, and
45% per 1% area; (H) 2 sigma. Data for (C) and (G) are from Haxel (1977).

to oligoclase amphibolites overprinted mineral assemblages transitional to the blueschist
facies (Dillon, 1976; Graham, 1975; Graham and Powell, 1984; Jacobson and Sorensen,
1986). Greenschist to amphibolite overprints of blueschists are common in metamorphic
belts (Ernst, 1973; Gray and Yardley, 1979; Holland and Richardson, 1979; Laird and
Albee, 1981a; Lister *et al.*, 1984; Matthews and Schliestedt, 1984; Trzcienski *et al.*,
1984; Yardley, 1982). In most cases, only the blueschists are thought to have formed
during subduction; the greenschists and amphibolites are generally attributed to recrystallization during subsequent uplift. This conclusion is based on thermal models which
predict that typical rates of erosional uplift are insufficient to prevent the high-P/T conditions established during subduction from being replaced by a more "normal" geotherm

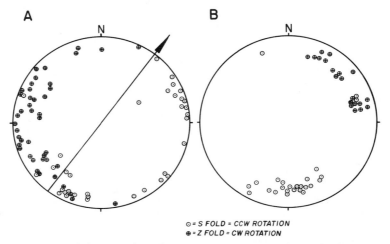

A   B

⊙ = S FOLD = CCW ROTATION
⊕ = Z FOLD = CW ROTATION

FIG. 35-3.   (A) Hansen (1967) analysis of style 2 folds from the Chocolate
Mountains thrust (Dillon *et al.*, in preparation). (B) Style 2 folds from one
subdomain in the northwestern Orocopia Mountains.

once subduction stops (Draper and Bone, 1981; England, 1978; England and Richardson, 1977; Rubie, 1984). The possibility of an origin during uplift has not previously been considered for the greenschists to oligoclase amphibolites of the POR schists. If they did form in this manner, the significance of the VCM thrust would have to be completely reevaluated. Alternatively, if the greenschists and lower amphibolites are primary high-*P/T* subduction rocks, then some special circumstance must have occurred to account for the lack of an uplift-related overprint.

PREMETAMORPHIC LITHOLOGIES

The POR schists consist of approximately 90% quartzofeldspathic and minor interbedded pelitic schist, 10% mafic schist, and less than 1% ferromanganiferous quartzite, marble, serpentinite, and talc-actinolite rock. Pervasive metamorphism and deformation have destroyed virtually all primary textures and structures.

Compositionally, the quartzofeldspathic schist is equivalent to graywacke (Haxel *et al.*, 1986). A sedimentary origin is confirmed by pervasive rhythmic compositional banding, presence of graphite, which is most abundant in the more micaceous layers and which presumably is derived from organic material, and rare occurrences of relict graded bedding.

The mafic schist occurs throughout the section. Abundances of minor and trace elements indicate that the parent rock was normal to enriched midocean ridge basalt (e.g., Fig. 35-4) (Dawson and Jacobson, 1986; Haxel *et al.*, 1986). The only exceptions are two samples from the Orocopia Mountains, which may have an intraplate origin. Well-preserved relict pillow structure occurs at one outcrop in the Sierra Pelona (Ehlig, 1981). The marbles and ferromanganiferous quartzites are almost invariably associated with mafic schist and are thought to be derived respectively from pelagic limestone and chert.

Some workers consider that the present interlayering of rock types represents a

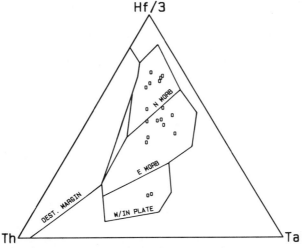

FIG. 35-4. Relative weight percents of Hafnium, Thorium, and Tantalum in mafic POR schists. N MORB, normal mid-ocean ridge basalt; E MORB, enriched MORB; W/IN PLATE, alkali within-plate basalt; DEST. MARGIN, destructive margin basalt (see Wood *et al.*, 1979).

deformed but more or less intact sedimentary/volcanic stratigraphic succession (Ehlig, 1981; Haxel and Dillon, 1978; Haxel *et al.*, 1986). In contrast, Sharry (1981) concluded that the protolith of the schist was a melange in which slivers of basalt, chert, and limestone from a subducting oceanic plate were tectonically interleaved with continentally derived graywacke.

STRUCTURE

Pelona-Orocopia-Rand Schists

The most obvious structure in the POR schists is a well-developed schistosity that is present in all rock types except serpentinite, a few massive bodies of metabasalt, and some relatively mica-poor quartzites. It is approximately parallel to the VCM thrust and mylonitic foliation in the upper plate. Orientation varies with attitude of the thrust but is on average horizontal. The schistosity is equally well developed both far from and close to the thrust.

Thin sections of quartzofeldspathic schist show very tight muscovite crenulation hinges and pseudomorphs of muscovite crenulations defined by graphite inclusions in albite porphyroblasts (Jacobson, 1980, 1983a, b; Sharry, 1981). These textures demonstrate that the present schistosity formed by crenulation and transposition of a preexisting one.

Lineation is defined by low-angle intersections between schistosity and compositional layering, alignment of elongated clots of minerals, streaking of muscovite on schistosity surfaces, elongated pressure shadows at the ends of porphyroblasts, and fine ribbing on schistosity surfaces. The lineations are generally of the type inferred to be parallel to the direction of maximum finite extension (Hobbs *et al.*, 1976).

Lineation is not as pronounced as schistosity. A weak correlation exists between strength of lineation and presence of isoclinal folds. There is no correlation between development of lineation and distance from the thrust.

Quartzites generally show a strong preferred orientation of *c*-axes (Harvill, 1969; Postlethwaite and Jacobson, 1987; Raleigh, 1958; Vargo, 1972). The shape of the pattern is closely tied to the presence or absence of lineation. Samples with lineation exhibit either single or crossed girdles. In both cases the line of intersection between girdles and schistosity is perpendicular to lineation. This is further evidence that the lineation is parallel to the extension direction (Lister and Hobbs, 1980; Sylvester and Christie, 1968). Where lineation is not present, *c*-axes define a small-circle distribution, the axis of which is perpendicular to foliation. The variation in type of pattern and its correlation with development of lineation implies that deformation in the schist is inhomogeneous, being a flattening strain in some areas but a plane strain in others.

As noted above, we have divided folds in the POR schists into two categories: style 1 and style 2. Style 1 folds are isoclinal and have axial-plane schistosity. Style 2 folds are open to tight and fold the main schistosity. Some have a secondary axial-plane crenulation cleavage. It is important to note that this classification is purely descriptive and, by itself, does not indicate when an individual fold was formed.

Style 1 folds show similar characteristics in all POR schists. They are highly attenuated and some have either or both limbs truncated by schistosity. Amplitudes range from centimeter scale to greater than 50 m. Reports of larger style 1 folds are rare (Vargo, 1972), although macroscopic isoclinal folds can be difficult to detect (Williams, 1967). Style 1 folds show extreme variation in abundance on an outcrop scale but no systematic variation in abundance with respect to distance from the thrust. They are synmetamorphic.

Several examples have been described of style 1 folds superimposed on older style 1 folds (Jacobson, 1980, 1983b; Sharry, 1981). The petrographic evidence for multiple development of schistosity may indicate that such isoclinally refolded folds are widespread but simply difficult to discern.

Axes of style 1 folds are parallel to lineation. In most areas, fold axes and lineations trend approximately northeast-southwest (Fig. 35-2). In the eastern San Gabriel Mountains they are oriented northwest-southeast.

Style 2 folds show less uniformity from area to area than do style 1 folds. We have studied the former in the San Gabriel, Rand, and Orocopia Mountains and will describe each area individually. The style 2 category includes the postschistosity folds used by Haxel and Dillon (1978) to infer thrusting direction in the Chocolate Mountains and vicinity.

The San Gabriel Mountains are divided into two subareas, the East Fork region southwest of the Punchbowl fault and the Blue Ridge area between the Punchbowl and San Andreas faults. In the East Fork region, all style 2 folds were found in the upper 1000 m of section and most are within the upper 700 m (the total exposed thickness of schist in this area is slightly greater than 4000 m). They are parasitic to the Narrows synform and their sense of asymmetry reverses across its axial surface. The style 2 folds have crenulated the style 1 schistosity, although only rarely does this cause them to split easily along their axial planes. The style 2 folds trend northwest-southeast, parallel to the style 1 folds (Fig. 35-2). Both sets of folds in this area formed during prograde metamorphism of greenschist facies (Jacobson, 1983a).

The schist of Blue Ridge is a sliver derived from the east end of the Sierra Pelona by right-lateral offset on the late Tertiary Punchbowl fault (Dibblee, 1968; Ehlig, 1968). Attitudes of foliation, lineation, and fold axes in this area are highly scattered due to young faulting. Peak metamorphic mineral assemblages are in the oligoclase-amphibolite facies (contemporaneous with style 1 folding). Many of the style 2 folds on Blue Ridge formed during retrogression to greenschist facies (replacement of garnet by chlorite and hornblende by actinolite).

Style 2 folds are uncommon in the Rand Mountains (Postlethwaite, 1983). Some of those plotted in Fig. 35-2 are probably synmetamorphic, but others are brittle in appearance and most likely related to young faulting. The style 2 folds are highly scattered in orientation. They occur at all structural levels within the Rand Schist.

Style 2 folds are scattered throughout the schist of the Orocopia Mountains but are most abundant in the northwest tip of the range. Figure 35-2H shows the orientations of style 2 folds from the entire range. The spread in trend is considerable. However, when separate plots are made for small domains (<1 km^2), it is generally found that axes of style 2 folds define two distinct clusters, one of which trends northeast-southwest to ENE-WSW, the other northwest-southeast to nearly north-south (e.g., Fig. 35-3B). Vergence of the northeast-southwest-trending folds varies from one domain to another. In contrast, the northwest-southeast-trending ones consistently verge northeast, similar to the vergence inferred by Haxel and Dillon (1978) for late movement on the Chocolate Mountains thrust. Apparently, style 2 folds in the Orocopia Mountains belong to two separate generations. Work in progress suggests that the northwest-southeast-trending folds are the youngest. Style 1 folds in the Orocopia Mountains formed during prograde metamorphism of epidote-amphibolite facies. Relations for the various style 2 folds have not yet been worked out.

Thrust Zone

As with the style 2 folds, the thrust zone above the POR schists is quite variable from one region to the next. The segment in the East Fork area of the San Gabriel Mountains was the first to be studied in detail (Ehlig, 1958). In that area, mylonites at the base of the upper plate range up to 1000 m in thickness. Mylonitic foliation is parallel to the thrust contact and to foliation in the underlying schist. Lineations are similar in nature and parallel to those in the schist. As in the schist, they are not well developed throughout. Folds range from relatively open to isoclinal. In some the mylonitic foliation is axial-planar, in others it is folded. Thus, the terms "style 1" and style 2," defined for the schists, can be applied equally well to the mylonites. Fold axes are parallel to lineation. During mylonitization, hornblende and Ca-bearing plagioclase in the parent rocks of the upper plate were retrograded to the greenschist assemblage of epidote, albite, chlorite, and actinolite.

In the Rand Mountains, upper-plate rocks are exposed above the Rand Schist in both the northeast and southwest ends of the range (Fig. 35-1). In the northeast, the schist is overlain by a minimum of 900 m of Johannesburg Gneiss. This unit consists of quartzofeldspathic gneiss, amphibolite (with or without garnet), marble, and quartzite. All the lithologies are locally cut by zones of mylonite that range from a few centimeters to a few tens of meters in thickness. The zones trend approximately east-west, parallel to the contact between the Johannesburg Gneiss and the Rand Schist. They occur throughout the gneiss.

In the southwest end of the range, the Rand Schist is overlain by the Atolia Quartz Monzonite. A several-meter-thick layer of highly laminated marble mylonite is present between the two in most areas. At some locations, the marble mylonite is accompanied by variably mylonitized quartzofeldspathic gneiss, amphibolite, and quartzite in lenses up to 50 m thick. The marble mylonite and associated lithologies appear to be derived from the Johannesburg Gneiss. The Atolia Quartz Monzonite is locally mylonitized within several meters of the marble mylonite. Elsewhere it has a uniformly massive igneous texture. At many locations, the contact zone between Rand Schist and Atolia Quartz Monzonite is overprinted by brittle faults and hydrothermal alteration.

In the Orocopia Mountains, the upper plate of the VCM thrust contains augen gneiss, granulite gneiss, and anorthosite to syenite of Precambrian age, as well as granitic rocks of Mesozoic age (Crowell, 1975; Crowell and Walker, 1962). Many of these units have correlatives in the upper plate of the San Gabriel Mountains. In the Orocopia Mountains, however, there is only a thin (generally a few tens of centimeters) zone of mylonitic gneiss at the contact between upper and lower plates. As in the Rand Mountains, the thin mylonite zone is commonly overprinted by brittle faulting and hydrothermal alteration.

METAMORPHISM

Facies

Metamorphism of the POR schists varies from transitional blueschist-greenschist facies to oligoclase-amphibolite facies, although most of the schist is of greenschist and albite-epidote amphibolite facies (Table 35-1). Definition of facies is after Graham and Powell (1984). Some previous workers have followed Miyashiro (1973) by lumping albite-epidote amphibolites and oligoclase amphibolites into a single epidote-amphibolite facies (Jacobson and Sorensen, 1986; Sharry, 1981). Metamorphic grade increases toward the upper plate (references cited in Table 35-1). Maximum grade at the contact with the upper plate ranges from greenschist to oligoclase-amphibolite facies.

Oligoclase amphibolites are generally restricted to within a few hundred meters of the upper plate. Transitional blueschist-greenschists (to be referred to as glaucophanic greenschists) occur only in the Rand Mountains. However, relict crossite, observed in mafic schist in several areas, indicates that this facies was formerly more widespread

TABLE 35-1. Metamorphic Facies of the POR Schists[a]

	gl-gs	gs	aea	oa	References
Chocolate Mountain		+	+	+	Dillon, 1976;
Orocopia Mountain			+		Dawson, unpublished data
Picacho–Peter Kane Mountain			+	+	Haxel, 1977
Rand Mountain	+	+	+	m	Jacobson and Sorensen, 1986
San Gabriel Mountain					Jacobson, 1980, 1983c
East Fork		+			
Blue Ridge			+	m	
Sierra Pelona		+	+	+	Graham and Powell, 1984
Tehachapi Mountain		+	+	+	Sharry, 1981

[a]"+" indicates facies occurs; "m" indicates assemblages of that facies observed in only a few samples. gl-gs, glaucophanic-greenschist; gs, greenschist; aea, albite-epidote amphibolite; oa, oligoclase-amphibolite.

(Dillon, 1976; Graham 1975; Graham and Powell, 1984). Crossite is common in meta-chert in some areas but does not coexist with epidote and is thus not indicative of blue-schist or glaucophanic-greenschist facies.

Quartzofeldspathic Schist

Albite, quartz, and celadonitic muscovite (phengite) in subequal proportions are the most abundant components of the quartzofeldspathic schist from glaucophanic-greenschist to albite-epidote amphibolite facies. At oligoclase-amphibolite facies, oligoclase to sodic andesine is present instead of, or in greater abundance than, albite. Epidote to clinozoisite (up to 10%) is virtually ubiquitous at the lower grades but is minor or absent in the oligoclase amphibolites. Small amounts of sphene occur in most samples; rutile is present in some albite-epidote amphibolites and oligoclase amphibolites. Finely disseminated graphite (generally minor but up to several percent) is characteristic of most quartzo-feldspathic schists.

Additionally, quartzofeldspathic schists of greenschist and glaucophanic-greenschist facies contain some combination of chlorite, stilpnomelane, actinolite, biotite, and micro-cline. Figure 35-5A shows these minerals on a schematic AFM diagram (Thompson, 1957) modified to include projection from albite and epidote in addition to the usual quartz and muscovite (cf. Brown, 1975; Jacobson, 1983c). Ideally, only one of the pairs actinolite-stilpnomelane and chlorite-biotite should be stable, although both have been observed within close proximity to each other (Jacobson, 1980). This probably is due to the fact that Al_2O_3 and Fe_2O_3, which are not differentiated in the projection, behave as separate components. The AFM minerals for the albite-epidote amphibolite and oligoclase-amphibolite facies are chlorite, garnet, hornblende, biotite, and microcline (Fig. 35-5B). Crossing tie lines exist within these facies, too, as the four-phase assemblage garnet-chlorite-biotite-hornblende occurs in many samples.

The relative abundances of the various AFM minerals are not the same for all areas. For example, in the Rand Mountains, chlorite is very abundant, garnet, biotite, and amphibole are of lesser importance, and microcline was not observed. In the San Gabriel Mountains, all the AFM minerals except microcline are common. In the Picacho-Peter Kane Mountain area, biotite and microcline are abundant, whereas chlorite, garnet, and amphibole are not. Taking into account variations in metamorphic grade within and between the different areas, the relations above imply that quartzofeldspathic schists in the Rand Mountains are relatively aluminous, whereas those in the Picacho area are rela-

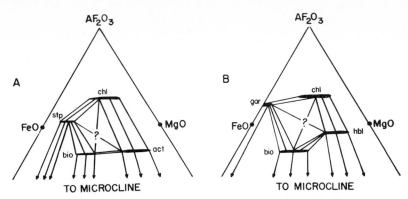

FIG. 35-5. Schematic AFM projections for quartzofeldspathic POR schists. (A) Glaucophanic-greenschist and greenschist facies. (B) Albite-epidote amphibolite and oligoclase-amphibolite facies. Projected from quartz, muscovite, albite, and epidote. AF203, $Al_2O_3 + Fe_2O_3$; act, actinolite; bio, biotite; chl, chlorite; gar, garnet; hbl, hornblende; stp, stilpnomelane.

tively alumina poor. Other minerals found in the quartzofeldspathic schist include calcite, magnetite, tourmaline, apatite, zircon, and allanite.

Mafic Schist

Mafic schists generally contain subequal amounts of albite, epidote to clinozoisite, chlorite, and amphibole. The amphibole is crossite at glaucophanic-greenschist facies, actinolite at greenschist facies, and hornblende at higher grades. Oligoclase to sodic andesine is present with or instead of albite in oligoclase amphibolites. Stilpnomelane is present in iron-rich samples at greenschist and glaucophanic-greenschist facies, garnet in iron-rich samples of albite-epidote amphibolite and oligoclase-amphibolite facies. Sphene is nearly ubiquitous at all but the highest grades. Rutile is present in many albite-epidote amphibolites and oligoclase amphibolites. Magnetite occurs in some glaucophanic green-schists and greenschists, ilmenite in some albite-epidote amphibolites and oligoclase amphibolites. Minor amounts of muscovite, biotite, and/or calcite occur in many samples at all metamorphic grades.

Figure 35-6 shows the compositions of chlorite, amphibole, stilpnomelane, and garnet on epidote-albite projections (Laird, 1980) for the greenschist and albite-epidote amphibolite facies in the San Gabriel Mountains. Tie lines are for the most part nonoverlapping, consistent with an overall approach to equilibrium. The increase of Al in amphibole (i.e., transition from actinolite to hornblende) with increase of metamorphic grade is evident, as is the restriction of stilpnomelane and garnet to Fe-rich bulk compositions. In addition, chlorite, and to some extent amphibole, has a higher maximum Fe content at greenschist facies than at albite-epidote amphibolite facies. This is consistent with the fact that garnet-bearing albite-epidote amphibolites are more common than stilpnomelane-bearing greenschists.

Conditions of Metamorphism

The occurrence of glaucophanic greenschists in the Rand Mountains and crossite relics in other areas is obvious evidence for high-P/T metamorphism of portions of the POR schists during at least some stage of their history. In addition, evidence exists which implies that

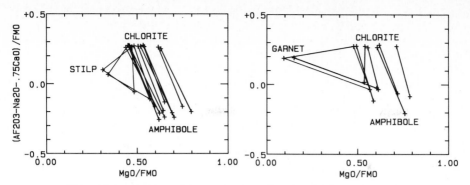

FIG. 35-6. Epidote-albite projections (Laird, 1980) for mafic rocks of greenschist (left) (East Fork) and albite-epidote amphibolite (right) (Blue Ridge) facies from the San Gabriel Mountains. AF203, $Al_2O_3 + Fe_2O_3$; FMO, FeO + MgO + MnO.

even the more extensive greenschist to oligoclase-amphibolite assemblages of the POR schists crystallized at relatively high P/T.

Many workers have noted that relative pressure of metamorphism affects the composition of actinolite and hornblende in mafic schist (Brown, 1974, 1977; Graham, 1974; Laird and Albee, 1981b; Leake, 1965; Raase, 1974). High-pressure metamorphism favors a high ratio of glaucophane substitution, $Na^{M4},(Al^{VI} + Fe^{3+} + Ti) \rightleftharpoons Ca, (Fe^{2+} + Mg + Mn)$, to tschermak, $(Al^{VI} + Fe^{3+} + Ti), Al^{IV} \rightleftharpoons (Fe^{2+} + Mg + Mn), Si$, plus edenite, $(Na + K)^A$, $Al^{IV} \rightleftharpoons Si$, substitution, as well as a high ratio of tschermak to edenite substitution. Laird and Albee (1981b) plotted compositional data for amphiboles from various metamorphic terranes of low- to high-pressure facies series on diagrams that reflect the foregoing substitutions (Fig. 35-7). Amphiboles from the POR schists are similar to those in the high-P/T Franciscan and Sanbagawa terranes.

Another factor generally considered indicative of high-P/T metamorphism is high Si content of muscovite (celadonite substitution) (Cipriani *et al.*, 1971; Ernst, 1963;

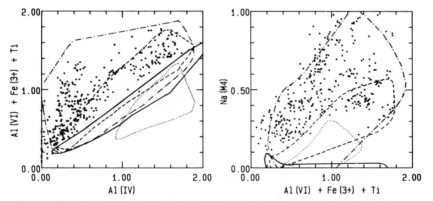

FIG. 35-7. Compositions of calcic-amphiboles in mafic POR schists. Ferric iron estimated according to Laird and Albee (1981a). Lined areas indicate the range of compositions for calcic amphiboles from (in order of increasing inferred P/T): The Abukuma terrane of Japan (dotted line), the Haast River Schist group of New Zealand (solid line), the Dalradian terrane of southwest Scotland (dashed line), and the Franciscan complex of California and the Sanbagawa Metamorphic belt of Japan (dot-dashed line) (after Laird and Albee, 1981b).

Guidotti and Sassi, 1976; Sorensen, 1986); Velde, 1965). Average and maximum Si contents are 3.4 and 3.5 atoms per 11 oxygens in the greenschist facies and 3.3 and 3.4 atoms p.f.u. in the albite-epidote amphibolite facies (calculated assuming all iron as FeO) (Jacobson, 1980). These are similar to values observed in the Franciscan Complex, the Sanbagawa metamorphic belt, and other high-pressure terranes.

The relations above are qualitative indicators of conditions of metamorphism. Quantitative estimates of P and T have been derived by Sharry (1981) for the Tehachapi Mountains and Graham and Powell (1984) for the Sierra Pelona. Sharry (1981) estimated temperatures of 526–631°C and pressures of 7.3–8.7 kb for oligoclase amphibolites adjacent to the upper plate using the garnet-biotite geothermometer and plagioclase-aluminosilicate-quartz-garnet and plagioclase-biotite-garnet-muscovite geobarometers. Graham and Powell (1984) estimated pressures of 10 ± 1 kb using the phengite geobarometer and 8-9 kb based on the jadeite content of clinopyroxene. Using the hornblende-garnet geothermometer, they derived a temperature of 480°C at the garnet isograd, 570°C at the oligoclase isograd, and 620–650°C directly adjacent to the upper plate.

Origin of the High-Pressure Metamorphism

Although the POR schists apparently were metamorphosed at relatively great depths, it is important to know whether this metamorphism occurred during subduction or whether it is an uplift-related overprint upon an even higher P/T metamorphism. Two lines of presently available evidence bear on this problem. The first deals with the similarity between the Rand Schist and a portion of the Catalina Schist of the southern California continental borderland (Fig. 35-1) (Jacobson and Sorensen, 1986; see also Sorensen, Chapter 36, this volume). The Catalina Schist is generally correlated with the Franciscan Complex of northern and central California (Bailey, 1941; Platt, 1975, 1976; Woodford, 1924). It consists of a blueschist-facies melange overlain along a thrust fault by a greenschist unit that is in turn overthrust by a nappe of amphibolite and ultramafic rock. The greenschist unit itself is zoned from glaucophanic-greenschist facies adjacent to the blueschist melange to oligoclase-amphibolite facies adjacent to the amphibolite unit. The inverted zonation within the greenschist unit is virtually identical in terms of mineral assemblages and amphibole compositions to the inverted metamorphic zonation within the Rand Schist (Jacobson and Sorensen, 1986). Platt (1975, 1976) concluded from the overall inverted zonation in the Catalina Schist and the similar radiometric ages of the blueschist and amphibolite units that metamorphism had occurred in a newly formed subduction zone due to influx of heat from hot, hanging-wall peridotite. The similarity of the Rand Schist to the greenschist unit of the Catalina Schist implies that a subduction origin is at least plausible for the greenschists to oligoclase amphibolites of the POR schists.

More direct evidence for an origin of the greenschists to oligoclase amphibolites during underthrusting rather than uplift comes from the compositions of muscovites in the San Gabriel Mountains (Jacobson, 1984). Older muscovites that crystallized during style 1 folding have lower Na (paragonite) contents and approximately the same or lower Si (celadonite) contents than newer muscovites that grew during and after style 2 folding (Fig. 35-8). The use of muscovite composition to determine metamorphic conditions is not straightforward (Guidotti and Sassi, 1976); however, Jacobson (1984) tentatively concluded that the higher Na contents of the new muscovites indicated that they grew at higher temperatures than the old muscovites. From the isopleths of maximum celadonite content in muscovite (Fig. 35-8) determined by Velde (1965), it can be concluded that if temperature rises, celadonite content will remain the same or increase only if there is a concomitant increase of pressure. That is, both P and T appear to have increased in the time between formation of style 1 and style 2 folds. This implies continued underthrusting even late in the deformational history.

It is not clear whether or not the conclusion above can be extended to the late structures associated with the Chocolate Mountains thrust. Style 2 folds in the San Gabriel Mountains are exactly parallel to style 1 folds. Style 2 folds used by Haxel and Dillon (1978) in the Chocolate Mountains and vicinity are variably oriented (Figs. 35-2G and 35-3A). In the Orocopia Mountains, variably oriented style 2 folds apparently belong to two separate generations, one of which is parallel to style 1 folds, the other of which is at a high angle to style 1 folds (Figs. 35-2H and 35-3B). We believe that until these complicated relations can be clarified, the tectonic significance of the NE vergence inferred for the Chocolate Mountains thrust must remain uncertain.

DISCUSSION

Evidence of Rapid Uplift of the POR Schists

Mineral assemblages associated with isoclinal folds and at least some postschistosity folds in the POR schists indicate a relatively high-P/T metamorphism. Limited evidence confirms previous assumptions that this metamorphism occurred during underthrusting (subduction) rather than uplift. This conclusion is important. As noted previously, typical rates of erosional uplift in orogenic belts apparently are not rapid enough to prevent overprinting of early high-P/T subduction assemblages during restoration of a normal geothermal gradient (Draper and Bone, 1981; England, 1978; England and Richardson, 1977; Rubie, 1984). The lack of such an overprint in the POR schists could be explained in several ways: (1) exceptionally rapid uplift upon cessation of subduction, (2) slow uplift during subduction due to underplating of subducted material (Ernst, 1977; Rubie, 1984), or (3) sufficiently long-lived subduction to "refrigerate" the upper plate so that a thermal overprint would be prevented even with normal rates of uplift (Peacock, Chapter 34, this volume).

Discriminating among the various possibilities requires dating of both the metamorphism and the uplift of the POR schists. Metamorphism of the Pelona and Orocopia schists has generally been considered to have occurred at 52–59 m.y.b.p. (Ehlig, 1981). Several new studies, however, imply that metamorphism may actually have occurred somewhat earlier (see also Haxel et al., 1985). For example, Dillon (1986) has recently contoured biotite K-Ar ages (Miller and Morton, 1980) in the upper plate of the Vincent thrust in the eastern San Gabriel Mountains. The contours were found to be approximately parallel to the thrust and to increase from 60 m.y. at the base of the upper plate to at least 74 m.y. at structurally high levels. This variation was considered to result from cooling of the upper plate at successively deeper levels during post-thrusting uplift. The 74-m.y. cooling date would thus place a minimum age on thrusting.

A very similar thermal history for the same area has been inferred by Mahaffie and Dokka (1986) based on fission-track dates of apatite, zircon, and sphene. Their data indicate a rapid cooling (from 500°C to 100°C) between 70 and 57 m.y.b.p., which they attributed to rapid, "Himalayan-style erosion uplift" during suturing of an exotic terrane to North America along the VCM thrust.

The work of Peacock (Chapter 34, this volume) indicates that, in general, pronounced cooling can be induced without rapid uplift simply by heat loss to the downgoing slab during prolonged subduction. However, the fact that Dillon (1986) noted the youngest cooling ages deepest in the structural section would seem to imply that there

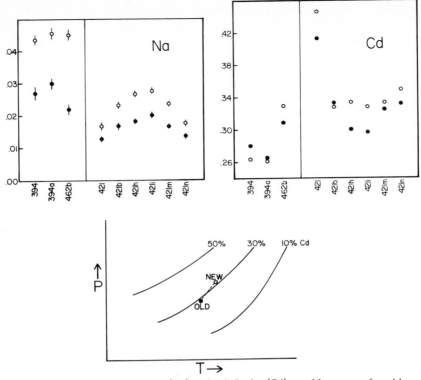

FIG. 35-8.   Moles sodium (Na) and celadonite (Cd) per 11 oxygens for old (solid circles) and new (open circles) muscovites in nine style 2 fold hinges from the San Gabriel Mountains (after Jacobson, 1984). Samples 394, 394a, and 462b are within 200 m (structural distance) of the Vincent thrust. The other six samples are from 1000 m below the thrust. Lower figure shows isopleths of maximum celadonite substitution in muscovite after Velde (1965).

must also have been substantial heat loss upward. This, in turn, suggests that rapid un-roofing *was* taking place.

Further evidence of substantial uplift at about this time is the recent discovery in the Picacho–Peter Kane Mountain area of a low-angle fault (Sortan fault) which locally truncates the Chocolate Mountains thrust (Haxel *et al.*, 1985). The Sortan fault has placed incipiently recrystallized Winterhaven Formation directly over Orocopia Schist. Contrasts in metamorphism between the Winterhaven Formation and Orocopia Schist indicate that the Sortan fault is a normal fault that has caused the excision of a crustal thickness probably on the order of 10 km. The Sortan fault is intruded by the Marcus Wash Granite, which has a K-Ar minimum age of 60 m.y. (Frost and Martin, 1983). This age relation is critical. It implies that at least some of the POR schists were at relatively shallow crustal levels by the end of the Cretaceous, just shortly after their presumed time of high-pressure metamorphism (Haxel *et al.*, 1985).

Several other observations may be relevant to the uplift history of the POR schists. Postlethwaite and Jacobson (1987) emphasized the general lack of deformation in the Atolia Quartz Monzonite of the upper plate of the Rand thrust in the southwestern Rand Mountains. This was contrasted with the pervasive isoclinal folding of the 3000m section of Rand Schist beneath the Atolia Quartz Monzonite, the presence of mylonites throughout the 900-m thickness of Johannesburg Gneiss exposed in the northeast end of the range, and the 1000m thickness of mylonites at the base of the upper plate in the San Gabriel Mountains. Postlethwaite and Jacobson (1987) also noted that the Atolia Quartz Monzonite is not typical of the deep crustal rocks that are expected to overlie the high-pressure Rand Schist. Such rocks *are* present in the Tehachapi Mountains (Sharry, 1981), which have been offset from the Rand Mountains by left-slip on the Garlock fault. Postlethwaite and Jacobson (1987) thus concluded that the Rand "thrust" in the southwestern Rand Mountains is actually a normal fault that has brought the lower-plate schist into contact with shallower rocks. The slices of Johannesburg Gneiss between the Rand Schist and Atolia Quartz Monzonite were interpreted as possible remnants of the true Rand thrust. A "late" origin for the southwestern Rand thrust has also been postulated by Silver *et al.* (1984), Nourse and Silver (1986), and Silver and Nourse (1986). Whether the reactivation of the Rand thrust occurred soon after the metamorphism of the Rand Schist or whether it is much younger (e.g., due to Tertiary detachment faulting, Frost and Martin, 1983) is not known.

As in the southwestern Rand Mountains, upper-plate rocks in the Orocopia Mountains are essentially undeformed by the thrust, even to within a meter of the Orocopia Schist. Haxel and Dillon (1978) suggested that wide variation in thickness of mylonites is a primary feature of the VCM thrust. We find it surprising that segments of the same thrust, that were at approximately the same temperature, have presumably similar offsets, and show similar extensive deformation of the lower plate, should exhibit such an exteme variation of deformation within the upper plate. Alternatively, we suggest that in some areas, the thrust zone has been disturbed by later movement.

It is of additional interest that Graham and Powell (1984), who found an increase of temperature in the Pelona Schist of the Sierra Pelona with approach to the mylonite zone, also analyzed two mylonites and obtained temperatures 40–50°C lower than from the directly subjacent schist. To explain this, Graham and Powell (1984) invoked retrograde exchange of Fe and Mg within the mylonites but not the schist. Alternatively, these relationships might be due to normal faulting of the mylonites onto relatively deeper schist along a fault slightly oblique to the original thrust zone.

In summary, the POR schists and VCM thrust present a major unresolved problem. Folds, composite foliations, and microstructures from the Chocolate Mountains thrust indicate northeast vergence. This movement direction implies that a major suture should lie inboard of the POR schist outcrops. Such a suture, however, has not been observed. We suggest that a way to resolve this conflict is if the NE vergence is due, not to the initial subduction of the POR schists, but rather to reactivation of the VCM thrust during later uplift. This would not only resolve the movement conflict, but might also explain the preservation of inverted, high-P metamorphic zonations in the schist and is consistent with observations of possible late movements on the VCM thrust.

Age Relations and Regional Correlations

The recent discovery of the pre-60-m.y. Sortan fault (Haxel *et al.*, 1985) and evidence regarding cooling ages of the upper plate (Dillon, 1986; Mahaffie and Dokka, 1986) imply that the VCM thrust may be slightly older than formerly thought. This might clear up a problem regarding the Rand Schist. The Rand Schist has generally been treated with the Pelona and Orocopia schists because of their similar lithology, metamorphism, structure, and setting beneath regional allochthons of continental basement (put more subjectively, the Rand Schist "looks" identical to the Pelona and Orocopia schists). The Rand Schist, however, is intruded by a postmetamorphic, 79-m.y.-old granodiorite (Silver and Nourse, 1986; see also Miller and Morton, 1977), which makes it older than the generally accepted age of metamorphism of the Pelona and Orocopia schists (52–59 m.y.). If the 52–59-m.y. dates are cooling ages, there might not be a conflict. Even if the age differences are real, we believe that the similarities among the various schist bodies are so striking that all must have formed by virtually identical processes (see also Burchfiel and Davis, 1981).

CONCLUSIONS

The POR schists exhibit a complex metamorphic history. Early glaucophanic greenschists have been overprinted by greenschists to oligoclase amphibolites. The structure of the schists is also complicated and includes multiple sets of isoclinal folds and multiple sets of open-to-tight folds. Exactly how many "phases" of deformation are represented by these various structures is not clear. A tentative interpretation is that all the isoclinal folds and some of the more open folds were produced by continuous deformation during subduction, with blueschist metamorphism occurring at the early stages and greenschist and amphibolite metamorphism at the later stages. Preservation of this subduction metamorphism could be due to refrigeration of the upper plate followed by slow uplift. Our preferred interpretation, however, is that cooling of the schists was related to major uplift either during (cf. Ernst, 1977; Rubie, 1984) or shortly after subduction. If so, the northeast vergence so commonly quoted for the POR schists may represent this later event rather than the initial subduction. It is not known whether the proposed uplift is due simply to underplating of additional subducted material, or whether it and features such as the Sortan fault represent a previously unrecognized period of late Mesozoic crustal extension.

ACKNOWLEDGMENTS

Funding for this work was provided by NSF grants EAR81-21210 and EAR83-19125. The manuscript was improved by comments from W. G. Ernst, G. B. Haxel, and R. M. Tosdal.

REFERENCES

Bailey, E. H., 1941, Mineralogy, petrology, and geology of Santa Catalina Island: Ph.D. thesis, Palo Alto, Calif., Stanford Univ., 193 p.

Bell, T. H., and Hammond, R. L., 1984, On the internal geometry of mylonite zones: J. Geology, v. 92, p. 667–686.

Brown, E. H., 1974, Comparison of the mineralogy and phase relations of blueschists from the North Cascades, Washington, and greenschists from Otago, New Zealand: Geol. Soc. America Bull., v. 85, p. 333–344.

―――, 1975, A petrogenetic grid for reactions producing biotite and other Al-Fe-Mg silicates in the greenschist facies: J. Petrology, v. 16, p. 258–271.

―――, 1977, The crossite content of a Ca-amphibole as a guide to pressure of metamorphism: J. Petrology, v. 18, p. 53–72.

Bryant, B., and Reed, J. C., 1969, Significance of lineation and minor folds near major thrust faults in the southern Appalachians and Norwegian Caledonides: Geol. Mag., v. 106, p. 412–429.

Burchfiel, B. C., and Davis, G. A., 1981, Mojave Desert and environs, *in* Ernst, W. G., ed., *The Geotectonic Development of California* (Rubey Vol. I): Englewood Cliffs, N.J., Prentice-Hall, p. 217–252.

Cipriani, C., Sassi, F. P., and Scolari, A., 1971, Metamorphic white micas: definition of paragenetic fields: *Schweiz. Mineral. Petrogr. Mitt.*, v. 51, p. 259–302.

Crowell, J. C., 1968, Movement histories of faults in the Transverse Ranges and speculations on the tectonic history of California, *in* Dickinson, W. R., and Grantz, A., eds., *Proceedings of Conference on Geologic Problems of San Andreas Fault System:* Stanford Univ. Publ. Geol. Sci., v. 11, p. 323–341.

——, 1975, Geologic sketch of the Orocopia Mountains, southeastern California, *in* Crowell, J. C., ed., *San Andreas Fault in Southern California:* Calif. Div. Mines Geology Spec. Rpt. 118, p. 99–110.

——, 1981, An outline of the tectonic history of southeastern California, *in* Ernst, W. G., ed., *The Geotectonic Development of California* (Rubey Vol I): Englewood Cliffs, N.J., Prentice Hall, p. 583–600.

——, and Walker, W. R., 1962, Anorthosite and related rocks along the San Andreas fault, southern California: *Calif. Univ. Publ. Geol. Sci.*, v. 40, p. 219–288.

Dawson, M. R., and Jacobson, C. E., 1986, Trace element geochemistry of the metabasites from the Pelona-Orocopia-Rand Schists, southern California: *Geol. Soc. America Abstr. Programs*, v. 18, p. 99.

Dibblee, T. W., 1968, Displacement on the San Andreas fault system in the San Gabriel, San Bernardino, and San Jacinto Mountains, southern California, *in* Dickinson, W. R., and Grantz, A., eds., *Proceedings of Conference on Geologic Problems of San Andreas Fault System:* Stanford Univ. Publ. Geol. Sci., v. 11, p. 260–278.

Dickinson, W. R., 1981, Plate tectonics and the continental margin of California, *in* Ernst, W. G., ed., *The Geotectonic Development of California* (Rubey Vol. I): Englewood Cliffs, N.J., Prentice Hall, p. 1–28.

Dillon, J. T., 1976, Geology of the Chocolate and Cargo Muchacho Mountains, southeasternmost California: Ph.D. thesis, Univ. California, Santa Barbara, Calif., 405 p.

——, 1986, Timing of thrusting and metamorphism along the Vincent-Chocolate Mountain thrust system, southern California: *Geol. Soc. America Abstr. with Programs*, v. 18, p. 101.

Draper, G., and Bone, R., 1981, Denudation rates, thermal evolution, and preservation of blueschist terrains: *J. Geology*, v. 79, p. 601–613.

Ehlig, P. L., 1958, The geology of the Mount Baldy region of the San Gabriel Mountains, California: Ph.D. thesis, Univ. California, Los Angeles, Calif., 195 p.

——, 1968, Causes of distribution of Pelona, Rand, and Orocopia Schists along the San Andreas and Garlock faults, *in* Dickinson, W. R., and Grantz, A., eds., *Proceedings of Conference on Geologic Problems of San Andreas Fault System:* Stanford Univ. Publ. Geol. Sci., v. 11, p. 294–306.

——, 1981, Origin and tectonic history of the basement terrane of the San Gabriel Mountains, central Transverse Ranges, *in* Ernst, W. G., ed., *The Geotectonic Development of California* (Rubey Vol. I): Englewood Cliffs, N.J., Prentice Hall, p. 253–283.

England, P. C., 1978, Some thermal considerations of the Alpine metamorphism — Past, present and future: *Tectonophysics*, v. 46, p. 21–40.

——, and Richardson, S. W., 1977, The influence of erosion upon the mineral facies of rocks from different metamorphic environments: *Geol. Soc. London J.*, v. 134, p. 201–213.

Ernst, W. G., 1963, Significance of phengitic micas from low-grade schists: *Amer. Mineralogist*, v. 48, p. 1357–1373.

_____, 1973, Interpretive synthesis of metamorphism in the Alps: *Geol. Soc. America Bull.*, v. 84, p. 2053–2078.

_____, 1977, Mineral parageneses and plate-tectonic settings of relatively high-pressure metamorphic belts: *Fortschr. Mineralogie*, v. 54, p. 192–222.

Escher, A., and Watterson, J., 1974, Stretching fabrics, folds, and crustal shortening: *Tectonophysics*, v. 22, p. 223–231.

Frost, E. G., and Martin, D. L., 1983, Overprint of Tertiary detachment deformation on the Mesozoic Orocopia Schist and Chocolate Mtns. thrust: *Geol. Soc. America Abstr. with Programs*, v. 15, p. 577.

Graham, C. M., 1974, Metabasite amphiboles of the Scottish Dalradian: *Contrib. Mineralogy Petrology*, v. 47, p. 165–185.

_____, 1975, Inverted metamorphic zonation and mineralogy of the Pelona Schist, Sierra Pelona, Transverse Ranges: *Geol. Soc. America Abstr. with Programs*, v. 7, p. 321–322.

_____, and England, P. C., 1976, Thermal regimes and regional metamorphism in the vicinity of overthrust faults — An example of shear heating and inverted metamorphic zonation from southern California: *Earth Planet. Sci. Lett.*, v. 31, p. 142–152.

_____, and Powell, R., 1984, A garnet-hornblende geothermometer: calibration, testing, and application to the Pelona Schist, southern California: *J. Met. Geology*, v. 2, p. 13–31.

Gray, J. R., and Yardley, B. W. D., 1979, A Caledonian blueschist from the Irish Dalradian: *Nature*, v. 278, p. 736–737.

Guidotti, C. V., and Saddi, F. P., 1976, Muscovite as a petrogenetic indicator mineral in pelitic schists: *Neues Jahrb. Mineralogie Abh.*, v. 127, p. 97–142.

Hamilton, W., 1978, Mesozoic tectonics of the western United States, *in* Howell, D. G., and McDougall, K. A., eds., *Mesozoic Paleogeography of the Western United States:* Pacific Section, Econ. Paleontologists Mineralogists, Pacific Coast Paleogeography Symp. 2, p. 33–70.

Hansen, E., 1967, Methods of deducing slip-line orientations from the geometry of folds: *Carnegie Inst. Wash. Year Book 65*, p. 387–405.

Harding, L. E., and Coney, P. J., 1985, The geology of the McCoy Mountains Formation, southeastern California and southwestern Arizona: *Geol. Soc. America Bull.*, v. 96, p. 755–769.

Harvill, L. L., 1969, Deformational history of the Pelona Schist, northwestern Los Angeles County, California: Ph.D. thesis, Univ. California, Los Angeles, Calif., 117 p.

Haxel, G. B., 1977, The Orocopia Schist and the Chocolate Mountain thrust, Picacho-Peter Kane Mountain area, southeasternmost California: Ph.D. thesis, Univ. California, Santa Barbara, Calif., 277 p.

_____, and Dillon, J., 1978, The Pelona-Orocopia Schist and Vincent-Chocolate Mountain thrust system, southern California, *in* Howell, D. G., and McDougall, K. A., eds., *Mesozoic Paleogeography of the Western United States:* Pacific Section, Soc. Econ. Paleontologists Mineralogists, Pacific Coast Paleogeography Symp. 2, p. 453–469.

_____, Tosdal, R. M., and Dillon, J. T., 1985, Tectonic setting and lithology of the Winterhaven Formation: A new Mesozoic stratigraphic unit in southeasternmost California and southwestern Arizona: *U.S. Geol. Survey Bull. 1599*, 19 p.

_____, Budahn, J. R., Fries, T. L., King, B. W., Taggart, J. E., and White, L. D., 1986, Protolith geochemistry of the Orocopia and Pelona Schists, southern California — Preliminary report: *Geol. Soc. America Abstr. with Programs*, v. 18, p. 115.

Hobbs, B. E., Means, W. D., and Williams, P. F., 1976, *An Outline of Structural Geology:* New York, Wiley, 571 p.

Holland, T. J. B., and Richardson, S. W., 1979, Amphibole zonation in metabasites as a guide to the evolution of metamorphic conditions: *Contrib. Mineralogy Petrology*, v. 70, p. 143–148.

Jacobson, C. E., 1980, Deformation and metamorphism of the Pelona Schist beneath the Vincent thrust, San Gabriel Mountains, California: Ph.D. thesis, Univ. California, Los Angeles, Calif., 231 p.

―――, 1983a, Structural geology of the Pelona Schist and Vincent thrust, San Gabriel Mountains, California: *Geol. Soc. America Bull.*, v. 94, p. 753–767.

―――, 1983b, Complex refolding history of the Pelona, Orocopia, and Rand Schists, southern California: *Geology*, v. 11, p. 583–586.

―――, 1983c, Relationship of deformation and metamorphism of the Pelona Schist to movement on the Vincent thrust, San Gabriel Mountains, southern California: Amer. J. Sci., v. 283, p. 587–604.

―――, 1984, Petrological evidence for the development of refolded folds during a single deformational event: *J. Struct. Geology*, v. 6, p. 563–570.

―――, and Sorensen, S. S., 1986, Amphibole compositions and metamorphic history of the Rand Schist and the greenschist unit of the Catalina Schist, southern California: *Contrib. Mineralogy Petrology*, v. 92, p. 308–315.

Laird, J., 1980, Phase equilibria in mafic schist from Vermont: *J. Petrology*, v. 21, p. 1–37.

―――, and Albee, A. L., 1981a, High-pressure metamorphism in mafic schist from northern Vermont: *Amer. J. Sci.*, v. 281, p. 97–126.

―――, 1981b, Pressure, temperature, and time indicators in mafic schist – Their application to reconstructing the polymetamorphic history of Vermont: *Amer. J. Sci.*, v. 281, p. 127–175.

Leake, B. E., 1965, The relationship between tetrahedral aluminum and the maximum possible octahedral aluminum in natural calciferous and subcalciferous amphiboles: *Amer. Mineralogist*, v. 50, p. 843–851.

Lister, G. S., and Hobbs, B. E., 1980, The simulation of fabric development during plastic deformation and its application to quartzite – The influence of deformation history: *J. Struct. Geology*, v. 2, p. 355–370.

―――, Banga, G., and Feenstra, A., 1984, Metamorphic core complexes of Cordilleran type in the Cyclades, Aegean Sea, Greece: *Geology*, v. 12, p. 221–225.

Mahaffie, M. J., and Dokka, R. K., 1986, Thermochronologic evidence for the age and cooling history of the upper plate of the Vincent thrust, California: *Geol. Soc. America Abstr. with Programs*, v. 18, p. 153.

Matthews, A., and Schliestedt, M., 1984, Evolution of the blueschist and greenschist facies rocks of Sifnos Cyclades, Greece: *Contrib. Mineralogy Petrology*, v. 88, 150–163.

Miller, F. K., and Morton, D. M., 1977, Comparison of granitic intrusions in the Pelona and Orocopia Schists, southern California: *U.S. Geol. Survey J. Res.*, v. 5, p. 643–649.

―――, and Morton, D. M., 1980, Potassium-argon geochronology of the eastern Transverse Ranges and southern Mojave Desert, southern California: *U.S. Geol. Survey Prof. Paper 1152*, 30. p.

Miyashiro, A., 1973, *Metamorphism and Metamorphic Belts:* New York, Wiley, 492 p.

Mukasa, S. B., Dillon, J. T., and Tosdal, R. M., 1984, A late Jurassic minimum age for the

Pelona-Orocopia Schist protolith, southern California: *Geol. Soc. America Abstr. with Programs*, v. 16, p. 323.

Nourse, J. A., and Silver, L. T., 1986, Structural and kinematic evolution of sheared rocks in the Rand "thrust" complex, northwest Mojave Desert, California: *Geol. Soc. America Abstr. with Programs*, v. 18, p. 165.

Platt, J. P., 1975, Metamorphic and deformational processes in the Franciscan Complex, California — Some insights from the Catalina Schist terrane: *Geol. Soc. America Bull.*, v. 86, p. 1337–1347.

____ , 1976, The petrology, structure, and geologic history of the Catalina Schist terrain, southern California: *Calif. Univ. Pub. Geol. Sci.*, v. 112, p. 1–111.

Postlethwaite, C. E., 1983, The structural geology of the western Rand Mountains, northwestern Mojave Desert, California: M.S. thesis, Iowa State Univ., Ames, Iowa, 91 p.

____ , and Jacobson, C. E., 1987, Early history and reactivation of the Rand thrust, southern California: *J. Struct. Geology*, v. 9, p. 195–205.

Raase, P., 1974, Al and Ti contents of hornblende, indicators of pressure and temperature of regional metamorphism: *Contrib. Mineralogy Petrology*, v. 45, p. 231–236.

Raleigh, C. B., Jr., 1958, Structure and petrology of a part of the Orocopia Schists: M.A. thesis, Claremont Grad. Sch., Claremont, Calif., 64 p.

Rubie, D. C., 1984, A thermal-tectonic model for high-pressure metamorphism and deformation in the Sesia Zone, Western Alps: *J. Geology*, v. 92, p. 21–36.

Sharry, J., 1981, The geology of the western Tehachapi Mountains, California: Ph.D. thesis, Massachusetts Inst. Technology, Cambridge, Mass., 215 p.

Silver, L. T., and Anderson, T. H., 1974, Possible left-lateral early to middle Mesozoic disruption of the southwestern North American craton margin: *Geol. Soc. America Abstr. with Programs*, v. 6, p. 955–956.

____ , and Nourse, J. A., 1986, The Rand Mountains "thrust" complex in comparison with the Vincent thrust–Pelona Schist relationship, southern California: *Geol. Soc. America Abstr. with Programs*, v. 18, p. 185.

____ , Sams, D. B., Bursik, M. I., Graymer, R. W., Nourse, J. A., Richards, M. A., and Salyards, S. L., 1984, Some observations on the tectonic history of the Rand Mountains, Mojave Desert, California: *Geol. Soc. America Abstr. with Programs*, v. 16, p. 333.

Simpson, C., 1986, Microstructural evidence for northeastward movement on the Vincent-Chocolate Mountains thrust system: *Geol. Soc. America Abstr. with Programs*, v. 18, p. 185.

Sorensen, S. S., 1986, Petrologic and geochemical comparison of the blueschist and greenschist units of the Catalina Schist terrane, southern California: *Geol. Soc. America Mem. 164*, p. 59–75.

Sylvester, A. G., and Christie, J. M., 1968, The origin of crossed-girdle orientations of optic axes in deformed quartzites: *J. Geology*, v. 76, p. 571–580.

Thompson, J. B., Jr., 1957, The graphical analysis of mineral assemblages in pelitic schists: *Amer. Mineralogist*, v. 42, p. 842–858.

Tosdal, R. M., Dillon, J. T., and Mukasa, S. B., 1984, Pelona–Orocopia Schist protolith: Accumulation in a middle Jurassic intra-arc basin: *Geol. Soc. America Abstr. with Programs*, v. 16, p. 323.

Trzcienski, W. E., Jr., Carmichael, D. M., and Helmstaedt, H., 1984, Zoned sodic amphibole: petrologic indicator of changing pressure and temperature during tectonism in the Bathurst Area, New Brunswick, Canada: *Contrib. Mineralogy Petrology*, v. 85, p. 311–320.

Vargo, J. M., 1972, Structural geology of a portion of the eastern Rand Mountains,

Kern and San Bernardino Counties: M.S. thesis, Univ. Southern California, Los Angeles, Calif., 117 p.

Vedder, J. G., Howell, D. G., and McLean, H., 1983, Stratigraphy, sedimentation, and tectonic accretion of exotic terranes, southern Coast Ranges, California, *in* Watkins, J. S., and Drake, C. L., eds., *Studies in Continental Margin Geology*: Amer. Assoc. Petrol. Geologists Mem. 34, p. 471–496.

Velde, B., 1965, Phengitic micas: Synthesis, stability and natural occurrence: *Amer. J. Sci.*, v. 263, p. 886–913.

Williams, P. F., 1967, Structural analysis of the Little Broken Hill area of New South Wales: *Geol. Soc. Australia J.*, v. 14, p. 317–332.

Wood, D. A., Joron, J.-L., and Treuil, M., 1979, A re-appraisal of the use of trace elements to classify and discriminate between magma series erupted in different tectonic settings: *Earth Planet. Sci. Lett.*, v. 45, p. 326–336.

Woodford, A. O., 1924, The Catalina metamorphic facies of the Franciscan Series: *Calif. Univ. Publ. Geol. Sci.*, v. 15, p. 49–68.

Yardley, B. W. D., 1982, The early metamorphic history of the Haast Schists and related rocks of New Zealand: *Contrib. Mineralogy Petrology*, v. 81, p. 317–327.

Yeats, R. S., 1968, Southern California structure, seafloor spreading, and history of the Pacific Basin: *Geol. Soc. America Bull.*, v. 79, p. 1693–1702.

Sorena Sorensen
Department of Mineral Sciences
National Museum of Natural History
Smithsonian Institution
Washington, D.C. 20560

36

TECTONOMETAMORPHIC SIGNIFICANCE OF THE BASEMENT ROCKS OF THE LOS ANGELES BASIN AND THE INNER CALIFORNIA CONTINENTAL BORDERLAND*

*This paper is dedicated to the memory of Alexander K. Baird, of Pomona College (1932–1985).

The Los Angeles Basin and the inner California Continental Borderland are underlain in part by the Catalina Schist, a relatively high-P/T terrane probably metamorphosed in the Cretaceous, and a group of Jurassic arc-like rocks metamorphosed (it at all) prior to the Late Cretaceous at relatively low-P/T conditions. The latter includes the Willows Plutonic Complex, and Santa Cruz Island Schist, the Santa Monica Formation, so-called "saussurite gabbros," and subsea and subsurface occurrences of amphibolite and greenschist.

The Catalina Schist consists of blueschist and relatively high-P greenschist and amphibolite facies rocks derived from graywacke, ocean-floor tholeiite, and ultramafic rocks. The arc-like basement rocks are slates, phyllites, and schists derived from graywacke-argillite (Santa Monica Formation) and basalt-andesite-rhyolite (Santa Cruz Island Schist) protoliths. Their greenschist to amphibolite-facies metamorphism occurred under low-P/T conditions and is in part related to contacts with granitoid plutons. The Willows Plutonic Complex is little affected by metamorphism; however, altered portions of the Willows Plutonic Complex are indistinguishable from the saussurite gabbros.

Estimates of metamorphic conditions, based on mineral assemblages and muscovite and amphibole mineral chemistries, delineate the high-P/T metamorphism of the Catalina Schist from the high-T/P metamorphic environment(s) of the arc-like rocks. The Catalina Schist apparently underwent subduction zone metamorphism in a significantly higher-T/P gradient than the Franciscan Complex of northern California. It may represent a short-lived subduction event initiated in the Cretaceous along the western Cordilleran margin. The arc-like rocks are similar to Jurassic rocks common in the western margins of the Sierra Nevada and Peninsular Ranges batholiths. Mineral assemblages of the Santa Cruz Island Schist are similar to those of relatively high-grade, arc-like protions of the Coast Range ophiolite. However, the schists and phyllites of the Santa Cruz Island Schist and the Santa Monica Formation were both deformed and metamorphosed, probably in a regime of arc-accretion. Metamorphic rocks that strongly resemble lithologies of the Franciscan Complex of northern California, and that are petrologically distinct from the Catalina Schist are apparently juxtaposed against arc-like rocks in the subsurface of the northern Channel Islands.

The Catalina Schist and the arc-like rocks are probably not portions of a coeval paired metamorphic belt. Their juxtaposition and distribution probably result from Cenozoic transcurrent tectonics.

INTRODUCTION

The California Continental Borderland consists of fault-controlled subsea ridges and basins (Emery, 1954; 1960). Many ridges are basement highs; two of their highest portions expose basement rocks on Santa Catalina and Santa Cruz Islands (Fig. 36-1). The on-land Los Angeles Basin displays several geologic features of the Borderland (Figs. 36-2 and 36-3). It is floored in part by the Catalina Schist, a metamorphic terrane that is otherwise restricted to the Borderland (Schoellhamer and Woodford, 1951; Woodford, 1960; Yeats, 1973). The basin fill conceals substantial (>3000 m) relief on the basement surface from its central deep to its margins (Woodford *et al.*, 1954; Yerkes *et al.*, 1965). The Los Angeles Basin and its faulted margins exhibit the two structural trends (northwest-southeast and east-west) characteristic of aligned faults, islands, ridges, and basins in the Borderland. The Los Angeles Basin and many basins in the Borderland probably formed in the Miocene as "pull-apart" basins at releasing bends in a transcurrent fault regime (Crowell, 1974).

Subsea and subsurface basement highs in the region have been relatively well-sampled

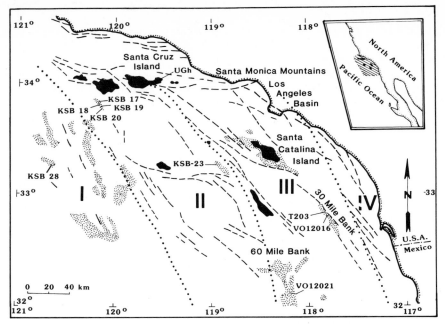

FIG. 36-1. Index map of subsea basement localities (stippled). Offshore faults are shown by dashed lines; heavy dotted lines separate basement terranes I–IV of Howell and Vedder (1981). Also shown: dredge haul localities KSB 17-20, KSB-23, KSB-28, T203, VO12016, and VO12021 and well Union Gherini 1 (UGh).

in 60 years of oil and gas drilling and 40 years of marine geological research. Samples from more than 100 oil wells to basement in the Los Angeles Basin were studied petrographically by Schoellhamer and Woodford (1951). Vedder *et al.* (1974, 1975, 1976a, b, 1977) collected and described samples of basement rocks from the California Continental Borderland.

Mesozoic basement rocks of the Los Angeles Basin and the inner California Continental Borderland are the Catalina Schist, a relatively high-P/T metamorphic complex, and a group of arc-like rocks metamorphosed (if at all) at relatively low-P/T conditions (Figs. 36-1, 36-2, and 36-3; Sorensen, 1985). The latter group includes the Willows Plutonic Complex, the Santa Cruz Island Schist, and the Santa Monica Formation, the "saussurite gabbros" described by Woodford (1925), Schoellhamer and Woodford (1951), and Platt (1976), the Sansiena schist of Yerkes (1972), and the Los Angeles Basin amphibolites of Sorensen (1985).

Previous petrotectonic interpretations of these rocks have relied on their field, structural, and age resemblance to Mesozoic terranes of northern California, supplemented by petrographic data and partial whole-rock analyses. Different proposals for the structural and tectonic evolution of the region have been based on different interpretations of the basement geology.

Howell and Vedder (1981) delineated four "terranes" (I–IV on Fig. 36-1), with contrasting basement-rock types and stratigraphic successions, in the California Continental Borderland. Crouch (1981) correlated the metagraywacke-dominated, mostly zeolite-facies but locally blueschist-facies rocks of terrane I with the Coastal Belt of the Franciscan Complex, and the basement rocks of terrane II with the Great Valley Sequence.

TECTONOMETAMORPHIC SIGNIFICANCE OF THE BASEMENT ROCKS

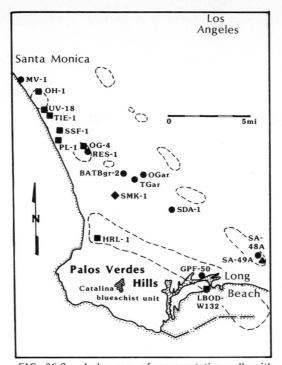

FIG. 36-2. Index map of representative wells with Catalina Schist as basement. Circles indicate the Catalina greenschist unit; squares, the Catalina blueschist unit; the diamond, ultramafic schist; and the triangle saussurite gabbro. Abbreviations: MV-1, Mobile Venice 1; OH-1, Ohio Howland 1; UV-18, Union Vidor 18; TIE-1, Texas Inglewood Extension 1-1; SSF-1, Standard Six Companies Fee 1; PL-1, Pauley Loftus 1; OG-4, Ohio Gough 4; Res-1, Republic El Segundo 1; BAT Bgr-2, British, American, and Texas Companies Bodger 2; OGar, Ohio Gardena 1; TGar, Texas Gardena 1; SMK-1, Shell McKenna 1; SDA-1, Standard Del Amo 1; HRL-1, Hunter-Devlin HRL-1; SA-48A, Shell Alamitos 48A; SA-49A, Shell Alamitos 49A; GPF-50, General Petroleum Company Ford-50; LBOD-W132, Long Beach Oil Development Company W-132. The surface exposure of the Catalina blueschist unit in George F. Canyon in the Palos Verdes Hills is shown by hatchures. Major oil fields, outlined in dashes, delineate the "Schist Ridge" trend and the Newport-Inglewood fault zone.

He included the locality SSW of Santa Rosa Island (KSB 17-20, Fig. 36-1) in his "outer belt" of Franciscan rocks. In contrast, Howell and Vedder (1981, Figs. 16-1 and 16-7) depicted this locality as a window of Franciscan Complex beneath terrane II basement. They described the basement rocks of terrane II as a composite of the Coast Range ophiolite and the Great Valley Sequence. They correlated the Willows Plutonic Complex, the Santa Cruz Island Schist, the Alamos Pluton of Santa Cruz Island, and "metagabbros" (saussurite gabbros) with the Coast Range ophiolite. The basement of terrane III is the Catalina Schist, long correlated with the Franciscan Complex (Bailey, 1941; Bailey *et al.*,

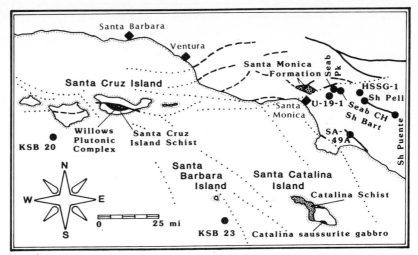

FIG. 36-3.   Index map, showing localities of Jurassic arc-like terranes. Dredge hauls KSB-20 and KSB-23 yielded rocks of probable arc affinities; saussurite gabbro occurs in well SA-49A. Well U-19-1 sampled amphibolite basement rock; well Shell Puente (Shell Puente A-3) bottomed in greenschist. Wells Seaboard Park 1-1 (Seab Pk), Seaboard Core Hole 5 (Seab CH), Shell Bartolo 1-1 (Sh Bart), Shell Pellisier 1 (Sh Pell), and Humble South San Gabriel 1 (HSSG-1) contain relatively high-grade Santa Monica Formation. HSSG-1 bottomed in amphibolite.

1964; Platt, 1975, 1976; Smith, 1897; Woodford, 1924). Howell and Vedder (1981) described the Peninsular Ranges-type basement rocks of terrane IV as a composite of Jurassic flysch and Jurassic and Cretaceous volcanic rocks and plutonic rocks.

The identification of the Willows Plutonic Complex, the Santa Cruz Island Schist, and the saussurite gabbros as dismembered Coast Range ophiolite (Hill, 1976; Hopson *et al.*, 1981; Platt and Stuart, 1974) suggested that a nappe of Coast Range ophiolite, now present only as "upper-plate remnants" on Santa Cruz and Santa Catalina Islands, had been mostly removed from a Franciscan-equivalent composite of the Catalina Schist and the Coastal Belt rocks of terrane I (Howell and Vedder, 1981). However, a composite "terrane" of the Willows Plutonic Complex, the Santa Cruz Island Schist, and the Santa Monica Formation was correlated by Jones *et al.* (1976) with Jurassic arc-like terranes of the western margins of the Sierra Nevada and Peninsular Ranges batholiths, implying that Cenozoic transcurrent faulting, accompanied by rotations, had juxtaposed relatively low-P/T arc-like terranes and the relatively high-P/T Catalina Schist (Crouch, 1979; Jones *et al.*, 1976; Kamerling and Luyendyk, 1979; Sorensen, 1985). The Coast Range ophiolite is now thought to contain arclike rocks (Sharp and Evarts, 1982; Shervais and Kimbrough, 1985), which renders either tectonic scenario equally probable.

Efforts to correlate the basement rocks of the California Continental Borderland with Mesozoic terranes of northern California have inevitably deemphasized many unique aspects of the former. The Catalina Schist and the arc-like rocks reflect Mesozoic subduction and arc-forming processes. The petrologic comparison that follows illustrates their contrasting protolith assemblages, as reflected by major, minor, and trace-element whole-rock chemistries, and metamorphic histories, seen by comparing mineral assemblages and mineral chemistries.

On Santa Catalina Island, the Catalina Schist consists of a structurally lowest blue-schist unit, overlain by a greenschist unit, both of which are overlain in turn by an amphibolite unit. All three are separated by low-angle faults (Platt, 1975, 1976). They do not share a metamorphic overprint, and were probably juxtaposed in a *P-T* regime that did not produce regional retrograde metamorphism (Sorensen, 1984a, 1986). K-Ar (91-109 m.y., Suppe and Armstrong, 1972) and U-Pb (112 m.y., Mattinson, 1986) data yield a narrow range of Cretaceous metamorphic ages for all units of the Catalina Schist. Garnet + diopside + hornblende-bearing blocks from the Catalina amphibolite unit yield 112-m.y. U-Pb ages on sphene, and are unique in the Franciscan Complex for their combination of high-grade metamorphism and relatively young ages (Mattinson, 1986).

The inverted metamorphic zonation of the Catalina Schist is interrupted by the low-angle faults which separate the three tectonic units, but apparently transcends the tectonic boundaries. The Catalina greenschist unit is itself inversely zoned. Structurally low in the Catalina greenschist unit, metabasites contain sodic amphibole (crossite); biotite + almandine-rich garnet-bearing metasediments and epidote amphibolites occur near the upper (greenschist unit-amphibolite unit) contact. Textural and zoning relationships of amphiboles suggest that the metabasites of the Catalina greenschist unit reflect an early inverted metamorphic zonation from epidote amphibolite to epidote (±lawsonite) + crossite-bearing ("glaucophanic greenschist") assemblages (Sorensen, 1984a, 1986). This zonation was overprinted by barroisite-, sodic actinolite-, or actinolite-bearing greenschist assemblages. The Catalina greenschist unit bears a striking petrologic resemblance to the Rand Schist occurrence of the Pelona Schist terrane (Jacobson and Sorensen, 1986). The greenschist unit is throughly recrystallized and penetratively deformed. Early isoclinal folds are isoclinally refolded; both sets of folds display axial planar schistosity. Still younger open folds, accompanied by crenulations, are sporadically developed throughout the unit (Platt, 1976).

The Catalina blueschist unit contains higher-*P/T* assemblages and differs structurally from the overlying greenschist unit. In some localities, the blueschist unit consists of blocks of metagraywacke, metashale, metaconglomerate, and quartz schist in a fine-grained schistose matrix. Rare garnet ± epidote-bearing high-grade blueschist and eclogite blocks that display lower-grade blueschist facies retrograde assemblages may be exotic blocks in this blueschist unit "melange." Structural geometries of the melange reflect localized deformation around blocks. Larger (>2 m) blocks are little deformed internally, but intensely so at their margins. Large (km^2) bodies of "coherent" metagraywacke underlie other areas of the Catalina blueschist unit. Glaucophane + lawsonite-bearing mineral assemblages are ubiquitous in metagraywacke, metashale, metaconglomerate, quartz schists, and metabasites. Jadeitic pyroxene is locally present in metagraywackes. Metasediments and greenstones typically retain relict sedimentary and igneous textures (Platt, 1976; Sorensen, 1984a).

The protoliths of the greenschist and blueschist units were basaltic rocks, clastic sediments, and cherts. The Catalina greenschist unit consists of ~50% metabasite schist, ~40% graywacke-composition grayschist, and ~10% Fe- and Mn-rich, locally piemontite-bearing quartz schist. Rare ultramafic lithologies are represented by tremolite-chlorite-fuchsite-chromite pods and lenses, typically intercalated with grayschist. Metagraywacke/metashale sequences comprise ~75% of the blueschist unit (Platt, 1976). The remainder consists of subequal amounts of metabasite lithologies (greenstone, blueschist, eclogite) and the ultramafic schist that comprises the melange matrix (Sorensen, 1984a). About 5% of the blueschist unit is metachert and Fe/Mn-rich quartz schist.

The protolith of the Catalina amphibolite unit consisted of ~70% ultramafic (serpentinite and ultramafic metasomatic schists) + gabbroic rocks, ~20% basaltic rocks, and ~10% sedimentary rocks (semipelitic schist and spessartine quartzites). A ~200-m-thick slab of zoisite + An_{05-15} plagioclase + magnesiohornblende-bearing amphibolites that display the major, minor, and trace-element characteristics of cumulate gabbros forms the lower part of the unit (Platt, 1976; Sorensen, 1984a). The metasediments are associated with the gabbroic amphibolites. They contain quartz + $An_{\sim08}$ plagioclase + paragonitic muscovite + alm-pyr-gross garnet ± biotite ± kyanite ± zoisite ± rutile ± graphite ± ilmenite ± magnetite ± chlorite. Both are overlain by a variably metasomatized, partially serpentinized ultramafic body that contains garnet + hornblende ± clinopyroxene-bearing blocks (Bailey, 1941; Platt, 1975, 1976; Sorensen, 1983, 1984a, b). Some of the blocks and most of the metasediments of the amphibolite unit are migmatitic (Sorensen *et al.*, 1985).

The Catalina Schist apparently underwent subduction zone metamorphism in a significantly higher-T/P gradient than northern California exposures of the Franciscan Complex (Platt, 1976; Sorensen, 1986). Platt (1975, 1976) proposed that the terrane reflected metamorphism in an inverted thermal gradient that developed below hot, hanging-wall peridotite in a newly-formed subduction zone. He suggested that the "cryptic source terrane" for high-grade eclogite and amphibolite blocks in the Franciscan Complex of the northern Coast Ranges might be an early-formed, accreted terrane similar to the Catalina Schist. However, the high-grade blocks of the Franciscan are the oldest (~150–160 m.y.) material in the Franciscan Complex (Mattinson, 1986; McDowell *et al.*, 1984; Suppe and Armstrong, 1972).

Relatively high-T, but nonetheless high-P/T, metamorphism is probably restricted to the early stages of subduction (Cloos, 1984; Platt, 1975, 1976). The Catalina Schist probably reflects the early stages of a Cretaceous subduction event (initiation of convergence) along the western Cordilleran margin.

ARC-LIKE ROCKS OF THE CALIFORNIA CONTINENTAL BORDERLAND AND THE LOS ANGELES BASIN

The southwesternmost Transverse Ranges region that includes Santa Cruz Island, the Santa Monica Mountains, and the margins of the Los Angeles Basin is underlain in part by Jurassic igneous and metamorphic rocks. The 162-m.y.-old cumulate pyroxene gabbros, hornblende gabbros, and hornblende diorites of the Willows Plutonic Complex, and the greenschist-facies metavolcanics of the Santa Cruz Island Schist occur as fault-bounded basement slivers south of the Santa Cruz Island fault on Santa Cruz Island (Figs. 36-1 and 36-3; Hill, 1976; Mattinson and Hill, 1976). The Santa Cruz Island Schist is intruded by a postmetamorphic leucotonalite body, the Alamos Pluton, that yielded a U-Pb zircon age of 141 ± 3 m.y. The late Oxfordian to early Kimmeridigan metagraywackes, metashales, and meta-argillites of the Santa Monica Formation are exposed in the Santa Monica Mountains (Criscione *et al.*, 1978; Hoots, 1931; Imlay, 1963; Jones *et al.*, 1976; Neuerberg, 1951). They are intruded by a granitoid pluton; metasediments and pluton are unconformably overlain by Upper Cretaceous sedimentary rocks.

Sorensen (1985) grouped saussurite gabbros exposed on Santa Catalina Island, and from well Shell Alamitos 49-A (Figs. 36-1, 36-2, and 36-3; Schoellhamer and Woodford, 1951), and altered plutonic rocks and greenschists from dredge haul KSB-23 (Figs. 36-1 and 36-3; Table 36-1; Vedder *et al.*, 1974, 1975) with the Willows Plutonic Complex and Santa Cruz Island Schist. Three samples of albite and oligoclase-epidote amphibolite, not exposed at the surface but found in oil wells Union-Signal-Texam U-19-1 and Humble South San Gabriel 1, and dredge haul KSB 20 (Figs. 36-1 and 36-3; Lamar, 1961; Yeats, 1973), along with the greenschist-facies metavolcanic basement rocks from wells in the Puente Hills (Fig. 36-3; Yerkes, 1972), are arc-like in their major, minor, and trace-element geochemistry (Sorensen, 1985).

Metamorphism of the arc-like rocks. Altered portions of the Willows Plutonic Complex, and the saussurite gabbros (Figs. 36-1, 36-2, and 36-3) retain igneous textures, yet contain greenschist-facies assemblages. Igneous plagioclase is replaced by albite + zoisite + clinozoisite; clinopyroxene is partially replaced by actinolitic amphibole + chlorite. Altered diorite from dredge KSB-23 (Vedder *et al.*, 1974) contains sericite in the plagioclase alteration assemblage and relict igneous hornblende rather than clinopyroxene (Table 36-1; Sorensen, 1984a).

The metavolcanic and metaplutonic rocks of the Santa Cruz Island Schist consist of schistose, fine-grained rocks, and coarser-grained rocks that retain igneous textures. Fine-grained metavolcanic rocks display relict igneous amphibole phenocrysts with actinolitic rims, albitized plagioclase phenocrysts, and radial pistacitic epidote pods after vesicles or amygdules (Hill 1976; Sorensen, 1984a, 1985). Mineral assemblages consist of albite + quartz + chlorite ± pistacitic epidote ± sphene ± actinolitic amphibole ± white mica ± magnetite ± hematite ± calcite.

The slates of the Santa Monica Formation locally retain relict sedimentary features. However, they are progressively transformed to spotted slates, phyllites, and schists near contacts with plutons. Slates and spotted slates consist of albite + quartz + white mica + carbonaceous material ± chlorite ± biotite ± cordierite ± chloritoid ± andalusite. Poikiloblastic cordierite may be replaced by aggregates of white mica + quartz ± chiastolitic andalusite. Neuerberg (1951) described chloritoid to occur both as late cross-cutting porphyroblasts overgrowing metamorphic foliations, and as coronas surrounding cordierite. Pelitic and semipelitic phyllites and schists of the Santa Monica Formation contain oligoclase + biotite + quartz + white mica + graphite ± K-feldspar ± cordierite ± sillimanite. Cordierite may be partially to entirely replaced by sericite or illite. Spotted slates, phyllites, and schists also occur in wells Seaboard Corehole 5, Seaboard Park 1, Shell Bartolo 1, and Humble South San Gabriel 1 from the northeastern margins of the Los Angeles Basin (Fig. 36-3). These rocks contain quartz + plagioclase ± biotite ± chlorite ± white mica ± cordierite ± sillimanite ± garnet ± staurolite, and are probably higher-grade lithologies of the Santa Monica Formation (Table 36-1; Sorensen, 1985).

The albite or oligoclase-epidote amphibolites found in oil wells Union-Signal-Texam U-19-1 and Humble South San Gabriel 1 and in dredge KSB-20 (Figs. 36-1 and 36-3) exhibit "pencil gneiss" textures in hand specimens and contain blue-green to pale olive-green hornblendic amphibole + albite or oligoclase ± pistacitic epidote ± sphene ± rutile ± magnetite ± ilmenite ± iron-poor epidote.

The greenschist-facies metavolcanic basement rocks of the Puente Hills (Yerkes, 1972; found in wells Shell Puente A-3, Shell Puente A-6-2, and Shell Menchego 6-2, all indicated by "Sh Puente" on Fig. 36-2) are schistose, yet exhibit pods of pistacitic epidote after vesicles or amygdules, as well as albite replacements of calcic plagioclase laths. They contain albite + epidote + chlorite + Fe-Ti oxides (Table 36-1). Yerkes (1972) reported rare actinolite and relict clinopyroxene from these samples.

The Willows Plutonic Complex, the Santa Cruz Island Schist, and the Santa Monica Formation are probably fragments of one or more Jurassic arc-like igneous and sedimentary terranes that were metamorphosed under relatively low-*P* greenschist to amphibolite

TABLE 36-1[a]   Mineral assemblages of subsea and subsurface samples from the California Continental Borderland and Los Angeles Basin

Sample	qtz	fld	chl	mus	law	epi	Ca-am	Na-am	bio	stp	gar	sph	gph	hem	mgt	apt	pyr	cal	pum	rut	Others
KSB 17A bs	x	a	x	x	x	–	–	x	–	–	–	x	–	–	–	x	–	–	–	–	–
KSB 17B gst	–	a	x	–	x	–	–	–	–	–	–	x	–	x	–	–	–	–	?	–	cpx
KSB 18A mgr	x	a	x	x	ia	–	–	x	–	–	–	–	x	x	x	–	–	–	–	–	–
KSB 18B sl	x	a, k	x	–	–	–	x	–	–	–	–	x	x	x	–	–	x	x	–	–	–
KSB 18C mgr	x	a	x	–	–	–	–	–	–	–	–	x	x	–	–	–	x	x	–	–	–
KSB 18D mgr	x	a	x	x	–	–	–	–	d	–	–	x	x	–	–	–	x	–	–	d	–
KSB 18E gst	–	a	x	–	–	x	–	–	–	x	–	x	–	s	x	x	x	x	–	–	–
KSB 19A hgbs	–	a	x	–	–	x	–	x	–	–	x	x	–	s	–	–	x	–	–	cs	–
KSB 19B hgbs	x	a	x	x	–	x	x	x	–	–	–	x	–	s	–	–	–	x	–	–	–
KSB 19C qs	x	–	x	x	–	–	–	–	–	–	–	x	–	x	–	–	–	–	–	–	omp
KSB 20A mch	x	–	–	x	–	d?	xc	xr	–	–	x	x	–	–	x	x	x	–	–	–	–
KSB 20C mgr	x	a	x	–	–	–	x	–	–	–	–	x	–	x	x	x	x	x	?	–	–
KSB 20D amp	–	a	x	x	–	ia	x	–	–	–	–	rr	–	–	x	x	–	–	–	x	–
KSB 20E hgbs?	–	a	–	x	–	ia	xc	xr	–	–	–	–	–	–	x	–	–	–	–	–	–
KSB 23A gs	x	a	–	–	–	x	x	–	–	–	–	x	–	–	x	–	–	x	–	–	–
KSB 23B mdi	–	a	x	x	–	x	x	–	–	–	–	–	–	–	x	–	–	–	–	–	–
KSB 28 ggs	–	a	x	–	–	x	–	x	–	–	–	x	–	s	x	–	x	x	–	–	–
T203B ggs	x	a	x	x	–	x	–	x	–	x	–	x	–	s	–	–	x	x	–	–	–
VO12016 ggs	x	a	x	x	–	x	x	x	–	–	–	x	–	–	–	–	–	–	–	–	–
VO12021 ggs	–	a	x	x	–	x	–	x	–	–	–	x	–	–	x	–	–	x	–	–	–
UGh1 gst	–	a	–	–	?	–	–	–	–	–	–	–	–	–	x	–	–	x	x	–	cpx?
MV-1 ggs	x	a	x	x	–	x	–	x	–	x	–	x	–	–	x	–	x	–	–	–	–
OH-1 gst	x	a	x	x	x	rt	–	x	–	–	–	x	–	–	x	–	x	x	–	–	–
UV-18 bs	–	a	x	–	x	–	–	x	–	–	–	x	–	–	–	x	x	x	–	–	–
TIE-1 hgbs	–	a	x	x	x	x	–	x	–	–	x	x	–	–	–	x	x	x	–	cs	–
SSF-1 msh	x	a	x	x	–	–	–	x	–	–	–	x	x	–	–	–	x	x	–	–	–
PL-1 msh	x	a	x	x	x	–	–	x	–	x	–	–	–	–	–	–	x	x	–	–	–
OG-4 gst	–	a	x	x	x	–	–	x	–	–	–	x	x	–	x	–	x	–	–	–	–
RES-1 gys	x	a	x	x	–	x	–	x	–	–	–	x	x	–	–	–	x	x	–	–	–
BAT Bgr2 gys	x	a	x	x	–	x	–	–	–	–	–	–	x	–	–	x	x	x	–	–	–
OGar gys	x	a	x	x	–	–	–	–	–	–	–	x	–	–	–	–	–	–	–	–	–
TGar gys	x	a	x	x	–	–	x	–	–	–	–	x	x	–	–	–	x	x	–	–	–
SMK-1 ums	–	–	x	–	–	–	–	–	–	–	–	–	–	–	x	–	–	–	–	–	–
SDA-1 gys	x	a	x	x	–	–	–	–	–	–	–	x	x	–	–	–	x	x	x	–	–
HRL-1 qs	x	a	x	x	–	–	–	x	–	–	–	x	–	–	x	–	x	x	x	–	–
SA-48A qs	x	–	x	x	–	–	x	–	–	–	x	x	–	–	–	–	x	–	–	–	–

Sample	qtz	fld	chl	mus	law	epi	Ca-am	Na-am	bio	stp	gar	sph	gph	hem	mgt	apt	pyr	cal	pum	rut
SA-49A sg	–	a	–	x	–	zo; cz	x	–	–	–	–	–	–	–	x	–	–	x	–	–
GPF-50 qs	x	–	x	x	–	rt	–	–	–	–	–	–	–	–	–	–	–	x	–	to
LBOD W132 gst	–	a	x	x	x	–	–	–	x	–	–	x	–	x	x	–	–	x	–	–
U-19-1 amp	–	o	x	x	x	zo; cz	x	–	x	–	–	x	–	–	–	x	–	–	–	–
SeabPk1 phy	x	o	x	x	–	–	–	x	x	x	x	–	x	–	x	–	–	–	–	–
SeabCH5 spsl	x	a	x	x	–	–	–	x	x	–	x	–	x	–	–	–	–	x	–	to
ShBart phy	x	a, k	x	x	–	–	–	x	x	x	x	–	–	–	–	–	–	–	d	to, il, sta
HSSG-1 sch	x	o	x	x	rt	–	–	x	x	–	x	–	–	–	x	–	–	–	–	il, zr
HSSG-1 amp	x	o	sc	x	–	x	–	–	x	–	x	s	–	x	–	–	–	–	–	–
ShPell grt	x	p, k	sc	s	–	hb	–	–	x	–	–	x	–	x	–	–	–	–	–	–
ShPuente SPA3 gs	–	a	x	x	–	x	–	–	x	–	–	x	–	–	–	x	–	–	–	–
ShSM 6-2 gs	–	a	x	x	–	x	–	–	x	–	–	x	–	x	x	–	–	x	–	–
ShSPA 6-2 gs	–	a	x	x	–	x	–	–	x	–	–	x	–	x	x	–	–	x	–	–

[a] qtz, quartz; fld, feldspar; chl, chlorite; mus, muscovite; law, lawsonite; epi, epidote; Ca-am, Ca-amphibole; Na-am, Na-amphibole; bio, biotite; stp, stilpnomelane; gar, garnet; sph, sphene; gph, graphite; hem, hematite; mgt, magnetite; apt, apatite; pyr, pyrite; cal, calcite; pum, pumpellyite; rut, rutile; a, albite; k, K-feldspar; p, plagioclase; ia, inclusion in albite; s, secondary; sc, sericite; rt, relict; d, detrital; zo, zoisite; cz, clinozoisite; c, core; r, rim; hb, hornblende; cs, as cores of sphene; rr, as rims on rutile; cpx, clinopyroxene; omp, omphacite; to, tourmaline; il, ilmenite; sta, staurolite; zr, zircon; bs, blueschist; gst, greenstone; mgr, metagraywacke; sl, slate; hgbs, high-grade blueschist; qs, quartz schist; mch, metachert; gs, greenschist; amp, amphibolite; mdi, altered diorite; ggs, glaucophanic greenschist; msh, metashale; gys, grayschist; ums, ultramafic schist; sg, saussurite gabbro; phy, phyllite; spsl, spotted slate; sch, schist; grt, granite.

facies, in part by thermal and hydrothermal effects of a Late Cretaceous silicic pluton (Sorensen, 1985). Greenschist- to amphibolite-facies basement rocks from the northern and northeastern margins of the Los Angeles Basin also resemble arc-like matabasites and metasediments metamorphosed under relatively low-P, water-rich conditions (Sorensen, 1985).

WHOLE-ROCK CHEMICAL CONTRASTS BETWEEN THE CATALINA SCHIST AND THE ARC-LIKE ROCKS

The whole-rock major, minor, and trace-element contents of most metabasites from the Catalina Schist and the arc-like rocks apparently reflect their protoliths. Despite relatively high Na/Na + Ca) ratios, probably acquired during alteration (Hill, 1976; Sorensen, 1984a, 1985, 1986). Catalina Schist metabasites plot in the ocean-floor field, and the arc-like rocks in the island arc field, of an A-F-M plot (Fig. 36-4; Miyashiro, 1974). TiO_2 contents of Catalina metabasites range from 0.5 to 3.2 wt %; SiO_2 contents range from 44 to 57 wt %. TiO_2 contents of metabasites from the Santa Cruz Island Schist range from 0.4 to 1.7 wt %; SiO_2 contents range from 54 to 74 wt %. TiO_2 contents of the Willows Plutonic Complex range from 0.3 to 0.6 wt %, and SiO_2 contents range from 48 to 55 wt % (all data from Sorensen, 1984a, 1985, 1986).

Some discriminant diagrams for magma types of metamorphic protoliths use those trace elements which are thought to be less mobile in metamorphic systems than major elements (Dungan *et al.*, 1983; Evans *et al.*, 1981; Pearce and Cann, 1973; Winchester and Floyd, 1976). The minor and trace-element contents of most metabasites from both the Catalina Schist and the Santa Cruz Island Schist apparently reflect ocean-floor tholeiite compositions in the former suite, and the arclike affinities of the latter group (Figs. 36-5 and 36-6). Unaltered rocks of the Willows Plutonic Complex and all of the Santa Cruz Island schists plot in a "volcanic arc basalt" field on a plot of Ti versus Cr; most of the Catalina Schist metabasites plot in the "ocean-floor basalt" field (Fig. 36-5; Pearce, 1975). REE (rare earth element) patterns of metabasites from the Catalina Schist are slightly depleted in L(light)REE relative to H(heavy)REE, a characteristic of ocean-floor tholeiites (Fig. 36-6; Lofgren *et al.*, 1981). The Santa Cruz Island Schist and the Los Angeles Basin amphibolites display slightly LREE-enriched REE patterns similar to those determined for the arc-like metavolcanics of the Smartville Complex (Fig. 36-6; Menzies *et al.*, 1980; Sorensen, 1985). Despite the potentially profound effect of metamorphic alteration, each terrane has seemingly retained a consistent goechemical signature: originally, the Catalina metabasites were a group of ocean-floor tholeiites, and the Santa Cruz Island Schist and Los Angeles Basin metabasites, an arc-like association of basalt-andesite-rhyolite.

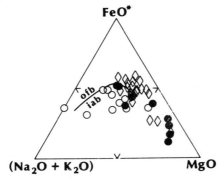

FIG. 36-4.   A-F-M plot of major-element, whole-rock analyses of metabasites from the Catalina Schist (diamond symbols) and the Santa Cruz Island Schist (open circles), and igneous rocks from the Willows Plutonic Complex (filled circles). Total iron = ferrous iron. Miyashiro's (1974) discriminant line separates ofb (ocean-floor basalt) from iab (island-arc basalt).

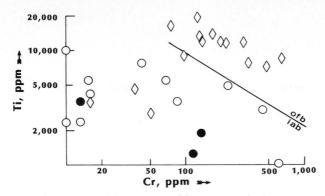

FIG. 36-5. Ti and Cr contents of basement rocks, in parts per million. Diamonds are metabasites from the Catalina Schist, open circles are metabasites from the Santa Cruz Island Schist, and filled circles are diorites and gabbros from the Willows Plutonic Complex. Pearce's (1975) empirical discriminant line separates ocean-floor basalt (ofb) and island-arc basalt (iab).

MINERAL CHEMISTRY OF MUSCOVITES AND AMPHIBOLES FROM THE BASEMENT TERRANES

Miyashiro (1973) proposed that metamorphic terranes could be classified in terms of pressure-dependent facies series, based on pressure-sensitive mineral parageneses. Greenschist-facies rocks from the Catalina Schist and the arc-like terrane(s) display mineral assemblages that are stable over a large pressure range (Ernst, 1976, 1979). However, the celadonite contents of muscovites (Frey *et al.*, 1983; Guidotti and Sassi, 1976; Jacobson, 1983; Massone and Schreyer, 1983; Saliot and Velde; 1982) and the glaucophane content of calcic amphiboles (Brown, 1974, 1977; Ernst, 1979; Graham, 1974; Holland and Richardson, 1979; Jacobson and Sorensen, 1986; Laird and Albee, 1981; Sorensen, 1986; Trzcienski *et al.*, 1984) both appear to increase with pressure and can be used to distinguish relatively low-P from high-P greenschists.

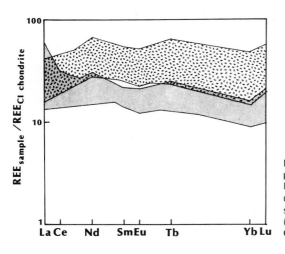

FIG. 36-6. Ranges of whole-rock REE patterns from metabasites of the Catalina blueschist, greenschist, and amphibolite units (random-dash pattern) and of greenschists from the Santa Cruz Island Schist (stipples). Data normalized to an average of CI chondrites (Ebihara *et al.*, 1982).

Muscovites. The Si-contents of muscovites from metasediments and metabasites of the blueschist and greenschist units of the Catalina Schist and from the arc-like greenschist and amphibolite facies metasediments and metabasites are plotted in Fig. 36-7; the data are from Sorensen (1984a, 1985, 1986).

Muscovites from metasediments and metabasites of the Catalina blueschist unit display significantly higher Si-contents than their counterparts in the Catalina greenschist unit (Fig. 36-7a and b). Phengitic muscovites have >3.0 Si in an 11-oxygen formula. This reflects a celadonitic substitution $[(Mg, Fe^{2+}) + Si = 2\ Al]$, favored by high-fluid pressure, low-temperature metamorphic conditions (Ernst, 1963; Guidotti, 1984; Guidotti and Sassi, 1976). Metabasites and metasediments within each of the blueschist and greenschist units of the Catalina Schist terrane contain muscovites with similar Si-contents (Fig. 36-7a and b). Semipelitic, migmatitic metasediments of the Catalina amphibolite unit contain coarse-grained, paragonitic muscovites (Si 3.08–3.16; Na 0.16–0.24 per 11-oxygen formula) that are partially replaced by fine-grained celadonitic muscovites (Sorensen, 1984a).

Muscovites from the Santa Monica Formation are Si-poor compared to those from the Santa Cruz Island Schist (Fig. 36-7c). Sorensen (1985) attributed the difference to a bulk compositional effect at relatively low pressures.

The formula $(Mg + Fe^{2+})$ contents of muscovites from the Catalina greenschist unit, the Catalina amphibolite unit, and the Santa Cruz Island Schist are generally in excess of a 1:1 correlation with Si (Figs. 36-8a and b), suggesting that some of their iron is present as Fe^{3+}.

Amphiboles. Calic amphiboles from the Catalina greenschist and amphibolite units exhibit higher total Al-contents and a greater degree of glaucophanic substitution than do calcic

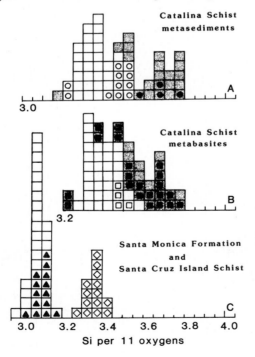

FIG. 36-7. Si-contents, per 11 oxygen formula, of muscovites. Analyses from (A) metasediments and (B) metabasites of the Catalina greenschist (unshaded) and blueschist (shaded) units are shown. Open (greenschist unit) and filled (blueschist unit) symbols indicate a subsea or subsurface locality. Analyses from metasediments from exposures (open squares) or well localities of the Santa Monica Formation (filled triangles), and from metabasites of Santa Cruz Island Schist (open diamond symbols), are shown in (C).

TECTONOMETAMORPHIC SIGNIFICANCE OF THE BASEMENT ROCKS

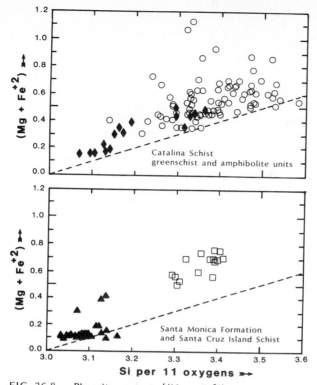

FIG. 36-8. Phengite contents [(Mg + Fe²⁺) versus Si per 11 oxygens] of muscovites from the Catalina Schist, the Santa Monica Formation, and the Santa Cruz Island Schist. The upper plot shows analyses from metasediments and metabasites of the Catalina greenschist unit (open circles) and metasediments of the amphibolite unit (filled diamonds). The lower plot shows analyses from exposures and subsurface localities (filled triangles) of the Santa Monica Formation and from the Santa Cruz Island Schist (open squares). The dashed line is a 1:1 correlation of (Mg + Fe²⁺) and Si in excess of 3.0.

amphiboles from the Santa Cruz Island Schist and the Los Angeles Basin amphibolites (Figs. 36-9 and 36-10; data from Sorensen, 1984a, 1985, 1986). This reflects the relatively high-*P* metamorphism of the Catalina greenschist and amphibolite units. The Catalina greenschist unit contains crossite-bearing glaucophanic greenschists. Sodic amphiboles from these lithologies plot outside the "Maximum Al VI" field, which applies only to calcic amphiboles (Fig. 36-9a). The empirical "5-kb" line was used by Raase (1974) to distinguish calcic amphiboles that occur in relatively high-pressure terranes from their lower-pressure counterparts.

Glaucophanic substitutions (sodic amphiboles plot above 1.34 Na in the M4-site on Fig. 36-10; Leake, 1978) also appear in sodic actinolites, barroisites, and winchites from the Catalina greenschist unit. The transition from actinolitic (greenschist-facies) to hornblendic (amphibolite-facies) amphiboles in each terrane is shown by a progressive increase in Al contents (Figs. 36-9 and 36-10).

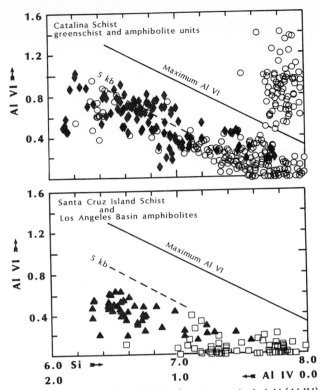

FIG. 36.9. Octahedral Al (Al VI) versus tetrahedral Al (Al IV) in a 23-oxygen amphibole formula. The Catalina greenschist (open circles) and amphibolite (filled diamonds) units are shown on the upper plot; analyses from the Santa Cruz Island Schist (open squares) and from amphibolites KSB-20, U-19-1, and HSSG-1 (filled triangles) are shown on the lower plot. Solid lines for "Maximum Al VI" and "5 kb" from Raase (1974).

METAMORPHIC *P-T* ESTIMATES FOR THE BASEMENT TERRANES

Multivariant, experimentally-determined phase equilibria illustrate *P-T* conditions for mineral assemblages of the Catalina Schist and the Santa Monica Formation (Fig. 36-11).

Catalina Schist. The three units of the Catalina Schist were each metamorphosed in a restricted temperature range at relatively high pressures (Fig. 36-11). The ubiquitous occurrence of the mineral assemblage lawsonite + glaucophane, the presence of jadeite forming at the expense of albite in some metagraywackes and metashales, and the paucity of chlorite in metagraywackes are all consistent with metamorphic $T \sim 200$–$400°C$ for the Catalina blueschist unit. A higher minimum T ($\sim 300°C$) is probably appropriate, since glaucophane + lawsonite + jadeite-bearing metagraywackes indicate relatively "high" temperatures as well as pressures of recrystallization (Ernst, 1965, 1971), and because zeolite-facies assemblages are absent. The celadonite contents of the muscovites suggest $P \sim 12$–15 kb at $T \sim 300°C$, although assemblages lacking jadeite could have formed at pressures as low as the minimum for the stability of glaucophane + lawsonite. The highest

TECTONOMETAMORPHIC SIGNIFICANCE OF THE BASEMENT ROCKS

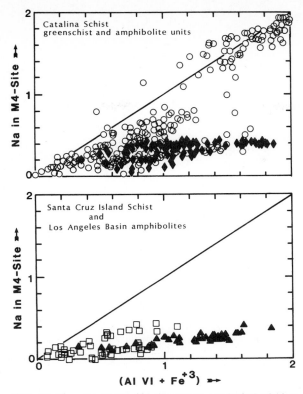

FIG. 36-10. Na in the M4-site versus octahedral trivalent cations Al Vi + Fe^{3+}) plotted for analyses of amphiboles from the Catalina amphibolite and greenschist units (upper diagram) and from the Los Angeles Basin amphibolites and the Santa Cruz Island Schist (lower diagram). The solid line shows a 1:1 (glaucophane substitution) correlation. Symbols as in Fig. 36-9.

temperature estimates (~520°C at $P \sim 12$ kb) for the blueschist unit are based on garnet-clinopyroxene geothermometry, combined with a pressure estimate from the jadeite content of omphacite in eclogite blocks from the blueschist unit melange (Sorensen, 1986).

The Catalina greenschist unit lacks lawsonite, jadeite, and aragonite, and was originally zoned from structurally low glaucophanic greenschists to structurally high epidote amphibolites. The absence of lawsonite and presence of pistacitic epidote + crossitic amphibole in the glaucophanic greenschists are evidence for $T \sim 350$–450°C at $P \sim 4$–7 kb (Fig. 36-11; Brown, 1974; Brown and Ghent, 1983). The epidote amphibolite facies rocks of the greenschist unit that contain almandine-rich garnet, ilmenite, or rutile rimmed with sphene, and hornblendic amphibole probably crystallized at $T \sim 450$–575°C and $P \sim 4$–7 kb (Sorensen, 1986). The mineral chemistry of amphiboles that rim and replace early crossites and hornblendes, and the celadonite contents of muscovites, suggest a relatively high-P greenschist-facies overprint.

Metamorphic P-T estimates for the Catalina amphibolite unit, constrained by the occurrence of kyanite in metasediments and migmatites, garnet-clinopyroxene geothermometry (Ellis and Green, 1979), the presence of assemblages such as anthophyllite + talc + quartz or enstatite + chlorite in ultramafic metasomatites, and fluid inclusion

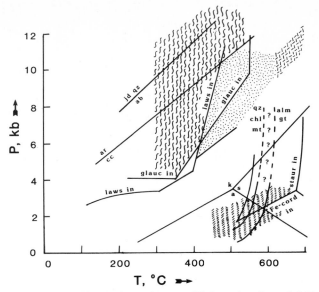

FIG. 36-11. Multivariant phase equilibria and estimated *P-T* conditions for metamorphism of the Catalina Schist terrane and the arc-like rocks. Abbreviations: jd, jadeite; qz, quartz; ab, albite; ar, aragonite; laws, lawsonite; glauc, glaucophane; chl, chlorite; mt, magnetite; alm gt, almandine garnet; staur, staurolite; Fe-cord, iron-rich cordierite; a, andalusite; k, kyanite; s, sillimanite. The reaction curves are: jd + qz = ab (Newton and Smith, 1967), ar = cc (Boettcher and Wyllie, 1968), laws in (Liou, 1971), glauc in (Maresch, 1977), staur in and Fe-cord in (Richardson, 1968), qz + chl + mt = alm gt (Hsu, 1968), and the aluminosilicates (Holdaway, 1971). Shaded areas are *P-T* estimates for the Catalina blueschist (s-shading), greenschist (stippled), and amphibolite (random-dashed) units; estimates for the Santa Monica Formation and the Santa Cruz Island Schist are herringbone-patterned.

data (Sorensen, 1984a; Sorensen *et al.*, 1985) yield $T \sim 600\text{--}750°C$ at $P \sim 8\text{--}12$ kb (Fig. 36-11).

Arc-like terrane(s). Experimentally determined phase relations for almandine, Fe-rich staurolite, chloritoid, and cordierite suggest metamorphic $T \sim 450\text{--}650°C$ at $P > 4$ kb for the higher-grade portions of the Santa Monica Formation (Fig. 36-11; Sorensen, 1985). Mineral assemblages of the Santa Monica Formation schists that are developed along pluton contacts include biotite + cordierite or biotite + sillimanite coexisting with quartz + muscovite + plagioclase, suggesting maximum $T \sim 550\text{--}600°C$ at $P \sim 3$ kb (Fig. 36-11). The *P-T* estimates are also appropriate for the metapelite from Humble South San Gabriel 1 (Fig. 36-3; Table 36-1) that contains staurolite in a reaction relation of the form: biotite + staurolite + H_2O = almandine garnet + chlorite, and for associated oligoclase-epidote amphibolite (Fig. 36-3; Table 36-1). The mineral assemblages of these rocks are typical of low-pressure greenschist (andalusite, chloritoid, chlorite) to epidote-amphibolite (andalusite, sillimanite, cordierite, almandine) facies metapelites (Miyashiro, 1973; Sorensen, 1985).

The Santa Cruz Island Schist is a low-pressure greenschist terrane. The Los Angeles Basin amphibolites are probably metabasite analogues to the higher-grade pelitic schists and phyllites of the Santa Monica Formation, formed by local heating of a metabasite protolith near pluton contacts (Sorensen, 1985). *P-T* conditions for the metabasite suite are comparable to those estimated for slates, phyllites, and schists of the Santa Monica Formation (Sorensen, 1985).

The Santa Monica Formation and Santa Cruz Island Schist resemble Jurassic basement-rock associations from the Sierra Nevada Foothills (Duffield and Sharp, 1975; Saleeby, 1981; Schweickert, 1981) as well as the prebatholithic, in part Jurassic, arclike metasedimentary and metavolcanic rocks of the western Peninsular Ranges (Gastil *et al.*, 1975; Hawkins, 1970; Larsen, 1948; Schoellhamer *et al.*, 1981). However, arc-like rocks with greenschist-facies mineral assemblages have also been reported from the Del Puerto locality of the Coast Range ophiolite (Evarts, 1977; Evarts and Schiffman, 1983). Evarts and Schiffman (1983) found pistacitic epidote + chlorite + albite + hematite + sphene + quartz ± calcite ± pyrite ± actinolite-bearing rocks in the deepest part of the Del Puerto volcanic section. The greenschists are successively overlain by pumpellyite and zeolite-facies rocks, and thus display "the typical ophiolite pattern of increasing grade, but generally decreasing intensity, of metamorphism with depth" (Evarts and Schiffman, 1983). Bauder and Liou (1979) reported mineral assemblages of chlorite + epidote + albite + actinolite from metamorphosed hornblende gabbro, diorite, and plagiogranite in a tectonic outlier of presumed Coast Range ophiolite from the Diablo Range. They suggested that "apparently these rocks were subjected only to the initial stages of greenschist-facies metamorphism." Neither investigation described schistose or phyllitic structures in metavolcanic or metaplutonic rocks. The Santa Cruz Island Schist and Santa Monica Formation, although similar in both protolith and metamorphic grade to both the Foothills and the Coast Ranges occurrences of Mesozoic arc-like rocks, were metamorphosed in a regime in which deformation accompanied local or regional heating. The Jurassic arc-like rocks of the California Continental Borderland probably represent portions of volcanic edifices deformed and metamorphosed by the emplacement of silicic plutons in an evolving arc.

PROBABLE FRANCISCAN BASEMENT ROCKS OF THE CALIFORNIA CONTINENTAL BORDERLAND

The Transverse Ranges structural trend (Figs. 36-1 and 36-3) contains two additional localities of probable Mesozoic basement, the basement rock recovered from well Union Gherini 1 (7370–7377 ft), drilled north of the Santa Cruz Island fault on Santa Cruz Island (Fig. 36-1), and the rocks in dredge hauls KSB 17-20, from a subsea ridge ~ 15 km SSW of Santa Rosa Island (Figs. 36-1 and 36-3).

Union Gherini 1 probably contains Franciscan rather than Catalina Schist basement. The rock is a greenstone that only locally (on the scale of a hand specimen) displays lawsonite and glaucophane, the latter typically as rims on relict clinopyroxene. It is cut by laumontite-bearing veins. Howell *et al.* (1976) reported a whole-rock K-Ar age determination made by Geochron Laboratories of 152 ± 8 m.y. for the so-called Gherini greenstone. Jones *et al.* (1976) observed pumpellyite and possibly prehnite in some of the six thin sections cut from the core. These authors reported that igneous textures are "almost completely obliterated by metamorphic recrystallization of lawsonite, actinolite, and glaucophane" in some portions of the rock. The reported age and the heterogeneous zeolite, prehnite-pumpellyite, and blueschist assemblages of the greenstone have no analogues in the Catalina greenstones studies by Platt (1976) and Sorensen (1984a; 1985).

Samples from KSB 17-20 are also unlike typical lithologies of the Catalina Schist.

Several samples of feebly recrystallized, detrital epidote-bearing metagraywacke apparently contain pumpellyite. A sample of green hornblende + epidote-bearing amphibolite is overprinted by sodic amphibole. Dredge hauls KSB 17-20 may in part represent a locality of Coastal Belt Franciscan (Crouch, 1981; Fig. 36-1, terrane I). The presence of arc-like amphibolite sample KSB 20D (Table 36-1; Figs. 36-1 and 36-3) suggests that the area may also be underlain by arc-like basement rocks.

SUMMARY

The Catalina Schist, apparently metamorphosed in the Cretaceous, consists of three tectonic units with distinct structural and metamorphic histories. The units are disposed in an inverted metamorphic gradient; a structually low blueschist unit is successively overlain by a greenschist and an amphibolite unit. Metamorphism of each unit occurred at high-P/T conditions. Metabasite protoliths were primarily ocean-floor tholeiites and cumulate gabbros; metasedimentary protoliths were primarily graywackes, shales, and cherts. The metamorphism probably reflects a P-T regime related to the initiation of subduction along some part of the Cordilleran margin during the Cretaceous.

One or more arc-like terranes are present in the California Continental Borderland. Exposed arc-like rocks include the Willows Plutonic Complex, the Santa Cruz Island Schist, and the Santa Monica Formation, which occur as fault-bounded slivers, mostly in the southwesternmost Transverse Ranges. The arc-like rocks are Jurassic in age; metamorphism is pre-Late Cretaceous, and in part >141 m.y. in age. The northern and northeastern margins of the Los Angeles Basin are underlain by slates, mafic and pelitic schists, and amphibolites that are also of probable arc origin. Low-P/T metamorphism, in part related to contacts with a granitoid pluton, displays a hydrothermal aspect. Metabasite protoliths were a basalt-andesite-rhyolite suite; metasediment protoliths were graywackes and argillites. Altered portions of the Willows Plutonic Complex are indistinguishable from saussurite gabbros exposed on Santa Catalina Island, and that occur in dredge KSB-23 and well Shell Alamitos 49A.

The arc-like rocks are similar to Jurassic terranes of the western margins of the Sierra Nevada and Peninsular Ranges batholiths. Metamorphic mineral assemblages of the Santa Cruz Island Schist are similar to those of relatively high-grade arc-like localities of the Coast Range ophiolite. However, the schistose and phyllitic nature of the Santa Cruz Island Schist and of the Santa Monica Formation suggest that these greenschist-facies rocks were dynamically recrystallized, perhaps in an arc-accretion process coeval with silicic plutonism.

In subsea and subsurface basement highs of the northern Channel Islands, the arc-like rocks are apparently juxtaposed against metamorphic rocks that strongly resemble lithologies of the Franciscan Complex of northern California, and that are distinct from the Catalina Schist.

The Catalina Schist terrane and the arc-like rocks were seemingly spatially and temporally unrelated when they acquired their geochemical properties and were metamorphosed. They do not comprise portions of a paired metamorphic belt. The juxtaposition and distribution of the two terranes is probably linked to Cenozoic transcurrent fault activity.

ACKNOWLEDGMENTS

Much of the data for this paper were obtained at U.C.L.A., with support from NSF grant EAR83-12702, a GSA Penrose grant, and a NSF Graduate Fellowship. Some samples were provided by J. E. Schoellhamer, J. G. Vedder, R. F. Yerkes, and D. G. Howell of the U.S. Geological Survey and by R. S. Yeats of Oregon State University. The late A. K. Baird of Pomona College, with characteristic generosity, devoted his time and XRF equipment for whole-rock, major-element analyses. Trace-element data were obtained at U.C.L.A. by instrumental neutron activation analysis with the able assistance of G. W. Kallemyen, F. T. Kyte, and J. N. Grossman. R. Alkaly made a large number (~800!) of high-quality thin sections and probe sections. R. E. Jones helped obtain the microprobe mineral analyses. This study was first suggested by A. O. Woodford of Pomona College, who offered substantial assistance and encouragement. W. G. Ernst enthusiastically supported the project. Discussions with J. N. Grossman, C. E. Jacobson, and J. S. Beard are gratefully acknowledged. J. S. Beard, R. F. Fudali, and P. Schiffman reviewed the manuscript. Their suggestions, although not always heeded, are certainly appreciated.

REFERENCES

Bailey, E. H., 1941, Mineralogy, petrology and geology of Santa Catalina Island: Ph.D. dissertation, Stanford Univ., Stanford, Calif., 193 p.

——, Irwin, W. P., and Jones, D. L., 1964, Franciscan and related rocks, and their significance in the geology of western California: *Calif. Div. Mines Geology Bull. 183*, 177 p.

Bauder, J. M., and Liou, J. G., 1979, Tectonic outlier of Great Valley sequence in Franciscan Terrain, Diablo Range, California: *Geol. Soc. America Bull.*, v. 90, p. 561–568.

Boettcher, A. L., and Wyllie, P. J., 1968, The calcite-aragonite transition measured in the system $CaO-CO_2-H_2O$: *J. Geology*, v. 76, p. 314–330.

Brown, E. H., 1974, Comparison of the mineralogy and phase relations of blueschists from the North Cascades, Washington, and greenschists from Otago, New Zealand: *Geol. Soc. America Bull.*, v. 85, p. 333–344.

——, 1977, The crossite content of Ca-amphibole as a guide to pressure of metamorphism: *J. Petrology*, v. 18, p. 53–72.

——, and Ghent, E. D., 1983, Mineralogy and phase relations in the blueschist facies of the Black Butte and Ball Rock areas, northern California Coast Ranges: *Amer. Mineralogist*, v. 68, p. 365–372.

Cloos, M. P., 1984, Flow melanges and the structural evolution of accretionary wedges: *Geol. Soc. America Spec. Paper 198*, p. 71–79.

Criscione, J. J., Davis, T. E., and Ehlig, P., 1978, The age and sedimentation/diagenesis for the Bedford Canyon Formation and the Santa Monica Formation in southern California — An Rb/Sr evaluation, *in* Howell, D. G., and McDougall, K. A., eds., *Mesozoic Paleogeography of the Western United States:* Pacific Section, Soc. Econ. Paleontologists Mineralogists, Pacific Coast Paleogeography Symp. 2, p. 385–396.

Crouch, J. K., 1979, Neogene tectonic evolution of the California continental borderland and western Transverse Ranges: *Geol. Soc. America, Bull.*, v. 90, pt. I, p. 338–345.

——, 1981, Northwest margin of California continental borderland — Marine geology and tectonic evolution: *Amer. Assoc. Petrol. Geologists Bull.*, p. 191–218.

Crowell, J. C., 1974, Origin of late Cenozoic basins in southern California, *in* Dickinson, W. R., ed., *Tectonics and Sedimentation:* Soc. Econ. Paleontologists Mineralogists Spec. Publ. 22, p. 190–204.

Duffield, W. A., and Sharp, R. V., 1975, Geology of the Sierra foothills melange and adjacent areas, Amador County, California: *U.S. Geol. Survey Prof. Paper 827*, 30 p.

Dungan, M. A., Vance, J. A., and Blanchard, D. P., 1983, Geochemistry of the Shuksan greenschists and blueschists, North Cascades, Washington — Variably fractionated and altered metabasalts of oceanic affinity: *Contrib. Mineralogy Petrology*, v. 82, p. 131–146.

Ebihara, M., Wolf, R., and Anders, E., 1982, Are Cl chondrites chemically fractionated? A trace element study: *Geochim. Cosmochim. Acta*, v. 46, p. 1849–1862.

Ellis, D. J., and Green, D. H., 1979, An experimental study of the effect of Ca upon garnet-clinopyroxene Fe-Mg exchange equilibria: *Contrib. Mineralogy Petrology*, v. 71, p. 13–22.

Emery, K. O., 1954, General geology of the offshore area, southern California, *in* Jahns, R. H., ed., *Geology of Southern California:* Calif. Div. Mines Geology Bull. 170, p. 107–111.

——, 1960, *The Sea off Southern California, a Modern Habitat of Petroleum:* New York, Wiley, 366 p.

Ernst, W. G., 1963, Significance of phengitic micas from low-grade schists: *Amer. Mineralogist*, v. 48, p. 1357–1373.

——, 1965, Mineral parageneses in Franciscan metamorphic rocks, Panoche Pass, California: *Geol. Soc. America Bull.*, v. 76, p. 879–914.

——, 1971, Petrologic reconnaisance of Franciscan metagraywackes from the Diablo Range, central California Coast Ranges: *J. Petrology* v. 12, p. 413–437.

——, 1976, *Petrologic Phase Equilibria:* San Francisco, W. H. Freeman, 333 p.

——, 1979, Coexisting sodic and calcic amphiboles from high-pressure metamorphic belts and the stability of barroisitic amphibole: *Mineral. Mag.*, v. 43, p. 269–278.

Evans, B. W., Trommsdorff, V., and Goles, G. G., 1981, Geochemistry of high-grade eclogites and metarodingites from the Central Alps: *Contrib. Mineralogy Petrology*, v. 76, p. 301–311.

Evarts, R. C., 1977, The geology and petrology of the Del Puerto ophiolite, Diablo Range, central California Coast Ranges, *in* Coleman, R. G., and Irwin, W. P., eds., *North American Ophiolites:* Oreg. Dept. Geology Min. Industries Bull. 95, p. 121–139.

——, and Schiffman, P., 1983, Submarine hydrothermal metamorphism of the Del Puerto ophiolite, California: *Amer. J. Sci.*, v. 283, p. 289–340.

Frey, M., Hunziker, J. C., Jager, E., and Stern, W. B., 1983, Regional distribution of white K-mica polymorphs and their phengite content in the Central Alps: *Contrib. Mineralogy Petrology*, v. 83, p. 185–197.

Gastil, R. G., Phillips, R. P., and Allison, E. C., 1975, Reconnaisance geology of the state of Baja California: *Geol. Soc. America Mem. 140*, 170 p.

Graham, C. M., 1974, Metabasite amphiboles of the Scottish Dalradian: *Contrib. Mineralogy Petrology*, v. 47, p. 165–185.

Guidotti, C. V., 1984, Micas in metamorphic rocks, *in* Bailey, S. W., ed., *Micas:* Min. Soc. America Rev. Mineralogy, v. 13, p. 357–456.

——, and Sassi, F. P., 1976, Muscovite as a petrogenetic indicator mineral in pelitic schists: *Neues Jahrb. Mineralogie Abh.*, v. 127, no. 2, p. 97–142.

Hawkins, J. W., 1970, Metamorphosed Late Jurassic andesites and dacites of the Tijuana-Tecate area, Baja California, *in Pacific Slope Geology of Northern Baja California and Adjacent Alta California:* Amer. Assoc. Petrol. Geologists, Pacific Section, Soc. Econ. Paleontologists Mineralogists, p. 25–29.

Hill, D. J., 1976, Geology of the Jurassic basement rocks, Santa Cruz Island, California, and correlation with other Mesozoic basement terranes in California, *in* Howell,

D. G., ed., *Aspects of the Geologic History of the California Continental Border-land:* Pacific Section, Amer. Assoc. Petrol. Geologists, Misc. Publ. 24, p. 16–46.

Holdaway, M. J., 1971, Stability of andalusite and the aluminum silicate phase diagram: *Amer. J. Sci.*, v. 271, p. 97–131.

Holland, T. J. B., and Richardson, S. W., 1979, Amphibole zonation in metabasites as a guide to the evolution of metamorphic conditions: *Contrib. Mineralogy Petrology*, v. 70, p. 143–148.

Hoots, H. W., 1931, Geology of the eastern part of the Santa Monica Mountains, Los Angeles County, California: *U.S. Geol. Survey Prof. Paper 165-C*, p. 83–134.

Hopson, C. A., Mattinson, J. M., and Pessagno, E. A., 1981, Coast Range Ophiolite, western California, *in* Ernst, W. G., ed., *The Geotectonic Development of California (Rubey Vol. I): Englewood Cliffs, N.J.*, Prentice-Hall, p. 418–510.

Howell, D. G., and Vedder, J. G., 1981, Structural implications of stratigraphic discontinuities across the southern California borderland, *in* Ernst, W. G., ed., *The Geotectonic Development of California* (Rubey Vol. I): Englewood Cliffs, N.J., Prentice-Hall, p. 535–558.

——, McLean, H., and Vedder, J. G., 1976, Cenozoic tectonism on Santa Cruz Island, *in* Howell, D. G., ed., *Aspects of the geologic history of the California Continental Borderland:* Pacific Section, Amer. Assoc. Petrol. Geologists, Misc. Publ. 24, p. 392–416.

Hsu, L. C., 1968, Selected phase relationships in the system Al-Mn-Fe-Si-O-H — A model for garnet equilibria: *J. Petrology*, v. 9, no. 1, p. 40–83.

Imlay, R. W., 1963, Jurassic fossils from southern California: *J. Paleontology*, v. 38, no. 3, p. 505–509.

Jacobson, C. E., 1983, Relationship of deformation and metamorphism of the Pelona Schist to movement on the Vincent thrust, San Gabriel Mountains, southern California: *Amer. J. Sci.*, v. 283, p. 587–604.

——, and Sorensen, S. S., 1986, Amphibole compositions and metamorphic history of two relatively high P/T, inversely zoned metamorphic terranes of southern California: *Contrib. Mineralogy Petrology*, v. 92, p. 308–315.

Jones, D. L., Blake, M. C., and Rangin, C., 1976, The four Jurassic belts of northern California and their significance to the geology of the southern California borderland, *in* Howell, D. G., ed., *Aspects of the Geologic History of the California Continental Borderland:* Pacific Section, Amer. Assoc. Petrol. Geologists, Misc. Publ. 24, p. 342–362.

Kamerling, M. J., and Luyendyk, B. P., 1979, Tectonic rotations of the Santa Monica Mountains region, western Transverse Ranges, California, suggested by paleomagnetic vectors: *Geol. Soc. America Bull.*, v. 90, pt. I, p. 331–337.

Laird, J., and Albee A. L., 1981, High-pressure metamorphism in mafic schist from northern Vermont: *Amer. J. Sci.*, v. 281, p. 97–126.

Lamar, D. L., 1961, Structural evolution of the northern margin of the Los Angeles Basin: Ph.D. dissertation, Univ. California, Los Angeles, Calif., 142 p.

Larsen, E. S., 1948, Batholith and associated rocks of Corona, Elsinore, and San Luis Rey quadrangles, southern California: *Geol. Soc. America Mem. 29*, 182 p.

Leake, B. E., 1978, Nomenclature of amphiboles: *Amer. Mineralogist*, v. 63, p. 1023–1052.

Liou, J. G., 1971, P-T stabilities of laumontite, wairakite, lawsonite, and related minerals in the system $CaAl_2Si_2O_8$-SiO_2-H_2O: *J. Petrology*, v. 12, p. 379–441.

Lofgren, G. E., *et al.*, 1981, Petrology and chemistry of terrestrial, lunar, and meteoritic basalts, *in* Kaula, W., ed., *Basaltic Volcanism of the Terrestrial Planets, Basaltic Volcanism Study Project:* Elmsford, N.Y., Pergamon, p. 132–157.

Maresch, W. V., 1977, Experimental studies on glaucophane — An analysis of present knowledge: *Tectonophysics*, v. 43, p. 109–125.

Massone, H.-J., and Schreyer, W., 1983, A new experimental phengite barometer and its application to a Variscan subduction zone at the southern margin of the rhenohercynicum: *Terra Cognita*, v. 3, no. 2-3, p. 187.

Mattinson, J., 1986, Geochronology of high pressure-low temperature Franciscan metabasites — A new approach using the U-Pb system: *Geol. Soc. America Mem. 164*, p. 95–105.

____, and Hill, D. J., 1976, Age of plutonic basement rocks, Santa Cruz Island, California, *in* Howell, D. G., ed., *Aspects of the Geologic History of the California Continental Borderland:* Pacific Section, Amer. Assoc. Petrol. Geologists Misc. Publ. 24, p. 53–58.

McDowell, F., Lehman, D. H., Gucwa, P. R., Fritz, D., and Maxwell, J. C., 1984, Glaucophane schists and ophiolites of the northern California Coast Ranges — Isotopic ages and their tectonic implications: *Geol. Soc. America Bull.*, v. 95, p. 1373–1382.

Menzies, M. A., Blanchard, D., and Xenophontos, C., 1980, Genesis of the Smartville arc-ophiolite, Sierra Nevada foothills, California: *Amer. J. Sci.*, v. 280-A, p. 329–344.

Miyashiro, A., 1973, *Metamorphism and Metamorphic Belts:* London, Allen & Unwin, 491 p.

____, 1974, Volcanic rock series in island arcs and active continental margins: *Amer. J. Sci.*, v. 274, p. 321–355.

Neuerberg, G. J., 1951, Petrology of the pre-Cretaceous rocks of the Santa Monica Mountains, California: Ph.D. dissertation, Univ. California, Los Angeles, Calif., 180 p.

Newton, R. C., and Smith, J. V., 1967, Investigations concerning the breakdown of albite at depth in the earth: *J. Geology*, v. 75, p. 268–286.

Pearce, J. A., 1975, Basalt geochemistry used to investigate past tectonic environment on Cyprus: *Tectonophysics*, v. 25, p. 41–67.

____, and Cann, J. R., 1973, Tectonic setting of basic volcanic rocks determined using trace element analysis: *Earth Planet. Sci. Lett.*, v. 19, p. 290–300.

Platt, J. P., 1975, Metamorphic and deformational processes in the Franciscan Complex, California — Some insights from the Catalina Schist terrain: *Geol. Soc. America Bull.*, v. 86, p. 1337–1347.

____, 1976, The petrology, structure, and geologic history of the Catalina Schist terrain, southern California: *Univ. Calif. Publ. Geol. Sci.*, v. 112, p. 1–111.

____, and Stuart, C. J., 1974, Newport-Inglewood fault zone, Los Angeles Basin — Discussion: *Amer. Assoc. Petrol. Geologists Bull.*, v. 58, p. 877–898.

Raase, P., 1974, Al and Ti contents of hornblende, indicators of pressure and temperature of regional metamorphism: *Contrib. Mineralogy Petrology*, v. 45, p. 231–236.

Richardson, S. W., 1968, Staurolite stability in part of the system Fe-Al-Si-O-H: *J. Petrology*, v. 9, p. 467–488.

Saleeby, J. B., 1981, Ocean floor accretion and volcanoplutonic arc evolution of the Mesozoic Sierra Nevada, *in* Ernst, W. G., ed., *The Geotectonic Development of California* (Rubey Vol. I): Englewood Cliffs, N.J., Prentice-Hall, p. 132–181.

Saliot, P., and Velde, B., 1982, Phengite compositions and post-nappe high-pressure metamorphism in the Pennine zone of the French Alps: Earth Planet. Sci. Lett., v. 57, p. 133–138.

Schoellhamer, J. E., and Woodford, A. O., 1951, The floor of the Los Angeles Basin, Los Angeles, Orange, and San Bernardino Counties, California: *U.S. Geol. Survey Oil Gas Inv. Map OM-117,* 2 p.

_____ , Vedder, J. G., Yerkes, R. F., and Kinney, D. M., 1981, Geology of the northern Santa Ana Mountains, California: *U.S. Geol. Survey Prof. Paper 420-D*, 107 p.

Schweickert, R. A., 1981, Tectonic evolution of the Sierra Nevada, *in* Ernst, W. G., ed., *The Geotectonic Development of California* (Rubey Vol. I): Englewood Cliffs, N.J., Prentice-Hall, 87–131.

Sharp, W. D., and Evarts, R. C., 1982, New constraints on the environment of formation of the Coast Range ophiolite at Del Puerto Canyon, California: *Geol. Soc. America Abstr. with Programs*, v. 14, no. 4, p. 233.

Shervais, J. W., and Kimbrough, D. L., 1985, Geochemical evidence for the tectonic setting of the Coast Range Ophiolite – A composite of island arc-oceanic crust terrane in western California: *Geology*, v. 13, p. 35–38.

Smith, W. S. T., 1897, The geology of Santa Catalina Island: *Proc. Calif. Acad. Sci.*, v. 1, no. 1, p. 1–70.

Sorensen, S. S., 1983, The formation of metasomatic rinds around amphibolite facies blocks, Catalina Schist terrane, southern California: *Geol. Soc. America Abstr. with Programs*, v. 15, no. 5, p. 436.

_____ , 1984a, Petrology of basement rocks of the California Continental Borderland and the Los Angeles Basin: Ph.D. dissertation, Univ. California, Los Angeles, Calif., 432 p.

_____ , 1984b, Trace element effects of eclogite/peridotite metasomatism, Catalina Schist terrane, southern California: *Geol. Soc. America Abstr. with Programs*, v. 16, no. 6, p. 663.

_____ , 1985, Petrologic evidence for Jurassic, island-arc-like basement rocks in the south-western Transverse Ranges and California Continental Borderland: *Geol. Soc. America Bull.*, v. 96, p. 997–1006.

_____ , 1986, Petrologic and geochemical comparison of the blueschist and greenschist units of the Catalina Schist terrane, southern California: *Geol. Soc. America Mem. 164*, p. 59–75.

_____ , Barton, M. D., and Ernst, W. G., 1985, Anatexis of garnet amphibolites from a sub-duction zone metamorphic terrane: *Geol. Soc. America Abstr. with Programs*, v. 17, no. 7, p. 722.

Suppe, J., and Armstrong, R. L., 1972, Potassium-argon dating of Franciscan metamorphic rocks: *Amer. J. Sci.*, v. 272, p. 217–233.

Trzcienski, W. E., Jr., Carmichael, D. M., and Helmstaedt, H., 1984, Zoned sodic amphibole – Petrologic indicator of changing pressure and temperature during tectonism in the Bathurst area, New Brunswick, Canada: *Contrib. Mineralogy Petrology*, v. 85, p. 311–320.

Vedder, J. G., Beyer, L. A., Junger, A., Moore, G. W., Roberts, A. E., Taylor, J. C., and Wagner, H. C., 1974, Preliminary report of the geology of the continental borderland of southern California: *U.S. Geol. Survey Misc. Field Studies Map MF-624*, 34 p., 9 maps.

_____ , Taylor, J. C., Wagner, H. C., and Junger, A., 1975, Seafloor bedrock patterns from the Patton Escarpment to the mainland shelf of southern California: Pacific Section, *Amer. Assoc. Petrol. Geologists Abstract.*, Long Beach, Calif.

_____ , Arnal, R. E., Bukry, D., and Barron, J. A., 1976a, Preliminary descriptions of pre-Quaternary samples, R/V Lee, March 1976, offshore southern California: *U.S. Geol. Survey Open File Rpt. 76-629*, 15 p.

_____ , Arnal, R. E., and Bukry, D., 1976b, Maps showing location of selected pre-Quaternary rock samples from the California Continental Borderland: *U.S. Geol. Survey Misc. Field Studies Maps MF-737*, 3 maps.

_____ , Crouch, J. K., Arnal, R. E., Bukry, D., Barron, J. A., and Lee-Wong, F., 1977, Descriptions of pre-Quaternary samples, R/V Ellen B. Scripps, September, 1976, Patton Ridge to Blake Knolls, California Continental Borderland: *U.S. Geol. Survey Open File Rpt. 77-474*, 19 p.

Winchester, J. A., and Floyd, P. A., 1976, Geochemical magma type discrimination — Application to altered and metamorphosed basic igneous rocks: *Earth Planet. Sci. Lett.*, v. 28, p. 459–469.

Woodford, A. O., 1924, The Catalina metamorphic facies of the Franciscan Series: *Univ. Calif. Publ. Geol. Sci.*, v. 15, p. 49–68.

_____ , 1925, The San Onofre Breccia — Its nature and origin: *Univ. Calif. Publ. Geol. Sci.*, v. 15, p. 159–280.

_____ , 1960, Bedrock patterns and strike-slip faulting in southwestern California: *Amer. J. Sci.*, v. 258-A, p. 400–417.

_____ , Schoellhamer, J. E., Vedder, J. G., and Yerkes, R. F., 1954, Geology of the Los Angeles Basin, *in* Jahns, R. H., ed.: *Geology of Southern California:* Calif. Div. Mines Geology Bull. 170, p. 65–83.

Yeats, R. S., 1973, Newport-Inglewood fault zone, Los Angeles Basin, California: *Amer. Assoc. Petrol. Geologists Bull.*, v. 57, p. 117–135.

Yerkes, R. F., 1972, Geology and oil resources of the western Puente Hills area, southern California: *U.S. Geol. Survey Prof. Paper 420-C*, 63 p.

_____ , McCullogh, T., Schoellhamer, J. E., and Vedder, J. G., 1965, Geology of the Los Angeles Basin, California — An introduction: *U.S. Geol. Survey Prof. Paper 420-A*, 56 p.

James M. Mattinson
Department of Geological Sciences
University of California
Santa Barbara, California 93106

37

CONSTRAINTS ON THE TIMING OF FRANCISCAN METAMORPHISM: GEOCHRONOLOGICAL APPROACHES AND THEIR LIMITATIONS

ABSTRACT

Metamorphic rocks of the Franciscan Complex were generated under conditions of high pressure and relatively low temperature during Jurassic and Cretaceous subduction and accretion. Determination of an accurate metamorphic chronology for the Franciscan Complex is critical to our understanding of the timing and nature of this subduction/accretion process. Dating of Franciscan metamorphic rocks by a variety of methods reveals a spread of apparent ages from about 160–165 m.y. to about 70–75 m.y., but there is considerable controversy over the interpretation of these ages. The oldest ages, determined chiefly on high-grade Franciscan tectonic blocks, clearly reflect initiation of subduction, as indicated by various types of thermal modeling. Ages of about 112 m.y. for high-grade amphibolites and blueschists of the Catalina schist terrane evidently record a separate subduction-initiation event. However, the wide spread of mostly younger ages for lower-grade, intact Franciscan metamorphics might reflect either separate metamorphic events in response to discrete pulses of subduction, random sampling of an age continuum produced by long-lived, steady-state subduction, partial or total resetting of older ages in response to one or more younger reheating events, or some combination. The key to resolving these problems lies in increased application of high-precision Ar-Ar, Rb-Sr, and U-Pb methods, plus an integration of the geochronology with thermal modeling for subduction/accretion systems.

INTRODUCTION

The Franciscan Complex of California is a diverse assemblage of sedimentary, igneous, and metamorphic rocks, primarily of oceanic affinities (Bailey *et al.*, 1964). The accretion of these rocks to North America, and their varying degrees of deformation and metamorphism, are related to Jurassic-Cretaceous convergence between the North American and paleo-Pacific plates (e.g., Dickinson, 1970, 1972; Ernst, 1970, 1974, 1977; Hamilton, 1969, 1978; Maxwell, 1974; Page, 1970a, b, 1981). Of particular interest here are the metamorphic rocks of the Franciscan Complex, which include blueschists and other high-pressure/low-temperature types. Blake *et al.* (Chapter 38, this volume) present an up-to-date review of the metamorphism and tectonic evolution of the Franciscan. The conditions necessary to form these characteristic high-pressure/low-temperature metamorphic rocks evidently obtain only in subduction zones during active subduction (e.g., see Ernst, 1977). Determination of precise and accurate ages for these metamorphic rocks has obvious importance for our understanding of the timing and nature of Jurassic-Cretaceous subduction in California. The purpose of this paper is to examine the present geochronological constraints on the timing of Franciscan metamorphism and the limitations on such constraints.

FRANCISCAN METAMORPHISM: TECTONIC SIGNIFICANCE

The Franciscan Complex of western North America, and similar complexes elsewhere in the world, have long been believed to be subduction/accretion assemblages; in other words, they represent collections of oceanic and trench deposits that have been accumulated, deformed, and locally carried to great depth in subduction complexes. The blueschists and related high-pressure/low-temperature metamorphic rocks of the Franciscan Complex are evidently the products of this subduction/accretion environment. Only in such an environment can the combination of high pressures (about 3 to more than 13 kb) and low temperatures (150 ± 50°C for low-grade blueschists; 500°C or more for higher-grade blueschists, eclogites, and amphibolites) be achieved simultaneously (e.g., Dickinson, 1970, 1972; Ernst, 1970, 1974, 1977; Hamilton, 1969, 1978; Page, 1970a, b).

The metamorphic rocks of the Franciscan Complex occur in two distinct tectonic settings. The highest-grade metamorphic rocks are found as exotic blocks, typically meters to tens of meters in diameter, set in a much-lower-grade matrix, usually mud-matrix melange, or to a lesser extent, serpentinite-matrix melange. Metamorphic rocks of variable, but usually lower grade than the highest-grade exotic blocks, occur as large, intact tracts of strata in which metamorphic zones can be traced with some consistency. The "intact" or "coherent" sequences of metamorphic rocks are generally interpreted to have been metamorphosed *in situ,* and thus to represent exposed portions of a fossil subduction/accretion complex. In contrast, the high-grade exotic blocks immersed in a lower-grade matrix clearly have been transported substantial distances from their place(s) of metamorphism.

Two outstanding problems concern the high-grade blocks: (1) do they represent a separate subduction event that is distinctly older than that represented by the "intact" metamorphic strata; and (2) by what mechanism have they been extracted from their site(s) of metamorphism, at great depth, and transported to the near-surface? The first problem is largely one of geochronologic interpretation, and will be considered in more detail later. With regard to the second problem, several mechanisms have been proposed for the sampling and rapid return from depth of the high-grade exotic blocks. Recent proposals include (1) major strike-slip faulting in the fore-arc region (Karig, 1980), (2) serpentinite diapirism (Carlson, 1981), and (3) circulation of pelitic melange material within the subduction zone (Cloos, 1982). Inasmuch as the exact tectonic mechanism for return of the blocks to the surface reflects on the detailed thermal evolution of the blocks, and this, in turn, bears on the behavior of the various geochronologic systems used to date the time of metamorphism, these three proposals are considered in more detail below.

Karig (1980) notes that strike-slip faults, related to oblique subduction, cut across the fore-arc region in some arc systems. Movement along these faults, some of which trends inboard from the trench, cutting into the accretionary prism, could result in the juxtaposition of relatively high-grade rocks from deep within the accretionary prism against unmetamorphosed rocks from near the trench walls. In fact, such movement could produce a kind of tectonic erosion, exposing relatively high-grade material on the trench slope. From this position, the high-grade material could be sloughed off into the trench to be incorporated in mud-matrix trench fill, or could be imbricated with low-grade rocks by thrusting near the trench (Karig, 1980, p. 31). Large-scale (several kilometers) vertical movements along major fore-arc strike-slip faults could account for juxtapositioning of higher-pressure metamorphic rocks with lower-grade or unmetamorphosed rocks, according to Karig.

Carlson (1981) proposes a very different mechanism for extracting high-grade metamorphic blocks from depth — diapiric serpentinite. Originally proposed over two decades ago (Coleman, 1961; Ernst, 1965), this model envisions the density-driven diapiric uprise of serpentinite from great depth. Entrained high-grade metamorphic blocks would be carried upward along with the serpentinite, to be recycled into surficial deposits and perhaps the trench by sedimentary processes, or to be inserted into lower-grade Franciscan at intermediate depths by tectonic processes. The presence of serpentinite protrusions within the Franciscan, some of which are associated with high-grade blocks, and of serpentinite-rich sedimentary rocks interlayered with "normal" Franciscan rocks, is a clear indication that this is a feasible mechanism, at least on a small scale.

The mechanisms proposed by Karig (1980) and Carlson (1981) for bringing high-grade exotic blocks to the surface and recycling them back into the trench both seem reasonable, mechanically. In particular, widespread evidence for the former, and in some cases, present association with serpentinite of the high-grade blocks from various parts of the Franciscan Complex favors the diapiric emplacement mechanism of Carlson (1981). However, while both the Karig and Carlson models can account for the initial exhumation of high-grade blocks, and their mixing with a lower-grade matrix, neither explains their subsequent reexhumation from depth in mud-matrix melange. This last stage in the tectonometamorphic history of the high-grade blocks is required by recent studies which reveal that the melange matrix itself has been subjected to high-pressure metamorphism, in common with the latest stage of high-pressure (retrograde) metamorphism of the enclosed high-grade blocks (e.g., Cloos, 1983; Moore, 1984).

In contrast to the above, the subduction-flowage model developed by Cloos (1982) appears to provide a comprehensive explanation for most aspects of the metamorphism and transport of high- and low-grade metamorphic rocks within the context of the overall tectonic and thermal evolution of subduction/accretion systems. According to this model (Fig. 37-1), the pelitic matrix of the melange within the accretionary prism and subduction zone behaves as a viscous fluid. The movement of the descending plate sets up a downward laminar flow within the matrix, and the narrowing and pinching out of the wedge with depth forces a return flow, or circulation. The circulation of the pelitic matrix provides a mechanism for plucking blocks of earlier-formed metamorphic rocks from the hanging wall, and a vehicle for returning such blocks to shallow levels. During circulation of the matrix, the high-density metamorphic blocks will have an additional downward velocity component, a "settling velocity." The implications of this are twofold: (1) blocks of different settling velocity will cross streamlines in the flow at different rates, and thus will tend to be widely dispersed, resulting in efficient mixing of block types within the melange; and (2) high-density blocks beyond some maximum size will sink through the upwelling apart of the melange and back into the descending limb while still at great depth within the wedge (e.g., Cloos, 1982, Fig. 6). These blocks will not return to the surface or near-surface environment, thus accounting for the fact that tectonic blocks are generally of modest size (e.g., 2–25 m). The predicted behavior is highly dependent on the exact parameters chosen for density contrast, subduction rate, wedge geometry, and in particular, matrix viscosity.

Cloos has recently (1985) extended his modeling to the thermal evolution of subduction/accretion systems, and implications for interpretation of ages. The thermal modeling suggests that, theoretically, a wide range of ages could be produced during the

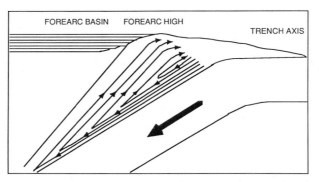

FIG. 37-1.   Tectonically driven circulation of accreted and subducted pelitic melange at convergent margins (after Cloos, 1982). See the text for a detailed discussion.

evolution of a single subduction system owing to (1) the long time interval over which various parts of the accretionary wedge reach peak temperatures, and (2) the similarly extended interval over which parts of the wedge remain at temperatures sufficiently high for continued recrystallization of metamorphic minerals, and/or continued diffusional losses of radiogenic daughter products (see Fig. 37-2 and the next section). Thus a wide range of ages might be expected from a single subduction event, particularly for lower-grade parts of the accretionary wedge (and thus the coherent metamorphic rocks?). Another point made clear by the thermal modeling of Cloos (1985) is that the ages of the highest-grade tectonic blocks, unless disturbed by later thermal events, should closely approximate the initiation of subduction (Fig. 37-2). A similar conclusion is drawn by Peacock (Chapter 34, this volume).

GEOCHRONOLOGY

Background

All geochronologic approaches to dating high-P/T rocks are based on the decay of a radio-active "parent" isotope to a stable "daughter" product (for example: $^{40}K \rightarrow ^{40}Ar$; $^{87}Sr \rightarrow ^{87}Rb$; $^{238}U \rightarrow ^{206}Pb$). By knowing the rate of decay of the parent isotope, by measuring the amounts of parent and daughter isotopes present in the system (either the total rock or pure separates of constituent minerals), and by correcting for any daughter isotope incorporated in the system at the time of its formation, we can calculate an age for the system. In practice, there are some minimum requirements for any of these dating methods. Both the parent and daughter isotopes must be present in sufficient abundance to permit precise and accurate measurements of their concentrations and isotopic compositions. Thus only certain minerals will be suitable for dating by any particular method. In the case of isochron dating methods, the suite of rocks or different minerals to be dated must have a range of parent/daughter ratios, to provide the spread of the isochron that is necessary for precision dating. Even when all these requirements are met, however, the event actually represented by the calculated age is subject to considerable interpretation, as discused below.

Uncertainties in interpreting ages (as opposed to analytical uncertainties) stem

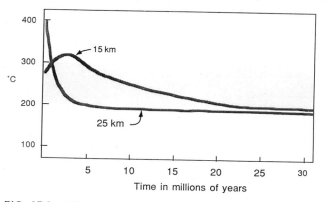

FIG. 37-2. Thermal evolution of metamorphic rocks within an accretion-subduction complex of the type shown in Fig. 37-1. Note the very different time-temperature trajectories for material buried at depths of 15 km versus 25 km (after Cloos, 1985). Implications for Franciscan geochronology are discussed in the text.

chiefly from the complexity of temperature-controlled diffusional losses of daughter product from the system (e.g., escape of radiogenic ^{40}Ar from a K-bearing mineral), or subsequent isotopic exchange of the daughter product among various parts of the system (e.g., isotopic homogenization of Sr among the phases of a rock). Such behavior can be understood generally in terms of the concept of "closure temperature" (e.g., Dodson, 1973, 1979), as shown in Fig. 37-3. According to this concept, each geochronologic system (e.g., Rb-Sr in biotite, or U-Pb in apatite) has a characteristic closure temperature. To a first approximation, the age calculated records the last time the system has dropped through this closure temperature. In more detail, three temperature-related domains can be recognized: (1) at relatively high temperatures, the rate of diffusional loss of the daughter product will exceed the rate of production by radioactive decay, resulting in little or no net accumulation of the daughter product within the system under consideration (region A in Fig. 37-3); (2) over a limited temperature range around the closure temperature, diffusional losses are still important, but are exceeded by production, so that some net accumulation of the daughter product takes place (region B in Fig. 37-3); and (3), at lower temperatures, diffusion effectively ceases, resulting in quantitative retention of all newly produced daughter products (region C in Fig. 37-3).

Applying the concept of closure temperature to Franciscan metamorphic rock and mineral systems reveals several possible types of behavior. In the simplest case of high-temperature metamorphism above the closure temperature, followed immediately by rapid cooling, the calculated age should approximate closely the actual time of metamorphism. Metamorphic crystallization of a mineral at a temperature below its closure temperature should also result in an "intact" geochronologic system, with the calculated age indicating the actual time of metamorphism. More typically, however, daughter products continue to be lost or to equilibrate isotopically after the peak of metamorphism owing to (1) prolonged slow cooling above the closure temperature, or later reheating through the closure temperature; (2) continued recrystallization at lower grades of metamorphism, below the original closure temperature; or (3) penetrative mechanical deformation promoting daughter-product losses below the simple thermal closure temperature. In these

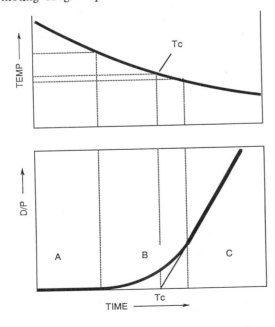

FIG. 37-3. Accumulation of radiogenic daughter isotopes and the concept of closure temperature (after Dodson, 1973). Closure temperature (Tc) is the extrapolated temperature corresponding to the apparent age of the system. This is given, to a first approximation, by the linear extrapolation of the D/P accumulation curve in part C of the diagram. See the text for further discussion.

cases, measured ages are only minimum ages for the time of metamorphism. Alternatively, incomplete metamorphic recrystallization, especially of metasedimentary rocks, could result in the incomplete expulsion of inherited daughter isotopes from the rock, or the incomplete isotopic homogenization of daughter isotopes among the mineral phases in the rock. In these cases, calculated ages could be anomalously high. At the present time, our qualitative, theoretical understanding of the concepts above greatly exceeds our ability to apply them in a rigorous, quantitative way to the metamorphic rocks of the Franciscan Complex. As a result, the significance of many published ages for Franciscan metamorphic rocks is the subject of considerable controversy.

Franciscan Metamorphic Ages

A variety of geochronological methods have been applied to the dating of Franciscan metamorphic rocks over the past two decades. Early work was based almost exclusively on the K-Ar method on whole rocks for fine-grained, lower-grade metamorphics, and on separated minerals for the coarse-grained, higher-grade metamorphics. The K-Ar method continues to be a mainstay in the geochronology of Franciscan metamorphic rocks, but recent years have seen the application of a number of other methods. In particular, Ar-Ar, U-Pb, and Rb-Sr have been used recently with very promising results, and the Nd-Sm method has some potential. A summary of methods and results follows.

As discussed above, previous geochronologic work on the blueschists and related metamorphic rocks of the Franciscan Complex has been chiefly by the K-Ar method. Regional studies by Coleman and Lanphere (1971) and Suppe and Armstrong (1972) provided the first detailed picture of Franciscan metamorphic ages. Coleman and Lanphere (1971) reported K-Ar mineral ages of high-grade blueschists, eclogites, and amphibolites that occur as exotic blocks in the Franciscan Complex. A summary of their results is included in Fig. 37-4. The data reveal, first, a tight clustering of the *oldest* mineral ages at around 150–155 m.y., predominantly from phengitic white mica and hornblende (note: all K-Ar ages from early work have been recalculated to reflect currently accepted decay constants). This age has been widely accepted as representing a major early (pre-Franciscan?) subduction event. Second, ages from coexisting minerals, for example, glaucophane, are typically lower. This has been interpreted as indicating the greater susceptibility of glaucophane to Ar loss, probably during later, retrograde metamorphic events.

K-Ar dates on Franciscan metamorphic rocks reported by Suppe and Armstrong (1972) include data for high-grade exotic blocks (mineral ages) as well as the lower-grade intact Franciscan metamorphic rocks (chiefly whole-rock ages). The ages for the high-grade exotic blocks corroborate those of Coleman and Lanphere (1971), with maximum ages of 150–155 m.y. on white mica and hornblende (Fig. 37-4), and significantly lower ages for glaucophane. Whole-rock ages for the lower-grade, intact Franciscan metamorphic rocks range widely, but are typically younger than the ages of the high-grade blocks, ranging down to about 70–75 m.y. (Fig. 37-4; and see Suppe and Armstrong, 1972).

More recent K-Ar studies include work by Suppe and Foland (1978), and McDowell *et al.* (1984). Ages for high-grade metamorphic blocks in melange reported by McDowell *et al.* range from 142 to 153 m.y. for white micas and actinolites, and 98 to 151 m.y. for glaucophanes. These data are in excellent agreement with the earlier results of Coleman and Lanphere (1971) and Suppe and Armstrong (1972) for the high-grade blocks (Fig. 37-4). Mineral and whole-rock ages for intact metamorphic strata, including the Goat Mountain Schist (Suppe and Foland, 1978) and the South Fork Mountain Schist (McDowell *et al.*, 1984), range from fully as old as the ages for the high-grade blocks (~150–160 m.y.) down to ~110 m.y. (Fig. 37-4). This spread in ages has been taken as an indication of a single, prolonged period of blueschist metamorphism, or, alternatively, that isotopic sys-

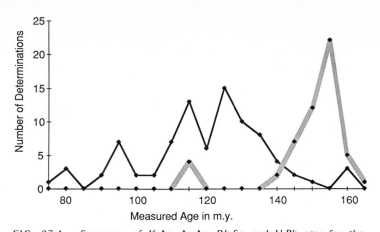

FIG. 37-4. Summary of K-Ar, Ar-Ar, Rb-Sr, and U-Pb ages for the Franciscan Complex and related rocks. Ages have been selected for reliability as follows: (1) for all ages, quoted 1σ errors must be 5 m.y. or less; and (2), for high-grade blocks, K-Ar ages have been restricted to those on white mica, hornblende, and actinolite. The Goat Mountain Schists of Suppe and Foland (1978) have been included in the high-grade group for this histogram. Ages of the high-grade blocks (wide, stippled line) cluster strongly around 150–165 m.y. The small cluster of ages at 100–115 m.y. includes K-Ar and U-Pb ages of high-grade garnet amphibolites from the Catalina Schist terrane. Lower-grade intact Franciscan rocks are represented by the narrow, solid line. Their ages range widely from as old as those for the high-grade blocks, down to 70–75 m.y., with a major peak between about 110 and 135 m.y. Sources of ages include Coleman and Lanphere (1971), Lanphere *et al.* (1978), Mattinson and Echeverria (1980), Mattinson (1981, 1986), McDowell *et al.* (1984), Nelson and DePaolo (1985), Suppe (1969), Suppe and Armstrong (1972), and Suppe and Foland (1978). See the text for further discussion.

tems originally recording ages similar to those well documented in the high-grade exotic blocks have been variably disturbed during subsequent thermal events ranging down to 120 m.y. or younger.

The conventional K-Ar dating method used in the studies discussed above has an inherent source of uncertainty, in that the K and Ar analyses require two separate splits of the sample. Any inhomogeneity in the split sample is a possible source of error. Fast-neutron irradiation of K-bearing samples results in the conversion, via an n,p reaction of ^{39}K to ^{39}Ar. This permits calculation of the age of the sample from the $^{40}Ar/^{39}Ar$ ratio of a single irradiated sample, without the need for a separate K determination. This "Ar/Ar" method has been applied to whole-rock samples of the South Fork Mountain Schist by Lanphere *et al.* (1978). The Ar/Ar ages cluster between 112 and 130 m.y., with a concentration at ~120 m.y. (Fig. 37-4), results that conflict somewhat with those of McDowell *et al.* (1984) discussed above. Lanphere *et al.* postulate that a later metamorphic event, totally separate from the one responsible for recrystallization of the high-grade blocks, produced the South Fork Mountain Schist. Unfortunately, Lanphere *et al.* (1978) used the Ar/Ar method only to obtain "total gas" ages. The most significant advantage of the $^{40}Ar/^{39}Ar$ method over conventional K/Ar dating is through the use of incremental Ar release via stepwise heating. This is discussed in more detail later.

The Rb/Sr method has been applied to the dating of a few Franciscan metamorphic rocks (Fig. 37-4). Peterman *et al.* (1967) report a Rb-Sr whole-rock isochron age of 112 ± 16 m.y. on metagraywackes from the Yolla Bolly area. This is in general agreement with

K-Ar ages from similar metagraywackes to the south. Rb-Sr mineral plus whole-rock isochrons are presented by Suppe and Foland (1978) for samples of the Goat Mountain Schist. Some of these samples yield high-precision ages in good agreement with K-Ar ages. For example, the Rb/Sr isochron age of a metabasalt, at 141 ± 2 m.y., is concordant with K/Ar ages of 141 ± 4 m.y. for white mica and 145 ± 5 m.y. for blue amphibole from the same sample. However, other isochrons determined for metasedimentary rocks yield apparent isochron ages that are significantly higher or lower than K/Ar mineral ages from the same samples. Recent Rb/Sr results on mineral pairs from veins and cavities in high-grade blocks reported by Nelson and DePaolo (1985) demonstrate the potential for high-precision dating by this method. Ages of 153.6 ± 2.4 m.y. on an aragonite-glaucophane pair, and 151.9 ± 0.4 m.y. on a white mica-zoisite pair presumably date blueschist retrograde metamorphism of older eclogite blocks.

The U-Pb isochron method has been applied to Franciscan metamorphic rocks only recently. Mattinson and Echeverria (1980) published a U-Pb isochron for metamorphic minerals from a large gabbroic sill that intruded an intact Franciscan sequence at Ortigalita Peak in the Diablo Range after the Franciscan had been deformed in an accretionary prism. The sill and surrounding Franciscan rocks were then both metamorphosed to blueschist grade. Based on zircon dates, the sill was emplaced about 95 m.y. ago, and, based on the mineral isochron, the blueschist metamorphism followed only about 3 m.y. later. Thus, the sill was intruded in the accretionary wedge just prior to subduction, according to Mattinson and Echeverria (1980). The 92-m.y. age of subduction (blueschist metamorphism) agrees well (Fig. 37-4) with K-Ar ages from intact graywacke sequences in the Diablo Range (Suppe and Armstrong, 1972), where the Ortigalita Peak gabbro is exposed. Mattinson (1986) reported U-Pb isochron ages of 162 ± 3 m.y. for blueschist-grade metabasalt from the Taliaferro complex in Northern California (isochron on sphene, lawsonite, and glaucophane), and 112.5 ± 1.1 m.y. for two garnet amphibolites from the Catalina Schist terrane (isochrons on sphene, garnet, hornblende, and apatite). Both of these results can be compared directly with K-Ar results from nearly identical samples. Suppe (1969) reported ages of ~154 ± 3 m.y. for white micas from the Taliaferro metabasalt. These ages are slightly but significantly younger than the U-Pb age, and are suggestive of minor Ar loss from the white mica during subsequent thermal events. Suppe and Armstrong (1972) reported a K-Ar age of 112 ± 3 m.y. for hornblende from garnet amphibolite of the Catalina Schist terrane, in excellent agreement with the U-Pb results.

Nd-Sm systematics have been studied for a number of Franciscan high-grade metamorphic blocks by Nelson and DePaolo[1] (1982). Results from separated minerals have yielded important evidence for metasomatic reactions and rare-earth mobility in these metamorphic rocks, and suggest MORB affinities for the protoliths of the metabasites. Unfortunately, the metasomatic disturbances of the Sm-Nd systematics of the individual minerals have thus far hampered the geochronologic applications of this method.

DISCUSSION

Geochronological studies over the past two decades have provided a good general picture of the timing of Franciscan metamorphism. Based on interpretation of abundant K-Ar and limited U-Pb, Rb-Sr, and Ar-Ar results (Fig. 37-4), the high-grade blocks appear to have been metamorphosed during a very limited period between about 150 and 165 m.y. In contrast, apparent ages of the lower-grade, "coherent" Franciscan metamorphics range widely. A few samples yield ages fully as old as those for the high-grade blocks, but most are younger, scattering down to 70–75 m.y. Clustering of K-Ar ages suggests the possibility

[1] Also B. K. Nelson and D. J. DePaolo, personal communications, 1984, 1985.

of a metamorphic event, or at least a peak, at 120–130 m.y. The high-grade metamorphic rocks of Santa Catalina apparently represent a distinct metamorphic event at about 112 m.y., some 40–50 m.y. younger than the ages of mainland Franciscan high-grade blocks. In fact, the age difference might be used to argue that the Catalina terrane should be excluded from the Franciscan *sensu stricto*. Metamorphism in the Diablo Range, particularly the Ortigalita Peak area, suggests another event at about 92 m.y. Finally, some K-Ar ages for low-grade whole-rock samples are as low as 70–75 m.y.

However, there is major controversy over the geologic/tectonic interpretation of these apparent peaks or events for the lower-grade rocks. Do they represent actual distinct events, perhaps related to separate cycles of subduction; are they merely samplings of a continuous, prolonged, single cycle of subduction-related metamorphism; or does much of the apparent range in ages result from the partial resetting of older ages in response to, say, a single much younger event? For example, Lanphere *et al.* (1978) have argued that the metamorphism of the intact, *in situ* South Fork Mountain Schist is a significantly younger event, perhaps totally unrelated to metamorphism of the high-grade blocks. On the other hand, McDowell *et al.* (1984) have argued that the apparently younger ages of the South Fork Mountain Schist are the result of partial resetting of older ages, and that a major gap in original crystallization age might not exist.

The difficulties in interpreting Franciscan metamorphic ages stem from two main sources: (1) by far the greatest number of Franciscan metamorphic ages have been determined by the conventional K-Ar method on minerals and fine-grained whole-rock samples. Our understanding of the complex responses of these systems to long-term cooling, to deformation, to continued and perhaps repeated low-grade retrograde recrystallization in the subduction/accretion environment, is simply inadequate; and (2) almost all published Franciscan geochronology has not been adequately integrated with models for the thermal and tectonic evolution of subduction zones. Indeed, such models are only recently becoming available (e.g., Cloos, 1985), and as discussed earlier, have profound implications for interpretation of ages. In fact, Cloos (1985) argues that the entire range of Franciscan metamorphic ages can be explained in terms of the thermal evolution of a single, prolonged subduction episode.

Future geochronological work can resolve at least some of these questions. To do so, however, it must follow different approaches. For example, the U-Pb isochron approach may be able to "see through" the low-grade thermal events that apparently have disturbed, at least to a small degree, most of the K-Ar ages (Mattinson, 1986). This method is demanding but should be more widely applied to critical samples in the future. High-precision Rb-Sr dating should also be expanded. Further, a dating method with great potential for contributing to our understanding of thermal histories and behavior of the K-Ar system in both mineral and rock systems, the ^{40}Ar-^{39}Ar method *with incremental Ar release*, should be used to supplement the conventional K-Ar and total-gas Ar-Ar methods. Incremental Ar release via stepwise heating permits the analysis of a series of apparent ages (an "age spectrum") from an individual sample. Successive increments of Ar, each extracted from the sample at a higher temperature than the preceding increment, presumably sample progressively more retentive sites within the mineral or rock analyzed. The total age spectrum for the sample can then, in many cases, reveal whether the system has lost Ar during postmetamorphic events, or has retained Ar from premetamorphic events. In some cases, valid ages can be obtained despite such complications. Finally, future dating studies should be designed not just to obtain crystallization ages for Franciscan metamorphism, but also to provide constraints on the overall thermal evolution of subduction zones.

Franciscan geochronological studies supported by the National Science Foundation (EAR82-05823). Drs. M. C. Blake, Jr., C. A. Hopson, and S. S. Sorensen have been particularly generous in their encouragement, advice, and in providing samples of U-Pb isotopic work.

REFERENCES

Bailey, E. H., Irwin, W. P., and Jones, D. L., 1964, Franciscan and related rocks, and their significance in the geology of western California: *Cal. Div. Mines and Geology, Bull.* 183, 171 p.

Carlson, C., 1981, Upwardly mobile melanges, serpentinite protrusions and transport of tectonic blocks in accretionary prisms: *Geol. Soc. America Abstr. with Programs*, v. 13, p. 48.

Cloos, M., 1982, Flow melanges — Numerical modeling and geological constraints on their origin in the Franciscan subduction complex, California: *Geol. Soc. America Bull.*, v. 93, p. 330–345.

——, 1983, Comparative study of melange matrix and metashales from the Franciscan subduction complex with the basal Great Valley sequence, California: *J. Geology*, v. 91, p. 291–306.

——, 1985, Thermal evolution of convergent plate margins — Thermal modeling and re-evaluation of isotopic Ar-ages for blueschists in the Franciscan complex of California: *Tectonics*, v. 4, p. 421–433.

Coleman, R. G., 1961, Jadeite deposits of the Clear Creek area New Idria district, San Benito County, California: *J. Petrology*, v. 2, p. 209–247.

Coleman, R. G., and Lanphere, M. A., 1971, Distribution and age of high-grade blueschists, associated eclogites, and amphibolites from Oregon and California: *Geol. Soc. America Bull.*, v. 82, p. 2397–2412.

Dickinson, W. R., 1970, Relation of andesites, granites, and derived sandstones to arc-trench tectonics: *Rev. Geophysics Space Physics*, v. 8, p. 813–860.

——, 1972, Evidence for plate tectonic regimes in the rock record: *Amer. J. Sci.*, v. 272, p. 551–576.

Dodson, M. H., 1973, Closure temperature in cooling geochronological and petrological systems: *Contrib. Mineralogy Petrology*, v. 40, p. 259–274.

——, 1979, Theory of cooling ages, *in* Jäger. E., and Hunziker, J. C., eds., *Lectures in Isotope Geology:* New York, Springer-Verlag, p. 194–202.

Ernst, W. G., 1965, Mineral parageneses in Franciscan metamorphic rocks, Panoche Pass, California: *Geol. Soc. America Bull.*, v. 76, p. 879–914.

——, 1970, Tectonic contact between the Franciscan melange and the Great Valley sequence, crustal expression of a Late Mesozoic Benioff zone: *J. Geophys. Res.*, v. 25, p. 886–901.

——, 1974, Metamorphism and ancient continental margins, *in* Burke, C. A., and Drake, C. L., eds., *The Geology of Continental Margins:* New York, Springer-Verlag, p. 907–919.

——, 1977, Mineral parageneses and plate tectonic settings of relatively high-pressure metamorphic belts: *Fortschr. Mineralogie*, v. 54, p. 192–222.

Hamilton, W. B., 1969, Mesozoic California and the underflow of the Pacific mantle: *Geol. Soc. America Bull.*, v. 80, p. 2409–2430.

——, 1978, Mesozoic tectonics of the western United States, *in* Howell, D. G., and McDougall, K. A., eds., *Mesozoic Paleogeography of the Western United States:* Pacific Section, Soc. Econ. Paleontologists Mineralogists, Pacific Coast Paleogeography Symp. 2, p. 33–70.

Karig, D. E., 1980, Material transport within accretionary prisms and the "knocker" problem: *J. Geology*, v. 88, p. 27–39.

Lanphere, M. A., Blake, M. C., Jr., and Irwin, W. P., 1978, Early Cretaceous metamorphic age of the South Fork Mountain Schist in the northern Coast Ranges of California: *Amer. J. Sci.*, v. 278, p. 798–815.

Mattinson, J. M., 1981, U-Pb systematics and geochronology of blueschists — Preliminary results [abstract] : (EOS) *Trans. Amer. Geophys. Un.*, v. 62, p. 1059.

——, 1986, Geochronology of high pressure–low temperature Franciscan metabasites — A new approach using the U-Pb system, *in* Evans, B. W., and Brown, E. H., eds., *Blueschists and Eclogites:* Geol. Soc. of America Mem. 164, p. 95–105.

——, and Echeverria, L. M., 1980, Ortigalita Peak gabbro, Franciscan complex — U-Pb ages of intrusion and high pressure-low temperature metamorphism: *Geology*, v. 8, p. 589–593.

Maxwell, J. A., 1974, Anatomy of an orogen: *Geol. Soc. America Bull.*, v. 85, p. 1195–1204.

McDowell, F. W., Lehman, D. H., Gucwa, P. R., Fritz, D., and Maxwell, J. C., 1984, Glaucophane schists and ophiolites of the northern California Coast Ranges — Isotopic ages and their tectonic implications: *Geol. Soc. America Bull.*, v. 95, p. 1373–1382.

Nelson, B. K., and DePaolo, D. J., 1982, Sr and Nd composition of Franciscan eclogite and blueschist — A sampling of subducted crust? (EOS) *Trans. Amer. Geophys. Un.*, v. 63, p. 1133.

——, and DePaolo, D. J., 1985, Isotopic investigation of metasomatism in subduction zones — The Franciscan complex, California: *Geol. Soc. America Abstr. with Programs*, v. 17, no. 7, p. 674–675.

Page, B. M., 1970a, Sur-Nacimiento fault zone of California — Continental margin tectonics: *Geol. Soc. America Bull.*, v. 81, p. 667–690.

——, 1970b, Time of completion of underthrusting of Franciscan beneath Great Valley rocks west of the Salinian block, California: *Geol. Soc. America Bull.*, v. 81, p. 2825–2834.

——, 1981, The southern Coast Ranges, *in* Ernst, W. G., ed., *The Geotectonic Development of California* (Rubey Vol. I): New York, Prentice-Hall, p. 329–417.

Peterman, Z. E., Hedge, C. E., Coleman, R. G., and Snavely. P. D., Jr., 1967, $^{87}Sr/^{86}Sr$ ratios in some eugeosynclinal sedimentary rocks and their bearing on the origin of granitic magma in orogenic belts: *Earth Planet. Sci. Lett.*, v. 2, p. 433–439.

Suppe, J., 1969, Times of metamorphism in the Franciscan terrain of the northern Coast Ranges, California: *Geol. Soc. America Bull.*, v. 80, p. 135–142.

——, and Armstrong, R. L., 1972, Potassium-argon dating of Franciscan metamorphic rocks: *Amer. J. Sci.*, v. 272, p. 217–233.

——, and Foland, K. A., 1978, The Goat Mountain Schists and Pacific Ridge complex — A redeformed but still-intact late Mesozoic schuppen complex, *in* Howell, D. G., and McDougall, K. A., eds., *Mesozoic Paleogeography of the Western United States:* Pacific Section, Soc. Econ. Paleontologists Mineralogists, Pacific Coast Paleogeography Symp. 2, p. 431–451.

M. C. Blake, Jr., A. S. Jayko, and R. J. McLaughlin
U. S. Geological Survey
Menlo Park, California 94025

M. B. Underwood
Department of Geology
University of Missouri
Columbia, Missouri 65211

38

METAMORPHIC AND TECTONIC EVOLUTION OF THE FRANCISCAN COMPLEX, NORTHERN CALIFORNIA

ABSTRACT

Franciscan rocks in northern California consist of three fault-bounded belts, the Eastern, Central and Coastal belts, which are further divided into tectonostratigraphic terranes, each having a different stratigraphy, structural state, and metamorphic history. The Eastern Franciscan belt consists of two terranes that were metamorphosed to the blueschist facies and which probably formed in an east-dipping subduction zone along the western edge of North America. The easternmost and structurally highest terrane of the Eastern Franciscan belt, the Pickett Peak terrane, consists of the South Fork Mountain Schist and the Valentine Spring Formation, which yield radiometric metamorphic ages that range from 110 to 143 m.y. The degree of textural reconstitution and the metamorphic mineral assemblages in the Pickett Peak terrane indicate an increase in metamorphic conditions from west to east. Estimated P-T conditions of this terrane range from approximately 250°C and 6 kb in the west to a maximum of about 345°C and 8 kb along the eastern margin.

The Yolla Bolly terrane lies west of and structurally below the Pickett Peak terrane. It contains fossils of Late Jurassic to mid-Cretaceous age. K/Ar radiometric ages on rocks and minerals range from about 90 m.y. to 115 m.y., although a Pb/U sphene age of 92 m.y. (mid-Cretaceous) is inferred to most closely approximate the time of blueschist-facies metamorphism. As in the Pickett Peak terrane, there is a general eastward increase in degree of reconstitution and metamorphic grade: the estimated P/T conditions range from 180°C and 6 kb in the west to 285°C and 9 kb in the east.

The Central Franciscan belt is a melange with a matrix of argillite, lithic graywacke, and greenish aquagene tuff. Radiolarians and megafossils within the matrix are of Late Jurassic to mid(?)-Cretaceous age. Pumpellyite is the dominant metamorphic mineral within the matrix graywacke. Vitrinite reflectance measurements on mudstone from both matrix and interleaved pumpellyite-bearing sandstone slabs give a range of mean values from 0.68% R_0 in the west to 1.69% R_0 in the east, indicating temperatures of 137–254°C. Pressures are poorly constrained, but the absence of lawsonite suggests pressures less than about 3 kb. Blocks of high-grade blueschist, eclogite, and amphibolite with metamorphic ages of about 160 m.y. are generally restricted to the Central Franciscan belt. Petrographic studies suggest that many of the blocks were initially amphibolite or garnet amphibolite, prior to retrograde metamorphism to their present blueschist and eclogite assemblages.

The Coastal Franciscan belt contains three terranes which, from west to east, are the King Range terrane (Late Cretaceous to Miocene), Coastal terrane (Late Cretaceous to Eocene), and Yager terrane (Paleocene and Eocene). Both the Yager and Coastal terranes are characterized by laumontite veins. Prehnite and datolite are also recognized locally in the Coastal terrane, particularly in sandstones with high percentages of volcanic detritus. The mean vitrinite reflectance values for Coastal and Yager terrane shales are from 0.40 to 1.78%, suggesting that temperatures ranged between 68 and 263°C.

Vitrinite and fluid inclusion data from the King Range terrane indicate temperatures up to 308°C and average about 250°C. Laumontite is typically present in graywacke, but local Late Cretaceous igneous and sedimentary rocks contain pumpellyite and prehnite. In addition, these rocks are overprinted by the hydrothermal vein assemblage quartz + calcite ± adularia ± fluorite ± numerous sulfide-bearing base and precious metals.

These structural and metamorphic data suggest that the Franciscan terranes formed during different plate tectonic regimes. From about 125 to 92 m.y. ago, plate convergence appears to have been roughly orthogonal with eastward subduction of an oceanic plate beneath North America and attendant blueschist-facies metamorphism. This was closely followed by large-scale faulting of the blueschists back to the surface and a major reorganization of the plates by about 85–90 m.y. ago that led to highly oblique, northerly-directed poleward convergence and righ-lateral transform faulting. We suggest that the cause of this reorganization was the arrival of a large oceanic plateau (Wrangellia?) that collided with the subduction zone. Transform faulting and northward dispersion of the previously accreted terranes continued until about 50 m.y. ago, when there was a return to more orthogonal convergence and accretion of the Coastal and Yager terranes. Finally, the northward-migrating Mendocino triple junction began to impinge on the continental margin a few million years ago and the King Range terrane was obducted onto the Coastal terrane as a result of convergence between the Pacific and North American plates.

INTRODUCTION

The Franciscan Complex of the California Coast Ranges is regarded as the type example of an accretionary complex that formed in a subduction zone (Bailey and Blake, 1969; Ernst, 1970; Hamilton, 1969). This paleotectonic interpretation is consistent with the general lithologic association, structural state, and metamorphic condition of the Franciscan Complex (Bailey et al., 1964; Blake et al., 1967; Ernst, 1984; Hamilton, 1978; Hsü, 1968). The high pressures and low temperatures inferred from the metamorphic mineral assemblages and a general eastward progression from low-grade laumontite-bearing rocks in the west to lawsonite and sodic amphibole-bearing assemblages in the east (Blake et al., 1967; Ernst, 1971a, 1980) are considered indicative criteria for a subduction zone environment. This systematic increase in metamorphic grade is consistent with eastward-directed subduction and differential uplift that exposes deeper levels of the accretionary prism from west to east (Ernst, 1971a). However, recent geologic studies in northern California indicate that the Franciscan Complex consists of a number of tectonostratigraphic terranes with different structural and metamorphic histories.

We describe herein the nature and timing of metamorphism within the Franciscan Complex of the northern Coast Ranges of California and relate metamorphic events to the regional tectonic history. These data are important because they allow an empirical check on the many models proposed for *P-T* regimes along accretionary continental margins (e.g., Cloos, 1985; Pavlis and Bruhn, 1983; Wang and Shi, 1984).

REGIONAL GEOLOGIC SETTING

Northern California (Fig. 38-1) includes portions of three major geologic and geographic provinces: (1) the Coast Ranges, consisting predominantly of Mesozoic and Cenozoic Franciscan rocks; (2) the Klamath Mountains, a complicated basement consisting of Paleozoic and Mesozoic oceanic and island arc terranes unconformably overlain by shallow to deep marine Lower Cretaceous (Valanginian and younger) sedimentary rocks; and (3) the Great Valley, with a composite basement composed predominantly of the Mesozoic Coast Range ophiolite (Middle Jurassic) on the west, and a granitic and metamorphic Sierran basement on the east (Paleozoic and Mesozoic), all overlain by marine deposits of the Great Valley sequence (Blake et al., 1985a).

All of the exposed contacts between the three geologic provinces of northern California are faults. The western boundary fault of the Klamath Mountains, the South Fork fault (Irwin et al., 1974), is considered an east-dipping thrust fault even though its present orientation appears to be very steep (Fig. 38-1). To the southeast, a zone of high-angle faults juxtaposes three distinct units: Franciscan Complex, Klamath basement with Cretaceous overlap assemblage, and the Great Valley sequence. This complicated region has been called the Yolla Bolly junction (Blake et al., 1984). Some of the faults display an apparent component of left-lateral strike-slip movement (Jones and Irwin,

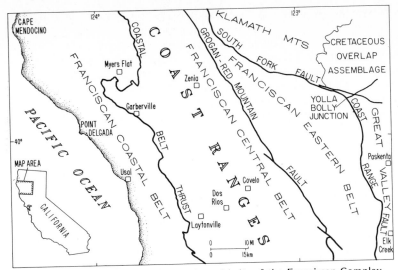

FIG. 38-1. Major geologic provinces and belts of the Franciscan Complex, northern California.

1971; Worrall, 1981) and evidently were active during the Cretaceous (Blake and Jones, 1981). Farther south, the faulted boundary between the Franciscan and the Coast Range ophiolite–Great Valley sequence composite unit is shown as the Coast Range fault (Fig. 38-1). This tectonic boundary was interpreted as a late Mesozoic subduction zone (Bailey *et al.*, 1970; Ernst, 1970); however, as pointed out by many workers, the Mesozoic Coast Range thrust is rarely, if ever, observed because of younger superposed faulting (see Page, 1981, p. 372–373, for a discussion of this problem). Although Cenozoic faulting has overprinted the tectonic boundaries, the continuity of the Franciscan rocks both north and south of the Yolla Bolly junction suggests that the Jurassic and older rocks of the Klamath Mountains and the Jurassic Coast Range ophiolite-basal part of the Great Valley sequence were juxtaposed prior to accretion of the Franciscan Complex.

FRANCISCAN GEOLOGY

Eastern Franciscan Belt

Pickett Peak terrane The rocks of the Eastern Franciscan belt have been divided into two terranes, Pickett Peak and Yolla Bolly (Fig. 38-2); both are regionally metamorphosed to the bluschist facies (Blake *et al.*, 1967; Jayko *et al.*, 1986; Suppe, 1973; Worrall, 1981). The easternmost and structurally highest Pickett Peak terrane consists of two units, the South Fork Mountain Schist, which includes the Chinquapin Metabasalt Member and the Valentine Spring Formation. These two units are separated by an east-dipping thrust fault (Worrall, 1981).

The South Fork Mountain Schist (Blake *et al.*, 1967) consists largely of quartz-veined mica schist (metamudstone), quartzofeldspathic schist, with minor metavolcanic rock and scarce metachert. All of the metavolcanic rocks are of basaltic composition and include both tholeiitic and alkalic varieties (Cashman *et al.*, 1986; Jayko *et al.*, 1986).

Some of the thinner metabasaltic layers appear to have formed as tuffs interbedded with the metasedimentary rocks; however, the larger bodies shown in Fig. 38-2 (mb = Chinquapin Metabasalt Member) include relict pillows and pillow breccias that are overlain depositionally by thin (1–5 m) beds of iron and manganese-rich metachert. At several localities, including North Yolla Bolly Mountain, Tomhead Mountain, and Black Butte (Fig. 38-2), the metachert if depositionally overlain by quartz-mica schist and at Black Butte, the metavolcanic rocks are overlain by metagraywacke. Worrall (1981) concluded that the Chinquapin Metabasalt Member was the basement to the South Fork Mountain Schist. This inference, combined with the tholeiitic to alkalic geochemistry of the basalts, suggests that the basement was probably a seamount or an oceanic island rather than oceanic crust. North of Thomes Creek (Fig. 38-2), the metaclastic rocks of the South Fork Mountain Schist are largely crumpled and quartz-veined mica schist (metamudstone); however west of Paskenta, there are minor interbeds of massive quartzo-feldspathic schist, derived from metagraywacke. South of Thomas Creek and, in particular, near Black Butte (Ghent, 1965), quartz-mica schist is uncommon and metagraywacke is the dominant lithology of the South Fork Mountain Schist.

The structurally lower Valentine Spring Formation (Worrall, 1981) consists largely of schistose to gneissic metagraywacke with scarce metavolcanic rocks and rare metachert. As in the South Fork Mountain Schist, the metavolcanic rocks are of basaltic composition and include both tholeiitic and alkalic varities (M. C. Blake and A. S. Jayko, unpublished data). Massive metabasalt up to 100 m thick, locally overlain by thin (1–2 m) metachert lenses, could represent oceanic crust upon which the Valentine Spring metaclastic sediments were deposited.

No fossils have been found in either the South Fork Mountain Schist or Valentine Spring Formation. Whole-rock K-Ar and Rb-Sr radiometric ages from both units range from 110 to 143 m.y. but cluster around 125 m.y. (Lanphere *et al.*, 1978; McDowell *et al.*, 1984; Suppe, 1973); these data suggest an Early Cretaceous age of metamorphism.

All of the rock units assigned to the Pickett Peak terrane contain evidence for three periods of penetrative deformation. The first period was characterized by segregation layering, and the second and third phases produced crenulation cleavages. Blueschist-facies conditions persisted during the first two deformation events.

Yolla Bolly terrane The Yolla Bolly terrane lies structurally below the Pickett Peak terrane and is divided into four thrust-bounded units. The structurally highest unit, the Chicago Rock melange, consists predominantly of metamorphosed, tectonized argillite and metagraywacke with intercalated thin-bedded radiolarian chert, minor greenstone, scarce diapiric(?) intrusions of serpentinite, and rare mafic blocks of amphibolite and blueschist. The chert and greenstone appear to be indigenous to the unit on the basis of stratigraphic relations (Blake and Jayko, 1983; Blake *et al.*, 1982a); however, the amphibolite, and mafic blueschist are found as tectonic blocks in or near the serpentinite.

Intermixed with the Chicago Rock melange are many slabs and thin thrust sheets of the Taliaferro Metamorphic Complex (Suppe, 1973; Suppe and Armstrong, 1972). These slabs are of slightly higher grade than the surrounding argillaceous matrix, but are considered to be part of the Yolla Bolly terrane because of lithologic and faunal similarities (Jayko *et al.*, 1986). The Taliaferro Metamorphic Complex, as defined in this study, includes the jadeite- and crossite-bearing metagraywackes and cherts of Suppe (1973) but excludes the older (approximately 160 m.y.) tectonic blocks of coarse-grained blueschist and amphibolite (Mattinson, 1981).

The metagraywacke of Hammerhorn Ridge, predominantly massive graywacke with a continuous horizon of interbedded radiolarian chert, structurally underlies the Chicago

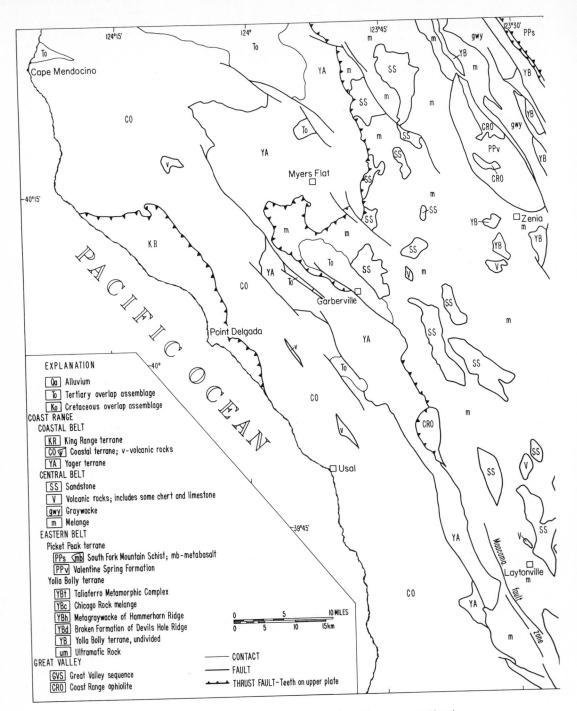

EXPLANATION

Qa	Alluvium
To	Tertiary overlap assemblage
Ko	Cretaceous overlap assemblage

COAST RANGE

COASTAL BELT

KR	King Range terrane
CO v	Coastal terrane; v-volcanic rocks
YA	Yager terrane

CENTRAL BELT

SS	Sandstone
V	Volcanic rocks; includes some chert and limestone
gwy	Graywacke
m	Melange

EASTERN BELT

Picket Peak terrane

| PPs mb | South Fork Mountain Schist; mb-metabasalt |
| PPv | Valentine Spring Formation |

Yolla Bolly terrane

YBt	Taliaferro Metamorphic Complex
YBc	Chicago Rock melange
YBh	Metagraywacke of Hammerhorn Ridge
YBd	Broken Formation of Devils Hole Ridge
YB	Yolla Bolly terrane, undivided
um	Ultramafic Rock

GREAT VALLEY

| GVS | Great Valley sequence |
| CRO | Coast Range ophiolite |

CONTACT
FAULT
THRUST FAULT-Teeth on upper plate

0   5   10 MILES
0   5   10   15km

FIG. 38-2. Geologic map of a portion of northern California.

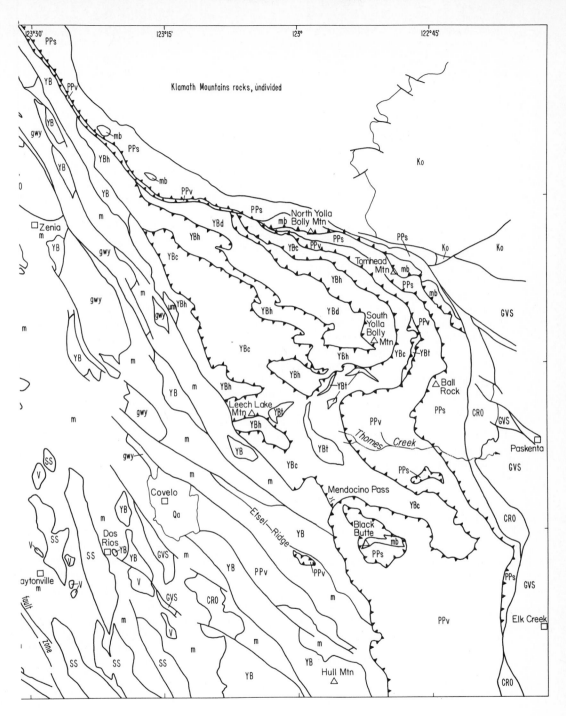

FIG. 38-2   Continued

Rock melange. Interbedded chert and graywacke of the Hammerhorn Ridge unit are locally intruded by Ti-rich alkalic gabbro sills and dikes, a lithologic association identical to that of the Taliaferro Metamorphic Complex.

The fourth unit of the Yolla Bolly terrane is the broken formation of Devils Hole Ridge. This unit consists of deformed graywacke and argillite (broken formation) and structurally underlies the Hammerhorn Ridge unit.

West of the terrane in the Central Franciscan belt (Fig. 38-2), several slabs and irregular bodies of fault-bounded low-grade metagraywacke and radiolarian chert are shown as the undivided Yolla Bolly terrane (YB) on Fig. 38-2.

In addition to the alkalic-titaniferous intrusive bodies, the Yolla Bolly terrane contains intrusive and extrusive keratophyres and quartz keratophyres. Chemical analyses of these igneous rocks indicate a bimodal SiO_2 content ranging from 44 to 50% in the alkalic gabbro and from 68 to 76% in the keratophyres and quartz keratophyres (Jayko *et al.*, 1986).

Radiolarian microfossils and scarce megafossils indicate a Late Jurassic (Tithonian) and Early Cretaceous (Valanginian) protolith age for most of the Yolla Bolly terrane. However, near Hull Mountain (Fig. 38-2), mid-Cretaceous (Cenomanian) *Inocerami* occur in Yolla Bolly shale and metagraywacke containing lawsonite and aragonite (Blake and Jones, 1974; Blake *et al.*, 1985a). K-Ar and Rb-Sr whole-rock metamorphic ages on metagraywacke from the Yolla Bolly terrane range from about 90 to 115 m.y. (Lanphere *et al.*, 1978; Suppe, 1973); but the presence of remnant detrital white mica suggest an inherited age in the metagraywacke. Therefore, the U/Pb sphene age of 92 m.y. from the Yolla Bolly terrane in the Diablo Range (Mattinson and Echeverria, 1980) may more accurately reflect the timing of metamorphism. Thus, it appears that sedimentation was terminated in the Cenomanian by a tectonic event that caused rapid high-pressure metamorphism. We infer that the Yolla Bolly terrane represents a collapsed continental-margin basin that formed in a transtensional or highly oblique convergent setting, and was subsequently subducted.

The Yolla Bolly terrane retains evidence for two phases of penetrative deformation that were coaxial with the second and third phases of deformation in the Pickett Peak terrane. The first phase of deformation in the Yolla Bolly terrane was also accompanied by blueschist-facies metamorphism.

Central Franciscan Belt

The Central Franciscan belt (Fig. 38-1) is largely a tectonic melange containing numerous blocks and slabs of resistant graywacke, greenstone, chert, limestone, ultramafic rock, and high-grade metamorphic rocks in a sheared matrix of argillite, lithic graywacke, and greenish radiolarian tuff. In the western part of the belt, the graywacke slabs (ss) (Fig. 38-2) are generally arkosic and contain scarce to abundant detrital potassium feldspar. Although similar to the Coastal belt sandstones, these rocks appear to be of Cretaceous age. Near the eastern margin of the belt, the graywacke slabs (gwy) are more quartzo-feldspathic, lack K-spar, and in the absence of other criteria (e.g., interbedded radiolarian chert), are very difficult to tell from low-grade Yolla Bolly metagraywacke.

Some of the resistant graywacke slabs west of Covelo (ss in Fig. 38-2) have yielded fossils of mid-Cretaceous (Albian) age, whereas most fossils from the sheared argillite matrix are Late Jurassic (Tithonian) to Early Cretaceous (Valanginian) in age. At another locality, south of Covelo, near Etsel Ridge, K. D. Berry (1982) described Cenomanian foraminifers that occur in limestone concretions. We have been unable to relocate Berry's locality, however, and no other fossils of this age are known from that area.

Limestones, in particular those near Laytonville, have been the object of much recent study. Both paleontologic and paleomagnetic data suggest the limestones were originally deposited at equatorial latitutes, between mid-Cretaceous (Albian) and Late Cretaceous (Coniacian) time (Alvarez *et al.*, 1980; Sliter, 1984; Tarduno *et al.*, 1986), and transported on an oceanic plate to the continental margin, where they were incorporated with the clastic rocks of the Central belt.

Structures within the Central belt have not been studied in detail within the map area. Well-exposed outcrops along streams show the characteristic boudinage pattern of slickensided shear surfaces. Where the matrix contains a high proportion of graywacke, the stratal disruption can be described as broken formation (Hsu, 1968). Where the matrix is largely argillite, block-in-matrix melange is more typical (Cowan, 1985). Although both tectonic and sedimentary processes have been invoked to produce the block-in-matrix character (Aalto, 1986; Aalto and Murphy, 1984; Cloos, 1985; Cowan, 1985), most workers agree that the melange formed within an accretionary wedge at the interface between an oceanic plate and the North American margin.

At least three sets of faults have contributed to, or have overprinted, the early-phase melange fabric of the Central belt. In the northern part of the transect area, a set of high-angle faults that trend ~N20°W truncates the Yolla Bolly terrane of the Eastern belt, and has transposed slabs of it, as well as other Eastern belt rocks (Pickett Peak terrane), Coast Range ophiolite, and Great Valley sequence as far north as southwest Oregon (Blake *et al.*, 1985b). The easternmost fault of this system is shown as the Grogan–Red Mountain fault on Fig. 38-1. Detailed mapping in the Pickett Peak quadrangle (Irwin *et al.*, 1974) suggests that these lateral faults dip about 60° to the east, implying oblique strike-slip faulting. Timing of this system of faulting is broadly constrained to postdate the age of the youngest fossils and blueschist-facies metamorphism in the Yolla Bolly terrane (Cenomanian, ≃90 m.y.) and to predate early Eocene (~50 m.y.) strata that overlap the displaced terrane fragments in southwest Oregon, although significant Neogene reactivation has undoubtedly occurred locally (Blake *et al.*, 1985b; Cashman *et al.*, 1986).

A second set of high- to low-angle faults, with reverse and right-slip components, trends about N30-50°W in the vicinity of Covelo (Fig. 38-2). This set of faults appears to truncate or disrupt the earlier N20°W trend. Similar-trending faults with prominent vertical slip components bound Miocene and Pliocene sedimentary rocks near Garberville.

A third set of faults trends ~N20-40°W, including the active Maacama fault zone (Fig. 38-2), which is parallel with the active Green Valley-Bartlett Springs–Lake Mountain fault zone to the south (DePolo and Ohlin, 1984; McLaughlin *et al.*, 1985a), and the active late Neogene Middle Mountain-Konocti Bay fault zone (Hearn *et al.*, in press).

Coastal Franciscan Belt

The westernmost and youngest Franciscan belt, the Coastal belt (Bailey and Irwin, 1959), consists of three mappable terranes, the Coastal, Yager, and King Range. Each terrane can be distinguished by differences in lithology, age, sandstone composition, and structural state (Bachman, 1982; Kramer, 1976; McLaughlin *et al.*, 1982, 1983a; Underwood, 1985). The Coastal belt is separated from the Central Franciscan belt by the Coastal Belt thrust (Jones *et al.*, 1978), an east-dipping structure locally folded into subvertical orientation and cut by younger high-angle faults.

The Coastal terrane is characterized by sandstone, mudstone, and conglomerate that range in age from Paleocene (~60-55 m.y.) to late Eocene (~40 m.y.). The Coastal terrane is highly disrupted by brittle fracturing, shearing, and folding, with the fabric varying from partially disrupted sandstone-rich rocks in the eastern part, to melange in the western exposures. More chaotic parts of the Coastal terrane contain rare far-traveled blocks of pillow basalt and pelagic foraminiferal limestone, which probably are accreted scraps of Late Cretaceous (Campanian to Maastrichtian) midocean seamounts or plateaus (Harbert *et al.*, 1984; Sliter, 1984; Sliter *et al.*, 1986); rare blocks of medium-grade blueschist (Type III, Coleman and Lee, 1963) also occur locally (McLaughlin *et al.*, 1982). The basalts that occur in the Coastal terrane are overlain in places by, or have intercalated with them, pelagic foraminiferal limestones. In addition to pelagic limestone, radiolarian and diatom-bearing chert is intercalated with basalts in the King Range terrane (McLaughlin *et al.*, 1982). In one critical area near Usal, manganiferous basalt and limestone are locally interbedded with arkosic sandstone and tachylitic tuffaceous argillite, suggesting (1) that the interbedded sandstones were tectonically transported an unknown distance with the basalt and pelagic limestone, and (2) that the active volcanic edifice from which the basalt and limestone were derived was near enough to the North American plate margin to receive terrigenous deposits, prior to tectonic interaction with that margin.

The Yager Formation is the sole component of the Yager terrane (McLaughlin, 1983; McLaughlin *et al.*, 1983b). Rocks of the Yager terrane consist largely of well-bedded, little sheared, but locally highly folded mudstone and sandstone, with thick lenses of polymict conglomerate. Sandstone in the Yager and Coastal terranes is characteristically arkosic, with abundant detrital K-spar; Yager rocks generally possess a uniformly arkosic to feldspathic composition compared to the much more variable arkosic to volcaniclastic compositional range of the Coastal terrane. Volcanic rock fragments in sandstones from both terranes are dominantly porphyritic to aphanitic and intermediate to silicic in composition, as is the abundant plutonic detritus, suggesting an arc-related derivation (Underwood and Bachman, 1986). A continental source is also suggested by the presence of detrital muscovite, garnet, epidote, biotite, minor white-mica schist and quartzite fragments, and the consistent presence of trace amounts of detrital shorlitic tourmaline.

In general, the Yager terrane is lithologically, structurally, and chronologically more homogeneous than the Coastal terrane. Shearing is prominent only locally, particularly near the western and eastern contacts with the Coastal terrane and Central belt; however, complex folding is widespread (Underwood, 1985). Dinoflagellates and foraminifers indicate that Yager rocks are no older than Paleocene and are as young as late Eocene. Thus, the Yager terrane overlaps, and, in part, is younger than the oldest (i.e., basalt and limestone) component of the Coastal terrane. Concretions from a Yager conglomerate along the Eel River, about 10 km northeast of Myers Flat (Fig. 38-2), have yielded a richly fossiliferous microfauna, including abundant well-preserved dinoflagellates of Paleocene age, identical to those described from the upper part of the Moreno Formation in central California (W. R. Evitt, written communication, 1985). The strata are considered to represent an uplifted and deformed slope or slope-basin deposit that accumulated above partly older, but largely coeval accreted slope and trench-slope deposits of the Coastal terrane (Bachman, 1978; Underwood, 1983, 1985).

Most of the deformation of the Coastal and Yager terranes predated deposition of the middle and upper Miocene Wildcat Group of Ogle (1953) (T on Fig. 38-2). The Central belt and Coastal belt were juxtaposed prior to deposition of the Wildcat Group (~15 m.y.) and after the deposition of the Yager (~38 m.y.); suturing was probably coincident with an Oligocene hiatus in the depositional record. North of Cape Mendocino, the Wildcat Group is locally imbricated with the Coastal terrane as the result of ongoing northeast-northwest-oriented convergence between the Pacific and North American plates at the Mendocino triple junction (McLaughlin *et al.*, 1983b; Ogle, 1953).

The youngest Franciscan terrane is the King Range terrane (McLaughlin *et al.*, 1982). The King Range terrane is a composite of two units (Point Delgada and King Peak that are stitched together across steep faults by cross-cutting adularia-bearing veins dated at 13.8 m.y. (McLaughlin *et al.*, 1985b). The Point Delgada unit consists of basaltic pillow flows, pillow breccias, and tuffs, intruded by diabase sills. This igneous section is overlain by folded and sheared, pumpellyite-bearing, quartz-rich to arkosic sandstone andargillite devoid of detrital K-spar. Melange zones containing rare blueschist blocks cut the igneous and overlying sedimentary rocks. Radiolarians from red argillite interbedded with pillow breccias are of Late Cretaceous (Campanian or Coniacian) age.

The King Peak unit is characterized by complexly folded and broken calcareous argillite and interbedded quartzofeldspathic to volcanoclastic sandstone, probably deposited in a lower-slope or inner-trench setting. These strata are associated locally with melange containing minor pods and lenses of pillow basalt and pelagic limestone, or basalt plus radiolarian and diatom-bearing chert (McLaughlin, 1983). Both the well-bedded pelagic and hemipelagic rocks and the melange blocks of pillow basalt-chert contain radiolarians and/or foraminifers of middle Miocene or younger age. A few poorly preserved foraminifers identified as Paleocene or Eocene in age are most likely reworked, although it is possible that part of the King Park subterrane could be Paleogene in age (McLaughlin *et al.*, 1982).

Aeromagnetic data (Griscom, 1980) indicate that the King Range terrane dips southwestward off the Coastal terrane, and thus was overthrust northeastward onto the North American margin in post-middle Miocene time. This accretion may have resulted from northeast-southwest-oriented compression and uplift associated with late Neogene converalthough it is possible that part of the King Peak unit could be Paleogene in age (McLaughlin *et al.*, 1982, 1983a). Accretion postdates the stitching of the King Peak and Point Delgada units at ∼14 m.y.

DESCRIPTION OF FRANCISCAN METAMORPHISM

Metamorphic Textural Zones

Earlier studies in northern California documented that metagraywacke of the Eastern Franciscan belt contains a progressive eastward increase in metamorphic textural grade, similar to that which had been described from New Zealand (Bishop, 1972; Blake *et al.*, 1967; Jayko *et al.*, 1986; Suppe, 1973; Worrall, 1981). Because each of these workers used slightly different criteria for defining textural zones, we here define our criteria, which are based entirely on field observations of graywacke that is medium-grained and in beds that are about o.5–2 m thick. Textural zone 1 (TZ-1) is defined as having no obvious cleavage in outcrop and no recognizable flatterning of the detrital grains under the hand lens. TZ-2A has a nonparallel, anastamosing cleavage in outcrop and an incipient flattening under the hand lens. TZ-2B has a planar-parallel cleavage in outcrop with a noticeable preferred orientation of the clastic grains under the hand lens. TZ-3A appears similar to TZ-2B in outcrop, but hand samples are noticeably banded with quartz-mica segregations up to 2 mm thick and several centimeters wide. TZ-3B is similar to TZ-3A; however, the segregation banding is greater than 2 mm.

The distribution of the textural zones are shown on Fig. 38-3. All of the rocks of

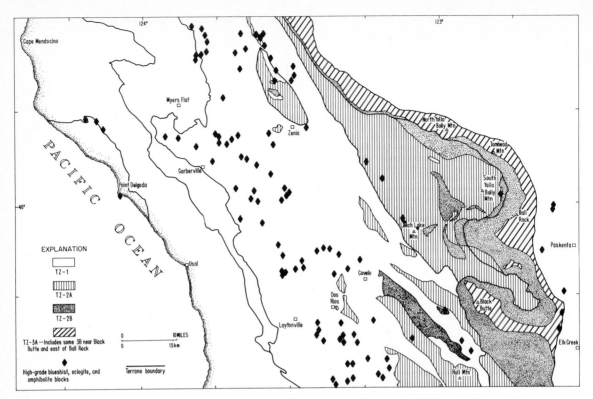

FIG. 38-3. Map showing metamorphic textural zones and high-grade blue-schists, amphibolites, and eclogites in a portion of northern California.

TZ-2A or higher grade are in the Eastern Franciscan belt or in outliers that are thought to have formed as part of the Eastern belt and were subsequently dispersed within the Central belt by strike-slip faults. Within both the Yolla Bolly and Pickett Peak terranes, there is an eastward increase in textural grade from TZ-1 to TZ-2B and from TZ-2A to TZ-3A, respectively. In addition, several large folds appear to be defined by the distribution of the textural zones.

Most metavolcanic rocks within a given textural zone are less deformed than nearby metagraywacke. In general, metavolcanic rocks of TZ-1 and 2A have no obvious schistosity in outcrop, and mesoscopic igneous textures and minerals are preserved. In TZ-2B, the metavolcanic rocks more commonly have a weak schistosity comparable to that of TZ-2A metagraywacke. In TZ-3A, metavolcanic rocks commonly show a strong schistosity and local segregation layering.

Mineral Zones

Metaigneous rocks

Eastern Franciscan Belt The distribution of metamorphic mineral assemblages in meta-igneous rocks of the Franciscan Complex show some important patterns, particularly in the Eastern Franciscan belt (Fig. 38-4). In the northeast area, between North Yolla Bolly Mountain and Tomhead Mountain (see Fig. 38-2), the metabasalts of the South Fork

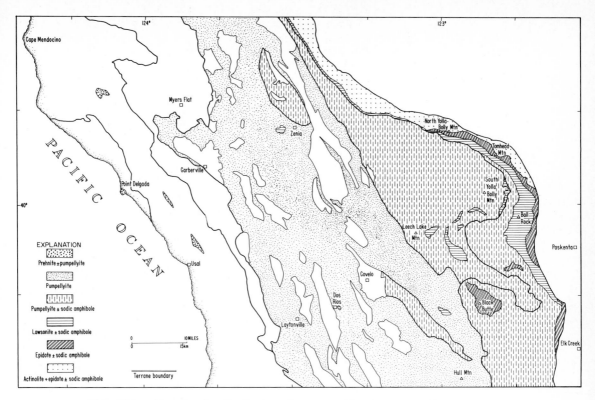

FIG. 38-4. Map showing distribution of metamorphic mineral assemblages in rocks of basaltic composition, northern California.

Mountain Schist, above the easternmost and structurally highest thrust (the Tomhead fault of Worrall, 1981), are characterized by greenish actinolite-albite-chlorite-epidote ± pumpellyite ± stilpnomelane layers that are intercalated with bluish layers containing sodic amphibole (typically crossite) + epidote. Chemical analyses indicate that the bluish metabasalts contain more total iron as well as more ferric iron, and less calcium than the greenish varieties; these differences in bulk-rock chemistry are believed to control the difference in the composition of the amphibole (Blake, 1965).

South of the Tomhead fault metabasalts of the South Fork Mountain Schist have been divided into lawsonite and epidote-bearing assemblages with distinct areal distributions (Fig. 38-4). The epidote-zone assemblage lies to the east of lawsonite-zone rocks. Sodic amphibole, Mg-pumpellyite, albite, and relict igneous clinopyroxene are also present. Accessory metamorphic minerals include acmitic-omphacitic pyroxene (which generally rims or replaces relict igneous clinopyroxene; Brown and Ghent, 1983), white mica, stilpnomelane, hematite, calcite or aragonite, quartz, opaque minerals, and sphene. Actinolite is rare and restricted to the Black Butte area (Ghent, 1965).

Metavolcanic rocks associated with TZ-2B and 3A metagraywacke of the Valentine Spring Formation commonly contain chlorite + albite + lawsonite + pumpellyite + blue amphibole and, more rarely, chlorite + albite + lawsonite + pumpellyite, or chlorite + albite + pumpellyite + sodic amphibole with accessory calcite, aragonite, celadonite, white mica, sphene, opaque minerals, and quartz. The assemblage chlorite + albite + pumpellyite + accessories is most common in TZ-2B.

Metabasalts of the Taliaferro Metamorphic Complex are characterized by the mineral assemblage lawsonite + sodic amphibole + chlorite + albite + pumpellyite, with accessory phases of quartz ± calcite ± aragonite ± omphacite ± sphene ± opaque minerals ± white mica. Both lawsonite and sodic amphibole are major phases. The lawsonite is coarser grained (0.1–0.3 mm) than in other metavolcanic rocks of the Yolla Bolly terrane (typically, 0.03–0.07 mm) or in metavolcanic rocks of the Valentine Spring Formation (0.05–0.1 mm). The rocks are either coarse grained, lacking foliation and relict igneous texture; fine grained and schistose with a poorly developed mineral segregation; or brecciated and partially annealed. Epidote is absent, but relict green amphibole cores in blue amphiboles occur in some samples. Although pumpellyite is fairly common, it appears to be a retrograde phase overprinting sodic amphibole and lawsonite, and also occurs in veins. Relict igneous pyroxene is rare.

The mineral assemblages in metabasites of the other Yolla Bolly terrane units (Chicago Rock melange, the metagraywacke of Hammerhorn Ridge, and the Devils Hole Ridge unit) are typically chlorite + albite + pumpellyite ± lawsonite. Trace amounts of incipient sodic amphibole rimming igneous green amphibole and/or relict pyroxene occur in about 30% of the samples; the sodic amphiboles are most common in coarse-grained gabbroic intrusive rocks. Relict igneous titaniferous clinopyroxene is preserved in many samples, and igneous(?) brown and green amphibole also is present locally. Pseudo-morphs by albite, calcite, and/or pumpellyite after plagioclase are common. Plagioclase in a few gabbroic samples contains intergrown lawsonite. Igneous textures commonly are well preserved and include typical hypidiomorphic granular, porphyritic, and extremely rare cumulate varieties. The metabasites are massive and lack a tectonite fabric.

Central Franciscan Belt Greenstones from the Central Franciscan melange belt range in grade from pumpellyite-chlorite-albite through pumpellyite-actinolite to completely recrystallized glaucophane and lawsonite-bearing metabasalt and keratophyre. Most, if not all, of the higher-grade blocks clearly are exotic with respect to the surrounding matrix. Nevertheless, many if not most, of the greenstone blocks, including those seen to be intercalated within the sedimentary rocks, are characterized by the mineral assemblage chlorite + pumpellyite + albite ± quartz ± calcite.

Coastal Franciscan Belt Greenstone is rare in the Coastal belt, but is present in rocks of the Coastal and King Range terranes. These basaltic rocks occur almost exclusively as sheared, transported blocks incorporated into tectonized sandstone and argillite. Our petrographic and X-ray diffraction data indicate that at least three metamorphic mineral assemblages are present in the basaltic rocks: (1) an early, moderate-temperature, low- to moderate-pressure prehnite-pumpellyite assemblage; (2) a later, lower-temperature, low- to moderate-pressure laumontite assemblage; and (3) near Point Delgada or even later, high-temperature, moderate-pressure hydrothermal vein assemblage, chlorite ± epidote ± prehnite. The latter mineral assemblage is probably associated with the earlier phases of the Miocene hydrothermal system that deposited base and precious metals and a potassic alteration mineral assemblage in veins that cut the entire King Range terrane.

Metasedimentary rocks

Eastern Franciscan belt The mineral assemblages in metasedimentary rocks of the South Fork Mountain Schist (Fig. 38-5) commonly include quartz + albite + chlorite + white mica + lawsonite ± calcite ± aragonite ± sodic amphibole ± stilpnomelane ± sphene ± hematite ± tourmaline. Fine-grained, pale sodic amphibole (glaucophane) occurs only in highly lithic or tuffaceous metagraywacke. Paragonite, detected by X-ray diffraction, is sporadically distributed with phengite (Jayko *et al.*, 1986). Jadeitic pyroxene and pumpell-yite were not observed in any samples of mica schist. Metacherts containing the assemblage ferrocrossite + quartz + magnetite ± stilpnomelane ± barite ± spessartine ± hematite crop out locally along the contact between the metavolcanic rocks and the mica schist. Spessartine in the metachert is widespread north of the Tomhead fault (Blake, 1965; Worrall, 1981); however, it is very rare to the south. A spessartine-in isograd may be present but has not been mapped, owing to the scarcity of metachert.

Metagraywacke in the Valentine Spring Formation is characterized by the presence of lawsonite, the absence of pumpellyite, and the local occurrence of glaucophane in lithic TZ-3A metagraywacke, where it occurs as incipient fibrous clusters intergrown in albite and chlorite. Mineral assemblages in rare metachert are quartz + stilpnomelane + hematite ± acmite ± ferrocrossite. As in the South Fork Mountain Schist, the chert is completely recrystallized and lacks any remnants of radiolarian tests.

Metagraywacke of the Taliaferro Metamorphic Complex is TZ-2B to 3A. The metasedimentary rocks are unusual in that they commonly contain jadeitic pyroxene in addition to the assemblage lawsonite ± pumpellyite ± sodic amphibole (Suppe, 1973). In fact, jadeitic pyroxene was one of the criteria Suppe employed for distinguishing the Taliaferro

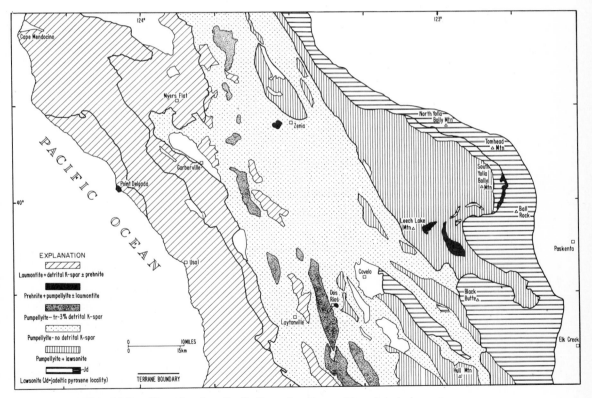

FIG. 38-5.   Map showing distribution of metamorphic minerals in metagraywacke, northern California.

Metamorphic Complex from surrounding rocks. The mineral assemblage is quartz + albite + chlorite ± pumpellyite ± jadeitic pyroxene ± lawsonite ± blue amphibole ± sphene ± calcite or aragonite. Abundant albite veins serve as a good field criterion for distinguishing the Taliaferro Metamorphic Complex from other units in the Yolla Bolly terrane.

Metagraywacke in other units of the Yolla Bolly terrane typically contains traces to minor amounts of lawsonite and pumpellyite, in addition to the ubiquitous assemblage quartz + albite + white mica + chlorite. Trace amounts of celadonite also occur. Accessory metamorphic minerals include sphene ± hematite ± calcite ± aragonite. Detrital grains of biotite, white mica, tourmaline, garnet, epidote, hornblende, and zircon, plus chert and volcanic fragments are all preserved in the metagraywacke. Detrital K-spar has not been found in any graywacke from the Yolla Bolly terrane; however, chessboard albite and detrital grains composed entirely of fine-grained sericite suggest that detrital K-spar was originally present but was not stable during the blueschist-facies metamorphism (Moore and Liou, 1979).

Graywacke from the Yolla Bolly terrane contains detrital grains of intergrown, coarse-grained (1–2 mm) lawsonite-white mica. Identical lawsonite has also been described in metagraywacke from the Diablo Range (K. E. Crawford, 1975; Ernst, 1971b) in rocks that we now correlate with the Yolla Bolly terrane (Blake *et al.*, 1982b). These observations indicate that a high-grade blueschist terrane was contributing a minor component of detritus to the Yolla Bolly basin during Late Jurassic and Early Cretaceous time.

Metachert in the Yolla Bolly terrane occasionally contains tiny (0.05 mm) crystals of acmitic pyroxene growing randomly in the groundmass. Radiolarians typically are recrystallized to a fine-grained quartz mosaic. Chert of the Taliaferro Metamorphic Complex is usually more metamorphosed; radiolarians are not preserved, and the groundmass recrystallized to a fine-grained (~0.05 mm) quartz mosaic with fairly well-segregated layers of ferrocrossite ± stilpnomelane, giving the rock a pronounced bluish color. Within these bluish cherts, however, it is sometimes possible to find pinkish to reddish zones from which radiolarians of Late Jurassic (Tithonian) age have been extracted.

Central Franciscan Belt Graywackes of the Central belt are typically pumpellyite-bearing. Characterizing the metamorphic state of the Central belt is complicated by the problem of distinguishing between the complex intercalation of matrix and blocks. Greenish TZ-1 graywackes that contain abundant detrital chert and volcanic rock fragments typically form discontinuous beds, lenses, and blocks in sheared argillite-tuff. These graywackes are characterized throughout the Central belt by the metamorphic mineral assemblage quartz + albite + chlorite + pumpellyite ± calcite ± white mica. Cloos (1983) identified lawsonite in the matrix of the Central belt, but we have not yet found lawsonite in graywacke that is clearly part of the melange matrix. Some of the slab sandstones (ss and gwy on Fig. 38-2) also contain abundant veins of laumontite in addition to fine-grained pumpellyite and prehnite (Fig. 38-5).

At least two slabs of TZ-2A metagraywacke within the Central belt contain jadeitic pyroxene in addition to lawsonite and pumpellyite. These slabs are located near Dos Rios and Zenia (Fig. 38-5) and are correlated with the graywacke of the Yolla Bolly terrane even though their degree of reconstitution and metamorphic grade would appear to be more typical of the Diablo Range than the predominantly, TZ-2B metagraywacke of the Taliaferro Metamorphic Complex of the northern Coast Ranges.

Coastal Franciscan Belt Laumontite ± calcite ± quartz is the principal metamorphic mineral assemblage of the Coastal terrane. The metamorphic mineral assemblage of the Yager terrane consists primarily of veins filled with laumontite ± calcite ± quartz, which occur most abundantly in sandstones along the tectonized western and eastern boundaries and only sparingly away from these boundaries. Laumontite veining is largely confined to sandstones (as opposed to argillites) and is especially prevalent in nearly all of the highly shattered sandstones of the Coastal terrane. Prehnite, datolite, and clinoptilolite have also been recognized in the Coastal terrane sandstone vein assemblages.

Sandstone of the King Range terrane contains quartz ± laumontite ± calcite in most places. However, in the Point Delgada unit of the King Range terrane, the sandstones contain pumpellyite and prehnite.

High-grade blocks Shown scattered throughout the Central belt and to a lesser extent along or near serpentinite-marked fault zones within other Franciscan terranes and in the Great Valley sequence (Fig. 38-3) are high-grade metamorphic blocks usually referred to as knockers (Berkland *et al.*, 1972). These blocks have been the subject of numerous detailed studies dealing largely with their mineralogy and age. Most blocks are a few meters to a few tens of meters across, roughly spheroidal, and contain a surrounding rind of talc, chlorite, and actinolite that formed by metasomatism when the block was encased in serpentinite (Bailey *et al.*, 1964; Coleman and Lee, 1963). Chemical analyses indicate that most of the blocks are of basaltic composition; however, some are clearly metachert and a few appear to be metagraywacke or metakeratophyre (R. G. Coleman and M. C. Blake, Jr., unpublished data).

Based on their mineralogy, the blocks have been called blueschist, eclogite, and amphibolite, with the blueschists further divided into garnet-bearing and garnet-free varieties (Type IV and Type III blueschists of Coleman and Lee, 1963). Recently, it has been shown that most of the high-grade blueschists and eclogites were initially foliated amphibolite and garnet amphibolite, prior to being converted to higher-pressure assemblages (Moore and Blake, 1986). For example, blueschists at the Laytonville quarry (Chesterman, 1966; Muir-Wood, 1982) contain glaucophane, lawsonite, deerite, howieite, and zussmanite, all growing across an earlier schistosity defined by green amphibole, white mica, garnet, epidote, and quartz. Radiometric K-Ar dating of some blocks has documented a Middle Jurassic age (~160 m.y.) for hornblende and white mica, with appreciably lower ages (~120 m.y.) for glaucophane (Coleman and Lanphere, 1971). On the other hand, other blocks have given concordant phengite and glaucophane ages of about 154 m.y. (Coleman and Lanphere, 1971). In addition, a single block of Type III metabasalt from near Leach Lake Mountain that shows no obvious early greenschist or amphibolite-facies event has given a concordant U/Pb mineral age of 162 m.y. (Mattinson, 1981, 1986). The 162-m.y. metamorphism is too old to be correlated with the 150-m.y. Nevadan orogeny (Harper and Wright, 1984), but the date does agree quite well with the timing of formation of the Coast Range ophiolite (Hopson *et al.*, 1981). As a speculative hypothesis, we suggest that blocks may have formed along the sole of the hot ophiolite as it was obducted onto the continental margin and that this amphibolite-greenschist terrane was subsequently transected by the Franciscan subduction zone and overprinted to eclogite or blueschist facies, dismembered, and carried back to the surface in serpentinite diapirs and thrust faults, and then subsequently dispersed along the North American margin through transcurrent movement within melange zones.

P/T conditions of metamorphism O^{18}/O^{16} data on quartz-magnetite pairs from metachert in the highest grade part of the South Fork Mountain Schist give estimated temperatures of about 330–345°C (James Drotleff, oral communication, 1983). Aragonite is stable in these rocks, suggesting maximum pressures of around 8 kb (W. A. Crawford and Hoersch, 1972; Johannes and Puhan, 1971). The lower-grade mineral assemblages seen in the Valentine Spring Formation, and the widespread occurrence of TZ-2B and even TZ-2A metagraywacke in these rocks, suggest that temperatures were probably as low as 250°C. Pressures constrained by the aragonite stability curves were about 6 kb.

Within the Yolla Bolly terrane, we used the technique of vitrinite reflectance (R_0)

to estimate peak paleotemperatures (Underwood and O'Leary, 1985). Temperature values (Fig. 38-6) are based on the empirical correlation of Price (1983), where $T = 302.97$ $\log_{10} R_0 + 187.33$; the analytical procedures and resulting data are described fully by Underwood and O'Leary (1985) and Underwood and Strong (1986). Mean R_0 values range from 1.00 to 1.99% in lawsonite and albite-bearing mudstone (TZ-2A), suggesting temperatures of about 187-275°C. Two samples of jadeitic pyroxene-bearing metagraywacke (TZ-2B) from the Taliaferro Metamorphic Complex gave mean R_0 values of 1.97 and 2.13%, suggesting temperatures of about 275 and 285°C. Using the stability curves for aragonite and jadeitic pyroxene (Jd_{80}), these temperatures give estimated pressures of about 5.7 and 6.7 kb.

In the Central belt, vitrinite measurements on mudstone from both the matrix and the tectonically interleaved pumpellyite-bearing sandstone slabs give mean R_0 values of 0.68-1.69%, suggesting temperatures of 137-254°C. On Fig. 38-7 we have plotted our vitrinite values plus some other published results from the same area (Larue, 1986). The data, while rather sparse, indicate that inferred paleotemperatures within the Central belt increase eastward toward the Yolla Bolly terrane. Based on the phengitic nature of fine-grained white mica, Cloos (1983) proposed that the Central belt matrix graywackes and shales were subjected to pressures on the order of 5-6 kb. A similar study of Central belt rocks near Crescent City (Aalto and Murphy, 1984), however, gave much lower phengite compositions. Based on the absence of lawsonite we suggest that pressures were less than about 3 kb (Liou et al., 1985).

In the Coastal belt, the Yager terrane has been sampled extensively for vitrinite (Fig. 38-7). Most R_0 values are around 0.6-0.8%, suggesting temperatures of 120-160°C. Near the Coastal belt thrust, however, there are several pronounced anomalies with values over 1.0%, suggesting temperatures as high as 240°C. These local anomalies are probably the result of shear heating along the thrust (Underwood and Strong, 1986). Most mudstones of the Coastal terrane show a similar range in reflectance (mean R_0 0.47-0.95%). For both the Yager and Coastal terranes a maximum pressure of about 3 kb is constrained by the laumontite to lawsonite reaction (Liou et al., 1985).

The King Range terrane is characterized by a prominent thermal anomaly (Fig. 38-7), with R_0 values reaching a maximum of 2.50% (~308°C) near Pt. Delgada. This vitrinite reflectance anomaly correlates well with the hydrothermal system initiated about 13.8 m.y. ago (McLaughlin et al., 1985b) and demonstrates that crustal heating occurred beneath the entire King Range terrane, but was confined only to the area of the King Range terrane.

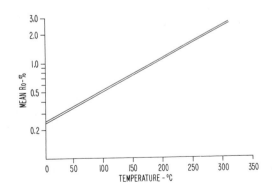

FIG. 38-6. Correlation between mean vitrinite reflectance and temperature established by Price (1983).

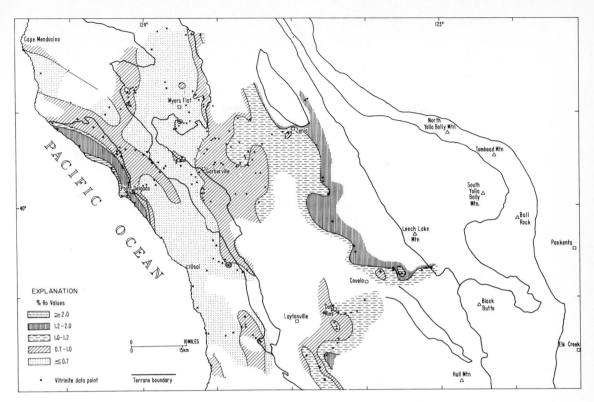

FIG. 38-7. Map showing contoured vitrinite values in metasedimentary rocks of the Franciscan Complex, northern California.

Plate Tectonic Models

Several different tectonic events are recorded in the Franciscan rocks of northern California. The oldest Franciscan rocks, the Pickett Peak terrane, apparently were subducted beneath the continental margin following the Nevadan orogeny. The timing of this event is poorly constrained but postdated the Nevadan orogeny at about 150 m.y.b.p. and may have occurred around 125 m.y. ago, as suggested by the radiometric ages. The Yolla Bolly terrane was deposited in Late Jurassic and Early Cretaceous time and subsequently subducted and metamorphosed to blueschist-facies conditions at about 92 m.y. ago (Mattinson and Echeverria, 1980). According to plate-motion models (Engebretson *et al.*, 1985), the Farallon plate was being rapidly subducted beneath North America at this time. Cretaceous plutons of the Sierra Nevada batholith were also being emplaced during the interval of about 120–80 m.y. b.p. (Chen and Moore, 1982) and provide further evidence for subduction at this time. Following subduction and high-P/T metamorphism, the rocks of both the Pickett Peak and Yolla Bolly terranes were involved in large-scale, west-directed overthrusts. We suggest that the cause of the thrusting was a collision between a large tectonic element such as Wrangellia, which from paleomagnetic data is thought to have been at the latitude of California during the mid-Cretaceous (Irving *et al.*, 1985).

During the latest Cretaceous-early Tertiary, the tectonic regime appears to have changed from orthogonal compression to north-directed oblique convergence with a large component of right-lateral strike-slip faulting associated with passing of the Kula-Farallon triple junction (Engebretson *et al.*, 1985). Extensional faulting which tectonically thinned the hanging wall to the Eastern Franciscan belt led to the farther unroofing of the blueschist-facies rocks, and is inferred to have accompanied this change in tectonic regime (Jayko *et al.*, 1987). Oblique convergence caused truncation of the western margin of the Eastern Franciscan belt and northward transport of slabs of Pickett Peak and Yolla Bolly terranes plus the Coast Range ophiolite and Great Valley sequence. These slabs were transported at least as far north as southwest Oregon along north- to northwest-trending high-angle faults that predate the San Andreas strike-slip regime. This period of oblique convergence appears to have strongly contributed to dispersion of large blocks as presently represented in the Central belt melange. The timing of oblique convergence postdates Yolla Bolly terrane metamorphism at about 92 m.y. ago, and predates an Eocene overlap deposit in southwest Oregon (~50 m.y.) that overlies the faulted boundary between the Eastern and Central belts (Blake *et al.*, 1985b). This event appears to correlate quite well with the cessation of granitic plutonism in the Sierra Nevada, and with the breakup of the Farallon plate into the Kula and Farallon plates, both about 85 m.y. ago (Engbretson *et al.*, 1985). According to their model, between about 85 and 74 m.y.p.b., an extremely high (~10 cm/yr) component of right-lateral tangential motion existed between the Kula plate and North America that decreased to about 5 cm/yr between 74 and 55 m.y. ago.

Finally, during the late Neogene, the Mendocino triple junction began to interact with the North American margin. The King Range terrane is interpreted as having been assembled at least 400 km to the south, along the Farallon–North American plate boundary prior to northward passage of the Mendocino triple junction at that latitude (McLaughlin *et al.*, 1985b). By this model, the King Range rocks were translated away from the central California margin and attached to the northeast corner of the Pacific plate as it propagated northward. At the time of this translation event about 14 m.y. ago, a shallow-level hydrothermal system developed from the rise of hot asthenosphere into the continuously opening, eastward extending slab window at the propagating end of the San Andreas transform (Dickinson and Snyder, 1979), a model which was further used by Lachenbruch and Sass (1980) to explain the heat flow regime of the San Andreas fault and Coast Ranges. The King Range terrane then moved northward with the Mendocino triple junction to its present position, where it accreted in late Neogene time.

SUMMARY

In summary, metamorphic and tectonic studies suggest that the Franciscan Complex in northern California records a complicated history of plate interactions, including subduction, collision, transcurrent faulting, and west-verging thrust faulting. We are not merely looking at an exhumed cross section of a subduction complex, but the complex history of at least 100 m.y. of plate convergence that was not at all times purely orthogonal to the margin, and that at times was destructive with respect to the margin rather than constructive (or accretionary).

This paper has benefited from critical reviews by W. G. Ernst, R. N. Brothers, and N. K. Huber. We also thank Bruce Rogers for drafting the figures and Mary Milan for her patient and careful work on the word processor.

REFERENCES

Aalto, K. R., 1986, Structural geology of the Franciscan Complex of the Crescent City area, northern California, in Abbote, P. L., ed., *Cretaceous Stratigraphy of Western North America:* Pacific Section, Soc. Econ. Paleontologists Mineralogists, v. 46, p. 197–209.

——, and Murphy, J. M., 1984, Franciscan Complex geology of the Crescent City area, northern California, in Black, M. C., Jr., ed., *Franciscan Geology of Northern California:* Pacific Section, Soc. Econ. Paleontologists Mineralogists, v. 43, p. 185–201.

Alvarez, Walter, Kent, D. V., Premoli Silver, Isabella, Schweickert, R. A., and Larson, Roger, 1980, Franciscan complex limestone deposited at 17° south paleolatitude: *Geol. Soc. America Bull.*, v. 91, p. 476–484.

Bachman, S. B., 1978, A Cretaceous and early Tertiary subduction complex, Mendocino Coast, northern California, in Howell, D. G., and McDougall, K. A., eds., *Mesozoic Paleogeography of the Western United States:* Pacific Section, Soc. Econ. Paleontologists Mineralogists, Pacific Coast Paleogeography Symp. 2, p. 419–430.

——, 1982, The Coastal Belt of the Franciscan — Youngest phase of northern California subduction, in Leggett, J. K., ed., *Trench and Forearc Sedimentation and Tectonics in Modern and Ancient Subduction Zones:* Geol. Soc. London Spec. Publ. 10, p. 401–417.

Bailey, E. H., and Blake, M. C., Jr., 1969, Tectonic development of western California during the late Mesozoic: *Geotektonika*, no. 3, p. 17–30, no. 4, p. 24–34.

——, and Irwin, W. P., 1959, K-feldspar content of Jurassic and Cretaceous graywacke of the northern Coast Ranges and Sacramento Valley, California: *Amer. Assoc. Petrol. Geologists Bull.*, v. 43, p. 2797–2802.

——, Irwin, W. P., and Jones, D. L., 1964, Franciscan and related rocks and their significance in the geology of western California: Calif. Div. Mines Geology Bull. 183, 177 p.

——, Blake, M. C., Jr., and Jones, D. C., 1970, On-land Mesozoic ocean crust in California and Coast Ranges, in *Geological Survey Research 1970:* U.S. Geol. Survey Prof. Paper 700-C, p. C70–C81.

Berkland, J. O., Raymond, L. A., Kramer, J. C., Moores, E. M., and O'Day, M., 1972, What is Franciscan? *Amer. Assoc. Petrol. Geologists Bull.*, v. 56, p. 2295–2302.

Berry, K. D., 1982, New age determinations in Franciscan limestone blocks, northern California [abstract]: *Amer. Assoc. Petrol. Geologists Bull.*, v. 66, no. 10, p. 1685.

Bishop, D. G., 1972, Progressive metamorphism from prehnite-pumpellyite to greenschist facies in the Dansey-Pass area, New Zealand: *Geol. Soc. America Bull.*, v. 83, p. 3177–3197.

Blake, M. C., Jr., 1965, Structure and petrology of low grade metamorphic rocks, blueschist facies — Yolla Bolly area, northern California: Ph.D. thesis, Stanford Univ., Stanford, Calif., 91 p.

____, and Jayko, A. S., 1983, Preliminary geologic map of the Yolla Bolly-Middle Eel Wilderness and adjacent roadless areas, northern California: *U.S. Geol. Survey Misc. Field Studies Map MF-1595-A*, scale 1:62,500.

____, and Jones, D. L., 1974, Origin of Franciscan melanges in northern California: *Soc. Econ. Paleontologists Mineralogists Spec. Paper 19*, p. 255–263.

____, and Jones, D. L., 1981, The Franciscan assemblage and related rocks in northern California, a reinterpretation, *in* Ernst, W. G., ed., *The Geotectonic Development of California* (Rubey Vol. I): Englewood Cliffs, N. J., Prentice-Hall, p. 306–328.

____, Irwin, W. P., and Coleman, R. G., 1967, Upside-down metamorphic zonation, blueschist facies, along a regional thrust in California and Oregon: *U.S. Geol. Survey Prof. Paper 575-C*, p. C1–C9.

____, Jayko, A. S., and Howell, D. G., 1982a, Sedimentation, metamorphism and tectonic accretion of the Franciscan assemblage of northern California, *in* Leggett, J. K., ed., *Trench and Forearc Sedimentation and Tectonics in Modern and Ancient Subduction Zones:* Geol. Soc. London Spec. Publ. 10, p. 433–438.

____, Howell, D. G., and Jones, D. L., 1982b, Tectonostratigraphic terrane map of California: *U.S. Geol. Survey Open File Rpt. 82-593*, scale 1:750,000.

____, Harwood, D. S., Helley, E. J., Irwin, W. P., Jayko, A. S., and Jones, D. L., 1984, Geologic map of the Red Bluff 1:100,000 quadrangle, California: *U.S. Geol. Survey Open File Rpt. 84-105*, 33 p., 1 map, scale 1:100,000.

____, Jayko, A. S., and McLaughlin, R. J., 1985a, Tectonostratigraphic terranes of northern California, *in* Howell, D. G., ed., *Tectonostratigraphic Terranes of the Circum-Pacific Region:* Circum-Pacific Council Energy Min. Resources, Earth Sci. Series, no. 1, p. 159–171.

____, Engebretson, D. C., Jayko, A. S., and Jones, D. L., 1985b, Tectonostratigraphic terranes in southwestern Oregon, *in* Howell, D. G., ed., *Tectonostratigraphic Terranes of the Circum-Pacific Region:* Circum-Pacific Council Energy Min. Resources, Earth Sci. Series, no. 1, p. 147–157.

Brown, E. H., and Ghent, E. D., 1983, Mineralogic and phase relations in the blueschist facies of the Black Butte and Ball Rock areas, northern California Coast Ranges: *Amer. Mineralogist*, v. 68, p. 365–372.

Cashman, S. M., Cashman, P. H., and Longshore, J. D., 1986, Deformational history and regional tectonic significance of the Redwood Creek schist, northwestern California: *Geol. Soc. America Bull.*, v. 97, p. 35–47.

Chen, J. H., and Moore, J. G., 1982, Uranium-lead isotopic ages from the Sierra Nevada batholith, California: *J. Geophys. Res.*, v. 87, p. 4761–4784.

Chesterman, C. W., 1966, Mineralogy of the Laytonville quarry, Mendocino County, California, *in* Bailey, E. H., ed., *Geology of Northern California:* Calif. Div. Mines Geology Bull. 190, p. 503–507.

Cloos, M., 1983, Comparative study of melange matrix and metashales from the Franciscan subduction complex with the basal Great Valley sequence, California: *J. Geology*, v. 91, p. 291–306.

____, 1985, Thermal evolution of convergent plate margins — Thermal modeling and reevaluation of isotopic Ar-ages for blueschists in the Franciscan Complex of California: *Tectonics*, v. 4, p. 421–434.

Coleman, R. G., and Lanphere, M. A., 1971, Distribution and age of high-grade blueschists, associated eclogites, and amphibolites from Oregon and California: *Geol. Soc. America Bull.*, v. 82, p. 2397–2412.

____, and Lee, D. E., 1963, Metamorphic aragonite in the glaucophane schists of Cazadero, California: *Amer. J. Sci.*, v. 260, p. 577–595.

Cowan, D. S., 1985, Structural styles in Mesozoic and Cenozoic melanges in the western Cordillera of North America: *Geol. Soc. America Bull.*, v. 96, p. 451–462.

Crawford, K. E., 1975, The geology of the Franciscan tectonic assemblage near Mount Hamilton, California: Ph.D. thesis, Univ. California, Los Angeles, Calif., 137 p.

Crawford, W. A., and Hoersch, A. L., 1972, Calcite-aragonite equilibrium from 50° to 150°C: *Amer. Mineralogist*, v. 57, p. 995–998.

DePolo, C. M., and Ohlin, H. N., 1984, The Bartlett Springs fault zone — An eastern member of the California plate boundary system: *Geol. Soc. America Abstr. with Programs*, v. 16, no. 6, p. 486.

Dickinson, W. R., and Snyder, W. S., 1979, Geometry of triple junctions related to San Andreas transform: *J. Geophys. Res.*, v. 84, p. 561–572.

Engebretson, D. C., Cox, Allan, and Gordon, R. G., 1985, Relative motions between oceanic and continental plates in the Pacific basin: *Geol. Soc. America Spec. Paper 206*, 59 p.

Ernst, W. G., 1970, Tectonic contact between the Franciscan melange and the Great Valley sequence — Crustal expression of a Late Mesozoic Benioff zone: *J. Geophys. Res.*, v. 75, p. 886–901.

——, 1971a, Metamorphic zonations on presumably subducted lithospheric plates from Japan, California, and the Alps: *Contrib. Mineralogy Petrology*, v. 34, p. 43–59.

——, 1971b, Petrologic reconnaissance of Franciscan metagraywackes from the Diablo Range, central California Coast Ranges: *J. Petrology*, v. 12, p. 413–437.

——, 1980, Mineral paragenesis in Franciscan metagraywackes of the Nacimiento block, a subduction complex of the southern California Coast Ranges: *J. Geophys. Res.*, v. 85, p. 7045–7055.

——, 1984, Phanerozoic continental accretion and the metamorphic evolution of northern and central California: *Tectonophysics*, v. 100, p. 287–320.

Ghent, E. D., 1965, Glaucophane-schist facies metamorphism in the Black Butte area, northern Coast Ranges, California: *Amer. J. Sci.*, v. 263, p. 385–400.

Griscom, Andrew, 1980, Aeromagnetic map and interpretation maps of the King Range and Chemise Mountain instant study areas, northern California: *U.S. Geol. Survey Misc. Field Studies Map MF-1196-B.*

Hamilton, Warren, 1969, Mesozoic California and the underflow of Pacific mantle: *Geol. Soc. America Bull.*, v. 80, p. 2409–2430.

——, 1978, Mesozoic tectonics of the western United States, *in* Howell, D. G., and McDougall, K. A., eds., *Mesozoic Paleogeography of the Western United States:* Pacific Section, Soc. Econ. Paleontologists Mineralogists, Pacific Coast Paleogeography Symp. 2, p. 33–70.

Harbert, W. P., McLaughlin, R. J., and Sliter, W. V., 1984, Paleomagnetic and tectonic interpretation of the Parkhurst Ridge limestone, Coastal belt Franciscan, northern California, *in* Blake, M. C., Jr., ed., *Franciscan Geology of Northern California:* Pacific Section, Soc. Econ. Paleontologists Mineralogists, v. 43, p. 175–183.

Harper, G. D., and Wright, J. E., 1984, Middle to Late Jurassic tectonic evolution of the Klamath Mountains, California-Oregon: *Tectonics*, v. 3, p. 759–772.

Hearn, B. C., Jr., McLaughlin, R. J., and Donnelly-Nolan, J. M., Tectonic Framework of the Clear Lake Basin, California, *in* Sims, J. D., ed., *Late Quaternary Climate, Tectonism, and Lake Sedimentation in Clear Lake, Northern California Coast Ranges:* Geol. Soc. America Spec. Paper, in press.

Hopson, C. A., Mattinson, J. M., and Pessagno, E. A., Jr., 1981, Coast Range ophiolite, western California, *in* Ernst, W. G., ed., *The Geotectonic Development of California* (Rubey Vol. I): Englewood Cliffs, N.J., Prentice-Hall, p. 418–510.

Hsü, K. J., 1968, Principles of melanges and their bearing on the Franciscan-Knoxville paradox: *Geol. Soc. America Bull.*, v. 79, p. 1063–1074.

Irving, E., Woodsworth, G., Wynne, P., and Morrison, A., 1985, Paleomagnetic evidence for displacement from the south of the Coast Plutonic Complex, British Columbia: *Can. J. Earth Sci.*, v. 22, p. 584–598.

Irwin, W. P., Wolfe, E. W., Blake, M. C., Jr., and Cunningham, C. G., Jr., 1974, Geologic map of the Pickett Peak quadrangle, Trinity County, California: *U.S. Geol. Survey Geol. Quad. Map GQ-1111*, scale 1:62,500.

Jayko, A. S., Blake, M. C., Jr., and Brothers, R. N., 1986, Blueschist metamorphism of the Eastern Franciscan belt, northern California, in Evans, B. W., and Brown, E. H., eds., *Blueschists and Eclogites:* Geol. Soc. America Mem. 164, p. 107–123.

Jayko, A. S., Blake, M. C., Jr., and Harmes, Tekla, 1987, Attenuation of the Coast Range Ophiolite by extensional faulting, and the nature of the Coast Range "thrust", California: *Tectonics*, v. 6, p. 475–488.

Johannes, W., and Puhan, D., 1971, The calcite-aragonite transition, reinvestigated: *Contrib. Mineralogy Petrology*, v. 31, p. 28–38.

Jones, D. L., and Irwin, W. P., 1971, Structural implications of an offset Early Cretaceous shoreline in northern California: *Geol. Soc. America Bull.*, v. 82, p. 815–822.

_____, Blake, M. C., Jr., Bailey, E. H., and McLaughlin, R. J., 1978, Distribution and character of Upper Mesozoic subduction complexes along the west coast of North America: *Tectonophysics*, v. 47, p. 207–222.

Kramer, J. C., 1976, Geology and tectonic implications of the Coastal Belt Franciscan, Ft. Bragg-Willits area, northern Coast Ranges, California: Ph.D. dissertation, Univ. California, Davis, Calif., 128 p.

Lachenbruch, A. H., and Sass, J. H., 1980, Heat flow and energetics of the San Andreas fault zone: *J. Geophys. Res.*, v. 85, p. 6185–6222.

Lanphere, M. A., Blake, M. C., Jr., and Irwin, W. P., 1978, Early Cretaceous metamorphic age of the South Fork Mountain Schist in the northern Coast Ranges of California: *Amer. J. Sci.*, v. 278, p. 798–815.

Larue, D. K., 1986, Organic matter in limestone and melange matrix from the Franciscan and Cedros subduction complexes, in Abbott, P. L., ed., *Cretaceous Stratigraphy of Western North America:* Pacific Section, Soc. Econ. Paleontologists Mineralogists, v. 46, p. 211–221.

Liou, J. G., Maruyama, S., and Cho, M., 1985, Phase equlibria and mineral parageneses of metabasites in low-grade metamorphism: *Mineral. Mag.*, v. 49, p. 321–333.

Mattinson, J. M., 1981, U-Pb systematics and geochronology of blueschists — Preliminary results: *Trans. Amer. Geophys. Un.*, v. 62, no. 45, p. 1059.

_____, 1986, Geochronology of high-pressure-low-temperature Franciscan metabasites — A new approach using the U-Pb system, in Evans, B. W., and Brown, E. H., eds., *Blueschists and Eclogites:* Geol. Soc. America Mem. 164, p. 95–105.

_____, and Echeverria, L. M., 1980, Ortigalita Peak gabbro, Franciscan complex — U-Pb date of intrusion and high-pressure-low temperature metamorphism: *Geology*, v. 8, p. 589–593.

McDowell, F. W., Lehman, D. H., Gucwa, P. R., Fritz, Deborah, and Maxwell, J. C., 1984, Glaucophane schists and ophiolites of the northern California Coast Ranges — Isotopic ages and their tectonic implications: *Geol. Soc. America Bull.*, v. 93, p. 595–605.

McLaughlin, R. J., 1983, Post-middle Miocene accretion of Franciscan rocks, northwestern California — Reply to "Discussion by S. G. Miller and K. R. Aalto": *Geol. Soc. America Bull.*, v. 94, p. 1028–1031.

_____, Kling, S. A., Poore, R. Z., McDougal, Kristin, and Beutner, E. C., 1982, Post-middle Miocene accretion of Franciscan rocks, northwestern California: *Geol. Soc. America Bull.*, v. 93, p. 595–605.

_____, Ellen, S. D., Miller, S. G., Lajoie, K. R., and Morrison, S. D., 1983a, Terrane boundary relations and tectonostratigraphic framework, south Eel River Basin, northwestern California: *Amer. Assoc. Petrol. Geologists 58th Ann. Meet. Pacific Section, Amer. Assoc. Petrol. Geologists, Soc. Explor. Geophyics, Soc. Econ. Paleontologists Mineralogists, Sacramento, Calif.*, p. 112–113.

_____, Ohlin, H. N., and Blome, C. D., 1983b, Tectonostratigraphic framework of the Franciscan assemblage and lower part of the Great Valley sequence in the Geysers-Clear Lake region, California: *Trans. Amer. Geophys. Un.*, v. 64, p. 868.

_____, Ohlin, H. N., Thomahlen, D. J., Jones, D. L., Miller, J. W., and Blome, C. D., 1985a, Geologic map and structural sections of the Little Indian Valley-Wilbur Springs geothermal area, northern Coast Range, California: *U.S. Geol. Survey Open File Rpt. 85-285.*

_____, Sorg, D. H., Morton, J. L., Theodore, T. G., Meyer, C. E., and Delevaux, M. H., 1985b, Paragenesis and tectonic significance of base and precious metal occurrences along the San Andreas fault at Pt. Delgada, California: *Econ. Geology*, v. 80, p. 344–359.

Moore, D. E., and Blake, M. C., Jr., 1986, Development of high-grade blueschist and eclogite assemblages in Franciscan exotic blocks from greenschist and amphibolite precursors: *Geol. Soc. America Abstr. with Programs*, v. 18, no. 1, p. 159.

_____, and Liou, J. G., 1979, Chessboard-twinned albite from Franciscan metaconglomerates of the Diablo Range, California: *Amer. Mineralogist*, v. 64, p. 329–336.

Muir-Wood, Robert, 1982, The Laytonville Quarry (Mendocino County, California) exotic block — Iron-rich blueschist-facies subduction-zone metamorphism: *Mineral. Mag.*, v. 45, p. 87–99.

Ogle, B. A., 1953, Geology of the Eel River Valley area, Humboldt County, California: *Calif. Div. Mines Geology Bull. 164*, 128 p.

Page, B. M., 1981, The southern Coast Ranges, *in* Ernst, W. G., ed., *The Geotectonic Development of California* (Rubey Vol. I): Englewood Cliffs, N.J., Prentice-Hall, p. 329–417.

Pavlis, T. L., and Bruhn, R. L., 1983, Deep-seated flow as a mechanism for the uplift of broad forearc ridges and its role in the exposure of high P/T metamorphic terranes: *Tectonics*, v. 2, p. 473–497.

Price, L. C., 1983, Geologic time as a parameter in organic metamorphism and vitrinite reflectance as an absolute paleogeothermometer: *J. Petrol. Geology*, v. 6, 5–38.

Sliter, W. V., 1984, Foraminifers from Cretaceous limestone of the Franciscan Complex, northern California, *in* Blake, M. C., Jr., ed., *Franciscan Geology of Northern California:* Pacific Section, Soc. Econ. Paleontologists Mineralogists, v. 43, p. 149–162.

_____, McLaughlin, R. J., Keller, Gerta, and Evitt, W. R., 1986, Paleogene accretion of Upper Cretaceous oceanic limestone in northern California: *Geology*, v. 14, p. 350–352.

Suppe, J., 1973, Geology of the Leech Lake Mountain-Ball Mountain region, California — A cross section of the northeastern Franciscan belt and its tectonic implications: *Univ. Calif. Publ. Geol. Sci.*, v. 107, 82 p.

_____, and Armstrong, R. L., 1972, Potassium-argon dating of Franciscan metamorphic rocks: *Amer. J. Sci.*, v. 272, p. 217–233.

Tarduno, J. A., McWilliams, M., Sliter, W. V., Cook, H. E., Blake, M. C., Jr., and Premoli-Silva, I., 1986, Southern hemisphere origin of the Cretaceous Laytonville Limestone of California: *Science*, v. 231, p. 1425–1428.

Underwood, M. B., 1983, Depositional setting of the Paleogene Yager Formation, northern Coast Ranges of California, *in* Larue, D., and Steel, R., eds., *Cenozoic Marine Sedimentation, Pacific Margin, U.S.A.:* Pacific Section, Soc. Econ. Paleontologists Mineralogists, p. 81–101.

―――, 1984, Franciscan and related rocks of southern Humboldt County, northern California Coast Ranges ― Analysis of structure, tectonics, sedimentary petrology, paleogeography, depositional history, and thermal maturity: Ph.D. dissertation, Cornell Univ., Ithaca, N.Y., 354 p.

―――, 1985, Sedimentology and hydrocarbon potential of the Yager structural complex, possible Paleogene source rocks in the Eel River basin, northern California: *Amer. Assoc. Petrol. Geologists Bull. 69*, p. 1088–1100.

―――, and Bachman, S. B., 1986, Sandstone petrofacies of the Yager Complex and the Franciscan Coastal belt, Paleogene of northern California: *Geol. Soc. America Bull.*, v. 97, p. 809–817.

―――, and O'Leary, J. D., 1985, Vitrinite reflectance and paleotemperature within Franciscan terranes of coastal northern California ― 38°45′N to 40°00′N: *U.S. Geol. Survey Open File Rpt. 85-663*, 32 p.

―――, and Strong R. H., 1986, Vitrinite reflectance and estimates of paleotemperature for Franciscan terranes of coastal northern California ― 40°00′N to 40°35′N: *U.S. Geol. Survey Open File Rpt. 86-258*, 41 p.

Wang, C. Y., and Shi, Y., 1984, On the thermal structure of subduction complexes ― A preliminary study: *J. Geophys. Res.*, v. 89, p. 7709–7718.

Worrall, D. M., 1981, Imbricate low-angle faulting in uppermost Franciscan rocks, south Yolla Bolly area, northern California: *Geol. Soc. America Bull.* v. 92, p. 703–729.

R. G. Coleman and C. E. Manning
Department of Geology
Stanford University
Stanford, California 94305

M. M. Donato
U.S. Geological Survey
Menlo Park, California 94025

N. Mortimer
New Zealand Geological Survey
Dunedin, New Zealand

L. B. Hill
Division of Natural Science
Chapman College
Orange, California 92666

39

TECTONIC AND REGIONAL METAMORPHIC FRAMEWORK OF THE KLAMATH MOUNTAINS AND ADJACENT COAST RANGES, CALIFORNIA AND OREGON

ABSTRACT

Six regional metamorphic events can be identified in the Klamath Mountains and adjacent Coast Ranges. These events, established by their age, mineral assemblages, areal distribution, and tectonic history, include:

1. Coffee Creek metamorphism, Middle to Late Devonian, low P/T
2. Fort Jones metamorphism, Late Triassic, high P/T
3. Siskiyou metamorphism, Middle Jurassic, low P/T
4. Condrey Mountain metamorphism, Middle Jurassic, moderate P/T
5. Nevadan metamorphism, Late Jurassic, low P/T
6. Franciscan metamorphism, Early to mid-Cretaceous, high P/T

All low- and high-P/T gradient series contain protoliths interpreted as fragments of island arcs, and oceanic or marginal basins of various ages. The low-P/T gradient series either overprint more than one terrane (Siskiyou, Nevadan) or are restricted to a single terrane (Coffee Creek). Low-P/T metamorphism is interpreted to have occurred at shallow levels during or immediately after tectonic imbrication in the vicinity of a spreading center (Nevadan) or island arc (Siskiyou, Coffee Creek) magmatic axis. Calc-alkaline plutons intruded Siskiyou and Nevadan metamorphic rocks immediately following regional metamorphism.

In contrast, the moderate- and high-P/T gradient series do not overprint major terrane boundaries and only rarely are intruded by postmetamorphic plutons. Metamorphism is generally thought to have occurred within the downgoing slab of a subduction zone, with postmetamorphic faulting juxtaposing these rocks against the lower-P/T facies rocks. Units metamorphosed during the Middle Jurassic Siskiyou and Condrey Mountain events may have been in a paired metamorphic belt relationship.

All six regional metamorphic events are preserved in thin, east-dipping thrust slices and attest to repeated telescoping of arc/subduction complexes outboard of the Eastern Klamath terrane from Paleozoic time to the present day.

INTRODUCTION

Since Irwin's (1966) regional reconnaissance, most ideas concerning the origin and history of the Klamath Mountains have been based on fieldwork conducted in the southern and western portions of the province (e.g. Ando *et al.*, 1983; G. A. Davis *et al.*, 1965; Harper, 1984; Harper and Wright, 1984; Irwin, 1972; Saleeby *et al.*, 1982; Wright, 1981). However, there have also been a number of detailed studies farther north and east. In this paper we incorporate these studies in a discussion of the tectonic and regional metamorphic framework of an area in the Klamath Mountains and adjacent Coast Ranges that are situated astride the California-Oregon border (Fig. 39-1). We emphasize the role of metamorphism and deformation, and how these processes can be used to supplement and modify previously published tectonic models. The time scale of Harland *et al.* (1982) is used throughout this paper, and old K-Ar ages have been recalculated using new decay constants to conform with this. The metamorphic geology of the contiguous region to the north and east is described by Kays *et al.* (Chapter 40, this volume).

TECTONIC FRAMEWORK

Irwin (1966) divided the Klamath Mountains into four major lithotectonic units, the eastern Klamath, central metamorphic, western Paleozoic and Triassic, and western

Jurassic belts. Portions of each of these belts are present in the area of our compilation map (Figs. 39-1 and 39-2). Although Jurassic radiolarians have subsequently been found in the western Paleozoic and Triassic belt (Irwin *et al.*, 1978), the original name is still used in the literature and is retained here. Belts, as used in this paper, are the largest units and can in some cases be subdivided into terranes.

Irwin (1972) introduced the term "terrane" to describe fault-bounded units differing from each other in lithology, age and structure, a scheme which has subsequently been adopted by Klamath geologists in naming and correlating map units on a regional scale. For example, the Hayfork terrane has been divided into eastern and western subterranes (Irwin, 1979; Wright, 1981, 1982) and the igneous part of Irwin's (1972) North Fork terrane has been given the status of a separate terrane, the Salmon River terrane, by Blake *et al.* (1972). In a strict stratigraphic hierarchy (e.g., C. H. Holland *et al.*, 1978) the Klamath terranes would be equivalent to fault-bounded, deformed, and metamorphosed formations and groups. In European Alpine terminology, they would probably be treated as nappes.

Figure 39-2 shows the major structural features of the north-central Klamath Mountains and adjacent Coast Ranges. The fundamental structural units (terranes) constitute a sequence of stacked thrust sheets, with the highest sheets containing rocks as old as Early Paleozoic and the lowest as young as Cretaceous (G. A. Davis *et al.*, 1978; Irwin, 1981). In this paper we have followed the terrane nomenclature of Blake *et al.* (1982) and Silberling *et al.* (1984), but have considerably modified the positions of some of their terrane boundaries in our compilation area.

Movement along major thrusts within the Klamath Mountains ceased by the end of the Jurassic (Irwin, 1981; Mortimer, 1984b); a Late Cretaceous marine overlap provides a firm lower age limit on deformational events. Juxtaposition of the Klamath province with Franciscan rocks of the Coast Ranges province took place along a major fault in Cretaceous time. High-angle faulting and broad warping of the thrust sheets occurred as the whole region was uplifted in the Late Tertiary.

Figure 39-2 provides a useful link between small-scale terrane maps (e.g., Blake *et al.*, 1983; Silberling *et al.*, 1984) and larger-scale maps of the Klamath Mountains (e.g., Coleman *et al.*, 1983; Donato *et al.*, 1982). Data used in the compilation of Figs. 39-2 and 39-3 are taken from a number of sources to which we refer throughout the text. Much of these data are already compiled on geologic maps at a scale of 1:250,000 by Smith *et al.* (1982) and Wagner and Saucedo (in press).

Tectonic Elements

Based primarily on geochronologic and metamorphic information, a number of regional metamorphic and penetrative deformational events have been recognized in the Klamath Mountains (Irwin, 1981). They are (1) the Coffee Creek event, Middle to Late Devonian age; (2) the Fort Jones event, latest Triassic age; (3) the Siskiyou event, latest Middle Jurassic age; (4) the Condrey Mountain event, Middle to Late Jurassic age; (5) the Nevadan event, Late Jurassic age; and (6) the Franciscan events, Early to mid-Cretaceous age. These events are discussed at length in the section Crustal Metamorphic Events.

Nonpenetrative deformation of the Klamath Mountains is of two kinds: high-angle, Cenozoic faults and low-angle Paleozoic and Mesozoic detachment faults (Fig. 39-2). Many of the older faults presently have steep dips, but after correction for Cenozoic deformation, their attitudes become low angle and mainly east dipping.

Paleozoic and Mesozoic faulting Most faults of this age are demonstrable thrusts. Four ages of thrust faulting can be distinguished (Fig. 39-2).

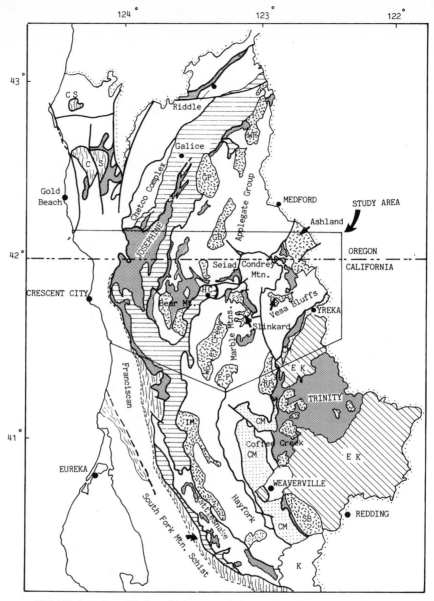

FIG. 39-1.   Simplified geologic map of the Klamath Mountains and Coast Ranges. Patterns as follows: heavy stipple, ultramafic rocks; dash, plutons; diagonal lines, Eastern Klamath belt; dot, Central Metamorphic terrane; horizontal lines, Western Klamath terrane; wavy lines, Pickett Peak terrane. Pluton abbreviations as follows: EP, English Peak; GP, Grants Pass; GB, Grayback; IM, Ironside Mountain; RP, Russian Peak; SB, Shasta Bally; SR, White Rock.

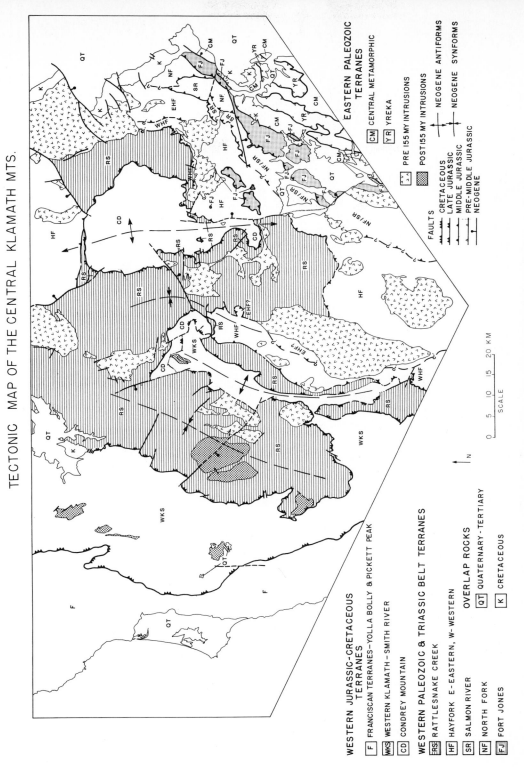

FIG. 39-2. Tectonic map of the Central Klamath Mountains.

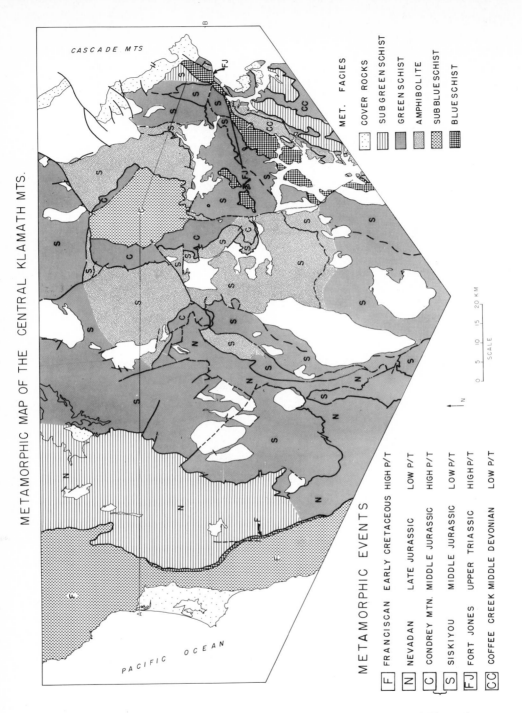

FIG. 39-3. Metamorphic facies map of the central Klamath Mountains.

1. Middle Jurassic or older: thrusts separating the Yreka and Central Metamorphic terranes from each other and the terranes to the west.

2. Middle Jurassic: the thrust separating the Fort Jones terrane and the North Fork terrane, and the faults that bound the North Fork, Salmon River, Eastern and Western Hayfork, and Rattlesnake Creek terranes. The latter faults were active just prior to and during the Siskiyou metamorphic event.

3. Late Jurassic: thrusts separating the Condrey Mountain terrane, Western Klamath terrane, and the structurally overlying Siskiyou metamorphic rocks.

4. Cretaceous: fault separating the terranes of the Klamath Mountains province from the Franciscan terranes to the west.

In addition to these interterrane faults (shown in Fig. 39-2), there are many intra-terrane faults that are too numerous to be shown on the map (e.g., see Donato *et al.*, 1982; Hotz, 1967). These faults juxtapose peridotite, schist, and volcanic and sedimentary rocks within the North Fork, Salmon River, eastern and western Hayfork, and Rattlesnake Creek terranes. This faulting has produced chaotic "block on block" tectonic melanges (Donato, 1985; Hill, 1985; Mortimer, 1985) within the terranes above and is considered to be in part prior to, and in part synchronous with, Siskiyou metamorphism. Despite this severe inter- and intraterrane faulting, the same five terranes can be recognized in the same structural order 100 km to the south.

The amount of movement along certain interterrane low-angle faults can be estimated using the klippe and fenster method, which yields the following minimum (east-west) displacements in our compilation area: Fort Jones terrane over structurally underlying terranes, 35 km; Hayfork terrane over Rattlesnake Creek terrane, 60 km; Rattlesnake Creek terrane over Western Klamath terrane, 45 km. This amounts to a minimum (east-west) shortening of 140 km or about 50% across the width of the Klamaths province in the compilation area.

Cenozoic faulting In contrast to the low-angle faults, many high-angle faults, despite their lesser displacement (1–2-km dip-slip at most), have marked topographic expression. Many faults formed in response to Neogene uplift in the area (Mortimer and Coleman, 1985) and are probably genetically related to the elongate north-south-trending flexures that are shown in Fig. 2. Minimum vertical uplift of 7 km has been estimated by strati-graphic reconstruction for one of these flexures, the Condrey Mountain dome (Mortimer and Coleman, 1985). Tertiary strike-slip faulting has affected Franciscan rocks north and south of our compilation area (Blake *et al.*, 1985b; Kelsey and Hagans, 1982), but is apparently absent within the Klamath Mountains province.

Description of Structural Units (Terranes)

Most Klamath terranes are long, thin, and traceable for 150 km or more along strike within the province (Irwin, 1981). The outcrop pattern and degree of structural coherence of most terranes in our compilation area is strongly a function of Jurassic and later dynamothermal events (Wright and Sharp, 1981).

As this is the first paper to synthesize tectonic information from the Klamath Mountains near the California-Oregon border, we provide brief descriptions of the terranes in this area (Fig. 39-2). In most cases they are northward extensions of terranes originally defined in, and described from, the southern Klamath Mountains. Criteria for the correlation of the rocks in the compilation area with these southern Klamath terranes include similar lithologies, structural order, paleontologic, and radiometric ages.

Yreka terrane The structurally highest terrane in the area of Fig. 39-2 also contains some of the oldest rocks in the Klamath Mountains. As we define it here, the Yreka terrane consists of olistostromal calcareous siltstone, quartzite, shale, and chert of the Moffett Creek Formation and the well-bedded Antelope Mountain Quartzite, which contains interbedded shale and chert (Lindsley-Griffin and Griffin, 1983; Potter *et al.*, 1977). The Moffett Creek Formation is dated as Ordovician or Silurian by radiolarians from a single locality (Irwin *et al.*, 1978), and the Antelope Mountain Quartzite is undated. These rocks are interpreted to be part of a lower Paleozoic siliciclastic submarine fan (Potter *et al.*, 1977).

Central Metamorphic terrane Hotz (1977) recognized that a thin strip of amphibole schist extending north to Yreka belonged to the Central Metamorphic terrane. We accept Cashman's (1980) correlation of the Duzel phyllite and Schulmeyer Gulch Sequence of Potter *et al.* (1977) with this terrane. There is also structural and paleontologic evidence that the strip of ultramafic rock that extends north to Yreka (Hotz, 1977) was deformed and metamorphosed along with Central Metamorphic terrane rocks in Devonian time (Cannat and Boudier, 1985; Cashman, 1980; Lindsley-Griffin and Griffin, 1983).

A wide range of lithologies is thus present in this terrane, including harzburgite, dunite, metagabbro (some showing cumulate textures), amphibolite, micaceous marble, phyllitic siltstone, chert, and micaceous quartzite (Cashman, 1980; Hotz, 1977). No fossils have been recovered from these rocks (nor from the Central Metamorphic terrane to the south), but some ^{40}Ar/^{39}Ar whole-rock ages of Duzel Phyllite are as old as 430–450 m.y. (Ordovician-Silurian boundary) and may reflect the age of detrital muscovite (Cashman, 1980). High-Ti metabasites (Salmon hornblende schist; G. A. Davis *et al.*, 1965; Hotz, 1977) suggest an oceanic origin; the rocks now constituting the Central Metamorphic terrane probably represent deformed and metamorphosed pre-Devonian oceanic crust and its sedimentary cover. Penetrative deformation of these rocks was synchronous with Devonian metamorphism (Cashman, 1980).

Fort Jones terrane Hotz (1977, 1970) correlated rocks in our compilation area with the Stuart Fork Formation of Davis and Lipman (1962), which is equivalent to the Fort Jones terrane of Blake *et al.* (1982). It consists predominantly of phyllitic quartzite (metachert), glaucophane, lawsonite and jadeite-bearing schist (metavolcanic rocks), and minor marble, metagabbro, serpentine, and eclogite (Borns, 1980, 1984; Hotz, 1977). In the Yreka area, three glaucophane schists have yielded latest Triassic (214–222 m.y.) K-Ar and ^{40}Ar/^{39}Ar white mica ages (Hotz *et al.*, 1977). Hotz interpreted the blueschists as isolated tectonic "knockers" but a detailed structural and petrologic study by Borns (1980) showed that all the lithologies of the Fort Jones terrane were metamorphosed to blueschist facies. The rocks of the Fort Jones terrane have been interpreted by Borns (1980) and Goodge (1985) to represent deformed and metamorphosed pre-Upper Triassic oceanic crust and sediments. West of the main outcrop area, several klippen of Fort Jones terrane rest on Hayfork terrane rocks (Borns, 1980; Hotz, 1977, 1979; Mortimer, 1984b).

North Fork terrane The North Fork terrane was first identified in the area of Fig. 39-2 by Mortimer (1984a). The terrane is internally faulted and consists of Upper Permian, Upper Triassic, and Lower Jurassic chert and argillite tectonically juxtaposed with units of amygdaloidal alkaline volcanic rocks. Limestones interbedded with the volcanic rocks have yielded Late Permian fossils of Tethyan faunal affinity at two localities (Elliott and Bostwick, 1973; C. H. Stevens, *in* Mortimer, 1984a).

Collectively, the rocks of the North Fork terrane record the development of Late Permian seamounts, deposition of Upper Permian to Upper Traissic open ocean hemipelagic sediments and deposition of Lower Jurassic near-continent sediments (Ando *et al.*, 1983; Irwin, 1972; Mortimer, 1984a; Wright, 1982).

Salmon River terrane Interfaulted nonamygdaloidal lava flows, pillows, diabase, and minor gabbro in the eastern part of the map area correlate with the Salmon River terrane of Blake *et al.* (1982). In the southern Klamath Mountains, this terrane was formerly defined as the ophiolitic part of the North Fork terrane (Ando *et al.*, 1983; Irwin, 1972; Wright, 1982). The lavas and diabase are compositionally uniform low-K subalkaline basalts for which a marginal basin origin has been proposed (Mortimer, 1984a). Ando *et al.* (1983) report a discordant U-Pb (zircon) age of 265–310 m.y. on a plagiogranite pod near a diabase-gabbro contact south of the map area.

Eastern Hayfork subterrane In the compilation area, the eastern Hayfork subterrane of Irwin (1979), equivalent to the eastern Hayfork terrane of Wright (1981, 1982), is petrologically identical to the North Fork terrane described above, except that (1) no fossils have been found (2) olistostromal chert-argillite-limestone breccias are present in the eastern Hayfork terrane, and (3) grade of metamorphism and degree of penetrative deformation are greater in eastern Hayfork terrane (Mortimer, 1984a). In the southern Klamath Mountains, Permian fossils of both Tethyan and Eastern Klamath faunal affinity have been described (Irwin, 1972; Irwin *et al.*, 1978; Miller and Wright, 1985) along with Paleozoic and Triassic conodonts in limestones (Irwin *et al.*, 1983) and Permian, Late Triassic, and Triassic or Jurassic radiolarians (Irwin *et al.*, 1982).

Western Hayfork subterrane This subterrane (equivalent to the western Hayfork terrane of Wright, 1981; 1982) is dominantly volcaniclastic and consists of fine- to medium-grained crystal (clinopyroxene and hornblende) and lithic wackes with interbedded siliceous argillite and marble. In the map area it is metamorphosed to the greenschist facies and transitional greenschist-amphibolite facies (Donato *et al.*, 1982; Hill, 1984; Mortimer, 1984a). In the southern Klamath Mountains, K-Ar ages on primary igneous hornblendes from less metamorphosed rocks give ages of 168–177 m.y. (Fahan, 1982). The western Hayfork terrane is interpreted as a Middle Jurassic island arc (Fahan, 1982; Wright, 1981).

Rattlesnake Creek terrane Rawson and Petersen (1982) correlated amphibolite facies rocks near Happy Camp with the relatively unmetamorphosed Rattlesnake Creek terrane in the southern Klamath Mountains. Gray and Peterson (1982) and Norman *et al.* (1983) have also extended the Rattlesnake Creek terranes into the Orleans and Preston Peak areas.

In the central part of our compilation area, the terrane consists of a tectonic melange of juxtaposed blocks (ranging in size from 10 m to 2 km) of metaperidotite, banded amphibolites, quartz-biotite schist, and marble (Donato, 1985; Donato *et al.*, 1982; Hill, 1984; Hotz, 1967). Blake *et al.* (1982) refer to this as the Marble Mountain terrane. Due to the strong metamorphic overprint, it is not always possible to identify protoliths unequivocally, and no fossils have been found.

In the southern Klamath Mountains, the Rattlesnake Creek terrane consists of a 193–207-m.y. mafic igneous complex constructed on serpentine matrix melange containing Lower Triassic and Lower Jurassic cherts (Irwin *et al.*, 1982; Wright, 1981). A mafic complex in the Preston Peak area (Snoke *et al.*, 1977), constructed on peridotite tectonite has a minimum age of 159 m.y. (Saleeby *et al.*, 1982). Hill (1985) reports a 172 ± 2-m.y. U-Pb zircon age on plagiogranite (dated by J. B. Saleeby) from a metamorphosed dike complex in the Rattlesnake Creek terrane near Seiad Valley. The rocks of the terrane thus record repeated earliest to Middle Jurassic development of oceanic crust transitional between interarc rift basins and island arc assemblages.

Condrey Mountain terrane The Condrey Mountain terrane of Blake *et al.* (1982) (Condrey

Mountain Schist of Hotz, 1967) can be divided into two lithologic units separated by a possible premetamorphic thrust fault (Donato *et al.*, 1982; Helper, 1985). The upper unit consists of a well-foliated basic schist with subordinate silicic and graphitic quartz-mica schists, whereas graphitic quartz-mica schist, metachert, glaucophane schist, and meta-peridotite comprise the lower unit (Coleman *et al.*, 1983; Donato *et al.*, 1980; Helper, 1985; Hotz, 1967, 1979). Rocks of the upper unit have been extended west to near Happy Camp, where they are in fault contact with rocks presumed to be part of the Western Klamath terrane (Hill, 1984, 1985). No fossils have been found.

The earliest deformation in both units is progressive and synmetamorphic, and based on similarities in orientation, Helper (1985) suggests that this deformational event is equivalent in the two units. A U-Pb zircon age of 170 ± 2 m.y. from a small metamorphosed and concordantly deformed diorite stock has been reported by Saleeby *et al.* (1984). Mapping indicates that the stock was intruded into greenschists and graphitic schists of the upper lithologic unit, not the lower unit as implied by Saleeby *et al.* (1984). The stock thus places minimum age constraints on the protolith of the upper unit and a maximum age limit on deformation in both units (Helper, 1985). Although some workers have correlated the entire Condrey Mountain terrane with the Upper Jurassic Galice Formation (described below; Irwin, 1981; Klein, 1977), such a correlation is precluded by the Middle Jurassic or older protolith age (Helper, 1985).

The rocks of the Condrey Mountain terrane probably formed in a Middle Jurassic or older arc-basin complex, perhaps similar to the Middle Jurassic Western Hayfork subterrane.

Western Jurassic terrane, Smith River subterrane Blake *et al.* (1985b) divide Irwin's (1966) western Jurassic belt into four subterranes (of the Western Klamath terrane). Each subterrane contains the Upper Jurassic Galice Formation but is in fault contact with its neighbors and contains different basement rocks. Only the Smith River subterrane is exposed in the area of Fig. 39-2.

The Smith River subterrane is composed of the Josephine Peridotite (Loney and Himmelberg, 1976) and overlying gabbro, diabase, dikes, and pillow lava, interpreted by Harper (1984) to be a cogenetic ophiolite suite. U-Pb zircon ages from two plagiogranites in the suite date the ophiolite at 157 ± 2 m.y. (Saleeby *et al.*, 1982). Included in the subterrane are distal turbidites and andesitic pyroclastics of the Oxfordian to Kimmeridgian Galice Formation that conformably overlie the ophiolite sequence. An interarc basin setting is interpreted for these Upper Jurassic rocks (Harper and Wright, 1984).

Franciscan assemblage (Pickett Peak and Yolla Bolly terranes) Upper Jurassic to Lower Cretaceous quartzose sandstones, shales, cherts, and basalts of the Franciscan assemblage and Dothan Formation immediately adjacent to the Western Klamath terrane are divided into the Pickett Peak and Yolla Bolly terranes by Blake *et al.* (1985a, b). In northern California there is a general increase in textural and metamorphic grade eastward, culminating in a narrow fault-bounded selvage of penetratively deformed glaucophane-bearing metamorphic rocks, the South Fork Mountain Schist, which Blake *et al.* (1982, 1985a) call the Pickett Peak terrane. This separates less metamorphosed Yolla Bolly terrane rocks from the Western Klamath terrane. North of our compilation area (Fig. 39-1), Blake *et al.* (1985b) include the Colebrooke Schist (Coleman, 1972) with the Pickett Peak terrane. Lying outboard of the Yolla Bolly terrane (but not exposed in the area of Fig. 39-2) are the melanges of the Central Belt Franciscan and the younger Coastal Belt Franciscan (Blake *et al.*, 1985a).

Tectonic Significance of Plutonic Rocks

Recent reviews of the geochronology of the plutonic rocks of the Klamath Mountains have been given by Harper and Wright (1984) and Irwin (1985). Three pulses of plutonism can be recognized in our compilation area. The oldest and volumetrically largest includes the Wooley Creek batholith and Vesa Bluffs, Ashland, Slinkard, English Peak, and Russian Peak plutons, which were intruded in the latest Middle Jurassic between 160 and 166 m.y. (Allen *et al.*, 1982; Harper and Wright, 1984; Hotz, 1971; Lanphere *et al.*, 1968). These collectively intrude the Rattlesnake Creek, eastern and western Hayfork, Salmon River, North Fork, Fort Jones, and Central Metamorphic terranes. Geologic and geophysical data suggest that these plutons are now rootless, having been truncated by the fault at the base of the Rattlesnake Creek terrane (Fig. 39-2; C. G. Barnes, 1982; G. C. Barnes *et al.*, 1986 a, b; Donato *et al.*, 1982; Jachens *et al.*, 1986).

Two volumetrically minor Late Jurassic plutonic events, represented by the Bear Mountain intrusive complex (149–153 m.y.) and the Lower Coon Mountain pluton (142–150 m.y.), are of pre-Nevadan and Nevadan age, respectively (Saleeby *et al.*, 1982). Geochemical, isotopic, and petrologic studies of Jurassic plutons in the Klamath Mountains reveal that they are most commonly quartz-saturated, calc-alkaline, have initial Rb/Sr ratios between 0.7028 and 0.7052 (Hotz, 1971; Masi *et al.*, 1981), and are therefore typical calc-alkaline arc rocks.

METAMORPHIC FRAMEWORK

The Klamath Mountains province displays a wide variety of metamorphic rocks of different facies and ages. In this paper we are principally concerned with major crustal regional metamorphic events. We do not discuss either contact metamorphism developed in aureoles around the numerous calc-alkaline plutons or ocean-floor metamorphism, which is demonstrable only in the Josephine ophiolite (Harper, 1984).

Facies Divisions

We show the distribution of five main crustal metamorphic facies in Figs. 39-3, 39-4, and 39-5. These are subgreenschist, greenschist, amphibolite, transitional blueschist-greenschist, and blueschist facies. The diagnostic mineral assemblages we have used to define these metamorphic facies are listed below.

1. Subgreenschist facies includes mineral assemblages characteristic of the zeolite, prehnite-pumpellyite, and pumpellite-actinolite facies. Minerals in poorly cleaved sedimentary rocks may include quartz, albite, chlorite, prehnite, ± white mica, illite-smectite. Nonschistose mafic and intermediate volcanics can contain prehnite, pumpellyite, epidote, laumontite ± quartz, albite, chlorite, carbonate, and sphene and preserve the primary igneous structures and minerals.

2. Greenschist facies rocks are characterized by strong foliations developed in sediments. Volcaniclastic rocks may also show well-developed fabrics, but volcanic and plutonic rocks are only slightly foliated, and retain original igneous structures. Albite, actinolite, epidote, chlorite, sphene, ± quartz, and calcite are the key minerals within mafic volcanics. Albite, muscovite, chlorite, calcite ± spessartine, biotite, epidote, and actinolite are common in quartz-bearing metasediments.

3. Amphibolite facies (including epidote-amphibolite facies) rocks are completely re-

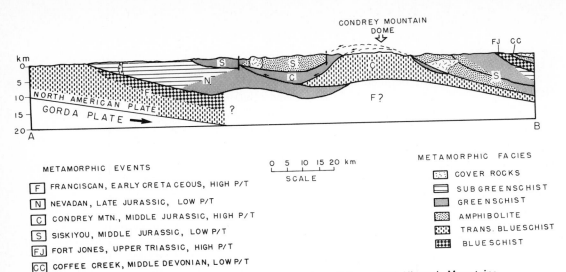

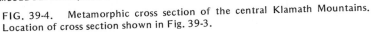

FIG. 39-4. Metamorphic cross section of the central Klamath Mountains.
Location of cross section shown in Fig. 39-3.

crystallized and exhibit well-developed foliations. Original igneous and sedimentary structures are generally eradicated. Mafic igneous rocks commonly contain the following key
minerals: plagioclase (An_{17-48}), brown hornblende, almandine, diopside ± biotite, rutile,
epidote and quartz. Quartzose schists can contain plagioclase ($>An_{15}$), microcline, biotite,
muscovite, almandine, staurolite, andalusite ± calcite, and actinolite.

4. Transitional blueschist-greenschist facies rocks may be completely recrystallized
and exhibit penetrative deformation or they may show incipient deformation. These
rocks usually contain only glaucophane or crossite in addition to typical greenschist facies
minerals. Pumpellyite, lawsonite, and/or aragonite are volumetrically minor and sporadically distributed. Jadeitic pyroxenes do not occur.

5. Blueschist facies rocks nearly always show multiple deformation and are commonly
foliated but original metabasite structures and textures may locally be preserved. Mafic
volcanic rocks contain lawsonite, glaucophane, actinolite, jadeite or omphacite, phengite,
chlorite ± quartz, pumpellyite, epidote, and aragonite. Common minerals in quartzose
metasediments include chlorite, albite or jadeite, phengite, lawsonite, glaucophane ±
stilpnomelane, epidote, and sphene.

Figure 39-6 shows pressure and temperature conditions that correspond approximately to the metamorphic facies discussed above. Contacts in the field between rocks of
different metamorphic facies are faults or are gradational contacts (Figs. 39-3 and 39-4).
Thermal metamorphism around the plutons is not shown in Fig. 39-3. Anatectic rocks
have not been observed at any structural level within the Klamaths.

Delineation of mineral reaction isograds in the Klamath Mountains is difficult
because much of the metamorphic information comes from metabasites which exhibit
continuous reactions during the transition from subgreenschist through greenschist and
amphibolite facies. The metasediments are silica-rich and alumina-poor; thus diagnostic
aluminosilicate index minerals are not widespread. Locally, however, apparent isograds
have been established (Hill, 1984; Mortimer, 1984a).

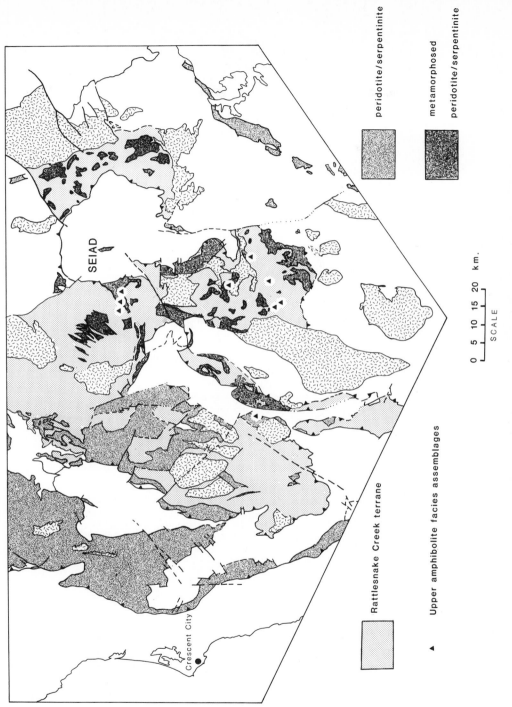

FIG. 39-5. Distribution map for the ultramafic rocks and upper amphibolite facies rocks in the central Klamath Mountains.

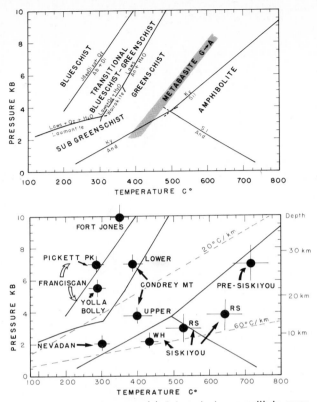

FIG. 39-6. *P-T* diagrams. (a) Selected phase equilbria representing index reactions for the five mapped metamorphic facies.
The lawsonite and aluminosilicate reactions are generated from
the equations and data of Helgeson *et al.* (1978). It is important
to note that these data predict that the reactions lawsonite =
anorthite + H_2O and lawsonite + quartz + H_2O = wairakite are
metastable with respect to zoisite + kyanite + quartz + H_2O.
The reaction Na-clinopyroxene ($Jd_{80}Di_{20}$) + quartz + diopside +
albite is taken from (T. J. B. Holland, 1983). Stipled area denotes
continuous reaction zone in mafic rocks changing from greenschist to amphibolite facies. (b) Dots indicate maximum estimated *P-T* conditions for the seven metamorphic events of the
central Klamath Mountains. Crosses on the dots for each event
show the range of possible errors in these estimates. Details and
sources of the data are given in the text. Abbreviations: WH,
Western Hayfork; RS, Rattlesnake Creek.

Crustal Metamorphic Events

Six distinct regional crustal metamorphic events of different ages can be distinguished in
the portion of the Klamaths shown in Fig. 3. These are (1) Coffee Creek metamorphism
of Middle to Late Devonian age; (2) Fort Jones metamorphism of Late Triassic age;
(3) Siskiyou metamorphism of Middle Jurassic age; (4) Condrey Mountain metamorphism,
also of Middle Jurassic age; (5) Nevadan metamorphism of Late Jurassic age; and (6)
Franciscan metamorphism of Early Cretaceous age. Recrystallization at these times also

coincided with major deformation. Details of these orogenic events are discussed below. The six metamorphic and deformational episodes are differentiated from one another based primarily on age of recrystallization or inferred contrasts in *P-T* conditions of prograde mineral assemblages.

Coffee Creek event This metamorphic and penetrative deformational event is restricted to rocks of the Central Metamorphic terrane (Figs. 39-2 and 39-3). We refer to it as the Coffee Creek event inasmuch as these metamorphic rocks were first described by G. A. Davis *et al.* (1965) from the Coffee Creek area in the central Klamath Mountains.

The mineral assemblages (Table 39-1) indicate that metamorphic grade extended from greenschist facies to epidote-amphibolite and (locally) amphibolite facies. Local strong retrogression of the high-grade assemblages is observed south of the area of Fig. 39-3 and may reflect Middle Jurassic overprinting during Siskiyou metamorphism (see below). Metabasite mineral assemblages including epidote, albite, and hornblende indicate

TABLE 39-1. Mineral Assemblages and Protoliths for the Central Metamorphic Belt[a]

Minerals	Protolith			
	Metabasite	Metasediment	Metaperidotite	Metacarbonate
Greenschist Facies				
Quartz		O		O
Plagioclase	O Ab	O Ab		
Mica		O		O
Chlorite	O		O	O
Antigorite			O	
Amphibole	O Act		O Trem	
Epidote	O	O		
Clinozoisite	O			
Sphene	O			
Calcite	O			O
Amphibolite Facies				
Quartz		O		
Plagioclase	O An$_{20-30}$	O An$_{20-35}$		
Muscovite		O		
Antigorite			O	
Ca-Amphibole	O Brown	O Hbd	O Parg	
Anthophyllite			O	
Olivine			O	
Epidote	O Al-rich	O		
Garnet	O	O		
Rutile	O			
Calcite	O	O		

[a]Ab, albite; Trem, tremolite; An, anorthite (subscripts denote mole fraction anorthite component in plagioclase); Hbd, hornblende; Parg, pargasite.
Sources of data: Cashman (1980), Hotz (1977), Holdaway (1965), Davis *et al.* (1965).

a moderate P/T gradient (Apted and Liou, 1983), which is in accordance with estimates of $T = 400$–$500°C$ and P = 3–5 kb based on phase equilibria and mineral compositions south of our compilation area (Peacock, 1985). Metaperidotites contain either a protected dry mantle assemblage of olivine + orthopyroxene + clinopyroxene or a low-grade antigorite + actinolite + chlorite assemblage; it is not clear that metaperidotites have been affected by amphibolite facies crustal metamorphism (Cashman, 1980).

K-Ar and $^{40}Ar/^{39}Ar$ whole-rock ages from amphibolite and siliceous schist in the area of Fig. 39-3 yield Devonian ages. Cashman (1980), in reviewing this and the data of Lanphere et al. (1968), concluded that recrystallization took place at about 370 ± 20 m.y. (Middle to Late Devonian).

The regional extent of Coffee Creek deformation and metamorphism indicates a major orogenic event. The Trinity peridotite is unlikely to have been the heat source for metamorphism because the age of the peridotite predates metamorphism by 70–100 m.y. (Peacock, 1985). Extensive zones of mylonite in the peridotite are also reported to overprint an older mantle fabric (Cannat and Boudier, 1985), with amphibolites near these zones showing concordant and synmetamorphic relations.

For the same reasons, interpretations involving asthenospheric heat flow seem equally unlikely. Pressure estimates (Peacock, 1985) indicate mid-crustal depths. Devonian plutonic rocks are unknown in the Central Metamorphic terrane but arc-related volcanics (including boninites), and comagmatic plutonic rocks dated at 400 ± 2 m.y. are present at the base of the Eastern Klamath terrane (Lapierre et al., 1985). We believe that the presence of amphibolite facies rocks with cooling ages slightly younger than structurally overlying arc-related igneous rocks is strong circumstantial evidence that Coffee Creek metamorphism occurred at moderate temperatures and pressures beneath the collapsed fore-arc region of a newly formed Middle Devonian intraoceanic island arc and basin complex.

Fort Jones event Hotz (1973) first described the blueschist facies mineral assemblages from the Klamath Mountains in rocks he correlated with the Stuart Fork Formation of Davis and Lipman (1962). Lawsonite and jadeite-bearing rocks are absent from the type locality, but there seems little doubt, on structural and lithologic grounds, that rocks in the two areas are correlative (Blake and Jayko, 1983; Borns, 1980; Goodge, 1985; Hotz, 1977). No high-pressure metamorphic overprinting of the adjacent Central Metamorphic terrane has been reported. $^{40}Ar/^{39}Ar$ ages (214–222 m.y.) on white micas separated from metabasalts near Yreka date the metamorphism as latest Triassic (Hotz et al., 1977).

Mineral assemblages (Table 39-2) are characteristic of blueschist facies metamorphism. Eclogites are present but rare (Borns, 1984). Where the Vesa Bluffs pluton intrudes the blueschists, glaucophane and lawsonite are replaced by albite and muscovite. Chemographic and phase relations for Fort Jones terrane rocks in the Yreka–Fort Jones area led Borns (1980) to conclude that the blueschists formed at pressures of 9–11 kb and temperatures of 300–400°C (Fig. 39-6b). The appearance of garnet + epidote in equilibrium with glaucophane and omphacite near the eastern contact with the Central Metamorphic terrane indicates increasing temperature in this up-structure direction.

Farther south, Goodge (1985), in a study of the transition from the lawsonite and jadeite-bearing assemblages near Yreka and Fort Jones to the greenschist facies assemblages in the type area of Davis and Lipman (1962), reported a geographically intermediate assemblage of glaucophane-clinozoisite and proposed that this progressive north-south mineralogical change reflects regionally shallower depths (4–8 kb) of Late Triassic metamorphism to the south.

In a plate tectonic context, the high pressures indicated by lawsonite and jadeite-bearing assemblages suggest a subduction-zone environment for Fort Jones metamor-

TABLE 39-2. Mineral Assemblages and Protoliths for the Fort Jones Blueschist[a, b]

Minerals	Protolith			
	Sediments	Mafic Igneous	Calcareous	Serpentine
Quartz	O		O	
Plagioclase		? Ab?		
Phengite	O	O	O	
Lawsonite	O	O		
Jadeite		O		
Chlorite		O		O
Pumpellyite		?		
Calcite-Aragonite			O	
Dolomite			O	
Omphacite		O $Di_{50}Ac_{20}Jd_{30}$		
Stilpnomelane	O			
Glaucophane	O	O		
Graphite	O			
Antigorite				O
Garnet		O High grade		
Epidote		O High grade		

[a]High-grade rocks found near basal thrust with the Central Metamorphic Belt.
[b]Ab, albite; Di, diopside; Ac, acmite; Jd, jadeite (subscripts denote mole fraction end-member components in omphacite).
Sources: Borns (1980), Hotz (1973).

phism and deformation (Borns, 1980; Ernst, 1984). Burchfiel and Davis (1981), Wright (1982), Goodge (1985), and in fact most commentators on Klamath geology implicitly regard the Fort Jones event as an integral part of a parautochthonous Permo-Triassic subduction complex in the Klamath Mountains. However, the lack of high-pressure overprinting of the Central Metamorphic terrane requires that the two terranes had to be juxtaposed after the Late Triassic blueschist event. Juxtaposition with the underlying North Fork, Salmon River, and Hayfork terranes could not have taken place until the Middle Jurassic inasmuch as the North Fork terrane contains Early Jurassic radiolarians (Irwin, *et al.*, 1978). Thus, although the Fort Jones terrane metamorphic rocks apparently formed in a subduction zone processes, it is not clear how this subduction zone was related to other Klamath terranes.

Siskiyou event Although it is the most geographically widespread tectono-metamorphic event in the Klamath Mountains, this Middle Jurassic deformation and metamorphism has received, at best, only cursory mention in synthesis articles to date (Burchfiel and Davis, 1981; Harper and Wright, 1984; Irwin, 1981; Saleeby *et al.*, 1982). We therefore describe below, in some detail, its features and significance.

Distribution Siskiyou metamorphism occurred both during and after widespread tectonic disruption and mutual juxtaposition of rocks in the North Fork, Salmon River, eastern and western Hayfork, and Rattlesnake Creek terranes (Donato, 1985; Hill, 1985; Hotz, 1979; Kays and Ferns, 1980; Mortimer, 1985). Recent detailed mapping and petrologic studies of small areas afford an opportunity to understand the broader distribution of metamorphism within these terranes.

Because the Upper Triassic Fort Jones blueschists do not show any later regional metamorphic overprint in the Yreka area (Borns, 1980; Hotz, 1977; Mortimer, 1985), the eastern limit of Siskiyou metamorphism is confined by the thrust fault that separates this terrane from the North Fork terrane (Figs. 39-2 and 39-3). Farther South, Goodge (1985) has demonstrated considerable greenschist overprinting, probably caused by Late Jurassic plutons. To the west, Siskiyou metamorphic rocks are truncated by the Orleans fault, a low-angle fault separating the Rattlesnake Creek terrane from the Western Klamath terrane (Jachens *et al.*, 1986). In the center of our compilation area, doming of Neogene age (Mortimer and Coleman, 1985) provides at least 7 km of structural relief and exposes the highest-grade Siskiyou metamorphic rocks at the base of the Rattlesnake Creek terrane which structurally overlie lower-grade rocks of the Condrey Mountain terrane (Hotz, 1979).

Mineral Assemblages On a regional scale, metamorphic grade increases down-structure. At deep structural levels (Rattlesnake Creek terrane) amphibolite facies assemblages are present with grade of metamorphism decreasing upward through greenschist facies in the Hayfork and Salmon River terranes to a pumpellyite-bearing subgreenschist facies assemblage in the North Fork terrane (Table 39-3). Metamorphic grade within the Rattlesnake Creek terrane also decreases in a southerly, westerly, and northwesterly direction away from the Happy Camp-Seiad Valley area (Fig. 39-3). Upper amphibolite and rare granulite assemblages reported from a few localities in the Rattlesnake Creek terrane represent a relict pre-Siskiyou metamorphism and are discussed below.

P-T Conditions Garnet-biotite geothermometry on rare andalusite-bearing metasediments in the Rattlesnake Creek terrane indicates temperatures of 480–560°C at pressures of <3.7 kb (Donato, 1985). Similar low pressures of metamorphism in metabasites of the Salmon River and Hayfork terranes are indicated by low Na(M4) content of calcic amphiboles in the buffering assemblage of Brown (1977a), and the presence of actinolite-calcic plagioclase assemblages at the greenschist-amphibolite transition (Hill, 1985; Mortimer, 1985). These assemblages imply pressures <3.5 kb at about 550°C (Fig. 39-6b; Apted and Liou, 1983). Absence of the pumpellyite-actinolite facies in the North Fork terrane is also evidence for a low-pressure series (Mortimer, 1985).

Extrapolation of the pumpellyite-out reaction of Nitsch (1971) and Liou *et al.* (1985) to these low pressures suggests that North Fork terrane rocks probably reached temperatures no hotter than 350°C. Estimates of maximum temperatures in the Rattlesnake Creek terrane are between 490 and 650°C, (based on metabasite mineral compositions; Hill, 1984). The presence of metamorphic clinopyroxene suggests still higher temperatures, but its presence may be controlled by whole-rock composition (Donato, 1985).

Age The age of Siskiyou metamorphism is fairly tightly bracketed. K-Ar hornblende ages from Rattlesnake Creek terrane amphibolites east of the Condrey Mountain window (149–151 m.y.; Lanphere *et al.*, 1968) are cooling ages and are now known to be too young. A minimum age of Siskiyou metamorphism and deformation is provided by the crystallization ages of plutons, such as the Wooley Creek batholith and Slinkard, Vesa Bluffs, and Ashland plutons, which cause thermal metamorphism of the Siskiyou regional metamorphic rocks (Allen *et al.*, 1982; C. G. Barnes, 1983; Hotz, 1979; Mortimer, 1984a). U-Pb zircon ages for the Wooley Creek and Slinkard plutons are 163 ± 3 m.y. and 162 ± 2 m.y., respectively (Allen *et al.*, 1982). The oldest K-Ar hornblende ages for the Ashland pluton

TABLE 39-3. Mineral Assemblages and Protoliths for Siskiyou Metamorphism[a]

Minerals	Sediments	Mafic Igneous	Calcareous Sediments	Ultramafic Igneous
		Protolith		
Amphibolite Facies				
Quartz	O	O	O	
Feldspar	O (An$_{30-50}$)	O (An$_{25-55}$)	O (Ksp)	
Amphibole	O (Hbd)	O (Hbd)	O (Trem)	Parg
Epidote		O		
Andalusite	O			
Sillimanite	O			
Chlorite	O		O	O
Garnet	O	O	O (Gross)	
Staurolite	O			
Calcite			O	
Pyroxene		O (Salite)	O (Diop)	O (Opx)
Olivine				O
Spinel			O	O
Rutile	O	O		
Wollastonite			O	
Biotite	O			
Muscovite	O			O
Anthophyllite				O
Talc				O
Greenschist Facies				
Quartz	O		O	
Plagioclase	O (Ab)	O (Ab)		
Amphibole		O (Act)		O (Trem)
Epidote		O		
Pumpellyite		O		
Chlorite	O	O		O
Calcite		O	O	
Antigorite				O
Sphene		O		
Muscovite		O		
Biotite	O	O		
Talc				O

[a]An, Anorthite (subscripts denote mole fraction anorthite component in plagioclase); Hbd, hornblende; Trem, tremolite; Parg, pargasite; Gross, grossular; Diop, diopside; Opx, orthpyroxene; Ab, albite; Act, actinolite.

Sources: Lieberman (1983), Donato (1985), Hill (1984), Kays and Ferns (1980), and Mortimer (1985).

are 170 ± 5 m.y. and 164 ± 5 m.y. and for the Vesa Bluffs pluton 164 ± 5 m.y. (Lanphere *et al.*, 1968). Thus, the minimum age of metamorphism and deformation is established at approximately 160–166 m.y. in various parts of the compilation area.

A maximum age on metamorphism is provided by the youngest protolith ages of rocks involved in the metamorphism. The North Fork terrane contains Pliensbachian radiolarian cherts (Irwin *et al.*, 1982; 194–200 m.y. on the time scale of Harland *et al.*, 1982). Plagiogranite from a dike complex in the Rattlesnake Creek terrane, metamorphosed to amphibolite facies during the Siskiyou event, has yielded a U-Pb zircon age of 172 ± 2 m.y. (work performed by J. B. Saleeby and quoted *in* Hill, 1985). Primary igneous hornblendes from unmetamorphosed western Hayfork terrane in the southern Klamath Mountains range in age from 168 to 177 m.y. (Fahan, 1982). In the area of Fig. 3, these rocks are metamorphosed to the greenschist-amphibolite facies (Mortimer, 1985). These ages bracket the maximum age of metamorphism at about 168–174 m.y. Siskiyou metamorphism and syn-metamorphic deformation occurred in a short period of time between 163 and 170 m.y. (i.e., latest Middle Jurassic).

Deformation associated with metamorphism As mentioned above, Siskiyou metamorphism was both preceded and accompanied by chaotic faulting of rocks within and between the North Fork, Salmon River, eastern and western Hayfork, and Rattlesnake Creek terranes. Prograde metamorphic minerals define lineations and flattening foliations that are respectively coaxial to and axial planar with isoclinal folds (Donato, 1985; Hill, 1984; Mortimer, 1985). Map-scale isoclinal folds can also be defined (Donato, 1985; Hill, 1984; Hotz, 1967; Kays and Ferns, 1980). The earliest recognizable deformation in the rocks is premetamorphic major faulting leading to the development of "block-on-block" melange. The style of this deformation and range in size of blocks is identical at all structural levels from the North Fork terrane down to the Rattlesnake Creek terrane (Mortimer, 1985). The synmetamorphic cleavage and lineation-forming events were likely continuous with this earlier inter- and intraterrane faulting as cleavage is commonly parallel to faults.

The Siskiyou event in the southern Klamath Mountains Metamorphic relations in the North Fork, Salmon River, eastern and western Hayfork, and Rattlesnake Creek terranes farther south are somewhat different. There, rocks of the Rattlesnake Creek terrane have not undergone the high-grade regional metamorphism that we describe from the northern Klamaths. This is demonstrated by mineral assemblages indicative of only greenschist and subgreenschist facies metamorphism (Wright, 1981) and the existence of unrecrystallized radiolarians in cherts (Irwin *et al.*, 1982). Locally observable metamorphic gradients are spatially related to plutons (Fahan, 1982). Amphibolite facies rocks (plagioclase + hornblende + quartz + sphene ± rutile) occur only as tectonic blocks in unmetamorphosed melange in the Rattlesnake Creek terrane (Wright, 1981; Wright and Wyld, 1985). As discussed in the next section, there is some evidence that an older metamorphic event, possibly equivalent to these blocks, may be preserved in parts of our compilation area.

Regional structural relations in Siskiyou terranes in the southern Klamath Mountains, however, appear identical to those in our compilation map area. In particular, a period of major interterrane thrust faulting in the southern Klamaths is constrained to have occurred between 165 and 169 m.y. (Fahan, 1982). The Siskiyou event in the southern Klamaths was thus primarily a deformational event with only low grades of regional metamorphism developed across the North Fork, Salmon River, eastern and western Hayfork, and Rattlesnake Creek terranes.

The transition between the northern high-grade and southern low-grade areas has been suggested south of Happy Camp, where Gray and Petersen (1982) observed a southward decrease in metamorphic grade within the Rattlesnake Creek terrane. A westward decrease in grade in the same rocks from Seiad Valley to Preston Peak and a northward decrease in grade from Seiad Valley to the Oregon border may also be present, as indicated by the regional distribution of metamorphic minerals.

Interpretation The Siskiyou event (~165-169 m.y.) was a major deformational and regional metamorphic event related to the imbrication and mutual juxtaposition (amalgamation) of the rocks of the North Fork, Salmon River, eastern and western Hayfork, and Rattlesnake Creek terranes. Pronounced vertical (decreasing up-structure) and lateral (decreasing to the south and west; present geographic reference frame) regional metamorphic gradients are recognizable. Mineral assemblages define relatively low-pressure metamorphic gradients.

Heat for this low-pressure event is considered to be derived from an island arc rather than a mid-ocean ridge because of the following: (1) involvement in the Siskiyou orogeny of volcanic and plutonic arc rocks (the western Hayfork terrane) that predate deformation and metamorphism by less than 10 m.y.; (2) intrusion of arc-related calc-alkaline plutons into deformed and regionally metamorphosed rocks within 5-10 m.y. of deformation and metamorphism; (3) absence of peridotites of Middle Jurassic age in or adjacent to the metamorphosed terranes; and (4) presence of low-angle, melangelike deformation of terranes prior to metamorphism, as expected beneath a fore-arc region.

Condrey Mountain event Condrey Mountain metamorphism is restricted to the rocks of the Condrey Mountain terrane; no overprinting of other terranes is seen. Mineral assemblages are given in Table 39-4.

Helper (1985) emphasized that within the Condrey Mountain terrane, deformation and metamorphism are progressive. In the lower structural levels, transitional blueschist-greenschist facies assemblages are present and he estimated pressures greater than 6 kb and peak temperatures between 360 and 410°C. Greenschist assemblages at higher structural levels indicate temperatures between 350 and 450°C and pressures between 2 and 7kb. A later epidote-amphibolite assemblage developed in the greenschist unit during thrusting beneath the Rattlesnake Creek terrane at pressures of 2-5 kb and temperatures around 500°C (Helper, 1985). As this recrystallization is restricted to an aureole below the thrust (Helper, 1985), it is not a regional event and is not shown on Figs. 3, 4 and 6b.

The age of Condrey Mountain metamorphism and the age of faulting involving the Condrey Mountain terrane are very poorly constrained. Available radiometric ages are as follows: K-Ar muscovite, 144 ± 3 m.y. (Lanphere *et al.*, 1968); K-Ar whole-rock, 158 ± 3 m.y. (Suppe and Armstrong, 1972); $^{40}Ar/^{39}Ar$ glaucophane 167 ± 12 m.y., $^{40}Ar/^{39}Ar$ actinolite, 125 ± 2 m.y., and $^{40}Ar/^{39}Ar$ muscovite 118 ± 2 m.y. (unpublished data quoted in Coleman *et al.*, 1983). As mentioned previously, a 170 ± 2 m.y. U-Pb zircon age of a deformed and metamorphosed diorite indicates that the sedimentary age of the upper unit must be older than 170 m.y. and the age of metamorphism and deformation younger than 170 m.y.

We consider the zircon and glaucophane ages to best constrain the time of Condrey Mountain metamorphism, the whole-rock age to be a partially reset age, the Lanphere *et al.* (1968) muscovite age to be a Nevadan muscovite and the two Cretaceous ages to possibly represent greenschist facies metamorphism synchronous with the blueschist facies Pickett Peak event. Alternatively, the Cretaceous ages may simply reflect prolonged cooling of the Klamath terranes (Helper, 1985; Mortimer, 1985).

There are no postmetamorphic plutons intruding the Condrey Mountain terrane to provide a good minimum age constraint on metamorphism and deformation. Based on a deformed contact near Happy Camp, Hill (1984, 1985) interpreted the already juxtaposed Condrey Mountain and Rattlesnake Creek terranes to share a deformation he correlated

TABLE 39-4. Mineral Assemblages and Protoliths for Condrey Mountain Metamorphism[a]

Minerals	Sediments (Clastic)	Mafic Igneous	Chert-iron-Stone	Ultramafic Igneous
Blueschist Facies				
Quartz	O		O	
Feldspar	O / Ab	O / Ab	O / Ab	
Amphibole		Cross + Act	O / Cross	O / Trem
Pyroxene		O / Acmite		
Garnet			O	
Epidote	O	O		
Chlorite	O	O	O	
Calcite			O	O
Mica	O		O	
Phengite		Pheng		
Stilpnomelane			O	
Sphene		O		
Graphite	O			
Antigorite				O
Brucite				O
Talc				O
Greenschist Facies				
Quartz	O	O	O	
Plagioclase	O / Ab	O / Ab	Ab	
Amphibole	O / Act	O / Act		O / Trem
Epidote	O	O		
Chlorite	O	O	O	O
Calcite		O		
Mica	O / Pheng		O / Pheng	
Sphene	O	O		
Garnet	O / Spess		O	
Stilpnomelane			O	
Antigorite				O
Brucite				O
Talc				O

[a] Ab, albite; corss, Crossite; Act, actinolite; Trem, tremolite; Pheng, phengite; Spess, spessartine.
Sources: Helper (1985), Coleman *et al.* (1983), Donato *et al.* (1982), and Hotz (1967).

with the Nevadan orogeny (see below). If this is the case, Condrey Mountain metamorphism is constrained to have taken place before about 150 m.y. The fault separating the Condrey Mountain and Rattlesnake Creek terranes provides no additional constraints on the age of metamorphism. However, faulting must postdate the 162 ± 2-m.y.-old Slinkard pluton, whose base is truncated by the fault (Allen *et al.*, 1982).

The presence of mineral assemblages intermediate between blueschist and greenschist facies within the Condrey Mountain terrane suggests to us that these rocks were metamorphosed in a subduction zone regime in latest Middle to earliest Late Jurassic time. Structural relations and the overlap in age of metamorphism between this low-T/P event and the high-T/P Siskiyou metamorphism suggest prior existence of a paired metamorphism (Helper, 1985).

Nevadan event The most recent and acceptable constraints on the age and distribution of the Nevadan orogeny in the Klamath Mountains have been presented by Harper and Wright (1984). They define the event as representing deformation and metamorphism of the western Klamath terrane between 145 and 150 m.y. Penetrative deformation is common in metasedimentary rocks of the Galice Formation but is barely discernible in the ophiolitic rocks. Nevadan age retrograde recrystallization is expressed in retrograde mineral assemblages in parts of the Rattlesnake Creek and Condrey Mountain terranes (Donato, 1985; Hill, 1985). However, at higher structural levels, Siskiyou rocks are deformed by open Nevadan folds; recrystallization is minor or absent (Donato, 1985; Hill, 1985; Mortimer, 1985). The Josephine ophiolite also shows an older, pre-Nevadan hydrothermal metamorphism, with grade increasing down-section within the ophiolite from subgreenschist facies in the volcanics to low-P amphibolite facies in the gabbros (Harper, 1984). Mineral assemblages associated with the Nevadan event are given in Table 39-5.

In a review of Nevadan metamorphism Hill (1985) interpreted Harper's (1980) results to indicate that metamorphism corresponded to pumpellyite-actinolite facies at inferred conditions of ~300°C and 2.5 kb. Greenschist facies conditions ($T = 350$–550°C, $P = 2$-7 kb) are indicated in the eastern part of the terrane (Fig. 39-6b).

A general eastward and southward increase in grade of metamorphism in the Galice

TABLE 39-5. Mineral Assemblages and Protoliths for Nevadan Metamorphism[a]

Minerals	Protolith		
	Sediments (Clastic)	Mafic Igneous	Ultramafic Igneous
	Greenschist – Subgreenschist Facies		
Quartz	○	○	
Feldspar	○	○	
	Ab	Ab	
Amphibole		○	
		Act	
Epidote		○	
Chlorite		○	○
Calcite		○	
White Mica	○	○	
	Ser	Musc	
Sphene	○	○	
Pyrite	○		
Serpentine			○
Prehnite		○	
Pumpellyite		○	
Brucite			○

[a]Ab, albite; Act, actinolite; Ser, sericite; Musc, muscovite.
Sources: Klein (1977), Gray (1985), Snoke (1977) and Hill (1984).

Formation in our compilation area is possibly related to syn- to post-Nevadan under-thrusting of the Western Klamath terrane beneath the still hot Rattlesnake Creek terrane. If this is the case, Nevadan metamorphism is technically "upside down"; however, no mineral zonation that suggests inverted gradients has been reported close to the thrust beneath the Rattlesnake Creek terrane (Gray, 1985; Klein, 1977).

Heat for Nevadan metamorphism was probably provided by rising hot asthenosphere under the collapsed Josephine/Galice interarc basin. This heat rose through the Western Jurassic terrane and into the Condrey Mountain and Rattlesnake Creek terranes, where it caused local recrystallization. This hypothesis is compatible with the tectonic interpretation of the Nevadan orogeny proposed by Harper and Wright (1984).

Franciscan events Franciscan age metamorphic rocks occur as coherent but allochthonous schist units in the Pickett Peak terrane (South Fork Mountain and Colebrooke Schists) adjacent to the surrounding clastic sedmentary rocks of the Yolla Bolly terrane, which show only incipient recrystallization. Deformation and the development of melanges has largely obscured original, premetamorphic relationships. Mineral assemblages are given in Table 39-6.

From assemblages in the South Fork Mountain Schist, Brown and Ghent (1983) estimate approximate metamorphic conditions as $T = 270$–$310°C$, $P = 7$ kb (Fig. 39-6b). Sporadic occurrence of aragonite and lawsonite in the Yolla Bolly terrane (Wiggins, 1980) indicate similar P-T conditions, but recrystallization and deformation are not pervasive. Pumpellyite coexists with chlorite and quartz rather than lawsonite and aragonite. Accord-to Brown (1977b), this assemblage may form at temperatures near $300°C$ at pressures <6 kb.

K-Ar and $^{40}Ar/^{39}Ar$ ages on various minerals form a continuum of Cretaceous ages that has not been easy to interpret. Lanphere *et al.* (1978) considered the age of South Fork Mountain Schist metamorphism to be between 115 and 120 m.y. McDowell *et al.* (1984) regarded it as older than 124 m.y. Both these estimates are in agreement with a 128 ± 18-m.y. Rb-Sr mineral isochron from the Colebrooke Schist (Coleman, 1972). Pickett Peak terrane blueschist facies metamorphism is thus probably Early Cretaceous in age.

No direct age measurements on Yolla Bolly terrane metamorphism have been made in northern California or Oregon, but Blake *et al.* (1985a) suggest that it may be as young as 90 m.y. based on fossils and radiometric ages established farther south.

Superimposed on the synmetamorphic ductile fabrics of the Yolla Bolly and Pickett Peak terranes is a brittle deformation related to postmetamorphic juxtaposition of the two terranes along a mid-Cretaceous thrust (Blake *et al.*, 1985a). Subsequent Late Cretaceous-early Tertiary strike-slip faulting has apparently dispersed Yolla Bolly and Pickett Peak terrane rocks northward in an area southwest of Fig. 39-2 (Kelsey and Hagans, 1982). Blake *et al.* (1985b) interpreted the Colebrooke Schist as such a dispersed remnant.

High-grade eclogite and gneissic blueschist blocks within melange of the Central Belt Franciscan (e.g., Moore, 1984) yield consistently older K-Ar mineral and whole-rock ages than other Franciscan metamorphic rocks. These are generally in the range 140–160 m.y. (Lanphere *et al.*, 1978; McDowell *et al.*, 1984). A U-Pb age of 162 ± 3 m.y. has been determined for a block (Mattinson, 1981), which also supports the view that the metamorphic age of these blocks predates the Nevadan orogeny and is broadly age equivalent to the Condrey Mountain and Siskiyou events.

The blueschist-facies mineral assemblages in the Yolla Bolly and Pickett Peak terranes are interpreted as having formed in Early Cretaceous subduction zones. The older eclogite blocks could represent higher-pressure fragments of the moderate-P/T Condrey Mountain event, remobilized and uplifted during the Early Cretaceous into the Franciscan subduction melange. Jayko *et al.* (1986) emphasize that there may be two distinct blueschist metamorphic events within the Franciscan: (1) Pickett Peak around 125 m.y., and (2) Yolla Bolly around 90 m.y.

Other Metamorphic events

Pre-Siskiyou Event in the Rattlesnake Creek Terrane Upper-amphibolite and rare granu-lite-facies metamorphic assemblages occur locally in rocks of ultramafic, rodingitic, calcareous, and mafic composition throughout the Rattlesnake Creek terrane in the northern Klamath Mountains (Fig. 39-5; Burton, 1982; Donato, 1985; Grover, 1984; Kays and Ferns, 1980; Lieberman and Rice, 1983; Lundquist, 1982; Medaris, 1975; Rawson, 1985). We interpret these rocks to be relics of an early, previously unrecognized tectonometamorphic event that is older than the Siskiyou event. Our interpretation is based on several lines of evidence. First, these rocks contain mineral assemblages which

TABLE 39-6. Mineral Assemblages and Protoliths for Franciscan Metamorphism[a]

Minerals	Sediments	Mafic Igneous	Chert
		Protolith	
		Blueschist (South Fork Mt., Pickett Pk., and Colebrooke Schist)	
Quartz	O		O
Feldspar	O Ab	O Ab	
Amphibole		O Cross + Act	O Rieb
Pyroxene			O Acmite
Garnet			O
Epidote		O	
Chlorite	O		
Aragonite	O	O	O
Mica	O Pheng		
Stilpnomelane		O	O
Sphene		O	
Lawsonite	O		
Deerite			O
Pumpellyite		O	
		Transitional Blueschist-Greenschist Facies (Dothan and Yolla Bolly)	
Quartz	O		O
Plagioclase	O Ab	O Ab	
Chlorite	O	O	
Mica	O		
Pumpellyite	O	O	
Lawsonite		O	

[a]Ab, albite; Act, actinolite; Cross, crossite; Rieb, riebeckite; Pheng, phengite.
Sources: Blake *et al.* (1985a, b), Coleman (1972) and Brown and Ghent (1983).

represent significantly higher temperatures and pressures than those found in contiguous rocks of the Rattlesnake Creek terrane (Donato, 1985; Gorman, 1985; Gray, 1985). Second, although transitional, progressive changes in mineral assemblages have been documented locally between these rocks and contiguous middle-amphibolite-facies rocks (Grover, 1984; Rawson, 1985), ambiguities exist because it is not clear in all cases that the higher- and lower-grade assemblages formed during the same event (e.g., at Scott Bar Mountain; Rawson, 1985). This suggests that the high-grade rocks may be blocks which were metamorphosed at high temperatures and pressures prior to incorporation in the Rattlesnake Creek terrane. Furthermore, many of these rocks display evidence for retrogression to middle-amphibolite facies, suggesting that the higher-temperature assemblages predate and were reequilibrated during the later Siskiyou metamorphic event. Finally, various geologic and structural evidence has led some workers independently to suggest an early tectonometamorphic event which predates the assembly and superimposed (Siskiyou) regional metamorphism of the Rattlesnake Creek terrane (e.g., Cannat, 1985; Gorman, 1985). High-grade amphibolite blocks which are clearly out of metamorphic context with respect to the surrounding rocks are reported by Gorman (1985) and by Gray (1985) in the low-grade part of the Rattlesnake Creek terrane in our compilation area by Wright (1981) and to the south by Wright and Wyld (1985).

The age of this pre-Siskiyou event is not known. However, sparse, anomalously old radiometric age determinations on metamorphic rocks throughout the Rattlesnake Creek terrane provide indirect evidence for their pre-Siskiyou age. For example, Saleeby *et al.* (1982) reported a reset 165 ± 3-m.y. age for retrograded amphibolite from the metamorphic basement complex of the Preston Peak ophiolite, suggesting that the true metamorphic age is even older. Hornblende from a 207-m.y. (U-Pb, zircon) metagabbro yields a K-Ar age of about 191 m.y. (Gray, 1985), which may reflect a thermal isotopic resetting at that time.

Representative mineral assemblages in the high-grade metaserpentinites include olivine + anthophyllite + chlorite, and olivine + enstatite + chlorite ± spinel (Donato, 1985; Grover, 1984; Rawson, 1985). Associated metarodingites contain calcite + zoisite + diopside + chlorite + spinel, and grossular + diopside + clintonite + zoisite + spinel (Rawson, 1985). Lieberman and Rice (1983) reported calcite + dolomite + diopside + olivine, and tremolite + calcite + quartz + diopside + K-feldspar in marbles near Seiad Valley. High-grade amphibolites contain brown hornblende + plagioclase (Gray, 1985).

Temperature estimates for rocks metamorphosed during the pre-Siskiyou metamorphic event range from about 610 to about 750°C based on olivine-spinel geothermometry on metaperidotites (Grover, 1984; Rawson, 1985); and calcite-dolomite geothermometry on marbles (Lieberman and Rice, 1983). Ultramafic rocks and associated metarodingites west of the Seiad ultramafic complex (Fig. 39-5) provide pressure estimates between 6 and 7 kb (Lieberman and Rice, 1983; Rawson, 1985). These conditions are in sharp contrast to the low-pressure, moderate-temperature conditions recorded by adjacent rocks in the Rattlesnake Creek terrane that were metamorphosed during the Siskiyou event (Fig. 39-6b).

Speculation on the tectonic significance of the pre-Siskiyou event is difficult because much of its signature has been obliterated by the subsequent Siskiyou and Nevadan events. Metamorphism and deformation in intraoceanic (fracture zone) settings have been proposed by Cannat (1985) and Gorman (1985). Intraoceanic thrusting, resulting in crustal thickening and imbrication of the Rattlesnake Creek terrane (Wright and Wyld, 1985) is also a possibility, and may have been a precursor to the subsequent Siskiyou metamorphic event. Future geochronological work on amphibolites may better define this early, somewhat cryptic metamorphic event.

Skookum Gulch Blueschists Paleozoic blueschists are found within the schist of Skookum Gulch below the Mallethead thrust within the Yreka terrane (Cotkin and Cotkin, 1985). The assemblage glaucophane + lawsonite + actinolite developed in metabasites. These rocks are considered to be the result of Paleozoic subduction, but inasmuch as they apparently occur as tectonic blocks in melange, their relation to other Paleozoic rocks in the Klamath Mountains remains enigmatic.

Eastern Klamath Terrane The Redding section of the Eastern Klamath terrane is notable among the Klamath terranes for its lack of tectonic imbrication and regionally distributed metamorphism. The nearly continuous 10-km-thick section of Devonian to Jurassic volcanic and sedimentary rocks shows no discernible metamorphic gradients. Low-grade (subgreenschist and lower greenschist facies) burial metamorphism has probably been operative in this island arc terrane throughout its 250-m.y.-long history.

GEOLOGIC SYNTHESIS

Pre-Jurassic

The Coffee Creek event has previously been related to Devonian westward thrusting of the Trinity Peridotite over the Central Metamorphic Belt (Davis *et al.*, 1965; Davis, 1968; Irwin, 1966; Peacock, 1985); increasing metamorphic grade toward the peridotite appears to substantiate this idea. We suggest that heat for metamorphism was somehow related to Early Devonian arc volcanism in the adjacent Eastern Klamath terrane. The complex array of small thrust plates in the Yreka terrane (including the Skookum Gulch melange) is possibly another manifestation of this Late Devonian arc-basin amalgamation (Lindsley-Griffin and Griffin, 1983).

Although we consider the Fort Jones deformation and metamorphism to record latest Triassic subduction (Ernst, 1984), this event cannot be linked directly with any other geologic event in the Klamath Mountains. The Fort Jones terrane was thrust under the Central Metamorphic terrane after blueschist facies metamorphism sometime in the Early or Middle Jurassic and was thrust over the underlying North Fork, Salmon River, and Hayfork terranes 50 m.y. after blueschist facies metamorphism in the Middle Jurassic.

From Permian to Late Triassic time, rocks now constituting the North Fork, Salmon River, and eastern Hayfork terranes were Panthalassic ocean floor, immature arcs, seamounts, and oceanic sediments (Fig. 39-7). In the Late Permian, these rocks were mostly in the Tethyan faunal province but also overlapped into an Eastern Klamath/Stikine/Quesnellia faunal realm. The North Fork and eastern Hayfork terrane cherts contain no detritus that suggests sedimentologic links with Permo-Triassic arcs; however, the associated clastics show volcanoclastic components (*cf.* Wright, 1982).

Early Jurassic

Primitive arc rocks of the Rattlesnake Creek terrane were erupted through and/or onto Upper Triassic oceanic rocks (Wright, 1981; Wright and Wyld, 1985). Intraoceanic deformation and granulite facies metamorphism involving ultramafic rocks of the Rattlesnake Creek Terrane may be of this age or older. Detritus derived from a continental volcanic arc was being received by North Fork terrane rocks (Mortimer, 1984a). Cherts in the eastern Hayfork and Rattlesnake Creek terranes may also be as young as Early Jurassic (Irwin *et al.*, 1982).

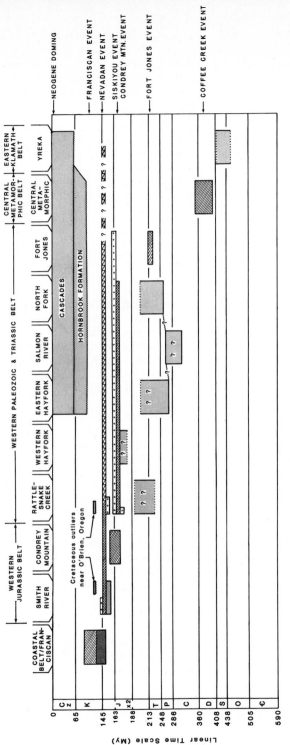

FIG. 39-7. Terrane-time diagram. Key to shading of boxes as follows: unpatterned, sedimentation and volcanism; diagonal line, regional metamorphism and penetrative deformation; zigzag, open folding; crosses, plutonism.

Middle Jurassic

Igneous activity continued in the Rattlesnake Creek terrane during this time interval, indicated by the dike complex near Happy Camp (Hill, 1985). Between 168 and 177 m.y., volcanic arc activity was initiated to form rocks now part of the western Hayfork terrane (Fahan, 1982). Harper and Wright (1984) considered this activity to be the earliest manifestation of continuous Middle and Late Jurassic arc activity between 177 and 159 m.y. However, on geochemical, structural, and petrologic grounds, it is possible to discriminate between (1) pre-Siskiyou, mildly alkaline volcanic arc rocks of the western Hayfork terrane and consanguineous Ironside Mountain batholithic rocks (Irwin, 1972) that are clearly cut by Siskiyou age faults (Fahan, 1982), and (2) post-Siskiyou calc-alkaline plutons such as the Wooley Creek batholith that cut across Siskiyou-age metamorphic rocks and interterrane faults.

A major deformational event resulted in the tectonic imbrication and chaotic juxtaposition of all the aforementioned Permian to Middle Jurassic oceanic and arc rocks (Fig. 39-7). As deformation continued, this thickened crust underwent regional metamorphism (the Siskiyou event). In latest Middle Jurassic time, these rocks were thrust under the Fort Jones terrane blueschists and intruded by post-Siskiyou calc-alkaline plutons (160–167-m.y. ages). This plutonic belt also intrudes the Central Metamorphic terrane, indicating amalgamation of the Siskiyou terranes with the rest of the Klamath Mountain terranes to the east by this time.

Pre-170-m.y. basinal sediments of the Condrey Mountain terrane were undergoing transitional greenschist-blueschist facies metamorphism at this time. A paired metamorphic belt interpretation between the Siskiyou and Condrey Mountain orogenies is tenable.

Late Jurassic

The Middle Jurassic plutonic axis moved west (Bear Mountain Complex) and intra-arc rifting opened a basin in which rocks of the Western Klamath terrane (including Josephine ophiolite and Galice Formation) formed (Harper and Wright, 1984). Collapse of this juvenile basin resulted in deformation and metamorphism of the Nevadan orogeny (Fig. 39-7). Limited post-Nevadan plutonism is represented by the Lower Coon Mountain pluton.

Cretaceous

Deformation and metamorphism of Pickett Peak terrane rocks took place in an Early Cretaceous subduction zone (Blake *et al.*, 1985a). Plutonic rocks of appropriate age in the Klamath Mountains are the Shasta Bally batholith, the White Rock pluton, and one in Shasta Valley. More voluminous Early Cretaceous plutonism, continuous with the Sierra Nevada and Idaho batholiths, is probably buried beneath the Casacades and Modoc Plateau (Fuis and Zucca, 1984), thus placing the Klamaths in a fore-arc setting from Cretaceous time to the present day.

K-Ar cooling ages in the range 117–125 m.y. are shown by some Condrey Mountain and Pickett Peak terrane rocks (Coleman *et al.*, 1983). Although structural evidence is lacking, these may also reflect a paired metamorphic belt, the bulk of whose low-P/T rocks may constitute the major thickness of the crust beneath the present-day Klamath Mountains.

Further (albeit tentative) evidence of tectonic and metamorphic links between the Klamath Mountains and Franciscan assemblage are found in Condrey Mountains–age eclogite blocks in Central Belt Franciscan melange. The structural geometry of these links is still a matter of speculation. It could be related to east-dipping thrusting, duplexing, and melange wedge recycling (Cloos, 1982). Alternatively, it could be related to east- and west-vergent structures produced during tectonic wedging of allochthonous Franciscan rocks under the Great Valley sediments, as new reflection profiles across the Great Valley–Franciscan contact seem to imply (Wentworth *et al.*, 1984).

Cenozoic

All Cenozoic structures in the Klamath Mountains and adjacent Coast Ranges are high-angle. Late Tertiary strike-slip faulting offset parts of the Western Klamath, Pickett Peak, and Yolla Bolly terranes by at least 300 km in a dextral sense (Blake *et al.*, 1985b; Kelsey and Hagans, 1982. Blake and Jayko, 1983). Blake (personal communication, 1986) has evidence that this strike-slip faulting affects Eocene units of the Tyee and Umpqua Formations in Oregon. Broad regional uplift-related warping and faulting of the Klamaths, Cascades, and northern Coast Ranges occurred in Neogene time. As much as 7 km of vertical uplift is implied for the Condrey Mountain dome (Mortimer and Coleman, 1985). The chemistry of spring waters in the area (I. Barnes *et al.*, 1981) suggests that metamorphic processes are still active today beneath northern California.

SUMMARY

Metamorphism in the Klamath Mountains is the result of large-scale tectonic events that relate directly to the processes of plate convergence. Two main types of metamorphism can be defined. (1) Low-pressure metamorphic facies series in which recrystallization is related to amalgamation and deformation of terranes in an arc environment; metamorphism and deformation generally predate calc-alkaline plutonism. (2) High-pressure metamorphic facies series that are structurally allochthonous with respect to surrounding terranes. Their formation in subduction zones under high-stress conditions produces multideformed masses characteristically devoid of cogenetic intrusive plutons and the low-strain rate at low temperature indicates slow subduction. Postmetamorphic intra-arc thrusting juxtaposes these high-P/T rocks against shallower-level metamorphic terranes and competely destroys premetamorphic tectonic configuration in the arc-subduction system.

ACKNOWLEDGMENTS

Support for this work has been provided by the U.S. Geological Survey (Coleman, Donato, Manning, Hill), National Science Foundation grant EAR82-13347, (Coleman and Mortimer), a Killam Postdoctoral Fellowship at the University of British Columbia (Mortimer), and the Shell and Chevron Companies Foundations Grants at Stanford (Manning and Hill). We also wish to acknowledge the geologic contributions of our numerous colleagues, without which this synthesis would not have been possible.

REFERENCES

Allen, C. A., Barnes, C. G., Kays, M. A., and Saleeby, J. B., 1982, Comagmatic nature of the Wooley Creek batholith and the Slinkard pluton and age constraints on tectonic and metamorphic events in the western Paleozoic and Triassic belt, Klamath Mountains, N. California: *Geol. Soc. America Abstr. with Programs*, v. 14, no. 4, p. 145.

Ando, C. J., Irwin, W. P., Jones, D. L., and Saleeby, J. B., 1983, The ophiolitic North Fork terrane in the Salmon River region, central Klamath Mountains, California: *Geol. Soc. America Bull.*, v. 94, p. 236–252.

Apted, M., and Liou, J. G., 1983, Phase relations among greenschist, epidote amphibolite and amphibolite in a basaltic system: *Amer. J. Sci.*, v. 283-A, p. 328–354.

Barnes, C. G., 1982, Geology and petrology of the Wooley Creek batholith, Klamath Mountains, northern California: Ph.D. thesis, Univ. Oregon, Eugene, Oreg., 214 p.

—— , 1983, Petrology and upward zonation of the Wooley Creek batholith, Klamath Mountains, California: *J. Petrology*, v. 24, p. 495–537.

—— , Rice, J. M., and Gribble, R. F., 1986a, Tilted plutons in the Klamath Mountains of California and Oregon: *J. Geophys. Res.*, v. 91, p. 6059–6072.

—— , Allen, C. M., and Saleeby, J. B., 1986b, Open- and closed-system characteristics of a tilted plutonic system, Klamath Mountains: *J. Geophys. Res.*, v. 91, p. 6073–6090.

Barnes, I., Kistler, R. W., Mariner, R. H., and Presser, T. S., 1981, Geochemical evidence on the nature of the basement rocks of the Sierra Nevada, California: *U.S. Geol. Survey Water-Supply Paper 2181*, 13 p.

Blake, M. C., Jr., and Jayko, A. S., 1983, Geologic map of the Yolla Bolly–Middle Eel Wilderness and adjacent roadless areas, northern California: *U.S. Geol. Survey Misc. Field Studies Map MF-1656-A*, scale 1:62,500.

—— , Howell, D. G., and Jones, D. L., 1982, Preliminary tectonostratigraphic terrane map of California: *U.S. Geol. Survey Open File Rpt. 82-593*, scale 1:750,000.

—— , Jayko, A. S., and McLaughlin, R. J., 1985a, Tectonostratigraphic terranes of the Northern Coast Ranges, California, *in* Howell, D. G., ed., *Tectonostratigraphic Terranes of the Circum-Pacific Region:* Circum-Pacific Council Energy Min. Resources, Earth Sci. Series, no. 1, p. 159–171.

—— , Engebretson, D. C., Jayko, A. S., and Jones, D. L., 1985b, Tectonostratigraphic terranes in southwest Oregon, *in* Howell, D. G., ed., *Tectonostratigraphic Terranes of the Circum-Pacific Region:* Circum-Pacific Council Energy Min. Resources, Earth Sci. Series, no. 1, p. 147–157.

Borns, D. J., 1980, Blueschist metamorphism of the Yreka–Fort Jones area, Klamath Mountains, Northern California: Ph.D. thesis, Univ. Washington, Seattle, Wash., 155 p.

—— , 1984, Eclogites in the Stuart Fork terrane, Klamath Mountains, California: *Geol. Soc. America Abstr. with Programs*, v. 16, no. 5, p. 271.

Brown, E. H., 1977a, The crossite content of Ca-amphibole as a guide to pressure of metamorphism: *J. Petrology*, v. 18, p. 53–72.

—— , 1977b, Phase equilibria among pumpellyite, lawsonite, epidote and associated minerals in low grade metamorphic rocks: *Contrib. Mineralogy Petrology*, v. 64, p. 123–136.

—— , and Ghent, E. D., 1983, Mineralogy and phase relations in the blueschist facies of the Black Butte and Ball Rock areas, northern California Coast Ranges: *Amer. Mineralogist*, v. 68, p. 365–372.

Burchfiel, B. C., and Davis, G. A., 1981, Triassic and Jurassic tectonic evolution of the Klamath Mountains-Sierra Nevada geologic terrane, *in* Ernst, W. G., ed., *The Geotectonic Development of California* (Rubey Vol. I): Englewood Cliffs, N.J., Prentice-Hall, p. 50–70.

Burton, W. C., 1982, Geology of the Bar Mountains, northern California: M.S. thesis, Univ. Oregon, Eugene, Oreg., 120 p.

Cannat, M., 1985, Tectonics of the Seiad massif, northern Klamath Mountains, California: *Geol. Soc. America Bull.*, v. 96, p. 15–26.

Cannat, M., and Boudier, F., 1985, Structural study of intra-oceanic thrusting in the Klamath Mountains, northern California — Implications on accretion geometry: *Tectonics*, v. 4, no. 5, p. 435–452; and correction in *Tectonics*, v. 4, no. 6, p. 597–601.

Cashman, S. M., 1980, Devonian metamorphic event in the northeastern Klamath Mountains, California: *Geol. Soc. America Bull.*, pt. I, v. 91, p. 453–459.

Cloos, M., 1982, Flow melanges — Numerical modeling and geologic constraints on their origin in the Franciscan subduction complex, California: *Geol. Soc. America Bull.*, v. 93, p. 330–345.

Coleman, R. G., 1972, The Colebrooke Schist of southern Oregon and its relation to the tectonic evolution of the region: *U.S. Geol. Survey Bull. 1339*, 61 p.

——, Helper, M. D., and Donato, M. M., 1983, Geologic map of the Condrey Mountain Roadless area, Siskiyou County, California: *U.S. Geol. Survey Misc. Field Studies Map MF-1540-A.*

Cotkin, M. L., and Cotkin, S. J., 1985, Structural style, metamorphism and tectonic implications of early Paleozoic blueschist, eastern klamath Mountains, California: *Geol. Soc. America Abstr. with Programs*, v. 17, no. 6, p. 349.

Davis, G. A., 1968, Westward thrust faulting in the south-central Klamath Mountains, California: *Geol. Soc. America Bull.*, v. 79, p. 911–934.

——, and Lipman, P. W., 1962, Revised structural sequence of pre-Cretaceous metamorphic rocks in the southern Klamath Mountains, California: *Geol. Soc. America Bull.*, v. 73, p. 1547–1552.

——, Holdaway, M. J., Lipman, P. W., and Romey, W. D., 1965, Structure, metamorphism and plutonism in the south-central Klamath Mountains, California: *Geol. Soc. America Bull.*, v. 76, no. 8, p. 933–966.

——, Monger, J. W. H., and Burchfiel, B. C., 1978, Mesozoic construction of the Cordilleran "collage," central British Columbia to central California, *in* Howell, D. G., and McDougall, K. A., eds., *Mesozoic Paleogeography of the Western United States:* Pacific Section, Soc. Econ. Paleontologists Mineralogists, Pacific Coast Paleogeography Symp. 2, p. 1–32.

Donato, M. M., 1985, Metamorphic and structural evolution of an ophiolitic tectonic melange, Marble Mountains, northern California: Ph.D. thesis, Stanford Univ., Stanford, Calif., 258 p.

——, Coleman, R. G., and Kays, M. A., 1980, Geology of the Condrey Mountain Schist, northern Klamath Mountains, California and Oregon: *Oreg. Geology*, v. 42, no. 7, p. 125–129.

——, Barnes, C. G., Coleman, R. G., Ernst, W. G., and Kays, M. A., 1982, Geologic map of the Marble Mountains Wilderness, Siskiyou County, California: *U.S. Geol. Survey Misc. Field Studies Map MF-1452-A.*

Elliot, M. A., and Bostwick, D. A., 1973, Occurrence of *Yabeina* in the Klamath Mountains, Siskiyou County, California: *Geol. Soc. America Abstr. with Programs*, v. 5, no. 1, p. 38.

Ernst, W. G., 1984, Californian blueschists, subduction, and the significance of tectonostratigrapic terranes: *Geology*, v. 12, p. 436–440.

Fahan, M. R., 1982, Geology and Geochronology of a part of the Hayfork terrane, Klamath Mountains, northern California: M.S. thesis, Univ. California, Berkeley, Calif., 127 p.

Fuis, G. S., and Zucca, J. J., 1984, A geologic cross section of northeastern California from seismic refraction results, *in* Nilsen, Tor H., ed., *Geology of the Upper Cre-*

taceous Hornbrook Formation, Oregon and California: Pacific Section, Soc. Econ. Paleontologists Mineralogists, v. 42, p. 203–209.

Goodge, J. W., 1985, Widespread blueschist assemblages in the Stuart Fork terrane, central Klamath Mountains, northern California: *Geol. Soc. America Abstr. with Programs, v. 17, p. 357.*

Gorman, C., 1985, Geology, chronology, and geochemistry of the Rattlesnake Creek terrane, west-central Klamath Mountains, California: M.S. thesis, Univ. Utah, Salt Lake City, Utah, 112 p.

Gray, G. C., 1985, Structural, geochronologic, and depositional history of the western Klamath Mountains, California and Oregon — Implications for the Early to Middle Mesozoic tectonic evolution of the western Cordillera: Ph.D. thesis, Univ. Texas, Austin, Tex., 161 p.

_____, and Petersen, S. W., 1982, Northward continuation of the Rattlesnake Creek terrane, north-central Klamath Mountains, California: *Geol. Soc. America Abstr. with Programs* v. 14, no. 4, p. 167.

Grover, T. W., 1984, Progressive metamorphism west of the Condrey Mountain Dome, north-central Klamath Mountains, northern California: M.S. thesis, Univ. Oregon, Eugene, Oreg., 129 p.

Harland, W. B., Cox, A. V., Llewellyn, P. G., Pickton, C. A. G., Smith, A. E., and Walters, R., 1982, *A Geologic Time Scale:* Cambridge, Cambridge University Press, 131 p.

Harper, G. D., 1984, The Josephine ophiolite, northwestern California: *Geol. Soc. America Bull.*, v. 95, p. 1009–1026.

_____, and Wright, J. E., 1984, Middle to Late Jurassic tectonic evolution of the Klamath Mountains, California-Oregon: *Tectonics*, v. 3, p. 759–772.

Helgeson, H. C., Delany, J. M., Nesbitt, H. W., and Bird, D. K., 1978, Summary and critique of the thermodynamic properties of rock-forming minerals: *Amer. J. Sci.*, v. 278-A, 229 p.

Helper, M. A., 1985, Structural, metamorphic and geochronologic constraints on the origin of the Condrey Mountain Schist, north central Klamath Mountains, northern California: Ph.D. thesis, Univ. Texas, Austin, Tex., 209 p.

_____, 1986, Deformation and high P/T metamorphism in the central part of the Condrey Mountain window, north-central Klamath Mountains, California and Oregon: *Geol. Soc. Amer. Mem. 164*, p. 125–141.

Hill, L. B., 1984, A tectonic and metamorphic history of the north-central Klamath Mountains, California: Ph.D. thesis, Stanford Univ. Stanford, Calif., 248 p.

_____, 1985, Metamorphic, deformational and temporal constraints on terrane assembly, northern Klamath Mountains, California, *in* Howell, D. G., ed., *Tectonostratigraphic Terranes of the Circum-Pacific Region:* Circum-Pacific Council Energy Min. Resources, Earth Sci. Series, no. 1, p. 173–186.

Holdaway, M. J., 1965, Basic regional metamorphic rocks in part of the Klamath Mountains, northern California: *Amer. Mineralogist*, v. 50, p. 953–977.

Holland, C. H., *et al.*, 1978, A guide to stratigraphical procedure: *Geol. Soc. London Spec. Rpt. no. 11*, 18 p.

Holland, T. J. B., 1983, The experimental determination of activities in disordered and short-range ordered jadeitic pyroxenes: *Contrib. Mineralogy Petrology*, v. 82, p. 214–220.

Hotz, P. E., 1967, Geologic map of the Condrey Mountain quadrangle, and parts of the Seiad Valley and Hornbrook quadrangles, California: *U.S. Geol. Survey Geol. Quad. Map GQ-618*, scale 1:62,500.

_____, 1971, Plutonic rocks of the Klamath Mountains, California and Oregon: *U.S. Geol. Survey Prof. Paper 684B*, 20 p.

_____, 1973, Blueschist, metamorphism in the Yreka–Fort Jones area, Klamath Mountains, California: *U.S. Geol. Survey J. Res.*, v. 1, p. 53–61.

_____, 1977, Geology of the Yreka quadrangle, Siskiyou County, California: *U.S. Geol. Survey Bull. 1436*, 72 p.

_____, 1979, Regional metamorphism in the Condrey Mountain Quadrangle, north-central Klamath Mountains, California: *U.S. Geol. Survey Prof. Paper 1086*, 25 p.

_____, Lanphere, M. A., and Swanson, D. A., 1977, Triassic blueschist from northern California and north-central Oregon: *Geology*, v. 5, no. 11, p. 659–663.

Irwin, W. P., 1966, Geology of the Klamath Mountains Province: *Calif. Div. Mines Geology Bull. 190*, p. 19–38.

_____, 1972, Terranes of the western Paleozoic and Triassic belt in the southern Klamath Mountains, California: *U.S. Geol. Survey Prof. Paper 800C*, p. C103–C111.

_____, 1979, Ophiolitic terranes of part of the western United States, *in International Atlas of Ophiolites:* Geol. Soc. America Maps Charts Series MC-33, sheet 1, scale 1:2,500,000.

_____, 1981, Tectonic accretion of the Klamath Mountains, *in* Ernst, W. G., ed., *The Geotectonic Development of California* (Rubey Vol. I): Englewood Cliffs, N. J., Prentice-Hall, p. 29–49.

_____, 1985, Age and tectonics of plutonic belts in accreted terranes of the Klamath Mountains, California and Oregon, *in* Howell, D. G., ed., *Tectonostratigraphic Terranes of the Circum-Pacific Region*, Circum-Pacific Council Energy Min. Resources, Earth Sci. Series, no. 1, p. 189–199.

_____, Jones, D. L., and Kaplan, T. A., 1978, Radiolarians from pre-Nevadan rocks of the Klamath Mountains, California and Oregon, *in* Howell, D. G., and McDougall, K. A., eds., *Mesozoic Paleogeography of the Western United States:* Pacific Section, Soc. Econ. Paleontologists Mineralogists, Pacific Coast Paleogeography Symp. 2, p. 303–310.

_____, Jones, D. L., and Blome, C. D., 1982, Map showing sampled radiolarian localities in the western Paleozoic and Triassic belt, Klamath Mountains, California: *U.S. Geol. Survey Misc. Field Studies Map MF-1399.*

_____, Wardlow, B. R., and Kaplan, T. A., 1983, Conodonts of the western Paleozoic and Triassic belt, Klamath Mountains, California and Oregon: *J. Paleontology*, v. 57, p. 1030–1039.

Jachens, R. C., Barnes, C. G., and Donato, M. M., 1986, Subsurface configuration of the Orleans fault — Implications for deformation in the western Klamath Mountains, California: *Geol. Soc. America Bull.*, v. 97, p. 388–395.

Jayko, A. S., Blake, M. C., and Brothers, R. N., 1986, Blueschist metamorphism of the Eastern Franciscan belt, northern California: *Geol. Soc. America Mem. 164*, p. 107–123.

Kays, M. A., and Ferns, M. L., 1980, Geologic field trip guide through the north-central Klamath Mountains: *Oreg. Geology*, v. 42, no. 2, p. 23–35.

Kelsey, H. M., and Hagans, D. K., 1982, Major right-lateral faulting in the Franciscan assemblage of northern California in late Tertiary time: *Geology*, v. 10, p. 387–391.

Klein, C. W., 1977, Thrust plates of the north-central Klamath Mountains near Happy Camp, California: *Calif. Div. Mines Geology Spec. Rpt. 129*, p. 23–26.

Lanphere, M. A., Irwin, W. P., and Hotz, P. E., 1968, Isotopic age of the Nevadan Orogeny and older plutonic and metamorphic events in the Klamath Mountains, California: *Geol. Soc. America Bull.*, v. 79, p. 1027–1052.

_____, Blake, M. C., Jr., and Irwin, W. P., 1978, Early Cretaceous metamorphic age of the South Fork Mountain Schist in the northern Coast Ranges of California: *Amer. J. Sci.*, v. 278, p. 798–815.

Lapierre, H., Albarede, F., Albers, J., Cabans, B., and Coulon, C., 1985, Early Devonian volcanism in the eastern Klamath Mountains, California — Evidence for an immature island arc: *Can. J. Earth Sci.*, v. 22, p. 214–226.

Lieberman, J. E., 1983 Petrology and petrogenesis of marble and peridotite, Seiad Ultramafic Complex, California: M.S. thesis, Univ. Oregon, Eugene, Oreg., 118 p.

———, and Rice, J. M., 1983, Prograde metamorphism of marble and peridotite in the Seiad Ultramafic Complex, California: *Geol. Soc. America Abstr. with Programs*, v. 15, no. 5, p. 436.

Lindsley-Griffin, N., and Griffin, J. R., 1983, The Trinity terrane — An Early Paleozoic microplate assemblage, *in* Stevens, C. H., *Pre-Jurassic Rocks in Western North American Suspect Terranes:* Pacific Section, Soc. Econ. Paleontologists Mineralogists, 141 p.

Liou, J. G., Maruyama, S., and Cho, M., 1985, Phase equilibria and mineral parageneses of metabasites in low-grade metamorphism: *Mineral. Mag.*, v. 49, p. 321–333.

Loney, R. A., and Himmelberg, G. R., 1976, Structure of the Vulcan Peak alpine-type peridotite, southwestern Oregon: *Geol. Soc. America Bull.*, v. 87, p. 259–274.

Lundquist, S. M., 1982, Deformation history of the ultramafic and associated rocks of the Seiad Complex, Seiad Valley, California: M.S. thesis, Univ. Washington, Seattle, Wash., 167 p.

Masi, U., O'Neil, J. R., and Kistler, R. W., 1981, Stable isotope systematics in Mesozoic granites of central and northern California and southwestern Oregon: *Contrib. Mineralogy Petrology*, v. 76, p. 116–126.

Mattinson, J. M., 1981, U-Pb systematics and geochronology of blueschists — Preliminary results: (EOS) *Trans. Amer. Geophys. Un.*, v. 62, no. 45, p. 1059.

McDowell, F. W., Lehman, D. H., Gucwa, P. R., Fritz, D., and Maxwell, J. C., 1984, Glaucophane schists and ophiolites of the northern California Coast Ranges — Isotopic ages and their tectonic implications: *Geol. Soc. America Bull.*, v. 95, p. 1373–1382.

Medaris, L. G., Jr., 1975, Coexisting spinel and silicates in alpine peridotites of the granulite facies: *Geochim. Cosmochim. Acta*, v. 39, p. 947–958.

Miller, M. M., and Wright, J. E., 1985, New biogeographic implications of a Permian Tethyan coral, Klamath Mountains, California: *Geol. Soc. America Abstr. with Programs*, v. 17, no. 7, p. 664–665.

Moore, D. E., 1984, Metamorphic history of a high-grade blueschist exotic block from the Franciscan complex, California: *J. Petrology*, v. 25, p. 126–150.

Mortimer, N., 1984a, Petrology and structure of Permian to Jurassic rocks near Yreka, Klamath Mountains, California: Ph.D. thesis, Stanford Univ., Stanford, Calif., 84 p.

———, 1984b, Geology of the Klamath Mountains near Yreka, California, *in* Nilsen, T. H., ed., *Geology of the Upper Cretaceous Hornbrook Formation, Oregon and California:* Pacific Section, Soc. Econ. Paleontologists Mineralogists, v. 42, p. 167–178.

———, 1985, Structural and metamorphic aspects of Middle Jurassic terrane juxtaposition, northeastern Klamath Mountains, California, *in* Howell, D. G., ed., *Tectonostratigraphic Terranes of the Circum-Pacific Region:* Circum-Pacific Council Energy Min. Resources, Earth Sci. Series, no. 1, p. 201–214.

———, and Coleman, R. G., 1985, A Neogene structural dome in the Klamath Mountains, California and Oregon: *Geology*, v. 13, p. 253–256.

Nitsch, K.-H., 1971, Die Niedrig-Temperaturgrenze des Anorthit-Stabilitatsfeldes: *Fortschr. Mineralogie*, v. 49, p. 34–36.

Norman, E. A., Gorman, C. M., Harper, G. D., and Wagner, D., 1983, Northern extension of the Rattlesnake Creek terrane: *Geol. Soc. America Abstr. with Programs*, v. 15, no. 6, p. 400.

Peacock, S. M., 1985, Inverted metamorphic gradient and fluid evolution of the central metamorphic belt, Klamath province, northern California: *Geol. Soc. America Abstr. with Programs*, v. 17, p. 400.

Potter, A. W., Hotz, P. E., and Rohr, D. M., 1977, Stratigraphy and inferred tectonic framework of Lower Paleozoic rocks in the eastern Klamath Mountains, northern California, *in* Stewart, J. H., Stevens, C. H., and Fritsche, A. E., eds., *Paleozoic Paleogeography of the Western United States:* Pacific Section, Soc. Econ. Paleontologists Mineralogists, Pacific Coast Paleogeography Symp. p. 421–440.

Rawson, S. A., 1985, Regional metamorphism of rodingites and related rocks from the north-central Klamath Mountains, California: Ph.D. thesis, Univ. Oregon, Eugene, Oreg., 235 p.

―― , and Petersen, S. W., 1982, Structural and lithologic equivalence of the Rattlesnake Creek terrane and high-grade rocks of the western Paleozoic and Triassic belt, north central Klamath Mountains, California: *Geol. Soc. America Abstr. with Programs*, v. 14, no. 4, p. 226.

Saleeby, J. B., Harper, G. D., Snoke, A. W., and Sharp, W. D., 1982, Time relations and structural-stratigraphic patterns in ophiolite accretion, west central Klamath Mountains, California: *J. Geophys. Res.*, v. 87, no. B5, p. 3831–3848.

―― , Blake, M. C., and Coleman, R. G., 1984, Pb/U zircon ages on thrust plates of the west central Klamath Mountains and Coast Ranges, northern California and southern Oregon: (EOS) *Trans. Amer. Geophys. Union*, v. 65, no. 45, p. 1147.

Silberling, N. J., Jones, D. L., Blake, M. C., Jr., and Howell, D. G., 1984, Lithotectonic terrane map of the North American Cordillera – Pt. C. *U.S. Geol. Survey Open File Rpt. 84-523*, scale 1:2,500,000.

Smith, J. G., Page, N. J., Johnson, M. G., Moring, B., and Gray, F., 1982, Preliminary geologic map of the Medford 1° X 2° quadrangle, Oregon and California: *U.S. Geol. Survey Open File Rpt. 82-955*, scale 1:250,000.

Snoke, A. W., Bowman, H. R., and Hebert, A. J., 1977, The Preston Peak ophiolite, Klamath Mountains, California, an immature island arc – Petrochemical evidence: *Calif. Div. Mines Geology Spec. Rpt. 129*, p. 67–79.

Suppe, J., and Armstrong, R. L., 1972, Potassium-argon dating of Franciscan metamorphic rocks: *Amer. J. Sci.*, v. 272, p. 217–233.

Wagner, D. L., and Saucedo, G., in press, *Geologic Map of California, Weed 1° X 2° sheet*, scale 1:250,000.

Wentworth, C. M., Blake, M. C. Jr., Jones, D. L., Walter, A. W., and Zoback, M. D., 1984, Tectonic wedging associated with emplacement of the Franciscan assemblage, California Coast Ranges, *in* Blake, M. C., Jr., ed., *Franciscan Geology of Northern California:* Pacific Section, Soc. Econ. Paleontologists Mineralogists, v. 43, p. 163–173.

Wiggins, B. D., 1980, Volcanism and sedimentation of a Late Jurassic to Early Cretaceous Franciscan terrane near Bosely Butte, southwestern Oregon: Ph.D. thesis, Univ. California, Berkeley, Calif., 290 p.

Wright, J. E., 1981, Geology and U-Pb geochronology of the Western Paleozoic and Triassic sub-province, Klamath Mountains, northern California; Ph.D. thesis, Univ. California, Santa Barbara, Calif., 300 p.

―― , 1982, Permo-Triassic Accretionary Subduction Complex, southwestern Klamath Mountains, northern California: *J. Geophys. Res.*, v. 87, no. B5, p. 3805–3818.

―― , and Sharp, W. D., 1981, The implication of regional tectonic correlations between the Klamath Mountain-Sierra Nevada provinces for the Late Paleozoic to Middle Mesozoic tectonic evolution of western North America: *Geol. Soc. America Abstr. with Programs*, v. 13, no. 7, p. 585.

—— , and Wyld, S. J., 1985, Multi-stage serpentinite matrix melange development — Rattlesnake Creek terrane, southwestern Klamath Mountains: *Geol. Soc. America Abstr. with Programs*, v. 17, no. 6, p. 419.

M. A. Kays
Department of Geological Sciences
University of Oregon
Eugene, Oregon 97405-1272

M. L. Ferns and H. C. Brooks
State of Oregon
Department of Geology and Mineral Industries
Baker, Oregon 97814

40

METAMORPHISM OF TRIASSIC-PALEOZOIC BELT ROCKS: A GUIDE TO FIELD AND PETROLOGIC RELATIONS IN THE OCEANIC MELANGE, KLAMATH AND BLUE MOUNTAINS, CALIFORNIA AND OREGON

ABSTRACT

In ophiolitic terranes of the Triassic-Paleozoic belt in the Klamath and Blue Mountains, the patterns of deformation and metamorphism correlate well with processes assumed to have operated during the formation and subsequent emplacement history of the ophiolitic rocks. A characteristic feature of each terrane is that high-temperature fabrics in the basal peridotites are among the oldest of the metamorphic structures. Subsequent strong modification of primary tectonite fabrics of ultramafic rocks and fabrics in the underlying metamorphic sequence is the result of widespread thrust faulting involving tectonic transport of peridotite over hydrothermally altered ocean-floor metabasic and metasedimentary rocks. Low-grade, hydrothermally altered metabasic rocks, with relict volcanic or hypabyssal structures and textures, are widespread in some terranes, and have ages that are about the same as the tectonite peridotite assemblages and fabrics. The age of the hydrothermally altered or spilitized metabasic assemblages in the Baker terrane in the Blue Mountains is older than regional 213–224-m.y. high-P/T metamorphism. A similar relationship between high-P/T metamorphism and hydrothermal alteration holds true for at least part of the North Fork and Salmon River terranes in the Klamath Mountains.

Superposition of metamorphic fabrics and incorporation of metamorphic tectonites into the melange of the ophiolitic terranes indicates that the melange component of each formed over an extended period, following or beginning with thrust faulting. Formation of the melange began later than formation of primary tectonite fabrics in the basal peridotites, and later than formation of the hydrothermal ocean-floor metamorphic assemblages, and probably continued through the regional high-P/T metamorphic event. High-P/T Norian (213–224 m.y.) metamorphism was followed by regional, high-T/P Jurassic "Nevadan" (150–175 m.y.) metamorphism. In the latter, assemblages in metapelitic rocks record a "geotherm" intermediate to those of the well-defined Buchan and Barrovian metamorphisms. Both regional events formed penetrative fabrics in the respective terranes where they dominate. High-P/T regional metamorphism is found in close association with low-grade, hydrothermally altered assemblages with abundant relict structures and fabrics that are older than about 225 m.y. in the Baker terrane in the Blue Mountains, and in the Fort Jones, North Fork, and Salmon River terranes in the Klamath Mountains. Regional, Jurassic, high-T/P metamorphism is not well developed in these terranes, but dominates in the Marble Mountains and Hayfork terranes, and in the May Creek Schist belt in the Klamath Mountains. There is no evidence of prior high-P/T metamorphism in these terranes.

INTRODUCTION

This paper reviews the character and chronology of metamorphism and deformation of Triassic-Paleozoic belt rocks in northern Californa and Oregon (Fig. 40-1), focusing on the oceanic and largely melanged part of the belt. The regional geology and metamorphic history of a more southerly portion of the Klamath Mountains is presented by Coleman *et al.*, (Chapter 39, this volume). A major objective of this paper is to help clarify the spatial and temporal patterns of metamorphic and fabric facies within melanged ophiolitic terranes. Therefore, the emphasis here is on field and petrologic relations, which provide important information about the timing of deformation and recrystallization. It is hoped that the patterns of deformation and metamorphism recognized in each of the several areas considered will add to an understanding of the overall pattern of evolution of metamorphic rocks within the Triassic-Paleozoic belt in northern California and Oregon.

KAYS, FERNS, BROOKS

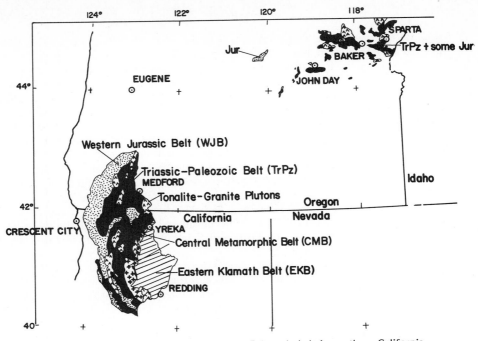

FIG. 40-1. Index map of the Triassic Paleozoic belt in northern California and Oregon.

REGIONAL GEOLOGIC SETTING OF OCEANIC TERRANES IN CALIFORNIA AND OREGON

Blue Mountains, Northeastern Oregon

The oceanic Baker terrane (Silberling *et al.* 1984) forms a belt with largely east-west trend (Fig. 40-2) and is bounded on the north and south, respectively, by the Wallowa–Seven Devils and Huntington volcanic arc terranes (Brooks, 1979). The Baker terrane is herein subdivided into the Elkhorn Ridge Argillite-dominated northern assemblage, serpentine-matrix melange-dominated southern assemblage, and the Burnt River Schist, based on characteristics described by Vallier *et al.* (1977), Brooks and Vallier (1978), and Brooks (1979). The Elkhorn Ridge-dominated terrane is composed mainly of supracrustal argillite, chert, and tuffaceous argillite. Less abundant are metamorphosed basaltic and keratophyric lavas and tuffs, and widely scattered limestone bodies (Brooks *et al.*, 1976; Gilluly, 1937). The Burnt River Schist and the Elkhorn Ridge Argillite have penetrative fabrics of two generations (Avé Lallèmant *et al.*, 1980; Oldow *et al.*, 1984). The two units have similar lithologies and small-scale structural features and may be equivalent (Ashley, 1967). The serpentine-matrix-dominated melange also includes coherent rock sequences of deeper-seated ophiolitic affinity, as well as all of the previously described lithologies in the Elkhorn Ridge Argillite and Burnt River Schist. Glaucophane schist blocks and amphibolite are also part of the melange.

Two rather large and partly dismembered ophiolitic sequences with apparent island-arc affinities (Mullen, 1983; Phelps, 1979) have been studied near the boundaries between oceanic and arc terranes. On the north, the Sparta complex (Phelps and Avé Lallèmant,

METAMORPHISM OF TRIASSIC-PALEOZOIC BELT ROCKS

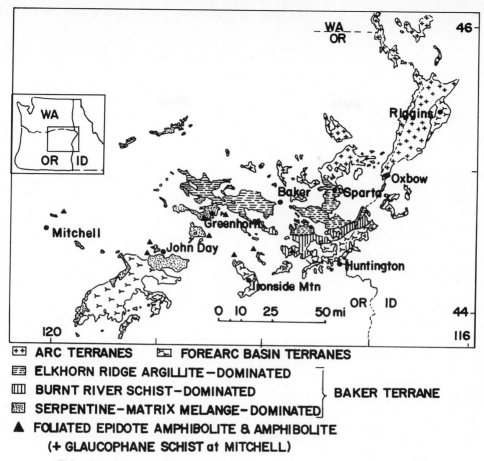

FIG. 40-2.   Paleozoic and Mesozoic terranes of the Blue Mountains.

1980; Prostka, 1962) consists in large proportion of gabbro and more silicic rocks, with only scarce amounts of serpentinized peridotite and pyroxenite. The Canyon Mountain ophiolite complex occurs to the south and is composed predominantly of peridotite and gabbro tectonite. The remainder of the complex consists of cumulate gabbroic rocks, and there are silicic differentiates, diabases, and basalts (Avé Lallèmant, 1976; C. E. Brown and Thayer, 1966; Thayer, 1956, 1963a; Thayer and Himmelberg, 1968). In the Canyon Mountain complex, the basal peridotite is separated from the underlying melange terrane by a thrust fault. The peridotite rests on tectonite amphibolite in a number of places where the contact with the underlying rocks is exposed. Smaller, fragmented peridotite bodies east of the complex are also in contact with garnet amphibolite, as for example in the Mine Ridge area north of Ironside Mountain (Fig. 40-2). Peridotite and underlying amphibolite are faulted against the melanged Baker terrane. The widespread distribution of peridotite and tectonite amphibolite within the predominantly broken or melanged Baker terrane suggests that the basal members of the ophiolite sequence contributed fragmental debris to the melange during emplacement of the ophiolite complex (Coleman, 1977; Nicolas and Le Pichon, 1980).

The relations at the base of the Sparta complex are less certain (Phelps and Avé

Lallèmant, 1980; Prostka, 1962). However, according to the regional tectonic setting as previously summarized, both the Sparta and Canyon Mountain ophiolites have a "right way up" stratigraphy that is constructional facing the respective arc terranes on the north and south, respectively. This would seem to require that the Sparta complex also either overlies the Baker terrane, or occurs at the base of the adjacent Wallowa–Seven Devils arc terrane. Geochemical characteristics suggest that the Sparta and Canyon Mountain complexes have island-arc affinities (Phelps, 1979; Phelps and Avé Lallèmant, 1980; Gerlach, et al., 1981).

Northernmost Klamath Mountains, Oregon

The northermost exposures of the Klamath Mountains geologic province are adjacent to, and partly covered by volcanic, volcaniclastic, and sedimentary rocks of the Tertiary, Western Cascades volcanic province (Peck et al., 1964). Here, too, Triassic-Paleozoic belt rocks are in west-directed thrust-fault contact with the underlying western Jurassic belt, Galice Formation. The thrust contact and the rocks on both sides are intruded by the ~141-m.y.-old White Rock pluton (Hotz, 1971). The initial mapping in this region by Diller and Kay (1924) recognized substantial areas of medium-grade metamorphosed rocks which they termed the May Creek Schist. Kays (1970) used the name May Creek Schist belt, produced a geologic map over part of the metamorphic belt (Fig. 40-3), and mapped metamorphic zone boundaries (Fig. 40-4). Plutonism, deformation by folding, thrusting, and low- to medium-grade regional metamorphism have obscured contact relations between the western Jurassic belt and Triassic-Paleozoic belt rocks in this area. The metasedimentary and smaller amount of metavolcanic greenschist facies rocks on the western margin of the study area (Fig. 40-3) were correlated with Galice Formation by Hotz (1971), apparently on the basis of structural position and lithology.

The upper-plate Triassic-Paleozoic belt rocks consist of bands of interlayered silicic and metapelitic schists tectonically interlayered with belts of mafic schist and gneiss, serpentinized ultramafic, and some gabbroic rocks. The interlayered rocks, their contacts, and penetrative fabrics are folded. Folding appears to be closely related in space and time to intrusion, but did not produce metamorphic fabrics in the granodiorite. Metamorphism in the belt ranges in grade from greenschist to middle amphibolite facies. There are no sharp discontinuities in the distribution of metamorphic assemblages, and the sequence is apparently prograde (Fig. 40-4).

Galice Formation silicic schists and slaty argillites are identifiable with confidence a few kilometers west of Tiller and Devils Flat, mostly just outside the mapped area of the May Creek Schist belt. Hotz (1971), however, draws the contact within the map area about 3 km west of Tiller. Galice Formation slates and phyllites are isoclinally folded (wavelengths in meters, amplitudes in tens of meters) and contain penetrative axial planar cleavage that transposes bedding in hinge areas of folds. However, cleavage and other penetrative fabrics are unfolded in the Galice, at least at the scale of the outcrop. This is in sharp contrast with the folded foliation in Triassic-Paleozoic belt schists.

North Central Klamath Mountains, California and Oregon

In this area the Triassic-Paleozoic belt is structurally above and in thrust fault contact with (1) a reentrant of the western Jurassic belt and Galice Formation (Smith River subterrane) on the west, and (2) the Condrey Mountain Schist in the central part (Fig. 40-5). The Condrey Mountain Schist is structurally above the western Jurassic terrane; there is some question whether it is an eastern facies of the western Jurassic belt (Mortimer, 1985; Mortimer and Coleman, 1984), or is a western facies of the Triassic-Paleozoic belt

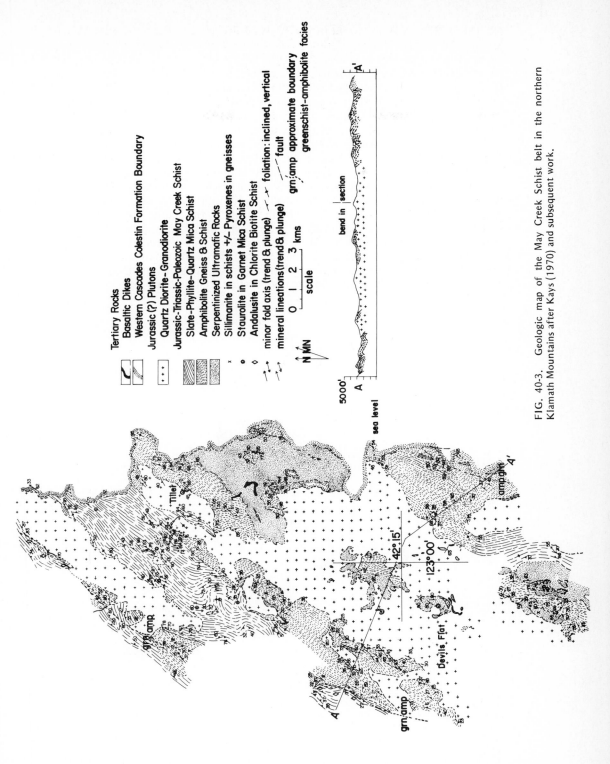

Tertiary Rocks
Basaltic Dikes
Western Cascades Colestin Formation Boundary
Jurassic (?) Plutons
    Quartz Diorite- Granodiorite
Jurassic-Triassic-Paleozoic May Creek Schist
    Slate-Phyllite-Quartz Mica Schist
    Amphibolite Gneiss 8 Schist
    Serpentinized Ultramafic Rocks
    Sillimanite in schists +/- Pyroxenes in gneisses
    Staurolite in Garnet Mica Schist
    Andalusite in Chlorite Biotite Schist
    minor fold axis (trend 8 plunge)   foliation: inclined, vertical
    mineral lineations (trend 8 plunge)   fault
                                        grn;amp approximate boundary
                                        greenschist-amphibolite facies

N MN

0 1 2 3 kms
    scale

FIG. 40-3. Geologic map of the May Creek Schist belt in the northern Klamath Mountains after Kays (1970) and subsequent work.

1103

(Helper, 1986). Oceanic terrane of the Triassic-Paleozoic belt has been termed the Marble Mountains terrane (Mortimer, 1985) and has characteristics similar to the Rattlesnake Creek terrane extended northward from the type location (Irwin, 1972) to the Klamath River near Happy Camp (Rawson and Petersen, 1982). The Marble Mountains terrane is greatly disrupted and melanged (block on block mostly), consisting of interlayered but disrupted sequences of silicic schists and amphibolites with minor lenses and layers of marble and quartzite, serpentinized ultramafic rocks (± gabbro), and discrete sheets and smaller blocks of amphibolite (± quartzite) with strong tectonite fabrics (Hotz, 1979). Near the base of the Marble Mountains sequence, most rocks have penetrative fabrics and low, middle, or upper amphibolite facies assemblages. These rocks are in thrust-faulted contact with the underlying Condrey Mountain Schist, composed of low-grade greenschist and transitional greenschist-blueschist-facies assemblages. With distance upward in the sequence, and generally radially away from contact with the Condrey Mountain Schist, the grade decreases. Structurally higher in the Marble Mountains terrane, the rocks have low- to low-medium-grade assemblages with locally preserved volcanic and sedimentary textures/fabrics and structures (pillows, filled vesicles, agglomeratic fragmental debris, sedimentary layering, pebbles, etc.). Peridotite in larger massifs and smaller tectonic slices occurs folded together with silicic schists and amphibolites. The central parts of some larger ultramafic bodies have textures, fabrics, and assemblages similar to

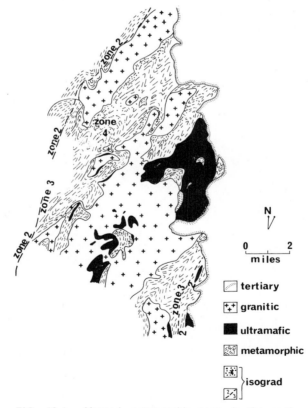

FIG. 40-4. Mapped metamorphic zones in the northern Klamath Mountains in the May Creek Schist belt after Kays (1970).

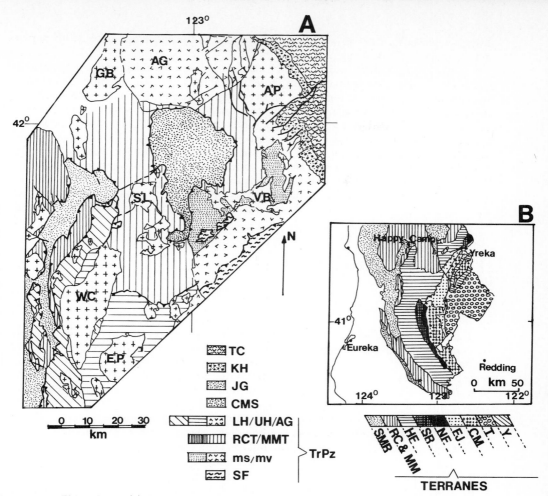

FIG. 40-5. (a). Geology of the north-central Klamath Mountains after
Barnes *et al.* (1986) from the compilation after Wagner and Saucedo (1985)
in California, Smith *et al.* (1982) and Kays and Ferns (1981) in Oregon. The
larger plutons are labeled according to informal name: Grayback (GB), Ash-
land (AP), Slinkard (SL), Vesa Bluffs (VB), Wooley Creek (WC), and English
Peak (EP). The Tertiary Colestin Formation (TC) and Cretaceous Hornbrook
Formation (KH) are cover sequences in the northeast corner of the map.
The Applegate terrane (AG) is tentatively correlated with the Lower and
Upper Hayfork terranes (LH and UH, respectively, after Wright, 1982). The
Jurassic Galice Formation (JG) and the Condrey Mountain Schist (CMS) have
similar lithologies, metamorphic age, and have the same structural relationship
to the overlying TrPz Marble Mountains (MMT), Rattlesnake Creek (RCT),
or Hayfork (LH, UH) terranes. The undivided, low-grade TrPz terranes are
shown simply as metasedimentary and metavolcanic (ms, mv respectively).
Klippe of Stuart Fork Formation with Norian high *P/T* metamorphic age rest
on TrPz units of younger Nevadan metamorphic age. (b). A terrane map after
Mortimer (1985) is based on the terminology and early correlations of Irwin
(1972) and Blake *et al.* (1982). Terranes are Smith River (SMR), Rattlesnake
Creek and Marble Mountains (RC & MM), Hayfork (HF), Salmon River (SR),
North Fork (NF), Fort Jones (FJ), Central Metamorphic (CM), Trinity (T),
and Yreka (Y). The sequence is ordered in the direction of structurally lowest
and generally youngest on the west to highest and oldest on the east.

those recognized in the Canyon Mountain complex and in other ophiolites (Christensen and Salisbury, 1975; Coleman, 1977).

Structurally highest in the Triassic-Paleozoic belt are flyschlike sequences of metagraywacke interlayered with metamorphosed volcaniclastic rocks, ash flows, and intermediate to silicic volcanic rocks. To the west and farther to the south of the Condrey Mountain Schist window (Fig. 40-5), the structurally higher rocks with arc-trench affinities have been correlated with the Hayfork subterrane (Irwin, 1972) by Mortimer (1985). To the north and northwest of the window, near Copper (Fig. 40-6), flyschlike, predominantly volcaniclastic rocks were named Applegate Group by F. G. Wells *et al.* (1940, 1949, 1956). The Applegate and Hayfork terranes are in thrust-fault contact with the Marble Mountains terrane. Structural relations among lithologies of the Marble Mountains terrane, the Condrey Mountain Schist, and Hayfork-Applegate terranes are summarized schematically in composite structure sections from five study areas in the north-central Klamath Mountains (Fig. 40-7).

AGE RELATIONS, PALEOZOIC-TRIASSIC OCEANIC TERRANE

Ages in the Blue Mountains

Tectonite gabbro resting on tectonite peridotite in the Canyon Mountain Complex is cut by hornblende-rich gabbro pegmatite. The gabbro yields an age of 262 ± 14 m.y. by Ar/Ar analysis on hornblende (Avé Lallèmant *et al.*, 1980). Vallier *et al.* (1977) also report K/Ar ages that are apparently the result of cooling over the interval 240–250 m.y. in pegmatitic dikes which cut the gabbro tectonite. Walker (1983) reports U/Pb zircon ages of three intrusive phases of the Canyon Mountain complex that range from 279 to 268 m.y. Because the cross-cutting rocks are not affected by the high temperature deformation, 262 m.y. may be taken as the minimum age for the high-temperature fabrics. Quartz diorites or plagiogranites which followed, or accompanied, later metamorphism of earliest basaltic dikes in gabbro or peridotite, have an apparent K/Ar cooling age of 210 m.y. Plagiogranite in the Sparta ophiolite has a similar cooling age of 213 m.y. by K/Ar analysis (Vallier *et al.*, 1977); Walker (1983) obtains crystallization ages by U/Pb on zircon from two separate intrusive phases that are 262 and 223 m.y.

Amphibolite gneiss occurs in the upper part of the oceanic melange terrane beneath the Canyon Mountain complex. Avé Lallèmant *et al.* (1980) report an Ar/Ar hornblende age of 255 m.y. for an amphibolite gneiss knocker in the melange at the base of the Canyon Mountain complex near Canyon City. Quartz diorite occurring in the melange of the Greenhorn-Granite area has an U/Pb zircon age of 243 m.y. (Brooks *et al.*, 1983). These ages may serve as a maximum value for melange formation inasmuch as cooling(?) ages by K/Ar analysis obtained for other blocks, including amphibolites, amphibolitized basaltic dikes that intrude peridotite, and quartz diorite cutting amphibolitized gabbro range from 186 to 234 m.y. (Vallier *et al.*, 1977). The ages suggest that perhaps melange formation continued for an extended period of time following emplacement of peridotite into and above ocean crust. Melange formation is also constrained by the age of the Late Triassic (Carnian, ~225–230 m.y.) Vester Formation and the Fields Creek Formation (Norian, ~208–225 m.y.). Both formations are depositional on the melange and contain fragmental debris from it.

The character and distribution of metamorphosed rocks in the Greenhorn melange (Hunt, 1985) and the Canyon Mountain complex, and the age relations of fragmental material as just discussed serve as the basis for a reconstructed premelange composite

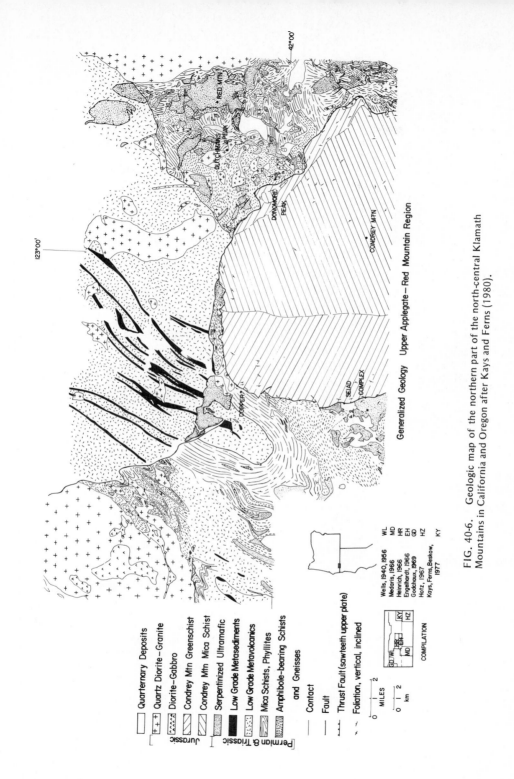

Quaternary Deposits] Quaternary

Quartz Diorite – Granite
Diorite – Gabbro
Condrey Mtn Greenschist
Condrey Mtn Mica Schist] Jurassic

Serpentinized Ultramafic
Low Grade Metasediments
Low Grade Metavolcanics
Mica Schists, Phyllites
Amphibole-bearing Schists and Gneisses] Permian & Triassic

— Contact
— Fault
⊥ Thrust Fault (sawteeth upper plate)
⟋ Foliation, vertical, inclined

0 1 2 MILES
0 1 2 km

Wells, 1940, 1956 .... WL
Medaris, 1966 .... MD
Heinrich, 1966 .... HR
Engelhardt, 1966 .... EH
Godchaux, 1969 .... GD
Holtz, 1967 .... HZ
Kays, Ferns, Beskow, 1977 .. KY

COMPILATION

Generalized Geology   Upper Applegate – Red Mountain Region

RED MTN
DUTCHMANS PEAK
DONOMORE PEAK
CONDREY MTN
COPPER
SEIAD COMPLEX

123°00'
42°00'

FIG. 40-6.   Geologic map of the northern part of the north-central Klamath Mountains in California and Oregon after Kays and Ferns (1980).

1107

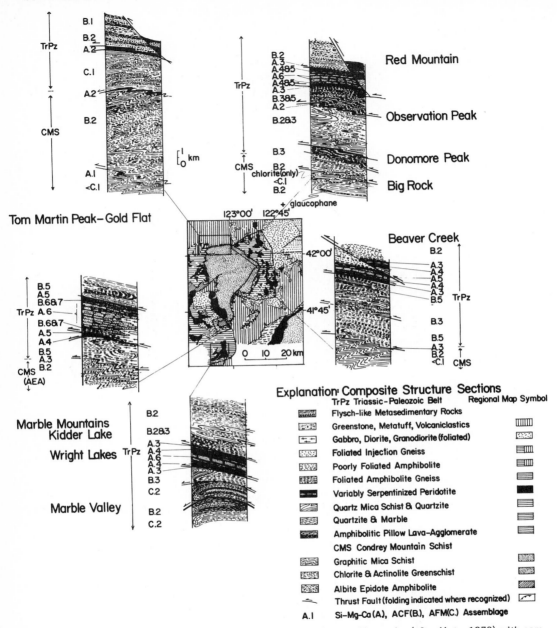

Upper Applegate

Tom Martin Peak – Gold Flat

**Marble Mountains
Kidder Lake**

Wright Lakes

Marble Valley

Red Mountain

Observation Peak

Donomore Peak

Big Rock

Beaver Creek

Explanation: Composite Structure Sections

TrPz Triassic-Paleozoic Belt   Regional Map Symbol

- Flysch-like Metasedimentary Rocks
- Greenstone, Metatuff, Volcaniclastics
- Gabbro, Diorite, Granodiorite (foliated)
- Foliated Injection Gneiss
- Poorly Foliated Amphibolite
- Foliated Amphibolite Gneiss
- Variably Serpentinized Peridotite
- Quartz Mica Schist & Quartzite
- Quartzite & Marble
- Amphibolitic Pillow Lava-Agglomerate

CMS Condrey Mountain Schist

- Graphitic Mica Schist
- Chlorite & Actinolite Greenschist
- Albite Epidote Amphibolite
- Thrust Fault (folding indicated where recognized)

A.I   Si-Mg-Ca (A.), ACF(B.), AFM(C.) Assemblage

FIG. 40-7.   Regional metamorphic map in the north-central Klamath Mountains (after Hotz, 1979) with composite sections compiled from geological relations in areas of more detailed mapping. Symbols in the composite structure sections refer to metamorphic grade in terms of (A) ultramafic, (B) mafic, and (C) pelitic mineral assemblages.

section of ocean crust and underlying upper mantle in the southern Baker terrane (Fig. 40-8). The reconstruction places the "reordered" Late Permian ocean floor above the tectonite peridotite-gabbro, cumulate gabbro, and plagiogranite-diabase of coherent ophiolite sections (Moores, 1982), or inverts the recognized structural sequence in northeastern Oregon.

Oldow *et al.* (1984) and Avé Lallèmant *et al.* (1980) correlate the north- to northeast-trending folds and penetrative axial planar cleavages in the Elkhorn Ridge Argillite and oceanic terrane with later compressional events (Late Triassic Carnian or Norian) associated with subduction that began in pre-Carnian. Subduction-related blueschists described by Hotz *et al.* (1977) have ages of 213–223 m.y. by K/Ar and $^{40}Ar/^{39}Ar$ analysis on micas. The blueschists occur interlayered with siliceous schists, thin-bedded quartzites, and marbles near Mitchell, and in isolated masses associated with fault-bounded Permian(?) rocks of various lithologies, as described previously in various locations in the serpentine-matrix-dominated melange (Mullen, 1978; Perkins, 1976). The latest regional deformation and metamorphism is associated with east-west-trending and north-northeast-trending structures and fabrics in the Permian-Triassic oceanic rocks in northeastern Oregon, forming apparently during Late Jurassic accretionary events (Avé Lallèmant *et al.*, 1980; Oldow *et al.*, 1984). The time of formation of latest penetrative fabrics in oceanic rocks is bracketed by unmetamorphosed and undeformed plutonic rocks of about 158 m.y. by K/Ar on hornblende in aplite, and 133 m.y. by Rb/Sr whole-rock isochron for the Bald Mountain Batholith that intrudes the Elkhorn Ridge Argillite and the melange (Armstrong *et al.*, 1977).

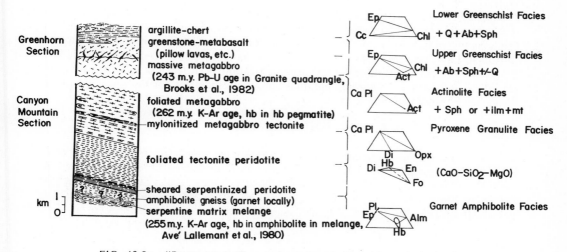

FIG. 40-8. "Reconstructed" premelange composite section of ocean crust and underlying upper mantle based on the character and distribution of the tectonic blocks in the Greenhorn melange and on distribution of tectonite gabbro-peridotite and tectonite amphibolite in the Canyon Mountain complex.

Ages in the Northern Klamath Mountains

Field data indicate that metamorphism of western Jurassic belt Galice Formation and Paleozoic-Triassic belt rocks occurred in the northern Klamath Mountains prior to final emplacement of the White Rock pluton and related bodies. Field data also support thrusting of Paleozoic-Triassic belt rocks over western Jurassic belt rocks, and the internal imbrication of the units prior to their final metamorphism. Petrographic studies indicate that penetrative metamorphic fabrics formed in Triassic-Paleozoic belt rocks earlier than porphyroblasts; assemblages of penetrative fabrics and porphyroblasts form the basis for the mapped prograde sequence (Fig. 40-3). Folding and formation of fabrics in western Jurassic belt rocks probably coincided with compression during convergence with Triassic-Paleozoic belt rocks.

K/Ar analyses of hornblendes in amphibolite gneiss intruded by quartz diorite near Tiller, and in amphibolitized metagabbro on Red Mountain south of Devils Flat yield ages of 130 ± 3 and 150 ± 11 m.y., respectively (Kays *et al.*, 1977). Hotz (1971) reports a K/Ar age of 141 m.y. on biotite from granodiorite of the White Rock pluton just south of the mapped area. The White Rock pluton, which is not metamorphosed or deformed, intrudes both Triassic-Paleozoic and western Jurassic belt rocks in the mapped area (Fig. 40-3). Allowing for the fact that the K/Ar system in most of the rocks in the area has been affected by plutonism, 150 m.y. is a minimum age for formation of penetrative fabrics and porphyroblasts in the May Creek Schist belt.

These data are essentially consistent with the findings of Saleeby *et al.* (1982) on juxtaposed western Jurassic belt Galice Formation, Josephine Peridotite, and Triassic-Paleozoic belt rocks, which all record regional Nevadan metamorphism-deformation in the Preston Peak area, California. In this area, near the border with Oregon, serpentine-matrix melange of the basal Triassic-Paleozoic belt contains widespread amphibolite tectonite with static textural overprint of retrograde mineral assemblages. K/Ar amphibole analysis yields an age of 165 m.y., which is clearly the minimum age of tectonite formation. Dacite and quartz keratophyre dikes in Galice flysch, and sheeted dikes in the Josephine Peridotite yield identical U/Pb zircon age of 151 m.y. Saleeby *et al.* (1982) also report closely comparable U/Pb zircon ages of 157 and 159 m.y. for quartz diorite and plagiogranite, respectively, in the Preston Peak complex, which lies structurally above the thrust fault and cuts rocks interpreted as oceanic basement to the Triassic-Paleozoic belt. These authors also indicate that Nevadan imbrication occurred just following dike emplacement into western Jurassic ophiolites and was responsible for the slaty cleavage in the Galice Formation at about 150 m.y.

Ages in the North-Central Klamath Mountains

Two separate regional metamorphic and deformational events are recorded in Triassic-Paleozoic belt rocks in the north-central Klamath Mountains. The earliest syntectonic event is associated with high-pressure lawsonite-glaucophane schist facies assemblages in the Stuart Fork Formation (Hotz, 1977; Hotz *et al.*, 1977). Metamorphism was apparently the result of the suturing of Triassic-Paleozoic belt rocks beneath the Central Metamorphic belt. The K/Ar and Ar/Ar ages of 214–222 m.y. on mica from the glaucophane schists are identical to those of the blueschists in the Blue Mountains, and suggest that there was widespread regional high P/T metamorphism at this time in the two areas. The two events may be the same in the Blue Mountains and Klamath Mountains. However, the evidence for rocks equivalent to the Devonian(?) Central Metamorphic belt has not been found or reported in northeastern Oregon or in the northern Sierra. Hamilton

(1969) and Ernst (1984) suggest that the belt, because of its discontinuous nature, represents a microcontinental fragment of uncertain affinity.

High-pressure lawsonite-glaucophane-bearing assemblages of Late Triassic age are apparently restricted to the structurally highest part of the Triassic-Paleozoic belt. The assemblages and fabrics correlative with those of the Stuart Fork blueschists have not been recognized or recorded in the Marble Mountains terrane, but have been mapped in Klippe that rest on it. Blueschist assemblages have been recognized in the upper part(?) of the North Fork antiform (Ando *et al.*, 1983). The North Fork terrane (Fig. 40-5) is part of Irwin's (1972) original subdivision of the Triassic-Paleozoic belt in the southwestern Klamath Mountains, and Ando *et al.* (1983) report U/Pb zircon ages of 265-310 m.y. for gabbro-plagiogranite in the ophiolite at the base of the North Fork terrane. However, the blueschists are interpreted to be olistoliths which overlie Upper Triassic North Fork cherts and argillites. Wright (1982) indicates that the metasedimentary rocks of the North Fork terrane are coherent beneath their thrusted contact with the overlying Central Metamorphic belt, and Ando *et al.* suggest that the blueschist olistoliths may be similar to those in the overlying Stuart Fork Formation.

If the Marble Mountains terrane (Fig. 40-5) and the Rattlesnake Creek terrane are equivalent, as Rawson and Petersen (1982) and Mortimer (1985) suggest, melange formation in both could have started as early as the subduction event which formed the Condrey Mountain transitional blueschist-greenschist-facies parageneses (Helper, 1986). A single concordant U/Pb zircon age of a deformed tonalite that intrudes Condrey Mountain greenschist near the southern margin of the window indicates a minimum depositional age of 170 ± 1 m.y. (Saleeby, *et al.*, 1984). This age is consistent with U/Pb zircon ages of 165-170 m.y. for compressive deformation-metamorphism of the entire Rattlesnake Creek terrane during juxtaposition against an active Middle Jurassic arc in the southwestern Klamath Mountains (Wright and Wyld, 1985). Gabbro intrusive into the melanged Rattlesnake Creek terrane, and the Late Triassic-Early Jurassic metavolcanic-metasedimentary arc-related cover sequence, has an U/Pb zircon age range of 192-212 m.y. (Wright and Wyld, 1985). Thus, an age of ~170 m.y. may be close to the minimum for formation of penetrative axial planar fabrics during syntectonic metamorphism of the Marble Mountains terrane. The assemblages and fabrics are associated with the earliest recognizable tight (and commonly recumbent) folds in the Triassic-Paleozoic schists and gneisses.

Fahan and Wright (1983) also report K/Ar metamorphic hornblende ages of 166-177 m.y. for the volcaniclastic rocks in the Triassic-Paleozoic Hayfork terrane in the southwestern Klamath Mountains. The age range of igneous and metamorphic rocks in these Triassic-Paleozoic units is consistent with an age of 172 m.y. for regional metamorphism of plagiogranite dikes in the Marble Mountains terrane near Seiad Valley, California (Hill, 1985). We have no isotopic data and few age data of any kind for the arc-related rocks of the Applegate terrane. However, early work of Diller (1906), and Diller and Kay (1924) yielded Late Triassic (Norian) fossil ages from the flysch-type sequences in the type location of the Upper Applegate River area near Copper (Fig. 40-6). The metavolcanic and metamorphosed volcaniclastic rocks are structurally coherent except near the faulted contact with the Marble Mountains terrane. Overall, the rocks have characteristics similar to the western Hayfork terrane (Wright, 1982).

A later period of folding and less pervasive recrystallization of the penetrative fabrics may be associated with intrusion of gabbroic to granodioritic plutons into the Marble Mountains, Hayfork, and Applegate terranes (Donato, 1985; Kays and Ferns, 1980). For example, ages of the plutonic rocks are 162 m.y. by U/Pb on zircon in the Slinkard pluton and Wooley Creek batholith (Allen *et al.*, 1982; Barnes, 1983), 147-170 m.y. and 150–164 m.y. by K/Ar on biotite and hornblende in the Ashland and Vesa Bluffs plutons, respectively (Hotz, 1971). Gabbro with an age of 164 m.y. intrudes metamorphosed rocks of the Fort Jones, Hayfork, Salmon River, and North Fork terranes (Hotz, 1967; Mortimer, 1985). Kays *et al.* (1977) report a K/Ar hornblende age of 144 m.y. for tight-folded amphibolite gneiss beneath the thrusted peridotite on Red Mountain northeast of the Condrey Mountain window (Fig. 40-6). The amphibolite, its penetrative fabric and early folds, the thrust fault, and overlying peridotite have all been folded by later mappable (F_3) folds with wavelengths measured in kilometers.

STRUCTURE-FABRIC RELATIONS IN THE PERIDOTITES WITH HIGH-TEMPERATURE ASSEMBLAGES IN THE BLUE AND KLAMATH MOUNTAINS

Fabrics in Early High-Temperature Assemblages in the Blue Mountains

The term "early" in reference to tectonite fabrics is constrained by radiometric ages and/or bracketing by structural-stratigraphic relations as discussed previously for the Canyon Mountain complex. The term "high temperature" refers to the aligned minerals (olivine, orthopyroxene, clinopyroxene, chrome-spinel, and plagioclase) that comprise the fabrics in the peridotites and gabbro. The fabrics and the rocks that contain them have been well documented by the work of Avé Lallèmant (1976), Thayer and Himmelberg (1968), and Thayer (1963a, b) in the Canyon Mountain complex. Here, the peridotitic and gabbroic rocks with tectonite fabrics occupy most of the lower two-thirds of a complex that is 19 km long, east to west, by 8 km wide north-south. A minimum age for the fabrics is given by the crosscutting 262-m.y.-old hornblende pegmatite dike. Mineral orientations of the kind found in the Canyon Mountain complex (Table 40-1) are commonly interpreted to form as a result of one of two mechanisms. Some evidence supports translation gliding along energetically favorable planes and directions in the minerals (Nicolas and Poirier, 1976; Nicolas *et al.*, 1980). Other evidence supports syntectonic recrystallization (Avé Lallèmant and Carter, 1970), possibly in the form of the pressure solution effect on minerals under stress (Kamb, 1961).

In the predominantly harzburgitic peridotite of the Canyon Mountain complex, the tectonite fabrics at the outcrop and in hand specimen consist of a foliation (S_{TP}) and a lineation (L_{TP}) defined by xenoblastic flattened and elongate spinel, respectively (Avé Lallèmant, 1976). Pyroxenes may have a similar orientation and can be used together with spinel or in its absence. Fabrics in the tectonite gabbro, S_{TG} and L_{TG}, are similar to those in the peridotite, but in the gabbro it is clear that foliations are axial planar and parallel to fold axes of tight to open folds not apparent in the peridotite. The terms "TP" and "TG" refer only to tectonite peridotite and tectonite gabbro, respectively. The fabric elements are clearly restricted to the complex and not the surrounding rocks. According to the mapping of Avé Lallèmant, the foliations in the peridotite and gabbro are of the same generation, and cross lithologic contacts at high angles.

Early High-Temperature Assemblages and Fabrics in Peridotites in the Klamath Mountains

The widespread occurrence of high-temperature fabrics and assemblages has been documented in certain of the peridotite bodies in the Klamath Mountains, including parts of the Trinity ultramafic sheet (Canat and Boudier, 1985; Goullaud, 1977), the Josephine

TABLE 40-1. Tectonite Peridotite Fabrics in the Klamath and Blue Mountains[a]

Rock Type	Foliation	Lineation	Mineral Orientation	Interpretation and Remarks
Canyon Mountain Complex				
Harzburgite, dunite, pyroxenite	S_{TP1}	L_{TP1}	Shape Opx, Cpx, Sp good; Ol microscopic Z Ol, Opx, Cpx form max. parallel to macroscopic lineation (chrome spinel); x Ol girdle normal to S_{TP1}	Flattened, chromite defines but also Opx
Gabbro	S_{TG1}	L_{TG1}	All about same as in peridotites	S_{TG1} is axial planar
Tom Martin Peak Ultramafic-Mafic Body				
Harzburgite dunite	S_{TP1-2}	L_{TP1-2}	Z Ol max. parallel to L_{TP1-2}; Z Opx girdle max. subparallel L_{TP1-2}; Z pargasite girdle subparallel to L_{TP1-2}; X, Y Ol about normal to S_{TP1-2}	S_{TP1-2}, L_{TP1-2} defined by flat, elongate spinel and pargasite +/– Opx
Serpentined harzburgite and hornblendite	S_{TP2}	L_{TP2}	Z Ol girdle max. about parallel to L_{TP1-2} but may be diffuse, but X Ol rotates parallel to L_{TP2} by slip	Border zone; S_{TP2}, L_{TP2} defined by flat and elongate pargasite parallel to mylonitic foliation
Wright Lakes				
Serpentinized harzburgite	S_{TP1-2}	L_{TP1-2}	Z Ol girdle max. subparallel to L_{TP1-2}; X, Y Ol at 90° but in S_{TP1-2}	Occurs only as tectonite blocks in serpentinized peridotite
Talc-olivine-magnesite	S_{1-2}	L_{1-2}	Z Ol point max. in S_{1-2} plane	Regional metamorphism

[a]Ta, talc; Antig, antigorite; Fo, forsterite; Ol, olivine; Parg, pargasite; Sp, spinel; Pl, plagioclase; Tr, tremolite; Anth, anthophyllite; [Gar = grossular (Gr), pyrope (Py), almandine (Alm)], garnet; Cpx, clinopyroxene; Opx, orthopyroxene; Chl, chlorite; Fa, fayalite; An, anorthite; V, vapor; Q, quartz; Di, diopside; En, enstatite.

peridotite (Loney and Himmelberg, 1976), and the Seiad Valley complex (Lundquist, 1983; Canat, 1985; Canat and Boudier, 1985). Regional studies in the same areas (Ando *et al.*, 1983; G. A. Davis *et al.*, 1965; Medaris, 1966; Ramp, 1975; Saleeby *et al.*, 1982) indicate that the high-temperature peridotite fabrics are of an earlier generation than the penetrative fabrics of the adjacent regionally metamorphosed rocks.

A large number of serpentinized peridotite bodies of varying size occur as dismembered fragments of ophiolite complexes distributed throughout the Marble Mountains terrane in approximately arcuate arrangement around and adjacent to the Condrey Mountain Schist window (Fig. 40-7). Most of the ultramafic bodies have tectonic contacts with the adjacent rocks, and their basal contacts are mapped thrust faults. It seems possible, because of nearly conformable contact relations along the thrusts, and the repetitive nature of the associated melanged lithologies in contact with the sheetlike peridotites, that emplacement was nappelike. However, large recumbent folds have not been defined through mapping. Emplacement and fragmentation of the peridotite was followed or accompanied (in later stages of emplacement) by intense flattening or compressional deformation that affected all the rocks. This deformation was responsible for the development of penetrative fabrics in the associated lithologies of the Marble Mountains terrane, and has largely obliterated evidence for the early high-temperature ductile deformation in the smaller peridotite bodies. In some of the larger peridotites (e.g., the Seiad Valley complex and the Tom Martin Peak ultramafic body), the harzburgite which predominates in the interiors has preserved assemblages and fabrics that are in sharp contrast with the subsequently deformed and highly serpentinized matrix at the margins of the bodies.

In the Tom Martin Peak ultramafic body (Fig. 40-7), chrome spinel is ubiquitous throughout the central part of the complex (Barrows, 1969), and its macroscopic elongation and flattening define lineation L_{TP1-2}, and foliation S_{TP1-2}. Orthopyroxene and pargasitic hornblende may also have measured, macroscopic elongation and flattening approximately the same as spinel. The effect of regional deformation-metamorphism on the Tom Martin Peak ultramafic body, according to Hanks (1981), has been to modify the earliest high-temperature tectonite fabrics. Therefore, the terms for lineation (L_{TP1-2}) and foliation (S_{TP1-2}) are used to indicate that earliest tectonite peridotite fabrics (subscript TP1) have been modified by later events (subscript TP1-2). Petrofabric measurements of preferred alignment of minerals in thin section are summarized in Table 40-1 and can be compared with those in the Canyon Mountain complex.

The Tom Martin Peak ultramafic body has a 100–200-m-wide, disturbed margin at its top and base, and a better developed schistosity which amplifies primary(?) layering of alternating gabbro, pyroxenite, and peridotite in the interior of the border zone (Barrows, 1969). The conclusion reached by Hanks (1981) is that perhaps the same deformation affecting the border zone may also have affected the fabrics throughout the body. Hanks shows, however, that, in contrast to the high-temperature olivine fabrics of the central zone, there is a range of disturbed olivine petrofabrics in the schistose border zones. This seems more consistent with the studies of Lundquist (1983) who finds that strong high-temperature olivine tectonite fabrics ([010] normal to S_T, [100] parallel to L_T, [001] in S_T; or [001] normal to S_T, [100] parallel to L_T in S_T, [010] in S_T) in the interior simply become more diffuse toward the margins of the Seiad Valley complex where regional assemblages and fabrics take over.

In the smaller, partially serpentinized sheetlike bodies, such as that of Red Mountain northeast of the Condrey Mountain Schist window, the peridotite is tectonically interlayered, folded together with, and metamorphosed to about the same grade as the surrounding schists and gneisses of the Marble Mountains terrane (Fig. 40-6). However, here, as well as in other serpentinized peridotites, there are lensoidal to blocklike masses with metamorphic mineral assemblages that are of higher grade, and in some places in sharp contrast to the surrounding ultramafic assemblage and the adjacent schists and gneisses. Taken together the blocks form sizable domains to several percent of the body. Ferns (1979) has shown, for example, that the high- and low-temperature assemblages are mappable, and that the boundaries between these assemblages were folded by later deformation. Locally, too, high-grade metamorphic blocks have layering and lineation recognizable from the flattening and elongation, respectively, of chrome spinel and in some places orthopyroxene. However, it is not clear here whether the chrome spinel fabrics reflect only the high-temperature, low-strain-rate episode, or if they have been modified by later events.

Blocky zones of peridotite with high-temperature assemblages are also recognized in the Wright Lakes peridotite in the Marble Mountains Wilderness (Table 40-1) (Chambers, 1983). The olivine petrofabrics are similar to those in the central zone of the Tom Martin Peak ultramafic body. In the Upper Wright Lakes body and in the Red Mountain peridotite sheet, preferred olivine orientations in the high-temperature assemblages cannot be related to foliation and lineation in the surrounding peridotite. This is because the isolated blocks with high-temperature assemblages have transposed fabrics with relicts of an earlier episode of recrystallization foreign to the surrounding medium-grade talc-olivine-magnesite peridotite.

TECTONITE AMPHIBOLITES AT THE SOLE OF PERIDOTITE BODIES IN THE OCEANIC MELANGE TERRANES OF THE PALEOZOIC-TRIASSIC BELT

The mafic tectonite amphibolites (TA) are generally fine- to medium-grained and have penetrative gneissosity (L_{TA}) and/or foliation (S_{TA}). They are conspicuous in some occurrences because of the sharp contrast in fabrics and/or grade of their metamorphic mineral assemblages compared to those of the underlying rocks. The penetrative foliation, comprised of flattened hornblende and plagioclase, may transpose an even earlier compositional banding. The mineral assemblage is hornblende-plagioclase (An_{25-50}) − epidote ± clinopyroxene ± garnet ± sphene or rutile. Such rocks are found to be tectonically attached to basal peridotite zones of most major ophiolites (Coleman, 1977; Moores, 1982) and have been the focus of several recent studies in the Klamath Mountains (Canat, 1985; Canat and Boudier, 1985) and elsewhere (Boudier and Coleman, 1981; Boudier et al., 1982; Ghent and Stout, 1981). The previously mentioned tectonite amphibolite in the melange south of Canyon City in the Blue Mountains has an age of 255 m.y. This is younger than that of the adjacent high-temperature tectonite peridotite in the Canyon Mountain complex, but is older than the high P/T metamorphism associated with other melange blocks in the surrounding Baker terrane.

In the Triassic-Paleozoic belt in northern California and in Oregon, the amphibolites have tectonic contacts with surrounding metasedimentary and metavolcanic rocks and the overlying or associated peridotite. Most commonly amphibolites occur tectonically interlayered and folded together with ultramafic rocks at the base of or in the lower part of peridotite sheet (e.g., Wright Lakes, Tom Martin Peak, Beaver Creek, Red Mountain, and the Canyon Mountain complex). Tectonite amphibolites occur probably as part of an imbricated thrust sequence, tectonically interlayered with schists and amphibolites of sedimentary affinity adjacent to and structurally above the Condrey Mountain Schist at the base of the Marble Mountains terrane. Because the amphibolites occur as fragments tectonically interleaved with other metamorphosed rocks (e.g., with interlayered metasedimentary schists and amphibolites), documentation through mapping may be difficult. Although the details of their history and origin are not always clear, mineral assemblages and fabrics of the tectonite amphibolites are distinctive and among the earliest after the high-temperature tectonite peridotites in the Klamath and Blue Mountains.

OTHER RELICT FABRIC FEATURES AND THEIR IMPORTANCE

Relict Fabrics in the Greenhorn Melange

Large areas of serpentine-matrix melange or megabreccia are exposed in the Mount Ireland–Greenhorn–Vinegar Hill areas in the western part of the Baker terrane (Fig. 40-2) in northeastern Oregon (Brooks and Vallier, 1978; Ferns and Brooks, 1983). Blocks in the Greenhorn area range from less than a meter to more than a kilometer across and are mainly gabbro, greenstone including pillow basalt, and chert-argillite. Lithologies of lesser abundance include diorite, quartz diorite, amphibolite, schist, and limestone. The chert-argillite is identified with the Permian (?) Elkhorn Ridge Argillite (Brooks *et al.*, 1983; Ferns *et al.*, 1983). Although the pillow basalts are not identified with a specific unit, such rocks have been mapped as part of the Elkhorn Ridge Argillite northeast of the Greenhorn area and eastward from Elkhorn Ridge. Some metabasalts have high TiO_2 contents and may correlate with Permian rocks (Olive Creek basalts of Ferns *et al.*, 1983; Mullen, 1978) based on conodonts (Dickenson and Thayer, 1978, p. 150) in interbedded ribbon cherts. The mapping shows that quartz diorite dikes (plagiogranites?) intrude the melange blocks including some parts of the cover sequence, but not the melange matrix, suggesting that dike emplacement preceded melange activity. In one such occurrence in the neighboring Granite quadrangle, Brooks *et al.* (1983) report a U/Pb zircon age of 243 m.y. for granodiorite cutting a block of metavolcanic and associated metasedimentary rocks in the melange. A cover sequence of sedimentary rocks with locally preserved depositional contacts on the melange contains clasts of melange material including serpentinite (Ferns *et al.*, 1983). However, the cover sequence may also occur as fragmental blocks within the melange, suggesting that the sedimentary sequence was being deposited during melange formation. The basinal sequence may be Paleozoic to Late Triassic in age.

The metamorphosed gabbroic and volcanic rocks in the melange have random relict igneous fabrics but low-grade assemblages, which are mainly prehnite-pumpellyite, pumpellyite-actinolite, or greenschist facies. These low-grade rocks do not have penetrative fabrics but are affected by a fracture cleavage tentatively correlated with the later of the two regional dynamothermal events (Hunt, 1985), i.e., Jurassic Nevadan orogenesis at ~150 m.y. (Avé Lallèmant *et al.*, 1980; Oldow *et al.*, 1984). Within the melange are blocks of amphibolite and glaucophane schist that have penetrative fabrics as do the blocks of Elkhorn Ridge Argillite, which are also cut by the fracture cleavage. In other areas, such as the more "coherent" parts of the Elkhorn Ridge Argillite to the north along Elkhorn Ridge, these rocks have two sets of penetrative fabrics, one identified with the ~220-m.y. Triassic subduction zone event, and the other with the Jurassic accretionary event (Avé Lallèmant *et al.*, 1980; Oldow *et al.*, 1984). This suggests that in the Greenhorn melange, the well-developed set of penetrative fabrics in the Elkhorn Ridge Argillite and in the glaucophane schist may be equivalent in age to those of the earlier "blueschist" event.

Thus, the picture that obtains from isotopic and structural studies in the Blue Mountains is that rocks in the melange have assemblages and fabrics representative of several metamorphic events (Avé Lallèmant, *et al.*, 1980; Vallier *et al.*, 1977). The age of hydrothermal metamorphism may not be much younger than the age of the relict igneous fabrics and assemblages in these Triassic-Paleozoic belt oceanic rocks. We also suggest that the age of the gabbro-basalt precursors to the melange blocks may be similar to the age determined for tectonite fabrics in gabbro and peridotite in bodies such as the Canyon Mountain ophiolite. Thus, the relict igneous fabrics and hydrothermal assemblages of the blocks in the Greenhorn melange may represent the earlier preorogenic, evolving ocean crust above the tectonite lithosphere. Later orogenesis beginning as early as 230(?) m.y. developed penetrative fabrics in the Elkhorn Ridge Argillite and associated metavolcanic rocks, with some nearer the suture zone developing blueschist-related assemblages. This activity was followed or accompanied by development of the melange approaching the suture zone with Triassic-Paleozoic arc terranes (Fig. 40-2). Many of the melanged rocks escaped formation of penetrative fabrics. Later Jurassic metamorphism was apparently short-lived and formed only a fracture cleavage in the rocks of the Greenhorn melange.

Relict Fabrics in the North-Central Klamath Mountains

Rocks of the Marble Mountains oceanic terrane and the adjacent structurally higher arc-related sequence (Hayfork and Applegate terranes) have locally well-preserved relict fabrics and structures. The relict fabrics differ from those in Triassic-Paleozoic belt rocks in the Blue Mountains in that metamorphic fabrics in the Marble Mountains terrane generally transpose the earlier features. This seems to be generally true for Hayfork rocks in the Marble Mountains, but arc-related rocks in the Upper Applegate area near Copper (Fig. 40-6) are feebly recrystallized and have relict assemblages, fabrics, and structures. For example, Heinrich (1966) documents the resemblance of primary features to flysch-type bedding sequences (Bouma, 1962), and to trachytic, pilotaxitic, or less commonly subophitic textures of silicic or intermediate metavolcanic rocks (andesites or keratophyres and quartz keratophyres).

Preservation of relict small-scale fabric features may be related in part to structural position within the Marble Mountains sequence. For example, we find that rocks structurally high in the Marble Mountains terrane have best preserved relict structures (e.g., in the Wright Lakes and Sky High Lakes areas of the Marble Mountains Wilderness above about 6200 ft elevation). The features include (mullioned) quartz clasts in the conglomerate, (stretched) pillows and agglomeratic clasts, and primary layering between marble and amphibolitized pillow lavas. Such features persist even though dynamothermal recrystallization and axial planar fabrics are also recognizable in the same rocks.

FABRICS AND STRUCTURES RELATED TO REGIONAL DYNAMOTHERMAL METAMORPHISM

Regional Fabrics in the Blue Mountains

The general character of the fabrics and their development during regional dynamothermal metamorphism in the Blue Mountains has been summarized by Oldow et al. (1984) and Avé Lallèmant et al., (1980). The early orogenic episode produced blueschists in meta-volcanic rocks with ages of about 225 m.y., and also generated penetrative cleavage-schistosity comprised of low-grade assemblages (muscovite-chlorite-quartz-bearing) in metaclastic rocks of the Elkhorn Ridge Argillite and the Burnt River Schist (Ashley, 1967). Avé Lallèmant et al. (1980) believe that this event may have been prolonged, starting in pre-Carnian (Late Triassic) time with the formation of subduction-related melange and continuing into the Early Jurassic.

The second major orogenic phase affected lower Oxfordian and older rocks (Brooks and Vallier, 1978), but deformation ceased before all terranes were intruded by granitic stocks and batholiths 160-95 m.y. ago (Armstrong et al., 1977). The later orogenesis and metamorphism was apparently of low grade and probably low pressure, creating penetrative fabrics in Elkhorn Ridge Argillite and Burnt River Schist. Regional metamorphism of this phase increases in grade eastward toward Riggins, Idaho (Hamilton, 1963). The impression gained through regional studies is that the later, or Nevadan phase had less of an effect on the melanged rocks (in terms of penetrative fabrics) than it did on more coherent parts of the Baker terrane (e.g., Elkhorn Ridge Argillite, Burnt River Schist). Possibly, too, because mineral assemblages in low-grade metaclastic rocks are about the same for low-pressure and high-pressure facies, it has been difficult to separate the two events (i.e., 225 m.y. versus 150 m.y.) on the basis of facies changes in northeastern Oregon. However, there are few published accounts of detailed studies of dynamothermal metamorphism in the Blue Mountains, and to date there has been little effort to illustrate distribution of metamorphic assemblages on the same scale as our understanding of tectonic or structural elements (e.g., Avé Lallèmant et al., 1980).

Regional Fabrics and Small-Scale Structural Features in the North-Central Klamath Mountains

In the Marble Mountains terrane interlayered amphibolites, metasedimentary schists, marbles, quartzites, some metagabbroic and metadioritic rocks, and tectonically interlayered serpentinized ultramafic rocks have penetrative foliation (S_{1-2}) and/or gneissose fabrics (L_{1-2}). We use the terminology S_{1-2}, etc., because there is evidence near the base of the Marble Mountains terrane for development of F_2 folds whose axes and planes are subparallel to S_1. The two are not easily distinguished. We exclude fabrics of tectonite peridotites (S_{TP}, L_{TP}) from this category because such fabrics are transposed or modified by and therefore earlier than penetrative (S_{1-2}, L_{1-2}) fabrics in Triassic-Paleozoic schists and gneisses. Fabrics such as those of modified tectonite peridotite (S_{TP1-2}, L_{TP1-2}) and tectonite amphibolite (S_{TA}, L_{TA}) have generally not been distinguished during detailed mapping, except by Hotz (1967). Such fabrics may represent the earliest stage of regional high-T/P metamorphism in the Marble Mountains, and in the May Creek Schist belt. The fabrics may be associated with early thrust faulting in the north-central Klamath Mountains. Thus, foliations (S_{1-2}) and gneissose (L_{1-2}) fabrics recognized in rocks of the Marble Mountains terrane are probably of several generations formed over a period of time. Schistosity or foliation composed of phyllosilicates and/or amphibole and in some places flattened quartz and feldspar is recognized in rocks with relict igneous or sedimentary structures. In such cases foliation is parallel to axial planes of tight, commonly isoclinal, and gently reclined to recumbent folds. Folds of this generation are generally recognizable at the scale of the outcrop but rarely on a regional scale.

In early folds, whether macroscopic or megascopic, compositional or primary layering is transposed in the hinges but has subparallel orientation to axial planar schistosity cleavage in the limbs. The folds are termed F_1 because primary layering and relict fabric features are folded and because there is no evidence of earlier folding. In many folds, wavelengths are measured in centimeters; amplitudes may be of the order of several times wavelengths, but generally the amplitude/wavelength ratio is low. Mineral elongation, especially amphibole, is generally parallel to F_1 fold axes where they can be measured. Quartz mullions and rods have the same orientation as F_{1-2} axes. Low- to medium-grade assemblages of the Marble Mountains terrane are largely syntectonic and have formed early and late during the M_1 crystallization stage which overlaps D_{1-2} deformation associated with F_{1-2} folding. Most porphyroblasts of S_{1-2} stage have elongation in the plane of schistosity. However, garnet, biotite, and hornblende may also have a late random growth stage that incorporates S_{1-2}.

Nearer the base of the Marble Mountains terrane and in the higher-grade rocks (e.g., at Gold Flat and nearby, on the west flank of Tom Martin Peak, and on Donomore Peak), tight recumbent secondary folds of S_1 foliation-cleavage, recognizable at the scale of an outcrop, have large amplitude/wavelength ratios, on the order of 10:20. In places, a new cleavage-foliation (S_2) is axial planar to the secondary folds and transposes the earlier (S_1). Locally within the secondary foliation, there are small rootless fold hinges a few centimeters wavelength at most, either greatly flattened or completely transposed by the new foliation. In the garnetiferous amphibolites at Gold Flat and on Tom Martin Peak, occurrence of boudin-like bodies (up to 35 cm long by 25 cm across in outcrop) preserve internally tight folds that are locally intrafolial, with compositional banding at high angle to the secondary foliation. The amplitude/wavelength ratio in the interior of the boudin is small, whereas externally it is large. The folds in the interior are clearly of earlier generation and have characteristics more like those structurally higher in the Marble Mountains terrane.

The structural features discussed above seem to support at least local prevalence of inhomogeneous strain near the base of the Marble Mountains terrane. This suggests that folds of earliest generation may have been flattened and deformed further by ductile flow in the higher-grade rocks following initial folding. In the higher-grade rocks, the earlier axial plane foliation (S_1) is partly transposed in the hinges of F_2 folds. Syntectonic recrystallization and generation of metamorphic foliation in the Marble Mountains terrane followed late Triassic convergence at ~220 m.y., and was complete prior to ~150–160 m.y. Voluminous calc-alkaline plutonism in the Marble Mountains terrane began at about 165–170 m.y. (Irwin, 1985) and overlapped regional high-T/P metamorphism. During the early period of metamorphism, ductile flow at the base of the Marble Mountains terrane was locally pronounced and apparently lasted longer than folding of comparable generation at structurally higher levels in the Marble Mountains terrane.

Broad, open, and mappable folds termed F_3 fold S_{1-2}, L_{1-2}, early thrust faults, and all other associated features including F_{1-2} folds where they can be observed. Examples of F_3 folds occur on the west flank of Tom Martin Peak; in the Red Mountain peridotite and the thrust plane that separates it from the underlying Marble Mountains schists and gneisses; in early thrust planes in the Beaver Creek melange (see the cross section in Fig. 40-9); and in the form of the broad synform in the Marble Mountains terrane west of Copper (Fig. 40-6).

Regional Fabrics in the Northern Klamath Mountains

Triassic-Paleozoic lithologies in the May Creek Schist belt are similar to those of the oceanic Marble Mountains terrane, and foliation-cleavage is defined in the same way. This does not imply, however, that the fabrics and structures here and in the north-central Klamath Mountains are more than broadly comparable. Although contacts between units are typically tectonic, units and their S_{1-2} cleavages-foliations are traceable for long distances and both appear to be continuous (Fig. 40-3). Penetrative foliation (S_{1-2}) in Triassic-Paleozoic metasedimentary rocks consists of parallel muscovite-biotite ± chlorite and subparallel layers of quartz and feldspar. In mafic rocks the foliation consists of flattened amphibole and chlorite and generally subparallel alternating layers that are plagioclase-richer and amphibole ± chlorite-richer. Evidence for early folds (F_{1-2}) is found in the form of intrafolial, or tight isoclinal folds of compositional layering, with transposed hinges in hand specimen and at the outcrop. F_{1-2} folds in outcrops of quartz-biotite-garnet (± staurolite) schist along the South Umpqua River about 5 km west of Tiller are sharply folded by F_3 folds.

Fabric studies of the metapelites indicates that porphyroblasts of staurolite and

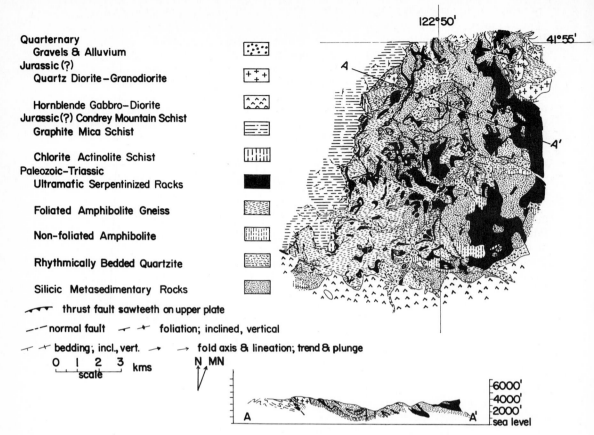

Legend (left column):

Quarternary
    Gravels & Alluvium
Jurassic (?)
    Quartz Diorite–Granodiorite

    Hornblende Gabbro–Diorite
Jurassic(?) Condrey Mountain Schist
    Graphite Mica Schist

    Chlorite Actinolite Schist
Paleozoic-Triassic
    Ultramafic Serpentinized Rocks

    Foliated Amphibolite Gneiss

    Non-foliated Amphibolite

    Rhythmically Bedded Quartzite

    Silicic Metasedimentary Rocks

thrust fault sawteeth on upper plate

normal fault   foliation; inclined, vertical

bedding; incl., vert.   fold axis & lineation; trend & plunge

0 1 2 3 kms
scale

N MN

6000'
4000'
2000'
sea level

FIG. 40-9.   Geologic map and cross section of the Beaver Creek melange (after Hotz, 1967).

andalusite incorporate penetrative S_{1-2} cleavage-schistosity. However, the incorporated schistosity shows curvature near margins of andalusite and staurolite porphyroblasts and connects to the external schistosity, which has been flattened and deformed about the partly rotated crystals. Amphibole porphyroblasts show similar relationships to surrounding flattened elongate amphibole-chlorite-plagioclase-epidote bands. The porphyroblasts are characteristically elongated parallel to the external schistosity. The interpretation is that growth of the porphyroblasts began late during development of S_{1-2}, and continued after this deformation. Later or continued deformation resulted in flattening of schistosity around the porphyroblasts. Fibrolitic sillimanite clusters may have preferred elongation in some sections but random orientation in others, with fibrolite growing on muscovite.

Schistosity or foliation in Triassic-Paleozoic rocks is broadly to sharply folded and there is local transposition of foliation in hinges of these later folds. The folds are termed F_3 and appear to be associated with intrusion of granodiorite into the May Creek Schist belt. S_{1-2} foliation-cleavage is folded or "molded" around the margins of smaller protrusions of granodiorite, but the plutonic rocks are not metamorphosed. The folds and deformation of a third generation are macroscopic or mappable features.

Penetrative cleavage in Galice Formation mudstones and graywackes is axial planar in well-defined folds observed in roadcuts, mostly to the west of the map area (Fig. 40-3).

In these cases the cleavages transpose bedding in the hinge areas of the folds, and the axial planar mineral assemblages are low-grade or greenschist facies. Galice cleavage-foliation is not folded, at least not by the same system that affected S_{1-2} foliation in the structurally overlying Triassic-Paleozoic rocks. The suggestion is that Galice folding and cleavage formed during Nevadan imbrication, and was probably associated with folding designated F_3 in the Triassic-Paleozoic belt.

METAMORPHISM IN THE KLAMATH AND BLUE MOUNTAINS

Pattern of Polymetamorphism

Regional studies in the Klamath and Blue Mountains suggest that the metamorphism in various tectonostratigraphic terranes helps characterize the amalgamation or accretionary events that formed the Triassic-Paleozoic belt. Furthermore, there is a recognizable pattern in the relative chronology of metamorphisms and deformations. In the oceanic or ophiolitic terranes of the Klamath Mountains, e.g., the Marble Mountains terrane, and in the May Creek Schist belt, the chronology of metamorphism is (1) high-temperature spinel lherzolite facies in peridotite ($\pm$ gabbro), (2) tremolite peridotite and upper amphibolite facies (amalgamation event) in peridotite and in the underlying metabasic and interlayered metasedimentary rocks, and (3) high-T/P regional metamorphism with well developed penetrative fabrics in rocks that are Bathonian or older, but probably younger than Carnian. Equilibria representative of the different phases are summarized in Fig. 40-10. Polymetamorphic assemblages in the five separate structure sections in the north-central Klamath Mountains are shown in Fig. 40-11. These assemblages can be compared with metamorphic assemblages in Fig. 40-12 shown according to specific occurrence in "Nevadan" metamorphic terranes.

In the ophiolitic Baker terrane of the Blue Mountains and in the Fort Jones, North Fork, and Salmon River terranes of the Klamath Mountains, high-temperature tectonite peridotites are earliest (1). However, age data suggest that widespread, shallow hydrothermal metamorphic mineral assemblages with directionless fabrics formed at nearly the same time. There is also evidence for mylonitization of the tectonite peridotites locally, and in their lower parts during amalgamation of the amphibolites to the base of the peridotite sheets (2) following the high-temperature and hydrothermal events. Regional metamorphism (3) was low grade and of high-P/T character (lawsonite-glaucophane schist in earliest Norian or older rocks); penetrative fabrics are of the same generation. The later high-T/P regional Bathonian (Middle Jurassic) or "Nevadan" metamorphism is not recognized as a well-developed regional event in these terranes. All metamorphic facies and types, except for hydrothermal, have penetrative fabrics, but (1) and (2) are found mostly on a local scale, depending on areal distribution of ultramafic rocks.

Preregional Allofacial Metamorphism: Tectonites Stage

Examination of allofacial tectonites, which include peridotites, pyroxenites, gabbros, and locally, some other kinds of basic rocks, indicates that a number of assemblages may prove to be useful in setting broad temperature and pressure limits of recrystallization (Fig. 40-10). Experimentally determined and calculated equilibria that correspond to assemblages identifiable in the field and/or petrographically are also reviewed in Table 40-2. The reactions for ultramafic rocks refer to end-member assemblages in the H_2O-Na_2O-CaO-MgO-Al_2O_3-SiO_2 system peridotites. Compositions of the phases in other rocks are also indicated in Table 40-2.

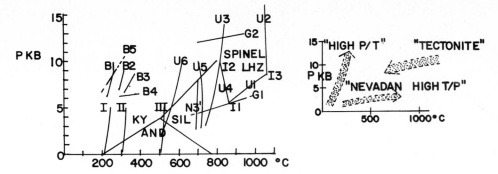

FIG. 40-10.   Equilibria in Triassic-Paleozoic belt rocks, with numbers keyed to the text and Table 40-2. Curves I, II, and III are after Elthon (1981) and represent the lower boundary of greenschist facies, the boundary between lower and upper greenschist facies, and the boundary between upper green-schist facies and lower actinolite facies, respectively for hydrothermal meta-morphism in the oceanic crust. Curve III also corresponds approximately to the equilibrium chloritoid + aluminosilicate = staurolite + quartz after Turner (1981). The schematic "paths of metamorphism" figured on the right approximate the sequence of equilibria encountered in metamorphic rocks as figured on the left. The path of "tectonites" (peridotites and amphibolites) is essentially a cooling path as indicated by the succession of assemblages and fabrics, and may converge at the lower temperature end with either "high P/T" (Norian) or "high T/P" (Nevadan) regional metamorphism.

Important observations are that aluminous and chrome spinels are common, and that plagioclase is generally absent in the tectonite peridotites-lherzolites in the Triassic-Paleozoic belt in the Klamath and Blue Mountains. Garnet, of course, has not been recorded in any of the peridotites, or in the two-pyroxene-plagioclase-bearing tectonite gabbros or granulites. Therefore, the pressure-dependent, univariant assemblages in peridotites and gabbros (e.g., U_1, G_1, and G_2 in Fig. 40-10; Table 40-2) set the rather broad pressure limits of more than about 6–8 kb, and less than about 12 kb, for the tectonite sections of the exposed ophiolites. The upper pressure limit is based on the combined equilibria in the system CaO-MgO-Al_2O_3-SiO_2 for diopside and enstatite with grossular-pyrope garnet, quartz, and anorthite (G_2) (Newton, 1983), and would be lowered by solution of almandine in garnet and albite in plagioclase. The higher tempera-ture limits are set by assemblages that become invariant and univariant on melting (I_3, U_2, Fig. 40-10). The invariant assemblage involves the phases orthopyroxene + clino-pyroxene + olivine + plagioclase + spinel + pargasite (+ liquid + vapor). The absence of plagioclase and presence of aluminous spinel in peridotites sets the pressure to higher values than the invariant, but above about 8 kb the high-temperature univariant assem-blage (U_2) pargasite + spinel + olivine + clinopyroxene (+ vapor + liquid) is not pressure sensitive. Hypothetically, the highest temperatures and pressures that could be recorded, if the assemblages are not affected by later reequilibration, involve the pargasite-bearing spinel lherzolites reported in parts of the Seiad Valley (Medaris, 1966) and Canyon Moun-tain ultramafic bodies (Mullen, 1983). These temperatures near the invariant I_3 (1050°C at 8 kb or so) are also compatible with generation of basic melts during deformation and high-temperature recrystallization.

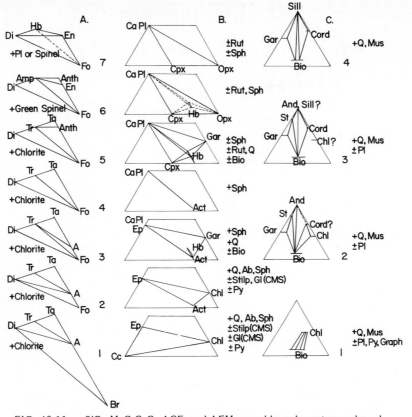

FIG. 40-11. SiO_2-MgO-CaO, ACF, and AFM assemblages in metamorphosed ultramafic, mafic, and pelitic rocks, respectively, in the Triassic-Paleozoic belt in the Klamath and Blue Mountains. Keyed to composite sections in Fig. 40-7.

Preregional Metamorphism: Amalgamation Stage of Tectonite Peridotites and Amphibolites

The experimentally determined "invariant" assemblages (I_1) amphibole + olivine + orthopyroxene + clinopyroxene + plagioclase + spinel, and (I_2) amphibole + olivine + orthopyroxene + clinopyroxene + chlorite + spinel may set the lower-temperature limits (about 850-875°C) for tectonite peridotites which have undergone a second recrystallization during mylonitization in the Klamath and Blue Mountains (Fig. 40-10). However, as fabric data suggest for peridotite bodies with remnants of such history, e.g., Canyon Mountain (Avé Lallèmant, 1976), Tom Martin Peak (Barrows, 1969; Hanks, 1981); Seiad Valley (Canat, 1985; Canat and Boudier, 1985; Lundquist, 1983; Medaris, 1966) and Wright Lakes (Chambers, 1983), their assemblages were disturbed during high-temperature

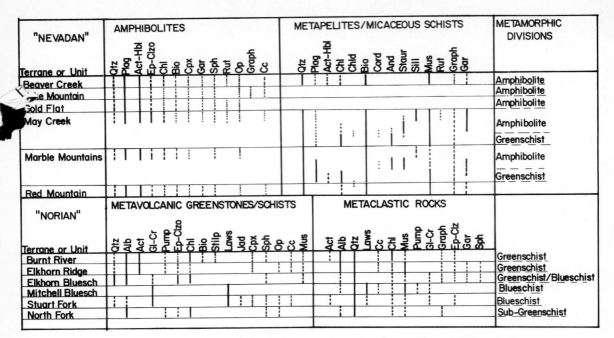

FIG. 40-12. Variation of metamorphic mineral assemblages in "Nevadan" and "Norian" metamorphic terranes in the Klamath and Blue mountains. "Nevadan" units are all part of the Marble Mountains terrane except for the May Creek Schist. The Burnt River Schist, Elkhorn Ridge Argillite, Elkhorn Ridge blueschist and Mitchell blueschist are all part of the Baker terrane. Data on mineral assemblages are from: Hotz (1967) in the Beaver Creek area; Hotz *et al.* (1977) and Hotz (1979) in the Stuart Fork Formation and Mitchell blueschist; Barrows (1969) in the Gold Flat amphibolite unit, and in the Lake Mountain metasedimentary unit; Donato *et al.* (1982), Donato (1985), and Kays (unpublished) in the Marble Mountains Wilderness and adjacent areas; Kays (1970) and Kays (unpublished) in the May Creek Schist; Kays and Ferns (1980), Ferns (1979), and Kays (unpublished) in the Red Mountain area; Ashley (1967), Mullen (1978), Perkins (1977), and Hunt (1985) in the Burnt River Schist, Elkhorn Ridge Argillite, and Elkhorn Ridge blueschist; Ando *et al.* (1983) and Mortimer (1985) in the North Fork terrane. Abbreviations are as follows: Qtz = quartz, Plag = plagioclase, Act-Hbl = actinolite-hornblende, Ep-Clzo = epidote-clinozoisite, Chl = chlorite, Bio = biotite, Cpx = clinopyroxene, Gar = garnet, Sph = sphene, Rut = rutile, Op = opaques; Graph = graphite, Cc = calcite, Chld = chloritoid, Cord = cordierite, And = andalusite, Staur = staurolite, Sill = sillimanite, Mus = muscovite, Alb = albite, Gl-Cr = glaucophane-crossite, Pump = pumpellyite, Stilp = stilpnomelane, Laws = lawsonite, Jad = jadeite. Bold solid lines represent major minerals, longer dashed lines represent minerals of lesser abundance, and shorter lines represent minor or accessory abundance.

conditions later than formation of primary tectonite fabrics. These conditions may have been imposed during an early deepseated phase of regional metamorphism at about 6–10 kb, or some amalgamation event involving thrust-fault displacement of the peridotite bodies into or onto oceanic crust. The latter has been termed "intraoceanic thrusting" by Nicolas and Le Pichon (1980) and Boudier *et al.* (1982). Nicolas and his co-workers suggest further that the finer-grained porphyroclastic fabric at the base of peridotites and in the immediately underlying amalgamated tectonite amphibolite (and other interlayered

METAMORPHISM OF TRIASSIC-PALEOZOIC BELT ROCKS

TABLE 40-2. Equilibria in Ultramafic and Some Associated
Mafic Rocks in the Klamath and Blue Mountains
Triassic-Paleozoic Oceanic Terranes[a]

Mineral Assemblage[b]	Symbol
Tr-Fo-Opx-Cpx-Sp-Pl-(V)	I_1
Tr-Fo-Opx-Cpx-Chl-Sp-(V)	I_2
Opx-Cpx-Sp-Pl-Fo	U_1
Gar (Gr_1Py_2)-Q-An-En	
Gar (Gr_2Py_1)-Q-An-Di	G_2
Chl-Opx-Sp-Fo-(V)	U_3
Tr-Fo-Opx-Cpx-Sp-(V)	U_4
An-Fa-Gar	G_1
("pseudounivariant" Gr-Alm exchange)	
Opx-Cpx-Sp-Pl-Fo-Parg-(V)-(L)	I_3
Parg-Sp-Cpx-Ol-(V)-(L)	U_2
Anth-Fo-En-(V)	U_5
Antig-Ta-Fo-(V)	U_6
Ta-Ol-Anth-(V)	U_7

[a] Symbols are keyed to Fig. 40-10.
[b] Ta, talc; Antig, antigorite; Fo, forsterite; Ol, olivine; Parg, pargasite;
Sp, spinel; Pl, plagioclase; Tr, tremolite; Anth, anthophyllite; [Gar =
grossular (Gr), pyrope (Py), almandine (Alm)], garnet; Cpx, clinopyroxene;
Opx, orthopyroxene; Chl, chlorite; Fa, fayalite; An, anorthite; V, vapor;
Q, quartz; Di, diopside; En, enstatite.

or tectonically intercalated rocks) is developed as a result of shear during the thrust
faulting event at temperatures in the range 800–1000°C.

In the Klamath and Blue Mountains, tectonic slivers and slices of oceanic rocks
juxtaposed with the high-temperature peridotites include siliceous metasedimentary
rocks, cherts, lentils of marble, and basic metavolcanic or intrusive rocks. Data on coexisting spinels and olivines in the Seiad Valley complex (Grover, 1984; Lieberman, 1983)
indicate that reequilibration occurred at temperatures between 700 and 800°C using
Engi's (1983) calibration coefficients for alpine-type peridotites. The work of Lieberman
(1983) provides apparent confirmation of reequilibration between peridotite and coexisting calcite-dolomite in the marble at the base of the Seiad Valley ultramafic complex. In
addition, the assemblages chlorite + orthopyroxene + spinel + olivine (U_3), and tremolite
+ olivine + orthopyroxene + clinopyroxene + spinel (U_4), and their intersection (I_2)
(Fig. 40-10) are all observed in the tectonite peridotites with modified, finer-grained
porphyroclastic or mylonitized fabrics. Since plagioclase is not recorded in the peridotites,
there is no evidence that conditions were attained that involved the assemblage of the
lower-pressure invariant (I_1). Thus, the following reactions may all be representative of
amalgamation-stage tectonite peridotites (Fig. 40-10):

$$\text{chlorite} = \text{orthopyroxene} + \text{forsterite} + \text{spinel} + 4H_2O \qquad (U_3)$$

$$\begin{aligned}\text{aluminous amphibole} + 3\text{forsterite} = 2.50\text{orthopyroxene} + 2\text{spinel} \\ + 2\text{clinopyroxene} + \text{vapor}\end{aligned} \qquad (U_4)$$

$$\text{calcic amphibole} + \text{forsterite} = 2.50\text{orthopyroxene} + \text{clinopyroxene} + \text{vapor} \qquad (U_4')$$

Reaction (U_3) studied by Jenkins (1981) serves to provide a frame of reference for use of the Fe-Mg olivine-spinel exchange geothermometer after Engi (1983). We can also use the two-pyroxene (Herzburg, 1978; Mori, 1977; P. R. A. Wells, 1977) and olivine-spinel geothermometers in the case of assemblage U_4, and the two-pyroxene geothermometer in the case of U_4'. A minimum temperature for amalgamation-stage metamorphism involving formation of tectonite peridotites is indicated by the reaction (Jenkins, 1983):

$$\text{anthophyllite} + \text{forsterite} = 9\text{enstatite} + H_2O \qquad (U_5)$$

The tectonite amphibolites, and interlayered or tectonically intercalated marbles, cherts, and garnetiferous quartz-mica schists which occur at the base of some porphyroclastic peridotites have fabrics and folds that are distinctive and of earlier generation than surrounding lower-grade metasedimentary and metavolcanic rocks. The amphibolites, which may be garnetiferous and/or clinopyroxene-bearing, are separated in turn by thrust faults from the lower-grade rocks. Hotz (1967) documents, through detailed mapping (Fig. 40-9) and petrographic studies, four structural sequences at the base of the Triassic-Paleozoic belt in the Beaver Creek area. The lowest structural zone, ordered in terms of structural superposition, contains amphibolite with the assemblage plagioclase + green-brown hornblende + sphene ± garnet ± clinopyroxene ± biotite. These amphibolites, interlayered and tectonically intercalated with siliceous schists, form the highest-grade part of the Triassic-Paleozoic sequence and are associated with tectonically dismembered and serpentinized peridotite bodies. Similar amphibolites are found at the base of the Tom Martin Peak, Seiad Valley, and Canyon Mountain peridotite bodies. Ghent and Stout (1981) have documented the presence of garnet-clinopyroxene-bearing amphibolites and associated quartzites, and other metasedimentary rocks beneath the Samail ophiolite.

Several geothermometers-geobarometers present in the amphibolites have been used with success in estimating conditions of their equilibration. Important in the amphibolites are temperature-dependent Fe-Mg exchange equilibria such as (Ellis and Green, 1979)

$$\tfrac{1}{3}\text{pyrope} + \text{hedenbergite} = \tfrac{1}{3}\text{almandine} + \text{diopside} \qquad (A)$$

This reaction is especially useful when coupled with more strongly pressure-dependent equilibria involving exchange of Ca, Mg, and Al in garnet-clinopyroxene (Newton and Perkins, 1981) in calc-silicate assemblages:

$$\text{anorthite} + \text{diopside} = \tfrac{1}{3}\text{pyrope} + \tfrac{2}{3}\text{grossular} + \text{quartz} \qquad (G_2)$$

Grover (1984) has also found that the equilibrium exchange of CaAl and NaSi in coexisting amphibole and plagioclase using the $\ln(X_{An}/X_{Ab})$ versus $\ln(Ca,M4/Na,M4)$, empirically calibrated plots of Spear (1981b) from the reaction

$$2\text{albite} + \text{tremolite} = 2\text{anorthite} + \text{glaucophane} \qquad (B)$$

was helpful in the overall assessment of metamorphic temperatures in the amphibolites surrounding and tectonically interlayered with the Seiad Valley ultramafic complex. Here temperature estimates for the amphibolites and associated metasedimentary rocks generally exceed 700°C and are less than about 850°C. These temperatures are generally consistent with those of the overlying mylonitized or finer-grained porphyroclastic spinel peridotites (Medaris, 1975).

Norian-Carnian High-*P/T* Regional Metamorphism and Associated Hydrothermal Metamorphism

Lawsonite-glaucophane schist facies metamorphism is known only in rocks that are earliest Norian or older in the Triassic-Paleozoic belt in the north-central Klamath Mountains and in the Blue Mountains. The most extensive distribution of such high-*P/T* metamorphic rocks is in the Stuart Fork Formation (G. A. Davis *et al.*, 1965; Hotz *et al.*, 1977) or Fort Jones Terrane (Mortimer, 1985). Olistostromal(?) blueschist similar in character and (apparently) in age to those in the Stuart Fork Formation also occur in the ophiolitic North Fork terrane, whose antiformal core (gabbro) has a U/Pb zircon age of 265–310 m.y. (Ando *et al.*, 1983). In the Blue Mountains, the new terrane subdivision shown in this report places the known blueschist-facies rocks in both the serpentine-matrix-dominated melange and in the Elkhorn Ridge Argillite-dominated part of the Baker Terrane. As a whole, the Baker terrane is dominated by metamorphism that is earliest Norian or older (Avé Lallèmant, *et al.*, 1980), but the character of the assemblages throughout the terrane is not well documented. What data as are available, however, show that earliest Norian metamorphism has more similarities than differences in terranes of the Klamath Mountains and Blue Mountains (see Fig. 40-12) and that lawsonite-glaucophane and jadeite-bearing blueschists have close association with feebly metamorphosed rocks and spilitized greenstones that also have relict assemblages (clinopyroxene and plagioclase) and fabrics. Overall, the greenstones have assemblages similar to those of hydrothermal or oceanic metamorphism (Elthon, 1981; Gass and Smewing, 1973). However, wherever glaucophane-lawsonite schists are found, structural and stratigraphic relationships among units are very complex. The solutions to such complexities appear to be crucial to an understanding of the high-*P/T* metamorphism recognized in these rocks.

In the Klamath Mountains, stratigraphic and isotopic data indicate that the North Fork and Salmon River terranes constitute parts of disrupted late Paleozoic oceanic crust (Ando *et al.*, 1983). According to Borns (1980), all components of the Fort Jones terrane were regionally metamorphosed to blueschist facies. The Fort Jones terrane has a close spatial relationship with the other two units (Mortimer, 1984), and the age of its metamorphism at 212–222 m.y. (Hotz *et al.*, 1977) suggests that it may be more closely equivalent to the other two terranes in terms of absolute age than it is to the Hayfork, Rattlesnake Creek, and Marble Mountains terranes of the Triassic-Paleozoic belt. The latter terranes, and the May Creek Schist belt, apparently were not subjected to high-*P/T* conditions of metamorphism, perhaps because of age and stratigraphic positioning. Mapping by Hotz (1979, 1977, 1967) demonstrates that the high-*P/T* Stuart Fork Formation overlies and is in thrust-faulted contact with Marble Mountains and Hayfork high-*T/P* terranes. This mapping also shows that high-*P/T* metamorphism in the Stuart Fork was earlier than the juxtaposition with Marble Mountains and Hayfork terranes. The juxtaposed Salmon River, North Fork, and Fort Jones terranes have a variety of low-grade assemblages (Table 40-3 and Fig. 40-12); relict igneous fabrics are recognizable, and metamorphic fabrics are poorly developed (Ando *et al.*, 1983; Burton, 1982; G. A. Davis *et al.*, 1965; Hotz, 1977; Mortimer, 1985). Regional correlation maps of Mortimer (1984, 1985) and the work of Hotz (1979), G. A. Davis *et al.* (1965), Ando *et al.* (1983), and Wright (1982) indicate that the high-*P/T* Fort Jones terrane, except for outlying klippe, is largely "inboard" or east of the North Fork and Salmon River terranes and has thrust-faulted contacts with the other Triassic-Paleozoic terranes and the Central Metamorphic, Eastern Klamath, and Trinity terranes (Fig. 40-5). The feebly metamorphosed North Fork and Salmon River terranes, with hydrothermal assemblages, would thus appear to occupy the same structural position with respect to the Fort Jones high-*P/T* terrane as do

Minerals	Pumpellyite-Prehnite-Actinolite	Glaucophane-Lawsonite	Greenschist	Actinolite
Glaucophane		xxxxx	xxx	
Lawsonite		xxxxx		
Sphene	xxxxx	xxxxx	xxxxx	
Aegerine		xxx		
Jadeite		xxxxx		
Muscovite	xxx	xxx	xxx	
Quartz	xxxxx	xxx	xxx	
Chlorite	xxxxx	xxx	xxxxx	
Stilpnomelane		xxx		
Garnet		xxx		
Actinolite			xxxxx	xxxxx
Plagioclase	xxx		Olig.	And.
Albite	xxxxx	xxxxx	xxxxx	
Cz-Ep	xxxxx		xxx	xxxxx
Magnetite	xxxxx		xxx	
Pumpellyite	xxx	xxx		
Calcite	xxx		xxx	
Prehnite	xxx			
Cpx	xxx			

[a]Stuart Fork Formation, Fort Jones Terrane, North Fork and Salmon River Terranes, Klamath Mountains; Elkhorn Ridge Argillite-Dominated, and Serpentine-Matrix Melange-Dominated Subterranes of the Baker Terrane, Blue Mountains.

[b]xxx represents major minerals commonly present; xxxxx represents additional minerals that may be present.

westernmost zeolite, and prehnite-pumpellyite facies terranes with respect to high-P/T Franciscan assemblages (Ernst *et al.*, 1970).

The basic metavolcanic rocks or greenstones of the North Fork and Salmon River terranes have been termed spilites (Hotz, 1979) and are characterized by coexistence of unaltered clinopyroxene, albite, and chlorite with veinlets of carbonate accompanied by prehnite and zeolites. Elsewhere, these rocks in the Yreka and Hornbrook quadrangles consist of sodic plagioclase (An_{15-20} to An_{30}), actinolitic amphibole, minor amounts of chlorite, biotite, clinozoisite-epidote, accessory sphene, and variable amounts of calcite; to a large extent original textures and structures are preserved. The assemblages in the greenstone metavolcanic rocks of the North Fork and Salmon River terranes in the north-central Klamath Mountains are similar to those described by Elthon (1981) as typical of hydrothermal oceanic metamorphism consisting of lower greenschist, upper greenschist, and lower actinolite facies. However, it seems important to clarify that the frame of reference is hydrothermal and that these assemblages according to Elthon's temperature estimates may not be equivalent to conventional greenschist facies assemblages that form under conditions of low water/rock ratio.

Metamorphosed rocks in the serpentine-matrix-dominated melange of the Baker terrane in the Blue Mountains consist of tectonic fragments of spilitic pillow basalts, other metavolcanic and related igneous and volcaniclastic rocks, chert-argillite, limestone or marble, glaucophane-lawsonite schist, glaucophanitic greenschist, pumpellyite-bearing greenschists, serpentinized harzburgite, altered gabbro, and occasional epidote-amphibolite and amphibolite. Except for the amphibolites, the assemblages are nearly identical to those described for the Stuart Fork Formation, the North Fork, and the Salmon River terranes. The assemblages in the mafic rock types (Table 40-3 and Fig. 40-12) keyed to ACF diagrams (Fig. 40-11) are lower greenschist facies [epidote + chlorite + calcite (+ quartz + albite + sphene)], upper greenschist facies [epidote + chlorite + actinolite (+ albite + sphene ± quartz)], and lower actinolite facies [plagioclase + actinolite (+ sphene)].

We suggest that the following hydration reactions (B_1)–(B_4) might be significant in the recrystallization of hydrothermal oceanic assemblages to lawsonite-glaucophane blueschists with increase in pressure (Fig. 40-10) during Norian-Carnian regional metamorphism (after E. H. Brown, 1977):

$$\text{epidote + chlorite + albite + quartz} + H_2O = \text{lawsonite + crossite + paragonite} \quad (B_1)$$

$$\text{pumpellyite + epidote + chlorite + albite + quartz} + H_2O = \text{lawsonite + crossite} \quad (B_2)$$

$$\text{epidote + actinolite + albite + chlorite} + H_2O = \text{crossite + pumpellyite + quartz} \quad (B_3)$$

$$\text{actinolite + hematite + albite + chlorite} + H_2O = \text{crossite + epidote + quartz} \quad (B_4)$$

Reactions (B_1)–(B_3) considered isobarically at 7 kb have successively higher temperatures (approximately 230, 280, and 320°C). The reactions represent the high-pressure boundary between hydrated hydrothermal oceanic assemblages and high-P/T lawsonite-glaucophane blueschists or transitional blueschist-greenschist-facies assemblages. Hydrated oceanic rocks on the left-hand sides of reactions (B_1)–(B_3) are lower and upper greenschist-facies assemblages, which on increase in pressure through loading or subduction react to form lawsonite-glaucophane blueschists, or transitional greenschists, depending on temperature. Reaction (B_4) is pressure dependent, and although of unusual oxidation assemblage, could serve as a geobarometer between upper greenschist facies and transitional blueschist-greenschist facies. Reactions (B_1)–(B_3) imply progressively increasing temperatures necessary to form lawsonite-glaucophane schists. However, hydrothermal assemblages, zoned according to depth in the vicinity of their heat source, could upon loading or subduction react discontinuously to produce lawsonite-glaucophane schist or pumpellyite-actinolite facies, depending on their temperature. This may explain in part why blueschist-facies rocks are found as discontinuous pods and layers surrounded by feebly recrystallized rocks of different assemblage in melanged parts of the Triassic-Paleozoic belt; i.e., cooling of hydrothermally altered oceanic rocks is slow compared to mass transfer during loading or subduction (Oxburgh and Turcotte, 1974). The reaction studied by Newton and Smith (1967),

$$\text{albite = jadeite + quartz} \quad (B_5)$$

can also be used to estimate pressure if the mixing of jadeite and clinopyroxene is assumed to be ideal with substitution of linked NaAl and CaMg units (i.e., ideal one-site substitution where $a_{Jd}^{Cpx} = X_{Na}$).

Middle Jurassic "Nevadan" Regional Metamorphism

Regionally developed "Nevadan" assemblages define a preliminary "geotherm" in Triassic-Paleozoic rocks in the north-central (Marble Mountains and Hayfork terranes) and northern Klamath Mountains (May Creek Schist belt). All the rocks have penetrative cleavage-foliation. Equivalent assemblages in terms of metamorphic age and facies series appear to be of limited areal distribution and are associated mainly with Jurassic plutonism in the Blue Mountains. We focus our attention here on Klamath Mountain parageneses because of the greater data base available, and because parageneses similar to those in the Klamath Mountains have not been studied or are lacking at the same scale in the Baker terrane in the Blue Mountains.

Work from many sources indicates that in any given part of the Triassic-Paleozoic Marble Mountains, Hayfork, or May Creek terranes of the Klamath Mountains, assemblages in pelitic, ultramafic, and mafic rocks are at uniformly consistent metamorphic grade. This suggests that the pattern of Jurassic metamorphism was not greatly displaced within a given terrane during or following Nevadan tectonic events even though large segments of these terranes are melanged or broken. In fact, metamorphic zones have been mapped in some areas, e.g., in the May Creek terrane and adjacent western Jurassic terrane (Kays, 1970), and in parts of the Marble Mountains terrane (Grover, 1984; Hotz, 1967; Klein, 1977; Mortimer, 1985). Progressive metamorphism seems well established to the west of the Condrey Mountain dome (Grover, 1984), and metamorphism generally increases in grade in the Triassic-Paleozoic belt rocks in all directions toward the dome. In all of these areas, however, the metamorphic zones have been partly disturbed or disrupted by faulting, and isograd boundaries may not be readily recognized. A thrust fault separates the amphibolite facies assemblages of the Marble Mountains terrane from the underlying transitional greenschist-blueschist facies rocks (Helper, 1986) of the Condrey Mountain Schist. In the Marble Mountains terrane, metamorphism increases in grade to upper amphibolite facies vertically downward but changes abruptly to blueschist-greenschist facies at the contact with rocks of the Condrey Mountain Schist.

"Nevadan" Assemblages in Metaperidotites

Nevadan assemblages form divariant zones in metabasic, metaperidotitic, and metapelitic rocks. There is widespread recognition in metaperidotites of talc, anthophyllite, and enstatite zones (Ferns, 1979; Grover, 1984; Kays and Ferns, 1980; Lieberman, 1983; Medaris *et al.*, 1980). A problem which hinders clear definition of each zone is the occurrence of serpentine as a retrogressive metamorphic product at all grades. Talc and clinopyroxene may have similar occurrence at the contact between metaperidotite and more siliceous rocks. Equilibrium assemblages in harzburgitic rocks in the three divariant zones have boundaries indicated by the following common assemblages: (1) olivine + antigorite + talc + chlorite + tremolite + oxide, (2) olivine + anthophyllite + chlorite + tremolite + oxide, (3a) olivine + chlorite + orthopyroxene + clinoamphibole + anthophyllite + aluminous or chrome spinel, and (3b) olivine + chlorite + orthopyroxene + clinoamphibole + aluminous spinel.

The first assemblage marks entrance into the talc-olivine zone in the compositional and stability field of clinoamphibole (tremolite) according to the reaction (Johannes, 1975)

$$\text{antigorite} = 4\text{talc} + 18\text{olivine} + 27\text{H}_2\text{O} \qquad (\text{U}_6)$$

The second assemblage with anthophyllite porphyroblasts to 10 cm long marks entrance into anthophyllite-olivine zone by the reaction (Hemley *et al.*, 1977)

$$9\text{talc} + 4\text{olivine} = 5\text{anthophyllite} + 4\text{H}_2\text{O} \qquad (\text{U}_7)$$

The third assemblage marks the first appearance of orthopyroxene by breakdown of (a) anthophyllite (Hemley *et al.*, 1977) (U_5), and (b) chlorite (Fawcett and Yoder, 1966; Jenkins, 1983) (U_3). The reactions are plotted in Fig. 40-10.

Equilibrium assemblages (U_5) and (U_3) are both found in the stability and compositional field that includes clinoamphibole, and as such probably define the uppermost part of the amphibolite facies. The fields of divariant stability and their boundaries for harzburgitic rocks are drawn in Fig. 40-10 based on measured phase compositions (Grover, 1984). Grover utilized ideal solution models for all the phases, and the effects on the phase fields in these Mg-rich rocks can be compared with those drawn by Evans (1977) for pure end-member minerals. Note, too, that in equilibrium assemblage (U_3), coexisting olivine and spinel also allows for application of Engi's (1983) geothermometer. Grover's traverse of olivine-spinel temperatures in ultramafic rocks of the Seiad Valley quadrangle, westward a distance of 7 km from the contact with the Condrey Mountain Schist, indicates reequilibration at generally lower temperatures than those of (U_5) or (U_3) for the phase compositions determined in those assemblages.

"Nevadan" Assemblages in Metapelitic Rocks

Nevadan regional metamorphism in the May Creek Schist belt of the northern Klamath Mountains produced three distinctive assemblages in metapelitic rocks. The assemblages form the basis for the mapped metamorphic zones (Kays, 1970) (Fig. 40-4 and 40-12). Important pelitic assemblages are (1) quartz + chlorite + biotite + muscovite + andalusite + albite + graphite + oxides, (2) quartz + biotite + muscovite + staurolite + plagiocase + tourmaline + graphite ± chlorite ± cordierite ± garnet ± oxides, and (3) quartz + biotite + muscovite + sillimanite + garnet + plagioclase + tourmaline ± staurolite ± graphite ± garnet ± oxides. The same or similar assemblages have been recognized in the Marble Mountains terrane of the north-central Klamath Mountains (Donato, 1985; Ferns, 1979; Hotz, 1979; Kays and Ferns, 1980; Mortimer, 1985) but their distribution in distinctive metamorphic zones has not been determined on a regional basis. Furthermore, there is indication that such zones are frequently tectonically disturbed (Hotz, 1979). Nevertheless, where assemblages of grade equivalent to (1)–(3) are documented in the adjacent ultramafic and mafic rocks of the north-central Klamath Mountains, metamorphism is generally consistent in all rock types. This suggests that during "Nevadan" regional metamorphism, ambient temperatures and pressures approached a path intermediate to that of classical Buchan and Barrovian metamorphism as described by Harte (1975) and Hudson (1980) in the Scottish Dalradian in northeastern Scotland.

The assemblages described above define the (1) andalusite-chlorite zone, (2) staurolite-biotite zone, and (3) sillimanite zone. The boundaries between divariant zones may be the continuous reactions as suggested by Harte (1975) and Hudson (1980):

$$\text{cordierite*} = \text{andalusite} + \text{biotite} \qquad (N_1)$$

$$\text{andalusite} + \text{biotite} = \text{staurolite} \qquad (N_2)$$

$$\text{staurolite} = \text{biotite} + \text{sillimanite} \qquad (N_3)$$

Cordierite is starred because of the difficulty in its recognition, but it has been identified in relatively fresh pelitic schists in association with andalusite adjacent to "Nevadan" plutons, e.g., near Wilderville, adjacent to the Merlin pluton west of Grants Pass, and in the Marble Mountains at the contact with the Wooley Creek Batholith (Barnes *et al.*, 1986). Note, however, that staurolite and garnet may also be produced discontinuously by any one of a number of possible reactions (Turner, 1981), e.g.,

$$\text{chlorite} + \text{andalusite} = \text{staurolite} + \text{biotite} \qquad (N_2')$$

$$\text{staurolite} + \text{quartz} = \text{garnet} + \text{sillimanite} \qquad (N_3')$$

The assemblages apparently require pressures no greater than about 5 kb maximum during highest temperatures of regional metamorphism. Otherwise, andalusite and cordierite would not occur, and the occurrence of staurolite and andalusite together in regionally metamorphosed rocks would be prohibited. Boundaries and phase fields for the metapelitic rocks are consistent with those calculated for ultramafic rocks. For example, breakdown of antigorite to form talc and olivine coincides approximately with the andalusite-staurolite (or the greenschist-amphibolite facies) boundary, and breakdown of talc and olivine to form anthophyllite is approximately in middle amphibolite facies in the stability field of staurolite (U_6 and U_7, Fig. 40-10). Breakdown of chlorite and anthophyllite (U_3 and U_5 respectively) in ultramafic rocks exceeds the stability of staurolite in the upper part of the sillimanite zone. Amphibolite facies assemblages, marked by the appearance of staurolite in metapelites, or talc + olivine in metaperidotites, are stable at temperatures beginning near 500–525°C. Regional metamorphism apparently culminated at temperatures of about 750°C.

Temperatures may be evaluated for the divariant assemblages indicated above from the balanced continuous reactions calculated by Thompson (1976a). All reactions employ some form of the relationship $\ln K = -[\Delta G^\circ_{1,T} + (P-1)\,\Delta V]/nRT$ utilizing the data base of Robie *et al*. (1978), the activity models of Thompson (1976a), and the measured equilibrium exchange relationships obtained from microprobe analysis of the minerals.

Pressure may be further constrained for the metapelites by the reaction suggested by Ghent (1976) and as reviewed by Newton and Haselton (1981) and Newton (1983) utilizing the equilibrium constant for the reaction

$$3\text{anorthite} = \text{grossular} + 2\text{aluminosilicate} + \text{quartz} \qquad (N_4)$$

In the interlayered calcareous assemblages, the following reaction (Newton, 1966) may also be useful:

$$4\text{clinozoisite} + \text{quartz} = \text{grossular} + 5\text{anorthite} + 2\text{water} \qquad (N_5)$$

"Nevadan" Assemblages in Mafic Rocks

The basis for a promising approach to an understanding of mafic mineral parageneses in progressive metamorphic sequences has been reviewed by Spear (1981a). The idea is that compositional changes involving amphibole and plagioclase may be systematic with change in metamorphic grade, and that the compositions of coexisting pairs may help define metamorphic grade. For example, Spear (1981b) shows that compositions of coexisting amphibole and plagioclase are statistically distinctive within most of the zonal subdivisions defined by aluminous minerals in the interlayered metapelites in New England metamorphic rocks. The distinction is based on the empirically examined exchange equilibrium

$$\text{NaSiCa}_{-1}\text{Al}_{-1}\ (\text{PLG}) = \text{NaAlCa}_{-1}\text{Si}_{-1}\ (\text{AMP}) \qquad (N_6)$$

where (PLG) is plagioclase and (AMP) is amphibole. Rewritten in terms of end-member minerals, the reaction is

$$2NaAlSi_3O_8 + Ca_2Mg_3Al_4Si_6O_{22}(OH)_2 = 2CaAl_2Si_2O_8 + Na_2Mg_3Al_2Si_8O_{22}(OH)_2 \quad (N_7)$$

A plot of $\ln(X_{An}/X_{Ab})$ versus $\ln(Ca,M4/Na,M4)$ shows systematic change in composition of coexisting plagioclase and amphibole, respectively, with changing temperature or grade. Very little work of this nature has been attempted in the Klamath Mountains until recently (Grover, 1984), but the results thus far have been encouraging.

The basis for subdivision of mafic amphibolitic gneisses into (1) greenschist, (2) epidote-amphibolite, and (3) amphibolite facies (Laird, 1982) is used here. The distribution of major minerals among the facies that agrees in general with "Nevadan" regionally metamorphosed mafic rocks in the Klamath Mountains is (1) actinolite + chlorite + epidote + albite (± quartz), (2) hornblende + albite (± oligoclase) + chlorite + epidote (± quartz), and (3) hornblende + plagioclase ($>An_{23}$) + epidote ± chlorite (Table 40-4, Fig. 12).

The greenschist and epidote-amphibolite facies data refer to the Condrey Mountain Schist and May Creek Schist in the north-central and northern Klamath Mountains, respectively, but without glaucophane and stilpnomelane in the latter. Epidote-amphibolite facies assemblages are apparently continuous or prograde with respect to greenschist facies in the Condrey Mountain Schist, and are largely concentrated at a structurally high position just below the thrust-faulted contact with the overlying amphibolite facies rocks of the Triassic-Paleozoic Marble Mountains terrane (see Fig. 1; Hotz, 1979). Amphibolite facies 2 is widespread in interlayered association with metasedimentary rocks over a large area in the Marble Mountains terrane in the north-central Klamath Mountains, and is also widespread in the May Creek Schist belt. In the north-central Klamath Mountains, amphibolite facies 2 of the Marble Mountains terrane is frequently in contact with epidote-amphibolite facies of the Condrey Mountain Schist. Massive, nonfoliated amphibolite facies 1 of the Marble Mountains terrane has fault-bound contacts with amphibolite facies 2. Amphibolite facies 3 is more restricted in areal distribution, occurring as tectonite amphibolite in close association with dismembered tectonite peridotite, as described previously. In the latter association, amphibolite facies 3 may be completely allofacial with respect to "Nevadan" metamorphism. Some occurrences, however, may be more closely associated as a contact facies with early and more mafic plutonic rocks.

Structural relationships between Condrey Mountain Schist and the Marble Mountains terrane may be crucial to an understanding of the inverted or "upside-down" prograde sequence with respect to the contact between these two terranes. For example, Barrows (1969) mapped gradational contacts between all facies, including the transitional greenschist-blueschist and epidote-amphibolite facies, and between these facies and the overlying Triassic-Paleozoic amphibolite facies rocks in the Gold Flat area and in the vicinity of Tom Martin Peak. Although foliations are all roughly parallel, they are of clearly different generation, as explained previously in the section on structural relationships between units in the north-central Klamath Mountains.

DISCUSSION

The Triassic-Paleozoic oceanic terranes considered here can be subdivided into (1) those that have high P/T character and Norian metamorphic ages, and (2) those that have high T/P "Nevadan," Middle (Bathonian) to Late Jurassic metamorphism. All these terranes, regardless of age, have relict early high-temperature peridotitic assemblages that were probably associated with solid flow in the upper mantle. This process was probably part of the normal evolution of an ocean plate as it grew and was transported landward, or

TABLE 40-4. Nevadan Assemblages in Mafic Rocks Except for Greenstone 1 and 2 Whose Affinity is Uncertain, but May Be North Fork or Salmon River, and Therefore "Pre-Nevadan" Assemblages

Minerals	Metamorphic Facies						
	Greenschist Facies[a]	Epidote-Amphibolite Facies[a]	Greenstone 1[b]	Greenstone 2[c]	Amphibolite Facies 1[d]	Amphibolite Facies 2[e]	Amphibolite Facies 3[f]
Albite	—	—	—				
Plagioclase				An_{15-20}	An_{25-35}	An_{30-35}	$>An_{40}$
Actinolite	—	—		—	—		
Blue-Green Hb		—			—	—	
Brown-Green Hb							—
Ep-Clzo	—	—	—	—	—	—	
Chlorite	—	—	—	—	—	—	
Almandine		—					—
Quartz	—	—	—	—	—	—	
Biotite	—	—	—	—	—	—	
Sphene	—	—	—	—	—	—	
Rutile	—	—				—	—
Clinopyroxene			"Relict"	"Relict"			
Muscovite	—	—	—				
Glaucophane	—	—					
Prehnite	—		—				
Calcite	—	—	—	—	—	—	
Magnetite	—	—	—	—	—	—	
Ilmenite	—	—	—			—	
Stilpnomelane	—						

[a] Greenschist facies and epidote-amphibolite facies refer mainly to the Condrey Mountain Schist; epidote-amphibolite facies rocks occur adjacent to the thrusted contact with overlying Triassic-Paleozoic amphibolite facies rocks. The two facies also refer to parts of the May Creek Schist belt. Assemblages are after Barrows (1969), Hotz (1979), Kays (1970), and Engelhardt (1966).
[b] Greenstone 1 refers to Hotz (1979) nonfoliated, feebly recrystallized "spilitic" metavolcanic rocks of the Triassic-Paleozoic belt.
[c] Greenstone 2 refers to Hotz (1979) nonfoliated but recrystallized metavolcanic rocks.
[d] Amphibolite 1 refers to Hotz (1979) nonfoliated, massive amphibolite.
[e] Amphibolite 2 refers to regional amphibolite facies of Barrows (1969) and zone 3 of Kays (1970).
[f] Amphibolite 3 refers to ultramafic and quartz diorite contact zones of Barrows (1969), and zone 4 of Kays (1970).

moved in the direction of marginal oceanic arcs. The peridotites also have less readily discernible relict, mylonitic fabrics and assemblages that transpose the earlier high-temperature tectonite fabrics and assemblages. These mylonitic events also involved ocean-floor rocks that were amalgamated to the base of the peridotite sheets during intra-oceanic thrusting. Such events may also have been responsible, at least in part, for formation of oceanic melange, and may have been the precursor to high-P/T events in pre-Late Triassic terranes, or high-T/P events beginning in the mid-Jurassic. We assume that earliest metamorphism in all Triassic-Paleozoic oceanic terranes involved partial melting in the peridotites and that these tectonites are residual. Assemblages and associations suggest that the tectonite peridotites started on the solidus at ~1050°C (U_2, Fig. 40-10) and cooled to temperatures as low as ~850°C (U_3 or U_4) during mylonitization associated with the amalgamation events.

The pre-Carnian terranes, such as the Baker, North Fork, and Fort Jones terranes, either contain regionally metamorphosed high-P/T assemblages with age of ~225 m.y., or low-grade hydrothermally metamorphosed assemblages mostly older than 225 m.y. The two metamorphic types may be gradational through pressure changes inasmuch as they all have overlapping temperature fields. If so, a major problem in explaining the discontinuous nature of blueschist facies in Triassic-Paleozoic terranes is overcome. That is, the thermal regime of each hydrothermal facies was approximately preserved as a relict of an ocean-floor regime, but rapid subduction and/or loading by intraoceanic thrusting resulted in a pressure increase and formed the appropriate facies without much change in temperature. The path of metamorphism is shown schematically for the oceanic rocks (Fig. 40-10). Oxburgh and Turcotte (1974) have shown that convection during plate-tectonic processes is rapid in comparison to conduction. Therefore, it is possible that for short time intervals (10–30 m.y.) between formation of hydrothermal assemblages and high-P/T assemblages, there would be little temperature change. Syntectonic deformations (D_1) are associated with orogeny during formation of high-P/T assemblages.

The younger or Bathonian high-T/P regional metamorphism in oceanic rocks is associated mainly with the Marble Mountains terrane and May Creek Schist belt in the Klamath Mountains. Here accretionary plutonism and crustal heating resulted in formation of andalusite, staurolite, and sillimanite in greenschist, lower amphibolite, and medium-high amphibolite facies, respectively. Penetrative fabrics in these rocks may have been associated with generation of recumbent nappe structures and loading during deformation D2 that also coincided with regional heating. Perhaps the nappe structures formed during intraoceanic thrusting in these terranes. The generally low pressures associated with this "Nevadan" or "accretionary" metamorphism require a thickness of about 15 km of oceanic crust, a thickness easily obtained through thrusting of ocean mantle over ocean-floor sediments and volcanic rocks. Perhaps, too, prograde metamorphism may have been associated with regional events that involved heating in conjunction with magmatic events by emplacement of Nevadan plutons into Marble Mountains terrane, Hayfork terrane, and the May Creek Schist belt. If so, accretion of the oceanic terrane to what is now western North America was an important part of "Nevadan" metamorphism. This final accretionary episode may explain the thrusting of the Marble Mountains terrane and the May Creek Schist belt over the Condrey Mountain Schist and Galice Formation, respectively. Plutonism and thrusting, followed by or accompanied by isoclinal, upright folding, would explain the chronology of deformation in the two terranes. The path of "Nevadan" metamorphism in these oceanic rocks is shown schematically in Fig. 40-10.

REFERENCES

Allen, C. A., Barnes, C. G., Kays, M. A., and Saleeby, J. B., 1982, Comagmatic nature of the Wooley Creek batholith and the Slinkard pluton and age constraints on tectonic and metamorphic events in the western Paleozoic and Triassic belt, Klamath Mountains, N. California: *Geol. Soc. America Abstr. with Programs*, v. 14, p. 145.

Ando, C. J., Irwin, W. P., Jones, D. L., and Saleeby, J. B., 1983, The ophiolitic North Fork terrane in the Salmon River region, central Klamath Mountains, California: *Geol. Soc. America Bull.*, v. 94, p. 236–252.

Armstrong, R. L., Taubeneck, W. H., and Hales, P. O., 1977, Rb-Sr and K-Ar geochronometry of Mesozoic granitic rocks and their Sr isotopic composition, Oregon, Washington, Idaho: *Geol. Soc. America Bull.*, v. 88, p. 397–411.

Ashley, R., 1967, Metamorphic petrology and structure of the Burnt River Canyon area, northeastern Oregon: Ph.D. thesis, Stanford Univ., Stanford, Calif., 193 p.

Ave Lallèmant, H. G., 1976, Structure of the Canyon Mountain (Oregon) ophiolite and its implication for sea floor spreading: *Geol. Soc. America Spec. Paper 173*, 49 p.

____, and Carter, N. L., 1970, Syntectonic recrystallization of olivine and modes of flow in the upper mantle: *Geol. Soc. America Bull.*, v. 81, p. 2203–2220.

____, Phelps, D. W., and Sutter, J. F., 1980, ^{40}Ar-^{39}Ar ages of some pre-Tertiary plutonic and metamorphic rocks of eastern Oregon and their tectonic relationships: *Geology*, v. 8, p. 371–374.

Barnes, C. G., 1983, Petrology and upward zonation of the Wooley Creek batholith, Klamath Mountains, California: *J. Petrology*, v. 24, p. 495–537.

____, Rice, J. M., and Gribble, R. F., 1986, Tilted plutons in the Klamath Mountains of California and Oregon: *J. Geophy. Res.*, v. 91, p. 6059–6072.

Barrows, A. G., Jr., 1969, Geology of the Hamburg–McGuffy Creek area, Siskiyou County, California, and petrology of the Tom Martin Peak ultramafic complex: Ph.D. thesis, Univ. California, Los Angeles, Calif., 301 p.

Borns, D. J., 1980, Blueschist metamorphism of the Yreka-Fort Jones area, Klamath Mountains, northern California: Ph.D. thesis, Univ. Washington, Seattle, Wash., 135 p.

Brooks, H. C., McIntyre, J. R., and Walker, G. W., 1976, Geology of the Oregon part of the Baker 1° by 2° quadrangle: *Oreg. Dept. Geol. Min. Industries Geol. Map Series GMS-7*.

Boudier, F., and Coleman, R. G., 1981, Cross section through the peridotite in the Semail ophiolite, southeastern Oman Mountains: *J. Geophys. Res.*, v. 86, p. 2573–2592.

____, Nicolas, A., and Bouchez, J. L., 1982, Kinematics of oceanic thrusting and subduction from basal section of ophiolites: *Nature*, v. 290, p. 825–828.

Bouma, A. H., 1962, *Sedimentology of Some Flysch Deposits:* New York, Elsevier, 168 p.

Brooks, H. C., 1979, Plate tectonics and the geologic history of the Blue Mountains: *Oreg. Geology*, v. 41, p. 71–80.

____, and Vallier, T. L., 1978, Mesozoic rocks and tectonic evolution of eastern Oregon and western Idaho, *in* Howell, D. G., and McDougall, K. A., eds., *Mesozoic Paleogeography of the Western United States:* Pacific Section, Soc. of Econ. Paleontologists Mineralogists, Pacific Coast Paleogeography Symp. 2, p. 133–146.

____, Ferns, M. L., Wheeler, G. R., and Avery, D. G., 1983, Geology and gold deposits map of the northeast corner of the Bates quadrangle, Baker and Grant Counties, Oregon: *Oreg. Dept. Geol. Min. Industries Geol. Map Series GMS-29*.

Brown, C. E., and Thayer, T. P., 1966, Geologic map of the Canyon City quadrangle, northeastern Oregon: *U.S. Geol. Survey Misc. Geol. Inv. Map I-447*, scale 1:250,000.

Brown, E. H., 1977, Phase equilibria among pumpellyite, lawsonite, epidote and associated minerals in low grade metamorphic rocks: *Contrib. Mineralogy Petrology*, v. 64, p. 123–136.

Burton, W. C., 1982, Geology of the Scott Bar Mountains, northern California: M.S. thesis, Univ. Oregon, Eugene, Oreg., 120 p.

Canat, M., 1985, Tectonics of the Seiad Massif, northern California: *Geol. Soc. America Bull.*, v. 96, p. 15–26.

——, and Boudier, F., 1985, Structural study of intra-oceanic thrusting in the Klamath Mountains, northern California — Implications on accretion geometry: *Tectonics*, v. 4, p. 435–452.

Chambers, J. M., 1983, The geology and structural petrology of ultramafic and associated rocks in the northeast Marble Mountain Wilderness, Klamath Mountains, northern California: M.S. thesis, Univ. Oregon, Eugene, Oreg., 146 p.

Christensen, N. I., and Salisbury, M. H., 1975, Structure and constitution of the lower oceanic crust: *Rev. Geophysics Space Physics*, v. 13, p. 87–137.

Coleman, R. G., 1977, *Ophiolites:* New York, Springer-Verlag, 229 p.

Davis, G. A., Holdaway, M. J., Lipman, P. W., and Romey, W. D., 1965, Structure, metamorphism and plutonism in the south-central Klamath Mountains, California: *Geol. Soc. America Bull.*, v. 76, p. 933–966.

Dickinson, W. R., and Thayer, T. P., 1978, Paleogeographic and Paleotectonic implications of Mesozoic stratigraphy and structure in the John Day inlier of central Oregon, *in* Howell, D. G., and McDougall, K. A., eds., *Mesozoic Paleogeography of the Western United States:* Pacific Section, Soc. Econ. Paleontologists Mineralogists, Pacific Coast Paleogeography Symp. 2, p. 147–161.

Diller, J. S., 1903, Klamath Mountain section, California: *Amer. J. Sci.*, v. 15, p. 342–362.

——, and Kay, G. F., 1924, Description of the Riddle quadrangle [Oregon]: *U.S. Geol. Survey Geol. Atlas, Folio 218*, 8 p.

Donato, M. M., 1985, Metamorphic and structural evolution of an ophiolitic tectonic melange, Marble Mountains, northern California: Ph.D. thesis, Stanford Univ., Stanford, Calif., 258 p.

——, Barnes, C. G., Coleman, R. G., Ernst, W. G., and Kays, M. A., 1982, Geologic map of the Marble Mountains Wilderness, Siskiyou County, California; *U.S. Geol. Survey Misc. Field Map MF-1452A*, scale 1:24,000.

Ellis, D. J., and Green, D. H., 1979, An experimental study of the effect of Ca upon garnet-clinopyroxene Fe-Mg exchange equilibria: *Contrib. Mineralogy Petrology*, v. 71, p. 13–22.

Elthon, D., 1981, Metamorphism in oceanic spreading centers, *in* Emiliani, Ceasare, *The Oceanic Lithosphere:* New York, Wiley, p. 285–303.

Engelhardt, C. L., 1966, The Paleozoic-Triassic contact in the Klamath Mountains, Jackson County, southwestern Oregon: M.S. thesis, Univ. Oregon, Eugene, Oreg., 98 p.

Engi, M., 1983, Equilibria involving Al-Cr spinel: Mg-Fe exchange with olivine — Experiments, thermodynamic analysis, and consequences for geothermometry: *Amer. J. Sci.*, v. 283-A, p. 29–71.

Ernst, W. G., 1984, California blueschists, subduction, and the significance of tectonostratigraphic terranes: *Geology*, v. 12, p. 436–440.

——, Seki, Y., Onuki, M. C., and Gilbert, M. C., 1970, Comparative study of low-grade metamorphism in California Coast Ranges and Outer Metamorphic Belt of Japan: *Geol. Soc. America Mem. 124*, 276 p.

Evans, B. W., 1977, Metamorphism of alpine peridotite and serpentinite: *Ann. Rev. Earth Planet. Sci.*, v. 5, p. 397–447.

Fahan, M. R., and Wright, J. E., 1983, Plutonism, volcanism, folding, regional metamorphism and thrust faulting — Contemporaneous aspects of a major middle Jurassic orogenic event within the Klamath Mountains, northern California: *Geol. Soc. America Abstr. with Programs*, v. 15, p. 272–273.

Fawcett, J. J., and Yoder, H. S., 1966, Phase relations in chlorites in the system MgO-Al_2O_3-SiO_2-H_2O: *Amer. Mineralogist*, v. 51, p. 353–380.

Ferns, M. L., 1979, The petrology and petrography of the Wrangle Gap-Red Mountain ultramafic body, Klamath Mountains, Oregon: M.S. thesis, Univ. Oregon, Eugene, Oreg., 124 p.

_____ , and Brooks, H. C., 1983, Serpentinite-matrix melanges in parts of the Blue Mountains of northeast Oregon: *Geol. Soc. America Abstr. with Programs*, v. 15, p. 371.

_____ , Brooks, H. C., and Avery, D. G., 1983, Geology and gold deposits map of the Greenhorn quadrangle, Baker and Grant Counties, Oregon: *Oreg. Dept. Geol. Min. Industries Geol. Map Series GMS-28*.

Gass, I. G., and Smewing, J. R., 1973, Intrusion, extrusion, and metamorphism at constructive margins — Evidence from the Troodos Massif, Cyprus: *Nature*, v. 242, p. 26–29.

Gerlach, D. C., Avé Lallèmant, H. G., and Leeman, W. P., 1981, An island arc origin for the Canyon Mountain ophiolite complex, eastern Oregon, U.S.A.: *Earth Planet. Sci. Lett.*, v. 53, p. 255–265.

Ghent, E. D., 1976, Application of activity-composition relations to displacement of a solid-solid equilibrium — Anorthite = grossular + kyanite + quartz, *in* Greenwood, H. J., ed., *Application of Thermodynamics to Petrology and Ore Deposits:* Mineral. Assoc. Canada, Short Course Handbook, v. 2, p. 99–108.

_____ , and Stout, M. Z., 1981, Metamorphism at the base of the Semail ophiolite southeastern Oman Mountains: *J. Geophys. Res.*, v. 86, p. 2557–2571.

Gilluly, J., 1937, Geology and mineral resources of the Baker quadrangle: *U.S. Geol. Survey Bull. 879*, 119 p.

Goullaud, L., 1977, Structure and petrology of the Trinity ultramafic complex, Klamath Mountains, northern California: Cordilleran Section, *Geol. Soc. America Field Guide*, 73rd Ann. Meet., p. 112–133.

Grover, T. W., 1984, Progressive metamorphism west of the Condrey Mountain dome, north-central Klamath Mountains, northern California: M.S. thesis, Univ. Oregon, Eugene, Oreg. 129 p.

Hamilton, W., 1963, Metamorphism in the Riggins region, western Idaho: *U.S. Geol. Survey Prof. Paper 436*, 95 p.

_____ , 1969, Mesozoic California and the underflow of Pacific mantle: *Geol. Soc. America Bull.*, v. 80, p. 2409–2430.

Hanks, C. L., 1981, The emplacement history of the Tom Martin ultramafic complex and associated metamorphic rocks, north-central Klamath Mountains, California: M.S. thesis, Univ. Washington, Seattle, Wash., 112 p.

Harte, B., 1975, Determination of a pelite petrogenetic grid for the eastern Scottish Dalradian: *Carnegie Inst. Wash. Yearbook 74*, p. 438–446.

Heinrich, M. A., 1966, The geology of the Applegate Group (Triassic) in the Kinney Mountain area, southwest Jackson County, Oregon: M.S. thesis, Univ. Oregon, Eugene, Oreg., 108 p.

Helper, M., 1986, Deformation and high P/T metamorphism in the central part of the Condrey Mountain Window, north-central Klamath Mountains, California and Oregon, *in* Evans, B. W., and Brown, E. H., eds., *Blueschists and Eclogites:* Geol. Soc. America Mem. 164, p. 125–141.

Herzberg, C. T., 1978, Pyroxene geothermometry and geobarometry — Experimental relations involving pyroxenes in the system CaO-MgO-Al$_2$O$_3$-SiO$_2$: *Geochim. Cosmochim. Acta.*, v. 42, p. 945–957.

Hill, L. G., 1985, Metamorphic, deformational, and temporal constraints on terrane assembly, northern Klamath Mountains, California, *in* Howell, D. G., ed., *Tectonostratigraphic Terranes of the Circum-Pacific Region:* Circum-Pacific Council Energy Min. Resources, Earth Sci. Series, no. 1, p. 173–186.

Hotz, P. E., 1967, Geologic map of the Condrey Mountain quadrangle, and parts of the Seiad Valley and Hornbrook quadrangle, California: *U. S. Geol. Survey Geol. Quad. Map GQ-618*, scale 1:62,500.

——, 1971, Plutonic rocks of the Klamath Mountains, California and Oregon: *U.S. Geol. Survey Prof. Paper 684B*, 20 p.

——, 1977, Geology of the Yreka quadrangle, Siskiyou County, California: *U. S. Geol. Survey Bull. 1436*, 72 p.

——, 1979, Regional metamorphism in the Condrey Mountain quadrangle, north-central Klamath Mountains, California: *U.S. Geol. Survey Prof. Paper 1086*, 25 p.

——, Lanphere, M. A., and Swanson, D. A., 1977, Triassic blueschist from northern California and north-central Oregon: *Geology*, v. 5, p. 659–663.

Hudson, N. F. C., 1980, Regional metamorphism of some Dalradian pelites in the Buchan area, N.E. Scotland: *Contrib. Mineralogy Petrology*, 73, p. 39–51.

Hunt, P. T., 1985, The metamorphic petrology and structural geology of the serpentinite-matrix melange in the Greenhorn Mountains, northeastern Oregon: M.S. thesis, Univ. Oregon, Eugene, Oreg., 127 p.

Irwin, W. P., 1972, Terranes of the western Paleozoic and Triassic belt in the southern Klamath Mountains, California: *U. S. Geol. Survey Prof. Paper 800C*, p. C103–C111.

——, 1985, Age and tectonics of plutonic belts in accreted terranes of the Klamath Mountains, California and Oregon: *in* Howell, D. G., ed., Tectonostratigraphic Terranes of the Circum-Pacific Region: Circum-Pacific Council Energy, Mine Resources, Earth Sci. Series, No. 1, p. 187–199.

Jenkins, D. M., 1981, Experimental phase relations of hydrous peridotites modelled in the system H$_2$O-CaO-MgO-Al$_2$O$_3$-SiO$_2$: *Contrib. Mineralogy Petrology*, v. 77, p. 166–176.

——, 1983, Stability and composition relations of calcic amphibole in ultramafic rocks: *Contrib. Mineralogy Petrology*, v. 83, p. 357–384.

Johannes, W., 1975, Zur Synthese und thermischen Stabilitat von Antigorit: *Fortschr. Mineralogie*, v. 53, p. 1–36.

Hemley, J. J., Montoya, J. W., and Shaw, D. R., 1977, Mineral equilibria in the MgO-SiO$_2$-H$_2$O system — II. Talc-antigorite-forsterite-anthophyllite-enstatite stability relations and some geologic implications in the system: *Amer. J. Sci.*, 277, p. 353–380.

Kamb, W. B., 1961, The thermodynamic theory of nonhydrostatically stressed solids: *J. Geophys. Res.*, v. 66, p. 259–271.

Kays, M. A., 1970, Mesozoic metamorphism, May Creek Schist belt, Klamath Mountains, Oregon: *Geol. Soc. America Bull.*, v. 81, p. 2743–2758.

——, and Ferns, M. L., 1980, Geologic field trip guide through the north-central Klamath Mountains: *Oreg. Geology*, v. 42, p. 23–35.

——, Ferns, M. L., and Beskow, L., 1977, Complementary metagabbros and peridotites in the northern Klamath Mountains, U.S.A., *in* Dick, H. G. B., ed., *Magma Genesis:* Oreg. Dept. Geol. Min. Industries Bull. 96, p. 91–107.

Klein, C. W., 1977, Thrust plates of the north-central Klamath Mountains near Happy Camp, California: *Calif. Div. Mines Geology Spec. Rpt. 129*, p. 23–26.

Laird, J., 1982, Amphiboles in metamorphosed basaltic rocks – Greenschist facies to amphibolite facies: *Mineral. Soc. America Rev.*, v. 9B, p. 113–137.

Lieberman, J. E., 1983, Petrology and petrogenesis of marble and peridotite Seiad Valley Complex, California: M.S. thesis, Univ. Oregon, Eugene, Oreg., 119 p.

Loney, R. A., and Himmelberg, G. R., 1976, Structure of the Vulcan Peak alpine-type peridotite, southwestern Oregon: *Geol. Soc. America Bull.*, v. 87, p. 259–274.

Lundquist, S. M., 1983, Deformation history of the ultramafic and associated metamorphic rocks of the Seiad Complex, Seiad Valley, California: M.S. thesis, Univ. Washington, Seattle, Wash., 166 p.

Medaris, L. G., Jr., 1966, Geology of the Seiad Valley area, Siskiyou County, California, and petrology of the Seiad ultramafic complex: Ph.D. thesis, Univ. California, Los Angeles, Calif., 212 p.

____ ,1975, Coexisting spinel and silicates in alpine peridotites of the granulite facies: *Geochim. Cosmochim. Acta*, v. 39, p. 947–958.

____ ,Walsh, J. L., Ferns, M. L., and Kays, M. A., 1980, Prograde metamorphism of serpentinites in the western Paleozoic and Triassic belt, Klamath Mountains Province: *Geol. Soc. America Abstr. with Programs*, v. 12, p. 120.

Moores, E. M., 1982, Origin and emplacement of ophiolites: *Rev. Geophysics Space Physics*, v. 20, p. 735–760.

Mori, T., 1977, Geothermometry of spinel lherzoirtes: *Contrib. Mineralogy Petrology*, v. 59, p. 261–279.

Mortimer, N., 1985, Structural and metamorphic aspects of Middle Jurassic terrane juxtaposition, northeastern Klamath Mountains, California, *in* Howell, D. G., ed., *Tectonostratigraphic Terranes of the Circum-Pacific Region:* Circum-Pacific Council Energy Min. Resources, Earth Sci. Series, no. 1, p. 201–214.

____ , and Coleman, R. G., 1984, A neogene structural dome in the Klamath Mountains, California and Oregon, *in* Nilsen, T. H., ed., *Geology of the Upper Cretaceous Hornbrook Formation, Oregon and California:* Pacific Section, Soc. Econ. Paleontologists Mineralogists, v. 42, p. 179–186.

Mullen, E. D., 1978, The geology of the Greenhorn Mountains, northeast Oregon: M.S. thesis, Oregon State Univ., Corvallis, Oreg., 372 p.

____ ,1983, Petrology of mafic rocks from the Canyon Mountain Ophiolite Complex, northeast Oregon: Rocky Mountain Section, *Geol. Soc. America Abstr. with Programs*, v. 15, p. 334.

Newton, R. C., 1966, Some calc-silicate equilibrium relations: *Amer. J. Sci.*, v. 264, p. 204–222.

____ ,1983, Geobarometry of high-grade metamorphic rocks: *Amer. J. Sci.*, v. 283–A, p. 1–28.

____ , and Haselton, H. T., 1981, Thermodynamics of garnet-plagioclase-Al_2SiO_5 geobarometer, *in* Newton, R. C., Navrotsky, A., and Wood, B. J., eds., *Thermodynamics of Minerals and Melts:* New York, Springer-Verlag, p. 131–147.

____ , and Perkins, D., 1981, Ancient granulite terrains – eight kbar metamorphism: (EOS) *Trans. Amer. Geophys. Un.*, v. 62, p. 420.

____ , and Smith, J. V., 1967, Investigations covering the breakdown of albite at depth in the earth: *J. Geology*, v. 75, p. 268–286.

Nicolas, A., and Le Pichon, X., 1980, Thrusting of young lithosphere in subduction zones with special reference to structures in ophiolitic peridotites: *Earth Planet. Sci. Lett.*, v. 46, p. 397–406.

——, and Poirier, J. P., 1976, *Crystalline Plasticity and Solid State Flow in Metamorphic Rocks:* Interscience, New York, 444 p.

——, Boudier, F., and Bouchez, J. L., 1980, Interpretation of peridotite structures from ophiolitic and oceanic environments: *Amer. J. Sci.*, v. 280A, p. 192–210.

Oldow, J. S., Avé Lallèmant, H. G., and Schmidt, W. J., 1984, Kinematics of plate convergence deduced from Mesozoic structures in the western Cordillera: *Tectonics*, v. 3, p. 201–227.

Oxburgh, E. R., and Turcotte, D. L., 1974, Thermal gradients and regional metamorphism in overthrust terranes with special reference to the eastern Alps: *Schweiz. Mineral. Petrogr. Mitt.*, v. 54, p. 641–662.

Peck, D. L., Griggs, A. B., Schlicker, H. G., Wells, F. G., and Dole, H. M., 1964, Geology of the central and northern parts of the western Cascade Range in Oregon: *U.S. Geol. Survey Prof. Paper 449*, 56 p.

Perkins, J. M., 1976, Geology of the Greenhorn quadrangle and the northwest portion of the Whitney quadrangle, Baker and Grant Counties, Oregon: M.S. thesis, Univ. Oregon, Eugene, Oreg., 96 p.

Phelps, D. W., 1979, Petrology, geochemistry, and origin of the Sparta quartz diorite-trondhjemite complex, northeastern Oregon, *in* Barker, F., ed., *Trondhjemites, Dacites, and Related Rocks:* New York, Elsevier, p. 547–579.

——, and Avé Lallèmant, H. G., 1980, The Sparta ophiolite complex, northeast Oregon — A plutonic equivalent to low-K_2O island arc volcanism: *Amer. J. Sci.*, v. 280A, p. 345–358.

Prostka, H. J., 1962, Geology of the Sparta quadrangle, Oregon: *Oreg. Dept. Geol. Min. Industries Map GMS-1*.

Ramp, L., 1975, Geology and mineral resources of the upper Chetco drainage area, Oregon: *Oreg. Dept. Geol. Min. Industries Bull.*, v. 88, 47 p.

Rawson, S. A., and Petersen, S. W., 1982, Structural and lithologic equivalence of the Rattlesnake Creek terrane and high-grade rocks of the western Paleozoic and Triassic belt, north-central Klamath Mountains, California: *Geol. Soc. America Abstr. with Programs*, v. 14, p. 226.

Robie, R. A., Hemingway, B. S., I. Fisher, J. R., 1978, Thermodynamic properties of minerals and related substances at 298.15K and 1 bar (10^5 pascals) pressure and at higher temperatures: *U.S. Geol. Survey Bull.*, 1452, 456 p.

Saleeby, J. B., Harper, G. D., Snoke, A. W., and Sharp, W. D., 1982, Time relations and structural-stratigraphic patterns in ophiolite accretion, west central Klamath Mountains, California: *J. Geophys. Res.*, v. 87, p. 3831–3848.

——, Blake, M. C., and Coleman, R. G., 1984, U/Pb zircon ages of thrust plates of the west-central Klamath Mountains and Coast Ranges, northern California and southern Oregon: *Trans. Amer. Geophys. Un.*, v. 65, p. 1147.

Silberling, N. J., Jones, D. L., Blake, M. C., Jr., and Howell, D. G., 1984, Lithotectonic terrane map of the North American Cordillera: *U.S. Geol. Survey Open File Rpt. 84-523C* scale 1:2,500,000, 43 p.

Spear, F. S., 1981a, An experimental study of hornblende stability and compositional variability in amphibolite: *Amer. J. Sci.*, v. 281, p. 697–734.

——, 1981b, NaSi = CaAl exchange equilibrium between plagioclase and amphibole: *Contrib. Mineralogy Petrology*, v. 72, p. 33–41.

Thayer, T. P., 1956, Preliminary geologic map of the John Day quadrangle, Oregon: *U.S. Geol. Survey Misc. Field Studies Map MF-51*, scale 1:62,500.

_____ , 1963a, The Canyon Mountain complex, Oregon, and the alpine mafic magma stem: *U.S. Geol. Survey Prof. Paper 475C*, p. C82–C85.

_____ , 1963, Flow-layering in alpine peridotite-gabbro complexes: *Mineral. Soc. America Spec. Paper 1*, p. 55–61.

_____ , and Himmelberg, G. R., 1968, Rock succession in the alpine type mafic complex at Canyon Mountain, Oregon: *Proc. 23rd Internat. Geol. Cong. Prague*, v. 1., p. 175–186.

Thompson, A. B., 1976a, Mineral reactions in pelitic rocks: I. Prediction of P-T-X (Fe-Mg) phase relations: *Amer. Jour. Sci.*, v. 276, p. 401–424.

_____ , 1976b, Mineral reactions in pelitic rocks: II. Calculations of some P-T-X (Fe-Mg) phase relations: *Amer. J. Sci.*, v. 276, p. 425–454.

Turner, F. J., 1981, *Metamorphic Petrology, Mineralogical, Field, and Tectonic Aspects*, 2nd Ed.: McGraw-Hill, New York, 524 p.

Vallier, T. L., Brooks, H. C., and Thayer, T. P., 1977, Paleozoic rocks of eastern Oregon and western Idaho, *in* Stewart, J. H., *et al.*, *Paleozoic Paleogeography of the Western United States:* Pacific Section, Soc. Econ. Paleontologists Mineralogists, Pacific Coast Paleogeography Symp. 1, p. 395–408.

Walker, N. W., 1983, Pre-Tertiary, tectonic evolution of northeastern Oregon and west-central Idaho — Constraints based on U/Pb ages of zircons: Rocky Mountain Section, *Geol. Soc. America, Abstr. with Programs*, v. 15, p. 371.

Wells, F. G., *et al.*, 1940, Preliminary geologic map of the Grants Pass quadrangle, Oregon: *Oreg. Dept. Geol. Min. Industries Map*, scale 1:96,000.

_____ , Hotz, P. E., and Cater, F. W., 1949, Preliminary description of the geology of the Kerby quadrangle, Oregon: *Oreg. Dept. Geol. Min. Industries Bull. 40*, 23 p.

_____ , *et al.*, 1956, Geology of the Medford quadrangle: *U.S. Geol. Survey Geol. Quad. Map GQ-89*, scale 1:96,000.

Wells, P. R. A., 1977, Pyroxene thermometry in simple and complex systems: *Contrib. Mineralogy Petrology*, v. 62, p. 129–139.

Wright, J. E., 1982, Permo-Triassic accretionary subduction complex, southwestern Klamath Mountains, northern California: *J. Geophys. Res.*, v. 87, p. 3805–3818.

_____ , and Wyld, S. J., 1985, Multi-stage serpentinite matrix melange development — Rattlesnake Creek terrane, southwestern Klamath Mountains: *Geol. Soc. America Abstr. with Programs*, v. 17, p. 419.

INDEX

Absaroka volcanic center, 355
Albion Mountains, northeastern Great Basin, 620-29
Alumina solubility in hornblende, Mojave-Sonoran Desert, 516-18
Aluminous metasomatic rocks, 488-89
Amphibolites, Farmington Canyon Complex, 435-37
Antelope Island, metamorphism on, 442
Arizona:
  differences between middle Tertiary and older fabrics, 495-96
  Mesozoic and Early Cenozoic metamorphism, 473-92
    controls of 489-92
    south-central Arizona, 481-83
    southeastern Arizona, 479-81
    western Arizona, 483-89
  middle Tertiary mylonitization and detachment faulting, 492-95
  thermal history of, 470-73
  See also Western Arizona.
"Autochthon", tectonic and metamorphic development of, 189-90
Autochthonous metasedimentary rocks, Peninsular Ranges, 908-9

Baker Lake blueschist, petrology, 200-201
Bare Mountain, Death Valley region, 724
Basaltic magmatism:
  Rio Grande rift, 93-95
  Snake River Plain and vicinity, 95-97
  southwest Nevada, 97-98
Basin and Range province, Paleomagnetic evidence for Sierra Nevada motion, 27
Beartooth Mountains:
  geochemistry and geochronology of Archean rocks 363-410
  eastern mountains, 387-88
  Late Archean rocks, 388-90
  northcentral mountains, 395
  northwestern mountains, 398-400
  older Archean rocks, 391-94
  southwestern mountains, 396-407
  isotopic systematics of Late Archean rocks, 390-91
  metamorphic petrology, 364-74

Jardin-South Snowy Block, 372-73
Lake Plateau area, 368-69
Long Lake and Broadwater River areas, 367-68
North Snowy Block (NSB), 369-72
Quad Creek and Hellroaring Plateau, 365-67
Stillwater Complex Hornfels aureole, 373
Yankee Jim Canyon, 372
Bear Valley Springs igneous suite, 872-76
  field/radiometric/petrographic data, 872-76
  gabbroids of Tunis Creek and Squirrel Spring, 875-76
  hypersthene tonalite of Bison Peak, 874-75
  tonalite of Bear Valley Springs, 873-74
Bedford Canyon Formation, 903-4
Bighorn Mountains, geochemistry and geochronology of Archean rocks, 401-2
Big Maria Mountains, 551-59
  age of Mesozoic metamorphism, 559
  general geology, 551
  geothermometry, 554-55
  isograds from siliceous dolomite 552-53
  peristerite gap, 556-58
  regional metamorphism, 551-52
  siliceous dolomites, 556
  Supai Formation, 558-59
Bitterroot batholith, post-Archean metamorphic and tectonic evolution, 347-49
Black Pine Mountains, northeastern Great Basin, 613-14
Blacktail Range, Wyoming, 375-76
Blue Mountains:
  ages, 1106-10
  fabrics in early high-temperature assemblages, 1112
  Middle Jurassic "Nevadan" regional metamorphism, 1129-30
  "Nevadan" assemblages, 1130-33
    in mafic rocks, 1132-33
    in metapelitic rocks, 1131-32
    in metaperidotites, 1130-31
  Norian-Carnian high-P-T regional metamorphism, 1127-29
  oceanic terranes, 1100-1102
    mafic tectonite amphibolites 1115-16
  polymetamorphism, pattern of, 1121

preregional allofacial metamorphism, 1121-27
  regional fabrics, 1117-18
Boulder batholith, post-Archean metamorphic and tectonic evolution, 353-54
British Columbia:
  Northwest Cascades, 197-209
  fault emplacement structures, 204-7
  petrology, 197-204
  phase relations, 203-4
  regional considerations, 207-9
  summary of ages, 202-3
  tectonic history, 207
Brittle deformation, Palm Canyon Complex, 922-23
Bullfrog Hills, Death Valley region 724

Calaveras Complex, 834-37
  age, 835-36
  correlation, 836-37
  lithology, 834-35
  origin, 836
Cascade River Schist, petrochemistry of, 221-23
Catalina Schist:
  metamorphic geology of, 1003-4
  upside-down metamorphism, 958-59
  whole-rock chemical contrasts between arc-like rocks and, 1008
  See also Los Angeles Basin.
Cenozoic model, northeast Great Basin, 639
Central Belt:
  Sierra Nevada (northern), 741-43
  chert-argillite unit, 741-42
  metamorphism, 747-49
  older ophiolitic rocks, 742
  younger volcanic arc complexes, 742-43
Central Metamorphic Belt (CMB), upside-down metamorphism, 959-60
Cheyenne Belt, Colorado, 413
Chilliwack Group, petrology, 198-200
Chocolate Mountain thrust, Mesozoic and Early Cenozoic metamorphism, 485-86
Church Mountain Thrust Plate, 187-88
  tectonic and metamorphic setting of, 187-88
Cimarron Mountains, Mexico, 451
Coffee Creek event, Klamath Mountains, 1075-76

Colorado:
  crustal accretion during
    Early Proterozoic, 423-25
    Middle Proterozoic, 425
  deformation and metamorphism,
    418-19
  geochronology, 419-22
    anorogenic assemblages,
      420-22
    orogenic assemblages, 419-20
  Proterozoic crustal rocks in,
    413-18
    buried basement of eastern
      Colorado, 418
    Cheyenne Belt and associated
      rocks, 413
    Front Range and Wet
      Mountain, 417-18
    Gunnison region, 413-16
    Needle Mountains, 416-17
    Salida area, 416
    Sierra Madre Range, 417
    Uinta Mountains, 418
  Rb-Sr ages, resetting of, 422-23
Comanche Point, gneiss units of,
  876-78
Condrey Mountain event,
  Klamath Mountains,
  1081-83
Condrey Mountain terrane,
  Klamath Mountains,
  1069-70
Continental arc magmatism:
  heat and water, 14-16
  lower crust, 13-14
  structure, 14
  upper and middle crust, 9-13
  wallrock deformation, 14
Continental borderland:
  arc-like rocks, 1004-8
    metamorphism of, 1005-8
    whole-rock chemical contrasts
      between Catalina Schist
      and, 1008
  metamorphic geology of
    Catalina Schist, 1003-4
  probable Franciscan basement
    rocks of, 1015-16
  See also Los Angeles Basin.
Crustal thickness, epidote-bearing
  plutons, 52-53
Crystalline basement, northeastern
  Great Basin, 633-35
Culbertson Lake allochthon, 768
Cultus Formation, petrology,
  198-200
Cuyumaca-Laguna Mountains
  shear zone, 915-18

Dear Peak unit, petrology, 201
Death Valley region:
  metamorphism in, 716-24

Bare Mountain, 724
Bullfrog Hills, 724
Funeral Mountains, 721-23
Panamint Mountains, 716-21
Tucki Mountain, 724
tectonics of metamorphism,
  724-31
nappes, 725-27
palinspastic reconstruction,
  727-28
Sierran batholith, 724
thermal environment during
  metamorphism, 728-31
Deep Creek Range, east-central
  Great Basin, 668, 673-74
Ductile deformation:
  Peninsular Ranges batholith
    Cuyumaca-Laguna Mountains
      shear zone, 915-18
    eastern Peninsular Ranges
      mylonite zone, 918-23
Duncan Peak allochthon, 768

Early Proterozoic:
  history of margin of Archean
    continent in Utah, 432-43
  isotopic data for Colorado,
    424-25
  petrologic and geochemical data
    of crustal accretion, 423-25
East-central Great Basin:
  Cretaceous age:
    Kern Mountains-Deep Creek
      Range, 668
    northern Snake Range,
      665-68
    Schell Creek Range, 662-64
    southern Snake Range, 665
  earliest metamorphism and
    deformation, 674-76
  geologic setting, 653-54
  Jurassic age, 655-62
    northern Snake Range,
      661-62
    Schell Creek Range, 657-59
    southern Snake Range,
      659-61
  muscovite-bearing plutons, 676
  Tertiary structures, 669-74
    Kern Mountains-Deep Creek
      Range, 673-74
    northern Snake Range 671-73
    southern Snake Range, 669-71
Eastern Belt:
  Sierra Nevada (northern), 744
    metamorphism, 750-52
Eastern Klamath terrane, 1087
Ecstall area, uplift rates, 54-56
Elbow Lake Formation, petrology,
  200
Eldorado Orthogneiss,
  petrochemistry of, 221

Eocene dike swarm, 354-55
Epidote-bearing plutons:
  Western Cordillera:
    crustal thickness, 52-53
    uplift rates, 53-60
Eureka thrust belt, Great Basin,
  585-86

Facer Formation, metamorphis
  in, 441
Farmington Canyon Complex,
  434-37
  age of rocks and metamorphic
    events, 437-38
  crustal history summary, 441
  metamorphic history, 438-39
  protoliths of, 434-37
    amphibolites, 435-37
    layered gneiss, 434-35
Fault emplacement structures,
  Washington and British
  Columbia, 197-209
Feather River Belt:
  Sierra Nevada (northern),
    743-44
    metamorphism, 749-50
Fort Jones event, Klamath
  Mountains, 1076-77
Fort Jones terrane, Klamath
  Mountains, 1068
Franciscan Complex, 1024-1033
  Central Franciscan belt,
    1042-43
  Coastal Franciscan belt,
    1043-45
  Eastern Franciscan belt,
    1038-42
    Pickett Peak terrane, 1038-39
    Yolla Bolly terrane, 1039-42
  geochronology of, 1027-1031
    background, 1027-1029
    Francisean metamorphic ages,
      1029-31
  metamorphism of, 1024-1027
    metamorphic textural zones,
      1045-46
    mineral zones, 1046-52
    plate tectonic models,
      1053-54
  regional geologic setting,
    1037-38
Franciscan events, Klamath
  Mountains, 1084
French Valley Formation,
  Peninsular Ranges, 904-5
Front Range, Colorado, 417-18
Funeral Mountains, 721-23

Gabriel Peak Orthogneiss,
  petrochemistry of,
  224-26
Gallatin Range, geochemistry and
  geochronology of Archean

rocks, 401
Gallatin Range, Wyoming, 374
Garnet-muscovite-plagioclase-
    biotite equilibria, Mojave-
    Sonoran Desert, 513-16
Glorieta Baldy, New Mexico, 453
Gold Butte area, Proterozoic
    high-grade metamorphism,
    528-31
Great Basin:
    Cenozoic extension,
        595-96
    foreland comparisons, 598-99
    kinematic model of foreland
        contraction, 596-98
    Mesozoic active margin tectonics,
        582-85
        foreland deformation, 585
        Mesozoic continental arc,
            582-84
        Mesozoic magmatism and
            metamorphism in
            foreland, 584
        tectonic attrition, 584-85
    passive margin collisions,
        579-80
    metamorphism during, 580-82
    passive margin formation,
        576-82
        edge of Precambrian North
            American, 578-79
        passive margin subsidence and
            sedimentation, 577-78
    Phanerozoic tectono-
        stratigraphy, 575-76
    Sevier and Eureka thrust belts,
        585-86
    thermal doming, model of,
        593-95
    Toiyabe uplift zone, 591-93
    Winnemucca deformation belt,
        586-91
    See also Northeastern Great
        Basin; East-central Great
        Basin.
Greenhorn melange, relict fabrics
    in, 1116-17
Grouse Creek Mountains, north-
    eastern Great Basin,
    620-29
Gunnison region, Colorado,
    413-16

Hayfork subterrane, Klamath
    Mountains, 1069
Hozameen Belt, northern
    Cascades, 190-91
Hydrocarbons, as product of
    continental arc
    magmatism, 8

Idaho batholith, as product of

continental arc
    magmatism, 12-13
Inverted metamorphic gradient
    (IMG), definition of,
    854-55
Iron Mountains, Mojave-Sonoran
    Desert, 507
Island arc juncture, west-central
    Idaho, 297-326
Isobaric and isothermal surfaces
    (metamorphic), northern
    and central New Mexico, 456
Isotope geochemistry:
    basaltic magmatism, 92-98
        Rio Grande rift, 93-95
        Snake River Plain and vicinity,
            95-97
        southwest Nevada, 97-98
    implications for
        structure/composition of
        deep continental
        lithosphere, 88-105
    isotopic systematics, 89-92
    silicic intrusive rocks, 98-105
        metaluminous granite, 101-3
        peraluminous granite, 99-101

Julian Schist, Peninsular Ranges
    904-5

Kern Mountains, east-central
    Great Basin, 668, 673-74
Ketchikan-Petersburg area, uplift
    rates, 54
Kettle Complex:
    evolution of, 286-89
    structure and metamorphism,
        284-86
Klamath Mountains, 1062-90
    ages in, 1110-12
    early high-temperature
        assemblages and fabrics in
        peridotites in, 1112-15
    geologic synthesis, 1087-90
        Cenozoic, 1090
        Cretaceous, 1089-90
        Early Jurassic, 1087
        Late Jurassic, 1089
        Middle Jurassic, 1089
        Pre-Jurassic, 1087
    metamorphic framework,
        1071-87
        crustal metamorphic events,
            1074-87
        facies divisions, 1071-73
    Middle Jurassic "Nevadan"
        regional metamorphism,
        1129-30
    Norian-Carnian high-P-T regional
        metamorphism, 1127-29
    oceanic terranes, 1102-1106
        mafic tectonite amphibolites

in, 1115-16
    polymetamorphism, pattern of,
        1121
    preregional allofacial
        metamorphism, 1121-27
    regional fabrics and small-scale
        structural features in,
        1118-20
    relict fabrics in, 1117
    tectonic framework, 1062-71
        Cenozoic faulting, 1067
        Paleozoic and Mesozoic
            faulting, 1063-67
        structural units (terranes),
            1067-70
        tectonic significance of
            plutonic rocks, 1071
Kootenay Arc:
    evolution of, 286-89
    structure and metamorphism,
        278-82

Lang sequence, 767-68
Laramide deformation, 20-23
Layered gneiss, Farmington
    Canyon Complex, 434-35
Los Angeles Basin:
    arc-like rocks, 1004-8
    metamorphism of, 1005-8
    whole-rock chemical contrasts
        between Catalina Schist
        and, 1008
    metamorphic geology of
        Catalina Schist, 1003-4
    metamorphic P-T estimates for
        basement terranes,
        1012-15
    mineral chemistry of muscovites
        and amphiboles from
        basement terranes, 1009-11
Lower crust, continental arc
    magmatism, 13-14

McCullough Range, Proterozoic
    high-grade metamorphism,
    531-33
Madison Range, geochemistry and
    geochronoly of Archean
    rocks, 400, 401
Mafic dikes, Bitterroot batholith,
    348-49
Mafic schist, Pelona, Orocopia, and
    Rand schists (POR suite),
    986
Magmatic systems of Western U.S.:
    accretionary wedge and fore-arc
        basin, 4
    Cenozoic crustal extension,
        23-30
        Basin and Range province,
            23-29
        Eocene extension in

Northwest, 28-29
new volcanic crust, 29-30
rifting and magmatism, 29
continental arc magmatism:
heat and water, 14-16
lower crust, 13-14
products of, 4-17
arc rocks east of Sierra
Nevada, 13
facies and equilibria, 6
hydrocarbons, 8
Idaho batholith, 12-13
Peninsular batholith, 12
Salinian block, 12
Sierra Nevada batholith,
10-12
uranium, 8
volcanic ash, 8
volcanism, 7-8
structure, 14
upper and middle crust, 9-13
wallrock deformation, 14
foreland thrust belt, 16-17
latest Cretaceous and early
Paleogene tectonics, 17-23
Laramide deformation, 20-23
Peninsular Ranges mylonites,
19-20
subcrustal erosion, 19
water release, 19
mantle rocks, 30
Mohorovicic Discontinuity,
30-31
Basin and Range Moho,
30
Neogene root of Sierra
Nevada, 31
plate motions, 3
subduction, mechanism of, 3-4
uplift of western interior, 31
Manzano Mountain New Mexico, 455
Marblemount Meta-Quartz
Diorite, petrochemistry
of, 219-221
Maria fold-and-thrust belt,
Mesozoic and Early
Cenozoic metamorphism,
486-88
Mesozoic active margin tectonics,
Great Basin, 582-85
Mesozoic contact metamorphism,
113-59
around single plutons, 145-51
correlations, 140-45
depth, 140-44
host composition, 145
intrusion composition, 144-45
distribution of, 117-45
in Arizona/New Mexico, 125-
125-26
in central/eastern Great Basin,
131-32

in Colorado, 125
in eastern Oregon/western
Idaho, 133-34
in Idaho batholith, 135-36
in Klamath Mountains,
132-33
magmatism and
metamorphism, synopsis
of, 118-25
in Mojave/southwestern
Arizona, 126-27
in Montana, 136-37
in northern Idaho/Washington,
137-40
Peninsular Range/Salinian
block, California, 127
in Sierra Nevada, 127-29
in western Great Basin, 129-31
diversity of, 113-17
relationship between Mesozoic
regional metamorphism
and, 151-58
P-T-time relationships, 151-54
regional igneous-related
metamorphism, model
for, 154-58
study problems, 158
field observations, 158
geochemistry, 158
petrology, 158
theory, 158
timing, 158
Mesozoic model, northeast Great
Basin, 638-39
Metaluminous granite, eastern
Great Basin, 101-3
Metamorphic petrology
(Wyoming/Montana),
363-78
Beartooth Mountains, 364-74
Jardine-South Snowy Block,
372-73
Lake Plateau area, 368-69
Long Lake and Broadwater
River areas, 367-68
North Snowy Block (NSB),
369-72
Quad Creek and Hellroaring
Plateau, 365-67
Stillwater Complex Hornfels
aureole, 373
summary of Archean geology
of, 373-74
Yankee Jim Canyon, 372
overview of western Wyoming
Province, 374-76
Blacktail Range, 375-76
northern Gallatin Range, 374
Spanish Peaks area, 374-75
Tobacco Root and Ruby
Ranges, 375
Methow-Pasayten and Hozameen

Belts, tectonic and
metamorphic of, 190-91
Methow-Pasayten Belt, northern
Cascades, 190-91
Middle Proterozoic, anorogenic
plutons, 425
Mineral zones:
Franciscan Complex:
high-grade blocks, 1051
metaigneous rocks, 1046-48
metasedimentary rocks,
1048-51
P/T conditions of
metamorphism, 1051-52
Miogeoclinal rocks:
Peninsular Ranges, 908-10
autochthonous meta-
sedimentary rocks, 908-9
parautochthonous metasedimentary rocks, 909-10
Mohorovicic Discontinuity, 30-31
Basin and Range Moho,
30
Neogene root of Sierra Nevada,
31
Mojave-Sonoran Desert:
detailed mineralogy, 508-12
amphibole, 511
biotite, 509-11
feldspars, 509
garnet, 512
muscovite, 509
geologic setting, 504-8
Iron Mountains, 507
Sacramento Mountains, 506-7
Santa Catalina Mountains,
507-8
Whipple Mountains, 506
geothermometry and
geobarometry, 512-19
alumina solubility in
hornblende, 516-18
garnet-muscovite-plagioclase
biotite equilibria, 513-16
muscovite-biotite-alkali
feldspar-quartz equilibria,
518-19
mylonitization, feldspar
thermometry, 518
Mule Mountains thrust, Mesozoic
and Early Cenozoic
metamorphism, 485-86
Muscovite-bearing plutons, 676
Muscovite-biotite-alkali feldspar-
quartz equilibria, Mojave-
Sonoran Desert, 518-19
Mylonitization, feldspar
thermometry, Mojave-
Sonoran Desert, 518

Nappes, Death Valley region,
725-27

Nd istopic data:
  isotope systematics, 89-92
  Proterozoic crust, 79-81
Needle Mountains, 416-17
"Nevadan" assemblages:
  Blue Mountains:
    in mafic rocks, 1132-33
    in metapelitic rocks, 1131-32
    in metaperidotites, 1130-31
Nevadan deformation,
        Sierra Nevada (northern),
        745-46
Nevadan event, Klamath
        Mountains, 1083-84
New Mexico (northern and central)
  metamorphic isothermalisobaric
        surfaces, 456
  metamorphic P-T path, 456-60
  metamorphism, 449-56
    Cimarron Mountains, 451
    Glorieta Baldy, 453
    Manzano Mountain, 455
    Pecos Greenstone Belt, 453
    Pedernal Hills, 455
    Picuris Range, 451-52
    Rincon Range, 452-53
    Rio Mora Uplift, 452
    Sandia Mountains, 453-55
    Taos Range, 451
    Truchas Range, 452
    Tusas Mountains, 449
    Zuni Mountains, 453
  regional geology, 447-48
  relations between
        metamorphism and
        deformation, 460-61
Northeastern Great Basin, 607-40
  areas of contrasting
        metamorphic grade,
        612-29
    Albion Mountains, 620-29
    Black Pine Mountains, 613-14
    Grouse Creek Mountains,
        620-29
    northern Pequop Mountains,
        617-20
    Pilot Range area, 614-17
    Raft River Mountains, 620-29
    Wood Hills, 617-20
  geologic framework, 609-11
  regional tectonic problems,
        629-36
  mylonitic rocks in
        metamorphic terranes,
        635-36
  P-T constraints for hinter-
        land metamorphism,
        637-43
  role of crystalline basement,
        633-35
  timing of metamorphism and
        deformation, 629-31

tectonic models, 636-40
  Cenozoic model, 639
  Mesozoic model, 638-39
  uncertainties about, 639-40
tectonic style of
        unmetamorphosed rocks,
        611-12
Northeastern Washington:
  regional overview, 274-76
  structural/metamorphic/plutonic
        evolution, 286-89
  structure and metamorphism,
        276-86
  tectonic history, 289-91
    Eocene, 291
    Late Cretaceous to Paleocene,
        291
    Middle Jurassic to
        Mid-Cretaceous, 290
    pre-Middle Jurassic, 290
  See also specific geographical
        areas.
Northern Cascades:
  tectonic and metamorphic
        development of, 180-91
    Methow-Pasayten and
        Hozameen Belts, 190-91
    Northwest Cascades Thrust
        System, 186-90
    Skagit Metamorphic Core,
        181-85
Northern Idaho:
  metamorphism/structure/
        tectonics of, 272-91
  regional overview, 274-76
  structural/metamorphic/plutonic
        evolution, 286-89
  structure and metamorphism, 276-86
  tectonic history, 289-91
    Eocene, 291
    Late Cretaceous to Paleocene, 291
    Middle Jurassic to mid-Cretaceous,
        290
    pre-Middle Jurassic, 290
  See also specific geographical areas.
Northern Sierra terrane, 765-85
  Jurassic deformation and
        metamorphism, 782-84
  Permian and Triassic deformation,
        779-81
  pre-Late Devonian deformation,
        769-72
  Shoo Fly Complex, 767-69
    Culbertson Lake allochthon, 768
    Duncan Peak allochthon, 768
    Lang sequence, 767-68
    Sierra City melange, 768-69
  Tahoe sequence, 778-79
  Taylorsville sequence, 772-78
    chert member of Peale Formation,
        776
    lower volcanic assemblage, 773-76

upper volcanic assemblage, 776-78
  Triassic and Jurassic rocks, 781-82
Northern Wyoming Province:
  geochemistry and geochronology of
        Archean rocks, 384-406
    Beartooth Mountains, 386-400
    Bighorn Mountains, 402-4
    northern Madison and Gallatin
        Ranges, 401
    Owl Creek Mountains, 404
    Southern Madison Range, 400
    Tobacco Root and Ruby Ranges,
        401-2
North Fork terrane, Klamath
        Mountains, 1068-69
Northwest Cascades:
  tectonic and metamorphic
        development of
        "Autochthon", 189-90
    Church Mountain Thrust Plate,
        187-88
    Shukasan Thrust Plate, 186-87
  Washington and British Columbia,
        197-209
    fault emplacement structures,
        204-7
    petrology, 197-204
    phase relations, 203-4
    regional considerations, 207-9
    summary of ages, 202-3
    tectonic history, 207
Northwest Cascades Thrust System,
        186-90

Okanogan Complex, geologic history
        of, 234-64
  core complex evolution models,
        261-64
  geologic relations, 239
  metamorphism, 247-59
    assemblages, 248-53
    conditions, 254-59
  radioisotopic age data, 259-61
  P-T evolution and deformation
        through time, 259-61
  regional tectonic setting, 236-38
  structures, 239-47
Old Woman-Piute Range, 541-51
  age of Mesozoic metamorphism, 559
  detailed study of Piute Mountains,
        544-51
  general geology, 541-42
  metamorphism, 542-44
  tectonic setting of, 539-41
  previous work, 541
Orocopia Schist:
  Mesozoic and Early Cenozoic
        metamorphism, 485
  Western Cordillera, relationship
        to, 955-58
Owl Creek Mountains, geochemistry
        and geochronoly of Archean

rocks, 404

Paleomagnetic studies:
   evidence for Sierra Nevada motion,
    27
   Skagit Crystalline Core (SCC),
    225-26
Palinspastic reconstruction, Death
   Valley region, 727-28
Palm Canyon Complex, 921-23
   brittle deformation, 922-23
   condition of metamorphism, 922
   "upper-plate" granitic rocks, 922
Panamint Mountains, Death Valley
   region, 716-21
Parautochthonous
   metasedimentary rocks,
   Peninsular Ranges, 909-10
Passive margin collisions, Great
   Basin, 579-80
Pastoria Creek, quartzofeldspathic
   gneiss of, 878-79
Pb istopic data:
   isotope systematics, 89-92
   Proterozic crust, 69-82
    for Arizona, 75-78
    implications of, 78-81
    for southeastern California,
     72-75
Peale Formation, Taylorsville
   sequence and, 776
Pecos Greenstone Belt, New
   Mexico, 453
Pedernal Hills, New Mexico, 455
Pelona, Orocopia, and Rand
   schists (POR suite),
   954-71, 977-92
   age relations and regional
    correlations, 991-92
   evidence of rapid uplift of,
    989-91
   high-pressure metamorphism,
    origin of, 988-99
   metamorphism, 984-89
    conditions of, 986-88
    facies, 984-85
    mafic schist, 986
    quartzofeldspathic schist,
     985-86
   premetamorphic lithologies,
    981-82
   structure, 982-84
   thrust zone, 983-84
Pelona Schist, and upside down
   metamorphism 955-58
Pelona Schist, See also Pelona,
   Orocopia, and Rand
   schists (POR suite);
   Western Cordillera.
Peninsular Ranges:
   Mesozoic contact
    metamorphism, 127

mylonite zone, 19-20, 918-23
   brittle deformation, 922-23
   Palm Canyon Complex,
    921-23
Peninsular Ranges batholith:
   metamorphic and tectonic
    evolution, 895-927
   major zones of ductile
    deformation, 915-23
   plutonic rocks, 910-13
   prebatholithic rocks, 896-910
   prebatholithic suture, 913-15
   regional setting, 896-98
   tectonic evolution of northern
    ranges, 923-27
   as product of continental arc
    magmatism, 12
Pequop Mountains, northeastern
   Great Basin, 617-20
Peraluminous granite, 99-101

Phanerozoic tectonostratigraphy,
   Great Basin, 575-76
Pickett Peak terrane, Klamath
   Mountains, 1070
Picuris Range, New Mexico, 451-52
Pilot Range, northeastern Great
   Basin, 614-17
Pioneer tectonic block,
   post-Archean metamorphic
   and tectonic evolution,
   351-52
Piute Mountains, detailed study
   of, 544-51
Post-Archean metamorphic and
   tectonic evolution, 333-56
   Archean basement, 333-34
   Bitterroot batholith, 347-49
    mafic dikes, 348-49
   central Plutonic and metamorphic
    zone (W. Montana, N.
    Idaho) 338-47
    batholith complex, 339-42
    deformation stages, 344-47
    eastern border fold-and-thrust
     belt, 347
    granite magma, 342-43
    regional metamorphism,
     deformation associated
     with, 343-44
   detachment blocks and thrust
    faulting, 349-54
   Boulder batholith, 353-54
   Pioneer tectonic block,
    351-52
   Sapphire tectonic block,
    350-51
   satellite plutons, 353
   diabase sils and dikes, 334-35
   early Mesozoic collision, 336-37
   early Tertiary events, 354-56
    Absaroka volcanic center,

     355
    Eocene dike swarm, 354-55
    Eocene granite and volcanic
     rocks, 355-56
    subduction, 354
   Lewis and Clark fault zone, 349
   Paleozoic deposition, 336
   post-Eocene tectonism, 356
   Precambrian metamorphism of
    Proterozoic section, 335
   Proterozoic deposition, 34
   western border deformed belt,
    337-38
   Salmon River arch, 337-38
Postsubduction thermal
   relaxation, inverted
   metamorphic gradients
   (IMGs), 970
Prebatholithic rocks:
   Peninsular Ranges batholith:
    miogeoclinal rocks, 908-10
    submarine fan deposits, 903-8
    volcanic and volcaniclastic
     rocks, 898-903
Pre-Late Devonian deformation,
   northern Sierra terrane,
   769-72
Pre-Nevadan deformation, Sierra
   Nevada (northern),
   744-45
Priest River Complex:
   evolution of, 286-89
   structure and metamorphism,
   278-82
Proterozic crust:
   Nd istopic data, 79-81
   Pb isotopic data, 69-82
    for Arizona, 75-78
    implications of, 78-81
    for southeastern California,
     72-75
Proterozoic high-grade
   metamorphism, 527-34
   Gold Butte area, 528-31
   McCullough Range, 531-33
   Whipple Mountains, 533-34
P-T constrains for hinterland
   metamorphism, north-
   eastern Great Basin,
   637-43
P-T path:
   northern and central New
    Mexico, 456-57
   changes in *P-T* during
    prophyroblast growth,
    459
   deformation and *P-T* path,
    461
   garnet zoning, 457
   plagioclase zoning, 457-59
   retrograde cooling, rate of,
    459-60

Quartzofeldspathic schist, 985-86
Quesnellia:
  evolution of, 286-89
  structure and metamorphism,
    283-84

Radioisotopic age data:
  Okanogan Complex, 259-61
    P-T evolution and
      deformation through
      time, 259-61
Raft River Mountains,
    northeastern Great Basin,
    620-29
Rattlesnake Creek terrane:
  Klamath Mountains, 1069
  pre-Siskiyou event in, 1085-86
Refrigeration, inverted
    metamorphic gradients
    (IMGs), 969-70
Rincon Range, New Mexico, 452-53
RioGrande rift, basaltic
    magmatism of, 93-95
Rio Mora Uplift, New Mexico, 452
Rocky Mountain Fold-and-Thrust
  Belt:
  evolution of, 286-89
  structure and metamorphism,
    276-78
Ruby Range, 375
  geochemistry and
    geochronology of Archean
    rocks, 401-2

Sacramento Mountains, Mojave-
  Sonoran Desert, 506-7
Salida area, Colorado, 416
Salinian block, 939-48
  compared to other terranes,
    947-48
  Mesozoic contact
    metamorphism, 127
  petrologic and structural frame-
    work, 941-43
    Sierra de Salinas schist,
      942-43
    southeastern sub-block rocks,
      941
    "Sur Series" rocks, 941-42
  as product of continental arc
    magmatism, 12
  similarities of metamorphic
    suites, 947
Salmon River suture:
  metamorphic rocks, structure
    of, 311-14
    age of deformation, 313-14
    eastern terrane, formation of,
      311-12
    western terrane formation of,
      312-13
  metamorphism adjacent to,

303-7
  age of metamorphism, 308-11
  comparison of metamorphism
    across suture, 307
  eastern terrane, 303-5
  western terrane, 305-7
  problematic features of, 324-25
  stratigraphy adjacent to,
    299-303
    eastern terrane, 299-301
    stratigraphic differences
      across, 302-3
    western terrane, 301-2
  suture zone plutons and
    superposed deformation,
    314-16
  tectonic model for, 321-23
    geometry of suture zone, 321
    sense of movement on suture
      zone, 322-23
    thermal history, 321-22
  tectonism outboard of, 318-20
    Late Jurassic tectonism, 320
    Late Triassic tectonism,
      319-20
    Upper Cretaceous deposits,
      320
  undeformed plutons, 316-18
    age of emplacement and
      cooling of, 316-17
    uplift history of eastern
      terrane, 317-18
Salmon River terrane, Klamath
    Mountains, 1069
Sandia Mountains, New Mexico,
    453-55
Santa Catalina Mountains,
    Mojave-Sonoran Desert,
    507-8
Sapphire tectonic block, post-
    Archean metamorphic and
    tectonic evolution, 350-51
Schell Creek Range, east-central
    Great Basin, 657-59,
    662--64
Sevier thrust belt, Great Basin,
    585-86
Shoo Fly Complex, 832
Shuskan Thrust Plate, tectonic
    and metamorphic
    development of, 186-87
Shuskan Metamorphic Suite,
    petrology, 201-2
Shuskan Metamorphic Suite,
    962-63
Sierra City melange, 768-69
Sierra de Salinas schist:
  Salinian block, 942-43
    metamorphic age of, 945
    provenance age of, 945-47
Sierra Madre Range, Colorado,
    417

Sierra Nevada batholith, Death
    Valley region, 10-12,
    724
Sierra Nevada (California region),
    824-54
  Calaveras Complex, 834-37
    age, 835-36
    correlation, 836-37
    lithology, 834-35
    origin, 836
  Foothills ophiolite terrane,
    838-44
    Early Mesozoic arc sequence,
      842-44
    origin, 840-41
    Paleozoic basement, 839-40
    pre-Jurassic tectonism, 841-42
  Foothills terranes, 844-49
    east-directed thrusting, 849
    epiclastic rocks, 844-45
    Middle Jurassic thrusting,
      845-49
  Middle to Late Jurassic arc,
    849-54
    intrusions, 850-51
    ophiolites, 851-52
    supracrustal sequences, 851
    syn-arc orogenesis, 852-54
  Paleozoic to Early Mesozoic
    continental borderland,
    830-33
    composite basement, 830-31
    Late Devonian to Permian
      marine arc, 831
    Lower Mesozoic marine
      strata, 832
    Shoo Fly Complex, 832
    western boundary of
      continental borderland,
      832-33
  parautochthonous North
    American basement,
    829-30
  Sullivan Creek terrane, 837-38
Sierra Nevada (northern), 738-59
  Central Belt, 741-43
    chert-argillite unit, 741-42
    older ophiolitic rocks, 742
    younger volcanic arc
      complexes, 742-43
  Feather River Belt, 743-44
  Eastern Belt, 744
  metamorphism, 746-53
    Central Belt, 747-49
    conditions of, 757-58
    contact metamorphism,
      752-53
    Eastern Belt, 750-52
    Feather River Belt, 749-50
    tectonic implications of,
      758-59
    time of, 753-57

Western Belt (Smartville
Complex), 746-47
structure, 744-46
Nevadan deformation, 745-46
pre-Nevadan deformation,
744-45
Western Belt, 739-41
Sierra Nevada (southernmost):
field/radiometric/petrographic
data, 869-79
Bear Valley Springs igneous
suite, 872-76
late- and postdeformational
intrusives, 879
metasedimentary framework
rocks, 871-72
Tehachapi Mountains gneiss
complex, 876-79
interrelations between igneous
and metamorphic
phenomena, 883-85
overview of isotopic data,
880-83
petrotectonic significance of
gneiss complex, 885-89
Silicic intrusive rocks, 98-105
metaluminous granites, 101-3
peraluminous granites, 99-101
Siskiyou event:
Klamath Mountains, 1077-81
age, 1078-80
distribution, 1077-78
interpretation, 1081
mineral assemblages, 1078
P-T conditions, 1078
in southern mountains,
1080-81
Skagit Crystalline Core (SCC):
composition of, 215
evolution of, 215-27
migmatitic rocks, 218
supracrustal rocks, 217-218
syntectonic intrusives, 219
Triassic intrusive age,
orthogneiss of, 218
paleomagnetic studies, 225-26
petrochemistry of type area,
219-26
Cascade River Schist, 221-23
Eldorado Orthogneiss, 221
Gabriel Peak Orthogneiss,
224-26
Marblemount Meta-Quartz
Diorite, 219-21
Skagit Gneiss, 223-24
Skagit Gneiss, petrochemistry of,
223-24
Skagit Metamorphic Core, 181-85
Skookum Gulch blueschists,
1087
Smith River subterrane, Klamath
Mountains, 1070

Snake Range:
northern range, 665-68
southern range, 659-61
Snake River Plain, basaltic
magmatism of, 95-97
Sonora fault, 803-4
Southeastern California
age of Mesozoic metamorphism,
559-60
Big Maria Mountains, 551-59
age of Mesozoic
metamorphism, 559
general geology, 551
geothermometry, 554-55
isograds from siliceous
dolomites, 552-53
peristerite gap, 556-58
regional metamorphism,
551-52
siliceous dolomites, 556
Supai Formation,
558-59
crustal anatexis, 560
heat transport by fluids,
evidence for, 562
Late Cretaceous regional
metamorphism,
implications of, 563
Old Woman-Piute Range,
541-51
age of Mesozoic
metamorphism, 559
detailed study of Piute
Mountains, 544-51
general geology, 541-42
metamorphism, 542-44
tectonic setting of, 539-41
previous work, 541
Southeastern Salinian sub-block:
941, 943-44
age of, 943-44
rocks of, 941
Southern Madison Range:
geochemistry and geochronology
of Archean rocks, 400
northern Wyoming Province,
400
South Fork Mountain schist,
960-62
Southwest Nevada, basaltic
magmatism of, 97-98
Spanish Peaks area, Wyoming,
374-75
Stillwater Complex hornfels
aureole, 373
Structural units (terranes):
Klamath Mountains, 1067-70
Central Metamorphic terrane,
1068
Condrey Mountain terrane,
1069-70
eastern Hayfork subterrane,

1069
Fort Jones terrane, 1068
Franciscan assemblage, 1070
North Fork terrane, 1068-69
Rattlesnake Creek terrane,
1069
Salmon River terrane, 1069
Smith River subterrane, 1070
Western Hayfork subterrane,
1069
Western Jurassic terrane, 1070
Yreka terrane, 1068
Subduction, mechanism of, 3-4
Submarine fan deposits:
Peninsular Ranges, 903-8
Bedford Canyon Formation,
903-4
conditions of metamorphism,
905-6
Julian Schist and French
Valley Formation, 904-5
transitional subzone, 906-8
Supai Formation, 558-59
"Sur Series":
Salinian block, 941-42, 944-45
metamorphic age of, 944
provenance age of, 944-45
rocks of, 941-42
Syn-arc orogenesis:
Sierra Nevada (California
region):
faulting, 853-54
penetrative deformation, 853
sedimentation, 852

Tahoe sequence, 778-79
Taos Range, New Mexico, 451
Taylorsville sequence, 772-78
chert member of Peale Formation, 776
lower volcanic assemblage,
773-76
upper volcanic assemblage,
776-78
Tehachapi Mountains gneiss
complex, 876-79,
gneiss units of Tejon Creek and
Comanche Point, 876-78
quartzofeldspathic gneiss of
Pastoria Creek, 878-79
Tejon Creek, gneiss units of,
876-78
Thermal model:
inverted metamorphic gradients
(IMGs)
heating upon descent, 968-69
postsubduction thermal
relaxation, 970
refrigeration, 969-70
Tobacco Root Range:
geochemistry and geochronology
of Archean rocks, 401-2

Wyoming, 375
Toiyabe uplift zone, Great
    Basin, 591-93
Truchas Range, New Mexico, 452
Tucki Mountain, Death Valley
    region, 724
Tusas Mountains, New Mexico, 449

Uinta Mountains, 418
Uplift rates:
    epidote-bearing plutons, 53-60
    Western Cordillera:
        Ecstall area, British Columbia,
            54-56
        implication of, 61
        Ketchikan-Petersburg area,
            southeast Alaska, 54
        Round Valley area, Idaho,
            53-54
        Tenpeak area, Washington,
            54
Upper/middle crust, continental
    arc magmatism, 9-13
"Upper-plate" granitic rocks, Palm
    Canyon Complex, 922
Uranium, as product of
 continental arc magmatism, 8
Utah;
    Farmington Canyon Complex,
        433-34
    age of rocks and metamorphic
        events, 437-38
    metamorphic history, 438-39
    protoliths of, 434-37
    margin of the Archean
        continent, 433
    quartz monzonite gneiss, age
        and origin of, 439-41

Vedder Complex, petrology, 200
Volcanism, as product of
    continental arc
    magmatism, 7-8
Volcano and volcaniclastic rocks,
    Peninsular Ranges, 898-903
Walker Lane Belt:
    Cenozoic structures, develop-
        ment of, 703-5
    characteristic structures of,
        695-99
    basin-range faults, 696-97
    large-scale extension/detach-
        ment faults/metamorphic
        core complexes, 697-99
    oroflexural folds, 697
    regional structural blocks,
        696
    strike-slip faults, 696
    definition of, 684-86
    Mesozoic contact
        metamorphism, 129-31

Mesozoic structures, develop-
    ment of, 699-702
structural blocks/fault
    patterns/displacements,
    686-95
    Carson section, 687-89
    Excelsior-Coaldale section,
        690-92
    Goldfield section, 694
    Inyo-Mono section, 692-94
    Lake Mead section, 695
    Pyramid Lake section, 686-87
    Spotted Range-Mine
        Mountain section, 694
    Spring Mountain section, 695
    Walker Lake sections, 690
Wallrock deformation, continental
    arc magmatism, 14
Washington:
    Northwest Cascades, 197-209
    fault emplacement structures,
        204-7
    petrology, 197-204
    phase relations, 203-4
    regional considerations, 207-9
    summary of ages, 202-3
    tectonic history, 207
Wells Creek Volcanics, petrology,
    201
Western Arizona:
    Mesozoic and Early Cenozoic
        metamorphism
    aluminous metasomatic rocks,
        488-89
    McCoy basin and Maria fold-
        and-thrust belt, 486-88
    Orocopia Schist and Chocolate
        Mountains thrust, 485
    upper plates of Chocolate
        Mountains and Mule
        Mountains thrusts, 485-86
Western Belt:
    Sierra Nevada (northern),
        739-41
    metamorphism, 746-47
Western Cordillera:
    environments of intrusion and
        inferred paleogeothermal
        gradients, 44-52
    Ecstall Pluton, British
        Columbia, 49-50
    Ketchikan-Petersburg area,
        Southeast Alaska, 51-52
    Round Valley Pluton, Idaho,
        45-48
    Tenpeak Area, Washington,
        48
    epidote-bearing plutons, crustal
        significance of, 52-60
    crustal thickness, 52-53
    uplift rates, 53-60
    inverted metamorphic gradients

(IMGs), 955-71
    Catalina Schist, 958-59
    Central Metamorphic Belt
        (CMB), 959-60
    Pelona-Orocopia schists,
        955-58
    Shuskan Suite, 962-63
    South Fork Mountain Schist,
        960-62
    summary of, 963
    thermal blocking of, 963-65
Western Great Basin, See Walker
    Lane Belt.
Western Sierra Nevada:
    Calveras Complex, 802-4
    age, 802
    igneous intrusions, 802-3
    lithology, 802
    Sonora fault, 803-4
    structural summary, 803
    Don Pedro terrane (Phyllite-
        Greenschist Belt), 804-6
    age, 804-5
    igneous intrusions, 805
    lithology, 804
    Melones fault, 805-6
    Foothills terrane, 806-8
    age, 807
    igneous intrusions, 807
    lithology, 806-7
    structural summary, 807-8
    metamorphism and deformation,
        808-11
    Mesozoic metamorphism,
        808-11
    Paleozoic metamorphism, 808
    regional setting, 791
    Shoo Fly Complex, 794-802
    age, 794-96
    Calaveras-Shoo Fly thrust,
        799-802
    igneous intrusions, 797
    lithology, 794
    structural summary, 797-99
    structural analysis, methods of,
        791-94
    conventional methods, 793
    problems of structural
        correlation, 791-93
Western Wyoming Province:
    evolution of, 376-77
    metamorphic petrology, 363-77,
    overview of, 374-76
Wet Mountains, 417-18
Whipple Mountains:
    Mojave-Sonoran Desert, 506
    Proterozoic high-grade
        metamorphism, 533-34
Winnemucca deformation belt,
    Great Basin, 586-91
Wood Hills, northeastern Great
    Basin, 617-20

Yellow Aster Complex, petrology,
    197-98
Yolla Bolly terrane, Klamath
    Mountains, 1070
Yreka terrane, Klamath
    Mountains, 1068

Zuni Mountains, New Mexico, 453